W9-BGE-078

HEATH ALGEBRA 2

AN INTEGRATED APPROACH

Teacher's Edition

Roland E. Larson

Timothy D. Kanold

Lee Stiff

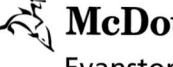

McDougal Littell
Evanston, Illinois Boston ◆ Dallas

Published simultaneously in Canada

Printed in the United States of America

International Standard Book Number: 0-669-43285-7

4 5 6 7 8 9 10 VHP 02 01 00 99

Authentic algebra
is algebra
that models the real world.
It's the all-new
ALGEBRA from D.C.Heath.

CONTENTS

Presenting algebra for all students

All of us learn and understand concepts better in context, and algebra is no exception. But over the last 20 years, most algebra texts have become so narrowly focused that they have lost track of the real purpose of algebra—to help us solve real-life problems. Algebra has been little more than the manipulation of symbols, producing symbolic solutions, with no larger context in sight.

ALGEBRA from D.C. Heath is designed to provide a context for the symbols of algebra. What you'll find in this new program is neither "old algebra" (however you might define that), nor some mysterious "new algebra." It is simply authentic algebra that makes sense— in the same way an architect's specifications suddenly make sense when examined next to a model of the design.

At every opportunity in this text, we demonstrate the usefulness and vitality of algebra—and in so doing we make it accessible to every student. We help each student set reasonable, reachable goals at the beginning of each lesson, and we provide continuing reasons for all students to develop confidence and a sense of accomplishment in their work.

ALGEBRA and the NCTM Standards

The NCTM Standards serve as the foundation of the new ALGEBRA from D.C. Heath, with its emphasis on

- Problem Solving
- Reasoning
- Communication
- Connections.

Look for the spirit of the Standards throughout the text.

Presenting algebra for the 21st Century— our distinguished author team

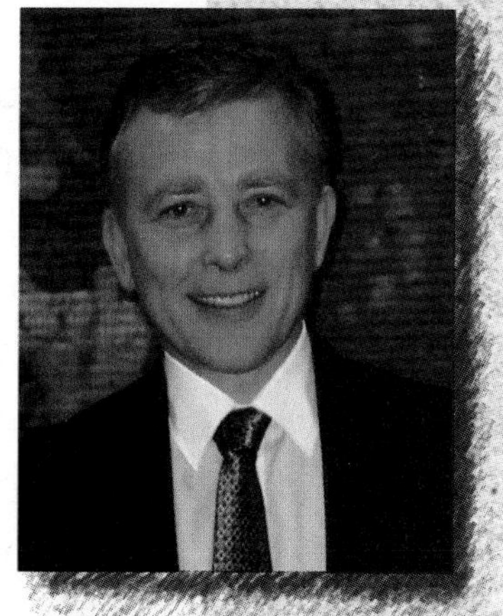

Roland E. Larson is Professor of Mathematics at the Behrend College of Pennsylvania State University in Erie. He is a member of NCTM and a highly successful author of many high school and college mathematics textbooks. Several of his D.C. Heath college titles have proven to be top choices for high school mathematics courses.

Timothy D. Kanold is Chairman of Mathematics and Science and a teacher at Adlai Stevenson High School in Prairie View, Illinois. He is the 1986 recipient of the Presidential Award for Excellence in Mathematics Teaching, as well as the 1991 recipient of the Outstanding Young Alumni Award from Illinois State University. A member of NCTM, he served on NCTM's *Professional Standards for Teaching Mathematics* Commission and is a member of the *Regional Services* Committee.

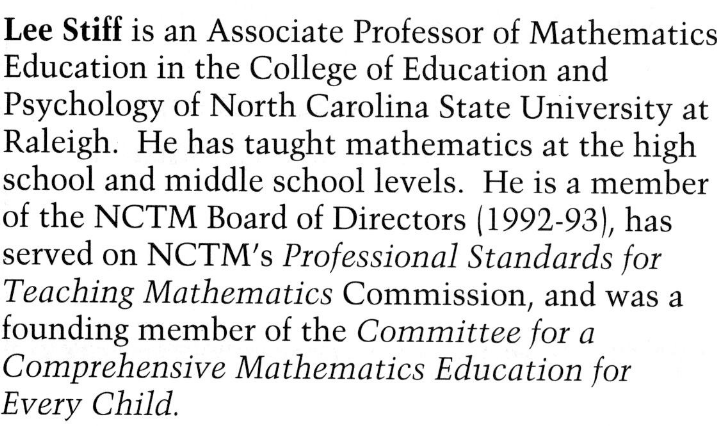

Lee Stiff is an Associate Professor of Mathematics Education in the College of Education and Psychology of North Carolina State University at Raleigh. He has taught mathematics at the high school and middle school levels. He is a member of the NCTM Board of Directors (1992-93), has served on NCTM's *Professional Standards for Teaching Mathematics* Commission, and was a founding member of the *Committee for a Comprehensive Mathematics Education for Every Child.*

What to look for in ALGEBRA from D.C.Heath

Real Life
Climatology

Real-life applications show the value of algebra. You don't have to search ALGEBRA for a subtle "orientation" to reality. You'll see it as solid grounding in every lesson, with exercise sets that blend interesting information with skill building and problem solving. *See pages T6, T18.*

Problem Solving
Draw a Graph

Problem solving is a continuing process in this program, not just an occasional topic. Students are asked to explain, justify, verify, interpret, draw and label—in short, to think critically all the time. *See page T8.*

Connections
Technology

Connections give algebra greater meaning. ALGEBRA from D.C. Heath helps you integrate such topics as geometry, statistics, trigonometry, probability, and matrix theory into your algebra course—together with topics from other disciplines, such as geography, history, economics, and the physical and biological sciences. *See page T10.*

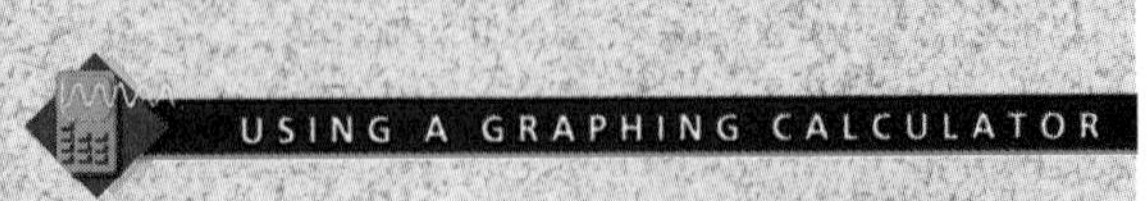

Technology is used to investigate and verify findings. Students are shown in every chapter how they can use a calculator or a computer to evaluate expressions, graph equations, draw scatter plots and best-fitting lines, and much more. Data analysis is presented in separate lessons and is integrated throughout the text. *See pages T12, T32.*

Formal and Alternative Assessments

Assessment gives a clear picture of individual needs. In ALGEBRA from D.C. Heath, both formal and alternative forms of assessment are natural components of the program, designed to help you meet the needs of all students. *See page T14.*

Communication is central to math activity. Lesson features in ALGEBRA encourage students to verbalize math concepts and share ideas with each other—to reinforce each others' understanding of algebra and emphasize the connections between other disciplines and algebra. *See page T16.*

Communicating about ALGEBRA

The text addresses a diverse student body as a matter of course—by referring naturally in problems and examples to students of varied interests and backgrounds, by presenting selected topics in historical perspective, by highlighting current career opportunities, and by giving teachers a range of ancillary materials. *See page T28.*

Milestones ORIGINS OF ALGEBRA

Career Interview

ALGEBRA provides cooperative learning opportunities. Mathematics should not be a solitary activity. It can be done and learned with others—through discussions, projects, activities, etc.—with students of different abilities, different backgrounds, different interests. *See margin, page 12.*

Communicating about ALGEBRA

Cooperative Learning

Extensive ancillaries help you reach all your students. You can enrich the new ALGEBRA with a variety of special materials designed to help you meet the needs of a diverse student population. *See page T33.*

COMPONENTS

To walk through a lesson from ALGEBRA 2, please turn to page T20.

A lesson that connects with life connects with students.

In developing this algebra text, the authors have kept uppermost in their minds the ways students learn—what builds interest in a topic, what makes an idea memorable.

Algebra instruction has for some time emphasized formal derivations—which ask a student to learn generalizations first, then to apply the results to particular cases. Applications, on the other hand, ask a student to learn particular, concrete cases first; then a general understanding follows. The authors have chosen to emphasize the applications of algebra in this text, feeling that moving from the need for algebra to the mechanics of algebra is what makes the most sense to students.

Derivations and proofs are provided in the Teacher's Edition, but the focus of the program is on the applications of algebra—its use in meaningful activities. Every lesson has at least one example that shows how the material in the lesson can be applied to a real-life situation.

These examples, and their corresponding exercises, comprise the most extensive, creative collection of mathematical applications ever assembled in an algebra series. The applications often use authentic data. This helps students validate what they are learning—a central theme of the NCTM Standards.

" We look at the hard data and see life in it. We're making it possible for students to do that, too."

—Rita Campanella, Executive Editor, ALGEBRA

Note: A separate Applications Handbook provides interesting additional background for applications in the text. Topics covered include astronomy, chemistry, physics, sports, economics, genetics, and music.

Look for meaning in this lesson.

While other texts discuss concepts as concepts, ALGEBRA from D.C. Heath discusses how a concept fits into the continuing story of the mathematics of real life. Compare various text treatments of inverse relations and functions, for instance. You'll find that the context and the logic of presentation in ALGEBRA are what makes the topic most meaningful.

Here "inverse relation" is defined in terms of the familiar "relation." A finite relation and its graph is used as an example.

3

Inverse Functions

...ld learn:
...ntify inverse
...e functions
...vo functions
...other
...e inverse
...situations

...learn it:
...y pairs of
...of inverse
...ple, you
...of the mea-
...of a
...umber of
...nber of
...of the
...les.

Goal 1 Finding Inverses

The **inverse** of a relation is the set of ordered pairs obtained by switching the coordinates of each ordered pair in the relation. Here is an example.

$\{(0,-3), (1,-1), (2, 1), (3, 3), (4, 5)\}$ *Original relation*

$\{(-3, 0), (-1, 1), (1, 2), (3, 3), (5, 4)\}$ *Inverse relation*

The graph of the inverse is the **reflection** of the graph of the original relation. The mirror of the reflection is the line $y = x$.

Graph of Original Relation

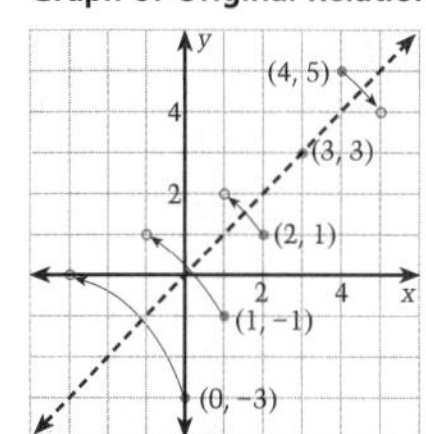

Graph of Inverse Relation

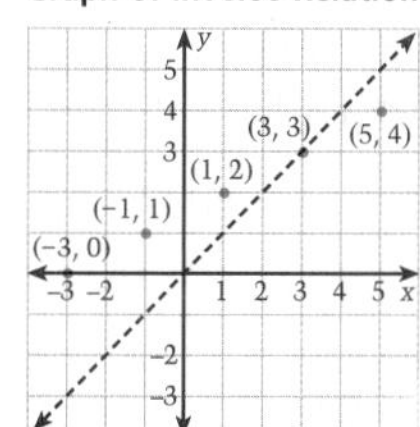

To find the inverse of a relation that is given by an equation in x and y, switch the roles of x and y in the equation.

Example 1 *Finding an Inverse Relation*

Find an equation for the inverse of the relation $y = 2x - 3$.

Solution Begin with the original equation. Then switch the roles of x and y.

$y = 2x - 3$ *Original equation*

$x = 2y - 3$ *Switch the roles of x and y to find an equation for the inverse relation.*

Be sure you see that these two equations are *not* equivalent. When finding an equation of an inverse relation, you are forming a *different* equation whose solutions reflect the switching of x and y. You can confirm this by sketching their graphs. ■

...2x − 3

...= 2x − 3 and ...eflections of ...e line y = x.

▪ Functions

Then students are shown how to find and graph an inverse relation with an infinite number of points.

After introducing inverse functions, four examples of skill development follow. The two shown here involve verifying inverse functions and applying them to a geometric formula.

Example 4 *Verifying Inverse Functions*

Verify that the functions from Example 2 are inverses of each other.

$f(x) = 3x - 1$ and $g(x) = \frac{1}{3}x + \frac{1}{3}$

Solution

$f(g(x)) = f\left(\frac{1}{3}x + \frac{1}{3}\right)$ *Substitute $\frac{1}{3}x + \frac{1}{3}$ for g(x).*

$= 3\left(\frac{1}{3}x + \frac{1}{3}\right) - 1$ *Apply f(□) = 3□ − 1.*

$= 3\left(\frac{1}{3}x\right) + 3\left(\frac{1}{3}\right) - 1$ *Distributive Property*

$= x + 1 - 1$ *Simplify.*

$= x$ *Simplify.*

$g(f(x)) = g(3x - 1)$ *Substitute 3x − 1 for f(x).*

$= \frac{1}{3}(3x - 1) + \frac{1}{3}$ *Apply g(□) = $\frac{1}{3}$□ + $\frac{1}{3}$.*

$= \frac{1}{3}(3x) - \frac{1}{3}(1) + \frac{1}{3}$ *Distributive Property*

$= x - \frac{1}{3} + \frac{1}{3}$ *Simplify.*

$= x$ *Simplify.*

Because both compositions produce x, you can conclude that the functions f and g are inverses of each other. ■

Goal 2 Using Inverse Functions in Real Life

A formula from geometry states that the sum, S (in degrees), of the angles of an n-sided polygon is S = 180n − 360.

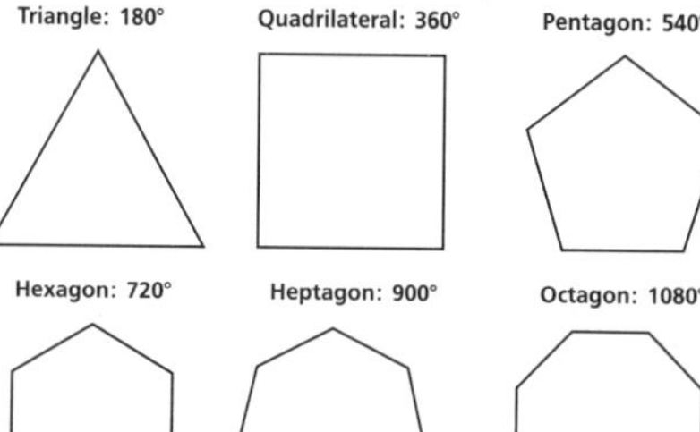

Example 5 *Rewriting a Formula*

The formula $S = 180n - 360$ gives the sum of the angles of an n-sided polygon, where n is the input and S the output. Solve the formula for n so that S is the input and n the output.

Solution Solve for n in terms of S.

$n = \frac{1}{180}S + 2$ *S is input, n is output.*

This "inverse" formula could be used to find the number of sides of a polygon when you are given the sum, S, of its angles. For example, a polygon whose interior angles have a sum of 1800° has $n = \frac{1}{180}(1800) + 2 = 12$ sides. ■

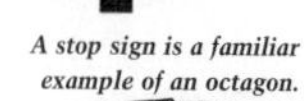

A stop sign is a familiar example of an octagon.

You could write the formulas in Example 5 as the inverse functions $f(x) = 180x - 360$ and $g(x) = \frac{1}{180}x + 2$. But when the variables have applications, letters that are easily associated with the quantities being measured are often used.

The Standards recommend...

Mathematics as Problem Solving

In grades 9-12, the mathematics curriculum should include the refinement and extension of methods of mathematical problem solving so that all students can—

- use, with increasing confidence, problem-solving approaches to investigate and understand mathematical content.
- apply integrated mathematical problem-solving strategies to solve problems from within and outside mathematics.
- recognize and formulate problems from situations within and outside mathematics.
- apply the process of mathematical modeling to real-world problem situations.

Increased attention to

▲ word problems with a variety of structures.
▲ everyday problems, applications, and open-ended questions.
▲ investigation and discussion of patterns, relationships, and problem-solving strategies.
▲ situations represented verbally, numerically, graphically, geometrically, or symbolically.

— from the NCTM Standards

Algebra is a way to solve problems—and to understand the world around us.

Algebra helps us solve problems efficiently. The real-life value of algebra as a problem-solving tool is a major theme stressed by Larson, Kanold, and Stiff throughout the new ALGEBRA.

Algebra is useful in real life.

Whenever students begin a new algebraic concept, the **Why you should learn it** feature in the Student Text explains how that concept will help them: that it's useful, for instance, when determining which video store offers the best deal or how much to charge for tickets to a fund-raiser. Because students encounter real-life problem solving in *every lesson*, they will come to see how valuable algebra can be as a problem-solving tool.

"The strength of ALGEBRA is the variety and quality of the word problems. Students are asked to apply concepts in a meaningful way. ALGEBRA encourages students to spend time working, thinking about and extending the problems."

— Jayne Fleener, ALGEBRA reviewer, University of Oklahoma, Norman, Oklahoma

Verbal modeling gets to the point of a problem.

Early in the text, ALGEBRA develops a problem-solving model. Students learn to develop a verbal model first—or "translate" a problem into their own words.

Once they have defined labels for the verbal model, they put together an equation, or algebraic model. Moving from the verbal to the algebraic helps students focus on the content of a problem.

Visual models broaden the picture.

To demonstrate the usefulness of algebra, D.C. Heath's ALGEBRA program connects linear equations—a difficult concept for many students—to real-life data. ALGEBRA also uses the power of new technology—the computer and the graphing calculator—to create meaningful visual representations of algebraic concepts. This visual approach gives students a firm grasp of concepts and strengthens reasoning skills.

Authentic data connect algebra to the real world.

Many problems and examples use real-life data and graphs to make algebraic concepts more meaningful to students—and to reinforce the role algebra plays in the real world. The use of authentic data also teaches students how to evaluate the information on graphs often found in newspapers and magazines.

Exercises build basic and critical-thinking skills.

Unusually rich and creative exercise sets blend basic skill building with critical thinking to strengthen students' problem-solving abilities. Students are continually asked to explain, justify, verify, interpret—in short, to think critically.

D.C. Heath's ALGEBRA provides...

- algebraic concepts set in real-life context.
- verbal modeling—to help students focus on what a problem is asking for.
- visual representations of algebraic concepts—for deeper understanding and stronger reasoning skills.
- exercises that strengthen students' skills *and* critical thinking.

"ALGEBRA enhances a teacher's role as guide, coach, and cheerleader. We try to help teachers nurture the problem-solving skills students already possess in order to build new problem-solving skills."

— *Roland E. Larson, ALGEBRA author*

The Standards recommend...

The mathematics curriculum should include investigation of the connections and interplay among various mathematical topics and their applications so that all students can—

- recognize equivalent representations of the same concept.
- relate procedures in one representation to procedures in an equivalent representation.
- use and value the connections among mathematical topics.
- use and value the connections between mathematics and other disciplines.

Increased attention to

▲ connections among math topics, among math and other curriculum areas, and between math and daily life.

▲ the relevance and value of mathematics in students' studies and lives so they view algebra as a whole rather than an isolated set of topics.

▲ the use of real-world problems to motivate and apply theory.

—from the NCTM Standards

Connections strengthen understanding of algebra.

This all-new program helps your students view algebra not as isolated collections of symbols, but as a vital body of knowledge with relevance to other math topics, to other academic disciplines, and to their own world. Seeing algebra's importance to their own lives makes studying it more meaningful and enjoyable.

Connections to other math topics

In this new program, concepts from geometry, statistics, probability, and other branches of mathematics are integrated with algebra so that students have an opportunity to recognize and understand algebraic principles.

ALGEBRA provides **Integrated Review** exercises to show students how previously learned concepts relate to the new concepts of a lesson.

Exploring Data lessons and exercises connect algebra to statistics, data organization, and finance.

Connections to other academic disciplines

In this new program, algebra is linked to many other disciplines, such as art, biology, geography, history, music, medicine, and business. This integration will expand your students' sense of the usefulness of algebra. They will see how algebraic concepts and procedures can be applied to problems arising in diverse areas.

Connections to real life

ALGEBRA presents mathematics in a relevant, meaningful context using real-life data and applications. Students value algebra as they understand the interrelationships that connect what they find in their text to their world. Not only are the real-life examples important to understanding algebra—they are also interesting and fun!

Each lesson of the text begins with a **Why you should learn it** feature that lets your students know—right from the start—the importance of what they are about to study.

Career Interview features in alternate chapters introduce students to the broad range of careers in which math is a useful tool.

The **Independent Practice** sections at the end of each lesson contain both skill-building exercises and real-life applications.

Real-Life Applications in ALGEBRA 2

Accounting 20, 22, 28, 807, 824
Advertising 127, 155, 214, 337, 451
Aerospace 351, 393, 720
Agriculture 22, 83, 112, 135, 163, 185, 451, 459, 580
Animal/Insect Studies 8, 23, 35, 42, 46, 48, 53, 167, 193, 323, 332, 382, 386, 388-389, 410, 425, 445, 451, 504, 507, 511, 741, 749, 786, 809, 827
Archaeology 238, 497
Architecture 106, 241, 249, 462, 476, 479
Arts 147, 220, 272, 356, 520, 603, 617, 688, 691, 721, 725, 794, 810
Astronomy 234, 350, 370, 505, 585, 588, 618, 628, 649, 720, 725
Banking 326, 422, 424, 437, 439, 450, 553, 560, 800
Biology 36, 46, 167, 366, 668, 741, 756, 784, 786, 827
Busines 26, 51, 83, 98, 124, 126, 142, 153, 155, 179, 184, 195, 214, 221, 234, 256, 290, 294-295, 307, 310, 351, 358, 365, 403-404, 430, 433, 447, 451, 468, 471, 483, 485-486, 490, 506, 523, 632, 634-635, 640, 807, 812, 824
Chemistry 14, 23, 40, 166, 212, 214, 411, 417, 424, 439, 451, 525, 530, 791
Communications 247, 267, 270, 310, 337, 580, 789, 799-800
Construction 25, 256, 707, 821-822
Consumer 8, 75, 82, 142, 148, 166, 184, 293, 307, 310, 317, 505, 546
Cryptography 198-199, 201
Earth Science 713, 751
Economics 166, 176, 195, 290, 295, 353
Education 13, 73, 176, 179, 296, 327, 330, 358, 394, 446, 793, 810, 825
Electronics 418, 526, 529
Engineering 393, 497, 531, 600, 690, 707
Entertainment 42, 98, 105, 141-142, 155, 218, 225, 290, 319, 330, 379, 445, 522, 538, 574, 627, 656, 698
Environmental Science 5, 28, 54, 133, 190, 355, 447, 525, 528, 566, 572, 751
Finance 22, 82, 91, 108, 126, 135, 139, 141-142, 174, 273, 287, 293-294, 303-304, 326, 352-353, 356, 401, 437, 439, 450, 468, 546, 555-560, 564, 642, 658-665, 668-669, 800, 807, 818, 824
Forensic Science 428
Fractals 261, 263
Game Theory 797, 802-803, 814, 816-818, 820-821
Geography 42, 66, 68, 75, 98, 192, 240, 281, 336, 506, 699, 714, 736, 742
Geology/Mining 68, 276, 415
Government 52, 89, 221, 269, 385, 424, 439, 492, 505, 798, 806, 825
History 75, 117, 505, 512
Home Economics 98, 372
Insurance 818, 821
Law Enforcement 76
Library Science 394
Mechanics 705, 726
Medical 49, 68, 214, 279, 311, 357, 379, 442, 446, 522, 801
Meteorology 7, 42, 88, 103, 303, 336, 383, 386, 410, 478, 543, 595, 739, 741, 806, 826
Military 492, 523
Music 53, 150, 162, 373, 525, 546, 554, 714, 733, 764
Nautical Science 363
Navigation 609, 718
Nutrition/Fitness 65, 91, 112, 184, 499
Oceanography 24, 48, 393, 408, 439
Optics 735, 762, 765
Photography 220, 417, 547
Physics 36, 91, 143, 231, 234, 254, 269, 337, 373, 393, 417-418, 424, 430, 478, 520, 528-531, 536, 564-565, 574, 575, 578, 580, 593, 612, 647, 649, 669, 707, 721, 733, 748, 768
Physiology 48, 98, 109, 388, 536, 742, 774
Population 13, 140, 166, 202, 269, 281, 319, 328-329, 331, 338, 431, 439, 461, 501, 505, 523, 550, 643, 741, 783, 787, 797, 825
Publishing 9, 11, 91, 149, 327, 461
Recreation 27, 160, 357
Retailing 8, 75, 80, 82, 91, 138, 142, 148, 166, 184, 268-269, 293, 461, 490, 499-500, 510, 625, 774, 819, 822
Sociology 50, 466
Sports 7, 27-28, 32, 48, 52-54, 96, 105, 112, 142, 150, 166, 178, 182, 208, 218, 221, 225, 234, 240, 249, 256, 303, 332, 365, 404, 425, 478, 500, 503, 539, 544, 586, 587, 602, 627, 656, 696, 698, 708, 717, 720, 792, 799-800, 809, 826
Testing 123, 417, 504, 511, 791
Time Management 329
Topography 3, 7, 681, 683-684, 711
Transportation 127, 135, 527, 565, 609, 720
Travel 162, 225, 232, 235, 357, 379, 815

D.C. Heath's ALGEBRA provides...

- **Integrated Review** exercises to relate previously learned concepts to new topics from the lesson.
- **Exploring Data** lessons and exercises to connect algebra to other disciplines.
- **Mathematical Models** and related exercises to apply materials found in each lesson to real-life situations.
- **Why You Should Learn It** features to let your students know the importance of what they are about to study.
- **Career Interview** features to acquaint students with the broad range of careers in which math is a useful tool.
- **Independent Practice** to promote understanding through skill-building exercises and real-life applications.
- **Real-life Themes** in Chapter Reviews to help students connect what they have just studied to the world around them.

"Modeling puts a mathematical overlay on the real world. It's attainable, reasonable, straightforward, and it fits with the topics at hand. Real-life connections make ALGEBRA meaningful for all students and promote success at all levels."

—Lee Stiff, ALGEBRA author

The Standards recommend...

- the integration of ideas from algebra and geometry, with graphic representation playing an important connecting role.
- the use of scientific calculators with graphing capabilities.
- the use of computers for demonstration purposes and for students to use in individual and group work.

Increased attention to

▲ visual representation of algebraic concepts.
▲ the use of calculators and computers as tools for learning and doing mathematics.
▲ the use of computer utilities to develop conceptual understanding.
▲ computer-based methods such as successive approximations and graphing utilities for solving equations and inequalities.

— from the NCTM Standards

Technology gives a visual dimension to algebra.

Seeing what the graph of an equation looks like helps students understand what that equation means. ALGEBRA emphasizes a visual approach to algebra—with the graphing calculator and the computer—because visual models of algebraic expressions make algebra more meaningful.

Technology helps students visualize algebra.

Graphing calculators and computers help students visualize statistics and data, linear equations, and other algebraic functions. When students use graphing calculators to graph equations, they are making visual models of algebraic expressions; they can readily see that an equation represents something. Graphing calculators let students create graphs quickly, too, so there's more time for "What if ... ?" questions—and for development of reasoning skills.

Technology is integrated throughout.

The graphing calculator is integrated into problem sets throughout the text, although it is not required. And in every chapter, the **Using a Graphing Calculator** feature shows students how to explore equations visually, practice data analysis techniques, and solve problems. The exercises in this feature are compatible with the newest TI, Casio and Sharp models. In the appendix, keystrokes are given for the TI-82, Casio fx-9700GE and Sharp EL-9300C along with a new application for use with these calculators.

D.C. Heath's ALGEBRA provides...

- a visual approach to algebra.
- optional but fully integrated use of the graphing calculator and computer.
- useful suggestions on when and how to use the graphing calculator.

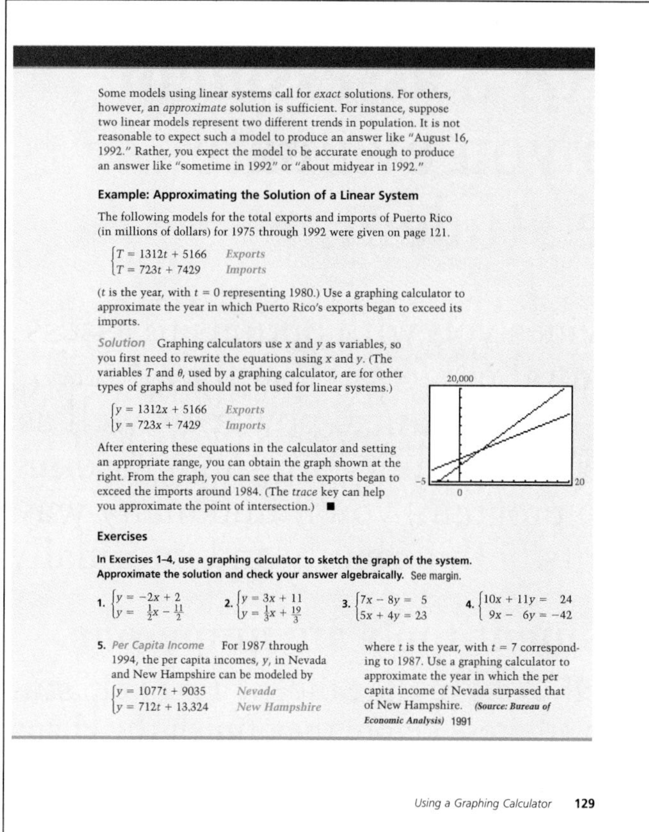

Some models using linear systems call for *exact* solutions. For others, however, an *approximate* solution is sufficient. For instance, suppose two linear models represent two different trends in population. It is not reasonable to expect such a model to produce an answer like "August 16, 1992." Rather, you expect the model to be accurate enough to produce an answer like "sometime in 1992" or "about midyear in 1992."

Example: Approximating the Solution of a Linear System

The following models for the total exports and imports of Puerto Rico (in millions of dollars) for 1975 through 1992 were given on page 121.

$$\begin{cases} T = 1312t + 5166 & \textit{Exports} \\ T = 723t + 7429 & \textit{Imports} \end{cases}$$

(t is the year, with $t = 0$ representing 1980.) Use a graphing calculator to approximate the year in which Puerto Rico's exports began to exceed its imports.

Solution Graphing calculators use x and y as variables, so you first need to rewrite the equations using x and y. (The variables T and θ, used by a graphing calculator, are for other types of graphs and should not be used for linear systems.)

$$\begin{cases} y = 1312x + 5166 & \textit{Exports} \\ y = 723x + 7429 & \textit{Imports} \end{cases}$$

After entering these equations in the calculator and setting an appropriate range, you can obtain the graph shown at the right. From the graph, you can see that the exports began to exceed the imports around 1984. (The *trace* key can help you approximate the point of intersection.) ■

Exercises

In Exercises 1–4, use a graphing calculator to sketch the graph of the system. Approximate the solution and check your answer algebraically. See margin.

1. $\begin{cases} y = -2x + 2 \\ y = \frac{1}{2}x - \frac{11}{2} \end{cases}$ **2.** $\begin{cases} y = 3x + 11 \\ y = \frac{1}{3}x + \frac{19}{3} \end{cases}$ **3.** $\begin{cases} 7x - 8y = 5 \\ 5x + 4y = 23 \end{cases}$ **4.** $\begin{cases} 10x + 11y = 24 \\ 9x - 6y = -42 \end{cases}$

5. *Per Capita Income* For 1987 through 1994, the per capita incomes, y, in Nevada and New Hampshire can be modeled by

$$\begin{cases} y = 1077t + 9035 & \textit{Nevada} \\ y = 712t + 13{,}324 & \textit{New Hampshire} \end{cases}$$

where t is the year, with $t = 7$ corresponding to 1987. Use a graphing calculator to approximate the year in which the per capita income of Nevada surpassed that of New Hampshire. ***(Source: Bureau of Economic Analysis)*** 1991

Using a Graphing Calculator **129**

Many lessons show students how to graph equations on a graphing calculator. Seeing visual representations of algebraic expressions makes algebraic concepts more meaningful to students.

Only the new ALGEBRA from D.C. Heath spells out the keystrokes so students learn the uses and nuances of graphing calculators quickly. The appendix pages 856–867 feature keystrokes for the TI-82, Casio fx-9700GE, and Sharp EL-9300C.

> "The Using a Graphing Calculator feature was wonderful! I've never seen anything written down for students to follow; it has always been 'This is what you do.' This feature gives a lot more freedom to explore."
>
> *— Deanna Mauldin,*
> *ALGEBRA field test teacher,*
> *Liberty Bell Middle School,*
> *Johnson City, TN*

ALGEBRA'S assessment tools help you evaluate the whole student.

ALGEBRA provides you with options to assess your students' total progress. You'll evaluate knowledge and mathematical power, as well as problem solving, communication, and reasoning. Throughout the program, you'll find many ways to assess students both formally and informally.

Formal assessment tools are numerous.

- **Mid-Chapter Self-Tests** in the Student Text help students assess their progress before they finish the chapter. Answers are provided for every exercise in the self-tests.
- **Mid-Chapter Tests**—provided for teachers in two forms—allow you to assess mid-chapter progress.
- **Chapter Tests** in the Student Text help students evaluate their understanding of chapter concepts.
- **Chapter Tests**—provided for teachers in three forms—allow you to assess end-of-chapter progress.
- **Independent Practice** in the Student Text reviews basic concepts directly correlated to lesson goals and lesson examples. Many of these homework exercises ask for explanations, comparisons, conclusions, and other responses that promote decision making.
- **A Computer Test Bank** allows you to generate customized tests by choosing from over 2,500 items to meet your students' individual needs.

The Standards recommend...

- Students should be given tasks that are challenging and multifaceted and that allow them to perform at their maximum level of ability.
- Assessment tools should not stress only one type of task or mode of response because this does not give an accurate indication of performance, nor does it allow students to show their individual capabilities.
- Assessment programs should have opportunities for students to show how well they have integrated their math knowledge by applying what they've learned in a larger context.
- Assessment must be more than testing—it must be a continuous, dynamic, and often informal process.

Increased attention to

▲ assessing the whole student.
▲ ongoing assessment programs.
▲ incorporating formal and alternative assessment programs.

— from the NCTM Standards

Alternative assessment allows students to demonstrate individual capabilities.

- **Communicating about Algebra**—a part of each lesson—lets you see how much your students understand as they talk about the lesson.
- **Guided Practice** helps you interact with students and judge their readiness to begin an assignment.
- **Integrated Review**, which draws on material from previous chapters relating to the current lesson, helps you measure retention and cumulative understanding. These exercises also contain *College Entrance Exam* questions that are relevant to the new material.
- **Exploration and Extension** exercises augment or extend material. Most of the exercises are appropriate for all students. In the Teacher's Edition, the exercises that present more challenge are starred.★
- **Teacher's Guide to Alternative Assessment** offers suggestions for evaluating communication skills (both oral and written), group work, and problem solving. It also includes forms for assessing student performance in these activities, and **Math Log Copymasters,** which give students opportunities to write about algebra procedures they are learning, do research, and work on open-ended problems.
- **Partner Quizzes** get students working in pairs to communicate ideas as you evaluate students' understanding.

"With ALGEBRA, teachers assess not just knowledge and mathematical power, but problem solving, communication, reasoning, and mathematical disposition—students' desire to know and do math with confidence. You assess the whole student."

—Timothy D. Kanold, ALGEBRA author

D.C. Heath's ALGEBRA provides...

an assessment program to evaluate the whole student in a continuous process with—

Formal Assessment

- Mid-Chapter Self-Tests
- Mid-Chapter Tests
- Chapter Tests
- Independent Practice
- Computer Test Bank

Alternative Assessment

- Communicating about Algebra
- Guided Practice
- Integrated Review
- Exploration and Extension
- Teacher's Guide to Alternative Assessment
- Math Log Copymasters
- Partner Quizzes

The Standards recommend...

The mathematics curriculum should include the continued development of language and symbolism to communicate mathematical ideas so that all students can—

- reflect upon and clarify their thinking about mathematical ideas and relationships.
- formulate mathematical definitions and express generalizations discovered through investigations.
- express mathematical ideas orally and in writing.
- read written presentations of mathematics with understanding.
- ask clarifying and extending questions related to mathematics they have read or heard about.

Increased attention to

▲ experience in listening to, reading about, writing about, speaking about, reflecting on, and demonstrating mathematical ideas.
▲ individual and small-group explorations that provide multiple opportunities for discussion, questioning, listening, and summarizing.
▲ acknowledging the merit of students' ideas and the importance of their own language in explaining their thinking.

—*from the NCTM Standards*

Communicating about algebra sharpens thinking.

Throughout the text, your students are given opportunities to exercise communication skills—in talking, listening, writing, representing, modeling, and reading. When students communicate about mathematical concepts, they refine their understanding. Their ability to verbalize demonstrates higher-order thinking skills in process.

Within the text, you'll find **Communicating about Algebra** features to help your students identify and share ideas relevant to *every* lesson. These exercises—designed for classroom discussion or for working in groups—offer opportunities for communication.

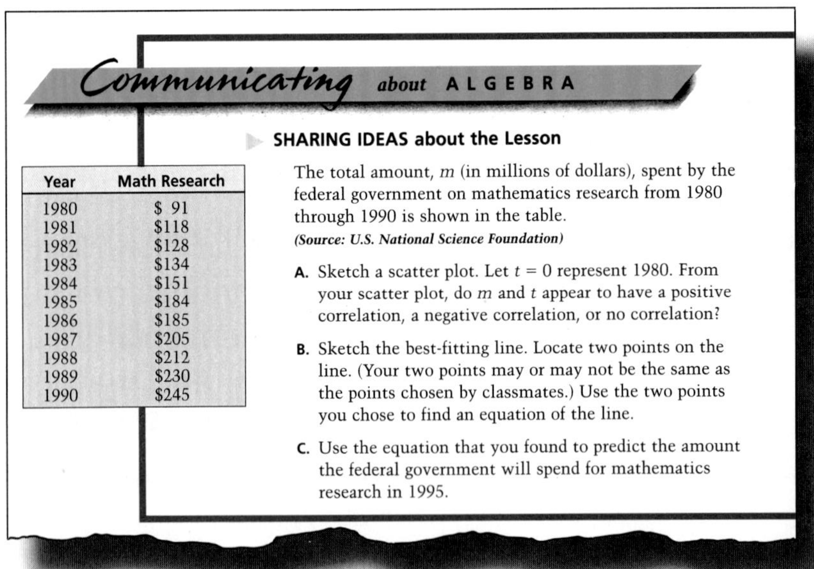

Communicating about ALGEBRA

SHARING IDEAS about the Lesson

Year	Math Research
1980	$ 91
1981	$118
1982	$128
1983	$134
1984	$151
1985	$184
1986	$185
1987	$205
1988	$212
1989	$230
1990	$245

The total amount, m (in millions of dollars), spent by the federal government on mathematics research from 1980 through 1990 is shown in the table.
(Source: U.S. National Science Foundation)

A. Sketch a scatter plot. Let $t = 0$ represent 1980. From your scatter plot, do m and t appear to have a positive correlation, a negative correlation, or no correlation?

B. Sketch the best-fitting line. Locate two points on the line. (Your two points may or may not be the same as the points chosen by classmates.) Use the two points you chose to find an equation of the line.

C. Use the equation that you found to predict the amount the federal government will spend for mathematics research in 1995.

An innovative approach to creative problem solving that promotes communication can be found in ALGEBRA's **Partner Quizzes.** Here, students work in pairs, verbalizing different ideas and listening to classmates justify their own answers. The partners confirm or reject ideas, connect others' thoughts with their own, and derive solutions based on teamwork and cooperation.

Math Logs provide opportunities for students to write their ideas about algebra. They investigate the algebra procedures they are learning, do research, and work on open-ended problem solving.

Communicating about Algebra, **Partner Quizzes,** and **Math Logs** foster active communication about algebra within your community of learners. These features also offer *you*, your students' mentor, the opportunity to monitor questions and responses.

"Communication in a classroom gets students talking about algebra. In concert, students come up with conclusions they couldn't come up with alone. When students communicate, they learn new, different perspectives—because problems don't always have just one 'right' solution."

—Lee Stiff, ALGEBRA author

D.C. Heath's ALGEBRA provides...

opportunities in every chapter to engage your students in meaningful communication—sharing, reflecting, and summarizing.

- **Communicating about Algebra** exercises are designed for classroom discussion or group work.
- **Partner Quizzes** make it possible for students to work together in pairs on algebra.
- **Math Logs** provide opportunities for students to write about algebra.

The Standards recommend...

- a core curriculum that provides a common body of mathematical ideas accessible to all students and that can be extended to meet the needs, interests, and performance levels of individual students or groups of students.
- that differences in background and ability be addressed by enrichment and extension of topics rather than by deletion.
- the use of a variety of instructional formats, such as small groups, individual exploration, peer instruction, whole-class discussion, and project work.

— from the NCTM Standards

ALGEBRA provides realistic ways to respond to diverse needs.

A basic premise of ALGEBRA is the belief that every student can succeed in algebra. It makes math concepts accessible to a wide range of students through an engaging writing style, carefully developed core concepts followed by opportunities for extension, emphasis on real-life applications of algebra, and ways to accommodate varied learning styles.

Thorough content coverage means success for all.

To reach students of all abilities, D.C. Heath's new ALGEBRA program puts the main concepts of algebra at the focus of instruction—followed by many opportunities in the exercise sets for challenge and extension. Each section of the exercise sets—Guided Practice, Independent Practice, Integrated Review, and Exploration and Extension—provides meaningful practice for every student.

Engaging writing involves students in the text.

Larson, Kanold, and Stiff write in a relaxed, encouraging style that speaks directly to the student. To get students even more involved in what they're learning, **What You Should Learn** and **Why You Should Learn It** features clearly spell out the objectives of the lesson and the usefulness of the material both in math class and in real life.

Real-life applications appeal to all.

Math problems and concepts that are based upon real-life situations give students a common basis for understanding. Because so many problems in ALGEBRA are set in real-life context, students of all abilities will find topics of interest—and will recognize the real-life value of the concepts they're learning.

ALGEBRA recognizes all learning styles.

To accommodate students of all ability levels and learning styles, ALGEBRA offers a range of activities—peer teaching, cooperative learning, and group activities. Students will find that working with others and sharing ideas are lifelong skills that go beyond algebra class.

D.C. Heath's ALGEBRA provides...

- content coverage that reaches all students and offers opportunities for extension.
- an engaging writing style, plus objectives that clarify the point of the lesson.
- real-life applications that facilitate student understanding.
- activities that recognize all abilities and learning styles.

"One of our goals in writing ALGEBRA was to show students that math is useful, interesting—and fun! Another goal was to provide enough real-life situations to give each student many opportunities to be an 'expert.' Maybe he or she has been to a location described in the text, or knows about motorcycling or music or baseball or dairy farming or raising collies—or any of the other real-life situations found in ALGEBRA"

— *Roland E. Larson, ALGEBRA author*

To walk through a lesson from ALGEBRA 2, please turn to page T20.

Relevance to real life

Each chapter begins with a two-page Opener, showing a table of contents for the chapter and a motivational example of the relevance of algebra to real life.

Linear equations and linear inequalities in two variables are common models of real-life situations. Ideas of slope and constant rates of change characterize these linear relationships.

The connection between geometry and algebra is strong and will be made throughout any discussion of linear equations and inequalities.

The Chapter Summary on page 114 provides you and the students with a synopsis of the chapter. It identifies key skills and concepts. You may want to have students look at the Chapter Summary as an overview before beginning the chapter.

CHAPTER 2

Linear Equations

LESSONS

Water-ski jumping is performed competitively in many parts of the country.

62

The concept of slope occurs in many applications of mathematics in real-life situations. The water-ski-jumping example is one among many. Others include the "pitch" of a roof of a house, the "grade" of the highway, and the "angle of descent" of an airplane.

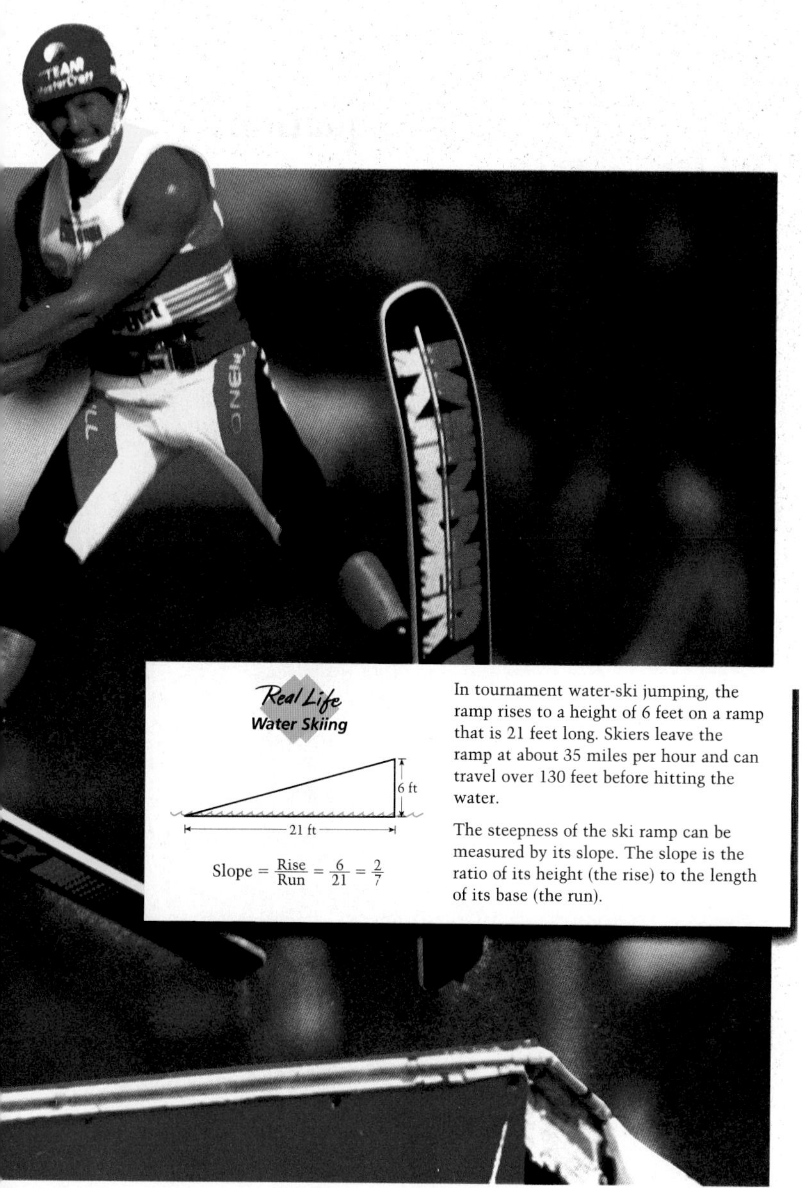

Real Life
Water Skiing

6 ft

21 ft

$$\text{Slope} = \frac{\text{Rise}}{\text{Run}} = \frac{6}{21} = \frac{2}{7}$$

In tournament water-ski jumping, the ramp rises to a height of 6 feet on a ramp that is 21 feet long. Skiers leave the ramp at about 35 miles per hour and can travel over 130 feet before hitting the water.

The steepness of the ski ramp can be measured by its slope. The slope is the ratio of its height (the rise) to the length of its base (the run).

63

Real data, not contrived

Many problems and examples in the new ALGEBRA from D.C. Heath use real data and diagrams to make algebraic concepts more meaningful to students. The applications reinforce the role algebra plays in the real world. Problems that use real data and newspaperlike graphics teach students how to interpret information in graphs they encounter in daily life.

What to Look For in Lesson 2.7*

- ▲ Stated objectives and purpose
- ▲ Problem of the Day
- ▲ Warm-Up Exercises
- ▲ Real-life examples
- ▲ Communicating about Algebra
- ▲ Assignment Guide
- ▲ Guided Practice
- ▲ Independent Practice
- ▲ Enrichment

*Lesson 2.7 is just a portion of Chapter 2.
The complete chapter, Linear Equations of Lines, begins on page 62 of this book.

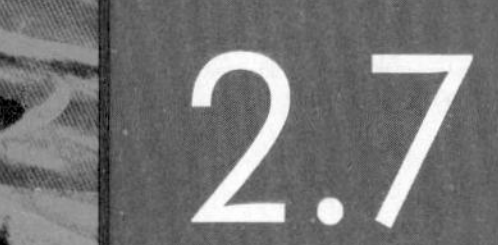

2.7 Exploring Data: Fitting a Line to Data

What you should learn:

Goal 1 How to fit a line to a set of data and to write an equation for the line

Goal 2 How to identify whether a set of data shows positive or negative correlation, or no correlation

Why you should learn it:

You can use the equation of a best-fitting line to investigate trends in data and to make predictions.

Goal 1 Fitting a Line to Data

In Lesson 2.4 you studied how to find an equation of a line that passes through *two* points. That type of problem always has only one correct solution—because two points determine exactly one line.

In this lesson, you will study problems that involve *several data points.* Usually, no single line passes through all the data points, so you try to find the line that best fits the data. We call this the **best-fitting line.** For instance, in the graph shown below, the line $y = -x + 3$ is the best-fitting line for the data points.

There are several ways to find the best-fitting line for a given set of data points. One way is to use a computer or a calculator and formulas from a branch of mathematics called *statistics.* In this lesson, however, a graphical approach will be used to *approximate* the best-fitting line. The basic steps of the graphical approach are as follows.

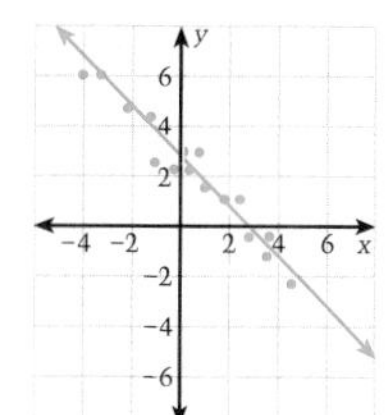

Approximating a Best-Fitting Line: Graphical Approach

To approximate the best-fitting line for a set of data points, use the following steps.

1. *Draw a Scatter Plot:* Carefully draw a scatter plot for the data.
2. *Sketch the Line:* Sketch the line that appears to most closely follow the pattern given by the points. There should be about as many points above the line as below the line.
3. *Locate Two Points on the Line:* Locate two points on the line, and approximate the *x*-coordinate and *y*-coordinate of each. (These two points do not have to be two of the original data points.)
4. *Find an Equation of the Line:* Use the technique described in Lesson 2.4 (page 80) to find an equation of the line that passes through the two points in step 3.

2.7 • *Exploring Data: Fitting a Line to Data* 107

Problem of the Day

A 12-inch square piece of wood is cut into two parts so that they cover a 9-inch by 16-inch hole. How was the cut made?

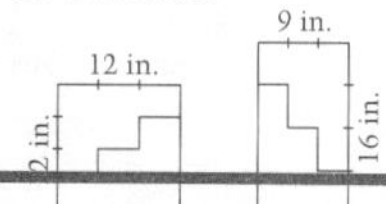

ORGANIZER

Warm-Up Exercises

Write the equation of the line that contains the given points.

1. (3, 8) and (−2, 1)
2. (0, −7) and (4, 9)
3. (1, −1) and (0, 3)

1. $y = \frac{7}{5}x + \frac{19}{5}$ 2. $y = 4x - 7$
3. $y = -4x + 3$

LESSON Notes

Example 1

Point out to students that the point (1.7, 1.5) is not a data point in the earnings-dividend list; it is a point found to be on the best-fitting line.

Be sure that students know that a "dividend" is usually a sum of money divided among the stockholders of a company.

107

Written for understanding

ALGEBRA is written in a relaxed, encouraging style that draws students into the text and helps them understand.

Organized for success

Your Teacher's Edition helps you organize each lesson. It gives you a Problem of the Day and Warm-Up Exercises to get students oriented for the lesson, and many point-of-use teaching suggestions.

Clearly stated objectives

Each lesson begins with a set of objectives that spell out the point of the lesson and the usefulness of the material both in math class and in real life.

Clear explanations

The title of an example gives students a quick overview of what the example is showing. Most steps in the examples have side comments that explain the solution steps. Notice Example 1 on this page.

Example 2

The vertical axis labeled "sleep hours" has been modified using a broken line so that the relevant information could be conveniently shown. Ask students to draw the graph without the broken line of the vertical axis.

Extra Example

Here is an additional example similar to **Examples 1–2.**

A local appliance store decided to monitor its sales of VCRs as related to their unit price. So they randomly chose 15 weeks out of the previous year. For those weeks, they obtained data on the number of VCRs that were sold and the unit sales price at which they were sold. These data can be found in the table below. Sketch a scatter plot, approximate the best-fitting line for the data, and write the equation of the best-fitting line.

Week	Unit Price (dollars) (P)	Number of VCRs (V)
1	350	45
2	360	35
3	365	30
4	325	54
5	395	25
6	400	18
7	330	35
8	335	50
9	365	45
10	390	20
11	375	45
12	350	50
13	390	19
14	400	20
15	425	17

108

Earnings (e)	Dividend (d)
1.67	1.73
1.73	1.46
1.77	1.48
1.79	1.42
1.84	1.63
1.90	1.60
1.92	1.83
1.97	1.46
1.99	1.56
1.99	1.67
2.00	1.72
2.00	1.65
2.00	1.86
2.23	1.74
2.23	1.56
2.25	1.80
2.38	2.20
2.48	1.60
2.55	2.35
2.56	2.00
2.58	1.80
2.69	2.46
2.74	2.10
2.77	1.74
2.79	2.30
3.02	1.90
3.26	1.78
3.32	2.62
3.38	2.51
3.45	2.81
3.54	2.28
3.70	2.50
3.79	2.76
4.12	2.40
4.40	2.96

Example 1 ***Approximating a Best-Fitting Line***

The 1990 earnings per share and dividends per share for 35 electric utility companies (in the central United States) are shown in the table. Approximate the best-fitting line for this data. Let e represent the earnings per share. Let d represent the dividend per share. What can you conclude from the result? ***(Source: The Value Line Investment Survey)***

Solution To begin, think of each of the data pairs as an *ordered pair.* For instance, the first pair of numbers in the table is represented by the ordered pair (1.67, 1.73). Next, draw a scatter plot for the ordered pairs. Then sketch the line that best fits the points, as shown below.

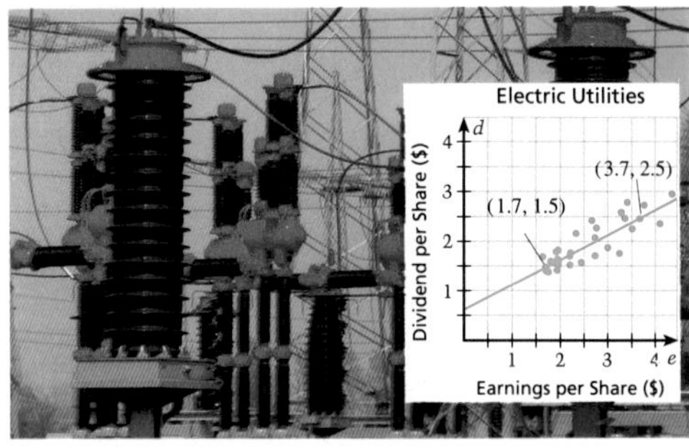

The next step is to find two points that lie on the best-fitting line. From the graph, you might choose the points (1.7, 1.5) and (3.7, 2.5). The slope of the line passing through these two points is

$$m = \frac{d_2 - d_1}{e_2 - e_1} = \frac{2.5 - 1.5}{3.7 - 1.7} = \frac{1.0}{2.0} = 0.5.$$

To find an equation of the line, use the point-slope form.

$d - d_1 = m(e - e_1)$	*Point-slope form*
$d - 1.5 = 0.5(e - 1.7)$	*Substitute values of m, e_1, d_1.*
$d = 0.5e - 0.85 + 1.5$	*Add 1.65 to both sides.*
$d = 0.5e + 0.65$	*Solve for d.*

From this equation, you can conclude two general results. As the utility companies earned more per share, they tended to pay a higher dividend per share. Another result is that the typical dividend was about half the earnings per share *plus* 65¢. ■

Normal human babies sleep more than half of their first year of life. A typical newborn sleeps about 16 hours a day, and a typical one-year-old sleeps about 13 hours a day.
(Source: Scientific American)

Connections **Physiology**

Age	Sleep	Age	Sleep
(yrs)	(hrs)	(yrs)	(hrs)
0.03	15.0	0.52	14.4
0.05	15.8	0.69	13.2
0.05	16.4	0.70	14.1
0.08	16.2	0.75	14.2
0.10	14.9	0.80	13.4
0.11	14.8	0.82	14.3
0.19	14.7	0.82	13.2
0.21	14.5	0.86	13.9
0.26	15.4	0.90	13.7
0.34	15.2	0.91	13.1
0.35	15.3	0.94	13.7
0.35	14.4	0.97	12.7
0.44	13.9	0.98	13.7
0.52	13.4	0.98	13.6

Example 2 ***Comparing Sleep to Age***

The data in the table shows the age, t (in years), and the number of hours, h, slept in a day by 28 infants who are less than one year old. Approximate the best-fitting line for this data.

Solution To begin, sketch a scatter plot as shown.

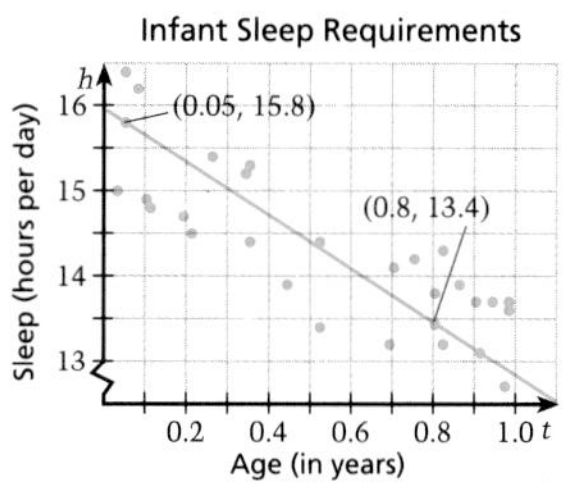

Next sketch the line that appears to fit the points best, and locate two points on the line: (0.05, 15.8) and (0.8, 13.4). The slope of the line passing through these two points is

$$m = \frac{h_2 - h_1}{t_2 - t_1} = \frac{13.4 - 15.8}{0.8 - 0.05} = \frac{-2.4}{0.75} = -3.2.$$

Use the point-slope form to find the equation of the line.

$h - h_1 = m(t - t_1)$	*Point-slope form*
$h - 15.8 = -3.2(t - 0.05)$	*Substitute values of m, t_1, h_1.*
$h = -3.2t + 0.16 + 15.8$	*Add 15.8 to both sides.*
$h = -3.2t + 15.96$	*Solve for h.* ■

Scatter Plot

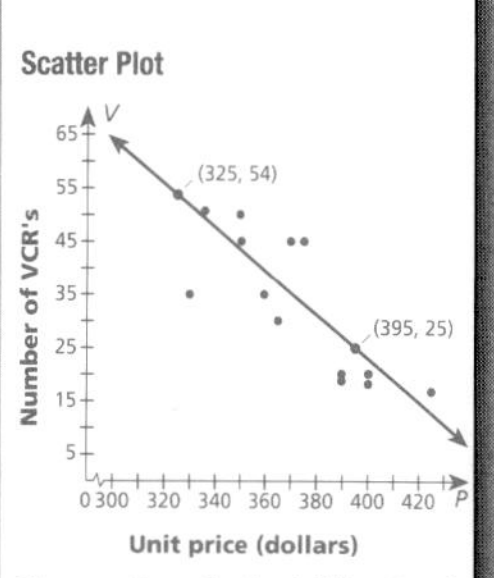

The equation of a best-fitting line is $V = -0.41P + 188.64$, where the points (395, 25) and (325, 54) were used to find the slope. Answers will vary.

Check Understanding

1. What is a scatter plot?
   a set of data points plotted on the coordinate plane
2. What are the steps in approximating best-fitting lines?
   See box on page 107.
3. For the same set of data, explain why best-fitting lines may not be identical.
   The line depends on the chosen points.
4. Describe the scatter plot of data that has a correlation coefficient of $r = 0$.
   The scatter plot cannot be approximated by a line of best fit.

Modeling with math

Real-life connections help students see math as a language for modeling the real world, not simply a language used to manipulate symbols.

Communicating about ALGEBRA

Answer

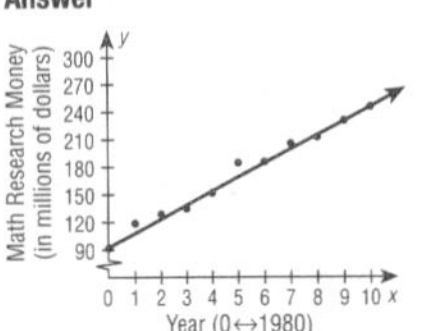

EXTEND *Communicating*

Ask students to identify real-life data that can be used to make scatter plots and best-fitting lines. Possible sources of data are school sports records, statistics from newspaper or magazine articles, or data from other classes, such as chemistry, economics, physics, biology, and history.

110

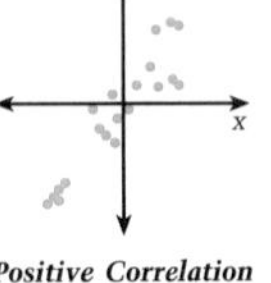

Positive Correlation

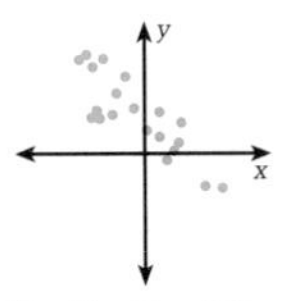

Negative Correlation

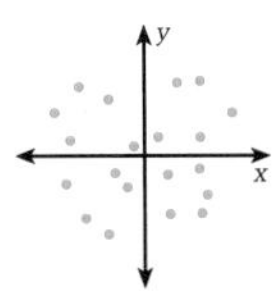

No Correlation

Goal 2 Positive and Negative Correlations

Three scatter plots are shown at the left. In the first, x and y have a **positive correlation,** which means that the points can be approximated by a line with a *positive slope.* In the second, x and y have a **negative correlation,** which means that the points can be approximated by a line with a *negative slope.* In the third, there is **no correlation** between x and y.

We have been talking about the best-fitting line. Sometimes, however, the best-fitting line isn't a very *good* fit. In statistics, the **correlation coefficient,** r, is used as a measure of how well a collection of data points can be modeled by a line. Correlation coefficients range between -1 and 1. The closer $|r|$ is to 1, the better the line fits the points. (There is more to the story than this—lines that are nearly horizontal have r-values that are close to 0. You can learn more about correlation in a statistics course.) The formula for r is

$$r = \frac{\text{Sum of } (x_i - \overline{x})(y_i - \overline{y})}{\sqrt{[\text{Sum of } (x_i - \overline{x})^2]\cdot[\text{Sum of } (y_i - \overline{y})^2]}}$$

where $\overline{x}$ is the average x-value and $\overline{y}$ is the average y-value.

Communicating about ALGEBRA

SHARING IDEAS about the Lesson

The total amount, m (in millions of dollars), spent by the federal government on mathematics research from 1980 through 1990 is shown in the table.

(Source: U.S. National Science Foundation)

Year	Math Research
1980	$ 91
1981	$118
1982	$128
1983	$134
1984	$151
1985	$184
1986	$185
1987	$205
1988	$212
1989	$230
1990	$245

A. Sketch a scatter plot. Let $t = 0$ represent 1980. From your scatter plot, do m and t appear to have a positive correlation, a negative correlation, or no correlation? positive correlation

B. Sketch the best-fitting line. Locate two points on the line. (Your two points may or may not be the same as the points chosen by classmates.) Use the two points you chose to find an equation of the line. $m = 15.125t + 91$

C. Use the equation that you found to predict the amount the federal government will spend for mathematics research in 1995. $317.88 million

Daily communication

Communicating about Algebra exercises in every lesson offer daily opportunities for students to share ideas about algebra.

EXERCISES

Guided Practice

CRITICAL THINKING about the Lesson

1. Do the points in the scatter plot at the right have a positive correlation or a negative correlation? Explain. See margin.
2. Find an equation of the line that you think best fits the points. $y = \frac{3}{4}x + 2$
3. Suppose you were given the shoe sizes, s, and the heights, h, of one hundred 25-year-old men. Do you think that s and h would have a positive correlation, a negative correlation, or no correlation? Explain. See margin.
4. Suppose you were given the incomes, i, and the percents of income, p, spent on food for one hundred families. Do you think i and p would have a positive correlation, a negative correlation, or no correlation? Explain. See margin.

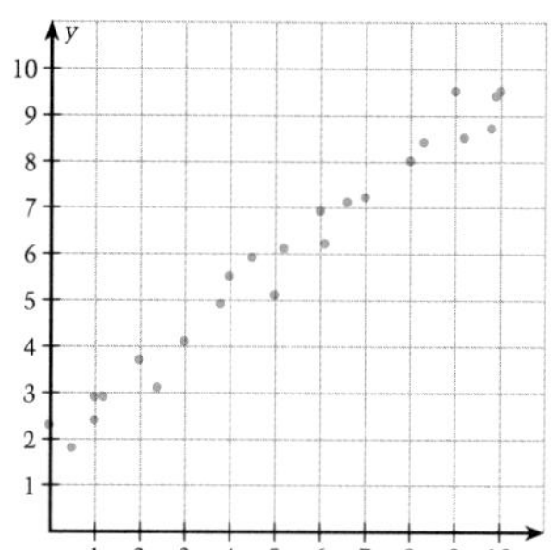

Independent Practice

In Exercises 5–8, state whether x and y have a positive correlation, a negative correlation, or no correlation.

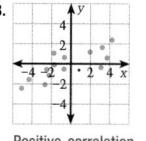

5. 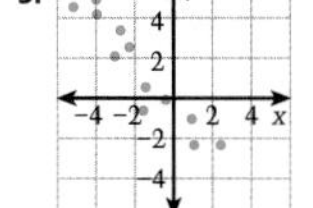
Negative correlation

6. 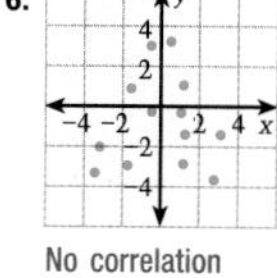
No correlation

7. 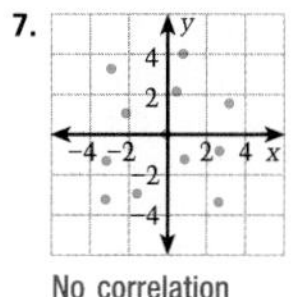
No correlation

8. Positive correlation

In Exercises 9–12, write an equation of the line that you think best fits the scatter plot. 9.–12. Equations vary.

9. 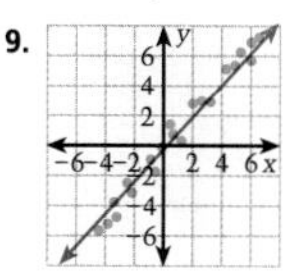
$y = \frac{5}{4}x + 1$

10. 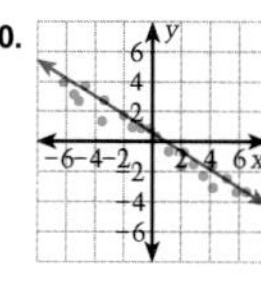
$y = -\frac{3}{5}x + \frac{2}{5}$

11. 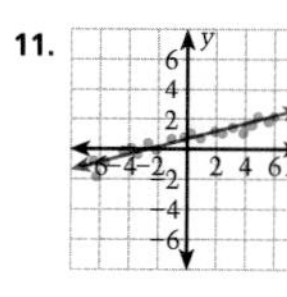
$y = \frac{1}{4}x + \frac{1}{2}$

12. 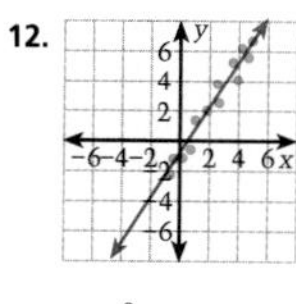
$y = \frac{3}{2}x - 1$

EXERCISE Notes

ASSIGNMENT GUIDE

Basic/Average: Ex. 1–4, 5–8, 9–13 odd, 16–19, 27–34

Above Average: Ex. 1–4, 5–8, 9–15 odd, 16–19, 27–36

Advanced: Ex. 1–4, 5–8, 9–15 odd, 16–18, 27–36

Selected Answers
Exercises 1–4, 5–25 odd

More Difficult Exercises
Exercises 27–36

Guided Practice

▶ **Ex. 4** There is no one "right" answer. Ask students to defend their answers.

Answers

1. Positive; the best-fitting line has a positive slope.
3. Positive correlation; taller men tend to have larger feet.
4. Negative correlation; the larger the income, the smaller the percent spent on food.

Independent Practice

▶ **Ex. 9–12** Remind students that they should select two points that lie on their approximated line of best fit.

Meeting individual needs

The Assignment Guide in each lesson of your Teacher's Edition helps you tailor exercises to the varying abilities of your students.

Plenty of practice

Guided Practice provides teacher-directed practice in basic algebra skills and gives teachers the opportunity to monitor students' comprehension of the material.

Varied homework assignments

Independent Practice exercises comprise the basic part of the homework assignment. They contain a balance of skill-building exercises and real-life applications. Many questions ask for explanations, comparisons, and other responses that teach decision making.

Answers

13., 14.

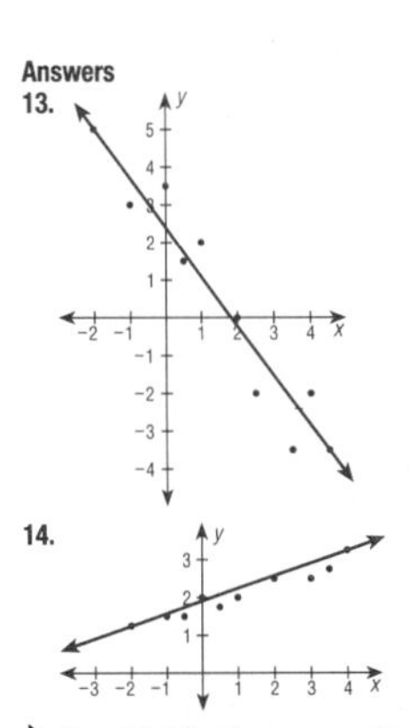

▶ **Ex. 15–18** These exercises provide real-life data models for finding the equation of the line. Ask students to use their models to predict future growth for low-fat ice cream consumption by evaluating them for the year 2000, when $t = 20$.

112

In Exercises 13 and 14, construct a scatter plot for the given data. Then find an equation of the line that you think best fits the data. See margin.

13.

x	-2	-1	0	0.5	1	2	2.5	3.5	4	4.5
y	5	3	3.5	1.5	2	0	-2	-3.5	-2	-3.5

$y = -\frac{4}{3}x + \frac{3}{2}$

13.–18. Equations vary.

14.

x	-2	-1	-0.5	0	0.5	1	2	3	3.5	4
y	1.25	1.5	1.5	2	1.75	2	2.5	2.5	2.75	3.25

$y = \frac{1}{4}x + \frac{7}{4}$

15. *Home Computers* The table lists the number of households, c (in millions), in the United States that owned computers between 1984 and 1991. Approximate the best-fitting line for this data. Let t represent the year, with $t = 4$ corresponding to 1984. If this pattern was to continue, how many households would own computers in 1993? ***(Source: Electronics Industry Association)*** $c = \frac{7}{2}t - 8$, ≈ 38 million

Year	1984	1985	1986	1987	1988	1989	1990	1991
Households	6.0	11.3	14.2	16.2	19.2	21.3	25.3	26.6

16. *Laying More Eggs* Between 1920 and 1990, improvements in poultry science produced hens that laid more eggs. The average number of eggs produced per hen per week for several years is shown in the graph. Find an equation of the line that you think best fits the data. ***(Source: United States Department of Agriculture)*** $y = \frac{1}{25}x + 2$

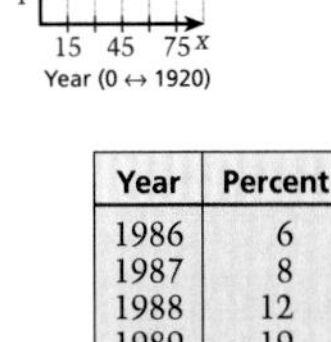

17. *Ice Cream* For 1986 through 1990, the percent, p, of low-calorie and/or low-fat ice cream sold in the United States increased as shown in the table. Approximate the best-fitting line for this data. (Let t represent the year, with $t = 6$ corresponding to 1986.) If this were to continue, what percent of ice cream sold in 1994 would be low-calorie and/or low-fat? ***(Source: A.C. Nielson)*** $p = \frac{13}{3}t - 20$, $\approx 41\%$

Year	Percent
1986	6
1987	8
1988	12
1989	19
1990	25

18. *100-Meter Sprint* A person who is running a 100-meter sprint was clocked several times during the sprint. The velocities, v (in meters per second), and the accelerations, a (in meters per second per second), are shown in the scatter plot at the right. Find an equation of the line that you think best fits this data. What can you conclude from the result? ***(Source: Mathematics in Sport ©1984 Halsted Press)***
$a = -\frac{2}{5}v + \frac{19}{5}$; the faster one runs, the harder it is to accelerate

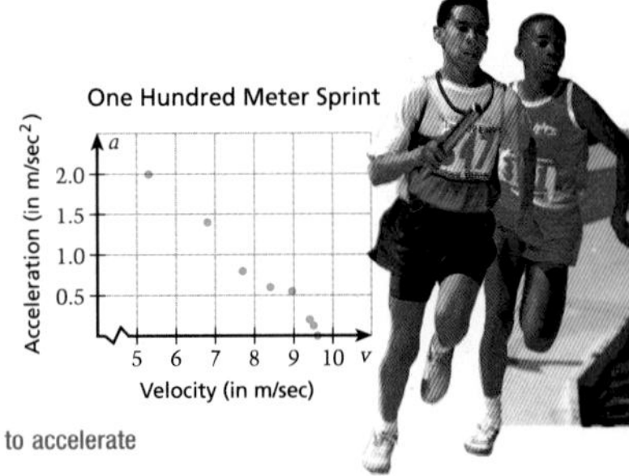

Showing algebra's value

Because students encounter real-life problems in every lesson, they will come to see how useful algebra can be as a problem-solving tool.

Problem solving process

ALGEBRA presents problem solving as a continuing process, not as a separate or single event.

Integrated Review

In Exercises 19–22, find the slope of the line containing the points.

19. (6, −9), (7, 7) 16 **20.** (6, 7), (−5, 5) $\frac{2}{11}$ **21.** (5, 0), (−5, 1) $-\frac{1}{10}$ **22.** (3, −3), (−2, 1) $-\frac{4}{5}$

In Exercises 23–26, find an equation of the line containing the points.

23. (7, 2), (8, 9) $y = 7x - 47$ **24.** (9, 4), (−8, 3) $y = \frac{1}{17}x + \frac{59}{17}$ **25.** (4, −8), (7, 5) $y = \frac{13}{3}x - \frac{76}{3}$ **26.** (2, 1), (7, −3) $y = -\frac{4}{5}x + \frac{13}{5}$

Exploration and Extension

Correlation Coefficient **In Exercises 27–30, match the correlation coefficient with the scatter plot. Explain your reasoning.** See margin.

27. $r = 0.93$ **28.** $r = 0.99$ **29.** $r = -0.64$ **30.** $r = -0.95$

a.

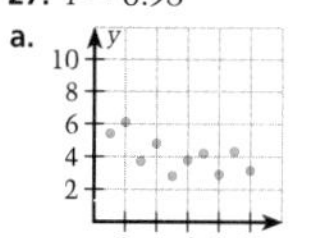

b.

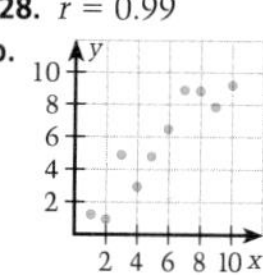

c.

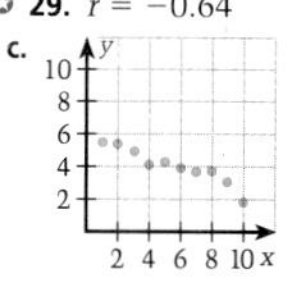

d.

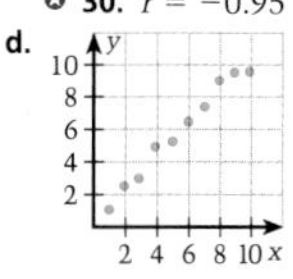

Correlation Coefficient **In Exercises 31–34, use the following information.**

The scatter plot at the right has five points. Find the correlation coefficient for these points by following the instructions in the exercises. See margin for 33.

31. Find $\bar{x}$, the mean (average) of the x-values. 20

32. Find $\bar{y}$, the mean of the y-values. 31.2

33. Complete the table and find the totals of the third, sixth, and seventh columns.

34. Use the formula given in the lesson to calculate the correlation coefficient, r. ≈0.99

x_i	$x_i - \bar{x}$	$(x_i - \bar{x})^2$	y_i	$y_i - \bar{y}$	$(y_i - \bar{y})^2$	$(x_i - \bar{x})(y_i - \bar{y})$
0	?	?	12	?	?	?
10	?	?	22	?	?	?
20	?	?	30	?	?	?
30	?	?	38	?	?	?
40	?	?	54	?	?	?

In Exercises 35 and 36, find the correlation coefficient, *r*, for the data.

35. (−5, 5), (−4, 3), (−2, 2), (−1, −1), (3, −5) ≈−0.98

36. (−20, −12), (−10, −7), (4, −3), (10, 0), (20, 5) ≈0.99

Ex. 27–30 These exercises should be assigned as a group.

Answers

27. b, slight positive correlation
28. d, strong positive correlation
29. a, slight negative correlation
30. c, strong negative correlation
33. 2nd column: −20, −10, 0, 10, 20; 3rd column: 400, 100, 0, 100, 400; 5th column: −19.2, −9.2, −1.2, 6.8, 22.8; 6th column: 368.64, 84.64, 1.44, 46.24, 519.84; 7th column: 384, 92, 0, 68, 456; totals: 1000, 1020.8, 1000

Enrichment Activity

This activity requires students to go beyond lesson goals.

TECHNOLOGY Extend the lesson by using the STAT feature of the graphics calculator. You can use this feature to enter the data points, display the scatter plot, and calculate the linear regression model $y = a + bx$ for the data, where a represents the y-intercept and b represents the slope. You can also find the correlation coefficient, r, for the data.

Consult the owner's manual that came with your graphics calculator for specific instructions for using the STAT feature. Then use the feature to check your answers to Exercises 13–18 and 34–36. How do you account for the differences? Students should understand that the graphics calculator uses a built-in series of mathematical instructions to find the line of best fit. The text suggests a more intuitive method based on data points.

Continuing review

Integrated Review exercises help students fit new material into the context of the bigger picture of algebra. The emphasis is on showing how previously learned concepts are related to the new concepts in the lesson. Samples of college entrance exam questions may be included in this section.

Enrichment resources

A wealth of enrichment activities are provided in the margins of the ALGEBRA Teacher's Edition. Writing is one of these activities. Students may be asked to discuss their thoughts in both narrative and symbolic form and to record their observations in a journal.

Extending the lesson

Exploration and Extension exercises ask all students to go beyond the lesson material in interesting ways.

Algebra—past, present, and future—is a language for everyone.

ALGEBRA seeks to make math appealing to students of all backgrounds and cultures—and to motivate all students to appreciate and succeed at algebra.

ALGEBRA provides ...

- representation of people of all backgrounds—not just in special features, but as an integral part of problems and applications.
- features that celebrate the contributions of both men and women, past and present, to the study of mathematics.

People of all walks of life are well represented.

People from all walks of life are an integral part of the settings of problems in ALGEBRA. The buildings of the Anasazi people of New Mexico and the desert etchings of the prehistoric Californians are as much a part of the book as, say, the Aztec calendar, the Great Wall of China, or the tomb of Tutankhamen.

"In ALGEBRA, the gamut of people represented is broader than usual—in special features, in the settings of problems, and in incidental learning situations."

— Lee Stiff, ALGEBRA author

Special features spotlight the role of math past and present.

Through anecdotes from the history of math, **Milestones** highlight the development of mathematics over the centuries, with time lines that show other important events of a period. By presenting men and women who find math a valuable tool in a range of careers, **Career Interviews** demonstrate that anyone can use and enjoy math.

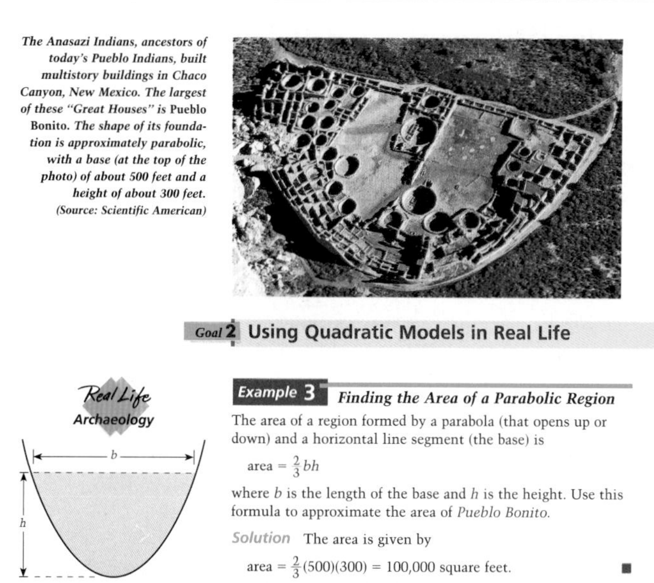

The Anasazi Indians, ancestors of today's Pueblo Indians, built multistory buildings in Chaco Canyon, New Mexico. The largest of these "Great Houses" is Pueblo Bonito. The shape of its foundation is approximately parabolic, with a base (at the top of the photo) of about 500 feet and a height of about 300 feet. (Source: Scientific American)

Goal 2 **Using Quadratic Models in Real Life**

Real Life Archaeology

Example 3 *Finding the Area of a Parabolic Region*

The area of a region formed by a parabola (that opens up or down) and a horizontal line segment (the base) is

area $= \frac{2}{3}bh$

where b is the length of the base and h is the height. Use this formula to approximate the area of *Pueblo Bonito.*

Solution The area is given by

area $= \frac{2}{3}(500)(300) = 100{,}000$ square feet.

ALGEBRA provides a wide range of settings and multicultural representations that give students the confidence that everyone can succeed in algebra.

Mixed **REVIEW**

1. Multiply: $\left(\frac{1}{2}\right)\left(\frac{2}{3}\right)\left(\frac{3}{4}\right)$. $\frac{1}{4}$
2. Add: $\frac{1}{2}+\frac{1}{3}$. $\frac{5}{6}$
3. Subtract: $\frac{1}{2}-\frac{1}{3}$. $\frac{1}{6}$
4. Divide: $\frac{1}{2}\div\frac{1}{3}$. $\frac{3}{2}$
5. What number is 14% of 35? 4.9
6. 17.6 is what percent of 80? 22%
7. Evaluate $2x^2-3x$ when $x=-1$. **(1.2)*** 5
8. Evaluate $2x^2\div 3x$ when $x=-1$. **(1.2)** $-\frac{2}{3}$
9. Multiply: $(1.25\times 10^2)(2.5\times 10^{-1})$. 31.25
10. Divide: $(1.25\times 10^2)\div(2.5\times 10^{-1})$. 500
11. Evaluate $-2x^3$ when $x=-2$. **(1.2)** 16
12. Evaluate $\left(\frac{1}{5}\right)^n$ when $n=4$. **(1.2)** $\frac{1}{625}$
13. Simplify: $4+6\div 2-1$. 6
14. Simplify: $-2\cdot 4-2$. **(1.2)** -10
15. Write in increasing order: $-2, 3, 0, -3$. **(1.1)** $-3, -2, 0, 1, 3$
16. Write in decreasing order: $-\frac{1}{2}, -\frac{1}{4}, -\frac{1}{3}, -\frac{1}{5}$. **(1.1)** $-\frac{1}{5}, -\frac{1}{4}, -\frac{1}{3}, -\frac{1}{2}$
17. Is -2 a solution of $4x^2+16=0$? No
18. Is 2 a solution of $x^1-1=0$? No
19. Write in exponential form: $2\cdot 2\cdot x\cdot x\cdot x$. **(1.2)** 2^2x^3
20. Simplify: $-3(x+4)+5(x-2)$. **(1.2)** $2x-22$

*Refers to lesson in which skill was taught. Exercises without lesson references review skills taught in previous mathematics courses.

Career Interview

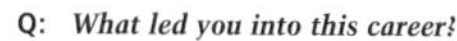

Architect

Heidi Johnson is an architect who designs buildings and the space in and around them. Her designs and drawings can range from an entire town square to the detail of a doorknob or hinge.

Q: ***What led you into this career?***

A: Before becoming an architect, I was teaching special needs students. The classrooms were often pretty depressing. That is how my interest in space and its impact on people developed. I wanted to design buildings and space that uplift or educate people rather than hinder them.

Q: ***Does having a math background help you on the job?***

A: Yes, it helps. I work with different scales to measure and draw both interior and exterior spaces. Architectural design has a lot of mathematical applications.

Q: ***What have you learned on the job?***

A: I have learned that culture often impacts the definition of space relationship. For example, in our culture a door has a specific size in relation to a room. In another country, however, that relationship may be very different. So it is necessary for us to be culturally sensitive when working on international projects.

1.2 ▪ Algebraic Expressions and Models 15

Career Interviews, which feature people of various backgrounds who use math on the job, demonstrate the utility of math in daily life.

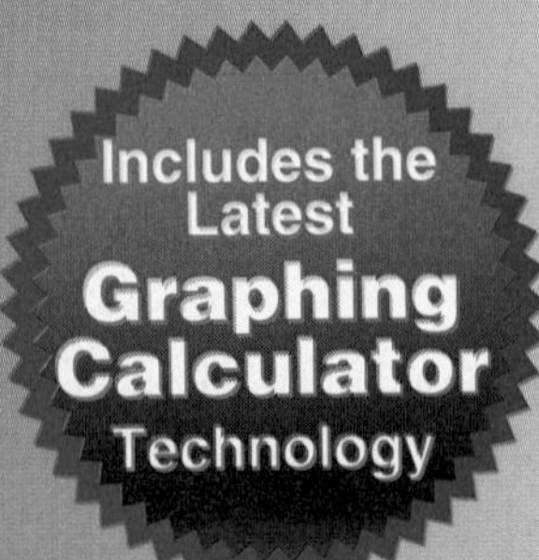

ALGEBRA makes learning and using technology easier.

In every chapter of ALGEBRA 1 and 2, there is a **Using a Graphing Calculator** feature in which students are shown how to use graphing calculators to evaluate expressions, draw scatter plots and best-fitting lines, graph equations, perform matrix operations, create visual representations of polynomial addition and subtraction, and much more.

> "Calculators and computers with appropriate software transform the mathematics classroom into a laboratory much like the environment in many science classes, where students use technology to investigate, conjecture, and verify their findings. In this setting, the teacher encourages experimentation and provides opportunities for students to summarize ideas and establish connections with previously studied topics."
>
> — *from the NCTM Standards*

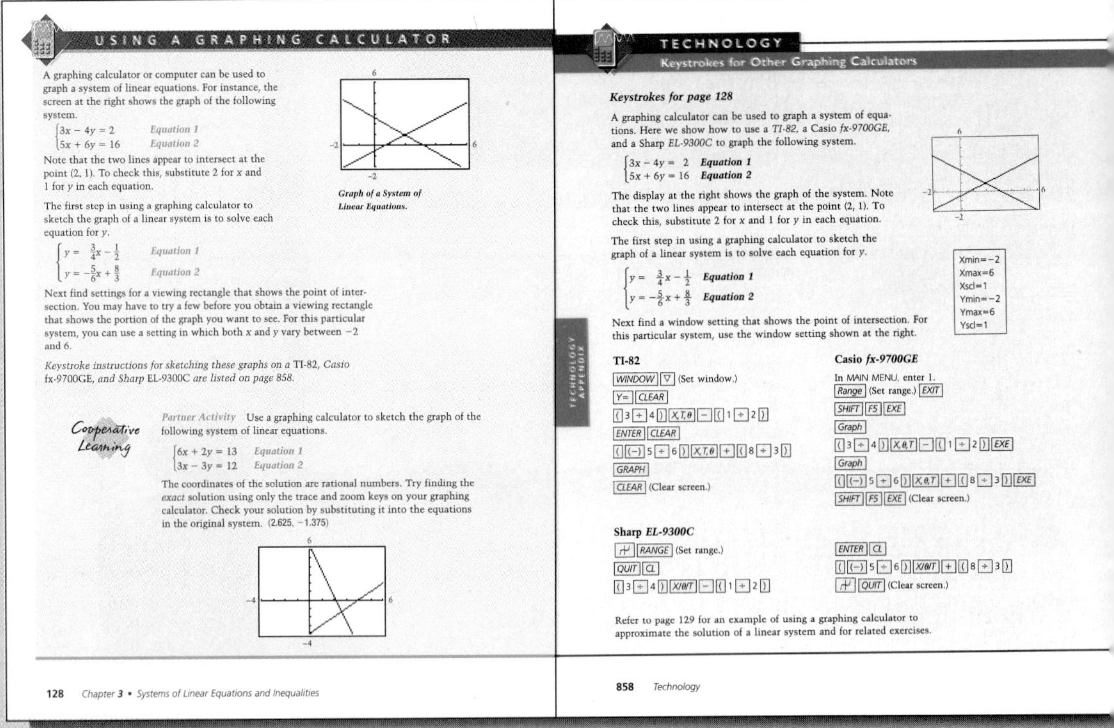

USING A GRAPHING CALCULATOR

A graphing calculator or computer can be used to graph a system of linear equations. For instance, the screen at the right shows the graph of the following system.

$$\begin{cases} 3x - 4y = 2 & \textit{Equation 1} \\ 5x + 6y = 16 & \textit{Equation 2} \end{cases}$$

Note that the two lines appear to intersect at the point (2, 1). To check this, substitute 2 for x and 1 for y in each equation.

Graph of a System of Linear Equations.

The first step in using a graphing calculator to sketch the graph of a linear system is to solve each equation for y.

$$\begin{cases} y = \frac{3}{4}x - \frac{1}{2} & \textit{Equation 1} \\ y = -\frac{5}{6}x + \frac{8}{3} & \textit{Equation 2} \end{cases}$$

Next find settings for a viewing rectangle that shows the point of intersection. You may have to try a few before you obtain a viewing rectangle that shows the portion of the graph you want to see. For this particular system, you can use a setting in which both x and y vary between -2 and 6.

Keystroke instructions for sketching these graphs on a TI-82, *Casio* fx-9700GE, *and Sharp* EL-9300C *are listed on page 858.*

Cooperative Learning

Partner Activity Use a graphing calculator to sketch the graph of the following system of linear equations.

$$\begin{cases} 6x + 2y = 13 & \textit{Equation 1} \\ 3x - 3y = 12 & \textit{Equation 2} \end{cases}$$

The coordinates of the solution are rational numbers. Try finding the *exact* solution using only the trace and zoom keys on your graphing calculator. Check your solution by substituting it into the equations in the original system. (2.625, −1.375)

128 Chapter **3** • *Systems of Linear Equations and Inequalities*

TECHNOLOGY

Keystrokes for Other Graphing Calculators

Keystrokes for page 128

A graphing calculator can be used to graph a system of equations. Here we show how to use a *TI-82*, a Casio *fx-9700GE*, and a Sharp *EL-9300C* to graph the following system.

$$\begin{cases} 3x - 4y = 2 & \textbf{Equation 1} \\ 5x + 6y = 16 & \textbf{Equation 2} \end{cases}$$

The display at the right shows the graph of the system. Note that the two lines appear to intersect at the point (2, 1). To check this, substitute 2 for x and 1 for y in each equation.

The first step in using a graphing calculator to sketch the graph of a linear system is to solve each equation for y.

$$\begin{cases} y = \frac{3}{4}x - \frac{1}{2} & \textbf{Equation 1} \\ y = -\frac{5}{6}x + \frac{8}{3} & \textbf{Equation 2} \end{cases}$$

Xmin=−2
Xmax=6
Xscl=1
Ymin=−2
Ymax=6
Yscl=1

Next find a window setting that shows the point of intersection. For this particular system, use the window setting shown at the right.

TI-82

[WINDOW] [▽] (Set window.)
[Y=] [CLEAR]
[(] 3 [÷] 4 [)] [X,T,θ] [−] [(] 1 [÷] 2 [)]
[ENTER] [CLEAR]
[(] [(−)] 5 [÷] 6 [)] [X,T,θ] [+] [(] 8 [÷] 3 [)]
[GRAPH]
[CLEAR] (Clear screen.)

Casio *fx-9700GE*

In MAIN MENU, enter 1.
[Range] (Set range.) [EXIT]
[SHIFT] [F5] [EXE]
[Graph]
[(] 3 [÷] 4 [)] [X,θ,T] [−] [(] 1 [÷] 2 [)] [EXE]
[Graph]
[(] [(−)] 5 [÷] 6 [)] [X,θ,T] [+] [(] 8 [÷] 3 [)] [EXE]
[SHIFT] [F5] [EXE] (Clear screen.)

Sharp *EL-9300C*

[⊬] [RANGE] (Set range.)
[QUIT] [CL]
[(] 3 [÷] 4 [)] [X/θ/T] [−] [(] 1 [÷] 2 [)]
[ENTER] [CL]
[(] [(−)] 5 [÷] 6 [)] [X/θ/T] [+] [(] 8 [÷] 3 [)]
[⊬] [QUIT] (Clear screen.)

Refer to page 129 for an example of using a graphing calculator to approximate the solution of a linear system and for related exercises.

858 *Technology*

Many lessons show students how to graph equations on a graphing calculator. Seeing visual representations of algebraic expressions makes algebraic concepts more meaningful to students.

In the **Using a Graphing Calculator** feature, students learn to use a graphing calculator to analyze data such as the data on the imports and exports of Puerto Rico given in this lesson. Some lessons provide a BASIC program that performs equivalent operations on a computer. In the appendix, keystrokes are given for the TI-82, Casio fx-9700GE and Sharp EL-9300C along with two new applications for use with these calculators.

ALGEBRA components give you a variety of options to support your teaching style.

❶ The **Teacher's Edition** provides you with extensive teaching support. This easy-to-use, extended-margin text helps define essential curriculum and instructional goals.
❷ **Color Transparencies for Real-Life Applications** provide 78 full-color visuals from the Student Text, such as maps, technical drawings, and statistical graphs, which facilitate discussion of examples and exercises.
❸ **Applications Handbook** provides more information about specific topics in the Student Text, such as astronomy, chemistry, physics, sports, economics, genetics, and music.
❹ **Cultural Diversity Extensions** offers projects for each chapter, including historical and contemporary investigations, library research, and interviews.
❺ **Algebra 2 Investigations for Performance Assessment** consists of eight long-term investigations providing rich opportunities for students to exercise mathematical thinking.
❻ **Technology: Using Calculators and Computers** consists of a Teacher's Guide and Copymasters for using graphing calculators, spreadsheets, and BASIC programs.
❼ **Technology Update** provides keystrokes for the TI-82, Casio fx-7700GE, Casio fx-9700GE, and Sharp EL-9300C graphing calculators to be used in conjunction with the Student Text and *Technology: Using Calculators and Computers.* New activities using these calculators are also included.
❽ **Interactive Real-Life Investigations** (CD-ROM and diskette versions) provides high-interest interactive investigations for solving real-world problems using rich data sets. One investigation is correlated to each chapter.
❾ **Formal Assessment** provides two forms of Mid-Chapter Test, three forms of Chapter Tests, and Practice for College Entrance Tests.
❿ **Alternative Assessment** suggests ways of using portfolios, oral presentations, communication skills, models, group projects, and problem-solving situations and includes forms for assessing student performance in these activities. Also includes Math Log copymasters that provide opportunities for students to write about algebra, to engage in research, and to solve open-ended problems.
⓫ **Computer Test Bank** (for Macintosh, Apple, or IBM) with graphic capability allows you to create customized tests by choosing from over 2,500 test items.
⓬ **Teaching Tools: Transparencies and Copymasters** include transparencies for number lines, coordinate grids, parabolas, rational functions, etc., plus copymasters for Problem of the Day Exercises, Warm-Up Exercises, and Answer Masters. The transparencies are in addition to the 78 full-color *Transparencies for Real-Life Applications.*
⓭ **Lesson Plans**
⓮ **Extra Practice Copymasters** offer a worksheet for every lesson with additional exercises like those found in the Student Text.
⓯ **Reteaching Copymasters** include teacher-directed and independent activities correlated to the Mid-Chapter Self-Tests and Chapter Tests.
⓰ **Complete Solutions Manual** includes step-by-step solutions for all exercises found in the Student Test.

Suggested Pacing Chart

*Full Course	*Basic Course
Chapter 1 **Review of Basic Algebra** 8 lessons 8 teaching days	Chapter 1 **Review of Basic Algebra** 8 lessons 9–10 teaching days
Chapter 2 **Linear Equations** 7 lessons 7–9 teaching days	Chapter 2 **Linear Equations** 7 lessons 9–12 teaching days
Chapter 3 **Systems of Linear Equations and Inequalities** 6 lessons 6–9 teaching days	Chapter 3 **Systems of Linear Equations and Inequalities** 6 lessons 10–13 teaching days
Chapter 4 **Matrices and Determinants** 7 lessons 7–8 teaching days	Chapter 4 **Matrices and Determinants** 6 lessons 10–14 teaching days
Chapter 5 **Quadratic Equations and Parabolas** 7 lessons 7–10 teaching days	Chapter 5 **Quadratic Equations and Parabolas** 7 lessons 10–13 teaching days
Chapter 6 **Functions** 7 lessons 7–10 teaching days	Chapter 6 **Functions** 7 lessons 11–15 teaching days
Chapter 7 **Powers, Roots, and Radicals** 6 lessons 6–8 teaching days	Chapter 7 **Powers, Roots, and Radicals** 6 lessons 11–14 teaching days
Chapter 8 **Exponential and Logarithmic Functions** 7 lessons 7–10 teaching days	Chapter 8 **Exponential and Logarithmic Functions** 6–7 lessons 12-14 teaching days
Chapter 9 **Polynomials and Polynomial Functions** 7 lessons 7–10 teaching days	Chapter 9 **Polynomials and Polynomial Functions** 6–7 lessons 12–14 teaching days
Chapter 10 **Rational Functions** 6 lessons 6–9 teaching days	Chapter 10 **Rational Functions** 5–6 lessons 10–13 teaching days
Chapter 11 **Other Quadratic Relations** 6 lessons 7–8 teaching days	Chapter 11 (Optional) **Other Quadratic Relations** 6 lessons 9 teaching days
Chapter 12 **Sequences and Series** 6 lessons 6–10 teaching days	Chapter 12 **Sequences and Series** 6 lessons 10-13 teaching days
Chapter 13 **Trigonometric Ratios and Functions** 6 lessons 8–11 teaching days Chapter 14 **Trigonometric Identities and Equations** 6 lessons 10–12 teaching days	To complete the coursework WITH TRIGONOMETRY, teach selected lessons from Chapters 13 and 14: Chapter 13 **Trigonometric Ratios and Functions** Chapter 14 **Trigonometric Graphs, Identities, and Equations**
Chapter 15 **Probability and Statistics** 6 lessons 11 teaching days	To complete the coursework WITHOUT TRIGONOMETRY, teach the entire chapter or selected lessons from: Chapter 15 **Probability and Statistics**
Total (average) 126 teaching days **Assessment** 45 days **Cumulative Reviews** 5 days **Total Course (average)** 176 days	**Total (average)** 131 teaching days **Assessment** 45 days **Cumulative Reviews** 5 days **Total Course (average)** 181 days

*See additional suggestions for pacing on page T35.

Suggestions for Pacing

Full Course:

Chapter 2 Allow extra time for Lesson 2.4 on writing linear equations, an important problem-solving tool, and for fitting a line to a scatter plot in Lesson 2.7.

Chapter 3 Allow extra time for developing the techniques for solving and using linear systems as shown in Lessons 3.2 and 3.6.

Chapter 4 Allow extra time for using inverse matrices as shown in Lessons 4.5.

Chapter 5 Allow extra time for developing the techniques for solving quadratic equations as shown in Lessons 5.2, 5.3, and 5.6.

Chapter 6 Allow extra time for introducing inverse functions in Lesson 6.3 and and for recursive functions in Lesson 6.6.

Chapter 7 Allow extra time for introducing rational exponents in Lessons 7.3 and graphing the square-root and cube-root functions as shown in Lesson 7.6.

Chapter 8 Allow extra time for the introduction of the natural base e in Lesson 8.5, and for the application of the properties of logarithms in solving equations as shown in Lessons 8.3 and 8.6.

Chapter 9 Allow extra time for factoring and dividing polynomials as shown in Lessons 9.3 and 9.4.

Chapter 10 Allow extra time for the development of the applications of rational functions such as variation, and complex fractions as shown in Lessons 10.1, 10.2, and 10.5.

Chapter 11 Allow extra time for translating and classifying conic sections as shown in Lessons 11.5 and 11.6.

Chapter 12 Allow extra time for developing arithmetic and geometric sequences and series as shown in Lessons 12.2 and 12.3, and for the applications of the binomial theorem as shown in Lesson 12.5.

Chapter 13 Allow extra time for introducing trigonometric functions as shown in Lessons 13.3 and 13.4.

Algebra 2 has been successfully taught as a two-year course in some schools.

To subdivide the course into two years, it is suggested that chapters 1–7 be taught in the first year. The second year can begin with several weeks of review of Chapters 1–7 before beginning the six new chapters for this course. These new chapters would be Chapters 8–12 and 15.

If this model seems to fit your situation, you may want to use some of these ideas for pacing:

- Spending extra time on the lessons in the first half of a chapter helps ensure that students understand the basic concepts.
- A full day spent on each graphing calculator lesson will ensure that all students can use this technology comfortably.
- Extra time can be spent on the cumulative reviews, in preparation for end-of-semester tests.

Basic Course:

Chapter 1 Allow extra time for Lesson 1.4, which introduces the problem-solving strategies that will be emphasized throughout the text.

Chapter 2 Allow extra time for important topics of solving, writing, and graphing linear equations as shown in Lessons 2.3, 2.4, 2.6, and 2.7.

Chapter 3 Allow extra time for developing the techniques of solving systems of linear equations as shown in Lessons 3.1, 3.2, 3.3, and 3.6, which are the focus of the chapter, and for introducing linear programming as shown in Lesson 3.5.

Chapter 4 Omitting Lesson 4.7 on Cramer's rule will allow extra time for introducing determinants as shown in Lesson 4.3. Lessons 4.5 and 4.6 on inverse and augmented matrices could be extended to 4 days each.

Chapter 5 Allow extra time for developing the techniques for solving and graphing quadratic equations as shown in Lessons 5.2, 5.3, 5.6, and 5.7. Lesson 5.6, which summarizes the various techniques, could be extended to 3 days.

Chapter 6 Allow extra time for developing and using functions as shown in Lessons 6.3–6.7. Lesson 6.6 on recursive functions could be extended to 3 days.

Chapter 7 Allow extra time for real-number and radical function development as shown in Lessons 7.2–7.6. Lesson 7.3 on n-th roots and Lesson 7.6 on graphs of radical functions could be extended to 3 days each.

Chapter 8 Omitting Lesson 8.7 on the logistics growth function will allow extra time for other exponential and logarithmic topics as shown in Lessons 8.2–8.6. Lesson 8.5 on natural logarithms and Lesson 8.6 on solving exponential and logarithmic equations could be extended to 3 days each.

Chapter 9 Omitting Lesson 9.7 on dispersion will allow extra time for introducing polynomial operations and graphs as shown in Lessons 9.2–9.6. Lesson 9.3 on factoring techniques could be extended to 3 days.

Chapter 10 Omitting Lesson 10.6 on the algebra of finance will allow extra time for introducing rational function operations and graphs as shown in Lessons 10.1, 10.2, 10.4, and 10.5. Lesson 10.2 on inverse and joint variation could be extended to 3 days.

Chapter 11 Omitting this chapter, which introduces quadratic relations and the conic sections, serves to balance the extra time allotted to the Basic Course in the earlier chapters.

Chapter 12 Allow extra time for the development of topics using sequences and series as shown in Lessons 12.2–12.6.

Chapter 13–15 If completing the Basic Course with trigonometry, use selected lessons from Chapter 13 and 14 which introduce selected topics from functional and right-triangle trigonometry. If completing the course without trigonometry, use some or all of the probability and statistics lessons from Chapter 15.

HEATH

ALGEBRA 2

AN INTEGRATED APPROACH

Roland E. Larson

Timothy D. Kanold

Lee Stiff

About the Cover

The cover shows mathematics used in aeronautics and air safety. The background is a photograph of an airplane above the clouds in a rainbow sky. The photo inset shows two air traffic controllers at work in an airport's air traffic control station. The equations $B = 10 \log_{10} \frac{I}{I_0}$ and $I = \frac{10}{r^2}$ relate the intensity of sound, I, to the level of sound, B, and the distance from the source of the sound, r. Knowing such relationships is important when designing safe working conditions for airport employees. A person's hearing could be permanently damaged if the level of sound is too great. The equation $S = 136.4\sqrt{p} + 4.5$ relates the indicated airspeed, S, of an airplane to the difference, p, between the static and dynamic pressures. Understanding such relationships is an important part of preparation for a career as an airplane pilot or an aeronautical engineer.

As you read the text, you will find information that is directly related to the cover.

Acknowledgments

Editorial Development Jane Bordzol, Rita Campanella, Anne M. Collier, Peter R. Devine, Tamara Trombetta Gorman, Marc Hurwitz, Albert Jacobson, Barbara M. Kelley, Savitri Kaur Khalsa, Pearl Ling, Marlys Mahajan, James O'Connell, George J. Summers, Folkert Van Karssen

Marketing/Advertising Jo DiGiustini, Phyllis D. Lindsay, Richard Ravich, Debbie R. Secrist

Design Robert Botsford, Pamela Daly, Leslie Dews, Robin Herr, Cynthia Maciel, Joan Williams

Production Sandra Easton, Lalia Nuzzolo, Mark Tricca

Author Acknowledgments

In Algebra 2, we have included numerous examples and exercises that use real-life data. This would not have been possible without the help of many people and organizations. Our wholehearted thanks goes to all for their time and effort.

We tried to use the data accurately—to give honest and unbiased portrayals of real-life situations. In the cases where models were fit to data, we used the least-squares method. In all cases, the square of the correlation coefficient, r^2, was at least 0.95. In most cases, it was 0.99 or greater.

(Acknowledgments continue following the Index.)

Published simultaneously in Canada

Printed in the United States of America

International Standard Book Number: 0-669-43394-2

1 2 3 4 5 6 7 8 9 10 VHP 02 01 00 99 98 97 96

Roland E. Larson is Professor of Mathematics at the Behrend College of Pennsylvania State University in Erie. He is a member of NCTM and a highly successful author of many high school and college mathematics textbooks. Several of his D.C. Heath college titles have proved to be top choices for senior high school mathematics courses.

Timothy D. Kanold is Chairman of Mathematics and Science and a teacher at Adlai Stevenson High School in Prairie View, Illinois. He is the 1986 recipient of the Presidential Award for Excellence in Mathematics Teaching, as well as the 1991 recipient of the Outstanding Young Alumni Award from Illinois State University. A member of NCTM, he served on NCTM's *Professional Standards for Teaching Mathematics* Commission and is a member of the Regional Services Committee.

Lee Stiff is an Associate Professor of Mathematics Education in the College of Education and Psychology of North Carolina State University at Raleigh. He has taught mathematics at the high school and middle school levels. He is a member of the NCTM Board of Directors, served on NCTM's *Professional Standards for Teaching Mathematics* Commission, and is a member of the *Committee for a Comprehensive Mathematics Education of Every Child.*

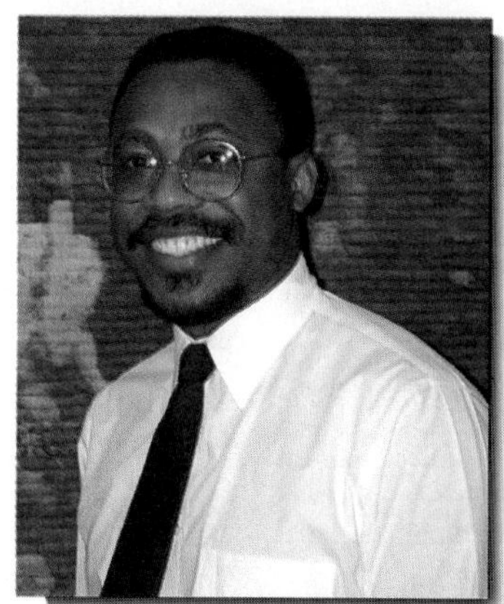

REVIEWERS AND CONTRIBUTORS

Linda Bailey
Putnam City Schools
Oklahoma City, OK

Pat Beck
Putnam City North High School
Oklahoma City, OK

Larry Demetrak
San Marin High School
Novato, CA

Charleen M. DeRidder
Knox County Schools
Knoxville, TN

Melanie Donnell
Russellville High School
Russellville, AR

Joseph P. Dooley
Upper Darby High School
Upper Darby, PA

David J. Enterlin
Upper Darby High School
Upper Darby, PA

M. Jayne Fleener
University of Oklahoma
Norman, OK

Gail Gonyo
Adirondack Community College
Queensbury, NY

Wanda Heflin
Russellville High School
Russellville, AR

Karen Humble
Western Oaks Junior High School
Oklahoma City, OK

Regina D. Kiczek
Ferris High School
Jersey City, NJ

Fran Kormes
Harry S. Truman High School
Levittown, PA

Chicha Lynch
Capuchino High School
San Bruno, CA

Elizabeth J. Manes
Harry S. Truman High School
Levittown, PA

Deanna Mauldin
Liberty Bell Middle School
Johnson City, TN

Terry McBroom
San Marin High School
Novato, CA

Dominic Mendalla
Capuchino High School
San Bruno, CA

Dennis H. Noe
School District of Waukesha
Waukesha, WI

Dean Orfanedes
Capuchino High School
San Bruno, CA

Jennifer J. Penley
Liberty Bell Middle School
Johnson City, TN

Susan Shumate
Edmond Public Schools
Edmond, OK

Sharon Siegel
Dodgen Middle School
Marietta, GA

Linda Tassano
San Marin High School
Novato, CA

John Wellington
Upper Darby High School
Upper Darby, PA

There are many ways to study algebra. After years of teaching and guiding students through algebra courses, we have compiled a list of learning ideas that will help you take responsibility for your own success. These ideas will take some time and effort—but they work!

Do the Homework: Learning algebra is like learning to play the piano or basketball. You cannot become skilled by just watching someone else do it. You must also do it yourself. When you start a new exercise set, understanding is much more important than speed. Your *ultimate* goal is to be able to solve the problems accurately and efficiently.

Prepare for Class: After completing your homework assignment, and before attending your next class, read the next portion of the text that is to be covered. This takes a lot of self-discipline, but it pays off.

Keep a Journal: Each day do a self-analysis in which you reflect on your learning of algebra. Ask yourself these questions: What were the major goals and tasks of the class? Why did I learn it? What procedures should I know to successfully do the problems? This type of reflection, combined with your completed attempt at the homework, will help to develop your confidence.

Participate in Class: As you are reading the text, communicating in class, or practicing problems, write down any questions you have about the material. Then ask your teacher.

Taking Notes: Take notes in class, especially on definitions, rules, and model examples. Then, as part of your homework, read through your notes—adding any explanations that are necessary to make your notes understandable to you. Highlight important aspects of the lesson, too.

Find a Study Partner: When you get stuck on a problem, it helps to be able to try to work with someone else. Even if you think you are giving more help than you are getting, you will find that teaching others is one of the best ways to learn.

Remember, math is not a spectator sport—it's a valuable tool you can use in everyday life! We believe that our book will help you to see, feel, and understand the wonderful value of Algebra.

ROLAND E. LARSON TIMOTHY D. KANOLD LEE STIFF

Real-Life Applications

Look through this list for things that interest you. Then find out how they are used with algebra.

"Modeling puts a mathematical overlay on the real world. It's attainable, reasonable, straight-forward, and it fits with the topics at hand. Real-world connections make ALGEBRA meaningful for all students and promotes success at all levels."

Lee Stiff Timothy D. Kanold Roland E. Larson

A Variety of Problem-Solving Approaches Throughout the program, **Algebra 2** provides many ways to look at problems in algebra: algebraically, graphically, using tables and charts, through real-life modeling, using reasoning, and using technology. Students learn to use algebra to solve problems in the real world.

Emphasis on Graphing **Algebra 2** emphasizes graphical representations of algebra. Throughout the text, students have many opportunities to make connections between algebraic, tabular, and graphical models. Lessons frequently explore algebraic concepts with the graphical techniques and formulas of coordinate geometry.

TABLE OF CONTENTS

Problem Solving in Every Lesson Chapter 1 reviews the key problem solving steps used throughout Algebra 1. Chapters 1–3 review the first seven chapters of Algebra 1 and introduce additional enriching models and applications.

CHAPTER 1 Review of Basic Algebra

Emphasis on Graphing and Coordinate Geometry Chapter 2 revisits the graphing techniques of quick graphs, coordinate geometry, and the power of graphing calculators introduced and developed in Algebra 1 and integrates them throughout Algebra 2.

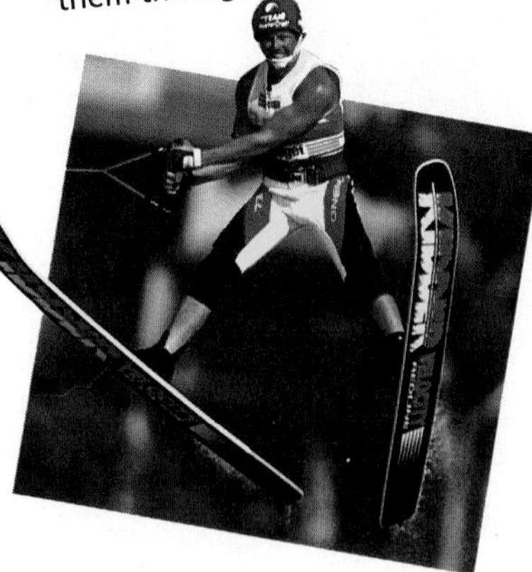

CHAPTER 2 Linear Equations

C Integrates Coordinate Geometry D Integrates Data Analysis
F Integrates Functions G Integrates Geometry

Integrating Technology
Algebra 2 integrates graphing calculator technology throughout the program in lessons, in exercises, and in the "Using a Graphing Calculator" feature — one per chapter — allowing you to continually explore, discover, and reinforce algebraic concepts.

CHAPTER 3 Systems of Linear Equations and Inequalities

CHAPTER 4 Matrices and Determinants

Visualizing Concepts: A Critical Problem-Solving tool.
Algebra 2 integrates a visual approach to the study of functions. Early on, students review the quick graphs and the techniques of coordinate geometry developed in **Algebra 1**. Students continue to explore equations visually first—by graphing them—then they solve them. This approach helps students grasp concepts and strengthens reasoning skills.

The Power of Technology
Since coordinate geometry is integrated throughout **Algebra 2**, so are graphing calculators. Introducing graphing calculators early encourages students to make connections between algebraic and graphical techniques and ultimately to use technology as a problem-solving tool.

Integrating Algebra and Geometry
In all chapters, algebra and geometry — both metric and coordinate geometry — are integrated together to strengthen important skills, develop concepts, and pave the way for future study of mathematics.

Emphasis on Mathematical Modeling
Algebra 2 is driven by the unifying idea of mathematical modeling and uses algebra to model data through the study of functions. Students explore mathematical modeling through a variety of functions: linear, exponential, quadratic, rational, general, polynomial, logarithmic, and trigonometric.

CHAPTER 5 Quadratic Equations and Parabolas

CHAPTER 6 Functions

C Integrates Coordinate Geometry D Integrates Data Analysis
F Integrates Functions G Integrates Geometry

Integrating Data Analysis
In Chapter 7, the "Exploring Data" lesson, as in most of the chapters in Algebra 2, integrates authentic data to model real life and to develop data analysis techniques.

CHAPTER 7

Powers, Roots, and Radicals

Modeling Algebra in Real Life
Algebra 2 uses an abundance of applications to model real life and to develop critical thinking. The use of real-life data and graphs makes algebraic concepts more meaningful to students—and reinforces the role algebra plays in the real world.

Modeling Algebra in Real Life
Algebra 2 uses an abundance of applications to model algebra in real life and to develop critical thinking. Chapter 8 explores the laws of exponents and follows with the study of exponential and logarithmic functions.

CHAPTER 8

Exponential and Logarithmic Functions

Connections Across Content Strands
Algebra 2 provides connections within mathematics—between data analysis, functions, algebra, geometry, and trigonometry—and to other disciplines, such as social sciences, physical sciences, music, and many more.

CHAPTER 9

Polynomials and Polynomial Functions

CHAPTER 10

Rational Functions

C Integrates Coordinate Geometry D Integrates Data Analysis
F Integrates Functions G Integrates Geometry

CHAPTER 11 Quadratic Relations

CHAPTER 12 Sequences and Series

Emphasis on Communication
Communicating about Algebra—a feature found in every lesson—offers opportunities for students to share their ideas about the lesson. Other vehicles for communicating and reasoning are found in the Teacher's Edition and Alternative Assessment the package.

Trigonometry
Chapters 13 and 14 introduce trigonometric functions — with special emphasis on modeling real-life applications using trigonometry.

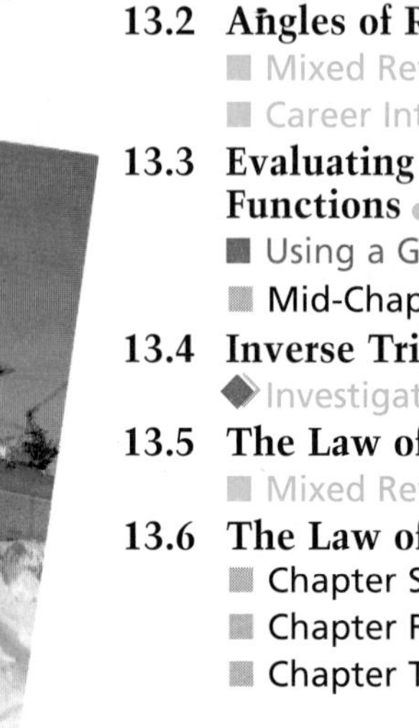

CHAPTER 13

Trigonometric Ratios and Functions

CHAPTER 14

Trigonometric Graphs, Identities, and Equations

Authentic Assessment
The program's assessment strand provides a variety of options for assessing students' total progress: their knowledge; their mathematical power; their problem-solving, communication, and reasoning skills; and ultimately their mathematical confidence.

C Integrates Coordinate Geometry D Integrates Data Analysis
F Integrates Functions G Integrates Geometry T Integrates Trigonometry

Probability
Chapter 15 explores the concepts of probability and statistics.

CHAPTER 15 Probability and Statistics

Technology Appendix
The **Technology Appendix** provides keystrokes for the *TI-82,* Casio *fx-9700GE,* and Sharp *EL-9300C* to be used in conjunction with the *Using the Graphing Calculator* feature as well as in the problem sets throughout the text. The appendix also features new applications for use with these calculators.

In the Teacher's Edition:

Interactive Technology

The following program is available for use with this textbook:

■ Larson Interactive Real-Life Investigations for Algebra 2

available on CD-ROM or diskette (Macintosh or Windows)

Chapter 1	Animal Voices and Hearing	**Chapter 2**	Working Men and Women
Chapter 3	Retail Sales of Companies	**Chapter 4**	Cryptograms
Chapter 5	Fractals	**Chapter 6**	Communication
Chapter 7	Growth of Animals	**Chapter 8**	Half-Life and Radioactivity
Chapter 9	Volume of a Box	**Chapter 10**	Air Resistance and Parachutes
Chapter 11	Comets	**Chapter 12**	Banking and Personal Finance
Chapter 13	Skyscrapers	**Chapter 14**	Meteorology
Chapter 15	Genes		

A distinctive feature of the text is its readability. Lessons provide clear and concise instruction. You may choose to build class presentations around them. Encourage students to read the text. It has been found that when you use the textbook lessons as the basis for your own lessons, students find it easier to read and follow classroom instructions.

Chapter 1 begins by reviewing the basic features of algebra. Stress that these same features will be used to help students interpret the concepts and skills that they will encounter in Algebra 2.

The Chapter Summary on page 56 provides you and the students with a synopsis of the chapter. It identifies key skills, concepts, and vocabulary. You may want to have students look at the Chapter Summary as an overview before beginning the chapter.

CHAPTER 1

Review of Basic Algebra

LESSONS

The National Aeronautics and Space Administration (NASA) is working on a replacement for the space shuttle. The current shuttle is propelled into orbit by expensive rockets that are used only once. NASA is designing a shuttle system whose rockets could be reused.

Real Life
Temperature Conversions

When the space shuttle leaves its orbit and enters Earth's atmosphere, the air friction produces tremendous heat. Temperatures on the wings can reach 2750°F. To write degrees Fahrenheit as degrees Celsius, you can use the formula $C = \frac{5}{9}(F - 32)$.

$$C = \frac{5}{9}(2750 - 32) = 1510^\circ$$

To write from degrees Celsius as degrees Fahrenheit, use the formula $F = \frac{9}{5}C + 32$.

$$F = \frac{9}{5}(1510) + 32 = 2750^\circ$$

An emphasis in this Algebra 2 text will be on making connections between mathematics and real-life situations and activities. The example given here is but one of the many that will be used in developing lessons, piquing student interest, evaluating student progress, and promoting student communication in algebra, especially as an answer to the question, Why do we have to study this?

Encourage students to ask questions about the mathematics of real-life situations found in their lessons. Questions may be asked during class or written in student math journals for consideration at a later time. For example, one of the many ways in which algebra is used is in establishing relationships between real-life measurements. The relationship between degrees Celsius and degrees Fahrenheit is modeled by $C = \frac{5}{9}(F - 32)$. This mathematical model may be used by students in their chemistry or biology classes.

Ask students to rewrite the following measurements in degrees Fahrenheit as degrees Celsius.

a. 0°F **b.** 98.6°F
c. 212°F **d.** 32°F

a. −18°C b. 37°C
c. 100°C d. 0°C

Students may wish to discuss the significance of the temperatures listed.

Review of Basic Algebra

PACING CHART

The daily Pacing Chart is meant to help you adjust your teaching pace. Students in the full course should finish the entire text by the end of the year. Students in the basic course are expected to complete the first twelve chapters. The Pacing Chart for each chapter contains suggestions for lessons that require more than one day and lessons that may be omitted for the basic course.

DAY	FULL COURSE	BASIC COURSE
1	1.1	1.1
2	1.2 & Using a Scientific Calculator or a Graphing Calculator	1.2 & Using a Scientific Calculator or a Graphing Calculator
3	1.3	1.3
4	1.4	1.4
5	Mid-Chapter Self-Test	1.4
6	1.5	Mid-Chapter Self-Test
7	1.6	1.5
8	1.7	1.6
9	1.8	1.7
10	Chapter Review	1.8
11	Chapter Test	Chapter Review
12		Chapter Test

CHAPTER ORGANIZATION

LESSON	PAGES	GOALS	MEETING THE NCTM STANDARDS
1.1	2–8	1. Use the real number line to graph and order real numbers 2. Use the properties of real numbers	Problem Solving, Communication, Reasoning, Connections, Structure
1.2	9–14	1. Evaluate an algebraic expression 2. Use algebraic expressions as models of real-life situations	Problem Solving, Communication, Connections, Geometry
Mixed Review	15	Review algebraic and arithmetic skills	
Career Interview	15	Architect	Connections
Using a Calculator	16–17	Evaluate expressions	Technology
1.3	18–23	1. Solve a linear equation 2. Use linear equations to answer questions about real life	Problem Solving, Communication, Connections
1.4	24–30	Use problem-solving strategies to solve real-life problems	Problem Solving, Communication, Connections
Mid-Chapter Self-Test	30	Diagnose student weaknesses and remediate with correlated Reteach worksheets	
1.5	31–36	Solve a literal equation for a given variable and evaluate it for specified values of the other variables	Problem Solving, Communication, Connections, Geometry
1.6	37–43	1. Solve simple and compound inequalities 2. Use inequalities to solve real-life problems	Problem Solving, Communication, Connections
Mixed Review	43	Review algebraic and arithmetic skills	
1.7	44–49	1. Solve absolute value equations and inequalities 2. Use absolute value inequalities in real-life settings	Problem Solving, Communication, Connections
1.8	50–55	Organize data by using tables and graphs	Problem Solving, Communication, Connections, Discrete Math
Chapter Summary	56	Restate for students what they have learned, why they learned it, and how it fits into the structure of algebra	Structure, Connections
Chapter Review	57–60	Review concepts and skills learned in the chapter	
Chapter Test	61	Diagnose student weaknesses and remediate with correlated Reteaching worksheets	

LESSON RESOURCES

MEETING INDIVIDUAL NEEDS

RETEACHING For students who need to spend more time on basics:

If a mid-chapter self-test or chapter test indicates a deficiency, teachers can help students with the appropriate ***Reteaching Copymaster.***

PRACTICE For students who need more practice:

Additional exercises like those in the Pupil's Edition are provided for each lesson in ***Extra Practice Copymasters.***

ENRICHMENT For enriching and broadening students' experiences:

Problem of the Day copymasters in ***Teaching Tools*** provide a daily opportunity to use logical reasoning, looking for a pattern, writing an equation, and other routine and non-routine problem-solving strategies.

Math Log copymasters in ***Alternative Assessment*** provide opportunities to report on investigations, research, and open-ended problems.

Technology: Using Calculators and Computers provides enriching activities with graphing and scientific calculators and computers.

The ***Applications Handbook*** provides additional information about the cross-curriculum topics such as astronomy, chemistry, physics, sports, economics, genetics, and music that are integrated into the Pupil's Edition.

LESSON	1.1	1.2	1.3	1.4	1.5	1.6	1.7	1.8
PAGES	2-8	9-14	18-23	24-29	31-36	37-43	44-49	50-55
Teaching Tools								
Transparencies	✓			✓		✓	✓	✓
Problem of the Day	✓	✓	✓	✓	✓	✓	✓	✓
Warm-up Exercises	✓	✓	✓	✓	✓	✓	✓	✓
Answer Masters	✓	✓	✓	✓	✓	✓	✓	✓
Extra Practice Copymasters	✓	✓	✓	✓	✓	✓	✓	✓
Reteaching Copymasters	Teacher-directed and independent activities tied to results on the Mid-Chapter Self-Tests and Chapter Tests							
Color Transparencies	✓	✓		✓	✓		✓	✓
Applications Handbook	Additional background information for many real-life applications							
Technology	Calculator and computer worksheets for appropriate lessons							
Complete Solutions Manual	✓	✓	✓	✓	✓	✓	✓	✓
Alternative Assessment	Assess student's ability to reason, analyze, solve problems, and communicate using mathematical language.							
Formal Assessment	Mid-Chapter Self-Tests, Chapter Tests, Cumulative Tests, and Practice for College Entrance Tests							
Computer Test Bank	Customized tests can be created by choosing from over 2500 items.							

INSIGHTS

1.1 Real Numbers and Number Operations

An understanding of algebra depends on understanding the real number system. In this lesson, students have an opportunity to review the components of the real number system and the rules and properties which govern it.

1.2 Algebraic Expressions and Models

Algebraic expressions are used to model real-life situations. Frequently, models result in equations that can be solved to provide information about the real-life situation. In this lesson, students review the use of algebraic models and how to evaluate them. Evaluation is an important component to solving equations.

1.3 Solving Linear Equations

Solving linear equations is a basic algebraic skill. In this lesson, students review the transformations that can be used to solve linear equations. Many real-life situations can be modeled by linear equations which, when solved, provide useful information.

1.4 Problem-Solving Strategies

Creating and using a mathematical model is a problem-solving strategy most characteristic of algebra. After listing some nonalgebraic problem-solving strategies, this lesson presents the key steps in creating and using an algebraic model.

1.5 Literal Equations and Formulas

Literal equations or formulas, such as the one in the chapter opener ($C = \frac{5}{9}(F - 32)$), occur frequently in the subject areas of physics, chemistry, statistics, and economics. Because of their common use, it is important to be able to solve these formulas for one of the two or more variables they contain. In this lesson, students learn how to solve literal equations.

1.6 Solving Linear Inequalities

Many real-life situations are modeled by linear inequalities rather than by equations. In this lesson, students review the transformations that can be used to solve both simple and compound linear inequalities.

1.7 Solving Absolute Value Equations and

Some real-life applications in chemistry, statistics, and engineering, for example, involve absolute-value equations or inequalities. In this lesson, students review the transformations used in solving absolute-value equations and inequalities. This review builds on the general methods for solving linear equations and inequalities.

1.8 Exploring Data: Tables and Charts

Much of the information that students receive in daily life is in the form of tables or graphs. This lesson emphasizes reading and organizing data that is presented in these forms. Tables and graphs thereby become tools with which students can explore data, detect patterns, and make inferences in problem-solving situations.

Problem of the Day

How many regular hexagons are in this figure? 27

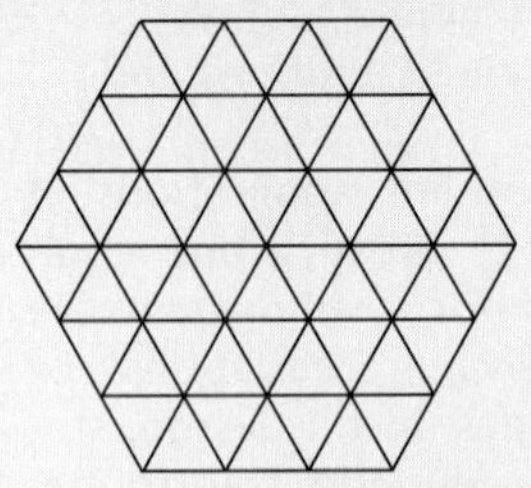

ORGANIZER

Warm-Up Exercises

Warm-Up Exercises are designed to get students ready quickly for the day's lesson.

1. Name as many different types of numbers as you can. Expect students to include counting numbers, whole numbers, integers, rational numbers, irrational numbers, real numbers, and complex numbers.
2. Identify the four basic operations in algebra. Provide examples of using these operations with the sets of numbers in Warm-Up Exercise 1. The four basic operations are addition, subtraction, multiplication, and division.

Lesson Resources

Teaching Tools
- Transparency: 1
- Problem of the Day: 1.1
- Warm-up Exercises: 1.1
- Answer Masters: 1.1

Extra Practice: 1.1

Color Transparency: 1

Applications Handbook: p. 52, 53

Technology Handbook: p. 1

LESSON Notes

Example 1

Remind students that it is not always possible to "accurately" graph irrational numbers on real number lines.

1.1 Real Numbers and Number Operations

What you should learn:

Goal 1 How to use the real number line to graph and order real numbers

Goal 2 How to use the properties of real numbers

Why you should learn it:

An understanding of real numbers and their properties is the basic building block to all of mathematics.

Goal 1 Using the Real Number Line

The numbers used in algebra are **real numbers.** They can be pictured as points on a horizontal line called the **real number line.** The point labeled *0* is the **origin.** Points to the left of 0 represent **negative numbers,** and points to the right of 0 represent **positive numbers.** Zero is neither positive nor negative.

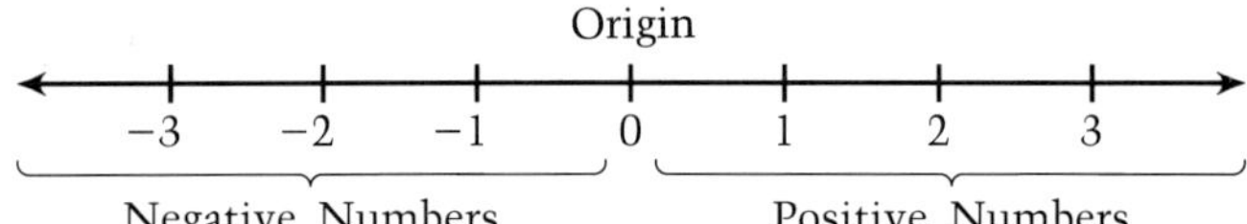

The scale marks on a number line are equally spaced and represent **integers.** An integer is either negative, zero, or positive.

$\{\ldots, -3, -2, -1, 0, 1, 2, 3, \ldots\}$ ***Integers***

Negative Zero Positive

Enclosing numbers in braces { } indicates that they form a **set.** The three dots on each side indicate that the list continues in each direction without end. Here are some other important **subsets** of real numbers.

Whole Numbers: 0, 1, 2, 3, . . .

Rational Numbers: Numbers such as $\frac{1}{2}$ and $-\frac{3}{1}$ that can be written as the ratio of two integers

Irrational Numbers: Real numbers such as $\sqrt{2}$ or π, that are not rational

The point on a number line that corresponds to a number is its **graph.** Graphing the number is called **plotting** the point.

Example 1 ***Graphing Numbers on a Number Line***

Sketch the graph of the real numbers $-\frac{2}{3}$, $\sqrt{2}$, and $\frac{5}{2}$.

Solution To begin, write each number in decimal form.

$-\frac{2}{3} \approx -0.67 \qquad \sqrt{2} \approx 1.41 \qquad \frac{5}{2} = 2.5$

−3 −2 −1 $-\frac{2}{3}$ 0 1 $\sqrt{2}$ 2 $\frac{5}{2}$ 3 ■

The real number line can be used to order real numbers.

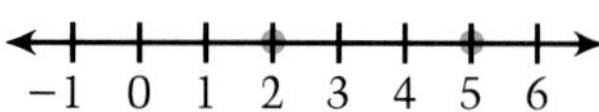

2 is less than 5

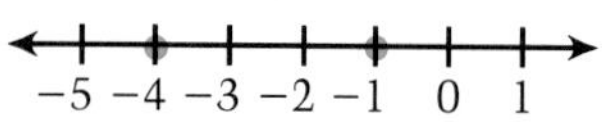

−1 is greater than −4

The **inequality symbols** $<$, $\leq$, $>$, and $\geq$ can be used to show the order of two numbers.

Verbal Statement	Inequality	Number Line Interpretation
2 is *less* than 5.	$2 < 5$	2 is to the *left* of 5.
−1 is *greater* than −4.	$-1 > -4$	−1 is to the *right* of −4.

Both statements, $2 < 5$ and $5 > 2$, give the same information.

Real Life **Topography**

Problem Solving **Draw a Diagram**

Example 2 *Ordering Real Numbers*

The lowest elevation in Mexico is −33 feet (33 feet below sea level) near Mexicali. The highest elevation is 18,701 feet at Orizaba. Order the following locations in *increasing* order according to their elevations.

Cancún (20)	Durango (6189)	Guadalajara (505)
La Paz (43)	Mexicali (−33)	Morelia (6386)
Monterrey (1765)	Orizaba (18,701)	Torreón (3740)

Solution Begin by plotting the elevations on a number line.

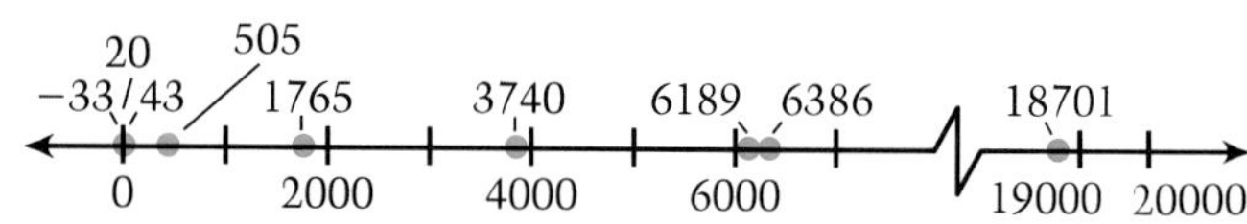

From the number line, you can see that the order is

Mexicali (−33)	Cancún (20)	La Paz (43)
Guadalajara (505)	Monterrey (1765)	Torreón (3740)
Durango (6189)	Morelia (6386)	Orizaba (18,701)

■

Two points on a number line that are the same distance from the origin but are on opposite sides of the origin are **opposites.** For instance, −4 and 4 are opposites. Zero is a special case and is its own opposite. In general, the opposite of a is $-a$ and the opposite of $-a$ is $-(-a) = a$.

1. If a is positive, then its opposite, $-a$, is negative.
2. The opposite of 0 is 0.
3. If a is negative, then its opposite, $-a$, is positive.

Example 2

Point out to students the break in the number line art. This type of graphic is a convenient way in which to show data on the real number line that is located far from most of the other data on the number line.

GOAL 2 MATH JOURNAL

You may want students to keep a mathematics journal in which important concepts and procedures are recorded. The properties of addition and multiplication, and the definitions of subtraction and division would make good journal entries. Consider the following questions.

a. Is the set of whole numbers closed under addition? That is, is the sum of two whole numbers also a whole number?
b. Is the set of whole numbers closed under subtraction? multiplication? division?
c. Repeat a and b above for the set of integers, and for the set of rational numbers.
d. Does the inverse property hold for the set of whole numbers? for the set of rational numbers? The set of whole numbers is closed under addition and multiplication. The set of integers is closed under addition, subtraction, and multiplication. The set of rational numbers is closed under all four operations with the exception of division by zero. For d: no; yes, with the exception of the inverse of 0.

Example 3

Students frequently ignore the fact that "the difference of" and "the quotient of" imply an order. Point out that "the difference of a and b" means $a - b$, in that order. And, "the quotient of a and b" means $a \div b$, in that order.

Example 4

Ask students to compute the number of barrels of oil that could be saved in a year operating at 95% of peak efficiency and express the result in decimal form and in scientific notation. Ask students which representation of the answer they prefer.
18,080,640 barrels per year, or approximately 1.81×10^7 barrels per year.

Extra Examples

Here are additional examples similar to **Examples 1–4.**

1. Sketch the graph of the following real numbers.
$-\frac{4}{5}, -\sqrt{5}, \pi, \frac{11}{4}$

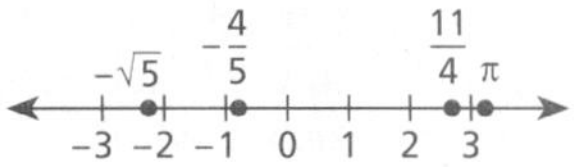

2. The following are temperature readings in degrees Celsius. Write them in decreasing order.
−13, 23.5, 0, −2, 18, 42.5, −8
42.5, 23.5, 18, 0, −2, −8, −13
3. Perform the indicated operations.
   a. the sum of −5 and −13
   b. the sum of 9 and −12
   c. the difference of −17 and −8
   d. the difference of −14 and 28
   e. the quotient of −56 and 8
   f. the quotient of 64 and 12
   g. the product of −23 and 7
   h. the product of 4 and 16

   a. −18 b. −3 c. −9
   d. −42 e. −7 f. $\frac{16}{3}$
   g. −161 h. 64
4. Suppose in Example 4, the California windmills operate at 75% of peak efficiency, but can only generate up to 1.35 million kilowatts per hour. How many barrels of oil could be saved in a year?
approximately 18 million barrels

Goal 2 Using Properties of Real Numbers

Four operations with real numbers are addition, subtraction, multiplication, and division. *Addition* and *multiplication* are the basic operations and are used to define the other two. Their properties are listed below.

Properties of Addition and Multiplication

Let a, b, and c be any real numbers.

Name of Property	Addition Properties	Multiplication Properties
1. Closure	$a + b$ is a real number.	ab is a real number.
2. Commutative	$a + b = b + a$	$ab = ba$
3. Associative	$(a + b) + c = a + (b + c)$	$(ab)c = a(bc)$
4. Identity	$a + 0 = a,\ 0 + a = a$	$a \cdot 1 = a,\ 1 \cdot a = a$
5. Inverse	$a + (-a) = 0$	$a \cdot \frac{1}{a} = 1,\ a \neq 0$

Properties Relating Addition and Multiplication

6. Left Distributive Property $a(b + c) = ab + ac$
Right Distributive Property $(a + b)c = ac + bc$

The **reciprocal** of a is $\frac{1}{a}$, provided a is not zero. Zero has no reciprocal.

Subtraction is defined as *adding the opposite,*

$$a - b = a + (-b) \qquad \textit{Definition of subtraction}$$

and *division* is defined as *multiplying by the reciprocal.*

$$\frac{a}{b} = a \cdot \frac{1}{b} \qquad \textit{Definition of division}$$

Example 3 *Operations with Real Numbers*

a. The **sum** of −12 and 14 is

$$-12 + 14 = 2.$$

b. The **difference** of 5 and −13 is

$$5 - (-13) = 5 + 13 = 18.$$

c. The **product** of −3 and −11 is

$$(-3)(-11) = 33.$$

d. The **quotient** of −42 and 14 is

$$\frac{-42}{14} = -3.$$

■

This "wind farm" near Palm Springs, California, consists of 4000 windmills that generate electricity. The states with the greatest wind power potential are shown in the table below.

Real Life
Conservation

States with the Greatest Wind Power Potential
(Percents represent potential of total electrical use in the United States.)

State	Percent
North Dakota	11%
Texas	11%
Kansas	10%
Montana	9%
South Dakota	9%
Nebraska	8%
Oklahoma	7%
Minnesota	7%
Wyoming	6%
Iowa	5%
Colorado	4%
New Mexico	4%

Example 4 *Operations with Real Numbers*

One barrel of oil can generate 545 kilowatt-hours of electricity. In 1990, the 17,000 windmills in California could generate up to 1.5 million kilowatt-hours per hour. At peak capacity, how many barrels of oil could be saved each hour? Operating at 75% of peak efficiency, how many barrels of oil could be saved in a year? ***(Source: U.S. Department of Energy)***

Solution

$$\text{Barrels of oil} = \frac{\text{Kilowatt-hours per hour (from the wind)}}{\text{545 kilowatt-hours per hour, per barrel}}$$

$$= \frac{1{,}500{,}000}{545}, \text{ or about 2752 barrels per hour}$$

Operating at 75% of peak capacity for a year, the savings would be the following.

$$(0.75)\left(\frac{2752 \text{ barrels}}{\text{hour}}\right)\left(\frac{24 \text{ hours}}{\text{day}}\right)\left(\frac{365 \text{ days}}{\text{year}}\right) \approx \frac{18 \times 10^6 \text{ barrels}}{\text{year}}$$ ■

Communicating about ALGEBRA

SHARING IDEAS about the Lesson

A. Four million 500-kilowatt windmills in the states listed in the table could generate 91% of the total electrical use. If this were done, how many windmills would be in North Dakota? (*Hint:* It is not 11% of 4 million.) ≈483,516

B. How many barrels of oil would be saved each hour if 4 million 500-kilowatt windmills were operating at 75% capacity? ≈2,752,294

Check Understanding

This feature is designed as a lesson quiz to assess students' understanding of the lesson goals.

1. Explain in words what the Closure Property states.
   See questions in Lesson Notes for Goal 2.
2. How can the number line be used to determine whether a given number is less than another?
   Numbers to the left are less than numbers to the right.
3. What numbers can be represented on the real number line?
   All real numbers; however, some real numbers are graphed by their decimal approximations.
4. Is there an associative property for the subtraction operation? Explain.
   No; for example $(4 - 3) - 2 \neq 4 - (3 - 2)$ because $-1 \neq 3$.

***Communicating* about ALGEBRA**

This feature is designed to give students an opportunity to share ideas with other students.

Point out that a 500-kilowatt windmill is one that produces 500 kilowatts of electricity each hour.

EXTEND *Communicating*

Ask students to determine how many barrels of oil would be saved in a year in North Dakota by the 500-kilowatt windmills operating at 75% of their peak capacity. $\approx 2.9 \times 10^9$ barrels
In New Mexico?
$\approx 1.1 \times 10^9$ barrels

EXERCISE Notes

ASSIGNMENT GUIDE

Basic/Average: Ex. 1–6, 7–33 odd, 34–37, 39–55 odd

Above Average: Ex. 1–6, 7–33 odd, 34–37, 39–55 odd

Advanced: Ex. 1–6, 7–33 odd, 34–37, 39–55 odd, 56

Selected Answers
Exercises 1–6, 7–53 odd

✪ More Difficult Exercises
Exercises 36, 55, 56

Guided Practice

▶ **Ex. 1–6** The Guided Practice exercises provide an opportunity for an in-class summary of essential concepts and skills from the lesson. You can use them as a final check of students' readiness and understanding.

The answers to all Guided Practice exercises will be found in the "Selected Answers" section of the student text.

▶ **Ex. 6** CHEMISTRY The factor analysis technique is similar to the "factor labeling" technique that students will encounter in chemistry.

Independent Practice

Each lesson is composed of two types of independent practice exercises based on levels of difficulty. One type of exercise presents challenging problems and extensions to the lesson Goals and lesson Examples. This type of exercise has been indicated in this Teacher's Edition by labeling each one with a star (✪).

The answers to the odd-numbered exercises of the Independent Practice and Integrated Review sections will be found in the "Selected Answers" section of the student text.

EXERCISES

Guided Practice

CRITICAL THINKING about the Lesson

1. Is zero a positive number, a negative number, or neither? Is zero an integer? Neither, yes
2. Give examples of real numbers that are whole numbers, integers, rational numbers, and irrational numbers. See page 2.
3. Which of the following is *false*? Explain. $\frac{1}{4}$ is not an integer.
   a. Every whole number is an integer.
   b. Every integer is a rational number.
   c. No integer is an irrational number.
   d. Every rational number is an integer.
4. Leon is 16 and Neesha is 18. Which of the following is *false*? Explain. On a number line, 16 is not to the right of 16.
   a. (Leon's age) < (Neesha's age)
   b. (Neesha's age) ≤ (Neesha's age)
   c. (Neesha's age) ≥ (Leon's age)
   d. (Leon's age) > (Leon's age)
5. Which of the following "associative properties" are true? Explain your answer. See below.
   a. *Addition:* $(a + b) + c = a + (b + c)$
   b. *Subtraction:* $(a - b) - c = a - (b - c)$
   c. *Multiplication:* $(a \cdot b) \cdot c = a \cdot (b \cdot c)$
   d. *Division:* $(a \div b) \div c = a \div (b \div c)$
6. *Factor Analysis* Find the unit of measure for the product. Explain your reasoning.
   $$\left(\frac{3 \text{ pounds}}{1 \text{ square inch}}\right)\left(\frac{144 \text{ square inches}}{1 \text{ square foot}}\right)\left(\frac{1 \text{ ton}}{2000 \text{ pounds}}\right)\left(\frac{25 \text{ dollars}}{1 \text{ ton}}\right)$$
   $\frac{\text{dollars}}{\text{square foot}}$ The other units cancel.

Independent Practice

In Exercises 7–12, plot the numbers on the real number line. Then decide which is greater. See margin for graphs.

7. $\frac{1}{2}$ and -3 $\frac{1}{2}$
8. 5 and $\frac{2}{3}$ 5
9. $\sqrt{3}$ and 1 $\sqrt{3}$
10. -6 and $\sqrt{2}$ $\sqrt{2}$
11. $-\frac{7}{3}$ and $\sqrt{5}$ $\sqrt{5}$
12. 0 and $-\sqrt{6}$ 0

In Exercises 13–16, write the numbers in decreasing order.

13. $1, -4, -\sqrt{3}, 9, \frac{1}{3}$ $9, 1, \frac{1}{3}, -\sqrt{3}, -4$
14. $\sqrt{13}, \frac{5}{2}, -\frac{7}{16}, 1, 10$ $10, \sqrt{13}, \frac{5}{2}, 1, -\frac{7}{16}$
15. $-\frac{5}{2}, -\sqrt{5}, 4, 0, \frac{3}{2}$ $4, \frac{3}{2}, 0, -\sqrt{5}, -\frac{5}{2}$
16. $-20, -0.5, -\sqrt{6}, 4.75, \frac{1}{3}$ $4.75, \frac{1}{3}, -0.5, -\sqrt{6}, -20$

In Exercises 17–24, state the property that is illustrated.

17. $(2 \cdot 4) \cdot 9 = 2(4 \cdot 9)$ Associative for ×
18. $6 \cdot 5 = 5 \cdot 6$ Commutative for ×
19. $3 + 1 = 1 + 3$ Commutative for +
20. $7 \cdot \frac{1}{7} = 1$ Inverse for ×
21. $4 + (-4) = 0$ Inverse for +
22. $(8 + 7) + 2 = 8 + (7 + 2)$ Associative for +
23. $2(3 + 12) = 2 \cdot 3 + 2 \cdot 12$ Left distributive
24. $9(1) = 9$ Identity for ×

5. b is not true because $(8 - 3) - 2 \neq 8 - (3 - 2)$, d is not true because $(12 \div 6) \div 2 \neq 12 \div (6 \div 2)$.

25. What is the sum of 64 and −5? 59

26. What is the sum of −14 and −7? −21

27. What is the difference of −18 and 6? −24

28. What is the difference of 4 and −7? 11

29. What is the product of 12 and 2? 24

30. What is the product of −16 and 3? −48

31. What is the quotient of 3 and $\frac{1}{2}$? 6

32. What is the quotient of 16 and $\frac{4}{3}$? 12

33. *Football Coaches* In 1990, the ten National Football League coaches with the most wins (in regular season games through 1990) are listed in alphabetical order. Order the coaches (in decreasing order) according to the number of wins. Who is the coach in the picture? Halas

Coach	Wins	Coach	Wins
Paul Brown	5 166	Tom Landry	3 250
Bud Grant	7 158	Chuck Noll	6 177
George Halas	1 319	Steve Owen	9 151
Chuck Knox	8 155	Don Shula	2 269
Earl Lambeau	4 226	Hank Stram	10 131

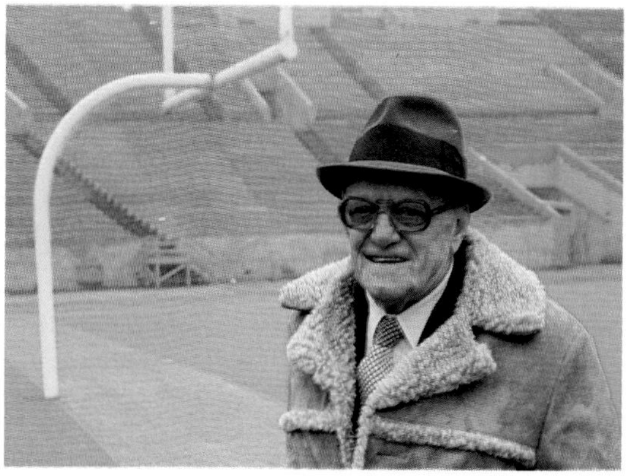

This NFL coach has had the most wins in regular season games.

34. *Low Temperatures of Cities* The lowest temperatures ever recorded in various cities are listed below. Order the cities in increasing order by their lowest temperatures.

City	Low Temp.	City	Low Temp.
Baltimore, MD	5 −7°F	Honolulu, HI	10 53°F
Cincinnati, OH	3 −25°F	Indianapolis, IN	4 −22°F
Columbia, SC	6 −1°F	Jackson, MS	8 2°F
Great Falls, MT	1 −43°F	Phoenix, AZ	9 17°F
Hartford, CT	2 −26°F	Seattle, WA	7 0°F

35. *A Close Call* The temperature of air drops about 3°F for each 1000-foot increase in altitude. About how much had the temperature dropped when Glaisher and Coxwell reached an altitude of 25,000 feet? If the temperature at sea level was 60°F, what was the temperature at 25,000 feet? 75°F; −15°F

In September 1862, James Glaisher, a meteorologist, and Henry Coxwell, a professional balloonist, went up in their hot-air balloon much higher than they had expected. At 25,000 feet, Glaisher became unconscious from lack of oxygen. To get the balloon to descend, Coxwell climbed up the rigging above the basket, but when he grasped the ice-cold hoop, his hands became too numb to pull the valve cord. Finally, he was able to pull the cord with his teeth. The balloon then began to descend, and both men made it safely back to Earth.

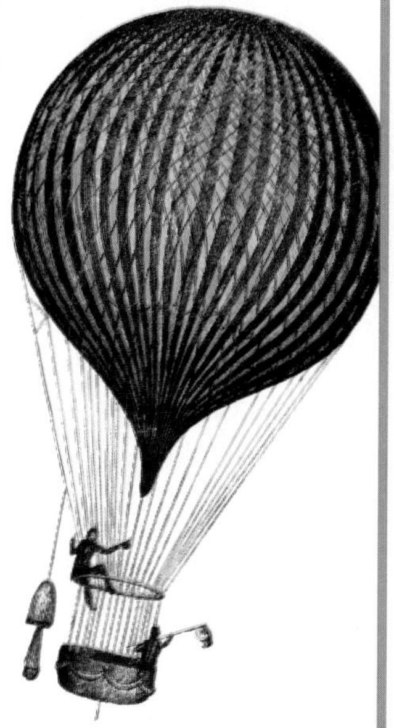

Answers

7.

8.

9.

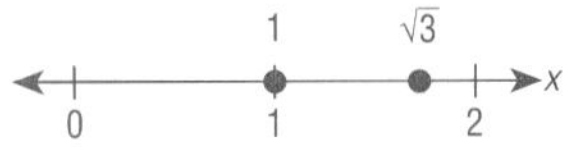

10.

11.

12.

Ex. 17–32 These exercises are outcomes of Goal 2 and are listed and modeled on page 4 of the lesson.

Ex. 33–36 The blue titles will appear throughout the text. These titles indicate applications to real life, connections to other mathematical topics, or technology.

These exercises present real-life connections to ordering numbers. This type of real-life modeling will be provided in every lesson.

Ex. 37–54 These Integrated Review exercises are found in every lesson. They are designed to review previous topics by connecting or integrating previously learned skills to the essential skills presented in the lesson.

▶ **Ex. 55–56** The Extension and Exploration exercises are not necessarily extremely difficult problems, nor are they only designated for gifted students. As in Exercises 55 and 56, this feature often provides an interesting extension of the lesson goals.

COOPERATIVE LEARNING
You may wish to use this UPC bar code application as a cooperative learning activity in class.

Enrichment Activities

These activities require students to go beyond lesson goals.

The following arrangement is called Pascal's triangle.

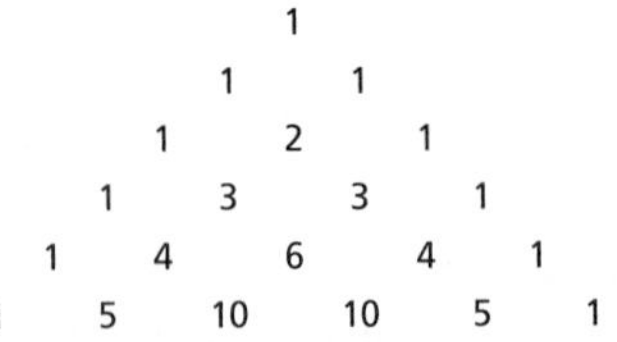

1. Look for a pattern and generate the next two rows.
Pattern: The triangle is bordered by ones. The sum of two adjacent numbers in one row is equal to the number in the next row between the numbers.
1 2 1 / 1 3 3 1
1, 6, 15, 20, 15, 6, 1; 1, 7, 21, 35, 35, 21, 7, 1

2. Describe any diagonal patterns you observe.
Sample answers: First diagonal numbers are all 1; Second diagonal numbers are consecutive integers 1, 2, 3, . . . ; Third diagonal number differences increase by 1 each time, that is, they differ by 2, 3, 4, 5, . . .

3. **RESEARCH** Use History of Mathematics texts in the library to find out more about Blaise Pascal and the uses of this triangle.

36. *Ouch!* The white-headed woodpecker's skull is specially designed to protect its brain from the force of continual pecking. This woodpecker pecks at a rate of 18 pecks per second. If you heard one pecking on a tree for 4 seconds, then 3 seconds, then 2 seconds, and then 4 seconds, how many times did it peck the tree? Show how the Distributive Property can be used to find the answer.
234; $18 \cdot 4 + 18 \cdot 3 + 18 \cdot 2 + 18 \cdot 4 = 18(4 + 3 + 2 + 4)$

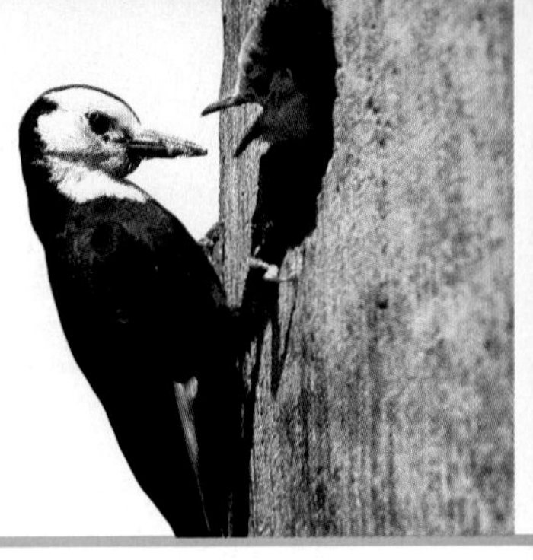

Integrated Review

In Exercises 37–40, decide whether the real number is rational or irrational.

37. 0.25 Rational **38.** $-\sqrt{3}$ Irrational **39.** $\sqrt{7}$ Irrational **40.** 13 Rational

In Exercises 41–44, place the correct inequality symbol between the numbers.

41. -2 [?] 1 < **42.** 2 [?] $\sqrt{6}$ < **43.** 3 [?] $\sqrt{10}$ < **44.** -5 [?] $\frac{7}{2}$ <

In Exercises 45–48, state the opposite of the number.

45. 14 -14 **46.** -6 6 **47.** $-\sqrt{5}$ $\sqrt{5}$ **48.** 19.9 -19.9

In Exercises 49–52, state the reciprocal of the number.

49. -31 $-\frac{1}{31}$ **50.** $\frac{10}{11}$ $\frac{11}{10}$ **51.** $-\frac{5}{7}$ $-\frac{7}{5}$ **52.** 14 $\frac{1}{14}$

In Exercises 53 and 54, find the product and its unit of measure.

53. $\left(\frac{50 \text{ miles}}{1 \text{ hour}}\right)\left(\frac{5280 \text{ feet}}{1 \text{ mile}}\right)\left(\frac{1 \text{ hour}}{3600 \text{ seconds}}\right)$
$73.\overline{3}$ feet per second

54. $\left(\frac{7 \text{ dollars}}{1 \text{ hour}}\right)\left(\frac{40 \text{ hours}}{1 \text{ week}}\right)\left(\frac{52 \text{ weeks}}{1 \text{ year}}\right)$
14,560 dollars per year

Exploration and Extension

UPC Codes **In Exercises 55 and 56, use the following information.**

Packaged products sold in the United States have a Universal Product Code (UPC) or bar code as shown at the right. The following algorithm is used on the first eleven digits and the result should equal the twelfth digit, called the check digit.

- Add the odd-numbered positions together. Multiply by 3.
- Add the even-numbered positions together.
- Add the results of steps 1 and 2.
- Subtract the result of step 3 from the next highest multiple of 10.

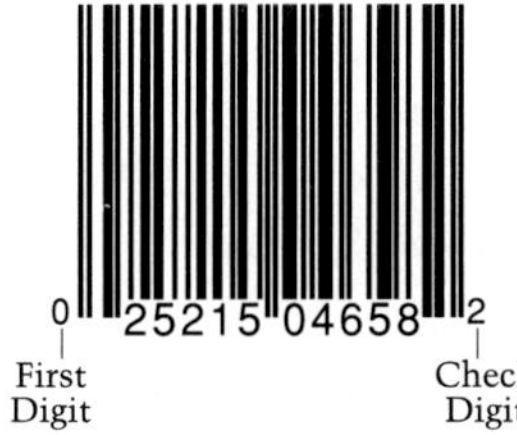

Using the algorithm on this UPC produces $(0 + 5 + 1 + 0 + 6 + 8)(3) + (2 + 2 + 5 + 4 + 5) = 78$. The next highest multiple of 10 is 80, so $80 - 78 = 2$, which is the check digit.

55. Does a UPC of 0 76737 20012 9 check? Explain.
Yes; $(0 + 6 + 3 + 2 + 0 + 2)3 + (7 + 7 + 7 + 0 + 1) = 61$, $70 - 61 = 9$

56. Does a UPC of 0 41800 48700 3 check? Explain.
No; $(0 + 1 + 0 + 4 + 7 + 0)3 + (4 + 8 + 0 + 8 + 0) = 56$, $60 - 56 = 4$

1.2 Algebraic Expressions and Models

What you should learn:

Goal 1 How to evaluate an algebraic expression

Goal 2 How to use algebraic expressions as models of real-life situations

Why you should learn it:

You can model many real-life situations with algebraic expressions, such as the average cost of children's books for the last ten years.

Goal 1 Evaluating Algebraic Expressions

A **variable** is a letter that is used to represent one or more numbers. Any number used to replace a variable is the **value of the variable.** An **algebraic expression** is a collection of numbers, variables, operations, and grouping symbols.

When the variables in an algebraic expression are replaced by numbers, you are **evaluating** the expression. The resulting number is called the **value of the expression.** To evaluate a variable expression, use the following *flowchart*.

Write algebraic expression. ⟶ Substitute values of variables. ⟶ Simplify.

The parts of an expression that form a sum, a product, or a quotient have special names.

Sum: $4x$ and 3 are the **terms** of $4x + 3$.
Product: 4 and x are the **factors** of $4x$.
Quotient: $3x$ is the **numerator** and 8 is the **denominator** of the fraction $\frac{3x}{8}$.

Example 1 shows how an algebraic expression can be used as a **model** of a real-life situation.

Real Life
Children's Books

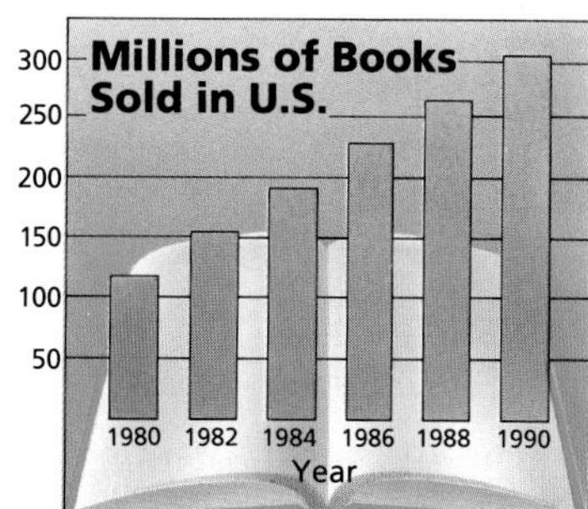

Example 1 *Algebraic Expressions as Models*

For 1980 through 1990, the number, B (in millions), of children's books that were sold in the United States can be modeled by $B = 18.7t + 119$ where $t = 0$ represents 1980. Use the model to approximate the number of children's books sold in 1985. ***(Source:* Book Industry Times)**

Solution If $t = 0$ represents 1980, then $t = 5$ represents 1985.

$B = 18.7t + 119$ *Model for number of books sold*

$B = 18.7(5) + 119$ *Substitute 5 for t.*

$B = 212.5$ *Simplify.*

About 212.5 million children's books were sold in 1985. ■

Problem of the Day

Each letter in this addition represents a different digit. Find the digits.

A H B C F	1 4 5 7 9
+ E B F A F	+ 8 5 9 1 9
A G G H F E	1 0 0 4 9 8

ORGANIZER

Warm-Up Exercises

1. Simplify the following expressions.
   a. $4(12) + 6$ 54
   b. $-5(-7) - 18$ 17
   c. 4^4 256
   d. $13(-8) + 3^5$ 139
   e. $(14 - 17)(23 + 2)$ −75
2. Let $m = 4$ and $n = -3$. Compute.
   a. $m - n$ 7
   b. $m + n$ 1
   c. $m \cdot n$ −12
   d. $\frac{n}{m}$ $-\frac{3}{4}$
   e. $(m - n)(6)$ 42
   f. $-5(2m)$ −40

Lesson Resources

Teaching Tools
Transparencies: 2, 3
Problem of the Day: 1.2
Warm-up Exercises: 1.2
Answer Masters: 1.2
Extra Practice: 1.2
Color Transparencies: 1, 2
Applications Handbook: p. 18, 19

LESSON Notes

Example 1

Use these questions to help students understand the example. "How many children's books were sold in 1987?" 249.9 million "How many more were sold in 1987 than in 1985?" 37.4 million more "How is the previous answer related to the coefficient of t?" It is twice the size of the coefficient.

Example 2

TECHNOLOGY Observe that most graphics calculators use the order of operations found in the text. If calculators are available, encourage students to check the results of the text using their calculators.

Example 3

The price, p, in $p = 0.2t + 3.3$ is in dollars. Ask students to identify the type of expression represented by S. It is the product of two binomials.

Point out to students that because t is to have values of 3, 4, 5, etc., you say that $t = 8$ corresponds to the year 1988. This technique for representing years such as 1983, 1984, 1985, and 1988, etc. by numbers such as 3, 4, 5, and 8 will be used throughout the text. You may find it worthwhile to spend some time on this notation. In order to check their understanding, have students discuss the reasons for its use.

Extra Examples

Here are additional examples similar to **Examples 1–3.**

1. Using the model of Example 1, determine how many children's books were sold in 1990. In 1986.
   306 million, 231.2 million
2. Evaluate each expression.
   a. $-6 + 3(7^2 - 12)$
   b. $-43 + 4 \cdot 12 - (2 + 4)$
   c. $(6 - 3)\left(-\frac{36}{12}\right)^3$
   a. 105 **b.** −1 **c.** −81
3. In Example 3, what were the total sales of children's books in 1983? $682.89 million What was the average price of a children's book in 1989? $5.10

Exponents are used to represent repeated factors in multiplication. For instance, the expression 2^5 represents the number that you obtain when 2 is used as a factor 5 times.

$$2^5 = \underbrace{2 \cdot 2 \cdot 2 \cdot 2 \cdot 2}_{\text{5 factors of 2}}$$

This expression is read as "2 to the 5th power." The number 2 is the **base,** and the number 5 is the **exponent.** The exponent represents the number of times the base is used as a factor. For *a number to the first power,* usually the exponent 1 is not written. For instance, 2^1 is written simply as 2.

An exponent applies only to the number, variable, or expression that is immediately to its left. In the expression $3x^4$, the base is x, not $3x$. In the expression $(3x)^4$, the base is $3x$.

An established **order of operations** helps avoid confusion when simplifying numerical expressions.

Order of Operations

1. First, do operations that occur within grouping symbols using Rules 2–4.
2. Next, evaluate powers.
3. Then do multiplications and divisions from left to right.
4. Finally, do additions and subtractions from left to right.

Example 2 *Evaluating Expressions Using Order of Operations*

a. $-4 + 2(-2 + 5)^2 = -4 + 2(3)^2$ *Add within parentheses.*
$= -4 + 2(9)$ *Evaluate the power.*
$= -4 + 18$ *Multiply.*
$= 14$ *Add.*

b. $(-4 + 2)(-2 + 5)^2 = (-2)(3)^2$ *Add within parentheses.*
$= (-2)(9)$ *Evaluate the power.*
$= -18$ *Multiply.*

c. $5 - 12 \div 3 - 7 = 5 - 4 - 7$ *Divide first.*
$= -6$ *Subtract from left to right.* ■

Over one half of the adults in the United States read at least one book for enjoyment each year. Reading books to children increases the likelihood that the children will read for enjoyment as adults. (Source: National Endowment for the Arts)

Goal 2 Writing Algebraic Expressions as Models

Real Life
Children's Books

Example 3 *Creating a Real-Life Model*

For 1980 through 1990, the average price, p, of a children's book can be modeled by $p = 0.2t + 3.3$ where $t = 0$ represents 1980. Use this model *and* the model given in Example 1 to find a model for the total sales (in millions of dollars) from children's books for 1980 through 1990. Then use the model to find the total sales in 1988.

Solution Let S represent yearly sales in millions of dollars.

Total yearly sales	=	Number sold each year	·	Average price per book each year

$$S = (18.7t + 119)(0.2t + 3.3)$$

To find the total sales in 1988, substitute 8 for t in this model.

$S = (18.7t + 119)(0.2t + 3.3)$ *Total sales model*

$= (18.7 \cdot 8 + 119)(0.2 \cdot 8 + 3.3)$ *Substitute 8 for t.*

$= (268.6)(4.9)$, or 1316.14 *Simplify.*

The total sales of children's books in 1988 was about $1316 million (or about $1.3 billion). ■

Communicating about ALGEBRA

SHARING IDEAS about the Lesson

Use a Model Use the model in Example 3 to construct a bar graph showing the total sales from children's books (in millions of dollars) for 1980 through 1990. See margin.

Check Understanding

1. What is an algebraic expression? See the top of page 9.
2. Why is the order of operations needed? See page 10.
3. Explain how algebraic models are used to represent real-life situations. See Example 3.
4. Distinguish between algebraic terms and factors. Terms are parts of indicated additions and subtractions; factors are parts of indicated multiplications and divisions.

Communicating about ALGEBRA

Answer

A.

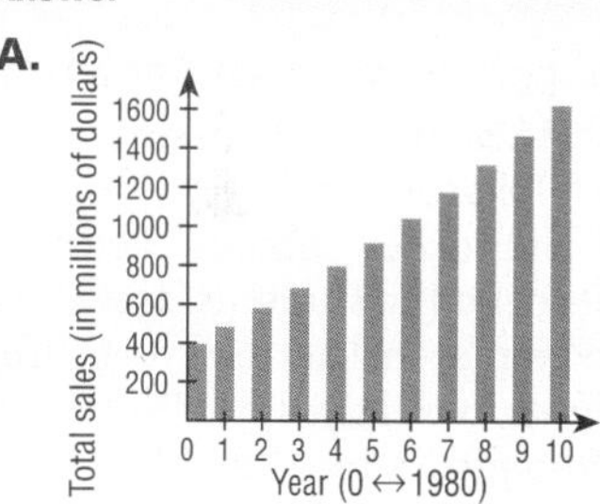

EXTEND ***Communicating***

Ask students to determine the number of children under the age of 13 in the United States in 1980 and in 1990. An almanac is a good place to get the information.

Then, ask students to explain how the change in population figures might have affected the sales of children's books from 1980 to 1990.

EXERCISE Notes

ASSIGNMENT GUIDE

Basic/Average: Ex. 1–6, 7–45 odd, 49–52, 54, 55–59 odd, 64–66

Above Average: Ex. 1–6, 7–49 odd, 49–52, 54, 55–63 odd, 64–67

Advanced: Ex. 1–6, 11–47 odd, 49–52, 53–63 odd, 64–67

Selected Answers
Exercises 1–6, 7–65 odd

Use **Mixed Review** as needed.

More Difficult Exercises
Exercises 48, 53, 54, 64, 67

Guided Practice

Ex. 1–6 COOPERATIVE LEARNING The Guided Practice exercises will tend to have higher-order "action" verbs to describe the students' tasks. For example, in Exercises 1–6, students are asked to "state an example," to "describe," to "identify," to "write a description," and to "explain." These exercises create an opportunity for students to work together and discuss their answers in a small-group setting.

Independent Practice

Ex. 11–12 and 17–20 Remind students to substitute the value of the variable *each* time it occurs.

Ex. 21–24 TECHNOLOGY Be sure to review specific calculator operations and parentheses use.

EXTEND Ex. 33 Ask students to discuss the difference between $(2a)^2$ and $2a^2$ when $a = 3$. $(2a)^2 = (2 \cdot 3)^2 = 6^2 = 36$ and $2a^2 = 2(3)^2 = 2(9) = 18$

Guided Practice

CRITICAL THINKING about the Lesson

1. Give an example of an algebraic expression for which the value of the expression is positive regardless of the value of the variable. Answers vary; x^2, $|x|$
2. Describe the difference between the terms of an expression and the factors of an expression. Terms are parts of an expression that are added, factors are multiplied.
3. Identify the terms of $6x^3 - 17x + 5$. $6x^3$, $-17x$, 5
4. Identify the factors of $10xy$. 10, x, y
5. Write a verbal description of the expression $(x^3 + 1)^2$. Square the sum of x cubed and 1.
6. Explain how the Order of Operations is used to evaluate $3 - 8^2 \div 4 + 1$. ($-$: 3rd, 8^2: 1st, $\div$: 2nd, $+$: 4th)

Independent Practice

In Exercises 7–16, evaluate the expression.

7. $x - 5$ when $x = 3$ -2
8. $7y + 2$ when $y = 5$ 37
9. $50a - 2b + 6$ when $a = 2$ and $b = 20$ 66
10. $16p - 32q + 5$ when $p = 2$ and $q = 1$ 5
11. $m + 2(m - 5)$ when $m = 25$ 65
12. $60n(2 - n)$ when $n = 3$ -180
13. $2x \div (13 - 9y)$ when $x = 14$ and $y = \frac{1}{3}$ 2.8
14. $(5 + y) \div (6x)$ when $x = 10$ and $y = 7$ $\frac{1}{5}$
15. $(20a - 1) \div b$ when $a = 2$ and $b = 13$ 3
16. $(60b \div 5) + c$ when $b = \frac{1}{2}$ and $c = 11$ 17

In Exercises 17–20, evaluate $12x \div (x - 1)$ for the given value of x.

17. $x = 7$ 14
18. $x = 0$ 0
19. $x = 3$ 18
20. $x = 10$ $\frac{40}{3}$

In Exercises 21–24, use a calculator to evaluate the expression. If necessary, round your result to two decimal places.

21. $65.1z - 0.4$ when $z = 1.2$ 77.72
22. $5.7 - 26.3z$ when $z = 80.0$ -2098.3
23. $15(0.5p + q)$ when $p = 31.5$ and $q = 0.8$ 248.25
24. $\frac{2.01p - q}{16.87q}$ when $p = 7.7$ and $q = 7.1$ 0.07

In Exercises 25–30, write the expression using exponents.

25. y squared y^2
26. seven cubed 7^3
27. x to the third power x^3
28. three to the rth power 3^r
29. $z \cdot z \cdot z \cdot z \cdot z \cdot z$ z^6
30. $16 \cdot 16 \cdot 16 \cdot 16$ 16^4

In Exercises 31–36, evaluate the expression.

31. s^2 when $s = 12$ 144
32. m^5 when $m = 2$ 32
33. $(2a)^2 + 4$ when $a = 3$ 40
34. $(4r)^3 + 10$ when $r = \frac{1}{2}$ 18
35. $11(b + d)^2$ when $b = 1$ and $d = 9$ 1100
36. $x^2 + y^2$ when $x = 5$ and $y = 6$ 61

In Exercises 37–44, evaluate the expression.

37. $6 - 3 + 1$ 4 **38.** $10 + 12 - 24$ −2 **39.** $20 \div 5 + 5 \cdot 4$ 24 **40.** $13 \cdot 4 - 2$ 50

41. $1 + 7 \cdot 4 - 5$ 24 **42.** $23 - 3 \cdot 6 \div 9$ 21 **43.** $4 - (6 \cdot 2)^2 + 2$ −138 **44.** $2 \cdot (13 - 9) \div 6$ $\frac{4}{3}$

Geometry **In Exercises 45–48, find an expression for the area of the figure. Then evaluate for the given value(s) of the variable(s).**

45. $n = 80$ $\frac{1}{2}n(10 + n)$, 3600

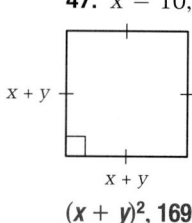

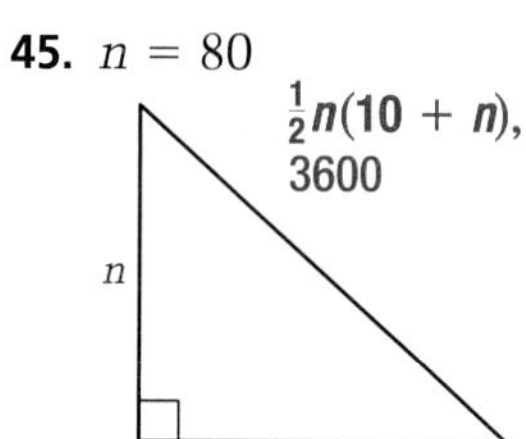

46. $a = 5, b = 4$

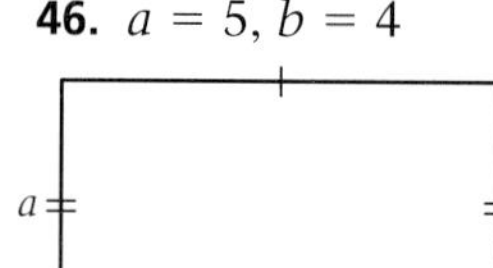

$a(a + b)$, 45

47. $x = 10, y = 3$

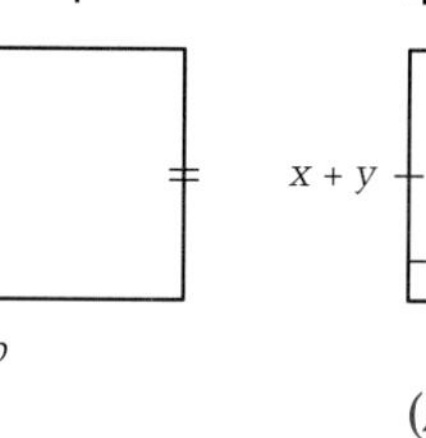

$(x + y)^2$, 169

48. $x = 9$

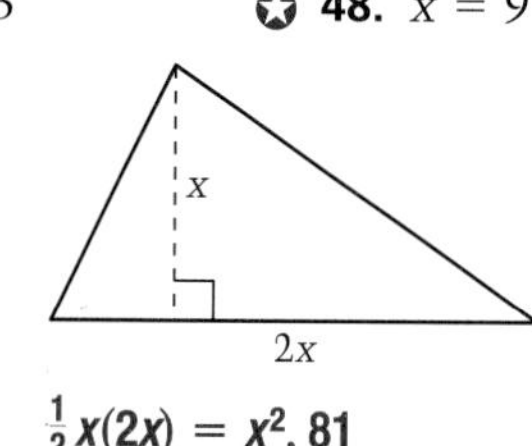

$\frac{1}{2}x(2x) = x^2$, 81

Average Salaries **In Exercises 49–52, use the following information.**

For 1980 through 1994, the average salary, A (in 1000's of dollars), of assistant principals at public high schools can be modeled by

$$A = 2.2t + 25$$

where $t = 0$ represents 1980. During the same years, the average salary, P, (in 1000's of dollars), of principals at public high schools followed the model

$$P = 2.6t + 29.$$

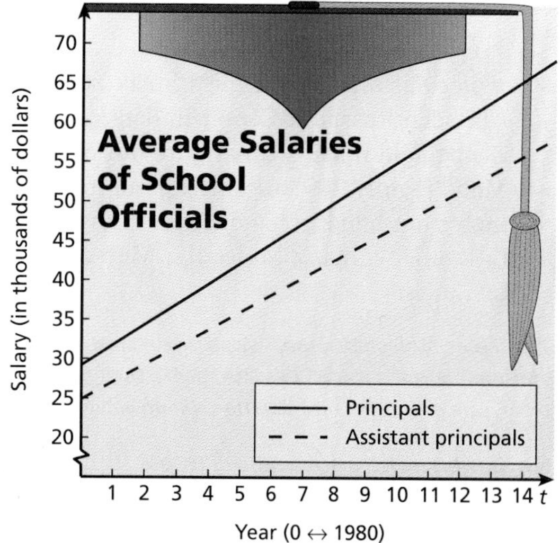

(Source: Educational Research Service)

49. Approximate a high school assistant principal's salary in 1992. $51,400

50. Approximate a high school principal's salary in 1992. $60,200

51. Write a model that gives the ratio of A to P.

52. Evaluate the model found in Exercise 51 when $t = 0$ and $t = 4$. Describe the results in words.

Population of Hawaii **In Exercises 53 and 54, use the following information.**

For 1980 through 1994, the population, P (in 1000's), of Hawaii can be modeled by

$$P = 15.5(t + 62)$$

where $t = 0$ represents 1980. *(Source: U.S. Bureau of the Census)*

53. What was the population in 1993? 1,162,500

54. What was the population increase from 1980 to 1993? 201,500

51. $\frac{A}{P} = \frac{2.2t + 25}{2.6t + 29}$ 52. $\frac{25}{29}, \frac{16.9}{19.7}$; the ratio is decreasing.

▶ **Ex. 45–48** You may need to review with students the formulas for the area of a rectangle and the area of a triangle.

▶ **Ex. 49–52** These exercises should be assigned as a group.

▶ Ex. 64 Review scientific notation with the students. Many of the real-life models studied in this text will express data in terms of scientific notation.

▶ Ex. 65–66 These College Entrance Samples will be integrated throughout the text.

Answer

67. No. If you start outside a 5-door room and you pass through all 5 doors, you will end inside the room. So, if you start outside all 3 of the 5-door rooms you must end inside all 3 rooms, which is impossible; if you start inside 1 of them, you must end inside both of the other 2, which is impossible.

Enrichment Activities

NUMBER SENSE Students can develop a number sense for powers of ten by generating their own concrete examples of what 10 million means.

In Exercise 64, you found that 10 million Molecular Men would be one inch wide. How big is 10 million?

1. Is your teacher 10 million seconds old?
   No; that is only 115.7 days
2. The circumference of the earth at the equator is 24,732 miles. Would 10 million people holding hands be able to circle the earth at the equator? The average arm span of an adult is about 1 yard.
   The earth at the equator would be 43,528,320 yards. So, 10 million people would not be able to circle the earth at the equator.
3. If you spent $100 a day, how long would it take you to spend $10,000,000?
   100,000 days or about 274 years
4. The average United States household used 1000 kilowatt hours of electricity each year for lighting. How many years would it take a household to use 10 million kilowatts? 10,000 years

Integrated Review

In Exercises 55–60, use a calculator to perform the given operations. If necessary, round your answers to two decimal places.

55. $19.5 \div 6.+$ 3.05

56. $210 \cdot 3.7$ 777

57. $65.16 \cdot 0.27$ 17.59

58. $0.14 \div 10.8$ 0.01

59. $8.05 \div 0.001 + 2$ 8052

60. $10.61 \cdot 5 - 2.3$ 50.75

Geometry **In Exercises 61–63, find the volume of the solid.**

61. Prism $V = lwh$ 36 cm³

62. Cone $V = \frac{1}{3}\pi r^2 h$ ≈20.94 in.³

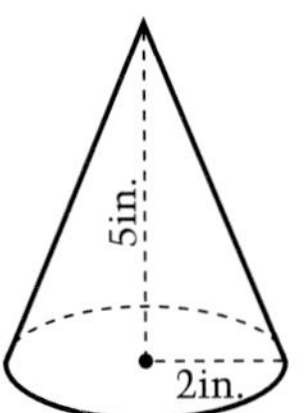

63. Sphere $V = \frac{4}{3}\pi r^3$ ≈113.10 ft³

64. *Molecular Man* During a break between experiments at the IBM Almaden Research Center in San Jose, California, scientist Peter Zeppenfeld made a tiny figure out of 28 molecules. His "Molecular Man" is only 200 billionths of an inch tall and 100 billionths of an inch from hand to hand. Express these dimensions in scientific notation. How many Molecular Men could stand side by side within one inch? 2×10^{-7} by 1×10^{-7}, 10^7 or 10,000,000

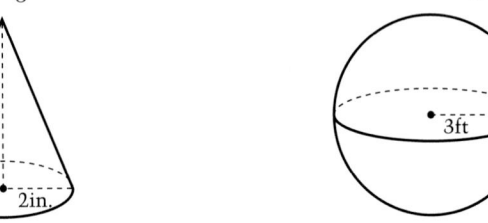

To create "Molecular Man," Zeppenfeld sprayed carbon monoxide gas onto a platinum surface cooled to −269°C. He then probed 28 carbon monoxide molecules, one at a time, into position. (Source:* Popular Science*)

65. *College Entrance Exam Sample* If on a certain day only 28 of 32 students were present for a gym class, what percent of the class was absent?
a. 4% **b.** 12.5% **c.** 25% **d.** 28% **e.** 87.5%

66. *College Entrance Exam Sample* Let $\boxed{x}$ represent the number of different pairs of positive integers whose product is x. For example, $\boxed{16} = 3$ because there are 3 different pairs of positive integers whose product is 16: 16×1, 8×2, and 4×4. What does $\boxed{36}$ equal?
a. 5 **b.** 6 **c.** 8 **d.** 10 **e.** 12

Exploration and Extension

67. *Going from Room to Room* The floor plan at the right shows five rooms. Some of the rooms have four doors and some have five doors. Is it possible to find an (unbroken) path through each door without going through any door twice? Explain your answer. No; see margin.

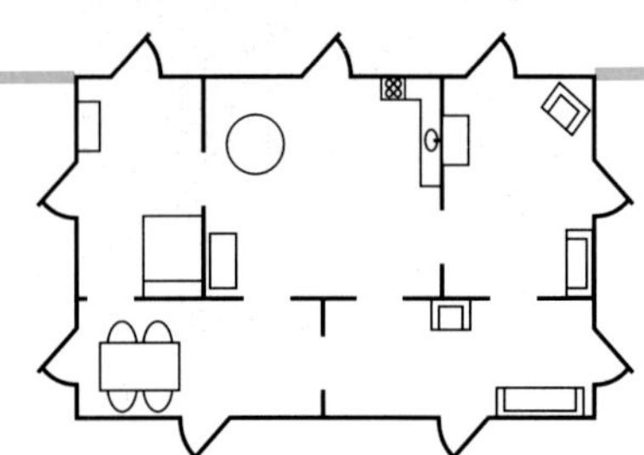

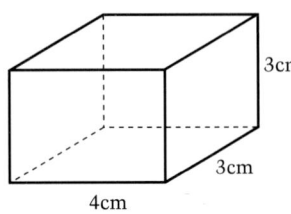

Mixed REVIEW

1. Multiply: $\left(\frac{1}{2}\right)\left(\frac{2}{3}\right)\left(\frac{3}{4}\right)$. $\frac{1}{4}$
2. Add: $\frac{1}{2} + \frac{1}{3}$. $\frac{5}{6}$
3. Subtract: $\frac{1}{2} - \frac{1}{3}$. $\frac{1}{6}$
4. Divide: $\frac{1}{2} \div \frac{1}{3}$. $\frac{3}{2}$
5. What number is 14% of 35? 4.9
6. 17.6 is what percent of 80? 22%
7. Evaluate $2x^2 - 3x$ when $x = -1$. **(1.2)*** 5
8. Evaluate $2x^2 \div 3x$ when $x = -1$. **(1.2)** $-\frac{2}{3}$
9. Multiply: $(1.25 \times 10^2)(2.5 \times 10^{-1})$. 31.25
10. Divide: $(1.25 \times 10^2) \div (2.5 \times 10^{-1})$. 500
11. Evaluate $-2x^3$ when $x = -2$. **(1.2)** 16
12. Evaluate $\left(\frac{1}{5}\right)^n$ when $n = 4$. **(1.2)** $\frac{1}{625}$
13. Simplify: $4 + 6 \div 2 - 1$. 6
14. Simplify: $-2 \cdot 4 - 2$. **(1.2)** −10
15. Write in increasing order: −2, 3, 0, −3. **(1.1)** −3, −2, 0, 1, 3
16. Write in decreasing order: $-\frac{1}{2}, -\frac{1}{4}, -\frac{1}{3}, -\frac{1}{5}$. **(1.1)** $-\frac{1}{5}, -\frac{1}{4}, -\frac{1}{3}, -\frac{1}{2}$
17. Is −2 a solution of $4x^2 + 16 = 0$? No
18. Is 2 a solution of $x^1 - 1 = 0$? No
19. Write in exponential form: $2 \cdot 2 \cdot x \cdot x \cdot x$. **(1.2)** 2^2x^3
20. Simplify: $-3(x + 4) + 5(x - 2)$. **(1.2)** $2x - 22$

*Refers to lesson in which skill was taught. Exercises without lesson references review skills taught in previous mathematics courses.

Career Interview

Architect

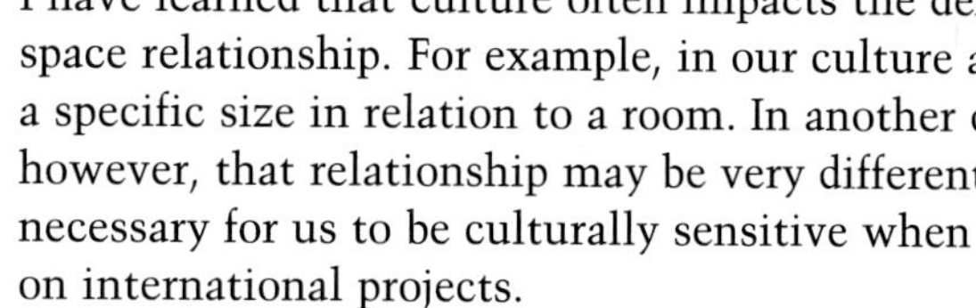

Heidi Johnson is an architect who designs buildings and the space in and around them. Her designs and drawings can range from an entire town square to the detail of a doorknob or hinge.

Q: ***What led you into this career?***

A: Before becoming an architect, I was teaching special needs students. The classrooms were often pretty depressing. That is how my interest in space and its impact on people developed. I wanted to design buildings and space that uplift or educate people rather than hinder them.

Q: ***Does having a math background help you on the job?***

A: Yes, it helps. I work with different scales to measure and draw both interior and exterior spaces. Architectural design has a lot of mathematical applications.

Q: ***What have you learned on the job?***

A: I have learned that culture often impacts the definition of space relationship. For example, in our culture a door has a specific size in relation to a room. In another country, however, that relationship may be very different. So it is necessary for us to be culturally sensitive when working on international projects.

5. **WRITING PROBLEMS** Obtain data from a newspaper or a weekly news magazine and write similar problems to share with your classmates.

Mixed Review exercises help students check their understanding of previous lessons and prior courses. Answers to the odd-numbered exercises will be found in the "Selected Answers" section of the student text.

Career Interview

Throughout the text, careers and individuals who use mathematics every day will be profiled. Having students interview local professionals who use mathematics at work is an excellent semester-long project. Students could share their findings with the entire class during a Career Day that they could have the responsibility of planning. A broad representation of people in various careers about which students can read has been identified. This collection of interviews should help convince all students of the importance of mathematics to their futures.

MULTICULTURAL *Winds of Change—A Magazine of American Indians* and *Ebony*, an African-American magazine, are two sources from which you can identify people using mathematics in their careers.

TECHNOLOGY Four types of technology will be featured throughout the text: Texas Instrument's TI-82 graphing calculator, Casio *fx*-9700GE graphing calculator, Sharp's EL-9300C graphing calculator, and BASIC computer programming. Technology can play an important role in learning algebra skills and concepts. In the textbook are uses of technology that enhance the study of algebra.

Encourage students to use their scientific or graphing calculators whenever possible. An advantage of using this technology is that more problems may be considered for a given exercise set. Another advantage is that students can spend more time on trying to understand the concepts and steps in procedures and less time on routine or tedious computations.

COOPERATIVE LEARNING Students with the same brand of calculator could work together in small groups to discuss possible key sequences.

USING A SCIENTIFIC CALCULATOR

This book includes many examples and exercises that are best done with a calculator. Most of these can be done with a scientific calculator *or* a graphing calculator. (Some require a graphing calculator.) Occasionally sample keystrokes for calculations will be given. These keystroke sequences, however, may not agree precisely with the steps required by *your* calculator. So be sure you are familiar with your own calculator.

Example 1 Evaluating Expressions on a Calculator

Expression	Keystrokes	Display	
a. $-4 - 5$	4 [+/−] [−] 5 [=]	−9	Scientific
$-4 - 5$	[(−)] 4 [−] 5 [*ENTER*]	−9	Graphing
b. $-3^2 + 4$	3 [x^2] [+/−] [+] 4 [=]	−5	Scientific
$-3^2 + 4$	[(−)] 3 [x^2] [+] 4 [*ENTER*]	−5	Graphing
c. $(-3)^2 + 4$	[(] 3 [+/−] [)] [x^2] [+] 4 [=]	13	Scientific
$(-3)^2 + 4$	[(] [(−)] 3 [)] [x^2] [+] 4 [*ENTER*]	13	Graphing ■

On a scientific calculator, notice the difference between the change sign key [+/−] and the subtraction key [−]. On a graphing calculator, the negation key [(−)] and the subtraction key [−] may not perform the same operations.

Example 2 Evaluating Expressions on a Calculator

Expression	Keystrokes	Display	
a. $24 \div 2^3$	24 [÷] 2 [y^x] 3 [=]	3	Scientific
$24 \div 2^3$	24 [÷] 2 [∧] 3 [*ENTER*]	3	Graphing
b. $(24 \div 2)^3$	[(] 24 [÷] 2 [)] [y^x] 3 [=]	1728	Scientific
$(24 \div 2)^3$	[(] 24 [÷] 2 [)] [∧] 3 [*ENTER*]	1728	Graphing
c. $\frac{5}{4 + 3 \cdot 2}$	5 [÷] [(] 4 [+] 3 [×] 2 [)] [=]	0.5	Scientific
$\frac{5}{4 + 3 \cdot 2}$	5 [÷] [(] 4 [+] 3 [×] 2 [)] [*ENTER*]	0.5	Graphing ■

Some scientific and most graphing calculators are programmable. This means that you can write and store a program or algorithm in the calculator, such as a program that would calculate the volume of a right circular cylinder. The formula for the volume is

$$V = \pi r^2 h$$

where r is the radius and h is the height.

Keystroke instructions for programming this formula into a *TI-82*, Casio *fx-9700GE*, and Sharp *EL-9300C* are listed on page 856.

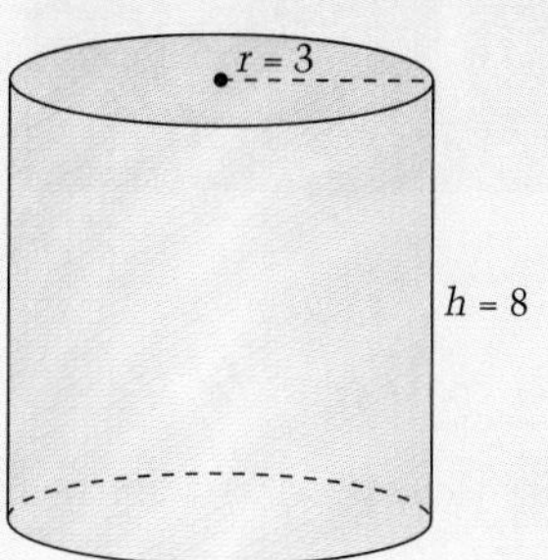

If you use one of these programs to find the volume of the cylinder shown at the right, you should get a volume of about 226.2 cubic units, as shown below.

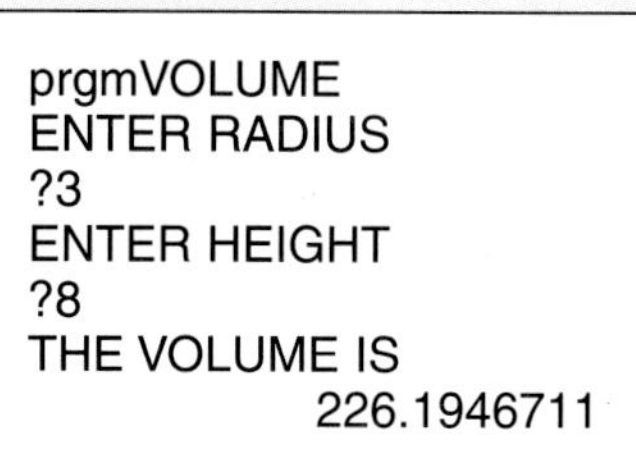

Exercises

In Exercises 1–4, write an expression that corresponds to the calculator keystrokes.

1. [(] [(–)] 4 [)] [∧] 2 [ENTER] $(-4)^2$
2. 7 [÷] [(] [(–)] 3 [–] 5 [)] [ENTER] $7 \div (-3 - 5)$
3. [(] 1 [+] 4 [)] [∧] [(–)] 1 [ENTER] $(1 + 4)^{-1}$
4. 3 [×] [(] 5 [–] 2 [)] [ENTER] $3(5 - 2)$

In Exercises 5–12, use a calculator to evaluate the expression. Round the result to three decimal places.

5. $3(5.3 - 4.1^2)^2$ 397.440
6. $(-3.7 - 5.2)^3$ −704.969
7. $(0.21 + 5.23)^{-1}$ 0.184
8. $\frac{4}{3}\pi(4^2)$ 67.021
9. $2075(1 + 0.65)^4$ 15,379.913
10. $\frac{9.2 - 4.5}{0.6}$ 7.833
11. $\frac{3.4}{-7.2 - 8.4}$ −0.218
12. $\frac{1 + 3(4^2)}{7.25}$ 6.759
13. Enter a program similar to those listed on page 856 on a graphing calculator. (See the calculator's manual to learn how to enter the program.) Then use the program to find the volume of several cylinders, including the one shown below.

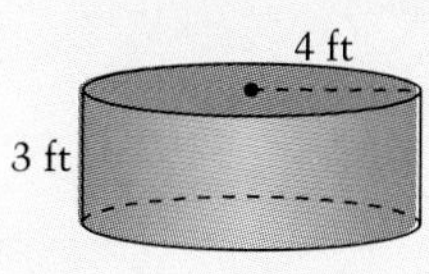

150.796 ft^3

14. Write a graphing calculator program that will evaluate the volume of a rectangular prism. Then use the program to find the volume of several prisms, including the one shown below.

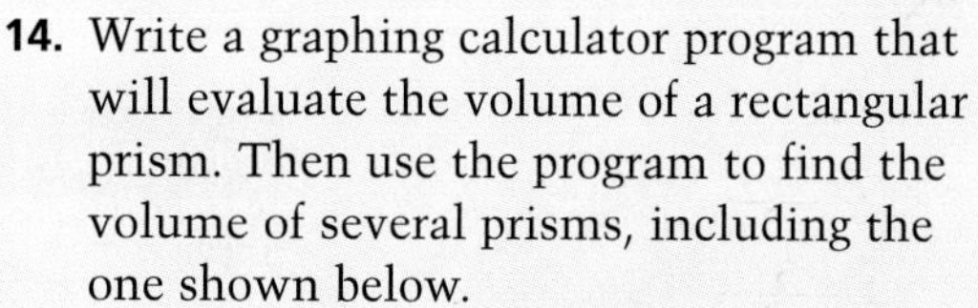

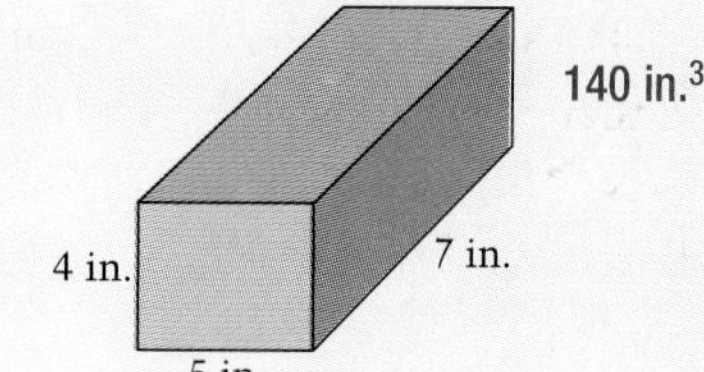

140 $in.^3$

Answer

14. :Disp "ENTER L"
:Input L
:Disp "ENTER W"
:Input W
:Disp "ENTER H"
:Input H
:L * W * H → V
:Disp "VOLUME"
:Disp V

Problem of the Day

A coin dealer knew that a lightweight counterfeit coin was mixed in with 8 rare coins of equal weight. She found it in just two weighings on a balance scale. How did she?
She placed 3 coins on each side. If they balanced, the counterfeit was one of the other 3. If they did not balance, the counterfeit was on the light side. In either case, she placed one coin from the 3 containing the counterfeit on each side. If they balanced, the counterfeit was the third coin. If they did not balance, it was the lighter coin.

ORGANIZER

Warm-Up Exercises

1. Combine.
   a. $8x + 3 - 4x - 12$
   b. $16(8 - 5x) - 42$
   c. $\frac{2}{3}(57 - 24x)$
   d. $5(6x + 7) + (-3)(7 - 8x)$
   a. $4x - 9$ b. $-80x + 86$
   c. $38 - 16x$ d. $54x + 14$
2. Solve.
   a. $5x = 20$ 4
   b. $-8x = 36$ -4.5
   c. $\frac{3}{4}x = 72$ 96
   d. $\frac{-4}{5}x = 2$ -2.5

Lesson Resources

Teaching Tools
Problem of the Day: 1.3
Warm-up Exercises: 1.3
Answer Masters: 1.3
Extra Practice: 1.3

1.3 Solving Linear Equations

What you should learn:

Goal 1 How to solve a linear equation

Goal 2 How to use linear equations to answer questions about real-life situations

Why you should learn it:

You can model many real-life situations with linear equations, such as your weekly pay, when you are working for an hourly wage.

Goal 1 Solving a Linear Equation

An **equation** is a statement that two expressions are equal. A **linear equation** in x is an equation in which x appears to the first power. A number is a **solution** of an equation if the statement is true when the number is substituted for the variable. For instance, 9 is a solution of the linear equation

$$2x - 5 = 13$$

because $2(9) - 5 = 13$ is a true statement. Two equations are **equivalent** if they have the same solutions. For instance, the equations $x - 4 = 1$ and $x = 5$ are equivalent because both have the number 5 as their only solution. The following transformations produce equivalent equations and can be used to **solve an equation.**

Transformations That Produce Equivalent Equations

1. *Add* the same number to both sides.
2. *Subtract* the same number from both sides.
3. *Multiply* both sides by the same nonzero number.
4. *Divide* both sides by the same nonzero number.
5. *Simplify* one or both sides.
6. *Interchange* the two sides.

Example 1 ***Solving a Linear Equation***

Solve $\frac{2}{5}x + 7 = 13$.

Solution This solution can be found by isolating the variable on one side of the equation.

$\frac{2}{5}x + 7 = 13$	*Rewrite original equation.*
$\frac{2}{5}x = 6$	*Subtract 7 from both sides.*
$x = \frac{5}{2}(6)$	*Multiply both sides by $\frac{5}{2}$.*
$x = 15$	*The solution is 15.*

Check this by substituting 15 into the original equation. ■

Example 2 ***Solving a Linear Equation***

Solve $8x + 15 = -4x + 51$.

Solution

$8x + 15 = -4x + 51$	*Rewrite original equation.*
$12x + 15 = 51$	*Add 4x to both sides.*
$12x = 36$	*Subtract 15 from both sides.*
$x = 3$	*Divide both sides by 12.*

Check: You can check the solution by substituting 3 for x in the original equation.

$8x + 15 = -4x + 51$	*Rewrite original equation.*
$8(3) + 15 \stackrel{?}{=} -4(3) + 51$	*Substitute 3 for x.*
$39 = 39$	*The solution checks.*

■

Example 3 ***Solving a Linear Equation***

Solve $15(1 - x) = -3(-x - 2)$.

Solution

$15(1 - x) = -3(-x - 2)$	*Rewrite original equation.*
$15 - 15x = 3x + 6$	*Distributive Property*
$15 = 18x + 6$	*Add 15x to both sides.*
$9 = 18x$	*Subtract 6 from both sides.*
$\frac{1}{2} = x$	*Divide both sides by 18.*

The solution is $\frac{1}{2}$. Check this in the original equation. ■

Example 4 ***Solving a Linear Equation***

$\frac{1}{3}(14x + 9) = 11 - 5(x - 3)$	*Original equation*
$\frac{1}{3}(14x + 9) = 11 - 5x + 15$	*Distributive Property*
$\frac{1}{3}(14x + 9) = 26 - 5x$	*Combine like terms.*
$14x + 9 = 78 - 15x$	*Multiply both sides by 3.*
$29x + 9 = 78$	*Add 15x to both sides.*
$29x = 69$	*Subtract 9 from both sides.*
$x = \frac{69}{29}$	*Divide both sides by 29.*

The solution is $\frac{69}{29}$. Check this in the original equation. ■

LESSON Notes

GOAL 1 Remind students that some equations may have no solution (such as $x = x + 1$) while others may have more than one (such as $x^2 = 4$).

Example 1

The most basic step in solving linear equations is isolating the variable. Then the unknown can usually be found by undoing the operation affecting the variable. Remind students to check all answers in the original equation.

Example 2

Again, an important first step is to isolate the variable, x, by combining x's on one side of the equation. Ask students if it matters on which side the unknown x's are collected.

Examples 3–4

Observe that it is necessary to simplify like terms on either side of the equation before the variable x is isolated.

Example 5

Discuss with students why the expression $t - 40$ is greater than or equal to 0. A good technique that helps students generate algebraic models is a "What if" approach to addressing the real-life situation. For example, what if a worker worked 20 hours in a week? Then, at \$7 per hour, the worker would earn wages equal to $\$7 \cdot 20$, or \$140. What if the worker worked 42 hours? Then there would be 2 hours of overtime. Consequently, wages would equal $\$7 \cdot 40$ plus 2 hours of overtime. But the overtime rate of pay is 1.5 times that of regular pay, or $1.5 \cdot \$7$. Using this type of reasoning, students will begin to understand how the verbal model of Example 5 was created.

Extra Examples

Here are additional examples similar to **Examples 1–5.**

1. Solve the equations.
   - a. $\frac{4}{5}x - 6 = 14$ — 25
   - b. $9x - 13 = 5x + 57$ — 17.5
   - c. $4(x - 3) = -2(9 - 8x)$ — $\frac{1}{2}$
   - d. $5(3x - 18) = 15 - 4(6x - 9)$ — $\frac{47}{13}$
2. In Example 5, how many hours must an employee work in one week to earn $311.00? 43 hours
   If no employee is allowed to work more than 60 hours in a single week, can an employee earn $479.50 in one week? Yes, by working 59 hours

Check Understanding

1. What is a solution of a linear equation?
   See the top of page 18.
2. What are equivalent equations? See the top of page 18.
3. Describe the steps required to solve the linear equation, $4(3x - 2) = -5(1 + x)$.
   See Example 3 for a solution model.
   $$\begin{aligned} 4(3x - 2) &= -5(1 + x) \\ 12x - 8 &= -5 - 5x \\ 17x - 8 &= -5 \\ 17x &= 3 \\ x &= \frac{3}{17} \end{aligned}$$

***Communicating* about ALGEBRA**

How many hours must an employee work in one week at $5.75 per hour to earn what a $6.50-per-hour worker makes in a 40-hour work week?
43.4783 hours or about 43.5 hours

Answer
Use the verbal model in Example 5 and the one-week earnings of 318.50 to write the equation:
$6.50(40) + 1.5(6.50)(t - 40) = 318.50$, and then solve the equation.

BakeryCorp in Miami, Florida, is a bakery that supplies baked goods to restaurants. The business was started in 1986 by three brothers: (clockwise from left) Luis, Juan, and José Lacal. During the first five years, sales approximately doubled each year. In 1991, BakeryCorp paid its employees between $4.50 and $7.00 per hour for the first 40 hours per week and time and a half for overtime.

(Source: Your Company Magazine)

Goal 2 Using Linear Equations to Solve Problems

Real Life
Accounting

Example 5 ***Finding the Number of Hours Worked***

Model the weekly earnings of a baker who earns $7 an hour at *BakeryCorp* and works *at least* 40 hours a week. How many hours would the baker work in a week to earn $332.50?

Solution

Verbal Model — Hourly wage · 40 hours + 1.5 · Hourly wage · Overtime hours

Labels
- Hourly wage = 7 (dollars)
- Total hours worked in week = t (hours)
- Overtime hours = $t - 40$ (hours)

Expression $7(40) + 1.5(7)(t - 40)$

The number of hours worked to earn $332.50 in a week is found by solving the following linear equation.

$$\begin{aligned} 332.50 &= 7(40) + 1.5(7)(t - 40) \\ 332.50 &= 280 + 10.5t - 420 \\ 472.50 &= 10.5t \\ 45 &= t \quad \text{(total hours worked in week)} \end{aligned}$$

***Communicating* about ALGEBRA**

SHARING IDEAS about the Lesson

Suppose the baker in Example 5 earns $6.50 per hour (regular pay rate). How many hours must the employee work in one week to earn $318.50? Explain. 46; see margin.

EXERCISES

Guided Practice

CRITICAL THINKING about the Lesson

In Exercises 1 and 2, decide whether the equations are equivalent.

1. $x - 5 = 5$ and $x = 0$ No
2. $4x = 2$ and $x = 0.5$ Yes
3. Which of the following transformations of the equation $-2x = 3 + 1$ does not produce an equivalent equation? Explain.
   a. Simplify the right side. **b.** Divide both sides by -2. **c.** Square both sides.
   -2 is a solution of the new equation but not of the original equation.
4. Is -4 the only solution of $2x + 9 = 1$? Explain your answer.
   Yes, $2x + 9 = 1$ is equivalent to $2x = -8$, which is equivalent to $x = -4$.

Independent Practice

In Exercises 5–10, describe the transformations you could use to solve the equation.

5. $x + 8 = -3$ Subtract 8.
6. $\frac{1}{7}x = 2$ Multiply by 7.
7. $-\frac{5}{2}x = 4$ Multiply by $-\frac{2}{5}$.
8. $x - 4 = 0$ Add 4.
9. $\frac{x}{3} = 2$ Multiply by 3.
10. $5 = -x$ Multiply by -1.

In Exercises 11–16, check whether the number is a solution of the equation.

11. $5 - t = 7, -2$ It is.
12. $9 + 5t = -6, 4$ It is not.
13. $-4x = 24, -8$ It is not.
14. $2y = -22, 10$ It is not.
15. $6x + 1 = 7, 1$ It is.
16. $3x - 11 = 1, 4$ It is.

In Exercises 17–22, solve the equation.

17. $16x = 8$ $\frac{1}{2}$
18. $x - 7 = 18$ 25
19. $8x - 1 = 15$ 2
20. $5x + 9 = 4$ -1
21. $6 - 2x = 5$ $\frac{1}{2}$
22. $10 + 7x = 24$ 2

In Exercises 23–40, solve the equation. Check your solution.

23. $13x = 52$ 4
24. $\frac{x}{2} = -5$ -10
25. $-\frac{x}{3} = 6$ -18
26. $-5x = 30$ -6
27. $\frac{1}{2}x - 12 = 4$ 32
28. $3x + 5 = 17$ 4
29. $2x + 3 = x - 7$ -10
30. $5x - 4 = 2x + 6$ $\frac{10}{3}$
31. $3(x - 2) = 6(5 + x)$ -12
32. $2(3 - x) = 16(x + 1)$ $-\frac{5}{9}$
33. $4(x + 5) = 6(2x - 1)$ $\frac{13}{4}$
34. $-2(4 + 3x) = 3(x - 2)$ $-\frac{2}{9}$
35. ✪ $3(x - 2) + 6 = 4(2 - x)$ $\frac{8}{7}$
36. ✪ $7(x + 1) = 1 - 2(5 - x)$ $-\frac{16}{5}$
37. ✪ $2(3x - 1) = 5 - (x + 3)$ $\frac{4}{7}$
38. ✪ $-4(3 + x) + 5 = 4(x + 3)$ $-\frac{19}{8}$
39. ✪ $-(x + 2) - 2x = -2(x + 1)$ 0
40. ✪ $-(x - 1) + 10 = -3(x - 2)$ $-\frac{5}{2}$

In Exercises 41 and 42, find the dimensions of the figure.

41. Area = 504
    36 by 14

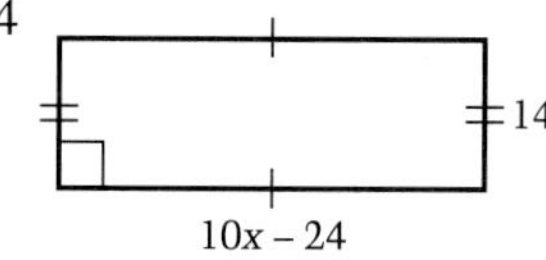

42. Perimeter = 23
    9, 8, 6

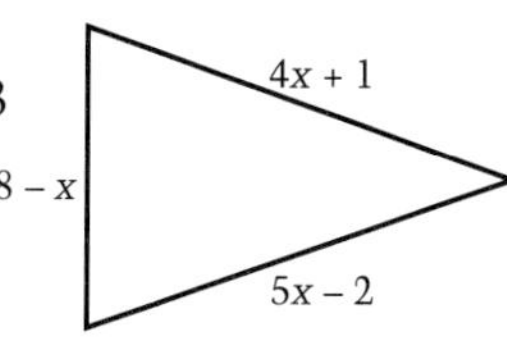

EXERCISE Notes

ASSIGNMENT GUIDE

Basic/Average: Ex. 1–4, 5–43 odd, 49, 63

Above Average: Ex. 1–4, 5–43 odd, 45–49, 63–65

Advanced: Ex. 1–4, 5–43 odd, 44–50, 63–65

Selected Answers
Exercises 1–4, 5–63 odd

✪ More Difficult Exercises
Exercises 35–40, 45–49, 65–66

Guided Practice

▶ **Ex. 3** Challenge students to provide an example of why squaring both sides of an equation may not produce equivalent equations. For example, $(x + 5) = 6$ has 1 as its only solution, and $(x + 5)^2 = 6^2$ has both 1 and -11 as solutions. So, they are not equivalent equations.

Independent Practice

▶ **Ex. 11–40** Encourage students to use a solution format similar to the solution models illustrated in Examples 1–4. This includes aligning the equal signs and stating reasons for each step in their solutions.

▶ **Ex. 41–42 GEOMETRY** These exercises should be a review from geometry.

Ex. 43–48 Students should use Example 5 as a solution model. Be sure to provide a similar model in class for student notes before assigning these exercises.

Ex. 45–48 These exercises should be assigned as a group.

Ex. 49 Refer students to the chapter opener, where the formula $F = \frac{9}{5}C + 32$ is provided.

Ex. 51–62 GROUP ACTIVITY These exercises review essential skills. You may wish to assign them separately to be completed in class so that students can correct any errors.

Ex. 63–64 These exercises are important reminders to students that numeration skills learned in class can transfer to the college entrance tests.

Enrichment Activities

These activities require students to go beyond lesson goals.

Practice your order-of-operation skills.

Expression trees can be used to diagram order of operations. For example, the following is an expression tree for $7 + 2 \times 3$.

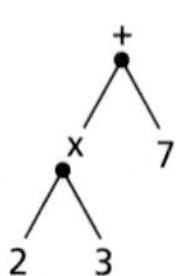

Note that operations are placed at a vertex and that you perform them from the bottom up.

43. *Summer Jobs* You have two summer jobs. In the first job, you work 40 hours a week and earn $6.25 an hour. The second one, you earn $5.50 an hour and can work as many hours as you want. If you want to earn $316 a week, how many hours must you work at your second job? 12

44. *Car Repair* The bill for the repair of your car was $415. The cost for parts was $265. The cost for labor was $25 per hour. How many hours did the repair work take? 6

75 Years of Tractors **In Exercises 45–48, use the following information.**

In 1917, a *Case* tractor was steam-powered. It used 20 chunks of hardwood per hour as a fuel and took $1\frac{1}{2}$ hours to start. The 1991 *Case Magnum 7140* is diesel-powered. It uses 11 gallons of fuel per hour and has a start-up time of 10 seconds. ***(Source: J. I. Case Corp)***

You can find some other comparisons between these two *Case* tractors by using the given clues.

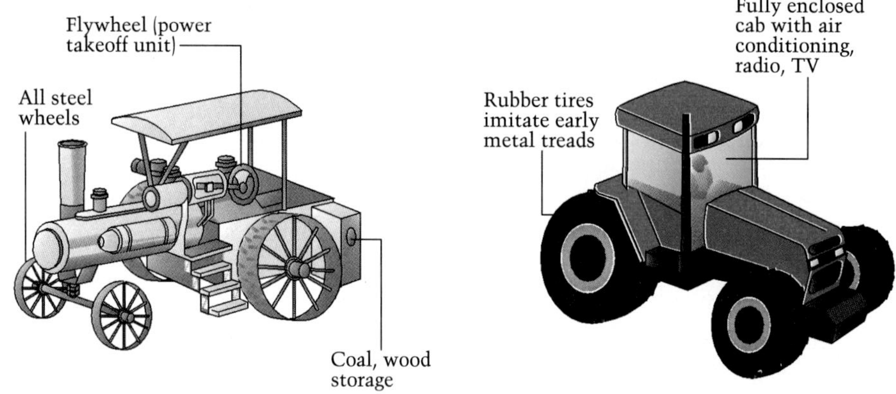

45. The 1991 tractor has a horsepower of 195, which is 5 less than 4 times the horsepower of the 1917 tractor. What was the horsepower of the 1917 tractor? 50

46. The maximum speed of the 1991 tractor is 19.6 miles per hour, which is 1.2 more than 8 times the maximum speed of the 1917 tractor. What was the maximum speed of the 1917 tractor? 2.3 mph

47. In 1917, the *Case* company made 65 tractors, which is 35 less than one hundredth of the *Magnum 7140* tractors made in 1991. How many were made in 1991? 10,000

48. The height of the rear wheels of the 1917 tractor is the same as the height of the front wheels of the 1991 tractor. The 1917 tractor's front wheels were 22 inches shorter than the rear wheels. The 1991 tractor's front wheels are 10 inches shorter than its rear wheels. The average wheel height for the 1991 model is 71 inches. Find the heights of the front and rear wheels of each model. 1917: 44 in., 66 in.; 1991: 66 in., 76 in.

49. *Dry Ice* Dry ice is solid carbon dioxide that changes to a gas at −78.5°C. What is the "melting point" of dry ice in degrees Fahrenheit? −109.3°F

50. *Hot Dog* Your dog's temperature is 38.4°C. Does it have a fever? Explain. No; 38.4°C = 101.12°F which is slightly below normal

The normal body temperature of a dog is 101.5°F.

Integrated Review

In Exercises 51–56, evaluate the expression.

51. $2(3 - 5x)$ when $x = 3$ −24

52. $-3(x + 3)$ when $x = -2$ −3

53. $5x - (x + 3)$ when $x = -5$ −23

54. $4 + (x - 4)$ when $x = 9$ 9

55. $7 + 3(2x - 9)$ when $x = -1$ −26

56. $13x - 2(x + 3)$ when $x = -5$ −61

In Exercises 57–62, simplify the expression.

57. $3x - 4(x - 3)$ $-x + 12$

58. $5(x - 4) + 3$ $5x - 17$

59. $9[x + 3(4 + x)] - 1$ $36x + 107$

60. $10 - [8(5 - x) + 2]$ $8x - 32$

61. $2 + [3(x + 4) - 10]$ $3x + 4$

62. $[2(5x + 3) + 4] - 3$ $10x + 7$

63. *College Entrance Exam Sample* If the following numbers are arranged from least to greatest, what will be their correct order? e

I. $\frac{9}{13}$ **II.** $\frac{13}{9}$ **III.** 70% **IV.** $\frac{1}{0.70}$

a. II, I, III, IV **b.** III, II, I, IV **c.** III, IV, I, II **d.** II, IV, III, I **e.** I, III, IV, II

64. *College Entrance Exam Sample* If $\frac{2}{3} + \frac{3}{4} + \frac{5}{6} + P = 3$, then $P = \boxed{?}$. b

a. $\frac{4}{3}$ **b.** $\frac{3}{4}$ **c.** $\frac{2}{3}$ **d.** $\frac{1}{2}$ **e.** $\frac{1}{3}$

Exploration and Extension

Round-Off Error **In Exercises 65 and 66, solve the equation in the following two ways. Explain why the two solutions are different.**

- Find the exact solution as a fraction. Then write it as a decimal, rounded to two decimal places.
- Write each coefficient as a decimal, rounded to two decimal places. Then solve the equation.

65. $\frac{1}{9}x = 20 + \frac{1}{10}x$ 1800, 2000; error caused by rounding decimal for $\frac{1}{9}$

66. $\frac{9}{20}x - \frac{4}{9}x = 60$ 10,800, 6000; error caused by rounding decimal for $\frac{4}{9}$

1. Make a tree diagram for these expressions.

a. $m(c - d) + n$

b. $(m + n)(c - d)$

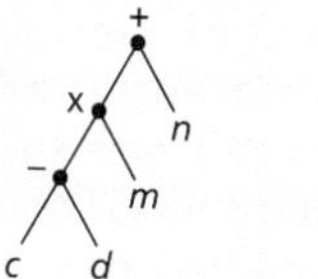

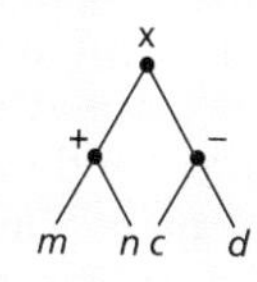

2. What expression is diagrammed by this tree?

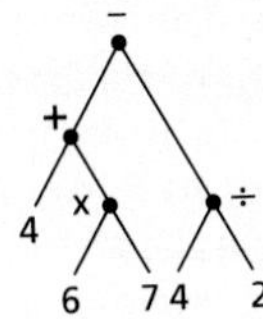

$4 + 6 \times 7 - 4 \div 2$

Challenge! As further practice of order-of-operation skills, you could have students play the game "krypto." Illustrate the game by writing 5 numbers and a krypto number, all between 1 and 15, on the board. For example, write 2, 5, 8, 11, and 14 and state that the krypto number is 10. Tell students to use each of the 5 numbers exactly once and any operation and grouping symbols to produce the "krypto" number. For example,

$(14 - 8) + (11 - 5) - 2 = 10$

You may wish to put a time limit of 3 minutes on the task. Once students understand the game, have them form teams and challenge each other.

Problem of the Day

A bucket contains 1 gallon of red paint, and a can contains 1 gallon of yellow paint. One pint of the yellow paint is placed in the bucket and mixed with the red paint. One pint of the mixture is returned to the can and mixed with the yellow paint. Is the amount of yellow paint in the bucket less than, the same as, or greater than the amount of red paint in the can?

The same as. As long as the containers begin and end with equal amounts of paint, as much has been moved from the bucket to the can as has been moved from the can to the bucket.

ORGANIZER

Warm-Up Exercises

Solve.

1. $8925 = 75x$
2. $679x + 34 = 235 - 125x$
3. $280.05 = 18.45x + 2(x - 13.35)$

1. 119 2. $\frac{1}{4}$ 3. 15

Lesson Resources

Teaching Tools
- Transparency: 4
- Problem of the Day: 1.4
- Warm-up Exercises: 1.4
- Answer Masters: 1.4

Extra Practice: 1.4
Color Transparency: 3

LESSON Notes

Example 1

Some students may wish to use the variable, d, to express the depth of the station. Encourage students to ask "What if" questions to motivate the selection of a verbal model. For example, What would the water pressure

(continued)

1.4 Problem-Solving Strategies

What you should learn:

Goal How to use problem-solving strategies to solve real-life problems

Why you should learn it:

You can use algebra to solve many real-life problems, such as those related to water pressure, building a railroad line, and starting your own business.

Goal Problem-Solving Strategies

In this lesson, you will review some of the most common strategies for solving real-life problems.

Problem-Solving Strategies

- **Guess** a reasonable solution. **Check** the guess. **Revise** the guess and continue until a correct solution is found.
- **Make a table** using the data from a problem. **Look for a pattern.** Use the pattern to complete the table and solve.
- **Draw a diagram** that shows the facts from the problem. Use the diagram to visualize the action of the problem. Use algebra to find a solution. Then check the solution.

Each of these strategies can be used alone or in combination with the strategy for creating algebraic models. The key steps in creating an algebraic model are as follows.

Write a verbal model. → Assign labels. → Write an algebraic model. → Solve the algebraic model. → Answer the question.

Real Life **Scuba Diving**

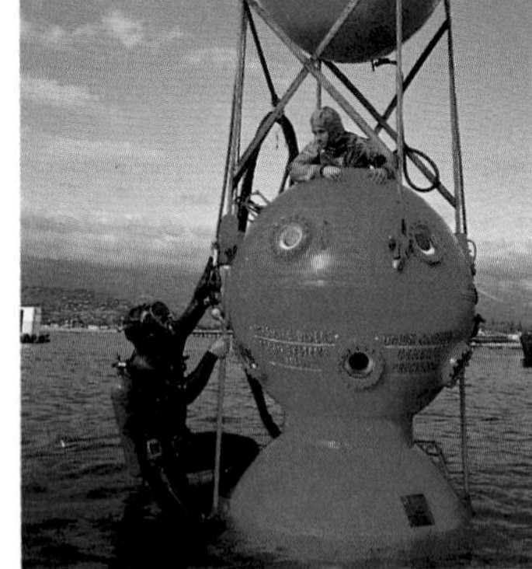

Water weighs 62.4 pounds per cubic foot. The water pressure at a depth, d (in feet), is 62.4d pounds per square foot.

Example 1 *Exploring Arctic Waters*

To study life in arctic waters, scientists worked in a submerged Sub-Igloo station in Resolute Bay in northern Canada. The water pressure at the floor of the station was 2184 pounds per square foot. How deep was the station floor?
(Source: National Geographic Society)

Solution

Verbal Model Water pressure = Weight of water per cubic foot · Depth of station

Labels
- Water pressure = 2184 (pounds per square foot)
- Weight of water = 62.4 (pounds per cubic foot)
- Depth of station = x (feet)

Equation
$2184 = 62.4x$ *Linear model*
$35 = x$ *Divide both sides by 62.4.*

The station floor was 35 feet deep. ■

In 1862, two railroad companies were given the rights to build a railroad connecting Omaha, Nebraska, with Sacramento, California. The Central Pacific Company began building eastward from Sacramento in 1863. Twenty-four months later the Union Pacific Company began building westward from Omaha.

Problem Solving
Draw a Diagram

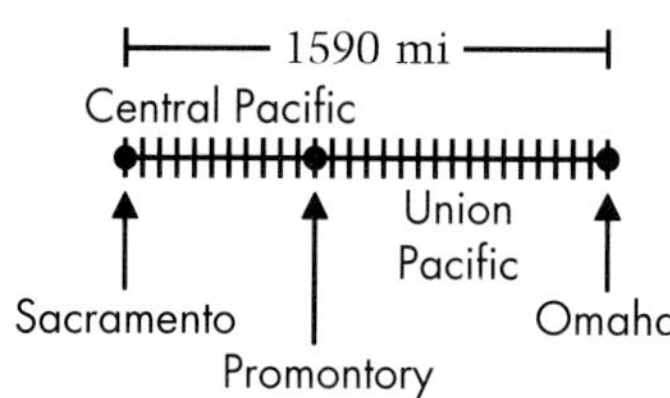

Example 2 *The First Transcontinental Railroad*

The Central Pacific Company averaged 8.75 miles of track per month. The Union Pacific Company averaged 20 miles of track per month. The photo shows the two companies meeting in Promontory, Utah, as the 1590 miles of track were completed. When was the photo taken? How many miles of track did each company build?

Solution

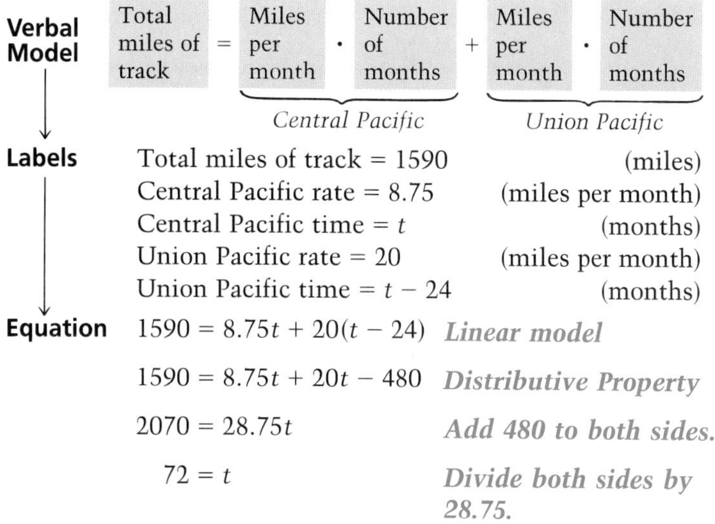

Verbal Model

Total miles of track	=	Miles per month	·	Number of months	+	Miles per month	·	Number of months
		Central Pacific				*Union Pacific*		

Labels

Total miles of track = 1590	(miles)
Central Pacific rate = 8.75	(miles per month)
Central Pacific time = t	(months)
Union Pacific rate = 20	(miles per month)
Union Pacific time = $t - 24$	(months)

Equation

$1590 = 8.75t + 20(t - 24)$ *Linear model*

$1590 = 8.75t + 20t - 480$ *Distributive Property*

$2070 = 28.75t$ *Add 480 to both sides.*

$72 = t$ *Divide both sides by 28.75.*

The construction took 72 months (6 years) from the time the Central Pacific Company began. Thus, the photo was taken in 1869. The number of miles built by each company is:

Central Pacific: (8.75 miles per month)(72 months) = 630 mi

Union Pacific: (20 miles per month)(48 months) = 960 mi ■

(continued)

be if the station were 10 feet deep? In answering such questions, students will discover the relationship between the water pressure, the weight of the water, and the depth d of the station.

Example 2

Discuss some reasons why the Union Pacific Company might have been able to build more track each month than the Central Pacific Company. (Perhaps Union Pacific used more men, or the terrain on which their track was built was less hostile.) Some students may wish to research the question and report to the class.

Example 3

If you receive 60% of the selling price, how many weeks must you work to earn a profit of $400? A little more than 7 weeks

Extra Examples

Here are additional examples similar to **Examples 1–3.**

1. You are driving to a nearby city that is 230 miles away. You believe you can average 50 miles per hour during your drive. How much time should you allocate for your trip?
   A verbal model of the drive is shown.

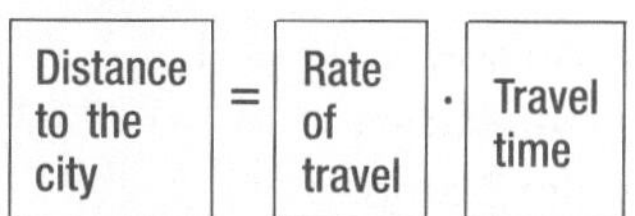

Distance to the city	=	Rate of travel	·	Travel time

   An equation of the drive is $230 = 50t$. The solution is $t = 4.6$ hours or 4 hours and 36 minutes.

2. In Example 2, when would the Union Pacific Company start building track if each company were to build the same amount of track and complete work at the same time?
   The Union Pacific Company should start about 51 months after the Central Pacific Company starts.

3. Suppose in Example 3 that your cost is \$0.75 per sculpture, and that it takes you about 3 hours to complete one. If you spend 20 hours a week working on the sculptures, how many weeks will you work to earn a profit of \$500? 14

Check Understanding

1. Identify the key steps in creating an algebraic model. See page 24.
2. What problem-solving strategies do you most frequently use? Explain. Answers vary. Encourage students to verbalize their individual problem-solving plans.

Communicating about ALGEBRA

Ask students to explore other ways in which the area of the house may be obtained. For instance, can the area of the large rectangle minus the area of the regions outside the house be used to compute the area of the house?

Answer

A. 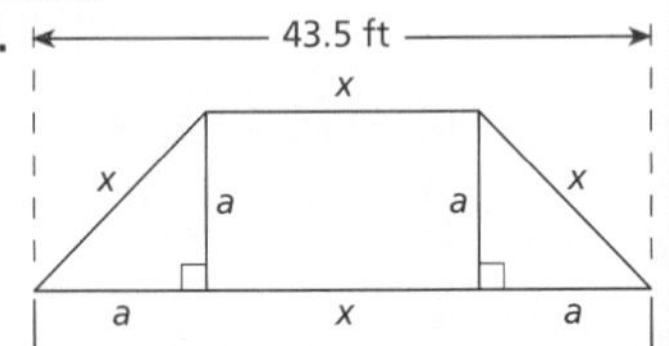

The exterior dining room wall is the hypotenuse of an isosceles right triangle, as indicated in the diagram. If each leg is a, then $a^2 + a^2 = x^2$ and $a = \frac{\sqrt{2}x}{2}$. Then the distance between opposite walls is $\left(\frac{\sqrt{2}x}{2} + x + \frac{\sqrt{2}x}{2}\right)$ ft.

Real Life **Arts and Crafts**

Example 3 ***Starting a Small Business***

You have started a small business making papier-mâché sculptures. Your cost per sculpture is only \$0.50. (The ingredients are water, flour, carpenter's glue, newspaper, wire, and paint.) Your pieces sell for \$12.50 at an arts-and-crafts shop, and you receive 50% of the selling price. Each sculpture takes about 2 hours to complete. If you spend 16 hours a week working on the sculptures, how many weeks will you work to earn a profit of \$400?

Solution Because you receive \$6.25 for each sculpture, your profit per sculpture is \$5.75. Working 16 hours a week, you can make 8 sculptures per week.

Verbal Model

Total profit = Profit per sculpture · Sculptures per week · Number of weeks

Labels

Total profit = 400	(dollars)
Profit per sculpture = 5.75	(dollars per sculpture)
Sculptures per week = 8	(sculptures per week)
Number of weeks = x	(weeks)

Equation

$400 = 5.75(8)(x)$ *Linear model*

$400 = 46x$ *Simplify*

$8.7 \approx x$ *Divide both sides by 46.*

It will take you about 9 weeks to make a profit of \$400. ■

Communicating about ALGEBRA

SHARING IDEAS about the Lesson

Use a Diagram The house plans at the left show a house whose shape is a regular octagon. (The 8 sides have the same length.)

A. Approximate the length, x (in feet), of each side of the house by solving the following linear equation. 18.0 ft, see margin.

$$43.5 = \frac{\sqrt{2}x}{2} + x + \frac{\sqrt{2}x}{2} \text{ or } 43.5 = \frac{2\sqrt{2}x + 2x}{2}$$

Explain how the Pythagorean Theorem can be used to obtain this equation. (What do you know about the green triangle in the diagram?)

B. Partition the house into two trapezoids and one rectangle. Then use these shapes to approximate the area of the house. 1570 ft²

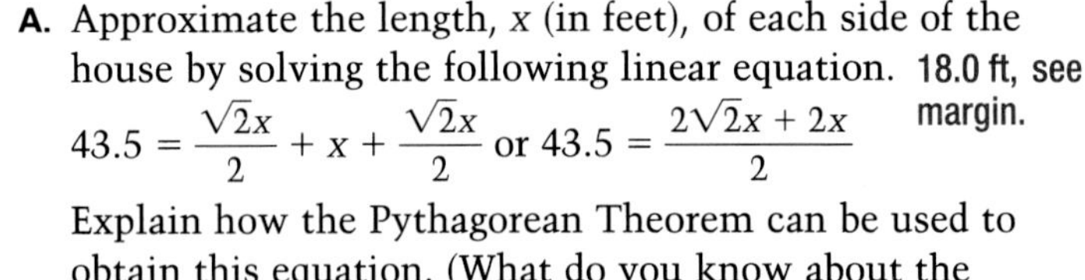

EXERCISES

Guided Practice

CRITICAL THINKING about the Lesson

Girl Scout Cookies **In Exercises 1–4, use the following information.**

Your sister is selling Girl Scout cookies for $2.60 a box. Your family bought 6 boxes. How many more boxes of cookies must she sell in order to collect $130? Use the following verbal model.

Boxes sold	·	Price per box	+	Boxes to be sold	·	Price per box	=	Total sales

1. Assign labels to each part of the verbal model. Indicate the units of measure. See margin.
2. Use the labels to translate the verbal model into an algebraic equation. $6(2.60) + x(2.60) = 130$
3. Solve the equation. 44
4. Answer the question. 44

Independent Practice

Sailing on the Seine **In Exercises 5–8, use the following information.**

You are on a boat traveling the navigable portion of the Seine River. The boat's average speed is 32 kilometers per hour. How long will the boat ride take? Use the following verbal model.

Distance traveled	=	Rate of boat	·	Time traveled

5. Assign labels for each part of the verbal model. See margin.
6. Use the labels to translate the verbal model into an algebraic equation. $547 = 32t$
7. Solve the equation. ≈ 17.1
8. Answer the question. ≈ 17.1 hours

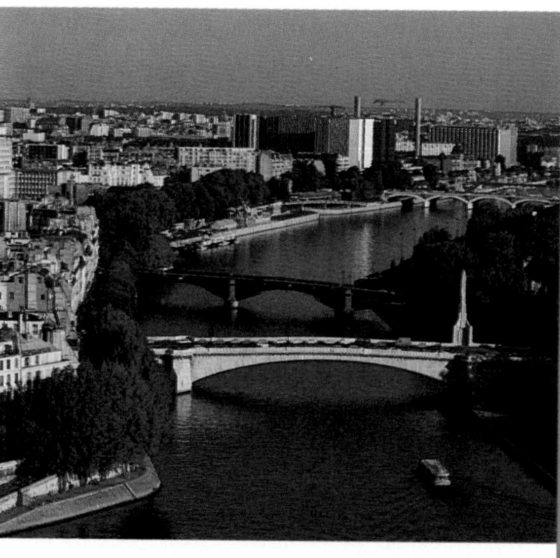

The Seine River flows under more than 30 bridges in Paris, France. The entire length of the river is 764 kilometers, but only 547 kilometers can accommodate boats.

EXERCISE Notes

ASSIGNMENT GUIDE

Basic/Average: Ex. 1–4, 5–8, 15–16, 18–24, 30–31

Above Average: Ex. 1–4, 5–8, 13–24, 31

Advanced: Ex. 1–4, 5–12, 15–24, 31

Selected Answers
Exercises 1–4, 5–29 odd

✪ More Difficult Exercises
Exercises 13–16, 31

Guided Practice

▶ **Ex. 1–4** COOPERATIVE LEARNING These exercises provide you with an opportunity to check on students' readiness for independent practice. You may wish to have students work in pairs to complete these exercises in order to discuss their strategies for creating an algebraic model.

Answer

1. Boxes sold = 6 (boxes)
   Price per box = 2.60 (dollars per box)
   Boxes to be sold = x (boxes)
   Total sales = 130 (dollars)

Independent Practice

▶ **Ex. 5–8 and 9–12** These exercises should be assigned as a group. They provide the verbal model, thus helping students to develop the remaining steps of the problem-solving process.

Answer

5. Distance traveled = 547 (km)
   Rate of boat = 32 (km per hour)
   Time traveled = t (hcurs)

Answer

9. Your rate = 5.5 (yards per second)
   Your time = t (seconds)
   Opponent's rate = 5 (yards per second)
   Opponent's time = $t + 16$ (seconds)

▶ **Ex. 13–14 and 15–16** These exercises should be assigned in pairs. Encourage students to follow the plan outlined in Examples 1–3 and begin by writing a verbal model and assigning labels.

MATH JOURNAL Students may wish to include the key problem-solving steps in their math journals. Several problem-solving strategies presented in the lesson should be recorded with brief explanations of them for future reference.

Relay Race **In Exercises 9–12, use the following information.**

You are the last runner in a relay race. The baton is passed to you 16 seconds after your opponents pass theirs. If you run 5.5 yards per second and tie your opponent, who runs 5 yards per second, how long did it take you to run your portion of the race? Use the following verbal model.

Your rate · Your time = Opponent's rate · Opponent's time

9. Assign labels to each part of the verbal model. **See margin.**

10. Use the labels to translate the verbal model into an algebraic equation. $5.5t = 5(t + 16)$

11. Solve the equation. **160**

12. Answer the question. **160 seconds**

Environmental Hazard **In Exercises 13 and 14, use the following information.**

You are attending a wedding in Cedar City, Utah. As the bride and groom prepare to go to the wedding reception, helium balloons are released in the air. The wind is blowing as shown on the map at the right. The radius of the blue circle is 1000 miles, and the radius of the green circle is 2000 miles. After 24 hours, the balloons fall to Earth due to the loss of helium.

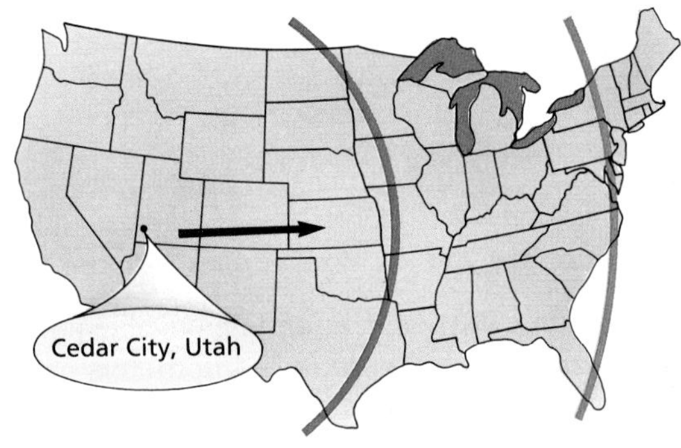

If a helium balloon travels as far as the ocean before falling to Earth, it can be mistaken as food by fish or sea mammals, who can die from ingesting it.

13. What wind speed will allow the balloons to reach the Atlantic Ocean? $83\frac{1}{3}$ **mph**

14. Suppose the balloons remain in the air for 36 hours. What wind speed will allow the balloons to reach the Atlantic Ocean? $55\frac{5}{9}$ **mph**

15. *Hang Gliding* You are taking hang gliding lessons. The cost of the first lesson is one and a half times the cost of each additional lesson. You spend \$260 for six lessons. How much did each lesson cost? **1st: \$60, others: \$40**

16. *Typing Papers* Some of your classmates ask you to type their English papers throughout a 10-week summer course. How much should you charge for each page if you want to earn enough to pay for the hang gliding classes in Exercise 15 and have \$50 left for spending money? You estimate that you can type 20 pages a week. **\$1.55**

17. *Density of Liquids* You are assigned a project in science class. You must use a hydrometer to measure the density of different liquids and then determine the specific gravity of the liquid. If you measure the density of ethanol to be 49 pounds per cubic foot, what is its specific gravity? ≈ 0.79

18. *Boyle's Law* The Irish scientist Robert Boyle (1627–1691) performed an experiment in which he varied the volume, V, and pressure, P, of a fixed quantity of air. (The temperature was kept constant.) Here are the actual data from the experiment. $VP = 1410$

V (in.^{3})	48	40	32	24
P (in. of Hg)	29.1	35.3	44.2	58.8

V (in.^{3})	20	16	12
P (in. of Hg)	70.7	87.2	117.5

The equation that relates V and P is an example of what is now called *Boyle's Law*. What is the equation?

The specific gravity of a liquid is the ratio of its density to the density of water (which is 62.4 pounds per cubic foot). Of two liquids, the one with the lower specific gravity will float on top of the other. The liquids in this photo (from top to bottom) are oil and vinegar.

Integrated Review

In Exercises 19–24, find the resulting unit of measure.

19. $\frac{\text{meters}}{\text{minutes}} \cdot \text{minutes}$ meters

20. $\frac{\text{hours}}{\text{week}} \cdot \text{weeks}$ hours

21. feet + feet feet

22. liters − liters liters

23. $\frac{\text{yards}}{\text{second}} \cdot \text{seconds} - \text{yards}$ yards

24. $\frac{\text{dollars}}{\text{hour}} \cdot \text{hours} + \text{dollars}$ dollars

In Exercises 25–30, solve the equation.

25. $6.7x = 93.1$ ≈ 13.9

26. $9.3 = 2.8y$ ≈ 3.3

27. $3y + 4.4 = 9.3y + 6$ ≈ -0.25

28. $4.1 - 5.3x = 12.9x$ ≈ 0.23

29. $3.1x = 7.5 - 1.9x$ 1.5

30. $6y - 4.2 = 12$ 2.7

Exploration and Extension

31. *You Be the Teacher* Create a real-life problem similar to those in Exercises 5–18. The problem should require a solver to use a linear equation. Pick a subject that is interesting to you. Use algebra to solve your problem. Answers vary.

▶ **Ex. 17–18** CONNECTIONS Throughout the text, exercises that represent examples of "connections" to other school courses or pastimes, such as science or sports, will be labeled. These exercises can provide an opportunity for students to share their knowledge with their classmates. For example, ask anyone who has ever used a hydrometer to describe its use to the class.

▶ **Ex. 19–24** These are important review questions for developing student confidence with labeling.

▶ **Ex. 31** PROJECT This could be used as an in-class project. Refer students to Example 3 for a solution model. Student projects could be collected and displayed. You might also save them to use as review later in the year.

Enrichment Activities

These activities require students to go beyond lesson goals.

Use an encyclopedia or other reference books to complete these activities.

1. **EXTEND EX. 18** Write a report on the Gas Laws as described by Robert Boyle and J.A.C. Charles. Include information about the scientists in your report.
2. While he was experimenting with hydrogen, Charles became interested in ballooning. On December 1, 1783, he put his principles to the test and made one of the earliest flights in a hydrogen balloon, over Paris. Write a report on the history of hot-air balloons and how they work.

Mid-Chapter SELF-TEST

Remind students to complete this test under test-like conditions.

The answers to the Mid-Chapter Self-Tests will be found in the "Selected Answers" section of the student text.

Mid-Chapter SELF-TEST

Take this test as you would take a test in class. The answers to the exercises are given in the back of the book.

1. Write the numbers in increasing order: $4, -2, -3, -\frac{3}{2}, 0, 3, \frac{1}{2}$.
(1.1) $-3, -2, -\frac{3}{2}, 0, \frac{1}{2}, 3, 4$

2. Plot -3 and -2 on the real number line. Which is greater? **(1.1)** -2

3. What is the product of $\frac{2}{3}$ and $-\frac{2}{3}$? **(1.1)** $-\frac{4}{9}$

4. What is the sum of $\frac{2}{3}$ and $-\frac{2}{3}$? **(1.1)** 0

5. Evaluate $6 - 4 \div 12 + 8 \div 6$. **(1.2)** 7

6. Evaluate $12 \cdot 2 - 2 \cdot 6$. **(1.2)** 12

In Exercises 7–10, evaluate the expression. (1.2)

7. $-3x^2 + 4x$ when $x = -2$ $\quad -20$

8. $-a(a - 3)$ when $a = 1$ $\quad 2$

9. $\dfrac{-2(y + 1)}{16 - 2y^2}$ when $y = 4$ $\quad \frac{5}{8}$

10. $\dfrac{ab - 1}{ab + 1}$ when $a = -2$ and $b = 2$ $\quad \frac{5}{3}$

In Exercises 11 and 12, evaluate the expression. Round the result to two decimal places. (1.2)

11. $1.34(x - 9.21)^2$ when $x = 4.13$. $\quad 34.58$

12. $\dfrac{1.67 - t}{4t + 1}$ when $t = 0.32$. $\quad 0.59$

In Exercises 13–16, solve the equation. (1.3)

13. $x - 6 = 3x + 2$ $\quad -4$

14. $3(a - 1) = 5 - a$ $\quad 2$

15. $7 = 7(2b + 5) - 6(b + 8)$ $\quad \frac{5}{2}$

16. $-2(4t - 7) = 3(t - 10)$ $\quad 4$

In Exercises 17–19, find the area of the blue region. (In Exercise 18, each of the yellow regions are quarter circles.) (1.2)

17. 9 cm^2

2cm
3cm
4cm

18. $\approx 86\text{ ft}^2$

10ft 10ft
10ft
10ft

19. 6 ft^2

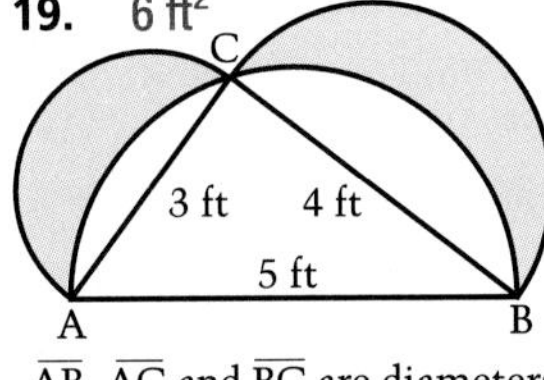

$\overline{AB}$, $\overline{AC}$ and $\overline{BC}$ are diameters

20. From 1980 through 1990, the prize money, P (in \$1000's), for the singles champions at the U.S. Tennis Open can be modeled by

$$P = 30.2t + 35.8$$

where $t = 0$ represents 1980. According to this model, when will the prize money be \$500,000? 1996
(Source: United States Tennis Association) **(1.4)**

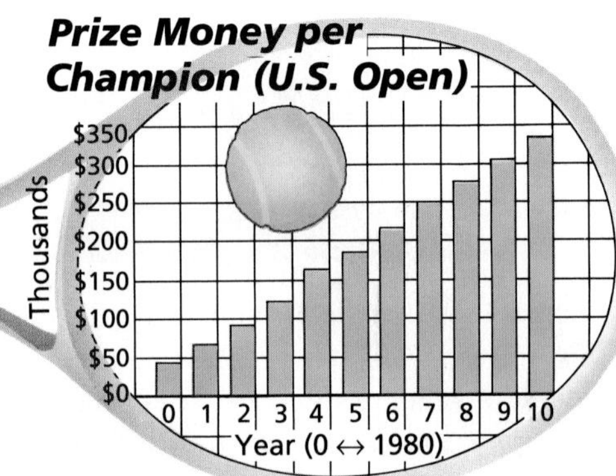

1.5 Literal Equations and Formulas

What you should learn:

Goal How to solve a literal equation for a specific variable, and how to evaluate it for specified values of the other variables

Why you should learn it:

To estimate a value in a formula, you can solve for that variable and substitute known values for the other variables.

Goal Solving a Literal Equation

A **literal equation** is an equation that has more than one variable. For instance, $5x + 2y = 7$ is a literal equation because it has two variables, x and y. The word *literal* comes from the Latin word for "letter." You can solve a literal equation for any one of its variables.

Solve for x	**Solve for y**
$5x + 2y = 7$	$5x + 2y = 7$
$5x = 7 - 2y$	$2y = 7 - 5x$
$x = \dfrac{7 - 2y}{5}$	$y = \dfrac{7 - 5x}{2}$

Example 1 *The Area of an Ellipse*

The area, A, of an ellipse is given by the formula $A = \pi ab$, where a and b are half the lengths of the major and minor axes, respectively.

a. Solve this formula for a.

b. Use the result to find the length of the major axis of an ellipse whose area is 15 square inches and whose minor axis has a length of 4 inches.

Solution

a. $A = \pi ab$ *Formula for area*

$\dfrac{A}{\pi b} = a$ *Divide both sides by πb.*

This gives the length of the semimajor axis in terms of the area and the length of the semiminor axis.

b. Substitute $A = 15$ and $b = 2$ into the formula for a.

$a = \dfrac{A}{\pi b}$ *Formula for a*

$= \dfrac{15}{\pi(2)}$ *Substitute 15 for A and 2 for b.*

≈ 2.39 *Simplify.*

The length of a is about 2.39 inches, so the major axis has a length of about 4.78. ■

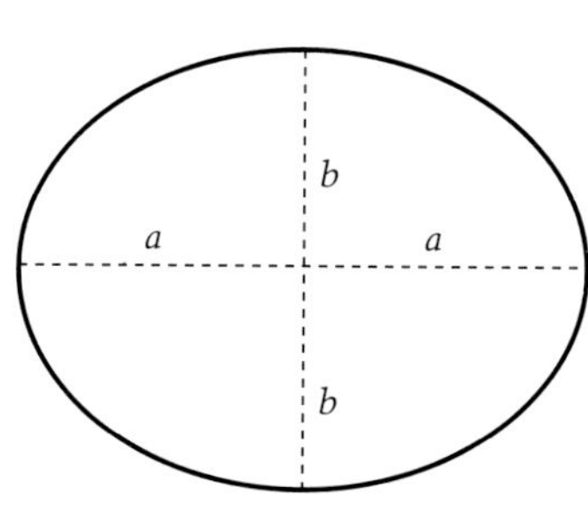

The longest chord of an ellipse is its major axis. The shortest chord through its center is its minor axis.

Problem of the Day

Determine the digits represented by the asterisks.

```
        * * *             193
* * ) * 9 * * *      99)19107
      * *                99
      * * *              920
      * * *              891
        2 * *            297
        * * *            297
```

ORGANIZER

Warm-Up Exercises

1. Solve.
   a. $3y + 4 = 56$ $y \approx 17.3$
   b. $96 = 3.14(6k)$ $k \approx 5.1$
   c. $36{,}144 = 45p$ $p = 803.2$
2. Solve for m, for each given value of n.
   a. $3m - 2n = 18$; $n = 4$
   b. $6n + 5m = 60$; $n = -2$
   c. $7n = 42 - 6m$; $n = 3$

   a. $m \approx 8.67$
   b. $m = 14.4$
   c. $m = 3.5$

Lesson Resources

Teaching Tools
- Problem of the Day: 1.5
- Warm-up Exercises: 1.5
- Answer Masters: 1.5

Extra Practice: 1.5

Color Transparency: 4

LESSON Notes

GOAL 1 Encourage students to justify the steps taken in solving the literal equation, $5x + 2y = 7$, for x and for y. Ask students to solve for both x and y in the literal equations that follow.

a. $6y - x = 15$

b. $8x + 3y = 48$

c. $-54 - 9y = 6x$

(continued)

(continued)

a. $x = 6y - 15,\ y = \frac{x + 15}{6}$

b. $x = 6 - \frac{3}{8}y,\ y = 16 - \frac{8}{3}x;$

c. $x = -9 - \frac{3}{2}y,\ y = -6 - \frac{2}{3}x$

Example 1

You may wish to observe that when the major axis equals the minor axis, the "ellipse" is a circle, and the formula for the area becomes $A = \pi a \cdot a = \pi a^2$.

EXTEND Example 2 Students may wish to explore whether a similar statistic can be found for approximating the number of wins in other sports such as football or basketball. Relevant data may be obtained from an almanac.

Example 3

Note that the smaller triangle (with base b_1 and height h_1) and the original triangle (with base b, and height h) are similar. Consequently, $\frac{b_1}{h_1} = \frac{b}{h}$. But $h_1 = \frac{2}{3}h$, so that $\frac{b_1 h}{h_1} = \frac{b_1}{\frac{2}{3}h} = \frac{b}{h}$. It follows that $b_1 = \frac{2}{3}h \cdot \frac{b}{h} = \frac{2}{3}b$; so, the area of the smaller triangle is $\frac{1}{2}\left(\frac{2}{3}b\right)\left(\frac{2}{3}h\right)$.

Extra Examples

Here are additional examples similar to **Examples 1–3.**

1. The area of a sector of a circle, A, is given by the formula $A = \frac{1}{2}rs$ where r is the radius of the circle and s is the arc length of the sector.
   a. Solve for s in this formula.
   b. Use the result from Part a. to find the arc length of a circle whose area is 57 square units and whose radius is 6 units.

a. $s = \frac{2A}{r}$

b. $s = 19$ units

The **Pythagorean Theorem** *of* **Baseball** *is a formula for approximating the ratio of the number of wins to the total games played. Let R be the number of runs scored during the season, A be the number of runs allowed the opponents, W be the number of games won, and T be the number of games played. These four variables are related by*

$$\frac{R^2}{R^2 + A^2} \approx \frac{W}{T}.$$

(Source: **Inside Sports***)*

Real Life
Sports Estimating

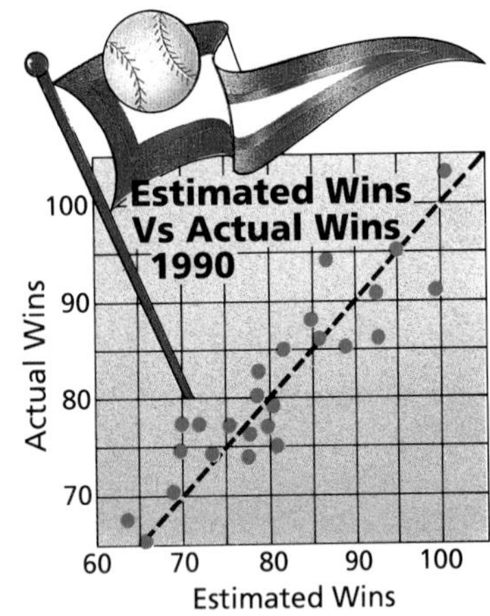

Example 2 *Approximating the Number of Wins in Baseball*

a. Solve the formula $\frac{R^2}{R^2 + A^2} \approx \frac{W}{T}$ for W.

b. In 1990, the Pittsburgh Pirates scored 733 runs and gave up 619 runs. How many of its 162 games would you estimate the team to have won?

Solution

a. $\frac{R^2}{R^2 + A^2} \approx \frac{W}{T}$ *Pythagorean Theorem of Baseball*

$\frac{TR^2}{R^2 + A^2} \approx W$ *Multiply both sides by T.*

b. Substitute $R = 733$, $A = 619$, and $T = 162$.

$$W \approx \frac{TR^2}{R^2 + A^2} = \frac{162(733^2)}{733^2 + 619^2} = 95$$

The formula would estimate the number of wins to be 95 (which, as it happens, is exactly the number of games the Pittsburgh Pirates won in 1990.) ■

The graph at the left shows the estimated number of wins for all 26 American League and National League baseball teams for 1990. For 15 of the teams, the formula gives a number of wins within three of the actual number.

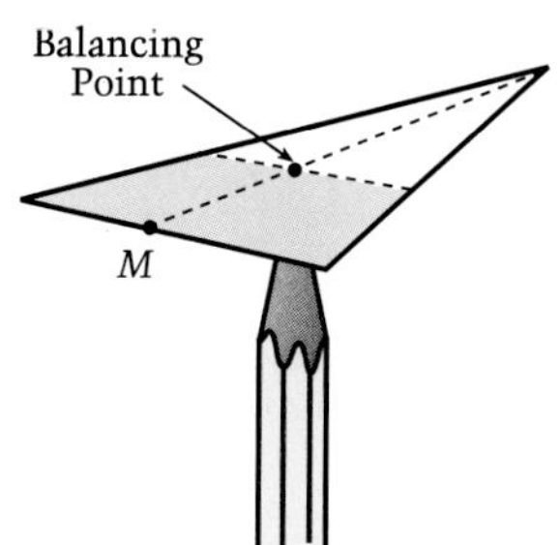

The **centroid** of a plane figure is the point at which the figure will balance. To find the balancing point of *any* triangular model, use the following steps.

1. Measure the height of the triangle, h, and draw a line that is parallel to the base at a height of $\frac{1}{3}h$.
2. Draw a line from the midpoint of the base to the opposite vertex.
3. The point at which the two lines intersect is the balancing point.

Connections **Geometry**

Example 3 *Finding a Formula for Area*

Find a formula for the area of the blue region. Is the area of the blue region half the total area of the triangle?

Solution

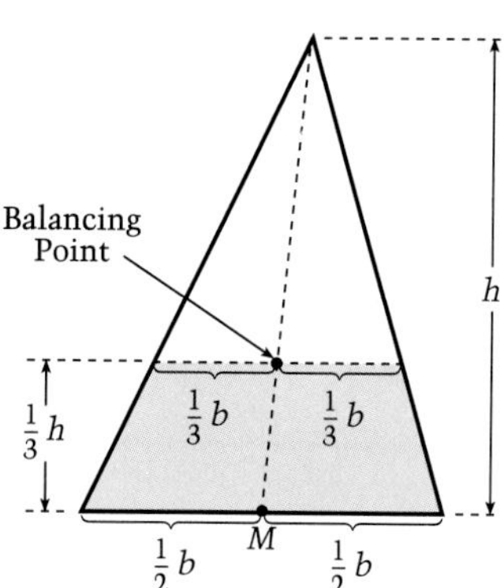

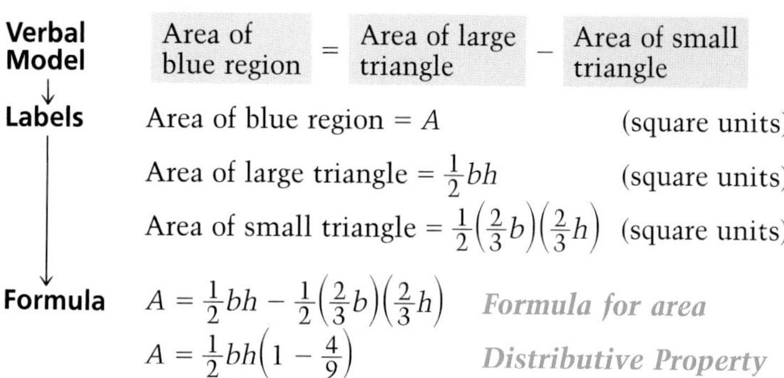

Verbal Model: Area of blue region = Area of large triangle − Area of small triangle

Labels:
Area of blue region = A (square units)
Area of large triangle = $\frac{1}{2}bh$ (square units)
Area of small triangle = $\frac{1}{2}\left(\frac{2}{3}b\right)\left(\frac{2}{3}h\right)$ (square units)

Formula:
$A = \frac{1}{2}bh - \frac{1}{2}\left(\frac{2}{3}b\right)\left(\frac{2}{3}h\right)$ *Formula for area*
$A = \frac{1}{2}bh\left(1 - \frac{4}{9}\right)$ *Distributive Property*
$A = \frac{1}{2}\left(\frac{5}{9}\right)bh$ *Simplify.*

Because the total area of the triangle is $\frac{1}{2}bh$, half its area is $\frac{1}{4}bh$. Thus, the blue area is more than half the total area. ■

Communicating about ALGEBRA

SHARING IDEAS about the Lesson

Cut a triangle out of a piece of cardboard. (Use a height of about 6 inches and a base of about 5 inches.) Use the steps listed above to find the balancing point of the triangle. Then test your result by balancing the triangle on the *eraser* of a pencil. (If you measured *very* carefully, the triangle could balance on the *point* of the pencil.) See margin.

2. In Example 2, solve the given formula for R^2. If your high school baseball team gave up 61 runs in its 20-game season and if the team won 12 games, how many runs would you estimate they earned during the season?
$R^2 \approx \frac{A^2W}{T - W}$; $A = 61$, $W = 12$, $T = 20$, and $R \approx 75$ runs
3. In Example 3, find the area, A, of the blue region for an equilateral triangle.
$A = \left(\frac{5\sqrt{3}}{36}\right)s^2$

Check Understanding

1. How are literal equations and algebraic formulas related? A formula is a general rule or principle. It is usually stated in terms of a literal equation.
2. Identify three or four common formulas. For each formula, solve for each of the variables in terms of the others.

Communicating about ALGEBRA

Answer
Cut the triangle into two pieces along the line through the balancing point and parallel to the base, then weigh each piece. The larger piece will weigh more.

EXTEND *Communicating*

Ask students to explore finding the centroid of various quadrilaterals. Begin with a square or a rectangle. Students should investigate the importance of diagonals in constructing the centroids.

EXERCISE Notes

ASSIGNMENT GUIDE

Basic/Average: Ex. 1–4, 5–10, 11–23 odd, 25–27, 21–39 odd

Above Average: Ex. 1–4, 5–10, 11–23 odd, 24–27, 31–39, 40–41

Advanced: Ex. 1–4, 5–12, 13–23 odd, 24–27, 31–39 odd, 40–41

Selected Answers
Exercises 1–4, 5–39 odd

More Difficult Exercises
Exercises 16, 19, 21–23

Guided Practice

Ex. 1–4 Formulas from geometry are good sources for literal equations. Use these exercises as an in-class assessment of student understanding. Students could copy each figure and label it using the variables.

Answers

1. a, Since the speed is not given, it is represented by a variable.
2. b: length of the side of the base, h: height
3. r: radius of the base, h: height
4. r: radius

Independent Practice

Ex. 5–10 These exercises should be assigned as a group. Students may not have been previously introduced to the formula for the area of an ellipse as $A = \pi ab$. Refer students to Example 1 for an explanation of the variables in this formula.

EXERCISES

Guided Practice

CRITICAL THINKING about the Lesson

1. Which of the following sentences can be written as a literal equation? Explain.
   a. The time it takes to drive 20 miles to the store is equal to 20 divided by the speed. **a, $t = \frac{20}{s}$**
   b. The height of the mast of the sailboat is 12.5 meters.

Geometry **In Exercises 2–4, match the formula for volume, V, with the figure. Describe the measurement represented by each variable (other than V).**

2. $V = \frac{1}{3}b^2h$ **b; see margin.**
3. $V = \frac{1}{3}\pi r^2h$ **c; see margin.**
4. $V = \frac{4}{3}\pi r^3$ **a; see margin.**

a. Sphere

b. Pyramid

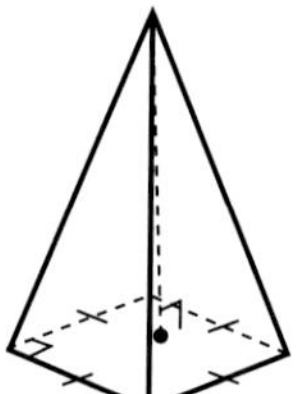

c. Cone

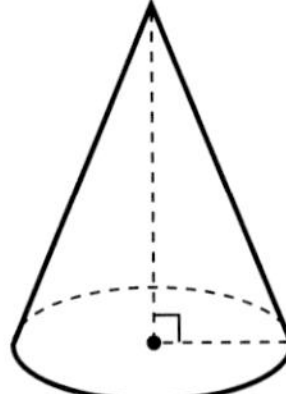

Independent Practice

Geometry **In Exercises 5–10, match the formula for area, A, with the figure. Describe the length represented by each variable (other than A).**

5. $A = \pi ab$ **b; see margin.**
6. $A = \pi r^2$ **e; see margin.**
7. $A = \frac{1}{2}bh$ **c; see margin.**
8. $A = LW$ **f; see margin.**
9. $A = bh$ **d; see margin.**
10. $A = \frac{1}{2}(b_1 + b_2)h$ **10. a; see margin.**

a. Trapezoid

b. Ellipse

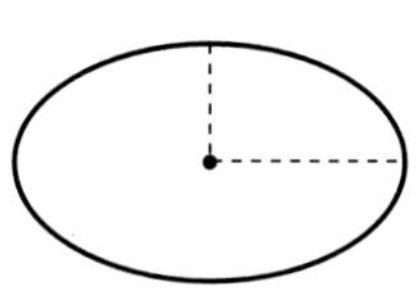

c. Triangle

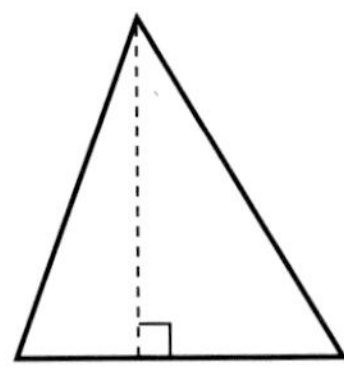

d. Parallelogram

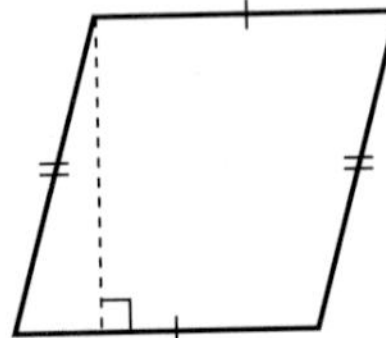

e. Circle

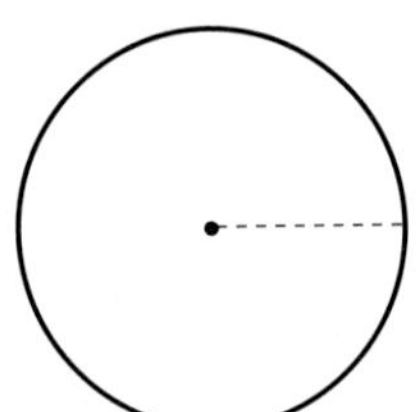

f. Rectangle

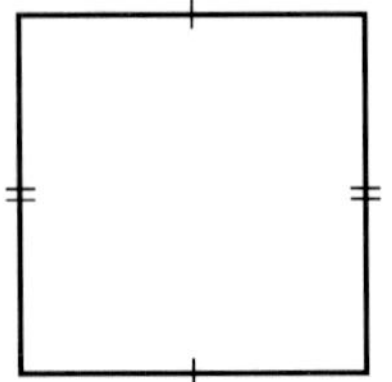

In Exercises 11–20, solve for the indicated variable.

11. *Volume of a right circular cylinder*
Solve for h: $V = \pi r^2 h$ $h = \dfrac{V}{\pi r^2}$

12. *Perimeter of a rectangle*
Solve for W: $P = 2L + 2W$ $W = \dfrac{P - 2L}{2}$

13. *Circumference of a circle*
Solve for r: $C = 2\pi r$ $r = \dfrac{C}{2\pi}$

14. *Volume of a general cone*
Solve for A: $V = \frac{1}{3}Ah$ $A = \dfrac{3V}{h}$

15. *Area of a trapezoid*
Solve for b_2: $A = \frac{1}{2}(b_1 + b_2)h$ $b_2 = \dfrac{2A}{h} - b_1$

16. *Discount*
Solve for L: $S = L - rL$ $L = \dfrac{S}{1 - r}$

17. *Conversion from Fahrenheit to Celsius*
Solve for F: $C = \frac{5}{9}(F - 32)$ $F = \frac{9}{5}C + 32$

18. *Profit*
Solve for C: $P = R - C$ $C = R - P$

19. *Investment at simple interest*
Solve for P: $A = P + Prt$ $P = \dfrac{A}{1 + rt}$

20. *Investment at compound interest*
Solve for P: $A = P(1 + r)^t$ $P = \dfrac{A}{(1 + r)^t}$

In Exercises 21–23, solve the formula for the indicated variable. Then evaluate the variable for the given values. (Include units of measure in your result.)

21. $p = \dfrac{A}{2\pi w}$; ≈ 1.8 cm

22. $h = \dfrac{S - 2\pi r^2}{2\pi r}$; ≈ 2.6 in.

21. Area of a circular ring: $A = 2\pi pw$. Solve for p. Find p given $A = 22\text{ cm}^2$ and $w = 2$ cm.

22. Surface area of a cylinder: $S = 2\pi rh + 2\pi r^2$. Solve for h. Find h given $S = 105\text{ in.}^2$ and $r = 3$ in.

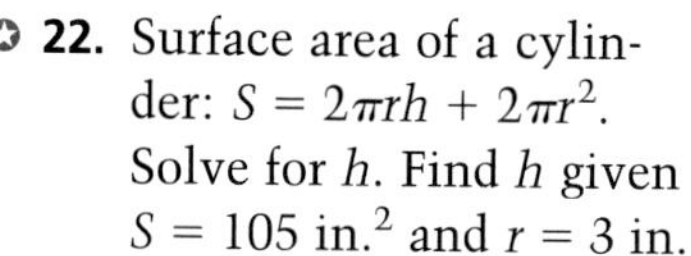

23. Perimeter of a pond: $P = 2\pi r + 2x$. Solve for r. Find r if $P = 75$ ft and $x = 20$ ft.
$r = \dfrac{P - 2x}{2\pi}$; ≈ 5.6 ft

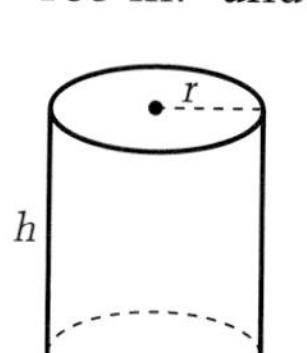

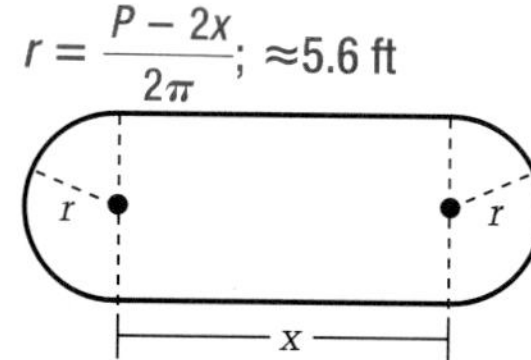

24. *Forager Honeybees* A forager honeybee spends about three weeks becoming accustomed to the immediate surroundings of its hive, and then the rest of its life collecting pollen and nectar. A forager's flight muscles can last only about 500 miles—after that the bee dies. The total number of miles, T, a forager can fly in its lifetime, L (in days), can be modeled by

$T = m(L - 21)$

where m is the number of miles it flies each day. A hardworking forager honeybee can fly about 55 miles each day. About how long does it live? ≈ 30 days

23. $b_1 = \dfrac{2A}{h} - b_2$; 25 ft

Answers

5. a: length of the major axis, b: length of the minor axis
6. r: radius
7. b: length of the base, h: height
8. L: length of a side, W: length of an adjacent side
9. b: length of the base, h: height
10. b_1: length of one parallel side, b_2: length of the other parallel side, h: height

▶ **Ex. 16 and 19** Students will need to apply the multiplicative identity property and the distributive property to express $L - rL$ as $(1 - r)L$ and $P + Prt$ as $(1 + rt)P$.

▶ **Ex. 24** It is important to help students identify each variable before substituting into the given formula.

▶ **Ex. 25–27** These exercises should be assigned as a group.

▶ **Ex. 40–41** Ask students if they remember how to simplify $\dfrac{4\pi r^2}{\frac{4}{3}\pi r^3}$. If they do, they should be able to express the formula as $Q = \frac{3}{r}$.

Enrichment Activities

These activities require students to go beyond lesson goals.

COOPERATIVE LEARNING

Encourage students to work together to research the data and complete these long-term projects.

1. **EXTEND EX. 25–27** Water conservation is everyone's responsibility. Most people leave the water running each time they brush their teeth. Put the stopper in the sink and measure how much water you use brushing your teeth; then estimate how much water you would use brushing your teeth in a week. Is there a way you think you can save water? Try it. Did it work? How much water would you save in a week? In a year?
2. Estimate the number of gallons of water used per person in your household for baths or showers each year. Use that information to estimate how much water everyone in your town uses per year for bathing and showering. If everyone in your household took a three-minute shower, how many gallons would be saved each year? Use that result to estimate how many gallons would be saved in the whole town each year.
3. If each person in the United States saved a gallon of water a day, how much water would be conserved nationally in a year? Assume that the population of the United States is about 250 million. ≈91.25 billion gallons of water

A Leaky Faucet **In Exercises 25–27, use the following information.**

Suppose that a leaky faucet continues to drip at the same rate for one year. Let v be the volume of water dripping from the faucet in t minutes and let V be the volume dripping in one year. Since there are about 525,600 minutes in a year, the three variables are related by the proportion

$$\frac{v}{t} = \frac{V}{525{,}600}.$$

25. You have a leaky faucet at home that fills an 8-ounce glass (0.03 gallon) in 12 minutes. How much water would drip from this faucet in one year? 1314 gal

26. An average bath uses 25 gallons of water. In one year, how many baths are "lost" through the leaky faucet? ≈52.56

27. An average 10-minute shower uses 50 gallons of water. In a year, how many showers are "lost"? 26.28

Integrated Review

In Exercises 28–33, solve the equation.

28. $3d + 16 = d - 4$ -10

29. $5 - x = 23 + 2x$ -6

30. $10(y - 1) = y + 4$ $\frac{14}{9}$

31. $p - 16 + 4 = 4(2 - p)$ 4

32. $x(2 - 6 \cdot 2) = 5x + 30$ -2

33. $12 = z(2 \cdot 6 \div 3)$ 3

In Exercises 34–39, evaluate the expression.

34. $a - 11b + 2$ when $a = 61$ and $b = 7$ -14

35. $\frac{1}{2}c + \frac{5}{2}d - 7$ when $c = 20$ and $d = 4$ 13

36. $15x + 8y$ when $x = \frac{1}{2}$ and $y = \frac{1}{3}$ $\frac{61}{6}$

37. $80\left(n - \frac{1}{4}m\right)$ when $n = 13$ and $m = 5$ 940

38. $\frac{1}{5}(p + q) - 7$ when $p = 5$ and $q = 3$ $-\frac{27}{5}$

39. $\frac{1}{10}\left(20g + \frac{1}{3}h\right)$ when $g = 6$ and $h = 6$ $\frac{61}{5}$

Exploration and Extension

Cells **In Exercises 40 and 41, use the following information.**

The surface-to-volume quotient, Q, of a living cell is related to the cell's ability to interact with its environment to obtain nourishment. If the surface area is small compared to its volume, the cell could not survive. Consider a spherical cell of radius r. Its surface-to-volume quotient is given by

$$Q = \frac{4\pi r^2}{\frac{4}{3}\pi r^3}.$$

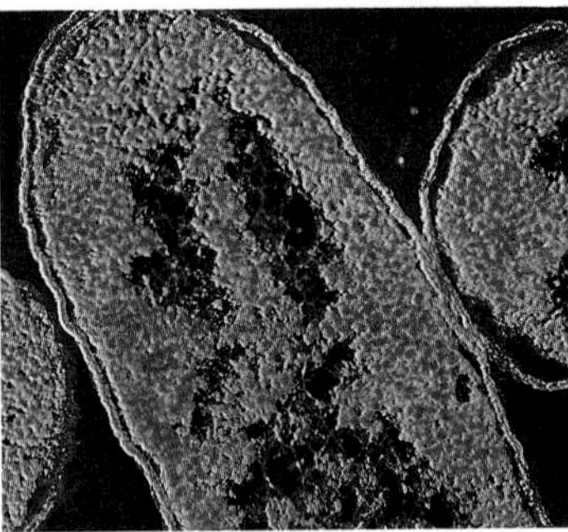

Escheichia coli

40. Measured in centimeters, a typical spherical cell's surface-to-volume quotient is 2400. What is the radius of such a cell? 0.00125 cm

41. Measured in centimeters, an *Escheichia coli* cell has a surface-to-volume ratio of 30,000. The cell is approximately spherical. What is its radius? (About 78,000 of these cells could fit within the period at the end of this sentence.) 0.0001 cm

1.6 Solving Linear Inequalities

What you should learn:

Goal 1 How to solve simple and compound inequalities

Goal 2 How to use inequalities to solve real-life problems

Why you should learn it:

You can model many real-life situations with inequalities, such as a range of frequencies for the human voice

Goal 1 Solving Inequalities

A **solution of an inequality** in one variable is any value of the variable that makes the inequality true. Most inequalities have many solutions. For instance, -2, 0, $\frac{1}{2}$, 0.872, and 1 are some of the many solutions of $x \le 1$.

The **graph** of a linear inequality in one variable is the graph on the real number line of all solutions of the inequality. For instance, here are the graphs of two inequalities. Notice that an open dot is used for $<$ or $>$ and a solid dot is used for $\le$ or $\ge$.

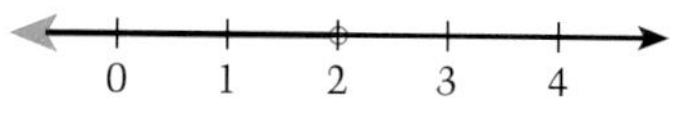
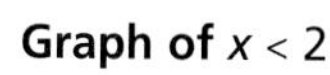
Graph of $x < 2$

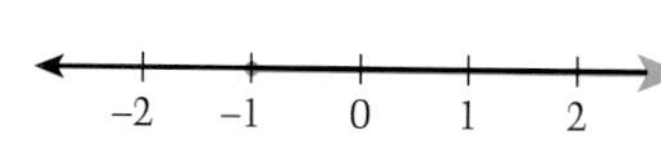
Graph of $x \ge -1$

1, 2, and 5 may be used.

a. $x \le -\frac{1}{3}$

b. $x < -1$

c. $x \ge 8$

3, 4, and 6 may not work all of the time.

LESSON INVESTIGATION

■ Investigating Solutions of Inequalities

Partner Activity Look at the six transformations listed on page 18. Discuss which of these transformations can be used to solve a linear inequality. Test your conclusions by solving the following inequalities

a. $3x + 5 \le -4$ **b.** $4 - 2x > 6$ **c.** $\frac{1}{2}(x - 2) \ge 3$

Do any of the transformations listed on page 18 *not* work with inequalities? Explain.

Study Tip

When solving a linear inequality, the goal is to isolate the variable on one side using transformations that produce ***equivalent*** *inequalities. In the transformations at the right, notice that when you multiply or divide both sides by a negative number, you must reverse the inequality sign. For instance, to reverse $<$, replace it with $>$.*

Transformations That Produce Equivalent Inequalities

1. *Add* the same number to both sides.
2. *Subtract* the same number from both sides.
3. *Multiply* both sides by the same *positive* number.
4. *Divide* both sides by the same *positive number.*
5. *Multiply* both sides by the same *negative* number and reverse the inequality.
6. *Divide* both sides by the same *negative* number and reverse the inequality.

Problem of the Day

Place the numbers 1 through 8 in the circles so that no two consecutive numbers are in circles connected by line segments, and so that $r > s$.

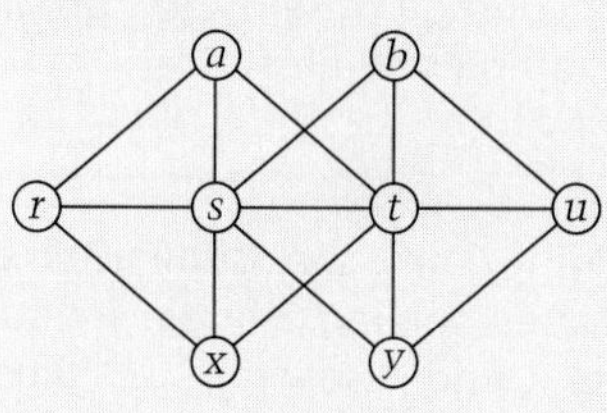

$s = 1$, $t = 8$, $u = 2$, $r = 7$ and $a = 3$, $b = 5$, $x = 4$, $y = 6$ or $a = 4$, $b = 6$, $x = 3$, $y = 5$

ORGANIZER

Warm-Up Exercises

1. Solve.
   a. $3m - 9 = 12$ $m = 7$
   b. $3x + 2 = 5x - 8$ $x = 5$
   c. $7 = 3y - 5$ $y = 4$
2. What is the definition of a solution to a linear equation? A number is a solution of a linear equation if the equation is true when the number is substituted for the variable.
3. What are equivalent linear equations? Equivalent linear equations are equations that have the same solution value(s).

Lesson Resources

Teaching Tools
- Transparency: 1
- Problem of the Day: 1.6
- Warm-up Exercises: 1.6
- Answer Masters: 1.6

Extra Practice: 1.6

Applications Handbook: p. 8, 9, 54, 55, 56

LESSON Notes

GOAL 1 Ask students to provide their own definitions for a solution of a linear inequality and for equivalent inequalities. Encourage them to provide examples to illustrate.

Example 1

Point out the similarity between solving a linear inequality and solving a linear equation. To emphasize the meaning of solution, encourage students to always check their answers.

Example 2

Observe that the solution to the inequality can be obtained without using transformation 6.

$$3x - 2 \le 5x - 3$$
$$3x + 1 \le 5x$$
$$1 \le 2x$$
$$\frac{1}{2} \le x \text{ or } x \ge \frac{1}{2}$$

Example 3

Some students may want to solve the given compound inequality as two inequalities.

$-4 \le 2t - 6$	and	$2t - 6 \le 12$
$2 \le 2t$	and	$2t \le 18$
$1 \le t$	and	$t \le 9$

The solution is still $1 \le t \le 9$.

Example 4

Observe that for compound inequalities joined by "or," you should solve each inequality separately. The solution of each inequality is joined by "or."

Example 5

Students may find it interesting to know that in degrees Celsius the boiling point is 100°C and the freezing point is 0°C. In degrees Fahrenheit, they are 212°F and 32°F, respectively.

Example 1 Solving a Simple Linear Inequality

Solve $3y - 8 < 10$.

Solution

$3y - 8 < 10$	*Rewrite original inequality.*
$3y < 18$	*Add 8 to both sides.*
$y < 6$	*Divide both sides by 3.*

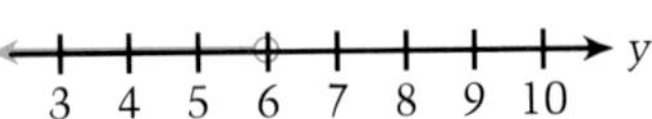

The solution is *all real numbers less than 6.* As a check, try several numbers that are less than 6 in the original inequality. (*Also,* try checking some numbers that are greater than or equal to 6 to see that they are *not* solutions of the original inequality.) The graph of this inequality is shown at the left. ■

Example 2 Solving a Simple Linear Inequality

Solve $3x - 2 \le 5x - 3$.

Solution

$3x - 2 \le 5x - 3$	*Rewrite original inequality.*
$3x \le 5x - 1$	*Add 2 to both sides.*
$-2x \le -1$	*Subtract 5x from both sides.*
$x \ge \frac{1}{2}$	*Divide both sides by −2 and reverse inequality.*

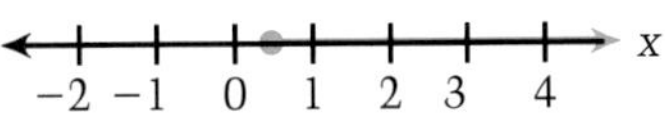

The solution is *all real numbers greater than or equal to* $\frac{1}{2}$. Check several numbers that are greater than or equal to $\frac{1}{2}$ in the original inequality. The graph of this inequality is shown at the left. ■

A **compound inequality** is two simple inequalities joined by "or" or "and." Here are two examples

Graph	Verbal Phrase	Inequality
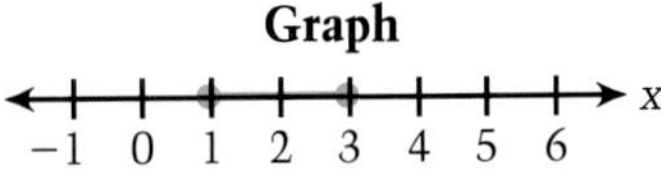	All real numbers that are greater than or equal to 1 *and* less than 3. (This inequality can also be written as $1 \le x$ *and* $x < 3$.)	$1 \le x < 3$
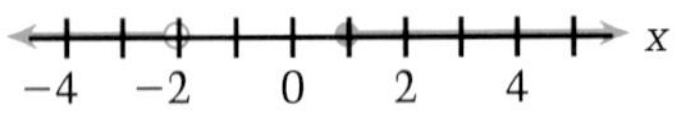	All real numbers that are less than −2 *or* greater than or equal to 1.	$x < -2$ *or* $x \ge 1$

The inequality $a < x < b$ is read as "*x* is *between a* and *b*." The inequality $a \le x \le b$ is read as "*x* is between *a* and *b*, *inclusive.*"

The transformations used to solve a compound linear inequality are the same as those used to solve a simple linear inequality.

Example 3 ***Solving a Compound Linear Inequality***

Solve $-4 \le 2t - 6 \le 12$.

Solution To solve this inequality, you must isolate the variable t between the two inequality signs.

$-4 \le 2t - 6 \le 12$	*Rewrite original inequality.*
$2 \le 2t \le 18$	*Add 6 to each expression.*
$1 \le t \le 9$	*Divide each expression by 2.*

Because t is between 1 and 9 inclusive, the solution is *all real numbers that are greater than or equal to 1 and less than or equal to 9.* Check several numbers between 1 and 9 in the original inequality. The graph is shown below.

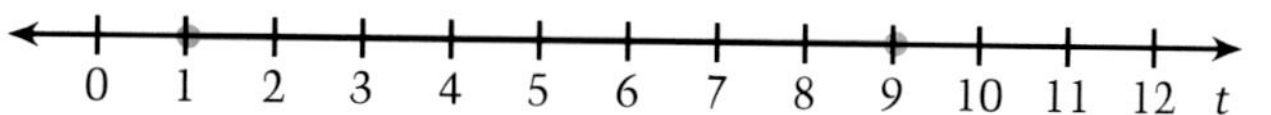

Example 4 ***Solving a Compound Linear Inequality***

Solve $4x + 3 < 7$ *or* $3x - 8 > 7$.

Solution A solution of this compound inequality is a solution of *either* of its simple parts, so you should solve each part separately.

Solution of First Inequality:

$4x + 3 < 7$	*Rewrite first inequality.*
$4x < 4$	*Subtract 3 from both sides.*
$x < 1$	*Divide both sides by 4.*

Solution of Second Inequality:

$3x - 8 > 7$	*Rewrite second inequality.*
$3x > 15$	*Add 8 to both sides.*
$x > 5$	*Divide both sides by 3.*

The solution is *all real numbers that are less than 1 or greater than 5.* The graph is shown below.

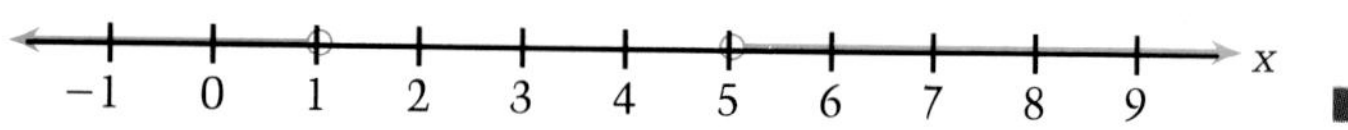

Extra Examples

Here are additional examples similar to **Examples 1–5.**

1. Solve each simple linear inequality.

a. $3x + 7 < 13$ $x < 2$

b. $63 - 15y \le 45$ $y \ge \frac{6}{5}$

c. $8x - 3 \le 23 - 5x$ $x \le 2$

d. $9 - 14y > 8y + 3$ $y < \frac{3}{11}$

2. Graph each solution in Exercise 1.

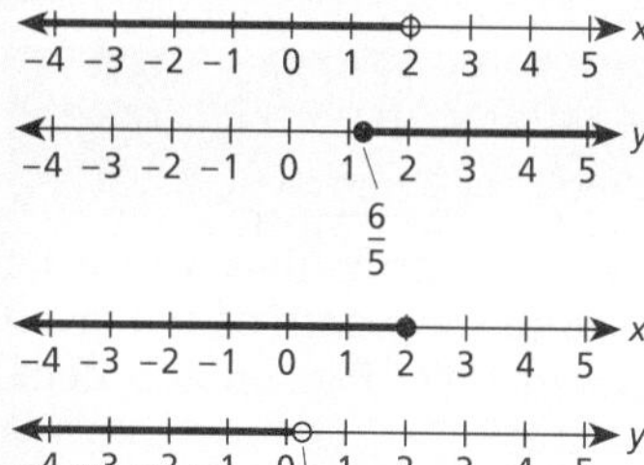

3. Solve each compound linear inequality.

a. $4 \le 4x - 12 \le 16$

b. $-8 < 7 - 3x < 10$

c. $5 - 9x < 14$ or $-4x - 17 > 3$

a. $4 \le x \le 7$

b. $-1 < x < 5$

c. $x > -1$ or $x < -5$

4. Graph each solution in Exercise 3.

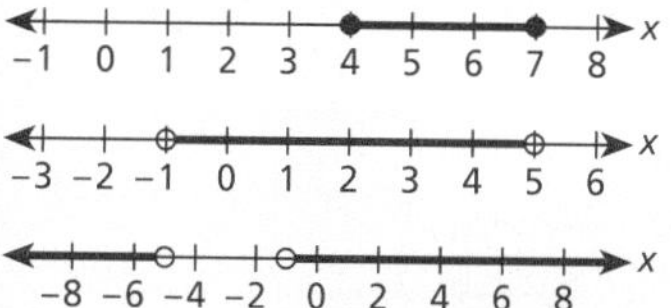

5. The area, A, of an ellipse with major axis of length 8 must be greater than 15 and less than 23 square units. Using the formula for the area of an ellipse found in Example 1 of Lesson 1.5, find the possible lengths for the minor axis.
$1.19 < b < 1.83$, and $2.38 <$ length of minor axis < 3.66

Check Understanding

1. What is a solution of a linear inequality? See the top of page 37.
2. Explain why the inequality sign is reversed when an inequality is multiplied by a negative number. For example, $6 > 5$ but $(-1)6 < (-1)5$.
3. Characterize the graph of a linear inequality in one variable. See page 37.

Communicating about ALGEBRA

Can transformations 5 or 6 be used to solve each of the given inequalities? Explain and demonstrate. Yes, you can combine terms so that the coefficients of the variable terms are always negative. Can transformations 5 and 6 always be avoided? Explain and demonstrate. Yes, you can combine terms so that the coefficients of the variable terms are always positive.

Answers

A.–D. Use the transformations on page 37 to solve the equations; use a solid dot for $\leq$ or $\geq$ and an open dot for $<$ or $>$ to graph the solutions.

A.

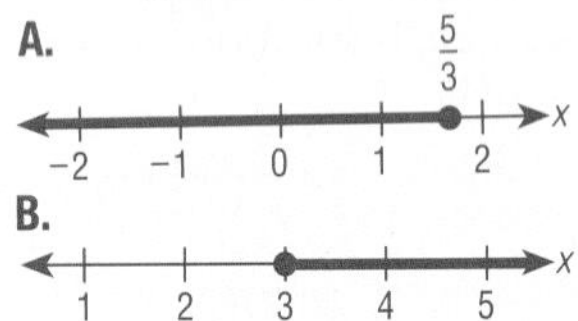

B.

x
1 2 3 4 5

C.

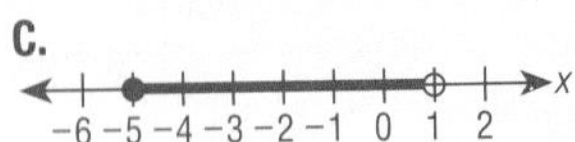

D.

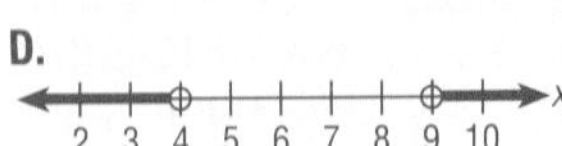

In most parts of the United States, people add antifreeze to automobile cooling systems. The antifreeze lowers the freezing point of the coolant and raises the boiling point. (The changes in freezing and boiling points depend on the amount and type of antifreeze that is added.)

Goal 2 Using Inequalities to Solve Problems

Example 5 *Finding the Freezing and Boiling Points*

You have added enough antifreeze to your car's cooling system to lower the freezing point to −35°C and raise the boiling point to 125°C. The coolant will remain a liquid as long as the temperature, C (in degrees Celsius), satisfies the inequality $-35 < C < 125$. Write this inequality in degrees Fahrenheit.

Solution Let F represent the temperature in degrees Fahrenheit and use the formula $C = \frac{5}{9}(F - 32)$.

$-35 < C < 125$	*Rewrite original inequality.*
$-35 < \frac{5}{9}(F - 32) < 125$	*Replace C by $\frac{5}{9}(F - 32)$.*
$-63 < F - 32 < 225$	*Multiply each expression by $\frac{9}{5}$.*
$-31 < F < 257$	*Add 32 to each expression.*

The coolant will remain liquid between −31°F and 257°F. ■

Communicating about ALGEBRA

SHARING IDEAS about the Lesson

Solve the following inequalities. Sketch the graph of each solution and explain your steps.

A. $3x + 5 \leq 10$ $x \leq \frac{5}{3}$

B. $-4x + 6 \leq 3x - 15$ $x \geq 3$

C. $-5 \leq 2x + 5 < 7$ $-5 \leq x < 1$

D. $3x - 4 < 8$ or $3x - 4 > 23$ $x < 4$ or $x > 9$

EXERCISES

Guided Practice

CRITICAL THINKING about the Lesson

In Exercises 1 and 2, write an inequality that corresponds to the phrase.

1. All real numbers less than or equal to -5 $x \le -5$
2. All real numbers less than 7 and greater than or equal to -2 $-2 \le x < 7$
3. *True or False?* Multiplying both sides of an inequality by the same number produces an equivalent inequality. Explain. False for multiplying by 0
4. Write a verbal phrase that describes $x > -1$. All real numbers greater than -1
5. Sketch the graph of $x < 3$. See margin.
6. Sketch the graph of $1 \le x < 5$. See margin.

Independent Practice

In Exercises 7–10, write an inequality for the verbal sentence.

7. x is less than 9. $x < 9$
8. 18 is less than or equal to a. $18 \le a$
9. x is greater than 4 and less than 6. $4 < x < 6$
10. y is greater than or equal to -5 and less than 0. $-5 \le y < 0$

In Exercises 11–14, match the inequality with its graph.

11. $x \ge 4$ c
12. $x \ge 4$ or $x < -4$ b
13. $-4 < x < 4$ d
14. $x < 4$ a

a. 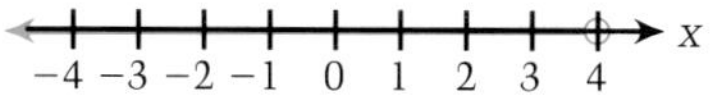

b. 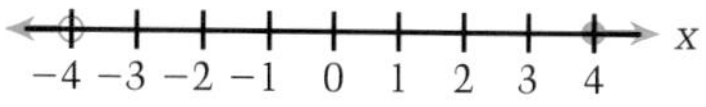

c. 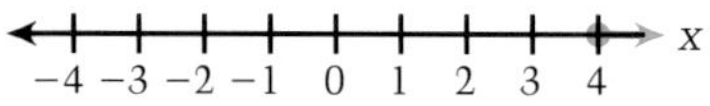

d. 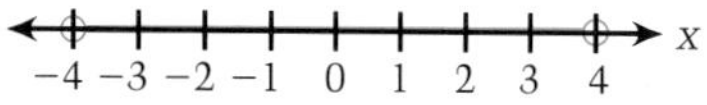

In Exercises 15–20, decide whether the number is a solution of the inequality.

15. $3y - 4 < 10$, 5 It is not.
16. $x + 4 > 9$, 5 It is not.
17. $7 - n \ge 3$, 4 It is.
18. $\frac{1}{2}x - 3 \le -3$, 4 It is not.
19. $-1 < 2x \le 8$, 4 It is.
20. $-4 < w + 4 < 6$, 0 It is.

In Exercises 21–32, solve the inequality.

21. $6t + 7 > 11$ $t > \frac{2}{3}$
22. $4 + x \le 19$ $x \le 15$
23. $9 - k < 4$ $k > 5$
24. $3 - 2x \ge 15$ $x \le -6$
25. $2h - 3 \ge 6 - h$ $h \ge 3$
26. $-x + 5 < 3x + 1$ $x > 1$
27. $-3 \le x - 3 \le 6$ $0 \le x \le 9$
28. $-14 < 2j + 3 \le 6$ $-\frac{17}{2} < j \le \frac{3}{2}$
29. $4 \le -3x - 1 \le 9$
30. $-3 \le -x - 5 \le 0$
31. $-1 < -2x + 1 \le 5$ $-2 \le x < 1$
32. $-7 \le 5y - 2 < 3$

29. $-\frac{10}{3} \le x \le -\frac{5}{3}$ 30. $-5 \le x \le -2$ 32. $-1 \le y < 1$ 34. $x \le 6$ or $x \ge 7$ 35. $x < 4$ or $x \ge 10$

In Exercises 33–38, solve the inequality. Then sketch its graph. See Additional Answers.

33. $4 \le r + 5 \le 5$ $-1 \le r \le 0$
34. $x - 1 \le 5$ or $x + 3 \ge 10$
35. $x - 4 < 0$ or $x - 6 \ge 4$
36. $-3 < 2x - 3 \le 17$ $0 < x \le 10$
37. $m + \frac{1}{2} \le -3$ or $m - 3 > -2$ $m \le -\frac{7}{2}$ or $m > 1$
38. $0 \le \frac{1}{3}x - 1 \le 3$ $3 \le x \le 12$

EXERCISE Notes

ASSIGNMENT GUIDE

Basic/Average: Ex. 1–6, 7–14, 17–37 odd, 39–43, 46–56

Above Average: Ex. 1–6, 9–14, 19–37 odd, 39–45, 52–57

Advanced: Ex. 1–6, 11–14, 21–37 odd, 39–45, 52–58

Selected Answers
Exercises 1–6, 7–55 odd

Use **Mixed Review** as needed.

More Difficult Exercises
Exercises 43–45, 56–58

Guided Practice

Ex. 1–4 Emphasize the connections between a compound inequality written as a conjunction and the intuitive notion of "betweenness." That is, in Exercise 2, $-2 \le x < 7$ indicates that x is -2 and x is between -2 and 7.

Answers

5. 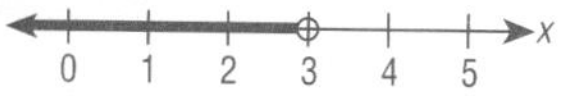

6. 

Independent Practice

Ex. 10 Ask students to express the solution using the word "between." y is -5 and y is between -5 and 0.

Ex. 21–38 Encourage students to use the format presented in Example 3.

Ex. 29–31 Remind students to reverse the directions of both inequality symbols to express the answer from least to greatest.

Ex. 39–45 These applications provide real-life data for inequality expressions. Students might wish to update the information using data available in newspapers or magazines.

Answers

40.

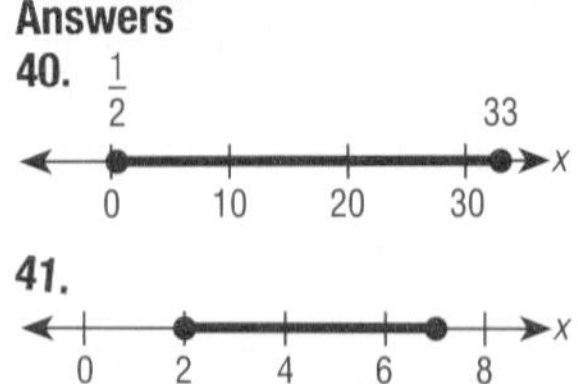

41.

45. If you scream, the bat will hear you only if the frequency of the scream is between 1000 and 1100 cycles per second. If the bat screams, you will hear the bat only if the frequency of the scream is between 10,000 and 20,000 cycles per second.

Ex. 43–45 These exercises should be assigned as a group.

Ex. 52–53 These exercises help to develop number sense.

The first Ferris wheel ever built was designed by the American mechanical engineer George Ferris (1859–1896) for use at the Columbian Exposition (World's Fair) in Chicago in 1893, and later used at the World's Fair in St. Louis in 1904. It had a diameter of 250 feet, and each of its 36 cars could hold 60 passengers.

39. *Ferris Wheels* Write an inequality that represents the number, n, of people on a single Ferris wheel. $0 \le n \le 2160$

40. *Centipedes* The longest species of centipede known measures about 33 centimeters. The shortest centipede recorded measures only about 0.5 centimeters. Write an inequality that represents the length of a centipede. Then graph the inequality. $0.5 \le x \le 33$, see margin.

41. *Eclipses* The greatest number of eclipses possible in a year is 7. This happened in 1935, when 5 solar and 2 lunar eclipses occurred. The smallest number of eclipses possible in a year is 2, both of which must be solar. Write an inequality that represents the number of eclipses in a year. Sketch the inequality. $2 \le x \le 7$, see margin.

42. *Entertainers* The highest-paid American entertainer in 1990 was Bill Cosby, who was paid \$55 million during that year. Write an inequality that describes the salary of any American entertainer in 1990. (Although Cosby makes his money in entertainment, he holds a doctorate in education.) $0 \le x \le 55{,}000{,}000$

Bats **In Exercises 43–45, use the following information.**

The range of a human's voice frequency, h (in cycles per second), is about $85 \le h \le 1100$. The range of a human's hearing frequency, H, is about $20 \le H \le 20{,}000$.

43. The relationship between a human's voice frequency and a bat's voice frequency, b, is $h = 85 + \frac{203}{22{,}000}(b - 10{,}000)$. Find the range of a bat's voice frequency. $10{,}000 \le b \le 120{,}000$

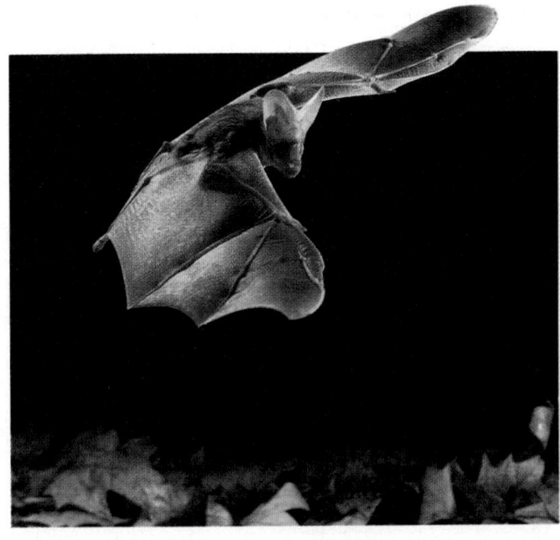

44. The relationship between a human's hearing frequency and a bat's hearing frequency, B, is $H = 20 + \frac{999}{5950}(B - 1000)$. Find the range of a bat's hearing frequency. $1000 \le B \le 120{,}000$

45. If a bat flies into your room and you scream, will it hear you? If you scare the bat and it screams, will you hear it? Explain. See margin.

Integrated Review

In Exercises 46–51, solve the equation.

46. $5 + 2x = 19$ 7

47. $-5x + 6 = 1 + 9x$ $\frac{5}{14}$

48. $3x - 9 = 2x - 12$ -3

49. $8 + x = -5x - 20$ $-\frac{14}{3}$

50. $6 - 2x = 7x - 4$ $\frac{10}{9}$

51. $4x - 13 = 5 - x$ $\frac{18}{5}$

52. *Powers of Ten* Which is the best estimate of the average annual income for an American family in 1990?
a. $\$3 \times 10^2$ **b.** $\$3 \times 10^3$ **c.** $\$3 \times 10^4$ **d.** $\$3 \times 10^5$

53. *Powers of Ten* Which is the best estimate of the diameter of an automobile tire?
a. 2.4×10^0 in. **b.** 2.4×10^1 in. **c.** 2.4×10^2 in. **d.** 2.4×10^3 in.

54. *College Entrance Exam Sample* If $6565 = 65(x + 1)$, then $x = \boxed{?}$.
a. 10 **b.** 11 **c.** 100 **d.** 101 **e.** 1001

55. *College Entrance Exam Sample* If $x = 2$ and $y = -1$, what is the value of $\frac{x - y}{x + y}$?
a. -3 **b.** -1 **c.** $\frac{1}{3}$ **d.** 1 **e.** 3

Exploration and Extension

In Exercises 56 and 57, write a compound inequality that is equivalent to the given inequality.

56. $x^2 < 25$ $-5 < x < 5$

57. $x^2 < 16$ $-4 < x < 4$

58. A number is in the **intersection** of two sets if it is in each set. Sketch the graphs of $2x - 5 > -9$ and $5x + 4 < 29$. Write a compound inequality that describes the intersection of the graphs. $-2 < x < 5$

Mixed REVIEW

1. What is one fifth of 120? 24

2. What is the ratio of 5 feet to 3 feet? $\frac{5}{3}$

3. Add: $-1.45 + 2.55$. 1.10

4. Multiply: $12^2 \cdot 12^3$. 12^5 or 248,832

5. What is the reciprocal of $-\frac{2}{3}$? **(1.1)** $-\frac{3}{2}$

6. What is the opposite of $-\frac{2}{3}$? **(1.1)** $\frac{2}{3}$

7. Write in scientific notation: 1,540,000. 1.54×10^6

8. Write in scientific notation: 0.0014. 1.4×10^{-3}

9. Is -4 a solution of $5 - (x + 2) = 3$? **(1.2)** No

10. Is 12 a solution of $-2(x - 3) - (x + 1) = 31$? **(1.2)** No

11. Solve $5(x - 2) = -2(x + 1)$. **(1.3)** $\frac{8}{7}$

12. Solve $-\frac{1}{2}(x + 1) = 5x - 6$. **(1.3)** 1

13. Find the volume: cube with 3-inch sides. **(1.5)** 27 in.^{3}

14. Find the area: circle with 2-inch radius. **(1.5)** 4π in.^{2} or ≈ 12.6 in.^{2}

15. Is -2 a solution of $-4x + 1 \le 9$? **(1.3)** Yes

16. Is -3 a solution of $-9x - 15 < 15$? **(1.3)** Yes

17. Solve and graph: $-2 \le -2x < 2$. **(1.6)**

18. Evaluate $\frac{2}{3}x$ when $x = \frac{3}{2}$. **(1.2)** 1

17. $-1 < x \le 1$, see margin.

Enrichment Activity

This activity requires students to go beyond lesson goals.

GEOMETRY Find x so that the area of the rectangle is less than 30. Show your solution steps.

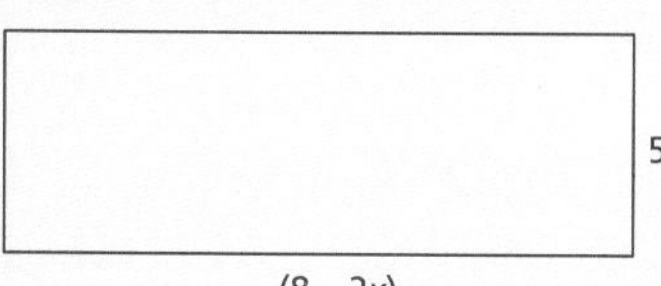

Most students will realize that $5(8 - 2x) < 30$ implies that $(8 - 2x) < 6$. But many will forget that since the side of the rectangle must be positive, $(8 - 2x) > 0$.

So, the solution is the conjunction of $5(8 - 2x) < 30$ and $(8 - 2x) > 0$.

$40 - 10x < 30$ $\quad -2x > -8$
$-10x < -10$ $\quad x < 4$
$x > 1$

So, $1 < x < 4$.

Answer

Mixed Review

17.

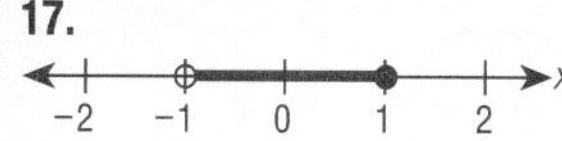

Problem of the Day

Biker A can bike around a circular track in 8 minutes, and biker B can do it in 6 minutes. If they start together, after how many minutes will biker B pass biker A? 24

ORGANIZER

Warm-Up Exercises

1. Solve each equation for x.
   a. $4 + 15x = 34$ 2
   b. $3 - 5x = 28$ -5
   c. $6x + 4 = 40$ 6
2. Solve each inequality for x.
   a. $7x - 3 > 11$ $x > 2$
   b. $14 - 8x < -18$ $x > 4$
   c. $9 + 4x < 7$ $x < -0.5$

Lesson Resources

Teaching Tools
- Transparency: 1
- Problem of the Day: 1.7
- Warm-up Exercises: 1.7
- Answer Masters: 1.7

Extra Practice: 1.7
Color Transparency: 5

LESSON Notes

Example 1

By using numerical examples, show students that the argument x of the absolute value expression, $|x| = a$, must either be positive or negative. A series of examples helps to make the point with students. For example,

$|x| = 5$ means either $x = 5$ or $x = -5$,

$|x| = 9$ means either $x = 9$ or $x = -9$,

(continued)

1.7 Solving Absolute Value Equations and Inequalities

What you should learn:

Goal 1 How to solve absolute value equations and inequalities

Goal 2 How to use absolute value inequalities in real-life settings

Why you should learn it:

You can model many problems, such as those dealing with a range of acceptable measurements, with absolute value equations and inequalities.

Goal 1 Solving Equations and Inequalities

The **absolute value** of a number, x, is the distance the number is from 0.

$$|x| = \begin{cases} x, & \text{if } x \text{ is positive.} \\ 0, & \text{if } x = 0. \\ -x, & \text{if } x \text{ is negative.} \end{cases}$$

Example

$|4| = 4$
$|0| = 0$
$|-2| = -(-2) = 2$

To solve an absolute value equation, remember that the expression inside the absolute value symbols can be either positive or negative. For instance, the solution of $|x| = 5$ is $x = 5$ or $x = -5$.

Example 1 *Solving an Absolute Value Equation*

Solve $|2x - 3| = 7$.

Solution

$\|2x - 3\| = 7$		*Rewrite original equation.*
$2x - 3 = 7$ or	$2x - 3 = -7$	*Expression can be 7 or −7.*
$2x = 10$ or	$2x = -4$	*Add 3 to both sides.*
$x = 5$ or	$x = -2$	*Divide both sides by 2.*

The equation has two solutions: 5 and -2. Check these solutions by substituting each into the original equation. ■

An absolute value *inequality* such as $|x - 2| < 4$ can be solved by rewriting it as a compound inequality.

Transformations of Absolute Value Inequalities

1. The inequality $|ax + b| < c$ means that $ax + b$ is *between* $-c$ and c. This is equivalent to $-c < ax + b < c$.
2. The inequality $|ax + b| > c$ means that $ax + b$ is *not between* $-c$ and c inclusive. This is equivalent to $ax + b < -c$ *or* $ax + b > c$.

In the first transformation, $<$ could be replaced by $\le$, adjusting inclusiveness. Similarly, in the second transformation, $>$ could be replaced by $\ge$.

Example 2 *Solving an Absolute Value Inequality*

Solve $|2x - 3| < 5$.

Solution

$\|2x - 3\| < 5$	*Rewrite original inequality.*
$-5 < 2x - 3 < 5$	*Write equivalent compound inequality.*
$-2 < 2x < 8$	*Add 3 to each expression.*
$-1 < x < 4$	*Divide each expression by 2.*

The solution is *all real numbers that are greater than −1* and *less than 4.* The graph is shown below.

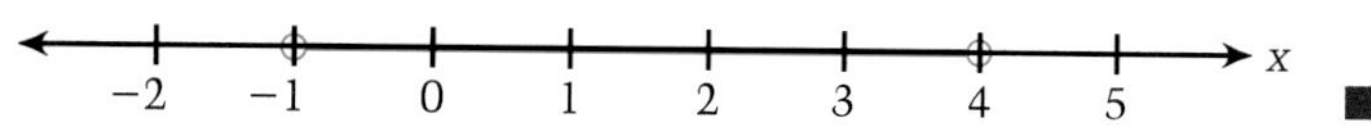

■

Example 3 *Solving an Absolute Value Inequality*

Solve $|3x + 4| \ge 10$.

Solution This absolute value inequality is equivalent to the following compound inequality.

$3x + 4 \le -10$ *or* $3x + 4 \ge 10$

Solve First Inequality:

$3x + 4 \le -10$	*Rewrite first inequality.*
$3x \le -14$	*Subtract 4 from both sides.*
$x \le -\frac{14}{3}$	*Divide both sides by 3.*

Solve Second Inequality:

$3x + 4 \ge 10$	*Rewrite first inequality.*
$3x \ge 6$	*Subtract 4 from both sides.*
$x \ge 2$	*Divide both sides by 3.*

The solution is *all real numbers that are less than or equal to* $-\frac{14}{3}$ or *greater than or equal to 2.* The graph is shown below.

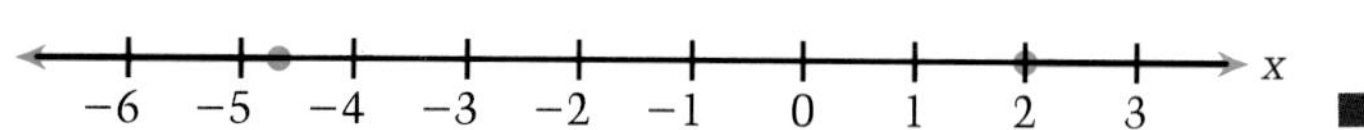

■

(continued)

$|x| = 25$ means either $x = 25$ or $x = -25$,

$|x| = 11$ means either $x = 11$ or $x = -11$.

In every instance, have students evaluate the absolute value of the two values; that is, for $|x| = 5$, ask students to evaluate both $|5|$ and $|-5|$.

Finally, stress that no matter what the argument is, it must be positive or negative. Consequently, $|x - 4| = 9$ means the argument, $x - 4$, must be positive 9 or negative 9.

Examples 2–3

Again use numerical examples to show students that $|x| < a$ means that x is between $-a$ and a, or $-a < x < a$; and that $|x| > a$ means x is not between $-a$ and a inclusive, but rather, $x < -a$ or $x > a$.

EXTEND Example 4

EXPERIMENT Select several objects of different weights. Ask students to guess the weights of each and describe the intervals in which their guesses fall. Guesses that are less than the weight of an object should be recorded using negative numbers. For example, if an object weighs 15 pounds and the guess is 13 pounds, then the guess is recorded as −2 off actual weight.

Extra Examples

Here are some exercises similar to **Examples 1–3.**

1. Solve.

a. $|3x + 4| = 10$
b. $|13 - 5x| = 2$
c. $|3x + 4| \le 8$
d. $|6 - 2x| > 14$

a. $x = -\frac{14}{3}$ or $x = 2$
b. $x = \frac{11}{5}$ or $x = 3$
c. $-4 \le x \le \frac{4}{3}$
d. $x < -4$ or $x > 10$

2. Graph the solutions to the absolute-value equations and inequalities in Exercise 1.

a.

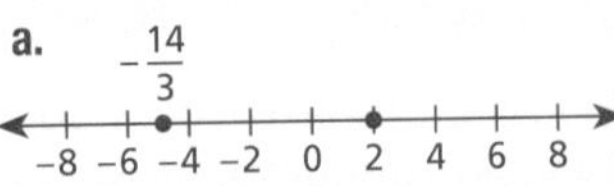

b.

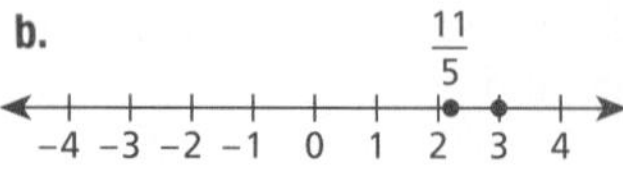

c.

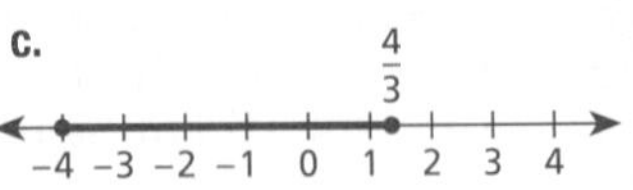

d.

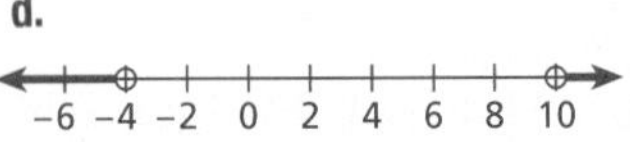

Check Understanding

1. Why must the absolute value of a number (or expression) always be positive? Absolute value is the measure of distance from 0 and distance is not negative.
2. What is the connection between solving an absolute-value equation involving linear expressions and solving linear equations? See Examples 1–3.
3. Provide a definition of the absolute value of x. See the top of page 44.

Communicating about ALGEBRA

One strategy is to graph the compound inequalities, identify the "center" of each graph, and determine how the absolute value inequality of each can be written based upon its graph.

Answers

A. Distance of x from zero is 3 or less.

B. Distance of x from zero is 2 or more.

(continued)

The pit organs of a rattlesnake can detect temperature changes that are as little as 0.005°F. By moving its head back and forth, a rattlesnake can detect warm prey, even in the dark.

Real Life
Biology

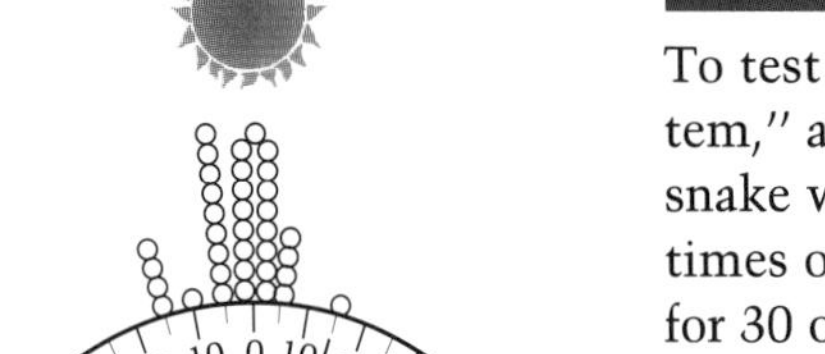

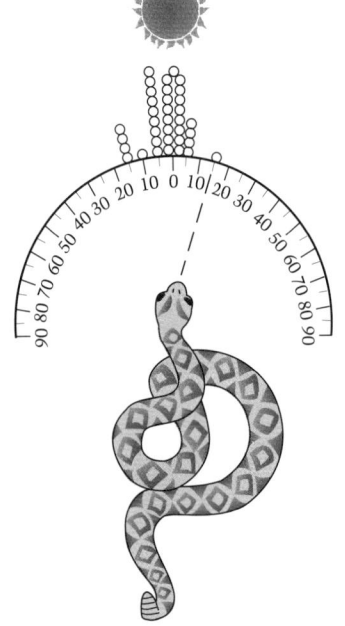

Goal 2 Using Absolute Value Inequalities

Example 4 *Creating a Model*

To test the accuracy of a rattlesnake's "pit-organ sensory system," a biologist blindfolded a rattlesnake and presented the snake with a warm "target." The snake was on-target 17 times out of 36. In fact, the snake was within 5° of the target for 30 of the strikes. Let A represent the number of degrees by which the snake is off-target. Then $A = 0$ represents a strike that is aimed directly at the target. Positive values of A represent strikes to the right of the target, and negative values of A represent strikes to the left of the target. Use the diagram at the left to write an absolute-value inequality that describes the interval in which the 36 strikes occurred.

Solution From the diagram, you can see that the snake was never off by more than 15° in either direction. This can be represented by the compound inequality $-15 \le A \le 15$.

As an absolute value inequality, the interval in which the strikes occurred can be represented by $|A| \le 15$. ■

Communicating about ALGEBRA

SHARING IDEAS about the Lesson

Write absolute value inequalities that are equivalent to the following compound inequalities. Explain your strategy.

A. $-3 \le x \le 3$ $|x| \le 3$ **B.** $x \le -2$ or $x \ge 2$

C. $0 < x < 6$ $|x - 3| < 3$ **D.** $x < 0$ or $x > 8$

See margin for explanation. **B.** $|x| \ge 2$ **D.** $|x - 4| > 4$

EXERCISES

Guided Practice

CRITICAL THINKING about the Lesson

1. The absolute value of a number cannot be negative. How, then, can the absolute value of a be $-a$? If a is negative, $-a$ is positive.
2. The absolute value of a can be defined as the distance from a to the origin on the real number line. Illustrate this definition with $a = -3$ and with $a = 4$. $|-3| = 3$, $|4| = 4$
3. The graph of the inequality $|x - 3| < 2$ can be described as *all real numbers that are within 2 units of 3.* Give a similar description of the graph of $|x - 4| < 1$. All real numbers that are within 1 unit of 4
4. The graph of the inequality $|y - 1| > 3$ can be described as *all real numbers that are greater than 3 units from 1.* Give a similar description of the graph of $|y + 2| > 4$. (Hint: $x + 2 = x - (-2)$.)
All real numbers that are more than 4 units from -2

Independent Practice

In Exercises 5–8, decide whether the number is a solution of the equation.

5. $|4x + 5| = 10, 2$ No
6. $|2x - 16| = 10, 3$ Yes
7. $|6 - 2w| = 2, 4$ Yes
8. $|\frac{1}{2}x + 4| = 8, 6$ No

In Exercises 9–12, transform the absolute value equation into two linear equations.

9. $|x - 10| = 17$ $x - 10 = 17$ or $x - 10 = -17$
10. $|7 - 2t| = 5$ $7 - 2t = 5$ or $7 - 2t = -5$
11. $|4x + 1| = \frac{1}{2}$ $4x + 1 = \frac{1}{2}$ or $4x + 1 = -\frac{1}{2}$
12. $|22k + 6| = 9$ $22k + 6 = 9$ or $22k + 6 = -9$

In Exercises 13–18, solve the equation.

13. $|9 + 2x| = 7$ $-1, -8$
14. $|4 - 6x| = 2$ $\frac{1}{3}, 1$
15. $|\frac{1}{4}m - 9| = 6$ $60, 12$
16. $|\frac{3}{5}x + 2| = 11$ $15, -\frac{65}{3}$
17. $|20 - 3x| = 7$ $\frac{13}{3}, 9$
18. $|20 - 9q| = 5$ $\frac{5}{3}, \frac{25}{9}$

In Exercises 19–22, transform the absolute value inequality into a compound inequality.

19. $|y + 5| < 3$ $-3 < y + 5 < 3$
20. $|6x + 7| \le 5$ $-5 \le 6x + 7 \le 5$
21. $|7 - 2h| \ge 9$ $7 - 2h \le -9$ or $7 - 2h \ge 9$
22. $|8 - x| > 25$ $8 - x < -25$ or $8 - x > 25$

In Exercises 23–30, solve the inequality.

23. $|d + 1| \le 6$ $-7 \le d \le 5$
24. $|11 - x| < 20$ $-9 < x < 31$
25. $|14 - 5x| > 8$ $x > \frac{22}{5}$ or $x < \frac{6}{5}$
26. $|18 + \frac{1}{2}x| \ge 10$ $x \le -56$ or $x \ge -16$
27. $|2x + 3| \ge 26$ $x \le -\frac{29}{2}$ or $x \ge \frac{23}{2}$
28. $|7r + 3| < 11$ $-2 < r < \frac{8}{7}$
29. $|19 + j| < 2$ $-21 < j < -17$
30. $|7 + 8x| > 5$ $x < -\frac{3}{2}$ or $x > -\frac{1}{4}$

In Exercises 31–34, sketch the graph of the inequality.

31. $\left|\frac{1}{4}p - \frac{1}{3}\right| \le \frac{1}{3}$
32. $|x + 1| \le 27$
33. $|x - 8| > 14$
34. $|6 - 5c| < 9$

(continued)

C. Distance of x from the midpoint, 3, is less than 3.

D. Distance of x from the midpoint, 4, is greater than 4.

EXERCISE Notes

ASSIGNMENT GUIDE

Basic/Average: Ex. 1–4, 5–33 odd, 35–36, 41–48, 58–60

Above Average: Ex. 1–4, 7–35 odd, 37–48, 58–62

Advanced: Ex. 1–4, 9–37, 38–48, 59–62

Selected Answers
Exercises 1–4, 5–59 odd

More Difficult Exercises
Exercises 37–38, 41–42, 61–62

Guided Practice

Ex. 1–4 Use these exercises to help students develop a better understanding of the concept of absolute value. In Exercises 3 and 4, you may wish to have students work in pairs to draw the graph.

Independent Practice

Ex. 13–18 Students should use a solution format similar to Example 1.

Ex. 33–34 Students should use a solution format similar to Examples 2 and 3.

Answers

31.

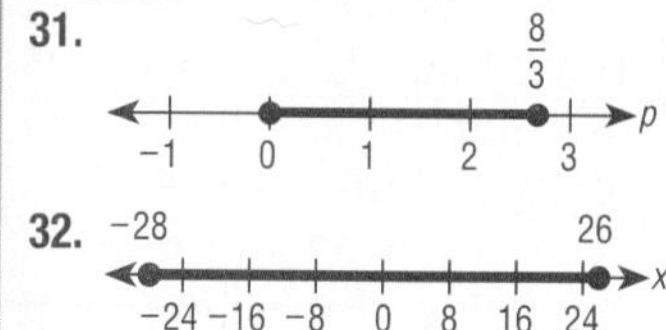

33.

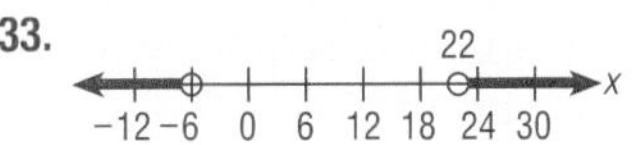

34.

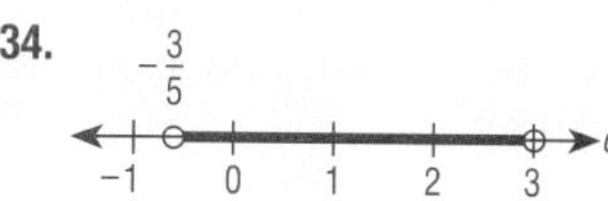

47

▶ **Ex. 35–42** To prepare students for these exercises, help them make the connection from "within" or "between" to the absolute-value inequality $|x - a| < b$. For example, $|x - 5| < 6$ creates a range around 5 that is 6 units in either direction. So, x must be between $5 - 6$, or -1, and $5 + 6$, or 11. Students need to build confidence in making the transition from the range to the absolute-value inequality, especially in Exercises 35–38 and 41–42.

▶ **Ex. 62** Students' answers will depend on whether they use the exact values provided in the table or approximate values from the chart.

Enrichment Activities

These activities require students to go beyond lesson goals.

An Alternate Approach
Boundary points are the solutions to the corresponding absolute-value equations.

The Boundary-Point Strategy
To solve absolute-value inequalities:

a. Solve the corresponding absolute-value equation to find the values of the boundary points.
b. Plot the boundary points on the number line. Use solid dots when the inequality was written with $\leq$ or $\geq$. Use open dots when the inequality was written with $<$ or $>$.
c. The boundary points divide the number line into three regions. Test a value from each region in the original inequality. If a value results in a true statement, the corresponding region is in the solution set.
d. Complete the graph and write the solution from the graph.

35. *Accuracy of Measurements* In woodshop class, you must cut several pieces of wood to within $\frac{3}{16}$ inch of the teacher's specifications. Let $(s - x)$ represent the difference between the specification, s, and the measured length, x, of a cut piece. Write an absolute value inequality that describes the values of x that are within specifications. One of the pieces of wood is specified to be $s = 5\frac{1}{8}$ inches. Describe the acceptable lengths for this piece. $|s - x| \leq \frac{3}{16}$, $4\frac{15}{16} \leq x \leq 5\frac{5}{16}$

36. *The Mile Run* The average time to run a mile for members of your gym class is 7 minutes 4 seconds. You want your time to be within 5 seconds of the class average. Rewrite the class average as seconds, then write and solve an absolute value inequality for the difference between your time, t, and the class average. $|t - 424| \leq 5$, $419 \leq t \leq 429$

Hand Measurements **In Exercises 37 and 38, use the following information.**

In a sampling conducted by the United States Air Force, the right-hand dimensions of 4000 Air Force men were measured. (The information could be used when designing control panels, keyboards, gloves, and so on.)

37. Ninety-five percent of the palm widths, p (in inches), were solutions of the inequality $|p - 3.49| \leq 0.26$. Sketch the graph of this inequality. See Additional Answers.

38. Ninety-nine percent of the palm widths were solutions of the inequality $|p - 3.49| \leq 0.37$. Sketch the graph of this inequality. How many of the 4000 men had palm widths that are solutions of the inequality $|p - 3.49| > 0.37$? See Additional Answers.

39. *Body Temperature* Physicians consider an adult's body temperature to be normal if it is between 97.6°F and 99.6°F. Write an absolute value inequality that describes this normal temperature range. $|t - 98.6| \leq 1.0$

40. *Green Ocean Plants* Green plants can live in the ocean only at depths of 0 to 100 feet. Find an absolute value inequality describing the possible depth range of green plants in an ocean. $|x - 50| \leq 50$

Show Collies **In Exercises 41 and 42, use the following information.**

The American Kennel Club had developed guidelines for judging the features of various breeds of dogs. For collies, the guidelines specify that the weight should be between 60 and 75 pounds for males and between 50 and 65 pounds for females.
(Source: American Kennel Club) $|x - 57.5| \leq 7.5$

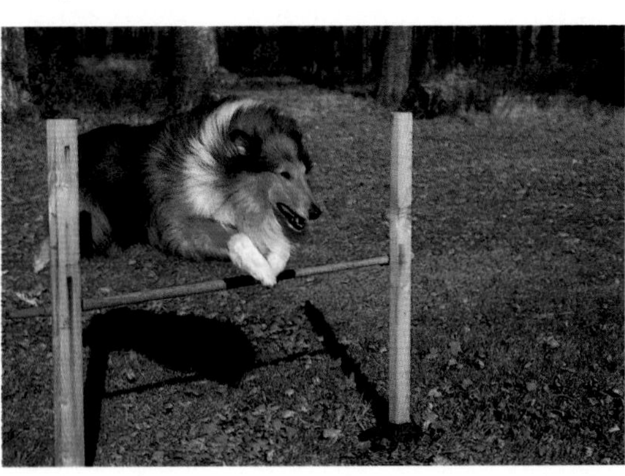

41. Write an absolute value inequality for the recommended weight of a female collie.

42. Write an absolute value inequality for the recommended weight of a male collie.
$|x - 67.5| \leq 7.5$

Integrated Review

In Exercises 43–48, match the inequality with its graph.

43. $-\frac{1}{2} < x \le \frac{3}{4}$ d

44. $1 \le 2x + 3 < 5$ a

45. $3 < x + 4 \le 5$ f

46. $-4 \le x - 7 \le -2$ e

47. $\frac{1}{3} < x + \frac{2}{3} \le \frac{7}{6}$ b

48. $-17 \le 17x \le 34$ c

a. 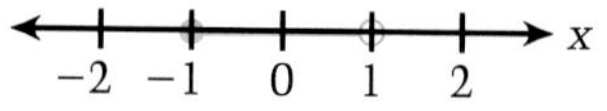

b. 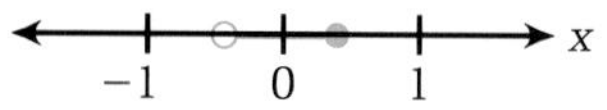

c. 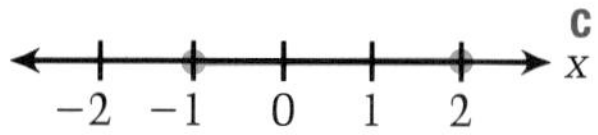

d. 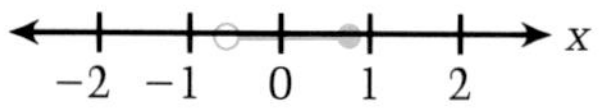

e. 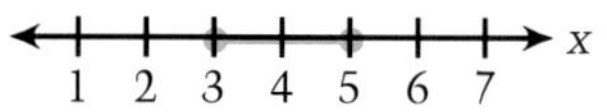

f. 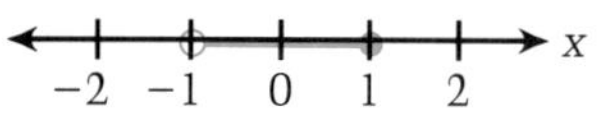

In Exercises 49–54, solve the equation.

49. $7x - \frac{1}{4} = 2$ $\frac{9}{28}$

50. $25 - x = 16x$ $\frac{25}{17}$

51. $\frac{10}{11}x + 5 = -\frac{5}{11}$ -6

52. $18x + \frac{1}{2} = 13x$ $-\frac{1}{10}$

53. $6 - 2x = 40x$ $\frac{1}{7}$

54. $35 + 3x = 10x$ 5

In Exercises 55–60, solve the inequality.

55. $-5x \le 65$ $x \ge -13$

56. $6x < 4 - 12x$ $x < \frac{2}{9}$

57. $10 > 6 - 2x$ $x > -2$

58. $13 \le 5x - 2$ $x \ge 3$

59. $\frac{1}{2} + \frac{3}{4}x < 10$ $x < \frac{38}{3}$

60. $-\frac{6}{5}x - \frac{1}{3} \ge \frac{2}{3}$ $x \le -\frac{5}{6}$

Exploration and Extension

American Doctors **In Exercises 61 and 62, use the following information.**

The table at the right shows the number, N, of practicing medical doctors per 100,000 people in each of the 50 states in 1988.
(Source: U.S. Bureau of Census)

AK-138	HI- 225	ME- 173	NJ- 234	SD- 138
AL- 151	IA- 145	MI- 180	NM-173	TN-187
AR-144	ID- 120	MN-212	NV- 158	TX- 169
AZ-191	IL- 210	MO-190	NY- 307	UT-177
CA-242	IN- 151	MS- 125	OH- 191	VA- 204
CO-202	KS- 169	MT-150	OK- 145	VT- 244
CT-296	KY- 161	NC-179	OR- 196	WA-206
DE-191	LA- 184	ND-167	PA- 227	WI- 181
FL- 203	MA-322	NE- 168	RI- 244	WV-164
GA-167	MD-325	NH-186	SC- 156	WY-134

61. The range of N within the 50 states is given by the compound inequality

$$120 \le N \le 325.$$

Write an absolute value inequality that is equivalent to this compound inequality.

62. Drop the five lowest and five highest values of N. Complete the following statement: *Eighty percent of the values of N can be described by the compound inequality*

$$\boxed{?} \le N \le \boxed{?}.$$

Write an absolute value inequality that is equivalent to the compound inequality. $|N - 194| \le 50$

61. $|N - 222.5| \le 102.5$

For example, here is a way to use this method to solve Exercise 25.

$$|14 - 5x| > 8$$

1. Write the corresponding absolute-value equation. Then solve it to find the boundary points.
   $|14 - 5x| = 8$; so, solve $14 - 5x = 8$ and $14 - 5x = -8$; the boundary points are $\frac{6}{5}$ and $\frac{22}{5}$.
2. Plot the boundary points on the number line. Since the original inequality was written with $>$, you need to use open dots.
   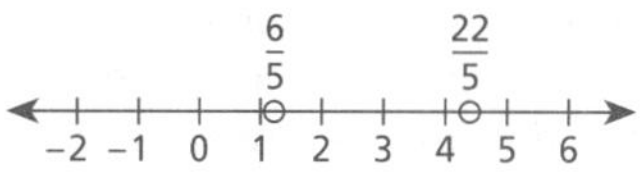

3. Define the three regions. Test a point in each region in the original inequality.
   Region 1: $x < \frac{6}{5}$; try 0.
   $|14 - 5(0)| > 8$
   $|14| > 8$ True
   Region 2: $\frac{6}{5} < x < \frac{22}{5}$; try 2.
   $|14 - 5(2)| > 8$
   $|14 - 10| > 8$
   $|4| > 8$ False
   Region 3: $x > \frac{22}{5}$; try 5.
   $|14 - 5(5)| > 8$
   $|14 - 25| > 8$
   $|-11| > 8$ True
4. Write and graph the solution.
   $x < \frac{6}{5}$ or $x > \frac{22}{5}$
   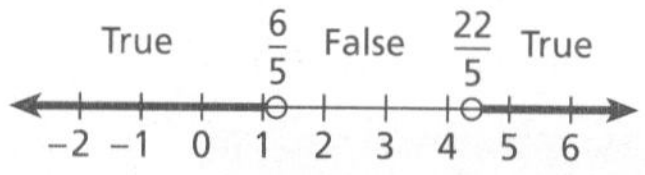

5. Check your solutions to Exercises 23–34 by using this technique.

Problem of the Day

How can you bring from the river exactly 6 quarts of water in one trip if the only containers you have are a 4-quart pail and a 9-quart pail?

9	9 5 5 1 1 0 9 6 6
4	0 4 0 4 0 1 1 4 0

ORGANIZER

Warm-Up Exercises

Order the following numbers from least to greatest. Indicate the frequency of occurrence for each number in the list.

1. 5, 8, 2, 9, 10, 3, 4, 6, 8, 7, 2, 2, 9, 4, 6, 9, 8, 11, 3, 6, 3, 8, 9
2. 34, 23, 45, 52, 36, 23, 45, 32, 43, 51, 32, 43, 50, 51, 26, 27, 31, 44, 43, 42, 43, 23, 45, 52, 38, 35, 33, 32

Number in list	Frequency
2	3
3	3
4	2
5	1
6	3
7	1
8	4
9	4
10	1
11	1

Number in list	Frequency
23	3
26	1
27	1
31	1
32	3
33	1
34	1
35	1
36	1
38	1
42	1
43	4
44	1
45	3
50	1
51	2
52	2

Lesson Resources

Teaching Tools
- Transparencies: 5, 6
- Problem of the Day: 1.8
- Warm-up Exercises: 1.8
- Answer Masters: 1.8

Extra Practice: 1.8
Color Transparencies: 6, 7, 8
Technology Handbook: p. 4

1.8 Exploring Data: Tables and Graphs

What you should learn:

Goal How to organize data by using tables and graphs

Why you should learn it:

Real-life problems can involve hundreds of pieces of information. Data-organization techniques help you to see patterns in the data.

Goal Organizing Data

The world we live in is becoming more and more complex. Almost every day of your life you will be given **data** that describe real-life situations. (The word *data* is plural, and it means information, facts, or numbers that describe something. The singular form, *datum,* is seldom used.)

A collection of data is easier to understand when it is organized in a table or in a graph. There is no "best way" to organize data, but there are many good ways. This lesson illustrates several types of tabular and graphic presentations: tables, frequency distributions, line plots, bar graphs, circle graphs, time-line graphs, and picture graphs.

Real Life **Family Size**

Example 1 *Making a Frequency Distribution and Line Plot*

Thirty-two students in a class were asked how many brothers and sisters they have (including half brothers, half sisters, stepbrothers, and stepsisters). The results were as follows.

0, 2, 0, 1, 1, 3, 4, 0, 0, 3, 2
1, 1, 6, 0, 3, 0, 2, 1, 2, 6, 1
3, 0, 2, 1, 5, 7, 1, 2, 0, 4

Construct a frequency distribution and a line plot for this data.

Solution In the frequency distribution, each tally mark represents one response. In the line plot, each × represents one response.

Number	Tally	Frequency
7	|	1
6	||	2
5	|	1
4	||	2
3	||||	4
2	~~||||~~ |	6
1	~~||||~~ |||	8
0	~~||||~~ |||	8

Frequency Distribution

	0	1	2	3	4	5	6	7
	××××××××	××××××××	××××××	××××	××	×	××	×

Line plot ■

The "size" of fast-food companies can be compared using either annual sales or the number of sales outlets.

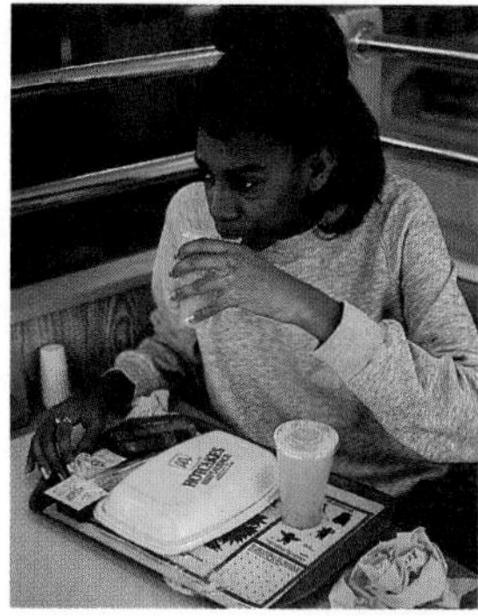

Connections
Fast-Food Sales

Example 2 *Making a Bar Graph and a Circle Graph*

The table shows the ten fast-food companies in the United States that had the greatest sales in 1990. Construct a double bar graph that shows the number of outlets and the sales of each company. Then construct a circle graph that shows each company's percent of the total amount sold by the ten companies. ***(Source: Restaurants and Institutions 400)***

Rank	Company	Number of Outlets	Sales ($ billions)
1.	McDonald's	11,162	17.3
2.	Burger King	6,051	5.7
3.	Kentucky Fried Chicken	7,945	5.3
4.	Pizza Hut	7,410	4.1
5.	Wendy's	3,755	3.0
6.	Hardee's	3,231	2.9
7.	Domino's Pizza	5,220	2.5
8.	Dairy Queen	5,206	2.2
9.	Taco Bell	3,125	2.1
10.	Arby's	2,247	1.3

Solution The double bar graph is shown below. The circle graph is shown at the left.

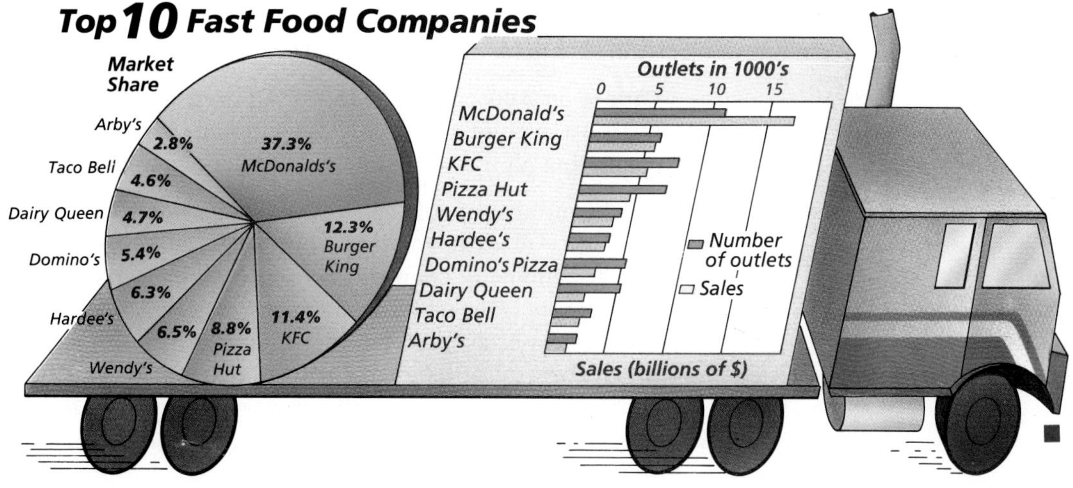

LESSON Notes

Example 1

A frequency distribution is one of the most basic forms of representing data. Students should order the data in the set to help keep track of the information.

Example 2

Ask students to discuss which representation conveys more information. Be sure students recognize that the double bar graph represents two different sets of data.

Example 3

Encourage students to identify other historical events that can be appropriately represented on a time line, such as events in the history of modern flight or events in the history of algebra or geometry.

Extra Examples

Here are additional examples similar to **Examples 1–2.**

1. Construct a frequency distribution and a line plot for the given data. Twenty-three students in a mathematics class were asked how often (on average) they went to the movies in a week during the summer. The results were as follows: 1, 2, 0, 1, 2, 2, 4, 1, 0, 1, 1, 3, 0, 2, 4, 2, 1, 1, 1, 2, 2, 0, 3,

Number	Tally	Frequency
0	||||	4
1	~~||||~~ |||	8
2	~~||||~~ ||	7
3	||	2
4	||	2

0	1	2	3	4
XXXX	XXXXXXXX	XXXXXXX	XX	XX

2. The table below shows the consumption of seven nations. Construct a bar graph and a circle graph of the data.

1989 Oil Consumption

Nation	Number of barrels (in millions)
1. Britain	634.0
2. Canada	643.1
3. France	677.4
4. Italy	708.1
5. Japan	1818.1
6. United States	6323.6
7. Germany	831.5

Check students' graphs.

Check Understanding

1. How are bar graphs and double bar graphs different? the same? See Example 2.
2. When would a circle graph be better to use than a bar graph? When the data involves percents that add up to 100%.
3. What is the relationship between a frequency distribution and a bar graph? The bar graph displays the frequency distribution.

Communicating about ALGEBRA

Answers

A. Double bar, to compare countries for numbers of two kinds of medals received.

B. Bar graph, to compare numbers of countries participating in various years.

C. Time line, to compare years when various sports became Olympic events.

D. Circle graph, to compare portions of the total number of medals won by each continent.

Example 3 *Creating a Time Line*

On August 18, 1920, the Nineteenth Amendment to the United States Constitution was ratified. This amendment gave women in each state the right to vote. Before the Nineteenth Amendment, each state was allowed to decide whether women could vote. Find the dates that women achieved full voting rights in each state. Then create a time line to show the results.

Solution Wyoming was the first state in 1869 to grant full voting rights to women. By the time South Dakota, Oklahoma, and Michigan granted voting rights in 1918, the women's suffrage movement had gained enough support to pass the Nineteenth Amendment. ■

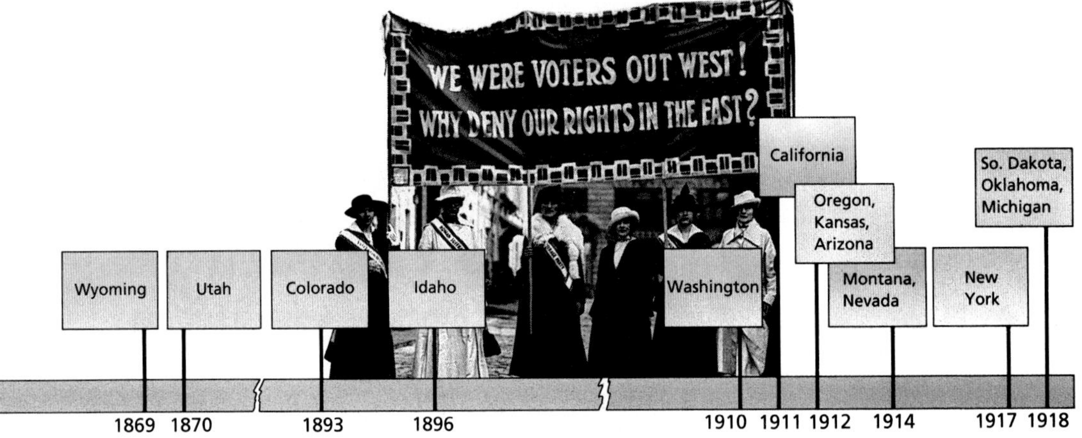

Communicating about ALGEBRA

SHARING IDEAS about the Lesson

Choose an Appropriate Graph For each set of data, choose the most appropriate graph—bar or double bar graph, circle graph, line plot, or time line. Justify your answer. See margin.

A. The number of gold and silver medals won by North and South American countries in the 1992 Olympics

B. The number of countries participating in each of the Olympic Games from 1960 to the present

C. The year each sport became an Olympic event

D. The percent of the medals that were won by participants from each of the seven continents

EXERCISES

Guided Practice

CRITICAL THINKING about the Lesson

1. Give an example of a collection of data that you could organize using a line plot. **Answers will vary.**

2. *Student Council Election* Which of the techniques discussed in this lesson do you think is best suited to organizing the results of a student council election? Explain. **Bar graph, see margin.**

3. *Favorite Music* Each of your classmates was asked to select their favorite type of music from these choices: classical, jazz, country, rock, and rap. How would you organize the results? **Bar graph**

4. *Pizza Prices* Which would give a better picture of the price of a cheeze pizza at pizza shops—a frequency distribution or a bar graph? Explain.
A bar graph because there will probably be 8 different prices

Independent Practice

5. *Fishing Trip* On each day of your three-week summer vacation in Alaska, you went fishing. At the end of each day, you recorded the number of legal-size fish you caught. The results are shown in the table at the right. Construct a frequency distribution and a line plot for this data. **See Additional Answers.**

S	M	T	W	T	F	S
5	7	4	5	2	3	4
3	4	6	2	7	6	1
5	1	3	6	2	4	3

6. *Band Practice* On each day except Sundays for two weeks, the band director recorded the attendance of his band. Construct a frequency distribution and line plot for the results. **See Additional Answers.**

M	T	W	T	F	S
299	298	300	300	297	300
300	299	295	296	300	299

✪ **7.** *Endangered Species* In 1990, 405 animal species were endangered in the United States. Construct a circle graph that shows each category's percent of the total number of endangered species. **See margin.**

(Source: U.S. Fish and Wildlife Service)

Category	Number
Amphibians/Reptiles	16
Arachnids/Insects	12
Birds	57
Clams/Snails	49
Fish	53
Mammals	36
Plants	182

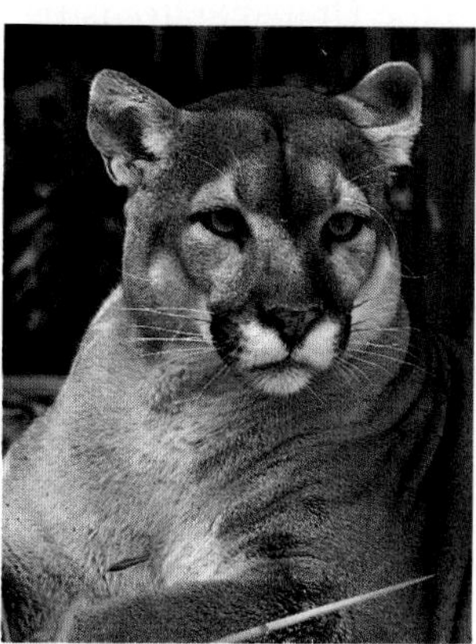

***The Florida panther is on the top-ten list of endangered species published by* Defenders of Wildlife.**

EXTEND *Communicating*
Select a team sport played at your school and determine its win-lose record for the past five or ten seasons. Construct the most appropriate graph in which to display the data. Do graphs differ by the sport selected?

EXERCISE Notes

ASSIGNMENT GUIDE

Basic/Average: Ex. 1–4, 5, 7–15, 17–23 odd

Above Average: Ex. 1–4, 5–15, 17–23 odd

Advanced: Ex. 1–4, 5–16, 21–23

Selected Answers
Exercises 1–4, 5–21 odd

✪ **More Difficult Exercises**
Exercises 7, 22, 23

Guided Practice

▶ **Ex. 1–4** These exercises can be used to summarize the data-organizing methods introduced in the lesson.

Answer

2. A bar graph, to compare candidates for numbers of votes received; or a circle graph, to compare candidates for parts of a total number of votes received.

Independent Practice

Ex. 5–6 These exercises are modeled in Example 1.

Ex. 7 Although Example 2 provides a solution model, students may need help in deciding how to translate the percent for each species to a sector in the circle graph.

Answers

7.

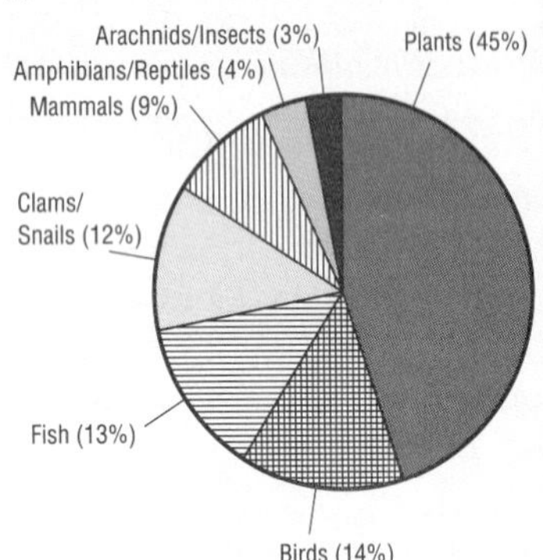

8.

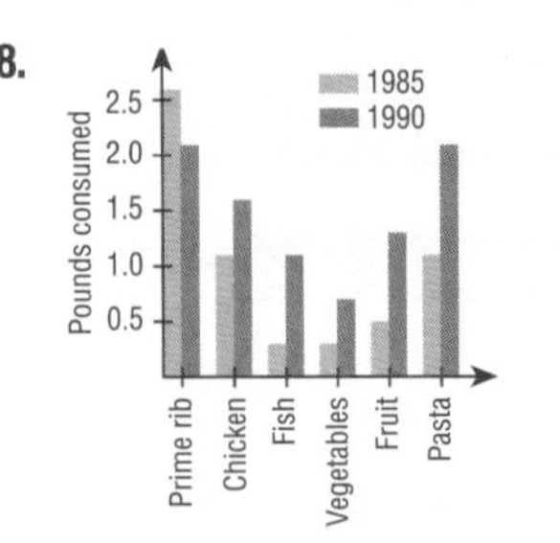

9.

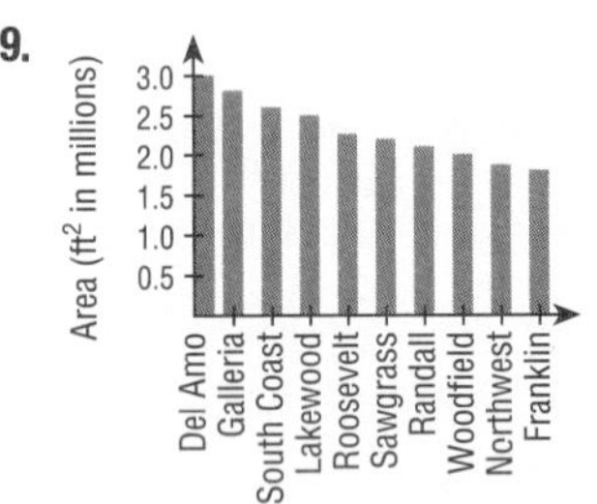

Ex. 9–12 These exercises should be assigned as a group.

Ex. 22–23 Students can use the Voting Rights model illustrated in Example 3 to display the data.

8. *You're Kidding, Aren't You?* In 1985, a typical daily serving for each New York Giants football player during training camp included about 2.6 pounds of prime rib, 1.1 pounds of chicken, 0.3 pound of fish, 0.3 pound of vegetables, 0.5 pound of fruit, and 1.1 pounds of pasta. In 1990, this menu changed to 2.1 pounds of prime rib, 1.6 pounds of chicken, 1.1 pounds of fish, 0.7 pound of vegetables, 1.3 pounds of fruit, and 2.1 pounds of pasta. Construct a double bar graph that shows the pounds of each type of food for 1985 and 1990. ***(Source: Dick Rossi, New York Giants)*** See margin.

United States Shopping Malls **In Exercises 9–12, use the following information.**

The list shows the ten largest shopping malls in the United States completed by 1990. ***(Source: International Council of Shopping Centers)***

Shopping Mall	Square Feet
Del Amo Fashion Center (Calif.)	3,000,000
The Galleria (Tex.)	2,808,000
South Coast Plaza & Town (Calif.)	2,600,000
Lakewood Center Mall (Calif.)	2,498,000
Roosevelt Field Mall (N.J.)	2,264,000
Sawgrass Mills (Fla.)	2,200,000
Randall Park Mall (Ohio)	2,097,000
Woodfield (Ill.)	2,000,000
Northwest Plaza (Mo.)	1,868,000
Franklin Mills (Penn.)	1,800,000

Retail space for shopping centers in 1990 was 4.5 billion square feet—18 sq. ft. for every person in the U.S.A.
(Source: International Council of Shopping Centers.)

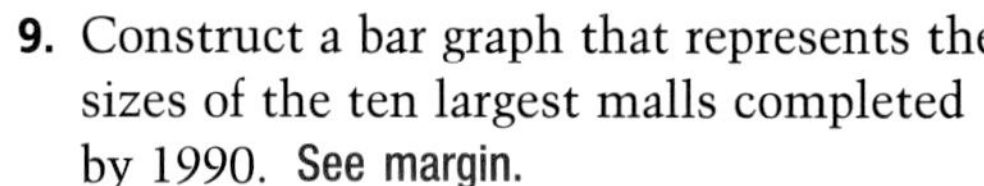

9. Construct a bar graph that represents the sizes of the ten largest malls completed by 1990. See margin.

10. What percent of the total shopping center space in the United States is represented by the ten shopping malls in the table? 0.5%

11. Use the graph at the right to estimate the number of shopping malls in the United States in 1970 and 1980. 11,000; 21,000

12. What was the average number of square feet in a shopping mall in the United States in 1990? ≈122,783

13. *Landfills in Vermont* In 1987, Vermont put 328,200 tons of solid waste into its 64 landfills. The types of material put into the landfills are shown in the circle graph. Use the graph to estimate the weight of the cardboard.
(Source: Vermont Agency of Natural Resources)
54,000 tons

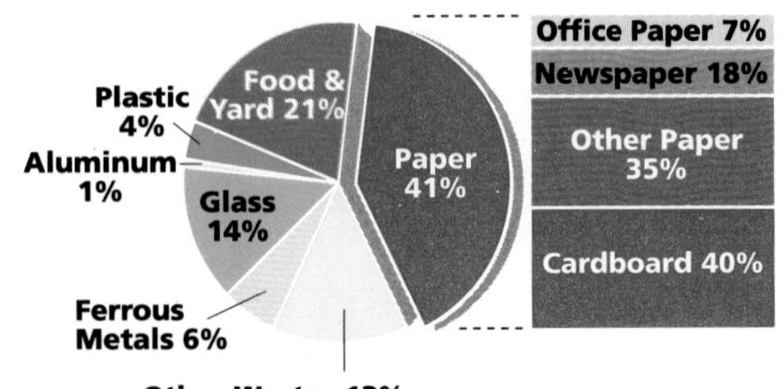

Integrated Review

House Wren Nest **In Exercises 14 and 15, use the following information.**

Wrens are very eccentric in their selection of building materials. A house wren's nest was taken apart for analysis, and the following was found among the nesting materials.
(Source:* America's Favorite Backyard Birds, *Simon and Schuster, 1983)

52 Hairpins	188 Nails
4 Thumbtacks	13 Staples
21 Pins	4 Pencil leads
6 Paper clips	52 Wires
1 Buckle	2 Hooks
3 Trash bag twists	2 Odds and ends

Of the 63 types of wrens in the world, 7 are native to the Eastern United States.
(Source:* Peterson's Field Guide of Eastern Birds*)

14. What percent of the items were nails? ≈54%

15. What percent of the items were staples? ≈3.7%

In Exercises 16–21, solve the inequality.

16. $9 + 3x \le 1 - x$ $x \le -2$

17. $1 + 4x > x - 5$ $x > -2$

18. $-12 < 3x - 1 \le 3$ $-\frac{11}{3} < x \le \frac{4}{3}$

19. $2 \le 5 + 7x \le 6$ $-\frac{3}{7} \le x \le \frac{1}{7}$

20. $|x + 4| \ge 4$ $x \le -8$ or $x \ge 0$

21. $|3 - x| < 21$ $-18 < x < 24$

Exploration and Extension

22. *The First Thirteen States* The Constitution of the United States was signed by representatives of the original thirteen colonies on September 17, 1787. Each colony, however, was not admitted into the union as a state until it had ratified (approved by vote) the Constitution. The dates of admission into the union are shown in the table. Create a time-line graph for this data.

23. *The Next Seven States* Use a reference source to find the dates of the next seven states that were admitted to the union: Vermont, Kentucky, Tennessee, Ohio, Louisiana, Indiana, and Mississippi. Make a time-line graph of these data. Did you use the same scale as in Exercise 22? See margin.

State	Date of Admission
1. Delaware	December 7, 1787
2. Pennsylvania	December 12, 1787
3. New Jersey	December 18, 1787
4. Georgia	January 2, 1788
5. Connecticut	January 9, 1788
6. Massachusetts	February 6, 1788
7. Maryland	April 28, 1788
8. South Carolina	May 23, 1788
9. New Hampshire	June 21, 1788
10. Virginia	June 25, 1788
11. New York	July 26, 1788
12. North Carolina	November 21, 1789
13. Rhode Island	May 29, 1790

Enrichment Activity

ALTERNATIVE ASSESSMENT This lesson provides an opportunity for a student research project. Encourage students to examine the lesson examples and applications for the types of data to research.

PROJECT Look for survey information in newspapers and magazines or compile sports data or population data from an encyclopedia or almanac. Organize the data using a method presented in the lesson.

Answers

22.

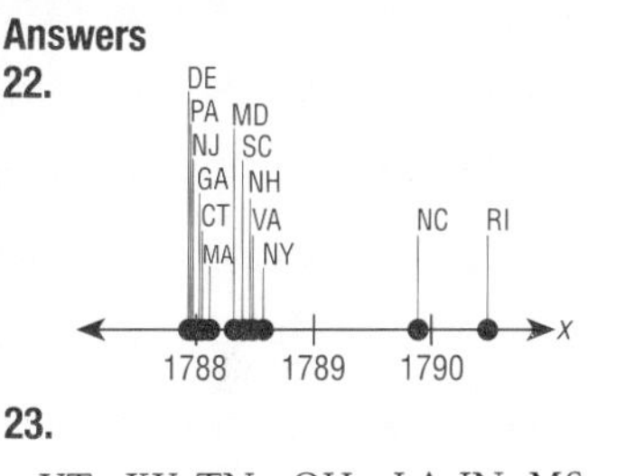

23.

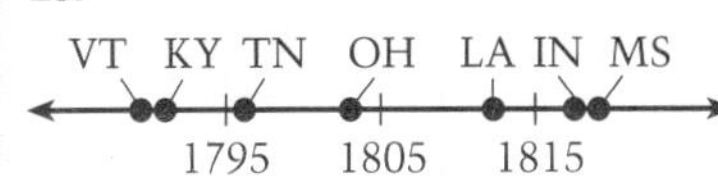

ORGANIZER

The Chapter Summary helps students organize the main ideas of the chapter. In this chapter, students reviewed many of the basic concepts and skills found in an Algebra 1 course. The connection between modeling real-life problems and solving the resulting algebraic models were also demonstrated.

Work with students to review the skills, strategies, and concepts of the chapter. The first day's homework assignment can be used as the basis for the second day of review.

COOPERATIVE LEARNING
Encourage students to study together. Emphasize the importance of "teaching" a classmate how to perform a skill or how to recall a strategy. When students work together, everyone wins. The students receiving help get additional instruction and the students giving help gain a deeper understanding of the skills and concepts involved.

Chapter SUMMARY

Review the chapter by asking students, "What did you learn? Why did you learn it? How does it fit with the other algebra skills and concepts you have learned?"

1

Chapter Summary

What did you learn?

Skills

1. Add, subtract, multiply, and divide real numbers. **(1.1, 1.2)**
2. Evaluate an algebraic expression. **(1.2)**
3. Solve
- a linear equation. **(1.3)**
- a literal equation for one of its variables. **(1.5)**
- an absolute value equation. **(1.7)**

4. Solve and graph
- a linear inequality. **(1.6)**
- an absolute value inequality. **(1.7)**

Strategies

5. Create algebraic models for real-life situations. **(1.1–1.7)**
6. Use problem-solving strategies to solve real-life problems. **(1.4)**

Exploring Data

7. Use tables and graphs to organize data. **(1.8)**

Why did you learn it?

The *primary* use of algebra is to create mathematical models for real-life situations that can be used to answer questions about those situations. As you continue your study of algebra, you should recognize that almost every facet of life uses some mathematics. Most people know that business, science, construction, engineering, and computer science use a lot of mathematics. But other fields such as history, art, music, politics, and psychology also use mathematics. For this book, we have chosen applications of mathematics from many different fields. We hope that we have chosen many that interest you.

How does it fit into the bigger picture of algebra?

This first chapter reviewed several basic skills and strategies from Algebra 1. From that course, you should remember that algebra is much more than just solving equations that contain letters (or variables). Other algebraic skills include extending the rules of arithmetic to form the rules of algebra, simplifying algebraic expressions, creating and using mathematical models in real-life settings, and organizing information with tables and graphs.

You practiced using these skills in this chapter, and you will continue expanding them throughout this course.

Chapter REVIEW

In Exercises 1–6, plot the numbers on the real number line. Then state which is greater. (1.1)

1. $\frac{2}{3}$ and $\frac{3}{4}$ $\frac{3}{4}$

2. $\frac{6}{7}$ and $\frac{5}{6}$ $\frac{6}{7}$

3. -4 and -3 -3

4. -5 and -5.1 -5

5. $\sqrt{3}$ and 1.7 $\sqrt{3}$

6. $-\sqrt{2}$ and -1.4 -1.4

In Exercises 7–12, write the numbers in increasing order. (1.1)

7. $-2, -0.5, 0, 0.5, -1.5, 1$ 7., 9.–11. See margin.

8. $-1.2, 0.6, -0.2, 1.4, 0.61, -0.1$

9. $\frac{1}{3}, -\frac{1}{2}, \frac{2}{3}, -\frac{1}{4}, \frac{2}{5}, -\frac{1}{8}$

10. $\frac{1}{3}, 0.33, \frac{3}{10}, \frac{3}{11}, 0.34, \frac{3}{8}$

11. $-\sqrt{2}, 1.41, -1.42, \sqrt{2}, -1.41, 1.42$

12. $\pi, \frac{22}{7}, 3.1416, \frac{355}{113}, 3.14, \frac{311}{99}$

8. $-1.2, -0.2, -0.1, 0.6, 0.61, 1.4$

12. $3.14, \frac{311}{99}, \pi, \frac{355}{113}, 3.1416, \frac{22}{7}$

In Exercises 13 and 14, find the product. (1.1)

13. $\left(\frac{80 \text{ kilometers}}{1 \text{ hour}}\right)\left(\frac{1000 \text{ meters}}{1 \text{ kilometer}}\right)\left(\frac{1 \text{ hour}}{60 \text{ minutes}}\right)\left(\frac{1 \text{ minute}}{60 \text{ seconds}}\right)$ $22\frac{2}{9}$ m per second

14. $\left(\frac{20 \text{ dollars}}{1 \text{ ton}}\right)\left(\frac{1 \text{ ton}}{2000 \text{ pounds}}\right)\left(\frac{1 \text{ pound}}{16 \text{ ounces}}\right)\left(\frac{100 \text{ cents}}{1 \text{ dollar}}\right)$ $\frac{1}{16}$ cent per ounce

In Exercises 15–20, evaluate the expression. (1.2)

15. $-2b^2$ when $b = -2$ -8

16. $7x - 8x^3$ when $x = -1$ 1

17. $5\sqrt{y} + 2$ when $y = 4$ 12

18. $|2x - 5|$ when $x = 2.1$ 0.8

19. $\frac{3ab^2 + 1}{3a^2b - 1}$ when $a = 2$ and $b = -2$ -1

20. $\frac{4x - 3y}{3x - 4y}$ when $x = 2$ and $y = 3$ $\frac{1}{6}$

In Exercises 21–24, evaluate the expression. (1.1)

21. $-3 - 6 \div 2 - 12$ -18

22. $7 - (3 \cdot 4)2 + 8$ -9

23. $12 - 3 \cdot 2 + 5$ 11

24. $-4 - 5 - 2 \cdot 3^2$ -27

In Exercises 25–28, find the area of the blue region. (1.2)

25. Triangle 14 cm^2

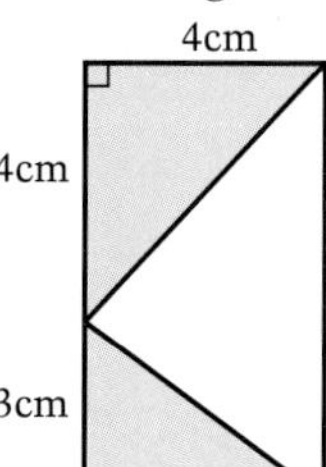

26. Circle $\approx 15.4 \text{ cm}^2$

7cm
4cm

27. Ellipse $\approx 6.0 \text{ cm}^2$

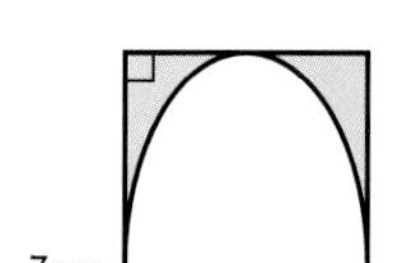

28. Trapezoid $15\frac{3}{8} \text{ cm}^2$

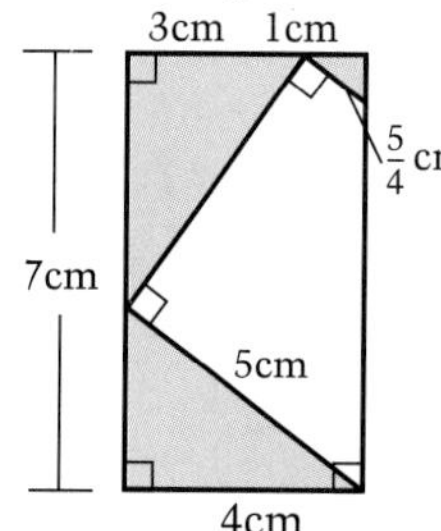

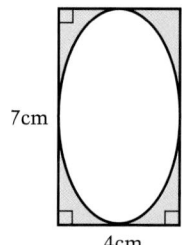

In Exercises 29–36, solve the equation. (1.3, 1.7)

29. $-5x + 3 = 18$ -3

30. $-2(x - 4) = 12$ -2

31. $3(x - 6) = 6x + 3$ -7

32. $-4x - 4 = 3(2 - x)$ -10

33. $|x| = 3$ $-3, 3$

34. $|x + 1| = 4$ $-5, 3$

35. $|2x - 3| = 5$ $-1, 4$

36. $-2|3x - 1| = -8$ $-1, \frac{5}{3}$

Chapter REVIEW

ASSIGNMENT GUIDE

Basic/Average: Ex. 1–43 odd, 45–50, 54–58 even, 60–62

Above Average: Ex. 1–43 odd, 51–59 odd, 60–63

Advanced: Ex. 1–43 odd, 51–59 odd, 60–65

✪ More Difficult Exercises
Exercises 54–65

▶ **Ex. 1–65** Since the Chapter Review exercises parallel the list of Skills, Strategies, and Exploring Data statements on page 56, tell students to use the Chapter Summary to identify the lesson in which a skill or concept was presented if they are uncertain how to do a given exercise. This makes the students categorize each task in the exercise set and locate relevant information in the chapter about the task, both of which are beneficial activities.

The answers to the odd-numbered exercises of the Chapter Review will be found in the "Selected Answers" section of the student text.

Answers

7. $-2, -1.5, -0.5, 0, 0.5, 1$

9. $-\frac{1}{2}, -\frac{1}{4}, -\frac{1}{8}, \frac{1}{3}, \frac{2}{5}, \frac{2}{3}$

10. $\frac{3}{11}, \frac{3}{10}, 0.33, \frac{1}{3}, 0.34, \frac{3}{8}$

11. $-1.42, -\sqrt{2}, -1.41, 1.41, \sqrt{2}, 1.42$

Answers

37.

38.

39.

40.

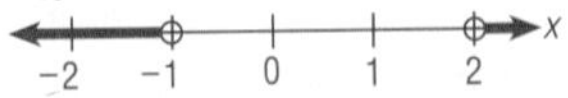

41.

42.

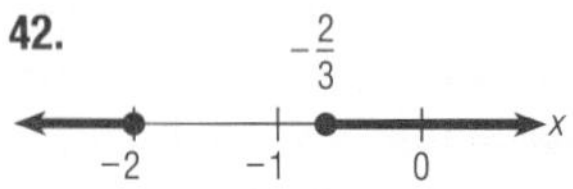

In Exercises 37–42, solve and graph the inequality. (1.7) See margin.

37. $2x - 3 < 5$ $x < 4$

38. $-3x + 4 \geq 2x + 19$ $x \leq -3$

39. $-3 \leq 2x + 1 < 5$ $-2 \leq x < 2$

40. $3x + 1 < -2$ or $3x + 1 > 7$ $x < -1$ or $x > 2$

41. $|2x - 5| < 9$ $-2 < x < 7$

42. $|3x + 4| \geq 2$ $x \leq -2$ or $x \geq -\frac{2}{3}$

Scouting Helicopter **In Exercises 43 and 44, use the following information.**

In 1991, Boeing Helicopters and Sikorsky Aircraft were awarded a $30 billion contract to build the United States Army's new scouting helicopter. Each two-person chopper will be equipped with night vision and will be able to fly 1250 nautical miles before refueling. The first of nearly 1300 helicopters is scheduled for production in 1998.
(Source:* Popular Science*)

43. A nautical mile is 6080 feet. How many (standard) miles will the new scouting helicopter be able to fly before refueling? ≈1439.4 mi

44. Approximate the average cost of each scouting helicopter. $23 million

Lettuce **In Exercises 45 and 46, use the following information.**

For 1970 through 1990, the average price, p (in dollars), of a pound of lettuce can be modeled by $p = 0.19 + 0.023t$, where $t = 0$ represents 1970. The total amount, L (in billions of pounds), of lettuce consumed in the United States can be modeled by $L = 4.5 + 0.05t$.
(Source:* U.S. Department of Agriculture*)

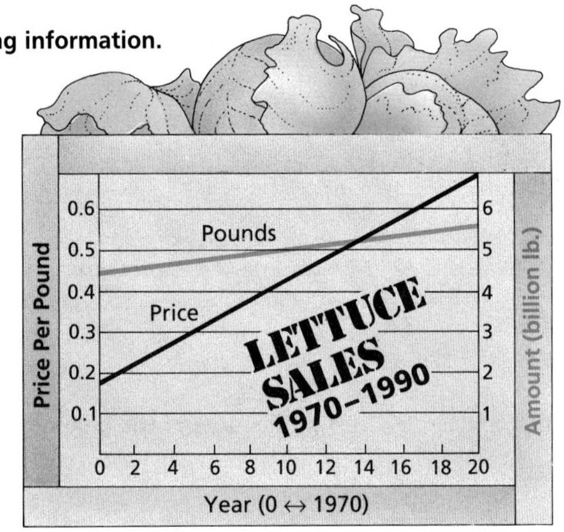

45. According to these models, during which year will the average price of lettuce be $0.765? 1995

46. Approximate the total sales in lettuce in the United States in 1990. ≈$3.58 billion

In Exercises 47–50, find the volume of the solid. (1.2)

47. Sphere ≈1436.76 cm³ **48.** Cone ≈56.55 ft³ **49.** Cylinder ≈384.85 m³ **50.** Rectangular Prism 24 ft³

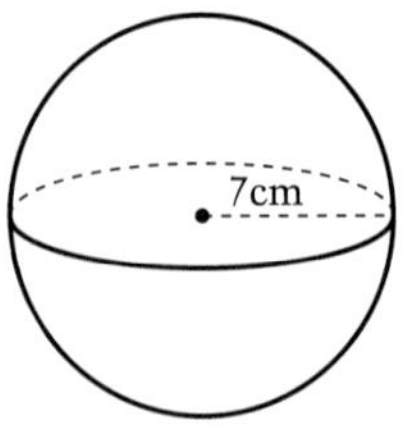

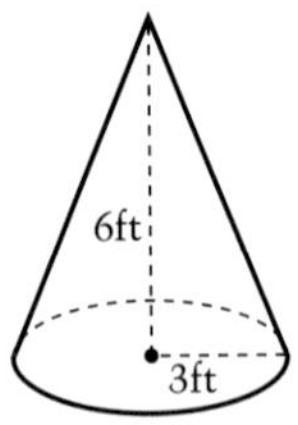

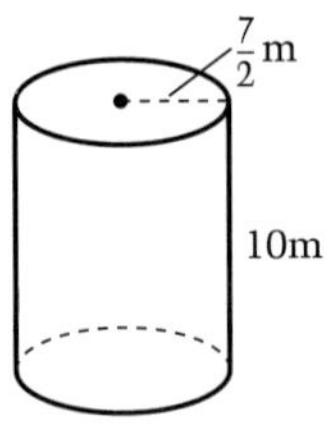

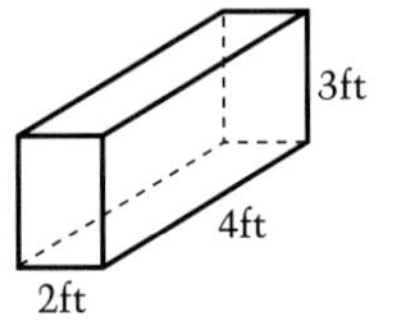

In Exercises 51 and 52, solve the literal equation for y. (1.5)

51. $7x + 3y = 9$ $y = -\frac{7}{3}x + 3$

52. $-3xy + 4y = 10$ $y = \dfrac{10}{-3x + 4}$

53. *Louisiana and Georgia* The range of recorded temperatures, L (in °F), in Louisiana is $-16 \le L \le 114$. The range of recorded temperatures, G, in Georgia is related to those in Louisiana by $L = G + 1$. What is the range of recorded temperatures in Georgia? $-17 \le G \le 113$

In 1994, 87 million full-time salaried and wage workers were employed in the United States.

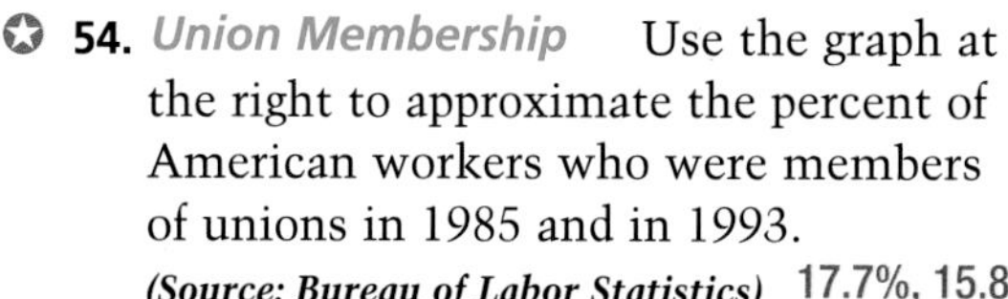

54. *Union Membership* Use the graph at the right to approximate the percent of American workers who were members of unions in 1985 and in 1993. *(Source: Bureau of Labor Statistics)* 17.7%, 15.8%

55. *Overtime* If you are paid \$12 an hour for up to 40 hours per week and time and a half for overtime, how many hours would you have to work in a week to earn \$624? Create a bar graph that shows your weekly pay for working overtime between 0 and 8 hours. 48, see margin.

56. *License Plate Colors* Use the graph at the right to create a circle graph that shows the colors used on automobile license plates in the United States. *(Source: USA Today)* See margin.

57. *Number of License Plates* In 1993, 10,613,000 vehicles were registered in New Mexico, Maryland, and Georgia. New Mexico had 3000 less than two fifths of the number in Maryland. Georgia had 52,000 less than 4 times the number in New Mexico. How many registered motor vehicles were in each of these three states? *(Source: U.S. Federal Highway Administration)* See margin.

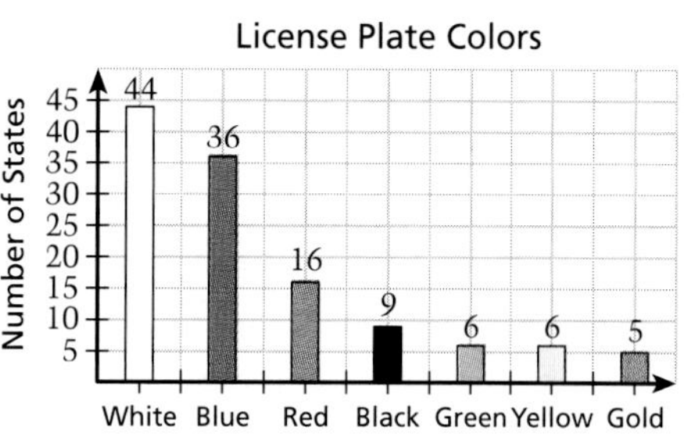

58. *Muscular Dystrophy Telethon* Use the graph below to estimate the amount of money raised by the Muscular Dystrophy Telethon in 1988. *(Source: Muscular Dystrophy Association)* \$41 million

59. *Muscular Dystrophy* Duchenne dystrophy is the most common form of muscular dystrophy. It is very rare in females (women who carry the defective gene rarely develop the disease), but it occurs in about one out of 3000 males. In 1992, about two million male babies were born in the United States. Approximately how many had Duchenne muscular dystrophy? 667

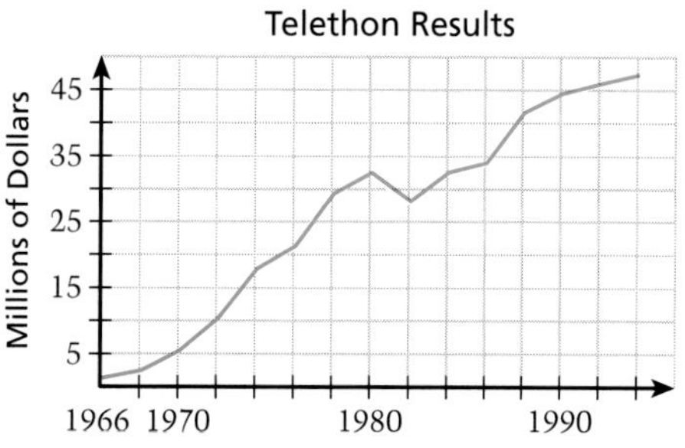

The Muscular Dystrophy Association Labor Day Telethon results.

55.

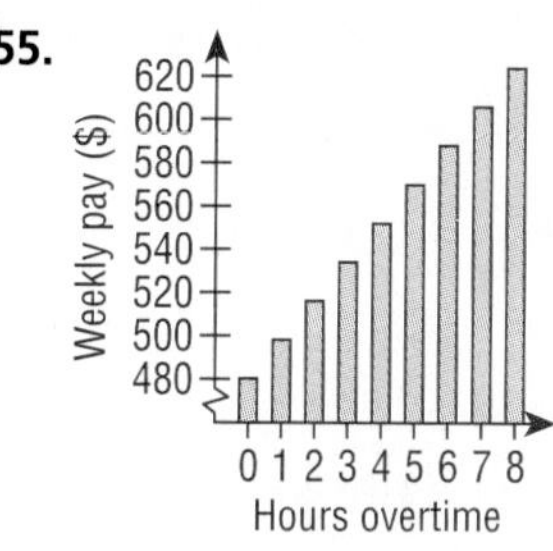

56.

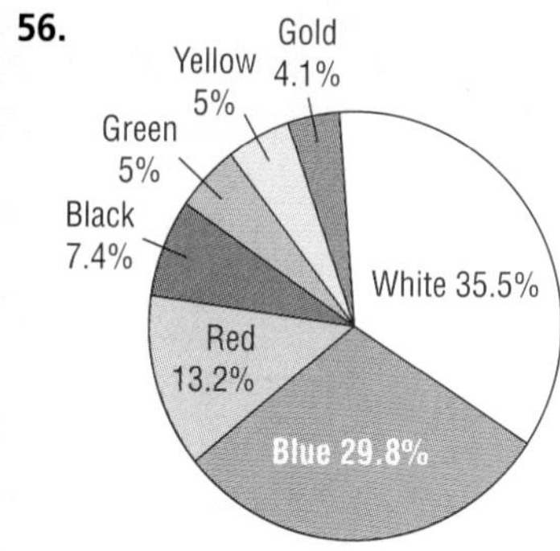

57. Georgia: 5,632,000; Maryland: 3,560,000; New Mexico: 1,421,000

Answers

Chapter Review

64.

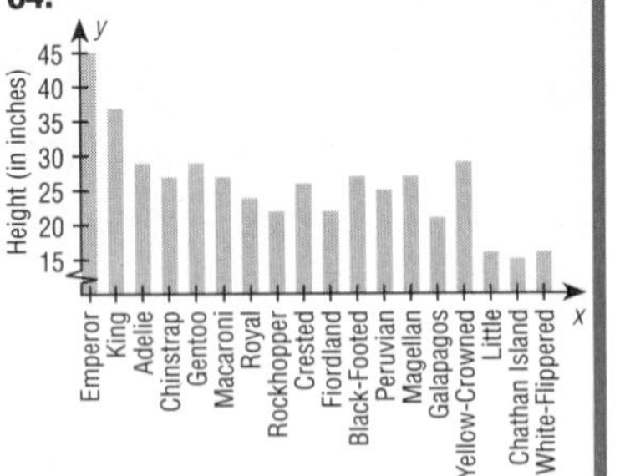

The Race to the South Pole **In Exercises 60–63, use the following information.**

In 1911, the Norwegian explorer Roald Amundsen and the English explorer Robert Scott both wanted to be the first human to reach the South Pole. Going by different routes (of about 900 miles each), the two explorers raced teams of men, dogs, and horses over the frozen land. Amundsen started on October 19 and averaged about 15.8 miles per day. Scott started on November 1 and averaged about 11.6 miles per day.

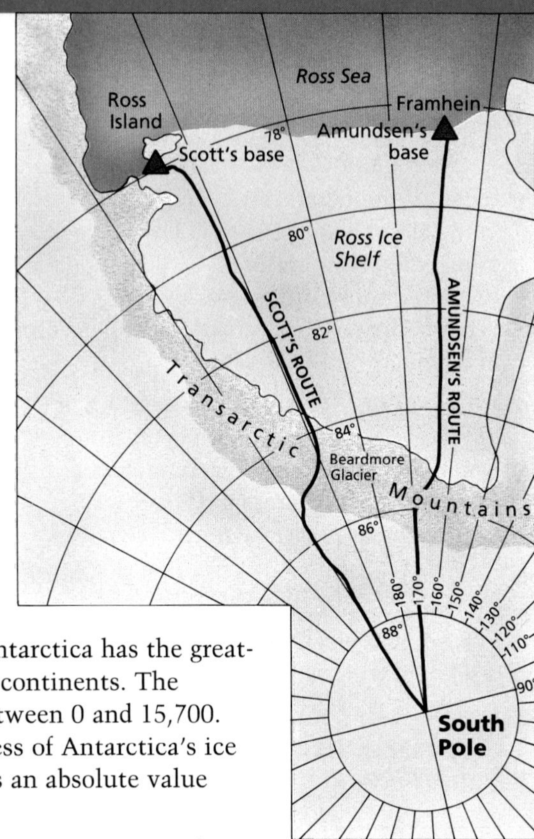

60. On which day did Amundsen's team reach the South Pole? **December 14, 1911**

61. On which day did Scott's team reach the South Pole? **January 17, 1912**

62. Because it is covered with a thick ice cap, Antarctica has the greatest average elevation of any of Earth's seven continents. The thickness, I (in feet), of the ice cap varies between 0 and 15,700. Write a compound inequality for the thickness of Antarctica's ice cap. Then, write the compound inequality as an absolute value inequality. **$0 < I < 15{,}700$; $|I - 7850| < 7850$**

63. The temperature, F (in degrees Fahrenheit), in Antarctica can be described by $-128.6 \le F \le 50$. Write an inequality for the temperature range in degrees Celsius. **$-89\frac{2}{9} \le C \le 10$**

64. *Penguins* The emperor penguin is the largest species of penguins—adults have heights of up to 45 inches. The heights (in inches) of the other species are king (37), Adelie (29), chinstrap (27), gentoo (29), macaroni (27), royal (24), rockhopper (22), crested (26), fiordland (22), black-footed (27), Peruvian (25), Magellan (27), Galápagos (21), yellow-crowned (29), little (16), Chathan Island little (15), and white-flippered (16). Create a bar graph to represent the heights of the 18 species. **64. See margin.**

65. *Kings and Emperors* Although king penguins are almost as tall as emperor penguins, the kings weigh only half of what the emperors weigh. Let K be the average weight (in pounds) of an adult king penguin and let E be the average weight (in pounds) of an adult emperor penguin. If $K + E = 99$, find K and E.

 65. $K = 33$, $E = 66$

Chapter TEST

1. Plot $-\frac{4}{3}$ and $-\frac{3}{4}$ on the real number line. Which is greater? **(1.1)** $-\frac{3}{4}$

2. Write the numbers in decreasing order: $-0.1, -0.6, 0.3, -0.5, 0.4$. **(1.1)** 0.4, 0.3, −0.1, −0.5, −0.6

3. Evaluate $8 - 4 \div 2 - 1$. 5

4. Simplify $3(x - 8) - 2(x + 5)$. $x - 34$

5. Evaluate $-2b^2 + 4ab$ when $a = 3$ and $b = -1$. −14

6. Evaluate $\frac{1}{4}x + \frac{1}{3}y$ when $x = 3$ and $y = 2$. $\frac{17}{12}$

In Exercises 7–10, solve the equation. (1.4, 1.7)

7. $2x + 19 = 5(x + 2)$ 3

8. $1.2n = 2.3n - 2.20$ 2

9. $|2x + 7| = 3$ −5, −2

10. $|5x + 11| = 9$ $-4, -\frac{2}{5}$

11. Solve the literal equation $5xy + 2x = 3$ for y. **(1.5)** $y = \frac{3 - 2x}{5x}$

12. Solve the literal equation $a^2b + 2b = 12$ for b. **(1.5)** $b = \frac{12}{a^2 + 2}$

In Exercises 13–16, solve and graph the inequality. (1.7) See margin.

13. $5x - 2 \le 7$ $x \le \frac{9}{5}$

14. $1 < 2x - 3 < 5$ $2 < x < 4$

15. $|1 - 2x| \le 3$ $-1 \le x \le 2$

16. $|-x + 3| > 4$ $x > 7$ or $x < -1$

In Exercises 17 and 18, find the area of the blue region. (1.2)

17.

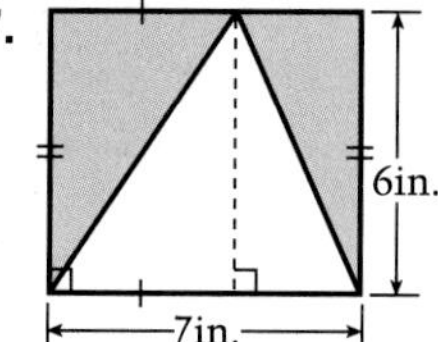

18.

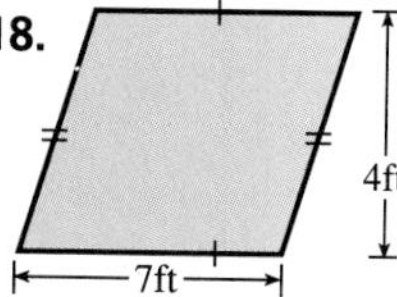

19. In 1990, the average sales tax percent in Connecticut, Illinois, Rhode Island, and Tennessee was 6%. The sales tax in Connecticut, Rhode Island, and Tennessee was, respectively, 2.5%, 1%, and 0.5% more than the sales tax in Illinois. What was the sales tax percent in each state? IL: 5%, CT: 7.5%, RI: 6%, TN: 5.5%

20. Of the Broadway shows running at the end of 1990, those that had the most performances were *Cats* (3427), *Les Misérables* (1527), *Phantom of the Opera* (1224), *Black and Blue* (805), *Grand Hotel* (473), *Gypsy* (468), *A Few Good Men* (465), *City of Angels* (439), *Aspects of Love* (305), and *Piano Lesson* (295). Create a bar graph to represent these data. *(Source: Variety)* See margin.

21. Let x represent the total number of performances of *Cats* at the end of 1991. In 1991, the show had at most 312 performances. Write an inequality that describes the possible values of x. $3427 \le x \le 3739$

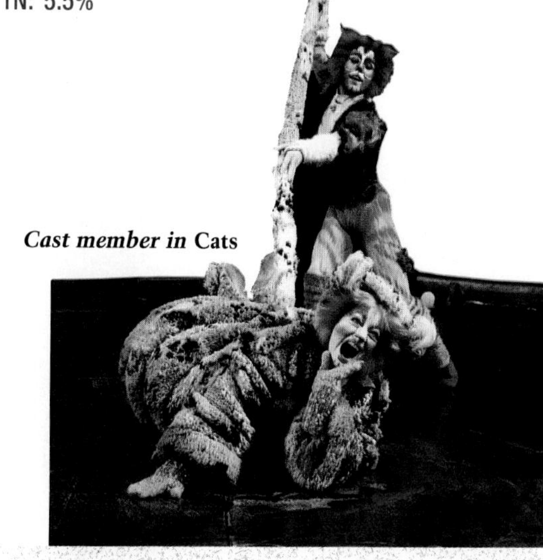

***Cast member in* Cats**

Chapter TEST

Taking the Chapter Test is a good way for students to prepare for your exam. Encourage students to take this test under test-like conditions.

Answers

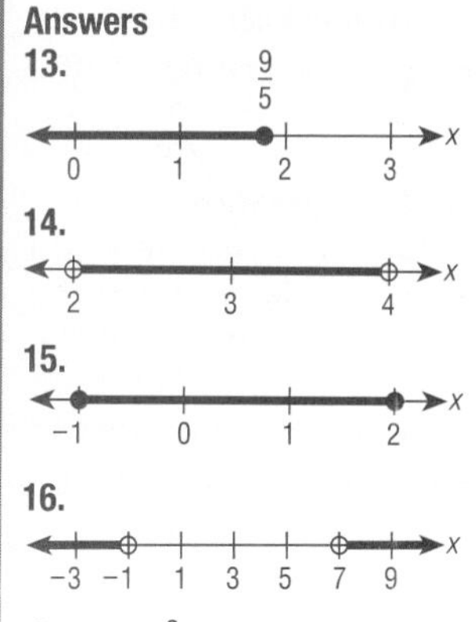

17. 21 in.^{2}

18. 28 ft^2

20.

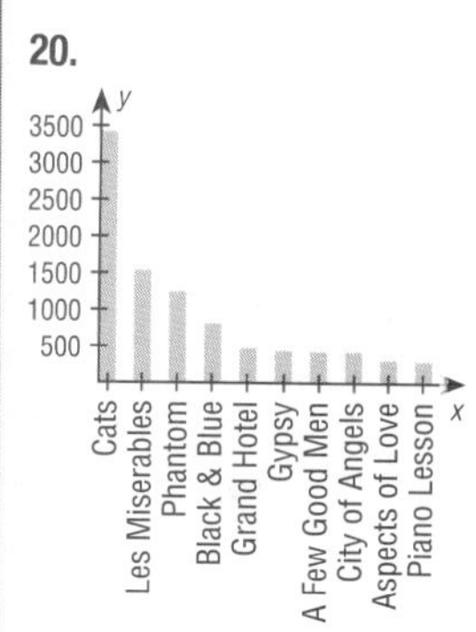

Linear equations and linear inequalities in two variables are common models of real-life situations. Ideas of slope and constant rates of change characterize these linear relationships.

The connection between geometry and algebra is strong and will be made throughout any discussion of linear equations and inequalities.

The Chapter Summary on page 114 provides you and the students with a synopsis of the chapter. It identifies key skills and concepts. You may want to have students look at the Chapter Summary as an overview before beginning the chapter.

CHAPTER 2

Linear Equations

LESSONS

Water-ski jumping is performed competitively in many parts of the country.

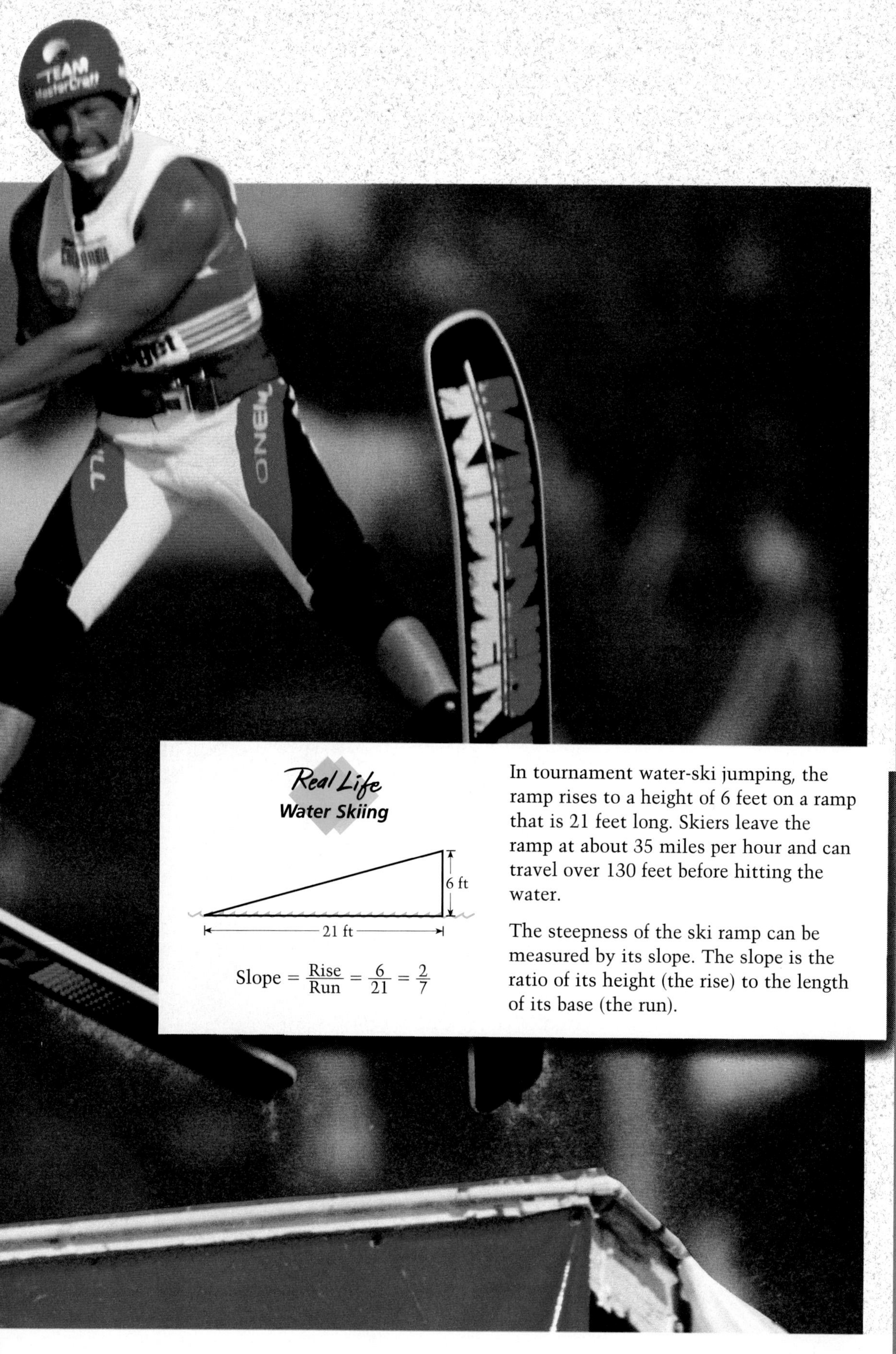

Real Life
Water Skiing

6 ft
21 ft

$$\text{Slope} = \frac{\text{Rise}}{\text{Run}} = \frac{6}{21} = \frac{2}{7}$$

In tournament water-ski jumping, the ramp rises to a height of 6 feet on a ramp that is 21 feet long. Skiers leave the ramp at about 35 miles per hour and can travel over 130 feet before hitting the water.

The steepness of the ski ramp can be measured by its slope. The slope is the ratio of its height (the rise) to the length of its base (the run).

The concept of slope occurs in many applications of mathematics in real-life situations. The water-ski-jumping example is one among many. Others include the "pitch" of a roof of a house, the "grade" of the highway, and the "angle of descent" of an airplane.

Linear Equations

PACING CHART

The daily Pacing Chart is meant to help you adjust your teaching pace. Students in the full course should finish the entire text by the end of the year. Students in the basic course are expected to complete the first twelve chapters. The Pacing Chart for each chapter contains suggestions for lessons that require more than one day and lessons that may be omitted for the basic course.

DAY	FULL COURSE	BASIC COURSE
1	2.1	2.1
2	2.2	2.2
3	2.3 & Using a Graphing Calculator	2.3 & Using a Graphing Calculator
4	2.4	2.4
5	Mid-Chapter Self-Test	2.4
6	2.5	Mid-Chapter Self-Test
7	2.6	2.5
8	2.7	2.6
9	Chapter Review	2.6
10	Chapter Test	2.7
11		Chapter Review
12		Chapter Test

CHAPTER ORGANIZATION

LESSON	PAGES	GOALS	MEETING THE NCTM STANDARDS
2.1	64–69	1. Sketch a graph using a table of values 2. Identify equations whose graphs are horizontal or vertical lines	Problem Solving, Communication, Connections
2.2	70–76	1. Find the slope of a line and identify parallel and perpendicular lines from their slopes 2. Interpret slope as a rate of change	Problem Solving, Communication, Connections, Geometry (algebraic), Calculus concepts
Mixed Review	77	Review algebraic and arithmetic skills	
Milestones	77	Mathematical symbols	Connections
2.3	78–83	1. Use intercepts to sketch a quick graph of a line 2. Use the slope-intercept form of a line to sketch a quick graph	Problem Solving, Communication, Connections
Using a Calculator	84–85	Graph an equation, and approximate the solution	Technology
2.4	86–92	1. Write equations of lines 2. Use equations of lines to solve real-life problems	Problem Solving, Communication, Connections, Calculus concepts
Mid-Chapter Self-Test	93	Diagnose student weaknesses and remediate with correlated Reteach worksheets	
2.5	94–100	1. Graph a linear inequality in two variables 2. Use a linear inequality in two variables to model real-life situations	Problem Solving, Communication, Connections, Geometry
Mixed Review	100	Review algebraic and arithmetic skills	
2.6	101–106	1. Graph absolute value equations 2. Use graphs of absolute value equations to answer questions about real-life situations	Problem Solving, Communication, Reasoning, Connections
2.7	107–113	1. Fit a line to a set of data and write an equation for the line 2. Identify whether a set of data shows positive or negative correlation, or no correlation	Problem Solving, Communication, Connections, Statistics, Discrete Math
Chapter Summary	114	Restate for students what they have learned, why they have learned it, and how it fits into the structure of algebra	Structure, Connections
Chapter Review	115–118	Review concepts and skills learned in the chapter	
Chapter Test	119	Diagnose student weaknesses and remediate with correlated Reteaching worksheets	

LESSON RESOURCES

MEETING INDIVIDUAL NEEDS

RETEACHING For students who need to spend more time on basics:

If a mid-chapter self-test or chapter test indicates a deficiency, teachers can help students with the appropriate ***Reteaching Copymaster.***

PRACTICE For students who need more practice:

Additional exercises like those in the Pupil's Edition are provided for each lesson in ***Extra Practice Copymasters.***

ENRICHMENT For enriching and broadening students' experiences:

Problem of the Day copymasters in ***Teaching Tools*** provide a daily opportunity to use logical reasoning, looking for a pattern, writing an equation, and other routine and non-routine problem-solving strategies.

Math Log copymasters in ***Alternative Assessment*** provide opportunities to report on investigations, research, and open-ended problems.

Technology: Using Calculators and Computers provides enriching activities with graphing and scientific calculators and computers.

The ***Applications Handbook*** provides additional information about the cross-curriculum topics such as astronomy, chemistry, physics, sports, economics, genetics, and music that are integrated into the Pupil's Edition.

LESSON	2.1	2.2	2.3	2.4	2.5	2.6	2.7
PAGES	64-69	70-76	78-83	86-92	94-100	101-106	107-113
Teaching Tools							
Transparencies	✓	✓	✓	✓	✓	✓	✓
Problem of the Day	✓	✓	✓	✓	✓	✓	✓
Warm-up Exercises	✓	✓	✓	✓	✓	✓	✓
Answer Masters	✓	✓	✓	✓	✓	✓	✓
Extra Practice Copymasters	✓	✓	✓	✓	✓	✓	✓
Reteaching Copymasters	Teacher-directed and independent activities tied to results on the Mid-Chapter Self-Tests and Chapter Tests						
Color Transparencies	✓	✓		✓			✓
Applications Handbook	Additional background information for many real-life applications						
Technology	Calculator and computer worksheets for appropriate lessons						
Complete Solutions Manual	✓	✓	✓	✓	✓	✓	✓
Alternative Assessment	Assess student's ability to reason, analyze, solve problems, and communicate using mathematical language.						
Formal Assessment	Mid-Chapter Self-Tests, Chapter Tests, Cumulative Tests, and Practice for College Entrance Tests						
Computer Test Bank	Customized tests can be created by choosing from over 2500 items.						

INSIGHTS

2.1 The Graph of an Equation

Many real-life situations can be modeled by linear equations in two variables. Graphs of linear equations are geometric representations of the relationship between the variables. Frequently, a "picture" of a relationship suggests strategies for handling a situation or solutions other than an algebraic one.

2.2 Slope and Rate of Change

The concept of slope is useful in describing many real-life phenomena, including marginal cost in economics, underwater pressure in oceanography, and product analysis in marketing. A critical interpretation of slope is rate of change, which has many applications in physics, chemistry, and economics. The widespread use of the concept of slope in real-life applications underlies the importance of this lesson to students.

2.3 Quick Graphs of Linear Equations

Frequently, because a sketch of a graph provides so much information about the relationship between variables in a linear equation, learning how to construct "quick graphs" is a useful algebraic skill. This skill is also useful as a check on the reasonableness of a more detailed graph.

2.4 Writing Equations of Lines

To express many real-life relationships algebraically, it is necessary to write equations of lines. Because the equation of a line may be generated from very different types of information, students should be able to recognize any of the three ways in which that information can be presented.

2.5 Graphing Linear Inequalities

The world of business, educational assessment, and engineering are three real-life contexts in which linear inequalities are used to model relationships. Students must learn how to solve linear inequalities to effectively manage information in these and other fields.

2.6 Graphs of Absolute-Value Equations

Absolute value equations may be used in measurement applications. Just as graphing a linear equation creates a picture of a linear relationship, knowing how to sketch absolute value equations presents an opportunity to picture an absolute-value relationship.

2.7 Exploring Data: Fitting a Line to Data

Many business and scientific concerns collect data on customers, merchandise, the physical attributes of materials, or the efficiency of a scientific process. To understand the information conveyed by such data, it is often necessary to employ the best-fitting-line technique discussed in this lesson.

Problem of the Day

The separate sections of rail on a railroad track are 30 ft long. As the train passes over the point where the sections are joined, there is an audible click. By what must the number of clicks in a minute be multiplied to find the speed of the train in miles per hour?

$\frac{15}{44}$

ORGANIZER

Warm-Up Exercises

1. Complete the tables by evaluating the corresponding expressions.

   a. Expression: $y = 4x - 5$

x-value	-3	-2	-1	0	1	2	3
y-value							

   b. Expression: $2x - 3y = 12$

x-value	-3	-2	-1	0	1	2	3
y-value							

   a. $-17, -13, -9, -5, -1, 3, 7$
   b. $-6, -\frac{16}{3}, -\frac{14}{3}, -4, -\frac{10}{3}, -\frac{8}{3}, -2$

2. Evaluate each expression at the indicated values of x and y.

   a. $5x + 2y;\ x = -1, y = 2$
   b. $y - 4x;\ x = 5, y = -4$
   c. $7x - 3y;\ x = 8, y = 0$
   a. -1, b. -24, c. 56

Lesson Resources

Teaching Tools
- Transparencies: 7, 8, 9
- Problem of the Day: 2.1
- Warm-up Exercises: 2.1
- Answer Masters: 2.1

Extra Practice: 2.1
Color Transparency: 10

2.1 The Graph of an Equation

What you should learn:

Goal 1 How to sketch a graph using a table of values

Goal 2 How to identify equations whose graphs are horizontal or vertical lines

Why you should learn it:

You can use a coordinate plane to visualize the relationship between two variables.

Goal 1 Sketching the Graph of an Equation

A **coordinate plane** is formed by two perpendicular real number lines. The horizontal line is the **horizontal axis,** or **x-axis,** and the vertical line is the **vertical axis,** or **y-axis.** The point at which the two lines meet is the **origin,** and the axes divide the plane into four parts called **quadrants.**

Each point in a coordinate plane corresponds to an **ordered pair** of real numbers. For an ordered pair, the first number is the **horizontal coordinate** and the second number is the **vertical coordinate.** For instance, the ordered pair $(4, -1)$ has a horizontal coordinate of 4 and a vertical coordinate of -1.

Horizontal coordinate → $(4, -1)$ ← Vertical coordinate

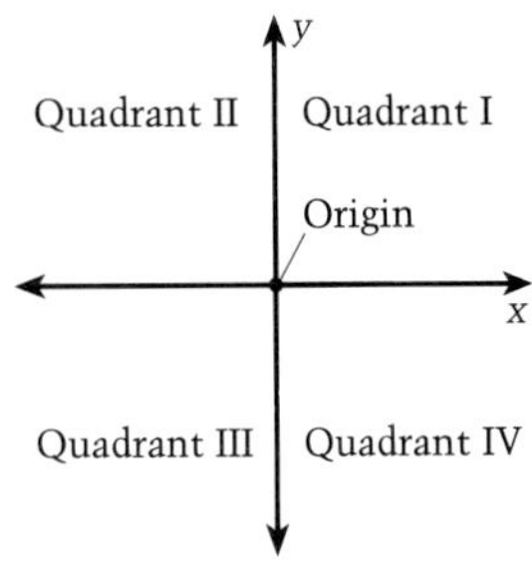

An ordered pair (x, y) is a **solution** of an equation in x and y if the equation is true when the values of x and y are substituted into the equation. The **graph** of an equation in x and y is the collection of *all* points (x, y) whose coordinates are **solutions** of the equation. For instance, since $(4, -1)$ is a solution of $y = -2x + 7$, you know that $(4, -1)$ is a point on the graph of $y = -2x + 7$. To sketch the graph of an equation, construct a *table of values,* graph enough solutions to recognize a pattern, and then connect the points.

Example 1 *Sketching the Graph of an Equation*

Sketch the graph of $y = -2x + 7$.

Solution Begin by constructing a table of values for $y = -2x + 7$.

Choose x.	-1	0	1	2	3	4	5
Evaluate y.	9	7	5	3	1	-1	-3

With this table of values, you have found seven solutions.

$$(-1, 9), (0, 7), (1, 5), (2, 3), (3, 1), (4, -1), (5, -3)$$

By plotting the points, you can see that they all lie on a line. The line through the points is the *graph* of the equation. ■

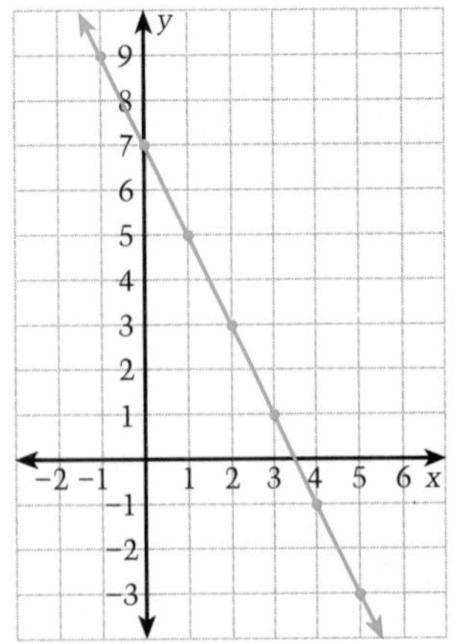

Goal 2 Identifying Equations of Lines

The point-plotting technique shown in Example 1 is useful when sketching the graph of an unfamiliar equation. Some equations, however, occur so often that you can learn to recognize their graphs and sketch a "quick graph." Two types of equations whose graphs are easy to recognize are equations of the form $x = a$ or $y = b$.

> **Equations of Horizontal and Vertical Lines**
>
> **1.** In a coordinate plane, the graph of $x = a$ is a **vertical line.**
>
> **2.** In a coordinate plane, the graph of $y = b$ is a **horizontal line.**

Example 2 *Sketching Horizontal and Vertical Lines*

Sketch the graphs of $x = 4$ and $y = -2$. Find the coordinates of the point of intersection of the two graphs.

Solution The graph of $x = 4$ is a vertical line and the graph of $y = -2$ is a horizontal line because the solutions must be of the following form.

Vertical line: $x = 4$ **Horizontal line: $y = -2$**

The point of intersection of the graphs must have an x-coordinate of 4 *and* a y-coordinate of -2. Thus the point is $(4, -2)$. ■

Annual Per Capita Consumption of Canned White Potatoes

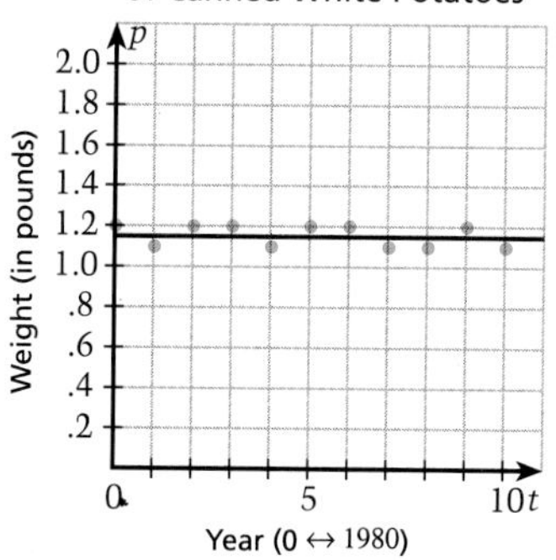

Example 3 *Creating a Model*

For 1980 through 1990, the per capita consumption, p (in pounds), of canned white potatoes in the United States is shown in the table. Write a model that approximates the number of pounds of canned white potatoes eaten per year by an American. Let $t = 0$ represent 1980.

(Source: U.S. Department of Agriculture)

Year, t	0	1	2	3	4	5	6	7	8	9	10
Pounds, p	1.2	1.1	1.2	1.2	1.1	1.2	1.2	1.1	1.1	1.2	1.1

Solution Each year the consumption was either 1.1 or 1.2 lb per person. Thus a reasonable model would be $p = 1.15$. ■

LESSON Notes

Example 1

Tables of values are frequently written vertically:

Choose x.	**Evaluate y.**

This is convenient because entries appear as coordinate pairs suitable for plotting.

Ask students why more than two or three points should be found and plotted. (Although it is sufficient to plot only two points, plotting several more provides a check because the points should form a straight line.)

Example 2

To help students understand the coordinates of vertical and horizontal lines, graph a horizontal (or vertical) line. Complete a table of values for the line by having students identify points that lie on the line.

Example 3

Ask students to identify characteristics of the graph that support the statement, "The consumption of canned white potatoes by Americans is constant." (The level slope of the horizontal line indicates no change in consumption.)

EXTEND Example 4 Ask students these questions.

a. How many acres are in a square mile? 640 acres

b. How long is the side of a square that has an area of 1 acre? (*Hint:* 1 mile = 5280 feet)
approximately 208.7 feet

Extra Examples

Here are additional examples similar to **Examples 1–2.**

1. Sketch the graph of each equation.
   a. $y = 4x + 6$
   b. $y = -3x + 8$
   c. $x = -5$
   d. $y = 2$

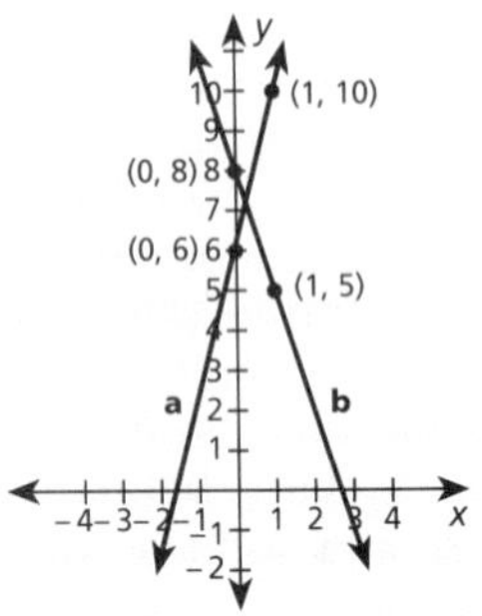

c. The graph is a vertical line.
d. The graph is a horizontal line.

Check Understanding

1. Which are better when graphing an equation: vertical tables of values or horizontal tables of values? Explain.
   Both are acceptable, but a vertical table lists points in coordinate form.
2. What are solutions of a linear equation?
   See page 64.
3. How can algebraic representations be transformed into geometric representations?
   by graphing

Communicating about ALGEBRA

EXTEND *Communicating*

Given a rectangle of dimensions 4 cm by 6 cm, describe all of the possible boundary lines of the rectangle in a coordinate plane in which one of the vertices of the rectangle is the origin (0, 0) and at least one of the sides of the rectangle lies on a

(continued)

The Homestead Act, passed by Congress in 1862, allowed Americans or new immigrants the right to obtain 160 acres of public land, called a* quarter, *because it formed a quarter of a square mile. In many of the midwestern states, the lines that marked adjoining square-mile* sections *became roads and county borders. The map at the right shows the county lines in Iowa. They were drawn so that no one would live more than one day's buggy ride from their county seat.

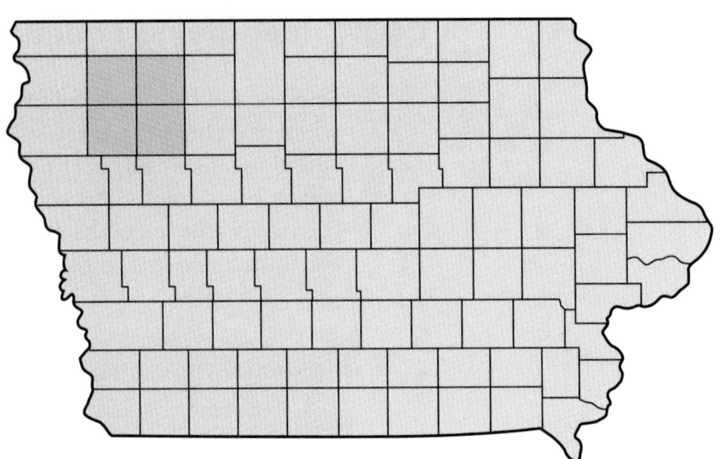

Connections **Geography**

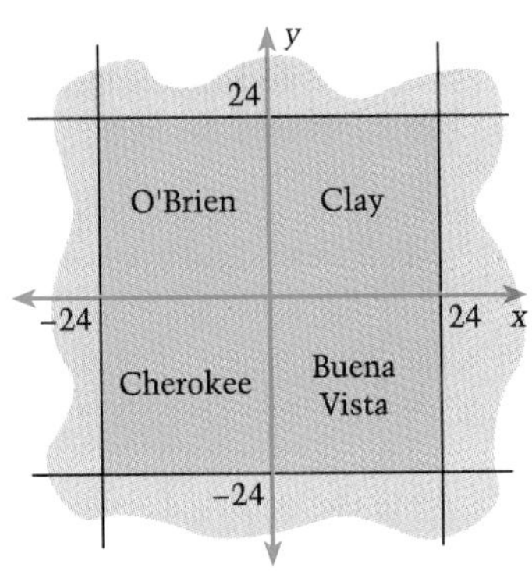

Example 4 *Finding the Boundary Lines of a County*

On the map above, a coordinate plane has been superimposed over four counties, O'Brien, Clay, Cherokee, and Buena Vista, in northwestern Iowa. Find equations that approximate the boundaries of these four counties. The scale on the coordinate plane is 1 mile to each unit. How many sections and quarters are in each of the four counties?

Solution Each of the four counties is approximately square, 24 miles by 24 miles. For the given coordinate plane, the north and south boundaries of the four counties are horizontal lines given by

$$y = 24, \ y = 0, \text{ and } y = -24.$$

The east and west boundaries are vertical lines given by

$$x = 24, \ x = 0, \text{ and } x = -24.$$

Each county contains about 576 sections. Since a section has four quarters, each county has 2304 quarters. ■

Communicating about ALGEBRA

SHARING IDEAS about the Lesson

Sketch the graph of each equation. See Additional Answers.

A. $y = -3$ **B.** $y = 2x - 4$ **C.** $y = -x + 3$

D. What is the equation of the x-axis? What is the equation of the y-axis? $y = 0$, $x = 0$

EXERCISES

Guided Practice

CRITICAL THINKING about the Lesson See Additional Answers for 1.–2.

1. Plot the point $(-1, 3)$ in a coordinate plane.
2. Plot the point $(6, -2)$ in a coordinate plane.
3. Construct a table of values for $y = 3x - 9$. Then use the table to sketch the graph of $y = 3x - 9$. Can you sketch the *entire* graph? Explain. See Additional Answers.
4. What does *quadrant* mean?
5. Explain how to sketch a quick graph of $x = -5$.
6. Explain how to sketch a quick graph of $y = 2$.

3. Tables vary.

x	−1	0	2	4
y	−12	−9	−3	3

4. One of the four parts into which a plane is divided by two perpendicular lines.
5. Draw a vertical line through the point −5 on the *x*-axis.
6. Draw a horizontal line through the point 2 on the *y*-axis.

Independent Practice

In Exercises 7–12, plot the point in a coordinate plane. In which quadrant does the point lie? (Points on an axis are not in any of the quadrants.)

7. $(5, -1)$ IV
8. $(4, 9)$ I
9. $(-6, -2)$ III
10. $(-3, 0)$ None
11. $(0, 5)$ None
12. $(-1, 2)$ II

In Exercises 13–18, find three points on the graph of the equation. (There are many such points.)

13. $y = x + 5$ (0, 5), (−2, 3), (1, 6)
14. $y = -6x - 8$ (0,−8), (−2, 4), (1,−14)
15. $y = 5x - 1$
16. $y = -8x + 3$ (0, 3), (−1, 11), (2,−13)
17. $y = -3x + 4$ (0, 4), (−2, 10), (1, 1)
18. $y = 9x + 2$

15. (0,−1), (−1,−6), (2, 9)
18. (0, 2), (1, 11), (−2,−16)

In Exercises 19–24, match the equation with its graph.

19. $x = 3$ e
20. $y = -x - 3$ d
21. $y = -2x + 1$ c
22. $y = x - 5$ a
23. $y = x + 1$ f
24. $y = 3$ b

a.
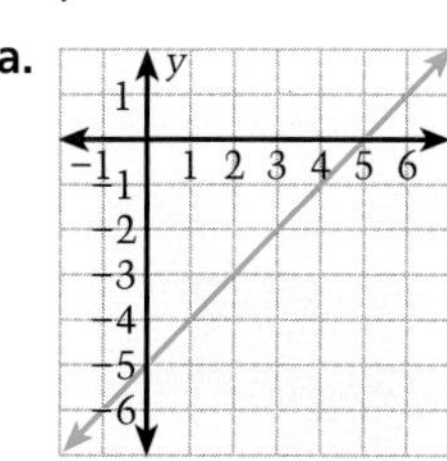

b.
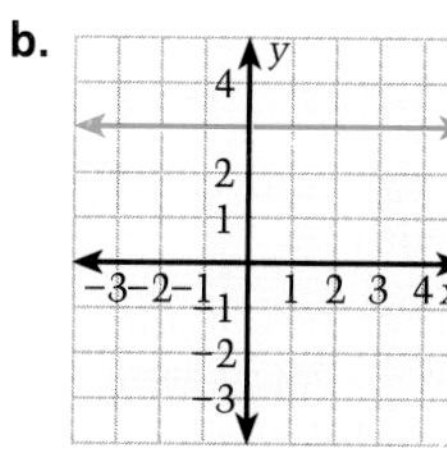

c.
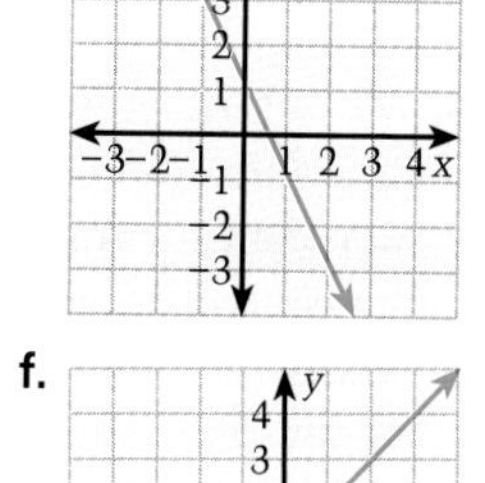

d.
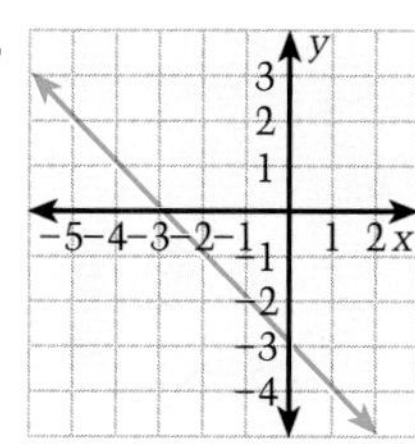

e.
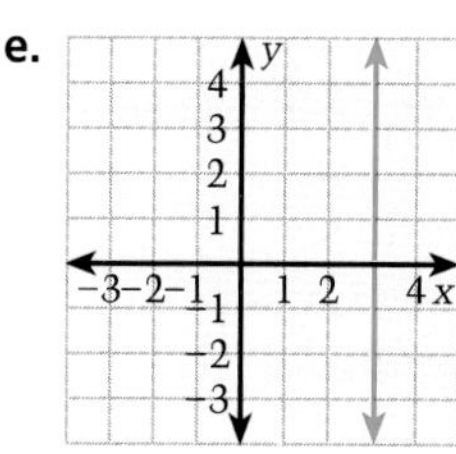

f.
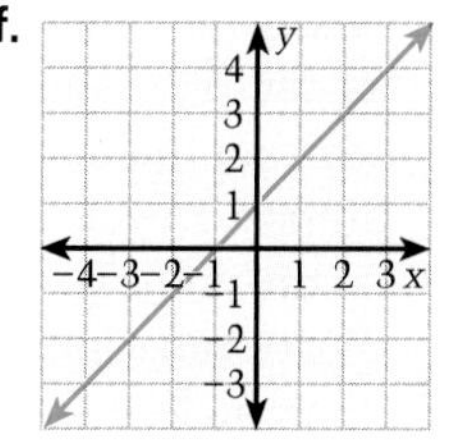

(continued)
coordinate axis. (*Hint:* There are 8 possibilities.)
$x = 4$, $x = 6$, $x = -4$, $x = -6$, $y = 4$, $y = 6$, $y = -4$, $y = -6$

EXERCISE Notes

ASSIGNMENT GUIDE

Basic/Average: Ex. 1–6, 7–12, 13–17 odd, 19–24, 25–45 odd, 47–48, 55–60

Above Average: Ex. 1–6, 7–24, 25–47 odd, 48, 57–61

Advanced: Ex. 1–6, 13–24, 25–47 odd, 48, 59–62

Selected Answers
Exercises 1–6, 7–59 odd

✪ More Difficult Exercise
Exercises 47, 48

Guided Practice

▶ **Ex. 1–6** These exercises summarize the essential concepts of the lesson.

EXTEND Ex. 5–6 Ask students for the coordinates of the point of intersection of $x = -5$ and $y = 2$. (−5, 2)

Independent Practice

▶ **Ex. 7–12** These exercises could all be plotted on the same coordinate plane.

▶ **Ex. 19–24** These exercises should be assigned as a group. Although linear equations are not formally discussed until Lessons 2.2 and 2.3, these exercises review students' intuitive understanding of them.

▶ **Ex. 25–32** Students can use Example 1 as a solution model. In Example 1, a table of values was used to sketch the graph. However, if students recall any other graphing methods from Algebra 1, such as using intercepts, they should be encouraged to use them.

▶ **Ex. 37–40** Ask students to state whether the line is horizontal or vertical.

▶ **Ex. 45** This exercise is similar to the horizontal-line model of Example 3 because the ratios of the services stay reasonably constant at a value of ≈ 0.25.

Answers

45.

t	5	6	7	8
r	0.245	0.250	0.253	0.256

9	10	11	12	13
0.248	0.246	0.247	0.249	0.243

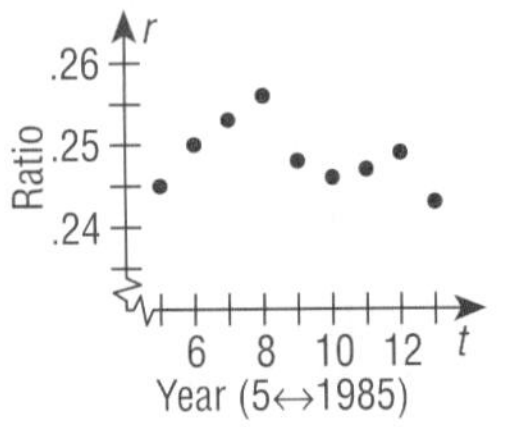

46. No, $x = 4$, $x = -8$, $y = 6$, $y = -2$

▶ **Ex. 47–48** HISTORY These exercises provide a historical connection. In Exercise 47, caution students to remember that each square unit on the number line is 50^2, or 2500, square feet.

▶ **Ex. 57–60** These exercises review literal equations skills that will be used later in the chapter for slope-intercept form.

In Exercises 25–32, sketch the graph of the equation. See Additional Answers.

25. $y = x - 5$ **26.** $y = -x + 8$ **27.** $y = 2x + 5$ **28.** $y = 4x - 3$

29. $y = -4x + 2$ **30.** $y = 3x + 1$ **31.** $y = 3x - 4$ **32.** $y = -2x - 4$

In Exercises 33–36, sketch both lines on the coordinate plane. Then find the coordinates of the point at which the two lines intersect. See Additional Answers.

33. $x = -2, y = 5$ $(-2, 5)$ **34.** $y = 0, x = -10$ $(-10, 0)$

35. $y = 1, x = 8$ $(8, 1)$ **36.** $x = 11, y = 6$ $(11, 6)$

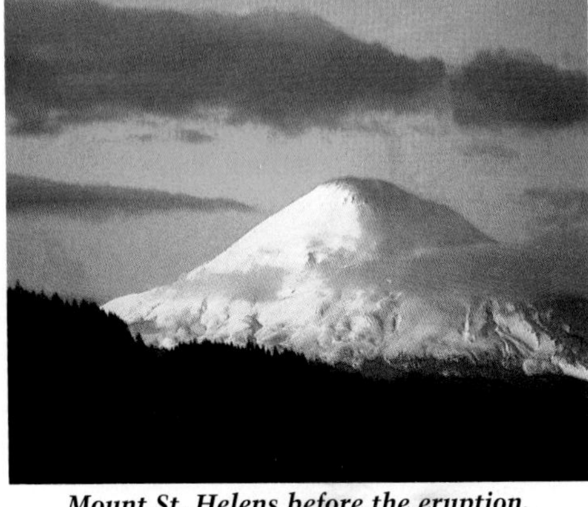
Mount St. Helens before the eruption.

In Exercises 37–40, write an equation for the line. See below.

37. The x-coordinate of each point on the line is -4.

38. The x-coordinate of each point on the line is 6.

39. The y-coordinate of each point on the line is 1.

40. The y-coordinate of each point on the line is -12.

In Exercises 41–44, write equations for the horizontal and vertical lines that pass through the point.

41. $(10, 0)$ $x = 10, y = 0$ **42.** $(-4, 15)$ $x = -4, y = 15$

43. $(-5, -3)$ $x = -5, y = -3$ **44.** $(6, -9)$ $x = 6, y = -9$

45. *Physician's Services* For 1985 through 1993, the average amount, p (in dollars), spent by an American on physician's services is shown in the table. The table also shows the average amount, h (in dollars), spent by an American on all health services. Let r represent the ratio of p to h, and let t represent the year, with $t = 5$ representing 1985. Calculate the value of r for each year, t. Graph your results, and write a model that approximates the relationship between r and t. ***(Source: U.S. Health Care Financing Administration)*** See margin; $r = 0.25$

Year, t	5	6	7	8	9	10	11	12	13
Physician's Services, p	231	256	282	310	329	356	373	398	411
Health Services, h	942	1023	1115	1212	1324	1447	1511	1597	1693

46. *Mount St. Helens* Mount St. Helens is a volcano in Washington that had been dormant for many years. In 1980, the volcano erupted, causing great damage for many miles around the volcano. The map at the right shows the affected region. Each unit on the coordinate plane that is superimposed over the map represents 2 miles. Is the point $(-4, 2)$ in the damaged region? Write equations for the sides of a rectangle that enclose the damaged area. See margin.

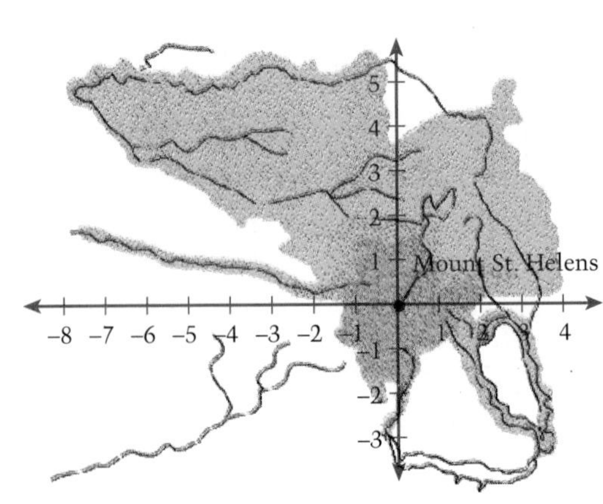

37. $x = -4$ **38.** $x = 6$ **39.** $y = 1$ **40.** $y = -12$

El Escorial **In Exercises 47 and 48, use the following information.**

El Escorial, a monumental creation of Philip II of Spain, was completed in 1584. It is located in the Guadarrama Mountains near Madrid. A combination church, monastery, palace, mausoleum, college, and library, El Escorial is almost a city in itself.

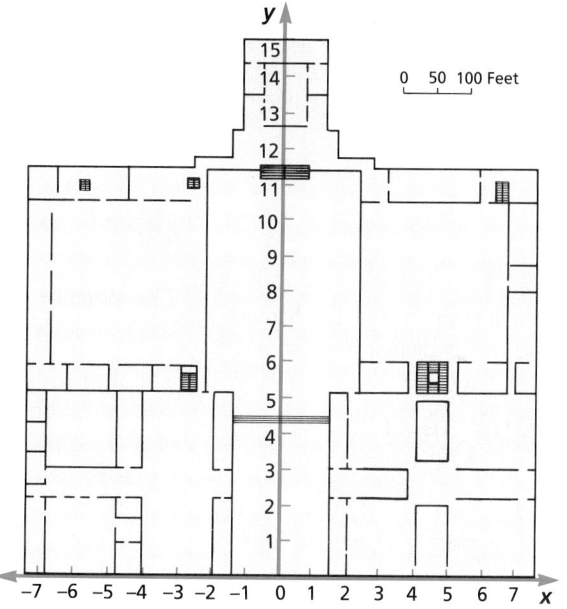

47. A coordinate plane has been superimposed over the floor plan of El Escorial so that the main gate lies at the origin. Each unit on the coordinate plane represents 50 feet. Approximate the number of square feet in El Escorial. 457,500

48. Philip II often joined the monks in their devotions in the High Choir located at (0, 7.25). Philip's apartment was located at (1.25, 12). About how many feet did Philip walk from his apartment to High Choir? Explain your reasoning. See below.

Integrated Review

49. Evaluate $-6x - 22$ when $x = 1$. -28

50. Evaluate $-14x + 8$ when $x = 2$. -20

In Exercises 51–54, find the value for y when x has the given value.

51. $y = x - 16$, $x = 9$ -7

52. $y = -2x + 8$, $x = -10$ 28

53. $y = 2x + 11$, $x = 0$ 11

54. $y = x - 20$, $x = -20$ -40

In Exercises 55 and 56, copy and complete the table.

55. $y = 3x - 5$

x	−3	−2	−1	0	1	2	3
y	?	?	?	?	?	?	?
	−14	−11	−8	−5	−2	1	4

56. $y = -8x + 11$

x	−3	−2	−1	0	1	2	3
y	?	?	?	?	?	?	?
	35	27	19	11	3	−5	−13

In Exercises 57–60, solve the equation for y.

57. $4x + y = 10$ $y = -4x + 10$

58. $x - 6y = 12$ $y = \frac{1}{6}x - 2$

59. $6x + 2y = 9$ $y = -3x + \frac{9}{2}$

60. $4x + 5y = 20$ $y = -\frac{4}{5}x + 4$

Exploration and Extension

In Exercises 61 and 62, find the point of intersection of the two lines.

61. $y = -5x + 9$ and $x = 12$ (12, −51)

62. $y = 3x - 14$ and $y = 4$ (6, 4)

48. He could not walk diagonally through the walls, so he would have to walk 1.25 units to the y-axis and 4.75 units along the y-axis to (0, 7.25); that is a walk of 6 units, or 300 ft.

Enrichment Activities

These activities require students to go beyond lesson goals.

Example 3 and Exercise 45 both illustrate data in which the ratio of the data values stays fairly constant. In Example 3, it is the ratio of potato consumption to population. In Exercise 45, it is the ratio of physician services (a subset of health services) to total health services.

1. Look back at Example 2 on page 51. Construct a bar graph of the ratio of average sales (in \$100,000) per number of outlets for each of the top ten fast-food restaurant chains.
   McDonalds: ≈ 15.5; Burger King: ≈ 9.4; KFC: ≈ 6.7; Pizza Hut: ≈ 5.5; Wendy's: ≈ 8; Hardee's: ≈ 9; Domino's: ≈ 4.8; Dairy Queen: ≈ 4.2; Taco Bell: ≈ 6.7; Arby's: ≈ 5.8

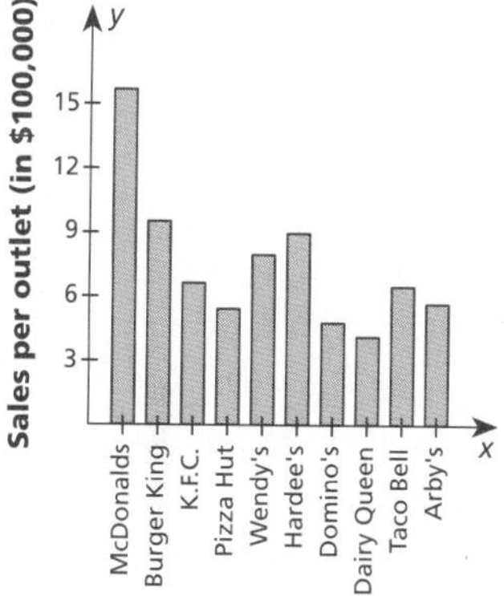

2. Are the ratios constant?
   No, they fluctuate with McDonalds being the greatest and Dairy Queen being the lowest.
3. What do the ratios indicate?
   The average sales depends on both the total sales and the number of outlets.
4. **RESEARCH** Look in newspapers or magazines and try to find data similar to the fast-food restaurant table. Compute the ratios and determine whether they are constant or not.

Problem of the Day

In a certain class, $\frac{1}{3}$ of the girls are $\frac{1}{5}$ of the students. What is the ratio of boys to girls in the class? $\frac{2}{3}$

ORGANIZER

Warm-Up Exercises

1. Evaluate each expression for the given values.
   a. $x - y$ when $x = -1$, $y = 5$
   b. $t - s$ when $s = -3$, $t = -12$
   c. $a - b$ when $a = 15$, $b = -4$

   a. −6, **b.** −9, **c.** 19
2. Evaluate the expression $\frac{x-y}{r-s}$ for the given values of x, y, r, and s.
   a. $x = 9$, $y = -3$, $r = 23$, $s = -1$
   b. $x = -7$, $y = 3$, $r = 8$, $s = 14$

   a. $\frac{1}{2}$, **b.** $\frac{5}{3}$
3. Complete the table for the given values of m.

	m	$\frac{1}{m}$	$-m$	$\frac{1}{-m}$
a.	1	1	−1	−1
b.	−2	$-\frac{1}{2}$	2	$\frac{1}{2}$
c.	$\frac{1}{4}$	4	$-\frac{1}{4}$	−4
d.	0	undefined	0	undefined

Lesson Resources

Teaching Tools
- Transparencies: 7, 8, 9
- Problem of the Day: 2.2
- Warm-up Exercises: 2.2
- Answer Masters: 2.2

Extra Practice: 2.2
Color Transparency: 11
Technology Handbook: p. 6

2.2 Slope and Rate of Change

What you should learn:

Goal 1 How to find the slope of a line and to identify parallel and perpendicular lines from their slopes

Goal 2 How to interpret slope as a rate of change

Why you should learn it:

You can model many real-life situations using slope—interpreting it either geometrically or as a rate of change.

Goal 1 Finding the Slope of a Line

The **slope** of a nonvertical line is the number of units the line rises (or falls) for each unit of horizontal change.

The line shown below at the left rises 2 units for each unit of horizontal change from left to right. Thus this line has a slope of 2. Two points on a line are all that is needed to determine its slope, which is represented by the letter "m."

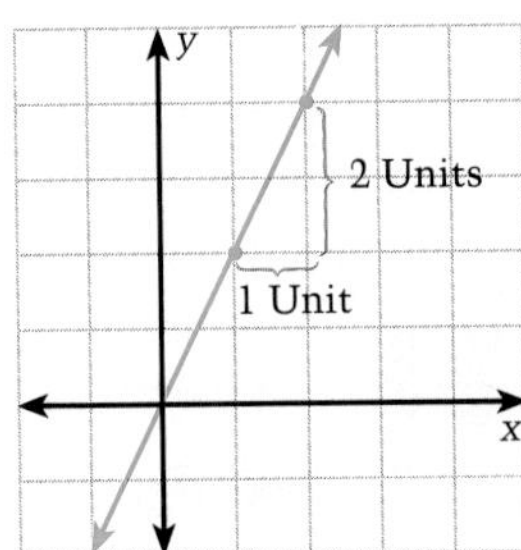

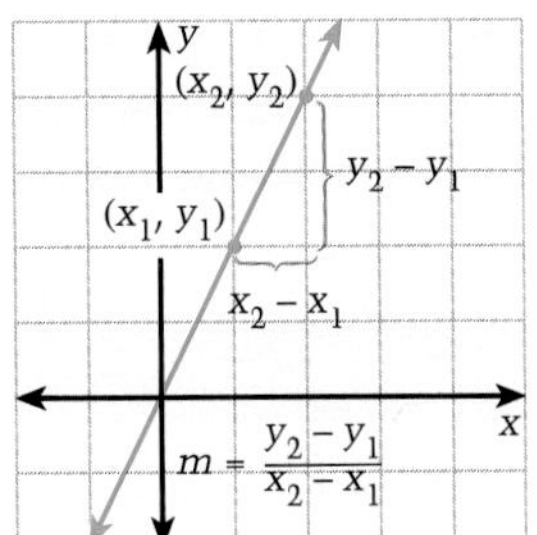

The Slope of a Line

The **slope,** m, of the nonvertical line passing through the points (x_1, y_1) and (x_2, y_2) is

$$m = \frac{y_2 - y_1}{x_2 - x_1}.$$

The numerator is read as "y sub 2 minus y sub 1" and is called the **rise.** The denominator is read as "x sub 2 minus x sub 1" and is called the **run.** The slope of a vertical line is not defined.

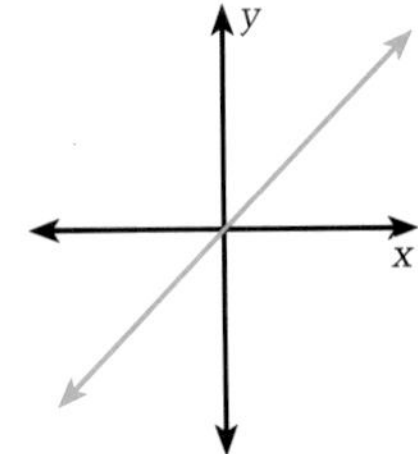

Line rises, positive slope.

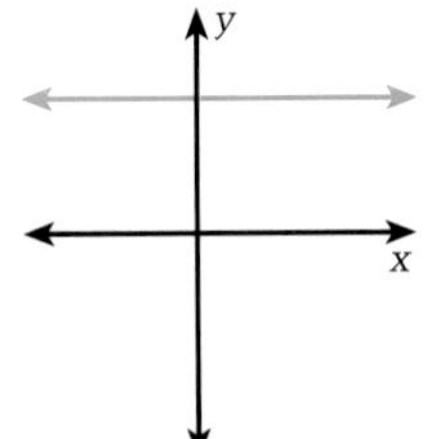

Line is horizontal, zero slope.

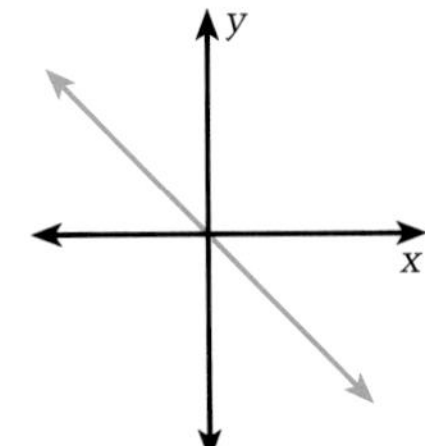

Line falls, negative slope.

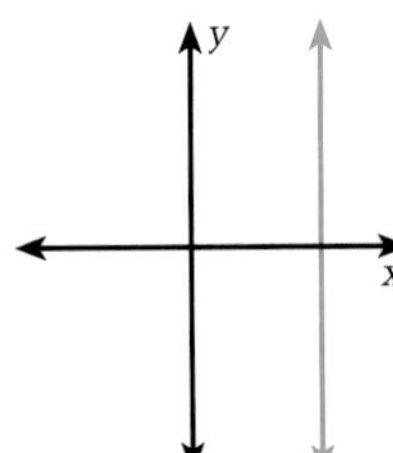

Line is vertical, undefined slope.

Example 1 *Finding the Slope of a Line*

Find the slope of the line containing (−3, 4) and (2, 1).

Solution Let $(x_1, y_1) = (-3, 4)$ and $(x_2, y_2) = (2, 1)$.

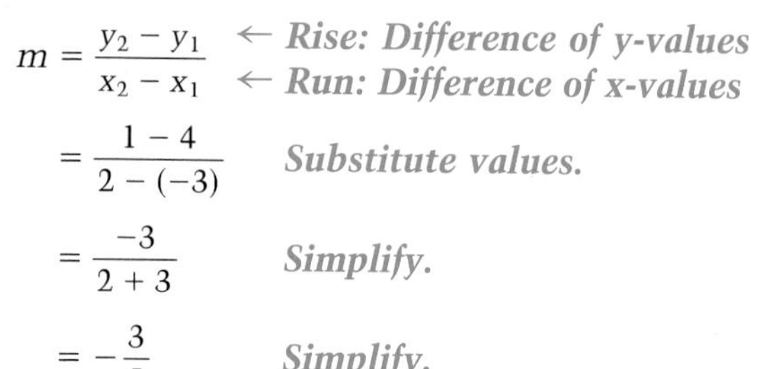

$$m = \frac{y_2 - y_1}{x_2 - x_1} \quad \begin{matrix} \leftarrow \textit{Rise: Difference of y-values} \\ \leftarrow \textit{Run: Difference of x-values} \end{matrix}$$

$$= \frac{1 - 4}{2 - (-3)} \qquad \textit{Substitute values.}$$

$$= \frac{-3}{2 + 3} \qquad \textit{Simplify.}$$

$$= -\frac{3}{5} \qquad \textit{Simplify.}$$

■

In Example 1, notice that the line falls from left to right, and the slope of the line is negative. This suggests one of the important uses of slope—to decide whether y decreases, increases, or is constant as x increases.

Classification of Lines by Slope

1. A line with positive slope *rises* from left to right. $(m > 0)$
2. A line with negative slope *falls* from left to right. $(m < 0)$
3. A line with slope zero *is horizontal.* $(m = 0)$
4. A line with undefined slope *is vertical.* (m is undefined.)

The slope of a line tells you more than whether the line rises, falls, or is horizontal. It also tells you the steepness of the line. For two lines with positive slopes, the line with the greater slope is steeper. For two lines with negative slopes, the line with the slope of greater absolute value is steeper.

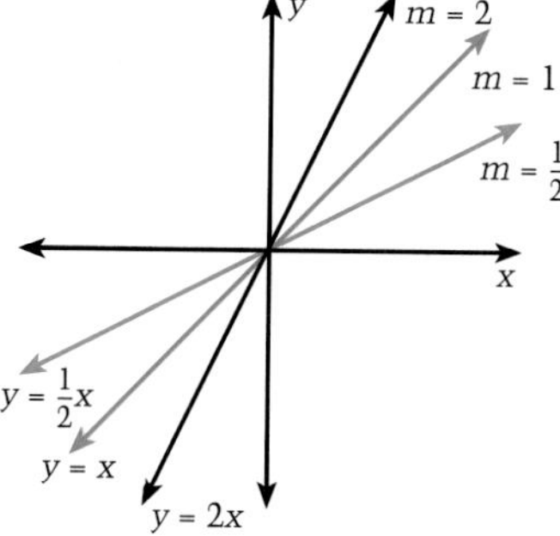

Lines with positive slopes

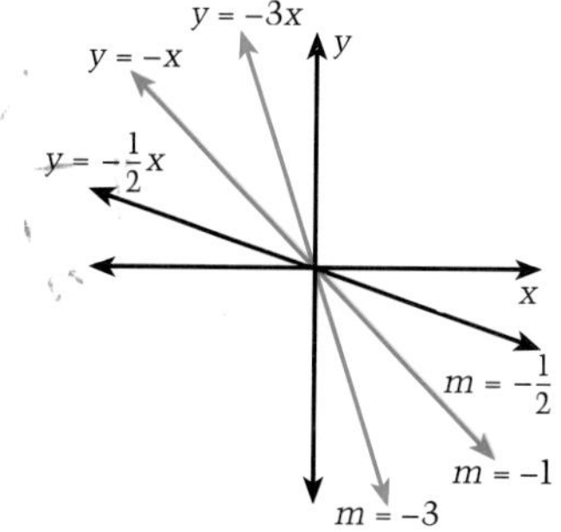

Lines with negative slopes

LESSON Notes

Example 1

The slope of a nonvertical line is given by the following formula.

$$m = \frac{y_2 - y_1}{x_2 - x_1}$$

Point out to students that points on a vertical line all have x-coordinates with the same value. For instance, for the vertical line $x = 5$, every point on the line has the form, $(5, y)$. Take the time to review what the subscripts of x and y represent. Often, students do not fully understand their role in distinguishing different number values.

If the formula for slope were used with vertical lines, the denominator of the formula would become 0. (Ask students why the denominator becomes 0.) Since division by 0 is undefined, the formula for the slope of a line is restricted to nonvertical lines.

Example 2

Clarify the meaning of "negative reciprocals "by showing students the transformation from one number value to the other.

Common-Error ALERT!

In **Example 1,** the order in which the coordinates are used in the formula for slope is important. Caution students that y_2 minus y_1 is the correct order of the y-coordinates and that x_2 minus x_1 is the correct order of the x-coordinates. However, point out to students that they may assign either point on the line as the (x_1, y_1) point.

Example 3

Discuss reasons why the number of women who receive doctorates in engineering is small. Allow students to suggest possible remedies. Stress the importance of mathematics in successfully completing any degree in engineering.

Extra Examples

Here are additional examples similar to **Examples 1–3.**

1. Find the slope of a line containing the following points,
   a. $(-6, 2)$ and $(4, 8)$
   b. $(12, 15)$ and $(16, 3)$
   a. $m = \frac{3}{5}$
   b. $m = -3$
2. Sketch the graph of the lines containing the points in Extra Example 1.

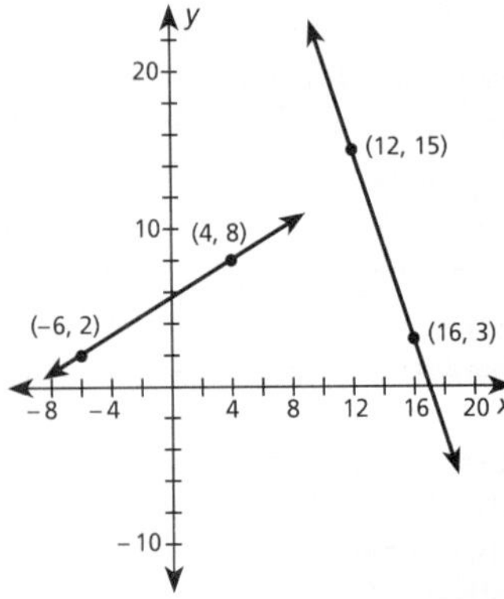

3. Decide whether the lines containing the given points are parallel, perpendicular, or neither.
   a. Line 1: $(4, 1)$ and $(7, 2)$
      Line 2: $(1, -2)$ and $(0, 1)$
   b. Line 1: $(3, 1)$ and $(-4, -13)$
      Line 2: $(-1, -8)$ and $(6, 5)$
   c. Line 1: $(-1, 9)$ and $(8, 36)$
      Line 2: $(2, -8)$ and $(9, 13)$
   a. perpendicular, **b.** neither, **c.** parallel
4. In 1975, women received 21.3% of all the doctoral degrees granted in the United States. In 1985, they received 34.1% of all doctoral degrees. Find the aver-

(continued)

Two lines in a plane are **parallel** if they do not intersect. Two lines in a plane are **perpendicular** if they intersect to form a right angle. Slope can be used to determine whether two (nonvertical) lines are parallel or perpendicular.

Slopes of Parallel and Perpendicular Lines

Consider two different, nonvertical lines with slopes of m_1 and m_2.

1. The lines are parallel if and only if they have the same slope.

   $m_1 = m_2$ *Parallel lines*

2. The lines are perpendicular if and only if their slopes are negative reciprocals of each other.

   $m_1 = -\frac{1}{m_2}$ *Perpendicular lines*

The phrase "if and only if" is used in mathematics as a way to write two **converse** statements as a single statement. For instance, the two statements in property 1 above are *if two lines are parallel, then they have the same slope* and its converse *if two lines have the same slope, then they are parallel.*

Example 2 *Deciding Whether Lines are Parallel or Perpendicular*

Decide whether the two lines are parallel, perpendicular, or neither.

	Line 1	Line 2
a.	Contains $(1, 4)$, $(2, 7)$	Contains $(-2, 1)$, $(1, 0)$
b.	Contains $(-2, 5)$, $(4, 3)$	Contains $(-1, 3)$, $(2, 2)$

Solution

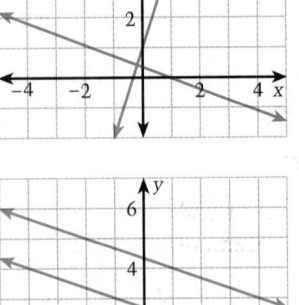

a. The slopes of the two lines are

$m_1 = \frac{7-4}{2-1} = 3$ and $m_2 = \frac{0-1}{1-(-2)} = -\frac{1}{3}$.

Because m_1 and m_2 are the negative reciprocals of each other, you can conclude that the lines are perpendicular. The graphs of these two lines are shown at the left.

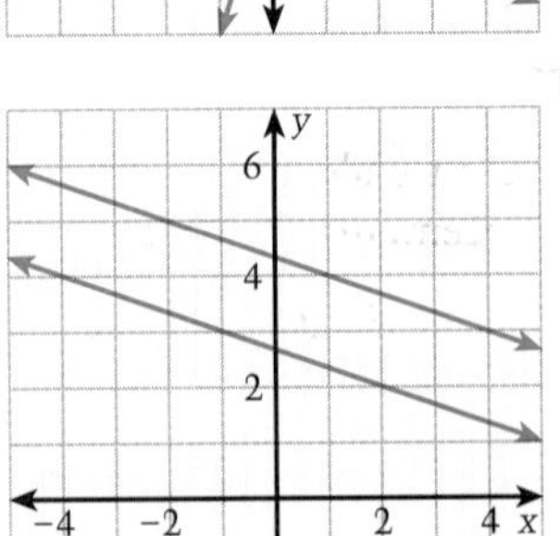

b. The slopes of the two lines are

$m_1 = \frac{3-5}{4-(-2)} = -\frac{1}{3}$ and $m_2 = \frac{2-3}{2-(-1)} = -\frac{1}{3}$.

Because $m_1 = m_2$, you can conclude that the lines are parallel. The graphs of these two lines are shown at the left. ■

Goal 2 Interpreting Slope as a Rate of Change

In real-life problems, slope is often used to describe a **constant rate of change** or an **average rate of change.** These rates include units of measure, such as miles per hour or dollars per year.

Connections **Education**

Example 3 *Slope as a Rate of Change*

In 1980, 3.5% of the doctorates in engineering were awarded to women. By 1990, 8.5% were earned by women. Find the average rate of change in the percent of engineering doctorates earned by women from 1980 to 1990. ***(Source: National Research Council)***

Solution Let p represent the percent of engineering doctorates who were women and let t represent the year. The two given data points are represented by (t_1, p_1) and (t_2, p_2).

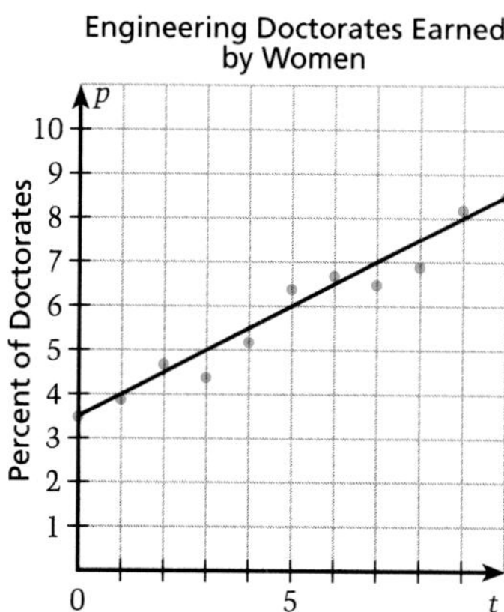

First year ↓, First percent ↓: $(t_1, p_1) = (1980, 3.5)$

Second year ↓, Second percent ↓: $(t_2, p_2) = (1990, 8.5)$

The formula for slope will give the average rate of change.

$$\text{Rate of change} = \frac{p_2 - p_1}{t_2 - t_1} \quad \leftarrow \textit{Difference of percents} \quad \leftarrow \textit{Difference of years}$$

$$= \frac{8.5 - 3.5}{1990 - 1980} \quad \textit{Substitute values.}$$

$$= \frac{5}{10}, \text{ or } 0.5 \quad \textit{Simplify.}$$

From 1980 through 1990, the *average rate of change* in the percent of engineering doctorates awarded to women was a half percent per year, although the exact changes varied slightly. ■

Communicating about ALGEBRA

SHARING IDEAS about the Lesson

Apply a Formula Use the formula for slope to write an average rate of change for each statement.

A. At 8:00 P.M., the temperature is 69°F. At 11:00 P.M., it is 53°F. $\frac{-16 \text{ degrees}}{3 \text{ hours}}$

B. At 4:00 P.M., an airplane takes off. By 4:45 P.M., it has traveled 340 miles. $\frac{68 \text{ miles}}{9 \text{ minutes}}$

age rate of change in the percent of doctorates earned by women from 1975 to 1985.
average rate of change was 1.28%

Check Understanding

1. What is the formula for the slope of a nonvertical line?
See page 70.
2. In what ways can a line be classified by its slope?
See page 71.
3. What is the relationship of the slopes of two nonvertical lines that are perpendicular?
They are negative reciprocals of each other.
4. How can an average rate of change be represented in algebra?
See Example 3.

Communicating about ALGEBRA

EXTEND *Communicating*

The converse of mathematical statements occur frequently in algebra and geometry. Caution students that not all converse statements are true. Ask students to construct the converse of the following statements and state whether they are true or false.

a. If lines are perpendicular, then their slopes are negative reciprocals of each other.

b. If $|x| = 0$, then $x = 0$.

c. For any real numbers a, b, and c, if $a = b$, then $a + c = b + c$.

a. If the slopes of lines are negative reciprocals of each other, then the lines are perpendicular. True.

b. If $x = 0$, then $|x| = 0$. True.

c. For any real numbers a, b, and c, if $a + c = b + c$, then $a = b$. True.

EXERCISE Notes

ASSIGNMENT GUIDE

Basic/Average: Ex. 1–4, 5–8, 11–21 odd, 27–30, 33–43 odd, 44, 55–59 odd

Above Average: Ex. 1–4, 5–10, 11–23 odd, 27–30, 33–43 odd, 44, 57–62

Advanced: Ex. 1–4, 5–8, 9–25 odd, 27–30, 31–41 odd, 42–44, 57–62

Selected Answers
Exercises 1–4, 5–57 odd

Use **Mixed Review as needed.**

More Difficult Exercises
Exercises 42, 44

Guided Practice

Ex. 1–4 **COOPERATIVE LEARNING** Have students work with partners to complete these exercises. Students should explain their thinking and reasoning to each other. To assess the students' understanding, you might wish to monitor their conversations.

Independent Practice

Ex. 5–8 These exercises should be assigned as a group.

Ex. 9–14 Remind students that a slope such as 4 can be thought of as $\frac{4}{1}$.

Ex. 15–26 Remind students that the slopes they determined should match the graph they plotted. Errors in signs are common when applying the slope formula. A graph should provide a visual check.

Ex. 27–30 These exercises should be assigned as a group.

Ex. 31–36 Refer students to both Example 2 and Guided Practice Exercise 4.

EXERCISES

Guided Practice

CRITICAL THINKING about the Lesson

1. Describe in your own words what is meant by the slope of a nonvertical line. Explain how your description relates to the definition of *slope*. **See below.**
2. Tell all you know about the slope of a line that falls from left to right, the slope of a line that rises from left to right, and the slope of a horizontal line. **negative; positive; 0**
3. Which line is steeper? **b**
   a. Line 1: through $(-4, 0)$ and $(2, 3)$
   b. Line 2: through $(0, 2)$ and $(1, 4)$
4. How can you decide, using slope, whether two nonvertical lines are parallel? perpendicular? **They are parallel if their slopes are equal. They are perpendicular if their slopes are negative reciprocals.**

Independent Practice

In Exercises 5–8, estimate the slope of the line if it exists.

5. 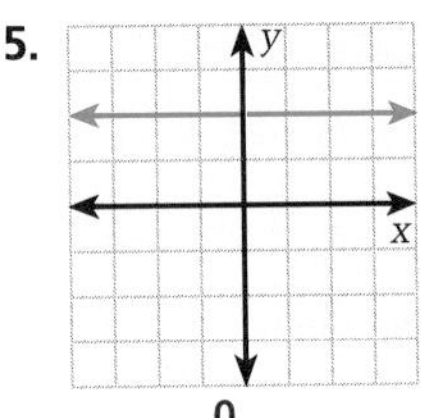

0

6. 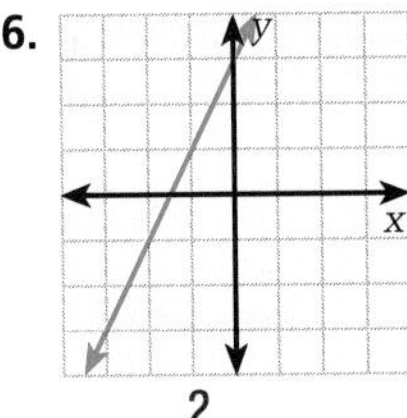

2

7. 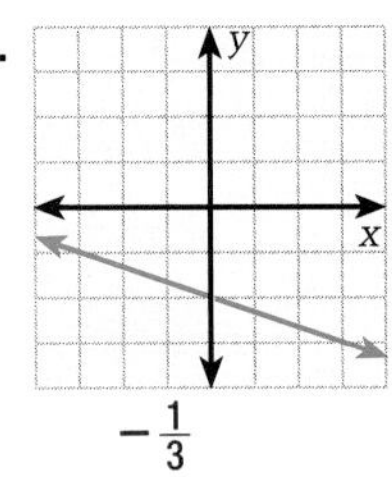

$-\frac{1}{3}$

8. 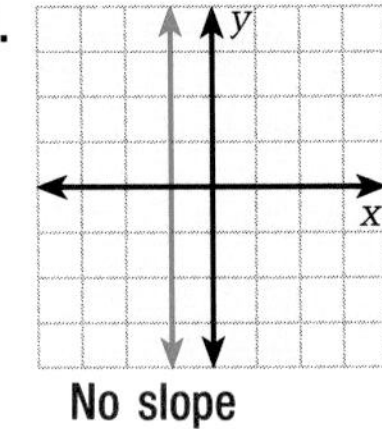

No slope

In Exercises 9–14, sketch the line containing the given point that has the given slope. See Additional Answers.

9. $(-3, 1)$, $m = 2$
10. $(10, 4)$, m is undefined.
11. $(4, 0)$, $m = -1$
12. $(-6, 1)$, $m = 4$
13. $(6, 5)$, $m = 0$
14. $(9, -12)$, $m = -\frac{1}{2}$

In Exercises 15–26, draw the line segment with the given endpoints. Then find the slope of the segment. See Additional Answers for graphs.

15. $(3, 1), (6, -4)$ $-\frac{5}{3}$
16. $(1, -5), (2, 7)$ 12
17. $(16, -3), (2, 8)$ $-\frac{11}{14}$
18. $(-12, -9), (1, -8)$ $\frac{1}{13}$
19. $(4, 5), (-1, 9)$ $-\frac{4}{5}$
20. $(3, -6), (2, 10)$ -16
21. $(4, 2), (-14, 3)$ $-\frac{1}{18}$
22. $(-4, 1), (-5, 2)$ -1
23. $(-2, 6), (10, 0)$ $-\frac{1}{2}$
24. $(-15, 8), \left(-9, \frac{1}{2}\right)$ $-\frac{5}{4}$
25. $\left(\frac{3}{4}, -8\right), (2, -5)$ $\frac{12}{5}$
26. $(11, 7), (12, -10)$ -17

In Exercises 27–30, match the pair of points with the description of the line containing the pair of points.

27. $(2, 3), (-15, 5)$ b
28. $(8, -12), (9, 2)$ c
29. $(-4, 7), (-4, 4)$ d
30. $(4, -3), (-20, -3)$ a

a. The line is horizontal.
b. The line falls to the right.
c. The line rises to the right.
d. The line is vertical.

1. The slope of a nonvertical line is a change in *y* divided by a corresponding change in *x*, which is equivalent to the rise or fall per unit of horizontal change, as the definition says.

In Exercises 31 and 32, decide which line is steeper.

31. a. Line 1 contains (−3, 4) and (1, 6). b
b. Line 2 contains (1, −5) and (6, 2).

32. a. Line 1 contains (6, 1) and (−4, 4). b
b. Line 2 contains (−2, 3) and (−1, −6).

In Exercises 33–36, decide whether the lines are parallel, perpendicular, or neither.

33. Line 1 contains (1, 7) and (−3, −5). Parallel
Line 2 contains (−6, −20) and (0, −2).

34. Line 1 contains (4, −4) and (−16, 1). Perpendicular
Line 2 contains (1, 5) and (5, 21).

35. Line 1 contains (3, −13) and (22, 4).
Line 2 contains (6, 2) and (−8, 9). Neither

36. Line 1 contains (20, 3) and (0, −7).
Line 2 contains (6, −4) and (12, −1). Parallel

In Exercises 37 and 38, find the average rate of change between the two points. List the units of measure.

37. (3, 2) and (7, 18)
x is measured in hours.
y is measured in dollars. 4 dollars per hour

38. (0, 3) and (3, 15)
x is measured in minutes.
y is measured in miles. 4 miles per minute

39. *Average Speed* At 10:00 A.M., you are 150 miles from home. At 4:00 P.M., you are 390 miles from home. Find your average speed. (Assume you are traveling directly away from home.) 40 mph

40. *Spending Your Paycheck* On the day after payday, you have $200. Eight days after payday, you have $46. At what average rate are you spending? $22 per day

41. *Leaning Tower of Pisa* When it was built, the Leaning Tower of Pisa in Italy was 180 feet tall. Since then, one side of the base has sunk 1 foot into the ground, causing the top of the tower to lean 16 feet off-center. Approximate the slope of the leaning side of the tower. 11.2

42. *Cheops's Great Pyramid* The Great Pyramid is the largest of the Egyptian pyramids. When it was built, it was 481 feet tall and had a square base with 755-foot sides. The pyramid has *two* different slopes—one along its sides and the other along its edges. Which slope is steeper? Explain. (*Hint:* You need the Pythagorean Theorem to find the slope along the pyramid's edges.)
Along its sides, because it has a shorter run.

Ex. 41, 43, and 44 Students need to translate the given information to ordered pairs and then evaluate using the slope formula.

Ex. 43–44 These exercises illustrate the changing demographic trends in the work force. You might use them to stimulate a class discussion as to the causes for the changes. For example, have students discuss why more women are in law enforcement professions and why more men are working as food servers.

Ex. 57 This exercise is an application of constant rates of change.

Ex. 61 This exercise connects slope to the equation of a line by using two points whose coordinates are solutions to the equation.

▶ **Ex. 62 GEOMETRY** This exercise is an application from analytic geometry.

Enrichment Activities

These activities require students to go beyond lesson goals.

EXTEND Ex. 43 and 44 The broken-line graph shows the percentage of double-income households—that is, households with both husband and wife employed outside the home for the years 1960–1990.

1. Find the average rate of change in the percent of double-income households from 1960–1990.
   average rate of change $= \frac{0.59 - 0.29}{30 - 0} = 0.01$

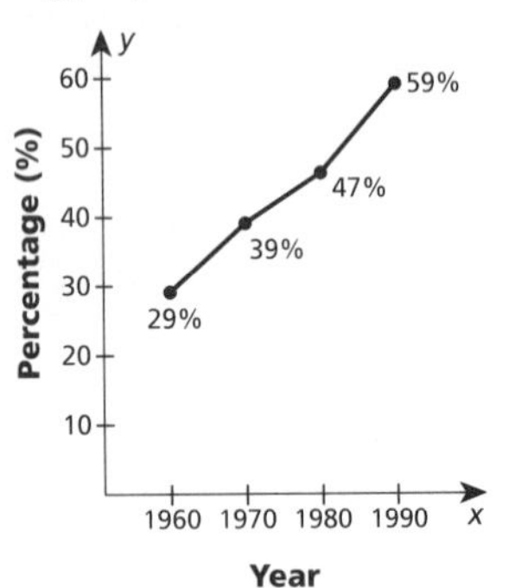

2. Assume that the trend continues at the same rate. Predict the percentage of double-income households by the year 2000.
   Evaluate student's prediction methods. Some students might recall that the average rate of change is a slope of a line and use the linear model $y = 0.01x + 0.29$ to predict a value of 0.69 or 69%.
3. **WRITING** Write a paragraph on the social implications of the data. Include in your paragraph a reason why you think the number of double-income households is on the increase.

43. *Law Enforcement* The percent of female law enforcement officers in the United States rose from 10.9 in 1986 to 16.0 in 1993. What was the average rate of change in the percent of women in law enforcement? ***(Source: U.S. Department of Labor)*** 0.73% per year

44. In 1986, 14.9% of all food servers in the United States were men. By 1994, this percent had risen to 21.4%. What was the average rate of change in the percent of male food servers? ***(Source: U.S. Bureau of Labor Statistics)*** ≈0.81% per year

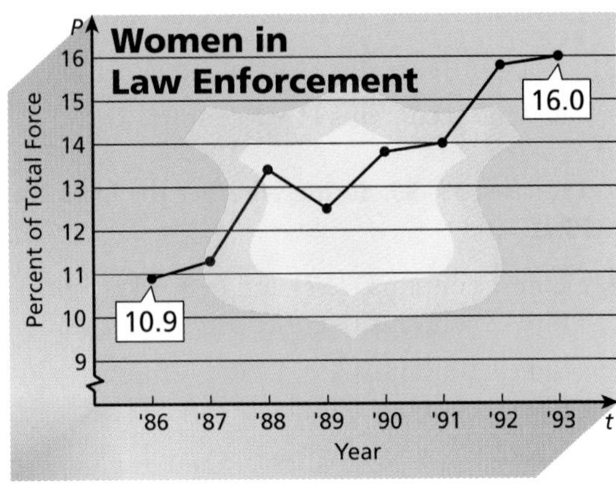

Integrated Review

In Exercises 45–50, plot the points in the coordinate plane. See Additional Answers.

45. (5, 1), (4, 5)

46. (1, 8), (7, −6)

47. $\left(-\frac{1}{2}, 3\right)$, (−14, 3)

48. (1, 0), (7, −1)

49. (−2, −2), (1, 0)

50. (3, 7), (−4, 5)

In Exercises 51–56, find two points on the graph of the equation. (There are many such points.)

51. $y = \frac{1}{4}x + 6$ (0, 6), (−4, 5)

52. $y = -\frac{1}{2}x - 4$ (0, −4), (−2, −3)

53. $y = 6x - 4$ (0, −4), (1,

54. $y = 3x + 4$ (0, 4), (−2, −2)

55. $y = -9x + 3$ (0, 3), (1, −6)

56. $y = -x + 8$ (0, 8), (2, 6

57. *College Entrance Exam Sample* From 9:00 A.M. to 2:00 P.M., the temperature rose at a constant rate from −14°F to +36°F. What was the temperature at noon? c

a. −4°F **b.** +6°F **c.** +16°F **d.** +26°F **e.** +31°F

58. *College Entrance Exam Sample* If 40,404 + x = 44,444, then 40,404 − 10x = [?]. c

a. −4.04 **b.** 0 **c.** 4 **d.** 4.04 **e.** 40.4

Exploration and Extension

59. Find x so that the line containing (1, 1) and (x, −2) will have a slope of 6. $\frac{1}{2}$

60. Find y so that the line containing (1, 1) and (4, y) will have a slope of −1. −2

61. The graph of the equation $3x + 5y = 30$ is a line. Explain how you could use the formula for slope to find the slope of the line.

61. Solve the equation for y, then find the coordinates of two points on the line to use in the slope formula.

62. A circle of radius 5 has its center at (0, 0). A line is tangent to the circle at (3, 4). What is the slope of the line? $-\frac{3}{4}$

Mixed REVIEW

4. $-\frac{4}{5}, -\frac{3}{4}, -\frac{2}{3}$ 14. $2 \cdot 2 \cdot 2 \cdot 3$ 15. $y = \frac{12}{3x^2 + 2}$

3. $11x - 24$ 1. 2.58×10^6

1. Write 2,580,000 in scientific notation.
2. 41.4 is 36% of what number? 115
3. Simplify: $-3(x - 2) + 4x + 5(2x - 6)$. **(1.2)**
4. Write in increasing order: $-\frac{2}{3}, -\frac{4}{5}, -\frac{3}{4}$.
5. Evaluate $3x^2$ when $x = 3$. **(1.2)** 27
6. Simplify: $\frac{1}{2}x - \frac{1}{3}x + \frac{1}{6}x$. **(1.2)** $\frac{1}{3}x$
7. Write the ratio of 27 to 54 in simplest form. **(1.1)** $\frac{1}{2}$
8. Subtract $3x - 9$ from $2x + 4$. **(1.2)** $-x + 13$
9. Divide: $(3.4 \times 10^{-1}) \div (1.7 \times 10^{-2})$. 20
10. Solve $3x - 4 = 5x + 6$. **(1.3)** -5
11. Solve $|2x - 7| = 3$. **(1.7)** 2, 5
12. Solve $2|x - 7| = 3$. **(1.7)** $5\frac{1}{2}, 8\frac{1}{2}$
13. Plot the points: $(-1, 2), (1, -3)$. **(2.1)** 13. See margin.
14. Write 24 as a product of prime numbers.
15. Solve $3x^2y + 2y = 12$ for y. **(1.5)**
16. Find the area of a circle of radius 4 feet.
17. Evaluate $3a^2b^3$ when $a = 2$ and $b = -3$. **(1.2)**
18. Write in exponential form: $2x \cdot 2x \cdot 2x$. **(1.2)**
19. Find the slope of the line through $(-3, 4)$ and $(2, -5)$. **(2.2)** $-\frac{9}{5}$
20. What is the slope of a horizontal line? **(2.2)** 0

16. ≈ 50.27 ft^2 17. -324 18. $(2x)^3$

Milestones MATHEMATICAL SYMBOLS

1500 — First use in print of + and – signs
1600 — Oughtred introduces multiplication sign (×)
1700 — Leibniz introduces multiplication symbol (•)

The language of algebra as it appears in this book evolved over the centuries. The first algebraic equations were written in words and diagrams. An equation written in 1545 as *cubus p̄ 6 rebus aequalis 20* is today written as $x^3 + 6x = 20$. Different symbols were introduced and adopted over time.

It is believed that our Hindu-Arabic numeral system was developed between the 4th and 7th centuries in India. Through commerce, conquest, and academic study, its use spread but only began to replace Roman numerals in Europe in the 13th century.

The minus and plus signs were first used in Germany in 1489. The plus sign (+) evolved from a shorthand notation of the Latin word *et*, meaning "and". The equal sign (=), the square root notation ($\sqrt{\ }$), and variables were introduced in the 1500's in Europe. In 1631, William Oughtred introduced the use of × to indicate multiplication. Gottfried Wilhelm von Leibniz thought it was easily confused with the variable x and proposed the dot symbol (•) in 1698.

1521 I□ e 32C° – 320 numeri
$x^2 + 32x = 320$

1525 Sit I℥ aequatus 12 ♃ – 36
$x^2 = 12x - 36$

1553 2 ♃ A + 2℥. aequata. 4335
$2xA + 2x^2 = 4{,}335$

1557 14.℞. + .15.℈ ═ 71.℈
$14x + 15 = 71$

1572 I. p. 8. Eguale à 20
$x^6 + 8x^3 = 20$

1585 3② + 4 egales à 2① + 4
$3x^2 + 4 = 2x + 4$

1631 aaa – 3 bba ═ + 2 ccc
$x^3 - 3b^2x = 2c^3$

Differences continue to exist in the written language of mathematics among countries. Find out what some of these differences are.

Answers may include that in many countries such as Mexico and Spain, decimals are written with a comma in place of the decimal point. In many Asian cultures, certain place value periods have more than 3 places.

Answers

Mixed Review

13.

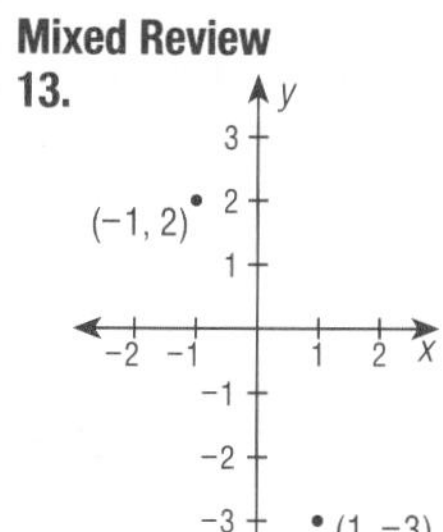

Milestones

Language Arts: *Critical Reading*

1. Find out some different ways in which fractions have been written.
   Answers may include fractions in which the denominator has been placed in the exponent position and fractions in which the positions of numerator and denominator have been switched.
2. Find out why in the late 13th century the city-state of Florence, Italy passed laws prohibiting the use of the newly introduced Hindu-Arabic numerals.
   It was believed that numerals like 0, 6, and 9 could too easily be interchanged by forgers.

Problem of the Day

If n is divided by 5, the remainder is 3. What is the remainder when $3n$ is divided by 5? 4

ORGANIZER

Warm-Up Exercises

1. Solve for y at the given values of x.
   a. $3x - 5y = 44;\ x = 3$
   b. $4y + 2x = 16;\ x = 0$
   c. $8x - 12 = 6y;\ x = 0$
   a. $y = -7$, b. $y = 4$, c. $y = -2$
2. Solve for x at the given values of y.
   a. $4y + 3x = 67;\ y = -1$
   b. $9x - 23y = 72;\ y = 0$
   c. $-3y + 18x = 48;\ y = 0$
   a. $x = \frac{71}{3}$, b. $x = 8$, c. $x = \frac{8}{3}$
3. Solve each equation for y in terms of x.
   a. $9x + 3y = 21$
   b. $-4x - 6y = 32$
   c. $8y - 3x = 48$
   a. $y = -3x + 7$, b. $y = -\frac{2}{3}x - \frac{16}{3}$, c. $y = \frac{3}{8}x + 6$

Lesson Resources

Teaching Tools
- Transparencies: 2, 3, 7, 8, 9
- Problem of the Day: 2.3
- Warm-up Exercises: 2.3
- Answer Masters: 2.3

Extra Practice: 2.3

LESSON Notes

Examples 1–2

Remind students that because two distinct points determine a straight line, the x- and y-intercepts are all that is needed to draw a quick graph. Students may need several opportunities to use the slope of a line to determine the second point when given the y-intercept.

2.3 Quick Graphs of Linear Equations

What you should learn:

Goal 1 How to use intercepts to sketch a quick graph of a line

Goal 2 How to use the slope-intercept form of a line to sketch a quick graph

Why you should learn it:

You can use graphs to see relationships between variables, such as the profits from the sale of two different items.

Goal 1 Using Intercepts to Sketch a Line

A **linear equation** in x and y is of the form $Ax + By = C$. This type of equation is called linear because its graph is always a line (unless both A and B are zero).

In Lesson 2.1, you sketched the graph of a linear equation by creating a table of values, plotting the corresponding points, and drawing the graph through the points. In this lesson, you will study two quicker ways to sketch such a graph.

One way is to find the points at which the graph intersects the x-axis and the y-axis.

Intercepts of a Line

Consider the line given by $Ax + By = C$.

1. The ***x*-intercept** of the line is the value of x when $y = 0$. To find the x-intercept, let $y = 0$ and solve for x in $Ax = C$.
2. The ***y*-intercept** of the line is the value of y when $x = 0$. To find the y-intercept, let $x = 0$ and solve for y in $By = C$.

Note: We can speak of the x-intercept as a or the point $(a, 0)$, and the y-intercept as b or the point $(0, b)$.

Example 1 *Sketching a Graph Using Intercepts*

Sketch the line given by $3x + 5y = 15$.

Solution First find the intercepts.

x-Intercept (Let $y = 0$.)	**y-Intercept (Let $x = 0$.)**
$3x + 5y = 15$	$3x + 5y = 15$
$3x + 5(0) = 15$	$3(0) + 5y = 15$
$x = 5$	$y = 3$
x-intercept is 5.	y-intercept is 3.

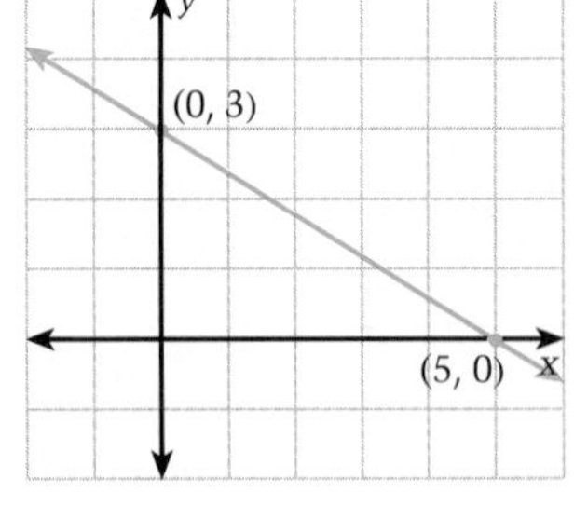

Quick graph using intercepts

To sketch the line, plot the intercepts, and draw a line through them. In the graph shown at the left, note that the x-intercept is $(5, 0)$ and the y-intercept is $(0, 3)$. ■

Goal 2 Using the Slope-Intercept Form

A second way to sketch a quick graph of a line requires that you write the equation in slope-intercept form.

Slope-Intercept Form of the Equation of a Line

The linear equation

Slope → m, y-Intercept → b

$$y = mx + b$$

is written in **slope-intercept form.** The slope of the line is m. The y-intercept is b.

To sketch a quick graph of a line, first write the equation in slope-intercept form (by solving for y). Next, graph the y-intercept. Finally, use the slope to find a second point on the line.

Example 2 *Using Slope and y-Intercept to Graph*

Sketch the graph of $x + 2y = 4$.

Solution Begin by solving the equation for y.

$x + 2y = 4$ *Original equation*

$2y = -x + 4$ *Subtract x from both sides.*

$y = \underbrace{-\frac{1}{2}}_{m}x + \underbrace{2}_{b}$ *Divide both sides by 2.*

You can sketch the line as follows. First, graph the y-intercept at (0, 2). Then locate a second point by moving 2 units to the right and 1 unit down. Finally, draw the line through the two points.

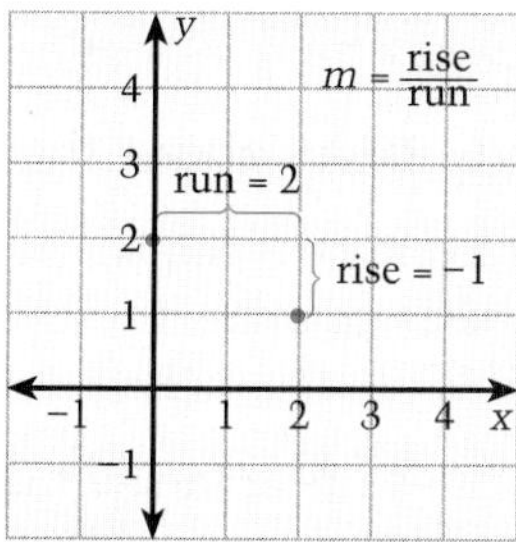

Plot y-intercept, then use the slope to find a second point.

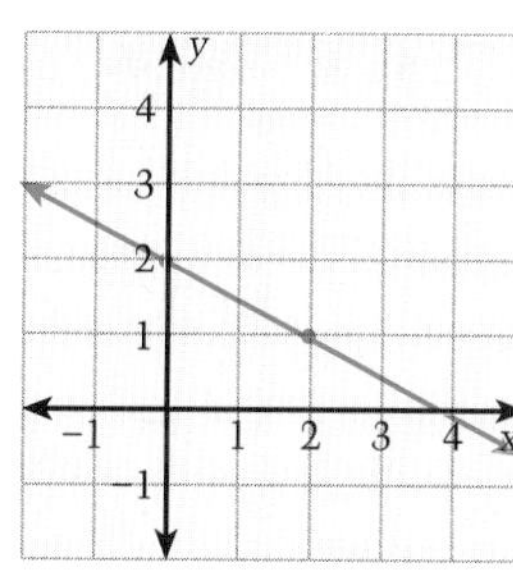

Draw the line containing the two points. ■

It is important for students to evaluate which quick graph technique to employ. For example, if the coefficient of y is different from 1, it may be easier to find the x- and y-intercepts by setting $y = 0$ and $x = 0$, respectively. However, if the coefficient of y is 1, then it is simpler to solve for y in terms of x.

Example 3

Ask students what the best technique is for constructing a quick graph of the relationship. Have students explain whether the line segment in the diagram continues in either direction.

Extra Examples

Here are additional examples similar to **Examples 1–3.**

1. Sketch a quick graph of each line. Indicate which technique was used and why.
   a. $4y - 3x = 24$
   b. $2y = 4x - 6$
   c. $12x + 8y = 16$
   d. $6x - y + 18 = 0$

   a., c.: Find the intercepts
   a. Intercepts: $(-8, 0)$ and $(0, 6)$
   c. Intercepts: $(\frac{4}{3}, 0)$ and $(0, 2)$
   b., d.: Solve for y in terms of x, and use the slope-intercept form
   b. $y = 2x - 3$; slope $= 2$, y-intercept $= -3$
   d. $y = 6x + 18$; slope $= 6$, y-intercept $= 18$
2. Suppose you are the manager of a department store in charge of clothing and jewelry. Based on retail prices, your markup on jewelry is 50% and your markup on clothes is 41.2%. Write a model that represents the different amounts of clothes and jewelry you could sell during a week to obtain a total markup of \$2000.

3. Sketch the graph of the linear model you find.
model: $0.5J + 0.412C = 2000$

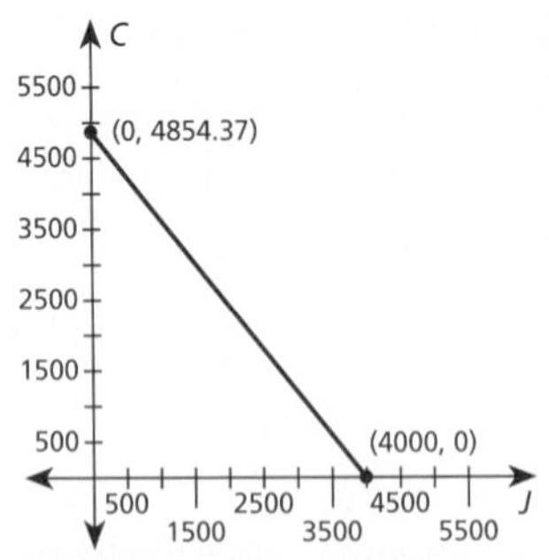

Check Understanding

1. Describe how the slope-intercept form of a linear equation is used to construct a quick graph.
   See Example 2.
2. How does one find the x-intercept and the y-intercept of a linear equation?
   See the box on page 78.
3. What is the general form of a linear equation?
   $Ax + By = C$

Communicating about ALGEBRA

Answer

B.

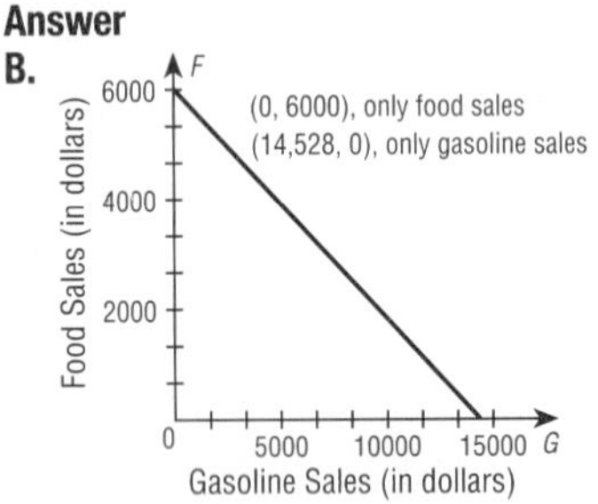

EXTEND *Communicating*

Ask students to explore the retail markup rates in their community on different types of merchandise. Combine data to obtain an average retail markup for the various types of merchandise researched.

Students could write linear models of various markups for the different types of merchandise and discuss the results.

Retail markup rates vary greatly. They depend on the store and the item that is being sold. For instance, typical retail markup rates are 9% for gasoline, 14% for new cars, 25% for groceries, 70% for clothes, and 100% for jewelry. These rates are based on cost; the same rates based on retail price would be 8.3%, 12.3%, 20%, 41.2%, and 50%.

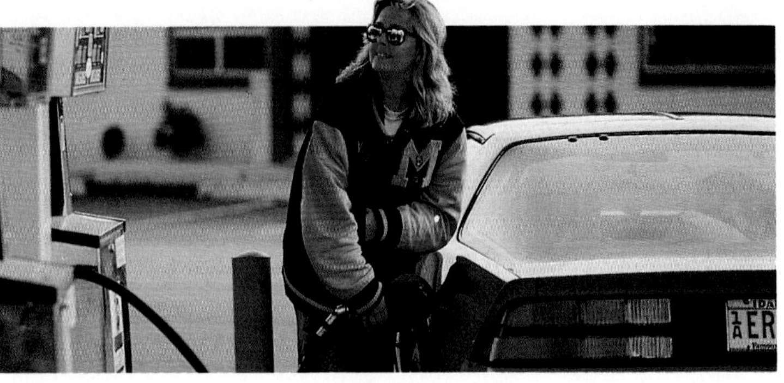

Connections
Retail Sales

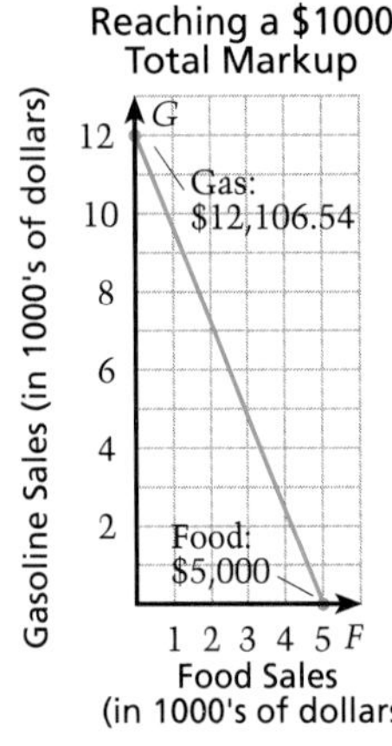

Example 3 *Creating a Linear Model*

You are the manager of a convenience store. Based on retail prices, your markup rate on gasoline is 8.3% and your average markup rate on food and other store items is 20%. Write a model that shows the amounts of gasoline and food you could sell during a week to obtain a total markup of $1000.

Solution

Verbal Model Markup rate · Gas sales + Markup rate · Food sales = Total markup

Labels

Markup rate for gas = 0.083	(% as decimal)
Gas sales = G	(dollars)
Markup rate for food = 0.2	(% as decimal)
Food sales = F	(dollars)
Total markup = 1000	(dollars)

Equation $0.083G + 0.2F = 1000$ *Linear model*

This equation has many solutions that can be seen from its graph. For instance, selling $5000 worth of food and no gasoline is one solution. Selling $12,106.54 worth of gasoline and no food is another solution. ■

Communicating about ALGEBRA

SHARING IDEAS about the Lesson

A. In Example 3, suppose you sold three times as much food as gasoline. If the store brought in a total markup of $1000, how much of each was sold?
   gas: $1464.13 food: $4392.39
B. Write a linear model that shows the amounts of gasoline and food that could be sold to bring in a weekly markup of $1200. Sketch the graph of the model. See margin.

B. $0.083G + 0.2F = 1200$

EXERCISES

Guided Practice

CRITICAL THINKING about the Lesson See margin for 1., 2., 5., and 6.

1. Why is an equation written in the form $Ax + By = C$ called a *linear* equation?
2. Describe the intercepts of a graph. How do you find the intercepts of a graph?
3. Find the intercepts of the line $3x - 6y = 12$. *x*-intercept: 4, *y*-intercept: −2
4. Find the slope and *y*-intercept of the line $y = -4x - 2$. −4, −2
5. Which of the two quick-graph techniques discussed in the lesson would you use to sketch the graph of $3x + 4y = 24$? Explain.
6. Which of the two quick-graph techniques discussed in the lesson would you use to sketch the graph of $y = -2x + 4$? Explain.

Independent Practice

In Exercises 7–12, find the *x*-intercept and *y*-intercept of the line.

7. $16x + 2y = 8$ $\frac{1}{2}$, 4
8. $-x + 7y = 21$ −21, 3
9. $3x - 8y = 9$ 3, $-\frac{9}{8}$
10. $5x - 10y = 10$ 2, −1
11. $21x + 7y = -7$ $-\frac{1}{3}$, −1
12. $-2x + 5y = 12$ −6, $\frac{12}{5}$

In Exercises 13–18, match the equation with its graph.

13. $5x + y = 10$ e
14. $-2x + 8y = 24$ f
15. $9x + 3y = 18$ c
16. $x - 2y = 5$ b
17. $2x + 8y = -24$ a
18. $3x - 6y = 18$ d

a.
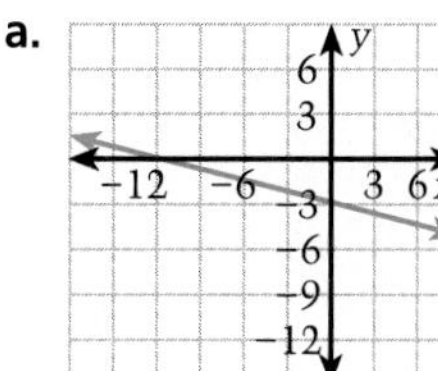

b.
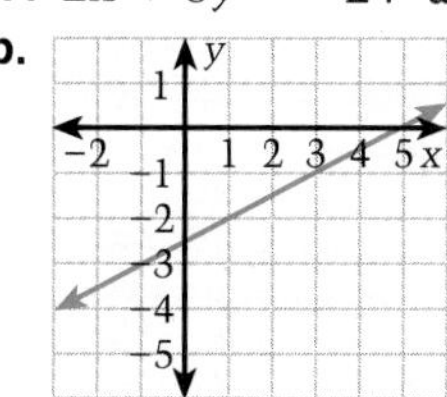

c.
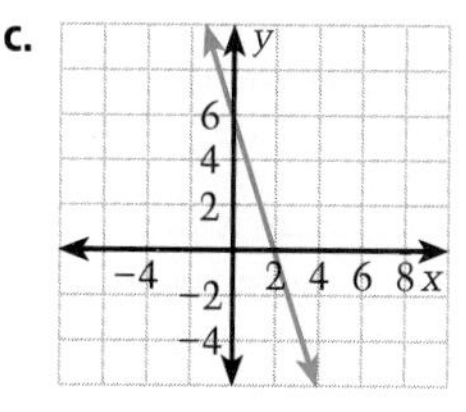

d.
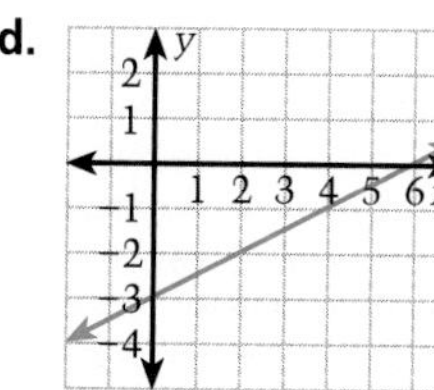

e.
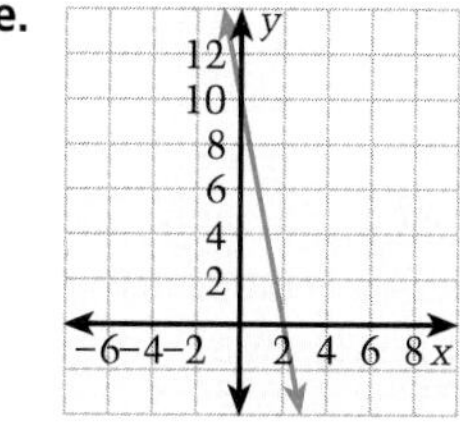

f.
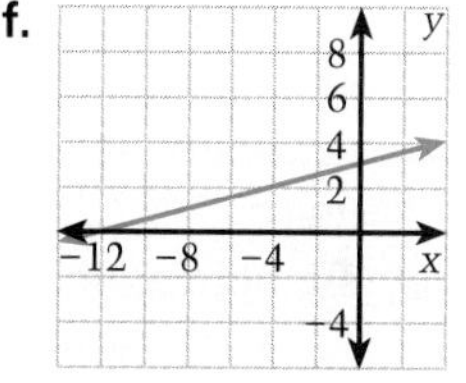

In Exercises 19–24, sketch the line with the given intercepts. (For one exercise, the intercepts do not determine a single line. Which is it and why?) See Additional Answers.

19. *x*-intercept: 1
    y-intercept: 4
20. *x*-intercept: −7
    y-intercept: −5
21. *x*-intercept: −2
    y-intercept: 5
22. *x*-intercept: 4
    y-intercept: −2
23. *x*-intercept: −3
    y-intercept: −1
24. *x*-intercept: 0
    y-intercept: 0
    See below.

24. These intercepts do not determine a single line; because they are the same point, the origin.

EXERCISE Notes

ASSIGNMENT GUIDE

Basic/Average: Ex. 1–6, 11, 13–18, 19–39 odd, 43–46, 47–59 odd, 67, 69, 70

Above Average: Ex. 1–6, 11–18, 23–39 odd, 43–47, 67–71

Advanced: Ex. 1–6, 13–18, 23–26, 31–43 odd, 44–46, 67–72

Selected Answers
Exercises 1–6, 7–65 odd

✪ More Difficult Exercises
Exercises 71, 72

Guided Practice

▶ **Ex. 1–6** These exercises require students to reflect on the important goals of the lesson. Require students to analyze the efficiency of a graphing method based on the form of the linear equation.

Answers

1. Because its graph is a line.
2. The *x*-intercept is the value of *x* when *y* is zero; the *y*-intercept is the value of *y* when *x* is zero. To find the *x*-intercept, let $y = 0$ and solve the equation of the line for *x*; to find the *y*-intercept, let $x = 0$ and solve for *y*.
5. Use intercepts, because 24 can be divided mentally by both 3 and 4.
6. Use slope and *y*-intercept, because the equation is already in slope-intercept form.

Independent Practice

▶ **Ex. 13–18** Assign these exercises as a group.

▶ **Ex. 24** Be sure students understand that at least two distinct points are necessary in order to draw a line.

Ex. 25–30 Encourage students to use the intercepts quick-graph technique illustrated in Example 1 on page 78.

Answers

25.

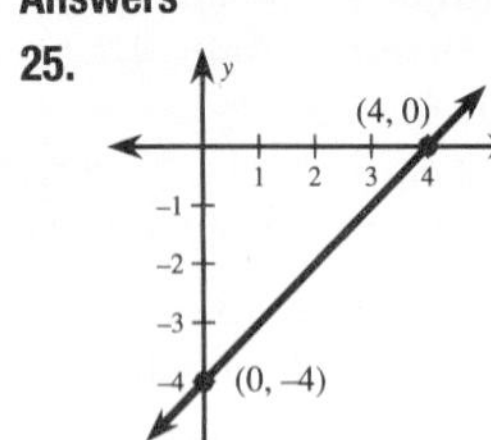

26.

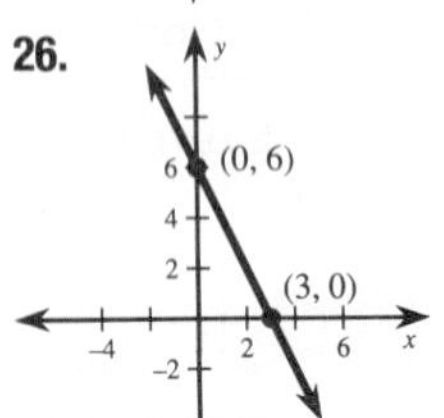

27.

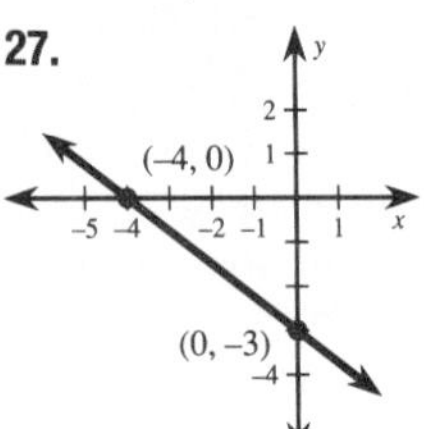

28.

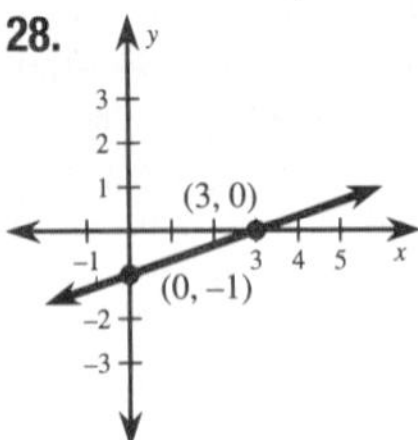

29.

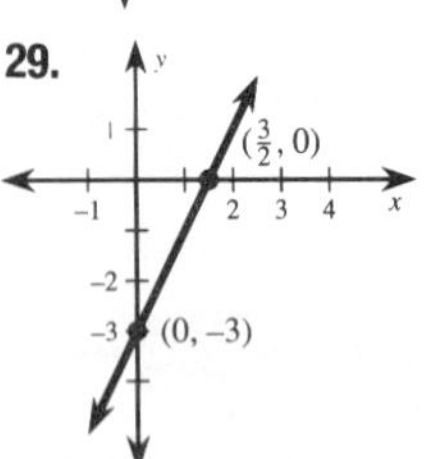

30.

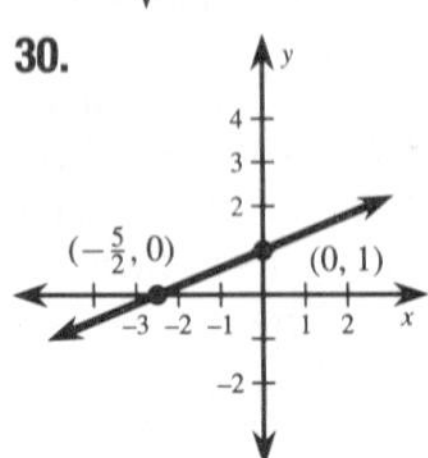

31. $y = -4x + \frac{1}{2}, -4, \frac{1}{2}$
32. $y = 2x - \frac{1}{2}, 2, -\frac{1}{2}$
33. $y = 13x + 2, 13, 2$
34. $y = -10x + 9, -10, 9$
35. $y = \frac{1}{2}x - 4, \frac{1}{2}, -4$

In Exercises 25–30, sketch the line. Label the coordinates of the x-intercepts and y-intercepts. See margins.

25. $x - y = 4$ **26.** $2x + y = 6$ **27.** $3x + 4y = -12$

28. $5x - 15y = 15$ **29.** $6x - 3y = 9$ **30.** $-2x + 5y = 5$

In Exercises 31–36, write the slope-intercept form of the equation. Identify the slope and y-intercept. See margin.

31. $8x + 2y = 1$ **32.** $-4x + 2y = -1$ **33.** $-13x + y = 2$

34. $10x + y = 9$ **35.** $-x + 2y = -8$ **36.** $-x + 3y = 15$

In Exercises 37–42, sketch the line using the slope and y-intercept.

37. $y = 4x + 4$ 37.–39. See margins. **38.** $y = -3x + 2$ **39.** $y = \frac{5}{6}x - 1$

40. $y = \frac{5}{2}x - 5$ 40.–42. See Additional Answers. **41.** $y = -\frac{1}{2}x + 6$ **42.** $y = -\frac{3}{4}x + 2$

43. *Ladder Safety* In the book *Fix It Fast, Fix It Right,* the following ladder-safety feature is given.

"Adjust the ladder until it is at least one quarter of the ladder height from the wall. For example, a 12-foot ladder should be at least 3 feet from the wall."

What is the maximum slope that this book is recommending for a ladder? Why do you think it is unsafe to set up a ladder with a steeper slope?

(Source:* Fix It Fast, Fix It Right, *©1991, Rodale Press) ≈3.9, the ladder is apt to tip back

44. *Fund-Raiser* The members of the marching band are selling pizzas and submarine sandwiches as a fund-raising project. Their goal is to make \$1000 profit. They will receive \$2.75 profit for every pizza sold and \$1.50 profit for every sub sold. Write a model that shows the different numbers of pizzas and subs the band members must sell to achieve their goal. $2.75p + 1.5s = 1000$

45. *Ticket Prices* Student tickets at your high school football game cost \$2 each. Non-student tickets cost \$5 each. The ticket sales at the homecoming game totaled \$6800. Write a model that shows the different numbers of students and adults who could have attended the game. $2s + 5n = 6800$

46. *Wheelchair Ramp* A recommended maximum slope for a wheelchair ramp is $\frac{1}{10}$. What does this mean? If a business was to install a wheelchair ramp that rose 16 inches (the height of two steps), what would be the minimum horizontal length of the ramp?
A rise of one foot in a run of 10 feet, 13 ft 4 in.

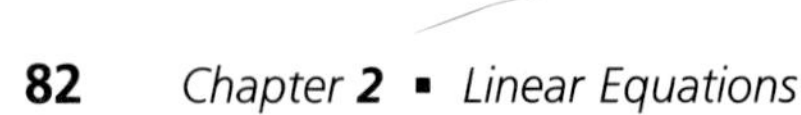

Integrated Review

47. Is $(1, -2)$ a solution of $-3x + 2y = 7$? No

48. Is $(5, 7)$ a solution of $6x - 4y = 2$? Yes

In Exercises 49–54, let $y = 0$ and solve for x.

49. $2x - y = 16$ 8

50. $-4x + 16y = 21$ $-5\frac{1}{4}$

51. $-6x + 10y = 15$ $-2\frac{1}{2}$

52. $3x - 5y = 15$ 5

53. $-x + 27y = 9$ -9

54. $5x + 12y = 144$ $28\frac{4}{5}$

In Exercises 55–60, solve for y.

55. $6x - 4y = 16$ $y = \frac{3}{2}x - 4$

56. $9x - 3y = 20$ $y = 3x - \frac{20}{3}$

57. $-5x + 3y = 9$ $y = \frac{5}{3}x + 3$

58. $-x + 5y = 15$ $y = \frac{1}{5}x + 3$

59. $x - 4y = 10$ $y = \frac{1}{4}x - \frac{5}{2}$

60. $2x - 8y = 5$ $y = \frac{1}{4}x - \frac{5}{8}$

In Exercises 61–66, sketch the line by creating a table of values. See Additional Answers.

61. $y = x - 5$

62. $y = -x + 7$

63. $y = 2x + 1$

64. $y = 4x - 9$

65. $y = -5x - 2$

66. $y = -8x - 9$

Exploration and Extension

67. *Herb Garden* You are planning an herb garden. The garden has 120 inches of "row space." The parsley seedlings each need 8 inches of row space and the garlic bulbs each need 6 inches of row space. Write a model that shows the numbers of parsley and garlic plants that will fit in your garden. If you want 6 parsley plants, how many garlic bulbs can you plant?

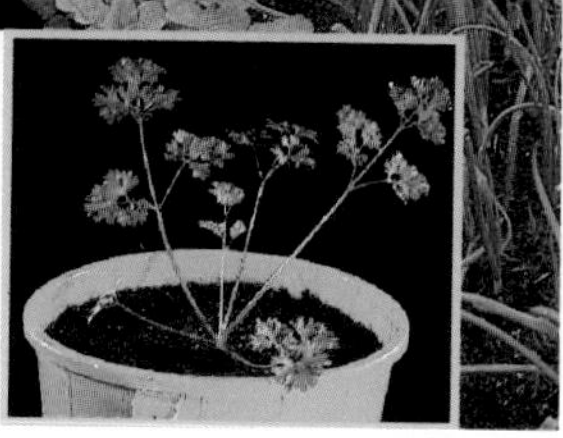

68. *Car Wash* You own a car wash. You charge \$6 for a wash and \$7 for a wash-and-wax. At the end of a busy day, total sales were \$2258. Write a model that shows the different number of car washes and wash-and-waxes that could have been sold during the day. If 170 cars got a wash-and-wax, how many got washes only?

69. What kind of line has no x-intercept? A horizontal line

70. What kind of line has no y-intercept? A vertical line

67. $8p + 6g = 120$, 12 **68.** $6h + 7w = 2258$, 178

In Exercises 71 and 72, write an equation for the line.

✪ **71.**

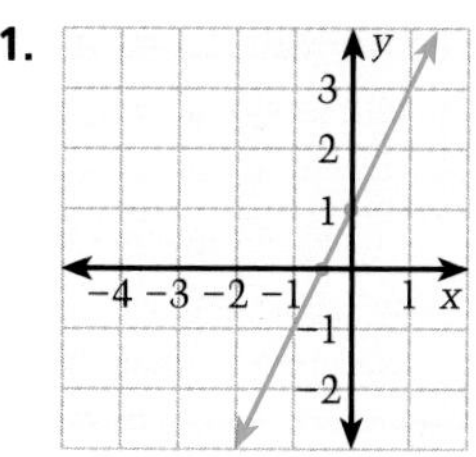

$y = 2x + 1$

✪ **72.**

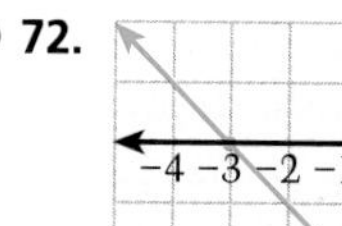

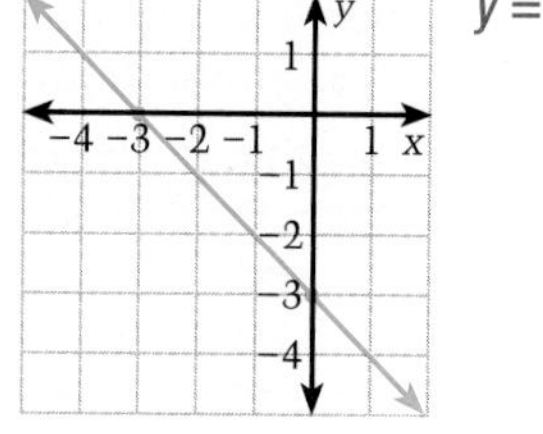

$y = -x - 3$

Answers

36. $y = \frac{1}{3}x + 5, \frac{1}{3}, 5$

37.

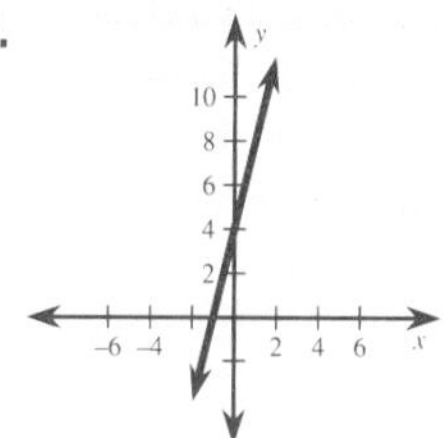

38.

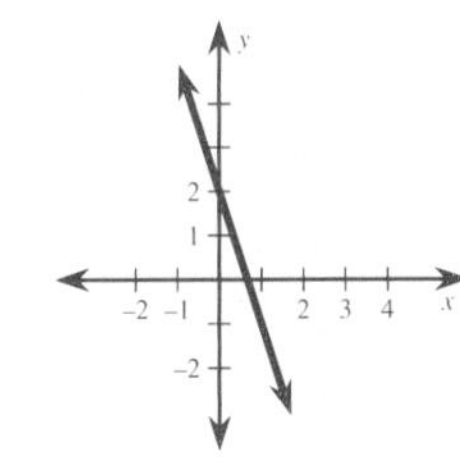

39.

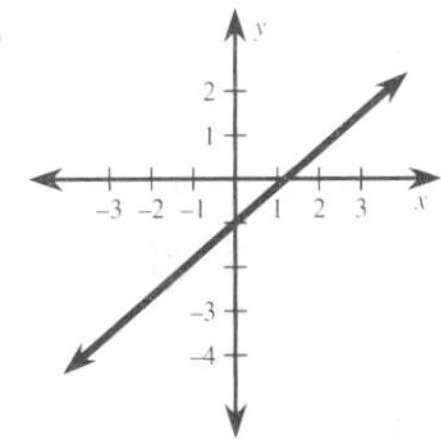

▶ **Ex. 37–42** Encourage students to use the slope-intercept quick-graph technique illustrated in Example 2 on page 79.

EXTEND Ex. 46 Have students discuss the special building requirements for handicap facilities.

▶ **Ex. 71–72** These exercises are a preview for Lesson 2.4.

Enrichment Activities

These activities require students to go beyond lesson goals.

WRITING PROBLEMS Write three problems similar to Exercises 43, 44, 67, and 68. Show your solutions on the back of your paper. Share your problems with your classmates.

ALTERNATIVE ASSESSMENT You could collect the problems and use them on your tests and quizzes, or you could save them for review at a later time.

TECHNOLOGY Observe that it is frequently necessary to solve for y in terms of x when you want to use the calculator.

A useful way to represent the viewing rectangle, or range, of a graph is given below.

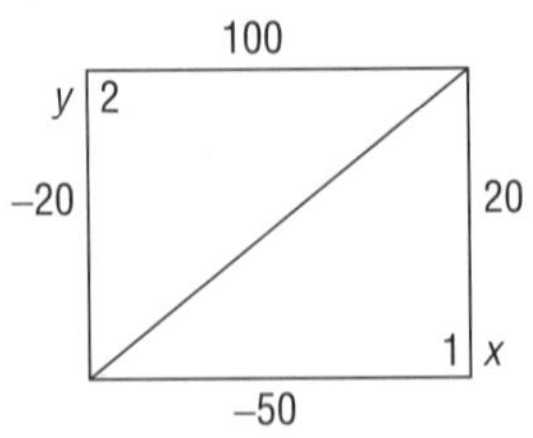

In this notation, Xmin = −20, Xmax = 20, Xscl = 1, Ymin = −50, Ymax = 100, Yscl = 2.

Alert students about the zooming feature found on graphing calculators and computer plotters. In particular, tell students that the cursor should be placed in the center of the region on which the user wishes to zoom.

Example 1

Approximating solutions of equations is facilitated by the use of the zoom-in feature of the graphing tool. (Check to see if the "box" feature is available on your graphing tool. This feature lets the user define the new viewing region graphically.) Repeated application of the zoom feature permits the user to obtain a solution to almost any degree of desired accuracy.

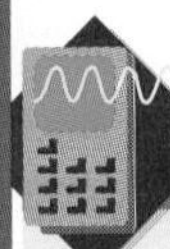

USING A GRAPHING CALCULATOR

A graphing calculator or computer can be used to sketch the graph of an equation. For instance, the screen at the right shows the graph of $x + 2y = 4$. Keystroke instructions for sketching this graph on a *TI-82*, Casio *fx-9700GE*, and Sharp *EL-9300C* are listed on page 857.

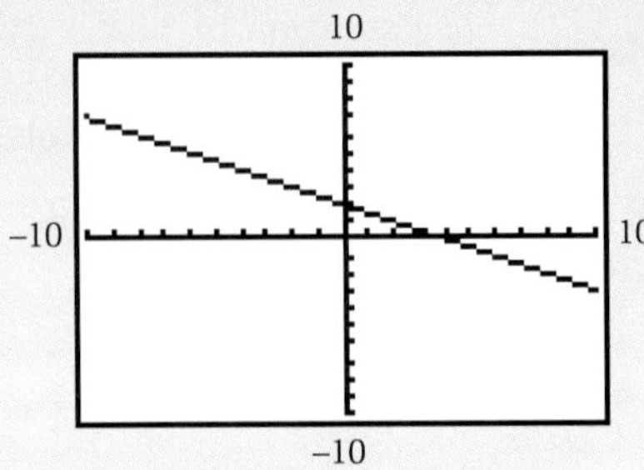

A standard viewing rectangle uses least x and y values of −10 and greatest x and y values of 10.

Using a graphing calculator to sketch a graph is fairly easy, but there are three ideas you must remember.

- Solve the equation for y in terms of x.
- Describe the viewing rectangle by entering the least and greatest x and y values and the scale (units per mark). This is called *setting the Window or Range.*
- Use parentheses if you are unsure of the calculator's order of operations.

After sketching the graph, you might want a better view of some part of the graph. For instance, if you want a closeup view of the point where the line $y = -\frac{1}{2}x + 2$ crosses the x-axis, you could enter new settings to produce a different viewing rectangle, as show below.

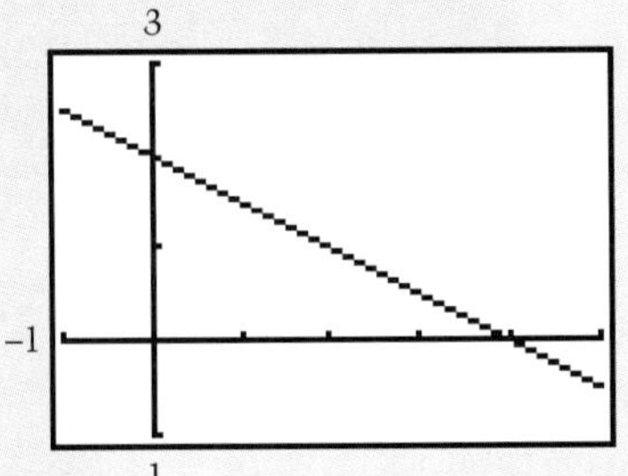

Cooperative Learning

Partner Activity Without showing your partner, enter an equation of the form

$$y = ax + b$$

into a graphing calculator. Choose an equation whose intercepts are both visible in a standard viewing rectangle. Then give the graphing calculator to your partner. His or her goal is to find the equation of the line. To do this, your partner is allowed to use only the trace key.

After your partner has found the equation of the line, switch roles and repeat the activity.

Changing the range setting to get a better view of a portion of a graph is called *zooming in* on the graph. Graphing calculators have a zoom key that will automatically change the setting to give you a close-up view of any part of the display. You can use the zoom feature to approximate the solution of an equation graphically.

Example 1 Approximating the Solution of an Equation

Use a graphing calculator to approximate the solution of

$$-340.5(x - 7.2) = 512.4(x + 13.7).$$

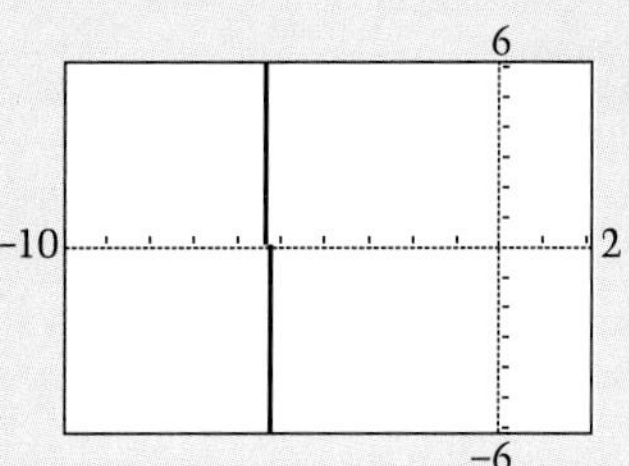

Solution To begin, rewrite the equation with 0 on one side.

$$-340.5(x - 7.2) - 512.4(x + 13.7) = 0$$

Then, set y equal to the nonzero side of the equation and sketch the graph of

$$y = -340.5(x - 7.2) - 512.4(x + 13.7).$$

The solution of the original equation is the x-intercept of the graph of this equation. Notice that the graph (shown above) is almost vertical. The x-intercept of the line is approximately -5.4. ■

Exercises

In Exercises 1–6, use a graphing calculator to sketch the graph of the equation. For 1–3, use the standard range (−10 to 10 by units of 1 for both x and y). For 4–6, use the indicated range. See Additional Answers.

1. $y = -3x + 5$

2. $y = 2x - 6$

3. $x - 3y = -2$

4. $y = 2x - 17$

RANGE
Xmin=−2 Ymin=−20
Xmax=12 Ymax=2
Xscl=1 Yscl=1

5. $y = -x + 30$

RANGE
Xmin=−5 Ymin=−5
Xmax=35 Ymax=35
Xscl=5 Yscl=5

6. $y = 0.2x + 1.8$

RANGE
Xmin=−10 Ymin=−1
Xmax=2 Ymax=3
Xscl=1 Yscl=1

In Exercises 7–10, use a graphing calculator to sketch the graph of the equation. Choose a range setting that makes both intercepts visible. See margin.

7. $y = 3.7x - 20$

8. $y = -1.54x + 63.4$

9. $y = -12.7x - 19.2$

10. $y = 1.4x + 208$

In Exercises 11 and 12, use a graphing calculator to approximate the solution.

11. $9.1(2x - 12.1) = 15.6x + 24.7$ ≈51.9

12. $1.4(2.1x - 5.8) = -0.2(x + 19.1)$ ≈1.4

Answers

7.

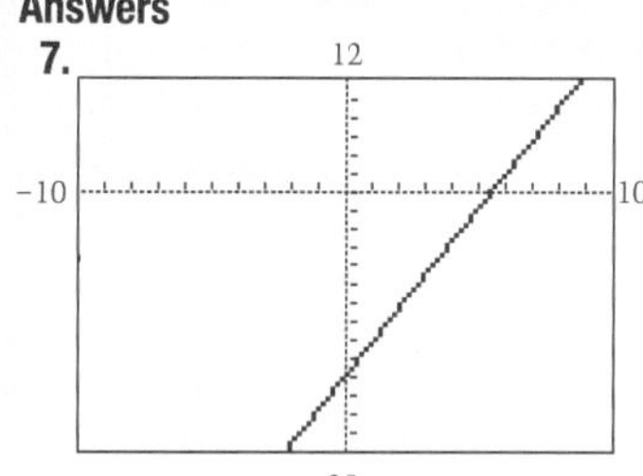

8.

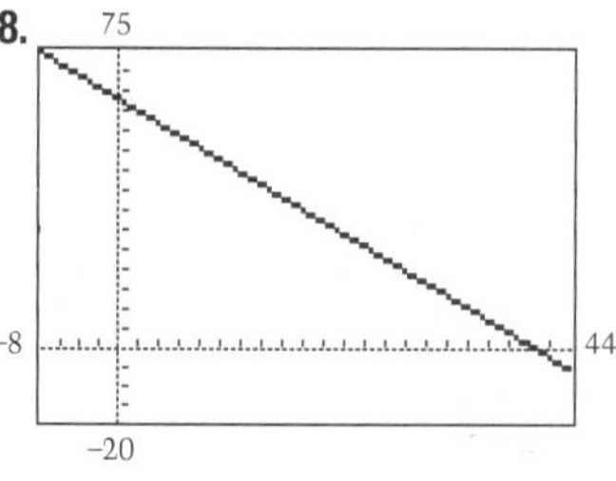

9.

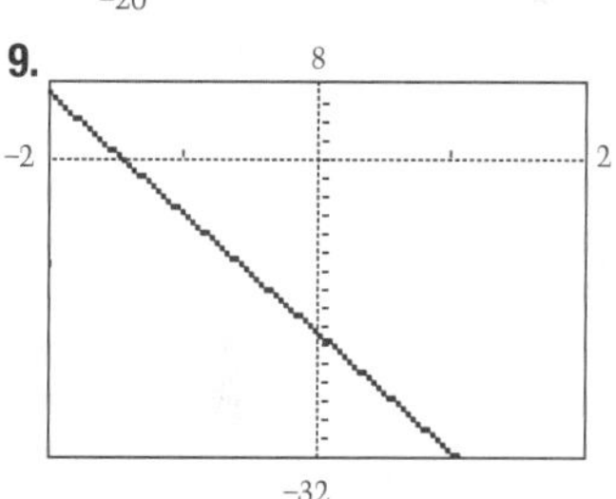

10.

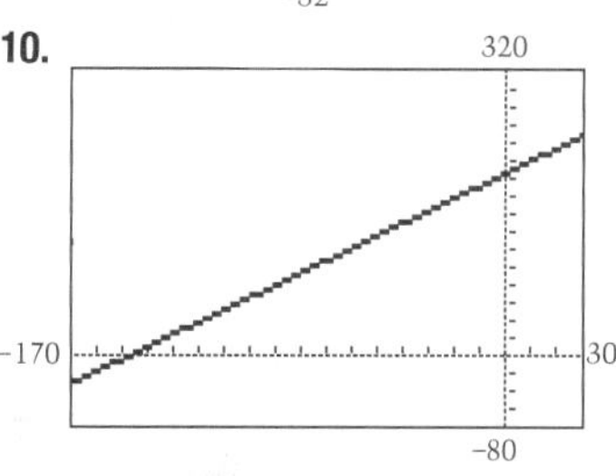

Problem of the Day

In a certain country, all the inhabitants are knights or knaves. The knights always tell the truth, and the knaves always lie. A stranger approaches A and B and asks what they are. A says, "At least one of us is a knave." What are they in fact? A is a knight, B is a knave.

ORGANIZER

Warm-Up Exercises

1. Determine the slope of each line by reading the rise and run from the graph.

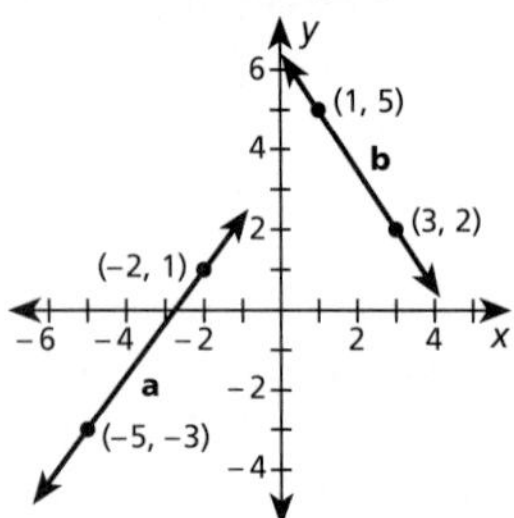

a. $m = \frac{4}{3}$, b. $m = -\frac{3}{2}$

2. Find the slope of the line containing the given points.
   a. $(-2, 1)$ and $(3, 4)$
   b. $(0, -4)$ and $(5, 7)$

a. $m = \frac{3}{5}$, b. $m = \frac{11}{5}$

3. Solve for y in terms of x.
   a. $6x + 3y = 12$
   b. $-4y + 12x = -10$

a. $y = -2x + 4$,
b. $y = 3x + 2.5$

Lesson Resources

Teaching Tools
- Transparencies: 7, 8, 9
- Problem of the Day: 2.4
- Warm-up Exercises: 2.4
- Answer Masters: 2.4

Extra Practice: 2.4
Color Transparencies: 11, 12
Applications Handbook: p. 41

2.4 Writing Equations of Lines

What you should learn:

Goal 1 How to write equations of lines

Goal 2 How to use equations of lines to solve real-life problems

Why you should learn it:

You can model many real-life situations with a linear equation, especially those involving direct variation.

Goal 1 Writing Equations of Lines

In Lesson 2.3, you learned to find the slope and y-intercept of a line whose equation is given. In this lesson, you will study the reverse process. That is, you will learn to write an equation of a line using one of the following: the slope and y-intercept of the line, the slope and a point on the line, or two points on the line.

Writing an Equation of a Line

1. Given the slope, m, and the y-intercept, b, use the equation

   $y = mx + b$. *Slope-intercept form*

2. Given the slope, m, and a point, (x_1, y_1), use the equation

   $y - y_1 = m(x - x_1)$. *Point-slope form*

3. Given two points, (x_1, y_1) and (x_2, y_2), use the formula

   $m = \frac{y_2 - y_1}{x_2 - x_1}$

   to find the slope, m. Then use the point-slope form with either of the given points to find an equation of the line.

Example 1 *Using the Slope-Intercept Form to Write an Equation*

Write an equation of the line shown at the left.

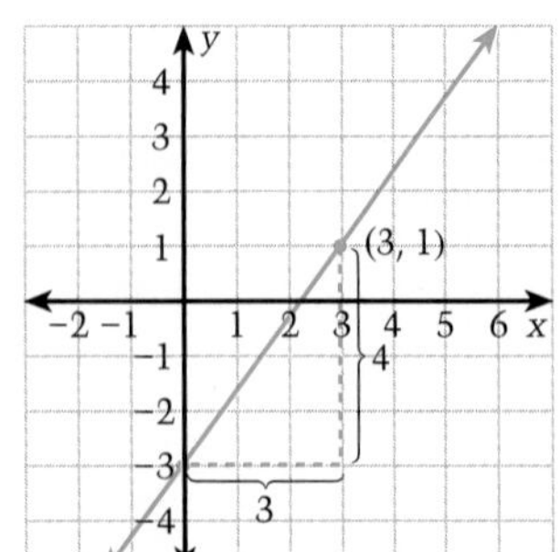

Solution Since the line intersects the y-axis at $(0, -3)$, the y-intercept, b, is -3. From the graph, you see that the slope is

$m = \frac{\text{rise}}{\text{run}} = \frac{4}{3}$.

Using the slope and y-intercept of the line, you can write

$y = mx + b$ *Slope-intercept form*

$y = \frac{4}{3}x - 3$ *Substitute $\frac{4}{3}$ for m and −3 for b.*

An equation of the line is $y = \frac{4}{3}x - 3$. ■

Example 2 Using the Point-Slope Form to Write an Equation

Write an equation of the line containing the point (1, 2) with a slope of $-\frac{1}{2}$.

Solution Using the point-slope form, with $(x_1, y_1) = (1, 2)$ and $m = -\frac{1}{2}$, you can write the following.

$$y - y_1 = m(x - x_1) \quad \textit{Point-slope form}$$

$$y - 2 = -\frac{1}{2}(x - 1) \quad \textit{Substitute for } m, x_1, \textit{ and } y_1.$$

Once you have used the point-slope form to find an equation, you can simplify your result to the slope-intercept form, as follows.

$$y - 2 = -\frac{1}{2}(x - 1) \quad \textit{Point-slope form}$$

$$y - 2 = -\frac{1}{2}x + \frac{1}{2} \quad \textit{Distributive Property}$$

$$y = -\frac{1}{2}x + \frac{5}{2} \quad \textit{Slope-intercept form}$$

Graphic Check Try sketching the graph of the line containing the point (1, 2) with a slope of $-\frac{1}{2}$. From your graph, does it appear that the y-intercept is $\frac{5}{2}$? ■

Remember that the point-slope form, $y - y_1 = m(x - x_1)$, has *two* minus signs. Be sure to account for these signs when the coordinates (x_1, y_1) involve negative numbers.

Example 3 Using Two Given Points to Write an Equation

Write an equation of the line shown at the left.

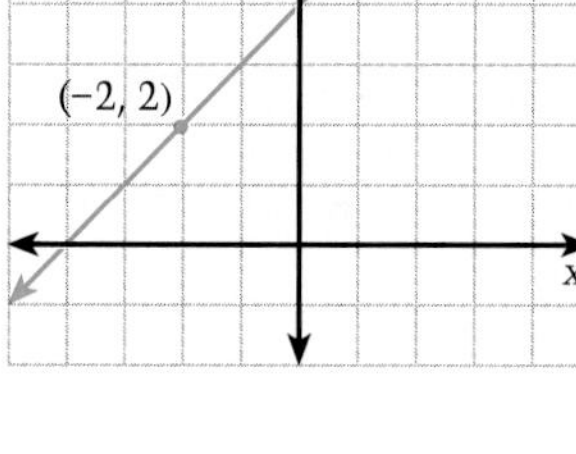

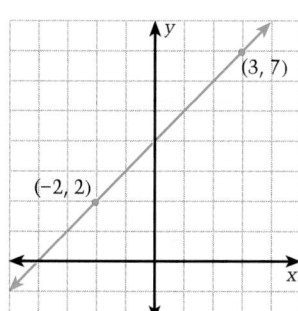

Solution Since the line passes through $(x_1, y_1) = (-2, 2)$ and $(x_2, y_2) = (3, 7)$, its slope is the following.

$$m = \frac{y_2 - y_1}{x_2 - x_1} = \frac{7 - 2}{3 - (-2)} = \frac{5}{5}, \text{ or } 1$$

To find an equation of the line, use the point-slope form and then simplify, as follows.

$$y - y_1 = m(x - x_1) \quad \textit{Point-slope form}$$

$$y - 2 = 1(x - (-2)) \quad \textit{Substitute for } m, x_1 \textit{ and } y_1.$$

$$y - 2 = x + 2 \quad \textit{Simplify.}$$

$$y = x + 4 \quad \textit{Slope-intercept form}$$

■

LESSON Notes

Example 1

The slope-intercept form is undoubtedly the most often used form of a linear equation. Therefore, it is important that students be able to express linear equations in this form. Students will see later that the slope-intercept form has a parallel form using function notation, $f(x) = mx + b$.

Example 2

The equation could be left in point-slope form. However, it is customary to "simplify" the point-slope form into the more common slope-intercept form. The slope-intercept is easier to use when sketching a graph of the linear equation.

Example 3

Students may indicate that the slope can be read from the graph as in Example 1 above. Acknowledge their response as correct. However, stress that it is important to use the technique described in this example because there are times when the rise and run cannot be easily read from the graph.

Extra Examples

Here are additional examples similar to **Examples 1–3.**

1. Use the slope-intercept form to write an equation for each line.

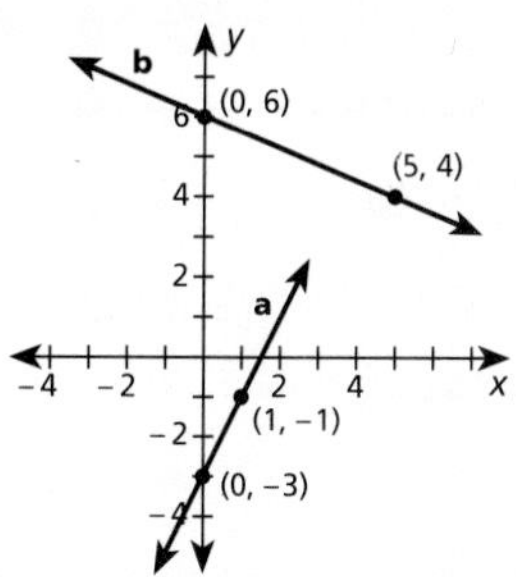

a. $y = 2x - 3$, **b.** $y = -\frac{2}{5}x + 6$

2. Use the point-slope form to write an equation for each line.

a. $m = -4$, $(3, 2)$

b. $m = \frac{4}{5}$, $(-2, -4)$

a. $y = -4x + 14$

b. $y = \frac{4}{5}x - \frac{12}{5}$

3. Write an equation of a line given two points on the line.

a. $(2.3, 18.9)$, $(-3.1, 2.7)$

b. $(-6.45, -5.37)$, $(2.6, 0.06)$

a. $y = 3x + 12$

b. $y = 0.6x - 1.5$

Example 4

Ask students to determine about how long hailstones that typically fall in their region of the country must remain in the clouds.
Answers will vary; students must first agree upon the size of a typical hailstone.

Example 5

Have students discuss the following proposition: By the year 2000, 2860 African-American women will hold public office.
Students should discuss whether the linear model may be used to predict the future.

88

Goal 2 Using Equations of Lines in Real Life

Two variable quantities that have the same rate or ratio, regardless of the values of the variables, have **direct variation.** For instance, if the ratio of y to x is k, then you can write $\frac{y}{x} = k$. By multiplying both sides of the equation by x, you obtain $y = kx$.

Direct Variation

The variables x and y **vary directly** if for a constant k

$$\frac{y}{x} = k \text{ or } y = kx.$$

The number k is the **constant of variation.**

Connections **Meteorology**

Example 4 *Creating a Linear Model for Direct Variation*

Hailstones are formed when frozen raindrops are caught in updrafts and carried into high clouds containing supercooled water droplets. The radius, r (in inches), of a hailstone varies directly with the time, t (in seconds), that the hailstone is in the high cloud. After a hailstone has been in a certain cloud for 10 seconds, the radius of the hailstone is 0.5 inch. Write a linear model that gives the radius of the hailstone in terms of the time, t.

Hailstones can be very large. For instance, on July 12, 1984, hailstones as big as tennis balls fell in a storm in Munich, Germany.

Solution

$r = kt$ *Model for direct variation*

$0.5 = k(10)$ *Substitute 0.5 for r, 10 for t.*

$0.05 = k$ *Divide both sides by 10.*

The model is $r = 0.05t$. (To grow to the size of a tennis ball, with a radius of about 1.25 inches, a hailstone would have to remain in the cloud for about 25 seconds.) ■

In 1973, Barbara Jordan became the first African-American woman from a southern state to serve in the United States Congress. The graph shows that between 1970 and 1990 the number of African-American women holding elective office increased at a linear rate.

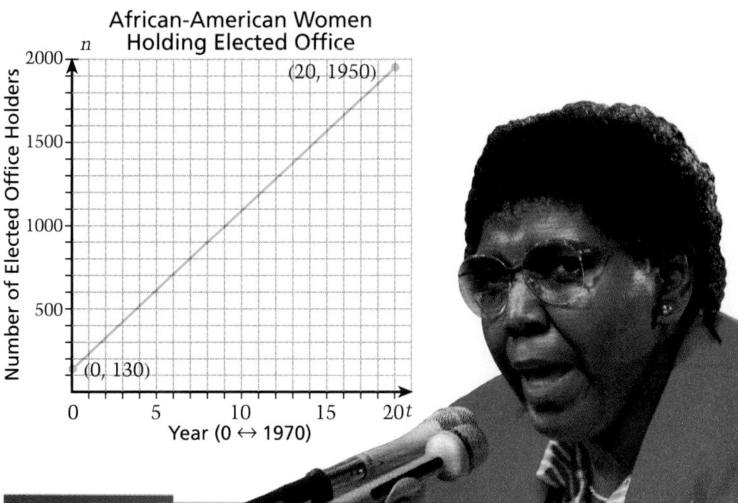

Real Life
Political Science

Example 5 *Writing a Linear Model*

In 1970, 130 African-American women held elected offices. By 1990, the number had increased to 1950. Write a linear model for the number, n, of African-American women who held elected offices between 1970 and 1990. (Let $t = 0$ represent 1970.) ***(Source: Joint Center for Political and Economic Studies)***

Solution You can find two points on the line.

In 1970, the value of n was 130: one point is (0, 130).

In 1990, the value of n was 1950: second point is (20, 1950).

The n-intercept, b, is 130. The slope of the line is

$$m = \frac{1950 - 130}{20 - 0} = \frac{1820}{20} = 91.$$

Using the slope-intercept form, the equation is $n = 91t + 130$.

Communicating about ALGEBRA

SHARING IDEAS about the Lesson

A. If x and y vary directly and are related by the linear equation $y = kx$, is (0, 0) a point on the graph? Explain. Yes, because $0 = k(0)$

B. What is the slope of the line? In direct variation problems, what is another name for the slope? k, constant of variation

C. Suppose that $y = 24$ when $x = 3$. What is the value of k? Explain your steps. 8; $\frac{y}{x} = k$, $\frac{24}{3} = 8$

Check Understanding

1. What is the most commonly used form of a linear equation?
   The slope-intercept form $y = mx + b$
2. Given the slope of a line, can the equation of the line be written? Explain.
   The slope alone would not determine a unique line.
3. Given the two points on a line, can the equation of the line be written?
   Yes, use-point-slope form.
4. Name two forms in which the equation of a line can be expressed.
   Slope-intercept form $y = mx + b$, standard form $Ax + By = C$.

Communicating about ALGEBRA

EXTEND *Communicating*
Many real-life situations represent examples of direct variation. For example, the water pressure on an object submerged in the ocean and the depth at which it is submerged, vary directly. Ask students to identify relationships studied in other classes (such as biology, chemistry, economics, etc.) that vary directly.

EXERCISE Notes

ASSIGNMENT GUIDE

Basic/Average: Ex. 1–4, 5–25 odd, 28–29, 33–37 odd, 38–41, 43–49 odd, 52

Above Average: Ex. 1–4, 9–25 odd, 27–30, 38–41, 43–51 odd, 52

Advanced: Ex. 1–4, 9–25 odd, 26–30, 38–41, 48–52

Selected Answers
Exercises 1–4, 5–49 odd

More Difficult Exercises
Exercises 26, 27, 30, 50, 51

Guided Practice

EXTEND Ex. 3 Present the intercept form of the equation of the line

$$\frac{x}{a} + \frac{y}{b} = 1$$

where a is the value of the x-intercept and b is the value of the y-intercept.

Answers

1. Use the point-slope form of the equation, $y - y_1 = m(x - x_1)$, where m is the slope and x_1 and y_1 are the coordinates of the given point.
2. Use the slope-intercept form of the equation, $y = mx + b$, where m is the slope and b is the y-intercept.
3. Use the coordinates of the intercepts to find the slope; then use the point-slope form of the equation with the slope just found and the coordinates of either of the two points.
4. Answers vary; for example, distance and miles per hour.

10. $y = 3x + 6$

EXERCISES

Guided Practice

CRITICAL THINKING about the Lesson See margin.

1. You are given the slope of a line and a point on the line. How can you find an equation of the line?
2. You are given the slope of a line and the y-intercept of the line. How can you find an equation of the line?
3. You are given the x-intercept and y-intercept of a line. How can you find an equation of the line?
4. Give a real-life example of two variable quantities that vary directly.

Independent Practice

In Exercises 5–10, write an equation of the line.

5. 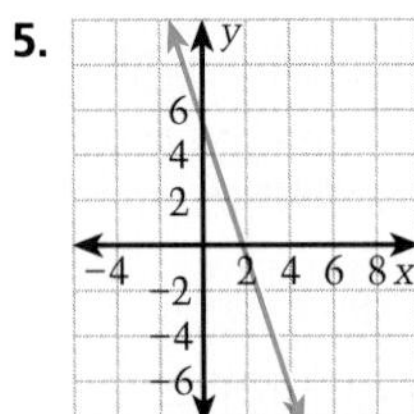

5. $y = -3x + 6$
6. $y = \frac{1}{2}x - 2$
7. $y = \frac{4}{5}x + 4$

6. 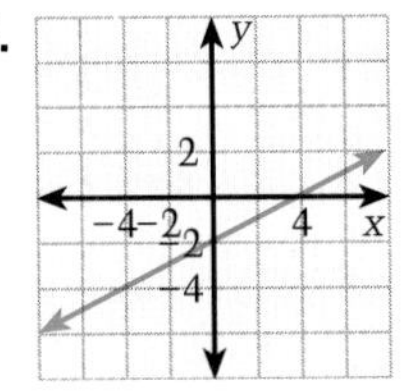

7. 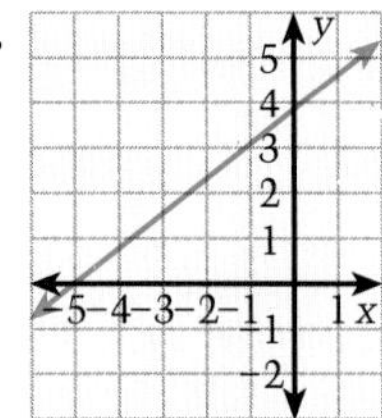

8. 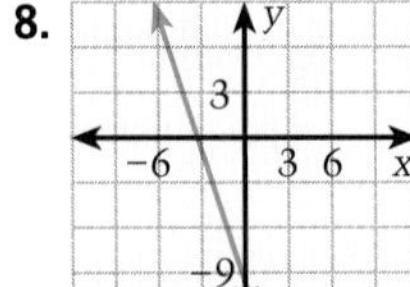

8. $y = -3x - 9$
9. $y = x + 10$
10. $y = 3x + 6$

9. 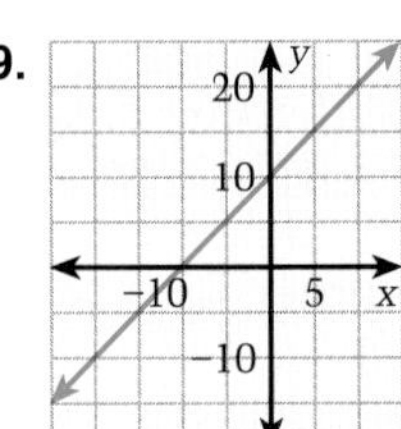

10. 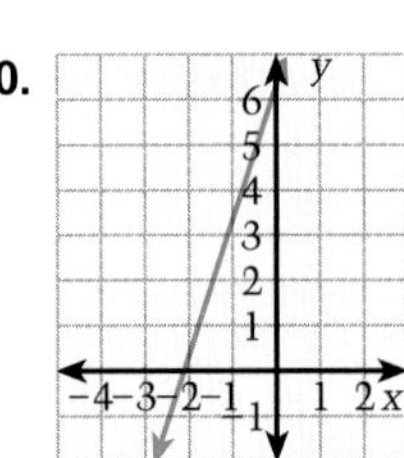

In Exercises 11–16, write an equation of the line containing the given point that has the given slope.

11. $(0, 3)$, $m = -\frac{1}{2}$ $y = -\frac{1}{2}x + 3$

12. $(4, 0)$, $m = 4$ $y = 4x - 16$

13. $(-8, 1)$, $m = 10$ $y = 10x + 81$

14. $(6, 7)$, $m = \frac{2}{3}$ $y = \frac{2}{3}x + 3$

15. $\left(\frac{1}{2}, -2\right)$, $m = -\frac{4}{3}$ $y = -\frac{4}{3}x - \frac{4}{3}$

16. $(3, -4)$, $m = \frac{2}{5}$ $y = \frac{2}{5}x - \frac{26}{5}$

In Exercises 17–19, write an equation of the line passing through the two points.

17. $(7, 6)$ and $(10, 15)$ $y = 3x - 15$

18. $(4, 11)$ and $(5, 9)$ $y = -2x + 19$

19. $(-9, 9)$ and $(0, 1)$ $y = -\frac{8}{9}x + 1$

In Exercises 20–25, the values of x and y vary directly and one pair of values for x and y are given. Write an equation that relates the variables.

20. $x = 2$, $y = 5$ $y = \frac{5}{2}x$

21. $x = -4$, $y = 6$ $y = -\frac{3}{2}x$

22. $x = -5$, $y = 10$ $y = -2x$

23. $x = 12$, $y = 3$ $y = \frac{1}{4}x$

24. $x = 0.1$, $y = 0.9$ $y = 9x$

25. $x = 1$, $y = 0.5$ $y = 0.5x$

26. $m = \frac{1}{2}h$; half your height; $\frac{34\frac{1}{2}}{69} = \frac{1}{2} = h$

26. *Mirror Length* To be able to see your complete reflection in a flat mirror that is hanging on a vertical wall, the mirror must have a minimum length of m inches. (Moving closer or farther back *doesn't* change the amount you see.) The value of m varies directly with your height, h (in inches). A person 69 inches tall requires a $34\frac{1}{2}$-inch mirror. Find a linear model for the minimum mirror length in terms of the person's height. What is the minimum length that you need to see your own full reflection? Explain. **See below.**

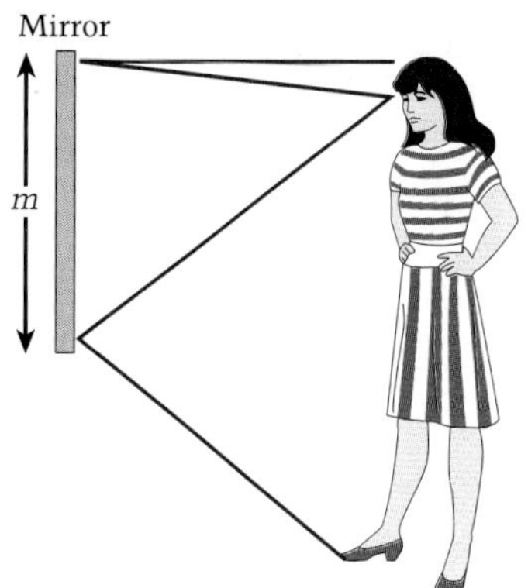

27. *Burning Calories* The number of Calories, c, a person burns and the time, t (in minutes), the person spends performing an activity vary directly. A 150-pound person can burn off 75 Calories (one cup of lowfat milk) by sitting in class for 50 minutes. Write a linear model that relates c and t. How long must a 150-pound person sit in class to burn off 545 Calories (a chicken pot pie)? **See below.**

28. *Bookstore Sales* For 1986 through 1991, retail sales of American bookstores increased at a rate that was approximately linear. In 1986, the retail sales were \$4.9 billion. In 1991, the retail sales were \$7.9 billion. Write a linear model for the retail sales, s (in billions of dollars), of American bookstores from 1986 to 1991. Let $t = 0$ represent 1986. **See below.**
(Source: American Booksellers Association)

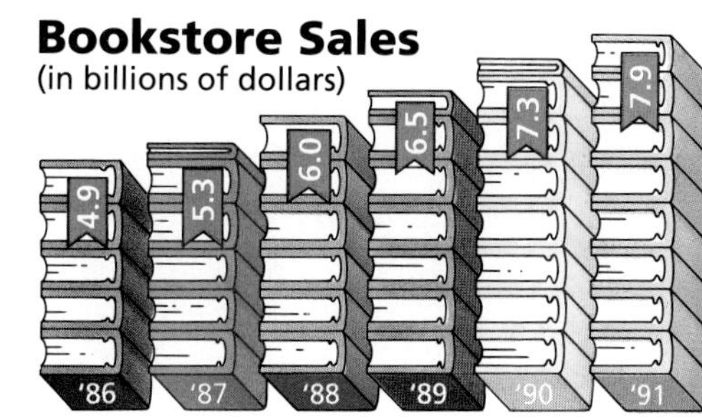

29. *Magazine Titles* From 1980 through 1991, the number, m, of magazines published in the United States increased at an approximately linear rate. In 1980, 1456 magazines were published and in 1991, 2276. Write a linear model for the number of magazines being published. (Let $t = 0$ represent 1980.) How many would you expect to be published in 1995? ***(Source: Wilkofsky Gruen Associates)*** **See below.**

Running the IRS **In Exercises 30 and 31, use the following information.**

The graph at the right shows the cost, C (in billions of dollars), to run the Internal Revenue Service from 1980 to 1990.
(Source: Internal Revenue Service)

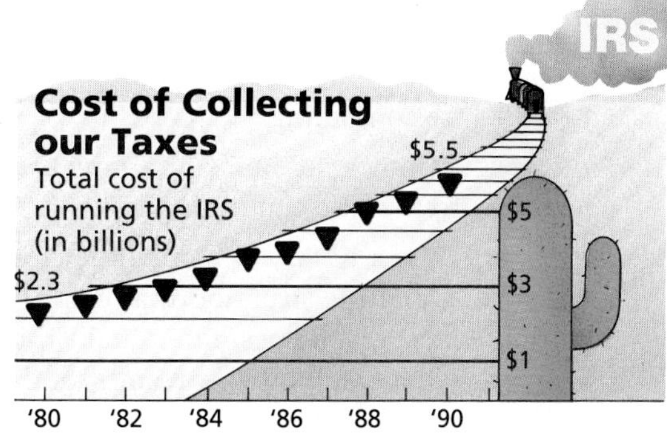

30. Write a linear model for the average cost. Let $t = 0$ represent 1980 and estimate the cost of running the IRS in 1995.

31. In 1989, the IRS collected about 9.5 trillion dollars. What percent of this was spent running the IRS? **≈0.055%**

30. $C = 0.32t + 2.3$, **\$7.1 billion**

27. $c = \frac{3}{2}t$; **6 hours,** $3\frac{1}{3}$ **minutes** **28.** $s = 0.6t + 4.9$ **29.** $m = \frac{820}{11}t + 1456$, **2574**

Independent Practice

▶ **Ex. 5–10** Encourage students to use the slope-intercept form as modeled in Example 2 .

EXTEND Ex. 11–16 You can derive the slope-intercept form for the equation of a line given its y-intercept $(0, b)$ and any point (x, y) on the line.

$$m = \frac{y-b}{x-0}$$
$$m = \frac{y-b}{x}$$
$$mx = y - b$$
$$mx + b = y$$
$$y = mx + b$$

▶ **Ex. 17–19** Encourage students to express their final answers in slope-intercept form.

▶ **Ex. 20–25** In direct variation models, the constant of variation is the slope and the y-intercept will always be 0.

EXTEND Ex. 28 Ask students to predict the bookstore sales in the year 2000 when $t = 14$ and interpret the results. **\$13.3 billion assuming an increasing population, an increasing production of books, and increasing book prices. If any of these variables change, then the linear model would be invalid.**

▶ **Ex. 29** TECHNOLOGY Use a graphics calculator to connect the visual to the symbolic by graphing the determined linear equation. The 1995 value ($t = 15$) can be verified either by substituting into the linear equation or by using the Trace feature of the graphics calculator.

Ex. 49 Assign this exercise for its geography connection.

Ex. 52 Assign this exercise as a class project. Ask your school librarian for a set of data resources for the class. Answers will vary.

Enrichment Activities

These activities require students to go beyond lesson goals.

In Example 5 and Exercise 28, you applied linear models to political science and bookstore sales.

1. Why were linear models appropriate for each real-life application?
   You were given two data points.
2. In each case, If the trends continue, what can be predicted for the year 2000?
   You should expect bookstore sales of \$13.3 billion and you should expect 2860 African-American women to hold public office.
3. Interpret your results; that is, state how this information impacts decisions about Afro-American women hoping to hold public office, or potential bookstore sales.
4. **WRITING** Write a paragraph on the limits of linear models. (*Hint:* Consider what happens to the model in the years preceding the given data and the years beyond the given data.)
   For many applications, the model is only accurate for the years listed because it was generated using those points.

Integrated Review

In Exercises 32–37, find the slope of the line containing the points.

32. $(1, 6), (6, 1)$ -1 **33.** $(3, 9), (0, -6)$ 5 **34.** $(8, -3), (7, 8)$ -11

35. $(2, 5), (9, 4)$ $-\frac{1}{7}$ **36.** $(6, 1), (-8, 2)$ $-\frac{1}{14}$ **37.** $(7, 2), (3, 2)$ 0

In Exercises 38–41, match the equation with its graph.

38. $y = -x + 4$ c **39.** $y = -\frac{1}{2}x - 6$ b **40.** $y = 2x - 6$ d **41.** $y = -2x + 3$ a

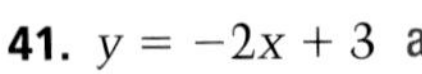

a.

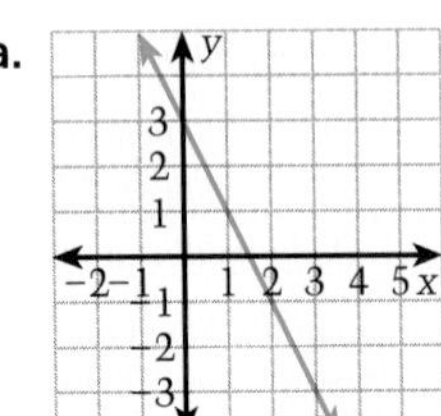

b.

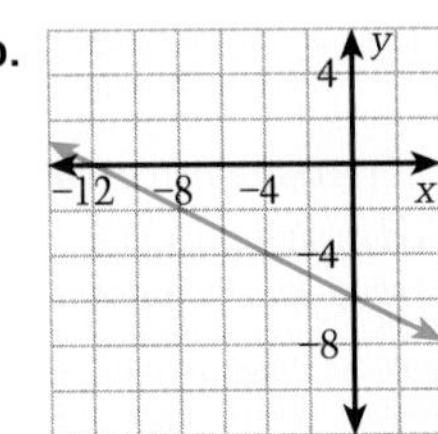

c.

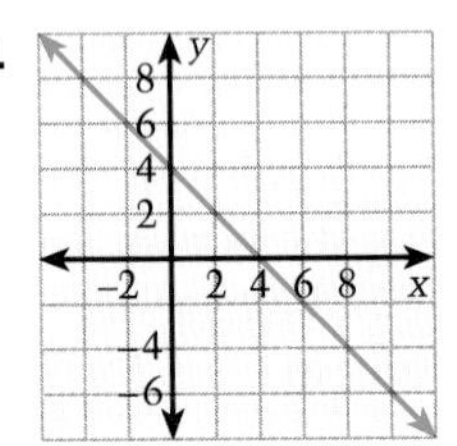

d.

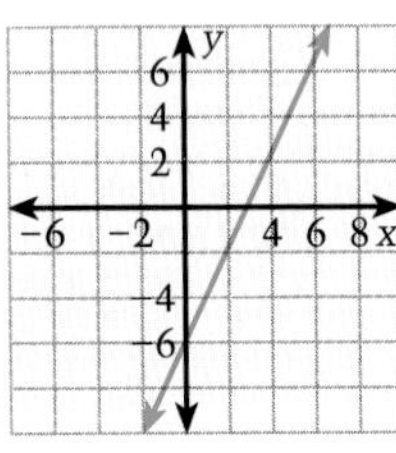

In Exercises 42–47, find the *x*-intercept and *y*-intercept of the line.

42. $4x + 2y = 7$ $\frac{7}{4}, \frac{7}{2}$ **43.** $-x + 8y = 21$

44. $9x - y = 18$ $2, -18$ **45.** $5x - 9y = 5$

46. $-12x + 8y = 72$ $-6, 9$ **47.** $-3x - 6y = 10$

48. *Powers of Ten* Which is the best estimate of the number of times your heart beats in an hour? a

a. 4×10^3 **b.** 4×10^4

c. 4×10^5 **d.** 4×10^6

49. *Powers of Ten* Which is the best estimate of the ratio of the area of Africa to the area of the United States? c

a. 3×10^{-2} **b.** 3×10^{-1}

c. 3×10^0 **d.** 3×10^1

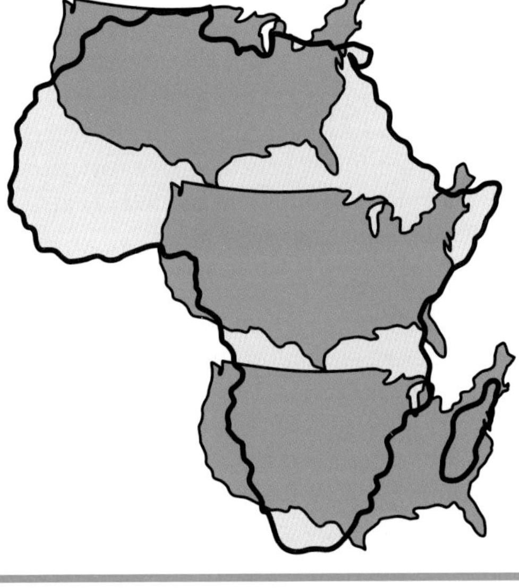

Exploration and Extension

50. Find an equation of the line containing the point $(4, 6)$ that is parallel to the line containing the points $(6, -6)$ and $(10, -4)$. $y = \frac{1}{2}x + 4$

51. Find an equation of the line containing the point $(1, -1)$ that is perpendicular to the line containing the points $(4, 2)$ and $(-4, 0)$. $y = -4x + 3$

52. *Research* Use a library or some other reference source to find an example of a real-life situation that can be modeled by a linear equation whose graph does not contain the origin. Describe the situation, sketch a graph of your results, and list the source of your information. See margin.

43. $-21, \frac{21}{8}$ **45.** $1, -\frac{5}{9}$ **47.** $-\frac{10}{3}, -\frac{5}{3}$

Mid-Chapter SELF-TEST

Take this test as you would take a test in class. The answers to the exercises are given in the back of the book.

1. Plot the points in a coordinate plane: $(-3, 5)$, $(4, -1)$, $(0, 2)$. **(2.1)** See margin.

2. Construct a table of values for $y = -4x + 12$. (Use $-3, -2, -1, 0, 1, 2, 3$ as x-values.) **(2.1)**

x	-3	-2	-1	0	1	2	3
y	24	20	16	12	8	4	0

In Exercises 3–5, find the slope of the line containing the points. (2.2)

3. $(2, 5), (4, 8)$ $\frac{3}{2}$

4. $(-3, 4), (2, -2)$ $-\frac{6}{5}$

5. $(-2, 5), (4, 5)$ 0

6. An airplane is flying from Owensboro, Kentucky, straight to Cincinnati, Ohio. At 1:40 P.M., the plane is 120 miles from Cincinnati. At 2:20 P.M., the airplane is 30 miles from Cincinnati. What is the average speed of the airplane? **(2.2)** 135 mph

In Exercises 7 and 8, state whether the lines are parallel, perpendicular, or neither. (2.2)

7. $2x - 4y = 8$, $-x + 2y = 4$ parallel

8. $2x - 4y = 8$, $2x + y = 4$ perpendicular

In Exercises 9 and 10, find the intercepts of the line. (2.3)

9. $-4x + 6y = 36$ x-intercept: -9, y-intercept: 6

10. $y = -5x + 20$ x-intercept: 4, y-intercept: 20

In Exercises 11 and 12, state the slope of the line and the y-intercept. (2.3)

11. $3x - 2y = 4x + 5$ $-\frac{1}{2}, -\frac{5}{2}$

12. $0 = 5x - 4y + 40$ $\frac{5}{4}, 10$

In Exercises 13 and 14, sketch the line. (2.3) See margin.

13. $4x + 2y = 9$

14. $y = -\frac{2}{3}x + 3$

15. Write an equation of the line containing $(4, -2)$ and $(-2, 0)$. **(2.4)** $y = -\frac{1}{3}x - \frac{2}{3}$

16. Write an equation of the line with a slope of $\frac{1}{4}$ and a y-intercept of 4. **(2.4)** $y = \frac{1}{4}x + 4$

17. Write an equation of the line containing $(3, -2)$ that has a slope of $\frac{1}{5}$. **(2.4)** $y = \frac{1}{5}x - \frac{13}{5}$

18. In 1980, Americans made an average of 2.5 phone calls per month to friends and relatives living more than 100 miles away. In 1990, the average was 4.1 phone calls per month. Estimate the average number of such phone calls per month for 1993. ***(Source: Roper Poll)*** **(2.2)** 4.58

19. The area, A (in square inches), of a rectangle varies directly with its width, w (in inches). When the width is 4 inches, the area is 12 square inches. Write an equation that relates A and w. **(2.4)**

20. Use the graph at the right to find the average rate of change per year in individual contributions to public television from 1981 to 1990. ***(Source: Public Broadcasting Service)*** **(2.2)** $2 per year

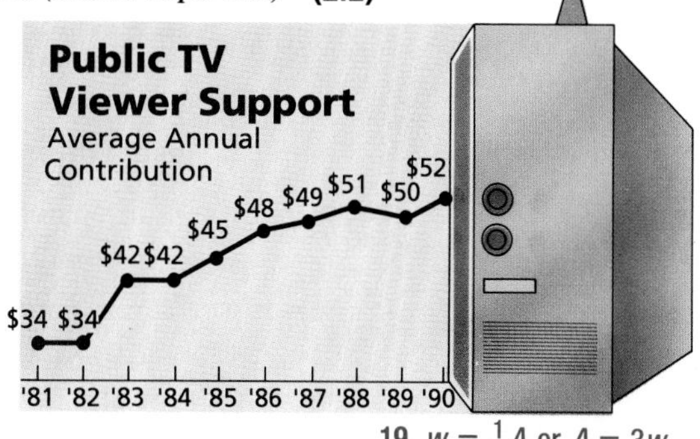

19. $w = \frac{1}{3}A$ or $A = 3w$

Answers

1.

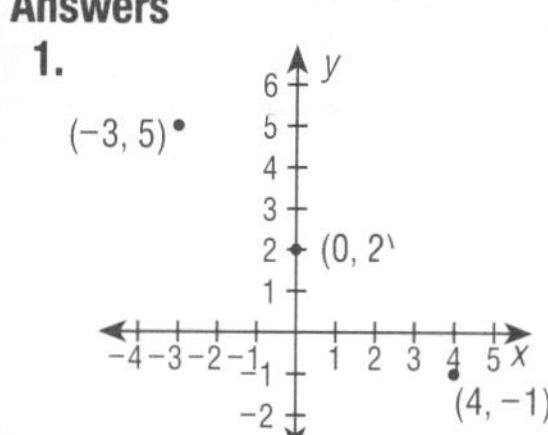

13.

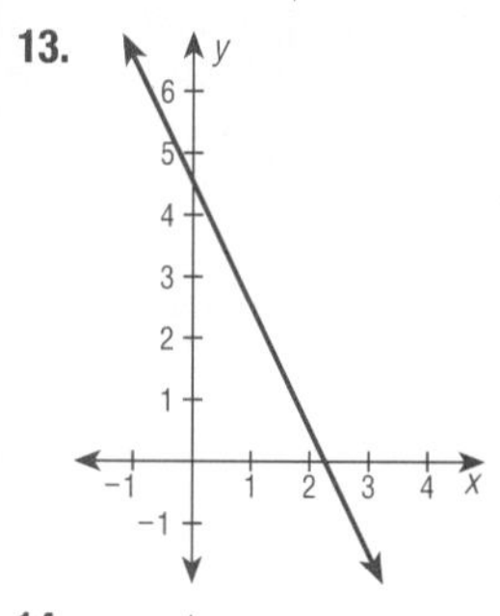

14.

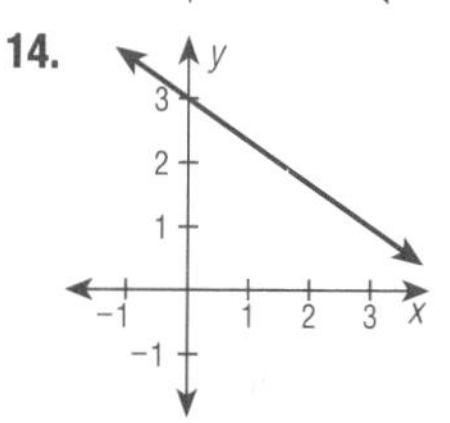

Problem of the Day

How many times will the digit 2 appear in numbering the pages of a book of 250 pages? 106

ORGANIZER

Warm-Up Exercises

1. Sketch the graphs of the following relationships.
   a. $x + y = 4$
   b. $3x - 5y = 2$
   c. $y = 3$
   d. $x = 8$

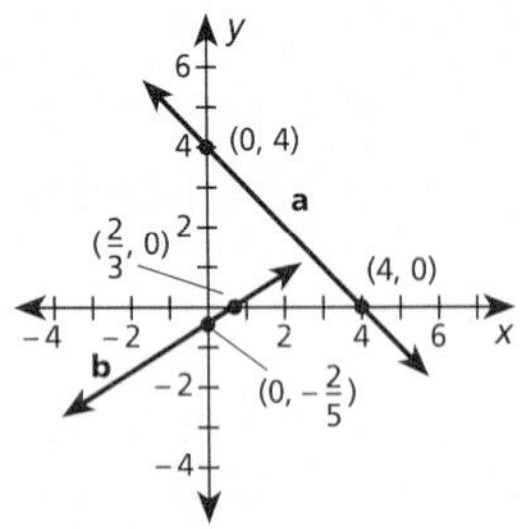

   c. The graph is a horizontal line through (0, 3).
   d. The graph is a vertical line through (8, 0).

2. Determine if the given points are solutions of the given linear equations.
   a. (3,−5), (−2, 13); $y = -4x + 7$
   b. (3,−1.5), (12, 9); $y = \frac{5}{6}x - 1$
   a. (3,−5), yes; (−2, 13), no;
   b. (3,−1.5), no; (12, 9), yes

Lesson Resources

Teaching Tools
- Transparencies: 7, 8, 10, 11
- Problem of the Day: 2.5
- Warm-up Exercises: 2.5
- Answer Masters: 2.5

Extra Practice: 2.5

2.5 Graphing Linear Inequalities in Two Variables

What you should learn:

Goal 1 How to graph a linear inequality in two variables

Goal 2 How to use a linear inequality in two variables to model real-life situations

Why you should learn it:

You can model many real-life problems with linear inequalities in two variables, such as the "safe surfacing" times and depths for scuba divers.

Goal 1 Graphing Inequalities

A **linear inequality** in x and y is an inequality that can be written in one of the following forms

$$ax + by < c,\ ax + by \le c,\ ax + by > c,\ ax + by \ge c$$

where a and b are not both 0. An ordered pair (x, y) is a **solution** of a linear inequality in x and y if the inequality is true when a and b are substituted for x and y, respectively. For instance, $(-2, 12)$ is a solution of $5x + y < 3$ because $5(-2) + 12 = 2$, which is less than 3.

Example 1 *Checking Solutions of Inequalities*

Check whether the ordered pairs are solutions of $3x + 2y \ge 2$.

a. (0, 0) **b.** (2,−1) **c.** (0, 2)

Solution

	(x, y)	Substitute	Conclusion
a.	(0, 0)	$3(0) + 2(0) = 0 < 2$	(0, 0) is not a solution.
b.	(2,−1)	$3(2) + 2(-1) = 4 \ge 2$	(2, −1) is a solution.
c.	(0, 2)	$3(0) + 2(2) = 4 \ge 2$	(0, 2) is a solution. ■

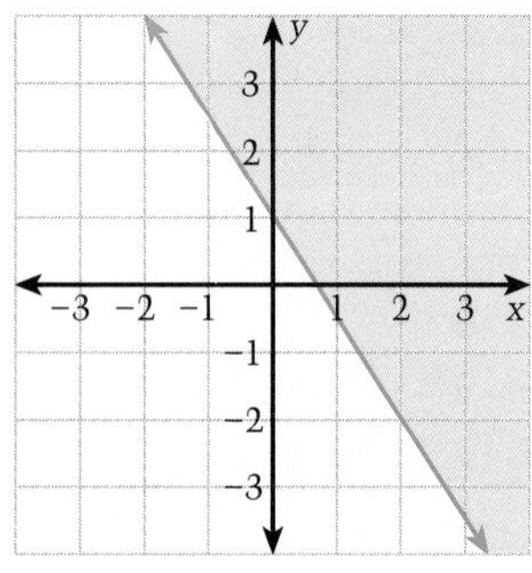

The **graph of an inequality** is the graph of the solutions of the inequality. For instance, the graph of $3x + 2y \ge 2$ is shown at the left. Every point in the shaded region is a solution of the inequality, and every other point in the plane is not.

To sketch the graph of a linear inequality in two variables, first sketch the line given by the corresponding equation. (Use a *dashed* line for inequalities with $<$ or $>$ and a *solid* line for inequalities with $\le$ or $\ge$.) This line separates the coordinate plane into two **half-planes.**

- In one of the half-planes, all points are solutions of the inequality.
- In the other half-plane, no point is a solution.

You can decide whether the points in an entire half-plane satisfy the inequality by testing *one* point in the half-plane.

Example 2 *Sketching the Graph of a Linear Inequality*

Sketch the graphs of the linear inequalities.

a. $y \geq 3$ **b.** $x < 2$

Solution

a. For $y \geq 3$, sketch the *(solid)* horizontal line

$y = 3.$ *Horizontal line*

Next test a point. The origin (0, 0) is *not* a solution and it lies below the line. Therefore, the graph of $y \geq 3$ is all points on or above the line $y = 3$.

b. For $x < 2$, sketch the *(dashed)* vertical line

$x = 2.$ *Vertical line*

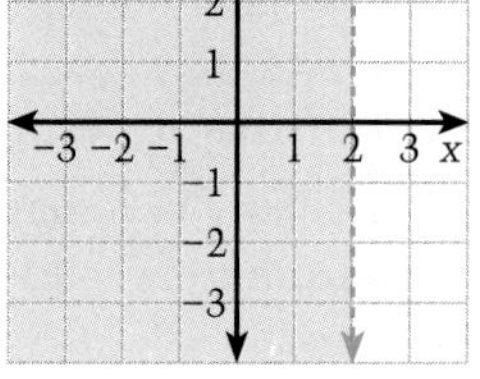

Next test a point. The origin (0, 0) *is* a solution, and it lies to the left of the line. Therefore, the graph of $x < 2$ is all points to the left of the line $x = 2$. ■

To sketch a line that is neither horizontal nor vertical, you may want to write the line in slope-intercept form first.

Example 3 *Writing in Slope-Intercept Form*

Sketch the graph of $-x + y < 5$.

Solution The corresponding equation is $-x + y = 5$. To sketch this line, first write the equation in slope-intercept form, $y = x + 5$.

Then sketch the (dashed) line that has a slope of 1 and a y-intercept of 5. The origin (0, 0) *is* a solution of $-x + y < 5$, and it lies below the line. Therefore, the graph is all points below the line. (Try checking another point below the line. Any one you choose will satisfy the inequality.)

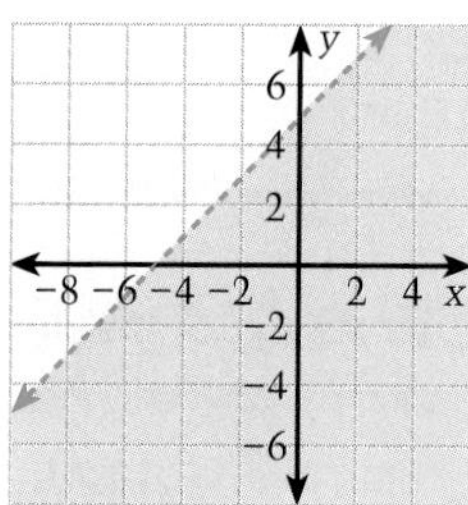

■

Notice that the origin is used as a test point for each inequality. You can use any point that is not on the line. The origin is often convenient because it is easy to evaluate expressions in which 0 is substituted for each variable.

LESSON Notes

Example 1

Discuss the similarity between a solution of a linear equation and a solution of a linear inequality. Ask students to construct a definition of solution that holds for either linear equations or inequalities.

Examples 2–3

To reinforce the test-point procedure, have students verify that various points on the coordinate plane are, or are not, solutions of the given linear inequalities.

Example 4

Does a diver with a depth and time described by the point (85, 45) need to decompress? Does a diver with a depth and time of (125, 12) need to decompress? yes; no

Extra Examples

Here are additional examples similar to **Examples 1–3.**

1. Which ordered pairs are solutions of the inequalities?

a. (3, 4), (−2, 0), (7, −1); $4x - 2y > 7$ (7, −1)

b. (−1, −1), (−6, 8), (11, 9); $6x + 5y \leq 13$ (−1, −1), (−6, 8)

2. Sketch the graph of the following inequalities.

a. $x \geq -3$ b. $y < 6$

c. $4x - 2y > 7$

a. The graph is the region to the right of the solid line $x = -3$.

b. The graph is the region below the dashed line $y = 6$.

c.

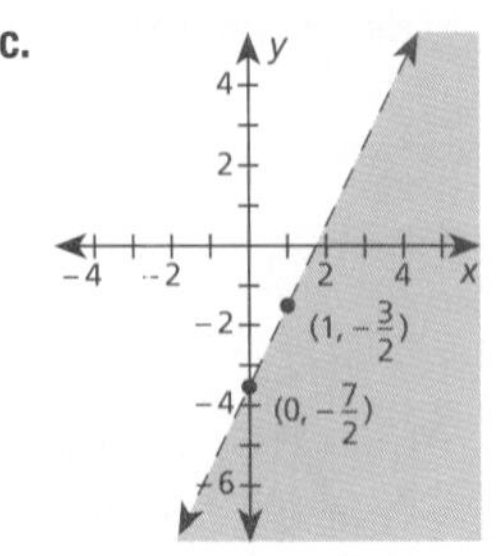

Check Understanding

1. What is a half-plane of a coordinate plane?
A region of the coordinate plane on either side of a line.
2. What is a solution of a linear inequality?
See the top of page 94.
3. What is the connection between graphing linear equations and linear inequalities?
See Examples 2 and 3.
4. In how many forms does a linear inequality exist?
See the top of page 94.

***Communicating* about ALGEBRA**

EXTEND *Communicating*
Ask students to develop and justify a procedure for graphing linear inequalities without using a test point. For example, for the inequality, $y < 3x - 6$, its graph is the region *below* the dashed line, $y = 3x - 6$. For $y \geq 3x - 6$, the graph is the region *above* and including the solid line. (Use a test point to check this assertion.)

The bends, or decompression sickness, is caused by a rapid decrease in pressure. Nitrogen bubbles form in body tissue and bloodstream, impairing the blood's circulation. To avoid the bends, scuba divers must carefully monitor the time they remain at certain depths.

Goal 2 Using Linear Inequalities in Real Life

Real Life
Scuba Diving

Example 4 *Using Linear Inequalities in Real Life*

The time, t (in minutes), that a scuba diver can remain at a depth, d (in feet), *without* having to decompress while surfacing is given by the following inequalities.

For depths between 60 and 90 feet: $t \leq 120 - d$

For depths between 90 and 130 feet: $t \leq 75 - \frac{1}{2}d$

Sketch the graph of these inequalities. Does a diver with a depth and time of (80, 50) need to decompress? How about a diver with a depth and time of (100, 20)?

(Source: U.S. Navy Standard Air Decompression Tables)

Time Limits for Scuba Divers

Decompression Required
Decompression Not Required
Time (in minutes): t — 10, 20, 30, 40, 50, 60
Depth (in feet): d — 20, 40, 60, 80, 100, 120

Solution The graph is shown at the left. The diver with a depth and time of (80, 50) was at a depth of 80 feet for 50 minutes. The maximum time allowed at a depth of 80 feet (without decompressing) is $120 - 80 = 40$ minutes. Therefore, this diver should use decompression stages when surfacing. The diver with a depth and time of (100, 20) does not need to decompress. Why? ■

Communicating about ALGEBRA

SHARING IDEAS about the Lesson See Additional Answers.

Sketch the graphs of the following. Explain your steps.

A. $3x \leq 4$ **B.** $-y \geq -2$ **C.** $2x - 3y < 6$

EXERCISES

Guided Practice

CRITICAL THINKING about the Lesson

1. Why is $ax + by < c$ called a *linear inequality*? Because its boundary is a line.
2. Is the ordered pair $\left(\frac{5}{6}, 0\right)$ a solution of $6x + y < 5$? Explain. No, $6\left(\frac{5}{6}\right) + 0$ is not less than 5.
3. Would you use a dashed line or a solid line for the graph of $ax + by < c$? for the graph of $ax + by \le c$? Explain. See below.
4. *True or False?* The graph of $y < 3x + 4$ is the set of all points in the coordinate plane that lie *below* the line $y = 3x + 4$. Explain. True, (0, 0) lies below the line and satisfies the inequality.

Independent Practice

In Exercises 5–12, check whether each ordered pair is a solution of the inequality.

5. $x + 2y \le -3$, $(0, 3)$, $(-5, 1)$ Only $(-5, 1)$ is.
6. $5x - y > 2$, $(-5, 0)$, $(5, 23)$ Neither is.
7. $12x + 4y \ge 3$, $(1, -3)$, $(0, 2)$ Only $(0, 2)$ is.
8. $\frac{1}{2}x - 7y \le 5$, $(20, 1)$, $(4, -2)$ Only $(20, 1)$ is.
9. $-8x - 3y < 5$, $(-1, 1)$, $(3, -9)$ Only $(3, -9)$ is.
10. $21x - 10y > 4$, $(2, 3)$, $(-1, 0)$ Only $(2, 3)$ is.
11. $-x - y \le -10$, $(-3, -7)$, $(5, 4)$ Neither is.
12. $4x + 5y < 6$, $\left(\frac{1}{2}, 1\right)$, $\left(\frac{1}{2}, -1\right)$ Only $\left(\frac{1}{2}, -1\right)$ is.

In Exercises 13–16, write an equation (in slope-intercept form) that corresponds to the given inequality.

13. $10x + 5y > 2$ $\quad y = -2x + \frac{2}{5}$
14. $13x - y < 6$ $\quad y = 13x - 6$
15. $-2x - 3y \le 1$ $\quad y = -\frac{2}{3}x - \frac{1}{3}$
16. $\frac{1}{2}x + \frac{3}{4}y \ge 4$ $\quad y = -\frac{2}{3}x + \frac{16}{3}$

In Exercises 17–24, sketch the graph of the inequality. See Additional Answers.

17. $x \le 4$
18. $-x > 3$
19. $-y < 50$
20. $y \le -\frac{1}{3}$
21. $12y > -6$
22. $2x \ge -\frac{5}{2}$
23. $7x \ge \frac{7}{3}$
24. $6y < 12$

In Exercises 25–28, match the inequality with its graph.

25. $-x + y > -2$ d
26. $-x + y \le 2$ b
27. $x + 2y > -4$ a
28. $-2x + y < 4$ c

a. 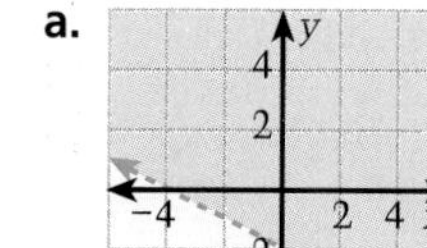

b. 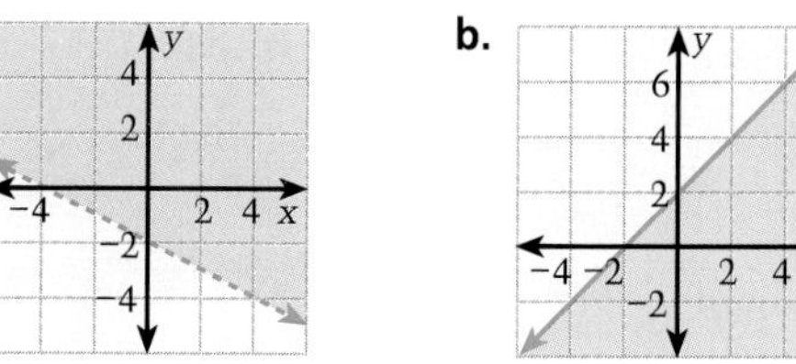

c. 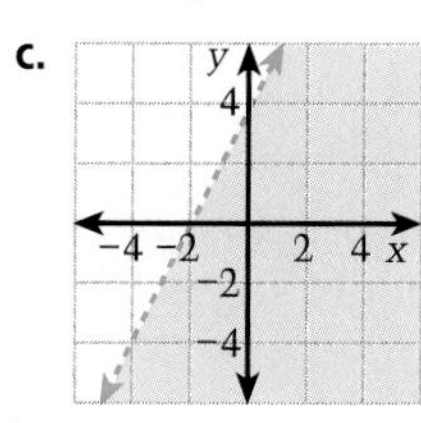

d. 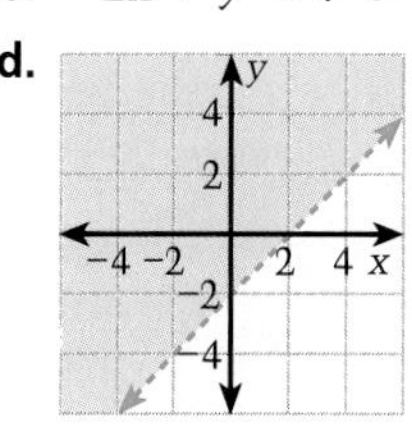

In Exercises 29–36, sketch the graph of the inequality. See Additional Answers.

29. $y < \frac{3}{4}x - 5$
30. $y \ge 2x + 10$
31. $4x + 12y > 3$
32. $-6x + y \ge 3$
33. $x - 5y \le 15$
34. $7x - 7y > -21$
35. $9x + 3y \le -1$
36. $2x + 4y < -8$

3. Dashed line, solid line; the solid line means the points on the line satisfy the inequality, the dashed line means the points on the line do not satisfy the inequality.

EXERCISE Notes

ASSIGNMENT GUIDE

Basic/Average: Ex. 1–4, 5–15 odd, 17–28, 29–39 odd, 40, 51–55, 59, 61

Above Average: Ex. 1–4, 7–15 odd, 17–28, 29–39 odd, 40–42, 51–55, 59, 61

Advanced: Ex. 1–4, 13–23 odd, 25–28, 29–37 odd, 38–42, 51–56, 59–63

Selected Answers
Exercises 1–4, 5–53 odd

Use **Mixed Review** as needed.

More Difficult Exercises
Exercises 41–42, 55–56, 61, 63

Guided Practice

Ex. 1–4 Use these exercises as an additional check for understanding before assigning the Independent Practice.

Independent Practice

Ex. 17–24 The special cases are modeled in Example 2.

Ex. 25–28 These exercises should be assigned as a group.

Answers

37. 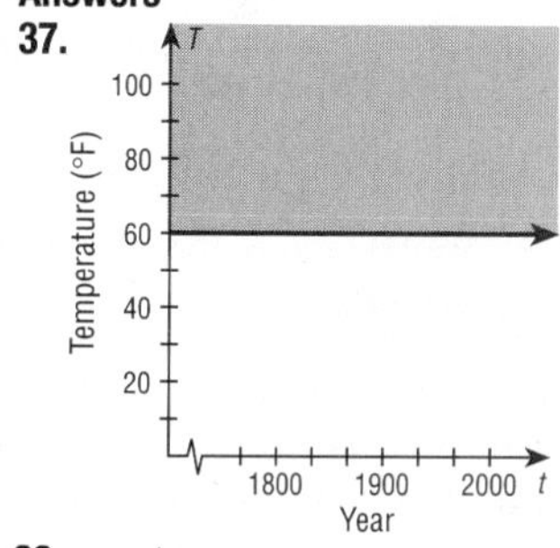

38. 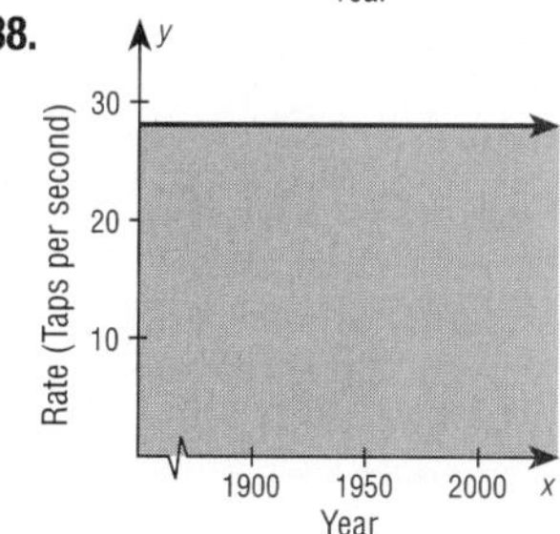

Ex. 39 Remind students that in this graph, the horizontal axis will be labeled p, and the vertical axis will be labeled t. Graphs usually show time along the horizontal axis.

Ex. 40 This is similar to the linear equality models of Lesson 2.4.

Ex. 41–42 In these exercises, help students translate the equations to ones they more easily recognize. For example: $y = -x + 220$ and $y = -0.75x + 165$.

Answers

39.

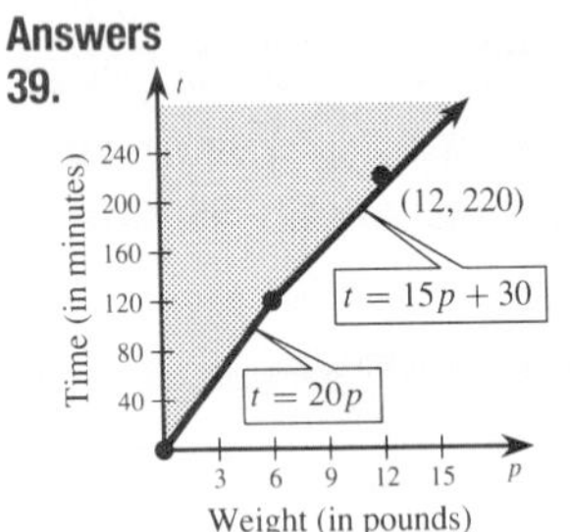

40.

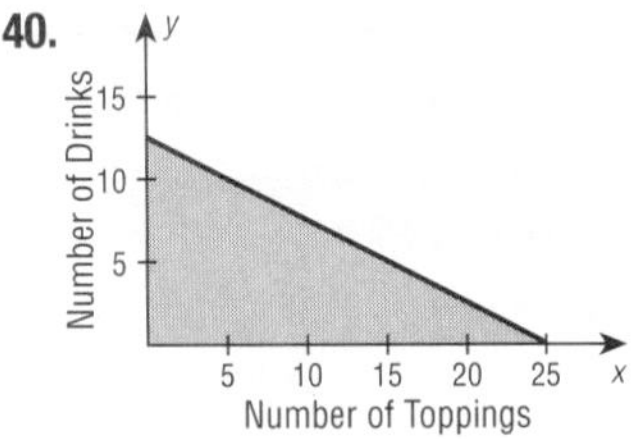

41.–42.

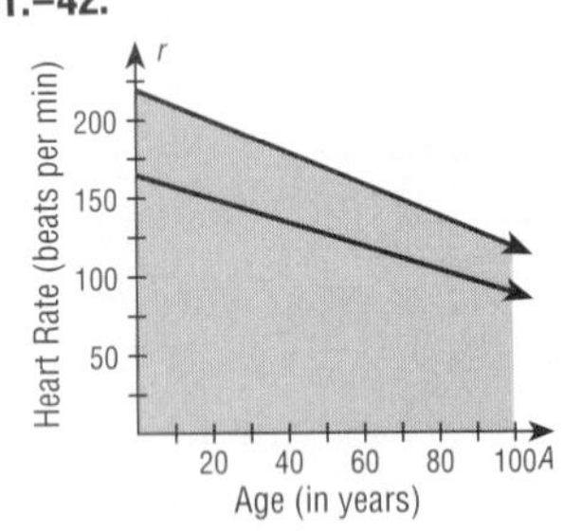

37. *San Juan* The lowest temperature ever recorded in San Juan, Puerto Rico, is 60°F. Write an inequality for T representing San Juan's recorded temperatures. Sketch the graph of the inequality in a coordinate plane. Use T on the vertical axis and t, representing years, on the horizontal axis. **See margin.**

38. *Tap Dancing* The fastest rate ever recorded for tap dancing in the United States is 28 taps per second by Michael Flatley of Palos Park, Illinois, on May 9, 1989. Let r represent the recorded rate (taps per second of a tap dancer). Write an inequality for r. Sketch the graph of the inequality in a coordinate plane. Use r on the vertical axis and t on the horizontal axis. ***(Source: Guinness Book of Records)*** **See margin.**

39. *Roasting a Turkey* The time, t (in minutes), that it takes at 350°F to roast a turkey weighing p pounds is given by these inequalities:

For a turkey up to 6 pounds: $t \geq 20p$

For a turkey over 6 pounds: $t \geq 15p + 30$ **(12, 220), yes**

Sketch the graphs of these inequalities. What are the coordinates for a 12-pound turkey that has been roasting for 3 hours and 40 minutes? Is this turkey fully cooked? **See margin.**

40. *Pizza and Soda Pop* You and some friends go out for pizza. Together you have \$26. You want to order two large pizzas with cheese at \$8 each. Each additional topping costs \$0.40, and each small soft drink costs \$0.80. Write an inequality that represents the number of toppings, x, and drinks, y, that your group can afford. Sketch the graph of the inequality. What are the coordinates for an order of 6 soft drinks and two large pizzas with cheese, each with 3 additional toppings? Is this a solution of the inequality? (Assume there is no sales tax.) **$x + 2y \leq 25$, (6, 6), yes. See margin.**

Getting a Workout **In Exercises 41 and 42, use the following information.**

The heart rate, r (in beats per minute), of a person in normal health is related to the person's age, A (in years). The relationship between r and A is given by the inequality $r \leq 220 - A$.

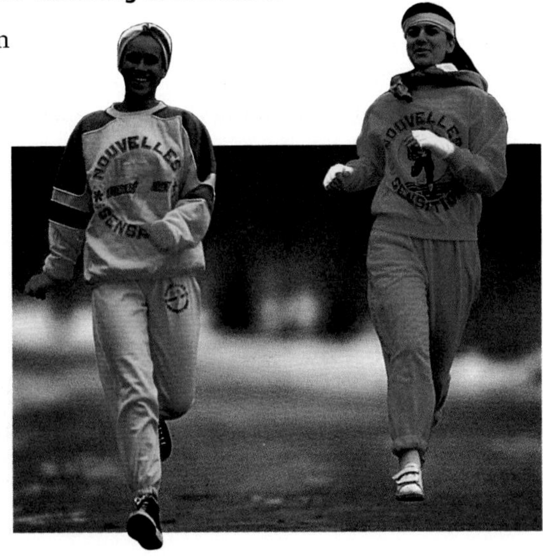

41. Sketch the graph of $r \leq 220 - A$. (Use A on the horizontal axis and use r on the vertical axis.) **See margin.**

42. Physiologists recommend that during a workout, a person should strive to increase his or her heart rate to 75% of the maximum rate for the person's age. Sketch the graph of $r = 0.75(220 - A)$ on the same coordinate plane used in Exercise 41. Write a short paragraph that describes how the graphs could be used. **See below.**

42. Locate your age on the lower line graph in order to read your corresponding normal heart rate; locate your age on the upper graph in order to read your corresponding recommended maximum heart rate after exercise.

Integrated Review

In Exercises 43–46, decide whether the ordered pair is a solution of the equation.

43. $3x - \frac{1}{2}y = 5$, $(-2, 0)$, $(6, 2)$ Neither is.

44. $x - 3y = 10$, $(0, 7)$, $(1, 2)$ Neither is.

45. $-11x + y = 1$, $(1, 10)$, $(4, 4)$ Neither is.

46. $-2x - 12y = 6$, $\left(5, \frac{1}{3}\right)$, $(-3, 0)$ Only $(-3, 0)$ is.

In Exercises 47–50, write the equation in slope-intercept form.

47. $14x + 2y = 1$ $y = -7x + \frac{1}{2}$

48. $6x - y = 4$ $y = 6x - 4$

49. $-9x - 3y = 12$ $y = -3x - 4$

50. $-4x + 16y = -2$ $y = \frac{1}{4}x - \frac{1}{8}$

In Exercises 51–54, sketch the graph of the equation. See Additional Answers.

51. $-x + y = 3$

52. $3x - y = -1$

53. $x + 2y = 6$

54. $-16x + 4y = -4$

Exploration and Extension

55. *Cutting Circles from a Square* You are cutting circles out of square pieces of sheet metal, 8 inches by 8 inches. How much of the sheet metal must you waste in each case if you cut 1, 4, 9, and 16 circles, as shown in the following diagram? The same: $\approx 13.7 \text{ in.}^2$

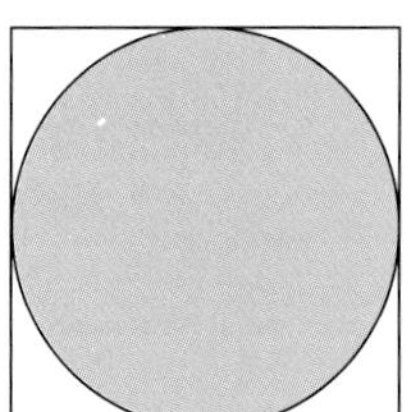

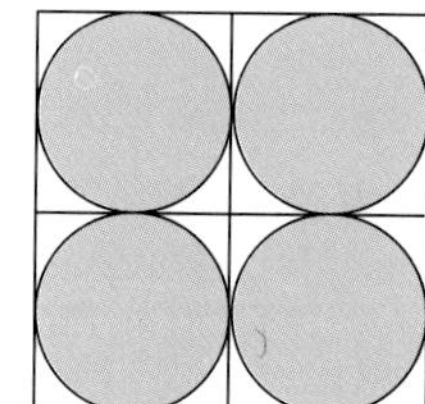

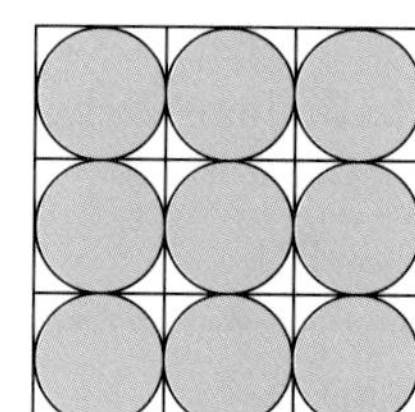

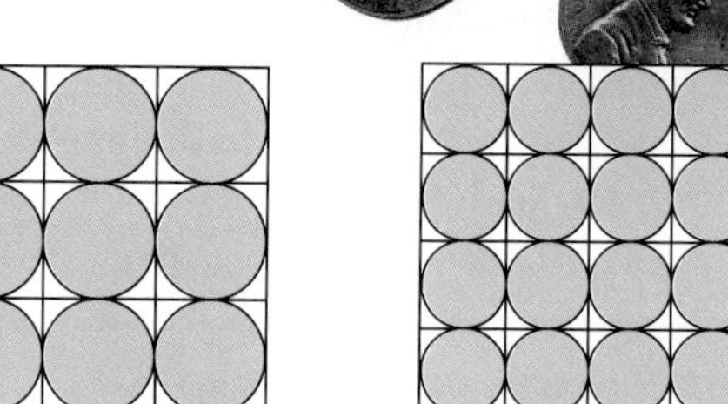

56. *Cutting More Circles from a Square* You are given a square piece of sheet metal that is $5\frac{1}{4}$ inches by $5\frac{1}{4}$ inches. You are asked to cut as many penny-size circles from the sheet as possible. How many can you cut? How much of the sheet metal must you waste? 52, $\approx 4.6 \text{ in.}^2$

TECHNOLOGY If students are using a graphics calculator, be sure to choose an appropriate scale for the graph in order to capture the intercepts in the viewing window.

EXTEND Ex. 41–42 Ask students to apply the equations and the graph to their age. For example, at age 16 the maximum heart rate is 204. Their aerobic workout heart rate should be 75% of 204, or 153. The point (16, 153) should be on the graph of the line from Exercise 42.

Ex. 51–54 These exercises review various graphing techniques.

Enrichment Activities

This activity requires students to go beyond lesson goals.

An Alternate Approach

The following program can be used on the TI-81 to compute the slope, *m*, of a line containing two points $A(x_1, y_1)$ and $B(x_2, y_2)$

```
Prgm # Slope
: ClrHome
: Disp "X1 = "
: Input A
: Disp "Y1 = "
: Input B
: Disp "X2 = "
: Input C
: Disp "Y2 = "
: Input D
: (D − B) → U
: (C − A) → L
: Disp "Slope Is"
: IF L = 0
: GOTO 1
: U/L → M
: Disp M
: End
: Lbl 1
: Disp "undefined"
```

Students with a Casio programmable graphing calculator should consult their manual for a similar program.

Use the program to verify the slopes you found in Exercises 32–37 on page 92.

Answers

61. $x < -1$, the solution of $2x + 2 < 0$ is all x-values of $y = 2x + 2$ that are below the x-axis.

62. $y \le -3x + 5$, $y \ge -3x - 4$, $y \le 2$, $y \ge -1$

63. The one below the line, test a point in either half-plane

Mixed Review

8.

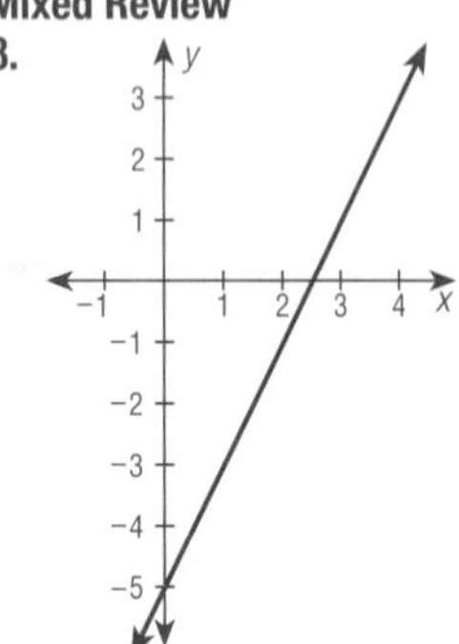

In Exercises 57–60, write the inequality whose graph is shown.

57. **58.**

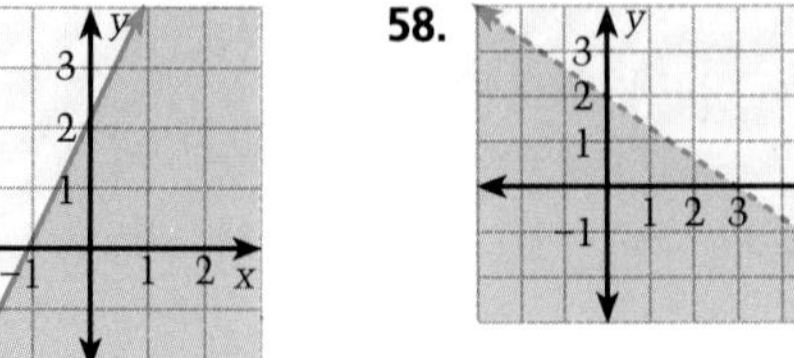

59.

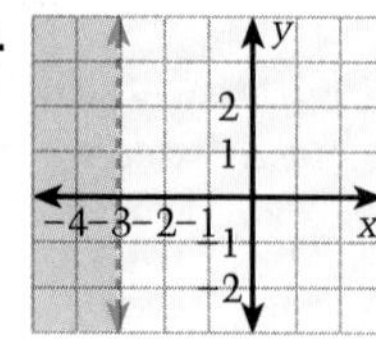

60.

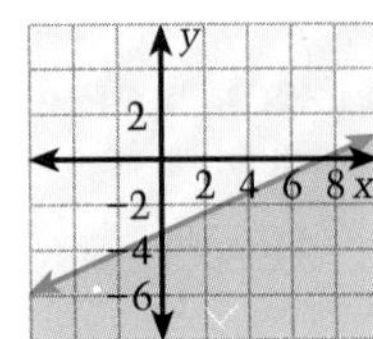

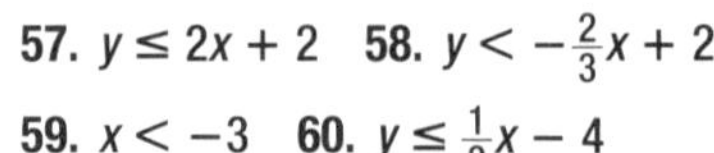

57. $y \le 2x + 2$ **58.** $y < -\frac{2}{3}x + 2$
59. $x < -3$ **60.** $y \le \frac{1}{2}x - 4$

61. For which values of x is the graph of $y = 2x + 2$ below the x-axis? Describe how you could use the graph of this line to graphically solve the inequality $2x + 2 < 0$. See margin.

62. Write four different inequalities whose intersection is the blue region shown at the right.

- Each point in the blue region must be a solution of all four inequalities.
- Each point that is not in the blue region must *not* be a solution of at least one of the four inequalities. See margin.

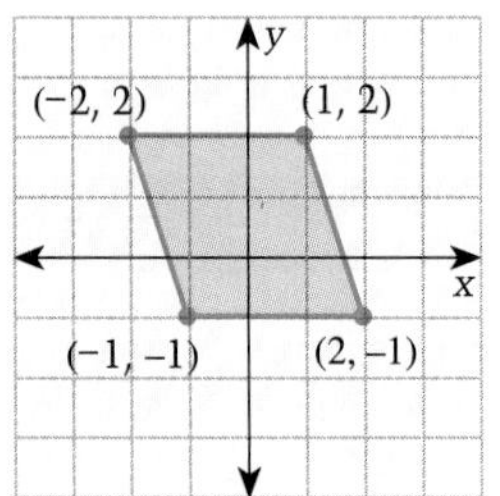

63. Graph the equation $y = 3x - 4$. For which half-plane is $y < 3x - 4$? Explain how you can determine which half-plane is the solution without graphing. See margin.

Mixed REVIEW

6. $-3x + 4$ **8.** See margin. **13.** $\frac{4}{2m^2 - m}$ **20.** $y = \frac{2}{5}x$

1. Write in decimal form: 6.13×10^{-3}. 0.00613

2. Divide $\frac{2}{3}$ by $\frac{4}{5}$. $\frac{5}{6}$

3. Write the reciprocal of $-\frac{3}{5}$. **(1.1)** $-\frac{5}{3}$

4. Write the opposite of $-\frac{3}{5}$. **(1.1)** $\frac{3}{5}$

5. Solve $3(x - 2) - 4x = -2(5 - x)$. **(1.3)** $\frac{4}{3}$

6. Simplify: $3(x - 2) - 4x + 2(5 - x)$. **(1.2)**

7. What is the ratio of 2 feet to 42 inches? $\frac{4}{7}$

8. Sketch the graph of $2x - y = 5$. **(2.1)**

9. Simplify: $\frac{5}{2}x - \frac{5}{4}x + x$. **(1.2)** $\frac{9}{4}x$

10. Divide $[9 - (-6)]$ by $[-2 - (-3)]$. **(1.1)** 15

11. Solve $|-3x + 4| = 12$. **(1.7)** $\frac{16}{3}, -\frac{8}{3}$

12. Solve $3x - 5 \le 2x - 7$. **(1.6)** $x \le -2$

13. Solve $8nm^2 - 4mn = 16$ for n. **(1.5)**

14. Thirty-four is what percent of eighty-five? 40%

15. Find the slope of the line $-5x + 3y = 8$. **(2.2)** $\frac{5}{3}$

16. Write in slope-intercept form: $4x - 12y = 18$. **(2.3)** $y = \frac{1}{3}x - \frac{3}{2}$

17. At a speed of 34 feet per second, how far will you travel in 1 minute? **(2.2)** 2040 ft

18. You traveled 72 miles in 84 minutes. What was your average speed (in miles per hour)? **(2.2)** ≈51.43 mph

19. Find an equation of the line passing through $(-4, 5)$ and $(3, -12)$. **(2.4)** $y = -\frac{17}{7}x - \frac{33}{7}$

20. x and y vary directly. When $x = 5$, $y = 2$. Find an equation relating x and y. **(2.4)**

100

2.6 Graphs of Absolute Value Equations

What you should learn:

Goal 1 How to graph absolute value equations

Goal 2 How to use graphs of absolute value equations to answer questions about real-life situations

Why you should learn it:

You can model some real-life images such as the Transamerica Pyramid in San Francisco, with graphs of absolute-value equations.

Goal 1 Graphing Absolute Value Equations

The absolute value of a number can be defined as follows.

$$|x| = \begin{cases} x, & \text{if } x > 0. \\ 0, & \text{if } x = 0. \\ -x, & \text{if } x < 0. \end{cases}$$

Example

$|2| = 2$

$|0| = 0$

$|-4| = -(-4) = 4$

In this lesson, you will learn to sketch the graph of absolute value equations. To begin, let's look at the graph of $y = |x|$. By constructing a table of values and plotting points, you can see that the graph is V-shaped and opens up. The corner point of the graph is its **vertex.** The vertex of this graph is (0, 0).

x	$y = \|x\|$
−4	4
−3	3
−2	2
−1	1
0	0
1	1
2	2
3	3
4	4

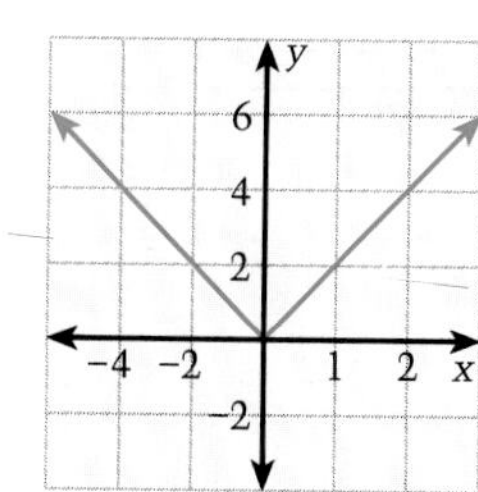

Other absolute value graphs in two variables are also V-shaped. Some open up and some open down. Here are some examples.

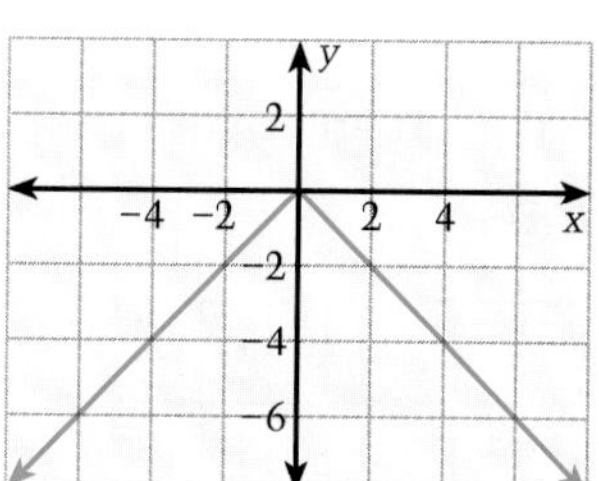

Graph of $y = -|x|$:
Vertex at (0, 0), opens down.

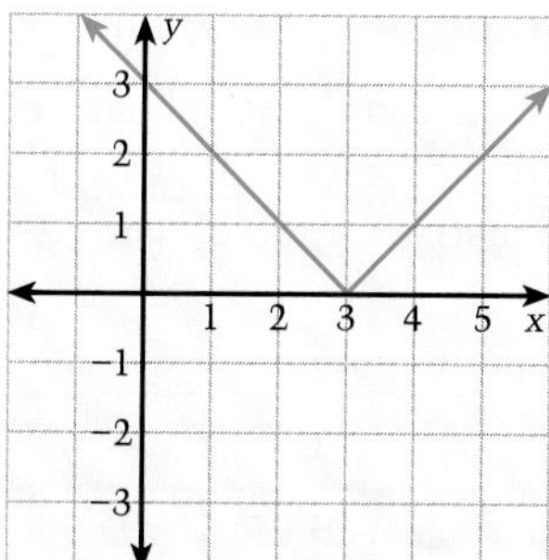

Graph of $y = |x - 3|$:
Vertex at (3, 0), opens up.

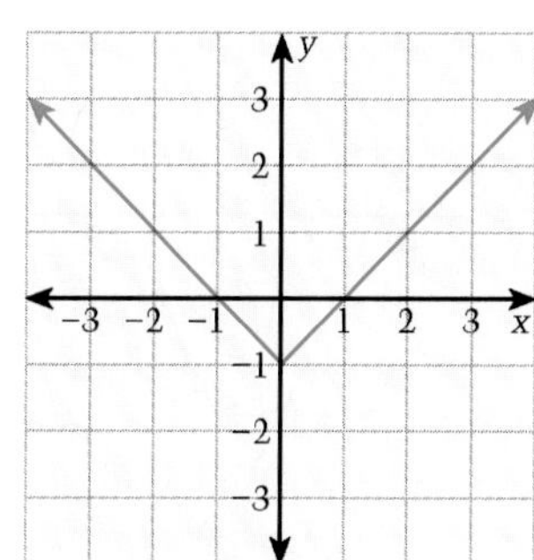

Graph of $y = |x| - 1$:
Vertex at (0, −1), opens up.

Problem of the Day

A dart board consists of only two circular areas: a nine-point bull's-eye surrounded by a four-point ring. Given an unlimited number of throws, what is the largest number that cannot be scored? 23

ORGANIZER

Warm-Up Exercises

1. Complete the table of values for the given expressions.

x	a. $2x - 8$	b. $4 - 5x$
−3	−14	19
−2	−12	14
−1	−10	9
0	−8	4
1	−6	−1
2	−4	−6
3	−2	−11

2. Sketch the graphs.

a. $y = 2x - 8$ **b.** $y = 4 - 5x$

Students' graphs should contain the points obtained in Warm-Up Exercise 1.

Lesson Resources

Teaching Tools
- Transparency: 7, 8, 12
- Problem of the Day: 2.6
- Warm-up Exercises: 2.6
- Answer Masters: 2.6

Extra Practice: 2.6

Color Transparency: 13

Technology Handbook: p. 9, 12

LESSON Notes

Example 1

Observe that the x- and y-intercepts may be found using techniques similar to those shown in Lesson 2.3.

Ask students if every graph of an absolute value equation has x- and y-intercepts, and whether the information is useful for graphing absolute value equations.

Example 2

Use the coordinates of the vertex, and x- and y-intercepts to graph the absolute value equation (vertex at $(-0.5, -4)$; x-intercepts at $(-2.5, 0)$ and $(1.5, 0)$; y-intercept at $(0, -3)$). Ask students why the vertex of the graph must be known.

Example 3

CONNECTIONS The approach to the problem uses ideas from calculus. Discuss with students why the area of the rectangle in Part a should represent the total rainfall. (Observe that total rainfall = (rate of rainfall) · (length of time of rainfall).)

To understand why the area of the triangle in Part b represents the total rainfall, ask students to approximate the area of the triangle using narrow rectangles as shown.

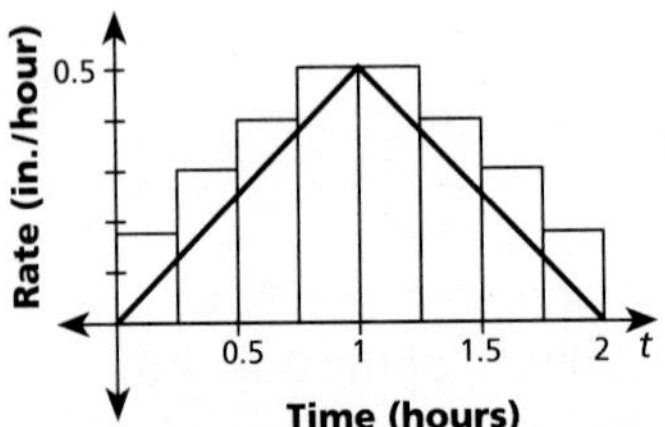

Why are these rectangles a reasonable estimate of the total rainfall? Are the same rectangles a reasonable estimate of the area of the triangle?
By using rectangles that have narrower bases, the approximation of the area is improved.

The following guidelines can help you sketch the graph of an absolute value equation.

Sketching the Graph of an Absolute Value Equation

To sketch the graph of $y = a|bx + c| + d$, use the following steps.

1. Find the x-coordinate of the vertex by finding the value of x for which $bx + c = 0$.
2. Construct a table of values, using the x-coordinate of the vertex, some x-values to its left, and some to its right.
3. Plot the points given in the table. Connect the points with a V-shaped graph, opening up if a is positive, down if a is negative.

Example 1 *Graphing an Absolute Value Equation*

Sketch the graph of $y = -|x - 2| + 3$.

Solution

1. The x-coordinate of the vertex is 2 because that is the value of x for which $x - 2 = 0$.
2. Construct a table, using x-values to the left and right of 2.

Vertex ↓

x	−1	0	1	2	3	4	5
$y = -\|x - 2\| + 3$	0	1	2	3	2	1	0

3. Plot the points. Notice that the vertex is at (2, 3). Draw a V-shaped graph (opening down) through the points. ■

Example 2 *Graphing an Absolute Value Equation*

Sketch the graph of $y = |2x + 1| - 4$.

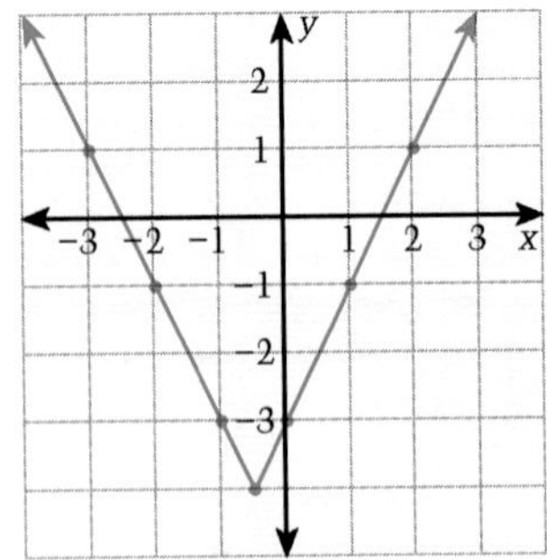

Solution

1. The x-coordinate of the vertex is $-\frac{1}{2}$ because that is the value of x for which $2x + 1 = 0$.
2. Construct a table, using x-values to the left and right of $-\frac{1}{2}$.

Vertex ↓

x	−3	−2	−1	$-\frac{1}{2}$	0	1	2
$y = \|2x + 1\| - 4$	1	−1	−3	−4	−3	−1	1

3. Plot the points. Notice that the vertex is at $\left(-\frac{1}{2}, -4\right)$. Then draw a V-shaped graph through the points. ■

Goal 2 Using Absolute Value Equations

Connections
Meteorology

Example 3 *Finding the Total Amount of Rainfall*

a. It is raining at a steady rate of $r = 0.5$ inch per hour for 2 hours. What is the total rainfall?

b. A rainstorm lasts for two hours, building up from a drizzle to a heavy rain, then dropping back to a drizzle. The rate, r, of the rain (in inches per hour) is given by

$$r = -0.5|t - 1| + 0.5; \quad 0 \le t \le 2$$

where t is the time in hours. What is the total rainfall?

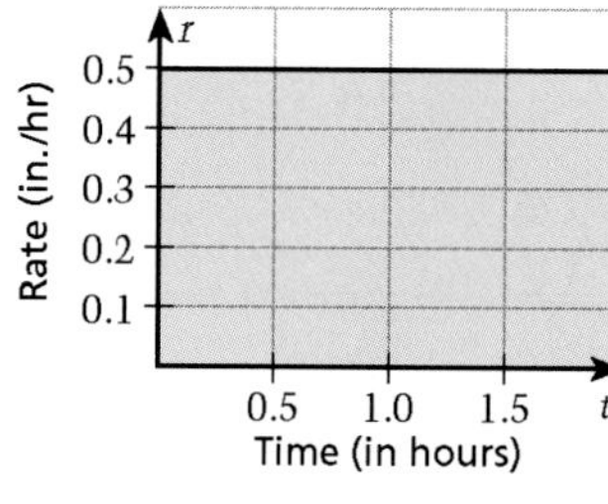

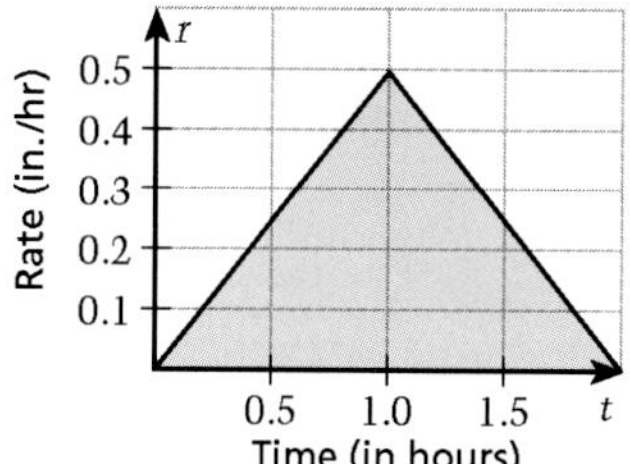

Solution In both cases, the total amount of rainfall can be found geometrically. In the first case, the rate is *constant,* and the total rainfall corresponds to the area of a rectangle. In the second case, the rate *varies,* and the total rainfall corresponds to the area of a triangle.

a. The total rainfall, T (in inches), is

$$T = \left(0.5\frac{\text{inches}}{\text{hour}}\right)(2 \text{ hours}) = 1 \text{ inch}.$$

Notice that this can be interpreted geometrically as the area of a rectangle that is 0.5 unit high and 2 units long.

b. In this case, the rate of the rain varies according to the equation $r = -0.5|t - 1| + 0.5$. To find the total rainfall, sketch the graph of this equation. Then find the area of the triangle formed by the x-axis and the graph.

$$T = \frac{1}{2}(\text{base})(\text{height}) = \frac{1}{2}(2)(0.5) = 0.5 \text{ inch}.$$

■

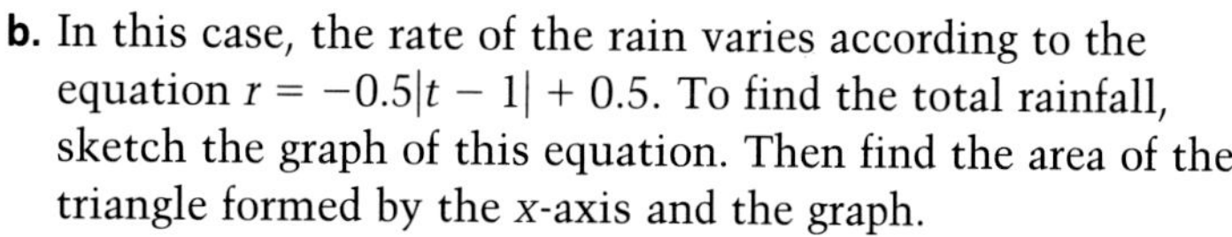

SHARING IDEAS about the Lesson

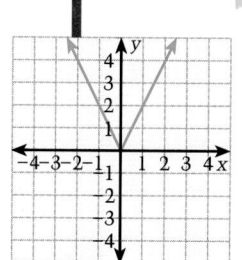

Compare Graphs Sketch the graph of each equation on the same coordinate plane. How does the shape of each compare to the graph of $y = |2x|$ shown at the left? How does the position of each compare to the graph at the left? Use translations and reflections in your comparisons.

A. $y = |2x - 4|$ Shifts 2 units to the right.

B. $y = |2x - 4| + 1$ Shifts 2 units to the right and 1 unit up.

C. $y = -|2x - 4| + 1$ Shifts 2 units to the right and reflects in x-axis.

See Additional Answers.

Extra Examples

Here are additional examples similar to **Examples 1–3.**

1. Sketch each graph.

a. $y = |x - 2|$

b. $y = -|x + 5|$

c. $y = |7x + 3| - 2$

d. $y = -2|3x - 4| + 1$

a. The graph is a horizontal shift of $y = |x|$ with vertex (2, 0).

b. The graph is a reflection about the x-axis and a horizontal shift of $y = |x|$ with vertex (−5, 0).

c. The graph has vertex $(-\frac{3}{7}, -2)$, x-intercepts of $(-\frac{5}{7}, 0)$ and $(-\frac{1}{7}, 0)$, and a y-intercept of (0, 1).

d. The graph has vertex $(\frac{4}{3}, 1)$, a y-intercept of (0, −7), and passes through $(\frac{8}{3}, -7)$.

2. Compare the area of the triangle with the sum of the areas of the rectangles. Describe how better approximations of the area of the triangle can be found using a collection of rectangles.

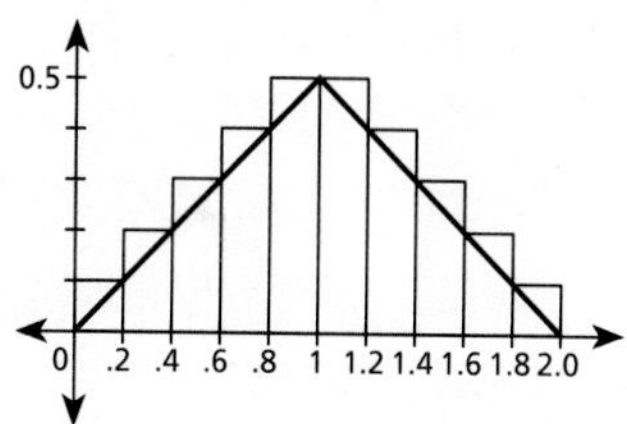

Increasing the number of rectangles by reducing the width of their bases will result in better approximations.

Check Understanding

1. What is the general shape of an absolute value equation?
   a V-shape
2. In the general form of an absolute value equation, $y = a|bx + c| + d$, why should $bx + c$ be set equal to zero?
   to find the x-coordinate of the vertex
3. What is the vertex of an absolute value equation?
   the corner point
4. How many points should be plotted to obtain the graph of an absolute value equation?
   at least two points on either side of the vertex

Communicating about ALGEBRA

Ask students to investigate the effects of constants on the graphs of absolute value equations by considering the differences in the graphs of $y = a|bx + c| + d$ for different values of a, b, c, and d.

EXERCISE Notes

ASSIGNMENT GUIDE

Basic/Average: Ex. 1–4, 5–8, 9–19 odd, 21–24, 27–39 odd, 44, 56–58

Above Average: Ex. 1–4, 5–8, 9–19 odd, 21–24, 27–41 odd, 45–48, 55–58

Advanced: Ex. 1–4, 5–8, 17–24, 27–41 odd, 45–48, 56–60

Selected Answers
Exercises 1–4, 5–55 odd

✪ More Difficult Exercises
Exercises 45–48, 57–60

EXERCISES

Guided Practice

CRITICAL THINKING about the Lesson

1. Explain in your own words why $|x|$ can equal either x, 0, or $-x$. Give an example of a value of x for which $|x| = -x$.
2. Which value (a, b, c, or d) determines whether the vertex of the graph of $y = a|bx + c| + d$ shows a maximum or a minimum value for y? Explain. See below.
3. What is the x-coordinate of the vertex of $y = |x - 5| + 1$? Which x-values would you use to construct a table of values?
4. Sketch the graph of $y = |x - 5| + 1$. See Additional Answers.

1. If x is positive or zero, its absolute value is the same as x. If x is negative, its absolute value is its opposite; for example, $|-7| = -(-7)$, or 7.
3. 5; 5 and values to the left and right of 5

Independent Practice

In Exercises 5–8, match the equation with its graph. Describe how the graph relates to the graph of $y = |x|$. See below.

5. $y = |x| + 4$
6. $y = |x| - 4$
7. $y = |x + 4|$
8. $y = |x - 4|$

a. 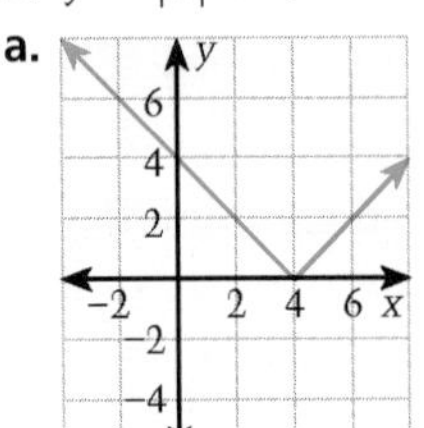

b. 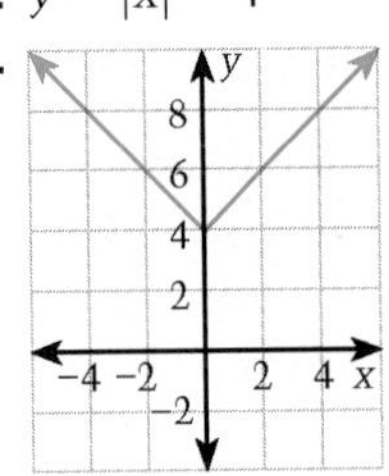

c. 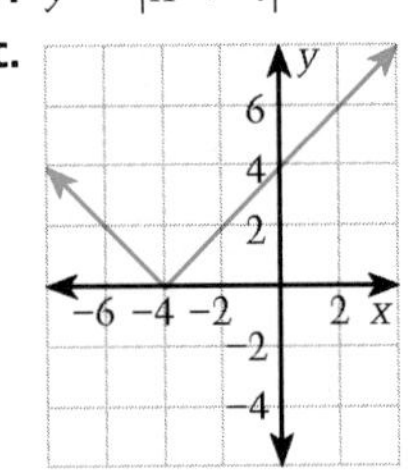

d. 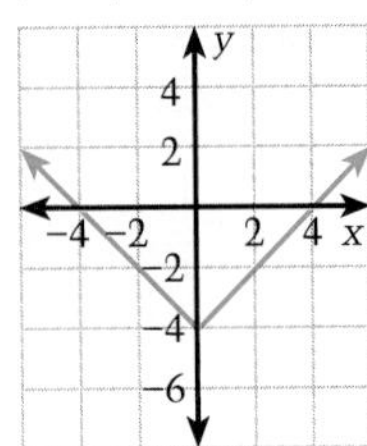

In Exercises 9–20, find the vertex of the graph.

9. $y = |x| + 1$ (0, 1)
10. $y = -|x| - 5$ (0, −5)
11. $y = |x - 1| + 3$ (1, 3)
12. $y = |x + 3| - 10$ (−3, −10)
13. $y = -|x - 2| - 12$ (2, −12)
14. $y = -|x + 14| - 1$
15. $y = |2x + 4| + 3$ (−2, 3)
16. $y = -2|4x - 1| + 16$ $\left(\frac{1}{4}, 16\right)$
17. $y = 3\left|\frac{1}{2}x - 3\right| + 6$
18. $y = -3\left|\frac{1}{9}x + 5\right| - 5$ (−45, −5)
19. $y = -2|5 - 2x| + 6$ $\left(\frac{5}{2}, 6\right)$
20. $y = -|16 - 6x| + 9$

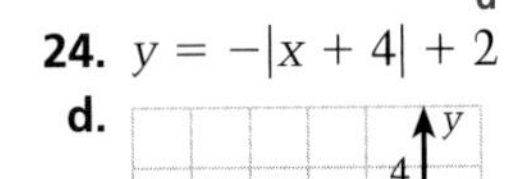

14. (−14, −1) 17. (6, 6) 20. $\left(\frac{8}{3}, 9\right)$

In Exercises 21–24, match the equation with its graph.

21. $y = -|x + 2| - 4$ b
22. $y = |x - 4| - 2$ c
23. $y = |x - 2| + 4$ a
24. $y = -|x + 4| + 2$ d

a. 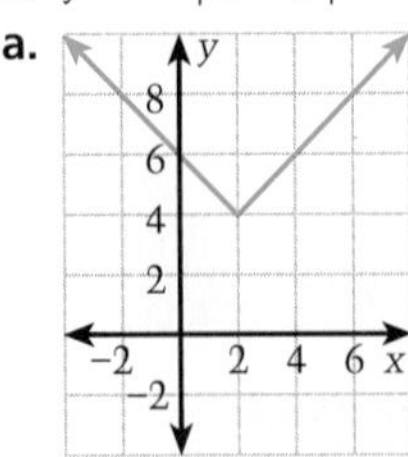

b. 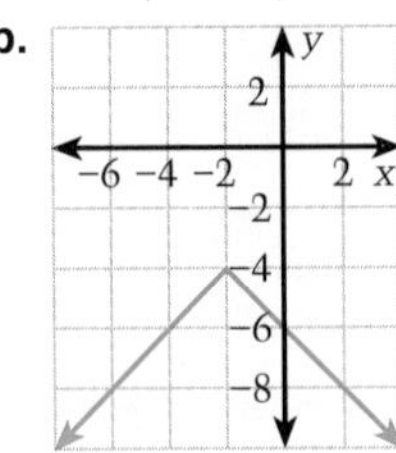

c. 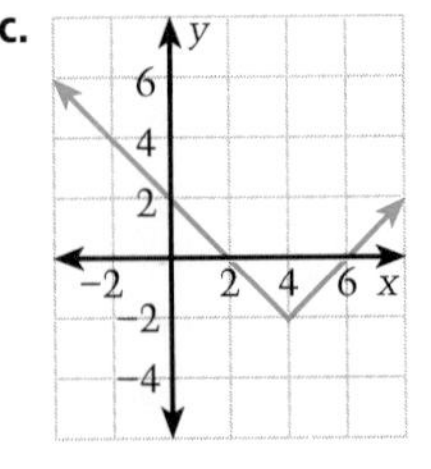

d. 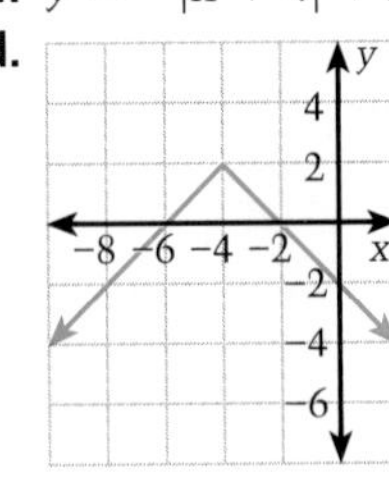

2. a; when a is positive the graph opens upward to show a minimum value for y, when a is negative the graph opens downward to show a maximum value for y

In Exercises 25–28, complete the table. Describe the pattern of symmetry.

25. $y = |x + 2|$ symmetric about $x = -2$

x	-5	-4	-3	-2	-1	0	1
y	?	?	?	?	?	?	?
	3	2	1	0	1	2	3

26. $y = |x| - 5$ symmetric about $x = 0$

x	-3	-2	-1	0	1	2	3
y	?	?	?	?	?	?	?
	-2	-3	-4	-5	-4	-3	-2

27. $y = 2|x - 1| + 3$ symmetric about $x = 1$

x	-2	-1	0	1	2	3	4
y	?	?	?	?	?	?	?
	9	7	5	3	5	7	9

28. $y = -|2x + 3| - 10$ symmetric about $x = -\frac{3}{2}$

x	$-\frac{9}{2}$	$-\frac{7}{2}$	$-\frac{5}{2}$	$-\frac{3}{2}$	$-\frac{1}{2}$	$\frac{1}{2}$	$\frac{3}{2}$
y	?	?	?	?	?	?	?
	-16	-14	-12	-10	-12	-14	-16

In Exercises 29–44, sketch the graph of the equation. See Additional Answers.

29. $y = |x - 5|$ **30.** $y = |x| - 5$ **31.** $y = |x| + 7$ **32.** $y = -|x - 8| + 3$

33. $y = -|x + 2| + 9$ **34.** $y = -|x - 8| - 9$ **35.** $y = |5 - x| + 4$ **36.** $y = -|10 - x| + 4$

37. $y = |2x| + 3$ **38.** $y = \left|\frac{1}{3}x\right| - 9$ **39.** $y = |3x - 1|$ **40.** $y = \left|\frac{1}{2}x + 6\right|$

41. $y = -4|2x + 1|$ **42.** $y = 2|x + 3| - 9$ **43.** $y = -2|9 - 2x| + 3$ **44.** $y = -|6 - 5x| - 4$

New Album Release **In Exercises 45–47, use the following information.** See Additional Answers.

A musical group's new album is released. Weekly sales, s (in thousands), increase steadily for a while and then decrease as given by the model

$$s = -2|t - 22| + 44$$

where t is the time in weeks.

The Recording Industry Association of America presents a **gold** *album to artists whose album sells over 500,000 copies. A* **platinum** *album is given for an album that sells over 1,000,000 copies.*

45. Sketch the graph of the sales model.

46. Estimate the total number of albums sold during the first 44 weeks. Did the album "go gold"? Did it "go platinum"? Explain.

47. Would the group have sold more albums during the 44 weeks if it had sold 20,000 albums per week? Give a geometric interpretation of your answer.

48. *Playing Billiards* You are trying to shoot the eight-ball into the corner pocket as shown. Imagine that a coordinate plane is superimposed over the pool table so that the points (10, 0) and (0, 5) are two corner pockets. The eight-ball follows the path given by $y = \frac{5}{4}|x - 6|$. Do you make your shot? Sketch a graph of the equation. At what point does the eight-ball bank off the first side of the table? Yes, (6, 0)
See Additional Answers.

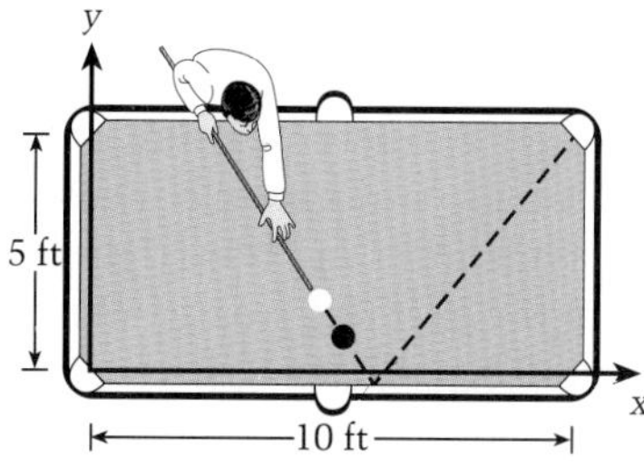

Guided Practice

▶ **Ex. 1–4** Use these exercises to check students' understanding of the essential concepts and skills of the lesson.

EXTEND Ex. 3 Discuss the vertical line symmetry of the graph.

Independent Practice

▶ **Ex. 5–8** Assign these exercises as a group. Ask students to describe each graph as a horizontal or vertical "shift" of the graph of $y = |x|$.

▶ **Ex. 15–19** Remind students that the x-coordinate of the vertex of $y = a|bx + c| + d$ occurs when $(bx + c) = 0$. This is not as obvious in these exercises as it was in Exercises 9–14.

▶ **Ex. 21–24** Assign these exercises as a group. Discuss the combination of vertical and horizontal graph shifts of $y = |x|$. For example, in Exercise 22, there is a horizontal shift 4 units to the right and a vertical shift 2 units down. You might wish to use transparencies of the graph of $y = |x|$ to demonstrate the shifts.

▶ **Ex. 29–44** Students should review the sketching algorithm before Example 1.

▶ **Ex. 45–47** These exercises should be assigned as a group. Students could review Example 3 before completing them.

5. b, shifts 4 units up **6.** d, shifts 4 units down **7.** c, shifts 4 units to the left **8.** a, shifts 4 units to the right

Ex. 58 Students can find the total distance by finding the area of the triangle formed by the graph and the x-axis.

Enrichment Activities

These activities require students to go beyond lesson goals.

1. Provide students with graphs of absolute value equations and have them write their equations.
   For example, given the following graph, students should realize that $a = 1$, $b = 1$, $c = -2$, and $d = -3$. So they would write the equation $y = |x + 2| - 3$.

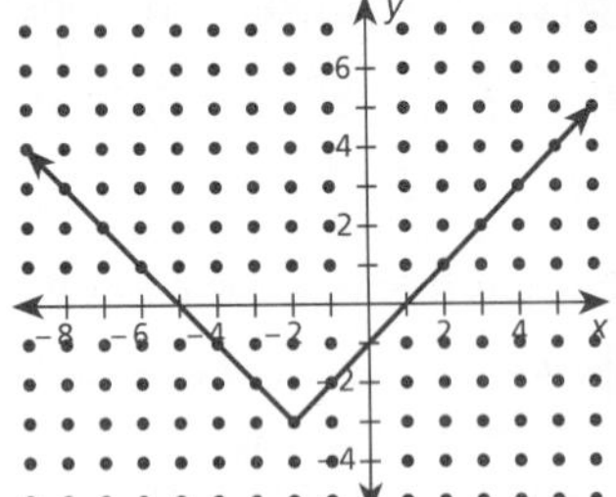

2. **MATH JOURNAL** Summarize, in your own words, the quick-graph techniques you have learned thus far. Include examples and information about how to use the technique and when it is applicable.

Integrated Review

In Exercises 49–52, the coordinates in the table of values are solutions of either a linear equation or an absolute value equation. Plot the points and sketch the graph. Then find an equation of the graph. See Additional Answers.

49.

x	0	1	2	3	4
y	1	0	-1	0	1

50.

x	-1	0	1	2	3
y	$-\frac{3}{2}$	$-\frac{1}{2}$	$\frac{1}{2}$	$\frac{3}{2}$	$\frac{5}{2}$

51.

x	-3	-2	-1	0	1
y	-4	-3	-2	-1	-2

52.

x	$\frac{1}{2}$	$\frac{3}{4}$	1	$\frac{5}{4}$	$\frac{3}{2}$
y	-2	-1	0	1	2

49. $y = |x - 2| - 1$ **50.** $y = x - \frac{1}{2}$ **51.** $y = -|x| - 1$ **52.** $y = 4x - 4$

In Exercises 53–56, evaluate the expression.

53. $2|x - 10| + 5$ when $x = -4$ 33

54. $6 - \frac{1}{2}|x - 5|$ when $x = -3$ 2

55. $10 + \frac{1}{4}|7 - x|$ when $x = 15$ 12

56. $3|x| + 4|1 - x|$ when $x = 6$ 38

Exploration and Extension

Hitting Red Lights **In Exercises 57 and 58, use the following information.**

When a red traffic light turns green, you accelerate at a constant rate to a speed of 60 feet per second (about 41 mph), then decelerate steadily to stop at the next traffic light. Your speed, r (in feet per second), is given by $r = -10|t - 6| + 60$, where t is the time in seconds.

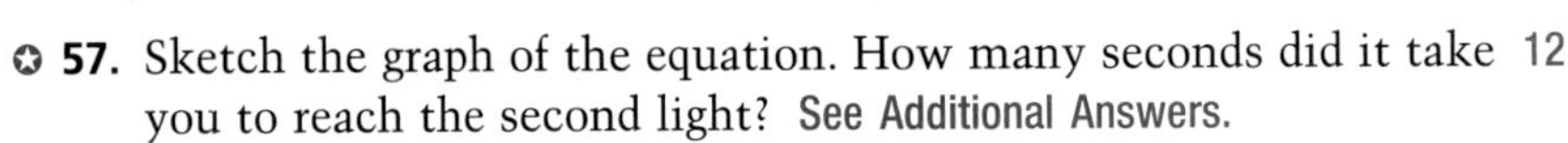

57. Sketch the graph of the equation. How many seconds did it take you to reach the second light? 12 See Additional Answers.

58. What is the distance between the two lights? (Your rate was not constant, but you can use a geometric model to find the distance.) 360 ft

Transamerica Pyramid **In Exercises 59 and 60, use the following information.**

The Transamerica Pyramid is an office building in San Francisco. It is about 850 feet tall and about 160 feet wide at its base.

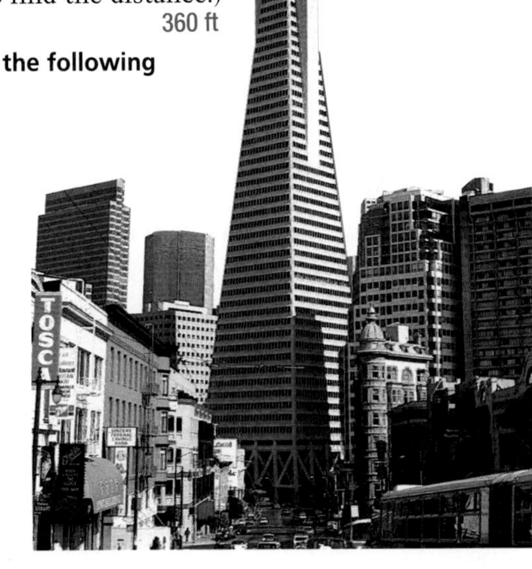

59. What is the slope of the sides of the building? $\frac{85}{8}$

60. Imagine that a coordinate plane has been superimposed over a head-on picture of the building. In the coordinate plane, each unit represents one foot, and the origin is at the center of the building's base. Write an absolute value equation whose graph is the V-shaped edges of the building.
$y = -\frac{85}{8}|x| + 850$

2.7 Exploring Data: Fitting a Line to Data

What you should learn:

Goal 1 How to fit a line to a set of data and to write an equation for the line

Goal 2 How to identify whether a set of data shows positive or negative correlation, or no correlation

Why you should learn it:

You can use the equation of a best-fitting line to investigate trends in data and to make predictions.

Goal 1 Fitting a Line to Data

In Lesson 2.4, you studied how to find an equation of the line that passes through *two* points. For this type of problem, there is always only one line—because two points determine exactly one line.

In this lesson, you will study problems that involve *several* data points. Usually there is no single line that passes through all the data points, so you try to find the line that best fits the data. This is called the **best-fitting line.** For instance, in the graph shown below, the line given by $y = -x + 3$ is the best-fitting line for the data points.

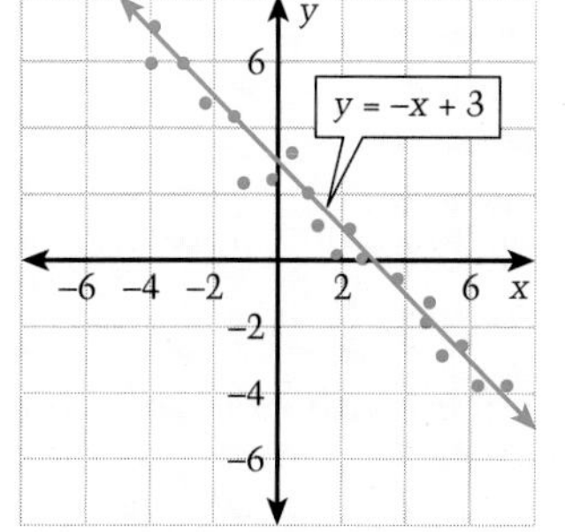

There are several ways to find the best-fitting line for a given set of data points whose graph suggests a linear relationship. One way is to use a graphing calculator or computer and formulas from a branch of mathematics called statistics. The statistical formula can *find* the best-fitting line. In this lesson, however, you will use a graphical approach, which *approximates* the best-fitting line.

LESSON INVESTIGATION

■ Investigating a Best-Fitting Line

Partner Activity With your partner, use the following steps to approximate a best-fitting line.

1. Carefully plot the following points on graph paper.
   $(-3, 3.4)$, $(-2, 3.8)$, $(-1, 4.2)$, $(0, 4.7)$, $(0, 5.1)$, $(1, 5.4)$, $(1, 5.8)$, $(2, 7.1)$, $(2, 7.4)$, $(3, 8.0)$, $(4, 9.1)$, $(4, 9.6)$, $(5, 10.2)$, $(6, 12.1)$, $(7, 14.0)$
2. Use a ruler to sketch the line that you think best approximates the data points.
3. Locate two points on the line and approximate the x-coordinate and y-coordinate of each point. (These don't have to be two of the original data points.)
4. Use the technique described in Lesson 2.4 to find an equation of the line that passes through the two points in Step 3. Answers will vary. $y = 1.1x + 5.28$

Problem of the Day

A 12-inch square piece of wood is cut into two parts so that they cover a 9-inch by 16-inch hole. How was the cut made?

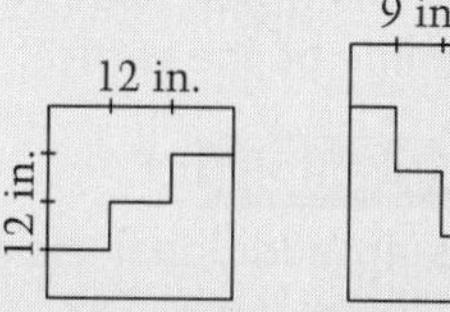

ORGANIZER

Warm-Up Exercises

Write the equation of the line that contains the given points.

1. (3, 8) and (−2, 1)
2. (0, −7) and (4, 9)
3. (1, −1) and (0, 3)

1. $y = \frac{7}{5}x + \frac{19}{5}$ 2. $y = 4x - 7$
3. $y = -4x + 3$

Lesson Resources

Teaching Tools
- Transparencies: 5, 7, 8, 9
- Problem of the Day: 2.7
- Warm-up Exercises: 2.7
- Answer Masters: 2.7

Extra Practice: 2.7
Color Transparencies: 14, 15
Technology Handbook: p. 14

LESSON Notes

Example 1

Point out to students that the point (1.7, 1.5) is not a data point in the earnings-dividend list; it is a point found to be on the best-fitting line.

Be sure that students know that a "dividend" is usually a sum of money divided among the stockholders of a company.

Example 2

The vertical axis labeled "sleep hours" has been modified using a broken line so that the relevant information could be conveniently shown. Ask students to draw the graph without the broken line of the vertical axis.

Extra Example

Here is an additional example similar to **Examples 1–2.**

A local appliance store decided to monitor its sales of VCRs as related to their unit price. So they randomly chose 15 weeks out of the previous year. For those weeks, they obtained data on the number of VCRs that were sold and the unit sales price at which they were sold. These data can be found in the table below. Sketch a scatter plot, approximate the best-fitting line for the data, and write the equation of the best-fitting line.

Week	Unit Price (dollars) (P)	Number of VCRs (V)
1	350	45
2	360	35
3	365	30
4	325	54
5	395	25
6	400	18
7	330	35
8	335	50
9	365	45
10	390	20
11	375	45
12	350	50
13	390	19
14	400	20
15	425	17

Earnings (e)	Dividend (d)
1.67	1.73
1.73	1.46
1.77	1.48
1.79	1.42
1.84	1.63
1.90	1.60
1.92	1.83
1.97	1.46
1.99	1.56
1.99	1.67
2.00	1.72
2.00	1.65
2.00	1.86
2.23	1.74
2.23	1.56
2.25	1.80
2.38	2.20
2.48	1.60
2.55	2.35
2.56	2.00
2.58	1.80
2.69	2.46
2.74	2.10
2.77	1.74
2.79	2.30
3.02	1.90
3.26	1.78
3.32	2.62
3.38	2.51
3.45	2.81
3.54	2.28
3.70	2.50
3.79	2.76
4.12	2.40
4.40	2.96

Example 1 *Approximating a Best-Fitting Line*

The 1990 earnings per share and dividends per share for 35 electric utility companies (in the central United States) are shown in the table. Approximate the best-fitting line for this data. Let e represent the earnings per share. Let d represent the dividend per share. What can you conclude from the result? ***(Source: The Value Line Investment Survey)***

Solution To begin, think of each of the data pairs as an *ordered pair.* For instance, the first pair of numbers in the table is represented by the ordered pair (1.67, 1.73). Next, draw a scatter plot for the ordered pairs. Then sketch the line that best fits the points, as shown below.

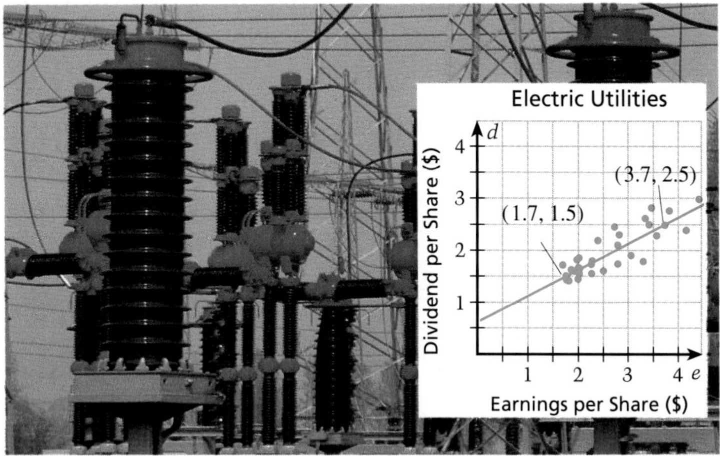

The next step is to find two points that lie on the best-fitting line. From the graph, you might choose the points (1.7, 1.5) and (3.7, 2.5). The slope of the line passing through these two points is

$$m = \frac{d_2 - d_1}{e_2 - e_1} = \frac{2.5 - 1.5}{3.7 - 1.7} = \frac{1.0}{2.0} = 0.5.$$

To find an equation of the line, use the point-slope form.

$$d - d_1 = m(e - e_1) \quad \textit{Point-slope form}$$

$$d - 1.5 = 0.5(e - 1.7) \quad \textit{Substitute values of } m, e_1, d_1.$$

$$d = 0.5e - 0.85 + 1.5 \quad \textit{Add 1.65 to both sides.}$$

$$d = 0.5e + 0.65 \quad \textit{Solve for } d.$$

From this equation, you can conclude two general results. As the utility companies earned more per share, they tended to pay a higher dividend per share. Another result is that the typical dividend was about half the earnings per share *plus* 65¢. ■

Normal human babies sleep more than half of their first year of life. A typical newborn sleeps about 16 hours a day, and a typical one-year-old sleeps about 13 hours a day.
(Source: Scientific American)

Connections
Physiology

Age	Sleep	Age	Sleep
(yrs)	(hrs)	(yrs)	(hrs)
0.03	15.0	0.52	14.4
0.05	15.8	0.69	13.2
0.05	16.4	0.70	14.1
0.08	16.2	0.75	14.2
0.10	14.9	0.80	13.4
0.11	14.8	0.82	14.3
0.19	14.7	0.82	13.2
0.21	14.5	0.86	13.9
0.26	15.4	0.90	13.7
0.34	15.2	0.91	13.1
0.35	15.3	0.94	13.7
0.35	14.4	0.97	12.7
0.44	13.9	0.98	13.7
0.52	13.4	0.98	13.6

Example 2 Comparing Sleep to Age

The data in the table shows the age, t (in years), and the number of hours, h, slept in a day by 28 infants who are less than one year old. Approximate the best-fitting line for this data.

Solution To begin, sketch a scatter plot as shown.

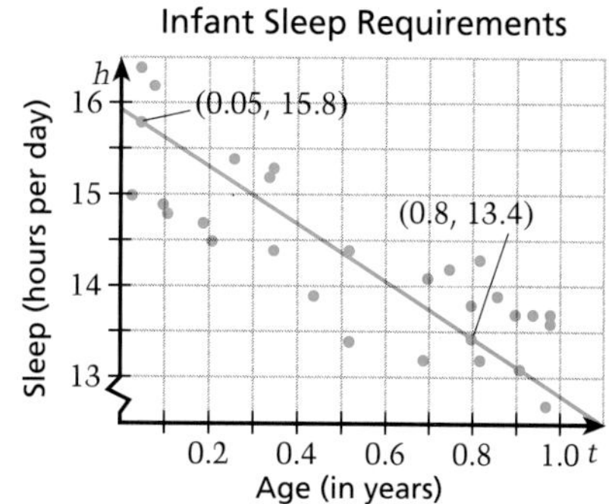

Next sketch the line that appears to fit the points best, and locate two points on the line: (0.05, 15.8) and (0.8, 13.4). The slope of the line passing through these two points is

$$m = \frac{h_2 - h_1}{t_2 - t_1} = \frac{13.4 - 15.8}{0.8 - 0.05} = \frac{-2.4}{0.75} = -3.2.$$

Use the point-slope form to find the equation of the line.

$h - h_1 = m(t - t_1)$	*Point-slope form*
$h - 15.8 = -3.2(t - 0.05)$	*Substitute values of m, t_1, h_1.*
$h = -3.2t + 0.16 + 15.8$	*Add 15.8 to both sides.*
$h = -3.2t + 15.96$	*Solve for h.* ■

Scatter Plot

Number of VCR's; Unit price (dollars); (325, 54); (395, 25)

The equation of a best-fitting line is $V = -0.41P + 188.64$, where the points (395, 25) and (325, 54) were used to find the slope. Answers will vary.

Check Understanding

1. What is a scatter plot?
   a set of data points plotted on the coordinate plane
2. What are the steps in approximating best-fitting lines?
   See box on page 107.
3. For the same set of data, explain why best-fitting lines may not be identical.
   The line depends on the chosen points.
4. Describe the scatter plot of data that has a correlation coefficient of $r = 0$.
   The scatter plot cannot be approximated by a line of best fit.

Communicating about ALGEBRA

Answer

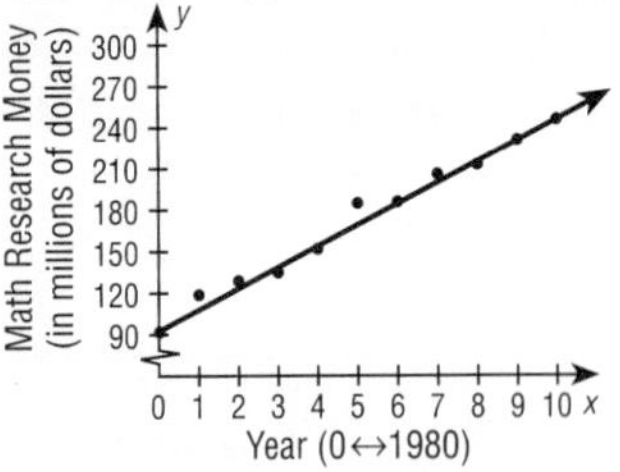

EXTEND *Communicating*

Ask students to identify real-life data that can be used to make scatter plots and best-fitting lines. Possible sources of data are school sports records, statistics from newspaper or magazine articles, or data from other classes, such as chemistry, economics, physics, biology, and history.

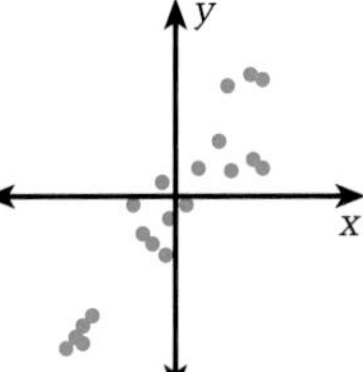

Positive Correlation

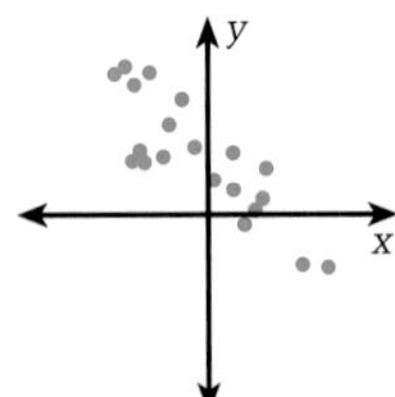

Negative Correlation

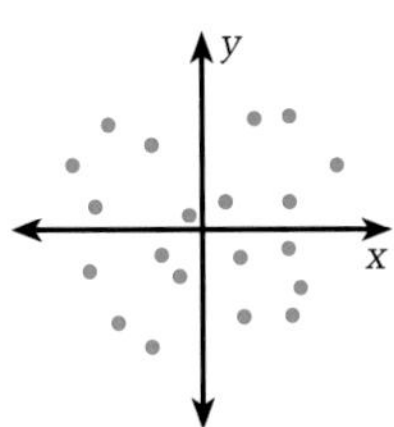

No Correlation

Goal 2 Positive and Negative Correlations

Three scatter plots are shown at the left. In the first, x and y have a **positive correlation,** which means that the points can be approximated by a line with a *positive slope.* In the second, x and y have a **negative correlation,** which means that the points can be approximated by a line with a *negative slope.* In the third, there is **no correlation** between x and y.

We have been talking about the best-fitting line. Sometimes, however, the best-fitting line isn't a very *good* fit. In statistics, the **correlation coefficient,** r, is used as a measure of how well a collection of data points can be modeled by a line. Correlation coefficients range between -1 and 1. The closer $|r|$ is to 1, the better the line fits the points. (There is more to the story than this—lines that are nearly horizontal have r-values that are close to 0. You can learn more about correlation in a statistics course.) The formula for r is

$$r = \frac{\text{Sum of } (x_i - \overline{x})(y_i - \overline{y})}{\sqrt{[\text{Sum of } (x_i - \overline{x})^2] \cdot [\text{Sum of } (y_i - \overline{y})^2]}}$$

where $\overline{x}$ is the average x-value and $\overline{y}$ is the average y-value.

Communicating about ALGEBRA

SHARING IDEAS about the Lesson

The total amount, m (in millions of dollars), spent by the federal government on mathematics research from 1980 through 1990 is shown in the table.
(Source: U.S. National Science Foundation)

Year	Math Research
1980	\$ 91
1981	\$118
1982	\$128
1983	\$134
1984	\$151
1985	\$184
1986	\$185
1987	\$205
1988	\$212
1989	\$230
1990	\$245

A. Sketch a scatter plot. Let $t = 0$ represent 1980. From your scatter plot, do m and t appear to have a positive correlation, a negative correlation, or no correlation? positive correlation

B. Sketch the best-fitting line. Locate two points on the line. (Your two points may or may not be the same as the points chosen by classmates.) Use the two points you chose to find an equation of the line. $m = 15.125t + 91$

C. Use the equation that you found to predict the amount the federal government will spend for mathematics research in 1995. \$317.88 million

EXERCISES

Guided Practice

CRITICAL THINKING about the Lesson

1. Do the points in the scatter plot at the right have a positive correlation or a negative correlation? Explain. See margin.
2. Find an equation of the line that you think best fits the points. $y = \frac{3}{4}x + 2$
3. Suppose you were given the shoe sizes, s, and the heights, h, of one hundred 25-year-old men. Do you think that s and h would have a positive correlation, a negative correlation, or no correlation? Explain. See margin.
4. Suppose you were given the incomes, i, and the percents of income, p, spent on food for one hundred families. Do you think i and p would have a positive correlation, a negative correlation, or no correlation? Explain. See margin.

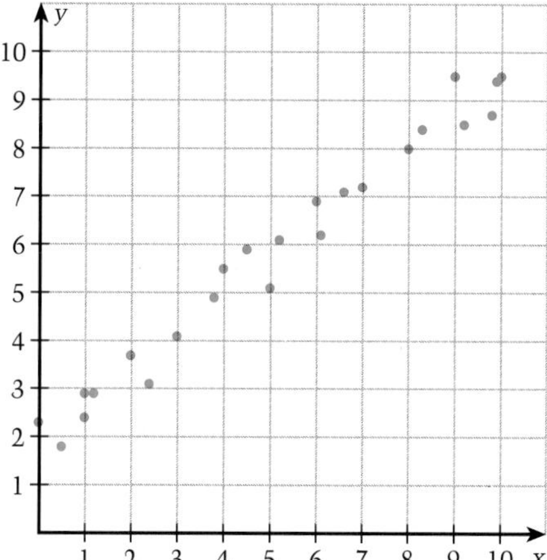

Independent Practice

In Exercises 5–8, state whether x and y have a positive correlation, a negative correlation, or no correlation.

5. 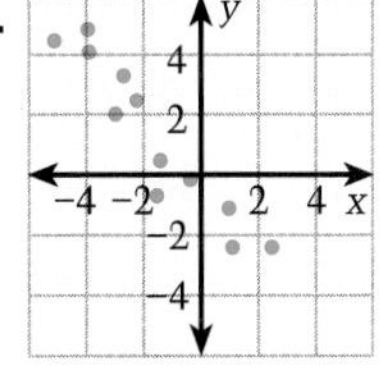

Negative correlation

6. 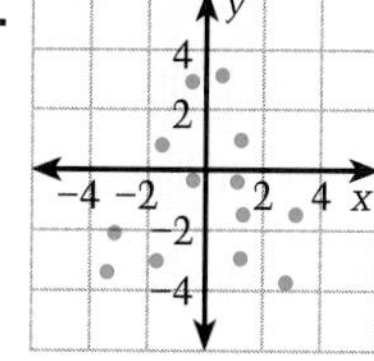 

No correlation

7. 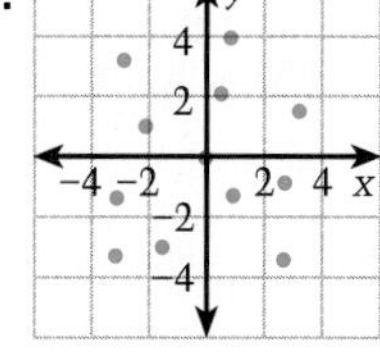 

No correlation

8. 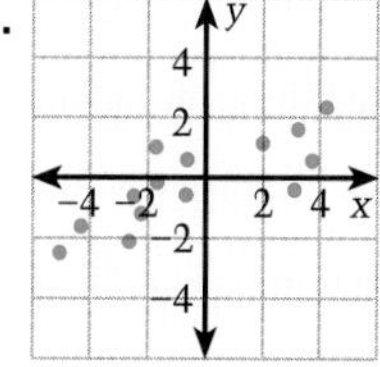

Positive correlation

In Exercises 9–12, write an equation of the line that you think best fits the scatter plot. 9.–12. Equations vary.

9. 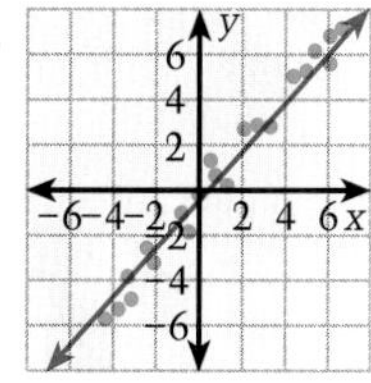

$y = \frac{5}{4}x - \frac{1}{2}$

10. 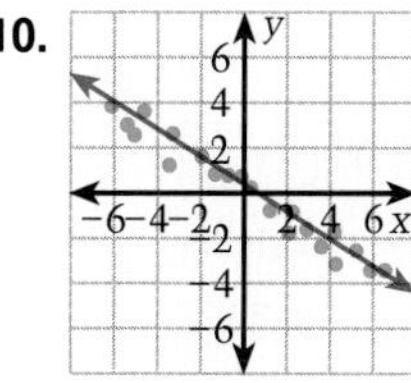 

$y = -\frac{3}{5}x + \frac{2}{5}$

11. 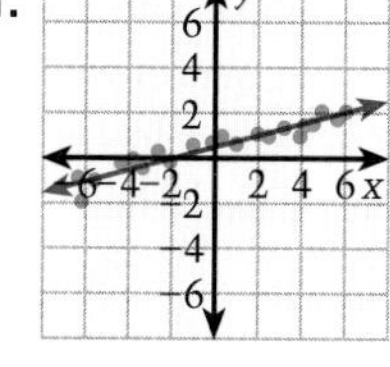

$y = \frac{1}{4}x + \frac{1}{2}$

12. 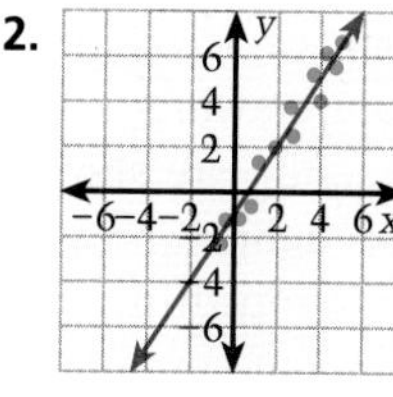

$y = \frac{3}{2}x - 1$

EXERCISE Notes

ASSIGNMENT GUIDE

Basic/Average: Ex. 1–4, 5–8, 9–13 odd, 16–19, 27–34

Above Average: Ex. 1–4, 5–8, 9–15 odd, 16–19, 27–36

Advanced: Ex. 1–4, 5–8, 9–15 odd, 16–18, 27–36

Selected Answers
Exercises 1–4, 5–25 odd

More Difficult Exercises
Exercises 27–36

Guided Practice

Ex. 4 There is no one "right" answer. Ask students to defend their answers.

Answers

1. Positive; the best-fitting line has a positive slope.
3. Positive correlation; taller men tend to have larger feet.
4. Negative correlation; the larger the income, the smaller the percent spent on food.

Independent Practice

Ex. 9–12 Remind students that they should select two points that lie on their approximated line of best fit.

Answers

13.

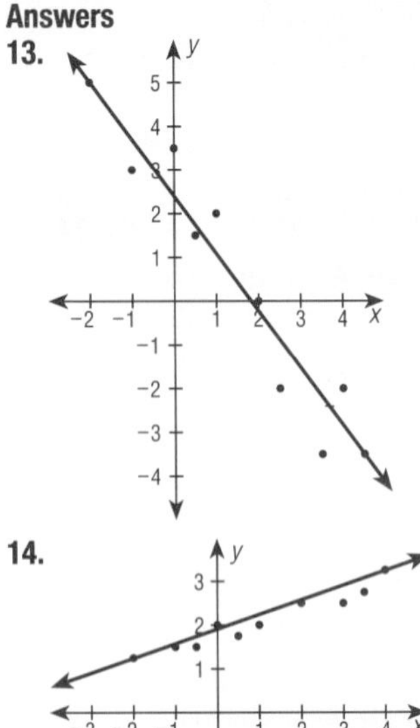

14.

▶ **Ex. 15–18** These exercises provide real-life data models for finding the equation of the line. Ask students to use their models to predict future growth for low-fat ice cream consumption by evaluating them for the year 2000, when $t = 20$.

In Exercises 13 and 14, construct a scatter plot for the given data. Then find an equation of the line that you think best fits the data. See margin.

13.

x	−2	−1	0	0.5	1	2	2.5	3.5	4	4.5
y	5	3	3.5	1.5	2	0	−2	−3.5	−2	−3.5

$y = -\frac{4}{3}x + \frac{5}{2}$

13.–18. Equations vary.

14.

x	−2	−1	−0.5	0	0.5	1	2	3	3.5	4
y	1.25	1.5	1.5	2	1.75	2	2.5	2.5	2.75	3.25

$y = \frac{1}{4}x + \frac{7}{4}$

15. *Home Computers* The table lists the number of households, c (in millions), in the United States that owned computers between 1984 and 1991. Approximate the best-fitting line for this data. Let t represent the year, with $t = 4$ corresponding to 1984. If this pattern was to continue, how many households would own computers in 1993? ***(Source: Electronics Industry Association)*** $c = \frac{7}{2}t - 8$, ≈38 million

Year	1984	1985	1986	1987	1988	1989	1990	1991
Households	6.0	11.3	14.2	16.2	19.2	21.3	25.3	26.6

16. *Laying More Eggs* Between 1920 and 1990, improvements in poultry science produced hens that laid more eggs. The average number of eggs produced per hen per week for several years is shown in the graph. Find an equation of the line that you think best fits the data. ***(Source: United States Department of Agriculture)*** $y = \frac{1}{25}x + 2$

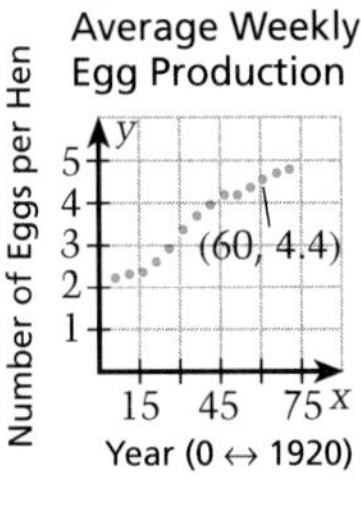

17. *Ice Cream* For 1986 through 1990, the percent, p, of low-calorie and/or low-fat ice cream sold in the United States increased as shown in the table. Approximate the best-fitting line for this data. (Let t represent the year, with $t = 6$ corresponding to 1986.) If this were to continue, what percent of ice cream sold in 1994 would be low-calorie and/or low-fat? ***(Source: A.C. Nielson)*** $p = \frac{13}{3}t - 20$, ≈41%

Year	Percent
1986	6
1987	8
1988	12
1989	19
1990	25

18. *100-Meter Sprint* A person who is running a 100-meter sprint was clocked several times during the sprint. The velocities, v (in meters per second), and the accelerations, a (in meters per second per second), are shown in the scatter plot at the right. Find an equation of the line that you think best fits this data. What can you conclude from the result? ***(Source: Mathematics in Sport ©1984 Halsted Press)***

$a = -\frac{2}{5}v + \frac{19}{5}$; the faster one runs, the harder it is to accelerate

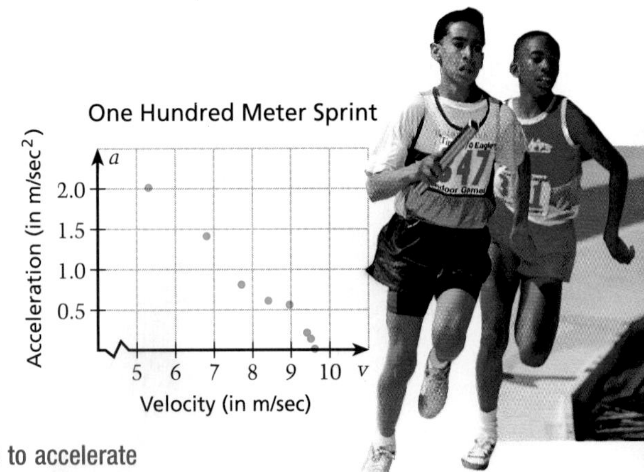

Integrated Review

In Exercises 19–22, find the slope of the line containing the points.

19. (6, −9), (7, 7) 16 **20.** (6, 7), (−5, 5) $\frac{2}{11}$ **21.** (5, 0), (−5, 1) $-\frac{1}{10}$ **22.** (3, −3), (−2, 1) $-\frac{4}{5}$

In Exercises 23–26, find an equation of the line containing the points.

23. (7, 2), (8, 9) $y = 7x - 47$

24. (9, 4), (−8, 3) $y = \frac{1}{17}x + \frac{59}{17}$

25. (4, −8), (7, 5) $y = \frac{13}{3}x - \frac{76}{3}$

26. (2, 1), (7, −3) $y = -\frac{4}{5}x + \frac{13}{5}$

Exploration and Extension

Correlation Coefficient **In Exercises 27–30, match the correlation coefficient with the scatter plot. Explain your reasoning.** See margin.

27. $r = 0.93$ **28.** $r = 0.99$ **29.** $r = -0.64$ **30.** $r = -0.95$

a.
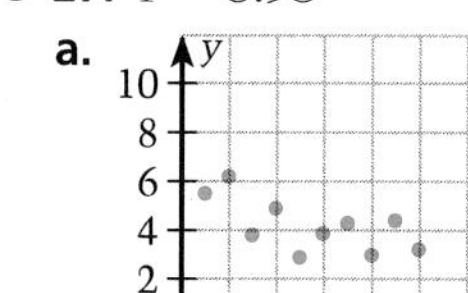

b.
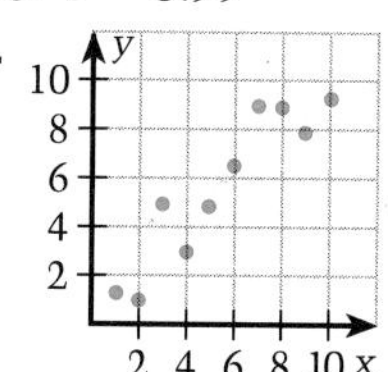

c.
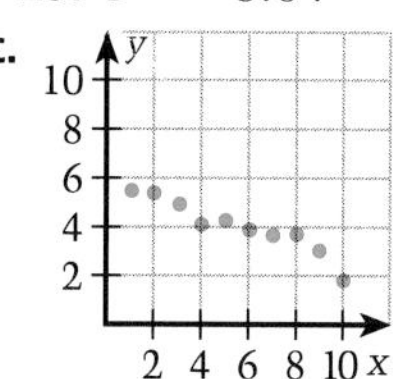

d.
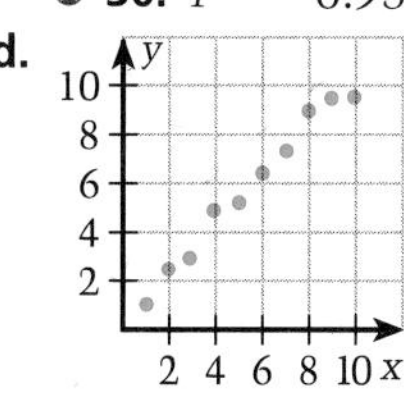

Correlation Coefficient **In Exercises 31–34, use the following information.**

The scatter plot at the right has five points. Find the correlation coefficient for these points by following the instructions in the exercises. See margin for 33.

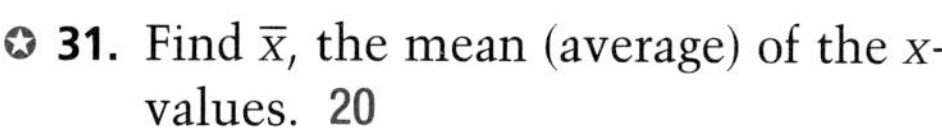

31. Find $\bar{x}$, the mean (average) of the x-values. 20

32. Find $\bar{y}$, the mean of the y-values. 31.2

33. Complete the table and find the totals of the third, sixth, and seventh columns.

34. Use the formula given in the lesson to calculate the correlation coefficient, r. ≈0.99

x_i	$x_i - \bar{x}$	$(x_i - \bar{x})^2$	y_i	$y_i - \bar{y}$	$(y_i - \bar{y})^2$	$(x_i - \bar{x})(y_i - \bar{y})$
0	?	?	12	?	?	?
10	?	?	22	?	?	?
20	?	?	30	?	?	?
30	?	?	38	?	?	?
40	?	?	54	?	?	?

In Exercises 35 and 36, find the correlation coefficient, *r*, for the data.

35. (−5, 5), (−4, 3), (−2, 2), (−1, −1), (3, −5) ≈−0.98

36. (−20, −12), (−10, −7), (4, −3), (10, 0), (20, 5) ≈0.99

▶ **Ex. 27–30** These exercises should be assigned as a group.

Answers

27. b, slight positive correlation
28. d, strong positive correlation
29. a, slight negative correlation
30. c, strong negative correlation
33. 2nd column: −20, −10, 0, 10, 20; 3rd column: 400, 100, 0, 100, 400; 5th column: −19.2, −9.2, −1.2, 6.8, 22.8; 6th column: 368.64, 84.64, 1.44, 46.24, 519.84; 7th column: 384, 92, 0, 68, 456; totals: 1000, 1020.8, 1000

Enrichment Activity

This activity requires students to go beyond lesson goals.

TECHNOLOGY Extend the lesson by using the STAT feature of the graphics calculator. You can use this feature to enter the data points, display the scatter plot, and calculate the linear regression model $y = a + bx$ for the data, where a represents the y-intercept and b represents the slope. You can also find the correlation coefficient, r, for the data.

Consult the owner's manual that came with your graphics calculator for specific instructions for using the STAT feature. Then use the feature to check your answers to Exercises 13–18 and 34–36. How do you account for the differences? Students should understand that the graphics calculator uses a built-in series of mathematical instructions to find the line of best fit. The text suggests a more intuitive method based on data points.

ORGANIZER

TIME SCHEDULE:
All levels, two days

The Chapter Summary helps students organize the main ideas of the chapter. In this chapter, the connection between modeling real-life problems and the resulting algebraic models was presented. Emphasis was placed upon graphs as a means for interpreting information.

Work with students to review the skills, strategies, and concepts of the chapter. The homework assignment can be used as the basis for the second day of review.

COOPERATIVE LEARNING
Encourage students to study together. Emphasize the importance of "teaching" a classmate how to perform a skill or how to recall a procedure. When students work together, everyone wins. The students receiving help get additional instruction and the students giving help gain a deeper understanding of the skills and concepts involved.

Chapter SUMMARY

Review the chapter by asking students, "What did you learn? Why did you learn it? How does it fit with the other algebra skills and concepts you have learned?"

2 Chapter Summary

What did you learn?

Skills

1. Plot points in a coordinate plane. **(2.1)**
2. Use a table of values to sketch the graph of an equation. **(2.1)**
3. Find the slope of a line
   - given two points on the line. **(2.2)**
   - given an equation of the line. **(2.3)**
4. Sketch the graph of a line
   - using intercepts. **(2.3)**
   - using the slope-intercept form. **(2.3)**
5. Write an equation of a line
   - given its slope and y-intercept. **(2.4)**
   - given its slope and a point on the line. **(2.4)**
   - given two points on the line. **(2.4)**
6. Sketch the graph of a linear inequality. **(2.5)**
7. Sketch the graph of an absolute value equation. **(2.6)**

Strategies

8. Use linear equations and inequalities to solve real-life problems. **(2.1–2.7)**

Exploring Data

9. Approximate the equation of the best-fitting line for related data. **(2.7)**

Why did you learn it?

Many relationships in real life are linear or approximately linear. By recognizing the relationship between two variables, you can learn more about the real-life situation. You can also make reasonable predictions of future trends. For instance, if a city's population has been increasing for the past 10 years at an approximately linear rate, you can find the average rate of change and use it to predict the population for the next 5 years. This use of linear models is important for businesses and city governments.

How does it fit into the bigger picture of algebra?

In Chapter 2, you studied linear equations and inequalities in two variables. A linear equation in x and y is an equation that can be written in the form $Ax + By = C$. In a coordinate plane, the graph of a linear equation is a line.

Using the coordinate plane to create geometric models of equations is an important part of algebra. Throughout this chapter, you translated geometric statements into algebraic statements, such as when you were given two points in a coordinate plane and found an equation of the line containing them. You also translated algebraic statements into geometric relationships, such as when you sketched the graph of an equation or inequality.

Chapter REVIEW

In Exercises 1–6, plot the points in a coordinate plane. In which quadrant does each point lie? (2.1) See Additional Answers.

1. $(5, 1), (-4, 0)$ I, none
2. $(2, -3), (1, -8)$ IV, IV
3. $(4, -6), (3, 8)$ IV, I
4. $(4, 1), (-4, 3)$ I, II
5. $(2, -2), (8, 6)$ IV, I
6. $(7, 9), (0, 2)$ I, none

In Exercises 7 and 8, find the average rate of change per year. (2.2)

7. A business had a profit of \$58,000 in 1989 and a profit of \$74,000 in 1993. **\$4,000 per year**

8. In 1790, each state was allowed one Congressman in the House of Representatives for every 33,000 citizens. In 1990, each Congressman or Congresswoman represented 555,000 citizens. **2610 citizens per year**

In Exercises 9–14, sketch the graph of the equation. (2.1) See margin.

9. $x = 4$
10. $y = -1$
11. $y = x - 5$
12. $y = -x + 9$
13. $y = -3x + 6$
14. $y = 2x + 7$

In Exercises 15–18, find the slope of the line containing the points. (2.2)

15. $(1, -3), (6, 7)$ 2
16. $(5, -3), (0, 12)$ -3
17. $(2, 2), (8, -1)$ $-\frac{1}{2}$
18. $(9, 1), (-5, 2)$ $-\frac{1}{14}$

In Exercises 19–22, find the *x*-intercept and *y*-intercept of the line. (2.3)

19. $2x - 5y = 20$ 10, -4
20. $6x - y = 36$ 6, -36
21. $-x + 4y = 16$ -16, 4
22. $-9y - 12x = 27$ $-\frac{9}{4}$, -3

In Exercises 23–28, write the equation in slope-intercept form. Then sketch the line. (2.3) See margin.

23. $x - y = 9$ $y = x - 9$
24. $2x - 10y = 15$ $y = \frac{1}{5}x - \frac{3}{2}$
25. $2x + 8y = 24$ $y = -\frac{1}{4}x + 3$
26. $-4x + 5y = 16$ $y = \frac{4}{5}x + \frac{16}{5}$
27. $-3x - 7y = 1$ $y = -\frac{3}{7}x - \frac{1}{7}$
28. $-x + 9y = 30$ $y = \frac{1}{9}x + \frac{10}{3}$

In Exercises 29–34, write an equation of the line containing the points. (2.4)

29. $(7, 0), (9, 6)$ $y = 3x - 21$
30. $(6, 5), (-7, 4)$ $y = \frac{1}{13}x + \frac{59}{13}$
31. $(3, -7), (8, 2)$ $y = \frac{9}{5}x - \frac{62}{5}$
32. $(0, 7), (5, 3)$ $y = -\frac{4}{5}x + 7$
33. $(8, 5), (-3, 2)$ $y = \frac{3}{11}x + \frac{31}{11}$
34. $(-8, 5), (9, 14)$ $y = \frac{9}{17}x + \frac{157}{17}$

In Exercises 35 and 36, *x* and *y* vary directly. Find an equation that relates *x* and *y*. (2.4)

35. $y = 10$ when $x = 1$ $y = 10x$
36. $y = 4$ when $x = 6$ $y = \frac{2}{3}x$

In Exercises 37–42, sketch the graph of the inequality. (2.5) See margin.

37. $2x < 6$
38. $4y \geq 9$
39. $y \geq x + 5$
40. $y < -x + 4$
41. $x + 8y < 1$
42. $3x + y < 9$

ASSIGNMENT GUIDE

Basic/Average: Ex. 1–45 odd, 47–55, 57–58, 62–64.

Above Average: Ex. 3–45 multiples of 3, 47–70

Advanced: Ex. 3–45 multiples of 3, 47–58, 64–70

✪ More Difficult Exercises
Exercises 65–70

▶ **Ex 1–70** The Chapter Review exercises include a reference to the lesson in which each skill or strategy was first presented. Students may use these references as a study aid if they are uncertain about how to do a given exercise in the Chapter Review.

Answers

9.

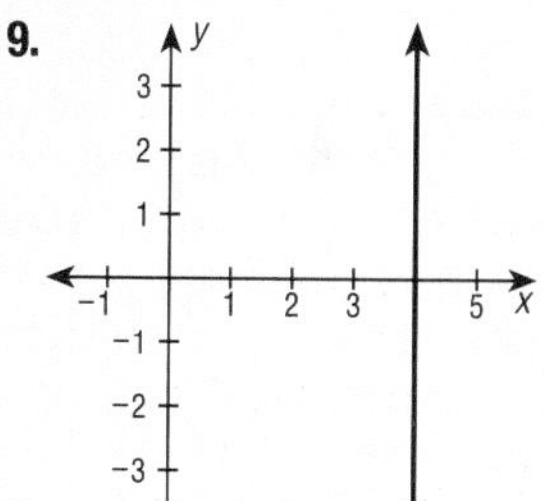

10.

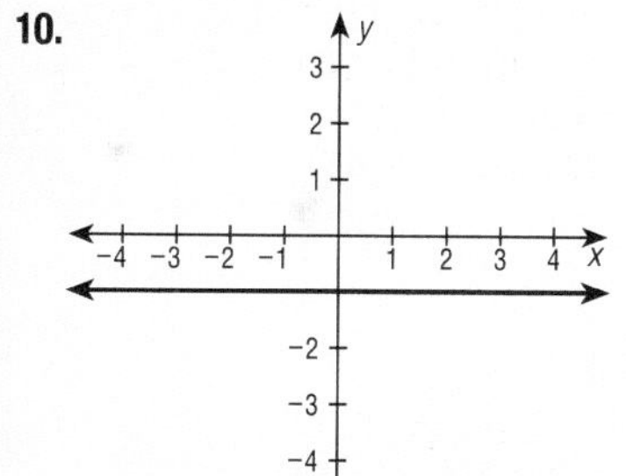

11.

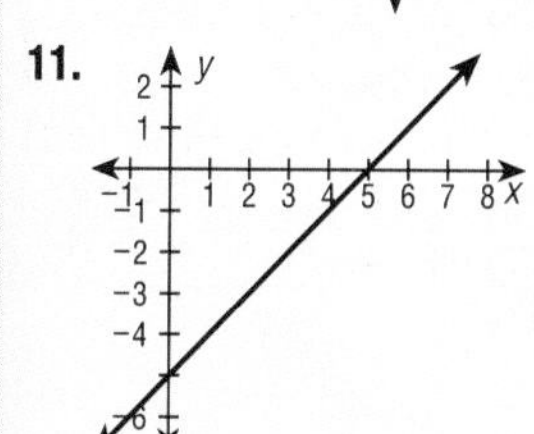

12.

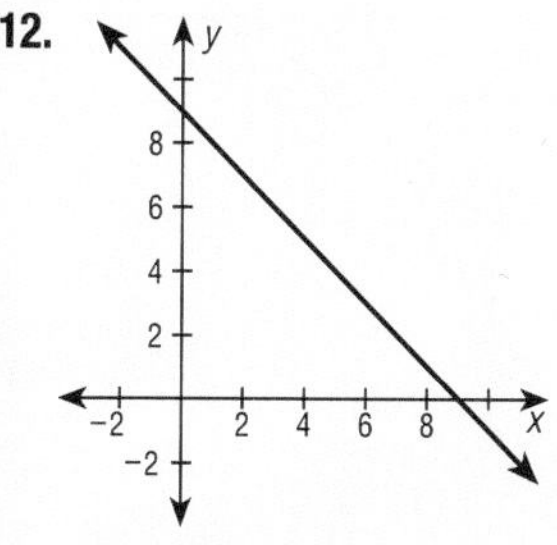

Answers

13. 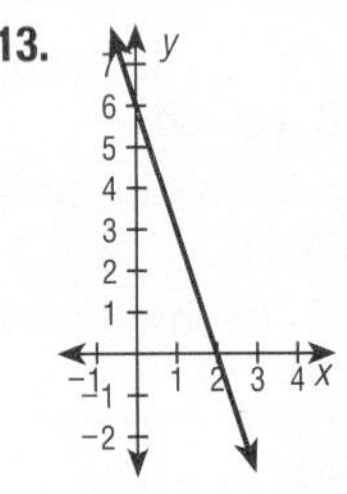

14. 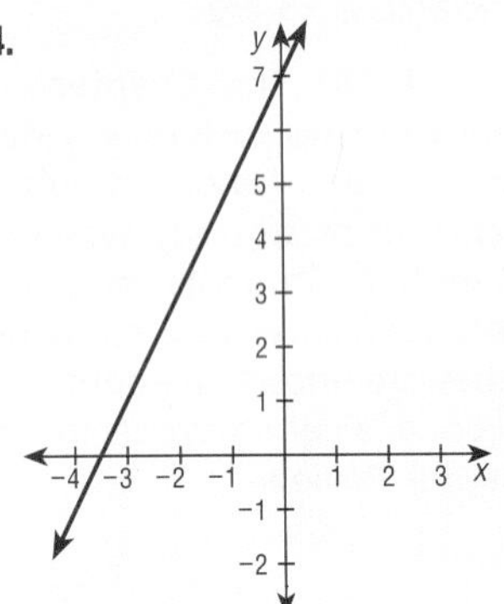

23. 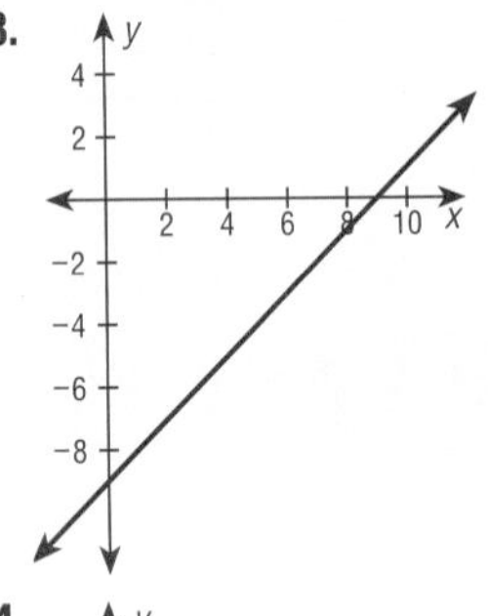

24. 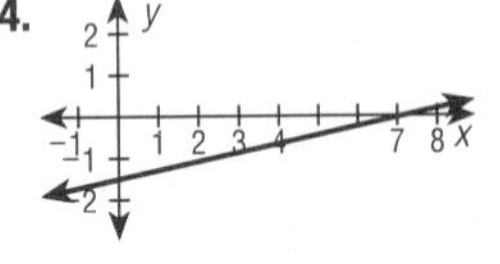

25. 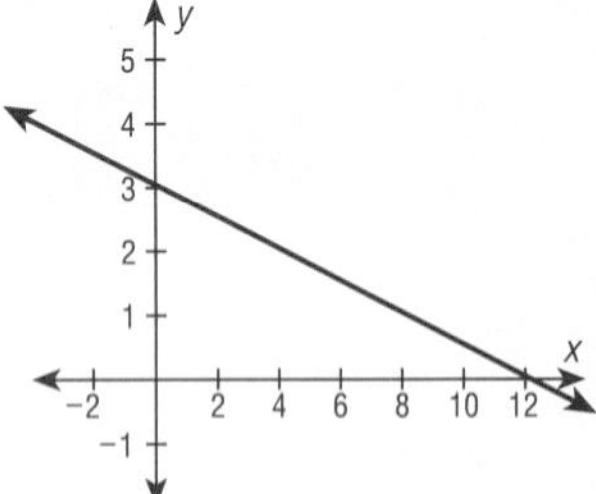

26. 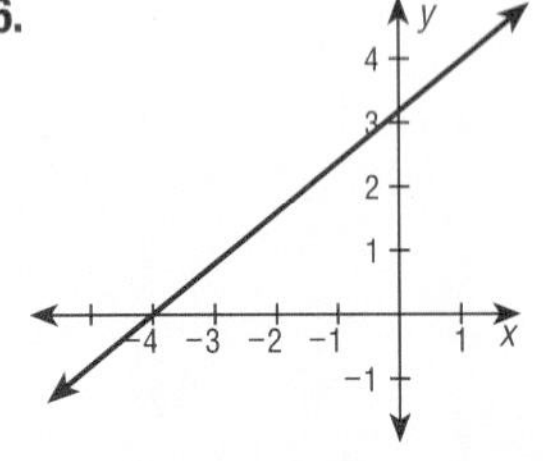

In Exercises 43–46, sketch the graph of the equation. (2.6) See margin.

43. $y = |3x| + 5$ **44.** $y = |x| + 3$ **45.** $y = |x - 4| + 3$ **46.** $y = |x + 7| - 12$

In Exercises 47–58, match the equation or inequality with its graph. (2.5, 2.6)

47. $y = -\frac{1}{2}x + 2$ e **48.** $y = \frac{1}{2}x - 2$ a **49.** $y = 2$ j **50.** $x = -2$ c

51. $y = |x + 4| - 1$ g **52.** $y = -|x + 4| + 1$ k **53.** $y = |x - 1| + 4$ b **54.** $x = -|x + 1| - 4$ d

55. $2x < 4$ h **56.** $3y > 6$ l **57.** $2x + 3y < 6$ f **58.** $2x - y > 4$ i

a.  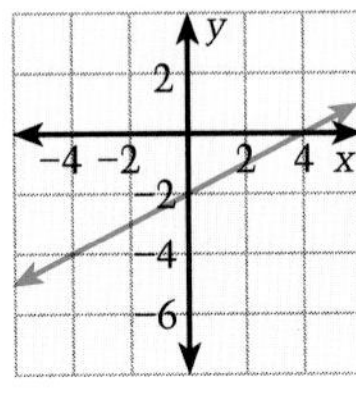

b. 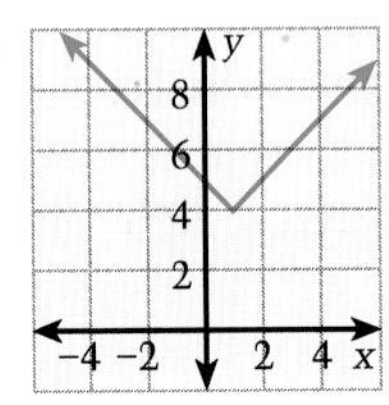

c. 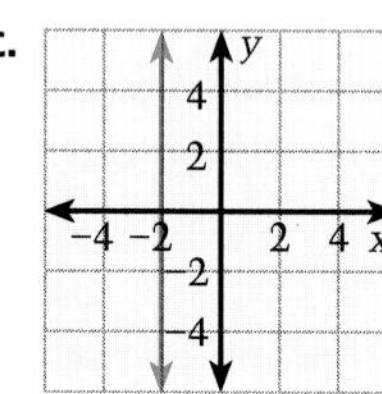

d. 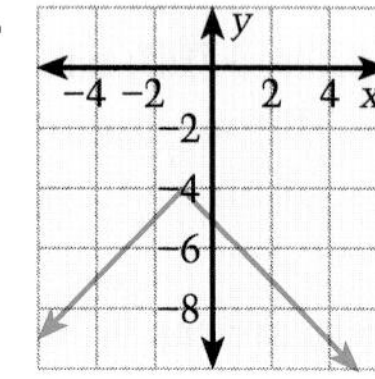

e. 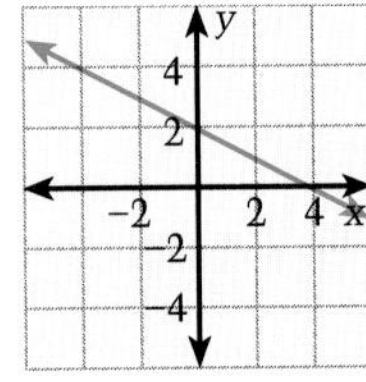

f. 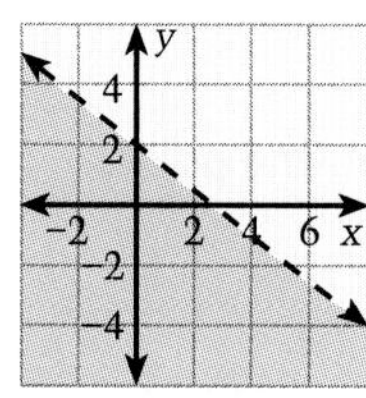

g. 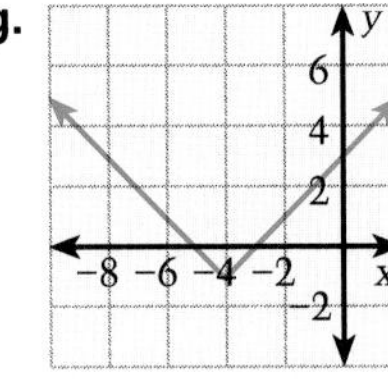

h. 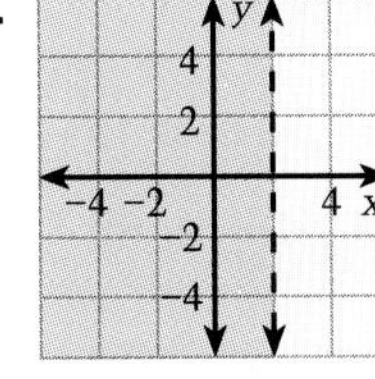

i. 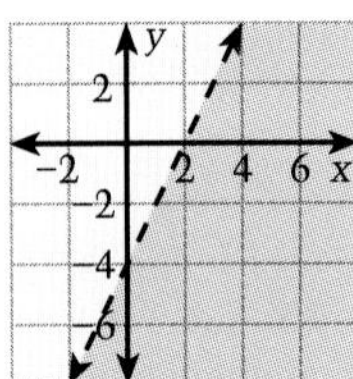

j. 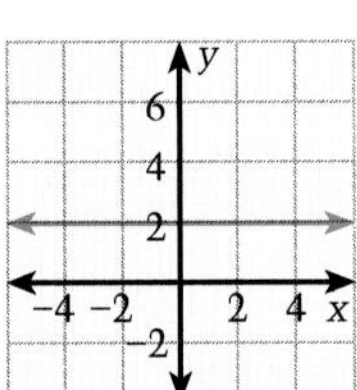

k. 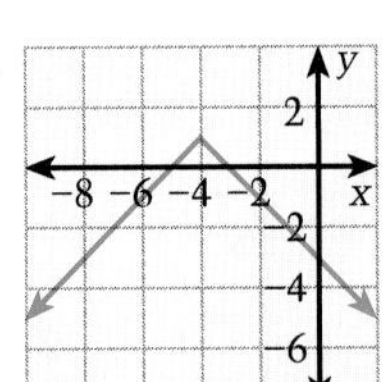

l. 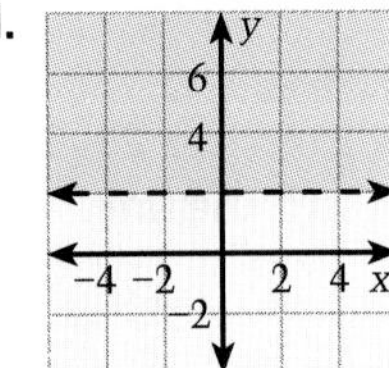

Medical School Applicants **In Exercises 59 and 60, use the following information.**

The graph at the right shows the number of applicants, a, and the number of students accepted, s, at American medical schools in the years from 1982 through 1990. Let t represent the year, with $t = 2$ corresponding to 1982. ***(Source: Association of American Medical Colleges)***

Getting Into Medical School

Thousands of Students: 0, 10, 20, 30, 40

36,000; 29,000; Applicants; Accepted; 17,000

Year: 1982, 1984, 1986, 1988, 1990

59. Find the average annual rate of change in the number of applications to medical schools between 1982 and 1990. −875 per year

60. Find an equation for the number of applicants accepted to medical schools per year between 1982 and 1990. $s = 17{,}000$

The Tomb of Tutankhamen **In Exercises 61–63, use the following information.**

Tutankhamen served as king of Egypt from the age of nine to about the age of eighteen. His underground 4-room tomb was discovered in 1922. The tomb contained over 5000 objects, including many art objects. In the diagram at the right, a coordinate plane has been superimposed over the tomb. Each unit in the coordinate plane represents 5 feet.

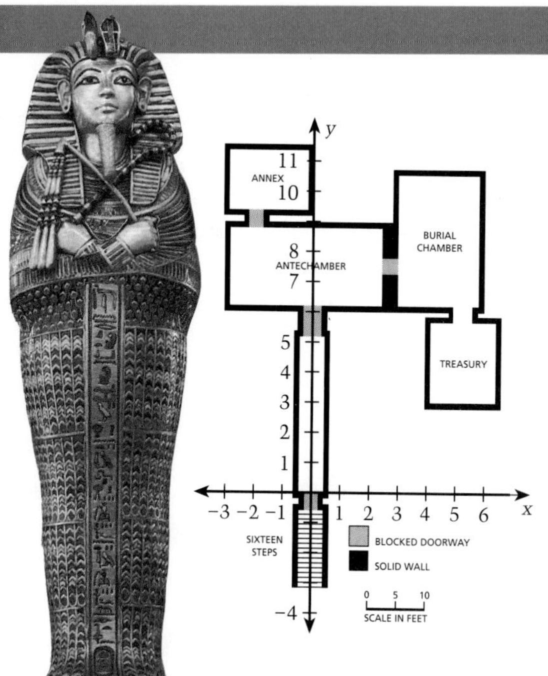

61. If you walk from the doorway at the foot of the steps to the point (4, 7.5), where are you standing? In the burial chamber

62. Approximate the area of the burial chamber. 337.5 ft^2

63. Each of the 16 steps are $\frac{3}{4}$-foot high and $\frac{3}{4}$-foot wide. Find the slope of the staircase. 1

Oxygen Use **In Exercises 64–67, use the following information.**

While at rest, a person in normal health inhales about 12.5 liters of oxygen per minute of which about 0.5 liter is absorbed by the blood through the person's lungs. The rest is dispelled when the person exhales. The amount, A (in liters), of oxygen used varies directly with the amount, I (in liters), that is inhaled.

64. Write an equation that relates the amount of oxygen absorbed, A, and the amount inhaled, I. $A = \frac{1}{25}I$

✪ **65.** The maximum amount of oxygen that a person's body is capable of using is about 3.5 liters per minute. (This would normally occur during heavy exercise.) At this rate, how much oxygen would be inhaled each minute?

✪ **66.** How much oxygen does your body absorb in a normal day? 720 liters

✪ **67.** During photosynthesis, 1 square foot of plant surface produces about 0.1 liter of oxygen per minute. Photosynthesis occurs only during daylight. Approximate the number of square feet of plant surface needed to replenish the oxygen that your body absorbs in 24 hours. Assume a day has 12 hours of sunlight. 10 ft^2

Plants produce oxygen during photosynthesis.

65. 87.5 liters

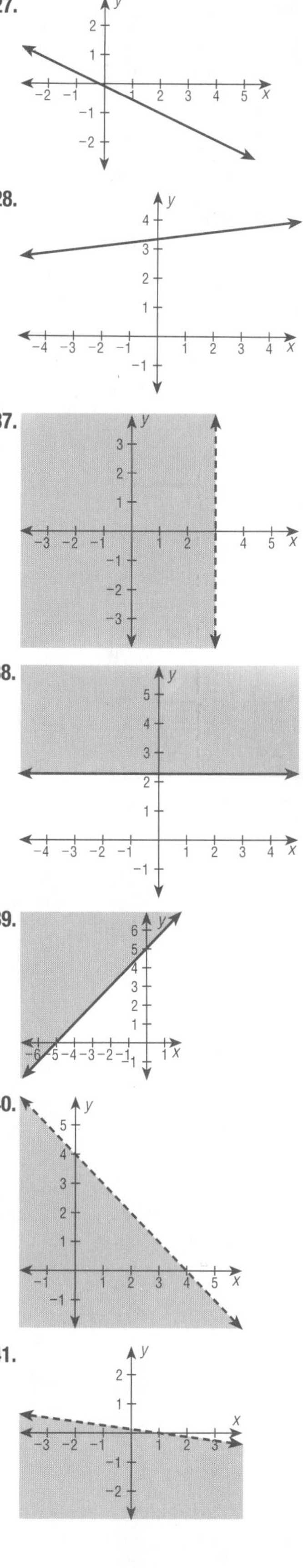

Answers
Chapter Review

42. 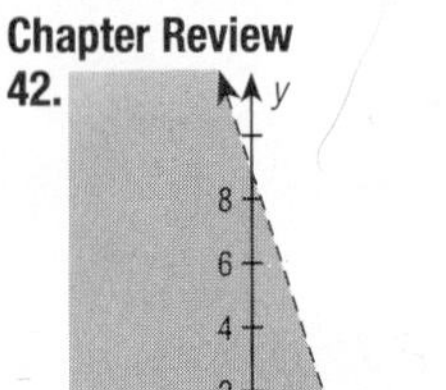
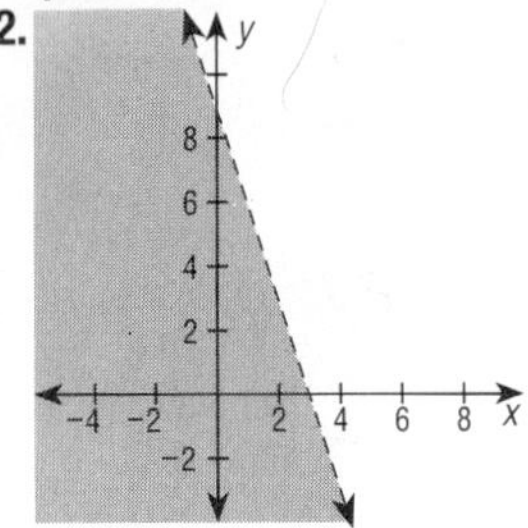

43. 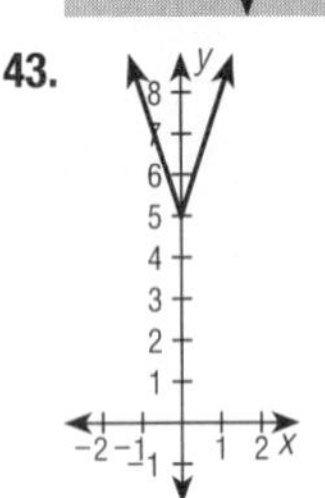

44. 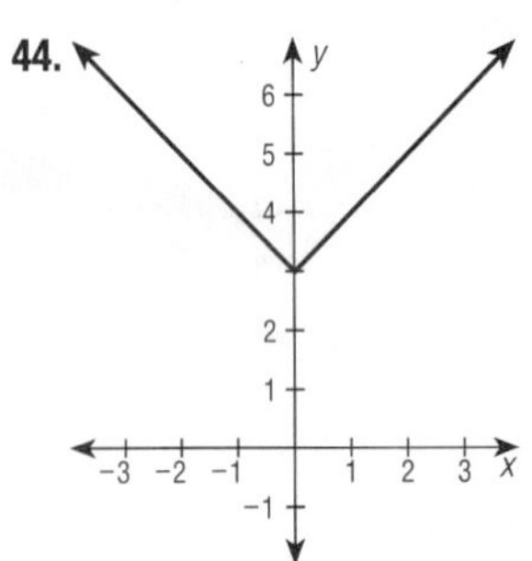

45. 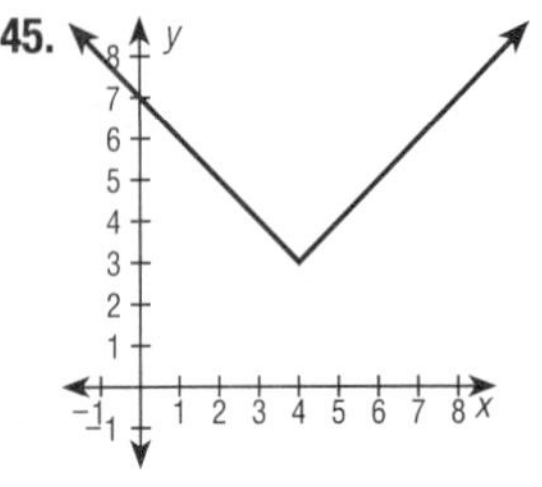

46. 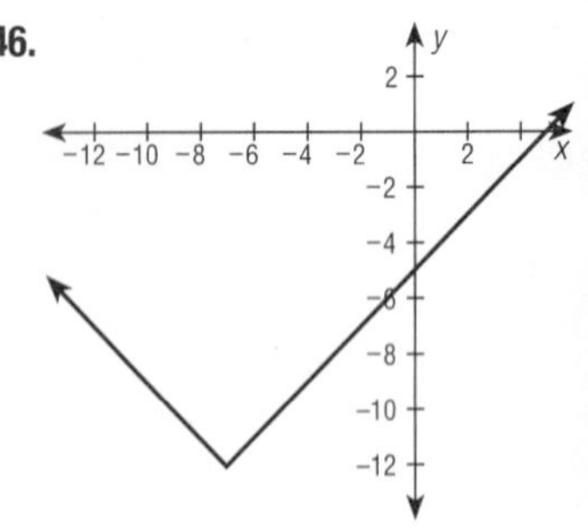

Low-Profile Tires **In Exercises 68–70, use the following information.**

Automobile tires sold in the United States have codes that give information about the tire. For instance, the meaning of P 205/60 R 13 85H is given at the right.

A 75% ratio of tire height to width is standard, 70% is low-profile, and 60% is ultralow-profile.

- Passenger car
- 205 mm ≈ cross section width
- 60% ratio of tire height to width
- Radial tire
- 13 inch wheel diameter
- 85 load index
- H speed rating

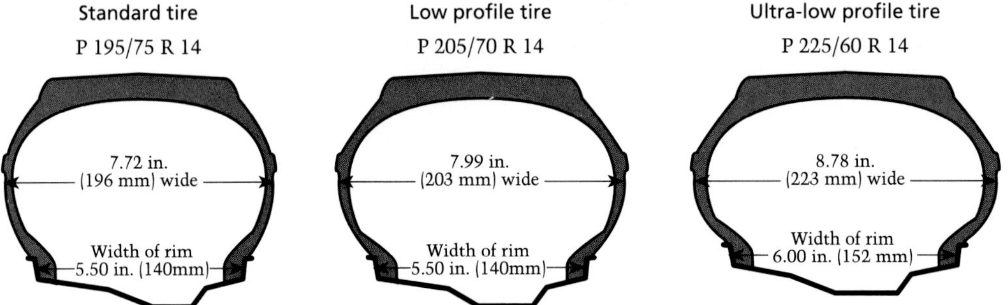

68. As recommended by *Motor Trend* magazine, the maximum width, W (in millimeters), of a tire for your car varies directly with the width, r (in millimeters), of the rim of one of the car's wheels. For a rim width of 152 millimeters, *Motor Trend* recommends a tire be no wider than 215 millimeters. Write an inequality that relates the maximum tire width, W, with the wheel width, r.

$W \le \frac{215}{152} r$, standard only

69. The speedometer on a car depends on the number of wheel revolutions per minute. Changing the size of the tires on a car can alter the accuracy of the car's speedometer. Let d be the diameter (in inches) of the tire. Let R be the number of revolutions per minute. Then the speed, s (in miles per hour), is given by the following.

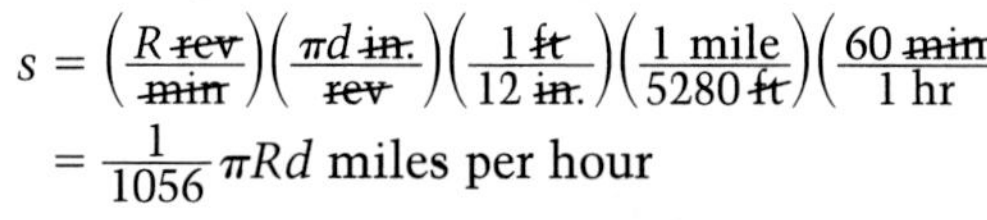

$$s = \left(\frac{R \text{ rev}}{\text{min}}\right)\left(\frac{\pi d \text{ in.}}{\text{rev}}\right)\left(\frac{1 \text{ ft}}{12 \text{ in.}}\right)\left(\frac{1 \text{ mile}}{5280 \text{ ft}}\right)\left(\frac{60 \text{ min}}{1 \text{ hr}}\right)$$
$$= \frac{1}{1056}\pi R d \text{ miles per hour}$$

The standard tire has a diameter of $d = 14 + 2(0.75)(7.72) \approx 25.6$ in. If your speedometer was set for this tire and you switched to the low-profile tire shown, would the readings on your speedometer become too high or too low? Explain.

Too high; the standard tire has a smaller diameter than the low-profile tire, so the speedometer would assume a greater distance was being traveled for each revolution of the tires.

70. The load index, I, used on tires can be used to find the maximum load, L, that should be placed on the tire as shown in the scatter plot. Is the relationship between I and L linear? Explain.
Yes, the points lie almost in a line.

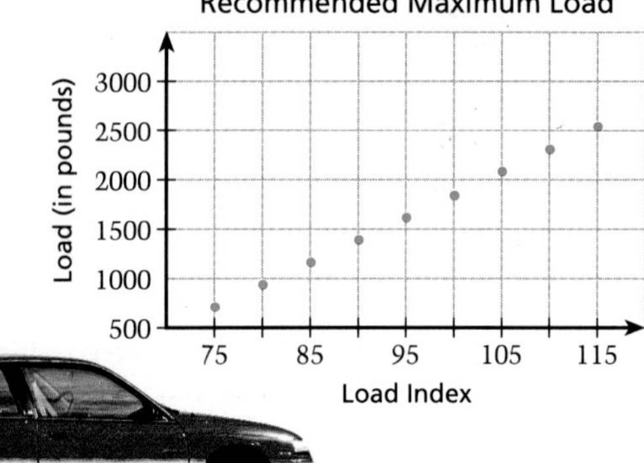

Chapter TEST

1. Find the slope of the line containing $(-1, 3)$ and $(-4, 5)$. **(2.2)** $-\frac{2}{3}$
2. At 3:10 P.M. a skydiver jumps from an airplane at an altitude of 15,000 feet. At 3:10:26 P.M., at an altitude of 4200 feet, the skydiver opens the parachute. What was the skydiver's average speed during the 26-second free-fall? **(2.2)** ≈415 ft per second
3. Sketch the graph of $-6x + 2y = 4$. See Additional Answers.
4. What is the slope of $2x + 4y = 15$? $-\frac{1}{2}$
5. What is the y-intercept of the line $-3x + 5y = 25$? **(2.3)** 5
6. What is the x-intercept of the line $8x - 7y = 16$? **(2.3)** 2

In Exercises 7 and 8, decide whether the lines are parallel, perpendicular, or neither. (2.2)

7. $-3x - 5y = 10$, $-3x + 5y = 10$ Neither
8. $-3x - 5y = 10$, $-5x + 3y = 10$ Perpendicular
9. Write an equation of the line containing $(5,-2)$ and $(7,-3)$. **(2.4)** $y = -\frac{1}{2}x + \frac{1}{2}$
10. Write an equation of the line that passes through $(1, 4)$ and is perpendicular to the line $y = -3x + 1$. **(2.4)** $y = \frac{1}{3}x + \frac{11}{3}$ 11.–14. See Additional Answers.
11. Sketch the graph of $4x - 7y \geq 28$. **(2.5)**
12. Sketch the graph of $y \leq -2$. **(2.5)**
13. Sketch the graph of $y = |x - 3| - 1$. **(2.6)**
14. Sketch the graph of $y = -2|x + 1| + 3$. **(2.6)**

In Exercises 15–18, match the inequality with its graph. (2.5)

15. $3x - 2y \leq 4$ d
16. $-3x + 2y \leq 4$ b
17. $3x + 2y \geq 4$ a
18. $3x - 2y \geq 4$ c

a.
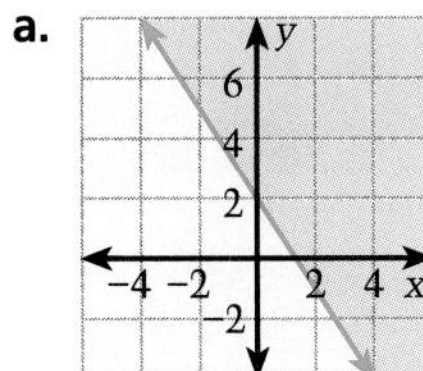

b.
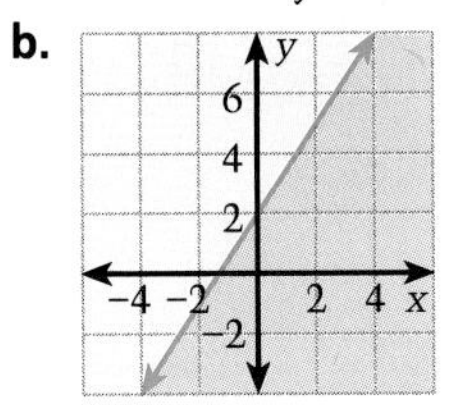

c.
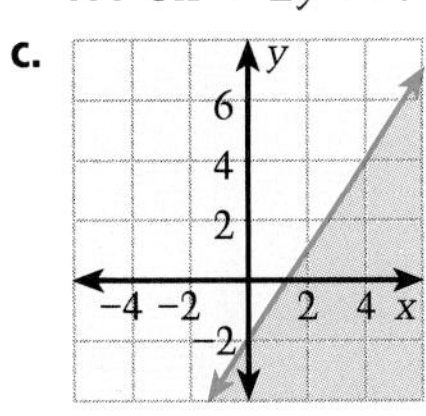

d.
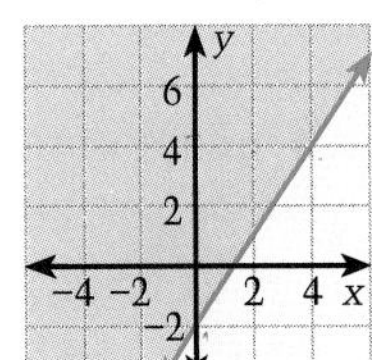

19. The height, b (in feet), that a basketball will bounce on a hardwood floor varies directly with the height, h (in feet), from which the ball is *dropped.* A basketball dropped from a height of 5 feet will bounce 2.8 feet. Write an equation that relates b and h. **(2.4)** $b = 0.56h$
20. You begin walking downtown. After 3 minutes you are $1\frac{1}{4}$ miles from your destination. Is this enough information to determine your average speed? Explain. **(2.2)** See below.
21. The scatter plot shows the number, s (in millions), of time-share owners of condominiums and houses worldwide for 1980 through 1990. Let t represent the year, with $t = 0$ corresponding to 1980. Write an equation of a line that you think best fits this data. ***(Source: American Resort and Residential Development Association)*** **(2.7)**
Answers vary, $s = \frac{1}{6}t + \frac{3}{25}$

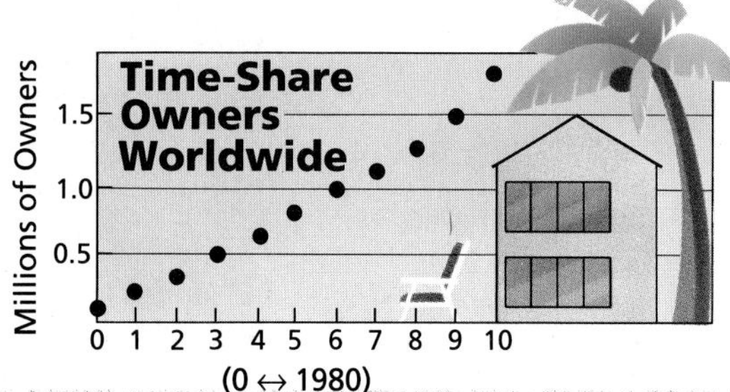

20. No, you need to know the distance walked.

The three basic techniques for solving systems of linear equations—graphing, substitution, and linear combinations—are presented in this chapter. The graphing technique has become a more robust approach to solving systems because of the use of computers and graphics calculators. Consequently, students must develop an awareness of when it is better to use one approach over another.

The Chapter Summary on page 165 provides you and the students with a synopsis of the chapter. It identifies key skills and concepts. You may want to have students look at the Chapter Summary as an overview before beginning the chapter.

CHAPTER 3

Systems of Linear Equations and Inequalities

LESSONS

The commercial harbor in San Juan, Puerto Rico, channels the country's exports to the rest of the world.

Real Life
Exports

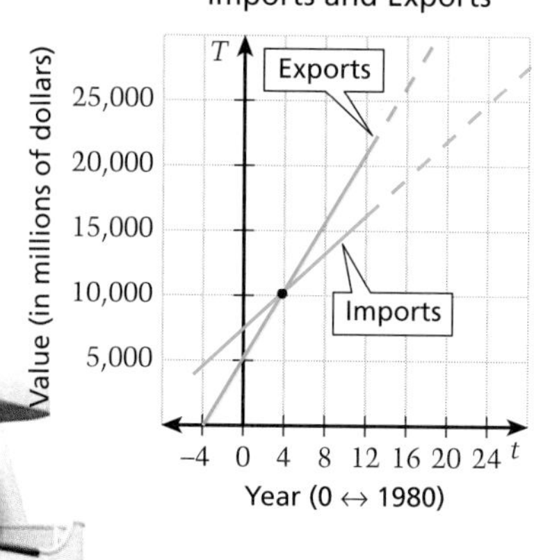

Puerto Rico exports many agricultural products, as well as clothing, medical supplies, petroleum products, and machinery.

For 1975 through 1992, Puerto Rico's total exports and imports, T (in millions of dollars), can be modeled by

$$\begin{cases} T = 1312t + 5166 & \text{Exports} \\ T = 723t + 7429 & \text{Imports} \end{cases}$$

where $t = 0$ corresponds to 1980. The graph of this *linear system* is two lines that intersect when $t = 3.8$. Up to about 1984, Puerto Rico had an unfavorable balance of trade, which means that it imported more than it exported. Since 1984, however, its balance of trade has been favorable, which means it has exported more than it has imported.

(Source: U.S. Bureau of the Census)

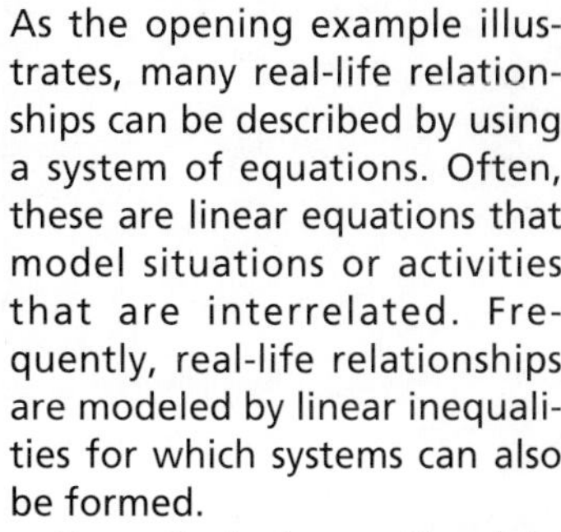

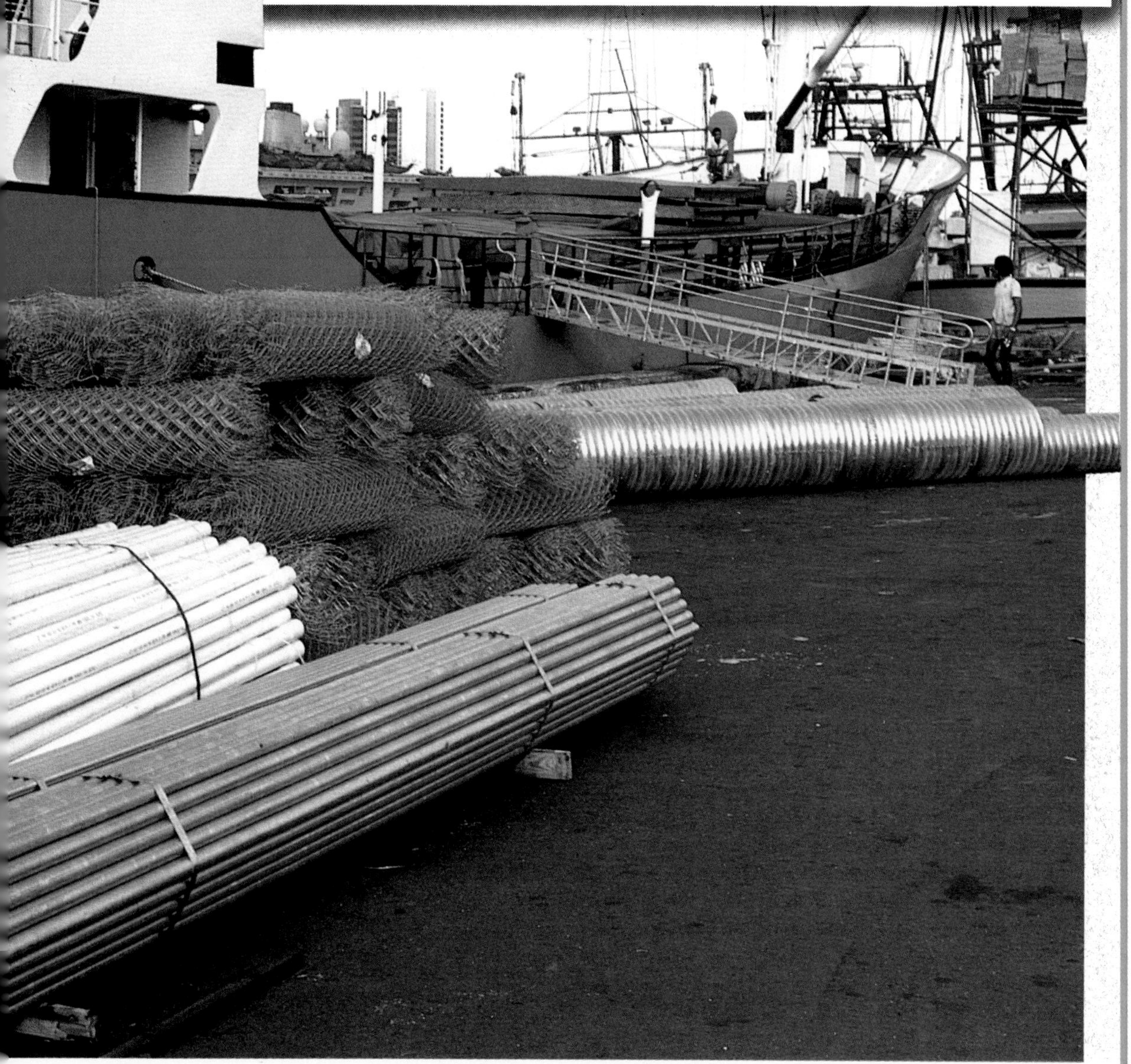

As the opening example illustrates, many real-life relationships can be described by using a system of equations. Often, these are linear equations that model situations or activities that are interrelated. Frequently, real-life relationships are modeled by linear inequalities for which systems can also be formed.

Have students use the data to answer these questions.

a. How much more (in millions of dollars) was exported by Puerto Rico in 1992 than was imported?
4805 million dollars

b. How does this compare to the export/import difference in 1980?
In 1980, 2263 million dollars more was imported than exported! Quite a dramatic turnaround.

Systems of Linear Equations and Inequalities

PACING CHART

The daily Pacing Chart is meant to help you adjust your teaching pace. Students in the full course should finish the entire text by the end of the year. Students in the basic course are expected to complete the first twelve chapters. The Pacing Chart for each chapter contains suggestions for lessons that require more than one day and lessons that may be omitted for the basic course.

DAY	FULL COURSE	BASIC COURSE
1	3.1 & Using a Graphing Calculator	3.1
2	3.2	3.1 & Using a Graphing Calculator
3	3.3	3.2
4	Mid-Chapter Self-Test	3.3
5	3.4	3.3
6	3.5	Mid-Chapter Self-Test
7	3.6	3.4
8	Chapter Review	3.5
9	Chapter Test	3.5
10	Cumulative Review 1–3	3.6
11		3.6
12		Chapter Review
13	Chapter Test	Chapter Test
14		Cumulative Review 1–3

CHAPTER ORGANIZATION

LESSON	PAGES	GOALS	MEETING THE NCTM STANDARDS
3.1	122–127	1. Graph and solve a system of linear equations 2. Use a system of linear equations to answer questions about a real-life situation	Problem Solving, Communication, Reasoning, Connections
Using a Calculator	128–129	Graph a linear system and approximate the solution	Technology
3.2	130–136	1. Use algebraic methods to solve a linear system 2. Use a linear system to answer questions about a real-life situation	Problem Solving, Communication, Reasoning, Connections
Mixed Review	137	Review algebraic and arithmetic skills	
Career Interview	137	Air Traffic Controller	Connections
3.3	138–143	Write and use linear systems to model real-life situations	Problem Solving, Communication, Connections, Statistics, Discrete Math
Mid-Chapter Self-Test	144	Diagnose student weaknesses and remediate with correlated Reteach worksheets	
3.4	145–150	1. Graph a system of linear inequalities to find the solution to the system 2. Use a system of linear inequalities to model a real-life situation	Problem Solving, Communication, Reasoning, Connections, Discrete Math
3.5	151–156	1. Solve a linear programming problem 2. Use linear programming to answer questions about real-life situations	Problem Solving, Communication, Connections, Discrete Math
Mixed Review	156	Review algebraic and arithmetic skills	
3.6	157–164	1. Solve a system of linear equations in three variables 2. Use a system in three variables to answer questions about real-life situations	Problem Solving, Communication, Connections
Chapter Summary	165	Restate for students what they have learned, why they have learned it, and how it fits into the structure of algebra	Structure, Connections
Chapter Review	166–168	Review concepts and skills learned in the chapter	
Chapter Test	169	Diagnose student weaknesses and remediate with correlated Reteaching worksheets	
Cumulative Review 1–3	170–171	Review concepts and skills from chapters 1–3	

LESSON RESOURCES

MEETING INDIVIDUAL NEEDS

RETEACHING For students who need to spend more time on basics:

If a mid-chapter self-test or chapter test indicates a deficiency, teachers can help students with the appropriate ***Reteaching Copymaster.***

PRACTICE For students who need more practice:

Additional exercises like those in the Pupil's Edition are provided for each lesson in ***Extra Practice Copymasters.***

ENRICHMENT For enriching and broadening students' experiences:

Problem of the Day copymasters in ***Teaching Tools*** provide a daily opportunity to use logical reasoning, looking for a pattern, writing an equation, and other routine and non-routine problem-solving strategies.

Math Log copymasters in ***Alternative Assessment*** provide opportunities to report on investigations, research, and open-ended problems.

Technology: Using Calculators and Computers provides enriching activities with graphing and scientific calculators and computers.

The ***Applications Handbook*** provides additional information about the cross-curriculum topics such as astronomy, chemistry, physics, sports, economics, genetics, and music that are integrated into the Pupil's Edition.

LESSON	7.1	7.2	7.3	7.4	7.5	7.6
PAGES	346-351	352-358	360-366	368-373	374-380	381-387
Teaching Tools						
Transparencies	✓	✓	✓	✓	✓	
Problem of the Day	✓	✓	✓	✓	✓	✓
Warm-up Exercises	✓	✓	✓	✓	✓	✓
Answer Masters	✓	✓	✓	✓	✓	✓
Extra Practice Copymasters	✓	✓	✓	✓	✓	✓
Reteaching Copymasters	Teacher-directed and independent activities tied to results on the Mid-Chapter Self-Tests and Chapter Tests					
Color Transparencies	✓	✓	✓	✓	✓	
Applications Handbook	Additional background information for many real-life applications					
Technology	Calculator and computer worksheets for appropriate lessons					
Complete Solutions Manual	✓	✓	✓	✓	✓	✓
Alternative Assessment	Assess student's ability to reason, analyze, solve problems, and communicate using mathematical language.					
Formal Assessment	Mid-Chapter Self-Tests, Chapter Tests, Cumulative Tests, and Practice for College Entrance Tests					
Computer Test Bank	Customized tests can be created by choosing from over 2500 items.					

INSIGHTS

3.1 Solving Linear Systems by Graphing

It is important for students to master the technique of graphing linear systems to provide them with a picture of the algebraic solution. With the increasing use of computers and scientific calculators, graphing has become an efficient way to solve systems of linear equations, particularly in real-life problems, such as those found in engineering or economics.

3.2 Solving Linear Systems Algebraically

While a graph can "picture" the approximate solution of a linear system, the most efficient way to find an exact solution is by algebraic means, using the techniques of substitution or linear combinations. These techniques are also important in other branches of mathematics, including calculus and linear algebra. Students may find it interesting that most computer applications that solve systems of linear equations use algebraic techniques.

3.3 Problem Solving Using Linear Systems

Solving real-life problems often involve systems of linear equations. In this lesson, students will have an opportunity to strengthen their understanding of systems and how to use systems to solve problems.

3.4 Solving Systems of Linear Inequalities

Systems of linear relationships are not necessarily equations. Many involve linear inequalities. Applications in economics, business, physics, calculus, and statistics often require solving systems of linear inequalities. An important class of problems in mathematics, known as optimization problems, relies heavily on such systems. (See Lesson 3.5.)

3.5 Exploring Data: Linear Programming

A common application of solving systems of linear inequalities is found in business problems of optimization and cost-effectiveness, in which the minimum or maximum value of a quantity (usually either cost or profit) is desired. Optimization problems are yet another way in which students investigate data.

3.6 Solving Systems of Linear Equations in Three Variables

Certain real-life situations are modeled by systems of three linear equations in three variables. In this lesson, students will be given general techniques for solving such systems, which are also applicable to systems of linear equations involving more than three variables. It should be pointed out to students that some of these more general techniques are employed in computer applications for solving systems of equations.

Problem of the Day

Right triangle ABC is inscribed in a quarter of a circle. What is the length of $\overline{AC}$? 10

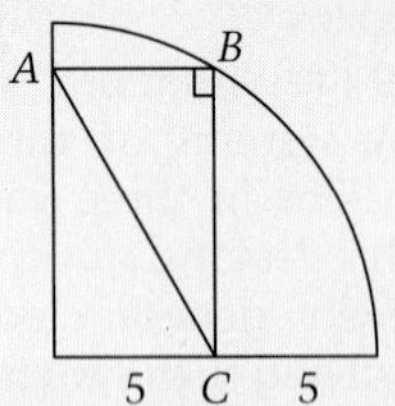

ORGANIZER

Warm-Up Exercises

1. Graph each linear equation.
   a. $y = 3x - 6$
   b. $4x + 5y = 10$
   c. $8 - 2y = 6x$
   a. The graph is a line with intercepts (2, 0) and (0, −6).
   b. The graph is a line with intercepts (2.5, 0) and (0, 2).
   c. The graph is a line with intercepts $(1\frac{1}{3}, 0)$ and (0, 4).
2. Explain what the solutions of the linear equation $y = 2x + 1$ are.
   Any ordered pair (a, b) for which $y = 2x + 1$ is a true statement when $x = a$ and $y = b$.

Lesson Resources

Teaching Tools
- Transparencies: 2, 3, 8, 9
- Problem of the Day: 3.1
- Warm-up Exercises: 3.1
- Answer Masters: 3.1

Extra Practice: 3.1
Color Transparencies: 18, 19
Technology Handbook: p. 17

LESSON Notes

Example 1

Carefully make the statement that (1, 3) is a solution of *each* linear equation in the system.

3.1 Solving Linear Systems by Graphing

What you should learn:

Goal 1 How to graph and solve a system of linear equations in two variables

Goal 2 How to use a system of linear equations in two variables to answer questions about a real-life situation

Why you should learn it:

You can use a system of linear equations as a model to find the number of test items answered correctly and incorrectly on a standardized test.

Goal 1 Graphing and Solving a System

A **system of two linear equations in two variables** x and y consists of two equations of the following form.

$$\begin{cases} Ax + By = C & \text{Equation 1} \\ Dx + Ey = F & \text{Equation 2} \end{cases}$$

A **solution** of a system of linear equations in two variables is an ordered pair (x, y) that is a solution of both equations.

A system of two linear equations in two variables can have exactly one solution, no solution, or infinitely many solutions. To see why this is true, consider the following graphs.

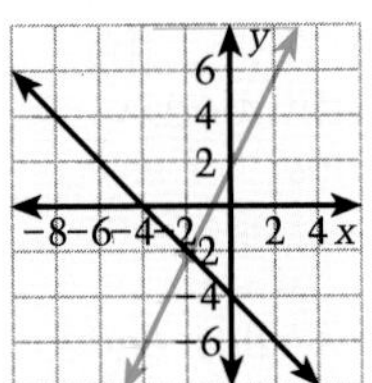

A system of linear equations whose graph is intersecting lines has exactly one solution.

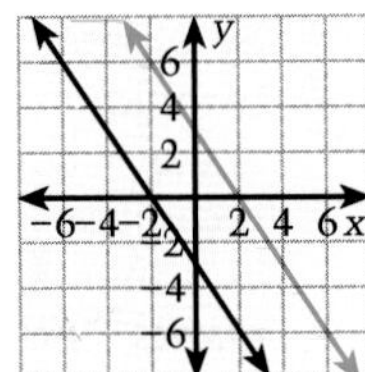

A system of linear equations whose graph is two parallel lines has no solution.

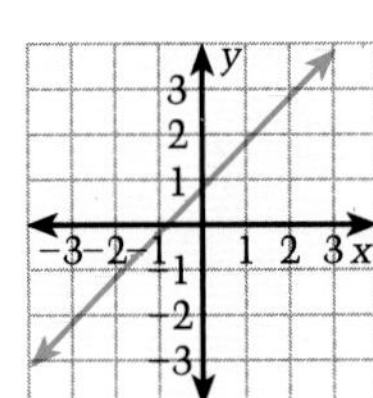

A system of linear equations whose graph is the same line has infinitely many solutions.

Example 1 *Checking a Solution*

Check that (1, 3) is a solution of the system.

$$\begin{cases} 5x - 3y = -4 & \text{Equation 1} \\ x + 2y = 7 & \text{Equation 2} \end{cases}$$

Solution To check (1, 3) as a solution, substitute 1 for x and 3 for y in each equation.

Equation 1 check: $5(1) - 3(3) = 5 - 9 = -4$

Equation 2 check: $(1) + 2(3) = 1 + 6 = 7$

To check the solution graphically, sketch a quick graph of each equation. The coordinates of the point of intersection will be the solution. ∎

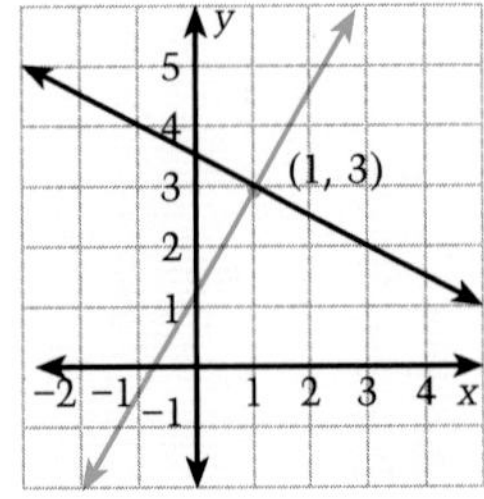

The **graph of a system** can be used to approximate the solution of the system. Be sure to check such a solution algebraically to ensure that the coordinates of the point are exact.

Example 2 *The Graph-and-Check Method*

Solve the system: $\begin{cases} 4x + 5y = -3 & \textit{Equation 1} \\ -x + \ \ y = \ \ 3 & \textit{Equation 2} \end{cases}$.

Solution To sketch the two lines, first write the equations in slope-intercept form.

$$\begin{cases} y = -\frac{4}{5}x - \frac{3}{5} & \textit{Equation 1} \\ y = \ \ \ \ \ \ \ x + 3 & \textit{Equation 2} \end{cases}$$

The two lines seem to intersect at $(-2, 1)$. To check, substitute -2 for x and 1 for y in *each* equation of the original system.

$$4(-2) + 5(1) \stackrel{?}{=} -3 \qquad\qquad -(-2) + 1 \stackrel{?}{=} 3$$
$$-3 = -3 \ \ \textit{(checks)} \qquad\qquad 3 = 3 \ \ \textit{(checks)} \quad ■$$

College Entrance Exam

Example 3 *Writing a Linear System as a Model*

On a college entrance exam, you answered 80 of the 85 questions. Each correct answer adds 1 point to your raw score, each unanswered question adds nothing, and each incorrect answer subtracts $\frac{1}{4}$ point. Your raw score was 70. How many questions did you answer correctly?

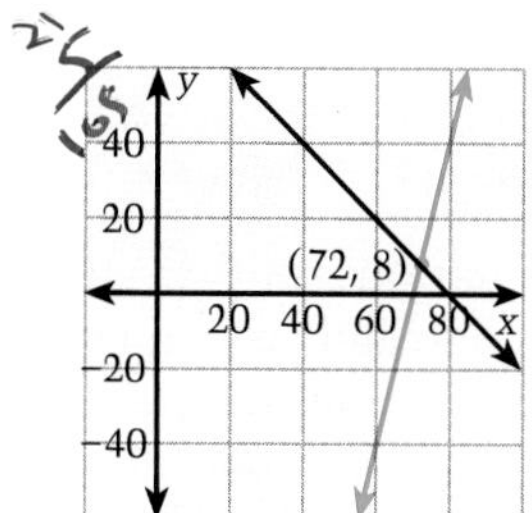

Solution

Verbal Model

Number correct + Number incorrect = Questions answered

1 point · Number correct + $-\frac{1}{4}$ point · Number incorrect = Raw score

Labels

Number correct $= x$	(Questions)
Number incorrect $= y$	(Questions)
Total answered $= 80$	(Questions)
Raw score $= 70$	(Points)

System

$$\begin{cases} x + \ \ y = 80 & \textit{Equation 1} \\ x - \frac{1}{4}y = 70 & \textit{Equation 2} \end{cases}$$

Sketch the graph of the system. The two lines appear to intersect at (72, 8). This solution *checks* with the original problem.

$$72 + 8 = 80 \qquad \textit{You answered 80 questions.}$$
$$1(72) - \left(\frac{1}{4}\right)(8) = 70 \qquad \textit{Your raw score was 70.}$$

Thus you answered 72 questions correctly. ■

Example 2

Note that it is not necessary to solve for y in each of the linear equations of the system in order to graph them. Some students may want to find the x- and y-intercepts in order to graph the equations.

Example 3

This is a good example of the use of a system of linear equations to solve a problem. A sketch of the graph of the system should be done on an appropriate coordinate system and estimates of points of intersection can be made. The tic marks on both the x- and y-axes should be scaled about 5 units apart or less. The intersection of the two lines should be at integer values because the number of questions answered correctly (or incorrectly) must be a whole number.

Example 4

It may be necessary to help students further in understanding the concept of "breaking even." Begin by asking what the cost is when $t = 0, t = 1, t = 10$, etc. What is the revenue when $t = 0$, $t = 1$, $t = 10$? Finally, ask students how the cost and revenue are related to the success of the business.

Extra Examples

Here are additional examples similar to **Examples 1–4.**

1. Check whether the given point is a solution of the system of linear equations.

   a. $(-2, 3)$; $\begin{cases} x - 6y = 20 \\ 3x + 2y = 0 \end{cases}$

   b. $(4, -9)$; $\begin{cases} 11x - 2y = 62 \\ 3x + y = 3 \end{cases}$

   a. no, b. yes

2. Solve each system of linear equations.

a. $\begin{cases} 3x + 8y = -25 \\ 9x - 2y = -23 \end{cases}$

b. $\begin{cases} 6x - 7y = 25, \\ 3x + 16y = -7 \end{cases}$

a. $(-3,-2)$, b. $(3,-1)$

3. Write a linear system as a model of the following situation, then solve.

Your school's star basketball player was excited about her play during a recent game. She remembered that she scored two 3–point shots, but could not recall how many free throws (worth 1 point each) and field goals (worth 2 points each) she scored. The scorekeeper said she had scored 20 times for 34 points. How many field goals did she make? How many free throws did she make?

The system becomes

$\begin{cases} x + y = 18 \\ 2x + y = 28 \end{cases}$

The solution is $x = 10$ field goals, $y = 8$ free throws.

Check Understanding

1. What is a solution of a system of linear equations?
See page 122.

2. A system of linear equations may have infinitely many solutions. Explain how this is possible.
The lines may be coincident.

3. After the solution of a system of linear equations is found graphically, why should the solution be checked algebraically?
Graphical answers are often approximations.

In 1992, there were about 20 million businesses in the United States. Of these, 15.5 million were owned by a sole proprietor, that is, a single person or family. The circle graph at the right shows the percents of these businesses in each revenue category. (Source: Internal Revenue Service)

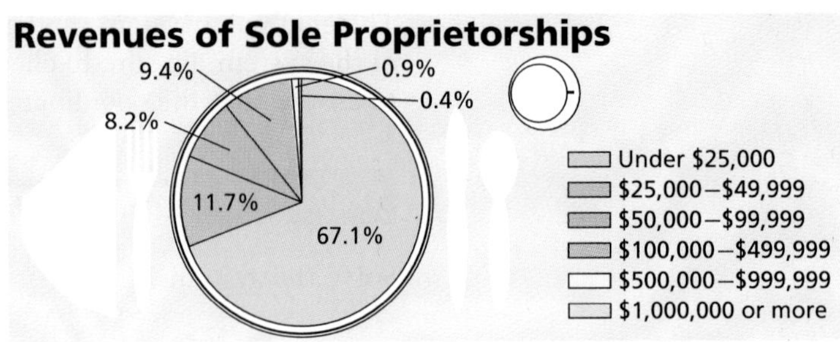

Goal 2 Using Linear Systems as Real-Life Models

Linear systems can be used to find the *break-even point* of a business. For a new business, the break-even point occurs when the *revenue* (the income from sales) is equal to the *cost* (the amount the business has spent).

Real Life **Business**

Example 4 *Finding the Break-Even Point*

You have purchased a flower shop. The previous owner's records for the past three years show the average monthly cost was \$5400. The average monthly revenue was \$8600. You paid \$153,600 for the shop. If monthly costs and revenues remain the same, how long will it take to break even?

Solution Let R represent the revenue you bring in during the first t months. A model for the revenue is $R = 8600t$.

Let C represent your cost, including purchase price, during the first t months. A model for the cost is $C = 5400t + 153{,}600$.

By sketching the graphs of both equations on the same coordinate plane, it appears that the revenue and cost equations intersect at 48 months. Substituting $t = 48$ in either equation will give the other coordinate, 412.8. Thus your break-even point occurs at 48 months, at which time you will have spent and taken in the same amount: \$412,800. ■

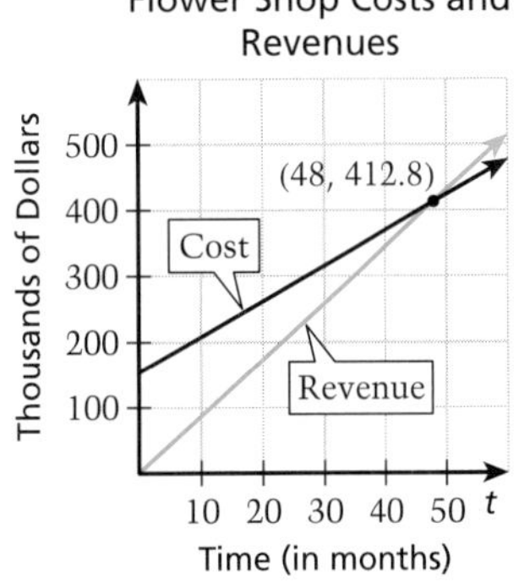

Communicating about ALGEBRA

SHARING IDEAS about the Lesson See Additional Answers.

Solve the linear systems graphically. Explain your steps.

A. $\begin{cases} 3x - 4y = 5 \\ -2x + y = -5 \end{cases}$ B. $\begin{cases} 3x - 4y = 5 \\ -3x + 4y = -5 \end{cases}$ C. $\begin{cases} 3x - 4y = 5 \\ -3x + 4y = 5 \end{cases}$

EXERCISES

Guided Practice

CRITICAL THINKING about the Lesson See Additional Answers for **1.–3.**

1. The graph of a system of two linear equations in two variables is a pair of lines. How can you use the graph to decide how many solutions the linear system has?

2. A linear system can have exactly one solution, no solution, or infinitely many solutions. Explain why a linear system in two variables cannot have exactly two solutions.

3. Sketch the graph of the following system and use the graph to find the solution. Check the solution algebraically.

$$\begin{cases} x + 5y = -1 \\ x + 4y = 0 \end{cases}$$

In Exercises 4–6, match the linear system with its graph. How many solutions does the system have?

4. $\begin{cases} 2x - y = -5 \\ x + 2y = 0 \end{cases}$ c, 1

5. 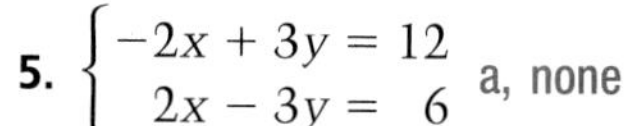 $\begin{cases} -2x + 3y = 12 \\ 2x - 3y = 6 \end{cases}$ a, none

6. $\begin{cases} 2x - y = 5 \\ -4x + 2y = -10 \end{cases}$ b, infinitely many

a.

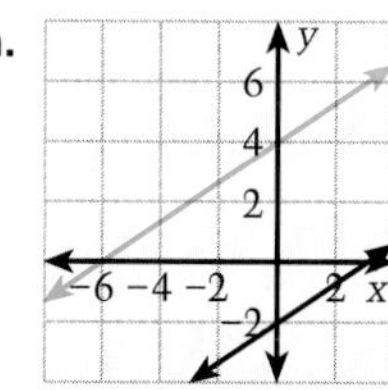

b.

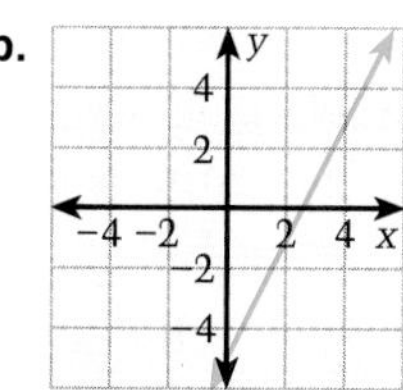

c.

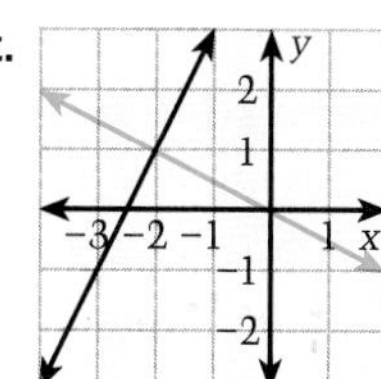

Independent Practice

In Exercises 7–12, how many solutions does the linear system have?

7. $\begin{cases} 24x - 9y = 3 \\ 8x - 3y = 1 \end{cases}$ Infinitely many

8. $\begin{cases} x + 4y = 9 \\ -2x + y = 0 \end{cases}$ 1

9. $\begin{cases} 2x + 8y = 16 \\ -3x + 6y = 30 \end{cases}$ 1

10. $\begin{cases} -x + 3y = 3 \\ 2x - 6y = 30 \end{cases}$ None

11. $\begin{cases} 2x + y = -1 \\ -4x - 2y = -5 \end{cases}$ None

12. $\begin{cases} -x + 2y = 16 \\ 4x - 8y = -64 \end{cases}$ Infinitely many

In Exercises 13–18, decide whether the ordered pair is a solution of the linear system. Use an algebraic check.

13. $(5, -2)$ It is.
$\begin{cases} 5x - y = 27 \\ -3x + 4y = -23 \end{cases}$

14. $(4, 0)$ It is not.
$\begin{cases} -x + 6y = 4 \\ 3x + y = 12 \end{cases}$

15. $(9, 1)$ It is not.
$\begin{cases} 2x - 9y = 9 \\ -x + 8y = 17 \end{cases}$

16. $(-3, 3)$ It is.
$\begin{cases} 2x - 4y = -18 \\ 4x - 2y = -18 \end{cases}$

17. $(-2, -7)$ It is.
$\begin{cases} 4x - 3y = 13 \\ 7x - y = -7 \end{cases}$

18. $(8, 9)$ It is not.
$\begin{cases} -2x + 4y = -20 \\ 3x + y = 33 \end{cases}$

Communicating about ALGEBRA

Consider the following system of linear equations. Discuss the problems of solving this system graphically (without the use of computers or calculators). Do not solve!

$$\begin{cases} \frac{153}{37}x - 4.9y = 1.51 \\ 3.8x + \frac{93.4}{6.1}y = 4.13 \end{cases}$$

The equations are difficult to graph because the coefficients contain decimals and fractions. Their intersection would probably not occur at an integral point on the graph.

If computers or graphics calculators are available, do the same problems identified above still exist? Explain.
No, see pages 128–129 that follow.

EXERCISE Notes

ASSIGNMENT GUIDE

Basic/Average: 1–6, 7–17 odd, 23–35 odd, 38–40, 42–45, 48, 51

Above Average: Ex. 1–6, 7–19 odd, 25–37 odd, 38–40, 42–45, 56–58

Advanced: Ex. 1–6, 7–19 odd, 25–37 odd, 38–44, 57–58

Selected Answers
Exercises 1–6, 7–55 odd

✪ More Difficult Exercises
Exercises 38–41

Guided Practice

Ex. 2 GEOMETRY Remind students of a theorem from geometry which states, "If two unique lines intersect, they intersect in exactly one point."

▶ **Ex. 3** **TECHNOLOGY** Students with graphics calculators may wish to write the equations in slope-intercept form, graph the equations, and use the trace key to approximate the values of the intersection point.

Independent Practice

▶ **Ex. 7–12** These exercises emphasize the use of slope in determining the answer.

▶ **Ex. 19–36** These should be completed on graph paper. If students have access to a graphics calculator, then encourage them to use it.

▶ **Ex. 40** Ask any students who have visited the memorial to describe it. In solving the problem, have students think of "The Wall" as being opened, so that it lies in one plane. Then the top of "The Wall" is similar to the top of a basement wall that is at ground level.

Review Example 4 as a solution model. Sketching a graph is a useful visual interpretation. The top stays level as you walk down the hill to the center. Thus, the two lines intersect at $(0, -10)$ and the height of the memorial is 10 feet. To find the length of each section, find the x-intercepts.

TECHNOLOGY A graphics calculator range setting of

X min: −30 Y min: −12
X max: 130 Y max: 2
X scl: 10 Y scl: 1
X res: 1

yields a "view" of the information.

▶ **Ex. 42–44** Although the lines are not exactly linear, they do provide an interesting application of intersection and timeline graphs.

In Exercises 19–24, decide whether the ordered pair is a solution of the linear system. Use a graphic check.

19. $(0, 3)$ It is not. $\begin{cases} x + 2y = 8 \\ 2x + y = -3 \end{cases}$

20. $(4, 4)$ It is. $\begin{cases} -x + 2y = 4 \\ -3x + 4y = 4 \end{cases}$

21. $(2, -1)$ It is. $\begin{cases} -x + 2y = -4 \\ y = -1 \end{cases}$

22. $\left(\frac{2}{3}, 0\right)$ It is not. $\begin{cases} -3x + 2y = 2 \\ 3x + 2y = -2 \end{cases}$

23. $(1, 4)$ It is. $\begin{cases} -5x + y = -1 \\ -3x + y = 1 \end{cases}$

24. $(3, 4)$ It is not. $\begin{cases} 3x + 4y = 12 \\ -4x + 3y = -3 \end{cases}$

In Exercises 25–36, sketch the graph of the linear system. Use your graph to estimate the solution. Check your estimate algebraically. See Additional Answers.

25. $\begin{cases} x + 2y = 0 \\ -x + y = -3 \end{cases}$ $(2, -1)$

26. $\begin{cases} -2x + y = 5 \\ x + y = 8 \end{cases}$ $(1, 7)$

27. $\begin{cases} 2x + y = 4 \\ x + y = 3 \end{cases}$ $(1, 2)$

28. $\begin{cases} 3x + 4y = -8 \\ -5x + y = -25 \end{cases}$ $(4, -5)$

29. $\begin{cases} 3x + y = 9 \\ -3x + y = 3 \end{cases}$ $(1, 6)$

30. $\begin{cases} 2x + y = 13 \\ x - y = 5 \end{cases}$ $(6, 1)$

31. $\begin{cases} y = \frac{1}{6}x - 2 \\ y = -\frac{1}{6}x + 2 \end{cases}$ $(12, 0)$

32. $\begin{cases} y = 3x \\ y = x + 2 \end{cases}$ $(1, 3)$

33. $\begin{cases} 2x + y = -5 \\ x - 2y = 0 \end{cases}$

34. $\begin{cases} x + 4y = -24 \\ x - 4y = 24 \end{cases}$ $(0, -6)$

35. $\begin{cases} x - 3y = -3 \\ 2x + y = 8 \end{cases}$ $(3, 2)$

36. $\begin{cases} x - 2y = 0 \\ x + 2y = 4 \end{cases}$ $(2, 1)$

33. $(-2, -1)$

37. *Investing in Two Funds* A total of \$4500 is invested in two funds paying 4% and 5% annual interest. The combined annual interest is \$210. How much of the \$4500 is invested in each fund? 4%: \$1500, 5%: \$3000

✪ **38.** *Exam Score* You answered 82 questions out of 85 on a college entrance exam. Your raw score was 67. Use the information in Example 3 to find the number of questions you answered correctly. 70

✪ **39.** *Break-Even Analysis* Your family is planning to open a restaurant. You need an initial investment of \$85,000. Each week your costs will be about \$7400. If your weekly revenue is \$8000, how many weeks will it take to break even? $141\frac{2}{3}$

✪ **40.** *Vietnam Veterans Memorial* "The Wall" in Washington, D.C., designed by Maya Ling Lin when she was a student at Yale University, has two vertical, triangular sections of black granite with a common side. The top of each section is level with the ground. The bottoms of the two sections can be modeled by the equations $y = \frac{2}{25}x - 10$ and $y = -\frac{5}{61}x - 10$, when the x-axis is superimposed on the top of the wall. Each unit in the coordinate system represents 1 foot. How deep is the memorial at the point where the two sections meet? How long is each section? See below.

The Vietnam Veterans Memorial honors all the men and women who served in the Vietnam War.

40. 10 ft; 125 ft, 122 ft

41. *Airport Runways* A small airport has two runways that intersect. (The runway in use depends on wind conditions.) A coordinate plane is superimposed over the airport, with the origin at the control tower. The endpoints of the center line of one runway are at $(-1, -6)$ and $(3, 2)$. The endpoints of the center line of the other runway are at $(-3, -4)$ and $(6, 2)$. Find equations for the lines that contain the runways. Then sketch the lines and find the point of intersection.
See Additional Answers.

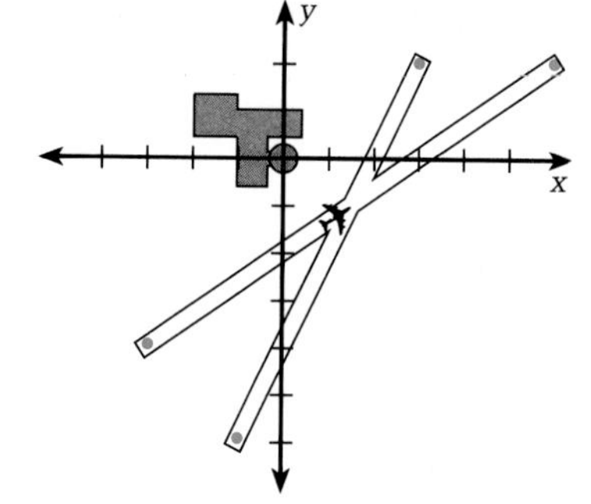

Advertising Expenditures **In Exercises 42–44, use the following information.**

The graph at the right shows advertising expenses in the United States as a percent of the gross national product. The dashed portions of the graphs are predictions.
(Source: American Demographics)

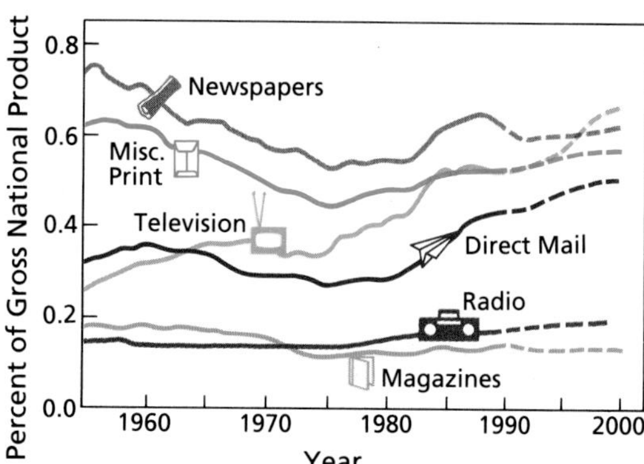

42. When did television advertising expenses surpass direct-mail advertising expenses? About 1963

43. When did radio advertising expenses surpass magazine advertising expenses? About 1972

44. When are television advertising expenses predicted to surpass newspaper advertising expenses? About 1996

Integrated Review

In Exercises 45–50, state whether the ordered pair is a solution of the equation. See below.

45. $2x + 3y = 9$, $(1, 4)$ **46.** $-x + 5y = 12$, $(3, 3)$ **47.** $16x + 2y = 12$, $(-2, 10)$

48. $\frac{1}{2}x + \frac{1}{6}y = 2$, $(10, -18)$ **49.** $-\frac{3}{4}x + y = -2$, $(2, -\frac{1}{2})$ **50.** $6x + 7y = -4$, $(-3, 2)$

In Exercises 51–53, rewrite the equation in slope-intercept form. See below.

51. $-7x + 2y = 6$ **52.** $3x + y = -10$ **53.** $4x - y = 13$

In Exercises 54–56, graph the equation. See Additional Answers.

54. $y = -4x - 5$ **55.** $y = \frac{1}{2}x + 4$ **56.** $2x + y = 10$

Exploration and Extension

In Exercises 57 and 58, describe a technique *other than graphing* that could be used to solve the linear system.

57. $\begin{cases} y = x + 2 \\ y = 4 \end{cases}$ Substitute 4 for y in the first equation of the system.

58. $\begin{cases} y = 3x - 6 \\ x = 2 \end{cases}$ Substitute 2 for x in the first equation of the system.

45. It is not. **46.** It is. **47.** It is not. **48.** It is. **49.** It is. **50.** It is. **51.** $y = \frac{7}{2}x + 3$

52. $y = -3x - 10$ **53.** $y = 4x - 13$

Enrichment Activities

These activities require students to go beyond lesson goals.

Sometimes the solution to a system of equations is not the solution to a problem. Consider the following.

Hare challenged Tortoise to a 100-meter rematch. Since Hare can run 50 meters in 7 seconds while Tortoise can run 20 meters in 5 seconds, Hare agreed to give Tortoise a 30-meter head start.

1. Assume that the animals run at a constant rate. Use quick-graph techniques and graph distance versus time for both animals on the same coordinate grid.
Hare: slope $= \frac{50}{7}$, y-intercept is 0; Tortoise: slope $= \frac{20}{5}$ or 4, y-intercept is 30.

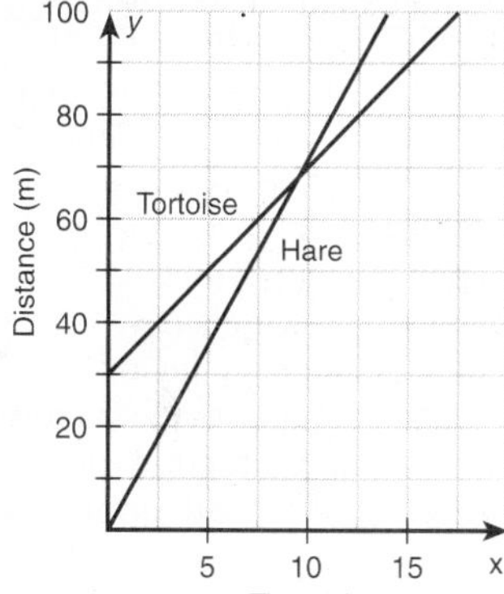

2. How can you tell from the graphs who won the race?
Hare arrives at 100 meters about three seconds before Tortoise, so Hare wins.

3. What does the intersection point tell you?
After 9 seconds, Hare caught up with Tortoise at about the 70-meter point.

4. COOPERATIVE LEARNING Work with a partner. Assume the rates stay constant. Devise ways for Tortoise to win. Justify your answers algebraically.

Sample answers: Tortoise could ask for a longer head start (any value greater than 44 meters); or Tortoise could request the same head start but shorten the race to less than 70 meters long.

TECHNOLOGY Points of intersection can be determined by using the trace feature of a graphing calculator. The zoom feature (or the box feature) can then be used to find the coordinates of the point of intersection to any degree of accuracy desired. (See the calculator manual for detailed instructions on the use of these features.)

Emphasize the importance of solving for y before using the graphing calculator to graph relationships. Point out that this is an example of the use of technology that requires "more algebra skills" not less!

Encourage students to use graphing calculators whenever possible in every lesson of the chapter. They provide a consistent visual interaction with the symbolic algebraic operations for systems of equations.

USING A GRAPHING CALCULATOR

A graphing calculator or computer can be used to graph a system of linear equations. For instance, the screen at the right shows the graph of the following system.

$$\begin{cases} 3x - 4y = 2 & \text{Equation 1} \\ 5x + 6y = 16 & \text{Equation 2} \end{cases}$$

Note that the two lines appear to intersect at the point (2, 1). To check this, substitute 2 for x and 1 for y in each equation.

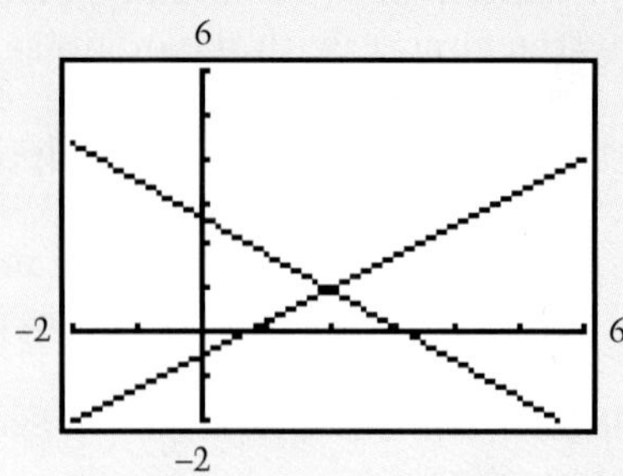

Graph of a System of Linear Equations.

The first step in using a graphing calculator to sketch the graph of a linear system is to solve each equation for y.

$$\begin{cases} y = \frac{3}{4}x - \frac{1}{2} & \text{Equation 1} \\ y = -\frac{5}{6}x + \frac{8}{3} & \text{Equation 2} \end{cases}$$

Next find settings for a viewing rectangle that shows the point of intersection. You may have to try a few before you obtain a viewing rectangle that shows the portion of the graph you want to see. For this particular system, you can use a setting in which both x and y vary between −2 and 6.

Keystroke instructions for sketching these graphs on a TI-82, *Casio* fx-9700GE, *and Sharp* EL-9300C *are listed on page 858.*

Cooperative Learning

Partner Activity Use a graphing calculator to sketch the graph of the following system of linear equations.

$$\begin{cases} 6x + 2y = 13 & \text{Equation 1} \\ 3x - 3y = 12 & \text{Equation 2} \end{cases}$$

The coordinates of the solution are rational numbers. Try finding the *exact* solution using only the trace and zoom keys on your graphing calculator. Check your solution by substituting it into the equations in the original system. (2.625, −1.375)

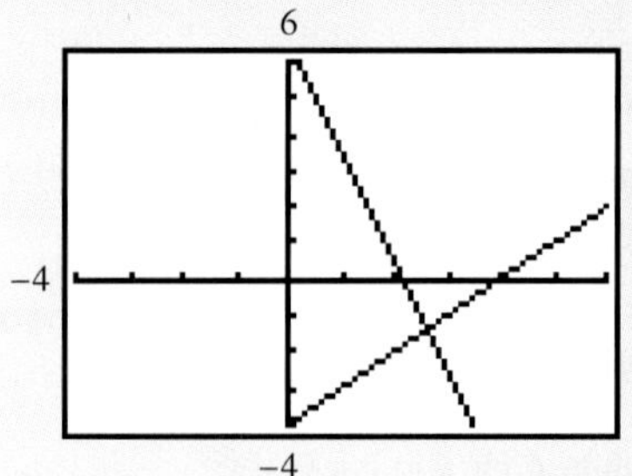

Some models using linear systems call for *exact* solutions. For others, however, an *approximate* solution is sufficient. For instance, suppose two linear models represent two different trends in population. It is not reasonable to expect such a model to produce an answer like "August 16, 1992." Rather, you expect the model to be accurate enough to produce an answer like "sometime in 1992" or "about midyear in 1992."

Example: Approximating the Solution of a Linear System

The following models for the total exports and imports of Puerto Rico (in millions of dollars) for 1975 through 1992 were given on page 121.

$$\begin{cases} T = 1312t + 5166 & \textit{Exports} \\ T = 723t + 7429 & \textit{Imports} \end{cases}$$

(t is the year, with $t = 0$ representing 1980.) Use a graphing calculator to approximate the year in which Puerto Rico's exports began to exceed its imports.

Solution Graphing calculators use x and y as variables, so you first need to rewrite the equations using x and y. (The variables T and θ, used by a graphing calculator, are for other types of graphs and should not be used for linear systems.)

$$\begin{cases} y = 1312x + 5166 & \textit{Exports} \\ y = 723x + 7429 & \textit{Imports} \end{cases}$$

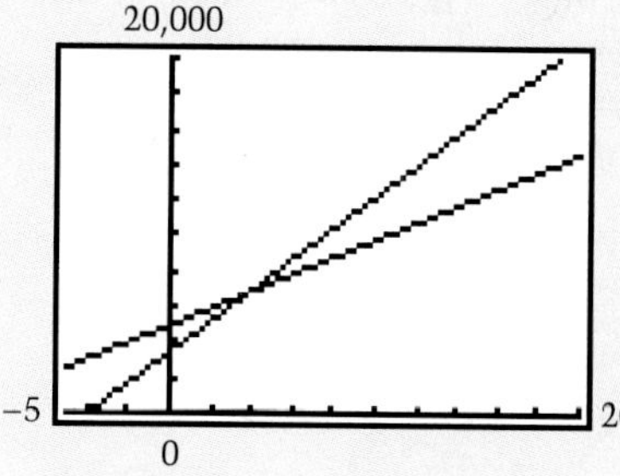

After entering these equations in the calculator and setting an appropriate range, you can obtain the graph shown at the right. From the graph, you can see that the exports began to exceed the imports around 1984. (The *trace* key can help you approximate the point of intersection.) ■

Exercises

In Exercises 1–4, use a graphing calculator to sketch the graph of the system. Approximate the solution and check your answer algebraically. See margin.

1. $\begin{cases} y = -2x + 2 \\ y = \frac{1}{2}x - \frac{11}{2} \end{cases}$ **2.** $\begin{cases} y = 3x + 11 \\ y = \frac{1}{3}x + \frac{19}{3} \end{cases}$ **3.** $\begin{cases} 7x - 8y = 5 \\ 5x + 4y = 23 \end{cases}$ **4.** $\begin{cases} 10x + 11y = 24 \\ 9x - 6y = -42 \end{cases}$

5. *Per Capita Income* For 1987 through 1994, the per capita incomes, y, in Nevada and New Hampshire can be modeled by

$$\begin{cases} y = 1077t + 9035 & \textit{Nevada} \\ y = 712t + 13{,}324 & \textit{New Hampshire} \end{cases}$$

where t is the year, with $t = 7$ corresponding to 1987. Use a graphing calculator to approximate the year in which the per capita income of Nevada surpassed that of New Hampshire. ***(Source: Bureau of Economic Analysis)*** 1991

Answers

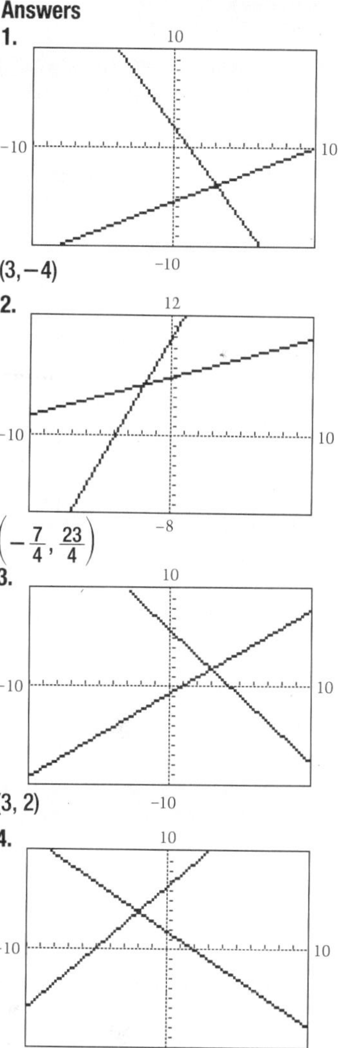

Problem of the Day

Use graph paper to draw squares with areas 2, 5, and 10.

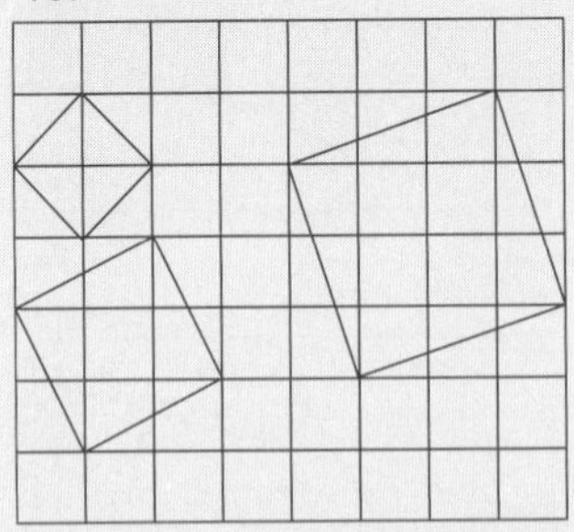

ORGANIZER

Warm-Up Exercises

1. Solve for x.
   a. $2x - (5x + 2) = 13$
   b. $3(9 - 11x) + 6x = 24$
   c. $4x - 7(x + 2) = -18$
   a. $x = -5$, b. $x = \frac{1}{9}$, c. $x = \frac{4}{3}$
2. Substitute the value of x into the given equation and solve for y.
   a. $y = 3x - 9$, $x = -2$
   b. $2x + 4y = 10$, $x = 1$
   c. $-3x - 2y = 16$, $x = -8$
   a. $y = -15$, b. $y = 2$, c. $y = 4$

Lesson Resources

Teaching Tools
  Transparencies: 7, 8, 9
  Problem of the Day: 3.2
  Warm-up Exercises: 3.2
  Answer Masters: 3.2
Extra Practice: 3.2
Color Transparencies: 20, 21

LESSON Notes

Example 1

To check the solution in the original system means evaluating each equation at the x- and y-values, -2 and 4.

(continued)

3.2 Solving Linear Systems Algebraically

What you should learn:

Goal 1 How to use algebraic methods to solve a linear system

Goal 2 How to use a linear system to answer questions about a real-life situation

Why you should learn it:

You can model many real-life situations with a linear system, such as the numbers of sheep and cattle grazing on national forest land during the last twenty years.

Goal 1 Using Algebraic Methods

In this lesson, you will study two algebraic methods for solving a linear system. The first is called **substitution.**

The Substitution Method for Solving a Linear System

1. *Solve* one of the equations for one of its variables.
2. *Substitute* this expression into the other equation and solve for the other variable.
3. *Substitute* this value in the revised first equation and solve.
4. *Check* the solution in each of the original equations.

Example 1 *The Substitution Method*

Solve the linear system.

$$\begin{cases} -2x + y = 8 & \text{Equation 1} \\ 3x + y = -2 & \text{Equation 2} \end{cases}$$

Solution Begin by solving for y in either of the equations.

$y = 2x + 8$ *Revised Equation 1*

Substitute this expression for y in Equation 2 and solve for x.

$3x + y = -2$ *Equation 2*

$3x + (2x + 8) = -2$ *Substitute $2x + 8$ for y.*

$5x = -10$ *Simplify.*

$x = -2$ *Solve for x.*

You now know that the value of x in the solution is -2. To find the value of y, substitute the value of x into the revised Equation 1.

$y = 2x + 8$ *Revised Equation 1*

$y = 2(-2) + 8$ *Substitute -2 for x.*

$y = 4$ *Solve for y.*

The solution is $(-2, 4)$. Check it in the original system. ■

When using the substitution method, you will obtain the same solution (x, y) whether you solve for y first or x first. Thus you should begin by solving for the variable that has a coefficient of 1.

Solve for x first here:

$$\begin{cases} 5x - 3y = 2 \\ x + 2y = 3 \end{cases}$$

Solve for y first here:

$$\begin{cases} -4x + y = 0 \\ 2x - 3y = -10 \end{cases}$$

If neither variable has a coefficient of 1, you can still use the substitution method by writing an equivalent equation that does have a coefficient of 1. In such cases, however, the **linear combination method** may be better.

Linear Combination Method for Solving a Linear System

1. *Arrange* the equations with like terms in columns.
2. *Obtain* coefficients for x (or y) that differ only in sign by multiplying each term of one or both equations by an appropriate number.
3. *Add* the equations and solve for the remaining variable.
4. *Substitute* the value obtained in step 3 into either of the original equations and solve for the other variable.
5. *Check* the solution in each of the original equations.

Example 2 *The Linear Combination Method*

Solve the linear system.

$$\begin{cases} 5x + 4y = 6 & \text{Equation 1} \\ -2x - 3y = -1 & \text{Equation 2} \end{cases}$$

Solution One way to obtain coefficients for y that differ only in sign is by multiplying the first equation by 3 and the second equation by 4. Add the results to obtain an equation that has only one variable, x.

$$\begin{aligned} 5x + 4y &= 6 \longrightarrow & 15x + 12y &= 18 && \text{Multiply by 3.} \\ -2x - 3y &= -1 \longrightarrow & -8x - 12y &= -4 && \text{Multiply by 4.} \\ & & 7x \quad &= 14 && \text{Add the equations.} \end{aligned}$$

Therefore $x = 2$. By substituting this value in the second equation, you can solve for y.

$$\begin{aligned} -2x - 3y &= -1 && \text{Equation 2} \\ -2(2) - 3y &= -1 && \text{Substitute 2 for } x. \\ y &= -1 && \text{Solve for } y. \end{aligned}$$

The solution is $(2, -1)$. Check it in the original system. ■

(continued)

$$\begin{aligned} -2x + y &= 8 \\ -2(-2) + 4 &\stackrel{?}{=} 8 \\ 4 + 4 &\stackrel{?}{=} 8 \\ 8 &= 8 ✓ \\ 3x + y &= -2 \\ 3(-2) + (4) &\stackrel{?}{=} -2 \\ -6 + 4 &\stackrel{?}{=} -2 \\ -2 &= -2 ✓ \end{aligned}$$

Ask students to solve the system using the substitution, $x = \frac{-y - 2}{3}$. Although the substitution would be a correct one, yielding the same solution as before, students will undoubtedly recognize the benefit of selecting the variable having the coefficient 1.

Example 2

Ask students to solve the system by eliminating the variable x instead of the variable y. (Note that this technique is sometimes called the elimination method.) Multiply Equation 1 by 2 and Equation 2 by 5 to obtain

$$\begin{cases} 10x + 8y = 12 \\ -10x - 15y = -5 \end{cases}$$

Example 3

There are systems of linear equations that have no solution or infinitely many solutions. Either of these possibilities is easily observed from a graph of the system, so suggest that students always make a quick sketch of the equations of the system (preferably by means of calculator or computer).

Example 4

Ask students why the substitution technique is a better method to use here than the linear combination technique. The process of multiplying either equation by −1 and eliminating y is virtually the same as substituting for y. If t is eliminated, the arithmetic is too cumbersome.

Extra Examples

Here are additional examples similar to **Examples 1–3.**

1. Solve by using the substitution method.

   a. $\begin{cases} 8x + y = 12 \\ -2x + 3y = 10 \end{cases}$ (1, 4)

   b. $\begin{cases} 5x - 3y = 2 \\ x + 2y = 3 \end{cases}$ (1, 1)

2. Solve by using the linear combination method.

   a. $\begin{cases} 3x - 5y = -6 \\ -2x + 7y = 4 \end{cases}$ (−2, 0)

   b. $\begin{cases} 11x + 7y = 9 \\ 6x + 7y = 24 \end{cases}$ (−3, 6)

3. Solve by using the most appropriate method.

   a. $\begin{cases} 6x + y = -5 \\ 4x - 3y = -7 \end{cases}$ (−1, 1)

   b. $\begin{cases} 12x + 8y = -8 \\ 3x + 4y = 2 \end{cases}$ (−2, 2)

   c. $\begin{cases} 9x + 12y = 3 \\ 3x + 4y = -2 \end{cases}$ no solution

   d. $\begin{cases} -2x - 4y = 2 \\ 10x + 20y = -10 \end{cases}$ infinitely many solutions

Check Understanding

1. Describe when it is better to use the substitution method instead of the linear combination method. See the top of page 131.
2. Does every system of linear equations have a solution? Explain. No; see Example 3.
3. Should solutions obtained by the substitution and linear combination methods be checked? Always
4. How can systems of linear equations that have no solution or infinitely many solutions be recognized algebraically? graphically? See Example 3 on page 132.

Guidelines for Solving a System of Linear Equations

To decide which method (graphing, substitution, or linear combination) to use, consider the following.

1. The graphing method is useful for approximating a solution, checking a solution for reasonableness, and for providing a visual model of the problem. (A graphing calculator is useful here.)
2. To find an exact solution, use either substitution or linear combinations.
3. For linear systems in which one variable has a coefficient of 1 or −1, substitution may be more efficient.
4. In other cases, the linear combination method is usually more efficient.

The linear systems in Examples 1 and 2 each have exactly one solution. Remember, however, that some linear systems have no solution and some have infinitely many solutions.

Example 3 *A Linear System with No Solution*

Solve the linear system.

$$\begin{cases} 3x - 2y = 5 & \text{Equation 1} \\ -6x + 4y = 7 & \text{Equation 2} \end{cases}$$

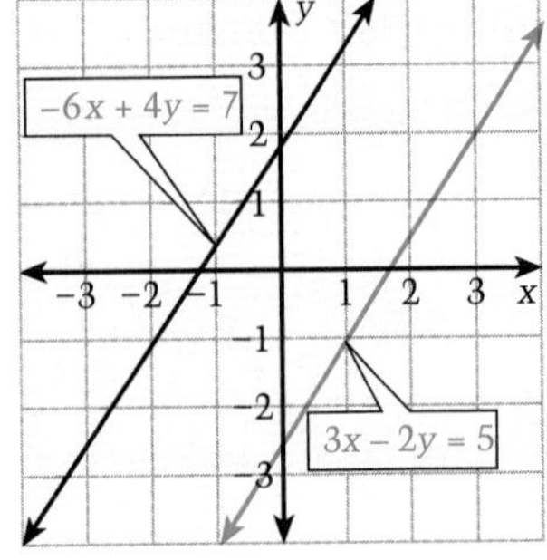

Solution You can use any of the three methods to solve the system. Using the graphing method, it appears that the two equations graph as parallel lines. Using either the substitution or linear combination method, you obtain a false "equation." For instance, the linear combination method produces the following.

$$\begin{array}{rcrl} 3x - 2y = 5 & \longrightarrow & 6x - 4y = 10 & \textit{Multiply by 2.} \\ -6x + 4y = 7 & \longrightarrow & -6x + 4y = 7 & \textit{Equation 2} \\ \hline & & 0 = 17 & \textit{Sum of equations} \end{array}$$

Because the statement $0 = 17$ is not true, you can conclude that there is no solution. ■

Using the substitution method or the linear combination method for a linear system with infinitely many solutions produces an equation such as $3 = 3$ or $0 = 0$. For instance, by adding the equations in the linear system

$$\begin{cases} 3x - 2y = 5 & \text{Equation 1} \\ -3x + 2y = -5 & \text{Equation 2} \end{cases}$$

you obtain $0 = 0$, which indicates that both equations graph as the same line. Every point on the line is a solution.

The federal government protects and manages about 300,000 square miles of national forest in the United States and Puerto Rico. (This is more than the combined areas of California and Nevada.) Some of the grass-covered areas of national forests are leased to ranchers for grazing cattle and sheep.

Goal 2 Using a Linear System as a Model

Real Life
Land Management

Example 4 Grazing in National Forests

For 1970 through 1990, the number, y (in 1000's), of cattle and sheep that grazed on national forest land can be modeled by the following equations. (Let $t = 0$ represent 1970.)

$y = 1600 - 11t$ *Cattle*

$y = 1650 - 30t$ *Sheep*

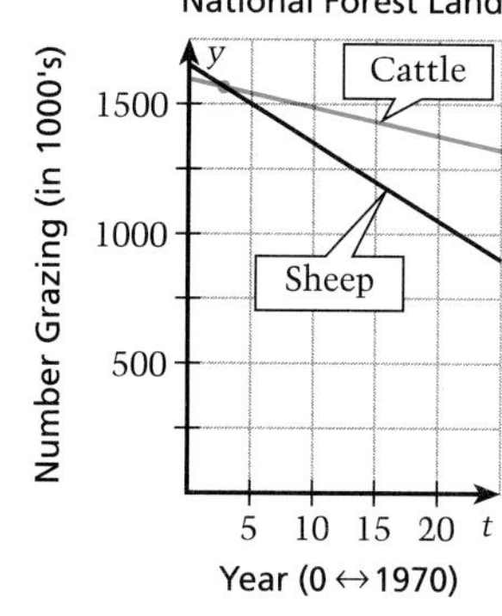

In 1970, more sheep than cattle were grazing in national forests. When did this change? ***(Source: National Forest Service)***

Solution This system can be solved using substitution.

$y = 1650 - 30t$ *Equation 2*

$(1600 - 11t) = 1650 - 30t$ *Substitute 1600 − 11t for y.*

$19t = 50$ *Subtract 1600 and add 30t.*

$t \approx 2.6$ *Solve for t.*

Thus according to this model, the number of grazing sheep fell below the number of grazing cattle around 1972 or 1973. ■

Communicating about ALGEBRA

SHARING IDEAS about the Lesson

Comparing Methods Sketch the graph of the linear system. Then use substitution and linear combinations to solve the system. Compare the merits of each method. See margin.

A. $\begin{cases} 4x - 3y = 2 \\ 5x + 2y = -1 \end{cases}$ **B.** $\begin{cases} 4x - y = 2 \\ 5x + 2y = 9 \end{cases}$

Communicating about ALGEBRA

EXTEND ***Communicating***
Discuss the following statement.

Exact solutions of systems of linear equations can be obtained by the graphing method if calculators or computers are used.

Graphical solutions are often approximations, and the approximations may be for numbers that are not rational. The lines for the equations may be nearly parallel, so that it is difficult to find an appropriate viewing window.

Answers

A.

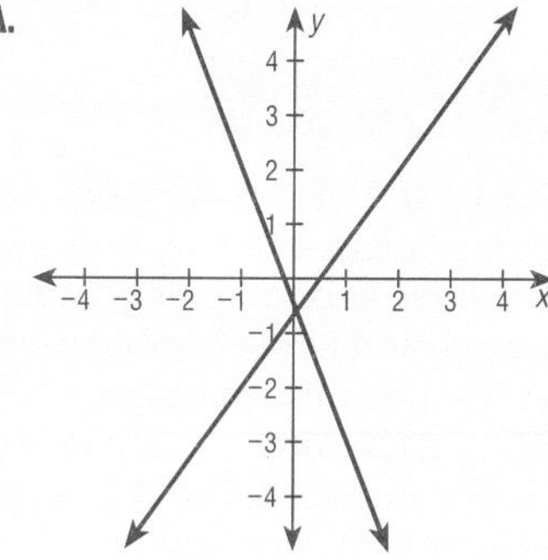

$\left(\frac{1}{23}, -\frac{14}{23}\right)$, the linear combination method is easiest to use.

B.

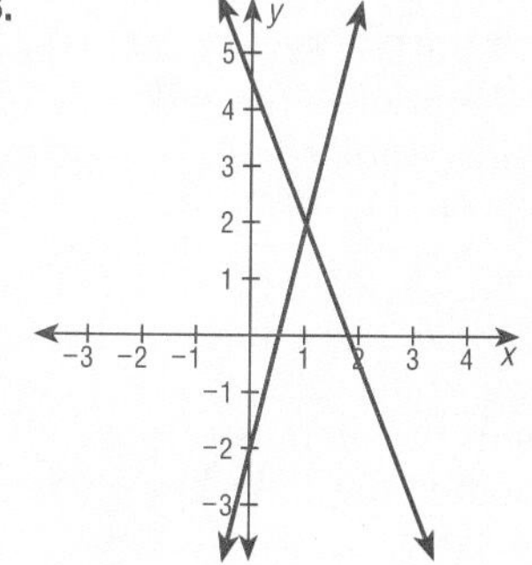

(1, 2), the substitution method is easiest to use.

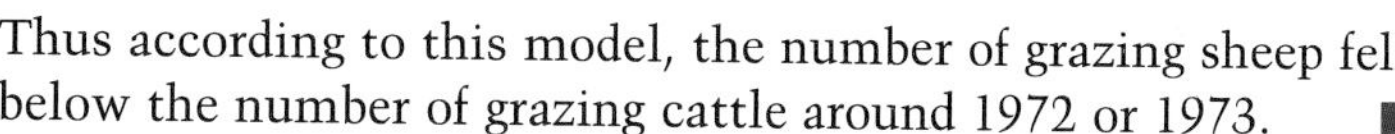

EXERCISE Notes

ASSIGNMENT GUIDE

Basic/Average: Ex. 1–6, 9–19 odd, 23–27 odd, 29–31, 37–45, 51–52, 54–56

Above Average: Ex. 1–6, 9–19 odd, 21–29 odd, 30–32, 48–56

Advanced: Ex. 1–6, 7–21 odd, 27, 29, 30–32, 49–56

Selected Answers
Exercises 1–6, 7–51 odd

Use **Mixed Review** as needed.

More Difficult Exercises
Exercises 31, 32, 53

Guided Practice

EXTEND Ex. 1–6 Have students discuss the strengths and weaknesses of graphing, substitution, and linear combinations as methods for solving systems of equations. Discussions should include when it is more efficient to use one method over another.

Independent Practice

EXTEND Ex. 19–24 Have students write the rationale for their choice of solution method.

Ex. 25–28 Students can use Example 3 as a solution model. They could also make judgements based on the slopes as to whether the lines are parallel, the same, or intersecting.

Ex. 32 Explain that the slopes for each equation represent the 15%, 28%, and 33% income tax brackets.

EXERCISES

Guided Practice

CRITICAL THINKING about the Lesson See Additional Answers for 1.–2.

1. Describe how you would use substitution to solve the system.

$$\begin{cases} 13x + 5y = 2 \\ x - 4y = 10 \end{cases}$$

2. Describe how you would use a linear combination to solve the system.

$$\begin{cases} 6x - 5y = 30 \\ -4x + 2y = 7 \end{cases}$$

3. Suppose you used substitution to solve a linear system and obtained the equation $-4 = -4$. What could you conclude? There are infinitely many solutions.

4. Suppose you used substitution to solve a linear system and obtained the result $12 = 0$. What could you conclude? There is no solution.

In Exercises 5 and 6, which method would you use to solve the linear system? Explain. See below.

5. $\begin{cases} 5x + 3y = -1 \\ -4x + 2y = 3 \end{cases}$

6. $\begin{cases} 5x + 3y = 1 \\ -4x + y = 6 \end{cases}$

Independent Practice

In Exercises 7–12, use substitution to solve the linear system.

7. $\begin{cases} 2x + y = 9 \\ 3x - 4y = 8 \end{cases}$ (4, 1)

8. $\begin{cases} 3x + 5y = 12 \\ x + 4y = 11 \end{cases}$ (−1, 3)

9. $\begin{cases} x - 9y = 25 \\ 6x - 5y = 3 \end{cases}$ (−2,−3)

10. $\begin{cases} 2x + y = -9 \\ 3x + 5y = 4 \end{cases}$ (−7, 5)

11. $\begin{cases} 2x + 3y = -6 \\ 3x + 2y = 25 \end{cases}$ $\left(\frac{87}{5}, \frac{-68}{5}\right)$

12. $\begin{cases} -x + 3y = 18 \\ 4x - 2y = 8 \end{cases}$ (6, 8)

In Exercises 13–18, use a linear combination to solve the system.

13. $\begin{cases} -2x + 7y = 10 \\ x - 3y = -3 \end{cases}$ (9, 4)

14. $\begin{cases} 2x - y = 2 \\ -5x + 4y = -2 \end{cases}$ (2, 2)

15. $\begin{cases} 3x + 11y = 4 \\ -2x - 6y = 0 \end{cases}$ (−6, 2)

16. $\begin{cases} -9x + 5y = 1 \\ 3x - 2y = 2 \end{cases}$ (−4,−7)

17. $\begin{cases} 7x + 20y = 11 \\ 3x + 10y = 5 \end{cases}$ $\left(1, \frac{1}{5}\right)$

18. $\begin{cases} 2x - y = 3 \\ 9x - 6y = 6 \end{cases}$ (4, 5)

In Exercises 19–24, sketch the graph of the system. Then use any convenient method to solve the system. See Additional Answers.

19. $\begin{cases} x - y = 10 \\ 3x - 2y = 25 \end{cases}$ (5,−5)

20. $\begin{cases} -7x + 5y = 0 \\ 14x - 8y = 2 \end{cases}$ $\left(\frac{5}{7}, 1\right)$

21. $\begin{cases} 4x + 3y = 1 \\ -3x - 6y = 3 \end{cases}$ (1,−1)

22. $\begin{cases} -4x - 10y = 12 \\ x + 5y = 2 \end{cases}$ (−8, 2)

23. $\begin{cases} 5x + 16y = 15 \\ -2x - 4y = 1 \end{cases}$ $\left(-\frac{19}{3}, \frac{35}{12}\right)$

24. $\begin{cases} 4x + y = 2 \\ 6x + 3y = 0 \end{cases}$ (1,−2)

In Exercises 25–28, how many solutions does the linear system have? Explain.

25. $\begin{cases} 3x + y = 12 \\ 2x + 2y = 4 \end{cases}$

26. $\begin{cases} 6x - y = 5 \\ 12x - 2y = 3 \end{cases}$

27. $\begin{cases} x - 2y = 1 \\ 3x - 6y = 2 \end{cases}$

28. $\begin{cases} 3x - y = -3 \\ -3x + y = 2 \end{cases}$

5. Linear combination; neither variable in either equation has a coefficient of 1.
6. Substitution; the second equation contains a variable with a coefficient of 1.

29. *World Food Production* For comparison purposes, the United Nations classifies countries as *developed* or *developing.* In 1990, the developed countries were Australia, Canada, all European countries, Israel, Japan, New Zealand, South Africa, the Soviet Union, and the United States. Based on an index for which *100* represents the food production in 1980, the food production, y, of developed and developing countries can be modeled by

$$\begin{cases} y = \frac{6}{5}t + 88 & \textit{Developed countries} \\ y = \frac{11}{5}t + 78 & \textit{Developing countries} \end{cases}$$

where $t = 0$ represents 1970. In 1970, food production was greater in developed countries. When did this change?
(Source: United Nations) 1980

This kibbutz in Israel grows peppers in greenhouses to conserve water.

30. *World Fuel Production* For 1980 through 1990, the coal production in the world, C (in millions of metric tons), and crude petroleum production, P (in millions of metric tons), can be modeled by the following equations. (Let $t = 0$ represent 1980.)

$C = 86.5t + 2728$ *Coal production*

$P = -27.5t + 2979$ *Petroleum production*

During which year were the production levels equal?
(Source: United Nations) 1982

31. *New York City Subway* The map shows a section of the New York City subway system. The portion of the track shown in blue lies approximately on the line given by $11x + 6y = 110$. From Central Park through most of the downtown area, the Avenue of the Americas is straight and lies on the line given by $-13x + 7y = 25$. (Each unit in the coordinate system represents a quarter mile.) Find the coordinates of the point at which the subway track goes under the Avenue of the Americas. (4, 11)

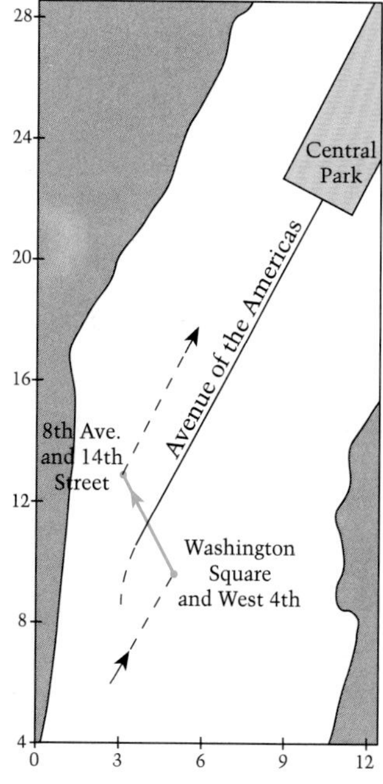

32. *Paying the IRS* In 1990, the federal income tax, I, for a single person whose taxable income, x, was less than \$97,620 was given by the following equations.

$I = 0.15x,$ *if $x \le A$.*

$I = 0.28x - 2528.5,$ *if $A \le x \le B$.*

$I = 0.33x - 4881,$ *if $B \le x \le 97{,}620$.*

At what taxable income level, A, did a person move from the 15% *tax bracket* to the 28% tax bracket? At what taxable income level, B, did a person move from the 28% tax bracket to the 33% tax bracket? Explain your answer.
See below.

25. 1, graphs have different slopes **26.–28.** none, derived equation is false
32. \$19,450; \$47,050; solved $0.15x = 0.28x - 2528.5$ and $0.28x - 2528.5 = 0.33x - 4881$

Enrichment Activities

These activities require students to go beyond lesson goals.

TECHNOLOGY Here is one way to use a graphics calculator to solve Exercise 30. Visually, the graph illustrates the declining production of petroleum as a natural resource, and the increasing production of coal.

1. In order to see the graph, you first need to locate an appropriate viewing window. Use the given range setting and graph the given data.

X min: 0	Y min: 2500
X max: 20	Y max: 3000
X scl: 10	Y scl: 100
X res: 1	

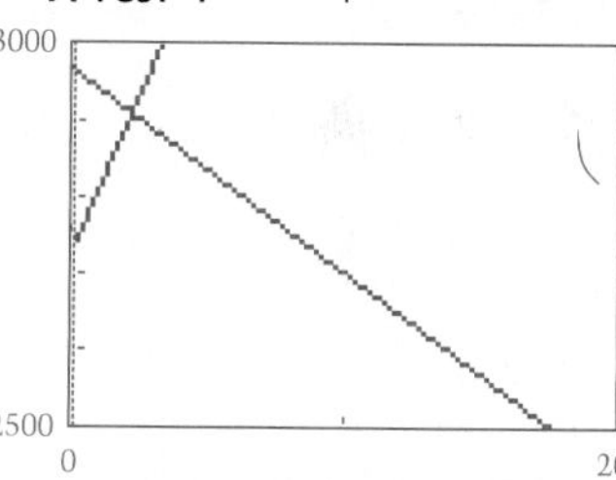

2. Since an exact answer is not necessary, use the trace key to find the point of intersection.
(2.10, 2910.10)
3. Interpret the solution.
At the beginning of 1982, the two production levels were equal at about 2910 million metric tons.
4. Use the graph to predict the production levels for both coal and oil in 1996 when $x = 16$. coal: 4112 million metric tons; petroleum: 2539 million metric tons.

Note: There is a growing scarcity of petroleum and natural gas. As time progresses, there will be greater reliance on coal.

▶ **Ex. 33–50** These exercises provide practice in essential skills.

▶ **Ex. 51–52** You may wish to use these exercises at the end of the class period for discussion. Compare the given data with your students' responses regarding their eating habits. Allow students to be creative regarding the justification of the data trends.

▶ **Ex. 54–56** By using slope concepts, students could view these exercises as mental math activities.

Answers

45. $y = -\frac{2}{9}x + 4$

46. $y = -\frac{4}{9}x + \frac{16}{9}$

47. $y = -\frac{16}{3}x + 8$

48. $y = \frac{1}{14}x - 2$

49. $y = \frac{1}{2}x + \frac{7}{5}$

50. $y = \frac{8}{9}x + \frac{16}{9}$

51. ≈ 0.37, ≈ 0.42, ≈ 0.42, ≈ 0.67; they increase

52. Younger people sharing living space are most likely not related and, thus, very likely to have variable work and play times; middle-aged people sharing living space are most likely related by marriage and have different work times; people 60 and older sharing living space are most likely not working and thus very likely to have all meals together.

53. $\left(\frac{8}{5}, \frac{11}{5}\right)$

54.–56. Answers vary.

54. $y = 2x$ 55. $y = -\frac{1}{2}x$

56. $2y = -x - 6$

Integrated Review

In Exercises 33–38, solve the equation.

33. $5x + 6(1 + x) = 17$ 1

34. $x - 8 - 5x = 10$ $-\frac{9}{2}$

35. $x - 10x + 4 = 10$ $-\frac{2}{3}$

36. $9 - 3x + x = 15$ -3

37. $7 - 4x + 3 + 2x = 6$ 2

38. $3x + 3 - 5x = 17$ -7

In Exercises 39–44, decide whether the ordered pair is a solution of the system.

39. $(4, -2)$ It is not.
$$\begin{cases} 2x - y = 6 \\ -x + 3y = 14 \end{cases}$$

40. $(1, 0)$ It is.
$$\begin{cases} 6x - 2y = 6 \\ -5x + 3y = -5 \end{cases}$$

41. $(1, 5)$ It is.
$$\begin{cases} x + y = 6 \\ 3x - y = -2 \end{cases}$$

42. $(-4, -7)$ It is not.
$$\begin{cases} -x - 7y = 46 \\ x + y = -11 \end{cases}$$

43. $(-3, -6)$ It is not.
$$\begin{cases} x - 6y = 33 \\ -2x - y = 0 \end{cases}$$

44. $(-6, 2)$ It is.
$$\begin{cases} 10x + 13y = -34 \\ -15x - 17y = 56 \end{cases}$$

In Exercises 45–50, solve for *y*. See margin.

45. $2x + 9y = 36$

46. $4x + 9y = 16$

47. $16x + 3y = 24$

48. $x - 14y = 28$

49. $-5x + 10y = 14$

50. $-8x + 9y = 16$

Eating Together **In Exercises 51 and 52, use the following information.** See margin.

The chart at the right shows the percents of Americans (18 years old and older) who eat breakfast with other people in their home and the percents who eat an evening meal with other people in their home. People who live alone are not included in these data.
(Source: Roper Organization)

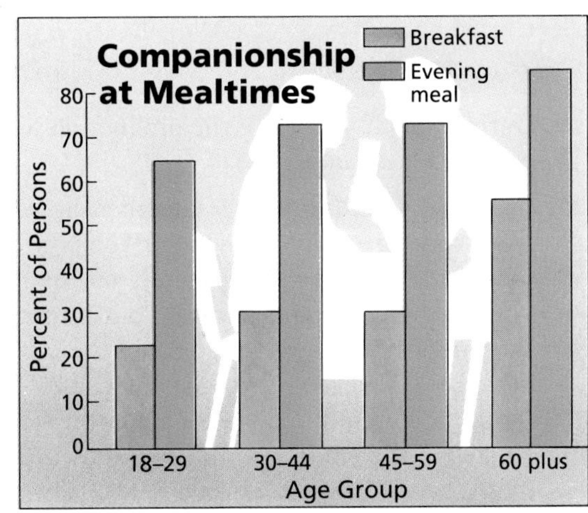

51. Find the ratio of the breakfast percent to the evening meal percent for each of the four age groups. What do you notice about the ratios?

52. How would you explain the differences in eating habits shown by the four age groups?

Exploration and Extension

53. Find the point of intersection of the line $y = 2x - 1$ and the line that passes through (4, 1) and is perpendicular to $y = 2x - 1$. See margin.

Creating Your Own Linear Systems **In Exercises 54–56, write a linear system that produces the given number of solutions. Use $y = -\frac{1}{2}x - 3$ as one of the equations.** See margin.

54. One solution

55. No solution

56. Infinitely many solutions

Mixed REVIEW

See margin.

1. Evaluate $3x - 2y$ when $x = -1$ and $y = 2$. **(1.2)**
2. Write in slope-intercept form: $-2x + 3y = 10$. **(2.3)**
3. Write in decimal form: 3.14×10^4.
4. Add: $\frac{2}{3} + \frac{1}{3} + \frac{4}{3}$.
5. Solve $-3(x - 2) = 2(5 - x)$. **(1.3)**
6. Solve $|2x - 5| = 3$. **(1.7)**
7. Sketch the graph of $y = 3x - 4$. **(2.3)**
8. Sketch the graph of $y = |x - 2| - 2$. **(2.6)**
9. Sketch the graph of $y \le -\frac{1}{2}x + 3$. **(2.5)**
10. Solve $-3x + 4 \le 2x + 24$. **(1.6)**
11. What is 49% of 135?
12. What is the reciprocal of $-\frac{1}{7}$?
13. Write *30* as the product of prime numbers.
14. For which type of line is slope not defined? **(2.2)**
15. Solve $3xy + 4y = 8$ for y. **(1.5)**
16. Find the volume of a cube with 3-foot edges.
17. Write an equation of the horizontal line through $(-5, 6)$. **(2.1)**
18. Write an equation of the vertical line through $(-5, 6)$. **(2.1)**
19. Write an equation of the line through $(2, -3)$ and $(-1, 0)$. **(2.4)**
20. What is the slope of a line that is perpendicular to $2x - 5y = 6$? **(2.2)**

Career Interview

Air Traffic Controller

Terry Gilbert is an air traffic controller at the control center in Chicago, which supervises the area that extends from Green Bay, Wisconsin, to Champagne, Illinois, to Detroit, Michigan, to Iowa City, Iowa. Over 2 million planes fly within this corridor annually!

Q: ***What do you do as a controller?***

A: O'Hare Airport, located within the center's airspace, is rated as an 80-plane-per-hour site. Controllers must balance the incoming load so that all 80 planes do not arrive within the first 5 minutes! We also keep track of the landing slots. That is, which slots are used, how often each is used, and whether each is used on a timely basis.

Q: ***How much of the actual math is done by radar?***

A: Airplanes on radar are tagged to show the plane's altitude, code number, direction, and speed.

Q: ***How much mental math and estimation do you use?***

A: Controllers often calculate distances, and climb and descent rates mentally. So good mental math and estimation skills are essential.

Answers

1. -7
2. $y = \frac{2}{3}x + \frac{10}{3}$
3. 31,400
4. $\frac{7}{3}$
5. -4
6. 4, 1
7.

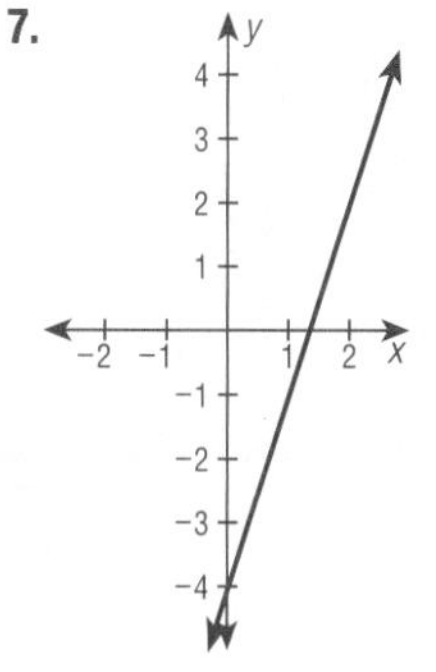

8.

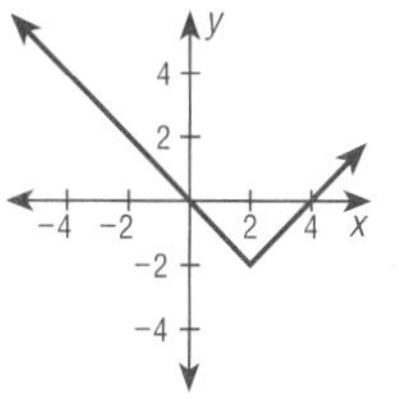

9.

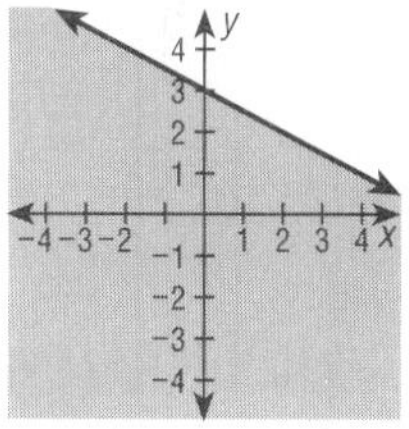

10. $x \ge -4$
11. 66.15
12. -7
13. $2 \cdot 3 \cdot 5$
14. Vertical
15. $y = \frac{8}{3x + 4}$
16. 27 ft^3
17. $y = 6$
18. $x = -5$
19. $y = -x - 1$
20. $-\frac{5}{2}$

Problem of the Day

Determine the digits represented by the asterisks.

**7*	1475
× *7*	× 677
*****	10325
***2*	10325
8*5*	8850
******	998575

ORGANIZER

Warm-Up Exercises

MATH JOURNAL Students might record their responses in their math journals in a special section on problem solving.

1. What is a verbal model of a real-life problem?
   A verbal model consists of phrases that describe (in an abbreviated way) the relationship between variables in a problem situation.
2. What is an algebraic model of a real-life problem?
   An algebraic model consists of variables (which replace the verbal phrases in a verbal model) that describe the relationship between variables in a problem situation.
3. Is the answer to a problem situation always the result obtained from solving the associated algebraic model?
   Not always. For example, see Example 2 on page 139.

Lesson Resources

Teaching Tools
- Transparencies: 4, 7, 8, 9
- Problem of the Day: 3.3
- Warm-up Exercises: 3.3
- Answer Masters: 3.3

Extra Practice: 3.3
Color Transparencies: 21, 22

LESSON Notes

Example 1

Point out to students that the variables c and b are more meaningful than x and y here.

3.3 Problem Solving Using Linear Systems

What you should learn:

Goal How to write and use linear systems to model real-life situations

Why you should learn it:

You can model situations with two unknown quantities using a system of linear equations, such as the premiums for liability and collision coverage for your automobile insurance.

Goal Linear Systems as Real-Life Models

You have already modeled some real-life situations with linear systems. Remember that the key steps are writing a *verbal model,* assigning *labels,* writing an *algebraic model, solving* the algebraic model, and *answering* the original question.

Write a verbal model. → Assign labels. → Write an algebraic model. → Solve the model. → Answer the question.

When you write a linear system to model a real-life problem, remember also that you have *three* methods for solving the system: graphing, substitution, and linear combination.

Example 1 *Finding the Price per Item*

You and several other people are helping your aunt move. At lunch, you volunteer to go out to buy sandwiches. You get 6 chicken sandwiches and 6 beef sandwiches for \$30. Later in the afternoon, everyone is hungry again. You go to the same shop and buy 4 chicken sandwiches and 8 beef sandwiches for \$30.60. What was the price of each type of sandwich? (Assume sales tax is included in the price.)

Solution

Verbal Model

$6 \cdot$ Price of chicken $+ 6 \cdot$ Price of beef $= 30.00$

$4 \cdot$ Price of chicken $+ 8 \cdot$ Price of beef $= 30.60$

Labels

Price of chicken sandwich $= c$ (dollars)

Price of beef sandwich $= b$ (dollars)

System

$$\begin{cases} 6c + 6b = 30 & \textit{Equation 1} \\ 4c + 8b = 30.6 & \textit{Equation 2} \end{cases}$$

Using the linear combination method, you can multiply the first equation by 2, multiply the second equation by -3, and add the results. The solution is \$2.35 per chicken sandwich and \$2.65 per beef sandwich. ■

Your automobile insurance premium depends on many factors: your age, where you live, the value of your car, the type of car, and your driving record. Of course, it also depends on the amount of coverage you obtain. At the right is a typical annual insurance premium schedule for a $15,000 car driven by a 16-year-old driver in 1991.

Liability, L (in $1000's)
Premium = $480 + 0.31L$
Collision Deductible, D (in dollars)
Premium = $692 - 0.49D$

Example 2 *Buying Automobile Insurance*

You are quoted a total annual premium of $1154 for automobile insurance. This premium gives liability coverage of L thousand dollars and collision coverage that has a deductible of D dollars. The insurance agent tells you that if you are willing to increase your collision deductible by $400, then you can get *five times* the liability coverage and still save $72 a year on your premium. What was the amount of the deductible and the amount of liability coverage in the first quote? (Use the premium schedule given above.)

Solution

First Quote:

$$\underbrace{480 + 0.31L}_{\text{Liability premium}} + \underbrace{692 - 0.49D}_{\text{Collision premium}} = 1154$$

Second Quote:

$$\underbrace{480 + 0.31(5L)}_{\text{Liability premium}} + \underbrace{692 - 0.49(D + 400)}_{\text{Collision premium}} = 1154 - 72$$

After simplifying these two equations, you can obtain the following linear system.

$$\begin{cases} 0.31L - 0.49D = -18 & \textit{Equation 1} \\ 1.55L - 0.49D = 106 & \textit{Equation 2} \end{cases}$$

Using the linear combination method (multiply the first equation by -1 and add the result to the second equation), you can determine that the solution is $L = 100$ and $D = 100$. Thus, the first quote was for liability coverage of $100,000 with a $100 collision deductible. The second was for liability coverage of $500,000 with a $500 collision deductible. By raising your deductible, you can afford to carry a higher liability coverage. ■

Example 2

Ask students if they know what liability and collision insurances are. Perhaps there are students in the class who maintain cars and must pay these types of insurances.

Liability insurance protects an insured driver from having to pay for damages inflicted upon another motorist's car.

Collision insurance protects the insured driver from having to pay for damages inflicted upon his or her own vehicle. Usually, collision insurance requires that the insured driver pay an amount toward the repair of the car. This payment is called a deductible.

Example 3

Remind students that exact answers are not required as the solution to some real-life situations. In comparing the populations of Illinois and Florida, an actual date, like May 26, 1974, is not necessary.

Extra Examples

Here are additional examples similar to **Examples 1–3.**

1. In Example 1, suppose on the second day of the move you again got sandwiches for everyone from the same shop as yesterday. At lunch you got 5 tuna sandwiches and 7 ham sandwiches for $31.95, and later in the day, you got 6 ham sandwiches, 4 tuna sandwiches, and 2 chicken sandwiches for $31.40. What was the price of each type of sandwich? (Assume that sales tax is included in the price and that the price of a chicken sandwich is the same as it was yesterday.)
   $2.40 for tuna sandwiches, $2.85 for ham sandwiches; the system of linear equations is
   $5t + 7h = 31.95$
   $4t + 6h = 26.70$

2. One Saturday, your sister left the house on foot, going to the movies. Ten minutes after she left, you remembered you had her money and set out on your bicycle to catch her! If your sister walks about 5 miles per hour and you can ride your bicycle about 15 miles per hour, how long will it take you to catch her if you take the same course she did? (*Hint:* Use the distance relationship, distance = rate · time, and note that you and your sister will travel the same distance when you catch up with her.)
The system of linear equations is
$d = 5(t + \frac{1}{6})$
$d = 15t$.
It will take 5 minutes to catch up with your sister.

Check Understanding

1. Identify the key steps in using an algebraic model to solve problems.
See the top of page 138.
2. What are the three methods for solving a system of linear equations?
Graphing, substitution, linear combinations

Communicating about ALGEBRA

Ask students to provide rationales for the processes used to estimate the revenues of Wal-Mart and K Mart.

Answers to Communicating

A. 1987, once Wal-Mart's revenues began increasing at a linear rate
B. \$43 billion, fit a line to the data and substituted for t in the equation of the line; \$58 billion, fit a line to the data and substituted for t in the equation of the line

Population Studies

Example 3 *Comparing Two Populations*

In 1980, the population of Illinois was 11,427,000, and the population of Florida was 10,195,000. In 1990, the population of Illinois was 11,467,000, and the population of Florida was 13,003,000. Over the ten-year period, both populations increased at a rate that was approximately linear. Estimate the year that the population of Florida surpassed the population of Illinois. *(Source: United States Census Bureau)*

Solution Let t represent the year, with $t = 0$ corresponding to 1980. The rate of change of Illinois's population was

$$\frac{(1990 \text{ population}) - (1980 \text{ population})}{10 - 0} = \frac{40{,}000}{10} = 4000 \text{ people/year.}$$

The rate of change of Florida's population was

$$\frac{(1990 \text{ population}) - (1980 \text{ population})}{10 - 0} = \frac{2{,}808{,}000}{10} = 280{,}800 \text{ people/year.}$$

Using the slope-intercept form, the linear models for the populations, P, of the two states are as follows.

$$\begin{cases} P = 4{,}000t + 11{,}427{,}000 & \textit{Illinois} \\ P = 280{,}800t + 10{,}195{,}000 & \textit{Florida} \end{cases}$$

Populations of Florida and Illinois

Population (millions)
15
10
5
P
Illinois
Florida
0 1 2 3 4 5 6 7 8 9 10 t
Year (0 ↔ 1980)

The graph of the system is shown at the left. From the graph, you can see that the population of Florida surpassed the population of Illinois around 1984 or 1985. ■

Communicating about ALGEBRA

SHARING IDEAS about the Lesson See margin.

The scatter plot at the left shows the annual revenue, y (in billions of dollars), for K Mart Corporation and Wal-Mart Stores for 1981 through 1990.
(Source: K Mart Corporation and Wal-Mart Stores)

Annual Sales

Sales (in billions of dollars)
40
35
30
25
20
15
10
5
y
K-Mart
Wal-Mart
1 2 3 4 5 6 7 8 9 10 t
Year (0 ↔ 1980)

A. In 1990, the total revenue for Wal-Mart was \$32.6 billion, and the total revenue for K Mart was \$32.1 billion. How early do you think you could have predicted that Wal-Mart's revenue would exceed K Mart's? Explain.

B. Estimate K Mart's revenue in 1995. Estimate Wal-Mart's revenue in 1995. Explain the process you used.

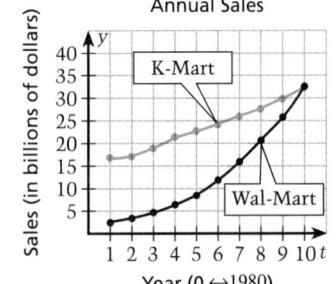

EXERCISES

Guided Practice

CRITICAL THINKING about the Lesson See margin.

Soccer Victory **In Exercises 1–4, use the following information.**

A soccer team bought ice-cream cones to celebrate a victory. The total cost of 12 double cones and 8 single cones was $17. A double cone cost $0.25 more than a single cone. What was the price of each type of cone?

1. Write a verbal model for this problem.
2. Assign labels to the verbal model.
3. Use the labels to write a linear system that represents the problem.
4. Solve the system and answer the question.

Independent Practice

5. *Forgotten Luggage* Your aunt and uncle have been visiting at your home. Five minutes after they drive away, you realize that they forgot their luggage. You happen to know that they drive slowly, so you get in your car and drive to catch up with them. Your average speed is 10 miles an hour faster than their average speed, and you catch up with them in 25 minutes. How fast did you drive? Use the following verbal model. 60 mph

Verbal Model [Your speed] = [Their speed] + [10 miles per hour]

[Your speed] · [Your time in hours] = [Their speed] · [Their time in hours]

6. *Auditing the Books* You run an accounting business that specializes in auditing (verifying accounting records). One of your auditors is working on the payroll records for a company with 75 employees. Some are part-time and some are full-time. After working for three days, your auditor tells you that the audit is completed for half of the full-time employees, but there are still 50 employee records to audit. After your auditor leaves your office, you wonder how many of the employees are full-time and how many are part-time. Use the following verbal model to find out.

Verbal Model [Number of full-time employees] + [Number of part-time employees] = [75] full-time: 50

$\frac{1}{2}$ [Number of full-time employees] + [Number of part-time employees] = [50] part-time: 25

7. *Tables and Chairs* You invited 56 people to your graduation party. You can afford to rent 5 tables, round and/or rectangular (each costing the same). Each round table can seat 8 people and each rectangular table can seat 12 people. How many round and rectangular tables should you rent? Round: 1, rectangular: 4

EXERCISE Notes

ASSIGNMENT GUIDE

Basic/Average: Ex. 1–4, 5–9 odd, 12, 13, 19–23 odd, 24, 25

Above Average: Ex. 1–4, 5–11 odd, 12–15, 23–25

Advanced: Ex. 1–4, 5–11 odd, 17–23 odd, 24, 25

Selected Answers
Exercises 1–4, 5–23 odd

✪ More Difficult Exercise
Exercise 11

Guided Practice

▶ **Ex. 1–4** Use these exercises as an in-class means of reviewing the problem-solving process one step at a time.

Answers

1. 12 · [Cost of double cone] + 8 · [Cost of single cone] = 17

[Cost of double cone] = [Cost of single cone] + 0.25

2. Cost of single cone = s (dollars)
Cost of double cone = d (dollars)

3. $\begin{cases} 12d + 8s = 17 \\ d = s + 0.25 \end{cases}$

4. $s = 0.70$, $d = 0.95$; single: $0.70, double: $0.95

Independent Practice

▶ **Ex. 5–10** Encourage students to use a format similar to Example 1.

Answers

8. Baseball: 250, football: 110

10. Single: 64, double: 196

13. 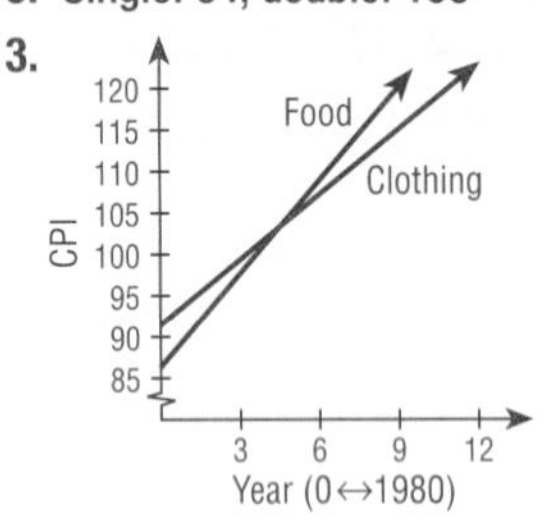

▶ **Ex. 11** You might want to review Example 2 before assigning this exercise.

▶ **Ex. 13–14** These applications are similar to the population model solved in Example 3. They can be used to promote interesting class discussions regarding both CPI and the increase in the number of U.S. households owning a microwave oven.

8. *Collecting Sports Cards* You collect baseball and football cards. Your uncle has an old collection of 360 cards that he gives to you. The collection has more baseball cards than football cards. In fact, it has 30 more baseball cards than *twice* the number of football cards. How many of each type are in your uncle's collection? See margin.

Collecting is one of the most common hobbies in America.

9. *Cable Television* Your family receives basic cable television and one movie channel for \$39 a month. Your neighbor receives basic cable and two movie channels for \$45.50. What is the monthly charge for basic cable? (Assume that each movie channel has the same monthly charge.) \$32.50

10. *Hotel Rates* A hotel has 260 rooms—some singles and some doubles. The singles cost \$35 and the doubles cost \$60. Because of a math teachers' convention, all of the hotel rooms are occupied. The sales for this night are \$14,000. How many of each type of room does the hotel have? See margin.

11. *Buying Automobile Insurance* You are quoted a total annual premium of \$1111.50 for automobile insurance. This premium gives a coverage of L thousand dollars for liability and collision coverage with a deductible of D dollars. The insurance agent tells you that if you are willing to double your collision deductible, you can increase your liability by \$500,000 and add only \$32.50 to your annual premium. What were the amounts of the liability and the deductible in the first quote? (Use the premium schedule given at the top of page 139.) Liability: \$200,000; deductible: \$250

12. *A Historical Puzzle* You can find the years of Napoleon's birth and death by solving the following system. How old was Napoleon when he crossed the Alps? 28 years old

$$\begin{cases} 3x - 2y = 1665 \\ -4x + 3y = -1613 \end{cases}$$

13. *Clothing and Food Costs* The Consumer Price Index (CPI) is used to measure the rate of inflation. From 1980 to 1988, the CPI for clothing increased from 90.9 to 115.4. (This means that clothing that cost \$90.90 in 1980 would have cost \$115.40 in 1988.) From 1980 to 1988, the CPI for food increased from 86.8 to 118.2. Use a graph to estimate the year that the CPI for clothing and food were the same. What does the graph tell you about the rate of inflation for these two items? See margin for graph, 1985, they increased

This etching is one interpretation of Napoleon crossing the Alps in 1797.

14. *Appliances* In 1978, 35% of U.S. households owned a dishwasher but only 8% owned a microwave oven. By 1993, 84% of the households owned a microwave oven and 45% owned a dishwasher. Use a graph to estimate the first year in which more homes owned a microwave oven than a dishwasher. ***(Source: Energy Information Administration)*** 1984; see margin for graph.

Integrated Review

In Exercises 15–17, solve the system by graphing. See margin for graphs.

15. $\begin{cases} x + 2y = 13 \\ -2x + y = 9 \end{cases}$ $(-1, 7)$

16. $\begin{cases} 3x - 2y = 3 \\ x - y = -1 \end{cases}$ $(5, 6)$

17. $\begin{cases} 2x + 3y = -12 \\ -3x + 2y = 5 \end{cases}$ $(-3, -2)$

In Exercises 18–20, solve the system by substitution. See margin.

18. $\begin{cases} x - 3y = 13 \\ 3x + 2y = 41 \end{cases}$

19. $\begin{cases} 3x + y = -1 \\ 4x + 5y = 6 \end{cases}$

20. $\begin{cases} 7x - 2y = 41 \\ 2x + y = 7 \end{cases}$

In Exercises 21–23, solve the system by using a linear combination. See margin.

21. $\begin{cases} 2x - 6y = -30 \\ x + 4y = 2 \end{cases}$

22. $\begin{cases} 3x + 2y = 6 \\ -5x - 4y = -8 \end{cases}$

23. $\begin{cases} 2x + 3y = -8 \\ 5x - 9y = 13 \end{cases}$

Exploration and Extension

Types of Apes **In Exercises 24 and 25, use the puzzle at the right, which contains the names of all four types of apes hidden within it.** See margin for 24.–25.

24. Find equations of the lines that contain the words *gorilla, chimpanzee,* and *orangutan.*

25. The name of the fourth type of ape lies on a line that intersects two of the lines found in Exercise 24. What is it?

```
 y
18 O C H Y E T O S B K A B R O N R Y O
   M B U K A C W P E L K T M R N I S C
16 P A N S S I H Y L O N E O Z E M N N
   K B N G E J L I N B B E S I R A L L
14 S B C I A U R E M P I P E Q T T S W
   A I M I D O O A N P D R T U U E A D
12 D M O R G T N E A Y A O G I R S P E
   K W I Y O I M A S T T N O P R T M L
10 H I S E R B B O O J A E Z A I U I D
   N O N K G R U B N R L E Y E K V B R
 8 I R E G T O T E O O U G H P E F H A
   A F E T K R U S T N A O A M L P O P
 6 J I P Z A O R C H I P R Z T G W C O
   I Q U S N O N I L L O Y M O N K E Y
 4 T E P B A B I G A B N S L A U R I E
   G R A P E A P E E W R I L A J O G N
 2 R S N A K E O H P L O H K B B B O N
   T E N A R O N G E T A N G I F L H O
     2   4   6   8  10  12  14  16  18 x
```


Answers

14.

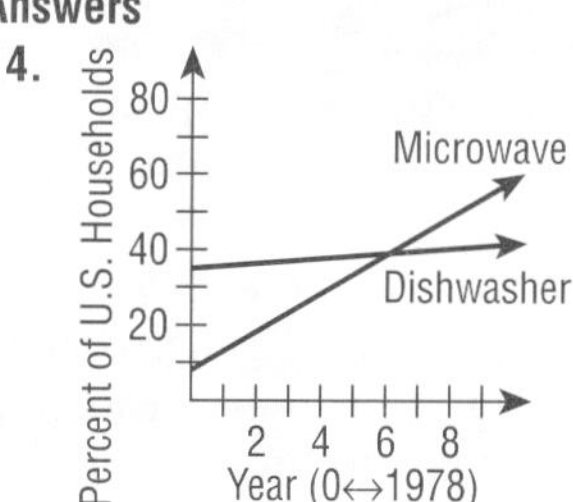

15.

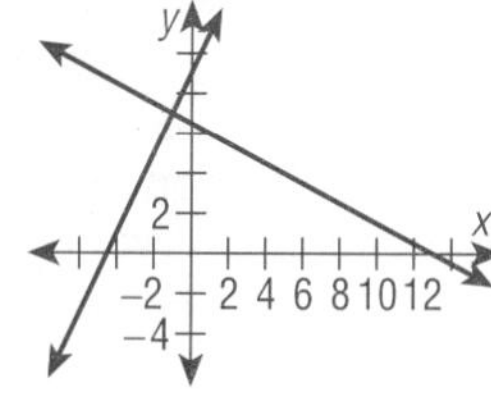

16.

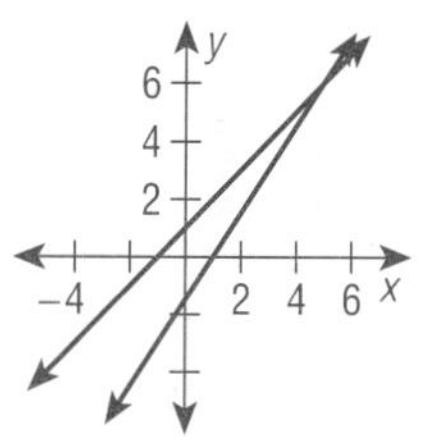

17.

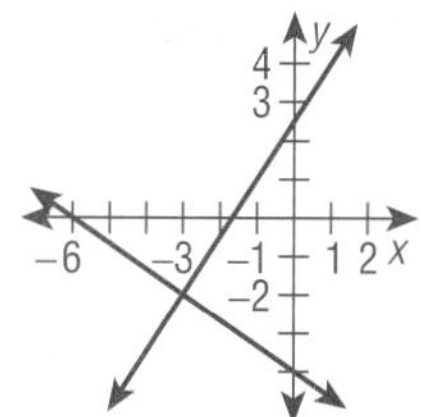

18. $\left(\frac{149}{11}, \frac{2}{11}\right)$ **19.** $(-1, 2)$

20. $(5, -3)$ **21.** $\left(-\frac{54}{7}, \frac{17}{7}\right)$

22. $(4, -3)$ **23.** $(-1, -2)$

▶ Ex. 24–25 COOPERATIVE LEARNING You may wish to use the exercises as an in-class partner quiz.

Answers

24. Gorilla: $y = x + 7$, chimpanzee: $y = -x + 23$, orangutan: $y = 1$

25. Gibbon

Enrichment Activities

These activities require students to go beyond lesson goals.

1. **WRITING PROBLEMS** Create your own word problems similar to those in Exercises 5–10 where the information is reasonable but fictitious. Include the solutions.
2. **COOPERATIVE LEARNING** Work with a partner. Research population data for communities in your area similar to those shown in Example 3. Assuming linear growth, develop algebraic models. Verify the accuracy of your models using current population statistics. How accurate were the linear models? Explain.

Answers

4.

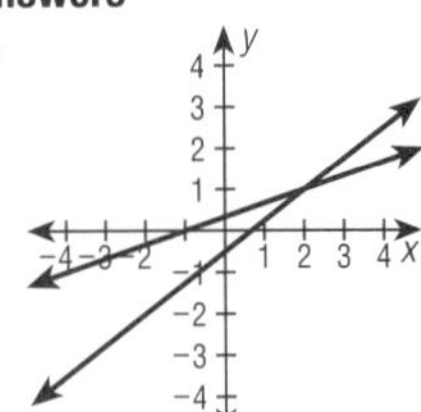

5.

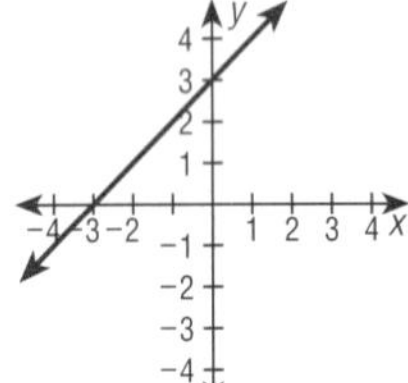

6.

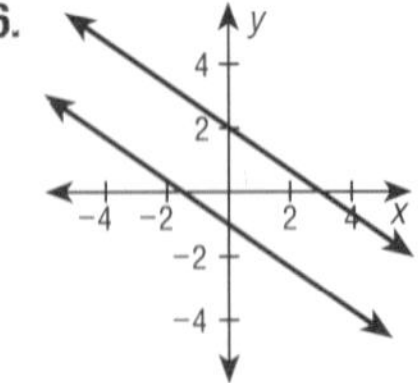

Mid-Chapter SELF-TEST

Take this test as you would take a test in class. The answers to the exercises are given in the back of the book.

In Exercises 1–3, match the linear system with its graph. How many solutions does the system have? (3.1)

1. $\begin{cases} 3x - 4y = 2 \\ -x + 3y = 1 \end{cases}$ a, one solution

2. $\begin{cases} -x + y = 3 \\ 2x - 2y = -6 \end{cases}$ c, many solutions

3. $\begin{cases} x + 2y = 3 \\ -2x - 4y = 3 \end{cases}$ b, no solutions

a.

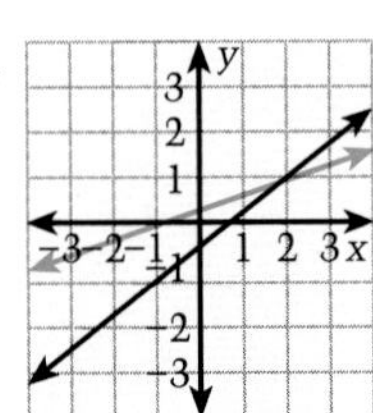

b.

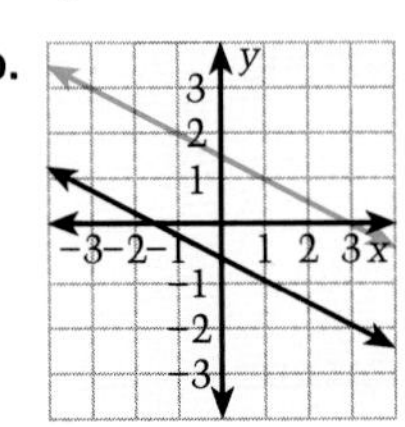

c.

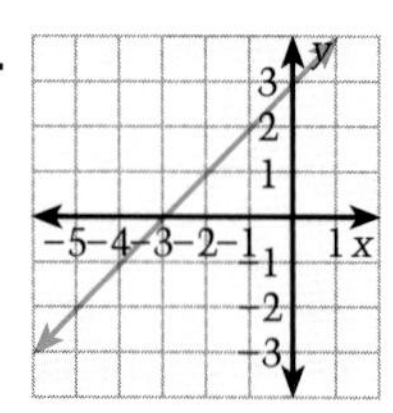

In Exercises 4–6, sketch the graph of the linear system. Then solve the system. (3.1) See margin.

4. $\begin{cases} 3x - 4y = 2 \\ -x + 3y = 1 \end{cases}$ (2, 1)

5. $\begin{cases} -x + y = 3 \\ 2x - 2y = -6 \end{cases}$ Any point on the line $-x + y = 3$

6. $\begin{cases} x + 2y = 3 \\ -2x - 4y = 3 \end{cases}$ no solution

In Exercises 7–9, use substitution to solve the system. (3.2)

7. $\begin{cases} 5x - 3y = 7 \\ x + 6y = -4 \end{cases}$ $\left(\frac{10}{11}, -\frac{9}{11}\right)$

8. $\begin{cases} 3x + 5y = 1 \\ -2x + y = 10 \end{cases}$ $\left(-\frac{49}{13}, \frac{32}{13}\right)$

9. $\begin{cases} 6x - 4y = 10 \\ -3x + 2y = 5 \end{cases}$ no solution

In Exercises 10–12, use a linear combination to solve the system. (3.2)

10. $\begin{cases} 2x + y = 12 \\ 3x - y = 13 \end{cases}$ (5, 2)

11. $\begin{cases} 5x - 3y = 18 \\ 4x - 6y = 0 \end{cases}$ (6, 4)

12. $\begin{cases} 3x - 4y = 10 \\ 4x - 5y = 13 \end{cases}$ (2, −1)

13. How can \$12,600 be split between two investments, one paying 6% annually and one paying 8% annually, so that the amounts of interest from the two investments are equal? **(3.1)** 6%: \$7200, 8%: \$5400

14. Halley's Comet appears about every 75 years. One appearance was the year Mark Twain was born. The next appearance was the year he died. The sum of these two years is 3745. In what year did Mark Twain die? **(3.3)** 1910

15. The measures of the two acute angles of a right triangle differ by 18 degrees. What are their measures? **(3.3)** 36°, 54°

16. A company purchases injection-mold equipment to make plastic knives, forks, and spoons. The cost of the equipment was \$82,800 and the cost of making a package is \$0.36. The company sells each package for \$1.08. How many packages must be sold for the company to break even? **(3.1)** 115,000

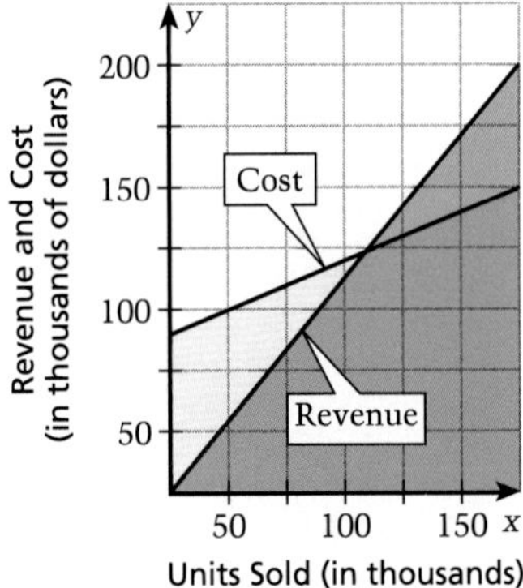

3.4 Solving Systems of Linear Inequalities

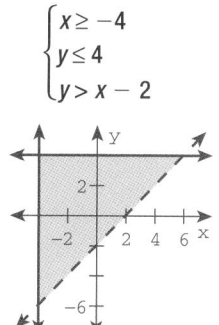

What you should learn:

Goal 1 How to graph a system of linear inequalities to find the solution of the system

Goal 2 How to use a system of linear inequalities to model a real-life situation

Why you should learn it:

You can model many real-life situations that involve area, such as regions of a map or a floor plan, with systems of linear inequalities.

$$\begin{cases} x \geq -4 \\ y \leq 4 \\ y > x - 2 \end{cases}$$

Goal 1 Graphing a System of Inequalities

A **solution of a system of linear inequalities** is an ordered pair that is a solution of each inequality in the system. The **graph of a system** is the graph of all solutions in the system.

LESSON INVESTIGATION

Investigating the Graph of a System

Partner Activity Three graphs of linear inequalities are shown below. Write the linear inequality for each graph. Then describe and sketch the graph of the system composed of the three inequalities.

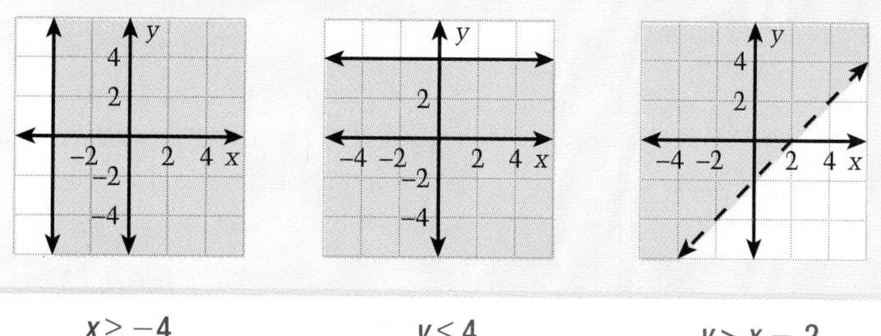

Example 1 *Graphing a System of Linear Inequalities*

Sketch the graph of the system of linear inequalities.

$$\begin{cases} x < 3 & \textit{Inequality 1} \\ y \geq -2 & \textit{Inequality 2} \\ y < x + 1 & \textit{Inequality 3} \end{cases}$$

Solution The graph of $x < 3$ is the half-plane lying to the left of the vertical line $x = 3$. The graph of $y \geq -2$ is the half-plane lying above the horizontal line $y = -2$. The graph of $y < x + 1$ is the half-plane lying below the diagonal line $y = x + 1$. The graph of the system is the **intersection** of these three half-planes. ■

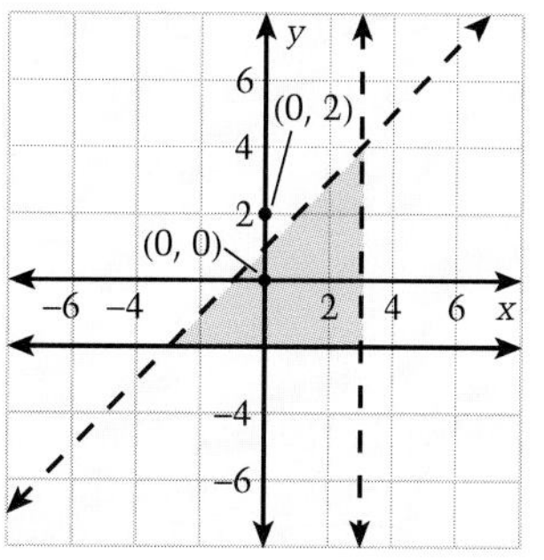

The point (0, 0) is in the graph of the system. The point (0, 2) is not.

After graphing a system of linear inequalities, you can find each **vertex** (corner point). For instance, the graph at the left has three vertices: $(-3, -2)$, $(3, 4)$, and $(3, -2)$.

Problem of the Day

Two fishermen wished to find the weight of a large fish. They set up a plank as a seesaw, placing it on a rock so that it would just balance when one stood on each end. Then they exchanged places, and the lighter man held the fish. This just brought the plank into balance again. The weights of the fishermen were 120 pounds and 150 pounds. What was the weight of the fish? $67\frac{1}{2}$ pounds

ORGANIZER

Warm-Up Exercises

1. Graph these lines.
   a. $y = 7x - 5$
   b. $x + 6y = 18$
   c. $8x = 12$
   a. The graph is the line with intercepts $(\frac{5}{7}, 0)$ and $(0, -5)$.
   b. The graph is the line with intercepts (18, 0) and (0, 3).
   c. The graph is the vertical line $x = 1.5$.
2. Graph the inequalities.
   a. $y \geq 7x - 5$ b. $8x \leq 12$
   c. $x + 6y < 18$ d. $y > -2$
   a. The graph is the region above the solid line $y = 7x - 5$.
   b. The graph is the region to the left of the solid line $x = 1.5$.
   c. The graph is the region below the dashed line $x + 6y = 18$.
   d. The graph is the region above the dashed line $y = -2$.

Lesson Resources

Teaching Tools
- Transparencies: 8, 10, 11
- Problem of the Day: 3.4
- Warm-up Exercises: 3.4
- Answer Masters: 3.4

Extra Practice: 3.4

Color Transparencies: 23, 24

LESSON Notes

Example 1

The vertices of the graph of the system are found by solving these three systems of linear equations:

$$\begin{cases} y = -2 \\ x = 3 \end{cases} \quad \begin{cases} y = -2 \\ y = x + 1 \end{cases}$$

$$\begin{cases} y = x + 1 \\ x = 3 \end{cases}$$

The solution of the first system is (3,−2); the solutions of the second and third are (−3,−2) and (3, 4).

Examples 2–3

Point out to students that the graphs in these examples illustrate how systems of linear inequalities can be used to graph familiar geometric regions, such as rectangles, squares, triangles, trapezoids, etc.

Extra Examples

Here are additional examples similar to **Examples 1–3.**

1. Sketch the graph of the given system.

a. $\begin{cases} y < 4 \\ y \geq -2x - 6 \\ x < 7 \end{cases}$ b. $\begin{cases} y > -1 \\ y \leq 3 \end{cases}$

c. $\begin{cases} y < 4x - 8 \\ y < -3x + 13 \\ y \geq 1 \end{cases}$

a.

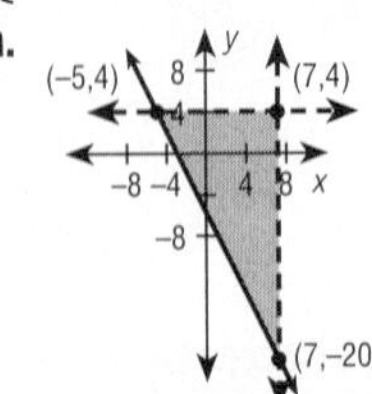

b.

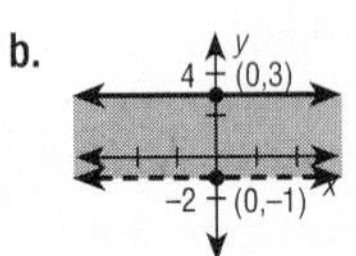

Here are some guidelines that may help you sketch the graph of a system of linear inequalities.

Graphing a System of Linear Inequalities

1. Sketch the line that corresponds to each inequality. (Use a dashed line for inequalities with < or > and a solid line for inequalities with ≤ or ≥.)
2. Lightly shade the half-plane that is the graph of each linear inequality. (Colored pencils may help you distinguish the different half-planes.)
3. The graph of the system is the intersection of each of the half-planes. (If you used colored pencils, it is the region that has been shaded with *every* color.)

Example 2 *Graphing a System of Linear Inequalities*

Sketch the graph of the system of linear inequalities.

$$\begin{cases} x \leq 3 & \textit{Inequality 1} \\ x > 1 & \textit{Inequality 2} \end{cases}$$

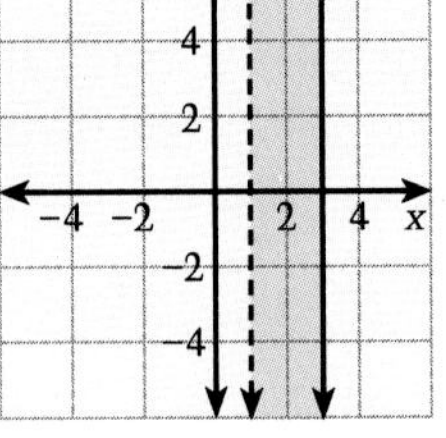

Solution The graph of the first inequality is the half-plane *to the left of and including* the vertical line $x = 3$. The graph of the second inequality is the half-plane *to the right* of the vertical line $x = 1$.

$x = 3$ *Right boundary*
$x = 1$ *Left boundary*

The graph of the system is the vertical strip *between* the two vertical lines (where $x \leq 3$ *and* $x > 1$) and includes the line on the right. (Note that this graph has no vertices.) ■

Example 3 *Graphing a System of Linear Inequalities*

Sketch the graph of the system of linear inequalities. Find the vertices of the graph.

$$\begin{cases} y \leq -\frac{1}{2}x + 2 & \textit{Inequality 1} \\ y \geq 0 & \textit{Inequality 2} \\ x \geq 0 & \textit{Inequality 3} \end{cases}$$

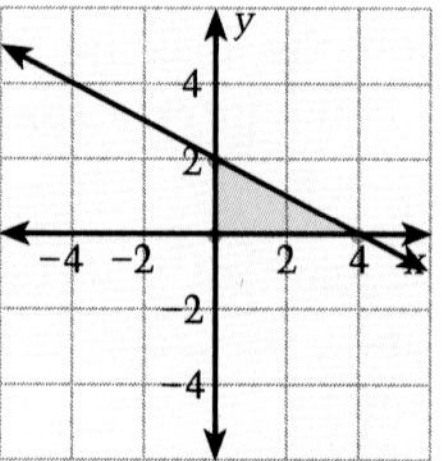

Solution The graph of the first inequality is the half-plane *below and including* the line $y = -\frac{1}{2}x + 2$. The graph of the second inequality is the half-plane above and including the x-axis, and the graph of the third inequality is the half-plane to the right of and including the y-axis. The vertices of the graph are (0, 0), (0, 2), and (4, 0). ■

Although major museums maintain security systems, some famous works of art have been stolen and/or damaged. In 1972, a person broke off several pieces of Michelangelo's beautiful marble sculpture, the Pietà. *Fortunately, it was able to be restored.*
(Source: National Geographic, *December 1985)*

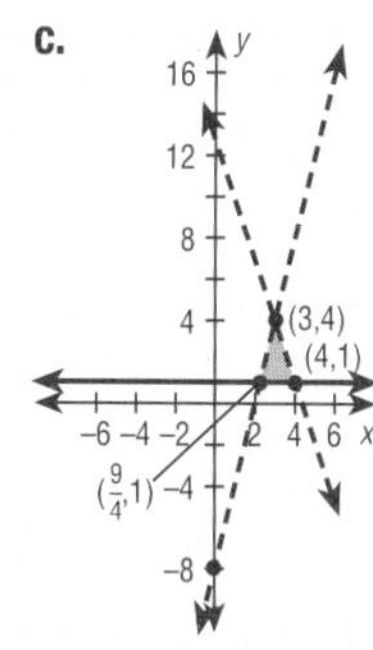

2. Find the corner points of each system described in Extra Example 1.
   a. $(-5, 4), (7, 4), (7, -20)$
   b. no corner points
   c. $(3, 4), (4, 1), (\frac{9}{4}, 1)$

Goal 2 Linear Inequalities as Real-Life Models

Real Life
Art Museums

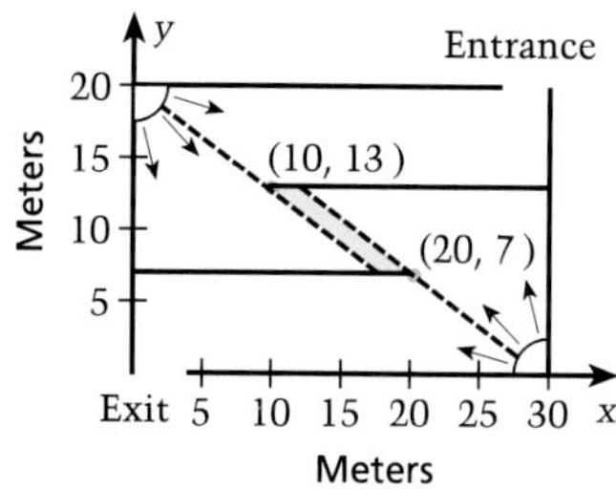

Each unit in the coordinate system represents 1 meter.

Example 4 *Writing a System of Linear Inequalities*

An art museum has installed an electronic surveillance system. In one hall, two cameras were installed at opposite corners, as shown in the diagram. To increase display space, however, the museum also built two temporary walls, each 20 meters long. Write a system of inequalities that describes the region of the room not visible to the cameras.

Solution The region that is not visible to either camera is shaded blue in the diagram. The line passing through the points (0, 20) and (10, 13) represents the left boundary of the region. The line passing through the points (30, 0) and (20, 7) represents the right boundary. The upper and lower boundaries are given by the lines $y = 7$ and $y = 13$. Using these equations, you can write the following system of inequalities.

$$\begin{cases} y \geq -\frac{7}{10}x + 20 & \textit{Left boundary} \\ y \leq -\frac{7}{10}x + 21 & \textit{Right boundary} \\ y \leq 13 & \textit{Upper boundary} \\ y \geq 7 & \textit{Lower boundary} \end{cases}$$

■

Communicating about ALGEBRA

SHARING IDEAS about the Lesson See Additional Answers.

Sketch the graph of the system of equations. Then mark and label the vertices of the region. Explain your steps.

A. $\begin{cases} x < 4 \\ y > 1 \\ y < x + 1 \end{cases}$ **B.** $\begin{cases} 4x + 3y \leq 12 \\ x \geq 0 \\ y \geq 0 \end{cases}$ **C.** $\begin{cases} x + 3y \geq 6 \\ 3x + 2y \leq 18 \\ y \leq 3 \end{cases}$

Communicating about ALGEBRA

EXTEND *Communicating*
Construct a system of linear inequalities whose solutions are regions that have the following areas (in square units): 1, 5, 12, and 16.
Answers will vary; the following system results in a rectangle with an area of 16 square units.

$$\begin{cases} x \leq 3 \\ x \geq 1 \\ y \geq -2 \\ y \leq 6 \end{cases}$$

Check Understanding

1. What is a solution of a system of linear inequalities?
   See the top of page 145.
2. Justify the statement, "To graph a linear inequality, you must first be able to graph a linear equation."
   The linear equation is a boundary or edge of the half-plane that is a solution.
3. How can colored pens help in graphing a system of linear inequalities?
   See the top of page 146.

EXERCISE Notes

ASSIGNMENT GUIDE

Basic/Average: Ex. 1–6, 7–12, 13–27 odd, 31–35 odd, 39–53 odd

Above Average: Ex. 1–6, 7–12, 15–37 odd, 43–53 odd, 54

Advanced: Ex. 1–6, 7–12, 17–37 odd, 38, 49–54

Selected Answers
Exercises 1–6, 7–51 odd

More Difficult Exercises
Exercises 32–38

Guided Practice

▶ **Ex. 1–6** Direct students to the graphing summary found on page 146.

Answer

1.

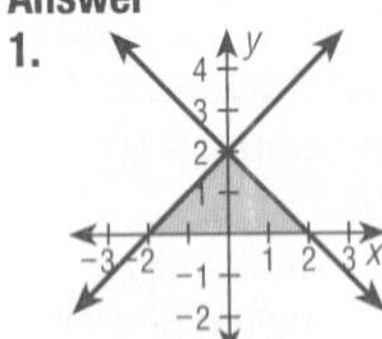

Independent Practice

▶ **Ex. 7–12** Assign these exercises as a group.

▶ **Ex. 25–30** TECHNOLOGY These exercises can be enhanced by the use of a graphics calculator. The trace feature can be used to locate the coordinates of the respective vertices. You might ask students to provide a copy of the graphs as part of their solutions.

EXERCISES

Guided Practice

▶ **CRITICAL THINKING about the Lesson**

1. Sketch the graph of the system of linear inequalities.

$$\begin{cases} y \le x + 2 \\ y \le -x + 2 \\ y \ge 0 \end{cases}$$ See margin.

In Exercises 2–5, decide whether the ordered pair is a solution of the system of linear inequalities. Justify your answer algebraically and graphically.

$$\begin{cases} x \ge -1 \\ y < 2 \\ y > 2x + 2 \end{cases}$$

2. $(-1, 0)$ It is not. **3.** $(0, 0)$ It is not. **4.** $(-1, 1)$ It is. **5.** $\left(-\frac{1}{2}, 1\right)$ It is not.

6. Explain how to find the vertices of the graph of a system of linear inequalities. Find the region that is the intersection of the half-planes; the vertices are the corner points of that region and are the intersections of the edges of half-planes.

Independent Practice

In Exercises 7–12, match the system of inequalities with its graph.

7. $\begin{cases} y > x \\ x > -3 \\ y \le 0 \end{cases}$ c

8. $\begin{cases} y \le 4 \\ y > -2 \end{cases}$ b

9. $\begin{cases} y < x \\ y > -3 \\ x \le 0 \end{cases}$ f

10. $\begin{cases} x \le 3 \\ y < 1 \\ y > -x + 1 \end{cases}$ e

11. $\begin{cases} y > -1 \\ x \ge -3 \\ y \le -x + 1 \end{cases}$ a

12. $\begin{cases} y > -4 \\ y \le 2 \end{cases}$ d

a.

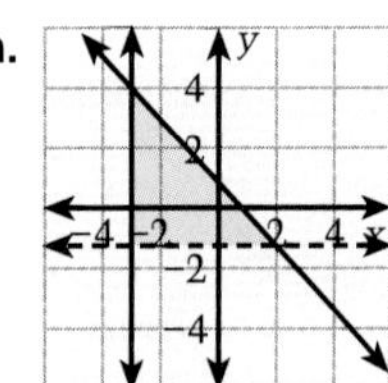

b.

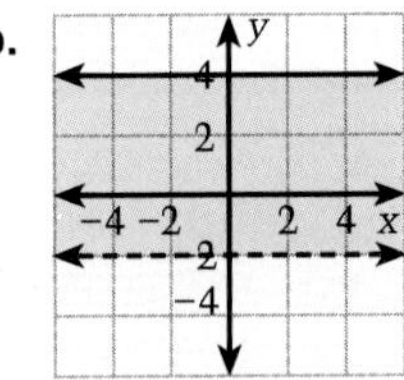

c.

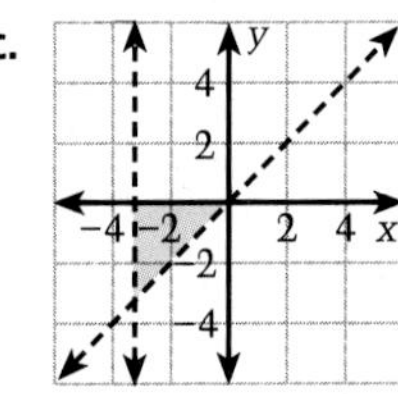

d.

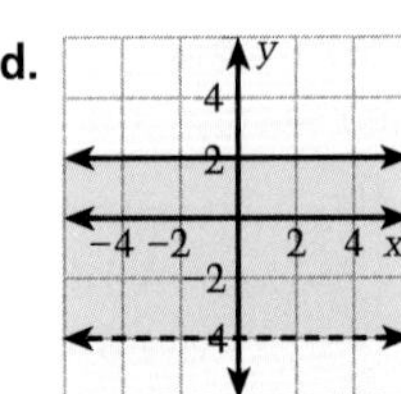

e.

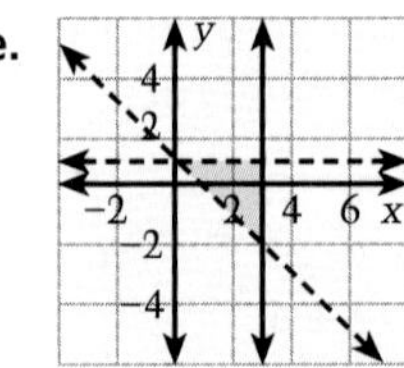

f.

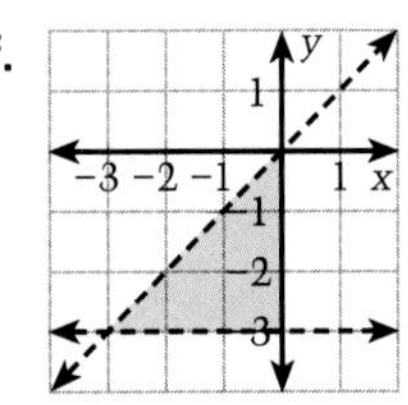

In Exercises 13–24, sketch the graph of the system of linear inequalities. See Additional Answers.

13. $\begin{cases} x < 3 \\ x > -2 \end{cases}$ 14. $\begin{cases} y > -1 \\ y \le 2 \end{cases}$ 15. $\begin{cases} y \le 4 \\ y > 1 \\ x \ge -3 \end{cases}$ 16. $\begin{cases} y < 4 \\ x > -4 \\ y > x \end{cases}$

17. $\begin{cases} y > -5 \\ x \le 2 \\ y \le x + 2 \end{cases}$ 18. $\begin{cases} y \le -1 \\ x \le 2 \\ y \le x + 2 \end{cases}$ 19. $\begin{cases} y \ge 2x - 3 \\ y > -5x - 8 \end{cases}$ 20. $\begin{cases} y \le x + 3 \\ y \le -\frac{1}{2}x - 1 \end{cases}$

21. $\begin{cases} x + y > 3 \\ 4x + y < 4 \end{cases}$ 22. $\begin{cases} 2x - y \ge 2 \\ 5x - y \ge 2 \end{cases}$ 23. $\begin{cases} x + y \ge -10 \\ x > -1 \\ y < 1 \end{cases}$ 24. $\begin{cases} 2x - y \le -1 \\ x - 2y \ge -8 \\ y \le 4 \end{cases}$

In Exercises 25–30, sketch the graph of the system of linear inequalities. Mark and label each vertex of the graph. See Additional Answers.

25. $\begin{cases} x - 3y \ge -1 \\ 4x - 2y \le 1 \\ x > -3 \end{cases}$ 26. $\begin{cases} 3x - 4y > -20 \\ 3x + 4y < 7 \\ y \ge -2 \end{cases}$ 27. $\begin{cases} 2x - 3y > -6 \\ 5x - 3y < 3 \\ x + 3y > -3 \end{cases}$

28. $\begin{cases} x - 4y > 0 \\ x + y \le 1 \\ x + 3y > -1 \end{cases}$ 29. $\begin{cases} x + y \le 4 \\ x + y \ge -1 \\ x - y \ge -2 \\ x - y \le 2 \end{cases}$ 30. $\begin{cases} y < 4 \\ y > -8 \\ 2x - y \ge -4 \\ 2x + y \le 6 \end{cases}$

In Exercises 31–34, write a system of linear inequalities for the region shown. See below.

31. 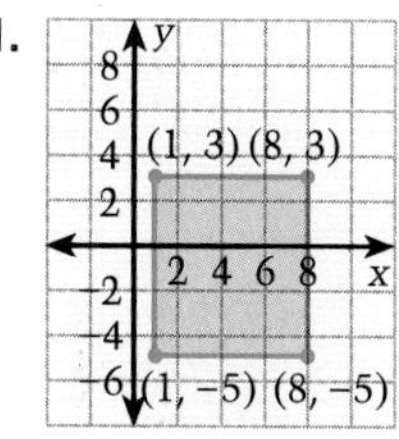

32. 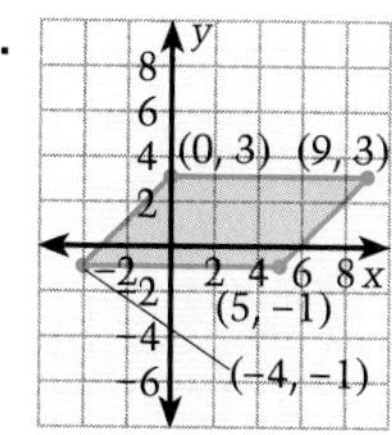

33. 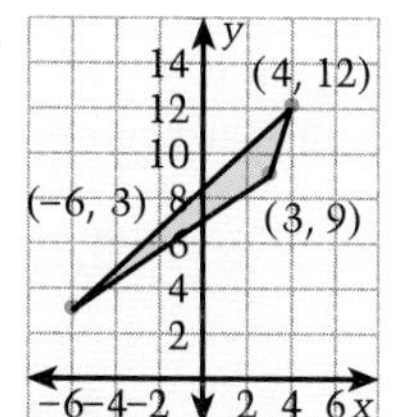

34. 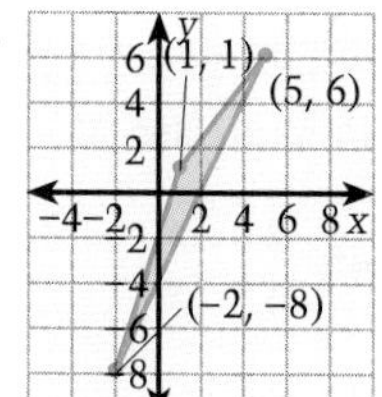

35. *Morning Snacks* You are a counselor at a summer day camp, and have been sent to buy drink boxes and boxes of raisins for the campers' morning snack. Each box of raisins costs \$0.10 and each juice box costs \$0.40. You have \$26, and you want to buy enough so that each camper will have at least one box of raisins and one juice box. There are at most 40 campers. Write a system of linear inequalities that shows the various numbers of boxes of raisins and juice boxes that you could buy. (You don't have to spend all the money.) Sketch a graph of the system. $x \ge 40$, $y \ge 40$, $0.1x + 0.4y \le 26$, see margin for graph.

36. *Research Paper Bibliography* You have been assigned a research paper. The guidelines specify that you must have at least ten sources in the bibliography, no more than three of which can be encyclopedias. Write a system of inequalities that describes the various number of encyclopedia sources and other sources that you could use. Graph this system. See margin.

31. $x \ge 1, x \le 8, y \le 3, y \ge -5$ 32. $y \le 3, y \ge -1, y \le x + 3, y \ge x - 6$

33. $y \le \frac{9}{10}x + \frac{42}{5}, y \ge 3x, y \ge \frac{2}{3}x + 7$ 34. $y \le \frac{5}{4}x - \frac{1}{4}, y \ge 2x - 4, y \le 3x - 2$

Ex. 31–34 COOPERATIVE LEARNING These exercises require extended student work time. You might suggest that students work with partners to share the task.

Ex. 35–36 These exercises are similar to the application modeled in Example 4.

Answers

35. 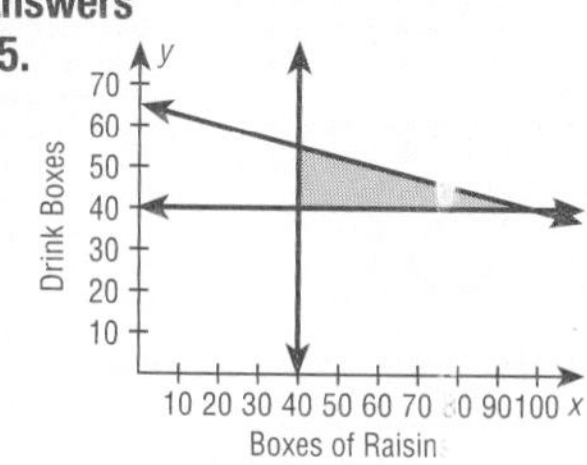

36. $x + y \ge 10$, $x \le 3$

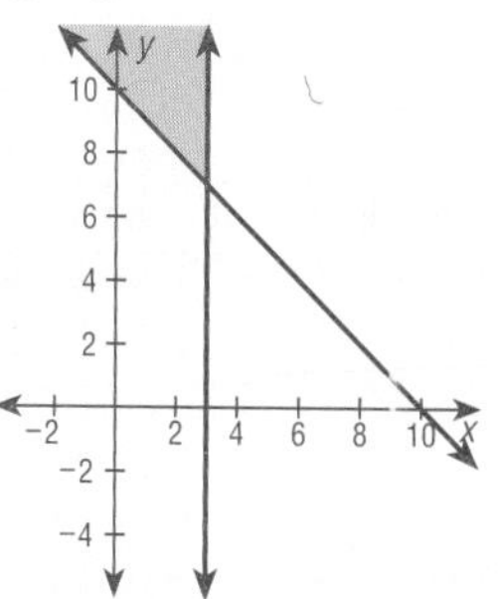

Ex. 53–54 TECHNOLOGY Graphics calculators have a shade feature that will graph linear inequalities. Have students consult their calculator manual for the proper key instructions.

Enrichment Activities

These activities require students to go beyond lesson goals.

1. **EXTEND** Example 4 Derive the left and right boundaries of the blue region.

Left boundary:

$m = \frac{(13-20)}{(10-0)} = -\frac{7}{10}$

y-intercept $= 20$

equation of line: $y = -\frac{7}{10}x + 20$

Right boundary:

$m = \frac{7-0}{20-30} = \frac{7}{-10} = -\frac{7}{10}$

point of line: (30, 0)

equation of line:

$y - 0 = -\frac{7}{10}(x-30)$

$y = -\frac{7}{10}x + 21$

2. Use the data in Example 4 to determine the percent of the room that cannot be viewed by the surveillance system.

total area $= (20 \cdot 30)\text{m}^2 = 600\ \text{m}^2$

To find the area, you need the corner points of the blue region, that is, the corner points associated with the systems

$\begin{cases} y = 13 \\ y = -\frac{7}{10}x + 21 \end{cases}$

and

$\begin{cases} y = 7 \\ y = -\frac{7}{10}x + 20 \end{cases}$

The corner points are $\left(\frac{80}{7}, 13\right)$ and $\left(\frac{130}{7}, 7\right)$, respectively. Knowing that the other corner points are (10, 13) and (20, 7), you can find that the base of the blue region is $\frac{10}{7}$ meters, and its height is 6 meters. So, the area of the blue region is $\left(\frac{10}{7}\right)(6)\text{m}^2 = \frac{60}{7}\text{m}^2$. It follows that the surveillance system cannot see about 1.4% of the room.

$\left[\frac{\frac{60}{7}}{600} = \frac{1}{70} \approx 0.014 = 1.4\%\right]$

37. *Swimming Safety* The diagram shows a cross section of a roped-off swimming area at a beach. Write a system of inequalities that describes the cross section. (Each unit in the coordinate system represents 1 foot.) $y \ge -\frac{1}{7}x,\ y \ge -10,\ x < 90,\ y \le 0$

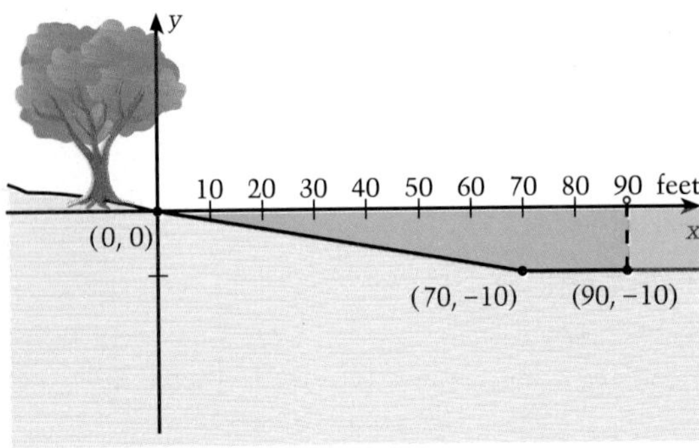

38. *A View of the Chorus* The diagram shows the chorus platform on a stage. Write a system of inequalities that describes the part of the audience that can see the full chorus. (Each unit in the coordinate system represents 1 meter.)
$y \ge -2x - 24,\ y \ge 2,\ y \ge 2x - 24,\ y \le 22$

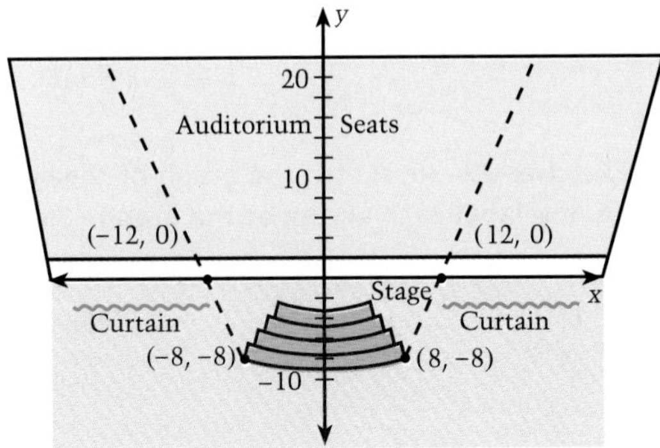

Integrated Review

In Exercises 39–44, sketch the graph of the inequality. See Additional Answers.

39. $y > -\frac{1}{2}x + 6$
40. $y < x - 56$
41. $y \le 5x - 1$
42. $y > 9 + 16x$
43. $9x - y > 16$
44. $4x + 2y \le 9$

In Exercises 45–50, solve the linear system.

45. $\begin{cases} x - 2y = 16 \\ -2x - y = -2 \end{cases}$ (4, −6)

46. $\begin{cases} 5x - 4y = -11 \\ -2x + y = 5 \end{cases}$ (−3, −1)

47. $\begin{cases} -4x + y = 5 \\ 3x - 2y = -15 \end{cases}$ (1, 9)

48. $\begin{cases} 3x - 5y = 5 \\ 4x + 3y = 26 \end{cases}$ (5, 2)

49. $\begin{cases} -2x - 3y = 3 \\ 3x + 7y = 13 \end{cases}$ (−12, 7)

50. $\begin{cases} 9x - 7y = 25 \\ -3x + 2y = -11 \end{cases}$ (9, 8)

51. *College Entrance Exam Sample* If $820 + R + S - 610 = 342$ and $R = 2S$, then $S =$ [?].
a. <u>44</u> b. 48 c. 132 d. 184 e. 192

52. *College Entrance Exam Sample* If $x + y = 8p$ and $x - y = 6q$, then $x =$ [?].
a. $7pq$ b. <u>$4p + 3q$</u> c. pq d. $4p - 3q$ e. $8p + 6q$

Exploration and Extension

Technology **In Exercises 53 and 54, use a graphing calculator to sketch the system of inequalities. Use the trace feature to estimate the vertex.** See Additional Answers.

53. $\begin{cases} y \le 9.2x + 12.1 \\ y > -4.1x - 2.0 \end{cases}$ (−1.1, 2.3)

54. $\begin{cases} y > 5.3x - 3.6 \\ y \le -2.1x + 18.8 \end{cases}$ (3.0, 12.4)

3.5 Exploring Data: Linear Programming

What you should learn:

Goal 1 How to solve a linear programming problem

Goal 2 How to use linear programming to answer questions about real-life situations

Why you should learn it:

You can model many business problems about maximizing costs or minimizing expenses using linear programming.

Goal 1 A Linear Programming Problem

Many real-life problems involve a process called **optimization,** which means finding the minimum or maximum value of some quantity. In this lesson, you will study one type of optimization process called **linear programming.**

A linear programming model consists of a system of linear inequalities called **constraints** and an **objective quantity** that can be minimized or maximized. To find the minimum or maximum value of the objective quantity, first find all the vertices of the graph of the constraint inequalities. Then, evaluate the objective quantity at each vertex. The least of these values is the minimum for the entire region and the greatest of these values is the maximum for the region (provided the objective quantity has a minimum and maximum value).

Example 1 *Solving a Linear Programming Problem*

Find the minimum value and maximum value of

$C = 4x + 5y$ *Objective quantity*

subject to the following constraints.

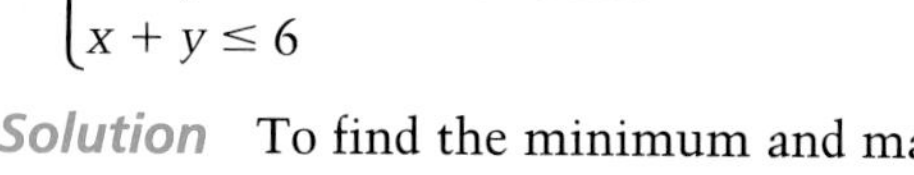

$$\begin{cases} x \geq 0 \\ y \geq 0 \\ x + y \leq 6 \end{cases}$$ *Constraints*

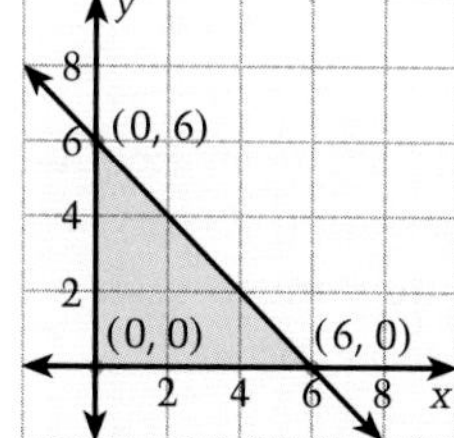

Solution To find the minimum and maximum values of C, evaluate $C = 4x + 5y$ at each of the three vertices of the graph.

At $(0, 0)$: $C = 4(0) + 5(0) = 0$ *Minimum value of C*

At $(6, 0)$: $C = 4(6) + 5(0) = 24$

At $(0, 6)$: $C = 4(0) + 5(6) = 30$ *Maximum value of C*

The minimum value of C is 0, when $x = 0$ and $y = 0$.
The maximum value of C is 30, when $x = 0$ and $y = 6$. ■

In Example 1, try evaluating C at the other feasible points in the graph of the constraints. No matter which point you choose, the value of C will be greater than or equal to 0 and less than or equal to 30.

Problem of the Day

A star is formed from two equilateral triangles. What is the ratio of the area of the star to the area of one of the two large triangles?
4:3

ORGANIZER

Warm-Up Exercises

1. Graph each system.
   a. $x > -3$, $y \geq -x - 2$, $y < 4$
   b. $x \leq 5$, $x \geq -2$, $y \leq 3$, $y \geq -6$

a.

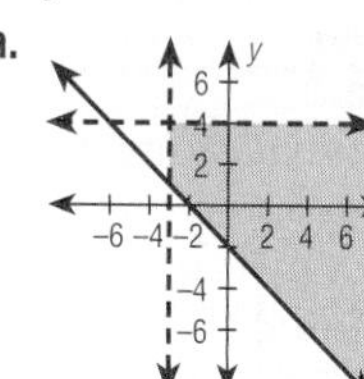

b.

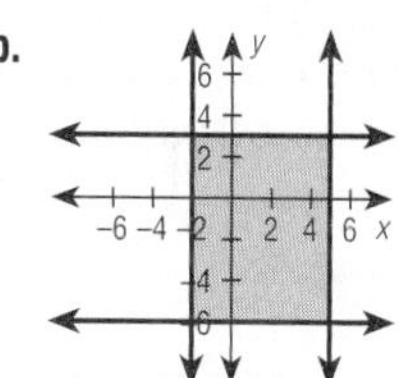

2. Evaluate each expression at the given point.
   a. $4y - 2x$, $(-1, 2)$ 10
   b. $7x + 11y$, $(3, -9)$ −78
   c. $-6x + 8y$, $(1, 5)$ 34

Lesson Resources

Teaching Tools
- Transparencies: 8, 10, 11
- Problem of the Day: 3.5
- Warm-up Exercises: 3.5
- Answer Masters: 3.5

Extra Practice: 3.5
Color Transparency: 25

LESSON Notes

Example 1

CONNECTIONS Linear programming is a branch of mathematics developed during World War II to cope with the complex task of shipping men and supplies to Europe from the United States.

Example 2

Objective quantities describe such relationships as those involving business profits or expenses, or productivity on assembly lines.

Stress the importance of knowing how to solve systems of linear equations in order to locate the vertices of a system of constraint inequalities.

Example 3

Observe for students that an unbounded region of a coordinate plane means that for every value of x (or y) there is always another value, x_0 (or y_0), such that $x < x_0$ (or $y < y_0$). Intuitively, an unbounded region is not contained within boundary lines.

Example 4

Take the time to show students how the vertices can be obtained from systems of two linear equations, derived from the collection of constraints. Begin by asking students how many such systems must be solved. (5) A useful approach for verifying the vertices is to have different groups of students determine the solution to one of the five systems of two linear equations. That is, share the work!

Example 2 *Solving a Linear Programming Problem*

Find the minimum value and maximum value of

$C = 3x + 4y$ *Objective quantity*

subject to the following constraints.

$$\begin{cases} x \geq 2 \\ x \leq 5 \\ y \geq 1 \\ y \leq 6 \end{cases}$$ *Constraints*

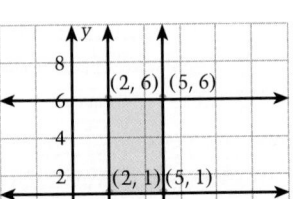

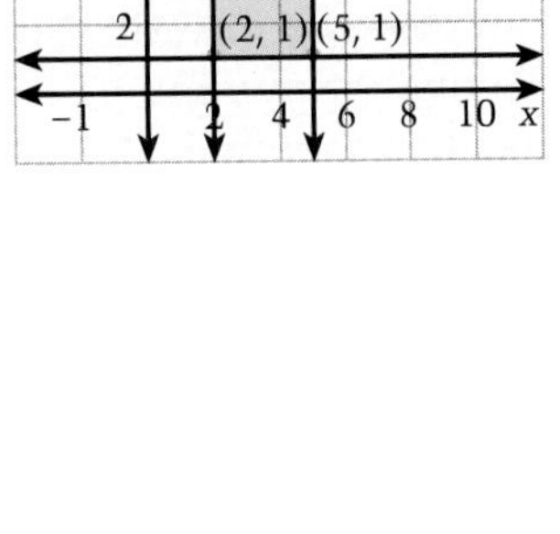

Solution The graph of the constraint inequalities is shown at the left. The four vertices are (2, 1), (2, 6), (5, 1), and (5, 6). To find the minimum and maximum values of C, evaluate $C = 3x + 4y$ at each of the four vertices.

At (2, 1): $C = 3(2) + 4(1) = 10$ *Minimum value of C*

At (2, 6): $C = 3(2) + 4(6) = 30$

At (5, 1): $C = 3(5) + 4(1) = 19$

At (5, 6): $C = 3(5) + 4(6) = 39$ *Maximum value of C*

The minimum value of C is 10. It occurs when $x = 2$ and $y = 1$. The maximum value of C is 39. It occurs when $x = 5$ and $y = 6$. ■

Example 3 *Solving a Linear Programming Problem*

Find the minimum value and maximum value of

$C = 6x + 7y$ *Objective quantity*

subject to the following constraints.

$$\begin{cases} x \geq 0 \\ y \geq 0 \\ 4x + 3y \geq 24 \\ x + 3y \geq 15 \end{cases}$$ *Constraints*

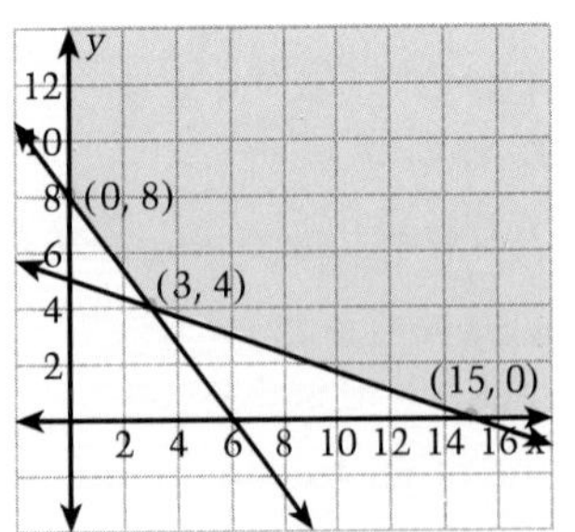

Solution The graph of the system of linear inequalities is shown at the left. The three vertices are (0, 8), (3, 4), and (15, 0). To find the minimum and maximum values of C, evaluate $C = 6x + 7y$ at each of the three vertices.

At (0, 8): $C = 6(0) + 7(8) = 56$

At (3, 4): $C = 6(3) + 7(4) = 46$ *Minimum value of C*

At (15, 0): $C = 6(15) + 7(0) = 90$

The minimum value of C is 46. It occurs when $x = 3$ and $y = 4$. There is no maximum value. (The graph of the constraints is unbounded.) ■

Goal 2 Modeling with Linear Programming

Real Life **Business**

Process A earns a profit of $80 per set and Process B earns a profit of $40 per set.

Example 4 *Finding the Maximum Profit*

You own a factory that makes metal patio sets using two processes. The hours of unskilled labor, machine time, and skilled labor *per patio set* are given in the table. You can use up to 4000 hours of unskilled labor, up to 1500 hours of machine time, and up to 2300 hours of skilled labor. How many patio sets should you make by each process to maximize profits?

	Assembly Hours	
	Process A	*Process B*
Unskilled labor	10	1
Machine time	1	3
Skilled labor	5	2

Solution Let a and b represent the number made with each process. To maximize profits, P, use the objective quantity

$P = 80a + 40b$. *Profit: $80 for Process A, $40 for Process B*

The constraints are as follows. The graph is at the left.

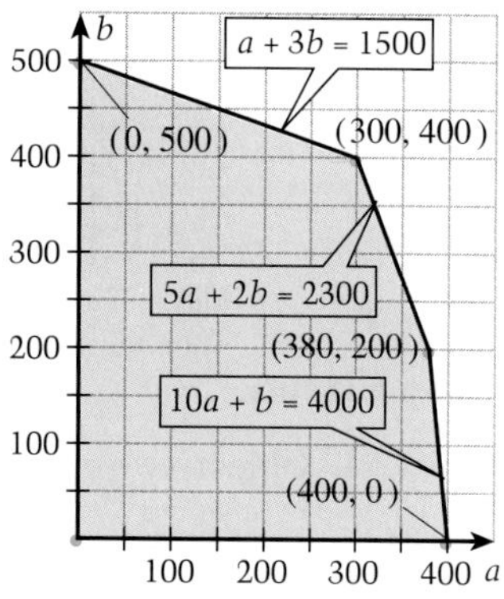

$$\begin{cases} 10a + b \le 4000 \\ a + 3b \le 1500 \\ 5a + 2b \le 2300 \\ a \ge 0 \\ b \ge 0 \end{cases}$$

Unskilled labor: Up to 4000 hours
Machine time: Up to 1500 hours
Skilled labor: Up to 2300 hours
Cannot produce a negative amount.
Cannot produce a negative amount.

The profit at each vertex of the graph is as follows.

At (0, 500):	$P = \$20{,}000$	At (400, 0):	$P = \$32{,}000$
At (300, 400):	$P = \$40{,}000$	At (0, 0):	$P = \$0$
At (380, 200):	$P = \$38{,}400$		

The maximum profit is obtained by making 300 patio sets with Process A and 400 patio sets with Process B. ■

Communicating about ALGEBRA

SHARING IDEAS about the Lesson

Vary the Conditions In Example 4, how many patio sets should you produce by each process if the profit per patio set is as follows?

A. Process A: $80
Process B: $30 **A: 380, B: 200**

B. Process A: $110
Process B: $10 **A: 400, B: 0**

Extra Examples

Here are additional examples similar to **Examples 1–3.**

1. Find the maximum and minimum values of the following objective quantities with given constraints.

a. $P = 5x + 6y$, subject to the constraints $x \ge 1$, $y \ge 0$, $3x + 2y \le 24$

b. $K = 7x + y$, subject to the constraints $x \ge 0$, $y \ge 0$, $2x + y \ge 11$, $x + 4y \ge 16$

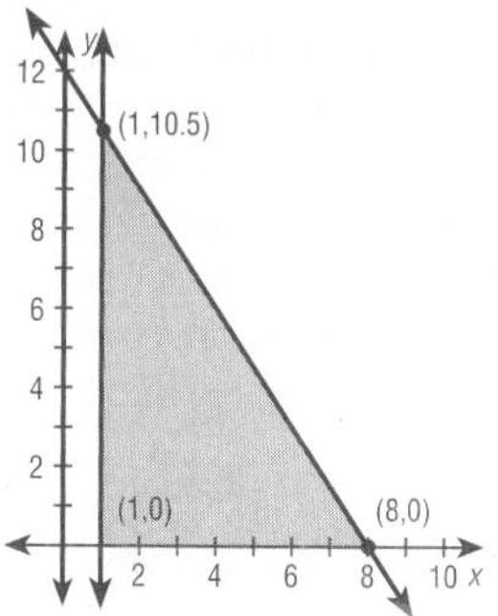

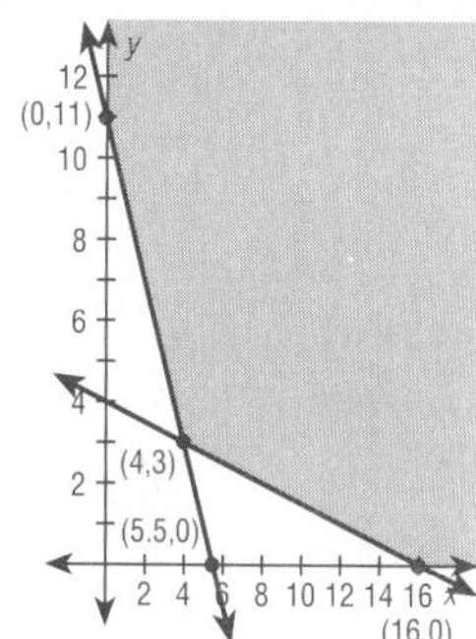

a. maximum value = 68, minimum value = 5;
b. maximum value = none, minimum value = 11

Check Understanding

1. What are optimization problems?
See the top of page 151.

2. Can the constraints of a linear-programming problem have inequalities involving only > or <?
Yes, the edges of the half-planes do not have to be within the region.

Communicating about ALGEBRA

Suppose in Example 4, the hours of unskilled labor, machine time, and skilled labor per patio set are those in the following table.

Assembly Hours	Process A	Process B
Unskilled labor	5	4
Machine time	5	1
Skilled labor	10	1

You can use up to 800 hours of unskilled labor, up to 350 hours of machine time, and up to 600 hours of skilled labor. How many patio sets should be made by each process to obtain the maximum profit?

$P = 80a + 40b$ with constraints: $5a + 4b \leq 800$, $5a + b \leq 350$, $10a + b \leq 600$, $a \geq 0$, $b \geq 0$; vertices: (0, 200), (40, 150), (50, 100), (60, 0), (0, 0)

The maximum profit is obtained by making 40 patio sets with Process A and 150 patio sets with Process B.

EXERCISES

Guided Practice

CRITICAL THINKING about the Lesson

1. How is the *objective quantity* used in a linear programming problem? How is the system of constraint inequalities used?
See margin, page 155.

2. In a linear programming problem, which ordered pairs should be tested to see whether they produce a minimum value?
The coordinates of the vertices

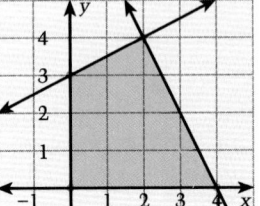

In Exercises 3 and 4, use the graph at the right.

3. What are the vertices of the region?
3. (4, 0), (2, 4) (0, 3), (0, 0)

4. What is the minimum and maximum value of the objective quantity $C = 5x + 7y$?
Minimum: 0; Maximum: 38

Independent Practice

In Exercises 5–8, you are given the graph of the constraint inequalities. What are the minimum and maximum values of the objective quantity?

5. $C = x - y$

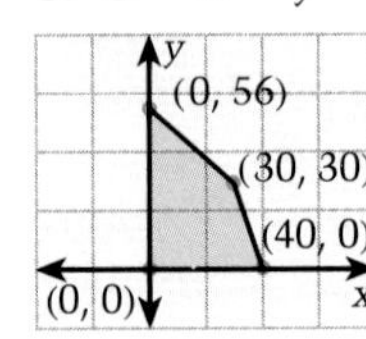

Min: 40
Max: 56

6. $C = x + 3y$

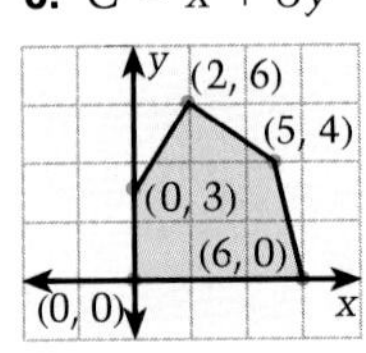

Min: 0
Max: 20

7. $C = 2x + 5y$

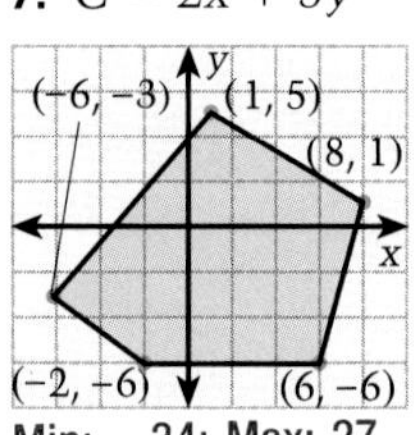

Min: −34; Max: 27

8. $C = 9x - 2y$

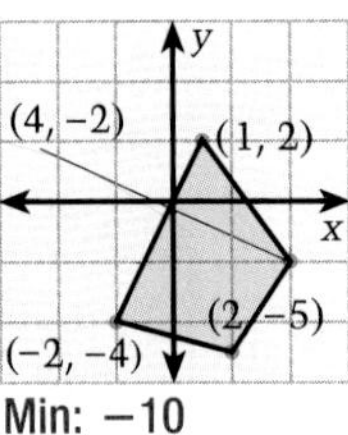

Min: −10
Max: 40

In Exercises 9–20, find the minimum and maximum values of the objective quantity.

9. **Objective quantity:**
$C = -2x + y$

Constraints:
$$\begin{cases} x \geq -5 \\ x \leq 4 \\ y \geq -1 \\ y \leq 3 \end{cases}$$
Min: −9
Max: 13

10. **Objective quantity:**
$C = 5x + 4y$

Constraints:
$$\begin{cases} x \leq -2 \\ x \geq -4 \\ y \geq 1 \\ y \leq 6 \end{cases}$$
Min: −16
Max: 14

11. **Objective quantity:**
$C = x + 4y$

Constraints:
$$\begin{cases} x \geq 0 \\ x \leq 12 \\ y \leq 10 \\ 2x + 3y \geq 24 \end{cases}$$
Min: 12
Max: 52

12. **Objective quantity:**
$C = 6x + 2y$ $\frac{3}{2}$,

Constraints:
$$\begin{cases} x \geq 0 \\ x \leq 5 \\ y \geq 0 \\ 4x - y \geq 1 \end{cases}$$
Min: $\frac{3}{2}$
Max: 68

13. **Objective quantity:**
$C = 2x + y$

Constraints:
$$\begin{cases} x \geq 0 \\ y \geq 2 \\ 2x + y \leq 10 \\ x - 3y \geq -3 \end{cases}$$
Min: 8
Max: 10

14. **Objective quantity:**
$C = 3x + y$

Constraints:
$$\begin{cases} x \geq 0 \\ y \geq 0 \\ 2x - 5y \geq 0 \\ 2x - 5y \leq 2 \end{cases}$$
Min: 0
No max

15. Objective quantity:
$C = 3x + 2y$ 0, 17

Constraints:

$$\begin{cases} x \geq 0 \\ y \geq 0 \\ x + 3y \leq 15 \\ 4x + y \leq 16 \end{cases}$$

16. Objective quantity:
$C = 6x + 4y$ 0, 52

Constraints:

$$\begin{cases} x \geq 0 \\ y \geq 0 \\ x + 6y \leq 30 \\ 6x + y \leq 40 \end{cases}$$

17. Objective quantity:
$C = 10x + 3y$ 12, $\frac{169}{3}$

Constraints:

$$\begin{cases} x \geq 0 \\ y \geq 0 \\ -x + y \geq 0 \\ 2x + y \geq 4 \\ 2x + y \leq 13 \end{cases}$$

18. Objective quantity:
$C = 5x + 2y$ 0, 55

Constraints:

$$\begin{cases} x \geq 0 \\ y \geq 0 \\ y \leq 8 \\ x + y \leq 14 \\ 5x + y \leq 50 \end{cases}$$

19. Objective quantity:
$C = 4x + 3y$ 6, 29

Constraints:

$$\begin{cases} x \geq 0 \\ y \geq 0 \\ 2x + 3y \geq 6 \\ 3x - 2y \leq 9 \\ x + 5y \leq 20 \end{cases}$$

20. Objective quantity:
$C = 3x + 6y$
8, no maximum

Constraints:

$$\begin{cases} x \geq 0 \\ y \geq 0 \\ x + 4y \geq 4 \\ x + y \geq 2 \end{cases}$$

21. *Finding the Maximum Profit* For each unit of a product made by Process A, you make \$50 profit. For each unit of a product made by Process B, you make \$55 profit. How many units should you make by each process to obtain a maximum profit? A: 300 units, B: 400 units

	Assembly Hours		
	Process A	*Process B*	*Maximum Hours*
Unskilled labor	1	6	2700
Machine time	3	1	1500
Skilled labor	1	1	700

✪ **22.** *Inventory Control* You are the assistant manager of an appliance store. Next month you will order two types of stereo systems. How many of each model should you order to minimize the cost?

- *Model A:* Your cost is \$300. Your profit is \$40. Model A: 60 units Model B: 40 units
- *Model B:* Your cost is \$400. Your profit is \$60.
- You expect a profit of at least \$4800.
- You expect to sell at least 100 units.

✪ **23.** *Magician* Your school has contracted with a professional magician to perform at the school. The school has guaranteed an attendance of at least 1000 and total ticket receipts of at least \$4800. The tickets are \$4 for students and \$6 for nonstudents, of which the magician receives \$2.50 and \$4.50, respectively. What is the minimum amount of money the magician could receive? \$3000

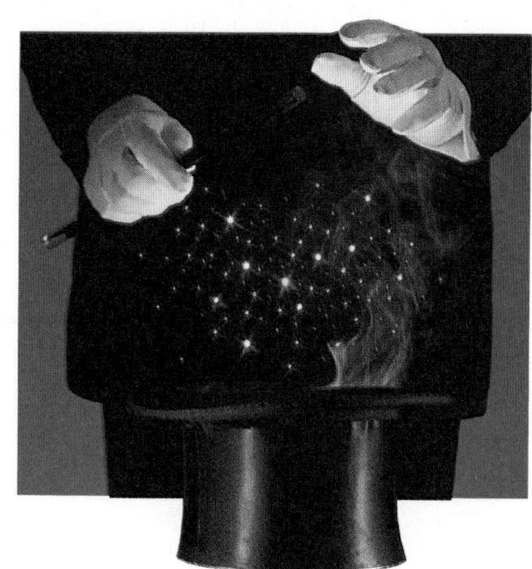

EXERCISE Notes

ASSIGNMENT GUIDE

Basic/Average: Ex. 1–4, 5–21 odd, 22, 23, 29–31

Above Average: Ex. 1–4, 7–21 odd, 22–25, 28–32

Advanced: Ex. 1–4, 7–21 odd, 22–24, 28–33

Selected Answers
Exercises 1–4, 5–31 odd

Use **Mixed Review** as needed.

✪ More Difficult Exercises
Exercises 22, 23, 32–33

Guided Practice

▶ **Ex. 1–4** The linear programming process contains many new terms such as optimization, objective quantity, and constraints. Use the Guided Practice exercises to review the terms.

Answers

1. It is evaluated at the vertices formed by the constraints; they define the region of possible answers.

Independent Practice

▶ **Ex. 9–20** Make sure students draw the graph of the region created by the system of constraints as part of the solution process.

▶ **Ex. 21–23** COOPERATIVE LEARNING These exercises require extended student work time. You might suggest that students work with partners and share the task.

EXTEND Ex. 32–33 Ask students if they notice a connection between one of the constraints and the objective quantity. For example, in Exercise 32, the objective quantity at $2x + 2y$ is a multiple of the constraint expression $x + y$.

Enrichment Activity

ALTERNATIVE ASSESSMENT Linear Programming is a part of a branch of mathematical study called "finite math" or "discrete math." Have students use library resources to find out more about its history and share their findings with their classmates.

RESEARCH Operations research is a branch of mathematics begun in the 20th century around the time of World War II. Investigate its origins and some of the individuals responsible for its growth, such as George Dantzig, who developed the so-called simplex method for solving linear programming problems.

WRITING Write a report on some aspect of the development of operations research and linear programming including the fields that make extensive use of operation-research techniques.

Answers

Mixed Review

13.

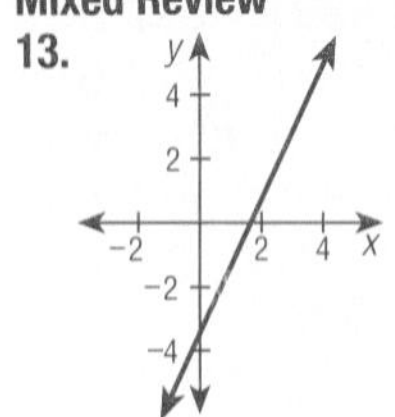

18.

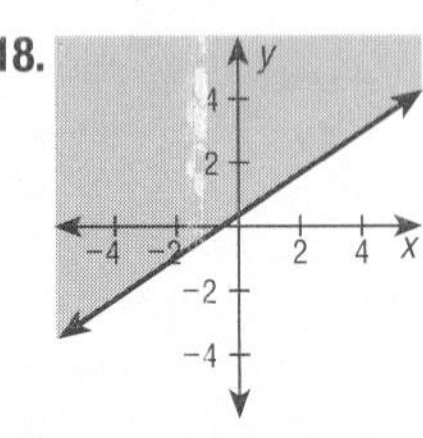

Integrated Review

In Exercises 24–29, sketch the graph of the inequality. See Additional Answers.

24. $2x - 4y > 7$ **25.** $-3x + 5y \leq 5$ **26.** $x + 9y \leq 27$

27. $2x - 7y > 3$ **28.** $12x + 16y \geq 33$ **29.** $5x + 4y \geq -15$

30. *Powers of Ten* Which is the best estimate for the area of Mississippi?
a. 4.8×10^3 mi² **b.** 4.8×10^4 mi² **c.** 4.8×10^5 mi² **d.** 4.8×10^6 mi²

31. *Powers of Ten* Which is the best estimate for the speed of a jet aircraft?
a. 4.8×10^0 mph **b.** 4.8×10^1 mph **c.** 4.8×10^2 mph **d.** 4.8×10^3 mph

Exploration and Extension

In Exercises 32 and 33, use the following information.

When the maximum (or minimum) occurs at two consecutive vertices in a linear programming problem, it occurs at any point on the line segment connecting the two vertices. Illustrate this result with each of the following.

32. Objective quantity: $C = 2x + 2y$

Constraints: $\begin{cases} y \leq 4 \\ x \leq 5 \\ x + y \leq 6 \end{cases}$

The maximum is 12 and occurs anywhere on the line segment connecting (2, 4) and (5, 1).

33. Objective quantity: $C = 5x - y$

Constraints: $\begin{cases} y \geq -1 \\ x \leq 3 \\ -5x + y \leq 4 \end{cases}$

The minimum is −4 and occurs anywhere on the line segment connecting (3, 19) and (−1, −1).

Mixed REVIEW

1. Simplify: $-(2x - 5) + 4(x + 3)$. **(1.2)** $2x + 17$

2. Simplify: $\frac{1}{4}(2x - 1) - \frac{3}{4}(2x + 1)$. **(1.2)** $-x - 1$

3. Rewrite 854 feet as miles. ≈0.162 mi

4. Rewrite 1.34 meters as centimeters. 134 cm

5. Evaluate: $\frac{1}{2}(3) - \frac{2}{3}(4)$. $-\frac{7}{6}$

6. Evaluate: $\frac{5}{6} + \frac{1}{2}$. $\frac{4}{3}$ or $1\frac{1}{3}$

7. Round to 3 decimal places: 4.04641. 4.046

8. Solve $5x - (x - 4) = -2(x + 4)$. **(1.3)** −2

9. Solve $2|x - 1| \geq 6$. **(1.7)** $x \geq 4$ or $x \leq -2$

10. Solve $-3(x - 4) \geq -2x + 5$. **(1.6)** $x \leq 7$

11. The sum of which two consecutive whole numbers is 37? **(1.3)** 18, 19

12. Find the area of a triangle whose height is 4 inches and whose base is 5 inches. 10 in.²

13. Sketch the graph of $2y - 4x + 7 = 0$. **(2.1)** See margin.

14. What is the opposite of $\frac{1}{3}$? **(1.1)** $-\frac{1}{3}$

15. 14.4 is 15% of what number? 96

16. Divide 1.4×10^3 by 2.8×10^4. 5×10^{-2}

17. Find the slope of the line $-3x + 5y = 2$. **(2.2)** $\frac{3}{5}$

18. Sketch the graph of $y \geq \frac{3}{4}x + \frac{1}{4}$. **(2.5)** See margin.

19. Solve $\begin{cases} 2x - 4y = 5 \\ -2x + 3y = -4 \end{cases}$. **(3.2)** $\left(\frac{1}{2}, -1\right)$

20. Solve $\begin{cases} -3x - 5y = -1 \\ 4x + 2y = 6 \end{cases}$. **(3.2)** (2, −1)

3.6 Solving Systems of Linear Equations in Three Variables

What you should learn:

Goal 1 How to solve a system of linear equations in three variables

Goal 2 How to use a linear system in three variables to answer questions about real-life situations

Why you should learn it:

You can model many real-life situations with a system of linear equations in three variables, such as data about three groups of people.

Goal 1 Solving a System in Three Variables

In this lesson, you will learn a technique for solving a **system of three linear equations** in three variables, such as these:

$$\begin{cases} x - 2y + 2z = 9 \\ -x + 3y = -4 \\ 2x - 5y + 3z = 16 \end{cases} \qquad \begin{cases} x - 2y + 2z = 9 \\ y + 2z = 5 \\ z = 3 \end{cases}$$

The linear system on the right is in **triangular form,** which means that its three equations follow a stair-step pattern, *and* each equation has 1 as the coefficient of its first nonzero term.

A **solution** of a linear system in x, y, and z is an **ordered triple** (x, y, z) that is a solution of each equation in the system. Finding the set of all solutions is called **solving** the system. A linear system in triangular form is easier to solve than a system that is not in that form as the next example shows.

Example 1 *Solving a Linear System That Is in Triangular Form*

Solve the system.

$$\begin{cases} x - 2y + 2z = 9 & \text{Equation 1} \\ y + 2z = 5 & \text{Equation 2} \\ z = 3 & \text{Equation 3} \end{cases}$$

Solution From Equation 3, you know the value of z. To solve for y, substitute 3 for z in Equation 2.

$$y + 2(3) = 5 \qquad \textit{Substitute 3 for z.}$$
$$y = -1 \qquad \textit{Solve for y.}$$

Next substitute -1 for y and 3 for z in Equation 1.

$$x - 2(-1) + 2(3) = 9 \qquad \textit{Substitute } -1 \textit{ for y and 3 for z.}$$
$$x = 1 \qquad \textit{Solve for x.}$$

The solution is $x = 1$, $y = -1$, and $z = 3$, which can be written as the *ordered triple* $(1, -1, 3)$. Check this solution by substituting into each equation in the original system.

Equation 1 check: $(1) - 2(-1) + 2(3) = 1 + 2 + 6 = 9$
Equation 2 check: $(-1) + 2(3) = -1 + 6 = 5$ ■

Problem of the Day

It took a train 45 seconds to pass through a 1320-foot tunnel, and 15 seconds to pass a watchman. How long was the train, and what was its speed in miles per hour? 660 feet, 30 mph

ORGANIZER

Warm-Up Exercises

1. Solve each system using the linear combination method.
   a. $\begin{cases} 3x + 2y = 1 \\ 5x - 6y = -17 \end{cases}$ $(-1, 2)$
   b. $\begin{cases} x + 7y = 10 \\ -4x - 2y = -20 \end{cases}$ $\left(\frac{60}{13}, \frac{10}{13}\right)$
2. Given a value of z, find x and y.
   a. $z = 1$, $\begin{cases} y + 2z = 3 \\ x - 4z = -2 \end{cases}$
   b. $z = -3$, $\begin{cases} 3y + z = 9 \\ -2x + 3z = 7 \end{cases}$
   c. $z = 2$, $\begin{cases} 5y + 2z = 19 \\ x + 2y - 3z = -7 \end{cases}$

   a. $x = 2, y = 1$; b. $x = -8, y = 4$; c. $x = -7, y = 3$

Lesson Resources

Teaching Tools
- Problem of the Day: 3.6
- Warm-up Exercises: 3.6
- Answer Masters: 3.6

Extra Practice: 3.6

LESSON Notes

Examples 1–2

Ask students to provide you with the sequence of steps and a rationale for solving a system in triangular form. This will help make the point that the

(continued)

(continued)
triangular form is desirable in solving systems of three linear equations. As always, emphasize the importance of checking solutions in the original system.

The Gaussian elimination method in Example 2, is the basis on which computer solutions are often found.

Example 3

Be certain students recognize that the variable eliminated in the first step must also be the variable eliminated in the second step. Another way of saying this is: all three equations in the system must be involved in order to eliminate a given variable. In this example, Equations 1 and 2 are used in Step 1, and Equations 2 and 3 are used in Step 2. (Equations 1 and 2 cannot be used in both Steps 1 and 2.)

How Many Solutions? The solutions to the two systems at the bottom of page 159 are presented here using the linear combinations method.

$$\begin{cases} x + y - 2z = 5 \\ x + 2y + z = 8 \\ 2x + 3y - z = 1 \end{cases}$$

$$\begin{array}{rl} x + y - 2z &= 5 \\ \underline{2x + 4y + 2z} &\underline{= 16} \\ 3x + 5y \quad &= 21 \end{array}$$

$$\begin{array}{rl} x + 2y + z &= 8 \\ \underline{2x + 3y - z} &\underline{= 1} \\ 3x + 5y \quad &= 9 \end{array}$$

$$\begin{cases} x + y - 2z = 5 \\ x + 2y + z = 8 \\ 2x + 3y - z = 13 \end{cases}$$

$$\begin{array}{rl} x + y - 2z &= 5 \\ \underline{2x + 4y + 2z} &\underline{= 16} \\ 3x + 5y \quad &= 21 \end{array}$$

$$\begin{array}{rl} x + 2y + z &= 8 \\ \underline{2x + 3y - z} &\underline{= 13} \\ 3x + 5y \quad &= 21 \end{array}$$

Two linear systems are **equivalent** if they have the same set of solutions. To solve a system that is not in triangular form, first rewrite it as an *equivalent* system in triangular form.

Equivalent Systems of Equations

Each of the following **row operations** on a system of linear equations produces an *equivalent* system of linear equations.

1. Interchange two equations.
2. Multiply one of the equations by a nonzero constant.
3. Add a multiple of one of the equations to another equation to replace the latter equation.

Rewriting a system of linear equations into triangular form usually involves a *chain* of equivalent systems, each of which is obtained by using one of the three basic row operations. This process is called **Gaussian elimination,** after the German mathematician Carl Friedrich Gauss (1777–1855).

Example 2 Using Elimination to Solve a System

Solve the system.

$$\begin{cases} x - 2y + 2z = 9 & \text{Equation 1} \\ -x + 3y = -4 & \text{Equation 2} \\ 2x - 5y + 3z = 16 & \text{Equation 3} \end{cases}$$

Solution Work from the upper left-hand corner. Save the x in the upper left-hand corner and eliminate the other x's from the first column.

$$\begin{cases} x - 2y + 2z = 9 \\ y + 2z = 5 \leftarrow & \text{Add the first equation to the second to obtain a new second equation.} \\ 2x - 5y + 3z = 16 \end{cases}$$

$$\begin{cases} x - 2y + 2z = 9 \\ y + 2z = 5 \\ -y - z = -2 \leftarrow & \text{Add } -2 \text{ times the first equation to the third to obtain a new third equation.} \end{cases}$$

All but the first x has been eliminated from the first column. Next you need to eliminate y from the third equation.

$$\begin{cases} x - 2y + 2z = 9 \\ y + 2z = 5 \\ z = 3 \leftarrow & \text{Add the second equation to the third to obtain a new third equation.} \end{cases}$$

The system is now in triangular form. In fact, it is the same triangular system we solved in Example 1. Thus the solution is $x = 1$, $y = -1$, and $z = 3$, or the ordered triple $(1, -1, 3)$. ■

A second way to solve a system of three linear equations in three variables is to use *linear combinations.*

Example 3 Using Linear Combinations

Solve the system.

$$\begin{cases} 3x + 2y + 4z = 11 & \text{Equation 1} \\ 2x - y + 3z = 4 & \text{Equation 2} \\ 5x - 3y + 5z = -1 & \text{Equation 3} \end{cases}$$

Solution The goal is to eliminate one of the variables in two of the given equations. This is done by adding multiples of two equations as demonstrated below.

$$\begin{array}{rl} 3x + 2y + 4z = 11 & \text{Add 2 times the second} \\ 4x - 2y + 6z = 8 & \text{equation to the first.} \\ \hline 7x \quad\quad + 10z = 19 & \text{New Equation 1} \end{array}$$

$$\begin{array}{rl} 5x - 3y + 5z = -1 & \text{Add } -3 \text{ times the second} \\ -6x + 3y - 9z = -12 & \text{equation to the third.} \\ \hline -x \quad\quad - 4z = -13 & \text{New Equation 2} \end{array}$$

This produces a system of linear equations in two variables.

$$\begin{cases} 7x + 10z = 19 & \text{New Equation 1} \\ -x - 4z = -13 & \text{New Equation 2} \end{cases}$$

Using any of the methods described in the previous sections, you solve this system and determine that $x = -3$ and $z = 4$. Finally, by substituting these values into any of the three original equations, you can solve for y.

$$\begin{array}{rl} 2x - y + 3z = 4 & \text{Equation 2} \\ 2(-3) - y + 3(4) = 4 & \text{Substitute } -3 \text{ for } x \text{ and } 4 \text{ for } z. \\ y = 2 & \text{Simplify and solve for } y. \end{array}$$

Therefore the solution is $x = -3$, $y = 2$, and $z = 4$, or the ordered triple $(-3, 2, 4)$. ■

You know that a linear system in two variables can have exactly one solution, no solution, or an infinite number of solutions. The same is true of a linear system in *three* variables. (A geometric interpretation of this is shown on page 164.)

The systems in Examples 1, 2, and 3 each have exactly one solution. Try solving the following systems. Can you see how to tell how many solutions the system has?

No Solution

$$\begin{cases} x + y - 2z = 5 \\ x + 2y + z = 8 \\ 2x + 3y - z = 1 \end{cases}$$

Many Solutions

$$\begin{cases} x + y - 2z = 5 \\ x + 2y + z = 8 \\ 2x + 3y - z = 13 \end{cases}$$

Observe that the resulting systems of two equations are

$$\begin{cases} 3x + 5y = 21 \\ 3x + 5y = 9 \end{cases} \quad \begin{cases} 3x + 5y = 21 \\ 3x + 5y = 21 \end{cases}$$

Any attempt to eliminate x also eliminates y. Adding the opposite of the second equation in each system to the first equation gives

$$\begin{array}{r} 3x + 5y = 21 \\ -3x - 5y = -9 \\ \hline 0 = 12 \end{array} \qquad \begin{array}{r} 3x + 5y = 21 \\ -3x - 5y = -21 \\ \hline 0 = 0 \end{array}$$

The result, $0 = 12$, is an untrue statement that indicates there is no solution to the first system. The statement, $0 = 0$, is a true statement that indicates there are infinitely many solutions.

Example 4

Some students may understand the following sequence of steps for simplifying the original system of linear equations better than multiplying Equations 1 and 2 by 2.5 and Equation 3 by 10.

Multiply each equation by 10 to obtain

$$\begin{cases} 4x + 8y + 8z = 4880 \\ 8x + 4y + 4z = 4120 \\ 5x + 7y + 8z = 4910 \end{cases}$$

Divide Equations 1 and 2 in this new system by 4 to get the second system of Example 4.

Extra Examples

Here are additional examples similar to **Examples 1–3.**

1. Solve by using the Gaussian elimination method.

   a. $\begin{aligned} 3x + 2y + 2z &= -1 \\ -3x + y - 3z &= 0 \\ x + 2y + z &= 0 \end{aligned}$

   b. $\begin{aligned} -x + y + 2z &= 2 \\ 2x + y - z &= -4 \\ x + 2y + 3z &= 2 \end{aligned}$

   a. $(-1, 0, 1)$, b. $(0, -2, 2)$

2. Solve by using the linear combinations method.

   a. $\begin{cases} 4x + 3y + z = 4 \\ 8x - 2y + 2z = 0 \\ -2x + y - z = 1 \end{cases}$

   b. $\begin{cases} -3x + 4y + 2z = 1 \\ x + 2y - z = 6 \\ 2x - y + 3z = 7 \end{cases}$

   c. $\begin{cases} x + y - 2z = 0 \\ 2y + 6z = 8 \\ x + 2y + z = 2 \end{cases}$

   a. $(0.5, 1, -1)$, b. $(3, 2, 1)$, c. no solution

Check Understanding

1. When is a system of linear equations said to be in triangular form? See the top of page 157.
2. What is a solution of a system of three linear equations? An ordered triple; see page 157.
3. What are row operations and how are they related to equivalent systems of linear equations? See page 158.

Goal 2 Using Linear Systems as Real-Life Models

Real Life
Recreation

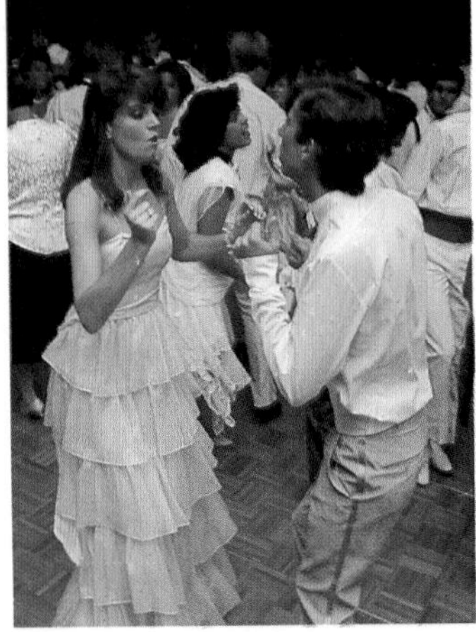

Emerson High School holds seven dances each year. Three of these are "open"—any student in Grades 10–12 can attend. The total attendance and the percent of each class that attended the open dances is shown in the table.

Example 4 *Finding Class Enrollments*

Use the information in the table. How many sophomores, juniors, and seniors attend Emerson High School?

	Sophomores	Juniors	Seniors	Total
Homecoming	40%	80%	80%	488
Winter Holiday	80%	40%	40%	412
Spring Fling	50%	70%	80%	491

Solution Let x, y, and z represent the number of sophomores, juniors, and seniors, respectively. The information in the table can be used to create a system of linear equations.

$$\begin{cases} 0.4x + 0.8y + 0.8z = 488 & \textit{Equation 1} \\ 0.8x + 0.4y + 0.4z = 412 & \textit{Equation 2} \\ 0.5x + 0.7y + 0.8z = 491 & \textit{Equation 3} \end{cases}$$

You can use elimination to solve this linear system. Begin by multiplying each term in the first and second equations by 2.5, and each term in the third equation by 10.

$$\begin{cases} x + 2y + 2z = 1220 \\ 2x + y + z = 1030 \\ 5x + 7y + 8z = 4910 \end{cases}$$

Now use elimination to write the system in triangular form.

$$\begin{cases} x + 2y + 2z = 1220 \\ y + z = 470 \\ z = 220 \end{cases}$$

Using substitution, you can find the solution to be $x = 280$ (sophomores), $y = 250$ (juniors), and $z = 220$ (seniors). ■

Communicating about ALGEBRA

SHARING IDEAS about the Lesson

Solve each system. Justify each step. See Additional Answers.

A. $\begin{cases} x - 3z = -2 \\ 3x + y - 2z = 5 \\ 2x + 2y + z = 4 \end{cases}$

B. $\begin{cases} x + y + z = 6 \\ 2x - y + z = 3 \\ 3x - z = 0 \end{cases}$

EXERCISES

Guided Practice

CRITICAL THINKING about the Lesson

In Exercises 1 and 2, solve the linear system.

1. $\begin{cases} 2x - y + 2z = 5 \\ y + z = 4 \\ z = 1 \end{cases}$ (3, 3, 1)

2. $\begin{cases} x + y - z = 4 \\ y + z = 4 \\ z = 2 \end{cases}$ (4, 2, 2)

3. Add -2 times Equation 1 to Equation 2. What effect does this operation have? See below.

$\begin{cases} x - 3y + 4z = 4 & \textbf{Equation 1} \\ 2x - 5y + 3z = -1 & \textbf{Equation 2} \\ 3y - 4z = -5 & \textbf{Equation 3} \end{cases}$

4. Add 3 times Equation 2 to Equation 3. What effect does this operation have? See below.

$\begin{cases} x - 2y - 2z = -4 & \textbf{Equation 1} \\ y + z = 3 & \textbf{Equation 2} \\ -3y - 4z = -13 & \textbf{Equation 3} \end{cases}$

In Exercises 5 and 6, write the system in triangular form. Triangular forms vary.

5. $\begin{cases} 2x + 10y + 2z = 6 \\ y + z = 1 \\ y + 2z = 3 \end{cases}$ $x + 5y + z = 3$, $y + z = 1$, $z = 2$

6. $\begin{cases} x + 5y - 3z = 2 \\ -x - 7y + 4z = -2 \\ 2x + 4y = 10 \end{cases}$ $x + 5y - 3z = 2$, $y - \frac{1}{2}z = 0$, $z = 2$

Independent Practice

In Exercises 7–12, solve the linear system.

7. $\begin{cases} x + 4y + z = 12 \\ y - 3z = -7 \\ z = 3 \end{cases}$ (1, 2, 3)

8. $\begin{cases} x + 2z = 30 \\ y + z = 12 \\ z = 1 \end{cases}$ (28, 11, 1)

9. $\begin{cases} x + y - 3z = -20 \\ -3y + 2z = 5 \\ z = 4 \end{cases}$ (−9, 1, 4)

10. $\begin{cases} x - y + z = 14 \\ y + z = 15 \\ z = 7 \end{cases}$ (15, 8, 7)

11. $\begin{cases} x + 2y - 5z = 10 \\ y + 2z = 2 \\ z = -1 \end{cases}$ (−3, 4, −1)

12. $\begin{cases} x + y - z = 17 \\ y + z = 1 \\ z = -3 \end{cases}$ (10, 4, −3)

In Exercises 13–16, perform the indicated row operation. Write the equivalent system. See Additional Answers.

13. Interchange Equations 1 and 3.

$\begin{cases} 4x + y - 3z = 19 & \textbf{Equation 1} \\ -2x + 2y - z = 12 & \textbf{Equation 2} \\ x + y - z = 39 & \textbf{Equation 3} \end{cases}$

14. Multiply Equation 1 by $\frac{1}{2}$.

$\begin{cases} 2x + 3y - z = 9 & \textbf{Equation 1} \\ -3x + y - 2z = 14 & \textbf{Equation 2} \\ 4x - 3y + 9z = 36 & \textbf{Equation 3} \end{cases}$

15. Add -2 times Equation 1 to Equation 2.

$\begin{cases} x - 3y + 4z = 24 & \textbf{Equation 1} \\ 2x + y - 2z = 30 & \textbf{Equation 2} \\ -3x - 2y + z = 46 & \textbf{Equation 3} \end{cases}$

16. Add 4 times Equation 1 to Equation 2.

$\begin{cases} x + 3y - 4z = 12 & \textbf{Equation 1} \\ -4x - y + 3z = 26 & \textbf{Equation 2} \\ 2x + y + z = 46 & \textbf{Equation 3} \end{cases}$

3. Eliminates the x-term. **4.** Eliminates the y-term.

Communicating about ALGEBRA

EXTEND *Communicating*

Solve the system by using the Gaussian elimination method. Justify each step.

$\begin{cases} 3x + y - z + w = -3 \\ x + 2y + 2z = 5 \\ -x - y + 2w = 2 \\ 2x + y + z + w = 2 \end{cases}$

$(x, y, z, w) = (-1, 1, 2, 1)$

EXERCISE Notes

ASSIGNMENT GUIDE

Basic/Average: Ex. 1–6, 9–21 odd, 27–35 odd, 41

Above Average: Ex. 1–6, 9–21 odd, 25–31 odd, 41, 42

Advanced: Ex. 1–6, 9–21 odd, 25–31 odd, 41–43

Selected Answers

Exercises 1–6, 7–41 odd

More Difficult Exercises

Exercises 21–31, 42–43

Guided Practice

Ex. 1–6 GROUP ACTIVITY Use the Guided Practice exercises as an in-class discussion of the steps for using the Gaussian elimination method.

Ex. 6 This exercise can be used to practice the linear combinations method outlined in Example 3 on page 159.

Independent Practice

Ex. 7–12 These equations are already in triangular form and should be solved by inspection.

Ex. 13–16 These exercises are designed to provide practice in row operations.

Ex. 17–20 These exercises are designed to provide practice in obtaining the triangular form.

Ex. 21–31 **COOPERATIVE LEARNING** These exercises require extended student work time. You might suggest that students work with partners and share the task.

Even though students may know other ways to solve the applications, encourage the student partners to practice the methods developed in the lesson. Refer them to Example 4 for a solution format.

Answers

29. String: 50, wind: 20, percussion: 8

30. Popcorn: \$3.15, granola: \$2.75, fruit: \$2.89 Solve the linear system:

$$\begin{cases} p + \frac{1}{2}g + \frac{1}{2}f = 5.97 \\ \frac{4}{3}p + \frac{1}{2}g + \frac{3}{2}f = 9.91 \\ \frac{1}{3}p + 2f = 6.83 \end{cases}$$

Ex. 42 In Chapter 4, students will study matrices and determinants. In this exercise, the determinant of the matrix for the system is 0. This verifies that the system has no solution. You might wish to informally discuss this "connection" to future learning.

In Exercises 17–20, write the system of linear equations in triangular form. Triangular forms may vary.

17. $\begin{cases} x - y + 2z = 15 \\ -2x + 2y - 3z = -25 \\ y + 2z = 9 \end{cases}$ $x - y + 2z = 15$, $y + 2z = 9$, $z = 5$

18. $\begin{cases} x - y + 2z = -1 \\ 2x - 2y + 2z = -8 \\ y + 3z = 15 \end{cases}$ $x - y + 2z = -1$, $y + 3z = 15$, $z = 3$

19. $\begin{cases} x - y - 3z = 1 \\ x - z = -4 \\ 3x = -15 \end{cases}$ $x - y - 3z = 1$, $y + 2z = -5$, $z = -1$

20. $\begin{cases} x + y - z = 0 \\ 2x + 2y - z = 1 \\ 3x + 4y = -2 \end{cases}$ $x + y - z = 0$, $y + 3z = -2$, $z = 1$

In Exercises 21–28, solve the system of equations. Check your solution.

21. $\begin{cases} x + 9y + z = 20 \\ x + 10y - 2z = 18 \\ 3x + 27y + 2z = 58 \end{cases}$ (−18, 4, 2)

22. $\begin{cases} x + 3y + 2z = 41 \\ 2x + 8y - 2z = 60 \\ -2x - 8y + z = -66 \end{cases}$ (8, 7, 6)

23. $\begin{cases} 2x - 4y + 2z = 16 \\ -2x + 5y + 2z = -34 \\ x - 2y + 2z = 4 \end{cases}$ (8,−2,−4)

24. $\begin{cases} 2x + 2y - 6z = -6 \\ x + y - 2z = 0 \\ -2x - y + 8z = 19 \end{cases}$ (−1, 7, 3)

25. $\begin{cases} -x + y - 3z = -4 \\ 3x - 2y + 8z = 14 \\ 2x - 2y + 5z = 7 \end{cases}$ (4, 3, 1)

26. $\begin{cases} -x + 2y + 4z = 10 \\ 2x - 3y - 7z = -13 \\ 2x - 4y - 6z = -24 \end{cases}$ (0, 9,−2)

27. $\begin{cases} 2x + 6y - 4z = 8 \\ 3x + 10y - 7z = 12 \\ -2x - 6y + 5z = -3 \end{cases}$ (−1, 5, 5)

28. $\begin{cases} 3x - 6y + 3z = 18 \\ 2x - 3y + 4z = 6 \\ 2x - 3y + 5z = 4 \end{cases}$ (4,−2,−2)

29. *School Orchestra* The table shows the percents of each section of the North High School orchestra that were chosen to participate in the city orchestra, the county orchestra, and the state orchestra. Thirty members of the city orchestra, 17 members of the county orchestra, and 10 members of the state orchestra are from North. How many members are in each section of North High's orchestra? See margin.

	Section		
	String	**Wind**	**Percussion**
City orchestra	40%	30%	50%
County orchestra	20%	25%	25%
State orchestra	10%	15%	25%

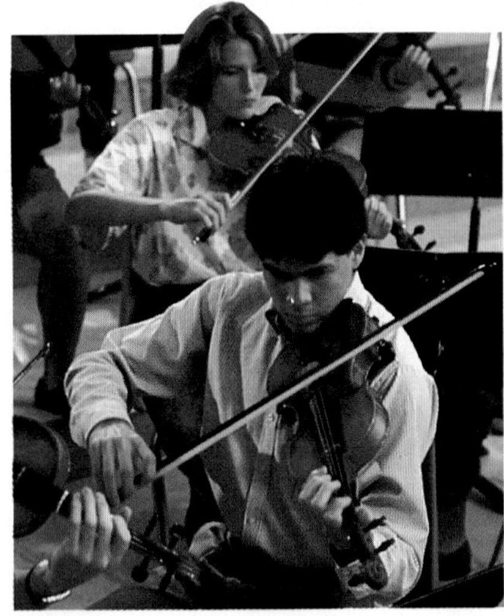

30. *It's Going to Be a Long Ride* To prepare for a field trip by bus, you and two friends buy some snacks. From the information at the right, can you determine the price per pound for popcorn, granola, and dried fruit? Explain. See margin.

Shopper	Popcorn	Granola	Dried Fruit	Total Price
You	1 lb	$\frac{1}{2}$ lb	$\frac{1}{2}$ lb	\$5.97
Greg	$1\frac{1}{3}$ lb	$\frac{1}{2}$ lb	$\frac{3}{2}$ lb	\$9.91
Sara	$\frac{1}{3}$ lb		2 lb	\$6.83

31. *County Fair* A dairy makes three types of cheddar cheese—mild, medium, and sharp—and sells the cheese in three booths at the county fair. At the beginning of one day, each booth was stocked with the same amount of each type of cheese but the total amount was different for each booth. By the end of the day, the dairy had sold 62 pounds of mild, 216 pounds of medium, and 162 pounds of sharp. The percents sold *of the amounts each booth was given* are shown in the table. How much of each type of cheddar was given to each booth?

	Booth 1	Booth 2	Booth 3
Mild	10%	20%	20%
Medium	30%	70%	80%
Sharp	20%	50%	70%

Mild: 180 lb, Medium: 140 lb, Sharp: 80 lb

Integrated Review

In Exercises 32–37, solve the linear system.

32. $\begin{cases} x + 3y = 16 \\ 9x - 5y = -16 \end{cases}$ (1, 5)

33. $\begin{cases} 6x - 5y = 11 \\ -9x + 6y = -15 \end{cases}$ (1, −1)

34. $\begin{cases} 2x - y = 32 \\ x + y = 4 \end{cases}$ (12, −8)

35. $\begin{cases} 5x - y = 28 \\ x + 3y = -4 \end{cases}$ (5, −3)

36. $\begin{cases} -3x + 2y = 24 \\ x - 4y = -28 \end{cases}$ (−4, 6)

37. $\begin{cases} -4x + 3y = 15 \\ -x + y = 6 \end{cases}$ (3, 9)

In Exercises 38–41, evaluate the expression.

38. $2x + 3y + 3z$ when $x = 2, y = -2, z = 5$ 13

39. $-3x + 2y + z$ when $x = -1, y = 8, z = 1$ 20

40. $x - 4y - z$ when $x = 4, y = -5, z = 6$ 18

41. $-5x + 2y - z$ when $x = -2, y = 3, z = -7$ 23

Exploration and Extension

42. Decide whether the linear system has exactly one solution, no solution, or an infinite number of solutions. Explain how you arrived at your decision.

$\begin{cases} x + 3y + z = 26 \\ y + z = 10 \\ x + y - z = 4 \end{cases}$

No solution; −1 times $x + y - z = 4$ added to $x + 3y + z = 26$ gives $2y + 2z = 22$ or $y + z = 11$, which contradicts $y + z = 10$

43. Which values should be given to a, b, and c so that the linear system has $(-1, 2, -3)$ as its only solution?

$\begin{cases} x + 2y - 3z = a \\ -x - y + z = b \\ 2x + 3y - 2z = c \end{cases}$

$a = 12$, $b = -4$, $c = 10$

Enrichment Activities

These activities require students to go beyond lesson goals.

1. The National Council of Teachers of Mathematics (NCTM) logo has been placed on a coordinate grid as shown. Determine the system of inequalities that define the "shield" (the interior of the logo).

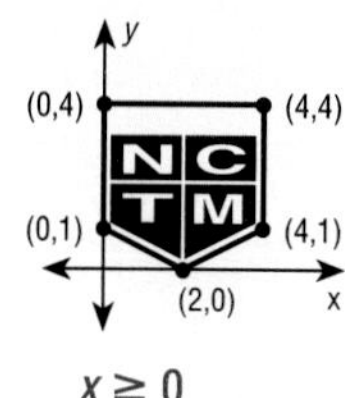

$x \geq 0$
$x \leq 4$
$y \leq 4$
$y \geq -\frac{1}{2}x + 1$
$y \geq \frac{1}{2}x - 1$

2. Design a logo for your school or for your homeroom. Superimpose it on a coordinate grid and write a system of inequalities to describe the region formed by the logo.

Note: You may wish to collect the suggested logos. Have students design and carry out a fair method for determining the best logo.

EXTEND *the Activity*

The corner of a room (or of a cardboard box) is a good physical model of a three-dimensional coordinate system in which $x > 0$, $y > 0$, and $z > 0$. Students should locate points in the coordinate system by positioning themselves in various (attainable) locations in the room. For example, in a classroom, locate the points

a. (0, 0, 0)
b. (4, 5, 0)
c. (7, 2, 0)
d. (4, 5, 2)
e. (7, 2, 5)

GRAPHING SYSTEMS IN THREE VARIABLES

Solutions of equations with three variables can be pictured with a **three-dimensional coordinate system.** To construct such a system, begin with the *xy*-coordinate plane in a horizontal position. Then draw the *z*-axis as a vertical line through the origin.

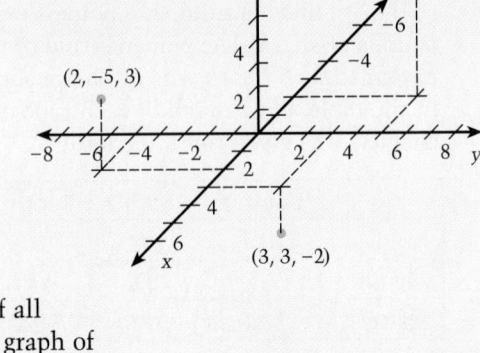

Every ordered triple (x, y, z) corresponds to a point in the three-dimensional coordinate system. For instance, the points corresponding to $(-2, 5, 4)$, $(2, -5, 3)$, and $(3, 3, -2)$ are shown at the right.

The **graph** of an equation in three variables consists of all points (x, y, z) that are solutions of the equation. The graph of a linear equation in three variables is a *plane.* Sketching graphs in a three-dimensional coordinate system is difficult because the sketch itself is only two-dimensional.

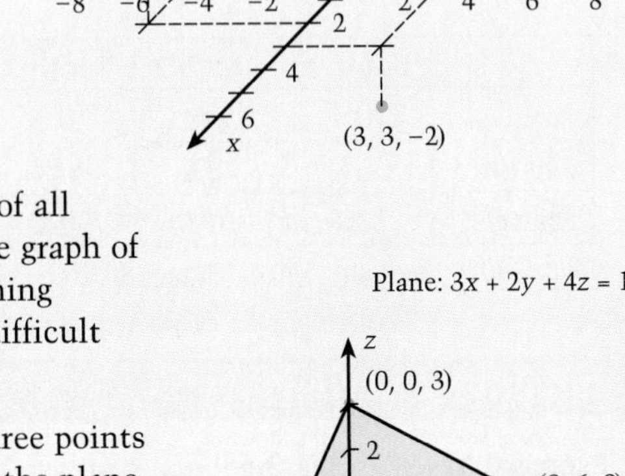

Plane: $3x + 2y + 4z = 12$

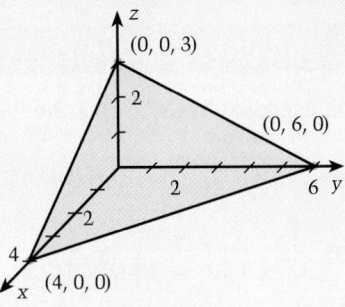

One technique for sketching a plane is to find the three points at which the plane intersects the axes. For instance, the plane given by

$$3x + 2y + 4z = 12$$

intersects the *x*-axis at the point (4, 0, 0), the *y*-axis at the point (0, 6, 0), and the *z*-axis at the point (0, 0, 3). By plotting these three points, connecting them with line segments, and shading the resulting triangular region, you can picture a portion of the graph, as shown at the right.

The graph of a system of three linear equations in three variables consists of *three* planes. When these planes intersect in a single point, the system has exactly one solution. When the three planes have no point in common, the system has no solution. When the three planes contain a common line, the system has infinitely many solutions.

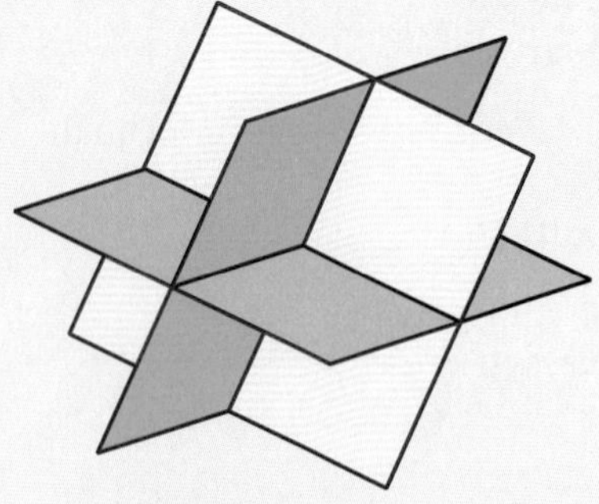

Exactly One Solution

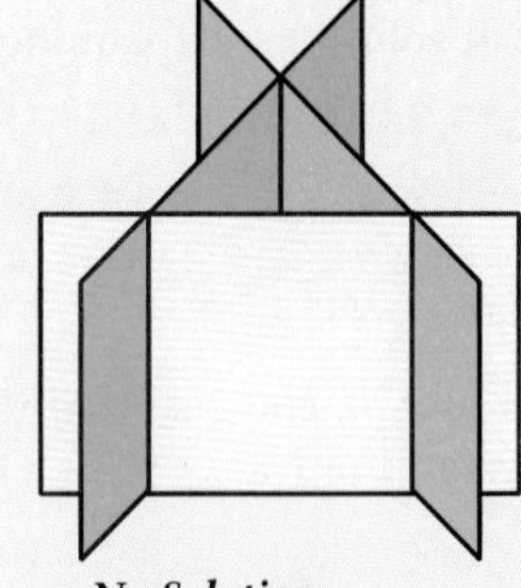

No Solution

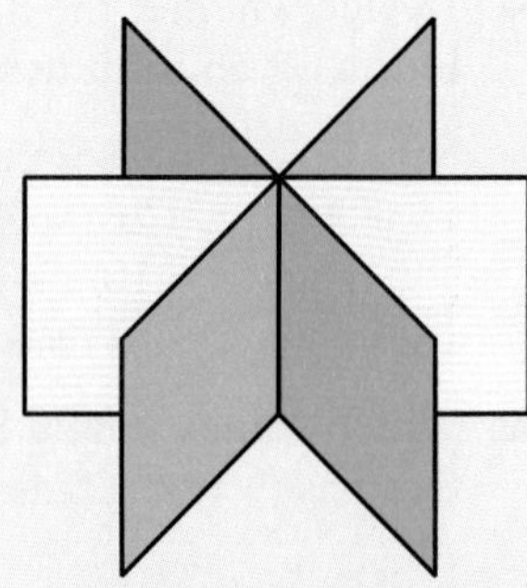

Infinitely Many Solutions

3

Chapter Summary

What did you learn?

Skills

1. Solve a system of linear equations in two variables
   - by graphing. **(3.1)**
   - by substitution. **(3.2)**
   - by linear combination. **(3.2)**
2. Solve a system of linear equations in three variables. **(3.6)**
3. Identify the number of solutions of a linear system. **(3.1, 3.6)**
4. Solve a system of linear inequalities. **(3.4)**
5. Solve a linear programming problem. **(3.5)**

Strategies

6. Use a system of linear equations to model a real-life situation. **(3.1–3.3)**
7. Use a system of linear inequalities to model a real-life situation. **(3.4)**

Exploring Data

8. Use linear programming to solve a real-life problem. **(3.5)**

Why did you learn it?

Systems of linear equations and systems of linear inequalities occur often as models of real-life situations. One common application is the use of a system of linear equations to compare the rate of change of two quantities. For instance, you might want to compare the total revenue for selling x units of a product with the total cost of selling y units of the product. Another common application is finding original data. For instance, you might want to find the total number of sophomores, juniors, and seniors when you know the dance attendance patterns.

Writing equations of lines and inequalities are the basic modeling skills you need for solving problems in this chapter.

How does it fit into the bigger picture of algebra?

In this chapter, you continued your study of linear equations and inequalities. Here, instead of working with a single equation or inequality, you worked with systems of two or more equations or inequalities.

You investigated the three basic techniques of solving a system of linear equations: graphing, substitution, and linear combinations. By now, you should recognize the advantages and disadvantages of each technique.

Most of this chapter dealt with linear systems in two variables. In real life, however, many applications of linear systems contain more than two variables. You explored such systems in the last lesson of the chapter.

ORGANIZER

TIME SCHEDULE: All levels, two days

The Chapter Summary helps students organize the main ideas of the chapter. In this chapter, the connection between modeling real-life problems and solving the resulting algebraic models was made.

Work with students to review the skills, strategies, and concepts of the chapter. The homework assignment can be used as the basis for the second day of review.

COOPERATIVE LEARNING

Encourage students to study together. Emphasize the importance of "teaching" a classmate how to perform a skill or how to recall a strategy. When students work together, everyone wins. The students receiving help get additional instruction and the students giving help gain a deeper understanding of the skills and concepts involved.

Chapter SUMMARY

Review the chapter by asking students, "What did you learn? Why did you learn it? How does it fit with the other algebra skills and concepts you have learned?"

ASSIGNMENT GUIDE

Basic/Average: Ex. 1–23 odd, 24, 26–27

Above Average: Ex. 1–23 odd, 24, 28–30

Advanced: Ex. 1–23 odd, 24, 25, 28–30

✪ More Difficult Exercises
Exercises 22–30

▶ **Ex. 1–30** The Chapter Review exercises include a reference to the lesson in which each skill or concept was presented. Students may use these references as a study aid if they are uncertain how to do a given exercise.

Answers

4.

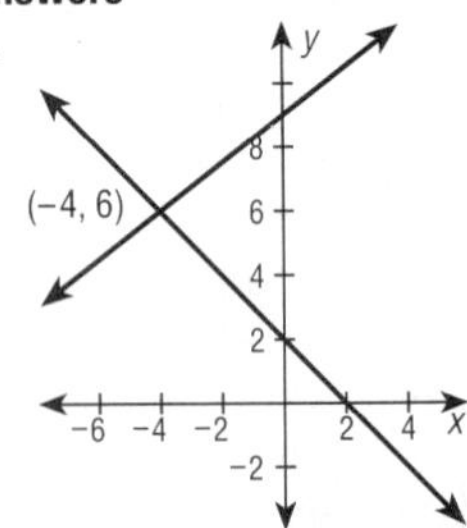

5.

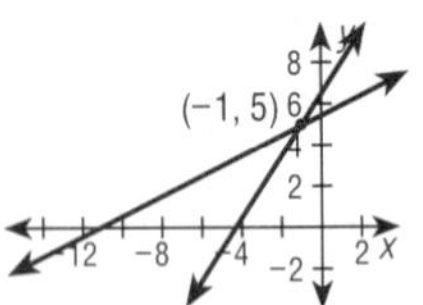

6.

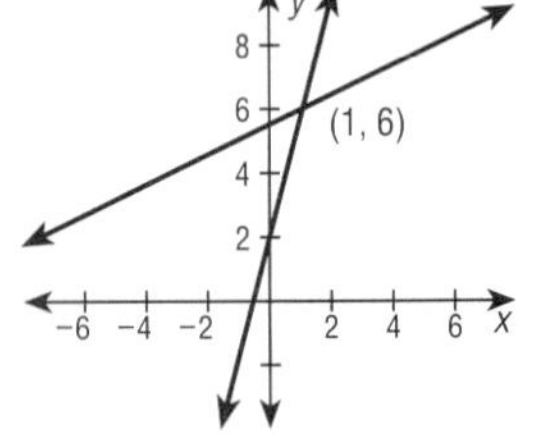

Chapter REVIEW

In Exercises 1–3, decide whether the ordered pair is a solution of the linear system. (3.1)

1. $(-9, 2)$ It is not.
$\begin{cases} -x + 2y = 5 \\ 2x - 4y = 8 \end{cases}$

2. $(1, 5)$ It is not.
$\begin{cases} 2x + 3y = 17 \\ -x + 2y = 11 \end{cases}$

3. $(-2, -1)$ It is
$\begin{cases} 7x + 2y = -16 \\ -6x + 9y = 3 \end{cases}$

In Exercises 4–6, sketch the graph of the linear system. Use the graph to approximate the solution. Then check your solution algebraically. (3.1) See margin.

4. $\begin{cases} x + y = 2 \\ -3x + 4y = 36 \end{cases}$ (−4, 6)

5. $\begin{cases} -x + 2y = 11 \\ 3x - 2y = -13 \end{cases}$ (−1, 5)

6. $\begin{cases} -4x + y = 2 \\ x - 2y = -11 \end{cases}$ (1, 6)

In Exercises 7–12, use substitution or a linear combination to solve the linear system. Then check your solution graphically. (3.2)

7. $\begin{cases} x - 4y = 20 \\ 2x + 5y = 1 \end{cases}$ (8, −3)

8. $\begin{cases} 9x - 5y = -30 \\ x + 3y = 18 \end{cases}$ (0, 6)

9. $\begin{cases} x + 3y = -2 \\ -3x + y = 6 \end{cases}$ (−2, 0)

10. $\begin{cases} -x + y = -14 \\ 2x - 3y = 33 \end{cases}$ (9, −5)

11. $\begin{cases} 2x + 3y = -7 \\ -4x - 5y = 13 \end{cases}$ (−2, −1)

12. $\begin{cases} 2x - 2y = -8 \\ 7x + 6y = 11 \end{cases}$ (−1, 3)

In Exercises 13–15, sketch the graph of the system of inequalities and find the vertices of the graph. (3.5) See margin.

13. $\begin{cases} x \geq 0 \\ y \geq 0 \\ x + 2y \leq 24 \\ 4x + y \leq 40 \end{cases}$ (0, 12), (8, 8), (10, 0), (0, 0)

14. $\begin{cases} x \geq 0 \\ y \geq 0 \\ -x + 2y \leq 8 \\ 3x + 2y \leq 24 \end{cases}$ (0, 4), (4, 6), (8, 0), (0, 0)

15. $\begin{cases} x \geq 0 \\ x \leq 5 \\ y \geq 0 \\ y \leq 3 \end{cases}$ (0, 3), (5, 3), (5, 0), (0, 0)

In Exercises 16–18, find the minimum value and maximum value of the objective quantity. (3.5)

16. Objective quantity:
$C = 5x + 2y$

Constraints:
$\begin{cases} x \geq 3 \\ y \geq 2 \\ 3x + 2y \leq 22 \end{cases}$ 19, 34

17. Objective quantity:
$C = 4x + 2y$

Constraints:
$\begin{cases} y \geq 0 \\ -3x + y \leq 4 \\ x + 5y \leq 180 \\ 16x - y \leq 288 \end{cases}$ $-\frac{16}{3}$, 144

18. Objective quantity:
$C = x + 4y$

Constraints:
$\begin{cases} x \geq 0 \\ y \geq 0 \\ -2x + y \leq 10 \\ 4x - y \leq 20 \\ 5x + 2y \leq 38 \end{cases}$ 0, 58

In Exercises 19–21, solve the linear system. (3.6)

19. $\begin{cases} x - 2y + z = 33 \\ y + 2z = 5 \\ z = 4 \end{cases}$ (23, −3, 4)

20. $\begin{cases} x + y + 3z = 20 \\ 2x + 4y + 5z = 30 \\ x + y + 4z = 26 \end{cases}$ (4, −2, 6)

21. $\begin{cases} 3x - 6y + z = 0 \\ 3x - 4y - 4z = 15 \\ -6x + 8y + 9z = -33 \end{cases}$ (1, 0, −3)

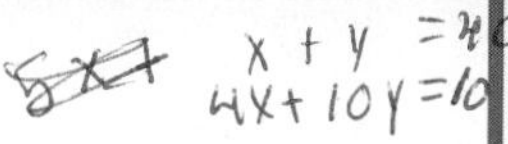

22. *Wearing Makeup* In a 1991 survey of American women, x percent said they wore makeup regularly and y percent said they wore makeup occasionally. The number who wore makeup regularly was 23% more than the number who wore makeup occasionally. What percent said they wore makeup regularly? What percent said they wore makeup occasionally? ***(Source: Roper Organization)*** **30%, 53%**

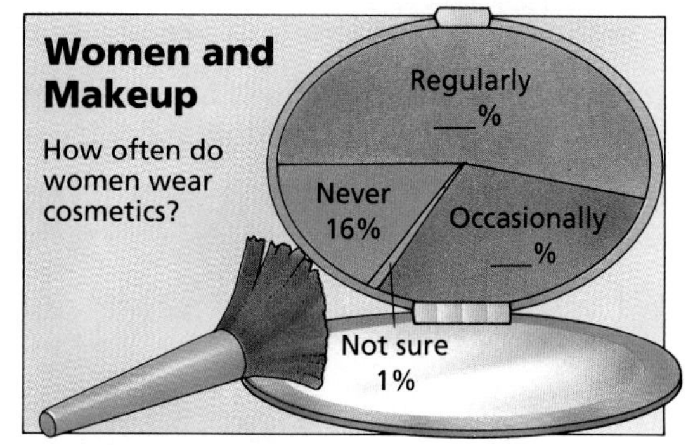

23. *Single-Parent Dads* The sum of the numbers of single-parent fathers in 1970, 1980, and 1990 is 2,433,000. The sum for 1980 and 1990 is 2,040,000. There were 1,350,000 single-parent fathers in 1990. Express this information as a system of three linear equations in three variables. (Let x, y, and z represent the number of single-parent fathers in the United States in 1970, 1980, and 1990, respectively.) Solve the system. ***(Source: U.S. Bureau of Census)*** **See margin.**

24. *Waiting to Remarry* From 1970 through 1988, the average number of years, R, that an American waited to remarry (after divorce or death of a spouse) is given by

$$\begin{cases} R = 0.067t + 2.9 & \textit{Women} \\ R = 0.072t + 2.3 & \textit{Men} \end{cases}$$

where t is the year, with $t = 0$ corresponding to 1970. Sketch the graph of this system. Do the two lines intersect between $t = 0$ and $t = 18$? ***(Source: National Center for Health Statistics)*** **No. See margin.**

25. *Why Ice Floats* Most liquids contract as they cool. This is true of pure water, however, only as it cools to 39°F. Between 39°F and 32°F, pure water expands as it cools. This is the reason ice floats—its density is less than the density of liquid water. As salt is added to water, it loses this property gradually. In fact, water that is more than 2.5% salt continues to contract as it cools until it finally freezes. (Icebergs float because they are broken off from glaciers, which are formed from fresh water.) Write a system of linear inequalities that models the temperature and saltiness of water that expands as it cools. **$s \geq 0$, $t \geq 0$, $4s + t \leq 39$, $\frac{6}{5}s + t \geq 32$**

Behavior of Salt Water

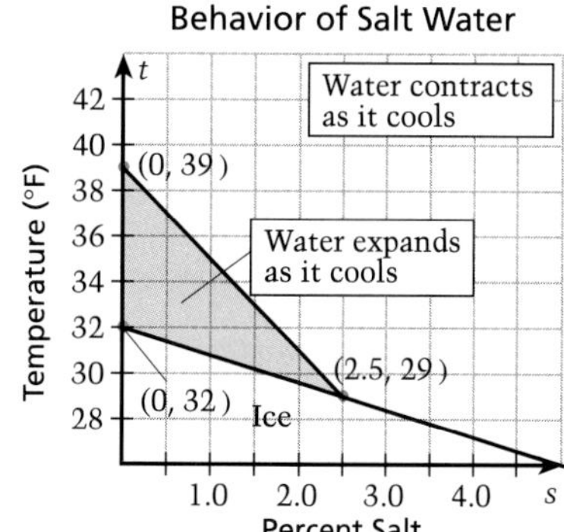

Answers

13.

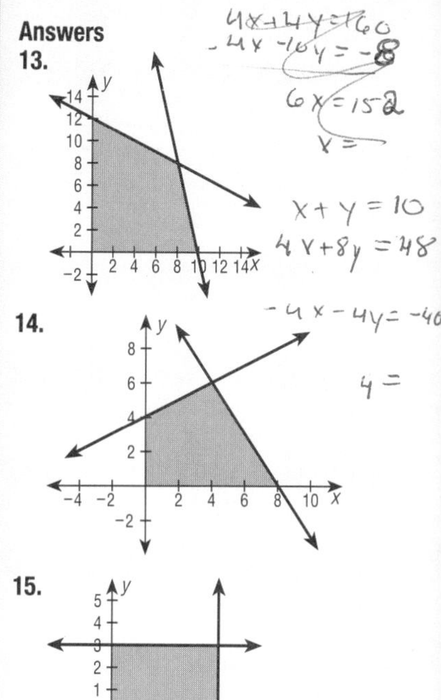

14.

15.

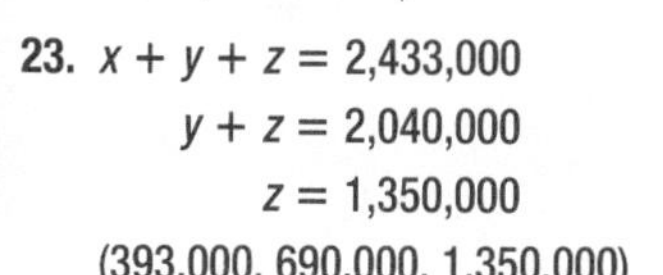

23. $x + y + z = 2{,}433{,}000$
$y + z = 2{,}040{,}000$
$z = 1{,}350{,}000$
(393,000, 690,000, 1,350,000)

24.

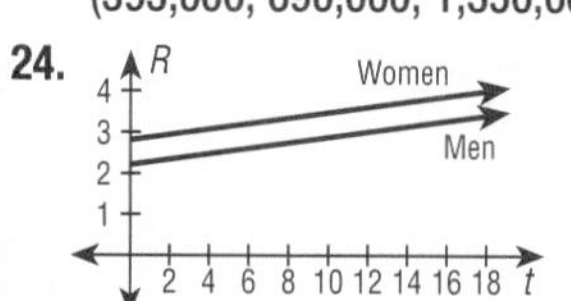

Answer

28.

Nucleus Radius (in millimeters)

Time (in months)

Pearl Necklace **In Exercises 26 and 27, use the following information.**

If the pearls shown at the right are strung to form a graduated pearl necklace, it would just fit within the graph of the system

$$\begin{cases} y \le \frac{1}{60}x + \frac{1}{10} \\ y \ge -\frac{1}{60}x - \frac{1}{10} \\ y \le -\frac{1}{60}x + \frac{3}{10} \\ y \ge \frac{1}{60}x - \frac{3}{10} \\ x \ge -3 \\ x \le 15 \end{cases}$$

where x and y are measured in inches.

26. How long is the necklace? 18 in.

27. Approximate the diameter of the largest pearl in the necklace. 0.4 in.

Growing a Pearl **In Exercises 28–30, use the following information.**

Pearls are grown by oysters and other mollusks that produce a lustrous substance called *nacre.* A pearl is formed when a foreign object is embedded in the mollusk. The mollusk reacts by slowly covering the object with layer upon layer of nacre. In a cultured oyster pearl, a spherical nucleus is embedded in an oyster. Most cultured oyster pearls are grown in Japan. The nuclei, however, come mainly from the nacre of freshwater mussels in Camden, Tennessee.

28. The radius, R (in millimeters), of a cultured pearl depends on the radius of the nucleus and the time, t (in months), the nucleus is left in the oyster. The relationship between R and t is indicated by:

$$\begin{cases} R = \frac{1}{30}t + 2 & \textit{Nucleus radius: 2 mm} \\ R = \frac{1}{30}t + 3 & \textit{Nucleus radius: 3 mm} \\ R = \frac{1}{30}t + 4 & \textit{Nucleus radius: 4 mm} \\ R = \frac{1}{30}t + 5 & \textit{Nucleus radius: 5 mm} \end{cases}$$

Sketch the graph of this system. Do the lines intersect? No

29. From the linear system given in Exercise 28, how much does a cultured pearl increase in radius each month? $\frac{1}{30}$ mm

30. Some natural pearls have grown in oysters for as long as 20 years. Estimate the diameter of such a pearl if its nucleus radius is 2 mm. 20 mm

Growing freshwater mussels in Tennessee

Chapter TEST

In Exercises 1–3, solve the linear system. Check your solution graphically. (3.1, 3.2)

1. $\begin{cases} y = -\frac{3}{2}x + \frac{7}{2} \\ y = \frac{5}{7}x - \frac{8}{7} \end{cases}$ $\left(\frac{65}{31}, \frac{11}{31}\right)$

2. $\begin{cases} x + 2y = 10 \\ 2x - 3y = -8 \end{cases}$ (2, 4)

3. $\begin{cases} 4x - 5y = 3 \\ 3x - 2y = 4 \end{cases}$ (2, 1)

In Exercises 4–6, sketch the graph of the system of inequalities. Label the vertices. (3.4) See margin.

4. $\begin{cases} x \geq 0 \\ y \geq 0 \\ x + y \leq 5 \end{cases}$

5. $\begin{cases} x \leq 5 \\ x \geq 1 \\ y \leq 7 \\ y \geq -2 \end{cases}$

6. $\begin{cases} x \geq 0 \\ y \geq 0 \\ 2x + y \leq 12 \\ -x + 4y \leq 20 \end{cases}$

In Exercises 7–9, find the minimum and maximum values of the objective quantity. (3.5)

7. Objective quantity: $C = 2x + 3y$

Constraints: $\begin{cases} x \geq 1 \\ y \geq 1 \\ x + y \leq 5 \end{cases}$ 5, 14

8. Objective quantity: $C = 8x + 7y$

Constraints: $\begin{cases} x \geq 0 \\ y \geq 0 \\ x + y \leq 7 \\ 4x + 3y \leq 24 \end{cases}$ 0, 52

9. Objective quantity: $C = 3x + 4y$

Constraints: $\begin{cases} x + y \geq 8 \\ x + y \leq 9 \\ x + 2y \leq 16 \\ 2x + y \leq 16 \end{cases}$ 24, 34

In Exercises 10–12, solve the linear system. (3.6)

10. $\begin{cases} x + 3y - 4z = 9 \\ y + 2z = 8 \\ z = 1 \end{cases}$ (−5, 6, 1)

11. $\begin{cases} x + 2y - 6z = 23 \\ x + 3y + z = 4 \\ 2x + 5y - 4z = 24 \end{cases}$ (1, 2, −3)

12. $\begin{cases} -x + 2y - 7z = -17 \\ 4x + 6y + 18z = 20 \\ 3x - y - 3z = -7 \end{cases}$ (−1, −2, 2)

13. Your hourly wage at a grocery store is greater after 6:00 P.M. than during the day. One week you work 18 daytime hours and 22 evening hours, and earn $264. The next week you work 30 daytime hours and 10 evening hours, and earn $240. What is the daytime hourly rate? What is the evening hourly rate? $5.50, $7.50

14. In 1990, the aquariums with the greatest attendance were the Sea Worlds in Florida, California, and Texas. The total attendance at these three was 10,200,000. Sea World of Florida had 300,000 more than Sea World of California. Sea World of California had 1,200,000 more than Sea World of Texas. What was the 1990 attendance of each? See margin.

(Source: American Association of Zoological Parks and Aquariums)

15. Write a system of linear inequalities to model the region shown at the right. See margin.

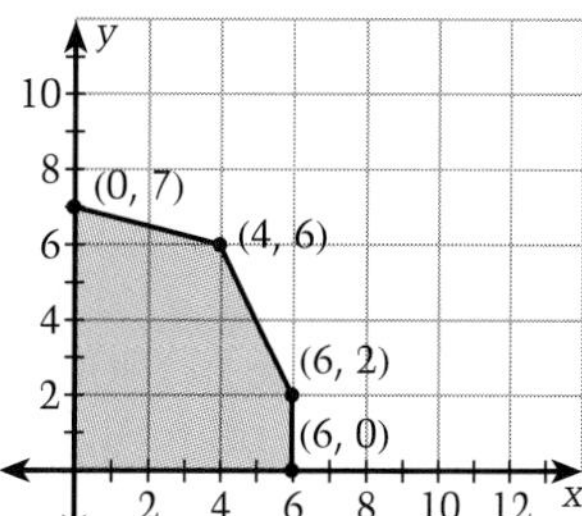

Answers

4.

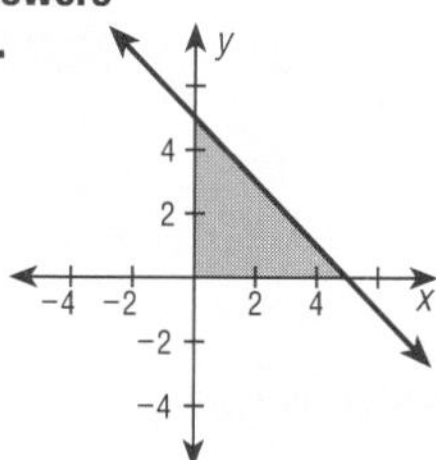

5.

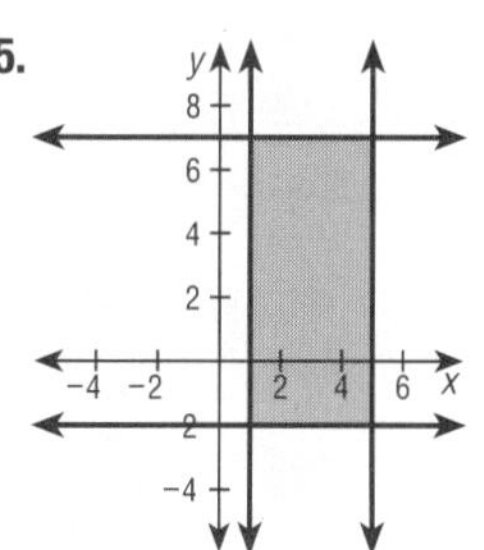

6.

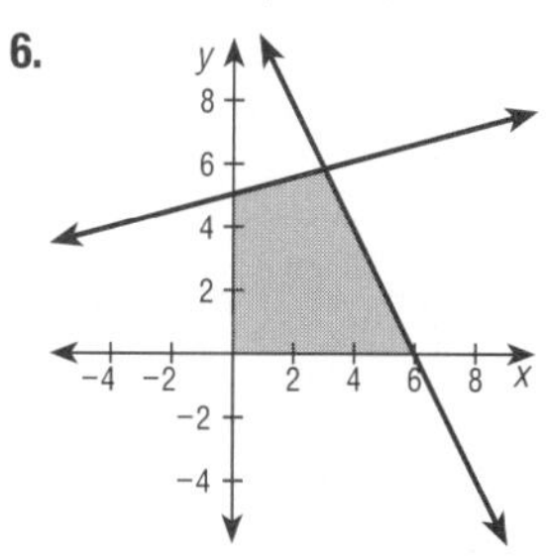

14. Sea World of California: 3,700,000
Sea World of Florida: 4,000,000
Sea World of Texas: 2,500,000

15. $y \leq -\frac{1}{4}x + 7$,
$y \leq -2x + 14$,
$x \leq 6$,
$y \geq 0$,
$x \geq 0$

13.

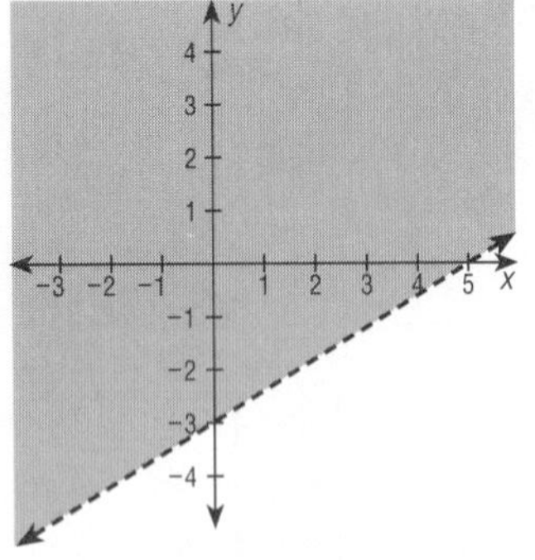

14.

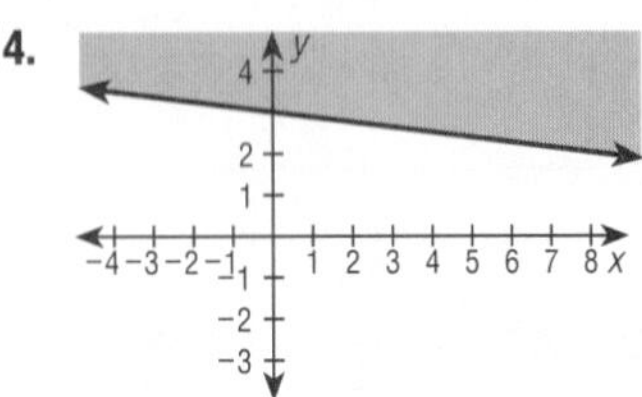

15.

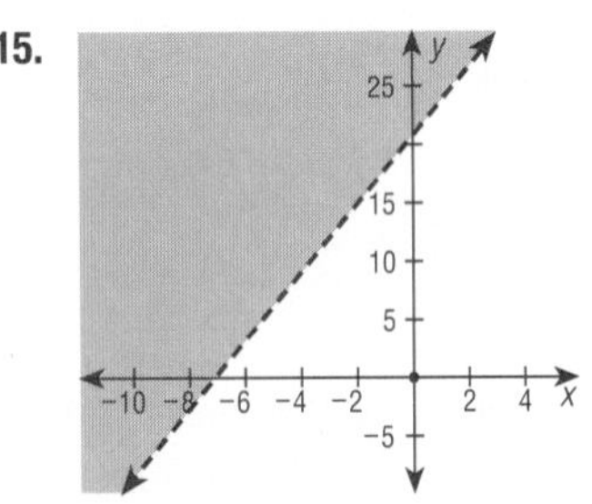

16.

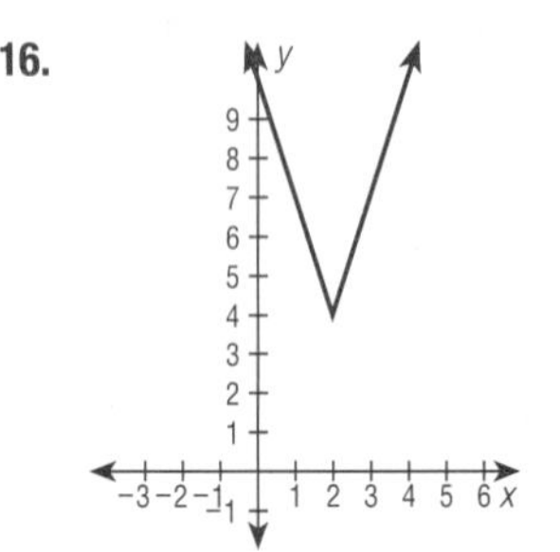

17.

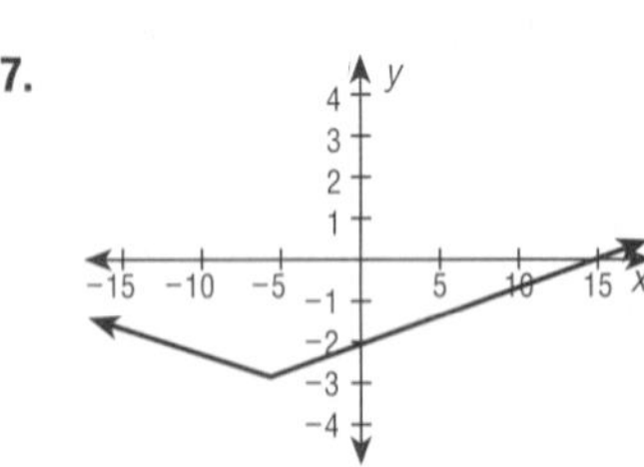

18.

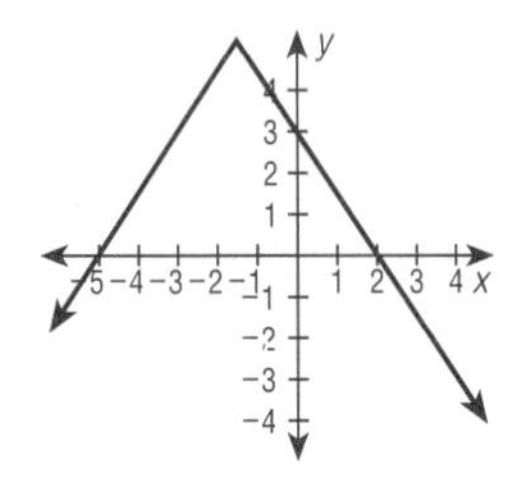

Cumulative REVIEW ▪ Chapters 1–3

In Exercises 1–4, solve the equation.

1. $3(x + 4) = (2 - x)7$ $\frac{1}{5}$
2. $\frac{1}{4}|x + 2| = 3$ $-14, 10$
3. $|3x - 4| = 15$ $-\frac{11}{3}, \frac{19}{3}$
4. $3x - 2(x + 6) = 13 - x$ $\frac{25}{2}$
5. *Simple Interest* Solve the formula $A = P + Prt$ for r. Then use the result to find the value of r when $A = \$575$, $P = \$500$, and $t = 3$ years. $r = \frac{A - P}{Pt}$, 0.05 or 5%

In Exercises 6–8, solve the inequality.

6. $4x - (2 + 3x) \geq 14$ $x \geq 16$
7. $\frac{1}{2}|2x - 3| \leq 13$ $-\frac{23}{3} \leq x \leq \frac{29}{2}$
8. $\left|\frac{1}{4}x + 3\right| > 5$ $x < -32$ or $x > 8$

In Exercises 9–12, write an equation of the indicated line. Then sketch its graph. See Additional Answers for graphs.

9. Slope: $\frac{2}{3}$, Point: $(0, 3)$ $2x - 3y = -9$
10. Slope: -4, y-intercept: -1 $4x + y = -1$
11. Two points: $(-3, 2)$, $(8, -1)$ $3x + 11y = 13$
12. Two points: $(-0.5, 6)$, $(0.25, 5)$ $4x + 3y = 16$

In Exercises 13–15, sketch the graph of the linear inequality. See margin.

13. $3x - 5y < 15$
14. $\frac{1}{4}x + 2y \geq 6$
15. $\frac{1}{3}y - x > 7$

In Exercises 16–18, sketch the graph of the absolute value equation. See margin.

16. $y = 3|x - 2| + 4$
17. $y = \frac{1}{7}|x + 6| - 3$
18. $y = -\left|\frac{3}{2}x + 2\right| + 5$

In Exercises 19–24, solve the system of equations.

19. $\begin{cases} 3x + 5y = 0 \\ y = -\frac{3}{2} \end{cases}$ $\left(\frac{5}{2}, -\frac{3}{2}\right)$
20. $\begin{cases} \frac{1}{4}x + \frac{2}{3}y = 7 \\ y = 4 \end{cases}$ $\left(\frac{52}{3}, 4\right)$
21. $\begin{cases} -4x + 8y = 4 \\ 3x + \frac{1}{2}y = 10 \end{cases}$ $(3, 2)$
22. $\begin{cases} 3x + 2y - 2z = 13 \\ 4y + 8z = 20 \\ 3z = 6 \end{cases}$ $(5, 1, 2)$
23. $\begin{cases} 8x - 2y + z = 32 \\ 6x + \frac{1}{2}y + \frac{1}{4}z = 26 \\ x + y - 5z = -14 \end{cases}$ $(4, 2, 4)$
24. $\begin{cases} -4x - 3y + 5z = 1 \\ 4x + 8y - 6z = 10 \\ 5x - y + 7z = -24 \end{cases}$ $(-3, 2, -1)$

In Exercises 25 and 26, sketch the graph of the system of linear inequalities. Label each vertex of the graph.

25. $\begin{cases} 2x + y \leq 4 \\ 5x + y \leq 8 \\ x \geq 0 \\ y \geq 0 \end{cases}$ See margin.
26. $\begin{cases} 4x - y > -6 \\ x + 2y < 6 \\ 4x - 3y < 9 \end{cases}$ See margin.

27. *Breakfast Calories* For breakfast, you had 1 cup of oatmeal, 1 cup of lowfat milk, and 1 cup of orange juice, for a total of 375 calories. Your brother had 2 cups of oatmeal, 2 cups of lowfat milk, and 1 cup of orange juice, for a total of 640 calories. Your sister had 1 cup of oatmeal, $\frac{3}{4}$ cup of lowfat milk, and $\frac{1}{2}$ cup of orange juice, for a total of 290 calories. How many calories are in 1 cup of oatmeal, 1 cup of lowfat milk, and 1 cup of orange juice? 145 calories, 120 calories, 110 calories

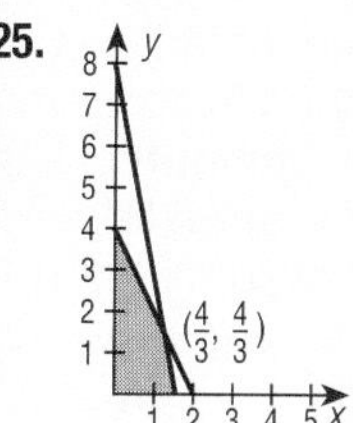

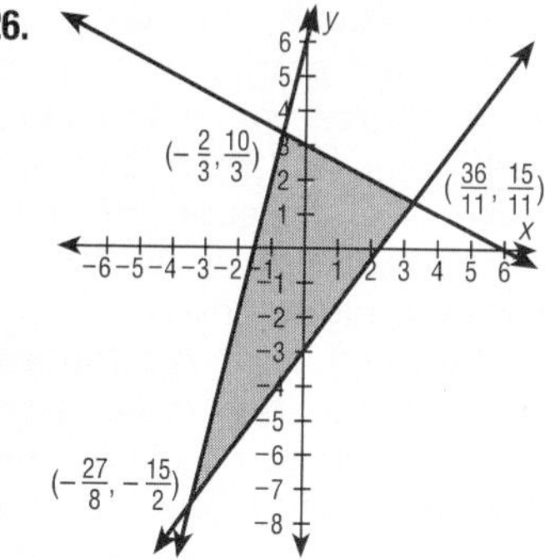

Hewlett-Packard **In Exercises 28–31, use the following information.**

Sales, S (in millions of dollars), for the Hewlett-Packard Company from 1980 through 1990 are given in the table.
(Source: Hewlett-Packard Company)

Year	1980	1981	1982	1983	1984	1985
Sales	\$3099	\$3578	\$4254	\$4710	\$6044	\$6505

Year	1986	1987	1988	1989	1990
Sales	\$7102	\$8090	\$9831	\$11,899	\$13,233

28. Sketch a scatter plot. Let $t = 0$ represent 1980. **See Additional Answers.**

29. Do S and t have a positive correlation, a negative correlation, or no correlation? **positive correlation**

30. Sketch the best-fitting line. Write an equation for the line. **$S = 1000t + 2000$, see Additional Answers.**

31. Estimate the 1994 sales for the Hewlett-Packard Company. **≈\$16,000 million**

32. *National Parks* The ten most popular national parks are listed in the table with their 1990 attendance. Construct a circle graph that shows each park's percentage of the total attendance. **See Additional Answers.**

National Park	Attendance	National Park	Attendance
Great Smokies, NC & TN	8,333,600	Yellowstone, WY	2,644,400
Acadia, ME	5,441,000	Rocky Mountain, CO	2,502,900
Grand Canyon, AZ	3,966,200	Zion, UT	1,998,900
Yosemite, CA	3,308,200	Shenandoah, VA	1,873,800
Olympic, WA	2,737,600	Glacier, MT	1,821,500

33. *Maximum Profit* Your company produces cabinets using two different processes. The number of assembly hours required for each process are listed in the table. You can use up to 3000 hours of machine time, up to 3600 hours of skilled labor, and up to 3600 hours of unskilled labor. The profit from Process A is \$50 per cabinet and the profit from Process B is \$70 per cabinet. How many cabinets should you make with each process to obtain a maximum profit? **Process A: 600, Process B: 1200**

	Assembly hours	
	Process A	Process B
Machine time	1	2
Skilled labor	2	2
Unskilled labor	3	1

The two most popular uses of matrices are found in computer software and computer systems. Computer spreadsheets, which are applications of matrices, are used to organize and manipulate data. Computer graphics, such as those commonly found in computer games, are possible because of how matrices are used to represent computer input and output, and because of how systems of equations can be solved using matrices.

The chapter offers a foundation in the study of matrix algebra, including adding and multiplying matrices, and solving systems of equations using matrices. For the first time, perhaps, students will encounter mathematical elements whose multiplication is not commutative.

The Chapter Summary on page 222 provides you and the students with a synopsis of the chapter. It identifies key skills and concepts. You may want to have students look at the Chapter Summary as an overview before beginning the chapter.

CHAPTER 4

Matrices and Determinants

LESSONS

New varieties of high-yield rice have helped reduce food shortages in Asia and the South Pacific. These farmers are harvesting rice in Bali, Indonesia.

Real Life
Food Production

Percent of Crop Grown in Each Country

	Wheat	Rice	Corn
China	16.6%	37.2%	13.6%
India	8.7%	18.9%	1.7%
C.I.S.	17.2%	0.5%	2.6%
U.S.	10.6%	1.3%	43.6%

During the mid 1960's, the population in developing nations was growing fast. Keeping the burgeoning population fed seemed an impossible task in these countries, until the introduction of high-yield varieties of wheat and rice. This produced what is known as the "green revolution." By the 1980's, some of these countries, such as China and India, were accumulating surpluses.

By 1990, China and India were two of the largest grain-producing countries in the world. The matrix at the left lists the percent of the world's wheat, rice, and corn produced in 1990 by the top four grain producers—China, the Commonwealth of Independent States (formerly the Soviet Union), India, and the United States.

Matrices provide a convenient way to organize data in tabular form. As mathematical objects, matrices can be added and multiplied. They are an important tool of mathematics, with applications in statistics, computer science, biology, and economics.

Have students use the matrix of the introduction to answer the following questions.

a. What percent of the total world production of corn comes from India? **1.7%**

b. What percent of the total world production of wheat comes from the United States? **10.6%**

c. What percent of the total world production of rice comes from China? **37.2%**

Matrices and Determinants

PACING CHART

The daily Pacing Chart is meant to help you adjust your teaching pace. Students in the full course should finish the entire text by the end of the year. Students in the basic course are expected to complete the first twelve chapters. The Pacing Chart for each chapter contains suggestions for lessons that require more than one day and lessons that may be omitted for the basic course.

DAY	FULL COURSE	BASIC COURSE
1	4.1	4.1
2	4.2	4.2
3	4.3 & Using a Graphing Calculator	4.3 & Using a Graphing Calculator
4	4.4	4.4
5	Mid-Chapter Self-Test	Mid-Chapter Self-Test
6	4.5	4.5
7	4.6	4.5
8	4.7	4.6
9	Chapter Review	4.6
10	Chapter Test	(4.7 Optional lesson)
11		(4.7 Optional lesson)
12		Chapter Review
13		Chapter Test

CHAPTER ORGANIZATION

LESSON	PAGES	GOALS	MEETING THE NCTM STANDARDS
4.1	174–179	1. Organize data into matrices and work with them, using addition, subtraction, and scalar multiplication 2. Use matrices in real-life settings	Problem Solving, Communication, Connections, Discrete Math
4.2	180–185	1. Multiply two matrices 2. Use matrix multiplication to answer questions about real-life situations	Problem Solving, Communication, Connections, Discrete Math, Structure
Mixed Review	186	Review algebraic and arithmetic skills	
Milestones	186	Arthur Cayley	Connections
4.3	187–193	1. Evaluate the determinant of a 2×2 or a 3×3 matrix 2. Use determinants to solve real-life problems	Problem Solving, Communication, Connections, Geometry (algebraic), Discrete Math
Using a Calculator	194–195	Perform operations with matrices	Technology
4.4	196–202	1. Find and use the inverse of a 2×2 matrix 2. Use inverse matrices in real-life settings	Problem Solving, Communication, Connections, Discrete Math, Structure
Mid-Chapter Self-Test	203	Diagnose student weaknesses and remediate with correlated Reteach worksheets	
4.5	204–209	1. Solve systems of linear equations using inverse matrices 2. Use systems of linear equations to solve real-life problems	Problem Solving, Communication, Connections, Discrete Math, Structure, Technology
Mixed Review	209	Review algebraic and arithmetic skills	
4.6	210–215	1. Use an augmented matrix to solve a system of linear equations 2. Use a system of linear equations to solve real-life problems	Problem Solving, Communication, Connections, Discrete Math, Structure
4.7	216–221	1. Use Cramer's rule to solve a system of linear equations 2. Use linear systems to solve real-life problems	Problem Solving, Communication, Connections, Discrete Math
Chapter Summary	222	Restate for students what they have learned, why they have learned it, and how it fits into the structure of algebra	Structure, Connections
Chapter Review	223–226	Review concepts and skills learned in the chapter	
Chapter Test	227	Diagnose student weaknesses and remediate with correlated Reteaching worksheets	

LESSON RESOURCES

MEETING INDIVIDUAL NEEDS

RETEACHING For students who need to spend more time on basics:

If a mid-chapter self-test or chapter test indicates a deficiency, teachers can help students with the appropriate ***Reteaching Copymaster.***

PRACTICE For students who need more practice:

Additional exercises like those in the Pupil's Edition are provided for each lesson in ***Extra Practice Copymasters.***

ENRICHMENT For enriching and broadening students' experiences:

Problem of the Day copymasters in ***Teaching Tools*** provide a daily opportunity to use logical reasoning, looking for a pattern, writing an equation, and other routine and non-routine problem-solving strategies.

Math Log copymasters in ***Alternative Assessment*** provide opportunities to report on investigations, research, and open-ended problems.

Technology: Using Calculators and Computers provides enriching activities with graphing and scientific calculators and computers.

The ***Applications Handbook*** provides additional information about the cross-curriculum topics such as astronomy, chemistry, physics, sports, economics, genetics, and music that are integrated into the Pupil's Edition.

LESSON	4.1	4.2	4.3	4.4	4.5	4.6	4.7
PAGES	174-179	180-185	187-193	196-202	204-209	210-215	216-221
Teaching Tools							
Transparencies							
Problem of the Day	✓	✓	✓	✓	✓	✓	✓
Warm-up Exercises	✓	✓	✓	✓	✓	✓	✓
Answer Masters	✓	✓	✓	✓	✓	✓	✓
Extra Practice Copymasters	✓	✓	✓	✓	✓	✓	✓
Reteaching Copymasters	Teacher-directed and independent activities tied to results on the Mid-Chapter Self-Tests and Chapter Tests						
Color Transparencies			✓				✓
Applications Handbook	Additional background information for many real-life applications						
Technology	Calculator and computer worksheets for appropriate lessons						
Complete Solutions Manual	✓	✓	✓	✓	✓	✓	✓
Alternative Assessment	Assess student's ability to reason, analyze, solve problems, and communicate using mathematical language.						
Formal Assessment	Mid-Chapter Self-Tests, Chapter Tests, Cumulative Tests, and Practice for College Entrance Tests						
Computer Test Bank	Customized tests can be created by choosing from over 2500 items.						

INSIGHTS

4.1 Exploring Data: Matrix Operations

The study of matrices begins with its most fundamental use—recording and organizing data. Students will see how to add and subtract matrices, and learn how to interpret the meaning of these operations in real-life terms.

4.2 Matrix Multiplication

Matrix multiplication can be used to solve many real-life problems involving tabular data, such as the total cost of business equipment, the annual number of subscribers of a magazine, or the yearly crop production in the United States by state. Matrix multiplication also provides students with an example of an operation that is not commutative. As such, it is an important example of a concept that may be counter-intuitive.

4.3 Determinants

Determinants are used to find the area of triangular regions, to find inverses of matrices, and to solve systems of equations. Determinants provide an efficient way to manipulate matrices and can also be used to characterize matrices.

4.4 Identity and Inverse Matrices

In the multiplication of real numbers, the product of a number and its inverse is the identity factor, 1. Similar relationships exist for some real-valued matrices. The collection of matrices for which these relationships hold is important (for instance, such matrices can be used to solve systems of equations). This lesson applies this concept for 2 x 2 matrices.

4.5 Solving Systems Using Inverse Matrices

Matrices offer students another technique for solving systems of linear equations. As described in the lesson, solving systems of linear equations will become analogous to solving linear equations in one variable. Students will see that matrices are a powerful tool for solving real-life problems.

4.6 Solving Systems Using Augmented Matrices

In this lesson, the Gaussian elimination method used to solve systems of linear equations will be extended to include an appropriate solution technique for a system of linear equations which has been rewritten as an augmented matrix. The techniques presented in this lesson add to the students growing list of ways to solve systems of equations.

4.7 Solving Systems Using Cramer's Rule

Many real-life situations are modeled by systems of linear equations. Consequently, much time has been spent demonstrating a variety of methods by which such systems can be solved. The final lesson of the chapter is another method known as Cramer's Rule. It uses determinants to solve systems of equations and is, therefore, a useful computational technique.

Problem of the Day

If $2^x + 2^x + 2^x + 2^x = 2^k$, express x in terms of k.
$x = k - 2$

ORGANIZER

Warm-Up Exercises

1. The following table lists data about the number of endangered species in the United States (U.S.) and foreign countries (F).
(*Source:* United States Interior Department, 1989)

Number of Endangered Species

	U.S. only	U.S. and F	F only
Mammals	32	19	241
Birds	61	15	145
Reptiles	8	7	59
Fishes	45	2	11
Plants	153	6	1

Use the table to answer the following questions.

a. What is the total number of endangered reptiles in the list?
b. What is the total number of endangered mammals in the list?
c. How many from all species in the list are only endangered in the United States?
d. How many from all species are on the endangered lists of both the United States and foreign countries?

a. 74, b. 292, c. 299, d. 49

Lesson Resources

Teaching Tools
Problem of the Day: 4.1
Warm-up Exercises: 4.1
Answer Masters: 4.1
Extra Practice: 4.1

4.1 Exploring Data: Matrix Operations

What you should learn:

Goal 1 How to organize data into matrices and to work with them, using addition, subtraction, and scalar multiplication

Goal 2 How to use matrices in real-life settings

Why you should learn it:

You can use matrices to organize many types of real-life data, such as income tax tables, nutrition charts, and accounting spreadsheets.

Goal 1 Organizing Data in a Matrix

A **matrix** is a rectangular arrangement of numbers in rows and columns. For instance, the matrix below has two rows and three columns. The **order** of this matrix is 2×3 (read "2 by 3"). A **row matrix** is a matrix that has only one row. A **column matrix** has only one column, and a **square matrix** has the same number of rows and columns.

$$\begin{bmatrix} 6 & 2 & -1 \\ -2 & 0 & 5 \end{bmatrix}$$

The numbers in a matrix are its **entries.** In the above matrix, the entry in the second row and third column is 5. A **zero matrix** is a matrix whose entries are all zero.

Two matrices (*matrices* is the plural of *matrix*) are equal if the entries in corresponding positions are equal.

$$\begin{bmatrix} 5 & 0 \\ -\frac{4}{4} & \frac{3}{4} \end{bmatrix} = \begin{bmatrix} 5 & 0 \\ -1 & 0.75 \end{bmatrix} \qquad \begin{bmatrix} -2 & 6 \\ 0 & -3 \end{bmatrix} \neq \begin{bmatrix} -2 & 6 \\ -3 & 0 \end{bmatrix}$$

Think of a matrix as a type of table used to organize data.

Real Life
Consumer Finance

Example 1 *Reading the Entries of a Matrix*

The monthly payment for a car loan depends on the annual interest rate, the amount of the loan, and the length of the loan. The matrix shows different monthly payments for a 12% annual interest rate. How much more interest would you pay for a 12% auto loan of \$15,000 if you took the loan for 5 years instead of 3 years?

Monthly Payment	36 Months	48 Months	60 Months
\$10,000	\$332.14	\$253.34	\$222.48
\$15,000	\$498.22	\$395.01	\$333.67
\$20,000	\$664.29	\$526.68	\$444.89

Solution For 3 years, you would make 36 payments of \$498.22 for a total of \$17,935.92. The total interest would be \$2935.92. For 5 years, you would make 60 payments of \$333.67 for a total of \$20,020.20. The total interest would be \$5020.20. Thus, you would pay \$2084.28 more in interest. ■

To add or subtract matrices, you simply add or subtract corresponding entries, as shown in Example 2.

Example 2 *Adding and Subtracting Matrices*

a. $\begin{bmatrix} 5 & -3 \\ -3 & 4 \\ 0 & 7 \end{bmatrix} + \begin{bmatrix} -2 & 1 \\ 3 & 0 \\ 4 & -3 \end{bmatrix} = \begin{bmatrix} 5+(-2) & -3+1 \\ -3+3 & 4+0 \\ 0+4 & 7+(-3) \end{bmatrix}$

$= \begin{bmatrix} 3 & -2 \\ 0 & 4 \\ 4 & 4 \end{bmatrix}$

b. $\begin{bmatrix} 8 & 3 \\ -4 & 0 \end{bmatrix} - \begin{bmatrix} 2 & -7 \\ -6 & 1 \end{bmatrix} = \begin{bmatrix} 8-2 & 3-(-7) \\ -4-(-6) & 0-1 \end{bmatrix}$

$= \begin{bmatrix} 6 & 10 \\ 2 & -1 \end{bmatrix}$ ■

You can add or subtract matrices only if they have the same orders. You cannot, for instance, add a matrix that has three rows to a matrix that has only two rows.

Matrix addition is associative and commutative. This means that three or more matrices can be added in any order. If A, B, and C are matrices of the same order, these properties can be stated as follows:

$(A + B) + C = A + (B + C)$ *Associative Property*

$A + B = B + A$ *Commutative Property*

In matrix algebra, a real number is often called a **scalar.** To *multiply a matrix by a scalar,* you multiply each entry in the matrix by the scalar. Multiplication of a matrix by a scalar obeys the Distributive Property.

$c(A + B) = cA + cB$ *Distributive Property (Addition)*

$c(A - B) = cA - cB$ *Distributive Property (Subtraction)*

Example 3 *Multiplying a Matrix by a Scalar*

a. $4\begin{bmatrix} -2 & 0 \\ 4 & -1 \end{bmatrix} = \begin{bmatrix} 4(-2) & 4(0) \\ 4(4) & 4(-1) \end{bmatrix}$

$= \begin{bmatrix} -8 & 0 \\ 16 & -4 \end{bmatrix}$

b. $-2\left(\begin{bmatrix} 1 & -2 \\ 0 & 3 \end{bmatrix} + \begin{bmatrix} -4 & 5 \\ 6 & -8 \end{bmatrix}\right) = -2\begin{bmatrix} -3 & 3 \\ 6 & -5 \end{bmatrix}$

$= \begin{bmatrix} 6 & -6 \\ -12 & 10 \end{bmatrix}$ ■

LESSON Notes

Example 1

Perhaps some students in the class are faced with a similar situation or are about to embark on purchasing a car. Emphasize the importance of knowing how much interest will be paid for borrowing money. Although students may see the 60-month payment plan as better because the monthly payments are smaller, the "better deal" is most certainly the more expensive one!

Example 2

You might have students use numerical examples to verify the associative and commutative properties for the addition of matrices.

Example 3

Bear in mind that scalar multiplication is different from the multiplication of matrices. Alert students that these two types of multiplication exist. Caution students to always refer to the operation as "scalar multiplication."

Example 4

Even though the 2-year private college is omitted from the data, it is reasonable to use the available data to calculate the averages.

Extra Examples

Here are additional examples similar to **Examples 1–4.**

1. In Example 1, how much more interest would you pay for a 12% auto loan of
   - **a.** $10,000 if you took the loan for 4 years instead of 3 years?
     $203.28
   - **b.** $20,000 if you took the loan for 5 years instead of 3 years?
     $2778.96

2. Perform the indicated operation.

a. $\begin{bmatrix} 0 & 2 \\ 2 & -5 \end{bmatrix} + \begin{bmatrix} -1 & 3 \\ -2 & 4 \end{bmatrix}$

b. $\begin{bmatrix} 11 & 5 \\ 6 & 9 \\ -5 & 2 \end{bmatrix} - \begin{bmatrix} -4 & 0 \\ 8 & 4 \\ -3 & 7 \end{bmatrix}$

c. $\begin{bmatrix} 12 & 0 \\ 3 & 3 \end{bmatrix} + \begin{bmatrix} -7 & 1 \\ 13 & 6 \end{bmatrix}$

d. $\begin{bmatrix} 8 & 4 \\ 4 & 9 \end{bmatrix} - \begin{bmatrix} 1 & 5 \\ 2 & 4 \end{bmatrix}$

e. $-3\begin{bmatrix} 0 & 3 \\ 1 & 3 \end{bmatrix} + \begin{bmatrix} -2 & 6 \\ 2 & 4 \end{bmatrix}$

f. $6\begin{bmatrix} 1 & 0 \\ 0 & 4 \end{bmatrix} + \begin{bmatrix} 2 & -2 \\ 2 & -1 \end{bmatrix}$

a. $\begin{bmatrix} -1 & 5 \\ 0 & -1 \end{bmatrix}$ b. $\begin{bmatrix} 15 & 5 \\ -2 & 5 \\ -2 & -5 \end{bmatrix}$

c. $\begin{bmatrix} 5 & 1 \\ 16 & 9 \end{bmatrix}$ d. $\begin{bmatrix} 7 & -1 \\ 2 & 5 \end{bmatrix}$

e. $\begin{bmatrix} -2 & -3 \\ -1 & -5 \end{bmatrix}$ f. $\begin{bmatrix} 8 & -2 \\ 2 & 23 \end{bmatrix}$

3. In Example 4, what was the average yearly cost of room-and-board at a college or university in 1985?
about $2650

Check Understanding

1. What is the order of a matrix?
The indication of the number of rows and columns
2. State the associative and commutative properties of matrix addition.
See the paragraph before Example 3.
3. What type of multiplication does the addition distributive property involve?
scalar multiplication
4. How many entries does an m by n matrix have? Explain.
mn entries; multiply the numbers of rows and columns

Goal 2 Using Matrices in Real-Life Settings

Real Life
Cooking

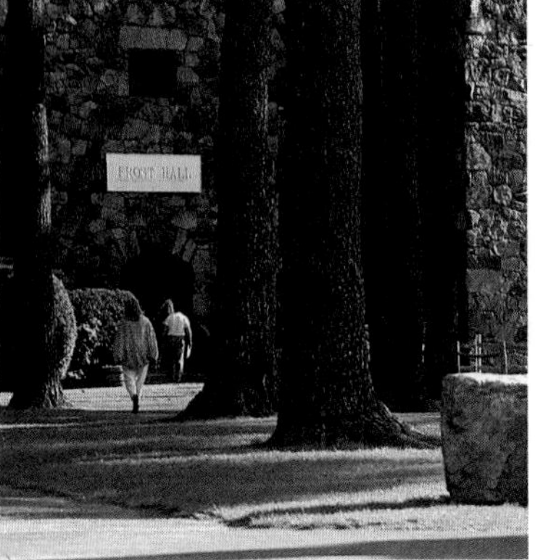

From 1991 to 1994, college costs at private 4-year universities increased by about 21% per year.

Example 4 *Finding the Total Cost of College*

The matrices below show the average yearly cost of tuition and room and board at colleges in the United States between 1991 and 1994. Use matrix addition to find a matrix showing the totals of these fees. ***(Source: U.S. National Center for Educational Statistics)***

Tuition	1991	1992	1993	1994
Public 4-Year University	2159	2410	2604	2822
Private 4-Year University	11379	12192	13055	13812
Public 4-Year College	1707	1933	2192	2368
Private 4-Year College	8389	9053	9533	10151
Public 2-Year College	824	937	1025	1114

Room and Board	1991	1992	1993	1994
Public 4-Year University	3425	3641	3838	3887
Private 4-Year University	5124	5587	5843	6215
Public 4-Year College	3296	3527	3548	3791
Private 4-Year College	3832	4136	4348	4551
Public 2-Year College	2644	2686	2774	2892

Solution The Total Charges matrix is as follows.

Total Charges	1991	1992	1993	1994
Public 4-Year University	5584	6051	6442	6709
Private 4-Year University	16503	17779	18898	20027
Public 4-Year College	5003	5460	5740	6159
Private 4-Year College	12221	13189	13881	14702
Public 2-Year College	3468	3623	3799	4006

■

Communicating about ALGEBRA

SHARING IDEAS about the Lesson See Additional Answers.

A. Use the following matrices and scalar to show that $c(A + B) = cA + cB$.

$$A = \begin{bmatrix} 4 & -5 \\ -2 & 8 \end{bmatrix}, \quad B = \begin{bmatrix} -3 & 7 \\ 8 & 2 \end{bmatrix}, \quad c = 2$$

B. Give a *convincing argument* that the general property $c(A + B) = cA + cB$ is true for *any* scalar c and *any* 2×2 matrices A and B.

EXERCISES

Guided Practice

CRITICAL THINKING about the Lesson

1. Give an example of a row matrix, a column matrix, a 4×3 matrix. See margin.
2. To add two matrices, what must be true of their orders? Must be the same
3. What is the order of $\begin{bmatrix} 10 & 4 & 2 & 4 \\ -2 & 0 & 1 & 7 \\ 3 & 1 & -9 & 6 \end{bmatrix}$? 3×4
4. Are the two matrices equal? Explain. $\begin{bmatrix} -6 & \frac{1}{2} \\ 0 & -2 \\ 1 & 5 \end{bmatrix} \stackrel{?}{=} \begin{bmatrix} -6 & 0.5 \\ 0 & 2 \\ 1 & 5 \end{bmatrix}$ No, $-2 \neq 2$
5. Add: $[2 \quad 1 \quad 3 \quad 5] + [0 \quad -4 \quad 3 \quad -2]$. $[2 \quad -3 \quad 6 \quad 3]$
6. Multiply the matrix by the scalar: $-2\begin{bmatrix} 4 & 6 \\ 1 & 0 \end{bmatrix}$. $\begin{bmatrix} -8 & -12 \\ -2 & 0 \end{bmatrix}$

Independent Practice

In Exercises 7–10, state the order of the matrix.

7. $\begin{bmatrix} 5 & -3 & 3 \\ 1 & 5 & 1 \\ 0 & 6 & -9 \end{bmatrix}$ 3×3

8. $\begin{bmatrix} 1 & -2 \\ 6 & 9 \\ 9 & 2 \\ 4 & 1 \end{bmatrix}$ 4×2

9. $\begin{bmatrix} 5 & 9 & -3 & 6 \\ 3 & 1 & 4 & 10 \end{bmatrix}$ 2×4

10. $\begin{bmatrix} 1 & 4 & 10 & 6 \\ 2 & 3 & -1 & 9 \\ -8 & 9 & -3 & -8 \end{bmatrix}$ 3×4

In Exercises 11–18, perform the indicated operation, if possible. If not possible, state the reason.

11. $\begin{bmatrix} 1 & 2 \\ -3 & 1 \end{bmatrix} + \begin{bmatrix} 1 & 5 \\ 2 & 0 \end{bmatrix}$ $\begin{bmatrix} 2 & 7 \\ -1 & 1 \end{bmatrix}$

12. $\begin{bmatrix} 4 & 2 \\ -9 & 5 \end{bmatrix} - \begin{bmatrix} 2 & 8 \\ 4 & 8 \end{bmatrix}$ $\begin{bmatrix} 2 & -6 \\ -13 & -3 \end{bmatrix}$

13. $\begin{bmatrix} 7 & -3 & 2 \\ 0 & 8 & 1 \\ 8 & 6 & -6 \end{bmatrix} + \begin{bmatrix} 1 & 0 & -5 \\ 0 & 8 & 7 \\ -7 & 9 & 6 \end{bmatrix}$ See margin.

14. $\begin{bmatrix} 1 & -6 \\ 9 & 7 \\ 5 & 6 \end{bmatrix} + \begin{bmatrix} 0 & -4 & 8 \\ 5 & 2 & 5 \\ 3 & 9 & -8 \end{bmatrix}$ Different number of columns

15. $\begin{bmatrix} 0 & -8 \\ 3 & 0 \\ -4 & 2 \end{bmatrix} - \begin{bmatrix} 5 & 4 \\ 2 & 5 \\ -7 & -5 \end{bmatrix}$ $\begin{bmatrix} -5 & -12 \\ 1 & -5 \\ 3 & 7 \end{bmatrix}$

16. $\begin{bmatrix} 2 & -1 & 7 \\ 5 & 8 & 2 \end{bmatrix} - \begin{bmatrix} 2 & 9 & 3 \\ -5 & 2 & -2 \end{bmatrix}$ See below.

17. $\begin{bmatrix} -6 & 8 & 5 \\ 4 & -4 & 7 \end{bmatrix} - \begin{bmatrix} 4 & 2 \\ 8 & -4 \\ 3 & 3 \end{bmatrix}$ Different numbers of rows and columns

18. $\begin{bmatrix} 9 & 1 & 6 \\ -5 & 0 & 9 \\ 2 & -2 & 3 \end{bmatrix} + \begin{bmatrix} 9 & 0 & 4 \\ -4 & 6 & 8 \\ 3 & -5 & -6 \end{bmatrix}$ See margin.

In Exercises 19–22, multiply the matrix by the scalar. See margin.

19. $-3\begin{bmatrix} 1 & -3 & 6 \\ 9 & -1 & 4 \end{bmatrix}$

20. $4\begin{bmatrix} 0 & -7 \\ 5 & 3 \end{bmatrix}$

21. $-8\begin{bmatrix} 0 & \frac{1}{2} \\ -2 & -1 \\ 4 & 3 \end{bmatrix}$

22. $2\begin{bmatrix} 1 & -1 & 0 \\ 6 & 3 & 1 \\ 3 & 2 & 9 \end{bmatrix}$

16. $\begin{bmatrix} 0 & -10 & 4 \\ 10 & 6 & 4 \end{bmatrix}$

Communicating about ALGEBRA

Ask students to provide convincing arguments that the associative and commutative properties hold for matrix addition. (*Note:* A general argument uses arbitrary real-valued entries, similar to those found in the matrix shown here, where a_{23} indicates that this entry lies in the 2nd row and the 3rd column.

$$\begin{bmatrix} a_{11} & a_{12} & a_{13} \\ a_{21} & a_{22} & a_{23} \\ a_{31} & a_{32} & a_{33} \\ a_{41} & a_{42} & a_{43} \end{bmatrix}$$

EXERCISE Notes

ASSIGNMENT GUIDE

Basic/Average: Ex. 1–6, 7–31 odd, 32–37, 39–45 odd

Above Average: Ex. 1–6, 7–31 odd, 37, 39–45 odd, 46

Advanced: Ex. 1–6, 7–33 odd, 34–37, 46

Selected Answers
Exercises 1–6, 7–45 odd

More Difficult Exercises
Exercises 33, 46

Guided Practice

Ex. 1–6 Be sure to review students' understanding of the order of a matrix.

Answers

1. Examples vary. $[1 \quad 5 \quad -3]$, $\begin{bmatrix} -1 \\ 0 \\ 2 \end{bmatrix}$, $\begin{bmatrix} 1 & 0 & -2 \\ 6 & 3 & 0 \\ -4 & 7 & 5 \\ 0 & 9 & 3 \end{bmatrix}$

13. $\begin{bmatrix} 8 & -3 & -3 \\ 0 & 16 & 8 \\ 1 & 15 & 0 \end{bmatrix}$

18. $\begin{bmatrix} 18 & 1 & 10 \\ -9 & 6 & 17 \\ 5 & -7 & -3 \end{bmatrix}$

Independent Practice

Answers

19. $\begin{bmatrix} -3 & 9 & -18 \\ -27 & 3 & -12 \end{bmatrix}$

20. $\begin{bmatrix} 0 & -28 \\ 20 & 12 \end{bmatrix}$

21. $\begin{bmatrix} 0 & -4 \\ 16 & 8 \\ -32 & -24 \end{bmatrix}$

22. $\begin{bmatrix} 2 & -2 & 0 \\ 12 & 6 & 2 \\ 6 & 4 & 18 \end{bmatrix}$

23. $\begin{bmatrix} 1 & 7 \\ -5 & 9 \end{bmatrix}$ **24.** $\begin{bmatrix} 1 & -1 \\ -9 & 4 \\ -5 & 0 \end{bmatrix}$

25. $\begin{bmatrix} -16 & 28 & 4 \\ 0 & -24 & 4 \\ 24 & -40 & -4 \end{bmatrix}$

26. $\begin{bmatrix} 24 & 20 \\ 30 & -30 \\ -6 & 66 \end{bmatrix}$

▶ **Ex. 32–37** These exercises require interpreting information in a matrix format. In Exercise 32, students who do not follow baseball may need to have the designated-hitter rule explained by you, or by another student. In the American League, the pitcher is not required to be a hitter. There is a designated hitter who stands in for the pitcher in the batting order. The rule is suspended in World Series play.

EXTEND Ex. 36–37 If you feel it is appropriate, you might share information about teachers' salaries in your school district for comparison.

In Exercises 23–27, simplify the expression. 23–26 See margin.

23. $\left(\begin{bmatrix} 4 & 0 \\ -1 & 5 \end{bmatrix} - \begin{bmatrix} 6 & -2 \\ 3 & 4 \end{bmatrix}\right) + \begin{bmatrix} 3 & 5 \\ -1 & 8 \end{bmatrix}$

24. $\begin{bmatrix} 3 & 2 \\ -2 & 0 \\ 1 & 6 \end{bmatrix} - \left(\begin{bmatrix} 9 & -1 \\ 2 & -3 \\ 3 & 8 \end{bmatrix} + \begin{bmatrix} -7 & 4 \\ 5 & -1 \\ 3 & -2 \end{bmatrix}\right)$

25. $-4\left(\begin{bmatrix} 5 & 1 & -3 \\ 4 & 0 & -1 \\ -3 & 9 & 6 \end{bmatrix} - \begin{bmatrix} 1 & 8 & -2 \\ 4 & -6 & 0 \\ 3 & -1 & 5 \end{bmatrix}\right)$

26. $6\begin{bmatrix} 6 & 4 \\ 9 & -3 \\ 5 & 11 \end{bmatrix} - 2\left(\begin{bmatrix} 1 & 0 \\ 5 & 3 \\ 9 & 4 \end{bmatrix} + \begin{bmatrix} 5 & 2 \\ 7 & 3 \\ 9 & -4 \end{bmatrix}\right)$

27. $\left(\begin{bmatrix} -2 & 0 & 5 \\ 5 & 1 & 9 \\ 7 & 2 & -1 \end{bmatrix} + 2\begin{bmatrix} 6 & -4 & 0 \\ 9 & 5 & 4 \\ -1 & 0 & 7 \end{bmatrix}\right) - \begin{bmatrix} 4 & 6 & 3 \\ -1 & 0 & 2 \\ 5 & 9 & 1 \end{bmatrix}$ $\begin{bmatrix} 6 & -14 & 2 \\ 24 & 11 & 15 \\ 0 & -7 & 12 \end{bmatrix}$

In Exercises 28 and 29, solve for *x*.

28. $\begin{bmatrix} 2x & 8 \\ 4 & -6 \end{bmatrix} + \begin{bmatrix} 3 & -2 \\ -1 & 5 \end{bmatrix} = \begin{bmatrix} 5 & 6 \\ 3 & -1 \end{bmatrix}$ 1

29. $\begin{bmatrix} 3 & 6 \\ 8 & -5 \end{bmatrix} - \begin{bmatrix} 8 & 1 \\ 10 & -4x \end{bmatrix} = \begin{bmatrix} -5 & 5 \\ -2 & 19 \end{bmatrix}$ 6

In Exercises 30 and 31, solve for *x*, *y*, and *z*.

30. $\begin{bmatrix} x \\ 0 \\ 0 \end{bmatrix} + \begin{bmatrix} 2y \\ y \\ 0 \end{bmatrix} + \begin{bmatrix} z \\ -z \\ 2z \end{bmatrix} = \begin{bmatrix} 5 \\ 3 \\ 6 \end{bmatrix}$ (−10, 6, 3)

31. $\begin{bmatrix} 2x \\ 0 \\ 0 \end{bmatrix} + \begin{bmatrix} -4y \\ 3y \\ 0 \end{bmatrix} + \begin{bmatrix} 2z \\ 6z \\ z \end{bmatrix} = \begin{bmatrix} 2 \\ 0 \\ -1 \end{bmatrix}$ (6, 2, −1)

32. *Designated Hitter Rule* The designated hitter rule in major league baseball applies only to the American League. The matrix below gives the percentages of the responses from the American League (AL) and National League (NL) fans when asked how they would treat the rule if they were league commissioners. From each league, 190 fans were surveyed. How many from each league said they would ban the designated hitter rule? How many from each league said they would expand it to the National League?

Ban: 137 from NL, 46 from AL; expand: 38 from NL, 101 from AL

Designated Hitter Rule?	NL Fans	AL Fans
Ban the rule.	72%	24%
Expand to NL.	20%	53%
Leave as is.	8%	23%

Dodger Stadium Video **In Exercises 33 and 34, use the following information.**

The video display at Dodger Stadium is made up of 3- by 3-inch modules. Each module contains 4 pixels. (A pixel is formed by 4 small cathode ray tubes.)

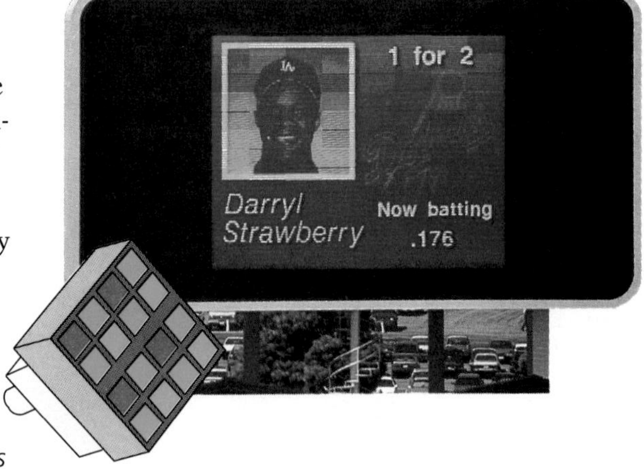

✪ **33.** The giant screen is rectangular: 25 feet by 30 feet. How many pixels are on the screen? 48,000

34. What is the order of the "pixel-matrix" shown? 4 × 4

35. *Sizes of Football Fields* The matrix shows the standard widths and lengths (in feet) of four types of football fields. What is the difference in the area of a college football field and an arena football field? **40,770 ft²**

Football Field Sizes	Width	Length
Arena Football	85	198
U.S. College Football	160	360
U.S. Professional Football	160	360
Canadian Football	195	450

Teachers' Salaries **In Exercises 36–37, use the following information.**

The matrices at the right list teachers' salaries in a certain Pennsylvania school district from the 1990–1991 school year to the 1993–1994 school year. The codes at the top of each matrix classify a teacher according to the number of graduate credits earned. The numbers to the left represent a teacher's current year of service to the district.

36. Suppose that in 1990, you accepted a teaching position in this school district. Your beginning code is A, but by the 1993–94 year you have earned enough graduate credits to be classified as a code B. What would be your total salary for your first four years of teaching? **$92,619**

37. Suppose that you had reached code B by your second year of teaching. What would be your total salary for your first four years of teaching? **$92,819**

1990–91	A	B	C
1	19,894	19,994	20,094
2	20,671	20,771	20,871
3	21,448	21,548	21,648
4	22,225	22,325	22,425

1991–92	A	B	C
1	20,525	20,625	20,725
2	21,375	21,475	21,575
3	22,225	22,325	22,425
4	23,075	24,025	23,275

1992–93	A	B	C
1	22,475	22,575	22,675
2	23,325	23,425	23,525
3	24,175	24,275	24,375
4	25,025	25,125	25,225

1993–94	A	B	C
1	24,525	24,625	24,725
2	25,375	25,475	25,575
3	26,225	26,325	26,425
4	27,075	27,175	27,275

Integrated Review

In Exercises 38–41, simplify the expression.

38. $10(5 - 1) - 5$ **35**

39. $6 + 2(1 - 6)$ **−4**

40. $4(-3 + 7) + 9$ **25**

41. $10 - (5 + 9)$ **−4**

42. What number is 24% of 700? **168**

43. What number is 86% of 250? **215**

44. 228 is what percent of 380? **60%**

45. 63 is what percent of 84? **75%**

Exploration and Extension

46. *Getting a Raise* The hourly wage of an employee at a restaurant depends on the employee's job description and the number of months the employee has worked. The matrix at the right shows the wage rates before the employees won a 5% wage hike. Write the new wage matrix. **See Additional Answers.**

Hourly Wage (in dollars)	Three Months	Six Months	Twelve Months
Regular Employee	5.00	5.10	5.35
Assistant Supervisor	5.75	5.85	6.10

Enrichment Activities

These activities require students to go beyond lesson goals.

In a recent survey, 74% of the National League baseball fans indicated they were opposed to the designated hitter rule of the American League.

1. Work with a partner and conduct a survey in your school regarding these questions on the designated hitter rule. To keep the survey unbiased, be sure to interview students who follow baseball and those who do not, American League fans and National League fans. What percent, by category, of the students surveyed think the rule
   a. should be banned from both leagues?
   b. should be expanded to the National League?
   c. should be retained in the American League?
   d. What percent have no opinion?

Use a matrix like the following to report your percentage data.

Question	AL Fan	NL Fan	Not a Fan
A			
B			
C			
D			

As a check on your survey, the sums in each column should add to 100%.

2. **WRITING** Write a paragraph interpreting the results of your survey.

Problem of the Day

A barrel of honey weighs 50 pounds. The same barrel with kerosene in it weighs 35 pounds. If honey is twice as heavy as kerosene, what is the weight of the empty barrel? 20 pounds

ORGANIZER

Warm-Up Exercises

1. Identify the order of each matrix in the pair.

a. $A = \begin{bmatrix} 2 & -1 \\ -1 & 2 \end{bmatrix}$, $B = \begin{bmatrix} 0 & 2 & 1 & 3 \\ -3 & 9 & -2 & -1 \end{bmatrix}$

b. $A = \begin{bmatrix} 7 & 2 & 1 \\ 2 & 1 & 4 \\ 3 & -9 & -2 \\ 1 & 0 & 1 \end{bmatrix}$, $B = \begin{bmatrix} 8 & 1 & 1 & 2 \\ -5 & -1 & 0 & -1 \end{bmatrix}$

a. A is 2×2, B is 2×4
b. A is 4×3, B is 2×4

2. In Warm-Up Exercise 1, which A matrix has the same number of columns as rows in the B matrix of the pair? In a., A has 2 columns, B has 2 rows.

Lesson Resources

Teaching Tools
Problem of the Day: 4.2
Warm-up Exercises: 4.2
Answer Masters: 4.2
Extra Practice: 4.2

4.2 Matrix Multiplication

What you should learn:

Goal 1 How to multiply two matrices

Goal 2 How to use matrix multiplication to answer questions about real-life situations

Why you should learn it:

You can model many real-life situations with matrix multiplication, such as data on a spreadsheet.

Goal 1 Multiplying Two Matrices

The **matrix multiplication** of A and B is defined only if the number of columns in A equals the number of rows in B. For instance, the product AB of the matrices below is defined since A has two columns (3×2) and B has two rows (2×2).

$$A = \begin{bmatrix} -1 & 3 \\ 4 & -2 \\ 5 & 0 \end{bmatrix} \quad \text{and} \quad B = \begin{bmatrix} -3 & 2 \\ -4 & 1 \end{bmatrix}$$

If A is an $m \times n$ matrix and B is an $n \times p$ matrix, then the product AB is an $m \times p$ matrix.

Example 1 Finding the Product of Two Matrices

Find the product: $\begin{bmatrix} -1 & 3 \\ 4 & -2 \\ 5 & 0 \end{bmatrix}\begin{bmatrix} -3 & 2 \\ -4 & 1 \end{bmatrix}$.

Solution Since A is a 3×2 matrix and B is a 2×2 matrix, the product AB is defined and is a 3×2 matrix. To find the entry in the first row and first column of AB, multiply corresponding entries in the first row of A and the first column of B. Then add.

$$(-1)(-3) + (3)(-4) = -9$$

$$\begin{bmatrix} -1 & 3 \\ 4 & -2 \\ 5 & 0 \end{bmatrix}\begin{bmatrix} -3 & 2 \\ -4 & 1 \end{bmatrix} = \begin{bmatrix} -9 & \square \\ \square & \square \\ \square & \square \end{bmatrix}$$

By continuing this pattern, you can obtain the following.

$$AB = \begin{bmatrix} -1 & 3 \\ 4 & -2 \\ 5 & 0 \end{bmatrix}\begin{bmatrix} -3 & 2 \\ -4 & 1 \end{bmatrix}$$

$$= \begin{bmatrix} (-1)(-3) + (3)(-4) & (-1)(2) + (3)(1) \\ (4)(-3) + (-2)(-4) & (4)(2) + (-2)(1) \\ (5)(-3) + (0)(-4) & (5)(2) + (0)(1) \end{bmatrix}$$

$$= \begin{bmatrix} -9 & 1 \\ -4 & 6 \\ -15 & 10 \end{bmatrix}$$

■

Be sure you understand that for the product of two matrices to be defined, the number of columns of the first matrix must equal the number of rows of the second matrix. That is, in the diagram below, the two blue numbers must match. Note also that the outside two numbers give the order of the product.

$$\underset{m \times n}{A} \cdot \underset{n \times p}{B} = \underset{m \times p}{AB}$$

equal (n and n)

order of AB (m and p)

Example 2 *Finding the Product of Two Matrices*

a. $\begin{bmatrix} 1 & 0 & 3 \\ 2 & -1 & -2 \end{bmatrix} \begin{bmatrix} -2 & 4 & 2 \\ 1 & 0 & 0 \\ -1 & 1 & -1 \end{bmatrix} = \begin{bmatrix} -5 & 7 & -1 \\ -3 & 6 & 6 \end{bmatrix}$

$2 \times 3 \qquad 3 \times 3 \qquad 2 \times 3$

b. $\begin{bmatrix} 3 & 4 \\ -2 & 5 \end{bmatrix} \begin{bmatrix} 1 & 0 \\ 0 & 1 \end{bmatrix} = \begin{bmatrix} 3 & 4 \\ -2 & 5 \end{bmatrix}$

$2 \times 2 \qquad 2 \times 2 \qquad 2 \times 2$

c. $\begin{bmatrix} 1 & -2 & -3 \end{bmatrix} \begin{bmatrix} 2 \\ -1 \\ 1 \end{bmatrix} = \begin{bmatrix} 1 \end{bmatrix}$

$1 \times 3 \qquad 3 \times 1 \qquad 1 \times 1$

d. $\begin{bmatrix} 2 \\ -1 \\ 1 \end{bmatrix} \begin{bmatrix} 1 & -2 & -3 \end{bmatrix} = \begin{bmatrix} 2 & -4 & -6 \\ -1 & 2 & 3 \\ 1 & -2 & -3 \end{bmatrix}$

$3 \times 1 \qquad 1 \times 3 \qquad 3 \times 3$ ■

Notice in Parts **c** and **d** of Example 2 that the products AB and BA are *not the same.* Matrix multiplication is not, in general, commutative. That is, for most matrices, $AB \neq BA$.

Properties of Matrix Multiplication

Let A, B, and C be matrices and let c be a scalar.

1. $A(BC) = (AB)C$ — *Associative Property*
2. $A(B + C) = AB + AC$ — *Left Distributive Property*
3. $(A + B)C = AC + BC$ — *Right Distributive Property*
4. $c(AB) = (cA)B = A(cB)$

LESSON Notes

Example 1

Be sure students understand how the entries in the product matrix are obtained. Specifically, the entry in the 2nd row, 3rd column of the product matrix comes from multiplying the 2nd row of matrix *A* with the 3rd column of matrix *B.* Ask students to identify how all of the entries of the product matrix are obtained.

Example 2

Emphasize that the commutative property of matrix multiplication fails because there is at least one example for which the commutative property does not hold. However, ask students to find other examples in which the commutative property of matrix multiplication fails. (Almost any choice of matrices should work!)

The "identity matrix" is a square matrix with 0's everywhere except on the main diagonal of the matrix. Have students multiply any square matrix by the identity matrix of the same order and make a conjecture about the results obtained.

Example 3

Observe that the Equipment matrix is nothing more than the table in which the information was originally presented. It is important for students to see that a matrix is another way in which to organize data.

The Cost matrix is constructed so that the order of its entries is identical to that of the Equipment matrix. Furthermore, the Cost matrix is constructed as a 3×1 matrix (instead of a 1×3 matrix) so that matrix multiplication can be used to determine the total cost per team.

EXTEND Example 3 Ask students if the same information about the total cost of equipment for each team can be obtained from a product of a 1×3 matrix with a 3×2 matrix. Have students construct the matrices and explain their results.

The same information can be obtained from the following matrices.

Cost	Bats	Balls	Gloves
Dollars	[21	4	30]

	Teams	
Equipment	Women's	Men's
Bats	12	15
Balls	45	38
Gloves	15	17

Extra Examples

Here are additional examples similar to **Examples 1–2.**

Find each product.

a. $\begin{bmatrix} -1 & 2 & -2 \\ -2 & 5 & 1 \\ 2 & 0 & -2 \end{bmatrix}\begin{bmatrix} 1 & 2 & 0 \\ 0 & 1 & 0 \\ 1 & -2 & 1 \end{bmatrix}$

b. $\begin{bmatrix} 1 & 2 \\ 6 & 1 \\ 0 & 2 \end{bmatrix}\begin{bmatrix} 1 & -1 & 1 & 2 \\ -2 & 2 & 3 & 0 \end{bmatrix}$

c. $\begin{bmatrix} 1 & 2 & -3 \\ 6 & 1 & 2 \\ 0 & 2 & -1 \end{bmatrix}\begin{bmatrix} 1 \\ -2 \\ 0 \end{bmatrix}$

a. $\begin{bmatrix} -3 & 4 & -2 \\ -1 & -1 & 1 \\ 0 & 8 & -2 \end{bmatrix}$

b. $\begin{bmatrix} -3 & 3 & 7 & 2 \\ 4 & -4 & 9 & 12 \\ -4 & 4 & 6 & 0 \end{bmatrix}$ c. $\begin{bmatrix} -3 \\ 4 \\ -4 \end{bmatrix}$

Check Understanding

1. Is matrix multiplication commutative? Explain. No, changing the order of the matrices to be multiplied can change the product matrix.
2. How does one decide if two matrices can be multiplied? Look at their orders. Any matrix of order $m \times n$ can multiply a matrix of order $n \times p$.

Goal 2 Using Matrix Multiplication in Real Life

Real Life
Softball Expenses

Example 3 *Finding the Total Cost*

Two softball teams submit equipment lists to their sponsors.

	Bats	Balls	Gloves
Women's Team	12	45	15
Men's Team	15	38	17

If a bat costs \$21, a ball costs \$4, and a glove costs \$30, use matrices to find the total cost of equipment for each team.

Solution To begin, write the equipment list and the cost per item in matrix form.

Equipment	Bats	Balls	Gloves
Women's Team	12	45	15
Men's Team	15	38	17

Cost	Dollars
Bats	21
Balls	4
Gloves	30

The total cost for each team can now be obtained by multiplying the equipment matrix by the cost matrix.

$$\begin{bmatrix} 12 & 45 & 15 \\ 15 & 38 & 17 \end{bmatrix}\begin{bmatrix} 21 \\ 4 \\ 30 \end{bmatrix} = \begin{bmatrix} 12(21) + 45(4) + 15(30) \\ 15(21) + 38(4) + 17(30) \end{bmatrix} = \begin{bmatrix} 882 \\ 977 \end{bmatrix}$$

The total cost of equipment for the women's team is \$822, and the total cost of equipment for the men's team is \$977. ■

Communicating about ALGEBRA

SHARING IDEAS about the Lesson See Additional Answers.

Two competing companies offer cable television to a city with 100,000 households. Company A has 25,000 subscribers and Company B has 30,000. (The other 45,000 households do not subscribe.) The percent changes in cable subscriptions each year are shown below. Explain how matrix multiplication can be used to find the number of subscribers each company will have in one year.

Percent Changes	From Company A	From Company B	From Nonsubscriber
To Company A	0.70	0.15	0.15
To Company B	0.20	0.80	0.15
Nonsubscriber	0.10	0.05	0.70

EXERCISES

Guided Practice

CRITICAL THINKING about the Lesson

1. Is matrix multiplication defined for the matrices shown? Explain your answer.

$$\begin{bmatrix} 2 & -1 & 5 \\ 1 & 0 & 4 \end{bmatrix} \text{ and } \begin{bmatrix} 0 & 6 \\ -1 & 4 \\ 1 & 2 \end{bmatrix}$$

Yes, the number of columns in the 1st matrix is equal to the number of rows in the second.

2. In the following matrix multiplication, find the missing entry.

$$\begin{bmatrix} 4 & 2 \\ 1 & 0 \\ 6 & 2 \\ -1 & 3 \end{bmatrix}\begin{bmatrix} 1 & -1 & 0 \\ 0 & 5 & 2 \end{bmatrix} = \begin{bmatrix} 4 & \boxed{?} & 4 \\ 1 & -1 & 0 \\ 6 & 4 & 4 \\ -1 & 16 & 6 \end{bmatrix}$$

6

3. A is a 6×1 matrix. B is a 1×2 matrix. Which of the matrix products is defined, AB or BA? Explain. See above.

3. AB, the number of columns in the 1st matrix is equal to the number of rows in the 2nd.

4. *True or False?* $\begin{bmatrix} 5 & 3 \\ -3 & 5 \end{bmatrix}\begin{bmatrix} 2 & 0 \\ 0 & 1 \end{bmatrix} = \begin{bmatrix} 2 & 0 \\ 0 & 1 \end{bmatrix}\begin{bmatrix} 5 & 3 \\ -3 & 5 \end{bmatrix}$ Explain.

False; $\begin{bmatrix} 10 & 3 \\ -6 & 5 \end{bmatrix} \neq \begin{bmatrix} 10 & 6 \\ -3 & 5 \end{bmatrix}$.

Independent Practice

In Exercises 5–12, state whether the matrix product AB is defined. If so, state the order of the product. See below.

5. Order of A: 2×3, Order of B: 3×1

6. Order of A: 1×4, Order of B: 4×5

7. Order of A: 3×2, Order of B: 3×5

8. Order of A: 5×5, Order of B: 5×4

9. Order of A: 5×4, Order of B: 4×2

10. Order of A: 3×2, Order of B: 3×4

11. Order of A: 8×3, Order of B: 3×4

12. Order of A: 5×1, Order of B: 1×3

In Exercises 13–22, find the product. If it is not defined, state the reason. 15.–17. See Additional Answers.

13. $\begin{bmatrix} 1 & 2 \\ -4 & 3 \end{bmatrix}\begin{bmatrix} 4 & 1 \\ 0 & -1 \end{bmatrix}$ $\begin{bmatrix} 4 & -1 \\ -16 & -7 \end{bmatrix}$

14. $\begin{bmatrix} 6 & 1 \\ 0 & 2 \end{bmatrix}\begin{bmatrix} 1 & 1 \\ -5 & 3 \end{bmatrix}$ $\begin{bmatrix} 1 & 9 \\ -10 & 6 \end{bmatrix}$

15. $\begin{bmatrix} 7 & 1 & 1 \\ 0 & 2 & 6 \\ 3 & -2 & 1 \end{bmatrix}\begin{bmatrix} 0 & 1 & 4 \\ 5 & -2 & 1 \\ -3 & 1 & -8 \end{bmatrix}$

16. $\begin{bmatrix} 3 & 0 & 3 \\ 1 & 2 & 1 \\ 2 & -1 & 0 \end{bmatrix}\begin{bmatrix} 1 & 0 & 0 \\ 4 & 1 & 1 \\ -5 & 0 & 2 \end{bmatrix}$

17. $\begin{bmatrix} 2 & 8 & 1 \\ 0 & 5 & -2 \end{bmatrix}\begin{bmatrix} 0 & 1 & -1 \\ 8 & 2 & 4 \end{bmatrix}$

18. $\begin{bmatrix} 5 & -2 \\ 0 & 2 \\ 1 & 1 \end{bmatrix}\begin{bmatrix} 3 & 0 \\ 6 & 2 \end{bmatrix}$ $\begin{bmatrix} 3 & -4 \\ 12 & 4 \\ 9 & 2 \end{bmatrix}$

19. $\begin{bmatrix} -3 & 2 & -5 \\ 4 & -6 & 3 \end{bmatrix}\begin{bmatrix} 4 \\ 1 \\ 0 \end{bmatrix}$ $\begin{bmatrix} -10 \\ 10 \end{bmatrix}$

20. $\begin{bmatrix} 0 & 1 & 0 \\ 6 & -3 & 1 \\ -1 & 4 & -2 \end{bmatrix}\begin{bmatrix} 6 & 7 & 1 \\ 2 & 10 & 5 \\ 1 & -10 & 9 \end{bmatrix}$

21. $\begin{bmatrix} 5 \\ 0 \\ 1 \end{bmatrix}\begin{bmatrix} 4 & 2 & -10 \end{bmatrix}$ $\begin{bmatrix} 20 & 10 & -50 \\ 0 & 0 & 0 \\ 4 & 2 & -10 \end{bmatrix}$

22. $\begin{bmatrix} 3 & 1 & 1 \end{bmatrix}\begin{bmatrix} 2 \\ -6 \\ 0 \end{bmatrix}$ $[0]$

20. $\begin{bmatrix} 2 & 10 & 5 \\ 31 & 2 & 0 \\ 0 & 53 & 1 \end{bmatrix}$

5. It is, 2×1 **6.** It is, 1×5 **7.** It is not. **8.** It is, 5×4 **9.** It is, 5×2 **10.** It is not. **11.** It is, 8×4 **12.** It is, 5×3

3. Why is there a left and right Distributive Property of matrix multiplication over addition?

Multiplication of matrices is not commutative. See Exercises 44–47.

Communicating about ALGEBRA

EXTEND *Communicating*
Use numerical examples to verify the Properties of Matrix Multiplication. That is, select matrices to demonstrate each of the properties shown on page 181.

How can the properties be shown to hold for any real-valued matrix of arbitrary order?

Use general matrix entries a_{ij}.

EXERCISE Notes

ASSIGNMENT GUIDE

Basic/Average: Ex. 1–4, 5–12, 13–31 odd, 34–43

Above Average: Ex. 1–4, 5–12, 13–33 odd, 35–46

Advanced: Ex. 1–4, 5–33 odd, 38–47

Selected Answers
Exercises 1–4, 5–43 odd

Use **Mixed Review** as needed.

More Difficult Exercises
Exercises 32, 33

Guided Practice

EXTEND Ex. 1–4 Have students generate examples illustrating why both a left and a right Distributive Property are needed.

Independent Practice

EXTEND Ex. 5–12 For each exercise, ask students if the matrix product *BA* exists.
The matrix product *BA* does not exist for any of the exercises.

▶ **Ex. 29–30** Students can save time by adding first.

▶ **Ex. 31–33** Encourage students to label the matrix entries as well as the final answers. For example, in **Exercise 32** students could show the following.

Why eat healthy?

To look better	60%	40%
To live longer	36%	53%
Don't eat healthy	4%	7%

$\cdot$

Women	8900
Men	7000

$=$

To look better	8140
To live longer	6914
Don't eat healthy	846

Answers

27. $\begin{bmatrix} -18 & -24 \\ -12 & -6 \end{bmatrix}$ 28. $\begin{bmatrix} -4 & 8 \\ 6 & 16 \end{bmatrix}$

31. $\begin{bmatrix} 3 \\ 1.25 \\ 4 \end{bmatrix} [6 \quad 4 \quad 3] = [35]$,
$\$35 \div 2 = \$17.50$

33. $\begin{bmatrix} 0.166 & 0.372 & 0.136 \\ 0.087 & 0.189 & 0.017 \\ 0.172 & 0.005 & 0.026 \\ 0.106 & 0.013 & 0.436 \end{bmatrix} \cdot \begin{bmatrix} 535{,}842 \\ 475{,}533 \\ 480{,}609 \end{bmatrix} =$

331,210.9	China
144,664.3	India
107,038.3	S.U.
272,526.7	U.S.

In Exercises 23 and 24, solve for *x* and *y*. 23. 2, −8 24. 3, 35

23. $\begin{bmatrix} -2 & 1 & 2 \\ 3 & 2 & 4 \\ 0 & -2 & 4 \end{bmatrix} \begin{bmatrix} 1 & 3 \\ x & 2 \\ 3 & -1 \end{bmatrix} = \begin{bmatrix} 6 & -6 \\ 19 & 9 \\ 8 & y \end{bmatrix}$

24. $\begin{bmatrix} 4 & 1 & 3 \\ -2 & x & 1 \end{bmatrix} \begin{bmatrix} 9 & -2 \\ 2 & 1 \\ -1 & 4 \end{bmatrix} = \begin{bmatrix} y & 5 \\ -13 & 11 \end{bmatrix}$

In Exercises 25–30, simplify the expression.

25. $2 \begin{bmatrix} 4 & 2 \\ 6 & -3 \end{bmatrix} \begin{bmatrix} 1 & -4 \\ 0 & 2 \end{bmatrix}$ $\begin{bmatrix} 8 & -24 \\ 12 & -60 \end{bmatrix}$

26. $-3 \begin{bmatrix} -5 & 3 \\ 1 & 2 \end{bmatrix} \begin{bmatrix} 0 & -3 \\ 2 & 4 \end{bmatrix}$ $\begin{bmatrix} -18 & -81 \\ -12 & -15 \end{bmatrix}$

27. $\begin{bmatrix} 6 & -4 \\ 0 & -2 \end{bmatrix} \begin{bmatrix} 2 & -3 \\ 3 & 1 \end{bmatrix} + \begin{bmatrix} 6 & -4 \\ 0 & -2 \end{bmatrix} \begin{bmatrix} -1 & 1 \\ 3 & 2 \end{bmatrix}$ See margin.

28. $\begin{bmatrix} 2 & -1 \\ 4 & 5 \end{bmatrix} \begin{bmatrix} 2 & 0 \\ -3 & 1 \end{bmatrix} + \begin{bmatrix} 2 & -1 \\ 4 & 5 \end{bmatrix} \begin{bmatrix} -3 & 4 \\ 5 & -1 \end{bmatrix}$ See margin.

29. $\begin{bmatrix} 1 & 3 & 4 \\ 0 & -3 & -5 \\ 2 & 5 & 1 \end{bmatrix} \left(\begin{bmatrix} 2 & 4 & 0 \\ -1 & 3 & 2 \\ 2 & -2 & 1 \end{bmatrix} + \begin{bmatrix} -3 & 1 & 9 \\ 8 & 3 & 4 \\ 4 & -3 & 0 \end{bmatrix} \right)$ $\begin{bmatrix} 44 & 3 & 31 \\ -51 & 7 & -23 \\ 39 & 35 & 49 \end{bmatrix}$

30. $\left(\begin{bmatrix} 4 & 2 & 1 \\ 1 & -1 & 3 \\ 0 & -3 & 2 \end{bmatrix} + \begin{bmatrix} -1 & 3 & 4 \\ 0 & -2 & 3 \\ 4 & 1 & 2 \end{bmatrix} \right) \begin{bmatrix} 0 & -5 & 4 \\ 1 & 3 & 9 \\ -3 & 1 & 3 \end{bmatrix}$ $\begin{bmatrix} -10 & 5 & 72 \\ -21 & -8 & -5 \\ -14 & -22 & 10 \end{bmatrix}$

31. *Roses, Carnations, and Lilies* You and your cousin decide to make two bouquets: one for your mother and one for your grandparents. The number of roses, carnations, and lilies that will be in your mother's and your grandparents' bouquets is shown in the matrix. Each rose costs \$3, each carnation costs \$1.25, and each lily costs \$4. Your cousin agrees to share the cost of your grandparents' bouquet. Show how matrix multiplication can be used to determine your cousin's share of the cost. See margin.

Number of Flowers	Roses	Carnations	Lilies
Mother's Bouquet	4	5	3
Grandparents' Bouquet	6	4	3

✪ **32.** *Eating Healthy* The matrix shows the results of a survey taken by *USA Today.* In the survey, people were asked why they tried to eat a healthful diet. The survey results show the percent of answers by 8900 women and 7000 men. Show how matrix multiplication can be used to find how many people answered in each of the three categories. See margin.

Why Eat Healthy?	Women	Men
To look better	0.6	0.4
To live longer	0.36	0.53
I don't eat healthy	0.04	0.07

33. *World Grain Production* The percents of the total 1990 world production of wheat, rice, and corn grown by the four countries that grow the most grain are given in the matrix on page 172. The total 1990 world production (in thousands of metric tons) of wheat, rice, and corn was 535,842, 475,533, and 480,609, respectively. Show how matrix multiplication can be used to find how many metric tons of the three grains were grown in each of the four countries. *(Source: United Nations)* See margin.

Integrated Review

In Exercises 34–37, state the order of the matrix. 34. 4 × 3 35. 3 × 3 36. 2 × 4 37. 2 × 5

34. $\begin{bmatrix} 5 & 6 & 1 \\ 4 & 2 & -4 \\ -1 & 5 & 0 \\ 0 & -2 & 3 \end{bmatrix}$ **35.** $\begin{bmatrix} 6 & -1 & 6 \\ 9 & 0 & 5 \\ -2 & 4 & 3 \end{bmatrix}$ **36.** $\begin{bmatrix} 1 & -4 & 3 & 4 \\ 9 & 6 & -3 & 7 \end{bmatrix}$ **37.** $\begin{bmatrix} 4 & -1 & 5 & -4 & 0 \\ 3 & 0 & -6 & 1 & 8 \end{bmatrix}$

In Exercises 38–43, state whether the operation is possible. If the operation is not possible, state the reason. If the operation is possible, state the order of the resulting matrix. See Additional Answers.

38. Add a 2 × 3 matrix to a 3 × 2 matrix.

39. Subtract a 4 × 4 matrix from a 4 × 2 matrix.

40. Add a 1 × 3 matrix to a 3 × 2 matrix.

41. Subtract a 2 × 3 matrix from a 2 × 3 matrix.

42. Multiply a 6 × 4 matrix by a 4 × 6 matrix.

43. Multiply a 1 × 3 matrix by a 3 × 4 matrix.

Exploration and Extension

In Exercises 44–47, use the given matrices and the scalar $c = 4$ to illustrate the property. See Additional Answers.

$$A = \begin{bmatrix} 1 & 3 \\ 9 & -1 \end{bmatrix} \quad B = \begin{bmatrix} 6 & 3 \\ 4 & 2 \end{bmatrix} \quad C = \begin{bmatrix} 2 & 0 \\ 4 & -2 \end{bmatrix}$$

44. *Associative Property:*
$A(BC) = (AB)C$

45. *Left Distributive Property:*
$A(B + C) = AB + AC$

46. *Right Distributive Property:*
$(A + B)C = AC + BC$

47. *Property of Scalar Multiplication:*
$c(AB) = (cA)B = A(cB)$

Ex. 44–47 TECHNOLOGY Matrix multiplication can be computed efficiently using a graphics calculator. This will be explored later in the chapter on page 194.

Enrichment Activities

These activities require students to go beyond lesson goals.

Powers of matrices, such as A^2 or A^3, are only defined for square matrices.

For example, if $A = \begin{bmatrix} 2 & 3 \\ -1 & 4 \end{bmatrix}$ then

$$A^2 = \begin{bmatrix} 2 & 3 \\ -1 & 4 \end{bmatrix} \cdot \begin{bmatrix} 2 & 3 \\ -1 & 4 \end{bmatrix} = \begin{bmatrix} 1 & 18 \\ -6 & 13 \end{bmatrix}$$

Let $A = \begin{bmatrix} 4 & 8 \\ -3 & 19 \end{bmatrix}$,

$B = \begin{bmatrix} 6 & 3 & 1 \\ -8 & 16 & 0 \\ -1 & 4 & 3 \end{bmatrix}$, and

$C = \begin{bmatrix} 6 & 4 & 0 \\ -1 & 3 & 9 \end{bmatrix}$

1. Find A^3. $\begin{bmatrix} -584 & 3432 \\ -1287 & 5851 \end{bmatrix}$

2. Find B^2. $\begin{bmatrix} 11 & 70 & 9 \\ -176 & 232 & -8 \\ -41 & 73 & 8 \end{bmatrix}$

3. Find C^2. It is undefined.

4. If $A = \begin{bmatrix} a & b \\ c & d \end{bmatrix}$, then $A^2 = ?$

$\begin{bmatrix} a^2 + bc & ab + bd \\ ca + dc & cb + d^2 \end{bmatrix}$

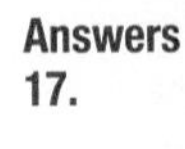

Answers

17.

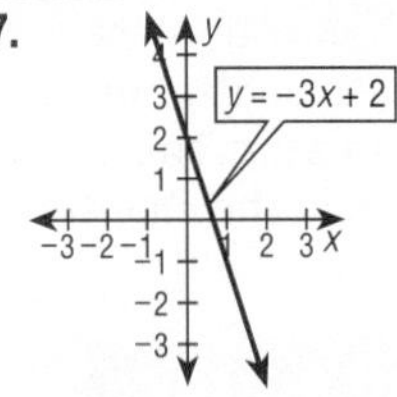

18.

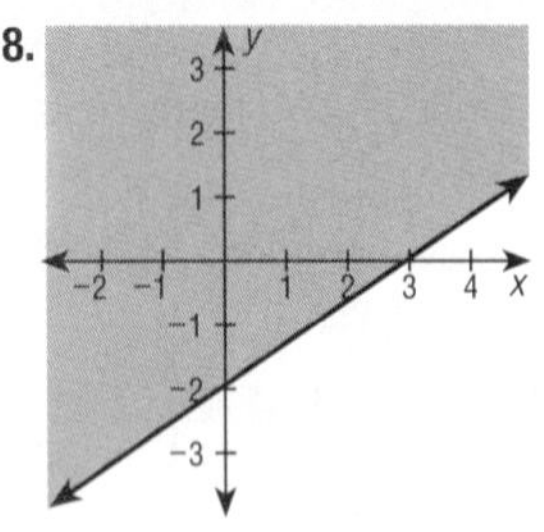

Milestones

Answer

quantum mechanics: the algebra of matrices is useful in describing quantum mechanics phenomena.

Language Arts: *Critical Reading*

Use your readings to answer these questions.

1. Arthur Cayley collaborated with another lawyer-mathematician, James Joseph Sylvester. Find out what they collaborated on.
   the theory of algebraic invariants
2. When did the University of Cambridge begin accepting women? 1873

Mixed REVIEW

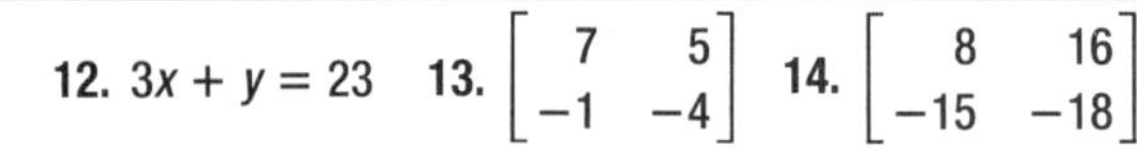

6. -20 **12.** $3x + y = 23$ **13.** $\begin{bmatrix} 7 & 5 \\ -1 & -4 \end{bmatrix}$ **14.** $\begin{bmatrix} 8 & 16 \\ -15 & -18 \end{bmatrix}$

1. Simplify $-2(3) + (-4)(-6) + 3(-5)$. **(1.1)** 3
2. Simplify $-3\left(\frac{2}{3}\right) + 4\left(\frac{1}{3}\right) - 2\left(-\frac{4}{3}\right)$. **(1.1)** 2
3. 39 is 150% of what number? 26
4. Write 3.24×10^{-3} in decimal form. 0.00324
5. Simplify $\frac{4}{5} - \frac{2}{3} + 4$. $4\frac{2}{15}$
6. Evaluate $-3x^2 + 4x$ when $x = -2$. **(1.2)**
7. What is the slope of the line $-4x + 6y = 5$? **(2.2)** $\frac{2}{3}$
8. Find the intercepts of the line $3y - x = 4$. **(2.3)** $\left(0, \frac{4}{3}\right)$, $(-4, 0)$
9. What is the reciprocal of $-\frac{3}{5}$? **(1.1)** $-\frac{5}{3}$
10. What is the opposite of $-\frac{3}{5}$? **(1.1)** $\frac{3}{5}$
11. Find the slope of the line through $(0, 2)$ and $(-3, 4)$. **(2.2)** $-\frac{2}{3}$
12. Find an equation of the line that passes through $(7, 2)$ with a slope of -3. **(2.4)**
13. Add: $\begin{bmatrix} 2 & -1 \\ -3 & 0 \end{bmatrix} + \begin{bmatrix} 5 & 6 \\ 2 & -4 \end{bmatrix}$. **(4.1)**
14. Multiply: $\begin{bmatrix} 2 & -1 \\ -3 & 0 \end{bmatrix}\begin{bmatrix} 5 & 6 \\ 2 & -4 \end{bmatrix}$. **(4.2)**
15. Solve $|-x + 4| < 3$. **(1.7)** $1 < x < 7$
16. Solve $-3(5 + x) = 2(9 - x)$. **(1.3)** -33
17. Sketch the graph of $y = -3x + 2$. **(2.3)**
18. Sketch the graph of $2x - 3y \le 6$. **(2.5)**

17. & 18. See margin.

19. Solve: $\begin{cases} 2x - 3y = 1 \\ -x - 7y = -11 \end{cases}$. **(3.2)** $\left(\frac{40}{17}, \frac{21}{17}\right)$
20. Solve the system: $\begin{cases} x + y = 34 \\ x - y = 2 \end{cases}$. **(3.2)** (18, 16)

Milestones ARTHUR CAYLEY

1820 1840 1860 1880 1900

Arthur Cayley born — Elected to chair at Cambridge — "The Collected Mathematical Papers" began publication

The invention of matrices is credited to Arthur Cayley. Born in England in 1821, Cayley grew up doing complex computations for amusement. He attended the University of Cambridge, studying languages and mathematics. Upon graduation in 1842, he received a 3-year appointment to Trinity College at Cambridge. He went on to study law and was admitted to the bar in 1849. While practicing law, Cayley continued his research in mathematics. Cayley returned to Cambridge in 1863 when he was elected to the new Sadlerian Chair of Pure Mathematics.

Cayley investigated many topics in modern mathematics. He developed matrix algebra, worked on group theory, and studied hyperspace. He was influential in getting Cambridge to accept women students. A prolific writer, Cayley authored more than 900 papers and notes that were published in 14 volumes from 1889 through 1898 in *The Collected Mathematical Papers.*

Arthur Cayley introduced matrices and their operations in an 1858 memoir.

Find out what mathematical field Werner Heisenberg developed. How did Cayley's work help him? See margin.

4.3 Determinants

What you should learn:

Goal 1 How to evaluate the determinant of a 2 × 2 or a 3 × 3 matrix

Goal 2 How to use determinants to solve real-life problems

Why you should learn it:

You can use the determinant of a matrix to answer real-life problems, such as finding the area of a triangle in a coordinate plane.

Goal 1 Evaluating Determinants

Recall from Lesson 4.1 that a *square* matrix has the same number of rows and columns. Associated with each square matrix is a real number called its **determinant.** In this lesson, you will learn to evaluate the determinant of a 2 × 2 matrix and a 3 × 3 matrix.

Determinant of a 2 × 2 Matrix

The **determinant** of the matrix $\begin{bmatrix} a & b \\ c & d \end{bmatrix}$ is $\begin{vmatrix} a & b \\ c & d \end{vmatrix} = ad - cb$.

The following diagram shows that the determinant of a 2 × 2 matrix is equal to the difference of the products of the entries on the two diagonals.

$$\begin{vmatrix} a & b \\ c & d \end{vmatrix} = ad - cb.$$

Example 1 *Evaluating the Determinant of a 2 × 2 Matrix*

Evaluate the determinant of each matrix.

a. $\begin{bmatrix} 2 & -3 \\ 1 & 4 \end{bmatrix}$ **b.** $\begin{bmatrix} -1 & 2 \\ 2 & -4 \end{bmatrix}$ **c.** $\begin{bmatrix} 1 & 3 \\ 2 & 5 \end{bmatrix}$

Solution

a. $\begin{vmatrix} 2 & -3 \\ 1 & 4 \end{vmatrix} = 2(4) - 1(-3) = 8 + 3 = 11$

b. $\begin{vmatrix} -1 & 2 \\ 2 & -4 \end{vmatrix} = (-1)(-4) - 2(2) = 4 - 4 = 0$

c. $\begin{vmatrix} 1 & 3 \\ 2 & 5 \end{vmatrix} = 1(5) - 2(3) = -1$ ■

Although vertical bars are used to denote both determinants and absolute value, notice in Example 1 that the determinant of a matrix can be positive, zero, or negative.

Problem of the Day

Plastic pipe is shipped in hexagonal bundles, as shown. How many pipes would be in a bundle with 9 on a side? 217

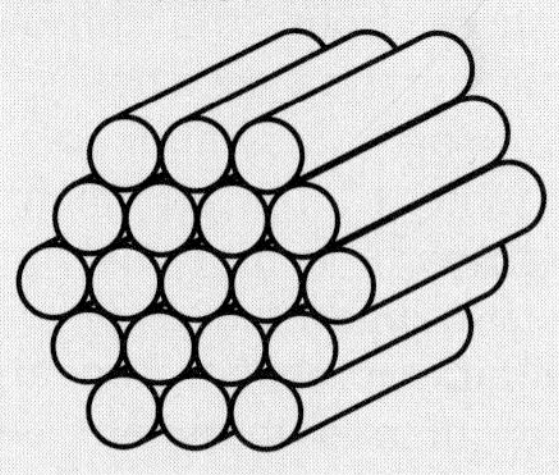

ORGANIZER

Warm-Up Exercises

1. Compute the following.
   - **a.** $2(3) - 8(-4)$ 38
   - **b.** $-5(6) + 3(9)$ −3
   - **c.** $7(-3) - (-2)(-5)$ −31
   - **d.** $2(6) - 3(4)$ 0
2. Give the order of each matrix.
   - **a.** $\begin{bmatrix} 2 & -1 & 0 & -5 & 4 & 3 \\ 0 & 2 & -1 & 3 & 3 & 1 \end{bmatrix}$ 2 × 6
   - **b.** $\begin{bmatrix} 2 & -1 & 0 \\ 0 & 8 & -1 \\ -6 & 2 & 3 \\ 1 & 3 & 1 \end{bmatrix}$ 4 × 3
   - **c.** $\begin{bmatrix} 2 & -1 & 0 \\ 0 & 8 & -1 \\ -6 & 2 & 3 \end{bmatrix}$ 3 × 3

Lesson Resources

Teaching Tools
- Transparencies: 2, 3
- Problem of the Day: 4.3
- Warm-up Exercises: 4.3
- Answer Masters: 4.3

Extra Practice: 4.3
Color Transparencies: 26, 27

The determinant of a 3×3 matrix is more difficult to evaluate than the determinant of a 2×2 matrix. Two methods are common. The first is **expansion by minors,** which expresses the determinant of a 3×3 matrix in terms of three 2×2 determinants.

Determinant of a 3 × 3 Matrix (Expansion by Minors)

The determinant of the matrix $\begin{bmatrix} a & b & c \\ d & e & f \\ g & h & i \end{bmatrix}$ is

$$\begin{vmatrix} a & b & c \\ d & e & f \\ g & h & i \end{vmatrix} = a\begin{vmatrix} a & b & c \\ d & e & f \\ g & h & i \end{vmatrix} - b\begin{vmatrix} a & b & c \\ d & e & f \\ g & h & i \end{vmatrix} + c\begin{vmatrix} a & b & c \\ d & e & f \\ g & h & i \end{vmatrix}$$

$$= a\begin{vmatrix} e & f \\ h & i \end{vmatrix} - b\begin{vmatrix} d & f \\ g & i \end{vmatrix} + c\begin{vmatrix} d & e \\ g & h \end{vmatrix}.$$

This pattern shows expansion by minors *along the first row.* Patterns for expanding along other rows or columns are similar. The *sign pattern* for expanding along the first or third row or column is "+ − +." The sign pattern for the second row or column is "− + −."

Example 2 *Evaluating the Determinant of a 3 × 3 Matrix*

Evaluate the determinant of $\begin{bmatrix} -1 & 1 & 2 \\ 0 & 2 & 3 \\ 3 & 4 & 2 \end{bmatrix}$.

Solution Expanding by minors along the *first* row, you obtain the following.

$$\begin{vmatrix} -1 & 1 & 2 \\ 0 & 2 & 3 \\ 3 & 4 & 2 \end{vmatrix} = (-1)\begin{vmatrix} 2 & 3 \\ 4 & 2 \end{vmatrix} - (1)\begin{vmatrix} 0 & 3 \\ 3 & 2 \end{vmatrix} + (2)\begin{vmatrix} 0 & 2 \\ 3 & 4 \end{vmatrix}$$

$$= (-1)(4 - 12) - (1)(0 - 9) + (2)(0 - 6)$$

$$= 8 + 9 - 12, \text{ or } 5$$ ■

In Example 2, expansion by minors along the *second* row would yield

$$\begin{vmatrix} -1 & 1 & 2 \\ 0 & 2 & 3 \\ 3 & 4 & 2 \end{vmatrix} = -(0)\begin{vmatrix} 1 & 2 \\ 4 & 2 \end{vmatrix} + (2)\begin{vmatrix} -1 & 2 \\ 3 & 2 \end{vmatrix} - (3)\begin{vmatrix} -1 & 1 \\ 3 & 4 \end{vmatrix}$$

$$= 0 - 16 + 21 = 5.$$

LESSON Notes

Example 1

The symbol used to denote a determinant can be distinguished from the absolute value notation by what is "inside" the | | symbol, or the "argument." The argument of an absolute value expression is a real-valued number or an algebraic expression. The argument of a determinant is an array of numbers or an array of expressions.

Example 2

Observe that the determinant is a fixed value for a given matrix. Ask students to verify that the determinant of the matrix is 5 if the expansion by minors is along the second column. The expansion is

$$-(1)\begin{vmatrix} 0 & 3 \\ 3 & 2 \end{vmatrix} + (2)\begin{vmatrix} -1 & 2 \\ 3 & 2 \end{vmatrix} - (4)\begin{vmatrix} -1 & 2 \\ 0 & 3 \end{vmatrix}$$

$$= -(1)(0 - 9) + (2)(-2 - 6) - (4)(-3 - 0)$$

$$= 9 - 16 + 12$$

$$= 5$$

Ask students why you might want to evaluate a determinant by expansion by minors along a row (or column) other than the first. (The expansion might be simpler if another row (or column) had a lot of zeros.)

Example 3

Observe that the diagonals method is equivalent to the expansion by minors along the first row. You may wish to demonstrate this (or ask students to derive the relationship; see Communicating about Algebra on page 190).

The second method for evaluating the determinant of a 3×3 matrix is called the **diagonals method.** To apply this method, it is easier to copy the first and second columns of the matrix to form fourth and fifth columns. The determinant is then obtained by adding the products of three diagonals and subtracting the products of three diagonals, as shown in the following diagram.

Second, subtract these products.

$$\begin{vmatrix} a & b & c \\ d & e & f \\ g & h & i \end{vmatrix} \begin{matrix} a & b \\ d & e \\ g & h \end{matrix} = aei + bfg + cdh - gec - hfa - idb$$

First, add these products.

Example 3 Evaluating the Determinant of a 3 × 3 Matrix

Evaluate the determinant of $\begin{bmatrix} 0 & 2 & 1 \\ 3 & -1 & 2 \\ 4 & -4 & 1 \end{bmatrix}$.

Solution

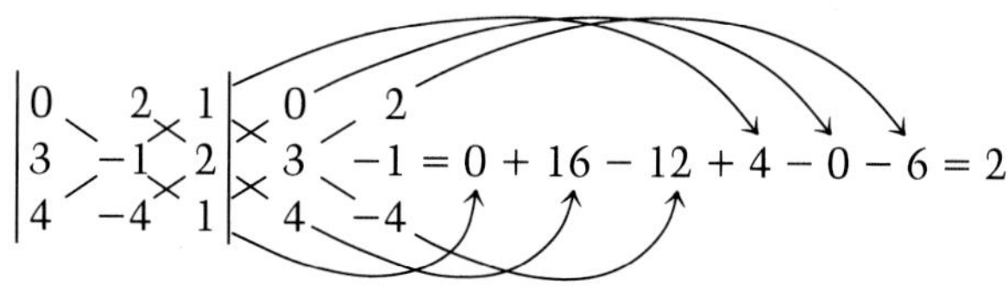

$$\begin{vmatrix} 0 & 2 & 1 \\ 3 & -1 & 2 \\ 4 & -4 & 1 \end{vmatrix} \begin{matrix} 0 & 2 \\ 3 & -1 \\ 4 & -4 \end{matrix} = 0 + 16 - 12 + 4 - 0 - 6 = 2$$

■

Goal 2 Using Determinants in Real-Life Situations

A determinant can be used to find the area of a triangle whose vertices are points in a coordinate plane.

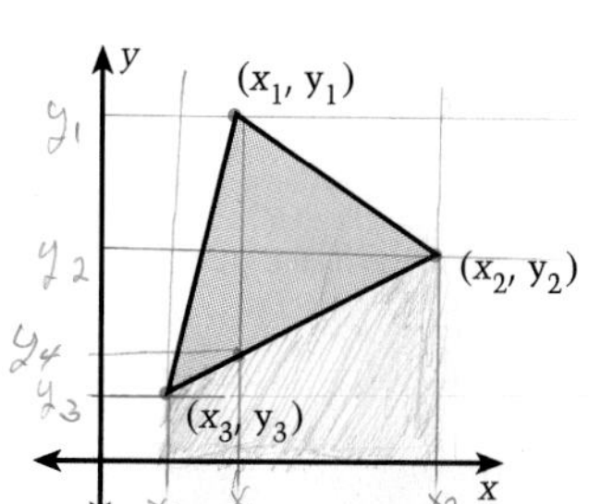

Area of a Triangle

The area of a triangle with vertices (x_1, y_1), (x_2, y_2), *and* (x_3, y_3) is given by

$$\text{area} = \pm\frac{1}{2}\begin{vmatrix} x_1 & y_1 & 1 \\ x_2 & y_2 & 1 \\ x_3 & y_3 & 1 \end{vmatrix}$$

where the symbol ($\pm$) indicates that the appropriate sign should be chosen to yield a positive value.

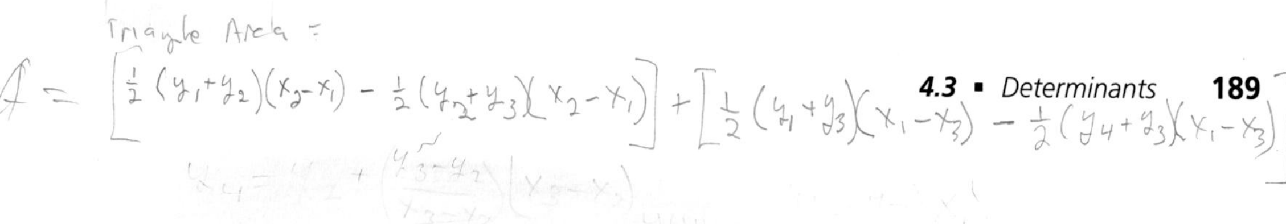

Example 4

Suppose a coordinate system were superimposed on the map at vertex *B*. Find the area of the triangular region using the resulting coordinate points for vertices *A*, *B*, and *C*. The vertices of the triangular regions are $(x_1, y_1) = (-10, 25)$, $(x_2, y_2) = (0, 0)$, $(x_3, y_3) = (18, 5)$. The area in square miles is

$$0.5\begin{vmatrix} -10 & 25 & 1 \\ 0 & 0 & 1 \\ 18 & 5 & 1 \end{vmatrix}$$

$= 0.5[-10(-5) - 25(-18) + 0]$
$= 0.5(500) = 250.$

Extra Examples

Here are additional examples similar to **Examples 1–3.**

1. Evaluate the determinant of each 2×2 matrix.
   a. $\begin{bmatrix} 3 & -3 \\ 2 & 5 \end{bmatrix}$ 21
   b. $\begin{bmatrix} -1 & 2 \\ 6 & 4 \end{bmatrix}$ −16
   c. $\begin{bmatrix} -7 & 9 \\ 3 & 0 \end{bmatrix}$ −27
   d. $\begin{bmatrix} 8 & 8 \\ -12 & 2 \end{bmatrix}$ 112
2. Evaluate the determinant of each 3×3 matrix using expansion by minors.
   a. $\begin{bmatrix} 3 & 4 & -3 \\ 2 & 0 & 5 \\ 0 & -2 & -3 \end{bmatrix}$ 66
   b. $\begin{bmatrix} -1 & 8 & 2 \\ 6 & 4 & 6 \\ -1 & 0 & 3 \end{bmatrix}$ −196
   c. $\begin{bmatrix} -7 & 1 & 9 \\ 3 & 0 & 1 \\ 0 & 5 & 4 \end{bmatrix}$ 158
   d. $\begin{bmatrix} 0 & 1 & 2 \\ -1 & 3 & 2 \\ 2 & 8 & 3 \end{bmatrix}$ −21
3. Evaluate the determinant of each 3×3 matrix in Exercise 2 above using the diagonals method.

Check Understanding

1. What is a square matrix?
   Any matrix of order $n \times n$.
2. Name the methods that can be used to evaluate determinants.
   Expansion of minors, diagonal method
3. Can determinants be used to find the area of triangular regions? Can they be used to find the area of quadrilaterals?
   Yes; see Goal 2 and Example 4; yes (See the Enrichment Activity that follows on page 193.)

***Communicating* about ALGEBRA**

Ask students to show that the diagonals method of evaluating determinants is equivalent to the expansion by minors along the first row.

Answers

A. $(2)\begin{vmatrix} 0 & 2 \\ 3 & 4 \end{vmatrix} - (3)\begin{vmatrix} -1 & 1 \\ 3 & 4 \end{vmatrix} + (2)\begin{vmatrix} -1 & 1 \\ 0 & 2 \end{vmatrix}$

$= 2(-6) - 3(-7) + 2(-2) = 5,$

$-4 + 9 + 0 - 12 + 12 - 0 = 5$

B. 0; by expanding by minors along that row, each product in the expansion becomes zero

Gypsy moths are native to France. In 1869, however, some were accidentally brought to the United States. Since then, the gypsy moth population has increased greatly and has done great damage to forests, especially in the northeastern United States.

Real Life
Forestry

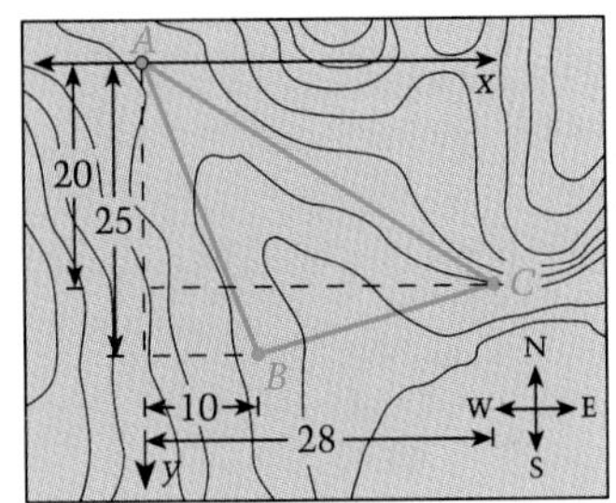

Example 4 ***Finding the Area of a Triangular Region***

A large region of forest has been infected with gypsy moths. The region is roughly triangular, as shown in the map at the left. From the northern vertex of the region, the distance to the other vertices is 20 miles south and 28 miles east (for vertex C), and 25 miles south and 10 miles east (for vertex B). Approximate the number of square miles in this region.

Solution Begin by superimposing a coordinate plane on the map. The vertices of the triangular region are $(x_1, y_1) = (0, 0)$, $(x_2, y_2) = (10, -25)$, and $(x_3, y_3) = (28, -20)$.

$$\text{Area} = \pm\frac{1}{2}\begin{vmatrix} x_1 & y_1 & 1 \\ x_2 & y_2 & 1 \\ x_3 & y_3 & 1 \end{vmatrix}$$

$$= \pm\frac{1}{2}\begin{vmatrix} 0 & 0 & 1 \\ 10 & -25 & 1 \\ 28 & -20 & 1 \end{vmatrix}$$

$$= \pm\frac{1}{2}\left(0\begin{vmatrix} -25 & 1 \\ -20 & 1 \end{vmatrix} - 0\begin{vmatrix} 10 & 1 \\ 28 & 1 \end{vmatrix} + 1\begin{vmatrix} 10 & -25 \\ 28 & -20 \end{vmatrix}\right)$$

$$= \pm\frac{1}{2}(0 - 0 + (1)(-200 + 700))$$

$$= 250 \text{ square miles.}$$

■

Communicating about ALGEBRA

SHARING IDEAS about the Lesson See margin.

A. Evaluate the determinant of the matrix in Example 2 by expanding by minors along the third column. Then evaluate the determinant using the diagonals method.

B. What is the value of the determinant of a matrix that has an entire row of zeros? Explain.

EXERCISES

Guided Practice

CRITICAL THINKING about the Lesson

1. Evaluate the determinant of $\begin{bmatrix} 0 & 1 \\ 6 & 2 \end{bmatrix}$. −6

2. Can two different matrices have the same determinant? If so, give an example. See Additional Answers.

3. Evaluate the determinant of the matrix in two ways: $\begin{bmatrix} 1 & -2 & 4 \\ 0 & 0 & -3 \\ 2 & 1 & -2 \end{bmatrix}$

   - Expand by minors along the first row.
   - Expand by minors along the second row.

   Which is easier? Why? See margin.

4. Use the diagonals method to evaluate. $\begin{vmatrix} 5 & -2 & -4 \\ 0 & 1 & -2 \\ 3 & -2 & 4 \end{vmatrix}$

   $20 + 12 + 0 - (-12) - (20) - 0 = 24$

Independent Practice

In Exercises 5–16, evaluate the determinant of the matrix.

5. $\begin{bmatrix} -4 & 2 \\ -6 & 3 \end{bmatrix}$ 0

6. $\begin{bmatrix} 9 & 0 \\ 4 & -3 \end{bmatrix}$ −27

7. $\begin{bmatrix} 8 & 3 \\ -1 & 2 \end{bmatrix}$ 19

8. $\begin{bmatrix} -6 & 11 \\ -6 & 5 \end{bmatrix}$ 36

9. $\begin{bmatrix} 1 & 7 \\ 2 & 9 \end{bmatrix}$ −5

10. $\begin{bmatrix} 1 & 4 \\ 6 & 8 \end{bmatrix}$ −16

11. $\begin{bmatrix} -4 & 1 \\ 7 & -2 \end{bmatrix}$ 1

12. $\begin{bmatrix} 5 & 0 \\ -2 & 5 \end{bmatrix}$ 25

13. $\begin{bmatrix} 3 & 4 \\ 2 & -1 \end{bmatrix}$ −11

14. $\begin{bmatrix} 9 & -3 \\ 6 & -2 \end{bmatrix}$ 0

15. $\begin{bmatrix} 0 & -3 \\ 4 & 1 \end{bmatrix}$ 12

16. $\begin{bmatrix} 12 & 6 \\ 3 & 1 \end{bmatrix}$ −6

In Exercises 17–22, evaluate the determinant of the matrix. (Expand by minors along the indicated row or column.)

17. Expand along row 1. $\begin{bmatrix} 5 & 0 & 2 \\ 1 & 4 & -1 \\ 3 & 2 & 0 \end{bmatrix}$ −10

18. Expand along row 3. $\begin{bmatrix} 10 & 4 & 2 \\ -8 & 1 & -1 \\ 2 & 3 & 2 \end{bmatrix}$ 54

19. Expand along row 2. $\begin{bmatrix} 6 & 2 & 1 \\ 5 & 6 & -3 \\ -1 & 2 & 2 \end{bmatrix}$ 110

20. Expand along column 3. $\begin{bmatrix} 1 & 2 & 1 \\ -3 & 3 & -1 \\ 2 & 1 & 0 \end{bmatrix}$ −12

21. Expand along column 2. $\begin{bmatrix} 10 & 0 & 1 \\ 6 & 0 & 4 \\ -1 & 5 & 5 \end{bmatrix}$ −170

22. Expand along column 1. $\begin{bmatrix} 1 & -4 & 3 \\ 0 & 10 & -12 \\ -1 & 2 & -2 \end{bmatrix}$ −14

In Exercises 23–26, use the diagonals method to evaluate the determinant of the matrix.

23. $\begin{bmatrix} 12 & 4 & -6 \\ 2 & 1 & 3 \\ 1 & 0 & -2 \end{bmatrix}$ 10

24. $\begin{bmatrix} 5 & -8 & 4 \\ 4 & 2 & 1 \\ 1 & 1 & 2 \end{bmatrix}$ 79

25. $\begin{bmatrix} 0 & 3 & 2 \\ 10 & 14 & 2 \\ 1 & -1 & 6 \end{bmatrix}$ −222

26. $\begin{bmatrix} 1 & 20 & 1 \\ 0 & 5 & 1 \\ 1 & 15 & -2 \end{bmatrix}$ −10

EXERCISE Notes

ASSIGNMENT GUIDE

Basic/Average: Ex. 1–4, 7–29 odd, 33, 37, 38, 40, 41–46

Above Average: Ex. 1–4, 7–31 odd, 37, 39, 40, 43–49

Advanced: Ex. 1–4, 7–31 odd, 37, 39, 40, 43–49

Selected Answers
Exercises 1–4, 5–45 odd

More Difficult Exercises
Exercises 39, 40, 47–49

Guided Practice

Ex. 1–4 Ask students to discuss and explain the method they prefer for evaluating the determinant of a 3 × 3 matrix.

TECHNOLOGY Students with graphics calculators might be encouraged to use the determinant keystroke to evaluate the determinants.

Answer

3. (1) $\begin{vmatrix} 0 & -3 \\ 1 & -2 \end{vmatrix} - (-2)\begin{vmatrix} 0 & -3 \\ 2 & -2 \end{vmatrix} + 4\begin{vmatrix} 0 & 0 \\ 2 & 1 \end{vmatrix}$

$= (1)(3) + (2)(6) + (4)(0)$
$= 15,$

$-(0)\begin{vmatrix} -2 & 4 \\ 1 & -2 \end{vmatrix} + (0)\begin{vmatrix} 1 & 4 \\ 2 & -2 \end{vmatrix} - (-3)\begin{vmatrix} 1 & -2 \\ 2 & 1 \end{vmatrix}$

$= (0)(0) + (0)(-10) + (3)(5)$
$= 15;$

expanding by minors along the second row, two products in the expansion become zero.

Independent Practice

▶ Ex. 17–26 and 33–40
COOPERATIVE LEARNING
These exercises will take students a long time to complete independently. You might encourage students to work together in pairs or in small groups to cut down on work time.

▶ Ex. 41–42 Remind students that $|AB|$ refers to the determinant of AB and not the absolute value of AB.

▶ Ex. 45–46 These exercises provide practice questions similar to those found on college entrance tests.

▶ Ex. 47–49 CONNECTIONS These exercises connect concepts from Chapter 2 on writing the equation of a line to determinants. You may wish to reserve these exercises to use as a 10-minute in-class partner activity.

Answers

27. The determinant of a 3 × 3 matrix with a column of zeros is 0.

41. $AB = \begin{bmatrix} 72 & 4 \\ 15 & -5 \end{bmatrix}$, $|AB| = -420$; $|A| = 21, |B| = -20$, $|A|\,|B| = -420$

42. $AB = \begin{bmatrix} 31 & 26 \\ 5 & 10 \end{bmatrix}$, $|AB| = 180$; $|A| = -20, |B| = -9$, $|A|\,|B| = 180$

47. 0; the points lie on a straight line

48. If $\begin{vmatrix} x_1 & y_1 & 1 \\ x_2 & y_2 & 1 \\ x_3 & y_3 & 1 \end{vmatrix} = 0$, then the points (x_1, y_1), (x_2, y_2), and (x_3, y_3) are collinear.

49. The value of the determinant of a 3 × 3 matrix is equivalent to the area of a triangle whose three vertices are part of the matrix; if the determinant is zero, then the three vertices lie on a straight line.
$y = -2x + 9$

27. Write a 3 × 3 matrix that has a row of zeros. Then evaluate its determinant. State a generalization about the determinant of a 3 × 3 matrix that has a column of zeros. See margin.

28. Write a 3 × 3 matrix that has two rows that are the same. Then evaluate its determinant. State a generalization about the determinant of a 3 × 3 matrix that has two rows that are the same. The determinant of a 3 × 3 matrix that has two rows that are the same is 0.

In Exercises 29–32, show that the determinant of the matrix is zero. Exercises 27 and 28 indicate two quick ways to recognize 3 × 3 matrices whose determinants are zero. State some other ways. See Additional Answers.

29. $\begin{bmatrix} 1 & 2 & 3 \\ 2 & 4 & 6 \\ -1 & 0 & 8 \end{bmatrix}$ **30.** $\begin{bmatrix} 4 & -5 & 0 \\ 3 & -1 & 0 \\ -2 & 7 & 0 \end{bmatrix}$ **31.** $\begin{bmatrix} 5 & 2 & 2 \\ -3 & -4 & -4 \\ 8 & 3 & 3 \end{bmatrix}$ **32.** $\begin{bmatrix} -2 & -6 & 3 \\ 1 & 3 & -4 \\ 3 & 9 & 1 \end{bmatrix}$

In Exercises 33–35, use a determinant to find the area of the triangle.

33.
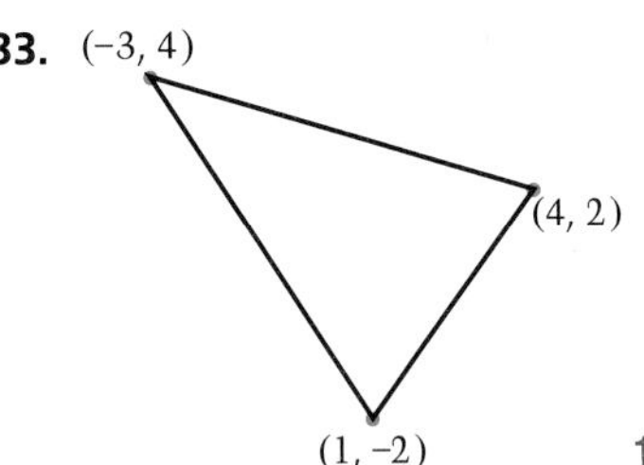

17

34.
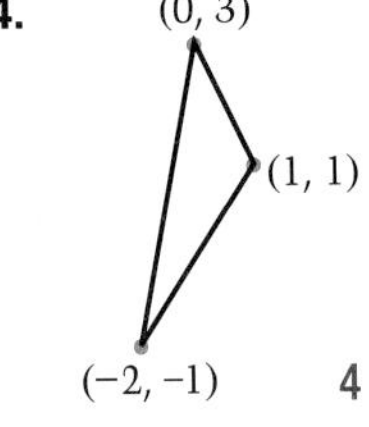

4

35.
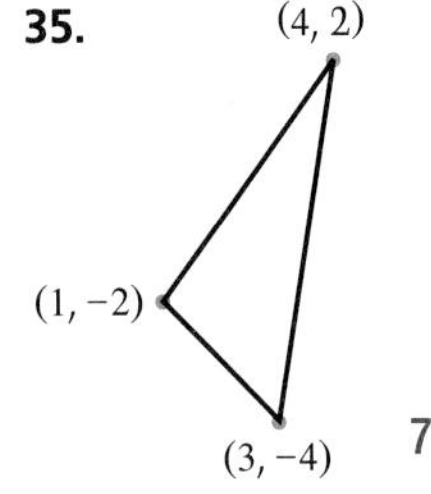

7

In Exercises 36–38, use two determinants to find the area of the region.

36.
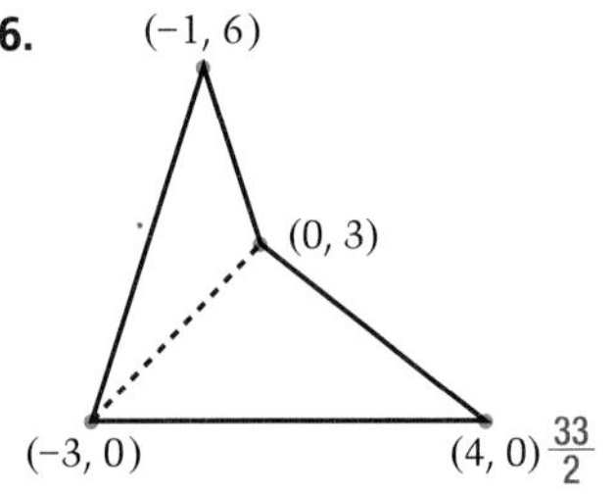

$\frac{33}{2}$

37.
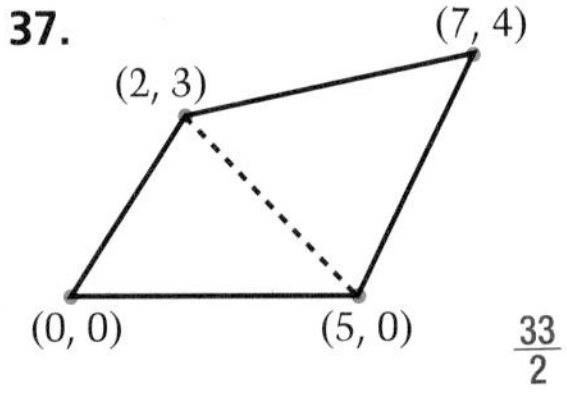

$\frac{33}{2}$

38.
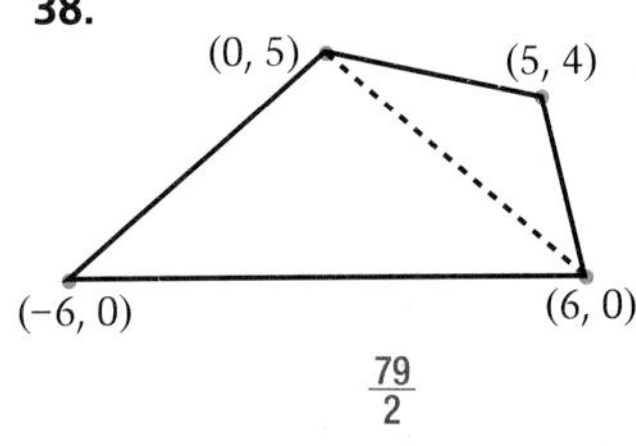

$\frac{79}{2}$

39. *The Bermuda Triangle* The Bermuda Triangle is a large triangular region in the Atlantic Ocean. Many ships and planes have been lost in this region. The triangle is formed by imaginary lines connecting Bermuda, Puerto Rico, and a point near Melbourne, Florida. Estimate the coordinates of these three locations on the map. Then use a determinant to estimate the number of square miles in the Bermuda Triangle. (*Hint:* Each unit on the map represents 47 miles.) 446,218 mi^2

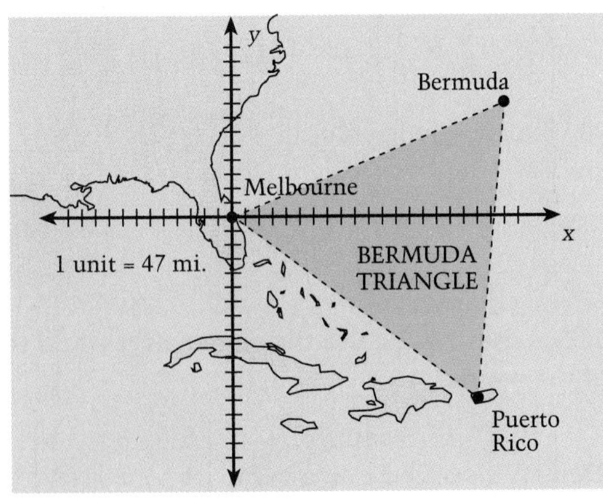

40. *Wetland Breeding Area* Black-necked stilts are birds that live throughout Florida and surrounding areas but breed mostly in the triangular region shown at the right. Each unit of the coordinate system represents a mile. Estimate the area of this region. 11,340 mi²

Integrated Review

In Exercises 41 and 42, illustrate the following property of determinants. "If *A* and *B* are square matrices, then $|AB| = |A||B|$." See margin.

41. $\begin{bmatrix} 5 & 6 \\ -1 & 3 \end{bmatrix} \begin{bmatrix} 6 & 2 \\ 7 & -1 \end{bmatrix}$

42. $\begin{bmatrix} -4 & 7 \\ 0 & 5 \end{bmatrix} \begin{bmatrix} -6 & -3 \\ 1 & 2 \end{bmatrix}$

43. *Powers of Ten* Which best estimates the distance between Virginia Beach, Virginia, and Malibu Beach, California? b
a. 3×10^2 miles **b.** 3×10^3 miles **c.** 3×10^4 miles **d.** 3×10^5 miles

44. *Powers of Ten* Which best estimates the length of a pencil? b
a. 160×10^{-2} cm **b.** 160×10^{-1} cm **c.** 160×10^{0} cm **d.** 160×10^{1} cm

45. *College Entrance Exam Sample* The symbol $\triangle$ represents one of the four basic operations of arithmetic. It has the following properties. a
- If a and b are real numbers, $a \triangle b = b \triangle a$.
- If a is a real number, then $a \triangle 0 = a$.

Which operation(s) must $\triangle$ represent?
a. + only **b.** × only **c.** + or × **d.** − **e.** ÷

46. *College Entrance Exam Sample* For all real numbers a, let ⓧ be defined by ⓧ $= x^2 - 1$. Which of the following is equal to the product of ③ and ④? b
a. ⑫ **b.** ⑪ **c.** ⑩ **d.** ⑨ **e.** ⑦

Exploration and Extension

See margin.

47. Use a determinant to find the area of the "triangle" whose vertices are $(-2, 4)$, $(4, 1)$, and $(8, -1)$. Based on your result, make a conjecture about the three points.

48. *Collinearity Test* Use the result of Exercise 47 to state a test that determines whether the points (x_1, y_1), (x_2, y_2), and (x_3, y_3) are collinear (lie on the same line).

49. *Determinant Form of Equation of a Line* Give a geometric argument why the equation at the right is an equation of the line through the points (x_1, y_1) and (x_2, y_2). Use this form to find an equation of the line containing $(2, 5)$ and $(6, -3)$.

$$\begin{vmatrix} x & y & 1 \\ x_1 & y_1 & 1 \\ x_2 & y_2 & 1 \end{vmatrix} = 0$$

Enrichment Activities

These activities require students to go beyond lesson goals.

The lesson presented a method for finding the area of a triangle by using determinants. The following algorithm can be used to find the area of any polygon, such as that shown in Exercise 37.

The "Shoelaces" Algorithm

A. List the ordered pairs of the vertices in clockwise order, repeating the initial vertex. (You may start at any vertex.)

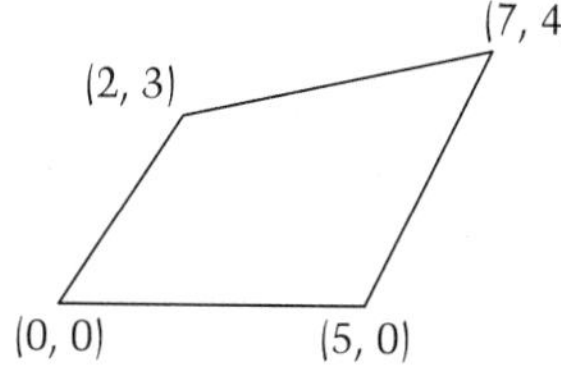

x	y	products
		21
		20
2	3	0
7	4	0
5	0	8
0	0	0
2	3	0
		0

B. Find the diagonal products as shown in both directions.

C. Find the sum of the "up" diagonals. 41

D. Find the sum of the "down" diagonals. 8

E. Find the absolute value of the difference. 33

F. The area of the enclosed figure is $\frac{1}{2}$ the absolute value of the difference found in Part E. What is the area? Compare the result to the answer you found by the determinant method.
$\frac{1}{2}$ of $|33| = 16\frac{1}{2}$. They are the same.

Check your answers to Exercises 33–40. Which method do you prefer? Why?

TECHNOLOGY Bear in mind that each entry of a matrix may not appear on screen when the number of rows and columns is large or the degree of decimal accuracy is more than one or two decimal places. However, entries can be viewed by using the appropriate directional arrows.

Answers

1. $\begin{bmatrix} 117 & -30 \\ -135 & 63 \\ 99 & 35 \end{bmatrix}$

2. $\begin{bmatrix} 17 & 32 & 29 \\ 28 & 55 & 46 \\ 12 & 24 & 19 \end{bmatrix}$

3. $\begin{bmatrix} 3 & 0 & 0 \\ 0 & 3 & 0 \\ 0 & 0 & 3 \end{bmatrix}$

6. *A:* 32,913.93, *B:* 45,300.98; yes, it increases less each year for each company

USING A GRAPHING CALCULATOR

Most graphing calculators can perform matrix algebra. That is, they can add, subtract, and multiply matrices, multiply a matrix by a scalar, find the determinant of a matrix, and find the inverse of a matrix. (The inverse of a matrix is introduced in Lesson 4.4.)

Keystroke instructions for entering and adding matrices with a *TI-82,* Casio *fx-9700GE,* and Sharp *EL-9300C* are listed on page 859.

Cooperative Learning

Partner Activity Enter the following matrix in a graphing calculator

$$A = \begin{bmatrix} 1 & 0.4 & 0.3 \\ 0 & 0.3 & 0.2 \\ 0 & 0.3 & 0.5 \end{bmatrix}$$

Then calculate the following powers of A by repeatedly multiplying A by itself.

$$A^1 = \begin{bmatrix} 1.0000 & 0.4000 & 0.3000 \\ 0.0000 & 0.3000 & 0.2000 \\ 0.0000 & 0.3000 & 0.5000 \end{bmatrix} \quad A^2 = \begin{bmatrix} 1.0000 & 0.6100 & 0.5300 \\ 0.0000 & 0.1500 & 0.1600 \\ 0.0000 & 0.2400 & 0.3100 \end{bmatrix}$$

$$A^3 = \begin{bmatrix} 1.0000 & 0.7420 & 0.6870 \\ 0.0000 & 0.0930 & 0.1100 \\ 0.0000 & 0.1650 & 0.2030 \end{bmatrix} \quad A^4 = \begin{bmatrix} 1.0000 & 0.8287 & 0.7919 \\ 0.0000 & 0.0609 & 0.0736 \\ 0.0000 & 0.1104 & 0.1345 \end{bmatrix}$$

$$A^5 = \begin{bmatrix} 1.0000 & 0.8862 & 0.8617 \\ 0.0000 & 0.0404 & 0.0490 \\ 0.0000 & 0.0735 & 0.0893 \end{bmatrix} \quad A^6 = \begin{bmatrix} 1.0000 & 0.9244 & 0.9081 \\ 0.0000 & 0.0268 & 0.0326 \\ 0.0000 & 0.0488 & 0.0594 \end{bmatrix}$$

$$A^7 = \begin{bmatrix} 1.0000 & 0.9497 & 0.9389 \\ 0.0000 & 0.0178 & 0.0216 \\ 0.0000 & 0.0325 & 0.0394 \end{bmatrix} \quad A^8 = \begin{bmatrix} 1.0000 & 0.9666 & 0.9594 \\ 0.0000 & 0.0118 & 0.0144 \\ 0.0000 & 0.0216 & 0.0262 \end{bmatrix}$$

With your partner, discuss any pattern you see. As the matrix A is raised to higher and higher powers, which matrix do the powers appear to be approaching? Explain how to use a graphing calculator to test your conjecture.

Matrix approaches $\begin{bmatrix} 1 & 1 & 1 \\ 0 & 0 & 0 \\ 0 & 0 & 0 \end{bmatrix}$.

Many matrices used to solve problems in engineering, business, and science have several rows and columns. Because it is so easy to make a mistake, people who work with such matrices almost always use technology to do the actual matrix calculations. Matrices have always been a powerful type of mathematical model—especially for problems that have a lot of data. With technology available to do the calculations, matrices have also become a *practical* type of mathematical model.

Example: Finding the Market Share

Two cable television companies compete in a city with 100,000 households. Company A has 25,000 subscribers and Company B has 30,000. (The other 45,000 households do not subscribe.) The percent changes in cable subscriptions each year are shown in the following **transition** matrix. How many subscribers will each company have in one year? In two years? (This is the same problem discussed on page 182.)

Percent Changes	From Company A	From Company B	From Nonsubscriber
To Company A	0.70	0.15	0.15
To Company B	0.20	0.80	0.15
To Nonsubscriber	0.10	0.05	0.70

Solution The matrix giving the present number of subscribers is shown at the right.

$$\begin{bmatrix} 25{,}000 \\ 30{,}000 \\ 45{,}000 \end{bmatrix} \begin{matrix} \textit{Company A} \\ \textit{Company B} \\ \textit{Nonsubscriber} \end{matrix}$$

To find the number of subscribers in one year, multiply the "subscriber matrix" by the transition matrix.

$$\begin{bmatrix} 0.70 & 0.15 & 0.15 \\ 0.20 & 0.80 & 0.15 \\ 0.10 & 0.05 & 0.70 \end{bmatrix} \begin{bmatrix} 25{,}000 \\ 30{,}000 \\ 45{,}000 \end{bmatrix} = \begin{bmatrix} 28{,}750 \\ 35{,}750 \\ 35{,}500 \end{bmatrix} \begin{matrix} \textit{Company A} \\ \textit{Company B} \\ \textit{Nonsubscriber} \end{matrix}$$

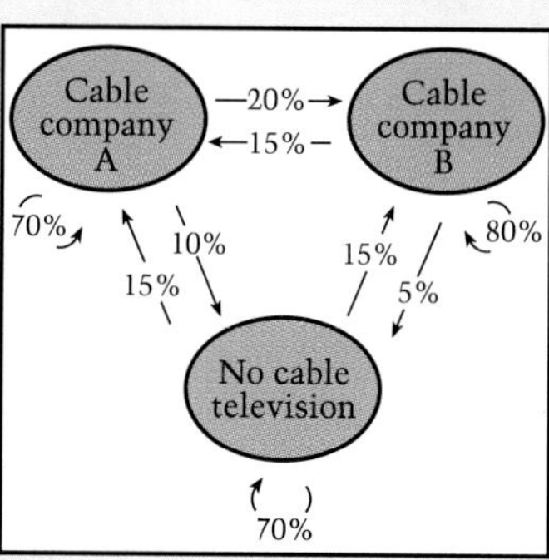

To find the number of subscribers in two years, multiply this new subscriber matrix by the transition matrix.

$$\begin{bmatrix} 0.70 & 0.15 & 0.15 \\ 0.20 & 0.80 & 0.15 \\ 0.10 & 0.05 & 0.70 \end{bmatrix} \begin{bmatrix} 28{,}750 \\ 35{,}750 \\ 35{,}500 \end{bmatrix} = \begin{bmatrix} 30{,}812.5 \\ 39{,}675.0 \\ 29{,}512.5 \end{bmatrix} \begin{matrix} \textit{Company A} \\ \textit{Company B} \\ \textit{Nonsubscriber} \end{matrix}$$ ■

Exercises

In Exercises 1–3, use a graphing calculator to multiply the matrices. See margin.

1. $\begin{bmatrix} 3 & -2 & 12 \\ 0 & -6 & -9 \\ 8 & -7 & 10 \end{bmatrix} \begin{bmatrix} 9 & 8 \\ 9 & -3 \\ 9 & -5 \end{bmatrix}$ **2.** $\begin{bmatrix} 2 & 0 & 5 \\ 1 & 3 & 7 \\ 0 & 3 & 2 \end{bmatrix} \begin{bmatrix} 1 & 1 & 2 \\ 2 & 4 & 3 \\ 3 & 6 & 5 \end{bmatrix}$ **3.** $\begin{bmatrix} -2 & 2 & 3 \\ 1 & -1 & 0 \\ 0 & 1 & 4 \end{bmatrix} \begin{bmatrix} -4 & -5 & 3 \\ -4 & -8 & 3 \\ 1 & 2 & 0 \end{bmatrix}$

In Exercises 4–5, use a graphing calculator to evaluate the determinant of the matrix.

4. $\begin{bmatrix} 2 & 4 & 7 \\ -3 & 5 & 8 \\ 10 & 2 & -9 \end{bmatrix}$ −302 **5.** $\begin{bmatrix} 8 & -7 & -7 \\ 2 & 5 & -5 \\ -8 & 7 & 7 \end{bmatrix}$ 0

6. *Market Share* In Example 1 above, how many subscribers will each company have in five years? Is the market stabilizing for each company? Explain. See margin.

Problem of the Day

Five sacks of grain are weighed two at a time in all possible combinations. The weights in pounds are 110, 112, 113, 114, 115, 116, 117, 118, 120, and 121. Find the weight of each sack. 54, 56, 58, 59, and 62 pounds

ORGANIZER

Warm-up Exercises

1. Find the determinant of each matrix.
   a. $A = \begin{bmatrix} 3 & 4 \\ 7 & 4 \end{bmatrix}$ −16
   b. $B = \begin{bmatrix} -2 & 3 \\ 0 & 1 \end{bmatrix}$ −2
   c. $C = \begin{bmatrix} -12 & 3 \\ 10 & 16 \end{bmatrix}$ −222
   d. $D = \begin{bmatrix} -8 & -7 \\ 9 & -5 \end{bmatrix}$ 103
2. Find each product.
   a. AB b. BA
   c. CD d. DC

a. $\begin{bmatrix} -6 & 13 \\ -14 & 25 \end{bmatrix}$ b. $\begin{bmatrix} 15 & 4 \\ 7 & 4 \end{bmatrix}$

c. $\begin{bmatrix} 123 & 69 \\ 64 & -150 \end{bmatrix}$

d. $\begin{bmatrix} 26 & -136 \\ -158 & -53 \end{bmatrix}$

Lesson Resources

Teaching Tools
  Problem of the Day: 4.4
  Warm-up Exercises: 4.4
  Answer Masters: 4.4
Extra Practice: 4.4
Technology Handbook: p. 19, 21

4.4 Identity and Inverse Matrices

What you should learn:

Goal 1 How to find and use the inverse of a 2 × 2 matrix

Goal 2 How to use inverse matrices in real-life settings

Why you should learn it:

You can use inverse matrices in many settings, such as in solving systems of linear equations and in decoding cryptograms.

Study Tip

*The **inverse** of a matrix A is written as A^{-1}. This notation is similar to that used for real numbers because a^{-1} is the multiplicative inverse of the nonzero real number a.*

Goal 1 Using the Inverse of a 2 × 2 Matrix

For real numbers, the number 1 is the multiplicative identity because for any real number a, $1 \cdot a = a$ and $a \cdot 1 = a$. For 2×2 matrices, the **multiplicative identity** is

$$I = \begin{bmatrix} 1 & 0 \\ 0 & 1 \end{bmatrix}.$$

Try multiplying any 2×2 matrix A by this identity. What do you obtain?

LESSON INVESTIGATION

Investigating Inverse Matrices

Partner Activity The real numbers 2 and $\frac{1}{2}$ are multiplicative inverses of each other because their product is equal to the multiplicative identity, 1. Which two of the following matrices are inverses of each other? Explain.

$$A = \begin{bmatrix} 3 & 1 \\ -5 & 2 \end{bmatrix} \qquad B = \begin{bmatrix} 3 & -1 \\ -5 & 2 \end{bmatrix} \qquad C = \begin{bmatrix} 2 & 1 \\ 5 & 3 \end{bmatrix}$$

B and C; $B \times C = C \times B = I$

The Inverse of a 2 × 2 Matrix

The inverse of the matrix $A = \begin{bmatrix} a & b \\ c & d \end{bmatrix}$ is

$$A^{-1} = \frac{1}{|A|}\begin{bmatrix} d & -b \\ -c & a \end{bmatrix} = \frac{1}{ad - cb}\begin{bmatrix} d & -b \\ -c & a \end{bmatrix}, \quad ad - cb \neq 0.$$

Example 1 *Finding the Inverse of a 2 × 2 Matrix*

Find the inverse of $A = \begin{bmatrix} 2 & 4 \\ 1 & 3 \end{bmatrix}$.

Solution

$$A^{-1} = \frac{1}{6-4}\begin{bmatrix} 3 & -4 \\ -1 & 2 \end{bmatrix} = \begin{bmatrix} \frac{3}{2} & -2 \\ -\frac{1}{2} & 1 \end{bmatrix}$$

Check this by showing that $AA^{-1} = I$ and $A^{-1}A = I$. ■

Example 2 Solving a Matrix Equation

Solve the matrix equation for the 2×2 matrix X.

$$\underbrace{\begin{bmatrix} 11 & 5 \\ 2 & 1 \end{bmatrix}}_{A} X = \underbrace{\begin{bmatrix} 4 & -1 \\ 2 & 0 \end{bmatrix}}_{B} \qquad \text{Matrix equation } AX = B$$

Solution Begin by finding the inverse of A.

$$A^{-1} = \frac{1}{11-10}\begin{bmatrix} 1 & -5 \\ -2 & 11 \end{bmatrix} = \begin{bmatrix} 1 & -5 \\ -2 & 11 \end{bmatrix}$$

To solve the equation for X, multiply both sides of the equation by A^{-1} *on the left.*

$$\begin{bmatrix} 1 & -5 \\ -2 & 11 \end{bmatrix}\begin{bmatrix} 11 & 5 \\ 2 & 1 \end{bmatrix} X = \begin{bmatrix} 1 & -5 \\ -2 & 11 \end{bmatrix}\begin{bmatrix} 4 & -1 \\ 2 & 0 \end{bmatrix} \qquad A^{-1}AX = A^{-1}B$$

$$\begin{bmatrix} 1 & 0 \\ 0 & 1 \end{bmatrix} X = \begin{bmatrix} -6 & -1 \\ 14 & 2 \end{bmatrix} \qquad IX = A^{-1}B$$

$$X = \begin{bmatrix} -6 & -1 \\ 14 & 2 \end{bmatrix} \qquad X = A^{-1}B$$

■

Some 2×2 matrices do not have an inverse. You can tell whether a matrix has an inverse by evaluating its determinant. If the determinant of A is zero, then A does not have an inverse. If the determinant is nonzero, then A has an inverse.

The inverse of a 3×3 matrix can be very difficult to compute. Of course, if you have a calculator that will compute inverse matrices, then the task is simplified.

Example 3 Verifying the Inverse of a 3 × 3 Matrix

Show that

$$B = \begin{bmatrix} 1 & -1 & 0 \\ 1 & 0 & -1 \\ 6 & -2 & -3 \end{bmatrix} \text{ is the inverse of } A = \begin{bmatrix} -2 & -3 & 1 \\ -3 & -3 & 1 \\ -2 & -4 & 1 \end{bmatrix}.$$

Solution

$$BA = \begin{bmatrix} 1 & -1 & 0 \\ 1 & 0 & -1 \\ 6 & -2 & -3 \end{bmatrix}\begin{bmatrix} -2 & -3 & 1 \\ -3 & -3 & 1 \\ -2 & -4 & 1 \end{bmatrix} = \begin{bmatrix} 1 & 0 & 0 \\ 0 & 1 & 0 \\ 0 & 0 & 1 \end{bmatrix}$$

$$AB = \begin{bmatrix} -2 & -3 & 1 \\ -3 & -3 & 1 \\ -2 & -4 & 1 \end{bmatrix}\begin{bmatrix} 1 & -1 & 0 \\ 1 & 0 & -1 \\ 6 & -2 & -3 \end{bmatrix} = \begin{bmatrix} 1 & 0 & 0 \\ 0 & 1 & 0 \\ 0 & 0 & 1 \end{bmatrix}$$

Because both products are equal to the 3×3 identity matrix, you can conclude that B is the inverse of A. ■

LESSON Notes

Example 1

Be sure to emphasize that the inverse of a 2×2 matrix, A, involves the determinant of matrix A. It is important for students to verify that the inverse matrix, A^{-1}, when multiplied with the matrix A, yields the identity matrix, I. For example,

$$\begin{bmatrix} 2 & 4 \\ 1 & 3 \end{bmatrix}\begin{bmatrix} \frac{3}{2} & -2 \\ -\frac{1}{2} & 1 \end{bmatrix} = \begin{bmatrix} 1 & 0 \\ 0 & 1 \end{bmatrix}$$

and

$$\begin{bmatrix} \frac{3}{2} & -2 \\ -\frac{1}{2} & 1 \end{bmatrix}\begin{bmatrix} 2 & 4 \\ 1 & 3 \end{bmatrix} = \begin{bmatrix} 1 & 0 \\ 0 & 1 \end{bmatrix}.$$

Example 2

Ask students why they must be so careful when multiplying expressions in a matrix equation. (The reason is that matrix multiplication is not commutative, and so, the order of multiplication may result in very different products. In the example, if you were multiplying $\begin{bmatrix} 4 & -1 \\ 2 & 0 \end{bmatrix}$ by $\begin{bmatrix} 1 & -5 \\ -2 & 11 \end{bmatrix}$ on the right, the product would have been $\begin{bmatrix} 6 & -31 \\ 2 & -10 \end{bmatrix}$ instead of $\begin{bmatrix} -6 & -1 \\ 14 & 2 \end{bmatrix}$.

Common-Error ALERT!

When solving matrix equations, as in **Example 2,** it is necessary to multiply both sides of the equation in the same manner. That is, if one side of a matrix equation is multiplied on the left, then the other side must also be multiplied on the left.

Example 3

Although computing the inverse matrix of a 3×3 matrix is tedious to do by paper and pencil methods, verifying that a 3×3 matrix is an inverse matrix is straight forward.

Examples 4–5

Students will find it fun to encode and decode their own messages using the encoding and decoding matrices presented in the examples. Have students exchange messages with classmates. This is a great way to practice using inverse matrices.

Extra Examples

Here are additional examples similar to **Examples 1–5.**

1. Find the inverse of each matrix.

a. $\begin{bmatrix} 2 & 3 \\ 2 & 4 \end{bmatrix}$ $\begin{bmatrix} 2 & -\frac{3}{2} \\ -1 & 1 \end{bmatrix}$

b. $\begin{bmatrix} 4 & 8 \\ -3 & -2 \end{bmatrix}$ $\begin{bmatrix} -\frac{1}{8} & -\frac{1}{2} \\ \frac{3}{16} & \frac{1}{4} \end{bmatrix}$

2. Solve each matrix equation for X.

a. $\begin{bmatrix} 1 & 2 \\ 0 & 2 \end{bmatrix} X = \begin{bmatrix} 1 & -5 \\ -2 & -6 \end{bmatrix}$

b. $\begin{bmatrix} 1 & -1 \\ -2 & 1 \end{bmatrix} X = \begin{bmatrix} -2 & -2 \\ 4 & 1 \end{bmatrix}$

a. $X = \begin{bmatrix} 3 & 1 \\ -1 & -3 \end{bmatrix}$;

b. $X = \begin{bmatrix} -2 & 1 \\ 0 & 3 \end{bmatrix}$

Goal 2 Using Inverse Matrices in Real Life

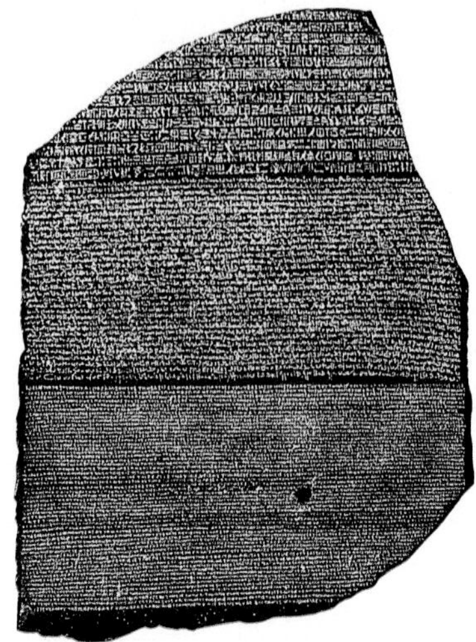

The text of the Rosetta Stone is written in Greek, Egyptian demotic, and hieroglyphics. Modern methods of cryptography were used to "break the code" of the hieroglyphs.

A *cryptogram* is a message written according to a secret code. (The Greek word *kryptos* means *"hidden."*) The following technique uses matrices to encode and decode messages.

To begin, assign a number to each letter in the alphabet (with 0 assigned to a blank space).

0 = __	6 = F	12 = L	18 = R	24 = X
1 = A	7 = G	13 = M	19 = S	25 = Y
2 = B	8 = H	14 = N	20 = T	26 = Z
3 = C	9 = I	15 = O	21 = U	
4 = D	10 = J	16 = P	22 = V	
5 = E	11 = K	17 = Q	23 = W	

The message is converted to numbers and partitioned into 1×2 *uncoded row matrices.* For instance, the message "STOP NOW" would be converted to the following row matrices.

[19 20]	[15 16]	[0 14]	[15 23]
S T	O P	__ N	O W

To encode a message, choose a 2×2 matrix A that has an inverse and multiply the uncoded row matrices (on the right) by A to obtain *coded row matrices.*

Real Life
Cryptography

Example 4 *Encoding a Message*

Use the matrix $A = \begin{bmatrix} -1 & 2 \\ 2 & -3 \end{bmatrix}$ to encode the message "STOP NOW."

Solution The coded row matrices are obtained by multiplying each of the uncoded row matrices (listed above) by the matrix A.

Uncoded Row Matrix	Encoding Matrix A		Coded Row Matrix
[19 20]	$\begin{bmatrix} -1 & 2 \\ 2 & -3 \end{bmatrix}$	=	[21 −22]
[15 16]	$\begin{bmatrix} -1 & 2 \\ 2 & -3 \end{bmatrix}$	=	[17 −18]
[0 14]	$\begin{bmatrix} -1 & 2 \\ 2 & -3 \end{bmatrix}$	=	[28 −42]
[15 23]	$\begin{bmatrix} -1 & 2 \\ 2 & -3 \end{bmatrix}$	=	[31 −39]

The coded message is 21, −22, 17, −18, 28, −42, 31, −39. ■

For those who do not know the matrix A, decoding the cryptogram in Example 4 is difficult. (When larger coding matrices are used, decoding is even more difficult.) But for an authorized receiver who knows the matrix A, decoding is simple. The receiver need only multiply the coded row matrices (on the right) by A^{-1} to retrieve the uncoded row matrices.

Problem Solving
Work Backward

Example 5 *Decoding a Message*

Use the inverse of $A = \begin{bmatrix} -1 & 2 \\ 2 & -3 \end{bmatrix}$ to decode the message

$-9, 23, -6, 16, 26, -39, 13, -12, 45, -65.$

Solution To decode the message, partition the message into groups of two numbers to form coded row matrices. Then multiply each coded row matrix by A^{-1} (on the right) to obtain the decoded row matrices.

Coded Row Matrix	Decoding Matrix A^{-1}		Decoded Row Matrix
$[-9 \quad 23]$	$\begin{bmatrix} 3 & 2 \\ 2 & 1 \end{bmatrix}$	=	$[19 \quad 5]$
$[-6 \quad 16]$	$\begin{bmatrix} 3 & 2 \\ 2 & 1 \end{bmatrix}$	=	$[14 \quad 4]$
$[26 \quad -39]$	$\begin{bmatrix} 3 & 2 \\ 2 & 1 \end{bmatrix}$	=	$[0 \quad 13]$
$[13 \quad -12]$	$\begin{bmatrix} 3 & 2 \\ 2 & 1 \end{bmatrix}$	=	$[15 \quad 14]$
$[45 \quad -65]$	$\begin{bmatrix} 3 & 2 \\ 2 & 1 \end{bmatrix}$	=	$[5 \quad 25]$

From the uncoded row matrices, read the message as follows.

[19 5]	[14 4]	[0 13]	[15 14]	[5 25]
S E	N D	— M	O N	E Y

■

Communicating about ALGEBRA

SHARING IDEAS about the Lesson

Use the inverse of $\begin{bmatrix} 3 & -4 \\ -2 & 3 \end{bmatrix}$ to decode the message.

$-4, 7, -23, 35, -26, 39, 15, -20, 31, -36, -38, 57, -21, 33, 20, -20, 75, -100$ BEAM ME UP SCOTTY

3. Verify that matrices A and B are inverses.

$A = \begin{bmatrix} \frac{3}{8} & \frac{1}{8} & 0 \\ \frac{1}{8} & -\frac{1}{8} & \frac{2}{8} \\ \frac{1}{8} & \frac{1}{8} & \frac{1}{8} \end{bmatrix},$

$B = \begin{bmatrix} 3 & 1 & -2 \\ -1 & -3 & 6 \\ -2 & 2 & 4 \end{bmatrix}$

$AB = I = BA$

Check Understanding

1. What is an identity matrix?
A square matrix with diagonal entries of 1 and all other entries zero.
2. Does every 2×2 matrix have an inverse? Explain.
No; if the determinant of a matrix is 0, the matrix does not have an inverse.
3. Describe how to find the inverse of a 2×2 matrix.
The inverse of $\begin{bmatrix} a & b \\ c & d \end{bmatrix}$ is $\frac{1}{ad - bc}\begin{bmatrix} d & -b \\ -c & a \end{bmatrix}$.

***Communicating* about ALGEBRA**

Use the 3×3 matrices of Example 3 in this lesson to devise a coding and decoding scheme similar to the one presented in Examples 4 and 5. Can other matrix operations be used in conjunction with inverses to encode and decode messages? Explain.
Yes, you could use addition, subtraction, and scalar multiplication.

EXERCISE Notes

ASSIGNMENT GUIDE

Basic/Average: Ex. 1–6, 7–29 odd, 35–39 odd, 40–42, 43–49 odd

Above Average: Ex. 1–6, 7–29 odd, 35, 37, 39–43, 47–52

Advanced: Ex. 1–6, 7–29 odd, 35, 37, 39–44, 47–52

Selected Answers
Exercises 1–6, 7–49 odd

✪ More Difficult Exercises
Exercises 39–42, 51, 52

Guided Practice

▶ **Ex. 3** Remind students to check both products, *AB* and *BA*.

▶ **Ex. 6** This exercise is developing a skill necessary for solving systems of equations, which will be presented in Lesson 4.5.

Answers

1. $\begin{bmatrix} 1 & 0 \\ 0 & 1 \end{bmatrix}$, $\begin{bmatrix} 1 & 0 & 0 \\ 0 & 1 & 0 \\ 0 & 0 & 1 \end{bmatrix}$

Independent Practice

▶ **Ex. 7–8** Remind students to check the product in both directions.

▶ **Ex. 17–22** These exercises are modeled by Example 2.

▶ **Ex. 23–26** Remind students that an inverse fails to exist if the determinant of the matrix is zero.

EXERCISES

Guided Practice

CRITICAL THINKING about the Lesson

1. What is the multiplicative identity for 2×2 matrices? For 3×3 matrices?

2. For two 2×2 matrices, A and B, to be inverses of each other, what must be true of AB and BA? **Both must equal *I*.**

3. Are A and B inverses of each other? Justify your answer. **Yes, *AB* = *I* and *BA* = *I***

$$A = \begin{bmatrix} 0 & -\frac{1}{2} \\ 2 & 5 \end{bmatrix} \text{ and } B = \begin{bmatrix} 5 & \frac{1}{2} \\ -2 & 0 \end{bmatrix}$$

4. Find the inverse of matrix A below.

$$A = \begin{bmatrix} -4 & 3 \\ -3 & 2 \end{bmatrix} \quad \begin{bmatrix} 2 & -3 \\ 3 & -4 \end{bmatrix}$$

5. Does $\begin{bmatrix} 6 & -3 \\ -2 & 1 \end{bmatrix}$ have an inverse? Explain. **No, its determinant is 0.**

6. How do you solve $\underbrace{\begin{bmatrix} 2 & 3 \\ 3 & 4 \end{bmatrix}}_{A} X = \underbrace{\begin{bmatrix} 1 & 3 \\ 2 & -1 \end{bmatrix}}_{B}$ for X?

Multiply both sides of the equation by A^{-1} on the left.

Independent Practice

In Exercises 7 and 8, decide whether the matrices are inverses of each other.

7. $\begin{bmatrix} 3 & 2 \\ 7 & 5 \end{bmatrix}$ and $\begin{bmatrix} 5 & -2 \\ -7 & 3 \end{bmatrix}$ **Yes**

8. $\begin{bmatrix} 9 & -2 \\ 5 & -1 \end{bmatrix}$ and $\begin{bmatrix} -1 & 2 \\ -5 & 9 \end{bmatrix}$ **Yes**

In Exercises 9–16, find the inverse of the matrix.

9. $\begin{bmatrix} 7 & -4 \\ -5 & 3 \end{bmatrix}$ $\begin{bmatrix} 3 & 4 \\ 5 & 7 \end{bmatrix}$

10. $\begin{bmatrix} 5 & 2 \\ 2 & 1 \end{bmatrix}$ $\begin{bmatrix} 1 & -2 \\ -2 & 5 \end{bmatrix}$

11. $\begin{bmatrix} 8 & 17 \\ -1 & -2 \end{bmatrix}$ $\begin{bmatrix} -2 & -17 \\ 1 & 8 \end{bmatrix}$

12. $\begin{bmatrix} -4 & -3 \\ 3 & 2 \end{bmatrix}$ **12.** $\begin{bmatrix} 2 & 3 \\ -3 & -4 \end{bmatrix}$

13. $\begin{bmatrix} 6 & -2 \\ 7 & -2 \end{bmatrix}$ $\begin{bmatrix} -1 & 1 \\ -\frac{7}{2} & 3 \end{bmatrix}$

14. $\begin{bmatrix} 3 & 2 \\ 4 & 1 \end{bmatrix}$ $\begin{bmatrix} -\frac{1}{5} & \frac{2}{5} \\ \frac{4}{5} & -\frac{3}{5} \end{bmatrix}$

15. $\begin{bmatrix} 11 & -5 \\ 3 & -1 \end{bmatrix}$ $\begin{bmatrix} -\frac{1}{4} & \frac{5}{4} \\ -\frac{3}{4} & \frac{11}{4} \end{bmatrix}$

16. $\begin{bmatrix} 7 & 4 \\ 3 & 2 \end{bmatrix}$ **See below.**

In Exercises 17–22, solve the matrix equation.

17. $\begin{bmatrix} -5 & -13 \\ 2 & 5 \end{bmatrix} X = \begin{bmatrix} 3 & 1 \\ -4 & 0 \end{bmatrix}$ $\begin{bmatrix} -37 & 5 \\ 14 & -2 \end{bmatrix}$

18. $\begin{bmatrix} 12 & 7 \\ 5 & 3 \end{bmatrix} X = \begin{bmatrix} 2 & -1 \\ 3 & 2 \end{bmatrix}$ $\begin{bmatrix} -15 & -17 \\ 26 & 29 \end{bmatrix}$

19. $\begin{bmatrix} 2 & 4 \\ 0 & 1 \end{bmatrix} X = \begin{bmatrix} 4 & 0 & 6 \\ 3 & -1 & 5 \end{bmatrix}$ $\begin{bmatrix} -4 & 2 & -7 \\ 3 & -1 & 5 \end{bmatrix}$

20. $\begin{bmatrix} -6 & -3 \\ 3 & 1 \end{bmatrix} X = \begin{bmatrix} 9 & 12 & 0 \\ -4 & 5 & -2 \end{bmatrix}$ **See margin.**

21. $\begin{bmatrix} 4 & 7 \\ 1 & 2 \end{bmatrix} X + \begin{bmatrix} 2 & 7 \\ -3 & 4 \end{bmatrix} = \begin{bmatrix} 6 & 2 \\ -2 & 3 \end{bmatrix}$ $\begin{bmatrix} 1 & -3 \\ 0 & 1 \end{bmatrix}$

22. $\begin{bmatrix} -7 & -9 \\ 4 & 5 \end{bmatrix} X + \begin{bmatrix} 3 & 4 \\ 4 & -3 \end{bmatrix} = \begin{bmatrix} 1 & 9 \\ 6 & -6 \end{bmatrix}$ **See margin.**

In Exercises 23–26, decide whether the matrix has an inverse. **See margin.**

23. $\begin{bmatrix} 8 & 2 & 0 \\ 4 & 1 & -1 \\ 0 & 0 & 4 \end{bmatrix}$

24. $\begin{bmatrix} 2 & -1 & 2 \\ 1 & 0 & -1 \\ 3 & 2 & -1 \end{bmatrix}$

25. $\begin{bmatrix} 5 & 1 & 0 \\ 2 & 3 & -2 \\ 1 & 3 & 1 \end{bmatrix}$

26. $\begin{bmatrix} 4 & -2 & 0 \\ 0 & 2 & 0 \\ 5 & 1 & 3 \end{bmatrix}$

16. $\begin{bmatrix} 1 & -2 \\ -\frac{3}{2} & \frac{7}{2} \end{bmatrix}$

In Exercises 27–30, decide whether the matrices are inverses of each other. See margin.

27. $\begin{bmatrix} 0 & 2 & -1 \\ 5 & 2 & 3 \\ 7 & 3 & 4 \end{bmatrix}$ and $\begin{bmatrix} -1 & -11 & 8 \\ 1 & 7 & -5 \\ 1 & 14 & -10 \end{bmatrix}$

28. $\begin{bmatrix} 11 & 2 & -8 \\ 4 & 1 & -3 \\ -8 & -1 & 6 \end{bmatrix}$ and $\begin{bmatrix} 3 & -4 & 2 \\ 0 & 2 & 1 \\ 4 & -5 & 3 \end{bmatrix}$

29. $\begin{bmatrix} 1 & 3 & 1 \\ 1 & 2 & 2 \\ 1 & 0 & 3 \end{bmatrix}$ and $\begin{bmatrix} 6 & -9 & 4 \\ -1 & 2 & -1 \\ -2 & 2 & -1 \end{bmatrix}$

30. $\begin{bmatrix} 10 & 2 & -25 \\ 4 & 1 & -10 \\ -9 & -2 & 23 \end{bmatrix}$ and $\begin{bmatrix} 3 & 4 & 5 \\ -2 & 5 & 0 \\ 1 & 2 & 2 \end{bmatrix}$

In Exercises 31–34, use a graphing calculator to find the inverse of the matrix. See Additional Answers.

31. $\begin{bmatrix} -3 & 4 & 9 \\ 1 & 0 & 0 \\ 5 & 3 & 7 \end{bmatrix}$

32. $\begin{bmatrix} -7 & 0 & -6 \\ -4 & 1 & 3 \\ 11 & -3 & -10 \end{bmatrix}$

33. $\begin{bmatrix} -2 & 1 & -1 \\ 2 & 0 & 4 \\ 0 & 2 & 5 \end{bmatrix}$

34. $\begin{bmatrix} 2 & 1 & -2 \\ 5 & 3 & 0 \\ 4 & 3 & 8 \end{bmatrix}$

Encoding Messages **In Exercises 35–38, use the matrix to encode the message.**

35. $A = \begin{bmatrix} 1 & -2 \\ -1 & 3 \end{bmatrix}$, BREAK A LEG −16, 50, 4, −7, 11, −22, 1, −2, 7, −9, 7, −14

36. $A = \begin{bmatrix} 1 & 2 \\ 1 & 3 \end{bmatrix}$, KNOCK ON WOOD 25, 64, 18, 39, 11, 22, 29, 72, 23, 69, 30, 75, 4, 8

37. $A = \begin{bmatrix} 2 & 3 \\ 1 & 2 \end{bmatrix}$, GET WELL SOON 19, 31, 40, 60, 51, 79, 36, 60, 19, 38, 45, 75, 28, 42

38. $A = \begin{bmatrix} 2 & -3 \\ -1 & 2 \end{bmatrix}$, COME TO DINNER −9, 21, 21, −29, −20, 40, 30, −45, −1, 6, 14, −14, −8, 21

Decoding Messages **In Exercises 39–42, use the following information.**

Primitive humans were skilled at making tools and jewelry from stone, bones, wood, and ivory. Use the inverse of

$$\begin{bmatrix} 5 & 2 \\ 3 & 1 \end{bmatrix}$$

to decode each message. Then match the message with an artifact made by primitive people in Europe. ***(Source: U.S. News and World Report)***

39. 111, 40, 129, 48, 125, 50, 95, 37, 82, 32, 47, 16, 100, 40 IVORY_PENDANT, b

40. 47, 16, 42, 15, 67, 24, 100, 40, 85, 33, 48, 17, 63, 25, 30, 11 ANCIENT_NECKLACE, c

41. 55, 19, 85, 33, 42, 14, 40, 15, 56, 20, 82, 29 BONE_NEEDLES, d

42. 57, 21, 119, 46, 24, 8, 120, 45, 112, 41 FISH_HOOKS, a

a.

b.

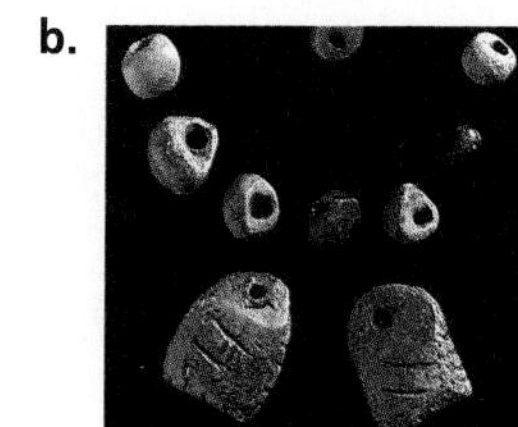

c.

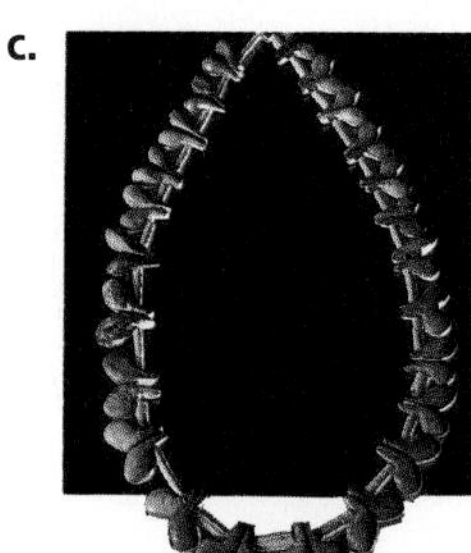

d.

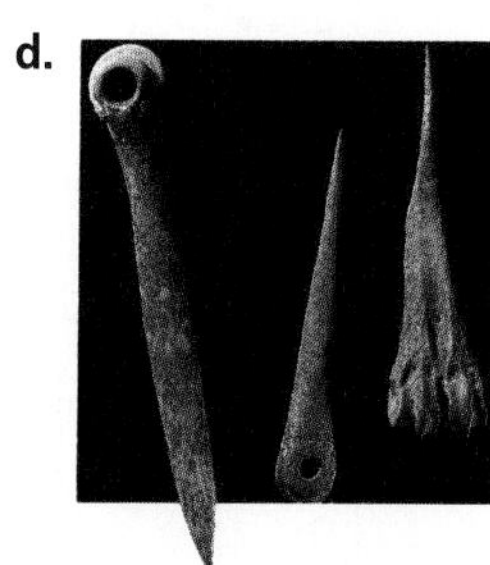

Ex. 27–30 Remind students to check the product in both directions.

Ex. 31–34 TECHNOLOGY Students might use their graphics calculator to find the inverse of a 3 × 3 matrix. Using a calculator is much more efficient than the laborious pencil-and-paper method. However, later in Exercise 51, students will be presented a paper-and-pencil algorithm that is simpler to use.

Ex. 35–38 These exercises are modeled by Example 4. Encourage students to show a similar format for their work. They will need the alphabet code above Example 4.

Ex. 39–42 These exercises should be assigned as a group. Refer students to Example 5 for a solution model.

Answers

20. $\begin{bmatrix} -1 & 9 & -2 \\ -1 & -22 & 4 \end{bmatrix}$

22. $\begin{bmatrix} 8 & -2 \\ -6 & 1 \end{bmatrix}$

23. It has not.

24. It has.

25. It has.

26. It has.

27.–28., 30. They are.

29. They are not.

Ex. 51–52 You may choose to use the exercises as in-class discussion problems. If possible, use an overhead graphics calculator to expedite the presentation process.

In Exercise 52, presenting students with the inverse of A will give you more time to discuss the population trends suggested by the exercise and you will spend less time doing the computations.

Enrichment Activities

These activities require students to go beyond lesson goals.

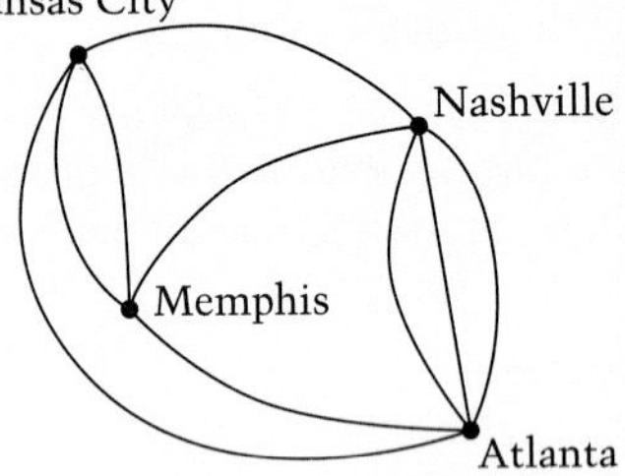

The number of direct flights between the four cities in the network is shown in this matrix.

From \ To	KC	Ns.	Mp.	At.
KC	0	1	2	1
Ns.	1	0	1	3
Mp.	2	1	0	1
At.	1	3	1	0

1. Give a reason why the diagonal entries from the upper left to the lower right are all zero.
   You cannot take a direct flight from a city to itself.
2. Why is the entry for flights from Atlanta to Nashville 3?
   There are three unique paths.

Integrated Review

In Exercises 43–46, simplify the matrix. 44, 46. See Additional Answers.

43. $-4\begin{bmatrix} 3 & 6 & 4 \\ 9 & 4 & 3 \\ -2 & 1 & 2 \end{bmatrix}$ $\begin{bmatrix} -12 & -24 & -16 \\ -36 & -16 & -12 \\ 8 & -4 & -8 \end{bmatrix}$

44. $\begin{bmatrix} 1 & 2 & 5 \\ 0 & 4 & 10 \\ -10 & 6 & 5 \end{bmatrix} - 2\begin{bmatrix} 4 & 3 & 1 \\ -1 & 1 & 7 \\ -5 & 6 & 4 \end{bmatrix}$

45. $\begin{bmatrix} 1 & 2 & 0 \\ 3 & 2 & -1 \\ -3 & -1 & 4 \end{bmatrix}\begin{bmatrix} 6 & 1 & -2 \\ 4 & 3 & 1 \\ 0 & 4 & -1 \end{bmatrix}$ $\begin{bmatrix} 14 & 7 & 0 \\ 26 & 5 & -3 \\ -22 & 10 & 1 \end{bmatrix}$

46. $\begin{bmatrix} 1 & -3 & 4 \\ 0 & 2 & -2 \\ 3 & 1 & 4 \end{bmatrix}\begin{bmatrix} -1 & 3 & 2 \\ -2 & 1 & -1 \\ 0 & 4 & 2 \end{bmatrix}$

In Exercises 47–50, evaluate the determinant of the matrix.

47. $\begin{bmatrix} 10 & 15 \\ 1 & 4 \end{bmatrix}$ 25

48. $\begin{bmatrix} -5 & -3 \\ 9 & 4 \end{bmatrix}$ 7

49. $\begin{bmatrix} 9 & 0 & 10 \\ 0 & 4 & 3 \\ -1 & 2 & 1 \end{bmatrix}$ 22

50. $\begin{bmatrix} 4 & 2 & 3 \\ -4 & 1 & 0 \\ 0 & 0 & 5 \end{bmatrix}$ 60

Exploration and Extension

51. The inverse of $A = \begin{bmatrix} a & b & c \\ d & e & f \\ g & h & i \end{bmatrix}$ is $A^{-1} = \dfrac{1}{|A|}\begin{bmatrix} \begin{vmatrix} e & f \\ h & i \end{vmatrix} & -\begin{vmatrix} b & c \\ h & i \end{vmatrix} & \begin{vmatrix} b & c \\ e & f \end{vmatrix} \\ -\begin{vmatrix} d & f \\ g & i \end{vmatrix} & \begin{vmatrix} a & c \\ g & i \end{vmatrix} & -\begin{vmatrix} a & c \\ d & f \end{vmatrix} \\ \begin{vmatrix} d & e \\ g & h \end{vmatrix} & -\begin{vmatrix} a & b \\ g & h \end{vmatrix} & \begin{vmatrix} a & b \\ d & e \end{vmatrix} \end{bmatrix}$.

Use this formula to find the inverse of $\begin{bmatrix} 1 & 0 & -2 \\ 0 & 2 & 4 \\ 2 & 1 & -1 \end{bmatrix}$. $\begin{bmatrix} -3 & -1 & 2 \\ 4 & \frac{3}{2} & -2 \\ -2 & -\frac{1}{2} & 1 \end{bmatrix}$

52. *City and Suburban Living* The following matrix equation describes the population changes in a city and its suburbs between 1980 and 1990. See below.

$$\underbrace{\begin{bmatrix} 0.8 & 0.1 \\ 0.2 & 0.9 \end{bmatrix}}_{A}\underbrace{\begin{bmatrix} x \\ y \end{bmatrix}}_{X} + \underbrace{\begin{bmatrix} 22{,}000 \\ 37{,}000 \end{bmatrix}}_{B} - \underbrace{\begin{bmatrix} 42{,}500 \\ 27{,}500 \end{bmatrix}}_{C} = \underbrace{\begin{bmatrix} 211{,}200 \\ 282{,}200 \end{bmatrix}}_{D} \begin{matrix} \leftarrow \text{City} \\ \leftarrow \text{Suburbs} \end{matrix}$$

The first row of each matrix refers to the city population, and the second row refers to the suburban population.

A: Transition matrix (shows population changes between city and suburb)
X: 1980 population matrix
B: New births and people who moved in from another area
C: Deaths and people who moved away to another area
D: 1990 population matrix

How many people lived in the city and its suburbs in 1980? Is the city losing population to the suburbs? Explain.

52. ≈258,943 and ≈245,457; yes, the city's population decreased and the suburb's population increased

Mid-Chapter SELF-TEST

Take this test as you would take a test in class. The answers to the exercises are given in the back of the book.

In Exercises 1–6, perform the matrix operations. (4.1, 4.2)

1. $\begin{bmatrix} 2 & -3 & 5 \\ 0 & 2 & -2 \end{bmatrix} + \begin{bmatrix} 4 & 5 & 0 \\ -2 & -3 & 8 \end{bmatrix}$ 1. $\begin{bmatrix} 6 & 2 & 5 \\ -2 & -1 & 6 \end{bmatrix}$

2. $\begin{bmatrix} 5 & -2 \\ 4 & 1 \end{bmatrix} - \begin{bmatrix} 0 & -2 \\ 4 & -2 \end{bmatrix}$ $\begin{bmatrix} 5 & 0 \\ 0 & 3 \end{bmatrix}$

3. $4\begin{bmatrix} 5 & -3 & 6 \\ 9 & -4 & 0 \\ -2 & 3 & 7 \end{bmatrix}$ $\begin{bmatrix} 20 & -12 & 24 \\ 36 & -16 & 0 \\ -8 & 12 & 28 \end{bmatrix}$

4. $-\frac{1}{2}\begin{bmatrix} 4 \\ 6 \\ -8 \end{bmatrix} + 2\begin{bmatrix} -3 \\ 4 \\ 0 \end{bmatrix}$ $\begin{bmatrix} -8 \\ 5 \\ 4 \end{bmatrix}$

5. $\begin{bmatrix} 2 & -1 & 3 \\ 2 & 4 & 0 \end{bmatrix}\begin{bmatrix} 1 & 0 \\ 9 & -3 \\ 2 & -5 \end{bmatrix}$ $\begin{bmatrix} -1 & -12 \\ 38 & -12 \end{bmatrix}$

6. $\begin{bmatrix} 1 & 0 & 0 \\ 0 & 1 & 0 \\ 0 & 0 & 1 \end{bmatrix}\begin{bmatrix} -2 & 7 & 9 \\ 4 & -2 & -7 \\ 9 & 8 & 2 \end{bmatrix}$ $\begin{bmatrix} -2 & 7 & 9 \\ 4 & -2 & -7 \\ 9 & 8 & 2 \end{bmatrix}$

7. Evaluate $\begin{vmatrix} 5 & 6 \\ -1 & 2 \end{vmatrix}$. **(4.3)** 16

8. Evaluate $\begin{vmatrix} -1 & 3 & 5 \\ 0 & 1 & 3 \\ -2 & 2 & 1 \end{vmatrix}$. **(4.3)** −3

9. Find the inverse of $\begin{bmatrix} -2 & 3 \\ -4 & 5 \end{bmatrix}$. **(4.4)**

$\begin{bmatrix} \frac{5}{2} & -\frac{3}{2} \\ 2 & -1 \end{bmatrix}$

10. Solve the equation. **(4.5)**

$\begin{bmatrix} 2 & -3 \\ 3 & -4 \end{bmatrix}\begin{bmatrix} x \\ y \end{bmatrix} = \begin{bmatrix} -8 \\ -9 \end{bmatrix}$ $\begin{bmatrix} 5 \\ 6 \end{bmatrix}$

11. Which is the inverse of $\begin{vmatrix} 1 & 2 & 1 \\ 2 & 1 & 1 \\ 0 & 1 & 1 \end{vmatrix}$? **(4.4)**
See margin.

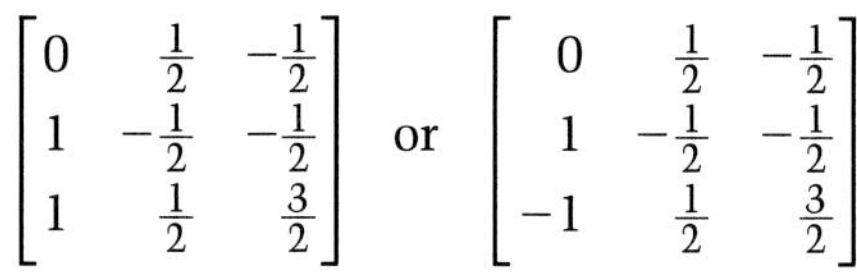

$\begin{bmatrix} 0 & \frac{1}{2} & -\frac{1}{2} \\ 1 & -\frac{1}{2} & -\frac{1}{2} \\ 1 & \frac{1}{2} & \frac{3}{2} \end{bmatrix}$ or $\begin{bmatrix} 0 & \frac{1}{2} & -\frac{1}{2} \\ 1 & -\frac{1}{2} & -\frac{1}{2} \\ -1 & \frac{1}{2} & \frac{3}{2} \end{bmatrix}$

12. Find the area of the triangle. **(4.3)** $\frac{41}{2}$

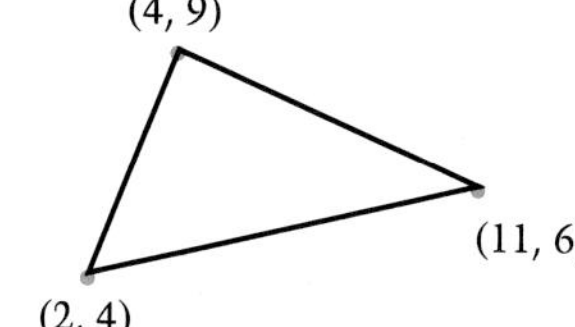

13. A company has two stores. The incomes and expenses (in dollars) for the two stores for the first two quarters of the year are shown in the matrices. Write a matrix that shows each store's profit for the first six months of the year. **(4.1)**

13., 14. See margin.

1st Quarter	Income	Expenses
Store 1	400,000	280,000
Store 2	320,000	240,000

2nd Quarter	Income	Expenses
Store 1	420,000	320,000
Store 2	340,000	250,000

14. A company keeps three types of sweatshirts in stock at each of its two stores. Use matrix multiplication to find a matrix that gives the inventory value of the sweatshirts at each store. **(4.2)**

Inventory	Type A	Type B	Type C
Store 1	25	32	34
Store 2	15	24	16

Type	Price (in dollars)
Type A	14
Type B	16
Type C	18

3. Square the matrix. Use the results to verify that squaring the matrix provides the number of two-flight routes between the cities. *Note:* Going from Kansas City to Nashville and from Nashville to Memphis is an example of a two-flight route.

$\begin{bmatrix} 6 & 5 & 2 & 5 \\ 5 & 11 & 5 & 2 \\ 2 & 5 & 6 & 5 \\ 5 & 2 & 5 & 11 \end{bmatrix}$

Answers

11. $\begin{bmatrix} 0 & \frac{1}{2} & -\frac{1}{2} \\ 1 & -\frac{1}{2} & -\frac{1}{2} \\ -1 & \frac{1}{2} & \frac{3}{2} \end{bmatrix}$

13.

	Profit ($)
Store 1	220,000
Store 2	170,000

14.

	Inventory Value ($)
Store 1	1474
Store 2	882

Problem of the Day

The product of the ages of a group of teenagers is 10,584,000. How many teenagers are in the group, and what is the sum of their ages?
6; 89

ORGANIZER

Warm-Up Exercises

1. Describe the steps used to solve the following linear equations.
   a. $x + 3 = 14$ b. $4x = 24$
   In each case, you "undo the operation" on the left side of the equation. In a., this means subtract 3 from both sides of the equation; in b., it means divide both sides of the equation by 4.
   Another way to describe the process is to say that you apply the "inverse operation" to both sides of the equation.
2. Solve each system of linear equations.
   a. $\begin{cases} 2x - 3y = -16 \\ 4x + 5y = 12 \end{cases}$ $(-2, 4)$
   b. $\begin{cases} x - 3y + 4z = 12 \\ 2x + y - z = -1 \\ x + 2y + 2z = 3 \end{cases}$
   $(1, -1, 2)$

Lesson Resources

Teaching Tools
- Problem of the Day: 4.5
- Warm-up Exercises: 4.5
- Answer Masters: 4.5

Extra Practice: 4.5

4.5 Solving Systems Using Inverse Matrices

What you should learn:

Goal 1 How to solve systems of linear equations using inverse matrices

Goal 2 How to use systems of linear equations to solve real-life problems

Why you should learn it:

You can use matrices to solve real-life problems that can be modeled with systems of linear equations.

Goal 1 Solving Systems Using Matrices

A system of linear equations can be written as a *single* matrix equation. Here is an example.

Linear System	Matrix Equation	
$\begin{cases} 2x - 3y = 10 \\ -5x + 8y = -26 \end{cases}$	$\underbrace{\begin{bmatrix} 2 & -3 \\ -5 & 8 \end{bmatrix}}_{A}\underbrace{\begin{bmatrix} x \\ y \end{bmatrix}}_{X} = \underbrace{\begin{bmatrix} 10 \\ -26 \end{bmatrix}}_{B}$	$AX = B$

The matrix A is the **coefficient matrix** of the system. X is the **matrix of variables,** and B is the **matrix of constants.**

Solution of a Linear System

Let $AX = B$ represent a system of linear equations. If the determinant of A is nonzero, then the linear system has exactly one solution, which is given by $X = A^{-1}B$.

Example 1 *Using an Inverse Matrix to Solve a Linear System*

Use an inverse matrix to solve the linear system.

$$\begin{cases} 2x - 3y = 10 \\ -5x + 8y = -26 \end{cases}$$

Solution Begin by writing the equation in matrix form.

$$\begin{bmatrix} 2 & -3 \\ -5 & 8 \end{bmatrix}\begin{bmatrix} x \\ y \end{bmatrix} = \begin{bmatrix} 10 \\ -26 \end{bmatrix}$$

Then, find the inverse of the coefficient matrix A.

$$A^{-1} = \frac{1}{16 - 15}\begin{bmatrix} 8 & 3 \\ 5 & 2 \end{bmatrix} = \begin{bmatrix} 8 & 3 \\ 5 & 2 \end{bmatrix}$$

Finally, multiply the matrix of constants by A^{-1} (on the left) to obtain the solution.

$$X = A^{-1}B = \begin{bmatrix} 8 & 3 \\ 5 & 2 \end{bmatrix}\begin{bmatrix} 10 \\ -26 \end{bmatrix} = \begin{bmatrix} 2 \\ -2 \end{bmatrix} = \begin{bmatrix} x \\ y \end{bmatrix}$$

The solution of the system is $x = 2$ and $y = -2$. ■

Using an inverse matrix to solve a linear system *by hand* is not particularly efficient, especially with linear systems that have more than two variables. If, however, you have access to a calculator or computer that does matrix algebra, then using inverse matrices is quite efficient.

Connections
Technology

Example 2 Using a Graphing Calculator to Solve a Linear System

Use a matrix equation and a graphing calculator to solve the linear system.

$$\begin{cases} 2x + 3y + z = -1 & \textit{Equation 1} \\ 3x + 3y + z = 1 & \textit{Equation 2} \\ 2x + 4y + z = -2 & \textit{Equation 3} \end{cases}$$

Solution The matrix equation that represents the system is

$$\begin{bmatrix} 2 & 3 & 1 \\ 3 & 3 & 1 \\ 2 & 4 & 1 \end{bmatrix}\begin{bmatrix} x \\ y \\ z \end{bmatrix} = \begin{bmatrix} -1 \\ 1 \\ -2 \end{bmatrix}.$$

Using a graphing calculator, you can find the inverse of the matrix A to be

$$A^{-1} = \begin{bmatrix} -1 & 1 & 0 \\ -1 & 0 & 1 \\ 6 & -2 & -3 \end{bmatrix}.$$

To find the solution, multiply B by A^{-1} (on the left).

$$X = A^{-1}B = \begin{bmatrix} -1 & 1 & 0 \\ -1 & 0 & 1 \\ 6 & -2 & -3 \end{bmatrix}\begin{bmatrix} -1 \\ 1 \\ -2 \end{bmatrix} = \begin{bmatrix} 2 \\ -1 \\ -2 \end{bmatrix} = \begin{bmatrix} x \\ y \\ z \end{bmatrix}$$

The solution is $x = 2$, $y = -1$, $z = -2$. ■

When you use a graphing calculator to solve the system of linear equations in Example 2, you don't have to display the matrix A^{-1} to find the solution. Simply enter the matrices A and B. Then, multiply B by A^{-1} to obtain the solution.

```
MATRIX[A] 3×3
[2  3  1]
[3  3  1]
[2  4  1]
```

Enter matrix A.

```
MATRIX[B] 3×1
[-1  ]
[1   ]
[-2  ]
```

Enter matrix B.

```
[A]-1 [B]
            [[2  ]
             [-1 ]
             [-2 ]]
```

Multiply B by A^{-1}.

When you use this method, remember that it only works if the matrix A has an inverse. If A does not have an inverse, then the system has either no solution or many solutions and you should use a different solution technique.

LESSON Notes

Example 1

Point out that the matrix equation, $AX = B$, can be solved much like a real-valued linear equation, $ax = b$, by multiplying by the inverse on both sides of the equation:

$$\begin{array}{ll} ax = b & AX = B \\ \frac{1}{a}ax = \frac{1}{a}b & (A^{-1})AX = (A^{-1})B \\ x = \frac{b}{a} & X = (A^{-1})B \end{array}$$

Notice that AX and B are both multiplied on the left by A^{-1}.

Example 2

Have students "check" the solution by solving the system by hand. They will be easily convinced about the efficiency with which graphing calculators solve systems of equations!

Example 3

Suppose the expected return for the stocks is as follows.

Stock	Expected Return
Stock X	10%
Stock Y	5%
Stock Z	9%

How much should be invested in each type of stock if the same investment strategy is used?

X: $6000, Y: $1500, Z: $22,500

Extra Examples

Here are additional examples similar to **Examples 1–3.**

1. Solve the systems of linear equations using inverse matrices.

a. $\begin{cases} 4x + 3y = -4 \\ 3x - y = -3 \end{cases}$ $(-1, 0)$

b. $\begin{cases} 2x - 7y = -3 \\ x + 5y = 7 \end{cases}$ $(2, 1)$

c. $\begin{cases} 3x + 2y - z = 11 \\ 2x - y + 2z = -3 \\ x + 3y - z = 8 \end{cases}$

$(2, 1, -3)$

2. In Example 3, suppose you have \$50,000 to invest using the same investment strategy except that you want the combined investments in Stocks Y and Z to be five times the amount invested in Stock X. How much should be invested in each type of stock?
X: \$8333.33, Y: \$16,666.67, Z: \$25,000.

Check Understanding

1. What is a coefficient matrix? a matrix of variables? a matrix of constants?
See page 204.
2. How is solving a real-valued linear equation similar to solving a matrix equation?
See Lesson Notes for Example 1.
3. Does every matrix equation have exactly one solution?
Yes, see page 204.

Communicating about ALGEBRA

The inverse of the coefficient matrix of a system of equations can also be found by solving a system of equations. For example, to find the inverse of $\begin{bmatrix} 5 & 2 \\ 4 & 3 \end{bmatrix}$ you could do the following. Since

$$\begin{bmatrix} 5 & 2 \\ 4 & 3 \end{bmatrix}\begin{bmatrix} a & b \\ c & d \end{bmatrix}=\begin{bmatrix} 1 & 0 \\ 0 & 1 \end{bmatrix},$$

solve for a, b, c, and d.

$$\begin{cases} 5a + 2c = 1 \\ 4a + 3c = 0 \end{cases} \quad \begin{cases} 5b + 2d = 0 \\ 4b + 3d = 1 \end{cases}$$

So, $a = \frac{3}{7}$, $b = -\frac{2}{7}$, $c = -\frac{4}{7}$, $d = \frac{5}{7}$.

Goal 2 Using a Linear System to Model Real Life

Real Life
Investment Strategy

Investors in the stock market often try to keep a diversified portfolio of stocks. This means that they will own some stocks that they believe will pay high returns, and others that are likely to pay average or even below-average returns. Typically stocks that have the potential for higher returns also are a greater risk.

Example 3 ***Dividing an Investment among Three Stocks***

You have \$30,000 to invest in the three stocks shown below. You want an average annual return of 9%. Because Stock X is a high-risk stock, you want the combined investment in Stock Y and Stock Z to be 4 times the amount invested in Stock X. How much should you invest in each type of stock?

Stock	Expected Return
Stock X	12%
Stock Y	9%
Stock Z	8%

Solution Let x, y, and z be the three investment amounts.

$$\begin{cases} x + y + z = 30{,}000 & \textit{Total investment: \$30,000} \\ 0.12x + 0.09y + 0.08z = 2{,}700 & \textit{Avg. return: 0.09(30,000)} \\ 4x - y - z = 0 & \textit{(y + z) is 4 times x.} \end{cases}$$

The coefficient matrix and matrix of constants are

$$A = \begin{bmatrix} 1 & 1 & 1 \\ 0.12 & 0.09 & 0.08 \\ 4 & -1 & -1 \end{bmatrix} \quad \text{and} \quad B = \begin{bmatrix} 30{,}000 \\ 2{,}700 \\ 0 \end{bmatrix}.$$

Using a graphing calculator, you can find the solution to be

$$X = A^{-1}B = \begin{bmatrix} 0.2 & 0 & 0.2 \\ -8.8 & 100 & -0.8 \\ 9.6 & -100 & 0.6 \end{bmatrix}\begin{bmatrix} 30{,}000 \\ 2{,}700 \\ 0 \end{bmatrix} = \begin{bmatrix} 6{,}000 \\ 6{,}000 \\ 18{,}000 \end{bmatrix}.$$

The solution is $x = \$6000$, $y = \$6000$, and $z = \$18{,}000$. ■

Communicating about ALGEBRA

SHARING IDEAS about the Lesson See Additional Answers.

Use an inverse matrix to solve each linear system. Solve the first by hand and use a graphing calculator to solve the second.

A. $\begin{cases} 5x + 2y = 21 \\ 4x + 3y = 35 \end{cases}$ (−1, 13)

B. $\begin{cases} x + 3y - 5z = -3 \\ x - 2y - z = 12 \\ x + 3y - z = 18 \end{cases}$ (19.65, 1.2, 5.25)

EXERCISES

Guided Practice

CRITICAL THINKING about the Lesson

1. Write the linear system $\begin{cases} 6x - y = 0 \\ x + 3y = 5 \end{cases}$ as a matrix equation. $\begin{bmatrix} 6 & -1 \\ 1 & 3 \end{bmatrix}\begin{bmatrix} x \\ y \end{bmatrix} = \begin{bmatrix} 0 \\ 5 \end{bmatrix}$

2. Is $\begin{bmatrix} 28 \\ -10 \\ -7 \end{bmatrix}$ a solution of $\begin{bmatrix} 3 & 7 & 2 \\ -2 & -5 & -1 \\ 0 & 1 & -2 \end{bmatrix}\begin{bmatrix} x \\ y \\ z \end{bmatrix} = \begin{bmatrix} 0 \\ 1 \\ 4 \end{bmatrix}$? Explain. See margin.

3. If $|A| \neq 0$, what is the solution to $AX = B$? $A^{-1}B$

4. Explain how to use an inverse matrix to solve $\begin{cases} -3x + y + 5z = 2 \\ x - z = 1. \\ 2y + 3z = 2 \end{cases}$ See margin.

Independent Practice

In Exercises 5–10, write the linear system as a matrix equation. See Additional Answers.

5. $\begin{cases} 2x - 4y = 7 \\ -3x + y = 12 \end{cases}$

6. $\begin{cases} x + 9y = 20 \\ 2x - 4y = 15 \end{cases}$

7. $\begin{cases} 2x - 5y = 12 \\ x - 3y = -3 \end{cases}$

8. $\begin{cases} x - 2y + 3z = 14 \\ 2x + y - 2z = 16 \\ -3x - 5y + 9z = 36 \end{cases}$

9. $\begin{cases} 3x - y + 4z = 16 \\ 2x + 4y - z = 10 \\ x - y + 3z = 31 \end{cases}$

10. $\begin{cases} x + y - z = 0 \\ 2x - z = 1 \\ y + z = 2 \end{cases}$

In Exercises 11–19, use an inverse matrix to solve the linear system.

11. $\begin{cases} 3x + y = 5 \\ 5x + 2y = 9 \end{cases}$ (1, 2)

12. $\begin{cases} x + y = 2 \\ 7x + 8y = 21 \end{cases}$ (−5, 7)

13. $\begin{cases} 2x + 7y = -53 \\ x + 3y = -22 \end{cases}$ (5, −9)

14. $\begin{cases} 8x + 5y = 38 \\ 3x + 2y = 14 \end{cases}$ (6, −2)

15. $\begin{cases} -2x - 3y = -26 \\ 3x + 4y = 36 \end{cases}$ (4, 6)

16. $\begin{cases} 5x - 7y = 54 \\ 2x - 4y = 30 \end{cases}$ (1, −7)

17. $\begin{cases} x + 2y = -9 \\ -2x - 3y = 14 \end{cases}$ (−1, −4)

18. $\begin{cases} -5x + 3y = -52 \\ 3x + 2y = 32 \end{cases}$ $\left(\frac{200}{19}, \frac{4}{19}\right)$

19. $\begin{cases} 3x - 7y = -16 \\ -2x + 4y = 8 \end{cases}$ (4, 4)

In Exercises 20–22, the inverse of the coefficient matrix is given. Use the inverse to solve the linear system.

20. $\begin{cases} -2x + y + z = 0 \\ x + 2z = 9 \\ x - 2y - 9z = -31 \end{cases}$

$A^{-1} = \begin{bmatrix} 4 & 7 & 2 \\ 11 & 17 & 5 \\ -2 & -3 & -1 \end{bmatrix}$ (1, −2, 4)

21. $\begin{cases} x + 9y + z = 20 \\ x + 10y - 2z = 18 \\ 3x + 27y + 2z = 58 \end{cases}$

$A^{-1} = \begin{bmatrix} -74 & -9 & 28 \\ 8 & 1 & -3 \\ 3 & 0 & -1 \end{bmatrix}$ (−18, 4, 2)

22. $\begin{cases} 2x - 4y + 2z = 16 \\ -2x + 5y + 2z = -34 \\ x - 2y + 2z = 4 \end{cases}$

$A^{-1} = \begin{bmatrix} 7 & 2 & -9 \\ 3 & 1 & -4 \\ -\frac{1}{2} & 0 & 1 \end{bmatrix}$ (8, −2, −4)

EXERCISE Notes

ASSIGNMENT GUIDE

Basic/Average: Ex. 1–4, 5–21 odd, 27–31 odd, 32, 33–39 odd

Above Average: Ex. 1–4, 5–31 odd, 32, 33–39 odd, 40

Advanced: Ex. 1–4, 9–27 odd, 29–32, 35, 39, 40

Selected Answers
Exercises 1–4, 5–37 odd

Use **Mixed Review** as needed.

✪ More Difficult Exercises
Exercises 29–32, 40

Guided Practice

▶ **Ex. 1–4** These exercise review students' understanding of the process of solving a system of linear equations by using inverse matrices.

Answers

2. Yes, $\begin{bmatrix} 3 & 7 & 2 \\ -2 & -5 & -1 \\ 0 & 1 & -2 \end{bmatrix}\begin{bmatrix} 28 \\ -10 \\ -7 \end{bmatrix} = \begin{bmatrix} 0 \\ 1 \\ 4 \end{bmatrix}$

4. $\begin{bmatrix} x \\ y \\ z \end{bmatrix} = \begin{bmatrix} -3 & 1 & 5 \\ 1 & 0 & -1 \\ 0 & 2 & 3 \end{bmatrix}^{-1}\begin{bmatrix} 2 \\ 1 \\ 2 \end{bmatrix}$

Independent Practice

▶ **Ex. 11–19** Encourage students to review Example 1.

▶ **Ex. 20–22** Encourage students to review Example 2. The inverse matrix is already provided for students, thus shortening the time needed to complete the exercises.

Ex. 23–28 TECHNOLOGY

These exercises should be assigned as independent practice only if students have access to a graphing calculator. If not, have students work on them in class in a cooperative group setting. Students should decide how to divide the work amongst themselves. You might also refer them to the algorithm presented in Exercise 51 on page 202. Beware that it is cumbersome and time-consuming.

Ex. 29–32 PROBLEM SOLVING These exercises are real-life models of a system of three equations with three unknowns. Refer students to Example 3 for a model for writing the system of equations in the appropriate form.

Answers

29. $e + c + s = 1162,$

$e = c + s,$

$s = 0.14(1162);$

operating expenses (e): $581 million,

facilities construction (c): $418.32 million,

surplus (s): $162.68 million

30. $s + a + i = 35,$

$s + a = \left(\frac{6}{7}\right)35,$

$s = \frac{26}{5}i;$

services (*s*): $26 billion,

agricultural products (*a*): $4 billion,

industrial goods (*i*): $5 billion

31. X: $5000, Y: $7500, Z: $7500

32. American Bowling Congress: 2.3 million,

Women's Bowling Congress: 2.2 million,

Young American's Bowling Alliance: 0.5 million

Graphing Calculator **In Exercises 23–28, use an inverse matrix and a graphing calculator to solve the linear system.**

23. $\begin{cases} 3x + 2y = 13 \\ 3x + 2y + z = 13 \\ 2x + y + 3z = 9 \end{cases}$ (5, −1, 0)

24. $\begin{cases} -x + y - 3z = -4 \\ 3x - 2y + 8z = 14 \\ 2x - 2y + 5z = 7 \end{cases}$ (4, 3, 1)

25. $\begin{cases} 2x + 6y - 4z = 8 \\ 3x + 10y - 7z = 12 \\ -2x - 6y + 5z = -3 \end{cases}$ (−1, 5, 5)

26. $\begin{cases} 2x + z = 2 \\ 5x - y + z = 5 \\ -x + 2y + 2z = 0 \end{cases}$ (2, 3, −2)

27. $\begin{cases} 5x + 4y + z = 14 \\ 5x + 2y = 3 \\ 2x + 5y + 2z = 24 \end{cases}$ (−1, 4, 3)

28. $\begin{cases} x + y - 2z = 5 \\ 2x + z = 2 \\ -9x - 2y + z = -1 \end{cases}$ (−15, 84, 32)

29. *Summer Olympics* The 1996 Summer Olympics were held in Atlanta, Georgia. In 1991, the planned budget for the event was $1162 million. The budget contained three categories: operating expenses, *e*, facilities construction, *c*, and surplus, *s*. The amount allowed for operating expenses was the same as the total amount allowed for facilities construction and surplus combined. The budget committee estimated that 14% of the total budget would be surplus. Write a linear system that represents this information. Solve the linear system to find the amount of money budgeted for each category. See margin.

30. *Gross State Product* The 1991 gross state product of Nebraska can be classified into: services, *s*, agricultural products, *a*, and industrial goods, *i*. The total gross state product was $35 billion. Services and agriculture made up sixth sevenths, and services produced five and one fifth times as much as industry. Write a linear system that represents this information. Solve the system to find the value of agricultural goods, industrial goods, and services in 1991. ***(Source: U.S. Bureau of Economic Analysis)*** See margin.

31. *Stock Investment* You have $20,000 to invest in three types of stocks. You expect the annual returns on Stock X, Stock Y, and Stock Z to be 12%, 10%, and 6%, respectively. You want the combined investment in Stock Y and Stock Z to be three times the amount invested in Stock X. How much should you invest in each type of stock to obtain an average return of 9%? See margin.

32. *Bowling* In 1993, the total membership in the American Bowling Congress, Women's Bowling Congress, and Young American Bowling Alliance was about 5 million. The American Bowling Congress and Women's Bowling Congress membership made up 90% of the total. The Young American Bowling Alliance and the Women's Bowling Congress made up 54% of the total. What was the membership in each bowling organization? See margin.

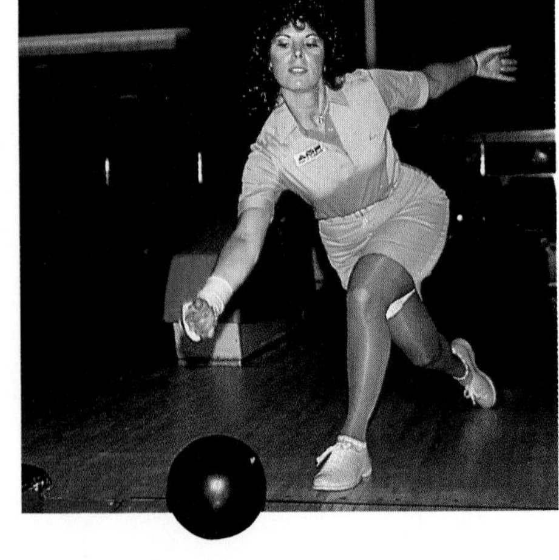

Integrated Review

In Exercises 33–36, solve the linear system by any convenient method.

33. $\begin{cases} 6x + 4y = 8 \\ 3x + 3y = 9 \end{cases}$ (−2, 5)

34. $\begin{cases} 4x - 2y = 0 \\ 2x - y = -1 \end{cases}$ No solution

35. $\begin{cases} x - 3y + z = 7 \\ y - z = 2 \\ x - 3y - z = 11 \end{cases}$ (9, 0, −2)

36. $\begin{cases} x + 3y + z = 1 \\ x + 5y + 2z = -1 \\ 2x + 6y + 3z = 2 \end{cases}$ (4, −1, 0)

In Exercises 37 and 38, solve the matrix equation.

37. $\begin{bmatrix} 0 & 1 \\ -1 & 4 \end{bmatrix}\begin{bmatrix} x \\ y \end{bmatrix} = \begin{bmatrix} 2 \\ 9 \end{bmatrix}$ $\begin{bmatrix} -1 \\ 2 \end{bmatrix}$

38. $\begin{bmatrix} 5 & 4 \\ 4 & 3 \end{bmatrix}\begin{bmatrix} x \\ y \end{bmatrix} = \begin{bmatrix} 7 \\ 6 \end{bmatrix}$ $\begin{bmatrix} 3 \\ -2 \end{bmatrix}$

Exploration and Extension

39. *Decoding a Message* Use the inverse of $A = \begin{bmatrix} 1 & 0 & 5 \\ 1 & 2 & 0 \\ 1 & 1 & 3 \end{bmatrix}$ to decode the message: 37, 45, 80, 16, 12, 56, 35, 17, 138. GOOD LUCK

40. Solve the linear system, using the given inverse of the coefficient matrix. (20, −1, −12, −8)

$$\begin{cases} x + 6y + 3z - 3w = 2 \\ 2x + 7y + z + 2w = 5 \\ x + 5y + 3z - 3w = 3 \\ -6y - 2z + 3w = 6 \end{cases} \qquad A^{-1} = \begin{bmatrix} 40 & -3 & -33 & 9 \\ 1 & 0 & -1 & 0 \\ -39 & 3 & 33 & -8 \\ -24 & 2 & 20 & -5 \end{bmatrix}$$

Mixed REVIEW

8.–12., 15.–18. See Additional Answers.

1. Solve $7x - 4 = -2x + 14$. **(1.3)** 2

2. Solve $\frac{1}{2}(-5x + 3) = \frac{3}{2}(-x + 2)$. **(1.3)** $-\frac{3}{2}$

3. Evaluate $y - 2z$ when $y = 1$ and $z = 2$. **(1.2)** −3

4. Find the slope of the line $3x - 4y = 9$. **(2.3)** $\frac{3}{4}$

5. Simplify $\frac{4}{5}(25) - \frac{3}{5}(30) + \frac{2}{5}(45)$. 20

6. Write in decimal form: 3.56×10^{-2}. 0.0356

7. Solve $3x + 4 \le 5x - 8$. $x \ge 6$

8. Evaluate $-3w^2 + 4w$ when $w = -2$. **(1.2)**

9. Sketch the graph of $y = -2x + 5$. **(2.1)**

10. Sketch the graph of $y \le \frac{1}{2}x - 2$. **(2.5)**

11. Sketch the graph of $|x - 3| \le 2$. **(2.6)**

12. Find the area of a circle of radius 4 feet.

13. Solve the system: $\begin{cases} 3x - 2y = 8 \\ -2x + 5y = 2 \end{cases}$ **(3.2)** (4, 2)

14. Solve the system: $\begin{cases} -x - 5y = 33 \\ 3x + 4y = -22 \end{cases}$ **(3.2)** (2, −7)

15. Add: $\begin{bmatrix} -2 & 5 & 3 \\ 0 & -3 & -4 \end{bmatrix} + \begin{bmatrix} -4 & -2 & -5 \\ 2 & -6 & 7 \end{bmatrix}$. **(4.1)**

16. Add: $\begin{bmatrix} -1 & 0 \\ 2 & -3 \end{bmatrix} + \begin{bmatrix} -2 & 4 \\ 1 & -7 \end{bmatrix}$. **(4.1)**

17. Multiply: $\begin{bmatrix} -9 & 4 \\ 7 & -3 \end{bmatrix}\begin{bmatrix} 3 & 4 \\ 7 & 9 \end{bmatrix}$. **(4.1)**

18. Multiply: $\begin{bmatrix} -8 & 4 \\ 4 & -2 \end{bmatrix}\begin{bmatrix} 1 & 2 \\ 2 & 4 \end{bmatrix}$. **(4.2)**

Ex. 39 TECHNOLOGY Finding A^{-1} using a graphics calculator will require less than 10 seconds of students' time.

Enrichment Activities

These activities require students to go beyond lesson goals.

1. What do you know about the following system of equations?

   $\begin{cases} 3x - y = -3 \\ -3x + y = 2 \end{cases}$

   The system of equations has no solution because the equations are inconsistent.

2. Describe what happens when the inverse matrix is used to solve a system of linear equations that has either no solution, or has infinitely many solutions. Provide examples of systems of linear equations to justify your explanations.

   For example, the coefficient matrix of the given system is

   $A = \begin{bmatrix} 3 & -1 \\ -3 & 1 \end{bmatrix}$

   Since $|A| = 0$, A^{-1} does not exist and you cannot use the inverse matrix technique.

   Similarly, $|A| = 0$ when the system of equations is dependent.

Problem of the Day

Suppose you have ten equal stacks of half-dollars. You know one stack is counterfeit. You also know the weight of a genuine half-dollar, and that a counterfeit weighs 1 gram more. What is the smallest number of weighings on a balance scale that is required in order to determine which stack is counterfeit?

One. Take one coin from stack 1, two from stack 2, three from stack 3, etc. You know what this group of coins would weigh if they were all genuine. The number of excess grams is the number of the counterfeit stack.

ORGANIZER

Warm-Up Exercises

Solve each system using the Gaussian elimination method.

1. $\begin{cases} 3x + 2y = 2 \\ 5x - 6y = 50 \end{cases}$

2. $\begin{cases} -x + 2y + 5z = -10 \\ 2x + y + 2z = 2 \\ x - 3y + z = -20 \end{cases}$

1. (4, −5), 2. (2, 6, −4)

Lesson Resources

Teaching Tools
- Problem of the Day: 4.6
- Warm-up Exercises: 4.6
- Answer Masters: 4.6

Extra Practice: 4.6

Technology Handbook: p. 23

4.6 Solving Systems Using Augmented Matrices

What you should learn:

Goal 1 How to use an augmented matrix to solve a system of linear equations

Goal 2 How to use a system of linear equations to solve real-life problems

Why you should learn it:

Solving linear systems with augmented matrices is the preferred computer technique for systems of linear equations with many variables and equations.

Goal 1 Using an Augmented Matrix

A matrix derived from a system of linear equations is the **augmented matrix** of the system.

Linear System	**Augmented Matrix**
$\begin{cases} x - 4y + 3z = 5 \\ -x + 3y - z = -3 \\ 2x \quad\quad - 4z = 6 \end{cases}$	$\left[\begin{array}{rrr:r} 1 & -4 & 3 & 5 \\ -1 & 3 & -1 & -3 \\ 2 & 0 & -4 & 6 \end{array}\right]$

The first three columns of the augmented matrix show the coefficients of x, y, and z in the linear system. (Note the use of 0 for the missing y-variable in the third equation.) The fourth column in the augmented matrix shows the constant terms in the linear system.

Recall from Lesson 3.6 the three operations that can be used on a system of linear equations in order to produce an equivalent system:

1. Interchange two equations.
2. Multiply an equation by a nonzero constant.
3. Add a multiple of an equation to another equation.

In matrix terminology, these three operations correspond to **elementary row operations.** An elementary row operation on an augmented matrix of a given linear system produces a new augmented matrix corresponding to a new (but equivalent) linear system. Two matrices are **row-equivalent** if one can be obtained from the other by a sequence of elementary row operations.

Elementary Row Operations

Any of the following elementary row operations performed on an augmented matrix will produce a matrix that is row-equivalent to the original matrix.

1. Interchange two rows.
2. Multiply a row by a nonzero constant.
3. Add a multiple of a row to another row.

In Lesson 3.6, you learned how to use elimination to solve a linear system—first write the system in triangular form, then use substitution to solve for each variable. With matrices, you use the same basic approach. One difference is that with matrices, you do not need to keep writing the variables.

Example 1 Using Elementary Row Operations to Solve a System

Linear System | **Augmented Matrix**

$$\begin{cases} x - 2y + 2z = 9 \\ -x + 3y = -4 \\ 2x - 5y + z = 10 \end{cases} \qquad \left[\begin{array}{ccc:c} 1 & -2 & 2 & 9 \\ -1 & 3 & 0 & -4 \\ 2 & -5 & 1 & 10 \end{array}\right]$$

Add the first equation to the second equation. | Add the first row to the second row ($R_1 + R_2$).

$$\begin{cases} x - 2y + 2z = 9 \\ y + 2z = 5 \\ 2x - 5y + z = 10 \end{cases} \qquad R_1 + R_2 \longrightarrow \left[\begin{array}{ccc:c} 1 & -2 & 2 & 9 \\ 0 & 1 & 2 & 5 \\ 2 & -5 & 1 & 10 \end{array}\right]$$

Add -2 times the first equation to the third equation. | Add -2 times the first row to the third row.

$$\begin{cases} x - 2y + 2z = 9 \\ y + 2z = 5 \\ -y - 3z = -8 \end{cases} \qquad -2R_1 + R_3 \longrightarrow \left[\begin{array}{ccc:c} 1 & -2 & 2 & 9 \\ 0 & 1 & 2 & 5 \\ 0 & -1 & -3 & -8 \end{array}\right]$$

Add the second equation to the third equation. | Add the second row to the third row.

$$\begin{cases} x - 2y + 2z = 9 \\ y + 2z = 5 \\ -z = -3 \end{cases} \qquad R_2 + R_3 \longrightarrow \left[\begin{array}{ccc:c} 1 & -2 & 2 & 9 \\ 0 & 1 & 2 & 5 \\ 0 & 0 & -1 & -3 \end{array}\right]$$

Multiply the third equation by -1. | Multiply the third row by -1.

$$\begin{cases} x - 2y + 2z = 9 \\ y + 2z = 5 \\ z = 3 \end{cases} \qquad -R_3 \longrightarrow \left[\begin{array}{ccc:c} 1 & -2 & 2 & 9 \\ 0 & 1 & 2 & 5 \\ 0 & 0 & 1 & 3 \end{array}\right]$$

At this point, you can use substitution to find that the solution is $x = 1$, $y = -1$, and $z = 3$. ■

Although elementary row operations are simple to perform, they involve a lot of arithmetic. Because it is easy to make a mistake, we suggest that you get in the habit of noting the elementary row operations performed in each step so that you can go back and check your work. The notation we use to denote each elementary row operation is to write an abbreviated version of the row operation to the left of the *row that has been changed.*

LESSON Notes

Example 1

Work the problems in parallel fashion as indicated in the text. This shows students that the "new" technique is, in fact, not so very new after all.

The common practice when using the Gaussian elimination method is to change the row "to which" another is added. Be careful not to say that two rows will be "added together," but rather, indicate which row will be added "to" the other.

Example 2

Wrought iron is resistant to corrosion, is tough, and can be hammered into thin layers without breaking. It is commonly used in fences, grating, and rivets.

Chromium is a grayish white metallic element that is very hard and highly resistant to corrosion. Steel is a strong iron alloy containing small amounts of carbon and various amounts of other metals such as chromium or nickel.

Cast iron is a strong but somewhat brittle metal formed by casting (that is, formed from a mold).

Extra Examples

Here are additional examples similar to **Examples 1–2.**

Solve each system using augmented matrices.

1. $\begin{cases} 3x + y = -5 \\ 4x - 2y = -20 \end{cases}$ $(-3, 4)$

2. $\begin{cases} x + y = -1 \\ 7x + 9y = -19 \end{cases}$ $(5, -6)$

3. $\begin{cases} -5x + 2y + z = -18 \\ 3x + y - 5z = 2 \\ 1x - 3y + 4z = 13 \end{cases}$ $(3, -2, 1)$

4. $\begin{cases} x + 6y + z = 3 \\ 11x - 15y + 9z = 31 \\ -8x + 3y + 12z = -4 \end{cases}$ $(2, 0, 1)$

Check Understanding

1. Describe how an augmented matrix is formed.
   See page 210.
2. How is solving systems of linear equations by the Gaussian elimination method similar to using augmented matrices?
   See Example 1.

Communicating about ALGEBRA

MATH JOURNAL Ask students to list in their math journals all the algebraic methods (up to the present) that can be used to solve systems of linear equations. Tell students to devise a scheme by which each method can be evaluated on such dimensions as: easy to use by hand, easy to use with technology, easy to explain how it works, etc.

Answers

A.–B. Elementary row operations vary.

A. $2R_1 + R_2 \to$ new R_2, $-\frac{1}{10}R_2$; $(-4, -3)$

B. $-3R_1 + R_2 \to$ new R_2, $-2R_1 + R_3 \to$ new R_3; $(1, 2, -1)$

Goal 2 Using Linear Systems in Real-Life Settings

Real Life **Metallurgy**

The properties of pure iron can be altered by adding other elements to form iron alloys. Elements that are often added are carbon, chromium, silicon, and nickel.

Example 2 *Solving a Mixture Problem*

Three iron *alloys* contain the following percents of carbon, chromium, and iron.

	Alloy X	Alloy Y	Alloy Z
Carbon	1%	1%	4%
Chromium	0%	15%	3%
Iron	99%	84%	93%

You have 15 tons of carbon, 39 tons of chromium, and 546 tons of iron. How much of each type of alloy can you make?

Solution Let x, y, and z be the amount of each iron alloy. You can model the situation with the following linear system.

$$\begin{cases} 0.01x + 0.01y + 0.04z = 15 & \text{Carbon} \\ \phantom{0.01x + {}} 0.15y + 0.03z = 39 & \text{Chromium} \\ 0.99x + 0.84y + 0.93z = 546 & \text{Iron} \end{cases}$$

The augmented matrix for this system is as follows.

$$\left[\begin{array}{ccc:c} 0.01 & 0.01 & 0.04 & 15 \\ 0 & 0.15 & 0.03 & 39 \\ 0.99 & 0.84 & 0.93 & 546 \end{array}\right]$$

After applying several elementary row operations, you obtain the following augmented matrix, which is in triangular form.

$$\left[\begin{array}{ccc:c} 1 & 1 & 4 & 1500 \\ 0 & 1 & \frac{1}{5} & 260 \\ 0 & 0 & 1 & 300 \end{array}\right] \longrightarrow \begin{cases} x + y + 4z = 1500 \\ y + \frac{1}{5}z = 260 \\ z = 300 \end{cases}$$

Using substitution, the solution is $x = 100$, $y = 200$, and $z = 300$. Thus, you can make 100 tons of Alloy X, 200 tons of Alloy Y, and 300 tons of Alloy Z. (Alloys X, Y, and Z are types of wrought iron, stainless steel, and cast iron, respectively.) ■

Communicating about ALGEBRA

SHARING IDEAS about the Lesson See margin.
Elementary row operations vary.

Use an augmented matrix to solve each linear system.

A. $\begin{cases} x - 3y = 5 \\ -2x - 4y = 20 \end{cases}$ **B.** $\begin{cases} x + y + 2z = 1 \\ 3x - y - z = 2 \\ 2x + 3y + 4z = 4 \end{cases}$

EXERCISES

Guided Practice

CRITICAL THINKING about the Lesson See page 210.

1. Describe the three elementary row operations that can be performed on an augmented matrix.
2. What is meant by saying that two augmented matrices are row-equivalent?
3. Are the following augmented matrices row-equivalent? Justify your answer. Yes; $2R_1 + R_2 \rightarrow$ new R_2, $-R_1 + R_3 \rightarrow$ new R_3

$$\left[\begin{array}{rrr:r} 1 & 2 & -2 & 8 \\ -2 & -3 & 7 & -2 \\ 1 & 2 & -1 & 2 \end{array}\right] \text{ and } \left[\begin{array}{rrr:r} 1 & 2 & -2 & 8 \\ 0 & 1 & 3 & 14 \\ 0 & 0 & 1 & -6 \end{array}\right]$$

4. What elementary row operation does $-3R_1 + R_3$ denote? Which row can it replace? Explain. −3 times the first row added to the third row; either Row 1 or Row 3

Independent Practice

In Exercises 5–7, write the augmented matrix for the linear system. See Additional Answers.

5. $\begin{cases} x + 3y - z = 16 \\ 4x - 6y = 9 \\ 2y - 3z = 12 \end{cases}$

6. $\begin{cases} -x - 3y + 2z = 8 \\ 2x + y - 3z = 9 \\ x - 3z = 10 \end{cases}$

7. $\begin{cases} 2x - y + 3z = 4 \\ -3x - z = 1 \\ x - 3y + z = 5 \end{cases}$

In Exercises 8–10, write the linear system that is represented by the augmented matrix. Then use substitution to solve the system. See margin.

8. $\left[\begin{array}{rrr:r} 1 & 1 & 2 & -5 \\ 0 & 1 & 5 & 2 \\ 0 & 0 & 1 & 4 \end{array}\right]$

9. $\left[\begin{array}{rrr:r} 1 & -6 & 4 & 7 \\ 0 & 1 & -8 & -2 \\ 0 & 0 & 1 & 0 \end{array}\right]$

10. $\left[\begin{array}{rrr:r} 1 & 2 & -2 & 1 \\ 0 & 1 & -3 & -8 \\ 0 & 0 & 1 & 6 \end{array}\right]$

In Exercises 11–19, use an augmented matrix to solve the linear system. See margin.

11. $\begin{cases} x + 3y + z = 3 \\ x + 5y + 5z = 1 \\ 2x + 6y + 3z = 8 \end{cases}$

12. $\begin{cases} 3x + 6y + 6z = 3 \\ x + 3y + 10z = -10 \\ x + 2y + 5z = -11 \end{cases}$

13. $\begin{cases} y - 5z = 15 \\ x + 2y - z = 7 \\ -3x - y + 2z = 10 \end{cases}$

14. $\begin{cases} 2x - 10y + 3z = -20 \\ x - 3y + 7z = 0 \\ x - 5y + z = -10 \end{cases}$

15. $\begin{cases} 2x + 4y + 5z = 5 \\ x + 3y + 3z = 2 \\ 2x + 4y + 4z = 2 \end{cases}$

16. $\begin{cases} x - y + 3z = 6 \\ x - 2y = 5 \\ 2x - 2y + 5z = 9 \end{cases}$

17. $\begin{cases} x + 2z = 4 \\ x + y + z = 6 \\ 3x + 3y + 4z = 28 \end{cases}$

18. $\begin{cases} -2x - 2y - 15z = 0 \\ x + 2y + 2z = 18 \\ 3x + 3y + 22z = 2 \end{cases}$

19. $\begin{cases} 2x + 10y = 28 \\ x + 3y + 4z = 22 \\ x + 5y - z = 10 \end{cases}$

EXERCISE Notes

ASSIGNMENT GUIDE

Basic/Average: Ex. 1–4, 7, 15–19 odd, 21–24, 34–35

Above Average: Ex. 1–4, 7, 15–19 odd, 20–23, 26, 35–37

Advanced: Ex. 1–4, 7, 15–19 odd, 20–23, 35–37

Selected Answers
Exercises 1–4, 5–35 odd

✪ More Difficult Exercises
Exercises 20–23, 36–37

Guided Practice

▶ **Ex. 1–4** These exercises review important information from the reading of the lesson. They require students to reflect and summarize the augmented matrix process.

Independent Practice

▶ **Ex. 8–10** Encourage students to notice that the matrices are in a "triangular" form.

Answers

8. (5, −18, 4) **9.** (−5, −2, 0)
10. (−7, 10, 6) **11.** (16, −5, 2)
12. (−33, 21, −4)
13. $\left(-\frac{13}{2}, \frac{35}{6}, -\frac{11}{6}\right)$
14. (15, 5, 0) **15.** (−1, −2, 3)
16. (−11, −8, 3) **17.** (−16, 12, 10)
18. (34, −4, −4) **19.** (−6, 4, 4)

▶ **Ex. 11–19** Each of these exercises will require ample time for students to complete. Students should use the format modeled by Example 1. Some students may be confused as to why they should perform a certain row operation or why certain row operations seem to undo previously obtained 1's and 0's. You may also wish to suggest the following strategy.

The Augmented Matrix Strategy for 3 × 3 Matrices

Follow these steps in the given order.

1. Obtain a 1 in row 1 and column 1, either by row exchange or by multiplying row 1 by a nonzero constant.
2. Obtain 0's in the rest of column 1 by using row operations with row 1.
3. Obtain a 1 in row 2 and column 2, by multiplying row 2 by a nonzero constant.
4. Obtain a 0 in row 3 column 2 by using row operations with row 2.
5. Obtain a 1 in row 3 and column 3, by multiplying row 3 by a nonzero constant.

▶ **Ex. 20–23** The solution to these exercises is modeled in Example 2. Exercises 22 and 23 are examples of problems that can be solved by using the augmented matrix technique. However, encourage students to practice the method, but to be aware of other methods of solution for systems of equations.

▶ **Ex. 37** TECHNOLOGY If students have access to graphics calculators, suggest they solve the exercise by using inverse matrices.

20. *Brochure Mailing* An art gallery is mailing brochures to 1960 people to promote the gallery's annual exhibition. The envelopes were addressed by hand with calligraphy (decorative handwriting). You began addressing envelopes early. Later Rosa started, and finally Rafael started. Together you and Rafael addressed 1680 envelopes and together Rosa and Rafael addressed 840 envelopes. How many did each of you address? You: 1120, Rosa: 280, Rafael: 560

21. *Toy Factory* You are working at a toy factory using two machines that can assemble plastic boats. You and two other employees, Jamie and Lan, take turns operating the machines. On Monday, Jamie and you operate the machines and produce 10,200 boats. On Tuesday, Jamie and Lan operate the machines and produce 10,100 boats. On Wednesday, Lan and you operate the machines and produce 9900 boats. The rates at which each of you work is the same each day. How many boats can each of you produce in a day? You: 5000, Jamie: 5200, Lan: 4900

22. *Dental Fillings* Dentists use mixtures of amalgam for silver fillings. The matrix shows the amounts of copper, mercury, and silver-tin that can be used to make two types of amalgam and one other mixture (which is not used in dental work). You have 320 grams of copper, 6140 grams of mercury, and 3540 grams of a silver-tin alloy. How much of each mixture can you make? I: 2000 g, II: 5000 g, III: 3000 g

Percent by Weight	I	II	III
Copper	0.02	0.02	0.06
Mercury	0.46	0.48	0.94
Silver-Tin	0.52	0.50	0.00

Very pure gold is mixed with other metals such as silver and copper to make jewelry.

23. *Gold Alloys* Gold jewelry is seldom made of pure gold because pure gold is soft and expensive. Instead, gold is mixed with other metals to produce a harder, less expensive gold alloy. The amount of gold (by weight) in an alloy is measured in karats. Anything made of 24-karat gold is 100% gold. An 18-karat gold mixture is 75% gold, and so on. Three gold alloys contain the percents of gold, copper, and silver shown in the matrix. You have 20,144 grams of gold, 766 grams of copper, and 1990 grams of silver. How much of each alloy can you make?

Percent by Weight	Alloy X	Alloy Y	Alloy Z
Gold	94%	92%	80%
Copper	4%	2%	4%
Silver	2%	6%	16%

X: 6600 g, Y: 7500 g, Z: 8800 g

Integrated Review

In Exercises 24–27, use substitution to solve the linear system.

24. $\begin{cases} x - y - 2z = 3 \\ 2y + 3z = 8 \\ z = 2 \end{cases}$ (8, 1, 2)

25. $\begin{cases} 2x + y + z = 10 \\ -y + 9z = 28 \\ z = 3 \end{cases}$ (4, −1, 3)

26. $\begin{cases} 2x - y - 3z = 16 \\ 3y + z = 8 \\ z = -7 \end{cases}$ (0, 5, −7)

27. $\begin{cases} 6x - 3y + 4z = -11 \\ 5y - z = 14 \\ z = 1 \end{cases}$ (−1, 3, 1)

In Exercises 28–31, perform the operation on the linear system.

28. Add −4 times the first equation to the second equation.

$\begin{cases} x - 2y + z = 7 \\ 4x + 2y - z = 3 \\ x + y - 5z = 13 \end{cases}$ $10y - 5z = -25$

29. Interchange the first and third equations.

$\begin{cases} 2x + y + z = -3 \\ 3x - 2y + 4z = 9 \\ x + 2y - 2z = -13 \end{cases}$ $\begin{cases} x + 2y - 2z = -13 \\ 3x - 2y + 4z = 9 \\ 2x + y + z = -3 \end{cases}$

30. Multiply the first equation by $-\frac{1}{3}$.

$\begin{cases} -3x + y - 2z = -11 \\ x - 3y + z = 5 \\ 2x + 2y - z = -2 \end{cases}$ $x - \frac{1}{3}y + \frac{2}{3}z = \frac{11}{3}$

31. Add 3 times the first equation to the second equation.

$\begin{cases} x - y + 2z = 10 \\ -3x + y - z = -11 \\ 2x - 2y + z = 11 \end{cases}$ $-2y + 5z = 19$

In Exercises 32–35, check whether the values of *x*, *y*, and *z* form a solution of the linear system.

32. $x = 1, y = -3, z = 2$

$\begin{cases} x - y + 3z = 10 \\ 2x + y - 2z = -5 \\ -x + 2y + z = -5 \end{cases}$ They do.

33. $x = 5, y = 0, z = -2$

$\begin{cases} -x + 2y + z = -7 \\ 3x - y + 2z = 11 \\ 4x + y - 3z = 26 \end{cases}$ They do.

34. $x = 4, y = 9, z = -1$

$\begin{cases} 2x + y - 3z = 18 \\ 4x - 2y + z = -5 \\ x - 3y - z = -23 \end{cases}$ They do not.

35. $x = -1, y = 4, z = 3$

$\begin{cases} -3x - y + 4z = 11 \\ 6x + y - z = -5 \\ x - 2y + 3z = 0 \end{cases}$ They do.

Exploration and Extension

36. The solution of the system represented by the augmented matrix is $x = -20$, $y = 24$, and $z = -16$. What is the value of a?

$$\begin{bmatrix} 1 & 2 & 1 & \vdots & 12 \\ 1 & 3 & a & \vdots & 4 \\ 3 & 6 & 4 & \vdots & 20 \end{bmatrix}$$ 3

37. Use an augmented matrix to solve the linear system.

$\begin{cases} a + 2b - 6c + d = 12 \\ -2a - 3b + 9c + d = -19 \\ a + 2b - 5c + 2d = 15 \\ 2a + 4b - 12c + 3d = 24 \end{cases}$ (2, 14, 3, 0)

Enrichment Activities

These activities require students to go beyond lesson goals.

TECHNOLOGY Graphics calculators can be instructed to compute row operations.

COOPERATIVE LEARNING Work in small groups. Consult the owner's manual for the brand of calculator you are using. Complete Exercises 11–19 by showing the solution steps, including the calculator key sequences, but using the calculator to perform the arithmetic computations.

Compare your solution steps and key sequences with a group of students using a different brand of calculator. How are the solutions the same? How are they different?

Problem of the Day

A man is $\frac{3}{8}$ of the way across a railroad bridge when he hears a train behind him. The trains speed is 60 mph. He concludes correctly that he has just enough time to run to either end of the bridge to prevent being hit by the train. How fast can he run? 15 mph

ORGANIZER

Warm-Up Exercises

1. Find the determinant of each 2×2 matrix.
   a. $\begin{bmatrix} 3 & 8 \\ 5 & 13 \end{bmatrix}$ -1
   b. $\begin{bmatrix} 4 & -6 \\ 5 & -7 \end{bmatrix}$ 2
2. Find the determinant of each 3×3 matrix.
   a. $\begin{bmatrix} 8 & 4 & 2 \\ 9 & -3 & 1 \\ 1 & -1 & 4 \end{bmatrix}$ -240
   b. $\begin{bmatrix} -2 & 3 & 12 \\ 3 & 0 & -5 \\ -3 & 1 & -1 \end{bmatrix}$ 80

Lesson Resources

Teaching Tools
- Problem of the Day: 4.7
- Warm-up Exercises: 4.7
- Answer Masters: 4.7

Extra Practice: 4.7
Color Transparenies: 28, 29

4.7 Solving Systems Using Cramer's Rule

What you should learn:

Goal 1 How to use Cramer's rule to solve a system of linear equations

Goal 2 How to use linear systems to solve real-life problems

Why you should learn it:

You can model many real-life problems with systems of linear equations, which can be solved using Cramer's rule.

Goal 1 Using Cramer's Rule

So far you have studied several methods for solving a linear system. In this lesson, you will learn one more method—called **Cramer's rule,** named after Gabriel Cramer (1704–1752). This rule uses determinants to express the solution of a linear system. To see how Cramer's rule works, consider the following linear system.

$$\begin{cases} ax + by = c \\ dx + ey = f \end{cases}$$

If $ae - bd$ is not zero, the solution of the system is

$$x = \frac{ce - fb}{ae - db} \text{ and } y = \frac{af - dc}{ae - db}.$$

The denominator for x and y is the determinant of the *coefficient matrix* of the system. The numerators are the determinants of the matrices formed by using the column of constants as replacements for the coefficients of either x or y. A similar pattern gives the solution of a linear system with three equations and three variables.

Cramer's Rule

Consider a linear system whose coefficient matrix has a determinant D. If $D \neq 0$, then the system has exactly one solution.

1. The solution of $\begin{cases} ax + by = c \\ dx + ey = f \end{cases}$ is $x = \dfrac{\begin{vmatrix} c & b \\ f & e \end{vmatrix}}{\begin{vmatrix} a & b \\ d & e \end{vmatrix}}$ and $y = \dfrac{\begin{vmatrix} a & c \\ d & f \end{vmatrix}}{\begin{vmatrix} a & b \\ d & e \end{vmatrix}}$.

2. The solution of $\begin{cases} ax + by + cz = d \\ ex + fy + gz = h \\ ix + jy + kz = l \end{cases}$ is

$$x = \frac{\begin{vmatrix} d & b & c \\ h & f & g \\ l & j & k \end{vmatrix}}{\begin{vmatrix} a & b & c \\ e & f & g \\ i & j & k \end{vmatrix}},\ y = \frac{\begin{vmatrix} a & d & c \\ e & h & g \\ i & l & k \end{vmatrix}}{\begin{vmatrix} a & b & c \\ e & f & g \\ i & j & k \end{vmatrix}},\text{ and } z = \frac{\begin{vmatrix} a & b & d \\ e & f & h \\ i & j & l \end{vmatrix}}{\begin{vmatrix} a & b & c \\ e & f & g \\ i & j & k \end{vmatrix}}.$$

Example 1 *Using Cramer's Rule for a 2 × 2 System*

Solve using Cramer's rule: $\begin{cases} 4x - 2y = 10 \\ 3x - 5y = 11 \end{cases}$.

Solution The determinant of the coefficient matrix is:

$$\begin{vmatrix} 4 & -2 \\ 3 & -5 \end{vmatrix} = -20 - (-6) = -14.$$

Since this determinant is not 0, you can apply Cramer's rule.

$$x = \frac{\begin{vmatrix} 10 & -2 \\ 11 & -5 \end{vmatrix}}{-14} = \frac{(-50) - (-22)}{-14} = \frac{-28}{-14} = 2$$

$$y = \frac{\begin{vmatrix} 4 & 10 \\ 3 & 11 \end{vmatrix}}{-14} = \frac{44 - 30}{-14} = \frac{14}{-14} = -1$$

The solution is $x = 2$ and $y = -1$. Check this solution in the original system. ■

Example 2 *Using Cramer's Rule for a 3 × 3 System*

Use Cramer's rule to solve $\begin{cases} -x + 2y - 3z = 1 \\ 2x + \qquad z = 0 \\ 3x - 4y + 4z = 2 \end{cases}$.

Solution The determinant of the coefficient matrix is

$$\left|\begin{matrix} -1 & 2 & -3 \\ 2 & 0 & 1 \\ 3 & -4 & 4 \end{matrix}\right| \begin{matrix} -1 & 2 \\ 2 & 0 \\ 3 & -4 \end{matrix} = 0 + 6 + 24 - 0 - 4 - 16 = 10.$$

Since this determinant is not 0, you can apply Cramer's rule.

$$x = \frac{\begin{vmatrix} 1 & 2 & -3 \\ 0 & 0 & 1 \\ 2 & -4 & 4 \end{vmatrix}}{10} = \frac{8}{10} = \frac{4}{5}$$

$$y = \frac{\begin{vmatrix} -1 & 1 & -3 \\ 2 & 0 & 1 \\ 3 & 2 & 4 \end{vmatrix}}{10} = \frac{-15}{10} = -\frac{3}{2}$$

$$z = \frac{\begin{vmatrix} -1 & 2 & 1 \\ 2 & 0 & 0 \\ 3 & -4 & 2 \end{vmatrix}}{10} = \frac{-16}{10} = -\frac{8}{5}$$

The solution is $x = \frac{4}{5}$, $y = -\frac{3}{2}$, and $z = -\frac{8}{5}$. Check this solution in the original system. ■

LESSON Notes

Examples 1–2

TECHNOLOGY Observe that Cramer's Rule can more easily be reduced to an algorithm (a series of steps leading to the desired result) than many of the other methods of solving systems of linear equations. This suggests that Cramer's Rule is a good technique to use with computers.

EXTEND Examples 1–2

An Alternate Approach

A Computer Program

Here is a computer program that uses Cramer's Rule to solve a system of equations given in the following form.

$$a_1x + b_1y = c_1$$
$$a_2x + b_2y = c_2$$

```
10 PRINT "ENTER THE VALUES
     OF A1, B1, and C1 SEPA-
     RATED BY COMMAS."
20 INPUT A1, B1, C1
30 PRINT "ENTER THE VALUES
     OF A2, B2, AND C2 SEPA-
     RATED BY COMMAS."
40 INPUT A2, B2, C2
50 D = A1 * B2 – A2 * B1
60 DX = C1 * B2 – C2 * B1
70 DY = A1 * C2 – A2 * C1
80 PRINT "D = "; D;", DX = ";
     DX;", DY = "; DY
90 IF D < > 0 THEN 150
100 IF DX = 0 AND DY = 0
     THEN 130
110 PRINT "THE SYSTEM HAS
     NO SOLUTION."
120 GOTO 160
130 PRINT "THE SYSTEM HAS
     INFINITELY MANY SOLU-
     TIONS."
140 GOTO 160
150 PRINT "THE SOLUTION IS
     (";DX / D;", ";DY / D;")."
160 END
```

Example 3

The combined male viewing audience can be found by subtraction; 109.75 − 47.38 = 62.37 (million). Note that the bar graph contains needed information about male viewers.

Extra Examples

Here are additional examples similar to **Examples 1–2.**

Solve each system using Cramer's Rule.

1. $\begin{cases} 8x + 5y = 2 \\ 2x - 4y = -10 \end{cases}$ (−1, 2)

2. $\begin{cases} 9x + 2y = -31 \\ 15x + 18y = 51 \end{cases}$ (−5, 7)

3. $\begin{cases} 5x + 2y + 2z = -14 \\ 7x - 13y - 4z = -58 \\ 17x + 23y + z = -21 \end{cases}$ (−4, 2, 1)

4. $\begin{cases} 38x - 19y + 13z = -51 \\ 4x + 16y - 21z = 53 \\ 13x + 18y - 11z = 47 \end{cases}$ (0, 2, −1)

Check Understanding

1. In general, what is the solution of a system of linear equations in x and y using Cramer's Rule? See page 216.
2. Why must the determinant of the coefficient matrix be nonzero when applying Cramer's Rule? The coefficient matrix is the denominator.

Communicating about ALGEBRA

EXTEND *Communicating*

If students have access to a computer, you might have them use the computer program shown on page 217 to solve these systems.

1. $\begin{cases} 5x + 2y = 1 \\ 2x - y = 4 \end{cases}$ (1, −2)

2. $\begin{cases} 5x - y = 8 \\ 10x - 2y = 16 \end{cases}$ many solutions

3. $\begin{cases} 3x - y = 2 \\ 6x - 2y = 5 \end{cases}$ no solutions

Goal 2 Using Linear Systems in Real Life

Real Life **Media Research**

Example 3 *World Series and Super Bowl Viewers*

The total viewing audience of the 1990 World Series and Super Bowl XXV was 109.75 million. The total female viewing audience was about 47.38 million. Determine the number of viewers for each of these sporting events.

(Source: Nielsen Media Research)

Solution Let x be the number of people who watched the 1990 World Series, and let y be the number of people who watched Super Bowl XXV. From the bar graph, you can write the following system.

$$\begin{cases} 0.41x + 0.44y = 47.38 \\ 0.59x + 0.56y = 62.37 \end{cases}$$

Using Cramer's rule, the solution of the system is as follows.

$$x = \frac{\begin{vmatrix} 47.38 & 0.44 \\ 62.37 & 0.56 \end{vmatrix}}{\begin{vmatrix} 0.41 & 0.44 \\ 0.59 & 0.56 \end{vmatrix}} = \frac{26.5328 - 27.4428}{0.2296 - 0.2596} = \frac{-0.91}{-0.03} \approx 30.33$$

$$y = \frac{\begin{vmatrix} 0.41 & 47.38 \\ 0.59 & 62.37 \end{vmatrix}}{\begin{vmatrix} 0.41 & 0.44 \\ 0.59 & 0.56 \end{vmatrix}} = \frac{25.5717 - 27.9542}{0.2296 - 0.2596} = \frac{-2.3825}{-0.03} \approx 79.42$$

Therefore, approximately 30.33 million people watched the 1990 World Series, and approximately 79.42 million people watched Super Bowl XXV. ■

Communicating about ALGEBRA

SHARING IDEAS about the Lesson

Use Cramer's rule to solve each linear system. See below.

A. $\begin{cases} 4x - 3y = -2 \\ 2x - 9y = 19 \end{cases}$

B. $\begin{cases} x + 3y = -20 \\ 5x - y = 12 \end{cases}$

C. $\begin{cases} 2x - 3y + 3z = 0 \\ x + 2y - 3z = 5 \\ 4x - 5y - 6z = 5 \end{cases}$

A. $\left(-\frac{5}{2}, -\frac{8}{3}\right)$ B. (1, −7) C. $\left(2, 1, -\frac{1}{3}\right)$

EXERCISES

Guided Practice

CRITICAL THINKING about the Lesson

1. To use Cramer's rule to solve a linear system, what must be true of the determinant of the coefficient matrix? Not equal to 0.

2. Complete the solution of $\begin{cases} x + 2y = 4 \\ 3x - y = 1 \end{cases}$. $x = \dfrac{\begin{vmatrix} & 2 \\ & -1 \end{vmatrix}}{\begin{vmatrix} 1 & 2 \\ 3 & -1 \end{vmatrix}}$ and $y = \dfrac{\begin{vmatrix} 1 & \\ 3 & \end{vmatrix}}{\begin{vmatrix} 1 & 2 \\ 3 & -1 \end{vmatrix}}$ 4, 1; 4, 1; $\left(\frac{6}{7}, \frac{11}{7}\right)$

3. Use Cramer's rule to solve $\begin{cases} 6x - 8y = 4 \\ 4x - 5y = -4 \end{cases}$. $(-26, -20)$

4. Complete the solution of $\begin{cases} 10x - y + 3z = 3 \\ 4x + y - z = 1. \\ 5x + 3y + 2z = -1 \end{cases}$ 3, 1, −1; 3, 1, −1; 3, 1, −1; $x = \frac{28}{84} = \frac{1}{3}$, $y = \frac{-56}{84} = -\frac{2}{3}$; $z = \frac{-28}{84} = -\frac{1}{3}$; $\left(\frac{1}{3}, -\frac{2}{3}, -\frac{1}{3}\right)$

$$x = \frac{\begin{vmatrix} & -1 & 3 \\ & 1 & -1 \\ & 3 & 2 \end{vmatrix}}{\begin{vmatrix} 10 & -1 & 3 \\ 4 & 1 & -1 \\ 5 & 3 & 2 \end{vmatrix}},\ y = \frac{\begin{vmatrix} 10 & & 3 \\ 4 & & -1 \\ 5 & & 2 \end{vmatrix}}{\begin{vmatrix} 10 & -1 & 3 \\ 4 & 1 & -1 \\ 5 & 3 & 2 \end{vmatrix}},\ \text{and } z = \frac{\begin{vmatrix} 10 & -1 & \\ 4 & 1 & \\ 5 & 3 & \end{vmatrix}}{\begin{vmatrix} 10 & -1 & 3 \\ 4 & 1 & -1 \\ 5 & 3 & 2 \end{vmatrix}}$$

Independent Practice

In Exercises 5–7, find the determinant of the coefficient matrix.

5. $\begin{cases} 4x + y = 19 \\ -3x + 3y = 10 \end{cases}$ 15

6. $\begin{cases} 5x - 3y = 6 \\ 2x + 7y = 18 \end{cases}$ 41

7. $\begin{cases} 3x + 4y + z = 17 \\ 2x + 3y + 2z = 15 \\ x + y = 4 \end{cases}$ 1

In Exercises 8–13, use Cramer's rule to solve the linear system.

8. $\begin{cases} 2x + y = 12 \\ 5x + 3y = 27 \end{cases}$ $(9, -6)$

9. $\begin{cases} x - 4y = 22 \\ 2x - 7y = 39 \end{cases}$ $(2, -5)$

10. $\begin{cases} x + 4y = 12 \\ 2x + 5y = 18 \end{cases}$ $(4, 2)$

11. $\begin{cases} 4x - y = -2 \\ -2x + y = 3 \end{cases}$ $\left(\frac{1}{2}, 4\right)$

12. $\begin{cases} x - 2y = 11 \\ 2x + 5y = -14 \end{cases}$ $(3, -4)$

13. $\begin{cases} 2x - 4y = 7 \\ -x + y = 1 \end{cases}$ $\left(-\frac{11}{2}, -\frac{9}{2}\right)$

In Exercises 14–16, use Cramer's rule to write the determinant form of the solution. (You do not need to evaluate the determinants.) See Additional Answers.

14. $\begin{cases} x - y + 10z = 2 \\ 3x + z = 4 \\ 7x + 2y + z = 0 \end{cases}$

15. $\begin{cases} 4x + 2z = 3 \\ -x + 2y + 5z = -1 \\ 3x + y - 7z = 10 \end{cases}$

16. $\begin{cases} 2x + 6y - z = 4 \\ 3x + y + 2z = -4 \\ 6x - y + 3z = 1 \end{cases}$

21. $\left(-\frac{2}{3}, -34, -12\right)$ 22. $(-1, 6, -2)$ 23. $\left(\frac{3}{4}, \frac{1}{4}, 0\right)$ 24. $\left(\frac{1}{2}, 3, -\frac{3}{2}\right)$ 25. $\left(\frac{1}{3}, 1, -\frac{2}{3}\right)$

26. $\left(2, \frac{1}{2}, -\frac{1}{2}\right)$ 27. $(5, 2, -1)$ 28. $(1, -1, 2)$ 29. $(2, 1\ 1)$ 30. $(-15, 1, 2)$ 31. $(3, -3, 0)$

EXERCISE Notes

ASSIGNMENT GUIDE

Basic/Average: Ex. 1–4, 5–25 odd, 26–31, 33–40, 43

Above Average: Ex. 1–4, 5–25 odd, 26–31, 33–35, 40–43

Advanced: Ex. 1–4, 9–25 odd, 26–35, 40–41

Selected Answers
Exercises 1–4, 5–41 odd

More Difficult Exercises
Exercises 26–35

Guided Practice

▶ **Ex. 1–4** These exercises provide students with practice at demonstrating another method for solving systems of equations.

You may wish to start a discussion about all the methods students know for solving systems of equations. These include graphing, substitution, using a graphics calculator or computer, Gaussian elimination, linear combinations, inverse matrices, augmented matrices, and Cramer's Rule.

Independent Practice

▶ **Ex. 14–16** Caution students that they are not required to find the actual solutions.

Answers

17. $(-1, 4, 3)$ 18. $(1, 1, -2)$
19. $(0, 5, 4)$ 20. $(2, 0, 1)$

Ex. 17–25 Remind students that if the determinant of the coefficient matrix is equal to 0, there is not exactly one solution; so Cramer's Rule would not apply. Students can find the determinants using their graphics calculators.

Ex. 26–31 Encourage students to choose any solution method, but to give an explanation as to why they chose a particular method. For example, since Exercise 30 is already in triangular form, Gaussian elimination may be the preferred method. Students with graphics calculators may choose to use inverse matrices, which would not be the method of choice of students without access to calculators.

Ex. 32–35 Encourage students to use Cramer's Rule to solve these exercises.

Answers

33. $x = \dfrac{\begin{vmatrix} 41.6 & 1 \\ 31.2 & -1 \end{vmatrix}}{\begin{vmatrix} 1 & 1 \\ 1 & -1 \end{vmatrix}} = \dfrac{-72.8}{-2} = 36.4$ million,

$y = \dfrac{\begin{vmatrix} 1 & 41.6 \\ 1 & 31.2 \end{vmatrix}}{\begin{vmatrix} 1 & 1 \\ 1 & -1 \end{vmatrix}} = \dfrac{-10.4}{-2} = 5.2$ million

color: 36.4 million, black-and-white: 5.2 million

34. $0.25x + 0.21y = 41$ million, $0.34x + 0.15y = 44$ million; (≈ 91 million, ≈ 87 million)

35. $0.18x + 0.10y + 0.02z = 200.40$, $0.10x + 0.12y + 0.03z = 154.50$, $0.16x + 0.26y + 0.16z = 361.80$; \$780 million, \$450 million, \$750 million

In Exercises 17–25, use Cramer's rule to solve the linear system. See page 219.

17. $\begin{cases} x + 3y - z = 8 \\ 2x - y + 2z = 0 \\ -3x + y - 3z = -2 \end{cases}$

18. $\begin{cases} x + 2z = -3 \\ x - y + 2z = -4 \\ 2x - 3y + z = -3 \end{cases}$

19. $\begin{cases} x + 2y - 3z = -2 \\ x - y + z = -1 \\ 3x + 4y - 4z = 4 \end{cases}$

20. $\begin{cases} x + 3y - z = 1 \\ -2x - 6y + z = -3 \\ 3x + 5y - 2z = 4 \end{cases}$

21. $\begin{cases} 3x + 2y - 5z = -10 \\ 6x - z = 8 \\ -y + 3z = -2 \end{cases}$

22. $\begin{cases} x + 2y + z = 9 \\ x + y + z = 3 \\ 5x - 2z = -1 \end{cases}$

23. $\begin{cases} 2x - 2y + 5z = 1 \\ -8x + z = -6 \\ x + y - 2z = 1 \end{cases}$

24. $\begin{cases} 4x + y + 6z = -4 \\ 2x + 2y + 4z = 1 \\ -x - y + z = -5 \end{cases}$

25. $\begin{cases} x + y - z = 2 \\ 6x + 4y + 3z = 4 \\ 3x + 6z = -3 \end{cases}$

In Exercises 26–31, use any convenient method to solve the linear system. See page 219. **Explain why you chose the method.** See below.

26. $\begin{cases} 3x + y + z = 6 \\ x - 4y + 2z = -1 \\ x - 3y + z = 0 \end{cases}$

27. $\begin{cases} x - 2y + 3z = -2 \\ -x + 3y - z = 2 \\ 2x - y + 2z = 6 \end{cases}$

28. $\begin{cases} x - y + 2z = 6 \\ -2x + 3y - z = -7 \\ 3x + 2y + 2z = 5 \end{cases}$

29. $\begin{cases} 2x + y - 3z = 2 \\ x - 2y + 4z = 4 \\ 3x - y + 2z = 7 \end{cases}$

30. $\begin{cases} x + 6y + 4z = -1 \\ y + 2z = 5 \\ z = 2 \end{cases}$

31. $\begin{cases} x = 3 \\ 2x + y = 3 \\ 3x + 2y + z = 3 \end{cases}$

32. *Western Painter* Charles Russell (1864–1926) is one of America's best-known western painters and sculptors. A self-taught artist, he started painting when he was x years old. During the next y years before he died, he completed over 4500 works of art that depict the life of the early settlers and Native Americans in the American and Canadian west. Find the values of x and y by solving the following linear system. (Use Cramer's rule.) How many years had Russell been painting when he completed the painting at the right? **(22, 40), 2**

$\begin{cases} x + y = \boxed{?} \\ 40x - 22y = 0 \end{cases}$

Captain Meriwether Meeting the Shoshones *by Charles Russell, 1888.*

33. *Camera Bugs* Americans take an average of about 41.6 million photographs each day. Of these, x are color photographs and y are black-and-white. The value of x is 31.2 million more than the value of y. What is the average number of color photographs taken each day? What is the average number of black-and-white photographs taken each day? Show how Cramer's rule can be used to answer the question. ***(Source: Photo Marketing Association International)*** See margin.

See margin.

✪ **34.** *Senators* Each state in the United States has two senators. In 1990, about 44 million Americans of voting age (voters and nonvoters) could name both senators from their state. About 41 million could name only one of their senators. Let x represent the number of Americans of voting age who were voters and let y represent the number who were nonvoters. Use the graph at the right to write a system of linear equations in x and y. Then use Cramer's rule to solve the system.

✪ **35.** *Shoe Purchases* The graph shows the percents (by age group) of the total amounts spent on three types of shoes in 1988. For instance, 18% of the total amount spent on gym shoes was spent by the 14–17 age group, 10% was spent by the 18–24 age group, and 16% was spent by the 25–34 age group. In 1988, the amount (in millions of dollars) spent on *all three* types of shoes is shown in the matrix. How many dollars worth of gym shoes, jogging shoes, and walking shoes were sold in 1988? See margin.

Amount ($ millions)	Spent on Shoes
14–17	200.4
18–24	154.5
25–34	361.8

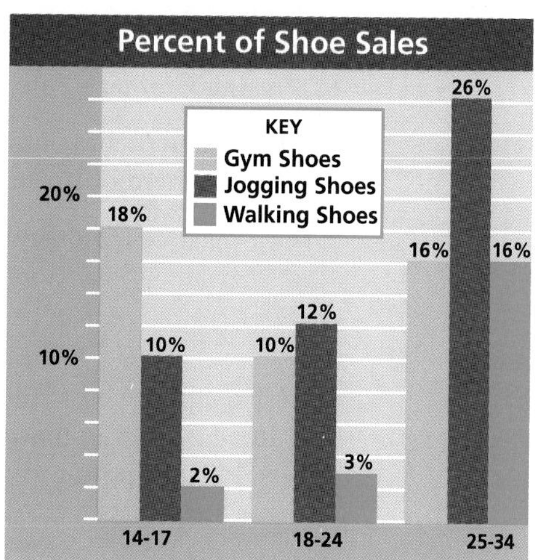

Integrated Review

In Exercises 36–39, evaluate the determinant of the matrix.

36. $\begin{bmatrix} 10 & 7 \\ 5 & 3 \end{bmatrix}$ −5

37. $\begin{bmatrix} 4 & 2 \\ 1 & 4 \end{bmatrix}$ 14

38. $\begin{bmatrix} 3 & -5 & 5 \\ 1 & 2 & -2 \\ 6 & 0 & 0 \end{bmatrix}$ 0

39. $\begin{bmatrix} 2 & 1 & 0 \\ 4 & 6 & -1 \\ 5 & 4 & -2 \end{bmatrix}$ −13

Top Spectator Sports **In Exercises 40 and 41, use the following information.**

In 1990, about 1240 sports fans were asked to choose their favorite sport to watch in person or on t.v. Of these fans, 35% said football, 16% said baseball, and 15% said basketball.

40. How many of the fans chose basketball? 186

41. How many of the fans chose football? 434

Exploration and Extension

42. Solve for x: $\begin{vmatrix} 11 & x \\ 7 & 6 \end{vmatrix} = 3$ 9

43. Solve for x: $\begin{vmatrix} 5 & 7 \\ x & -6 \end{vmatrix} = -2$ −4

EXTEND Ex. 32–35

CONNECTIONS Use the exercises to promote in-class discussion on their topics. For example, how many of your students can name their senator? How does your class compare with the national average presented in Exercise 34? Similar comparisons could be made with the data of Exercises 33 and 35.

Enrichment Activity

This activity requires students to go beyond lesson goals.

WRITING Write a summary essay of all the methods for solving systems of equations you have learned in both Chapters 3 and 4. Include sample problems that can be solved using the methods.

Discuss the strengths and weaknesses of each method. Indicate your preferred methods and explain why they are your methods of choice.

ORGANIZER

TIME SCHEDULE All levels, two days

The Chapter Summary helps students organize the main ideas of the chapter. In this chapter, topics addressing each of the basic features of algebra were studied. An extensive look at how systems of linear equations can be solved was made and an important area of mathematics, the algebra of matrices, was introduced.

Work with students to review the skills, strategies, and concepts of the chapter. The first day's homework assignment can be used as the basis for the second day of review.

COOPERATIVE LEARNING Encourage students to study together. Emphasize the importance of "teaching" a classmate how to perform a skill or how to recall a strategy. When students work together, everyone wins. The students receiving help get additional instruction and the students giving help gain a deeper understanding of the skills and concepts involved.

Chapter SUMMARY

Review the chapter by asking students, "What did you learn? Why did you learn it? How does it fit with the other algebra skills and concepts you have learned?"

4 Chapter Summary

What did you learn?

Skills

1. Perform matrix operations.
   - Add and subtract matrices. **(4.1)**
   - Multiply a matrix by a scalar. **(4.1)**
   - Multiply two matrices. **(4.2)**
2. Find the determinant of a square matrix. **(4.3)**
3. Find the inverse of a square matrix. **(4.4)**
4. Solve a linear system
   - using an inverse matrix. **(4.5)**
   - using an augmented matrix. **(4.6)**
   - using Cramer's rule. **(4.7)**

Strategies

5. Use matrices and determinants to model real-life situations. **(4.1–4.4)**
6. Use a system of linear equations to model a real-life situation. **(4.5–4.7)**

Exploring Data

7. Use matrices to organize data. **(4.1)**

Why did you learn it?

In this chapter, you studied two basic uses of matrices. One is to organize real-life data. Often matrices used for this can be added, subtracted, or multiplied by another matrix or a scalar to help answer questions about real-life situations.

The second basic use of matrices and determinants is to solve a linear system. In this chapter, you studied three ways of solving a linear system: using the inverse matrix, using an augmented matrix, and using Cramer's rule.

How does it fit into the bigger picture of algebra?

Two important uses of a matrix are to organize data and to solve a system of linear equations. You learned to use matrices in both of these ways in this chapter.

Just like real numbers, matrices can be added or multiplied. All matrices have opposites (so you can subtract), and some matrices have inverses. The rules for operating with matrices form part of *the algebra of matrices.* Some of the properties of this algebra are similar to the properties in the algebra of real numbers. Others, however, are not. For example, in Lesson 4.2, you saw that the product of two matrices depends on the order in which the two matrices are multiplied. In other words, matrix multiplication is *not* commutative.

In Exercises 1–8, perform the matrix operation(s). (4.1)

1. $\begin{bmatrix} 0 & 1 & -5 \\ 4 & 1 & 6 \end{bmatrix} + \begin{bmatrix} 10 & 3 & 11 \\ -2 & 8 & 3 \end{bmatrix}$ $\begin{bmatrix} 10 & 4 & 6 \\ 2 & 9 & 9 \end{bmatrix}$

2. $\begin{bmatrix} 15 & 4 \\ 3 & 12 \end{bmatrix} - \begin{bmatrix} 0 & 9 \\ 2 & 7 \end{bmatrix}$ $\begin{bmatrix} 15 & -5 \\ 1 & 5 \end{bmatrix}$

3. $\begin{bmatrix} 5 & 1 & 10 \\ -1 & 0 & 0 \\ 2 & 3 & 4 \end{bmatrix} - \begin{bmatrix} 6 & 7 & 3 \\ 0 & 14 & 6 \\ 1 & -1 & 2 \end{bmatrix}$ $\begin{bmatrix} -1 & -6 & 7 \\ -1 & -14 & -6 \\ 1 & 4 & 2 \end{bmatrix}$

4. $\begin{bmatrix} 6 & 10 \\ 9 & 6 \\ 4 & -1 \end{bmatrix} + \begin{bmatrix} 2 & 1 \\ 0 & 7 \\ 4 & 7 \end{bmatrix}$ $\begin{bmatrix} 8 & 11 \\ 9 & 13 \\ 8 & 6 \end{bmatrix}$

5. $3\begin{bmatrix} 4 & 6 & -1 \\ 10 & -5 & 2 \\ 2 & 11 & 1 \end{bmatrix}$ $\begin{bmatrix} 12 & 18 & -3 \\ 30 & -15 & 6 \\ 6 & 33 & 3 \end{bmatrix}$

6. $8\begin{bmatrix} 3 & 2 & 0 & 1 \\ 5 & 3 & 3 & 6 \\ -9 & 4 & 2 & 0 \end{bmatrix}$ $\begin{bmatrix} 24 & 16 & 0 & 8 \\ 40 & 24 & 24 & 48 \\ -72 & 32 & 16 & 0 \end{bmatrix}$

7. $-2\left(\begin{bmatrix} 6 & 4 \\ 0 & 3 \end{bmatrix} - \begin{bmatrix} 5 & 10 \\ 1 & 3 \end{bmatrix}\right)$ $\begin{bmatrix} -2 & 12 \\ 2 & 0 \end{bmatrix}$

8. $-3\left(\begin{bmatrix} 4 & 7 \\ 1 & -2 \end{bmatrix} + \begin{bmatrix} 3 & -5 \\ 6 & 0 \end{bmatrix}\right)$ $\begin{bmatrix} -21 & -6 \\ -21 & 6 \end{bmatrix}$

In Exercises 9–14, write the product as a single matrix. (4.2)

9. $\begin{bmatrix} 1 & 0 \\ 4 & 9 \end{bmatrix}\begin{bmatrix} -1 & 1 \\ 3 & 2 \end{bmatrix}$ $\begin{bmatrix} -1 & 1 \\ 23 & 22 \end{bmatrix}$

10. $\begin{bmatrix} 2 & 15 \\ 3 & 10 \end{bmatrix}\begin{bmatrix} -5 & 12 \\ 1 & 0 \end{bmatrix}$ $\begin{bmatrix} 5 & 24 \\ -5 & 36 \end{bmatrix}$

11. $\begin{bmatrix} 6 & 6 & 0 \\ 1 & -1 & 5 \end{bmatrix}\begin{bmatrix} -6 & 1 & 4 \\ 5 & -2 & 1 \\ 3 & -8 & 0 \end{bmatrix}$ $\begin{bmatrix} -6 & -6 & 30 \\ 4 & -37 & 3 \end{bmatrix}$

12. $\begin{bmatrix} 4 & -1 & 3 \\ 7 & 10 & -3 \\ 1 & 2 & -5 \end{bmatrix}\begin{bmatrix} 0 & 1 \\ 9 & 2 \\ 12 & 0 \end{bmatrix}$ $\begin{bmatrix} 27 & 2 \\ 54 & 27 \\ -42 & 5 \end{bmatrix}$

13. $\begin{bmatrix} 10 & 2 & 1 & 5 \end{bmatrix}\begin{bmatrix} 1 \\ 0 \\ -2 \\ 3 \end{bmatrix}$ [23]

14. $\begin{bmatrix} 6 & 2 & 1 & -3 \end{bmatrix}\begin{bmatrix} 0 \\ 5 \\ 4 \\ 3 \end{bmatrix}$ [5]

In Exercises 15–22, evaluate the determinant of the matrix. (4.3)

15. $\begin{bmatrix} 6 & -3 \\ 2 & 1 \end{bmatrix}$ 12

16. $\begin{bmatrix} -2 & -6 \\ 1 & 4 \end{bmatrix}$ −2

17. $\begin{bmatrix} 9 & 1 \\ -3 & 2 \end{bmatrix}$ 21

18. $\begin{bmatrix} 4 & 1 \\ 0 & 6 \end{bmatrix}$ 24

19. $\begin{bmatrix} 2 & 1 & 5 \\ -1 & 6 & 3 \\ 2 & -4 & 2 \end{bmatrix}$ 16

20. $\begin{bmatrix} -3 & 1 & 0 \\ 2 & -1 & 1 \\ 0 & 3 & 4 \end{bmatrix}$ 13

21. $\begin{bmatrix} 2 & -3 & 4 \\ 0 & 1 & -2 \\ 1 & 2 & -3 \end{bmatrix}$ 4

22. $\begin{bmatrix} 3 & -2 & 1 \\ 4 & -2 & 1 \\ 0 & 3 & -2 \end{bmatrix}$ −1

In Exercises 23 and 24, use a determinant to find the area of the triangle. (4.3)

23.
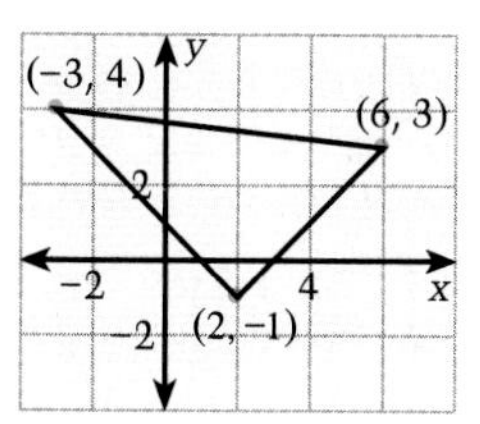

20

24.
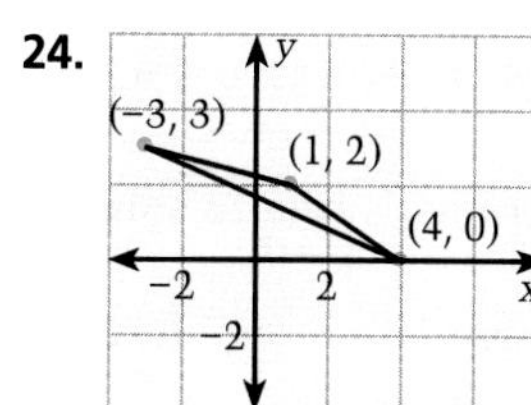

$\frac{5}{2}$

Chapter REVIEW

ASSIGNMENT GUIDE

Basic/Average: Ex. 1–41 odd, 42, 45, 48, 49, 52–54

Above Average: Ex. 1–41 odd, 42, 45, 48, 49, 51–54

Advanced: Ex. 5–41 odd, 42, 45, 48, 49, 51–58

✪ More Difficult Exercises
Exercises 52–58

▶ **Exercises 1–58** The Chapter Review exercises include a reference to the lesson in which each skill or concept was presented. Students may use these references as a study guide if they are uncertain how to do a given exercise.

In Exercises 25–30, find the inverse of the matrix. (4.4)

25. $\begin{bmatrix} 7 & 3 \\ 5 & 2 \end{bmatrix}$ $\begin{bmatrix} -2 & 3 \\ 5 & -7 \end{bmatrix}$

26. $\begin{bmatrix} 2 & 3 \\ 7 & 11 \end{bmatrix}$ $\begin{bmatrix} 11 & -3 \\ -7 & 2 \end{bmatrix}$

27. $\begin{bmatrix} 2 & 2 \\ 1 & 3 \end{bmatrix}$ $\begin{bmatrix} \frac{3}{4} & -\frac{1}{2} \\ -\frac{1}{4} & \frac{1}{2} \end{bmatrix}$

28. $\begin{bmatrix} 2 & -2 \\ 4 & -3 \end{bmatrix}$ $\begin{bmatrix} -\frac{3}{2} & 1 \\ -2 & 1 \end{bmatrix}$

29. $\begin{bmatrix} 8 & -3 \\ 4 & -2 \end{bmatrix}$ $\begin{bmatrix} \frac{1}{2} & -\frac{3}{4} \\ 1 & -2 \end{bmatrix}$

30. $\begin{bmatrix} 6 & 10 \\ 2 & 4 \end{bmatrix}$ $\begin{bmatrix} 1 & -\frac{5}{2} \\ -\frac{1}{2} & \frac{3}{2} \end{bmatrix}$

In Exercises 31 and 32, decide whether the matrices are inverses of each other. (4.4)

31. $\begin{bmatrix} 1 & -1 & -3 \\ 5 & 2 & 1 \\ -3 & -1 & 0 \end{bmatrix}, \begin{bmatrix} 1 & 3 & 5 \\ -3 & -9 & -16 \\ 1 & 4 & 7 \end{bmatrix}$ They are.

32. $\begin{bmatrix} -2 & 0 & -3 \\ -7 & 1 & -11 \\ -3 & 0 & -4 \end{bmatrix}, \begin{bmatrix} -4 & 0 & 3 \\ 5 & -1 & -1 \\ 3 & 0 & -2 \end{bmatrix}$ They are not.

In Exercises 33–36, solve the matrix equation. (4.5)

33. $\begin{bmatrix} 3 & -1 \\ 5 & 2 \end{bmatrix}\begin{bmatrix} x \\ y \end{bmatrix} = \begin{bmatrix} -3 \\ 2 \end{bmatrix}$

33. $\begin{bmatrix} -\frac{4}{11} \\ \frac{21}{11} \end{bmatrix}$

34. $\begin{bmatrix} 5 & 6 \\ 4 & 5 \end{bmatrix}\begin{bmatrix} x \\ y \end{bmatrix} = \begin{bmatrix} 1 \\ -5 \end{bmatrix}$ $\begin{bmatrix} 35 \\ -29 \end{bmatrix}$

35. $\begin{bmatrix} -1 & -1 \\ 4 & 2 \end{bmatrix}\begin{bmatrix} x \\ y \end{bmatrix} + \begin{bmatrix} -1 \\ 2 \end{bmatrix} = \begin{bmatrix} -5 \\ 0 \end{bmatrix}$ $\begin{bmatrix} -5 \\ 9 \end{bmatrix}$

36. $\begin{bmatrix} 0 & 1 \\ 2 & 4 \end{bmatrix}\begin{bmatrix} x \\ y \end{bmatrix} - \begin{bmatrix} 10 \\ -3 \end{bmatrix} = \begin{bmatrix} -8 \\ 5 \end{bmatrix}$ $\begin{bmatrix} -3 \\ 2 \end{bmatrix}$

In Exercises 37 and 38, use the matrix $\begin{bmatrix} 4 & 0 \\ -1 & 2 \end{bmatrix}$ to encode the message.

37. MAKE A WISH
51, 2, 39, 10, −1, 2, −23, 46, 17, 38, 32, 0

38. SPECIAL DELIVERY
60, 32, 17, 6, 35, 2, 48, 0, 11, 10, 39, 18, 83, 10, 47, 50

In Exercises 39 and 40, use the inverse of $\begin{bmatrix} 2 & -1 \\ 3 & -1 \end{bmatrix}$ to decode the message. (4.5)

39. 44, −15, 3, −1, 80, −32, 39, −17, 3, −1, 12, −4, 77, −26 AN APPLE A DAY

40. 53, −24, 40, −18, 24, −8, 46, −17, 32, −16, 15, −7, 98, −39 SEND HELP FAST

In Exercises 41–43, use an inverse matrix to solve the linear system. (Use a graphing calculator for Exercises 42 and 43.) (4.5)

41. $\begin{cases} x - 3y = 10 \\ 2x + 5y = -2 \end{cases}$ (4, −2)

42. $\begin{cases} x + 4y + 2z = 1 \\ -x + 5y + 2z = 3 \\ 4x + z = -5 \end{cases}$ (3, 8, −17)

43. $\begin{cases} -5x + 4y + 2z = 2 \\ y - 3z = -6 \\ -4x + 3y + 2z = 1 \end{cases}$ (20, 21, 9)

In Exercises 44–46, use an augmented matrix to solve the linear system. (4.6)

44. $\begin{cases} x - y - 5z = 1 \\ 2x - y - 6z = 2 \\ 2x - 2y - 9z = -1 \end{cases}$ (−2, 12, −3)

45. $\begin{cases} x + 3y - 5z = 2 \\ x + 4y - 9z = 3 \\ 2x + 6y - 9z = 6 \end{cases}$ (−15, 9, 2)

46. $\begin{cases} x - y - 4z = 3 \\ -x + 3y - z = -1 \\ x - y + 5z = 3 \end{cases}$ (4, 1, 0)

In Exercises 47–49, use Cramer's rule to solve the linear system. (4.6)

47. $\begin{cases} 2x - 3y = -17 \\ 3x + 4y = 0 \end{cases}$ (−4, 3)

48. $\begin{cases} x + 5y = -16 \\ 5x - 2y = -4 \end{cases}$ $\left(-\frac{52}{27}, -\frac{76}{27}\right)$

49. $\begin{cases} x + 2y + z = -2 \\ 2x - y - 3z = 1 \\ -3x + y - 2z = -7 \end{cases}$ (1, −2, 1)

50. *Rod Woodson* In 1989, Rod Woodson led the National Football League on kick-off returns with 982 yards in 36 carries. Let x represent the total return yardages for Woodson during the first half of the 1989 season, and let y represent his total return yardages during the second half of the season. Use an inverse matrix to solve the following linear system for x and y.

$$\begin{cases} x + y = 982 \\ 3x - 2y = 461 \end{cases}$$ (485, 497)

Rod Woodson

51. *Stock Investment* You have \$18,000 to invest in three types of stocks. You expect the annual returns on Stock X, Stock Y, and Stock Z to be 10%, 9%, and 6%. You want the combined investment in Stock Y and Stock Z to be twice the amount invested in Stock X. How much should you invest in each type of stock to obtain an average return of 8%? See margin.

In Exercises 52 and 53, find the area of the blue region. (The area of the rectangle is 24 square units.)

52. $\frac{29}{2}$ units^2

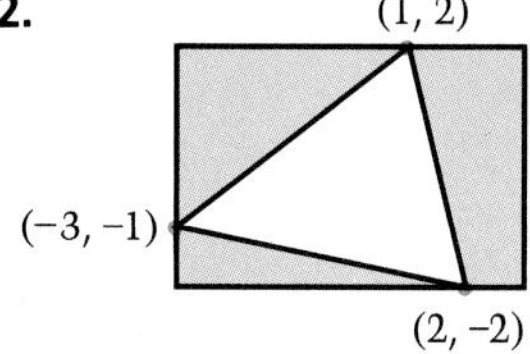

53. 15 units^2

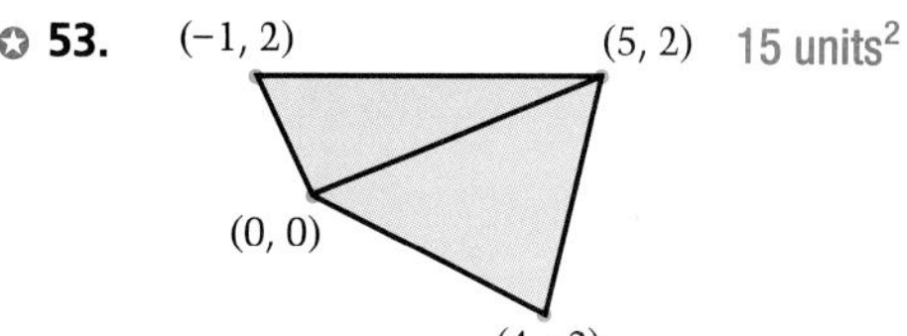

54. *Summer Vacations* More summer vacations were taken by Americans in 1991 than in 1988, but the vacations were shorter. The percents of vacations lasting 1 night, 2–3 nights, and 4 nights or more for the two summers are shown in the bar graph at the right. In 1988 and 1991 combined, a total of 53.9 million summer vacations lasted 1 night, and a total of 130.9 million lasted 2–3 nights. How many summer vacations were taken in 1988? In 1991?

(Source: U.S. Travel Data Center)

≈162.7 million, ≈184.4 million

56. Denmark France Indonesia
Norway Panama Thailand
Iceland Netherlands Chile

57. $\begin{bmatrix} 2 & 3 & 8 \\ 1 & 7 & 6 \\ 5 & 9 & 4 \end{bmatrix}$

58. Italy Senegal Guinea
Ivory Coast Colombia Chad
Hungary Cameroon Mali

Flags of the World **In Exercises 55–58, use the following information.**

Each of the following matrices contains the flags of nine countries. Your goal is to match each country to its flag.

Matrix A

1. Indonesia
2. Chile
3. Norway
4. Iceland
5. Thailand
6. Netherlands
7. France
8. Denmark
9. Panama

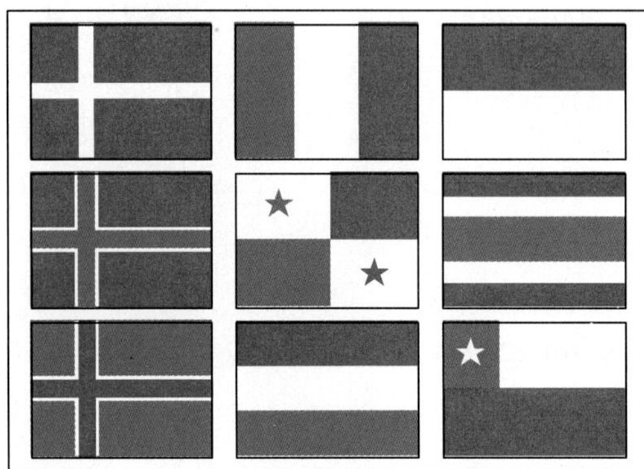

Matrix B

1. Ivory Coast
2. Italy
3. Senegal
4. Mali
5. Hungary
6. Chad
7. Colombia
8. Guinea
9. Cameroon

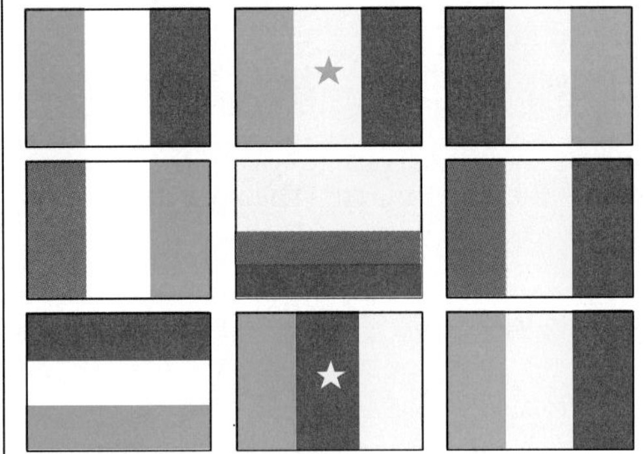

55. Solve the matrix equation

$$\begin{bmatrix} 1 & 1 & 2 \\ 3 & 1 & 0 \\ -2 & 0 & 3 \end{bmatrix}\begin{bmatrix} a & b & c \\ d & e & f \\ g & h & i \end{bmatrix} = \begin{bmatrix} 19 & 28 & 10 \\ 27 & 30 & 8 \\ -4 & 4 & 4 \end{bmatrix}.$$

56. Use the clues given by the solution of Exercise 55 to match the countries to the flags in Matrix A.

57. Solve the matrix equation

$$\begin{bmatrix} 1 & 1 & 2 \\ 3 & 1 & 0 \\ -2 & 0 & 3 \end{bmatrix}\begin{bmatrix} a & b & c \\ d & e & f \\ g & h & i \end{bmatrix} = \begin{bmatrix} 13 & 28 & 22 \\ 7 & 16 & 30 \\ 11 & 21 & -4 \end{bmatrix}.$$

58. Use the clues given by the solution of Exercise 57 to match the countries to the flags in Matrix B.

Chapter TEST

In Exercises 1–4, perform the matrix operation(s). **(4.1, 4.2)** See margin.

1. $2\left(\begin{bmatrix} 1 & 5 & -6 \\ 2 & 0 & 3 \end{bmatrix} - \begin{bmatrix} 6 & 2 & -1 \\ 9 & 9 & 5 \end{bmatrix}\right)$

2. $-2\left(\begin{bmatrix} 1 & 12 \\ -6 & -7 \end{bmatrix} + \begin{bmatrix} 4 & -8 \\ 2 & 4 \end{bmatrix}\right)$

3. $\begin{bmatrix} 3 & 1 & 0 \\ -1 & 9 & -4 \\ 2 & 3 & 0 \end{bmatrix}\begin{bmatrix} -1 \\ 4 \\ 2 \end{bmatrix}$

4. $\begin{bmatrix} 0 & 1 & 0 \\ 2 & -1 & 1 \\ 0 & 2 & -1 \end{bmatrix}\begin{bmatrix} -1 & 2 & 0 \\ 4 & 6 & 0 \\ 1 & 0 & 1 \end{bmatrix}$

In Exercises 5–7, evaluate the determinant of the matrix *and* find the inverse matrix. **(4.3)**

5. $\begin{bmatrix} 7 & -2 \\ -3 & 4 \end{bmatrix}$ 22

6. $\begin{bmatrix} -6 & 2 \\ 24 & -8 \end{bmatrix}$ 0

7. $\begin{bmatrix} -2 & -1 \\ 1 & -1 \end{bmatrix}$ 3

In Exercises 8–10, evaluate the determinant of the matrix. Which two of the three matrices are inverses of each other? **(4.4)** Matrices in Exercises 8 and 9

8. $\begin{bmatrix} 4 & 10 & -1 \\ 11 & 28 & -4 \\ -6 & -15 & 2 \end{bmatrix}$ 1

9. $\begin{bmatrix} -4 & -5 & -12 \\ 2 & 2 & 5 \\ 3 & 0 & 2 \end{bmatrix}$ 1

10. $\begin{bmatrix} -2 & -1 & 3 \\ -2 & 4 & 3 \\ 0 & 1 & -1 \end{bmatrix}$ 10

In Exercises 11–16, use the indicated method to solve the linear system. (*Hint:* For Exercise 14, use the information from Exercises 8–10.) **(4.5, 4.6)**

11. Inverse Matrix (−4, 2)
$\begin{cases} 3x + 2y = -8 \\ -2x + 5y = 18 \end{cases}$

12. Augmented Matrix (5, −4)
$\begin{cases} x - 3y = 17 \\ 9x + 6y = 21 \end{cases}$

13. Cramer's Rule (5, 2)
$\begin{cases} -4x + 5y = -10 \\ 5x - 6y = 13 \end{cases}$

14. Inverse Matrix (19, −9, −11)
$\begin{cases} 4x + 10y - z = -3 \\ 11x + 28y - 4z = 1 \\ -6x - 15y + 2z = -1 \end{cases}$

15. Augmented Matrix (8, −6, −1)
$\begin{cases} x + y - 2z = 4 \\ -2x - 3y + 4z = -2 \\ x + y - 3z = 5 \end{cases}$

16. Cramer's Rule $\left(\frac{3}{2}, \frac{1}{2}, 1\right)$
$\begin{cases} x + y = 2 \\ 2y - z = 0 \\ -x - y + z = -1 \end{cases}$

17. Use the inverse of $\begin{bmatrix} 4 & 3 \\ 1 & 1 \end{bmatrix}$ to decode the message. **(4.4)** JUST SAY NO

61, 51, 96, 77, 19, 19, 29, 28, 14, 14, 60, 45

18. Write a 3 × 3 matrix that has no entries that are zero, but whose determinant is zero. **(4.5)** See margin.

19. You have 296 ounces of tin, 271 ounces of lead, and 5 ounces of silver. How much of Alloy X, Alloy Y, and Alloy Z can you make? **(4.5)**

Percents	Alloy X	Alloy Y	Alloy Z
Tin	20%	10%	80%
Lead	78%	88%	20%
Silver	2%	2%	0%

Alloys X, Y, and Z are types of soft solder.
X: 134 oz, Y: 116 oz, Z: 322 oz

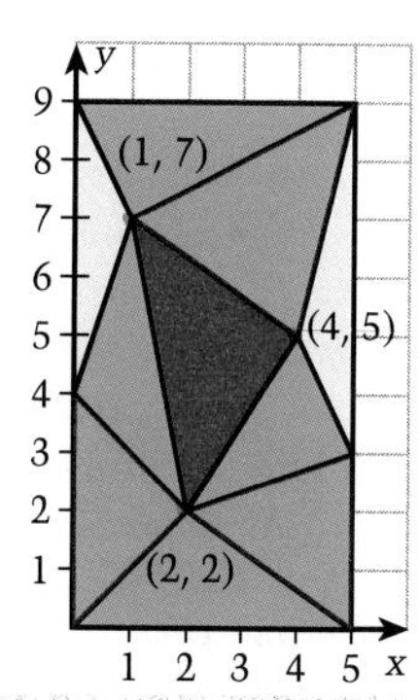

20. Find the area of the red triangle in the stained glass. **(4.3)** $\frac{13}{2}$

Answers

1. $\begin{bmatrix} -10 & 6 & -10 \\ -14 & -18 & -4 \end{bmatrix}$

2. $\begin{bmatrix} -10 & -8 \\ 8 & 6 \end{bmatrix}$

3. $\begin{bmatrix} 1 \\ 29 \\ 10 \end{bmatrix}$

4. $\begin{bmatrix} 4 & 6 & 0 \\ -5 & -2 & 1 \\ 7 & 12 & -1 \end{bmatrix}$

18. Answers may vary.

$\begin{bmatrix} 1 & 1 & 1 \\ 1 & 1 & 1 \\ 1 & 1 & 1 \end{bmatrix}$

Any 3 × 3 matrix where any two rows or columns are the same.

Techniques for solving quadratic equations will be introduced in this chapter. The geometric interpretation of a quadratic equation in two variables, the parabola, will also be presented.

The Quadratic Formula will be derived by completing the square, and complex numbers will be introduced in order to solve such equations as $x^2 + 1 = 0$.

The Chapter Summary on page 277 provides you and the students with a synopsis of the chapter. It identifies key skills and concepts. You may want to have students look at the Chapter Summary as an overview before beginning the chapter.

CHAPTER 5

Quadratic Equations and Parabolas

LESSONS

The Brooklyn Bridge connects two boroughs of New York City: Manhattan and Brooklyn. Designed by the American engineer John Roebling, it was completed in 1883. At the time, the Brooklyn Bridge was the longest suspension bridge in the world.

A quadratic equation can be used to model the shape of the suspension cables of the Brooklyn Bridge. This is but one example of real-life situations that can be represented using quadratic relationships. Other situations involve falling bodies, distances between cities on a map, and areas of regions.

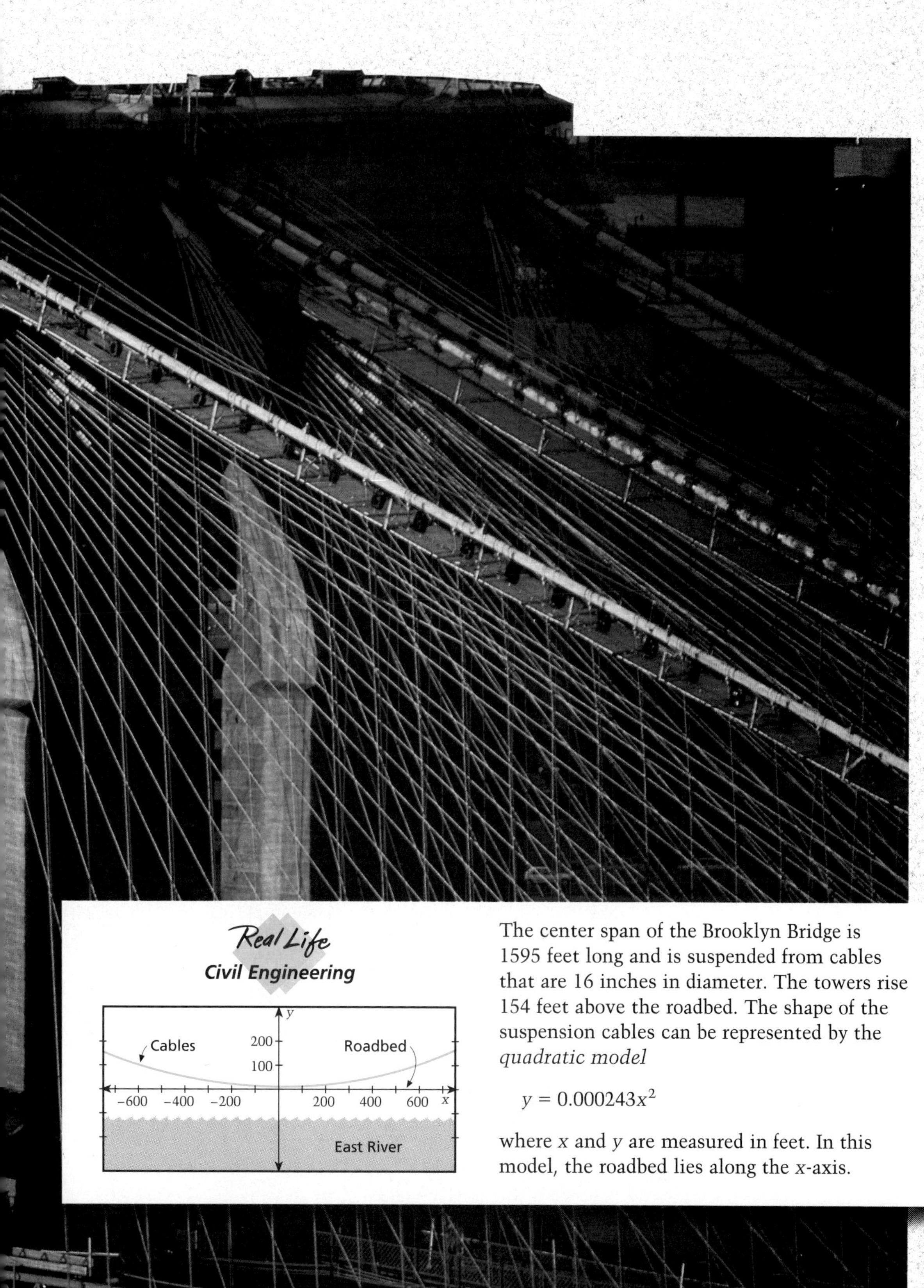

The center span of the Brooklyn Bridge is 1595 feet long and is suspended from cables that are 16 inches in diameter. The towers rise 154 feet above the roadbed. The shape of the suspension cables can be represented by the *quadratic model*

$$y = 0.000243x^2$$

where x and y are measured in feet. In this model, the roadbed lies along the x-axis.

Quadratic Equations and Parabolas

PACING CHART

The daily Pacing Chart is meant to help you adjust your teaching pace. Students in the full course should finish the entire text by the end of the year. Students in the basic course are expected to complete the first twelve chapters. The Pacing Chart for each chapter contains suggestions for lessons that require more than one day and lessons that may be omitted for the basic course.

DAY	FULL COURSE	BASIC COURSE
1	5.1	5.1
2	5.2 & Using a Graphing Calculator	5.2
3	5.3	5.2 & Using a Graphing Calculator
4	Mid-Chapter Self-Test	5.3
5	5.4	5.3
6	5.5	Mid-Chapter Self-Test
7	5.6	5.4
8	5.7	5.5
9	Chapter Review	5.6
10	Chapter Test	5.6
11		5.7
12		Chapter Review
13		Chapter Test

CHAPTER ORGANIZATION

LESSON	PAGES	GOALS	MEETING THE NCTM STANDARDS
5.1	230–235	1. Solve a quadratic equation by finding square roots 2. Use quadratic equations as models of real-life situations	Problem Solving, Communication, Connections, Geometry (synthetic)
5.2	236–241	1. Graph a quadratic equation 2. Use the graph of a quadratic equation in a real-life setting	Problem Solving, Communication, Connections, Geometry (algebraic)
Using a Calculator	242–243	Approximating the solutions of a quadratic equation	Technology
Mixed Review	244	Review algebraic and arithmetic skills	
Career Interview	244	Epidemiologist	Connections
5.3	245–250	1. Solve a quadratic equation by completing the square 2. Use "completing the square" to solve real-life problems	Problem Solving, Communication, Connections, Geometry (algebraic)
Mid-Chapter Self-Test	251	Diagnose student weaknesses and remediate with correlated Reteach worksheets	
5.4	252–257	1. Use the quadratic formula to solve quadratic equations 2. Use quadratic models in real-life settings	Problem Solving, Communication, Connections
5.5	258–264	1. Identify, add, subtract, and multiply imaginary and complex numbers 2. Plot complex numbers in the complex plane	Problem Solving, Communication, Connections, Structure
Mixed Review	263	Review algebraic and arithmetic skills	
5.6	264–270	1. Solve any quadratic equation (even ones that have complex number solutions) 2. Use complex numbers and programmable calculators to solve real-life problems	Problem Solving, Communication, Connections, Structure, Technology
5.7	271–276	1. Sketch the graph of a quadratic inequality 2. Use quadratic inequalities in real-life modeling	Problem Solving, Communication, Connections
Chapter Summary	277	Restate for students what they have learned, why they have learned it, and how it fits into the structure of algebra	Structure, Connections
Chapter Review	278–280	Review concepts and skills learned in the chapter	
Chapter Test	281	Diagnose student weaknesses and remediate with correlated Reteaching worksheets	

LESSON RESOURCES

MEETING INDIVIDUAL NEEDS

RETEACHING For students who need to spend more time on basics:

If a mid-chapter self-test or chapter test indicates a deficiency, teachers can help students with the appropriate ***Reteaching Copymaster.***

PRACTICE For students who need more practice:

Additional exercises like those in the Pupil's Edition are provided for each lesson in ***Extra Practice Copymasters.***

ENRICHMENT For enriching and broadening students' experiences:

Problem of the Day copymasters in ***Teaching Tools*** provide a daily opportunity to use logical reasoning, looking for a pattern, writing an equation, and other routine and non-routine problem-solving strategies.

Math Log copymasters in ***Alternative Assessment*** provide opportunities to report on investigations, research, and open-ended problems.

Technology: Using Calculators and Computers provides enriching activities with graphing and scientific calculators and computers.

The ***Applications Handbook*** provides additional information about the cross-curriculum topics such as astronomy, chemistry, physics, sports, economics, genetics, and music that are integrated into the Pupil's Edition.

LESSON	5.1	5.2	5.3	5.4	5.5	5.6	5.7
PAGES	230-235	236-241	245-250	252-257	258-264	265-270	271-276
Teaching Tools							
Transparencies		✓	✓	✓			✓
Problem of the Day	✓	✓	✓	✓	✓	✓	✓
Warm-up Exercises	✓	✓	✓	✓	✓	✓	✓
Answer Masters	✓	✓	✓	✓	✓	✓	✓
Extra Practice Copymasters	✓	✓	✓	✓	✓	✓	✓
Reteaching Copymasters	Teacher-directed and independent activities tied to results on the Mid-Chapter Self-Tests and Chapter Tests						
Color Transparencies		✓		✓	✓		✓
Applications Handbook	Additional background information for many real-life applications						
Technology	Calculator and computer worksheets for appropriate lessons						
Complete Solutions Manual	✓	✓	✓	✓	✓	✓	✓
Alternative Assessment	Assess student's ability to reason, analyze, solve problems, and communicate using mathematical language.						
Formal Assessment	Mid-Chapter Self-Tests, Chapter Tests, Cumulative Tests, and Practice for College Entrance Tests						
Computer Test Bank	Customized tests can be created by choosing from over 2500 items.						

INSIGHTS

5.1 Solving Quadratic Equations by Finding Square Roots

Several techniques for solving quadratic equations will be presented in this chapter. Perhaps, the most straightforward technique for simple quadratic equations is solving by finding square roots. The technique can frequently be completed using mental mathematics instead of paper and pencil. Because many real-life problems can be represented by quadratic models, it is important to have a collection of strategies for solving quadratic equations.

5.2 Parabolas: Graphs of Quadratic Equations

Once again emphasizing the strong connection between algebra and geometry, this lesson employs graphs as a way to picture quadratic relationships. The graph of a quadratic equation in two variables is a parabola. Geometric models often provide additional insights and perspectives into the behavior of algebraic relationships.

5.3 Completing the Square

Completing-the-square is a useful technique for solving quadratic equations in one variable, but perhaps its most important use is in deriving the Quadratic Formula, a formula that generates the solutions of any quadratic equation.

5.4 The Quadratic Formula

The quadratic formula is used to find the solutions of any quadratic equation. Because it can always be used, it is, undoubtedly, the most useful way to solve quadratic equations. In this lesson, students apply the quadratic formula in a variety of ways. In the *Handbook of Mathematical Connections* on page 846, students are shown how to complete the square to derive the quadratic formula.

5.5 Complex Numbers

Complex numbers were first defined by the sixteenth century Italian mathematician, Rafael Bombelli. Much later, in 1843, the Irish mathematician William Rowan Hamilton extended the concept of complex numbers to a new set of numbers called "quaternions," for which the first noncommutative algebra was created. In fact, complex numbers play an important role in the study of algebra.

5.6 Solving Any Quadratic Equation

Many real-life situations can be modeled by quadratic equations. The quadratic formula can now be used to find the solutions of any quadratic equation, including those in which the discriminant is negative. Since it is such a powerful problem-solving tool, students should take time to memorize the quadratic formula.

5.7 Graphs of Quadratic Inequalities

Real-life applications of quadratic inequalities can be found in economics, business, physics, and probability computations. The procedure used in this lesson to graph quadratic inequalities is similar to the procedure presented earlier for graphing linear inequalities. Students will find it interesting to see how personalized logos may be created from combinations of quadratic inequalities.

Problem of the Day

Find the length of the edge of the smallest cubical box that will hold a 10-foot rod.
$\frac{10\sqrt{3}}{3}$ ft, or about 5.77 ft

ORGANIZER

Warm-Up Exercises

1. Compute the following square roots to two decimal places. (A calculator may be needed to evaluate some of the square roots.)
   a. $\sqrt{49}$ b. $\sqrt{169}$
   c. $\sqrt{225}$ d. $\sqrt{324}$
   e. $\sqrt{12}$ f. $\sqrt{30}$
   g. $\sqrt{125}$ h. $\sqrt{784}$
   a. 7, b. 13, c. 15, d. 18, e. 3.46, f. 5.48, g. 11.18, h. 28
2. Solve for x.
   a. $4x + 12 = 40$ 7
   b. $-3x - 13 = 62$ −25
   c. $126 + 14x = 0$ −9
   d. $23 - 8x = 159$ −17

Lesson Resources

Teaching Tools
Problem of the Day: 5.1
Warm-up Exercises: 5.1
Answer Masters: 5.1
Extra Practice: 5.1
Applications Handbook: p. 42–51

Common-Error ALERT!

In **Example 1,** be sure students recognize the difference between finding the real number values that satisfy the statement, $x^2 = 7$, versus evaluating the square root of 7, that is, $\sqrt{7}$. In the former case, there are two possible answers, $+\sqrt{7}$ and $-\sqrt{7}$. In the latter case, there is only one result, $+\sqrt{7}$.

5.1 Solving Quadratic Equations by Finding Square Roots

What you should learn:

Goal 1 How to solve a quadratic equation by finding square roots

Goal 2 How to use quadratic equations as models of real-life situations

Why you should learn it:

You can model many real-life situations with quadratic equations, such as the height of an object as it falls.

Goal 1 Solving a Quadratic Equation

A **quadratic equation** in x is an equation that can be written in the **standard form**

$$ax^2 + bx + c = 0$$

where a, b, and c are real numbers ($a \neq 0$). The number a is the **leading coefficient.** In this lesson, you will solve quadratic equations of the form $ax^2 + c = 0$.

Recall from Algebra 1 that all positive real numbers have two square roots: the **positive square root** and the **negative square root.** Square roots are written with the **radical symbol** $\sqrt{\ }$. (The number inside a radical symbol is the **radicand.**)

Solving $x^2 = d$ by Finding Square Roots

1. If d is positive, then $x^2 = d$ has two solutions: $x = \pm\sqrt{d}$.
2. The equation $x^2 = 0$ has one solution: $x = 0$.
3. If d is negative, then $x^2 = d$ has no *real-number* solution.

(The case in which d is negative will be discussed in Lesson 5.5. There you will learn about solutions that are not real numbers.)

The symbol $\pm\sqrt{d}$ is read: "plus or minus the square root of d."

Example 1 *Solving Quadratic Equations*

Solve $3x^2 = 21$.

Solution

$3x^2 = 21$ *Rewrite original equation.*

$x^2 = 7$ *Divide both sides by 3.*

$x = \pm\sqrt{7}$ *Find square roots.*

The solutions are $\sqrt{7}$ and $-\sqrt{7}$. Check these in the original equation. To do that, you need to use the properties that $(\sqrt{d})(\sqrt{d}) = d$ and $(-\sqrt{d})(-\sqrt{d}) = d$. ■

Connections
Physics

Goal 2 Using Quadratic Equations in Real Life

When an object (with little air resistance) is dropped, its speed continually *increases.* The height, h (in feet), of such an object is given by the model

$$h = -16t^2 + s \quad \textit{Falling-Object Model}$$

where s is the height from which the object was dropped and t is the number of seconds it has fallen. This model works on Earth, but not on planets with different gravity.

Problem Solving
Make a Table

Example 2 *Finding the Time a Falling Object Takes*

At 9:55 A.M. on Saturday, July 28, 1945, a terrible airplane accident occurred. A B-25 bomber crashed into the 78th and 79th floors of the Empire State Building in New York City. Hundreds of pieces of debris fell 975 feet to the streets below. How much time after hearing the crash did the people on the street have to get out of the way?

This accident killed 14 people. The toll would have been greater, but it was a drizzly Saturday morning and most of the building's 15,000 office workers were at home.

Solution Because the debris dropped from a height of 975 feet, the model for its height is $h = -16t^2 + 975$. The table gives the heights at different times in seconds.

Time, t	0	1	2	3	4	5	6	7	7.8
Height, h	975	959	911	831	719	575	399	191	2

From the table, you can see that the debris would take about 8 seconds to hit the ground. Because the sound of the crash would take about 1 second to reach the ground (sound travels at about 1100 feet per second), you can conclude that people had about 7 seconds to get out of the way of the falling debris.

Another way to determine how long it took the debris to hit the ground is to solve the quadratic equation for the time that gives a height of $h = 0$ feet.

$$-16t^2 + 975 = h \quad \textit{Falling-object model}$$

$$-16t^2 + 975 = 0 \quad \textit{Substitute 0 for h.}$$

$$975 = 16t^2 \quad \textit{Add } 16t^2 \textit{ to both sides.}$$

$$\frac{975}{16} = t^2 \quad \textit{Divide both sides by 16.}$$

$$\sqrt{\frac{975}{16}} = t \quad \textit{Find the positive square root.}$$

$$7.8 \approx t \quad \textit{Use a calculator.}$$

Since it took about 1 second to hear the crash, you can conclude that the people on the street had about 7 seconds to get out of the way. ■

LESSON Notes

Example 1

The expression $\sqrt{7}$ would indicate the principal or positive square root of seven.

Example 2

PROBLEM SOLVING An important feature of solving problems is illustrated in this example. Namely, there are at least two ways that a solution can be reached—analyzing a table and finding an algebraic solution. Either method is correct, and either should be accepted as valid. In fact, you may wish to ask students which approach they understand better and which approach is more efficient.

Point out that the final answer is not the number value obtained in the table or the number obtained from solving the quadratic model. It was necessary to use the additional information about how long it took the sound of the crash to reach the streets. Observe that the algebraic solution to the quadratic equation is ±7.8, but that real-life considerations require that the solution be positive.

Example 3

Remind students that if $d = rt$ then $t = \frac{d}{r}$, where d is distance, r is speed, and t is time.

Have students read the map and identify the two possible routes to be followed on the trip. Point out that the two routes very nearly form a right triangle and so it is appropriate to use the Pythagorean Theorem to obtain the information. (Describing the routes as a right triangle is another example of modeling real-life situations with algebra.)

Extra Examples

Here are additional examples similar to **Examples 1–3.**

1. Solve.
   - **a.** $5x^2 = 180$ ± 6
   - **b.** $8x^2 = 40$ $\pm\sqrt{5}$
   - **c.** $4x^2 - 76 = 0$ $\pm\sqrt{19}$
   - **d.** $7x^2 = 42$ $\pm\sqrt{6}$
2. Suppose in Example 2, that debris fell 1265 feet to the streets below. How long would it take the debris to hit the streets? From which floor did the debris fall?
   time $\approx$ 8.9 seconds; $\frac{1265 \text{ ft}}{12.5 \text{ ft per floor}} \approx 101$ floor
3. You remember (in Example 3) that Fond du Lac is being repaired along a 5-mile stretch near Capitol Drive. You figure that you can average 35 miles per hour along this portion of the trip. Is Fond du Lac to Menomonee still the quickest route?
   Yes, the trip will now take about 19 minutes.

Check Understanding

1. What is the standard form of a quadratic equation? An equation in the form $ax^2 + bx + c = 0$, $a \neq 0$
2. How is finding the solution of $x^2 = 4$ different from evaluating $\sqrt{4}$? There are two solution values for $x^2 = 4$, 2 and -2, while $\sqrt{4} = 2$.
3. What is the Falling Object Model? $h = -16t^2 + s$
4. What does the model $d = rt$ represent?
   See Lesson Notes for Example 3.

Example 3 *Using the Pythagorean Theorem*

You are in downtown Milwaukee and want to drive to Menomonee Falls. You are considering two options. You could drive 9 miles north on Interstate 43 and then 9 miles west on Brown Deer, *or* you could drive northwest on Fond du Lac Avenue. You can average 55 miles per hour on Interstate 43, then 40 miles per hour on Brown Deer, and 45 miles per hour on Fond du Lac. How much time would each route take?

Solution The time it would take on I-43 and Brown Deer is:

$$\text{time} = \underbrace{\frac{9 \text{ miles}}{55 \text{ miles per hour}}}_{\text{Interstate 43}} + \underbrace{\frac{9 \text{ miles}}{40 \text{ miles per hour}}}_{\text{Brown Deer}}$$

$$\approx 0.164 + 0.225 = 0.389 \text{ hours} \approx 23 \text{ minutes.}$$

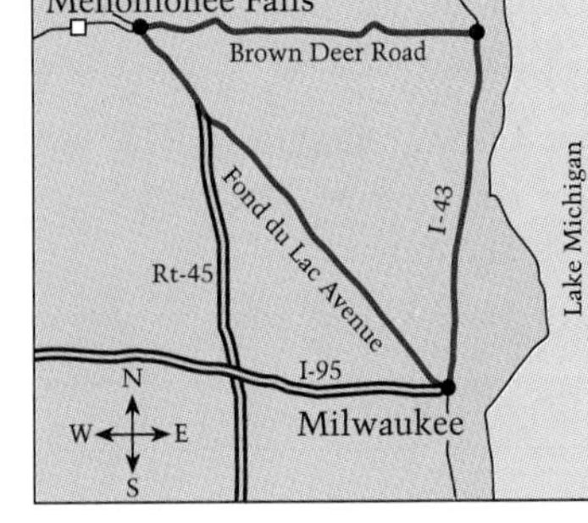

To find the distance, c (in miles), on Fond du Lac Avenue, use the Pythagorean Theorem.

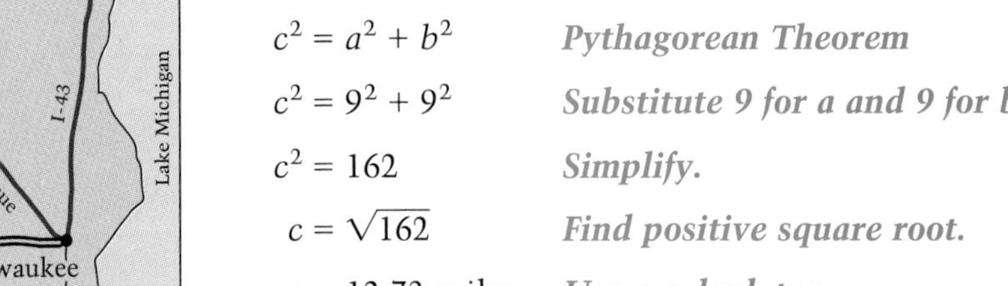

$c^2 = a^2 + b^2$ *Pythagorean Theorem*

$c^2 = 9^2 + 9^2$ *Substitute 9 for a and 9 for b.*

$c^2 = 162$ *Simplify.*

$c = \sqrt{162}$ *Find positive square root.*

$c \approx 12.73$ miles *Use a calculator.*

The time it will take you on Fond du Lac Avenue is

$$\text{time} = \frac{12.73 \text{ miles}}{45 \text{ miles per hour}} \approx 0.28 \text{ hours} \approx 17 \text{ minutes.}$$

■

Communicating about ALGEBRA

SHARING IDEAS about the Lesson

Use a Geometry Theorem Use a radical to write the *exact* length of x. Then write approximate lengths correct to the nearest 0.01 cm. **A.** $\sqrt{52}$ cm, 7.21 cm **B.** $\sqrt{109}$ cm, 10.44 cm

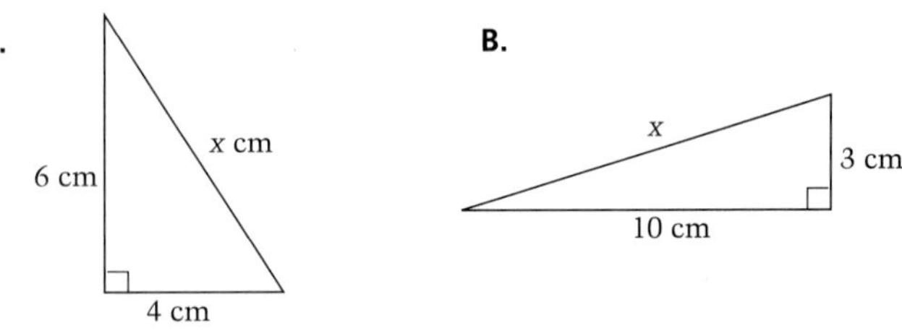

EXERCISES

Guided Practice

CRITICAL THINKING about the Lesson

1. Write the quadratic equation $3 - 2x = -4x^2$ in standard form. What is its leading coefficient? $4x^2 - 2x + 3 = 0$, 4

In Exercises 2–4, solve the equation. If the equation has no real-number solutions, explain why.

2. $2x^2 + 4 = 0$
3. $x^2 + 3 = 3$ 0
4. $3x^2 = 12$ ± 2

No real number solutions; the square of a number cannot be negative.

Independent Practice

In Exercises 5–16, solve the equation.

5. $2x^2 = 32$ ± 4
6. $3x^2 = 27$ ± 3
7. $\frac{1}{3}x^2 = 4$ $\pm\sqrt{12}$
8. $\frac{2}{3}x^2 = 6$ ± 3
9. $4x^2 - 5 = 9$ $\pm\frac{\sqrt{14}}{2}$
10. $6x^2 + 3 = 16$ $\pm\sqrt{\frac{13}{6}}$
11. $-x^2 + 9 = 2x^2 - 6$ $\pm\sqrt{5}$
12. $-x^2 + 2 = x^2 + 1$ $\pm\sqrt{\frac{1}{2}}$
13. $3x^2 - 7 = 2(x^2 + 3)$
14. $\frac{1}{2}x^2 - 5 = 13$ ± 6
15. $\frac{2}{3}x^2 + 6 = 18$ $\pm\sqrt{18}$
16. $\frac{1}{4}x^2 + 1 = 33$

13. $\pm\sqrt{13}$ 16. $\pm\sqrt{128}$

In Exercises 17–22, solve the equation. Round the solutions to two decimal places.

17. $7x^2 + 4 = 12$ ± 1.07
18. $x^2 - 7 = 3$ ± 3.16
19. $3x^2 - 5 = 9$ ± 2.16
20. $-\frac{1}{2}x^2 + 5 = 2$ ± 2.45
21. $2x^2 + 3 = 10$ ± 1.87
22. $5x^2 - 8 = 1$ ± 1.34

✪ 23. *Geometry* The surface area of a cube is 326 square meters. How long is each edge? ≈ 7.37 m

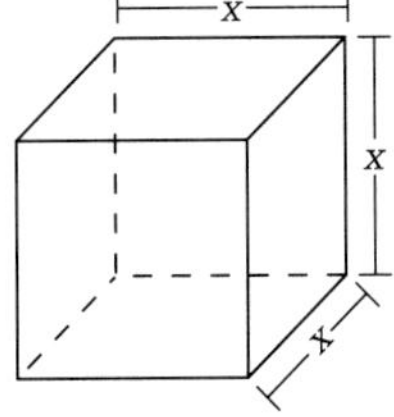

24. *Geometry* The surface area of a sphere is 576π square inches. Find the radius. 12 in.

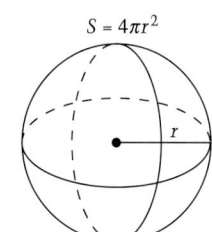

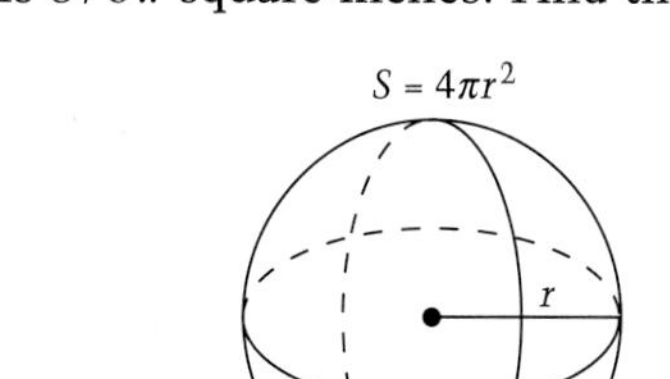

Geometry **In Exercises 25 and 26, use the Pythagorean Theorem to find *x*.**

25. 4

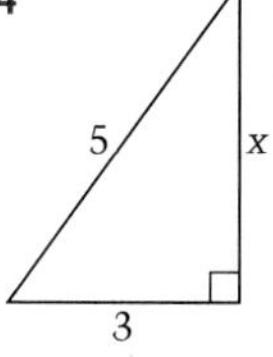

26. 40

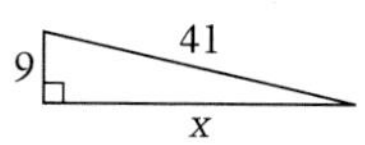

Communicating about ALGEBRA

Have students discuss the circumstances under which they might need the exact form rather than the approximate form.

EXERCISE Notes

ASSIGNMENT GUIDE

Basic/Average: Ex. 1–4, 5–9 odd, 17–27 odd, 29–33, 35–39 odd, 49–53

Above Average: Ex. 1–4, 11–27 odd, 29–35, 49–53

Advanced: Ex. 1–4, 11–29 odd, 30–36, 49–53

Selected Answers
Exercises 1–4, 5–49 odd

✪ More Difficult Exercises
Exercises 23, 29, 30, 51–53

Guided Practice

▶ **Ex. 1–4** The Guided Practice exercises can be used to summarize the lesson.

EXTEND Ex. 1–4 COOPERATIVE LEARNING Ask student pairs to state two things they remember about quadratic equations from Algebra 1.

Independent Practice

▶ **Ex. 5–16** Students should use the "square-root" method as modeled by Example 1 because there is no middle term.

Ex. 17–26 TECHNOLOGY Encourage students to use a scientific calculator to approximate the solutions.

Ex. 27–28 Students will need to use the height model provided on page 231.

Ex. 29–30 Students can use Example 3 as a solution model.

Ex. 31–34 MATH JOURNAL These applications are the first in a series of real-life applications and questions that can be modeled by using quadratic models. Students might wish to include a section on quadratic modeling in their math journals.

Ex. 49–50 NUMBER SENSE These exercises provide practice in scientific notation.

Ex. 51–53 These exercises explore a visual connection of parabolas with quadratic equations.

Falling Object **In Exercises 27 and 28, find the time it takes an object to hit the ground when it is dropped from a height of *s*.**

27. $s = 25$ feet 1.25 seconds

28. $s = 64$ feet 2 seconds

Falling on Mars **In Exercises 29 and 30, use the following information.**

The height, h (in feet), of a falling object on Mars after t seconds is given by

$$h = -6t^2 + s.$$

29. Suppose one of the rocks scooped up by the Viking I space probe fell back to Mars' surface from a height of 72 feet. About how long would it take to reach the surface? $\approx$3.5 seconds

30. If the same rock was dropped from the same height on Earth, how long would it take to reach Earth's surface? $\approx$2.1 seconds

On July 20, 1976, the Viking I space probe landed on Mars and scooped up samples of rock. This photo shows Viking I orbiting Mars.

31. *Looking at the Horizon* At sea, the distance, d (in miles), that you can see to the horizon depends on your height. If your eyes are h feet above sea level, then the approximate distance you can see to the horizon is given by

$$d^2 = \frac{3}{2}h.$$

Solve this equation for d. A 6-foot person is standing on the seashore. About how far out to sea can the person see? 3 miles

32. *Period of a Pendulum* The *period* of a pendulum is the time the pendulum takes to swing back and forth. The period, t (in seconds), of a pendulum of length d (in inches) is given by

$$d = 9.78t^2.$$

The longest pendulum in the world is part of a clock on a building in Tokyo, Japan. This pendulum is about 74 feet (888 inches) long. What is the period of this pendulum?
(Source: Guinness Book of World Records) $\approx$9.53 seconds

33. *PGA Prize Money* From 1970 to 1990, the prize money, P (in dollars), at the Professional Golf Association Championship can be approximated by the model $P = 2875t^2 + 200{,}000$ where t is the year, with $t = 0$ corresponding to 1970. During which year was the prize money first over \$1,350,000? 1990
(Source: Professional Golf Association)

34. *Company Profit* From 1980 to 1991, the profit, P (in dollars), of a company could be modeled by $P = 9400t^2 + 525{,}000$ where $t = 0$ represents 1980. According to this model, when would the company have a profit of \$2,640,000? 1995

35. *Stair Steps* Find the distance between the edges of two consecutive steps in the stairs shown below. Approximate the length of the handrail shown above the stairs. ≈11.3 in., ≈56.6 in.

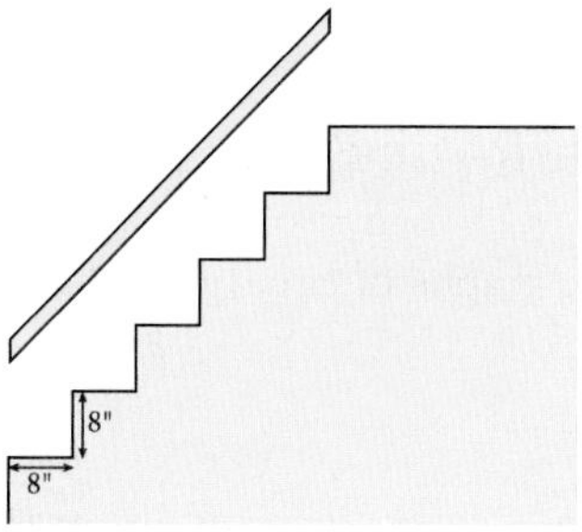

36. *Taking a Trip* You are driving from Deerfield to Campbellsport in Ohio. Highway 14 is under construction, so you drive 5.5 miles north on Route 225, then 5.8 miles west on Interstate 76. How many miles longer is this route than taking Highway 14? ≈3.3 miles

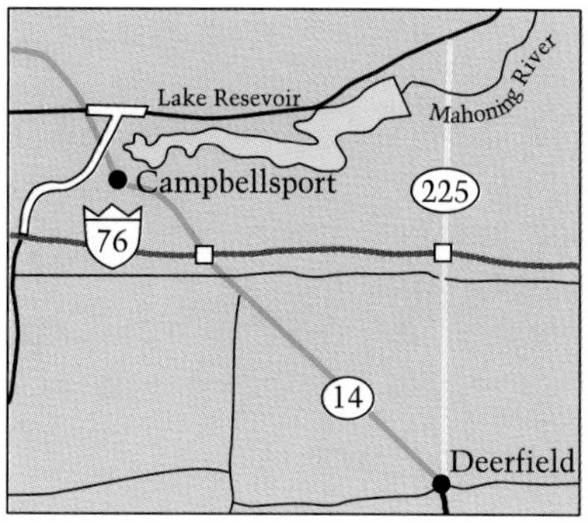

Integrated Review

In Exercises 37–42, evaluate the expression.

37. $3x^2 + 9x - 4$ when $x = 3$ 50 **38.** $x^2 - 5x + 12$ when $x = 4$ 8 **39.** $\sqrt{x} - 12$ when $x = 49$ −5

40. $x - \sqrt{x} + 5$ when $x = 4$ 7 **41.** $\frac{1}{2}x^2 - x + 1$ when $x = 1$ $\frac{1}{2}$ **42.** $-\frac{1}{3}x^2 + 6x - 1$ when $x = 12$ 23

In Exercises 43–48, solve the equation.

43. $5x + 3 = 18$ 3 **44.** $-x + 4 = 13$ −9 **45.** $-\frac{1}{3}x - 7 = 1$ −24

46. $\frac{1}{2}x - 5 = 17$ 44 **47.** $-4x + 11 = -9$ 5 **48.** $9x - 5 = 13$ 2

49. *Powers of Ten* Which best estimates the weight of a house cat? c
a. 1.0×10^{-1} lb **b.** 1.0×10^{0} lb **c.** 1.0×10^{1} lb **d.** 1.0×10^{2} lb

50. *Powers of Ten* Which best estimates the volume of an automobile fuel tank? b
a. 1.5×10^{0} gal **b.** 1.5×10^{1} gal **c.** 1.5×10^{2} gal **d.** 1.5×10^{3} gal

Exploration and Extension

✪ **51.** Complete the table and plot the points on a coordinate system.

x	−3	−1	0	1	3
$y = -x^2 + 1$	?	?	?	?	?

−8, 0, 1, 0, −8 See margin.

✪ **52.** Use the graph to approximate the solutions of $-x^2 + 1 = -3$. 2, −2

✪ **53.** Confirm your results by solving the equation algebraically. $x^2 = 4$, $x = \pm 2$

Enrichment Activities

These activities require students to go beyond lesson goals.

1. Find both the simplest exact value (no perfect *n*th power in the radicand) and approximate decimal value (to the nearest hundredth) for each expression.

a. $\sqrt{18}$ $3\sqrt{2} \approx 4.24$
b. $\sqrt{180}$ $6\sqrt{5} \approx 13.42$
c. $\sqrt{1800}$ $30\sqrt{2} \approx 42.43$
d. $\sqrt{18000}$ $60\sqrt{5} \approx 134.16$
e. $\sqrt{1008}$ $12\sqrt{7} \approx 31.75$
f. $\sqrt{10080}$ $12\sqrt{70} \approx 100.40$

2. **NUMBER SENSE** Look for patterns in the answers from Exercise 1. Apply the patterns to the answers in Parts e and f and use mental math to simplify and approximate.

a. $\sqrt{0.1008}$ **b.** $\sqrt{1.008}$
c. $\sqrt{10.08}$ **d.** $\sqrt{100.8}$

a. $1008 \div 10000 = 0.1008$, so $\sqrt{0.1008} = \sqrt{1008} \div \sqrt{10000} = 12\sqrt{7} \div 100 = 0.12\sqrt{7} \approx 0.3175$
b. $10080 \div 10000 = 1.008$; $0.12\sqrt{70} \approx 1.004$
c. $1008 \div 100 = 10.08$; $1.2\sqrt{7} \approx 3.1749$
d. $10080 \div 100 = 100.8$; $1.2\sqrt{70} \approx 10.04$

Answer
51.

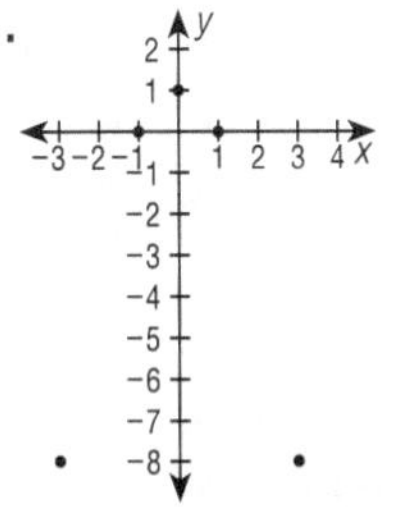

Problem of the Day

Find a number that is one greater than its own reciprocal.

$\frac{1+\sqrt{5}}{2} \approx 1.618$ or $\frac{1-\sqrt{5}}{2} \approx -0.618$

ORGANIZER

Warm-Up Exercises

1. Complete the table of values for the indicated relationships. Use the tables to plot the resulting points on a coordinate plane.

a. $y = 2x + 1$

x	−3	−2	−1	0	1	2	3
y	−5	−3	−1	1	3	5	7

b. $y = 4 - 3x$

x	−3	−2	−1	0	1	2	3
y	13	10	7	4	1	−2	−5

a.

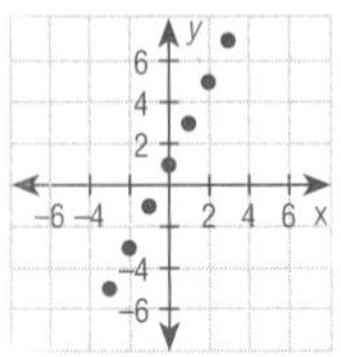

b. Students should graph the points shown in the table.

2. Evaluate the expression, $-\frac{b}{2a}$, at the given values of a and b.

a	1	−3	2	8	−5
b	−6	7	−6	4	15
$-\frac{b}{2a}$	3	$\frac{7}{6}$	$\frac{3}{2}$	$-\frac{1}{4}$	$\frac{3}{2}$

Lesson Resources

Teaching Tools
- Transparencies: 2, 3, 7, 8, 13
- Problem of the Day: 5.2
- Warm-up Exercises: 5.2
- Answer Masters: 5.2

Extra Practice: 5.2

Color Transparencies: 32, 33

Technology Handbook: p. 25, 28

5.2 Parabolas: Graphs of Quadratic Equations

What you should learn:

Goal 1 How to graph a quadratic equation

Goal 2 How to use the graph of a quadratic equation in a real-life setting

Why you should learn it:

You can use the graph of a quadratic equation to model real-life shapes, such as the foundation of *Pueblo Bonita* in Chaco Canyon, New Mexico.

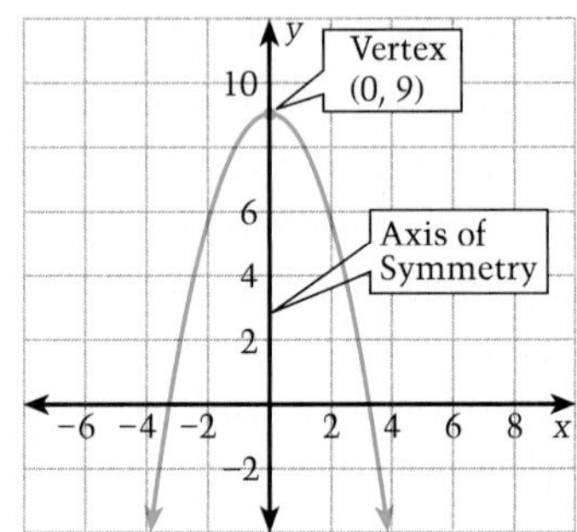

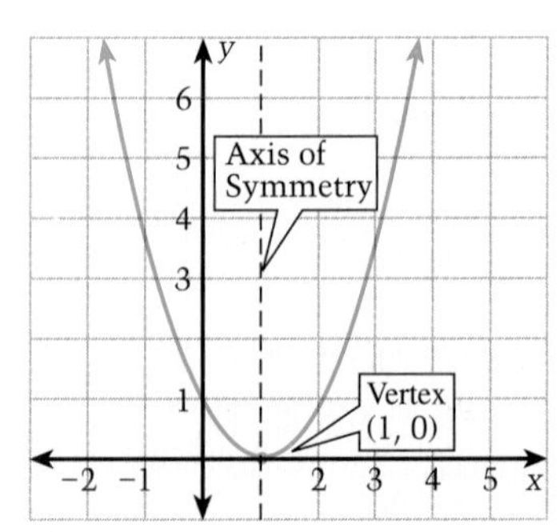

Goal 1 Graphing a Quadratic Equation

In this lesson, you will study the graphs of equations of the form

$$y = ax^2 + bx + c, \quad a \neq 0.$$

This equation is quadratic (second degree) in x and linear (first degree) in y. For simplicity, however, we will call the equation a **quadratic equation.**

The graph of a quadratic equation is U-shaped and is called a **parabola.** If the leading coefficient a is negative, the parabola opens *down.* If the leading coefficient is positive, the parabola opens *up.* On a parabola that opens up, the lowest point is the **vertex.** On a parabola that opens down, the highest point is the vertex. The vertical line passing through the vertex is the **axis of symmetry.** This line divides a parabola into two symmetrical parts, which are mirror images of each other. Here are two examples of quadratic equations.

$y = -x^2 + 9$	$y = x^2 - 2x + 1$
Leading coefficient: $a = -1$	Leading coefficient: $a = 1$
Parabola opens down.	Parabola opens up.
Vertex: (0, 9)	Vertex: (1, 0)
Axis of symmetry: $x = 0$	Axis of symmetry: $x = 1$

Graph of a Quadratic Equation

The graph of $y = ax^2 + bx + c$ is a parabola.

1. If a is positive, the parabola opens up. If a is negative, the parabola opens down.
2. The x-coordinate of the vertex is $-\frac{b}{2a}$.
3. The axis of symmetry is the vertical line $x = -\frac{b}{2a}$.

To sketch the graph of a quadratic equation, first find the x-coordinate of the vertex. Then construct a table of values, using x-values to the left and right of the vertex. Finally, plot the points and connect them with a U-shaped graph.

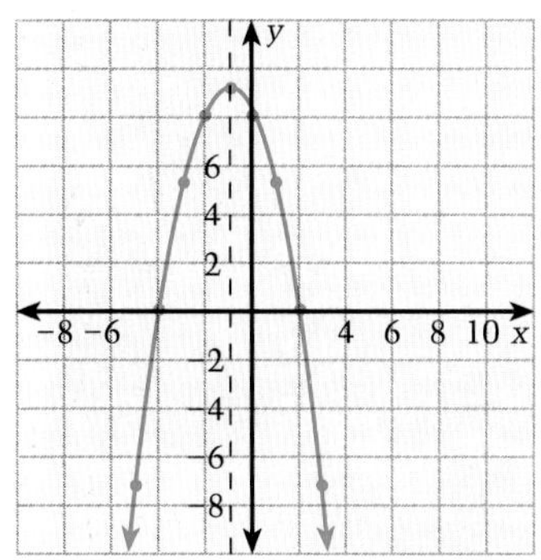

Example 1 Sketching the Graph of a Quadratic Equation

Sketch the graph of $y = -x^2 - 2x + 8$.

Solution In the standard form $ax^2 + bx + c$, the coefficients are $a = -1$, $b = -2$, and $c = 8$. The x-coordinate of the vertex is

$$-\frac{b}{2a} = -\frac{-2}{2(-1)} = -1. \quad \textit{x-coordinate of vertex}$$

Using x-values to the left and right of this x-value, construct a table of values.

Vertex ↓ (at $x = -1$)

x	−5	−4	−3	−2	−1	0	1	2	3
$y = -x^2 - 2x + 8$	−7	0	5	8	9	8	5	0	−7

The vertex is $(-1, 9)$ and the axis of symmetry is $x = -1$. Plot the points given in the table and connect them with a parabola that opens down. ■

The graph in Example 1 has x-intercepts at $(-4, 0)$ and $(2, 0)$. The y-intercept occurs at $(0, 8)$.

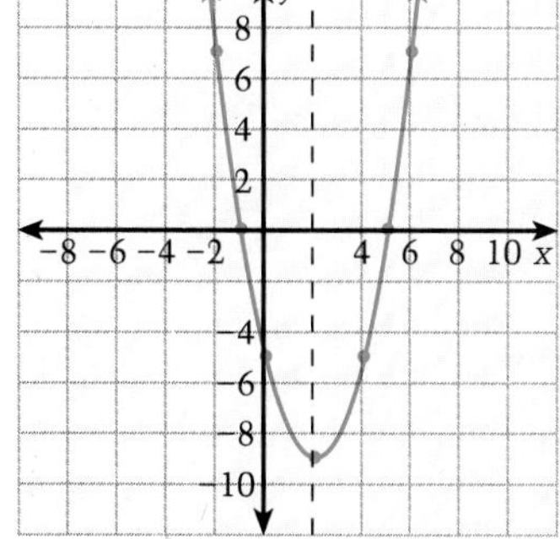

Example 2 Sketching the Graph of a Quadratic Equation

Sketch the graph of $y = x^2 - 4x - 5$.

Solution For this quadratic equation, $a = 1$, $b = -4$, and $c = -5$. The x-coordinate of the vertex is

$$-\frac{b}{2a} = -\frac{-4}{2(1)} = 2. \quad \textit{x-coordinate of vertex}$$

Using x-values to the left and right of this x-value, construct a table of values.

Vertex ↓ (at $x = 2$)

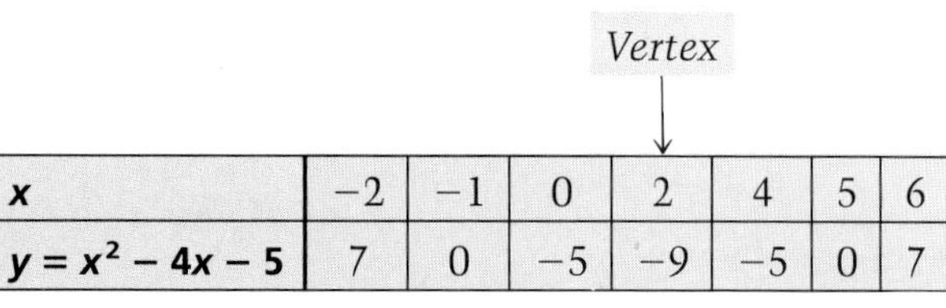

x	−2	−1	0	2	4	5	6
$y = x^2 - 4x - 5$	7	0	−5	−9	−5	0	7

The vertex is $(2, -9)$ and the axis of symmetry is $x = 2$. Plot the points given in the table and connect them with a parabola that opens up. ■

The graph in Example 2 has x-intercepts at $(-1, 0)$ and $(5, 0)$. The y-intercept occurs at $(0, -5)$. You can verify this from the graph.

LESSON Notes

Example 1–2

Students should memorize the formula for the x-coordinate of the vertex of the parabola for the quadratic equation in standard form. Emphasize the importance of plotting points on either side of the vertex to obtain a good sketch of the parabola.

You may wish to observe that a sketch of the parabola can often be made from the x- and y-intercepts of the curve. The y-intercept can be "read" from a quadratic equation in standard form; it is the constant term, c. The x-intercepts are found by solving the quadratic equation for x when y is set to zero. (More techniques for solving quadratic equations follow.)

EXTEND Example 3 Have students use the formula area $= \frac{2}{3}bh$ to determine the area of the shaded region.

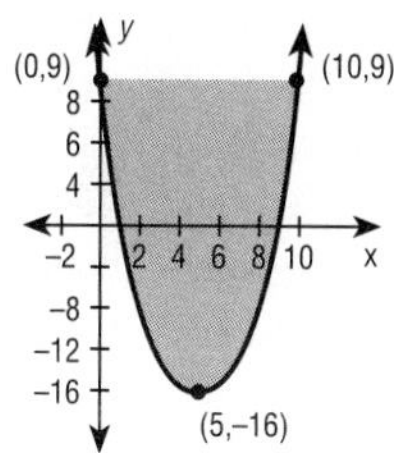

166.67 cm^2

Extra Examples

Here are additional examples similar to **Examples 1–2.**

1. Complete a table of values for each quadratic equation by first finding the vertex of the parabola.
   a. $y = x^2 + 4x - 12$
   b. $y = -x^2 + 6x + 7$

   Tables vary.

   a. Vertex ↓ (at $x = -2$)

x	−8	−6	−3	−2	−1	2	4
y	20	0	−15	−16	−15	0	20

   b. Vertex ↓ (at $x = 3$)

x	−1	0	2	3	4	6	7
y	0	7	15	16	15	7	0

2. Sketch the graph of each quadratic equation from Exercise 1.
   a. Graph opens upward.
   b. Graph opens downward.

Check Understanding

1. What is the standard form of a quadratic equation?
   $y = ax^2 + bx + c;\ a \neq 0$
2. If the leading coefficient of a quadratic equation in standard form is negative, in which direction does the parabola open?
   Downward
3. What is the equation of the axis of symmetry of a parabola given by a quadratic equation in standard form?
   $x = -\frac{b}{2a}$

The Anasazi Indians, ancestors of today's Pueblo Indians, built multistory buildings in Chaco Canyon, New Mexico. The largest of these "Great Houses" is Pueblo Bonito. *The shape of its foundation is approximately parabolic, with a base (at the top of the photo) of about 500 feet and a height of about 300 feet. (Source: Scientific American)*

Goal 2 Using Quadratic Models in Real Life

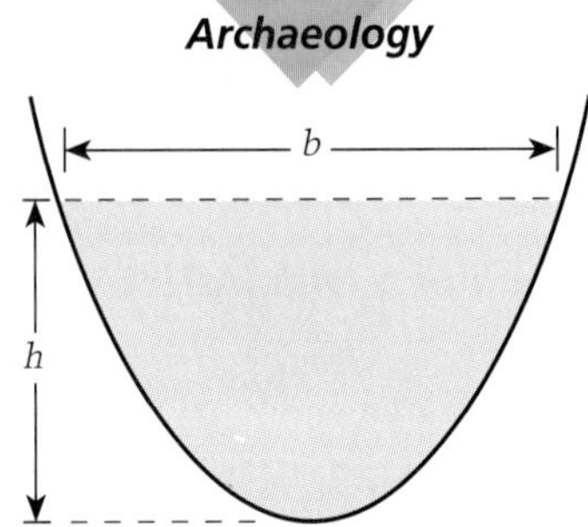

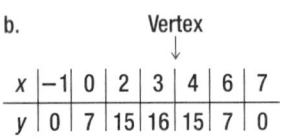

Example 3 *Finding the Area of a Parabolic Region*

The area of a region formed by a parabola (that opens up or down) and a horizontal line segment (the base) is

$$\text{area} = \frac{2}{3}bh$$

where b is the length of the base and h is the height. Use this formula to approximate the area of *Pueblo Bonito.*

Solution The area is given by

$$\text{area} = \frac{2}{3}(500)(300) = 100{,}000 \text{ square feet.}$$ ■

Communicating about ALGEBRA

SHARING IDEAS about the Lesson

The parabola shown at the left is given by

$$y = 0.0048x^2 - 300.$$

A. Find the x-intercepts and y-intercepts of the parabola. A. (250, 0), (−250, 0); (0, −300)

B. Find the area of the shaded region. 100,000 square feet

C. A drawing of *Pueblo Bonito* has been superimposed over the parabolic area. Do you think the approximation found in Example 3 is too large, too small, or just about right? Explain. about right

EXERCISES

Guided Practice

CRITICAL THINKING about the Lesson

1. Describe the graph of $y = ax^2 + bx + c$. What is the name of this type of graph? U-shaped parabola
2. Does the graph of $y = 3x^2 - x - 2$ open up or down? Explain. 2. Up; the leading coefficient, 3, is positive.
3. Find the coordinates of the vertex of the parabola given by $y = -x^2 + 4x$. Explain your steps. 3. (2, 4); $-\frac{b}{2a} = -\frac{4}{2(-1)} = 2$, $y = -x^2 + 4x = -2^2 + 4 \cdot 2 = 4$
4. The table shows several x-values and y-values for a quadratic equation. Explain how you can use the table to find the x-intercepts and y-intercepts of the graph of the equation. An x-intercept is the value of x when $y = 0$. A y-intercept is the value of y when $x = 0$.

x	−2	−1	0	1	2	3	4
y	−8	−3	0	1	0	−3	−8

Independent Practice

In Exercises 5–10, decide whether the parabola opens up or down.

5. $y = -x^2 + 4x - 2$ Down
6. $y = 2 - 10x^2$ Down
7. $y = -6 + x^2 - 5x$ Up
8. $y = 2x^2 - 5x - 6$ Up
9. $y = -3x + 2 + 5x^2$ Up
10. $y = 6 - x^2 + 7x$ Down

In Exercises 11–16, find the vertex and axis of symmetry of the parabola. See margin.

11. $y = -2x^2 + 5x - 10$
12. $y = 3x^2 + 6x - 2$
13. $y = 4x^2 - 2x + 1$
14. $y = -x^2 - 2x - 3$
15. $y = 10x^2 - 20x - 15$
16. $y = 7x^2 + 2x - 10$

In Exercises 17–22, complete the table. Then use the table to find the x- and y-intercepts of the graph.

17. $y = -x^2 - 2x + 3$ x-intercept: (−3, 0), (1, 0); y-intercept: (0, 3)

x	−4	−3	−2	−1	0	1	2
y	?	?	?	?	?	?	?
	−5	0	3	4	3	0	−5

18. $y = x^2 - 4$ x-intercept: (−2, 0), (2, 0); y-intercept: (0, −4)

x	−3	−2	−1	0	1	2	3
y	?	?	?	?	?	?	?
	5	0	−3	−4	−3	0	5

19. $y = 3x^2 - 12$ x-intercept: (−2, 0), (2, 0); y-intercept: (0, −12)

x	−3	−2	−1	0	1	2	3
y	?	?	?	?	?	?	?
	15	0	−9	−12	−9	0	15

20. $y = -x^2 - 10x - 24$ x-intercept: (−6, 0),(−4, 0); y-intercept: (0, −24)

x	−6	−5	−4	−3	−2	−1	0
y	?	?	?	?	?	?	?
	0	1	0	−3	−8	−15	−24

21. $y = x^2 - 8x + 12$ x-intercept: (2, 0), (6, 0); y-intercept: (0, 12)

x	0	1	2	3	4	5	6	7
y	?	?	?	?	?	?	?	?
	12	5	0	−3	−4	−3	0	5

22. $y = -x^2 + 4x + 12$ x-intercept: (−2, 0), (6, 0); y-intercept: (0, 12)

x	−4	−2	0	2	4	6	8
y	?	?	?	?	?	?	?
	−20	0	12	16	12	0	−20

Communicating about ALGEBRA

Ask students to graph the following quadratic equations and find the vertex of each parabola using the formula: x-coordinate of vertex $= -\frac{b}{2a}$. Have them look for a pattern that might indicate another way to determine the vertex of a parabola.

a. $y = (x - 2)^2 + 3$ (2, 3)
b. $y = (x - 6)^2 - 2$ (6, −2)
c. $y = (x + 3)^2 + 1$ (−3, 1)
d. $y = (x + 4)^2 - 5$ (−4, −5)

Note that, for any quadratic equation written in the form, $y = (x - h)^2 + k$, the coordinates of the vertex are (h, k).

EXERCISE Notes

ASSIGNMENT GUIDE

Basic/Average: Ex. 1–4, 5–19 odd, 23–28, 35–41 odd, 51–59 odd, 60–61

Above Average: Ex. 1–4, 7–21 odd, 23–28, 33–41 odd, 44–46, 57–62

Advanced: Ex. 1–4, 9–21 odd, 23–28, 33–39 odd, 43–46, 59–62

Selected Answers
Exercises 1–4, 5–59 odd

Use **Mixed Review** as needed.

More Difficult Exercises
Exercises 43–46, 60

Guided Practice

Ex. 1–4 GROUP ACTIVITY Use these exercises as an in-class activity to verify students' understanding of the lesson concepts.

Independent Practice

▶ **Ex. 5–10** Remind students to focus on the coefficient of the x^2 term.

EXTEND Ex. 5–16 Have students decide when the vertex provides the maximum and when it provides the minimum. Have them explain their decision.

Answers

11. $\left(\frac{5}{4}, -\frac{55}{8}\right), x = \frac{5}{4}$
12. $(-1, -5), x = -1$
13. $\left(\frac{1}{4}, \frac{3}{4}\right), x = \frac{1}{4}$
14. $(-1, -2), x = -1$
15. $(1, -25), x = 1$
16. $\left(-\frac{1}{7}, -\frac{71}{7}\right), x = -\frac{1}{7}$

▶ **Ex. 23–28** Assign these exercises as a group.

▶ **Ex. 29–40** Throughout the chapter, there are exercises that connect the visual behaviour of the graph of a quadratic equation and the symbolic operation of solving a quadratic equation. Students with a graphics calculator should be encouraged to use it. However, there are times when you do not need an exact graph, so students should develop "quick-graph" techniques for graphing parabolas similar to those for graphing linear equations.

One graphing technique for paper-and-pencil graphs is to begin by finding the vertex. Then, by examining the coefficients and understanding the symmetry of a parabola, you can generate a table of values as modeled in Examples 1 and 2 on page 237.

▶ **Ex. 41–42** Remind students of the formula given in Example 3, area $= \frac{2}{3}bh$.

In Exercises 23–28, match the equation with its graph.

23. $y = x^2 - 2$ b
24. $y = x^2 + 4x - 1$ f
25. $y = -x^2 - 4x - 3$ e

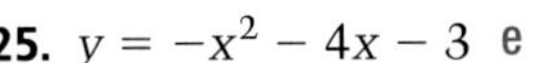

26. $y = x^2 + 2$ c
27. $y = -x^2 + 4x + 1$ a
28. $y = -x^2 + 2$ d

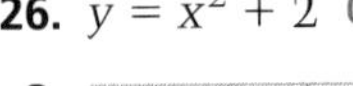

a.
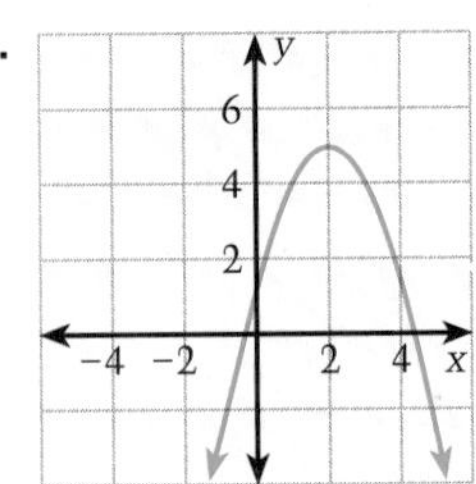

b.
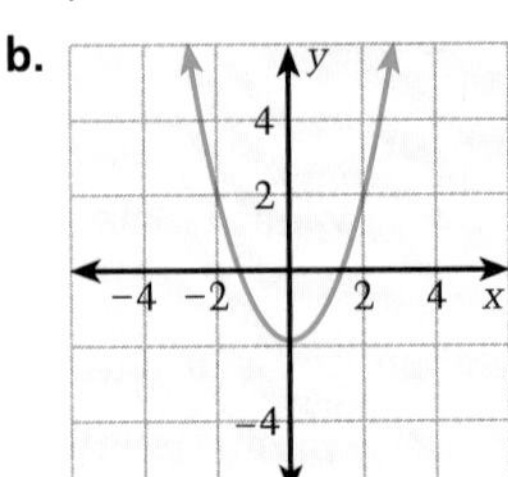

c.
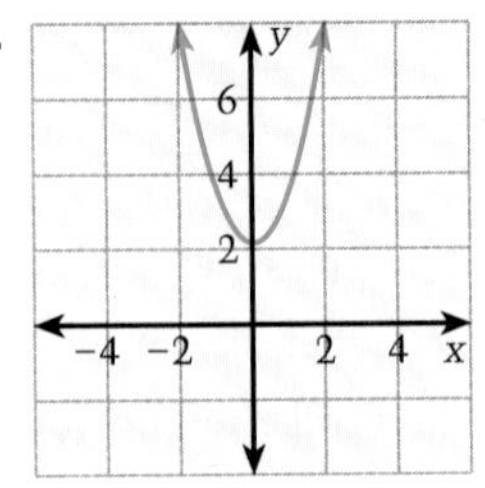

d.
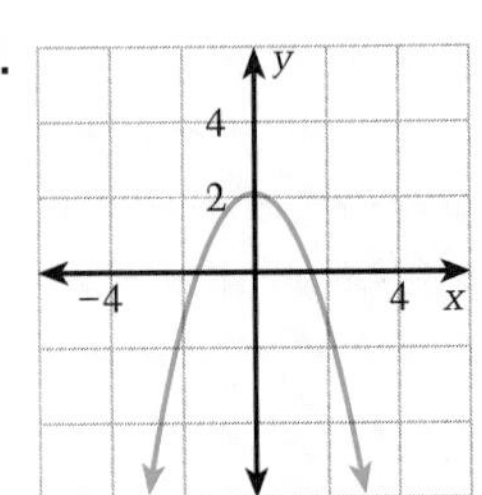

e.
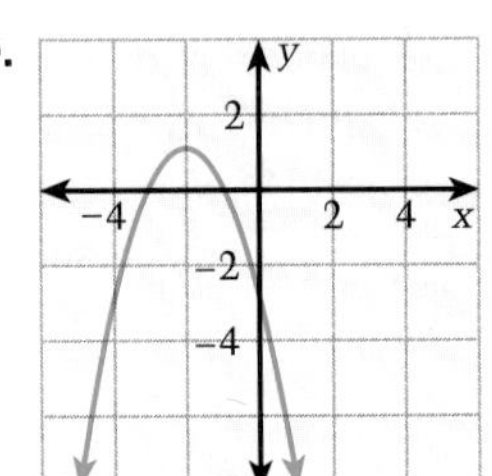

f.
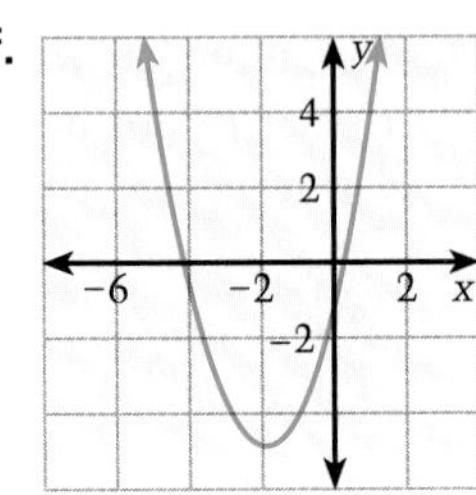

In Exercises 29–40, sketch the graph of the equation. See Additional Answers.

29. $y = x^2 - 2x + 6$
30. $y = x^2 + 8x + 16$
31. $y = -x^2 + 6x - 9$
32. $y = -x^2 - 2x - 3$
33. $y = x^2 + 14x + 45$
34. $y = x^2 - 4x + 7$
35. $y = -\frac{1}{4}x^2 + 4$
36. $y = \frac{1}{4}x^2 - 4$
37. $y = -\frac{1}{4}x^2 + x + 3$
38. $y = -\frac{1}{6}x^2 + 6$
39. $y = -\frac{1}{3}x^2 + 3$
40. $y = -\frac{1}{3}x^2 + 4x - 9$

41. *Area of Mongolia* The map below shows the country of Mongolia. Each tic mark on the coordinate system represents 60 miles. Use the map to estimate the area of Mongolia. Explain your procedure. See below.

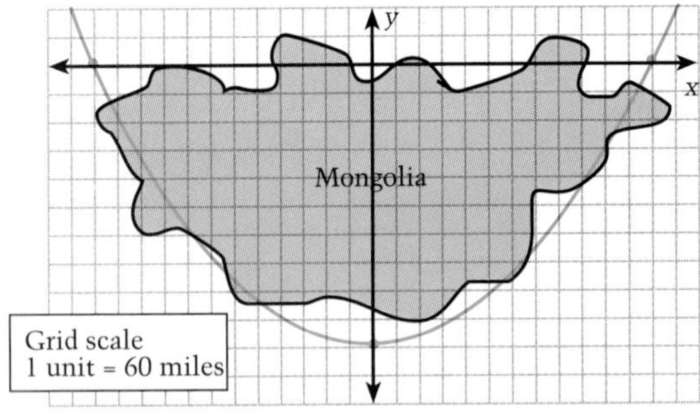

42. *Area of a Lake* The map below shows the tip of Lake Michigan superimposed on a coordinate system. Each tic mark represents about 12.5 miles. Use the map to estimate the area of the lake south of Milwaukee and Grand Haven. Explain. See below.

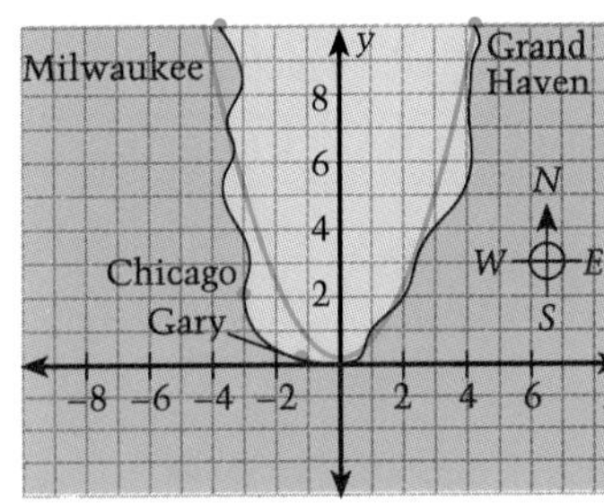

43. *Kicking a Football* A football player kicked a 41-yard punt. The path of the ball was modeled by $y = -0.035x^2 + 1.4x + 1$, where x and y are measured in yards. What was the maximum height of the ball? The player kicked the football toward midfield from the 18 yard line. Over which yard line was the ball when it was at its maximum height? 15 yd; 38-yard line

41. $576{,}000$ mi^2; area $= \frac{2}{3}bh = \frac{2}{3}(24 \times 60)(10 \times 60)$

42. $8333\frac{1}{3}$ mi^2; area $= \frac{2}{3}bh = \frac{2}{3}(8 \times 12.5)(10 \times 12.5)$

Golden Gate Bridge **In Exercises 44–46, use the following information.**

The shape of each suspension cable on the Golden Gate Bridge (in San Francisco) can be approximated by either of the models

$$y = \frac{1}{9000}x^2 + 5$$

or

$$y = \frac{1}{9000}(x - 2100)^2 + 5$$

where x and y are measured in feet. (The roadbed of the bridge lies on the x-axis.)

44. Sketch the graph of each model. See Additional Answers.

45. In each model, what are the coordinates of the lowest point on the cable? 1st: (0, 5), 2nd: (2100, 5)

46. The span between the two towers is 4200 feet. How high above the roadbed are the cables connected to the towers? 495 ft

Integrated Review

In Exercises 47–52, sketch the graph. See Additional Answers.

47. $y = -2x + 3$ **48.** $y = 10x - 5$ **49.** $y = 5x - 7$

50. $y = -6x + 3$ **51.** $2x + 5y = 10$ **52.** $x - 2y = -8$

In Exercises 53–58, solve the equation.

53. $-3x^2 = -9$ $\pm\sqrt{3}$ **54.** $2x^2 + 5 = 9$ $\pm\sqrt{2}$ **55.** $6x^2 + 3 = 3$ 0

56. $4x^2 = 0$ 0 **57.** $\frac{1}{2}x^2 - 7 = 1$ ±4 **58.** $\frac{1}{7}x^2 - 2 = 5$ ±7

59. *College Entrance Exam Sample* In a right triangle, the ratio of the legs is 1 to 2. The area of the triangle is 25 square units. What is the length of the hypotenuse? b

a. $\sqrt{5}$ **b.** $5\sqrt{5}$ **c.** $5\sqrt{3}$ **d.** $10\sqrt{3}$ **e.** $25\sqrt{5}$

60. *College Entrance Exam Sample* What is the total length of the line segments in the drawing at the right? c

a. 10 **b.** $\sqrt{17} + \sqrt{6}$ **c.** $2(\sqrt{13} + \sqrt{5})$ **d.** $2(\sqrt{15} + \sqrt{7})$

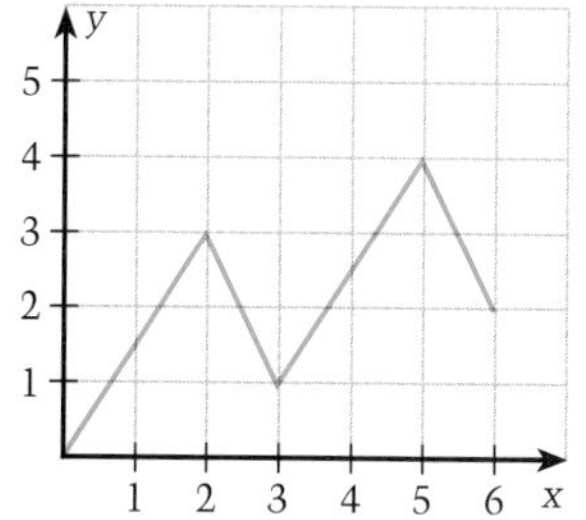

Exploration and Extension

In Exercises 61 and 62, sketch the graphs of the two equations. Then find the area of the region that is bounded by the two graphs. See Additional Answers.

61. $y = 0$ and $y = x^2 - 9$ 36 **62.** $y = 0$ and $y = -x^2 + 36$ 288

Ex. 46 Refer students to the chapter opener on page 228. You may wish to have students discuss possible increments and labels for the x- and y-axes before assigning this exercise.

Ex. 59–60 These college-entrance-test samples are an application of concepts from Lesson 5.1.

Ex. 61–62 These area-between-two-curves problems are a precursor to the study of integration in calculus.

Enrichment Activities

TECHNOLOGY You may wish to use these activities as an in-class review prior to the graphics calculator feature that follows on pages 242 and 243.

1. **EXTEND Ex. 44–46** Graph $y = \frac{1}{9000}x^2 + 5$ in a standard viewing window.

   X min = −10 Y min = −10
   X max = 10 Y max = 10
   X scl = 1 Y scl = 1
   X res = 1

   Describe what you see. Explain why you see it.
   A horizontal line appears across the screen because the viewing window is too small.

2. **COOPERATIVE LEARNING** Work with a partner. Try some other viewing windows. Which window gives you a better view of the cable? State reasons for your choice.

Sample answer:

The x-axis represents the road bed. The span between towers is 4200 feet; the lowest point occurs halfway between the towers. The y-values represent the heights of the cable connections, which are between 0 and 500 feet.

X min = −2200 Y min = 0
X max = 2200 Y max = 500
X scl = 100 Y scl = 100
X res = 1

(continued)

(continued)

3. Graph $y = \frac{1}{9000}x^2 + 5$ in your viewing window. Use Trace to verify that when $x = 2100$, $y = 495$.

4. How would you describe the graph $y = \frac{1}{9000}(x - 2100)^2 + 5$? How would you adjust the viewing rectangle?
The graph is a horizontal shift 2100 feet to the right, so X min = −100 and X max = 4300.

USING A GRAPHING CALCULATOR

You can use the zoom feature of a graphing calculator to approximate the solutions of an equation to any desired accuracy. The four graphs below show four steps in approximating the positive solution of

$$x^2 - 5 = 0.$$

Here, the approximation is $x \approx 2.237$. (The exact solution is $x = \sqrt{5}$.) By repeated zooming, you can obtain whatever accuracy you need.

1.

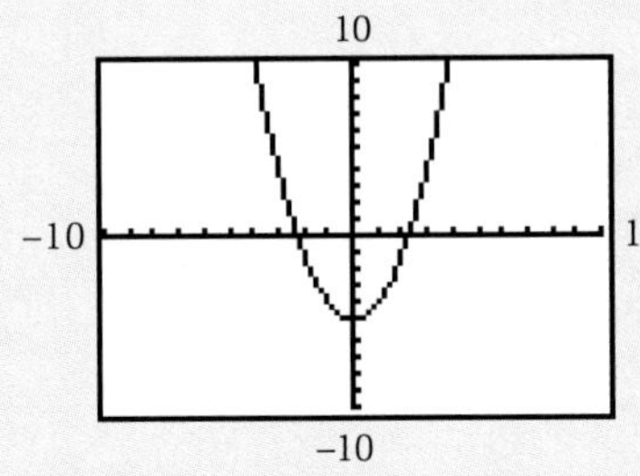

Sketch the graph of $y = x^2 - 5$.

2.

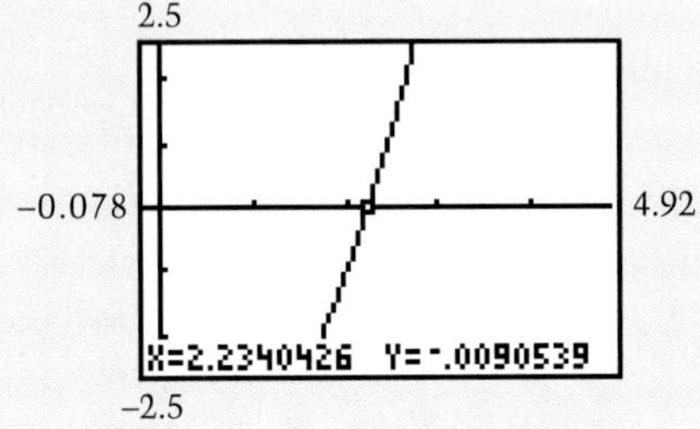

Zoom once and Trace to get a better view of the positive x-intercept. (The viewing rectangle will change automatically.)

3.

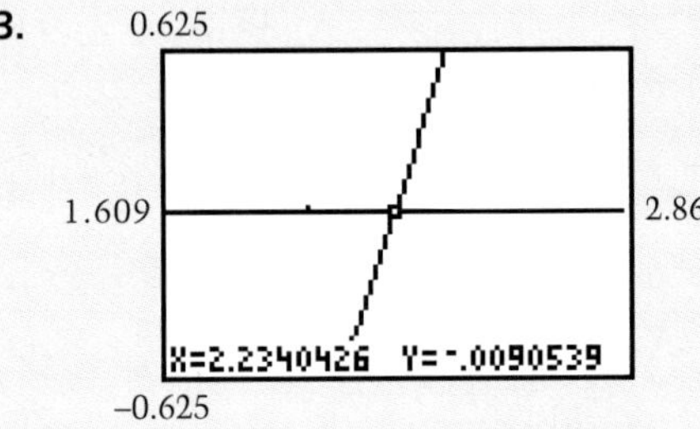

Zoom a second time and Trace to get an even better view. Use the cursor keys to determine that the x-intercept is about 2.24.

4.

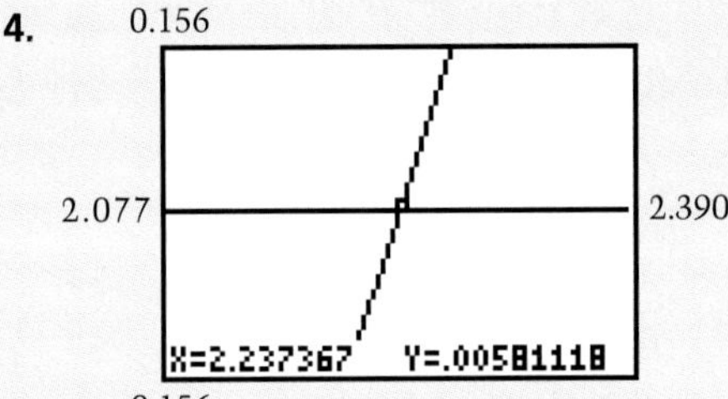

Zoom a third time. Use the trace feature to find the y-value that is closest to 0. The corresponding x-value, which is 2.237, is your approximation.

Many equations can be solved by find the x-intercepts of a related equation in x and y. For instance, the above screens show how to solve the equation

$x^2 - 5 = 0$ *Equation in x*

by finding the x-intercepts of the graph of

$y = x^2 - 5.$ *Equation in x and y*

You will encounter this technique again in Chapters 8, 10, and 14.

Example: Using a Graphing Calculator to Approximate a Solution

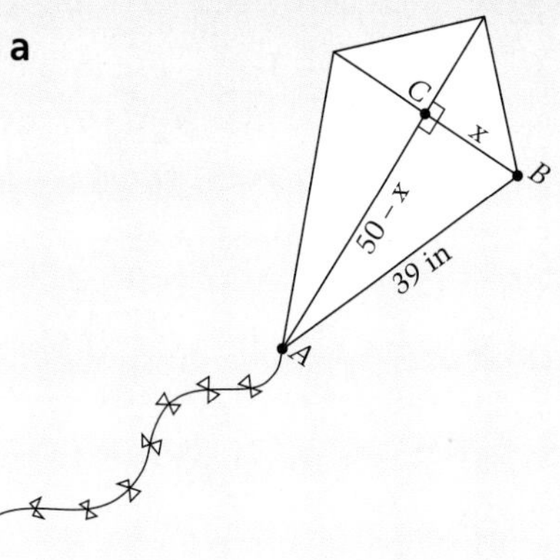

On the kite shown at the right, the length of $\overline{AB}$ is 39 inches. The combined length of $\overline{AC}$ and $\overline{CB}$ is 50 inches. How long are $\overline{AC}$ and $\overline{CB}$?

Solution Let x represent the length of $\overline{CB}$. Because the combined length of $\overline{AC}$ and $\overline{CB}$ is 50 inches, it follows that the length of $\overline{AC}$ is $50 - x$. Using the Pythagorean Theorem, you can write the following equation.

$$x^2 + (50 - x)^2 = 39^2$$
$$x^2 + 2500 - 100x + x^2 = 1521$$
$$2x^2 - 100x + 979 = 0$$

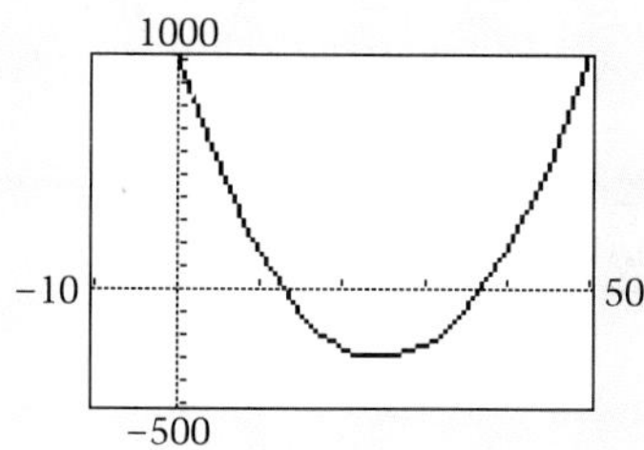

To solve this equation with a graphing calculator, sketch the graph of $y = 2x^2 - 100x + 979$ and approximate the x-intercepts of the graph. Using the zoom feature as described on page 242, you can approximate the x-intercepts to be 13.36 and 36.64. Since $BC < AC$, the length of $\overline{CB}$ is approximately 13.36 inches, and the length of $\overline{AC}$ is approximately 36.64 inches. ■

Exercises

In Exercises 1–8, use a graphing calculator to approximate the solutions of the equation to one decimal place.

1. $3x^2 - 5x - 7 = 0$ 2.6, −0.9

2. $-x^2 - 6x + 12 = 0$ 1.6, −7.6

3. $\frac{1}{2}x^2 + 12x + 5 = 0$ −23.6, −0.4

4. $\frac{4}{5}x^2 - 7x - 9 = 0$ 9.9, −1.1

5. $-0.3x^2 - 4.2x + 8.1 = 0$ 1.7, −15.7

6. $4.2x^2 + 5.6x + 1.3 = 0$ 1.0, −0.3

7. $15x^2 + 15.4x - 6.8 = 0$ −1.4, 0.3

8. $0.4x^2 + 1.2x - 3.2 = 0$ −4.7, 1.7

Geometry **In Exercises 9 and 10, solve for *x*, as in the example.**

9. ≈10.9

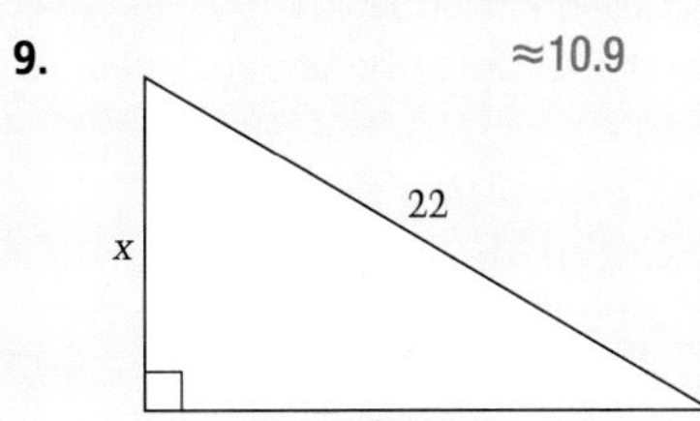

10. No solution

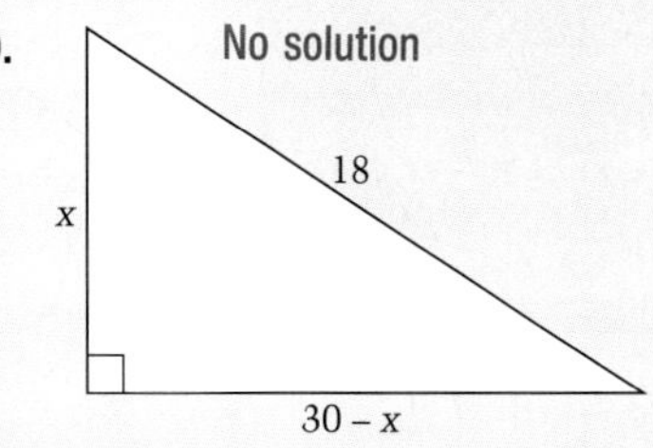

TECHNOLOGY Many find the "box" feature of graphing calculators to be an easy-to-use enhancement of the Zoom function. Refer to your user's manual for details.

Observe that for practical applications of quadratic equations, approximate solutions are usually sufficient. Accordingly, using a calculator or computer to solve quadratic equations becomes a powerful tool.

Answers

13.

14.

−6 −5 −4 −3 −2 −1 0 1 2 3 4 5 x

Mixed REVIEW

1. Simplify: $2(x^2 - x) - 3(2x - 7)$. **(1.2)** $2x^2 - 8x + 21$
2. Find the intercepts of $3x - 7y = 21$. **(2.3)** x: 7, y: −3
3. Solve $|x - 7| < 3$. **(1.7)** $4 < x < 10$
4. Solve $-3|x - 4| \le -6$. **(1.7)** $x \le 2$ or $x \ge 6$
5. What is 15% of 36? 5.4
6. Multiply: $\frac{2}{3} \cdot \frac{9}{16}$. $\frac{3}{8}$
7. Find the slope of $3x - 4y = 2x - 10$. **(2.2)** $\frac{1}{4}$
8. Solve $-3(x - 5) + (4x - 5) = 4x - 2$. **(1.3)** 4
9. Solve $(x + 3)^2 < 4$. **(1.6)** $-5 < x < -1$
10. Solve $|2x - 5| = 9$. **(1.7)** −2, 7
11. Solve for y: $2xy + 3y = x$. **(1.5)** $\frac{x}{2x + 3}$
12. Evaluate: $(3 \cdot 4 - 2) - (3^2 - 8)$. **(1.2)** 9
13. Sketch the graph of $-4 < x < 1$. **(1.6)**
14. Sketch the graph of $x < -2$ or $x > 3$. **(1.6)**

13, 14. See margin.

15. Solve $|x + 7| \le 2$. **(1.7)** $-9 \le x \le -5$
16. Solve $\begin{cases} 2x - 3y = 20 \\ 5x + 2y = -7 \end{cases}$ **(3.2)** (1, −6)
17. Evaluate $3\begin{bmatrix} 2 & 5 \\ -1 & 3 \end{bmatrix}$. **(4.1)** $\begin{bmatrix} 6 & 15 \\ -3 & 9 \end{bmatrix}$
18. Find the determinant of $\begin{bmatrix} 2 & 5 \\ -1 & 3 \end{bmatrix}$. **(4.3)** 11
19. What is the radius of a circle with a circumference of 24 inches? ≈3.82 in.
20. An object dropped from 49 feet will hit the ground in how many seconds? **(5.1)**

20. 1.75 sec

Career Interview

Epidemiologist

Nancy Dreyer is an epidemiologist who specializes in studies relating to the causes of human diseases. "Part of my job is to use math and logic to determine the appropriate research needed to address specific concerns."

Q: ***What math did you take in school?***

A: In high school, I took Algebra 1, Algebra 2, and Geometry. In graduate school, I took statistics courses; I especially remember biostatistics where I learned how to evaluate data and make comparisons using statistical tools, such as means and medians.

Q: ***How do you use math on the job?***

A: I use math to calculate disease rates among different groups, and then apply rate ratios for comparisons. Or in drug related studies, I use math to describe the differences in side effects between groups.

Q: ***How much mental math and estimation do you use?***

A: I use mental math and estimation all the time. You can't just assume that answers from computers are always correct. You have to check the accuracy of each answer. I use mental math and estimation to check myself, so I can find mistakes before anyone else does.

5.3 Completing the Square

What you should learn:

Goal 1 How to solve a quadratic equation by completing the square

Goal 2 How to use "completing the square" to solve real-life problems

Why you should learn it:

You can use quadratic equations to solve many real-life problems, such as finding the dimensions of some rectangles.

Goal 1 Solving by Completing the Square

In this lesson, you will solve quadratic equations by **completing the square.** Before doing that, however, let's look at some examples of *expanding a binomial* by squaring it.

Square of Binomial		Perfect-Square Trinomial
$(x + 3)^2$	=	$x^2 + 6x + 9$
$(x - 4)^2$	=	$x^2 - 8x + 16$

In each case, the constant term of the trinomial is the square of half of the coefficient of the trinomial's x-term.

Perfect-Square Trinomial	Square of Binomial
$x^2 + bx + \left(\frac{b}{2}\right)^2$ (half of b)^2	$\left(x + \frac{b}{2}\right)^2$

Thus, to make the expression $x^2 + bx$ a perfect square, you must add $\left(\frac{1}{2}b\right)^2$ to the expression.

When completing the square to solve a quadratic equation, remember that you *must preserve the equality.* When you add a constant to one side of the equation, be sure to add the same constant to the other side of the equation.

Example 1 *Completing the Square: Leading Coefficient Is 1*

Solve $x^2 - 4x + 2 = 0$ by completing the square.

Solution

$x^2 - 4x + 2 = 0$ *Rewrite original equation.*

$x^2 - 4x = -2$ *Subtract 2 from both sides.*

$x^2 - 4x + (-2)^2 = -2 + 4$ *Add $(-2)^2 = 4$ to both sides.*

(half of -4)^2

$(x - 2)^2 = 2$ *Binomial squared*

$x - 2 = \pm\sqrt{2}$ *Take square roots.*

$x = 2 \pm \sqrt{2}$ *Add 2 to both sides.*

The equation has two solutions: $2 + \sqrt{2}$ and $2 - \sqrt{2}$. ■

Problem of the Day

A woman paid \$2.25 for some apples, bananas, and pears. The apples cost 10¢ each, the pears 20¢ each, and the bananas 25¢ each. The number of apples was equal to the number of pears and bananas combined. How many of each did she buy?

7 apples, 4 pears, 3 bananas

ORGANIZER

Warm-Up Exercises

1. Expand each binomial.
   a. $(x + 2)^2$ $x^2 + 4x + 4$
   b. $(x - 5)^2$ $x^2 - 10x + 25$
2. Factor each trinomial.
   a. $x^2 + 16x + 64$ $(x + 8)^2$
   b. $x^2 - 12x + 36$ $(x - 6)^2$

Lesson Resources

Teaching Tools
- Transparencies: 7, 8, 13
- Problem of the Day: 5.3
- Warm-up Exercises: 5.3
- Answer Masters: 5.3

Extra Practice: 5.3

Technology Handbook: p. 31

LESSON Notes

Example 1

The solutions $2 + \sqrt{2}$ and $2 - \sqrt{2}$ are exact solutions. In many real-life applications, the decimal solutions, 3.41 and 0.59 (accurate to two decimal places), would probably be sufficient.

Example 2

A convenient way in which to represent the range, or viewing rectangle, of a graphics calculator or computer is given below. For example, the range presented with the graph of the quadratic equation in this example would be written as

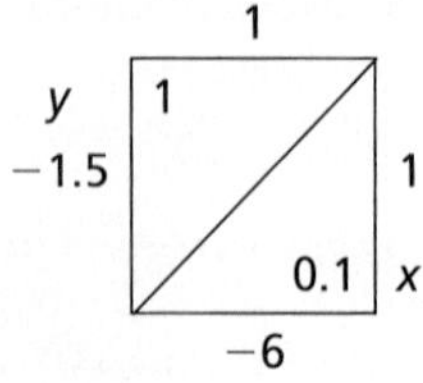

Example 3

Be sure students understand that the width of the picture is 48 in. because it is 12 in. wider than the length of 36 in. The extraneous solution, −48, cannot be transformed, somehow, into the "other answer."

Extra Examples

Here are additional examples similar to **Examples 1–3.**

1. Solve by completing the square.

a. $x^2 + 6x - 16 = 0$

b. $x^2 - 2x - 8 = 0$

c. $x^2 + 8x + 10 = 0$

a. $-8, 2$; **b.** $-2, 4$;

c. $-4 \pm \sqrt{6}$

2. Solve by completing the square.

a. $2x^2 + 6x - 16 = 0$

b. $3x^2 - 2x - 8 = 0$

c. $4x^2 + 8x + 1 = 0$

a. $\frac{-3 \pm \sqrt{41}}{2}$; **b.** $-\frac{4}{3}, 2$;

c. $\frac{-2 \pm \sqrt{3}}{2}$

If the leading coefficient of the quadratic is not 1, you should divide both sides of the equation by this coefficient *before* completing the square.

Example 2 Completing the Square: Leading Coefficient Is Not 1

Solve $4x^2 + 2x - 5 = 0$ by completing the square.

Solution

$4x^2 + 2x - 5 = 0$	*Rewrite original equation.*
$4x^2 + 2x = 5$	*Add 5 to both sides.*
$x^2 + \frac{1}{2}x = \frac{5}{4}$	*Divide both sides by 4.*
$x^2 + \frac{1}{2}x + \left(\frac{1}{4}\right)^2 = \frac{5}{4} + \frac{1}{16}$ $\left(\text{half of } \frac{1}{2}\right)^2$	*Add $\left(\frac{1}{4}\right)^2 = \frac{1}{16}$ to both sides.*
$\left(x + \frac{1}{4}\right)^2 = \frac{21}{16}$	*Binomial squared*
$x + \frac{1}{4} = \pm\frac{\sqrt{21}}{4}$	*Find square roots.*
$x = -\frac{1}{4} \pm \frac{\sqrt{21}}{4}$	*Solve for x.*

The equation has two solutions: $-\frac{1}{4} + \frac{\sqrt{21}}{4}$ and $-\frac{1}{4} - \frac{\sqrt{21}}{4}$. ■

Graph of $y = 4x^2 + 2x - 5$

Checking solutions that have fractions and square roots is cumbersome. A graphing check is usually more efficient. For instance, in Example 2, the solutions of $4x^2 + 2x - 5 = 0$ are

$$-\frac{1}{4} + \frac{\sqrt{21}}{4} \approx 0.9 \quad \text{and} \quad -\frac{1}{4} - \frac{\sqrt{21}}{4} \approx -1.4.$$

From the graph of $y = 4x^2 + 2x - 5$ at the left, you can see that the x-intercepts are about -1.4 and 0.9.

Connections
Technology

A graphing check is especially efficient with a graphing calculator. The graph below shows the display screen from a graphing calculator. Note that the scale on the x-axis has been set with tick marks that are 0.1 units apart.

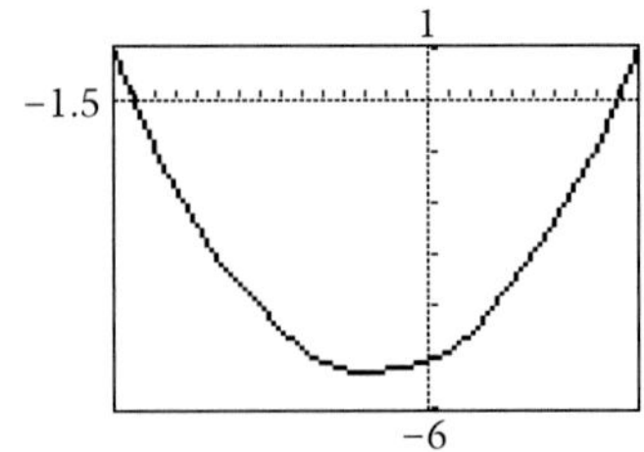

RANGE
Xmin=−1.5
Xmax=1
Xscl=.1
Ymin=−6
Ymax=1
Yscl=1

Goal 2 Using Completing-the-Square in Real Life

Real Life
Television

Example 3 *Finding the Dimensions of a Rectangle*

Television screens are usually measured by the length of the diagonal. The oversized television at the left has a 60-inch diagonal. The screen is 12 inches wider than it is high. Find the dimensions of the screen.

Solution Use the Pythagorean Theorem to write a model.

Verbal Model Square of height + Square of width = Square of diagonal

Labels
Height = x (inches)
Length = $x + 12$ (inches)
Diagonal = 60 (inches)

Equation

$x^2 + (x + 12)^2 = 60^2$ *Pythagorean Theorem*

$x^2 + x^2 + 24x + 144 = 3600$ *Expand the binomial.*

$2x^2 + 24x = 3456$ *Subtract from both sides.*

$x^2 + 12x = 1728$ *Divide both sides by 2.*

$x^2 + 12x + (6)^2 = 1728 + 36$ *Add $6^2 = 36$ to both sides.*

(half of 12)^2

$(x + 6)^2 = 1764$ *Binomial squared*

$x + 6 = \pm 42$ *Find square roots.*

$x = -6 \pm 42$ *Solve for x.*

$x = 36$ or -48 *Simplify.*

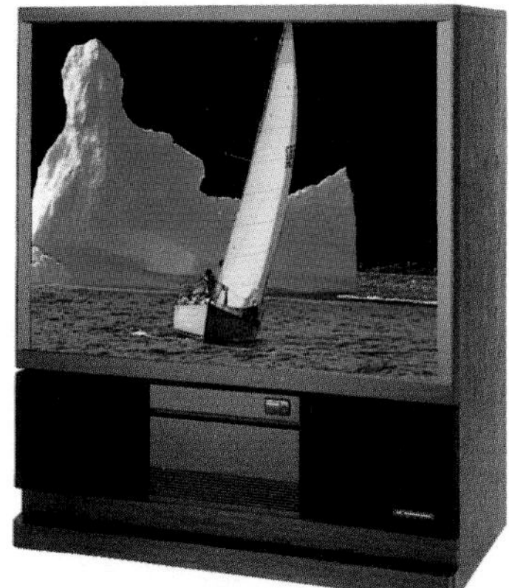

Choosing the positive solution, you can conclude that the height of the picture is 36 inches. The width of the picture is 48 inches (because it is 12 inches wider than it is high). ■

Communicating about ALGEBRA

SHARING IDEAS about the Lesson

Solve the equation by completing the square. Explain your steps *and* use a graphic check. See Additional Answers.

A. $x^2 - 8x + 10 = 0$ **B.** $-2x^2 - 8x - 6 = 0$

3. Suppose the cabinet you built to hold your television is 20 inches tall and 30 inches wide. The 31-inch television set you have been considering is 8 inches wider than it is tall. Would the television fit in your cabinet?
The dimensions of the television set are approximately 26 inches wide and 18 inches tall. It will fit in the cabinet you built.

Check Understanding

1. What is a perfect square trinomial?
See the top of page 245.
2. What is an efficient way to check solutions (of quadratic equations) that have square roots?
See the notes below Example 2 on page 246.
3. Is the process of completing the square the same for quadratic equations with (a) leading coefficients of 1 and (b) leading coefficients not equal to 1? Explain.
Yes; see Examples 1 and 2.

Communicating about ALGEBRA

Ask students to solve these equations by completing the square.

a. $x^2 + 4x = 0$ 0, -4
b. $4x^2 + 6x = 0$ 0, $-\frac{3}{2}$
c. $ax^2 + bx = 0$ 0, $-\frac{b}{a}$

EXERCISE Notes

ASSIGNMENT GUIDE

Basic/Average: Ex. 1–6, 9–19 odd, 23–29 odd, 33–45 odd

Above Average: Ex. 1–6, 11–21 odd, 27–35 odd, 45–51 odd, 52–55

Advanced: Ex. 1–6, 13–27 odd, 28–32, 50–55

Selected Answers

Exercises 1–6, 7–49 odd

✪ More Difficult Exercises

Exercises 19–24, 28–32

Guided Practice

▶ **Ex. 1–5** These exercises provide practice for each of the steps in the completing-the-square process.

▶ **Ex. 6** The graphic-check model is shown on page 246 below Example 2.

Independent Practice

▶ **Ex. 13–24** At this time, students have practiced two techniques that can be used to solve *any* quadratic equation—graphing and completing the square. A third technique, using the quadratic formula, will be reviewed in Lesson 5.4.

In past years, there was an emphasis on factoring in solving quadratic equations. Since the number of quadratic equations that can be solved by factoring is very limited, this text has chosen to introduce factoring as a mental-math technique to use with polynomials in Chapter 9.

EXERCISES

Guided Practice

CRITICAL THINKING about the Lesson

1. Which of the following is a perfect-square trinomial? Explain.

a. $x^2 - 4x + 4$ **b.** $x^2 - 6x + 6$ **a.** 4 is the square of half of −4

2. Write the perfect-square trinomial $x^2 - 8x + 16$ as the square of a binomial. $(x - 4)^2$

3. For the equations at the right, describe each step used to complete the square. What is the missing value in the *completed-square form?*

$$x^2 + 8x + 2 = 0$$
$$x^2 + 8x = -2$$
$$x^2 + 8x + (4)^2 = -2 + 16$$
$$(x + \boxed{?})^2 = 14$$

Subtract 2 from both sides, add the square of half of 8 to both sides; 4

4. What term must be added to $x^2 - 10x$ to complete the square? 25

5. Write $2x^2 + 6x = 1$ in completed-square form. Explain your steps. See Additional Answers.

6. Describe two methods for checking the solutions of a quadratic equation. See Additional Answers.

Independent Practice

In Exercises 7–12, write the trinomial as the square of a binomial.

7. $x^2 - 2x + 1$ $(x - 1)^2$ **8.** $x^2 + 24x + 144$ $(x + 12)^2$ **9.** $x^2 + 14x + 49$ $(x + 7)^2$

10. $x^2 + \frac{4}{3}x + \frac{4}{9}$ $\left(x + \frac{2}{3}\right)^2$ **11.** $x^2 - x + \frac{1}{4}$ $\left(x - \frac{1}{2}\right)^2$ **12.** $x^2 - 12x + 36$ $(x - 6)^2$

In Exercises 13–18, solve the equation by completing the square.

13. $x^2 + 10x + 24 = 0$ −6, −4 **14.** $x^2 - 4x + 3 = 0$ 1, 3 **15.** $x^2 - 2x - 15 = 0$ −3, 5

16. $x^2 + 4x - 1 = 0$ $-2 \pm \sqrt{5}$ **17.** $x^2 + 6x - 4 = 0$ $-3 \pm \sqrt{13}$ **18.** $x^2 - 2x - 5 = 0$ $1 \pm \sqrt{6}$

In Exercises 19–24, solve the equation. Check your results graphically. See Additional Answers.

✪ **19.** $2x^2 - 6x - 5 = 0$ $\frac{3}{2} \pm \frac{\sqrt{19}}{2}$ ✪ **20.** $-3x^2 + 3x + 4 = 0$ $\frac{1}{2} \pm \sqrt{\frac{19}{12}}$ ✪ **21.** $3x^2 + 9x + 1 = 0$

✪ **22.** $2x^2 - 10x - 6 = 0$ $\frac{5}{2} \pm \frac{\sqrt{37}}{2}$ ✪ **23.** $-5x^2 - 5x + 9 = 0$ $-\frac{1}{2} \pm \sqrt{\frac{41}{20}}$ ✪ **24.** $4x^2 - 16x - 10 = 0$ $2 \pm \frac{\sqrt{26}}{2}$

21. $-\frac{3}{2} \pm \sqrt{\frac{23}{12}}$

Geometry **In Exercises 25 and 26, find the dimensions of the figure.**

25. Triangle Area = 12 cm^2 4 cm, 6 cm **26.** Rectangle Area = 160 ft^2 20 ft, 8 ft

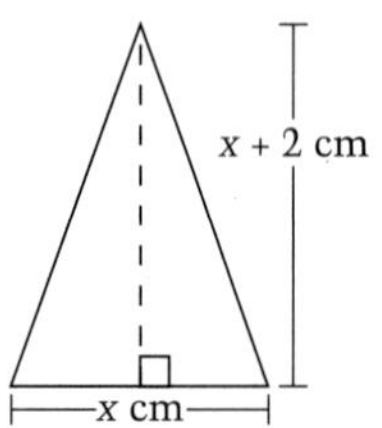

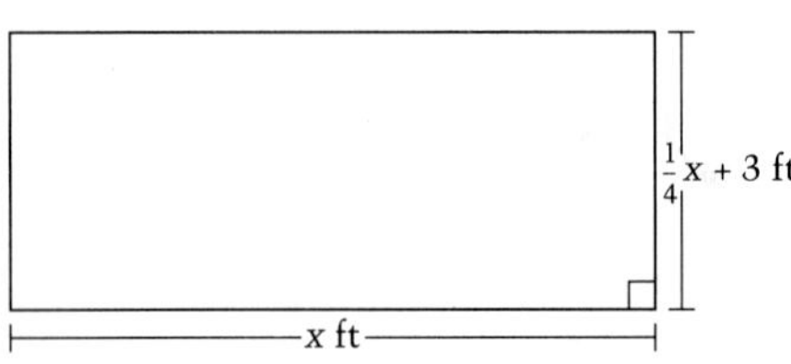

27. *Cutting across the Lawn* On the sidewalk, the walk from the high school to the junior high school is 400 meters. By cutting across the lawn, the walking distance is shortened to 300 meters. How long is each part of the L-shaped sidewalk? **≈271 m, ≈129 m**

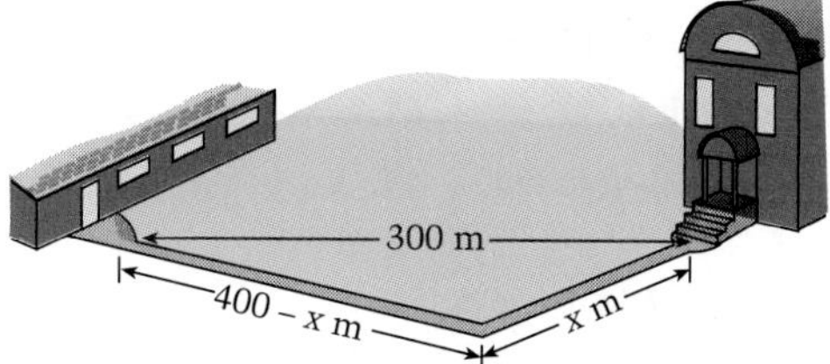

28. *More Grass to Mow* Your home is built on a square lot. To add more space to your yard, you purchase an additional 4 feet along the side of the property. The area of the lot is now 9600 square feet. What are the dimensions of the new lot? **96 ft by 100 ft**

29. *Pulling a Boat into the Dock* A windlass is used to pull a boat to the dock. The rope is attached to the boat at a point 15 feet below the level of the windlass. Find the distance from the boat to the dock when the rope is 75 feet. **≈73.5 ft**

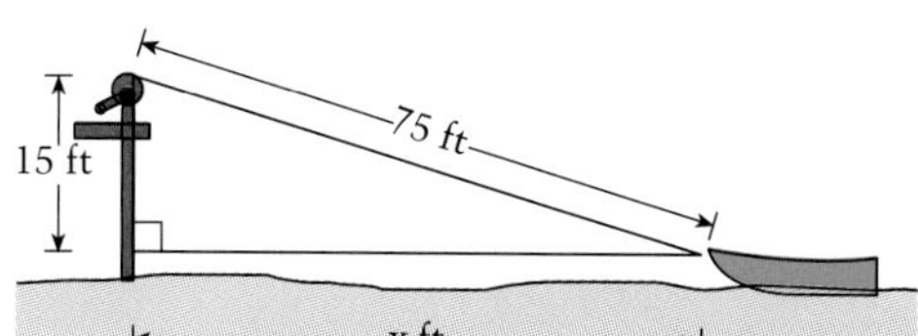

30. *Fencing in a Corral* You have 200 feet of fencing to enclose two adjacent rectangular corrals. The total area of the enclosed region is 1400 feet. What are the dimensions of each corral? (The corrals are the same size.) **35 ft by 20 ft or 15 ft by $46\frac{2}{3}$ ft**

Gateway Arch **In Exercises 31 and 32, use the following information.**

The Gateway Arch in St. Louis, Missouri, was designed in the shape of a *catenary* (a U-shaped curve that resembles a parabola). The outer edge of the arch can be approximated by the equation

$$y = \frac{-2}{315}x^2 + \frac{92}{21}x - \frac{880}{7}$$

where x and y are measured in feet.

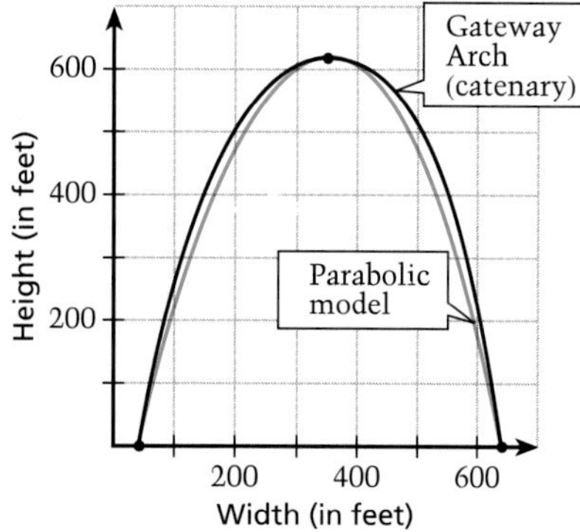

31. Find the x-intercepts of the graph. How wide is the arch? **30 and 660, 630 ft**

32. How high is the arch? How does the parabolic model differ from the arch's actual shape? **≈630 ft, parabola is narrower**

Ex. 13–18 The leading coefficient is always 1.

Ex. 19–24 The leading coefficient is not 1.

Ex. 27–32 TECHNOLOGY Encourage students to use a solution format similar to that modeled in Example 3. Those students with graphics calculators should use the Zoom and Trace features.

Ex. 51–55 These exercises connect the completing-the-square process to the completed-square form of a quadratic equation. You may wish to use them as an in-class discusssion of another quick-graph technique for graphing quadratic equations. This completed-square form is ideal for graphing because the vertex of the parabola can be read from the equation.

In Exercises 53–55, students can study the effects of the other coefficents in the completed-square form to discover that each graph is a horizontal or vertical shift of the basic graph $y = x^2$.

a. $y = (x - 1)^2$: a horizontal shift 1 unit to the right

b. $y = (x)^2 + 3$: a vertical shift 3 units up

c. $y = (x - 1)^2 + 3$: a horizontal shift 1 unit to the right and a vertical shift 3 units up

Enrichment Activities

These activities require students to go beyond lesson goals.

You now know three methods of graphing a quadratic equation such as $y = x^2 + 10x + 28$.

a. The vertex/symmetry method from Lesson 5.2.

b. The completed-square form method from this lesson

c. The graphics calculator technique

1. Graph the equation
$y = x^2 + 10x + 28$
by each method.

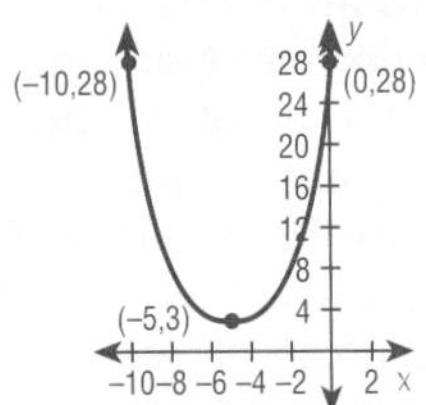

2. Which method did you prefer? Why? Which method was most efficient? Why?

3. **WRITING** Write a paragraph describing when each method would probably be more efficient than the other methods.
For example, the graph of an equation like $y = x^2 + 18$ is just a vertical shift of the graph of $y = x^2$, so a quick-graph is possible. The graph of an equation like $y = 0.18x^2 - 1.6x - 1.2$ is best graphed using a graphics calculator.

Integrated Review

In Exercises 33–38, expand the binomial.

33. $(x + 6)^2$ $x^2 + 12x + 36$
34. $(x - 1)^2$ $x^2 - 2x + 1$
35. $\left(x - \frac{1}{3}\right)^2$ $x^2 - \frac{2}{3}x + \frac{1}{9}$
36. $(x + 5)^2$ $x^2 + 10x + 25$
37. $(2x - 1)^2$ $4x^2 - 4x + 1$
38. $(3x - 2)^2$ $9x^2 - 12x + 4$

In Exercises 39–44, decide whether the numbers are solutions of the equation.

39. $x^2 + 8x + 16 = 0, -4$ It is.
40. $x^2 - 4x + 4 = 0, 2$ It is.
41. $x^2 - 6x + 9 = 0, -3$ It is not.
42. $x^2 - 18x + 81 = 0, 9$ It is.
43. $x^2 - \frac{3}{2}x + \frac{9}{16} = 0, \frac{3}{2}$ It is not.
44. $x^2 + \frac{2}{9}x + \frac{1}{81} = 0, \frac{1}{9}$ It is not.

In Exercises 45–50, sketch the graph of the equation. See Additional Answers.

45. $y = x^2 - 4x + 8$
46. $y = x^2 - 6x + 14$
47. $y = -x^2 - 8x - 17$
48. $y = x^2 + 10x + 28$
49. $y = -x^2 - 2x + 2$
50. $y = x^2 - 14x + 47$

Exploration and Extension

Finding the Vertex of a Parabola **In Exercises 51 and 52, use the following information.**

The equation $y = a(x - h)^2 + k$ is written in *completed-square form.* In this form, the coordinates of the vertex of the graph are (h, k). Completing the square can be used to write a quadratic equation in completed-square form. When you do this, you complete the square by *adding and subtracting* the same quantity. Here is an example.

$y = x^2 - 6x + 11$ *Standard form*

$y = [x^2 - 6x + (-3)^2] + 11 - (-3)^2$ *Add and subtract $(-3)^2$ from the right side.*

$y = (x - 3)^2 + 2$ *Completed-square form*

51. Write $y = x^2 - 2x + 5$ in completed-square form. Then find the coordinates of the vertex of the graph. $y = (x - 1)^2 + 4$, $(1, 4)$

52. Write $y = x^2 + 4x + 5$ in completed-square form. Then find the coordinates of the vertex of the graph. $y = (x + 2)^2 + 1$, $(-2, 1)$

In Exercises 53–55, match the equation with its graph.

53. $y = (x - 1)^2 + 3$ c
54. $y = (x - 1)^2$ a
55. $y = (x - 1)^2 - 3$ b

a.

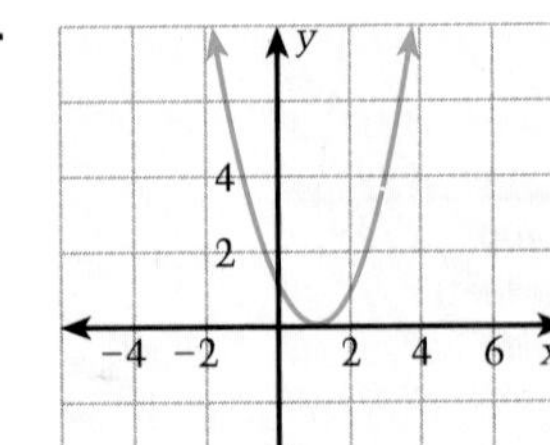

b.

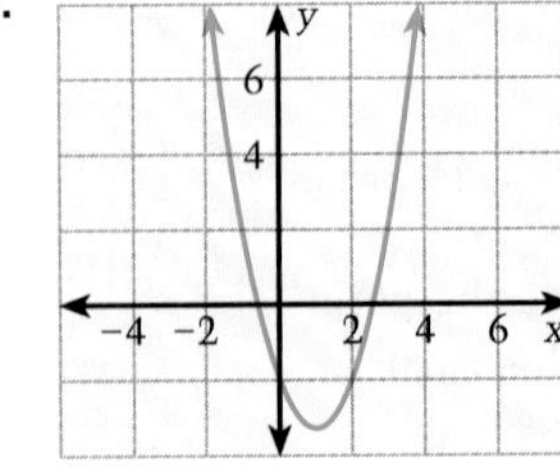

c.

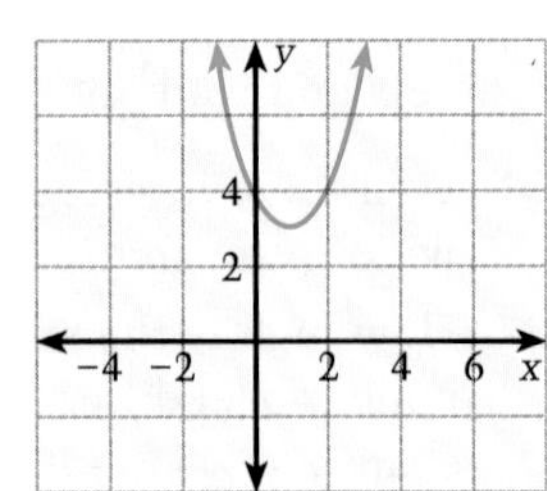

Mid-Chapter SELF-TEST

Take this test as you would take a test in class. The answers to the exercises are given in the back of the book.

In Exercises 1–3, solve the equation. (5.1)

1. $4x^2 = 28$ $\pm\sqrt{7}$

2. $\frac{2}{5}x^2 = 6$ $\pm\sqrt{15}$

3. $x^2 - 2x + 3 = 8 - 2x$ $\pm\sqrt{5}$

In Exercises 4–6, solve the equation by completing the square. (5.3)

4. $x^2 + 8x + 3 = 0$ $-4 \pm \sqrt{13}$

5. $x^2 - 10x - 4 = 0$ $5 \pm \sqrt{29}$

6. $3x^2 + 4x - 1 = 0$ $-\frac{2}{3} \pm \frac{\sqrt{7}}{3}$

In Exercises 7–9, find the vertex and the axis of symmetry of the parabola. (5.2, 5.3)

7. $y = 2x^2 - 4x + 3$ Vertex: $(1, 1)$, axis: $x = 1$

8. $y = 6x^2 + 3x$ Vertex: $\left(-\frac{1}{4}, -\frac{3}{8}\right)$, axis: $x = -\frac{1}{4}$

9. $y = (x - 7)^2 + 5$ Vertex: $(7, 5)$; axis: $x = 7$

In Exercises 10–12, sketch the graph of the parabola. (5.2, 5.3) See margin.

10. $y = x^2 + 4x + 7$

11. $y = (x + 2)^2 - 5$

12. $y = -\frac{1}{3}x^2 - 4x - 8$

13. What must be added to $x^2 + 8x$ to make the expression a perfect square trinomial? **(5.3)** 16

14. Rewrite $y = x^2 + 12x + 29$ in completed square form. **(5.3)** $y = (x + 6)^2 - 7$

15. The area of a circle is 20π square inches. To the nearest hundredth, what is the radius? **(5.1)** 4.47 in.

16. A skydiver jumps from a plane at an altitude of 7000 feet. What is the approximate altitude of the diver after 15 seconds of free fall? **(5.1)** 3400 ft

17. A ladder is leaning against a building. The base of it is 6 feet from the base of the building, and the top of it touches the building at a point that is 21 feet above the ground. How long is the ladder? **(5.1)** 21.8 ft

18. The parabola $y = x^2 + 6x$ has x-intercepts at $(-6, 0)$ and $(0, 0)$. What is the area between the parabola and the x-axis? **(5.2)** 36 square units

19. The period, t (in seconds), of a pendulum is given by $d = 9.78t^2$, where d is the length of the pendulum in inches. A grandfather clock with a 60-inch pendulum is running fast. The clock is adjusted by lengthening the pendulum $\frac{1}{2}$ inch. To the nearest hundredth of a second, how much has the period changed? **(5.1)** 0.01 sec.

20. When looking at the ocean, the distance, d (in miles), you can see to the horizon is given by $d^2 = \frac{3}{2}h$, where h is the height of the viewer (in feet above sea level). To the nearest mile, how far can a sailor see by climbing a 60 foot mast? **(5.1)** $\approx$9.5 mi

Answers

10.

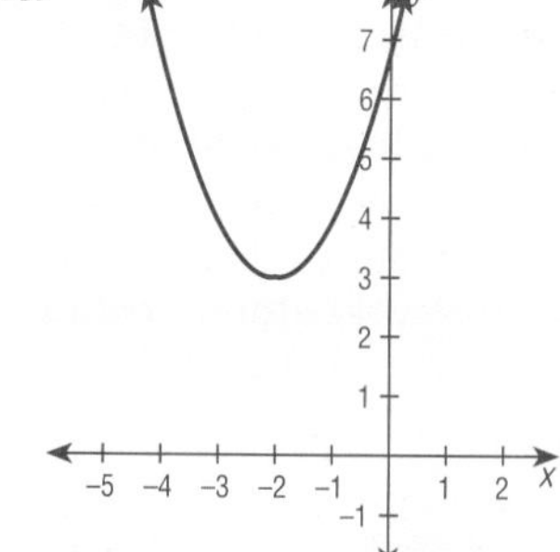

11.

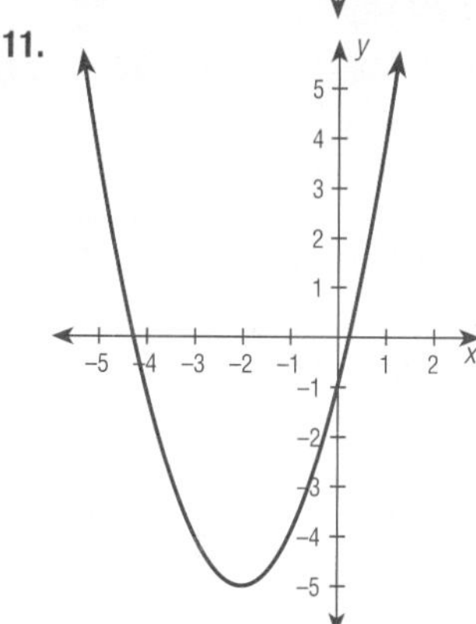

12.

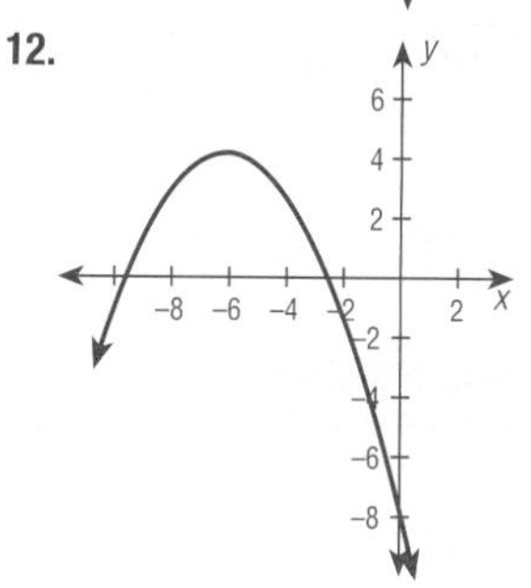

Problem of the Day

A fly lands on a vertex of a solid 3-inch cube, and walks along edges without re-tracing any steps. What is the greatest distance it could travel before arriving at a vertex a second time?
24 inches

ORGANIZER

Warm-Up Exercises

1. Rewrite each quadratic equation in standard form.
   a. $x^2 + 2x = 8$
   b. $4x^2 - 6x = -12$
   c. $5x^2 = 9 - 11x$
   a. $x^2 + 2x - 8 = 0$
   b. $4x^2 - 6x + 12 = 0$
   c. $5x^2 + 11x - 9 = 0$
2. For each equation in Exercise 1, evaluate $b^2 - 4ac$, where a, b, and c are real numbers in the standard form of the quadratic equation.
   a. 36, b. −156, c. 301

Lesson Resources

Teaching Tools
- Transparencies: 7, 8, 13
- Problem of the Day: 5.4
- Warm-up Exercises: 5.4
- Answer Masters: 5.4

Extra Practice: 5.4
Color Transparencies: 33, 34
Applications Handbook: p. 46–48
Technology Handbook: p. 33

LESSON Notes

Example 1

Make the connection that the solutions of a quadratic equation are the x-intercepts of the graph of the same quadratic equation.

5.4 The Quadratic Formula

What you should learn:

Goal 1 How to use the quadratic formula to solve quadratic equations

Goal 2 How to use quadratic models in real-life settings

Why you should learn it:

You can use quadratic equations to model many real-life situations, such as the height of an object as it falls.

Goal 1 Using the Quadratic Formula

In Lesson 5.3, you solved quadratic equations by completing the square for *each problem*. However, completing the square *once* for the standard quadratic equation $ax^2 + bx + c = 0$ produces a formula for the solutions of any quadratic equation. The formula for the solutions is called the quadratic formula. **The derivation of the quadratic formula is on page 846.**

The Quadratic Formula

Let a, b, and c be real numbers such that $a \neq 0$. The solutions of the quadratic equation $ax^2 + bx + c = 0$ are

$$x = \frac{-b \pm \sqrt{b^2 - 4ac}}{2a}.$$

Remember this formula: *The opposite of b, plus or minus the square root of b squared minus 4ac, all divided by 2a.*

Remember that *before* you can apply the quadratic formula, you must first write the equation in standard form, $ax^2 + bx + c = 0$.

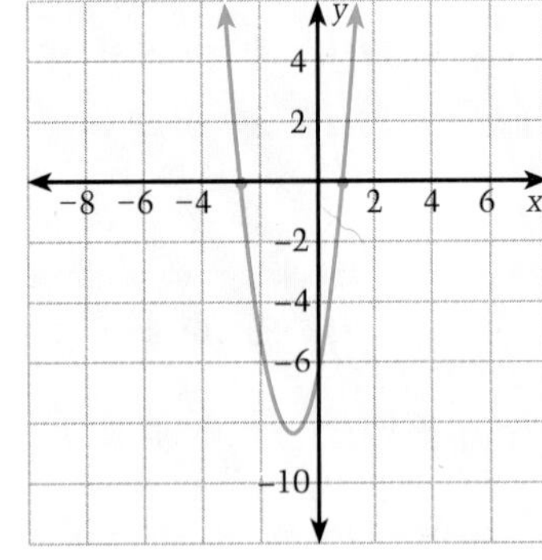

Graph of $y = 3x^2 + 5x - 6$. The x-intercepts are approximately 0.81 and −2.47.

Example 1 *Writing in Standard Form First*

Solve $3x^2 + 5x = 6$.

Solution

$3x^2 + 5x = 6$ — *Rewrite original equation.*

$3x^2 + 5x - 6 = 0$ — *$a = 3$, $b = 5$, $c = -6$*

$x = \frac{-5 \pm \sqrt{5^2 - 4(3)(-6)}}{2(3)}$ — *Quadratic formula*

$x = \frac{-5 \pm \sqrt{97}}{6}$ — *Simplify.*

The equation has two solutions.

$$x = \frac{-5 + \sqrt{97}}{6} \approx 0.81 \quad \text{and} \quad x = \frac{-5 - \sqrt{97}}{6} \approx -2.47$$

■

The expression $b^2 - 4ac$ under the radical in the quadratic formula is the **discriminant** of $ax^2 + bx + c$.

Discriminant

$$x = \frac{-b \pm \sqrt{b^2 - 4ac}}{2a} \quad \textit{Quadratic formula}$$

The discriminant of a quadratic equation indicates how many real-number solutions the equation has.

The Number of Real-Number Solutions of a Quadratic Equation

Consider the quadratic equation $ax^2 + bx + c = 0$.

1. If $b^2 - 4ac$ is positive, then the equation has two solutions.
2. If $b^2 - 4ac$ is zero, then the equation has one solution.
3. If $b^2 - 4ac$ is negative, then the equation has no *real-number* solution. (The equation does have two solutions that are not real numbers. You will study this case in Lesson 5.6.)

Example 2 *Finding the Number of Solutions*

$ax^2 + bx + c = 0$	Discriminant $b^2 - 4ac$	Number of Real-Number Solutions
a. $x^2 - 4x + 5 = 0$	$(-4)^2 - 4(1)(5) = -4$	None
b. $x^2 - 4x + 4 = 0$	$(-4)^2 - 4(1)(4) = 0$	One
c. $x^2 - 4x + 3 = 0$	$(-4)^2 - 4(1)(3) = 4$	Two

Using the quadratic formula, the solution to Part **b** is

$$x = \frac{4 \pm \sqrt{(-4)^2 - 4(1)(4)}}{2(1)} = \frac{4 \pm 0}{2} = 2.$$

Similarly, you can use the quadratic formula to find the solutions in Part **c** to be 1 and 3. ■

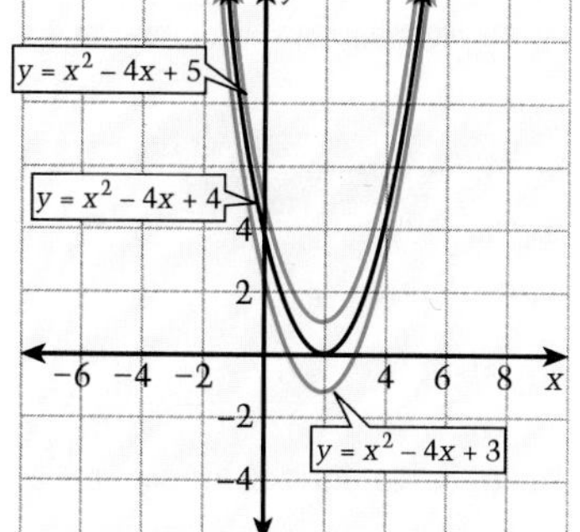

In Example 2, note that the number of solutions can be changed by just changing the value of c. A graph can help you see why this occurs. By changing the value of c, you can move the graph of $y = x^2 - 4x + c$ up or down in the coordinate plane. If the graph is moved too high, it won't have an x-intercept and the equation $x^2 - 4x + c = 0$ won't have a real-number solution.

$y = x^2 - 4x + 5$ *Graph above x-axis (no x-intercepts)*

$y = x^2 - 4x + 4$ *Graph touches x-axis (one x-intercept)*

$y = x^2 - 4x + 3$ *Graph crosses x-axis (two x-intercepts)*

Example 2

Observe that the discriminant indicates how many x-intercepts the graph of the quadratic equation has. In particular, the quadratic in Part a has none; in Part b, there is one; and in Part c, there are two x-intercepts.

EXTEND Example 3 Suppose our baton twirler wishes to match the record set by Donald Garcia and can launch the baton at 6 feet above the ground at a velocity of 40 feet per second or at 5 feet above the ground at a velocity of 45 feet per second. If the baton is caught when it falls back to a height of 5 feet, which release height should be selected?
The release height of 5 feet gives the baton twirler more time to make ten spins.

Extra Examples

Here are additional examples similar to **Examples 1–2.**

1. Solve.
   - **a.** $5x^2 + 3x - 4 = 0$
   - **b.** $3x^2 - 7x + 3 = 0$
   - **c.** $-2x^2 + 8x + 12 = 0$

   a. $\frac{-3 \pm \sqrt{89}}{10}$
   b. $\frac{7 \pm \sqrt{13}}{6}$
   c. $2 \pm \sqrt{10}$
2. Use the discriminant to determine the number of real-valued solutions of the following quadratic equations.
   - **a.** $6x^2 + 3x + 4 = 0$ none
   - **b.** $2x^2 - x + 4 = 0$ none
   - **c.** $x^2 = -6x - 9$ one
   - **d.** $3x^2 + 5x - 6 = 0$ two

Example 3

The value $t = 0.03$ is disregarded because it represents the time that the baton reaches a height of 6 feet on its *ascent*.

Check Understanding

1. Give the quadratic formula. $x = \frac{-b \pm \sqrt{b^2 - 4ac}}{2a}$
2. What is the relationship between the solutions of a quadratic equation and the x-intercepts of its graph? See notes below Example 2 on page 253.
3. Why is it important that quadratic equations be written in standard form before applying the formula? In order to determine the correct values of a, b, and c

Communicating about ALGEBRA

The derivation of the quadratic formula by completing the square of the general quadratic equation $ax^2 + bx + c = 0$ is straightforward (although a bit messy). Many students may enjoy the challenge of the derivation while others may want you to provide the steps so they finish what you started. Recognize that there will be some students who will be happy just to let you do it all. You should decide which approach is best for your class.

A sample derivation will be found on page 846 of the student text in the *Handbook of Mathematical Connections.*

Answers

A. No solution: $c > 16$, one: $c = 16$, two: $c < 16$

B. Number of x-intercepts is the number of solutions.

Connections
Physics

Goal 2 Using Quadratic Models in Real Life

In Lesson 5.1, you studied the model for the height (in feet) of a falling object that is *dropped.* For an object that is *thrown* down or up, the model needs an extra term. Problems that use either of these models are **vertical motion** problems.

Vertical Motion Models	$h = -16t^2 + s$	Height after object is *dropped*
	$h = -16t^2 + vt + s$	Height after object is *thrown*
Labels	h = height	(feet)
	t = time in motion	(seconds)
	s = initial height (when $t = 0$)	(feet)
	v = initial velocity (when $t = 0$)	(feet per second)

Be sure you understand that v is the velocity, not the speed. Velocity can be positive (moving up), negative (moving down), or zero (not moving). Speed is the absolute value of velocity.

Real Life
Baton Tossing

When the baton twirler tosses the baton into the air, its initial velocity is 30 feet per second.

Example 3 *Vertical Motion Problem*

When the baton twirler releases the baton into the air, the twirler's hand is 5 feet above the ground. The twirler will catch the baton when it falls back to a height of 6 feet. How many seconds is the baton in the air?

Solution When the baton is thrown up, its initial velocity is $v = 30$ feet per second. The initial height is $s = 5$ feet. The twirler will catch the baton when the height is 6 feet.

$h = -16t^2 + 30t + 5$ *Vertical motion model*

$6 = -16t^2 + 30t + 5$ *Substitute 6 for h.*

$16t^2 - 30t + 1 = 0$ *a = 16, b = −30, c = 1*

$t = \frac{30 \pm \sqrt{836}}{32}$ *Quadratic formula*

$t \approx 1.84 \text{ or } 0.03$

Thus the twirler has less than 2 seconds to catch the baton. ■

Communicating about ALGEBRA

SHARING IDEAS about the Lesson See margin.

A. Find values of c so that the equation $x^2 - 8x + c = 0$ will have no solution, one solution, and two solutions.

B. Give a graphical interpretation of your Part A answers.

EXERCISES

Guided Practice

CRITICAL THINKING about the Lesson

1. State the quadratic formula *in words.* See Additional Answers.
2. What is the discriminant of $ax^2 + bx + c = 0$? How is the discriminant related to the number of solutions of $ax^2 + bx + c = 0$? See Additional Answers.
3. What are the missing numbers in the solutions of the equation? How many solutions does the equation have? Explain your answer. −10, 2, −5; two, the discriminant is positive.

$$2x^2 + 10x - 5 = 0 \quad x = \frac{\boxed{?} \pm \sqrt{100 - 4\boxed{?}\cdot\boxed{?}}}{4}$$

4. Compare the two vertical motion models, $h = -16t^2 + s$ and $h = -16t^2 + vt + s$. See Additional Answers.

Independent Practice

In Exercises 5–8, write the equation in standard form.

5. $5 - 3x^2 = 7x + x^2$ $4x^2 + 7x - 5 = 0$
6. $4x^2 - 10 = -13x - 7x$ $4x^2 + 20x - 10 = 0$
7. $12x + 1 = 6 - 30x^2$ $30x^2 + 12x - 5 = 0$
8. $2x^2 + x - 3 = 4x^2 - 5$ $2x^2 - x - 2 = 0$

In Exercises 9–14, state the discriminant of the quadratic equation.

9. $3x^2 - 8x + 12 = 0$ −80
10. $6x^2 + 9x - 4 = 0$ 177
11. $15x^2 - 10x + 1 = 0$ 40
12. $4x^2 + x - 20 = 0$ 321
13. $3x^2 - 6x + 3 = 0$ 0
14. $5x^2 - 12x + 10 = 0$ −56

In Exercises 15–20, use the discriminant to determine the number of real solutions of the equation.

15. $x^2 + x + 1 = 0$ None
16. $3x^2 - 6x + 1 = 0$ 2
17. $20x^2 - 10x + 5 = 0$ None
18. $-2x^2 - 5x - 6 = 0$ None
19. $-7x^2 - 8x - 10 = 0$ None
20. $-2x^2 + 3x - 1 = 0$ 2

In Exercises 21–30, use the quadratic formula to solve the equation.

21. $-5x^2 - 15x + 10 = 0$ $-\frac{3}{2} \pm \frac{\sqrt{425}}{10}$
22. $4x^2 - 6x + 2 = 0$ $1, \frac{1}{2}$
23. $-9x - 1 = -9x^2$ $\frac{1}{2} \pm \frac{\sqrt{117}}{18}$
24. $10x^2 - 5x + 5 = -25x$ $-1 \pm \frac{\sqrt{200}}{20}$
25. $-13x^2 - 9x + 4 = 5 + 2x^2 - x$ $-\frac{1}{3}, -\frac{1}{5}$
26. $7x^2 + x = 12x^2 - 3x + 2$ No real solution.
27. $11x^2 + 4 = 2 - 7x^2 + 11x$ No real solution.
28. $x - x^2 = 1 - 6x^2$ $-\frac{1}{10} \pm \frac{\sqrt{21}}{10}$
29. $4x - 2 = -3x^2 + 3x$ $-1, \frac{2}{3}$
30. $12x^2 - 9x + 1 = 10x^2 + 10x + 7$ $\frac{19}{4} \pm \frac{\sqrt{409}}{4}$

In Exercises 31–34, solve the equation. Round to two decimal places.

31. $5.1x^2 - 0.33x - 0.1 = 0$ 0.18, −0.11
32. $6.3x^2 + 10.08x + 0.99 = 0$ −0.11, −1.49
33. $3.2x^2 - 0.01 = 15.1 + 3.3x$ 2.75, −1.72
34. $17.02x - 3x^2 = 4.01x^2 - 1.5$ 2.51, −0.09

EXERCISE Notes

ASSIGNMENT GUIDE

Basic/Average: Ex. 1–4, 7–27 odd, 29–41 odd, 49–53 odd, 57–60

Above Average: Ex. 1–4, 7–27 odd, 29–41 odd, 44–48, 54–60

Advanced: Ex. 1–4, 7–27 odd, 29–41 odd, 44–48, 53–60

Selected Answers
Exercises 41, 43, 55, 56

More Difficult Exercises
Exercises 41, 43, 55, 56

Guided Practice

Ex. 2 Be sure to check that students understand that "no solutions" really means "no real solutions." These nonreal solutions (complex numbers) will be introduced in Lessons 5.5 and 5.6.

Ex. 4 Consider the differences between objects that are dropped versus objects that are thrown.

Independent Practice

EXTEND Ex. 9–20 Ask students to describe the number of times that the related graph would cross the *x*-axis. (See the bottom of page 253.)

Ex. 21–30 Be sure that students express the equations in standard form. As a check, ask them to state the values of *a*, *b*, and *c* for each equation.

Common-Error ALERT!

In **Exercise 2,** students often mistake the discriminant to be $\sqrt{b^2 - 4ac}$, not $b^2 - 4ac$.

▶ **Ex. 31–34 and 41–44 TECHNOLOGY** Students should use scientific calculators in applying the quadratic formula. In Lesson 5.6, a graphing calculator program for the quadratic formula will be provided.

▶ **Ex. 35–38** A solution model is provided below Example 2 on page 253.

EXTEND Ex. 41–44 Have students state some reasons why the trends in cellular phones and recreational boats are occurring.

EXTEND Ex. 45–54 Students now have four methods that can be used to solve any quadratic equation. Ask students to state which method they used to solve each of the exercises and explain why.

Answers

35. $c < 9, c = 9, c > 9$
36. $c < 36, c = 36, c > 36$
37. $c < 16, c = 16, c > 16$
38. $c < 1, c = 1, c > 1$

Enrichment Activities

These activities require students to go beyond lesson goals.

An Alternate Approach
Another Graphing Technique
Consider Exercise 58. The CD model is a quadratic model and the albums model is linear. The year is which the sales are equal occurs when $4.3t^2 - 9.8t - 42.7 = -28.8t + 301$.

(continued)

In Exercises 35–38, find values of *c* such that the equation has two real-number solutions, one real-number solution, and no real-number solution. See margin.

35. $x^2 - 6x + c = 0$ **36.** $x^2 - 12x + c = 0$ **37.** $x^2 + 8x + c = 0$ **38.** $x^2 + 2x + c = 0$

Geometry **In Exercises 39 and 40, use the quadratic formula to find the dimensions of the figure.**

39. Rectangle Area $= 58.14$ in.^{2} 5.1 in., 11.4 in.

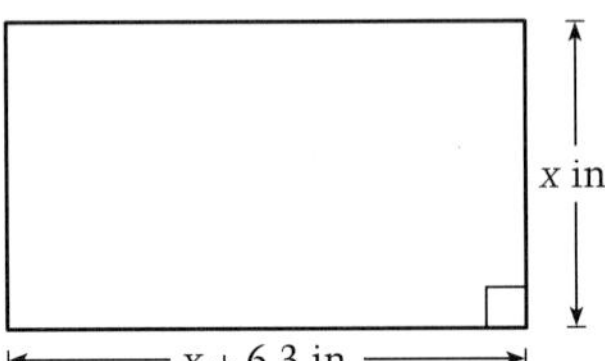

40. Ellipse Area $= 9.02$ (area of ellipse is πab) ≈ 0.99, ≈ 2.89

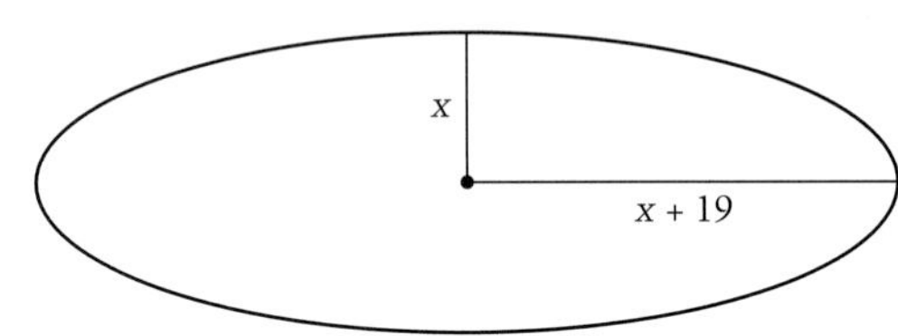

✪ **41.** *Hitting Baseballs* You are hitting baseballs. When tossing the ball into the air, your hand is 5 feet above the ground. You hit the ball when it falls back to a height of 4 feet. If you toss the ball with an initial velocity of 25 feet per second, how much time will pass before you hit the ball? ≈ 1.6 seconds

42. *Cellular Phones* For 1987 through 1994, the number of cellular phone subscribers, *s* (in millions), in the United States can be approximated by the model

$$s = 0.52t^2 - 7.94t + 31.87$$

where $t = 7$ represents 1987. In which year did cellular phone companies have about 7.5 million subscribers? ***(Source: Cellular Telecommunications Industry Association)*** 1991

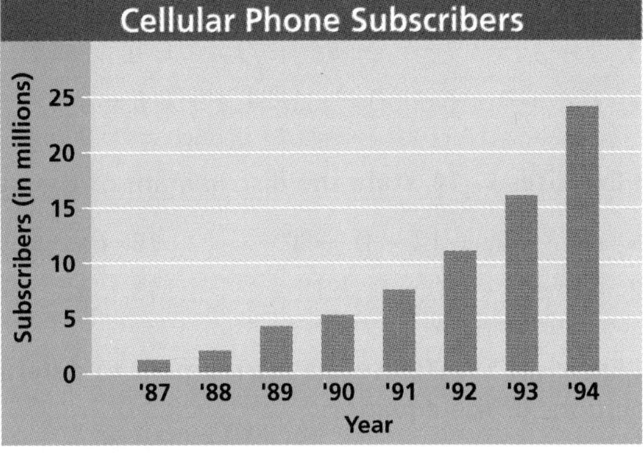

✪ **43.** *Conveyor Belt* Gravel is falling 4 feet from one conveyor belt to another. Because of the way the conveyor belt is constructed, the gravel is given a slight downward velocity (-1.4 feet per second). How long does it take each piece of gravel to fall from the upper belt to the lower belt? ≈ 0.46 seconds

44. *Boat Owners* For 1983 to 1993, the number, *b* (in millions), of recreational boats owned in the United States can be approximated by the model

$$b = -0.01t^2 + 0.59t + 11.28$$

where $t = 3$ represents 1983. In what year were 15.8 million recreational boats owned? ***(Source: National Marine Manufacturers Association)*** 1989

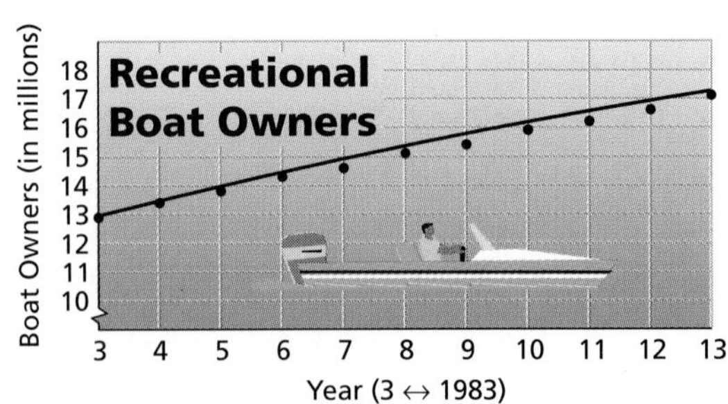

Integrated Review

In Exercises 45–54, solve the equation using the most convenient method. Round to the nearest hundredth.

45. $6x^2 - x + 1 = x^2 - x + 4$ ±0.77

46. $2x^2 + 5x - 9 = -10$ −2.28, −0.22

47. $4.5x^2 - 3x = 0.33 - 2.1x^2$ 0.55, −0.09

48. $16x - 3x^2 - 7 = -6x^2 + 10x$ 0.83, −2.83

49. $-5x^2 - 4 = x^2 - 10x$ 0.67, 1

50. $-10x^2 + 2x - 5 = -6 - 6x^2 + 2x$ −0.5, 0.5

51. $6.7x^2 + 30x - 0.04 = 1.7$ 0.06, −4.53

52. $-5x^2 + 21.2x = 10.2x - 6.01$ 2.65, −0.45

53. $3x^2 + 5x - 10 = 11 + 5x$ ±2.65

54. $-2x^2 + 2x - 2 = -9x^2 - 5x$ 0.23, −1.23

55. *College Entrance Exam Sample* In the figure, the diameter of the small circle is AC and the diameter of the large circle is AD. If $AB = BC = CD$, what is the ratio of the area of the shaded region to the area of the smaller circle? d

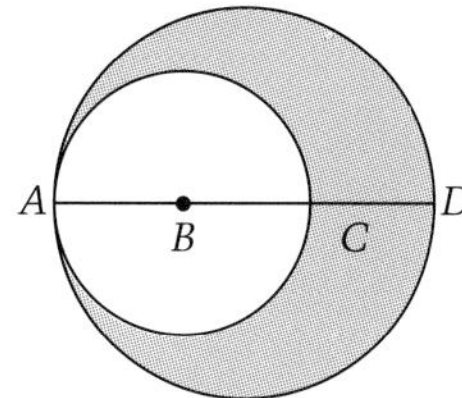

a. 1 : 1 **b.** 3 : 2 **c.** 4 : 3
d. 5 : 4 **e.** 9 : 4

56. *College Entrance Exam Sample* In the figure, square $PQRS$ is divided into four smaller squares. If the area of the shaded region is 3, what is the area of $PQRS$? a

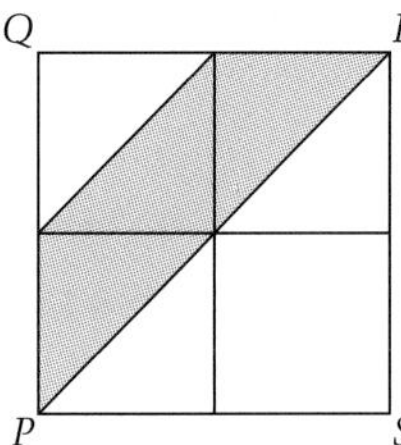

a. 8 **b.** 7 **c.** 6 **d.** 5 **e.** 1

Exploration and Extension

Compact Disks and Records **In Exercises 57 and 58, use the following information.**

For 1986 through 1990, the number of compact disks, C (in millions), and the number of record albums, R (in millions), sold in the United States can be approximated by the models $C = 4.3t^2 - 9.8t - 42.7$ and $R = -28.8t + 301$ where $t = 6$ represents 1986.
(Source: Recording Industry Association of America)

57. How many compact disks were sold in 1986? How many record albums were sold in 1986? 53.3 million, 128.2 million

58. During which year was the number of compact disks sold approximately equal to the number of record albums sold? 1987

In Exercises 59 and 60, sketch the graph of each equation. Use your graph to approximate the points at which the two graphs intersect. Describe an algebraic procedure for finding the points of intersection. See Additional Answers for graphs.

59. $\begin{cases} x + 3y = 11 \\ x^2 - 3y = -5 \end{cases}$
$(2, 3), \left(-3, \frac{14}{3}\right)$
Add the two equations, solve for the one variable, substitute for the one variable to find the others.

60. $\begin{cases} x + 2y = -7 \\ -x - y^2 = -17 \end{cases}$
(1, −4), (−19, 6)

To solve the equation algebraically using any of the methods presented thus far, you would have to use algebra to express the quadratic equation in standard form or in completed-square form. As an alternative, consider the following graphing technique.

Let y_1 be the left-hand side of the equation and y_2 be the right-hand side. If you graph both y_1 and y_2 on the same coordinate grid, you can find their point(s) of intersection. To find the year in which the sales are equal, look for the point(s) where the parabola $y_1 = 4.3t^2 - 9.8t - 42.7$ intersects the line $y_2 = -28.8t + 301$.

COOPERATIVE LEARNING

1. Decide on an appropriate viewing rectangle for the graphics calculator and graph the two equations.
X min = −10 Y min = 0
X max = 10 Y max = 300
X scl = 1 Y scl = 10

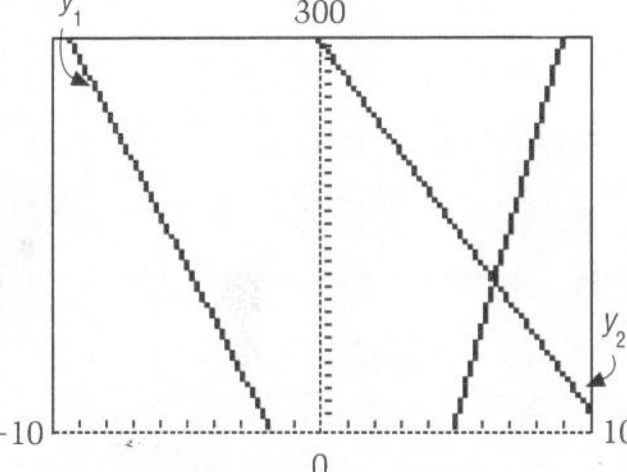

2. Use the Trace function to determine the intersection point. Interpret your results.
The graphs intersect when $t \approx 7.00$. Since $t = 6$ corresponds to 1986, the sales were equal in 1987.

Problem of the Day

List 12 points in the coordinate plane exactly 5 units from the origin.
(5, 0), (4, 3), (3, 4), (0, 5), (−3, 4), (−4, 3), (−5, 0), (−4,−3), (−3,−4), (0,−5), (3,−4), (4,−3)

ORGANIZER

Warm-Up Exercises

1. Solve.
   a. $x^2 - 3 = 0$ $\pm\sqrt{3}$
   b. $3x^2 - 27 = 0$ ± 3
   c. $2x^2 + 8 = 18$ $\pm\sqrt{5}$
   d. $-2x^2 + 15 = 0$ $\pm\sqrt{\frac{15}{2}}$
2. Combine like terms.
   a. $(3x^2 + 4x) + (8x - 13)$
   b. $(4x - 6x^2) - (3x^2 + 10x)$
   c. $(4x - 8 + 2x^2) + (12x - 4x^2)$
   a. $3x^2 + 12x - 13$
   b. $-9x^2 - 6x$
   c. $-2x^2 + 16x - 8$
3. Multiply.
   a. $(2x + 5)(3x - 4)$
   b. $3(10 - 4x + 2x^2)$
   c. $(7 - 5x)(3 + 4x)$
   a. $6x^2 + 7x - 20$
   b. $30 - 12x + 6x^2$
   c. $21 + 13x - 20x^2$

Lesson Resources

Teaching Tools
  Transparencies: 7, 8
  Problem of the Day: 5.5
  Warm-up Exercises: 5.5
  Answer Masters: 5.5
Extra Practice: 5.5
Color Transparencies: 34, 35
Applications Handbook: p. 26–28

5.5 Complex Numbers

What you should learn:

Goal 1 How to identify, add, subtract, and multiply imaginary and complex numbers

Goal 2 How to plot complex numbers in the complex plane

Why you should learn it:

You can use the complex plane to illustrate relationships between complex numbers, such as in fractal geometry.

Goal 1 Imaginary and Complex Numbers

You have seen that some quadratic equations have no real-number solutions. For instance, the quadratic equation $x^2 + 1 = 0$ has no real-number solution because no real number x can be squared to produce -1. To overcome this problem, mathematicians created an expanded system of numbers using the **imaginary unit** i, defined as

$i = \sqrt{-1}$. *Imaginary unit i*

Note that $i^2 = -1$. The imaginary unit i can be used to write the square root of *any* negative number.

Imaginary Numbers

If a is a positive real number, then $\sqrt{-a}$ is an **imaginary number** and

$\sqrt{-a} = i\sqrt{a}$. *Imaginary number*

The imaginary number $i\sqrt{a}$ has the property that $(i\sqrt{a})^2 = -a$.

Example 1 Writing Imaginary Numbers

Use the imaginary unit i to write the solutions of $x^2 = -3$.

Solution

$x^2 = -3$ *Rewrite original equation.*

$x = \pm\sqrt{-3}$ *Take square root.*

$x = \pm i\sqrt{3}$ *Write with imaginary unit.*

The two solutions are $i\sqrt{3}$ and $-i\sqrt{3}$.

Check:

$x^2 = -3$	$x^2 = -3$
$(i\sqrt{3})^2 \stackrel{?}{=} -3$	$(-i\sqrt{3})^2 \stackrel{?}{=} -3$
$-3 = -3$ ***Check***	$-3 = -3$ ***Check***

■

A **complex number** is a number that is the sum of a real number and an imaginary number. Each complex number can be written in the **standard form,** $a + bi$.

Complex Numbers

For any real numbers a and b, the number

$$a + bi$$

is a **complex number.** If $b = 0$, then the complex number is a real number. Two complex numbers $a + bi$ and $c + di$ are **equal** if $a = c$ and $b = d$.

To add (or subtract) two complex numbers, add (or subtract) the real and imaginary parts of the numbers.

Sum: $(a + bi) + (c + di) = (a + c) + (b + d)i$

Difference: $(a + bi) - (c + di) = (a - c) + (b - d)i$

Example 2 *Adding and Subtracting Complex Numbers*

a. $(3 - i) + (2 + 3i) = 3 - i + 2 + 3i$
$= 3 + 2 - i + 3i$
$= (3 + 2) + (-1 + 3)i$
$= 5 + 2i$

b. $2i + (-4 - 2i) = 2i - 4 - 2i$
$= -4 + 2i - 2i$
$= -4$

c. $3 - (-2 + 3i) + (-5 + i) = 3 + 2 - 3i - 5 + i$
$= 3 + 2 - 5 - 3i + i$
$= 0 - 2i$
$= -2i$ ■

The following properties of real numbers hold for complex numbers as well.

Associative Properties of Addition and Multiplication
Commutative Properties of Addition and Multiplication
Distributive Property of Multiplication over Addition

LESSON Notes

Example 1

Imaginary numbers may be written in two ways: "$i\sqrt{3}$" or "$\sqrt{3}\ i$." The latter is often abandoned because of the possible confusion about whether "i" is under the radical sign or not. Use care when expressing imaginary numbers in the second manner.

Example 2

Adding and subtracting complex numbers is similar to adding and subtracting other algebraic expressions. Consequently, students already have the necessary skills with which to add and subtract complex numbers, that is, "add (or subtract) like terms."

EXTEND Example 3 PROBLEM SOLVING It may be interesting to record the first four powers of the imaginary unit, i. Ask students to continue raising i to successive powers. What patterns can be seen?

$i = \sqrt{-1}$
$i^2 = \sqrt{-1} \cdot \sqrt{-1} = -1$
$i^3 = -1 \cdot i = -i$
$i^4 = -i \cdot i = -i^2 = -(-1) = 1$

Pattern: For all integers n, $i^{4n+1} = i$, $i^{4n+2} = i^2 = -1$, $i^{4n+3} = -i$, and $i^{4n} = 1$.

Example 4

Complex planes and the related vector treatment of complex numbers are frequently used in the study of physics. A complex number can be seen as a vector (a directed line segment) as pic-

(continued)

Example 3 Multiplying Complex Numbers

a. $i \cdot -3i = -3i^2 = -3(-1) = 3$

b. $i(4 + 3i) = 4i + 3i^2$ *Distributive Property*

$= 4i + 3(-1)$ $i^2 = -1$

$= -3 + 4i$ *Standard form*

c. $(2 - i)(4 + 3i) = (2 - i)(4) + (2 - i)(3i)$

$= 8 - 4i + 6i - 3i^2$

$= 8 - 4i + 6i - (-3)$

$= 8 - 4i + 6i + 3$

$= 11 + 2i$ ■

Goal 2 The Complex Number Plane

Most applications of complex numbers are either theoretical or very technical, and are therefore not appropriate for this text. However, to give you some idea of how complex numbers can be used, consider their role in **fractal geometry.**

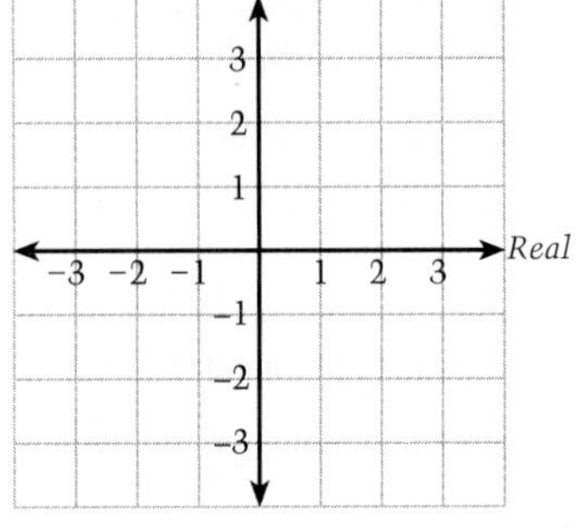

Just as every real number corresponds to a point on the real line, every complex number can be made to correspond to a point in a plane called the **complex plane.** In the complex plane, the horizontal axis is the **real axis,** and the vertical axis is the **imaginary axis.**

Don't worry about confusing the complex plane with the standard coordinate plane. This lesson is the *only* one in the book that uses complex planes—*all* other coordinate planes in the book are standard coordinate planes whose points are ordered pairs of *real numbers.*

Example 4 Plotting Complex Numbers in the Complex Plane

Plot the complex numbers in the complex plane.

a. $2 - 3i$ **b.** $-1 + 2i$ **c.** $2i$

Solution

a. To plot the complex number $2 - 3i$, move (from the origin) two units to the right on the real axis and 3 units down.

b. To plot the complex number $-1 + 2i$, move (from the origin) one unit to the left and 2 units up.

c. To plot the complex number $2i$, move (from the origin) two units up. ■

(continued)

tured below. A vector has both magnitude and direction, and is, therefore, a good way to represent a displacement, a force, an acceleration, or a velocity.

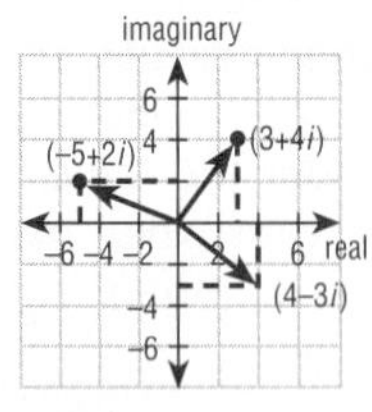

Example 5

Ask students to identify other complex numbers that are (and that are not) in the Mandelbrot set.

For example, $-\frac{1}{2}i$ (and $1 - i$)

Extra Examples

Here are additional examples similar to **Examples 1–4.**

1. Solve. Use imaginary numbers to express nonreal solutions.
   a. $2x^2 + 14 = 0$ $\pm i\sqrt{7}$
   b. $x^2 + 9 = 0$ $\pm 3i$
2. Combine.
   a. $(6 + 2i) - (5 - 4i)$
   b. $(12 - 7i) + (10 + 9i)$
   c. $(3 + 8i) - (6 - 15i) + (2 - 4i)$

   a. $1 + 6i$, b. $22 + 2i$, c. $-1 + 19i$
3. Multiply.
   a. $3i \cdot -5i$ 15
   b. $6i(4 - 2i)$ $12 + 24i$
   c. $(6 + 2i)(5 - 4i)$ $38 - 14i$
   d. $(12 - 7i)(10 + 9i)$ $183 + 38i$
4. Plot these complex numbers in the same complex plane.
   a. $5 - 2i$ b. $4i$ c. $-3 - 5i$

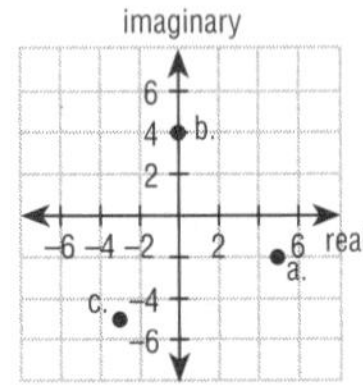

Ashley Reiter of Charlotte, North Carolina, was the first-place winner of the 1991 Westinghouse Science Talent Search. Her winning project was on fractals.

People who understand fractal geometry can use the complex plane to draw stunning pictures, called **fractals.** The most famous such picture is called the **Mandelbrot set,** after the Polish-born mathematician Benoit Mandelbrot.

In the picture below, the Mandelbrot set is the black region in the complex plane. Each point in one of the colored regions is *not* in the set. (The reason for assigning colors to points not in the set is too complicated to mention here.)

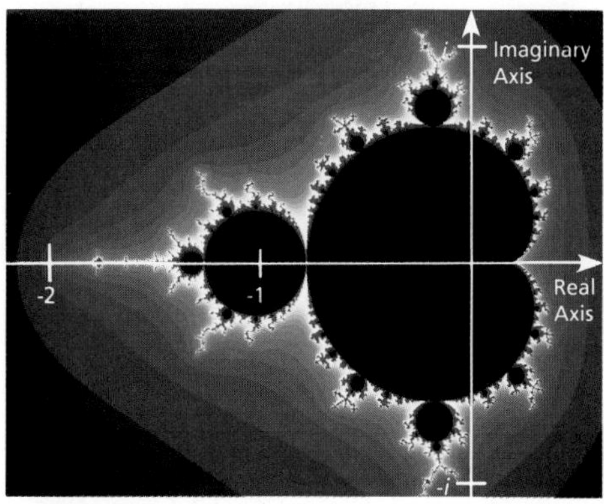

Connections
Fractal Geometry

Example 5 *The Mandelbrot Set*

Decide whether the complex number is in the Mandelbrot set.

a. -1 **b.** $\frac{1}{2}i$ **c.** $1 + i$

Solution

a. The point that represents -1 is on the real axis, one unit to the left of the origin. This number *is* in the Mandelbrot set.

b. The point that represents $\frac{1}{2}i$ is on the imaginary axis, one-half unit above the origin. It *is also* in the Mandelbrot set.

c. The point that represents $1 + i$ is one unit to the right and one unit above the origin. It is *not* in the Mandelbrot set. ■

Communicating about ALGEBRA

SHARING IDEAS about the Lesson See margin.

A. Is a real number also a complex number? Explain.

B. The rule for adding two complex numbers is sometimes called the parallelogram rule. Graph the complex numbers $2 + i$ and $1 + 3i$ in the complex plane. Then graph their sum. How are the three points and the origin related to a parallelogram?

Check Understanding

1. What is the standard form of a complex number?
   a number in the form $a + bi$, where a and b are real
2. Which properties of real numbers also hold for complex numbers?
   See the bottom of page 259.
3. Can the quadratic equation, $x^2 + 5 = 0$, be solved in the set of real numbers? in the set of complex numbers?
   No; yes

Communicating about ALGEBRA

Answers

A. Yes, a real number can be written as $a + bi$ with $b = 0$.

B.

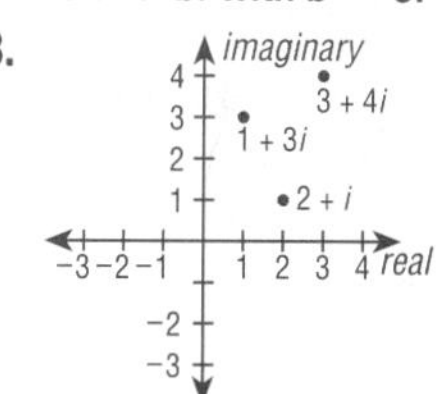

They are the vertices of a parallelogram.

EXTEND *Communicating*

Use the concept of complex numbers as vectors (directed line segments) to model addition and subtraction of complex numbers. (Below are examples of how complex numbers are represented as vectors in the complex plane.)

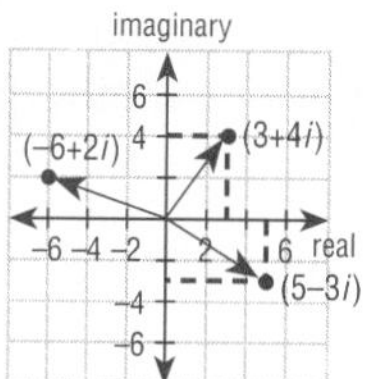

EXERCISE Notes

ASSIGNMENT GUIDE

Basic/Average: Ex. 1–8, 13–23 odd, 29–39 odd, 49–57 odd, 65, 69–72, 85–88

Above Average: Ex. 1–8, 13–23 odd, 31–41 odd, 51–61 odd, 66, 69–72, 85–88

Advanced: Ex. 1–8, 15–25 odd, 31–41 odd, 51–61 odd, 68–72, 85–88

Selected Answers
Exercises 1–8, 9–83 odd

Use **Mixed Review** as needed.

More Difficult Exercises
Exercises 85–88

Guided Practice

Ex. 1–8 Use these exercises as a check for understanding of complex-number operations.

Answers

7. Yes, $(-i\sqrt{2})^2 + 2 = i^2 \cdot 2 + 2 = -2 + 2 = 0$

8.

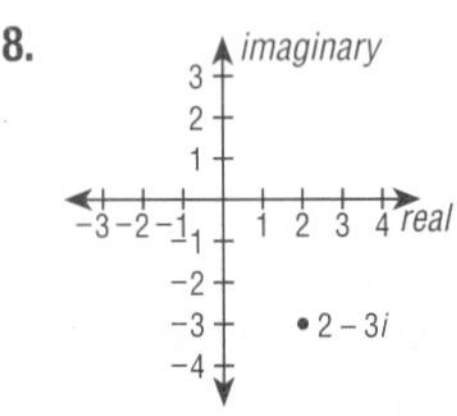

Independent Practice

Ex. 27–32 The equations do not have a linear, or middle, term. They can be solved by using the square-root method of Lesson 5.1.

EXERCISES

Guided Practice

CRITICAL THINKING about the Lesson

1. Does the equation $x^2 = -1$ have a real-number solution? Does it have solutions that are not real numbers? Explain. See margin. No, yes; the solutions, $\pm i$, are not real numbers.
2. Rewrite $\sqrt{-4}$ using the imaginary unit i. $2i$

In Exercises 3–6, perform the indicated operation.

3. $(i\sqrt{2})^2$ -2
4. $(4 + 3i) + (-2 - 2i)$ $2 + i$
5. $i(-2 + 2i)$ $-2 - 2i$
6. $(1 - i)(4 + 2i)$ $6 - 2i$
7. Is $-i\sqrt{2}$ a solution of $x^2 + 2 = 0$? Justify your answer. See margin.
8. Plot the complex number $2 - 3i$ in the complex plane. See margin.

Independent Practice

In Exercises 9–14, write the number using the imaginary unit *i*.

9. $\sqrt{-9}$ $3i$
10. $\sqrt{-4}$ $2i$
11. $-\sqrt{-25}$ $-5i$
12. $-\sqrt{-1}$ $-i$
13. $\sqrt{-5}$ $i\sqrt{5}$
14. $-\sqrt{-15}$ $-i\sqrt{15}$

In Exercises 15–20, simplify the expression.

15. $2i^2$ -2
16. $(2i)^2$ -4
17. $-i^2$ 1
18. $(\sqrt{-6})^2$ -6
19. $(\sqrt{-10})^2$ -10
20. $(i\sqrt{3})^2$ -3

In Exercises 21–26, decide whether the complex number is a solution of the equation.

21. $x^2 = -9,\ 3i$ It is.
22. $x^2 = -14,\ 14i$ It is not.
23. $x^2 + 9 = 8,\ -i$ It is.
24. $x^2 - 4 = 2,\ i\sqrt{2}$ It is not.
25. $x^2 + 10 = 3,\ 7i$ It is not.
26. $x^2 + 5 = -5,\ -i\sqrt{10}$ It is.

In Exercises 27–32, solve the equation.

27. $x^2 = -4$ $\pm 2i$
28. $x^2 = -5$ $\pm i\sqrt{5}$
29. $-3x^2 + 1 = 7$ $\pm i\sqrt{2}$
30. $4x^2 + 15 = 3$ $\pm i\sqrt{3}$
31. $6x^2 + 5 = 2x^2 + 1$ $\pm i$
32. $x^2 = 7x^2 + 1$ $\pm i\sqrt{\frac{1}{6}}$

In Exercises 33–44, perform the indicated operations.

33. $(3 + 2i) + (9 + i)$ $12 + 3i$
34. $(9 + 2i) + (3 - i)$ $12 + i$
35. $(-4 + i) + (i - 4)$ $-8 + 2i$
36. $(1 + i) + (7 - 4i)$ $8 - 3i$
37. $(5 - 3i) - (-1 + i)$ $6 - 4i$
38. $(2 - i) - (5 + i)$ $-3 - 2i$
39. $(1 - 3i) - (8 + i)$ $-7 - 4i$
40. $(-4 + 6i) - (-3 + 12i)$
41. $(30 - i) - (18 + 6i) + 3i^2$
42. $(3 - 36i) + (7 + 28i) + i$
43. $-i + (9 - 2i) - (5 + 7i)$
44. $(8 - i) + (5 + 13i) - 2i^2$

40. $-1 - 6i$ 41. $9 - 7i$ 42. $10 - 7i$ 43. $4 - 10i$ 44. $15 + 12i$

In Exercises 45–56, perform the indicated multiplication.

45. $i(1 + i)$ $-1 + i$

46. $i(2 - i)$ $1 + 2i$

47. $-i(5 + 2i)$ $2 - 5i$

48. $3i(1 - 2i)$ $6 + 3i$

49. $2i(6 + i)$ $-2 + 12i$

50. $-8i(6 + i)$ $8 - 48i$

51. $(3 + i)(2 + i)$ $5 + 5i$

52. $(2 + 7i)(4 - 2i)$ $22 + 24i$

53. $(1 + 4i)(2 - i)$ $6 + 7i$

54. $(9 - 10i)(-8 + 3i)$ $-42 + 107i$

55. $(-7 - i)(3 + 2i)$ $-19 - 17i$

56. $(-14 + 6i)(1 - i)$ $-8 + 20i$

In Exercises 57–62, simplify the expression.

57. $4i(1 + i) + (6 + 2i)$ $2 + 6i$

58. $(6 + 6i) - 3i^2(1 - 2i)$ 9

59. $i^2[(1 + 4i) - (2 - 3i)]$

60. $i(7 - 3i)(2 + 10i)$ $-64 + 44i$

61. $10i(2 - 2i)(5 + i)$ $80 + 120i$

62. $3i[2i(1 + i) - (2 - 10i)]$

59. $1 - 7i$ 62. $-36 - 12i$

In Exercises 63–68, graph the complex number in the complex plane. See Additional Answers.

63. $3 - i$

64. $4 + 2i$

65. $-2 + 5i$

66. $-10 - 4i$

67. $-3i$

68. $2 - 12i$

Fractals **In Exercises 69–72, use the following information.**

Trace the Mandelbrot set on page 261 on a piece of paper. Then evaluate the expressions and mark the locations of the results on your tracing. This shows where each enlargement (pictured below) is found. The colors of the enlargements are not necessarily those used on page 261. See Additional Answers.

69. *Cos Antenna* $(6.2 - 1.2i) - (7.6 - 1.2i)$ $-1.4 + 0i$

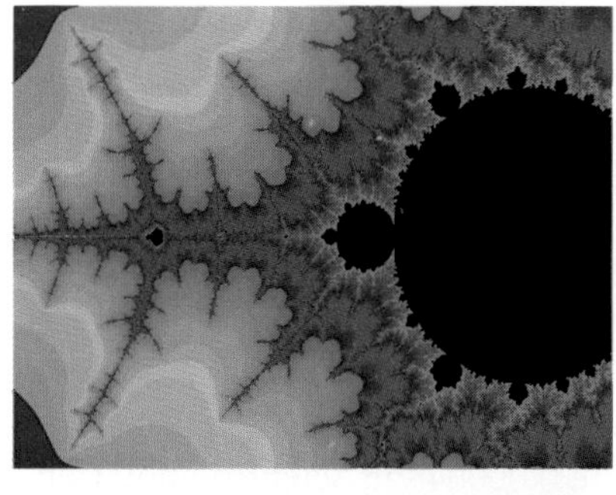

70. *Star Factory* $-\frac{1}{170}(3 + 5i)(8 + 15i)$ $\frac{3}{10} - \frac{1}{2}i$

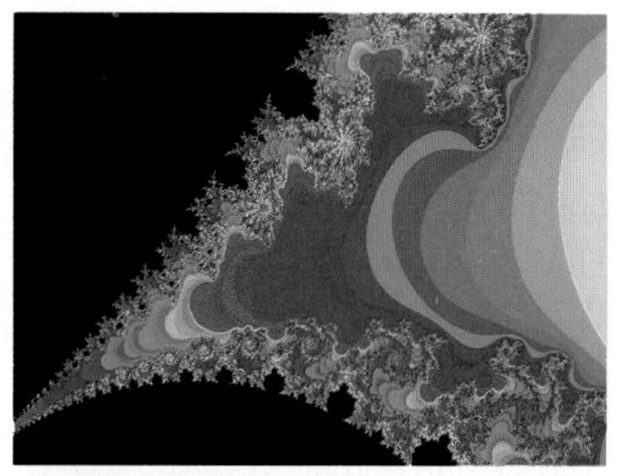

71. *Lightning* $(1 + 0.5i)(0.396 + 0.822i)$ $-0.015 + 1.02i$

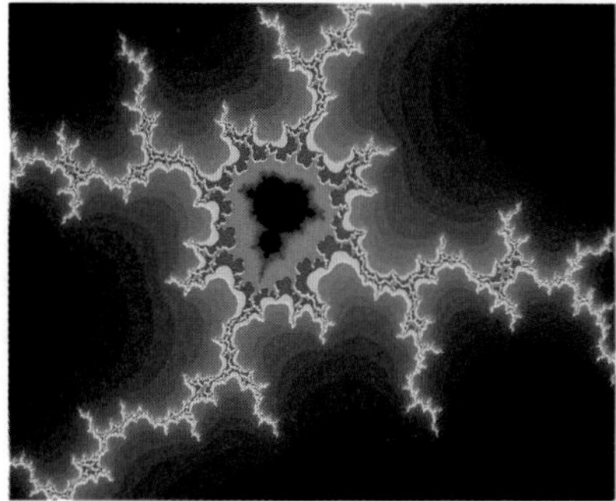

72. *Spanish Lace* $2i(0.1 + 0.36i)$ $-0.72 + 0.2i$

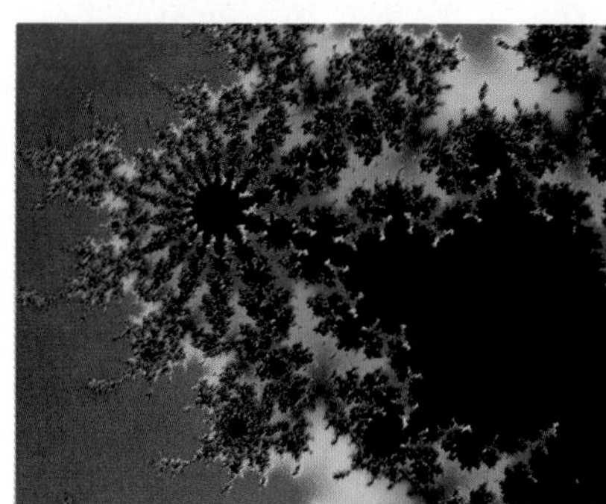

▶ **Ex. 45–56** Encourage students to follow the solution model provided by Example 3 and give their answers in $a + bi$ form.

▶ **Ex. 63–68** Students can use Example 4 as a solution model.

Enrichment Activity

PROJECT Students have an opportunity to learn about a new branch of mathematics, the development of fractal geometry.

EXTEND Ex. 69–72 Write a two-page report on fractals. You can use resources such as the library, your school's computer science or geometry teacher, or other reference books. In this report, include the history of fractals, some examples of fractals, and the connection of fractals to the Mandelbrot set.

Benoit Mandelbrot established the set in 1980 while he was conducting research for IBM.

Some examples of fractals occurring in nature include oak tree development, ferns, broccoli, and clouds.

Fractals can also be used to describe shapes with self-similarity such as the Sierpinski Triangle. (See the Enrichment Activity that follows in Lesson 5.6.)

▶ **Ex. 79–84** You can use this review of real-number solutions of quadratic equations as preparation for Lesson 5.6.

▶ **Ex. 85–88** The powers of i can be thought of as a mod 4 system. (See the Lesson Notes for Example 3.) Since $1001 = 1000 + 1 = 4(250) + 1$; $i^{1001} = i^1 = i$.

Mixed Review Answers

12. $a = \pm\sqrt{\dfrac{b+2}{bc-1}}$

13.

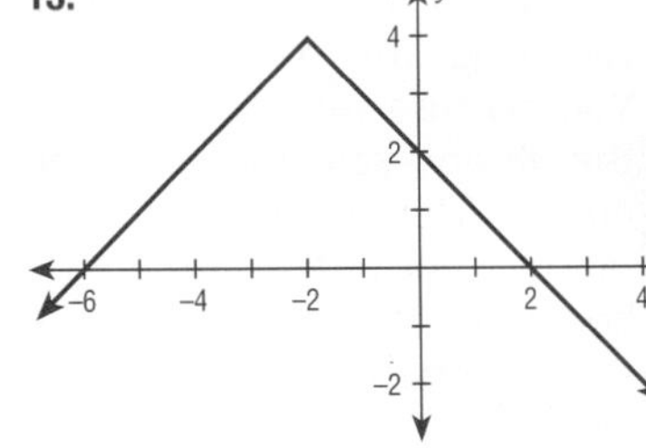

Integrated Review

In Exercises 73–78, simplify the expression.

73. $-3(2x + 3)$ $-6x - 9$

74. $4(-3a + b)$ $-12a + 4b$

75. $\frac{1}{2}s(s - 3)$ $\frac{1}{2}s^2 - \frac{3}{2}s$

76. $-\frac{1}{4}t(2t - 8)$ $-\frac{1}{2}t^2 + 2t$

77. $x(-3x + 1)$ $-3x^2 + x$

78. $-2y(2y + 7)$ $-4y^2 - 14y$

In Exercises 79–84, solve the equation.

79. $x^2 - 2x = 1$ $1 \pm \sqrt{2}$

80. $3x^2 - 5x + 2 = 0$ $\frac{2}{3}, 1$

81. $16x^2 = 4$ $\pm\frac{1}{2}$

82. $3x - 5x^2 = -7$ $\frac{3}{10} \pm \frac{\sqrt{149}}{10}$

83. $30x^2 - 10x - 1 = 0$ $\frac{1}{6} \pm \frac{\sqrt{220}}{60}$

84. $-8x^2 - 2x + 5 = 1$ $-\frac{1}{8} \pm \frac{\sqrt{33}}{8}$

Exploration and Extension

Powers of i **In Exercises 85–88, raise the imaginary unit i to various powers, then find and use a pattern.**

✪ 85. Complete the table.

i	i^2	i^3	i^4	i^5	i^6	i^7	i^8
i	-1	$-i$	?	?	?	?	?

1 i -1 $-i$ 1

✪ 86. Describe the pattern in the second row of the table. 1 and i alternate and alternate pairs are negative.

✪ 87. Use the pattern to evaluate i^{26}. -1

✪ 88. Use the pattern to evaluate i^{52}. 1

Mixed REVIEW

2. $x^2 + 4x + 4$ 5. $10\frac{2}{3}$ square units 7. 5.32×10^6 17. 64 ft 18. ≈ 8.94

1. Solve $7x^2 = 28$. **(5.1)** ± 2

2. Write $(x + 2)^2$ as a trinomial. **(5.3)**

3. Find the y-intercept of $y = 3x^2 + 8$. **(5.2)** $(0, 8)$

4. Find the x-intercept of $y = \frac{2}{3}x - 4$. **(2.3)** 6

5. Find the area of the region between the graphs of $y = x^2 - 4$ and $y = 0$. **(5.2)**

6. Divide: $\frac{4}{9} \div \frac{8}{27}$. $1\frac{1}{2}$

7. Write 5,320,000 in scientific notation.

8. Write 0.000048 in scientific notation. 4.8×10^{-5}

9. Solve $|3x - 2| > 5$. **(1.7)** $x < -1$ or $x > \frac{7}{3}$

10. 12 is 6% of what number? 200

11. Solve $\frac{3}{2}x^2 + 8 = \frac{26}{3}$. **(5.1)** $\pm\frac{2}{3}$

12. Solve for a: $a^2bc - a^2 - b = 2$. **(1.5)**

13. Sketch the graph of $y = -|x + 2| + 4$. **(2.6)** 12, 13. See margin.

14. Find the vertex of $y = (x - 5)^2 - 7$. **(5.2)** $(5, -7)$

15. Write $y = x^2 + 4x - 7$ in completed square form. **(5.3)** $y = (x + 2)^2 - 11$

16. Find the axis of symmetry of $y = 5x^2 - 9x + 4$. **(5.2)** $x = \frac{9}{10}$

17. A stone dropped into a well hits the water after 2 sec. How deep is the well?

18. In order of length, the sides of a right triangle measure 8, x, and 12. Find x. **(5.1)**

19. Multiply: $\begin{bmatrix} 3 & 1 \\ 0 & 2 \end{bmatrix}\begin{bmatrix} 1 & 3 \\ -2 & 4 \end{bmatrix}$. **(4.2)** $\begin{bmatrix} 1 & 13 \\ -4 & 8 \end{bmatrix}$

20. Find the inverse of $\begin{bmatrix} 2 & -4 \\ 1 & -1 \end{bmatrix}$. **(4.3)** $\begin{bmatrix} -\frac{1}{2} & 2 \\ -\frac{1}{2} & 1 \end{bmatrix}$

5.6 Solving Any Quadratic Equation

What you should learn:

Goal 1 How to solve any quadratic equation (even ones that have complex-number solutions)

Goal 2 How to use complex numbers and programmable calculators to solve real-life problems

Why you should learn it:

You can solve *any* quadratic equation by using complex numbers. In real-life settings, these solutions must be interpreted to fit the context.

Goal 1 Solving Any Quadratic Equation

In Lesson 5.4, you learned that if the discriminant of the quadratic equation $ax^2 + bx + c = 0$ is negative, then the equation has no *real-number* solution, but it does have two *complex-number* solutions.

You can find the complex solutions in the same way you find real solutions—by using the quadratic formula. For instance, the solutions of $x^2 + 2x + 2 = 0$ are as follows.

$$x = \frac{-2 \pm \sqrt{2^2 - 4(1)(2)}}{2(1)} \quad \textit{Quadratic formula}$$

$$= \frac{-2 \pm \sqrt{-4}}{2} \quad \textit{Simplify.}$$

$$= \frac{-2 \pm 2i}{2} \quad \textit{Write with imaginary unit.}$$

$$= \frac{2(-1 \pm i)}{2} \quad \textit{Distributive Property}$$

$$= -1 \pm i \quad \textit{Divide common factor.}$$

The two solutions are $-1 + i$ and $-1 - i$. You can check this solution by substituting it into the original equation. For instance, you can check $-1 + i$ as follows.

$$(-1 + i)^2 + 2(-1 + i) + 2 = 1 - 2i - 1 - 2 + 2i + 2 = 0$$

Example 1 Finding Complex-Number Solutions

Solve $x^2 - 3x + \frac{5}{2} = 0$.

Solution

$$x = \frac{3 \pm \sqrt{(-3)^2 - 4(1)\left(\frac{5}{2}\right)}}{2(1)} \quad \textit{Quadratic Formula}$$

$$= \frac{3 \pm \sqrt{-1}}{2} \quad \textit{Simplify.}$$

$$= \frac{3 \pm i}{2} \quad \textit{Write with imaginary unit.}$$

$$= \frac{3}{2} \pm \frac{1}{2}i \quad \textit{Write in standard form.}$$

The two solutions are $\frac{3}{2} + \frac{1}{2}i$ and $\frac{3}{2} - \frac{1}{2}i$ ■

Problem of the Day

Each face of a number of identical wooden cubes is to be painted red or blue. How many cubes can be produced that are distinguishable? 10

ORGANIZER

Warm-Up Exercises

1. Write the quadratic formula. $x = \frac{-b \pm \sqrt{b^2 - 4ac}}{2a}$
2. Compute the following.
   a. $\sqrt{36 - 4(4)(5)}$
   b. $\sqrt{49 - 4(-3)(-6)}$
   c. $\sqrt{8^2 - 4(12)(-8)}$
   d. $\sqrt{4^2 - 4(2)(5)}$
   a. $i\sqrt{44} = 2i\sqrt{11}$
   b. $i\sqrt{23}$
   c. $\sqrt{448} = 8\sqrt{7}$
   d. $i\sqrt{24} = 2i\sqrt{6}$

Lesson Resources

Teaching Tools
- Problem of the Day: 5.6
- Warm-up Exercises: 5.6
- Answer Masters: 5.6

Extra Practice: 5.6

LESSON Notes

Example 1

The graph of the quadratic equation has no (real-valued) x-intercepts because the solutions are complex-valued. Ask students to verify this by actually graphing the quadratic equation.

Example 2

Point out that if the solutions to the equation will be used in other algebraic expressions, exact (real-valued) solutions may be preferred.

Example 3

Ask students to determine the maximum amount of sales from 1970 through 1990.

about 85.3 million dollars

Extra Examples

Here are additional examples similar to **Examples 1–2.**

1. Find the exact solutions of the following.
   a. $x^2 - 5x + 8 = 0$
      $\frac{5}{2} \pm \frac{i\sqrt{7}}{2}$
   b. $2x^2 = 6x - 12$
      $\frac{3}{2} \pm \frac{i\sqrt{15}}{2}$
   c. $3x^2 + 8x - 14 = 0$
      $\frac{-4}{3} \pm \frac{\sqrt{58}}{3}$
2. TECHNOLOGY Use a programmable calculator or computer to solve.
   a. $4x^2 + 8x - 3 = 0$
   b. $5x^2 = 16x - 3$
   c. $8x^2 + 23x + 15 = 0$

   a. $-2.323, 0.323$; b. $0.2, 3$; c. $-1.875, -1$

Check Understanding

1. What is the solution of a quadratic equation? Replacement values for the variable that result in true statements.
2. Can the quadratic formula always be used to solve quadratic equations? Yes
3. How are complex-number solutions graphed on the real-valued coordinate plane? They cannot be graphed on the real-valued coordinate plane.

Goal 2 Using Technology to Solve Equations

You can program a graphing calculator to find the real solutions of a quadratic equation. The keystroke instructions for doing this on a *TI-82,* Casio *fx-9700GE,* and Sharp *EL-9300C* are listed on page 861.

LESSON INVESTIGATION

Investigating Real Solutions of Quadratics

Partner Activity Enter one of the programs listed on page 861 in a graphing calculator. The program is written to solve any quadratic equation of the form

$ax^2 + bx + c = 0.$ *Quadratic equation*

After entering the program, use it to find the real solutions of the following equations. In each case, give a graphical interpretation of the number of real solutions.

$x^2 + 6x + 3 = 0,\ x^2 + 6x + 9 = 0,\ x^2 + 6x + 15 = 0$

≈ -5.45, ≈ -0.55; -3; no real solution.
The number of real solutions is the number of x-intercepts of the graph.

Connections
Technology

Example 2 *Finding Real Solutions*

Use one of the programs listed on page 861 to solve the equations.

a. $3x^2 - 5x + 1 = 0$ **b.** $3x^2 - 5x + 2 = 0$ **c.** $3x^2 - 5x + 3 = 0$

Solution

a. Running the program with $A = 3$, $B = -5$, and $C = 1$ produces two real solutions that are approximately 1.434 and 0.232. Using the quadratic formula, the *exact* solutions are $\frac{5}{6} \pm \frac{\sqrt{13}}{6}$.

b. Running the program with $A = 3$, $B = -5$, and $C = 2$ produces two real solutions that are 1 and approximately 0.667. The exact solutions are 1 and $\frac{2}{3}$.

c. Running the program with $A = 3$, $B = -5$, and $C = 3$ produces the display "No Real Solution." In this case the solutions are not real numbers. Using the quadratic formula, you can find the solutions to be $\frac{5}{6} \pm \frac{\sqrt{11}}{6}i$. ■

During the 1960's and 1970's, most homes in the United States had television antennas. With the introduction of cable television, however, fewer homes needed a television antenna.

Real Life
Television

Example 3 *Interpreting Complex Solutions*

For 1970 through 1990, the sales, A (in millions of dollars), of television and FM antennas in the United States is given by

$$A = -0.2t^2 + 3.5t + 70$$

where t represents the year, with $t = 0$ corresponding to 1970. According to this model, during which year did the sales of antennas reach \$95 million?

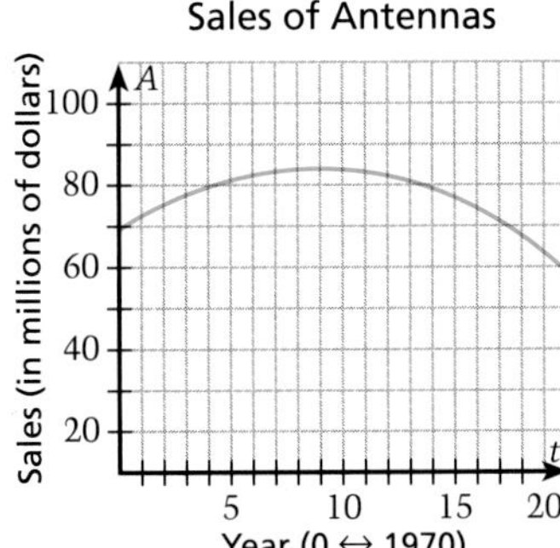

Solution

$$A = -0.2t^2 + 3.5t + 70$$

$$95 = -0.2t^2 + 3.5t + 70$$

$$0.2t^2 - 3.5t + 25 = 0$$

$$t = \frac{3.5 \pm \sqrt{(3.5)^2 - 4(0.2)(25)}}{2(0.2)}$$

$$t = \frac{3.5 \pm \sqrt{-7.75}}{0.4}$$

Because the discriminant is negative, the equation has no real solutions. In no year did antenna sales reach \$95 million. This conclusion is supported by the graph at the left. Note that the graph of $A = -0.2t^2 + 3.5t + 70$ is a parabola that opens down. The maximum sales occurred around 1979. ■

Communicating about ALGEBRA

SHARING IDEAS about the Lesson

Find an equation of a parabola with the given description.

A. The parabola opens down and has no x-intercepts.

B. The parabola opens up and has no x-intercepts.

Answers will vary; samples are given. **A.** $y = -x^2 + x - 1$ **B.** $y = x^2 + x + 1$

Communicating about ALGEBRA

EXTEND *Communicating*

COOPERATIVE LEARNING

Work with a partner. Indicate how to modify the calcualtor program used in Example 2 so that the calculator will also find complex solutions. Test your program on a graphing calculator.

A sample answer for the TI-82 follows. Students need to realize that if $D < 0$, then the solutions are conjugate pairs of the form $a + bi$ and $a - bi$. The calculator program can be adapted to give the real values of a and b.

At Lbl 1, input the following:

```
: – B/(2A)→W
:√ (–D)/(2A)→Z
:Disp "SOLUTIONS ARE"
:Disp "W + ZI AND"
:Disp "W–ZI WHERE"
:Disp "W = "
:Disp W
:Disp "Z = "
:Disp Z
```

EXERCISE Notes

ASSIGNMENT GUIDE

Basic/Average: Ex. 1–4, 9–19 odd, 23–27 odd, 29–33 odd, 34, 39, 44, 45, 49

Above Average: Ex. 1–4, 9–19 odd, 23–27 odd, 29–34, 39, 41, 47, 51

Advanced: Ex. 1–4, 9–19 odd, 23–27 odd, 29–34, 37, 43, 49, 51

Selected Answers
Exercises 1–4, 5–49 odd

More Difficult Exercises
Exercises 29–34, 51

Guided Practice

Ex. 1 Have students actually graph the three equations simultaneously. Graph **a** is above the x-axis, graph **b** is on the x-axis, and graph **c** intersects the x-axis twice. Since only the constant term changes, the graphs are all vertical shifts of each other.

Ex. 4 This exercise is similar to Example 3 on page 267. If $S = 64$ there is no real solution. Graphically, the parabola $S = -\frac{3}{4}t^2 + 6t + 51$ never reaches a value of 64.

Independent Practice

EXTEND Ex. 5–22 Discuss with students the fact that complex solutions always exist in conjugate pairs—that is, if $a + bi$ is a solution, then $a - bi$ is also a solution.

EXERCISES

Guided Practice

CRITICAL THINKING about the Lesson

1. One of the following equations has two real solutions, one has one real solution, and one has no real solutions. Which is which? **c, b, a**
   a. $4x^2 - 8x + 5 = 0$ b. $4x^2 - 8x + 4 = 0$ c. $4x^2 - 8x + 3 = 0$
2. Write $\frac{4 \pm \sqrt{-2}}{6}$ in standard complex-number form. $\frac{2}{3} \pm \frac{\sqrt{2}}{6}i$
3. Solve $2x^2 + 6x + 5 = 0$. $-\frac{3}{2} \pm \frac{1}{2}i$
4. *New Style of Jeans* The monthly sales, S (in thousands of pairs), of a new style of jeans can be modeled by

   $S = -\frac{3}{4}t^2 + 6t + 51$

   where $t = 0$ corresponds to the first month of sales. In which month were 64 thousand pairs of jeans sold? Explain your answer algebraically and graphically. See below.

Sales of Jeans

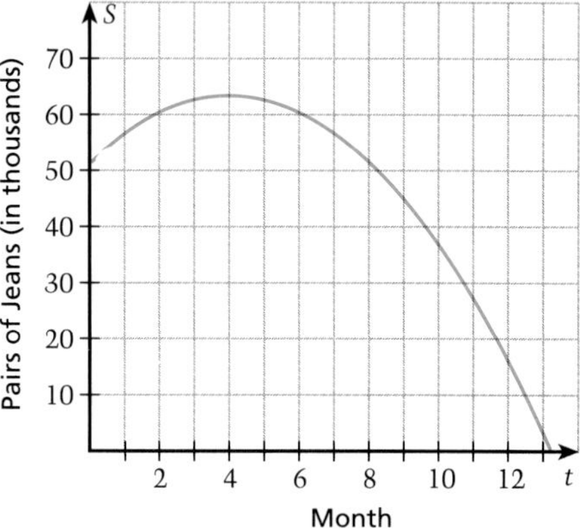

Independent Practice

In Exercises 5–10, decide whether the complex number is a solution of the equation.

5. $x^2 - 2x + 2 = 0, 1 + i$ It is.
6. $x^2 - 4x + 5 = 0, 2 + i$ It is.
7. $x^2 - 2x + 5 = 0, 1 + 2i$ It is.
8. $x^2 - 2x + 5 = 0, 1 - 2i$ It is.
9. $x^2 - 2x + 10 = 0, 1 + 3i$ It is.
10. $x^2 - 2x + 10 = 0, 1 - 3i$ It is.

In Exercises 11–22, solve the equation. See Additional Answers.

11. $-x^2 + 5x - 9 = 0$
12. $6x^2 - 7x + 3 = 0$
13. $x^2 + 2x + 7 = 0$
14. $-5x^2 + 10x - 6 = 0$
15. $-4x^2 + 6x - 3 = 0$
16. $x^2 - x + 1 = 0$
17. $-3x^2 - 5x - 3 = 0$
18. $2x^2 + 12x + 19 = 0$
19. $16x - x^2 = -4 - 22x^2$
20. $x + 2x^2 + 1 = -1 - x$
21. $13x^2 - 5 = x + 10x^2 - 8$
22. $7x^2 - 4 - x = 6x^2 - 10$

Technology **In Exercises 23–28, use a programmable calculator to find all real solutions of the equation.**

23. $16x^2 - 10x + 1 = 0$ 0.5, 0.125
24. $3x^2 + 7x - 2 = 0$ ≈0.2657, −2.591
25. $4x^2 + 5x + 11 = 0$ No real solutions.
26. $-9x^2 + 8x - 1 = 0$ ≈0.150, 0.738
27. $8x^2 - 6x + 6 = 0$ No real solutions.
28. $-14x^2 - 12x - 4 = 0$ No real solutions.

4. No month, the discriminant is negative and (4, 63) is the highest point on the graph.

29. *Fencing the Yard* A friend's family has built a fence around three sides of their property. (See figure.) In total, they used 550 feet of fencing. By their calculations, the lot is one acre (43,560 square feet). Is this correct? Explain your answer. **No**

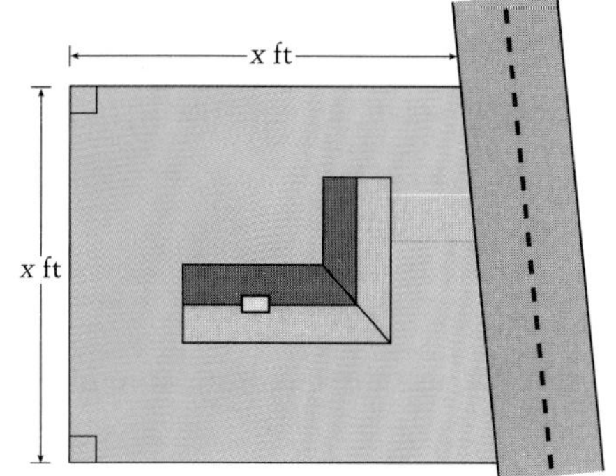

30. *More Fencing* You have 100 feet of fencing. Do you have enough fencing to enclose a rectangular region whose area is 630 square feet? How about a circular region whose area is 630 square feet? Explain. **No, yes**

31. *High School Enrollment* Randall Curtain High School's enrollment, E, decreased from 1955 to 1975. In 1975, the enrollment began to increase. For 1955 through 1995, the enrollment can be modeled by $E = 0.37t^2 - 14.8t + 460$ where $t = 5$ corresponds to 1955. In what year was enrollment down to 300 students? Explain your answer algebraically and graphically.

29.–33. See Additional Answers.

32. *Grocery Sales* In 1975, your uncle opened a neighborhood grocery store. For 1975 through 1995, his annual sales, S (in millions of dollars), could be modeled by $S = -0.015t^2 + 0.36t + 0.5$ where $t = 5$ represents 1975. During which years were the annual sales $2,300,000? During which year were the annual sales $2,750,000? Explain your answers algebraically and graphically.

33. *United States Postal Service Revenue* For 1980 through 1994, the postal service's revenue, R (in millions of dollars), for fourth-class mail can be approximated by the model

$$R = 5.16t^2 - 36.68t + 822.51$$

where $t = 0$ represents 1980. Construct a table that shows the revenue for each year. From the table, did the annual revenue drop as low as $750 million from 1980 through 1994? What happens when you try to solve the equation $750 = 5.16t^2 - 36.68t + 822.51$?

(Source: U.S. Postal Service)

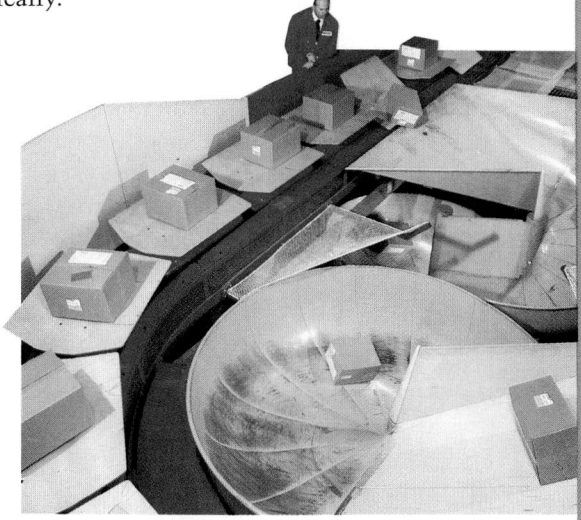

34. *Sale of Electricity* For 1970 through 1992, the sale of electricity, S (in billions of dollars), in the United States can be modeled by the equation

$$S = -0.22t^2 + 9.38t + 105.21$$

where $t = 0$ represents 1980. In what year did electricity sales reach $150 billion? ***(Source: U.S. Energy Information Administration)*** **1985**

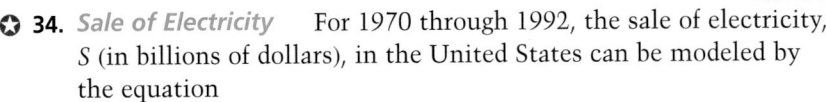

▶ **Ex. 23–28** Students should use the programs provided on page 861 (and the adaptation shown on page 267) to complete these exercises. If students do not have access to a graphing calculator or a computer, omit these exercises from the assignment guides for all levels.

EXTEND **Ex. 33–34** COOPERATIVE LEARNING You may wish to have groups of students solve these real-life applications in more than one way. One group could solve the problems algebraically, while another group could use a graphing calculator and solve them graphically. Students could discuss advantages and disadvantages of each method.

▶ **Ex. 51** Test each value of d either by graphing or by using the quadratic-formula program.

Enrichment Activities

This activity requires students to go beyond lesson goals.

The TI-82 program on pages 269 and 270 will generate a Sierpinski gasket, the result of self-similarity transformations of random points that take place in the plane. The more chaotic and random the transformations appear, the more ordered their results.

Work with a partner. Run the program. Describe the results. (Students with Casio calculators will have to make slight adjustments.)

```
PROGRAM: SRPINSKI
:ClrDraw
:1→Xmin
:3→Xmax
:1→Ymin
:3→Ymax
:2rand + 1 → X
:2rand + 1 → Y
:PT-On (X,Y)
:Lbl 1
:iPart (3rand) + 1 → A
: If A = 1
```

(continued)

(continued)

```
:Goto 2
:If A = 2
:Goto 3
:If A = 3
:Goto 4
:Lbl 5
:PT-On (X, Y)
:Goto 1
:Lbl 2
:(1 + X)/2 → X
:(3 + Y)/2 → Y
:Goto 5
:Lbl 3
:(3 + X)/2 → X
:(3 + Y)/2 → Y
:Goto 5
:Lbl 4
:(3 + X)/2 → X
:(1 + Y)/2 → Y
:Goto 5
```

Run the program.
To Stop the program, press ON while running.

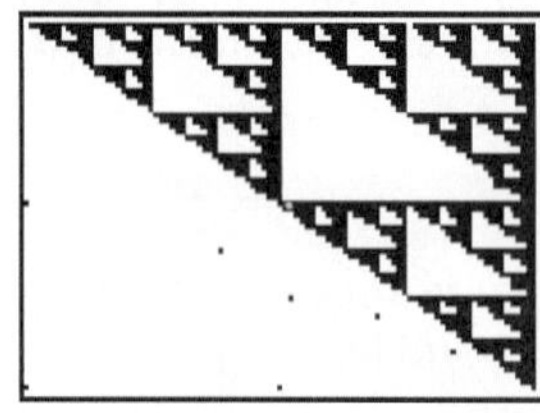

The program produces a series of nested triangles of varying sizes.

Integrated Review

In Exercises 35–40, solve the equation.

35. $-10x^2 + 16 = 30$ $\pm i\sqrt{\frac{7}{5}}$

36. $5 - x^2 = 10 + 4x^2$ $\pm i$

37. $2x^2 + 9 = -2x^2$ $\pm \frac{3}{2}i$

38. $7x^2 + 56 = 3x^2 - 8$ $\pm 4i$

39. $x^2 - 17 = 13 - x^2$ $\pm\sqrt{15}$

40. $-3x^2 - 7 = -3$ $\pm \frac{2}{\sqrt{3}}i$

In Exercises 41–44, use the quadratic formula to solve the equation.

41. $x^2 - 3x + x = 0$ $0, 2$

42. $-2x^2 + 12x - x = 0$ $0, 5\frac{1}{2}$

43. $-3x^2 + x + 10 = 0$ $2, -1\frac{2}{3}$

44. $15x - 6x^2 = 12x - 14x^2 - 2$ $-\frac{3}{16} \pm \frac{\sqrt{55}}{16}i$

In Exercises 45–50, sketch the graph of the equation. See Additional Answers.

45. $y = 2x^2 + x + 5$

46. $y = x^2 - 3x + 1$

47. $y = 6x^2 - x + 2$

48. $y = 7x^2 + x - 2$

49. $y = x^2 + 5x - 4$

50. $y = -8x^2 + 10x - 5$

Exploration and Extension

51. *Morse Code* For some values of d, the equation

$$2x^2 - 16.8x + 37.64 = d$$

has at least one real solution. For other values of d, the equation has no real solution. Complete the following tables by placing a dot (•) if the equation has at least one real solution and a dash (▬) if the equation has no real solution. After completing the tables, use the International Morse Code at the right to find the five-letter word that is being spelled.

d	4.2	10.1	13.5	6.3
Code	?	?	?	?

••••

d	6.0	5.0	0.0
Code	?	?	?

••▬

d	11.25	−1.0	12.0
Code	?	?	?

•▬•

d	7.2	2.2	8.0
Code	?	?	?

•▬•

d	1.6	10.5	−7.2	2.0
Code	?	?	?	?

▬•▬▬

51. ••••, ••▬, •▬•, •▬•, ▬•▬▬ HURRY

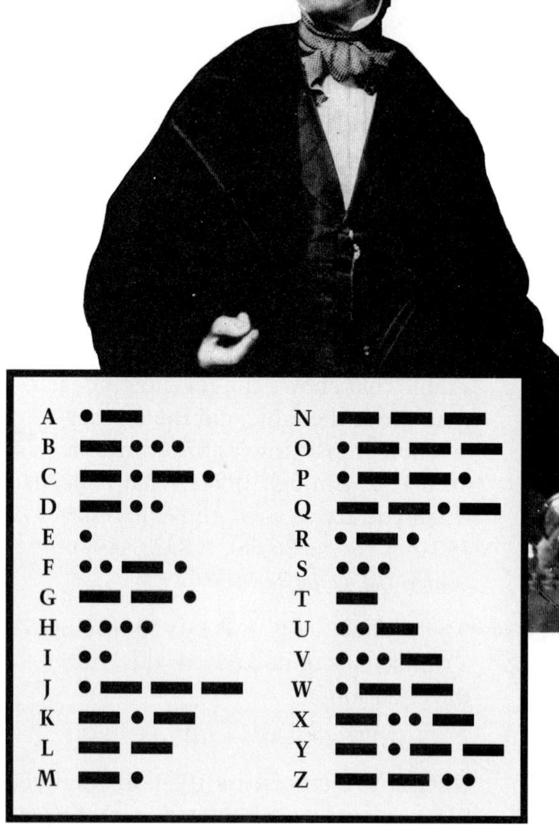

5.7 Graphs of Quadratic Inequalities

What you should learn:

Goal 1 How to sketch the graph of a quadratic inequality

Goal 2 How to use quadratic inequalities in real-life modeling

Why you should learn it:

You can model many real-life situations with quadratic equations, such as some of the unusual shapes found in corporate logos.

Goal 1 The Graph of a Quadratic Inequality

In this lesson, you will study the following four types of **quadratic inequalities.**

$y < ax^2 + bx + c \qquad y \le ax^2 + bx + c$

$y > ax^2 + bx + c \qquad y \ge ax^2 + bx + c$

The **graph** of any such inequality consists of the graph of all solutions (x, y) of the inequality. The steps used to sketch the graph of a quadratic inequality are similar to those used to sketch the graph of a linear inequality. (See Lesson 2.5.)

Sketching the Graph of a Quadratic Inequality

1. Sketch the graph of the parabola $y = ax^2 + bx + c$. (Use a *dashed* parabola for inequalities with $<$ or $>$ and a *solid* parabola for inequalities with $\le$ or $\ge$.)
2. Test a point inside the **U**-shape and one outside the **U**-shape.
3. Only one of the test points will be a solution. Shade the region that contains this test point.

Example 1 *Sketching the Graph of a Quadratic Inequality*

Sketch the graph of $y < 3x^2 - 6x$.

Solution The parabola $y = 3x^2 - 6x$ has a vertex at $(1, -3)$. Its dashed graph is shown at the left. Choose a point inside the **U**-shape, say $(1, 0)$, and a point outside the **U**-shape, say $(3, 0)$.

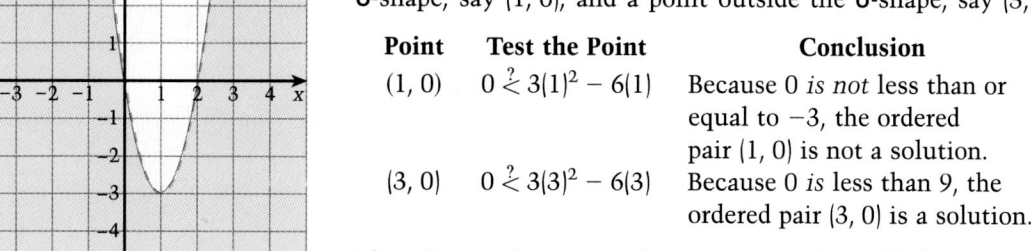

Point	Test the Point	Conclusion
$(1, 0)$	$0 \stackrel{?}{<} 3(1)^2 - 6(1)$	Because 0 *is not* less than or equal to -3, the ordered pair $(1, 0)$ is not a solution.
$(3, 0)$	$0 \stackrel{?}{<} 3(3)^2 - 6(3)$	Because 0 *is* less than 9, the ordered pair $(3, 0)$ is a solution.

Thus the graph must be the region outside the **U**-shape. ■

Problem of the Day

Solve. $|x - 1| \le |x - 2|$

$x \le \frac{3}{2}$

ORGANIZER

Warm-Up Exercises

1. Sketch each graph.
   a. $y = 3x^2 - 8x$
   b. $y = -4x^2 + 4x + 12$
   a. The graph is a parabola equivalent to $y = 3x^2$ with vertex $\left(\frac{4}{3}, -\frac{16}{3}\right)$ passing through $(0, 0)$ and $\left(\frac{8}{3}, 0\right)$.
   b. The graph is a parabola equivalent to $y = -4x^2$ with vertex $\left(\frac{1}{2}, 13\right)$ passing through $(0, 12)$, $(1, 12)$, $(-1, 4)$, and $(2, 4)$.
2. Sketch the graph of the following.
   a. $y \le 4x - 8$
   b. $y > 12 - 3x$
   a. The graph is the region below the solid line $y = 4x - 8$.
   b. The graph is the region above the dashed line $y = 12 - 3x$.

Lesson Resources

Teaching Tools
- Transparencies: 7, 8, 13
- Problem of the Day: 5.7
- Warm-up Exercises: 5.7
- Answer Masters: 5.7

Extra Practice: 5.7

Color Transparency: 36

LESSON Notes

Example 1

Observe that other test points provide the same information. For example, test the points (0, 2) (inside the parabola) and (−2, 0) (outside the parabola).

(continued)

(continued)

Point (0, 2)

Test

$2 < 3(0)^2 - 6(0)$

$2 < 0$ False

So, (0, 2) is not a solution.

Point (–2, 0)

Test

$0 < 3(-2)^2 - 6(-2)$

$0 < 24$ True

So, (–2, 0) is a solution.

Example 2

The intersections, points (–1, 1) and (2, 4), of the two inequalities may be easily found by setting the inequalities equal to each other.

$x^2 = -x^2 + 2x + 4$

$2x^2 - 2x - 4 = 0$

$x^2 - x - 2 = 0$

$(x - 2)(x + 1) = 0.$

Solving for x gives $x = 2$ and $x = -1$. By evaluating either quadratic equation, say $y = x^2$, you can determine the corresponding y-coordinates.

EXTEND Example 3 What is the take-home pay of someone with four deductions and a gross monthly income of (a) \$2500, or (b) \$4500? (c) What can you conclude if someone with a monthly gross income of \$3500 has a take-home pay of \$2950?

a. \$1968.75, **b.** \$3318.75,

c. You can conclude that the person claims more than four deductions.

Extra Examples

Here are additional examples similar to **Examples 1–2.**

1. Sketch each graph.
   - **a.** $y \le 3x^2 + 5x$
   - **b.** $y > x^2 + 6x - 4$
   - **a.** The graph is the region outside the parabola and including the parabola with vertex $\left(-\frac{5}{6}, -\frac{25}{12}\right)$.
   - **b.** The graph is the region inside the parabola, not including the parabola, with vertex (–3, –13) and intercepts (–6.606, 0) and (0.606, 0).

272

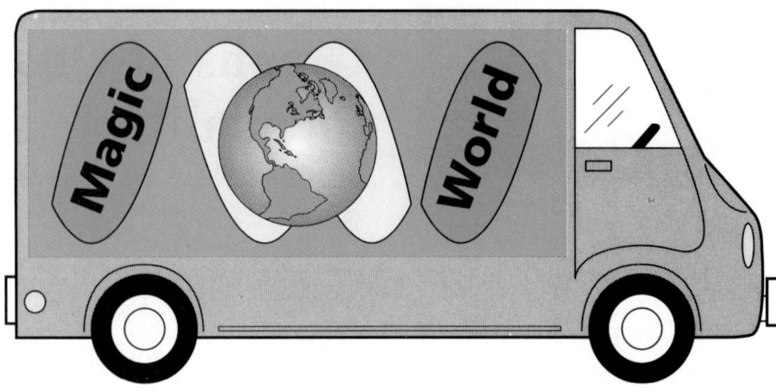

***Many organizations have a logo (short for logotype) that is designed to be easily recognized. For example, the logo for the company that publishes this book is shown at the bottom of the spine on the outside cover of the book. The name of the company is* D. C. Heath.**

Goal 2 Using Quadratic Inequalities in Real Life

Real Life **Design**

Example 2 *Sketching the Graph of an Inequality*

You are using a computer to create a logo for a company called Magic World. Each of the four green and yellow regions can be described as the *intersection* of the graphs of two quadratic inequalities. Create one of these by sketching the graphs of the following inequalities.

$$y \ge x^2 \quad \text{and} \quad y \le -x^2 + 2x + 4$$

Solution The parabola $y = x^2$ opens up and has its vertex at the origin. The graph of the inequality

$$y \ge x^2$$

is the region on and inside the parabola. The parabola $y = -x^2 + 2x + 4$ opens down and has its vertex at (1, 5). The graph of the inequality

$$y \le -x^2 + 2x + 4$$

is the region on and inside the parabola.

You can see from the coordinate plane at the left that the intersection of these two graphs is a region that has the shape of the green regions in the logo for Magic World. ■

Sketch the graphs of the following inequalities. Does the intersection of these two graphs look like one of the green regions in the logo or does it look like one of the yellow regions in the logo?

$$y \ge x^2 + 4x + 4 \quad \text{and} \quad y \le -x^2 - 6x + 4$$

Real Life
Salary and Wages

Example 3 ***Take-Home Pay***

In 1990, the monthly take-home pay, y (in dollars), for a person claiming four *income-tax deductions* can be modeled by

$$y = -0.000025x^2 + 0.85x$$

where x represents the monthly gross pay (in dollars). Describe the deductions claimed by a person whose take-home pay is given by one of the following inequalities.
(Source: Hooper International Accounting Software, based on a state and local income-tax rate of 3.1%)

$$y > -0.000025x^2 + 0.85x$$

$$y < -0.000025x^2 + 0.85x$$

Solution The graphs of the two inequalities are shown below. For the graph at the left, the monthly take-home pay is greater than that of a person who claims four deductions, so the person must be claiming *more than four deductions.* For the graph at the right, the monthly take-home pay is less than that of a person who claims four deductions, so the person must be claiming *less than four deductions.* ■

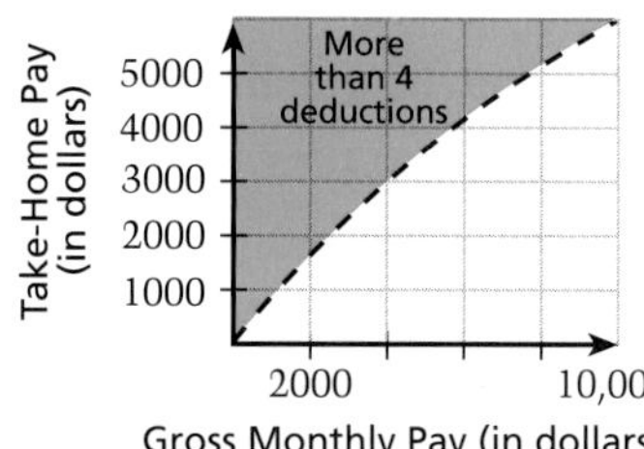

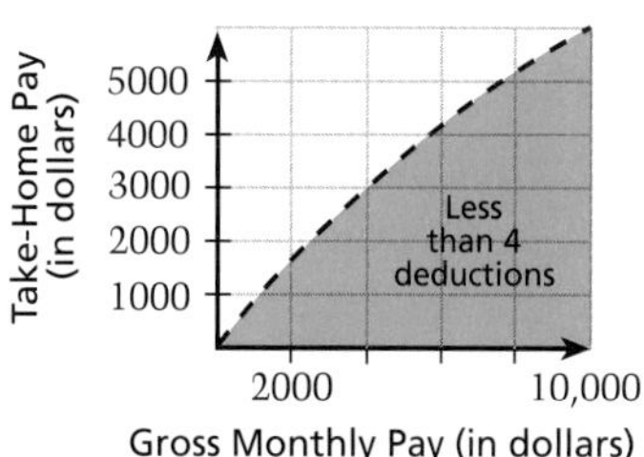

Communicating about ALGEBRA

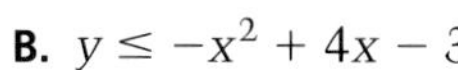

SHARING IDEAS about the Lesson

Match each inequality with its graph. Explain your choices.

A. $y \leq -x^2 + 2x$

B. $y \leq -x^2 + 4x - 3$

C. $y \leq -x^2 + 6x - 8$

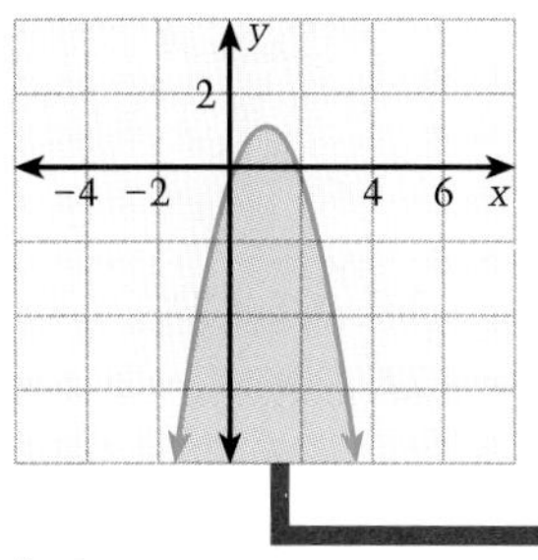

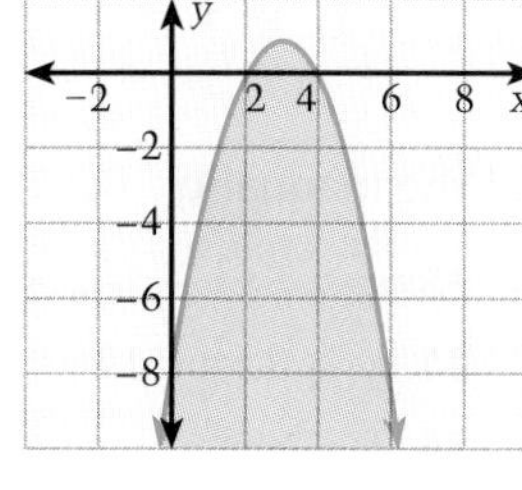

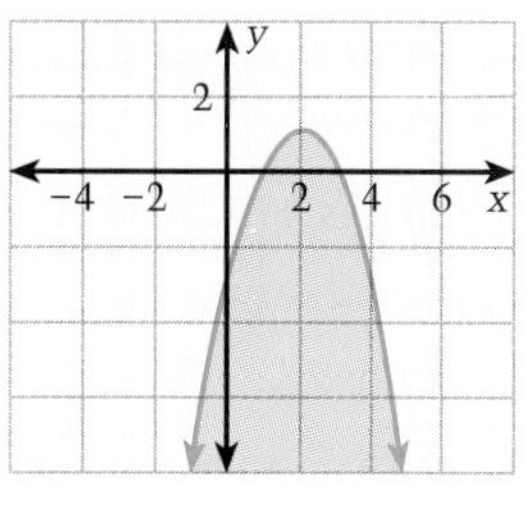

A. I **B. III** **C. II**

2. Sketch the intersection of the graphs of the quadratic inequalities.

$y < -3x^2 + 9x - 3$

$y > 2x^2 - 8x + 4$

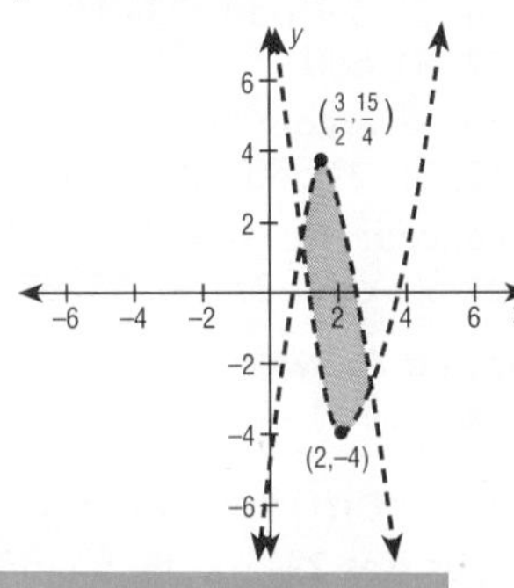

Check Understanding

1. What is a solution of a quadratic inequality?
Replacement sets of ordered pairs (x, y) that result in true statements when substituted for the variables in the inequalities.

2. Describe the graph of a quadratic inequality.
See the box on page 271.

Communicating about ALGEBRA

Have students discuss the similarities between graphing linear inequalities and graphing quadratic inequalities. The procedures are identical. First you graph the boundaries (lines or parabolas) using a dashed boundary (line or parabola) for $<$ and $>$ and a solid boundary (line or parabola) for $\leq$ and $\geq$. The boundaries divide the coordinate plane into regions. Test points in each region. Shade the regions that contain the solutions.

EXERCISES

Guided Practice

CRITICAL THINKING about the Lesson

In Exercises 1–4, use the inequality $y \le 4x^2 - 3x + 1$.

1. What are the coordinates of the lowest point in the graph? $\left(\frac{3}{8}, \frac{7}{16}\right)$
2. Sketch the graph of the parabola $y = 4x^2 - 3x + 1$. **2., 4.** See margin.
3. Test two points in the coordinate plane—one inside the parabola and one outside. Test points will vary.
4. Use the result of your test in Exercise 3 to shade the graph of the inequality.

Independent Practice

In Exercises 5–10, determine whether the ordered pair is a solution of the inequality.

5. $y > -x^2 + x + 4$, $(1, 6)$ It is.
6. $y \ge -x^2 - 5x + 3$, $(2, -10)$ It is.
7. $y \le 3x^2 - x - 1$, $(-1, 5)$ It is not.
8. $y < 4x^2 - x + 5$, $(0, 5)$ It is not.
9. $y \ge -\frac{1}{2}x^2 + 2x - 4$, $(4, -2)$ It is.
10. $y \le -x^2 + 9x + 7$, $(-1, 3)$ It is not.

In Exercises 11–16, match the inequality with its graph.

11. $y \le -x^2 + 3x + 5$ c
12. $y > -x^2 - 6x - 2$ b
13. $y > -2x^2 + 3x + 5$ f
14. $y \ge 2x^2 + 3x - 5$ a
15. $y \le \frac{1}{2}x^2 + 3x + 2$ d
16. $y > x^2 - 3x - 5$ e

a.
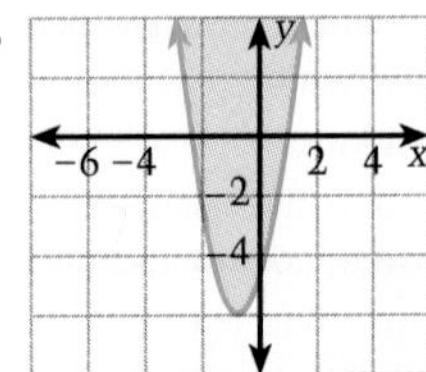

b.
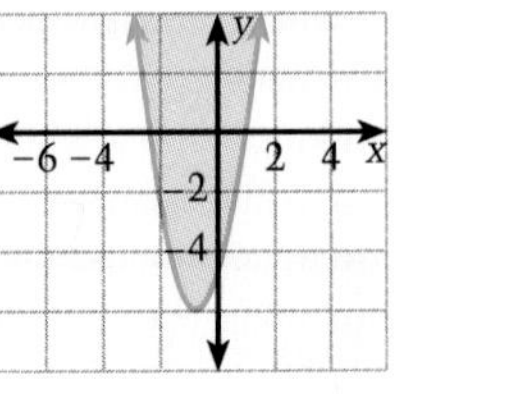

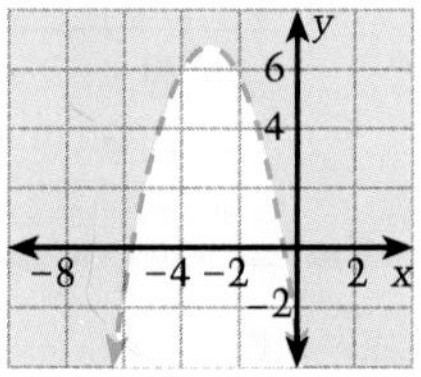

c.

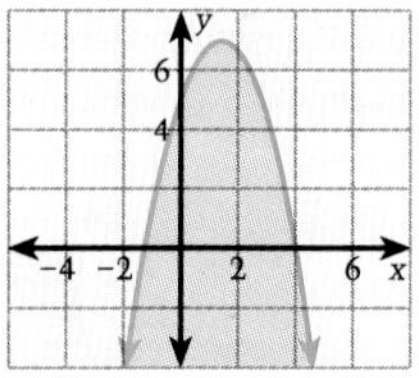

d.
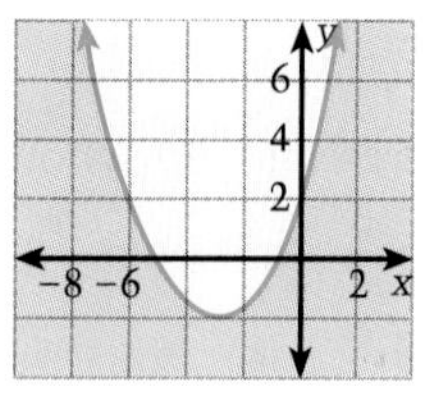

e.
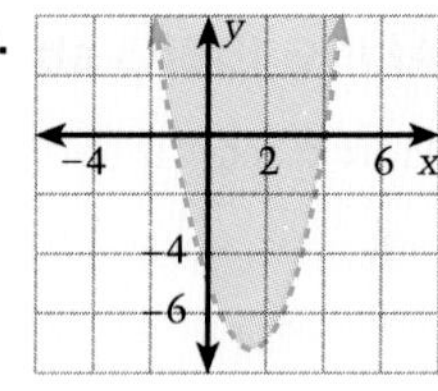

f.
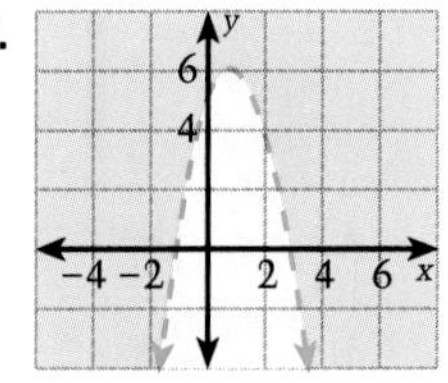

In Exercises 17–28, sketch the graph of the inequality. See Additional Answers.

17. $y > x^2 - x + 10$
18. $y \ge 5x^2 + 10x + 7$
19. $y \le -2x^2 + 4x - 1$
20. $y < 3x^2 - 18x + 2$
21. $y > 3x^2 + 12x$
22. $y \ge x^2 - 20x + 90$
23. $y \ge 6x^2 + 36x + 50$
24. $y < -x^2 + 14x - 7$
25. $y \ge -2x^2 + 4x + 8$
26. $y < -3x^2 + 30x - 50$
27. $y < -x^2 + 6x - 2$
28. $y \le 4x^2 + 16x$

EXERCISE Notes

ASSIGNMENT GUIDE

Basic/Average: Ex. 1–4, 9, 11–16, 23–29 odd, 33–37 odd, 47–51 odd

Above Average: Ex. 1–4, 7, 11–16, 21–29 odd, 33–36, 54–56

Advanced: Ex. 1–4, 9, 11–16, 21–29 odd, 32–36, 55–56

Selected Answers
Exercises 1–4, 5–53, odd

More Difficult Exercises
Exercises 29–34, 38, 55–56

Guided Practice

Ex. 2 Review with students all the available methods of graphing. They can locate the vertex by using the formula $x = -\frac{b}{2a}$, or they can write the quadratic equation in completed-square form to find the vertex, or they can use the Shade feature of a graphics calculator.

Answers
2.–4.
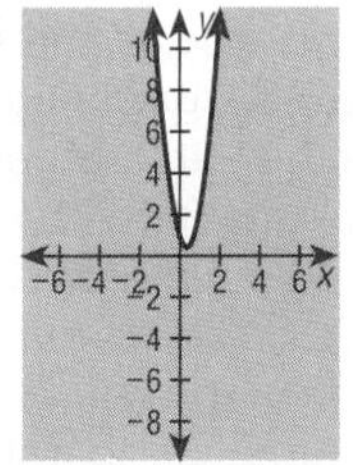

Independent Practice

Ex. 11–16 Assign these exercises as a group. Students can use the x-intercepts as an intuitive check.

In Exercises 29–34, sketch the intersection of the graphs of the inequalities. See Additional Answers.

29. $y > x^2 - 6x + 9$
$y < -x^2 + 6x - 3$

30. $y \geq 3x^2 - 12x + 12$
$y \leq -x^2 + 4x$

31. $y \geq x^2 - 2x + 2$
$y < x^2 - 8x + 20$

32. $y \geq 2x^2$
$y \leq -2x^2 - 8x + 2$

33. $y > x^2 + 2x - 4$
$y > x^2 - 4x - 1$

34. $y < -3x^2 + 6x + 7$
$y \leq -2x^2 + 16x - 22$

35. *The Sleepy Hare* You are using a computer to create a logo for The Sleepy Hare restaurant. The logo you have designed is shown at the right. Sketch the intersections of the graphs of the inequalities. Which region represents the hare's right ear? Which region represents the hare's left ear? **b, a,** see Additional Answers.

a. $y \leq -0.4x^2 + 8x - 30$
$y \geq 0.4x^2 - 12x + 90$

b. $y \leq -0.4x^2 + 8x - 30$
$y \geq 0.4x^2 - 4x + 10$

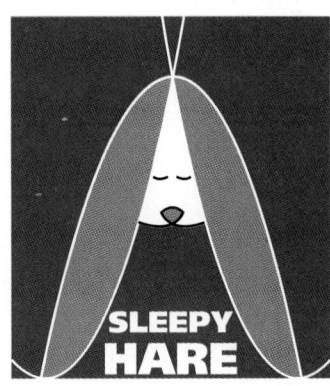

36. *The Fish Bowl* You are using a computer to create a logo for an aquarium supplies store called The Fish Bowl. The logo is the intersection of the graphs of *eight* inequalities.

$y \leq x^2 + 10x + 25.25$	$x \leq 6$
$y \geq -x^2 - 10x - 25.25$	$x \geq -6$
$y \leq x^2 - 10x + 25.25$	$y \leq 0.25x^2$
$y \geq -x^2 + 10x - 25.25$	$y \geq -0.25x^2$

Sketch the graphs of these inequalities and label each of the borders with its equation. What are the coordinates of the upper and lower points of the fins of each fish? See Additional Answers.

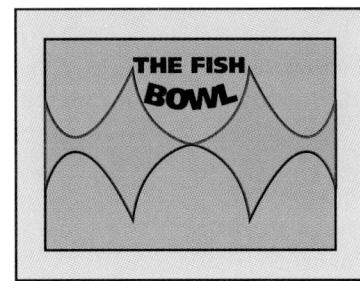

37. *United States Water Use* For 1940 through 1980, the daily water use, W (in billions of gallons), in the United States followed the quadratic model

$$W = 0.063t^2 + 5.69t + 133.24$$

where $t = 0$ represents 1940. Suppose this model had been used to predict the water use for 1980 through 1990. The actual use in 1985 was 400 billion gallons. Was this less or more than the predicted consumption for 1985? Explain. ***(Source: Water Resources Activities in the United States)***
Less; the predicted use was 516.865 billion gallons.

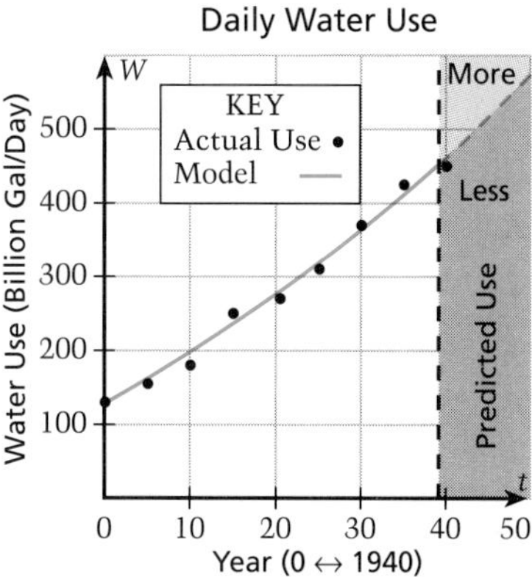

▶ **Ex. 17–34** Students can graph the quadratic equation by locating the vertex and a few points on either side of the vertex.

EXTEND Ex. 17–34 You might wish to show students that the vertex of the equation $y = ax^2 + bx + c$ is the point $\left(-\frac{b}{2a}, c - \frac{b^2}{4a}\right)$.
You already know the x-coordinate for the vertex is $x = -\frac{b}{2a}$.
Substitute to find the y-coordinate.

$$y = ax^2 + bx + c$$
$$y = a\left(-\frac{b}{2a}\right)^2 + b\left(-\frac{b}{2a}\right) + c$$
$$= \frac{b^2}{4a} - \frac{b^2}{2a} + c$$
$$= \frac{b^2 - 2b^2}{4a} + c$$
$$= -\frac{b^2}{4a} + c$$
$$= c - \frac{b^2}{4a}$$

▶ **Ex. 35–36** TECHNOLOGY Students should use graphics calculators to solve these problems.

EXTEND Ex. 37 Ask students why they think that the 1985 consumption was less than the 1940-through-1980 consumption.

▶ **Ex. 55–56** Remind students that they need to locate the *y*-coordinate of the vertex. See the pattern developed in the Exercise Notes extension for Exercises 17–34.

Enrichment Activity

This activity requires students to go beyond lesson goals.

GROUP ACTIVITY Work in a group. Use linear inequalities, or quadratic inequalities, or combinations of linear and quadratic inequalities to design personalized logos. Personalized logos might be a fancy way of writing your initials, or your school's emblem; or, you can copy a company logo.

38. *Chicago Fire* A coordinate system has been superimposed on the map at the right, with (0, 0) at the fire's origin. Which of the inequalities below, used with $y \geq -\frac{9}{5}x + 9$ and $y \geq 7$, approximates the north side region of Chicago that was burned in the fire? d

a. $y \geq x^2 - 2x + 7$
b. $y \geq -\frac{3}{4}x^2 + 3x + 20$
c. $y \leq x^2 - 2x + 7$
d. $y \leq -\frac{3}{4}x^2 + 3x + 20$

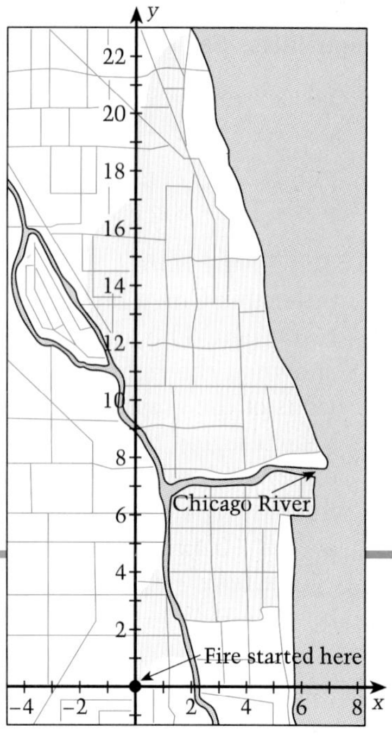

In October 1871, the Great Chicago Fire started on the southwest side of the city, raced north, and crossed the river to the north side. The fire covered the area shaded on the map.

Integrated Review

In Exercises 39–42, sketch the graph of the equation.

39. $y = 2x^2 - 8x + 12$
40. $y = x^2 - 12x + 39$
41. $y = -6x^2 + 1$
42. $y = 5x^2 + 10x + 7$

See Additional Answers.

In Exercises 43–48, decide whether the ordered pair is a solution of the inequality.

43. $y > x + 1$, $(-2, -3)$ It is not.
44. $y > -2x + 4$, $(4, 1)$ It is.
45. $y \leq -x - 3$, $(1, 1)$ It is not.
46. $y < 2x + 3$, $(-3, -1)$ It is not.
47. $y \geq \frac{1}{2}x - 2$, $(1, 3)$ It is.
48. $y \leq -\frac{1}{4}x - 1$, $(5, 0)$ It is not.

In Exercises 49–54, sketch the graph of the inequality. See Additional Answers.

49. $y \leq 10x - 5$
50. $y < \frac{1}{2}x - 8$
51. $y > -x + 3$
52. $y \leq 2x - 5$
53. $y \leq \frac{1}{4}x - 4$
54. $y \geq x + 11$

Exploration and Extension

Geometry **In Exercises 55 and 56, find the area of the intersection of the graphs of the inequalities. (Use the formula $A = \frac{2}{3}bh$.)**

55. $y \geq \frac{1}{9}x^2 + \frac{4}{3}x + 1$ 32
$y \leq 1$

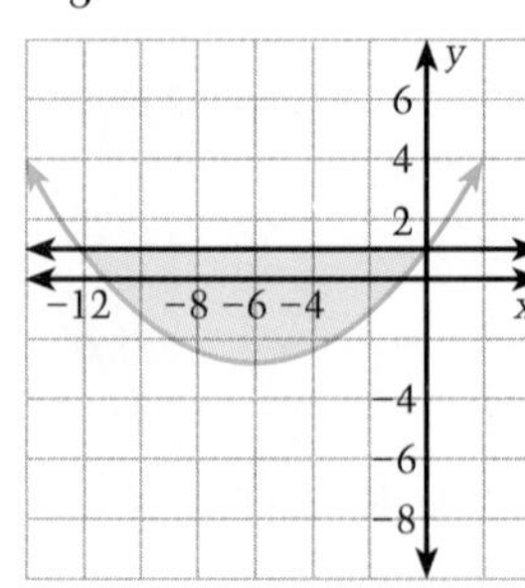

56. $y \leq -\frac{1}{15}x^2 + 2x + 5$ 300
$y \geq 5$

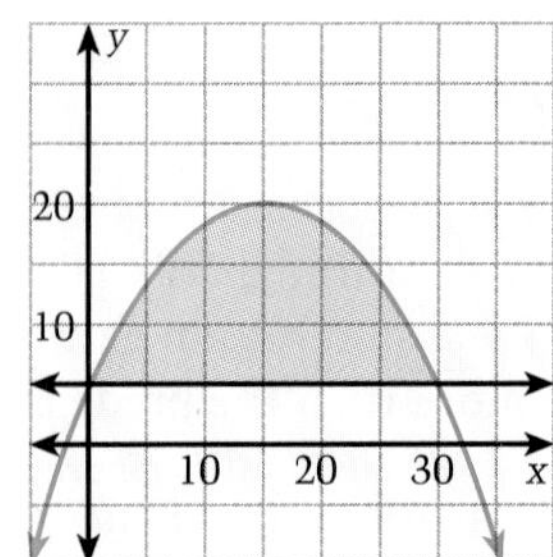

5 Chapter Summary

What did you learn?

Skills

1. Solve a quadratic equation
   - by finding square roots. **(5.1)**
   - by completing the square. **(5.3)**
   - by using the quadratic formula. **(5.4, 5.6)**
2. Check a solution of a quadratic equation
   - algebraically. **(5.1, 5.4, 5.6)**
   - graphically. **(5.3)**
3. Use the discriminant to find the number of solutions of a quadratic equation. **(5.4, 5.6)**
4. Sketch the graph of a quadratic equation in two variables. **(5.2)**
   - Find the vertex of a parabola. **(5.2)**
   - Find the intercepts of a parabola. **(5.2)**
5. Add, subtract, and multiply complex numbers. **(5.5)**
   - Plot complex numbers in the complex plane. **(5.5)**
   - Write complex solutions in standard form. **(5.6)**

Strategies

6. Use quadratic equations to model real-life situations. **(5.1–5.7)**

Why did you learn it?

One of your goals in Algebra 2 is to become familiar with different types of equations and inequalities. You have now studied three types of equations in one and two variables: linear equations and inequalities, absolute value equations and inequalities, and quadratic equations and inequalities.

In this chapter, you studied examples of real-life situations that can be modeled with quadratic equations. For instance, you studied falling-object problems, parabolic-shaped regions, and quantities that are related to time by a quadratic model.

How does it fit into the bigger picture of algebra?

In this chapter, you studied quadratic equations in one variable and in two variables. A quadratic equation in one variable contains the square of the variable. You solved these equations by finding square roots, by completing the square, and by using the Quadratic Formula.

The graph of a quadratic equation in two variables, such as $y = x^2 - 4$, is a parabola. Solving equations such as these often relies on the algebra-geometry connection between equations in one and two variables. For instance, the equation $x^2 - 4 = 0$ has two solutions, which correspond to the two x-intercepts of the graph of $y = x^2 - 4$.

ORGANIZER

TIME SCHEDULE: All levels, two days.

The Chapter Summary helps students organize the main ideas of the chapter. In this chapter, a variety of techniques for solving quadratic equations and inequalities were studied. Many examples of real-life situations that can be modeled by quadratic equations were presented.

Work with students to review the skills, strategies, and concepts of the chapter. The first day's homework assignment can be used as the basis for the second day of review.

COOPERATIVE LEARNING Encourage students to study together. Emphasize the importance of "teaching" a classmate how to perform a skill or how to recall a procedure. When students work together, everyone wins. The students receiving help get additional instruction and the students giving help gain a deeper understanding of the skills and concepts involved.

Chapter SUMMARY

Review the chapter by asking students, "What did you learn? Why did you learn it? How does it fit with the other algebra skills and concepts you have learned?"

Chapter REVIEW

ASSIGNMENT GUIDE

Basic/Average: Ex. 3–54 multiples of 3, 55, 57–58, 60–61

Above Average: Ex. 3–54 multiples of 3, 55, 57–58, 62–64

Advanced: Ex. 3–54 multiples of 3, 56–58, 62–64

More Difficult Exercises
Exercises 55–64

Ex. 1–66 The Chapter Review exercises include a reference to the lesson in which a skill or strategy was first presented. Students should use these references as study aids if they are uncertain about how to do a given exercise in the Chapter Review.

Answers

19. $-\frac{5}{18} \pm \frac{\sqrt{61}}{18}$

20. $-\frac{3}{4} \pm \frac{\sqrt{148}}{8}$

21. $-1, 2$ 22. $-\frac{4}{3} \pm \frac{\sqrt{232}}{12}$

23. $\frac{15}{22} \pm \frac{\sqrt{812}}{44}$

24. $-\frac{10}{33} \pm \frac{\sqrt{2512}}{66}$

29. $-\frac{1}{3} \pm \frac{\sqrt{28}}{6}$

30. No real solution

43. $-\frac{3}{4} \pm \frac{\sqrt{23}}{4}i$

44. $\frac{1}{12} \pm \frac{\sqrt{23}}{12}i$

45. $-\frac{7}{6} \pm \frac{\sqrt{71}}{6}i$

46. $-\frac{5}{12} \pm \frac{\sqrt{23}}{12}i$

47. $-3 \pm \sqrt{11}$ 48. -1

49.

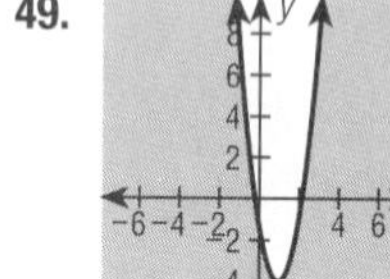

Chapter REVIEW

In Exercises 1–3, solve the equation. (5.1)

1. $\frac{1}{3}x^2 = 12$ ± 6
2. $-x^2 - 7 = -56$ ± 7
3. $3 - x^2 = 5x^2 - 9$ $\pm\sqrt{2}$

In Exercises 4–9, sketch the graph of the equation. (5.2) See Additional Answers.

4. $y = x^2 + 4x - 7$
5. $y = 5x^2 + 20x - 18$
6. $y = 2x^2 + 6x - 1$
7. $y = 4x^2 - 32x + 20$
8. $y = -x^2 + x - 5$
9. $y = x^2 - 3x + 2$

In Exercises 10–15, solve the equation by completing the square. (5.3)

10. $x^2 + 4x - 1 = 0$ $-2 \pm \sqrt{5}$
11. $x^2 + 3x - 2 = 0$ $-\frac{3}{2} \pm \frac{\sqrt{17}}{2}$
12. $2x^2 + 4x + 1 = 3$ $-1 \pm \sqrt{2}$
13. $3x^2 - 6x + 2 = 5$ $1 \pm \sqrt{2}$
14. $3x^2 - x + 7 = 4 - 4x$ 14. $-\frac{1}{2} \pm \frac{\sqrt{3}}{2}i$
15. $4x^2 + 10x + 1 = -3 - 6x$ $-2 \pm \sqrt{3}$

In Exercises 16–18, state the number of real solutions of the equation. (5.4)

16. $5x^2 + x - 4 = 0$ 2
17. $3x^2 + 6x + 3 = 0$ 1
18. $-x^2 + 7x - 2 = 0$ 2

In Exercises 19–24, solve the equation by using the quadratic formula. (5.4) See margin.

19. $1 - 6x^2 = 5x + 3x^2$
20. $4x^2 - 2 = -6x + 5$
21. $3x + 2 - 10x^2 = 2x - 9x^2$
22. $16x + 6x^2 + 5 = 4$
23. $2.2x^2 - 3x + 1 = 0.9$
24. $5.6 - x = 4 + x + 3.3x^2$

In Exercises 25–30, find the real solutions by using the most convenient method. (5.3, 5.4)

25. $1.4x^2 + 0.2x - 1 = 0$
26. $4x^2 + 5 = 9$ ± 1
27. $x^2 - 2x + 1 = 0$ 1
28. $x^2 + 5x - 7 = 0$ $-\frac{5}{2} \pm \frac{\sqrt{53}}{2}$
29. $3x^2 + 2x - 2 = 0$ See margin.
30. $-6x^2 + 2x - 0.3 = 0$ $\pm\frac{1}{3}$

25. $-\frac{1}{14} \pm \frac{\sqrt{564}}{28}$

In Exercises 31–39, perform the indicated operation(s). (5.5)

31. $(4 + i) + (-6 - 3i)$ $-2 - 2i$
32. $(-3 + i) + (11 - 8i)$ $8 - 7i$
33. $(1 + 3i) - (4 - i)$ $-3 + 4i$
34. $(7 + 6i) - (15 + 9i)$ $-8 - 3i$
35. $-5(3 + i) - (1 - 2i)$ $-16 - 3i$
36. $6i(2 + i) + (4 - 2i)$
37. $(2 + 3i)(1 + i)$ $-1 + 5i$
38. $(1 - 5i)(-7 - 10i)$ $-57 + 25i$
39. $(1 + 2i)(3 + i) + i$ $1 + 8i$

36. $-2 + 10i$

In Exercises 40–42, graph the complex number in the complex plane. (5.5) See Additional Answers.

40. $2 - 3i$
41. $-3 - 2i$
42. $1 + 9i$

In Exercises 43–48, solve the equation. (Give complex solutions in standard form.) (5.6) See margin.

43. $2x^2 + 3x + 4 = 0$
44. $6x^2 - x + 1 = 0$
45. $3x^2 + 7x = -10$
46. $-6x^2 - 5x = 2$
47. $5x^2 - 2 = 4x^2 - 6x$
48. $3x^2 = 2x^2 - 2x - 1$

In Exercises 49–54, sketch the graph of the inequality. (5.7) See margin.

49. $y \leq 3x^2 - 6x - 1$
50. $y < 2x^2 - 4x - 4$
51. $y \geq x^2 - 4x + 4$
52. $y \leq \frac{1}{2}x^2 + 4x - 2$
53. $y > 2x^2 - 8x + 11$
54. $y > 5x^2 + \frac{2}{5}x + 4$

55. *Truck Off the Road* Two tow trucks have attached cables to another truck that has gone through a guardrail. Find the length of each tow truck's cable.
≈53.9 ft and 67.1 ft

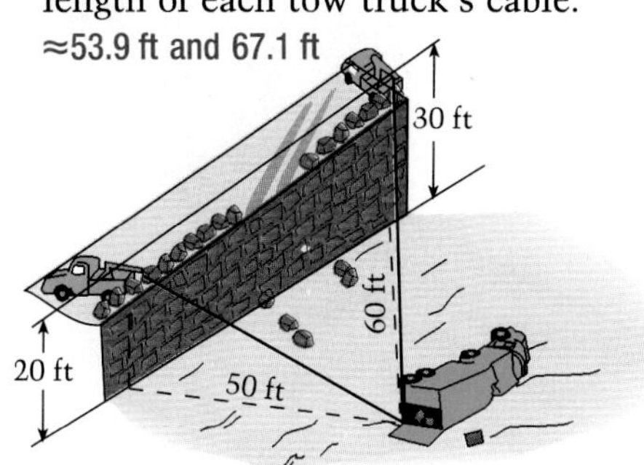

56. *Pickup Lift* The end of the beam on the pickup is x feet above the ground. Find this height by solving the equation $(x-4)^2 + 18^2 = 22^2$. ≈16.6 ft

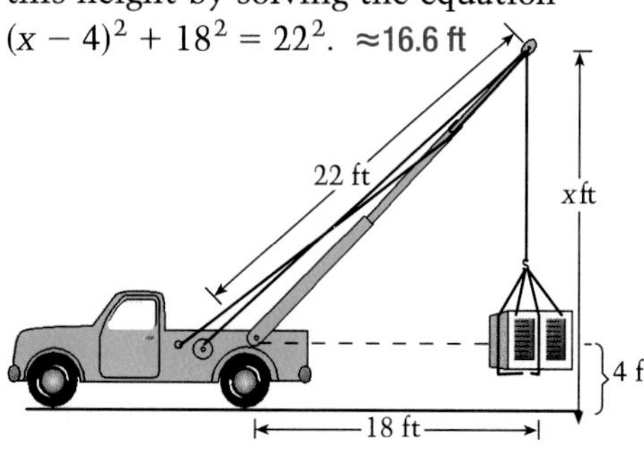

Medicaid Payments **In Exercises 57 and 58, use the following information.**

For 1970 through 1990, the total amount, M (in billions of dollars), spent on Medicaid can be modeled by

$$M = 0.13t^2 + 0.58t + 5.8 \quad \textit{Federal share}$$
$$M = 0.06t^2 + 0.26t + 2.5 \quad \textit{State share}$$

where $t = 0$ represents 1970.
(Source: National Association of State Budget Officers)

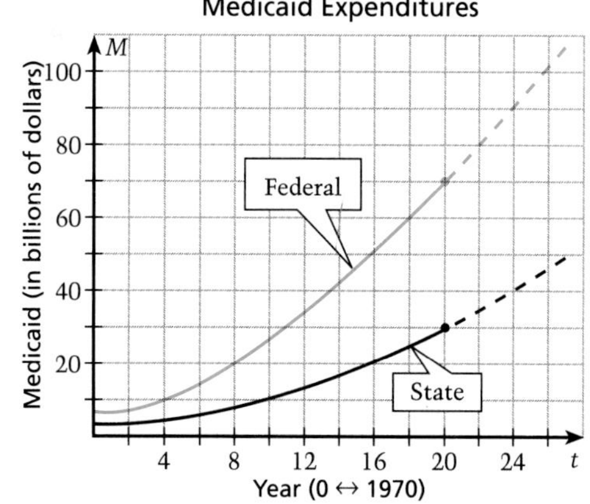

57. During which year was the federal government's share of Medicaid payments approximately \$28 billion? Explain your answer algebraically and graphically. See margin.

58. In 1990, a review commission of the National Association of State Budget Offices projected the 1995 Medicaid spending by the federal government to be \$155 billion and by the state governments to be \$67 billion. Did the review commission use the quadratic models given above, or another model to make the projections? Explain. See margin.

59. *Contact Lens* The shape and thickness of a contact lens depends on the vision correction that is needed. The intersection of the following inequalities describes a typical contact lens cross section. (x and y are measured in millimeters.) Sketch the region. How wide is the contact lens? How thick is it?
12 mm, 0.36 mm, see margin.

$y \le -0.065x^2 + 2.34$
$y \ge -0.055x^2 + 1.98$

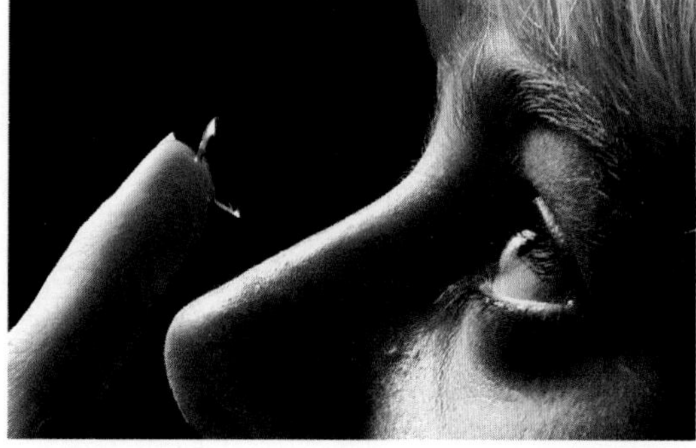

50.

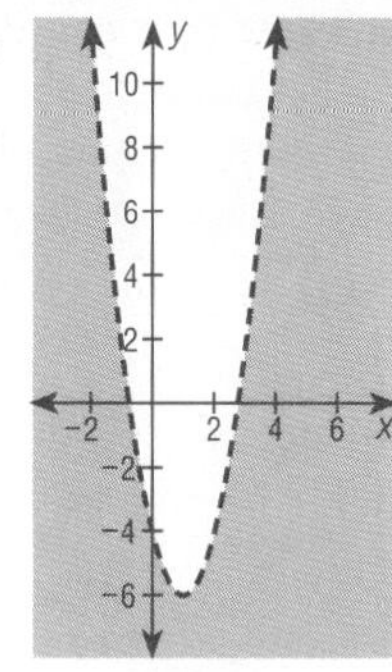

51.

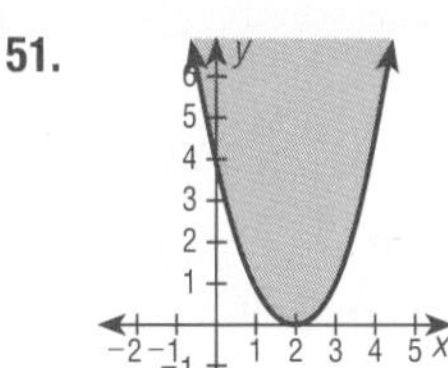

52.

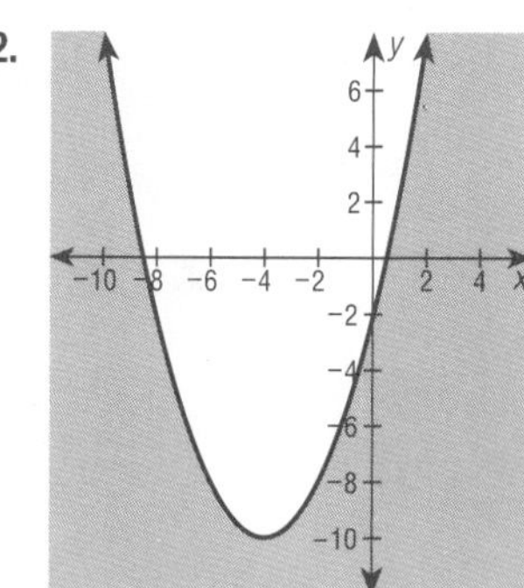

53.

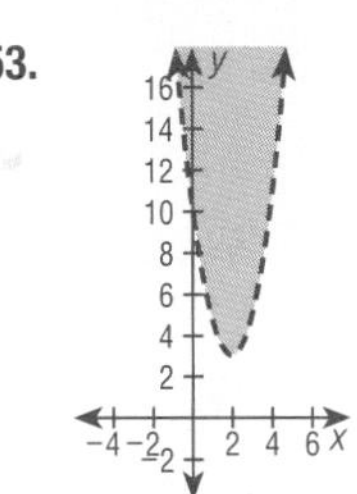

54.

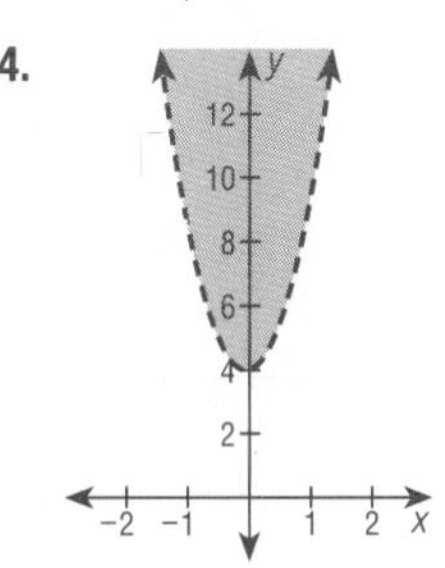

57. 1981; when $M = 28$, $t \approx 11$; the point (11, 28) is on the graph.

58. Some other models; these models yield $\approx$ \$102 billion and $\approx$ \$47 billion

59.

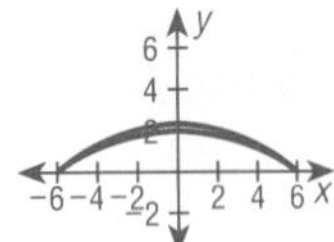

Population of Madagascar **In Exercises 60 and 61, use the following information.**

For 1970 through 1990, the population, P (in thousands), of Madagascar can be modeled by

$$P = 6.4t^2 + 135t + 6750$$

where $t = 0$ represents 1970.
(Source: United Nations)

The central part of Madagascar was once covered with rain forest. When the forest was destroyed, millions of cubic feet of soil eroded.

60. During which year was the population of Madagascar 8,740,000? 1980

61. According to this model, when will the population of Madagascar reach 16,560,000? 2000

Rain Forests of Madagascar **In Exercises 62 and 63, use the following information.**

Madagascar has an area of about 226,660 square miles. The number of square miles (in thousands) of rain forest in 1955 and in 1985 can be found by solving the equation

$$x^2 - 96x + 2048 = 0.$$

(Source: National Geographic)

62. What percent of Madagascar's area was rain forest in 1955? What percent was rain forest in 1985? 28%; 14%

63. The cross section of a hill is modeled by a quadratic equation, as shown in the diagram. How wide is the hill? How high is it? What is its volume? (In the diagram, x and y are measured in feet.) 500 ft; 156.25 ft; 26 million ft^3

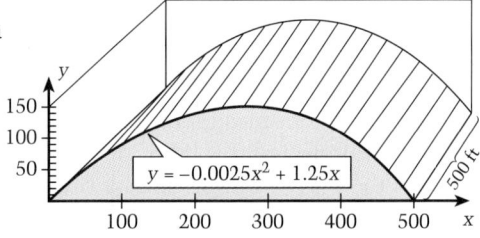

$y = -0.0025x^2 + 1.25x$

64. *Jumping from Branch to Branch* One branch on a tree is 6 feet directly above another. A lemur on the lower branch jumps up toward the higher branch with an initial velocity of 18 feet per second. The lemur is 18 inches tall. Will it reach the higher branch? Explain.

Many of Madagascar's forest animals, such as this ring-tailed lemur, are endangered because of the dwindling forest areas.

Yes; the discriminant of $6 = -16t^2 + 18t + 1\frac{1}{2}$ is positive.

280

Chapter TEST

In Exercises 1 and 2, solve the equation by completing the square. **(5.3)**

1. $x^2 + 6x - 3 = 0$ $-3 \pm 2\sqrt{3}$

2. $2x^2 - 8x + 3 = 0$ $2 \pm \sqrt{\frac{5}{2}}$

In Exercises 3 and 4, solve the equation by the quadratic formula. **(5.4, 5.6)**

3. $3x^2 - 6x + 2 = 0$ $1 \pm \frac{\sqrt{12}}{6}$

4. $5x^2 - 2x + 1 = 0$ $\frac{1}{5} \pm \frac{2}{5}i$

In Exercises 5–10, simplify the expression. **(5.5)**

5. $(2 + 7i) + (3 - 6i)$ $5 + i$

6. $(3 + 4i) - (1 - 3i)$ $2 + 7i$

7. $i(4i - 5)$ $-4 - 5i$

8. $(2 + 3i)(3 - 4i)$ $18 + i$

9. i^{42} -1

10. $i^{15} \cdot i^{20}$ $-i$

In Exercises 11 and 12, state the number of real solutions of the equation. **(5.4)**

11. $3x^2 - 5x + 3 = 0$ None

12. $9x^2 - 6x + 1 = 0$ One

In Exercises 13–16, sketch the graph of the equation or inequality. **(5.2, 5.3, 5.7)** See margin.

13. $y = 2x^2 - 8x + 5$

14. $y = (x + 3)^2 - 2$

15. $y > x^2 + 2x - 3$

16. $y \geq -x^2 + 6x - 7$

17. Find the value of the discriminant of $5x^2 - 6x + 2 = 0$. **(5.4)** -4

18. For what values of c will $2x^2 - 9x + c = 0$ have two real solutions? **(5.4)** $c < 10\frac{1}{8}$

19. Find the axis of symmetry of the graph of $y = 2x^2 - 10x + 7$. **(5.2)** $x = 2\frac{1}{2}$

20. Write $y = x^2 + 8x + 7$ in completed square form. **(5.3)** $y = (x + 4)^2 - 9$

21. A ball was thrown upward from the top of a tower at a velocity of 28 feet per second. The ball hit the ground at the base of the tower 4 seconds later. How high is the tower? **(5.4)** 144 ft

22. For 1980 through 1995, the annual sales, S (in thousands of dollars), of a photography studio can be modeled by $S = -\frac{1}{8}t^2 + 3t + 42$, where $t = 0$ represents 1980. During which year did the studio have sales of \$59,500? During which year did the studio have sales of \$65,500? **(5.4, 5.6)** See margin.

23. A rectangular stage in a park is being decorated for a Fourth of July celebration. The stage has an area of 450 square feet. Is 80 feet of skirting material enough to create a skirt that covers all four sides of the stage? **(5.6)** No

24. You are hanging a rectangular mirror in your room so that one diagonal of the mirror is parallel to the base of the wall. The diagonal of the mirror is 15 inches long. The width of the mirror is three inches shorter than its length. What are the dimensions of the mirror? **(5.3)** 9 in. × 12 in.

Chapter Test Answers

13.

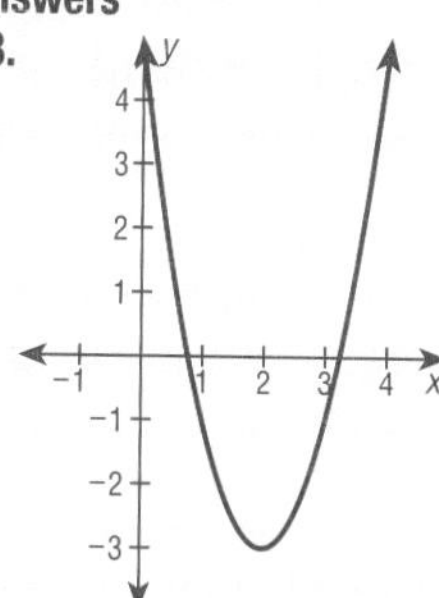

14.

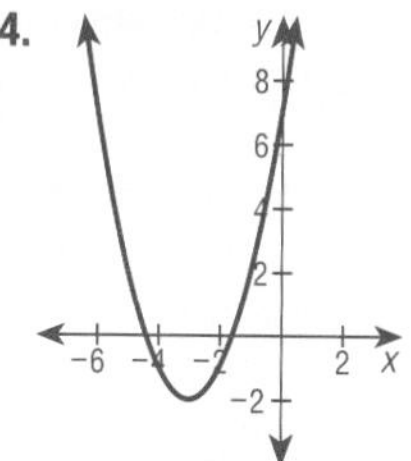

15.

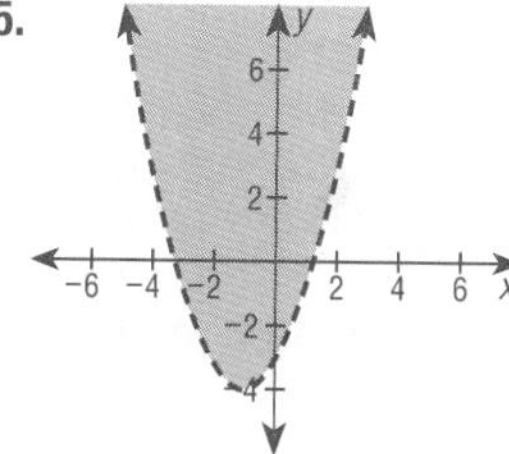

16.

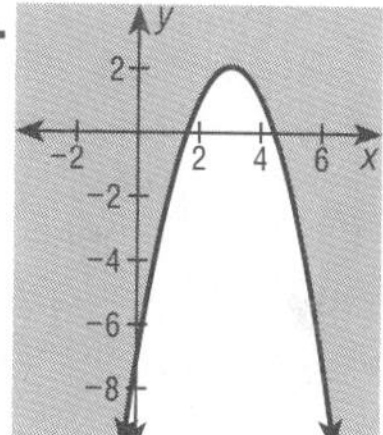

22. 1990 and 1994, sales never reached \$65,500.

The concept of function is one of the most important mathematical ideas students will study. Functions are used in essentially every branch of mathematics because they are an efficient and powerful way in which to organize and manage a variety of mathematical concepts and relationships.

Indeed, functions, as a mathematical class of objects, can be studied and categorized to make them even more useful as problem-solving tools. In this chapter some of the ways in which functions are used to organize and manage real-life situations will be introduced.

The Chapter Summary on page 333 provides you and the students with a synopsis of the chapter. It identifies key skills and concepts. You may want to have students look at the Chapter Summary as an overview before beginning the chapter.

CHAPTER 6

Functions

LESSONS

Whether a musical tone is high-pitched or low-pitched depends on the frequency of the sound wave. In a pipe organ, such as the one in Milwaukee, Wisconsin, shown here, the pitch is determined by the length of the pipe that is sounded.

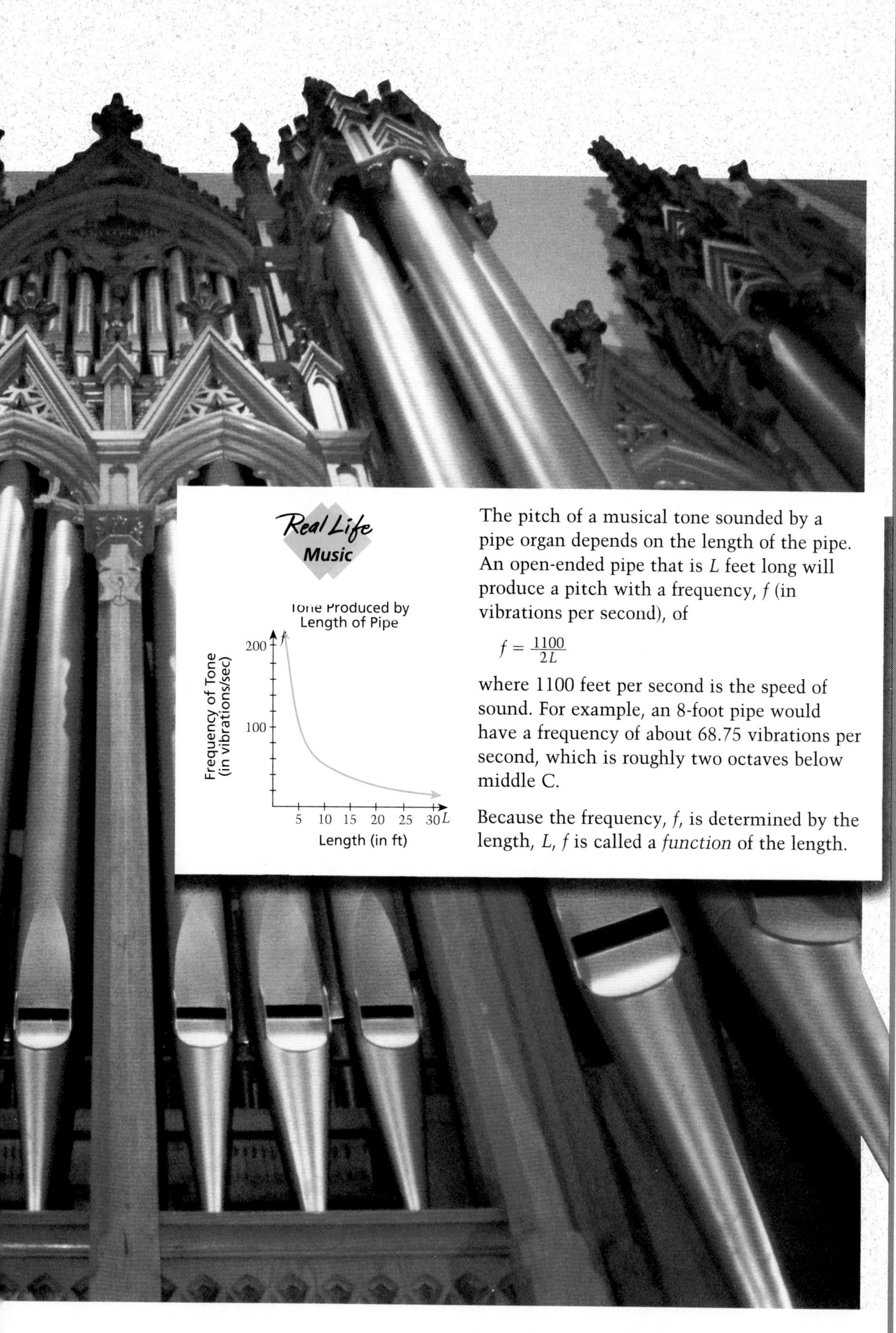

The pitch of a musical tone sounded by a pipe organ depends on the length of the pipe. An open-ended pipe that is L feet long will produce a pitch with a frequency, f (in vibrations per second), of

$$f = \frac{1100}{2L}$$

where 1100 feet per second is the speed of sound. For example, an 8-foot pipe would have a frequency of about 68.75 vibrations per second, which is roughly two octaves below middle C.

Because the frequency, f, is determined by the length, L, f is called a *function* of the length.

As the opening example illustrates, many real-life situations involve relationships in which one variable quantity depends upon another. There are many relationships that occur in real life, called functions, in which the dependency is well-defined. In fact, many mathematical concepts can be expressed in terms of such functional relationships.

Functions

PACING CHART

The daily Pacing Chart is meant to help you adjust your teaching pace. Students in the full course should finish the entire text by the end of the year. Students in the basic course are expected to complete the first twelve chapters. The Pacing Chart for each chapter contains suggestions for lessons that require more than one day and lessons that may be omitted for the basic course.

DAY	FULL COURSE	BASIC COURSE
1	6.1	6.1
2	6.2	6.2
3	6.3	6.3
4	Mid-Chapter Self-Test	6.3
5	6.4 & Using a Graphing Calculator	Mid-Chapter Self-Test
6	6.5	6.4
7	6.6	6.4 & Using a Graphing Calculator
8	6.7	6.5
9	Chapter Review	6.5
10	Chapter Test	6.6
11	Cumulative Review 1–6	6.6
12		6.7
13		Chapter Review
14		Chapter Test
15		Cumulative Review 1–6

CHAPTER ORGANIZATION

LESSON	PAGES	GOALS	MEETING THE NCTM STANDARDS
6.1	284–290	1. Identify a relation and a function 2. Identify real-life relations that are functions	Problem Solving, Communication, Reasoning, Connections, Functions, Geometry
6.2	291–296	1. Perform operations with functions 2. Use function operations in real-life situations	Problem Solving, Communication, Reasoning, Connections, Functions, Structure
Mixed Review	297	Review algebraic and arithmetic skills	
Milestones	297	Fractals	Connections
6.3	298–304	1. Identify inverse relations and inverse functions and verify that two functions are inverses of each other 2. Use inverse functions in real-life situations	Problem Solving, Communication, Reasoning, Connections, Functions
Mid-Chapter Self-Test	305	Diagnose student weaknesses and remediate with correlated Reteach worksheets	
6.4	306–311	1. Use special functions, such as compound functions and step functions 2. Use these functions in real-life situations	Problem Solving, Communication, Connections, Functions
Using a Calculator	312–313	Verify inverse functions by using reflections	Technology
6.5	314–320	1. Use translations and reflections to sketch the graph of a function. 2. Use transformations of graphs of functions in real-life settings	Problem Solving, Communication, Connections, Functions, Geometry (algebraic)
Mixed Review	320	Review algebraic and arithmetic skills	
6.6	321–326	1. Identify and classify recursive functions 2. Use recursive functions as real-life models	Problem Solving, Communication, Reasoning, Connections, Functions, Discrete Math, Structure
6.7	327–332	1. Find the mean, median, mode, and quartiles of a set of numbers 2. Use measures of central tendency in real-life situations	Problem Solving, Communication, Connections, Statistics, Discrete Math, Structure
Chapter Summary	333	Restate for students what they have learned, why they have learned it, and how it fits into the structure of algebra	Structure, Connections
Chapter Review	334–338	Review concepts and skills learned in the chapter	
Chapter Test	339	Diagnose student weaknesses and remediate with correlated Reteaching worksheets	
Cumulative Review 1–6	340–343	Review concepts and skills from chapters 1–6	

LESSON RESOURCES

MEETING INDIVIDUAL NEEDS

RETEACHING For students who need to spend more time on basics:

If a mid-chapter self-test or chapter test indicates a deficiency, teachers can help students with the appropriate ***Reteaching Copymaster.***

PRACTICE For students who need more practice:

Additional exercises like those in the Pupil's Edition are provided for each lesson in ***Extra Practice Copymasters.***

ENRICHMENT For enriching and broadening students' experiences:

Problem of the Day copymasters in ***Teaching Tools*** provide a daily opportunity to use logical reasoning, looking for a pattern, writing an equation, and other routine and non-routine problem-solving strategies.

Math Log copymasters in ***Alternative Assessment*** provide opportunities to report on investigations, research, and open-ended problems.

Technology: Using Calculators and Computers provides enriching activities with graphing and scientific calculators and computers.

The ***Applications Handbook*** provides additional information about the cross-curriculum topics such as astronomy, chemistry, physics, sports, economics, genetics, and music that are integrated into the Pupil's Edition.

LESSON	6.1	6.2	6.3	6.4	6.5	6.6	6.7
PAGES	284-290	291-296	298-304	306-311	314-320	321-326	327-332
Teaching Tools							
Transparencies			✓	✓	✓		✓
Problem of the Day	✓	✓	✓	✓	✓	✓	✓
Warm-up Exercises	✓	✓	✓	✓	✓	✓	✓
Answer Masters	✓	✓	✓	✓	✓	✓	✓
Extra Practice Copymasters	✓	✓	✓	✓	✓	✓	✓
Reteaching Copymasters	Teacher-directed and independent activities tied to results on the Mid-Chapter Self-Tests and Chapter Tests						
Color Transparencies		✓		✓		✓	✓
Applications Handbook	Additional background information for many real-life applications						
Technology	Calculator and computer worksheets for appropriate lessons						
Complete Solutions Manual	✓	✓	✓	✓	✓	✓	✓
Alternative Assessment	Assess student's ability to reason, analyze, solve problems, and communicate using mathematical language.						
Formal Assessment	Mid-Chapter Self-Tests, Chapter Tests, Cumulative Tests, and Practice for College Entrance Tests						
Computer Test Bank	Customized tests can be created by choosing from over 2500 items.						

INSIGHTS

6.1 Relations and Functions
Functions are powerful mathematical tools for describing, analyzing, and organizing real-life information. Because of this, students should learn to identify those mathematical concepts and relationships that are functions, and to express them in appropriate terms.

6.2 Function Operations
Function operations can be used to represent real-life situations in fields such as physics, economics, and engineering. As a tool, function operations can be used to analyze and interpret more complex mathematical relationships. (Calculus is a good example of a mathematical branch of study that uses function operations.)

6.3 Inverse Functions
Many functions have inverses that are themselves interesting and important in real-life applications. Among these functions are the exponential and trigonometric functions which we will encounter later in more detail.

6.4 Special Functions
Real-life situations are often described by compound functions or step functions. For example, the cost (in dollars) of mailing a letter as a function of its weight (in ounces) is a step function. In this lesson, students study two more types of functions that they may encounter in real-life applications.

6.5 Transformations of Graphs of Functions
In this lesson, the connection between algebra and geometry again becomes obvious. An understanding of transformations applied to graphs provides an efficient and convenient way to sketch the graphs of functions without the use of technology. It also enables the users of technology to quickly check the reasonableness of graphs produced by calculators and computers.

6.6 Recursive Functions and Finite Differences
We continue to explore different types of functions that have applications in real-life situations. Recursive functions can be used to model phenomena that frequently occur in nature, in such contexts as biology, computer science and physics. In mathematics, recursive functions frequently occur in the study of series and sequences (which we will see later in the text).

6.7 Exploring Data: Measures of Central Tendency
This chapter ends with a discussion of measures of central tendency: means, medians, and modes. Although each measure represents something "typical" about the data, under scrutiny each measure describes something different about the data. Therefore, it is necessary to focus on what distinguishes these "averages" from each other.

Problem of the Day

A cardboard square lies on an identical square, covering part of it. A vertex of the first square is at the center of the second square. What fraction of the second square is covered? $\frac{1}{4}$

ORGANIZER

Warm-Up Exercises

1. Evaluate each expression for the given set of values of x.
   a. $2x + 3$, $\{-2, -1, 0, 1, 2\}$
   b. $3x^2 - 4x + 6$, $\{-5, -3, 0, 1, 2\}$
   a. $\{-1, 1, 3, 5, 7\}$
   b. $\{101, 45, 6, 5, 10\}$
2. Plot the following sets of ordered pairs on the coordinate plane.
   a. $\{(-2, -1), (-1, 1), (0, 3), (1, 5), (2, 7)\}$
   b. $\{(-5, 10), (-3, 4), (0, 6), (1, 5), (2, 10)\}$
   Check students' graphs of ordered pairs.
3. Evaluate the following expressions.
   a. $-(4)^2 + 3(4) - 5$ $\quad -9$
   b. $5(-3)^2 - 6(-3) + 4$ $\quad 67$
   c. $7(3)^2 - 4(3) - 37$ $\quad 14$

Lesson Resources

Teaching Tools
- Transparencies: 7, 8
- Problem of the Day: 6.1
- Warm-up Exercises: 6.1
- Answer Masters: 6.1

Extra Practice: 6.1

6.1 Relations and Functions

What you should learn:

Goal 1 How to identify a relation and a function

Goal 2 How to identify real-life relations that are functions

Why you should learn it:

You can use functions to describe many relationships in science and social studies, such as the life expectancy of currency.

Goal 1 Identifying Relations and Functions

A **relation** between two variables x and y is a set of ordered pairs. The x-values are the **inputs,** or the **domain,** and the y-values are the **outputs,** or the **range.** Here is an example.

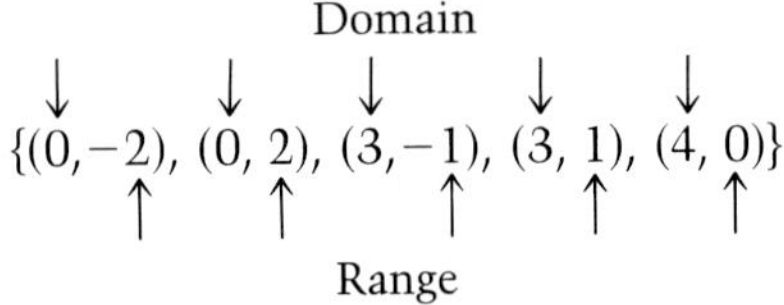

For this relation, the domain is the set $\{0, 3, 4\}$, and the range is the set $\{-2, 2, -1, 1, 0\}$.

A relation may be described by a set of ordered pairs, a graph, a mapping diagram, a table, an equation, or a verbal sentence.

Example 1 *Describing a Relation*

The above relation was described as a set of ordered pairs. Here are some other ways to describe the same relation.

a. Graph

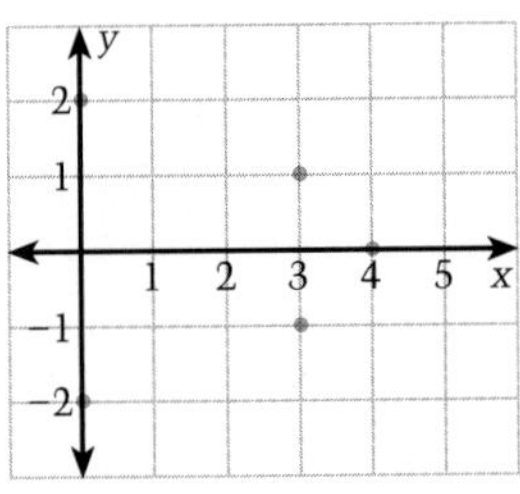

b. Mapping Diagram

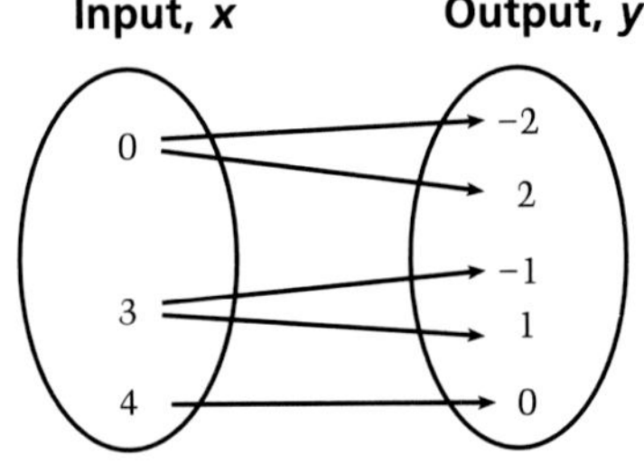

c. Table

Input, x	0	0	3	3	4
Output, y	−2	2	−1	1	0

d. Equation $\quad y^2 = 4 - x$, where $x = 0, 3, 4$.

You can find the y-values that correspond to a given x-value by substituting the x-value in the equation. For instance, when $x = 0$, $y = \pm 2$. ■

A relation between x and y is a **function** of x if each value of x corresponds to *exactly one* value of y. From this definition, you can see that the relation in Example 1 is *not* a function of x because two y-values may correspond to the same x-value. For instance, when $x = 0$, y has the values of -2 and 2.

Example 2 *Identifying Functions*

a. The relation

$$\{(0,-5), (0, 5), (3,-4), (3, 4), (4,-3), (4, 3), (5, 0)\}$$

describes seven points that lie on a circle of radius 5. This relation is *not* a function because two y-values correspond to each of the x-values, 0, 3, and 4. For instance, when $x = 3$, y has the values of -4 and 4. The domain and range of the relation are as follows.

$\{0, 3, 4, 5\}$ *Domain: Set of x-values*

$\{-5, 5, -4, 4, -3, 3, 0\}$ *Range: Set of y-values*

b. The relation

$$\{(-5, 0), (-4, 3), (-3, 4), (0, 5), (3, 4), (4, 3), (5, 0)\}$$

also describes seven points that lie on a circle of radius 5. This relation, however, *is* a function because each x-value corresponds to exactly one y-value. The domain and range are as follows.

$\{-5, -4, -3, 0, 3, 4, 5\}$ *Domain: Set of x-values*

$\{0, 3, 4, 5\}$ *Range: Set of y-values* ■

The graph of a relation can help you see whether the relation is a function. If you can find one vertical line that passes through two points of the relation, then the relation is *not* a function.

Graph of Example 2a

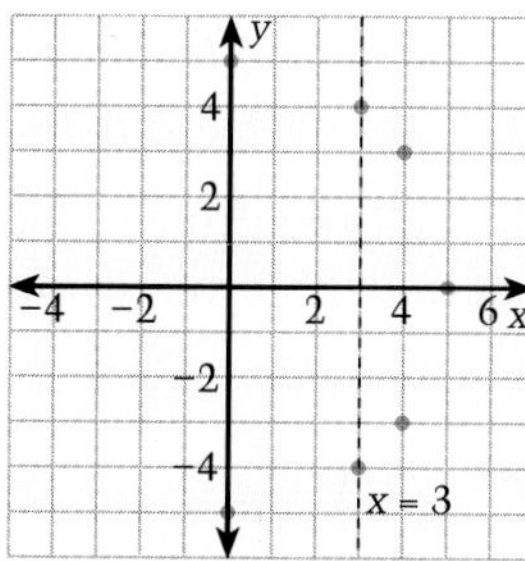

Relation is not a function because at least one vertical line contains two points.

Graph of Example 2b

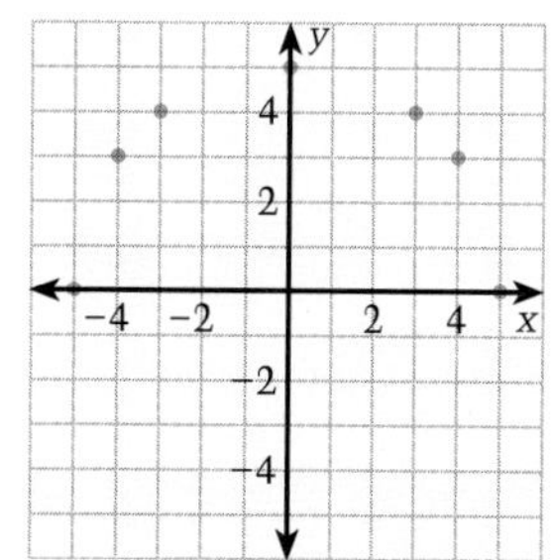

Relation is a function because no two points lie on any one vertical line.

LESSON Notes

Example 1

Observe that the relation presented in the example is not a function. Call students' attention to the fact that there are numbers in the domain that are assigned more than one value in the range.

Example 2

Just as every square is a quadrilateral, every function is a relation. Be sure students understand the connection between relations and functions. The "vertical line test" of a function should be demonstrated graphically. It provides a useful connection between algebraic and geometric interpretations of algebraic concepts.

Example 3

TECHNOLOGY If students have graphics calculators, expressions such as $(-1)^2 - 3(-1) + 5$ can be entered on the calculator as written and can be easily evaluated. Be sure that students understand that the notation $f(x)$ does not represent the multiplication of f times x.

Example 4

Ask students why the average life span of United States currency is a function of (or depends upon) the value of the currency. Point out that the converse statement does not make sense—that is, when the table is rewritten as follows:

Life span, x	1.5	2	3	4	5	9
Value y	\$1	\$5	\$10	\$20	\$50	\$100

This mathematical relationship does represent a function.

(continued)

(continued)

However, you would not say that the value of the currency is a function of its life span.

In the strict mathematical sense, ask students if earnings are a function of dividends. No, they are not.

Extra Examples

Here are additional examples similar to **Examples 1–3.**

1. Describe the given relations in two other ways.
   a. {(0, −1), (−3, 4), (4, 2), (2, 0)}
   b. {(−4, −12), (3, 2), (−2, −8), (5, 6)}

   The relations could be shown as mapping diagrams, graphically, as tables, in equation form, or as verbal sentences. For example, relation 1 could be shown using the following diagram:

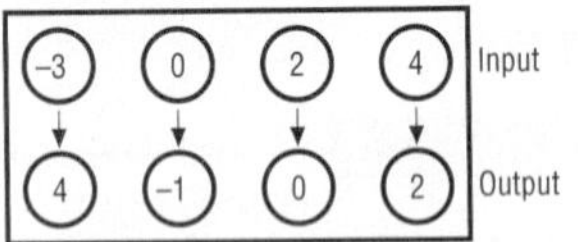

   Relation b. could be expressed in equation form as $y = 2x - 4$.

2. Which graphs display functions?

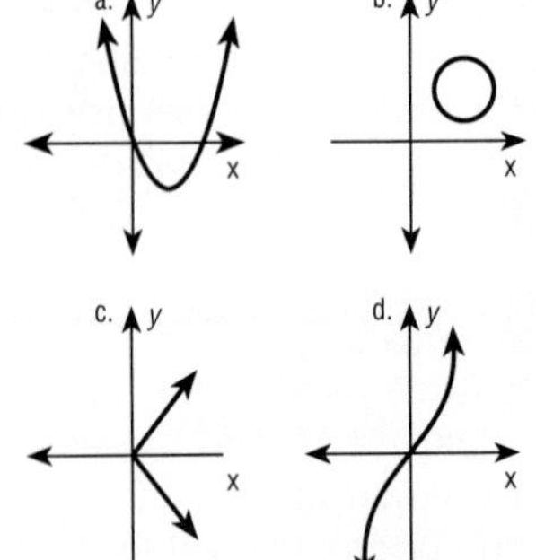

   a. a function, b. not a function, c. not a function, d. a function

3. Evaluate the following functions at the given x-values.
   a. $f(x) = 4x + 2$, $x = 3$
   b. $g(x) = 2x^2 - 4x - 3$, $x = -4$
   c. $h(x) = -3x^2 + 8x + 12$, $x = 2$

   a. $f(3) = 14$, b. $g(-4) = 45$, c. $h(2) = 16$

Functions are often described by equations. For example, the equation $y = 4x - 3$ describes y as a function of x. By giving this function the name "f", you can use **function notation.**

Input	**Output**	**Equation**
x	$f(x)$	$f(x) = 4x - 3$
		input (x) output ($4x - 3$)

The symbol $f(x)$ is read as the "value of f at x," or simply "f of x." Note that $f(x)$ is another name for y.

To find $f(x)$ for a given value of x is to **evaluate** the function f by substituting an input value into the equation and simplifying. When an equation is used to describe a function, the domain is the set of all x-values that make sense in the equation.

x-value	**Function value, $f(x) = 4x - 3$**
$x = 0$	$f(0) = 4 \cdot 0 - 3 = 0 - 3 = -3$
$x = 1$	$f(1) = 4 \cdot 1 - 3 = 4 - 3 = 1$
$x = 2$	$f(2) = 4 \cdot 2 - 3 = 8 - 3 = 5$

Up to this point in the text, parentheses have been used to represent multiplication. The function notation $f(x)$ is a *different* use of parentheses. Remember that $f(x)$ means "the value of f at x." It does not mean "f times x."

Example 3 Evaluating a Function

Evaluate the function $f(x) = -x^2 - 3x + 5$ at the given x-values.

a. $x = -1$ **b.** $x = 4$

Solution

a. To find the value of $f(x)$ when $x = -1$, substitute -1 for x in the equation that describes the function.

$f(x) = -x^2 - 3x + 5$ *Equation for f*

$f(-1) = -(-1)^2 - 3(-1) + 5$ *Substitute −1 for x.*

$= -1 + 3 + 5$, or 7 *Simplify.*

b. To find the value of $f(x)$ when $x = 4$, substitute 4 for x in the equation that describes the function.

$f(x) = -x^2 - 3x + 5$ *Equation for f*

$f(4) = -(4)^2 - 3(4) + 5$ *Substitute 4 for x.*

$= -16 - 12 + 5$, or -23 *Simplify.* ■

Although f is often used as the name of a function, other letters such as g or h can also be used.

Goal 2 Identifying Functions in Real Life

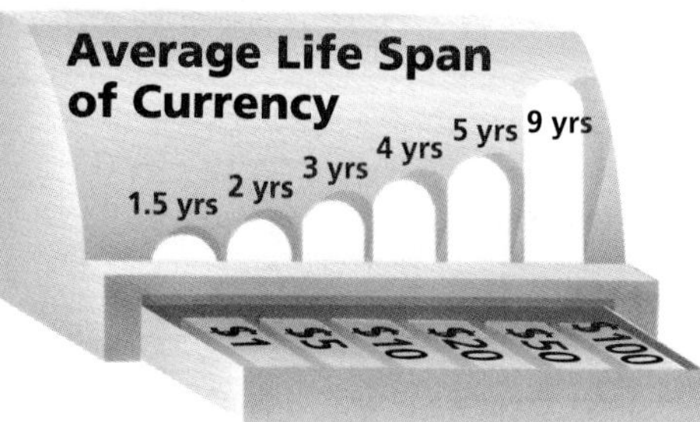

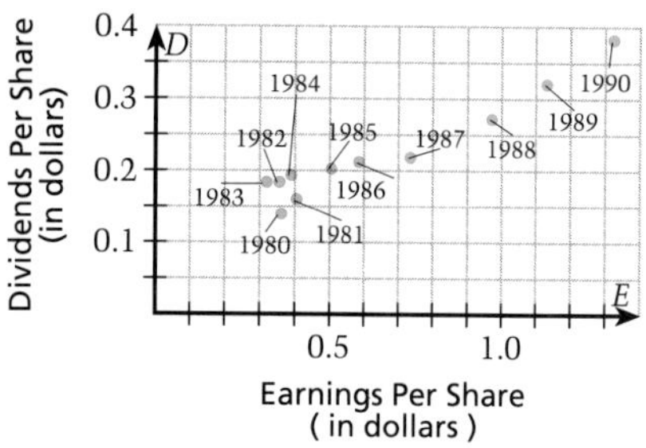

Real Life **Currency**

Example 4 *Identifying Relations That Are Functions*

a. The average life span of United States currency is a function of the value of the currency. For instance, a $1 bill has an average life span of 1.5 years, a $5 bill has an average life span of 2 years, etc. ***(Source: Bureau of Engraving and Printing)***

Value, *x*	$1	$5	$10	$20	$50	$100
Life Span, *y*	1.5 yrs	2 yrs	3 yrs	4 yrs	5 yrs	9 yrs

Real Life **Business**

b. For 1980–1985, the dividend per share was not a function of the earnings per share for PepsiCo, Inc. In 1980 and in 1982, the company earned $0.36 per share, but the dividend was $0.14 in 1980 and $0.18 in 1982. ***(Source: PepsiCo, Inc.)***

Earnings, *E*	$0.36	$0.40	$0.36	$0.33	$0.38	$0.50
Dividend, *D*	$0.14	$0.16	$0.18	$0.18	$0.19	$0.20

Communicating about ALGEBRA

SHARING IDEAS about the Lesson See below.

Does the mapping diagram represent a function? Explain.

A.

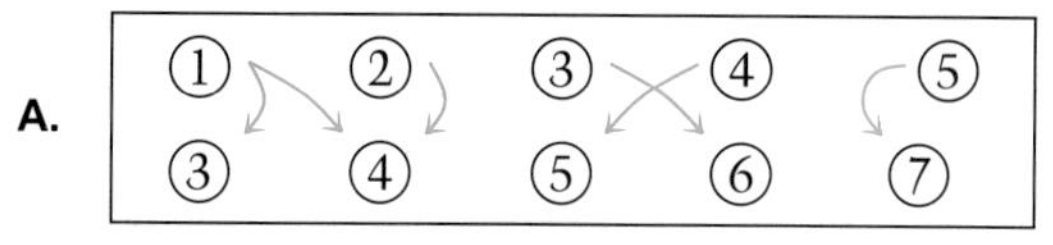

← *Input*

← *Output*

B.

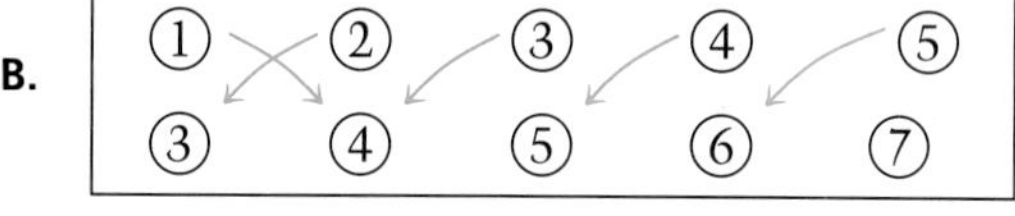

← *Input*

← *Output*

A. No, one input corresponds to more than one output.
B. Yes, each input corresponds to exactly one output.

Check Understanding

1. List the ways in which a relation or function may be represented.
   For example: ordered pairs, mapping diagrams, graphs, tables, equations, verbal sentences
2. How does a relation differ from a function?
   A function is a relation. A relation is a function if each input value corresponds to exactly one output value.
3. How can a graph of a relation be used to decide if the relation is a function?
   See the bottom of page 285.
4. How are evaluating an expression and evaluating a function alike?
   See Example 3.

Communicating about ALGEBRA

Functions may also be thought of as "rules of association." For example, the function $f(x) = x^2$ associates each x-value with its square.

EXTEND *Communicating*
For each set of the following ordered pairs, ask students to discover a rule of association that describes the relationship between coordinates of each ordered pair in a set.

a. {(−2, −4), (−1, −1), (0, 2), (1, 5), (2, 8)}
b. {(−2, 8), (−1, 2), (0, 0), (1, 2), (2, 8)}
c. {(−2, 2), (−1, 1), (0, 0), (1, 1), (2, 2)}
d. {(−2, 1), (−1, −2), (0, −3), (1, −2), (2, 1)}

Answers may vary.
a. $f(x) = 3x + 2$, **b.** $f(x) = 2x^2$, **c.** $f(x) = |x|$, **d.** $f(x) = x^2 - 3$

EXERCISE Notes

ASSIGNMENT GUIDE

Basic/Average: Ex. 1–4, 5–19 odd, 21, 22, 23–35 odd, 37–39, 45

Above Average: Ex. 1–4, 5–21 odd, 22, 23–35 odd, 36–38, 45, 47

Advanced: Ex. 1–4, 5–21 odd, 23–35 odd, 36–38, 45, 47

Selected Answers
Exercises 1–4, 5–45 odd

More Difficult Exercise
Exercise 47

Guided Practice

Ex. 1–4 GROUP ACTIVITY
Use these exercises in class as a summary review of the different formats used to express the relationships. Help students to establish the differences between a relation and a function.

Answers

2. A relation is a function if each input corresponds to exactly one output.
3. Yes, there is no vertical line that passes through two points of the relation when it is graphed.

Independent Practice

Ex. 5–20 These exercises represent some of the different formats used to express relations.

EXERCISES

Guided Practice

CRITICAL THINKING about the Lesson

1. Do the mapping diagram and graph describe the same relation? Explain.

Yes, they both describe the same ordered pairs: (−3, 0), (−2, 1), (−1, 2), (−1,−2), (0,−2), (1,−2), (2, 3).

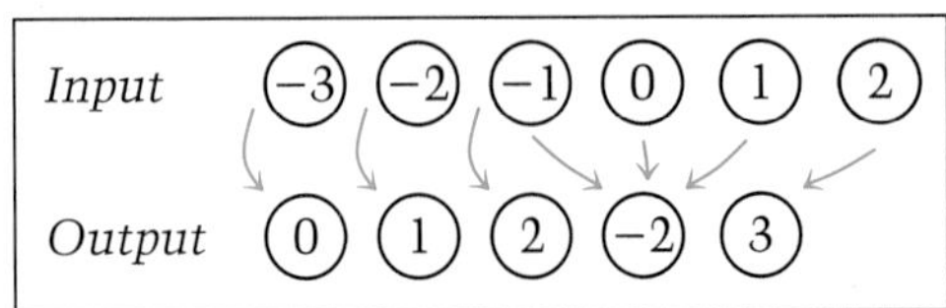

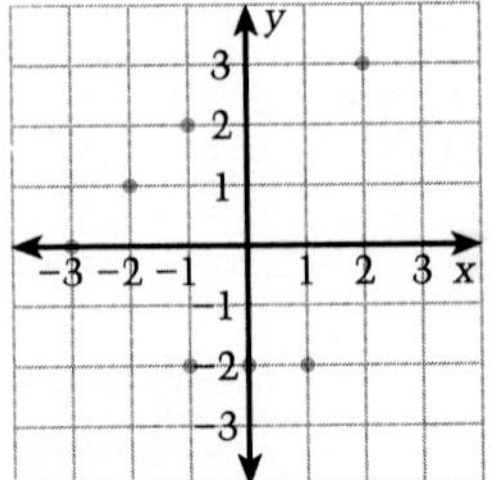

2. Describe in your own words which types of relations are functions. See margin.

3. Is the relation described by the table a function? Why or why not? Justify your answer graphically. See margin.

Input, *x*	−3	−2	−1	0	1	2	3	4
Output, *y*	2	1	−2	−2	−1	0	2	2

4. Which set of ordered pairs describes a function? Explain.

$A = \{(-2, 2), (0, 5), (1, 6), (1, 7), (2, -1), (3, 2)\}$
$B = \{(0, 1), (2, -1), (3, 2), (4, 2), (5, 3), (-5, 1)\}$

B; in *A* the *x*-value 1 corresponds to the *y*-values 6 and 7, and in *B* each *x*-value corresponds to exactly one *y*-value.

Independent Practice

In Exercises 5–8, find the domain and range of the relation. See below.

5. $\{(1, 5), (-2, 8), (0, 4), (-1, 5), (2, 8)\}$

6. $\left\{\left(-\frac{1}{2}, 0\right), (4, 3), (4, -3), (0, 1), (0, -1)\right\}$

7. $\{(4, 1), (7, 2), (7, -2), (3, 0), (4, -1)\}$

8. $\{(0, -1), (1, 2), (-1, 0), (-2, 5), (2, 9)\}$

In Exercises 9–12, sketch a graph of the relation. See margin.

9.
Input: 3, −3, 0, 4, −4
Output: 0, 9, −7

10.
Input: 1, −1, 0, −3, 3
Output: 1, −1, 17

11.
Input: −3, −4, 0
Output: 1, −1, 0, 2, −2

12.
Input: 1, 3, −5
Output: 2, −2, 0, 4, −4

5. $\{-2, -1, 0, 1, 2\}$, $\{4, 5, 8\}$
6. $\left\{-\frac{1}{2}, 0, 4\right\}$, $\{-3, -1, 0, 1, 3\}$
7. $\{3, 4, 7\}$, $\{-2, -1, 0, 1, 2\}$
8. $\{-2, -1, 0, 1, 2\}$, $\{-1, 0, 2, 5, 9\}$

In Exercises 13–16, use a table to describe the relation. See Additional Answers.

13. 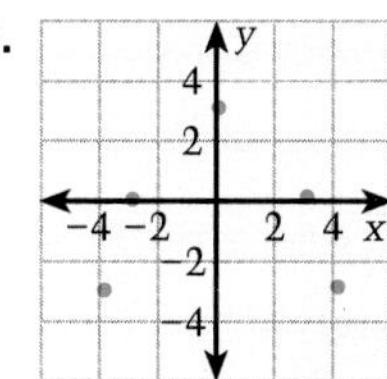

14. 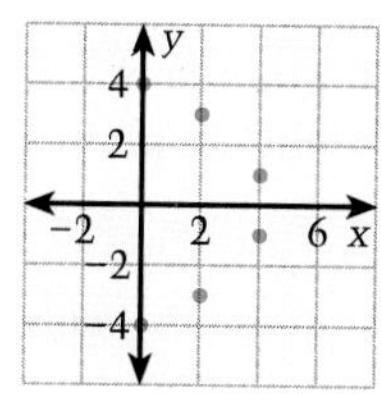

15. 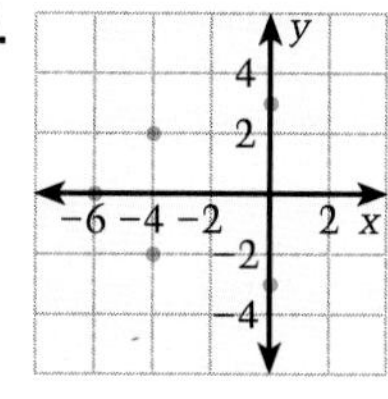

16. 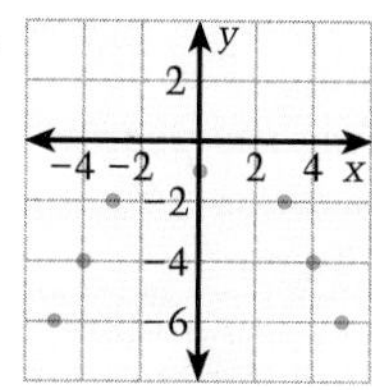

In Exercises 17–20, use a mapping diagram to describe the relation. See Additional Answers.

17. 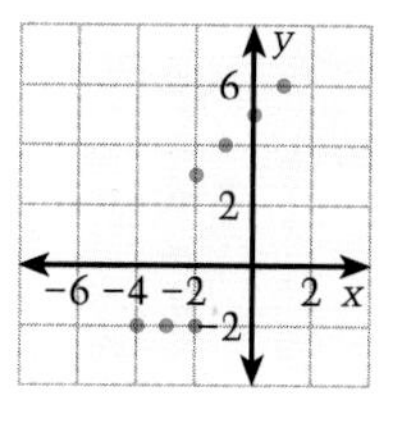

18. 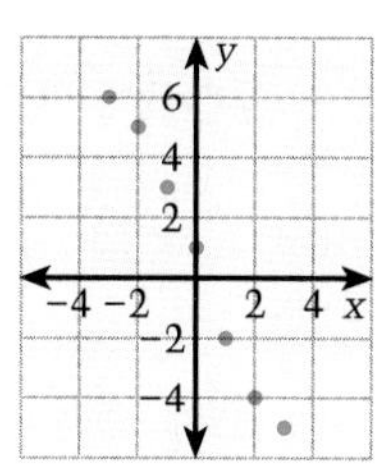

19. 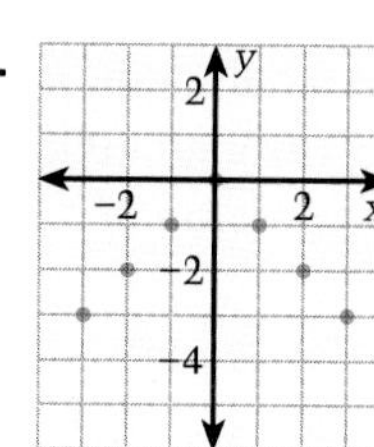

20. 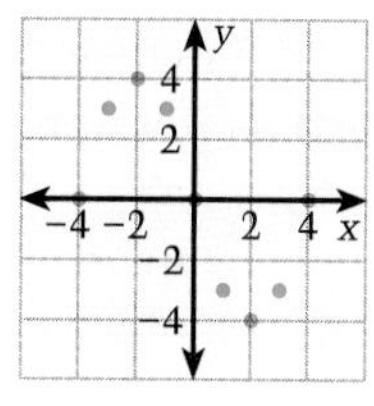

21. Which of the relations in Exercises 13–16 are functions? Those in Exercises 13 and 16.

22. Which of the relations in Exercises 17–20 are functions? Those in Exercises 18–20.

In Exercises 23–26, determine whether the relation is a function.

23. $\{(0, -5), (1, 3), (2, 2), (0, 4), (-5, 6), (3, 4)\}$ It is not.

24. $\{(-1, 6), (-1, 3), (2, 4), (0, 3), (1, 2), (3, 1)\}$ It is not.

25. $\left\{\left(-10, \frac{1}{2}\right), (-6, 1), \left(-4, \frac{3}{2}\right), (-2, 2), (2, 0), (0, 1)\right\}$ It is.

26. $\{(0, -5), (5, 4), (6, 6), (1, -5), (7, 7), (2, 3)\}$ It is.

In Exercises 27–34, find the indicated value of *f*(*x*).

27. $f(x) = x - 9$, $f(3)$ -6

28. $f(x) = -x^2 + 3$, $f(-1)$ 2

29. $f(x) = x^2 - 3x + 2$, $f(-2)$ 12

30. $f(x) = 3x^2 - x - 1$, $f(4)$ 43

31. $f(x) = \frac{1}{2}x^2 - 4x$, $f(1)$ $-3\frac{1}{2}$

32. $f(x) = \frac{-2}{3}x^2 - x + 5$, $f(6)$ -25

33. $f(x) = x^2 - 2x - 4$, $f\left(\frac{1}{2}\right)$ $-4\frac{3}{4}$

34. $f(x) = -x^2 + 4x + 3$, $f\left(-\frac{1}{2}\right)$ $\frac{3}{4}$

35. *Geometry* The volume of a cube with edges of length x is given by the function $f(x) = x^3$. Find $f(5)$. Explain what $f(5)$ represents.

36. *Geometry* The volume of a regular tetrahedron with edge length e is given by the function $f(e) = \frac{\sqrt{2}}{12}e^3$. Find $f(2)$. Explain what $f(2)$ represents. $\frac{2}{3}\sqrt{2}$, the volume of a regular tetrahedron with edge length 2.

35. 125, the volume of the cube with an edge of length 5.

Cubes and tetrahedrons are only two possible shapes for dice.

Answers

9. 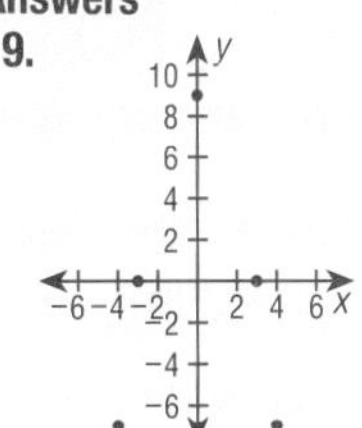

10. 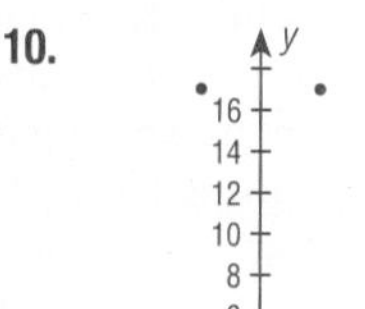

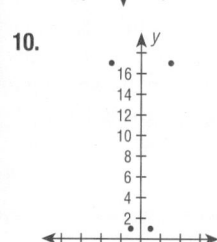

11. 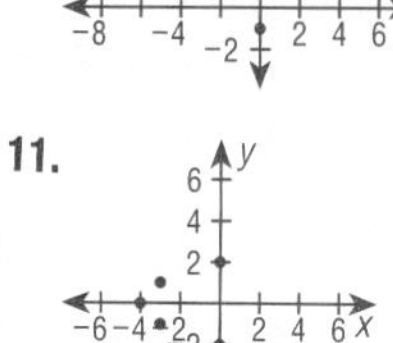

12. 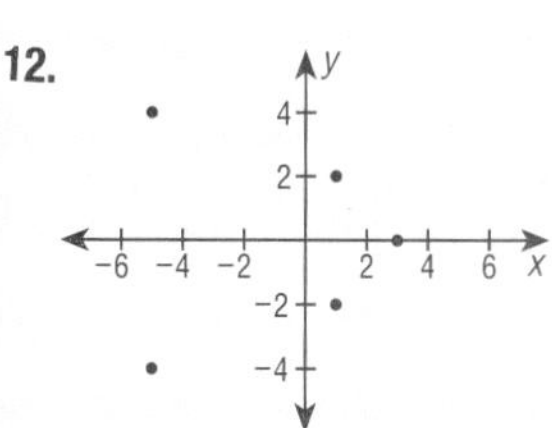

Ex. 27–36 Refer students to Example 3 for a solution model.

EXTEND Ex. 38 Ask students what condition would have been necessary for the number of movies not to have been a function of the year. Why isn't this possible? There cannot be two different numbers of new releases in a given year.

Ex. 39–42 Explain to students that quadratic models of the form shown in these exercises generate relations (sets of ordered pairs) that are functions.

Enrichment Activity

This activity requires students to go beyond lesson goals.

RESEARCH The notation $f(x)$, read as "f of x," is attributed to Leonard Euler (pronounced "oiler"), a Swiss mathematician (1707–1780).

Write a short essay on Euler and his accomplishments. Include at least one formula from geometry that bears his name. The most famous formula is Euler's theorem for polyhedrons (solid figures bounded by polygons). For any simple polyhedron, $V - E + F = 2$, where V is the number of vertices, E the number of edges, and F the number of faces.

37. *Peanut Production* For 1984 through 1991, the peanut production, P (in billions of pounds), in the United States is shown in the table. The table also shows the average yield, y (in pounds per acre). Was the peanut production a function of average yield? Explain. Yes, each y-value corresponds to exactly one P-value.

Year	1984	1985	1986	1987	1988	1989	1990	1991
Average Yield, y	2880	2805	2400	2335	2440	2440	1895	2596
Production, P	4.44	4.13	3.71	3.60	4.00	4.00	3.62	5.09

38. *New Movie Releases* The graph at the right shows the number of new major movies released from 1985 through 1990. Is the number of movies a function of the year? Explain.

Yes, each year corresponds to exactly one number of movies.

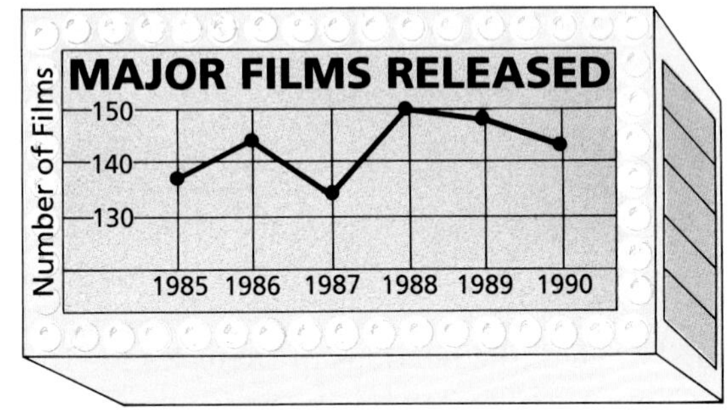

Integrated Review

In Exercises 39–42, complete the table.

39. $y = 7x^2 + 4x - 10$ 41 10 −7 −10 1 26

x	−3	−2	−1	0	1	2
y	?	?	?	?	?	?

40. $y = x^2 + \frac{1}{2}x + 6$ 9 $6\frac{1}{2}$ 6 $7\frac{1}{2}$ 11 $16\frac{1}{2}$ 24

x	−2	−1	0	1	2	3	4
y	?	?	?	?	?	?	?

41. $y = -x^2 + 5x + 5$ 5 9 11 11 9 5 −1

x	0	1	2	3	4	5	6
y	?	?	?	?	?	?	?

42. $y = 3x^2 + 10x - 2$

x	−5	−4	−3	−2	−1	0	1
y	?	?	?	?	?	?	?

23 6 −5 −10 −9 −2 11

In Exercises 43–46, sketch the graph of the equation. See Additional Answers.

43. $y = x^2 - 2x + 6$ **44.** $y = x^2 + 10x + 15$ **45.** $y = 2x^2 - 4x - 1$ **46.** $y = -4x^2 - 16x - 15$

Exploration and Extension

47. Let y be a prime factor of x (y is a factor of x and also a prime number), where x and y are positive integers. Draw a mapping diagram for the inputs and outputs below. Then decide whether the relation is a function.

Input x: 2, 3, 4, 5, 6, 7, 8

Output y: 2, 3, 4, 5, 6, 7, 8

2 4 8 6 3 5 7
2 3 4 5 6 7

It is not.

6.2 Function Operations

What you should learn:

Goal 1 How to perform operations with functions

Goal 2 How to use function operations in real-life situations

Why you should learn it:

You can model many real-life situations with function operations, such as the application of successive discounts.

Goal 1 Performing Function Operations

So far in the book, you have performed addition and multiplication with real numbers, matrices, and complex numbers. These operations can also be performed with functions. Here are some examples that use the functions

$f(x) = 3x$ and $g(x) = x - 5.$

The domain for both f and g is the set of all real numbers. When an equation is used to describe a function, the domain is the set of all x-values that make sense in the equation.

Operation	Resulting Function	Domain
Addition	$f(x) + g(x) = 3x + (x - 5)$ $= 4x - 5$	All real numbers
Subtraction	$f(x) - g(x) = 3x - (x - 5)$ $= 2x + 5$	All real numbers
Multiplication	$f(x) \cdot g(x) = 3x(x - 5)$ $= 3x^2 - 15x$	All real numbers
Division	$f(x) \div g(x) = \dfrac{3x}{x - 5}$	All real numbers except $x = 5$

Note that the domain of the quotient of f and g does not include 5, because that would produce a denominator of zero.

Example 1 *Finding the Domain of a Function*

a. The domain of $f(x) = -3x^2 + 4x + 5$ is the set of all real numbers.

b. The domain of $g(t) = \sqrt{t}$ is the set of nonnegative real numbers, because the square root of a negative number is not a real number. (Unless stated otherwise, you can assume that function values are real numbers.)

c. The domain of

$$h(x) = \frac{3}{x - 4}$$

is the set of all real numbers *except* 4. The number 4 is not in the domain because the denominator cannot be zero. ■

Problem of the Day

A man has 90¢ in quarters, dimes, and nickels. Half the coins are nickels, and a fourth of them are dimes. How many of each does he have?
2 quarters, 2 dimes, and 4 nickels.

ORGANIZER

Warm-Up Exercises

1. Evaluate $f(x)$ for the indicated values.
   a. $f(x) = 3x + 6$; $f(-3)$, $f(9)$
   b. $f(x) = 7x - 2$; $f(4)$, $f(a)$
   c. $f(x) = 8x^2 - 5x$; $f(-1)$, $f(2)$
   d. $f(x) = 2x^2 + 3x - 16$; $f(-6)$, $f(m)$
   a. $f(-3) = -3$, $f(9) = 33$
   b. $f(4) = 26$, $f(a) = 7a - 2$
   c. $f(-1) = 13$, $f(2) = 22$
   d. $f(-6) = 38$, $f(m) = 2m^2 + 3m - 16$
2. Solve.
   a. $x + 3 = 0$ $x = -3$
   b. $2x - 4 = 0$ $x = 2$
   c. $2x - 6 > 0$ $x > 3$
   d. $8 - 12x > 0$ $x < \frac{2}{3}$

Lesson Resources

Teaching Tools
  Problem of the Day: 6.2
  Warm-up Exercises: 6.2
  Answer Masters: 6.2
Extra Practice: 6.2
Color Transparency: 37
Technology Handbook: p. 35

LESSON Notes

Example 1

Only "real-valued functions," that is, functions whose range values consist only of real numbers, will be considered, unless otherwise noted.

(continued)

(continued)

You may wish to ask students to identify the range of each function in the example.

a. all real numbers less than or equal to $6\frac{1}{3}$

b. all nonnegative real numbers

c. all nonzero real numbers

Examples 2–3

The composition of two functions does not have a numerical counterpart. Students should evaluate composite functions and compare $f(g(a))$ with $f(x)$ evaluated at $g(a)$. For example, $f(g(x)) = 8 - 4x$. Therefore, $f(g(2)) = 8 - 4(2) = 0$. This should be compared to $f(x) = 4x$ evaluated at $g(2) = 2 - 2 = 0$, or $f(0) = 4(0) = 0$.

You may wish to introduce the common notation for the composition of functions. If f and g are functions, then $f(g(x))$ can be written $f \circ g$ (read as "f composed with g") and $g \circ f$ would be equivalent to $g(f(x))$.

EXTEND Examples 2–3 Ask students to find these composites in the two ways described above.

a. $f(g(-5))$ b. $f(g(3))$

c. $g(f(-8))$ d. $g(f(0))$

a. 28, b. −4, c. 34, d. 2

Example 4

Explain why a 12% discount is the same as 88% of the original cost.

Extra Examples

Here are additional examples similar to **Examples 1–3.**

1. Find the domain of each function.
   a. $f(x) = 4x^2 + 3x - 8$
   b. $f(x) = \dfrac{6}{2x - 10}$
   c. $f(x) = \sqrt{x - 3}$

   a. all real numbers
   b. all real numbers except $x = 5$
   c. all real numbers greater than or equal to 3

A fifth operation that is performed with functions is called **composition.**

The Composition of Two Functions

The **composition** of the function f with the function g is given by

$$h(x) = f(g(x)).$$

A good way to find the composition of two functions is to use three steps, *working from the inside out.* For instance, if $f(x) = 4x + 2$ and $g(x) = x - 3$, then the composition of f with g, $f(g(x))$, can be found as follows.

1. Substitute formula for $g(x)$. $\quad f(g(x)) = f(\overbrace{x - 3}^{g(x)})$
2. Apply formula for $f(x)$. $\quad f(g(x)) = \overbrace{4(x - 3) + 2}^{f(\square) = 4\square + 2}$
3. Simplify. $\quad f(g(x)) = 4x - 10$

Example 2 ***Finding the Composition of Functions***

Find the composition of $f(x) = 4x$ with $g(x) = 2 - x$.

Solution

$$f(g(x)) = f(\overbrace{2 - x}^{g(x)}) \qquad \textit{Substitute } 2 - x \textit{ for } g(x).$$
$$= 4(2 - x) \qquad \textit{Apply } f(\square) = 4\square.$$
$$= 8 - 4x \qquad \textit{Simplify.}$$

■

Example 3 ***Finding the Composition of Functions***

Find the composition of $g(x) = 2 - x$ with $f(x) = 4x$.

Solution

$$g(f(x)) = g(\overbrace{4x}^{f(x)}) \qquad \textit{Substitute } 4x \textit{ for } f(x).$$
$$= 2 - (4x) \qquad \textit{Apply } g(\square) = 2 - \square.$$
$$= 2 - 4x \qquad \textit{Simplify.}$$

■

Notice in Examples 2 and 3 that the composition of f with g is not the same as the composition of g with f.

Goal 2 Using Function Operations in Real Life

Real Life
Consumer Finance

Car dealers often offer two types of incentives to buyers. One is a factory rebate, which is a dollar amount that is subtracted from the price of the car. The other is a percent discount that is taken off the price of the car.

Example 4 *Comparing Composition Orders*

The regular price of a certain new car is \$15,800. The dealership advertised a factory rebate of \$1500 *and* a 12% discount. Compare the sale price obtained by subtracting the rebate first, then taking the discount, with the sale price obtained by taking the discount first, then subtracting the rebate.

Solution Using function notation, where x is the price of the car, the price after rebate, $f(x)$, and the price after discount, $g(x)$, can be represented by

$f(x) = x - 1500$ *Rebate of \$1500*

$g(x) = 0.88x$ *Discount of 12%*

Rebate First: If you subtract the rebate first, the sale price is given by the composition of g with f.

$g(f(15{,}800)) = g(15{,}800 - 1500)$ *Subtract the rebate first.*

$= g(14{,}300)$ *Simplify.*

$= 0.88 \cdot 14{,}300$ *Take discount.*

$= \$12{,}584$ *Simplify.*

Discount First: If you take the discount first, the sale price is given by the composition of f with g.

$f(g(15{,}800)) = f(0.88 \cdot 15{,}800)$ *Take the discount first.*

$= f(13{,}904)$ *Simplify.*

$= 13{,}904 - 1500$ *Subtract rebate.*

$= \$12{,}404$ *Simplify.*

Thus given an option, you should take the discount first. ■

Communicating about ALGEBRA

SHARING IDEAS about the Lesson See Additional Answers.

A car dealership is offering a rebate of \$1200 *and* a discount of 15% on the purchase of some new cars. Represent the sales price after each type of discount by writing a function for each. Then use composition of functions to express each sale price. What is the difference in the cost of a new car whose regular price is \$18,400?

2. Find the composition of $f(x)$ with $g(x)$, and $g(x)$ with $f(x)$.

a. $f(x) = 2x + 3$, $g(x) = 8 - x$

b. $f(x) = \frac{4}{3x + 6}$, $g(x) = 2x$

c. $f(x) = 6x - 2$, $g(x) = 1 - x$

a. $f(g(x)) = 19 - 2x$, $g(f(x)) = 5 - 2x$

b. $f(g(x)) = \frac{2}{3x + 3}$, $g(f(x)) = \frac{8}{3x + 6}$

c. $f(g(x)) = 4 - 6x$, $g(f(x)) = 3 - 6x$

Check Understanding

1. What is the domain of a function $f(x)$?
   The set of all x-values that make sense
2. Describe how to find the composition of two functions.
   See the top of page 292 and Examples 2 and 3.
3. In general, does $f(g(x)) = g(f(x))$?
   No

Communicating about ALGEBRA

Ask students, "Under what conditions would the order of the rebate and discount not matter?"
The order is unimportant only if there is no discount (0% discount) or if there is no rebate!

EXTEND *Communicating*
Ask students to construct one or more functions $f(x)$ and $g(x)$ such that $f(g(x)) = g(f(x))$.
Answers will vary. One pair that works is $f(x) = x$ and $g(x) = \frac{1}{x}$; another is $f(x) = x - 1$ and $g(x) = x + 1$.

EXERCISE Notes

ASSIGNMENT GUIDE

Basic/Average: Ex. 1–4, 5–12, 15–23 odd, 31, 33, 34, 37–41, 53–58

Above Average: Ex. 1–4, 5–12, 15–23 odd, 33, 34, 35–43, 55–59

Advanced: Ex. 1–4, 5–12, 15–23, 33, 35–43, 56–60

Selected Answers
Exercises 1–4, 5–57 odd

Use **Mixed Review** as needed.

More Difficult Exercises
Exercises 35–43, 56–60

Guided Practice

Ex. 2 Ask students why the domain is restricted.

Independent Practice

Ex. 5–12 The graphs can be used to support the notion of domain as the set of x-values and range as the set of y-values or $f(x)$-values.

Answers

5. All real numbers, all real numbers
6. All real numbers, all real numbers $\geq -\frac{1}{4}$
7. All real numbers, all real numbers $\geq \frac{7}{8}$
8. All real numbers, all real numbers ≤ 1
9. All real numbers ≥ 3, all real numbers ≥ 0
10. All real numbers ≥ -1, all real numbers ≥ 0
11. All real numbers except -1, all real numbers except 0
12. All real numbers except -3, all real numbers except 1.

EXERCISES

Guided Practice

CRITICAL THINKING about the Lesson

1. What is the domain of $f(x) = 6x - 21$? See below.
2. What is the domain of $f(x) = \frac{3x + 2}{x - 5}$? See below.
3. Name the five function operations discussed in this lesson. See below.
4. Consider the functions $f(x) = x + 1$ and $g(x) = 3x$. Find the compositions $f(g(x))$ and $g(f(x))$. Are they the same? $f(g(x)) = 3x + 1$, $g(f(x)) = 3x + 3$; no

Independent Practice

In Exercises 5–12, find the domain and range of the function. See margin.

5. $f(x) = x + 1$

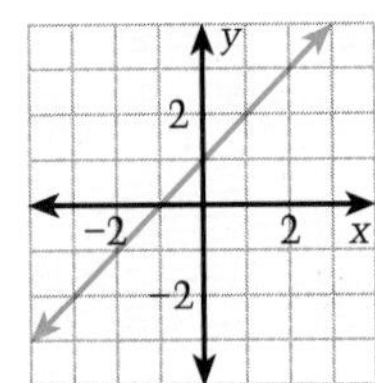

6. $f(x) = x^2 - x$

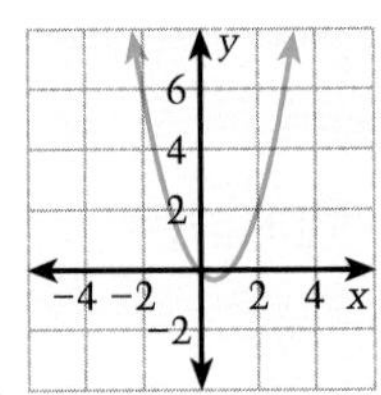

7. $f(x) = 2x^2 - x + 1$

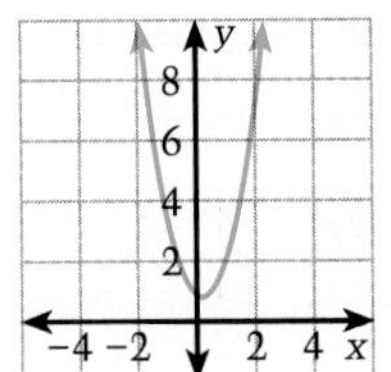

8. $f(x) = 1 - x^2$

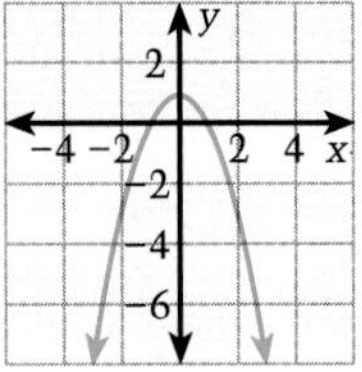

9. $f(x) = \sqrt{x - 3}$

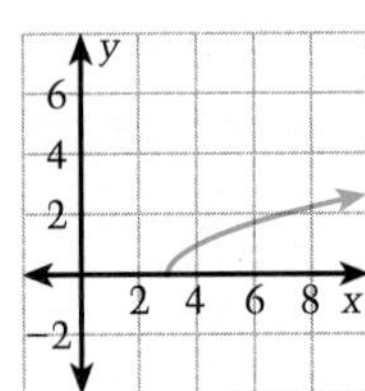

10. $f(x) = \sqrt{x + 1}$

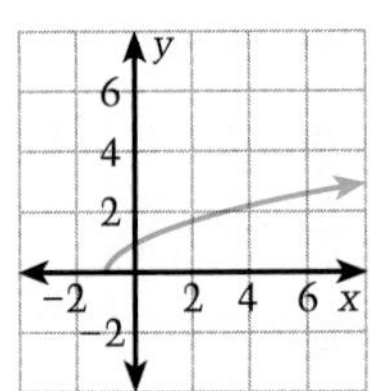

11. $f(x) = \frac{4}{x + 1}$

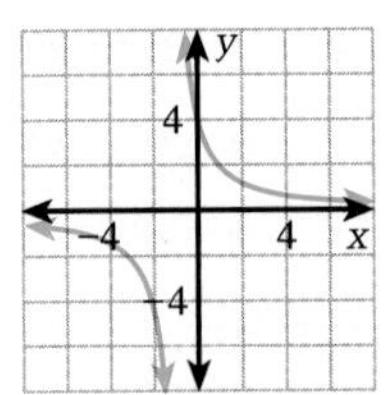

12. $f(x) = \frac{x - 3}{x + 3}$

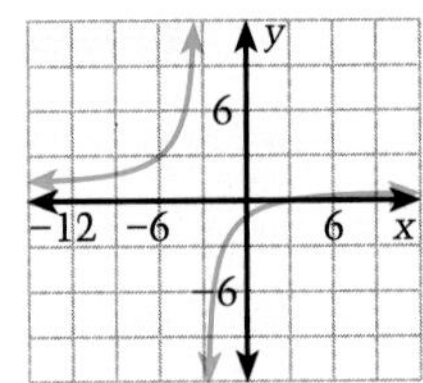

In Exercises 13–20, let $f(x) = x + 1$ and $g(x) = 2x$. Write and simplify an equation for $h(x)$.

13. $h(x) = f(x) + g(x)$ $h(x) = 3x + 1$
14. $h(x) = f(x) - g(x)$ $h(x) = -x + 1$
15. $h(x) = 2f(x) - g(x)$ $h(x) = 2$
16. $h(x) = \frac{1}{2}g(x) + f(x)$ $h(x) = 2x + 1$
17. $h(x) = f(x) \cdot g(x)$ $h(x) = 2x^2 + 2x$
18. $h(x) = g(x) \cdot f(x)$ $h(x) = 2x^2 + 2x$
19. $h(x) = g(x) \div f(x)$ $h(x) = \frac{2x}{x + 1}$
20. $h(x) = f(x) \div g(x)$ $h(x) = \frac{x + 1}{2x}$

In Exercises 21–26, find $f(g(x))$ and $g(f(x))$. See Additional Answers.

21. $f(x) = x + 1$ and $g(x) = 2x$
22. $f(x) = 2x + 1$ and $g(x) = x - 3$
23. $f(x) = x^2$ and $g(x) = x - 1$
24. $f(x) = x^2 - 1$ and $g(x) = x + 2$
25. $f(x) = x - 3$ and $g(x) = x + 3$
26. $f(x) = -x^2 - 1$ and $g(x) = x + 5$

1. All real numbers 2. All real numbers except $x = 5$
3. Addition, subtraction, multiplication, division, and composition

In Exercises 27–34, let $f(x) = 2x - 8$ and $g(x) = x^2$. Write an equation for $h(x)$. Then describe the domain of h. See Additional Answers.

27. $h(x) = 2f(x)$	**28.** $h(x) = f(x) + g(x)$	**29.** $h(x) = g(x) - f(x)$	**30.** $h(x) = 6g(x)$
31. $h(x) = f(x) \div g(x)$	**32.** $h(x) = f(x) \cdot g(x)$	**33.** $h(x) = f(g(x))$	**34.** $h(x) = g(f(x))$

35. *Winter Coat Sale* You have a coupon for $50 off the price of a winter coat. When you arrive at the store, you find that the coats are on sale for 20% off. Use function notation to describe the cost with your coupon and the cost with the 20% discount. $f(x) = x - 50$, $g(x) = 0.8x$

36. Would you pay less for the coat in Exercise 35 if you used your coupon after the discount? Explain. Yes, $10 less

37. *Sales Bonus* You are a sales representative for a clothing manufacturer. You are paid an annual salary plus a bonus of 2% of your sales *over* $200,000. Consider two functions: $f(x) = x - 200{,}000$ and $g(x) = 0.02x$. If x is greater than $200,000, which of the following represents your bonus? Explain.

a. $f(g(x))$ **b.** $g(f(x))$ b, you must subtract first

38. See margin.

38. *A Bridge over Troubled Waters* You are standing on a bridge over a calm pond and drop a pebble, causing ripples of concentric circles in the water. The radius (in feet) of the outer ripple is given by $r(t) = 0.6t$, where t is time in seconds after the pebble hits the water. The area of a circle is given by the function $A(r) = \pi r^2$. Find an equation for the composite function $A(r(t))$. What is the input and output of this composite function?

39. *Weekly Production Cost* The weekly cost of producing x units in a manufacturing process is given by the function $C(x) = 60x + 750$. The number of units produced in t hours during a week is given by $x(t) = 50t$, $0 \le t \le 40$. Find, simplify, and interpret $C(x(t))$.

Dogs and Cats **In Exercises 40 and 41, use the following information.**

For 1982 through 1990, the numbers of dogs, d (in millions), and cats, c (in millions), in the United States can be modeled by the following functions. (Let $t = 2$ represent 1982.)

Dogs: $d(t) = 0.875t + 46.25$

Cats: $c(t) = 2.5t + 38$

(Source: Market Research Corporation of America)

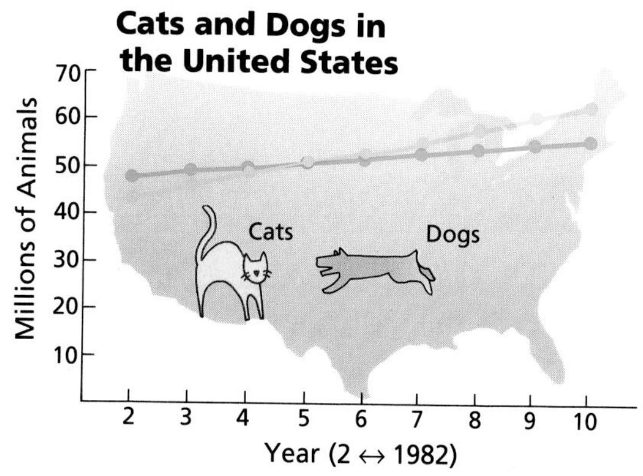

40. How many dogs were there in 1982? 48 million

41. Find, simplify, and interpret the sum of $d(t)$ and $c(t)$. See margin.

EXTEND Ex. 5–12 Ask students to explain the necessity for the restrictions placed on the domains. For example, in Exercise 9, the square-root function is restricted to real numbers, so the domain for x is restricted to values greater than or equal to 3.

EXTEND Ex. 21–26 Does $f(g(x)) = g(f(x))$ for any of these exercises?
Only for Exercise 25.

▶ **Ex. 33–34** Finding the domain for the function composition is simplified since both $f(x)$ and $g(x)$ are defined for all real numbers.

▶ **Ex. 35–39** Encourage students to solve these exercises using the format illustrated in Example 4.

Answers

38. $A(r(t)) = 0.36\pi t^2$; real numbers ≥ 0

39. $C(x(t)) = 60(50t) + 750 = 3000t + 750$, $0 \le t \le 40$; cost: $750 to $120,750

▶ **Ex. 40–41** These exercises should be assigned as a pair. They are examples of real-life applications of linear functions.

Answer

41. $(0.875t + 46.25) + (2.5t + 38) = 3.375t + 84.25$; total number of dogs and cats in the U.S. in a given year.

EXTEND Ex. 40–41 Ask students to evaluate $d(t)$ and $c(t)$ for a given year such as 1996 ($t = 14$) and predict the number of dogs and cats.
dogs: 58.5 million, cats: 73 million.

Also, you could ask students to find the year in which the number of cats exceeded the number of dogs.
Somewhere in the fifth year or sometime in 1985.

▶ **Ex. 42–43** Assign these exercises as a pair.

▶ **Ex. 60** Encourage students to consider simple functions such as $f(x) = -x$ and $g(x) = x^2$ since students have not been introduced to inverse functions.

Enrichment Activities

These activities require students to go beyond lesson goals.

EXTEND Ex. 13–20 Consider the linear functions $f(x) = x + 1$ and $g(x) = 2x$.

1. Evaluate the following for integer values $-3 \le x \le 3$.
   a. $f(x)$ b. $g(x)$
   c. $f(x) + g(x)$ d. $f(x) - g(x)$
   e. $f(x) \cdot g(x)$ f. $f(x) \div g(x)$
   Look for a pattern. Did you have to evaluate the functions every time?
   a. −2, −1, 0, 1, 2, 3, 4
   b. −6, −4, −2, 0, 2, 4, 6
   c. −8, −5, −2, 1, 4, 7, 10
   d. 4, 3, 2, 1, 0, −1, −2
   e. 12, 4, 0, 0, 4, 12, 24
   f. $\frac{1}{3}, \frac{1}{4}, 0, -, 1, \frac{3}{4}, \frac{2}{3}$
   Pattern: To evaluate the functions in c.–f. you can add, subtract, multiply, or divide the $f(x)$- and $g(x)$- values.
2. Use the pattern to graph $f(x)$, $g(x)$, $f(x) + g(x)$, and $f(x) - g(x)$ on the same coordinate grid. Use the graphs to check your answers to Exercises 13 and 14.

TECHNOLOGY If you have a graphics calculator, you can graph the four functions simultaneously.

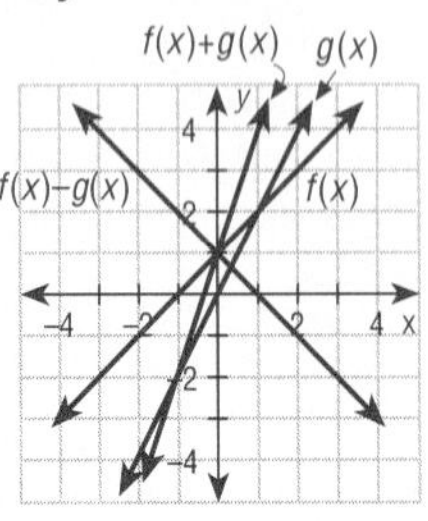

Extending the School Year **In Exercises 42 and 43, use the following information.**

Polls were taken in 1982 and in 1991 to determine whether American adults think the school year should be extended. The results can be represented by the following functions, where the function values are percents in decimal form. (Let $t = 2$ represent 1982.)

Favor: $f(t) = 0.0156t + 0.3389$

Oppose: $g(t) = -0.0122t + 0.5544$

Don't Know: $h(t) = -0.0034t + 0.1067$

(Source: Gallup Poll)

Should the School Year Be Extended?

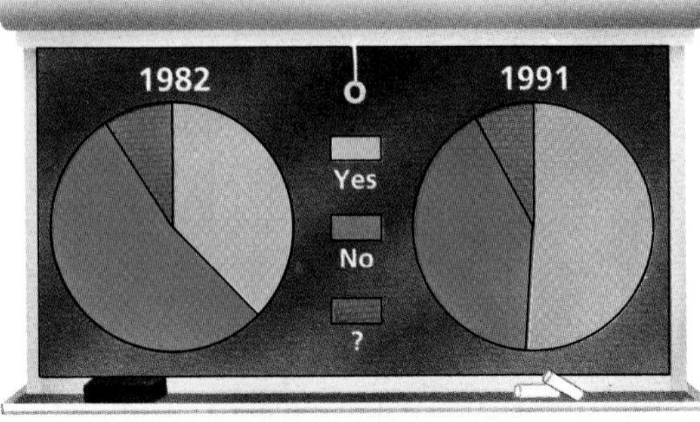

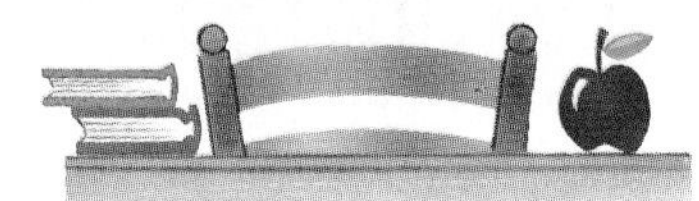

42. What percent favored extending the school year in 1982? In 1991? 37.01%, 51.05%

43. Find and interpret the sum of $f(t)$, $g(t)$, and $h(t)$. 1, represents 100% of people polled

Integrated Review

In Exercises 44–55, let $f(x) = x - 4$ and $g(x) = -x + 6$. Find the indicated value.

44. $f(3) + g(3)$ 2
45. $f(1) - g(1)$ −8
46. $-3f(3)$ 3
47. $10 + g(6)$ 10
48. $f(2) \cdot g(2)$ −8
49. $f(5) \cdot g(5)$ 1
50. $f(-4) \div g(-4)$ $-\frac{4}{5}$
51. $f(4) \div g(4)$ 0
52. $f(g(2))$ 0
53. $g(f(0))$ 10
54. $g(f(1))$ 9
55. $f(g(-3))$ 5

56. *College Entrance Exam Sample* Fill in the box with either <, >, or =. =

Percent of increase when a \$10.00 price is increased by \$1.00	?	Percent of increase when a \$15.00 price is increased by \$1.50

57. *College Entrance Exam Sample* In a certain army post, 30% of the enlistees are from New York State, and 10% of these are from New York City. What percent of the enlistees in the post are from New York City? c
a. 0.03% **b.** 0.3% **c.** 3% **d.** 13% **e.** 20%

Exploration and Extension

58. Find $f(x)$ and $g(x)$ such that $f(g(x)) = (x + 1)^2$. $f(x) = x^2$, $g(x) = x + 1$

59. Find $f(x)$ and $g(x)$ such that $g(f(x)) = \sqrt{x - 5}$. $f(x) = x - 5$, $g(x) = \sqrt{x}$

60. For most functions f and g that are different, the composite functions $f(g(x))$ and $g(f(x))$ are also different. Find two functions f and g, such that f and g are different, but $f(g(x))$ and $g(f(x))$ are the same. See below.

60. Answers vary. $f(x) = x^2$, $g(x) = \sqrt{x}$; $f(x) = x + 3$, $g(x) = x - 3$

Mixed REVIEW

1. $\{-3, -2, -1, 0, 1, 2\}$ 7. $4a + 10b + 3c$ 8, 9. See margin.
11. x: $-3, 2$, $y = -6$ 17. $y = \frac{3}{4}x - \frac{1}{2}$

1. Find all integers such that $-4 < x < 3$.
2. Add: $\frac{2}{3} + \frac{3}{4}$. $1\frac{5}{12}$
3. Simplify: $(3.6 \times 10^6) \div (1.2 \times 10^3)$. 3000
4. Solve $(x - 5)^2 = 16$. 1, 9
5. Sketch the graph of $y = (x + 3)^2 - 4$. **(5.3)** See Additional Answers.
6. Find the y-intercept: $y = 3x^2 - 7x + 2$. **(2.3)** 2
7. Add: $(3a + 4b - c) + (a + 6b + 4c)$.
8. Sketch the graph of $y = \frac{2}{5}x - 3$. **(2.1)**
9. Solve for x: $ax + by = cx$. **(1.5)**
10. Find the vertex of $y = 3|x - 2| + 7$. **(2.6)** (2, 7)
11. Find the intercepts of $y = x^2 + x - 6$. **(5.3)**
12. Has $2x^2 - 4x - 5 = 0$ any real solutions? **(5.4)** Yes
13. Solve $x^2 = -49$. **(5.5)** $\pm 7i$
14. Solve $9x^2 = 49$. **(5.1)** $\pm\frac{7}{3}$
15. Simplify: $(0.03)(8.2)$. 0.246
16. Write the reciprocal of $\frac{5}{16}$. $\frac{16}{5}$
17. Write an equation of the line parallel to $y = \frac{3}{4}x + 7$, through (2, 1). **(2.2)**
18. Find the area of the triangle with vertices at (0, 4), (4, 6), and (2, 1). **(4.3)** 8
19. Solve $\begin{bmatrix} 2 & 1 \\ 0 & -1 \end{bmatrix} X = \begin{bmatrix} 7 & -4 \\ -1 & 0 \end{bmatrix}$. **(4.4)** See margin.
20. Write $y = x^2 + 8x + 13$ in completed square form. **(5.3)** $y = (x + 4)^2 - 3$

Milestones FRACTALS

700 500 300 100 B.C. A.D. 100 300 500 700 900 1100 1300 1500 1700 1900 2100

Euclidean Geometry — Fractal Geometry

The literal translation of the word *geometry* is *earth measurement.* Euclidean geometry measures, or describes quantitatively, many of the shapes found in the natural world, such as circles, spheres, and cones. However, it doesn't describe shapes in nature such as clouds, coastlines, and mountain ranges. Until very recently it was thought that such shapes could not be described in mathematical terms.

In the 1970's, Benoit Mandelbrot discovered that these shapes do have a pattern in their irregularity. He found these shapes are made up of smaller parts that are scaled-down versions of the shape itself. These parts are made up of yet smaller parts that again resemble the shape and so on through numerous reductions. Once the smallest shape that repeats is quantified, computers are helpful in generating the larger image. Mandelbrot coined the term *fractal* for these shapes and for the geometry used to describe them.

By R. F. Voss; from the Fractal Geometry of Nature *by Benoit B. Mandelbrot, 1982*

The study of fractals involves the concept of chaos. Find out what this is.

Chaos describes behavior of major proportions that arises from a seemingly minor behavior. An example would be a landslide resulting from the dislodging of a single pebble.

3. **EXTEND EX. 27–34** Let $f(x) = 2x - 8$ and $g(x) = x^2$. Graph $f(x)$, $g(x)$, $f(x) + g(x)$, and $f(x) - g(x)$ on the same coordinate grid. Does the pattern still hold when the functions are not linear? Yes

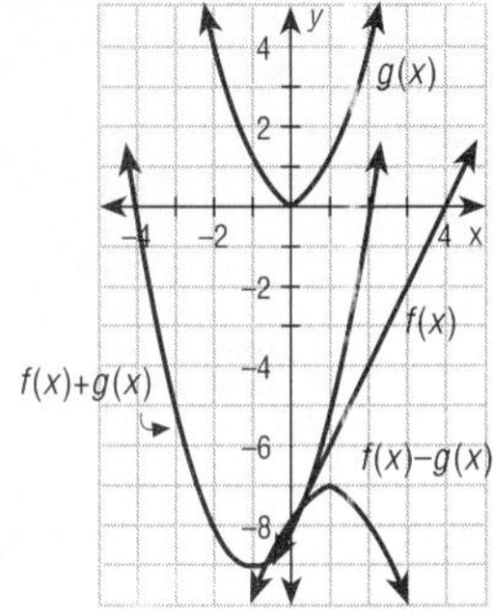

Answers

8.

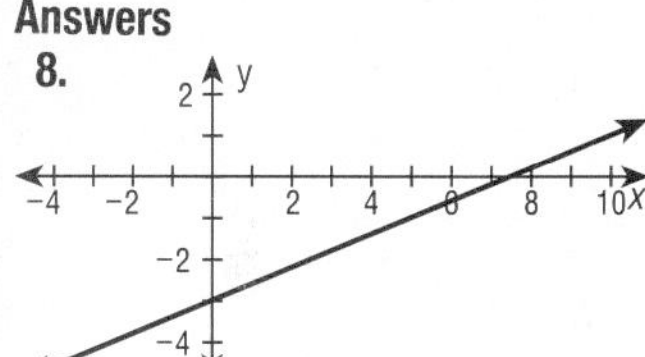

9. $x = \frac{by}{c - a}$

19. $\begin{bmatrix} 3 & -2 \\ 1 & 0 \end{bmatrix}$

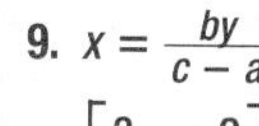

Milestones

Language Arts: *Critical Reading*

1. The words *fractal* and *fraction* come from the same Latin root word. What is similar about the meanings of the two words?
   Both words have to do with part/whole relationships
2. About how many years elapsed between the development of Euclidean geometry and the development of fractal geometry?
   Accept answers between 2200 and 2400 years.

Problem of the Day

$\overline{AB}$ is the side of a square and the diameter of a semicircle. A smaller square is inscribed in the semicircle. The area of the smaller square is what fraction of the area of the larger square? $\frac{1}{5}$

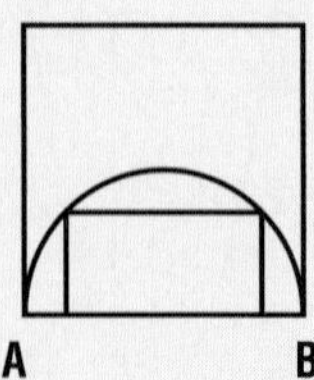

ORGANIZER

Warm-Up Exercises

1. Express y in terms of x.
   a. $3y + 4x - 9 = 0$
   b. $-2y - 12x + 8 = 0$
   c. $6x = 4y - 24$
   d. $x = 2y + 14$

   a. $y = -\frac{4}{3}x + 3$,
   b. $y = -6x + 4$,
   c. $y = \frac{3}{2}x + 6$,
   d. $y = \frac{1}{2}x - 7$
2. Evaluate $f(g(x))$ for the indicated values of x.
   a. $f(x) = 4x + 2$, $g(x) = x - 3$, $x = -3$
   b. $f(x) = 8x - 12$, $g(x) = 2x$, $x = 4$

   a. -22, b. 52

Lesson Resources

Teaching Tools
- Transparencies: 7, 8, 9, 13
- Problem of the Day: 6.3
- Warm-up Exercises: 6.3
- Answer Masters: 6.3

Extra Practice: 6.3
Applications Handbook: p. 54

6.3 Inverse Functions

What you should learn:

Goal 1 How to identify inverse relations and inverse functions and to verify that two functions are inverses of each other

Goal 2 How to use inverse functions in real-life situations

Why you should learn it:

You can model many pairs of situations with pairs of inverse functions. For example, you could find the sum of the measures of the angles of a polygon given the number of sides, or find the number of sides given the sum of the measures of the angles.

Goal 1 Finding Inverses

The **inverse** of a relation is the set of ordered pairs obtained by switching the coordinates of each ordered pair in the relation. Here is an example.

$\{(0, -3), (1, -1), (2, 1), (3, 3), (4, 5)\}$ *Original relation*

$\{(-3, 0), (-1, 1), (1, 2), (3, 3), (5, 4)\}$ *Inverse relation*

The graph of the inverse is the **reflection** of the graph of the original relation. The mirror of the reflection is the line $y = x$.

Graph of Original Relation

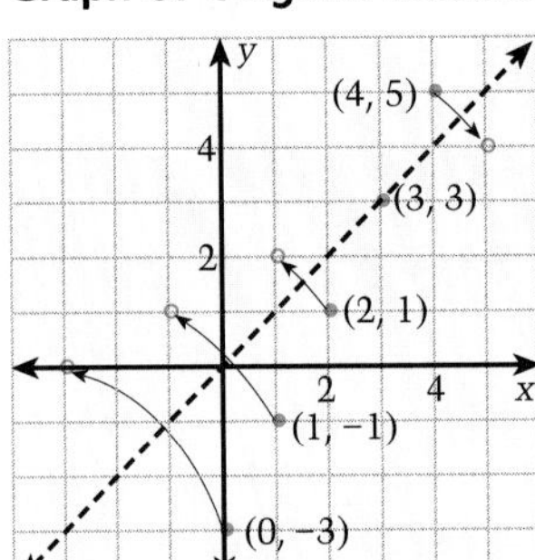

Graph of Inverse Relation

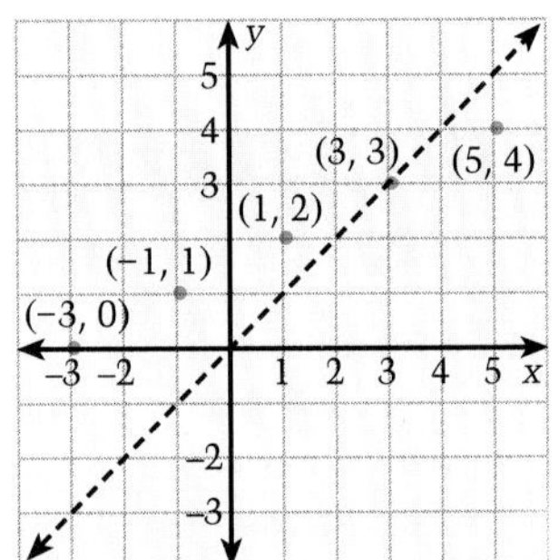

To find the inverse of a relation that is given by an equation in x and y, switch the roles of x and y in the equation.

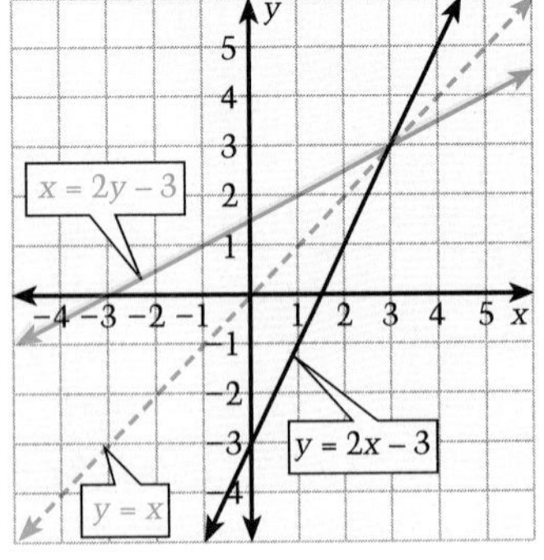

The graphs of $y = 2x - 3$ and $x = 2y - 3$ are reflections of each other in the line $y = x$.

Example 1 *Finding an Inverse Relation*

Find an equation for the inverse of the relation $y = 2x - 3$.

Solution Begin with the original equation. Then switch the roles of x and y.

$y = 2x - 3$ *Original equation*

$x = 2y - 3$ *Switch the roles of x and y to find an equation for the inverse relation.*

Be sure you see that these two equations are *not* equivalent. When finding an equation of an inverse relation, you are forming a *different* equation whose solutions reflect the switching of x and y. You can confirm this by sketching their graphs. ■

The inverse of a function is found in the same way you find the inverse of a relation (because a function is simply a special type of relation). It is important to realize, however, that the inverse of a function may not itself be a function.

Example 2 Finding the Inverse of a Function

Find the inverse of the function $f(x) = 3x - 1$. Is the inverse a function of x?

Solution To begin, replace $f(x)$ by y. Then switch x and y in the equation.

$f(x) = 3x - 1$ *Original function*

$y = 3x - 1$ *Replace f(x) by y.*

$x = 3y - 1$ *Switch x and y to obtain inverse.*

Thus the inverse relation is given by the linear equation $x = 3y - 1$. To check whether this equation describes a function of x, sketch its graph—it helps to write the equation in slope-intercept form, $y = \frac{1}{3}x + \frac{1}{3}$. Because no vertical line intersects the graph twice, you can conclude that the inverse relation is itself a function. In such cases, function notation is usually used to describe the inverse.

$g(x) = \frac{1}{3}x + \frac{1}{3}$ *Inverse function* ■

Example 3 An Inverse Relation That Is Not a Function

Find the inverse of the function $f(x) = x^2 + 1$. Is the inverse a function of x?

Solution

$f(x) = x^2 + 1$ *Original function*

$y = x^2 + 1$ *Replace f(x) by y.*

$x = y^2 + 1$ *Switch x and y to obtain inverse.*

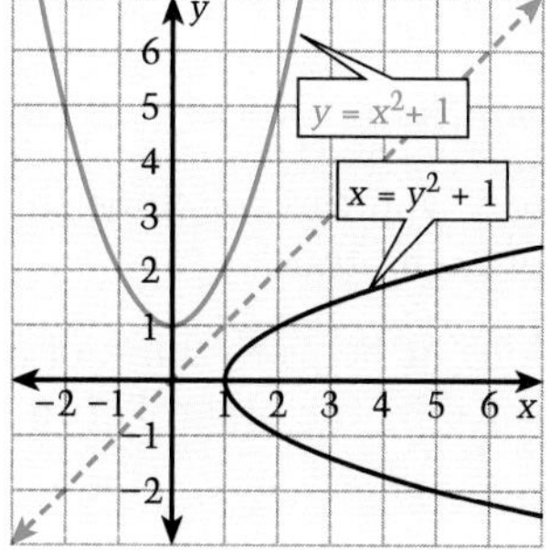

Thus the inverse relation is given by the quadratic equation $x = y^2 + 1$. By constructing a table of values, you can graph this equation, as shown at the left. Notice that the graph does not pass the vertical line test for functions. Thus this inverse relation is *not* a function, and you cannot use function notation to represent the inverse relation. ■

The graph of a function can tell you whether the inverse is also a function. If no *horizontal line* intersects the graph of f twice, then the inverse of f is itself a function. Why?

LESSON Notes

Example 1

Convince students (or have them convince themselves) that $x = 2y - 3$ is the inverse of $y = 2x - 3$ by finding sets of ordered pairs that demonstrate the relationship between the two equations. For example, a set of ordered pairs associated with $y = 2x - 3$ is $\{(-3,-9), (-2,-7), (-1,-5), (0,-3), (1,-1), (2, 1)\}$. Evaluating $x = 2y - 3$ for x by setting y equal to the former x-values gives the following set of ordered pairs. $\{(-9,-3), (-7,-2), (-5,-1), (-3, 0), (-1, 1), (1, 2)\}$.

An Alternate Approach
Using Technology
Graphically, the inverse of a linear relation can easily be shown to be the mirror image of the original relation. If students have a graphics calculator or a computer graphing program, have them graph the original equation, the inverse relation, and the equation $y = x$ simultaneously.

Examples 2–3

Common notation for the inverse function of $f(x)$ is $f^{-1}(x)$. One disadvantage of using this notation is that many students think that the "-1" is an exponent. Consider the differences among $f^{-1}(3)$, $(f(3))^{-1}$, and $f(3^{-1})$ when $f(x) = 3x - 1$.

$f^{-1}(3) = \frac{4}{3}$, $(f(3))^{-1} = \frac{1}{8}$, $f(3^{-1}) = f\left(\frac{1}{3}\right) = 0$.

Example 4

Students may find it easier to first verify that $f(x)$ and $g(x)$ are inverses by using fixed values of x. For example, show that $f(g(5)) = 5$ and that $g(f(5)) = 5$.

Example 5

It is easy to distinguish between the original function, $S = 180n - 360$, and its inverse, the function $n = \frac{1}{180}S + 2$, because of the way in which the variables are used. However, it is common practice to use the same variable, often x, when expressing a function, f, and its inverse, g. Help students see that, although the variables have been renamed, the relationships expressed by the functions are as they should be.

Extra Examples

Here are additional examples similar to **Examples 1–4.**

1. Find the inverse relation of each function. Indicate whether the inverse relation is a function or not.
   a. $y = 4x + 8$
   b. $f(x) = 7 - 5x$
   c. $f(x) = 3x^2 + 2x$

   a. $x = 4y + 8$, a function;
   b. $x = 7 - 5y$, a function;
   c. $x = 3y^2 + 2y$, not a function
2. Determine whether the pair of functions are inverses of each other or not.
   a. $f(x) = 2x + 4$ and $g(x) = \frac{1}{2}x - 2$
   b. $f(x) = \frac{1}{3}x + 2$ and $g(x) = 3x - 2$
   c. $f(x) = 12 + 4x$ and $g(x) = \frac{1}{4}x - 3$

   a. and **c.** are inverses of each other, **b.** are not inverses

If g is the inverse of f, then it is also true that f is the inverse of g. This means the functions f and g are inverses of each other. You can verify that two functions defined by equations are inverses of each other by using composition of functions.

Using Composition to Verify Inverse Functions

If the functions f and g are inverses of each other, then

$f(g(x)) = x$ and $g(f(x)) = x$.

Example 4 *Verifying Inverse Functions*

Verify that the functions from Example 2 are inverses of each other.

$f(x) = 3x - 1$ and $g(x) = \frac{1}{3}x + \frac{1}{3}$

Solution

$f(g(x)) = f\left(\frac{1}{3}x + \frac{1}{3}\right)$	*Substitute $\frac{1}{3}x + \frac{1}{3}$ for g(x).*
$= 3\left(\frac{1}{3}x + \frac{1}{3}\right) - 1$	*Apply f(□) = 3□ − 1.*
$= 3\left(\frac{1}{3}x\right) + 3\left(\frac{1}{3}\right) - 1$	*Distributive Property*
$= x + 1 - 1$	*Simplify.*
$= x$	*Simplify.*
$g(f(x)) = g(3x - 1)$	*Substitute 3x − 1 for f(x).*
$= \frac{1}{3}(3x - 1) + \frac{1}{3}$	*Apply g(□) = $\frac{1}{3}$□ + $\frac{1}{3}$.*
$= \frac{1}{3}(3x) - \frac{1}{3}(1) + \frac{1}{3}$	*Distributive Property*
$= x - \frac{1}{3} + \frac{1}{3}$	*Simplify.*
$= x$	*Simplify.*

Because both compositions produce x, you can conclude that the functions f and g are inverses of each other. ■

When working with inverse functions, it helps to think of one as "undoing" the operations performed by the other. For instance, the functions

$f(x) = 4x$ and $g(x) = \frac{1}{4}x$

are inverses because if you multiply x by 4 and then multiply the result by $\frac{1}{4}$, you come back to x. Similarly, the functions $f(x) = x + 5$ and $g(x) = x - 5$ are inverses. The first function adds 5 to x, and the second function subtracts 5, which brings you back to x.

Goal 2 Using Inverse Functions in Real Life

A formula from geometry states that the sum, S (in degrees), of the angles of an n-sided polygon is $S = 180n - 360$.

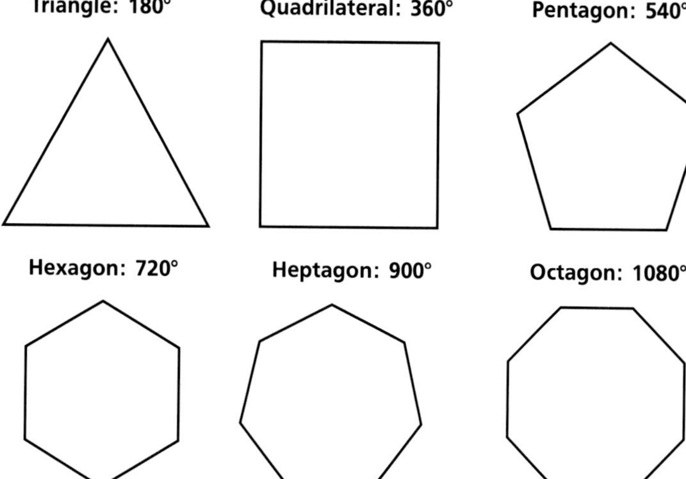

Connections
Geometry

A stop sign is a familiar example of an octagon.

Example 5 *Rewriting a Formula*

The formula $S = 180n - 360$ gives the sum of the angles of an n-sided polygon, where n is the input and S the output. Solve the formula for n so that S is the input and n the output.

Solution Solve for n in terms of S.

$n = \frac{1}{180}S + 2$ *S is input, n is output.*

This "inverse" formula could be used to find the number of sides of a polygon when you are given the sum, S, of its angles. For example, a polygon whose interior angles have a sum of 1800° has $n = \frac{1}{180}(1800) + 2 = 12$ sides. ■

You could write the formulas in Example 5 as the inverse functions $f(x) = 180x - 360$ and $g(x) = \frac{1}{180}x + 2$. But when the variables have applications, letters that are easily associated with the quantities being measured are often used.

Communicating about ALGEBRA

SHARING IDEAS about the Lesson See Additional Answers.

Find the inverse of the function. Is the inverse itself a function? Use a graph to help explain your answer.

A. $f(x) = \frac{2}{3}x - 4$ **B.** $f(x) = \frac{1}{2}x^2 + 3$

Check Understanding

1. Is the inverse of a function also a function?
Not necessarily
2. How can the graph of the original function be used to decide whether the inverse relation will also be a function?
If no horizontal line intersects the graph of f in more than one point, then the inverse of f will be a function.
3. What does it mean for functions to be inverses of each other?
See the top of page 300.

Communicating about ALGEBRA

EXTEND *Communicating*
Have students determine if the inverses of the following functions are functions. If the inverses are not functions, have students describe how they can be modified to become functions.

a. $f(x) = x^2$ **b.** $f(x) = x^2 + 2x$
c. $f(x) = 3x^2$ **d.** $f(x) = 2x^2 + 4$

Generally, you need to restrict the domain and the range of the inverse relation so that there is only one output value for each possible input value. For example, in Part a, the inverse is not a function. To make the inverse a function, you need to restrict the domain of the inverse so that x is nonnegative and you need to restrict the inverse, $g(x) = \pm\sqrt{x}$, to either $g(x) = +\sqrt{x}$ or $g(x) = -\sqrt{x}$.

None of the inverses in Parts b–d is a functions.

b. $g(x) = \sqrt{x+1} - 1$ or $g(x) = -\sqrt{x+1} - 1, x \geq -1$

c. $g(x) = \frac{\sqrt{3x}}{3}$ or $g(x) = -\frac{\sqrt{3x}}{3}$, $x \geq 0$

d. $g(x) = \frac{\sqrt{2(x-4)}}{2}$ or $g(x) = \frac{\sqrt{2(x-4)}}{2}$, $x \geq 4$

EXERCISE Notes

ASSIGNMENT GUIDE

Basic/Average: Ex. 1–6, 10–14, 21–37 odd, 41–45 odd, 46, 51–54, 57–58

Above Average: Ex. 1–6, 10–14, 21–37 odd, 41–46, 51–54, 58–60

Advanced: Ex. 1–6, 10–14, 21–37, 41–46, 51–54, 59–62

Selected Answers
Exercises 1–6, 7–57 odd

More Difficult Exercises
Exercises 41–44, 59–62

Guided Practice

Ex. 1–6 Use these exercises to help students connect the algebraic and geometric processes of finding an inverse function. In Exercise 4, $f(x)$ has an inverse relation that is not a function.

Answers

1. Switch the coordinates of each ordered pair, for example: $\{(-3, 4), (-1, 2)\} \rightarrow \{(4, -3), (2, -1)\}$
2. Switch x and y, for example: $y = 3x - 1 \rightarrow x = 3y - 1$
3. $x = \frac{1}{2}y - 4$ or $g(x) = 2x + 8$, yes

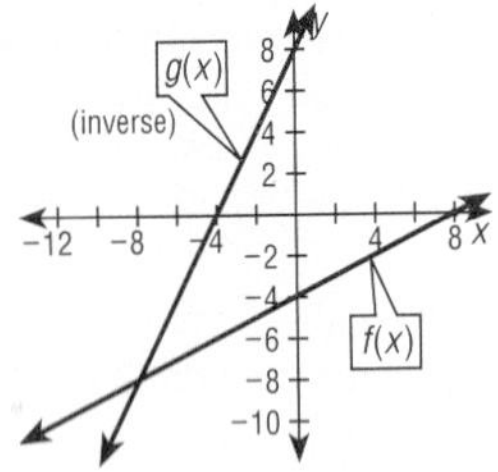

4. No, its graph does not pass the vertical line test.
5. They are reflections of each other in the line $y = x$.
6. $f(g(x)) = 2\left(\frac{1}{2}x - 2\right) + 4 = x$, $g(f(x)) = \frac{1}{2}(2x + 4) - 2 = x$

EXERCISES

Guided Practice

CRITICAL THINKING about the Lesson See margin.

1. Describe how to find the inverse relation of a relation given by a set of ordered pairs. Give an example.
2. Describe how to find the inverse relation of a relation given by an equation in x and y. Give an example.
3. Find an equation that represents the inverse of the function $f(x) = \frac{1}{2}x - 4$. Sketch the graphs of f and its inverse. Is the inverse of f a function?
4. The graphs of $f(x) = -|x| - 1$ and the inverse of f are shown at the right. Is the inverse of f a function of x? Explain.
5. How is the graph of a relation related to the graph of its inverse?
6. Verify that $f(x) = 2x + 4$ and $g(x) = \frac{1}{2}x - 2$ are inverses of each other.

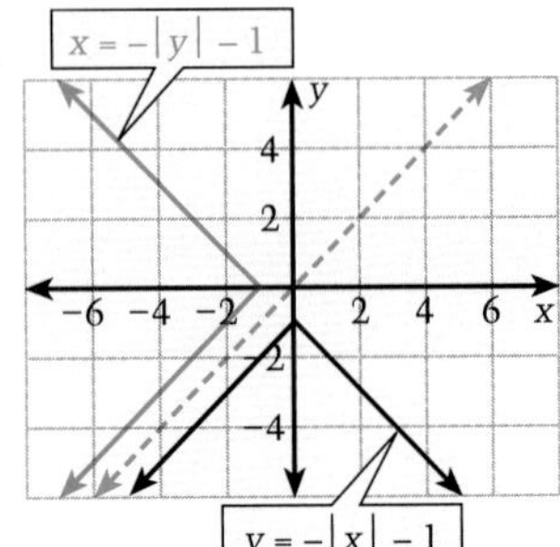

Independent Practice

In Exercises 7–10, find the inverse of the relation. See below.

7. $\{(1, 5), (4, -3), (1, 9), (0, -1)\}$
8. $\{(0, -4), (-4, 3), (3, 2), (2, 1)\}$
9. $\{(-3, 1), (-2, 2), (-1, 0), (3, 3)\}$
10. $\{(1, -4), (3, -2), (0, -6), (-3, 2)\}$

In Exercises 11–14, match the graph with the graph of its inverse.

11. 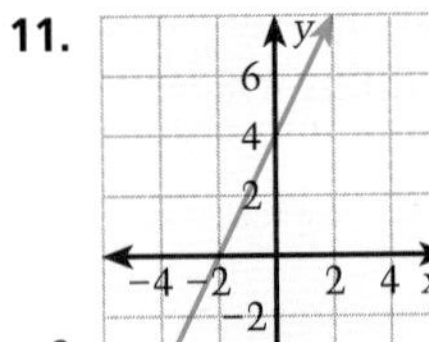 c

12. 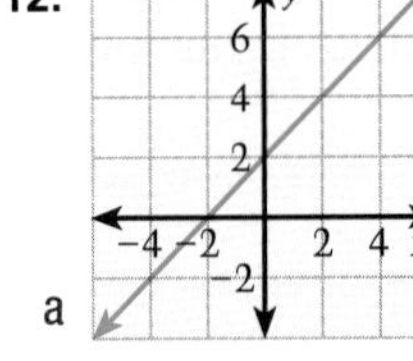 a

13. 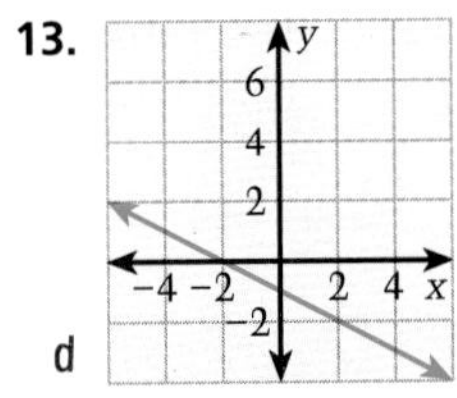 d

14. 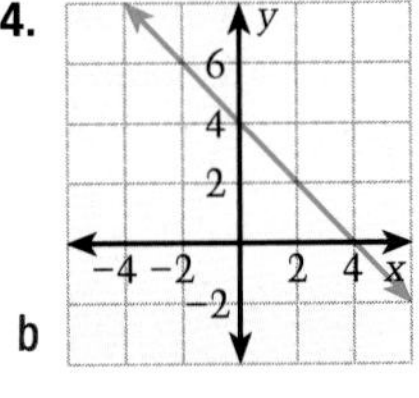 b

a. 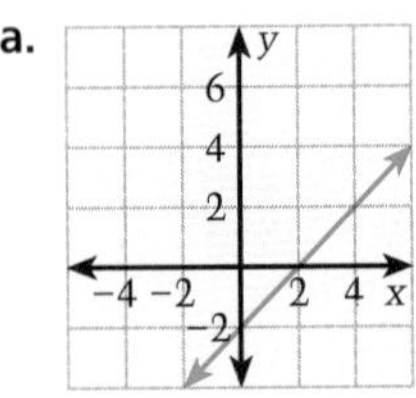

b. 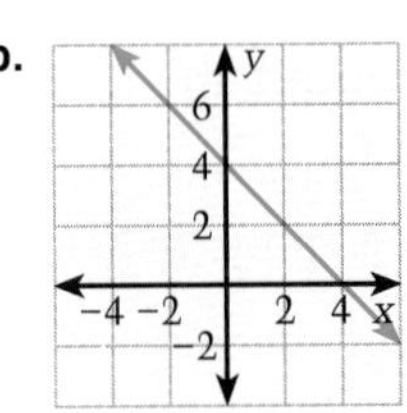

c. 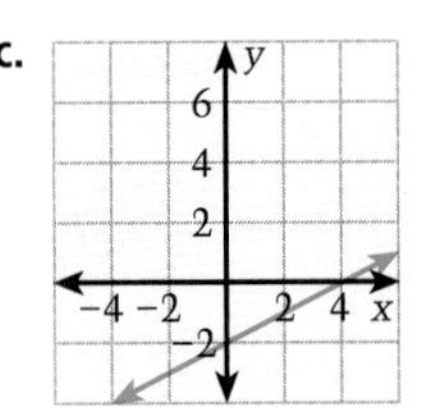

d. 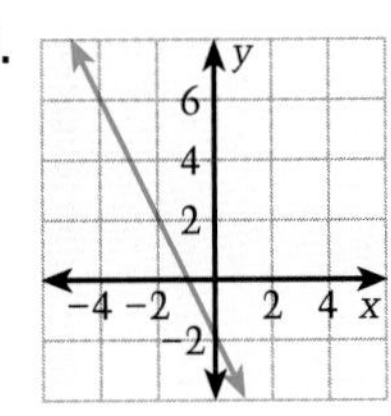

7. $\{(5, 1), (-3, 4), (9, 1), (-1, 0)\}$ 8. $\{(-4, 0), (3, -4), (2, 3), (1, 2)\}$ 9. $\{(1, -3), (2, -2), (0, -1), (3, 3)\}$
10. $\{(-4, 1), (-2, 3), (-6, 0), (2, -3)\}$

In Exercises 15–22, write an equation for the inverse of the relation. See below.

15. $y = -3x + 5$ 16. $y = 2x - 3$ 17. $y = x + 4$ 18. $y = -5x + 9$
19. $y = 12x - 6$ 20. $y = -13x + 6$ 21. $y = 4x - 1$ 22. $y = 9x - 14$

In Exercises 23–30, sketch the function and its inverse in the same coordinate plane. Is the inverse a function of x? See Additional Answers.

23. $f(x) = x + 3$ Yes 24. $f(x) = -x + 2$ Yes 25. $f(x) = 3x + 4$ Yes 26. $f(x) = 2x^2 + 4$ No
27. $f(x) = x^2 + 1$ No 28. $f(x) = -x^2 + 3$ No 29. $f(x) = -x^2 - 4$ No 30. $f(x) = \frac{1}{2}x + 9$ Yes

In Exercises 31–34, sketch the graph of the function. Use the graph of f to decide whether the inverse of f is a function of x. See Additional Answers.

31. $f(x) = x + 1$ Yes 32. $f(x) = -2x^2 - 1$ No 33. $f(x) = x^2 - 2$ No 34. $f(x) = |x| - 1$ No

In Exercises 35–40, verify that f and g are inverses of each other. See margin.

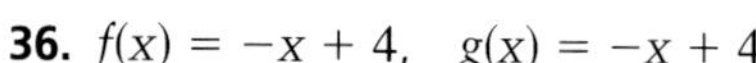

35. $f(x) = x + 9, \quad g(x) = x - 9$ 36. $f(x) = -x + 4, \quad g(x) = -x + 4$
37. $f(x) = 2x - 1, \quad g(x) = \frac{1}{2}x + \frac{1}{2}$ 38. $f(x) = \frac{1}{2}x + 2, \quad g(x) = 2x - 4$
39. $f(x) = -3x + \frac{1}{2}, \quad g(x) = -\frac{1}{3}x + \frac{1}{6}$ 40. $f(x) = -4x + 8, \quad g(x) = -\frac{1}{4}x + 2$

41. *Temperature Conversion* The formula to convert temperature in degrees Celsius to temperature in degrees Fahrenheit is

$F = \frac{9}{5}C + 32.$ $C = \frac{5}{9}(F - 32)$

For this formula, C is the input and F is the output. Rewrite the formula so that F is the input and C is the output.

42. *Geometry* The height h of an equilateral triangle with sides of length s is

$h = \frac{\sqrt{3}s}{2}.$ $s = \frac{2h}{\sqrt{3}}$

Rewrite the formula so that the height is the input and the length of the sides is the output.

43. *Bowling Handicap* You belong to a bowling league in which each bowler's handicap, H, is determined by his or her average, A, according to the following formula.

$H = 0.8(200 - A)$ $A = 200 - \frac{H}{0.8}$, 160

(If the bowler's average is over 200, the handicap is 0.) Rewrite this formula so that the handicap is the input and the average is the output. If your handicap is 32, what is your average?

44. *State Sales Tax* In a state that has a 6% sales tax, the total cost, C, of a taxable item that has a price of p is $C = 0.06p + p$. Rewrite this formula so that C is the input and p is the output. $p = \frac{C}{1.06}$

15. $x = -3y + 5$ 16. $x = 2y - 3$ 17. $x = y + 4$ 18. $x = -5y + 9$ 19. $x = 12y - 6$
20. $x = -13y + 6$ 21. $x = 4y - 1$ 22. $x = 9y - 14$

Independent Practice

▶ **Ex. 11–14** Assign these exercises as a group. Be sure students understand the idea of a reflection in the line $y = x$.

▶ **Ex. 15–22** Students should use Example 1 on page 298 as a solution model for these exercises.

▶ **Ex. 23–34** Students should use Examples 2 and 3 on page 299 as a solution model for these exercises.

▶ **Ex. 35–40** Encourage students to show that $f(g(x)) = g(f(x)) = x$.

Answers

35. $f(g(x)) = (x - 9) + 9 = x$, $g(f(x)) = (x + 9) - 9 = x$

36. $f(g(x)) = (-x + 4) + 4 = x$, $g(f(x)) = (-x + 4) + 4 = x$

37. $f(g(x)) = 2\left(\frac{1}{2}x + \frac{1}{2}\right) - 1 = x$, $g(f(x)) = \frac{1}{2}(2x - 1) + \frac{1}{2} = x$

38. $f(g(x)) = \frac{1}{2}(2x - 4) + 2 = x$, $g(f(x)) = 2\left(\frac{1}{2}x + 2\right) - 4 = x$

39. $f(g(x)) = -3\left(-\frac{1}{3}x + \frac{1}{6}\right) + \frac{1}{2} = x$, $g(f(x)) = -\frac{1}{3}\left(-3x + \frac{1}{2}\right) + \frac{1}{6} = x$

40. $f(g(x)) = -4\left(-\frac{1}{4}x + 2\right) + 8 = x$, $g(f(x)) = -\frac{1}{4}(-4x + 8) + 2 = x$

▶ **Ex. 41–44** These exercises connect the rewriting of a literal equation or formula, so that the dependent and independent variables are interchanged, to finding the inverse relation for the literal equation or formula.

EXTEND Ex. 45–46 Ask students to translate their states' Tax Freedom Day to a corresponding month and day. For example, April 29 would be Alabama's day in a nonleap year.

Answers

51. Yes

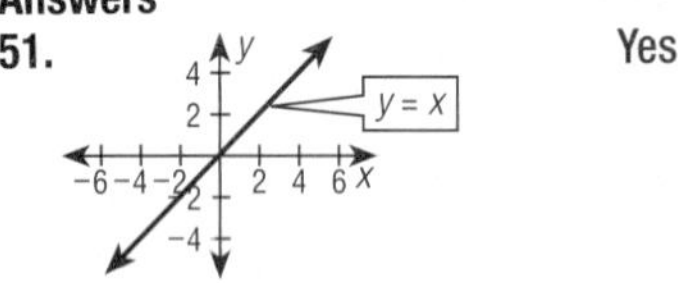

52. Yes

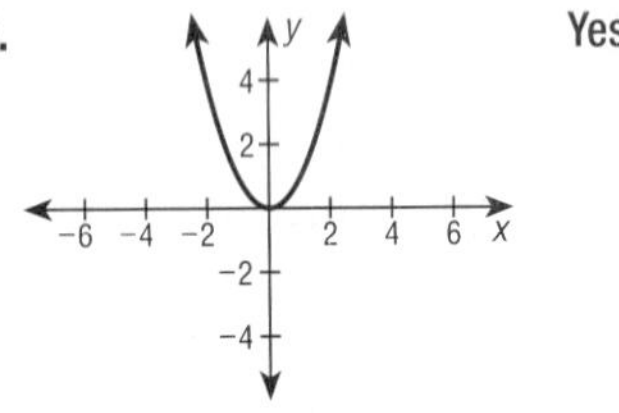

53. No

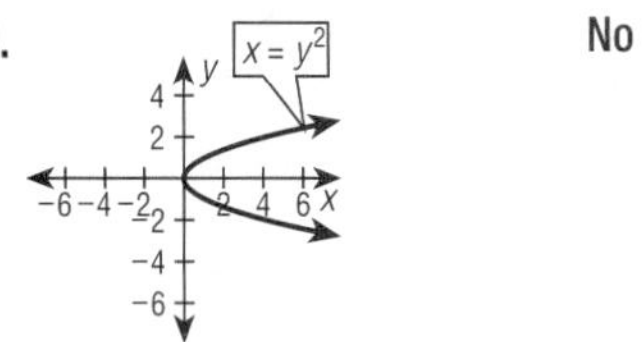

54. No

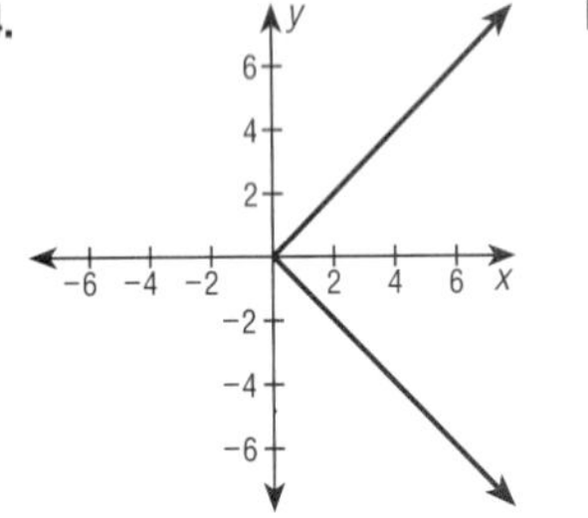

▶ **Ex. 59–60** Students should justify their solutions graphically.

Answers

59. Yes; $x = y^2$ is not a function, but $y = x^2$ is.

60. True; the inverse is $y = 2 - x$, since $2 - (2 - x) = x$

Enrichment Activities

These activities require students to go beyond lesson goals.

A function is a *one-to-one function* if each element of the domain is associated with a unique element of the range,

(continued)

Tax-Freedom Day **In Exercises 45 and 46, use the following information.**

Tax-Freedom Day is the day of the year that the average taxpayer will have earned enough money to pay his or her total federal, state, and local taxes. The table gives the number of the Tax-Freedom Day in each state (and the District of Columbia) in 1990. ***(Source: Tax Foundation)***

AK–129	FL–122	LA–122	NC–122	OK–125	VA–120
AL–119	GA–119	MA–128	ND–117	OR–115	VT–120
AR–115	HI–137	MD–132	NE–115	PA–126	WA–125
AZ–124	IA–116	ME–123	NH–109	RI–127	WI–123
CA–124	ID–114	MI–123	NJ–129	SC–123	WV–120
CO–117	IL–125	MN–128	NM–130	SD–109	WY–120
CT–132	IN–117	MO–121	NV–123	TN–119	
DC–143	KS–122	MS–110	NY–143	TX–118	
DE–138	KY–119	MT–116	OH–124	UT–118	

45. Let T represent the total federal, state, and local taxes for the average taxpayer in Tennessee. Let I represent the taxpayer's total 1990 income. Which of the following equations describes the relationship between T and I? a

a. $T = \frac{119}{365}I$ **b.** $I = \frac{119}{365}T$

46. Explain the differences in the numbers in the table. State and local taxes vary.

Integrated Review

In Exercises 47–50, write the equation in slope-intercept form. See below.

47. $x = 2y - 3$ **48.** $x = 9y + 18$ **49.** $x = y + 2$ **50.** $x = -2y - 6$

In Exercises 51–54, sketch the graph of the equation. Is *y* a function of *x*?

51. $y = x$ **52.** $y = x^2$ **53.** $x = y^2$ **54.** $x = |y|$

In Exercises 55–58, find *f*(*g*(*x*)). See below.

55. $f(x) = \frac{1}{2}x,\ \ g(x) = 3x + 2$

56. $f(x) = x + 1,\ \ g(x) = 2x - 6$

57. $f(x) = x^2 - 6,\ \ g(x) = -10x$

58. $f(x) = -5x + 1,\ \ g(x) = 2x$

Exploration and Extension

59. Is it possible for a relation that is not a function to have an inverse that is a function? Why or why not? See margin.

60. True or False? The function $f(x) = 2 - x$ is its own inverse. Explain.

61. *Geometry* Find the area of the polygon formed by the graphs of $y = 4 - x$, $f(x) = 8 - 2x$, and the inverse of f. $\frac{8}{3}$

62. *Geometry* Find the area of the polygon formed by the graphs of $y = 2 - x$, $y = 6 - x$, $f(x) = x + 2$, and the inverse of f. 8

47. $y = \frac{1}{2}x + \frac{3}{2}$ **48.** $y = \frac{1}{9}x - 2$ **49.** $y = x - 2$ **50.** $y = -\frac{1}{2}x - 3$

55. $f(g(x)) = \frac{3}{2}x + 1$ **56.** $f(g(x)) = 2x - 5$ **57.** $f(g(x)) = 100x^2 - 6$ **58.** $f(g(x)) = -10x + 1$

Mid-Chapter SELF-TEST

Take this test as you would take a test in class. The answers to the exercises are given in the back of the book.

1. Express the relation $\{(2, 3), (4, 7), (4, 12), (5, 3), (6, 8)\}$ as a mapping. **(6.1)** See Additional Answers.
2. Sketch the graph of $\{(0, -4), (1, 0), (2, 1), (3, 5), (4, 1)\}$. **(6.1)** See Additional Answers.
3. Which is *not* the graph of a function? **(6.1)** b

a. 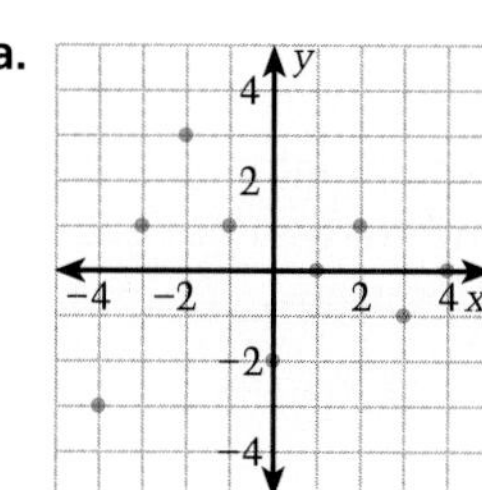

b. 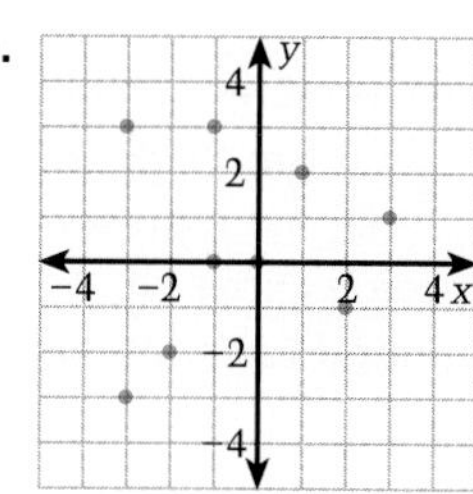

c. 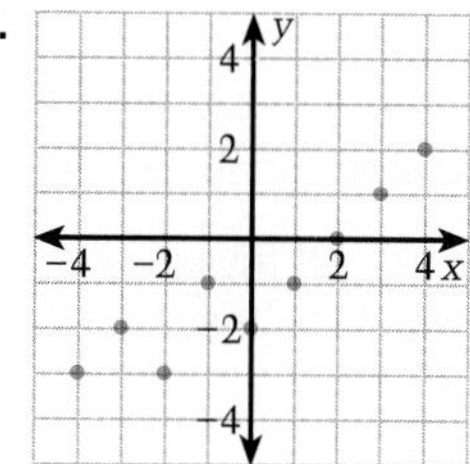

4. Given $f(x) = 2x^2 + 5x + 2$, find $f(-3)$. **(6.1)** 5
5. Is $\{(-2, 1), (-1, 2), (-1, 3), (0, 4), (3, 5)\}$ a function? Explain. **(6.1)** No, -1 maps onto both 2 and 3.
6. Find the domain and range of $\{(-4, 1), (-3, 2), (-2, 3), (-1, 4), (0, 5)\}$. **(6.1)** See below.

In Exercises 7–9, find the domain of the function. (6.2)

7. $f(x) = \frac{x-5}{x+2}$ All real numbers except -2
8. $h(x) = 4x^2$ All real numbers
9. $k(x) = \sqrt{x-7}$ All real numbers ≥ 7

In Exercises 10–12, find the range of the function. (6.2)

10. $k(x) = x^2 + 9$ All real numbers ≥ 9
11. $f(x) = (x+5)^2 + 3$ All real numbers ≥ 3
12. $h(x) = \sqrt{x+16}$ All real numbers ≥ 0

In Exercises 13–15, use $f(x) = x^2$ and $g(x) = 4x - 6$ to find $h(x)$. (6.2, 6.3) See below.

13. $h(x) = f(x) - g(x)$
14. $h(x) = f(g(x))$
15. $h(x)$ is the inverse of $g(x)$.

In Exercises 16–18, use the following information. With a coupon, if you purchase one spaghetti dinner at the regular price, *x*, you can purchase the second dinner at half price. (6.2)

16. Write a function, $P(x)$, for the price of the two spaghetti dinners. $P(x) = 1.5x$
17. Write a function, $C(x)$, that represents the total amount spent for the two dinners. In the total price, include a 15% tip (based on the full regular price of two dinners), and a 7% sales tax. $C(x) = 1.905x$
18. Evaluate the function in Exercise 17 when x is \$5. \$9.53

In Exercises 19 and 20, use the following information. After giving a 75-point test, a teacher adjusts the test scores by adding 7 points to each student's score. After adding the 7 points, each student is given a percent score (based on 75 possible points). (6.3)

19. Write a function that inputs the original point score and outputs the percent score.
20. Find the inverse of the function in Exercise 19. Use this inverse to find the original score of a student whose percent score is 89%. $g(x) = \frac{75x}{100} - 7$, 60

19. $f(x) = \left(\frac{x+7}{75}\right)(100)$

6. Domain: $\{-4, -3, -2, -1, 0\}$, Range: $\{1, 2, 3, 4, 5\}$
13. $h(x) = x^2 - 4x + 6$
14. $h(x) = 16x^2 - 48x + 36$
15. $h(x) = \frac{1}{4}x + \frac{3}{2}$

(continued)

and each element of the range is associated with a unique element of the domain.

1. Are the following functions one-to-one functions?
   a. $f(x) = \{(-2, 1), (-1, 1), (0, 1), (1, 2)\}$
   b. $f(x) = x^2, x \geq 0$
   c. $f(x) = \frac{3}{4}x + 12$
   d. $f(x) = \{(1, 1), (2, 8), (3, 27), (4, 64)\}$

   a. is not one-to-one; b., c., and d., are one-to-one.
2. Find the inverses of the functions shown in Activity 1. What do you notice about the functions that have inverses that are functions?
   a. $g(x) = \{(1, -2), (1, -1), (1, 0), (2, 1)\}$; not a function
   b. $g(x) = \sqrt{x}, x \geq 0$; a function
   c. $g(x) = \frac{4}{3}x - 16$; a function
   d. $g(x) = \{(1, 1), (8, 2), (27, 3), (64, 4)\}$; a function.

   b.–d. They are one-to-one functions.

Problem of the Day

A 7 is written at the end of a two-digit number, increasing its value by 700. Find the original number.
77

ORGANIZER

Warm-Up Exercises

1. Graph the pair of functions using the same coordinate plane.

   $f(x) = 3x - 2$ and $g(x) = x + 2$

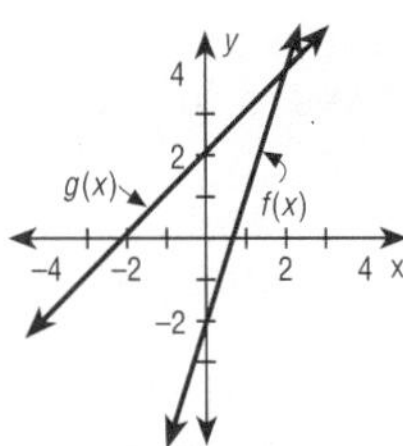

2. Graph the pair of functions using the same coordinate plane, subject to the indicated constraints.

   For all $x \geq 2$, $f(x) = 3x - 2$ and $g(x) = x + 2$.

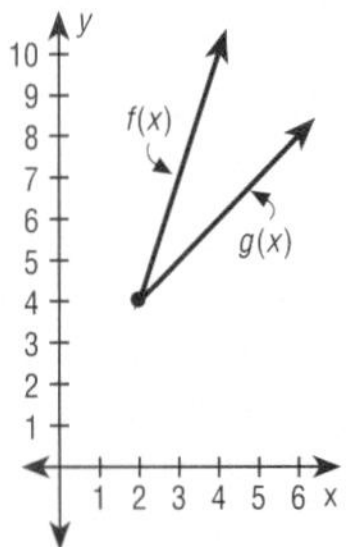

Lesson Resources

Teaching Tools
- Transparencies: 2, 3, 7, 8, 12, 13
- Problem of the Day: 6.4
- Warm-up Exercises: 6.4
- Answer Masters: 6.4

Extra Practice: 6.4
Color Transparency: 37

6.4 Special Functions

What you should learn:

Goal 1 How to use special functions, such as compound functions and step functions

Goal 2 How to use these functions in real-life situations

Why you should learn it:

You can model many situations with compound functions, such as the changing costs of mobile homes in the United States.

Goal 1 Using Special Functions

In the first three lessons of this chapter, a single equation represented each function. In many real-life problems, however, functions may be represented by a combination of equations. Such functions are called **compound functions.** For example, the compound function given by

$$f(x) = \begin{cases} 2x - 1, & x \leq 1 \\ x^2 - 2x + 1, & x > 1 \end{cases}$$

is defined by two equations. One equation gives the values of $f(x)$ when x is less than or equal to 1, and the other equation gives the values of $f(x)$ when x is greater than 1.

Example 1 *Using a Compound Function*

Sketch the graph of the compound function. What is $f(-2)$? What is $f(3)$?

$$f(x) = \begin{cases} -x^2 + 2, & x < 0 \\ x + 2, & x \geq 0 \end{cases}$$

Solution To the left of the y-axis, the graph of the function is a parabola that opens down and whose vertex is (0, 2). To the right of the y-axis, the graph is a line that has a slope of 1 and a y-intercept of 2. To sketch the graph of the function, you can lightly draw both graphs. Then darken the portion of the graph that represents the function, as shown at the left.

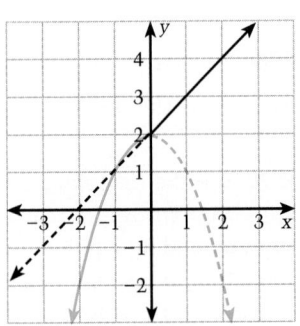

To find the value of the function when $x = -2$, use the first equation.

$$f(-2) = -(-2)^2 + 2 = -4 + 2 = -2$$

To find the value of the function when $x = 3$, use the second equation.

$$f(3) = 3 + 2 = 5$$

In Example 1, the left and right portions of the graph of the compound function connect at the point (0, 2). Try sketching the graph of the compound function listed at the top of the page. Do the left and right portions of its graph connect? ■

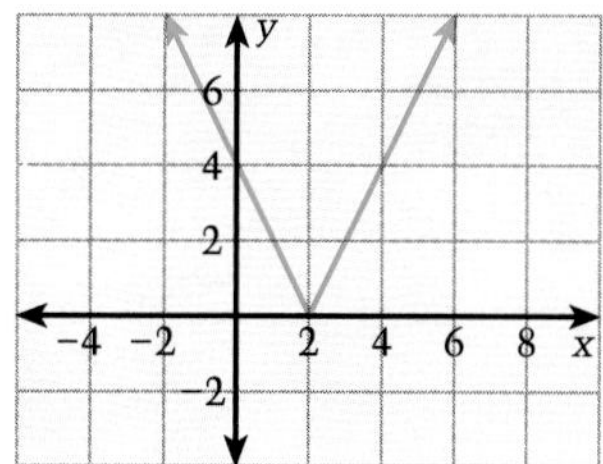

Example 2 Writing an Absolute Value Function as a Compound Function

Write the *absolute value function* $f(x) = |2x - 4|$ as a compound function.

Solution For $x < 2$, the values of $2x - 4$ are negative, and for $x > 2$, the values of $2x - 4$ are positive. Thus, the function could be written as follows.

$$f(x) = \begin{cases} -(2x - 4), & x < 2 \\ 2x - 4, & x \geq 2 \end{cases}$$

The graph of the function is shown at the left. ■

Some compound functions are defined by many different equations. One special type is called a **step function** because its graph looks like stair steps.

Real Life
Postage Rates

Example 3 Sketching the Graph of a Step Function

The cost, C, of using an overnight delivery service to send a package depends on the weight, x (in ounces), of the package. The following step function gives the cost for packages that weigh less than 1 pound. Sketch the graph of this function.

$$C = \begin{cases} 8.50, & 0 < x < 2 \\ 9.75, & 2 \leq x < 4 \\ 12.00, & 4 \leq x < 6 \\ 13.25, & 6 \leq x < 8 \\ 15.50, & 8 \leq x < 10 \\ 16.75, & 10 \leq x < 12 \\ 18.00, & 12 \leq x < 14 \\ 19.25, & 14 \leq x < 16 \end{cases}$$

Cost of Overnight Delivery

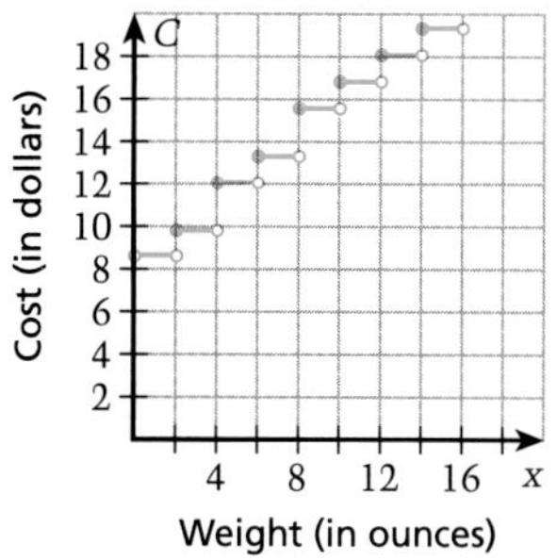

Solution The graph consists of horizontal line segments. The endpoints of the line segments are represented by either open dots or solid dots. For instance, the first line segment has an open dot on the left because you cannot send a package that weighs 0 ounces. The first line segment also has an open dot on the right because a 2-ounce package does *not* cost \$8.50, but \$9.75 (which is symbolized by the solid dot on the left of the second line segment). ■

LESSON Notes

Example 1

An Alternate Approach
One way to determine if the left and right portions of the graph of a compound function *connect* involves solving algebraic equations.

For example, consider the equations shown at the top of page 306.

$$f(x) = \begin{cases} 2x - 1, & x \leq 1 \\ x^2 - 2x + 1, & x > 1 \end{cases}$$

Set the two equations of the compound function equal to each other and solve for x.

Since $2x - 1 = x^2 - 2x + 1$, $x = 2 \pm \sqrt{2}$.

The solution, if it exists, satisfies each equation of the compound definition, and therefore, indicates where the graphs of the compound function connect. If the value of x at which the compound function changes from one equation to the other is a solution value, then the left and right portions of the graph of the compound function are connected.

Since the solution $x = 2 \pm \sqrt{2} \neq 1$, the value of the compound function at which the function changes from one equation to the other, the graph of this compound function is not connected.

Example 2

Ask students to express the absolute value function $f(x) = |3x + 2|$ as a compound function. How is the x-value at which the compound function changes from one equation to the other determined?

$$f(x) = \begin{cases} -(3x + 2), & x < -\frac{2}{3} \\ 3x + 2, & x \geq -\frac{2}{3} \end{cases}$$

The x-value is found by solving for x in $3x + 2 = 0$. (The equal sign may go with either $<$ or $>$.)

Example 3

Ask students if you would still have a compound function if there were solid dots throughout instead of the open dots as presented. You would no longer have a function because one weight could have two prices associated with it.

Example 4

(a) According to the model, what was the average cost of a new mobile home in 1994?
(b) What would have been the cost of a new mobile home in 1994 if the linear model employed between 1975 and 1980 had been used?

a. \$32,774; b. \$44,800

Extra Examples

Here are additional examples similar to **Examples 1–3.**

1. Evaluate the function for x when x is equal to -2, 2, 8, and 12.

   a. $f(x) = \begin{cases} x + 2, & x \geq 6 \\ 3x - 4, & x < 6 \end{cases}$

   b. $f(x) = \begin{cases} -4, & x \geq 2 \\ x^2 - 8, & x < 2 \end{cases}$

   c. $f(x) = \begin{cases} 3, & 0 \leq x < 4 \\ 4, & 4 \leq x < 10 \\ 6, & 10 \leq x \end{cases}$

   a. $f(-2) = -10$, $f(2) = 2$, $f(8) = 10$, $f(12) = 14$;

   b. $f(-2) = -4$, $f(2) = -4$; $f(8) = -4$, $f(12) = -4$;

   c. $f(-2) =$ undefined, $f(2) = 3$, $f(8) = 4$, $f(12) = 6$

2. Graph each function in Exercise 1 above. Check student's graphs.

3. Write the absolute value functions $f(x) = |12 - 3x|$ and $g(x) = -|2x + 1|$ as compound functions.

In 1993, the United States had approximately 106,611,000 housing units. Of these, approximately 7,072,000 were mobile homes.
This mobile home park is in California. (Source: U.S. Bureau of Census, Current Housing Reports)

Goal 2 Using Special Functions in Real Life

Real Life
Housing Cost

Example 4 *The Price of a Mobile Home*

For 1975 through 1994, the average cost, C, of a new mobile home in the United States can be modeled by the compound function

$$C = \begin{cases} 19{,}600 + 1800t, & -5 \leq t < 0 \\ 19{,}600 + 297t + 46t^2, & 0 \leq t \leq 14 \end{cases}$$

where $t = 0$ represents 1980. Sketch the graph of this compound function. Use the graph to describe the change in the average cost of a mobile home.
(Source: U.S. Bureau of Census Construction Reports)

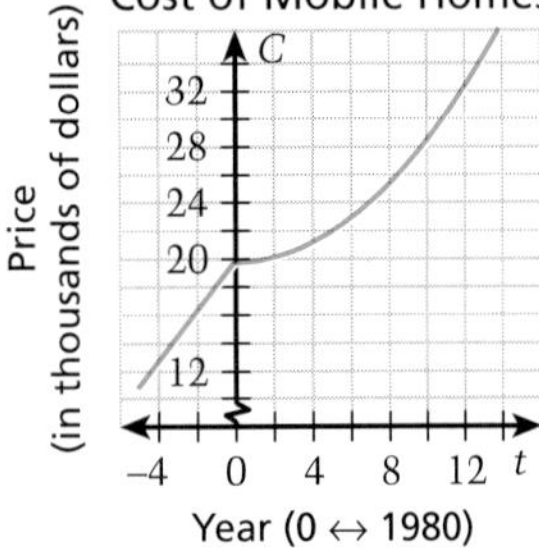

Solution To the left of the y-axis, the graph is a line, and to the right of the y-axis, the graph is a parabola. From the graph, you can see that between 1975 and 1980, the average cost of a mobile home was increasing at a constant rate. Between 1980 and 1994, however, the average cost began increasing according to a quadratic model. ■

Communicating about ALGEBRA

SHARING IDEAS about the Lesson See Additional Answers.

Sketch the graph of each function. Explain the steps you used.

A. $f(x) = \begin{cases} -2x^2 + 4, & x \leq 0 \\ 2x^2 + 4, & x > 0 \end{cases}$ **B.** $f(x) = \begin{cases} 4, & 0 \leq x < 3 \\ 5, & 3 \leq x < 6 \\ 6, & 6 \leq x < 9 \end{cases}$

EXERCISES

Guided Practice

CRITICAL THINKING about the Lesson See Additional Answers.

1. Give an example of a compound function. Draw its graph.

In Exercises 2 and 3, consider the function $f(x) = |10 - 2x|$.

2. Write $f(x)$ as a compound function.
3. Sketch the graph of f. What is $f(3)$? What is $f(6)$?
4. Give an example of a step function. Draw its graph.

Independent Practice

In Exercises 5–8, evaluate the function for the given value of x.

$$f(x) = \begin{cases} -x^2 - 2x, & x < -1 \\ x + 2, & x \geq -1 \end{cases}$$

5. $f(-3)$ −3
6. $f(0)$ 2
7. $f(-1)$ 1
8. $f(3)$ 5

In Exercises 9–12, evaluate the function for the given value of x.

$$g(x) = \begin{cases} -2x - 1, & x \leq 1 \\ -x^2 + 3x - 5, & x > 1 \end{cases}$$

9. $g(1)$ −3
10. $g(-2)$ 3
11. $g(4)$ −9
12. $g(0)$ −1

In Exercises 13–18, sketch the graph of the function. See Additional Answers.

13. $f(x) = \begin{cases} x + 5, & x < -2 \\ x^2 + 2x + 3, & x \geq -2 \end{cases}$
14. $g(x) = \begin{cases} 2x, & x \leq 2 \\ -2x^2 + 3x + 6, & x > 2 \end{cases}$
15. $h(x) = \begin{cases} \frac{1}{4}x - 2, & x \leq 4 \\ x^2 - x - 13, & x > 4 \end{cases}$
16. $f(x) = \begin{cases} x^2 - 4x + 5, & x < 3 \\ -2x + 8, & x \geq 3 \end{cases}$
17. $f(x) = \begin{cases} x^2 + 4x - 20, & x < 3 \\ -x + 4, & x \geq 3 \end{cases}$
18. $f(x) = \begin{cases} \frac{1}{2}x^2 - x + 7, & x \leq -4 \\ 2x + 11, & x > -4 \end{cases}$

In Exercises 19–22, match the equation with its graph.

19. $f(x) = |x - 6|$ d
20. $f(x) = |-x + 4|$ a
21. $f(x) = |-2x - 8|$ b
22. $f(x) = \left|\frac{1}{2}x + 4\right|$ c

a. 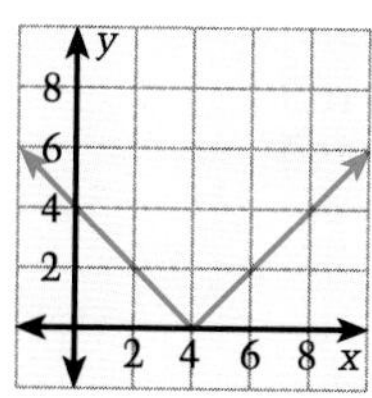

b. 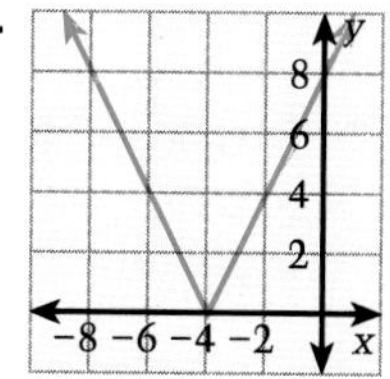

c. 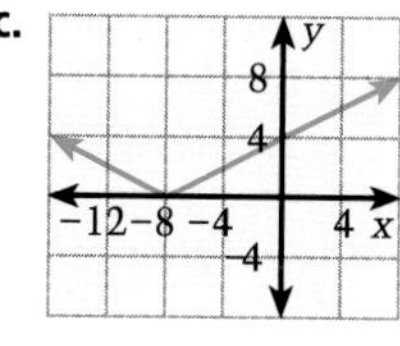

d. 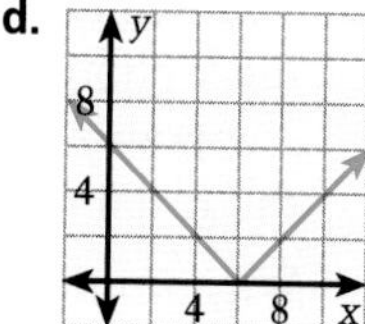

$$f(x) = \begin{cases} -(12 - 3x), & x < 4 \\ 12 - 3x, & x \geq 4 \end{cases}$$

$$g(x) = \begin{cases} 2x + 1, & x \leq -0.5 \\ -(2x + 1), & x > -0.5 \end{cases}$$

Check Understanding

1. Why do "step functions" have this name?
   The graphs look like stair steps.
2. Can every absolute value function be rewritten as a compound function? Explain.
   Yes, even "trivial" functions like $f(x) = |3|$ can be rewritten in compound form.
3. Can either equation in a compound function be used to evaluate the compound function at a given x-value? Explain.
   No, only the equations defined for that value of x.

Communicating about ALGEBRA

Do the graphs in Part A and B connect?
A: yes; B. no

EXTEND *Communicating*
Decide, without graphing, whether the following compound functions have graphs with left and right portions connected. (See the Alternate Approach on pages 306 and 307.)

a. $f(x) = \begin{cases} x - 3, & x \geq 3 \\ x^2 - 8, & x < 3 \end{cases}$

b. $f(x) = \begin{cases} 3, & x \geq 3 \\ x^2 - 6, & x < 3 \end{cases}$

a. not connected; $x - 3 = x^2 - 8$, $x = \frac{1}{2} \pm \frac{\sqrt{21}}{2}$

b. $x^2 - 6 = 3$, $x = \pm 3$; connected at $x = 3$.

EXERCISE Notes

ASSIGNMENT GUIDE

Basic/Average: Ex. 1–4, 5–15 odd, 19–22, 27–33 odd, 34, 37–38, 45–50

Above Average: Ex.1–4, 5–15 odd, 19–22, 25–31 odd, 33–38, 43, 51

Advanced: Ex. 1–4, 5–17 odd, 19–23, 31–38, 50–51

Selected Answers
Exercises 1–4, 5–49 odd

More Difficult Exercises
Exercises 23–28, 32, 33, 37–38, 51

Guided Practice

Ex. 2–3 You may need to illustrate other examples of writing compound functions. Many students will not understand why $f(x) = |10 - 2x|$ can be rewritten as

$$f(x) = \begin{cases} 2x - 10, & x < 5 \\ 10 - 2x, & x \geq 5 \end{cases}$$

Why does the interval change at $x = 5$?
At $x = 5$, the value of $(10 - 2x)$ is 0.

Independent Practice

Ex. 13–18 Students may want to use a dotted line or curve for the portion of the graph that is not being used. See Example 2 for a solution model.

TECHNOLOGY Sometimes a compound function is called an interval function. Graphics calculators will graph interval or compound functions. Have students consult the calculator manual for specific directions.

Ex. 19–22 Assign these exercises as a group. Help students to locate the turning points for the graphs, that is, the values of x for which $f(x)$ has its minimum of 0.

In Exercises 23–28, write the absolute value function as a compound function. See Additional Answers.

23. $f(x) = |x + 2|$ **24.** $f(x) = |-x + 5|$ **25.** $f(x) = |-2x - 6|$

26. $f(x) = \left|\frac{1}{2}x - 3\right|$ **27.** $f(x) = \left|\frac{1}{3}x - 1\right|$ **28.** $f(x) = |-3x + 6|$

In Exercises 29–31, sketch the graph of the step function. See Additional Answers.

29. $y = \begin{cases} 4, & -2 \leq x < 3 \\ 6, & 3 \leq x < 5 \\ 9, & 5 \leq x < 6 \\ 12, & 6 \leq x < 10 \end{cases}$ **30.** $y = \begin{cases} 10, & 0 \leq x < 1 \\ 4, & 1 \leq x < 2 \\ 1, & 2 \leq x < 3 \\ -3, & 3 \leq x < 4 \end{cases}$ **31.** $y = \begin{cases} 4.6, & -3 \leq x < -1 \\ 2.1, & -1 \leq x < 2 \\ 0, & 2 \leq x < 3 \\ -3.1, & 3 \leq x < 5 \end{cases}$

32. *Summer Job* You are working at a summer job that pays time and a half for overtime. That is, if you work more than 40 hours per week, your hourly wage for the extra hours is 1.5 times the normal hourly wage of \$5.50. Write a compound function that gives your weekly pay, P, in terms of the number of hours, x, you work. See below.

33. *Silk-screen T-Shirts* A silk-screen shop has the following price schedule for silk-screen T-shirts.

- An initial charge of \$10 to create the silk-screen.
- \$8.50 per shirt for orders of 25 or fewer shirts.
- \$7.75 per shirt for orders of more than 25 shirts.

Write a compound function that gives the cost, C, for an order of x shirts. Sketch the graph of this function.

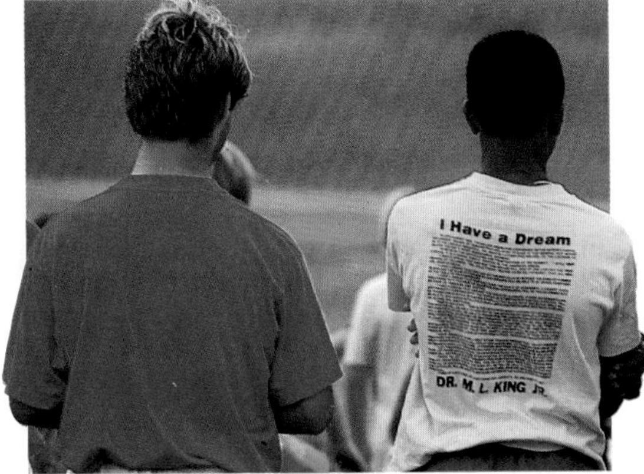

33. See Additional Answers.

34. *Telephone Call* The cost, C, of a t-minute telephone call from Bogalusa, Louisiana, to Biloxi, Mississippi, is given by the following step function. (The rates are based on an evening call in 1991.) Sketch the graph of this function. What is the cost of a call that lasts 20 minutes and 35 seconds?

$$C = \begin{cases} 0.28, & 0 < t \leq 1 \\ 0.28 + 1(0.18), & 1 < t \leq 2 \\ 0.28 + 2(0.18), & 2 < t \leq 3 \\ 0.28 + 3(0.18), & 3 < t \leq 4 \\ 0.28 + 4(0.18), & 4 < t \leq 5 \\ \vdots & \vdots \end{cases}$$

\$3.88 See Additional Answers.

35. *Photocopy Rates* The cost, C, of making x photocopies is given by the compound function at the right. Sketch the graph of this function. See Additional Answers.

$$C = \begin{cases} 0.10x, & 0 < x \leq 25 \\ 0.08x, & 25 < x \leq 100 \\ 0.06x, & 100 < x \leq 250 \\ 0.04x, & 250 < x \end{cases}$$

36. Why would it not make sense to order 240 photocopies with this rate schedule? See below.

32. $P = \begin{cases} 5.5x, & x \leq 40 \\ 5.5(40) + 1.5(5.5)(x - 40), & x > 40 \end{cases}$

36. 240 would cost \$14.40, 251 would cost \$10.04.

Health and Research **In Exercises 37 and 38, match the compound function given in the exercise with its graph. Then answer the questions.**

37. For 1970 through 1990, the total health expenses, H (in billions of dollars), in the United States can be modeled by

$$H = \begin{cases} t^2 + 6.4t + 76, & 0 \le t < 10 \\ 36.5t - 125, & 10 \le t \le 20 \end{cases}$$

where $t = 0$ represents 1970. What were the expenses in 1975? in 1985?

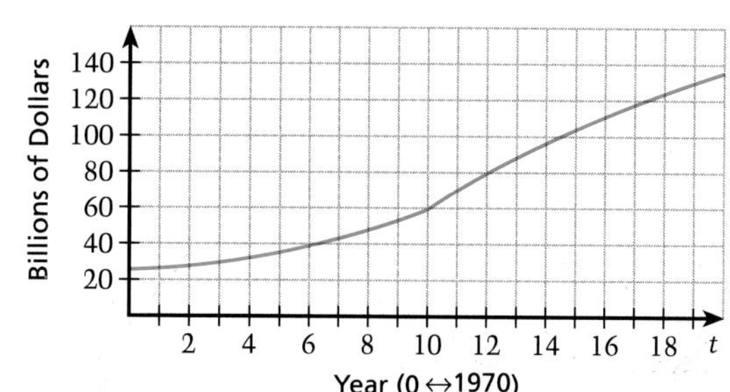

38. For 1970 through 1990, the total amount, R (in billions of dollars), spent on research and development in the United States can be modeled by

$$R = \begin{cases} 0.31t^2 + 0.25t + 26.5, & 0 \le t < 10 \\ -0.26t^2 + 15.3t - 67, & 10 \le t \le 20 \end{cases}$$

where $t = 0$ represents 1970. How much was spent on research and development in 1970? In 1980? In 1990?

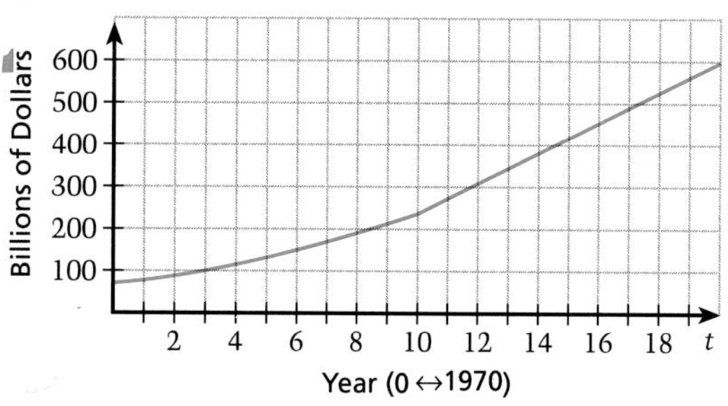

37. Second graph; $133 billion, $422.5 billion

38. First graph; $26.5 billion, $60 billion, $135 billion

Integrated Review

In Exercises 39–44, sketch the graph of the equation. See Additional Answers.

39. $y = -\frac{1}{2}x + 6$ **40.** $y = 2x + 9$ **41.** $y = x^2 + 6x - 9$

42. $y = -2x^2 - x + 4$ **43.** $y = -\frac{1}{2}x^2 + 4x + 10$ **44.** $y = 3x^2 - x + 9$

In Exercises 45–50, find the indicated value of *f*(*x*).

45. $f(x) = x - 2,\quad f(3)$ 1 **46.** $f(x) = -\frac{1}{2}x + 3,\quad f(4)$ 1 **47.** $f(x) = x^2 - x,\quad f(-1)$ 2

48. $f(x) = x^2 - 5x,\quad f(5)$ 0 **49.** $f(x) = 3x^2 - x - 7,\quad f(-5)$ 73 **50.** $f(x) = 3x^2 + 4x - 6,\quad f(-3)$ 9

Exploration and Extension

51. *Writing a Compound Function* In the coordinate plane at the right, the equations for the graphs are as follows.

A: $y = -\frac{1}{4}x^2 + 2x + 4$
B: $y = 2x - 12$
C: $y = -2x + 36$
D: $y = -\frac{1}{4}x^2 + 10x - 92$

Write a compound function that describes the entire graph. See Additional Answers.

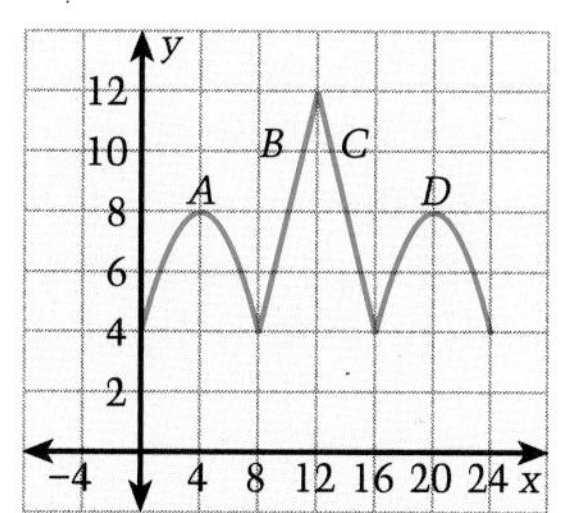

EXTEND Ex. 37–38 **SOCIAL STUDIES** Use these exercises for an in-class discussion regarding health and research costs. What is the implication of the trend in funding for research and development as defined by the given model when $t \ge 20$?
Since the coefficient of the t^2-term is negative, research funding will reach a maximum, then will eventually decrease.

Enrichment Activities

These activities require students to go beyond lesson goals.

Many real-life applications are best defined by using compound functions. Consider the following data.

Over the years, Americans have become more interested in health and nutrition. The following chart lists the percent of adults who indicate that they consume certain foods and beverages because they are considered to be low in calories.

Year	Percent
1978	25%
1984	40%
1986	45%
1989	51%
1991	55%

1. Graph the ordered pairs (year, percent). Let $t = 0$ correspond to 1970.

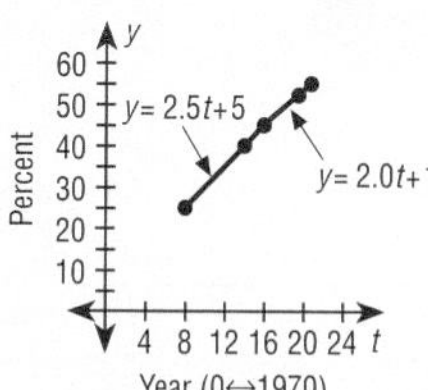

2. Create a compound function consisting of linear models that describe the data from 1978 to 1986 and from 1986 through 1991.

$$f(t) = \begin{cases} 2.5t + 5, & 8 \le t < 16 \\ 2.0t + 13, & 16 \le t \le 21 \end{cases}$$

TECHNOLOGY To be sure students understand that the ratio of screen height to screen width must be 2 to 3 to avoid graph distortion, ask students to provide several viewing rectangles with the appropriate ratio. (Use the special notation developed earlier to represent the viewing screens.)

Possible square settings are as follows.

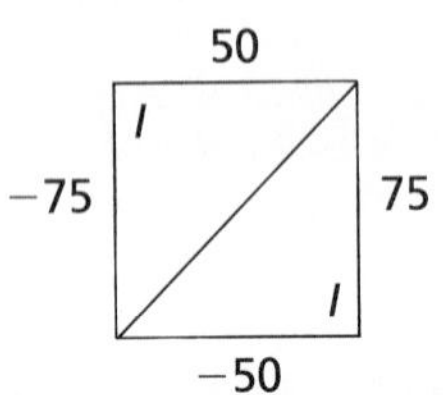

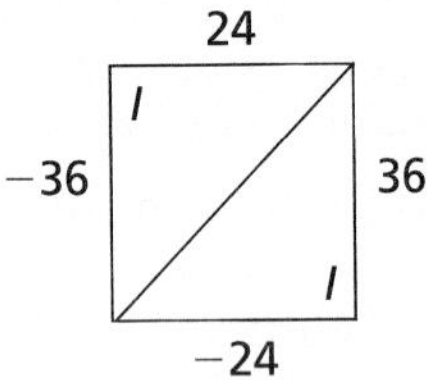

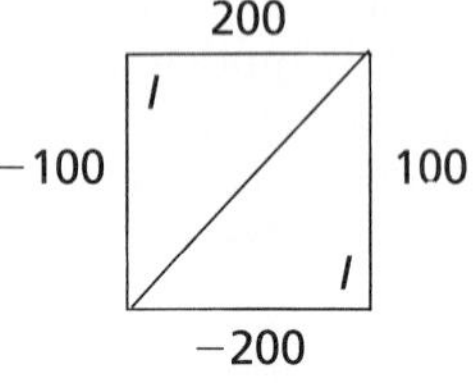

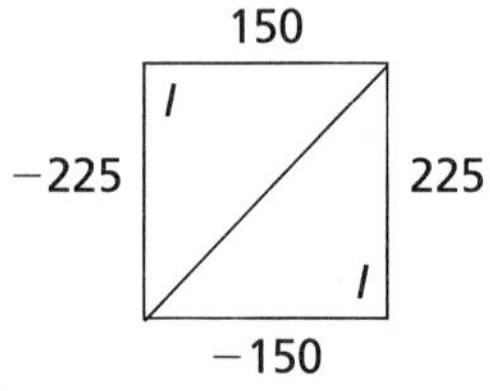

USING A GRAPHING CALCULATOR

After finding an equation that represents the inverse of a function, you can use a graphing calculator to verify that the graph of the inverse is a reflection (in the line $y = x$) of the graph of the original function. For instance, to verify graphically that the functions

$$f(x) = 2x + 3 \quad \text{and} \quad g(x) = \frac{1}{2}(x - 3)$$

are inverses of each other, you can sketch the graph of f, the graph of g, and the graph of $y = x$, all on the same screen. The graph of g should be the mirror image (in the line $y = x$) of the graph of f.

Because the display screens on most graphing calculators are not square, you will not obtain a true reflection with a standard setting. You need to use a **square setting,** in which the ratio of (Ymax − Ymin) to (Xmax − Xmin) is equal to the ratio of the screen height to the screen width. On most graphing calculators, this ratio is two thirds.

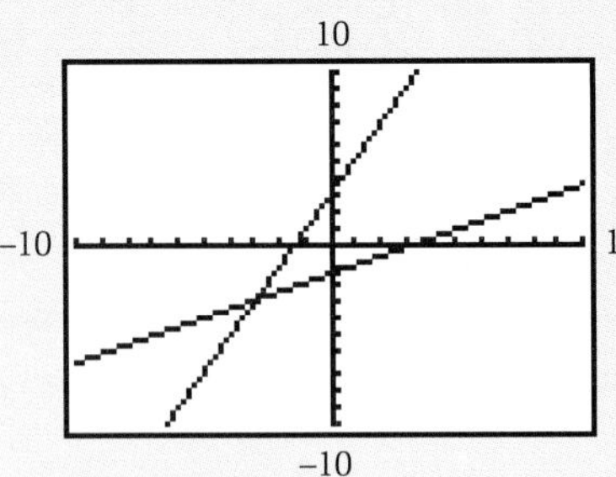

Standard setting: Graph is distorted.

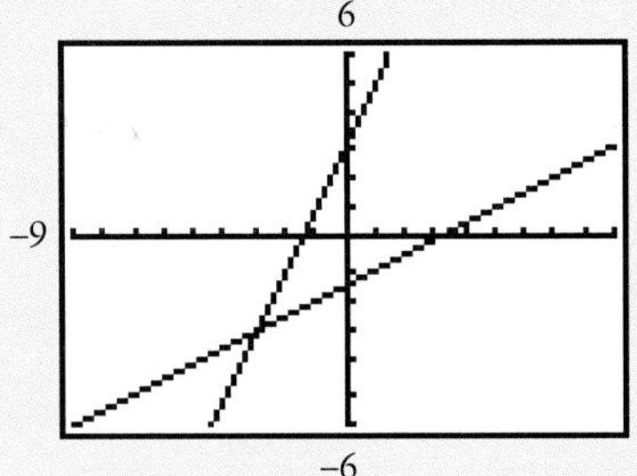

Square setting: Graph is not distorted.

Another case for which it is helpful to use a square setting is to sketch graphs of two perpendicular lines.

Example 1: Using a Square Setting for Perpendicular Lines

Sketch the graphs of $y = -3x + 12$ and $y = \frac{1}{3}x + 2$. Choose a viewing rectangle that shows the two lines intersecting at right angles.

Solution If you sketch the two lines with a standard setting, you can see that they intersect at the point (3, 3). The angles formed by the two lines at the point, however, do not appear to be right angles. (Try this with your own graphing calculator.) To eliminate the distortion in the graph, change to a square setting, such as $-9 \le x \le 9$ and $-6 \le y \le 6$. For this setting, there is no distortion, and the lines are shown to be perpendicular. ■

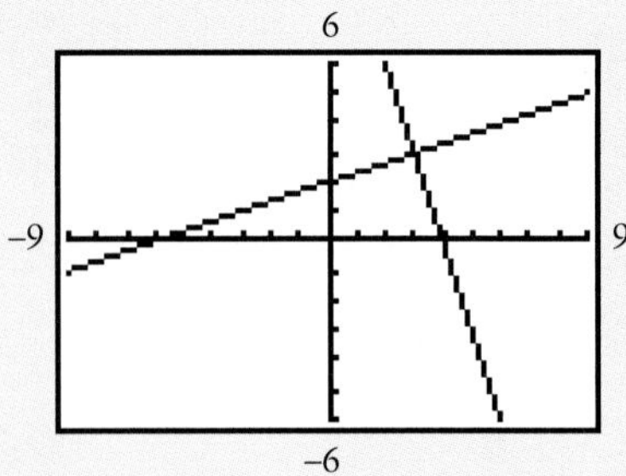

Example 2: Sketching the Graph of a Circle

Sketch the graphs of the functions

$$f(x) = \sqrt{36 - x^2} \qquad \text{and} \qquad g(x) = -\sqrt{36 - x^2}$$

on the same coordinate plane. What recognizable shape do the graphs form?

Solution Begin by changing the range to the square setting given by $-9 \le x \le 9$ and $-6 \le y \le 6$. Then sketch the graphs of both functions. As shown at the right, the two graphs form the upper and lower portions of a circle. The center of the circle is the origin, and the radius of the circle is 6.

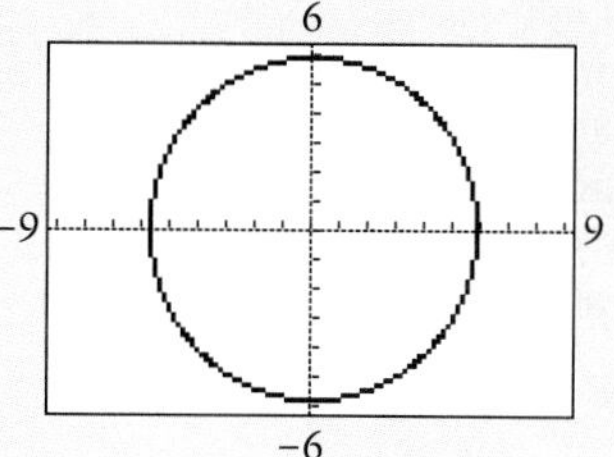

Notice that the entire circle could not be represented by a single function, because it does not pass the vertical line test. The domain and range of each function are as follows.

Domain of f: $-6 \le x \le 6$ *Range of f:* $0 \le y \le 6$
Domain of g: $-6 \le x \le 6$ *Range of g:* $-6 \le y \le 0$ ■

Exercises

In Exercises 1–6, find the inverse of the function. Then use a graphing calculator to verify that the graph of the inverse is a reflection (in the line y = x) of the graph of f. Use a square setting. See margin.

1. $f(x) = 4x - 9$
2. $f(x) = -\frac{1}{2}x + 5$
3. $f(x) = \frac{2}{3}x - 6$
4. $f(x) = -3x + 8$
5. $f(x) = -2.5x + 5$
6. $f(x) = 0.35x - 4$
7. Find an equation of the line that passes through the point $(2, -3)$ and is perpendicular to the line $4x - 5y = 8$. Then use a graphing calculator to sketch the graph of each line. Use a square setting. $y = \frac{-5}{4}x - \frac{1}{2}$
8. Find an equation of the line that passes through the point $(-1, 4)$ and is perpendicular to the line $-2x + 3y = 4$. Then use a graphing calculator to sketch the graph of each line. Use a square setting. $y = -\frac{3}{2}x + \frac{5}{2}$

In Exercises 9–12, sketch the graphs of both functions on the same display screen. (Use a square setting.) State the domain and range of each function. Then use a formula from geometry to find the area of the region enclosed by the two graphs. See margin.

9. $f(x) = \sqrt{25 - x^2}$, $g(x) = -\sqrt{25 - x^2}$
10. $f(x) = \frac{2}{3}\sqrt{81 - x^2}$, $g(x) = -\frac{2}{3}\sqrt{81 - x^2}$
11. $f(x) = 6 - |x|$, $-6 \le x \le 6$
    $g(x) = -6 + |x|$, $-6 \le x \le 6$
12. $f(x) = 3 + \frac{1}{2}x - \frac{1}{2}|x|$, $-6 \le x \le 6$
    $g(x) = -3 + \frac{1}{2}x + \frac{1}{2}|x|$, $-6 \le x \le 6$

Answers

1. $g(x) = \frac{1}{4}x + \frac{9}{4}$
2. $g(x) = -2x + 10$
3. $g(x) = \frac{3}{2}x + 9$
4. $g(x) = -\frac{1}{3}x + \frac{8}{3}$
5. $g(x) = -\frac{2}{5}x + 2$
6. $g(x) = \frac{20}{7}x + \frac{80}{7}$
9. $-5 \le x \le 5$, $0 \le y \le 5$; $-5 \le x \le 5$, $-5 \le y \le 0$; 25π

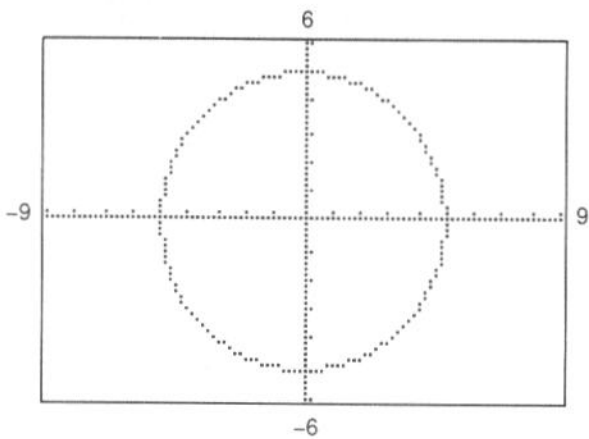

10. $-9 \le x \le 9$, $0 \le y \le 6$; $-9 \le x \le 9$, $-6 \le y \le 0$; 54π

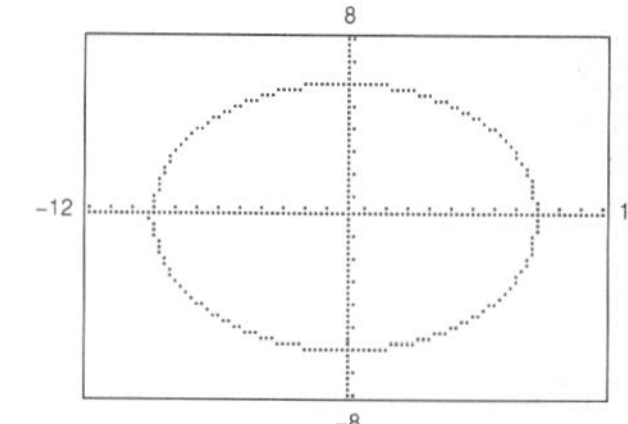

11. $-6 \le x \le 6$, $0 \le y \le 6$; $-6 \le x \le 6$, $-6 \le y \le 0$; 72

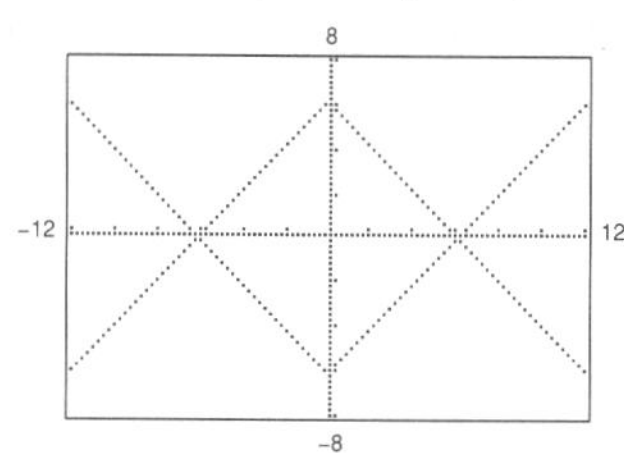

12. $-6 \le x \le 6$, $-3 \le y \le 3$; $-6 \le x \le 6$, $-3 \le y \le 3$; 36

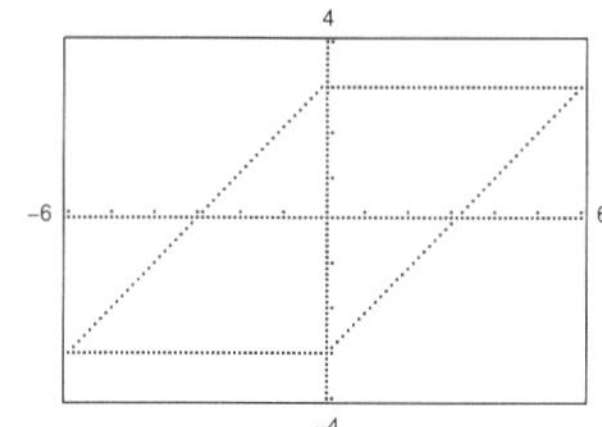

Problem of the Day

One hundred people spent \$100 to enter the fairgrounds. Tickets cost \$10 per man, \$3 per woman, and 50¢ per child. How many children entered the fairgrounds? 94

ORGANIZER

Warm-Up Exercises

1. Graph the following functions.
   a. $f(x) = x - 1$
   b. $g(x) = x^2 - 3$
   c. $h(x) = |x - 1|$

a.

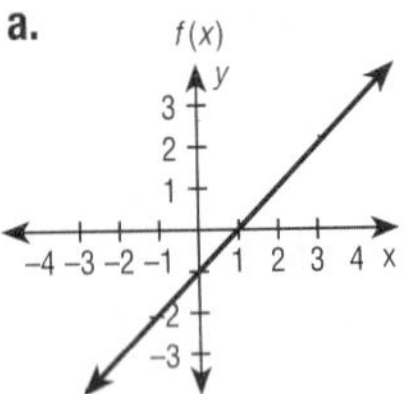

b.

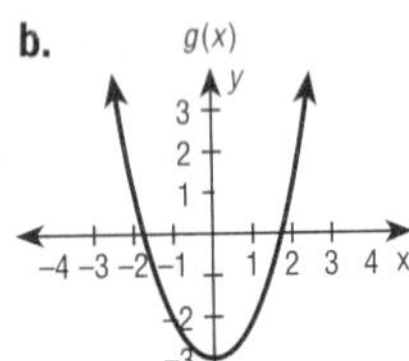

c.

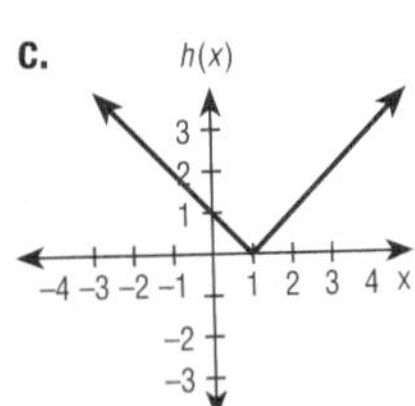

Lesson Resources

Teaching Tools
- Transparencies: 7, 8, 12, 13
- Problem of the Day: 6.5
- Warm-up Exercises: 6.5
- Answer Masters: 6.5

Extra Practice: 6.5

Technology Handbook: p. 37

6.5 Transformations of Graphs of Functions

What you should learn:

Goal 1 How to use translations and reflections to sketch the graph of a function

Goal 2 How to use transformations of graphs of functions in real-life settings

Why you should learn it:

Translations and reflections provide a quick way of sketching graphs of functions.

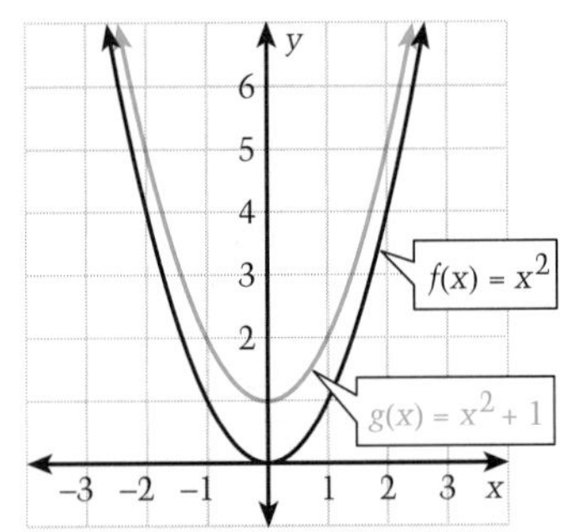

Shift the graph of f up 1 unit to get the graph of g.

Goal 1 Using Translations and Reflections

Many functions have graphs that are **geometric transformations** of the graphs of simpler functions. For instance, the graph of $g(x) = x^2 + 1$ is a shift of the graph of $f(x) = x^2$ *upward* by one unit. In function notation, g and f are related by

$g(x) = x^2 + 1 = f(x) + 1.$ *Upward shift of 1*

LESSON INVESTIGATION

Investigating Horizontal and Vertical Shifts

Partner Activity Use a graphing calculator to sketch the graph of $f(x) = x^2$ using a standard viewing rectangle. Then with your partner, write equations for the following graphs.

a. Graph of g: Shift the graph of f up 2 units.
b. Graph of g: Shift the graph of f down 3 units.
c. Graph of g: Shift the graph of f left 1 unit.
d. Graph of g: Shift the graph of f right 4 units.

In general, discuss how you can alter an equation so that its graph is shifted horizontally or vertically.

a. $g(x) = x^2 + 2$ b. $g(x) = x^2 - 3$ c. $g(x) = (x + 1)^2$
d. $g(x) = (x - 4)^2$

The list below summarizes the various **horizontal shifts** and **vertical shifts** (or translations) of the graphs of functions.

Vertical and Horizontal Shifts

Let c be a positive real number. **Vertical** and **horizontal shifts** of the graph of f are represented as follows.

1. Vertical shift of c units *upward*: $g(x) = f(x) + c$
2. Vertical shift of c units *downward*: $g(x) = f(x) - c$
3. Horizontal shift of c units *to the right*: $g(x) = f(x - c)$
4. Horizontal shift of c units *to the left*: $g(x) = f(x + c)$

Example 1 Shifts of the Graphs of Functions

Use the graph of $f(x) = x^2$ to sketch the graph of each function.

a. $g(x) = x^2 - 2$

b. $h(x) = (x + 3)^2$

Solution

a. To sketch the graph of $g(x) = x^2 - 2$, shift the graph of $f(x) = x^2$ *downward* by 2 units.

b. To sketch the graph of $h(x) = (x + 3)^2$, shift the graph of $f(x) = x^2$ *to the left* by 3 units.

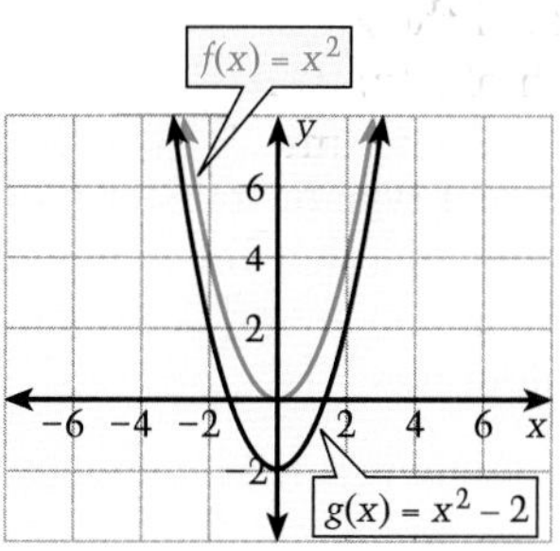

a. ***Shift downward of 2 units***

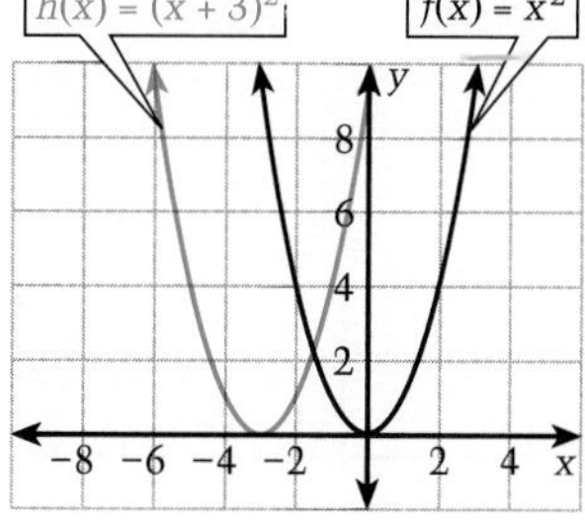

b. ***Shift left of 3 units***

Some graphs can be obtained from a *combination* of vertical and horizontal shifts, as shown in the next example. ■

Example 2 Shifts of the Graphs of Functions

Use the graph of $f(x) = |x|$ to sketch the graph of each function.

a. $g(x) = |x - 3| + 1$

b. $h(x) = |x + 2| - 3$

Solution

a. To sketch the graph of $g(x) = |x - 3| + 1$, shift the graph of $f(x) = |x|$ by 3 units *to the right* and 1 unit *upward.*

b. To sketch the graph of $h(x) = |x + 2| - 3$, shift the graph of $f(x) = |x|$ by 2 units *to the left* and 3 units *downward.*

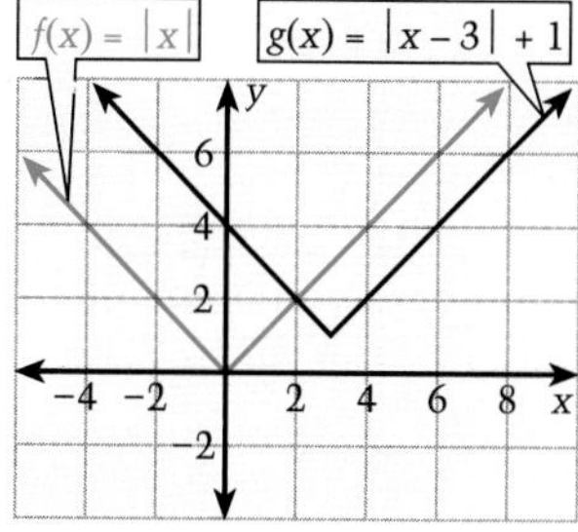

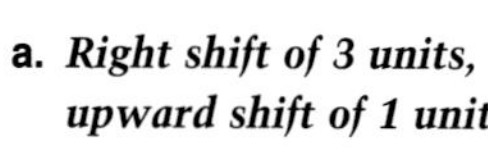

a. ***Right shift of 3 units, upward shift of 1 unit***

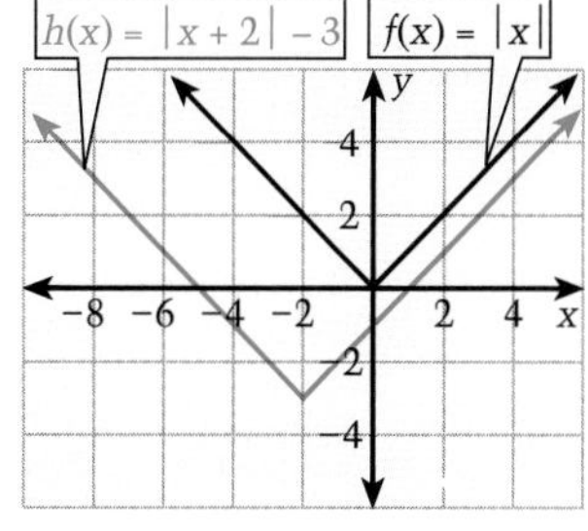

b. ***Left shift of 2 units, downward shift of 3 units*** ■

LESSON Notes

GOAL 1 The list of vertical and horizontal shifts will become familiar to students if they have the opportunity to discuss the effects of constants on the basic functions (for example, $f(x) = x^2$, $f(x) = |x|$, and $f(x) = x$).

MATH JOURNAL Students should record these shifts in the graphing section of their math journals.

Example 1

Develop the contrast between graphing the given function, such as $g(x) = x^2 - 2$, and applying a vertical shift to $f(x) = x^2$, by using two coordinate planes to represent the basic function on one and then moving it, followed by graphing the given function on the other coordinate plane. Compare the results, observing that the graphs are identical. Repeated uses of this procedure will allow students to construct their own list of vertical and horizontal shifts.

Example 2

It may be helpful to show the intermediate graphs of the given functions. That is, for the function $g(x) = |x - 3| + 1$, it is a good idea to show the graphs in the following sequence:

$g(x) = |x|$
$g(x) = |x - 3|$
$g(x) = |x - 3| + 1$.

Example 3

Another way to indicate the reflection of a function is to fold the paper (on which the graph is displayed) along the x-axis. Imagine that the original graph is drawn with charcoal, a substance that will leave a trace on the folded side of the paper. The image left on the folded side of the paper will be the reflection of the original function.

An Alternate Approach
Using Technology
The graphics calculator and a computer graphing program are both ideal tools for demonstrating the relationships among several functions and the basic function from which they are derived. A successful technique that emphasizes the effects of constants on the basic function utilizes the calculator's or the computer's capacity to graph several functions in sequence. For example, graph these four functions.

$y_1 = x^2$ $\quad$ $y_2 = x^2 + 2$
$y_3 = x^2 + 4$ $\quad$ $y_4 = x^2 + 8$

Each function is graphed, one after the other, and each graph remains on the screen. Students can actually see the effects of the constants on the basic function.

Changing the basic function, say to $f(x) = |x|$, (but keeping the same constants that are added to the basic function) emphasizes that the effect of adding a constant to a function is the same in every case!

A similar approach should be used for each type of vertical and horizontal shift, and reflection in the x-axis.

Example 4

Point out that the curve itself (and the information that it conveys) will remain unchanged. Consequently, the amount of money spent in 1980 remains at $20.9 billion. On the coordinate plane, you are required to display this information at $t = 0$; hence, a shift to the left of its original position on the coordinate plane.

The reflection of the hill and tree is seen in the lake.

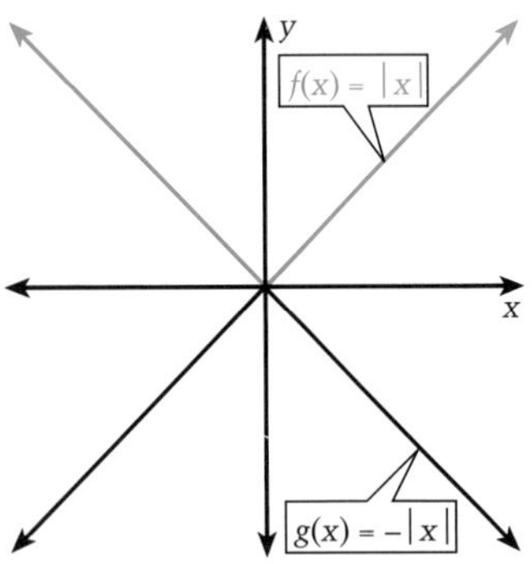

A second type of transformation is a **reflection.** For instance, if you imagine that the x-axis represents a mirror, then the graph of $g(x) = -|x|$ is the mirror image (or reflection) of the graph of $f(x) = |x|$.

Reflection in the x-Axis
If $g(x) = -f(x)$, then the graph of g is a reflection in the x-axis of the graph of f.

Example 3 *Reflections of the Graphs of Functions*

Use the graph of $f(x) = x^2$ to sketch the graph of each function.
a. $g(x) = -x^2$ **b.** $h(x) = -(x - 3)^2$

Solution
a. To sketch the graph of $g(x) = -x^2$, reflect the graph of $f(x) = x^2$ in the x-axis.
b. To sketch the graph of $h(x) = -(x - 3)^2$, first shift the graph of $f(x) = x^2$ by 3 units *to the right.* Then reflect the result in the x-axis.

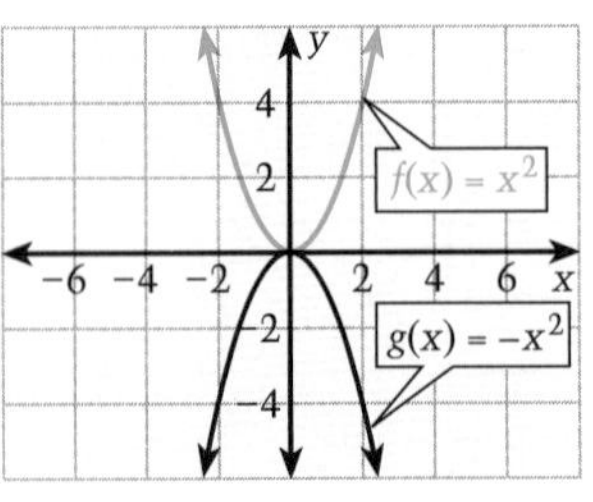

a. ***Reflection in the x-axis***

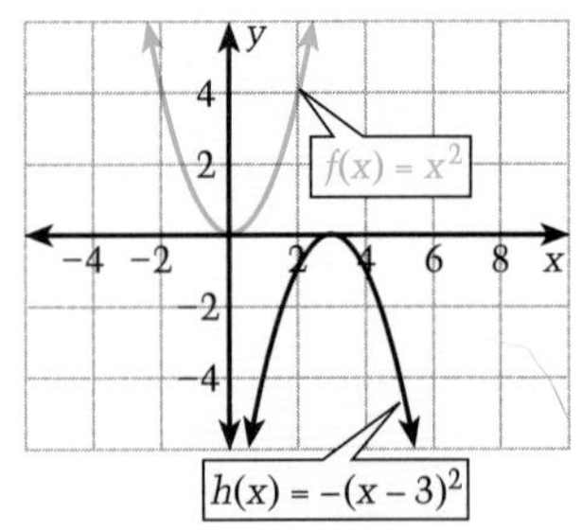

b. ***Right shift of 3 units, reflection in the x-axis*** ■

Goal 2 Using Transformation of Graphs

Real Life
Consumer Spending

Example 4 *Shifting a Graph*

For 1970 through 1990, the amount of money, $f(t)$ (in billions of dollars), that Americans spent on radios, televisions, recordings, and musical instruments can be modeled by

$$f(t) = 8.4 + 0.125t^2$$

where $t = 0$ represents 1970. Rewrite this model so that $t = 0$ represents 1980. ***(Source: U.S. Bureau of Economic Analysis)***

Solution In the formula for f, $t = 10$ represents 1980. You want to obtain a formula in which $t = 0$ represents 1980. Therefore, you need to shift the graph 10 units *to the left.* The resulting model is

$$g(t) = 8.4 + 0.125(t + 10)^2. \quad \textit{t = 0 represents 1980.}$$

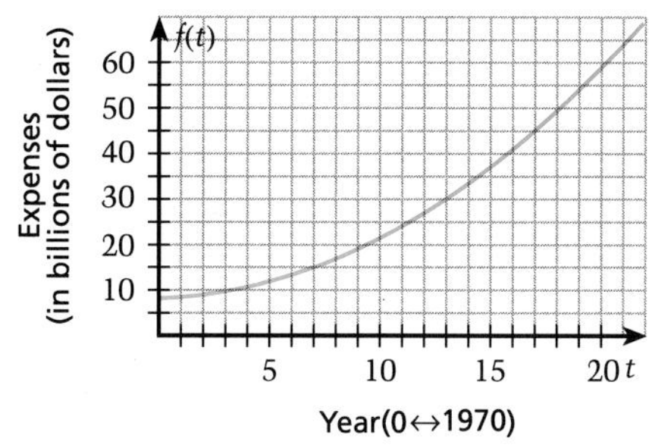

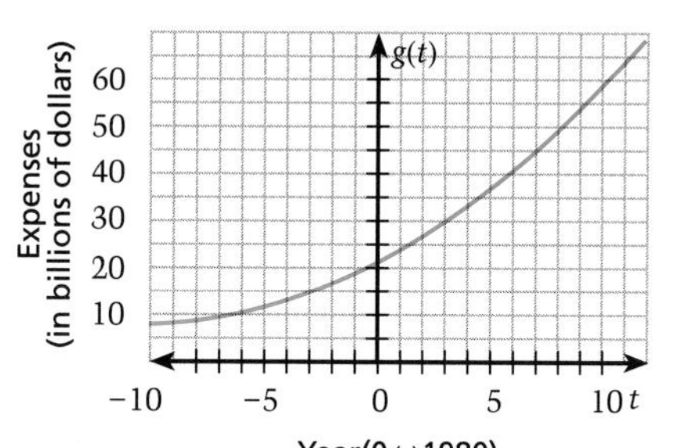

Communicating about ALGEBRA

SHARING IDEAS about the Lesson See below.

Each of the graphs represents a transformation of the graph of $f(x) = x^2$. Find an equation for each graph. Explain your reasoning.

A.

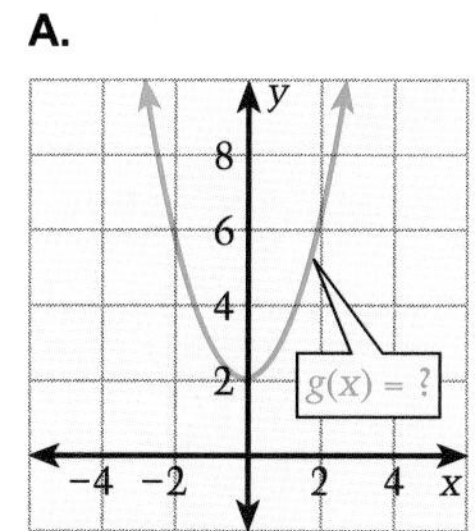

B.

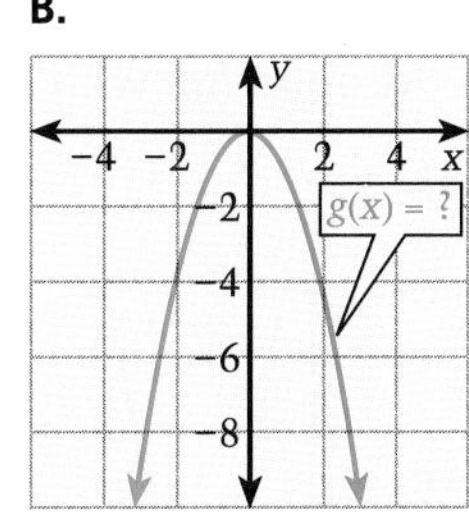

C.

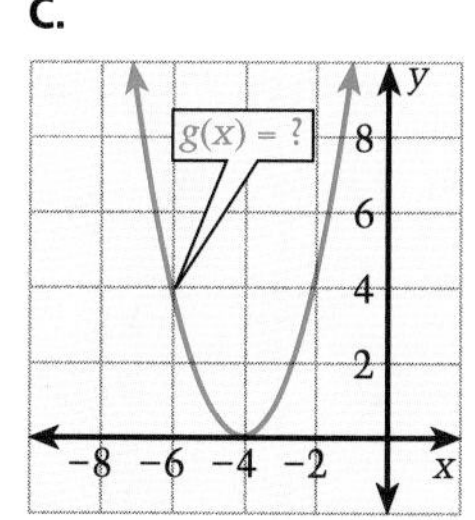

A. $g(x) = x^2 + 2$, vertical shift 2 units up **B.** $g(x) = -x^2$, reflection
C. $g(x) = (x + 4)^2$, horizontal shift 4 units to the left

Extra Examples

Here are additional examples similar to **Examples 1–3.**

1. Use the appropriate basic graphs to sketch the graphs of the following functions.
   a. $g(x) = x^2 + 3$
   b. $h(x) = -(x - 5)^2$
   c. $p(x) = (x + 7)^2 - 2$
   d. $g(x) = |x + 3|$
   e. $h(x) = |x - 6| - 4$
   f. $p(x) = -|x + 2|$
   a.–c. The shifts are all applied to the function $y = x^2$, vertex (0, 0)
   a. vertical shift up 3 units, vertex (0, 3)
   b. a reflection about the x-axis and a horizontal shift 5 units to the right , vertex (5, 0)
   c. horizontal shift 7 units to the left and a vertical shift 2 units down, vertex (−7, −2)
   d.–f. The shifts are all applied to the function $y = |x|$, vertex (0, 0).

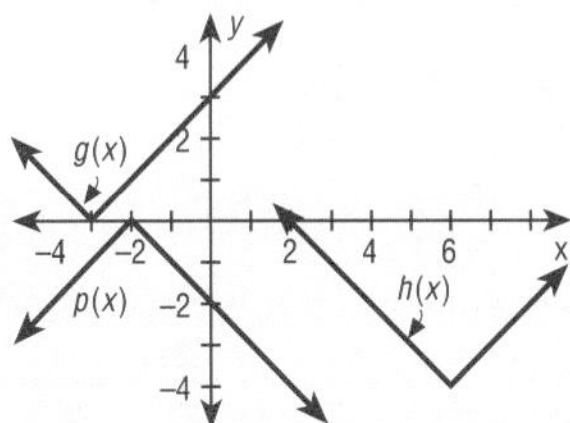

2. Write the equation of the function, $g(x)$, that is the result of shifting the basic function, $f(x) = |x|$, 7 units to the left and 2 units down.
   $g(x) = |x + 7| - 2$

Check Understanding

1. What is the reflection in the x-axis of the graph of a function?
   See page 316.
2. Describe the steps in graphing a function of the form $g(x) = f(x - c) + d$, where the basic graph of $f(x)$ is well-known.
   See page 314.

Communicating about ALGEBRA

EXTEND *Communicating*

Up to this point, students have applied the rules for horizontal and vertical shifts outlined in this lesson to linear, absolute-value, and quadratic graphs. Do they apply to higher-order graphs?

Consider the function $f(x) = x^3$. Show that the techniques of vertical and horizontal shifts apply for this basic function.

COOPERATIVE LEARNING

Specifically, have groups of students graph one or more of the following functions and describe the shifts.

a. $g(x) = x^3 - 2$
b. $h(x) = x^3 + 4$
c. $g(x) = (x - 3)^3$
d. $h(x) = -(x + 1)^3$
e. $g(x) = (x + 1)^3 - 7$
f. $h(x) = -(x - 8)^3 + 3$

EXERCISE Notes

ASSIGNMENT GUIDE

Basic/Average: Ex. 1–8, 13–19 odd, 21–24, 33–41 odd, 43–49, 57–58

Above Average: Ex. 1–8, 13–19 odd, 21–24, 31–43 odd, 42–50, 57–58

Advanced: Ex. 1–8, 13–19 odd, 21–24, 31–43 odd, 42–50, 57–60

Selected Answers
Exercises 1–8, 9–57 odd

Use **Mixed Review** as needed.

More Difficult Exercises
Exercises 51–56, 59, 60

EXERCISES

Guided Practice

CRITICAL THINKING about the Lesson See below for 1.–2., Additional Answers for 3.–4.

1. Describe the four types of shifts of the graph of a function.
2. How are the graphs of $f(x)$ and $g = -f(x)$ related?
3. Use a shift to sketch the graph of $h(x) = (x + 2)^2$.
4. Use a reflection to sketch the graph of $g(x) = -|x|$.

In Exercises 5–8, match the function with its graph.

5. $f(x) = |x - 3|$ d
6. $f(x) = |x| + 2$ a
7. $f(x) = -(x - 2)^2$ c
8. $f(x) = (x + 2)^2 - 2$ b

a. 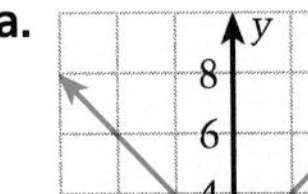
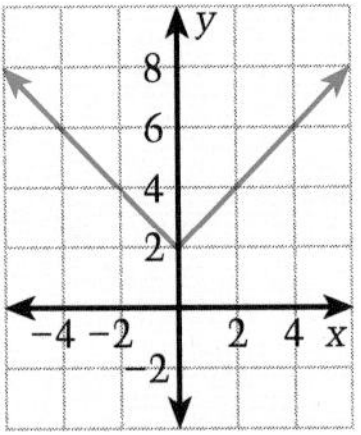

b. 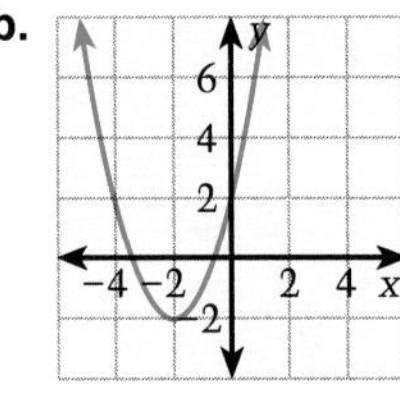

c. 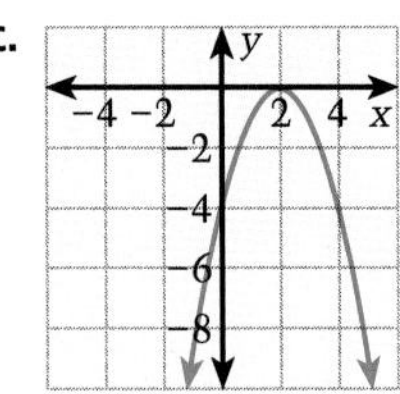

d. 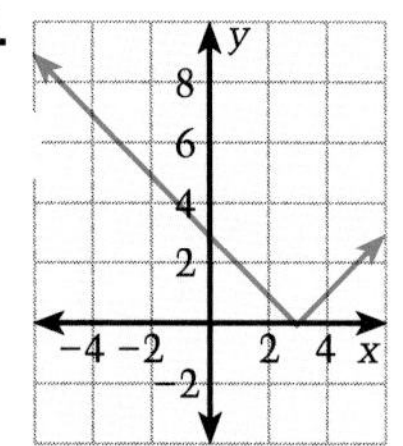

Independent Practice

In Exercises 9–16, compare the graph of *g* to the graph of $f(x) = x^2$. See Additional Answers.

9. $g(x) = x^2 + 3$
10. $g(x) = x^2 + 9$
11. $g(x) = x^2 - 1$
12. $g(x) = x^2 - 3$
13. $g(x) = (x - 8)^2$
14. $g(x) = (x + 4)^2$
15. $g(x) = (x + 5)^2$
16. $g(x) = (x - 7)^2$

In Exercises 17–20, show how *f* and *g* are related by writing an equation of the form $g(x) = f(\boxed{?})$ or $g(x) = f(x) + \boxed{?}$. See Additional Answers.

17. $f(x) = x^2$
 $g(x) = (x + 1)^2$
18. $f(x) = |x|$
 $g(x) = |x - 9|$
19. $f(x) = |x|$
 $g(x) = |x| - 3$
20. $f(x) = x^2$
 $g(x) = x^2 + 4$

In Exercises 21–24, match the equation with its graph.

21. $y = (x - 4)^2$ b
22. $y = x^2 + 4$ d
23. $y = |x| + 4$ c
24. $y = |x + 4|$ a

a. 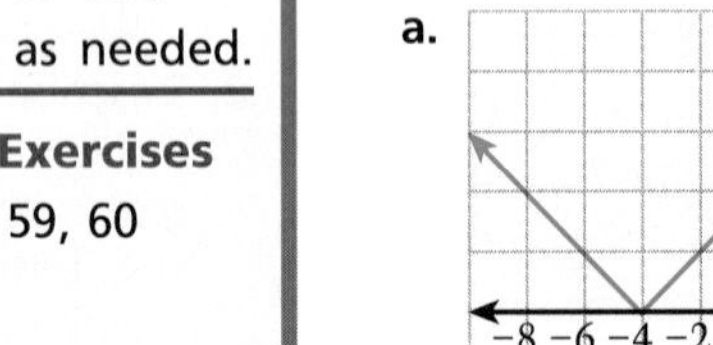

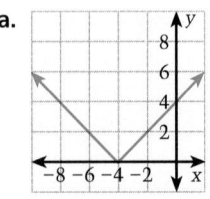

b. 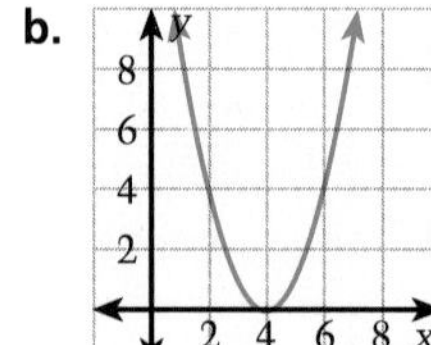

c. 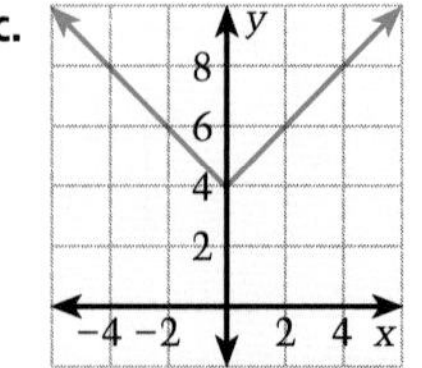

d. 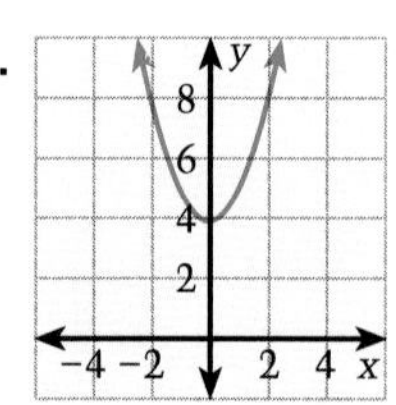

1. vertical up, vertical down, horizontal to the right, horizontal to the left
2. They are reflections of each other.

In Exercises 25–42, sketch the graph of the function. See Additional Answers.

25. $f(x) = |x| + 1$
26. $f(x) = |x + 2|$
27. $f(x) = x^2 + 2$
28. $f(x) = |x + 5| + 1$
29. $f(x) = |x - 2| + 5$
30. $f(x) = (x + 3)^2$
31. $f(x) = (x - 5)^2 + 4$
32. $f(x) = |x - 4| - 3$
33. $f(x) = |x + 6| - 3$
34. $f(x) = -|x|$
35. $f(x) = -(x - 1)^2$
36. $f(x) = -(x^2 + 1)$
37. $f(x) = -(x + 5)^2$
38. $f(x) = -(x + 3)^2 + 1$
39. $f(x) = (x - 2)^2 + 4$
40. $f(x) = -|x + 3| - 3$
41. $f(x) = (x + 7)^2 - 1$
42. $f(x) = -|x - 1| + 4$

In Exercises 43–46, match the equation with its graph.

43. $y = -|x - 3|$ d
44. $y = -|x - 2|$ b
45. $y = -(x + 2)^2$ c
46. $y = -(x - 3)^2$ a

a. 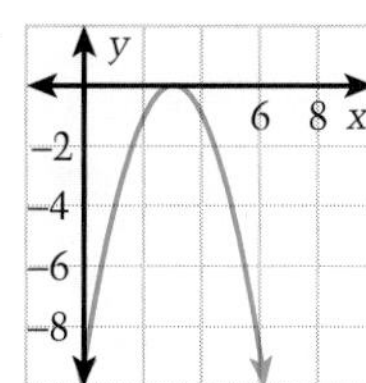

b. 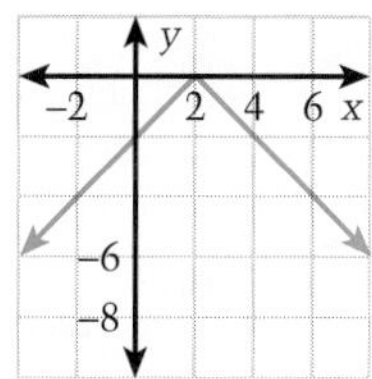

c. 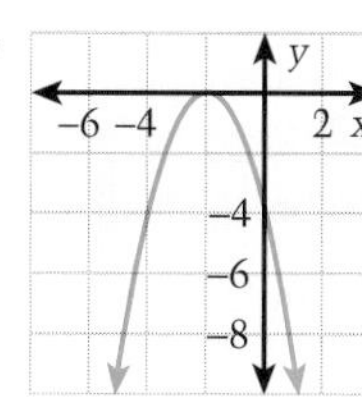

d. 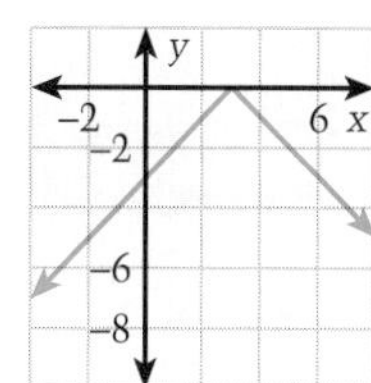

Cable Television Systems **In Exercises 47–49, use the following information.**

For 1970 through 1994, the number of cable television systems, $C(t)$, in the United States can be modeled by

$$C(t) = 14t^2 + 63.2t + 2500$$

where $t = 0$ represents 1970.
(Source: Television and Cable Factbook) See below.

47. Rewrite the model so that $t = 0$ represents 1980. Describe the shift from the original model to this one.
48. Rewrite the model so that $t = 0$ represents 1990. Describe the shift from the original model to this one.
49. Approximate the average number of subscribers of cable television in 1993.

50. *Civilian Population of United States* For 1950 through 1994, the civilian population, P (in thousands), in the United States can be modeled by

$$P = -8t^2 + 2767t + 150{,}000$$

where $t = 0$ represents 1950. Rewrite this model so that $t = 0$ represents 1970.
(Source: U.S. Bureau of the Census) See below.

Cable television began in the United States in the 1950's. The cables are connected to large antennas that receive the television signals. In 1994, there were 60.5 million cable television subscribers in the United States.

Guided Practice

▶ **Ex. 1–8** Students should review the quick-graph techniques from Chapters 1–5 before completing these exercises. Be sure to assign Exercises 5–8 as a group.

Independent Practice

▶ **Ex. 9–16** Students should "compare" by describing the type of shift.

▶ **Ex. 25–42** **TECHNOLOGY** Although students can use mental math and quick-graph techniques to complete these graphs, you might wish to have a group of students use graphics calculators to complete the graphs. Compare the two types of graphs as a check.

▶ **Ex. 47–49** Assign these exercises as a group. Students should refer to Example 4 for a solution model.

EXTEND Ex. 47–49 Use the data as the basis for an in-class discussion of cable television. What percent of students have cable? Have students estimate how much of their viewing time is spent watching cable stations rather than locally available stations. Students could also discuss whether they think cable TV is worth the monthly cost.

RESEARCH Update the data given for Exercises 47–49. How accurate were the models?

▶ **Ex. 51–56** Encourage students to rewrite the quadratic equations in completed-square form so as to use the quick-graph techniques outlined in Chapter 5.

47. $C(t) = 14(t + 10)^2 + 63.2(t + 10) + 2500$, a horizontal shift 10 units to the right
48. $C(t) = 14(t + 20)^2 + 63.2(t + 20) + 2500$, a horizontal shift 20 units to the right
49. 11,360 50. $P = -8(t + 20)^2 + 2767(t + 20) + 150{,}000$

Enrichment Activities

These activities require students to go beyond lesson goals.

EXTEND Ex. 38, 40, and 42 Consider horizontal and vertical shifts in relation to reflections in the x-axis. The order in which the shifts and the reflection are listed and performed is important.

1. In Exercise 38, the function is $f(x) = -(x + 3)^2 + 1$. Does this represent a horizontal shift of 3 units to the left, followed by a vertical shift of 1 unit up, followed by reflection in the x-axis?
   No, it represents a horizontal shift of 3 units to the left, followed by a reflection in the x-axis, followed by a vertical shift of 1 unit up.
2. Consider the following graphs. Which graphs are reflections of each other in the x-axis? In the y-axis?

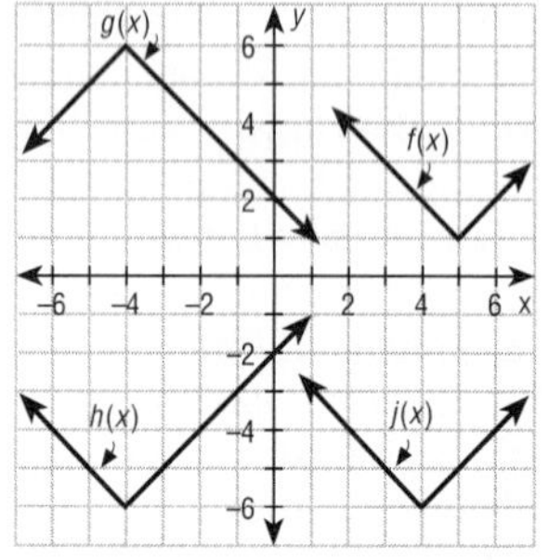

   Reflections in the x-axis: $g(x)$ and $h(x)$
   Reflections in the y-axis: $h(x)$ and $j(x)$
3. Write the equations for each of the functions shown in Activity 2.
   $f(x) = |x - 5| + 1$
   $g(x) = -|x + 4| + 6$
   $h(x) = |x + 4| - 6$
   $j(x) = |x - 4| - 6$

Integrated Review

In Exercises 51–56, sketch the graph of the function. See Additional Answers.

51. $f(x) = x^2 + x - 4$
52. $f(x) = 2x^2 - 8x + 1$
53. $f(x) = 3x^2 + 6x + 5$
54. $f(x) = x^2 - 4x + 3$
55. $f(x) = 6x^2 - 24x + 10$
56. $f(x) = x^2 - 2x + 5$

57. *Powers of Ten* Which best estimates the 1990 federal income tax for a single person who had no dependents and earned $20,000? d
    a. $\$2.2 \times 10^0$ **b.** $\$2.2 \times 10^1$ **c.** $\$2.2 \times 10^2$ **d.** $\$2.2 \times 10^3$
58. *Powers of Ten* Which best estimates the weight of a newborn child? c
    a. 1.1×10^0 oz **b.** 1.1×10^1 oz **c.** 1.1×10^2 oz **d.** 1.1×10^3 oz

Exploration and Extension

More about Reflections **In Exercises 59 and 60, use the following information.**

The graph of $y = f(-x)$ is a reflection of the graph of f in the y-axis. Use this fact to sketch the graph of the function. See Additional Answers.

59. $g(x) = (-x + 2)^2$
60. $g(x) = (-x - 1)^2$

Mixed REVIEW

7., 8. See Additional Answers.

1. Subtract: $\frac{4}{5} - \frac{2}{3}$. $\frac{2}{15}$
2. Simplify: $(4.8 \times 10^4)(1.6 \times 10^{-2})$. 7.68×10^2
3. Solve $|x - 2| > 4$. **(1.7)** $x < -2$ or $x > 6$
4. Solve for x: $ax^2 + b = y$. **(1.5)** $\pm\sqrt{\frac{y-b}{a}}$
5. Find the intercepts of $y = (x - 3)^2 - 5$. **(2.3)** x: $3 \pm \sqrt{5}$, y: 4
6. Simplify: $(3a^2b^4)(4a^3b)$. $12a^5b^5$
7. Sketch the graph of $5x + 2y + 4 = 0$. **(2.1)**
8. Sketch the graph of $y = x^2 + 4x - 1$. **(5.2)**
9. Subtract: $(4x - 2y + 3) - (x - 8y + 2)$. $3x + 6y + 1$
10. Write $y = 4x^2 - 8x + 3$ in completed square form. **(5.3)** $y = 4(x - 1)^2 - 1$

In Exercises 11–16, use $f(x) = 2x$ and $g(x) = 3x - 9$ to write an equation for $h(x)$.

11. $h(x) = \frac{1}{3}g(x) + f(x)$ **(6.2)** $h(x) = 3x - 3$
12. $h(x) = f(x) - g(x)$ **(6.2)** $h(x) = -x + 9$
13. $h(x) = f(x) \cdot g(x)$ **(6.2)** $h(x) = 6x^2 - 18x$
14. $h(x) = g(f(x))$ **(6.2)** $h(x) = 6x - 9$
15. $h(x)$ is the inverse of $g(x)$. **(6.3)**
16. $h(x) = f(g(x))$ **(6.2)** $h(x) = 6x - 18$
17. Simplify: $\frac{8x^2y^4}{12x^5y} \cdot \frac{2y^3}{3x^3}$
18. Write an equation of the line perpendicular to $y = -\frac{2}{3}x - 11$ through $(-1, 3)$. **(2.2)**
19. With what velocity must an arrow be shot straight up, to return in 8 sec? **(5.4)** 128 ft/sec
20. Find the area of the region enclosed between $y = 9$ and $y = x^2 - 6x + 9$. **(5.2)** 36

15. $h(x) = \frac{x + 9}{3}$
18. $y = \frac{3}{2}x + 4\frac{1}{2}$

6.6 Recursive Functions and Finite Differences

What you should learn:

Goal 1 How to identify and classify recursive functions

Goal 2 How to use recursive functions as real-life models

Why you should learn it:

You can model many real-life situations with recursive functions, such as Fibonacci's famous rabbit problem.

Goal 1 Identifying Recursive Functions

A **recursive function** is a function whose domain is the set of nonnegative integers (or sometimes the set of positive integers). To indicate that the domain is this set and not the set of real numbers, n is usually used as the variable.

A recursive function can be defined by stating the value of the function at 0 and by giving an equation for the value of the function at n in terms of the value of the function at $n - 1$.

A well-known example of a recursive function is the **factorial function.** It is defined as follows. (The exclamation mark is read "factorial.")

$0! = 1$ — *State the value of n! at n = 0.*

$n! = n \cdot (n - 1)!$ — *Define n! in terms of (n − 1)!.*

To evaluate a function that is defined recursively, first evaluate the function at 0, then at 1, then at 2, and so on.

$0! = 1$ — *The value of 0! is given.*

$1! = 1 \cdot 0! = 1 \cdot 1 = 1$ — *Use 0! to find 1!.*

$2! = 2 \cdot 1! = 2 \cdot 1 = 2$ — *Use 1! to find 2!.*

$3! = 3 \cdot 2! = 3 \cdot 2 = 6$ — *Use 2! to find 3!.*

$4! = 4 \cdot 3! = 4 \cdot 6 = 24$ — *Use 3! to find 4!.*

Example 1 *Evaluating a Recursive Function*

Use the following definition to find the value of $f(4)$.

$f(1) = 3$ and $f(n) = f(n - 1) + n$

Solution

$f(1) = 3$ — *The value of f(1) is given.*

$f(2) = f(1) + 2 = 3 + 2 = 5$ — *Use f(1) to find f(2).*

$f(3) = f(2) + 3 = 5 + 3 = 8$ — *Use f(2) to find f(3).*

$f(4) = f(3) + 4 = 8 + 4 = 12$ — *Use f(3) to find f(4).* ■

Problem of the Day

A man and his grandson have the same birthday. For six consecutive birthdays the man's age was an integral multiple of his grandson's age. How old was each on the sixth consecutive birthday?
6 and 66

ORGANIZER

Warm-Up Exercises

1. Evaluate $f(x) = 2x^2 + x + 3$ for the given values of x.
   a. $x = 0$ **b.** $x = 1$ **c.** $x = 2$
   d. $x = 3$ **e.** $x = 4$ **f.** $x = 5$
   a. 3, b. 6, c. 13, d. 24, e. 39, f. 58
2. The domain of the function, $f(x)$, is the set {0, 1, 2, 3, 4, . . . }. Continue the following pattern to evaluate $f(x)$ for the indicated values:
   $f(0) = 2$
   $f(1) = 5$
   $f(2) = 2 + 5 = 7$
   $f(3) = 5 + 7 = 12$
   $f(4) = 7 + 12 = 19$
   a. $f(5) = ?$ $f(5) = 31$
   b. $f(6) = ?$ $f(6) = 50$
   c. $f(7) = ?$ $f(7) = 81$
   d. $f(8) = ?$ $f(8) = 131$

Lesson Resources

Teaching Tools
- Problem of the Day: 6.6
- Warm-up Exercises: 6.6
- Answer Masters: 6.6

Extra Practice: 6.6
Color Transparency: 36

LESSON Notes

Example 1

Note that the first term of this function occurs when $n = 1$. It is common for a recursive function to start at $n = 1$ instead of at $n = 0$.

Example 2

Ask students to examine the differences in the following function.

n	$f(n)$
1	2
2	16
3	54
4	128
5	250
6	432
7	686

Differences

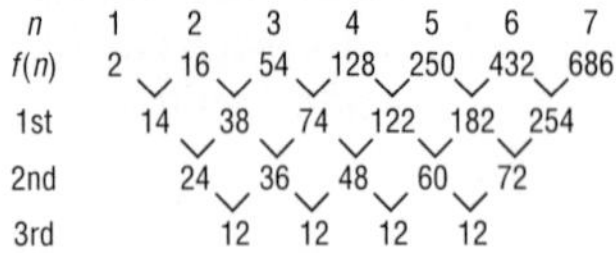

The diagram indicates that the function is cubic. (In fact, the function is $f(x) = 2x^3$; see Communicating about Algebra.)

Example 3

Ask students to evaluate $f(8)$ and $f(12)$.
$f(8) = 34$, $f(12) = 233$

Extra Examples

Here are additional examples similar to **Examples 1–2.**

1. Use the given recursive definitions to find $f(5)$.
   a. $f(0) = 2$, $f(n) = f(n - 1) + 1$
   b. $f(0) = -1$, $f(1) = 3$, $f(n) = f(n - 1) \cdot f(n - 2)$
   c. $f(0) = 2$, $f(n) = f(n - 1)!$
   a. $f(5) = 7$, b. $f(5) = -243$, c. $f(5) = 2$
2. Find the quadratic model whose values are $f(1) = 2$, $f(2) = 13$, and $f(3) = 30$.
   $f(n) = 3n^2 + 2n - 3$

The **first differences** of a recursive function are found by subtracting consecutive values of the function. The **second differences** are found by subtracting consecutive first differences. The first and second differences of the function in Example 1 are as follows.

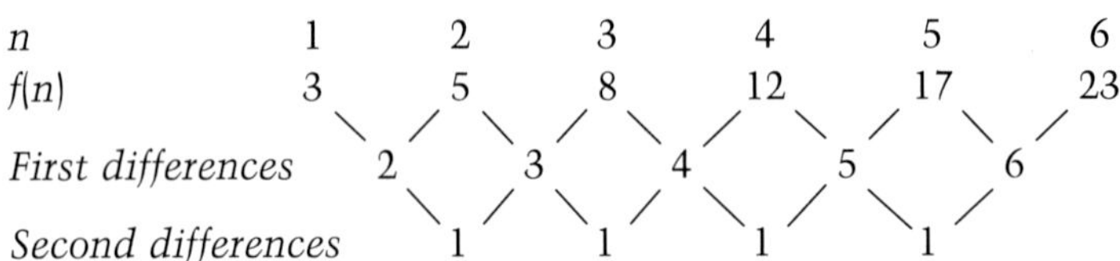

For this function, the second differences are all the same nonzero number. When this happens, the function has a quadratic model. (If the first differences are all the same number, then the function has a linear model.) The next example shows you how to find the quadratic model for the function given in Example 1.

Example 2 *Finding a Quadratic Model*

Find a quadratic model whose values are $f(1) = 3$, $f(2) = 5$, and $f(3) = 8$.

Solution You know the model has the form

$$f(n) = an^2 + bn + c.$$

By substituting 1, 2, and 3 for n, you can obtain a system of three linear equations in three variables.

$f(1) = a(1^2) + b(1) + c = 3$ *Substitute 1 for n.*

$f(2) = a(2^2) + b(2) + c = 5$ *Substitute 2 for n.*

$f(3) = a(3^2) + b(3) + c = 8$ *Substitute 3 for n.*

You now have a system of three equations in a, b, and c.

$$\begin{cases} a + b + c = 3 & \text{Equation 1} \\ 4a + 2b + c = 5 & \text{Equation 2} \\ 9a + 3b + c = 8 & \text{Equation 3} \end{cases}$$

Using the techniques discussed in Chapters 3 and 4, you can find the solution to be $a = \frac{1}{2}$, $b = \frac{1}{2}$, and $c = 2$. Thus a quadratic model is

$$f(n) = \frac{1}{2}n^2 + \frac{1}{2}n + 2.$$

Try checking the values of $f(1)$, $f(2)$, and $f(3)$. ■

Notice that the constant second differences were used to conclude that the function in Example 1 *has* a quadratic model. To *find* the model, three values of n were used to create a system of three linear equations.

Goal 2 Using Recursive Functions in Real Life

The numbers given by the recursive function in Example 3 are called Fibonacci numbers. They are named after the Italian mathematician Leonardo of Pisa (c. 1170–1250), also called Fibonacci.

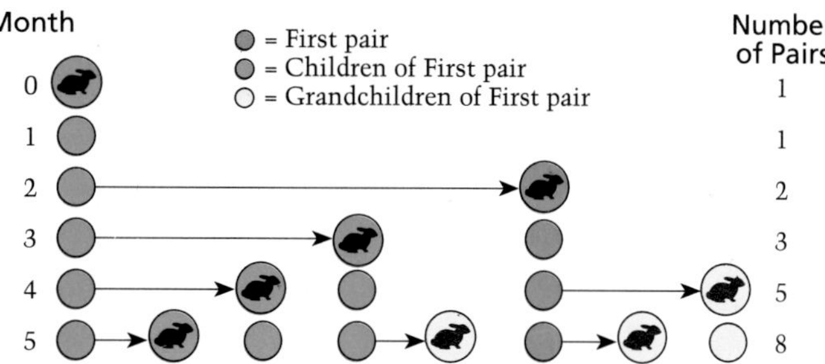

Problem Solving
Look for a Pattern

Example 3 *Using a Recursive Function*

Fibonacci proposed this rabbit problem: *Begin with one pair of newborn rabbits. Suppose that when a pair of rabbits is two months old, they begin to produce a new pair of rabbits each month. If none of the rabbits ever die, how many pairs of rabbits will there be in the nth month?*

Solution The above diagram shows the number of pairs of rabbits in months 0 through 5. Fibonacci answered the question with a recursive function, in which $f(n)$ is the number of pairs in the nth month.

$$f(0) = 1 \qquad f(1) = 1$$
$$f(n) = f(n-2) + f(n-1)$$

The following calculations confirm the definition.

$f(0) = 1$ *The value of f(0) is given.*

$f(1) = 1$ *The value of f(1) is given.*

$f(2) = 1 + 1 = 2$ *Use f(0) + f(1) to find f(2).*

$f(3) = 1 + 2 = 3$ *Use f(1) + f(2) to find f(3).*

$f(4) = 2 + 3 = 5$ *Use f(2) + f(3) to find f(4).*

$f(5) = 3 + 5 = 8$ *Use f(3) + f(4) to find f(5).* ■

Communicating about ALGEBRA

SHARING IDEAS about the Lesson See margin.

Find the Fibonacci numbers through $f(9)$. Find the first and second differences for these ten numbers. Do they have a linear or quadratic model? Explain.

Check Understanding

1. Evaluate 7! 5040
2. Must recursive function always begin with a value of $f(0)$? No, see Lesson Notes for Example 1.
3. How can finite differences be used to determine the algebraic model of a recursive function? See page 322.

Communicating about ALGEBRA

Find the appropriate model for the following differences.

n	$f(n)$
1	2
2	16
3	54
4	128
5	250
6	432
7	686

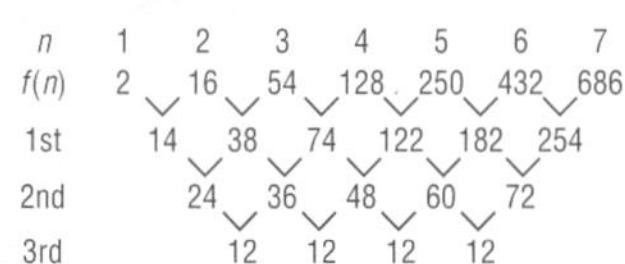

The model is cubic, $f(n) = 2n^3$.

Answers

Fibonacci numbers: 1, 1, 2, 3, 5, 8, 13, 21, 34, 55

First differences: 0, 1, 1, 2, 3, 5, 8, 13, 21; Second differences: 1, 0, 1, 1, 2, 3, 5, 8; no, neither the first nor the second differences are all the same.

EXERCISE Notes

ASSIGNMENT GUIDE

Basic/Average: Ex. 1–6, 11–21 odd, 29–37 odd, 41–47 odd, 55, 58, 60, 71

Above Average: Ex. 1–6, 11–21 odd, 27–37 odd, 45–55 odd, 57–60, 72–74

Advanced: Ex. 1–6, 11–21 odd, 27–37 odd, 45–55 odd, 57–60, 72–74

Selected Answers
Exercises 1–6, 7–71 odd

More Difficult Exercises
Exercises 57–60, 73–74

Guided Practice

Ex. 1 Ask students to explain the term "recursive" in their own words. *Note:* A dictionary defines a recursive procedure as "a procedure that can repeat itself indefinitely or until a specific condition is met."

Ex. 5–6 These exercises help students connect finite differences to quadratic models.

Independent Practice

EXTEND Ex. 19–24 Ask students to identify the linear models.

Answers

19. 3, −2, 6, 3, 13, 12, 24; −5, 8, −3, 10, −1, 12

20. 1, 2, 5, 26, 677, 458330, 210066388901; 1, 3, 21, 651, 457653, 210065930571

21. 0, 3, 6, 9, 12, 15, 18; 3, 3, 3, 3, 3, 3

22. 2, 0, 3, 1, 4, 2, 5; −2, 3, −2, 3, −2, 3

23. −2, −2, −7, −9, −16, −20, −29; 0, −5, −2, −7, −4, −9

24. −1, 4, −1, 4, −1, 4, −1; 5, −5, 5, −5, 5, −5

EXERCISES

Guided Practice

CRITICAL THINKING about the Lesson

1. Give an example of a recursive function. Answers vary. $f(1) = 2$, $f(n) = f(n-1) + 4$

2. Find the value of $f(5)$. 11
$f(0) = 1$
$f(n) = f(n-1) + n - 1$

3. Find the value of $f(7)$. 21
$f(1) = 1$
$f(2) = 2$
$f(n) = f(n-2) + f(n-1)$

4. Find the first and second differences for the following values of $f(n)$. −1, 1, 3, 5, 7; 2, 2, 2, 2

n	0	1	2	3	4	5
f(n)	3	2	3	6	11	18

5. Can the function whose values are given in Exercise 4 be described with a linear or quadratic model? Explain. Yes, a quadratic model, the second differences are all the same.

6. Does the following function have a linear or quadratic model? Explain. Quadratic model, the second differences are all the same
$f(0) = 0$
$f(n) = f(n-1) + 3n$

Independent Practice

In Exercises 7–18, find the first five values of the recursive function.

7. $f(0) = 1$ 1, 3, 5, 7, 9
$f(n) = f(n-1) + 2$

8. $f(1) = 4$ 4, 8, 13, 19, 26
$f(n) = n + f(n-1) + 2$

9. $f(0) = 0$ 0, 1, 5, 14, 30
$f(n) = f(n-1) + n^2$

10. $f(0) = -3$
$f(n) = |f(n-1)| - 10$

10. −3, −7, −3, −7, −3

11. $f(1) = 2$ See below.
$f(n) = [f(n-1)]^2 + 1$

12. $f(0) = 7$ 7, −6, 10, −1, 17
$f(n) = n^2 - f(n-1)$

13. $f(1) = 10$ See below.
$f(n) = 4f(n-1)$

14. $f(0) = 3$ See below.
$f(n) = [f(n-1)]^2 - 2$

15. $f(0) = 4$ 4, 0, 10, 8, 20
$f(n) = n^2 + 3n - f(n-1)$

16. $f(0) = 4$ 4, 2, −2, −4, −2
$f(1) = 2$
$f(n) = f(n-1) - f(n-2)$

17. $f(0) = 0$ 0, 2, 2, 6, 10
$f(1) = 2$
$f(n) = f(n-1) + 2f(n-2)$

18. $f(1) = 1$ 1, 2, 2, 4, 8
$f(2) = 2$
$f(n) = f(n-1) \cdot f(n-2)$

In Exercises 19–24, find the first seven values of the function. Then use the values to find the first differences of the function. See margin.

19. $f(0) = 3$
$f(n) = n^2 - f(n-1)$

20. $f(0) = 1$
$f(n) = [f(n-1)]^2 + 1$

21. $f(1) = 0$
$f(n) = f(n-1) + 3$

22. $f(1) = 2$
$f(n) = n - f(n-1)$

23. $f(1) = -2$
$f(n) = |f(n-1)| - n^2$

24. $f(3) = -1$
$f(n) = -f(n-1) + 3$

11. 2, 5, 26, 677, 458330 **13.** 10, 40, 160, 640, 2560 **14.** 3, 7, 47, 2207, 4870847

In Exercises 25–30, find the first seven values of the function. Then use the values to calculate the second differences of the function. See margin.

25. $f(1) = 3$
$f(n) = f(n-1) - n$

26. $f(0) = 1$
$f(n) = n^2 - f(n-1)$

27. $f(0) = 0$
$f(n) = -3[f(n-1)] + 5$

28. $f(1) = 0$
$f(n) = f(n-1) + (n-1)^2$

29. $f(3) = 9$
$f(n) = f(n-1) + n^2 - n$

30. $f(2) = -3$
$f(n) = -2f(n-1)$

In Exercises 31–36, state whether the function has a linear or quadratic model or neither.

31. $f(0) = 0$ Quadratic
$f(n) = f(n-1) + n$

32. $f(0) = 2$ Neither
$f(n) = [f(n-1)]^2$

33. $f(1) = 2$ Linear
$f(n) = f(n-1) + 2$

34. $f(1) = 0$ Quadratic
$f(n) = f(n-1) + 2n$

35. $f(0) = 1$ Neither
$f(n) = f(n-1) + n^2$

36. $f(0) = 0$ Linear
$f(n) = f(n-1) - 1$

In Exercises 37–42, find a linear model that has the indicated values.

37. $f(1) = 1, \quad f(2) = 3$ $f(n) = 2n - 1$

38. $f(0) = 4, \quad f(2) = 6$ $f(n) = n + 4$

39. $f(4) = 1, \quad f(1) = 7$ $f(n) = -2n + 9$

40. $f(2) = 2, \quad f(4) = 1$ $f(n) = -\frac{1}{2}n + 3$

41. $f(1) = 1, \quad f(2) = 6$ $f(n) = 5n - 4$

42. $f(1) = 6, \quad f(4) = 15$ $f(n) = 3n + 3$

In Exercises 43–50, find a quadratic model that has the indicated values. See margin.

43. $f(0) = 3, \quad f(1) = 3, \quad f(4) = 15$

44. $f(0) = 7, \quad f(1) = 6, \quad f(3) = 10$

45. $f(0) = -3, \quad f(2) = 1, \quad f(4) = 9$

46. $f(0) = 3, \quad f(2) = 0, \quad f(6) = 36$

47. $f(0) = 4, \quad f(1) = -2, \quad f(2) = -14$

48. $f(0) = 4, \quad f(1) = 2, \quad f(2) = -2$

49. $f(0) = -3, \quad f(1) = 4, \quad f(3) = 30$

50. $f(0) = 1, \quad f(1) = 3, \quad f(2) = 15$

In Exercises 51–56, find a linear or quadratic model for the function.

51. $f(0) = 2$ $f(n) = 2n + 2$
$f(n) = 2 + f(n-1)$

52. $f(1) = 2$ $f(n) = 3n - 1$
$f(n) = f(n-1) + 3$

53. $f(1) = 0$ $f(n) = 2n^2 + 2n - 4$
$f(n) = f(n-1) + 4n$

54. $f(1) = 3$ $f(n) = 5n - 2$
$f(n) = 5 + f(n-1)$

55. $f(0) = 0$ $f(n) = -n^2 - n$
$f(n) = f(n-1) - 2n$

56. $f(1) = 1$
$f(n) = f(n-1) + 3n$
$f(n) = \frac{3}{2}n^2 + \frac{3}{2}n - 2$

57. *Diagonals of a Polygon* The number of diagonals, $f(n)$, of an n-sided polygon is given by the following function. Write the first six values of the function. Find a quadratic model for the function.

$f(3) = 0$ 0, 2, 5, 9, 14, 20; $f(n) = \frac{1}{2}n^2 - \frac{3}{2}n$
$f(n) = f(n-1) + n - 2$

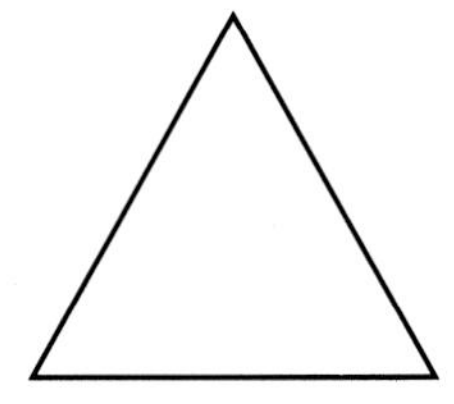
3 Sides

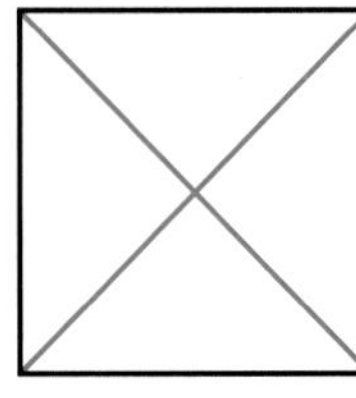
4 Sides

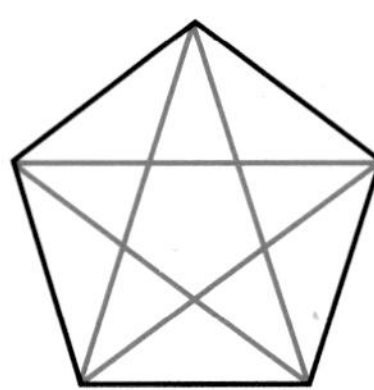
5 Sides

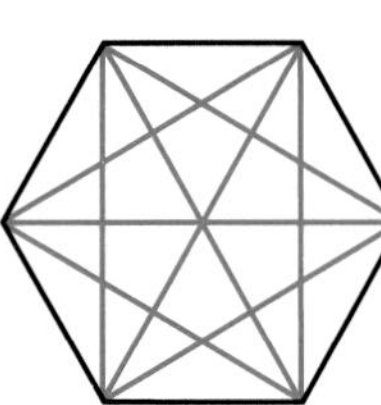
6 Sides

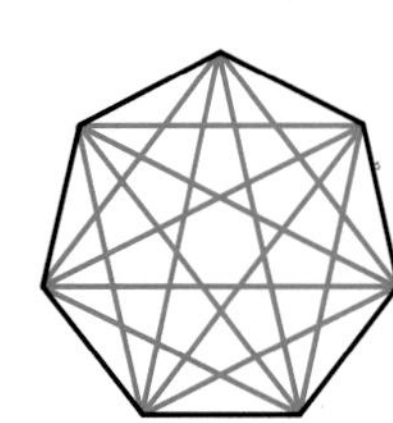
7 Sides

Answers

25. 3, 1, −2, −6, −11, −17, −24; −1, −1, −1, −1, −1

26. 1, 0, 4, 5, 11, 14, 22; 5, −3, 5, −3, 5

27. 0, 5, −10, 35, −100, 305, −910; −20, 60, −180, 540, −1620

28. 0, 1, 5, 14, 30, 55, 91; 3, 5, 7, 9, 11

29. 9, 21, 41, 71, 113, 169, 241; 8, 10, 12, 14, 16

30. −3, 6, −12, 24, −48, 96, −192; −27, 54, −108, 216, −432

▶ **Ex. 31–36** Remind students to test the first and second finite differences.

▶ **Ex. 43–58** Students can use Example 2 on page 322 as a solution model.

Answers

43. $f(n) = n^2 - n + 3$

44. $f(n) = n^2 - 2n + 7$

45. $f(n) = \frac{1}{2}n^2 + n - 3$

46. $f(n) = \frac{7}{4}n^2 - 5n + 3$

47. $f(n) = -3n^2 - 3n + 4$

48. $f(n) = -n^2 - n + 4$

49. $f(n) = 2n^2 + 5n - 3$

50. $f(n) = 5n^2 - 3n + 1$

Answer

59. \$20, \$40.10, \$60.30, \$80.60, \$101.00, \$121.51, \$142.12, \$162.83, \$183.64, \$204.56, \$225.58, \$246.71; quadratic model

▶ **Ex. 73–74** The solution should be given in a recursive-function form, not as a linear or quadratic model.

Enrichment Activities

These activities require students to go beyond lesson goals.

THE GOLDEN RATIO The ancient Greeks felt that the rectangle that is most pleasing to the eye is one in which the ratio of length to width equals the ratio of the sum of its length and width to its length.

1. Express the golden ratio in terms of l and w.
   $\frac{l}{w} = \frac{l+w}{l}$
2. The golden ratio is a solution of the equation $l^2 - l - 1 = 0$. Express the ratio in an exact form and as a decimal accurate to 4 decimal places.
   $\frac{1+\sqrt{5}}{2} \approx 1.6180$
3. In Example 3 on page 323, you studied the Fibonacci sequence. Examine the ratios of consecutive terms $\frac{f_{n+1}}{f_n}$ in the Fibonacci sequence. What do you notice?
   The ratios $\frac{8}{5}, \frac{13}{8}, \frac{21}{13}$, etc. are close to the golden ratio.
4. **WRITING** Write a two paragraph report on the occurrences of the golden ratio in art, architecture, or nature. For example, the Greek Parthenon, the paintings of Leonardo daVinci, the seed spirals in a sunflower, or the florets in a daisy.

58. *Triangular Numbers* The diagram at the right shows the first four triangular numbers. The nth triangular number is given by the following function. Write the first eight values of the function. Decide whether the function has a linear or quadratic model. If it does, find the model.

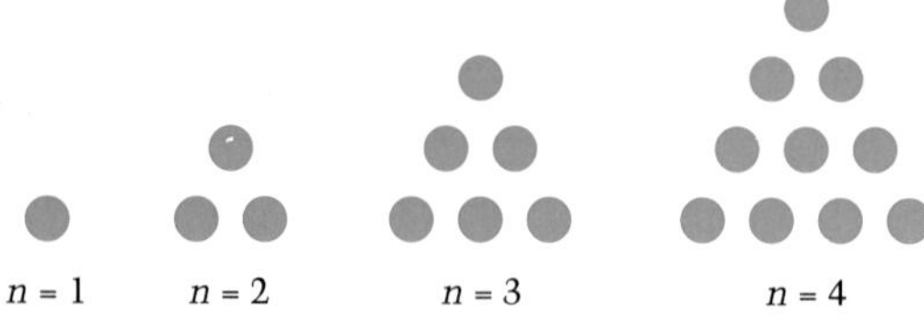

$f(1) = 1$ Quadratic, $f(n) = \frac{1}{2}n^2 + \frac{1}{2}n$
$f(n) = f(n-1) + n$

59. *Savings* Your savings account earns 0.005% monthly (6% annually). Each month you deposit \$20. The balance at the beginning of the nth month is given by the following recursive function.

$f(1) = 20$
$f(n) = 20 + 1.005f(n-1)$

Find the balances at the beginning of each of the first 12 months. Can this function be represented by a linear model or a quadratic model? See margin.

60. The values of the following recursive function form better and better approximations of a simple fraction. What is the fraction? $\frac{1}{3}$

$f(0) = \frac{3}{10}$
$f(n) = f(n-1) + \left(\frac{1}{10}\right)^n\left(\frac{3}{10}\right)$

Integrated Review

In Exercises 61–66, evaluate the function as indicated.

61. $f(x) = x - 3,\ f(-6)$ -9
62. $f(x) = 3x - 2,\ f(-1)$ -5
63. $f(x) = x^2 + 9,\ f(-1)$ 10
64. $f(x) = -2x^2 + 5,\ f(8)$ -123
65. $f(x) = |x + 3|,\ f(4)$ 7
66. $f(x) = (x - 3)^2 + 7,\ f(-3)$ 43

In Exercises 67–72, solve the system of equations. 70. $\left(\frac{15}{11}, -\frac{6}{11}, -\frac{20}{11}\right)$

67. $\begin{cases} x - 3y = 13 \\ 3x + y = -1 \end{cases}$ $(1, -4)$

68. $\begin{cases} 2x - 3y = 61 \\ -x + 5y = -83 \end{cases}$ $(8, -15)$

69. $\begin{cases} \frac{1}{2}x - y = -5 \\ -x + 3y = -5 \end{cases}$ $(-40, -15)$

70. $\begin{cases} x + y + z = -1 \\ 2x - y + 4z = -4 \\ 5x + 3y - z = 7 \end{cases}$

71. $\begin{cases} -x + 3y + 4z = 5 \\ x - z = 1 \\ 2x - y + 2z = 15 \end{cases}$ $(4, -1, 3)$

72. $\begin{cases} -x - 2y + 4z = 13 \\ 2x + y - z = 1 \\ x - y + 2z = 8 \end{cases}$ $(1, 5, 6)$

Exploration and Extension

In Exercises 73 and 74, find a recursive function whose values agree with those given in the table. (There is more than one correct solution.) See below.

73.

n	0	1	2	3	4	5
$f(n)$	3	5	7	9	11	13

74.

n	0	1	2	3	4	5
$f(n)$	3	8	15	24	35	48

73. Answers vary. $f(0) = 3,\ f(n) = f(n-1) + 2$ 74. Answers vary. $f(0) = 3,\ f(n) = f(n-1) + 2n + 3$

6.7 Exploring Data: Measures of Central Tendency

What you should learn:

Goal 1 How to find the mean, median, mode, and quartiles of a set of numbers

Goal 2 How to use measures of central tendency to answer questions about real-life situations

Why you should learn it:

You can use the mean, median, or mode of a set of data to represent a "typical" number of the data.

Goal 1 Measures of Central Tendency

There are three commonly used **measures of central tendency** for a collection of numbers.

1. The **mean,** or average, of n numbers is the sum of the numbers divided by n.
2. The **median** of n numbers is the middle number when the numbers are written in order. (If n is even, the median is the average of the two middle numbers.)
3. The **mode** of n numbers is the number that occurs most frequently.

These three measures can be different, *or* they can all be the same. For example, the numbers 1, 2, 2, 3, 4, 4, 4, 5, 6, 6, 7 have a mean of $\frac{44}{11} = 4$, a median of 4, and a mode of 4.

Many collections of numbers that are taken from real life have bar graphs that are *bell-shaped* (or nearly bell-shaped). For such collections, the mean, median, and mode are approximately equal.

Connections **Reading**

Word Lengths in "A Visit From St. Nicholas"

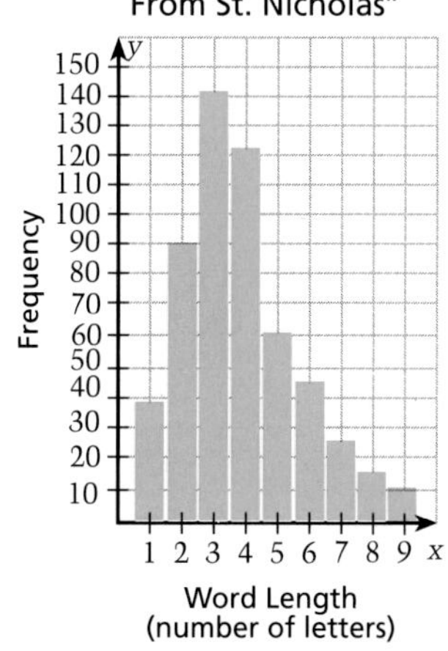

Example 1 *Comparing Central Tendencies*

One of the ways that the reading level of material is assessed is to count the letters in each word and find the average word length of the material. The number of words of different lengths in the poem "A Visit from St. Nicholas," by Clement Clarke Moore, is shown in the bar graph at the left. Find the mean, median, and mode of these data.

Solution The poem has 544 words. To find the mean word length, multiply each word length by its *frequency* and divide the result by 544.

$$\text{mean} = \frac{38(1) + 89(2) + 141(3) + \cdots + 15(8) + 10(9)}{544}$$

$$= \frac{2076}{544}, \text{ or about } 3.8.$$

The median word length is 4 because the middle two word lengths (numbers 272 and 273) are both 4. The word length mode is 3 because there are more 3 letter words than words of any other length. ■

Problem of the Day

Find the ratio of the area of the small square to the area of the large square. 1:2

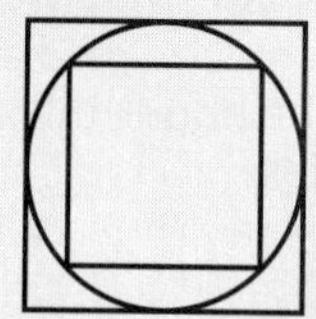

ORGANIZER

Warm-Up Exercises

1. Find the average of the following set of real numbers: {4, 2, 8, 1, 3, 2, 4, 2, 5, 7, 4, 8, 2, 1, 3, 5, 6, 3, 7, 3, 4, 1, 8, 7, 7, 7, 4, 3, 1, 2, 4, 5, 2, 6, 1, 7, 3}
   The average is 4.1, to the nearest tenth.
2. Order the following numbers from smallest to largest:
   32.0, 27.2, 27.1, 27.3, 26.5, 26.3, 23.5, 22.3, 25.2, 22.7, 28.0, 26.2, 25.2, 30.3, 25.9, 26.3, 26.2, 26.3, 29.4, 22.6, 24.8, 25.2, 26.5, 26.0, 25.5, 29.8, 27.4, 25.2, 27.3, 26.4, 25.3, 23.7, 29.8, 25.2, 24.3, 26.0, 27.2, 24.8, 23.7, 28.4, 23.1, 27.3, 27.6, 25.6, 29.6, 37.3, 24.4, 25.3, 25.6, 26.2, 25.5, 29.4
   See Example 2 in the lesson.

Lesson Resources

Teaching Tools
- Transparencies: 1, 5
- Problem of the Day: 6.7
- Warm-up Exercises: 6.7
- Answer Masters: 6.7

Extra Practice: 6.7
Color Transparencies: 38, 39
Technology Handbook: p. 40

LESSON Notes

Example 1

Ask students if they would expect the mean, median, and mode to be about the same in this example.

Answers may vary. They would probably be somewhat different because the shape of the bar graph does not closely resemble a bell-shaped curve. See the bell-shaped curve below.

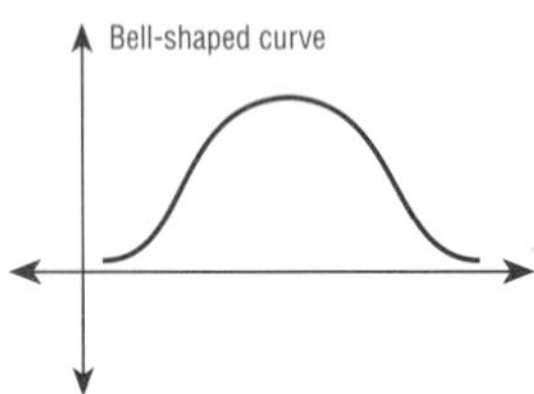

Example 2

Observe that in each case, the first quartile, the median, and the third quartile are found by taking the average of the two middle numbers from the appropriate list of numbers. For example, the third quartile is found by finding the average of the 13th and 14th entries in the upper half of the collection of numbers.

$$\frac{27.3 + 27.3}{2} = 27.3$$

Point out how the box-and-whisker plot is used to indicate the distribution of numbers in the collection. The range of values from 22.3 to 37.3 is recorded in the box-and-whisker plot with the three quartiles. Observe how the "whiskers" (the line segments on either end of the box) represent the distances of the extreme values from the quartile values.

Example 3

Ask students if there is any information in the table that surprises them. Discuss their observations in class.

The median of an ordered collection of numbers roughly partitions the collection into two halves: numbers below the median and numbers above the median. The **first quartile** is the median of the lower half. The **second quartile** is the median of the entire collection. The **third quartile** is the median of the upper half. Quartiles roughly partition an ordered collection into four **quarters.**

Example 2 *Finding the Quartiles of a Collection of Numbers*

The percent of the population of each state (including the District of Columbia and Puerto Rico) that is 18 years old or younger is given in the table. Find the first, second, and third quartiles for these data.

AK—32.0	FL—22.7	LA—29.4	NC—25.2	OK—27.2	UT—37.3
AL—27.2	GA—28.0	MA—22.6	ND—27.3	OR—24.8	VA—24.4
AR—27.1	HI—26.2	MD—24.8	NE—26.4	PA—23.7	VT—25.3
AZ—27.3	IA—25.2	ME—25.2	NH—25.3	PR—28.4	WA—25.6
CA—26.5	ID—30.3	MI—26.5	NJ—23.7	RI—23.1	WI—26.2
CO—26.3	IL—25.9	MN—26.0	NM—29.8	SC—27.3	WV—25.5
CT—23.5	IN—26.3	MO—25.5	NV—25.2	SD—27.6	WY—29.4
DC—22.3	KS—26.2	MS—29.8	NY—24.3	TN—25.6	
DE—25.2	KY—26.3	MT—27.4	OH—26.0	TX—29.6	

Solution There are 52 numbers in the collection. Begin by ordering the numbers. Then form the quartiles by breaking the numbers into 4 groups of 13 each.

First quarter: 22.3, 22.6, 22.7, 23.1, 23.5, 23.7, 23.7, 24.3, 24.4, 24.8, 24.8, 25.2, 25.2

Second quarter: 25.2, 25.2, 25.2, 25.3, 25.3, 25.5, 25.5, 25.6, 25.6, 25.9, 26.0, 26.0, 26.2

Third quarter: 26.2, 26.2, 26.3, 26.3, 26.3, 26.4, 26.5, 26.5, 27.1, 27.2, 27.2, 27.3, 27.3

Fourth quarter: 27.3, 27.4, 27.6, 28.0, 28.4, 29.4, 29.4, 29.6, 29.8, 29.8, 30.3, 32.0, 37.3

From the above grouping, you can see that the first quartile is 25.2, the second quartile (or median) is 26.2, and the third quartile is 27.3.

A **box-and-whisker plot** can be used to present a graphic summary of the information found in Example 2.

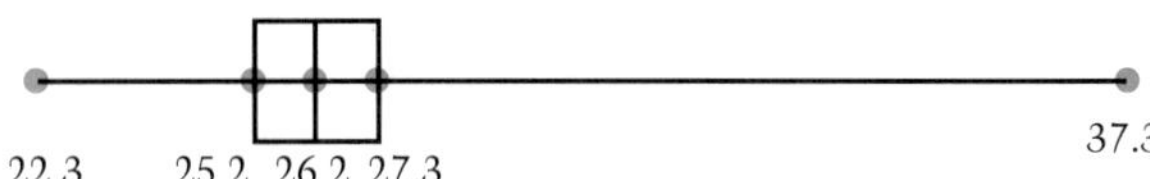

■

Goal 2 Using Measures of Central Tendency

Scheduled Time	12 to 17 yr	18 to 29 yr	Free Time	12 to 17 yr	18 to 29 yr
Classes	18.1	4.1	Visiting, social	4.4	7.8
Homework	3.1	3.4	Conversation	2.6	3.1
Work-related	2.8	29.0	Correspondence	0.4	0.4
Cleaning, laundry	2.2	3.2	Organizations, meetings	1.2	0.5
Yard and garden	2.4	1.4	Church, synagogue	1.0	0.5
Shopping	2.4	2.4	Travel	2.5	2.8
Travel (shopping)	1.5	1.9	Sports and sports events	4.1	2.1
Services	0.5	0.7	Walking outdoors	2.0	1.9
Child care	1.2	1.8	Games and hobbies	3.9	1.6
Adult care	0.6	0.6	Other recreation	1.0	0.5
Meals	9.0	9.1	Cultural events	0.3	0.4
Bathing, dressing	6.4	6.8	Movies	0.6	0.7
Sleep	62.6	55.9	Radios, records	0.9	0.7
Naps	1.3	1.7	Television	17.7	14.2
Travel (school/work)	3.4	2.4	Reading	1.3	1.9
Other	5.8	3.4	Relaxing, thinking	0.8	1.1

Real Life
Time Management

Example 3 *Interpreting Measures of Central Tendency*

The data in the above table show the mean hours per week that two groups of young Americans spend on their activities. The "12 to 17 yr" and "18 to 29 yr" categories represents *unmarried* Americans between the given ages. Which activities show significantly different uses of time between the two groups? How could you explain the differences? ***(Source: American Demographics, 1991)***

Solution Two clear differences are the times spent in class and at work. This seems reasonable because most Americans between the ages of 12 and 17 attend school each year, whereas most Americans between the ages of 18 and 29 attend school for only a portion of those years. ■

Communicating about ALGEBRA

SHARING IDEAS about the Lesson See below.

What are some other comparisons that you can make for Americans aged 12-to-17 and 18-to-29 from the table above?

Answers vary. Americans aged 18-to-29 use more time to wash and dress than Americans aged 12-to-17.

Extra Examples

Here are additional examples similar to **Examples 1–2.**

1. Find the mean, median, and mode of the collection of data in Example 2 of this lesson. mean = 26.46, median = 26.2, mode = 25.2
2. In Exercise 1 of the Warm-Up Exercises, find the mean, median, and mode. Construct a box-and-whisker plot of the data. mean = 4.11, median = 4, mode = 2, 3, 4, 7 (there are four numbers that occur with equal frequency!)

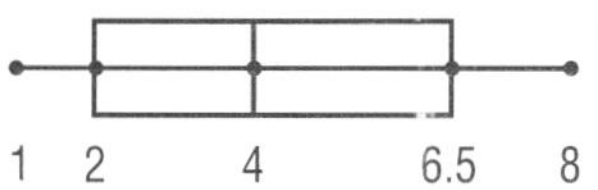

Check Understanding

1. Must the mean, median, and mode of a collection of numbers always be different? no
2. What is wrong with a word like "typical" or "average" when describing a collection of data? There is more than one average or measure of central tendency.
3. What does a bell-shaped curve look like? See the Lesson Notes for Example 1.

Communicating about ALGEBRA

GROUP ACTIVITY Collect data from your class and present the results using box-and-whisker plots. You should decide what data you want to collect, how the data should be collected, and how it should be reported. Some possible data you may wish to collect are:

1. Information about student activities on a Saturday night.
2. The top three occupations that students foresee in their future.
3. The number of siblings in each student's family group.

EXERCISE Notes

ASSIGNMENT GUIDE

Basic/Average: Ex. 1–4, 5–9, 10–12, 13–14, 17–20

Above Average: Ex. 1–4, 5–20

Advanced: Ex. 1–4, 5–20

Selected Answers

Exercises 1–4, 5–17 odd

More Difficult Exercises

Exercises 12, 14, 16, 19, 20

Guided Practice

Ex. 1–4 Use these exercises to summarize the terminology of the lesson. Note that sometimes the first and third quartiles are referred to as the lower and upper quartiles, respectively.

Answer

4.

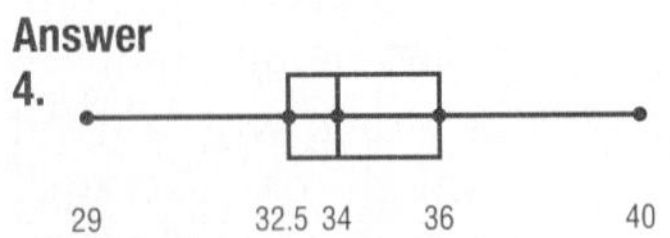

Independent Practice

Ex. 5–9 The measures of central tendency and the box-and-whisker plots will help students analyze the trends in the data. You may wish to use these exercises for in-class discussions regarding the trends.

Answers

5. Preamble: ≈5.15, Green Eggs and Ham: ≈3.04; the Preamble has a higher reading level.
6. ≈96.14 million, 96 million
7. ≈\$331.67, \$340, \$400

EXERCISES

Guided Practice

CRITICAL THINKING about the Lesson

1. Define the mean, median, and mode of a set of n numbers in your own words. See page 327.

In Exercises 2–4, use the following set of numbers.

{34, 35, 40, 31, 34, 34, 31, 38, 29, 35, 35, 32, 37, 33, 34, 37, 37, 35, 34, 32}

2. Find the mean, median, and mode of the data. 34.35, 34, 34
3. Find the first, second, and third quartiles of the data. 32.5, 34, 36
4. Construct a box-and-whisker plot for the numbers. See margin.

Independent Practice

5. *Reading Levels* The Preamble to the Constitution of the United States of America contains 52 words. The blue bars in the bar graph represent the number of words of different lengths in the Preamble. The green bars represent the number of words of different lengths (of the first 52 words) in *Green Eggs and Ham* by Dr. Seuss. Find the mean word length of each set of 52 words. Compare the reading levels of the two passages. See margin.

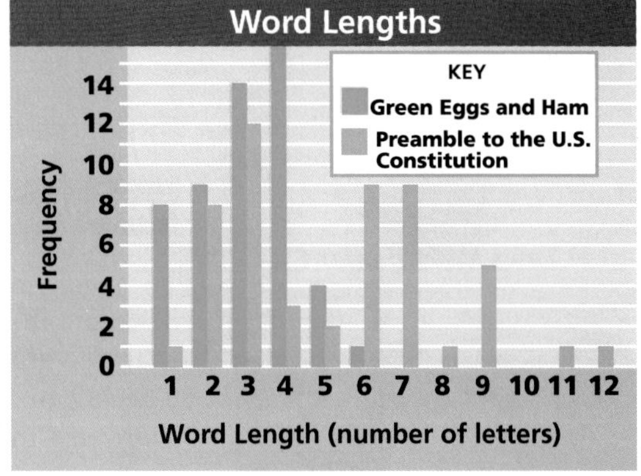

6. *Prime-Time Viewers* The numbers (in millions) of Americans who watched television during prime time in the fall of 1990 is shown in the graph at the right. What was the mean number of viewers per night? What was the median number of viewers per night? See margin.
(Source: Nielsen Media Research)

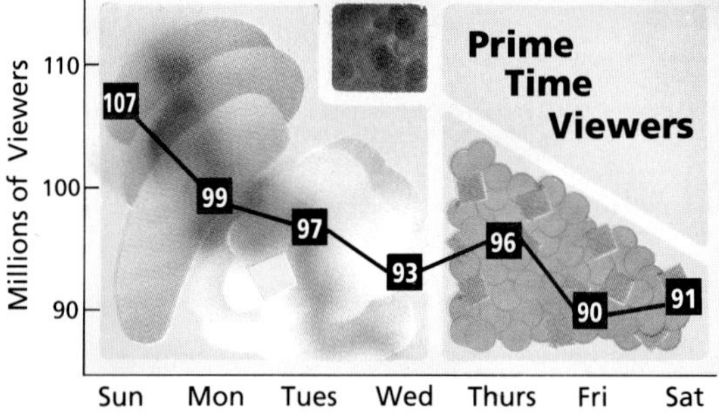

7. *Wedding Cake* The table below shows the prices in nine regions of the United States for a wedding cake that serves 300. Find the mean, median, and mode of these prices. *(Source: How to Have a Big Wedding on a Small Budget, 1990, Writer's Digest Books)* See margin.

Region	Cost	Region	Cost	Region	Cost
Northwestern	\$340	Central Mountain	\$325	Northeastern (urban)	\$400
California (urban)	400	Midwestern	255	Northeastern (rural)	350
California (rural)	225	Southwestern	400	Southeastern	290

8. *Birthday Months* The table shows numbers (in millions) of living Americans who were born in each of the 12 months. Construct a box-and-whisker plot. See margin.

Month	Number	Month	Number
January	21.2	July	20.4
February	20.4	August	21.2
March	22.2	September	22.8
April	20.0	October	17.4
May	21.4	November	19.0
June	20.0	December	20.4

9. *Greeting Cards* The table shows a greeting card company's monthly sales (in thousands of dollars) for one year. Construct a box-and-whisker plot and a bar graph for the data. Which do you think gives a better picture of the monthly sales? Why? See margin.

Month	Sales	Month	Sales
January	260	July	262
February	284	August	262
March	275	September	279
April	272	October	281
May	310	November	276
June	288	December	300

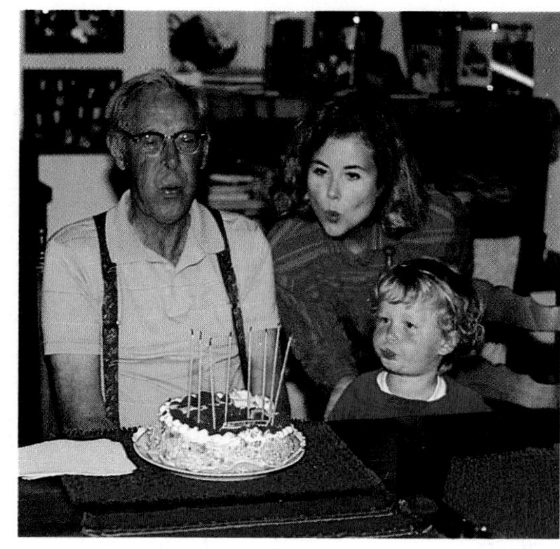

Super Bowl Scores **In Exercises 10–12, use the following information.**

The table shows the final score of each Super Bowl game from 1967 to 1990. (1967 was the year of the first Super Bowl.)

Year	1967	1968	1969	1970	1971	1972	1973	1974
Score	35–10	33–14	16–7	23–7	16–13	24–3	14–7	24–7

Year	1975	1976	1977	1978	1979	1980	1981	1982
Score	16–6	21–17	32–14	27–10	35–31	31–19	27–10	26–21

Year	1983	1984	1985	1986	1987	1988	1989	1990
Score	27–17	38–9	38–16	46–10	39–20	42–10	20–16	55–10

10. Find the margin of victory for each Super Bowl. Then find the mean, median, and modes of the margins of victory. 25, 19, etc.; ≈16.7, 17, 4 and 17

11. Find the first, second, and third quartiles for the margins of victory. 8, 17, 21.5

12. Construct two box-and-whisker plots: one for the winning scores and one for the losing scores. See margin.

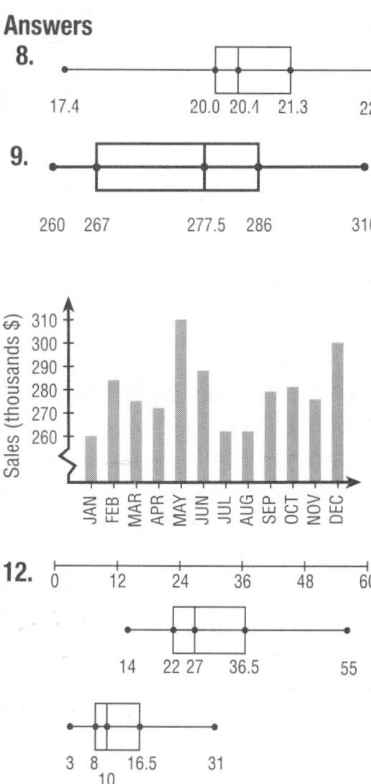

Ex. 10–12 **COOPERATIVE LEARNING** Have students work with partners to complete these exercises in class. Students can share the work. For example, one student could do the winning-score plot while the other completes the losing-score plot.

Answers

14. 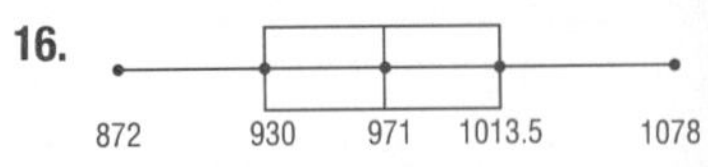

16.

872 930 971 1013.5 1078

▶ **Ex. 17–18** These college entrance test samples relate to this lesson.

EXTEND Ex. 19–20 Ask students to create a data set where the mean is distorted by a few extreme values. For example, the selling price for 10 houses in a town is about $120,000 each; two houses in town are selling for $3,000,000 each. How is the mean selling price distorted?

Enrichment Activities

These activities require students to go beyond lesson goals.

COOPERATIVE LEARNING
Work with a partner. Collect sets of data from newspapers or other resources. (You could also collect data by designing and conducting a survey on some topic that interests you.) Create a set of questions using measures of central tendency for your data set. Share your questions with your classmates.

Sports Injuries **In Exercises 13 and 14, use the following information.**

The following table shows the percent of sports participants who are injured each year. ***(Source: Consumer Products Safety Commission)***

Sport	Percent	Sport	Percent
Archery	1.4%	Racquetball	1.0%
Baseball	9.8	Roller-skating	1.9
Basketball	5.4	Snow-skiing	1.1
Bicycling	2.4	Soccer	4.2
Football	9.6	Swimming	0.3
Golf	0.4	Tennis	0.4
Ice-skating	0.5	Water-skiing	0.4

13. Find the mean and median of the data. ≈2.8%, 1.25%

✪ **14.** Construct a box-and-whisker plot for the data.

Polar Bears **In Exercises 15 and 16, use the following information.**

The following list shows the weights (in pounds) of 20 adult male polar bears that were weighed by a Norwegian scientific team.

952 937 932 1062 925 964 954
980 1022 872 1030 992 1005 928
925 1078 978 1026 1002 894

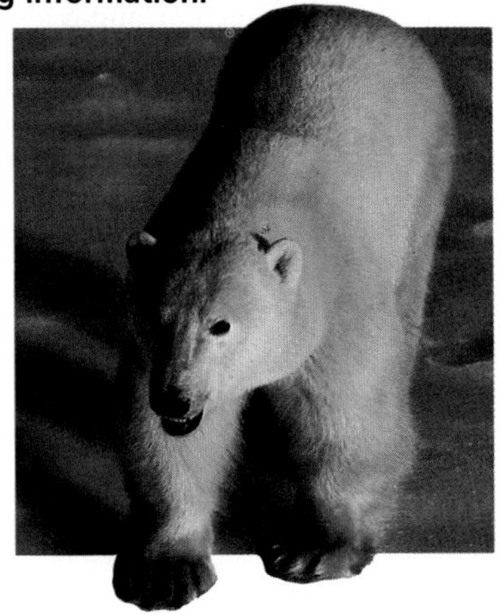

15. Which measure of central tendency do you think best describes the average weight of an adult male polar bear? Explain. See below.

✪ **16.** Construct a box-and-whisker plot of the data. See margin.

Integrated Review

College Entrance Exam Samples **In Exercises 17 and 18, use the following information.**

The graph at the right compares the test scores of five students on two different tests.

Test Score Correlations

Score on Test 2: 20, 40, 60, 80, 100
Score on Test 1: 20, 40, 60, 80, 100
A, B, C, D, E

17. For which student was the change in scores from test I to test II the greatest? a

a. *A* **b.** *B* **c.** *C* **d.** *D* **e.** *E*

18. What was the mean of the scores of the five students on test II? c

a. 60 **b.** 65 **c.** 68 **d.** 70 **e.** 72

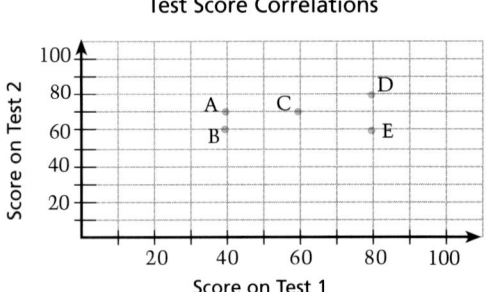

Exploration and Extension

✪ **19.** *Prices* Create a list of 8 prices so that the mean is $20 and the median is $18. See below.

✪ **20.** *Days* Create a list of data so that the mean is 5 days and the median is 6 days. See below.

15. Either the mean (972.9) or the median (971) describes the average weight well; the mode (925) is too low.
19. Answers vary; example: $10, $12, $14, $15, $21, $24, $26, $38 **20.** Answers vary; example: 2, 2, 6, 7, and 8 days

6 Chapter Summary

What did you learn?

Skills

1. Identify relations and functions. **(6.1)**
- Use function notation. **(6.1)**

2. Perform operations with functions. **(6.2)**

3. Find the inverse of a relation. **(6.3)**
- Decide whether an inverse relation is a function. **(6.3)**

4. Evaluate compound functions and step functions. **(6.4)**

5. Use translations and reflections to graph functions. **(6.5)**

6. Evaluate recursive functions. **(6.6)**
- Use first and second differences to decide whether a recursive function has a linear or quadratic model. **(6.6)**

Strategies

7. Use functions to model real-life situations. **(6.1–6.6)**

Exploring Data

8. Find measures of central tendency for data. **(6.7)**

9. Sketch box-and-whisker plots to represent data. **(6.8)**

Why did you learn it?

One of the primary uses of algebra is to create and use models of real-life situations. Often the model is a function. For instance, the equation $V = \frac{4}{3}\pi r^3$ shows how the volume of a sphere is related to the radius of the sphere. This equation describes the volume, V, as a function of the radius, r.

As you study the remaining chapters of the book, you will encounter many references to functions. Remember that a function is simply a special type of relation. Moreover, most functions are specified by equations such as $f(x) = 3x - 4$ or $g(x) = x^2 - 4x + 1$.

How does it fit into the bigger picture of algebra?

This chapter explored functions. A function is a special type of relationship between two variables in which each input value corresponds to exactly one output value. For instance, the area of a circle is a function of its radius. You can think of the radius of a circle as input and the area of the circle as output. If someone gives you the radius, r, of a circle, you know the area is $A = \pi r^2$.

Now you should be able to use function notation, perform operations with functions, and sketch graphs of functions.

The last lesson in the chapter discussed measures of central tendency and box-and-whisker plots.

ORGANIZER

TIME SCHEDULE: All levels, two days

The Chapter Summary helps students organize the main ideas of the chapter. In this chapter, the concept of function was introduced and developed. Special functions were also presented and connections to real-life applications were made.

Work with students to review the skills, strategies, and concepts of the chapter. The first day's homework assignment can be used as the basis for the second day of review.

COOPERATIVE LEARNING Encourage students to study together. Emphasize the importance of "teaching" a classmate how to perform a skill or how to recall a strategy. When students work together, everyone wins. The students receiving help get additional instruction and the students giving help gain a deeper understanding of the skills and concepts involved.

Chapter SUMMARY

Review the chapter by asking students, "What did you learn? Why did you learn it? How does it fit with the other algebra skills and concepts you have learned?"

Chapter REVIEW

ASSIGNMENT GUIDE

Basic/Average: Ex. 3–57 multiples of 3, 59–65 odd, 67–76, 78–90 multiples of 3, 92

Above Average: Ex. 3–57 multiples of 3, 59–65 odd, 67–76, 78–79, 85–91 odd, 92

Advanced: Ex. 3–57 multiples of 3, 59–65 odd, 67–76, 78–79, 85–91 odd, 92

More-Difficult Exercises
Exercises 77–92

Ex. 1–92 Since the Chapter Review exercises parallel the list of Skills, Strategies, and Exploring Data statements on page 333, tell students to use the Chapter Summary to identify the lesson in which a skill or concept was presented if they are uncertain how to do a given exercise. This makes the students categorize each task in the exercise set and locate relevant information in the chapter about the task, both of which are beneficial activities.

Answers

11. All real numbers except 1
12. All real numbers ≤ 0
13. All real numbers ≥ 0
14. All real numbers except 3
15. All real numbers
16. All real numbers except 2
17. $h(x) = 3x - 5$
18. $h(x) = x - 3$
19. $h(x) = -x - 1$
20. $h(x) = 3x - 4$
21. $h(x) = 2x^2 - 6x + 4$
22. $h(x) = \dfrac{2x - 4}{x - 1}$
23. $h(x) = \dfrac{x - 1}{2x - 4}$
24. $h(x) = 2x - 6$
25. $h(x) = 2x - 5$
26. $x = 11y + 22$
27. $x = \frac{1}{2}(y - 10)$
28. $x = \frac{1}{3}(-6y + 2)$

Chapter REVIEW

In Exercises 1–4, state whether the relation is a function. (6.1)

1. 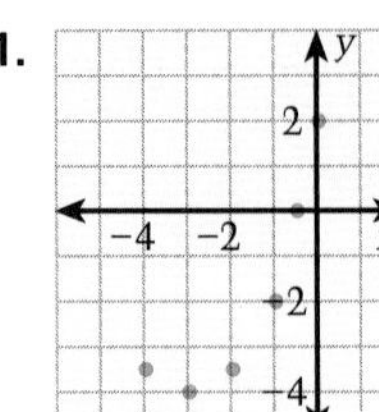
It is.

2. 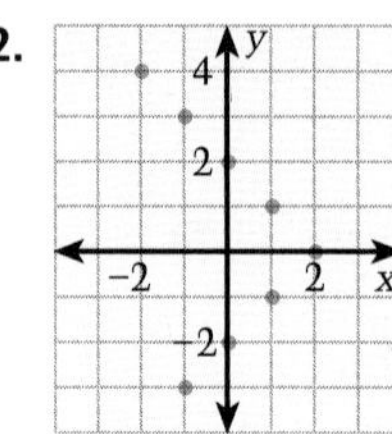
It is not.

3. 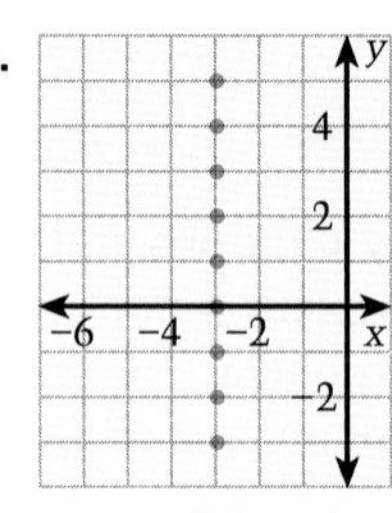
It is not.

4. 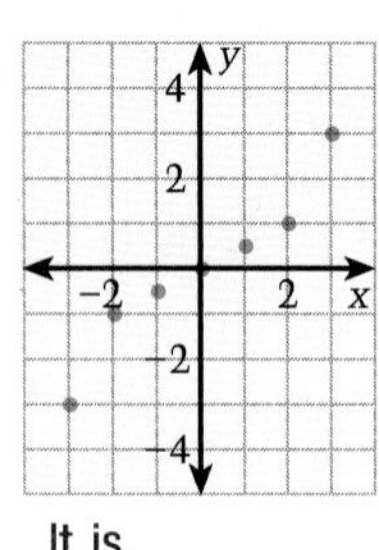
It is.

In Exercises 5–10, find the indicated value of $f(x)$. (6.1)

5. $f(x) = 10x + 2$, $f(-1)$ -8

6. $f(x) = -x^2 - 5$, $f(-3)$ -14

7. $f(x) = -6x^2 + 13$, $f(-2)$ -11

8. $f(x) = 2x^2 + \frac{1}{2}x - 1$, $f(4)$ 33

9. $f(x) = \frac{1}{3}x^2 - 3x + 1$, $f(6)$ -5

10. $f(x) = -x^2 - 4x + 3$, $f(10)$ -137

In Exercises 11–16, find the domain of the function. (6.2) See margin.

11. $f(x) = \dfrac{10x - 4}{x - 1}$

12. $f(x) = \sqrt{-x}$

13. $f(x) = \dfrac{\sqrt{x}}{5}$

14. $f(x) = \dfrac{x^2}{3 - x}$

15. $f(x) = -16x^2 + 2$

16. $f(x) = \dfrac{x^2 + 3}{x - 2}$

In Exercises 17–25, let $f(x) = 2x - 4$ and $g(x) = x - 1$. Write and simplify an equation for $h(x)$. (6.2) See margin.

17. $h(x) = f(x) + g(x)$

18. $h(x) = f(x) - g(x)$

19. $h(x) = f(x) - 3g(x)$

20. $h(x) = \frac{1}{2}f(x) + 2g(x)$

21. $h(x) = f(x) \cdot g(x)$

22. $h(x) = f(x) \div g(x)$

23. $h(x) = g(x) \div f(x)$

24. $h(x) = f(g(x))$

25. $h(x) = g(f(x))$

In Exercises 26–28, find the inverse of the function. (6.3) See margin.

26. $f(x) = 11x + 22$

27. $f(x) = \frac{1}{2}(x - 10)$

28. $f(x) = \frac{1}{3}(-6x + 2)$

In Exercises 29–34, sketch the graph of the function and its inverse on the same coordinate plane. (6.3) See margin.

29. $f(x) = x^2 - 1$

30. $f(x) = \frac{1}{2}x - 4$

31. $f(x) = x - 10$

32. $f(x) = 2x + 1$

33. $f(x) = -2x + 7$

34. $f(x) = -x^2$

In Exercises 35–38, verify that the two functions are inverses of each other. (6.3) See margin.

35. $f(x) = 2x - 5$, $g(x) = \frac{1}{2}x + \frac{5}{2}$

36. $f(x) = -\frac{1}{2}x + 7$, $g(x) = -2x + 14$

37. $f(x) = 3x + 6$, $g(x) = \frac{1}{3}x - 2$

38. $f(x) = \frac{1}{3}x + 8$, $g(x) = 3x - 24$

39. Evaluate $f(3)$. **(6.4)**

$$f(x) = \begin{cases} -3x^2 + 10, & x < 3 \\ x - 20, & x \geq 3 \end{cases}$$ −17

40. Evaluate $f(-2)$. **(6.4)**

$$f(x) = \begin{cases} x^2 + 2x - 1, & x \leq 1 \\ 3 - x, & x > 1 \end{cases}$$ −1

In Exercises 41–46, sketch the graph of the function. (6.4) See margin.

41. $y = \left|\frac{1}{2}x + 3\right|$

42. $y = |3x - 6|$

43. $f(x) = \begin{cases} 2x + 2, & x \leq 2 \\ \frac{1}{4}x^2 + x + 1, & x > 2 \end{cases}$

44. $f(x) = \begin{cases} x^2 - 6, & x < 0 \\ \frac{1}{2}x - 6, & x \geq 0 \end{cases}$

45. $f(x) = \begin{cases} -3, & 0 \leq x < 2 \\ -1, & 2 < x \leq 4 \\ 0, & 4 < x \leq 6 \\ 3, & 6 < x \leq 8 \end{cases}$

46. $f(x) = \begin{cases} -1, & -2 \leq x \leq 4 \\ 2, & 4 < x \leq 6 \\ 4, & 6 < x \leq 20 \\ 6, & 20 < x \leq 30 \end{cases}$

In Exercises 47–58, use shifts and reflections to sketch the graph of the function. (6.5) See Additional Answers.

47. $y = x^2 - 6$
48. $y = (x + 7)^2$
49. $y = |x + 10|$
50. $y = |x| - 7$
51. $y = x^2 + 12$
52. $y = -x^2 + 1$
53. $y = -(x - 2)^2$
54. $y = -|x - 4|$
55. $y = -|x| + 9$
56. $y = (x - 2)^2 + 3$
57. $y = -|x + 3| - 2$
58. $y = -(x - 1)^2 + 1$

In Exercises 59–62, find the first five values of the recursive function. (6.6)

59. $f(1) = 4$ 4, 12, 21, 31, 42
$f(n) = f(n - 1) + n + 6$

60. $f(1) = -3$ −3, −1, −3, −1, −3
$f(n) = |f(n - 1)| - 4$

61. $f(0) = 1$ 1, 2, 5, 26, 677
$f(n) = [f(n - 1)]^2 + 1$

62. $f(2) = 2$ 2, 2, 2, 2, 2
$f(n) = 6 - [f(n - 1)]^2$

In Exercises 63–66, match the function with its graph. Then find a linear or quadratic model for the function. (6.6)

63. $f(0) = 0$ c, $f(n) = 2n$
$f(n) = f(n - 1) + 2$

64. $f(1) = 0$ a, $f(n) = \frac{1}{2}n^2 + \frac{1}{2}n - 1$
$f(n) = f(n - 1) + n$

65. $f(0) = 10$ d, $f(n) = -\frac{1}{2}n^2 - \frac{1}{2}n + 10$
$f(n) = f(n - 1) - n$

66. $f(1) = 12$ b, $f(n) = -3n + 15$
$f(n) = f(n - 1) - 3$

a. 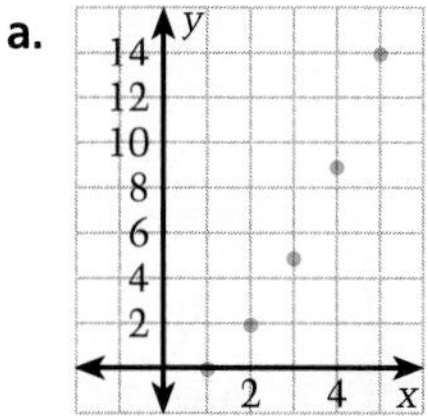

b. 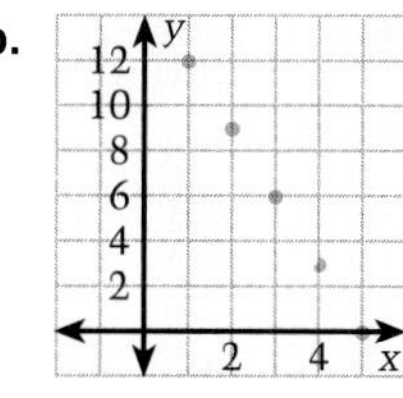

c. 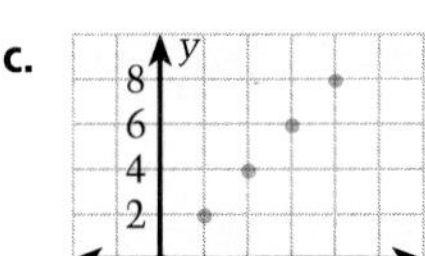

d. 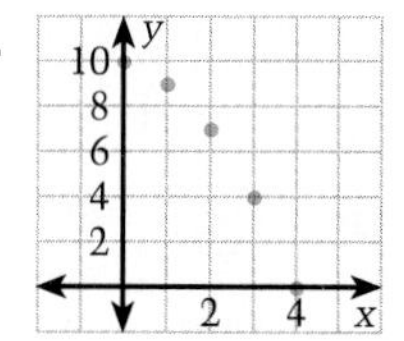

Answers

29 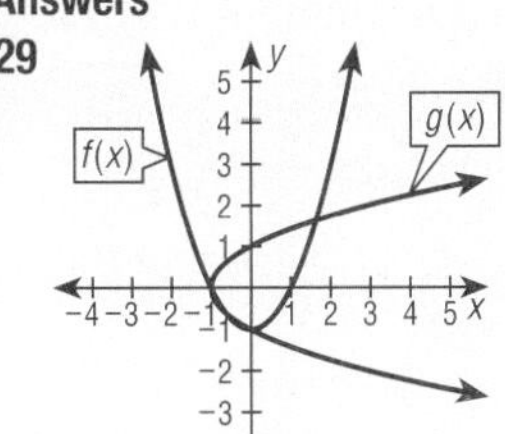

30. 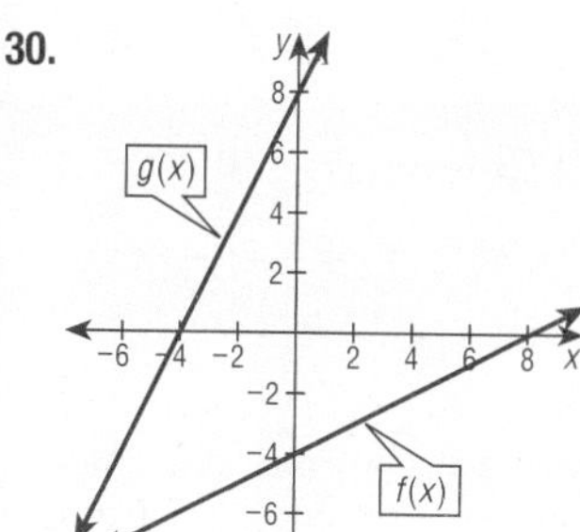

31. 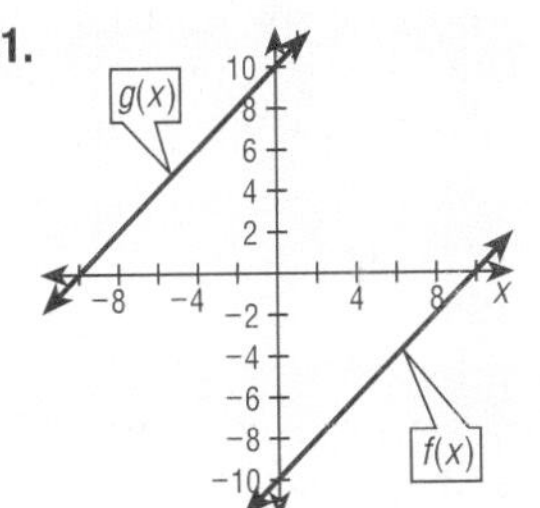

32. 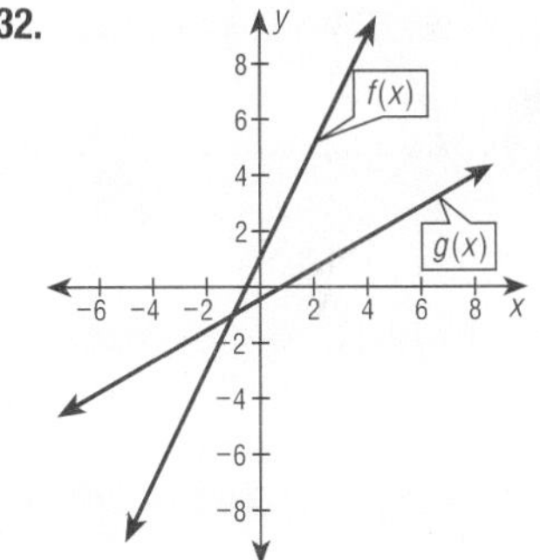

33. 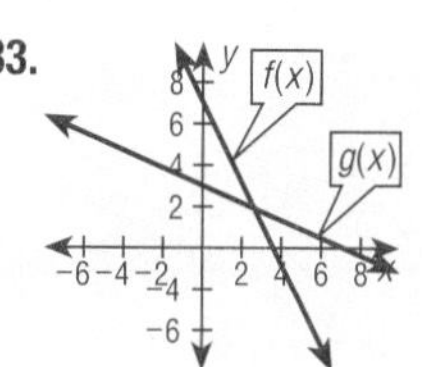

34. 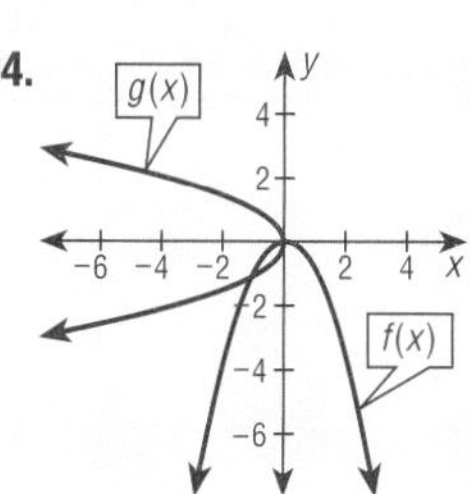

Answers

35. $f(g(x)) = 2\left(\frac{1}{2}x + \frac{5}{2}\right) - 5 = x,$
$g(f(x)) = \frac{1}{2}(2x - 5) + \frac{5}{2} = x$

36. $f(g(x)) = -\frac{1}{2}(-2x + 14) + 7 = x,$
$g(f(x)) = -2\left(-\frac{1}{2}x + 7\right) + 14 = x$

37. $f(g(x)) = 3\left(\frac{1}{3}x - 2\right) + 6 = x,$
$g(f(x)) = \frac{1}{3}(3x + 6) - 2 = x$

38. $f(g(x)) = \frac{1}{3}(3x - 24) + 8 = x,$
$g(f(x)) = 3\left(\frac{1}{3}x + 8\right) - 24 = x$

41.

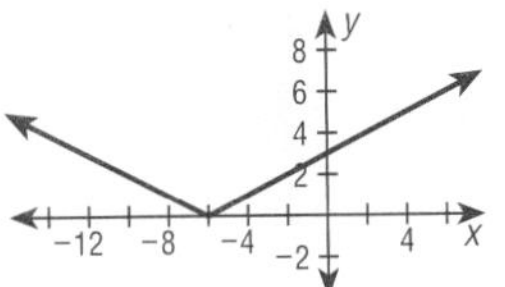

42.

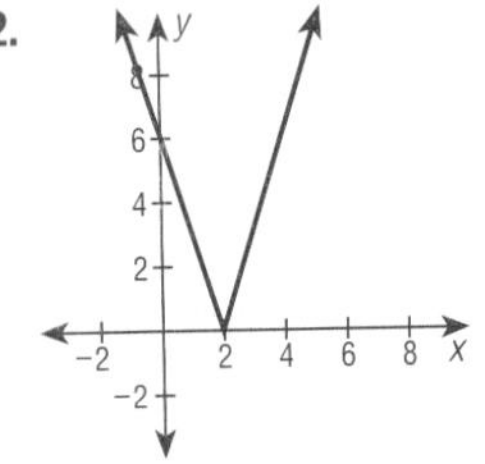

43.

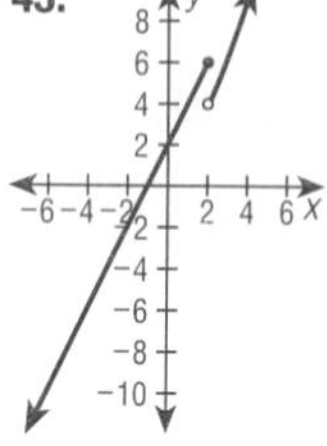

44.

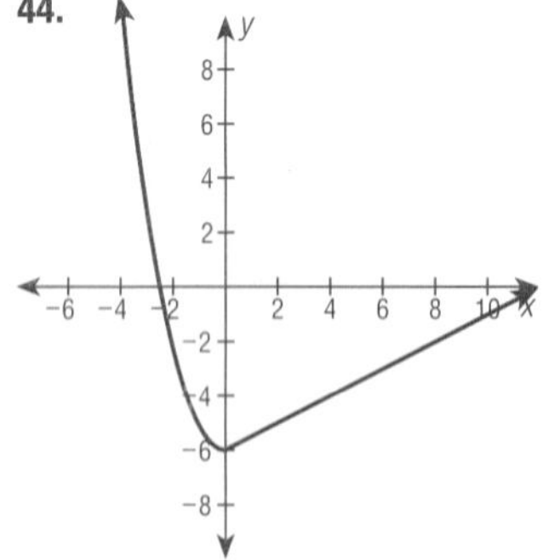

In Exercises 67–74, match the function with its graph. (6.4, 6.5)

67. $f(x) = -2x + 1$ e

68. $f(x) = (x - 1)^2 + 1$ a

69. $f(x) = |x - 1| + 1$ c

70. $f(x) = -(x - 2)^2$ h

71. $f(x) = \begin{cases} -2x + 1, & x < 0 \\ x + 1, & x \geq 0 \end{cases}$ b

72. $f(x) = \begin{cases} x^2, & x < 0 \\ -x^2, & x \geq 0 \end{cases}$ f

73. $f(x) = \begin{cases} 2, & 0 \leq x < 1 \\ 3, & 1 \leq x \leq 2 \end{cases}$ g

74. $f(x) = \begin{cases} 2, & 0 \leq x \leq 1 \\ 3, & 1 < x \leq 2 \end{cases}$ d

a.

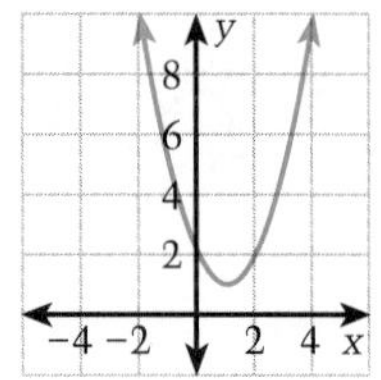

b.

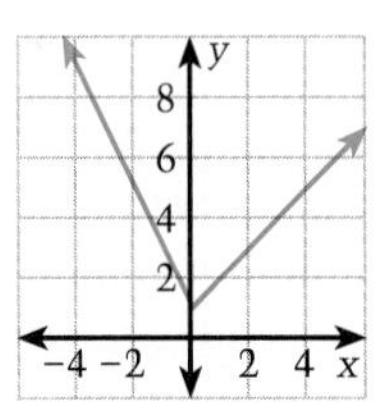

c.

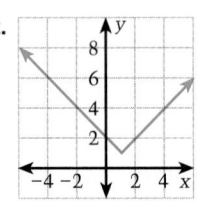

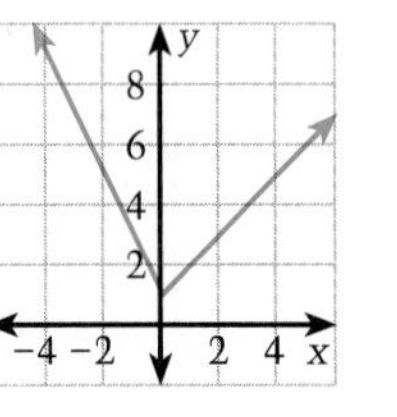

d.

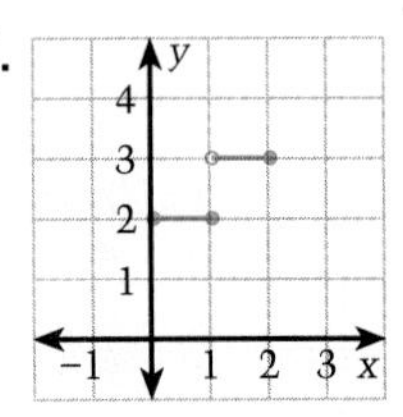

e.

f.

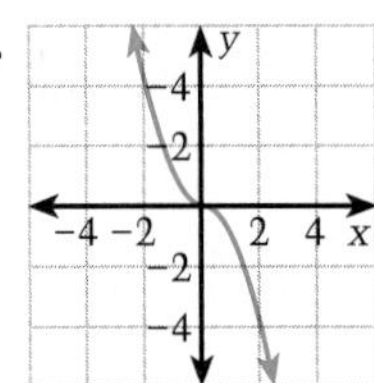

g.

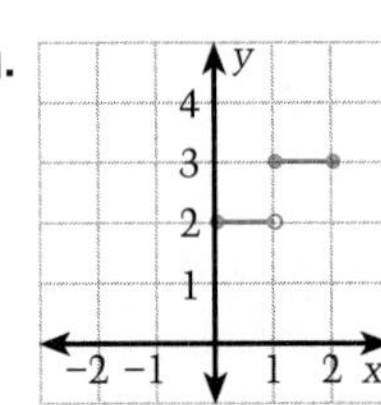

h.

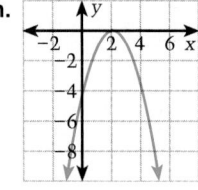

Temperatures in Phoenix **In Exercises 75 and 76, use the following information. (6.7)**

For each month of the year, the average daily high temperature in degrees Fahrenheit in Phoenix, Arizona, is shown in the table.

Month	Temperature	Month	Temperature	Month	Temperature
January	52.3	May	77.0	September	84.6
February	56.1	June	86.5	October	92.3
March	60.6	July	92.3	November	60.6
April	68.0	August	89.9	December	53.5

75. Find the mean, median, and mode of these data. $\approx$72.8, 72.5, 60.6 and 92.3

76. Find the first, second, and third quartiles for these data. 58.35, 72.5, 88.2

77. *Geometry* The radius, $r(x)$, of a circle inscribed in an equilateral triangle with sides of length x is given by the function

$$r(x) = \frac{\sqrt{3}}{6}x.$$

Evaluate $r(18)$. Explain in words what the value $r(18)$ represents. Explain how the formula for $r(x)$ can be derived. $3\sqrt{3}$, $r(18)$ is the radius of a circle inscribed in an equilateral triangle with a side of length 18.

78. *Sled Riding* A person's velocity, v (in meters per second), while sliding down a hill on a sled depends on several things, such as the slope of the hill, the type of snow, and the type of sled. The following model approximates the velocity of the person sliding down the hill shown at the right.

$$v = \sqrt{19.6h}$$

In this model, h is the change in height (in meters) measured from the top of the hill. Rewrite this equation so that velocity is the input and height is the output. What is the value of h when the velocity is 14 meters per second? $h = \frac{v^2}{19.6}$, 10

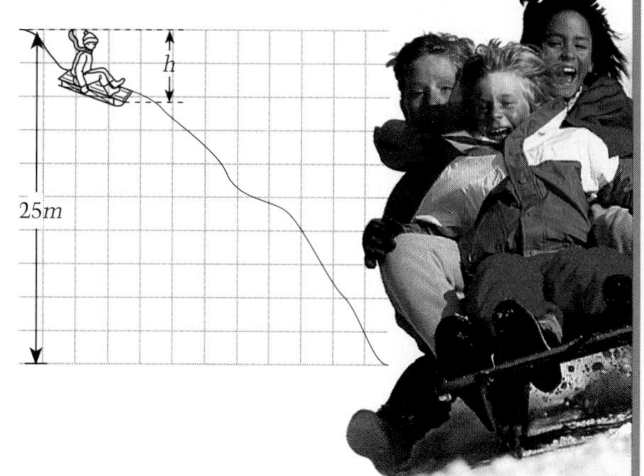

79. *Call or Write?* Over the past half century, telephone calls have far exceeded letters as a means of personal communication in the United States. From 1952 through 1988, the average number, P, of phone calls made in a year, can be approximated by

$$P = 1.2t^2 + 430 \quad P = 1.2(t + 28)^2 + 430$$

where $t = 0$ represents 1952. Rewrite this model so that $t = 0$ represents 1980.

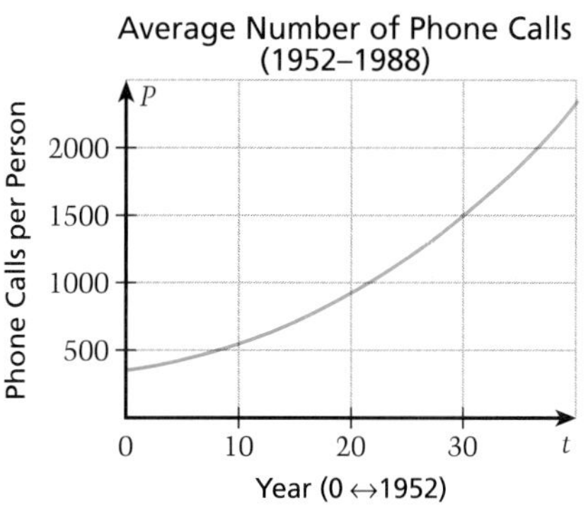

Local Television Ads **In Exercises 80–83, use the following information.**

For 1950 through 1990, the total expenses, T (in millions of dollars), for local television advertisements can be modeled by

$$T = \begin{cases} 225t + 2450, & 0 \le t \le 18 \\ 930t - 10{,}240, & 18 < t \le 25 \\ 2324t - 45{,}090, & 25 < t \le 32 \\ 3900t - 95{,}522, & 32 < t \le 40 \end{cases}$$

where $t = 0$ represents 1950.

(Source: McCann-Erickson, Inc.) See below.

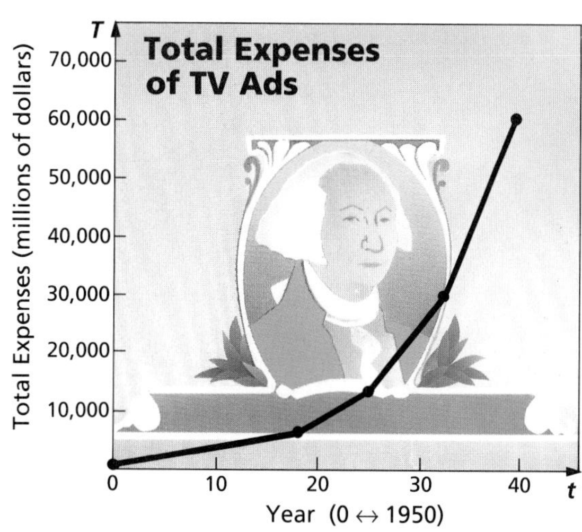

80. Approximate the total expense for 1960.

81. Approximate the total expense for 1970.

82. Approximate the total expense for 1980.

83. Approximate the total expense for 1990.

80. \$4700 million **81.** \$8360 million **82.** \$24,630 million **83.** \$60,478 million

Answers

45.

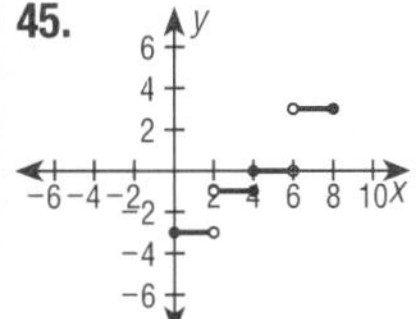

46.

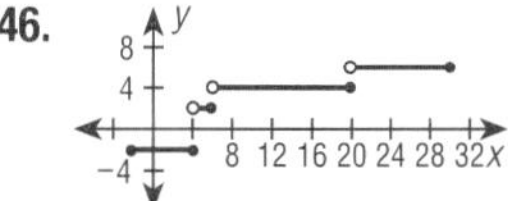

In Exercises 84–91, use the following information.

The map shows the center of population in the contiguous United States for each of the census years from 1790 through 1990. For each exercise, use the following functions.

$f(x) = 800 + 300x$, $g(x) = 990 - 200x$ *h(x) is the inverse of g(x)*

Find the year that the center of population was at (or near) the indicated town. Each exercise represents a town from a different state. Name the state.

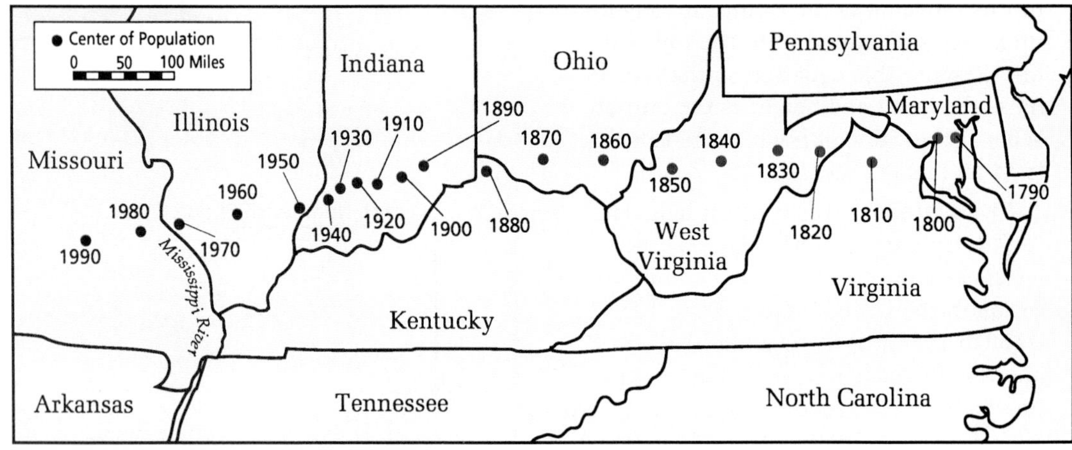

84. Mascoutah: $\frac{1}{10}f(3) + 1800$ **1970, Illinois**

85. Clarksburg: $-8 \cdot \frac{f(5)}{g(5)}$ **1840, West Virginia**

86. Middleburg: $f(4) - g(4)$ **1810, Virginia**

87. Davisville: $g(f(-4)) - 79{,}000$ **1990, Missouri**

88. Hebron: $4080 + f(g(5))$ **1880, Kentucky**

89. Poole's Island: $f(0) + g(0)$ **1790, Maryland**

90. Chillicothe: $\frac{1}{100}f(2) \cdot g(2) - 6400$ **1860, Ohio**

91. Whitehall: $384h(-10)$ **1920, Indiana**

92. Which of the compound functions has a graph that most resembles the roof of a covered wagon? **c**

a. $f(x) = \begin{cases} \frac{1}{3}|x - 9| + 5, & 6 \le x < 12 \\ \frac{1}{3}|x - 15| + 5, & 12 \le x \le 18 \end{cases}$

b. $g(x) = \begin{cases} 6, & 6 \le x < 9 \\ 5, & 9 \le x < 15 \\ 6, & 15 \le x \le 18 \end{cases}$

c. $h(x) = \begin{cases} \frac{1}{9}x^2 - 2x + 14, & 6 \le x < 12 \\ \frac{1}{9}x^2 - \frac{10}{3}x + 30, & 12 \le x \le 18 \end{cases}$

Chapter TEST

1. Which, if any, of the following are graphs of functions? **(6.1)** a, c

a. 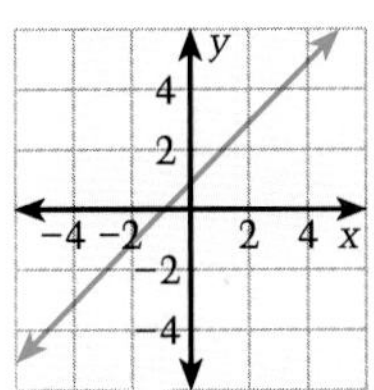

b. 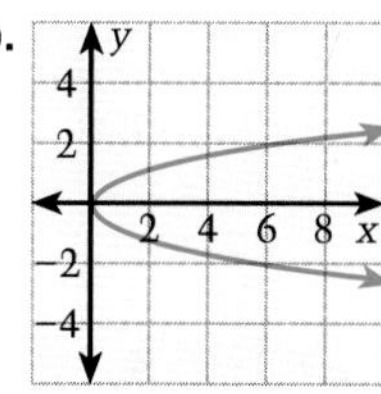

c. 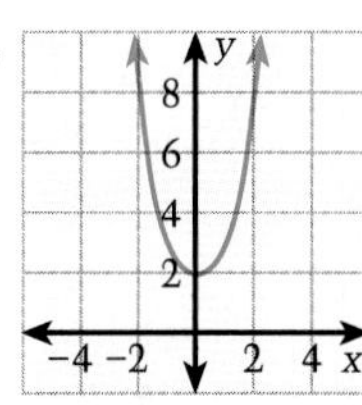

In Exercises 2–4, find the domain and range of the function. **(6.2)** See margin.

2. $f(x) = \sqrt{x + 2}$ 3. $f(x) = \frac{4}{x - 3}$ 4. $f(x) = x^2 + 3$

5. Find the inverse of $\{(-4,-1), (-3, 0), (-2, 1), (0, 2)\}$. **(6.3)** See margin.

6. Use $f(x) = x^2 - 7$ and $g(x) = |x| + 2$ to write the equation for $g(f(x))$. **(6.2)** $|x^2 - 7| + 2$

In Exercises 7–9, sketch the inverse of the function. Is the inverse a function? **(6.3)** See margin.

7. $f(x) = \frac{1}{2}x + 2$ 8. $g(x) = |x - 2| + 3$ 9. $h(x) = (x - 2)^2 - 1$

10. Given that $f(x)$ and $g(x)$ are inverse functions, find $f(g(7))$. **(6.3)** 7

11. Write $f(x) = |3x - 12|$ as a compound function. **(6.4)** $f(x) = \begin{cases} 3x - 12, & \text{if } x \geq 4 \\ -3x + 12, & \text{if } x < 4 \end{cases}$

12. Sketch the graph of $f(x) = \begin{cases} |x + 2| - 2, & \text{if } x < 0 \\ -x^2 + 4x, & \text{if } x \geq 0 \end{cases}$. **(6.4)** See Additional Answers.

13. Given $f(x) = x^2$, write an equation $g(x)$ where $g(x)$ is the graph of $f(x)$ shifted 4 units to the right and down 8 units. **(6.5)** $g(x) = (x - 4)^2 - 8$

14. Given $f(1) = 2$ and $f(n) = 2f(n - 1) - 1$, find $f(6)$. **(6.6)** 33

In Exercises 15 and 16, write a linear or quadratic model for the function. **(6.6)**

15.

n	1	2	3	4	5
***f*(*n*)**	−2	−1	1	4	8

15. $f(n) = \frac{1}{2}n^2 - \frac{1}{2}n - 2$

16.

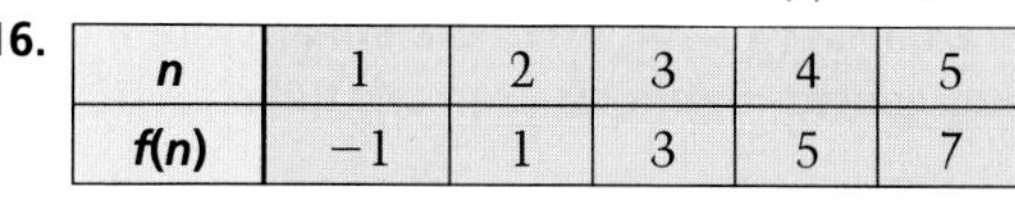

n	1	2	3	4	5
***f*(*n*)**	−1	1	3	5	7

16. $f(n) = 2n - 3$

17. Lead poisonings in Massachusetts children under 6 years of age from 1985 through 1990 can be modeled by $f(t) = 0.3t^2 - 53.8t + 2415$, where $f(t)$ is the number of cases found per 1000 people and $t = 85$ represents 1985. Using this model, about how many cases per 1000 were found in 1990? *(Source:* **Boston Globe***)* **(6.1)** 3

18. For 1990, the ratio of per capita highway miles traveled to rail miles traveled is given for several countries. Compute the mean, first with and then without, the ratio for the United States. *(Source:* **World Monitor***)* **(6.7)** 19.9, 7.4

Austria	7	France	7	Poland	1	Sweden	10
Belgium	8	Italy	7	Portugal	5	United States	158
Britain	10	Netherlands	9	Spain	6	West Germany	11

Answers

2. Domain: all reals greater than or equal to −2; Range: all non-negative reals
3. Domain: all reals except 3; Range: all reals except 0
4. Domain: all reals; Range: all reals greater than or equal to 3
5. $\{(-1,-4), (0,-3), (1,-2), (2, 0)\}$
7. 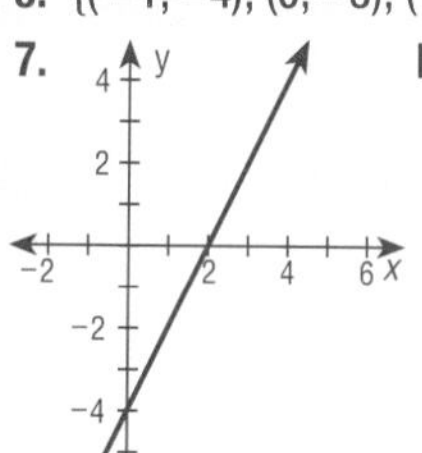 It is.
8. 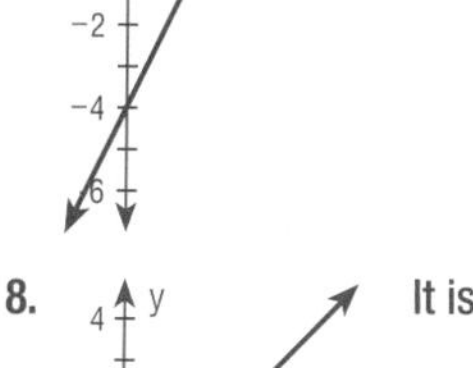 It is not.
9. 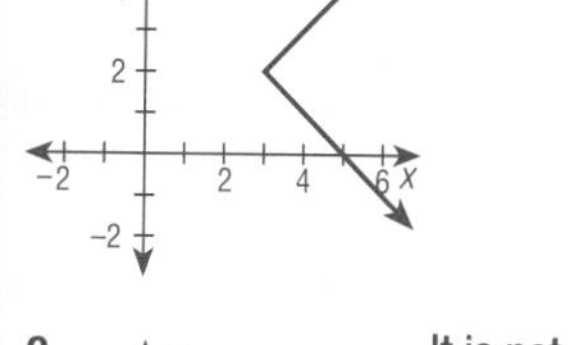 It is not.

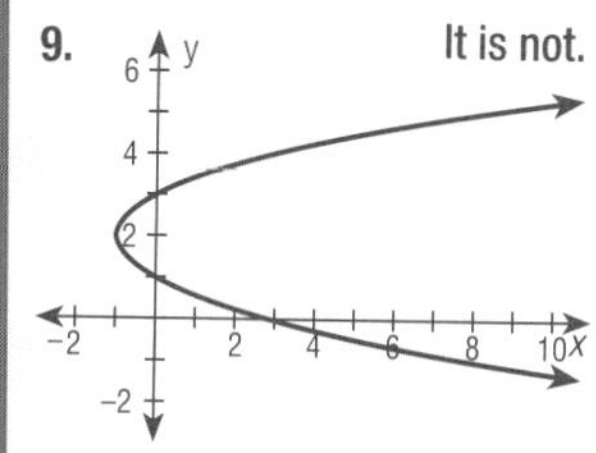

23.

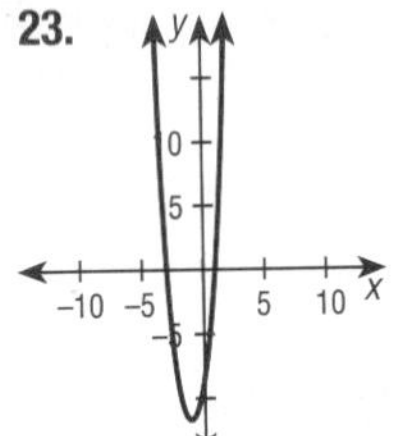

24.

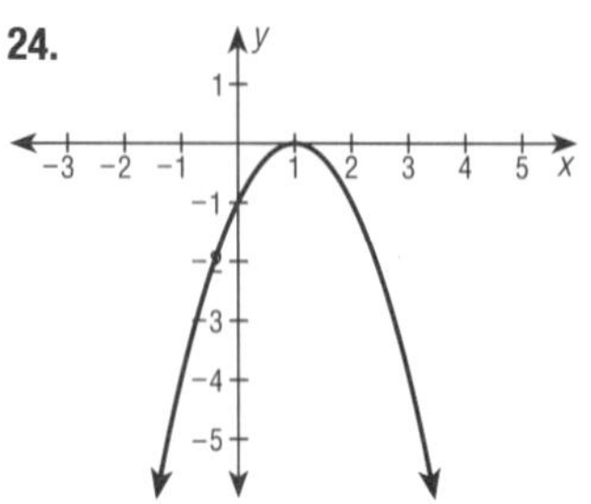

25.

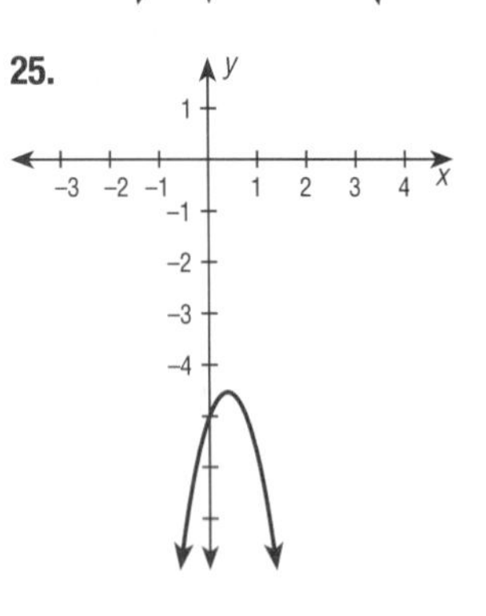

Cumulative REVIEW ▪ *Chapters* 1–6

In Exercises 1–4, perform the indicated matrix operation.

1. $5\begin{bmatrix}3&-4&\frac{1}{2}\\-6&7&2\end{bmatrix}+\begin{bmatrix}-8&22&6\\\frac{3}{2}&-10&3\end{bmatrix}$

2. $\begin{bmatrix}0&17\\8&4\\-10&5\end{bmatrix}-\frac{1}{3}\begin{bmatrix}6&7\\3&-2\\-5&21\end{bmatrix}$ $\begin{bmatrix}-2&\frac{44}{3}\\7&\frac{14}{3}\\-\frac{25}{3}&-2\end{bmatrix}$

3. $\begin{bmatrix}5&1\\6&0\\8&3\end{bmatrix}\begin{bmatrix}2&1\\-4&7\end{bmatrix}$ $\begin{bmatrix}6&12\\12&6\\4&29\end{bmatrix}$

1. $\begin{bmatrix}7&2&\frac{17}{2}\\-\frac{57}{2}&25&13\end{bmatrix}$

4. $\begin{bmatrix}7&5&0\\1&2&-4\\-8&7&1\end{bmatrix}\begin{bmatrix}-3\\2\\6\end{bmatrix}$ $\begin{bmatrix}-11\\-23\\44\end{bmatrix}$

In Exercises 5 and 6, evaluate the determinant of the matrix. Then find the inverse of the matrix.

5. $\begin{bmatrix}4&7\\2&1\end{bmatrix}$ -10, $\begin{bmatrix}-\frac{1}{10}&\frac{7}{10}\\\frac{1}{5}&-\frac{2}{5}\end{bmatrix}$

6. $\begin{bmatrix}3&-2\\1&0\end{bmatrix}$ 2, $\begin{bmatrix}0&1\\-\frac{1}{2}&\frac{3}{2}\end{bmatrix}$

In Exercises 7 and 8, evaluate the determinant of the matrix.

7. $\begin{bmatrix}-3&11&6\\4&-2&7\\5&0&1\end{bmatrix}$ 407

8. $\begin{bmatrix}-1&1&2\\6&2&5\\-3&8&-9\end{bmatrix}$ 205

In Exercises 9 and 10, decide whether the matrices are inverses of each other.

9. $\begin{bmatrix}0&2&-3\\5&8&4\\2&3&2\end{bmatrix}$ and $\begin{bmatrix}-4&13&-32\\2&-6&15\\1&-4&10\end{bmatrix}$ Yes

10. $\begin{bmatrix}11&8&-28\\3&2&-7\\-7&-5&18\end{bmatrix}$ and $\begin{bmatrix}-1&4&0\\5&-2&7\\1&1&2\end{bmatrix}$ Yes

In Exercises 11–14, solve the linear system using an inverse matrix, an augmented matrix, or Cramer's Rule.

11. $\begin{cases}3x-2y=-7\\-5x+y=14\end{cases}$ $(-3,-1)$

12. $\begin{cases}4x+5y=11\\-3x+2y=32\end{cases}$ $(-6, 7)$

13. $\begin{cases}3x+4z=-1\\-3x+2y-z=-6\\x+4y+2z=-9\end{cases}$ $(1,-2,-1)$

14. $\begin{cases}5x-y+6z=-3\\-2x+2y-z=0\\4x+3y+2z=7\end{cases}$ $(2, 1,-2)$

In Exercises 15–18, solve the quadratic equation.

15. $x^2+7x+42=-5x+6$ -6

16. $4x^2-121=0$ $-\frac{11}{2}, \frac{11}{2}$

17. $3x^2+2x-10=0$ $-\frac{1}{3}\pm\frac{\sqrt{124}}{6}$

18. $5x^2+5=-5x^2+11x+11$ $-\frac{2}{5}, \frac{3}{2}$

In Exercises 19–22, use the discriminant to find the number of real solutions of the quadratic equation.

19. $3x^2+4x-10=0$ Two real solutions

20. $-2x^2+3x-4=0$ No real solutions

21. $2x^2-4x+2=0$ One real solution

22. $5x^2+4x+4=0$ No real solutions

In Exercises 23–25, sketch the graph of the quadratic equation. See margin.

23. $y = 3x^2 + 6x - 9$ **24.** $y = -x^2 + 2x - 1$ **25.** $y = -4x^2 + 3x - 5$

In Exercises 26–28, solve the quadratic equation.

26. $\frac{1}{2}x^2 - 3x + 5 = 0$ $3 \pm i$ **27.** $-3x^2 + 4x - 3 = 0$ $\frac{2}{3} \pm \frac{\sqrt{20}}{6}i$ **28.** $x^2 + x + 2 = 0$ $\frac{-1 \pm i\sqrt{7}}{2}$

In Exercises 29–32, perform the indicated operations.

29. $3i(4 - 3i) + i^2(1 + 2i)$ $8 + 10i$ **30.** $(3 - i) + 2(4 + 2i)$ $11 + 3i$

31. $4i(7 - 2i)(2 - i)$ $44 + 48i$ **32.** $(3i)^2 - 6i(2 - 3i)$ $-27 - 12i$

In Exercises 33 and 34, sketch the graph of the quadratic inequality.

33. $y \geq 3x^2 + 20x + 12$ See margin. **34.** $y < -x^2 + 6x - 8$ See margin.

In Exercises 35–38, match the function with its graph.

35. $f(x) = |x - 3| + 3$ c **36.** $f(x) = \begin{cases} x + 3, & x > -3 \\ -x - 3, & x \leq -3 \end{cases}$ a

37. $f(x) = -(x - 2)^2 + 4$ b **38.** $f(x) = \begin{cases} 6x + 1, & x \geq 0 \\ -2x^2 + 1, & x < 0 \end{cases}$ d

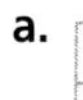

a.

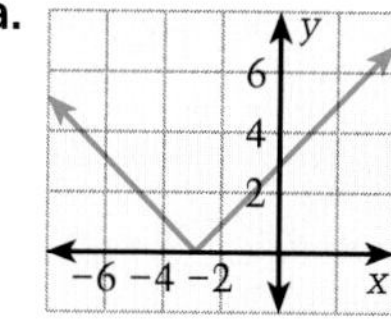

b.

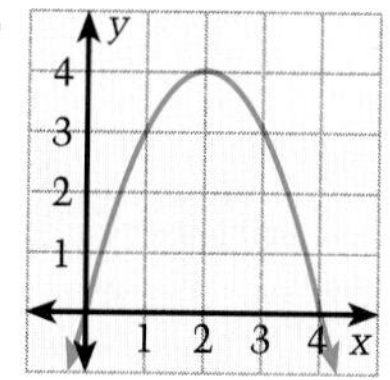

c.

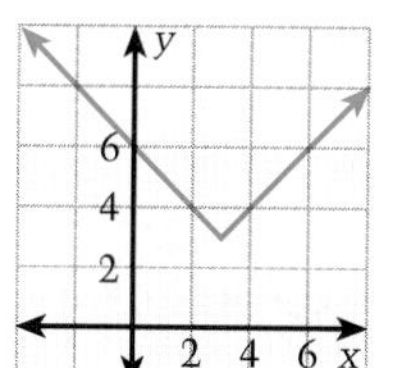

d.

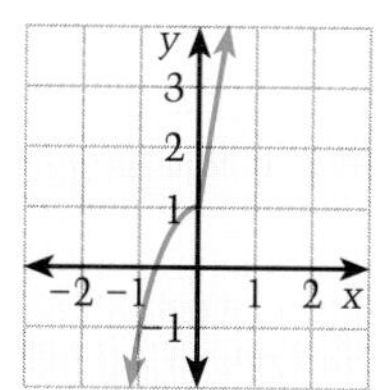

In Exercises 39–42, let $f(x) = 3x + 3$ and $g(x) = -4x + 1$. Write and simplify an equation for $h(x)$.

39. $h(x) = \frac{1}{3}f(x) - g(x)$ $h(x) = 5x$ **40.** $h(x) = f(x) + g(x)$ $h(x) = -x + 4$

41. $h(x) = f(x) \cdot g(x)$ $h(x) = -12x^2 - 9x + 3$ **42.** $h(x) = f(g(x))$ $h(x) = -12x + 6$

In Exercises 43 and 44, sketch the graph of the function and its inverse on the same coordinate plane.

43. $f(x) = \frac{1}{4}x - 6$ See margin. **44.** $f(x) = 2x + 3$ See margin.

In Exercises 45 and 46, find the first five values of the recursive function.

45. $f(2) = 4$ 4, 2, 0, −2, −4
$f(n) = f(n - 1) - 2$

46. $f(1) = 1$ 1, 6, 41, 1686, 2,842,601
$f(n) = [f(n - 1)]^2 + 5$

In Exercises 47–50, evaluate the factorial.

47. 5! 120 **48.** 6! 720 **49.** 7! 5040 **50.** 8! 40,320

Answers

33.

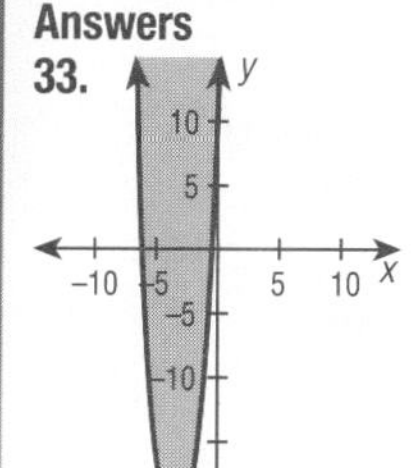

34.

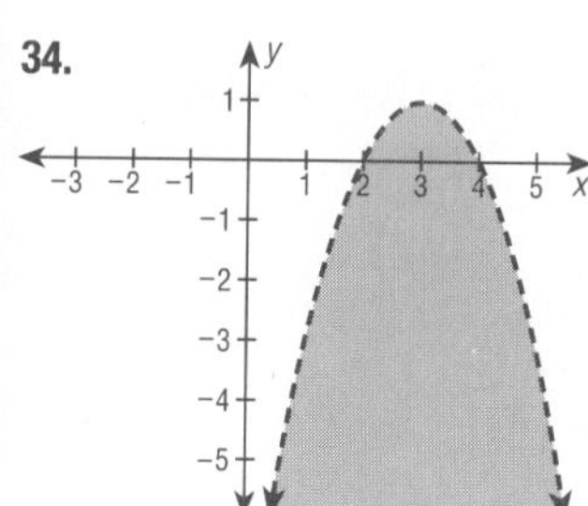

43.

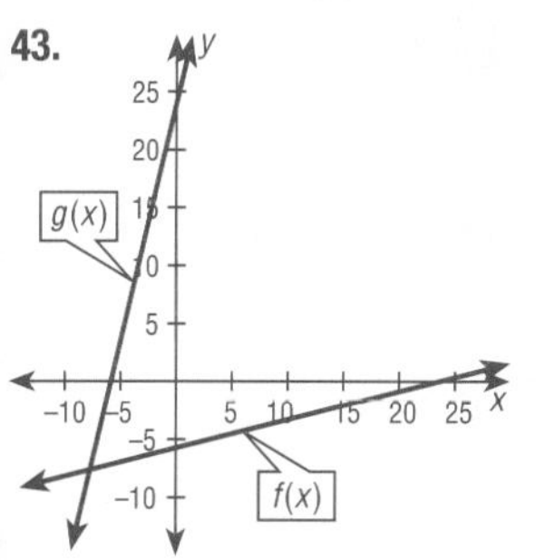

44.

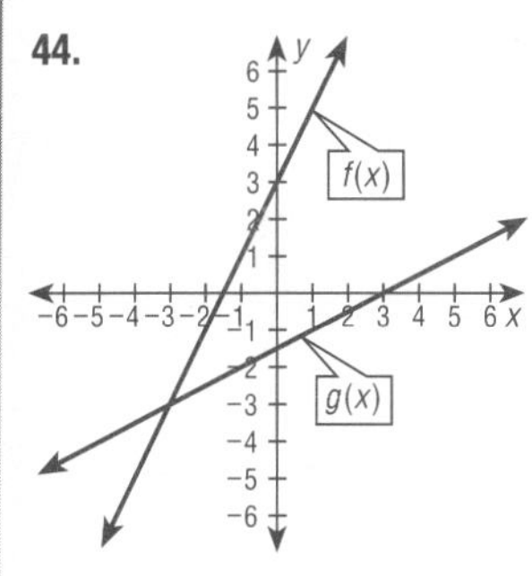

341

51. *High-School Memories* Five hundred high-school students and 500 parents of high-school students were asked to name the most memorable part of high school. The matrix gives the percent (in decimal form) of those who thought friendship, sports, or classes were memorable. Show how matrix multiplication can be used to find the total number of people who answered that friendship, sports, or classes were the most memorable. ***(Source: 1991 Harris Poll)*** See below.

Which is Most Memorable?

	Friendship	Sports/Activities	Teachers/Classes
Students	0.34	0.28	0.13
Parents	0.18	0.33	0.15

52. *Library Book Sale* The library is having its annual book sale. Pat spent \$7.00 on books, Chris spent \$5.50, and Mike spent \$4.50. The matrix below shows the number of books each one bought. Use an augmented matrix to determine the prices that the library charged for books.

Books

	Price A	Price B	Price C
Pat	1	2	3
Chris	2	3	1
Mike	3	0	2

Price A: 50¢, Price B: \$1, Price C: \$1.50

The Great Wall of China **In Exercises 53–56, use the following information.**

The height of the Great Wall of China varies from 15 feet to 39 feet.

53. If you dropped a ball from the maximum height of the wall, how many seconds would it take to reach the ground? $\approx$1.6 sec

54. If another ball was dropped from the minimum height, how long would it take to reach the ground? $\approx$1 sec

55. Someone finds the second ball at the base of the wall and tosses it straight up from a height of 4 feet with an initial velocity of 16 feet per second. Does the ball reach the minimum height of the wall? No

56. If the ball is tossed at twice the velocity of that in Exercise 55, does it reach the minimum height? Yes

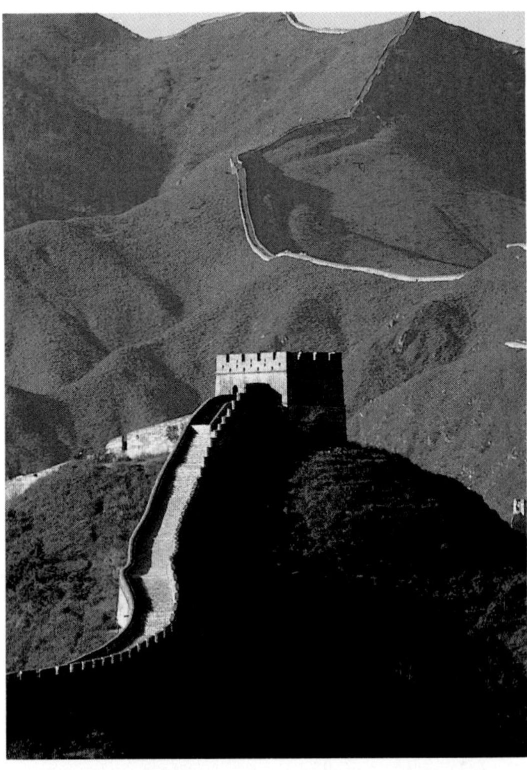

The Great Wall of China has a main-line length of 2,150 miles with spurs and branches totaling 1,780 miles. Erected by hand, it is the longest structure ever built.

51. $[500 \quad 500]\begin{bmatrix} 0.34 & 0.28 & 0.13 \\ 0.18 & 0.33 & 0.15 \end{bmatrix} = [260 \quad 305 \quad 140]$

Cotton Crops **In Exercises 57 and 58, use the following information.**

The table gives the 1995 cotton yield (in pounds per acre) for twelve states. *(Source: Agricultural Statistics Board)*

Alabama	382
Arizona	1006
Arkansas	637
California	843
Georgia	635
Louisiana	614
Mississippi	624
Missouri	559
North Carolina	504
Oklahoma	190
Tennessee	531
Texas	378

All parts of a cotton plant are useful. The lint is used to make fiber. The cottonseed provides oil. The stalks and leaves are plowed under to improve soil structure.

57. What is the average yield per acre for these states? 575.25

58. Construct a box-and-whisker plot to represent the data. See Additional Answers.

Farm Foods **In Exercises 59–62, use the following information.**

For 1980 through 1993, the amount, $f(t)$ (in billions of dollars), American consumers spent on farm foods can be modeled by

$$f(t) = \begin{cases} 15t + 268, & 0 \le t < 7 \\ -1.4t^2 + 48t + 105.6, & 7 \le t \le 13 \end{cases}$$

where $t = 0$ represents 1980. *(Source: U.S. Department of Agriculture)*

59. How much did consumers spend on farm foods in 1985? \$343 billion

60. How much did consumers spend on farm foods in 1992? \$480 billion

61. Sketch a graph of the function f. See margin.

62. During which year was \$373 billion spent on farm foods? Which equation did you use to obtain the answer? Could you have used the other equation? 1987; The second one; Yes

Farm Land **In Exercise 63, use the following information.**

For 1980 through 1994, the total amount (in millions of acres) of farm land in the United States can be modeled by

$$f(t) = 0.14t^2 - 6.76t + 1040$$

where $t = 0$ represents 1980. *(Source: U.S. Department of Agriculture)*

63. Rewrite this model so that $t = 0$ represents 1990. $f(t) = 0.14(t + 10)^2 - 6.76(t + 10) + 1040$

Answer

61.

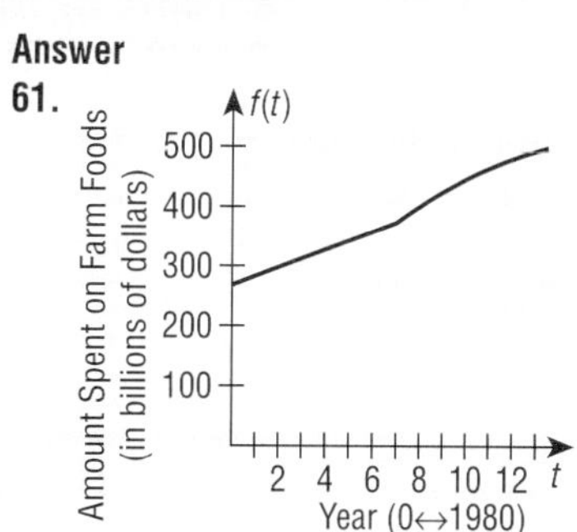

The rules that govern the manipulation of integer exponents are extended to rational exponents. The connection between rational exponents and radicals is used to develop rules for the manipulation of radicals as well. The properties of exponents are then used to solve radical equations. Finally, graphs of radical functions are explored.

The Chapter Summary on page 390 provides you and the students with a synopsis of the chapter. It identifies key skills, concepts, and vocabulary. You may want to have students look at the Chapter Summary as an overview before beginning the chapter.

CHAPTER 7

Powers, Roots, and Radicals

LESSONS

The scene shown on this page is from the prize-winning computer-animated film Red's Dream. *Storage of computer graphic images requires a tremendous amount of computer memory.*

Real Life
Computer Graphics

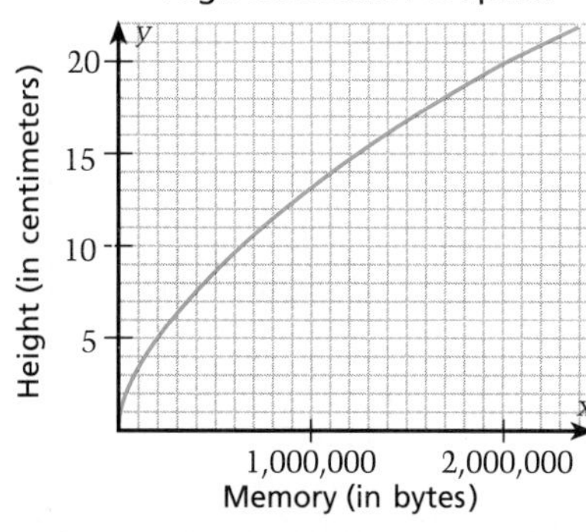

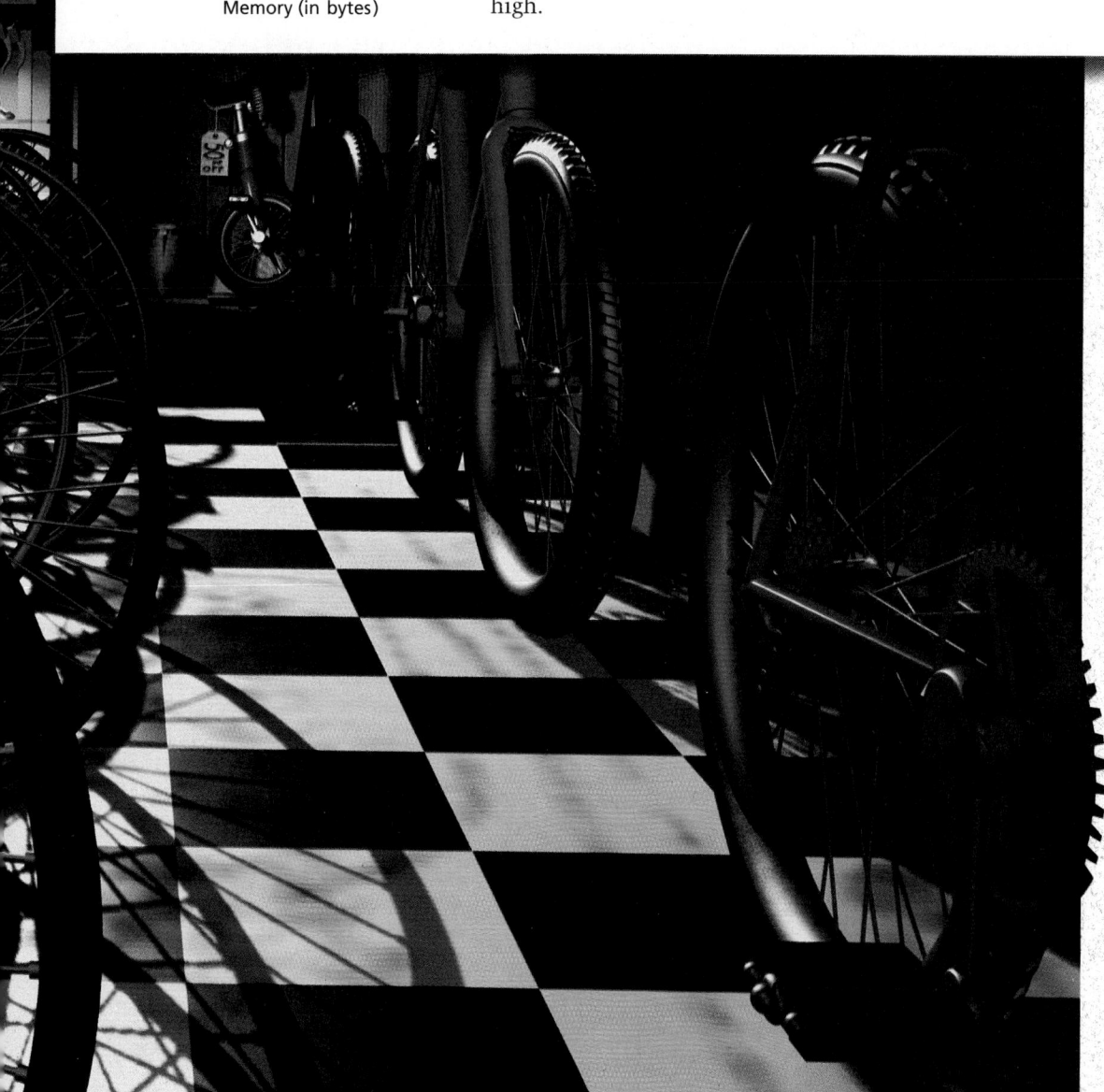

High resolution graphics, which have 36 pixels per centimeter and use 3 bytes of memory per pixel, need the following number of bytes to store a picture that is y centimeters high and $\frac{4}{3}y$ centimeters wide.

$$x = \underbrace{(36)(y)}_{\text{Number of pixels high}} \cdot \underbrace{(36)\left(\frac{4}{3}y\right)}_{\text{Number of pixels wide}} \cdot \underbrace{(3)}_{\text{Bytes per pixel}} = 5184y^2 \text{ bytes}$$

Solving this equation for y in terms of x produces

$$y = \frac{1}{72}\sqrt{x}.$$

From the graph, you can see that half a megabyte of memory (500,000 bytes) can store a picture that is about 9.8 centimeters high. With a full megabyte of memory, you can store a picture that is about 13.9 centimeters high.

Relationships do not always involve only whole numbers. Often, rational numbers are used to describe algebraic relationships, and frequently, rational exponents are needed. Rational exponents can be expressed as "roots" or "radical" expressions, as in the number of bytes needed to store a computer image. In order to answer questions about the memory a computer needs in order to produce a picture of given height, it is necessary to be able to solve the "radical equation," $y = \frac{1}{72}\sqrt{x}$, for its unknown, x.

Powers, Roots, and Radicals

PACING CHART

The daily Pacing Chart is meant to help you adjust your teaching pace. Students in the full course should finish the entire text by the end of the year. Students in the basic course are expected to complete the first twelve chapters. The Pacing Chart for each chapter contains suggestions for lessons that require more than one day and lessons that may be omitted for the basic course.

DAY	FULL COURSE	BASIC COURSE
1	7.1	7.1
2	7.2	7.2
3	7.3	7.2
4	Mid-Chapter Self-Test	7.3
5	7.4	7.3
6	7.5	Mid-Chapter Self-Test
7	7.6 & Using a Graphing Calculator	7.4
8	Chapter Review	7.4
9	Chapter Test	7.5
10		7.5
11		7.6
12		7.6 & Using a Graphing Calculator
13		Chapter Review
14		Chapter Test

CHAPTER ORGANIZATION

LESSON	PAGES	GOALS	MEETING THE NCTM STANDARDS
7.1	346–351	1. Use properties of exponents to evaluate and simplify exponential expressions 2. Use powers as models in real-life problems	Problem Solving, Communication, Reasoning, Connections, Geometry, Structure
7.2	352–358	1. Use the compound interest formula and the exponential growth and decay formulas 2. Solve real-life problems modeled by exponential growth formulas	Problem Solving, Communication, Connections
Mixed Review	359	Review algebraic and arithmetic skills	
Career Interview	359	Contractor	Connections
7.3	360–366	1. Evaluate *n*th roots of real numbers using radical notation and rational exponent notation 2. Use *n*th roots to solve real-life problems	Problem Solving, Communication, Connections, Geometry, Technology
Mid-Chapter Self-Test	367	Diagnose student weaknesses and remediate with correlated Reteach worksheets	
7.4	368–373	1. Use properties of roots to evaluate and simplify expressions containing radicals and rational exponents 2. Use properties of roots to solve real-life problems	Problem Solving, Communication, Connections
7.5	374–380	1. Solve equations that have radicals and rational exponents 2. Use radical equations to solve real-life problems	Problem Solving, Communication, Connections, Geometry (algebraic)
Mixed Review	380	Review algebraic and arithmetic skills	
7.6	381–387	1. Graph square root and cube root functions 2. Use square root and cube root functions to solve real-life problems	Problem Solving, Communication, Connections, Functions, Geometry (algebraic)
Using a Calculator	388–389	Graph and use equations of the form $y = x^n$	Technology
Chapter Summary	390	Restate for students what they have learned, why they have learned it, and how it fits into the structure of algebra	Structure, Connections
Chapter Review	391–394	Review concepts and skills learned in the chapter	
Chapter Test	395	Diagnose student weaknesses and remediate with correlated Reteaching worksheets	

LESSON RESOURCES

MEETING INDIVIDUAL NEEDS

RETEACHING For students who need to spend more time on basics:

If a mid-chapter self-test or chapter test indicates a deficiency, teachers can help students with the appropriate ***Reteaching Copymaster.***

PRACTICE For students who need more practice:

Additional exercises like those in the Pupil's Edition are provided for each lesson in ***Extra Practice Copymasters.***

ENRICHMENT For enriching and broadening students' experiences:

Problem of the Day copymasters in ***Teaching Tools*** provide a daily opportunity to use logical reasoning, looking for a pattern, writing an equation, and other routine and non-routine problem-solving strategies.

Math Log copymasters in ***Alternative Assessment*** provide opportunities to report on investigations, research, and open-ended problems.

Technology: Using Calculators and Computers provides enriching activities with graphing and scientific calculators and computers.

The ***Applications Handbook*** provides additional information about the cross-curriculum topics such as astronomy, chemistry, physics, sports, economics, genetics, and music that are integrated into the Pupil's Edition.

LESSON	7.1	7.2	7.3	7.4	7.5	7.6
PAGES	346-351	352-358	360-366	368-373	374-380	381-387
Teaching Tools						
Transparencies		✓				✓
Problem of the Day	✓	✓	✓	✓	✓	✓
Warm-up Exercises	✓	✓	✓	✓	✓	✓
Answer Masters	✓	✓	✓	✓	✓	✓
Extra Practice Copymasters	✓	✓	✓	✓	✓	✓
Reteaching Copymasters	Teacher-directed and independent activities tied to results on the Mid-Chapter Self-Tests and Chapter Tests					
Color Transparencies		✓		✓	✓	✓
Applications Handbook	Additional background information for many real-life applications					
Technology	Calculator and computer worksheets for appropriate lessons					
Complete Solutions Manual	✓	✓	✓	✓	✓	✓
Alternative Assessment	Assess student's ability to reason, analyze, solve problems, and communicate using mathematical language.					
Formal Assessment	Mid-Chapter Self-Tests, Chapter Tests, Cumulative Tests, and Practice for College Entrance Tests					
Computer Test Bank	Customized tests can be created by choosing from over 2500 items.					

INSIGHTS

7.1 Properties of Exponents

Algebraic relationships and models often contain exponents, including negative and zero exponents. Students must be able to manipulate such algebraic expressions using the properties of exponents. Understanding the properties of exponents will be helpful when solving radical equations.

7.2 Exploring Data: Compound Interest and Exponential Growth

Situations modeled by exponential growth or decay (or depreciation) are commonplace in the real world. A familiar application of exponential growth is compound interest. In this lesson, examples about compound interest and other real-life applications will be presented. Students have the opportunity to explore data in several interesting contexts.

7.3 *n*th Roots and Rational Exponents

In this lesson, the connection between roots and rational exponents will be examined. Students will learn to rewrite the *n*th root as the exponent $\frac{1}{n}$, and to manipulate and evaluate expressions containing rational exponents. Real-life applications involving rational exponents occur in economics, business, and engineering.

7.4 Properties of Roots of Real Numbers

The properties developed in Lesson 7.1 for integer exponents also are true for rational exponents. Young's Modulus (Exercises 69, 70) and Musical Notes (Exercises 73, 74) are typical of real-life contexts that can be modeled with expressions containing rational exponents.

7.5 Solving Radical Equations

Real-life contexts that can be modeled by equations containing radicals are found in physics, economics, and engineering. Solving equations containing radicals uses knowledge about solving linear and quadratic equations. Consequently, students must maintain and extend their problem-solving techniques in this lesson.

7.6 Graphing Square Root and Cube Root Functions

This lesson emphasizes the value of geometry in picturing algebraic models that contain radicals. The more common radical expressions involve square roots and cube roots. Such relationships can be expressed as functions, and therefore can be graphed on a coordinate plane as shown in the examples.

Problem of the Day

In a certain country, everyone is either a knight or a knave. The knights always tell the truth, and the knaves always lie. A stranger approaches A, B, and C, and asks what they are. A says, "All of us are knaves." B says, "Exactly one of us is a knight." What are they in fact?

B is a knight, A and C are knaves.

ORGANIZER

Warm-Up Exercises

Indicate the number of times each factor is used in the following expressions.

a. $x^3x^2y^4xy$

b. $ab^5a^3a^2b^2$

c. rs^7tr^2st

There are a. 6 factors of x and 5 factors of y; b. 6 factors of a and 7 factors of b; c. 3 factors of r, 8 of s, and 2 of t.

Lesson Resources

Teaching Tools
- Transparencies: 7, 8
- Problem of the Day: 7.1
- Warm-up Exercises: 7.1
- Answer Masters: 7.1

Extra Practice: 7.1

Applications Handbook: p. 4

LESSON Notes

GOAL 1 It is a good idea to illustrate each of the Properties of Exponents using integers for the exponents, m and n. For example, Property 2 can be illustrated by comparing $(a^3)^2 = a^3 \cdot a^3 = a^{3+3} = a^6$ and $a^{3\cdot 2} = a^6$.

(continued)

7.1 Properties of Exponents

What you should learn:

Goal 1 How to use properties of exponents to evaluate and simplify exponential expressions

Goal 2 How to use powers as models in real-life problems

Why you should learn it:

You can use models containing powers for many real-life situations, such as volume measurements.

Goal 1 Using Properties of Exponents

Recall that the expression a^n, when n is a positive integer, represents the number that you obtain when a is used as a factor n times. The number a is the **base,** and the number n is the **exponent.** The expression a^n is called a **power** and is read as "a to the nth power."

LESSON INVESTIGATION

Investigating Operations with Powers

Partner Activity Here are two examples of multiplying and dividing powers.

Multiplying Powers

$$2^4 \cdot 2^2 = 2 \cdot 2 \cdot 2 \cdot 2 \cdot 2 \cdot 2 = 2^6$$

Dividing Powers

$$\frac{3^5}{3^3} = \frac{3 \cdot 3 \cdot \cancel{3} \cdot \cancel{3} \cdot \cancel{3}}{\cancel{3} \cdot \cancel{3} \cdot \cancel{3}} = 3^2$$

With your partner, write several other examples of multiplying and dividing powers. Then describe a rule for multiplying powers and a rule for dividing powers. Check that your rules fit each of your examples.

To multiply, add the exponents. To divide, subtract the exponents.

Study Tip

Rather than simply trying to memorize the rules at the right, we suggest that you try to see that each rule makes sense. For instance, the investigation above shows how to make sense of the rules for the product and quotient of powers.

Properties of Exponents

Let a and b be real numbers and let m and n be integers.

1. $a^m \cdot a^n = a^{m+n}$ — *Product of Powers Property*
2. $(a^m)^n = a^{m \cdot n}$ — *Power of a Power Property*
3. $(ab)^m = a^mb^m$ — *Power of a Product Property*
4. $a^{-n} = \frac{1}{a^n}$, $a \neq 0$ — *Negative Power Property*
5. $a^0 = 1$, $a \neq 0$ — *Zero Power Property*
6. $\frac{a^m}{a^n} = a^{m-n}$, $a \neq 0$ — *Quotient of Powers Property*
7. $\left(\frac{a}{b}\right)^m = \frac{a^m}{b^m}$, $b \neq 0$ — *Power of a Quotient Property*

The properties of exponents may be used to simplify both real numbers and variable expressions.

Example 1 *Using Properties of Exponents*

a. $(-3)(-3)^5 = (-3)^{1+5}$
$= (-3)^6$
$= 729$

b. $[(-2)^3]^3 = (-2)^{3 \cdot 3}$
$= (-2)^9$
$= -512$

c. $(3^2xy)^4 = (3^2)^4 \cdot x^4 \cdot y^4$
$= 3^8 \cdot x^4 \cdot y^4$
$= 6561x^4y^4$

d. $4^{-3} \cdot 4^3 = 4^{-3+3}$
$= 4^0$
$= 1$ ■

Expressions involving negative powers may be rewritten with positive powers by using the Negative Power Property.

Example 2 *Using Properties of Exponents*

a. $(3^{-2})^2 = 3^{-2 \cdot 2}$
$= 3^{-4}$
$= \frac{1}{3^4}$
$= \frac{1}{81}$

b. $x^3 \cdot \frac{1}{x^5} = \frac{x^3}{x^5}$
$= x^{3-5}$
$= x^{-2}$
$= \frac{1}{x^2}$

c. $\frac{6^2 \cdot 6}{6^5} = \frac{6^3}{6^5}$
$= 6^{-2}$
$= \frac{1}{6^2}$
$= \frac{1}{36}$

d. $\left(\frac{2}{3}\right)^{-2} = \frac{2^{-2}}{3^{-2}}$
$= \frac{3^2}{2^2}$
$= \frac{9}{4}$ ■

Example 3 *Simplifying an Expression*

Simplify $\frac{4x^3y}{10x^2} \cdot \frac{5x^3y}{y^3}$.

Solution

$\frac{4x^3y}{10x^2} \cdot \frac{5x^3y}{y^3} = \frac{(4x^3y)(5x^3y)}{(10x^2)(y^3)}$ *Multiply fractions.*

$= \frac{20x^6y^2}{10x^2y^3}$ *Product of Powers Property*

$= 2x^4y^{-1}$ *Quotient of Powers Property*

$= \frac{2x^4}{y}$ *Write with positive exponents.* ■

(continued)

You may wish to ask your students to illustrate the validity of the other properties.

The following pattern can be used to explain Property 4.

$a^4 = a \cdot a \cdot a \cdot a$
Divide by a. $a^3 = a \cdot a \cdot a$
Divide by a. $a^2 = a \cdot a$
Divide by a. $a^1 = a$
Divide by a. $a^0 = 1$
Divide by a. $a^{-1} = \frac{1}{a}$
Divide by a. $a^{-2} = \frac{1}{a^2}$

Examples 1–2

Ask students to identify the properties used to evaluate the expressions in these examples. Can more than one set of properties be used to simplify a given expression? Explain.

Example 3

Emphasize that the properties of exponents can only be applied to expressions that have the same base. Consequently, the expression x^3y^2 cannot be simplified further.

Observe that the solution is written using a positive exponent for y. It is customary to write solutions using positive exponents.

Example 4

Recall for students that the volume of a sphere of radius r is $V_S = \frac{4}{3}\pi r^3$, and that the volume of a cylinder of radius r and height h is $V_C = \pi r^2 h$.

Observe that the number of marbles, N, is given by $\frac{\text{container volume}}{\text{volume of one marble}} \cdot 0.524$.

Extra Examples

Here are additional examples similar to **Examples 1–3.**

1. Evaluate the following.
   a. $(-4)(-4)^3$ b. $[(-3)^2]^3$
   c. $(3^2x^2y)^2$ d. $5^{-4} \cdot 5^3$
   e. $(2^{-3})^2$ f. $m^7 \cdot \dfrac{1}{m^4}$
   g. $\dfrac{8^3 \cdot 8^5}{8^9}$ h. $\left(\dfrac{5}{6}\right)^{-3}$

   a. 256, b. 729, c. $81x^4y^2$, d. $\frac{1}{5}$, e. $\frac{1}{64}$, f. m^3, g. $\frac{1}{8}$, h. $\frac{216}{125}$
2. Simplify the following expressions.
   a. $\dfrac{5x^4y^3}{8x^5} \cdot \dfrac{3x^3y^5}{6y^4}$ $\dfrac{5x^2y^4}{16}$

   b. $\dfrac{2x^6y^4}{6x^3} \cdot \dfrac{4x^2y^3}{12y^5}$ $\dfrac{x^5y^2}{9}$

Check Understanding

1. When multiplying powers, what is the importance of "same base"?
   See Lesson Notes for Example 3.
2. Do the Properties of Exponents hold for decimals such as 5.6 and 3.45?
   Yes, the properties stated on page 346 are valid for all real numbers *a* and *b*.

Communicating about ALGEBRA

COOPERATIVE LEARNING

Have students work with a partner and use integers to illustrate that the Properties of Exponents in the beginning of this lesson hold true.

Goal 2 Using Powers to Solve Real-Life Problems

Real Life **Estimation**

Example 4 *Comparing Two Volumes*

A store is having a contest to guess the number of regular-sized playing marbles in a large jar. The jar is cylindrical with a radius of 6 inches and a height of 15 inches. How many marbles would you guess are in the jar?

Solution To make your guess, you use some playing marbles and a cylindrical container that you have at home. First, you measure the radius of a marble to be $\frac{3}{10}$ inch, which means that each marble has a volume of

$$\frac{4}{3}\pi\left(\frac{3}{10}\right)^3 \text{ cubic inches.}$$

Next, you count the number of marbles that will fit into your cylindrical container that has a radius of $\frac{3}{2}$ inches and a height of $\frac{11}{2}$ inches. You find that the container holds 180 marbles. Thus, the ratio of marble volume to container volume is

$$\frac{\text{Marble volume}}{\text{Container volume}} = \frac{180 \cdot \frac{4}{3}\pi\left(\frac{3}{10}\right)^3}{\pi\left(\frac{3}{2}\right)^2\left(\frac{11}{2}\right)} = \frac{144}{275} \approx 52.4\%.$$

Finally, you guess that 52.4% of the volume of the large contest jar is filled with marbles.

$$\text{Number of marbles} \approx \frac{(0.524)(15\pi 6^2)}{\frac{4}{3}\pi\left(\frac{3}{10}\right)^3} \approx 7860$$

Your guess is that 7860 marbles are in the contest jar. ■

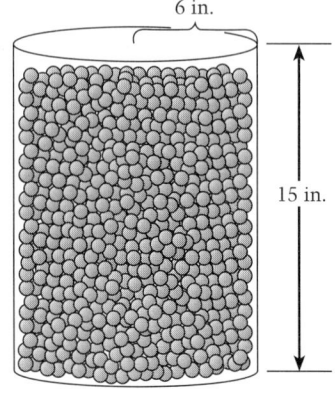

Communicating about ALGEBRA

SHARING IDEAS about the Lesson See Additional Answers.

A. Use the properties of exponents to help complete the missing steps in Example 4. That is, show that

$$\frac{180 \cdot \frac{4}{3}\pi\left(\frac{3}{10}\right)^3}{\pi\left(\frac{3}{2}\right)^2\left(\frac{11}{2}\right)} = \frac{144}{275}.$$

B. If 1000 marbles will fill a cube that is 6 inches on each edge, approximately how many marbles will fill a cube that is 12 inches on each edge? Explain your reasoning.

EXERCISES

Guided Practice

CRITICAL THINKING about the Lesson

1. How do you read the expression $\frac{x^3}{y^4}$? "x to the third power divided by y to the fourth power"

In Exercises 2–5, simplify the expression. State each property of exponents used. See below.

2. $2a^2 \cdot a^6$
3. $\frac{a^{-3}}{a}$
4. $(a^2)^3$
5. $\left(\frac{1}{a}\right)^{-2}$

6. Use a calculator to evaluate $(2.15)^7$. Round your result to two decimal places. 212.36

Independent Practice

In Exercises 7–10, identify the base and the exponent of the expression.

7. 3^4 3; 4
8. $\left(\frac{1}{4}\right)^3$ $\frac{1}{4}$; 3
9. x^3 x; 3
10. 2^{-3} 2; −3

In Exercises 11–38, simplify the expression.

11. $(2)^3(2)^5$ 256
12. $(-3)^2(-3)^1$ −27
13. $\left(\frac{1}{2}\right)^2\left(\frac{1}{2}\right)^{-2}$ 1
14. $3\left(\frac{2}{3}\right)^3\left(\frac{3}{2}\right)^2$ 2
15. $x^4 \cdot x^{-2}$ x^2
16. $3y^2 \cdot y^2$ $3y^4$
17. $(4^3)^2$ 4096
18. $(6^2)^{-2}$ $\frac{1}{1296}$
19. $(x^{-3})^5$ $\frac{1}{x^{15}}$
20. $(y^4)^3$ y^{12}
21. $(x^3y)^4$ $x^{12}y^4$
22. $(2xy^3)^{-2}$
23. $(-3x^2)^2$ $9x^4$
24. $(x^2y^2)^{-1}$ $\frac{1}{x^2y^2}$
25. $4x^{-1}y$ $\frac{4y}{x}$
26. $xy^{-2}x$ $\frac{x^2}{y^2}$
27. $3x^0y^4$ $3y^4$
28. $-2x^{-2}y^0$ $-\frac{2}{x^2}$
29. $\frac{4^3}{4^1}$ 16
30. $\frac{x^3}{x^{-1}}$ x^4
31. $\frac{x^{-3}y}{xy^{-2}}$ $\frac{y^3}{x^4}$
32. $\frac{2x^2y^7}{6xy^{-1}}$ $\frac{xy^8}{3}$
33. $\frac{6x}{5y} \cdot \frac{y^2x^{-2}}{x^3}$ $\frac{6y}{5x^4}$
34. $\frac{-3x^5}{x^{13}} \cdot \frac{2x^{10}y}{15y^2}$
35. $\frac{-12xy}{7x^4} \cdot \frac{21x^5y^2}{4y}$ $-9x^2y^2$
36. $\frac{xy^9}{3y^{-2}} \cdot \frac{-7y}{21x^5}$ $-\frac{y^{12}}{9x^4}$
37. $\frac{y^{10}}{2x^3} \cdot \frac{20x^{14}}{xy^6}$ $10x^{10}y^4$
38. $\frac{5x^{-2}}{3x} \cdot \frac{2y^3}{x^{10}}$

22. $\frac{1}{4x^2y^6}$ 34. $-\frac{2x^2}{5y}$ 38. $\frac{10y^3}{3x^{13}}$

In Exercises 39–47, use properties of exponents to solve for x.

Sample:
$$2^x2^4 = 2^2$$
$$2^{x+4} = 2^2$$
$$x + 4 = 2$$
$$x = -2$$

39. $4^x4^2 = 4^5$ 3
40. $\frac{a^2}{a^x} = a^{-3}$ 5
41. $x^{-1} = \frac{1}{4}$ 4
42. $x^{-2} = 9$ $-\frac{1}{3}, \frac{1}{3}$
43. $\frac{x^2}{3} \cdot \frac{1}{x} = 4$ 12
44. $\frac{x^6}{2x^{-2}} \cdot \frac{10}{x^7} = 20$ 4
45. $\frac{x^{16}y^2}{2x^{12}y} \cdot \frac{3}{x^3y} = \frac{9}{2}$ 3
46. $\frac{12x^5a^2}{2x^4} \cdot \frac{2a}{3a^2} = -12$ $-\frac{3}{a}$
47. $\frac{2kx^4}{5k^2x} \cdot 25kx^{-1} = \frac{1}{10}$ $\frac{1}{10}, -\frac{1}{10}$

2. $2a^8$; Product of Powers Property
3. $\frac{1}{a^4}$; Negative Power Property, Product of Powers Property
4. a^6; Power of a Power Property
5. a^2; Power of a Quotient Property, Negative Power Property

EXERCISE Notes

ASSIGNMENT GUIDE

Basic/Average: Ex. 1–6, 9–27 odd, 43–49 odd, 50, 51, 54–56, 60–62

Above Average: Ex. 1–6, 9–31 odd, 37–43 odd, 48–56, 60–62

Advanced: Ex. 1–6, 9–31 odd, 35–43 odd, 48–56, 60–63

Selected Answers
Exercises 1–6, 7–61 odd

More Difficult Exercises
Exercises 50–53, 62–63

Guided Practice

Ex. 2–5 Remind students to express the final answers using only positive exponents.

Ex. 6 TECHNOLOGY One way to examine the reasonableness of a calculator answer is to have students estimate the answer before using the calculator.

Independent Practice

Ex. 11–38 Remind students to express their final answers using only positive exponents.

Ex. 39–47 Review with students the intuitive idea that $a^x = a^y$ if and only if $x = y$.

Common-Error ALERT!

In **Exercises 31–38**, be sure students understand the difference between expressions like $-2x^{-2} = \frac{-2}{x^2}$ and $(-2x)^{-2} = \frac{1}{(-2x)^2}$. Some students will write $-2x^{-2}$ as $\frac{1}{-2x^2}$ or even as $\frac{1}{4x^2}$.

Ex. 51 Students should use the formula $\frac{T^2}{R^3} = k$ and let $k = 1$. They will need to convert their T-value answer by multiplying by 365.25.

Ex. 54–55 TECHNOLOGY These exercises provide students with practice in using calculators to compute with exponents. Exponential growth models such as $M = (19)(1.2)^t$ will be studied in the next lesson.

48. *Geometry* Find an expression for the area of the rectangle. $4x^8$

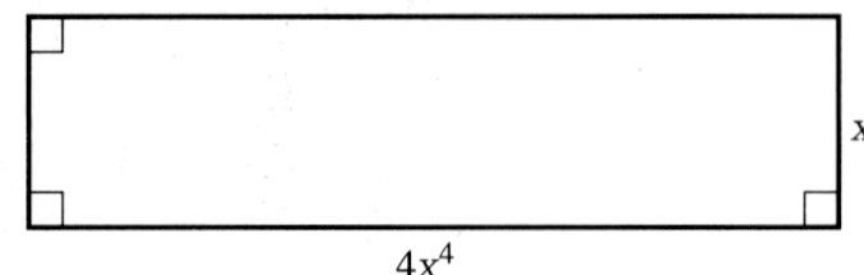

49. *Geometry* Find an expression for the area of the circle. $36\pi x^8$

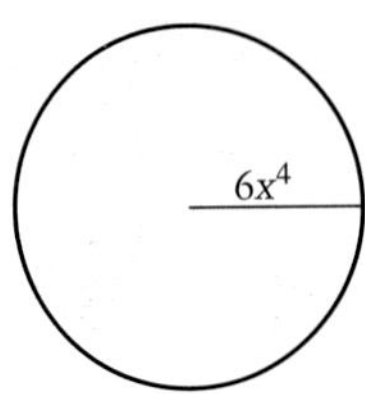

50. *Kepler's Third Law* In 1619, Johannes Kepler, a German astronomer, discovered that the period, T (in years), of each planet in our solar system is related to the planet's mean distance, R (in astronomical units), from the sun by the equation Each planet: ≈1.00, yes

$$\frac{T^2}{R^3} = k.$$

Test Kepler's equation for the nine planets in our solar system, using the table at the right. (Astronomical units relate the other planets' periods and mean distances to Earth's period and mean distance.) Do you get approximately the same value for k for each planet?

Planet	T	R
Mercury	0.241	0.387
Venus	0.615	0.723
Earth	1.000	1.000
Mars	1.881	1.523
Jupiter	11.861	5.203
Saturn	29.457	9.541
Uranus	84.008	19.190
Neptune	164.784	30.086
Pluto	248.350	39.507

51. *Discovering a Planet* Suppose that you discovered a planet whose mean distance from the sun is 48.125 astronomical units. Use the results of Exercise 50 to find the period of this planet in days. (The period of Earth is 365.25 days.) ≈121,940 days

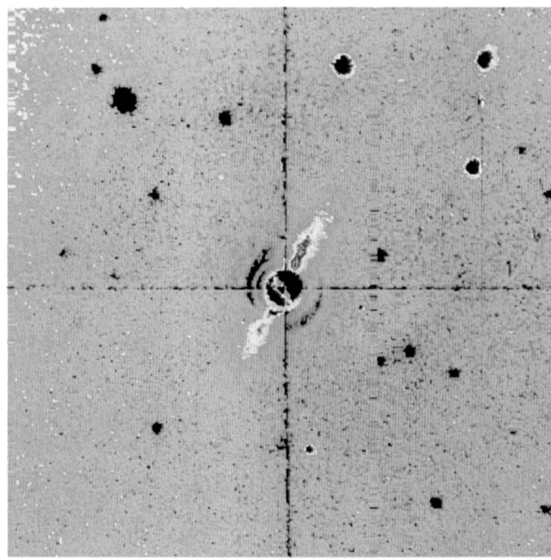

The first visual evidence of a solar system other than our own was recorded in 1984 by the Las Campanas Observatory in Chile.

52. *Pluto* Find the ratio of Pluto's mean distance (from the sun) to Mercury's mean distance. $\frac{39.507}{0.387} \approx 102.1$

53. *Pluto* If you draw a diagram of our solar system in which Mercury's mean distance from the sun is represented by 1 inch, how many inches would be needed to represent Pluto's mean distance from the sun? 102.1 in.

54. National Park Spending For 1985 through 1994, the total expenditures, M (in millions of dollars), for all the national parks in the United States can be modeled by

$$M = 575(1.066)^t$$ \$791.5 million, $\approx$ \$1319.8 million

where $t = 5$ represents 1985. What was the total expenditure in 1985? In 1993? *(Source: U.S. National Park Service)*

55. Aerospace Sales For 1980 through 1993, new orders, O (in billions of dollars), in the aerospace industry can be modeled by

$$O = \begin{cases} 68.7(1.1)^t, & 0 \le t < 10 \\ 386.87 - 24.03t, & 10 \le t \le 13 \end{cases}$$

where $t = 0$ represents 1980. What was the value of new orders in 1988? In 1990? *(Source: U.S. Bureau of Census)* $\approx$ \$147.26 billion, $\approx$ \$146.57 billion

A National Park Service range guides students in a nature studies program.

Integrated Review

In Exercises 56–59, evaluate the expression.

56. x^4y^2 when $x = 2$ and $y = 3$ 144

57. $2x^3 + 6y^5$ when $x = 3$ and $y = -1$ 48

58. $x^4 - xy - y^2$ when $x = -5$ and $y = 4$ 629

59. $3x^2 + 4xy - 2y^2$ when $x = 4$ and $y = -10$ −312

60. Powers of Ten Which best estimates the number of hours that a person sleeps in a year? c

a. 3.0×10^1 b. 3.0×10^2 c. 3.0×10^3 d. 3.0×10^4

61. Powers of Ten Which best estimates the number of cubic inches in a 12-ounce soda pop can? a

a. 2.2×10^1 b. 2.2×10^2 c. 2.2×10^3 d. 2.2×10^4

Exploration and Extension

62. Geometry When the radius of a sphere is doubled, how much is the volume increased? Justify your answer algebraically. 8 times, $(2r)^3 = 8r^3$

63. Geometry The volume of a circular cone is $V = \frac{1}{3}\pi r^2 h$, where r is the radius of the base and h is the height. Write a formula for the volume of a circular cone whose height is equal to the diameter of its base. (Use the letters V and r in your formula.) Use the formula to find the volume of a cone whose height and diameter are each 10 centimeters. $V = \frac{2}{3}\pi r^3$, $\frac{250\pi}{3}$

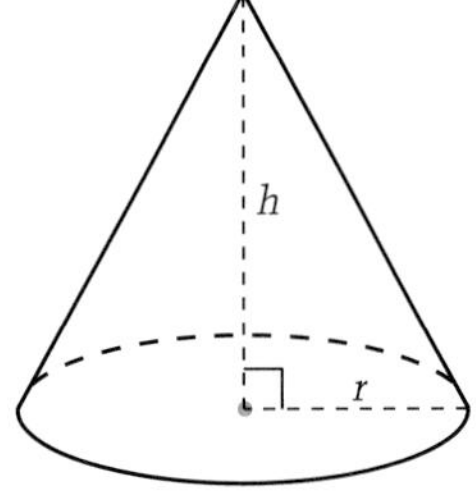

▶ **Ex. 62–63** These exercises connect algebra skills to geometric concepts. Remind students that the volume of a sphere of radius r is $V = \frac{4}{3}\pi r^3$.

Enrichment Activities

These activities require students to go beyond lesson goals.

1. List the first 9 powers of 2.
   $2^1 = 2$ $2^2 = 4$ $2^3 = 8$
   $2^4 = 16$ $2^5 = 32$ $2^6 = 64$
   $2^7 = 128$ $2^8 = 256$ $2^9 = 512$
2. Look at the ending digits. Do you notice a repeating pattern?
   The last digits end in 2, 4, 8, and 6 repeatedly in that order.
3. Use your pattern to predict the last digit of 2^{17} and 2^{2003}.
   17 is 4(4) + 1; so, 2^{17} ends in 2; 2003 is 4(500) + 3; so, 2^{2003} ends in 8.
4. Can you find a pattern in the ending digits of the powers of 3? the powers of 4? the powers of 5? Use your patterns to predict the ending digits of 3^{15}, 4^{15}, and 5^{15}. (You may want to use a calculator to check your answers.)
   Powers of 3: 3, 9, 7, 1, 3, 9, etc.; so, 3^{15} ends in 7. Powers of 4: 4, 6, 4, 6, etc.; so, 4^{15} ends in 4. Powers of 5 all end in 5; so, 5^{15} ends in 5.

Problem of the Day

A fly is in one corner of a ceiling of a 9×12 foot room with an 8-foot ceiling. A spider is on the floor in the farthest corner of the room. How far must the spider crawl to reach the fly? About 20.8 ft

ORGANIZER

Warm-Up Exercises

1. **TECHNOLOGY** Use a calculator to evaluate the following to two decimal places.
   a. $\left[4 + \frac{3.8}{5}\right]^2$ 22.66
   b. $\left[1 + \frac{4.23}{16}\right]^4$ 2.56
2. Construct a table of values and graph the relationship $y = 2^x$.

x	-2	-1	0	1	2	3
y	$\frac{1}{4}$	$\frac{1}{2}$	1	2	4	8

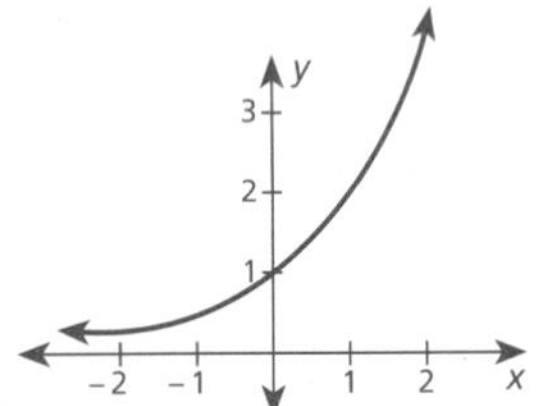

Lesson Resources

Teaching Tools
- Transparencies: 5, 7, 8, 14, 15
- Problem of the Day: 7.2
- Warm-up Exercises: 7.2
- Answer Masters: 7.2

Extra Practice: 7.2
Color Transparency: 41
Technology Handbook: p. 42, 45

7.2 Exploring Data: Compound Interest and Exponential Growth

What you should learn:

Goal 1 How to use the formulas for compound interest and exponential growth and decay formulas

Goal 2 How to solve real-life problems modeled by exponential growth formulas

Why you should learn it:

You can use exponential models for many problems, such as those measuring acidity with the pH scale.

Goal 1 Using Exponential Formulas

Simple interest is paid only on the initial *principal*. *Compound interest* is paid on the initial principal and on previously earned interest.

Compound Interest

Consider an initial principal of P deposited in an account that pays interest at an annual rate of r, compounded n times per year. The balance, A, in the account after t years is

$$A = P\left(1 + \frac{r}{n}\right)^{nt}.$$

Example 1 *Finding the Balance in an Account*

Assume \$1000 is deposited in an account that pays 8% annual interest compounded quarterly. What is the balance after one year?

Solution The number of compoundings per year is $n = 4$. This means that the account earns $\frac{r}{n} = 2\%$ interest each quarter. You can find the balance after one year in two ways. One way is to find the balance at the end of each quarter.

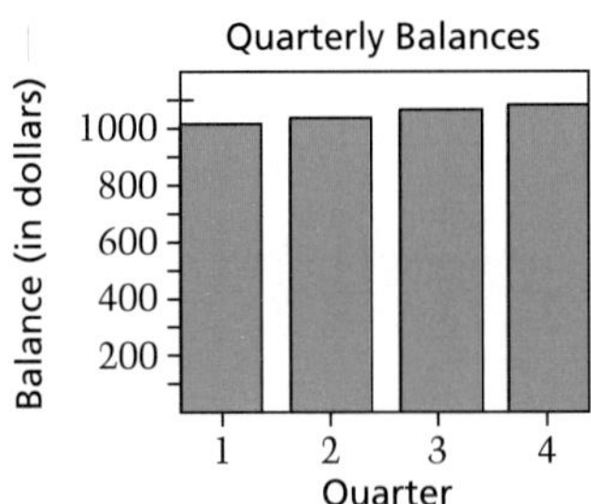

Finding the Balance after Each Quarter:

$1000.00(1.02) = \$1020.00$ *1st Quarter Balance*

$1020.00(1.02) = \$1040.40$ *2nd Quarter Balance*

$1040.40(1.02) = \$1061.21$ *3rd Quarter Balance*

$1061.21(1.02) = \$1082.43$ *4th Quarter Balance*

Another way is to apply the compound interest formula:

$$A = 1000\left(1 + \frac{0.08}{4}\right)^{4 \cdot 1} = 1000(1.02)^4 = \$1082.43$$

The balance at the end of one year is \$1082.43. Notice that if the account paid simple interest, the balance at the end of one year would be $1000(1.08) = \$1080.00$. ■

The formula for compound interest shows you four ways to increase the balance, A, in a savings account: increase the initial principal P, increase the interest rate r, increase the time t, cr increase the number of compoundings per year n.

Example 2 ***Comparing Quarterly and Monthly Compounding***

Assume \$5000 is deposited in an account that pays 6% annual interest. Find the balance after 25 years if the interest is

a. compounded quarterly. **b.** compounded monthly.

Solution

a. If the interest is compounded quarterly, the balance at the end of 25 years would be

$$A = 5000\left(1 + \frac{0.06}{4}\right)^{4 \cdot 25} \qquad P = 5000,\ r = 0.06,\ n = 4,\ t = 25$$

$$= 5000(1.015)^{100} \qquad \textit{Simplify.}$$

$$= \$22{,}160.23 \qquad \textit{Use a calculator.}$$

b. If the interest is compounded monthly, the balance at the end of 25 years would be

$$A = 5000\left(1 + \frac{0.06}{12}\right)^{12 \cdot 25} \qquad P = 5000,\ r = 0.06,\ n = 12,\ t = 25$$

$$= 5000(1.005)^{300} \qquad \textit{Simplify.}$$

$$= \$22{,}324.85 \qquad \textit{Use a calculator.}$$

Notice that more frequent compoundings produce a greater balance. ■

The formula for compound interest can be used to find the value of other quantities that increase by the same percent over equal time intervals.

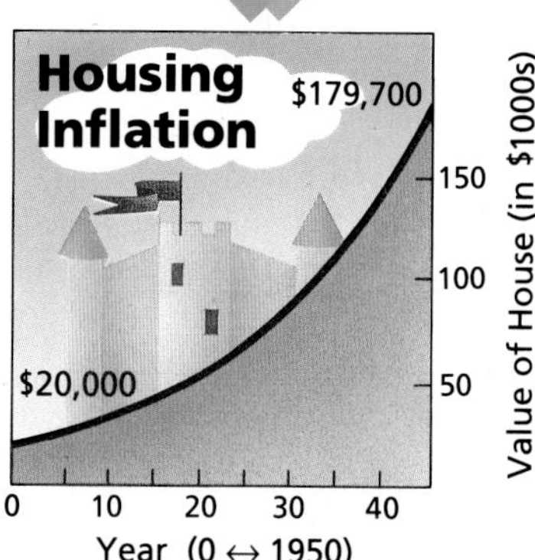

Example 3 ***Finding the Value of a House***

You have inherited a house that was purchased for \$20,000 in 1950. It is now 1995, and the value of the house increased by approximately 5% each year. What is the value of the house now?

Solution Using $P = 20{,}000$, $r = 0.05$, $n = 1$, and $t = 45$, you can apply the formula for compound interest to obtain a 1995 value of

$$A = 20{,}000(1 + 0.05)^{45}$$

$$= 20{,}000(1.05)^{45}$$

$$\approx \$179{,}700.$$

The house is now (in 1995) worth about \$179,700. ■

LESSON Notes

Example 1

Ask students to indicate the frequency of compounding for accounts compounded (a) quarterly, (b) semiannually, (c) monthly, and (d) daily.

four times per year; two times per year; 12 times per year; 365 times per year. (*Note*: Some banking formulas assume a year is 360 days.)

Example 2

Ask students to explain why more frequent compounding produces a greater balance over a finite period of time.

Example 3

Suppose the value of the house only increased by approximately 3.5% each year. What is the value of the house in 1995?

about \$94,047

Example 4

TECHNOLOGY You might want to ask students with graphics calculators to sketch graphs like the following:

a. $y = 2(0.25)^x$ and $y = 5(0.1)^x$

b. $y = 2(1.25)^x$ and $y = 5(1.1)^x$

What patterns exist among these relationships and those given in the example?

When the base is less than 1, the model is of exponential decay; when the base is greater than 1, the model is of exponential growth. So, the graphs in Part a are decreasing and represent decay, while the graphs in Part b are increasing and represent growth.

Extra Examples

Here are additional examples similar to **Examples 1–4.**

1. The amount of \$500 is deposited into an account that pays 9.5% compounded monthly. What is the balance in the account after 3 years? \$664.14
2. Suppose you have \$250 to invest. The bank near your home pays 8.2% compounded semiannually, and the bank near school pays 8% compounded quarterly. In which account should you invest your money? Explain why.
   Invest at 8.2% compounded semiannually because it has a higher yield.
3. Your collection of baseball cards cost you \$150 in 1988. In 1994, the value of the collection had increased by approximately 7.5% each year. What is the value of the cards in 1994? \$231.50
4. Classify each exponential model as either exponential growth or exponential decay.
   a. $y = 5^x$ **b.** $y = (0.4)^x$
   c. $y = 8(0.25)^x$ **d.** $y = 4(2.5)^x$
   a., d. growth; **b., c.** decay

Check Understanding

1. State the compound interest formula.
   See page 352; $A = P(1 + \frac{r}{n})^{nt}$
2. Describe the graph of an exponential decay model.
   See Example 4; the values of y decrease as the values of x increase.
3. Identify two different procedures that can be used to find the compound interest of a given deposit. See Example 1.

Goal 2 Modeling Exponential Growth and Decay

Compound interest is an example of exponential growth.

Exponential Growth and Decay

Let a and C be real numbers, with $C > 0$. The models for **exponential growth** and **exponential decay** are as follows.

$y = Ca^x$ *Exponential growth model, $a > 1$*

$y = Ca^x$ *Exponential decay model, $0 < a < 1$*

Example 4 *Classifying Exponential Models*

Classify the model as exponential growth or exponential decay.

a. $y = 3\left(\frac{1}{2}\right)^x$ **b.** $y = 2\left(\frac{3}{2}\right)^x$

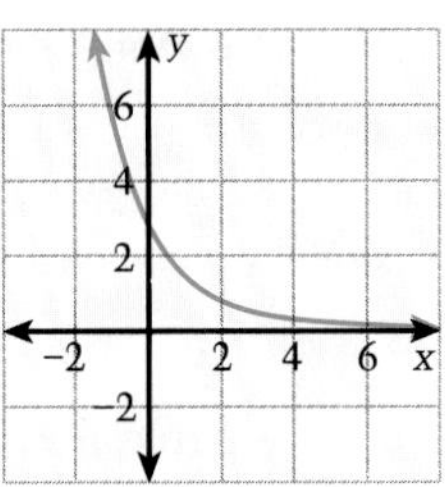

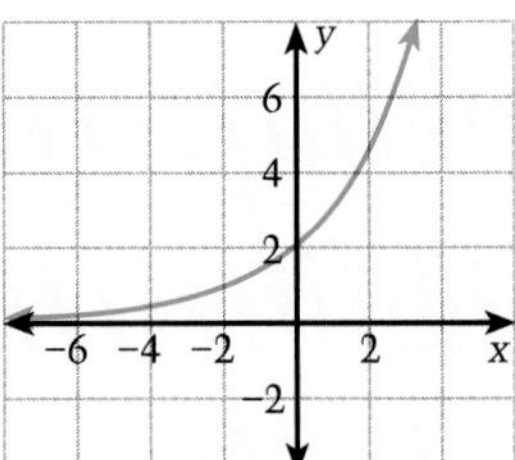

Solution

a. Because the base, $a = \frac{1}{2}$, is less than 1, this model represents exponential decay. Notice from the graph that the values of y decrease as the values of x increase.

b. Because the base, $a = \frac{3}{2}$, is greater than 1, this model represents exponential growth. Notice from the graph that the values of y increase as the values of x increase. ■

To sketch the graph of an exponential growth or decay model, construct a table of values, graph the points, and then connect the points with a curve. For instance, the following tables could be used to sketch the graphs shown in Example 4.

a.

x	-2	-1	0	1	2
$y = 3\left(\frac{1}{2}\right)^x$	12	6	3	$\frac{3}{2}$	$\frac{3}{4}$

b.

x	-2	-1	0	1	2
$y = 2\left(\frac{3}{2}\right)^x$	$\frac{8}{9}$	$\frac{4}{3}$	2	3	$\frac{9}{2}$

The map at the right shows the **pH** *levels of rain in the United States. The acid in the rain is the result of pollutants in the air—especially those from burning gasoline, coal, and oil. The brown and purple areas on the map indicate regions with low alkalinity in the surface water. Because an acid can neutralize alkalinity, these areas are very susceptible to harm from acid rain.*

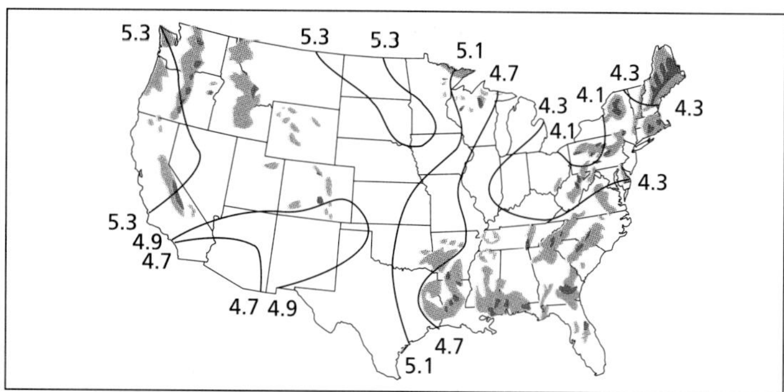

Example 5 *Using an Exponential Model*

To test whether a solution is a *base* (such as baking soda) or an *acid* (such as vinegar), chemists have devised a test called the pH test. The test measures the concentration of hydrogen ions (in moles per liter) in the solution.

$$\text{Concentration of hydrogen ions} = \left(\frac{1}{10}\right)^{\text{pH}}$$

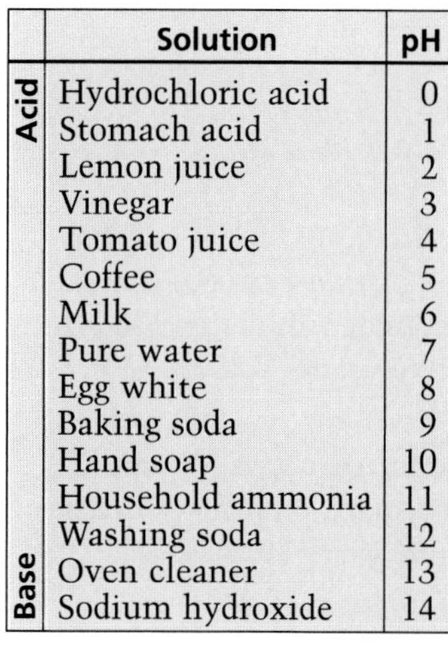

	Solution	pH
Acid	Hydrochloric acid	0
	Stomach acid	1
	Lemon juice	2
	Vinegar	3
	Tomato juice	4
	Coffee	5
	Milk	6
	Pure water	7
	Egg white	8
	Baking soda	9
	Hand soap	10
	Household ammonia	11
	Washing soda	12
	Oven cleaner	13
Base	Sodium hydroxide	14

Pure water has a pH of 7. Solutions with pH values greater than 7 are bases and solutions with pH values less than 7 are acids. Normal rain has a pH value of about 5.5. Which parts of the United States have rain that is at least 10 times the acidity of normal rain? ***(Source:* Scientific American*)***

Solution A pH of 4.5 is ten times more acidic than a pH of 5.5. To see this, you can write the following ratio.

$$\frac{4.5 \text{ pH}}{5.5 \text{ pH}} = \frac{\left(\frac{1}{10}\right)^{4.5}}{\left(\frac{1}{10}\right)^{5.5}} = \left(\frac{1}{10}\right)^{-1} = \frac{1}{\frac{1}{10}} = 10$$

The northeastern portion of the United States has rain with pH values that are 4.5 or lower, which means that its rain has at least 10 times the acidity of normal rain. ■

Example 5

Of course, the "base" referred to here is not the base of an exponential expression!

One mole of an element indicates that its mass in grams numerically equals its atomic weight. The "atomic weight" is a number representing the weight of one atom of an element compared to a standard, usually carbon at a value of 12.

EXTEND Example 5 **CHEMISTRY** Students could collect rain water and ask their chemistry teacher for a method of identifying the pH levels of the rain in their region of the United States. Compare the experimental values with those of normal rain.

Communicating *about* **ALGEBRA**

In Example 1, have students verify that the two procedures that were used to find the compound interest produce equivalent values. That is, have them show that finding the balance after each quarter of one year results in the same final balance as that found by using the formula when the time is one year.

Communicating *about* ALGEBRA

SHARING IDEAS about the Lesson

Which will yield the greater balance? Explain your answer.

A, because use of $A = P\left(1 + \frac{r}{n}\right)^{nt}$ shows it will

A. \$1000 deposited for 2 years in an account with interest compounded quarterly at an annual rate of 6.25%. **\$1132.05**

B. \$1000 deposited for 2 years in an account with interest compounded monthly at an annual rate of 6.15%. **\$1130.53**

EXERCISE Notes

ASSIGNMENT GUIDE

Basic/Average: Ex. 1–6, 7–17 odd, 19–22, 27–31 odd, 35–39

Above Average: Ex. 1–6, 7–17 odd, 19–22, 25–31 odd, 32, 37–39

Advanced: Ex. 1–6, 11–19 odd, 20–23, 27–32, 37–39

Selected Answers
Exercises 1–6, 7–37 odd

Use **Mixed Review** as needed.

More Difficult Exercises
Exercises 14, 29–32, 37–39

Guided Practice

Ex. 1–6 Use these exercises as an in-class summary of the lesson skills before assigning the independent practice. Check that students correctly use their calculators when applying the compound interest formula.

Independent Practice

Ex. 7–13 Remind students to use the compound interest formula.

Ex. 14 PROBLEM SOLVING Students can use a calculator and a guess-and-check strategy to narrow down the possible answer. For example:

$53{,}900{,}000 = 50(1 + r)^{98}$

Simplify first.

$1{,}078{,}000 = (1 + r)^{98}$

Try $r = 0.10$.

$(1.10)^{98} \approx 11{,}389$ Too small.

Try $r = 0.20$

$(1.20)^{98} \approx 57{,}512{,}482$ Too big.

Try 0.15 and 0.16.

$(1.15)^{98} \approx 887{,}950$ and $(1.16)^{98} \approx 2{,}074{,}354$.

So the answer is between 15% and 16%. It is closer to 15%.

EXERCISES

Guided Practice

CRITICAL THINKING about the Lesson

1. What is the difference between simple and compound interest? See page 352.
2. *Comparing Balances* Which will yield the larger balance in one year? Explain. b; it is compounded more frequently
   a. \$3000 at 6% annual interest, compounded twice a year
   b. \$3000 at 6% annual interest, compounded four times a year

2a. $3000\left(1 + \frac{0.06}{2}\right)^{2 \cdot 1} = 3000(1.03)^2 = 3182.07$

2b. $3000\left(1 + \frac{0.06}{4}\right)^{4 \cdot 1} = 3000(1.015)^4 = 3184.09$

3. *Balance in an Account* \$2000 is deposited in an account that pays 8% annual interest, compounded monthly. What is the balance in five years? \$2979.69
4. *Balance in an Account* \$2000 is deposited in an account that pays 8% annual interest, compounded annually. What is the balance in five years? \$2938.66
5. Is $y = 5\left(\frac{6}{5}\right)^x$ a model for exponential growth or decay? Explain. Growth, $\frac{6}{5} > 1$
6. Is $y = \frac{1}{4}\left(\frac{3}{4}\right)^x$ a model for exponential growth or decay? Explain. Decay, $\frac{3}{4} < 1$

Independent Practice

7. *Balance in an Account* \$800 is deposited in an account that pays 9% annual interest, compounded annually. Find the balance after four years. \$1129.27
8. *Balance in an Account* \$100 is deposited in an account that pays 6% annual interest, compounded annually. Find the balance after five years. \$133.82
9. *Balance in an Account* \$2250 is deposited in an account that pays 6% annual interest, compounded quarterly. Find the balance after ten years. \$4081.54
10. *Balance in an Account* \$1800 is deposited in an account that pays 5% annual interest, compounded monthly. Find the balance after one year. \$1892.09
11. *How Much to Deposit?* How much must you deposit in an account that pays 6.25% interest, compounded annually, to have a balance of \$700 after two years? \$620.07
12. *How Much to Deposit?* How much must you deposit in an account that pays 8% annual interest, compounded monthly, to have a balance of \$1000 after one year? \$923.36
13. *Inheritance* You have inherited an emerald ring that had an appraised value of \$2400 in 1960. It is now 1996, and the appraised value of the ring has increased by approximately 6% each year. What is the value now? \$19,553.40
14. *Fine Art* Vincent Van Gogh's painting "Irises" was auctioned for \$53.9 million dollars in 1987. Suppose that it sold for \$50 in 1889, when it was painted. Assuming exponential growth, by what percent did its value increase each year? ≈15.23%

In Exercises 15–18, state whether the function is a model of exponential growth or exponential decay.

15. $y = \frac{1}{3}(2)^x$ Growth
16. $y = 6\left(\frac{1}{5}\right)^x$ Decay
17. $A = 100(1.05)^x$ Growth
18. $y = 0.05(10)^x$ Growth

In Exercises 19–22, match the function with its graph.

19. $y = \frac{3}{2}(5)^x$ c

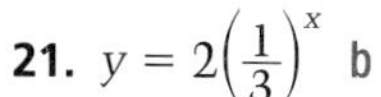

20. $y = 5\left(\frac{4}{3}\right)^x$ a

21. $y = 2\left(\frac{1}{3}\right)^x$ b

22. $y = 3(0.75)^x$ d

a.

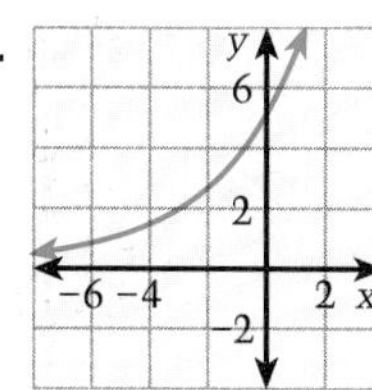

b.

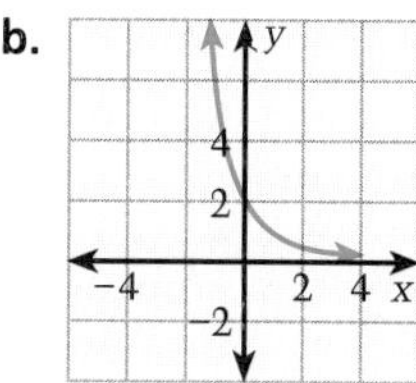

c.

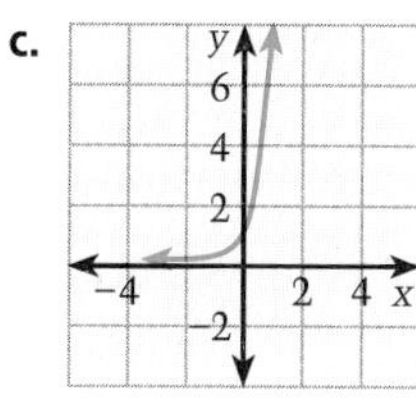

d.

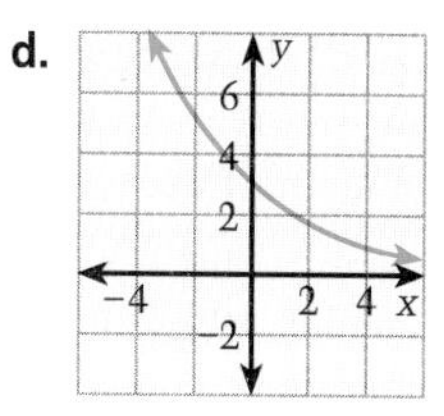

In Exercises 23–28, copy and complete the table of values and sketch the graph of the exponential function. See Additional Answers.

23. $y = 4\left(\frac{1}{2}\right)^x$

x	−3	−2	−1	0	1	2
y	32	16	8	4	2	1

24. $y = 2\left(\frac{1}{4}\right)^x$

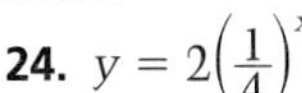

x	−3	−2	−1	0	1	2
y	128	32	8	2	$\frac{1}{2}$	$\frac{1}{8}$

25. $y = \frac{1}{3}(3)^x$

x	−3	−2	−1	0	1	2	3
y	$\frac{1}{81}$	$\frac{1}{27}$	$\frac{1}{9}$	$\frac{1}{3}$	1	3	9

26. $y = 10(2)^x$

x	−3	−2	−1	0	1	2	3
y	$\frac{5}{4}$	$\frac{5}{2}$	5	10	20	40	80

27. $A = 2000(1.02)^{4t}$

t	1	2	3	4	5	6
A						

28. $A = 2000(1.08)^t$

t	1	2	3	4	5	6
A						

29. *Recreational Books* Of the total amounts of money spent on recreation for 1970 through 1990, the percent, P, spent on books and maps can be modeled by

$$P = 2.94(1.07)^t$$

where $t = 0$ represents 1970. Sketch the graph of this model. See margin.
(Source: U.S. Bureau of Economic Analysis)

30. *Wheelchairs* For 1980 through 1990, the total amount, E (in billions of dollars), spent on health appliances such as eyeglasses, hearing aids, and wheelchairs can be approximated by the model

$$E = 4.9(1.1)^t$$

where $t = 0$ represents 1980. Sketch the graph of this model.
(Source: U.S. Health Care Financing Administration)
See margin.

Modern lightweight wheelchairs give greater freedom to physically challenged people.

▶ **Ex. 19–28** Refer students to Example 4 for a solution model. Assign Exercises 19–22 as a group.

▶ **Ex. 29–32** TECHNOLOGY Encourage students to use graphics calculators to sketch the graphs.

Answers

29.

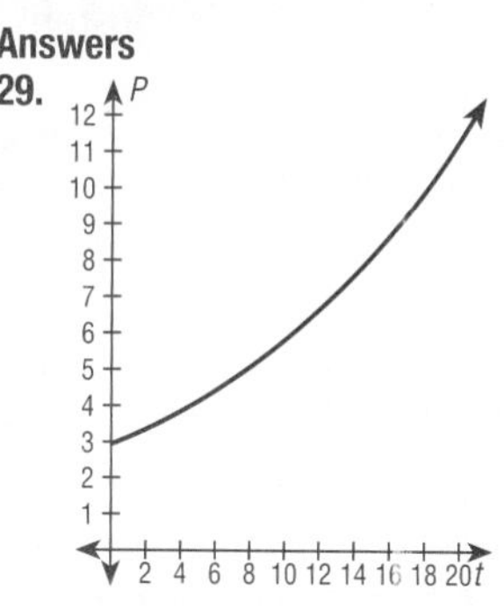

30.

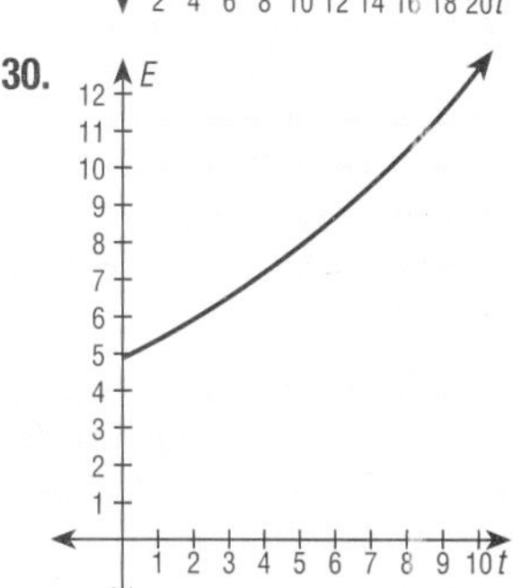

▶ **Ex. 31–32** Be sure students understand the differences between the three types of models, that is, linear, exponential, and quadratic models.

EXTEND Ex. 31–32 SOCIAL STUDIES You may wish to use these exercises to motivate an in-class discussion of salaries and inflation.

Ex. 37 Emphasize that this periodic growth problem assumes that the percent increase in each time period is constant.

Answers

31. Yes, the inflation graph is below the salary graph.

32. No, the average wage graph is below the inflation graph.

39. 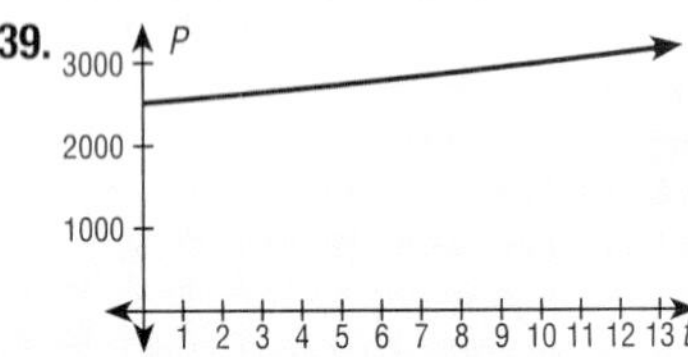

Enrichment Activities

These activities require students to go beyond lesson goals.

1. **EXTEND EX. 38–39** Use an almanac to investigate the population data for another country over a 10- to 15-year period. Use the research to write a linear, quadratic, or exponential model that best approximates the data. Use your model to predict the 1996 population for that country.
2. **CONNECTIONS** Discuss the reasonableness of your predicted value in terms of the social, natural, and political factors that contributed to the population increases or decreases in the previous years.

31. *Teachers' Salaries* For 1980 through 1990, the average salary, S (in thousands of dollars), of public school teachers in the United States can be modeled by $S = 1.5t + 16$ ($t = 0$ represents 1980). During that time, the average inflation rate was 6.2% per year. Did the average salary keep up with inflation? Explain. Yes. See margin.

32. *Miners' Wages* For 1980 to 1990, the average hourly wage, M (in dollars), of miners in the United States can be modeled by $M = -0.045t^2 + 0.8t + 9.2$ where $t = 0$ represents 1980. During that period of time, the average rate of inflation was 6.2% per year. Did the average wage keep up with inflation? Explain. No. See margin.

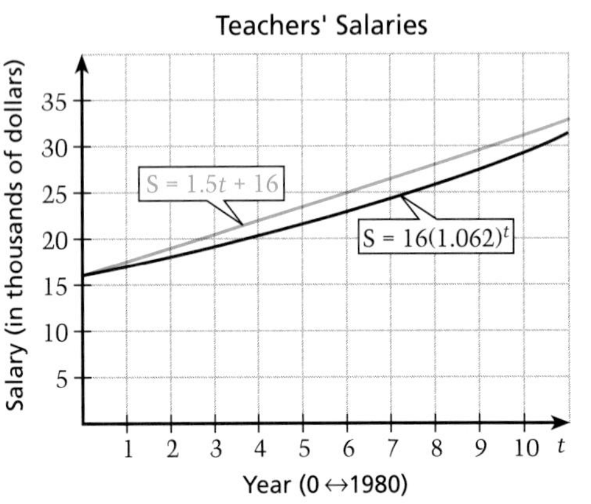

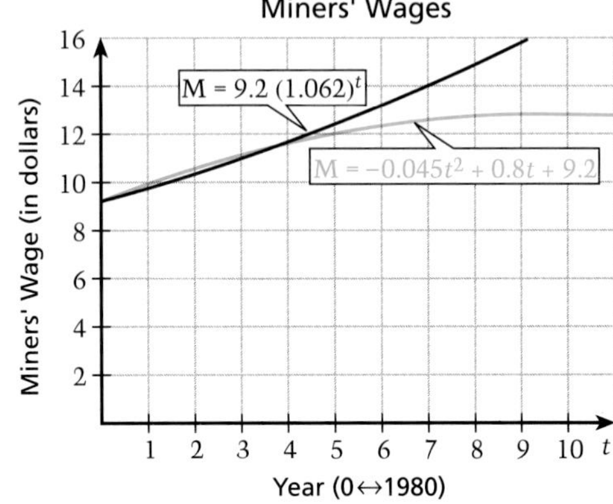

Integrated Review

In Exercises 33–36, evaluate the expression.

33. $6\left(\frac{4}{5}\right)^3$ $\frac{384}{125}$

34. $\frac{1}{3}\left(\frac{1}{2}\right)^7$ $\frac{1}{384}$

35. $250(1.06)^{15}$ ≈ 599.14

36. $1024\left(\left(\frac{1}{2}\right)^2\right)^4$ 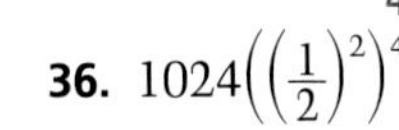 4

37. *College Entrance Exam Sample* Suppose that the average cost of a new home increased from 1969 to 1978 by the same percent that it increased from 1960 to 1969. Use the graph at the right to approximate the average cost of a new home in 1978. a

a. \$35,000 b. \$31,000 c. \$28,000
d. \$25,000 e. \$24,000

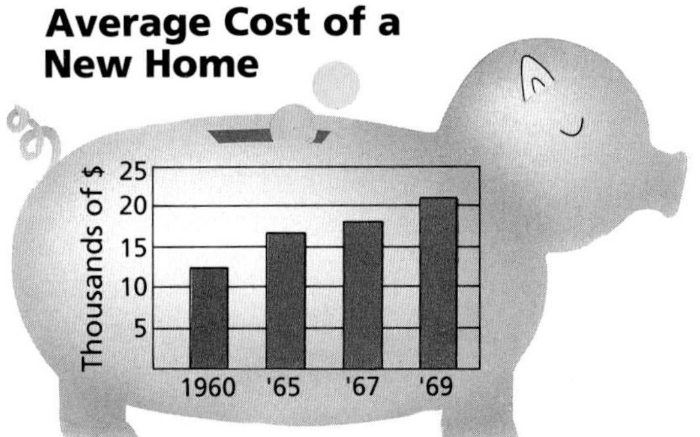

Exploration and Extension

38. *Population Growth* In 1980, the population of Shady Bridge was 2500. Each year between 1980 and 1993, the population increased by 1.8%. Write a model for the population, P, in year, t, where $t = 0$ represents 1980. $P = 2500(1.018)^t$

39. *Population Growth* Sketch the graph of the model in Exercise 38. During which year did the population surpass 3000? 1991. See margin.

Answer
6.

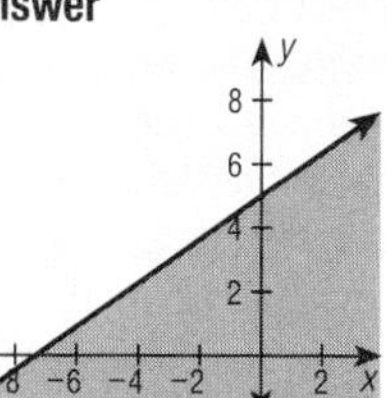

Mixed REVIEW

5. x: 7, y: 3 **6.** See margin.
9. All real numbers > 4 **10.** $y = \frac{3}{2}x - 6$ **16.** $\frac{1}{2}(h^2 + 3h)$

1. 24 is what percent of 120? **20%**

2. Solve $|x - 7| > 2$. **(1.6)** $x < 5$ or $x > 9$

3. Solve $|2x - 4| = 8$. **(1.7)** $-2, 6$

4. In which quadrant is $(-3, 8)$? **(2.1)** II

5. Find the intercepts of $\frac{x}{7} + \frac{y}{3} = 1$. **(2.3)**

6. Sketch the graph of $y \leq \frac{2}{3}x + 5$. **(2.5)**

7. Solve $2(3x - 5) - (2x - 3) = x + 2$. **(1.3)** 3

8. Find the intercepts of $y = -2|x - 4| + 6$. **(2.6)** x: 1, 7, y: -2

9. Find the domain of $f(x) = \frac{3}{\sqrt{x - 4}}$. **(6.2)**

10. Write the inverse of $y = \frac{2}{3}x + 4$. **(6.3)**

11. Simplify: $4.7 \times 10^3 + 8.9 \times 10^3$. 1.36×10^4

12. Simplify: $(2 \times 10^{-2})(6 \times 10^4)$. 1.2×10^3

13. Find the vertex of $y = 4|x - 3| + 8$. **(2.6)** (3, 8)

14. If $f(1) = -1$ and $f(n + 1) = [f(n)]^2 - 3$, find $f(4)$. **(6.6)** -2

15. Write the equation for the line passing through $(2, -4)$ and $(3, 1)$. $y = 5x - 14$

16. State the area of a triangle whose base is 3 units greater than its height, h. **(1.2)**

17. Solve $\begin{cases} 2x + 3y = 6 \\ 4x + 4y = \frac{10}{3} \end{cases}$ **(3.2)** $\left(-3\frac{1}{2}, 4\frac{1}{3}\right)$

18. For what values of c will $y = 3x^2 - 7x + c$ have no x-intercepts? **(5.4)** $c > 4\frac{1}{12}$

Career Interview

Contractor

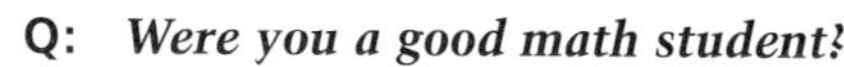

James Simko is a general contractor. His work involves the design, construction, and renovation of residential buildings, and he oversees all aspects of running the business.

Q: ***Were you a good math student?***

A: I didn't think I was a particularly good math student. I didn't expect to use much math in my career, because at that time I was more interested in the humanities.

Q: ***What math skill do you now apply in your work?***

A: I enjoyed geometry, but after college, I hardly used it. When I started this job, the geometric formulas I learned in school came to life. I can use the formulas instead of making guesses!

Q: ***Is it important to know other math skills?***

A: I do cost estimates when bidding on a project. These estimates are legally binding, so there isn't any room for errors. Even when I use a calculator or a computer program, I still need to understand the calculations and to have a sense of what is reasonable.

Q: ***How do you know if you are successful in your job?***

A: I know I'm successful when the company is profitable. That comes from a combination of good management and communication skills.

Problem of the Day

A cube is painted red, and then cut into 64 little cubes, all of the same size. How many of the little cubes have one red face, two red faces, three red faces, four red faces, and no red faces?
1 red face: 24; 2 red faces: 24
3 red faces: 8; 4 red faces: 0
no red faces: 8

ORGANIZER

Warm-Up Exercises

1. TECHNOLOGY Use a calculator to evaluate the following to two decimal places.
   a. $\sqrt{8}$ 2.83
   b. $12\sqrt{6}$ 29.39
2. Simplify.
   a. $(-3)^4$ 81
   b. 4^3 64

Lesson Resources

Teaching Tools
Problem of the Day: 7.3
Warm-up Exercises: 7.3
Answer Masters: 7.3
Extra Practice: 7.3

LESSON Notes

Examples 1–2

It is important to be able to write exponential expressions in both radical and rational-exponent forms. Ask students to rewrite each expression in Examples 1 and 2 in its alternate form.

7.3 *n*th Roots and Rational Exponents

What you should learn:

Goal 1 How to evaluate *n*th roots of real numbers using radical notation and rational exponent notation

Goal 2 How to use *n*th roots to solve real-life problems

Why you should learn it:

You can model many situations with *n*th roots, such as the lengths of the sides of regular polyhedrons.

Goal 1 Evaluating *n*th Roots

You know how to find the square roots of a number. For instance, $\pm\sqrt{5}$ are square roots of 5 because $(\sqrt{5})^2 = 5$ and $(-\sqrt{5})^2 = 5$.

In this lesson, you will study other types of roots of numbers: cube roots, fourth roots, fifth roots, and so on. For instance, 2 is a cube root of 8 because $2^3 = 8$. Similarly, -3 and 3 are fourth roots of 81 because $(-3)^4 = 81$ and $3^4 = 81$.

_n_th Roots of a Real Number

Let n be an integer greater than 1. If $b^n = a$, then b is an nth root of a. (Notice below that nth roots can be written using either radicals or rational exponents.)

1. If n is *odd*, then a has one nth root.
   $b = \sqrt[n]{a}$ *Radical notation*
   $b = a^{1/n}$ *Rational exponent notation*
2. If n is *even* and a is positive, then a has two nth roots.
   $b = \pm\sqrt[n]{a}$ *Radical notation*
   $b = \pm a^{1/n}$ *Rational exponent notation*

Example 1 *Using Radical Notation for* ***n*th Roots**

a. $\sqrt[3]{64} = 4$ because $4^3 = 64$.

b. $\sqrt[4]{81} = 3$ because $(3^4) = 81$.

c. $\sqrt[5]{-32} = -2$ because $(-2)^5 = -32$. ■

Example 2 *Using Rational Exponent Notation*

a. $16^{1/2} = 4$ because $4^2 = 16$.

b. $(-27)^{1/3} = -3$ because $(-3)^3 = -27$.

c. $625^{1/4} = 5$ because $5^4 = 625$. ■

Most calculators have a square root key but do not have keys for other roots. Thus, when using a calculator to approximate an nth root, you should first rewrite the nth root using a rational exponent. Then use the calculator's power key.

Connections Technology

Example 3 Approximating Roots with a Calculator

To approximate $\sqrt[3]{5}$, write the expression as $5^{1/3}$. Then use the power key as follows.

Expression	Keystrokes	Display	
$5^{1/3}$	5 [y^x] [(] 1 [÷] 3 [)] [=]	1.70998	*Scientific*
$5^{1/3}$	5 [∧] [(] 1 [÷] 3 [)] [=]	1.70998	*Graphing*

Rounded to two decimal places, the value is $\sqrt[3]{5} \approx 1.71$. ■

A rational exponent doesn't have to be of the form $\frac{1}{n}$. Rational numbers such as $\frac{3}{2}$ and $-\frac{1}{2}$ can also be used as exponents.

Rational Exponents

Let $a^{1/n}$ be an nth root of a and let m be a positive integer.

1. $a^{m/n} = (a^{1/n})^m$

2. $a^{-1/n} = \dfrac{1}{a^{1/n}},\ a \neq 0$

3. $a^{-m/n} = \dfrac{1}{a^{m/n}} = \dfrac{1}{(a^{1/n})^m},\ a \neq 0$

Example 4 Evaluating Expressions with Rational Exponents

a. $4^{3/2} = (4^{1/2})^3 = 2^3 = 8$

b. $9^{-1/2} = \dfrac{1}{9^{1/2}} = \dfrac{1}{3}$

c. $8^{2/3} = (8^{1/3})^2 = 2^2 = 4$

d. $32^{-3/5} = \dfrac{1}{32^{3/5}}$

$= \dfrac{1}{(32^{1/5})^3} = \dfrac{1}{2^3} = \dfrac{1}{8}$ ■

Example 5 Using a Calculator to Evaluate Rational Exponents

a. $6^{3/4} \approx 3.834$ *Use a calculator.*

b. $8^{-1/5} \approx 0.660$ *Use a calculator.* ■

Example 3

TECHNOLOGY To be sure students' calculators use the same key sequences, have them use their calculators to evaluate the following.

a. $64^{\frac{1}{3}}$ **b.** $32^{\frac{1}{5}}$
c. $8^{0.4}$ **d.** $158^{0.2}$

a. 4, **b.** 2, **c.** 2.30, **d.** 2.75

Examples 4–5

The connection between radicals and rational exponents enables the Properties of Rational Exponents to be rewritten using radicals. For example,

$$a^{\frac{m}{n}} = (a^{\frac{1}{n}})^m$$
$$\downarrow \qquad \downarrow$$
$$\sqrt[n]{a^m} = (\sqrt[n]{a})^m.$$

EXTEND Example 5 Consider the graph of the function $y = 2^x$. (See Warm-Up Exercise 2 on page 352.) Since the domain of x includes all real numbers, ask students if expressions like 2^{π} or $2^{\sqrt{2}}$ have meaning. They can use their calculators to show that properties for exponents of real numbers are not limited to just rational exponents.

Example 6

Density, d, is defined as the mass, M, per unit volume, V, of a substance, or $d = \frac{M}{V}$. Solving for the volume, V, you get $V = \frac{M}{d}$.

Example 7

Remind students that $\sqrt[3]{350}$ is computed using rational exponents. That is, $\sqrt[3]{350} = (350)^{\frac{1}{3}}$.

Extra Examples

Here are additional examples similar to **Examples 1–5.**

1. Evaluate.
   a. $\sqrt[3]{27}$ b. $\sqrt[5]{243}$
   c. $\sqrt[4]{256}$ d. $\sqrt{169}$
   a. 3, b. 3, c. 4, d. 13
2. Evaluate the following to two decimal places.
   a. $-32^{\frac{1}{5}}$ b. $64^{\frac{1}{4}}$
   c. $54^{\frac{2}{5}}$ d. $41^{\frac{3}{5}}$
   e. $125^{-\frac{2}{3}}$ f. $8^{-\frac{4}{5}}$
   a. −2, b. 2.83, c. 4.93, d. 9.28, e. 0.04, f. 0.19

Check Understanding

1. If b is the nth root of a, what equation states the relationship between b and a?
   $b = \sqrt[n]{a}$ or $b = a^{\frac{1}{n}}$
2. Describe how a calculator is used to compute the nth root of a number.
   See the top of page 361.
3. Does every radical expression have a counterpart written with rational exponents?
   Yes; if the radical expression is defined, it has a counterpart.

***A polyhedron** is a geometric solid whose faces are polygons. A **regular polyhedron** has sides that are congruent regular polygons that make equal angles with each other. There are only five such polyhedrons—the five **Platonic solids**. Their names and the approximate formulas for their volumes in terms of the length of an edge a are given at the right.*

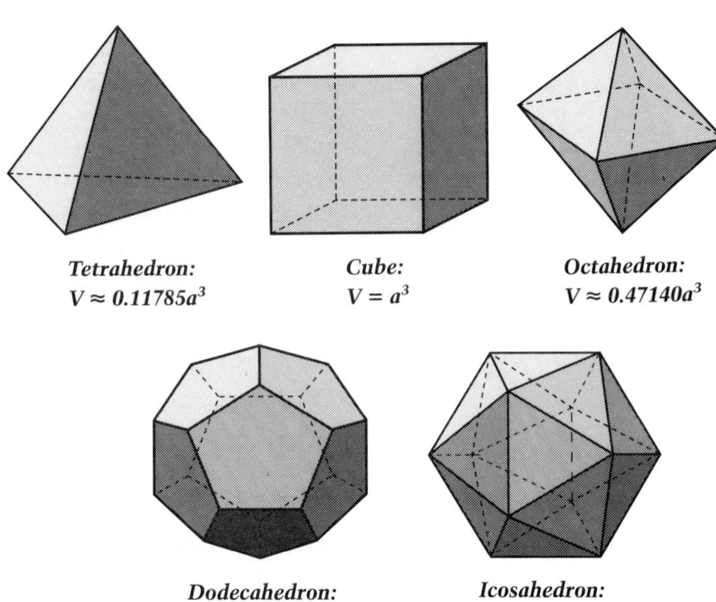

Goal 2 Using nth Roots in Real-Life Problems

Example 6 Finding the Length of an Edge of a Polyhedron

The uncut diamond shown at the left weighs 616 carats. It has the shape of an octahedron. Approximate the length of each edge of this diamond. (A 1-carat diamond has a mass of $\frac{1}{5}$ gram.)

The density of diamond is 3.513 grams per cubic centimeter.

Solution The mass of the diamond in grams is $\frac{1}{5}(616)$, or about 123.2. Since the density of diamond is 3.513 grams per cubic centimeter, the volume of the diamond is

$$\text{volume} = \frac{\text{mass}}{\text{density}} = \frac{123.2 \text{ grams}}{3.513 \text{ g/cm}^3} \approx 35.07 \text{ cubic centimeters.}$$

Using the formula for the volume of an octahedron, you can solve for the length of an edge as follows.

$V \approx 0.47140a^3$	*Volume of octahedron*
$35.07 \approx 0.47140a^3$	*Substitute 35.07 for V.*
$74.40 \approx a^3$	*Divide both sides by 0.47140.*
$\sqrt[3]{74.40} \approx a$	*Find cube root.*
$4.2 \approx a$	*Use a calculator.*

Each edge of the diamond is about $\sqrt[3]{74.40} \approx 4.2$ centimeters. ■

The Olympias *was built in 1987. It is a reconstruction of a trireme (a Greek galley ship). The people shown are volunteers who operated the ship's triple set of oars.*

Real Life
Nautical Science

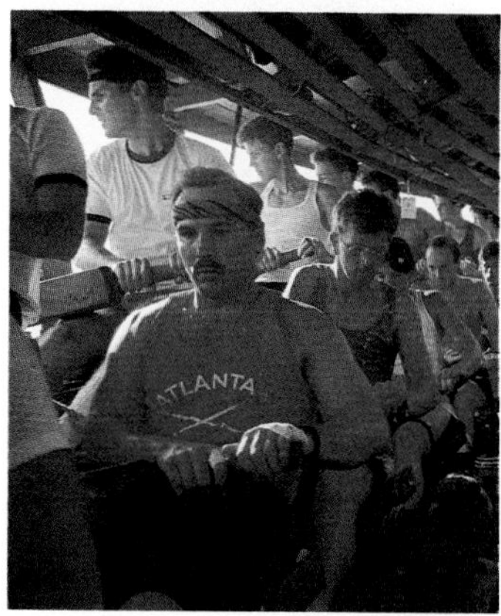

Example 7 *Finding the Speed of a Ship*

The speed, s (in knots), of the *Olympias,* was found to be related to the power, P (in kilowatts), generated by the rowers by the model $P = 0.03s^3$. The volunteer crew of the *Olympias* was able to generate a maximum power of 10.5 kilowatts. What was its greatest speed? ***(Source:*** **Scientific American*)***

Solution

$P = 0.03s^3$	*Cubic model*
$10.5 = 0.03s^3$	*Substitute 10.5 for P.*
$350 = s^3$	*Divide both sides by 0.03.*
$\sqrt[3]{350} \approx s$	*Find cube root.*
$7 \approx s$	*Use a calculator.*

The greatest speed attained by the *Olympias* was about 7 knots (about 8 miles per hour). ■

Communicating about ALGEBRA

SHARING IDEAS about the Lesson

A. Find the length of an edge of a cube that has a volume of 14 cubic feet. ≈ 2.41 ft

B. Find the length of an edge of a tetrahedron that has a volume of 8 cubic inches. ≈ 4.08 in.

C. If the volunteer crew of the *Olympias* could have matched the power output of the crew of the U.S. Navy light racing cutters (21.8 kilowatts), how fast could the *Olympias* have traveled? ≈ 8.99 knots

Communicating about ALGEBRA

EXTEND *Communicating*

1. Have students find the length of an edge of a dodecahedron that has a volume of 21 cubic meters. $21 \approx 7.66312a^3$, ≈ 1.4 meters
2. Find the length of an edge of an icosahedron that has a volume of 116 cubic centimeters. $116 \approx 2.18169a^3$, ≈ 3.76 cm
3. **EXTEND Example 7** How fast can the volunteer crew of the *Olympias* row if they can generate 25 kilowatts of power? $25 = 0.03s^3$, ≈ 9.4 knots

EXERCISE Notes

ASSIGNMENT GUIDE

Basic/Average: Ex. 1–6, 7–31 odd, 39–49 odd, 59–67 odd, 78

Above Average: Ex. 1–6, 7–33 odd, 39–49 odd, 57–65 odd, 66–68, 77–82

Advanced: Ex. 1–6, 9–35 odd, 41–49 odd, 57–65 odd, 66–68, 77–82

Selected Answers
Exercises 1–6, 7–79 odd

More Difficult Exercises
Exercises 63, 64, 67, 68, 81, 82

Guided Practice

Ex. 5–6 TECHNOLOGY These exercises are examples of an appropriate use of calculators in computations. Remind students that the formula for the volume of a sphere is $\frac{4}{3}\pi r^3$.

Independent Practice

Ex. 19–24 Remind students that even roots lead to two solutions.

EXERCISES

Guided Practice

CRITICAL THINKING about the Lesson

1. Is 4 a cube root of 64? Is -4? Explain. Yes, $4^3 = 64$; no, $(-4)^3 \neq 64$
2. Is 4 a square root of 16? Is -4? Explain. Yes, $4^2 = 16$; yes, $(-4)^2 = 16$
3. Evaluate $\sqrt[3]{125}$ and $\sqrt[4]{10,000}$. 5, 10
4. Evaluate $(-125)^{1/3}$ and $\sqrt[4]{-10,000}$. -5, does not exist
5. *Geometry* The Unisphere, which was the symbol of the 1964 World's Fair in New York City, has a volume of approximately 1,440,000 cubic feet. Approximate the radius of the Unisphere. ≈ 70 ft
6. Evaluate $27^{-2/3}$ without a calculator. Check your result with a calculator. $\frac{1}{9}$

Unisphere, Flushing Meadows Park, New York

Independent Practice

In Exercises 7–12, rewrite the expression using rational exponents.

7. $\sqrt[5]{12}$ $12^{1/5}$
8. $\sqrt[3]{9}$ $9^{1/3}$
9. $\sqrt[9]{16}$ $16^{1/9}$
10. $\sqrt[4]{29}$ $29^{1/4}$
11. $\sqrt[4]{2}$ $2^{1/4}$
12. $\sqrt[7]{3}$ $3^{1/7}$

In Exercises 13–18, rewrite the expression using radical notation.

13. $4^{1/3}$ $\sqrt[3]{4}$
14. $5^{1/7}$ $\sqrt[7]{5}$
15. $10^{1/6}$ $\sqrt[6]{10}$
16. $7^{1/3}$ $\sqrt[3]{7}$
17. $3^{1/4}$ $\sqrt[4]{3}$
18. $6^{1/4}$ $\sqrt[4]{6}$

In Exercises 19–24, use radical notation to write all of the indicated roots.

19. Cube root(s) of 3 $\sqrt[3]{3}$
20. Fifth root(s) of 14 $\sqrt[5]{14}$
21. Sixth root(s) of 9 $\pm\sqrt[6]{9}$
22. Ninth root(s) of 216 $\sqrt[9]{216}$
23. Fourth root(s) of 10 $\pm\sqrt[4]{10}$
24. Tenth root(s) of 110 $\pm\sqrt[10]{110}$

Technology **In Exercises 25–42, evaluate the expression using a calculator. Round the result to two decimal places, when appropriate.**

25. $\sqrt[3]{125}$ 5
26. $\sqrt[3]{729}$ 9
27. $\sqrt[5]{-32,768}$ -8
28. $\sqrt[4]{81}$ 3
29. $\sqrt[4]{256}$ 4
30. $\sqrt[3]{216}$ 6
31. $15,625^{1/6}$ 5
32. $2401^{1/4}$ 7
33. $-243^{1/5}$ -3
34. $-512^{1/3}$ -8
35. $1024^{1/5}$ 4
36. $1024^{1/10}$ 2
37. $\sqrt[4]{9}$ ≈ 1.73
38. $\sqrt[5]{108}$ ≈ 2.55
39. $\sqrt[5]{118}$ ≈ 2.60
40. $\sqrt[7]{210}$ ≈ 2.15
41. $\sqrt[6]{6}$ ≈ 1.35
42. $\sqrt[3]{-96}$ ≈ -4.58

In Exercises 43–54, evaluate the expression without using a calculator.

43. $16^{-3/2}$ $\frac{1}{64}$
44. $4^{7/2}$ 128
45. $81^{3/4}$ 27
46. $64^{-2/3}$ $\frac{1}{16}$
47. $125^{2/3}$ 25
48. $-8^{5/3}$ −32
49. $-27^{4/3}$ −81
50. $9^{3/2}$ 27
51. $8^{2/3}$ 4
52. $25^{3/2}$ 125
53. $100^{3/2}$ 1000
54. $256^{3/4}$ 64

Technology **In Exercises 55–60, use a calculator to evaluate the expression. Round the result to three decimal places.**

55. $4^{2/3}$ ≈2.520
56. $8^{3/8}$ ≈2.181
57. $10^{-1/3}$ ≈0.464
58. $9^{-1/4}$ ≈0.577
59. $16^{-4/5}$ ≈0.109
60. $26^{-3/4}$ ≈0.087

61. *Geometry* Find the length of an edge of an icosahedron that has a volume of 21 cubic centimeters. ≈2.13 cm

62. *Geometry* Find the length of an edge of a dodecahedron that has a volume of 30 cubic feet. 1.58 ft

63. *Lawn Bowling* The target ball in lawn bowling is called the jack. It weighs about 283 grams and is made of wood whose density is 2.11 grams per cubic centimeter. What is the radius of a jack? ≈3.18 cm

64. *Water Balloon* You fill a spherical balloon with one pint of water. What is the radius of the water balloon? (*Hint:* 1 pint = 28.875 cubic inches.) ≈1.90 in.

Lawn bowling is an ancient game in which players roll balls at a white target ball. About 20 nations have lawn bowling associations.

65. *Spectator Sports* For 1980 through 1990, the admission receipts, S (in billions of dollars), for spectator sports in the United States can be modeled by

$$S = 2.84 + 0.014t^3$$

where $t = 0$ represents 1985. In what year were the receipts approximately \$3.2 billion? 1988

66. *Cement Production* For 1980 through 1990, the world production of cement, C (in millions of metric tons), can be modeled by

$$C = 868 + \frac{1}{2}t^3$$

where $t = 0$ represents 1980. In what year were about 868 million metric tons of cement produced? 900 million metric tons? 1980, 1984

Concrete is used in most construction projects, such as La Grand Arche in Paris.

Ex. 43–54 These exercises should be completed using mental math.

Ex. 55–60 By contrast with Ex. 43–54, these exercises are an appropriate use of calculators.

Ex. 61–62 Students should use the polyhedron volumes shown at the top of page 362.

Ex. 63–64 These exercises are modeled in Example 6 on page 362. Remind students to first use the formula, volume = $\frac{\text{mass}}{\text{density}}$, to find the volume and then use the formula, $V = \pi r^3$, to find r.

Ex. 65–66 TECHNOLOGY Encourage students to use calculators to evaluate the cube root.

▶ **Ex. 77–80** Remind students of the difference between solving the equation $x^2 = 4$ and evaluating the expression $\sqrt{4}$. The equation $x^2 = 4$ has two solutions, 2 and −2, while $\sqrt{4}$ simplifies to the principal root 2.

▶ **Ex. 81–82** GROUP ACTIVITY Explore these exercises in class. Since factoring has not been formally discussed yet in this text, you can use the Distributive Property to write the equation $x^7 - 5x^2 = 0$ as $x^2(x^5 - 5) = 0$. Ask students if they remember the Zero Product Property from Algebra 1. Either $x^2 = 0$ or $x^5 - 5 = 0$; so, $x = 0$ or $x = \sqrt[5]{5}$.

Enrichment Activities

These activities require students to go beyond lesson goals.

1. Use a calculator. Find 2^k when $k = \frac{n}{n+1} = \frac{1}{2}, \frac{2}{3}, \frac{3}{4}, \frac{4}{5}, \frac{5}{6}, \frac{6}{7}, \frac{7}{8}$.
   1.414; 1.587; 1.682; 1.741; 1.782; 1.811; 1.834
2. What do you think happens to the values of 2^k, where $k = \frac{n}{n+1}$, as n becomes very large?
   As n becomes large, the fraction $\frac{n}{n+1}$ approaches 1; so the values of 2^k approach $2^1 = 2$.

67. *Douglas Fir* The volume, V (in cubic feet), of wood in the main trunk of a Douglas fir can be approximated by the model $V = 50r^3$, where r is the radius of the trunk at the base of the tree. Assume that the trunk of a Douglas fir has the shape of a tall, slender right circular cone. What is the ratio of the tree's height to its radius? $\frac{150}{\pi}$

68. *Doyle Log Rule* The Doyle Log Rule is a formula for approximating the number of board feet, B, in a circular log that is L feet long and has a radius of r inches.

$$B \approx \left(\frac{r-2}{2}\right)^2 L$$

What was the approximate radius of a 20-foot log that produced 5000 board feet of lumber? 33.6 in.

Douglas firs produce more lumber than any other type of North American tree.

Lumber mills use different log-sawing patterns when cutting large logs than when cutting small logs, as shown in the diagrams at the right. The patterns are designed to produce as many board feet as possible from a log. One board foot is the amount of wood in a 1-inch thick piece of lumber that is 12 inches square.

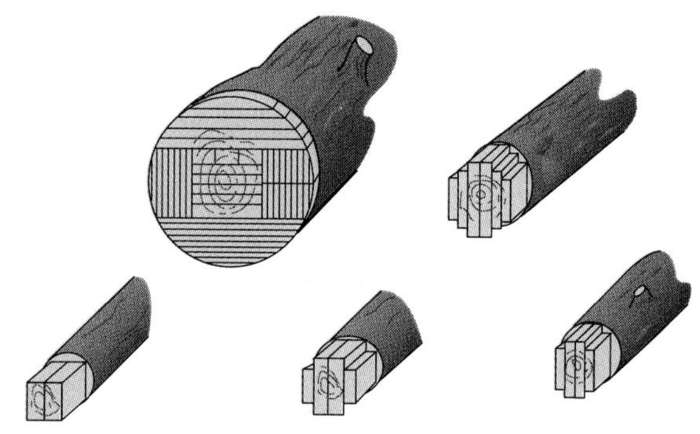

Integrated Review

In Exercises 69–72, simplify the expression.

Sample: $\sqrt{72} = \sqrt{36} \cdot \sqrt{2} = 6\sqrt{2}$

69. $\sqrt{20}$ $2\sqrt{5}$ **70.** $\sqrt{180}$ $6\sqrt{5}$ **71.** $\sqrt{54}$ $3\sqrt{6}$ **72.** $\sqrt{216}$ $6\sqrt{6}$

In Exercises 73–76, evaluate the expression.

73. $4x^6$ when $x = 2$ 256 **74.** x^4 when $x = 5$ 625 **75.** 4^x when $x = 3$ 64 **76.** 10^{-x} when $x = 2$ $\frac{1}{100}$

In Exercises 77–80, solve the equation for x.

77. $25 = 16x^2$ $\pm\frac{5}{4}$ **78.** $13x^2 = 39$ $\pm\sqrt{3}$ **79.** $2x^2 + 5 = 13$ ± 2 **80.** $512 = 2x^2$ ± 16

Exploration and Extension

In Exercises 81 and 82, solve the equation for x.

81. $x^7 - 5x^2 = 0$ $0, \sqrt[5]{5}$ **82.** $x^4 - 9x = 0$ $0, \sqrt[3]{9}$

Mid-Chapter SELF-TEST

Take this test as you would take a test in class. The answers to the exercises are given in the back of the book.

In Exercises 1–3, simplify the expression. (7.1)

1. $\frac{a^4b^0c^{-3}}{a^2b^{-7}c}$ $\frac{a^2b^7}{c^4}$

2. $\left(\frac{a^2}{b^3}\right)^5 \cdot \left(\frac{b}{a^5}\right)^2$ $\frac{1}{b^{13}}$

3. $\frac{(a^2b^3)^4}{(ab^{-2})^{-3}}$ $a^{11}b^6$

In Exercises 4–6, evaluate the expression. (7.3)

4. $(-8)^{2/3}$ 4

5. $81^{-3/4}$ $\frac{1}{27}$

6. $\sqrt[4]{16}$ 2

In Exercises 7–9, rewrite using only positive exponents. (7.1)

7. $\frac{15^{-2}3^{-3}}{5^{-2}}$ $\frac{1}{3^5}$

8. $2(6^{-2})(3)(4^2)$ $\frac{2^3}{3}$

9. $\frac{2^{-3}}{8^{-2}}$ 8^1

In Exercises 10 and 11, copy and complete the table of values. (7.2)

10.

x	-2	-1	0	1	2
$y = 4\left(\frac{2}{3}\right)^x$	?	?	?	?	?

9 6 4 $\frac{8}{3}$ $\frac{16}{9}$

11.

x	-2	-1	0	1	2
$y = 3(4^x)$	?	?	?	?	?

$\frac{3}{16}$ $\frac{3}{4}$ 3 12 48

12. Rewrite $3^{1/2}$ using radical notation. **(7.3)** $\sqrt{3}$

13. Rewrite $\sqrt[3]{7}$ using rational exponents. **(7.3)** $7^{\frac{1}{3}}$

14. Use radical notation to write all fourth roots of 90. **(7.3)** $\pm\sqrt[4]{90}$

15. In 1980, the population of a town was 12,000. If the population increases by 4% each year, what would the population be in 1995? **(7.2)** ≈21,611

16. \$2000 is deposited in an account that earns 6% annual interest, compounded monthly. Find the balance after 10 years. **(7.2)** \$3,638.79

17. How much money must be deposited in an account paying 8% annual interest, compounded quarterly, to have \$3585 in the account after 11 years? **(7.2)** \$1499.97

18. 100 grams of radioactive actinium decays according to the model $A = 100(2)^{-0.05t}$ where A is the amount (in grams) and t is the time (in years). Find the amount remaining after 5 years. How long does it take for half of the actinium to decay? **(7.2)** ≈84.09 grams, 20 yr

19. For 1980 through 1990, home sales of personal computers can be modeled by the equation $y = 0.05(1.37)^t + 2.6$, where y is sales in millions and $t = 0$ represents 1980. Using the model, estimate the sales for 1995. *(Source:* USA Today) **(7.2)** \$8.2 million

20. In the early 1400s, a plague took the lives of about one-third of the population of Europe. Had the plague not occurred, the estimated world population, y (in billions), for the years between 1400 and 1900 can be modeled by $y = 0.003(2.05)^t + 0.514$ where $t = 4$ represents 1400, $t = 5$ represents 1500, and so on. Using the model, estimate the world population in 1900, had the plague not occurred. **(7.2)** ≈2.43 billion

Problem of the Day

A square barn with sides x units long has a horse tethered outside at one corner, by a rope $2x$ units long. How large is the area in which the horse can graze?

$3\frac{1}{2}\pi x^2$

ORGANIZER

TIME SCHEDULE: All levels, two days

Warm-Up Exercises

1. Simplify.
   a. $a^3 \cdot a^4$ — a^7
   b. $(a^5)^3$ — a^{15}
   c. $(2a \cdot 5b)^4$ — $10{,}000a^4b^4$
   d. $\frac{a^8}{a^6}$ — a^2
2. Evaluate.
   a. $3^2 \cdot 5^3$ — 1125
   b. $4^{-3} \cdot 6^3$ — $\frac{27}{8}$ or 3.375
   c. $(\frac{9}{12})^3$ — $\frac{27}{64}$ or ≈ 0.422
   d. $3^4 \cdot 2^3$ — 648

Lesson Resources

Teaching Tools
- Problem of the Day: 7.4
- Warm-up Exercises: 7.4
- Answer Masters: 7.4

Extra Practice: 7.4
Color Transparencies: 41, 42
Applications Handbook: p. 56

LESSON Notes

Example 1

Ask students to identify which properties are used in each part.

7.4 Properties of Roots of Real Numbers

What you should learn:

Goal 1 How to use properties of roots to evaluate and simplify expressions containing radicals and rational exponents

Goal 2 How to use properties of roots to solve real-life problems

Why you should learn it:

You can model many problems with expressions containing roots, such as ratios of areas or volumes.

Goal 1 Using Properties of Roots

You can use the definition of rational exponents to show that the properties of exponents presented in Lesson 7.1 also apply to these exponents.

Properties of Roots

Let a and b be real numbers and let m and n be rational numbers.

Property	Example
1. $a^m \cdot a^n = a^{m+n}$	$2^{1/2} \cdot 2^{3/2} = 2^{(1/2\,+\,3/2)} = 2^2 = 4$
2. $(a^m)^n = a^{m \cdot n}$	$(4^{3/2})^2 = 4^{(3/2 \cdot 2)} = 4^3 = 64$
3. $(a \cdot b)^m = a^m \cdot b^m$	$(4 \cdot 9)^{1/2} = 4^{1/2} \cdot 9^{1/2} = 2 \cdot 3 = 6$
4. $\frac{a^m}{a^n} = a^{m-n},\ a \neq 0$	$\frac{5^{3/2}}{5^{1/2}} = 5^{(3/2\,-\,1/2)} = 5^1 = 5$
5. $\left(\frac{a}{b}\right)^m = \frac{a^m}{b^m},\ b \neq 0$	$\left(\frac{27}{8}\right)^{1/3} = \frac{27^{1/3}}{8^{1/3}} = \frac{3}{2}$
6. $a^{-m} = \frac{1}{a^m},\ a \neq 0$	$4^{-1/2} = \frac{1}{4^{1/2}} = \frac{1}{2}$

For rational exponents, the third and fifth properties above are often shown in radical notation.

$\sqrt[m]{a \cdot b} = \sqrt[m]{a} \cdot \sqrt[m]{b}$ *Product Property*

$\sqrt[m]{\frac{a}{b}} = \frac{\sqrt[m]{a}}{\sqrt[m]{b}}$ *Quotient Property*

Example 1 *Using Properties of Roots*

a. $3^{1/2} \cdot 3^{1/4} = 3^{3/4}$

b. $\frac{1}{2^{-1/2}} = 2^{1/2}$

c. $\frac{2}{2^{1/3}} = \frac{2^1}{2^{1/3}} = 2^{2/3}$

d. $(3^{1/2} \cdot 2^{1/3})^2 = (3^{1/2})^2 \cdot (2^{1/3})^2 = 3 \cdot 2^{2/3}$

e. $2^{1/3} \cdot 4^{1/3} = (2 \cdot 4)^{1/3} = 8^{1/3} = 2$

f. $(5^4 \cdot 3^4)^{1/4} = ((5 \cdot 3)^4)^{1/4} = (15^4)^{1/4} = 15$ ■

Example 2 Simplifying Radical Expressions

a. $\sqrt[3]{3} \cdot \sqrt[3]{9} = \sqrt[3]{3 \cdot 9} = \sqrt[3]{27} = 3$

b. $\dfrac{\sqrt{50}}{\sqrt{2}} = \sqrt{\dfrac{50}{2}} = \sqrt{25} = 5$

c. $\sqrt[3]{4} \cdot \sqrt{4} = 4^{1/3} \cdot 4^{1/2} = 4^{(1/3+1/2)} = 4^{5/6}$ ■

Applying the properties of roots to expressions involving variables can be difficult—especially with even roots such as $\sqrt{x^2}$ and $\sqrt[4]{x^4}$. In these cases, an absolute value sign is necessary to ensure that the result will be positive when the variable is negative.

$\sqrt[n]{x^n} = x$ *n is odd.*

$\sqrt[n]{x^n} = |x|$ *n is even.*

Example 3 Simplifying Expressions Involving Variables

a. $\sqrt[3]{8x^3} = \sqrt[3]{8}\sqrt[3]{x^3} = 2x$

b. $\sqrt{4x^2} = \sqrt{4}\sqrt{x^2} = 2|x|$

c. $x^{1/2} \cdot x^{3/2} = x^2$ ■

In the radical expression $c\sqrt[n]{a}$, the number a is the **radicand,** and the number n is the **index.** Two radical expressions are **like radicals** if they have the same index and the same radicand. For instance, $\sqrt[3]{2}$ and $4\sqrt[3]{2}$ are like radicals. To add or subtract like radicals, use the Distributive Property.

Sum: $\sqrt[3]{2} + 4\sqrt[3]{2} = (1 + 4)\sqrt[3]{2} = 5\sqrt[3]{2}$

Difference: $\sqrt[3]{2} - 4\sqrt[3]{2} = (1 - 4)\sqrt[3]{2} = -3\sqrt[3]{2}$

Example 4 Adding and Subtracting Roots and Radicals

a. $\sqrt[4]{5} + 6\sqrt[4]{5} = 7\sqrt[4]{5}$

b. $5(4^{1/3}) - 2(4^{1/3}) = 3(4^{1/3})$

c. $\sqrt{8} + \sqrt{2} = \sqrt{4 \cdot 2} + \sqrt{2}$
$= \sqrt{4} \cdot \sqrt{2} + \sqrt{2}$
$= 2\sqrt{2} + \sqrt{2}$
$= 3\sqrt{2}$ ■

Example 2

Point out to students that the Product and Quotient Properties of radicals are used in this example. Notice, however, that the properties are used by reading them from right to left, that is, $\sqrt[m]{a} \cdot \sqrt[m]{b} = \sqrt[m]{a \cdot b}$ and $\dfrac{\sqrt[m]{a}}{\sqrt[m]{b}} = \sqrt[m]{\dfrac{a}{b}}$.

Example 3

Observe the use of the absolute value when working with even roots. Ask students to explain why the absolute value is needed. (The principal root of an even root must be positive. Taking the absolute value of x guarantees that it will be.)

Example 4

Adding and subtracting roots and radicals imply the use of exact forms. Remind students that radical terms are alike when the roots and radicands are alike. For example, it can be shown that the following terms cannot be combined.

$\sqrt[3]{54} + \sqrt[3]{81}$
$= \sqrt[3]{27 \cdot 2} + \sqrt[3]{27 \cdot 3}$
$= \sqrt[3]{27} \cdot \sqrt[3]{2} + \sqrt[3]{27} \cdot \sqrt[3]{3}$
$= 3\sqrt[3]{2} + 3\sqrt[3]{3}$

Extra Examples

Here are additional examples similar to **Examples 1–4.**

1. Simplify.
   a. $5^{\frac{1}{3}} \cdot 5^{\frac{1}{4}}$ $5^{\frac{7}{12}}$
   b. $\dfrac{1}{4^{-\frac{3}{2}}}$ 8
   c. $(2^{\frac{1}{5}} \cdot 2^{\frac{1}{3}})^{15}$ 256
   d. $6^{\frac{1}{3}} \cdot 108^{\frac{1}{3}}$ $6\sqrt[3]{3}$
2. Simplify.
   a. $\sqrt[3]{4} \cdot \sqrt[3]{54}$ 6
   b. $\dfrac{\sqrt[4]{8}}{\sqrt[4]{128}}$ $\dfrac{1}{2}$
   c. $\sqrt{8x^4}$ $2x^2\sqrt{2}$
   d. $\sqrt[3]{32x^6}$ $2x^2\sqrt[3]{4}$

3. Combine.

a. $5\sqrt{3} - 2\sqrt{27}$ $-\sqrt{3}$

b. $12\sqrt[4]{80} + 18\sqrt[4]{5}$ $42\sqrt[4]{5}$

c. $3(6^{\frac{2}{3}}) - 6^{\frac{2}{3}}$ $2(6^{\frac{2}{3}})$

d. $\sqrt[3]{192} + \sqrt[3]{375}$ $9\sqrt[3]{3}$

Example 5

The solution illustrates the use of the properties of roots and radicals. By using a calculator, students can verify that they could have found each volume without simplifying first. The answer will still be the same. However, finding each volume without simplifying requires a great number of keystrokes, which can lead to possible keystroke errors. It is good practice to always simplify first.

Check Understanding

1. Describe the attributes that make two radical expressions alike.
Same index and same radicand
2. Are there any properties of roots and radicals that hold for integers but not for rational numbers? No
3. Describe how the Distributive Property is used to add and subtract radical expressions.
See Example 4.

Communicating about ALGEBRA

Ask students to guess and check which ratio of balls is closest to 1; closest to 0.5.
The ratio of the volume of a tennis ball to the volume of a baseball is about 0.831. The ratio of the volume of a Ping-Pong ball to the volume of a billiard ball is about 0.528.

Goal 2 Using Properties of Radicals in Real Life

Real Life Astronomy

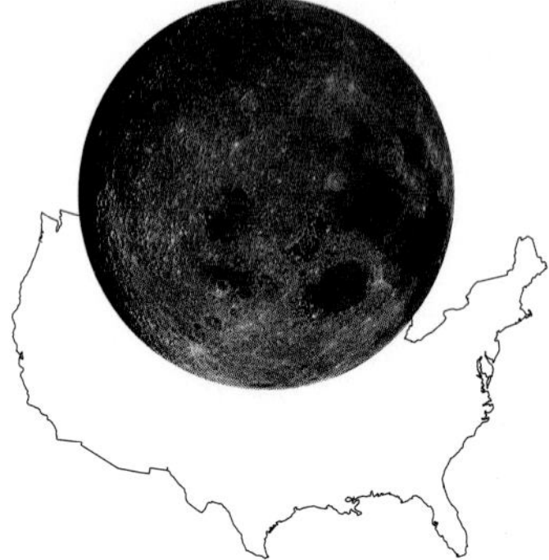

The surface area of the moon is 4,668,868π square miles. The surface area of Earth is 62,835,320π square miles. If the moon were placed on top of the United States, its diameter would span more than three-fourths of the distance between the east and west coasts.

Example 5 *Finding the Volume of a Sphere*

Find the ratio of the volume of the moon to the volume of Earth.

Solution The surface area and volume of a sphere are given by $S = 4\pi r^2$ and $V = \frac{4}{3}\pi r^3$, where r is the radius. Begin by solving for each radius.

Moon Radius

$$S = 4\pi r^2$$
$$4{,}668{,}868\pi = 4\pi r^2$$
$$1{,}167{,}217 = r^2$$
$$(1{,}167{,}217)^{1/2} = r$$

Earth Radius

$$S = 4\pi r^2$$
$$62{,}835{,}320\pi = 4\pi r^2$$
$$15{,}708{,}830 = r^2$$
$$(15{,}708{,}830)^{1/2} = r$$

The ratio of the volume of the moon to the volume of Earth is as follows.

$$\frac{\text{Volume of moon}}{\text{Volume of Earth}} = \frac{\frac{4}{3}\pi((1{,}167{,}217)^{1/2})^3}{\frac{4}{3}\pi((15{,}708{,}830)^{1/2})^3} = \frac{(1{,}167{,}217)^{3/2}}{(15{,}708{,}830)^{3/2}} = \left(\frac{1{,}167{,}217}{15{,}708{,}830}\right)^{3/2} \approx 0.02$$

Thus the volume of the moon is about 2% or $\frac{1}{50}$ the volume of Earth. ■

Communicating about ALGEBRA

SHARING IDEAS about the Lesson

Use the given surface areas to find the indicated ratios.

Ping-Pong Ball:	2.24π in.^{2}	Tennis Ball:	6.28π in.^{2}
Golf Ball:	2.84π in.^{2}	Baseball:	7.56π in.^{2}
Billiard Ball:	4.24π in.^{2}	Softball:	17.12π in.^{2}

A. Find the ratio of the volume of a Ping-Pong ball to the volume of a billiard ball. ≈0.38

B. Find the ratio of the volume of a tennis ball to the volume of a golf ball. ≈3.29

C. Find the ratio of the volume of a baseball to the volume of a softball. ≈0.29

EXERCISES

Guided Practice

CRITICAL THINKING about the Lesson

In Exercises 1–4, simplify the expression.

1. $\sqrt[6]{4}\sqrt[3]{4}$ 2
2. $8^{3/2}\cdot 2^{3/2}$ 64
3. $\frac{x}{x^{1/5}}$ $x^{4/5}$
4. $\frac{\sqrt[3]{24}}{\sqrt[3]{3}}$ 2

5. Are $2\sqrt[3]{3}$ and $2\sqrt[3]{4}$ like radicals? Explain. No; 3 ≠ 4 (The radicands are not equal.)

6. In the expression $10\sqrt[4]{7}$, identify the radical, the radicand, and the index. $\sqrt[4]{7}$, 7, 4

Independent Practice

In Exercises 7–42, simplify the expression.

7. $(3)^{1/4}\cdot(3)^{3/4}$ 3
8. $\sqrt[3]{4}\cdot\sqrt[3]{16}$ 4
9. $\sqrt[4]{20}\cdot\sqrt[4]{\frac{4}{5}}$ 2
10. $(-9)^{1/7}(-9)^{3/7}$ $(-9)^{4/7}$
11. $\left(\frac{1}{3}\right)^{1/5}\cdot\left(\frac{1}{3}\right)^{3/5}$ $\left(\frac{1}{3}\right)^{4/5}$
12. $(2^{1/2})^{2/3}$ $2^{1/3}$
13. $(4^{1/3})^{9/4}$ $4^{3/4}$
14. $-(8^{5/4})^{6/5}$ $-8^{3/2}$
15. $-(6^{4/3})^{3/7}$ $-6^{4/7}$
16. $\left(\frac{2}{3}\right)^{5/3}\cdot\left(\frac{2}{3}\right)^{1/3}$ $\frac{4}{9}$
17. $(1296x)^{1/4}$ $6x^{1/4}$
18. $\frac{4^{1/5}}{-4^{3/5}}$ $-4^{-2/5}$
19. $(243y)^{1/5}$ $3y^{1/5}$
20. $\left(\frac{96}{16}\right)^3$ 216
21. $\frac{2^{1/5}}{2^{6/5}}$ $\frac{1}{2}$
22. $\sqrt[7]{128x^7y}$ $2x\sqrt[7]{y}$
23. $\left(\frac{100}{64}\right)^3$ $\frac{15625}{4096}$
24. $(32x)^{1/5}$ $2x^{1/5}$
25. $\frac{9^{2/3}}{9^{1/6}}$ 3
26. $\frac{12^{10/8}}{12^{-3/8}}$ $12^{13/8}$
27. $\sqrt[4]{256x}$ $4\sqrt[4]{x}$
28. $\sqrt[3]{512x}$ $8\sqrt[3]{x}$
29. $\sqrt[5]{\frac{7776}{243}}$ 2
30. $\sqrt[3]{32x}\cdot\sqrt[3]{2}$ $4\sqrt[3]{x}$
31. $\frac{1}{6^{-1/2}}$ $\sqrt{6}$
32. $\frac{1}{x^{-5/4}}$ $x^{5/4}$
33. $(4^{1/4}\cdot4^{1/3})^2$ $4^{7/6}$
34. $(6^{2/3}\cdot6^{1/4})^{12/11}$ 6
35. $\frac{\sqrt[3]{648}}{\sqrt[3]{3}}$ 6
36. $\frac{\sqrt[5]{2048}}{\sqrt[5]{2}}$ 4
37. $\sqrt[4]{6x^4}$ $|x|\sqrt[4]{6}$
38. $\sqrt[3]{27xy^3}$ $3y\sqrt[3]{x}$
39. $\sqrt[5]{32x^5}$ $2x$
40. $\sqrt[6]{2x^6}$ $|x|\sqrt[6]{2}$
41. $\sqrt[3]{2x^6}$ $x^2\sqrt[3]{2}$
42. $\sqrt[3]{8x^9}$ $2x^3$

Technology **In Exercises 43–54, use a calculator to evaluate the expression. Then simplify the expression and use a calculator to evaluate the simplified expression. Compare your results.** 43.–54. Results are the same.

43. $\sqrt[3]{14}\cdot\sqrt[3]{13}$ ≈5.67
44. $\sqrt[8]{3}\cdot\sqrt[8]{5}$ ≈1.40
45. $6^{1/4}\cdot6^{4/5}$ ≈6.56
46. $(5^{6/7})^{1/2}$ ≈1.99
47. $(6^{-3})^{5/3}$ ≈0.000129
48. $3^{1/8}\cdot3^{3/5}$ ≈2.22
49. $(6^{1/4}\cdot6^{4/5})^{1/7}$ ≈1.31
50. $(2^{5/7}\cdot2^{4/3})^{3/5}$
51. $(9^{1/2}\cdot9^{2/3})^{1/6}$ ≈1.53
52. $(5^{1/3}\cdot5^{2/5})^{3/2}$ ≈5.87
53. $\sqrt[5]{11\cdot14}$ ≈2.74
54. $\sqrt[5]{3\cdot160}$ ≈3.44

50. ≈2.34

In Exercises 55–66, simplify the expression (if possible).

55. $\sqrt[4]{5}+4\sqrt[4]{5}$ $5\sqrt[4]{5}$
56. $\sqrt[7]{7}+6\sqrt[7]{7}$ $7\sqrt[7]{7}$
57. $4\sqrt[9]{6}-6\sqrt[9]{6}$ $-2\sqrt[9]{6}$
58. $-3\sqrt[5]{2}+10\sqrt[5]{2}$
59. $\sqrt[4]{768}-3\sqrt[4]{3}$ $\sqrt[4]{3}$
60. $\sqrt[3]{6}+\sqrt[3]{48}$ $3\sqrt[3]{6}$
61. $\sqrt[3]{270}+2\sqrt[3]{10}$ $5\sqrt[3]{10}$
62. $\sqrt[3]{14}-\sqrt[3]{112}$
63. $\sqrt[5]{1701}+4\sqrt[5]{7}$ $7\sqrt[5]{7}$
64. $2\sqrt[4]{176}+5\sqrt[4]{11}$ $9\sqrt[4]{11}$
65. $\sqrt[5]{160}-\sqrt[5]{1215}$ $-\sqrt[5]{5}$
66. $\sqrt[3]{375}+\sqrt[3]{81}$

58. $7\sqrt[5]{2}$ 62. $-\sqrt[3]{14}$ 66. $8\sqrt[3]{3}$

EXERCISE Notes

ASSIGNMENT GUIDE

Basic/Average: Ex. 1–6, 9–25 odd, 39–47 odd, 63–65, 67, 69–72, 75–89 odd

Above Average: Ex. 1–6, 13–27 odd, 39–49 odd, 61–65, 67–74, 75–89 odd, 90–92

Advanced: Ex. 1–6, 13–29 odd, 41–65 odd, 67–74, 75–89 odd, 90–92

Selected Exercises
Exercises 1–6, 7–89 odd

✪ More Difficult Exercises
Exercises 67–74, 91, 92

Guided Practice

Ex. 1–6 Use the Guided Practice exercises to help students develop strategies for simplifying radical expressions.

You may wish to review the simplest form of a radical expression. A radical expression is in simplest form if:

- the radicand does not contain a fraction;
- the radicand has no factor under the radical which possesses the root indicated by the index;
- no radicals appear in the denominator.

So, $\sqrt{12}$ is not in simplest form; nor is $\sqrt{\frac{5}{4}}$.

EXTEND Ex. 5 Ask students when like radicands are necessary.

Independent Practice

Note: These exercises could be assigned over a two-day period. You may assign Exercises 1–66 on the first day and Exercises 67–92 on the second day.

▶ **Ex. 7–42** Encourage students to use the properties to simplify these expressions. Students can use the radical form or they can convert to rational exponent notation first.

Note that expressions with rational exponents such as $\frac{6}{5}$ or $\frac{9}{4}$ can also be simplified by transforming the exponents to mixed numbers. For example, $25^{\frac{3}{2}}$ can be expressed as $25^{1+\frac{1}{2}} = (25^1)(25^{\frac{1}{2}}) = (25)(\pm 5) = \pm 125$.

EXTEND Ex. 17–42 Ask students if there are any restrictions on the domains of variables in these exercises so they can check whether absolute values are needed in their simplifications. For example, absolute values are not needed in Exercises 17 and 27 because *x* is restricted to nonnegative values. In Exercise 32, absolute values are not needed because *x* is restricted to positive values. In Exercises 37 and 40, absolute values are necessary because there are no restrictions on *x*.

▶ **Ex. 43–54** **COOPERATIVE LEARNING** You may wish to have students work with partners to complete these exercises as an in-class activity.

▶ **Ex. 55–66** Remind students that they can only add or subtract like radicals.

▶ **Ex. 69–70** Assign these exercises as a pair.

67. *Geometry* The areas of the circles below are 15 square centimeters and 20 square centimeters. Find the ratio of the radius of the small circle to the radius of the large circle. $\frac{\sqrt{3}}{2}$

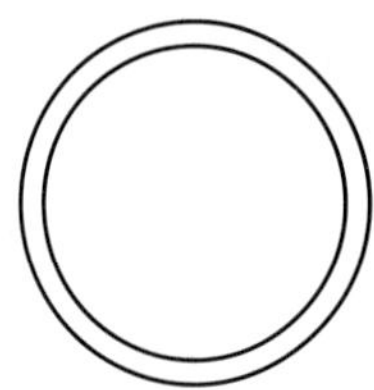

68. *Geometry* The volume of the sphere and the circular cylinder shown below are each 4 cubic feet. What is the ratio of the radius of the sphere to the radius of the circular cylinder? $\sqrt[3]{\frac{9}{4}}$

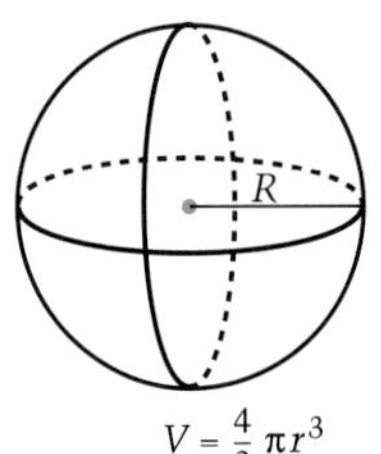

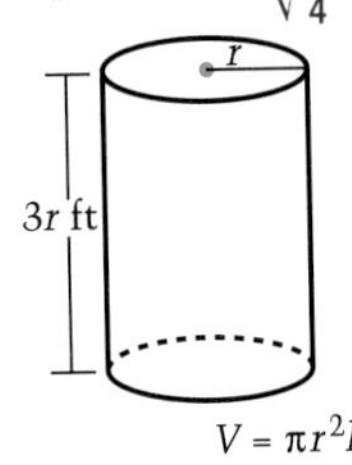

Strength-to-Weight Efficiency **In Exercises 69 and 70, use the following information.**

Young's Modulus Young's Modulus, named after the English scientist Thomas Young (1773–1829), is a measure of flexibility. Materials with a large Young's modulus are stiff. Materials with a small Young's modulus are flexible. The table gives the Young's modulus and density (in grams per cubic centimeter) of seven materials. The *strength-to-weight efficiency* of a column is $\frac{1}{d}\sqrt{E}$ and of a panel is $\frac{1}{d}\sqrt[3]{E}$.

Material	Steel	Aluminum	Brick	Bone	Concrete	Oak	Spruce
Young's Modulus, *E*	210,000	73,000	21,000	18,000	15,000	12,000	10,000
Density, *d*	7.8	2.8	3.0	3.0	2.5	0.65	0.35

69. Which of the materials has the greatest strength-to-weight efficiency as a column? Which has the least? **Spruce, bone**

70. Which of the materials has the greatest strength-to-weight efficiency as a panel? Which has the least? **Spruce, steel**

71. *Quilts* Find a radical expression that represents the total perimeter of the red triangles in the quilt at the right. Simplify the expression. **See below.**

72. *Quilts* Find a radical expression that represents the total perimeter of the blue triangles. Simplify the expression. **See below.**

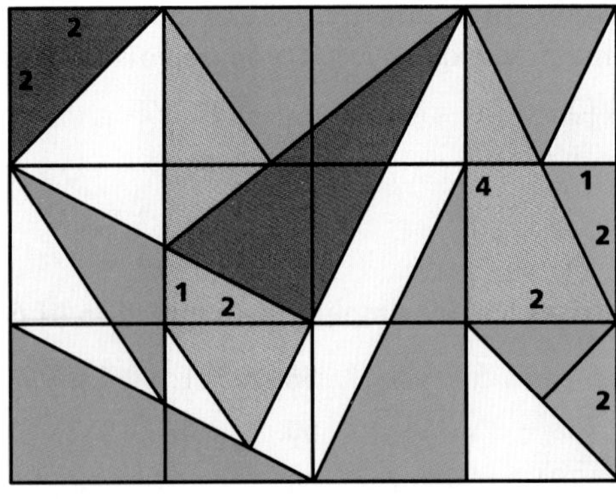

71. $9 + 2\sqrt{2} + 3\sqrt{5}$ **72.** $5 + 2\sqrt{2} + \sqrt{5}$

Musical Notes **In Exercises 73 and 74, use the following information.**

The musical note A-440 (above middle C) has a frequency of 440 vibrations per second. The frequency, F, of any note can be found by the equation

$$F = 440 \cdot 2^{n/12}$$

where n represents the number of black and white keys below or above the given note and A-440.

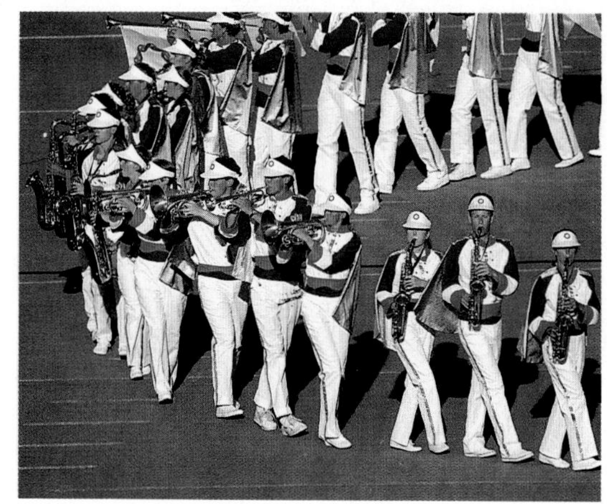

73. Approximate the lowest and highest frequencies of a trumpet. See below.

74. Describe the pattern of the frequencies of successive notes of the same letter. Frequencies double going to the right.

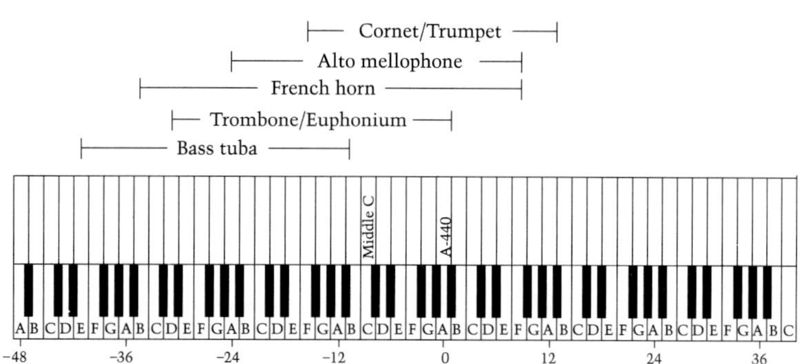

Integrated Review

In Exercises 75–82, simplify the expression.

75. $x^4 \cdot x^7$ x^{11}

76. $x^4 x^{-8}$ $\frac{1}{x^4}$

77. $(x^3)^5$ x^{15}

78. $\frac{x}{x^{-10}}$ x^{11}

79. $x^{-3}x^2$ $\frac{1}{x}$

80. $(x^7)^8$ x^{56}

81. $\frac{x^2}{x^9}$ $\frac{1}{x^7}$

82. $x^2 \cdot x^5$ x^7

In Exercises 83–90, evaluate the expression.

83. $(-3)^{1/3}$ ≈ -1.44

84. $2^{1/4}$ ≈ 1.19

85. $16^{1/6}$ ≈ 1.59

86. $27^{2/3}$ 9

87. $\sqrt[3]{200}$ ≈ 5.85

88. $\sqrt[3]{56}$ ≈ 3.83

89. $\sqrt[4]{14}$ ≈ 1.93

90. $\sqrt[4]{625}$ 5

Exploration and Extension

In Exercises 91 and 92, solve for *a* and *b*.

91. $\frac{3^a \cdot 5^b}{5^{1/2} \cdot 3^{5/6}} = 5^{1/6} 3^{1/3}$ $a = \frac{7}{6}$, $b = \frac{2}{3}$

92. $\frac{2^b (6^{1/2} \cdot 6^a)^4}{2^5} = \frac{6^{10}}{2^6}$ $a = 2$, $b = -1$

73. $\approx$ 175 vibrations per second, $\approx$932 vibrations per second

▶ **Ex. 73–74** Assign these exercises as a pair. If possible, you might ask the music teacher or the band director to demonstrate these instruments to the class.

▶ **Ex. 91–92** Remind students that for all positive real numbers $a \neq 1$, then $a^x = a^y$ if and only if $x = y$.

Enrichment Activity

This activity requires students to go beyond lesson goals.

MATH JOURNAL Following the Cooperative Learning discussion, students might record their personal responses in their math journals.

COOPERATIVE LEARNING
Work with a partner. Using the properties of powers and roots, develop strategies that can be used to simplify the various types of numerical and variable expressions in both radical form and rational-exponent form. Create your own problems similar to Exercises 7–42 to illustrate how your strategies correspond to the type of algebraic expression to be simplified.

Problem of the Day

If one wiggle equals two waggles and one waggle equals five woggles, how many wiggles are there in a wiggle plus a waggle plus a woggle? 1.6

ORGANIZER

Warm-Up Exercises

1. Solve.
   a. $4x + 12 = 16$ $\quad x = 1$
   b. $5 - 3x = 50$ $\quad x = -15$
   c. $3x^2 - 23 = 25$ $\quad x = \pm 4$
   d. $89 - 6x^2 = -205$ $\quad x = \pm 7$
2. Determine whether the given value of x is a solution to the given equation.
   a. $x = 1;\ 5x - 18 = -23$
   b. $x = 15;\ 28 + 6x = 8x - 2$
   c. $x = -2\sqrt{10};\ 4x^2 - 55 = 105$

   a. not a solution, b. solution, c. solution

Lesson Resources

Teaching Tools
- Transparencies: 7, 8
- Problem of the Day: 7.5
- Warm-up Exercises: 7.5
- Answer Masters: 7.5

Extra Practice: 7.5
Color Transparency: 42
Technology Handbook: p. 48

Common-Error ALERT!

In **Examples 1–2,** students frequently treat any radical expression as if it were a square-root radical. Consequently, when told to "eliminate the radical" to obtain a linear or quadratic expression, students will square the given expression without thinking!

7.5 Solving Radical Equations

What you should learn:

Goal 1 How to solve equations that have radicals and rational exponents

Goal 2 How to use radical equations to solve real-life problems

Why you should learn it:

You can model many real-life situations with radical expressions, such as the distance between two points.

Goal 1 Solving a Radical Equation

To solve a **radical equation**—one that contains radicals or rational exponents—you try to eliminate the radicals and obtain a linear or quadratic equation. Then you can solve the new equation using the standard procedures. The following property plays a key role.

> **Raising Both Sides of an Equation to the nth Power**
> If $a = b$, then $a^n = b^n$.

Before raising both sides of an equation to the nth power, you must isolate the radical expression on one side of the equation.

Example 1 *Solving a Radical Equation*

Solve $\sqrt[3]{x} - 2 = 0$.

Solution

$\sqrt[3]{x} - 2 = 0$	*Rewrite original equation.*
$\sqrt[3]{x} = 2$	*Add 2 to both sides.*
$(\sqrt[3]{x})^3 = 2^3$	*Cube both sides.*
$x = 8$	*Simplify.*

The solution is 8. Check this in the original equation. ■

Example 2 *Solving an Equation Containing Exponents*

Solve: $x^{3/2} = 27$.

$(x^{3/2})^{2/3} = 27^{2/3}$	*Raise both sides to $\frac{2}{3}$ power.*
$x = (27^{1/3})^2$	*Apply properties of exponents.*
$x = 9$	*Simplify.*

The solution is 9. Check this in the original equation. ■

Raising both sides of an equation to the nth power may introduce **extraneous** (or false) **solutions.** So when you use this procedure, it is critical that you check each solution in the *original* equation. For instance, squaring both sides of the equation

$$\sqrt{x} = -1$$

produces $x = 1$. This result, however, is extraneous. The original equation has no real-number solution.

Example 3 Solving a Radical Equation

Solve $\sqrt{3x - 8} + 1 = 3$.

Solution

$$\sqrt{3x - 8} + 1 = 3 \quad \textit{Rewrite original equation.}$$
$$\sqrt{3x - 8} = 2 \quad \textit{Subtract 1 from both sides.}$$
$$(\sqrt{3x - 8})^2 = 2^2 \quad \textit{Square both sides.}$$
$$3x - 8 = 4 \quad \textit{Simplify.}$$
$$3x = 12 \quad \textit{Add 8 to both sides.}$$
$$x = 4 \quad \textit{Divide both sides by 3.}$$

Check

$$\sqrt{3x - 8} + 1 = 3 \quad \textit{Rewrite original equation.}$$
$$\sqrt{3(4) - 8} \stackrel{?}{=} 2 \quad \textit{Substitute 4 for x.}$$
$$\sqrt{4} \stackrel{?}{=} 2 \quad \textit{Simplify.}$$
$$2 = 2 \quad \textit{Solution checks.}$$

The solution is 4. ■

The **geometric mean** of a and b is $\sqrt{ab}$. For instance, the geometric mean of 2 and 18 is $\sqrt{2 \cdot 18} = \sqrt{36} = 6$.

Example 4 Using the Geometric Mean

The geometric mean of 4 and a is 20. What is a?

Solution

$$\text{Geometric mean} = \sqrt{a \cdot 4}$$
$$20 = \sqrt{4a}$$
$$20^2 = (\sqrt{4a})^2$$
$$400 = 4a$$
$$100 = a$$

Since $a = 100$, the geometric mean of 4 and 100 is 20. ■

LESSON Notes

Example 1

Caution students to first identify the index or exponent of an expression. Once the index or exponent has been identified, select another exponent that will undo it. That is, raise the radical expression to a power such that the result is a linear or quadratic expression.

Example 2

The given exponent is $\frac{3}{2}$. The product of $\frac{2}{3}$ and $\frac{3}{2}$ is 1; consequently, the radical expression can be rewritten as a linear expression if both sides of the equation are raised to a power of $\frac{2}{3}$.

EXTEND Examples 1–2 Remember, always check the solutions in the original equations. Ask students to check the solutions.

Example 3

Stress the importance of isolating the radical expression on one side of the equation. Ask students what results if both sides of the original equation are squared before isolating the radical. (The radical is not eliminated from the resulting equation.)

Example 4

The geometric mean of 6 and b is 18. What is b? The geometric mean of c and 16 is 8. What is c? $b = 54$, $c = 4$

EXTEND Example 4 The geometric mean of n numbers is $\sqrt[n]{a_1 \cdot a_2 \cdot a_3 \cdots a_n}$. If 2, 6, and d have a geometric mean of 4, what is d? $\frac{16}{3}$

Example 5

Observe that the given order of the two points $(0,-3)$ and $(x, 4)$ is unimportant. That is, if (x_1, y_1) had been $(x, 4)$ and (x_2, y_2) had been $(0,-3)$, the outcome would be the same. You may want to have students verify this assertion.

Extra Examples

Here are additional examples similar to **Examples 1–5.**

1. Solve.

a. $8\sqrt[3]{x} - 24 = 0$ $x = 27$

b. $\sqrt[4]{x} - 3 = 0$ $x = 81$

c. $x^{\frac{4}{3}} = 256$ $x = 64$

d. $3x^{\frac{2}{5}} - 12 = 0$ $x = 32$

2. Solve.

a. $\sqrt{4x + 8} - 3 = 1$ $x = 2$

b. $\sqrt[3]{15 - 7x} + 15 = 17$ $x = 1$

3. Find the distance between the given points.

a. $(-2, 1)$ and $(5, 7)$
$d = \sqrt{85}$

b. $(-3,-11)$ and $(8, 4)$
$d = \sqrt{346}$

4. If the distance between $(-2, 12)$ and $(x, 6)$ is $\sqrt{37}$, what is the value of x?
$x = -1$ or $x = -3$

Radicals are also used to find the distance between two points in a coordinate plane. Suppose you were asked to find the distance between the points $A(2, 2)$ and $B(6, 5)$. How would you do it? One way is to locate point $C(6, 2)$ so that $\triangle ABC$ is a right triangle. Then find the lengths of the two legs and use the Pythagorean Theorem.

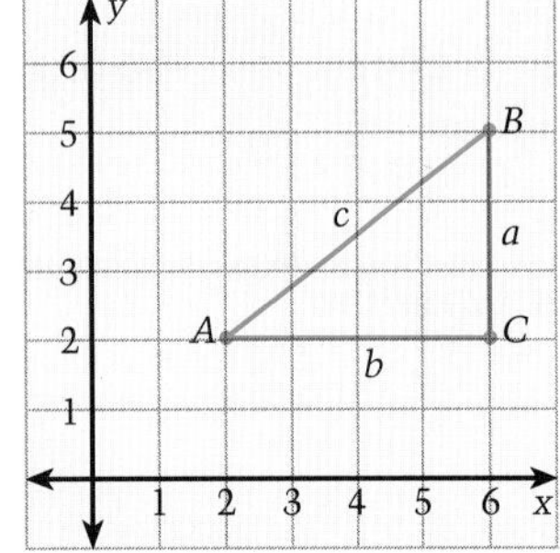

$a = 5 - 2 = 3$	*Length of $\overline{BC}$*
$b = 6 - 2 = 4$	*Length of $\overline{AC}$*
$c = \sqrt{a^2 + b^2}$	*Pythagorean Theorem*
$c = \sqrt{3^2 + 4^2}$	*Substitute $a = 3$ and $b = 4$.*
$c = \sqrt{25}$	*Simplify.*
$c = 5$	*Simplify.*

By performing this process with two general points (x_1, y_1) and (x_2, y_2), you can obtain the Distance Formula.

The Distance Formula

The distance d between the points (x_1, y_1) and (x_2, y_2) is

$$d = \sqrt{(x_2 - x_1)^2 + (y_2 - y_1)^2}.$$

Example 5 ***Using the Distance Formula***

The distance between $(0,-3)$ and $(x, 4)$ is $\sqrt{65}$. What is x?

Solution Let $(x_1, y_1) = (0,-3)$ and $(x_2, y_2) = (x, 4)$.

$$\sqrt{65} = \sqrt{(0 - x)^2 + (-3 - 4)^2}$$
$$\sqrt{65} = \sqrt{x^2 + 49}$$
$$65 = x^2 + 49$$
$$16 = x^2$$
$$\pm 4 = x$$

There are two solutions: $x = -4$ and $x = 4$. A graphic check is shown below.

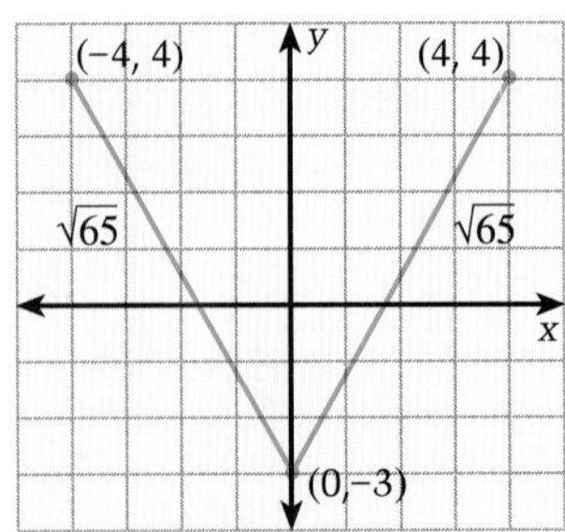

■

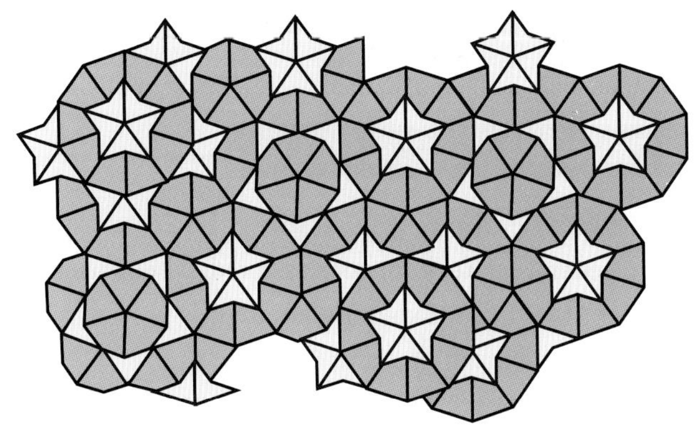

This tile pattern is made with Penrose's two simple shapes—one shaped like a kite and the other like an arrowhead. Although these tiles can be used to cover the plane, every such covering is nonrepeating!

Goal 2 Solving Radical Equations in Real Life

Connections **Geometry**

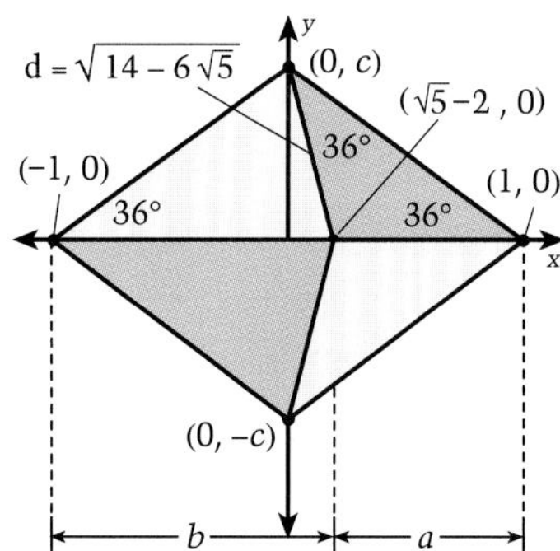

Example 6 *Using the Distance Formula*

In the early 1970's, Roger Penrose, an English physicist, discovered two tiling shapes whose dimensions involve the *golden ratio.* In the diagram at the left, find the value of c.

Solution Let $(x_1, y_1) = (0, c)$ and let $(x_2, y_2) = (\sqrt{5} - 2, 0)$. Using the Distance Formula, you can solve for c as follows.

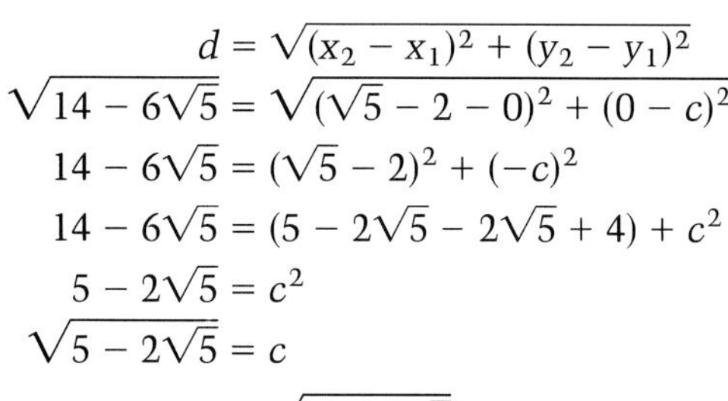

$$d = \sqrt{(x_2 - x_1)^2 + (y_2 - y_1)^2}$$
$$\sqrt{14 - 6\sqrt{5}} = \sqrt{(\sqrt{5} - 2 - 0)^2 + (0 - c)^2}$$
$$14 - 6\sqrt{5} = (\sqrt{5} - 2)^2 + (-c)^2$$
$$14 - 6\sqrt{5} = (5 - 2\sqrt{5} - 2\sqrt{5} + 4) + c^2$$
$$5 - 2\sqrt{5} = c^2$$
$$\sqrt{5 - 2\sqrt{5}} = c$$

The value of c is $\sqrt{5 - 2\sqrt{5}}$. ■

Communicating about ALGEBRA

SHARING IDEAS about the Lesson See Additional Answers.

A. The **golden ratio** is $\frac{\sqrt{5} + 1}{2}$. In the diagram, show that the ratio of b to a is the golden ratio.

B. The golden ratio can also be written as $\frac{2}{\sqrt{5} - 1}$. Show that this is equivalent to the form in Part **A** by multiplying the numerator and denominator by $(\sqrt{5} + 1)$ and simplifying.

Example 6

A tiling pattern is called "nonrepeating" if there is no translation that will allow a tracing of the pattern to be moved so that it exactly matches the original pattern. The value of c, $\sqrt{5 - 2\sqrt{5}}$, equals the tangent of 36°. Students will study trigonometric functions in Chapter 13.

Check Understanding

1. What is a "radical equation"?
   one that contains radicals or rational exponents
2. Show that the following statement is not always true: "If $a^n = b^n$, then $a = b$."
   For example, let $a = 2$, $b = -2$, and let n be an even number.
3. What is an "extraneous solution"?
   a false solution; see the top of page 375
4. What important theorem is used to derive the Distance Formula?
   Pythagorean Theorem

Communicating about ALGEBRA

Have students examine the following geometric sequences to determine a connection between geometric means and the terms of geometric sequences. (If *i, j,* and *k* are three consecutive terms of a geometric sequence, then *j* is the geometric mean of *i* and *k*.)

a. 2, 4, 8, 16, 32, . . .
b. 3, 6, 12, 24, 48, . . .
c. 3, 12, 48, 192, 768, . . .
d. 1, 3, 9, 27, 81, 243, . . .
e. 1.5, 3.15, 6.615, 13.8915, . . .

EXERCISE Notes

ASSIGNMENT GUIDE

Basic/Average: Ex. 1–4, 9–27 odd, 45, 47, 51–54, 55–63 odd, 64

Above Average: Ex. 1–4, 9–27 odd, 41–47, 51–54, 55–63 odd, 64

Advanced: Ex. 1–4, 9–29 odd, 37–47 odd, 50–54, 63–65

Selected Answers
Exercises 1–4, 5–63 odd

Use **Mixed Review** as needed.

✪ More Difficult Exercises
Exercises 39, 40, 46–52

Guided Practice

▶ **Ex. 1–4** Use these exercises as an in-class summary of the lesson concepts. Have students state a general strategy for solving a radical equation that contains either a variable under a radical sign or a variable with a rational (non-integral) exponent. Students should recognize that the strategies that work with equations in radical form are the same as the strategies that can be used with equations in rational-exponent form.

Independent Practice

▶ **Ex. 11–40** Emphasize that all solutions must be checked in the original equations, not in the transformed equations. Note that the equations in Exercises 19, 20, 29, and 34 have no solutions.

▶ **Ex. 46–48** Remember that $(3\sqrt{3})^2$ is equivalent to $(3)^2(\sqrt{3})^2$, or $9 \cdot 3 = 27$.

EXERCISES

Guided Practice

CRITICAL THINKING about the Lesson Except for 3. See Additional Answers.

1. Describe a general strategy for solving a radical equation.
2. Solve $\sqrt[3]{2x + 10} + 1 = 3$. Explain your steps.
3. Find the geometric mean of 3 and 6. $\sqrt{18}$
4. Find the distance between (6, 1) and (−2, 5) in two ways: use the Distance Formula and use the Pythagorean Theorem.

Independent Practice

In Exercises 5–10, check whether the value is a solution of the equation. 5.–9. The value is a solution. 10. The value is not a solution.

5. $\sqrt{x} - 4 = 4, \quad x = 64$
6. $2(x - 3)^{1/2} = 12, \quad x = 39$
7. $(x + 4)^{3/2} - 60 = 4, \quad x = 12$
8. $4(2x + 1)^{2/3} = 36, \quad x = 13$
9. $\sqrt[3]{4x} + 9 = 3, \quad x = -54$
10. $2\sqrt[4]{2x + 1} + 2 = 0, \quad x = 0$

In Exercises 11–40, solve the equation. Check for extraneous solutions. 29. and 34. No solution

11. $\sqrt{x} = 20$ 400
12. $x^{1/2} = 5$ 25
13. $y^{1/3} - 7 = 0$ 343
14. $x^{2/3} + 13 = 17$ −8, 8
15. $\sqrt[4]{2x} + 2 = 6$ 128
16. $2x^{3/4} = 54$ 81
17. $\sqrt{a + 100} = 25$ 525
18. $\sqrt[3]{b + 12} = 5$ 113
19. $(x - 2)^{3/2} = -8$ No solution
20. $(8x)^{1/2} + 6 = 0$ No solution
21. $(3y + 5)^{1/2} - 3 = 4$ $\frac{44}{3}$
22. $\sqrt[3]{5z - 4} + 7 = 10$ $\frac{31}{5}$
23. $\sqrt[4]{x + 2} + 9 = 14$ 623
24. $2(x + 4)^{2/3} = 8$ −12, 4
25. $\sqrt{x^2 + 5} = x + 3$ $-\frac{2}{3}$
26. $\sqrt{x^2 - 4} = x - 2$ 2
27. $\sqrt{x + 3} = \sqrt{2x - 1}$ 4
28. $\sqrt{3t + 1} = \sqrt{t + 15}$ 7
29. $\sqrt{3y - 5} - 3\sqrt{y} = 0$
30. $\sqrt{2u + 10} - 2\sqrt{u} = 0$ 5
31. $\sqrt[3]{3x - 4} = \sqrt[3]{x + 10}$ 7
32. $2\sqrt[3]{10 - 3x} = \sqrt[3]{2 - x}$ $\frac{78}{23}$
33. $\sqrt[3]{2x + 15} - \sqrt[3]{x} = 0$ −15
34. $\sqrt[4]{2x} + \sqrt[4]{x + 3} = 0$
35. $\sqrt{2x} = x - 4$ 8
36. $\sqrt{x} = x - 6$ 9
37. $\sqrt{8x + 1} = x + 2$ 1, 3
38. $\sqrt{3x + 7} = x + 3$ −2, −1
39. ✪ $\sqrt{2t + 3} = 3 - \sqrt{2t}$ $\frac{1}{2}$
40. ✪ $\sqrt{x + 3} - \sqrt{x - 1} = 1$ $\frac{13}{4}$

In Exercises 41 and 42, find *x*.

41. The geometric mean of 2 and x is $3\sqrt{6}$. 27
42. The geometric mean of x and 6 is $4\sqrt{3}$. 8

In Exercises 43–45, find the distance between the two points.

43. (0, 1), (1, 5) $\sqrt{17}$
44. (−2, 6), (3, 6) 5
45. (3, 7), (9, −1) 10

In Exercises 46–48, find all values of *x* for which the distance between the given points is *d*.

46. ✪ (5, −8), (x, −11) 2, 8 $d = 3\sqrt{2}$
47. ✪ (−3, 2), (−10, x) 0, 4 $d = \sqrt{53}$
48. ✪ (3, x), (5, 7) −1, 15 $d = 2\sqrt{17}$

49. *Geometry* In a right triangle, a line segment is drawn from the vertex of the right angle, perpendicular to the hypotenuse. Let h represent the length of the segment and let x and y represent the lengths of the two parts of the hypotenuse. One of these lengths is the geometric mean of the other two lengths. Which is it? Explain. See Additional Answers.

50. *Geometry* In the diagram at the right, the rectangle whose sides have lengths x and 6 is similar to the rectangle whose sides have lengths 24 and x. Which of the three lengths (6, 24, and x) is the geometric mean of the other two lengths? Explain your answer and then use the result to solve for x. See below.

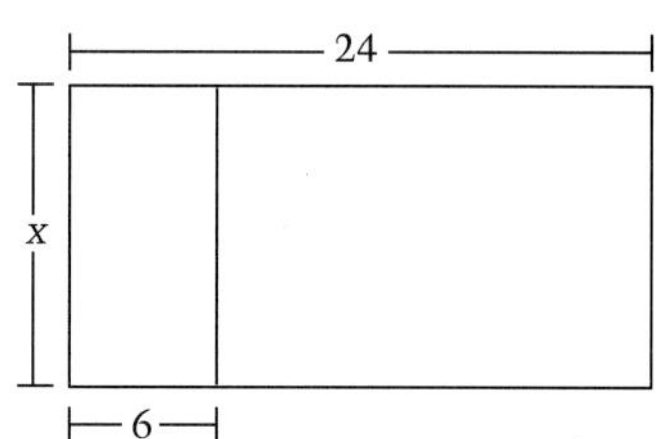

Wisconsin **In Exercises 51 and 52, use the following information.**

A grid is superimposed on the map of Wisconsin at the right. Each unit of the grid represents 10.5 miles. **51.** ≈166, ≈44 minutes

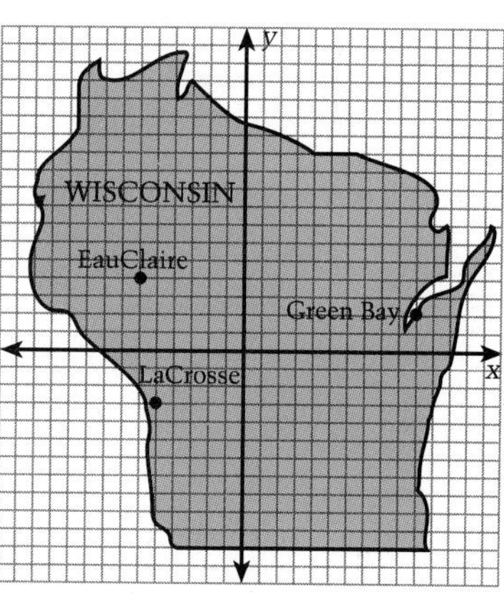

51. Approximate the distance in miles between La Crosse and Green Bay. How long would a flight from La Crosse to Green Bay take traveling at 225 m.p.h.?

52. Approximate the distance in miles between Eau Claire and La Crosse. What is the minimum time necessary to walk from Eau Claire to La Crosse walking at a rate of 5 miles per hour? ≈ 74, ≈15 hours

53. *Public Television* For 1970 through 1990, the average number, h, of broadcast hours per year per public television station can be modeled by $h = \sqrt{-91{,}150t^2 + 3{,}187{,}200t + 4{,}255{,}600}$ where $t = 0$ represents 1970. In which year was the average equal to 5420 hours? 1982

54. *Women in Medicine* For 1980 through 1990, the percent, p, of Doctor of Medicine (MD) degrees earned each year by women can be modeled by

$$p = \sqrt{3.3t^2 + 50.7t + 528.4}$$

where t represents the year with $t = 0$ corresponding to 1980. In what year were 28% of the MD degrees earned by women? 1984

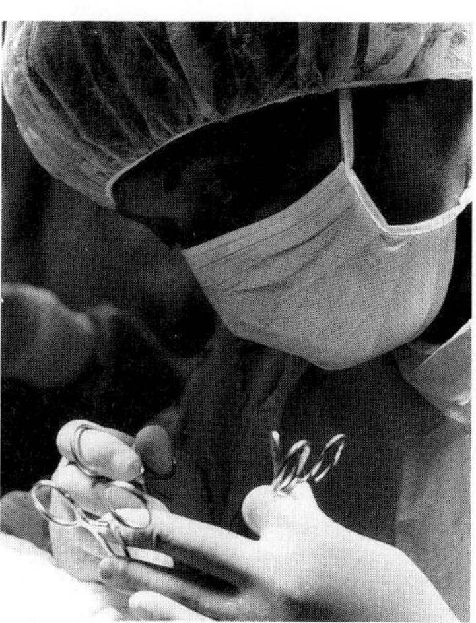

In 1981, Dr. Alexa Canady became the first African-American woman to become a neurosurgeon in the United States. She now directs neurosurgery at Children's Hospital of Michigan in Detroit.

Ex. 49 You may wish to use this exercise in class. Have students draw a 3-4-5 right triangle on graph paper and measure the values of x, y, and h to verify that $h = \sqrt{xy}$.

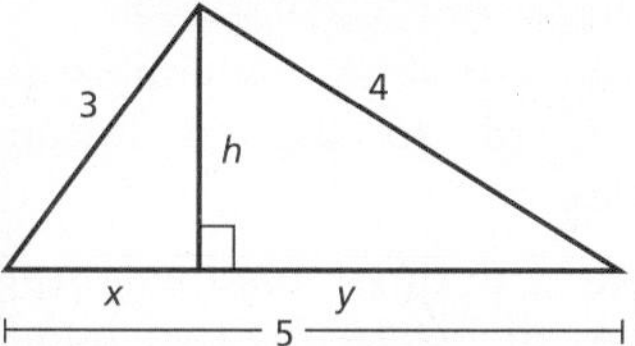

For a 3-4-5 right triangle, $h = 2.4$, $x = 3.2$, and $y = 1.8$.

Ex. 50 GEOMETRY You might need to review similar figures with students. Ask them to redraw the rectangles as shown.

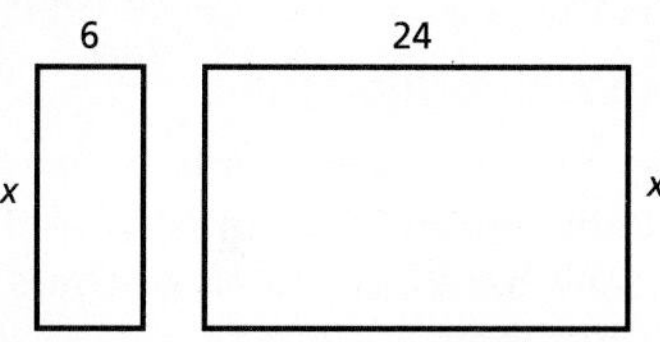

If they are similar, then the ratios of corresponding parts are equal. If $\frac{6}{x} = \frac{x}{24}$, then $x^2 = 6 \cdot 24$ and $x = \sqrt{6 \cdot 24}$, and x is the geometric mean of 6 and 24.

Ex. 53–54 Students need to use the Quadratic Formula to solve for t.

EXTEND Ex. 53–54 Have students predict the values for h and p in the year 2000. Ask students what their answers indicate for the future.
h: ≈4223, Public TV hours decreasing; p: ≈53.5%, MD's earned by women increasing

50. x is the geometric mean of 6 and 24. Corresponding sides are proportional, so $\frac{6}{x} = \frac{x}{24}$, $x = 12$

EXTEND Ex. 65–66 Ask students to determine another method for verifying collinearity. Use the points to write the equations of the lines that contain them.

Enrichment Activities

These activities require students to go beyond lesson goals.

Here are three mathematical types of means for two numbers a and b.

The arithmetic mean of a and b is the average of their sum, $\frac{a+b}{2}$. The geometric mean is the square root of their product, $\sqrt{ab}$, and the harmonic mean is the ratio of twice their product to their sum, $\frac{2ab}{a+b}$.

1. Find the arithmetic mean, the geometric mean, and the harmonic mean of these numbers.
   a. 4 and 9 6.5; 6; 5.5
   b. 5 and 15 10; 8.7; 7.5
   c. 6 and 8 7.0; 6.9; 6.9
   d. 7 and 100 53.5; 26.5; 13.1
2. Examine your answers. What do you notice about the relative values of the means? If $a \neq b$, then the arithmetic mean is greater than the geometric mean which is greater than the harmonic mean.
3. **ALTERNATIVE ASSESSMENT** Write a one- or two-paragraph report on finding and using the harmonic mean.
   The harmonic mean of a set of n numbers can also be found by dividing n by the sum of the reciprocals of the numbers. The harmonic mean of a and b is $\frac{2}{\frac{1}{a}+\frac{1}{b}}$ or $\frac{2ab}{a+b}$.

Integrated Review

In Exercises 55–62, simplify the expression.

55. $[(3x)^4]^{1/2}$ $9x^2$
56. $[(4x)^{1/2}]^{-3}$ $\frac{1}{8}x^{-3/2}$
57. $(\sqrt[5]{6x^3})^3$ $216^{1/5}x^{9/5}$
58. $(64)^{1/3} \cdot (64)^{1/3}$ 16
59. $(-8)^{2/3}$ 4
60. $(\sqrt[6]{64x^7})^{-1}$ $\frac{1}{2}x^{-7/6}$
61. $(13)^{3/2} \cdot (13)^{-5/2}$ $\frac{1}{13}$
62. $(\sqrt[3]{27x})^4$ $81x^{4/3}$

63. *College Extrance Exam Sample* In the diagram, ABC is a right triangle with $AB = 6$ and $AC = 7$. What does BC equal? b

a. 1 **b.** $\sqrt{13}$ **c.** 6 **d.** $\sqrt{39}$ **e.** $\sqrt{78}$

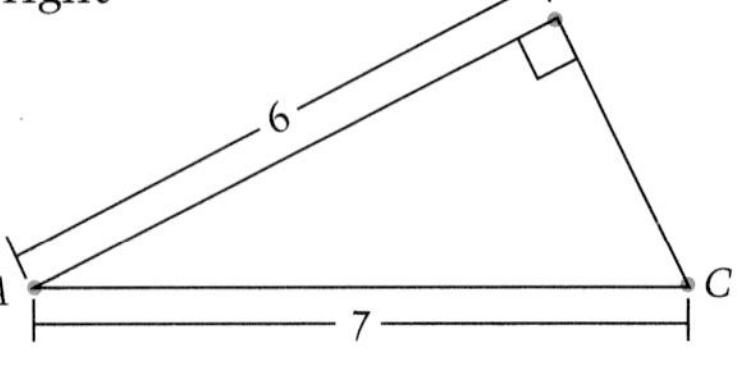

64. *College Entrance Exam Sample* The lengths of the sides of a triangle are 6 in., 8 in., and 10 in. A rectangle equal in area to that of the triangle has a width of 3 in. What is the perimeter of the rectangle (in in.)? c

a. 11 **b.** 16 **c.** 22 **d.** 24 **e.** 30

Exploration and Extension

Collinear Points **In Exercises 65 and 66, use the following information.**

Three points are collinear if the sum of the distances between two pairs of the points is equal to the distance between the third pair. Use this to help you decide whether the three points lie on a line.

65. (0, 5), (1, 7), (2, 9) They do.
66. (0, 2), (−2, 3), (2, −1) They do not.

Mixed REVIEW

2. $y = \frac{1}{3}x + \frac{1}{3}$ 17. $f(x) = \frac{1}{2}x^2 + \frac{3}{2}x - 2$

1. For $f(x) = x^3$, find $f(f(x))$. **(6.2)** x^9
2. Find the inverse of $g(x) = 3x - 1$. **(6.3)**
3. Find $\frac{3}{7}$ of 56. 24
4. $52\frac{1}{2}$ is $\frac{5}{8}$ of what? 84
5. Given $f(x) = 4^{-x}$, find $f(-2)$. **(7.1)** 16
6. Simplify: $(3x^2y^3)^4$. **(7.1)** $81x^8y^{12}$
7. 3×10^{-3} is 6% of what? 0.05
8. Simplify: $\left(\frac{1}{2}\right)^{-4}$. **(7.1)** 16
9. Simplify: $\frac{8x^3y^{-5}z^3}{27xyz^5}$. **(7.1)** $\frac{8x^2}{27y^6z^2}$
10. Simplify: $\frac{(1762.4)^0x^{-5}}{x^{-8}}$. **(7.1)** x^3
11. Write $\left(\frac{1}{3}\right)^4$ using negative exponents. **(7.1)** 3^{-4}
12. Sketch the graph of $y \leq -|x+4| + 3$. **(5.7)** See Additional Answers.
13. Find the vertex of $y = 2x^2 - 8x + 5$. **(5.2)** (2, −3)
14. Find the vertex of $y = |x-2| + 1$. **(6.5)** (2, 1)
15. Sketch the graph of $y = (-x-3)^2$. **(6.5)**
16. Sketch the graph of $y = -|x-2| - 3$. **(6.5)**
15., 16. See Additional Answers.
17. Write a model for the function containing (−1, −3), (0, −2), (1, 0), and (2, 3). **(6.6)**
18. Solve $\begin{bmatrix} 2 & 3 \\ 1 & 0 \end{bmatrix} X = \begin{bmatrix} 1 & 6 \\ -1 & 0 \end{bmatrix}$. **(4.4)** $\begin{bmatrix} -1 & 0 \\ 1 & 2 \end{bmatrix}$

7.6 Graphing Square Root and Cube Root Functions

What you should learn:

Goal 1 How to graph square root and cube root functions

Goal 2 How to use square root and cube root functions to solve real-life problems

Why you should learn it:

You can model many real-life situations with square root functions, such as the Beaufort scale.

Goal 1 Graphing Radical Functions

In this lesson, you will study graphs of functions of the form

$$f(x) = \sqrt{x - a} + b \quad \text{and} \quad f(x) = \sqrt[3]{x - a} + b.$$

Each of these functions has graphs that are translations of the following two graphs.

Square Root: $f(x) = \sqrt{x}$

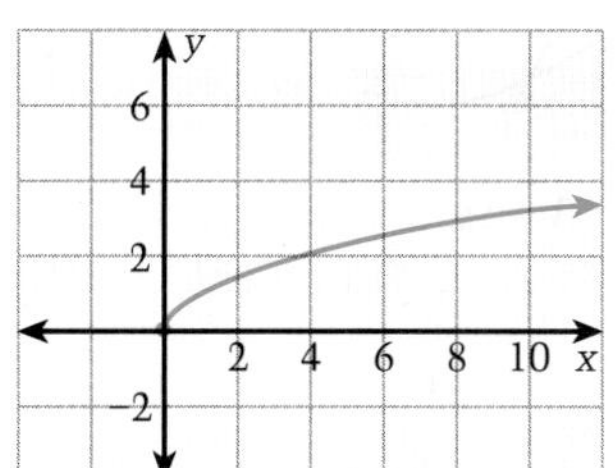

Domain: $x \geq 0$
Range: $y \geq 0$

Cube Root: $f(x) = \sqrt[3]{x}$

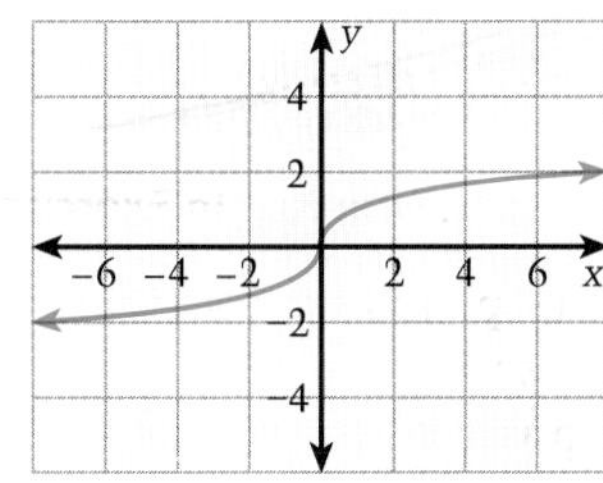

Domain: All real numbers
Range: All real numbers

Example 1 *Graphing a Square Root Function*

Sketch the graphs of $g(x) = \sqrt{x - 3}$ and $h(x) = -\sqrt{x - 3}$ on the same coordinate plane.

Solution To sketch the graph of g, draw the graph of $f(x) = \sqrt{x}$ three units to the right of the origin. To sketch the graph of h, reflect the graph of g in the x-axis. A table of values can help you locate several points on each graph.

x	3	4	5	6	7
$g(x) = \sqrt{x - 3}$	0	1	$\sqrt{2} \approx 1.4$	$\sqrt{3} \approx 1.7$	2
$h(x) = -\sqrt{x - 3}$	0	−1	$-\sqrt{2} \approx -1.4$	$-\sqrt{3} \approx -1.7$	−2

The domain of both g and h is $x \geq 3$. The range of g is $y \geq 0$, and the range of h is $y \leq 0$. ■

Notice that the two graphs together form a parabola. The parabola can be represented by the single equation $y^2 = x - 3$, but this equation does not define y as a function of x. (Why?)

Problem of the Day

99 girls and 1 boy are in a mathematical lecture hall. How many girls must leave so that the percent of girls becomes 98%? 50

ORGANIZER

Warm-Up Exercises

1. Graph each pair of functions on the same coordinate plane.
   a. $f(x) = |x|$ and $g(x) = |x - 2|$
   b. $f(x) = x^2$ and $g(x) = x^2 + 3$

a.

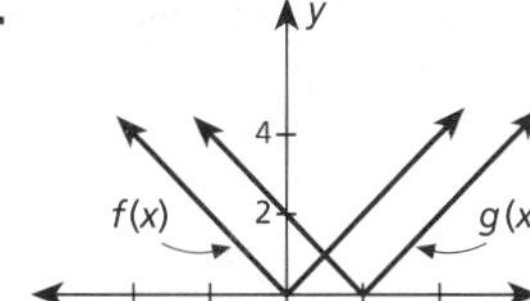

b.

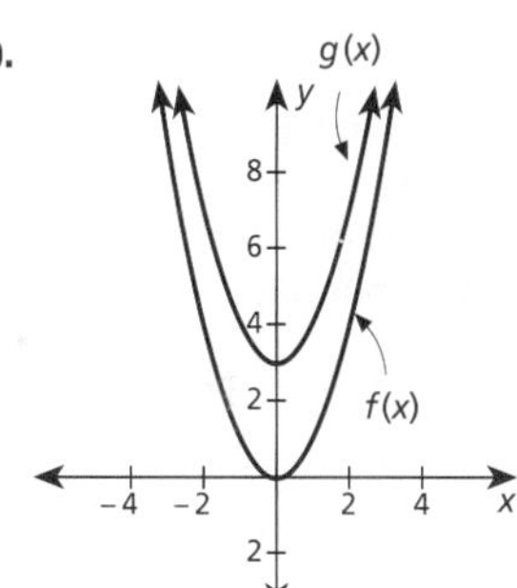

2. Use a calculator to complete the following table for $f(x) = \sqrt{x}$.

x	−3	−1	0	1	2	3	4	5
$f(x)$	undefined		0	1	1.4	1.7	2	2.2

Lesson Resources

Teaching Tools
- Transparencies: 2, 3, 7, 8, 13, 16
- Problem of the Day: 7.6
- Warm-up Exercises: 7.6
- Answer Masters: 7.6

Extra Practice: 7.6

Color Transparency: 43

LESSON Notes

Example 1

The domain of the square root function is the set of all x-values for which the radicand of the function is greater than or equal to zero. In this example, it is all x-values such that $x - 3 \geq 0$, or $x \geq 3$. Notice that the graph of a function is helpful in determining the range of the function.

Example 2

The domain of a cube root function is all real numbers, because the radicand of a cube root function may be negative.

Examples 3–4

Emphasize that cube root and square root functions, in particular, and algebraic functions, in general, can be used to describe biological and natural phenomena.

Extra Examples

Here are additional examples similar to **Examples 1–2.**

1. Sketch each set of functions on the same coordinate plane.
   - **a.** $f(x) = \sqrt{x}$, $g(x) = \sqrt{x+3}$, $h(x) = \sqrt{x-6} + 5$
   - **b.** $f(x) = \sqrt[3]{x}$, $g(x) = \sqrt[3]{x-4}$, $h(x) = \sqrt[3]{x+1} - 5$

   a. Shift $f(x)$ 3 units left to get $g(x)$; shift $f(x)$ 6 units right and 5 units up to get $h(x)$.

   b. Shift $f(x)$ 4 units right to get $g(x)$; shift $f(x)$ 1 unit left and 5 units down to get $h(x)$.

Example 2 *Sketching the Graph of a Cube Root Function*

Sketch the graph of $g(x) = \sqrt[3]{x+2} + 1$.

Solution To sketch the graph of g, draw the graph of $f(x) = \sqrt[3]{x}$ shifted two units to the left and one unit up. A table of values can help you locate several points on the graph.

x	-4	-3	-2	-1	0	1
$g(x) = \sqrt[3]{x+2} + 1$	-0.26	0	1	2	2.26	2.44

Each of the domain and range of g is all real numbers. ■

Goal 2 Using Square and Cube Root Functions

Real Life **Zoology**

Example 3 *Graphing a Cube Root Function*

The shoulder height, h (in feet), of a male African elephant can be modeled by the equation

$$h = \sqrt[3]{13t} + 3, \quad 0 \leq t$$

where t is the age of the elephant in years. Sketch the graph of this equation.

Solution To sketch the graph, construct a table of values, graph the points, and connect the points with a smooth curve.

t	0	1	3	10	15	40
$h = \sqrt[3]{13t} + 3$	3	5.4	6.4	8	8.8	11

The graph is shown below. ■

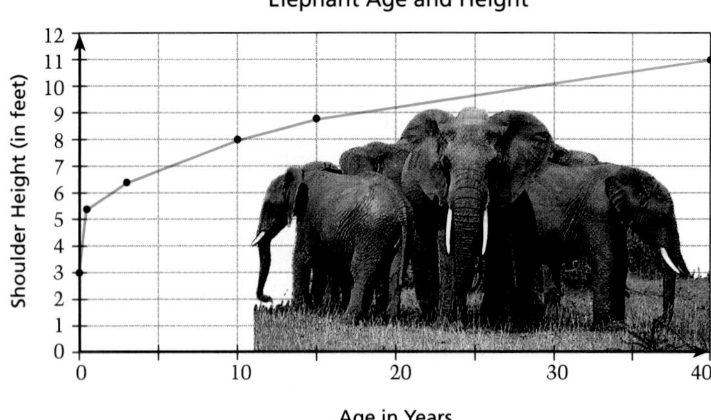

Elephants grow slowly and are among the longest lived animals. A newborn male African elephant measures about 3 feet high at the shoulders. A fully-grown male (about 40 years old) measures about 11 feet high at the shoulders.

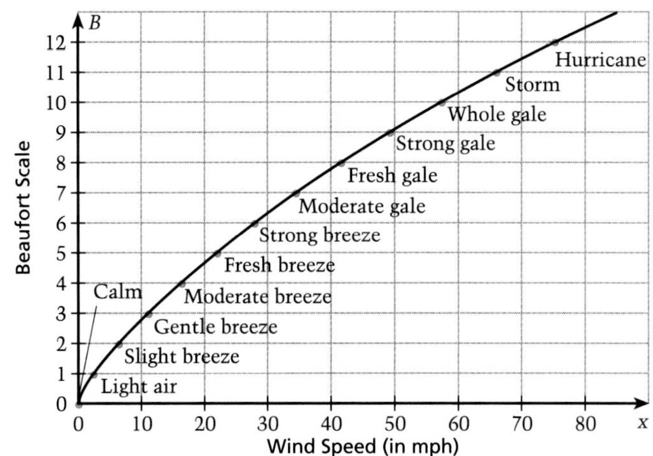

Beaufort Scale and Wind Effects

0 Smoke rises vertically.
1 Smoke shows wind direction.
2 Wind felt on face.
3 Leaves move, flags extend.
4 Paper, small branches move.
5 Small trees sway, flags ripple.
6 Large branches sway, flags beat.
7 Large trees sway, walking is difficult.
8 Twigs break, walking is hindered.
9 Slight roof damage
10 Severe damage, trees uprooted
11 Widespread damage
12 Devastation

The* Beaufort scale *was devised by Francis Beaufort in 1805 to measure wind speeds. The scale is numbered from 0 to 12, and represents wind speeds in the open, 33 feet above ground.

Example 4 *Graphing a Square Root Function*

The Beaufort scale, B, can be modeled by the function

$$B = 1.9\sqrt{x + 8} - 5.4$$

where x is the speed of the wind in miles per hour. Sketch the graph of this function.

Solution To sketch the graph, construct a table of values, graph the points, and connect the points with a smooth curve.

Speed, x	0	3.30	7.15	11.55	16.50	22.00	28.05
Scale, B	0	1	2	3	4	5	6

Speed, x	34.65	41.80	49.50	57.75	66.55	75.90
Scale, B	7	8	9	10	11	12

The graph is shown above. ■

Communicating about ALGEBRA

SHARING IDEAS about the Lesson See Additional Answers.

Sketch the graphs of the functions. Explain the procedure you used.

A. $f(x) = \sqrt{x - 5} - 2$ **B.** $f(x) = \sqrt[3]{x + 2} + 3$

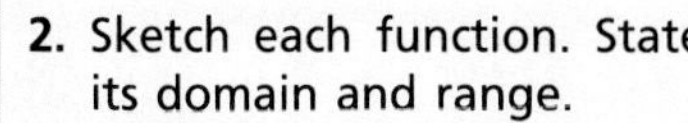

2. Sketch each function. State its domain and range.
   a. $f(x) = \sqrt[3]{8x - 3}$
   b. $g(x) = \sqrt{12 - 4x} - 5$

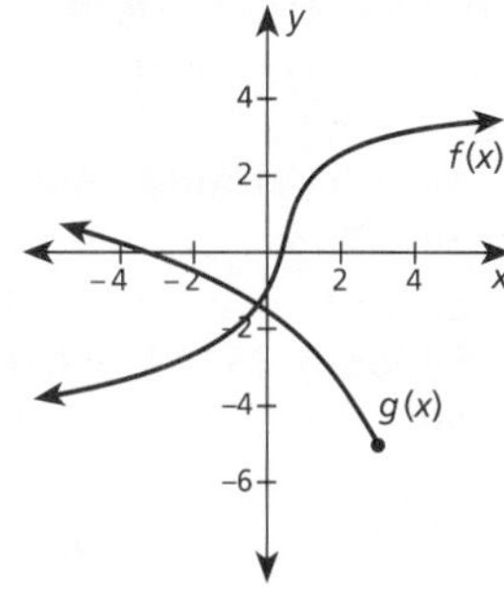

a. domain and range: all real numbers
b. domain: all real numbers $x \le 3$, range: all real numbers $y \ge -5$

Check Understanding

1. What is a translation of a graph?
   a horizontal or vertical shift
2. What is the effect of the constant "a" on $f(x) = \sqrt{x}$ if $g(x) = \sqrt{x - a}$?
   a horizontal translation $|a|$ units to the left or right
3. What is the effect of the constant "b" on $f(x) = \sqrt{x}$ if $g(x) = \sqrt{x} + b$?
   a vertical translation $|b|$ units up or down

***Communicating* about ALGEBRA**

Find the domain and range of each function. Does the sketch of each function assist in obtaining this information?

A. domain: $x \ge 5$, range: $y \ge -2$;
B. domain and range: all real numbers. Yes

EXERCISE Notes

ASSIGNMENT GUIDE

Basic/Average: Ex. 1–4, 9, 11–16, 21–31 odd, 35–42, 51–56

Above Average: Ex. 1–4, 9, 11–16, 19–31 odd, 35–42, 51–56, 59–60

Advanced: Ex. 1–4, 9, 11–16, 19–33 odd, 35–40, 43, 44, 51–60

Selected Answers
Exercises 1–4, 9–55 odd

✪ More Difficult Exercises
Exercises 31, 34, 40, 57–60

Guided Practice

▶ **Ex. 1–2** Emphasize the difference in the domain and range of the two functions.

▶ **Ex. 3–4** Review quick-graph techniques. Translate the original graph two units to the left and then reflect the graph in the x-axis in order to visualize the domain and range.

Answers

1. b, a; only the cube root function has negative roots.
4. 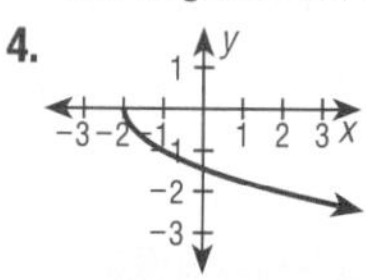
5. g is a reflection of f in the x-axis translated to the left 16 units.
6. g is a translation of f up 15 units.
7. g is a reflection of f in the x-axis translated down 9 units.
8. g is a reflection of f in the x-axis translated to the left 2 units.
9. g is a translation of f to the right 10 units and down 1 unit.
10. g is a translation of f to the left 7 units and down 4 units.

EXERCISES

Guided Practice

CRITICAL THINKING about the Lesson

1. Which of the graphs at the right is the graph of a square root function and which is the graph of a cube root function? Explain. See margin.

a. 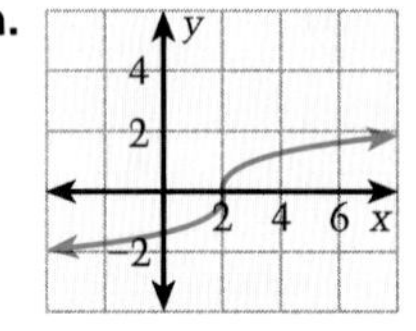

b.

2. Compare the graph of $g(x) = \sqrt[3]{x - 10}$ to the graph of $f(x) = \sqrt[3]{x}$. g is a translation of f to the right 10 units.
3. Copy and complete the table. What are the domain and range of the function? $x \geq -2$, $y \leq 0$

x	-2	-1	0	1	2
$y = -\sqrt{x+2}$	?	?	?	?	?

0 $\quad -1$ $\quad -\sqrt{2}$ $\quad -\sqrt{3}$ $\quad -2$

4. Sketch the graph of the function in Exercise 3. See margin.

Independent Practice

In Exercises 5–10, compare the graph of g to the graph of f. See margin.

5. $f(x) = \sqrt{x}$ and $g(x) = -\sqrt{x + 16}$
6. $f(x) = \sqrt{x}$ and $g(x) = \sqrt{x} + 15$
7. $f(x) = \sqrt[3]{x}$ and $g(x) = -\sqrt[3]{x} - 9$
8. $f(x) = \sqrt[3]{x}$ and $g(x) = -\sqrt[3]{x + 2}$
9. $f(x) = \sqrt{x}$ and $g(x) = \sqrt{x - 10} - 1$
10. $f(x) = \sqrt[3]{x}$ and $g(x) = \sqrt[3]{x + 7} - 4$

In Exercises 11–16, match the function with its graph.

11. $f(x) = \sqrt[3]{x} - 1$ b
12. $f(x) = \sqrt[3]{x - 3} - 1$ f
13. $f(x) = \sqrt{x + 3} + 1$ a
14. $f(x) = -\sqrt[3]{x} - 1$ c
15. $f(x) = \sqrt{x} - 1$ e
16. $f(x) = \sqrt{x - 3}$ d

a. 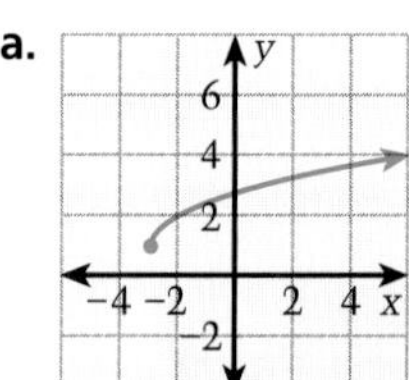

b. 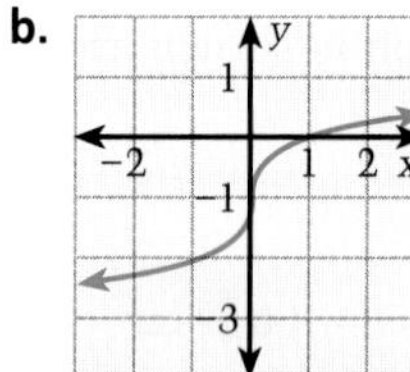

c. 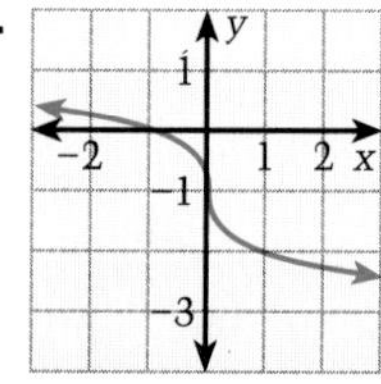

d. 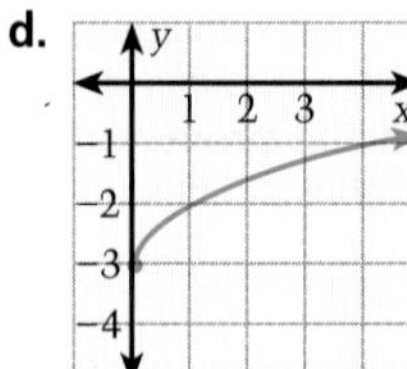

e. 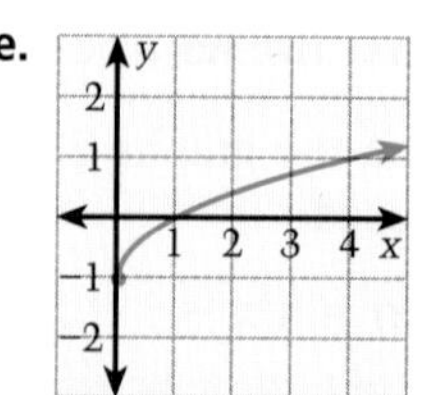

f. 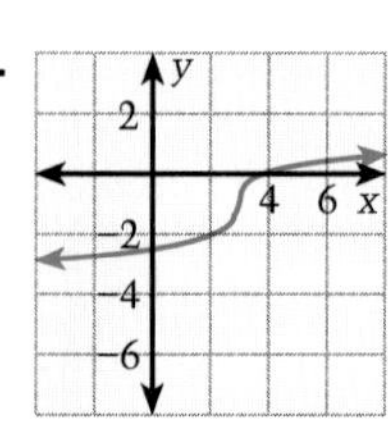

In Exercises 17–34, sketch the graph of the function. State the domain and range of the function. See Additional Answers.

17. $f(x) = \sqrt{x-1}$
18. $g(x) = \sqrt{x+2}$
19. $f(x) = \sqrt[3]{x-5}$
20. $g(x) = (x+1)^{1/3}$
21. $g(x) = \sqrt[3]{x} + 1$
22. $h(x) = \sqrt{x} + 4$
23. $f(x) = (x-1)^{1/2} + 7$
24. $f(x) = \sqrt[3]{x-2} + 4$
25. $h(x) = -\sqrt[3]{x} + 4$
26. $f(x) = \sqrt[3]{x-6} + 2$
27. $f(x) = -\sqrt{x+5}$
28. $g(x) = \sqrt{x+2} + 5$
29. $f(x) = -x^{1/2} - 3$
30. $f(x) = -\sqrt[3]{x-1}$
31. ✪ $h(x) = (x-10)^{1/3} + 2$
32. $f(x) = -\sqrt{x-1} + 1$
33. $g(x) = 2\sqrt{x+4}$
34. ✪ $f(x) = 3(x-2)^{1/3}$

Congressional Aides **In Exercises 35–38, use the following information.**

Each member of the United States Congress has a staff of congressional aides. For 1930 through 1990, the number of aides, A, assigned to members of the House of Representatives can be modeled by the function

$$A = 1220\sqrt[3]{t-42} + 4900.$$

where $t = 0$ represents 1930.

35. In what year were about 4900 aides assigned to members of the House of Representatives? 1972
36. In what year were about 7340 aides assigned to members of the House of Representatives? 1980
37. Sketch the graph of the function $A = 1220\sqrt[3]{t-42} + 4900.$ See Additional Answers
38. The House of Representatives has 435 members. Write a function that gives the average number of congressional aides per representative. Sketch the graph of this function. See Additional Answers.

Members of Congress appoint young people (between 14 and 18 years old) to work as congressional pages. Pages attend a special high school from 6:30 A.M. to 9:45 A.M.

39. *Geometry* Write a function that gives the radius, r, of a circle in terms of the circle's area, A. Sketch the graph of this function.

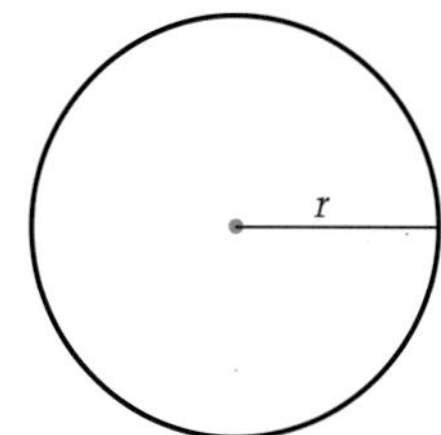

$r = \sqrt{\frac{A}{\pi}}$ See Additional Answers.

40. ✪ *Geometry* Write a function that gives the radius, r, of a sphere in terms of the sphere's volume, V. Sketch the graph of this function.

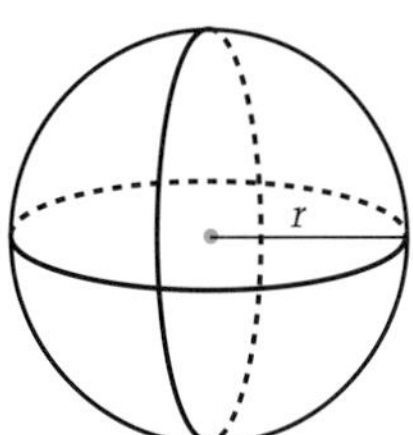

$r = \sqrt[3]{\frac{3V}{4\pi}}$ See Additional Answers.

Independent Practice

▶ **Ex. 5–10** Ask students to write a sentence description similar to the one modeled in Example 1.

▶ **Ex. 11–16** Assign these exercises as a group. Ask students to list their choices and justify them.

▶ **Ex. 17–34** **TECHNOLOGY** Students should complete these exercises using quick-graph techniques. However, you might wish to have some students use graphics calculators to graph the equations as a check.

▶ **Ex. 35–38** Assign these exercises as a group.

▶ **Ex. 40** Remind students that even though a cube-root function has all real numbers as its domain and range, *r* and *V* in this function are restricted to positive real numbers.

▶ **Ex. 41–44** Assign these exercises as a group. These models relate square-root and cube-root functions to nature.

▶ **Ex. 51–56** These exercises should be assigned as a review of the quick-graph techniques for each type of function.

▶ **Ex. 59–60** TECHNOLOGY Display the graphs by using an overhead graphics calculator or overhead computer and ask students to "guess" the equation.

Enrichment Activities

These activities require students to go beyond lesson goals.

If a relation is defined by an equation in x and y, the inverse of that relation can be defined by interchanging x and y in the original relation. So, the relations $y = x^2$ and $x = y^2$ are inverse relations.

1. Graph both relations on the same coordinate grid.

Answer

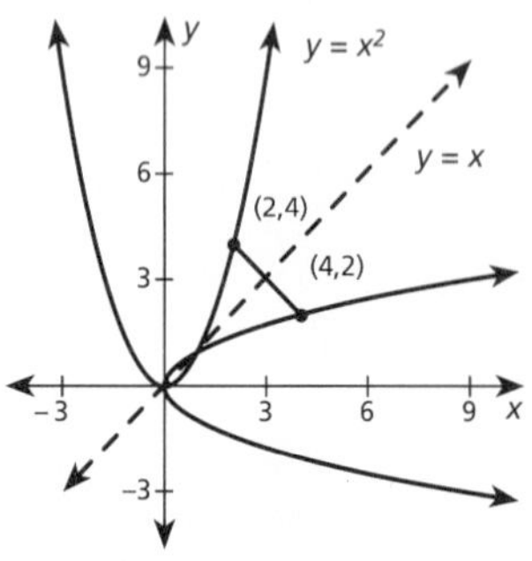

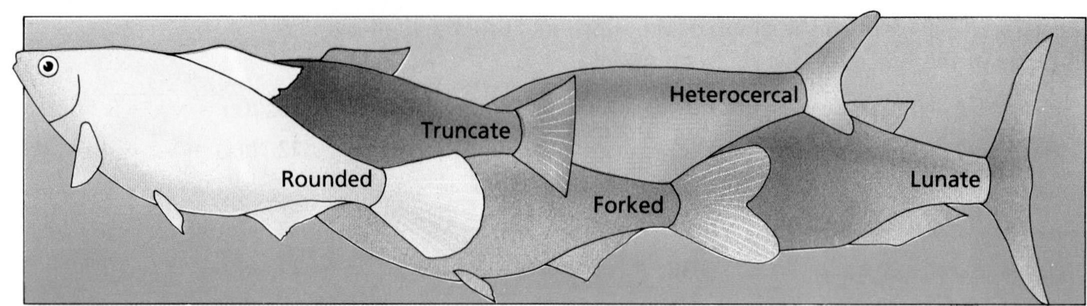

The aspect ratio of a fish's tail fin depends on the shape of the fin. (The aspect ratio is defined to be the ratio of the square of the fin height to the fin area.) The truncate tail fin, like that of salmon or bass, has an aspect ratio of 3. The forked tail fin, like that of herring or yellowtail, has an aspect ratio of 5.

Fish Tail Fins **In Exercises 41 and 42, use the following information.**

The tail fin height, h, of a fish is related to the tail fin area, A, and the aspect ratio, r, by the model

$h = \sqrt{rA}$. **41.–42.** See Additional Answers for graphs.

41. Write a function that gives the tail fin height of a salmon in terms of the tail fin area. Then sketch the graph of the function. $h = \sqrt{3A}$

42. Write a function that gives the tail fin height of a herring in terms of the tail fin area. Then sketch the graph of the function. $h = \sqrt{5A}$

Storms at Sea **In Exercises 43 and 44, use the following information.**

The *fetch*, f (in nautical miles), of the wind at sea is the distance over which the wind is blowing. The minimum fetch required to create a "fully developed" storm is related to the wind speed, S (in knots), by the model

$S = 4.6\sqrt[3]{f} + 16.$

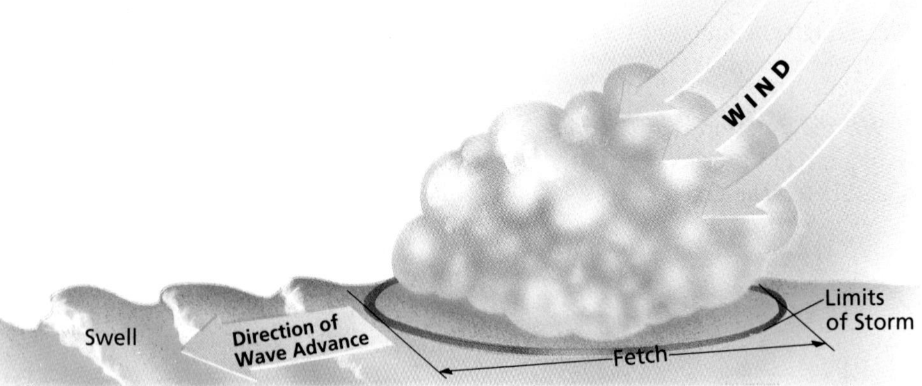

43. Sketch the graph of this function. See Additional Answers.

44. If the wind is blowing at 25 knots, what is the minimum fetch required to create a fully developed storm? ≈144.5 nautical miles

Integrated Review

In Exercises 45–50, evaluate the expression.

45. $\sqrt[3]{x}$ when $x = 125$ 5 **46.** $\sqrt{x - 3}$ when $x = 52$ 7 **47.** $\sqrt{x^2 + x - 29}$ when $x = 10$ 9

48. $\sqrt[3]{x^2 + 28}$ when $x = -6$ 4 **49.** $\sqrt[4]{x - 4} - 6$ when $x = 20$ −4 **50.** $\sqrt{3x - 15} + \sqrt[3]{x}$ when $x = 8$ 5

In Exercises 51–56, match the function with its graph.

51. $f(x) = |x + 4|$ f **52.** $f(x) = |x| - 4$ b **53.** $f(x) = x^2 - 4$ e

54. $f(x) = \sqrt[3]{x + 4}$ c **55.** $f(x) = \sqrt{x} - 4$ d **56.** $f(x) = (x - 4)^2$ a

a. 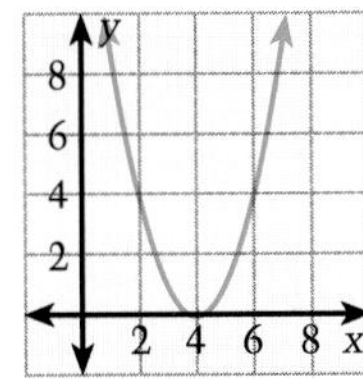

b. 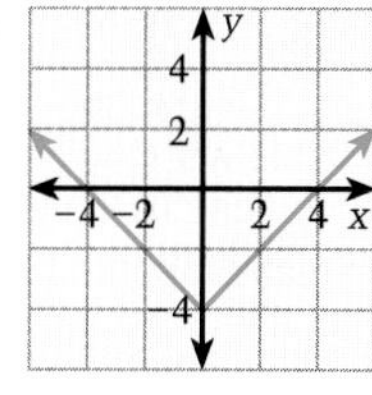

c. 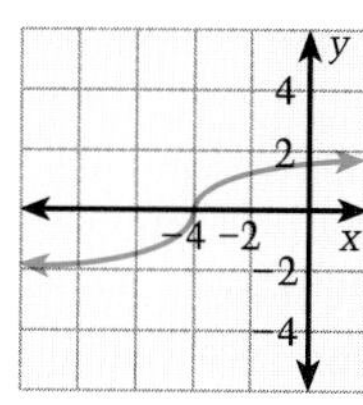

d. 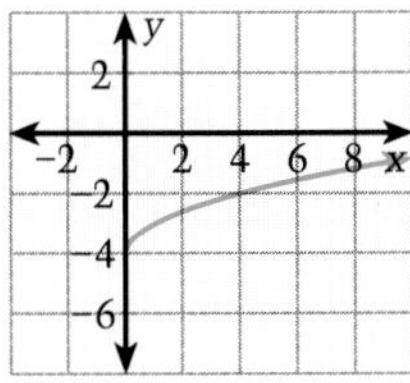

e. 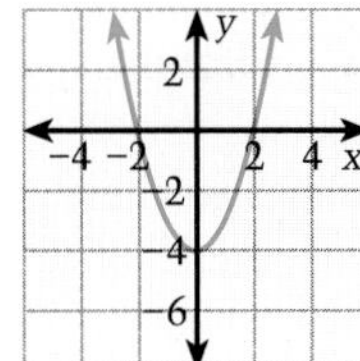

f. 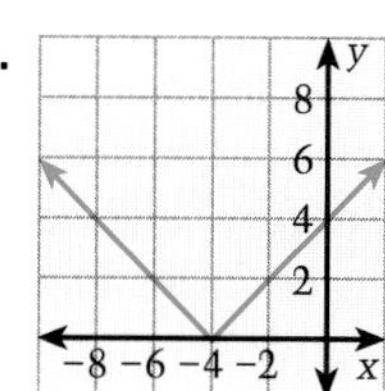

Exploration and Extension

Reflections in the y-Axis **In Exercises 57 and 58, use the following information.** See Additional Answers.

The graph of $y = f(-x)$ is a reflection of the graph of f in the y-axis. Use this result to sketch the graph of the function. State the domain and range of the function.

57. $g(x) = \sqrt{-x} + 1$ **58.** $h(x) = \sqrt[3]{-x} - 1$

In Exercises 59 and 60, write an equation for the function whose graph is given.

59. 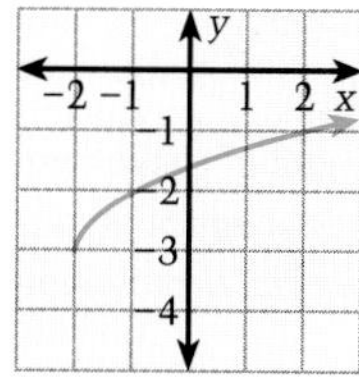 $f(x) = \sqrt{x + 2} - 3$

60. 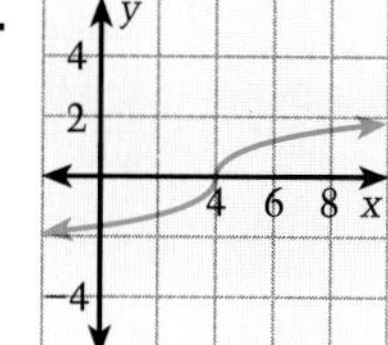 $f(x) = \sqrt[3]{x - 4}$

2. Draw the line $y = x$ on the same coordinate grid as your graphs in the previous activity. Fold your grid on the line $y = x$. What do you notice? The graphs of $y = x^2$ and $x = y^2$ are reflections of each other in the line $y = x$.

3. Graph the square-root function $f(x) = \sqrt{x - 2} + 4$. Graph its *inverse function* by using the line $y = x$. Be sure to consider the domains and ranges of the functions.

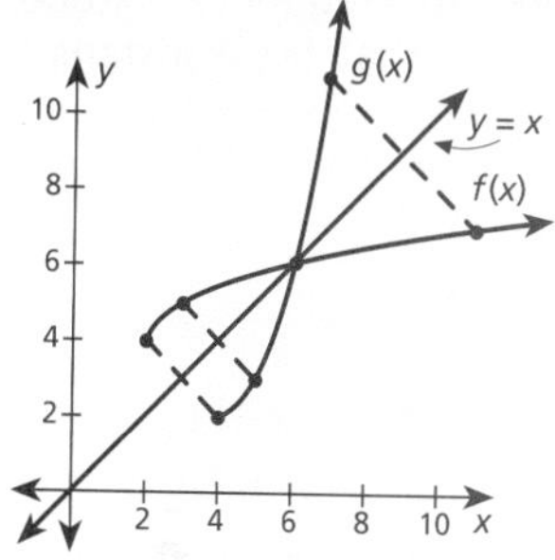

The square-root function has its domain restricted to $x \geq 2$. Its range is restricted to the principal square root. So, the graph of the inverse function is restricted to only one branch of the parabola, that is, the branch defined when $x \geq 4$.

4. What kind of a function is the inverse of a square-root function? a quadratic function

5. Write a defining equation for $g(x)$. Remember, the domain of x is restricted. $g(x) = (x - 4)^2 + 2,\ x \geq 4.$

EXTEND *the Activity*

TECHNOLOGY If graphics calculators are available, Lesson 7.6 can be revisited to demonstrate the relationship of the basic square root and cube root functions to their translations.

By graphing several functions in sequence, the effect of a given constant on the graph of the function can be observed. For example, by graphing the following functions in sequence, students can see that the constant "b" (in $f(x) = \sqrt{x - a} + b$) shifts the basic graph in either the "up" or "down" direction.

A. **1.** $f(x) = \sqrt{x}$

2. $g(x) = \sqrt{x} + 2$

3. $h(x) = \sqrt{x} + 4$

4. $j(x) = \sqrt{x} + 10$

B. **1.** $f(x) = \sqrt{x}$

2. $g(x) = \sqrt{x} - 1$

3. $h(x) = \sqrt{x} - 3$

4. $j(x) = \sqrt{x} - 8$

USING A GRAPHING CALCULATOR

A graphing calculator can be used to investigate the graphs of functions of the form

$$y = x^n.$$

For instance, to produce the screen shown at the right, a graphing calculator was used to sketch the first-quadrant portions of the graphs of the following functions.

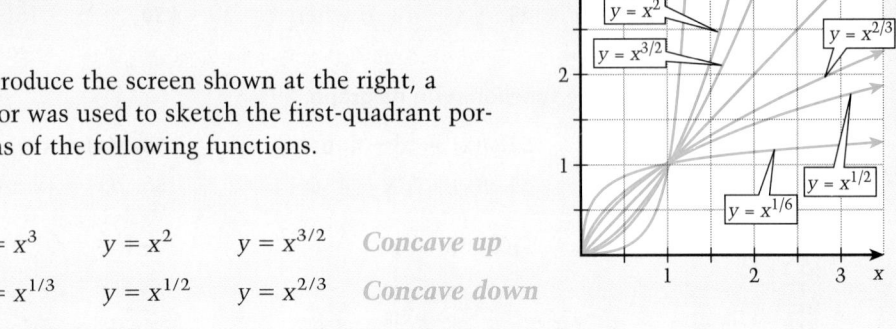

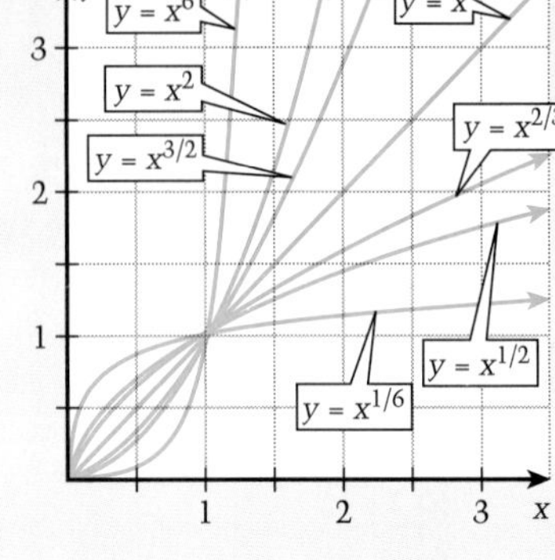

$y = x^1$

$y = x^6$	$y = x^3$	$y = x^2$	$y = x^{3/2}$	*Concave up*
$y = x^{1/6}$	$y = x^{1/3}$	$y = x^{1/2}$	$y = x^{2/3}$	*Concave down*

Notice that if $n = 1$, the graph of $y = x^n$ is a line that makes an angle of 45° with the x-axis. If $n > 1$, the graph is **concave up.** If $0 < n < 1$, the graph is **concave down.**

Example 1: Sketching the Graph of a Model

The chest circumferences, C (in inches), of five different species of adult primates (tamarins, squirrel monkeys, vervet monkeys, macaques, and baboons) were compared to their weights, W (in pounds). From 32 sets of measurements, the relationship between C and W was modeled by

$$C = 5W^{37/100}.$$

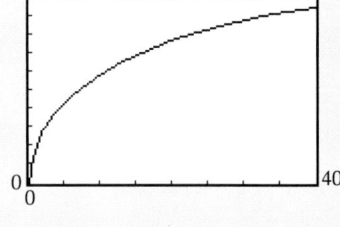

Sketch the graph of this model for weights between 0 pounds and 40 pounds. ***(Source:* Scientific American*)***

Solution Using a graphing calculator, you can obtain the graph shown at the right. ■

Macaques are a type of rhesus monkey. The Golden Macaques, as shown, are a rare coloring for these monkeys.

Answer

5.

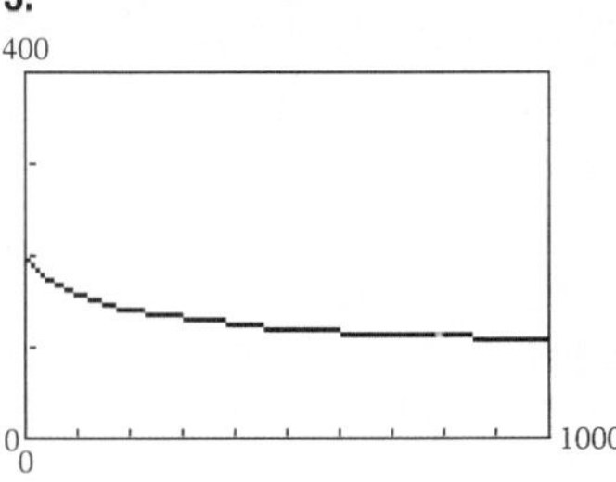

Example 2: Finding a Mathematical Model

Most graphing calculators are capable of fitting power models of the form $y = ax^b$ to data. The table below lists the length, x (in inches), and diameter, y (in inches), of several sizes of common nails. Find a power model for these data.

Length, x	1.00	2.00	3.00	4.00	5.00	6.00
Diameter, y	0.070	0.111	0.146	0.176	0.204	0.231

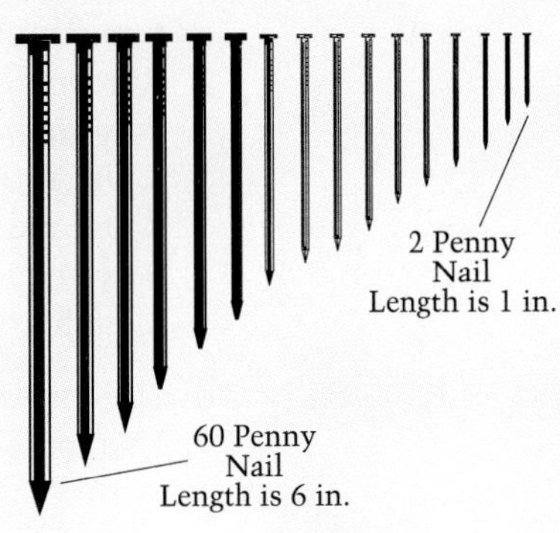

Solution Begin by entering the data into a graphing calculator. Keystrokes for doing this on a *TI-82*, Casio *fx-9700GE*, and Sharp *EL-9300C* are listed on page 862. After running the power regression program, you should obtain

$$a \approx 0.07 \quad \text{and} \quad b \approx 0.67.$$

Because $\frac{2}{3} \approx 0.67$, you can write the model as follows.

$$y = 0.07x^{2/3} \quad \blacksquare$$

Exercises

In Exercises 1–4, sketch the first-quadrant portion of the graph of each function. State whether each graph is concave up or concave down.

1.–4. Concave up, concave down. See Additional Answers.

1. $y = x^5, y = x^{1/5}$ **2.** $y = x^4, y = x^{1/4}$ **3.** $y = x^{4/3}$ $y = x^{3/4}$ **4.** $y = x^{6/5}, y = x^{5/6}$

5. *Galloping Speed of Animals* Four-legged animals run with two types of motion: trotting and galloping. An animal that is trotting has at least one foot on the ground at all times. When an animal is galloping, sometimes all four feet are off the ground. The number of strides per minute at which an animal breaks from a trot to a gallop depends on the animal's weight. Use the table to find an equation of the form $y = ax^b$ that relates the lowest galloping speed, y (in strides per minute), with the weight of the animal, x (in pounds). Use a graphing calculator to graph the model.

Weight, x	25	35	50	75	500	1000
Galloping Speed, y	191.5	182.7	173.8	164.2	125.9	114.2

$y = 300.7x^{-0.14}$ See margin.

ORGANIZER

TIME SCHEDULE: All levels, two days

The Chapter Summary helps students organize the main ideas of the chapter. In this chapter, properties of exponents were extended to rational numbers, students learned how to solve radical equations, and square root and cube root functions were introduced. An important application of exponents, compound interest, was also presented.

Work with students to review the skills, strategies, and concepts of the chapter. The first day's homework assignment can be used as the basis for the second day of review.

COOPERATIVE LEARNING
Encourage students to study together. Emphasize the importance of "teaching" a classmate how to perform a skill or how to recall a strategy. When students work together, everyone wins. The students receiving help get additional instruction and the students giving help gain a deeper understanding of the skills and concepts involved.

Chapter SUMMARY

Review the chapter by asking students, "What did you learn? Why did you learn it? How does it fit with the other algebra skills and concepts you have learned?"

7

Chapter Summary

What did you learn?

Skills

1. Use properties of exponents
   - to perform operations with powers. **(7.1, 7.4)**
   - to simplify expressions. **(7.1, 7.4)**
2. Recognize exponential growth and decay models. **(7.2)**
3. Find the nth roots of a number. **(7.3)**
   - Use rational exponent notation. **(7.3)**
   - Use radical notation. **(7.3)**
4. Solve equations that contain radicals or roots. **(7.5)**
5. Sketch graphs of square root functions and cube root functions. **(7.6)**

Strategies

6. Use exponential growth and decay in real-life problems. **(7.2)**
7. Use roots and radicals in real-life problems. **(7.3–7.5)**

Exploring Data

8. Solve compound interest problems. **(7.2)**

Why did you learn it?

Many models of real-life situations involve exponents and roots. For instance, the length of the edge of a cube is the cube root of the volume of the cube.

In this chapter, you studied whole number exponents, integer exponents, and rational exponents. The ability to use all these exponents is an important skill—one that you will need as long as you study and use mathematics.

How does it fit into the bigger picture of algebra?

The square root of 2 can be written with the familiar radical notation as $\sqrt{2}$, or it can be written as $2^{1/2}$. The second notation uses a *rational exponent.* The advantage of the rational exponent notation is that the standard properties of exponents can be used to simplify radicals or perform operations involving radicals.

The chapter began by reviewing properties of integer exponents. In Lessons 7.3 and 7.4, these properties were extended to rational exponents. In the last two lessons, you learned how to solve *radical equations* and sketch the graphs of *radical functions.*

Chapter REVIEW

In Exercises 1–16, simplify the expression. **(7.1)**

1. $(5)^2(5)^6$ 390,625 **2.** $(3)^5(3)^7$ 531,441 **3.** $x^{-5} \cdot x^4$ $\frac{1}{x}$ **4.** $y^3 \cdot y^{-6}$ $\frac{1}{y^3}$

5. $(y^4)^{-3}$ $\frac{1}{y^{12}}$ **6.** $(10^{-4})^6$ $\frac{1}{10^{24}}$ **7.** $(6x)^2$ $36x^2$ **8.** $(14x)^3$ $2744x^3$

9. $3x^0y$ $3y$ **10.** $\frac{x^{-4}}{x^{-6}}$ x^2 **11.** $\frac{x^5}{x^7}$ $\frac{1}{x^2}$ **12.** $4x^4y^0$ $4x^4$

13. $\frac{y^6}{6x^4} \cdot \frac{18x}{x^2y}$ $\frac{3y^5}{x^5}$ **14.** $\frac{9x^2y}{y^3} \cdot \frac{xy^3}{3x}$ $3x^2y$ **15.** $\frac{4y^2}{x^2} \cdot \frac{x^3}{2y^4}$ $\frac{2x}{y^2}$ **16.** $\frac{5x^2}{y^{-2}} \cdot \frac{1}{25xy}$ $\frac{xy}{5}$

17. *Balance in an Account* A principal of \$500 is deposited in an account that pays 7% interest, compounded annually. Find the balance after 5 years. **(7.2)** \$701.28

18. *Balance in an Account* A principal of \$800 is deposited in an account that pays 6% interest, compounded quarterly. Find the balance after 8 years. **(7.2)** \$1288.26

19. *How Much to Deposit?* How much must you deposit in an account that pays 8% interest, compounded monthly, to have a balance of \$1000 after 10 years? **(7.2)** \$450.53

20. *How Much to Deposit?* How much must you deposit in an account that pays 6% interest, compounded quarterly, to have a balance of \$2000 after 20 years? **(7.2)** \$607.79

In Exercises 21–24, complete the table. **(7.2)**

21.

x	−2	−1	0	1	2
$y = 2\left(\frac{1}{2}\right)^x$	?	?	?	?	?

22.

x	−2	−1	0	1	2
$y = \frac{1}{2}(4)^x$	?	?	?	?	?

23.

x	−2	−1	0	1	2
$y = \frac{1}{3}\left(\frac{3}{2}\right)^x$	?	?	?	?	?

24.

x	−2	−1	0	1	2
$y = -2\left(\frac{1}{3}\right)^x$	?	?	?	?	?

21. 8 4 2 1 $\frac{1}{2}$ **22.** $\frac{1}{32}$ $\frac{1}{8}$ $\frac{1}{2}$ 2 8 **23.** $\frac{4}{27}$ $\frac{2}{9}$ $\frac{1}{3}$ $\frac{1}{2}$ $\frac{3}{4}$ **24.** −18 −6 −2 $-\frac{2}{3}$ $-\frac{2}{9}$

In Exercises 25–32, evaluate the expression without using a calculator. **(7.4)**

25. $\sqrt[4]{625}$ 5 **26.** $(243)^{1/5}$ 3 **27.** $(729)^{1/6}$ 3 **28.** $\sqrt[3]{64}$ 4

29. $(512)^{-1/3}$ $\frac{1}{8}$ **30.** $(27)^{-1/3}$ $\frac{1}{3}$ **31.** $\sqrt[4]{16}$ 2 **32.** $\sqrt[5]{-32}$ −2

In Exercises 33–44, simplify the expression. **(7.4)**

33. $(5)^{1/4}(5)^{-9/4}$ $\frac{1}{25}$ **34.** $((5)^{1/3})^{3/4}$ $5^{1/4}$ **35.** $\sqrt[3]{16} \cdot \sqrt[3]{\frac{1}{4}}$ $\sqrt[3]{4}$ **36.** $(81x)^{1/4}$ $3x^{1/4}$

37. $\left(\frac{200}{8}\right)^{1/3}$ $25^{1/3}$ **38.** $\frac{2^{1/9}}{2^{5/6}}$ $\frac{1}{2^{13/18}}$ **39.** $(6^{1/2} \cdot 2^{1/3})^6$ 864 **40.** $\sqrt[6]{4x^6}$ $|x|\sqrt[6]{4}$

41. $3\sqrt[3]{6} - 9\sqrt[3]{6}$ $-6\sqrt[3]{6}$ **42.** $5\sqrt[3]{17} + 4\sqrt[3]{17}$ $9\sqrt[3]{17}$ **43.** $\sqrt[4]{7} + 2\sqrt[4]{1792}$ $9\sqrt[4]{7}$ **44.** $4\sqrt[5]{3} + \sqrt[5]{729}$ $7\sqrt[5]{3}$

Chapter REVIEW

ASSIGNMENT GUIDE

Basic/Average: Ex. 1–57 odd, 59–64, 66–81 multiples of 3, 86–87

Above Average: Ex. 1–57 odd, 59–64, 66–81 multiples of 3, 87–88

Advanced: Ex. 1–57 odd, 59–64, 66–75 multiples of 3, 78–88 even

✪ More Difficult Exercises
Exercises 80–81, 86–88

▶ **Ex. 1–88** The Chapter Review exercises include a reference to the lesson in which a skill or concept was presented. Students may use these references as a study aid if they are uncertain how to do a given exercise.

Answers

65.

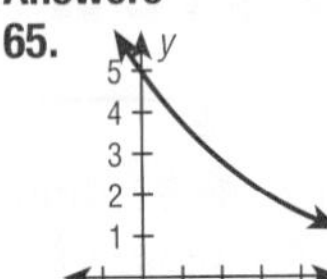

66.

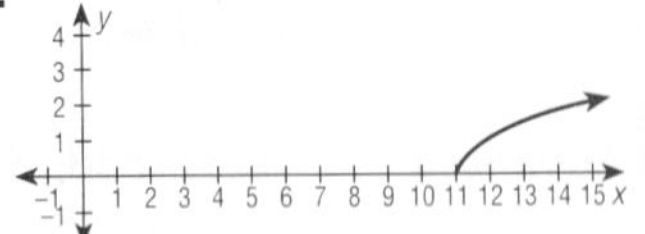

67.

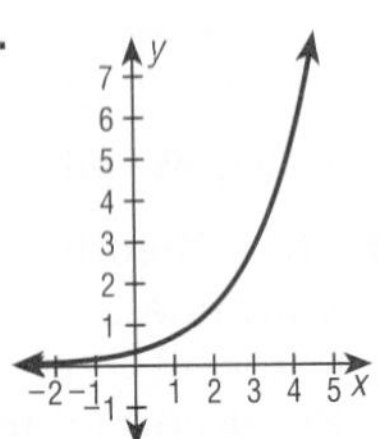

68.

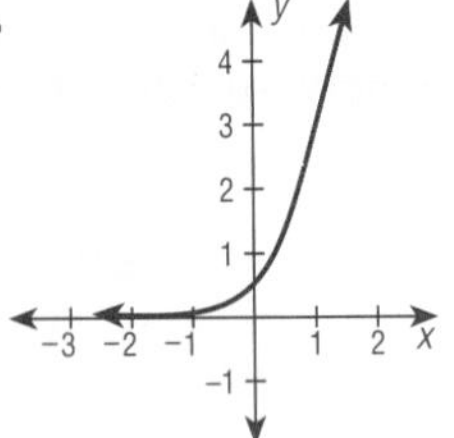

69.

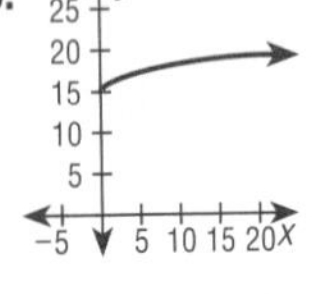

70.

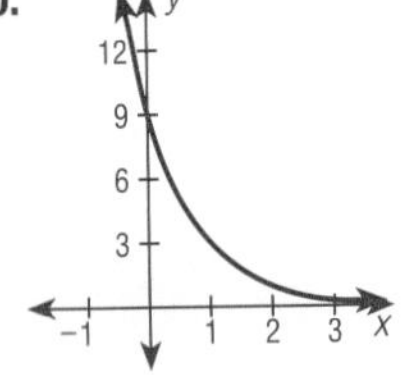

In Exercises 45–52, solve the equation if possible. (7.5)

45. $\sqrt{x} + 3 = 4$ 1

46. $\sqrt{x} - 9 = 1$ 100

47. $3(x - 4)^{1/2} + 5 = 11$ 8

48. $2\sqrt{x + 5} + 5 = 7$ −4

49. $-4\sqrt{2x - 1} - 10 = 6$ No solution

50. $-2(3x + 4)^{1/2} - 3 = 21$ No solution

51. $\sqrt[3]{x - 4} + 6 = 4$ −4

52. $2(x + 3)^{1/3} - 5 = 1$ 24

Geometric Mean **In Exercises 53 and 54, find *x*. (7.5)**

53. The geometric mean of 8 and x is 16. 32

54. The geometric mean of 14 and x is 28. 56

In Exercises 55–58, find all values of *x* for which the distance between the given points is *d*. (7.5)

55. $(3, x), (-2, 4);\ d = 6$ $4 \pm \sqrt{11}$

56. $(10, x), (2, 4);\ d = 10$ −2, 10

57. $(6, 10), (1, x);\ d = 13$ −2, 22

58. $(8, 12), (x, 4);\ d = 17$ −7, 23

In Exercises 59–64, match the function with its graph. (7.2, 7.6)

59. $f(x) = 3\left(\frac{1}{2}\right)^x$ c

60. $f(x) = \sqrt{x - 7}$ f

61. $f(x) = \sqrt[3]{x + 7}$ e

62. $f(x) = \frac{1}{2}(2)^x$ a

63. $f(x) = \sqrt{x} + 2$ b

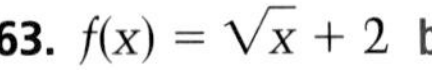

64. $f(x) = \sqrt[3]{x} - 2$ d

a.

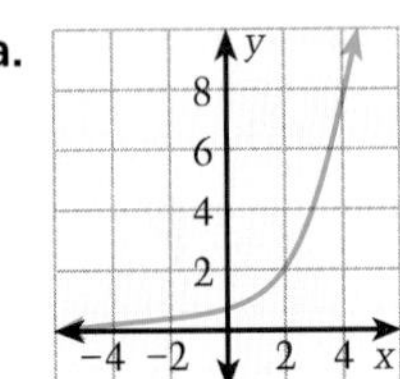

b.

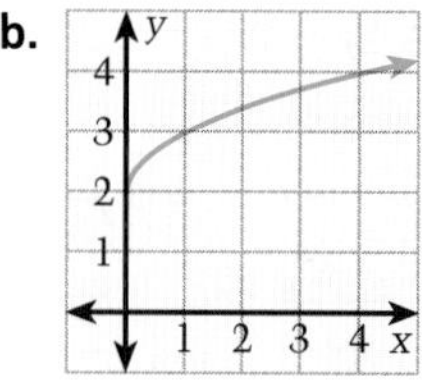

c.

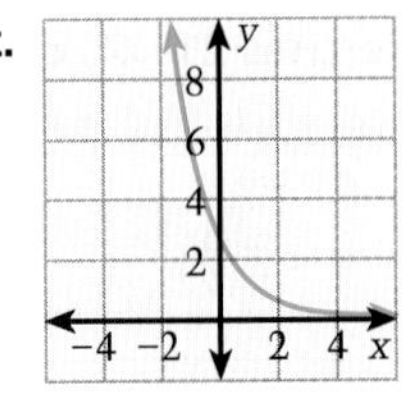

d.

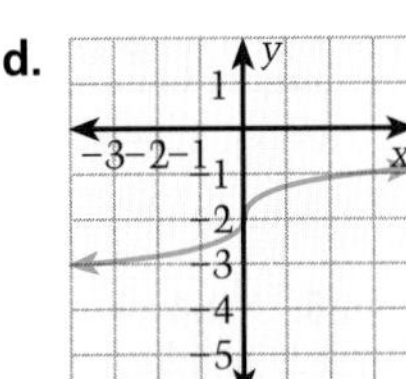

e.

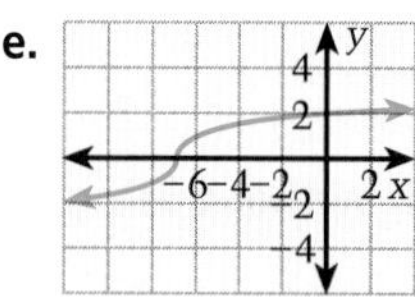

f.

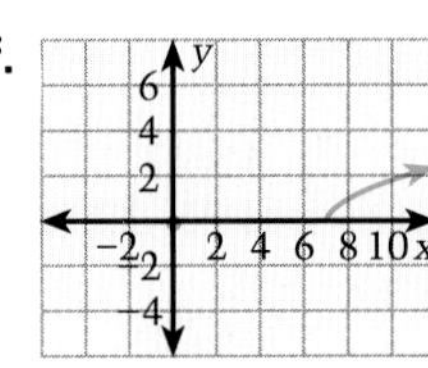

In Exercises 65–76, sketch the graph of the function. (7.2, 7.6) See margin.

65. $f(x) = 5\left(\frac{3}{4}\right)^x$

66. $g(x) = (x - 11)^{1/2}$

67. $h(x) = \frac{1}{3}(2)^x$

68. $f(x) = \frac{1}{2}(6)^x$

69. $f(x) = \sqrt{x} + 15$

70. $g(x) = 9\left(\frac{1}{3}\right)^x$

71. $g(x) = (x - 7)^{1/3}$

72. $f(x) = \sqrt[3]{x} + 16$

73. $f(x) = \sqrt{x - 9} + 3$

74. $h(x) = \sqrt[3]{x + 4} - 9$

75. $h(x) = (x + 2)^2 - 4$

76. $h(x) = -(x - 3)^2 + 9$

77. R&D For 1970 through 1990, the average cost of research and development, C (in thousands of dollars per scientist or engineer), can be modeled by $C = 61 + 0.33t^2 - 12.88(2.7)^{-t}$ where $t = 0$ represents 1970. Approximate the average cost in 1980.
(Source: U.S. National Science Foundation) $94,000

78. Fish Catch For 1970 through 1990, the total amount of fish caught, F (in millions of metric tons), in the world can be modeled by $F = 67 + 0.005t^3$, where t represents the year with $t = 0$ corresponding to 1970. In what year were approximately 72 million metric tons of fish caught? ***(Source: Statistical Office of the United Nations)*** 1980

79. Car Advertising For 1970 through 1990, the amount, A (in millions of dollars), spent on automotive advertising in the United States can be modeled by $A = 0.52(t^{5/2}) + 90$ where $t = 0$ represents 1970. In what year was approximately $471 million spent on automotive advertising? 1984

Airspeed **In Exercises 80 and 81, use the following information.**

Airplane speeds are measured in three different ways: *indicated airspeed, true airspeed,* and *ground speed.* The indicated airspeed is the airspeed given by an instrument called an airspeed indicator. The true airspeed is the speed of the airplane relative to the wind. The ground speed is the speed of the airplane relative to the ground. (For instance, if a plane were flying at a true airspeed of 200 knots into a headwind of 50 knots, then the ground speed would be 150 knots.)

80. The indicated airspeed, S (in knots), of an airplane is given by an airspeed indicator that measures the difference, p (in inches of mercury), between the static and dynamic pressures. The relationship between S and p is given by the model

$$S = 136.4\sqrt{p} + 4.5.$$

Construct a table of values for this function and sketch its graph. See margin.

81. An airplane pilot can approximate the true airspeed, T (in knots), by

$$T = \left(1 + \frac{A}{50{,}000}\right)S$$

where A is the altitude (in feet) and S is the indicated airspeed (in knots). A plane is flying at an altitude of 20,000 feet and the differential pressure is 2.05 (inches of mercury). What is its true airspeed?
≈ 279.7 knots

Pilots do not usually calculate their airspeed, but they should know how to estimate it in emergencies.

Answers

71.–73.

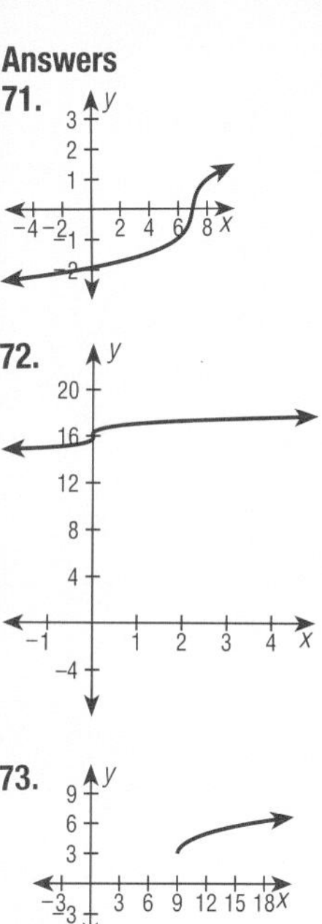

74.

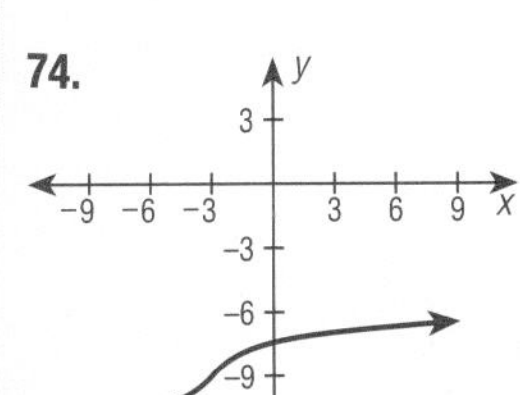

75.

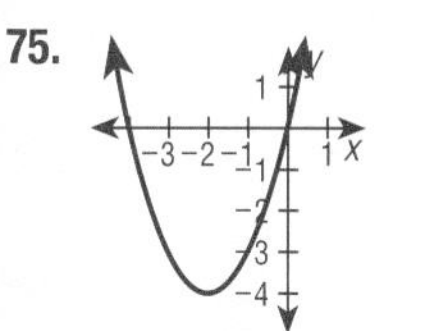

76.

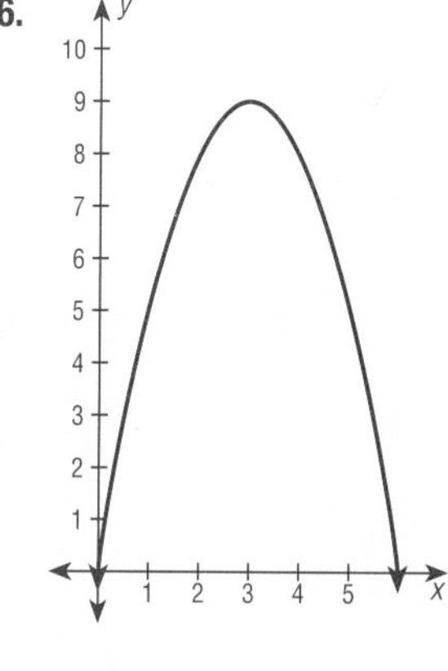

80.

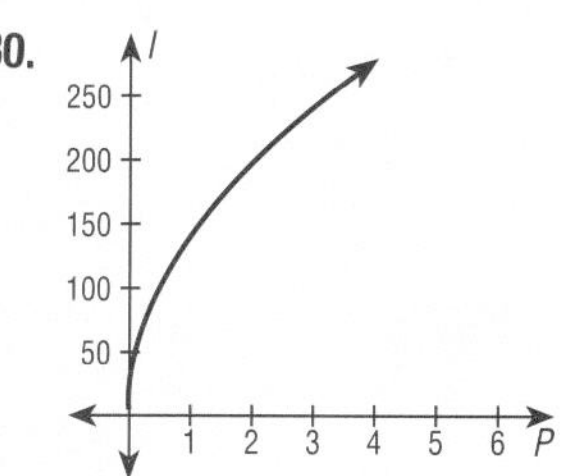

Library Statistics **In Exercises 82–85, simplify the expression to find the indicated library statistic. (The statistics refer to college and university libraries in the United States in 1990.)**

82. *Total Number of Libraries:*

$$\frac{10^6}{2^4} \cdot \frac{2^3}{10^2} \quad 5000$$

83. *Total Number of Employees:*

$$2000\sqrt[3]{24389} \quad 58{,}000$$

84. *Total Expenses (Millions of Dollars):*

$$\left(\frac{4 \cdot 25^3}{36^{-1}}\right)^{1/2} \quad \$1500 \text{ million}$$

85. *Total Number of Books (in Millions):*

$$\frac{\sqrt[3]{8^{11}}}{\sqrt[3]{8^2}} + \sqrt[3]{8} \quad 514 \text{ million}$$

86. *Largest Public Library* The ten-story Harold Washington Library Center opened in 1991 in Chicago, Illinois. If all the bookshelves in this library were laid end to end, they would stretch from Chicago to Belvidere, Illinois. How many miles would they stretch? (Each unit on the coordinate axes represents about 3.3 mi.) ≈68 mi.

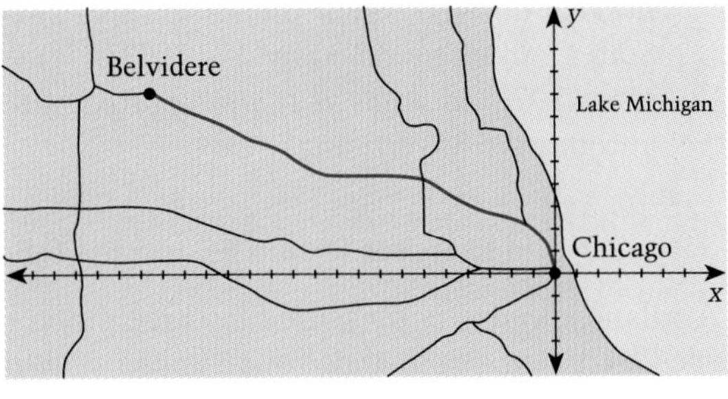

87. *Overdue Book* According to the *Guinness Book of World Records,* the record for an overdue library book was set in 1955 when a book that had been borrowed from the Sidney Sussex College Library in Cambridge, England, was returned. The book had been borrowed for x years where x is a solution of

$$\sqrt{\frac{x}{2}} = 12.$$

When was the book borrowed? 1667

88. *Masters Degrees* For 1975 through 1990, the number of library and archival science masters degrees, L, earned in the United States can be modeled by

$$L = 670\sqrt{280 - 31.6t + t^2}$$

where $t = 5$ represents 1975. In which year (prior to 1990) were approximately 5360 masters degrees in library and archival science earned? 1980

The Harold Washington Library Center in Chicago, Illinois.

Chapter TEST

1. The geometric mean of x and 18 is $6\sqrt{3}$. Find x. **(7.5)** 6
2. Find the distance between $(4, -6)$ and $(1, 2)$. **(7.5)** $\sqrt{73}$
3. A circle has a center at $(-4, 2)$ and a radius of $5\sqrt{2}$. Find all possible values of x such that $(3, x)$ is a point on the circle. **(7.5)** 1, 3
4. Simplify: $\left(\frac{x^2y}{y^{-4}}\right)^2 \cdot \frac{x^0y^{-6}}{y^2}$. **(7.1)** x^4y^2
5. Sketch the graph of $y = \sqrt{x+4} - 2$. **(7.6)** See margin.
6. Sketch the graph of $y = \sqrt[3]{x-2}$. **(7.6)** See margin.

In Exercises 7–12, simplify the expression. (7.3)

7. $\sqrt[3]{27x^3}$ $3x$
8. $2^{3/4} \cdot 2^{1/2}$ $2\sqrt[4]{2}$
9. $(8^{1/2})^{2/3}$ 2
10. $\sqrt[5]{64x^{10}}$ $2x^2\sqrt[5]{2}$
11. $3\sqrt[4]{2} + 5\sqrt[4]{2}$ $8\sqrt[4]{2}$
12. $\sqrt[3]{28} + \sqrt[3]{224}$ $3\sqrt[3]{28}$

In Exercises 13–18, solve the equation. (7.5)

13. $\sqrt{x^2+7} = x + 1$ 3
14. $\sqrt{x^2+16} = \sqrt{x^2 - 5x + 1}$ -3
15. $\sqrt{11 - x} - 2\sqrt{x+9} = 0$ -5
16. $\sqrt{x+4} = x - 8$ 12
17. $\sqrt[3]{2x} = \sqrt[3]{x-4}$ -4
18. $x^{2/3} = 9$ ± 27

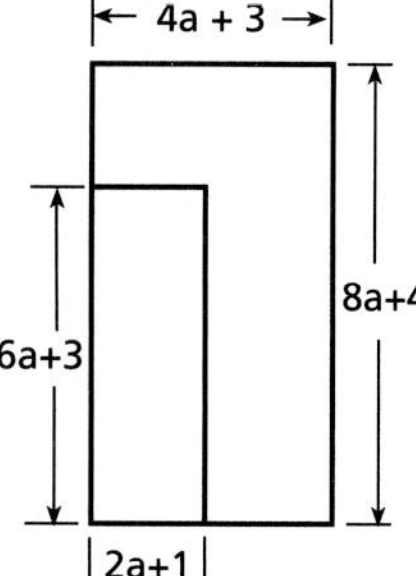

19. In the figure at the right, x is the geometric mean of the lengths of the shorter side of the small rectangle and the longer side of the large rectangle. Find x. **(7.5)** $4a + 2$
20. The geometric mean of two numbers equals one half the first number. If the first number is 8, what is the second number? **(7.5)** 2
21. The surface area of Venus is 5×10^8 square kilometers and the surface area of Mars is 1.43×10^8 square kilometers. Find the ratio of the surface area of Venus to the surface area of Mars. **(7.4)** $\approx \frac{3.5}{1}$
22. \$4000 is deposited in an account that earns 6% annual interest, compounded monthly. What is the balance after 8 years? **(7.2)** \$6456.57
23. A 70,000-gallon swimming pool is leaking water at the rate of 10% per week. Write a model for the volume of water, W (in gallons), in the pool after t weeks. If none of the water is replaced, how much will remain after 4 weeks? **(7.2)** $W = 70{,}000(0.9)^t$, 45,927 gal
24. In 1990, the sales, y (in billions of dollars), of portable computers were predicted to increase, based on the model $y = 6(1.25)^t - 3.75$, where $t = 0$ represents 1990. What are the predicted sales for the year 2000? *(Source: USA Today)* **(7.2)** $\approx$52.1 billion

Answers

5.

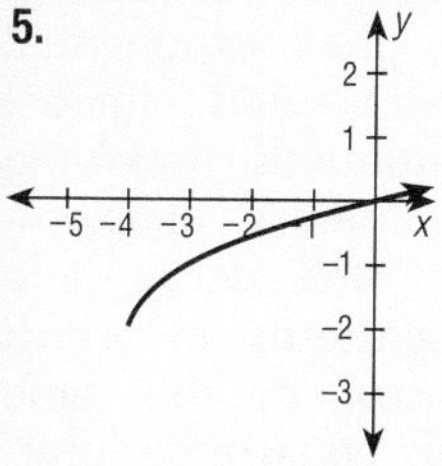

6.

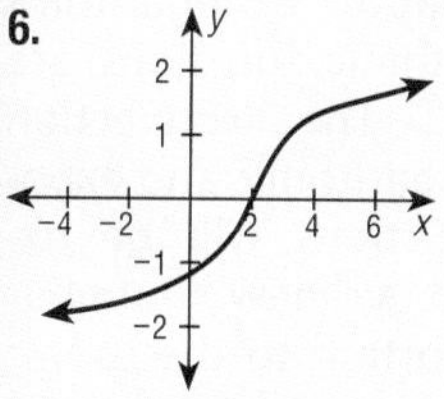

Students have already seen, in Chapter 7, that exponential functions can be used to model real-life situations involving compound interest, exponential growth and decay, and "growth" patterns in which increases occur by the same percent over equal time intervals. In Chapter 8, logarithms and logarithmic functions are introduced. The connection between logarithmic and exponential functions will be explored and a "new" number that is important to the use of these functions will be identified!

The Chapter Summary on page 448 provides you and the students with a synopsis of the chapter. It identifies key skills and concepts. You may want to have students look at the Chapter Summary as an overview before beginning the chapter.

CHAPTER 8

Exponential and Logarithmic Functions

LESSONS

Monteca is a jazz fusion band from Toronto, Canada. They use a combination of standard and electronic instruments.

Real Life
Music

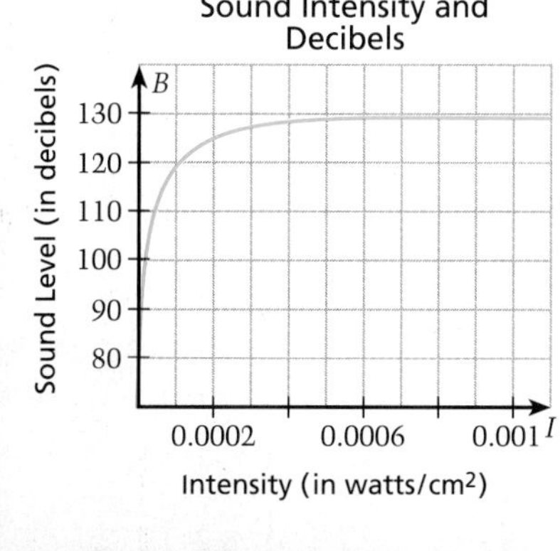

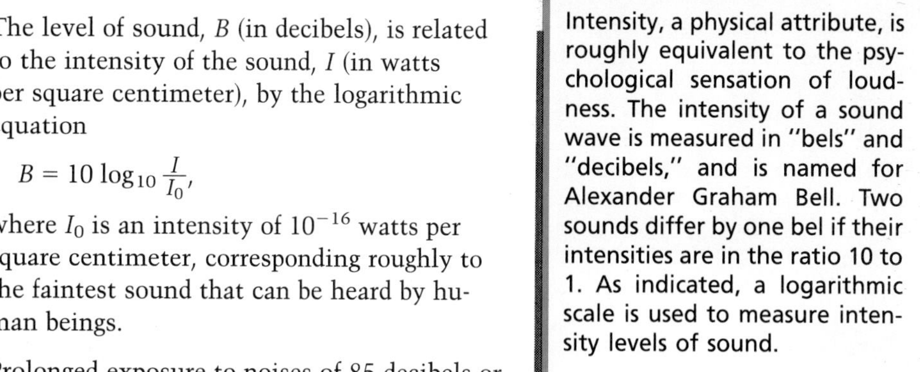

The level of sound, B (in decibels), is related to the intensity of the sound, I (in watts per square centimeter), by the logarithmic equation

$$B = 10 \log_{10} \frac{I}{I_0},$$

where I_0 is an intensity of 10^{-16} watts per square centimeter, corresponding roughly to the faintest sound that can be heard by human beings.

Prolonged exposure to noises of 85 decibels or greater can cause hearing loss. Above 100 decibels, even brief exposure can be damaging.

(Source: **Time** ***Magazine)***

Intensity, a physical attribute, is roughly equivalent to the psychological sensation of loudness. The intensity of a sound wave is measured in "bels" and "decibels," and is named for Alexander Graham Bell. Two sounds differ by one bel if their intensities are in the ratio 10 to 1. As indicated, a logarithmic scale is used to measure intensity levels of sound.

Exponential and Logarithmic Functions

PACING CHART

The daily Pacing Chart is meant to help you adjust your teaching pace. Students in the full course should finish the entire text by the end of the year. Students in the basic course are expected to complete the first twelve chapters. The Pacing Chart for each chapter contains suggestions for lessons that require more than one day and lessons that may be omitted for the basic course.

DAY	FULL COURSE	BASIC COURSE
1	8.1	8.1
2	8.2	8.2
3	8.3	8.2
4	Mid-Chapter Self-Test	8.3
5	8.4 & Using a Graphing Calculator	8.3
6	8.5	Mid-Chapter Self-Test
7	8.6	8.4
8	8.7 & Using a Graphing Calculator	8.4 & Using a Graphing Calculator
9	Chapter Review	8.5
10	Chapter Test	8.5
11		8.6
12		8.6
13		8.7 & Using a Graphing Calculator
14		Chapter Review
15		Chapter Test

CHAPTER ORGANIZATION

LESSON	PAGES	GOALS	MEETING THE NCTM STANDARDS
8.1	398–404	1. Graph exponential functions and evaluate exponential expressions 2. Use exponential functions as models for real-life situations	Problem Solving, Communication, Connections, Functions, Geometry (algebraic), Structure
8.2	405–411	1. Evaluate logarithmic expressions and graph logarithmic functions 2. Use logarithms in real-life situations	Problem Solving, Communication, Connections, Functions, Structure
Mixed Review	412	Review algebraic and arithmetic skills	
Milestones	412	Logarithms	Connections
8.3	413–418	1. Use properties of logarithms 2. Use logarithms to solve real-life problems	Problem Solving, Communication, Connections, Structure
Mid-Chapter Self-Test	419	Diagnose student weaknesses and remediate with correlated Reteach worksheets	
8.4	420–425	1. Use the number e as the base of an exponential function 2. Use the natural base e in real-life problems	Problem Solving, Communication, Connections, Reasoning, Functions, Calculus concepts, Structure
8.5	426–431	1. Evaluate natural logarithmic expressions and graph natural logarithmic functions 2. Use natural logarithmic functions as real-life models	Problem Solving, Communication, Reasoning, Connections, Functions, Calculus concepts, Geometry (algebraic)
Using a Calculator	432–433	Fit logarithmic and exponential models to data	Technology
8.6	434–440	1. Solve exponential and logarithmic equations 2. Use exponential and logarithmic equations to answer questions about real-life	Problem Solving, Communication, Connections, Geometry (algebraic), Structure
Mixed Review	440	Review algebraic and arithmetic skills	
8.7	441–446	1. Graph a logistics growth function 2. Use logistics growth functions to answer questions about real-life situations	Problem Solving, Communication, Connections, Functions
Using a Calculator	447	Solve logarithmic and exponential equations	Technology
Chapter Summary	448	Restate for students what they have learned, why they have learned it, and how it fits into the structure of algebra	Structure, Connections
Chapter Review	449–452	Review concepts and skills learned in the chapter	
Chapter Test	453	Diagnose student weaknesses and remediate with correlated Reteaching worksheets	

LESSON RESOURCES

MEETING INDIVIDUAL NEEDS

RETEACHING For students who need to spend more time on basics:

If a mid-chapter self-test or chapter test indicates a deficiency, teachers can help students with the appropriate ***Reteaching Copymaster.***

PRACTICE For students who need more practice:

Additional exercises like those in the Pupil's Edition are provided for each lesson in ***Extra Practice Copymasters.***

ENRICHMENT For enriching and broadening students' experiences:

Problem of the Day copymasters in ***Teaching Tools*** provide a daily opportunity to use logical reasoning, looking for a pattern, writing an equation, and other routine and non-routine problem-solving strategies.

Math Log copymasters in ***Alternative Assessment*** provide opportunities to report on investigations, research, and open-ended problems.

Technology: Using Calculators and Computers provides enriching activities with graphing and scientific calculators and computers.

The ***Applications Handbook*** provides additional information about the cross-curriculum topics such as astronomy, chemistry, physics, sports, economics, genetics, and music that are integrated into the Pupil's Edition.

LESSON	8.1	8.2	8.3	8.4	8.5	8.6	8.7
PAGES	398-404	405-411	413-418	420-425	426-431	434-440	441-446
Teaching Tools							
Transparencies	✓	✓		✓	✓		
Problem of the Day	✓	✓	✓	✓	✓	✓	✓
Warm-up Exercises	✓	✓	✓	✓	✓	✓	✓
Answer Masters	✓	✓	✓	✓	✓	✓	✓
Extra Practice Copymasters	✓	✓	✓	✓	✓	✓	✓
Reteaching Copymasters	Teacher-directed and independent activities tied to results on the Mid-Chapter Self-Tests and Chapter Tests						
Color Transparencies	✓				✓	✓	✓
Applications Handbook	Additional background information for many real-life applications						
Technology	Calculator and computer worksheets for appropriate lessons						
Complete Solutions Manual	✓	✓	✓	✓	✓	✓	✓
Alternative Assessment	Assess student's ability to reason, analyze, solve problems, and communicate using mathematical language.						
Formal Assessment	Mid-Chapter Self-Tests, Chapter Tests, Cumulative Tests, and Practice for College Entrance Tests						
Computer Test Bank	Customized tests can be created by choosing from over 2500 items.						

INSIGHTS

8.1 Exponential Functions
In this lesson, students continue to use vertical and horizontal shifts and reflections in the *x*-axis to graph complex functions derived from the basic exponential and logarithmic functions. They apply previously-learned concepts and techniques to investigate the relationship between logarithmic and exponential functions. In this way, students gain greater ownership of the mathematics that is developed.

8.2 Logarithmic Functions
Logarithmic functions and exponential functions are closely related to each other. In fact, they are inverse relations. Students must understand this relationship to use either effectively in solving problems. Logarithms occur in the physical and biological sciences, in education, in statistics, and in economics.

8.3 Properties of Logarithms
In modeling the real world, an understanding of the properties of logarithms enables us to simplify relationships containing exponents and logarithms, and are extremely useful for computational purposes.

8.4 The Natural Base *e*
Engineering and business applications of logarithms often involve the use of a special number, "e." In this lesson, the exponential function, base *e*, will be explored.

8.5 Natural Logarithms
Natural logarithms are logarithms with base *e*. For those models of real-life phenomena that involve natural logarithms, students should be able to evaluate and graph such functions. Happily, the properties of common logarithms, and the methods of evaluating and graphing common logarithms apply also to natural logarithms.

8.6 Solving Exponential and Logarithmic Equations
As a necessary accompaniment to business and engineering models that involve logarithms and exponents, solving exponential and logarithmic equations require some strategies that are unique to these special relationships. This lesson explores those strategies.

8.7 Exploring Data: Logistics Growth Functions
In this lesson, students add yet another useful function to their collection of possible models. Logistics growth models occur frequently in the description of "organic models" found in chemistry, biology, and environmental studies. Logistics growth models are a useful application of exponential relationships.

Problem of the Day

Find the dimensions of the triangle that must be cut from the corners of a square with side s to form a regular octagon.
Isosceles right triangle with legs $\frac{2s - s\sqrt{2}}{2}$ and hypotenuse $s(\sqrt{2} - 1)$

ORGANIZER

Warm-Up Exercises

Use horizontal and vertical shifts, and reflections of $f(x) = x^2$ or $g(x) = |x|$ to sketch the following graphs.

a. $j(x) = -x^2$

b. $k(x) = (x + 2)^2 - 3$

c. $m(x) = -|x| - 2$

d. $n(x) = |x - 4| + 2$

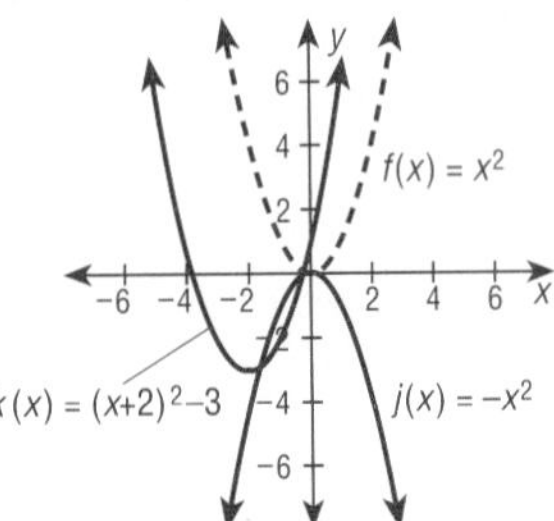

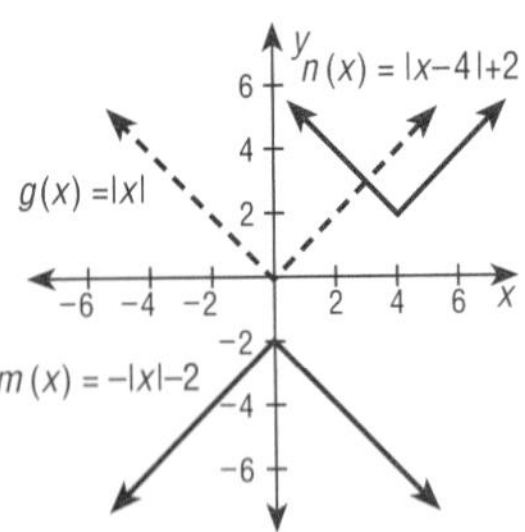

Lesson Resources

Teaching Tools
- Transparencies: 7, 8, 14, 15
- Problem of the Day: 8.1
- Warm-up Exercises: 8.1
- Answer Masters: 8.1

Extra Practice: 8.1
Color Transparency: 44

8.1 Exponential Functions

What you should learn:

Goal 1 How to graph exponential functions and evaluate exponential expressions

Goal 2 How to use exponential functions as models for real-life problems

Why you should learn it:

You can model many real-life problems with exponential models, such as the depreciating value of a car as it ages.

Goal 1 Graphing Exponential Functions

In Lesson 7.2, you studied the graphs of basic exponential functions of the form $f(x) = Ca^x$. If $0 < a < 1$, then the graph represents exponential decay. If $a > 1$, then the graph represents exponential growth.

Exponential Decay: $f(x) = \left(\frac{1}{2}\right)^x$ **Exponential Growth:** $f(x) = 3^x$

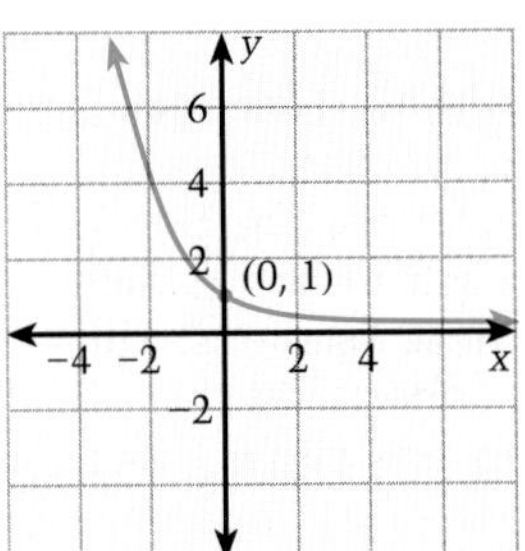

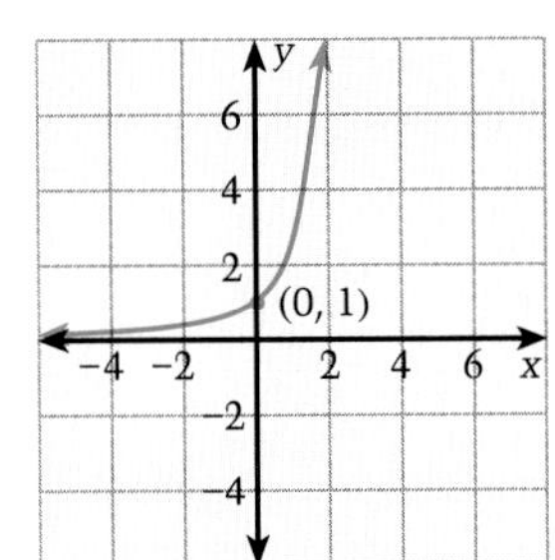

In this lesson, you will study vertical shifts, horizontal shifts, and reflections of these two basic types of graphs.

Example 1 *Using Vertical Shifts to Sketch Graphs*

Sketch the graph of each function.

a. $g(x) = \left(\frac{1}{2}\right)^x + 1$

b. $h(x) = 3^x - 2$

Solution

a. To sketch the graph of g, shift the graph of $f(x) = \left(\frac{1}{2}\right)^x$ upward by 1 unit.

b. To sketch the graph of h, shift the graph of $f(x) = 3^x$ downward by 2 units.

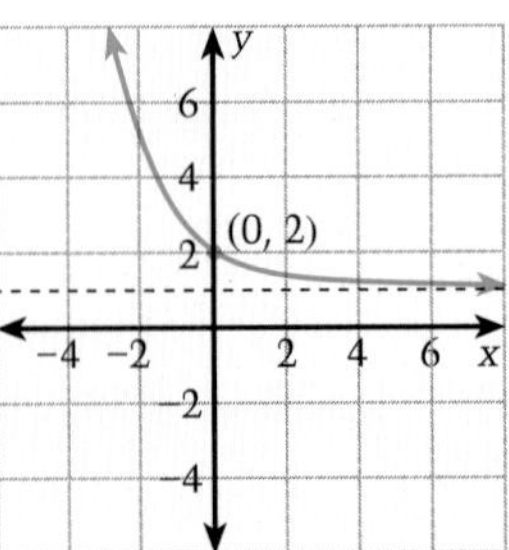

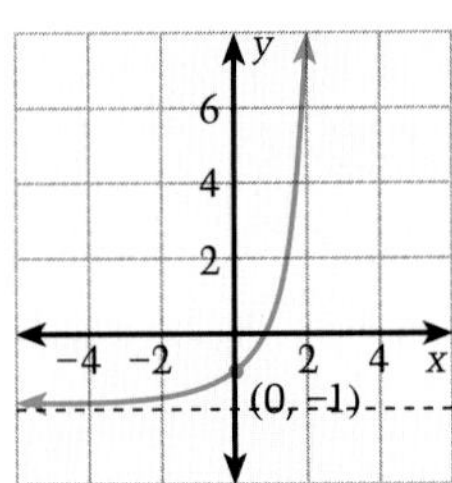

Example 2 Using Horizontal Shifts

Sketch the graph of each function.

a. $g(x) = 2^{x+3}$

b. $h(x) = 2^{x-2}$

Solution

a. To sketch the graph of g, shift the graph of $f(x) = 2^x$ to the left 3 units.

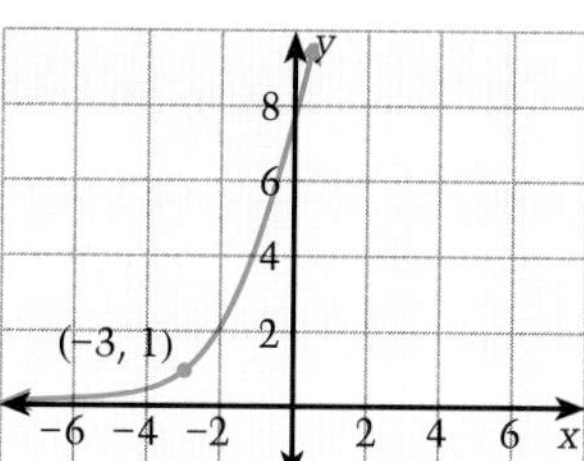

b. To sketch the graph of h, shift the graph of $f(x) = 2^x$ to the right 2 units.

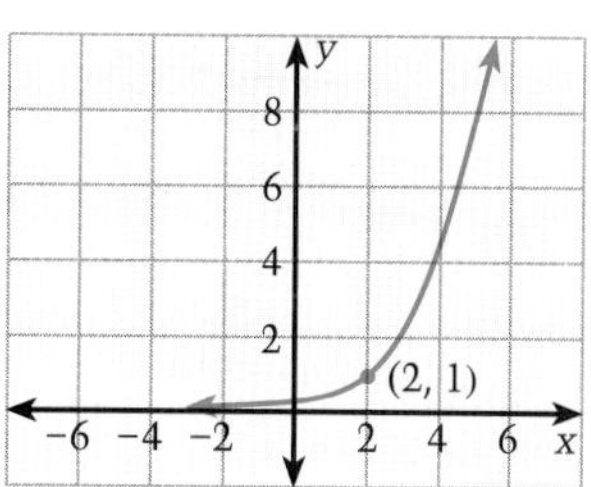

Each graph in Examples 1 and 2 shows a single vertical shift or a single horizontal shift. Example 3 shows how combinations of shifts and reflections can be used to sketch the graph of a function.

Example 3 Using Reflection and Shifts

Sketch the graph of each function.

a. $g(x) = -\left(\frac{1}{2}\right)^x + 4$

b. $h(x) = -2^{x-2} + 2$

Solution

a. To sketch the graph of g, reflect the graph of $f(x) = \left(\frac{1}{2}\right)^x$ in the x-axis. Then shift the graph upward 4 units.

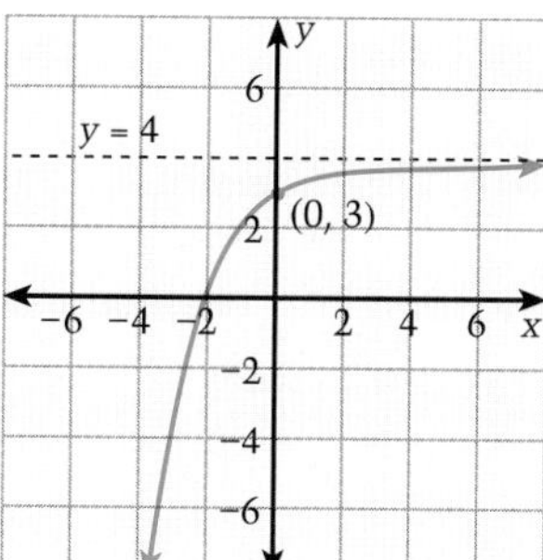

b. To sketch the graph of h, shift the graph of $f(x) = 2^x$ 2 units to the right. Then reflect the graph in the x-axis and shift 2 units upward.

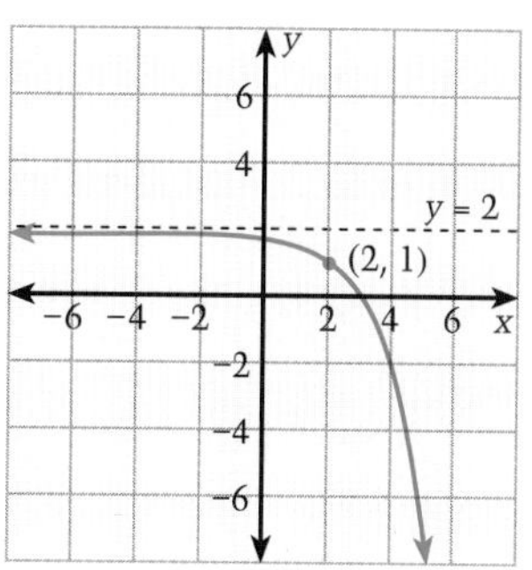

LESSON Notes

The effects of constants on the graphs of basic algebraic functions are embodied in important techniques for producing quick graphs of more complex algebraic functions.

Examples 1–3

Students have used horizontal and vertical shifts, and reflections to graph other algebraic functions in previous lessons. Ask students to state conjectures about the effects of constants a, b, and c for any algebraic function $f(x)$. That is, how is $g(x)$ related to $f(x)$ when

(a) $g(x) = f(x) + c$?
(b) $g(x) = f(x + b)$?
(c) $g(x) = af(x)$?

(a) a vertical shift $|c|$ units up or down; (b) a horizontal shift $|b|$ units left or right; (c) an enlargement (for $a > 1$); a shrinking (for $0 < a < 1$); and for $a < 0$, a reflection in the x-axis.

Encourage students to test their conjectures using unfamiliar functions, such as $f(x) = x^3$, $f(x) = x^4$, and $f(x) = \frac{1}{x}$.

In Examples 1 and 3, the broken line that bounds the graphs is called an *asymptote.* In Example 2, the x-axis is the asymptote.

Example 4

TECHNOLOGY Emphasize that the most practical method for evaluating powers with irrational exponents is to use computers or calculators.

Point out that the fraction $\frac{1}{3}$ is the "limit" of the infinite sequence of rational numbers, 0.3, 0.33, 0.333, 0.3333, 0.33333, ...

Example 5

Remind students that situations involving the same percent increase over equal time intervals can be represented by the exponential growth model. In a similar way, situations involving the same percent reduction each year can be represented by the exponential decay model. See Lesson 7.2.

Extra Examples

Here are additional examples similar to **Examples 1–5.**

1. Sketch the graph of each function on the same coordinate grid.

   a. $f(x) = 2^x$ and $p(x) = 2^{x+3} - 2$

   b. $g(x) = \left(\frac{1}{5}\right)^x$ and $p(x) = -\left(\frac{1}{5}\right)^{x+2} + 6$

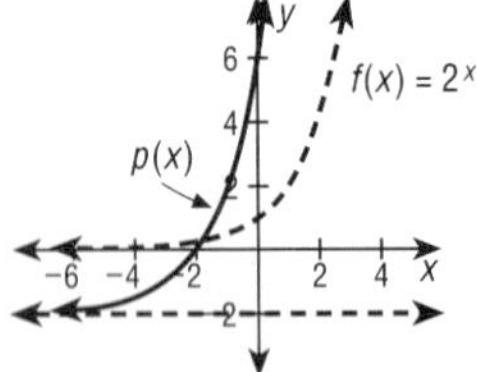

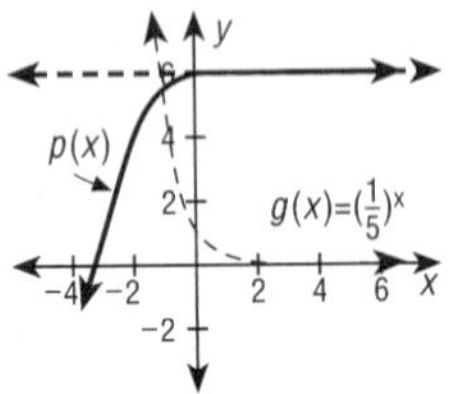

Study Tip

Here are some definitions of powers that you have already studied.

Positive Integer Exponent:

$a^n = a \cdot a \cdot a \cdots a$ (n factors)

Zero Exponent:

$a^0 = 1$

Rational Exponent:

$a^{1/n} = \sqrt[n]{a}$

Negative Exponent:

$a^{-n} = \frac{1}{a^n}$

When the graph of an exponential function is sketched over the entire x-axis, you are assuming that *any* real number can be used as an exponent. What, however, does it mean to raise a number to an irrational power, such as $\sqrt{2}$?

LESSON INVESTIGATION

■ Investigating Irrational Exponents

Partner Activity Use a calculator to evaluate the powers. Round your results to five decimal places.

$3^{14/10}$	4.65554
$3^{141/100}$	4.70697
$3^{1414/1000}$	4.72770
$3^{14,142/10,000}$	4.72873
$3^{141,421/100,000}$	4.72879
$3^{1,414,214/1,000,000}$	4.72881

Each of these powers has a rational exponent, which means that it can be defined by

$$a^{m/n} = (\sqrt[n]{a})^m.$$

With your partner, discuss how you can use these powers to define the power $3^{\sqrt{2}}$, which has an irrational exponent.

Example 4 *Evaluate Powers That Have Irrational Exponents*

Use a calculator to evaluate the power. Round the result to 3 decimal places.

a. $3^{\sqrt{2}}$ **b.** $3^{-\pi}$

Solution

Keystrokes	Display	Calculator
a. 3 [x^y] 2 [√] [=]	4.72880	*Scientific*
3 [∧] [√] 2 [ENTER]	4.72880	*Graphing*

Rounding to 3 decimal places, you obtain $3^{\sqrt{2}} \approx 4.729$.

Keystrokes	Display	Calculator
b. 3 [x^y] π [+/−] [=]	0.03170	*Scientific*
3 [∧] [(−)] π [ENTER]	0.03170	*Graphing*

Rounding to 3 decimal places, you obtain $3^{-\pi} \approx 0.032$. ■

Goal 2 Using Exponential Models in Real Life

Real Life **Automobiles**

Example 5 *Exponential Decay Model*

Suppose that you purchased a new car for $20,000 in 1990. If the value of the car decreases by 16% each year to 84% of its previous value, what will the car be worth in 1996? In 1998?

Solution One way to solve this problem is to create an exponential decay model. Let V represent the value of the car and let $t = 0$ represent 1990. Using the exponential decay model, you can write the following.

$V = Ca^t$ *Exponential decay model*

$= \text{Initial Value}\,(0.84)^t$

$= 20{,}000(0.84)^t$

With this model, you can determine that the value of the car in 1996 is

$V = 20{,}000(0.84)^6 \approx \$7026$ *Value in 1996*

and the value in 1998 is

$V = 20{,}000(0.84)^8 \approx \$4958.$ *Value in 1998*

The graph of $V = 20{,}000(0.84)^t$ is shown below.

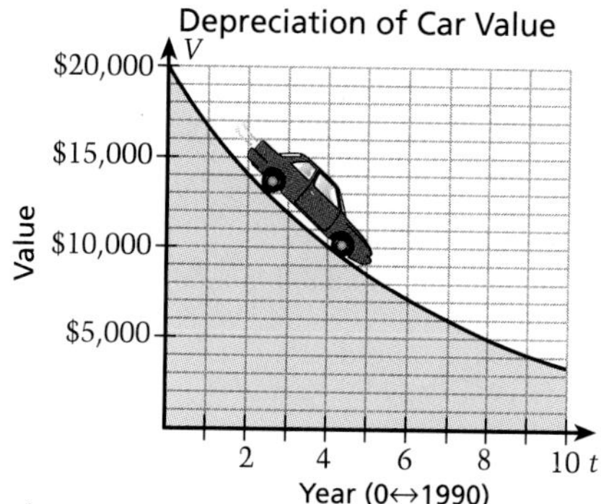

■

Communicating about ALGEBRA

SHARING IDEAS about the Lesson See Additional Answers.

Sketch the graph of the function. Explain your steps.

A. $f(x) = 2^{x-2} - 1$ **B.** $f(x) = -3^{x+1} + 4$

2. **TECHNOLOGY** Use a calculator to evaluate each expression to three decimal places.
   a. $5^{\sqrt{3}}$ 16.242
   b. 4^{π} 77.880
   c. $2^{\sqrt{5}}$ 4.711
   d. $3^{-\sqrt{2}}$ 0.211
3. Suppose you purchased a new car for $15,000 in 1992. If the value of the car decreases by 10% each year to 90% of its previous value, what will the car be worth in 1999?
   about $7,174

Check Understanding

1. True or False? The rules for rational exponents hold for irrational exponents. True
2. True or False? Exponential growth and exponential decay are both examples of exponential functions. True
3. How are the effects of constants (such as horizontal and vertical shifts) on exponential functions related to the effects of constants on other functions, such as the function $f(x) = x^2$? They are similar; see Warm-Up Exercises.
4. What is the limit of this sequence: 0.6, 0.66, 0.666, 0.6666, 0.66666, 0.666666, . . . ? the fraction $\frac{2}{3}$

Communicating about ALGEBRA

MATH JOURNAL Have students construct a flowchart of steps that describes how to graph the effects of horizontal and vertical shifts, and reflections in the x-axis, for exponential functions, $f(x) = Ca^x$, for any real number a. Students should include these charts in their math journals in a section reserved for quick-graph techniques.

See Lesson Notes for Examples 1–3. Students should adapt the rules given in the Lesson Notes for Examples 1–3.

EXERCISE Notes

ASSIGNMENT GUIDE

Basic/Average: Ex. 1–4, 7–11 odd, 13–20, 27–39 odd, 46, 47–51 odd, 65–66

Above Average: Ex. 1–4, 7–11 odd, 13–20, 27–39 odd, 45–52, 53–60, 65–66

Advanced: Ex. 1–4, 7–11 odd, 13–20, 21–29 odd, 35–51 odd, 52, 65–72

Selected Answers
Exercises 1–4, 5–65 odd

More Difficult Exercises
Exercises 45–52, 67–72

Guided Practice

EXTEND Ex. 2 To prepare students for logarithms and review the properties of exponents, have them compare the functions $f(x) = 5^x$ and $g(x) = 3^x$ with the functions $h(x) = \left(\frac{1}{5}\right)^x$ and $j(x) = \left(\frac{1}{3}\right)^x$ and look for a pattern. (See the Enrichment Activities that follow for other examples of using the properties of exponents.) Since $\frac{1}{5} = 5^{-1}$, $h(x) = ((5^{-1})^x) = 5^{-x} = f(-x)$. Similarly, $j(x) = g(-x)$. In general, if $f(x) = a^x = y$, then $g(x) = \left(\frac{1}{a}\right)^x = a^{-x} = \frac{1}{y} = f(-x)$. Note also that the graphs of the functions are reflections of each other in the y-axis.

Answers

1. The graph of g is the graph of f shifted 3 units down.
2. The graph of g is the graph of f reflected in the x-axis and shifted 2 units to the left.

Independent Practice

Ex. 5–12 Students should state their answers in terms of reflections of the graph of $f(x)$ and vertical and horizontal shifts of the graph of $f(x)$.

EXERCISES

Guided Practice

CRITICAL THINKING about the Lesson

1. Compare the graphs of $f(x) = 5^x$ and $g(x) = 5^x - 3$. **See margin.**
2. Compare the graphs of $f(x) = 3^x$ and $g(x) = -3^{x+2}$. **See margin.**
3. Use a calculator to evaluate $5^{\sqrt{3}}$. Round the result to 3 decimal places. **16.242**
4. Write an expression that could be the limit of the following sequence. **2^{π}**

   $2^{3.1}, 2^{3.14}, 2^{3.142}, 2^{3.1416}, 2^{3.14159}, 2^{3.141593}, \ldots$

Independent Practice

In Exercises 5–12, compare the graphs of f and g. See Additional Answers.

5. $f(x) = \left(\frac{1}{2}\right)^x$, $g(x) = \left(\frac{1}{2}\right)^x - 2$
6. $f(x) = 4^x$, $g(x) = 4^{x-1}$
7. $f(x) = 5^x$, $g(x) = 5^x + 6$
8. $f(x) = \left(\frac{1}{5}\right)^x$, $g(x) = \left(\frac{1}{5}\right)^x + 10$
9. $f(x) = \left(\frac{1}{3}\right)^x$, $g(x) = -\left(\frac{1}{3}\right)^{x+4}$
10. $f(x) = 2^x$, $g(x) = -2^x + 9$
11. $f(x) = 8^x$, $g(x) = 8^{x-4} + 5$
12. $f(x) = 10^x$, $g(x) = 10^{x+2} - 2$

In Exercises 13–20, match the function with its graph.

13. $f(x) = \left(\frac{1}{2}\right)^x - 4$ h
14. $f(x) = 4^{x-1}$ d
15. $f(x) = \left(\frac{1}{2}\right)^{x-4}$ g
16. $f(x) = 4^x + 1$ b
17. $f(x) = (4)^{x+1}$ f
18. $f(x) = \left(\frac{1}{2}\right)^{x+4}$ a
19. $f(x) = (4)^x - 1$ c
20. $f(x) = \left(\frac{1}{2}\right)^x + 4$ e

a. 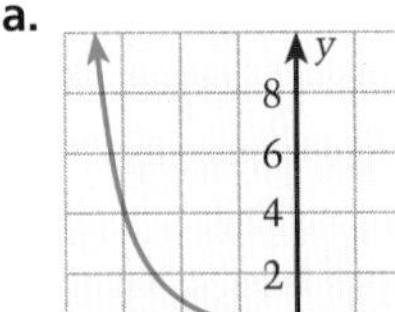

b. 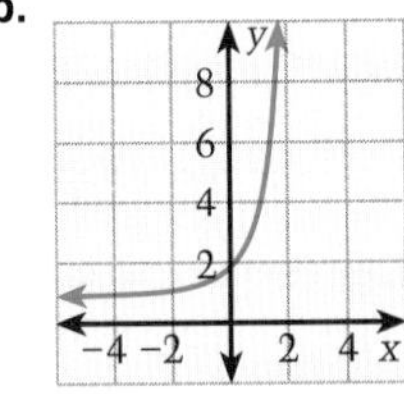

c. 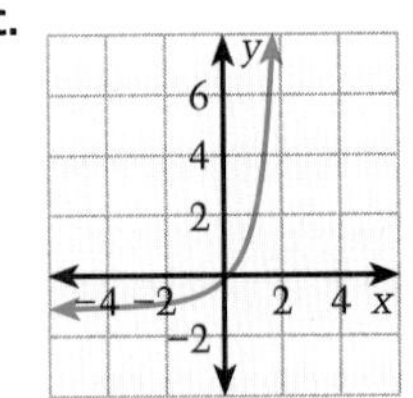

d. 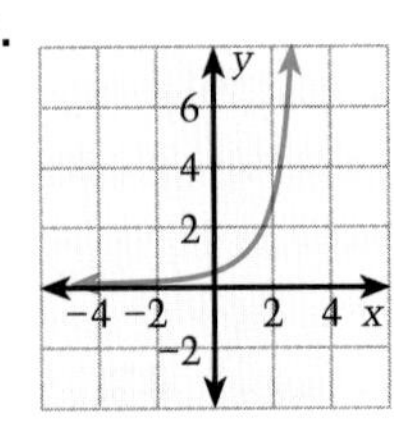

e. 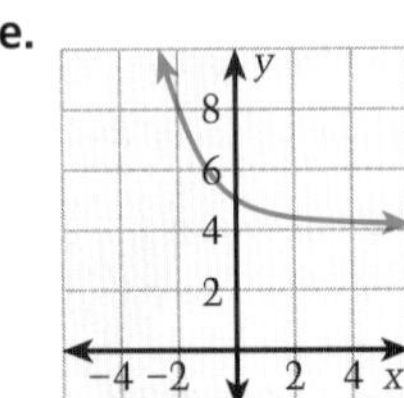

f. 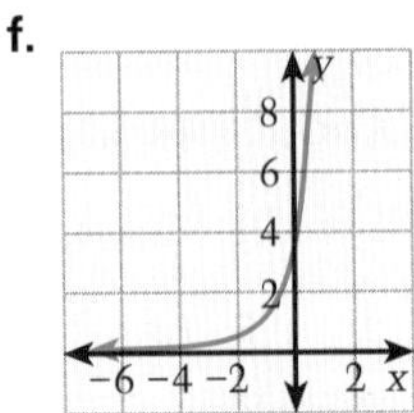

g. 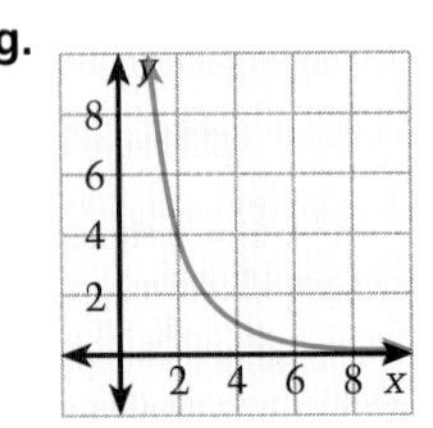

h. 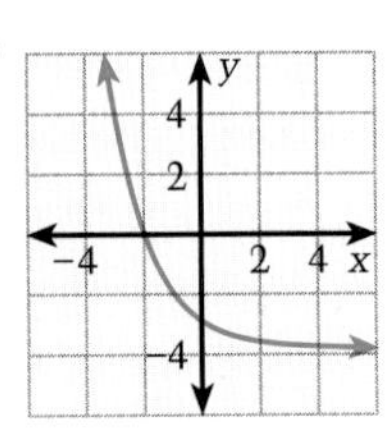

In Exercises 21–36, sketch the graph of the function. See Additional Answers.

21. $f(x) = \left(\frac{2}{3}\right)^x + 2$
22. $f(x) = 2^{x-5}$
23. $f(x) = \left(\frac{1}{4}\right)^{x-1}$
24. $f(x) = 2^x - 5$
25. $f(x) = 5^{x+3}$
26. $f(x) = 4^{x-4}$
27. $f(x) = 5^x - 4$
28. $f(x) = \left(\frac{3}{4}\right)^x + 1$
29. $f(x) = -\left(\frac{1}{3}\right)^x$
30. $f(x) = -6^x$
31. $f(x) = -2^{x+2}$
32. $f(x) = -\left(\frac{1}{2}\right)^x + 4$
33. $f(x) = -\left(\frac{1}{2}\right)^{x-3} + 7$
34. $f(x) = -2^{x-8} + 1$
35. $f(x) = -5^{x+1} - 2$
36. $f(x) = 4^{x+3} - 3$

Technology **In Exercises 37–44, use a calculator to evaluate the expression. Round your result to three decimal places.**

37. $5^{\sqrt{5}}$ 36.555
38. $21^{\sqrt{11}}$ 24,283.165
39. 6^{π} 278.378
40. $7^{-\pi}$ 0.002
41. $2^{\sqrt{3}}$ 3.322
42. $13^{2\pi}$ 9,979,585.521
43. $4^{\pi/3}$ 4.270
44. $18^{\sqrt{6}}$ 1187.892

45. *Technology* In the following sequence of powers, which power is the first to approximate $4^{\sqrt{3}}$ accurately to 5 decimal places? e

a. $4^{1.732}$ b. $4^{1.7321}$ c. $4^{1.73205}$ d. $4^{1.732051}$ e. $4^{1.7320508}$

46. *Technology* In the following sequence of powers, which power is the first to approximate 6^{π} accurately to 1 decimal places? d

a. $6^{3.1}$ b. $6^{3.14}$ c. $6^{3.142}$ d. $6^{3.1416}$ e. $6^{3.14159}$

In Exercises 47 and 48, write an exponential model that describes the situation.

47. *Increase in Value* In 1985, you bought a sculpture for \$380. Each year, t, the value, v, of the sculpture increases by 8%. $v = 380(1.08)^t$ where $t = 0$ represents 1985.

48. *Decrease in Value* In 1990, you bought a television for \$600. Each year, t, the value, v, of the television decreases by 6%. $v = 600(0.94)^t$ where $t = 0$ represents 1990.

49. *Automobile Depreciation* If the value of a new \$19,000 car that you bought in 1991 decreases by 14% each year, what will the car be worth in 1995? In 1998? Sketch the graph of the exponential decay model. See Additional Answers.

Elvis Presley has been awarded more gold records by the recording industry than any other individual musician.

50. *Business Revenue* In 1980, your business had a revenue of \$30,000. Each year after that, the revenue increased by 15%.What was the revenue in 1990? In 1992? Sketch the graph of the exponential growth model. See Additional Answers.

51. *Record Albums* Record albums increased in popularity until about 1980. For 1985 through 1994, however, the number, n (in millions), of record albums sold each year can be modeled by

$$n = 817.75(0.732)^t$$

where $t = 5$ represents 1985. How many were sold in 1990? What might explain why the sale of record albums decreased so rapidly after 1985? *(Source: Recording Industry Association of America)* $\approx$ 36 million, tape cassettes became more popular.

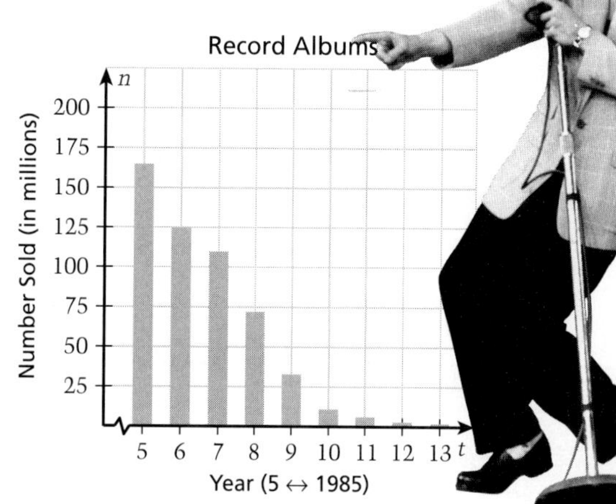

▶ **Ex. 13–20** Assign these exercises as a group. Be sure that students understand the visual difference between functions with bases that are between 0 and 1 (decay) and functions with bases greater than 1 (growth).

▶ **Ex. 21–36** Students should use the quick-graph techniques illustrated in Examples 1–3. Be sure they follow this order: (1) horizontal shift; (2) reflection in the x-axis; (3) vertical shift.

TECHNOLOGY To provide a visual check, you might wish to have some students use graphing calculators to graph the functions.

▶ **Ex. 37–46** Students can use Example 4 on page 400 as a solution model.

▶ **Ex. 47–48** Remind students to use 1.08 as the base in $V = Ca^t$, not 0.08.

EXTEND Ex. 51 To understand the role of asymptotes, ask students whether the model can be used to find a year in which record albums would no longer exist.
No, the expression $817.75(0.732)^t$ approaches 0 but never reaches it. For example, in the year 2030, when $t = 50$, only 137 record albums would be sold.

EXTEND **Ex. 52** Have students use the model to predict the number of boats in the year 2000. ≈23.6 million

▶ **Ex. 67–72** These exercises can be used to introduce the natural base, *e*, which will be completely presented in Lessons 8.4 and 8.5 . Encourage students with graphics calculators to graph $f(x) = e^x$.

Enrichment Activities

COOPERATIVE LEARNING

Have students work with partners to complete these activities.

1. Use a guess-and-check strategy with a calculator (or the TRACE function of a graphics calculator) to find the solution for this equation.

   $5^x = 2$ $x \approx 0.4307$

2. Show how to use the properties of exponents and the solution to Activity 1 to solve these equations. Check by using a calculator.

   a. $5^k = 4$ **b.** $25^n = 2$

   c. $\left(\frac{1}{5}\right)^w = \frac{1}{8}$ **d.** $125^z = 10$

   a. $5^k = 2^2 \approx (5^{0.4307})^2 = 5^{0.8614}$; so $k \approx 0.8614$.

   b. $25^n = (5^2)^n = 5^{2n} \approx 5^{0.4307}$; so $2n \approx 0.4307$ and $n \approx 0.2153$.

   c. $\left(\frac{1}{5}\right)^w = (5^{-1})^w = 5^{-w}$; $\frac{1}{8} = \left(\frac{1}{2}\right)^3 = (2^{-1})^3 = 2^{-3}$; so $5^{-w} = 2^{-3} \approx (5^{0.4307})^{-3} = (5^{-1.2921})$; so $-w \approx -1.2921$ and $w = 1.2921$.

   d. $125^z = (5^3)^z = 5^{3z}$; $10 = 2 \cdot 5$; so $5^{3z} = 2 \cdot 5 \approx (5^{0.4307}) \cdot 5^1 = (5^{1.4307})$; so $3z \approx 1.4307$ and $z \approx 0.4769$.

52. *Recreational Boats* From 1970 through 1990, the number of recreational boats, B (in millions), in the United States can be modeled by $B = 6.65(1.04)^t + 2$, where $t = 0$ represents 1970. Sketch the graph of this function. How many recreational boats were owned in the United States in 1984? ***(Source: National Marine Manufacturers Association)*** ≈13.5 million; see Additional Answers.

Integrated Review

In Exercises 53–60, sketch the graph of the function. See Additional Answers.

53. $f(x) = (x + 8)^2$ **54.** $f(x) = -x^2 + 2$ **55.** $f(x) = |x| + 3$ **56.** $f(x) = |x + 2|$

57. $f(x) = \sqrt{x - 9}$ **58.** $f(x) = \sqrt{x} + 9$ **59.** $f(x) = \sqrt[3]{x} + 4$ **60.** $f(x) = \sqrt[3]{x + 1}$

In Exercises 61–64, decide whether the function represents exponential growth or exponential decay. See Additional Answers.

61. $f(x) = \left(\frac{1}{4}\right)^x$ **62.** $f(x) = (6)^x$

63. $f(x) = \left(\frac{3}{2}\right)^x$ **64.** $f(x) = \left(\frac{2}{3}\right)^x$

65. *Powers of Ten* Which best approximates the length of an F-18 Hornet fighter airplane? c

a. 5.6×10^{-1} ft **b.** 5.6×10^0 ft
c. 5.6×10^1 ft **d.** 5.6×10^2 ft

66. *Powers of Ten* Which best approximates the length of an aircraft carrier? c

a. 10.9×10^0 ft **b.** 10.9×10^1 ft
c. 10.9×10^2 ft **d.** 10.9×10^3 ft

Exploration and Extension

The Natural Base **In Exercises 67–70, use the following information.**

The number $e \approx 2.71828$ ~~is~~ called the natural base. Use this approximation of e to evaluate the powers in Exercises 67–70. Round the results to two decimal places.

67. 5^e 79.43 **68.** e^{-4} 0.02 **69.** $(e)^{-\frac{1}{3}}$ 0.72 **70.** $\left(\frac{1}{2}\right)^e$ 0.15

71. *The Natural Exponential Function* Copy and complete the table using $e \approx 2.71828$. Round your answers to two decimal places.

x	−3	−2	−1	0	1	2	3
e^x	?	?	?	?	?	?	?

0.05 0.14 0.37 1 2.72 7.39 20.09

72. *The Natural Exponential Function* Use the result of Exercise 71 to sketch the graph of $f(x) = e^x$. See Additional Answers.

61. Exponential decay **62.** Exponential growth **63.** Exponential growth **64.** Exponential decay

8.2 Logarithmic Functions

What you should learn:

Goal 1 How to evaluate logarithmic expressions and graph logarithmic functions

Goal 2 How to use logarithms in real-life situations

Why you should learn it:

You can model many real-life situations with logarithmic models, such as the slope of a beach relative to the size of the sand particles.

Goal 1 Evaluating Logarithmic Expressions

You know that $2^2 = 4$ and $2^3 = 8$, but for what value of y does

$2^y = 6$?

Because $2^2 < 6 < 2^3$, you would expect the solution to be between 2 and 3. To answer this question exactly, mathematicians defined *logarithms.* In terms of a logarithm, $y = \log_2 6 \approx 2.585$.

Definition of Logarithm to Base *a*

Let a and x be positive numbers, $a \neq 1$. The **logarithm of x with base a** is denoted by $\log_a x$ and is defined as follows.

$$\log_a x = y \quad \text{if and only if} \quad a^y = x$$

The expression $\log_a x$ is read as the "log base a of x." The function $f(x) = \log_a x$ is the **logarithmic function with base *a*.**

This definition tells you that the equations $\log_a x = y$ and $a^y = x$ are equivalent. The first is in logarithmic form, and the second is in exponential form.

Example 1 *Rewriting Exponential and Logarithmic Equations*

	Logarithmic Form	Exponential Form
a.	$\log_2 16 = 4$	$2^4 = 16$
b.	$\log_{10} 10 = 1$	$10^1 = 10$
c.	$\log_3 1 = 0$	$3^0 = 1$
d.	$\log_{10} 0.1 = -1$	$10^{-1} = 0.1$
e.	$\log_2 6 \approx 2.585$	$2^{2.585} \approx 6$ ■

Be sure you see that a logarithm is defined as an exponent. Thus, to evaluate the expression $\log_3 9$, you should ask the question "3 to what power gives 9?" Because the answer is 2, it follows that $\log_3 9 = 2$.

Problem of the Day

Solve for x. $(x - 4)^{x-1} = 1$
1, 5

ORGANIZER

TIME SCHEDULE: All levels, two days

Warm-Up Exercises

1. Solve for x.
   - **a.** $2^x = 16$ $\quad x = 4$
   - **b.** $3^x = 27$ $\quad x = 3$
   - **c.** $5^x = \frac{1}{25}$ $\quad x = -2$
   - **d.** $(-2)^x = -8$ $\quad x = 3$
2. Evaluate the following.
   - **a.** 3^{-3} $\quad \frac{1}{27}$
   - **b.** 5^3 $\quad 125$
   - **c.** $4^{2.345}$ $\quad 25.813$
   - **d.** 10^0 $\quad 1$
   - **e.** 10^{-2} $\quad 0.01$
   - **f.** 10^3 $\quad 1000$

Lesson Resources

Teaching Tools
- Transparencies: 8, 17, 18
- Problem of the Day: 8.2
- Warm-up Exercises: 8.2
- Answer Masters: 8.2

Extra Practice: 8.2
Applications Handbook: p. 21

LESSON Notes

Note: Since logarithms are a new topic for your students, introduce Examples 1–3 and Exercises 1–5 and 7–58 on the first day and Examples 4–6 and Exercises 6 and 59–93 on the second day.

Example 1

It is important to understand the inverse relationship between the exponential and logarithmic functions. Recall that one way to determine the inverse of a function is to interchange the variables in the defining equation.

So, the base-2 exponential function $y = 2^x$ has an inverse defined by the equation $x = 2^y$. This inverse is called the base-2 logarithmic function and can be written as $y = \log_2 x$.

Example 2

Ask students if there is any base, b, such that $\log_b(-1)$ is defined.
There is no such base b. Remember, in the definition of logarithms, the base must be a positive number $b > 0$, $b \neq 1$; so b raised to any real-number power would always be positive.

Example 3

Logarithms were invented in 1614 by Scottish mathematician John Napier. On the advice of the English mathematician Henry Briggs, Napier's logarithmic tables were altered in 1615 to have the base 10, and so "common logarithms" were invented.

Interestingly enough, although logarithms are usually defined as exponents, logarithms were invented before noninteger exponents were used. See Milestones in the pupil text on page 412.

Example 4

Given $f(x) = 3^x$ and $g(x) = \log_3 x$, ask students to evaluate the following.

a. $f(g(27))$ b. $g(f(2))$
a. 27, b. 2

Example 2 *Evaluating Logarithms*

Evaluate the logarithms.

a. $\log_4 16$ **b.** $\log_5 1$
c. $\log_4 2$ **d.** $\log_3(-1)$

Solution In each case, note the question asked.

a. 4 to what power gives 16?

$$4^2 = 16 \longrightarrow \log_4 16 = 2$$

b. 5 to what power gives 1?

$$5^0 = 1 \longrightarrow \log_5 1 = 0$$

c. 4 to what power gives 2?

$$4^{\frac{1}{2}} = 2 \longrightarrow \log_4 2 = \frac{1}{2}$$

d. There is no power of 3 that gives -1. (Think of the graph of $y = 3^x$.) Therefore, $\log_3(-1)$ is undefined. ■

There are some special logarithm values that you should learn to recognize.

Special Logarithm Values

Let a and x be positive real numbers such that $a \neq 1$.

1. $\log_a 1 = 0$ because $a^0 = 1$.
2. $\log_a a = 1$ because $a^1 = a$.
3. $\log_a a^x = x$ because $a^x = a^x$.

The logarithmic function with base 10 is called the **common logarithmic function.** On most scientific calculators, this function can be evaluated with the key [*log*]. To use a calculator to evaluate logarithms to other bases, you can use the **change of base formula.**

$$\log_a x = \frac{\log_{10} x}{\log_{10} a} \qquad \textit{Change of Base Formula}$$

Example 3 *Using a Calculator to Evaluate a Logarithm*

Evaluate $\log_2 6$.

Solution Using the change of base formula, you can write

$$\log_2 6 = \frac{\log_{10} 6}{\log_{10} 2} \approx \frac{0.778}{0.301} \approx 2.585. \quad ■$$

By definition, the functions $f(x) = a^x$ and $g(x) = \log_a x$ are inverses of each other. This means the graph of $g(x) = \log_a x$ is a reflection of the graph of $f(x) = a^x$ in the line $y = x$.

Example 4 Sketching the Graph of a Logarithmic Function

Sketch the graph of $g(x) = \log_2 x$.

Solution One way to sketch the graph is to construct a table of values, plot the points, and connect the points with a smooth curve.

	Without Calculator					With Calculator		
x	$\frac{1}{4}$	$\frac{1}{2}$	1	2	4	3	5	6
$\log_2 x$	−2	−1	0	1	2	1.58	2.32	2.58

Another way to sketch the graph of g is to reflect the graph of $f(x) = 2^x$ in the line $y = x$. The graph is shown at the left. ■

From the graph of $g(x) = \log_2 x$, you can see that the domain of the function is the set of positive numbers and the range is the set of all real numbers.

Example 5 Sketching the Graph of a Logarithmic Function

Sketch the graph of $f(x) = \log_{10} x$.

Solution

	Without Calculator				With Calculator		
x	$\frac{1}{100}$	$\frac{1}{10}$	1	10	2	5	8
$\log_{10} x$	−2	−1	0	1	0.301	0.699	0.903

The graph is shown below. Notice how slowly the graph rises for $x > 1$. On this graph, you would need to move out to $x = 1000$ before the graph rises to $y = 3$.

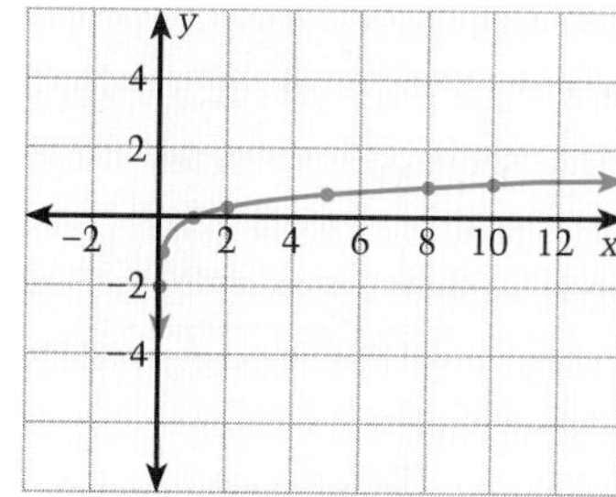

■

Example 5

Because the graph of a logarithmic function rises so slowly, logarithms are ideal for expressing numbers of great magnitude.

Example 6

What is the slope of a beach in which the sand particles are 4 millimeters in diameter?
about 0.23 or a rise of 23 meters for a run of 100 meters

Extra Examples

Here are additional examples similar to **Examples 1–6.**

1. Evaluate without the use of a calculator.
   - **a.** $\log_3 81$ 4
   - **b.** $\log_5 125$ 3
   - **c.** $\log_4 256$ 4
   - **d.** $\log_2 \frac{1}{32}$ −5
2. Evaluate to three decimal places using a calculator as needed.
   - **a.** $\log_2 10$ 3.322
   - **b.** $\log_2 2$ 1
   - **c.** $\log_3 10$ 2.096
   - **d.** $\log_6 216$ 3
   - **e.** $\log_5 12$ 1.544
   - **f.** $\log_5 1$ 0
3. Sketch each graph.
   - **a.** $f(x) = \log_5 x$
   - **b.** $g(x) = \log_3 x$

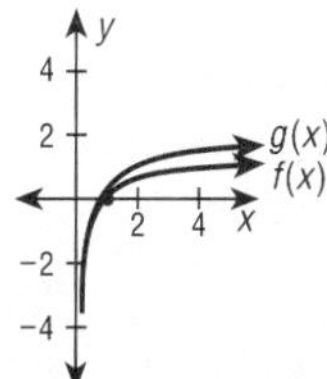

4. Find the slope of a beach if its sand is regarded by oceanographers as "fine"; as "very fine."
   fine sand: $d = 0.125$, slope is about 0.05; very fine sand: $d = 0.0625$, slope is about 0.02

Check Understanding

1. What is the definition of the logarithm of x with base a? See page 405.
2. Explain the statement, "A logarithm is an exponent." See Lesson Notes for Example 1; $\log_a x = y$ implies that $a^y = x$.
3. Explain why the logarithm of a negative integer is undefined. See Lesson Notes for Example 2.
4. Can any positive real number be the base of a logarithm? Explain. No, $b \neq 1$.

COOPERATIVE LEARNING

Have students work with a partner and use graphics calculators to complete the following activity.

Let $f(x) = \log_{10} x$. Ask students to determine the effects of constants on the graph of $f(x)$. In particular, students should substitute both positive and negative values for the constants and then describe the graphs for the functions $g(x)$, $h(x)$, and $j(x)$ in terms of $f(x)$.

a. $g(x) = f(x) + c$, for any real number c
b. $h(x) = f(x + b)$, for any real number b
c. $j(x) = -f(x)$

a. The graph of $g(x)$ represents a vertical shift (translation) of the graph of $f(x)$ $|c|$ units up or down.
b. The graph of $h(x)$ represents a horizontal shift (translation) of the graph of $f(x)$ $|b|$ units left or right.
c. The graph of $j(x)$ is a reflection in the x-axis of the graph of $f(x)$.

The slope of a beach depends on the size of its sand particles. If the sand is very coarse, the slope of the beach will be steeper than if the sand is fine. The photo shows a coral sand beach in St. Thomas, Virgin Islands. Because the sand is fine, the beach has a gentle slope.

Goal 2 Using Logarithms in Real-Life Problems

Real Life **Oceanography**

Diameter (in mm)	Type of Sand Particle
4	Pebble
2	Granule
1	Very coarse sand
0.5	Coarse sand
0.25	Medium sand
0.125	Fine sand
0.0625	Very fine sand

Example 6 *Evaluating a Logarithmic Function*

The slope, s, of a beach is related to the average diameter, d (in millimeters), of the sand particles on the beach by the model

$$s = 0.159 + 0.118 \log_{10} d.$$

Find the slope of the beach if the average diameter of the sand is 0.25 millimeter.

Solution If $d = 0.25$, then the slope of the beach is

$s = 0.159 + 0.118 \log_{10} 0.25$ *Substitute 0.25 for d.*

$\approx 0.159 + 0.118(-0.602)$ *Use a calculator.*

$\approx 0.09.$

The slope of the beach is about 0.09. This is a gentle slope that has a rise of only 9 meters for a run of 100 meters. ■

Communicating about ALGEBRA

SHARING IDEAS about the Lesson See below.

Match the function with its graph. Explain your reasoning.

A. $f(x) = \log_3 x$ **B.** $f(x) = \log_4 x$ **C.** $f(x) = \log_5 x$

1.

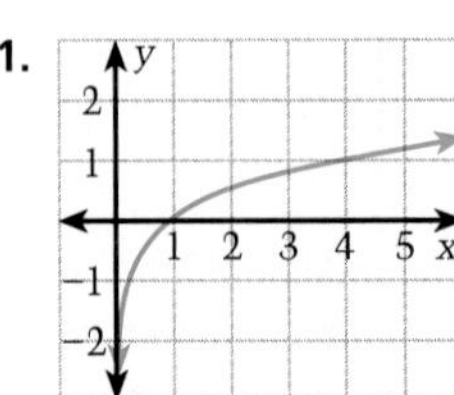

2.

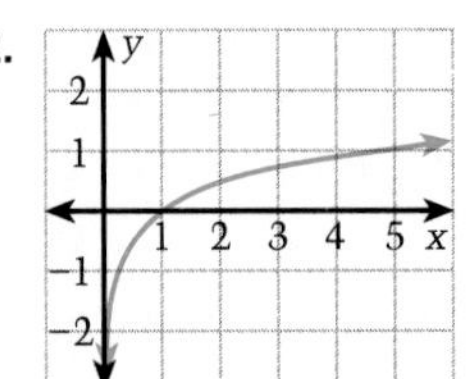

3.

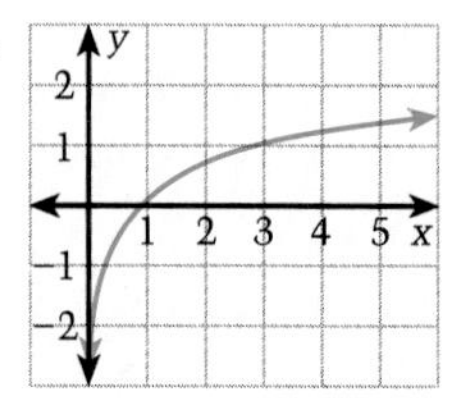

A. 3, the graph includes (3, 1) B. 1, the graph includes (4, 1) C. 2, the graph includes (5, 1)

EXERCISES

Guided Practice

CRITICAL THINKING about the Lesson

1. Write the expression symbolically, "logarithm of 3 with base 2." $\log_2 3$
2. Rewrite $\log_3 x = 1.7$ in exponential form. $3^{1.7} = x$
3. Rewrite $4^x = 8$ in logarithmic form. $\log_4 8 = x$
4. Evaluate $\log_{25} 5$ without a calculator. $\frac{1}{2}$
5. Use the change of base formula to evaluate $\log_6 7$ with a calculator. ≈ 1.086
6. In the coordinate plane at the right, identify the graph of $f(x) = 5^x$ and the graph of $g(x) = \log_5 x$. At left: f; at right: g

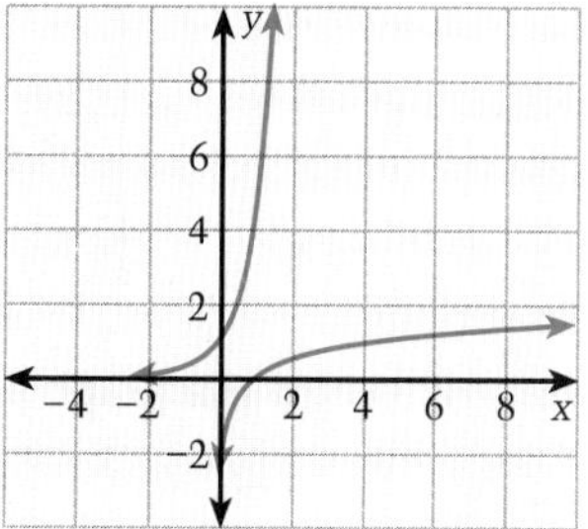

Independent Practice

In Exercises 7–14, write the equation in exponential form.

7. $\log_3 27 = 3$ $3^3 = 27$
8. $\log_4 256 = 4$ $4^4 = 256$
9. $\log_6 36 = 2$ $6^2 = 36$
10. $\log_6 216 = 3$ $6^3 = 216$
11. $\log_2 32 = 5$ $2^5 = 32$
12. $\log_8 64 = 2$ $8^2 = 64$
13. $\log_5 125 = 3$ $5^3 = 125$
14. $\log_1 1 = 3$ $1^3 = 1$

In Exercises 15–22, write the equation in logarithmic form. See margin.

15. $3^5 = 243$
16. $10^{0.12} \approx 1.318$
17. $4^{-1} = \frac{1}{4}$
18. $16^{-\frac{1}{2}} = \frac{1}{4}$
19. $4^3 = 64$
20. $13^2 = 169$
21. $9^{\frac{2}{3}} \approx 4.327$
22. $6^4 = 1296$

In Exercises 23–34, evaluate the expression without using a calculator.

23. $\log_5 25$ 2
24. $\log_9 9^{-0.56}$ -0.56
25. $\log_{12} 1$ 0
26. $\log_9 81$ 2
27. $\log_4 4^{\frac{5}{6}}$ $\frac{5}{6}$
28. $\log_4 2$ $\frac{1}{2}$
29. $\log_{\frac{1}{2}} \frac{1}{2}$ 1
30. $\log_5(-25)$
31. $\log_2 \frac{1}{8}$ -3
32. $\log_9 \frac{1}{3}$ $-\frac{1}{2}$
33. $\log_4(-4)$ Undefined
34. $\log_{10} 10$ 1

30. Undefined

In Exercises 35–38, estimate a range of values for the expression. Then evaluate using the change of base formula. See below.

35. $\log_4 5$
36. $\log_2 5$
37. $\log_6 3$
38. $\log_3 10$

In Exercises 39–46, use a calculator to evaluate the expression.

39. $\log_2 9$ 3.170
40. $\log_4 7$ 1.404
41. $\log_{1.4} 10$ 6.843
42. $\log_5 2$ 0.431
43. $\log_3 20$ 2.727
44. $\log_{0.8} 10$ -10.319
45. $\log_6 -3$ No solution
46. $\log_{\frac{1}{4}} 6$ -1.292

In Exercises 47–58, solve the equation for *x*.

47. $\log_3 81 = x$ 4
48. $\log_2 256 = x$ 8
49. $\log_6 x = -1$ $\frac{1}{6}$
50. $\log_x 343 = 3$ 7
51. $\log_3 x = 5$ 243
52. $\log_4 x = 2$ 16
53. $\log_x 64 = 3$ 4
54. $\log_x 243 = 5$ 3
55. $\log_x 256 = 8$ 2
56. $\log_5 x = 1$ 5
57. $\log_2(x + 1) = 1$ 1
58. $\log_5(x - 4) = 0$ 5

35. Between 1 and 2, ≈ 1.161
36. Between 2 and 3, ≈ 2.322
37. Between 0 and 1, ≈ 0.613
38. Between 2 and 3, ≈ 2.096

EXERCISE Notes

ASSIGNMENT GUIDE

Day 1

Basic/Average: Ex. 1–5, 7–29 odd, 35–53 odd

Above Average: Ex. 1–5, 11–33 odd, 37–38, 41–55 odd

Advanced: Ex. 1–5, 13–33 odd, 37–38, 43–57 odd

Day 2

Basic/Average: Ex. 6, 59–60, 63–66, 67, 69, 75, 79–83, 87–90

Above Average: Ex. 6, 61–66, 71–77 odd, 79–83, 88–91

Advanced: Ex. 6, 61–66, 71–77 odd, 79–83, 91–93

Selected Answers
Exercises 1–6, 7–89 odd

Use **Mixed Review** as needed.

✪ More Difficult Exercises
Exercises 79–83, 91–93

Guided Practice

Ex. 1–6 Use these Guided Practice exercises in class to check for students' understanding of the connection between the exponential and logarithmic functions. The ability to write expressions in both logarithmic and exponential form is an essential prerequisite skill for simplifying logarithmic expressions and solving logarithmic equations.

Answers

15. $\log_3 243 = 5$
16. $\log_{10} 1.318 \approx 0.12$
17. $\log_4 \frac{1}{4} = -1$
18. $\log_{16} \frac{1}{4} = -\frac{1}{2}$
19. $\log_4 64 = 3$
20. $\log_{13} 169 = 2$
21. $\log_9 4.327 \approx \frac{2}{3}$
22. $\log_6 1296 = 4$

Independent Practice

▶ **Ex. 35–38 TECHNOLOGY** Students should use a calculator to evaluate each expression.

▶ **Ex. 47–58** Remind students to express the equations in exponential form before solving them.

▶ **Ex. 63–78** Students can read Example 5 on page 407 if they need help in solving these problems. Encourage students to use graphics calculators with Exercises 67–78.

▶ **Ex. 82–83 COOPERATIVE LEARNING** You may wish to have students work with a partner to complete these exercises as an in-class activity. Help students understand that since $C = 1$ is assigned to living things, the concentration of C in fossils which are no longer living is less than 1.

EXTEND Ex. 82–83 Have students estimate the concentration of carbon-14 atoms in the samples taken from the Indian etchings near Blythe, California.
0.88 (where $t \approx 1100$)

▶ **Ex. 91** This formula is a more generalized change-of-base formula. Students should come to realize that $\log_a b$ and $\log_b a$ are reciprocals of each other.

In Exercises 59–62, find the inverse of the function. See below.

59. $f(x) = \log_6 x$ **60.** $f(x) = 8^x$ **61.** $f(x) = \left(\frac{1}{3}\right)^x$ **62.** $f(x) = \log_{\frac{1}{2}} x$

In Exercises 63–66, match the function with its graph.

63. $f(x) = 1 + \log_2 x$ c **64.** $f(x) = \log_2 x$ a **65.** $f(x) = 1 + \log_4 x$ d **66.** $f(x) = \log_4 x$ b

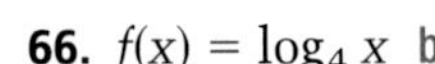

a.

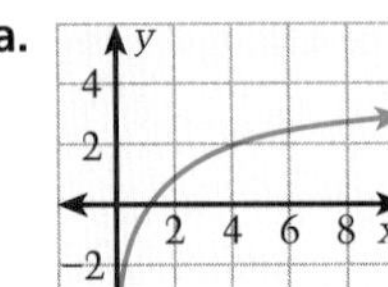

b.

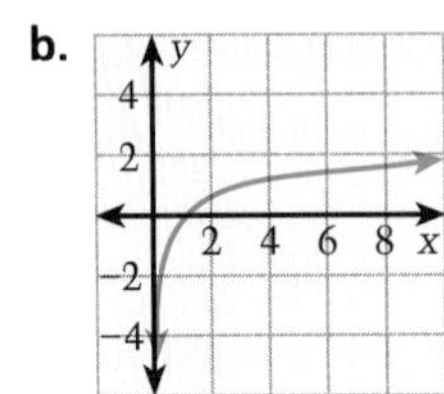

c.

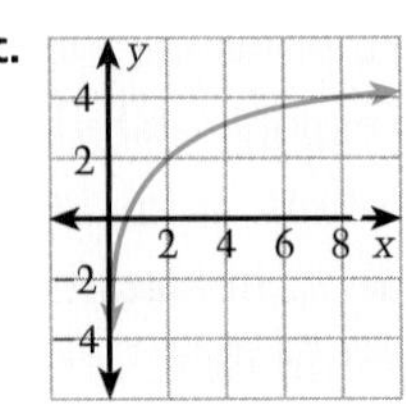

d.

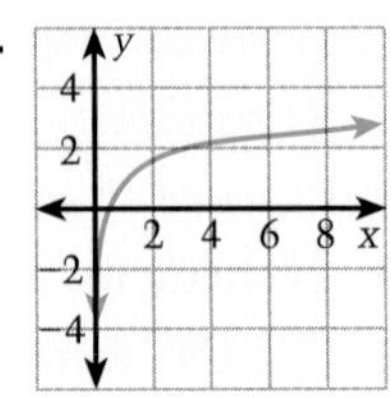

In Exercises 67–78, sketch the graph of the function. See Additional Answers.

67. $f(x) = \log_5 x$ **68.** $f(x) = \log_3 x$ **69.** $f(x) = 1 + \log_2 x$ **70.** $f(x) = 2 + \log_{10} x$

71. $f(x) = \log_8 x$ **72.** $f(x) = -1 + \log_3 x$ **73.** $f(x) = \log_{10} 10x$ **74.** $f(x) = \log_8 2x$

75. $f(x) = \log_2(x - 1)$ **76.** $f(x) = \log_{10}(x - 2)$ **77.** $f(x) = \log_3(x + 2)$ **78.** $f(x) = \log_4(x + 3)$

79. *American Elk* The antler spread, a (in inches), and shoulder height, h (in inches), of an adult male elk are related approximately by the model

$$h = 116 \log_{10}(a + 40) - 176.$$

Approximate the shoulder height of a male American elk. 56 in.

Male American elks grow antlers with a spread of about 5 feet.

80. *Tornadoes* Most tornadoes last less than an hour and travel less than 20 miles. The speed of the wind, S (in miles per hour), near the center of the tornado is related to the distance the tornado travels, d (in miles), by the model

$$S = 93 \log_{10} d + 65.$$

On March 18, 1925, a large tornado struck Missouri, Illinois, and Indiana, covering a distance of 220 miles. Approximate the speed near the center of this tornado. 283 mph

81. *Horn of Gabriel* A portion of the three-dimensional mathematical figure called the horn of Gabriel is shown at the right. The area of the cross section (in the coordinate plane) of the horn is given by

$$A \approx \frac{2}{\log_{10} 2.71828}.$$

Approximate this area. 4.6

59. $g(x) = 6^x$ **60.** $g(x) = \log_8 x$ **61.** $g(x) = \log_{\frac{1}{3}} x$ **62.** $g(x) = \left(\frac{1}{2}\right)^x$

Carbon Dating **In Exercises 82 and 83, use the following information.**

Sunlight produces radioactive carbon (carbon-14), which is absorbed by living plants and animals. Once the plant or animal dies, it stops absorbing carbon-14. Using carbon-14's half-life of 5700 years, scientists can approximate the age of plant or animal fossils by measuring the fossil's concentration, C, of carbon-14 atoms. For living things, C is given a value of 1. Thus, C is related to the number of years, t (since the plant or animal died), by the model $t = 5700 \log_{\frac{1}{2}} C$.

82. For a particular plant fossil, the concentration of carbon-14 atoms is one tenth of what it was when the plant was alive. About how long ago did the plant die? 18,935 years

83. A particular animal fossil has $\frac{1}{8}$ of the carbon-14 atoms that it had when the animal was alive. How long ago did the animal die? 17,100 years

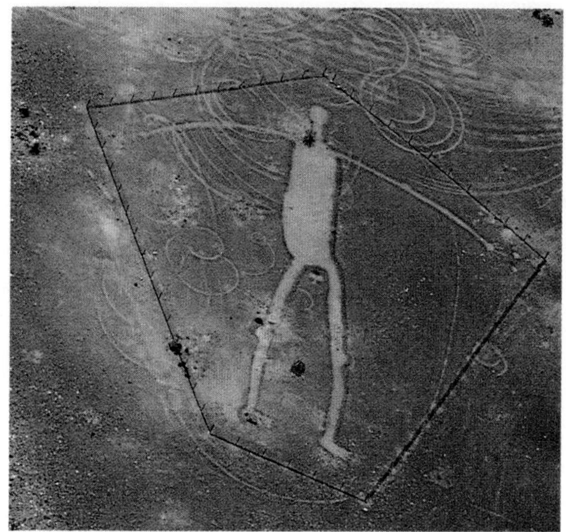

Hundreds of figures, like this 170-foot giant, were etched in the desert near Blythe, California, by prehistoric natives of California. In 1952, historians concluded that the figures were created between the sixteenth and nineteenth centuries. Later, carbon dating of organic matter that grew on the gravel showed that they were created about A.D. 900.

Integrated Review

In Exercises 84–88, sketch the graph of the function. See Additional Answers.

84. $y = 3^x$ **85.** $y = \left(\frac{1}{3}\right)^x$ **86.** $y = 4^x - 3$ **87.** $y = 4^{x-2}$ **88.** $y = \left(\frac{1}{2}\right)^x + 3$

89. *College Entrance Exam Sample* A baseball team has won 40 games out of 60 played. It has 32 more games to play. How many of these must the team win to obtain a 75% win record for the year? b

a. 26 **b.** 29 **c.** 28 **d.** 30 **e.** 32

90. *College Entrance Exam Sample* The enrollment at a university was 3000 in 1950 and 12,000 in 1975. By what percent did the enrollment increase? c

a. 125% **b.** 25% **c.** 300% **d.** 400% **e.** 3%

Exploration and Extension

See Additional Answers.

91. *Change of Base* Use the formula $\log_a x = \dfrac{\log_b x}{\log_b a}$ to verify that $\log_a b = \dfrac{1}{\log_b a}$.

In Exercises 92 and 93, verify the equation without using a calculator.

92. $\log_2 4 = \dfrac{1}{\log_4 2}$ $2 = \dfrac{1}{\frac{1}{2}}$

93. $\log_5 125 = \dfrac{1}{\log_{125} 5}$ $3 = \dfrac{1}{\frac{1}{3}}$

Enrichment Activities

These activities require students to go beyond lesson goals.

1. TECHNOLOGY To graph a function like $y = \log_2 x$ using a graphics calculator, you use the change-of-base formula to graph $y = \dfrac{\log_{10} x}{\log_{10} 2}$. Graph the functions y_1, y_2, and y_3 on the same coordinate grid for the following range:

X min = −5 Y min = −6
X max = 5 Y max = 4
X scl = 1 Y scl = 1
X res = 1

$y_1 = \log_2 x$, $y_2 = \log_{\frac{1}{2}} x$, $y_3 = \log_2 x^2$

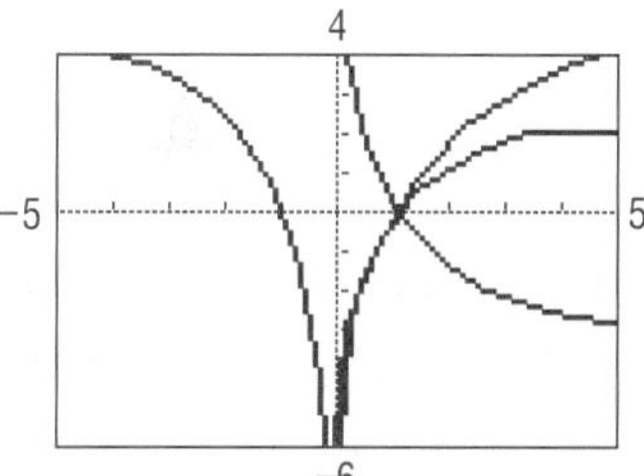

2. What point is contained in all three graphs? Why?
The point (1, 0), since $\log_b 1 = 0$ for any base b.

3. What do you notice about the graphs of y_1 and y_2?
They are reflections of each other in the x-axis, so $\log_2 x = -\log_{\frac{1}{2}} x$.

4. Show algebraically that the expressions $\log_a x$ and $-\log_{\frac{1}{a}} x$ are equivalent.
If $\log_a x = y$, then $a^y = x$. Since $a = \left(\frac{1}{a}\right)^{-1}$, $a^y = \left(\left(\frac{1}{a}\right)^{-1}\right)^y = \left(\frac{1}{a}\right)^{-y} = x$. Therefore, $\log_{\frac{1}{a}} x = -y$ and $-\log_{\frac{1}{a}} x = y$. So, $\log_a x$ and $-\log_{\frac{1}{a}} x$ are equivalent.

5. Why is y_3 defined for $x < 0$? Compare the graphs of y_1 and y_3 for $x > 1$. What do you notice?
The square of a negative number is positive; the y-coordinates of y_3 are double the y-coordinates of y_1.

Mixed Review Answer

16.

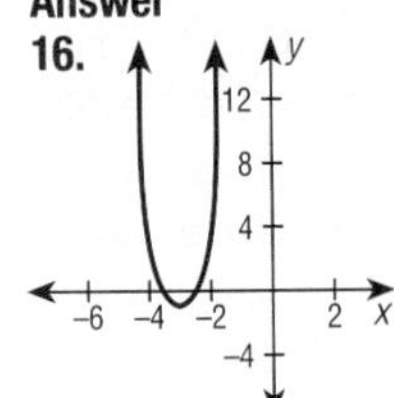

Mixed REVIEW

1. Simplify: $(a^2b^0c^4)(a^{-5}b^{-3}c^{-2})$. **(7.1)** $\frac{c^2}{a^3b^3}$
2. Simplify: $\sqrt{x^2}$. **(7.4)** $|x|$
3. Simplify: $\sqrt[3]{-27}$. **(7.4)** -3
4. Simplify: $\sqrt{50a^2b^5}$. **(7.4)** $5|a|b^2\sqrt{2b}$
5. Simplify: $(3^{2/3})^{3/5}$. **(7.4)** $3^{2/5}$
6. Simplify: $\sqrt{3} \cdot \sqrt{6}$. **(7.4)** $3\sqrt{2}$
7. Simplify: $(3i^4)(2i^3)$. **(5.5)** $-6i$
8. Solve $\sqrt{5x + 1} - 2 = 4$. **(7.5)** 7
9. In which quadrant is $(5, -2)$? **(2.6)** IV
10. Find the slope of $3x - 4y = -2$. **(2.2)** $\frac{3}{4}$
11. Find the vertex of $y = 2|x - 7| + 5$. **(2.5)** (7, 5)
12. Find the distance from (3, 5) to (6, 1). **(7.5)** 5
13. For $f(1) = -1$ and $f(n + 1) = 2(-1)^n f(n)$, find $f(4)$. **(6.6)** -8
14. For $f(x) = x - 3$ and $g(x) = x^2 + 2$, find $g(f(x))$. **(6.2)** $x^2 - 6x + 11$
15. Find the geometric mean of 4 and 16. **(7.5)** 8
16. Sketch the graph of $y = 4(x + 3)^2 - 1$. **(5.2)** See margin.
17. Simplify: $\frac{\sqrt[5]{64}}{\sqrt[5]{2}}$. **(7.4)** 2
18. Find the domain of $y = \frac{2}{\sqrt{x - 3}}$. **(6.2)** $x > 3$
19. \$1200 saved at 8% compounded quarterly will be how much in 6 years? **\$1930.12**
20. Write an equation of the line passing through $(2, -5)$ and $(-1, -3)$. **(2.4)** $y = -\frac{2}{3}x - \frac{11}{3}$

Milestones LOGARITHMS

1600 | 1620 | 1640 | 1660 | 1680 | 1700

Napier introduces logarithms — Briggs publishes common logarithm tables — Recognition of relationship between logarithms and exponents

As the field of mathematics grew through the centuries, so did the need for computational tools. In 1614, a Scotsman, John Napier, introduced logarithms. Logarithms made it possible to use addition and subtraction in place of multiplication and division. Pierre-Simon Laplace said the invention of logarithms, "by shortening the labors, doubled the life of the astronomer."

John Napier

Henry Briggs, an English geometry professor, was excited by Napier's work and traveled to Scotland to meet him. They decided that a convenient logarithm system would be based on the number 10. From their visit, came the system of common logarithms. In 1624, Briggs published a book of logarithms of the numbers between 1 and 20,000 and from 90,000 to 100,000.

The invention of logarithms was remarkable in many ways. It made decimals, which were slowly coming into use, a necessity. Any discussion of logarithms is certain to include exponents, but their relationship was not recognized until 1685.

Find out who developed logarithms independently in Switzerland shortly after Napier developed them in Scotland.

Jobst Bürgi, a watchmaker

Milestones

Language Arts: *Critical Reading*

1. First tables and now calculators are used to solve logarithmic equations. What other instrument has been used to solve logarithmic equations?
   slide rule
2. Why do you think it made sense to base a system of logarithms on the number 10?
   It made sense because the Hindu-Arabic number system, the one we use, is a base-10 system.

8.3 Properties of Logarithms

What you should learn:

Goal 1 How to use properties of logarithms

Goal 2 How to use logarithms to solve real-life problems

Why you should learn it:

You can model many real-life situations with logarithms, such as the intensity of earthquakes as measured by the Richter scale.

Goal 1 Using Properties of Logarithms

Since the logarithm function with base a is the inverse of the exponential function with base a, it makes sense that each property of exponents should have a corresponding property of logarithms. For instance, the exponential property $a^0 = 1$ has the corresponding logarithmic property $\log_a 1 = 0$.

In this lesson, you will study the properties of logarithms that correspond to the following three properties of exponents.

1. $a^n a^m = a^{n+m}$ **2.** $\dfrac{a^n}{a^m} = a^{n-m}$ **3.** $(a^n)^m = a^{nm}$

Properties of Logarithms

Let a, u, and v be positive numbers such that $a \neq 1$, and let n be any real number.

1. $\log_a(uv) = \log_a u + \log_a v$ *Product Property*

2. $\log_a \dfrac{u}{v} = \log_a u - \log_a v$ *Quotient Property*

3. $\log_a u^n = n \log_a u$ *Power Property*

Note that there is no general property of logarithms that can be used to simplify $\log_a(u + v)$. For instance, $\log_a(u + v)$ does not equal $\log_a u + \log_a v$.

Example 1 *Demonstrating Properties of Logarithms*

Use $\log_{10} 2 \approx 0.301$ and $\log_{10} 3 \approx 0.477$ to approximate the following.

a. $\log_{10} \frac{2}{3}$ **b.** $\log_{10} 6$ **c.** $\log_{10} 9$

Solution

a. $\log_{10} \frac{2}{3} = \log_{10} 2 - \log_{10} 3 \approx 0.301 - 0.477 = -0.176$

b. $\log_{10} 6 = \log_{10}(2 \cdot 3) = \log_{10} 2 + \log_{10} 3$
$\approx 0.301 + 0.477 = 0.778$

c. $\log_{10} 9 = \log_{10}(3^2) = 2 \log_{10} 3 \approx 2(0.477) = 0.954$ ■

Problem of the Day

If the area and perimeter of an equilateral triangle have the same numerical value, what is the radius of the inscribed circle?
2

ORGANIZER

TIME SCHEDULE: All levels, two days

Warm-Up Exercises

1. Use the properties of exponents to simplify.
   - a. $4^3 \cdot 4^2$ 1024
   - b. $2^4 \cdot 2^5$ 512
   - c. $\frac{3^6}{3^4}$ 9
   - d. $\frac{7^2}{7^5}$ $\frac{1}{343}$
   - e. $(6^2)^2$ 1296
2. Use a calculator to evaluate the following to three decimal places.
   - a. $\log_{10} 2$ 0.301
   - b. $\log_{10} 3$ 0.477
   - c. $\log_{10} 5$ 0.699
   - d. $\log_{10} 7$ 0.845

Lesson Resources

Teaching Tools
- Problem of the Day: 8.3
- Warm-up Exercises: 8.3
- Answer Masters: 8.3

Extra Practice: 8.3

Applications Handbook: p. 34, 35, 57

LESSON Notes

Note: Since computing with logarithms is a new topic for your students, introduce Goal 1 (Examples 1–5 and Exercises 1–38) on the first day and Goal 2 (Example 6 and Exercises 39–74) on the second day.

Example 1

You might have students verify that $\log_a (u + v)$ does not equal $\log_a u + \log_a v$ by evaluating such logarithms as $\log_{10} 5$ in these two ways.

$\log_{10} (2 + 3) = \log_{10} 5 \approx 0.699$. But, $\log_{10} 2 + \log_{10} 3 \approx 0.301 + 0.477 = 0.778$.

Examples 2–5

Expanding and condensing logarithmic expressions is useful in solving logarithmic equations. Suppose, in Example 3, that $\log_{10} x - \log_{10} 3 = 2$; then, from the condensed expression, $\log_{10} \left(\frac{x}{3}\right) = 2$, and $\frac{x}{3} = 100$, so $x = 300$.

In Example 2, you can solve $\log_{10} 7x^3 = 2.345$ for x by using an expanded expression.

$$\begin{aligned} \log_{10} 7x^3 &= 2.345 \\ \log_{10} 7 + 3 \log_{10} x &= 2.345 \\ 3 \log_{10} x &\approx 1.5 \\ \log_{10} x &\approx 0.5 \\ x &\approx 3.162 \end{aligned}$$

Example 6

Suppose an earthquake in San Diego, California is half the intensity of the 1989 earthquake in San Francisco. What would it measure on the Richter scale? $R \approx 6.8$

Extra Examples

Here are additional examples similar to **Examples 1–5.**

1. Evaluate each expression using the values of $\log_{10} 2$, $\log_{10} 3$, $\log_{10} 5$, and $\log_{10} 7$ obtained in Warm-up Exercise 2.
   - **a.** $\log_{10} 35$ 1.544
   - **b.** $\log_{10} \left(\frac{3}{5}\right)$ −0.222
   - **c.** $\log_{10} 147$ 2.167
   - **d.** $\log_{10} 125$ 2.097

When using the properties of logarithms, you may find that it helps to remember verbal forms of the properties. For instance, the verbal form of $\log_a(uv) = \log_a u + \log_a v$ is "The log of a product is the sum of the logs of the factors." Similarly, the verbal form of $\log_a \frac{u}{v} = \log_a u - \log_a v$ is "The log of a quotient is the difference of the logs of the numerator and denominator."

Example 2 *Rewriting the Logarithm of a Product*

Use the properties of logarithms to rewrite $\log_{10} 7x^3$.

Solution

$$\begin{aligned} \log_{10} 7x^3 &= \log_{10} 7 + \log_{10} x^3 && \textit{Product Property} \\ &= \log_{10} 7 + 3 \log_{10} x && \textit{Power Property} \end{aligned}$$ ■

When a logarithmic expression is rewritten as in Example 2, it is called expanding the expression. The reverse procedure is demonstrated in Example 3 and is called condensing a logarithmic expression.

Example 3 *Condensing a Logarithmic Expression*

Condense $\log_{10} x - \log_{10} 3$.

Solution

$$\log_{10} x - \log_{10} 3 = \log_{10} \frac{x}{3} \quad \textit{Quotient Property}$$ ■

Example 4 *Expanding a Logarithmic Expression*

Expand $\log_2 3xy^2$.

Solution

$$\begin{aligned} \log_2 3xy^2 &= \log_2 3 + \log_2 x + \log_2 y^2 && \textit{Product Property} \\ &= \log_2 3 + \log_2 x + 2 \log_2 y && \textit{Power Property} \end{aligned}$$ ■

Example 5 *Condensing a Logarithmic Expression*

Condense $\log_{10} 2 - 2 \log_{10} x$.

Solution

$$\begin{aligned} \log_{10} 2 - 2 \log_{10} x &= \log_{10} 2 - \log_{10} x^2 && \textit{Power Property} \\ &= \log_{10} \frac{2}{x^2} && \textit{Quotient Property} \end{aligned}$$ ■

The Richter scale is used to measure the magnitude of an earthquake. The magnitude, R, of an earthquake of intensity I is given by R = $\log_{10}$ I, where the intensity is a measure of the wave energy of an earthquake per unit of area. The photo at the right shows the devastation caused by the San Francisco Earthquake of 1989, which measured 7.1 on the Richter scale.

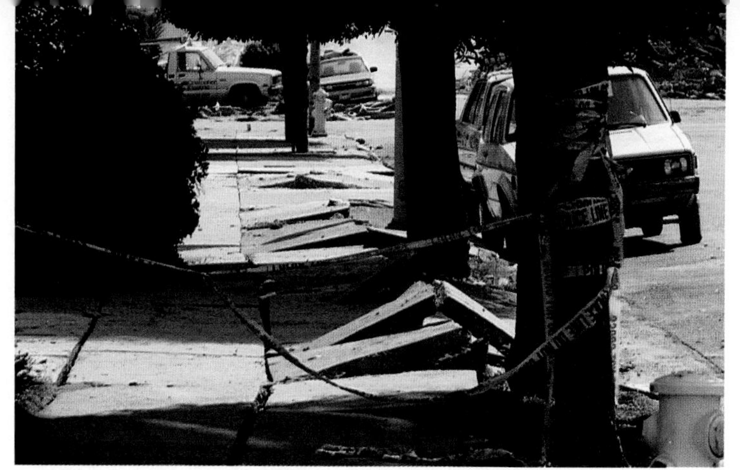

Goal 2 Using Logarithms to Solve Problems

Example 6 *Using Properties of Logarithms*

Find the intensity per unit of area for the earthquakes listed.

a. Great San Francisco Earthquake in 1906, $R = 8.3$

b. San Francisco Bay Area Earthquake in 1989, $R = 7.1$

Solution

a. Using the formula $R = \log_{10} I$ and $R = 8.3$, you can write $8.3 = \log_{10} I$ or

$$I = 10^{8.3} \approx 199{,}526{,}000.$$

b. Using $R = 7.1$, you can write $7.1 = \log_{10} I$ or

$$I = 10^{7.1} \approx 12{,}589{,}000.$$

Note that an increase of 1.2 units on the Richter scale (from 7.1 to 8.3) represents an intensity change by a factor of

$$\frac{199{,}526{,}000}{12{,}589{,}000} \approx 16.$$

In other words, the earthquake in 1906 had an intensity about 16 times greater than that of the one in 1989. ■

Communicating about ALGEBRA

SHARING IDEAS about the Lesson

The energy, E (in ergs), of an earthquake is related to its magnitude, R (on the Richter scale), by the equation $11.4 + 1.5R = \log_{10} E$. Find the energy of the two San Francisco earthquakes in Example 6. See below.

The 1906 earthquake was $\approx 7.08 \times 10^{23}$ ergs; the 1989 earthquake was $\approx 1.12 \times 10^{22}$ ergs

2. Expand the logarithmic expressions.

a. $\log_{10} 5mn$

b. $\log_5 8x^3$

c. $\log_2 \left(\frac{3y^2}{4}\right)$

a. $\log_{10} 5 + \log_{10} m + \log_{10} n$
b. $\log_5 8 + 3 \log_5 x$
c. $\log_2 3 + 2 \log_2 y - \log_2 4$

3. Condense the following logarithmic expressions.

a. $5 \log_5 x$

b. $\log_5 7 + 3 \log_5 t$

c. $3 \log_2 x - (\log_2 4 + \log_2 y)$

a. $\log_5 x^5$, b. $\log_5 7t^3$, c. $\log_2 \left(\frac{x^3}{4y}\right)$

Check Understanding

1. If the logarithms of prime numbers are known, explain how the logarithms of positive whole numbers can be determined.
Every whole number greater than 1 is either prime or composite. Composite numbers can be expressed in terms of prime numbers.

2. Distinguish between condensing and expanding logarithmic expressions.
See Examples 3–5.

Communicating about ALGEBRA

You might wish to show students a derivation of the change-of-base formula.

$$\log_a x = \frac{\log_{10} x}{\log_{10} a}$$

Use the properties of logarithms. Begin by noting that:
If $y = \log_a x$, then $a^y = x$.
Take the logarithm of both sides:
$\log_{10} a^y = \log_{10} x$.
Apply the Power Property of logarithms:
$y \cdot \log_{10} a = \log_{10} x$.
Solve for y:
$$y = \frac{\log_{10} x}{\log_{10} a}.$$
But $y = \log_a x$, so substitute:
$$\log_a x = \frac{\log_{10} x}{\log_{10} a}.$$

EXERCISE Notes

ASSIGNMENT GUIDE

Day 1

Basic/Average: Ex. 1–4, 5–37 odd

Above Average: Ex. 1–4, 5, 7, 15–37 odd

Advanced: Ex. 1–4, 5, 7, 15–33 odd

Day 2

Basic/Average: Ex. 39–43, 50–56, 57–67 odd, 68, 73–74

Above Average: Ex.39–43, 47–56, 61, 63, 68–74

Advanced: Ex. 39–43, 47–56, 62, 64, 68–74

Selected Answers
Exercises 1–4, 5–69 odd

✪ More Difficult Exercises
Exercises 47–56, 71–74

Guided Practice

▶ **Ex. 1–4** Use these Guided Practice exercises in class to emphasize the appropriate use of the properties of exponents. You may wish to direct the students' attention to proofs of the properties of logarithms in the *Handbook of Mathematical Connections* on page 847.

Independent Practice

▶ **Ex. 9–38** Students can use Examples 2–5 as solution models.

Answers

9. $\log_2 3 + \log_2 x$
10. $\log_8 16 + \log_8 x$
11. $\log_{10} 2 + 3\log_{10} x$
12. $3\log_3 x$
13. $\log_4 6 - \log_4 5$
14. $2\log_5 3$

416

EXERCISES

Guided Practice

CRITICAL THINKING about the Lesson

1. Which is equivalent to $\log_{10}\left(\frac{5}{7}\right)^2$? Explain. a, because of the quotient and power properties of logarithms

a. $2(\log_{10} 5 - \log_{10} 7)$ **b.** $\dfrac{2\log_{10} 5}{\log_{10} 7}$

2. Use a property of logarithms to help evaluate $\log_3 27^5$. 15

3. Use $\log_2 5 \approx 2.322$ and $\log_2 3 \approx 1.585$ to approximate $\log_2 15$. 3.907

4. Which is equivalent to $\log_6(2x^2 + 5)$? Explain.
a. $\log_6 2x^2 + \log_6 5$ **b.** $\log_6 2x^2 \cdot \log_6 5$ **c.** Neither **a** nor **b** is equivalent.
c, logarithm properties refer to products and quotients, not sums.

Independent Practice

In Exercises 5–8, use $\log_{10} 3 \approx 0.477$ and $\log_{10} 12 \approx 1.079$ to approximate the value of the expression. Use a calculator to verify your results.

5. $\log_{10} 4$ 0.602 **6.** $\log_{10} \frac{1}{4}$ −0.602 **7.** $\log_{10} 36$ 1.556 **8.** $\log_{10} 144$ 2.158

In Exercises 9–24, expand the expression. See margin.

9. $\log_2 3x$ **10.** $\log_8 16x$ **11.** $\log_{10} 2x^3$ **12.** $\log_3 x^3$
13. $\log_4 \frac{6}{5}$ **14.** $\log_5 9$ **15.** $\log_6 \frac{10}{3}$ **16.** $\log_3 6xy$
17. $\log_{10} 7x^3yz$ **18.** $\log_6 36x^2$ **19.** $\log_2 x^{1/2}y^3$ **20.** $\log_3 12^{2/3}x^7$
21. $\log_{10} \sqrt{x}$ **22.** $\log_2 \sqrt[3]{x}$ **23.** $\log_{10} \sqrt[3]{x^2}$ **24.** $\log_2 \sqrt{4x}$

In Exercises 25–38, condense the expression.

25. $\log_5 6 - \log_5 4$ $\log_5 \frac{3}{2}$
26. $\log_3 13 + \log_3 3$ $\log_3 39$
27. $2\log_{10} x + \log_{10} 5$ $\log_{10} 5x^2$
28. $5\log_4 12 - 5\log_4 2$ $\log_4 7776$
29. $3\log_3 19 - 3\log_3 38$ $\log_3 \frac{1}{8}$
30. $\log_7 48 - 4\log_7 2$ $\log_7 3$
31. $\log_{10} 8 + \log_{10} x + 2\log_{10} y$ $\log_{10} 8xy^2$
32. $\log_{10} 6 - 3\log_{10} \frac{1}{3}$ $\log_{10} 162$
33. $6\log_8 2 + 2\log_8 x + 2\log_8 y$ $\log_8 64x^2y^2$
34. $\log_3 2 + \frac{1}{2}\log_3 y$ $\log_3 2y^{1/2}$
35. $10\log_{10} x + \frac{2}{3}\log_{10} 64$ $\log_{10} 16x^{10}$
36. $2\log_{10} 9 + 5\log_{10} x + \log_{10} \frac{1}{3}$ $\log_{10} 27x^5$
37. $2(\log_5 18 - \log_5 3) + \frac{1}{2}\log_5 \frac{1}{16}$ $\log_5 9$
38. $\frac{1}{3}\log_4 27 - \left(2\log_4 6 - \frac{1}{2}\log_4 81\right)$ $\log_4 \frac{3}{4}$

In Exercises 39–46, use properties of logarithms to solve for *x*.

39. $\log_4 2 - \log_4 x = \log_4 \frac{2}{3}$ 3
40. $\log_3 6 = \log_3 3 + \log_3 x$ 2
41. $\log_4 5 = \log_4 10 - \log_4 x$ 2
42. $\log_3 3 = \log_3 x - \log_3 3$ 9
43. $\log_3 8 = x\log_3 2$ 3
44. $\log_{10} 16 = x\log_{10} 2$ 4
45. $\log_2 x = \frac{1}{2}\log_2 4$ 2
46. $\frac{1}{3}\log_4 x = \log_4 4$ 64

Molecular Transport **In Exercises 47–49, use the following information.**

The energy, E (in kilocalories per gram molecule), required to transport a substance from the outside to the inside of a living cell is given by

$$E = 1.4(\log_{10} C_2 - \log_{10} C_1),$$

where C_1 is the concentration of the substance outside the cell and C_2 is the concentration inside the cell.

47. Condense the given equation. $E = \log_{10}\left(\frac{C_2}{C_1}\right)^{1.4}$

48. The concentration of a particular substance inside a cell is twice the concentration outside the cell. How much energy is required to transport the substance from outside to inside the cell? ≈0.42 kilocalories per gram molecule

49. The concentration of a particular substance inside a cell is six times the concentration outside the cell. How much energy is required to transport the substance from outside to inside the cell? ≈1.09 kilocalories per gram molecule

50. *Human-Memory Model* Students in a ninth grade class were given an exam on French vocabulary words. During the next two years, the same students were retested several times. The average score, S, of the class can be modeled by

$$S = 90 - \log_{10}(t + 1)^{15},$$

where t is the time in months since the original exam. Expand this model. What was the average score after 1 year? $S = 90 - 15\log_{10}(t + 1)$, ≈73

Photography **In Exercises 51 and 52, use the following information.** See margin.

The f-stops on a 35-millimeter camera control the amount of light that enters a camera. Let s be a measure of the amount of light that strikes the film and let f be the f-stop. Then s and f are related by the model

$$s = \log_2 f^2.$$

51. The table shows the first eight f-stops on a 35-millimeter camera. Copy and complete the table and describe the pattern.

f	1.414	2.000	2.828	4.000
s	?	?	?	?

f	5.657	8.000	11.314	16.000
s	?	?	?	?

52. Many 35-millimeter cameras have nine f-stops. What do you think the ninth f-stop is? Explain your reasoning.

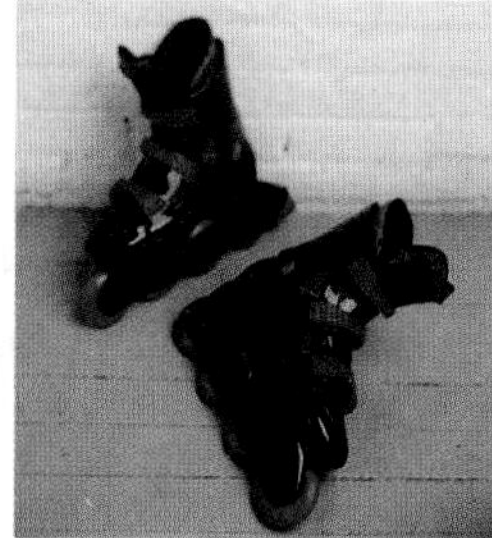

The two top photos were taken with an f-stop of 2. With that setting, the camera could not be focused on both near and far subjects. The bottom photo was taken with an f-stop of 16. With that setting, all figures appear in focus.

Answers
15. $\log_6 10 - \log_6 3$
16. $\log_3 6 + \log_3 x + \log_3 y$
17. $\log_{10} 7 + 3\log_{10} x + \log_{10} y + \log_{10} z$
18. $2\log_6 6 + 2\log_6 x$
19. $\frac{1}{2}\log_2 x + 3\log_2 y$
20. $\frac{2}{3}\log_3 12 + 7\log_3 x$
21. $\frac{1}{2}\log_{10} x$
22. $\frac{1}{3}\log_2 x$
23. $\frac{2}{3}\log_{10} x$
24. $1 + \frac{1}{2}\log_2 x$

Ex. 39–46 Explain to students that $\log_a u = \log_a v$ if and only if $u = v$. For example, in Exercise 41, students should condense the right side of the equation to obtain $\log_4 5 = \log_4\left(\frac{10}{x}\right)$, which implies that $5 = \frac{10}{x}$ and $x = 2$.

Ex. 47–49 Assign these exercises as a group. Help students understand that twice the amount means the ratio of C_2 to C_1 is 2 to 1.

Ex. 51–52 If possible, have a student photographer show the class a 35-mm camera and explain the various uses of the *f*-stop feature.

Answers
51. 1.000, 2, 3.000, 4, 5.000, 6, 7.000, 8; pattern for s: consecutive integers
52. ≈22.627; for the ninth *f*-stop, $s = 9$, $9 = \log_2 f^2$, and $f \approx 22.627$

Ex. 65–69 TECHNOLOGY Students should use a calculator to evaluate.

EXTEND Ex. 71–72 TECHNOLOGY Ask students to graph both the condensed versions and the expanded versions to check that they are equal. For example, in Exercise 71, students would graph these equations:
$y_1 = \log_{10} 2x - \log_{10} 2$ and
$y_2 = \log_{10} x$.

Answers

71.

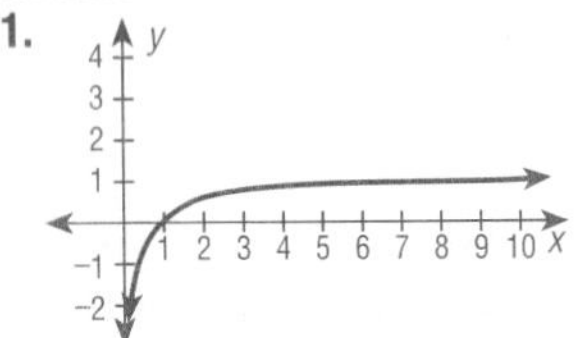

72.

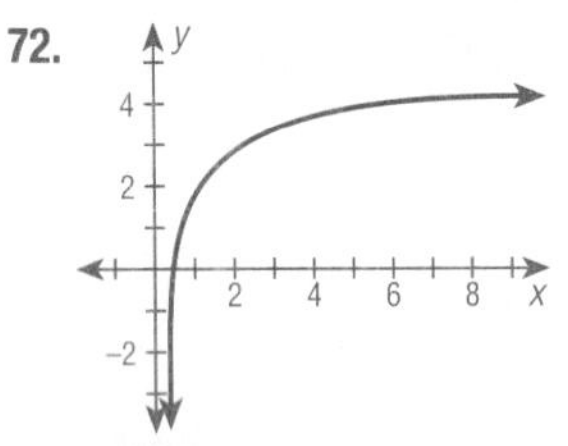

73. False;
left side: $\log_8 6 - \log_8 3 + \log_8 2$,
right side: $\log_8 6 - \log_8 3 - \log_8 2$

74. True;
left side: $2(\log_5 4) - 2(\log_5 3)$
right side: $2(\log_5 4) - 2(\log_5 3)$

Enrichment Activity

This activity requires students to go beyond lesson goals.

For about 300 years, prior to the technological advances of inexpensive calculators and computers, logarithm computing machines called slide rules were used by students, scientists, and mathematicians to make the cumbersome tasks of multiplication, division, raising to powers, and finding roots easier.

Intensity of Sound **In Exercises 53–56, use the following information.** See below.

The level of sound, B (in decibels), is related to the intensity of the sound, I (in watts per square centimeter), by the model

$$B = 10 \log_{10} \frac{I}{I_0},$$

where I_0 is an intensity of 10^{-16} watts per square centimeter, corresponding roughly to the faintest sound that can be heard by humans.

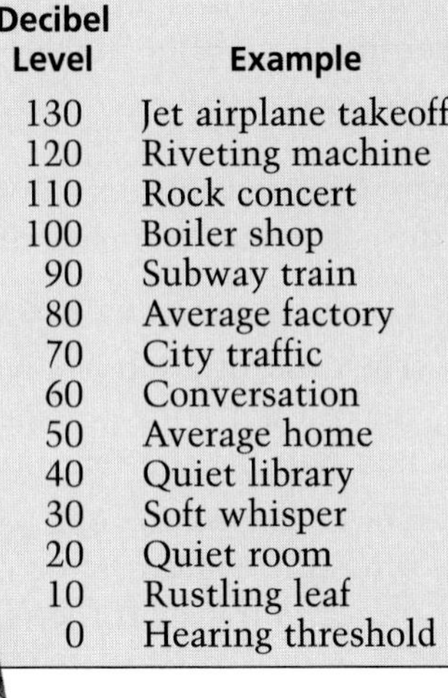

Decibel Level	Example
130	Jet airplane takeoff
120	Riveting machine
110	Rock concert
100	Boiler shop
90	Subway train
80	Average factory
70	City traffic
60	Conversation
50	Average home
40	Quiet library
30	Soft whisper
20	Quiet room
10	Rustling leaf
0	Hearing threshold

53. What is the sound intensity in an average home?

54. What is the sound intensity of city traffic?

55. Sound levels above 120 decibels can cause pain. What is the intensity for such a sound?

56. What is the intensity of a jet airplane takeoff?

Integrated Review

In Exercises 57–64, evaluate the expression without a calculator.

57. $\log_3 243$ 5

58. $\log_2 16$ 4

59. $\log_4 16$ 2

60. $\log_3 27$ 3

61. $\log_4 256$ 4

62. $\log_5 125$ 3

63. $\log_2 32$ 5

64. $\log_4 1024$ 5

In Exercises 65–70, evaluate the expression.

65. $\log_{10} x$ when $x = 5$ ≈0.700

66. $\log_{10} 2x$ when $x = 2$ ≈0.602

67. $3 \log_{10}\left(\frac{1}{2}x\right)$ when $x = 6$ ≈1.431

68. $\log_6 x$ when $x = 9$ ≈1.23

69. $\log_3 3x$ when $x = 2$ ≈1.631

70. $\log_4 x^2$ when $x = 4$ 2

Exploration and Extension

In Exercises 71 and 72, sketch the graph of the function. 71.–74. See margin.

71. $y = \log_{10} 2x - \log_{10} 2$

72. $y = \log_4 8x + \log_4 2x$

73. True or False? Explain. $(\log_8 6 - \log_8 3) + \log_8 2 = \log_8 6 - (\log_8 3 + \log_8 2)$

74. True or False? Explain. $2(\log_5 4 - \log_5 3) = \log_5 4^2 - \log_5 3^2$

53. 10^{-11} watts per cm^2 54. 10^{-9} watts per cm^2 55. 10^{-4} watts per cm^2 56. 10^{-3} watts per cm^2

Mid-Chapter SELF-TEST

Take this test as you would take a test in class. The answers to the exercises are given in the back of the book.

In Exercises 1–3, sketch the graph of the function. (8.1) See margin.

1. $h(x) = \left(\frac{1}{3}\right)^x - 4$ **2.** $h(x) = 5^x + 7$ **3.** $h(x) = 4^{x-2} + 1$

In Exercises 4–6, write the equation in exponential form. (8.2)

4. $\log_2 8 = 3$ $2^3 = 8$ **5.** $\log_5 5 = 1$ $5^1 = 5$ **6.** $\log_7 1 = 0$ $7^0 = 1$

In Exercises 7–9, write the equation in logarithmic form. (8.2)

7. $6^0 = 1$ $\log_6 1 = 0$ **8.** $7^1 = 7$ $\log_7 7 = 1$ **9.** $4^3 = 64$ $\log_4 64 = 3$

In Exercises 10–15, evaluate the expression, if possible. (8.2)

10. $\log_2 32$ 5 **11.** $\log_3 0$ Not defined **12.** $\log_3 \frac{1}{27}$ -3

13. $\log_9 9$ 1 **14.** $\log_2(-8)$ Not defined **15.** $\log_{10} 17$ 1.230

In Exercises 16–18, expand the expression. (8.3)

16. $\log_{10} 6x$ $\log_{10} 6 + \log_{10} x$ **17.** $\log_{10} \frac{3}{a}$ $\log_{10} 3 - \log_{10} a$ **18.** $\log_{10} x^4$ $4 \log_{10} x$

In Exercises 19–21, condense the expression. (8.3)

19. $\log_{10} 3 + \log_{10} x$ $\log_{10} 3x$ **20.** $\log_{10} 7 - \log_{10} b$ $\log_{10} \frac{7}{b}$ **21.** $4 \log_{10} y$ $\log_{10} y^4$

In Exercises 22–27, use the properties of logarithms to solve for *x*. (8.2, 8.3)

22. $\log_4 x = 3$ 64 **23.** $\log_x 64 = 6$ 2 **24.** $\log_3(x - 2) = 4$ 83

25. $\log_2 x + \log_2 5 = 4$ $3\frac{1}{5}$ **26.** $x \log_4 8 = 3$ 2 **27.** $\frac{1}{3} \log_9 x = \frac{1}{2}$ 27

28. Sketch the graph of $g(x) = \log_{10} x$. **(8.2)** See margin.

29. Given $\log_5 x = 2$, evaluate $\log_x 5$. **(8.2)** $\frac{1}{2}$

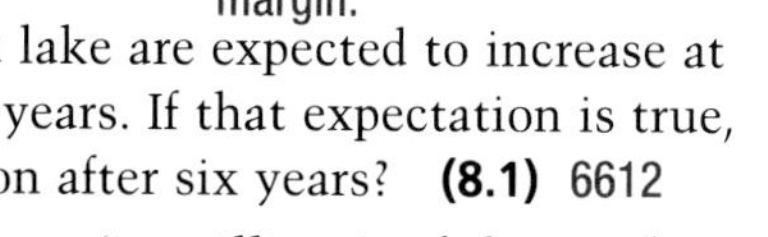

30. 1000 trout fry introduced into a lake are expected to increase at the rate of 37% per year for six years. If that expectation is true, what will be the trout population after six years? **(8.1)** 6612

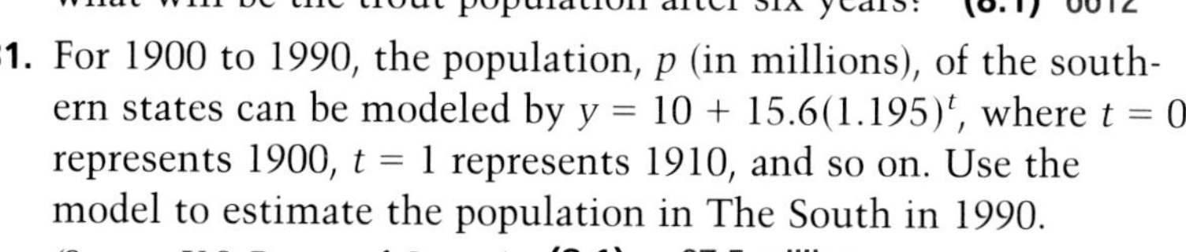

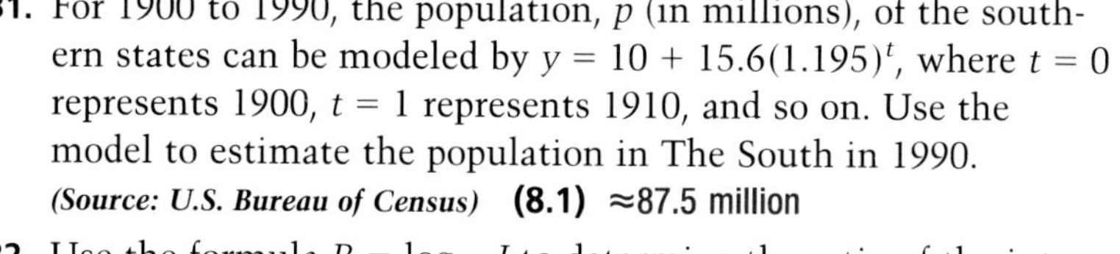

31. For 1900 to 1990, the population, *p* (in millions), of the southern states can be modeled by $y = 10 + 15.6(1.195)^t$, where $t = 0$ represents 1900, $t = 1$ represents 1910, and so on. Use the model to estimate the population in The South in 1990. ***(Source: U.S. Bureau of Census)*** **(8.1)** ≈87.5 million

32. Use the formula $R = \log_{10} I$ to determine the ratio of the intensity of the Mexico City aftershock to the intensity of the earthquake. **(8.3)** ≈0.158

On September 19, 1985, an earthquake measuring R = 8.1 shook Mexico City. One aftershock measured R = 7.3 on the Richter scale.

WRITING Write a one- or two-paragraph report on the history of these logarithmic computing machines and how they worked.

The first type of slide rule was Napier's bones, tables of logarithms that could be arranged on cylinders to complete multiplications. Circular slide rules were devised by English mathematicians Edmund Gunter in 1620 and William Oughtred in 1622. Rectangular slide rules were made by Newton in 1675 and John Robertson in 1775.

Answers

1.

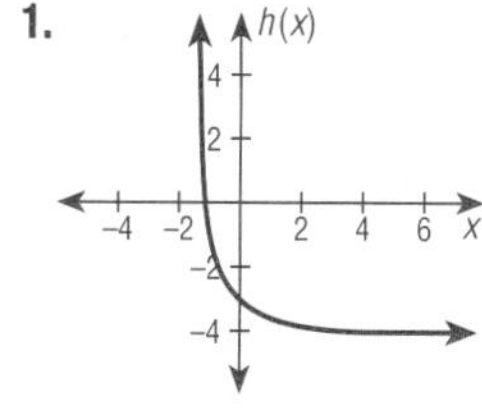

2.

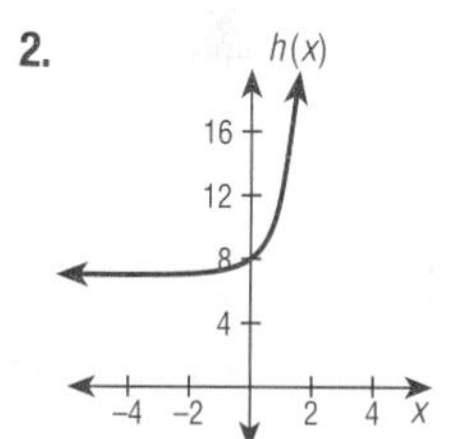

3.

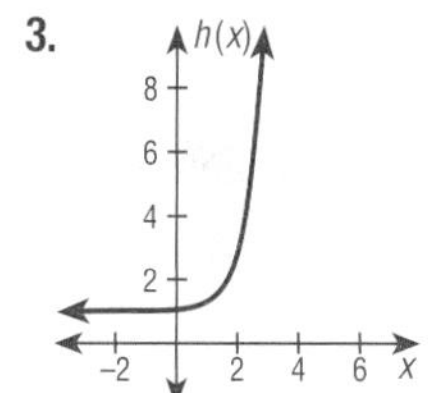

28.

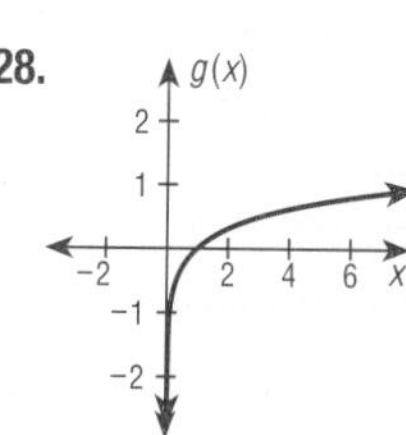

Problem of the Day

Suppose a 6-foot-tall person could walk around the earth. How much farther would the top of the person's head travel than the person's feet? 12π feet

ORGANIZER

Warm-Up Exercises

1. Sketch the graph of each function.
   a. $f(x) = 3(0.25)^x$
   b. $g(x) = 2(2.1)^x$

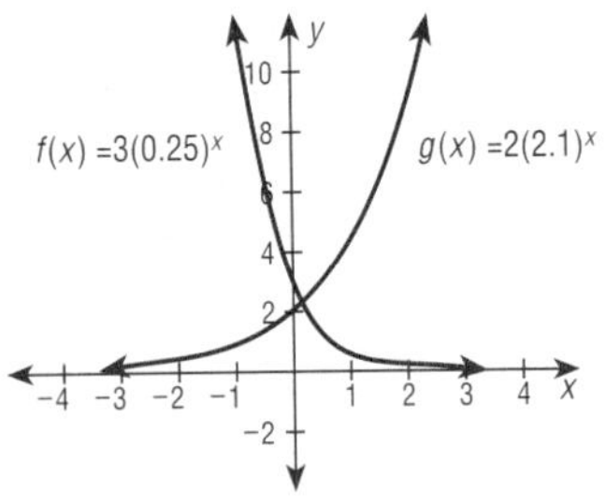

2. Identify each function in Warm-up Exercise 1 as an example of exponential growth or exponential decay.
   $f(x)$ models exponential decay; $g(x)$ models exponential growth.

Lesson Resources

Teaching Tools
- Transparencies: 7, 8, 19
- Problem of the Day: 8.4
- Warm-up Exercises: 8.4
- Answer Masters: 8.4

Extra Practice: 8.4

Applications Handbook: p. 21

8.4 The Natural Base *e*

What you should learn:

Goal 1 How to use the number e as a base of an exponential function

Goal 2 How to use the natural base e in real-life problems

Why you should learn it:

Many business and scientific models use the natural base e, such as the formula for compounding interest continually.

Goal 1 Using the Natural Base e

Much of the history of mathematics is marked by the discovery of special types of numbers such as the counting numbers, zero, negative numbers, π, and imaginary numbers.

In this lesson, you will study one of the most famous numbers of modern times. Like π and i, the number is denoted by a letter. The number is called the **natural base *e*,** or the **Euler number,** after its discoverer, Leonhard Euler (1707–1783).

> **The Natural Base e**
>
> The natural base e is defined to be
>
> $$e = \frac{1}{0!} + \frac{1}{1!} + \frac{1}{2!} + \frac{1}{3!} + \frac{1}{4!} + \frac{1}{5!} + \cdots$$
> $$= 1 + 1 + \frac{1}{2} + \frac{1}{6} + \frac{1}{24} + \frac{1}{120} + \cdots$$
> $$\approx 2.718281828459.$$
>
> The number e is irrational—its decimal representation does not terminate or follow a repeating pattern.

A function of the form $f(x) = Ce^{bx}$ is a natural-base exponential function, or a base-e exponential function. To evaluate this function, you need a calculator. For instance, to evaluate e^2 on a graphing calculator, you can enter

[2nd] [e^x] 2 [ENTER]. *Display is 7.389056099.*

Example 1 *The Graph of a Base e Function*

Sketch the graph of $f(x) = e^x$.

Solution You can begin by constructing a table of values.

x	−2	−1.5	−1	−0.5	0	0.5	1	1.5	2
e^x	0.14	0.22	0.37	0.61	1	1.65	2.72	4.48	7.39

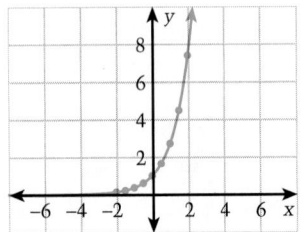

By plotting points and connecting them with a smooth curve, you can obtain the graph shown at the left. ■

Example 2 *Simplifying Natural Base Expressions*

a. $e^2 \cdot e^3 = e^{2+3} = e^5$

b. $\dfrac{6e^2}{2e} = 3e^{2-1} = 3e$

c. $(2e^{-1})^2 = 2^2e^{-2} = 4e^{-2} = \dfrac{4}{e^2}$ ■

The graph of the function $f(x) = Ce^{bx}$ represents exponential growth if b is positive and exponential decay if b is negative.

Example 3 *Graphs of Natural Base Exponential Functions*

Sketch the graph of each function.

a. $f(x) = 0.5e^{0.6x}$

b. $g(x) = 0.5e^{-0.6x}$

Solution

a. The graph of f is an example of exponential growth. The domain of the function is all real numbers, and the range is all positive numbers.

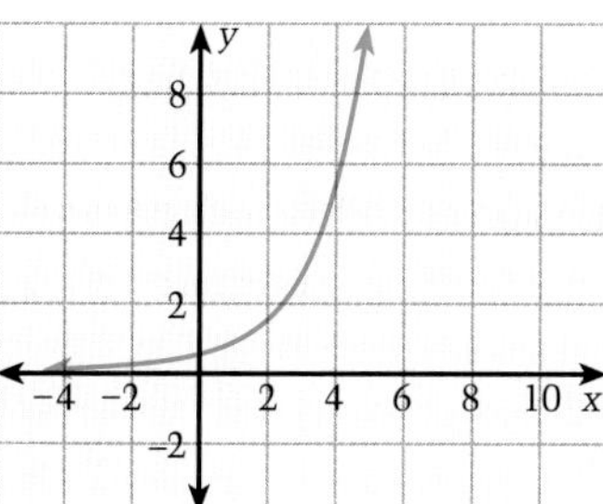

b. The graph of g is an example of exponential decay. The domain of the function is all real numbers, and the range is all positive numbers.

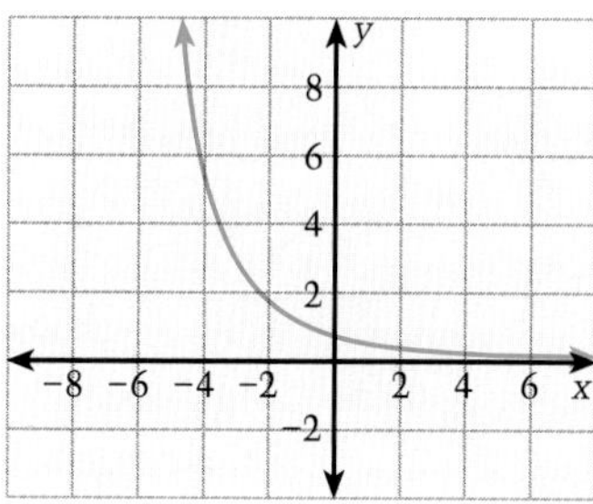

■

The x-axis is a horizontal asymptote of the graph of both functions in Example 3 because if you move far to the left on the graph of $f(x) = 0.5e^{0.6x}$ or far to the right on the graph of $g(x) = 0.5e^{-0.6x}$, you get closer and closer to the x-axis. For instance, when $x = 25$, the value of $g(x)$ is

$$g(25) = 0.5e^{-0.6(25)}$$
$$\approx 0.00000015.$$

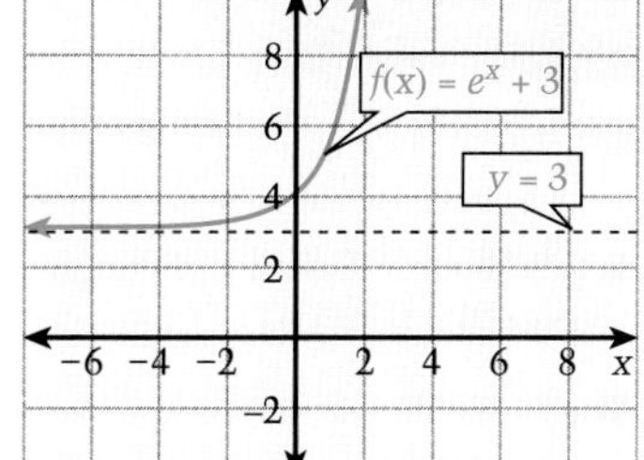

In general, the graph of $f(x) = Ce^{bx} + d$ has the line $y = d$ as its horizontal asymptote. For instance, the graph of $f(x) = e^x + 3$ is shown at the left. Its horizontal asymptote is the line $y = 3$.

LESSON Notes

The number e is defined as the infinite sum, $1 + \frac{1}{1!} + \frac{1}{2!} + \frac{1}{3!} + \frac{1}{4!} + \ldots + \frac{1}{k!} + \ldots$ and, in calculus, is often defined as the following limit:

$$\lim_{n\to\infty}\left(1 + \frac{1}{n}\right)^n.$$

Example 1

Observe that, although e is a special number, the base-e exponential function has a graph similar to other exponential growth functions.

Example 2

Note that the same properties of exponents hold for expressions involving base e as for expressions involving any other base a.

Example 3

For any real number b, $e^{bx} = (e^b)^x$. If b is positive, ask students to substitute values for b and use their calculators to verify that $(e^b) > 1$. If b is negative, verify that $0 < (e^b) < 1$.

Example 4

Since e can be defined as

$$e = \lim_{n\to\infty}\left(1 + \frac{1}{n}\right)^n,$$

you can derive the formula for continuously compounded interest, $A = Pe^{rt}$, by finding the limit, as n approaches ∞, of the expression $A = P\left(1 + \frac{r}{n}\right)^{nt}$. Since you can write $P\left(1 + \frac{r}{n}\right)^{nt}$ as $P\left(\left(1 + \frac{r}{n}\right)^{\frac{n}{r}}\right)^{rt}$, and r is very small compared to n, as n approaches ∞, the limit of the expression $\left(1 + \frac{r}{n}\right)^{\frac{n}{r}}$ is e; so you can arrive at the formula $A = Pe^{rt}$.

Extra Examples

Here are additional examples similar to **Examples 1–4.**

1. Simplify the following.
   a. $e^3 \cdot e^{-5}$ $\quad \frac{1}{e^2}$
   b. $\frac{18e^6}{3e^2}$ $\quad 6e^4$
   c. $(4e^8 \cdot 3e^{-5})^3$ $\quad 1728e^9$
2. Graph each function.
   a. $f(x) = 4e^{0.25x}$
   b. $g(x) = 0.1e^{-2x}$

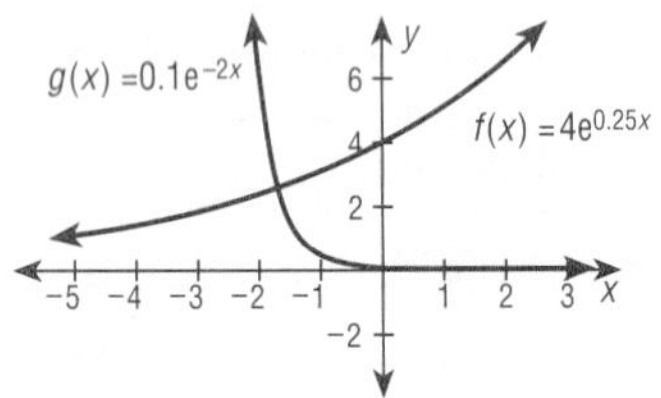

3. For a deposit of \$5,000 into an account that pays 8.5%, find the balance after 5 years if the interest is compounded a. semiannually or b. continuously. By how much is compounding interest continuously a better deal?
   a. \$7,581.07,
   b. \$7,647.95, continuously compounding is a better deal by \$66.88

Check Understanding

1. How is the number e defined? Is e a rational or irrational number?
   See page 420; irrational
2. For what values of b will the base-e exponential function, $f(x) = Ce^{bx}$, model exponential growth?
   $b > 0$
3. What is an asymptote?
   a line that the graph of a function approaches but never reaches
4. What is the frequency of compounding for interest compounded continuously?
   roughly equivalent to compounding every second

Goal 2 Using the Natural Base e in Real Life

In Lesson 7.2, you studied the formula for the value of an investment in which interest is compounded n times per year.

$$A = P\left(1 + \frac{r}{n}\right)^{nt} \qquad \textit{Compounded n times per year}$$

In this formula, A is the balance, P is the initial principal, r is the annual interest rate, n is the number of times the interest is compounded in a year, and t is the number of years. Another common investment formula for compound interest is

$$A = Pe^{rt}. \qquad \textit{Compounded continuously}$$

Here the interest is **continuously compounded interest,** and it is roughly equivalent to compounding every second.

Real Life **Banking**

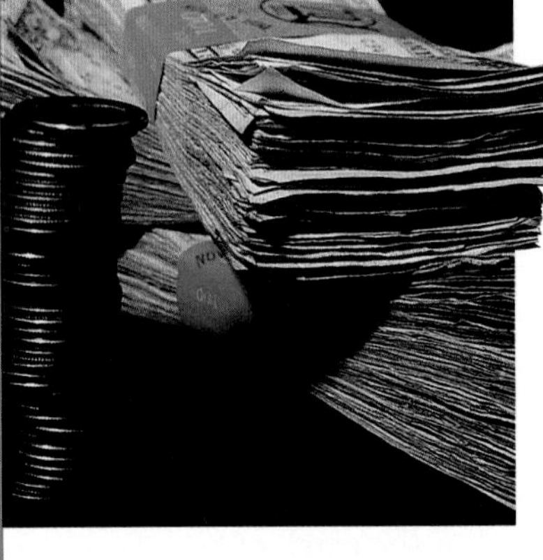

Example 4 *Finding Continuously Compounded Interest*

You deposit \$10,000 in an account that pays 6% interest. Find the balance after 10 years if the interest is compounded
a. quarterly. **b.** continuously.

Solution

a. Compounding quarterly yields a balance of

$$A = 10{,}000\left(1 + \frac{0.06}{4}\right)^{4(10)} = 10{,}000(1.015)^{40} \approx \$18{,}140.18.$$

b. Compounding continuously yields a balance of

$$A = 10{,}000e^{0.06(10)} = 10{,}000e^{0.6} \approx \$18{,}221.19.$$

Thus by compounding continuously, you would obtain a balance that is \$81.01 greater than compounding quarterly. This is a characteristic of continuous compounding—it produces a balance that is greater than or equal to the balance obtained by compounding n times per year. ∎

Communicating about ALGEBRA

SHARING IDEAS about the Lesson See Additional Answers.

Sketch the graph of the function. Classify the graph as showing exponential growth or exponential decay.

A. $f(x) = 0.2e^{0.4x}$ **B.** $f(x) = 0.3e^{-0.2x}$

EXERCISES

Guided Practice

CRITICAL THINKING about the Lesson

1. **True or False?** $e = \frac{271801}{99990}$. Explain your reasoning. False, e is irrational.

2. Simplify $\frac{5e^3}{20e^{-2}}$. Check your result with a calculator. $\frac{e^5}{4}$

3. Sketch the graph of $f(x) = \frac{1}{2}e^{3x}$. Is this function an example of exponential growth or exponential decay? See Additional Answers.

4. State the horizontal asymptote of the graph shown at the right. (The function is $f(x) = 2e^x - 2$.) $y = -2$

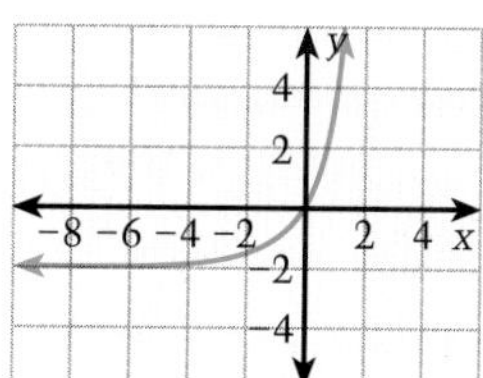

Independent Practice

In Exercises 5–12, use a calculator to evaluate the expression. Round your answer to three decimal places.

5. e^5 148.413
6. e^3 20.086
7. $e^{-3/4}$ 0.472
8. $e^{1.2}$ 3.320
9. $e^{1/2}$ 1.649
10. $e^{1/3}$ 1.396
11. $e^{3.2}$ 24.533
12. $e^{-1/2}$ 0.607

In Exercises 13–24, simplify the expression.

13. $e^2 \cdot e^4$ e^6
14. $3e^{-2}e^6$ $3e^4$
15. $e^x \cdot 4e^{2x+1}$ $4e^{3x+1}$
16. $\frac{e^x}{2e}$ $\frac{1}{2}e^{x-1}$
17. $\frac{4e^x}{e^{4x}}$ $\frac{4}{e^{3x}}$
18. $e^6 \cdot e^x \cdot e^{-2x}$ e^{6-x}
19. $(3e^{-3x})^{-1}$ $\frac{e^{3x}}{3}$
20. $\left(\frac{1}{2}e^{-3}\right)^3$ $\frac{1}{8e^9}$
21. $(2e^{5x})^2$ $4e^{10x}$
22. $(10e^{0.5x})^{-4}$ $\frac{1}{10{,}000e^{2x}}$
23. $\sqrt{4e^2x}$ $2e\sqrt{x}$
24. $\sqrt[3]{64e^6x}$ $4e^2\sqrt[3]{x}$

In Exercises 25–28, decide whether the function is an example of exponential growth or exponential decay. See below.

25. $f(x) = 3e^{2x}$
26. $f(x) = e^{-4x}$
27. $f(x) = 4e^{-2x}$
28. $f(x) = \frac{1}{9}e^{8x}$

In Exercises 29–32, match the equation with its graph.

29. $f(x) = e^{-2x} + 1$ c
30. $f(x) = 2e^{-x} - 3$ b
31. $f(x) = e^{-3x} - 1$ d
32. $f(x) = 2e^x - 3$ a

a.
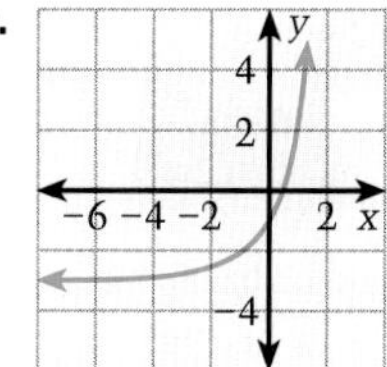

b.
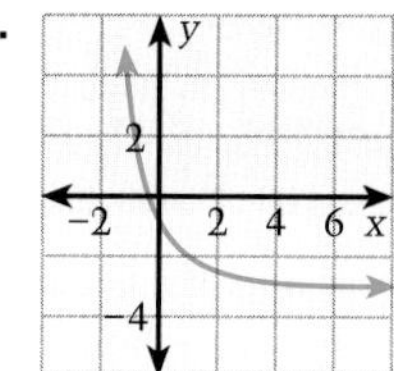

c.
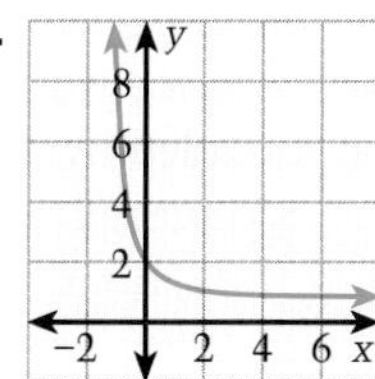

d.
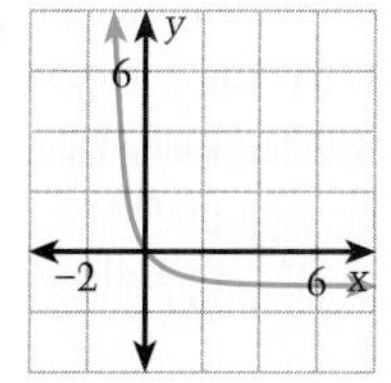

Communicating about ALGEBRA

EXTEND *Communicating*
Have students describe the effects of constants on the graph of the base-e exponential function. In particular, have students describe the graphs of these functions in terms of $f(x) = e^x$.

a. $g(x) = e^x + c$, for any constant c.
vertical shift; $|c|$ units up or down

b. $h(x) = e^{x+b}$, for any constant b.
horizontal shift; $|b|$ units left or right

c. $j(x) = -e^x$
reflection in the x-axis

EXERCISE Notes

ASSIGNMENT GUIDE

Basic/Average: Ex. 1–4, 9–19 odd, 25, 27, 29–32, 33, 41, 45–51 odd, 52, 58–60

Above Average: Ex. 1–4, 9–21 odd, 25–32, 39–41, 45–52, 58–60

Advanced: Ex. 1–4, 9–21 odd, 25–32, 37–41 odd, 45–52, 58–62

Selected Answers
Exercises 1–4, 5–59 odd

More Difficult Exercises
Exercises 33–44, 49–52, 61–62

Guided Practice

▶ **Ex. 1–4** TECHNOLOGY
Students with graphics calculators could use them to develop an understanding of the graphs for the general base-e function $y = ce^{bx}$. Note that the constant c is the y-intercept since e^{bx} is 1 when $x = 0$.

25. Exponential growth 26. Exponential decay 28. Exponential growth

Independent Practice

▶ **Ex. 5–12** **TECHNOLOGY** Students may need to review calculator key sequences. You might suggest that students who have the same type of calculator work together to verify the proper key sequences.

EXTEND Ex. 25–32 **WRITING** Have students write a verification for their answer choices.

▶ **Ex. 33–44** Encourage students to use the graphing techniques suggested in Example 3 on page 421. Some students might use a graphics calculator to provide a visual check.

▶ **Ex. 45–48** Students should use Example 4 as a solution model.

Answers

47. Quarterly: \$4930.86, continuously: \$4946.16; continuously compounded interest pays \$15.30 more

48. The effect is the same as if the interest was compounded once a year at 6.45%.

EXTEND Ex. 52 Use the formula to estimate the amount the Postal Service spent in 1990. If the population of the United States in 1990 was estimated to be 250 million, then how much was spent per person? $\approx$\$42,136.16 million; $\approx$\$168.54

Answers

57. a. $g(x)$, **b.** $h(x)$, **c.** $f(x)$ **d.** $k(x)$; the larger the base, the steeper the curve

58. a. $g(x)$ and $k(x)$, **b.** $h(x)$, **c.** $f(x)$; $h(x) = \left(\frac{1}{e}\right)^x$ and $k(x) = \left(\frac{1}{3}\right)^x$, and the larger the base, the steeper the curve

In Exercises 33–44, sketch the graph of the function. See Additional Answers.

33. $f(x) = 2e^x$ **34.** $f(x) = \frac{1}{2}e^x$ **35.** $f(x) = e^{-x}$ **36.** $f(x) = e^{-4x}$

37. $f(x) = 1.3e^{-0.5x}$ **38.** $f(x) = 3e^{2x} + 2$ **39.** $f(x) = \frac{2}{3}e^x - 1$ **40.** $f(x) = 0.1e^{-x}$

41. $f(x) = 2e^{-1.4x} + 3$ **42.** $f(x) = e^{-3.5x}$ **43.** $f(x) = 0.1e^{2x} - 4$ **44.** $f(x) = \frac{4}{3}e^x + 1$

45. *Continuous Compounding* \$600 is deposited in an account that pays 7% annual interest, compounded continuously. What is the balance after 5 years? **\$851.44**

46. *Continuous Compounding* \$1250 is deposited in an account that pays 6.5% annual interest, compounded continuously. What is the balance after 8 years? **\$2102.53**

47. *Comparing Types of Compounding* \$3000 is deposited in an account that pays 5% annual interest. Compare the balance at the end of 10 years for continuous compounding of interest with the balance for quarterly compounding. **See margin.**

Our Savings Accounts Pay 6.25% Interest Compounded Continuously

Effective Yield: 6.45%

48. *Effective Yield* The sign at the right is in a bank lobby. What is meant by the effective yield of a savings account? **See margin.**

49. *Air Pressure* The air pressure, P, at sea level is about 14.7 pounds per square inch. As the altitude, h (in feet above sea level), increases, the air pressure decreases. The relationship between air pressure and altitude can be modeled by

$P = 14.7e^{-0.00004h}$. $\approx$**4.6 lb per in.^{2}, answers vary**

Mount Everest in Tibet and Nepal rises to a height of 29,108 feet above sea level. What is the air pressure at the peak of Mount Everest? What is the ratio of the air pressure at your school to the air pressure at the peak of Mount Everest?

50. *Radioactive Decay* One hundred grams of radium is stored in a container. The amount of radium, R (in grams), present after t years is given by $R = 100e^{-0.00043t}$ How much of the radium will remain after 10,000 years? $\approx$**1.36 g**

51. *Radioactive Decay* Sketch the graph of the model in Exercise 50.

52. *Postal Service Expenses* The amount of money, M (in millions of dollars), spent by the United States Postal Service from 1975 through 1990 can be modeled by $M = 8679e^{0.079t}$, where $t = 5$ represents 1975. Estimate the amount the United States Postal Service spent in 1986. **\$30719.77 million**

Pierre and Marie Curie discovered radium. They jointly won a Nobel prize in physics in 1903.

51. See Additional Answers.

Integrated Review

In Exercises 53–56, simplify the expression.

53. $5^{2x}(5^{-x})$ 5^x

54. $\frac{4^{5x}}{4^{2x}}$ 4^{3x}

55. $(4^{2x})^5$ 4^{10x}

56. $\left(\frac{2^x}{3^x}\right)^{-1}$ $\left(\frac{3}{2}\right)^x$

57. *Comparing Exponential Growth Curves* Match the function with its graph. Explain your reasoning. **See margin.**

$f(x) = \left(\frac{5}{2}\right)^x$

$g(x) = 3^x$

$h(x) = e^x$

$k(x) = 2^x$

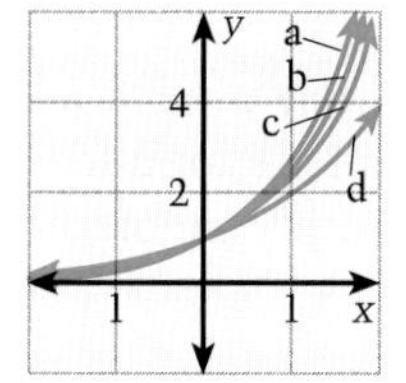

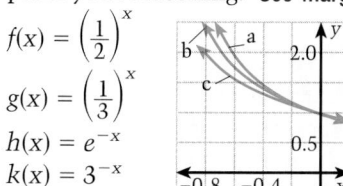

58. *Comparing Exponential Decay Curves* Match the function with its graph. Explain your reasoning. **See margin.**

$f(x) = \left(\frac{1}{2}\right)^x$

$g(x) = \left(\frac{1}{3}\right)^x$

$h(x) = e^{-x}$

$k(x) = 3^{-x}$

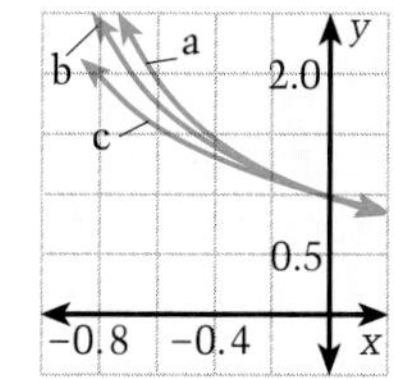

59. *Alligators* Let A be the number of pounds of alligator meat sold in Louisiana from 1980 to 1990. Let $t = 0$ represent 1980. Compare the values given by $A = 100{,}835(1.2755)^t$ to the actual data represented by the graph. **See Additional Answers**

60. *Minor League Attendance* Let A be the average season attendance per team in minor league baseball from 1955 to 1990. Let $t = 0$ represent 1970. Compare the model $A = 93{,}285(1.01473)^t$ to the actual data represented by the graph.

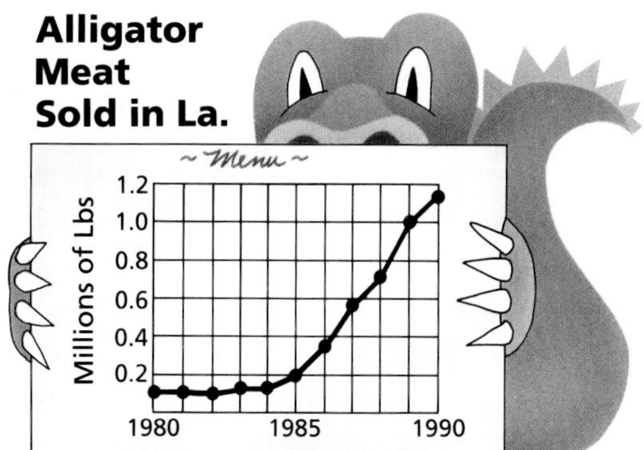

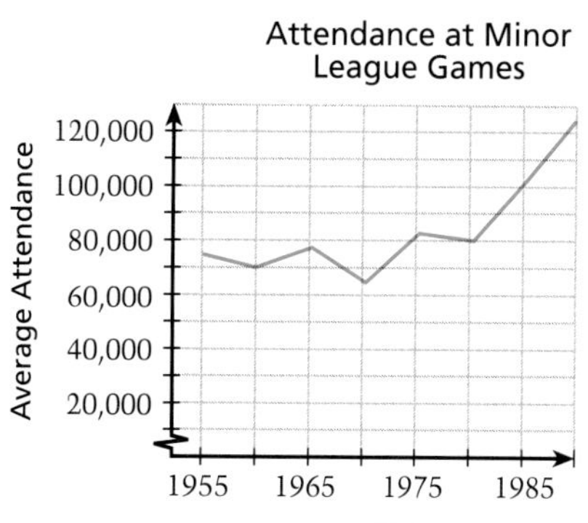

See Additional Answers.

Exploration and Extension

61. Find the first several values of the recursive function. What number is being approached by the values of the recursive function?

$f(0) = \frac{1}{0!}$, $f(n) = f(n-1) + \frac{1}{n!}$ $1, 2, 2\frac{1}{2}, 2\frac{2}{3}, 2\frac{17}{24}, 2\frac{43}{60}$; e

62. Complete the table. The function values appear to be getting closer and closer to what number?

n	10^0	10^1	10^2	10^3	10^4	10^5	10^6
$f(n) = \left(1 + \frac{1}{n}\right)^n$	?	?	?	?	?	?	?

2, ≈2.593742, ≈2.704814, ≈2.716924, ≈2.718146, ≈2.718268, ≈2.718280; e

▶ **Ex. 59–60** Students might observe how closely these exponential graphs approximate the real data by evaluating the models for various t-values.

EXTEND Ex. 61 COOPERATIVE LEARNING Pairs of students might evaluate e^x by finding the sum of the series

$1 + \left(\frac{x}{1!}\right) + \left(\frac{x^2}{2!}\right) + \left(\frac{x^3}{3!}\right) + \left(\frac{x^4}{4!}\right) + \ldots + \left(\frac{x^k}{k!}\right) + \ldots$

Enrichment Activity

This activity requires students to go beyond lesson goals.

WRITING The number e is called Euler's number. Write a two-paragraph report on the connection between Euler and the number e.

Students should find that Euler was the first to recognize that e was an exact number. He estimated its value by using the formula from Exercise 61. He was only 21 at the time. Later, he discovered the limit formula

$$e = \lim_{n\to\infty}\left(1 + \frac{1}{n}\right)^n$$

and this continued-fraction formula for e:

$$e = 2 + \cfrac{1}{1 + \cfrac{1}{2 + \cfrac{2}{3 + \cfrac{3}{4 + \cfrac{4}{\ddots}}}}}$$

e is called Euler's number because Euler established this famous identity that relates the five most important numbers of mathematics:

$e^{i\pi} + 1 = 0.$

Problem of the Day

Simplify.

$\dfrac{2^{n+4} - 2(2^n)}{2(2^{n+3})}$ $\frac{7}{8}$

ORGANIZER

TIME SCHEDULE: All levels, two days

Warm-Up Exercises

1. Evaluate the following to three decimal places using the change-of-base formula of Lesson 8.2.
   a. $\log_3 45$ 3.465
   b. $\log_5 6$ 1.113
   c. $\log_\pi 15$ 2.366
   d. $\log_{\sqrt{2}} 8$ 6
2. Condense the following logarithmic expressions.
   a. $3(\log_{10} x - \log_{10} y)$
   b. $\log_{10} 3 + 2\log_{10} t$
   a. $\log_{10}\left(\frac{x}{y}\right)^3$, **b.** $\log_{10} 3t^2$
3. Expand the following logarithmic expressions.
   a. $\log_{10}\left(\frac{2x^5}{3}\right)$ **b.** $\log_{10}\left(\frac{x^3}{y^4}\right)$
   a. $\log_{10} 2 + 5\log_{10} x - \log_{10} 3$
   b. $3\log_{10} x - 4\log_{10} y$

Lesson Resources

Teaching Tools
- Transparencies: 2, 3, 7, 8, 19
- Problem of the Day: 8.5
- Warm-up Exercises: 8.5
- Answer Masters: 8.5

Extra Practice: 8.5
Color Transparency: 44
Technology Handbook: p. 50

LESSON Notes

Note: Since computing with natural logarithms is a new topic for your students, introduce Goal 1 (Examples 1–4 and Exercises 1–3 and 5–44) on the

(continued)

8.5 Natural Logarithms

What you should learn:

Goal 1 How to evaluate natural logarithmic expressions and graph natural logarithmic functions

Goal 2 How to use natural logarithmic functions as real-life models

Why you should learn it:

You can model many real-life situations with natural logarithmic functions, such as the time an object takes to cool to the temperature of a new environment.

Goal 1 Graphing Natural Log Functions

In Lesson 8.4, the number e was used as a base for an exponent. It is also used as a base for a logarithm, called the **natural base logarithm,** or simply the **natural logarithm.**

Natural Logarithm

If x is a positive real number, then the natural logarithm of x is denoted by

$\log_e x$ or $\ln x$.

(The second notation is more common.) A function given by $f(x) = a + \ln bx$ is called a natural logarithmic function.

Many calculators have a special key for evaluating natural logarithms. For example, to evaluate ln 2 on a graphing calculator, you can use the keystrokes

[ln] 2 [ENTER]. *Display is 0.6931471806.*

Example 1 Graphing the Natural Log Function

Sketch the graph of $f(x) = \ln x$.

Solution You can begin by constructing a table of values.

x	0.25	0.5	1	2	3	4	5	6	7
$\ln x$	−1.39	−0.69	0	0.69	1.10	1.39	1.61	1.79	1.95

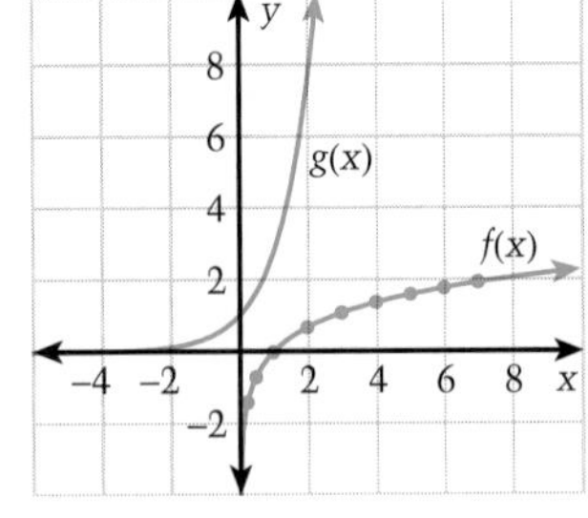

By plotting points and connecting them with a smooth curve, you can obtain the graph of $f(x)$ shown at the left. The domain of the function is the positive real numbers. The range is all real numbers. ■

The functions $f(x) = \ln x$ and $g(x) = e^x$ are inverses of each other. This can be verified by observing that the graph of f is the reflection of the graph of g in the line $y = x$. In such a reflection, the x-axis (the horizontal asymptote of the graph of g) becomes the y-axis (the vertical asymptote of the graph of f).

Example 2 Using Logarithmic Properties to Evaluate Natural Logarithms

a. $\ln 1 = 0$ because $e^0 = 1$ **b.** $\ln e = 1$ because $e^1 = e$

c. $\ln e^2 = 2$ **d.** $\ln e^{-3} = -3$ ■

Example 3 Expanding and Condensing Natural Logarithms

a. $\ln 3x = \ln 3 + \ln x$

b. $\ln x^3y = \ln x^3 + \ln y = 3 \ln x + \ln y$

c. $\ln x - \ln 2 = \ln \frac{x}{2}$ ■

Example 4 Sketching the Graph of a Natural Logarithmic Function

Sketch the graph of $f(x) = 3 - \ln(x - 2)$.

Solution The domain of the function is all real numbers that are greater than 2. Thus you can begin by constructing a table of values that uses several x-values greater than 2.

x	2.25	2.5	3	4	5	6	7	8
$3 - \ln(x - 2)$	4.39	3.69	3	2.31	1.90	1.61	1.39	1.21

By plotting points and connecting them with a smooth curve, you obtain the graph shown at the left. The range of the function is all real numbers. The vertical line $x = 2$ is a vertical asymptote of the graph. ■

In Lesson 8.2, you used a change-of-base formula to evaluate logarithms in bases other than 10 that used common logarithms. A similar formula exists for natural logarithms.

$$\log_a x = \frac{\ln x}{\ln a} \quad \textit{Change-of-base formula}$$

Example 5 Using the Change-of-Base Formula

Use a calculator to evaluate $\log_3 12$.

Solution Using natural logarithms, you can write

$$\log_3 12 = \frac{\ln 12}{\ln 3} \approx \frac{2.4849}{1.0986} \approx 2.262.$$

You can check the result by raising 3 to the 2.262 power. The result should be approximately 12. ■

(continued)

first day and Goal 2 (Examples 5 and 6 and Exercises 4 and 45–70) on the second day.

Examples 1–3

Note that the graph of $f(x) = \ln x$ is similar to all the other logarithmic functions seen in earlier lessons. Emphasize the importance of plotting points to determine what the basic graph of a function is.

Observe that the same properties that hold for logarithms in general hold for natural logarithms as well.

Example 4

Ask students if they can provide other rationales for why the graph of $f(x) = 3 - \ln(x - 2)$ is sketched as it is shown. (Remind students that the constants in a functional equation produce horizontal and vertical shifts and reflections in the x-axis.)

Example 5

Ask students whether the value of $\log_3 12$ should be the same regardless of which change-of-base formula (base 10 or base e) is used. Encourage students to evaluate several expressions, if needed, to form a speculation. Both change-of-base formulas give approximately the same results.

Extra Examples

Here are additional examples similar to **Examples 1–5.**

1. Condense the following.
   a. $3 \ln x + \ln 5$
   b. $\ln x - (2 \ln 4 + \ln y)$
   c. $\ln x + 4 \ln y - \ln 5$
   a. $\ln 5x^3$, **b.** $\ln \frac{x}{16y}$, **c.** $\ln \frac{xy^4}{5}$
2. Expand the following.
   a. $\ln x^3$ **b.** $\ln 4xy^2$
   c. $\ln \frac{5t}{3x^2}$
   a. $3 \ln x$, **b.** $\ln 4 + \ln x + 2 \ln y$, **c.** $\ln 5 + \ln t - (\ln 3 + 2 \ln x)$

4. Evaluate the following to three decimal places using the change-of-base formula for natural logarithms.
   a. $\log_4 13$ 1.850
   b. $\log_7 123$ 2.473
   c. $\log_5 18$ 1.796
   d. $\log_2 78$ 6.285

Example 6

You may want to perform the computations of the problem for students. (See Communicating About Algebra in the pupil text.) Point out that T_1 is a temperature reading made by the coroner. It follows that

(a) $kt = \ln\left[\dfrac{85.7 - 70}{98.6 - 70}\right]$,
$kt \approx \ln(0.54895) \approx -0.6$

(b) $k(t + 0.5) = \ln\left[\dfrac{82.8 - 70}{98.6 - 70}\right]$,
$k(t + 0.5) \approx \ln(0.44755) \approx -0.8$.

Solve the system

$$\begin{cases} kt = -0.6 \\ k(t + 0.5) = -0.8 \end{cases}$$

by the substitution method.

$kt + 0.5k = -0.8$ or
$-0.6 + 0.5k = -0.8$.

From this last equation:

$0.5k = -0.8 + 0.6 = -0.2$

$k = \dfrac{-0.2}{0.5} = -0.4$.

Solving for t gives

$-0.4t = -0.6$, or $t = \dfrac{-0.6}{-0.4}$,

$t = 1.5$.

Check Understanding

1. How can graphs be used to show that $f(x) = \ln x$ and $g(x) = e^x$ are inverses of each other?
   The graphs are reflections in the line $y = x$.
2. Does $\dfrac{\ln x}{\ln a} = \ln\left(\dfrac{x}{a}\right)$? Explain.
   No; $\ln\left(\dfrac{x}{a}\right) = \ln x - \ln a$

Forensic scientists work in crime laboratories of police departments. Modern crime laboratories use computers and scientific analysis to examine clues.

Goal 2 Using Natural Logarithms in Real Life

Real Life
Forensic Science

Example 6 *Estimating the Time of Death*

At 8:30 A.M., a coroner was called to the home of a person who had died during the night. In order to estimate the time of death, the coroner took the person's body temperature twice. At 9:00 A.M. the temperature was 85.7°F and at 9:30 A.M. the temperature was 82.8°F. The room temperature stayed constant at 70°F. Find the approximate time of death.

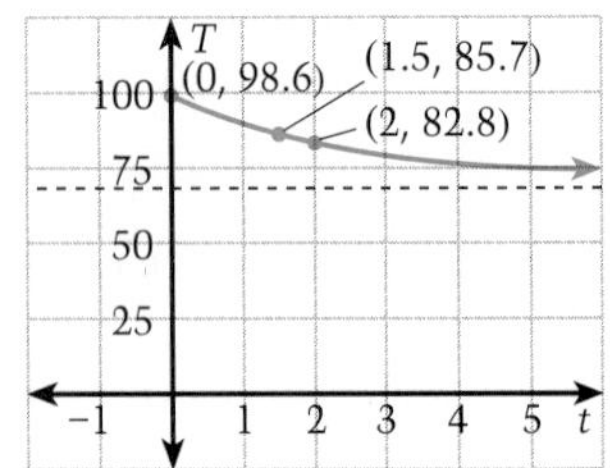

Solution To estimate the time of death, the coroner used Newton's law of cooling

$$kt = \ln \frac{T - S}{T_0 - S}, \quad \textit{Newton's law of cooling}$$

where k is a constant, S is the temperature of the surrounding air, and t is the time it takes for the body temperature to cool from $T_0 = 98.6$ to T. From the first temperature reading, the coroner could obtain $kt \approx -0.6$. From the second temperature reading, the coroner could obtain $k\left(t + \frac{1}{2}\right) \approx -0.8$. Solving these two equations for k and t produces $k = -0.4$ and $t = 1.5$. Therefore, the coroner could approximate that the first temperature reading took place 1.5 hours after death, which implies that the death occurred at about 7:30 A.M. ■

Communicating about ALGEBRA

SHARING IDEAS about the Lesson See Additional Answers.

Show how you can use the following equations to conclude that $k \approx -0.4$ and $t \approx 1.5$ in Example 6.

1st Temperature Reading	2nd Temperature Reading
$kt = \ln \dfrac{85.7 - 70}{98.6 - 70}$	$k\left(t + \frac{1}{2}\right) = \ln \dfrac{82.8 - 70}{98.6 - 70}$

EXERCISES

Guided Practice

CRITICAL THINKING about the Lesson

1. Explain why $\ln e = 1$.
2. Explain why $\ln e^6 = 6$.
3. Sketch the graph of $f(x) = -\ln(x)$. What is the domain of the function? What is the range? How is its graph related to the graph of $f(x) = \ln x$ at the right?
4. Explain how to use natural logarithms to evaluate $\log_6 10$.
$\log_6 10 = \frac{\ln 10}{\ln 6}$

1. Let $\ln e = \log_e e = x$, then $e^x = e$ and $x = 1$.
2. Let $\ln e^6 = \log_e e^6 = x$. Then $e^x = e^6$ and $x = 6$.

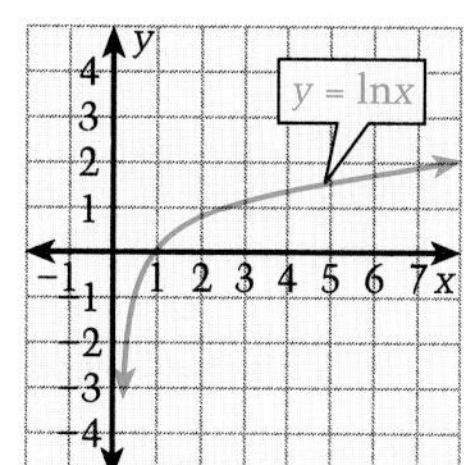

3. $x > 0$, all real numbers See Additional Answers.

Independent Practice

In Exercises 5–7, evaluate the expression without using a calculator.

5. $\ln e^4$ 4
6. $\ln e^{-2}$ −2
7. $\ln\left(\frac{1}{e^3}\right)$ −3

In Exercises 8–16, use a calculator to evaluate the expression.

8. $\ln 1.4$ ≈0.3365
9. $\ln 15$ ≈2.7081
10. $2 \ln 3$ ≈2.1972
11. $3.2 \ln 8$ ≈6.6542
12. $4 \ln 6 + 5$ ≈12.1670
13. $2 - 5 \ln \frac{1}{2}$ ≈5.4657
14. $\ln 6 - \ln 9$ ≈−0.4055
15. $\ln 2 + \ln 16$ ≈3.4657
16. $6 \ln 3.3 + 2 \ln 1.5$ ≈7.9745

In Exercises 17–22, expand the expression.

17. $\ln \frac{8}{6x}$ $\ln 8 - \ln 6 - \ln x$
18. $\ln 3xy^2$ $\ln 3 + \ln x + 2 \ln y$
19. $\ln 16x^2$ $\ln 16 + 2 \ln x$
20. $\ln \frac{2xy}{x^2}$ $\ln 2 + \ln y - \ln x$
21. $\ln \frac{2y^7}{x^3}$ $\ln 2 + 7 \ln y - 3 \ln x$
22. $\ln 32x^5y$ $\ln 32 + 5 \ln x + \ln y$

In Exercises 23–28, condense the expression.

23. $\ln 16 - \ln 4$ $\ln 4$
24. $\ln 20 + 2 \ln \frac{1}{2} + \ln x$ $\ln 5x$
25. $3 \ln 5 - (\ln 15 - \ln 3)$ $\ln 25$
26. $4 \ln 3 - \ln 9$ $\ln 9$
27. $3 \ln x + \ln y + \ln 10$ $\ln 10x^3y$
28. $2(\ln 2 - \ln x) + (\ln x - \ln 4)$ $-\ln x$

In Exercises 29–32, match the function with its graph.

29. $f(x) = \ln(x + 1)$ a
30. $f(x) = 4 - \ln(x + 4)$ d
31. $f(x) = \ln\left(x - \frac{3}{2}\right)$ c
32. $f(x) = -\frac{3}{2} \ln x$ b

a.
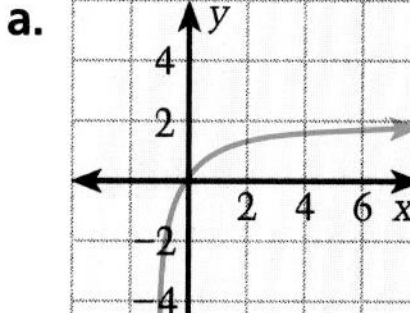

b.
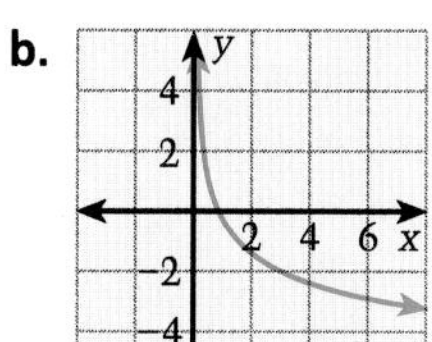

c.
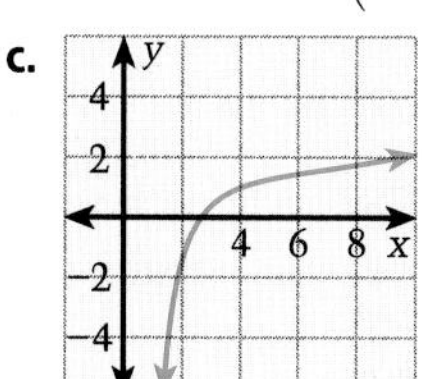

d.
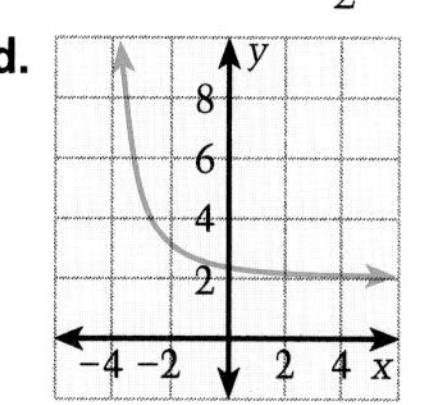

Communicating about ALGEBRA

Answers
See the Lesson Notes for Example 6.

EXTEND Communicating COOPERATIVE LEARNING
Have students work in groups to complete the following project. Have students take a cup of hot water and record its temperature and the temperature of the surrounding air at regular intervals as the water is cooling. They can use the data and Newton's law of cooling to determine the constant k for the water. Groups of students can compare the values they obtained.

EXERCISE Notes

ASSIGNMENT GUIDE

Day 1

Basic/Average: Ex. 1–3, 11–19 odd, 23, 29–32, 37–43

Above Average: Ex. 1–3, 11–21 odd, 27–32, 37–43 odd

Advanced: Ex. 1–3, 11–21 odd, 27–32, 37–43 odd

Day 2

Basic/Average: Ex. 4, 45–49 odd, 51–53, 55–60, 61–67 odd

Above Average: Ex. 4, 45–49 odd, 51–56, 59–69

Advanced: Ex. 4, 45–49 odd, 51–56, 59–69 odd, 70

Selected Answers
Exercises 1–4, 5–67 odd

✪ More Difficult Exercises
Exercises 33–44, 51–56, 69–70

Guided Practice

▶ **Ex. 1–4** Use these Guided Practice exercises as an in-class summary review of the lesson topics. Be sure students understand that the notation $\ln x$ is equivalent to the notation $\log_e x$.

Independent Practice

▶ **Ex. 17–28** Students can use Example 3 as a solution model.

▶ **Ex. 29–32** Assign these exercises as a group.

▶ **Ex. 33–44** TECHNOLOGY Students can use Example 4 as a solution model. Encourage students to use graphics calculators or computers if they have access to them.

▶ **Ex. 51–53** Assign these exercises as a group. Students can use Example 6 as a solution model.

EXTEND Ex. 55 Use the given formula to predict the amount spent on salary and wages for the year 2000. $t = 20$; ≈\$514 million

EXTEND Ex. 56 Have students predict the median age for the current year. Ask your students to guess whether you are above the median age or below the median age.

▶ **Ex. 69–70** In a logarithmic spiral, the distances between the loops form a geometric sequence. Students who are interested might find other examples of spirals and present them to the class.

In Exercises 33–44, sketch the graph of the function. See Additional Answers.

33. $f(x) = -\ln x$ **34.** $f(x) = -2 \ln x$ **35.** $f(x) = 5 + \ln x$

36. $f(x) = \ln x - 4$ **37.** $f(x) = \ln(x + 9)$ **38.** $f(x) = -\ln(x - 2)$

39. $f(x) = -10 - \ln(x + 4)$ **40.** $f(x) = \ln(x + 2) - 1$ **41.** $f(x) = 1 + \ln(x + 6)$

42. $f(x) = 2 - \ln(x - 1)$ **43.** $f(x) = \frac{4}{3} \ln(x - 1)$ **44.** $f(x) = \frac{5}{2} \ln(x + 2)$

In Exercises 45–50, use natural logarithms to rewrite the expression. Then use a calculator to evaluate the expression. Round the result to three decimal places.

45. $\log_2 9$ $\frac{\ln 9}{\ln 2} \approx 3.170$ **46.** $\log_2 180$ $\frac{\ln 180}{\ln 2} \approx 7.492$ **47.** $\log_5 4.2$ $\frac{\ln 4.2}{\ln 5} \approx 0.892$

48. $\log_5 30$ $\frac{\ln 30}{\ln 5} \approx 2.113$ **49.** $\log_8 \sqrt{3}$ $\frac{\ln \sqrt{3}}{\ln 8} \approx 0.264$ **50.** $\log_2 1.5$ $\frac{\ln 1.5}{\ln 2} \approx 0.585$

Making Ice Cubes **In Exercises 51–53, use the following information.**

You place a tray of 60°F water into a freezer that is set at 0°F. The water cools according to Newton's law of cooling

$$kt = \ln \frac{T - S}{T_0 - S},$$

where T is the temperature (in °F), t is the number of hours the tray is in the freezer, S is the temperature of the surrounding air, and T_0 is the original temperature of the water.

51. If the water freezes in 4 hours, what is the constant k? (Hint: Water freezes at 32°F.) ≈−0.157

52. Suppose you lower the temperature in the freezer to −10°F. At this temperature, how long will it take for the ice cubes to form? ≈3.25 hours

53. Suppose the initial temperature of the water is 50°F. If the freezer temperature is 0°F, how long will it take for the ice cubes to form? ≈2.84 hours

54. *Thawing Steaks* A package of frozen steaks is taken out of the freezer and left at room temperature. The temperature, T (in °F), of the package is related to the time, t (in hours), by the model

$$t = -3.8 \ln \frac{T - 70}{24 - 70}.$$ ≈5.8 hours

How long will it take the package to reach a temperature of 60°F?

55. *National Park Salaries* For 1981 through 1990, the amount, A (in millions of dollars), spent on salaries and wages for National Park Service employees can be modeled by

$$A = 286.52 + 7.25t + 27.51 \ln t,$$

where $t = 1$ represents 1981. About how much was spent on salaries and wages in 1984? ≈\$354 million

National Park Service Ranger

56. *Median Age of Americans* For 1980 through 2030, the median age, M (in years), of Americans is predicted to follow the population model

$$M = 19.17 + 0.107t + 4.028 \ln t,$$

where $t = 10$ represents 1980. According to this model, will the median age be increasing or decreasing? Explain.

Increasing, t and $\ln t$ are increasing functions of t

Integrated Review

In Exercises 57–60, identify the asymptote of the graph of the function.

57. $f(x) = \ln(x + 5)$ $x = -5$

58. $f(x) = e^x + 3$ $y = 3$

59. $f(x) = e^x - 2$ $y = -2$

60. $f(x) = 2 - \ln(x - 1)$ $x = 1$

In Exercises 61–64, expand the expression. See below.

61. $\log_{10}\left(\frac{x}{y}\right)^2$

62. $\log_4 21x^4$

63. $\log_6 81y^2$

64. $\log_2 \frac{16}{x^3y}$

In Exercises 65–68, condense the expression.

65. $\log_3 6 + \log_3 \frac{1}{2}$ 1

66. $\log_2 3 - \log_2 5$ $\log_2 \frac{3}{5}$

67. $5(\log_{10} 4 - \log_{10} 2)$ $\log_{10} 32$

68. $2 \log_{10} x + 3 \log_{10} y$ $\log_{10} x^2y^3$

Exploration and Extension

Logarithmic Spiral **In Exercises 69 and 70, use the following information.**

The shape of the seashell fossil shown at the right is a logarithmic spiral. Consider a point, P at (1, 0), on the positive x-axis. To form the spiral, the point moves counterclockwise around the origin. The distance, r, between the point and the origin is given by

$$r = e^{0.001745\theta},$$

where θ (the Greek letter *theta*) is the number of degrees through which the point has spiraled.

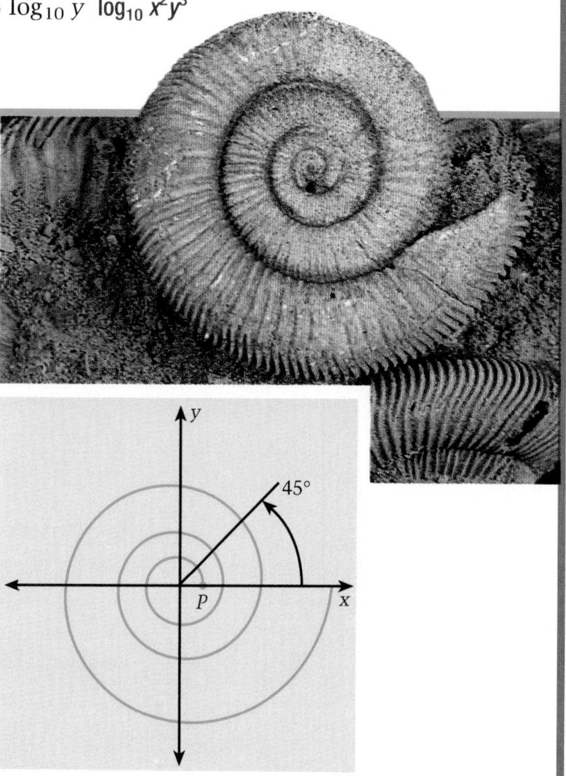

69. What is the distance between the point and the origin when the point has spiraled through 45°? $\approx$1.08 units

70. Write the given equation in logarithmic form. Use the result to find the number of degrees that correspond to an r-value of 2. $\ln r = 0.001745\theta$, $\approx$397°

Enrichment Activities

These activities require that students go beyond lesson goals.

After students complete the Graphing Calculator feature on pages 432–433, present them with other data such as the following.

TECHNOLOGY Use your graphics calculator and write an appropriate model relating the year to each of the four sets of data that follow. What kind of model (linear, quadratic, etc.) is each one?

1.

Year	Women in labor force	Women in military
1973	39%	3%
1975	40%	7.5%
1980	42%	9%
1985	44%	10%
1990	45%	11%

Let $t = 0$ correspond to 1970.

2. Home smoke detectors were introduced in 1969. In 1980, 50% of American homes were estimated to have smoke detectors. That percentage grew to 70% in 1985 and to 86% in 1990. In 1970, the death rate due to fire and burns per 100,000 deaths was 3.3. In the years 1980, 1985, and 1990, the death rates fell to 2.6, 2.1, and 1.7, respectively.

Let $t = 0$ correspond to 1969.

Women in labor force: linear, $y = 38.18 + 0.36x$; $r \approx 0.99$.
Women in military: logarithmic, $y = -0.075 + 3.815 \ln x$; $r \approx 0.95$.

Home smoke detectors: logarithmic, $y = -83.40 + 55.53 \ln x$; $r \approx 0.9996$.
Death rate due to burns: linear, $y = 3.412 - 0.081x$; $r \approx -0.997$.

61. $2 \log_{10} x - 2 \log_{10} y$ **62.** $\log_4 21 + 4 \log_4 x$ **63.** $\log_6 81 + 2 \log_6 y$ **64.** $4 - 3 \log_2 x - \log_2 y$

Example 1

Observe that the values of the amounts increase by about the same percent over equal time intervals. This is an indication that an exponential model may best represent these data. (A linear model is not appropriate since the increase does not appear to be constant over equal time intervals.)

Example 2

In contrast to Example 1, the ratio of consecutive years required to obtain the specified balances approaches 1, which indicates that a logarithmic model may best describe the relationship.

Note: The graphing calculator may place restrictions on data points for some models.

Model	Restriction
Logarithmic:	All x values > 0
Exponential:	All y values > 0
Power:	All x and y values > 0

EXTEND the Activity

Have students use an almanac to find examples of real-life data and determine the appropriate model for the data. (See the Enrichment Activities on page 431 for samples.) Some examples of data to consider are the consumer price index; the cost of a postage stamp; the car manufacturing rates of the United States and Japan; and the population increases and decreases of cities and countries.

USING A GRAPHING CALCULATOR

A graphing calculator can be used to fit exponential and logarithmic models of the form

$$y = a(b^x) \quad \text{and} \quad y = a + b \ln x$$

to data. For details on the actual keystrokes, consult the user's manual that accompanies your graphing calculator.

Example 1: Finding an Exponential Model

For 1950 through 1990, the total amount, y (in billions of dollars), of life insurance in the United States is shown in the table. Use a graphing calculator to find an exponential model that approximates these data.

(Source: American Council of Life Insurance)

Year	1950	1955	1960	1965	1970	1975	1980	1985	1990
Amount	234	373	586	901	1402	2314	3541	6053	9198

Solution Let $x = 0$ represent 1950. Then enter the following coordinates into the statistical data bank of the calculator.

(0, 234), (5, 373), (10, 586), (15, 901), (20, 1402)
(25, 2314), (30, 3541), (35, 6053), (40, 9198)

After running the exponential regression program, the display should read $a \approx 231.5$ and $b \approx 1.0963$. (The correlation of $r \approx 0.9998$ tells you that the fit is very good.) Thus a model for these data is

$$y = 231.5(1.0963)^x.$$

The graph below compares the actual data with the model. ■

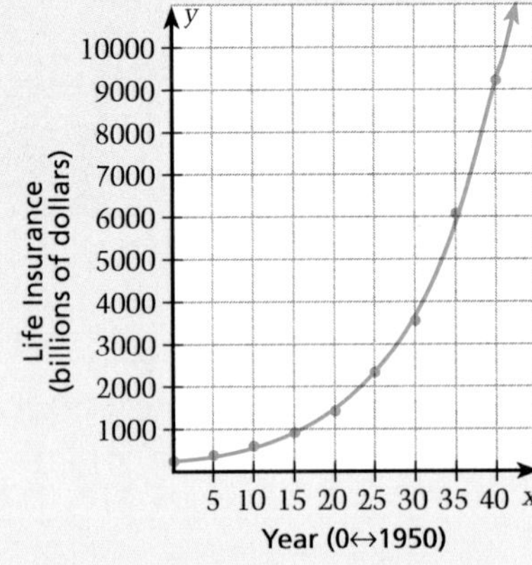

Example 2: Finding a Logarithmic Model

A savings account contains \$1000, which earns 5% annual interest, compounded continuously. The number of years, y, the money must be left in the account for it to grow to a balance of x dollars is given in the table. Find a logarithmic model for these data.

Balance, x	1000	2000	3000	4000	5000	6000	7000
Years, y	0	13.86	21.97	27.73	32.19	35.84	38.92

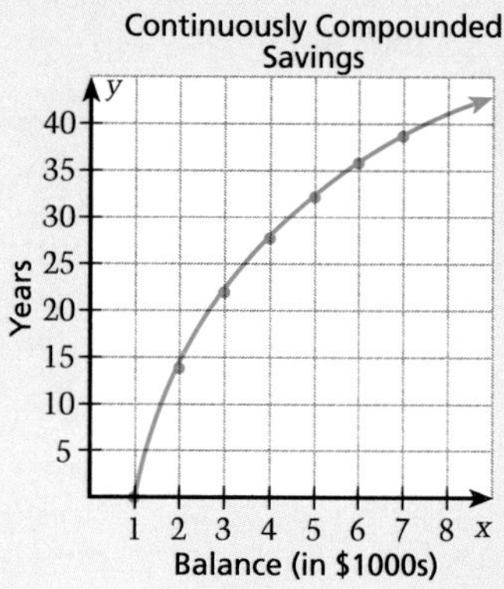

Solution To find a logarithmic model for these data, enter the seven ordered pairs from the table into the statistical data bank of the calculator. Then run the "logarithmic regression" program. The display should read $a \approx -138.2$ and $b \approx 20$. (The correlation of $r \approx 1$ tells you that the fit is nearly perfect.) Thus the model for these data is $y = -138.2 + 20 \ln x$. The graph at the right compares the actual data with the model. ■

Exercises

In Exercises 1–4, sketch a scatter plot of the data. Then find an exponential or logarithmic model that approximates the data. See Additional Answers.

1.

x	1	2	3	4	5	6	7	8	9	10
y	0.58	0.66	0.76	0.87	1.01	1.16	1.33	1.53	1.76	2.02

$y = 0.5(1.1496)^x$

2.

x	1	2	3	4	5	6	7	8	9	10
y	2.42	2.93	3.54	4.29	5.19	6.28	7.59	9.19	11.12	13.45

$y = 2.0(1.2099)^x$

3.

x	1	2	3	4	5	6	7	8	9	10
y	2.00	2.35	2.55	2.69	2.80	2.90	2.97	3.04	3.10	3.15

$y = 2.0 + 0.5 \ln x$

4.

x	1	2	3	4	5	6	7	8	9	10
y	5.00	4.58	4.34	4.17	4.03	3.92	3.83	3.75	3.68	3.62

$y = 5.0 - 0.6 \ln x$

5. *Liz Claiborne, Inc.* The table shows the annual sales, y (in millions of dollars), for Liz Claiborne, Inc., from 1981 through 1990. Sketch a scatter plot of the data. (Let $x = 1$ represent 1981.). Then find an exponential or a logarithmic model that approximates the data. Use your model to estimate the 1991 sales. ***(Source: Liz Claiborne, Inc.)*** See Additional Answers for graph.

Year	1981	1982	1983	1984	1985	1986	1987	1988	1989	1990
Sales, y	116.8	165.7	228.7	391.3	556.6	813.5	1053.3	1184.2	1410.7	1728.9

5. $y = 100.5(1.3607)^x \approx \$2975.4$ million

Problem of the Day

Determine the digit represented by each asterisk.

**	11
× **	× 91
1*	11
**	99
****	1001

ORGANIZER

TIME SCHEDULE: All levels, two days

Warm-Up Exercises

1. Evaluate the following to three decimal places.
   a. $\log_3 8$ 1.893
   b. $\log_5 17$ 1.760
   c. $\log_2 16$ 4
   d. $\log_3 81$ 4
   e. $\log_2 2^5$ 5
2. Solve for x in each of the following equations.
   a. $3^x = 2187$ $x = 7$
   b. $7^x = 2401$ $x = 4$
   c. $\log_7 7^x = 3$ $x = 3$
   d. $\log_2 x^4 = 4$ $x = 2$

Lesson Resources

Teaching Tools
- Problem of the Day: 8.6
- Warm-up Exercises: 8.6
- Answer Masters: 8.6

Extra Practice: 8.6
Color Transparency: 45
Applications Handbook: p. 21
Technology Handbook: p. 52

LESSON Notes

Note: Since solving exponential and logarithmic equations is a new topic for your students, introduce Goal 1 (Examples 1–5 and Exercises 1–46) on the first day and Goal 2 (Example 6 and Exercises 47–63) on the second day.

8.6 Solving Exponential and Logarithmic Equations

What you should learn:

Goal 1 How to solve exponential and logarithmic equations

Goal 2 How to use exponential and logarithmic equations to answer questions about real life

Why you should learn it:

You can solve an exponential equation to solve many real-life problems, such as the length of time your money must be invested to double your principal.

Goal 1 Solving Equations

Consider the exponential equation $2^x = 32$. To solve this equation, you can rewrite it as

$$2^x = 2^5,$$

from which you can conclude that $x = 5$. Suppose, however, that you were asked to solve the equation $2^x = 7$. To solve for x in this equation, you can take the logarithm with base 2 of both sides to obtain the following.

$2^x = 7$	*Rewrite original equation.*
$\log_2 2^x = \log_2 7$	*Take log with base 2 of both sides.*
$x = \log_2 7$	*Simplify.*

The last step uses one of the inverse properties of exponents and logarithms.

Inverse Properties of Exponents and Logarithms

Let a be a positive real number, $a \neq 1$.

	Base a	**Base e**
1.	$\log_a a^x = x$	$\ln e^x = x$
2.	$a^{\log_a x} = x$	$e^{\ln x} = x$

Example 1 *Solving an Exponential Equation*

Solve $e^x = 72$.

Solution

$e^x = 72$	*Rewrite original equation.*
$\ln e^x = \ln 72$	*Take ln of both sides.*
$x = \ln 72$	*Simplify.*
$x \approx 4.277$	*Use a calculator.*

The solution is $x = \ln 72$. Check this in the original equation. ■

Example 2 Solving an Exponential Equation

Solve $10^x + 5 = 60$.

Solution

$10^x + 5 = 60$	*Rewrite original equation.*
$10^x = 55$	*Subtract 5 from both sides.*
$\log_{10} 10^x = \log_{10} 55$	*Take log of both sides.*
$x = \log_{10} 55$	*Simplify.*
$x \approx 1.740$	*Use a calculator.*

The solution is $x = \log_{10} 55$. Check this in the original equation. ■

Example 3 Solving an Exponential Equation

Solve $4e^{2x} = 5$.

Solution

$4e^{2x} = 5$	*Rewrite original equation.*
$e^{2x} = \frac{5}{4}$	*Divide both sides by 4.*
$\ln e^{2x} = \ln \frac{5}{4}$	*Take ln of both sides.*
$2x = \ln \frac{5}{4}$	*Simplify.*
$x = \frac{1}{2} \ln \frac{5}{4}$	*Divide both sides by 2.*
$x \approx 0.112$	*Use a calculator.*

The solution is $x = \frac{1}{2} \ln \frac{5}{4}$. Check this in the original equation. ■

To solve a logarithmic equation such as

$\ln x = 3$ *Logarithmic form*

write the equation in exponential form as follows.

$e^{\ln x} = e^3$ *Exponentiate both sides.*

$x = e^3$ *Exponential form*

This procedure is called exponentiating both sides of an equation.

Example 1

Take time to explain the inverse properties of exponents and logarithms. Stress the logarithm definition in Lesson 8.2:

$\log_a x = y$ if and only if $a^y = x$.

It follows from the definition that for $M = a^x$, $\log_a M = x$ if and only if $a^x = M$. But $M = a^x$; so the first inverse property of logarithms holds.

Ask students to demonstrate the second inverse property.

Example 2

Ask students why you wouldn't take the logarithm, base 10, of the original left side of the equation. (There is no property of logarithms for simplifying sums.)

If the original equation had been $2^x + 5 = 60$, the solution would be similar.

$2^x + 5 = 60$

$2^x = 55$

Take the log of both sides, simplify, and solve.

$\log_{10} 2^x = \log_{10} 55$

$x \log_{10} 2 = \log_{10} 55$

$x = \dfrac{\log_{10} 55}{\log_{10} 2}$

$x \approx \dfrac{1.7404}{0.3010} \approx 5.78$

Example 3

Here is a check of this example problem.

$4e^{2x} = 5$

$4e^{2[(\frac{1}{2})\ln(\frac{5}{4})]} \stackrel{?}{=} 5$

$4e^{\ln(\frac{5}{4})} \stackrel{?}{=} 5$

$4\left(\frac{5}{4}\right) \stackrel{?}{=} 5$

$5 = 5$ It checks.

Example 4

Why is it necessary first to subtract 5 from both sides of the original equation?

Example 5

Ask students if there is another sequence of steps that can be used to obtain a solution. For example:

$2 \log_{10} 3x = 4$	Original equation
$\log_{10} (3x)^2 = 4$	Power property
$\log_{10} 9x^2 = 4$	Simplify.
$10^{\log_{10} 9x^2} = 10^4$	Exponentiate.
$9x^2 = 10{,}000$	Simplify.
$x^2 = \frac{10{,}000}{9}$	Divide both sides by 9.
$x^2 \approx 1111.111$	Simplify.
$x \approx +\sqrt{1111.11}$	Take the square root of both sides.
$x \approx 33.333$	Simplify.

Why must the x-value in the next-to-last step be positive?
Logarithms are only defined for positive values of x.

Example 6

Be sure students recognize that the value $A = 1000$ represents a doubling of the initial investment of \$500. Ask students if it would take the same amount of time for \$800 to double as for \$1100 to double.
yes

Extra Examples

Here are additional examples similar to **Examples 1–6.**

1. Solve the following exponential equations.
   - **a.** $5^x = 56$ $\quad x \approx 2.501$
   - **b.** $e^{3x} = 124$ $\quad x \approx 1.607$
   - **c.** $10^x - 4 = 34$ $\quad x \approx 1.580$
   - **d.** $12e^{3x-2} = 8$ $\quad x \approx 0.532$

Example 4 Solving a Logarithmic Equation

Solve $5 + 2 \ln x = 4$.

Solution

$5 + 2 \ln x = 4$	*Rewrite original equation.*
$2 \ln x = -1$	*Subtract 5 from both sides.*
$\ln x = -\frac{1}{2}$	*Divide both sides by 2.*
$e^{\ln x} = e^{-1/2}$	*Exponentiate both sides.*
$x = e^{-1/2}$	*Simplify.*
$x \approx 0.607$	*Use a calculator.*

The solution is $x = e^{-1/2}$. Check this in the original equation. ■

Example 5 Solving a Logarithmic Equation

Solve $2 \log_{10} 3x = 4$.

Solution

$2 \log_{10} 3x = 4$	*Rewrite original equation.*
$\log_{10} 3x = 2$	*Divide both sides by 2.*
$10^{\log_{10} 3x} = 10^2$	*Exponentiate both sides.*
$3x = 100$	*Simplify.*
$x = \frac{100}{3}$	*Divide both sides by 3.*
$x \approx 33.333$	*Simplify.*

The solution is $x = \frac{100}{3}$. Check this solution in the original equation. ■

When solving exponential or logarithmic equations, the following properties are useful. Can you see where these properties were used in the examples in this lesson?

Properties of Exponential and Logarithmic Equations

Let a be a positive real number.

1. $a^x = a^y$ if and only if $x = y$.
2. $\log_a x = \log_a y$ if and only if $x = y$.

Goal 2 Using Exponential Equations in Real Life

Real Life
Finance

Example 6 Doubling Time

You have deposited \$500 in an account that pays 6.75% interest, compounded continuously. How long will it take for your money to double?

Solution Using the formula for continuous compounding, you can write

$$A = Pe^{rt} = 500e^{0.0675t}.$$

To find the time required for the balance to double, let $A = 1000$ and solve the resulting equation for t.

$500e^{0.0675t} = 1000$	*Let A = 1000.*
$e^{0.0675t} = 2$	*Divide both sides by 500.*
$\ln e^{0.0675t} = \ln 2$	*Take ln of both sides.*
$0.0675t = \ln 2$	*Simplify.*
$t = \frac{1}{0.0675} \ln 2$	*Divide both sides by 0.0675.*
$t \approx 10.27$	*Use a calculator.*

The balance in the account will double after approximately 10.27 years. This result is shown visually in the graph below.

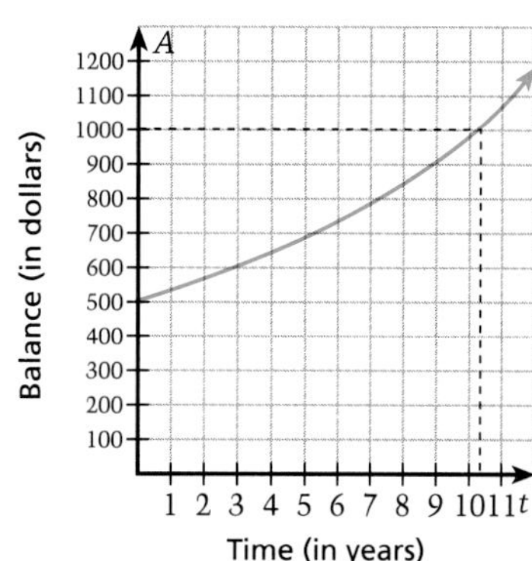

Interest rates paid by banks and other savings institutions have varied considerably over the past 20 years. Moreover, interest rates depend on the type of savings account. Typically, passbook accounts pay lower interest rates than certificates of deposit (which require committing a deposit for a specified period of time).

Communicating about ALGEBRA

SHARING IDEAS about the Lesson See margin.

Solve the following equations. Explain your steps.

A. $4 + 2(3^{2x}) = 12$ **B.** $-3 + 2\ln(x - 3) = 2$

2. Solve the following logarithmic equations.
   a. $13 - \log_{10} x = 17$
   b. $15 \ln 8x = 38$
   c. $\ln x + \ln 2 = 7$

   a. $x = 0.0001$, b. $x \approx 1.574$, c. $x \approx 548.317$
3. How long will it take an investment to triple in an account that pays 8.5%, compounded continuously?
   about 12.92 years

Check Understanding

1. What are logarithmic equations? Exponential equations?
   See Examples 4 and 5; see Examples 1–3
2. Describe the process called "exponentiating."
   Writing an equation in exponential form

Communicating about ALGEBRA

Answers

A. ≈ 0.631; subtract 4 from both sides, divide both sides by 2, take log with base 3 of both sides, simplify, divide both sides by 2, use a calculator

B. ≈ 15.18; add 3 to both sides, divide both sides by 2, exponentiate both sides, simplify, add 3 to both sides, use a calculator

MATH JOURNAL Have students include a flowchart of the steps needed to solve exponential equations and a flowchart of the steps needed to solve logarithmic equations in their math journals.

Then have them compare the flowcharts for similarities. Can one flowchart be designed that handles both types of equations?

The steps of both flowcharts are similar. For example, (1) Rewrite the original equation. (2) Isolate the variable expression. (3) Take the log (or ln) of both sides; or, exponentiate both sides. (4) Simplify to solve for the variable. (5) Use a calculator to evaluate.

EXERCISE Notes

ASSIGNMENT GUIDE

Day 1

Basic/Average: Ex. 1–4, 9–21 odd, 37–45 odd

Above Average: Ex. 1–4, 9–23 odd, 35–45 odd

Advanced: Ex. 1–4, 9–23 odd, 35–45 odd

Day 2

Basic/Average: Ex. 47–59 odd, 60–63

Above Average: Ex.47–51 odd, 52–54, 60–63

Advanced: Ex. 47–51 odd, 52–54, 60–63

Selected Answers
Exercises 1–4, 5–61 odd

Use **Mixed Review** as needed.

✪ More Difficult Exercises
Exercises 47–54, 63

Guided Practice

▶ **Ex. 1** Remind students to use the change-of-base formula to evaluate $\log_2 10$.

▶ **Ex. 3** Another strategy for solving the equation $\log_3 4x = 6$ is to rewrite it in its equivalent exponential form $3^6 = 4x$.

▶ **Ex. 4** If $\ln x = 0.5$ is written as $\log_e x = 0.5$, then $x = e^{0.5}$ might seem more obvious.

Independent Practice

▶ **Ex. 11–28** Remind students that they should isolate the exponent term and apply the log or the ln to both sides of the equation.

EXERCISES

Guided Practice

CRITICAL THINKING about the Lesson

1. Explain how to solve $2^x = 10$ by taking the log of both sides of the equation. $x = \log_2 10 \approx 3.322$
2. Explain how to solve $\log_{10} x = 2.4$ by exponentiating both sides of the equation. $\log_{10} x = 2.4;\ 10^{\log_{10} x} = 10^{2.4};\ x \approx 251.19$
3. Which is the correct set of steps in solving $\log_3 4x = 6$? Explain. b; you need to use 3, not e, as the base

   a. $\log_3 4x = 6$
   $e^{\log_3 4x} = e^6$
   $4x = e^6$
   $x = \frac{1}{4}e^6$

   b. $\log_3 4x = 6$
   $3^{\log_3 4x} = 3^6$
   $4x = 3^6$
   $x = \frac{1}{4}(3)^6$
4. Solve $\ln x = 0.5$. Explain your steps. $x = e^{0.5} \approx 1.649$

Independent Practice

In Exercises 5–10, decide whether the *x*-value is a solution of the equation.

5. $\ln x = 31,\ x = e^{31}$ It is.
6. $\log_5 \frac{1}{2}x = 17,\ x = 2e^{17}$ It is not.
7. $3 - \log_8 3x = 1,\ x = \frac{64}{3}$ It is.
8. $\ln 5x = 4,\ x = \frac{1}{4}e^5$ It is not.
9. $2e^x = 4,\ x = \ln 2$ It is.
10. $e^x + 2 = 18,\ x = \log_2 16$ It is not.

In Exercises 11–28, solve the exponential equation. 25. ≈ -0.164

11. $e^x = 12$ ≈ 2.485
12. $2^x = 1.5$ ≈ 0.585
13. $4^x - 5 = 3$ 1.5
14. $10^{2x} + 3 = 8$ ≈ 0.349
15. $6e^{-x} = 3$ ≈ 0.693
16. $\frac{1}{2}e^{3x} = 20$ ≈ 1.230
17. $5(2)^{3x} - 4 = 13$ ≈ 0.589
18. $1.2e^{-5x} + 2.6 = 3$ ≈ 0.220
19. $-12e^{-x} + 8 = 7$ ≈ 2.485
20. $\frac{2}{3}(3)^{4x} + \frac{1}{3} = 4$ ≈ 0.388
21. $6^{0.1x} + 5 = 10$ ≈ 8.982
22. $\left(\frac{1}{4}\right)^x - \frac{3}{5} = 1$ ≈ -0.339
23. $\frac{1}{2}(8)^{2x} + 2 = 9$ ≈ 0.635
24. $\frac{1}{4}(3)^{-x} - 18 = 18$ ≈ -4.524
25. $10^{-12x} + 6 = 100$
26. $12 + 3(3)^{0.1x} = 16$ ≈ 2.619
27. $4 - 2e^x = -23$ ≈ 2.603
28. $-16 + 0.2(10)^x = 35$ ≈ 2.407

In Exercises 29–46, solve the logarithmic equation. 37. ≈ -33.115 43. ≈ 49.471

29. $\ln x = 4$ ≈ 54.598
30. $\log_2 x = -1$ $\frac{1}{2}$
31. $\log_{10} 2x = 1.5$ ≈ 15.811
32. $\ln \frac{1}{2}x = \frac{1}{4}$ ≈ 2.568
33. $4 \log_3 x = 28$ 2187
34. $16 \ln x = 30$ ≈ 6.521
35. $1 - 2 \ln x = -4$ ≈ 12.182
36. $\frac{1}{3} \log_2 x + 5 = 7$ 64
37. $2 \ln(-x) + 7 = 14$
38. $8 + 9 \log_{10} x = 3$ ≈ 0.278
39. $3 \log_5 5x = 9$ 25
40. $\frac{1}{2} \log_6 16x = 3$ 2916
41. $15 + 2 \log_2 x = 31$ 256
42. $6 \ln 4x - 1 = 14$ ≈ 3.046
43. $-5 + 2 \ln 3x = 5$
44. $2 \log_2 2x = 19$ ≈ 362.039
45. $10 \ln 100x - 3 = 117$ ≈ 1627.548
46. $6.5 \log_5 3x = 20$ ≈ 47.158

47. *Compound Interest* You deposit \$1000 in an account that pays 5% annual interest, compounded continuously. How long will it take for the balance to reach \$1500? $\approx$**8.11 years**

48. *Compound Interest* You deposit \$2000 in an account that pays 7.5% interest, compounded quarterly. How long will it take for the balance to reach \$2400? $\approx$**2.45 years**

49. *Doubling Time* You deposit \$200 in an account that pays 7% annual interest, compounded continuously. How long will it take for the balance to double? $\approx$**9.90 years**

50. *Tripling Time* You deposit \$600 in an account that pays 6% annual interest, compounded monthly. How long will it take for the balance to triple? $\approx$**18.36 years**

51. *Municipal Waste* For 1960 through 1986, the amount of municipal waste, W (in pounds per person per day), processed for energy recovery in the United States can be approximated by the equation $W = 0.007e^{0.0051t^2}$, where $t = 0$ represents 1960. In what year did the amount of waste reach 0.22 pound per person? **1986**

52. *Population of Colonial America* The first permanent colony in America was established in Jamestown, Virginia, in 1607. For 1610 through 1780, the population, P (in thousands), in colonial America can be modeled by

$$P = 242.4e^{0.00008t^2} - 244,$$

where $t = 10$ represents 1610. When was the population about 345 thousand? **1705**

53. *Oceanography* Oceanographers use the density, d (in grams per cubic centimeter), of seawater to obtain information about the circulation of water masses and the rates at which waters of different densities mix. For water with a salinity of 30%, the water temperature, T (in degrees Celsius), is related to the density by

$$T = 7.9 \ln(1.0245 - d) + 61.84.$$

Find the densities of the subantarctic water and the antarctic bottom water shown in the diagram.

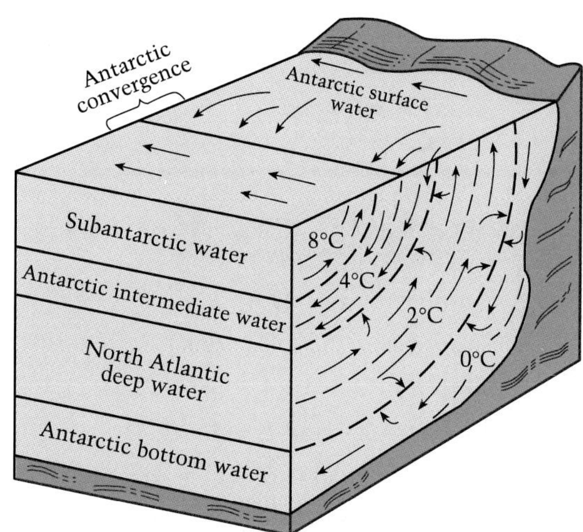

This cross section shows complex currents at various depths in the South Atlantic Ocean off Antarctica.

53. 1.0234 g per cm^3; 1.0241 g per cm^3

54. *Muon Decay* A muon is an elementary particle that is similar to an electron, but much heavier. Muons are unstable—they quickly decay to form electrons and other particles. In an experiment conducted in 1943, the number of muon decays, m (of an original 5000 muons), was related to the time T (in microseconds), by the model $T = 15.7 - 2.48 \ln m$. How many decays were recorded when T was equal to 2.5 microseconds? $\approx$**204.9**

▶ Ex. 29–46 Students can solve the equations either by the method modeled in Examples 4 and 5, or by rewriting the equations in the equivalent exponential form shown in Lesson 8.2.

▶ Ex. 47–52 The variables are in the exponent. Remind students that they must take the log or the ln of both sides.

TECHNOLOGY Students with graphics calculators could graph the functions and use the Trace function to locate the t-values.

Ex. 63 Use the graphs to discuss the differences in the domains of the functions $f(x)$, $g(x)$, and $h(x)$. If students graph $h(x)$ first, only the first-quadrant values for $y = x$ are displayed because the domain for $h(x)$ is $x > 0$.

Enrichment Activities

These activities require students to go beyond lesson goals.

1. Speculate why either the change-of-base formula for common logarithms or the change-of-base formula for natural logarithms may be used to evaluate such expressions as $\log_a M$. In other words, show that

$$\frac{\log_{10} M}{\log_{10} a} = \log_a M = \frac{\ln M}{\ln a}.$$

$$\frac{\log_{10} M}{\log_{10} a} = \frac{\frac{\ln M}{\ln 10}}{\frac{\ln a}{\ln 10}}$$

$$= \frac{\ln M}{\ln 10} \cdot \frac{\ln 10}{\ln a} = \frac{\ln M}{\ln a}$$

2. Use the operation of applying a logarithm to both sides of an equation to prove the change-of-base formula. That is, show that for positive numbers a and b, $a \neq 1$ and $b \neq 1$, $\log_a x = \dfrac{\log_b x}{\log_b a}$.

If $\log_a x = y$, then $a^y = x$.

Take the log, base b, of both sides.

If $a^y = x$, then $\log_b a^y = \log_b x$.

Apply the power rule.

$$y \log_b a = \log_b x$$

$$y = \frac{\log_b x}{\log_b a}.$$

Substitute $\log_a x$ for y.

$$\log_a x = \frac{\log_b x}{\log_b a}.$$

Integrated Review

In Exercises 55–60, evaluate the expression for the given value of x.

55. $6 \ln 2x$, $x = 10$ ≈ 17.974

56. $\frac{1}{2} \log_2(-x)$, $x = -5$ ≈ 1.161

57. $6 + e^{4x}$, $x = \frac{2}{3}$ ≈ 20.392

58. $21 - 5 \log_{10} x$, $x = 16$ ≈ 14.979

59. $0.05 \log_3 6x$, $x = 12$ ≈ 0.195

60. $10e^{-x}$, $x = 5$ ≈ 0.067

Your Gasoline Dollar **In Exercises 61 and 62, use the following information.**

The circle graph at the right shows what happens to each dollar spent on gasoline in the United States. ***(Source: Computer Petroleum)***

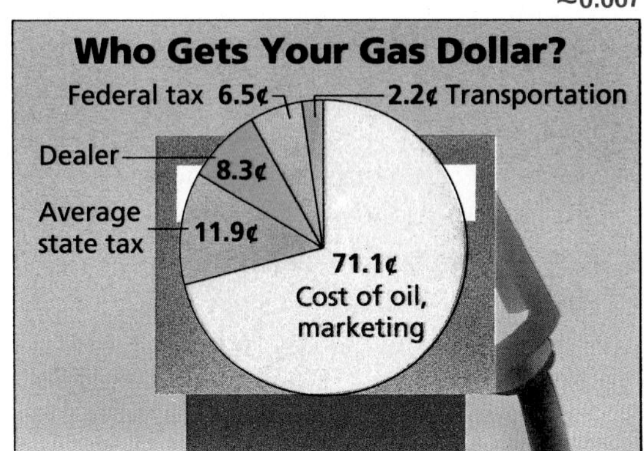

61. Suppose you spend \$16 for gasoline. How much of that goes for federal tax? \$1.04

62. Suppose gasoline costs \$1.39 per gallon. How much of the per-gallon price goes to the gasoline dealer? $\approx$\$0.115

Exploration and Extension

63. *Technology* Use a graphing calculator to sketch the graphs of $f(x) = x$, $g(x) = \ln e^x$, and $h(x) = e^{\ln x}$ on the same coordinate plane. What can you conclude about each function? See Additional Answers.

Mixed REVIEW

8. $6a^2 + 11a - 10$ 12., 15. See Additional Answers.

1. Simplify: $\sqrt[3]{x^3}$. **(7.4)** x

2. Simplify: $\sqrt[5]{64}$. **(7.4)** $2\sqrt[5]{2}$

3. Simplify: $\sqrt[3]{24a^6b^{14}}$. **(7.4)** $2a^2b^4\sqrt[3]{3b^2}$

4. Simplify: $\sqrt[3]{9} \cdot \sqrt[3]{15}$. **(7.4)** $3\sqrt[3]{5}$

5. Simplify: $(2 + 3i)(4 - 5i)$. **(5.5)** $23 + 2i$

6. Simplify: $(16^{1/3})^{3/4}$. **(7.4)** 2

7. For $f(x) = 2^x$, find $f(-1)$. **(6.1)** $\frac{1}{2}$

8. Write $(3a - 2)(2a + 5)$ as a trinomial.

9. Write the inverse of $f(x) = 3x - 1$. **(6.3)** $g(x) = \frac{1}{3}(x + 1)$

10. Find the vertex of $y = 3(x + 4)^2 - 7$. **(5.2)** $(-4, -7)$

11. Solve $\sqrt[3]{7a + 1} - 4 = 0$. **(7.5)** 9

12. Sketch the graph of $y < \frac{3}{4}x - 5$. **(2.5)**

13. Find the geometric mean of $2a^5$ and $8a^3$. **(7.5)** $4a^4$

14. Find the distance from $(-4, -1)$ to $(2, -3)$. **(7.5)** $2\sqrt{10}$

15. Sketch the graph of $|x - 5| < 3$. **(1.7)**

16. Solve $2x^2 - 3x + 5 = 0$. **(5.4)** $\frac{3 \pm i\sqrt{31}}{4}$

17. Simplify: $\frac{a^2b^{-3}}{c^0b^{-2}} \cdot \left(\frac{a^{-2}c}{b^{-3}}\right)^4$. **(7.1)** $\frac{b^{11}c^4}{a^6}$

18. Find the domain of $y = \frac{3}{x^2 + 4}$. **(6.2)** All reals

19. \$4000 saved at 6% compounded monthly is how much in 5 years? **(7.2)**

20. A rock dropped from a 400-ft bridge hits the water in how many seconds? **(5.1)**

19. \$5395.40 20. 5 sec

8.7 Exploring Data: Logistics Growth Functions

What you should learn:

Goal 1 How to graph a logistics growth function

Goal 2 How to use logistics growth functions to answer questions about real-life situations

Why you should learn it:

You can model many types of growth using logistics functions, such as the spread of an infectious disease in a small environment.

Goal 1 Graphing Logistics Growth Functions

Four common types of mathematical models involving exponential or logarithmic functions are as follows.

1. Exponential Growth: $y = a^x$, $1 < a$ $\quad$ $y = e^{bx}$, $0 < b$
2. Exponential Decay: $y = a^x$, $0 < a < 1$ $\quad$ $y = e^{-bx}$, $0 < b$
3. Logarithmic: $y = \log_{10} bx$ $\quad$ $y = \ln bx$
4. Logistics Growth: $y = \dfrac{1}{1 + be^{-cx}}$

You have already studied the first three models. In this lesson, you will study the fourth model.

Logistics growth models are used to represent populations or quantities whose rate of growth changes—from an increasingly rapid growth to a decreasingly rapid growth. An example would be a bacteria culture allowed to grow initially under ideal conditions, followed by less favorable conditions that inhibit growth.

Example 1 *Graphing a Logistics Growth Function*

Sketch the graph of $f(x) = \dfrac{1}{1 + e^{-x}}$.

Solution As usual, begin by using a calculator to construct a table of values.

x	−3	−2	−1	0	1	2	3
$f(x)$	0.05	0.12	0.27	0.50	0.73	0.88	0.95

Plot the points given by the table of values and connect the points with a smooth curve to obtain the graph at the left.

Connections **Technology**

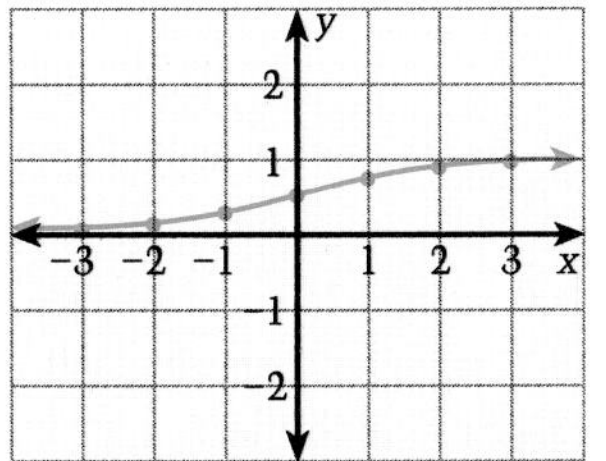

The domain of the function is all real numbers. The range of the function is $0 < y < 1$. The x-axis is a horizontal asymptote of the graph (to the left), and the line $y = 1$ is a horizontal asymptote of the graph (to the right). ■

In Example 1, notice that the graph is rising more and more rapidly until it reaches the y-axis. At that point, the graph continues to rise, but at a slower and slower rate. Eventually, for large values of x, the graph is nearly horizontal.

Problem of the Day

Tweedledum and Tweedledee are twins who cannot be distinguished by appearance. One lies on Mondays, Tuesdays, and Wednesdays, and tells the truth on other days. The other lies on Thursdays, Fridays, and Saturdays, and tells the truth on other days. Once Alice in Wonderland met the two, and the first one said, "I'm Tweedledum." The second said, "I'm Tweedledee." Which is really which?
They are what they claim to be. (It is Sunday.)

ORGANIZER

Warm-Up Exercises

1. Make a table of values of the functions $f(x) = 2^{-x}$ and $g(x) = \dfrac{1}{1 + 2^{-x}}$ and graph the functions on the same coordinate grid.

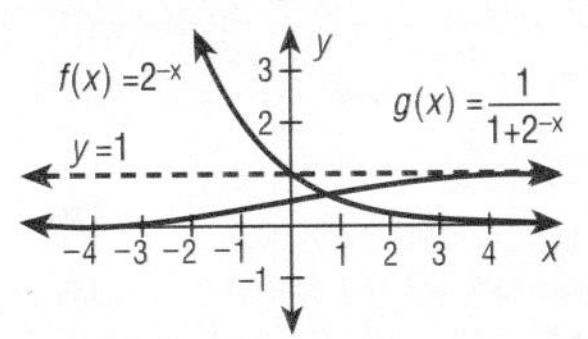

2. Does the function $f(x)$ have any asymptotes? Does the function $g(x)$?
$f(x)$: asymptote at $y = 0$; $g(x)$: asymptotes appear to be at $y = 0$ and $y = 1$.

Lesson Resources

Teaching Tools
- Transparencies: 7, 8
- Problem of the Day: 8.7
- Warm-up Exercises: 8.7
- Answer Masters: 8.7

Extra Practice: 8.7
Color Transparency: 45

LESSON Notes

Example 1

Ask students to compare and contrast the graph of $g(x)$ of the Warm-up Exercises with that of this example.

Example 2

Ask your class if the 54 students in this example who are infected with the virus over a five-day period represent a large enough number of students to classify the outbreak as an epidemic. If not, ask your class to decide what an epidemic level might be, and then determine the number of days that would have to pass before that many students became infected. For example, if 500 students infected with the flu virus represent an epidemic level of infection, it would take about 8 days to reach that number.

Example 3

Remind students that the median of a collection of data is the number such that half the data is above it and half is below it.

Extra Examples

Here are additional examples similar to **Examples 1–3.**

1. Sketch the graph of

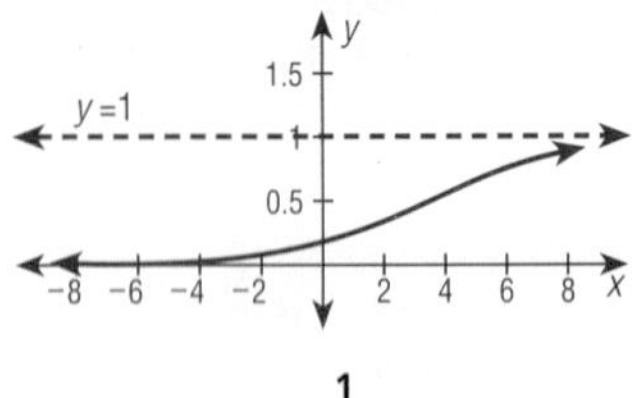

$$g(x) = \frac{1}{1 + 5e^{-0.43x}}$$

2. Evaluate the following at the indicated value.

   a. $y = \dfrac{5000}{1 + 4999e^{-0.83x}}$, $x = 8$

   b. $A = \dfrac{26 + 3.1}{1 + 53e^{-0.56x}}$, $x = 15$

   a. $y \approx 663.67$, b. $A \approx 28.76$

Epidemics are a major threat to human, animal, and plant populations. The worst in recent times was the global influenza epidemic of 1918 and 1919. About 20 million people died worldwide, including about one-half million people in the United States.

Goal 2 Using Logistics Growth Models

Real Life
Epidemiology

Example 2 *Spread of a Virus*

On a college campus of 5000 students, one student returned from vacation with a contagious three-day flu virus. The spread of the virus through the student body can be modeled by

$$y = \frac{5000}{1 + 4999e^{-0.8t}},$$

where y is the total number infected after t days. How many students will have been infected by the end of five days? Sketch a graph of this function.

Solution After five days, the number infected is

$$y = \frac{5000}{1 + 4999e^{-0.8t}} \quad \text{Logistics growth model}$$

$$= \frac{5000}{1 + 4999e^{-0.8(5)}} \quad \text{Substitute 5 for } t.$$

$$= \frac{5000}{1 + 4999e^{-4}} \quad \text{Simplify.}$$

$$\approx 54. \quad \text{Use a calculator.}$$

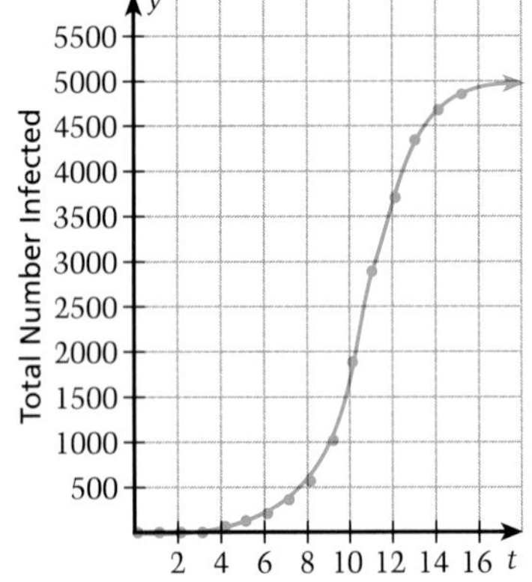

In a similar way, you can compute other values of y, as shown in the table. By plotting the values and connecting them with a smooth curve, you obtain the graph shown at the left.

t	0	1	2	3	4	5	6	7
y	1	2	5	11	24	54	119	257

t	8	9	10	11	12	13	14	15
y	537	1057	1868	2851	3735	4340	4680	4851

■

Real Life
Social Science

Example 3 *Median Age of First Marriages*

For 1970 through 1990, the median age, A, of Americans at their first marriage can be modeled by

$$A = 20.4 + \frac{3.1}{1 + 43e^{-0.36t}} \quad \textit{Women}$$

$$A = 22.4 + \frac{3.3}{1 + 63e^{-0.36t}} \quad \textit{Men}$$

where $t = 0$ represents 1970. Find the median age of women and men at their first marriage in **(a)** 1970, **(b)** 1980, and **(c)** 1990. ***(Source: U.S. Center for Health Statistics)***

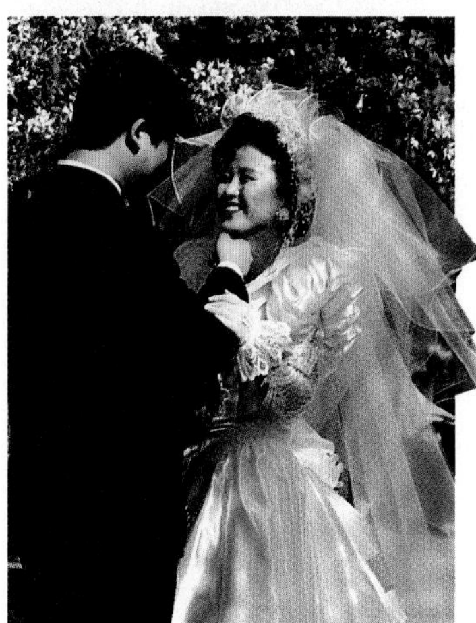

The ages at which women and men marry vary from time to time and from country to country. From 1970 to 1990, the median age of American women and men at their first marriage increased by about 3 years.

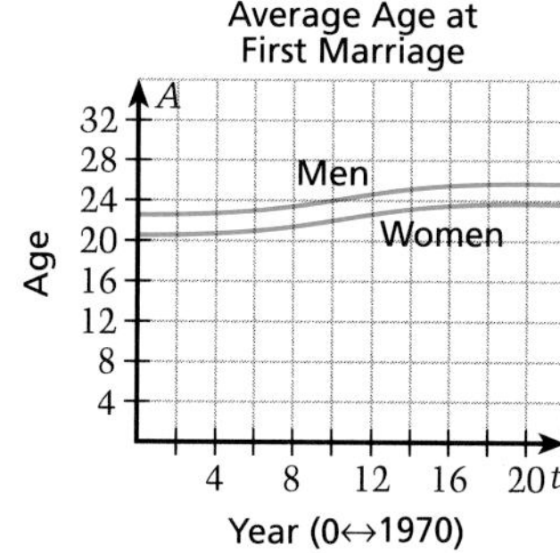

Solution

Women | **Men**

a. 1970:

$20.4 + \frac{3.1}{1 + 43e^{-0.36(0)}} \approx 20.5$ $\qquad$ $22.4 + \frac{3.3}{1 + 63e^{-0.36(0)}} \approx 22.5$

b. 1980:

$20.4 + \frac{3.1}{1 + 43e^{-0.36(10)}} \approx 21.8$ $\qquad$ $22.4 + \frac{3.3}{1 + 63e^{-0.36(10)}} \approx 23.6$

c. 1990:

$20.4 + \frac{3.1}{1 + 43e^{-0.36(20)}} \approx 23.4$ $\qquad$ $22.4 + \frac{3.3}{1 + 63e^{-0.36(20)}} \approx 25.6$ ■

Communicating about ALGEBRA

SHARING IDEAS about the Lesson

Sketch the graphs of the logistics growth functions. Explain your steps. What are the horizontal asymptotes of each graph? See Additional Answers for graphs.

A. $f(x) = \frac{2}{1 + e^{-0.5x}}$ **B.** $g(x) = \frac{5}{1 + e^{-0.2x}}$

A.–B. Make a table of values, plot the points, then connect the points with a smooth curve.

Check Understanding

1. Characterize the behavior of the graphs of the following mathematical models:
See the top of page 441.
   a. exponential growth
   b. exponential decay
   c. logarithmic
   d. logistics growth
2. What approach is used to graph functions that have not been graphed previously?
See Example 1.

Communicating about ALGEBRA

EXTEND *Communicating*
COOPERATIVE LEARNING
Have groups of students use graphics calculators and examine the effects of constants on the logistics growth model, $f(x) = \frac{1}{1 + e^{-x}}$. In particular, for real number values b and c, they should describe the effects of these constants in the following functions in relationship to $f(x)$ as a vertical shift, a horizontal shift, and a reflection in the x-axis.

a. $g(x) = c + \frac{1}{1 + e^{-x}}$

b. $h(x) = \frac{1}{1 + e^{-x+b}}$

c. $j(x) = \frac{-1}{1 + e^{-x}}$

EXERCISE Notes

ASSIGNMENT GUIDE

Basic/Average: Ex. 1–4, 5–14, 23, 25, 27–32, 33–38

Above Average: Ex. 1–4, 5–14, 21–27 odd, 28–39

Advanced: Ex. 1–4, 5–14, 21–27 odd, 28–40

Selected Answers
Exercises 1–4, 5–37 odd

More Difficult Exercises
Exercises 27–32, 39–40

Guided Practice

Ex. 1–4 You can use these exercises to explore students' intuitive ideas about asymptotes as limits. As x becomes very large, the expression $4e^{-x}$ approaches 0; so $f(x)$ approaches $\frac{4}{1+0}$, or 4. As x becomes very large negatively, the expression $4e^{-x}$ becomes very large; so $f(x)$ approaches 0.

TECHNOLOGY You can ask students with graphics calculators to graph the function for very large positive values and very small negative values to verify the limits.

Independent Practice

Ex. 11–14 WRITING Ask students to write a one- or two-sentence justification for their choices.

Ex. 15–26 TECHNOLOGY Students can either use Example 1 as a solution model or use a graphics calculator or computer graphing program.

Answers
15.

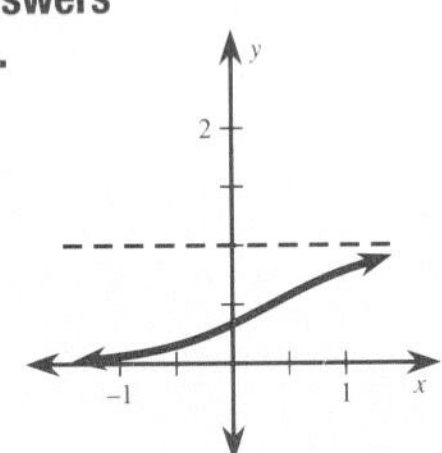

(continued)

EXERCISES

Guided Practice

CRITICAL THINKING about the Lesson

In Exercises 1–4, use the function $f(x) = \frac{4}{1 + 4e^{-x}}$.

1. Complete the table of values for the function. 0.049 0.131 0.337 0.800 1.618 2.595 3.336

x	−3	−2	−1	0	1	2	3
f(x)	?	?	?	?	?	?	?

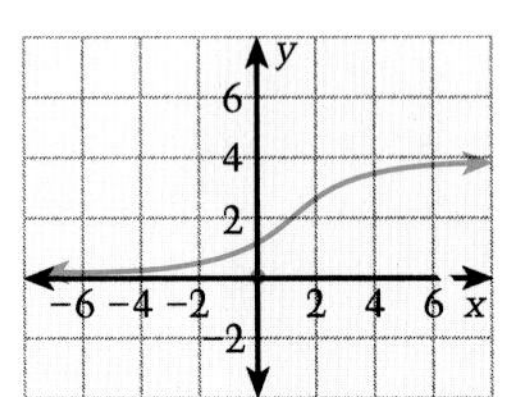

2. What kind of mathematical model is the function? 2. logistics growth
3. Estimate the horizontal asymptotes of the graph. 3. $y = 0$ and $y = 4$
4. What are the function's domain and range? 4. All real numbers; $0 < y < 4$

Independent Practice

In Exercises 5–10, identify the type of mathematical model.

5. $f(x) = \ln 2x$ logarithmic
6. $f(x) = e^{-0.11x}$ exponential decay
7. $f(x) = \frac{10}{1 + 6e^{-0.5x}}$ logistics growth
8. $f(x) = 4 \log_2(0.5x)$ logarithmic
9. $f(x) = 2000(1.05)^x$ exponential growth
10. $f(x) = 2.04e^{0.8x}$ exponential growth

In Exercises 11–14, match the function with its graph.

11. $f(x) = \frac{2}{1 + e^{-2x}}$ a
12. $f(x) = \frac{3}{1 + e^{-x}}$ c
13. $f(x) = \frac{5}{1 + e^{-0.5x}}$ b
14. $f(x) = \frac{3}{1 + e^{-x}} + 1$ d

a. 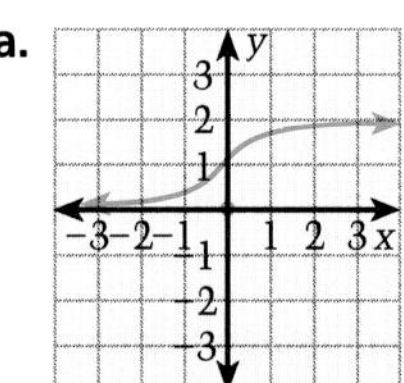

b. 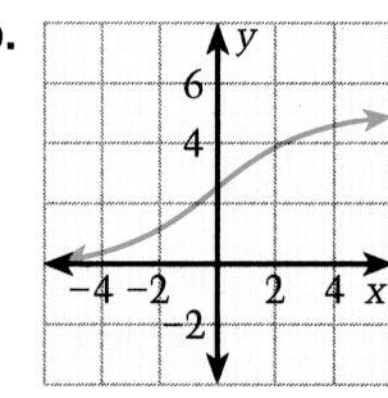

c. 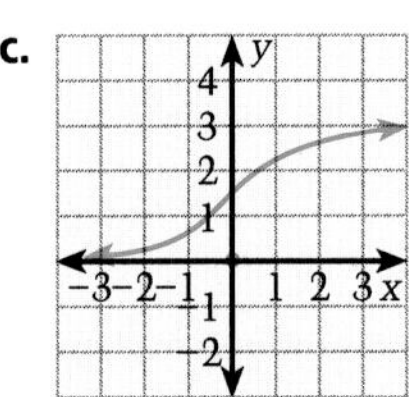

d. 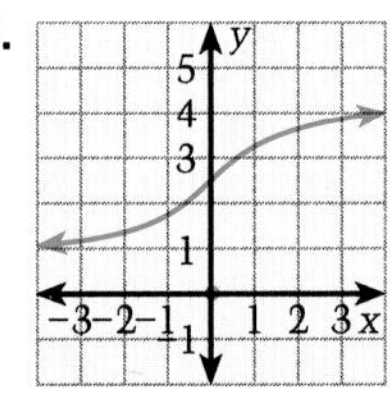

In Exercises 15–26, sketch the graph. Identify the horizontal asymptotes.

15. $f(x) = \frac{1}{1 + 2e^{-2x}}$ $y = 0, y = 1$
16. $f(x) = \frac{1}{1 + 0.8e^{-x}}$ $y = 0, y = 1$
17. $f(x) = \frac{5}{1 + e^{-10x}}$ $y = 0, y = 5$
18. $f(x) = \frac{4}{1 + 3e^{-3x}}$ $y = 0, y = 4$
19. $f(x) = \frac{4}{1 + 0.05e^{-1.2x}}$ $y = 0, y = 4$
20. $f(x) = \frac{2}{1 + 2e^{-x}}$
21. $f(x) = \frac{10}{1 + 4e^{-x}} - 4$
22. $f(x) = \frac{3}{1 + e^{-8x}} + 1$ $y = 1, y = 4$
23. $f(x) = \frac{8}{1 + e^{-1.02x}} + 4$
24. $f(x) = \frac{4}{1 + 5e^{-4x}} - 2$
25. $f(x) = \frac{6}{1 + 0.8e^{-2x}} - 6$
26. $f(x) = \frac{12}{1 + 2e^{-x}} + 1$

15.–21. See margins for graphs. 22.–26. See Additional Answers for graphs.

27. *Stocking a Lake* A lake has been stocked with 500 trout. During the next several years, the trout population, P, in the lake can be modeled by

$$P = \frac{10{,}000}{1 + 19e^{-0.2t}},$$

10,000 trout; the denominator must be at least 1.

where t is the time in years. Sketch the graph of this model. What is the maximum number of trout that the lake can support? Explain.

28. *Endangered Species* One hundred animals of an endangered species have been released into a game preserve that has a carrying capacity of 1000 animals. The population, P, of the herd can be modeled by

$$P = \frac{1000}{1 + 9e^{-0.16t}},$$

where t is the time in years. What is the population after 5 years? ≈198 animals

29. *Discontinued Advertising* In 1980, the manufacturer of Lotus Egg Rolls decided to discontinue all advertising of the product. After that, the annual sales, S (in dollars), can be modeled by

$$S = \frac{500{,}000}{1 + 0.6e^{0.05t}},$$

where $t = 0$ represents 1980. Did the sales increase or decrease after 1980? Approximate the sales in 1992. Decrease; $238,860.58

30. *Videocassette Prices* For 1980 through 1990, the average price, p (in dollars), of a prerecorded videocassette in the United States can be modeled by

$$p = 65.5 - \frac{49.5}{1 + 127e^{-1.175t}},$$

where $t = 0$ represents 1980. Sketch the graph of this function. Explain why the average price was decreasing during the 1980's. ***(Source: Paul Kagan Associates, Inc.)*** See Additional Answers.

31. *Videocassette Sales* For 1980 through 1990, the number, x (in thousands), of prerecorded videocassettes sold in the United States can be modeled by

$$x = 1630 + \frac{82{,}100}{1 + 1506e^{-1.22t}},$$

where $t = 0$ represents 1980. Use the model given in Exercise 30 to find the total revenue (in $1000s) for prerecorded videocassettes in 1980, 1985, and 1990.
(Source: Paul Kagan Associates, Inc.)

≈$110,000 thousand, ≈$590,000 thousand, ≈$1,300,000 thousand

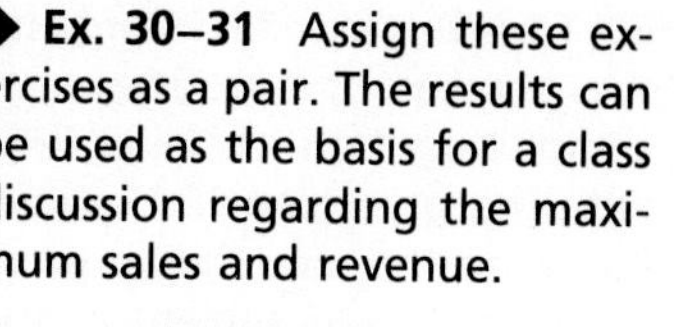

Ex. 30–31 Assign these exercises as a pair. The results can be used as the basis for a class discussion regarding the maximum sales and revenue.

Answers *(continued)*

16.

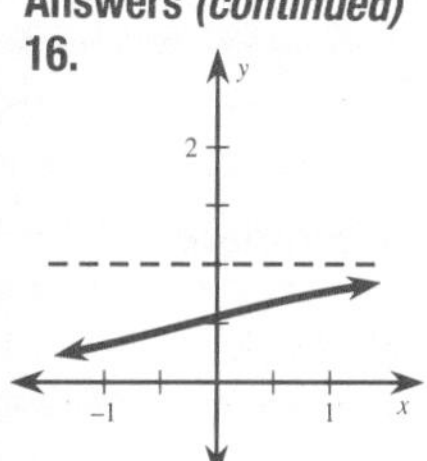

17.

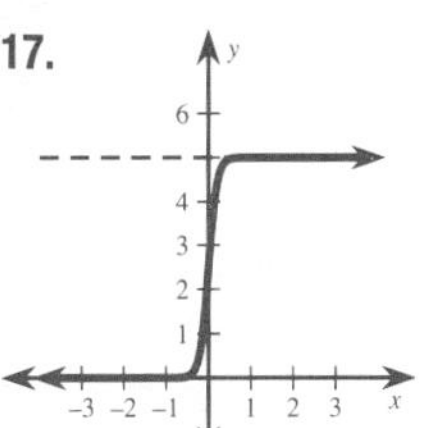

18.

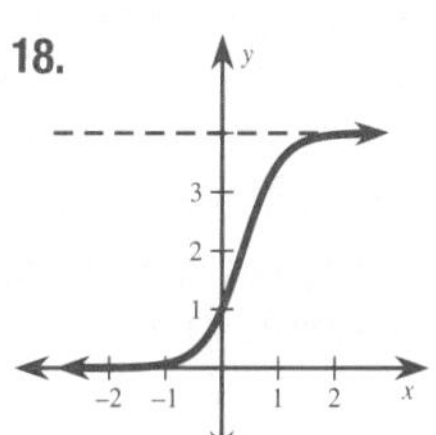

19.

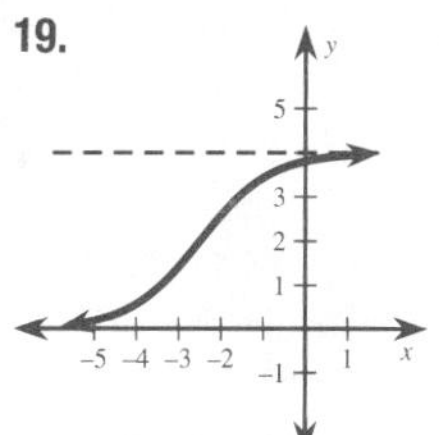

20.

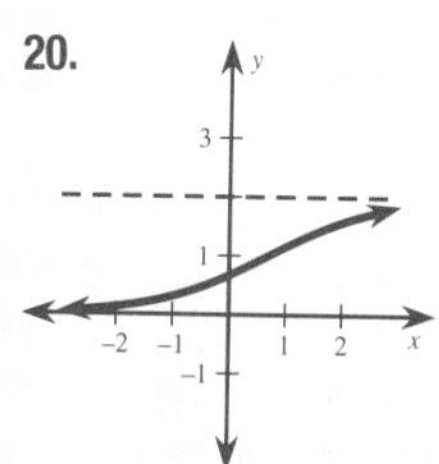

21.

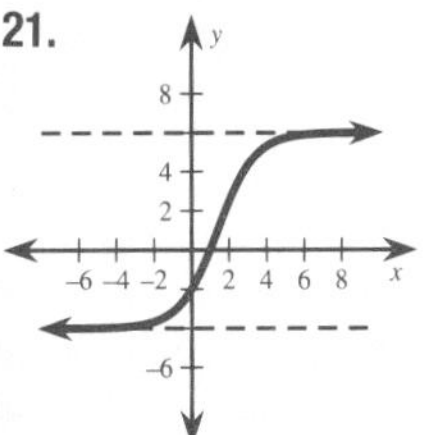

EXTEND Ex. 32 Compare the percent of Americans not completing high school in 1974 with the predicted 1995 value. Based on the model, will there always be some percent of the population not completing high school?
yes

How does this compare to the 1995 value?
1974: 33.8%; 1995: 23.5%

Answers
33. c; all real numbers, $y > -3$
34. d; $x > -3$, all real numbers
35. b; all real numbers, $y > 3$
36. a; all real numbers, $0 < y < 2$

Ex. 40 Discuss why the maximum population of 11.14 billion is realistic. Can the population keep growing or should its growth rate gradually slow down?

Enrichment Activity

This activity requires students to go beyond lesson goals.

COOPERATIVE LEARNING
Work with a partner or in a small group. List each of the four types of mathematical models discussed in this lesson. Describe an appropriate problem that uses each model, draw a sketch of the model, and explain the trends in the data that would occur in the model.
Refer to page 441. An exponential growth model is appropriate when the data is rapidly increasing with no maximum; an exponential decay model is appropriate when the data is rapidly decreasing toward 0; a logarithmic growth model is appropriate when the data is increasing slowly with no maximum; a logistics growth model is appropriate when the rate of growth changes from increasing rapidly to decreasing rapidly and moves toward a maximum.

32. *High School Graduates* For 1980 through 1990, the percent, P, of Americans who had not completed high school can be modeled by

$$P = 33.8 - \frac{10.3}{1 + 25.8e^{-0.833t}},$$

where $t = 0$ represents 1980. What percentage of Americans did not complete high school in 1988? Did the percentage increase or decrease during the 1980's? ≈23.8%, decrease

Integrated Review

In Exercises 33–36, match the function with its graph. State the domain and range. See margin for domain and range.

33. $f(x) = 2e^x - 3$ c  **34.** $f(x) = 2\ln(x + 3)$ d  **35.** $f(x) = -2^x + 3$ b **36.** $f(x) = \frac{2}{1 + 2e^{-1.5x}}$ a

a. 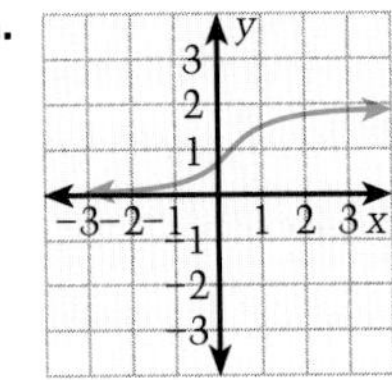

b. 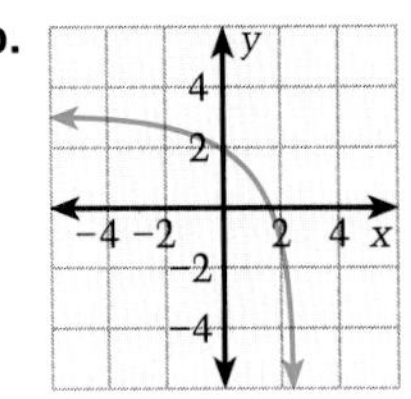

c. 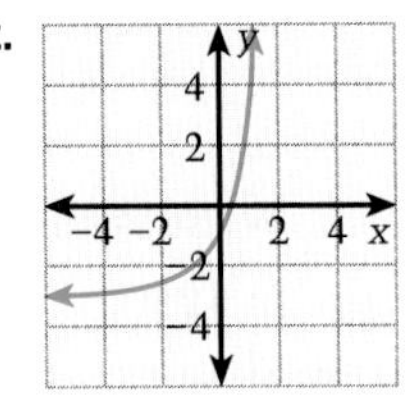

d. 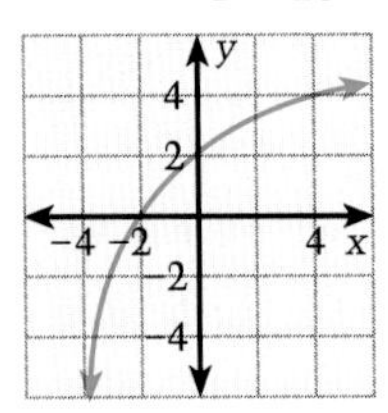

37. *College Entrance Exam Sample* Which is the best approximation of $5 - \sqrt{32.076} + 1.00017^3$?
a. 9 **b.** 7 **c.** 5 **d.** 3 **e.** 0

38. *College Entrance Exam Sample* Which fraction is greater than $\frac{3}{4}$?
a. $\frac{35}{71}$ **b.** $\frac{13}{20}$ **c.** $\frac{71}{101}$ **d.** $\frac{19}{24}$ **e.** $\frac{15}{20}$

Exploration and Extension

39. *Flu Epidemic* In a small town, one person has a flu virus. The virus' spread through the population is modeled by

$y = \frac{1000}{1 + 990e^{-0.7t}}$, where y is the total number infected after t days.

In how many days will 818 people be infected with the virus? ≈12 days

40. *World Population* The graph at the right shows the population, P (in billions), of the world as projected by the Population Reference Bureau. The bureau's projection can be modeled by

$$P = \frac{11.14}{1 + 1.101e^{-0.051t}},$$

where $t = 0$ represents 1990. *If* this model is accurate, what is the maximum world population? 11.14 billion

Projected World Population

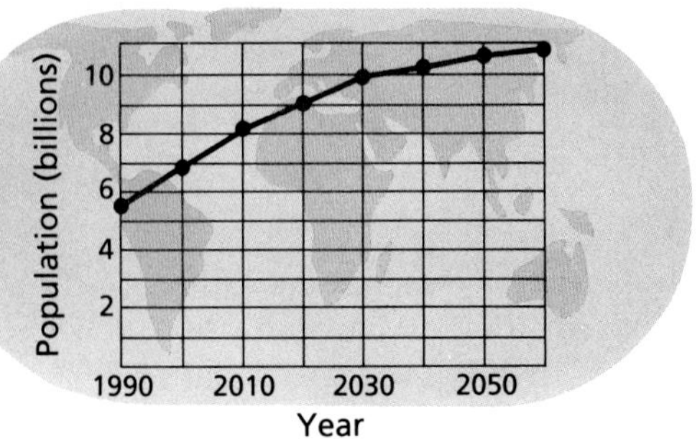

USING A GRAPHING CALCULATOR

In Lesson 8.6, you studied how to solve exponential and logarithmic equations by hand. The strategies discussed in that lesson work well for equations such as $2e^{-x} + 5 = 7$ or $\log_{10}(3x - 2) = 8$. Equations that have more than one type of expression, however, cannot usually be solved in such a straightforward manner. For instance, it is not easy to solve the equation $\ln x = x - 2$ by hand. To solve such equations, we usually rely on technology such as a computer or a graphing calculator.

Example: Using a Graphing Calculator to Solve an Equation

For 1980 through 2030, the median age, M, of Americans is predicted to follow the model

$$M = 19.17 + 0.107t + 4.028 \ln t,$$

where $t = 10$ represents 1980. (See Exercise 56, Lesson 8.5.) In 1980 the median age was about 30 (half of all Americans were 30 years old or younger). By 1990, the median age had risen to about 33. According to this model, when will the median age be 40?

Solution To answer this question, solve the equation

$$40 = 19.17 + 0.107t + 4.028 \ln t.$$

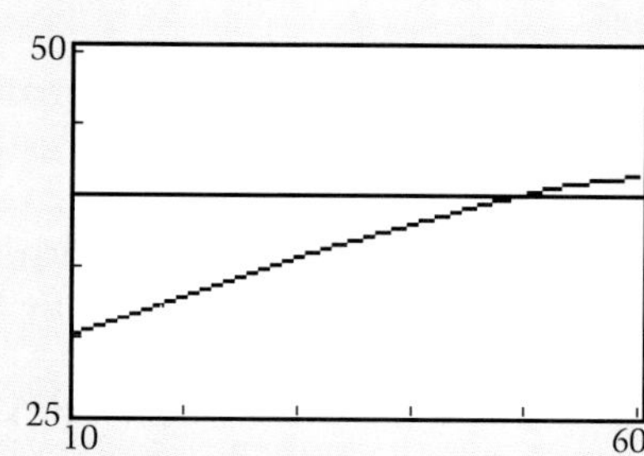

On a graphing calculator, you can sketch the graphs of $y = 40$ and $y = 19.17 + 0.107x + 4.028 \ln x$ and find the point at which the two graphs intersect. From the calculator display shown at the right, you can see that the two graphs intersect when x is about 49. Thus, the median age will be 40 in about 2019. ■

Exercises

In Exercises 1–6, use a graphing calculator to approximate the solution(s) of the equation. (Some equations may have no solutions.)

1. $\ln x = x - 2$ 0.16, 3.15

2. $\log_{10} x = x^2 - 3$ 1.8, 0.0015

3. $2 \ln x = -x + 4$ 2.3

4. $e^x = -x + 3$ 0.8

5. $2^x = x^2$ −0.8, 2, 4

6. $e^x = 2 + \ln x$ No solution

7. *National Park Salaries* For 1981 through 1990, the amount, A (in millions of dollars), spent on salaries and wages for National Park Service employees can be modeled by $A = 286.52 + 7.25t + 27.51 \ln t$, where $t = 1$ represents 1981. (See Exercise 55, Lesson 8.5.) During which year was the amount spent on salaries and wages about \$367 million? 1985

TECHNOLOGY It is important to note that the graphing technique of solving equations is very powerful if computer and calculator technologies are available.

EXTEND *the Activity*
Ask students how the following relationship is related to the model used in Example 1:

$y = -20.03 + 0.107t + 4.028 \ln t.$

How can this model be used to answer the question posed in Example 1?
Since $-20.03 = 40 - 19.17$, the models are equivalent for $M = 40$; let $y = 0$.

ORGANIZER

TIME SCHEDULE: All levels, two days

The Chapter Summary helps students organize the main ideas of the chapter. In this chapter, we explored exponential and logarithmic relationships and functions. The irrational number, e, was introduced, and techniques for solving exponential and logarithmic equations were also presented. Finally, the logistics growth model was introduced.

Work with students to review the skills, strategies, and concepts of the chapter. The first day's homework assignment can be used as the basis for the second day of review.

COOPERATIVE LEARNING

Encourage students to study together. Emphasize the importance of "teaching" a classmate how to perform a skill or how to recall a procedure. When students work together, everyone wins. The students receiving help get additional instruction and the students giving help gain a deeper understanding of the skills and concepts involved.

Chapter SUMMARY

Review the chapter by asking students, "What did you learn? Why did you learn it? How does it fit with the other algebra skills and concepts you have learned?"

8 Chapter Summary

What did you learn?

Skills

1. Evaluate and graph an exponential function
   - with base a. **(8.1)**
   - with natural base e. **(8.4)**
2. Evaluate and graph a logarithmic function
   - with base a. **(8.2)**
   - with natural base e. **(8.5)**
3. Use properties of logarithms. **(8.3, 8.5)**
4. Solve exponential or logarithmic equations. **(8.6)**

Strategies

5. Use exponential models to solve real-life problems. **(8.1–8.7)**
6. Use logarithmic models to solve real-life problems. **(8.2–8.6)**

Exploring Data

7. Use logistics growth functions as real-life models. **(8.7)**

Why did you learn it?

Many quantities in real life are related exponentially. For instance, in savings accounts in which the interest is compounded continuously, the balances are related to the time by the exponential equation $A = Pe^{rt}$. Other quantities in real life such as the decibel level and intensity of sound are related by logarithmic equations.

Exponential functions and logarithmic functions are inverses of each other. They form an important part of advanced mathematics—especially calculus and probability.

How does it fit into the bigger picture of algebra?

In this chapter, you studied exponential functions and their inverses, called logarithmic functions. Exponential functions are used in real-life situations to model growth that increases by the same percent over equal periods of time. For instance, if you receive a 5% increase in salary each year, your salary could be modeled by an exponential function. (Remember from Chapter 2 that quantities that increase by the same amount over equal periods of time are modeled by linear, not exponential, functions.)

Any positive number other than 1 can be used as a base for an exponential or a logarithmic function. There is, however, a special number that is often used. This important number, called e, is used in the last four lessons of the chapter.

Chapter REVIEW

In Exercises 1–9, use a calculator to evaluate the expression. Round your result to three decimal places. (8.1)

1. 8^{π} 687.291
2. $2^{-\sqrt{3}}$ 0.301
3. $\log_6 10$ 1.285
4. $\left(\frac{1}{2}\right)^{\pi}$ 0.113
5. $\ln 4$ 1.386
6. $\ln 23$ 3.135
7. $e^{4.2}$ 66.686
8. $e^{-2.1}$ 0.122
9. π^{e} 22.459

In Exercises 10–15, evaluate the expression without using a calculator. (8.2)

10. $\log_2 16$ 4
11. $\log_4 64$ 3
12. $\log_2(-1)$ Undefined
13. $\log_2 \frac{1}{8}$ −3
14. $\log_3 \frac{1}{9}$ −2
15. $\log_6 1$ 0

In Exercises 16–18, write the equation in exponential form. (8.2)

16. $\log_5 125 = 3$ $5^3 = 125$
17. $\log_3 81 = 4$ $3^4 = 81$
18. $\log_4 1024 = 5$ $4^5 = 1024$

In Exercises 19–21, write the equation in logarithmic form. (8.2)

19. $2^5 = 32$ $\log_2 32 = 5$
20. $6^3 = 216$ $\log_6 216 = 3$
21. $25^{-1/2} = \frac{1}{5}$ $\log_{25} \frac{1}{5} = -\frac{1}{2}$

In Exercises 22–27, simplify the expression. (8.4)

22. $e^{x-3} \cdot e^{3x}$ e^{4x-3}
23. $4e^{x^2} \cdot 2e^{2x+1}$ $8e^{x^2+2x+1}$
24. $\frac{5e^{2x}}{e}$ $5e^{2x-1}$
25. $\frac{e^{-3}}{3e^{2x}}$ $\frac{1}{3}e^{-2x-3}$
26. $(2e^{-4x})^2$ $\frac{4}{e^{8x}}$
27. $\left(6e^{20x} \cdot \frac{1}{2}e^{-10x}\right)^2$ $9e^{20x}$

In Exercises 28–33, expand the expression. (8.5)

28. $\log_{10} 5x^3$ $\log_{10} 5 + 3\log_{10} x$
29. $\ln\left(\frac{2x}{x^2}\right)^3$ $\ln 8 - 3\ln x$
30. $\ln \frac{7x}{3}$ $\ln 7 + \ln x - \ln 3$
31. $\log_3 6xy^2$ $\log_3 6 + \log_3 x + 2\log_3 y$
32. $\log_2 \frac{4x}{3}$ $\log_2 4 + \log_2 x - \log_2 3$
33. $\ln \frac{x^5y^{-2}}{2y}$ $5\ln x - \ln 2 - 3\ln y$

In Exercises 34–39, condense the expression. (8.5) 37.–39. See margin.

34. $2\ln 3 - \ln 5$ $\ln \frac{9}{5}$
35. $\log_2 x + \frac{1}{2}\log_2 9$ $\log_2 3x$
36. $\log_4 3 + 3\log_4 2$ $\log_4 24$
37. $\frac{1}{3}\log_6 8 + \log_6 x + 2\log_6 y$
38. $\frac{1}{2}\ln 4 + 2(\ln 6 - \ln 2)$
39. $-\frac{3}{2}(\log_{10} x + 2\log_{10} 2)$

In Exercises 40–48, solve the equation. (8.6)

40. $\log_4 x = 3$ 64
41. $\ln x = 5$ ≈ 148.413
42. $\log_x 256 = 4$ 4
43. $10^x = 4.3$ ≈ 0.633
44. $e^x = 6$ ≈ 1.791
45. $\log_x 729 = 3$ 9
46. $3e^{-x} - 4 = 9$ ≈ -1.466
47. $10 + 0.1e^{3x} = 18$ ≈ 1.461
48. $2(3)^{2x} = 5$ ≈ 0.417

In Exercises 49–54, identify the type of mathematical model. (8.1, 8.2, 8.7)

49. $f(x) = \frac{66}{1 + e^{2x}}$ logistics growth
50. $f(x) = e^{-4x}$ exponential decay
51. $f(x) = \ln(x + 3)$ logarithmic
52. $f(x) = \log_{10} x + 4$ logarithmic
53. $f(x) = 10\left(\frac{1}{2}\right)^x$ exponential decay
54. $f(x) = 4 + \frac{1}{1 + e^x}$ logistics growth

Chapter REVIEW

ASSIGNMENT GUIDE

Basic/Average: Ex.3–48 multiples of 3, 49–56, 57–78 multiples of 3, 79, 84.

Above Average: Ex.3–48 multiples of 3, 49–56, 57–78 multiples of 3, 79, 81–83, 85–94.

Advanced: Ex.3–48 multiples of 3, 49–56, 57–78 multiples of 3, 79, 81–83, 85–94.

More Difficult Exercises
Exercises 79–84

Ex. 1–94 The Chapter Review exercises include a reference to the lesson in which a skill or concept was presented. Students may use the reference as a study aid if they are uncertain how to do a given exercise in the Chapter Review.

Answers

37. $\log_6 2xy^2$

38. $\ln 18$

39. $\log_{10} \frac{1}{8} x^{-\frac{3}{2}}$

63.

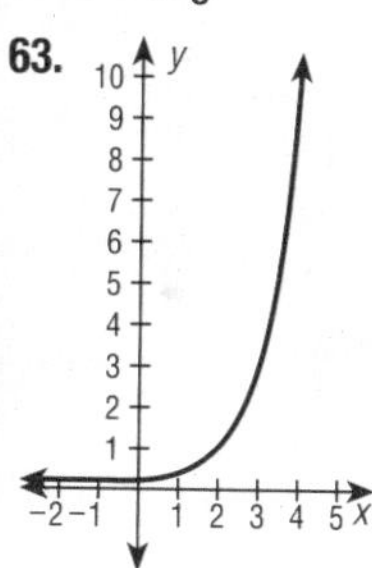

Answers

64.

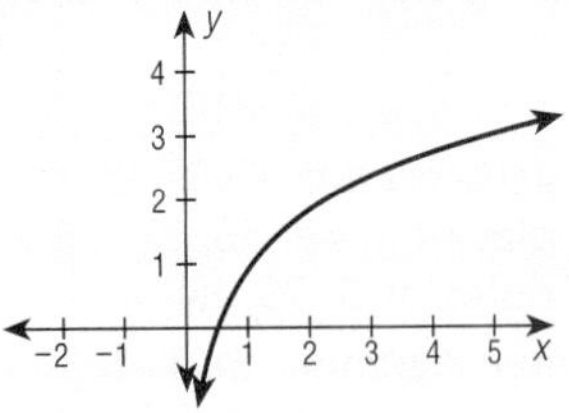

65.

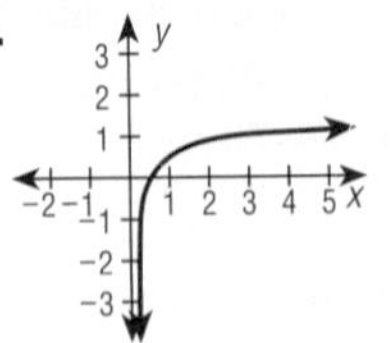

66.

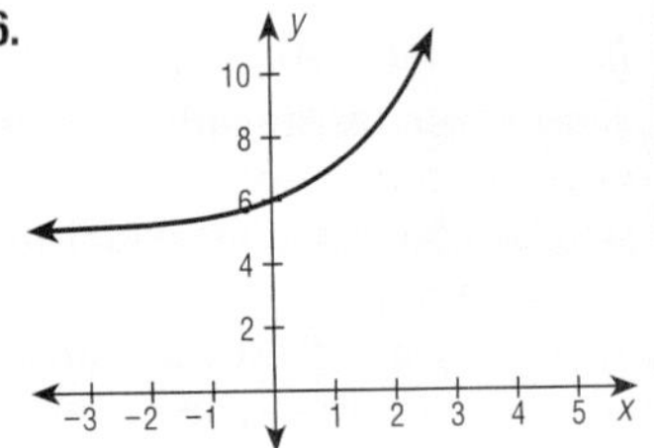

67.

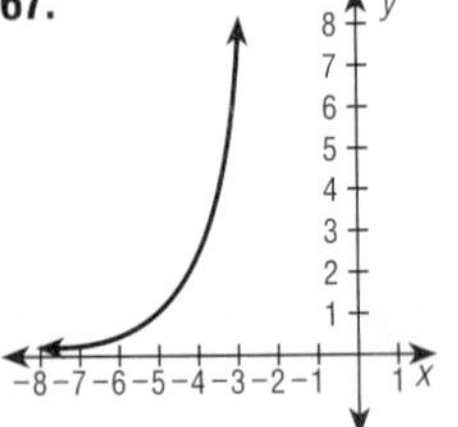

68.

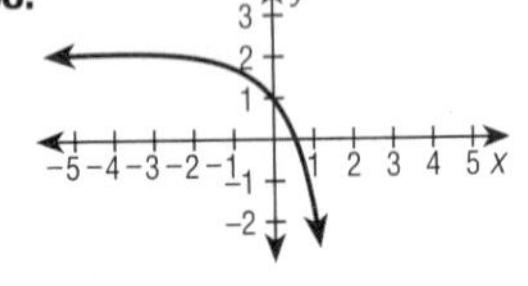

69.

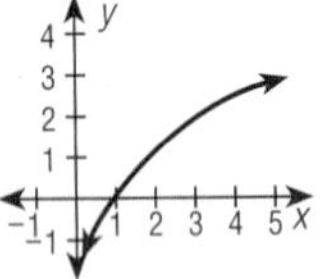

70.

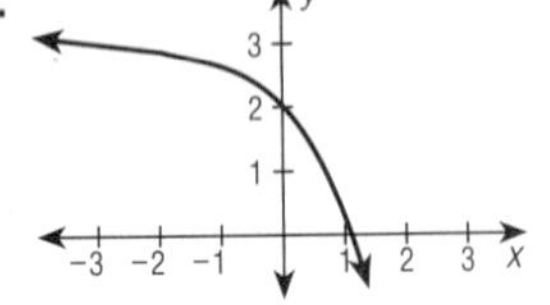

In Exercises 55–62, match the function with its graph. State the domain and range. (8.1, 8.2, 8.7) See below for domain and range.

55. $f(x) = 2e^x - 1$ b

56. $f(x) = -\ln x + 2$ h

57. $f(x) = -2^x + 4$ a

58. $f(x) = \dfrac{4}{1 + e^{3x}} - 2$ g

59. $f(x) = \ln(x + 3)$ f

60. $f(x) = \log_3 2x$ d

61. $f(x) = \dfrac{2}{1 + e^{-2x}}$ c

62. $f(x) = e^{x-1}$ e

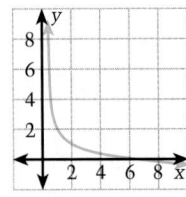

a.

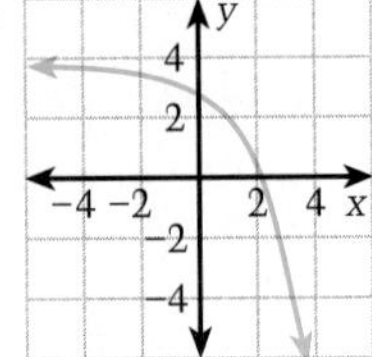

b.

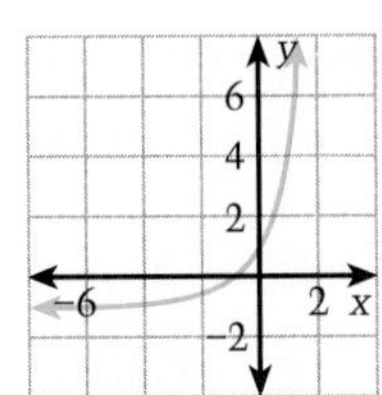

c.

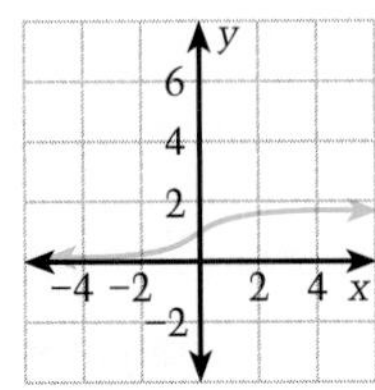

d.

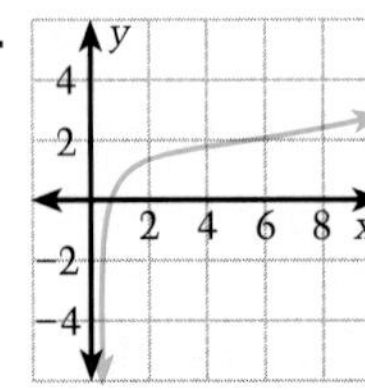

e.

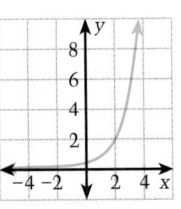

f.

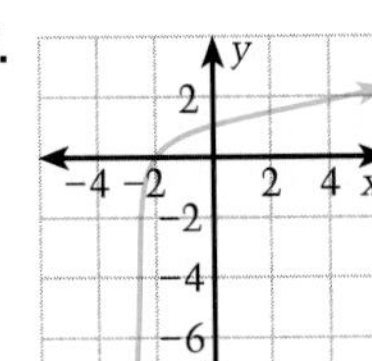

g.

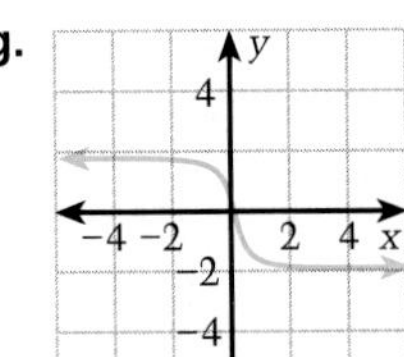

h.

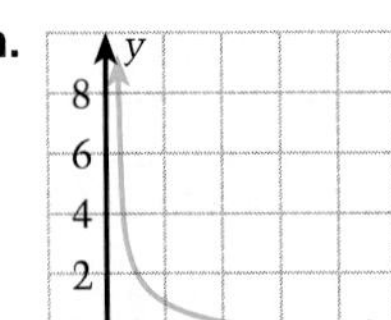

In Exercises 63–74, sketch the graph of the function. (8.1, 8.2, 8.7) See margin.

63. $f(x) = 3^{x-2}$

64. $f(x) = \log_3(2x)$

65. $f(x) = \log_{10}(4x)$

66. $f(x) = 2^x + 5$

67. $f(x) = e^{x+5}$

68. $f(x) = -4^x + 2$

69. $f(x) = 3 \log_5 x$

70. $f(x) = -e^x + 3$

71. $f(x) = 3 \ln x + 4$

72. $f(x) = 2e^x - 3$

73. $f(x) = \dfrac{3}{1 + e^{-0.5x}} - 1$

74. $f(x) = \dfrac{2}{1 + e^{2x}} + 1$

75. *Compound Interest* You deposit \$1500 in an account that pays 6.5% annual interest, compounded continuously. Find the balance after 10 years. \$2873.31

76. *Compound Interest* You deposit \$300 in an account that pays 7% annual interest, compounded continuously. Find the balance after 20 years. \$1216.56

77. *Doubling Time* You deposit \$500 in an account that pays 6% annual interest, compounded continuously. How long before your balance doubles? Would a larger deposit double in a shorter time? Explain. ≈11.5 years, no, see margin.

78. *Compound Interest* You deposit \$2500 in an account that pays 7.5% annual interest, compounded quarterly. How long will it take for your balance to become \$3000? Would a larger deposit increase by \$500 in a shorter time? Explain. ≈2.5 years, yes, see margin.

55. All real numbers, $y > -1$ **56.** $x > 0$, all real numbers **57.** All real numbers, $y < 4$

58. All real numbers, $-2 < y < 2$ **59.** $x > -3$, all real numbers **60.** $x > 0$, all real numbers

61. All real numbers, $0 < y < 2$ **62.** All real numbers, $y > 0$

79. *Chemical Reaction* The rate, R, of many chemical reactions can be modeled by $R = ke^{-4727/(T+273)}$, where T is the temperature in degrees Celsius and k is a constant. The chemical reactions that take place when a cake is baked can be approximated by this model. Find the ratio of the chemical reaction rates of batter placed in a 200°C oven to batter placed in a 150°C oven. ≈3.26

80. *Deer Herd* The state Parks and Wildlife Department releases 100 deer into a wilderness area. The population, P, of the herd can be modeled by

$$P = \frac{500}{1 + 4e^{-0.36t}},$$

where t is measured in years. Find the population after 5 years. What is the maximum number of deer that the wilderness area can support? ≈301, 500 deer

Fruits and Vegetables **In Exercises 81–83, use the following information.**

The pH of a material is a measure of its acidity or alkalinity. A material with a pH of 7 is neutral. A material with a pH less than 7 is an acid. The pH of a material is given by

$$pH = \log_{10} \frac{1}{H},$$

where H is the concentration of hydrogen ions in moles per liter. The pH values of several fruits and vegetables are shown in the table.

Fruit	pH	Fruit	pH
Apples	3.1	Lemons	2.3
Asparagus	5.6	Oranges	3.5
Cabbage	5.3	Pears	3.8
Cherries	3.6	Potatoes	5.8
Corn	6.3	Pumpkins	5.0
Grapes	4.0	Tomatoes	4.2

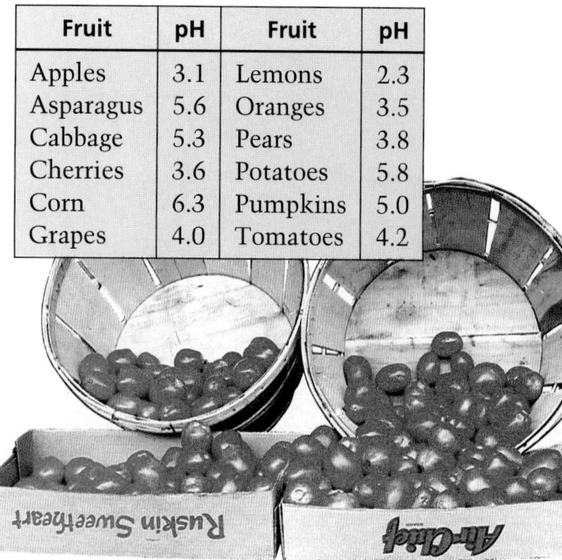

81. What is the concentration of hydrogen ions in apples? ≈0.00079 moles per liter

82. What is the concentration of hydrogen ions in pumpkins? ≈0.00001 moles per liter

83. What is the ratio of the concentration of hydrogen ions in lemons to the concentration of hydrogen ions in grapes? ≈50 : 1

84. *Car Advertising Expenses* For 1970 through 1990, the amount of money, A (in billions of dollars), spent by new car dealerships on advertising their product line can be modeled by

$$A = 0.11e^{0.2t} + 0.3,$$

where $t = 0$ represents 1970. In what year was about \$4.3 billion spent by new car dealerships? 1988

Answers

71.

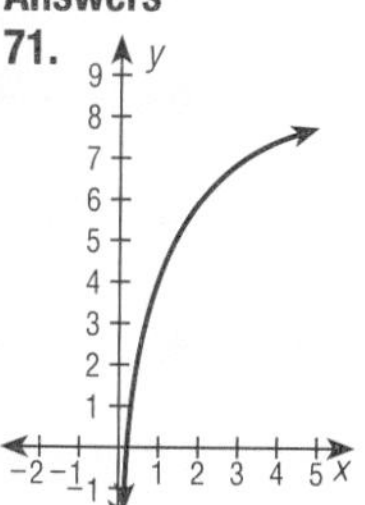

72.

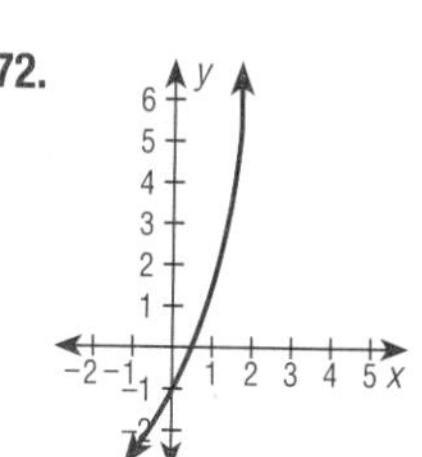

73.

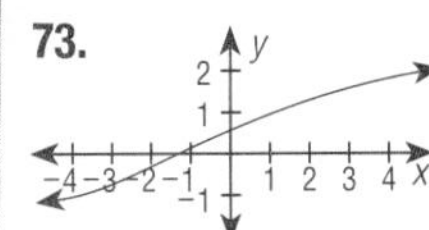

74.

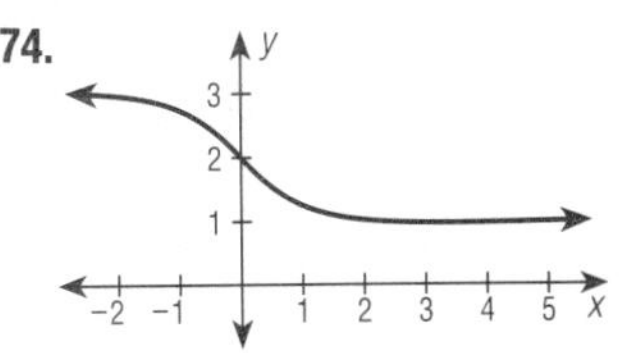

77. ≈11.55 years; no, $\frac{A}{P} = 2$ in each case

78. ≈2.45 years; yes, $\log \frac{A}{P}$ would be smaller and result in a smaller t

The Stone of the Fifth Sun **In Exercises 85–94, use the following information.**

This Stone of the Fifth Sun, uncovered in 1790 in Mexico City, is a depiction of the Aztec cosmos. In each exercise, evaluate the function at $x = 2$. Then arrange the function values in ascending order to find the animal or object that is depicted in each of the ten day signs numbered below. (*Hint:* Some of the functions are easier to evaluate if they are first simplified.)

85. *Deer:* $d(x) = 1 + \dfrac{1}{1 + e^{3x}}$ ≈1.002, 4th

86. *Eagle:* $e(x) = \dfrac{e^x}{12e^{-2x}}$ ≈33.619, 9th

87. *Flower:* $f(x) = 6^x \cdot 6^4$ 46,656, 10th

88. *Grass:* $g(x) = \dfrac{3}{1 + e^{-2x}}$ ≈2.946, 7th

89. *Jaguar:* $j(x) = (5e^{-2x})^{-1}$ ≈10.920, 8th

90. *Lizard:* $l(x) = \ln 4 - 2 \ln x$ 0, 2nd

91. *Monkey:* $m(x) = e^{-4x} \cdot e^9$ ≈2.718, 6th

92. *Rabbit:* $r(x) = \log_5 5 + \log_5 3x$

93. *Serpent:* $s(x) = \log_4 3x - \log_4(x + 2)$

94. *Wind:* $w(x) = 2 \log_2 x - \log_2 8$

92. ≈2.113, 5th 93. ≈0.292, 3rd 94. −1, 1st

Chapter TEST

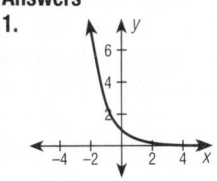

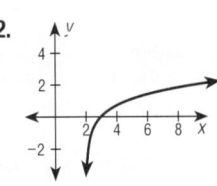

In Exercises 1–3, sketch the graph of the function. **(8.4, 8.5, 8.7)** See margin.

1. $f(x) = e^{-x}$

2. $f(x) = \ln(x - 2)$

3. $f(x) = \dfrac{1}{1 + e^{-x}}$

In Exercises 4–6, simplify the expression. **(8.4)**

4. $(2e^{-4})(3e^{2})$ $\dfrac{6}{e^2}$

5. $\dfrac{-6e^x}{2e^2}$ $-3e^{x-2}$

6. $\sqrt{4e^6x^2}$ $2e^3|x|$

In Exercises 7–9, evaluate the expression. If appropriate, round your result to three decimal places. **(8.2, 8.5)**

7. $\ln e^3$ 3

8. $\log_3 \frac{1}{9}$ -2

9. $\log_6 \sqrt{5}$ ≈ 0.449

In Exercises 10–12, expand the expression. **(8.3, 8.5)**

10. $\ln \dfrac{7x}{2}$ $\ln 7 + \ln x - \ln 2$

11. $\log_{10} 4x^2$ $\log_{10} 4 + 2\log_{10} x$

12. $\ln \dfrac{3x}{4y^2}$ $\ln 3 + \ln x - \ln 4 - 2\ln y$

In Exercises 13–15, condense the expression. **(8.3, 8.5)**

13. $\log_4 24 - \log_4 6$ $\log_4 4 = 1$

14. $\ln 7 + \ln 8$ $\ln 56$

15. $3\ln x - 2\ln y$ $\ln \dfrac{x^3}{y^2}$

In Exercises 16–18, solve the equation. Round your result to three decimal places. **(8.6)**

16. $3e^{5x} = 8$ ≈ 0.196

17. $3\ln 5x = 27$ ≈ 1620.617

18. $\log_5 3x = 4$ 208.333

19. Write $\ln 5 = x$ in exponential form. **(8.6)** $e^x = 5$

20. Write $e^x = 17$ in logarithmic form. **(8.6)** $\ln 17 = x$

21. \$4000 is deposited in an account that earns 7% annual interest, compounded continuously. Find the balance at the end of 5 years. **(8.4)** \$5676.27

22. Money is deposited in an account that earns 7% annual interest, compounded continuously. How long will it take for the balance to double? **(8.4)** About 10 years

23. For 1980 through 1990, the sales, S (in thousands of dollars), of electronic keyboards can be modeled by

$$y = \frac{70}{1 + 0.67e^{-2.2(t-6)}} + 45$$

where $t = 0$ represents 1980. Use the model to estimate the sales of electronic keyboards in 1989. ***(Source: USA Today)*** $\approx$ \$115,000

24. For 1978 through 1988, the sales, S (in thousands of units), of VCRs can be modeled by

$$y = 0.14(1.78)^t$$

where $t = 1$ represents 1978. Use the model to estimate the year in which 8000 units were sold. ***(Source: Electronic Market Data Book)*** 1984

25. The ice trays of a freezer are filled with water at 53°F. The freezer maintains a temperature of 20°F. According to Newton's Law of Cooling, the water temperature, T, is related to the time, t (in hours), by the equation

$$kt = \ln\frac{T - 20}{53 - 20}.$$

After 1 hour, the water temperature in the ice trays is 40°F. Use the fact that $T = 40$ when $t = 1$ to find how long it takes the water to freeze. **(8.5)** About 2 hours after being put in the freezer

Answers

1.

2.

3.

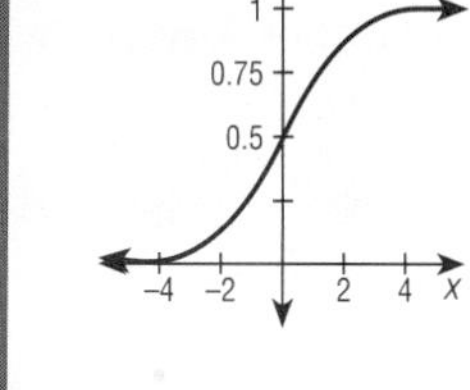

A number of classic problems of algebra will be considered in this chapter. Much of the work done in algebra over the past few centuries has involved polynomials. Students should pay close attention to the relationships between the zeros and factors of a polynomial function and the zeros, factors, and solutions of polynomial equations.

The Chapter Summary on page 508 provides you and the students with a synopsis of the chapter. It identifies key skills and concepts. You may want to have students look at the Chapter Summary as an overview before beginning the chapter.

CHAPTER 9

Polynomials and Polynomial Functions

LESSONS

Rectangular prisms are among the most common structures in our environment, as this photograph of Oahu, Hawaii shows. Yet they give rise to surprising relationships, such as the one described on the opposite page.

Connections
Geometry

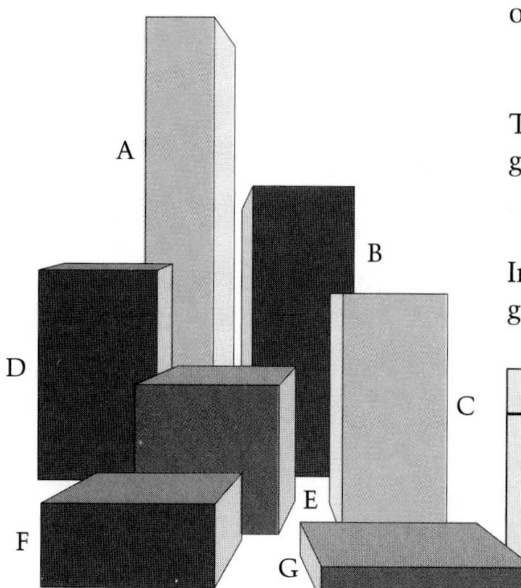

Each of the boxes shown has a square base. If a side of the base is x inches, then the height of the box was made to be $h = 54x^{-1} - \frac{1}{2}x$ inches. This implies that the surface area, S, of each box is the same!

$$S = 2x^2 + 4xh = 2x^2 + 4x\left(54x^{-1} - \tfrac{1}{2}x\right) = 216$$

The volumes of the boxes, however, differ greatly. The volume, V, of each is given by

$$V = x^2\left(54x^{-1} - \tfrac{1}{2}x\right) = 54x - \tfrac{1}{2}x^3.$$

In Lesson 9.2, you will be asked to sketch the graph of this *polynomial equation.*

	x	h	S	V
A.	3.0	16.50	216	148.5
B.	4.0	11.50	216	184.0
C.	4.5	9.75	216	197.4
D.	5.0	8.30	216	207.5
E.	6.0	6.00	216	216.0
F.	8.0	2.75	216	176.0
G.	9.0	1.50	216	121.5

Polynomial relationships are useful in describing many real-life situations. In fact, logarithmic and exponential relationships can often be approximated by polynomial expressions, which are conducive to easier computational formulas.

Ask students to show that the surface area, S, of each box illustrated here is a constant value, 216.

$$\begin{aligned} S &= 2x^2 + 4xh \\ &= 2x^2 + 4x(54x^{-1} - 0.5x) \\ &= 2x^2 + 4 \cdot 54 \cdot x \cdot x^{-1} \\ &\quad - 4 \cdot 0.5 \cdot x \cdot x \\ &= 2x^2 + 216 - 2x^2 \\ &= 216 \end{aligned}$$

Polynomials and Polynomial Functions

PACING CHART

The daily Pacing Chart is meant to help you adjust your teaching pace. Students in the full course should finish the entire text by the end of the year. Students in the basic course are expected to complete the first twelve chapters. The Pacing Chart for each chapter contains suggestions for lessons that require more than one day and lessons that may be omitted for the basic course.

DAY	FULL COURSE	BASIC COURSE
1	9.1	9.1
2	9.2 & Using a Graphing Calculator	9.2
3	9.3	9.2 & Using a Graphing Calculator
4	9.4	9.3
5	Mid-Chapter Self-Test	9.3
6	9.5	9.4
7	9.6	9.4
8	9.7	Mid-Chapter Self-Test
9	Chapter Review	9.5
10	Chapter Test	9.5
11	Cumulative Review 7–9	9.6
12		9.6
13		9.7
14		Chapter Review
15		Chapter Test
16		Cumulative Review 7–9

CHAPTER ORGANIZATION

LESSON	PAGES	GOALS	MEETING THE NCTM STANDARDS
9.1	456–462	1. Add, subtract, and multiply polynomials 2. Use polynomial operations to solve real-life problems	Problem Solving, Communication, Connections
9.2	463–469	1. Sketch the graph of a polynomial function 2. Use polynomial functions as models of real-life situations	Problem Solving, Communication, Connections, Functions, Geometry (algebraic), Calculus concepts
Using a Calculator	470–471	Sketch and compare the graphs of polynomial functions	Technology
Mixed Review	472	Review algebraic and arithmetic skills	
Career Interview	472	Administrator	Connections
9.3	473–479	1. Factor polynomial expressions and equations 2. Use factoring to solve real-life problems	Problem Solving, Communication, Reasoning, Connections
9.4	480–486	1. Divide polynomials using long division and synthetic division and relate the quotient to the Remainder Theorem and the Factor Theorem 2. Use polynomial division in real-life problems	Problem Solving, Communication, Connections
Mid-Chapter Self-Test	487	Diagnose student weaknesses and remediate with correlated Reteach worksheets	
9.5	488–493	1. Find the rational zeros of a polynomial function 2. Use the zeros of a polynomial function to solve real-life problems	Problem Solving, Communication, Reasoning, Connections, Functions
Mixed Review	493	Review algebraic and arithmetic skills	
9.6	494–500	1. Look at zeros of polynomial functions, factors of polynomials, and solutions of polynomial equations to see their connections 2. Use polynomials to solve real-life problems	Problem Solving, Communication, Reasoning, Connections, Functions, Structure
9.7	501–507	1. Find the range of a collection of numbers 2. Find the standard deviation of a collection of numbers	Problem Solving, Communication, Connections, Statistics, Discrete Math
Chapter Summary	508	Restate for students what they have learned, why they have learned it, and how it fits into the structure of algebra	Structure, Connections
Chapter Review	509–512	Review concepts and skills learned in the chapter	
Chapter Test	513	Diagnose student weaknesses and remediate with correlated Reteaching worksheets	
Cumulative Review 7–9	514–515	Review concepts and skills from chapters 7–9	

LESSON RESOURCES

MEETING INDIVIDUAL NEEDS

RETEACHING For students who need to spend more time on basics:

If a mid-chapter self-test or chapter test indicates a deficiency, teachers can help students with the appropriate ***Reteaching Copymaster.***

PRACTICE For students who need more practice:

Additional exercises like those in the Pupil's Edition are provided for each lesson in ***Extra Practice Copymasters.***

ENRICHMENT For enriching and broadening students' experiences:

Problem of the Day copymasters in ***Teaching Tools*** provide a daily opportunity to use logical reasoning, looking for a pattern, writing an equation, and other routine and non-routine problem-solving strategies.

Math Log copymasters in ***Alternative Assessment*** provide opportunities to report on investigations, research, and open-ended problems.

Technology: Using Calculators and Computers provides enriching activities with graphing and scientific calculators and computers.

The ***Applications Handbook*** provides additional information about the cross-curriculum topics such as astronomy, chemistry, physics, sports, economics, genetics, and music that are integrated into the Pupil's Edition.

LESSON	9.1	9.2	9.3	9.4	9.5	9.6	9.7
PAGES	456-462	463-469	473-479	480-486	488-493	494-500	501-507
Teaching Tools							
Transparencies	✓	✓					✓
Problem of the Day	✓	✓	✓	✓	✓	✓	✓
Warm-up Exercises	✓	✓	✓	✓	✓	✓	✓
Answer Masters	✓	✓	✓	✓	✓	✓	✓
Extra Practice Copymasters	✓	✓	✓	✓	✓	✓	✓
Reteaching Copymasters	Teacher-directed and independent activities tied to results on the Mid-Chapter Self-Tests and Chapter Tests						
Color Transparencies	✓	✓	✓	✓	✓		✓
Applications Handbook	Additional background information for many real-life applications						
Technology	Calculator and computer worksheets for appropriate lessons						
Complete Solutions Manual	✓	✓	✓	✓	✓	✓	✓
Alternative Assessment	Assess student's ability to reason, analyze, solve problems, and communicate using mathematical language.						
Formal Assessment	Mid-Chapter Self-Tests, Chapter Tests, Cumulative Tests, and Practice for College Entrance Tests						
Computer Test Bank	Customized tests can be created by choosing from over 2500 items.						

INSIGHTS

9.1 Operations with Polynomials

To effectively manipulate polynomial expressions in solving real-life problems, students must be able to add, subtract, multiply, and divide the "terms" that are the "building blocks" of polynomials. Much of this lesson is review from Algebra 1, but provides a common starting point for the chapter.

9.2 Graphs of Polynomial Functions

Linear and quadratic functions are examples of polynomial functions that students have already studied. In this lesson, students extend their study of polynomial functions to polynomials of the third degree or higher. The effects of constants on polynomial functions, and consequently, the transformation techniques that can be applied, are the same as for other algebraic functions explored in earlier lessons.

9.3 Factoring Polynomials and Solving Polynomial Equations

Factoring polynomial expressions, where possible, may be useful in solving polynomial equations. Furthermore, an understanding of the factoring of polynomials is necessary to appreciate some of the important results (or theorems) of algebra, including the identification of roots of polynomials.

9.4 Polynomial Division, Factors, and Remainders

Factoring polynomial expressions into other polynomial expressions of lesser degree with integer coefficients is not always possible. The general long division algorithm is useful for dealing with polynomials that cannot otherwise be factored. Since many polynomial division problems contain divisors of the first degree, synthetic division of polynomials is an important labor-saving algorithm for students to learn.

9.5 Finding Rational Roots

Not all solutions to a polynomial equation ("zeros" of the polynomial *function*) are simple integers. In general, all real solutions are either rational or irrational. For polynomials of the third degree or higher, no convenient formula comparable to the quadratic formula exists. Other techniques for solving these equations are required. Students explore one of these techniques in this lesson.

9.6 Connections: Zeros, Factors, and Solutions

How are the zeros, factors, and solutions of polynomial functions related to each other? The answer to this question is developed in this lesson. The connection among these algebraic concepts helps students to make multiple representations of an important branch of algebra, solving polynomial equations. Multiple representations, in turn, increases students' mathematical power to understand real-life situations.

9.7 Exploring Data: Measures of Dispersion

The information obtained from measures of central tendency (means, medians, and modes) is often not adequate to convey the true relationship among the data collected from a real-life situation. Variations in the measures of central tendency for a collection of data may suggest very different interpretations of the information collected. Measures of dispersion are used to account for variation within a collection of data.

Problem of the Day

Eight squares of paper, all the same size, are placed on a table, overlapping as shown. The top sheet, numbered 1, is completely visible, while the others are only partly visible. Number the squares in order from the top layer to the bottom.

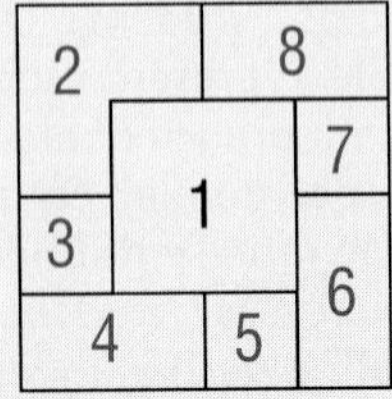

ORGANIZER

Warm-Up Exercises

1. Simplify.
   a. $3x^2 \cdot 5x^4$ $15x^6$
   b. $\frac{45x^8}{20x^6}$ $\frac{9x^2}{4}$
   c. $(2x^3)^4$ $16x^{12}$
2. Find the volume of a box with the following dimensions: width = 7.8 cm, length = 14.5 cm, height = 20 cm.
   2262 cm^3

Lesson Resources

Teaching Tools
- Transparency: 22
- Problem of the Day: 9.1
- Warm-up Exercises: 9.1
- Answer Masters: 9.1

Extra Practice: 9.1
Color Transparency: 47

9.1 Operations with Polynomials

What you should learn:

Goal 1 How to add, subtract, and multiply polynomials

Goal 2 How to use polynomial operations to solve real-life problems

Why you should learn it:

You can combine two polynomial models to get a new polynomial model, such as using the difference between consumer spending for an item and the cost of its raw materials to find the cost of processing, packaging, and marketing the item.

Goal 1 Adding, Subtracting, and Multiplying

An expression whose terms are of the form $a_k x^k$, where k is a nonnegative integer, is a *polynomial in x*. The real number a_k is the **coefficient** of x^k, and the integer k is the **degree** of $a_k x^k$. For instance, the degree of a_2x^2 is two, the degree of a_1x is one, and the degree of the **constant term** a_0 is zero. (The coefficient a_k is read as "a sub k.")

Definition of Polynomial in x

A polynomial in x is an expression of the form

$$a_nx^n + \cdots + a_2x^2 + a_1x + a_0.$$

If $a_n \neq 0$, then the polynomial is of **degree** n, and the number a_n is called the **leading coefficient.** A polynomial that is written with descending powers of x is in **standard form.** Polynomials with one, two, or three terms are called **monomials, binomials,** or **trinomials,** respectively.

To add or subtract polynomials, add or subtract the coefficients of terms with the same degree. You can use a horizontal or vertical format.

Example 1 *Adding and Subtracting Polynomials*

a. $(2x^2 + 3x) - (3x^2 + x - 4) = 2x^2 + 3x - 3x^2 - x + 4$
$= -x^2 + 2x + 4$

b.
$$\begin{array}{r} 4x^3 + 5x^2 - 9x - 3 \\ + \; 5x^3 - 2x^2 \qquad\;\; + 2 \\ \hline 9x^3 + 3x^2 - 9x - 1 \end{array}$$
■

A common mistake is to forget to change signs correctly when subtracting one polynomial from another. Here is an example.

Wrong Signs

$$(x^2 - 3x + 4) - (x^2 + 5x - 1) \neq x^2 - 3x + 4 - x^2 + 5x - 1$$

A basic type of polynomial multiplication is that of multiplying a polynomial by a monomial. Here is an example.

$$\begin{aligned}(6x)(3x^2 + 4x - 1) &= (6x)(3x^2) + (6x)(4x) - (6x)(1)\\ &= 18x^3 + 24x^2 - 6x\end{aligned}$$

To multiply two binomials, you can use both left and right Distributive Properties.

$$\begin{aligned}(2x + 3)(5x - 1) &= 2x(5x - 1) + 3(5x - 1)\\ &= (2x)(5x) - (2x)(1) + (3)(5x) - (3)(1)\\ &= 10x^2 - 2x + 15x - 3\end{aligned}$$

Product of First terms	Product of Outer terms	Product of Inner terms	Product of Last terms
$10x^2$	$-2x$	$15x$	-3

$$= 10x^2 + 13x - 3$$

This technique for mentally multiplying two binomials in a single step is called using the **FOIL pattern.**

To multiply two polynomials that have three or more terms, remember that *each term of one polynomial must be multiplied by each term of the other polynomial.* This can be done using either a vertical or horizontal format.

Example 2 Multiplying Polynomials (Vertical Format)

Multiply: $(x - 3)(4 + 2x - x^2)$.

Solution Begin by writing each polynomial in standard form.

$$\begin{array}{rl}-x^2 + 2x + 4 & \textit{Standard form}\\ \times \qquad\quad x - 3 & \textit{Standard form}\\ \hline 3x^2 - 6x - 12 & \longleftarrow -3(-x^2 + 2x + 4)\\ -x^3 + 2x^2 + 4x \qquad & \longleftarrow x(-x^2 + 2x + 4)\\ \hline -x^3 + 5x^2 - 2x - 12 & \textit{Simplified form}\end{array}$$

■

Example 3 Multiplying Polynomials (Horizontal Format)

Multiply: $(x - 3)(3x^2 - 2x - 4)$.

Solution

$$\begin{aligned}(x - 3)(3x^2 - 2x - 4) &= (x - 3)(3x^2) - (x - 3)(2x) - (x - 3)(4)\\ &= 3x^3 - 9x^2 - 2x^2 + 6x - 4x + 12\\ &= 3x^3 - 11x^2 + 2x + 12\end{aligned}$$

■

Try using a vertical format and compare your results.

LESSON Notes

Example 1

Most students prefer to add and subtract polynomials written in vertical form because it is easier to identify like terms. Encourage students who need to, to rewrite expressions given in the horizontal format in the vertical format.

Examples 2–3

Emphasize that the Distributive Properties are essential to multiplying polynomial expressions. Have students select the format that works best for them. Make sure that students understand that neither format is better than the other; student preference determines which approach is better.

Example 4

Observe that the model, V, may be initially evaluated in the form,

$V = (x + 2)(x - 2)(2x + 1)$.

In this case,

$$V = (5 + 2)(5 - 2)(2 \cdot 5 + 1) = 7 \cdot 3 \cdot 11 = 231.$$

Example 5

Ask students if the graphs of the models found in this example are straight lines. If not, why do the graphs appear to be straight lines or linear relationships?

Although the lines may appear to be straight, each relationship found in this example is a quadratic relationship. The graphs appear to be straight lines because the domain over which these relationships are drawn is restricted to a small subset of the original functions.

Extra Examples

Here are additional examples similar to **Examples 1–4.**

1. Combine the following.
   a. $(5x^2 - 4x - 8) + (2x^2 + 3x + 3)$
   b. $(14x^3 + 8x^2 + 4x + 2) - (6x^3 + 4x^2 - 7)$

   a. $7x^2 - x - 5$
   b. $8x^3 + 4x^2 + 4x + 9$
2. Multiply.
   a. $8x(4 - 3x + 2x^2)$
   b. $(5x + 3)(8x - 4)$
   c. $(2x - 7)(3x^2 + x - 3)$
   d. $(2x + 3)(2x - 3)$
   e. $(5 - 6x)^2$
   f. $(2x + 9)^3$
   g. $(4x + 5)(2x - 1)(3x)$

   a. $32x - 24x^2 + 16x^3$
   b. $40x^2 + 4x - 12$
   c. $6x^3 - 19x^2 - 13x + 21$
   d. $4x^2 - 9$
   e. $25 - 60x + 36x^2$
   f. $8x^3 + 108x^2 + 486x + 729$
   g. $24x^3 + 18x^2 - 15x$

Some binomial products occur so frequently that it is worth memorizing their special product patterns. You can verify each of these products by multiplying. Here is an example.

$$\begin{aligned}(u - v)^3 &= (u - v)(u - v)^2 \\ &= (u - v)(u^2 - 2uv + v^2) \\ &= u^3 - 3u^2v + 3uv^2 - v^3\end{aligned}$$

Special Product Patterns

	Example
Sum and Difference	
$(u + v)(u - v) = u^2 - v^2$	$(x + 4)(x - 4) = x^2 - 16$
Square of a Binomial	
$(u + v)^2 = u^2 + 2uv + v^2$	$(x + 3)^2 = x^2 + 6x + 9$
$(u - v)^2 = u^2 - 2uv + v^2$	$(3x - 2)^2 = 9x^2 - 12x + 4$
Cube of a Binomial	
$(u + v)^3 = u^3 + 3u^2v + 3uv^2 + v^3$	$(x + 2)^3 = x^3 + 6x^2 + 12x + 8$
$(u - v)^3 = u^3 - 3u^2v + 3uv^2 - v^3$	$(x - 1)^3 = x^3 - 3x^2 + 3x - 1$

Connections
Geometry

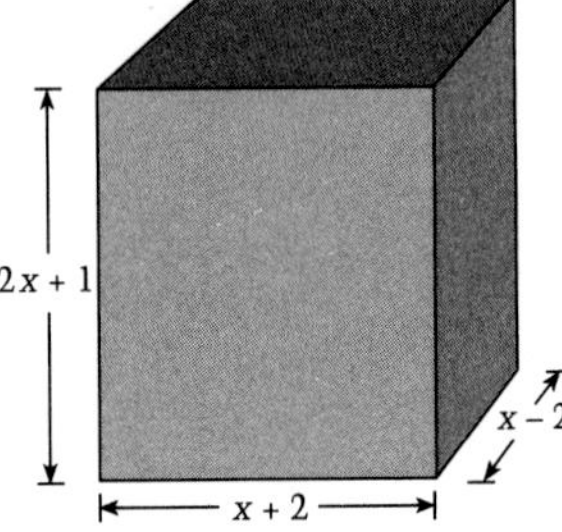

Example 4 ***A Polynomial Model for Volume***

A rectangular box has sides whose lengths (in inches) are $(x + 2)$, $(x - 2)$, and $(2x + 1)$. Write a polynomial model, in standard form, for the volume of the box. Then use the model to find the volume when x is equal to 5 inches.

Solution

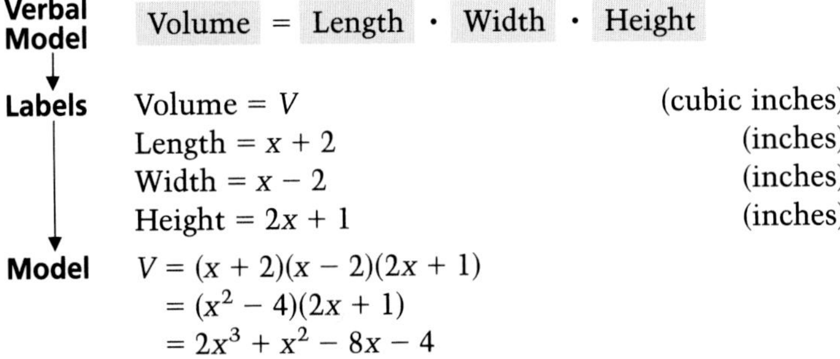

Verbal Model: Volume = Length · Width · Height

Labels:
Volume = V (cubic inches)
Length = $x + 2$ (inches)
Width = $x - 2$ (inches)
Height = $2x + 1$ (inches)

Model:
$$\begin{aligned}V &= (x + 2)(x - 2)(2x + 1) \\ &= (x^2 - 4)(2x + 1) \\ &= 2x^3 + x^2 - 8x - 4\end{aligned}$$

When x is 5 inches, the volume is

$$\begin{aligned}V &= 2(5^3) + 5^2 - 8(5) - 4 \\ &= 250 + 25 - 40 - 4 \\ &= 231 \text{ cubic inches.}\end{aligned}$$

Check: When $x = 5$ inches, the box has dimensions 7 inches by 3 inches by 11 inches. The volume is

(7 inches)(3 inches)(11 inches) = 231 cubic inches. ■

The world's major sources of food are barley, cassava, corn, millet, oats, potato, rice, rye, sorghum, soybeans, sweet potato, and wheat. Some people consume these foods directly. Americans, however, tend to prefer ready-to-eat forms (pasta, bread) or converted forms (meats, eggs). This preference, as you see in Example 5, is costly.

Goal 2 Using Polynomials as Real-Life Models

Real Life
Agriculture

Example 5 *Consumer Expenses for Farm Foods*

For 1980 through 1990, the total amount, A (in billions of dollars), spent by Americans for farm foods can be modeled by

$$A = -0.035t^2 + 15.9t + 267.6$$

where $t = 0$ represents 1980. The amount, F (in billions of dollars), paid to farmers can be modeled by

$$F = 0.156t^2 + 0.475t + 81.5.$$

Find a model that represents the amount that was spent on processing, packaging, and marketing the food.

Solution

Let M represent the amount (in billions of dollars) spent on processing, packaging, and marketing the food. To find a model for M, subtract the amount paid to farmers from the total amount.

$$\begin{aligned} M &= A - F \\ &= (-0.035t^2 + 15.9t + 267.6) - (0.156t^2 + 0.475t + 81.5) \\ &= -0.191t^2 + 15.425t + 186.1 \end{aligned}$$

The graphs of all three models are shown at the left. ■

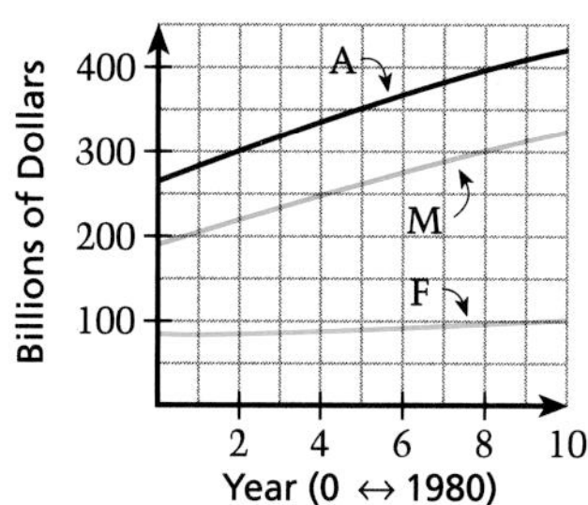

Communicating about ALGEBRA

SHARING IDEAS about the Lesson See margin.

Perform the indicated polynomial operations. Explain your steps.

A. $(3x^3 - 4x + 5) + (3x^2 + 6x)$ **B.** $(2x - 5)(3x - 9)$
C. $(-x^2 + x - 4) - (2x^2 + 3x - 4)$ **D.** $(x - 5)(x + 5)(2x - 1)$

Check Understanding

1. What do the letters in "FOIL" represent? See page 457.
2. Expand the expression $(u + v)^2$. $u^2 + 2uv + v^2$
3. Which computational format do you prefer when adding, subtracting, or multiplying polynomials? Provide a justification. Answers vary.
4. Define what a polynomial expression is. See the box on page 456.

Communicating about ALGEBRA

Answers

A. $3x^3 + 3x^2 + 2x + 5$, add the numerical coefficients of like terms
B. $6x^2 - 33x + 45$, use the distributive property
C. $-3x^2 - 2x$, change the signs of the second polynomial and add the numerical coefficients of like terms
D. $2x^3 - x^2 - 50x + 25$, use the distributive property

EXTEND *Communicating*

Have students perform the indicated polynomial operations using the special product patterns and explain their steps.

a. $(2x + 3 - y)^3$ *Hint:* Represent $2x + 3$ by u.
$[(2x + 3) - y]^3 =$
$(2x + 3)^3 - 3(2x + 3)^2y +$
$3(2x + 3)y^2 - y^3$
Evaluate $(2x + 3)^3$ and $(2x + 3)^2$ first; then substitute and simplify to arrive at the expression:
$8x^3 + 36x^2 + 54x + 27 - 12x^2y - 36xy - 27y + 6xy^2 + 9y^2 - y^3$.

b. $(4x - 3 + y)(4x + 3 - y) =$
$[4x - (3 - y)][4x + (3 - y)] =$
$16x^2 - 9 + 6y - y^2$ or
$16x^2 - y^2 + 6y - 9$

EXERCISE Notes

ASSIGNMENT GUIDE

Basic/Average: Ex. 1–8, 9–29 odd, 41–53 odd, 60–62, 65–71, 78

Above Average: Ex. 1–8, 9–31 odd, 41–53 odd, 59–62, 69–71, 76–78

Advanced: Ex. 1–8, 9–31 odd, 41–53 odd, 59–64, 69–71, 77–80

Selected Answers
Exercises 1–8, 9–75 odd

More Difficult Exercises
Exercises 58–64, 77–80

Guided Practice

Ex. 1–8 Use the Guided Practice exercises to summarize the key terms and processes in the lesson.

Ex. 5, 6, and 8 Be sure students understand that $(a + b)^n \neq a^n + b^n$.

Independent Practice

Ex. 9–16 Students should refer to the opening statements of the lesson on page 456.

Ex. 24–34 Be sure that students understand that in subtraction, all terms in the quantity preceded by a negative sign get subtracted. They can use Part a of Example 1 as a solution model.

Ex. 37–48 Encourage students to use either the vertical or the horizontal format modeled by Examples 2 and 3.

EXERCISES

Guided Practice

CRITICAL THINKING about the Lesson

1. Identify the degree, the leading coefficient, and the other coefficients of $-3x^3 + x^2 + 1$. 3; −3, 1, 1
2. Find two second-degree polynomials whose sum is a first-degree polynomial. Answers vary; $-3x^2 + x$ and $3x^2$
3. Use the FOIL pattern to multiply: $(7x - 1)(2x + 3)$. $14x^2 + 19x - 3$
4. Use a vertical format to multiply: $(6x + 1)(2x^2 - 3x + 2)$. $12x^3 - 16x^2 + 9x + 2$

In Exercises 5–8, write the polynomial in standard form.

5. $(x + 2)^2$ $x^2 + 4x + 4$
6. $(x - 1)^3$ $x^3 - 3x^2 + 3x - 1$
7. $(2x - 1)(2x + 1)$ $4x^2 - 1$
8. $(x - 3)^2$ $x^2 - 6x + 9$

Independent Practice

In Exercises 9–16, state whether the expression is a polynomial. See below.

9. $2x - x^{-1}$
10. $14x + 3$
11. $\sqrt{6x^2 + 1}$
12. 9
13. $\dfrac{2}{x^2 + 1}$
14. 2^x
15. $(x - 2)^2$
16. $\log_{10}(x^2 + 3)$

In Exercises 17–22, write the polynomial in standard form. Then state its degree. See Additional Answers.

17. $14 + x + 13x^2$
18. $\frac{2}{3}x - \frac{5}{3}x^2$
19. $-1 + x^2 + 6x^3$
20. $x^3 + 5x - 2x^2 + 2$
21. $-3x^2 + 3 - x^3$
22. $-4x^3 + 6x^2 - 19x + 18$

In Exercises 23–36, perform the indicated operation. See Additional Answers.

23. $(6x^2 + 1) + (5x^2 - 4)$
24. $(2x^3 + 11x + 2) - (x^3 - 2x + 7)$
25. $(x^2 - 3x + 3) - (x^2 + x - 1)$
26. $(14 - 16x) + (10x - 5)$
27. $(8x^3 - 1) - (20x^3 + 2x^2 - x - 5)$
28. $6x - (22x + 3 - 36x^2 + x^3)$
29. $(4x^2 - 15x + 16) + (2x - 20)$
30. $(7x^3 - 2 + x^2 + 13x) - (4x^3 + 10)$
31. $(-3x^3 + 4x - 9) - (2x^3 + x^2 - x)$
32. $(6x^2 - 18x + 3) - (14x^2 - 12x + 9)$
33. $(15 - 10x^3 - 2x^2 + x) - (x^2 + 7x)$
34. $(50x - 3) - (8x^3 + 9x^2 + 2x + 4)$
35. $(4x - 33 + 9x^2) + (20x^3 - 19x + 3)$
36. $(12x^3 - 5x^2 - 70x + 1) + (-17x^3 + 56x)$

In Exercises 37–48, perform the indicated operations. See Additional Answers.

37. $x(x^2 + 9x - 5)$
38. $12x^2(x - 8)$
39. $-2x(x + 4)$
40. $2x(3x^2 - x + 6)$
41. $(x - 2)(x - 4)$
42. $(x + 8)(x - 1)$
43. $(x + 3)(x^2 - x + 2)$
44. $(x + 9)(x^2 - 6x + 4)$
45. $(2x - 1)(3x^3 - x + 3)$
46. $(6x + 2)(2x^2 + x + 1)$
47. $(x + 9)(x^2 - 2x + 6)$
48. $(2x - 3)(4x^2 - 3x + 3)$

9. Not a polynomial
10. Polynomial
11. Not a polynomial
12. Polynomial
13. Not a polynomial
14. Not a polynomial
15. Polynomial
16. Not a polynomial

In Exercises 49–57, write the polynomial in standard form.

49. $(x-9)(x+9)$ x^2-81

50. $(x+2)(x-2)$ x^2-4

51. $(x+5)^2$ $x^2+10x+25$

52. $(x-3)^2$ x^2-6x+9

53. $(x-4)^3$ $x^3-12x^2+48x-64$

54. $(x+6)^3$ 54. $x^3+18x^2+108x+216$

55. $(x+1)^3$ x^3+3x^2+3x+1

56. $(3x+4)^2$ $9x^2+24x+16$

57. $(2x-1)^2$ 57. $4x^2-4x+1$

In Exercises 58–60, find the area of the blue region.

58.

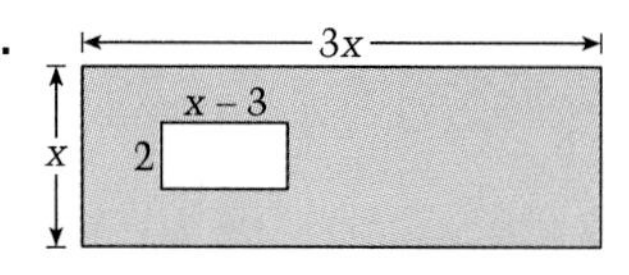

$3x^2-2x+6$

59.

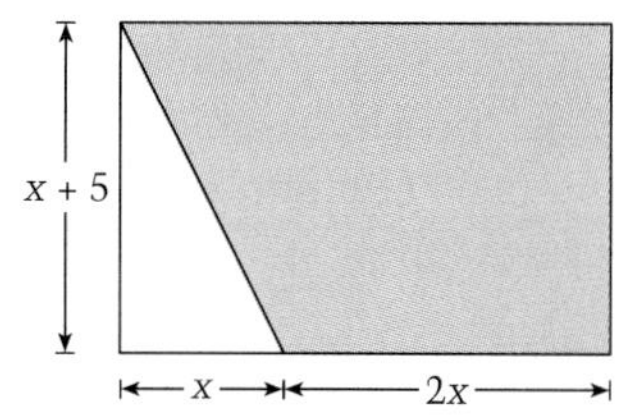

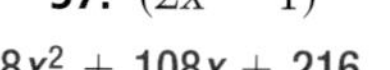

$\frac{5}{2}x^2+\frac{25}{2}x$

60.

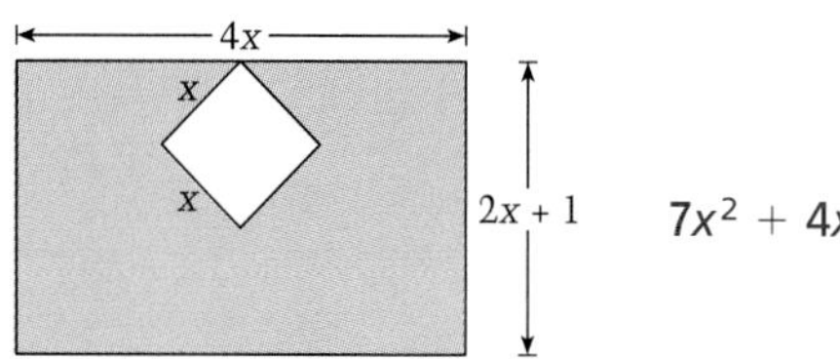

$7x^2+4x$

61. *Age Distribution* For 1980 through 1994, the population, P (in thousands), of the United States can be modeled by

$$P = 2348t + 226{,}427$$

where $t = 0$ represents 1980. The number, S (in thousands), of people 65 years or older can be modeled by

$$S = -3.53t^2 + 595t + 25{,}584.$$

Write a model that represents the number, y (in thousands), of people under age 65. How many people were under age 65 in 1990?
(Source: U.S. Bureau of Census) See margin.

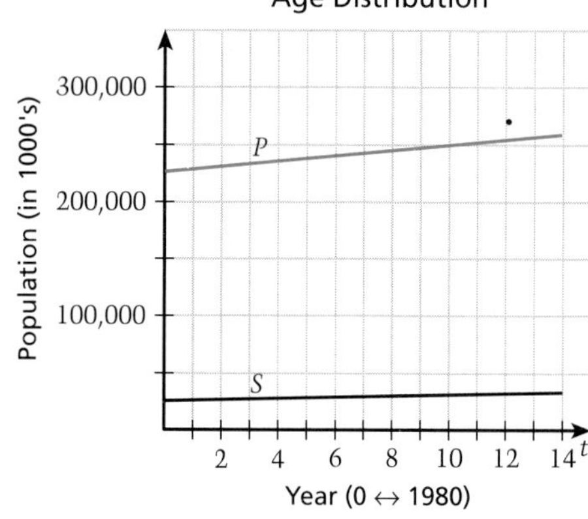

62. *Hardback Book Revenue* For 1985 through 1993, the average price for a hardback book, h (in dollars), can be modeled by

$$h = -0.195t^2 + 5.13t + 10.41$$

where $t = 5$ represents 1985. The total number, x (in millions), of hardback books sold in the United States can be modeled by

$$x = -0.316t^2 + 5.49t + 588.$$

Write a model that represents the total revenue, R (in millions of dollars), received from hardback books. What was the revenue in 1991?
(Source: Book Industry Study Group) See margin.

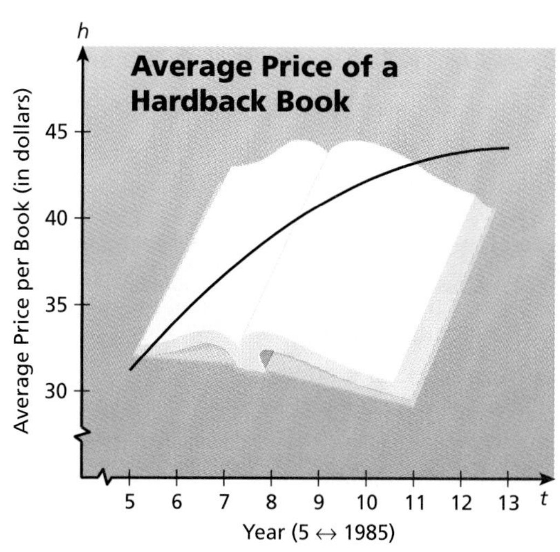

Ex. 49–57 Caution students that expressions such as $(x-3)^2$ are not equivalent to x^2-9.

Ex. 58–60 CONNECTIONS These exercises connect polynomial operations to geometry.

Answers

61. $y = 3.53t^2 + 1753t + 200{,}843$; 218,726 thousand

62. $R = (-0.195t^2 + 5.13t + 10.41) \cdot (-0.316t^2 + 5.49t + 588)$; \$26,386.10973 million

Ex. 62 Remind students that the revenue will be $h \cdot x$, the price per book times the number of books sold.

EXTEND Ex. 62 Ask students to predict the 1995 revenue. Is the answer reasonable?
\$26,058.39 million; answers vary, encourage discussion

▶ **Ex. 63** Remind students that the formula for the area of a trapezoid that has bases b_1 and b_2 and height h is $A = \frac{1}{2}h(b_1 + b_2)$.

▶ **Ex. 70** Help students visualize the size of 1 cubic foot as a box, 1 foot on a side.

Answers

71. $7y^2 + 3y + 3$
72. $8x^2 + 3x - 5$
73. $-x + 5$
74. $-10x^3 + 48x$
75. $-x^3 + x^2 + 3x + 4$
76. $-6y^3 + y^2 + 9y$

Enrichment Activities

These activities require students to go beyond lesson goals.

1. Multiply to show that each pattern is true.
$x^2 - 1 = (x - 1)(x + 1)$
$x^3 - 1 = (x - 1)(x^2 + x + 1)$
$x^4 - 1 = (x - 1)(x^3 + x^2 + x + 1)$
2. How do you think you would express $x^5 - 1$ and $x^6 - 1$ as the product of $(x - 1)$ and another polynomial? Multiply to check your answer.
$(x - 1)(x^4 + x^3 + x^2 + x + 1)$; $(x - 1)(x^5 + x^4 + x^3 + x^2 + x + 1)$
3. In general, how can the expression $x^n - 1$ be expressed as the product of $(x - 1)$ and another polynomial?
$(x - 1)(x^{n-1} + x^{n-2} + \ldots + x^2 + x + 1)$

John Hancock Center **In Exercises 63 and 64, use the following information.**

Each of the four sides of the 100-story John Hancock Center in Chicago, Illinois, are shaped like trapezoids. The height of the building is 28 feet less than 7 times the length of the base along Michigan Avenue. The length of the roof on the Michigan Avenue side of the building is 65 feet shorter than the length of the base below it.

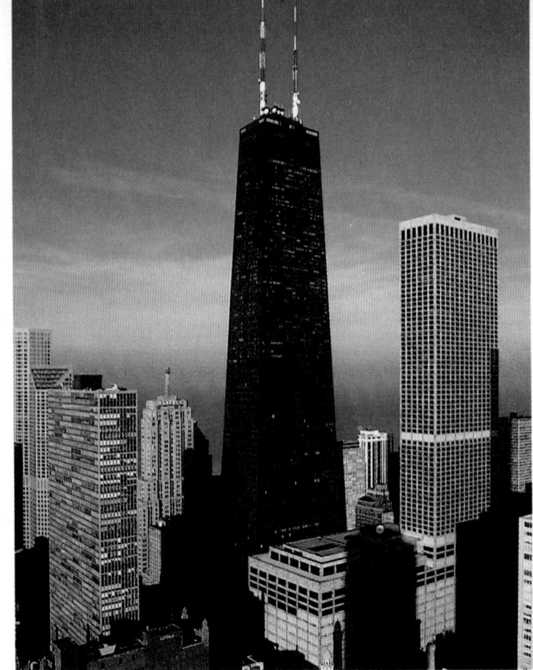

63. The area of the side of the John Hancock Center that faces Michigan Avenue is 149,327.5 square feet. What is the length of the base along Michigan Avenue? How tall is the building? 165 ft, 1127 ft

64. The area of the rectangular roof is 16,000 square feet. What are its dimensions? 160 ft by 100 ft

Integrated Review

In Exercises 65–68, simplify the expression.

65. $e^{-2} \cdot e^4$ e^2 **66.** $(-2)^{x+1}(-2)^3$ $(-2)^{x+4}$ **67.** $3^x \cdot 3^2$ 3^{x+2} **68.** $e^9 \cdot e^{-9}$ 1

69. *Powers of Ten* Which best approximates the volume of a stack of 12 *National Geographic* magazines? c
a. 2.6×10^0 in.^{3} **b.** 2.6×10^1 in.^{3} **c.** 2.6×10^2 in.^{3} **d.** 2.6×10^3 in.^{3}

70. *Powers of Ten* Which best approximates the volume of the interior of an automobile? a
a. 1.4×10^2 ft^3 **b.** 1.4×10^3 ft^3 **c.** 1.4×10^4 ft^3 **d.** 1.4×10^5 ft^3

In Exercises 71–76, simplify the expression. See margin.

71. $6y^2 + 3y + 3 + y^2$ **72.** $17x^2 - 5 + 3x - 9x^2$ **73.** $x^2 + 2 - (x - 3 + x^2)$
74. $2(6x - 3x^3) - 4(x^3 - 9x)$ **75.** $x^3 + x^2 - 2x^3 + 3x + 4$ **76.** $-10y^3 - 3y + 12y + y^2 + 4y^3$

Exploration and Extension

77. What must you add to $3x^2 - x + 5$ to obtain $-x^2 - 4x + 1$? $-4x^2 - 3x - 4$

78. What must you subtract from $x^3 + 3x - 2$ to obtain $x^3 - x^2 + 1$? $x^2 + 3x - 3$

In Exercises 79 and 80, solve for *a*, *b*, and *c*.

79. $(4x^2 + bx + 9) - (ax^2 + 3x - 2) = 2x^2 + x + c$ 2, 4, 11

80. $3x(ax^2 - 5) + (bx + cx^2) = x^3 + 2x^2 - x$ $\frac{1}{3}$, 14, 2

9.2 Graphs of Polynomial Functions

What you should learn:

Goal 1 How to sketch the graph of a polynomial function

Goal 2 How to use polynomial functions as models of real-life situations

Why you should learn it:

You can use a graph to help you visualize the relationships in a polynomial model, such as the model for the amounts of charitable contributions.

Goal 1 Sketching a Polynomial Function

A function of the form

$$f(x) = a_n x^n + a_{n-1}x^{n-1} + \cdots + a_1 x + a_0, \qquad a_n \neq 0$$

is a **polynomial function** of degree n. You know how to sketch the graph of a polynomial function of degree 0, 1, or 2.

Function	Degree	Graph
$f(x) = a$	0	Horizontal line
$f(x) = ax + b$	1	Line of slope a
$f(x) = ax^2 + bx + c$	2	Parabola

The graphs of polynomial functions of degree greater than 2 are more difficult to sketch, as the graphs below illustrate. In this lesson, however, you will learn to recognize some of their basic features. With these features, point-plotting, and intercepts, you will be able to make reasonably accurate sketches by hand. Of course, if you have a graphing calculator or graphing software for a computer, then the task is easier.

Features of Graphs of Polynomial Functions

1. The graph of a polynomial function is **continuous.** This means that the graph has no breaks—you could sketch the graph without lifting your pencil from the paper.
2. The graph of a polynomial function has only smooth turns. A function of degree n has *at most* $n - 1$ turns.
3. If the leading coefficient of the polynomial function is positive, then the graph rises to the right. If the leading coefficient is negative, then the graph falls to the right.

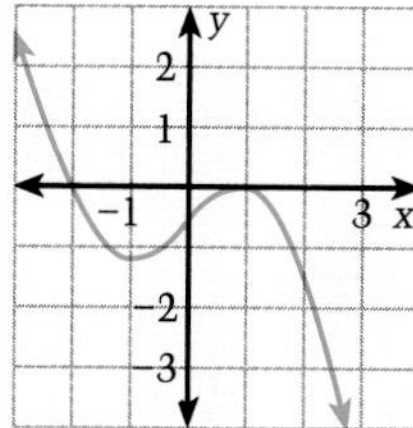

$f(x) = -\frac{1}{3}x^3 + x - \frac{2}{3}$

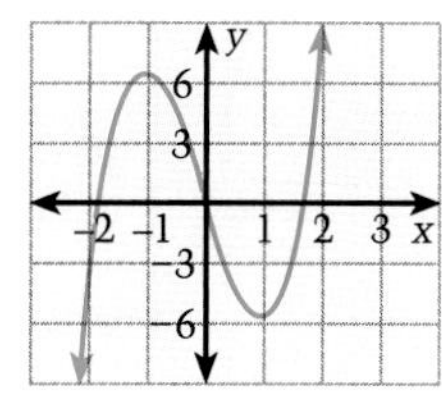

$f(x) = 3x^3 - 9x + 1$

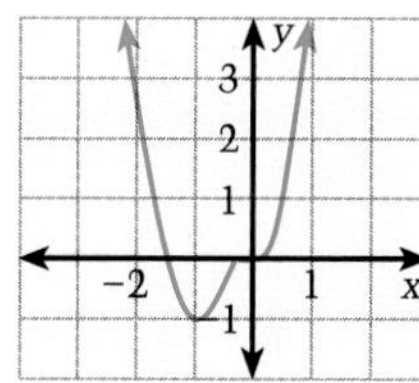

$f(x) = 3x^4 + 4x^3$

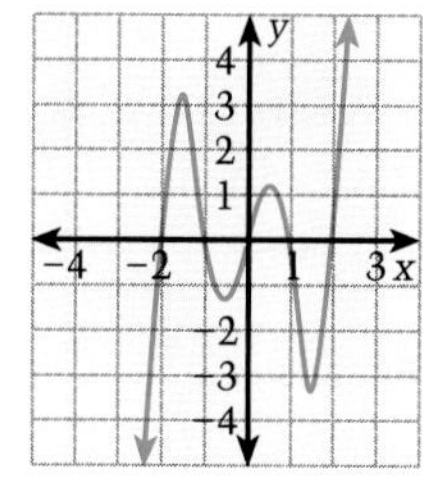

$f(x) = x^5 - 5x^3 + 4x$

Problem of the Day

Solve for x.

$\left(\frac{1}{3}\right)^{-1} \cdot 9^{x-1} = \left(\frac{1}{9}\right)^{2x-1}$ $\frac{1}{2}$

ORGANIZER

Warm-Up Exercises

1. Sketch these functions.
   a. $f(x) = 3x^2 + 2x - 1$
   b. $g(x) = 5x + 3$

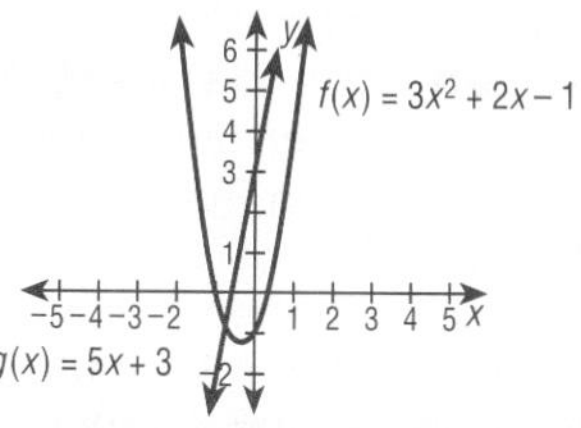

2. Make a table of values for $-3 \le x \le 3$ and sketch the graphs of these functions.
   $f(x) = 3x^3 - 9x + 1$
   $g(x) = x^5 - 5x^3 + 4x$

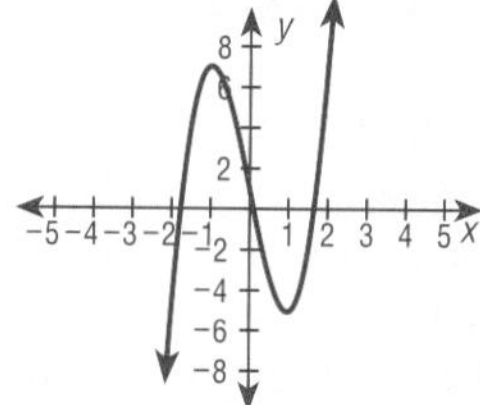

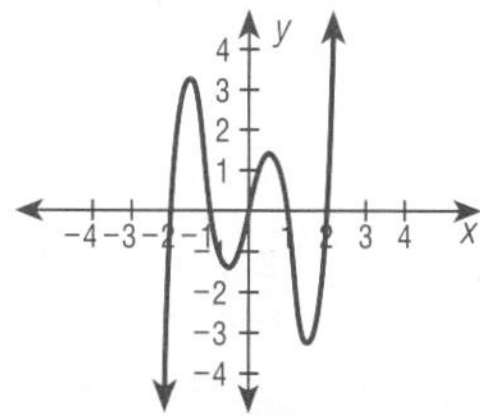

Lesson Resources

Teaching Tools
- Transparencies: 2, 3, 7, 8, 20
- Problem of the Day: 9.2
- Warm-up Exercises: 9.2
- Answer Masters: 9.2

Extra Practice: 9.2

Color Transparency: 48

Technology Handbook: p. 54

LESSON Notes

Example 1

Students should recognize the effects of constants on these functions because they are the same as the effects on quadratic, exponential, and logarithmic functions explored in earlier lessons.

Students may find it helpful to review the effects of constants on the basic quadratic function, $f(x) = x^2$.

Be attentive to students who may not understand the general expression for a polynomial function of degree n. In particular, students may not understand the role of the subscripts in the statement

$f(x) = a_n x^n + a_{n-1}x^{n-1} + \dots .$

Help students recognize that the subscripts are just counters for the number of terms occurring in the expression.

Example 2

Ask students to speculate whether a cubic function, such as $f(x)$, could have had fewer, or more, zeros than three. (Later, the Fundamental Theorem of Algebra can be used to answer this question.)

Example 3

It may be necessary to construct tabular values for these functions so that they can be sketched and the assertions made about them can be verified.

The polynomial functions that have the simplest graphs are the monomial functions $f(x) = a_n x^n$. When n is *even* the graph is similar to the graph of $f(x) = x^2$. When n is *odd* the graph is similar to the graph of $f(x) = x^3$. Moreover, the greater the value of n, the flatter the graph of a monomial is on the interval $-1 \le x \le 1$.

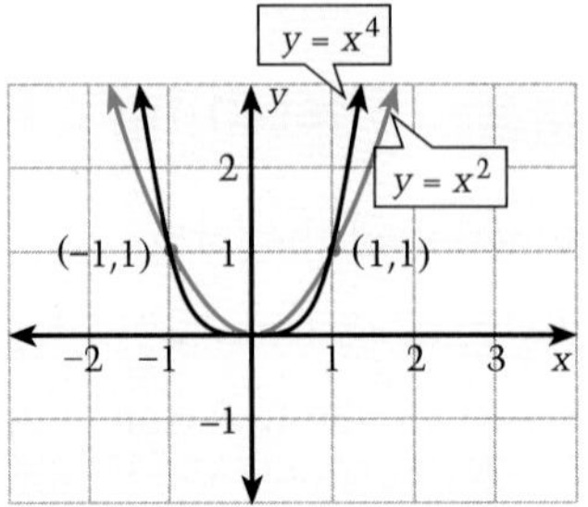

For n even, the graph of y = x^n is tangent to the x-axis at the origin.

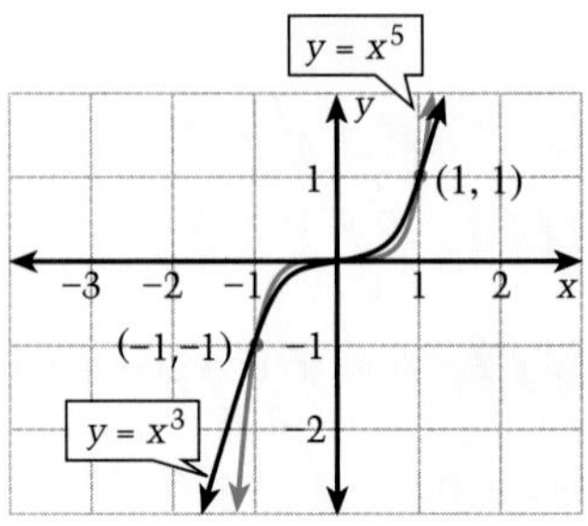

For n odd, the graph of y = x^n crosses the x-axis at the origin.

Example 1 *Sketching Transformations of Monomial Functions*

a. ***Reflection*** To sketch the graph of $g(x) = -x^5$ reflect the graph of $f(x) = x^5$ in the x-axis.

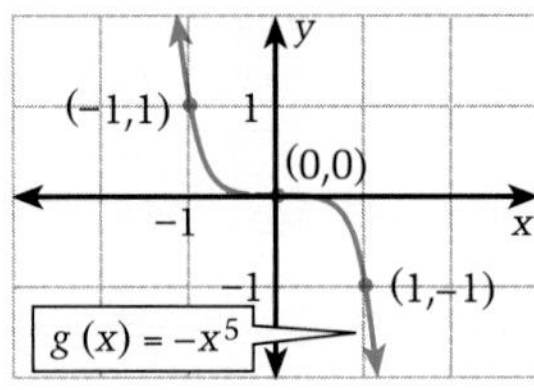

b. ***Vertical shift*** To sketch the graph of $g(x) = x^4 + 1$ shift the graph of $f(x) = x^4$ up one unit.

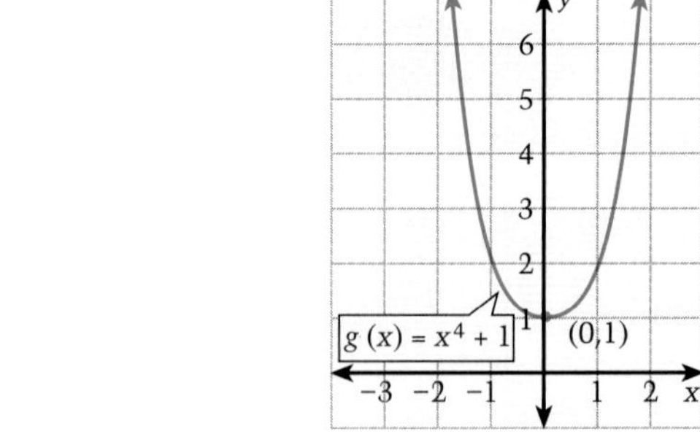

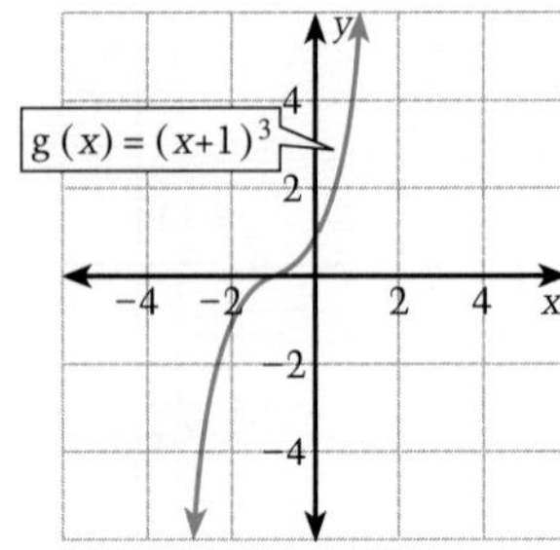

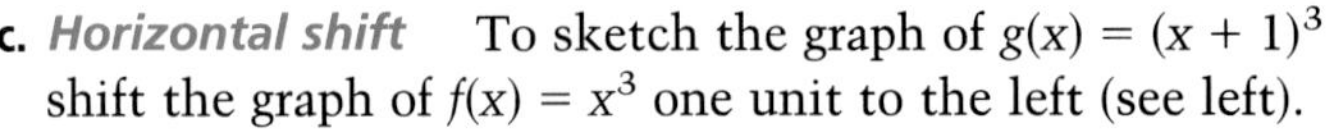

c. ***Horizontal shift*** To sketch the graph of $g(x) = (x + 1)^3$ shift the graph of $f(x) = x^3$ one unit to the left (see left). ■

The y-intercept of the graph of a function occurs when $x = 0$. The x-intercepts are the **zeros** of the function. At an x-intercept, the value of the function is zero.

Example 2 *Sketching a Cubic Polynomial Function*

Sketch the graph of $f(x) = x^3 - 9x$.

Solution Because the leading coefficient is positive, you know that the graph rises to the right. Also, because the degree of the polynomial function is 3, you know that the graph can have at most two turns. Using these general observations with the values given in the table, you can sketch the graph as shown at the left. Notice that the function has three zeros: -3, 0, and 3.

x	-4	-3	-2	-1	0	1	2	3	4
$f(x)$	-28	0	10	8	0	-8	-10	0	28

■

If the degree of a polynomial function is even, then its graph has the same behavior to the left and right. For instance, the graph of $f(x) = x^4$ rises to the right and rises to the left.

If the degree is odd, the graph has opposite behaviors to the right and left. For instance, the graph of $f(x) = -x^3$ falls to the right and rises to the left.

Example 3 *Identifying the Left and Right Behavior*

a. Because the leading coefficient of $f(x) = -x^3 + 4x$ is negative, the graph falls to the right. Because the degree is odd, the graph rises to the left.

b. Because the leading coefficient of $f(x) = x^4 - 5x^2 + 4$ is positive, the graph rises to the right. Because the degree is even, the graph rises to the left.

c. Because the leading coefficient of $f(x) = x^5 - x$ is positive, the graph rises to the right. Because the degree is odd, the graph falls to the left.

a.

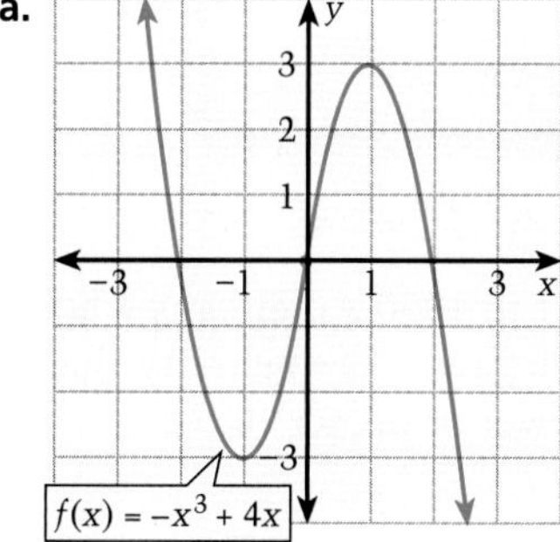

b.

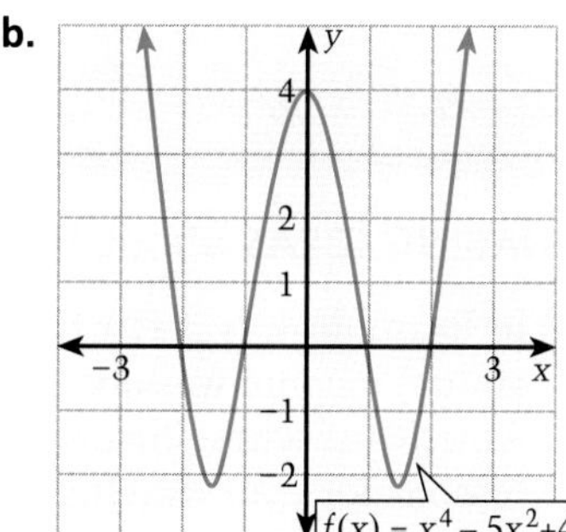

c.

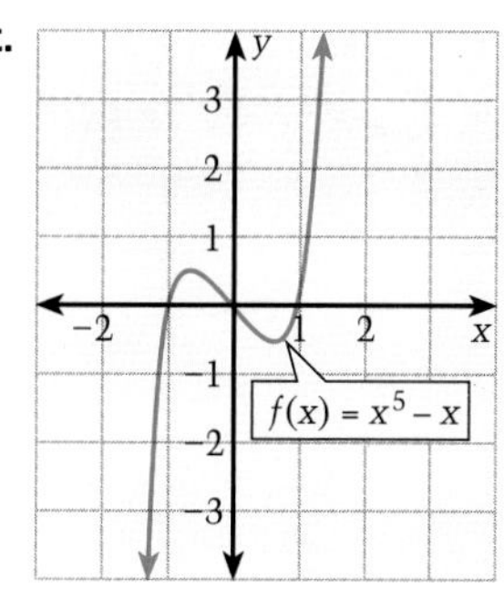

■

Extra Examples

Here are additional examples similar to **Examples 1–3.**

1. Given the graphs of $y_1 = x^4$ and $y_2 = x^7$, how would you sketch these graphs?

a. $f(x) = (x - 2)^4$

b. $g(x) = (x + 3)^7$

c. $h(x) = -x^4 + 2$

d. $j(x) = -(x - 5)^7 - 3$

a. $f(x)$: Shift the graph of y_1 2 units to the right.

b. $g(x)$: Shift the graph of y_2 3 units to the left.

c. $h(x)$: First, reflect the graph of y_1 in the x-axis, then shift it up 2 units.

d. $j(x)$: First, shift the graph of y_2 5 units to the right; then reflect it in the x-axis; then shift it down 3 units.

2. Describe the left and right behavior of the graph of each function.

a. $f(x) = x^4 + x^2 - 5x$

b. $g(x) = -x^5 + 3x^4 - 9x + 3$

c. $h(x) = 3x^3 - 6x^2 + 12$

d. $j(x) = -5x^6 - 3x^4 + 8$

a. The graph rises to the left and to the right.

b. The graph rises to the left and falls to the right.

c. The graph falls to the left and rises to the right.

d. The graph falls to the left and to the right.

Example 4

Ask students to explain why the two very different-looking graphs reflect the same information about charitable contributions.

Check Understanding

1. How many terms can there be in a polynomial expression of degree 5 (that has been simplified)? 6
2. How many turns can the graph of a polynomial function of degree 5 have? at most 4
3. Draw a sketch of the general quadratic, cubic, and quartic (degree 4) functions. See page 464. The graph of a quadratic function is U-shaped; the graph of a cubic function resembles a sideways S; the graph of a quartic function resembles a rounded W.

***Communicating* about ALGEBRA**

EXTEND *Communicating*

Have students describe a characteristic of the graphs of cubic and quartic (degree 4) functions with the largest number of turns.

The functions cross the x-axis, at most, 3 and 4 times, respectively.

Then ask them to devise a method for constructing cubic and quartic functions whose graphs have the largest number of possible turns.

For a cubic function, you would need 3 unequal linear x-factors (zeros of the function):

$f(x) = (x - a)(x - b)(x - c)$.

For a quartic function, you would need 4 unequal x-factors:

$g(x) = (x - a)(x - b)(x - c)(x - d)$.

Goal 2 Using Polynomial Functions in Real Life

Real Life **Social Science**

Example 4 *Charitable Contributions*

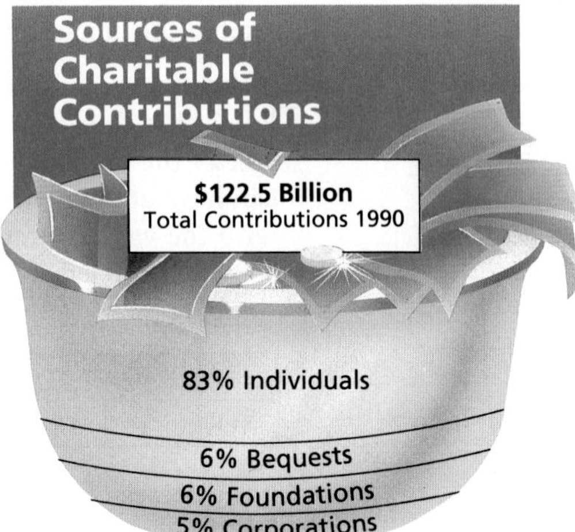

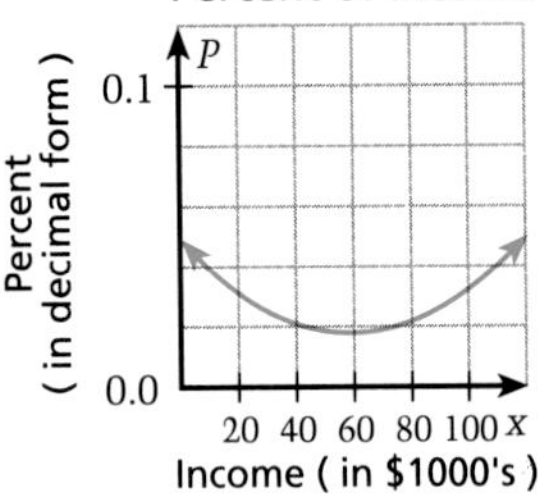

Amount of Charitable Contributions

A
6000
5000
4000
3000
2000
1000
Amount Given to Charity (in dollars)
20 40 60 80 100 x
Income (in $1000's)

In 1990, the average percent, P (in decimal form), of income that Americans gave to charitable organizations was related to their income, x (in thousands of dollars), by the model

$$P = 0.000014x^2 - 0.001529x + 0.05855, \qquad 5 \le x \le 100.$$

Use this model to write a model for the average amount given. ***(Source:* Giving and Volunteering in the United States*)***

Solution To find the average amount, A, that an individual gave to charity, you can multiply the average percent-of-income model by the income ($1000x$) to obtain

$$A = 1000x(0.000014x^2 - 0.001529x + 0.05855)$$
$$= 0.014x^3 - 1.529x^2 + 58.55x, \qquad 5 \le x \le 100.$$

The graphs of both models are shown at the left. ■

Communicating about ALGEBRA

SHARING IDEAS about the Lesson See below.

In Example 4, is it fair to say that in 1990 Americans whose annual income was about $55,000 gave less to charity than Americans whose income was lower or higher? Give reasons for your conclusion.

No; they gave more money to charity than Americans whose income was less, even though they gave a smaller percent of their income.

EXERCISES

Guided Practice

CRITICAL THINKING about the Lesson

1. Explain what is meant by a continuous graph. See below.
2. Name a feature of the graph of $f(x) = |x|$ that is *not* shared by graphs of polynomial functions. See below.
3. Does the graph of $f(x) = 13x^4 - 3x$ rise or fall to the right? How can you tell? The graph rises to the right, the leading coefficient is positive.
4. Which is the graph of $f(x) = 2x^5 - 2$? Explain.

a.

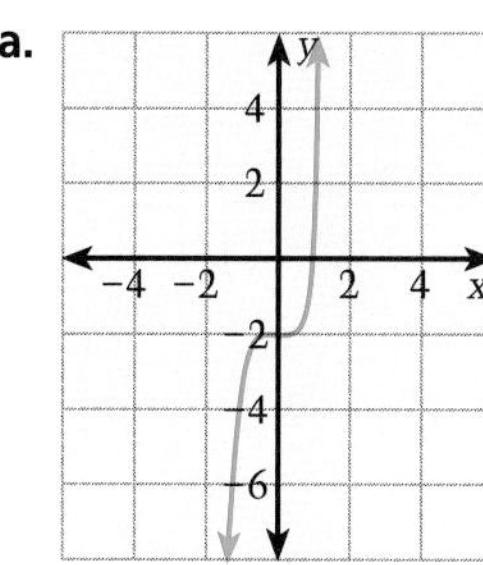

b.

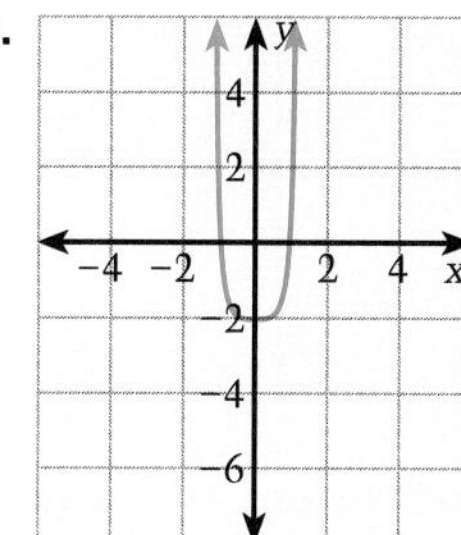

a, the graph rises to the right since the leading coefficient is positive.

5. How is the graph of $g(x) = x^4 + 3$ related to the graph of $f(x) = x^4$? The graph of g is the graph of f shifted up 3 units.
6. Find the zeros of $f(x) = x^2 + 2x - 15$. $-5, 3$

Independent Practice

In Exercises 7–10, describe the transformation of *g* that would produce *f*. See Additional Answers.

7. $f(x) = (x - 5)^3$ $g(x) = x^3$
8. $f(x) = x^4 + 3$ $g(x) = x^4$
9. $f(x) = -x^6$ $g(x) = x^6$
10. $f(x) = (x + 1)^5$ $g(x) = x^5$

In Exercises 11–14, state the maximum number of turns in the graph.

11. $f(x) = 2x^5 - x + 6$ 4
12. $f(x) = \frac{1}{2}x^3 + 1$ 2
13. $f(x) = x^4 + 9x - 1$ 3
14. $f(x) = -2x^6 + 3x + 7$ 5

In Exercises 15–24, describe the left and right behaviors of the graph. See Additional Answers.

15. $f(x) = -5x^4$
16. $f(x) = -x^2$
17. $f(x) = 2x^2$
18. $f(x) = 10x^3$
19. $f(x) = -x^6 + 2x^3 - x$
20. $f(x) = \frac{1}{2}x^5 + 4x^2$
21. $f(x) = x^5 - 4x^4 + 4x^2 + 3$
22. $f(x) = -x^5 + 3x^3 + x$
23. $f(x) = 6x^7 - 3x^3 + 2x - 3$
24. $f(x) = 3x^6 - x - 11$

1. It is a graph that you can sketch without lifting your pencil from the paper.
2. It does not have only smooth turns; it is shaped like a V.

EXERCISE Notes

ASSIGNMENT GUIDE

Basic/Average: Ex. 1–6, 7–19 odd, 25–30, 37–43, 51–55 odd, 56, 63–64

Above Average: Ex. 1–6, 7–19 odd, 25–30, 35–45, 51–56, 63–64

Advanced: Ex. 1–6, 7–19 odd, 25–30, 35–45, 51–56, 63–68

Selected Answers
Exercises 1–6, 7–63 odd

Use **Mixed Review** as needed.

More Difficult Exercises
Exercises 52–56, 64–68

Guided Practice

▶ **Ex. 1–6** COOPERATIVE LEARNING Use the Guided Practice to review the major concepts in the lesson. Encourage students to work in pairs to discuss their answers to the exercises.

Independent Practice

▶ **Ex. 7–10** Students should describe the transformations in terms of shifts and reflections in the *x*-axis.

▶ **Ex. 11–14** Students may need to review the features of the graphs of polynomial functions listed on page 463, especially feature 2.

▶ **Ex. 15–24** Students should review Example 3 on page 465 before doing these exercises.

▶ **Ex. 25–30 WRITING** Assign these exercises as a group. Ask students not only to choose the letter, but to include a one-sentence justification of their choice. For example, in Exercise 27, the answer is e. Since the leading coefficient is negative and the degree is odd, the graph rises to the left and falls to the right with, at most, two turns.

▶ **Ex. 31–48** Be sure students understand they can use the basic graphs of $g(x) = x^3$ (or x^4 or x^5) as starting points for the transformations.

▶ **Ex. 49–51 TECHNOLOGY** Encourage students to locate the zeros by using the Quadratic Formula or by using the Trace function of a graphics calculator. For example, in Exercise 49, students could graph $y_1 = x^2 + 9x - 4$ and $y_2 = 0$. This will allow them to "skip down" to the x-axis on the calculator graph and locate the zeros for $f(x)$.

Answers

52.

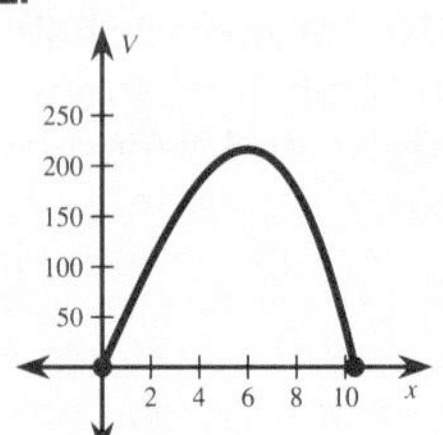

53.

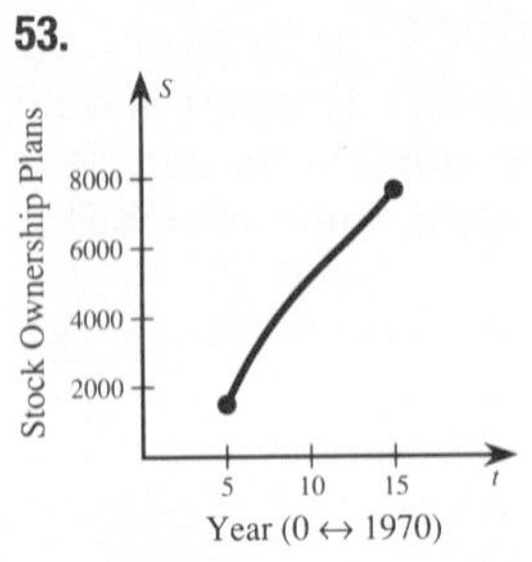

In Exercises 25–30, match the function with its graph.

25. $f(x) = x^3 - 5x$ a

26. $f(x) = -x^4 + x + 4$ f

27. $f(x) = -x^3 + 2x^2$ e

28. $f(x) = -x^5 + 3x^3 + 1$ d

29. $f(x) = x^5 - 3x^3 + 2$ c

30. $f(x) = x^4 - 4x^2 + x$ b

a.

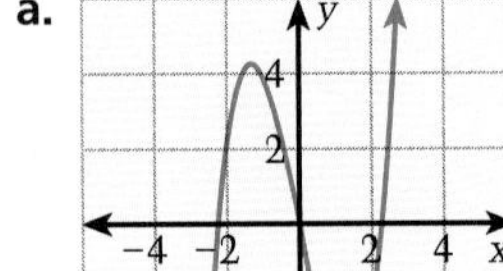

b.

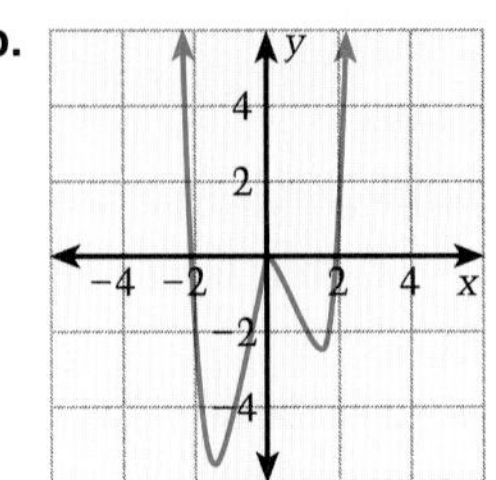

c.

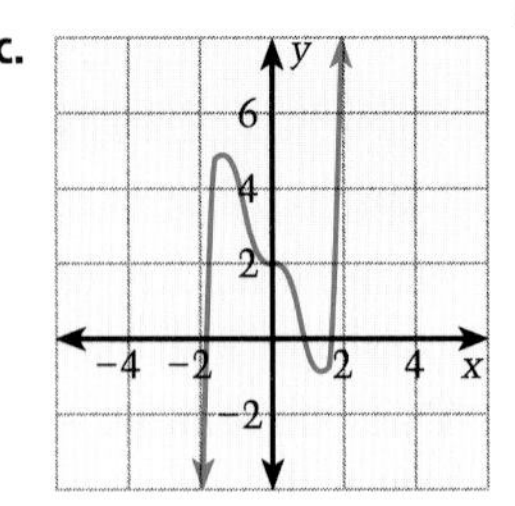

d.

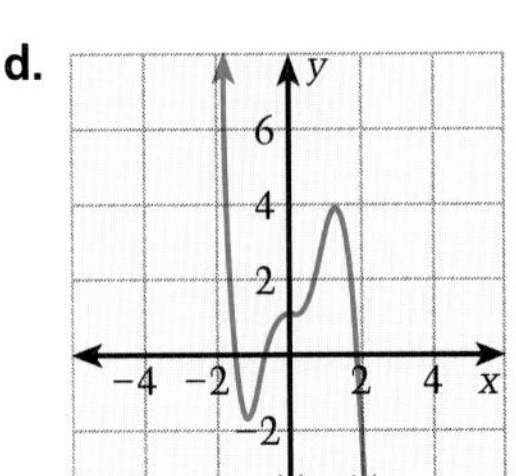

e.

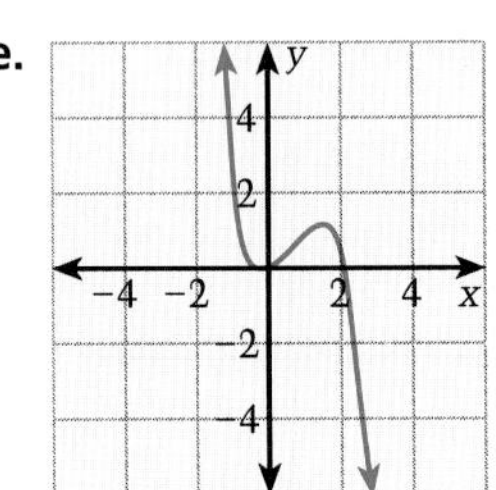

f.

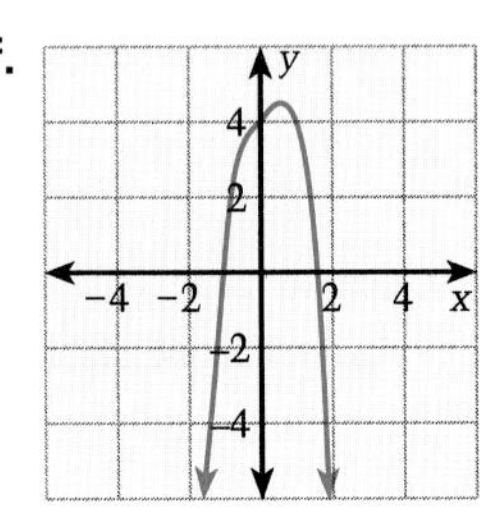

In Exercises 31–48, sketch the graph of the function. (Use a graphing calculator for Exercises 43–48.) See Additional Answers.

31. $f(x) = -x^3$

32. $f(x) = -x^4$

33. $f(x) = x^5 + 1$

34. $f(x) = x^4 - 2$

35. $f(x) = (x + 2)^4$

36. $f(x) = (x - 4)^3$

37. $f(x) = -(x - 3)^3$

38. $f(x) = -(x + 2)^5$

39. $f(x) = (x + 2)^4 - 6$

40. $f(x) = (x + 1)^3 + 3$

41. $f(x) = -x^5 - 2$

42. $f(x) = -x^4 + 8$

43. $f(x) = -x^3 + 4x^2$

44. $f(x) = x^5 + x^2 - 10x - 6$

45. $f(x) = x^3 - 8x^2 + x - 1$

46. $f(x) = x^5 + x$

47. $f(x) = x^5 + 3x^3 - x$

48. $f(x) = x^4 - 2x + 6$

In Exercises 49–51, find the zeros of the function.

49. $f(x) = x^2 + 9x - 4$ **49.** ≈0.42, ≈−9.42

50. $f(x) = 2x^2 - 2x - 1$ ≈1.37, ≈−0.37

51. $f(x) = -2x^2 - 3x + 6$ −2.64, ≈1.14

52. *Maximum Volume* The volume of each of the boxes on page 455 is given by $V = 54x - \frac{1}{2}x^3$, where x is the length of each side of the base. Sketch the graph of this function. Which box has the greatest volume? The box with a base side of 6 in.

53. *Stock Ownership Plans* From 1975 through 1990, the number of employee stock ownership plans, S, in the United States can be modeled by $S = 4.2t^3 - 148t^2 + 2210t - 6380$, where $t = 5$ represents 1975. Sketch the graph of this function. How many plans were there in 1982? ≈6086 See margin.

54. *Clothes Cleaning Services* From 1980 through 1990, the annual revenue, R (in millions of dollars), for laundry and dry cleaning can be modeled by $R = 12.2(t^4 - 16.3t^3 + 86t^2 - 92t + 754)$ where $t = 0$ represents 1980. Sketch the graph of this function. What was the revenue in 1988? $15,523.28 million

55. *Making a Box* An open box is to be made from a 10-in. by 12-in. piece of cardboard by cutting x-in. squares from each corner and folding up the sides. Write a function giving the volume of the box in terms of x. Sketch the graph of the function and use it to approximate the value of x that produces the greatest volume.
$V = x(12 - 2x)(10 - 2x)$ 1.81 in. See Additional Answers.

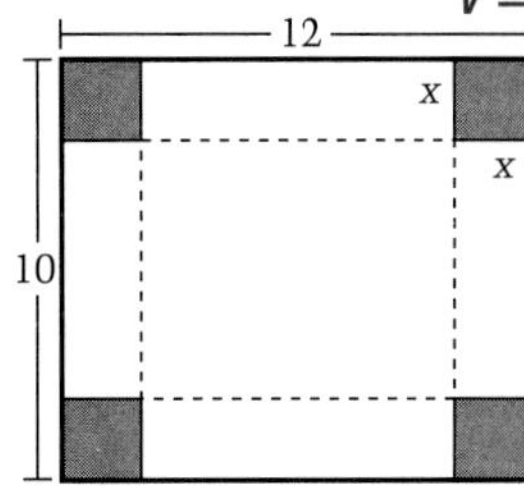

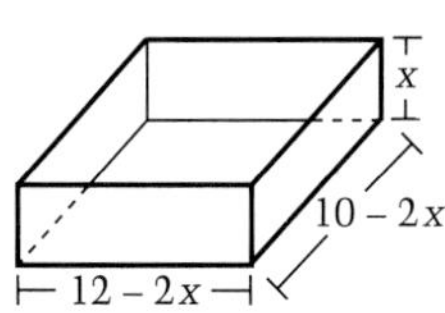

56. *Putting a Lid on It* Using a 10-in. by 12-in. piece of cardboard, you construct a box with a lid, as shown. Write a function that gives the volume in terms of x. Sketch the graph and use it to approximate the value of x that produces the greatest volume. $V = x(6 - x)(10 - 2x)$ 1.81 in. See Additional Answers

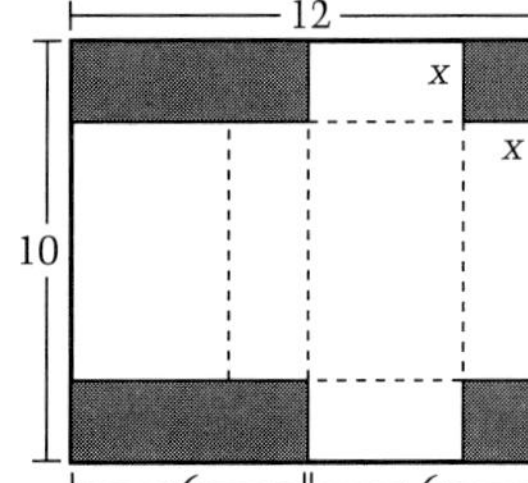

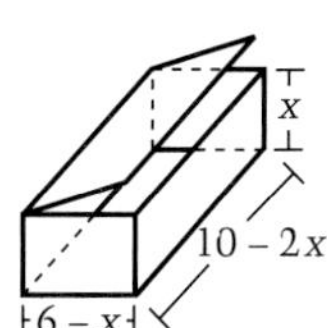

Integrated Review

In Exercises 57—62, sketch the graph of the function. See Additional Answers.

57. $f(x) = 6$

58. $f(x) = -x + 2$

59. $f(x) = \frac{1}{2}x - 1$

60. $f(x) = x^2 - 2x + 6$

61. $f(x) = -x^2 + 4x - 3$

62. $f(x) = -2x^2 - x + 1$

63. *College Entrance Exam Sample* In the equation $y = x^2 + rx - 3$, for what value of r will $y = 11$ when $x = 2$? b

a. 6 **b.** 5 **c.** 4 **d.** $3\frac{1}{2}$ **e.** 0

64. *College Entrance Exam Sample* If a cubic inch of metal weighs 2 pounds, a cubic foot of the same metal weighs how many pounds? e

a. 8 **b.** 24 **c.** 96 **d.** 288 **e.** 3456

Exploration and Extension

In Exercises 65–68, find a polynomial function that has only the given zeros. See below.

65. 2, −1

66. 0, 1, 2

67. 1, −3, −1, 0

68. 2, −2, 3, −3

65. $f(x) = (x - 2)(x + 1)$ **66.** $f(x) = x(x - 1)(x - 2)$ **67.** $f(x) = (x - 1)(x + 3)(x + 1)x$
68. $f(x) = (x - 2)(x + 2)(x - 3)(x + 3)$

Ex. 55–56 Remind students of the restrictions on the value of x. For example, in Exercise 55, $0 < x < 5$. Students with graphics calculators could graph the function to find the greatest V-value.

Enrichment Activities

This activity requires students to go beyond lesson goals.

EXTEND Ex. 55–56
A PROJECT

1. Get an open rectangular box such as a shoe box, a cigar box, or a mailer box. Measure its length, width, and height in inches. Calculate its volume in cubic inches.
2. Cut down the seams of the box to create a two-dimensional model as shown in Exercise 55 or 56. Trace the flattened box on paper and draw the "squares" that were cut from each corner. Measure the outline of the box, including the corners, to find the length, l, and width, w, of the original piece of cardboard from which the box was cut. Assuming that the squares have side x, write an algebraic model for the volume, V, of the box in terms of x.
   Students need to substitute their values of l and w in the model $V = x(l - 2x)(w - 2x)$.
3. Graph this equation. Find the x-value that maximizes the expression and determine whether it matches the actual value as measured on the box. Did the box manufacturer maximize the volume for the given surface area of your box? Explain.

The "viewing rectangle" or "viewing window" is referred to as the range of values in which the basic shape of the graph of a function can be viewed. Finding the dimensions of the viewing rectangle of the graph of a function is not always simple. It may require a brief search for range values in which the characteristics of the function are taken into consideration as the dimensions of the viewing rectangle are selected and checked.

The following display is a convenient way in which to indicate the viewing rectangle for a given function. The display

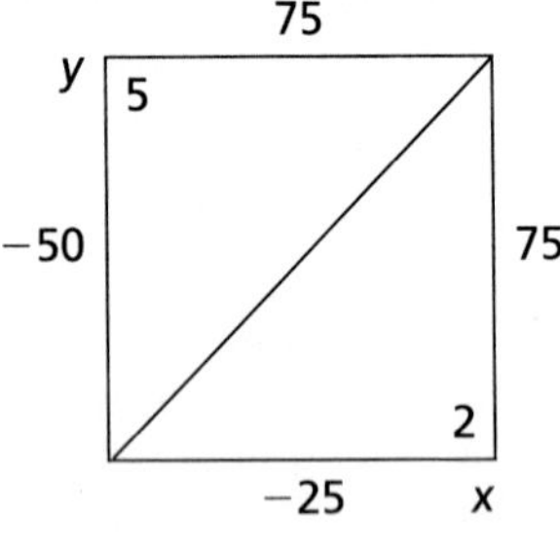

indicates that the x-values in the range are from -50 to 75, that the y-values go from -25 to 75, and that the tic marks on the x- and y-axes are 2 and 5 units apart, respectively.

Example 2

Numbers that satisfy a polynomial equation with integer coefficients are called *algebraic numbers.* Numbers that are not algebraic, such as π, e, log 2, and $2^{\sqrt{2}}$, are called *transcendental numbers.* Point out that there are polynomial functions that can be used to approximate functions like $f(x) = e^x$.

USING A GRAPHING CALCULATOR

Sketching the graph of a polynomial function with a graphing calculator can be tricky. The problem is in finding screen boundaries that will display the basic shape of the graph. For instance, both of the screens below show a portion of the graph of $f(x) = 4x^3 + 2x^2 - x + 12$. The screen on the left, however, is not as good a view for seeing the basic characteristics of the graph as the screen on the right.

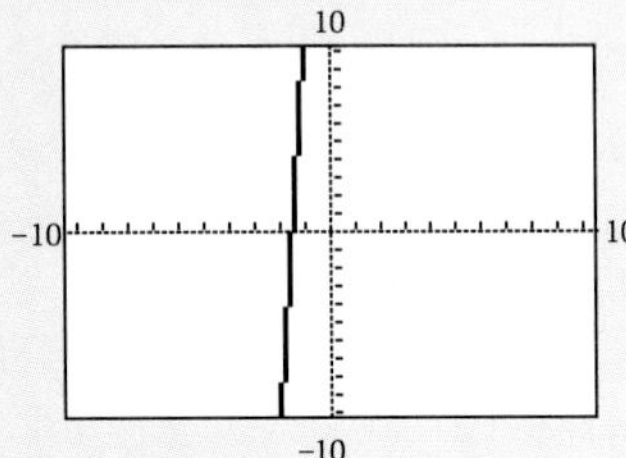

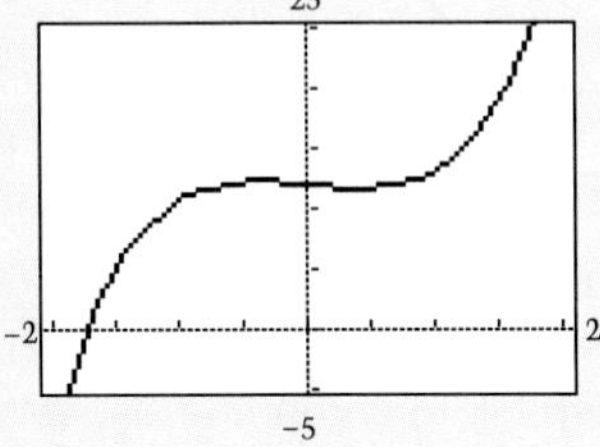

Example 1: Sketching the Graph of a Polynomial Function

For 1970 through 1990, the amount, P (in millions of dollars), spent for public elementary and secondary schools can be modeled by

$$P = -0.2t^4 + 15t^3 - 53t^2 + 4667t + 40{,}724$$

where $t = 0$ represents 1970. For private schools, the total amount, R (in millions of dollars), can be modeled by

$$R = -0.2t^4 + 6t^3 - 19t^2 + 261t + 2508.$$

Find bounds for x and y that will show the relevant portions of the graphs of both models.

Solution Because the models represent the years from 1970 through 1990, you can set the horizontal bounds to be

$0 \le t \le 20.$ *Horizontal bounds*

When $t = 20$, the value of P is 200,864 and the value of R is 16,128. Because it is reasonable to assume that the expenses were increasing, you can set the vertical bounds to be

$0 \le y \le 210{,}000.$ *Vertical bounds*

The graph is shown at the right. ■

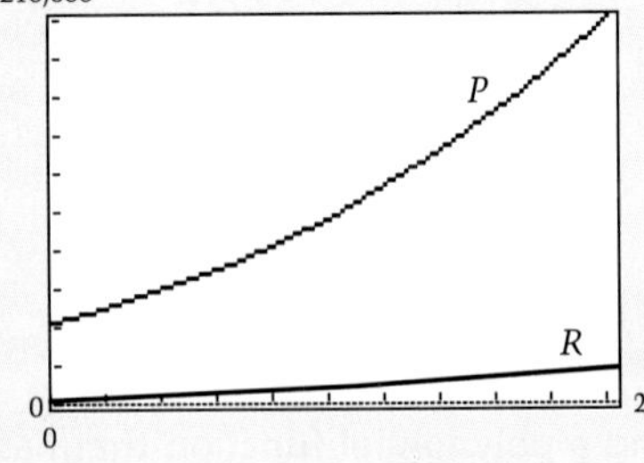

One of the reasons that polynomial functions are important is that they can be used as approximations for other types of functions. (If you go on to study calculus, you will encounter this use of polynomials.)

Example 2: Comparing the Graphs of Two Functions

The graph of the polynomial function

$$f(x) = 1 + x + \frac{1}{2}x^2 + \frac{1}{6}x^3 + \frac{1}{24}x^4 + \frac{1}{120}x^5 + \frac{1}{720}x^6$$

can be used as an approximation of the graph of $g(x) = e^x$. Find bounds for x and y for which the graphs of f and g appear to be the same.

Solution Using trial and error, you can find the graphs to be close to each other over the interval $-2.5 \le x \le 2.5$. ■

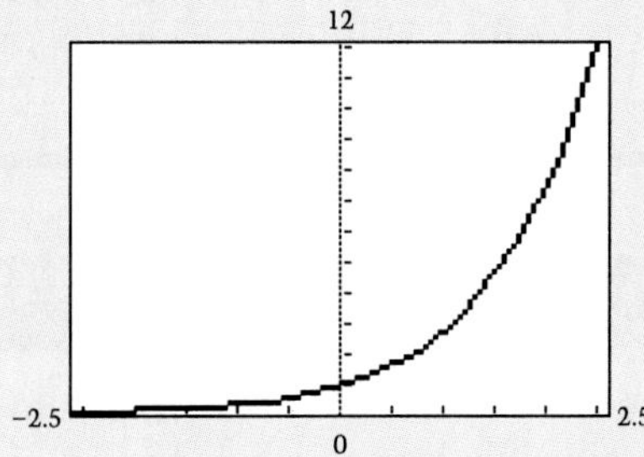

Exercises

In Exercises 1–6, find bounds for x and y that display the basic characteristics of the graph of the polynomial function. 1.–6., 8., 9. Bounds vary. See margin.

1. $f(x) = x^3 + 6x^2 - 11x + 3$
2. $f(x) = -x^3 + 25x^2 + 4$
3. $f(x) = x^4 - 5x^2 + 6$
4. $f(x) = -x^4 - 3x^3 + x^2 - x + 5$
5. $f(x) = x^5 - 10x^4 + 35x^3 - 50x^2 + 24x$
6. $f(x) = -x^5 + 5x^3 - 4x + 10$
7. Sketch the graph of $f(x) = (x - 1)(x - 2)(x - 3)$. What are the zeros? See margin.
8. Find bounds for x and y for which the graphs of f and g appear the same.

   $$f(x) = 1 - x^2 + x^4 - x^6 + x^8 - x^{10} \quad \text{and} \quad g(x) = \frac{1}{x^2 + 1}$$
9. *Company Profit* For 1980 through 1992, the profit, P (in millions of dollars), for a company can be modeled by

   $$P = -t^3 + 10t^2 + 3t + 204$$

   where $t = 0$ represents 1980. Find appropriate bounds for x and y for the graph of this function. For which year (between 1980 and 1992) was the profit the greatest?

Answers

1.–6. Answers may vary. Samples are given.

1. $-10 \le x \le 5, -5 \le y \le 90$
2. $-10 \le x \le 30, -25 \le y \le 2500$
3. $-3 \le x \le 3, -1 \le y \le 7$
4. $-4 \le x \le 2, -10 \le y \le 25$
5. $-1 \le x \le 5, -5 \le y \le 4$
6. $-3 \le x \le 3, -5 \le y \le 15$
7. 1, 2, 3

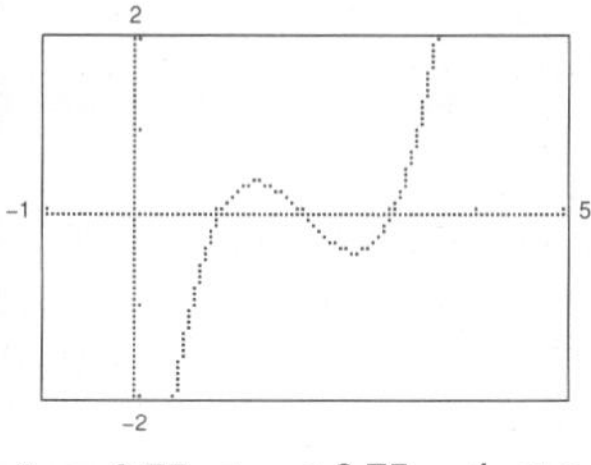

8. $-0.75 \le x \le 0.75, -1 \le y \le 2$
9. Answers may vary; sample— $0 \le t \le 15, 0 \le P \le 400$; 1987

Answers

1.

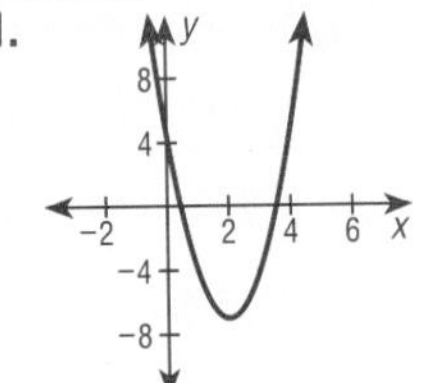

9.

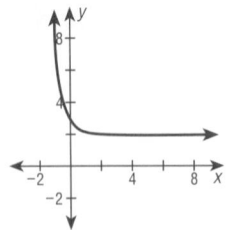

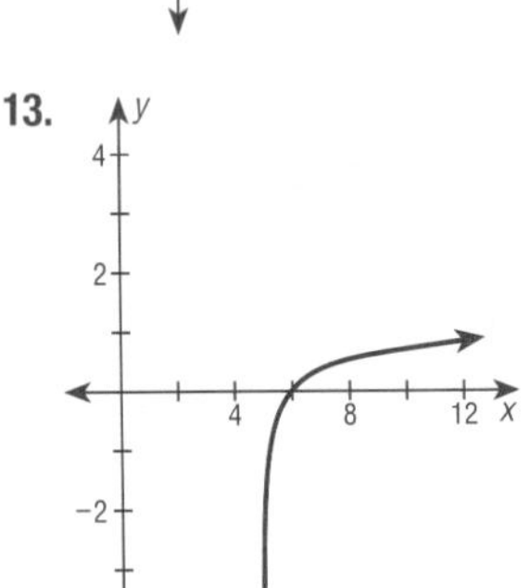

13.

Mixed REVIEW

4. $\frac{1 \pm i\sqrt{2}}{3}$ **8.** -2 **10.** None

1. Sketch $f(x) = 3x^2 - 12x + 5$. **(5.2)** See margin.

2. Solve $|3x - 4| = 6$. **(1.7)** $-\frac{2}{3}, 3\frac{1}{3}$

3. Subtract: $\frac{4}{5} - \frac{2}{7}$. $\frac{18}{35}$

4. Solve $3x^2 - 2x + 1 = 0$. **(5.6)**

5. Simplify: $(2 + 3i)^2$. **(5.5)** $-5 + 12i$

6. Simplify: $\sqrt{48a^2b^4c^7}$. **(7.4)** $4|a|b^2c^3\sqrt{3c}$

7. Evaluate $16^{3/4}$. **(7.3)** 8

8. Find the y-intercept of $f(x) = 5^x - 3$. **(8.1)**

9. Sketch $f(x) = \left(\frac{1}{7}\right)^x + 2$. **(8.1)** See margin.

10. Find the x-intercept of $f(x) = e^{x+3}$. **(8.1)**

11. Find the y-intercept of $y = \frac{1}{1 + e^{-x}}$. **(8.7)** $\frac{1}{2}$

12. Find the geometric mean of 4 and 27. **(7.5)** $6\sqrt{3}$

13. Sketch $f(x) = \log_{10}(x - 5)$. **(8.2)** See margin.

14. Evaluate $e^{\ln 4}$. **(8.6)** 4

15. Expand $\ln \frac{8x}{3}$. **(8.3)** $\ln 8 + \ln x - \ln 3$

16. Solve: $\log_7 x = 5$. **(8.2)** 7^5

17. Find the distance from $(2,-5)$ to $(-6,-2)$. **(7.5)** $\sqrt{73}$

18. \$2000 at 8% compounded continuously is how much in 10 years? **(8.4)** \$4451.08

19. Write an equation of the line through $(4,-5)$ perpendicular to $y = -\frac{2}{3}x + 4$. **(2.4)** $y = \frac{3}{2}x - 11$

20. Is the graph through $(1,-3)$, $(2,-1)$, $(3, 2)$, and $(4, 3)$ linear, quadratic, or neither? **(6.6)** Neither

Career Interview

Administrator

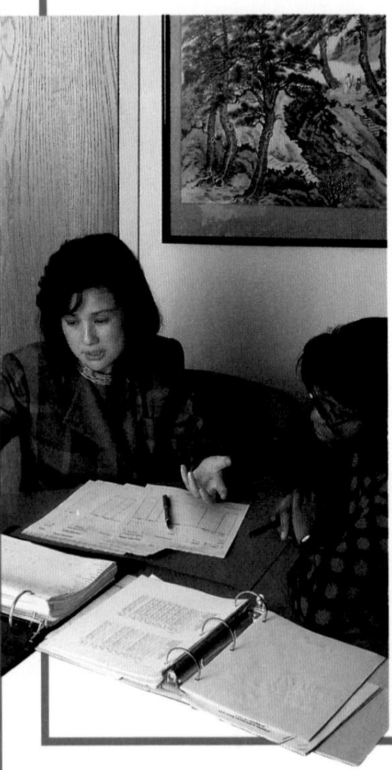

Nancy Hsu is the administrator of a nursing home. "I am like a mayor. I am responsible for the whole facility, from keeping budgets to planning menus to conducting fire drills."

Q: ***What math did you take in school?***

A: I love math. I took algebra, trigonometry, and calculus in high school and more calculus in college.

Q: ***Does having a math background help you on the job?***

A: All the time. Recently I was figuring out how much cleaning solution to order each month. I had to consider the quantity needed per gallon per room per month and also the storage space available.

Q: ***How has new technology changed your job experiences?***

A: Computer spreadsheet programs save me the most time. I can plan and respond to budget changes instantly. For example, if there is a 10% budget cut, the spreadsheet shows me the impact immediately.

Q: ***Are mental math and estimation strategies helpful?***

A: Mental math and estimation allow me to respond to a situation or question quickly and with reasonable accuracy.

Q: ***What would you like to tell kids about math?***

A: Think of mathematics as a language, like learning French or Spanish. To be good at it, you need to practice.

9.3 Factoring Polynomials and Solving Polynomial Equations

What you should learn:

Goal 1 How to factor polynomial expressions and equations

Goal 2 How to use factoring to solve real-life problems

Why you should learn it:

You can model many real-life situations with polynomials that can be factored, such as the model for the velocity that a basketball player must achieve to make a slam dunk.

Goal 1 Factoring Expressions and Equations

LESSON INVESTIGATION

Investigating Factoring Patterns

Partner Activity From Algebra 1, remember that **factoring** a polynomial means to write it as the product of polynomials of lesser degrees. With your partner, use the factoring techniques you learned in Algebra 1 to factor each of the following. In each case, describe the factoring pattern.

1. Common monomial $3x^2 + 6x$ $3x(x + 2)$
2. Difference of Squares $4x^2 - 9$ $(2x + 3)(2x - 3)$
3. Perfect-Square Trinomial $x^2 - 8x + 16$ $(x - 4)^2$
4. Factorable Trinomial $x^2 - 4x + 3$ $(x - 1)(x - 3)$

Example 1 *Factoring Polynomials Completely*

a. $3x - 12x^3 = 3x(1 - 4x^2)$ *Common monomial*

$= 3x(1 - 2x)(1 + 2x)$ *Difference of squares*

b. $16x^4 - 81 = (4x^2 - 9)(4x^2 + 9)$ *Difference of squares*

$= (2x - 3)(2x + 3)(4x^2 + 9)$ *Difference of squares*

c. $2x^3 + 12x^2 + 18x = 2x(x^2 + 6x + 9)$ *Common monomial*

$= 2x(x + 3)^2$ *Perfect-square trinomial*

d. $6x^2 + 21x + 9 = 3(2x^2 + 7x + 3)$ *Common monomial*

$= 3(2x + 1)(x + 3)$ *Factorable trinomial*

e. $x^4 - 4x^2 + 3 = (x^2 - 1)(x^2 - 3)$ *Factorable trinomial*

$= (x - 1)(x + 1)(x^2 - 3)$ *Difference of squares* ■

Study Tip

A polynomial with integer coefficients is completely factored with respect to the integers if it cannot be further factored using integer coefficients.

Problem of the Day

A container of 650 doses of medicine has 5 mL of medicine removed. When $\frac{1}{5}$ of the remaining doses are removed, only 512 doses are left. If each dose has the same number of mL, how many mL are in a dose? 0.5

ORGANIZER

TIME SCHEDULE: All levels, two days

Warm-Up Exercises

Factor each expression.

a. $12x + 8xy$
b. $18x^3 - 42x^2$
c. $9 - 16m^2$
d. $81x^2 - 49y^2$
e. $25x^2 + 30x + 9$
f. $14 + 52x - 16x^2$

a. $4x(3 + 2y)$
b. $6x^2(3x - 7)$
c. $(3 - 4m)(3 + 4m)$
d. $(9x - 7y)(9x + 7y)$
e. $(5x + 3)^2$
f. $(7 - 2x)(2 + 8x)$

Lesson Resources

Teaching Tools
- Problem of the Day: 9.3
- Warm-up Exercises: 9.3
- Answer Masters: 9.3

Extra Practice: 9.3
Color Transparencies: 48, 49
Applications Handbook: p. 46–48
Technology Handbook: p. 57

LESSON Notes

Note: You should present factoring expressions (Examples 1–3, along with Exercises 1–5 and 7–50) on the first day, and solving polynomial equations and real-life applications (Ex-

(continued)

(continued)
amples 4 and 5, along with Exercises 6 and 51–83) on the second day.

Example 2

It is important that students be able to identify the factors of each term that is cubed in a given expression. For example, in Part b, students should be able to state that neither $3x^3$ nor 192 is a perfect cube involving integral values. However, upon factoring out the common monomial factor, 3, the remaining terms of the other factor, x^3 and 64, are both perfect cubes for which u and v, in the special factoring patterns, can be assigned the values x and 4, respectively. The pattern can then be used to factor the expression.

Example 3

One indicator that the factor-by-grouping pattern should be used is that there are four terms in the polynomial expression.

Example 5

Perform a "reality check." Some of your students may already know Michael Jordan's height in feet and inches. Determine what Michael Jordan's height is (or is believed to be; about 6 feet 6.5 inches) and then ask your students to determine how long a meter must be in inches. ≈39.25 in.

Common-Error ALERT!

Many students will be tempted to divide both sides of the equation $3x^4 = 48x^2$ by $3x^2$ in **Example 4.** However, if this is done, one of the solutions, 0, will be omitted. Caution students to write the equation in standard form, factor, and apply the Zero-Product Property.

None of the polynomials in Example 1 can be factored further without using fractions, irrational numbers, or imaginary numbers.

$2x - 3 = 2\left(x - \frac{3}{2}\right)$ *Requires fractions.*

$x^2 - 3 = (x - \sqrt{3})(x + \sqrt{3})$ *Requires irrational numbers.*

$4x^2 + 9 = (2x - 3i)(2x + 3i)$ *Requires imaginary numbers.*

Special Factoring Patterns

Sum of Two Cubes

$u^3 + v^3 = (u + v)(u^2 - uv + v^2)$

Example: $x^3 + 8 = (x + 2)(x^2 - 2x + 4)$

Difference of Two Cubes

$u^3 - v^3 = (u - v)(u^2 + uv + v^2)$

Example: $27x^3 - 1 = (3x - 1)(9x^2 + 3x + 1)$

Factoring by Grouping

$au^3 + acu^2 + bu + bc = au^2(u + c) + b(u + c) = (au^2 + b)(u + c)$

Example: $x^3 + 2x^2 + 3x + 6 = x^2(x + 2) + 3(x + 2)$
$= (x^2 + 3)(x + 2)$

Example 2 *Factoring the Sum or Difference of Cubes*

Factor the polynomials.

a. $x^3 - 27$ **b.** $3x^3 + 192$

Solution

a. $x^3 - 27 = x^3 - 3^3$ *Difference of cubes*
$= (x - 3)(x^2 + 3x + 9)$

b. $3x^3 + 192 = 3(x^3 + 4^3)$ *Sum of cubes*
$= 3(x + 4)(x^2 - 4x + 16)$ ■

Example 3 *Factoring by Grouping*

Factor the polynomials.

a. $4x^3 - 6x^2 + 10x - 15$ **b.** $x^3 - 2x^2 - 4x + 8$

Solution

a. $4x^3 - 6x^2 + 10x - 15 = 2x^2(2x - 3) + 5(2x - 3)$
$= (2x^2 + 5)(2x - 3)$

b. $x^3 - 2x^2 - 4x + 8 = (x^3 - 2x^2) - (4x - 8)$
$= x^2(x - 2) - 4(x - 2)$
$= (x - 2)(x^2 - 4)$
$= (x - 2)(x - 2)(x + 2)$
$= (x - 2)^2(x + 2)$ ■

The **Zero Product Property** states that if the product of two factors is zero, then one (or both) of the factors must be zero.

Zero Product Property

Let a and b be real numbers. If $ab = 0$, then $a = 0$ or $b = 0$.

This property connects factoring to solving equations. For instance, to solve the equation

$$(2x + 1)(x - 4) = 0$$

you can use the Zero Product Property to conclude that either $(2x + 1)$ or $(x - 4)$ must be zero. Setting $2x + 1$ equal to zero produces the solution $x = -\frac{1}{2}$. Setting $x - 4$ equal to zero produces the solution $x = 4$.

The Zero Product Property can be generalized to include equations with more than two factors, as shown in Example 4.

Example 4 *Solving a Polynomial Equation by Factoring*

Solve $3x^4 = 48x^2$.

Solution The basic approach is to write the polynomial equation in standard form first (with zero on the right side of the equation), factor the left side, and then set each factor equal to zero.

$$3x^4 = 48x^2 \quad \textit{Given equation}$$

$$3x^4 - 48x^2 = 0 \quad \textit{Collect terms on left side.}$$

$$3x^2(x^2 - 16) = 0 \quad \textit{Common monomial factor}$$

$$3x^2(x + 4)(x - 4) = 0 \quad \textit{Difference of two squares}$$

$$\left.\begin{array}{l} 3x^2 = 0 \longrightarrow x = 0 \\ x + 4 = 0 \longrightarrow x = -4 \\ x - 4 = 0 \longrightarrow x = 4 \end{array}\right\} \textit{Zero Product Property}$$

The equation has three solutions: 0, -4, and 4. Check these solutions in the original equation. ■

When solving the equation shown in Example 4, some people like to divide both sides of the equation by 3. This does help simplify the equation a little. Remember, however, that you cannot divide both sides of an equation by a variable such as $3x^2$. Doing this would "lose" $x = 0$ as one of the solutions.

Extra Examples

Here are additional examples similar to **Examples 1–4.**

1. Factor completely.

a. $108x^4 - 12$

b. $7x^2 + 28x + 21$

c. $8x^3 - 48x^2 + 72x$

d. $x^4 - 3x^2 - 4$

a. $12(3x^2 - 1)(3x^2 + 1)$

b. $7(x + 1)(x + 3)$

c. $8x(x - 3)^2$

d. $(x - 2)(x + 2)(x^2 + 1)$

2. Factor completely.

a. $2x^3 + x^2 - 18x - 9$

b. $8x^6 - 125$

c. $54x^3 + 128$

d. $24x^3 + 8x^2 + 12x + 4$

e. $30x^3 + 18x^2 - 5x - 3$

a. $(2x + 1)(x - 3)(x + 3)$

b. $(2x^2 - 5)(4x^4 + 10x^2 + 25)$

c. $2(3x + 4)(9x^2 - 12x + 16)$

d. $4(2x^2 + 1)(3x + 1)$

e. $(6x^2 - 1)(5x + 3)$

3. Solve each equation.

a. $2(5 - x)(3x - 4)(x + 2) = 0$

b. $6x^5 = 54x^3$

a. $5, \frac{4}{3}, -2$

b. $0, -3, 3$

Check Understanding

1. Describe what it means to factor a polynomial.
   To write it as the product of polynomials of lesser degree
2. State the special factoring pattern for the difference of two cubes.
   $u^3 - v^3 = (u - v)(u^2 + uv + v^2)$
3. Can factoring by grouping be used with trinomials?
   No; see Example 3. You need at least 4 terms.
4. Does the Zero-Product Property hold for three or more factors?
   Yes; see the Extend Communicating activity that follows.

Communicating about ALGEBRA

EXTEND *Communicating*
COOPERATIVE LEARNING

Ask students to work together to show the following. Given the Zero-Product Property, show that for real numbers *a*, *b*, *c*, and *d*, if $abcd = 0$ then either $a = 0$, or $b = 0$, or $c = 0$, or $d = 0$.

(*Hint:* Use the Associative Property of Multiplication to demonstrate that the product of four factors can be viewed as a product of two factors.)

Sample answer:
Let $abcd = (ab)(cd)$.
If $abcd = 0$, then $ab = 0$ or $cd = 0$.
If $ab = 0$, then $a = 0$ or $b = 0$.
If $cd = 0$, then $c = 0$ or $d = 0$.
So, if $abcd = (ab)(cd) = 0$, then $a = 0$, or $b = 0$, or $c = 0$, or $d = 0$.

Goal 2 Using Factoring in Real-Life Problems

Real Life **Architecture**

World Trade Center, New York City

Example 5 *A Trip to the World Trade Center*

Each of the twin towers of the World Trade Center in New York City has a volume of 1,695,744 cubic meters. One of the tour guides at the World Trade Center, who is a Michael Jordan fan, likes to tell tour groups the following riddle. From these clues, can you find the dimensions of each tower?

> *Each tower of the World Trade Center is 207 times as tall as Michael Jordan. The base of each tower is square. To find the length of each side of the base, subtract 1 meter from Michael Jordan's height and multiply the result by 64.*

Solution Let x represent Michael Jordan's height (in meters).

Verbal Model Volume = Height · Length · Width

Labels

Volume	$= 1{,}695{,}744$	(cubic meters)
Height	$= 207x$	(meters)
Length	$= 64(x - 1)$	(meters)
Width	$= 64(x - 1)$	(meters)

Equation $1{,}695{,}744 = 207x \cdot 64(x - 1) \cdot 64(x - 1)$

You can solve this equation as follows.

$207x(64^2)(x - 1)^2 = 1{,}695{,}744$	*Model for volume*
$x(x - 1)^2 = 2$	*Divide both sides by $207(64^2)$.*
$x^3 - 2x^2 + x - 2 = 0$	*Standard form*
$x^2(x - 2) + (x - 2) = 0$	*Factor by grouping.*
$(x - 2)(x^2 + 1) = 0$	*Factored form*

From the factored form, you can see that $x = 2$. So Michael Jordan's height is 2 meters. Thus each tower is 414 meters high and has a base that is 64 meters square. ■

Communicating about ALGEBRA

SHARING IDEAS about the Lesson See Additional Answers.

Solve the equations by factoring. Explain your steps.

A. $x^3 - 2x^2 = 9x - 18$ **B.** $x^4 - 29x^2 + 100 = 0$

EXERCISES

Guided Practice

CRITICAL THINKING about the Lesson

1. What does it mean for a polynomial with integer coefficients to be completely factored with respect to the integers? Give an example of such a second-degree polynomial that cannot be further factored with respect to the integers. It cannot be factored further using integer coefficients; example: $x^2 + 1$

In Exercises 2–5, factor completely with respect to the integers. See below.

2. $81x^4 - 1$ **3.** $16x^3 - 64x$ **4.** $1 - 64x^3$ **5.** $2x^3 - 8x^2 + 3x - 12$

6. What is wrong with the following solution steps?

$2x^4 - 18x^2 = 0$	*Original equation*
$2x^2(x^2 - 9) = 0$	*Common monomial factor*
$2x^2(x + 3)(x - 3) = 0$	*Difference of squares*
$(x + 3)(x - 3) = 0$	*Divide both sides by $2x^2$.*
$x + 3 = 0 \longrightarrow x = -3$	*Zero-Product Property*
$x - 3 = 0 \longrightarrow x = 3$	*Zero-Product Property*

The Zero-Product Property must be used for each factor; the fourth step should be: $2x^2 = 0 \longrightarrow x = 0$.

Independent Practice

In Exercises 7–14, find the missing factor.

7. $10x^4 - 40x^2 = (\ ?\ )(x + 2)(x - 2)$ $10x^2$

8. $8x^3 - 18x = 2x(2x + 3)(\ ?\ )$ $2x - 3$

9. $6x^3 + 9x = 3x(\ ?\ )$ $2x^2 + 3$

10. $18x^3 - 6x^2 + 3x = (\ ?\ )(6x^2 - 2x + 1)$ $3x$

11. $3x^3 + 24 = 3(\ ?\ )(x^2 - 2x + 4)$ $x + 2$

12. $16x^3 - 54 = 2(2x - 3)(\ ?\ )$ $4x^2 + 6x + 9$

13. $3x^3 + 12x^2 + 2x + 8 = (\ ?\ )(x + 4)$ $3x^2 + 2$

14. $14x^3 + 7x^2 + 10x + 5 = (7x^2 + 5)(\ ?\ )$ $2x + 1$

In Exercises 15–50, factor completely with respect to the integers. See Additional Answers.

15. $9x^2 - 4$ **16.** $3x^2 - 48$ **17.** $x^3 - 8$ **18.** $x^3 + 64$

19. $216x^3 + 1$ **20.** $125x^3 - 1$ **21.** $200x^2 - 50$ **22.** $32x^3 - 4$

23. $128x^3 - 2$ **24.** $8x^3 - 64$ **25.** $3x^3 + 81$ **26.** $40x^3 - 5$

27. $x^3 + x^2 + x + 1$ **28.** $30x^3 + 40x^2 + 3x + 4$ **29.** $x^3 + 2x^2 + 5x + 10$

30. $x^3 - 2x^2 + 4x - 8$ **31.** $9x^3 + 18x^2 + 7x + 14$ **32.** $-2x^3 - 4x^2 - 3x - 6$

33. $2x^3 + 4x^2 + 4x + 8$ **34.** $18x^3 + 30x^2 + 3x + 5$ **35.** $2x^3 - 2x^2 + 5x - 5$

36. $2x^3 + 3x^2 - 32x - 48$ **37.** $5x^3 - 20x^2 + 3x - 12$ **38.** $18x^3 - 2x^2 + 27x - 3$

39. $7x^3 + 14x^2 + 7x$ **40.** $3x^2 - 24x + 48$ **41.** $2x^3 - 4x^2 - 3x + 6$

42. $6x^3 - 18x^2 - 2x + 6$ **43.** $3x^3 - 24$ **44.** $x^3 + 125$

45. $3x^4 - 300x^2$ **46.** $28x^3 - 7x$ **47.** $3x^4 + 3x^3 + 6x^2 + 6x$

48. $x^4 + 12x^3 + 4x^2 + 48x$ **49.** $10x^3 - 20x^2 - 2x + 4$ **50.** $18x^3 - 9x^2 - 18x + 9$

2. $(9x^2 + 1)(3x - 1)(3x + 1)$ **3.** $16x(x - 2)(x + 2)$ **4.** $(1 - 4x)(1 + 4x + 16x^2)$ **5.** $(2x^2 + 3)(x - 4)$

EXERCISE Notes

ASSIGNMENT GUIDE

Day 1

Basic/Average: Ex. 1–5, 7–11 odd, 15–45 odd

Above Average: Ex. 1–5, 11–49 odd

Advanced: Ex. 1–5, 11–49 odd

Day 2

Basic/Average: Ex. 6, 51–55 odd, 62–67, 69–77 odd

Above Average: Ex. 6, 51–57 odd, 62–67, 69–81 odd

Advanced: Ex. 6, 51–61 odd, 62–68, 69–75 odd, 79–83

Selected Answers

Exercises 1–6, 7–77 odd

✪ More Difficult Exercises

Exercises 63–68, 79–83

Guided Practice

Ex. 2–5 Ask students to describe the factoring technique being used. For example, in Exercise 2, it is the "difference of two squares technique." Factoring by grouping and factoring the sums and differences of two cubes may be new techniques for students.

Independent Practice

Ex. 15–50 Remind students to look for common monomial factors first.

Ex. 27–38 Help students to observe that these expressions are third-degree polynomials with four terms so the grouping technique modeled in Example 3 would apply.

▶ **Ex. 51–62** Students should consider these exercises as examples of a mental math approach to solving polynomial equations.

▶ **Ex. 63–65** Assign these exercises as a group. Remind students that 5′7″ ≈ 5.583 ft. Ask a student who plays basketball to explain what hang time is: the total time in the air. When $h = 0 = -16t^2 + 10.77t$, the equation can be solved by factoring.

▶ **Ex. 66** When the piece of equipment hits the ground, $h = 0$.

▶ **Ex. 68** Refer students to Example 5 for a solution model.

▶ **Ex. 81–82** Suggest that students use factoring by grouping and the Quadratic Formula to solve the equations.

Enrichment Activities

These activities require students to go beyond the lesson goals.

While it is true that calculators have produced a dramatic change in how difficult computations are accomplished, it is not true that calculator usage has made algebraic techniques like factoring obsolete. The following activity should illustrate that factoring can be used as an efficient mental-math technique for solving many quadratic equations.

In Exercises 51–62, find all real-number solutions.

51. $x^2 - 3x = 0$ 0, 3

52. $2x^3 - 6x^2 = 0$ 0, 3

53. $x^2 + 8x + 16 = 0$ −4

54. $x^2 - 4x + 4 = 0$ 2

55. $x^3 - 15x^2 + 75x - 125 = 0$ 5

56. $x^3 + 3x^2 + 3x + 1 = 0$

57. $2x^2 - x = 3$ $\frac{3}{2}, -1$

58. $x^2 + 3x = 10$ −5, 2

59. $x^3 + 3x^2 - 2x - 6 = 0$

60. $x^4 + 7x^3 - 8x - 56 = 0$ 2, −7

61. $x^3 + 2x^2 - x = 2$ −2, ±1

62. $x^3 - x^2 + 2x = 2$ 1

56. −1

59. $\pm\sqrt{2}, -3$

Making a Slam Dunk **In Exercises 63–65, use the following information.**

A basketball hoop is 10 feet above the playing floor. To make a "slam dunk," a player must jump high enough to reach $10\frac{1}{2}$ feet. The November, 1991 issue of *Popular Mechanics* analyzed the mechanics of a slam dunk as follows.

"If you take a person's height, H (in feet), subtract the distance from the top of the head to the shoulder (about 1 foot), and then add half of the result (about one arm's length), you obtain the formula $R = \frac{3}{2}(H - 1)$, where R is the person's standing reach in feet."

Using this model, the initial velocity, v (in feet per second), at which a person must jump to make a slam dunk is given by

$$\frac{21}{2} - \frac{3}{2}(H - 1) = \frac{1}{64}v^2$$

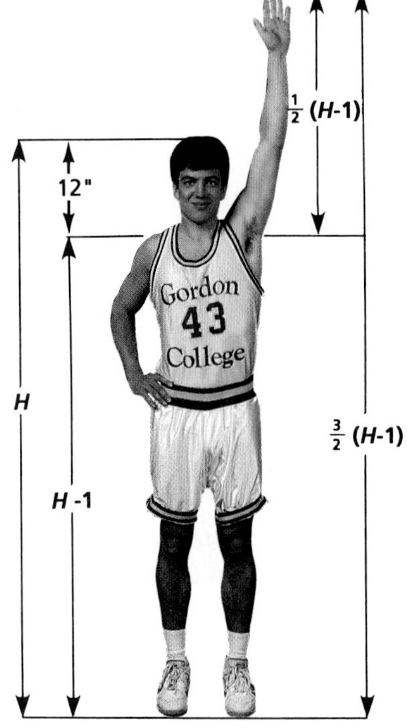

63. Copy and complete the table. See below.

Height, H	5.5	6.0	6.5	7.0	7.5
Reach, R	?	?	?	?	?
Velocity, v	?	?	?	?	?

64. *Popular Mechanics* states that "perhaps the most amazing dunker in professional basketball is 5′7″ Spud Webb, of the Atlanta Hawks." With what initial velocity must Webb jump to make a slam dunk? ≈15.23 ft per second

65. Let h (in feet) be the distance between a player's shoes and the floor and let v be the initial velocity in feet per second. Then h and v are related by the equation $h = -16t^2 + vt$. What is Spud Webb's minimum hang time—the time that he must be in the air to make a slam dunk? ≈0.95 second

66. *Weather Balloon* A heavy piece of equipment falls from a weather balloon that is 1600 feet above the ground. Let h be the height (in feet) above ground of the piece of equipment. Use factoring and the vertical motion model $h = -16t^2 + 1600$ to find the time, t (in seconds), that the piece of equipment takes to hit the ground. 10 seconds

63. 6.75, 7.5, 8.25, 9, 9.75; ≈15.49, ≈13.86, 12, ≈9.80, ≈6.93

67. *Room Dimensions* A rectangular room has an area of 160 square feet. The length of the room is 6 feet more than its width. Find the dimensions of the room. 16 ft by 10 ft

68. *Big Ben* Big Ben is the name of a famous bell in a clock tower in London and also the name of its clock tower. Let x represent one-third the diameter of Big Ben, the bell. The volume of Big Ben, the clock tower (not counting the belfry) is 336,000 cubic feet. What are the dimensions of the clock tower if it has a square base? What is the diameter of the bell at its widest place?
40 ft by 40 ft by 210 ft, 9 ft

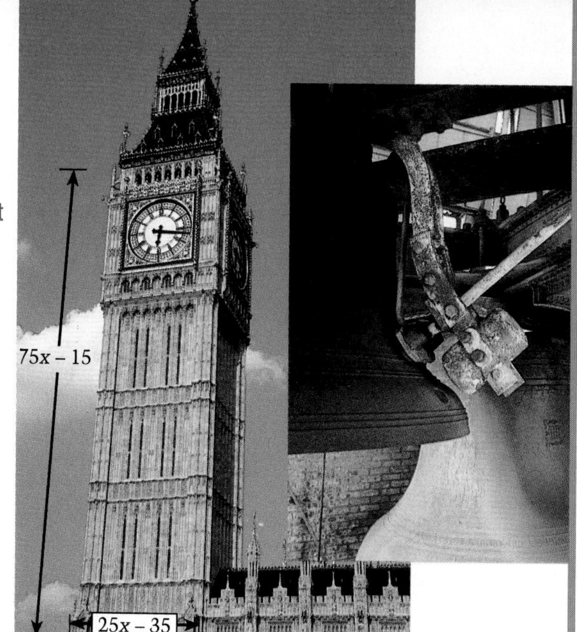

Integrated Review

In Exercises 69–74, write in standard polynomial form. See Additional Answers.

69. $(x + 4)^2$
70. $(x - 2)^2$
71. $(x - 3)^3$
72. $(x + 1)^3$
73. $(x + 9)^3$
74. $(x - 5)^3$

In Exercises 75–78, find all real and imaginary solutions.

75. $2x^2 + 2x - 60 = 0$ $-6, 5$
76. $x^2 + x + 1 = 0$ $-\frac{1}{2} \pm \frac{\sqrt{3}}{2}i$
77. $4x^2 + x - 3 = 0$ $\frac{3}{4}, -1$
78. $3x^3 - 18x = 0$ $0, \pm\sqrt{6}$

Exploration and Extension

In Exercises 79 and 80, find all integers *b* such that the trinomial can be factored using integer coefficients.

79. $x^2 + bx + 15$ $\pm 16, \pm 8$
80. $x^2 + bx - 12$ $\pm 11, \pm 4, \pm 1$

In Exercises 81 and 82, find all real and imaginary solutions of the equation.

81. $9x^5 - 4x^3 + 9x^2 - 4 = 0$ $-1, \pm\frac{2}{3}, \frac{1}{2} \pm \frac{\sqrt{3}}{2}i$
82. $x^5 - 4x^3 + 8x^2 - 32 = 0$ $\pm 2, 1 \pm \sqrt{3}i$
83. *Geometry* The cube at the right is formed by four solids: I, II, III, and IV. Explain how the figure can be used as a geometric factoring model for the *difference of two cubes* factoring pattern.

$$a^3 - b^3 = (a - b)(a^2 + ab + b^2)$$
$$= (a - b)a^2 + (a - b)ab + (a - b)b^2$$

See Additional Answers.

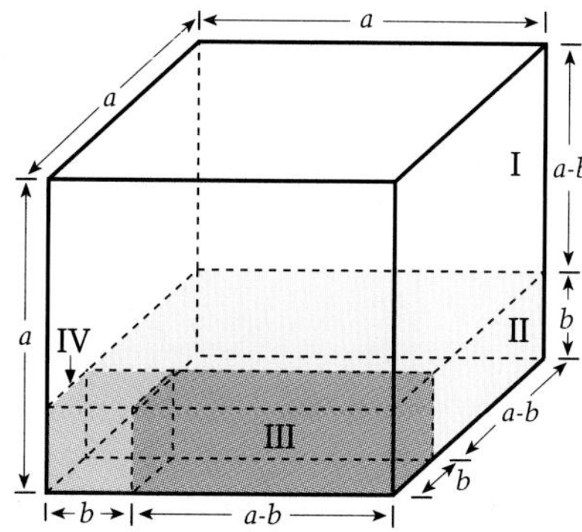

1. COOPERATIVE LEARNING Work with a partner. You have 10 minutes to solve these equations. One student in the pair is to solve the equations by factoring without using a calculator. The other student must use a calculator and solve the equations using the quadratic formula.
   a. $x^2 + 6x + 5 = 0$
   b. $x^2 - 3x + 1 = 0$
   c. $x^2 - 4x + 4 = 0$
   d. $2x^2 - 7x - 9 = 0$
   e. $x^2 + 14x - 51 = 0$
   f. $20x^2 - 33x - 221 = 0$
   g. $15x^2 - 6x + 25 = 0$

   a., c., d., e., and f. are factorable
   a. $(x + 5)(x + 1) = 0; -5, -1$
   c. $(x - 2)^2 = 0; 2$
   d. $(2x - 9)(x + 1) = 0; \frac{9}{2}, -1$
   e. $(x + 17)(x - 3) = 0; -17, 3$
   f. $(5x + 13)(4x - 17) = 0; -\frac{13}{5}, \frac{17}{4}$

   b. and g. are not factorable
   b. $\frac{3}{2} \pm \frac{\sqrt{5}}{2}$
   g. imaginary roots
2. Which equations did you find easy to solve by your method? Which equations did you find awkward to solve by your method? Which were impossible to solve by your method?
3. WRITING Compare the two methods of solving quadratic equations with your partner. Write a one-paragraph report on your observations.

Problem of the Day

When new, a rubber tire has an outside diameter of 24 inches. If the tire wears down 0.1 inch, about how many more times will the wheel go around in a mile? About 7

ORGANIZER

Warm-Up Exercises

1. Show the steps involved in this long division.

$$\begin{array}{r} 131 \leftarrow \text{quotient} \\ \text{divisor} \rightarrow 372\overline{)49,023} \leftarrow \text{dividend} \\ -37\,2 \\ 11\,82 \\ -11\,16 \\ 663 \\ -372 \\ 291 \leftarrow \text{remainder} \end{array}$$

What is the quotient? the remainder? the divisor? the dividend?

2. Perform the indicated operation and simplify.

a. $\frac{2x^5}{x^2}$ **b.** $\frac{56x^3}{8x}$

c. $\frac{-4x^6}{2x^5}$ **d.** $4x^2(2x^2 - 3x)$

e. $3x^3(-5x^4)$

f. $6x^3(x^2 + 2x - 9)$

a. $2x^3$, **b.** $7x^2$, **c.** $-2x$

d. $8x^4 - 12x^3$, **e.** $-15x^7$

f. $6x^5 + 12x^4 - 54x^3$

Lesson Resources

Teaching Tools

Problem of the Day: 9.4

Warm-up Exercises: 9.4

Answer Masters: 9.4

Extra Practice: 9.4

Color Transparencies: 49, 50, 51

Technology Handbook: p. 59

9.4 Polynomial Division, Factors, and Remainders

What you should learn:

Goal 1 How to divide polynomials using long division and synthetic division and relating the quotient to the Remainder Theorem and the Factor Theorem

Goal 2 How to use polynomial division in real-life problems

Why you should learn it:

You can factor many polynomial models to find other solutions, such as alternate prices that will give the same amount of profit.

Goal 1 Dividing Polynomials

There are two algorithms for polynomial division: *long division* and *synthetic division.* Of the two, synthetic division is quicker, but long division is more general—it will work for divisors of any degree. (Synthetic division works *only* for divisors of the form $x \pm k$.) An example of how long division can be used to divide $x^2 + 3x + 5$ by $x + 1$ is shown.

$$\begin{array}{r} x + 2 \leftarrow \text{Quotient} \\ \text{Divisor} \rightarrow x + 1\overline{)x^2 + 3x + 5} \leftarrow \text{Dividend} \\ \underline{x^2 + x} \\ 2x + 5 \\ \underline{2x + 2} \\ 3 \leftarrow \text{Remainder} \end{array}$$

You can write this result as follows.

$$\overbrace{\frac{x^2 + 3x + 5}{\underbrace{x + 1}_{\text{Divisor}}}}^{\text{Dividend}} = \overbrace{x + 2}^{\text{Quotient}} + \overbrace{\frac{3}{\underbrace{x + 1}_{\text{Divisor}}}}^{\text{Remainder}}$$

If there is a zero remainder, then the divisor *divides evenly* into the dividend.

Example 1 ***Long Division of Polynomials***

Divide $2x^4 + 4x^3 - 5x^2 + 2x - 3$ by $x^2 + 2x - 3$.

Solution

$$\begin{array}{r} 2x^2 + 1 \\ x^2 + 2x - 3\overline{)2x^4 + 4x^3 - 5x^2 + 2x - 3} \\ \underline{2x^4 + 4x^3 - 6x^2} \\ x^2 + 2x - 3 \\ \underline{x^2 + 2x - 3} \\ 0 \end{array}$$

This result can be written as follows.

$$\frac{2x^4 + 4x^3 - 5x^2 + 2x - 3}{x^2 + 2x - 3} = 2x^2 + 1$$

■

When the divisor is of the form $x \pm k$, an abbreviated form of long division, called **synthetic division,** can be used. The pattern for synthetic division of a cubic polynomial is summarized below.

Synthetic Division (of a Cubic Polynomial)

To divide $ax^3 + bx^2 + cx + d$ by $x - k$, use the following pattern. (The pattern for higher-degree polynomials is similar.)

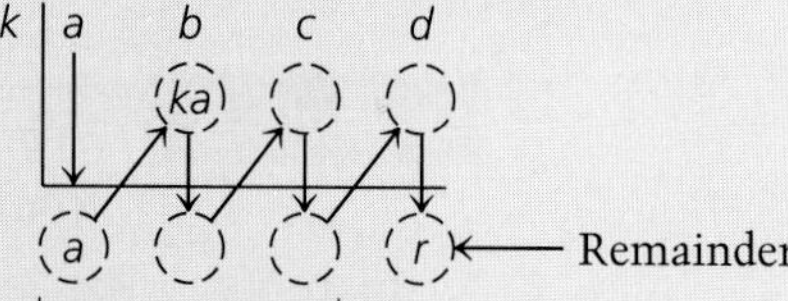

Coefficients of Quotient (in decreasing order)

Vertical Pattern: Add terms.

Diagonal Pattern: Multiply by k.

Note: k can be positive or negative. For instance, if you divide by $(x + 2) = (x - (-2))$, then $k = -2$.

Example 2 *Using Synthetic Division*

Use synthetic division to divide $x^4 - 10x^2 + 2x + 3$ by $x - 3$.

Solution When setting up the synthetic division array, you must include a zero for each "missing" term in the dividend.

$$\begin{array}{r|rrrrr} 3 & 1 & 0 & -10 & 2 & 3 \\ & & 3 & 9 & -3 & -3 \\ \hline & 1 & 3 & -1 & -1 & 0 \end{array} \leftarrow \textit{Remainder: } 0$$

Quotient: $x^3 + 3x^2 - x - 1$

In fractional form, you can write the result as follows.

$$\frac{x^4 - 10x^2 + 2x + 3}{x - 3} = x^3 + 3x^2 - x - 1$$

You can check this result by multiplying. That is,

$$(x - 3)(x^3 + 3x^2 - x - 1) = x^4 - 10x^2 + 2x + 3.$$

The use of long division to obtain the same result is shown at the left. ■

$$\begin{array}{r} x^3 + 3x^2 - x - 1 \\ x - 3\overline{)x^4 \qquad - 10x^2 + 2x + 3} \\ \underline{x^4 - 3x^3} \qquad\qquad\qquad \\ 3x^3 - 10x^2 \qquad\quad \\ \underline{3x^3 - 9x^2} \qquad\quad \\ -x^2 + 2x \quad \\ \underline{-x^2 + 3x} \quad \\ -x + 3 \\ \underline{-x + 3} \\ 0 \end{array}$$

Be sure you see that synthetic division works *only* for divisors of the form $x - k$. You cannot, for example, use synthetic division to divide a polynomial by a quadratic such as $x^2 + 4$. For that, you would have to use long division.

LESSON Notes

Example 1

Point out that each successive (partial) product (such as $2x^4 + 4x^3 - 6x^2$ and $x^2 + 2x - 3$, in this example) must be subtracted from the dividend. The process is sometimes described as changing the sign of the product and adding the product to the dividend.

Example 2

Ask students to explain the connection between long division and synthetic division, using this example. [Students should observe that the leading term of each successive (partial) dividend is always eliminated by the selection of the appropriate (partial) quotient.]

Example 3

Ask students to verify that $f(-2) = -9$.

$3(-2)^3 + 8(-2)^2 + 5(-2) - 7$
$= 3(-8) + 8(4) + 5(-2) - 7$
$= -24 + 32 - 10 - 7$
$= -9$

Be sure students understand that the factor indicated by the Remainder Theorem is $(x + 2)$.

Example 4

Stress that a factor of a polynomial function, $f(x)$, has the form $(x - a)$ where $f(a) = 0$. The real-valued constant may be positive or negative.

Extra Examples

Here are additional examples similar to **Examples 1–4.**

1. Divide using long division.

$2x + 1\overline{)2x^4 + x^3 - 2x^2 + 9x + 5}$

$x^3 - x + 5$

2. Use synthetic division to divide.
   a. $3x^5 - 8x^3 + 2x^2 - x + 1$ by $x + 2$
   b. $x^4 - 2x^3 - 31x - 4$ by $x - 4$

   a. $3x^4 - 6x^3 + 4x^2 - 6x + 11 + \frac{-21}{x+2}$

   b. $x^3 + 2x^2 + 8x + 1$

3. Use the Remainder Theorem to evaluate the function for (a) $x = 5$ and (b) $x = -2$. $f(x) = 3x^4 - 18x^3 + 20x^2 - 24x$

   a. $f(5) = 5$, b. $f(-2) = 320$

4. Given that $(x + 2)$ is a factor of $h(x)$, use the Factor Theorem to factor the polynomial $h(x) = 2x^3 - x^2 - 13x - 6$.

−2	2	−1	−13	−6
		−4	10	6
	2	−5	−3	0

So,
$h(x) = (x + 2)(2x^2 - 5x - 3)$
$= (x + 2)(x - 3)(2x + 1)$.

Example 5

Point out that, although fewer units of the product are sold, each unit costs more than when 1,600,000 units were sold; so the same profit can be obtained selling fewer units.

The synthetic division used in this example is given here.

1.6	−5	0	25	−19.52
		−8	−12.8	19.52
	−5	−8	12.2	0

Ask students to verify the use of the Quadratic Formula in solving the equation $-5x^2 - 8x + 12.2 = 0$.

The remainder obtained when dividing has an interesting connection to polynomial functions.

Remainder Theorem

If a polynomial $f(x)$ is divided by $(x - k)$, then the remainder is $r = f(k)$.

Example 3 Evaluating a Polynomial by the Remainder Theorem

Evaluate $f(x) = 3x^3 + 8x^2 + 5x - 7$ at $x = -2$.

Solution Begin by using synthetic division.

−2	3	8	5	−7
		−6	−4	−2
	3	2	1	−9

Because the remainder is -9, you can conclude from the Remainder Theorem that $f(-2) = -9$. ■

Factor Theorem

A polynomial $f(x)$ has a factor $(x - k)$ if and only if $f(k) = 0$.

Example 4 Using Synthetic Division to Find Factors of a Polynomial

Given that $f(x) = 2x^3 + 11x^2 + 18x + 9$ has $x = -3$ as a zero, factor $f(x)$.

Solution Because $f(-3) = 0$, you know that $(x - (-3))$, which is equal to $(x + 3)$, is a factor of $f(x)$. Using synthetic division produces the following.

−3	2	11	18	9
		−6	−15	−9
	2	5	3	0

Thus you can write

$$2x^3 + 11x^2 + 18x + 9 = (x + 3)(2x^2 + 5x + 3)$$
$$= (x + 3)(2x + 3)(x + 1).$$

The complete factorization is $(x + 3)(2x + 3)(x + 1)$. ■

Goal 2 Using Polynomial Division in Real Life

In business, the demand function gives the price per unit as a function of the number of units sold. The demand function whose graph is shown at the immediate right is $p = 40 - 5x^2$, where p is in dollars and x is measured in millions of units. Note that to increase sales, the price must decrease. The revenue, R, (in millions of dollars) is given by $R = xp = x(40 - 5x^2)$.

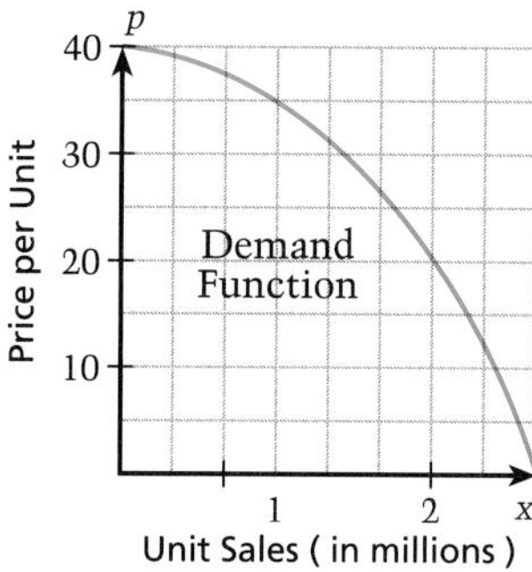

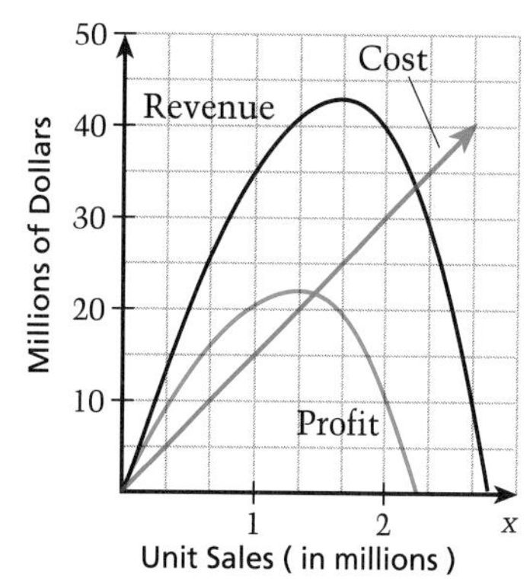

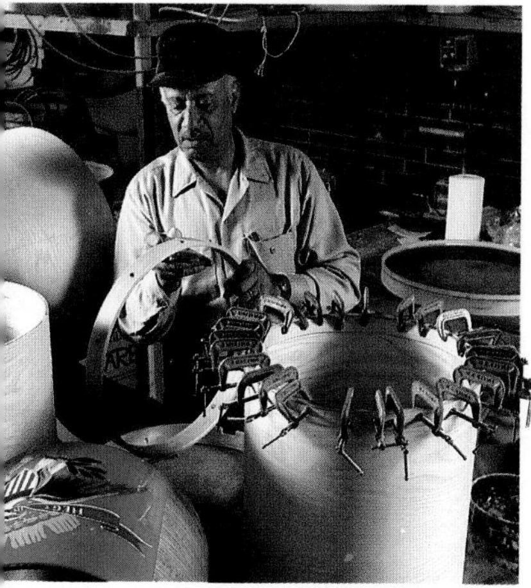

Economic models are used by both large and small businesses, such as this company in Essex, Conn., that handcrafts drums.

Example 5 *Making a Profit*

The cost of producing the product whose demand function is shown above is \$15 per unit. The total cost, C (in millions of dollars), of producing x million units is 15x. The total profit, P (in millions of dollars), for selling these x million units is

$$P = R - C = x(40 - 5x^2) - 15x = -5x^3 + 25x.$$

By selling 1,600,000 units ($x = 1.6$) at \$27.20 each, the company made a profit of \$19,520,000 ($P = 19.52$). Find a smaller unit sales that would have produced the same profit.

Solution Let $P = 19.52$ and solve the resulting equation.

$$-5x^3 + 25x = 19.52 \quad \text{or} \quad -5x^3 + 25x - 19.52 = 0$$

Because one solution is $x = 1.6$, you know that $(x - 1.6)$ is a factor of the left side of the equation. Using synthetic division, you can factor the left side to obtain

$$(x - 1.6)(-5x^2 - 8x + 12.2) = 0.$$

Using the Quadratic Formula, solve $-5x^2 - 8x + 12.2 = 0$. The company could have made the same profit by selling approximately 955,000 units ($x \approx 0.955$) at \$39.09 each. ■

Communicating about ALGEBRA

SHARING IDEAS about the Lesson ≈\$2.2 million, ≈\$32

In Example 5, approximate the maximum profit that the company could have made. What price per unit should the company charge to obtain a maximum profit? ≈\$32 per week

Check Understanding

1. What does it mean for a divisor to divide a dividend *evenly* in a polynomial division problem?
   There is a zero remainder.
2. Describe the steps in a synthetic division problem.
   See page 481.
3. State the Remainder Theorem and tell how it can be used to evaluate functions.
   See page 482 and Example 3.

Communicating about ALGEBRA

COOPERATIVE LEARNING

Have students work together in small groups to prove the Factor Theorem. That is, show the following.

a. If $f(k) = 0$ then $(x - k)$ is a factor of $f(x)$.

b. If $(x - k)$ is a factor of $f(x)$ then $f(k) = 0$.

Hints: **a.** Use the Remainder Theorem to express $f(x)$ divided by $(x - k)$.

b. Express $f(x)$ as a product with the factor $(x - k)$.

Proofs:

a. The Remainder Theorem implies that there is a quotient polynomial, $q(x)$, such that $f(x) = (x - k)q(x) + f(k)$. But if $f(k) = 0$, then $f(x) = (x - k)q(x)$ and $(x - k)$ is a factor of $f(x)$.

b. If $(x - k)$ is a factor of $f(x)$, then there is a quotient polynomial, $q(x)$, such that $f(x) = (x - k)q(x)$. Evaluate $f(k)$.
$f(k) = (k - k)q(x) = 0 \cdot q(x) = 0.$

EXERCISE Notes

ASSIGNMENT GUIDE

Basic/Average: Ex. 1–6, 9–15 odd, 23–33 odd, 39, 43–49 odd, 51–57, 58, 61

Above Average: Ex. 1–6, 9–27 odd, 33–39 odd, 43–50, 57, 58, 61

Advanced: Ex. 1–6, 9–27 odd, 33–39 odd, 43–50, 57–61

Selected Answers
Exercises 1–6, 7–57 odd

More Difficult Exercises
Exercises 43–50, 59–61

Guided Practice

Ex. 1–6 Use these Guided Practice exercises to review the terms and techniques of the lesson. In Exercises 5 and 6, remind students that since $f(1) = 0$, the binomial $(x - 1)$ must be a factor of the polynomial $2x^3 + 6x^2 - 8$.

Independent Practice

Ex. 19–30 Remind students that they must write a 0 for each missing term in the dividend. For example, in Exercise 25, the expression $x^2 + 9$ is missing a linear, or x-, term.

Ex. 7–30 "Fractional form" means to write the remainder as a fraction.

EXERCISES

Guided Practice

CRITICAL THINKING about the Lesson

1. Write the result of the long division at the right in fractional form. $2x^2 + 3x - 1 + \frac{1}{x-1}$

$$\begin{array}{r} 2x^2 + 3x - 1 \\ x - 1\overline{)2x^3 + x^2 - 4x + 2} \\ \underline{2x^3 - 2x^2} \qquad\qquad \\ 3x^2 - 4x \qquad \\ \underline{3x^2 - 3x} \qquad \\ -x + 2 \\ \underline{-x + 1} \\ 1 \end{array}$$

2. Does $x^2 - 3$ divide evenly into

$4x^4 + x^3 - 12x^2 + 3x$?

Justify your answer. No, there is a remainder of $6x$.

3. What kind of divisor is required for synthetic division? Of the form $x \pm k$

4. Write the polynomial divisor, dividend, and quotient that the synthetic division below represents.

$$\begin{array}{r|rrrr} 1 & 2 & 6 & 0 & -8 \\ & & 2 & 8 & 8 \\ \hline & 2 & 8 & 8 & 0 \end{array}$$

$(2x^3 + 6x^2 - 8) \div (x - 1) = 2x^2 + 8x + 8$

5. Evaluate $f(x) = 2x^3 + 6x^2 - 8$ at $x = 1$. 0

6. Factor $2x^3 + 6x^2 - 8$ completely. $2(x - 1)(x + 2)^2$

Independent Practice

In Exercises 7–18, use long division. Write the result in fractional form. See Additional Answers.

7. $(x^2 + 7x - 2) \div (x - 2)$
8. $(2x^2 + x - 1) \div (x + 4)$
9. $(3x^2 + 9x + 1) \div (x + 3)$
10. $(x^2 - 5x + 4) \div (x - 1)$
11. $(4x^4 + 5) \div (x^2 + 1)$
12. $(x^3 + 6x^2 - 5x + 20) \div (x^2 + 5)$
13. $(6x^4 + 2x^3 + 5x^2 + 3x) \div (3x^2 + x - 2)$
14. $(10x^3 + 27x^2 + 14x + 5) \div (x^2 + 2x)$
15. $(2x^3 + 3x^2 - 4x - 7) \div (x^2 - 2)$
16. $(2x^4 + 2x^3 - 10x - 9) \div (x^3 + x^2 - 5)$
17. $(5x^4 + 18x^3 + 10x^2 + 3x) \div (x^2 + 3x)$
18. $(x^4 + 3x^3 - 5x^2 - 8x + 6) \div (x^2 + 2x - 6)$

In Exercises 19–30, use synthetic division. Write the result in fractional form. See Additional Answers.

19. $(x^3 - 7x + 6) \div (x - 2)$
20. $(x^3 - 28x - 48) \div (x + 4)$
21. $(4x^2 + 3x - 3) \div (x + 1)$
22. $(7x^2 + 4x + 3) \div (x + 2)$
23. $(2x^2 + 7x + 8) \div (x - 2)$
24. $(4x^2 - 6x) \div (x - 2)$
25. $(x^2 + 9) \div (x + 3)$
26. $(x^2 + 3) \div (x + 3)$
27. $(10x^4 + 5x^3 + 4x^2 - 9) \div (x + 1)$
28. $(x^4 - 2x^3 - 70x + 20) \div (x - 5)$
29. $(2x^4 - 6x^3 + x^2 - 3x - 3) \div (x - 3)$
30. $(4x^4 + 5x^3 + 2x^2 - 1) \div (x + 1)$

In Exercises 31–36, use the Remainder Theorem to evaluate the function.

31. $f(x) = 4x^2 - 10x - 21$ at $x = 5$ 29

32. $f(x) = 6x^3 - x^2 - 16x$ at $x = 2$ 12

33. $f(x) = 2x^2 - 2x + x + 1$ at $x = 4$ 29

34. $f(x) = 4x^4 + x^2 + 6x + 2$ at $x = -1$ 1

35. $f(x) = 5x^4 + 2x^3 - 20x - 6$ at $x = 2$ 50

36. $f(x) = x^2 - 26x + 7$ at $x = 5$ −98

In Exercises 37–42, find the missing factors. See margin.

37. $x^3 - 8x^2 + 4x + 48 = (x - 4)(\ ?\ )(\ ?\ )$

38. $2x^3 - 14x^2 - 56x - 40 = (x - 10)(\ ?\ )(\ ?\ )$

39. $x^3 + 2x^2 - 5x - 6 = (x + 3)(\ ?\ )(\ ?\ )$

40. $x^3 - 13x^2 + 24x + 108 = (x + 2)(\ ?\ )(\ ?\ )$

41. $6x^3 - 24x^2 - 42x + 60 = 2(x - 5)(\ ?\ )(\ ?\ )$

42. $x^4 + 14x^3 + 51x^2 + 54x = x(x + 9)(\ ?\ )(\ ?\ )$

Geometry **In Exercises 43–45, you are given an expression for the volume of the solid shown. Find an expression for the missing dimension.**

✪ **43.** $V = 3x^3 + 8x^2 - 45x - 50$

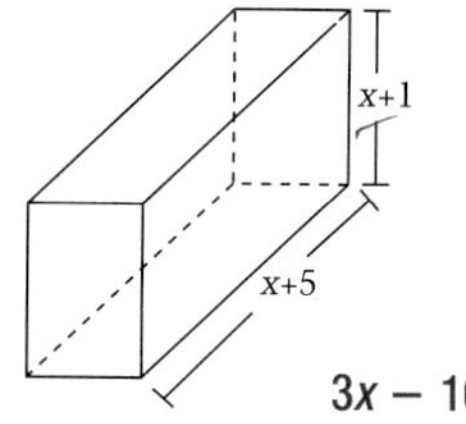

3x − 10

✪ **44.** $V = x^3 + 18x^2 + 80x + 96$

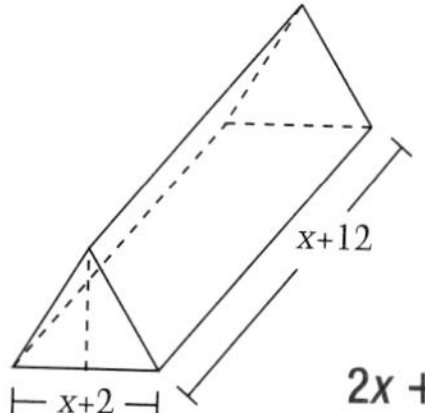

2x + 8

✪ **45.** $V = 2x^3 + 17x^2 + 40x + 25$

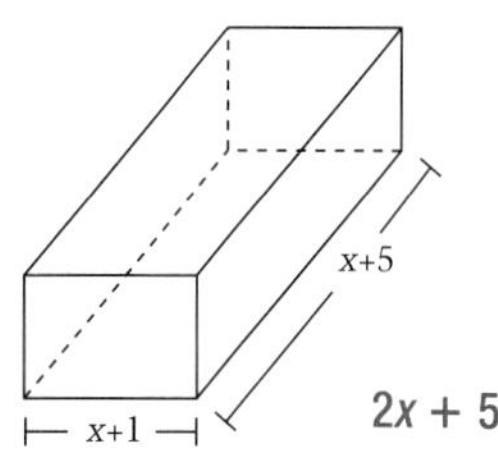

2x + 5

Points of Intersection **In Exercises 46–48, find all points of intersection of the two graphs. One intersection occurs at $x = 1$.**

✪ **46.**

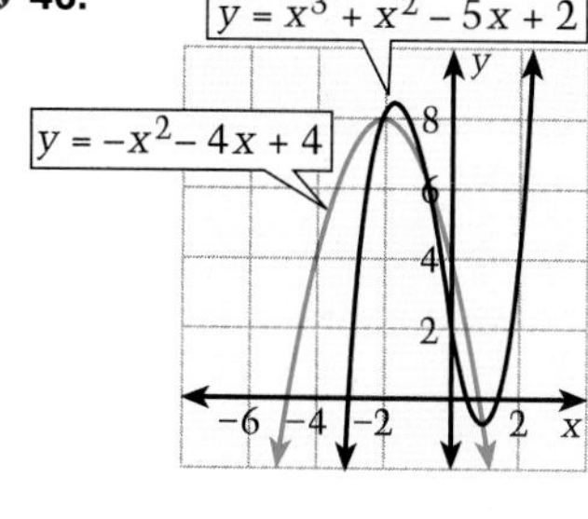

(1, −1), (−1, 7), (−2, 8)

✪ **47.**

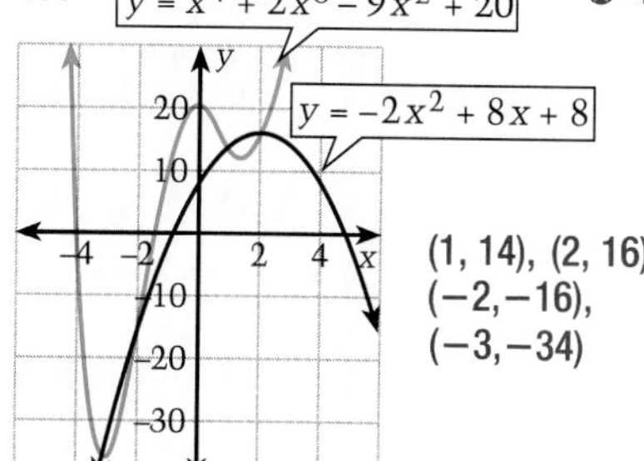

(1, 14), (2, 16), (−2, −16), (−3, −34)

✪ **48.**

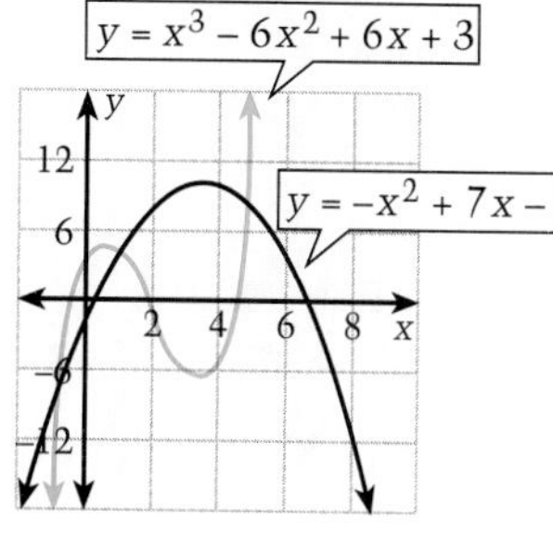

(1, 4), (5, 8), (−1, −10)

✪ **49.** *Company Profit* The demand function for a type of camera is given by the model $p = 100 - 8x^2$ where p is measured in dollars and x is measured in millions of units. The production cost is \$25 per camera. A sale of 2.5 million cameras produced a profit of \$62.5 million. What other amount of cameras could the company sell to make the same profit?
≈0.92 million

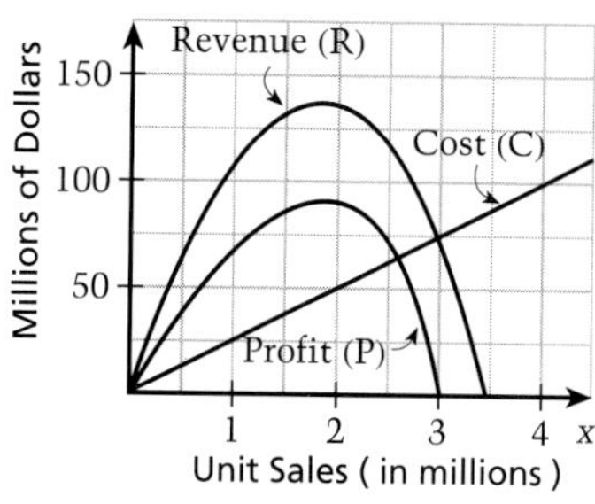

▶ **Ex. 31–36** Students should use synthetic division to evaluate $f(x)$. They can verify the solution by using substitution. Ask students which method they think is more efficient. Ask students which method they prefer to use. Why? (Note that even though synthetic division seems more efficient, some students will probably prefer to use substitution because they have always done so before and are more comfortable with it.)

▶ **Ex. 37–42** Remind students to use the synthetic division value determined by $x - h$. Thus, a factor of $(x + 4)$ indicates division by -4, not 4.

Answers

37. $(x - 6), (x + 2)$

38. $2, (x + 2), (x + 1)$

39. $(x - 2), (x + 1)$

40. $(x - 9), (x - 6)$

41. $3, (x + 2), (x - 1)$

42. $(x + 3), (x + 2)$

▶ **Ex. 46–48** Students can find the points of intersection by setting the two equations equal to each other, simplifying, and factoring.

TECHNOLOGY Suggest that students with graphics calculators use the Trace function to find the points. Have students compare the solutions as a check.

▶ **Ex. 49–50** Students should refer to Example 5 on page 483 as a solution model.

Answers

51. $3(x + 3)(x - 3)$
52. $2(x + 3)(x^2 - 3x + 9)$
53. $2(x - 10)(x + 1)$
54. $9(1 + 2x)(1 - 2x)$
55. $3(4x^2 + 1)(x - 1)$
56. $2(x^2 + 6)(x + 2)$

Enrichment Activities

These activities require students to go beyond lesson goals.

1. Build a cubic polynomial function by choosing any three zeros and multiplying their corresponding linear factors.
   For example, for zeros −5, 2, and 6, you could build the function $f(x) = (x + 5)(x - 2)(x - 6) = x^3 - 3x^2 - 28x + 60$.
2. **TECHNOLOGY** Use a graphics calculator and graph both the factored form and the expanded form of your function on the same coordinate grid. What do you notice?
   The graphs should coincide. This should reinforce the concept that the chosen zeros are solutions of the function.

50. *Company Profit* Suppose the company described in Example 5 on page 483 had sold 200,000 units. How much profit did the company make? Would any fewer unit sales produce the same profit? Would any larger unit sales produce the same profit? Explain. $4.96 million; no, yes, 2.13 million would produce the same profit

Integrated Review

In Exercises 51–56, factor completely. See margin.

51. $3x^2 - 27$ **52.** $2x^3 + 54$ **53.** $2x^2 - 18x - 20$

54. $9 - 36x^2$ **55.** $12x^3 - 12x^2 + 3x - 3$ **56.** $2x^3 + 4x^2 + 12x + 24$

57. *Eating Pasta* In 1990, the retail sales of pasta in the United States totaled about \$1.3 billion. How much was spent on spaghetti and linguini? How much was spent on noodles? ***(Source: National Pasta Association)***

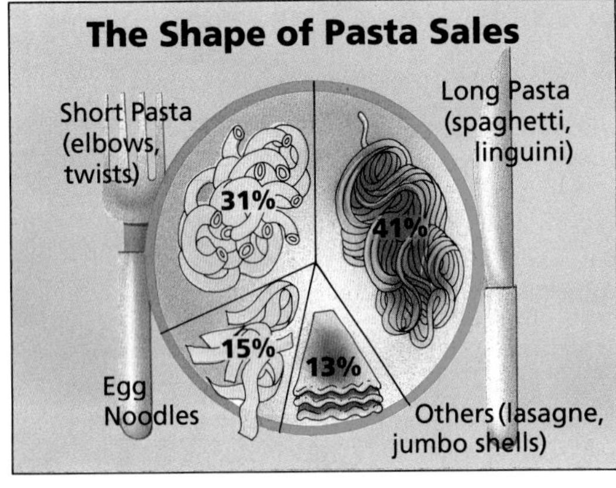

$533 million, $195 million

58. *Eating Salad* 1000 people were asked how often they eat salad. Of those surveyed, how many eat salad once or twice a day? Less than once a month? ***(Source: The Association for Dressings and Sauces)***

200, 40

Exploration and Extension

In Exercises 59 and 60, factor completely. 59. $(x + 2)(x - 4)(x - 3)(x + 3)$ 60. $(x - 1)(x - 2)(x - 4)^2$

59. $x^4 - 2x^3 - 17x^2 + 18x + 72$ **60.** $x^4 - 11x^3 + 42x^2 - 64x + 32$

61. *Counterfeit Coins* You are given ten stacks of silver dollars. Each stack contains ten coins. Nine of the stacks are completely genuine and one is completely counterfeit. You are told that a genuine silver dollar weighs 26.7 grams and a counterfeit silver dollar weighs 27.7 grams. You are given a scale and are allowed to weigh any number of coins, but you are allowed *only one* weighing! How many coins should you weigh to be sure of identifying the counterfeit stack?
55

Mid-Chapter SELF-TEST

Take this test as you would take a test in class. The answers to the exercises are given in the back of the book.

1. $3x^3 + x^2 - 7x + 3$

In Exercises 1–4, perform the indicated operation. (9.1) 2. $7x^4 + 5x^3 + 7x - 2$

1. $(3x^3 - 4x^2 + 7) + (5x^2 - 7x - 4)$

2. $(8x^4 - 3x^2 + 7x) - (x^4 - 5x^3 - 3x^2 + 2)$

3. $3x^2(2x^3 - 4x^2 + 5)$ $6x^5 - 12x^4 + 15x^2$

4. $(3x - 7)(2x + 1)$ $6x^2 - 11x - 7$

In Exercises 5 and 6, sketch the graph of the function. (9.2)

5. $f(x) = x^3$ See margin.

6. $f(x) = (x - 2)^4$ See margin.

In Exercises 7–10, completely factor the polynomial with respect to the integers. (9.3)

7. $16x^2 - 9$ $(4x - 3)(4x + 3)$

8. $6x^3 - 21x^2$ $3x^2(2x - 7)$

9. $4x^2 - 20x + 25$ $(2x - 5)^2$

10. $8x^3 - 125$ $(2x - 5)(4x^2 + 10x + 25)$

In Exercises 11 and 12, find all real-number solutions. (9.3)

11. $12x^2 + 7x = 10$ $-\frac{5}{4}, \frac{2}{3}$

12. $16x^4 - 81 = 0$ $\pm\frac{3}{2}$

In Exercises 13 and 14, use long division to write the quotient. (9.4)

13. $(10x^2 - 11x - 6) \div (2x - 3)$ $5x + 2$

14. $(12x^4 - 5x^2 - 3) \div (3x^2 + 1)$ $4x^2 - 3$

In Exercises 15 and 16, use synthetic division to write the quotient. (9.4) 15. See margin.

15. $(3x^4 + 12x^3 - 5x^2 - 18x + 8) \div (x + 4)$

16. $(x^3 - 2x^2 - 9) \div (x - 3)$ $x^2 + x + 3$

17. Use the remainder theorem to evaluate $f(x) = 3x^3 - 7x^2 + 4x - 2$ when $x = -2$. **(9.4)** -62

18. Completely factor $x^3 - 5x^2 - 2x + 24$, given that $(x - 4)$ is a factor. **(9.4)** $(x - 4)(x - 3)(x + 2)$

19. For 1910 through 1990, the number, $f(t)$, of representatives from New England in the House of Representatives can be modeled by $f(t) = -1.17t + 32.9$, where $t = 1$ represents 1910, $t = 2$ represents 1920, and so on. The number of representatives from California can be modeled by $g(t) = 4.9t + 5.2$. Write a linear function that models the number of representatives from New England and California combined. *(Source:* **Boston Globe)** **(9.1)** $h(t) = 3.73t + 38.1$

20. The playhouse shown at the right has a total volume of $9x^3 + 46x^2 + 59x + 6$ cubic feet. Find the missing dimension of the playhouse. **(9.4)** $x + 3$

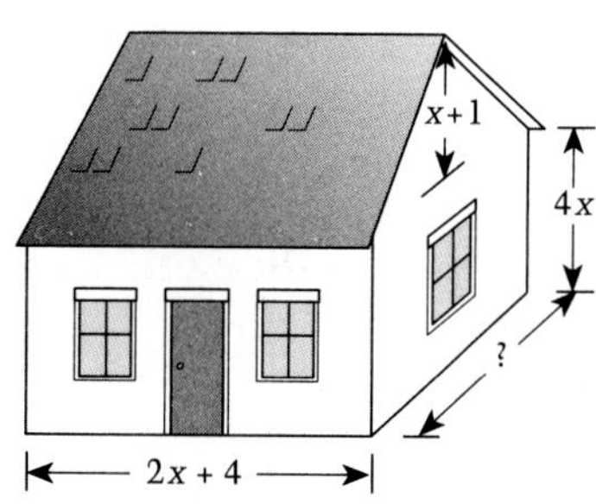

21. The demand function of a product is given by $p = 20 - 2x^2$, where p is the unit price in dollars and x is measured in millions of units. The cost of producing the product is $8 per unit. Write a polynomial for the total profit, P (in millions of dollars), of selling x million units. **(9.1)** $-2x^3 + 12x$

Answers

5.

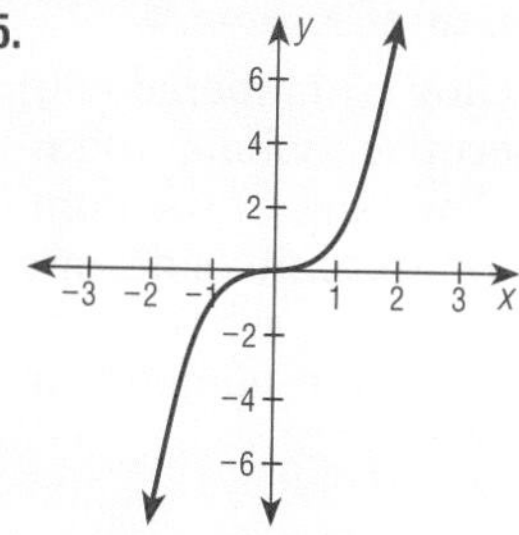

6.

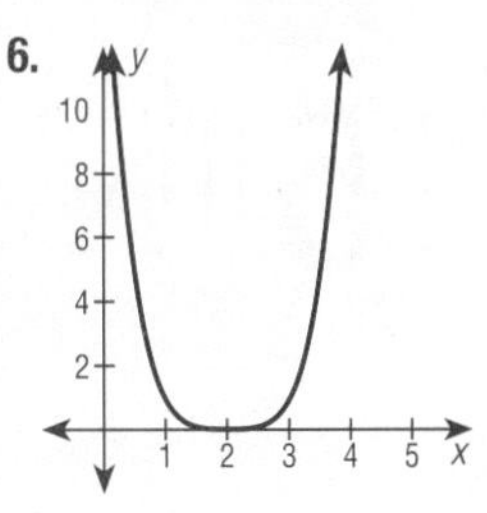

15. $3x^3 - 5x + 2$

Problem of the Day

This cube is numbered with consecutive whole numbers. The sums of the numbers on opposite sides are all the same. What is the sum of all the numbers on the cube?
81

ORGANIZER

Warm-Up Exercises

Use the Factor Theorem and synthetic division to determine which values of k make $(x - k)$ a factor of the function $f(x) = x^5 - 5x^3 + 4x$.

a. $k = 1$ **b.** $k = 3$ **c.** $k = -2$

a. $(x - 1)$ is a factor, b. $(x - 3)$ is not a factor, c. $(x + 2)$ is a factor.

Lesson Resources

Teaching Tools
- Transparency: 8
- Problem of the Day: 9.5
- Warm-up Exercises: 9.5
- Answer Masters: 9.5

Extra Practice: 9.5
Color Transparencies: 51, 52
Technology Handbook: p. 61

LESSON Notes

Example 1

Until students have a complete understanding of the abbreviated notation frequently used to represent the possible rational zeros, it is often helpful to have students list all the possible rational zeros of a polynomial. In Example 1, the possible rational zeros occur at the values −6, −3, −2, −1, 1, 2, 3, 6.

9.5 Finding Rational Zeros

What you should learn:

Goal 1 How to find the rational zeros of a polynomial function

Goal 2 How to use the zeros of a polynomial function to solve real-life problems

Why you should learn it:

You can model many real-life situations with polynomial models that have rational zeros, such as the model for the actual and projected number of home computers in the United States.

Goal 1 Using the Rational-Zero Test

The real-number zeros of a polynomial function are either rational or irrational. (Remember that a rational number is one that can be written as the quotient of two integers.) For instance, the function $f(x) = x^2 - 3x - 4$ has two rational zeros: -1 and 4. On the other hand, the function $f(x) = x^2 - 2$ has no rational zeros—its two zeros, $\pm\sqrt{2}$, are both irrational.

In this lesson, you will learn to use the Rational-Zero Test to compile a list that will include all rational zeros of a polynomial function with integer coefficients. By testing the numbers in the list, you can find the zeros that are rational.

The Rational-Zero Test

Consider a polynomial $f(x) = a_nx^n + \cdots + a_1x + a_0$ with *integer* coefficients. Every rational zero of f has the form

$$\frac{p}{q} = \frac{\text{Factor of constant term, } a_0}{\text{Factor of leading coefficient, } a_n}.$$

Example 1 Using the Rational-Zero Test

Find the rational zeros of $f(x) = x^4 - x^3 - 5x^2 + 3x + 6$.

Solution The leading coefficient is 1, and the constant term is 6. Thus the rational zeros are of the following form.

$$\text{Possible rational zeros: } \frac{\pm 1, \pm 2, \pm 3, \pm 6}{\pm 1}$$

Using synthetic division and the Remainder Theorem, you can determine that two of the *possible* rational zeros are *actual* zeros: $x = -1$ and $x = 2$. This implies that $(x + 1)$ and $(x - 2)$ are factors of $f(x)$.

$$f(x) = (x + 1)(x - 2)(x^2 - 3)$$

From this factorization, you can see that f has only two rational zeros. (The other two zeros, $\pm\sqrt{3}$, are irrational.) The graph of f is shown at the left. ■

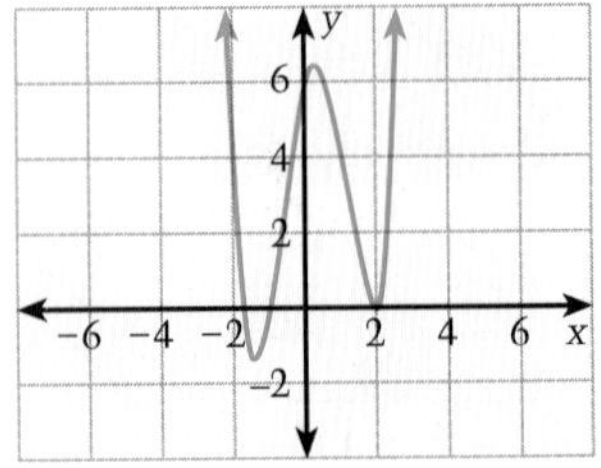

In Example 1, synthetic division was used to test each rational zero. Remember that when you find a zero, you can also find the factorization associated with the zero by the Factor Theorem. Here is what we mean.

$$\begin{array}{r|rrrrr}
-1 & 1 & -1 & -5 & 3 & 6 \\
 & & -1 & 2 & 3 & -6 \\ \hline
 & 1 & -2 & -3 & 6 & 0
\end{array}$$

Test −1 as a zero.

$(x + 1)(x^3 - 2x^2 - 3x + 6)$

$$\begin{array}{r|rrrr}
2 & 1 & -2 & -3 & 6 \\
 & & 2 & 0 & -6 \\ \hline
 & 1 & 0 & -3 & 0
\end{array}$$

Test 2 as a zero of $x^3 - 2x^2 - 3x + 6$.

$(x + 1)(x - 2)(x^2 - 3)$

When the leading coefficient is not 1, the list of possible rational zeros can increase dramatically. In such cases the search can be shortened by making a rough sketch of the function—either by hand or on a graphing calculator.

Finding the first zero is often the hardest part. After that, the search is simplified by using the lower-degree polynomial obtained in synthetic division.

Example 2 Using the Rational-Zero Test

Find all the real zeros of the function

$$f(x) = 10x^3 - 15x^2 - 16x + 12.$$

Solution Since the leading coefficient is 10 and the constant term is 12, you have a long list of possible rational zeros.

$$\textit{Possible rational zeros: } \frac{\pm 1, \pm 2, \pm 3, \pm 4, \pm 6, \pm 12}{\pm 1, \pm 2, \pm 5, \pm 10}$$

With so many possibilities (32, in fact), it is worth your time to sketch the graph of the function. From the graph, it looks like three reasonable choices would be $x = -\frac{6}{5}$, $x = \frac{1}{2}$, and $x = 2$. Testing these by synthetic division shows that only $x = 2$ works.

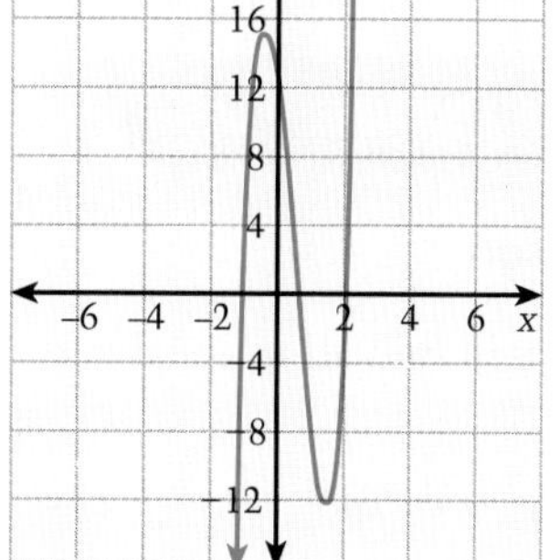

$$\begin{array}{r|rrrr}
2 & 10 & -15 & -16 & 12 \\
 & & 20 & 10 & -12 \\ \hline
 & 10 & 5 & -6 & 0
\end{array}$$

From this result, a factored form is

$$f(x) = (x - 2)(10x^2 + 5x - 6).$$

Using the Quadratic Formula, you can determine that the other two zeros are irrational numbers.

$$x = \frac{-5 + \sqrt{265}}{20} \approx 0.5639 \quad \text{and} \quad x = \frac{-5 - \sqrt{265}}{20} \approx -1.0639$$

■

Example 2

From the graph of $f(x)$, you can approximate where the curve crosses the x-axis (that is, where the zeros are). This information helps to narrow the investigation of the possible rational zeros. In particular, you can easily determine that $f(-2) < 0$ and $f(-1) > 0$, so $f(x)$ crosses the x-axis between −2 and −1. Similarly, $f(x)$ crosses the x-axis between 0 and 1. When you discover that $f(2) = 0$, use synthetic division to obtain the resulting factorization of $f(x)$.

Example 3

From the Rational-Zero Test, the possible roots of the function are −175, −105, −75, −35, −25, −21, −15, −7, −5, −3, −1, 1, 3, 5, 7, 15, 21, 25, 35, 75, 105, and 175. A rough sketch of $f(x)$ indicates that the zero is positive and less than 10 so only the values 1, 3, 5, and 7 should be checked.

Extra Examples

Here are additional examples similar to **Examples 1–2.**

1. Use the Rational-Zero Test to find the rational zeros of each function.
   a. $f(x) = 2x^3 + x^2 + 2x + 1$
   b. $g(x) = 2x^4 + x^3 + 3x - 18$
   a. $x = -\frac{1}{2}$, **b.** $x = -2, \frac{3}{2}$
2. Use the Rational-Zero Test to find the real-valued zeros of each function.
   a. $f(x) = 15x^4 - 2x^3 - 46x^2 + 6x + 3$
   b. $g(x) = 2x^3 + x^2 + 10x + 5$
   a. $-\frac{1}{5}, \frac{1}{3}, \pm\sqrt{3}$;
   b. $-\frac{1}{2}$; other roots are imaginary

Check Understanding

1. If the leading coefficient and the constant term of $f(x)$ are 3 and 4, respectively, how many possible rational zeros must be considered to find the zeros of $f(x)$?
   12; $\pm 1, \pm 2, \pm 4, \pm\frac{1}{3}, \pm\frac{2}{3}, \pm\frac{4}{3}$
2. Describe when it is advisable to sketch a graph of $f(x)$ when using the Rational-Zero Test.
   When the leading coefficient is not 1

Communicating *about* ALGEBRA

Have students graph the functions to check their solutions. For example, in graph A, the function crosses the x-axis at $x = -2$, $x = -1$, and $x = 3$. In graph B, the graph crosses the x-axis at one point between 1 and 2, so it has one real zero. By the Rational Root Theorem, the zero is not rational. Using the Trace feature of a graphics calculator, you can estimate the real root as ≈ 1.286.

EXTEND *Communicating*
COOPERATIVE LEARNING
Have students work with a partner to show that numbers such as $\sqrt{7}$ or $\sqrt[3]{7}$ are irrational by examining the solutions to the quadratic equation $x^2 - 7 = 0$ or the equation $x^3 - 7 = 0$. By the Rational Root Theorem, the possible rational roots are ± 1 and ± 7. Testing will show that none of these values are roots of the equations. Thus, $\sqrt{7}$ and $\sqrt[3]{7}$ are real numbers that are not rational and must be irrational.

Goal 2 Using the Rational-Zero Test in Real Life

Real Life
Computer Purchases

The first home computer, the Altair, was introduced in 1975. It did not sell well, however, because it came as a kit. In 1977, two American students, Steve Wozniak and Steve Jobs (shown above) founded the highly successful Apple Computer Company. Since then, millions of Americans have purchased home computers.

Example 3 *Solving a Polynomial Equation*

For 1985 through 1995, the actual and projected number, C (in millions), of home computers sold in the United States can be modeled by

$$C = 0.0092(t^3 + 8t^2 + 40t + 400)$$

where $t = 0$ represents 1990. During which year are 8.51 million computers projected to be sold? ***(Source: Dataquest)***

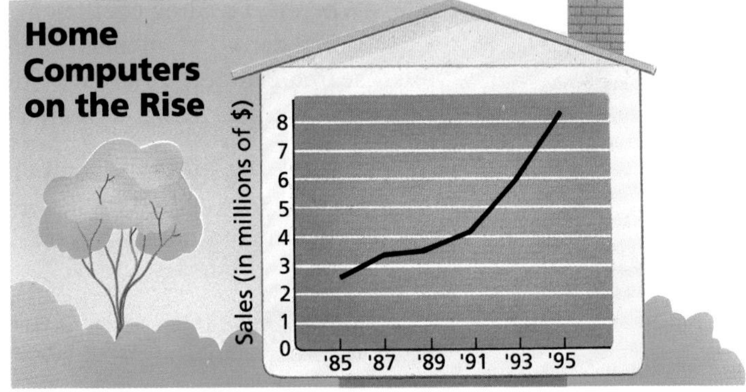

Solution Let $C = 8.51$ and solve the resulting equation.

$$0.0092(t^3 + 8t^2 + 40t + 400) = 8.51$$
$$t^3 + 8t^2 + 40t + 400 = 925$$
$$t^3 + 8t^2 + 40t - 525 = 0$$

From the Rational-Zero Test, you can determine that 5 is a solution. To check, substitute $t = 5$ into the original model.

$$C = 0.0092(5^3 + 8(5^2) + 40(5) + 400) = 8.51$$

Thus the sales are projected to be 8.51 million in 1995. ■

Communicating *about* ALGEBRA

SHARING IDEAS about the Lesson

Use the Rational-Zero Test to help find all rational zeros of the function.

A. $f(x) = x^3 - 7x - 6$ −2, −1, 3

B. $f(x) = 3x^3 - 9x^2 + 27x - 27$ None

EXERCISES

Guided Practice

CRITICAL THINKING about the Lesson

1. List the possible rational zeros given by the Rational-Zero Test for $f(x) = 3x^3 - 6x^2 + 7x - 2$. $\frac{\pm 1, \pm 2}{\pm 1, \pm 3}$
2. Factor $x^3 - 3x^2 - 6x + 8$ completely. $(x-1)(x-4)(x+2)$
3. Factor $x^3 + 4x^2 - x - 4$ completely. $(x-1)(x+1)(x+4)$

In Exercises 4–6, find all real zeros of the polynomial function.

4. $y = 4x^3 - 12x^2 - x + 15$

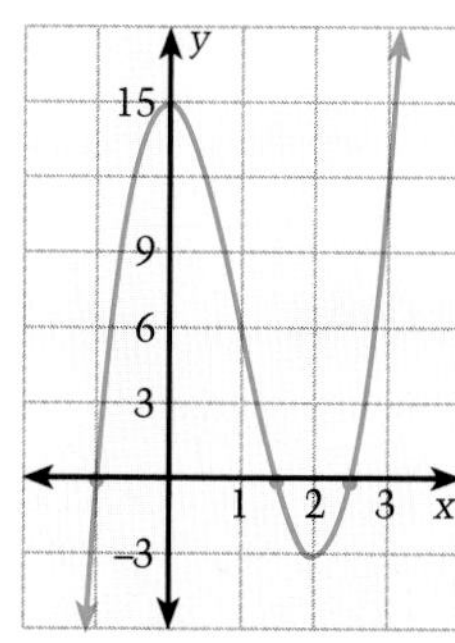

$-1, \frac{3}{2}, \frac{5}{2}$

5. $y = -4x^3 + 15x^2 - 8x - 3$

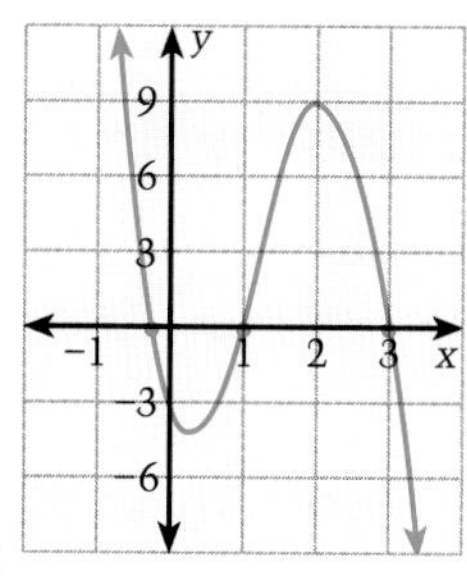

$-\frac{1}{4}, 1, 3$

6. $y = -3x^3 + 20x^2 - 36x + 16$

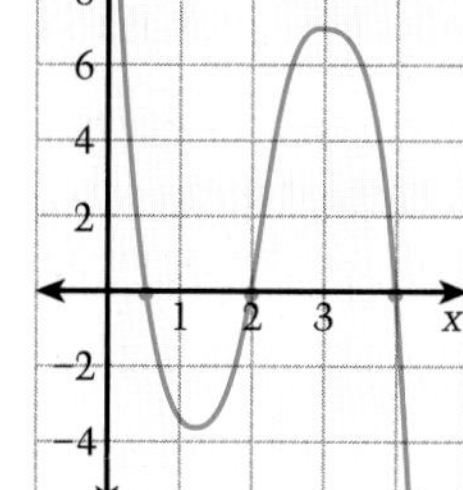

$\frac{2}{3}, 2, 4$

Independent Practice

In Exercises 7–10, list the possible rational zeros given by the Rational-Zero Test.

7. $f(x) = x^3 - 2x^2 + x - 3$ $\frac{\pm 1, \pm 3}{\pm 1}$
8. $g(x) = 2x^3 + 5x^2 - 6x - 10$ $\frac{\pm 1, \pm 2, \pm 5, \pm 10}{\pm 1, \pm 2}$
9. $f(x) = 2x^5 + x^2 + 16$ $\frac{\pm 1, \pm 2, \pm 4, \pm 8, \pm 16}{\pm 1, \pm 2}$
10. $h(x) = 3x^3 - 5x - 24$ $\frac{\pm 1, \pm 2, \pm 3, \pm 4, \pm 6, \pm 8, \pm 12, \pm 24}{\pm 1, \pm 3}$

In Exercises 11–20, find all rational zeros of the function.

11. $g(x) = x^3 + 2x^2 - 11x - 12$ $-4, -1, 3$
12. $f(x) = x^3 + x^2 - 10x + 8$ $-4, 1, 2$
13. $f(x) = x^3 - 8x^2 - 23x + 30$ $-3, 1, 10$
14. $g(x) = x^3 - 9x^2 + 15x + 25$ $-1, 5$
15. $f(x) = x^3 - 7x^2 + 2x + 40$ $-2, 4, 5$
16. $g(x) = x^3 - x^2 - 21x + 45$ $-5, 3$
17. $f(x) = 2x^3 + 4x^2 - 2x - 4$ $-2, \pm 1$
18. $g(x) = 3x^3 + 12x^2 + 3x - 18$ $-3, -2, 1$
19. $h(x) = 2x^3 - 5x^2 - 14x + 8$ $-2, \frac{1}{2}, 4$
20. $f(x) = 3x^3 - 3x + x^2 - 1$ $\pm 1, -\frac{1}{3}$

In Exercises 21–30, find all real zeros of the function.

27. $-8, -1, \pm\sqrt{3}$ 28. $-2, -\frac{1}{5}, \pm\sqrt{10}$

21. $g(x) = x^3 + x^2 - 2x - 2$ $-1, \pm\sqrt{2}$
22. $f(x) = x^3 + 6x^2 - 6x - 36$ $-6, \pm\sqrt{6}$
23. $h(x) = x^3 + 2x^2 - 9x - 18$ $\pm 3, -2$
24. $f(x) = x^3 + 9x^2 - 4x - 36$ $\pm 2, -9$
25. $g(x) = 2x^3 - x^2 - 32x + 16$ $\pm 4, \frac{1}{2}$
26. $f(x) = 2x^3 + x^2 - 50x - 25$ $\pm 5, -\frac{1}{2}$
27. ✪ $h(x) = x^4 + 9x^3 + 5x^2 - 27x - 24$
28. ✪ $g(x) = 5x^4 + 11x^3 - 48x^2 - 110x - 20$
29. ✪ $f(x) = x^5 + 5x^4 + 2x^3 - 22x^2 - 35x - 15$ $-3, -1, \pm\sqrt{5}$
30. ✪ $f(x) = x^5 - x^4 - 7x^3 + 7x^2 + 12x - 12$ $1, \pm 2, \pm\sqrt{3}$

EXERCISE Notes

ASSIGNMENT GUIDE

Basic/Average: Ex. 1–6, 9–17 odd, 27–31 odd, 32, 34, 43–47 odd, 51–52

Above Average: Ex. 1–6, 9–15 odd, 23–31 odd, 32–35, 45–51 odd, 52, 53

Advanced: Ex. 1–6, 9–15 odd, 23–31 odd, 32–34, 49–53, 55, 56

Selected Answers
Exercises 1–6, 7–51 odd

Use **Mixed Review** as needed.

✪ **More Difficult Exercises**
Exercises 27–34, 39, 40, 53–56

Guided Practice

▶ **Ex. 1–3** COOPERATIVE LEARNING Seeking the rational zeros of a polynomial function presents an ideal opportunity for students to work in small groups where the task of identifying the zeros can be shared among the members of the group. Discrepancies over which values are zeros can be discussed and resolved in the group.

▶ **Ex. 4–6** Encourage students to use the graphs to visualize possible values for the zeros.

Independent Practice

▶ **Ex. 7–10** Students can list the possible rational zeros as in Example 2 or they can list each value separately.

▶ **Ex. 11–20** Encourage students to use synthetic division to find the first root.

▶ **Ex. 21–30** TECHNOLOGY Have students use graphics calculators to test for possible rational zeros.

▶ **Ex. 31–34** A solution model is provided in Example 3 on page 490.

▶ **Ex. 35–40** Students who do not use graphing calculators should be reminded to review the graphing techniques presented in Lesson 9.2.

Answers

41. $x^3 - x^2 + 3x + 4$
42. $x^2 + 4x - 3$
43. $x^2 - x + 1$
44. $x^3 - x - 1$
45. $x^5 - x^2 + 3$
46. $x^4 - x + 5$

31. *Portable Personal Computers* For 1990 through 2000, the predicted sales, S (in millions of dollars), of portable personal computers in the United States can be modeled by

$$S = -2.6t^4 + 80t^3 - 219t^2 + 1710t + 3000$$

where $t = 0$ represents 1990. In what year are the sales predicted to be approximately $52,200 million? 2000

32. *Defense-Related Jobs* In 1991, the United States Department of Defense announced that it was beginning to cut back military and defense-related jobs. The planned cutback was to take place over a five-year period. The number of jobs remaining, J (in thousands), can be modeled by

$$J = 12t^3 - 100t^2 + 5100$$

where $t = 1$ represents 1991. In what year was the number of jobs remaining projected to be 4,100,000? 1995

33. *United States Exports* For 1980 through 1994, the total exports, E (in billions of dollars), can be modeled by

$$E = -0.21t^3 + 6.32t^2 - 27.9t + 239$$

where $t = 0$ represents 1980. In what year were the total exports about $312.76 billion? ***(Source: U.S. Bureau of the Census)*** 1988

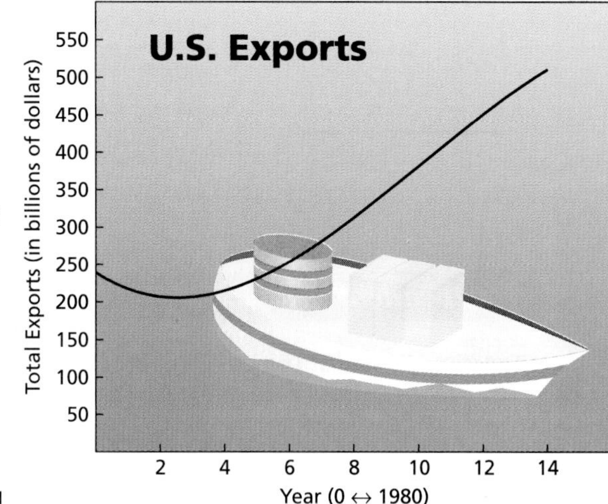

34. *Education Donations* For 1983 through 1993, the amount of private donations, D (in millions of dollars), allocated to education can be modeled by

$$D = -2.98t^3 + 92t^2 + 6159$$

where $t = 3$ represents 1983. In what year did about $14,258 million of charitable contributions go to education? ***(Source: AAFRC Trust for Philanthropy)*** 1992

Integrated Review

In Exercises 35–40, sketch the graph of the function. See Additional Answers.

35. $g(x) = x^4 + 9$ **36.** $f(x) = x^2 - 3x + 7$ **37.** $h(x) = x^3 - x^2 + 2x - 4$

38. $h(x) = (x + 4)^3 + 2$ **39.** $g(x) = x^4 - 2x^3 + x - 3$ **40.** $f(x) = x^4 - 2x^3 - x - 3$

In Exercises 41–46, perform the indicated division. See margin.

41. $(x^4 - 2x^3 + 4x^2 + x - 4) \div (x - 1)$ **42.** $(x^3 + 6x^2 + 5x - 6) \div (x + 2)$

43. $(x^3 - 4x^2 + 4x - 3) \div (x - 3)$ **44.** $(x^4 + 3x^3 - x^2 - 4x - 3) \div (x + 3)$

45. $(x^6 + x^5 - x^3 - x^2 + 3x + 3) \div (x + 1)$ **46.** $(x^5 - 2x^4 - x^2 + 7x - 10) \div (x - 2)$

In Exercises 47–50, one of the zeros of the polynomial is $x = 2$. Use this fact to completely factor the polynomial.

47. $x^3 - 2x^2 - x + 2$ $(x - 2)(x + 1)(x - 1)$

48. $x^3 + 3x^2 - 4x - 12$ $(x - 2)(x + 2)(x + 3)$

49. $x^3 - 4x^2 + x + 6$ $(x - 2)(x - 3)(x + 1)$

50. $x^3 + 4x^2 - 39x + 54$ $(x - 2)(x + 9)(x - 3)$

51. *Powers of Ten* Which best estimates the length (in millimeters) of the world's smallest bird as shown at the right? b

a. 5.5×10^0 **b.** 5.5×10^1
c. 5.5×10^2 **d.** 5.5×10^3

52. *Powers of Ten* Which best estimates the weight (in ounces) of a new "Number 2" pencil? c

a. 2.5×10^{-2} **b.** 2.5×10^{-1}
c. 2.5×10^0 **d.** 2.5×10^1

Exploration and Extension

In Exercises 53–56, find a polynomial function that has only the given zeros. See below.

53. $2, \pm\sqrt{3}$

54. $\pm\sqrt{2}, 1, 2$

55. $0, 1, \pm i$

56. $1, 2, 3, 4$

Mixed REVIEW

3. $(3x - 5)(5x - 2)$ 5. $\pm 3, \pm 3i$ 13. $\ln \frac{x^5\sqrt{y}}{2}$
6. See Additional Answers. 16. $y = \pm\sqrt{e^{x+2}}$

1. Multiply: $(2x - 3)(4x + 5)$. **(9.1)** $8x^2 - 2x - 15$

2. Factor $4x^2 - 25$. **(9.3)** $(2x - 5)(2x + 5)$

3. Factor $15x^2 - 31x + 10$. **(9.3)**

4. Sketch $f(x) = (x - 4)(x - 1)(x + 3)$. **(9.2)** See Additional Answers.

5. Solve $x^4 - 81 = 0$. **(9.3)**

6. Sketch $f(x) = 2^{x-3}$. **(8.1)**

7. Sketch $f(x) = \ln(x - 3)$. **(8.2)** See Additional Answers.

8. Write $e^{x-6} = 8$ in logarithmic form. **(8.4)** $\ln 8 = x - 6$

9. Evaluate $\log_3 27$. **(8.2)** 3

10. Evaluate $\ln e^9$. **(8.6)** 9

11. Find the y-intercept of $y = \log_2(x + 8)$. **(8.2)** 3

12. Find the y-intercept of $y = -3 + \log_4 x$. **(8.2)** There is none.

13. Condense $5 \ln x + \frac{1}{2} \ln y - \ln 2$. **(8.3)**

14. Solve $\log_{10}(x - 3) = 2$. **(8.2)** 103

15. Solve $e^{3x-4} = 7$. **(8.6)** $\frac{\ln 7 + 4}{3}$, or ≈ 1.982

16. Solve for y: $2 \ln y - x = 2$. **(8.6)**

17. Find the y-intercept of $f(x) = 5x^4 + 2x^3 - 7x + 8$. **(2.3)** 8

18. Find the slope of the line through $(-4, -1)$ and $(2, -8)$. **(2.2)** $-\frac{7}{6}$

19. Solve $\begin{cases} 3x + 2y = 1 \\ 7x + 3y = -2 \end{cases}$. **(3.2)** $\left(-\frac{7}{5}, \frac{13}{5}\right)$

20. Sketch the solution of $-4 < 2x + 5 < 2$. **(2.5)** See Additional Answers.

53. $f(x) = (x - 2)(x^2 - 3)$ 54. $f(x) = (x - 1)(x - 2)(x^2 - 2)$ 55. $f(x) = x(x - 1)(x^2 + 1)$
56. $f(x) = (x - 1)(x - 2)(x - 3)(x - 4)$

Enrichment Activities

These activities require students to go beyond lesson goals.

Not all roots, or zeros, of polynomial functions are rational. Not all polynomial functions have integer coefficients.

1. Determine polynomial functions in factored form that have the following zeros.
   a. a cubic polynomial $f(x)$ with zeros $-\sqrt{7}, -2, 6$
   b. a quartic function $g(x)$ with zeros $\frac{1}{2}, \frac{3}{4}, \pm i$

Answers vary.
a. $f(x) = (x + \sqrt{7})(x + 2)(x - 6)$
b. $g(x) = \left(x - \frac{1}{2}\right)\left(x - \frac{3}{4}\right)(x^2 + 1)$

2. TECHNOLOGY Write the functions you found in Activity 1 in expanded form and use a graphics calculator to graph them. How are the zeros of the function related to the x-intercepts of the graphs?

$f(x) = x^3 - (4 - \sqrt{7})x^2 - 4(3 + \sqrt{7})x - 12\sqrt{7}$

$g(x) = x^4 - \frac{5}{4}x^3 + \frac{11}{8}x^2 - \frac{5}{4}x + \frac{3}{8}$

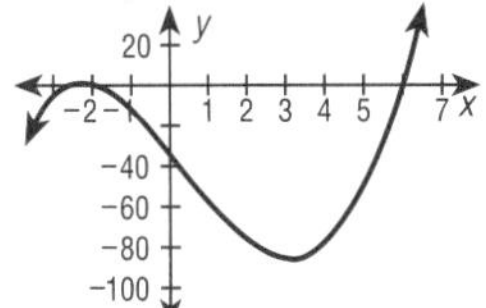

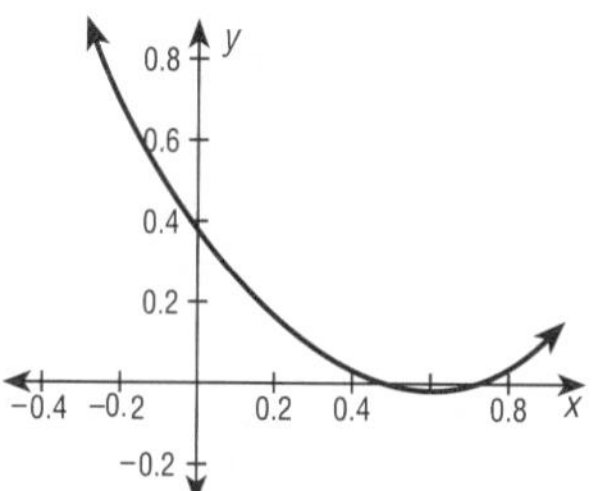

Problem of the Day

In an international competition among 24 students, three languages were spoken: French, Spanish, and English. Eight of the students who spoke English did not speak Spanish. Eight of the students who spoke Spanish did not speak English. Of the 8 students who spoke French, 3 spoke only French. Six students spoke only English, and 1 student spoke all three languages. How many spoke Spanish and English, but not French?
4

ORGANIZER

Warm-Up Exercises

1. Find the zeros of the function.
$f(x) = 3x^3 - 10x^2 - 23x - 10$
$-1, -\frac{2}{3}, 5$
2. Find the factors of the polynomial.
$3x^3 - 10x^2 - 23x - 10$
$(x + 1), (3x + 2), (x - 5)$
3. Evaluate $f(-1)$, $f\left(-\frac{2}{3}\right)$, and $f(5)$.
$f(x) = 3x^3 - 10x^2 - 23x - 10$
$f(-1) = 0$, $f\left(-\frac{2}{3}\right) = 0$, and $f(5) = 0$.

Lesson Resources

Teaching Tools
 Problem of the Day: 9.6
 Warm-up Exercises: 9.6
 Answer Masters: 9.6
Extra Practice: 9.6
Technology Handbook: p. 65

9.6 Connections: Zeros, Factors, and Solutions

What you should learn:

Goal 1 How to look at zeros of polynomial functions, factors of polynomials, and solutions of polynomial equations to see their connections

Goal 2 How to use polynomials to solve real-life problems

Why you should learn it:

You can use polynomial models for many real-life situations, such as the model for the volume of the blocks of concrete found in the remains of Caesarea harbor.

Goal 1 Viewing Polynomials in Three Ways

In this chapter, we have talked about polynomials, polynomial functions, and polynomial equations. These are closely related.

Zeros, Factors, and Solutions

Let $f(x) = a_nx^n + \cdots + a_1x + a_0$ be a polynomial function. The following statements are equivalent.

1. k is a zero of the polynomial function f.
2. $(x - k)$ is a factor of the polynomial $f(x)$.
3. k is a solution of the polynomial equation $f(x) = 0$.

These three statements are equivalent for *any real* or *complex* number. For *real* numbers, a fourth equivalent statement can be added: $(k, 0)$ is an x-intercept of the graph of f.

Example 1 *Zeros, Factors, Solutions, and x-Intercepts*

a. The polynomial function $f(x) = x^3 - x^2 - 4x + 4$ has three zeros: $-2, 1, 2$.

$$f(-2) = (-2)^3 - (-2)^2 - 4(-2) + 4 = -8 - 4 + 8 + 4 = 0$$
$$f(1) = (1)^3 - (1)^2 - 4(1) + 4 = 1 - 1 - 4 + 4 = 0$$
$$f(2) = (2)^3 - (2)^2 - 4(2) + 4 = 8 - 4 - 8 + 4 = 0$$

b. The polynomial $x^3 - x^2 - 4x + 4$ factors as the product of three linear factors.

$$x^3 - x^2 - 4x + 4 = (x + 2)(x - 1)(x - 2)$$

c. The polynomial equation $x^3 - x^2 - 4x + 4 = 0$ has three solutions: -2, 1, and 2. (Check them!)

d. Because all three zeros of f are real, they show up as x-intercepts of the graph of f.

$(-2, 0)$, $(1, 0)$, and $(2, 0)$

(Complex zeros of a polynomial function do not show up as x-intercepts of the graph of the function.) ■

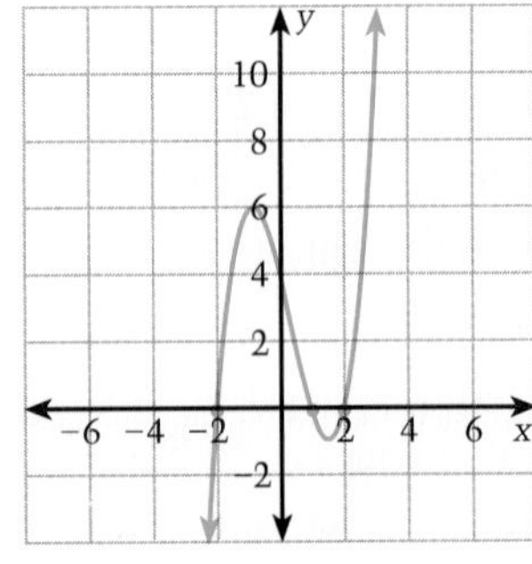

The following theorem is called the **Fundamental Theorem of Algebra.** Its beauty lies in the simplicity of its statement and its strong, logical appeal.

Carl Friederich Gauss

> **The Fundamental Theorem of Algebra**
> Counting complex and repeated solutions, an nth-degree polynomial equation has *exactly n* solutions.

Note: The Fundamental Theorem of Algebra is often stated as "Any polynomial equation of degree 1 or greater has at least one complex number solution." This version is equivalent to the one stated above. (Remember that each real number is also a complex number.)

The Fundamental Theorem of Algebra was proved by the great German mathematician Carl Friederich Gauss (1777–1855) when he was only 22 years old. The proof of this theorem marked the end of an era for algebra. The theorem told mathematicians that the set of complex numbers contains all possible solutions of polynomial equations. After 1799, it was unnecessary (at least for solving polynomial equations) to hunt for new types of numbers.

Example 2 *Demonstrating the Fundamental Theorem of Algebra*

a. The first-degree polynomial equation $x - 4 = 0$ has exactly *one* solution: $x = 4$.

b. Counting repeated solutions, the second-degree polynomial equation

$$x^2 - 6x + 9 = 0$$
$$(x - 3)(x - 3) = 0$$

has exactly *two* solutions: $x = 3$ and $x = 3$.

c. The third-degree polynomial equation

$$x^3 + 4x = 0$$
$$x(x - 2i)(x + 2i) = 0$$

has exactly *three* solutions: $x = 0$, $x = 2i$, and $x = -2i$.

d. The fourth-degree polynomial equation

$$x^4 - 1 = 0$$
$$(x - 1)(x + 1)(x - i)(x + i) = 0$$

has exactly *four* solutions: $x = 1$, $x = -1$, $x = i$, and $x = -i$. ■

LESSON Notes

Example 1

Point out that the factorization of the function

$f(x) = (x + 2)(x - 1)(x - 2)$

can be used to quickly evaluate $f(x)$ at -2, 1, and 2 in Part c of this example. Have students indicate why this form of $f(x)$ is especially easy to evaluate.

Example 2

Observe that the Fundamental Theorem of Algebra does not indicate how to find the solutions of an nth-degree polynomial, only that there are exactly n solutions (perhaps counting some of them more than once as in part b of this example). The techniques developed in this chapter remain useful approaches to finding solutions of polynomial equations.

Example 3

Remind students that a quick sketch of the function $f(x)$ helps to narrow down the number of possible zeros before using synthetic division.

Ask students to multiply the following complex conjugates. Will the product always be a real number?

a. $3i$ and $-3i$

b. $2 + i$ and $2 - i$

c. $-5 + 8i$ and $-5 - 8i$

a. 9, **b.** 5, **c.** 89; yes

Example 4

There are 28 possible solutions! A sketch of the equation will be useful in narrowing the choices from the list. Ask students to provide a rationale for how they would narrow the list of possible solutions.

Extra Examples

Here are additional examples similar to **Examples 1–3.**

1. Suppose the zeros of $f(x)$, a polynomial function of degree 4, are -3, 5, i, and $-i$.
   a. What are the factors of the polynomial expression that defines $f(x)$?
   b. What are the x-intercepts of the graph of $f(x)$?
   c. Evaluate $f(i)$ and $f(-3)$.
   a. $(x + 3)$, $(x - 5)$, $(x - i)$, and $(x + i)$
   b. $(-3, 0)$ and $(5, 0)$
   c. $f(i) = 0$ and $f(-3) = 0$
2. Find the zeros of these polynomial functions.
   a. $f(x) = 3x^3 - 4x^2 + 75x - 100$
   b. $g(x) = x^4 + 5x^2 + 4$
   a. $\frac{4}{3}, -5i, 5i$ b. $\pm i, \pm 2i$

Check Understanding

1. State the Fundamental Theorem of Algebra.
   See page 495.
2. Is every zero of a polynomial function also an x-intercept? Explain.
   No, complex zeros are not x-intercepts.
3. Is every solution of a polynomial equation complex-valued?
   Yes, see page 495.
4. Is the Linear Factorization Theorem equivalent to the Fundamental Theorem of Algebra?
   Yes, see page 496.

Because each solution of a polynomial equation corresponds to a linear factor of the polynomial, the Fundamental Theorem of Algebra can also be stated in terms of factors of a polynomial.

Linear Factorization Theorem

A polynomial of degree n has exactly n linear factors

$$f(x) = a_n(x - k_1)(x - k_2)\cdots(x - k_n)$$

where $k_1, k_2, \ldots, k_n$ are complex numbers and a_n is the leading coefficient of $f(x)$.

Example 3 *Finding the Zeros of a Polynomial Function*

Write $f(x) = x^5 + x^3 + 2x^2 - 12x + 8$ as the product of linear factors and list all of its zeros.

Solution The possible rational zeros are ± 1, ± 2, ± 4, and ± 8. Synthetic division produces the following.

1	1	0	1	2	−12	8	
		1	1	2	4	−8	
	1	1	2	4	−8	0	*1 is a zero.*

1	1	1	2	4	−8	
		1	2	4	8	
	1	2	4	8	0	*1 is a repeated zero.*

−2	1	2	4	8	
		−2	0	−8	
	1	0	4	0	*−2 is a zero.*

From the results of the synthetic division, you can write

$$f(x) = (x - 1)(x - 1)(x + 2)(x^2 + 4)$$
$$= (x - 1)(x - 1)(x + 2)(x - 2i)(x + 2i).$$

This factorization gives the following five zeros:

$$1, 1, -2, 2i, \text{ and } -2i.$$

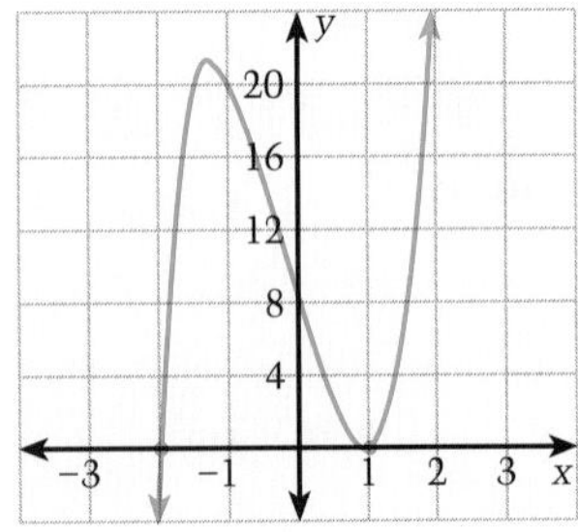

Note from the graph of f that the *real* zeros are the only ones that appear as x-intercepts. ■

In Example 3, the zeros $2i$ and $-2i$ are **complex conjugates.** The complex zeros of a polynomial function with *real* coefficients always occur in complex conjugate pairs. That is, if $a + bi$ is a solution, then $a - bi$ must also be a solution.

The building of the port of Caesarea, around 30 B.C., was the first known use of hydraulic concrete—concrete that will set up in water. Caesarea was located in Israel, which at the time was a province of Rome. Recent archeological finds include the large block of hydraulic concrete shown.

Goal 2 Using Polynomials in Real-Life Problems

Real Life
Engineering

Example 4 *Solving a Polynomial Equation*

In 1980, archaeologists at the ruins of Caesarea discovered a huge hydraulic concrete block. The block had a volume of 330 cubic yards. (A typical concrete driveway to a house would use about 5 cubic yards.) The dimensions of the block were x yards by $5x + 1$ yards by $7x + 1$ yards. How high was the block?

Solution

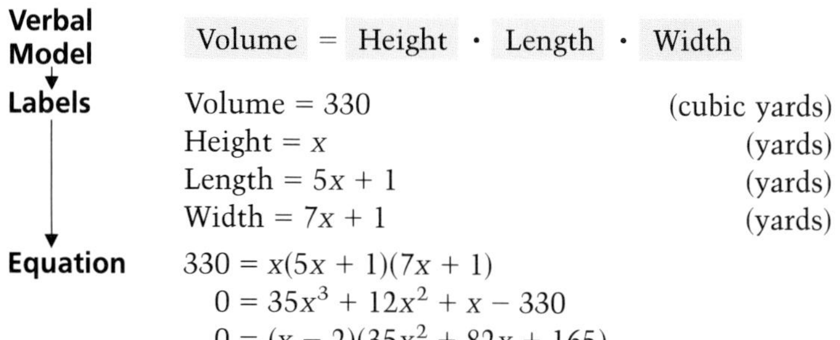

Verbal Model Volume = Height · Length · Width

Labels

Volume = 330	(cubic yards)
Height = x	(yards)
Length = $5x + 1$	(yards)
Width = $7x + 1$	(yards)

Equation

$$330 = x(5x + 1)(7x + 1)$$
$$0 = 35x^3 + 12x^2 + x - 330$$
$$0 = (x - 2)(35x^2 + 82x + 165)$$

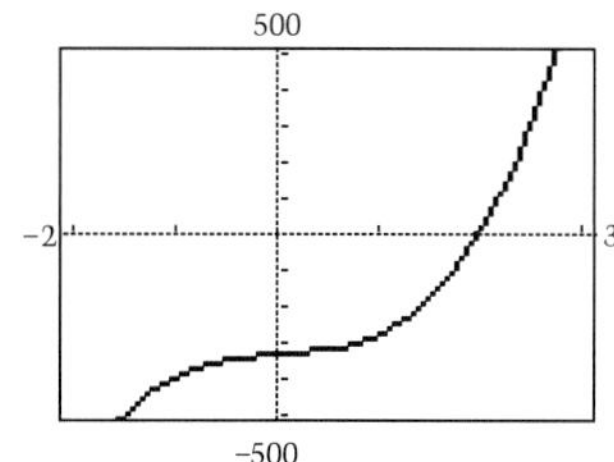

The equation has only one real solution: $x = 2$. Thus the block was 2 yards high. The dimensions of the block were 2 by 11 by 15, which produces a volume of 330 cubic yards. ■

Communicating about ALGEBRA

SHARING IDEAS about the Lesson

A graphing calculator was used to sketch the graph of the equation in Example 4, $f(x) = 35x^3 + 12x^2 + x - 330$. What are the scales on the x-axis and y-axis? Are all of the x-intercepts shown in the graph? Explain.

1, 100; yes, two roots are not real

Communicating about ALGEBRA

EXTEND Communicating
COOPERATIVE LEARNING
Have students work in small groups to investigate the relationship between zeros and x-intercepts for the type of polynomial functions presented.

1. Describe the graph of the function $g(x) = x^4 + 5x^2 + 4$ in terms of its zeros and x-intercepts.
   The graph does not intersect the x-axis. There are no real zeros.
   Are there values of x for which $g(x)$ is sometimes negative and sometimes positive?
   No; for all real numbers x, $g(x)$ is always greater than or equal to 4.
2. Describe the graph of the function $h(x) = x^4 - 2x^3 - 4$ in terms of its zeros and x-intercepts.
   The graph intersects the x-axis between $x = -2$ and $x = -1$ and between $x = 2$ and $x = 3$, so there are two real zeros.
   Are there values of x for which $h(x)$ is sometimes negative and sometimes positive?
   yes
3. Have students speculate on the existence of x-intercepts when $f(x)$ has the same sign for all values of x in the domain. They can test their conjectures by constructing relevant examples of polynomial functions.

Note: If $f(x)$ is a polynomial function, and r and s are real numbers such that $f(r)$ and $f(s)$ are real-valued and are not both positive nor both negative, then there is at least one real-valued zero of $f(x)$ between r and s.

EXERCISE Notes

ASSIGNMENT GUIDE

Basic/Average: Ex. 1–5, 7–15 odd, 25–35 odd, 36–43, 51–55

Above Average: Ex. 1–5, 7–21 odd, 25–35 odd, 36–47, 54–56, 58

Advanced: Ex. 1–5, 7–21 odd, 25–35 odd, 36–46, 54–58

Selected Answers
Exercises 1–5, 7–55 odd

More Difficult Exercises
Exercises 36–45, 56–58

Guided Practice

▶ **Ex. 3** Students will find no rational zeros.

EXTEND Ex. 3 Since $f(x)$ can be factored into the product $(x^2 + 4)(x^2 - 3)$, you might wish to explore with students a factoring pattern for the sum of two squares, as follows. Since $(a + bi)(a - bi) = a^2 - b^2i^2$ $= a^2 - b^2(-1) = a^2 + b^2$, the sum of two squares can be factored over the complex number system. For example,
$f(x) = (x^2 + 4)(x^2 - 3)$
$= (x \pm 2i)(x \pm \sqrt{3})$
and $x = -2i$, or $x = 2i$, or $x = -\sqrt{3}$, or $x = \sqrt{3}$.

Independent Practice

▶ **Ex. 6–11** Ask students to describe two methods for deciding whether the given x-value is a zero. For example, students could evaluate $f(x)$ at x, graph $f(x)$, and check whether the given x-value is an x-intercept, or students could use synthetic division and check for a 0 remainder.

▶ **Ex. 12–19** Remind students that they can use the Linear Factorization Theorem to "build" their polynomial function.

EXERCISES

Guided Practice

CRITICAL THINKING about the Lesson

1. Two of the zeros of $f(x) = x^3 - 6x^2 - 16x + 96$ are 4 and -4. Explain why the third zero must also be a real number. Using only the constant of the polynomial, what must the other zero be? Explain. See below.
2. How many solutions does the equation $x^5 + x - 3 = 0$ have? Explain. A fifth degree equation has 5 solutions.
3. Find the zeros of $f(x) = x^4 + x^2 - 12$. $\pm 2i, \pm\sqrt{3}$
4. State the Linear Factorization Theorem in your own words. See page 496.
5. At the right is the graph of $f(x) = 2x^4 + 2x^3 - 9x^2 - 2x + 7$. How many real zeros does the graph have? How many are complex? 4, 0

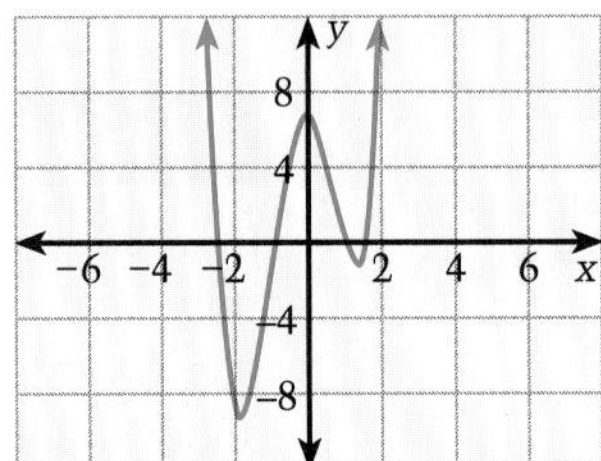

Independent Practice

In Exercises 6–11, decide whether the given x-value is a zero of the function.

6. $f(x) = x^3 - x^2 + 3x - 3,\ x = 1$ It is.
7. $f(x) = x^3 + 4x^2 + 3x + 12,\ x = 3$ It is not.
8. $f(x) = x^4 - x + 4,\ x = 0$ It is not.
9. $f(x) = x^3 + 3x^2 + x + 3,\ x = -3$ It is.
10. $f(x) = x^3 - 2x^2 + 4x - 8,\ x = 2i$ It is.
11. $f(x) = x^3 - 2x^2 + x - 2,\ x = -i$ It is.

In Exercises 12–19, write a polynomial function that has the given zeros and has a leading coefficient of 1. See Additional Answers.

12. 2, 1, 5
13. 1, −3, 4
14. −6, 4, 2
15. 6, −2, 2
16. 5, i, $-i$
17. $3i$, $-3i$, 4
18. 3, −3, $2i$, $-2i$
19. i, $-i$, $2i$, $-2i$

In Exercises 20–23, write a polynomial function whose graph has the given x-intercepts and has a leading coefficient of 1. See Additional Answers.

20. (4, 0), (1, 0)
21. (−2, 0), (5, 0)
22. (−3, 0), (2, 0)
23. (9, 0), (−1, 0)

In Exercises 24–35, write the polynomial as a product of linear factors. See Additional Answers.

24. $x^4 + x^3 - 19x^2 + 11x + 30$
25. $x^4 + 3x^3 - 2x^2 - 12x - 8$
26. $x^3 - x^2 - 2x$
27. $x^3 + 5x^2 - 9x - 45$
28. $x^4 - 5x^2 - 36$
29. $x^3 - x^2 + 9x - 9$
30. $x^3 - x^2 + 16x - 16$
31. $x^3 + 3x^2 + 25x + 75$
32. $x^4 - x^3 - 5x^2 - x - 6$
33. $x^4 - x^3 + 2x^2 - 4x - 8$
34. $2x^4 - 7x^3 - 27x^2 + 63x + 81$
35. $x^3 - 2x^2 + 25x - 50$

1. Nonreal roots can occur only in pairs; 6; $(-r_1)(-r_2)(-r_3) = 96$, $r_3 = 6$

36. *Harvard Step Test* A group of people were given the Harvard Step Test to determine their physical fitness. Each person was asked to step up onto a one-foot box, then step down, repeatedly, for five minutes. At the end of the test, each person was given a score, S, determined by the rate at which their pulse returned to normal. The scores of the people in the group were found to be related to the amounts of their blood hemoglobin, x (in grams per 100 milliliters of blood), by the model $S = -0.015x^3 + 0.6x^2 - 2.4x + 19$. Approximate the amount of hemoglobin in the blood of a person who scored 40 on the Harvard Step Test. See margin.

37. *Grocery Store Revenue* A neighborhood grocery store has been open for 25 years. During that time, the store's annual revenue, R (in millions of dollars), can be modeled by

$$R = \frac{1}{10{,}000}(-t^4 + 12t^3 - 77t^2 + 600t + 13{,}650)$$

where $t = 0$ represents the store's first year. In what years did the revenue reach $1.5 million? In the fourth and tenth years

Silkworm Puzzle **In Exercises 38–43, find the sum of the real zeros of the function. Then use the sum to complete the fact about silkworms.**

When the casing of a silkworm's cocoon is unraveled, a single unbroken thread of silk results.

38. Adult silkworm length (in inches): $f(x) = x^4 - 3x^3 - 27x^2 - 3x - 28$ 3, 3 in.

39. Number of times a silkworm sheds its skin: $f(x) = 2x^3 - 8x^2 + x - 4$ 4, 4 times

40. Number of breathing vents on a silkworm: $f(x) = x^3 - 9x^2 + 4x - 36$ 9, 9 vents

41. Number of species of wild silkworms (in hundreds):

$f(x) = x^4 - 5x^3 + 3x^2 - 45x - 54$ 5, 500 species

42. Average silkworm life span (in days):

$f(x) = x^4 - 26x^3 + 168x^2 - 262x + 119$ 26, 26 days

43. Ratio of adult-to-birth weight (in 1000's):

$f(x) = x^4 - 10x^3 + 28x^2 - 40x + 96$ 10, 10,000

Ex. 36–37 Encourage students to sketch the graphs of the functions to help determine the solutions.

Answer

36. 10 grams per 100 milliliters of blood

Ex. 38–43 Assign these exercises as an in-class group activity.

Alternative Assessment

When students complete the exercises, ask each group to write a strategy that describes how they found the zeros.

▶ **Ex. 44** Students can graph S and look for a y-value of 664, or they could graph $S - 664$ and look for a zero.

EXTEND Ex. 45 What are the predicted sales for 1995? Is the answer reasonable?
$3160.5 million, no

▶ **Ex. 58** Ask students why $f(x)$, a fifth-degree polynomial in x, has only 4 zeros.
−1 is a double root.

Enrichment Activity

This activity requires students to go beyond lesson goals.

PROJECT Prior to the invention of negative numbers and imaginary numbers, the idea of the Fundamental Theorem of Algebra would have been difficult. The Hindus by the year 1100 had a vague idea that quadratic equations had two roots. In 1545, the Italian Girolamo Cardano realized that cubic equations should have three roots. Carl Friedrich Gauss (1777–1855) actually gave four proofs of the Fundamental Theorem of Algebra in his lifetime. He gave the first proof in 1797 when he was 20 years old and the last proof when he was 70 years old.

Research the history of polynomial equations and their roots in some history-of-mathematics texts. Write a report on some other mathematicians and their techniques for solving polynomial equations.

Ski Equipment **In Exercises 44 and 45, use the following information.**

For 1980 through 1990, the sales, S (in millions of dollars), of snow skiing equipment can be modeled by

$S = 0.42t^4 - 9.7t^3 + 72.3t^2 - 134t + 378,$

where $t = 0$ represents 1980.

44. Use a graphing calculator to sketch the graph. Was there a year in which sales were about $664 million? Explain.

45. Was there a year in which sales were about $280 million? Explain.
See below.

44. Yes; for $t = 7$, $S \approx \$664$ million

Integrated Review

In Exercises 46–53, find all the zeros of the function.

46. $f(x) = x^3 - 7x^2 + 7x + 15$ −1, 3, 5

47. $f(x) = 2x^3 + 3x^2 - 3x - 2$ $-2, -\frac{1}{2}, 1$

48. $f(x) = x^3 - 4x^2 - 15x + 18$ −3, 1, 6

49. $f(x) = x^3 + 13x^2 + 48x + 36$ −6, −1

50. $f(x) = 2x^3 - 15x^2 - 47x - 30$ $-\frac{3}{2}, -1, 10$

51. $f(x) = x^4 + 4x^3 + 3x^2 - 4x - 4$ −2, ±1

52. $f(x) = x^4 + 3x^3 - 18x^2 - 32x + 96$ −4, 2, 3

53. $f(x) = x^4 - 15x^2 - 10x + 24$ −3, −2, 1, 4

54. *College Entrance Exam Sample* If $z = -x$ and $x \neq 0$, what are all values of y for which $(x + y)^2 + (y + z)^2 = 2x^2$? a
a. 0 **b.** 0, 1 **c.** −1, 0, 1 **d.** All positive numbers
e. There are no values of y for which the equation is true.

55. *College Entrance Exam Sample* A *perfect square* is an integer whose square root is an integer. The numbers 16, 25, and 36 are perfect squares. If a and b are perfect squares, which of the following is *not* necessarily a perfect square? e
a. $25a$ **b.** a^2 **c.** a^3 **d.** ab **e.** $a - b$

Exploration and Extension

56. Write a third-degree polynomial, two of whose zeros are 1 and i.

57. Write a fourth-degree polynomial, two of whose zeros are $2i$ and $-3i$.

58. The graph of

$f(x) = x^5 - 2x^3 - 2x^2 - 3x - 2$

is shown at the right. Find all the zeros of the function. −1, 2, $\pm i$

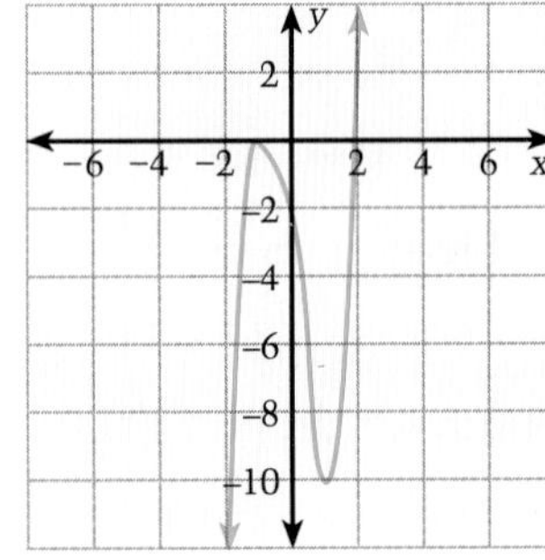

56. The simplest is $x^3 - x^2 + x - 1$

57. The simplest is $x^4 + 13x^2 + 36$

45. No, the minimum value of the function is ≈$305 million.

9.7 Exploring Data: Measures of Dispersion

What you should learn:

Goal 1 How to find the range of a collection of numbers

Goal 2 How to find the standard deviation of a collection of numbers

Why you should learn it:

You can use the range and standard deviation to describe many collections of data, such as the quality control data for the color brightness of paper.

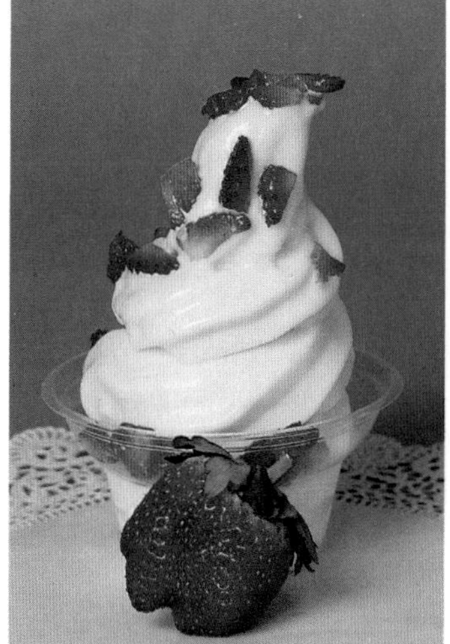

Goal 1 Finding the Range

You already know several ways to display a collection of data, a frequency distribution, a bar graph, or a box-and-whisker plot. You also know how to find the mean, median, mode, and quartiles for a collection of numbers. In this lesson, you will study two **measures of dispersion** of a collection of numbers: the *range* and the *standard deviation.* Each of these gives you an idea of how much the numbers in the collection vary.

The **range** of a collection of numbers is the difference between the largest and smallest values in the collection. For instance, the range of the collection 2, 3, 4, 6, 9 is $9 - 2 = 7$.

Example 1 *Finding the Range of a Set of Numbers*

Three employees at a frozen yogurt store were each asked by their manager to fill a dozen 8-ounce cups of yogurt. The manager weighed each of the cups and recorded the following results. Which employee(s) had the most consistent record?

Sean	Eric	Jody
6.5, 7.0, 9.0	8.0, 7.0, 10.0	7.0, 8.0, 8.0
9.0, 8.0, 7.0	11.0, 7.5, 5.0	7.5, 8.0, 8.5
7.5, 9.5, 9.0	9.5, 5.0, 10.5	10.0, 8.0, 7.5
9.5, 7.0, 7.0	9.0, 7.5, 6.0	8.0, 8.0, 7.5

Solution One way to answer the question is to calculate the range for each set of measurements.

Sean's range: $9.5 - 6.5 = 3$ ounces
Eric's range: $11.0 - 5.0 = 6$ ounces
Jody's range: $10.0 - 7.0 = 3$ ounces

Judging from the ranges of each, Sean's and Jody's performances were more consistent than Eric's. ■

In Example 1, notice that the mean of each set of measurements was 8 ounces. (Try checking this.) Their abilities to consistently fill cups with 8 ounces of yogurt, however, varied. From this example, you can see that two collections of numbers with the same mean can have very different dispersion patterns.

Problem of the Day

Each of the four arcs is a semicircle of radius 2, and A is the midpoint of both arcs on which it lies. Find the area of the figure. 16

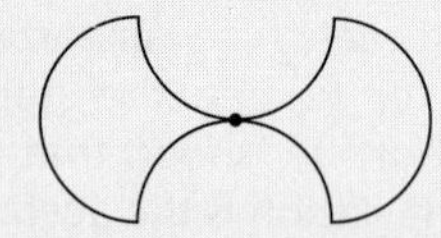

ORGANIZER

Warm-Up Exercises

1. Find the mean of these data. {9.4, 9.0, 1.4, 5.1, 4.0, 7.3, 0.0, 3.3, 9.9, 2.0, 7.9, 9.5, 2.2, 3.6, 0.0, 9.3, 1.0, 0.0, 5.4, 8.5}
   mean $= \frac{98.8}{20} = 4.94$
2. Evaluate each expression to three decimal places.
   a. $\sqrt{[1.2 - 3]^2 + [3.4 - 4]^2}$
   b. $\sqrt{\frac{2^2 + (-6)^2 + (-2.3)^2 + 9^2}{8}}$
   a. 1.897, b. 3.973

Lesson Resources

Teaching Tools
Transparency: 5
Problem of the Day: 9.7
Warm-up Exercises: 9.7
Answer Masters: 9.7
Extra Practice: 9.7
Color Transparencies: 52, 53
Applications Handbook: p. 11
Technology Handbook: p. 68

LESSON Notes

Example 1

Although Eric gave his customers an average of 8 ounces of yogurt, explain why the manager should be concerned about Eric's performance. (One possible reason for concern is that customers may feel

(continued)

(continued)
cheated if the difference in the amount of yogurt received varies greatly from customer to customer.)

Example 2

Ask students to analyze the steps for finding the standard deviation to show that standard deviation is the average of the distances of the data points from the mean.

Example 3

Observe how the computed standard deviation value is used to partition the collection of data. It is important to recognize that most of the data points lie within 1 standard deviation of the mean and that practically all of the data points lie within 2 standard deviations of the mean.

Extra Examples

Here are additional examples similar to **Examples 1–3.**

1. Find the range for this data. {9.4, 9.0, 1.4, 5.1, 4.0, 7.3, 0.0, 3.3, 9.9, 2.0, 7.9, 9.5, 2.2, 3.6, 0.0, 9.3, 1.0, 0.0, 5.4, 8.5}
   9.9
2. **TECHNOLOGY** Use a calculator to find the standard deviation of the data in Extra Example 1.
   ≈3.54

Check Understanding

1. What is the range of a collection of data?
   One measure of dispersion; see page 501.
2. What is the standard deviation of a collection of data?
   Another measure of dispersion; see page 502.
3. Why is a measure of dispersion needed to fully understand a collection of data?
   To give you an idea of how much the numbers in the collection vary.

Goal 2 Finding the Standard Deviation

The range of a collection of numbers is only a simple measure of its dispersion. In Example 1, for instance, the range was not able to distinguish between Sean's and Jody's performance. To obtain a more sophisticated measure of dispersion, statisticians use the *standard deviation.*

Standard Deviation of a Collection of Numbers

The **standard deviation** of $x_1, x_2, x_3, \ldots, x_n$ is

$$s = \sqrt{\frac{(x_1 - \bar{x})^2 + (x_2 - \bar{x})^2 + \cdots + (x_n - \bar{x})^2}{n}}$$

where $\bar{x}$ is the mean of $x_1, x_2, \ldots,$ and x_n.

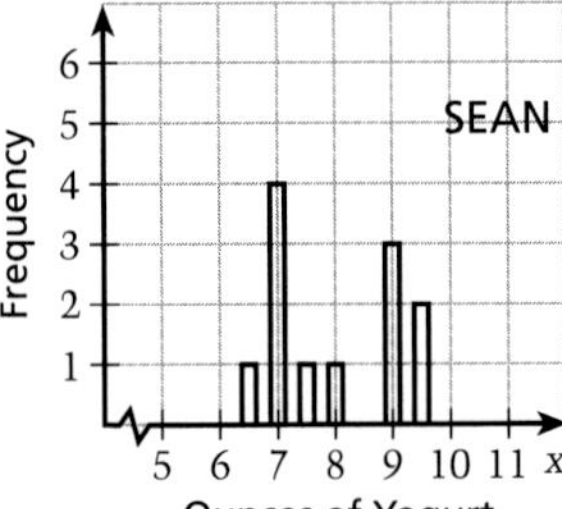

ERIC
Frequency
Ounces of Yogurt

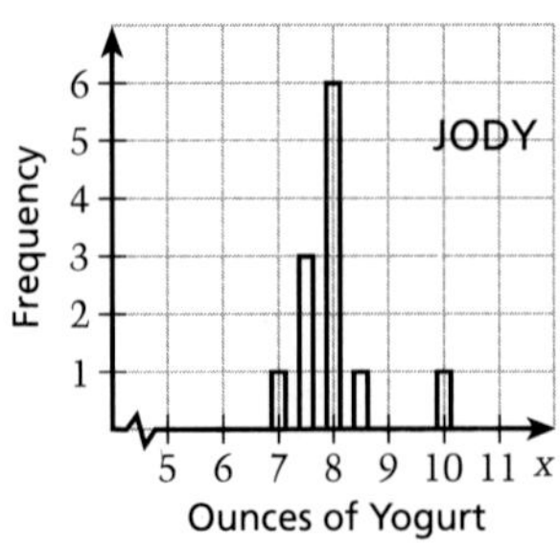

Example 2 *Finding the Standard Deviation of a Collection of Numbers*

The standard deviations for the sets of measurements given in Example 1 are as follows. (The mean for each is $\bar{x} = 8$.)

$$\text{Sean: } s = \sqrt{\frac{(6.5 - 8)^2 + (7.0 - 8)^2 + \cdots + (7.0 - 8)^2}{12}} = \sqrt{\frac{14}{12}} \approx 1.080$$

$$\text{Eric: } s = \sqrt{\frac{(8.0 - 8)^2 + (7.0 - 8)^2 + \cdots + (6.0 - 8)^2}{12}} = \sqrt{\frac{46}{12}} \approx 1.958$$

$$\text{Jody: } s = \sqrt{\frac{(7.0 - 8)^2 + (8.0 - 8)^2 + \cdots + (7.5 - 8)^2}{12}} = \sqrt{\frac{6}{12}} \approx 0.707$$

A larger standard deviation tells you that the numbers in the collection are more "spread out"—the dispersion is greater. A smaller standard deviation tells you that the numbers in the collection are more tightly grouped. Thus, using standard deviation as a measure, the employee with the most consistent record was Jody. The bar graphs at the left confirm that Eric's measurements are more spread out than Sean's, and Sean's are more spread out than Jody's. ■

Calculating the standard deviation using the formula given in its definition can be tedious (and may involve significant round-off error). The following alternate formula is easier to use. This formula is *equivalent* to the formula on page 502.

Alternate Formula for Standard Deviation

$$s = \sqrt{\frac{x_1^2 + x_2^2 + \cdots + x_n^2}{n} - \bar{x}^2}$$

Example 3 ***Finding Standard Deviation with the Alternate Formula***

The heights (in inches) of 20 members of a high school football team are given below. Find the mean and standard deviation of the heights.

64.0, 66.0, 66.5, 68.0, 68.5, 70.0, 70.0, 70.5, 71.0, 71.5
71.5, 72.0, 72.0, 72.5, 73.0, 73.0, 74.0, 74.5, 76.0, 78.0

Height of 20 Football Players

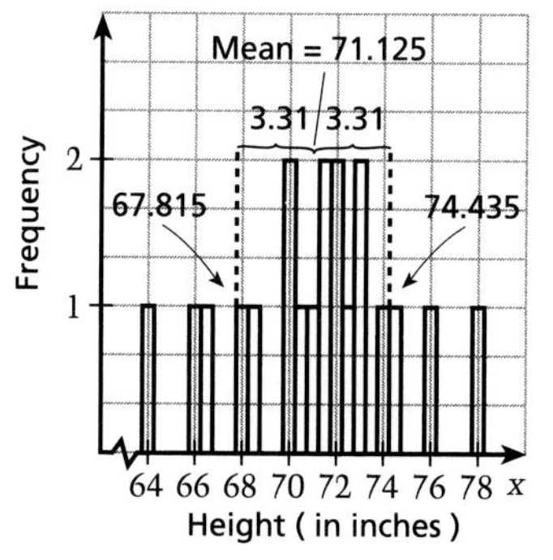

Solution The mean height for the team is

$$\bar{x} = \frac{64.0 + 66.0 + \cdots + 76.0 + 78.0}{20} = \frac{1422.5}{20} = 71.125 \text{ inches.}$$

Using the alternate formula, the standard deviation is

$$s = \sqrt{\frac{64.0^2 + 66.0^2 + \cdots + 76.0^2 + 78.0^2}{20} - (71.125)^2}$$

$$\approx \sqrt{5069.7375 - 5058.7656} \approx 3.31 \text{ inches.}$$

Note in the bar graph that most of the heights lie within 1 standard deviation of the mean. That is, most of the heights lie between 67.815 and 74.435. ■

Communicating about ALGEBRA

SHARING IDEAS about the Lesson

Find the range and standard deviation of each collection.

A. The weights (in pounds) of ten giant pandas: 302, 310, 331, 298, 348, 305, 314, 284, 321, 337 64, ≈18.47

B. The age (in years) of ten giant pandas: 3.4, 4.2, 8.6, 5.1, 3.6, 2.8, 7.1, 4.4, 5.2, 5.6 5.8, ≈1.68

Communicating about ALGEBRA

EXTEND *the Activity*

COOPERATIVE LEARNING

Have students work in small groups to show that the alternate formula for standard deviation, given on page 503, is equivalent to the formula of the definition, given on page 502. Consider a collection of three data points x_1, x_2, and x_3 with mean $\bar{x} = \frac{x_1 + x_2 + x_3}{3}$ and transform the expression

$$\frac{(x_1 - \bar{x})^2 + (x_2 - \bar{x})^2 + (x_3 - \bar{x})^2}{3}$$

into the expression

$$\frac{x_1^2 + x_2^2 + x_3^2}{3} - \bar{x}^2.$$

$$\frac{(x_1 - \bar{x})^2 + (x_2 - \bar{x})^2 + (x_3 - \bar{x})^2}{3} =$$

$$\frac{x_1^2 - 2x_1\bar{x} + \bar{x}^2 + \ldots - 2x_3\bar{x} + \bar{x}^2}{3}$$

$$= \frac{x_1^2 + x_2^2 + x_3^2}{3} + \frac{-2x_1\bar{x} - 2x_2\bar{x} - 2x_3\bar{x}}{3} + \frac{3\bar{x}^2}{3}$$

$$= \frac{x_1^2 + x_2^2 + x_3^2}{3} + \frac{-2\bar{x}(x_1 + x_2 + x_3)}{3} + \bar{x}^2$$

$$= \frac{x_1^2 + x_2^2 + x_3^2}{3} - 2\bar{x}(\bar{x}) + \bar{x}^2$$

$$= \frac{x_1^2 + x_2^2 + x_3^2}{3} - 2\bar{x}^2 + \bar{x}^2$$

$$= \frac{x_1^2 + x_2^2 + x_3^2}{3} - \bar{x}^2$$

EXERCISE Notes

ASSIGNMENT GUIDE

Basic/Average: Ex. 1–4, 5–9 odd, 11–17, 21–26

Above Average: Ex. 1–4, 5–9 odd, 10–21, 24–26

Advanced: Ex. 1–4, 5–9 odd, 10–21, 24–25

Selected Answers
Exercises 1–4, 5–23 odd

✪ More Difficult Exercises
Exercises 13–20, 25

Guided Practice

▶ **Ex. 1** Discuss with the class the implications of a greater standard deviation.
See Example 2.

▶ **Ex. 2** Ask students how you can tell which plot has the greater "interquartile" range.
By the length of the box

▶ **Ex. 3–4 TECHNOLOGY** Use the Stats feature of a graphing calculator to provide a visual interpretation for these exercises. Entering the data efficiently yields a standard deviation of 2.65 for Exercise 3a and 3.84 for Exercise 3b. Consult the manuals for instructions.

Answers

1. Example with range 3: 4, 4, 5, 7. Example with range 7: 1, 5, 6, 8. The collection with range 7
2. a, 3 and 10 are farther apart than 5 and 11
3. a, the data in a are not spread as far apart as the data in b

Independent Practice

▶ **Ex. 5–12** Encourage students to use either the alternate formula provided on page 503, or the standard deviation feature on their scientific or graphing calculators.

504

EXERCISES

Guided Practice

CRITICAL THINKING about the Lesson

1. Give an example of two collections of four numbers, each with a mean of 5. Choose the numbers so that one collection has a range of 3 and the other has a range of 7. Which has the greater standard deviation? See margin.

2. The following box-and-whisker plots represent two collections of numbers. Which collection has the greater range? Explain. See margin.

a.
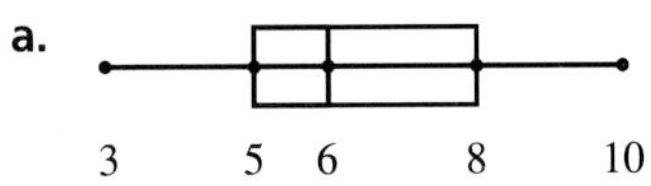

b.
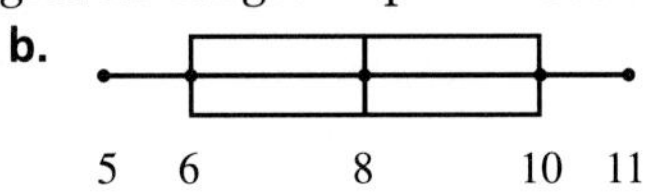

3. *Test Scores* The following bar graphs represent two collections of test scores. Which collection has the smaller standard deviation? Explain. See margin.

a.
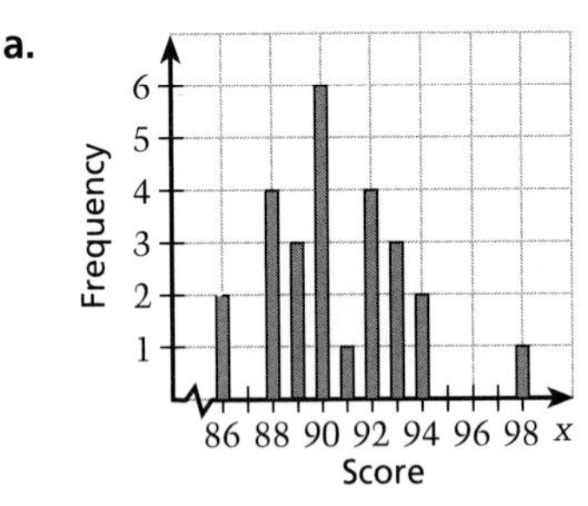

b.
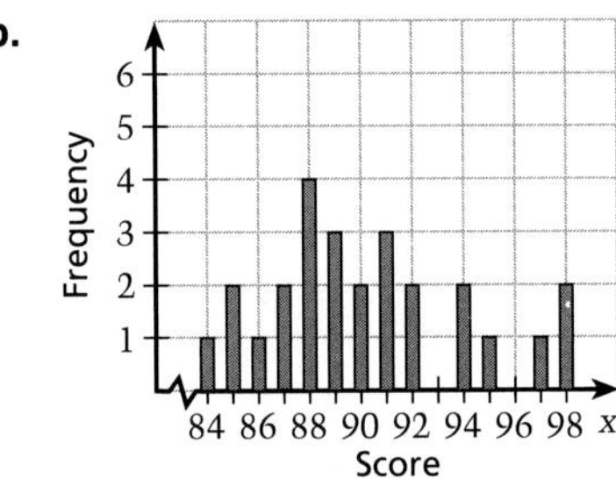

4. *Test Scores* Find the standard deviation of the collection of test scores you chose in Exercise 3. If you received a score of 98 on the test, would your score have been within one standard deviation of the mean score? ≈2.65, no

Independent Practice

In Exercises 5–8, find the range and standard deviation of the data.

5. 5.6, 5.6, 5.7, 5.8, 5.8, 5.8, 5.9, 6.1 0.5, ≈0.15
6. 100, 156, 158, 159, 162, 165, 170, 190 90, ≈23.98
7. 0, 0.1, 0.1, 0.3, 0.6, 0.6, 0.7, 0.8 0.8, ≈0.29
8. 201, 201, 203, 205, 208, 210, 211, 220 19, ≈5.98
9. *Squirrel Tails* The average tail lengths (in inches) of ten types of squirrels are shown in the list. Find the range and standard deviation of these lengths. 10, ≈3.37

Western gray squirrel	12	Apache fox squirrel	11
Tassel-eared squirrel	8.5	Red squirrel	6
Eastern gray squirrel	10	Chickaree	5
Arizona gray squirrel	12	Southern flying squirrel	4
Eastern fox squirrel	14	Northern flying squirrel	5

10. *Brothers and Sisters* A class of 28 students took a survey of the number of brothers and sisters each student had. Find the range and standard deviation of the results. 6, ≈1.31

0 0 0 0 0 1 1 1 1 1 1 1 1 2
2 2 2 2 2 2 2 2 2 3 3 3 4 6

11. *Grocery Shopping* Over the past year, your parents went grocery shopping once every two weeks. The costs of the groceries were as follows. What is the range of these costs? What is the mean? $48.25, ≈$195.64

$210.55	$178.60	$197.22	$181.23	$172.45	$201.74	$211.38	$199.71	$193.30
$202.01	$190.64	$180.13	$180.95	$192.76	$219.82	$220.70	$192.02	$194.40
$195.06	$202.36	$185.12	$198.46	$194.62	$204.28	$198.54	$188.60	

12. *Moons of the Planets* The numbers of known moons for the nine planets in our solar system are shown in the table. Find the range and standard deviation of these data. 17, ≈6.99

Mercury	0	Mars	2	Uranus	15
Venus	0	Jupiter	16	Neptune	8
Earth	1	Saturn	17	Pluto	1

Each of the moons in our solar system has been named—except for our moon. It is simply called* the moon. *For instance, the names of the 17 moons of Saturn are Atlas, Calypso, Dione, Enceladus, Epimetheus, Helene, Hyperion, Janus, Japetus, Mimas, Pandora, Phoebe, Prometheus, Rhea, Telesto, Tethys, and Titan.

President's Ages **In Exercises 13–16, use the following information.**

The ages of the first 40 Presidents of the United States at the time of their inaugurations are given in the list.

42, 43, 46, 47, 48, 49, 50, 51, 51, 51, 51, 51, 52, 52
54, 54, 54, 54, 55, 55, 55, 55, 56, 56, 56, 57, 57, 57
57, 58, 60, 61, 61, 61, 62, 64, 64, 65, 68, 69

13. What is the range of the ages? 27

14. What is the mean of the ages? 55.225

15. The sum of the squares of the ages is 126,703. Use this to find the standard deviation of the ages. ≈10.85

16. Ronald Reagan was the oldest to be inaugurated as president. John F. Kennedy, at age 43, was the youngest to be *elected.* Use a library or some other reference source to find who was the youngest person to become President. Theodore Roosevelt

Ex. 12 Ask students to compute the mean. Ask them if it is a good indicator of the "average" number of moons.
$6.\overline{6}$, no

Ex. 13–16 Assign these exercises as a group. Ask students if Kennedy was within one standard deviation of the mean.
no

Ex. 16 Vice president Theodore Roosevelt became president on September 6, 1901 after William McKinley was assassinated.

▶ **Ex. 17** Ask students to explain how the standard deviation is connected to consistency of color. Would the manufacturer desire a low or a high standard deviation?
The lower the standard deviation, the greater the consistency of color; low

17. *Color Brightness of Paper* A company that manufactures paper pulls ten samples from one roll of paper and ten samples from another roll. The color brightness of each sample is measured with a spectrophotometer. The company's goal is to produce rolls of paper that have the same color brightness throughout the roll. In practice, however, absolute consistency is almost impossible. Judging from the samples, which roll has the more consistent color? Roll #2

Roll #1: 81.35, 81.24, 81.76, 81.64, 81.77, 81.85, 81.70, 83.24, 83.42, 82.49
Roll #2: 81.73, 81.80, 81.60, 81.69, 81.75, 81.20, 81.75, 81.30, 81.95, 81.44

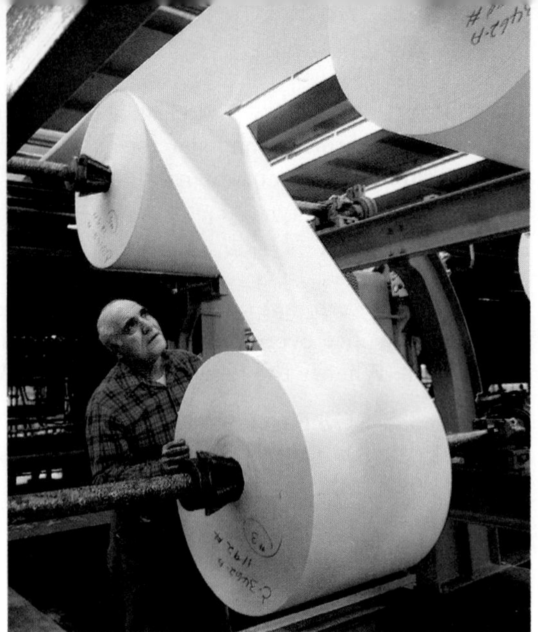

Rain in New Mexico **In Exercises 18 and 19, use the following information.**

The tables list the average number of days with rain or snow in Albuquerque and Roswell, New Mexico. Find the mean and standard deviation of each collection. Which data set is more consistent?

18. *Albuquerque:* 5, ≈2.08

Month	Jan	Feb	Mar	Apr	May	Jun	Jul	Aug	Sept	Oct	Nov	Dec
Days	4	4	4	4	4	4	9	10	5	5	3	4

19. *Roswell:* 4, 2; Roswell

Month	Jan	Feb	Mar	Apr	May	Jun	Jul	Aug	Sept	Oct	Nov	Dec
Days	3	4	2	3	4	4	9	7	4	4	2	2

20. *Log Length* The lengths (in feet) of 30 logs are listed below. What is the mean of their lengths? What percent of the logs have lengths that differ from the mean by more than one standard deviation?

25, 26, 26, 27, 27, 28, 28, 28, 29, 29, 30, 30, 30, 30, 30, 31, 31, 31, 31, 32, 32, 32, 32, 32, 33, 34, 34, 35, 35, 36
≈30.47 ft, $33\frac{1}{3}$%

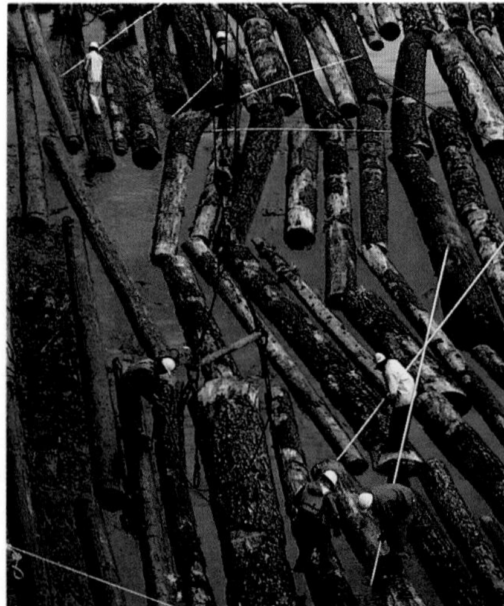

Hundreds of logs are shown on their way to a paper pulp mill in Oregon.

Integrated Review

21. *College Entrance Exam Sample* In a family of five, the heights of the members are 5 feet 1 inch, 5 feet 7 inches, 5 feet 2 inches, 5 feet, and 4 feet 7 inches. What is the average height? c

a. 4′4″ **b.** 5′ **c.** 5′1″ **d.** 5′2″ **e.** 5′3″

22. *College Entrance Exam Sample* A recipe for a cake called for $1\frac{1}{4}$ cups of milk and 3 cups of flour. With this recipe, a cake is baked using 12 cups of flour. How many cups of milk are required? c

a. 3 **b.** 4 **c.** 5 **d.** 6 **e.** 7

23. *Postage Prices* Although the price of a first-class stamp rose four cents in 1991, it was still cheaper to mail a letter in the United States than in other major industrialized countries. What was the mean cost of a first-class stamp in the five countries below? *(Source: U.S. Postal Service)* 46.2¢

Postage Prices

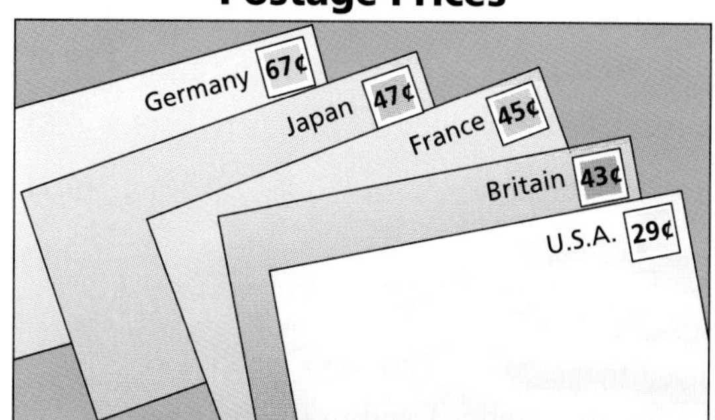

24. *Overseas Travel* In 1993, about 44 million Americans who traveled abroad spent about 40.5 billion dollars. Use the information below to approximate the average amount spent by each traveler to Canada, to Mexico, and overseas. *(Source: U.S. Travel and Tourism Administration)*

Canada: ≈ \$306.82; Mexico ≈ \$351.94; Overseas: ≈ \$1840.91

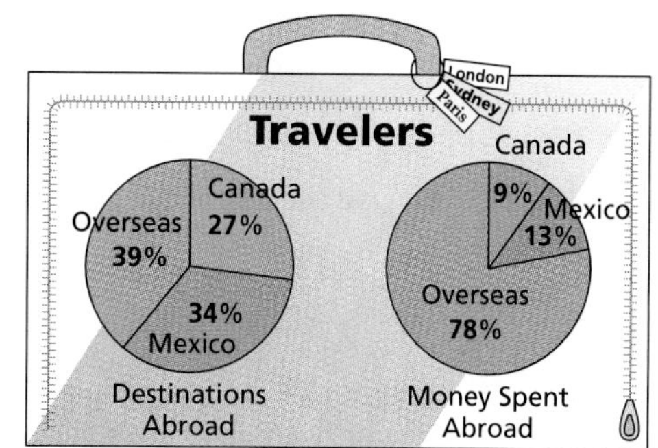

Exploration and Extension

25. *Household Sizes* The following list gives the average number of people per household in 1994 for each of the 50 states. Use a calculator or a computer to find the mean and standard deviation of these data. 2.6328, ≈ 0.117

AL	2.61	GA	2.67	ME	2.54	NE	2.56	OH	2.59	TX	2.75
AK	2.81	HI	2.99	MD	2.67	NV	2.56	OK	2.56	UT	3.13
AZ	2.66	ID	2.75	MA	2.57	NH	2.61	OR	2.53	VT	2.54
AR	2.58	IL	2.66	MI	2.65	NJ	2.72	PA	2.57	VA	2.60
CA	2.83	IN	2.59	MN	2.60	NM	2.77	RI	2.57	WA	2.56
CO	2.52	IA	2.52	MS	2.74	NY	2.64	SC	2.66	WV	2.53
CT	2.60	KS	2.56	MO	2.56	NC	2.55	SD	2.63	WI	2.62
DE	2.59	KY	2.59	MT	2.58	ND	2.54	TN	2.57	WY	2.62
FL	2.50	LA	2.72								

Which states have household sizes that differ from the mean by more than one standard deviation? By more than two standard deviations? See margin.

▸ **Ex. 25–26** Encourage students to use a statistics program on a calculator to compute the mean and standard deviation. Remind students that the difference from the mean could be *exactly* one standard deviation below or above the mean. Thus, students should seek values below 2.5148 or above 2.7508.

EXTEND Ex. 25 Ask students to find the household average for their states and compare it to the mean and standard deviation of the United States.

Answer

25. AK, CA, FL, HI, ID, NM, TX, UT; HI, UT

Enrichment Activity

This activity requires students to go beyond the lesson goals.

EXTEND *the Lesson*

Consult the reference librarian in your school for assistance with using reference books like almanacs or the State Book of Lists. Examine the sources to look for data sets similar to those found in the Independent Practice exercises. Interpret the data. Find the mean and standard deviation for your data set to support your interpretation. What interpretation can you put on points outside of one standard deviation for your data set?

ORGANIZER

TIME SCHEDULE: All levels, two days

The Chapter Summary helps students organize the main ideas of the chapter. In this chapter, we explored polynomials and polynomial functions. The connection among zeros, factors, and solutions to polynomial equations was examined, and we continued our investigation of ways to organize and interpret data.

Work with students to review the skills, strategies, and concepts of the chapter. The first day's homework assignment can be used as the basis for the second day of review.

COOPERATIVE LEARNING
Encourage students to study together. Emphasize the importance of "teaching" a classmate how to perform a skill or how to recall a strategy. When students work together, everyone wins. The students receiving help get additional instruction and the students giving help gain a deeper understanding of the skills and concepts involved.

Chapter SUMMARY

Review the chapter by asking students "What did you learn? Why did you learn it? How does it fit with the other algebra skills and concepts you have learned?"

9 Chapter Summary

What did you learn?

Skills

1. Add, subtract, and multiply polynomials. **(9.1)**
   - Use special product patterns to multiply polynomials. **(9.1)**
2. Sketch the graph of a polynomial function. **(9.2)**
   - Identify basic features of polynomial graphs. **(9.2)**
3. Factor a polynomial.
   - Use special factoring patterns. **(9.3)**
   - Use the Rational-Zero Test. **(9.5)**
4. Divide polynomials.
   - Use long division. **(9.4)**
   - Use synthetic division. **(9.4)**
5. Solve polynomial equations. **(9.3, 9.5, 9.6)**
6. Find zeros of polynomial functions. **(9.2–9.6)**

Strategies

7. Use polynomial models to solve real-life problems. **(9.1–9.6)**

Exploring Data

8. Use measures of dispersion to describe data. **(9.7)**

Why did you learn it?

Polynomial models have been used by people for many hundreds of years—longer than any other type of model. In Algebra 1 and in the first part of Algebra 2, you studied many uses of polynomial models of degree 0, 1, and 2. In this chapter, you have seen that polynomials of degree 3 or more can also be used to model a variety of real-life situations.

One way to create real-life models is to add, subtract, or multiply other models. For instance, to create a model for the profit, P, you can subtract the model for total cost, C, from the model for total revenue, R.

How does it fit into the bigger picture of algebra?

In this chapter, you studied the graphs of *polynomial functions.* You also learned how to find the x-intercepts of the graph of a polynomial function by finding the real-number *zeros* of the function. Knowing how to find the real zeros of a polynomial function, you also know how to *find the real solutions of a polynomial equation.* The two problems are essentially the same.

The problem of finding the solutions of a polynomial equation is *the* most classic problem in all of algebra. Because of that, some of the results in this chapter are theoretical in nature. For instance, the chapter contains the *Fundamental Theorem of Algebra.*

Chapter REVIEW

In Exercises 1–14, perform the indicated operation. (9.1) See Additional Answers.

1. $(4x^3 + 2x^2 + 1) + (x^2 + x - 5)$
2. $(x^4 - x + 3) + (x^3 - x^2 - 3x + 4)$
3. $(9x^2 - 3x - 4) - (x^2 + 2x - 1)$
4. $(-3x^3 + 2x^2 - x + 1) - (x^3 + x + 3)$
5. $2x(x^2 + 3x - 4)$
6. $x(x^4 + 2x^2 - 3)$
7. $(x + 6)(x - 6)$
8. $(x + 14)(x - 14)$
9. $(x - 3)(x^2 + x - 7)$
10. $(x + 1)(x^2 + 3x + 1)$
11. $(x + 9)^2$
12. $(x - 10)^2$
13. $(x - 7)^3$
14. $(x + 5)^3$

In Exercises 15–20, match the function with its graph. (9.2)

15. $f(x) = (x + 1)^3 - 6$ b
16. $f(x) = x^4 - 3x^3 + x - 2$ d
17. $f(x) = x^5 + 2x^2 - 3x + 1$ a
18. $f(x) = x^3 + 2x - 1$ c
19. $f(x) = -x^4 + 2x^3 + 5$ f
20. $f(x) = -\frac{1}{4}(x + 1)^5 + 2$ e

a.
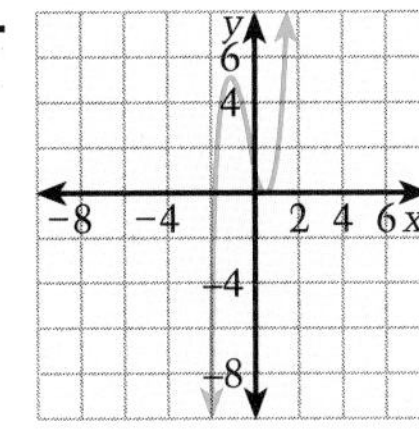
b.
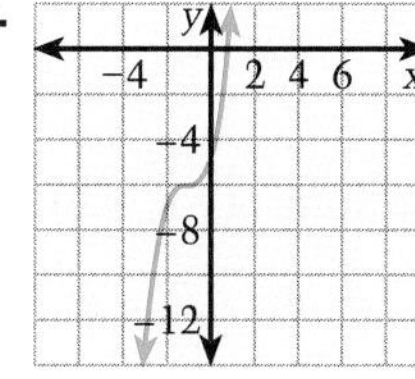
c.
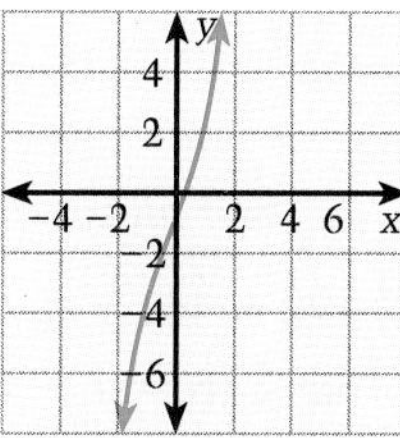
d.
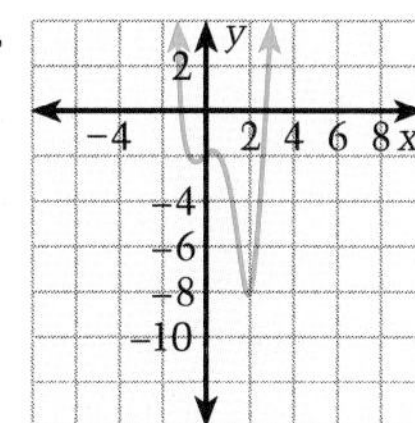
e.
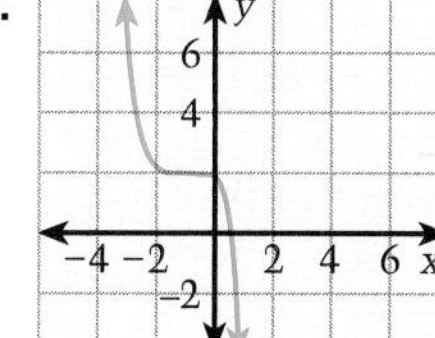
f.
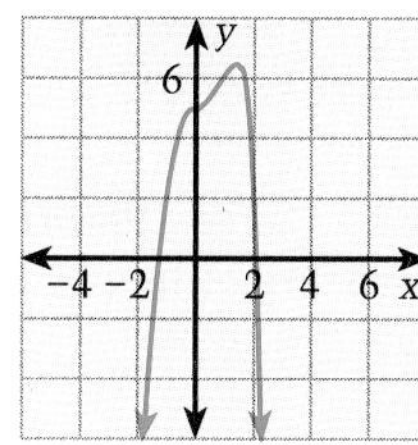

In Exercises 21–32, sketch the graph of the function. (9.2) See Additional Answers.

21. $f(x) = -x^3 + 5$
22. $f(x) = (x - 7)^3 + 1$
23. $f(x) = (x + 2)^3$
24. $f(x) = -x^3 + 4$
25. $f(x) = (x - 1)^4$
26. $f(x) = x^4 - 3$
27. $f(x) = -(x + 3)^4 + 5$
28. $f(x) = -(x + 6)^3$
29. $f(x) = x^5 + 6$
30. $f(x) = -(x - 2)^5$
31. $f(x) = -x^3 + 4x + 1$
32. $f(x) = x^4 - 5x + 4$

In Exercises 33–44, completely factor the polynomial with respect to the integers. (9.3) See Additional Answers.

33. $16x^2 - 144$
34. $4x^2 - 49$
35. $125x^3 - 8$
36. $64x^3 + 343$
37. $x^2 - 9x - 22$
38. $x^2 + 8x - 9$
39. $2x^2 - 5x + 2$
40. $3x^2 - 11x - 20$
41. $x^3 - x^2 + 5x - 5$
42. $x^3 - 3x^2 - 5x + 15$
43. $2x^3 - 3x^2 + 4x - 6$
44. $2x^3 + x^2 + 2x + 1$

Chapter REVIEW

ASSIGNMENT GUIDE

Basic/Average: Ex.3–12 multiples of 3, 15–20, 24–72 multiples of 3, 73–77

Above Average: Ex.3–12 multiples of 3, 15–20, 24–72 multiples of 3, 74–79

Advanced: Ex.3–12 multiples of 3, 15–20, 24–72 multiples of 3, 75–78, 79–89 odd

✪ More Difficult Exercises
Exercises 65–72, 78–89

▶ **Ex. 1–89** The Chapter Review exercises include a reference to the lesson in which a skill or concept was presented. Students may use this reference as a study aid if they are uncertain how to do a given exercise in the Chapter Review.

Answers

53. $x^2 - 7x + 6$
54. $x^2 - 6x - 7$
55. $2x^2 + x + 6$
56. $x^3 + 6x^2 + 33x + 156 + \frac{776}{x - 5}$
57. $x^4 - x + 3 - \frac{3}{x + 3}$
58. $x^3 + 6x^2 + 5x + 2 + \frac{1}{x - 1}$
59. $2x^3 + 5x^2 + 5x + 6 - \frac{3}{x - 1}$
60. $2x^4 - 10x^3 + 50x^2 - 256x + 1280 - \frac{6405}{x + 5}$
61. $3x^2 - 13x + 64 - \frac{316}{x + 5}$
62. $-2x^5 - 6x^4 - 15x^3 - 46x^2 - 138x - 413 - \frac{1241}{x - 3}$
63. $x^5 - 2x^4 + 4x^3 - 5x^2 + 4x - 8 + \frac{23}{x + 2}$
64. $6x^2 + 22x + 79 + \frac{322}{x - 4}$

In Exercises 45–52, find all real solutions of the equation. (9.3)

45. $x^2 - 6x = 27$ $-3, 9$
46. $x^2 + 9x - 10 = 0$ $-10, 1$
47. $x^3 + 3x^2 - x - 3 = 0$ $-3, \pm 1$
48. $x^3 - 4x^2 - 4x + 16 = 0$ $4, \pm 2$
49. $x^3 + x^2 - 16x = 16$ $-1, \pm 4$
50. $x^3 + 6x^2 - 25x = 150$ $-6, \pm 5$
51. $2x^3 - 3x^2 - 18x + 27 = 0$ $\frac{3}{2}, \pm 3$
52. $4x^3 - 8x^2 - x + 2 = 0$ $2, \pm\frac{1}{2}$

In Exercises 53–56, divide using long division. (9.4) See margin.

53. $(x^3 + 2x^2 - 57x + 54) \div (x + 9)$
54. $(x^3 - 13x^2 + 35x + 49) \div (x - 7)$
55. $(2x^3 - x^2 + 5x - 6) \div (x - 1)$
56. $(x^4 + x^3 + 3x^2 - 9x - 4) \div (x - 5)$

In Exercises 57–64, divide using synthetic division. (9.4) See margin.

57. $(x^5 + 3x^4 - x^2 + 6) \div (x + 3)$
58. $(x^4 + 5x^3 - x^2 - 3x - 1) \div (x - 1)$
59. $(2x^4 + 3x^3 + x - 9) \div (x - 1)$
60. $(2x^5 - 6x^2 - 5) \div (x + 5)$
61. $(3x^3 + 2x^2 - x + 4) \div (x + 5)$
62. $(-2x^6 + 3x^4 - x^3 + x - 2) \div (x - 3)$
63. $(x^6 + 3x^3 - 6x^2 + 7) \div (x + 2)$
64. $(6x^3 - 2x^2 - 9x + 6) \div (x - 4)$

In Exercises 65–72, find *all* solutions of the equation. (9.5, 9.6)

65. $x^3 - 9x^2 + 49x - 441 = 0$ $9, \pm 7i$
66. $x^4 + x^3 - x^2 + x - 2 = 0$ $-2, 1, \pm i$
67. $x^3 - x^2 + 16x - 16 = 0$ $1, \pm 4i$
68. $2x^3 - 3x^2 + 2x - 3 = 0$ $\frac{3}{2}, \pm i$
69. $x^4 + 2x^3 + x^2 + 8x - 12 = 0$ $-3, 1, \pm 2i$
70. $x^4 - 8x^3 - 8x^2 - 8x - 9 = 0$ $-1, 9, \pm i$
71. $2x^4 - 3x^3 + 7x^2 - 9x + 3 = 0$ $1, \frac{1}{2}, \pm i\sqrt{3}$
72. $3x^3 - 2x^2 + 3x = 2$ $\frac{2}{3}, \pm i$

73. *Grocery Sales* For 1980 through 1990, the number, G (in thousands), of grocery stores in the United States can be modeled by

$$G = 0.09t^2 - 1.6t + 172.5$$

where $t = 0$ represents 1980. The average sales, S (in thousands of dollars), per grocery store can be modeled by

$$S = -0.5t^2 + 90t + 1190.$$ $T = (0.09t^2 - 1.6t + 172.5)(-0.5t^2 + 90t + 1190)$, $310733.88 *million*

Write a model that represents the total sales, T (in millions of dollars), for grocery stores in the United States. Use the model to approximate the total sales in 1988. ***(Source: U.S. Department of Agriculture)***

74. *Foreign Investment in the United States* For 1970 through 1990, the total foreign investment, I (in millions of dollars), in the United States can be modeled by

$$I = 13t^4 - 414t^3 + 4958t^2 - 13453t + 13435$$

where $t = 0$ represents 1970. In what year was the investment $327,913 million? ***(Source: U.S. Bureau of Economic Analysis)***
1988

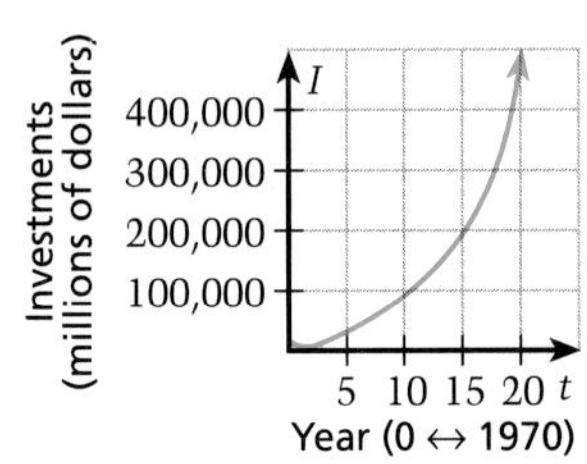

75. *Advertising* For 1970 through 1990, the total amount, T (in millions of dollars), spent on advertising in the United States can be modeled by

$$T = 257t^2 + 1016t + 19{,}100$$

where $t = 0$ represents 1970. The total amount, L (in millions of dollars), spent on local advertising can be modeled by

$$L = 114t^2 + 434t + 8450.$$

Write a model that represents the total amount, N (in millions of dollars), spent on national advertising. Use the model to approximate the amount spent on national advertising in 1986.
(Source: McCann-Erickson, Inc.) See margin.

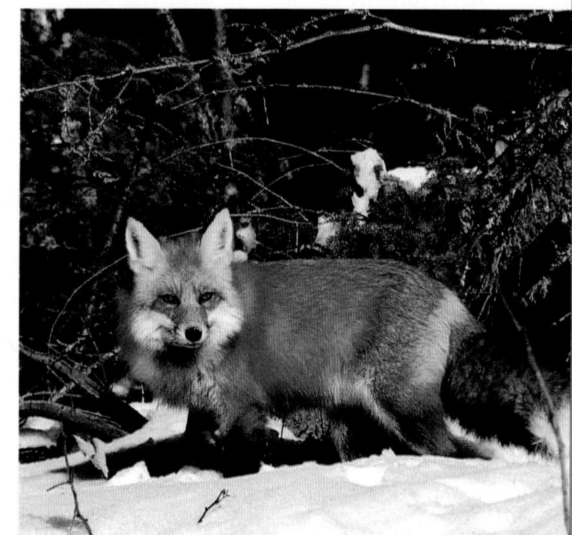

76. *Red Foxes* The weights (in pounds) of twenty red foxes is given below. Find the mean, range, and standard deviation for these data. 9.25, 3, ≈0.86

8.5	9.0	11.0	9.5	10.0
10.5	8.0	9.0	9.0	10.5
9.5	8.5	8.0	10.0	9.0
10.0	9.0	9.5	8.0	8.5

77. *Red Foxes* What percent of the weights in Exercise 76 lie within one standard deviation of the mean weight? 70%

78. *Market Testing* To test consumer reaction to a new type of potato chip dip, people in grocery stores were asked to taste the dip and then check the box by the picture that best describes their reaction to the dip. Each of the pictures was assigned a number from -3 to 3. The numbers at the right show the reactions of 25 people to the chip dip. Find the mean and standard deviation of these data. 0.52, ≈1.72

+2,	−1,	+3,	+1,	+3
+1,	0,	+1,	−3,	−1
+1,	+1,	+1,	0,	0
−2,	+3,	0,	+2,	+1
−1,	+3,	+2,	−3,	−1

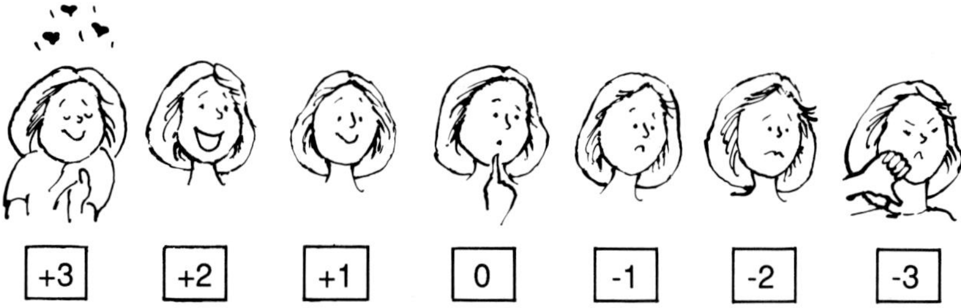

Answer

75. $N = 143t^2 + 582t + 10{,}650$; \$56,570 million

▶ **Ex. 79** Ben Field II collected all kinds of memorabilia from the 1889 centennial celebration of Washington's first inauguration. The box contained newspapers, magazines, and programs about the centennial, as well as centennial medals. Also included were Field family hymn books, photographs, letters, and journals.

Answer
79. 3, 15 in. by 8 in. by 19 in.

79. *One-Hundred-Year-Old Box* On April 30, 1989, Benjamin P. Field opened a box that had been in his family since 1889. His father had told him that the box had been sealed on April 30, 1889, and that it contained memorabilia of the centennial of George Washington's first inaugural. The box had a volume of 2280 cubic inches. The box's dimensions were $5x$ inches by $2(x + 1)$ inches by $6x + 1$ inches. What is x? What were the dimensions of the box? (To see what was in the box, look at the September 1989 issue of *National Geographic.*) See margin.

Ten Decades in America **In Exercises 80–89, each of the events, people, or objects is associated with a decade between 1890 and 1990. To find the decade, find the only real zero of the function. Then use the chart below to match the decade with the event, person(s), organization, activity, or object.**

a. 1890–1899	**b.** 1900–1909	**c.** 1910–1919	**d.** 1920–1929
e. 1930–1939	**f.** 1940–1949	**g.** 1950–1959	**h.** 1960–1969
i. 1970–1979	**j.** 1980–1989		

80. The Supremes: h, 1960's
$f(x) = 2x^3 - 16x^2 + x - 8$

81. Wright Brothers: b, 1900's
$f(x) = 2x^3 - x^2 - 4x - 4$

82. Shirley Temple: e, 1930's
$f(x) = x^3 - x^2 - 12x - 40$

83. Compact Discs: j, 1980's
$f(x) = x^3 - 10x^2 + x - 10$

84. United Nations: f, 1940's
$f(x) = x^3 - 6x^2 + 3x - 18$

85. Richie Valens: g, 1950's
$f(x) = 4x^3 - 28x^2 + 3x - 21$

86. World War I: c, 1910's
$f(x) = 3x^3 - 9x^2 + 7x - 21$

87. United States Bicentennial: i, 1970's
$f(x) = 3x^3 - 27x^2 + 4x - 36$

88. Chicago World's Fair: a, 1890's
$f(x) = x^3 - x^2 + 9x - 9$

89. Charleston (the dance): d, 1920's
$f(x) = 2x^3 - 8x^2 + x - 4$

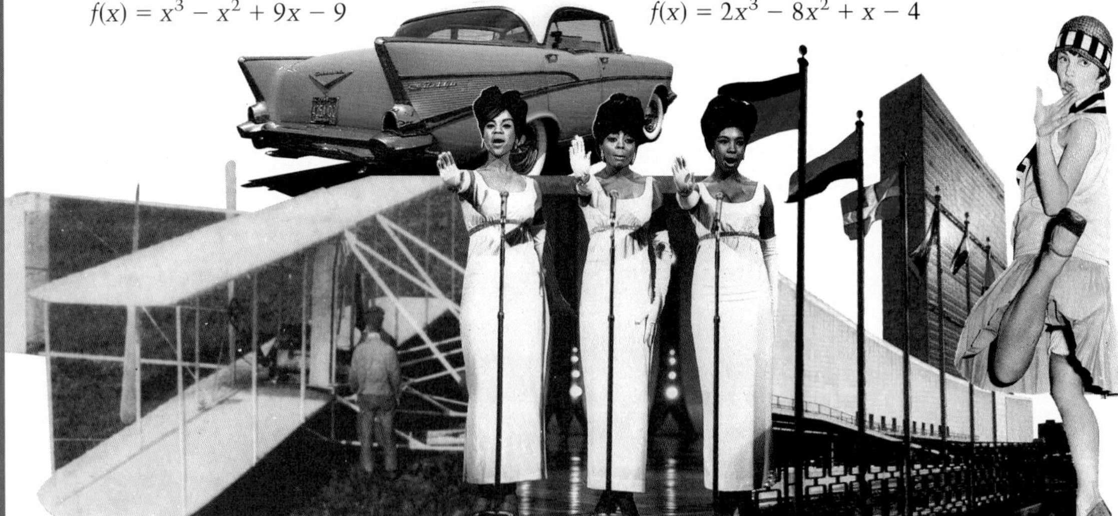

About the Photo
(From left to right:)
The Wright brothers and their plane, a 1957 Chevrolet Impala, the Supremes singing their hit "Stop in the Name of Love," the United Nations building, a woman doing the Charleston in a 1926 London stage review.

Chapter TEST

In Exercises 1 and 2, perform the indicated operation. (9.1)

1. $(4x^3 - 3x^2 + 5) - (x^3 - 2x^2 + x - 7)$ $3x^3 - x^2 - x + 12$

2. $(2x - 7)^2$ $4x^2 - 28x + 49$

In Exercises 3 and 4, sketch the graph of the function. (9.2)

3. $f(x) = x^4 - 2$ See margin.

4. $f(x) = (x - 5)^3$ See margin.

In Exercises 5 and 6, solve the equation. (9.3)

5. $x^3 - 10x^2 + 25x = 0$ 0, 5

6. $x^4 - 3x^3 + x - 3 = 0$ $-1, 3, \frac{1 \pm i\sqrt{3}}{2}$

In Exercises 7 and 8, completely factor the polynomial with respect to the integers. (9.3)

7. $2x^3 - 2x^2 - 12x$ $2x(x + 2)(x - 3)$

8. $x^4 - 3x^2 - 4$ $(x - 2)(x + 2)(x^2 + 1)$

9. Determine how many x-intercepts the graph of $f(x) = x^3 - x^2 - 3x + 3$ has. **(9.6)** Three

10. Write an equation of a polynomial function that has a leading coefficient of 1 and has $2i$, $-2i$, and 5 as its only zeros. **(9.6)** $f(x) = x^3 - 5x^2 + 4x - 20$

11. Find $f(4)$ for $f(x) = 5x^4 - 16x^3 - 14x^2 - 11x + 11$. **(9.4)** -1

12. Use long division to divide $(6x^3 - 17x^2 - 5x + 6)$ by $(2x - 1)$. **(9.4)** $3x^2 - 7x - 6$

13. Use synthetic division to divide $(12x^3 + 31x^2 - 17x - 6)$ by $(x + 3)$. **(9.4)** $12x^2 - 5x - 2$

14. Find the x-intercepts of the graph of $f(x) = x^3 - 19x + 30$. **(9.5)** $-5, 2, 3$

15. Find all zeros of $f(x) = x^3 - 4x^2 + 9x - 36$. (Hint: $3i$ is a solution.) **(9.6)** $-3i, 3i, 4$

16. Use the rational zero test to list all *possible* rational zeros of $f(x) = 5x^3 + 19x^2 + 22x + 8$. **(9.5)** $\pm 1, \pm 2, \pm 4, \pm 8, \pm\frac{1}{5}, \pm\frac{2}{5}, \pm\frac{4}{5}, \pm\frac{8}{5}$

17. Which of the possible rational zeros listed in Exercise 16 are actual zeros of $f(x) = 5x^3 + 19x^2 + 22x + 8$? **(9.5)** $-1, -2, -\frac{4}{5}$

18. The circle graph at the right shows the approximate number of calories per serving for eight types of pies. Find the mean and the standard deviation of the eight numbers. (Calories are based on a one-eighth serving of a 9-inch pie.) 304, 56.6

A Piece Of The Pie

How many calories in your slice of pie? Check the "pie chart" for your favorites

Pumpkin 241
Pecan 431
Coconut Custard 268
Raisin 319
Apple 302
Mincemeat 320
Sweet Potato 243
Cherry 308

In Exercises 19 and 20, use the following information.

The average number of miles driven annually by men and by women can be modeled by

$f(t) = 3.14t^2 + 177.52t + 11{,}660.61$ **Men**

$g(t) = 13.72t^2 - 74.30t + 5421.10$ **Women**

where $t = 0$ represents 1970. ***(Source: U.S. Department of Transportation)***

19. In which year did men drive 15,586 miles? 1987

20. In which year did women drive 5470 miles? 1976

Answers

3.

4.

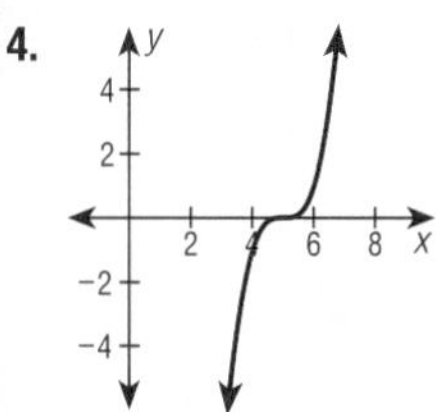

Answers

12.

13.

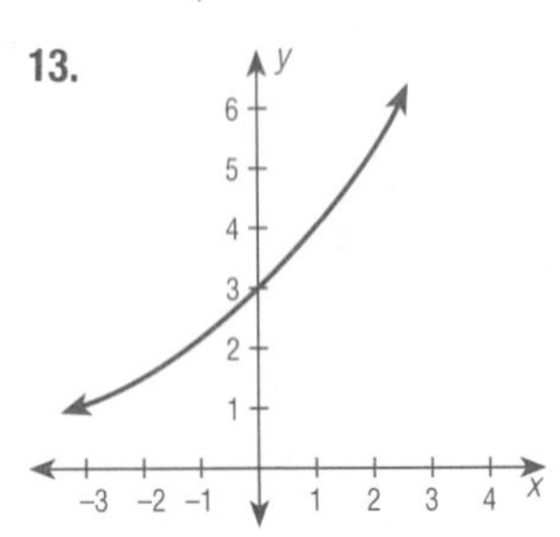

14.

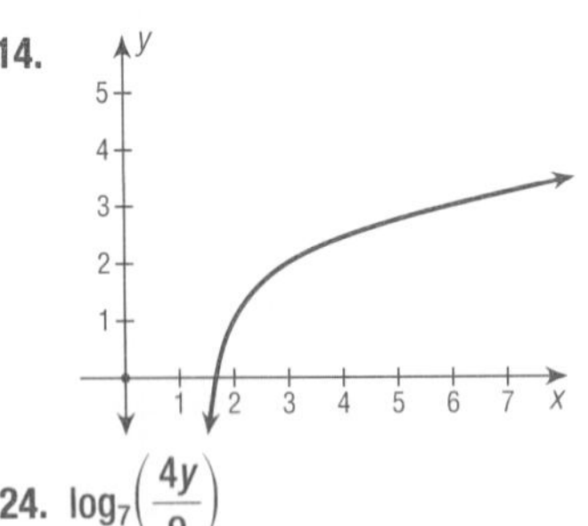

24. $\log_7\left(\frac{4y}{9}\right)$

Cumulative REVIEW ■ Chapters 7–9

In Exercises 1–3, simplify the expression.

1. $3\sqrt[5]{5} + 4\sqrt[5]{5}$ $7\sqrt[5]{5}$
2. $7\sqrt[3]{9} - 2\sqrt[3]{72}$ $3\sqrt[3]{9}$
3. $-4\sqrt[4]{405} + \sqrt[4]{80}$ $-10\sqrt[4]{5}$
4. Find the geometric mean of 45 and 125. 75
5. The geometric mean of 27 and x is 9. What's x? 3

In Exercises 6–11, solve the equation.

6. $\sqrt{x} - 2 = 7$ 81
7. $3x^{1/2} + 4 = 14$ $\frac{100}{9}$
8. $-2\sqrt{x+5} + 6 = -10$ 59
9. $\frac{1}{2}(x-6)^{1/3} + 7 = 6$ -2
10. $5\sqrt[3]{x+2} - 6 = -16$ -10
11. $3(2x+1)^{1/2} - 4 = 5$ 4

In Exercises 12–14, sketch the graph of the function. 12.–14. See margin.

12. $f(x) = 2\left(\frac{1}{4}\right)^x$
13. $f(x) = 3\left(\frac{4}{3}\right)^x$
14. $f(x) = \sqrt{x-2} + 1$

In Exercises 15–20, solve the equation.

15. $\log_2 x = 3$ 8
16. $\ln x = 10$ $\approx 22{,}026.466$
17. $3e^x = 8$ ≈ 0.981
18. $10^x = 4$ ≈ 0.602
19. $6 + \frac{1}{2}e^{2x} = 9$ ≈ 0.896
20. $\frac{1}{4}(3)^x = \frac{3}{4}$ 1

In Exercises 21–23, expand the expression.

21. $\log_{10} 3x^2y^3$ $\log_{10} 3 + 2\log_{10} x + 3\log_{10} y$
22. $\ln 4xy^2$ $\ln 4 + \ln x + 2\ln y$
23. $\ln\left(\frac{6x}{5y}\right)$ $\ln 6 + \ln x - \ln 5 - \ln y$

In Exercises 24–26, condense the expression.

24. $\log_7 4 + \log_7 y - 2\log_7 3$ See margin.
25. $\frac{1}{4}\ln 3 - \ln 4$ $\ln\frac{3^{1/4}}{4}$
26. $\frac{3}{2}(\log_2 x + \log_2 y)$ $\log_2(xy)^{3/2}$

In Exercises 27–32, sketch the graph of the function. 27.–32. See Additional Answers.

27. $f(x) = 1 + \ln(x-2)$
28. $f(x) = e^{x+2} - 4$
29. $f(x) = \log_{10}(4x-1)$
30. $f(x) = 3\left(\frac{1}{2}\right)^x + 2$
31. $f(x) = (x+1)^3 + 2$
32. $f(x) = x^4 - 6$

In Exercises 33–38, perform the indicated operation.

33. $(4x^2 + 3x - 1) + (5x^3 + 2x^2 + 5)$
34. $(9x + 3) + (4x^4 - x^3 + 2x + 6)$
35. $(3x^3 + 2x + 7) - (4x^2 - 3x + 2)$
36. $(9x^4 - x^2 + 6) - (x^3 + 5x + 5)$
37. $(2x+1)(x^3 + 2x - 3)$ $2x^4 + x^3 + 4x^2 - 4x - 3$
38. $(3x-1)^3$ $27x^3 - 27x^2 + 9x - 1$

33. $5x^3 + 6x^2 + 3x + 4$ 34. $4x^4 - x^3 + 11x + 9$ 35. $3x^3 - 4x^2 + 5x + 5$ 36. $9x^4 - x^3 - x^2 - 5x + 1$

In Exercises 39 and 40, use long division. In Exercises 41 and 42 use synthetic division.

39. $(x^3 + 3x + 2) \div (x - 7)$ See below.
40. $(x^4 + 5x^2 - x - 5) \div (x + 1)$ See below.
41. $(2x^3 + 4x^2 - 3x + 1) \div (x - 5)$ See below.
42. $(-6x^4 + 3x^3 - x + 6) \div (x + 2)$ See below.

In Exercises 43–46, find all solutions of the equation.

43. $2x^3 - 3x^2 - 5x + 6 = 0$ $-\frac{3}{2}, 1, 2$
44. $x^4 - 12x^2 + 27 = 0$ $\pm 3, \pm\sqrt{3}$
45. $2x^4 + x^3 - 8x^2 - x + 6 = 0$ $-2, -1, 1, \frac{3}{2}$
46. $x^4 + 3x^3 - 2x^2 + 6x - 8 = 0$ $-4, 1, \pm\sqrt{2}i$

39. $x^2 + 7x + 52 + \frac{366}{x-7}$ 40. $x^3 - x^2 + 6x - 7 + \frac{2}{x+1}$ 41. $2x^2 + 14x + 67 + \frac{336}{x-5}$

42. $-6x^3 + 15x^2 - 30x + 59 - \frac{112}{x+2}$

Depreciation **In Exercises 47 and 48, use the following information.**

You are an accountant for a company that has just purchased a delivery truck for \$30,000. For tax purposes, the truck is to be depreciated using the *double-declining balances method.* With this method, each year the "book-value" of the truck drops by $\frac{2}{n}$ of the previous year's value, where n is the number of years. If the useful life of the truck is estimated to be 5 years, then the book value, V, in year t is given by

$$V = 30{,}000 \cdot \left(1 - \frac{2}{5}\right)^t.$$

47. What is the book value of the delivery truck after 2 years? **\$10,800**

48. When is the book value of the truck equal to \$3880? **After 4 years**

Tennessee **In Exercises 49 and 50, use the grid superimposed on the map of Tennessee. Each unit of the grid represents 8 miles.**

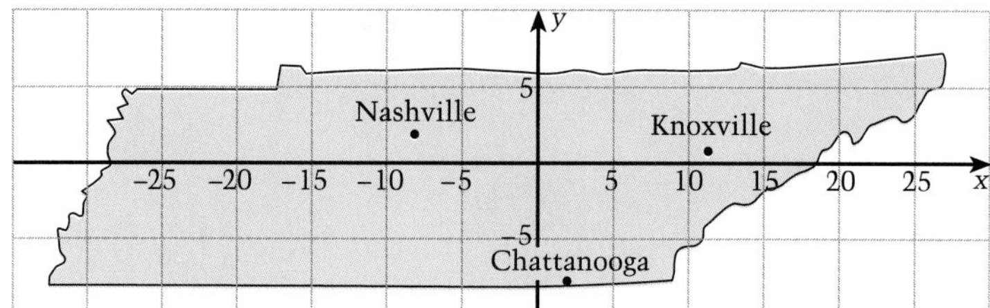

49. You are riding in a "bike-a-thon" fundraiser from Knoxville to Nashville, Tennessee. Your pledges total \$12 per mile. How much money will you raise by traveling the entire distance? **≈\$1922**

50. Approximately how much closer is Chattanooga to Knoxville than Nashville is to Chattanooga? **≈5.5 mi**

Acid and Base **In Exercises 51 and 52, use the formula for pH on page 451.**

51. The pH of cola is 3. What is its concentration of hydrogen ions? **0.001 mole per liter**

52. Hydrochloric acid secreted by the stomach after a meal has a hydrogen ion concentration of 0.0012 moles per liter. What is the pH of hydrochloric acid? **2.92**

Major League Baseball **In Exercises 53 and 54, use the following information.**

For 1982 through 1990, the revenue, R (in millions of dollars), and expenses, E (in millions of dollars), for major league baseball can be modeled by $R = 117t + 113$ and $E = 8t^2 - 29t + 633$ where $t = 3$ represents 1983.

53. In what year did major league baseball have expenses of about \$1020 million? **1989**

54. In what year did major league baseball have a loss of about 44 million dollars? **1984**

Many real-life situations can be described by using rational expressions. Numerous examples in physics, economics, and social studies will be shown in the lesson applications. Recognizing various types of problems that utilize rational expressions is a primary objective of this chapter. A secondary objective is the development of techniques that students can use effectively to solve problems.

The Chapter Summary on page 561 provides you and the students with a synopsis of the chapter. It identifies key skills, concepts, and vocabulary. You may want to have students look at the Chapter Summary as an overview before beginning the chapter.

CHAPTER 10

Rational Functions

LESSONS

Fire fighters, such as these in New York City, must handle water that is under great pressure. The speed of this pressurized water in their hoses is inversely proportional to the diameter of the hose.

Real Life
Fire Fighting

Speed of Water Flowing at a Rate of 2 ft³/sec

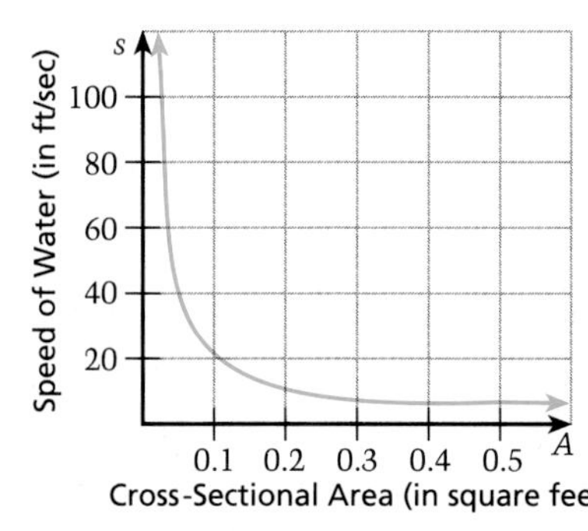

A fire department may use different size hoses for fighting fires. When the water flow remains constant, water flows at a faster speed in a narrower hose than in a wider one. The speed of water, s (in feet per second), varies inversely with the cross-sectional area, A (in square feet), of the hose.

$$s = \frac{k}{A}$$

If a hose has a cross-sectional area of 0.02 square feet, and the speed of water flowing through the hose is 100 feet per second, you can conclude that $k = 2$ cubic feet per second. This means that using the same pump, a hose with a cross-sectional area of 0.05 square feet will flow at a speed of 40 feet per second.

Hydrodynamics is the branch of physics that studies the behavior of liquids that are in motion. For example, The Principle of Continuity in Liquid Flow states that the velocity of a liquid flowing through a pipe increases as the cross-sectional area of the pipe decreases, and decreases as the cross-sectional area of the pipe increases.

On a given day, a fire department uses the same pump to put out two fires. The rate of water is given by the rational function

$$r = \frac{1250}{A},$$

where r is the rate of water in gallons per minute, and A is the cross-sectional area of the hose in square inches. Does this function obey the Principal of Continuity in Liquid Flow? Yes. As the cross-sectional area of the hose increases, the velocity of the water decreases, and as the cross-sectional area of the hose decreases, the velocity of the water increases.

During the morning fire, the fire department used a hose with a cross-sectional area of 5 square inches. What was the velocity of the water? 250 gallons per minute

During the evening fire, the velocity of the water was 100 gallons per minute. What was the cross-sectional area of the hose? 12.5 square inches

Rational Functions

PACING CHART

The daily Pacing Chart is meant to help you adjust your teaching pace. Students in the full course should finish the entire text by the end of the year. Students in the basic course are expected to complete the first twelve chapters. The Pacing Chart for each chapter contains suggestions for lessons that require more than one day and lessons that may be omitted for the basic course.

DAY	FULL COURSE	BASIC COURSE
1	10.1 & Using a Graphing Calculator	10.1
2	10.2	10.1 & Using a Graphing Calculator
3	10.3	10.2
4	Mid-Chapter Self-Test	10.2
5	10.4	10.3
6	10.5	Mid-Chapter Self-Test
7	10.6	10.4
8	Chapter Review	10.4
9	Chapter Test	10.5
10		10.6
11		Chapter Review
12		Chapter Test

CHAPTER ORGANIZATION

LESSON	PAGES	GOALS	MEETING THE NCTM STANDARDS
10.1	518–523	1. Graph a rational function using asymptotes 2. Use rational functions as real-life models	Problem Solving, Communication, Connections, Functions
Using a Calculator	524–525	Graph rational functions that have asymptotes	Technology
10.2	526–531	1. Create and use real-life models using inverse variation 2. Create and use models of real-life situations that use joint variation	Problem Solving, Communication, Connections, Calculus concepts
Mixed Review	532	Review algebraic and arithmetic skills	
Milestones	532	Abel and Galois	Connections
10.3	533–539	1. Multiply and divide rational expressions, writing the result in simplest form 2. Use rational expressions as real-life models	Problem Solving, Communication, Reasoning, Connections, Geometry (synthetic)
Mid-Chapter Self-Test	540	Diagnose student weaknesses and remediate with correlated Reteach worksheets	
10.4	541–547	1. Solve equations that contain rational expressions 2. Use rational equations to solve real-life problems	Problem Solving, Communication, Connections
10.5	548–554	1. Add and subtract rational expressions and simplify complex fractions 2. Use rational expressions as models of real-life situations	Problem Solving, Communication, Connections
Mixed Review	554	Review algebraic and arithmetic skills	
10.6	555–560	1. Construct an amortization table for a loan 2. Find the monthly payment for an installment loan	Problem Solving, Communication, Connections, Discrete Math
Chapter Summary	561	Restate for students what they have learned, why they have learned it, and how it fits into the structure of algebra	Structure, Connections
Chapter Review	562–566	Review concepts and skills learned in the chapter	
Chapter Test	567	Diagnose student weaknesses and remediate with correlated Reteaching worksheets	

LESSON RESOURCES

MEETING INDIVIDUAL NEEDS

RETEACHING For students who need to spend more time on basics:

If a mid-chapter test or chapter test indicates a deficiency, teachers can help students with the appropriate ***Reteaching Copymaster.***

PRACTICE For students who need more practice:

Additional exercises similar to those in the Pupil's Edition are provided for each lesson in ***Extra Practice Copymasters.***

ENRICHMENT For enriching and broadening students' experiences:

Problem of the Day copymasters in ***Teaching Tools*** provide a daily opportunity to use logical reasoning, looking for a pattern, writing an equation, and other routine and non-routine problem-solving strategies.

Math Log copymasters in ***Alternative Assessment*** provide opportunities to report on investigations, research, and open-ended problems.

Technology: Using Calculators and Computers provides enriching activities with graphing and scientific calculators and computers.

The ***Applications Handbook*** provides additional information about the cross-curriculum topics such as astronomy, chemistry, physics, sports, economics, genetics, and music that are integrated into the Pupil's Edition.

LESSON	10.1	10.2	10.3	10.4	10.5	10.6
PAGES	518-523	526-531	533-539	541-547	548-554	555-560
Teaching Tools						
Transparencies	✓			✓		
Problem of the Day	✓	✓	✓	✓	✓	✓
Warm-up Exercises	✓	✓	✓	✓	✓	✓
Answer Masters	✓	✓	✓	✓	✓	✓
Extra Practice Copymasters	✓	✓	✓	✓	✓	✓
Reteaching Copymasters	Teacher-directed and independent activities tied to results on the Mid-Chapter Self-Tests and Chapter Tests					
Color Transparencies	✓	✓		✓		✓
Applications Handbook	Additional background information for many real-life applications					
Technology	Calculator and computer worksheets for appropriate lessons					
Complete Solutions Manual	✓	✓	✓	✓	✓	✓
Alternative Assessment	Assess student's ability to reason, analyze, solve problems, and communicate using mathematical language.					
Formal Assessment	Mid-Chapter Self-Tests, Chapter Tests, Cumulative Tests, and Practice for College Entrance Tests					
Computer Test Bank	Customized tests can be created by choosing from over 2500 items.					

INSIGHTS

10.1 Graphs of Rational Functions

In preparing to graph rational functions, students must be more than usually selective about their choice of *x*-values. They are reminded of the meaning of restricted values, asymptotes, and the relationship between a graph and its asymptotes. Once again, students learn the importance of a visual representation in helping to understand concepts such as restricted domain, discontinuity, limiting values, etc.

10.2 Inverse and Joint Variation

A common example of a simple rational function is the relationship $f(x) = \frac{k}{x}$, with the vertical and horizontal axes as asymptotes. This function models the relationship of inverse variation. The lesson first deals with real-life applications of inverse variation. As a reminder that variation in real life is not always a simple direct variation, as in lesson 2.4, nor a simple inverse variation as $f(x) = \frac{k}{x}$, the student is also introduced to joint variation.

10.3 Multiplying and Dividing Rational Expressions

Using rational expressions as models can be much less formidable if the expressions can first be simplified. This is where the students' skill at factoring can be applied, since common factors in the numerator and the denominator can be divided out (with the necessary cautions regarding restricted values and avoiding a very common error). Students are then walked through the multiplication and division of rational expressions.

10.4 Solving Rational Equations

To the multiplication and division skills taught in lesson 10.3, this lesson adds the addition and subtraction of simple rational expressions, as a means to solving rational equations. The lesson deals with the simplest kind of rational equation, where cross-multiplication can be applied, and also with the equation where a common denominator must be applied. The students are cautioned about extraneous solutions, and are encouraged to use the quick-graph method for checking solutions. This is applied in an interesting way at the end of Example 4.

10.5 Addition, Subtraction, and Complex Fractions

The fractional sums and differences presented in the preceding lesson contained relatively simple rational expressions. In this lesson, students learn to manipulate more complex expressions, including complex fractions. Example 4, and Exercises 41 through 44, show how real-life models can be complex enough to require this type of manipulation.

10.6 Exploring Data: The Algebra of Finance

This is one of two lessons (the other is 12.6) dealing with a topic that deeply affects the financial "health" of the average citizen. Students are presented with the tabular data relating to monthly installment payments on a loan such as a mortgage, and with the algebraic model that generates the data. They are introduced to the relationship between the term of a loan, the interest rate, and the total interest paid. Throughout the exercises, students are encouraged to do some "comparison shopping" for car loans and mortgages.

Problem of the Day

How can a square cake be cut into six pieces so that all the pieces have the same volume and the same amount of frosting?

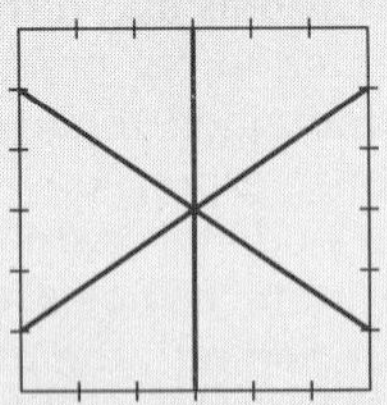

ORGANIZER

Warm-Up Exercises

1. Evaluate each fraction for the given values of x.

$\frac{x-2}{x-4}, \frac{x+2}{x^2+x-20}$

a. $x = 2$ b. $x = -2$

c. $x = \frac{2}{3}$ d. $x = -5$

a. $0, -\frac{2}{7}$ b. $\frac{2}{3}, 0$

c. $\frac{2}{5}, -\frac{12}{85}$ d. $\frac{7}{9}$, undefined

2. Divide.

a. $24x^2 - 32x$ by $8x$

b. $(3x^2 + 4x - 14)$ by $(x + 1)$

c. $(18x^3 - 3x^2 + 6)$ by $(2x - 3)$

a. $3x - 4$

b. $(3x + 1) + \frac{-15}{x+1}$

c. $9x^2 + 12x + 18 + \frac{60}{2x-3}$

Lesson Resources

Teaching Tools

Transparencies: 2, 3, 7, 8, 21

Problem of the Day: 10.1

Warm-up Exercises: 10.1

Answer Masters: 10.1

Extra Practice: 10.1

Color Transparency: 54

10.1 Graphs of Rational Functions

What you should learn:

Goal 1 How to graph a rational function using asymptotes

Goal 2 How to use rational functions as real-life models

Why you should learn it:

You can model many real-life situations with rational functions, such as average costs over time.

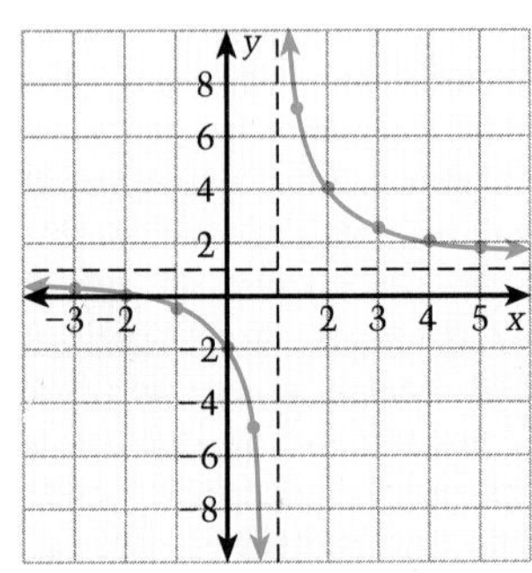

Goal 1 Graphing a Rational Function

A **rational function** is a function of the form

$$f(x) = \frac{p(x)}{q(x)}, \qquad q(x) \neq 0$$

where $p(x)$ and $q(x)$ are polynomials in x. The *domain* of a rational function consists of the values of x for which the denominator $q(x)$ is not zero. For instance, the domain of

$$f(x) = \frac{x+2}{x-1}$$

is all real numbers except $x = 1$. When sketching the graph of a rational function, you should give special attention to the shape of the graph near x-values that are not in the domain.

Example 1 *Sketching the Graph of a Rational Function*

Sketch the graph of $f(x) = \frac{x+2}{x-1}$.

Solution Begin by noticing that the domain is all real numbers except $x = 1$. Next construct a table of values, including x-values that are close to 1 on the left *and* on the right.

x-Values to the Left of 1

x	−3	−2	−1	0	0.5	0.9
f(x)	0.25	0	−0.5	−2	−5	−29

x-Values to the Right of 1

x	1.1	1.5	2	3	4	5
f(x)	31	7	4	2.5	2	1.75

Plot the points to the left of 1 and connect them with a smooth curve. Do the same for the points to the right of 1. *Do not* connect the two portions of the graph, which are called **branches.** ■

For the graph of the function in Example 1, as x approaches 1 from the left, the values of $f(x)$ approach negative infinity. As x approaches 1 from the right, the values of $f(x)$ approach positive infinity.

An **asymptote** of a graph is a line to which the graph becomes arbitrarily close as $|x|$ or $|y|$ increases without bound. In other words, if a graph has an asymptote, then it is possible to move far enough from the origin so that there is almost no difference between the graph and the asymptote.

The graph in Example 1 has two asymptotes: the line $x = 1$ is a **vertical asymptote,** and the line $y = 1$ is a **horizontal asymptote.** Here are some other examples.

Graph of $f(x) = \dfrac{1}{x^2 + 1}$
Horizontal asymptote: $y = 0$
Vertical asymptote: None

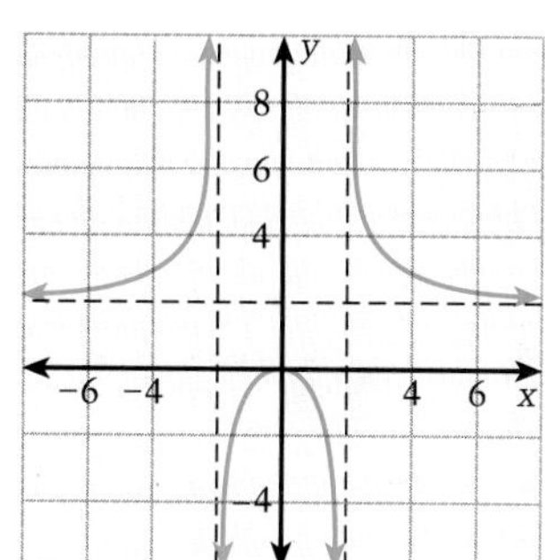

Graph of $f(x) = \dfrac{2x^2}{x^2 - 4}$
Horizontal asymptote: $y = 2$
Vertical asymptote: $x = 2, x = -2$

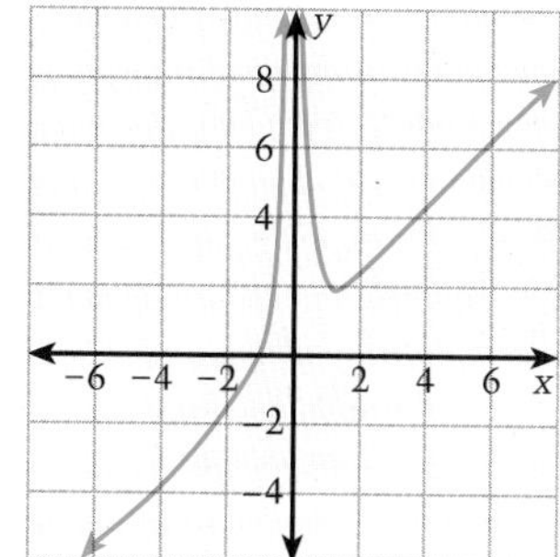

Graph of $f(x) = \dfrac{x^3 + 1}{x^2}$
Horizontal asymptote: None
Vertical asymptote: $x = 0$

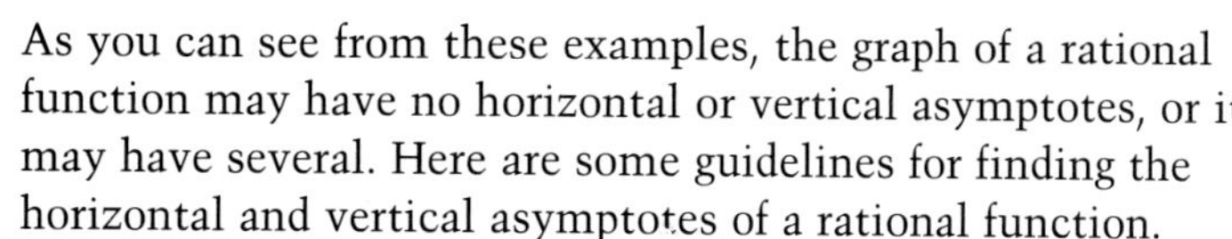

As you can see from these examples, the graph of a rational function may have no horizontal or vertical asymptotes, or it may have several. Here are some guidelines for finding the horizontal and vertical asymptotes of a rational function.

Horizontal and Vertical Asymptotes

Let $f(x) = \dfrac{p(x)}{q(x)}$ where $p(x)$ and $q(x)$ have *no common factors.*

1. The graph of f has a vertical asymptote at *each* real zero of $q(x)$.
2. The graph of f has, at most, one horizontal asymptote.
   - If the degree of $p(x)$ is less than the degree of $q(x)$, then the line $y = 0$ is a horizontal asymptote.
   - If the degree of $p(x)$ is equal to the degree of $q(x)$, then the line $y = \dfrac{a}{b}$ is a horizontal asymptote, where a is the leading coefficient of $p(x)$ and b is the leading coefficient of $q(x)$.
   - If the degree of $p(x)$ is greater than the degree of $q(x)$, then the graph has no horizontal asymptote.

Try applying these guidelines to the three graphs shown above.

LESSON Notes

Example 1

Stress to students the importance of finding the domains of rational functions. Point out how the graph represents the domain geometrically.

Example 2

GOAL 2 This example gives a real-life situation whose model is a rational function. Ask students to determine whether the average cost will decrease indefinitely. The average cost is bounded below by the asymptote $y = 3.25$. Therefore, the average cost will never be less than \$3.25.

Extra Examples

Here are additional examples similar to **Examples 1–2.**

1. Sketch the graph of each function. Include the asymptotes on the graphs.

   a. $f(x) = \dfrac{x - 2}{x + 2}$

   b. $g(x) = \dfrac{2(x^2 - 9)}{x^2 - 4}$

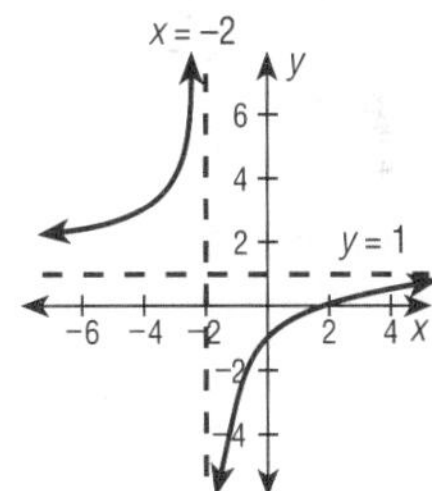

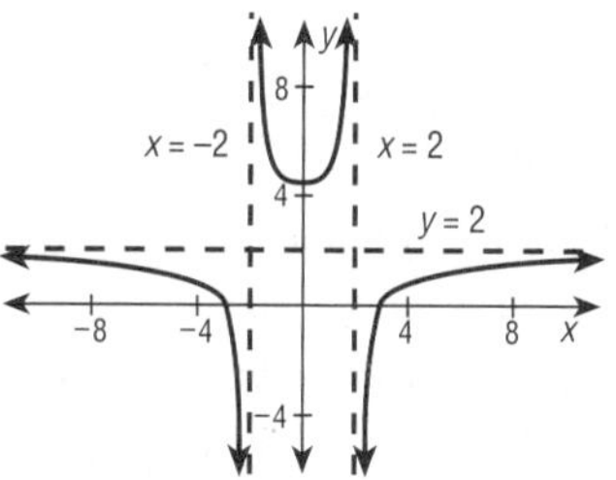

2. In Example 2, suppose that the one-time charges are reduced to \$775. Write a model that represents the average cost per calendar.
   $A = \dfrac{3.25x + 775}{x}$

Check Understanding

1. What is a rational function? See the top of page 518.
2. How do you find the domain of a rational function? See the top of page 518.
3. What is the vertical asymptote of the graph of the model that represents the average cost per calendar in Example 2? $x = 0$

Communicating about ALGEBRA

Making a table of (x, y) values can, of course, be used to help in graphing rational functions. However, students can gain a greater understanding of rational functions, needed for calculus, by studying the guidelines for horizontal and vertical asymptotes as listed on page 519. If you have not already done so, be sure to have students work together in small groups to apply these guidelines to the three graphs shown at the top of page 519 and to those of Extra Example 1. For example, consider the function $g(x) = \frac{2(x^2 - 9)}{x^2 - 4}$. Let $p(x) = 2(x^2 - 9)$ and let $q(x) = x^2 - 4$. $p(x)$ and $q(x)$ have no common factors.

(1) The real zeros of $q(x)$ occur at $x = 2$ and $x = -2$ so the graph of g has vertical asymptotes at $x = 2$ and $x = -2$.

(2) The graph of $g(x)$ can have, at most, one horizontal asymptote. $p(x)$ and $q(x)$ have the same degree, so the horizontal asymptote has the form $y = \frac{a}{b}$, where a and b are the leading coefficients of $p(x)$ and $q(x)$, respectively. So, the horizontal asymptote of $g(x)$ is $y = \frac{a}{b} = \frac{2}{1}$, or $y = 2$.

The work of the Dutch artist M. C. Escher (1889–1972) has intrigued many people. Escher's drawings involve repeated patterns and optical illusions. For instance, the Escher piece at the right, Drawing Hands, *shows the impossible, but fun to look at, illusion of two hands drawing each other.*

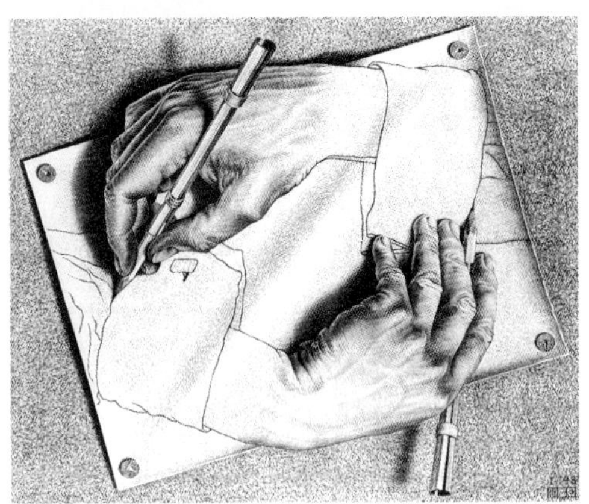

Goal 2 Using Rational Functions in Real Life

Example 2 *Finding the Average Cost*

As a fund-raising project, your math club is publishing a calendar with an Escher drawing on the cover. The cost of photography, typesetting, printing, and using the Escher drawing is \$850. In addition to these one-time charges, the *unit cost* of printing each calendar is \$3.25. Let x represent the number of calendars that your club has printed. Write a model that represents the average cost per calendar.

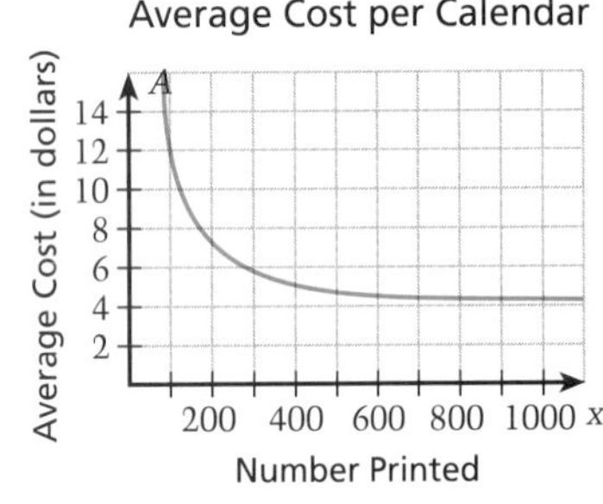

Solution The total cost, C, of printing x calendars is $C = 3.25x + 850$. The average cost, A, per calendar for printing x calendars is given by

$$A = \frac{3.25x + 850}{x}.$$

From the graph at the left, notice that the average cost decreases as the number of calendars increases. ■

Communicating about ALGEBRA

SHARING IDEAS about the Lesson See below.

A. Use the model in Example 2. Copy and complete the table.

Number of calendars, x	10	100	1000	10,000
Average cost, A (in dollars)	?	?	?	?

B. What is the horizontal asymptote of the graph shown in Example 2? What is the real-life significance of this asymptote?

A. 88.25, 11.75, 4.1, 3.335 B. $y = 3.25$, the club needs to charge at least \$3.25 per calendar.

EXERCISES

Guided Practice

CRITICAL THINKING about the Lesson

1. Explain what is meant by "the domain of a rational function." See page 518.

2. Complete the table for $f(x) = \frac{x^2+1}{2x^2-4}$. Then sketch the graph of f. ≈0.607, 1.25, ≈3.179, −37, −1, −0.25, −1, −37, ≈3.179, 1.25, ≈0.607 See Additional Answers.

x	−4	−2	−1.6	−1.4	−1	0	1	1.4	1.6	2	4
f(x)	?	?	?	?	?	?	?	?	?	?	?

3. Give an example of a rational function whose graph has two vertical asymptotes: $x = 1$ and $x = 5$. $f(x) = \frac{1}{x^2 - 6x + 5}$

In Exercises 4–6, identify all horizontal and vertical asymptotes of the graph of the function.

4. $f(x) = \frac{10x^2+3}{5x^2}$ $y = 2, x = 0$

5. $f(x) = \frac{3x^4-2}{2x-1}$ $x = \frac{1}{2}$

6. $f(x) = \frac{4x^2}{8x^3-1}$ $y = 0, x = \frac{1}{2}$

Independent Practice

In Exercises 7–12, state the domain of the function. 7.–12. All real numbers except:

7. $f(x) = \frac{3x}{x+16}$ $x = -16$

8. $f(x) = \frac{9}{2x-4}$ $x = 2$

9. $f(x) = \frac{x+4}{x^2+2x+1}$ $x = -1$

10. $f(x) = \frac{x+6}{(x+1)(x-8)}$ $x = -1$ and $x = 8$

11. $f(x) = \frac{1+x}{1-x}$ $x = 1$

12. $f(x) = \frac{2}{x^2-4x-12}$ 12. $x = 6$ and $x = -2$

13. Complete the table for $f(x) = \frac{x-2}{x-3}$.

x	2	2.5	2.75	3.25	3.5	4
f(x)	?	?	?	?	?	?
	0	−1	−3	5	3	2

14. Complete the table for $f(x) = \frac{x+2}{x-2}$.

x	1	1.5	1.75	2.25	2.5	3
f(x)	?	?	?	?	?	?
	−3	−7	−15	17	9	5

In Exercises 15–26, identify all horizontal and vertical asymptotes of the graph of the function. See margin.

15. $f(x) = \frac{1}{x-5}$

16. $f(x) = \frac{x}{x^2-9}$

17. $f(x) = \frac{7x}{x^3+1}$

18. $f(x) = \frac{3x^2-1}{x^3}$

19. $f(x) = \frac{6x^2+3}{x-1}$

20. $f(x) = \frac{5x^3-4}{x^2+4x-5}$

21. $f(x) = \frac{3x^3+30}{2x^3}$

22. $f(x) = \frac{2x^2+x-9}{3x^2-12}$

23. $f(x) = \frac{12x^4+10x-3}{3x^4}$

24. $f(x) = \frac{13x^4+x^2}{6x+3}$

✪ **25.** $f(x) = \frac{4x^2}{x^3-x^2-2x}$

✪ **26.** $f(x) = \frac{11x}{5x^3+40}$

EXERCISE Notes

ASSIGNMENT GUIDE

Basic/Average: Ex. 1–6, 9–21 odd, 27–30, 31–37 odd, 43–45, 53–55

Above Average: 1–6, 9–23 odd, 26–30, 31–37 odd, 43–46, 53–55

Advanced: 1–6, 9–23 odd, 26–30, 31–37 odd, 43–46, 53–57

Selected Answers
Exercises 1–6, 7–55 odd

✪ More Difficult Exercises
25, 26, 43–46, 56, 57

Guided Practice

▶ **Ex. 1** The denominator of a rational expression cannot be zero.

▶ **Ex. 2** Ask students why $f(-x) = f(x)$ for the values given for x in the table.

Independent Practice

▶ **Ex. 13–14** Discuss the behavior of the graph of $f(x)$ as it approaches its vertical asymptote.

▶ **Ex. 15–26** Refer students to page 519 of the lesson. For students who plan to take Precalculus or Calculus, obtaining information about asymptotes is helpful in curve sketching.

Answers

15. $y = 0; x = 5$
16. $y = 0; x = 3, x = -3$
17. $y = 0; x = -1$
18. $y = 0; x = 0$
19. $x = 1$
20. $x = -5, x = 1$
21. $y = \frac{3}{2}; x = 0$
22. $y = \frac{2}{3}; x = 2, x = -2$
23. $y = 4; x = 0$
24. $x = -\frac{1}{2}$
25. $y = 0; x = -1, x = 2, x = 0$
26. $y = 0; x = -2$

Ex. 27–30 Assign these exercises as a group. Help students focus on the location of the vertical asymptotes of the graph of each function.

WRITING Have students give a one-sentence justification for their answers.

Ex. 31–42 Students should locate any horizontal or vertical asymptotes, construct an appropriate table of values, and then sketch the graph. Students can also use graphics calculators. If students use graphics calculators, they should use the dot mode. "Using a Graphing Calculator" on page 524 shows students how to make use of this mode when graphing rational functions.

Ex. 43–44 Assign these exercises as a pair.

Ex. 45 TECHNOLOGY This is an important question. Students should use a graphics calculator to approximate the hospital costs in 1990 ($t = 20$). about $586 per day

The graph of this model has a horizontal asymptote of $y = 0$ and no vertical asymptotes. Using the range

Xmin = −50
Xmax = 100
Xscl = 10
Ymin = −50
Ymax = 600
Yscl = 50

students can "visualize" the tendency of the data beyond 1990. In this case the model is appropriate only for 1970–1990 and should not be used to predict future trends.

In Exercises 27–30, match the function with its graph.

27. $f(x) = \dfrac{x+3}{x-2}$ b

28. $f(x) = \dfrac{-8}{x^2-4}$ a

29. $f(x) = \dfrac{3}{x+2}$ d

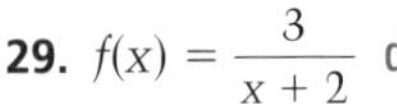

30. $f(x) = \dfrac{x-3}{2x+4}$ c

a.

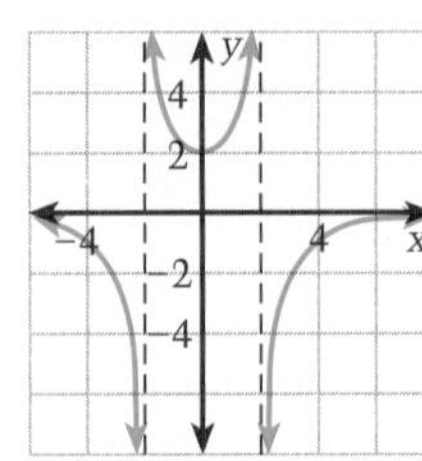

b.

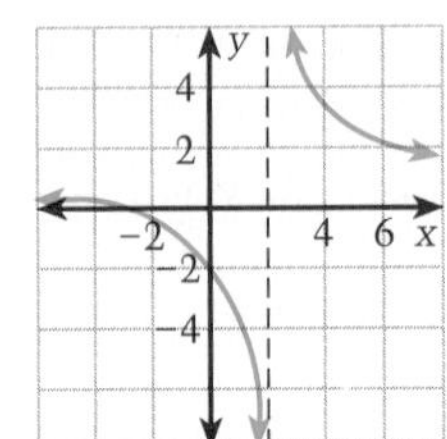

c.

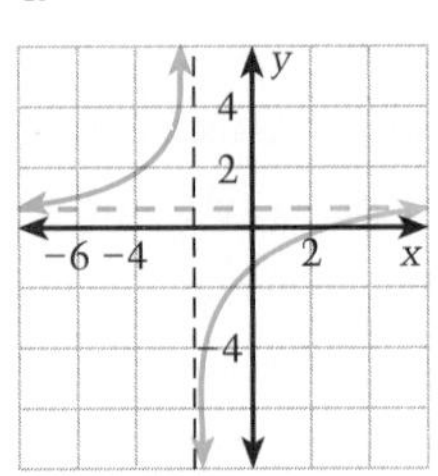

d.

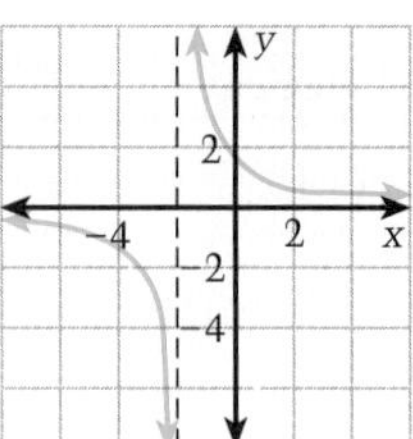

In Exercises 31–42, sketch the graph of the function. See Additional Answers.

31. $f(x) = \dfrac{2x-6}{x+4}$

32. $f(x) = \dfrac{-5}{x+9}$

33. $f(x) = \dfrac{3}{4x+10}$

34. $f(x) = \dfrac{5x+1}{x^2-1}$

35. $f(x) = \dfrac{3x^2+4x+4}{x^2-5x-6}$

36. $f(x) = \dfrac{-3x^2}{2x+6}$

37. $f(x) = \dfrac{3x^3+1}{4x^3-32}$

38. $f(x) = \dfrac{5-x}{4x^2-3x-1}$

39. $f(x) = \dfrac{x^2-6x-9}{x^3+27}$

40. $f(x) = \dfrac{-2x^2}{x^2-9}$

41. $f(x) = \dfrac{x^2-10x+24}{3x}$

42. $f(x) = \dfrac{x^3+5x^2-1}{x^2-4x}$

Renting Videos **In Exercises 43 and 44, use the following information.** 43. See Additional Answers.

You have purchased a VCR for $180. You also joined a video rental store where you can rent movies for $3 each.

43. Write a model that represents your average cost per movie (including the price of the VCR). Sketch the graph of this model. What is the horizontal asymptote and what does it represent?

44. Suppose that the cost of admission to a movie theater is $6.00. How many videos must you rent to make your average cost per video less than $6.00? More than 60

45. *Hospital Costs* For 1970 to 1990, the average daily cost, C (in dollars), per patient at a community hospital can be modeled by

$$C = \frac{746{,}676 + 16{,}557t}{10{,}000 - 864t + 22.8t^2}$$

where $t = 0$ represents 1970. Sketch the graph of this model. Would you expect this model to continue to represent patient costs through the year 2010? Explain.

No, it is unlikely that costs will decrease to the level of 1970 in 2010. See Additional Answers for graph.

In 1990, Americans rented about 6 million movie videos each day, about twice the number of tickets sold at movie theaters.

47. $6x^2 - 24x + 105 - \dfrac{424}{x+4}$ **48.** $9x^2 + 7x + 8 + \dfrac{6}{x-1}$ **49.** $2x^3 - 5x^2 + 4x - 4 + \dfrac{13}{x+1}$

46. *Hotel Revenue* For 1970 through 1990, the total revenue, R (in billions of dollars), from hotels and motels can be modeled by

$$R = \frac{3200 + 270t}{1000 - 33t}$$

where $t = 0$ represents 1970. Sketch the graph of this model. Would you expect this model to continue to represent hotel and motel revenue through the year 2000? Explain. No, the revenue would have to become abnormally high, since the graph has a vertical asymptote near the year 2000. See Additional Answers for graph.

Integrated Review

In Exercises 47–50, perform the indicated division. See below.

47. $(6x^3 + 9x - 4) \div (x + 4)$

48. $(9x^3 - 2x^2 + x - 2) \div (x - 1)$

49. $(2x^4 - 3x^3 - x^2 + 9) \div (x + 1)$

50. $(x^5 - 6x^3 - 2x^2 + 5x - 6) \div (x + 3)$

In Exercises 51–53, completely factor the polynomial with respect to the integers. See below.

51. $x^4 + 2x^3 + 6x^2 + 18x - 27$

52. $2x^3 - x^2 - 2x + 1$

53. $3x^4 - 9x^2 - 12$

54. *Women Military Pilots* Find the ratio of the total number of women military pilots (in the Air Force, Army, Navy, and Marines in 1991) to the total number of men military pilots. Use the graph at the right.

55. *Women Executives* In 1991, women made up only 2.6% of the top corporate officers at Fortune 500 companies. Find the ratio of the number of women who were top corporate officers to the number of men who were top corporate officers. See margin.

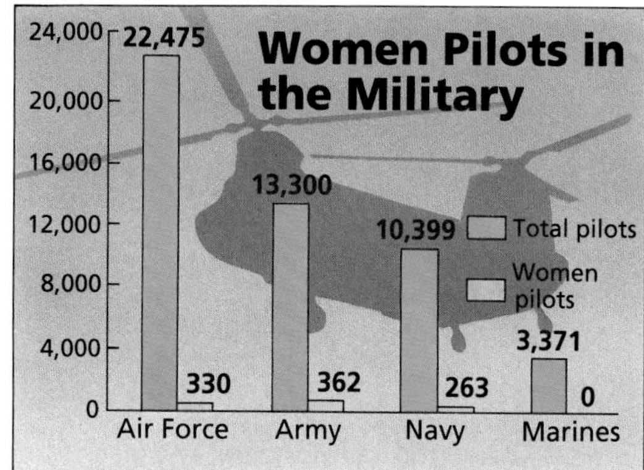

54. See margin.

Exploration and Extension

Slant Asymptotes **In Exercises 56 and 57, use the information below to find the slant asymptote of the graph of the function. Then sketch the graph of the function and the slant asymptote.** See Additional Answers.

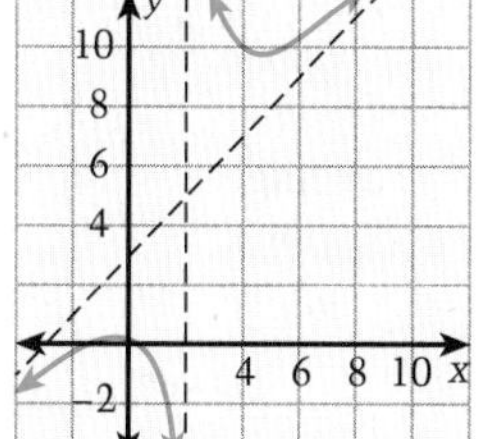

If the degree of the numerator of a *rational function* is exactly one more than the degree of the denominator, the graph of the function has a slant asymptote. For example, the graph of

$$f(x) = \frac{x^2 + x}{x - 2} = x + 3 + \frac{6}{x - 2}$$

has the line $y = x + 3$ as a slant asymptote.

56. $f(x) = \dfrac{x^2 - x - 2}{x - 1}$

57. $f(x) = \dfrac{x^2 + 5x + 8}{x + 3}$

50. $x^4 - 3x^3 + 3x^2 - 11x + 38 - \dfrac{120}{x + 3}$ **51.** $(x - 1)(x + 3)(x^2 + 9)$

52. $(2x - 1)(x + 1)(x - 1)$ **53.** $3(x + 2)(x - 2)(x^2 + 1)$

Ex. 46 Encourage students to use graphics calculators. Note that in the year 2000 (t = 30), S is approaching a vertical asymptote.

$t = \dfrac{1000}{33} \approx 30.30$

Ex. 54–55 CONNECTIONS Use these exercises to discuss the changing roles of women in the military and work force.

Answers

54. $\dfrac{191}{9718}$

55. $\dfrac{13}{487}$

Enrichment Activities

These activities require students to go beyond lesson goals.

1. Graph $f(x) = \dfrac{x^2 - 4}{x - 2}$ using a graphics calculator. Using a standard window with the Trace feature, trace the graph to $x = 2$. What happens?

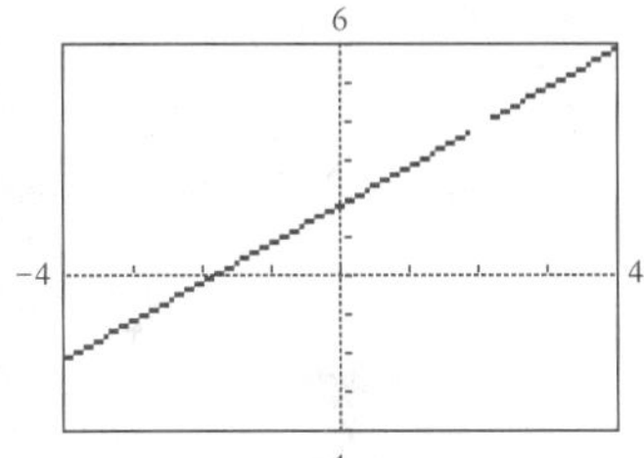

The cursor disappears and the coordinates shown are $x = 2$, $y =$ (space).

2. Why is there a "break" in the line at $x = 2$? Why is the graph a line without a vertical asymptote at $x = 2$? If x is "close to" 2, what is the y value "close to"?
$f(x)$ factors to $f(x) = \dfrac{(x + 2)(x - 2)}{(x - 2)}$. If the factor $(x - 2)$ is divided out, $f(x)$ is equivalent to the function $g(x) = x + 2$, which is linear. As x gets close to 2, y gets close to 4.

USING A GRAPHING CALCULATOR

Most graphing calculators have two graphing modes: a *connected mode* and a *dot mode.* The connected mode works well for graphs of functions that are continuous (have no holes or breaks). The connected mode does not, however, work well for the graphs of many rational functions because they often have two or more disconnected branches. To correct this problem, change the calculator to dot mode.

In the graphs shown below, notice that the graph on the left has a vertical line at approximately $x = 1$. This line is *not* part of the graph—it is simply the calculator's attempt at connecting the two branches of the graph.

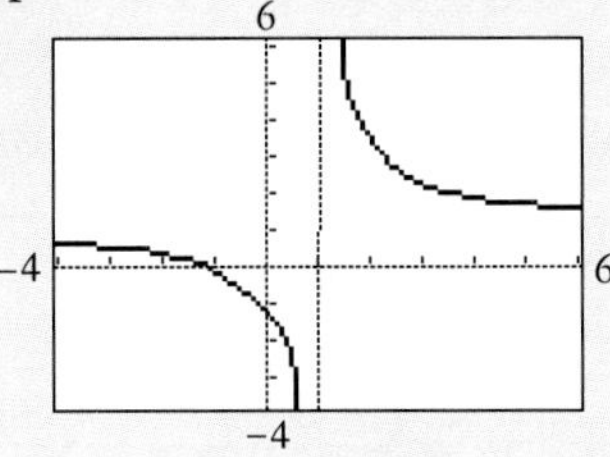

Connected Mode

Graph of $f(x) = \dfrac{x+1}{x-1}$

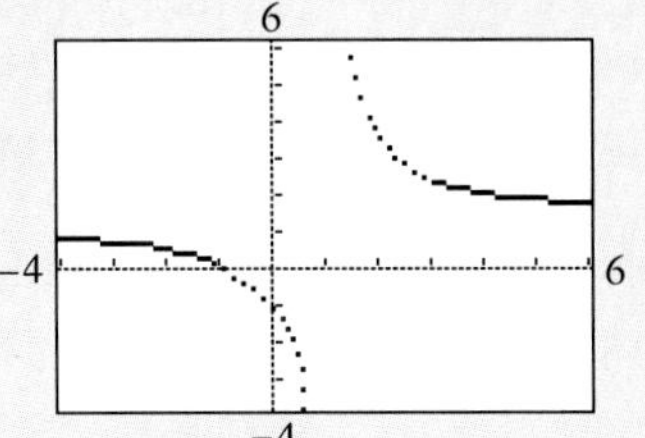

Dot Mode

Graph of $f(x) = \dfrac{x+1}{x-1}$

Example 1: Investigating Asymptotic Behavior

Some people think that the graph of a rational function cannot cross its horizontal asymptote. This, however, is not true. Use a graphing calculator to sketch the graph of

$$f(x) = \frac{2x^2 - 3x + 5}{x^2 + 1}$$

and its horizontal asymptote $y = 2$ on the screen. Find the x-value at which the graph crosses its horizontal asymptote.

Solution From the screen below, it appears that the graph crosses the horizontal line $y = 2$ when $x = 1$. You can confirm this by substituting $x = 1$ in the function.

$$f(1) = \frac{2(1^2) - 3(1) + 5}{1^2 + 1} = \frac{4}{2} = 2 \quad \blacksquare$$

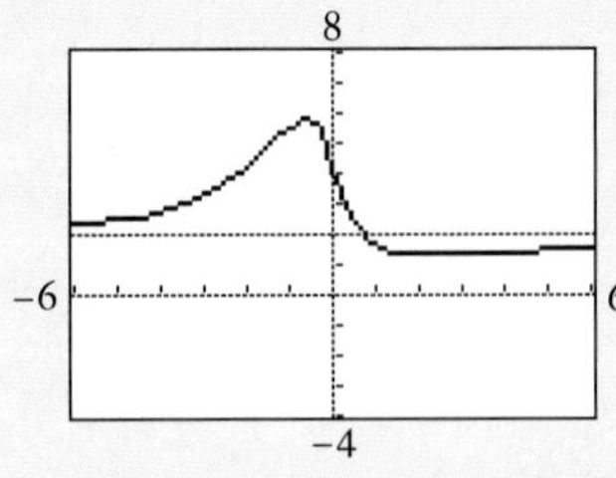

Example 2: Pollution Level of a Pond

Some organic waste has fallen into a pond. Part of the decomposition process includes oxidation, whereby oxygen that is dissolved in the pond water is combined with decomposing material. Let $L = 1$ represent the normal oxygen level in the pond and let t represent the number of weeks after the waste is dumped. The oxygen level in the pond can be modeled by

$$L = \frac{t^2 - t + 1}{t^2 + 1}.$$

Sketch the graph of this model and use the graph to explain how the oxygen level changed during the 15 weeks after the waste was dumped.

Oxygen Level of Pond

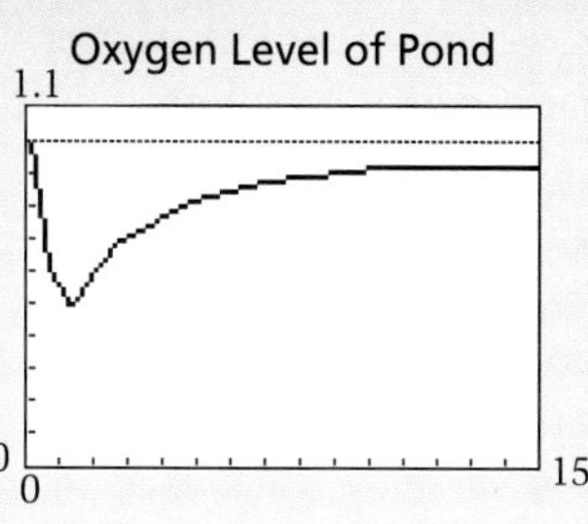

Solution The screen at the right shows the graph of the model and the line $L = 1$ (the normal oxygen level). From the graph, you can see that the oxygen dropped to 50% of its normal level after 1 week. Then, during the next several weeks, the oxygen level gradually returned to normal. By the end of the 15th week, the oxygen level had reached 93% of normal. ■

Exercises

In Exercises 1–6, find ranges for x and y so that the calculator screen displays the basic characteristics of the graph of the rational function. See margin.

1. $f(x) = \dfrac{x-5}{x+1}$ **2.** $f(x) = \dfrac{2x-5}{x-4}$ **3.** $f(x) = \dfrac{x^2-1}{x^2-4}$

4. $f(x) = \dfrac{2x^2-7x+5}{x^2+2x+1}$ **5.** $f(x) = \dfrac{x^2-5x}{2x^2+1}$ **6.** $f(x) = \dfrac{x^3-1}{x^3+1}$

7. Use a graphing calculator to sketch the graphs of f and g on the same screen. (Use a range of $-5 \le x \le 5$ and $-4 \le y \le 20$.)

$f(x) = x^2$ and $g(x) = \dfrac{x^3 + x^2 + 1}{x + 1}$

Use long division to rewrite $g(x)$. Then use the result to explain why the two graphs are close to each other for large values of $|x|$.

8. *Sale of Long-Playing Albums* For 1975 to 1990, the number, N (in millions), of long-playing albums sold in the United States can be approximated by the model

$$L = \frac{1000 - 74.41t + 1.45t^2}{2.94 - 0.336t + 0.0125t^2}$$

where $t = 5$ represents 1975. Sketch the graph of this function and use the graph to find the year that the sales peaked.

(Source: Recording Industry Association of America) 1985 See margin.

Answers

1. $-10 \le x \le 10$, $-10 \le y \le 10$
2. $-10 \le x \le 10$, $-10 \le y \le 10$
3. $-10 \le x \le 10$, $-5 \le y \le 5$
4. $-15 \le x \le 15$, $-1 \le y \le 20$
5. $-10 \le x \le 10$, $-3 \le y \le 3$
6. $-10 \le x \le 10$, $-5 \le y \le 5$

8.

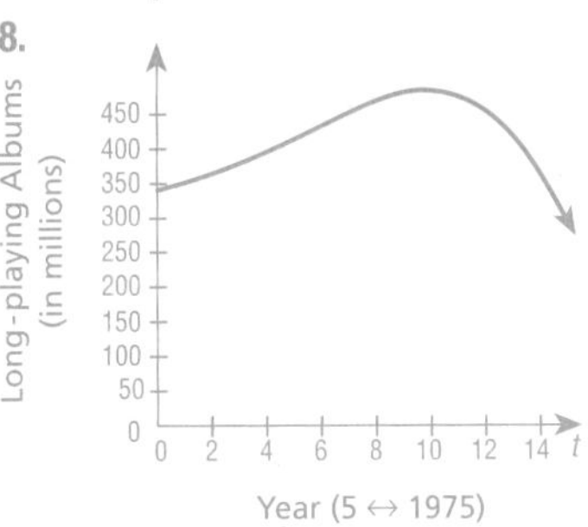

7. $g(x) = x^2 + \dfrac{1}{x+1}$ When $|x|$ is large, $\dfrac{1}{x+1} \to 0$; so the graphs are close to each other when $|x|$ is large.

If 24 karat gold is 100% pure, what percent pure is 14 karat?

$58\frac{1}{3}\%$

ORGANIZER

Warm-Up Exercises

1. Solve for the unknown.
   a. $45x = 75$ b. $18y = 90$
   c. $\frac{60}{x} = 48$ d. $36 = \frac{78}{y}$

   a. $\frac{5}{3}$, b. 5, c. $\frac{5}{4}$, d. $\frac{13}{6}$
2. You are driving by car to see your aunt who lives 40 miles away. Complete the table showing how many hours it will take you to arrive at the indicated speeds (or rates) in mph. Describe any patterns that you observe.

Distance, d	40	40	40	40	40
Speed, r	20	30	40	50	60
Time, t					

$t = 2, \frac{4}{3}, 1, \frac{4}{5}, \frac{2}{3}$; as the speed increases, the time decreases.

Lesson Resources

Teaching Tools
- Problem of the Day: 10.2
- Warm-up Exercises: 10.2
- Answer Masters: 10.2

Extra Practice: 10.2
Color Transparency: 55
Applications Handbook: p. 57

LESSON Notes

GOAL 1 The table in Warm-Up Exercise 2 displays the inverse variation, $rt = 40$, or $t = \frac{40}{r}$.

10.2 Inverse and Joint Variation

What you should learn:

Goal 1 How to create and use real-life models using inverse variation

Goal 2 How to create and use models of real-life situations that use joint variation

Why you should learn it:

You can model many real-life relationships with joint or inverse variation, such as sound intensity and distance from its source.

Connections
Electronics

Goal 1 Using Inverse Variation

In Lesson 2.4, you studied *direct* variation. Recall that two variable quantities x and y vary directly if, for a constant k,

$$\frac{y}{x} = k \quad \text{or} \quad y = kx. \quad \textit{Direct variation}$$

In this lesson, you will study two other types of variation: *inverse variation* and *joint variation.*

Inverse Variation

The variables x and y **vary inversely** if, for a constant k,

$$yx = k \quad \text{or} \quad y = \frac{k}{x}. \quad \textit{Inverse variation}$$

The number k is the **constant of variation.**

Example 1 *Using Inverse Variation*

The resistance, R, of wire varies inversely with its cross-sectional area, A. A 500-centimeter length of copper wire with a radius of 0.2 centimeter has a resistance of 0.025 ohm. Find the resistance of a copper wire of the same length whose radius is 0.1 centimeter.

Solution Since the wire has a radius of 0.2 centimeter, it has a cross-sectional area of $A = \pi r^2 = \pi(0.2)^2 = 0.04\pi \text{ cm}^2$.

$$RA = k \quad \textit{Model for inverse variation}$$

$$(0.025)(0.04\pi) = k \quad \textit{Substitute 0.025 for R; } 0.04\pi \textit{ for A.}$$

$$0.001\pi = k \quad \textit{Simplify.}$$

Thus an equation that relates R and A is

$$RA = 0.001\pi \quad \text{or} \quad R = \frac{0.001\pi}{A}.$$

The resistance of the piece of copper wire whose radius is 0.1 centimeter is

$$R = \frac{0.001\pi}{(0.1)^2\pi} = \frac{0.001\pi}{0.01\pi} = 0.1 \text{ ohm.}$$

■

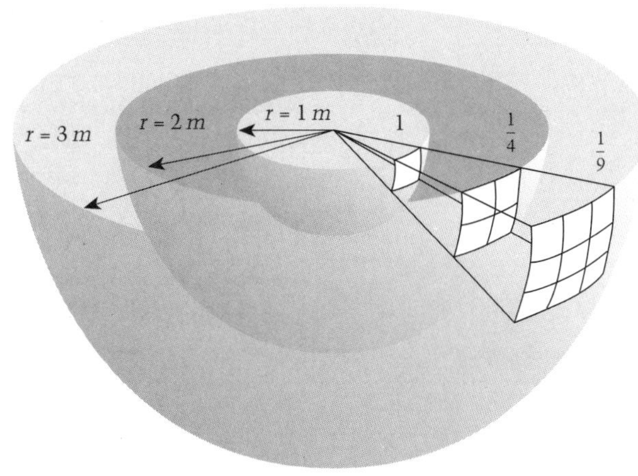

Intensity of sound is measured in watts per square centimeter (or watts per square meter). The intensity of a sound varies inversely with the square of its distance from the source. The diagram suggests why the intensity, at a distance of 3 meters, is one-ninth the intensity at a distance of 1 meter.

Connections
Airport Safety

Airport workers wear ear protectors to prevent ear damage from the intensity of jet engine noise.

Example 2 *Using Inverse Variation*

At a distance of 1 meter, the intensity of jet engine noise is 10 watts per square meter. An airport cargo worker is 15 meters from the jet engine. What is the sound intensity at 15 meters?

Solution Let I represent the intensity (in watts per square meter) and let r represent the distance (in meters). From the information given above, you know that I varies inversely as r^2. Thus you can write the following model.

$Ir^2 = k$ *Model for inverse variation*

$(10)(1) = k$ *Substitute 10 for I and 1 for r^2.*

$10 = k$ *Simplify.*

Thus an equation that relates I and r^2 is

$$Ir^2 = 10$$
$$I = \frac{10}{r^2}.$$

At a distance of $r = 15$ meters, the intensity is

$$I = \frac{10}{15^2}$$
$$\approx 0.044 \text{ watt per square meter.}$$

This intensity corresponds to about 126 decibels—above the level that can cause damage. (To rewrite this intensity as decibels, see page 418.) ■

Examples 1–2

Observe that if two variables vary inversely, their product is constant. Such variables are said to vary inversely because as one value increases, the other value decreases.

Emphasize the connections to real-life situations. Ask students if knowing the constant of variation is useful in solving similar problems.

Example 3

Students may need to see more steps to help them write equations of inverse and joint variation. For example, after solving for k, students may need to be reminded to rewrite the equation using this new constant, and then use this new equation as a formula to solve related problems. Horsepower is still commonly used as a unit of power in some industries. One horsepower is equivalent to 746 Watts.

EXTEND Example 3 RESEARCH Students may want to find out about alternate forms of energy. The 1991 *Alternate Energy Sourcebook* and an encyclopedia are good places to begin.

Extra Examples

Here are additional examples similar to **Examples 1–3.**

1. The volume of a gas V at a constant temperature varies inversely with its pressure, P. The pressure acting on 15 m^3 of air is raised from 1 atmosphere to 2 atmospheres while the temperature is kept constant. Find the volume of the air after the pressure is applied. 7.5 m^3

2. The weight, w, of a body varies inversely as the square of its distance, d, from the center of the earth. On the surface of the earth (4000 miles from the center), a certain body weighs 125 pounds. How much would it weigh 300 miles above the earth? ≈ 108 pounds
3. The power in watts of an electrical circuit varies jointly as the resistance and the square of the current. For a 600-watt microwave oven that draws a current of 5.0 amperes, the resistance is 24 ohms. What is the resistance of a 200-watt refrigerator that draws a current of 1.7 amperes? ≈ 69 ohms

Check Understanding

Define inverse variation and joint variation.
See the top of pages 526 and 528.

Communicating about ALGEBRA

EXTEND *Communicating*
Many real-life applications of mathematics result in equations that combine joint and inverse variation (called *combined* variation). Discuss the following situation with your students:

Heat loss, h, per hour through a glass window of a house on a cold day varies directly as the difference, d, between the inside and outside temperatures and the area, A, of the window, and inversely as the thickness, w, of the window. Have the students write the model for this variation.
$h = k\frac{dA}{w}$

Goal 2 Using Joint Variation

Joint variation occurs when a quantity varies directly as the product of *two or more* other quantities. For instance, if $z = kxy$, then z varies jointly with the product of x and y.

Real Life
Solar Power

Example 3 *Using Joint Variation*

The horsepower, h, needed for a water pump varies jointly with the well depth, d (in feet), and the rate, r (in gallons per minute). A 55-horsepower motor can pump water from a depth of 150 feet at a rate of 990 gallons per minute. The 1991 *Alternative Energy Sourcebook* has the following description of a pump that runs on solar energy. Estimate the horsepower of this pump.

> *For your deep-well pumping needs (well depths to 1000 feet at rates of $2\frac{3}{4}$ gallons per minute), we offer the Solarjack Pump water pumping system.*

Solution

$h = krd$ — *Joint variation model*

$55 = k(990)(150)$ — *Let $h = 55$, $r = 990$, and $d = 150$.*

$\frac{1}{2700} = k$ — *Solve for k.*

Thus an equation that relates h, r, and d is

$$h = \frac{1}{2700}rd.$$

For a rate of $r = 2.75$ gallons per minute and a depth of $d = 1000$ feet, you would need a motor that has

$$h = \frac{1}{2700}(2.75)(1000)$$
$$\approx 1 \text{ horsepower.}$$

■

Communicating about ALGEBRA

SHARING IDEAS about the Lesson

The Sunraycer, a solar-powered car, has 9500 solar cells to generate electricity for a motor that can deliver up to 4 horsepower. How many solar cells would you estimate are in the Solarjack Pump? To answer the question, did you use direct or inverse variation? Explain.

2375, direct variation, the number of solar cells varies directly with horsepower.

EXERCISES

Guided Practice

CRITICAL THINKING about the Lesson

In Exercises 1–4, do x and y vary directly or inversely?

1. $xy = \frac{1}{2}$ Inversely

2. $x = \frac{9}{y}$ Inversely

3. $\frac{x}{y} = 5$ Directly

4. $y = 3x$ Directly

Distance, Rate, and Time **In Exercises 5 and 6, use the formula $d = rt$, where d is the distance in miles, r is the rate in miles per hour, and t is the time in hours.**

5. A car travels a distance of 30 miles. Do the rate and time vary directly or inversely? Explain. Inversely; rt is a constant.

6. A car travels for 30 minutes. Do the distance and rate vary directly or inversely? Explain. Directly; $\frac{d}{r}$ is a constant.

Independent Practice

In Exercises 7–9, the variables x and y vary inversely. Use the given pair of values to find an equation that relates the variables.

7. $x = 3, y = -2$ $yx = -6$

8. $x = 4, y = 6$ $yx = 24$

9. $x = 5, y = 1$ $yx = 5$

In Exercises 10–12, the variable z varies jointly with the product of x and y. Use the given values to find an equation that relates the variables. See margin.

10. $x = 3, y = 8, z = 2$

11. $x = -5, y = 2, z = \frac{3}{4}$

12. $x = 1, y = \frac{1}{2}, z = 4$

In Exercises 13–15, find an equation that relates the variables.

13. *Typing Time* The time, t (in minutes), you need to type a 2000-word paper varies inversely with the rate, r (in words per minute), that you can type. When typing 40 words per minute, you can finish in 50 minutes. $rt = 2000$

14. *Pressure* The pressure, P (in pounds per square inch), that a 160-pound person's shoe exerts on the ground when walking varies inversely with the area, A (in square inches), of the sole of the shoe. When the shoes have a sole area of 40 square inches, the pressure is 4 pounds per square inch. $PA = 160$

15. *Electrical Resistance* The resistance, R (in ohms), of a wire varies directly with the length, L (in centimeters), of the wire, and inversely with the cross-sectional area, A (in square centimeters), of the wire. A 500-centimeter piece of copper wire with a radius of 0.2 centimeter has a resistance of 0.025 ohm. $R = \frac{\pi L}{500{,}000A}$

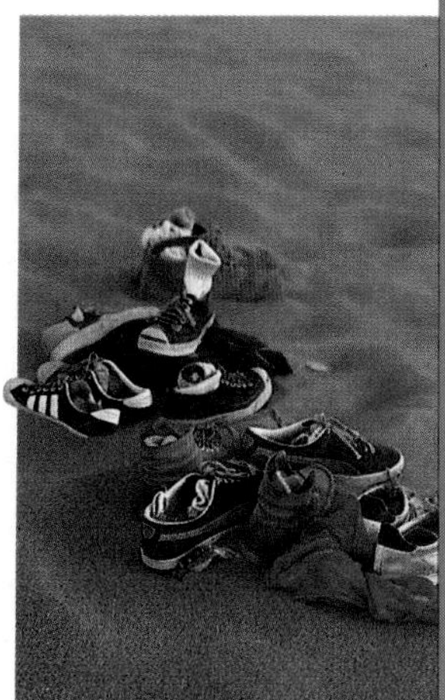

EXERCISE Notes

ASSIGNMENT GUIDE

Basic/Average: Ex. 1–6, 7–15 odd, 17, 18, 22–24, 30–35

Above Average: Ex. 1–6, 7–19 odd, 20–24, 30–35

Advanced: Ex. 1–6, 7–19 odd, 20–24, 30–35

Selected Answers
Exercises 1–6, 7–31 odd

Use **Mixed Review** as needed.

More Difficult Exercises
Exercises 16–23, 32–35

Guided Practice

Ex. 1–6 Use these exercises to highlight the differences between inverse and direct variation. Extend the Guided Practice by modeling any of the Independent Practice exercises from 16–19.

Independent Practice

Ex. 10–12 Refer students to the top of page 528 for the meaning of joint variation.

Answers

10. $z = \frac{1}{12}xy$

11. $z = -\frac{3}{40}xy$

12. $z = 8xy$

Ex. 13–15 Remind students that the main task is to write an equation by first finding the appropriate value of the constant of variation. Examples 1 and 3 on pages 526 and 528 provide solution models for these exercises.

▶ **Ex. 16–19** Encourage students to use a solution format similar to those given in Examples 2 and 3 on pages 527 and 528.

▶ **Ex. 16** From ancient times, the Strait of Messina has been part of sea lore. It contained two obstacles for sailors. The whirlpool, called Charybdis, was on the Sicilian side of the Strait, while the sea monster Scylla, half fish, half woman, with dogs and serpents growing out of her waist, resided in a cave at the top of a sheer rock on the Italian coast. Three classic heros of antiquity, Jason, Odysseus, and Aeneas, were forced to "go between Scylla and Charybdis" in their epic quests.

The Strait of Messina is also the site of the famous "Fata Morgana" mirage.

▶ **Ex. 20–21 and 22–23** Assign these exercises in pairs.

EXTEND Ex. 22–23 Ask students to investigate the Joule as a unit of measurement. They can ask a physics teacher to provide information.

A Joule is a unit of heat in the metric system (1 BTU, British Thermal Unit, is approximately 1055 Joules).

▶ **Ex. 32–35** The equation of variation for these exercises is $P = \frac{WD^2}{L}$.

16. *Speed of a Whirlpool* A naturally-occurring whirlpool in the Strait of Messina, a channel between Sicily and the Italian mainland, is about 6 feet across at its center, and is said to be large enough to swallow small fishing boats. The speed, s (in feet per second), of the water in the whirlpool varies inversely with the radius, r (in feet). If the water speed is 2.5 feet per second at a radius of 30 feet, what is the speed of the water at a radius of 3 feet? 25 ft/sec

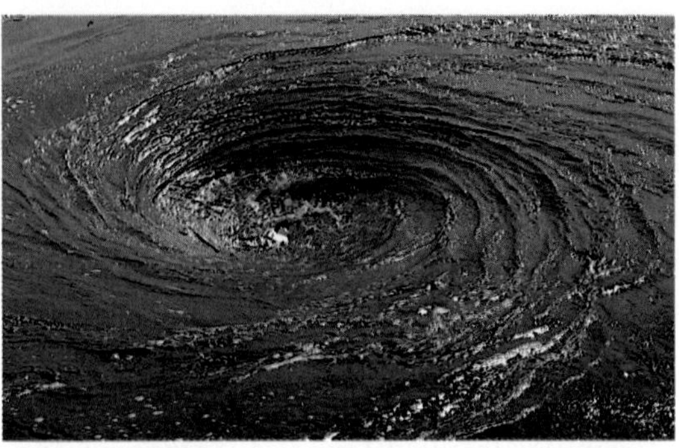

Many whirlpools occur naturally. This one is on the lake side of the Rance barrage in northern France.

17. *Seesaw* To balance a seesaw, the distance, d (in feet), a person is from the fulcrum is inversely proportional to his or her weight, w (in pounds). Roger, who weighs 120 pounds, is sitting 6 feet away from the fulcrum. Ellen weighs 108 pounds. How far from the fulcrum must she sit to balance the seesaw? $6\frac{2}{3}$ ft

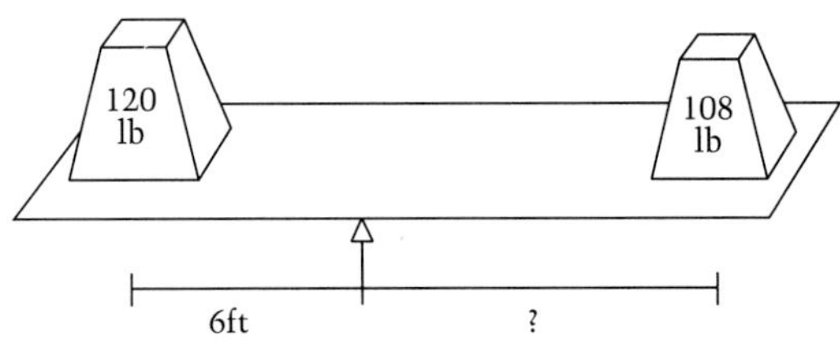

18. *Chemistry* The temperature, T (in degrees Kelvin), of an enclosed gas varies jointly with the product of the volume, V (in cubic centimeters), and the pressure, P (in kilograms per square centimeter). The temperature of a gas is 294°K when the volume is 8000 cubic centimeters and the pressure is 0.75 kilogram per square centimeter. What is the temperature when the volume is 7000 cubic centimeters and the pressure is 0.87 kilogram per square centimeter? 298.41°K

19. *Simple Interest* The simple interest, I (in dollars), for a savings account is jointly proportional to the product of the time, t (in years), and the principal, P (in dollars). After three months, the interest on a principal of \$5000 is \$106.25. What is the interest after nine months? \$318.75

Light Illumination **In Exercises 20 and 21, use the following information.**

The illumination, I (in luxes), of a surface varies inversely with the square of the distance, d (in meters), from the light source to the surface. One meter away from a certain light, the illumination is 750 luxes. (One lux is equal to 0.093 footcandle.) $Id^2 = 750$

20. Write an equation that relates I and d.

21. Find the illumination at a distance of 2 meters. 187.5 luxes

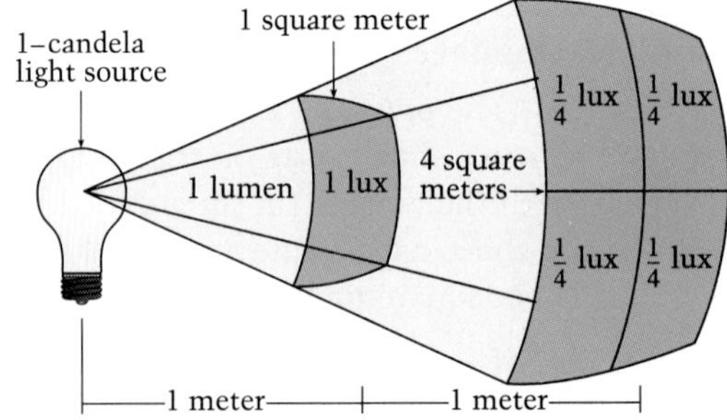

A 200-watt incandescent bulb in a desk lamp provides about 750 luxes (70 footcandles) at the desk's surface.

Lifting an Object **In Exercises 22 and 23, use the following information.**

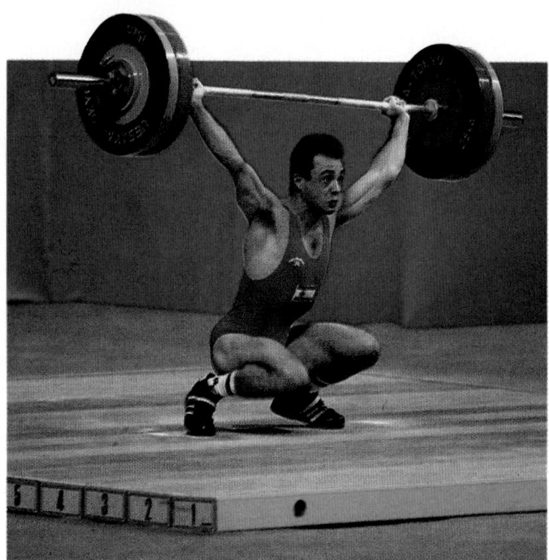

The work, W (in joules), done when lifting an object is jointly proportional to the product of the mass, m (in kilograms), of the object and the height, h (in meters), that the object is lifted. The work done when a 120-kilogram object is lifted 1.8 meters above the ground is 2116.8 joules.

22. Write an equation that relates W, m, and h. $W = 9.8mh$

23. How much work is done when lifting a 100-kilogram object 1.5 meters above the ground? 1470 joules

Integrated Review

In Exercises 24–29, sketch the graph of the rational function. See Additional Answers.

24. $f(x) = \dfrac{x-5}{2x-4}$

25. $f(x) = \dfrac{-2}{x+1}$

26. $f(x) = \dfrac{6}{x-9}$

27. $f(x) = \dfrac{3x-4}{x+2}$

28. $f(x) = \dfrac{x+2}{x^2+x-6}$

29. $f(x) = \dfrac{4x-1}{x^2-9}$

30. *Powers of Ten* Which best estimates the voltage of common electrical house current? d

a. 1.1×10^{-1} **b.** 1.1×10^{0}
c. 1.1×10^{1} **d.** 1.1×10^{2}

31. *Powers of Ten* Which best estimates the number of people who have been president of the United States? c

a. 4.0×10^{-1} **b.** 4.0×10^{0}
c. 4.0×10^{1} **d.** 4.0×10^{2}

Exploration and Extension

Engineering **In Exercises 32–35, use the following information.**

The load, P (in pounds), that can be safely supported by a horizontal beam varies jointly as the product of the width, W (in feet), of the beam and the square of its depth, D (in feet), and inversely as its length, L (in feet).

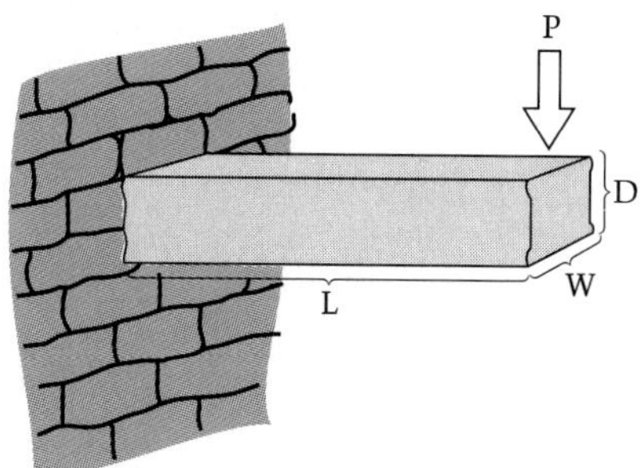

32. How does P change when the width and length of the beam are doubled? P does not change.

33. How does P change when the width and depth of the beam are doubled? P is 8 times as heavy.

34. How does P change when all three of the dimensions are doubled? P is 4 times as heavy.

35. How does P change when the depth of the beam is halved? P is $\frac{1}{4}$ as heavy.

Enrichment Activities

These activities require students to go beyond lesson goals.

CONNECTIONS This activity will help students connect the concepts learned in this lesson to the concepts learned in Lesson 10.1.

1. Sketch the graphs for $y = \dfrac{k}{x}$ and $y = \dfrac{k}{x^2}$ with $k > 0$.

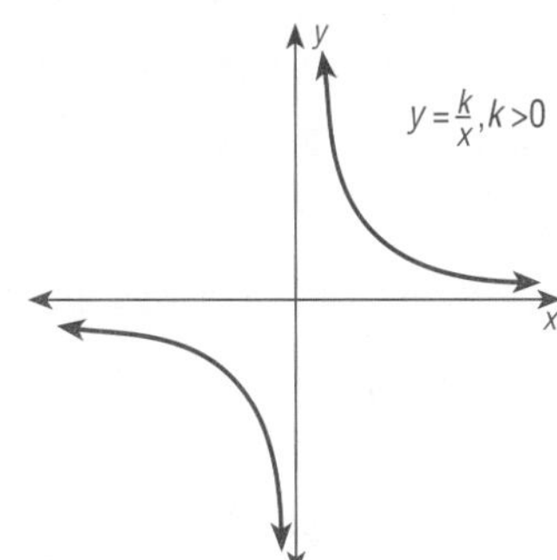

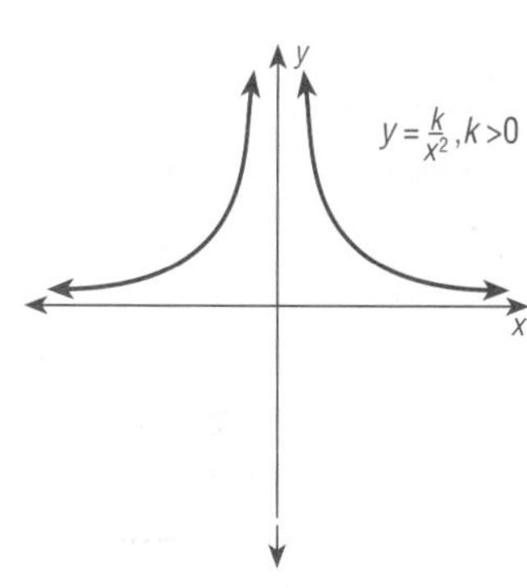

2. In Exercise 14, $P = \dfrac{160}{A}$ where P is the pressure of a 160-pound person's shoe and A is the area in square inches of the sole for the shoe. Does the graph of $y = \dfrac{160}{x}$ correctly depict the relationship given in this exercise?

Yes. If $y = \dfrac{160}{x}$, then $xy = 160$.

Milestones

Language Arts: *Critical Reading*

1. Find out who defined a general abstract group. When was it defined? (Arthur Cayley; 1854)
2. Find out why the significance of Galois' work was not recognized at first. (Possible answers may include: The originality of his work made it difficult for his contemporaries to understand and appreciate it; Galois was killed in a duel in 1932, so he was not alive to discuss and promote his ideas.)
3. Which of the following was also published in 1831: Edgar Allen Poe's *Poems,* Charles Dickens' *Oliver Twist,* Edith Wharton's *Ethan Frome*? (Edgar Allen Poe's *Poems*)

Mixed REVIEW

1. $3x^3 - 7x$ 5.–8. See Additional Answers. 13. $(3x + 2)(2x - 5)$ 17. See below.

1. Simplify: $(2x^3 - 3x) + (x^3 - 4x)$. **(9.1)**
2. Simplify: $(3x - 4)(2x + 1)$. **(9.1)** $6x^2 - 5x - 4$
3. Evaluate $\log_3 81^4$. **(8.2)** 16
4. Evaluate $27^{-2/3}$. **(7.3)** $\frac{1}{9}$
5. Sketch $f(x) = (x - 3)^5 + 1$. **(9.2)**
6. Sketch $f(x) = e^{x-2} - 4$. **(8.1)**
7. Sketch $y = (x - 5)^2 + 3$. **(5.3)**
8. Sketch $y < 3|x - 2| - 5$. **(2.6)**
9. Solve $|2x + 5| < 3$. **(1.7)** $-4 < x < -1$
10. Solve $\sqrt{3x - 2} = 5$. **(7.5)** 9
11. Solve $|x - 5| > 2$. **(1.7)** $x < 3$ or $x > 7$
12. Solve $x^{2/3} = 36$. **(7.5)** 216
13. Factor $6x^2 - 11x - 10$. **(9.3)**
14. Find the intercepts of $\frac{x}{8} - \frac{y}{2} = 1$. **(2.3)** x-int.: 8, y-int.: -2
15. What is 108% of \$300? 324
16. Divide $\frac{3}{8}$ by $\frac{15}{28}$. $\frac{7}{10}$
17. Find a quadratic function containing $(1,-3)$, $(2,-1)$, $(3, 2)$, and $(4, 6)$. **(6.6)**
18. Find the slope of the line through $(2,-7)$ and $(-5,-1)$. **(2.2)** $-\frac{6}{7}$
19. Given $f(1) = 2$ and $f(n) = f(n - 1) - 3$, find $f(4)$. **(6.6)** -7
20. For what values of c has $f(x) = 3x^2 - 4x - c$ only one x-intercept? **(5.4)** $-\frac{4}{3}$

Milestones ABEL AND GALOIS

1822 1823 1824 1825 1826 1827 1828 1829 1830 1831 1832 1833 1834 1835

1824: Abel's proof published — 1831: Galois' proof published

Abel

Galois

In the exploration of algebra over the centuries, general solutions were found for second-, third-, and fourth-degree polynomial equations. Attempts to find an algebraic solution for the fifth-degree polynomial equation, however, met with failure time and again.

In the early 1800's, Niels Henrik Abel thought he had found the solution for a general fifth-degree equation, but later discovered it contained an error. In 1824, he published a proof that showed that for any degree greater than four, the general polynomial equation could not be solved algebraically. Another young mathematician, Évariste Galois, published a proof in 1831 in which he elaborated on Abel's work to show how to determine whether or not a given equation can be solved algebraically. Of significance in both men's work was the concept of a *group.* Subsequent research in group theory gave mathematicians new ways to explore and describe algebraic structures. The work of Abel and Galois marked the beginning of the modern theory of equations.

Which of these events also took place in 1831: Peru's secession from Columbia, Belgium's separation from the Netherlands, or John Quincy Adams' inauguration as president of the United States?

Belgium's separation from the Netherlands.

17. $f(x) = \frac{1}{2}x^2 + \frac{1}{2}x - 4$

10.3 Multiplying and Dividing Rational Expressions

What you should learn:

Goal 1 How to multiply and divide rational expressions, writing the result in simplest form

Goal 2 How to use rational expressions as real-life models

Why you should learn it:

You can model many real-life situations with rational models, such as the ratio of the surface area of the volume of biological species.

Goal 1 Working with Rational Expressions

A rational expression is in **simplified** (or **simplest**) **form** if its numerator and denominator have no common factors (other than ± 1). To simplify a rational expression, apply the following property.

Simplifying Fractions

Let a, b, and c be nonzero real numbers.

$\frac{a\not{c}}{b\not{c}} = \frac{a}{b}$ *Divide out common factor of c.*

LESSON INVESTIGATION

Investigating Simplification Strategies

Partner Activity The simplification property listed above allows you to divide out factors, not terms. With your partner, simplify the following. Discuss your strategy.

a. $\frac{x^2 + 4x + 4}{x + 2}$ **b.** $\frac{x - 3}{x^2 - 4x + 3}$ **c.** $\frac{x^2 - 12x}{12 - x}$

a. $x + 2$ b. $\frac{1}{x - 1}$ c. $-x$

Study Tip

Simplifying a rational expression usually requires two steps. First, factor the numerator and the denominator. Then, divide out any factors that are common to both numerator and denominator.

Example 1 ***Simplifying a Rational Expression***

Simplify: $\frac{x^2 - 4x - 12}{x^2 - 4}$.

Solution

$\frac{x^2 - 4x - 12}{x^2 - 4} = \frac{(x + 2)(x - 6)}{(x + 2)(x - 2)}$ *Factor numerator and denominator.*

$= \frac{\cancel{(x + 2)}(x - 6)}{\cancel{(x + 2)}(x - 2)}$ *Divide out common factor.*

$= \frac{x - 6}{x - 2}$ *Simplified form* ■

Problem of the Day

Three tennis balls snugly fit in a can. What is the ratio of the volume of the balls to the volume of the space in the can that is around them?

2:1

ORGANIZER

Warm-Up Exercises

1. Simplify. Identify the greatest common factor of the numerator and denominator for each fraction.

a. $\frac{45}{60}$ **b.** $\frac{36}{252}$

c. $\frac{360}{375}$ **d.** $\frac{48}{54}$

a. $\frac{3}{4}$, 15 b. $\frac{1}{7}$, 36

c. $\frac{24}{25}$, 15 d. $\frac{8}{9}$, 6

2. Factor.

a. $x^2 - 2x$

b. $x^2 - 3x - 10$

c. $6x^2 - 5x - 4$

d. $4x^2 + 9x - 9$

a. $x(x - 2)$

b. $(x + 2)(x - 5)$

c. $(2x + 1)(3x - 4)$

d. $(4x - 3)(x + 3)$

3. Evaluate.

a. $\frac{3x - 4}{x - 3}$ at $x = 2$

b. $\frac{(x + 8)(2x - 5)}{x}$ at $x = -1$

c. $\frac{12(x^2 - 2x)}{x - 4x^2}$ at $x = 3$

a. -2, b. 49, c. $-\frac{12}{11}$

Lesson Resources

Teaching Tools

Problem of the Day: 10.3

Warm-up Exercises: 10.3

Answer Masters: 10.3

Extra Practice: 10.3

LESSON Notes

Examples 1–6

In each example, factors are cancelled before factored expressions are multiplied. This reduces the amount of work required to complete the computation.

Examples 4–6

Be sure students recognize that the "invert-and-multiply" technique applies to rational expressions as well.

EXTEND Example 7 Students may want to use the geometric model to evaluate the ratio for other animals.

Extra Examples

Here are additional examples similar to **Examples 1–7.**

1. Simplify. $\dfrac{8x^3 - 2x^2}{4x^2 - x}$

$2x$

2. Perform the indicated operations.

a. $\dfrac{8x^3}{26xy} \cdot \dfrac{14y^4}{12x^2}$

b. $\dfrac{2x^2 - x - 3}{x^2 + x} \cdot \dfrac{x^2 + x}{2x + 3}$

c. $\dfrac{3x - 1}{2x^3 - 2x} \div \dfrac{x}{2x^3 - 2x^2}$

d. $\dfrac{x^2 + 4x + 4}{x + 3} \div (x + 2)$

e. $(x^2 - 1) \div \dfrac{x - 1}{x + 5}$

a. $\dfrac{14y^3}{39}$ b. $\dfrac{(2x - 3)(x + 1)}{2x + 3}$

c. $\dfrac{3x - 1}{x + 1}$ d. $\dfrac{x + 2}{x + 3}$

e. $(x + 1)(x + 5)$

Multiplying rational expressions is similar to multiplying numerical fractions. In both cases, you multiply numerators, multiply denominators, and write the new fraction in simplest form.

$$\frac{a}{b} \cdot \frac{c}{d} = \frac{ac}{bd}$$

To divide one rational expression by another, multiply the first by the reciprocal of the second.

$$\frac{a}{b} \div \frac{c}{d} = \frac{a}{b} \cdot \frac{d}{c}$$

Example 2 Multiplying Rational Expressions

Multiply: $\dfrac{5x^2y}{2xy^3} \cdot \dfrac{6x^3y^2}{10y}$.

Solution

$$\frac{5x^2y}{2xy^3} \cdot \frac{6x^3y^2}{10y} = \frac{30x^5y^3}{20xy^4}$$ *Multiply numerators and denominators.*

$$= \frac{3 \cdot \cancel{10} \cdot \cancel{x} \cdot x^4 \cdot \cancel{y^3}}{2 \cdot \cancel{10} \cdot \cancel{x} \cdot y \cdot \cancel{y^3}}$$ *Factor and divide out common factors.*

$$= \frac{3x^4}{2y}$$ *Simplified form* ■

Example 3 Multiplying Rational Expressions

Multiply: $\dfrac{4x^2 - 4x}{x^2 + 2x - 3} \cdot \dfrac{x^2 + x - 6}{4x}$.

Solution

$$\frac{4x^2 - 4x}{x^2 + 2x - 3} \cdot \frac{x^2 + x - 6}{4x}$$

$$= \frac{4x(x - 1)(x + 3)(x - 2)}{(x - 1)(x + 3)(4x)}$$ *Multiply numerators and denominators.*

$$= \frac{\cancel{4x}\cancel{(x - 1)}\cancel{(x + 3)}(x - 2)}{\cancel{(x - 1)}\cancel{(x + 3)}\cancel{(4x)}}$$ *Factor and divide out common factors.*

$$= x - 2$$ *Simplified form* ■

The procedure for multiplying fractions can be extended to factors that are not all in fractional form. To do this, rewrite the nonfractional expression as a fraction with a denominator of 1. Here is a simple example.

$$\frac{x + 3}{x - 2} \cdot (5x) = \frac{x + 3}{x - 2} \cdot \frac{5x}{1}$$

$$= \frac{(x + 3)(5x)}{x - 2}$$

$$= \frac{5x(x + 3)}{x - 2}$$

Example 4 Dividing Rational Expressions

Divide: $\frac{2x}{3x-12} \div \frac{x^2-2x}{x^2-6x+8}$.

Solution

$$\frac{2x}{3x-12} \div \frac{x^2-2x}{x^2-6x+8} = \frac{2x}{3x-12} \cdot \frac{x^2-6x+8}{x^2-2x}$$
$$= \frac{(2x)(x-2)(x-4)}{(3)(x-4)(x)(x-2)}$$
$$= \frac{(2x)\cancel{(x-2)}\cancel{(x-4)}}{(3)\cancel{(x-4)}(x)\cancel{(x-2)}}$$
$$= \frac{2}{3}$$

■

Example 5 Dividing a Rational Expression by a Polynomial

Divide: $\frac{x^2-3x+2}{4x^2} \div (x-1)$.

Solution

$$\frac{x^2-3x+2}{4x^2} \div (x-1) = \frac{x^2-3x+2}{4x^2} \cdot \frac{1}{x-1}$$
$$= \frac{(x-1)(x-2)}{4x^2(x-1)}$$
$$= \frac{\cancel{(x-1)}(x-2)}{4x^2\cancel{(x-1)}}$$
$$= \frac{x-2}{4x^2}$$

■

Example 6 Dividing a Polynomial by a Rational Expression

Divide: $(2x-3) \div \frac{4x^2-9}{x+5}$.

Solution

$$(2x-3) \div \frac{4x^2-9}{x+5} = \frac{2x-3}{1} \cdot \frac{x+5}{4x^2-9}$$
$$= \frac{(2x-3)(x+5)}{(2x-3)(2x+3)}$$
$$= \frac{\cancel{(2x-3)}(x+5)}{\cancel{(2x-3)}(2x+3)}$$
$$= \frac{x+5}{2x+3}$$

■

3. The number of canoes, C, in the United States from 1980 to 1990 can be modeled by the linear equation $C = 100{,}000(t + 13)$ where $t = 0$ represents 1980. The total number of recreational boats, B, can be modeled by the linear equation $B = 400{,}000(t + 30)$.

a. Find a model for the ratio of canoes to boats.
$\frac{100000(t+13)}{400000(t+30)}$

b. Use the model to decide whether canoeing is becoming more or less popular compared to the total use of recreational boats in the years from 1980 to 1990.
Answers will vary. The ratios increase in the ten-year span, which shows that canoeing is becoming more popular.

Common-Error ALERT!

In **Examples 1–6,** the Simplifying Fractions property is also known as the Cancelling property. Only factors that occur in both the numerator and denominator can be divided out or cancelled using this property. Stress that factors are multiplied (or divided). Terms (which are added or subtracted) cannot be cancelled.

Check Understanding

1. When is a rational expression in simplified (or simplest) form?
   See the top of page 533.
2. How is multiplying or dividing rational expressions similar to multiplying or dividing numerical fractions?
   See the top of page 534.
3. At what point in the multiplication or division computation should rational expressions be reduced?
   Answers depend on personal preference. Students might reduce after each step to simplify the solution process.
4. Explain why it is important to know how to factor when multiplying or dividing rational expressions.
   It is easier to divide out common factors when the expressions are in factored form.

Communicating about ALGEBRA

Help students discuss the connection between the surface-area-to-volume ratio and air resistance when falling through the air. Consider a marble and a feather of equal weights. The ratio of surface area to volume of a marble is less than the ratio of surface area to volume of a feather.

According to the theoretical falling-body model, if there were no air resistance, and you dropped the marble and the feather at the same time, they both would strike the ground at the same time. In air, however, the marble, because of its compact shape, falls in a straight path and strikes the ground ahead of the feather. The feather's larger surface area makes it more susceptible to air resistance. So, buoyed up by the air, it floats in a zigzag path. Since the distance it travels is greater, it strikes the ground after the marble.

The geometric model at the right can be used to find an algebraic model for the ratio of the surface area to the volume for a person or animal. The model is created from six rectangular boxes. The dimensions, surface area, S, and volume, V, of each box are shown at the right.

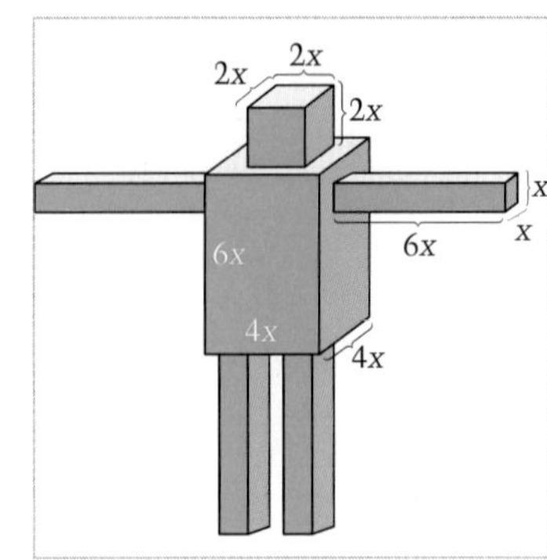

Each arm and leg:

$S = x^2 + 4(6x^2) = 25x^2$

$V = 6x^3$

Head:

$S = 5(4x^2) = 20x^2$

$V = 8x^3$

Trunk:

$S = 2(16x^2) + 4(24x^2) - 4x^2 - 4x^2 = 120x^2$

$V = 96x^3$

Goal 2 Using Rational Expressions in Real Life

Connections
Physics

The greater the ratio of an animal's surface area to its volume, the greater its air resistance when falling through the air.

Example 7 Comparing Surface Area to Volume

Find a rational expression that represents the ratio of the surface area to the volume of the geometric model shown above. Simplify the expression. Then evaluate the simplified expression for several values of x, where x is measured in feet.

Solution

$$\frac{\text{Surface area}}{\text{Volume}} = \frac{4(25x^2) + 20x^2 + 120x^2}{4(6x^3) + 8x^3 + 96x^3}$$

$$= \frac{240x^2}{128x^3}, \quad \text{or} \quad \frac{15}{8x}$$

In the following table, notice that small animals have large ratios and large animals have small ratios.

Animal	Mouse	Squirrel	Cat	Human	Elephant
x	$\frac{1}{24}$	$\frac{1}{12}$	$\frac{1}{6}$	$\frac{2}{5}$	2
Ratio	45	22.5	11.25	4.7	0.9

■

Communicating about ALGEBRA

SHARING IDEAS about the Lesson See below.

A. A mouse and a human each fall 30 feet. Why is the mouse less likely to be hurt?

B. Why do parachutes allow humans to fall from great heights without injury?

A. A mouse has a greater ratio of surface area to volume, and, so, has a greater air resistance. B. A parachute has a very great ratio of surface area to volume and, so, has a very great air resistance.

EXERCISES

Guided Practice

CRITICAL THINKING about the Lesson

In Exercises 1 and 2, simplify the expression, if possible.

1. $\dfrac{3x^2}{3x^3 + 21x}$ $\dfrac{x}{x^2+7}$

2. $\dfrac{x^2 + 10x - 4}{x^2 + 10x}$ Not possible

3. Multiply: $\dfrac{x^2 - x - 2}{3x^2} \cdot \dfrac{3x - 12}{x^2 - 6x + 8}$. $\dfrac{x+1}{x^2}$

4. Divide: $\dfrac{5x^2 + 10x}{x^2 - x - 6} \div \dfrac{15x^3 + 45x^2}{x^2 - 9}$. $\dfrac{1}{3x}$

Independent Practice

In Exercises 5–8, simplify the expression.

5. $\dfrac{x^2 + 6x + 9}{x^2 - 9}$ $\dfrac{x+3}{x-3}$

6. $\dfrac{x^2 - 3x + 2}{x^2 + 5x - 6}$ $\dfrac{x-2}{x+6}$

7. $\dfrac{x^2 - 2x - 3}{x^2 - 7x + 12}$ $\dfrac{x+1}{x-4}$

8. $\dfrac{x - 2}{x^3 - x^2 - 4}$ $\dfrac{1}{x^2+x+2}$

In Exercises 9–20, multiply and simplify.

9. $\dfrac{3xy^3}{x^3y} \cdot \dfrac{y}{6x}$ $\dfrac{y^3}{2x^3}$

10. $\dfrac{120x^5}{y^4} \cdot \dfrac{xy}{5x^3}$ $\dfrac{24x^3}{y^3}$

11. $\dfrac{16x^3}{5y^9} \cdot \dfrac{x^3y^7}{80xy^2}$ $\dfrac{x^5}{25y^4}$

12. $\dfrac{x^{10}y^4}{33x^4} \cdot \dfrac{39x^5}{4y^{10}}$ $\dfrac{13x^{11}}{44y^6}$

13. $\dfrac{2x^2 - 10}{x + 1} \cdot \dfrac{x - 4}{4x^2 - 20}$ $\dfrac{x-4}{2(x+1)}$

14. $\dfrac{x^2 + 3x}{x^2 + 6x + 8} \cdot \dfrac{x^2 + x - 2}{4x^3 + 12x^2}$ $\dfrac{x-1}{4x(x+4)}$

15. $\dfrac{x - 3}{2x - 8} \cdot \dfrac{6x^2 - 96}{x^2 - 9}$ $\dfrac{3(x+4)}{x+3}$

16. $\dfrac{x^2 - 5x - 6}{4x^5} \cdot \dfrac{x + 2}{x^2 + 3x + 2}$ $\dfrac{x-6}{4x^5}$

17. $\dfrac{x^2 + 6x - 7}{x^4 + 8x^3 + 7x^2} \cdot 3x^2$ $\dfrac{3(x-1)}{x+1}$

18. $(x + 2) \cdot \dfrac{x^2 - 9}{x^2 - x - 6}$ $x + 3$

19. $(x^2 + 8x + 16) \cdot \dfrac{16x^2 - 64}{x^2 - 16}$ $\dfrac{16(x+4)(x^2-4)}{(x-4)}$

20. $\dfrac{2x^2 - 2}{x^2 - 6x - 7} \cdot (x^2 - 10x + 21)$ $2(x-1)(x-3)$

In Exercises 21–32, divide and simplify.

21. $\dfrac{x^2 + 4x - 5}{2x^2} \div \dfrac{x - 1}{4x}$ $\dfrac{2(x+5)}{x}$

22. $\dfrac{2x^2 - 16x}{x^2 - 9x + 8} \div \dfrac{2x}{5x - 5}$ 5

23. $\dfrac{x^2 + 8x + 16}{x + 2} \div \dfrac{x^2 + 6x + 8}{x^2 - 4}$ $\dfrac{(x+4)(x-2)}{x+2}$

24. $\dfrac{x^2 - 8x - 20}{15x^4 - 150x^3} \div \dfrac{x^2 - 9x}{x^2 - 7x - 18}$ $\dfrac{(x+2)^2}{15x^4}$

25. $\dfrac{x - 3}{x^2 - 5x - 14} \div \dfrac{x^2 - x - 6}{x - 7}$ $\dfrac{1}{(x+2)^2}$

26. $\dfrac{x^2 + 3x - 4}{x^2 - 4x + 3} \div \dfrac{x + 4}{x - 3}$ 1

27. $\dfrac{x^2 + 10x + 24}{3x^2 + 3x} \div (x + 6)$ $\dfrac{x+4}{3x(x+1)}$

28. $\dfrac{x^2 - 14x + 48}{x^2 - 6x} \div (3x - 24)$ $\dfrac{1}{3x}$

29. $(x + 4) \div \dfrac{x^2 - 6x - 40}{x^2 + 3x}$ $\dfrac{x(x+3)}{x-10}$

30. $(x + 1) \div \dfrac{x^2 + 2x + 1}{x^2 + 13x + 12}$ $x + 12$

31. $(x^2 + 10x + 25) \div \dfrac{x^2 + 12x + 35}{x^2 - 8x - 105}$ $(x+5)(x-15)$

32. $(x^2 + 6x - 16) \div \dfrac{2x^2 + 16x}{x + 2}$ $\dfrac{(x+2)(x-2)}{2x}$

EXERCISE Notes

ASSIGNMENT GUIDE

Basic/Average: Ex. 1–4, 7–15 odd, 25–33 odd, 42–46, 57–59

Above Average: Ex. 1–4, 7–17 odd, 27–35 odd, 42–46, 50, 57–59

Advanced: Ex. 1–4, 9–19 odd, 27–37 odd, 42–46, 52, 57–60

Selected Answers
Exercises 1–4, 5–57 odd

More Difficult Exercises
Exercises 39–46, 59, 60

Guided Practice

▶ **Ex. 1–4** Students should understand that they must factor each expression first. Exercise 2 is a good test for this. Discuss with students why $\dfrac{x^2 + 10x - 4}{x^2 + 10x}$ does not simplify as follows: $\dfrac{x^2 + 10x}{x^2 + 10x} - 4 = 1 - 4 = -3$.

Independent Practice

▶ **Ex. 5–32** Remind students to factor first.

▶ **Ex. 21–32** When dividing fractions, make sure that students find the reciprocal of the second fraction only.

▶ **Ex. 27** Students may need to be reminded that the reciprocal of $(x + 6)$ is $\dfrac{1}{x+6}$.

Answers

34. $\dfrac{(x+5)(x+1)}{x-4}$

35. $3(x+2)$

36. $\dfrac{2(x+4)}{3x}$

37. $(x+4)(x+3)$

38. $\dfrac{2(x+2)}{x+6}$

▶ **Ex. 43** The total amount, A, is given by the product of the average cost per snow tire, C, and the number of snow tires bought by the shop, T, or $A = CT$.

▶ **Ex. 44** The average number, x, of cars admitted per month is given by the revenue, R, from admissions divided by the average price, p, per car, or $x = \frac{R}{p}$. Encourage students to use a calculator when finding factors of large numbers such as 131,124. One such factor is 3122.

In Exercises 33–40, perform the operations and simplify. 34.–38. See margin.

33. $\dfrac{x-11}{2x+10} \div \dfrac{x^2-8x-33}{x+5}$ $\quad \dfrac{1}{2(x+3)}$

34. $\dfrac{x^2+3x-10}{x-2} \cdot \dfrac{x^2+3x-10}{x-2} \cdot \dfrac{x+1}{x^2+x-20}$

35. $\dfrac{x^2+12x+20}{4x^2-9} \cdot \dfrac{6x^3-9x^2}{x^3+10x^2} \cdot (2x+3)$

36. $\left[\dfrac{x^2+x-20}{x+1} \div \dfrac{33x^2-132x}{16x+16}\right] \div \dfrac{8x+40}{11x+44}$

37. $(x^2+x-30) \div \dfrac{x^2-11x+30}{x^2+7x+12} \cdot \dfrac{x-6}{x+6}$

38. $\dfrac{x^2-5x-14}{x^3-6x^2-7x} \cdot (x^2-4x-5) \div \dfrac{x^2+x-30}{2x}$

✪ **39.** $\left[\dfrac{x^2+11x}{x-2} \div (3x^2+6x)\right] \cdot \dfrac{x^2-4}{x+11}$ $\quad \dfrac{1}{3}$

✪ **40.** $(x^2+15x) \div \left[\dfrac{x^2+2x-8}{x-2} \cdot \dfrac{x^2+15x}{x+4}\right]$ $\quad 1$

Geometry **In Exercises 41 and 42, find the ratio of the area of the blue region to the total area. Write your result in simplified form.**

✪ **41.** $\dfrac{7}{6x(x+5)}$

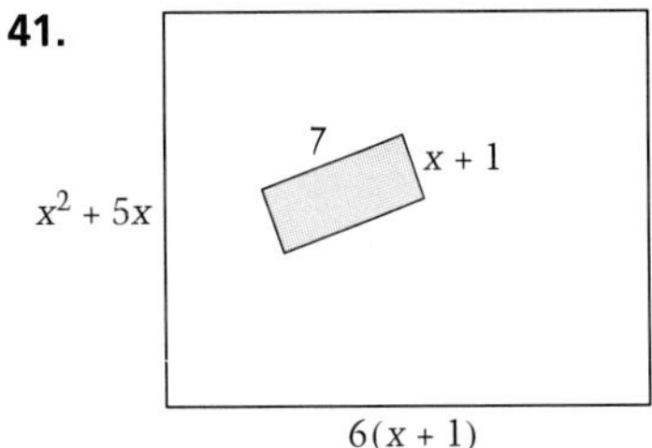

✪ **42.** $\dfrac{x}{3(x+3)}$

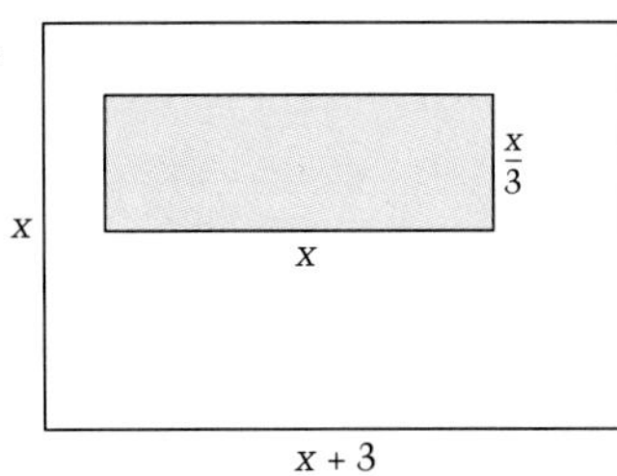

✪ **43.** *Cost of Tires* For 1980 through 1992, Alfredo's Autobody Shop spent an average of C dollars per snow tire per year, where

$$C = \frac{t+75}{1-0.05t}$$

and $t = 0$ represents 1980. The number, T, of snow tires bought per year by the shop can be modeled by

$$T = \frac{500(t+20)}{t+75}. \quad A = \frac{500(t+20)}{1-0.05t}; \ \$23{,}333.33$$

Write a model for the total amount, A, that the shop spent each year on snow tires. How much did the shop spend in 1988?

✪ **44.** *Drive-In Movies* For 1980 through 1992, the average monthly revenue, R (in dollars), from admissions at the Sunset Drive-In Theater can be modeled by

$$R = \frac{131{,}124 + 3122t}{26-t}$$

where $t = 0$ represents 1980. The average price, p (in dollars), per car can be modeled by

$$p = \frac{294+7t}{100-5t}. \quad x = \frac{2230(20-t)}{26-t}, \ \approx 1394$$

Write a model for the average number, x, of cars admitted per month to the theater. How many cars were admitted in a month of 1990?

Swimming Pools **In Exercises 45 and 46, use the following information.**

You are considering buying a swimming pool and have narrowed the choices to two—one that is circular and one that is rectangular. The rectangular pool's width is three times its depth. Its length is 6 feet more than its width. The circular pool has a diameter that is twice the width of the rectangular pool, and it is 2 feet deeper.

45. Find the ratio of the circular pool's volume to the rectangular pool's volume. π

46. The volume of the rectangular pool is 2592 cubic feet. How many gallons of water are needed to fill the circular pool if 1 gallon ≈ 0.134 ft^3? $\approx$60,769

Integrated Review

In Exercises 47–52, factor the expression completely with respect to the integers. See below.

47. $x^3 - 9x^2 + 5x - 45$ **48.** $x^3 + 2x^2 + 2x + 4$ **49.** $x^3 - 4x^2 + 9x - 36$

50. $x^3 + x^2 - 16x - 16$ **51.** $2x^3 - 3x^2 + 2x - 3$ **52.** $3x^3 + 6x^2 - 3x - 6$

In Exercises 53–56, simplify the expression.

53. $\frac{24x^2y^3}{8x^5y}$ $\frac{3y^2}{x^3}$ **54.** $\frac{3xy^4}{2x^6y}$ $\frac{3y^3}{2x^5}$ **55.** $\frac{-9y^2}{3x^3y}$ $\frac{-3y}{x^3}$ **56.** $\frac{2x^{-1}y}{x^4y^2}$ $\frac{2}{x^5y}$

57. *College Entrance Exam Sample* If $\frac{a}{b} = \frac{5}{9}$ and $\frac{b}{c} = \frac{3}{5}$, then $\frac{a}{c} = \boxed{?}$. c

a. $\frac{1}{9}$ **b.** $\frac{1}{5}$ **c.** $\frac{1}{3}$ **d.** $\frac{5}{9}$ **e.** $\frac{3}{5}$

58. *College Entrance Exam Sample* $[(2x^2y^3)^2]^3 = \boxed{?}$ d

a. $4x^4y^6$ **b.** $12x^4y^6$ **c.** $64x^4y^6$ **d.** $64x^{12}y^{18}$ **e.** $64x^{64}y^{216}$

Exploration and Extension

59. Find two pairs of rational expressions with first-degree numerators and first-degree denominators whose product is

$\frac{2x^2 + 11x - 6}{3x^2 + x - 2}$. $\frac{2x-1}{3x-2}$ and $\frac{x+6}{x+1}$, $\frac{2x-1}{x+1}$ and $\frac{x+6}{3x-2}$

60. *Technology* Use a graphing calculator to compare the graphs of the following two functions. What is the domain of each function? Can you find a range setting that shows a difference in the graphs? Explain. All real numbers except $x = -2$, all real numbers; no

$$f(x) = \frac{x^2 + 3x + 2}{x + 2} \quad \text{and} \quad g(x) = x + 1$$

▶ **Ex. 45** The volume of a circular pool is given by $v = \pi r^2 h$, the volume of a cylinder of radius, r, and height, h.

Enrichment Activity

This activity requires students to go beyond lesson goals.

Common errors usually occur in computational problems when there is a misunderstanding of the conditions of the problem, or when there is a misapplication of a technique or strategy. For example, here is an example of a common student error that occurs in simplifying rational expressions.

Wendell claims that $\frac{x^2 - 9}{x - 3} = x + 3$, since $\frac{x^2}{x} = x$ and $\frac{-9}{-3} = 3$. Though Wendell came up with the correct answer in this one case, his reasoning is faulty. If he tried to apply the same reasoning to the expression $\frac{x^2 - 8}{x - 2}$, he would simplify *wrongly* to get $x + 4$.

WRITING PROBLEMS Write some other examples of faulty reasoning that occur as common errors in simplifying rational or fractional expressions. Share your examples with your classmates.

Answers may vary. The example above shows a misapplication of a strategy. One of the most common misunderstandings involving operations with rational expressions occurs when students see two or more fractions in an expression or equation. Some students will automatically find their least common denominator first, whether they need to or not.

MATH JOURNAL You may wish to have students include some of the most common errors in their math journals for further reference.

47. $(x^2 + 5)(x - 9)$ **48.** $(x^2 + 2)(x + 2)$ **49.** $(x^2 + 9)(x - 4)$
50. $(x + 4)(x - 4)(x + 1)$ **51.** $(x^2 + 1)(2x - 3)$ **52.** $3(x + 1)(x - 1)(x + 2)$

Mid-Chapter SELF-TEST

Take this test as you would take a test in class. The answers to the exercises are given in the back of the book.

In Exercises 1–6, sketch the graph of the function. (10.1) See Additional Answers.

1. $f(x) = \dfrac{1}{x+2}$ **2.** $f(x) = \dfrac{2}{x^2+1}$ **3.** $f(x) = \dfrac{1}{x^2-1}$

4. $f(x) = \dfrac{x}{x+4}$ **5.** $f(x) = \dfrac{x^2}{x+1}$ **6.** $f(x) = \dfrac{x^2+2x-3}{x-1}$

In Exercises 7–10, simplify the rational expression. (10.3)

7. $\dfrac{x^2-2x-8}{x^2-16}$ $\dfrac{x+2}{x+4}$ **8.** $\dfrac{x^2-4x+4}{2x^2-7x+6}$ $\dfrac{x-2}{2x-3}$

9. $\dfrac{6x^6-486x^2}{9x^5+9x^4-108x^3}$ $\dfrac{2(x^2+9)(x+3)}{3x(x+4)}$ **10.** $\dfrac{8x^2-2x-15}{10x^2-17x+3}$ $\dfrac{4x+5}{5x-1}$

In Exercises 11 and 12, multiply and simplify. (10.3)

11. $\dfrac{2x^3-2x^2-12x}{x-4} \cdot \dfrac{x^2-5x+4}{4x^3+4x^2-8x}$ $\dfrac{x-3}{2}$ **12.** $\dfrac{8x^2-32}{x^3+x^2} \cdot \dfrac{x^5-x^4-2x^3}{4x^2-16x+16}$ $2x(x+2)$

In Exercises 13–16, divide and simplify. (10.3)

13. $\dfrac{x^3-3x^2}{3x+6} \div \dfrac{x^3-8x^2+15x}{6x^2-18x-60}$ $2x$ **14.** $\dfrac{6x^3+4x^2-16x}{3x^2+3x-6} \div \dfrac{45x-60}{24x^3+8x^2-32x}$ $\dfrac{16x^2(3x+4)}{45}$

15. $\dfrac{x^2+2x-3}{3x^2-x-2} \div (x+3)$ $\dfrac{1}{3x+2}$ **16.** $(30x^3-6x^2) \div \dfrac{15x^2-3x}{4x^2+4x-24}$ $8x(x+3)(x-2)$

17. The variables x and y vary inversely. Given that $x = 2$ when $y = -5$, write an equation that relates x and y. **(10.2)** $y = \frac{-10}{x}$

18. The variable x varies jointly with the product of y and z. Given that $x = 4$ when $y = -1$ and $z = 6$, write an equation that relates the variables. **(10.2)** $x = -\frac{2}{3}yz$

19. Consider two people who weigh W_1 pounds and W_2 pounds and are sitting d_1 feet and d_2 feet, respectively, from the center of a see-saw. To balance the see-saw, the people must adjust their distances so that $d_1W_1 = d_2W_2$. Rodriguez, who weighs 80 pounds, is sitting 6 feet from the center. Pearl weighs 96 pounds. To balance the see-saw, where should she sit? **(10.2)** 5 feet from the center

20. For 1956 through 1991, the average cost per year, A, of possessing a United States passport can be modeled by

$$A = \frac{18.5t^2 - 310.8t + 5452.2}{-1.2t^2 + 87.2t + 1000}$$

where $t = 0$ represents 1956. Approximate the average cost per year of possessing a United States passport in 1990. **(10.1)** \$6.31

10.4 Solving Rational Equations

What you should learn:

Goal 1 How to solve equations that contain rational expressions

Goal 2 How to use rational equations to solve real-life problems

Why you should learn it:

Many real-life problems can be modeled by rational equations, such as the year that a rational model for prize money reached a given level.

Goal 1 Solving Rational Equations

In this lesson, you will learn two basic methods for solving equations that contain rational expressions. The first method works for any rational equation; the second works only for equations in which each side is a single fraction.

The **least common multiple** of two (or more) polynomials is the simplest polynomial that is a multiple of each of the original polynomials. The **least common denominator** of two or more fractions is the least common multiple of the denominators of the fractions.

Example 1 *Least Common Multiple of Polynomials*

	Polynomials	Least Common Multiple
a.	x and $x - 4$	$x \cdot (x-4) = x(x-4)$
b.	x^2, $2x$, and $x(x-3)$	$2 \cdot x \cdot x \cdot (x-3) = 2x^2(x-3)$
c.	$x^2 - 4$ and $x + 2$	$(x+2) \cdot (x-2)$
d.	$x^2 - x - 6$ and $x^2 - 2x - 8$	$(x+2)(x-3)(x-4)$

To solve a rational equation, multiply each term on both sides of the equation by the least common denominator of the terms. Simplify, and solve the resulting polynomial equation.

Example 2 *Solving a Rational Equation*

Solve $\frac{2}{x} + \frac{5}{3} = \frac{7}{x}$.

Solution The least common denominator is $3x$.

$\frac{2}{x} + \frac{5}{3} = \frac{7}{x}$ *Rewrite original equation.*

$3x\left(\frac{2}{x} + \frac{5}{3}\right) = 3x \cdot \frac{7}{x}$ *Multiply both sides by 3x.*

$6 + 5x = 21$ *Simplify.*

$5x = 15$ *Subtract 6x from both sides.*

$x = 3$ *Divide both sides by 5.*

The solution is 3. Check this in the original equation. (Or use a graphing calculator to check that the graphs of "each side" intersect when $x = 3$.)

Problem of the Day

Four brown cows and three black cows give as much milk in five days as three brown cows and five black cows give in four days. Which kind of cow is the better milker?

Black

ORGANIZER

Warm-Up Exercises

1. Solve for x.
   a. $\frac{12}{8} + \frac{2}{3} = x$
   b. $\frac{25}{x} = \frac{40}{16}$
   c. $\frac{2}{3} + \frac{4}{5} - \frac{2}{8} = x$

   a. $\frac{13}{6}$, b. 10, c. $\frac{73}{60}$
2. Solve for x.
   a. $x^2 - 2x - 35 = 0$
   b. $3x - 14 = 5(x + 2)$
   c. $4x^2 + 12x = -9$

   a. $x = -5$, $x = 7$ b. $x = -12$

   c. $x = -\frac{3}{2}$

Lesson Resources

Teaching Tools
- Transparency: 8
- Problem of the Day: 10.4
- Warm-up Exercises: 10.4
- Answer Masters: 10.4

Extra Practice: 10.4

Color Transparency: 56

Technology Handbook: p. 71

LESSON Notes

Example 1

The skill of finding the least common multiple of polynomials is important since it has a variety of applications. Emphasize that leaving the least common multiple in factored form is especially helpful when solving rational equations.

Examples 2–4

Each example results in an equation that can be solved using techniques learned in previous lessons. Point out that it is important to check the solutions of a rational equation by using the original equation.

EXTEND Examples 3–4 Ask students to provide the rationale for each step in the solution.

Example 5

Point out that, in this case, cross multiplying and multiplying both sides of the equation by $4(x + 3)$ achieves the same result.

Examples 6–7

TECHNOLOGY Students should be encouraged to graph the model given in each example on a graphics calculator or computer. In Example 6, for example, have students approximate the average monthly high temperatures in the summer months. 89.88°, 92.48°, 91.35°

Ask the students if the model provides reasonable information for temperatures in Mississippi. Yes

Students may want to investigate similar temperature information for their local areas.

Common-Error ALERT!

Cross multiplying does not always give the same result as multiplying by the least common denominator. For example, consider solving the equation $\frac{1}{x^2 - x} = \frac{1}{2x - 2}$. Cross multiplying results in a quadratic equation with roots of 2 and 1; the root 1 is extraneous. Multiplying both sides by the LCD, $2x(x - 1)$, results in a linear equation with a root of 2.

Multiplying both sides of an equation by the least common denominator may introduce extraneous solutions.

Example 3 *An Extraneous Solution*

Solve $\frac{5x}{x-2} = 7 + \frac{10}{x-2}$.

Solution The least common denominator is $x - 2$.

$$\frac{5x}{x-2} = 7 + \frac{10}{x-2}$$
$$(x-2)\cdot\frac{5x}{x-2} = (x-2)(7) + (x-2)\cdot\frac{10}{x-2}$$
$$5x = 7(x-2) + 10$$
$$5x = 7x - 4$$
$$4 = 2x$$
$$x = 2$$

At this point, the solution appears to be 2. After checking 2 in the original equation, however, you can conclude that this "trial solution" is extraneous. Thus, the original equation has no solution. ■

$y = \frac{5x}{x-2}$

$y = 7 + \frac{10}{x-2}$

If you have a graphing calculator, try using it to perform a graphic check. Note that the graphs of the two equations do not intersect.

Example 4 *An Equation That Has Two Solutions*

Solve $\frac{3x}{x+1} = \frac{12}{x^2-1} + 2$.

Solution The least common denominator is $(x + 1)(x - 1) = x^2 - 1$.

$$\frac{3x}{x+1} = \frac{12}{x^2-1} + 2$$
$$(x^2-1)\cdot\frac{3x}{x+1} = (x^2-1)\cdot\frac{12}{x^2-1} + (x^2-1)\cdot 2$$
$$(x-1)(3x) = 12 + 2(x^2-1)$$
$$3x^2 - 3x = 12 + 2x^2 - 2$$
$$x^2 - 3x - 10 = 0$$
$$(x+2)(x-5) = 0$$
$$x + 2 = 0 \longrightarrow x = -2$$
$$x - 5 = 0 \longrightarrow x = 5$$

The solutions are −2 and 5. Check these solutions in the original equation. ■

Cross multiplying can be used only for equations that have *exactly one* fraction on each side of the equation.

Example 5 *Solving a Rational Equation by Cross Multiplying*

Solve $\frac{7}{x+3} = \frac{x}{4}$.

Solution

$\frac{7}{x+3} = \frac{x}{4}$	*Rewrite original equation.*
$7(4) = x(x+3)$	*Cross multiply.*
$28 = x^2 + 3x$	*Simplify.*
$0 = x^2 + 3x - 28$	*Standard form*
$0 = (x+7)(x-4)$	*Factor.*
$x = -7$ or 4	*Zero Product Property*

The solutions are -7 and 4. Check these solutions in the original equation. ■

Goal 2 Using Rational Equations in Real Life

Real Life **Meteorology**

Example 6 *A Temperature Model*

The average monthly high temperature in Jackson, Mississippi, can be modeled by

$$T = \frac{-191(t-30)}{t^2 - 16.5t + 114}$$

where T is measured in degrees Fahrenheit and $t = 1, 2, \ldots, 12$ represents the months of the year. During which month is the average monthly high temperature equal to 57.3°?

Solution

$$\frac{-191(t-30)}{t^2 - 16.5t + 114} = 57.3$$

$$-191(t-30) = 57.3(t^2 - 16.5t + 114)$$

$$-\frac{10}{3}(t-30) = t^2 - 16.5t + 114$$

$$-10t + 300 = 3t^2 - 49.5t + 342$$

$$0 = 3t^2 - 39.5t + 42$$

$$0 = (3t - 3.5)(t - 12)$$

The solutions are 1.17 (January) and 12 (December). The graph at the left confirms these solutions. ■

Average Monthly High Temperatures, Jackson, Mississippi

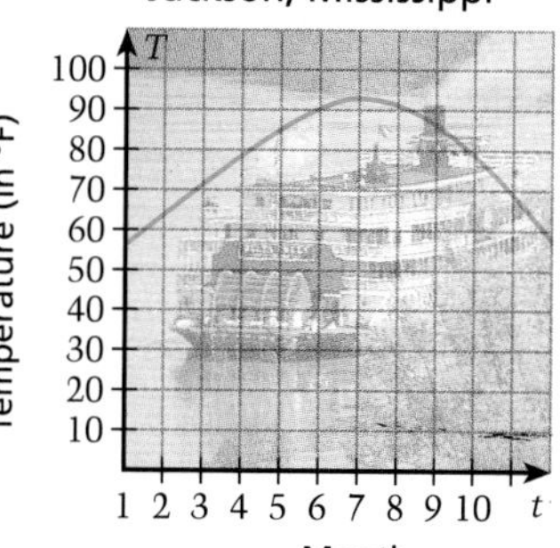

Extra Examples

Here are additional examples similar to **Examples 1–7.**

1. Find the least common multiple.
   a. x^2 and $x^3 - 1$
   b. x^4, $6x^2$, and $x(2x - 1)$
   c. $x^2 + 5x + 6$ and $x + 3$
   d. $x^2 - 9$ and $x^2 - 5x - 24$
   a. $x^2(x^3 - 1)$
   b. $6x^4(2x - 1)$
   c. $(x + 2)(x + 3)$
   d. $(x + 3)(x - 3)(x - 8)$
2. Solve each equation.
   a. $\frac{5}{x} + 2 = \frac{6}{x}$
   b. $\frac{3}{x(x+1)} + \frac{3}{x+1} = -\frac{1}{x}$
   c. $\frac{2}{x-2} = \frac{x}{x-2} + \frac{6}{x^2 + 3x - 10}$
   d. $\frac{2}{x-4} = \frac{x+5}{5}$
   a. $\frac{1}{2}$
   b. No solution
   c. $-4, 1$
   d. $5, -6$
3. Using the model in Example 6, during which month is the average monthly high temperature equal to 91.3°?
   The solutions are 8.01 (August) and 6.39 (June). Check the graph on the pupil page.
4. Using the model in Example 7, during which year was the total prize money approximately $11 million?
   1985

Check Understanding

1. What is the least common multiple of two (or more) polynomials?
   See the top of page 541.
2. How do you solve a rational equation?
   See the middle of page 541.
3. When can cross multiplying be used to solve a rational equation?
   See the top of page 543.

Communicating about ALGEBRA

An Alternate Approach: *Using Technology*

Students may sketch the graph of each function on a graphics calculator and use the Trace function to approximate where the zeros of each function are.

When finding the zeros, you may want to point out that the zeros of a rational function occur only when the numerator of the fraction is equal to zero. For example, in Part A, if $f(x) = 0$, then $2x - 15 = 0$, or $x = \frac{15}{2}$ or 7.5. In Part B, discuss why this rational function has no vertical asymptotes.
This occurs since the denominator function, $x^2 + 1$, has no real zeros.

The photo shows a contestant in the saddle-bronc event at the Indian rodeo in Gallup, New Mexico. Many rodeo events are dangerous. Fortunately, this rider escaped serious injury.

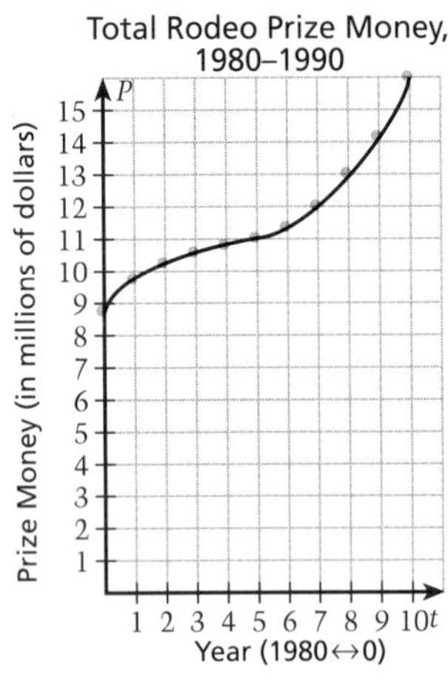

Example 7 *Solving a Rational Equation*

For 1980 through 1990, the total prize money, P (in millions of dollars), at Professional Rodeo Cowboys Association events can be modeled by

$$P = \frac{69(t^2 + 3t + 30)}{-t^3 + 16t^2 - 15t + 240}$$

where $t = 0$ represents 1980. During which year was the total prize money approximately \$16 million?

Solution You could try to solve this equation algebraically. In doing that, however, you will have to solve a third-degree polynomial equation (which, you know from Chapter 9, can be difficult). In the context of this problem, a graphical approach is much better. After sketching the graph of the equation, as shown at the left, you can see that the total prize money was approximately \$16 million in 1990 (when $t = 10$). ■

Communicating about ALGEBRA

SHARING IDEAS about the Lesson

Use the graph to help find the zeros of the function. 5, −1

A. $f(x) = \frac{2x - 15}{x - 3}$ $\frac{15}{2}$

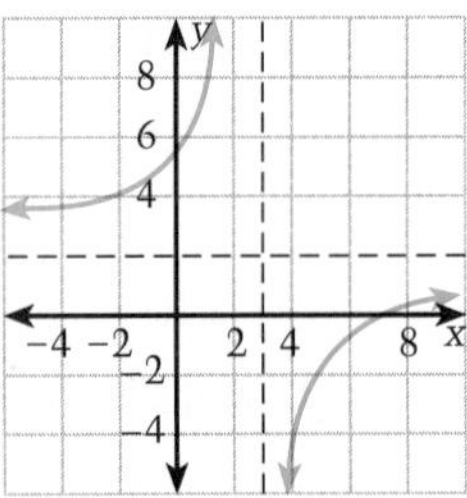

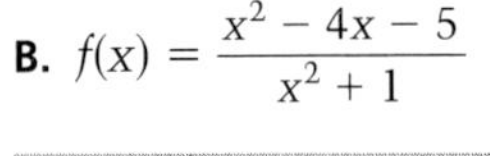

B. $f(x) = \frac{x^2 - 4x - 5}{x^2 + 1}$

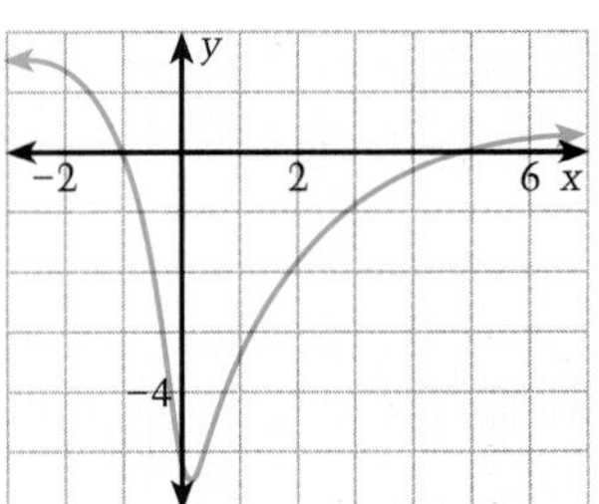

EXERCISES

Guided Practice

CRITICAL THINKING about the Lesson

1. Describe two methods that can be used to solve a rational equation. Which method is more general? Explain.
2. Find the least common multiple of x, $x^2 + 5x + 6$, and $x + 3$.
3. Use the least common denominator method to solve $\frac{3x}{x-5} = \frac{4-x}{x-5} - \frac{4}{x}$.
4. Use the cross-multiplication method to solve $\frac{x}{x^2-10} = \frac{3}{2x+1}$.

Independent Practice

In Exercises 5–10, find the least common multiple.

5. $(x^2 - 9), (x - 3), 3$
6. $(x + 5), x^2, x$
7. $4x, x^3, (x - 1)$
8. $1, 3x, x(x + 1)$
9. $x^2, 2x(x - 2), (x^2 - 4)$
10. $(x - 4), (x^2 - 16), x$

In Exercises 11–16, determine whether the *x*-value is a solution of the equation.

11. $\frac{3x-2}{x+4} = \frac{2x}{x+3}, -1$
12. $\frac{x}{2x+1} = \frac{5}{4-x}, -1$
13. $\frac{3x-1}{x-2} + 3 = \frac{x}{x-2}, -2$
14. $\frac{2x}{x+3} = 1 - \frac{6}{x+3}, -3$
15. $\frac{3}{x(x-3)} + \frac{4}{x} = \frac{1}{x-3}, 3$
16. $\frac{x}{x-5} + 4 = \frac{1}{x+3}, 4$

In Exercises 17–24, find the zeros of the function.

17. $f(x) = \frac{x+2}{x-2}$
18. $f(x) = \frac{2x}{x+4}$
19. $f(x) = \frac{6}{x+5}$
20. $f(x) = \frac{x-1}{2x-4}$

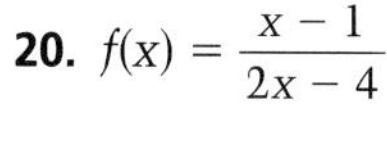

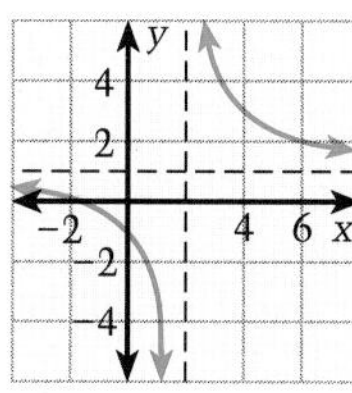

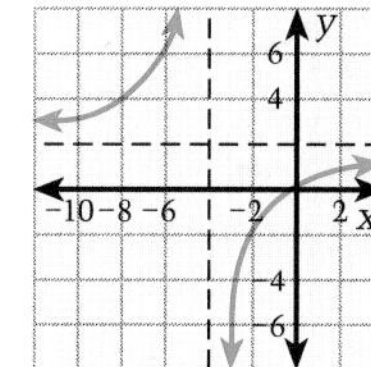

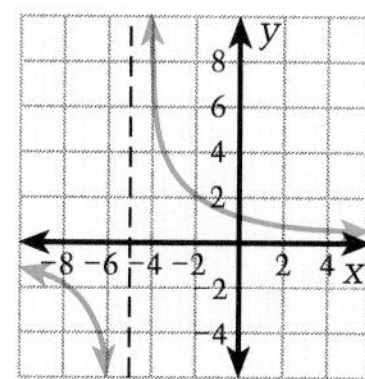

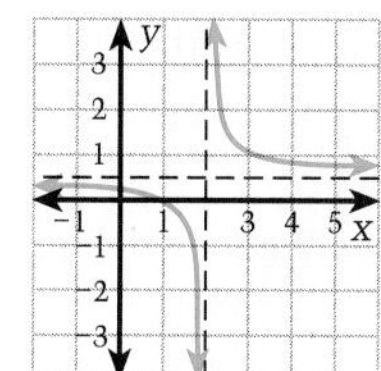

21. $f(x) = \frac{x^2-4}{x^2+4x+3}$
22. $f(x) = \frac{x^2-4x-21}{x+3}$
23. $f(x) = \frac{2x^2+9x-5}{2x^2-1}$
24. $f(x) = \frac{x^2+8x+12}{4x+1}$

EXERCISE Notes

ASSIGNMENT GUIDE

Basic/Average: Ex. 1–4, 7, 11–23 odd, 31–41 odd, 49, 51–60

Above Average: Ex. 1–4, 7–23 odd, 33–43 odd, 49–60

Advanced: Ex. 1–4, 7–23 odd, 33–43 odd, 49–61

Selected Answers
Exercises 1–4, 5–59 odd

More Difficult Exercises
Exercises 45–52, 61, 62

Guided Practice

Ex. 1–4 COOPERATIVE LEARNING These exercises can be used as a summary for the lesson and can provide in-class practice of the methods demonstrated for solving rational equations. Have students discuss their answers with their cooperative-learning partners.

Ex. 3–4 Remind students to check for extraneous solutions.

Answers

1. Multiplying by the least common denominator and cross multiplying; multiplying by the least common denominator, it works for any rational equation.
5. $3(x + 3)(x - 3)$
6. $x^2(x + 5)$
7. $4x^3(x - 1)$
8. $3x(x + 1)$
9. $2x^2(x + 2)(x - 2)$
10. $x(x + 4)(x - 4)$

Independent Practice

Ex. 14 Students should be able to tell that -3 is not a solution to the equation by inspection because -3 is not in the domain of one of the rational functions that make up the equation. Students should refer to page 518 for a review of the concept of the domain of a rational function.

▶ **Ex. 25–48** Students may be encouraged to use the cross-multiplying property when appropriate. Examples 2–4 provide models for solving these equations.

TECHNOLOGY Students with graphics calculators can verify solutions by graphing the functions on each side of the equation and locating the intersection(s).

▶ **Ex. 49–50** A solution model for these exercises is provided by Example 2 on page 520 of Lesson 10.1.

▶ **Ex. 53–56** These exercises should be assigned as a group.

▶ **Ex. 57–60** **CONNECTIONS** Use these exercises to stimulate a class discussion comparing local prices of color prints with the prices given.

In Exercises 25–48, solve the equation. Check each solution. 30. $\frac{5}{7}$, 3 33., 34. No solution

25. $\frac{10}{x+3} + \frac{10}{3} = 6$ $\frac{3}{4}$

26. $\frac{6}{x} - \frac{7x}{5} = \frac{x}{10}$ ±2

27. $\frac{-1}{x-3} = \frac{x-4}{x^2-27}$ $-\frac{3}{2}, 5$

28. $\frac{x}{2x+7} = \frac{x-5}{x-1}$ −5, 7

29. $\frac{6x}{x+4} + 4 = \frac{2x+2}{x-1}$ $-\frac{3}{2}, 2$

30. $\frac{7x+1}{2x+5} + 1 = \frac{10x-3}{3x}$

31. $\frac{2x}{5} = \frac{x^2-5x}{5x}$ −5

32. $\frac{x}{x^2-8} = \frac{2}{x}$ ±4

33. $\frac{10}{x^2-2x} + \frac{4}{x} = \frac{5}{x-2}$

34. $\frac{2(x+7)}{x+4} - 2 = \frac{2x+20}{2x+8}$

35. $\frac{2}{6x+5} - \frac{3}{4(6x+5)} = \frac{1}{28}$ 5

36. $3\left(\frac{1}{x} + 4\right) = 2 + \frac{4}{3x}$

37. $\frac{-2}{x-1} = \frac{x-8}{x+1}$ 2, 5

38. $\frac{8(x-1)}{x^2-4} = \frac{4}{x-2}$ 4

39. $\frac{x^2+2x+2}{x-1} = \frac{2x+3}{x-1}$

40. $\frac{2x}{4-x} = \frac{x^2}{x-4}$ −2, 0

41. $\frac{2x}{x+2} = \frac{1}{x^2-4} + 1$ 1, 3

42. $\frac{6}{x-1} + \frac{2x}{x-2} = 2$ $\frac{8}{5}$

43. $\frac{8}{2x+4} - \frac{3x+1}{3x^2+2} = \frac{2}{x+2}$ $\frac{1}{3}, 2$

44. $\frac{3x}{x+1} = \frac{12}{x^2-1} + 2$ −2, 5

36. $-\frac{1}{6}$ 39. −1

✪ **45.** $\frac{2x^2-5}{x^2-4} + \frac{6}{x+2} = \frac{4x-7}{x-2}$ $1, \frac{3}{2}$

✪ **46.** $\frac{2}{x-10} - \frac{3}{x-2} = \frac{6}{x^2-12x+20}$ 20

✪ **47.** $\frac{10x^2+x-2}{2x^2-9x-18} = \frac{4x}{x-6} + \frac{2x-3}{2x+3}$ 5

✪ **48.** $\frac{13x+16}{4x-1} + \frac{8x^2+12x+14}{4x^2+11x-3} = \frac{24x-8}{x+3}$ $-\frac{9}{25}, 2$

✪ **49.** *Average Cost* A greeting card manufacturer can produce a dozen cards for $6.50. If the initial investment by the company was $60,000, how many dozen cards must be produced before the average cost per dozen falls to $11.50? **12,000**

✪ **50.** *Average Cost* You invest $40,000 to start a nacho stand in a shopping mall. You can make each basket of nachos for $0.70. How many baskets must you sell before your average cost per basket is $1.50? **50,000**

Keyboards **In Exercises 51 and 52, use the following information.**

For 1980 through 1990, the number, P, of conventional pianos sold in the United States can be modeled by

$$P = 223{,}720 - 9480t$$

where $t = 0$ represents 1980. The number, K, of electronic pianos and keyboards sold can be modeled by

$$K = 27{,}820 + 8710t.$$

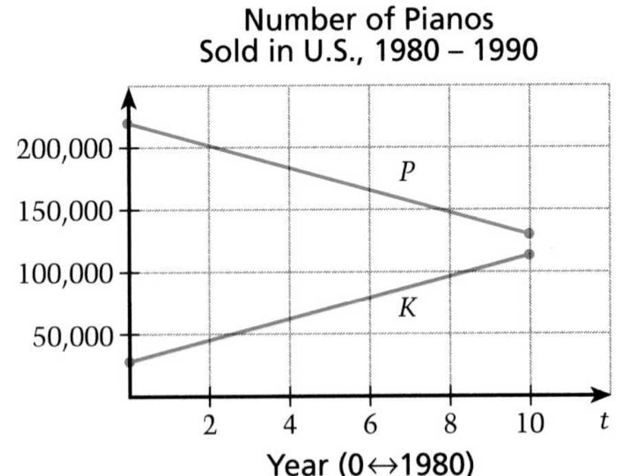

✪ **51.** In what year was the ratio of conventional pianos sold to electronic pianos and keyboards sold equal to 3? **1984**

✪ **52.** In what year does it appear that the number of electronic pianos and keyboards sold will surpass the number of conventional pianos sold? **1991**

Integrated Review

In Exercises 53–56, match the function with its graph.

53. $f(x) = \dfrac{5}{x-6}$ d

54. $f(x) = \dfrac{6}{x+5}$ a

55. $f(x) = \dfrac{6x}{x-5}$ c

56. $f(x) = \dfrac{2x}{x+6}$ b

a.

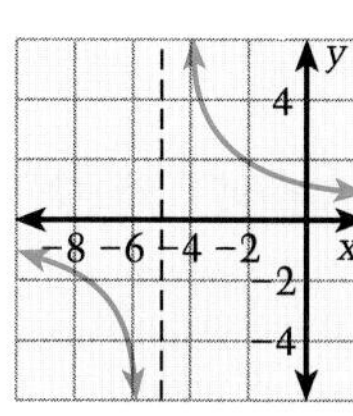

b.

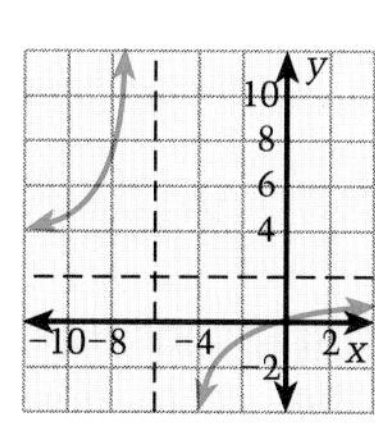

c.

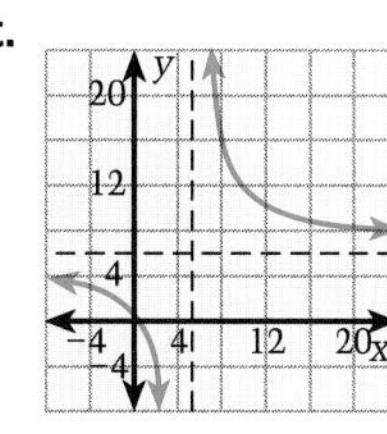

d.

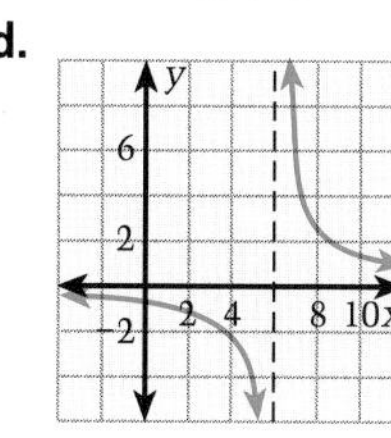

Photography **In Exercises 57–60, use the following information.**

The graph below shows the total amount spent by Americans on photofinishing, film, cameras, and other photographic equipment.

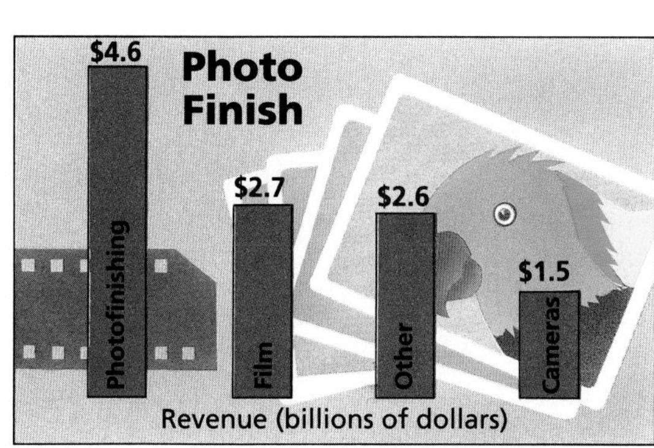

Professional wildlife photographers, such as Steven Maka shown above, go to great lengths to photograph wildlife in their natural habitats.

57. What percent of the amount spent by amateur photographers was spent on photofinishing? ≈40%

58. What percent of the amount spent by amateur photographers was spent on film? ≈24%

59. Consumers shot about 15.2 billion color prints in 1990, and paid an average of \$0.29 for each one printed. How much did consumers spend on color prints? ≈\$4.4 billion

60. What percent of the photofinishing revenue comes from the sales of color prints? ≈96%

Exploration and Extension

In Exercises 61 and 62, solve the equation.

✪ **61.** $\dfrac{6x}{x+2} - \dfrac{4x^2-7x+25}{x^2-4} = \dfrac{x-5}{x}$ $-4, -1, 5$

✪ **62.** $\dfrac{x}{x+2} - \dfrac{x+6}{2x} = \dfrac{-2(2x+3)}{x^2+3x+2}$ 3

Enrichment Activities

These activities require students to go beyond lesson goals.

The height and width of the image in a camera depends on the distance from lens to film, the height and width of the object, and the distance from object to lens. This relationship is given by the proportion $\frac{i}{b} = \frac{o}{a}$ where i = the height or width of the image, o = the height or width of the object, b = the image-to-lens distance, and a = the object-to-lens distance.

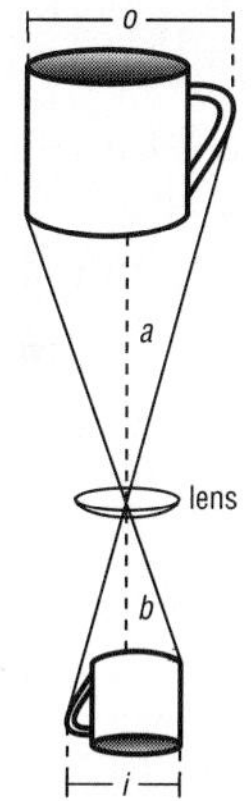

1. Solve this equation for i in terms of a, b, and o.
   $i = \frac{bo}{a}$
2. Solve this equation for a in terms of i, b, and o.
   $a = \frac{bo}{i}$
3. How tall is the image of an object 50 meters tall and 80 meters from the lens, if the image-to-lens distance is 50 mm?
   31.25 mm
4. How far from the lens is an object that has an image 30 mm wide, if the image-to-lens distance is 60 mm and the object is 100 meters wide?
   200 meters

Problem of the Day

The Golden Ratio, $\frac{1+\sqrt{5}}{2}$, is often represented by the Greek letter ϕ (phi). It is a number one greater than its own reciprocal. That is, $\phi = \frac{1}{\phi} + 1$. Find the surface area of a gold brick whose dimensions are $\phi \times 1 \times \phi^{-1}$. Express your answer as a multiple of ϕ.
4ϕ

ORGANIZER

Warm-Up Exercises

1. Find the least common multiple.
   a. $5a$, a^2b, $15b^2$
   b. $x - 1$, $x^2 - x$
   c. $x^2 - xy$, $y^2 - xy$
   a. $15a^2b^2$
   b. $x^2 - x$
   c. $xy^2 - x^2y$
2. Simplify.
   a. $\frac{7}{8} + \frac{1}{8}$ b. $\frac{5}{6} - \frac{2}{9}$ c. $\frac{\frac{2}{3}}{\frac{5}{6}}$
   a. 1, b. $\frac{11}{18}$ c. $\frac{4}{5}$

Lesson Resources

Teaching Tools
  Problem of the Day: 10.5
  Warm-up Exercises: 10.5
  Answer Masters: 10.5
Extra Practice: 10.5

10.5 Addition, Subtraction, and Complex Fractions

What you should learn:

Goal 1 How to add and subtract rational expressions and simplify complex fractions

Goal 2 How to use rational expressions as models of real-life situations

Why you should learn it:

You can model many real-life situations with the sum, difference, or ratio of rational functions, such as the difference between the marriage and divorce rates in the United States.

Goal 1 Working with Rational Expressions

As with numerical fractions, the procedure used to add (or subtract) two rational expressions depends upon whether the expressions have *like* or *unlike* denominators.

To add (or subtract) two rational expressions with *like* denominators, simply add (or subtract) their numerators and place the result over the common denominator. For example:

$$\frac{4}{3x} + \frac{5}{3x} = \frac{4+5}{3x} = \frac{9}{3x} = \frac{3}{x} \quad \textit{Add with like denominators.}$$

$$\frac{2x}{x+3} - \frac{4}{x+3} = \frac{2x-4}{x+3} \quad \textit{Subtract with like denominators.}$$

To add or subtract rational expressions with *unlike* denominators, first find the least common denominator of the original rational expressions. Then rewrite each expression as an equivalent rational expression using the least common denominator and proceed as with fractions with like denominators.

Example 1 ***Adding and Subtracting Rational Expressions***

a.
$$\frac{7}{6x} + \frac{5}{8x} = \frac{7(4)}{6x(4)} + \frac{5(3)}{8x(3)} \quad \textit{Rewrite fractions using least common denominator of 24x.}$$
$$= \frac{28}{24x} + \frac{15}{24x} \quad \textit{Simplify.}$$
$$= \frac{28+15}{24x} \quad \textit{Add fractions.}$$
$$= \frac{43}{24x} \quad \textit{Simplify.}$$

b.
$$\frac{3}{x-3} - \frac{5}{x+2} = \frac{3(x+2)}{(x-3)(x+2)} - \frac{5(x-3)}{(x-3)(x+2)}$$
$$= \frac{3x+6}{(x-3)(x+2)} - \frac{5x-15}{(x-3)(x+2)}$$
$$= \frac{(3x+6)-(5x-15)}{(x-3)(x+2)}$$
$$= \frac{3x+6-5x+15}{(x-3)(x+2)} = \frac{-2x+21}{(x-3)(x+2)}$$

A **complex fraction** is a fraction that contains a fraction in its numerator or denominator. Complex fractions may have numerators or denominators that are the sum or difference of fractions.

To simplify a complex fraction, write its numerator and its denominator as single fractions. Then divide by multiplying by the reciprocal of the denominator.

Example 2 *Simplifying a Complex Fraction*

Simplify $\dfrac{\left(\frac{x}{4}+\frac{3}{2}\right)}{\left(2-\frac{3}{x}\right)}$.

Solution

$$\frac{\left(\frac{x}{4}+\frac{3}{2}\right)}{\left(2-\frac{3}{x}\right)} = \frac{\left(\frac{x}{4}+\frac{6}{4}\right)}{\left(\frac{2x}{x}-\frac{3}{x}\right)}$$ *Find least common denominators.*

$$= \frac{\left(\frac{x+6}{4}\right)}{\left(\frac{2x-3}{x}\right)}$$ *Combine fractions in numerator and in denominator.*

$$= \frac{x+6}{4} \cdot \frac{x}{2x-3}$$ *Multiply by the reciprocal.*

$$= \frac{x(x+6)}{4(2x-3)}$$ *Simplify.* ■

Another way to simplify a complex fraction is to multiply the numerator and denominator by the least common denominator of *every* fraction in the numerator and denominator.

Example 3 *Simplifying a Complex Fraction*

Simplify $\dfrac{\left(\frac{2}{x+2}\right)}{\left(\frac{1}{x+2}+\frac{2}{x}\right)}$.

Solution

$$\frac{\left(\frac{2}{x+2}\right)}{\left(\frac{1}{x+2}+\frac{2}{x}\right)} = \frac{\left(\frac{2}{x+2}\right)(x)(x+2)}{\frac{1}{x+2}(x)(x+2)+\frac{2}{x}(x)(x+2)}$$

$$= \frac{2x}{x+2(x+2)}$$

$$= \frac{2x}{3x+4}$$ ■

LESSON Notes

Example 1

Stress to the students that when adding and subtracting rational expressions with unlike denominators, the first step is to find the least common denominator as in Lesson 10.4. Remind students that when renaming the fractions they must multiply both the numerator and the denominator by the same number or expression. This will assure that each fraction is multiplied by a form of 1.

Example 2

One way to simplify a complex fraction is to simplify the numerator and denominator first and then divide the fraction in the numerator by the fraction in the denominator.

Example 3

When using the technique used in this example, stress to the student that every term in the numerator and denominator must be multiplied by the LCD.

Example 4

When using real-life models, it is sometimes easier to simplify expressions like 9.6 − 4.4, but not to add or subtract algebraic fractions.

TECHNOLOGY If a graphics calculator or computer is used to graph the model for R, there is no need to combine the two rational expressions first.

Extra Examples

Here are additional examples similar to **Examples 1–4.**

1. Add or subtract the rational expressions.

   a. $\frac{2}{3x} + \frac{5}{6x}$

   b. $\frac{3}{x+2} - \frac{8}{x-2}$

   a. $\frac{3}{2x}$, b. $\frac{-5x-22}{(x+2)(x-2)}$

2. Simplify each complex fraction.

   a. $\dfrac{\frac{1}{x} + \frac{3}{2x}}{\frac{1}{3x} + \frac{3}{4x}}$ b. $\dfrac{\frac{1}{x+2}}{6 + \frac{4}{x}}$

   a. $\frac{30}{13}$, b. $\frac{x}{(x+2)(6x+4)}$

3. Using the models given in Example 4, what is the marriage rate in 1975? What is the divorce rate in 1975? What is the difference, R, of marriages and divorces (per 1000 people per year) in 1975? What is the value of R in 1985?

   10.09; 4.9; 5.19; 5.35

Check Understanding

1. How do you add or subtract two rational expressions with like denominators? How do you add or subtract rational expressions with unlike denominators?
   See page 548.
2. What is a complex fraction?
   See page 549.
3. What are two ways of simplifying a complex fraction?
   See page 549.

Communicating about ALGEBRA

Discuss these activities with the class to prepare them for Exercises 8–22.

Marriage (and divorce) rates are often given in the number of marriages (or divorces) per 1000 people. Since 1900, the marriage rate has remained relatively constant—the year with the lowest rate was 1958 (a rate of 8.4) and the year with the highest rate was 1945 (a rate of 12.2). The divorce rate, however, has increased greatly since 1900—from a low of 0.7 in 1900 to a high of 5.3 around 1980.

Goal 2 Using Rational Expressions in Real Life

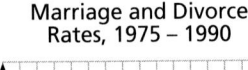

Real Life **Social Trends**

Example 4 *Subtracting Rational Expressions*

For 1975 through 1990, the marriage rate, M (per 1000 people), and the divorce rate, D (per 1000 people), can be modeled by

$$M = 9.6 + \frac{-2t + 48}{t^2 - 25.2t + 178} \quad \textit{Marriage rate}$$

$$D = 4.4 + \frac{2t + 13}{t^2 - 17t + 106} \quad \textit{Divorce rate}$$

where $t = 5$ represents 1975. Find a model that gives the difference, R, of marriages and divorces (per 1000 people per year). (You do not need to simplify the model.)

Marriage and Divorce Rates, 1975 – 1990

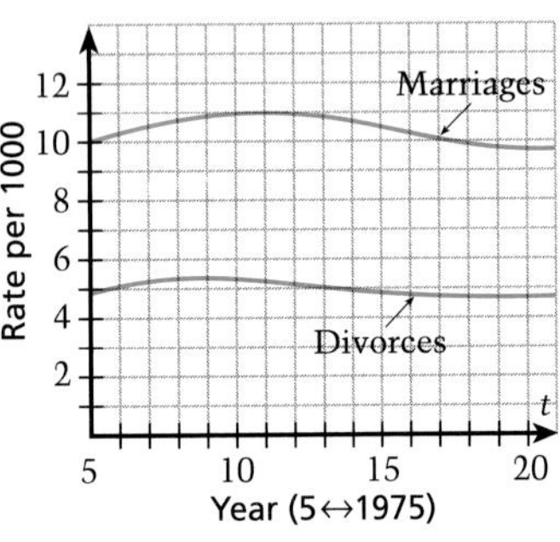

Solution You can find the model for R by subtracting the models for M and D.

$$R = M - D$$

$$= \left(9.6 + \frac{-2t + 48}{t^2 - 25.2t + 178}\right) - \left(4.4 + \frac{2t + 13}{t^2 - 17t + 106}\right)$$

$$= 5.2 + \frac{-2t + 48}{t^2 - 25.2t + 178} - \frac{2t + 13}{t^2 - 17t + 106}$$

■

Communicating about ALGEBRA

SHARING IDEAS about the Lesson

Perform the indicated operation and simplify. Explain your steps.

A. $\frac{2x+1}{x+3} - \frac{x+4}{x-5}$ **B.** $\frac{5}{x} - \frac{x}{3} + \frac{1}{x-1}$

Rewrite fractions using least common denominator, simplify, add or subtract fractions.

A. $\frac{x^2 - 16x - 17}{(x+3)(x-5)}$ B. $\frac{-x^3 + x^2 + 18x - 15}{3x(x-1)}$

EXERCISES

Guided Practice

CRITICAL THINKING about the Lesson

1. State the least common denominator of $\frac{3}{2x}$, $\frac{7}{x}$, and $\frac{x}{x+4}$. Add the expressions. See margin.

2. Which is correct? Which is incorrect? Explain. See margin.

   a. $\dfrac{\left(\frac{1}{x} - 4\right)}{\left(\frac{5}{3} + \frac{3}{2x}\right)} = \dfrac{\left(\frac{1-4x}{x}\right)}{\left(\frac{10x+9}{6x}\right)}$

   b. $\dfrac{\left(\frac{1}{x} - 4\right)}{\left(\frac{5}{3} + \frac{3}{2x}\right)} = \left(\frac{1}{x} - 4\right) \cdot \left(\frac{3}{5} + \frac{2x}{3}\right)$

In Exercises 3 and 4, simplify $\dfrac{\left(\frac{x}{5} + \frac{4}{x}\right)}{\left(\frac{6}{5x} + 1\right)}$ using the indicated method. 3., 4. $\frac{x^2 + 20}{5x + 6}$

3. Combine the numerator and denominator into single fractions, then divide.

4. Multiply the numerator and denominator by the least common denominator of all fractions, then simplify.

Independent Practice

In Exercises 5–7, state the least common denominator.

5. $\frac{18}{2(x+1)}, \frac{5}{2x}$ $2x(x + 1)$

6. $\frac{2x+3}{9x^2-1}, \frac{2}{x}, \frac{8x}{3x+1}$ $x(9x^2 - 1)$

7. $\frac{1}{x(x-5)}, \frac{32}{x^2-3x-10}$ $x(x - 5)(x + 2)$

In Exercises 8–22, perform the operations and simplify. See Additional Answers.

8. $\frac{4}{2x^2} + \frac{1}{3x}$

9. $\frac{-2}{7x} - \frac{5}{4x}$

10. $\frac{11}{3(x-5)} - \frac{x+1}{3x}$

11. $\frac{4x+1}{x^2-4} - \frac{3}{x-2}$

12. $\frac{4}{7x} - \frac{3x^2+x-5}{14x+21}$

13. $\frac{3x^2}{x^2-x-30} - \frac{3x+5}{x-6}$

14. $\frac{2x^2}{3x+5} - \frac{14}{x+7}$

15. $\frac{12x^2-x+9}{3x+33} - \frac{16}{x+11}$

16. $\frac{1-3x}{x-6} + \frac{2}{2x+1}$

17. $\frac{4x}{x+1} + \frac{5}{2x-3} - \frac{4}{x}$

18. $\frac{2}{x+2} - \frac{x}{3} - \frac{5}{x-2}$

19. $\frac{-4x^2+2}{x^2+9x-10} + \frac{3}{x+10}$

20. $\frac{x^2+x-3}{x^2-12x+32} + \frac{3x}{x-8}$

21. $\frac{3-4x}{x^2+3x-10} - \frac{2}{x+5}$

22. $\frac{10x}{3x^2-3} + \frac{4}{x-1} + \frac{5}{6x}$

EXERCISE Notes

ASSIGNMENT GUIDE

Basic/Average: Ex. 1–4, 5, 9–17 odd, 29–35 odd, 41–44, 47–57 odd, 59

Above average: Ex. 1–4, 5–17 odd, 27–37 odd, 41–45, 56–59

Advanced: Ex. 1–4, 7–19 odd, 27–37 odd, 41–45, 56–59

Selected Answers
Exercises 1–4, 5–57 odd

Use **Mixed Review** as needed.

More Difficult Exercises
Exercises 32–34, 59

Guided Practice

Ex. 1 Ask students to explain why a common denominator is necessary for adding the expressions.

Answers

1. $2x(x + 4), \frac{2x^2 + 17x + 68}{2x(x + 4)}$

2. *a, b,* the numerator must be multiplied by the reciprocal of the sum of the fractions in the denominator (not by the sum of the reciprocals)

Ex. 3–4 Ask students to discuss which method they think is more efficient.

Independent Practice

Ex. 5–22 Remind students to factor the denominators as an aid in determining the LCD.

▶ Ex. 23–34 Encourage students to try both methods illustrated by Examples 2 and 3 on page 549. Remind students to use a similar solution format for showing their work.

▶ Ex. 35–40 Help students solve these equations by modeling a solution for them.

▶ Ex. 41–42 Assign these exercises as a pair.

▶ Ex. 41 In this exercise $I = M + B$. Graphics calculators can be used to obtain the values in 1985 ($t = 15$). Students can graph $y_1 = M$, $y_2 = B$, and $y_3 = y_1 + y_2$ to "view" the graph resulting in the sum of the two ratios of polynomial expressions.

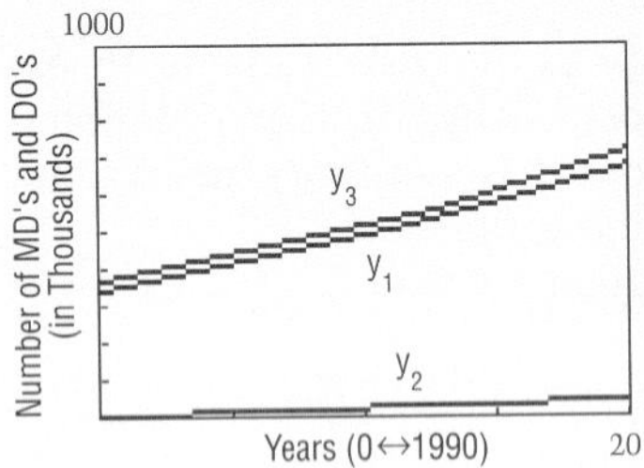

In Exercises 23–34, simplify the complex fraction.

23. $\dfrac{\left(\dfrac{x}{3} - 4\right)}{\left(5 + \dfrac{1}{x}\right)}$ $\quad \dfrac{x(x-12)}{3(5x+1)}$

24. $\dfrac{\left(\dfrac{4}{x} + 2\right)}{\left(\dfrac{1}{2x} - 8\right)}$ $\quad \dfrac{4(2+x)}{1-16x}$

25. $\dfrac{\left(\dfrac{3}{x^2} + \dfrac{1}{x}\right)}{\left(2 - \dfrac{4}{5x}\right)}$ $\quad \dfrac{5(x+3)}{10x^2-4x}$

26. $\dfrac{\left(16 - \dfrac{1}{x^2}\right)}{\left(\dfrac{1}{4x^2} - 4\right)}$ $\quad -4$

27. $\dfrac{\left(\dfrac{10}{x+1}\right)}{\left(\dfrac{1}{2} + \dfrac{3}{x+1}\right)}$ $\quad \dfrac{20}{x+7}$

28. $\dfrac{\left(\dfrac{2}{3x+15}\right)}{\left(\dfrac{2}{x+5} + \dfrac{1}{4x+20}\right)}$ $\quad \dfrac{8}{27}$

29. $\dfrac{\left(\dfrac{1}{2x+1} - \dfrac{3}{4(2x+1)}\right)}{\left(\dfrac{x}{2x+1}\right)}$ $\quad \dfrac{1}{4x}$

30. $\dfrac{\left(\dfrac{2}{x^2-1} + \dfrac{1}{x+1}\right)}{\left(\dfrac{1}{12x^2-3}\right)}$ $\quad \dfrac{3(4x^3-1)}{x-1}$

31. $\dfrac{\left(\dfrac{1}{x+1} + \dfrac{1}{2}\right)}{\left(\dfrac{3}{2x^2+4x+2}\right)}$ $\quad$ 31. $\dfrac{(x+3)(x+1)}{3}$

✪ 32. $\dfrac{\left(\dfrac{x}{x-3} - \dfrac{2}{3}\right)}{\left(\dfrac{10}{3x} + \dfrac{x^2}{x-3}\right)}$ $\quad \dfrac{x(x+6)}{3x^3+10x-30}$

✪ 33. $\dfrac{\left(\dfrac{1}{2x} - \dfrac{6}{x+5}\right)}{\left(\dfrac{x}{x-5} + \dfrac{1}{x}\right)}$ $\quad \dfrac{(-11x+5)(x-5)}{2(x+5)(x^2+x-5)}$

✪ 34. $\dfrac{\left(\dfrac{4}{x^2-9} + \dfrac{2}{x-3}\right)}{\left(\dfrac{1}{x+3} + \dfrac{1}{x-3}\right)}$ $\quad \dfrac{x+5}{x}$

In Exercises 35–40, solve the equation.

35. $\dfrac{\left(\dfrac{4}{x} + \dfrac{6}{x+1}\right)}{\left(\dfrac{3}{x+1} - 5\right)} = -1$ $\quad 2$

36. $\dfrac{\left(\dfrac{3}{x-2} + 2\right)}{\left(\dfrac{2x-3}{x-2} + \dfrac{2x}{3}\right)} = 1$ $\quad 3, -1$

37. $\dfrac{\left(x - \dfrac{5x+3}{x-1}\right)}{\left(\dfrac{2x}{2x-2} - \dfrac{x}{2}\right)} = -2$ $\quad -1$

38. $\dfrac{\left(\dfrac{3x}{x-1} - \dfrac{4}{x}\right)}{\left(\dfrac{1}{2x} - \dfrac{1}{x}\right)} = 16$ $\quad \dfrac{2}{3}, -2$

39. $\dfrac{\left(\dfrac{x}{x-2} + \dfrac{7}{x}\right)}{\left(6 - \dfrac{12}{x}\right)} = -1$ $\quad \dfrac{10}{7}, 1$

40. $\dfrac{\left(\dfrac{x+4}{5} - 3\right)}{\left(\dfrac{1}{2} + \dfrac{x+4}{10}\right)} = -2$ $\quad 1$

Doctors **In Exercises 41 and 42, use the following information.**

For 1970 to 1990, the number of doctors of medicine, M (in thousands), in the United States can be approximated by

$$M = \frac{28{,}390 + 693t}{85 - t},$$

where $t = 0$ represents 1970. The number of doctors of osteopathy, B (in thousands), can be approximated by

$$B = \frac{776 - 12t}{55 - 2t}.$$

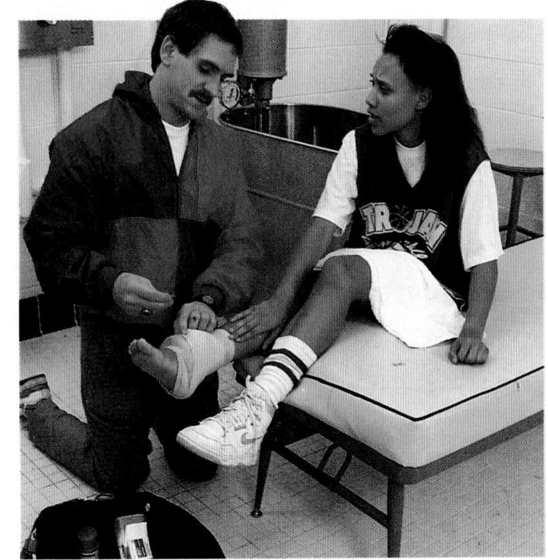

41. Write an expression for the total number, I, of doctors of medicine (MD) and doctors of osteopathy (DO). Simplify the result. See below.

42. How many MDs and DOs did the United States have in 1985?
≈554,000, ≈24,000

41. $I = \dfrac{-1374t^2 - 20{,}461t + 1{,}627{,}410}{(85-t)(55-2t)}$

Banks **In Exercises 43 and 44, use the following information.**

For 1970 to 1990, the number of commercial banks, T, in the United States can be modeled by

$$T = \frac{17{,}510(41 + 5t)}{20 + t}$$

where $t = 0$ represents 1970. The number, M, of commercial banks that were members of the Federal Reserve System can be modeled by

$$M = \frac{174{,}720(148 + 5t)}{1174 - t}.$$ 43. See margin.

43. Write an equation that represents the number, N, of commercial banks that were not members of the Federal Reserve System. Do not simplify the result.

44. How many banks were not members in 1988? ≈24,392

The Federal Reserve System, nicknamed the Fed, *is an independent agency of the federal government. The* Fed *manages the nation's supply of money and credit and provides services to commercial banks.*

Integrated Review

In Exercises 45–50, find the least common multiple. See margin.

45. $x - 7,\ x^2,\ x(x + 7)$

46. $x + 2,\ x^2 - 4,\ x$

47. $x^2 - 1,\ x + 1,\ x$

48. $x - 6,\ x(x - 6),\ x^3$

49. $x,\ x^2(x + 1),\ x + 1$

50. $9 - x,\ x^2,\ x + 1$

In Exercises 51–56, solve the equation.

51. $\frac{5}{x - 2} = \frac{3x + 1}{x - 1}$ $\frac{1}{3}$, 3

52. $\frac{11 + x}{x - 1} = \frac{-6}{x + 4}$ −19, −2

53. $\frac{3x + 2}{x} + \frac{3x}{x + 1} = 6$ $\frac{2}{3}$

54. $\frac{-6}{x - 2} + \frac{6}{x - 1} = -1$ −1, 4

55. $\frac{x + 2}{x - 6} + \frac{4}{x - 4} = -3$ 5, 2

56. $\frac{x + 5}{x - 2} + \frac{x}{x - 6} = 5$ $\frac{10}{3}$, 9

57. *Powers of Ten* Which best estimates the length of the tail of the Parson's chameleon shown in the photo? (The twig is about one-half inch thick.) c

a. 5.0×10^{-2} in.
b. 5.0×10^{-1} in.
c. 5.0×10^{0} in.
d. 5.0×10^{1} in.

58. *Powers of Ten* Which best estimates the average mileage of an American automobile (in miles per gallon) in 1990? b

a. 2.0×10^{0}
b. 2.0×10^{1}
c. 2.0×10^{2}
d. 2.0×10^{3}

▶ **Ex. 43–44** Assign these exercises as a pair. Encourage students to use graphics calculators to help visualize the solutions and data trends.

Answers

43. $N = \frac{17{,}500(41 + 5t)}{20 + t} - \frac{174{,}720(148 + 5t)}{1174 - t}$

45. $x^2(x - 7)(x + 7)$

46. $x(x + 2)(x - 2)$

47. $x(x + 1)(x - 1)$

48. $x^3(x - 6)$

49. $x^2(x + 1)$

50. $x^2(9 - x)(x + 1)$

▶ **Ex. 59** COOPERATIVE LEARNING This exercise can be used as a small-group activity. Each small group should write a paragraph giving the reasoning for their solution.

Enrichment Activities

These activities require students to go beyond lesson goals.

1. Evaluate each of the following complex fractions using a calculator.

$$1+\frac{1}{2} \qquad 1.5$$

$$1+\cfrac{1}{2+\cfrac{1}{2}} \qquad 1.400$$

$$1+\cfrac{1}{2+\cfrac{1}{2+\cfrac{1}{2}}} \qquad 1.417$$

$$1+\cfrac{1}{2+\cfrac{1}{2+\cfrac{1}{2+\cfrac{1}{2}}}} \qquad 1.414$$

2. If this process is continued, what number does this complex fraction approximate? $\sqrt{2} = 1.41421\ldots$

3. Evaluate each of the continued fractions. What special number do these continued fractions approximate? (*Hint:* Euler derived the following continued fractions and showed that this special number must be irrational.)

a. $$2+\cfrac{1}{1+\cfrac{1}{2+\cfrac{2}{3+\cfrac{3}{4+\cfrac{4}{\cdots}}}}}$$
≈ 2.718

b. $$1+\cfrac{2}{1+\cfrac{1}{6+\cfrac{1}{10+\cfrac{1}{14+\cdots}}}}$$
≈ 2.7183

The special number, e, is sometimes called Euler's number.

Answer Mixed Review

19. $\begin{bmatrix} 0 & 0 \\ 1 & 1 \end{bmatrix}$

Exploration and Extension

59. *Country Music Records* Bob's and Helen's favorite pastime is trying to stump each other with puzzles. Can you solve the puzzle?

(Puzzle is from aha! Insight, Martin Gardner, copyright © Scientific American, Inc.)

***Bob:** Do you still have your country and western records?*

***Helen:** No, I gave half of them, and half a record more, to Suzy.*

***Helen:** Then I gave half of what was left, and half a record more, to Joe.*

***Helen:** That left me with just one record, which I'll give to you. . . .*

***Helen:** . . . If you can tell me how many I began with.*

Bob was puzzled. Suddenly he had an aha! and answered Helen's question.
Yes, 7 records

Mixed REVIEW

3.–8. See Additional Answers.
10. $(3x - 2)(3x + 2)(9x^2 + 4)$
11. $\frac{7 \pm \sqrt{37}}{6}$

1. Simplify: $(5x - 2)^2$. **(9.1)** $25x^2 - 20x + 4$
2. Simplify: $(8x^3 - 3x^2) - (2x^3 - 4x^2)$. **(9.1)** $6x^3 + x^2$
3. Sketch $f(x) = (x + 4)^6 - 3$. **(9.2)**
4. Sketch $f(x) = \log_{10}(x - 2)$. **(8.2)**
5. Sketch $|3x - 7| > 4$. **(2.6)**
6. Sketch $f(x) = x^2 + 6x + 4$. **(5.2)**
7. Sketch $f(x) = |x - 4| - 2$. **(2.6)**
8. Sketch $f(x) = -3(x - 5)^2 + 2$. **(5.3)**
9. Factor $49x^2 - 28x + 4$. **(9.3)** $(7x - 2)^2$
10. Factor $81x^4 - 16$. **(9.3)**
11. Solve $3x^2 - 7x + 1 = 0$. **(5.4)** $\frac{9 \pm \sqrt{37}}{6}$
12. Solve $3x^3 - 18x^2 + 27x = 0$. **(9.3)** 0, 3
13. In which quadrant is $(-2, -7)$? **(2.1)** III
14. Find 140% of 60. 84
15. Find the y-intercept of $y = 3|x - 8| + 4$. **(2.3)** 28
16. Find the distance from $(-5, 1)$ to $(4, 6)$. **(7.5)** $\sqrt{106}$
17. The longest sides of a right triangle are 8 ft and 6 ft. Find the third side. **(5.1)** $2\sqrt{7}$
18. \$300 saved at 6% compounded monthly for four years is how much? **(7.2)** \$381.15
19. Solve $\begin{bmatrix} 2 & -1 \\ 1 & 0 \end{bmatrix} X = \begin{bmatrix} -1 & -1 \\ 0 & 0 \end{bmatrix}$. See margin.
20. Find the area of the triangle with vertices $(-3, 5)$, $(4, 4)$ and $(1, -8)$. **(4.3)** 43.5 units^2

10.6 Exploring Data: The Algebra of Finance

What you should learn:

Goal 1 How to construct an amortization table for a loan

Goal 2 How to find the monthly payment for an installment loan

Why you should learn it:

You will repay most of your consumer loans with monthly installments. Knowing how these installments are computed will ensure that you are not overcharged.

Goal 1 Constructing an Amortization Table

When you receive a loan, you usually are expected to repay it. Most loans are not repaid with a single payment. Instead they are repaid over a period of time with payments called **installments.** This type of loan is an **installment loan,** and the process of repaying the loan is called **amortization.**

The period of time over which the loan is paid is the **term** of the loan. The amount you receive is the **principal.**

Example 1 *Constructing an Amortization Table*

You borrow $1600.00 and make 11 monthly payments of $144.41 each and one final payment of $144.44. The annual percentage rate is 15%, compounded monthly. Construct an amortization table for this loan. (Assume that repayment begins one month after the loan is obtained.)

Solution Because the annual percentage rate is 15%, the interest for the first month is $I = Prt = 1600(0.15)\left(\frac{1}{12}\right) = \$20.00$. This means $20 of the first payment is interest and the remaining $124.41 reduces the principal. The balance after the first payment is 1600.00 − 124.41 = $1475.59.

Payment Number	Balance before Payment	Payment	Interest Payment	Principal Payment	Balance after Payment
1	$1600.00	$144.41	$20.00	$124.41	$1475.59
2	$1475.59	$144.41	$18.44	$125.97	$1349.62
3	$1349.62	$144.41	$16.87	$127.54	$1222.08
4	$1222.08	$144.41	$15.28	$129.13	$1092.95
5	$1092.95	$144.41	$13.66	$130.75	$ 962.20
6	$ 962.20	$144.41	$12.03	$132.38	$ 829.82
7	$ 829.82	$144.41	$10.37	$134.04	$ 695.78
8	$ 695.78	$144.41	$ 8.70	$135.71	$ 560.07
9	$ 560.07	$144.41	$ 7.00	$137.41	$ 422.66
10	$ 422.66	$144.41	$ 5.28	$139.13	$ 283.53
11	$ 283.53	$144.41	$ 3.54	$140.87	$ 142.66
12	$ 142.66	$144.44	$ 1.78	$142.66	$ 0.00

■

Problem of the Day

Three men and their wives were given $5400. The wives together received $2400. Sue has $200 more than Jan, and Ellen had $200 more than Sue. Lou got half the amount his wife got. Rob got the same amount as his wife, and Matt got twice the amount his wife got. Who is married to whom?

Sue and Lou, Ellen and Matt, Jan and Rob

ORGANIZER

Warm-Up Exercises

1. Convert to decimal form.
   - a. 24% 0.24
   - b. 13% 0.13
   - c. 0.12% 0.0012
   - d. 436% 4.36
   - e. 35.7% 0.357
   - f. $\frac{45}{75}$ 0.6
2. Solve.
   - a. $2000 is deposited in an account that pays 8% annual interest. What will the balance be in ten years?
   - b. How much must you deposit in an account that pays 6.25% interest, compounded yearly, to have a balance of $1400 after two years?

   a. $4317.84, b. $1240.14

Lesson Resources

Teaching Tools
- Problem of the Day: 10.6
- Warm-up Exercises: 10.6
- Answer Masters: 10.6

Extra Practice: 10.6
Color Transparency: 57
Technology Handbook: p. 73, 75

LESSON Notes

Example 1

Remind students of the formula $I = Prt$ where $P =$ the principal (the amount of money borrowed), $r =$ the annual percentage rate at which the money is borrowed, and $t =$ the amount of time in years for which the money is borrowed. Notice that $t = \frac{1}{12}$ since the loan will be payed back over a period of one year in 12 monthly payments.

Example 2

The total interest is given by the number of payments multiplied by the monthly payment minus the original amount of the loan.

Example 3

Notice that $10,976.88 was paid in the first year, yet only $500.52 was paid on the principal. Discuss this situation with the students.

Extra Examples

Here are additional examples similar to **Examples 1–3.**

1. A $75,000 mortgage was obtained for a small condominium at 15.75% for 25 years. The monthly payment is $1004.47.
   a. What part of the first monthly payment goes to interest?
   $984.38
   b. What is the total interest paid on the mortgage?
   $226,341
2. A loan of $62,500 at 15.75% is obtained for 35 years.
   a. Find the monthly payment.
   $823.76
   b. Find the total interest.
   $283,453.34

Goal 2 Finding a Monthly Payment

Monthly Installment Payments

Consider a monthly installment loan with a principal of P, an annual interest rate of r (in decimal form), and a term of t years. The monthly interest rate is $i = \frac{1}{12}r$ and the monthly payment is

$$M = P\left[\frac{i}{1 - \left(\frac{1}{1+i}\right)^{12t}}\right].$$

From this formula you can see how the monthly payment was calculated for the loan described in Example 1. Using $P = 1600$, $r = 0.15$, $t = 1$, and $i = \frac{1}{12}(0.15) = 0.0125$, you obtain

$$M = (1600)\left[\frac{0.0125}{1 - \left(\frac{1}{1 + 0.0125}\right)^{12}}\right] = \$144.41.$$

Because payments are rounded, the final payment will usually differ from the other payments by a few cents.

When obtaining an installment loan, borrowers often decrease the amount of each installment by extending the term of the loan, but this can greatly increase the total interest paid.

Real Life
Home Financing

Example 2 *Comparing Installments with Total Interest Payments*

You are considering a home mortgage for $100,000 with an annual percentage rate of 10.5% and monthly installments. Find the monthly payment for 20 years and 30 years. What is the total interest paid for each term?

Solution

a. Using $P = 100{,}000$, $r = 0.105$, and $t = 20$, you can find the monthly installment to be $998.38. The total interest paid over the 20-year period is

$240(998.38) - 100{,}000.00 = \$139{,}611.20.$

b. Using $P = 100{,}000$, $r = 0.105$, and $t = 30$, you can find the monthly installment to be $914.74. The total interest paid over the 30-year period is

$360(914.74) - 100{,}000.00 = \$229{,}306.40.$

Increasing the term from 20 to 30 years decreased the monthly payment by only $83.64, but increased the total interest by $89,695.20! ■

For installment loans over long terms, a large portion of each installment during the early years is the interest. Only a small portion goes toward reducing the principal.

Example 3 Comparing the Principal Payment with the Interest Payment

Construct the first 12 months of an amortization table for a home mortgage of $100,000 at 10.5% for 30 years (with monthly payments). How much of the first year's payments are applied against the principal and how much is interest?

Solution The monthly payment for this home mortgage is $914.74, and the amortization table for the first 12 months is shown below.

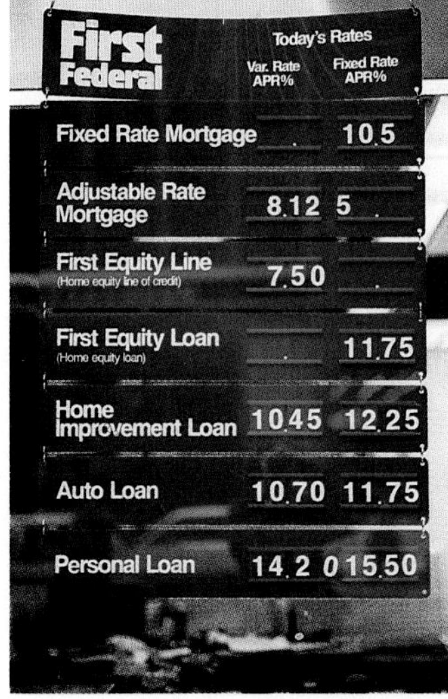

Payment Number	Balance before Payment	Payment	Interest Payment	Principal Payment	Balance after Payment
1	$100,000.00	$914.74	$875.00	$39.74	$99,960.26
2	$ 99,960.26	$914.74	$874.65	$40.09	$99,920.17
3	$ 99,920.17	$914.74	$874.30	$40.44	$99,879.73
4	$ 99,879.73	$914.74	$873.95	$40.79	$99,838.94
5	$ 99,838.94	$914.74	$873.59	$41.15	$99,797.79
6	$ 99,797.79	$914.74	$873.23	$41.51	$99,756.28
7	$ 99,756.28	$914.74	$872.87	$41.87	$99,714.41
8	$ 99,714.41	$914.74	$872.50	$42.24	$99,672.17
9	$ 99,672.17	$914.74	$872.13	$42.61	$99,629.56
10	$ 99,629.56	$914.74	$871.76	$42.98	$99,586.58
11	$ 99,586.58	$914.74	$871.38	$43.36	$99,543.22
12	$ 99,543.22	$914.74	$871.00	$43.74	$99,499.48

The principal was reduced only $100,000.00 − $99,499.48 = $500.52 at the end of the first year. The total payments for the first year were 12($914.74) = $10,976.88; the portion that went toward interest was $10,976.88 − $500.52 = $10,476.36. ■

Communicating about ALGEBRA

SHARING IDEAS about the Lesson

You are considering a car loan for $12,500. Compare the monthly payments and total interest paid for each loan.

A. A 3-year loan with an 8% annual interest rate.
B. A 3-year loan with a 12% annual interest rate.
C. A 3-year loan with an 18% annual interest rate.

A. $391.70; $1601.20 B. $415.18; $2446.48 C. $451.90; $3768.40

Check Understanding

1. Define amortization.
See page 555.

2. You wish to make lower payments on an installment loan. What happens to the term of the loan? What happens to the total interest payment of the loan?
The term of the loan increases. The total interest payment will increase.

Communicating about ALGEBRA

EXTEND *Communicating*
COOPERATIVE LEARNING
Have groups of students investigate some purchases they wish to make—for example, buying a new or used car or stereo equipment. Students may wish to compare purchases made with bank credit cards, store credit plans, or bank loans. Students may get information from local banks, newspaper ads, and radio and television announcements. Each group can present information on interest rates, monthly payments, total interest, or finance charges.

EXERCISE Notes

ASSIGNMENT GUIDE

Basic/Average: Ex. 1–4, 7, 11, 15, 19, 21–24, 26, 27, 29, 31–34

Above average: Ex. 1–4, 5–19 odd, 21–25, 27–33 odd, 35

Advanced: Ex. 1–4, 5–19 odd, 21–27, 29–35 odd, 36

Selected Answers
Exercises 1–4, 5–33 odd

More Difficult Exercises
Exercises 25–30, 35, 36

Guided Practice

Ex. 1–4 These exercises can be used to check understanding of the formula for monthly installment payments on page 556 and to generate the first entry of the amortization table. Make sure that students understand the terminology. M is the monthly payment, $P = 2000$, $r = 0.12$. Thus $i = \frac{1}{12}(0.12)$ or 0.01 and $t = 1$. Help students to evaluate the rational expression accurately using their calculators. They can find the amount for the first month's interest by using the formula $I = Prt$ with $r = 0.12$ and $t = \frac{1}{12}$.

Independent Practice

Ex. 5–8 TECHNOLOGY These exercises are similar to Example 1 on page 555. These are excellent examples to solve using computer spreadsheets.

Ex. 9–12 Use the monthly payment formula on page 556. See the Enrichment Activity that follows for a graphics calculator program that can be used to evaluate the monthly payment formula.

EXERCISES

Guided Practice

CRITICAL THINKING about the Lesson

Monthly Payment **In Exercises 1–4, consider a \$2000 monthly installment loan with an annual interest rate of 12% and a term of one year.**

1. What is the monthly payment? **\$177.70**
2. How much of the first month's payment goes toward interest? **\$20**
3. How much of the first month's payment goes toward principal? **\$157.70**
4. What is the balance after the first month's payment? **\$1842.30**

Independent Practice

Amortization Table **In Exercises 5–8, construct an amortization table.** See Additional Answers.

5. You borrow \$1200 at an annual interest rate of $10\frac{1}{2}\%$ to be repaid in eight monthly payments of \$139.23 and one final payment of \$139.27. (*Hint:* $t = \frac{3}{4}$ year.)
6. You borrow \$750 at an annual interest rate of 14% to be repaid in eight monthly payments of \$88.27 and one final payment of \$88.25.
7. You borrow \$5000 at an annual interest rate of 8% to be repaid in 11 monthly payments of \$434.94 and one final payment of \$434.97.
8. You borrow \$400 at an annual interest rate of 15% to be repaid in five monthly payments of \$69.61 and one final payment of \$69.63.

Monthly Payment **In Exercises 9–12, find the monthly payment.**

9. A home mortgage for \$80,000, with an annual interest rate of 11%, is taken out for a 15-year term. **\$909.28**
10. A business loan for \$20,000, with an annual interest rate of 12.5%, is taken out for a 10-year term. **\$292.75**
11. You borrow \$75,000 for 20 years at an annual interest rate of 9%. **\$674.79**
12. Your home mortgage is for \$50,000, with an annual interest rate of 10%, over a 25-year term. **\$454.35**

Home Mortgage **In Exercises 13–16, use the table of monthly payments to find the total interest payment on a \$100,000 home mortgage.** See below.

13. 8% interest, 20 years
14. 8% interest, 30 years
15. 12% interest, 20 years
16. 9% interest, 15 years

	8%	9%	10%	11%	12%
15 years	955.65	1014.27	1074.61	1136.60	1200.17
20 years	836.44	899.73	965.02	1032.19	1101.09
25 years	771.82	839.20	908.70	980.11	1053.22
30 years	733.76	804.62	877.57	952.32	1028.61

13. \$100,745.60 **14.** \$164,153.60 **15.** \$164,261.60 **16.** \$82,568.60

Total Interest **In Exercises 17–20, find the total interest payment.**

17. An installment loan for \$2000 to be paid in 12 monthly payments with an annual interest rate of 13% \$143.56

18. An installment loan for \$14,000 to be paid in 18 monthly payments with an annual interest rate of 11% (*Hint:* $t = \frac{3}{2}$ years.) \$1250.68

19. An installment loan for \$5500 to be paid in 6 monthly payments with an annual interest rate of 12.5% \$202.28

20. An installment loan for \$8000 to be paid in 9 monthly payments with an annual interest rate of 9.5% \$319.96

Comparing Two Repayment Plans **In Exercises 21–24, consider a \$2000 loan with an annual interest rate of 12%.**

21. What is the monthly payment for a term of 12 months? \$177.70

22. What is the monthly payment for a term of 18 months? \$121.96

23. Find the total interest and the total payment for each term. 23. \$132.40, \$2132.40; \$195.28, \$2195.28

24. What is one advantage of each repayment plan? 1st: lower amount of interest, 2nd: lower monthly payment

25. *Fitting a Loan to Your Budget* You take out a used car loan for \$3500 with an annual interest rate of 14%. You want to repay the loan as soon as possible, but you can afford monthly payments of, at most, \$100. Which of the following terms should you choose? Explain. e See margin.

a. 24 months **b.** 30 months **c.** 36 months **d.** 42 months **e.** 48 months

26. *Fitting a Loan to Your Budget* You take out a loan for \$2600 with an annual interest rate of 18%. You want to repay the loan as soon as possible, but you can afford monthly payments of, at most, \$100. Which of the following terms should you choose? Explain. d See margin.

a. 6 months **b.** 12 months **c.** 24 months **d.** 36 months **e.** 48 months

27. *Comparing Two Repayment Plans* You borrow \$2000 and have two repayment options. Which has the lesser total payment? b

a. One lending source offers the loan for 24 monthly payments at an annual interest rate of 14%.

b. Another lending source offers the loan for 18 monthly payments at an annual interest rate of 16%.

Most monthly payments for personal loans are paid with checks, which are processed at Federal Reserve Banks with machines such as those shown above.

28. *Comparing Two Repayment Plans* You borrow \$20,000 and have two repayment options. Which has the lesser total payment? a

a. One lending source offers the loan for 24 monthly payments at an annual interest rate of 8%.

b. Another lending source offers the loan for 36 monthly payments at an annual interest rate of 6%.

Ex. 13–16 Students may be surprised by the high amount of interest paid on a loan over the entire time period. Remind the students to multiply the amount by the number of years and by 12, the number of months per year.

Ex. 17–20 Refer students to Example 2 for a model for solving these problems.

Ex. 21–24 Assign these exercises as a group. Discuss the merits of lower total interest versus lower monthly payment, and "fitting the monthly budget."

Answers

25. e; only e has a monthly payment of at most \$100.

26. d; only d and e have monthly payments of at most \$100, but d has a lower amount of interest than e.

Ex. 27–30 Refer students to Example 3 on page 557 for a model for solving these problems.

▶ **Ex. 35–36** Solve the formula for monthly payment for *P*.

Enrichment Activities

These activities require students to go beyond lesson goals.

An Alternate Approach: *Using Technology*

The following "Mortgage Payment" program can be used on a TI-82 graphing calculator.

```
PROGRAM:MORTGAGE
:Clr Home
:Disp "MORTGAGE AMOUNT"
:Input P
:Disp "INTEREST RATE"
:Input I
:Disp "NUMBER OF YEARS"
:Input T
:.011→ R
:PR/(12(1−(1+R/12)^−12T)) →M
:Disp "MONTHLY PAYMENT"
:Disp M
```

Use the program to assist you in completing Exercises 9–12 and 17–30.

For example, to find *M* in Exercise 9, execute the program as follows.

MORTGAGE AMOUNT
?80000
INTEREST RATE
?11
NUMBER OF YEARS
?15
MONTHLY PAYMENT:
909.28

29. *Home Mortgage* You have taken out a home mortgage for \$90,000 at 9.5% annual interest for 20 years. You make monthly payments for one year. How much of the total monthly payments made during the first year was interest? What is the balance at the end of the year? **\$8,482.17, \$88,415.13**

30. *Home Mortgage* You have taken out a home mortgage for \$120,000 at 11% annual interest for 30 years. You make monthly payments for one year. How much of the total monthly payments made during the first year was interest? What is the balance at the end of the year? **\$13,173.30, \$119,459.82**

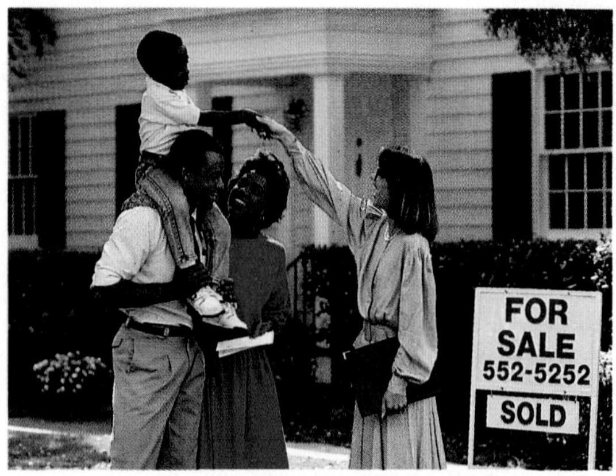

Most people need to obtain a large mortgage to purchase a home. A wise shopper will determine how much of each payment will be applied to the principal.

Integrated Review

31. *Color Television* You purchased a new color television set. The total price, including a 6% sales tax, was \$663.03. What was the price of the television? **\$625.50**

32. *Bicycle* You purchased a new bike. The total price, including a 7% sales tax, was \$437.70. What was the price of the bike? **\$409.07**

33. *Savings* In 1991, about 70 million Americans were in the 18-to-34 age group. Using the graph at the right, approximate how many of them saved between \$50 and \$100 a month. **≈7 million**

34. *Savings* Suppose that each "saver" in Exercise 33 saved \$75 a month. How much did the group save in 1991? **≈6.3 billion**

Percent of Age Group in Each Savings Category

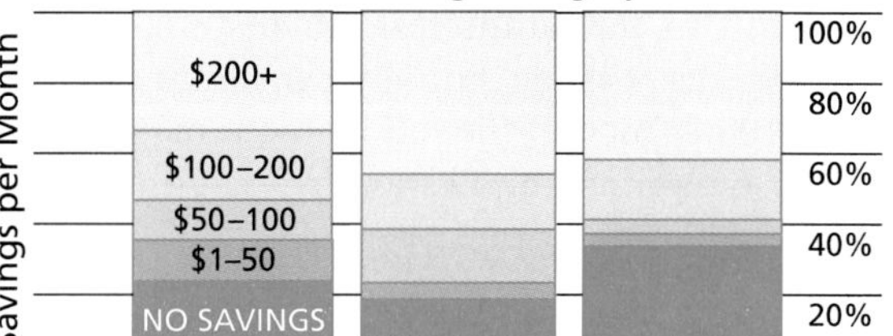

Source: The Survey of Consumer Financial Behavior, conducted by the Gallup Organization and Bank Advertising News

Exploration and Extension

35. *Finding the Principal* You are paying \$36.10 each month on a loan with an annual interest rate of 11.5% for a term of 6 years. What was the original amount of the loan? **\$1871.32**

36. *Finding the Principal* You are paying \$77.71 each month on a loan with an annual interest rate of 9.5%. If the original amount of the loan was \$3700, what is the term of your loan? **5 years**

10 Chapter Summary

What did you learn?

Skills

1. Sketch the graph of a rational function. **(10.1)**
   - Find the horizontal and vertical asymptotes. **(10.1)**
2. Perform operations with rational expressions.
   - Multiply and divide rational expressions. **(10.3)**
   - Add and subtract rational expressions. **(10.5)**
3. Solve rational equations. **(10.4)**
4. Simplify complex fractions. **(10.5)**

Strategies

5. Use rational models to solve real-life problems. **(10.1–10.5)**
6. Solve inverse variation problems. **(10.2)**
7. Solve joint variation problems. **(10.2)**

Exploring Data

8. Find monthly payments for installment loans. **(10.6)**
9. Construct an amortization table for an installment loan. **(10.6)**

Why did you learn it?

The primary characteristic that distinguishes rational functions from polynomial functions is that graphs of rational functions can have horizontal or vertical asymptotes. They can also have discontinuities. (No polynomial function has a horizontal or vertical asymptote or a discontinuity.)

Because rational functions can have horizontal or vertical asymptotes, they can be used to model real-life situations that involve asymptotic behavior. For instance, the average cost of a product can asymptotically approach the unit cost as more and more units are produced.

How does it fit into the bigger picture of algebra?

In this chapter, you studied rational expressions and rational functions. By now you should be able to sketch the graph of a rational function. You have seen that the graphs of many rational functions have horizontal and vertical asymptotes.

You also learned to multiply, divide, add, and subtract rational expressions and solve rational equations. Because a rational expression is the ratio of two polynomials, this chapter provided additional practice with the polynomial operations that you studied in Chapter 9.

ORGANIZER

TIME SCHEDULE: All levels, two days

The Chapter Summary helps students organize the main ideas of the chapter. In this chapter, the operations of rational functions were developed. These operations were utilized to solve rational equations and variation problems. An important consumer application, installment loans, was also presented.

Work with students to review the skills, strategies, and concepts of the chapter. The first day's homework assignment can be used as the basis for the second day of review.

COOPERATIVE LEARNING
Encourage students to study together. Emphasize the importance of "teaching" a classmate how to perform a skill or how to recall a strategy. When students work together, everyone wins. The students receiving help get additional instruction and the students giving help gain a deeper understanding of the skills and concepts involved.

Chapter SUMMARY

Review the chapter by asking students, "What did you learn? Why did you learn it? How does it fit with the other algebra skills and concepts you have learned?"

Chapter REVIEW

ASSIGNMENT GUIDE

Basic/Average: Ex. 1–5 odd, 6–10, 12–60 multiples of 3, 67, 73–79 odd, 86–89

Above Average: Ex. 1–5 odd, 6–10, 12–63 multiples of 3, 67, 69, 73–81 odd, 83–85

Advanced: Ex. 1–5 odd, 6–10, 12–63 multiples of 3, 69, 71–81 odd, 83–85

More Difficult Exercises

Exercises 63–72, 78–85

Ex. 1–89 The Chapter Review exercises includes a reference to the lesson in which a skill or concept was presented. Students should use these references as a study aid if they are uncertain how to do a given exercise.

Chapter REVIEW

In Exercises 1–6, find the horizontal and vertical asymptotes, if any, of the graph of the function. 4. $y = 3$; $x = 0$, $x = -1 \pm \sqrt{7}$

1. $f(x) = \dfrac{2}{x^2}$ $y = 0$; $x = 0$
2. $f(x) = \dfrac{1}{x+4}$ $y = 0$; $x = -4$
3. $f(x) = \dfrac{3x^2}{4x^2 - 4}$ $y = \frac{3}{4}$; $x = -1$, $x = 1$
4. $f(x) = \dfrac{3x^3}{x^3 + 2x^2 - 6x}$
5. $f(x) = \dfrac{5x^2 - 6x + 1}{x - 3}$ $x = 3$
6. $f(x) = \dfrac{3x^3 - 7}{4x + 6}$ $x = -\frac{3}{2}$

In Exercises 7–10, match the function with its graph.

7. $f(x) = \dfrac{-4}{2 - x}$ c
8. $f(x) = \dfrac{x - 5}{x + 3}$ a
9. $f(x) = \dfrac{-2}{x^2 - 1}$ d
10. $f(x) = \dfrac{10}{x^2 - 4x - 5}$ b

a. 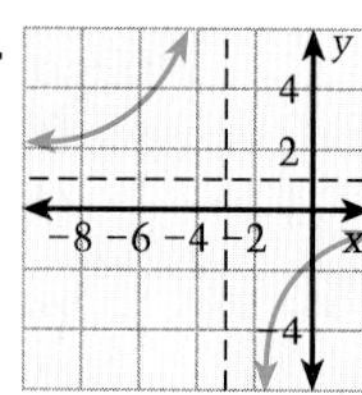

b. 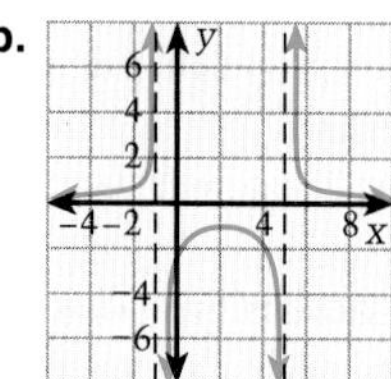

c. 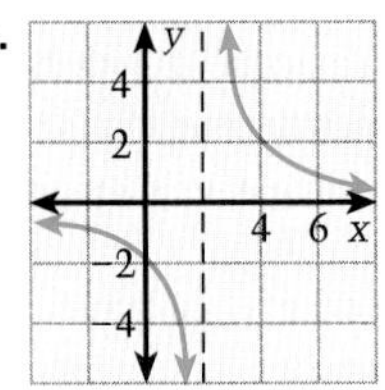

d. 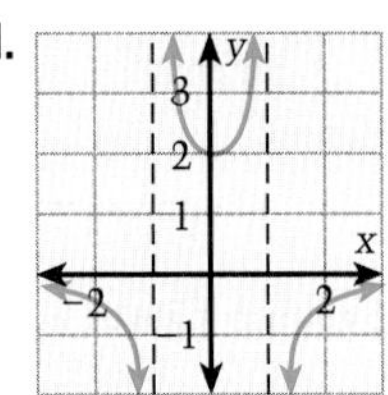

In Exercises 11–22, sketch the graph of the function. See Additional Answers.

11. $f(x) = \dfrac{1}{x^2}$
12. $f(x) = -\dfrac{1}{x}$
13. $f(x) = \dfrac{2}{x - 5}$
14. $f(x) = \dfrac{3}{4 + 3x}$
15. $f(x) = \dfrac{-6x}{x + 2}$
16. $f(x) = \dfrac{3x^2 + 1}{x^2 - 1}$
17. $f(x) = \dfrac{3x^2}{x^2 + 4x + 4}$
18. $f(x) = \dfrac{4x^2 - 1}{x^2 + 8x + 7}$
19. $f(x) = \dfrac{2x^3 - x + 10}{3x^2 + 6x}$
20. $f(x) = \dfrac{4x^4 + x^3 + x - 4}{x^3 - 27}$
21. $f(x) = \dfrac{2x^4 + 1}{x^3 + 64}$
22. $f(x) = \dfrac{5x^2 + 6x - 7}{x - 2}$

In Exercises 23 and 24, the variables *x* and *y* vary inversely. Use the given pair of *x*- and *y*-values to find an equation that relates the variables.

23. $x = 1$, $y = 5$ $xy = 5$
24. $x = 4$, $y = 3$ $xy = 12$

In Exercises 25 and 26, the variable *z* varies jointly with the product of *x* and *y*. Use the given values to find an equation that relates the variables.

25. $z = 4$, $x = 1$, $y = 12$ $z = \frac{1}{3}xy$
26. $z = 6$, $x = 6$, $y = 8$ $z = \frac{1}{8}xy$

In Exercises 27 and 28, state whether *x* and *y* vary directly or inversely.

27.

x	1	2	3	4	5
y	60	30	20	15	12

Inversely

28. 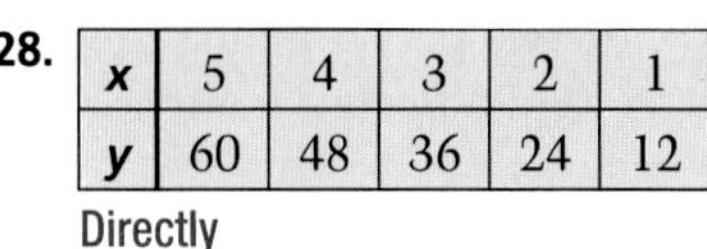

x	5	4	3	2	1
y	60	48	36	24	12

Directly

In Exercises 29–32, simplify the expression. See margin.

29. $\frac{x^2 - 3x - 10}{x^2 - 4}$ 30. $\frac{x - 2}{x^2 + 7x - 18}$ 31. $\frac{x + 5}{x^2 - 4x - 45}$ 32. $\frac{x^2 - 2x - 8}{x + 2}$

In Exercises 33–44, perform the operations and simplify.

33. $\frac{24x^3y^2}{5xy} \cdot \frac{10x^2y^4}{3x^4y}$ $16y^4$

34. $\frac{9xy^5}{2x^2y} \cdot \frac{6x^{-1}y}{x^3y^2}$ $\frac{27y^3}{x^5}$

35. $\frac{x^3 + 3x^2 - 10x}{3x^2 + 9x - 30} \cdot (3x - 30)$ $x(x - 10)$

36. $\frac{x^2 - 3x}{4x^2 - 8x} \cdot (4x^2 - 16)$ $(x - 3)(x + 2)$

37. $\frac{x^2 - 2x - 3}{x + 1} \div \frac{x^2 + x - 12}{x^2 - 1}$ $\frac{(x + 1)(x - 1)}{x + 4}$

38. $\frac{x - 4}{x^4} \div \frac{x^2 + 6x - 40}{x^4 + 9x^3 - 10x^2}$ $\frac{x - 1}{x^2}$

39. $\frac{4x^3 - 40x^2}{2x + 10} \div \frac{x^3 - 9x^2 - 10x}{x + 5}$ $\frac{2x}{x + 1}$

40. $\frac{5x^4 - 50x^3}{x^2 - 3x - 70} \div (5x^3 + 25x^2)$ $\frac{x}{(x + 7)(x + 5)}$

41. $\frac{2x}{x + 3} \cdot \frac{x^2 - 9}{5} \div \frac{x - 3}{4}$ $\frac{8x}{5}$

42. $5x \div \frac{1}{x - 6} \cdot \frac{x^2 - 9}{x}$ $5(x - 6)(x^2 - 9)$

43. $\frac{x^2 - 2x - 8}{x - 3} \div \frac{x + 2}{x - 3} \cdot \frac{x}{x - 4}$ x

44. $\frac{x - 1}{x^2 + 5x + 6} \cdot \frac{x^2 + x - 2}{3} \div \frac{x - 1}{3x}$ $\frac{x(x - 1)}{x + 3}$

In Exercises 45–54, solve the equation. Check your solutions.

45. $\frac{x}{x - 1} = \frac{2x + 10}{x + 11}$ $-2, 5$

46. $\frac{2}{3x + 6} = \frac{x + 2}{x^2 - 10}$ $-8, -4$

47. $\frac{x + 3}{x} - 1 = \frac{1}{x - 1}$ $\frac{3}{2}$

48. $\frac{x + 6}{-x} + \frac{2x + 1}{x + 3} = 1$ $-\frac{18}{11}$

49. $\frac{x + 8}{x - 4} + 3 = \frac{6(x - 2)}{x - 4}$ No solution

50. $\frac{x}{x + 2} - 2 = \frac{-9}{2x - 3}$ $-3, 5$

51. $\frac{2}{x - 2} - \frac{2x}{3} = \frac{x - 3}{3}$ $0, 3$

52. $\frac{x + 6}{x} + 3 = \frac{x + 5}{x - 1}$ $-1, 2$

53. $\frac{2x + 4}{x} + \frac{x + 2}{x - 2} = \frac{3x}{x - 2}$ 4

54. $\frac{3x + 2}{x + 1} = 2 - \frac{2x + 3}{x + 1}$ No solution

In Exercises 55–66, perform the operations and simplify.

55. $\frac{6}{x^2 - 4} + \frac{x}{x + 4}$ $\frac{x^3 + 2x + 24}{(x^2 - 4)(x + 4)}$

56. $\frac{x - 1}{x + 4} + \frac{3}{x(x - 4)}$ $\frac{x^3 - 5x^2 + 7x + 12}{x(x + 4)(x - 4)}$

57. $\frac{5x - 1}{x + 3} + \frac{9}{x(x + 3)}$ $\frac{5x^2 - x + 9}{x(x + 3)}$

58. $\frac{5}{x^2(x - 2)} + \frac{x}{x - 2}$ $\frac{5 + x^3}{x^2(x - 2)}$

59. $\frac{x}{x - 1} - \frac{4}{x + 3}$ $\frac{x^2 - x + 4}{(x - 1)(x + 3)}$

60. $\frac{x + 5}{x - 5} - \frac{3}{x + 5}$ $\frac{x^2 + 7x + 40}{(x - 5)(x + 5)}$

61. $\frac{x}{x^2 - 16} - \frac{x - 2}{x(x + 4)}$ $\frac{6x - 8}{x(x + 4)(x - 4)}$

62. $\frac{x}{x - 7} - \frac{16}{x^2 - 49}$ $\frac{x^2 + 7x - 16}{(x + 7)(x - 7)}$

✪ 63. $\frac{x - 5}{x^2 - x - 2} - \frac{3}{2} + \frac{4x + 1}{x + 1}$ $\frac{5x^2 - 9x - 8}{2(x - 2)(x + 1)}$

✪ 64. $\frac{x - 2}{5x(x - 1)} + \frac{1}{x - 1} - \frac{3x + 2}{x^2 + 4x - 5}$

✪ 65. $\frac{x + 3}{4x - 3} - 2\left(\frac{x}{x + 2} + \frac{5x}{4x^2 + 5x - 6}\right)$

✪ 66. $3\left(\frac{10}{x - 4} - \frac{x + 3}{2x + 1}\right) + \frac{x^2 - 5}{2x^2 - 7x - 4}$

64. $\frac{-9x^2 + 18x - 10}{5x(x - 1)(x + 5)}$ 65. $\frac{-7x^2 + x + 6}{(4x - 3)(x + 2)}$ 66. $\frac{-2x^2 + 63x + 61}{(x - 4)(2x + 1)}$

Answers

29. $\frac{x - 5}{x - 2}$

30. $\frac{1}{x + 9}$

31. $\frac{1}{x - 9}$

32. $x - 4$

Answers

67. $\dfrac{(-x^2+5x-5)(x+1)}{(x^2+6x+6)(x-1)}$

68. $\dfrac{(x^2+4x+27)(x+1)}{3(-4x-5)}$

69. $\dfrac{x(x^2+2x+16)}{(-x^2+5x-10)(x+2)}$

70. $\dfrac{x(x+2)(x^2+x-4)}{(x-1)(x^3-2x^2-3x-2)}$

71. $\dfrac{-9x(2x-1)}{(x+4)(x^2+2x-30)}$

72. $\dfrac{5(2x^2+x-12)}{x(x^2+4x+15)}$

In Exercises 67–72, simplify the complex fraction. See below.

67. $\dfrac{\left(\dfrac{5}{x}-\dfrac{x}{x-1}\right)}{\left(\dfrac{x}{x+1}+\dfrac{6}{x}\right)}$

68. $\dfrac{\left(\dfrac{4+x}{3}+\dfrac{9}{x}\right)}{\left(\dfrac{1}{x+1}-\dfrac{5}{x}\right)}$

69. $\dfrac{\left(\dfrac{x-2}{x+2}+\dfrac{6}{x-2}\right)}{\left(\dfrac{5}{x}-\dfrac{x}{x-2}\right)}$

70. $\dfrac{\left(\dfrac{4}{x}+\dfrac{x-3}{x-1}\right)}{\left(\dfrac{x-1}{x+2}-\dfrac{x+1}{x^2}\right)}$

71. $\dfrac{\left(\dfrac{x-5}{x+4}-\dfrac{x+4}{x-5}\right)}{\left(\dfrac{1}{x-5}+\dfrac{6+x}{x}\right)}$

72. $\dfrac{\left(\dfrac{x-3}{x}+\dfrac{x}{x+4}\right)}{\left(\dfrac{3}{x+4}+\dfrac{x}{5}\right)}$

Geometry **In Exercises 73 and 74, find an expression for the ratio of the area of the blue region to the total area. Then simplify the expression.**

73.

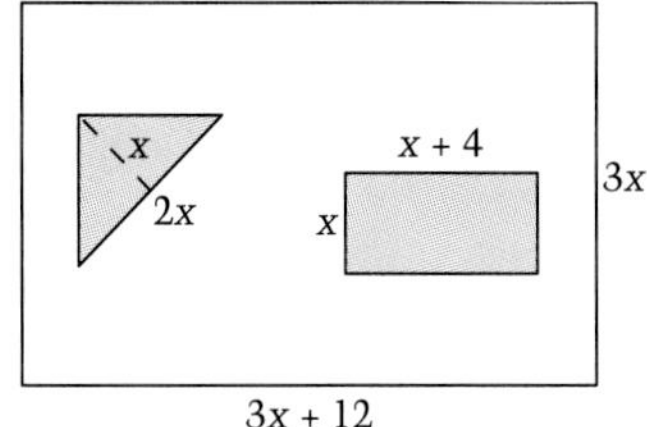

74.

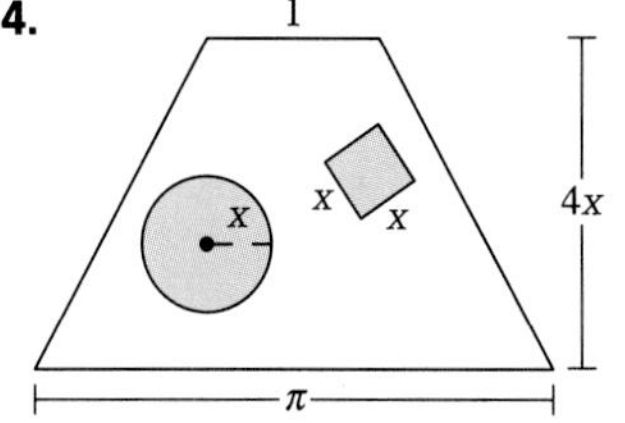

73. $\dfrac{2x^2+4x}{9x^2+36x}=\dfrac{2x+4}{9x+36}$

74. $\dfrac{(1+\pi)x^2}{2x(1+\pi)}=\dfrac{x}{2}$

75. *Amortization* You borrow \$900 at a 10% annual interest rate. To repay the loan, you make 11 monthly payments of \$79.12 and a final payment of \$79.16. Construct an amortization table for this loan. See Additional Answers.

76. *Home Mortgage* You have a home mortgage of \$85,000 with an annual interest rate of 11%. What is the monthly payment for a 20-year term? What is the total interest paid over the 20 years? \$877.36; \$125,566.40

77. *Principal Payment* You borrow \$15,000 at 8% annual interest for 10 years. How much of the first year's monthly installment payments are used to reduce the principal? \$1020.77

78. *Worm and Wheel Gears* For the worm and wheel gears shown at the right, the distance, D, on the perimeter of the wheel that the wheel gear rotates varies jointly with the product of the distance, P, between each tooth on the wheel gear and the number, N, of threads on the worm gear. A wheel gear whose teeth are 0.8 centimeter apart rotates 4 centimeters each time the 5-thread worm gear rotates once. How far would a wheel gear whose teeth are 0.5 centimeter apart rotate each time a 4-thread worm gear rotates once? 2 cm

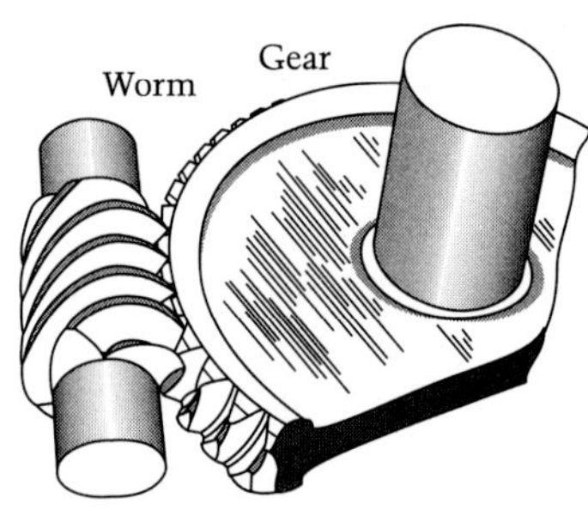

79. *Gravitational Force* The gravitational force, F, between two bodies varies jointly with the product of their masses, m_1 and m_2, and inversely with the square of the distance, r, between the two bodies. For instance, the gravitational force between the moon and Earth is 1.985×10^{20} kilogram-meters per second squared. Use the following values to find an equation that relates F, m_1, m_2, and r.

Mass of moon: $m_1 = 7.354 \times 10^{22}$ kilograms
Mass of Earth:
$m_2 = 5.979 \times 10^{24}$ kilograms
Distance between moon and Earth:
$r = 3.844 \times 10^{8}$ meters

Earth, as photographed by the Apollo II mission while on the moon.

79. $F = \dfrac{6.67 \times 10^{-11} m_1 m_2}{r^2}$

80. *Commercial Airline Revenue* For 1985 through 1993, the commercial airline revenue from passengers, P (in millions of dollars), and from other sources, R (in millions of dollars), can be modeled by

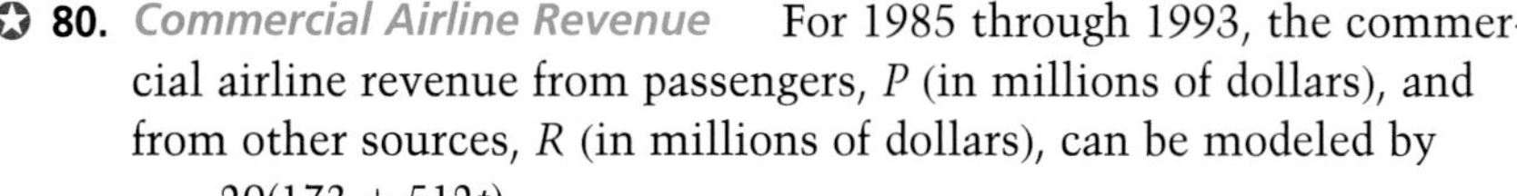

$$P = \frac{20(173 + 512t)}{1 + 0.089t} \quad \text{and} \quad R = 46.7t^3 - 1398t^2 + 13{,}792t - 34{,}995$$

where $t = 5$ represents 1985. Sketch the graphs of these two models. Which type of revenue increased more from 1985 to 1993, passenger or other sources? ***(Source: Air Transport Association of America)***

80. Passenger, see margin for graphs.

81. *Railroad Salaries* For 1985 through 1993, the average annual salary, S (in dollars), for railroad employees can be modeled by

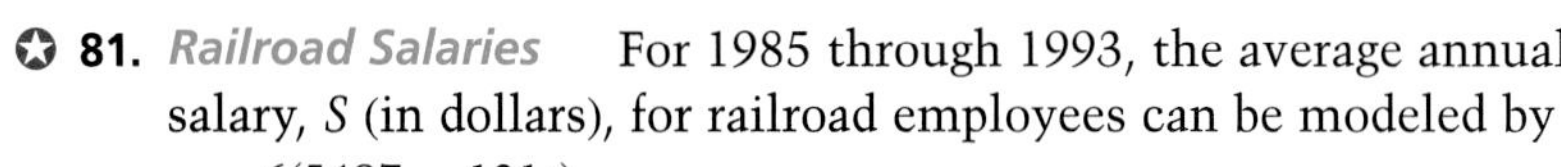

$$S = \frac{6(5407 - 121t)}{1 - 0.038t}$$

where $t = 5$ represents 1985. The number of railroad employees, N (in thousands), can be modeled by

$$N = \frac{126t + 878}{t}.$$

Write an expression for the total salary (in thousands of dollars) paid to railroad employees. What was the total salary paid in 1991? ***(Source: Association of American Railroads)***

81. $\dfrac{6(5407 - 121t)(126t + 878)}{t(1 - 0.038t)}$; \$8,648,607.31 thousand

82. *Dairy Production* For 1985 through 1994, the average amount of milk, M (in thousands of pounds per cow), produced by American dairy cows can be modeled by

$$M = \frac{11{,}525 + 223t}{1000 - 6.6t}$$

where $t = 5$ represents 1985. In what year was the average amount of milk about 15,000 pounds per cow? ***(Source: U.S. Department of Agriculture)*** 1990

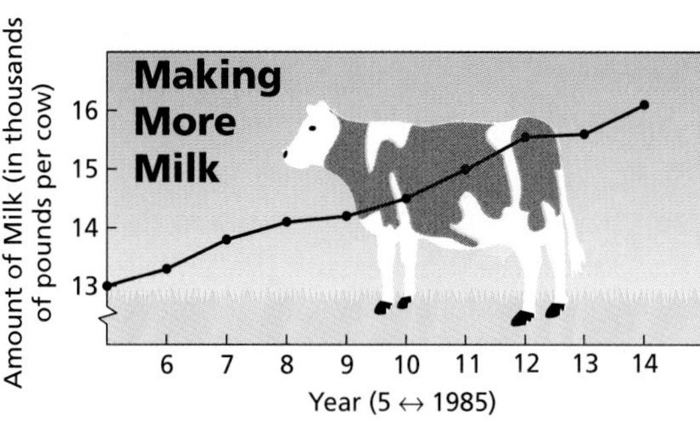

Answer

80.

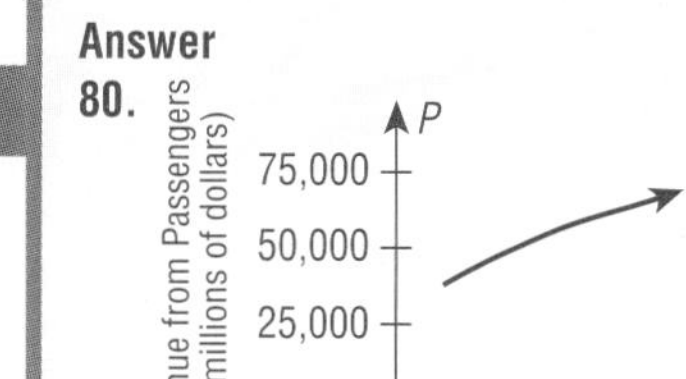

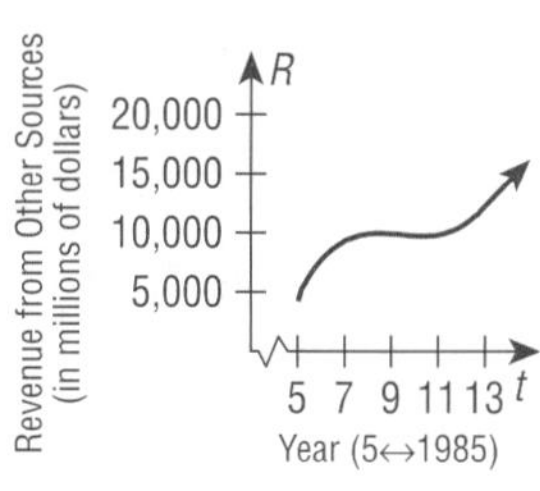

Answers

86. $\dfrac{860}{3(t + 330)}$

87. $\dfrac{t^3(40 - t)}{18{,}750(t + 330)}$

88. $\dfrac{4(28 + 5t)(40 - t)}{(1250 - 33t)(t + 330)}$

89. $\dfrac{5000}{3(51 - t)} - \dfrac{215}{40 - t}$, more and more nonrecycled paper is being used

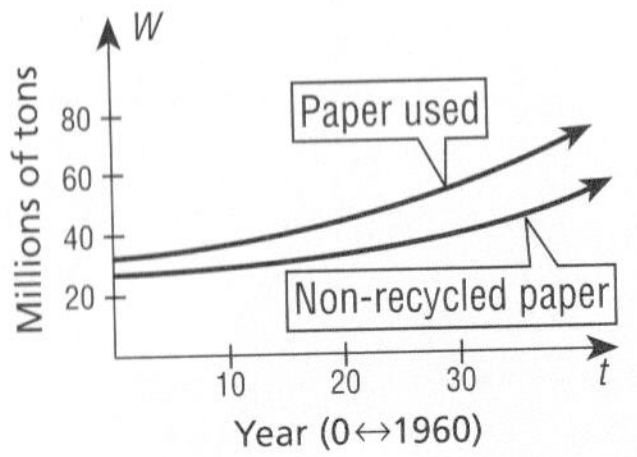

Recycled Paper, Glass, and Aluminum **In Exercises 83–85, use the following information.**

For 1960 through 1990, the amount (in millions of tons) of paper, P, aluminum, A, and glass, G, that was recycled in the United States can be modeled by

$$P = \frac{215}{40 - t}, \qquad A = \frac{t^3}{25{,}000}, \qquad \text{and} \qquad G = \frac{84 + 15t}{1250 - 33t}$$

where $t = 0$ represents 1960.

83. Sketch the graph of each model. See Additional Answers.

84. In what year did Americans recycle about 16.5 million tons of paper? 1987

85. In what year did Americans recycle about 0.1 million tons of glass? Was this the same year that 0.1 million tons of aluminum were recycled? 1962, no

More About Recycling **In Exercises 86–88, use the following information *and* the information given in Exercises 83–85.** See below.

For 1960 through 1990, one model for the total amount of material, T (in millions of tons), that was recycled in the United States can be written

$$T = \frac{3(t + 330)}{4(40 - t)}$$

where $t = 0$ represents 1960.

86. Of the total recycled materials, write an expression that represents the decimal percent (by weight) that was paper.

87. Of the total recycled materials, write an expression that represents the decimal percent (by weight) that was aluminum.

88. Of the total recycled materials, write an expression that represents the decimal percent (by weight) that was glass.

89. *Nonrecycled Paper* For 1960 through 1990, the amount of paper, W (in millions of tons) used in the United States can be modeled by

$$W = \frac{5000}{3(51 - t)}$$

where $t = 0$ represents 1960. Write an expression for the amount of paper products used and *not* recycled. Sketch the graph of W and the model for nonrecycled paper on the same coordinate system. What can you conclude from the graphs?
See Additional Answers.

Chapter TEST

In Exercises 1 and 2, sketch the graph of the equation. (10.1) See Additional Answers.

1. $y = \dfrac{3}{x-2}$

2. $y = \dfrac{4}{x^2+4}$

3. The variables x and y vary inversely. When $x = 2$, $y = 4$. Write an equation that relates x and y. **(10.2)** $xy = 8$

4. The variable x varies jointly with the product of y and z. When $x = 1$ and $y = 2$, $z = 8$. Write an equation that relates x, y, and z. **(10.2)** $x = \frac{1}{16}yz$

In Exercises 5–10, perform the operation and simplify. (10.3, 10.5)

5. $\dfrac{x^2-4}{x+3} \cdot \dfrac{x^2+4x+3}{2x-4}$ $\frac{1}{2}(x+1)(x+2)$

6. $\dfrac{4x-8}{x^2-3x+2} \div \dfrac{3x-6}{x-1}$ $\dfrac{4}{3(x-2)}$

7. $\dfrac{3}{5b} + \dfrac{2}{15b}$ $\dfrac{11}{15b}$

8. $\dfrac{5}{x+1} + \dfrac{x-7}{x-2}$ $\dfrac{x^2-x-17}{(x+1)(x-2)}$

9. $\dfrac{x+1}{3y} - \dfrac{x+2}{6y}$ $\dfrac{x}{6y}$

10. $\dfrac{x-1}{x-2} - \dfrac{x-4}{x+1}$ 10. $\dfrac{6x-9}{(x-2)(x+1)}$

In Exercises 11–13, solve the equation. (10.4, 10.5)

11. $\dfrac{x-3}{8} = \dfrac{2x}{3x+7}$ $-1, 7$

12. $\dfrac{x}{x+3} + \dfrac{2}{x} = \dfrac{5}{2x}$ $-1, \frac{3}{2}$

13. $\dfrac{\left(3 - \dfrac{2}{x+1}\right)}{\left(3 + \dfrac{3}{x}\right)} = 2$ $-3, -\frac{2}{3}$

In Exercises 14 and 15, find the least common multiple. (10.4)

14. x^2, $x(x-3)$, $(x-3)^2$ $x^2(x-3)^2$

15. $x^4 - 3x^3 - 4x^2$, $x^3 - x^2 - 2x$ $x^2(x+1)(x-2)(x-4)$

16. Simplify the complex fraction $\dfrac{\left(1 + \dfrac{3}{x}\right)}{\left(2 - \dfrac{5}{x^2}\right)}$. **(10.5)** $\dfrac{x^2+3x}{2x^2-5}$

17. You are starting a small beekeeping business. You spend \$500 for the equipment and the bees. You figure that it will cost \$1.25 per pound to collect, clean, bottle, and label the honey. How many pounds must you produce before your average cost per pound is \$1.79? **(10.1)** 926 lbs

18. From 1970 through 1990, the per capita consumption of red meat, m (in pounds), and boneless chicken, c (in pounds), in the United States can be modeled by $m = -4.4t + 132$ and $c = 0.8t^2 - 1.6t + 37$, where $t = 0$ represents 1970, $t = 1$ represents 1975, and so on. According to these models, in what year was the ratio of red meat consumption to chicken consumption 3 to 1? ***(Source: USA Today)*** **(10.4)** 1985

In Exercises 19 and 20, you take out a 20-year loan for \$100,000 at 6% annual interest. You will repay the loan with monthly installments. (10.6)

19. Calculate the amount of each monthly installment. \$716.43

20. If you continue making payments each month for 20 years, how much interest will you have paid? \$71,943.20

Answers

1.

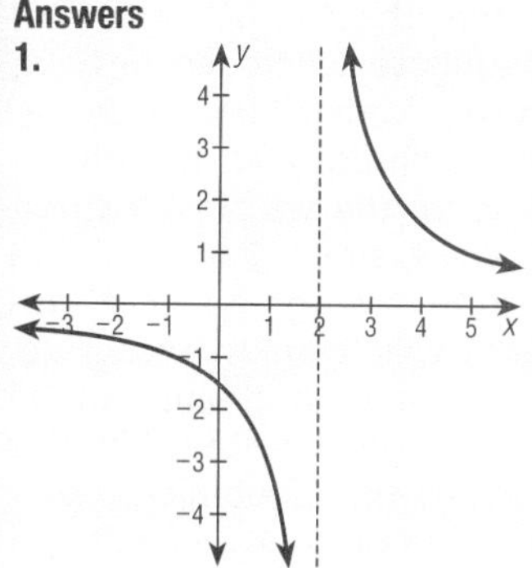

2.

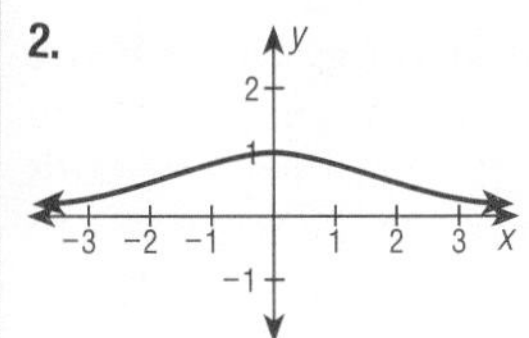

In this chapter, students will investigate "conic sections" or, simply, "conics." Parabolas, circles, ellipses, and hyperbolas make up the set of conics that we will examine. Each conic has a distinctive graph that students will learn to recognize. Although each graph can be defined geometrically, in this chapter, the algebraic properties of these four relations will be defined and explored.

The Chapter Summary on page 614 provides you and the students with a synopsis of the chapter. It identifies key skills and concepts. You may want to have students look at the Chapter Summary as an overview before beginning the chapter.

CHAPTER 11

Quadratic Relations

LESSONS

The West Wacker Building in Chicago follows the curve of the Chicago River. It has 36 stories and is 468 feet tall.

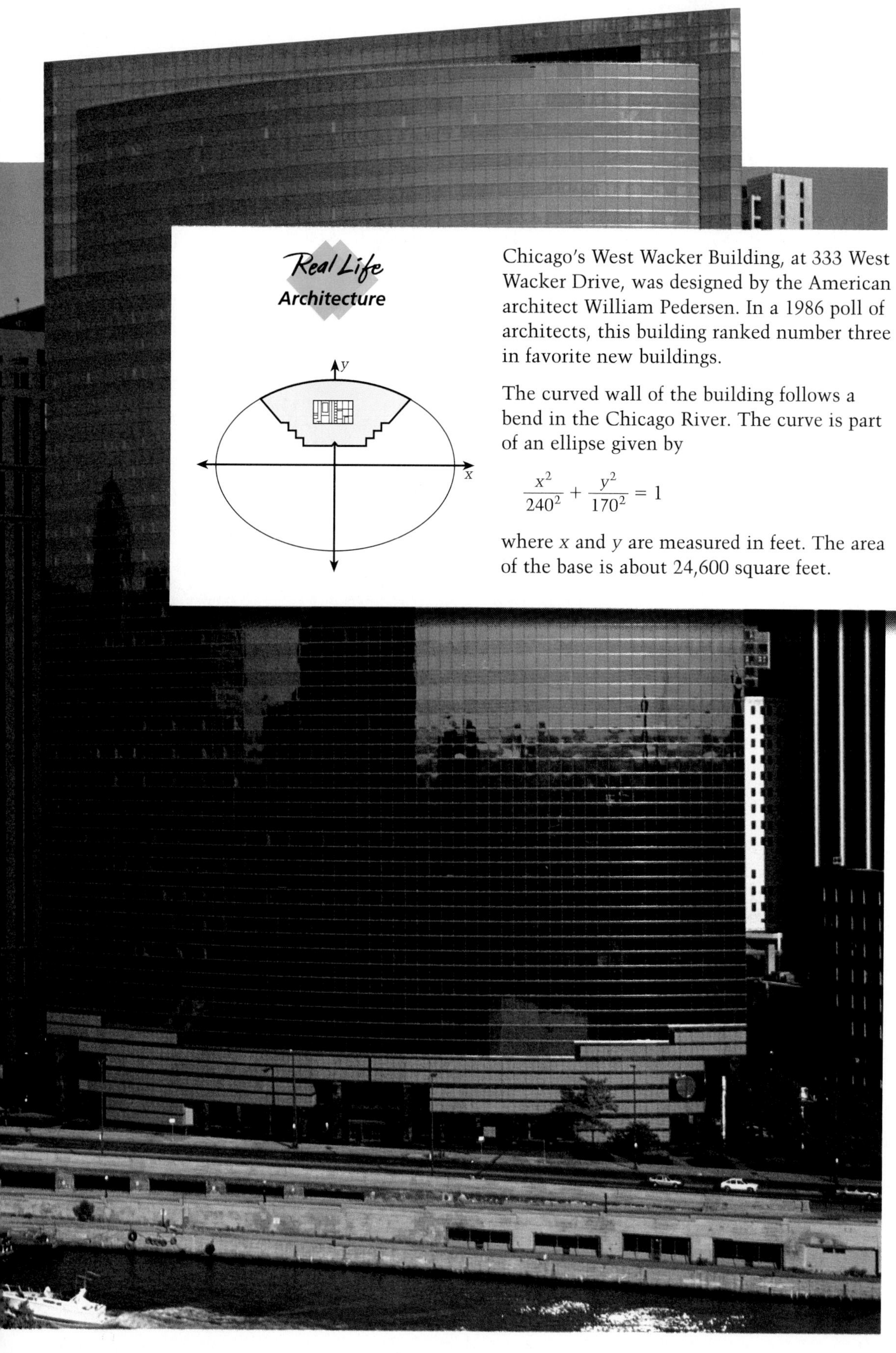

Real Life
Architecture

Chicago's West Wacker Building, at 333 West Wacker Drive, was designed by the American architect William Pedersen. In a 1986 poll of architects, this building ranked number three in favorite new buildings.

The curved wall of the building follows a bend in the Chicago River. The curve is part of an ellipse given by

$$\frac{x^2}{240^2} + \frac{y^2}{170^2} = 1$$

where x and y are measured in feet. The area of the base is about 24,600 square feet.

Many real-life relationships are not modeled by functions. In particular, not all quadratic relationships are represented by quadratic functions of the form $f(x) = ax^2 + bx + c$, where a, b, and c are real-valued, and a is non-zero. This chapter investigates quadratic relations of the general form, $Ax^2 + Bxy + Cy^2 + Dx + Ey + F = 0$, where A and B are not both zero, and A, B, C, D, E, and F are real-valued.

The relation that describes the curved wall of Chicago's West Wacker building is an ellipse. Ask students to show that the equation $\frac{x^2}{240^2} + \frac{y^2}{170^2} = 1$ can be rewritten in the general form given above.
$170^2x^2 + 240^2y^2 - (170^2)(240^2) = 0$; $A = 170^2$, $B = 0$, $C = 240^2$, $D = 0$, $E = 0$, $F = -(170^2)(240^2)$

Other Quadratic Relations

PACING CHART

The daily Pacing Chart is meant to help you adjust your teaching pace. Students in the full course should finish the entire text by the end of the year. Students in the basic course are expected to complete the first twelve chapters. The Pacing Chart for each chapter contains suggestions for lessons that require more than one day and lessons that may be omitted for the basic course.

DAY	FULL COURSE	BASIC COURSE
1	11.1	11.1
2	11.2	11.2
3	11.3	11.3
4	Mid-Chapter Self-Test	Mid-Chapter Self-Test
5	11.4	11.4
6	11.5 & Using a Graphing Calculator	11.4
7	11.6	11.5
8	11.6	11.5 & Using a Graphing Calculator
9	Chapter Review	11.6
10	Chapter Test	11.6
11		Chapter Review
12		Chapter Test

CHAPTER ORGANIZATION

LESSON	PAGES	GOALS	MEETING THE NCTM STANDARDS
11.1	570–575	1. Write an equation of a parabola and sketch its graph 2. Use parabolas to solve real-life problems	Problem Solving, Communication, Connections, Geometry (algebraic)
11.2	576–581	1. Write an equation of a circle and sketch its graph 2. Use equations and graphs of circles in real-life problems	Problem Solving, Communication, Reasoning, Connections, Geometry (algebraic)
Mixed Review	582	Review algebraic and arithmetic skills	
Career Interview	582	Sales Coordinator	Connections
11.3	583–588	1. Write an equation of an ellipse and sketch its graph 2. Use ellipses to solve real-life problems	Problem Solving, Communication, Connections, Geometry (algebraic)
Mid-Chapter Self-Test	589	Diagnose student weaknesses and remediate with correlated Reteach worksheets	
11.4	590–596	1. Write the equation of a hyperbola and graph it, using asymptotes 2. Use hyperbolas to solve real-life problems	Problem Solving, Communication, Connections, Geometry (algebraic)
11.5	597–603	1. Write an equation of a parabola with a vertex at (h, k) and an equation of a circle, an ellipse, or a hyperbola with its center at (h, k) 2. Use translations to solve real-life problems	Problem Solving, Communication, Connections, Geometry (algebraic perspective)
Mixed Review	603	Review algebraic and arithmetic skills	
Using a Calculator	604–605	Graph conic sections	Technology
11.6	606–613	1. Classify a conic from its general equation 2. Use equations of conics to solve mathematical and real-life problems	Problem Solving, Communication, Reasoning, Connections, Geometry (algebraic)
Chapter Summary	614	Restate for students what they have learned, why they have learned it, and how it fits into the structure of algebra	Structure, Connections
Chapter Review	615–618	Review concepts and skills learned in the chapter	
Chapter Test	619	Diagnose student weaknesses and remediate with correlated Reteaching worksheets	

LESSON RESOURCES

MEETING INDIVIDUAL NEEDS

RETEACHING For students who need to spend more time on basics:

If a mid-chapter self-test or chapter test indicates a deficiency, teachers can help students with the appropriate ***Reteaching Copymaster.***

PRACTICE For students who need more practice:

Additional exercises like those in the Pupil's Edition are provided for each lesson in ***Extra Practice Copymasters.***

ENRICHMENT For enriching and broadening students' experiences:

Problem of the Day copymasters in ***Teaching Tools*** provide a daily opportunity to use logical reasoning, looking for a pattern, writing an equation, and other routine and non-routine problem-solving strategies.

Math Log copymasters in ***Alternative Assessment*** provide opportunities to report on investigations, research, and open-ended problems.

Technology: Using Calculators and Computers provides enriching activities with graphing and scientific calculators and computers.

The ***Applications Handbook*** provides additional information about the cross-curriculum topics such as astronomy, chemistry, physics, sports, economics, genetics, and music that are integrated into the Pupil's Edition.

LESSON	11.1	11.2	11.3	11.4	11.5	11.6
PAGES	570-575	576-581	583-588	590-596	597-603	606-613
Teaching Tools						
Transparencies	✓					✓
Problem of the Day	✓	✓	✓	✓	✓	✓
Warm-up Exercises	✓	✓	✓	✓	✓	✓
Answer Masters	✓	✓	✓	✓	✓	✓
Extra Practice Copymasters	✓	✓	✓	✓	✓	✓
Reteaching Copymasters	Teacher-directed and independent activities tied to results on the Mid-Chapter Self-Tests and Chapter Tests					
Color Transparencies	✓	✓		✓	✓	✓
Applications Handbook	Additional background information for many real-life applications					
Technology	Calculator and computer worksheets for appropriate lessons					
Complete Solutions Manual	✓	✓	✓	✓	✓	✓
Alternative Assessment	Assess student's ability to reason, analyze, solve problems, and communicate using mathematical language.					
Formal Assessment	Mid-Chapter Self-Tests, Chapter Tests, Cumulative Tests, and Practice for College Entrance Tests					
Computer Test Bank	Customized tests can be created by choosing from over 2500 items.					

INSIGHTS

11.1 Parabolas

A parabola is a quadratic relation that can be defined as a function. Students have already worked extensively with quadratic functions in earlier chapters. In this lesson, students explore a geometric interpretation of parabolas and real-life quadratic models for which the geometric interpretation is most appropriate.

11.2 Circles and Points of Intersection

The geometric interpretation of circles is the most commonly known definition. The algebraic definition of a circle, however, is more useful in solving real-life problems. In this lesson, students will explore the algebraic representation of circles and the connection to points of intersection.

11.3 Ellipses

In this lesson, students examine quadratic relationships modeled by the ellipse. Ellipses are important geometric figures used in such diverse areas as building construction and astronomy.

11.4 Hyperbolas

The last quadratic relationship that students explore is the hyperbola. Its definition is similar to that of the ellipse. Like the ellipse, it has foci, vertices, and a center. However, it has only one transverse axis, and therefore, only one pair of vertices. Upon the completion of this lesson, students will have a fundamental understanding of all those quadratic relationships which, in the next lesson, will be referred to as conic sections.

11.5 Translations of Conics

Although it is common for real-life applications of conic sections to be centered at the origin (or in the case of parabolas, to have vertices at the origin), many models must deal with situations in which the centers (or vertices for parabolas) are not located at the origin. No investigation of conic sections would be complete without this more general approach to the subject.

11.6 Classifying Conics

Until now, students have used informal means to decide which conic section is the graph of a given equation. Each conic has an equation that can be written in the form $Ax^2 + Bxy + Cy^2 + Dx + Ey + F = 0$ (where A and C cannot *both* be zero). Note how a "discriminant" exists that can be used to classify conic sections. This is a particularly useful indicator, since some second-degree equations are not conics, or have axes that are neither horizontal nor vertical.

Problem of the Day

$\frac{2}{3}$ of the men in town are married. $\frac{3}{5}$ of the women in town are married. What fraction of the men and women in town are not married?

$\frac{7}{19}$

ORGANIZER

Warm-Up Exercises

1. Solve for p.
   a. $\frac{4}{3}y = 4py$ b. $28x = 4px$
   c. $-36y = 4py$
   d. $-80x = 4px$
   e. $\frac{2}{3}y = 4py$ f. $\frac{8}{5}x = 4px$

   a. $p = \frac{1}{3}$, b. $p = 7$, c. $p = -9$, d. $p = -20$, e. $p = \frac{1}{6}$, f. $p = \frac{2}{5}$
2. Write the equation of each straight line described below.
   a. a horizontal line 3 units below the x-axis
   b. a vertical line 6 units to the right of the y-axis
   c. a vertical line containing the point $(-4, 2)$
   d. a horizontal line passing through the point $(-12, 7)$

   a. $y = -3$, b. $x = 6$, c. $x = -4$, d. $y = 7$

Lesson Resources

Teaching Tools
- Transparencies: 7, 8, 13
- Problem of the Day: 11.1
- Warm-up Exercises: 11.1
- Answer Masters: 11.1

Extra Practice: 11.1
Color Transparency: 58

11.1 Parabolas

What you should learn:

Goal 1 How to write an equation of a parabola and sketch its graph

Goal 2 How to use parabolas to solve real-life problems

Why you should learn it:

You can model many real-life objects, such as solar collectors, with parabolic models.

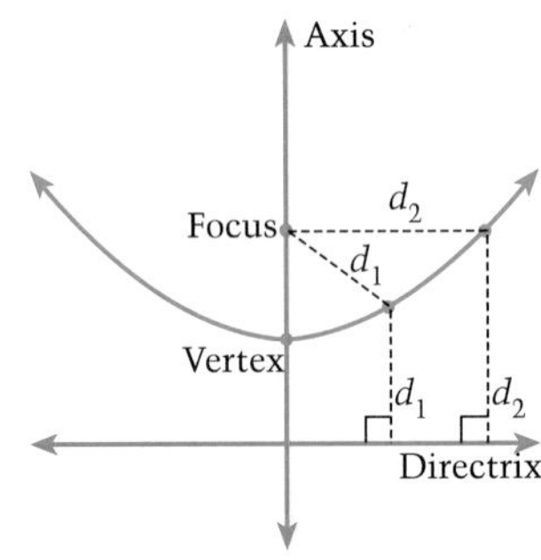

Goal 1 Writing an Equation of a Parabola

You already know that the graph of a quadratic function is a *parabola.* In this lesson, you will study properties of parabolas.

A **parabola** is the set of all points (x, y) that are equidistant from a fixed line, the **directrix,** and a fixed point, the **focus.** The midpoint of the segment from the focus perpendicular to the directrix is the **vertex.** The line passing through the focus and the vertex is the **axis.**

Standard Equation of a Parabola (Vertex at Origin)

The **standard form of the equation of a parabola** with a vertex at (0, 0) is as follows.

Equation	Focus	Directrix	
$x^2 = 4py$	$(0, p)$	$y = -p$	*Vertical axis* $(x = 0)$
$y^2 = 4px$	$(p, 0)$	$x = -p$	*Horizontal axis* $(y = 0)$

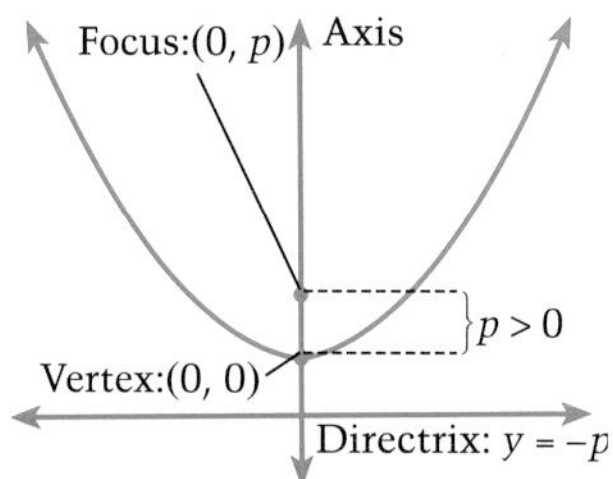

Vertical axis: $p > 0$

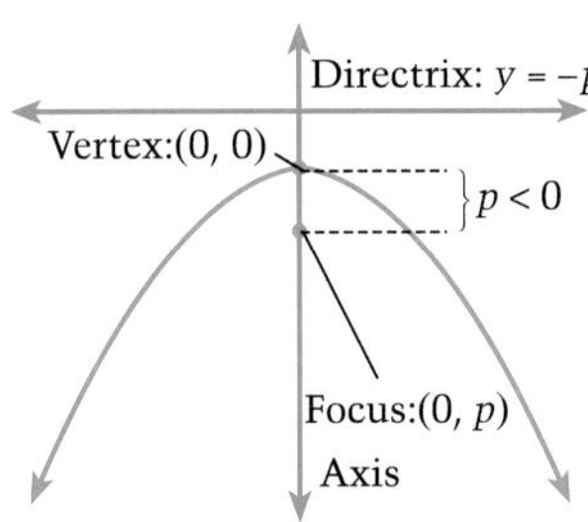

Vertical axis: $p < 0$

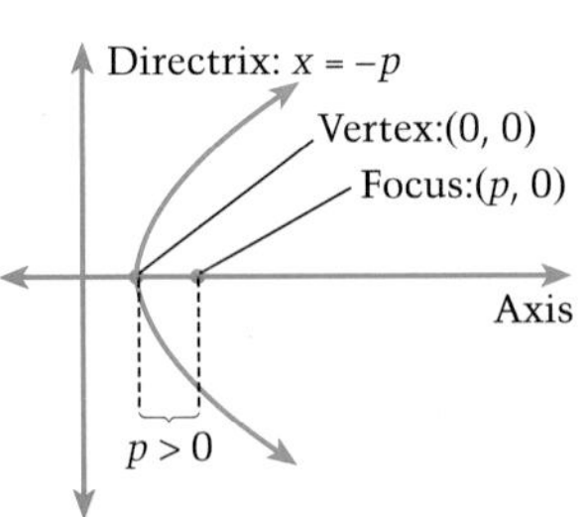

Horizontal axis: $p > 0$

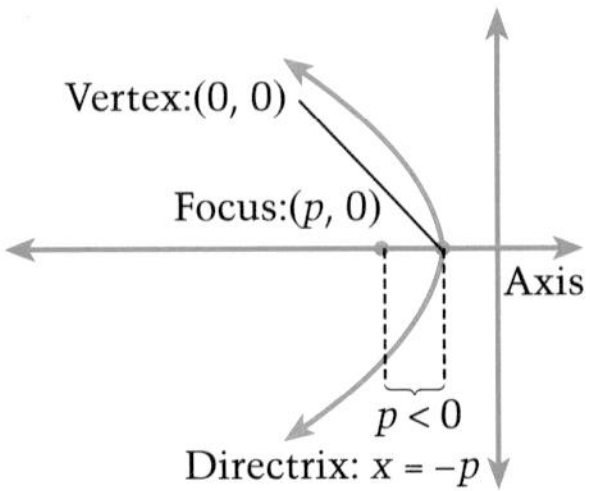

Horizontal axis: $p < 0$

The four ways a parabola can be oriented in the coordinate plane when its axis is horizontal or vertical are shown at right.

Example 1 *Finding the Focus and Directrix*

Find the focus and the directrix of the parabola $y = -2x^2$.

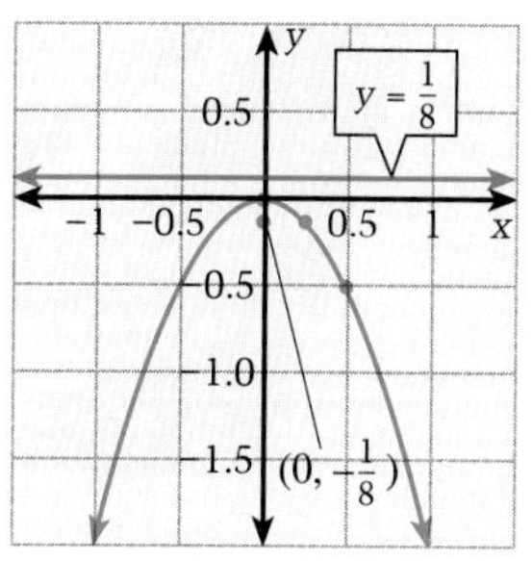

Solution Since the x-term is squared, you know that the axis is vertical, and the standard form is

$x^2 = 4py$. *Standard form, vertical axis*

You can write the given equation in this form as follows.

$-2x^2 = y$ *Given equation*

$x^2 = -\frac{1}{2}y$ *Divide both sides by −2.*

$x^2 = 4\left(-\frac{1}{8}\right)y$ *Standard form*

Thus $p = -\frac{1}{8}$. Since p is negative, the parabola opens downward, the focus of the parabola is $(0, p) = \left(0, -\frac{1}{8}\right)$, and the directrix is $y = -p = \frac{1}{8}$. Note that the focus is $\frac{1}{8}$ unit from the vertex, $(0, 0)$. The directrix is also $\frac{1}{8}$ unit from the vertex, but on the side opposite the focus. The directrix is the horizontal line through the point $\left(0, \frac{1}{8}\right)$. ■

Example 2 *Writing an Equation and Sketching a Graph*

Write the standard form of the equation of the parabola with vertex at the origin and focus at $(2, 0)$. Sketch the parabola.

Solution The axis of the parabola is horizontal, through $(0, 0)$, and the focus is at $(p, 0) = (2, 0)$. The standard form of the equation is as follows.

$y^2 = 4px$ *Standard form, horizontal axis*

Because $p = 2$, the equation is

$y^2 = 4(2)x$ *Substitute 2 for p.*

$y^2 = 8x$. *Standard form*

To sketch the parabola, it helps to write the equation as $x = \frac{1}{8}y^2$ and construct a table of values. From the table, you can sketch the graph shown at the left.

y	−3	−2	−1	0	1	2	3
x	$\frac{9}{8}$	$\frac{1}{2}$	$\frac{1}{8}$	0	$\frac{1}{8}$	$\frac{1}{2}$	$\frac{9}{8}$

■

Be sure you understand that the term *parabola* is a technical term used in mathematics and does not simply refer to any U-shaped curve. For instance, the graph of $y = x^4$ is a U-shaped curve but is *not* a parabola.

LESSON Notes

Example 1

Emphasize that both equations, $y = -2x^2$ and $x^2 = 4\left(-\frac{1}{8}\right)y$, represent the same relationship between x and y, and have the same graph. Techniques for graphing the parabola, learned previously, are still appropriate.

Example 2

Be sure students understand that the focus at $(2, 0)$ indicates that the axis of the parabola is the x-axis and is, therefore, a horizontal axis. Ask students to identify the equation of the directrix ($x = -2$).

Extra Examples

Here are additional examples similar to **Examples 1–2.**

1. Find the focus and directrix of the given parabolas.

a. $y = 4x^2$ **b.** $y = -6x^2$
c. $x = -3y^2$ **d.** $x = 8y^2$

a. $\left(0, \frac{1}{16}\right), y = -\frac{1}{16}$,
b. $\left(0, -\frac{1}{24}\right), y = \frac{1}{24}$,
c. $\left(-\frac{1}{12}, 0\right), x = \frac{1}{12}$,
d. $\left(\frac{1}{32}, 0\right), x = -\frac{1}{32}$

2. Write the standard form of the equation of the indicated parabola with its vertex at the origin.

a. focus at $(-4, 0)$
b. directrix: $y = 6$
c. directrix: $x = -1$
d. focus at $(0, 3)$

a. $y^2 = -16x$ b. $x^2 = -24y$
c. $y^2 = 4x$ d. $x^2 = 12y$

3. Sketch the parabolas in Extra Example 2.

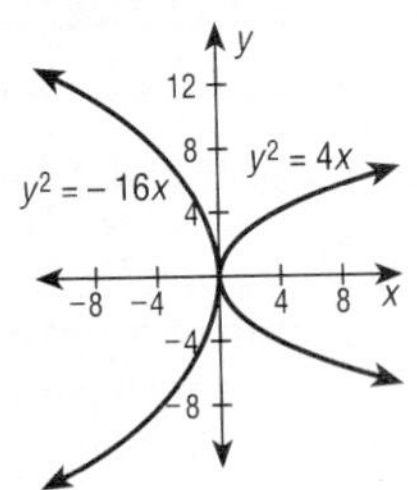

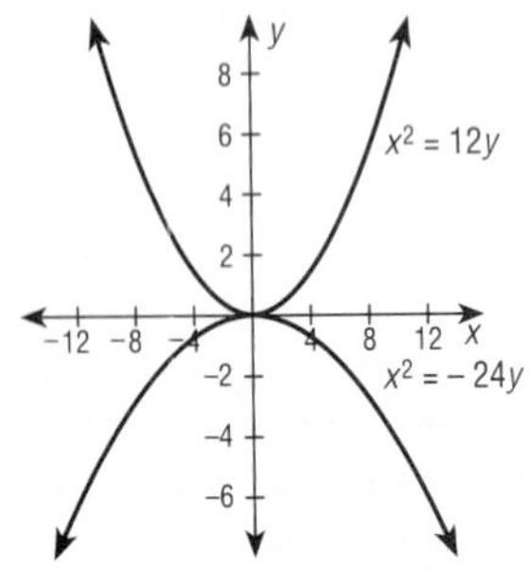

Check Understanding

1. Describe the set of points that define a parabola.
See page 570.
2. In how many ways can a parabola be oriented when its axis is either horizontal or vertical?
4; see the graphs on page 570
3. Is the function $y = x^4$ a model of a parabola?
No
4. What is the standard form of the equation of a parabola?
$x^2 = 4py$ or $y^2 = 4px$

Communicating about ALGEBRA

COOPERATIVE LEARNING
Have students work in small groups to verify that each parabola in Parts A and B consists of points (x, y) that are equidistant from its directrix and focus. (*Hints:*
1. Use the Distance Formula $d = \sqrt{(x_2 - x_1)^2 + (y_2 - y_1)^2}$ to find the distance between the points (x_1, y_1) and (x_2, y_2) in a coordinate plane.
2. An arbitrary point on the parabola of Part A is $(x, 2x^2)$.)

The world's largest solar-thermal complex, Luz International, is located in California's Mojave Desert. The parabolic mirrors reflect the sun's rays on tubes filled with oil. The heated oil is then used to boil water, which sends steam to a turbine. There are 1.5 million mirrors in the complex. A crew of 20 people works at night to keep the mirrors clean.

Goal 2 Using Parabolas for Real-Life Problems

Real Life
Solar Energy

Example 3 *Farming the Sun*

Cross sections of parabolic mirrors at the solar-thermal complex can be modeled by the equation

$$\frac{1}{25}x^2 = y$$

where x and y are measured in feet. The oil-filled heating tube is located at the focus of the parabola. How high above the vertex of the mirror is the heating tube?

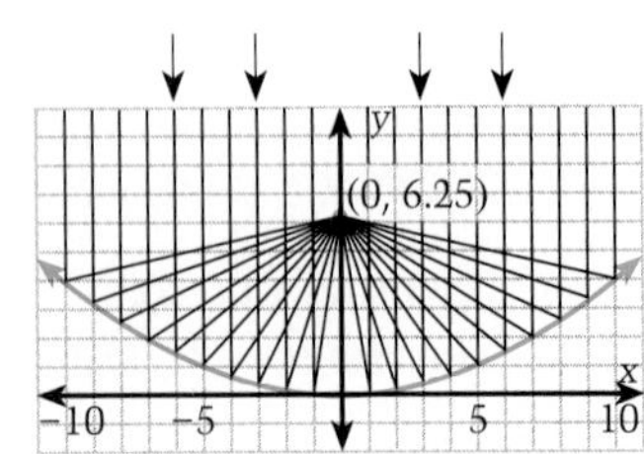

In a parabolic reflector, all incoming rays that are parallel to the parabola's axis are directed to the focus.

Solution

$$\frac{1}{25}x^2 = y \quad \textit{Given equation}$$

$$x^2 = 25y \quad \textit{Multiply both sides by 25.}$$

$$x^2 = 4\left(\frac{25}{4}\right)y \quad \textit{Standard form}$$

Thus $p = \frac{25}{4} = 6.25$. Since p is positive, the parabola opens upward, and the focus of the parabola is 6.25 feet above the vertex. ■

Communicating about ALGEBRA

SHARING IDEAS about the Lesson See below.

Find the focus and directrix of the parabola. Explain your steps.

A. $y = 2x^2$ $\left(0, \frac{1}{8}\right), -\frac{1}{8}$ **B.** $y = -6x^2$ $\left(0, -\frac{1}{24}\right), \frac{1}{24}$

C. $x = \frac{1}{2}y^2$ $\left(\frac{1}{2}, 0\right), -\frac{1}{2}$ **D.** $x = -12y^2$ $\left(-\frac{1}{48}, 0\right), \frac{1}{48}$

A–D. Isolate the second-degree variable, then set the coefficient of the first-degree variable equal to $4p$ and solve for p; the coordinates of the focus are either $(0, p)$ or $(p, 0)$, the directrix is either $x = -p$ or $y = -p$.

EXERCISES

Guided Practice

CRITICAL THINKING about the Lesson

1. Describe in words the location of a parabola's vertex relative to its focus and directrix. The vertex is the midpoint of the distance between the focus and the directrix.

2. Does the parabola $y^2 = x$ have a horizontal or vertical axis? Horizontal

3. Write the equation of the directrix of the parabola $x^2 = 4y$. $y = -1$

4. Find the focus of the parabola $y = -\frac{1}{8}x^2$. $(0, -2)$

5. In which direction does the parabola with vertex at (0, 0) and focus at (0, 3) open? Opens up

6. Write an equation of the parabola in Exercise 5. Sketch the parabola. $x^2 = 12y$ See Additional Answers.

Independent Practice

In Exercises 7–14, write the equation of the parabola in standard form. See Additional Answers.

7. $y = 5x^2$ **8.** $y = -3x^2$ **9.** $y + 2x^2 = 0$ **10.** $x = 9y^2$

11. $3y^2 = 2x$ **12.** $\frac{1}{9}y^2 = 2x$ **13.** $14y^2 - x = 0$ **14.** $2y^2 + x = 0$

In Exercises 15–18, decide whether the parabola has a vertical or horizontal axis. See below.

15. $y = -7x^2$ **16.** $-2x^2 = 5y$ **17.** $3y^2 = 8x$ **18.** $x = 9y^2$

In Exercises 19–30, find the focus and the directrix of the parabola. See Additional Answers.

19. $2x^2 = -y$ **20.** $4y^2 = x$ **21.** $x^2 = 8y$ **22.** $y^2 = 20x$

23. $y^2 = -16x$ **24.** $x^2 = -36y$ **25.** $x^2 + 12y = 0$ **26.** $-80y + 2x^2 = 0$

27. $3y^2 - 12x = 0$ **28.** $3x^2 - 4y = 0$ **29.** $-24y + x^2 = 0$ **30.** $-4x + 9y^2 = 0$

In Exercises 31–38, write the standard form of the equation of the parabola with vertex at the origin and the given focus.

31. $(5, 0)$ $y^2 = 4(5)x$ **32.** $(-2, 0)$ $y^2 = 4(-2)x$ **33.** $(-3, 0)$ $y^2 = 4(-3)x$ **34.** $(0, 2)$ $x^2 = 4(2)y$

35. $(-4, 0)$ $y^2 = 4(-4)x$ **36.** $(2, 0)$ $y^2 = 4(2)x$ **37.** $(0, -1)$ $x^2 = 4(-1)y$ **38.** $(0, -3)$ $x^2 = 4(-3)y$

In Exercises 39–42, match the equation with its graph.

39. $y^2 = 12x$ c **40.** $x^2 = -12y$ d **41.** $x^2 = 12y$ b **42.** $y^2 = -12x$ a

a.

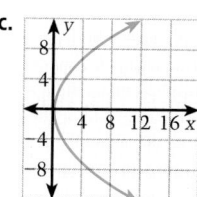

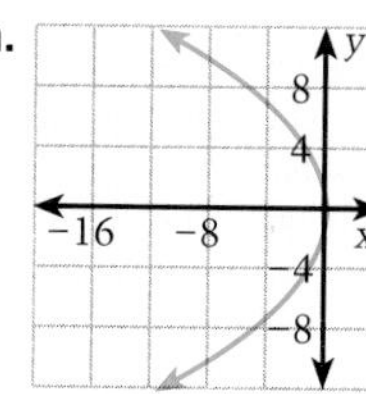

b. y 16 12 8 4 −8 −4 4 8 x

c. y 8 4 4 8 12 16 x −4 −8

d.

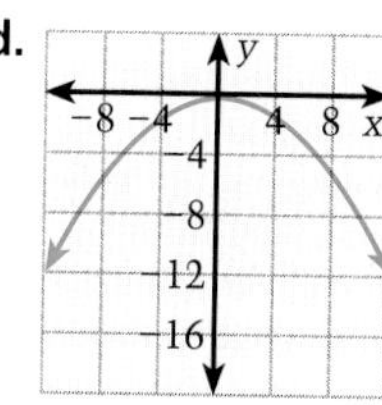

15. Vertical **16.** Vertical **17.** Horizontal **18.** Horizontal

EXERCISE Notes

ASSIGNMENT GUIDE

Basic/Average: Ex. 1–6, 13–23 odd, 29–37 odd, 39–42, 47–55 odd, 57, 67–69

Above Average: Ex. 1–6, 13–23 odd, 29–37 odd, 39–42, 45–55 odd, 56–59, 67–70

Advanced: Ex. 1–6, 13–23 odd, 29–37 odd, 39–42, 45–55 odd, 56–60, 67–72

Selected Answers
Exercises 1–6, 7–67 odd

More Difficult Exercises
Exercises 57–60, 69–72

Guided Practice

▶ **Ex. 1–6** Students have worked with parabolas before in the study of quadratic functions. This lesson introduces new terms such as *focus* and *directrix.* Use the Guided Practice exercises to check whether students can write the equation of a parabola and sketch its graph given its focus and directrix.

Independent Practice

EXTEND Ex. 7–14 Ask students to indicate which parabolas represent functions. The parabolas in Exercises 7–9 are functions.

▶ **Ex. 19–30** Students should use Example 1 as a solution model.

▶ **Ex. 31–38** Students should use Example 2 as a solution model.

▶ **Ex. 39–42** Assign these exercises as a group. Ask students to interpret the effect on the graph of the sign of p. When the sign is positive, the graph opens up or to the right; when the sign is negative, the graph opens down or to the left.

▶ **Ex. 43–54** Remind students that it helps to solve for the linear term. For example, in Exercise 49, write the equation as $y = -\frac{1}{20}x^2$. In Exercise 52, write the equation as $x = -\frac{1}{20}y^2$.

▶ **Ex. 55–56** Students should use Example 3 as a solution model.

Answers

55. Signals hit the parabolic dish and reflect into the focus.

57. The distance from F to the gold bar is the same as the distance from the gold bar to the directrix.

59. The distance from the pole to the zipper clasp is equal to the distance from tab A to the zipper clasp. Tab A is the focus and the pole is the directrix.

60. The sun's rays focus on one point of the candle.

▶ **Ex. 69–72 TECHNOLOGY** Students with graphics calculators can use the Trace function to check the solutions.

In Exercises 43–54, sketch the parabola. See Additional Answers.

43. $y^2 = 16x$ **44.** $x^2 = -6y$ **45.** $y^2 = -2x$ **46.** $y^2 = 36x$

47. $x^2 = 8y$ **48.** $y^2 = -10x$ **49.** $x^2 = -20y$ **50.** $x^2 = 14y$

51. $x^2 - 18y = 0$ **52.** $x + \frac{1}{20}y^2 = 0$ **53.** $x - \frac{1}{8}y^2 = 0$ **54.** $5x^2 = 4y$

55. *Television Antenna Dish* Cross sections of television dish antennas have parabolic shapes. For the dish below, the receiver is located at the focus and is 4 feet above the vertex. Find an equation of a cross section of the dish. (Assume the vertex is at the origin.) Explain how the antenna works. $x^2 = 16y$ See margin.

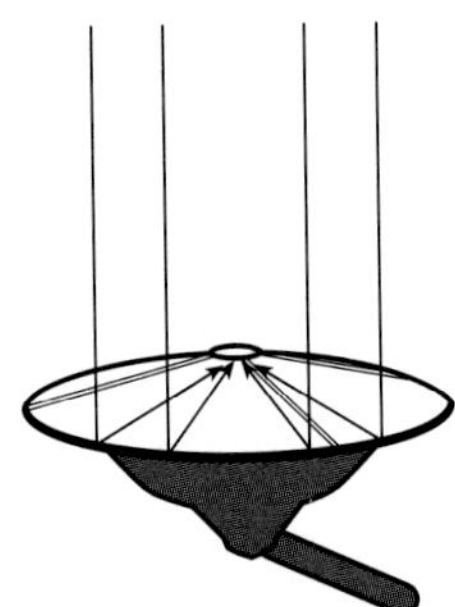

56. *Automobile Headlights* The headlights on a car work opposite to the way in which the television dish antenna works. The headlight bulb is at the focus of a parabolic reflector behind it. The bulb is 1.5 inches from the vertex of the reflector. Find an equation of a cross section of the reflector. The reflector is 6 inches wide. How deep is it? $y^2 = 6x$; 1.5 in.

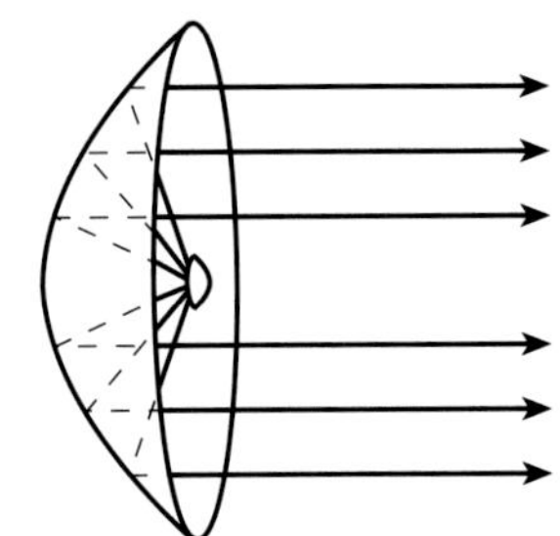

57. *Gold Rush* Each second in the *Gold Rush* video game, a gold bar is released from point F in any direction, except straight up. At the same time, a gold rusher is released from some point at the bottom of the screen. The gold rusher moves straight up at the same speed as the gold bar. The path G represents all possible locations at which the gold rusher intersects the gold bar. Explain why G is a parabola. If the point F is four units from the bottom of the screen, what is the equation of G? $x^2 = 8y$; See margin.

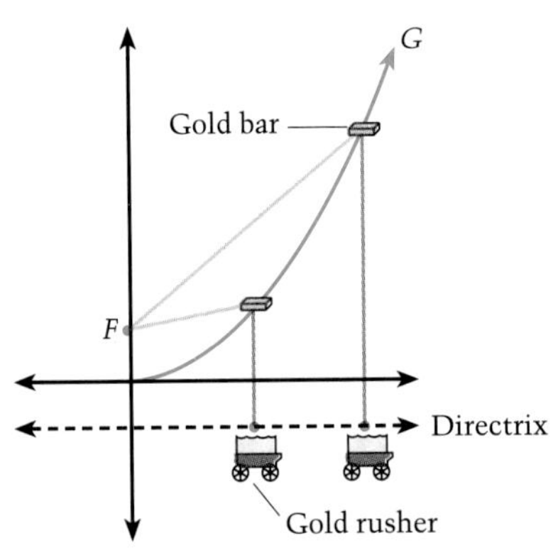

58. *Storage Building* A storage building for rock salt (for melting ice on roads) has the shape of a *paraboloid*—vertical cross sections are parabolas and horizontal cross sections are circles. An equation of a vertical cross section of the building is $y = -\frac{1}{12}x^2$. If the building is 27 feet high, how much rock salt will it hold? (The volume of a paraboloid is $V = \frac{1}{2}\pi r^2 h$, where r is the radius of the base and h is the height.) 4374π ft^3

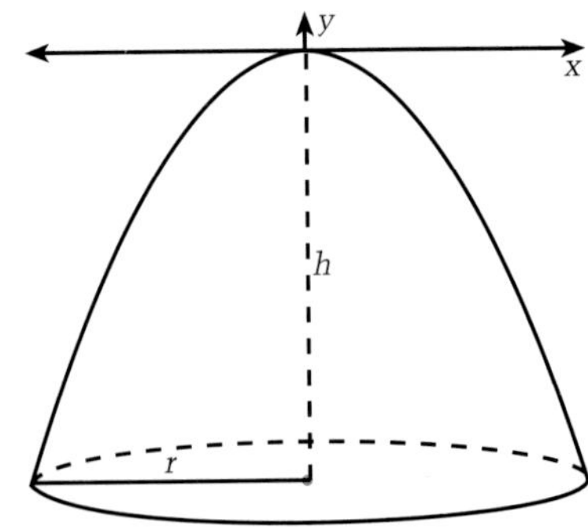

59. *Unzipping a Zipper* In the picture below, a thumbtack holds tab A fixed at the top of the zipper. Tab B glides along the pole without friction. Explain why the zipper traces out a parabola. Where are the focus and directrix? See margin.

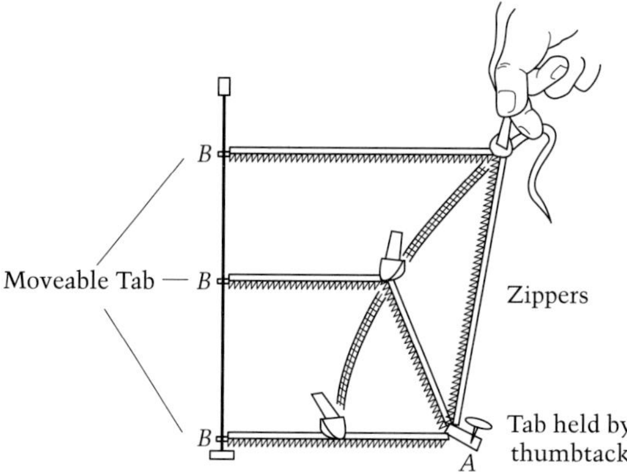

60. *Burning Mirror* The drawing below was taken from a book that was printed in the 17th century. In the drawing, the rays of the sun are shown lighting a candle. Explain why the sun's rays can generate enough heat to do this. See margin.

Integrated Review

In Exercises 61–66, sketch the graph of the function. See Additional Answers.

61. $f(x) = (x + 1)^2$
62. $f(x) = x^2 + 3$
63. $f(x) = x^2 - 4$
64. $f(x) = (x + 2)^2 - 4$
65. $f(x) = (x - 1)^2 + 3$
66. $f(x) = x^2 + 1$

67. *College Entrance Exam Sample* A rectangular box with a square base contains 6 cubic feet. The height of the box is 18 inches. How many feet are in each edge of the base? **b**

a. 1 **b.** 2 **c.** $\sqrt{3}$ **d.** $\frac{\sqrt{3}}{3}$ **e.** 4

68. *College Entrance Exam Sample* The surface area of a cube is 150 square feet. What is the volume of the cube in cubic feet? **d**

a. 30 **b.** 50 **c.** 100 **d.** 125 **e.** 150

Exploration and Extension

In Exercises 69–72, find the points of intersection of the graphs of *f* and *g* by setting *f*(*x*) equal to *g*(*x*) and solving the resulting equation for *x*.

69. $f(x) = 2x^2$ $(-3, 18), (2, 8)$
$g(x) = -2x + 12$

70. $f(x) = 5x^2$ $(-2, 20), (-1, 5)$
$g(x) = -15x - 10$

71. $f(x) = -3x^2$ $(1, -3), (2, -12)$
$g(x) = -9x + 6$

72. $f(x) = -\frac{1}{2}x^2$ $(2, -2), (6, -18)$
$g(x) = -4x + 6$

Enrichment Activity

This activity requires students to go beyond lesson goals.

TECHNOLOGY The graphics calculator is designed to graph functions in the form $y = f(x)$. Parabolas such as $y^2 = 36x$ (Exercise 46) are not functions; they are relations. In order to graph relations using a graphics calculator, you must transform them into forms the calculator will accept.

$$y^2 = 36x$$

Solve for *y*.

$$y = \pm\sqrt{36x} = \pm 6\sqrt{x}$$

There are two possible *y*-values for every *x*-value. So, you still do not have the equation of the parabola in a form that the calculator will accept. A parabola has a line of symmetry that goes through its vertex, so you can separate the parabola into "halves." To graph the entire parabola, you need to graph each half on the same grid. Thus, to graph the entire parabola, $y^2 = 36x$, you need to graph its top half, $y_1 = 6\sqrt{x}$, and then graph its bottom half, $y_2 = -6\sqrt{x}$.

What equations would you graph for Exercises 43, 45, 48, 52, and 53?

43. $y_1 = 4\sqrt{x}$, $y_2 = -4\sqrt{x}$
45. $y_1 = \sqrt{-2x}$, $y_2 = -\sqrt{-2x}$
48. $y_1 = \sqrt{-10x}$, $y_2 = -\sqrt{-10x}$
52. $y_1 = 2\sqrt{-5x}$, $y_2 = -2\sqrt{-5x}$
53. $y_1 = 2\sqrt{2x}$, $y_2 = -2\sqrt{2x}$

(If you have access to a graphics calculator, graph your equations to verify your answers.)

Problem of the Day

Suppose that the surface of the earth is smooth and spherical and that the distance around the equator is 25,000 miles. A steel band is made to fit tightly around the earth at the equator, then the band is cut and a piece of band 18 feet long in inserted. What will be the gap, all the way around, between the band and the earth's surface?

About 3 feet

ORGANIZER

Warm-Up Exercises

1. Find the distance between the given points in a coordinate plane.
   a. $(8,-3)$ and $(-1,-3)$
   b. $(4, 5)$ and $(4, 11)$
   c. $(0, 0)$ and $(5, 12)$
   d. $(-2, 7)$ and $(2,-5)$

   a. 9 b. 6
   c. 13 d. ≈ 12.65
2. Solve each system of equations by substitution.
   a. $\begin{cases} 2x + y = 3 \\ x - y = 3 \end{cases}$
   b. $\begin{cases} 3x + 4y = -3 \\ 2x + y = -7 \end{cases}$

   a. $(2,-1)$, b. $(-5, 3)$

Lesson Resources

Teaching Tools
- Transparencies: 7, 8
- Problem of the Day: 11.2
- Warm-up Exercises: 11.2
- Answer Masters: 11.2

Extra Practice: 11.2
Color Transparencies: 58, 59
Applications Handbook: p. 57
Technology Handbook: p. 77

11.2 Circles and Points of Intersection

What you should learn:

Goal 1 How to write an equation of a circle and sketch its graph

Goal 2 How to use equations and graphs of circles in real-life problems

Why you should learn it:

You can model many real-life situations with circular models, such as finding where two sounds are of equal intensity.

Goal 1 Writing an Equation of a Circle

A **circle** is the set of all points (x, y) that are equidistant from a fixed point, called the **center** of the circle. The distance, r, between the center and any point (x, y) on the circle is the **radius.**

The distance formula can be used to obtain an equation of the circle whose center is the origin and whose radius is r. Because the distance between any point on the circle, (x, y), and the center, $(0, 0)$, is r, you can write the following.

$$\sqrt{(x-0)^2 + (y-0)^2} = r \qquad \textit{Distance formula}$$

$$(x-0)^2 + (y-0)^2 = r^2 \qquad \textit{Square both sides.}$$

$$x^2 + y^2 = r^2 \qquad \textit{Simplify.}$$

> **Standard Equation of a Circle (Center at Origin)**
>
> The **standard form of the equation of a circle** with center at $(0, 0)$ and radius of r is
>
> $x^2 + y^2 = r^2.$

Example 1 *Writing an Equation of a Circle*

The point $(2, 5)$ is on a circle whose center is the origin. Write the standard form of the equation of the circle.

Solution Because the point $(2, 5)$ is on the circle, the radius of the circle must be the distance between the center and the point $(2, 5)$. Using the distance formula, the radius is

$$r = \sqrt{(2-0)^2 + (5-0)^2} = \sqrt{4+25} = \sqrt{29}.$$

Since the radius is $\sqrt{29}$, use the standard form to write the equation.

$$x^2 + y^2 = r^2 \qquad \textit{Standard form}$$

$$x^2 + y^2 = (\sqrt{29})^2 \qquad \textit{Substitute } \sqrt{29} \textit{ for r.}$$

$$x^2 + y^2 = 29 \qquad \textit{Simplify.}$$

■

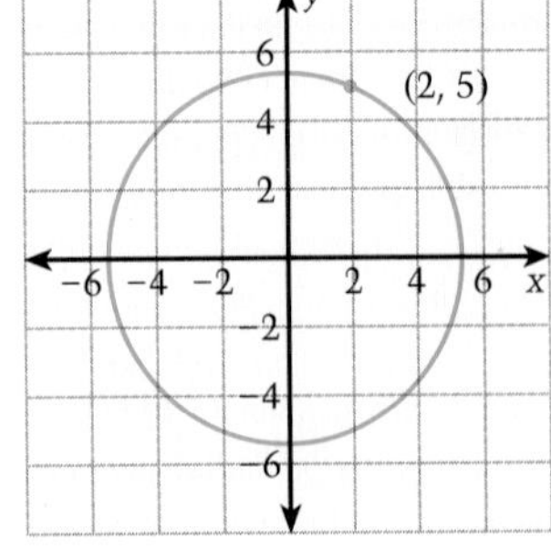

Goal 2 Using Equations and Graphs of Circles

A **point of intersection** of two graphs is a point that lies on both graphs. Two graphs can have no point of intersection, one point, or two or more points.

LESSON INVESTIGATION

■ Investigating Points of Intersection

Partner Activity Find values of r so that the circle $x^2 + y^2 = r^2$ and the line $y = -x + 2$ have the indicated number of points of intersection. If it is not possible, explain why.

a. No points **b.** 1 point **c.** 2 points **d.** 3 points

a. $r < \sqrt{2}$ b. $r = \sqrt{2}$ c. $r > \sqrt{2}$

d. Not possible, a line and circle can only intersect at at most 2 points.

Study Tip

To find any points of intersection of the graphs of two equations, you can use substitution, as shown in Example 2. The general strategy is to solve for one of the variables in one equation and substitute for that variable in the other equation.

Example 2 *Finding Points of Intersection*

Find the points of intersection of the graphs of

$x^2 + y^2 = 8$ and $y = \frac{1}{2}x + 1$.

Solution To find the points of intersection, substitute $\frac{1}{2}x + 1$ for y in the equation of the circle.

$x^2 + y^2 = 8$	*Equation of circle*
$x^2 + \left(\frac{1}{2}x + 1\right)^2 = 8$	*Substitute — for y.*
$x^2 + \frac{1}{4}x^2 + x + 1 = 8$	*Simplify.*
$\frac{5}{4}x^2 + x - 7 = 0$	*Simplify.*
$5x^2 + 4x - 28 = 0$	*Multiply both sides by 4.*
$(x - 2)(5x + 14) = 0$	*Factor.*
$x = 2, -\frac{14}{5}$	*Zero Product Property*

You now know the x-coordinates of the points of intersection. To find the y-coordinates, substitute $x = 2$ and $x = -\frac{14}{5}$ into the linear equation $y = \frac{1}{2}x + 1$ to conclude that the points of intersection are $(2, 2)$ and $\left(-\frac{14}{5}, -\frac{2}{5}\right)$. A graphical check, at the left, helps verify the result. ■

LESSON Notes

Example 1

Ask students to verify whether the points (2,−5), (5,−2), and (4,−4) are on the circle.
The points (2,−5) and (5,−2) are on the circle; (4,−4) is not.

Example 2

Ask students why you substitute the x-coordinates into the linear equation to obtain the y-coordinates of the points of intersection. The linear equation is easier to solve, but either relation could be used to get the y-coordinates.

Example 3

Students may wish to verify that the intensities of speakers A and B are the same at any point on the circle. Suggest that they investigate the point on the x-axis between speakers A and B. The intensity level from each speaker at the point $(\sqrt{200}, 0)$ is approximately $\frac{I}{17.16}$. Have students simplify the initial equation given in the example.
From the initial equation,
$(\sqrt{(x-20)^2 + (y-0)^2})^2$
$= 2(\sqrt{(x-10)^2 + (y-0)^2})^2$.
Square both sides as indicated.
$(x-20)^2 + (y-0)^2$
$= 2((x-10)^2 + (y-0)^2)$, or
$x^2 - 40x + 400 + y^2$
$= 2(x^2 - 20x + 100 + y^2)$.
This simplifies to
$x^2 - 40x + 400 + y^2$
$= 2x^2 - 40x + 200 + 2y^2$.
Combining like terms,
$x^2 + y^2 = 200$.

Extra Examples

Here are additional examples similar to **Examples 1–2**.

1. Write an equation of the circle with its center at the origin and the given point on the circle.
   a. (4, 5) b. (−11, 13)
   a. $x^2 + y^2 = 41$
   b. $x^2 + y^2 = 290$

2. Find the points of intersection of the graphs of the following equations.
   a. $x^2 + y^2 = 25$ and $y = \frac{2}{3}x + 2$
   b. $x^2 + y^2 = 13$ and $y = x + 1$

a. (3, 4) and $\left(-\frac{63}{13}, -\frac{16}{13}\right)$;
b. (−3, −2) and (2, 3).

Check Understanding

1. Write the standard form of the equation for a circle with its center at the origin and a radius of k. $x^2 + y^2 = k^2$
2. Explain how the Distance Formula is used to write the equation of a circle given a point on the circle whose center is the origin. See the top of page 576.
3. If two graphs intersect, how many points of intersection must there be? At least one

Communicating about ALGEBRA

EXTEND *Communicating*
Ask students to select a point and verify that the intensity level from each speaker in Example 3 is the same at any point on the circle.

Also have them select a point to justify that Speaker A is louder than Speaker B at any point inside the circle.
Consider the point (−10, −10) on the circle. Substituting shows that the intensity from Speaker A = $\frac{I}{500}$ = the intensity from Speaker B.
Consider the point (−1, −2) inside the circle. Substituting shows that the intensity from Speaker A = $\frac{I}{125}$. The intensity from Speaker B = $\frac{2I}{445}$. $\frac{I}{125} > \frac{2I}{445}$

Two electronic speakers are in a room. One is 10 feet from the center of the room and the other is 20 feet from the center, as shown at the right. The speaker that is 20 feet from the center produces twice the intensity as the speaker that is 10 feet from the center.

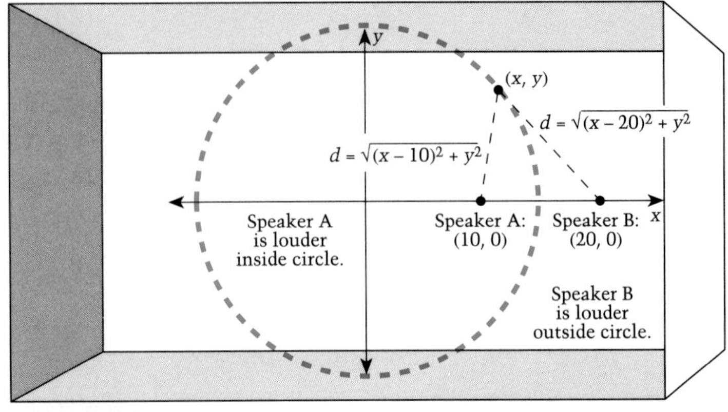

Real Life **Acoustics**

Example 3 *Writing an Equation of a Circle*

Find all points in the room at which the sound from Speaker A has the same intensity as the sound from Speaker B. (Use a two-dimensional model.)

Solution Let I represent the power of Speaker A at (10, 0). Then $2I$ represents the power of Speaker B at (20, 0). Knowing that the intensity of sound at a point varies inversely as the square of the distance between the point and the source (see page 527), you can conclude that the intensity at a point (x, y) is as follows.

$$\frac{I}{(\sqrt{(x-10)^2 + (y-0)^2})^2} \quad \textit{Intensity from Speaker A}$$

$$\frac{2I}{(\sqrt{(x-20)^2 + (y-0)^2})^2} \quad \textit{Intensity from Speaker B}$$

The points at which these two intensities are equal are given by

$$\frac{I}{(\sqrt{(x-10)^2 + (y-0)^2})^2} = \frac{2I}{(\sqrt{(x-20)^2 + (y-0)^2})^2}.$$

Simplifying this equation produces $x^2 + y^2 = 200$. Thus at any point on the circle $x^2 + y^2 = 200$, you would hear sound of equal intensity from the two speakers. ■

Communicating about ALGEBRA

SHARING IDEAS about the Lesson

Complete the "missing steps" in Example 3 by showing that the following equation simplifies to $x^2 + y^2 = 200$.

$$\frac{I}{(\sqrt{(x-10)^2 + (y-0)^2})^2} = \frac{2I}{(\sqrt{(x-20)^2 + (y-0)^2})^2}$$

See Additional Answers.

EXERCISES

Guided Practice

CRITICAL THINKING about the Lesson

1. What is the radius of the circle $x^2 + y^2 = 1$? Where is the center? 1, (0, 0)
2. Write the standard form of the equation of the circle that passes through $(6, -4)$ and whose center is the origin. Explain the strategy you used. $x^2 + y^2 = 52$, $r^2 = (6 - 0)^2 + (-4 - 0)^2 = 52$
3. Sketch the graphs of $x^2 + y^2 = 10$ and $y = 2x + 1$ in the same coordinate plane. Are there any points of intersection? Yes, 2. See Additional Answers for graphs.
4. Give an algebraic verification of your answer to Exercise 3. See Additional Answers.

Independent Practice

In Exercises 5–8, write the standard form of the equation of the circle with the given radius and whose center is the origin. See below.

5. $r = 4$
6. $r = \frac{1}{2}$
7. $r = \sqrt{3}$
8. $r = 1 + \sqrt{2}$

In Exercises 9–12, write the standard form of the equation of the circle that passes through the given point and whose center is the origin. See below.

9. $(-2, -4)$
10. $(0, 6)$
11. $(-5, 1)$
12. $(-3, -1)$

In Exercises 13–16, match the equation with its graph.

13. $x^2 + y^2 = 16$ b
14. $x^2 + y^2 = 5$ d
15. $x^2 + y^2 = 4$ a
16. $x^2 + y^2 = 25$ c

a. 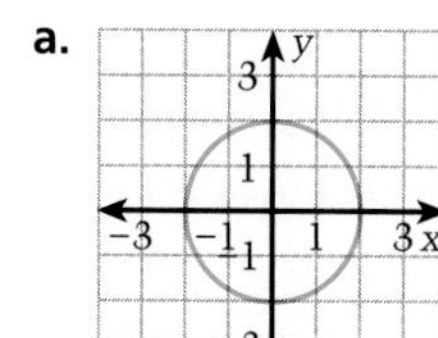

b. 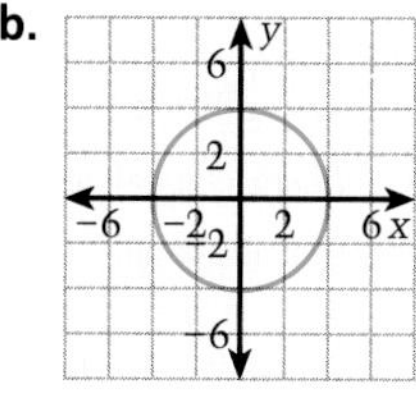

c. 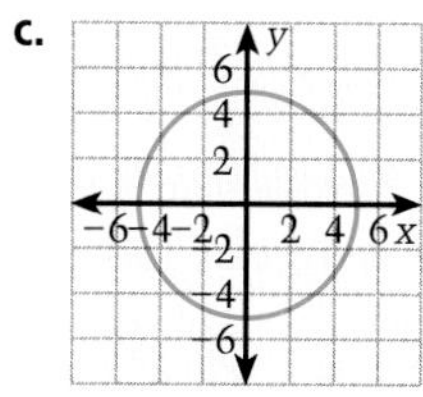

d. 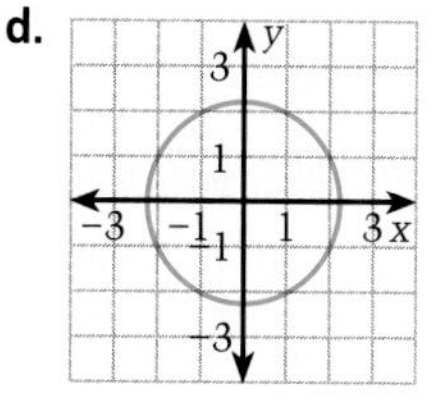

In Exercises 17–28, sketch the graph of the equation. See Additional Answers.

17. $x^2 + y^2 = 2$
18. $x^2 + y^2 = 7$
19. $x^2 + y = 0$
20. $x^2 + y^2 = 9$
21. $x^2 + y^2 = 36$
22. $x^2 + y^2 = 49$
23. $x^2 + y^2 = 8$
24. $4x^2 + y = 0$
25. $6x^2 + 6y^2 = 144$
26. $7x^2 + 7y^2 = 252$
27. $20x^2 + 20y^2 = 400$
28. $3x^2 + 3y^2 = 48$

In Exercises 29–36, find the points of intersection, if any, of the graphs. See margin.

29. $x^2 + y^2 = 1$, $x + y = -1$
30. $x^2 + y^2 = 100$, $x + y = 0$
31. $x^2 + y^2 = 100$, $x + y = 14$
32. $x^2 + y^2 = 9$, $x - 3y = 3$
33. $x^2 + y^2 = 4$, $x - y = 1$
34. $x^2 = -12y$, $2x - y = 15$
35. $x^2 = 8y$, $y = -x$
36. $y^2 = 16x$, $4x - y = -24$

5. $x^2 + y^2 = 16$ 6. $x^2 + y^2 = \frac{1}{4}$ 7. $x^2 + y^2 = 3$ 8. $x^2 + y^2 = 3 + 2\sqrt{2}$ 9. $x^2 + y^2 = 20$
10. $x^2 + y^2 = 36$ 11. $x^2 + y^2 = 26$ 12. $x^2 + y^2 = 10$

EXERCISE Notes

ASSIGNMENT GUIDE

Basic/Average: Ex. 1–4, 7, 11, 13–16, 21–31 odd, 38, 42, 53, 55–60

Above Average: Ex. 1–4, 7, 9, 13–16, 19–33 odd, 37–41, 53–61

Advanced: Ex.1–4, 7, 9, 13–16, 19–33 odd, 37–42, 54–61

Selected Answers
Exercises 1–4, 5–55 odd

Use **Mixed Review** as needed.

✪ More Difficult Exercises
Exercises 39–42, 57–61

Guided Practice

▶ **Ex. 3–4** Use these exercises to connect the visual and the algebraic. The graphs need only be quick sketches that verify the intersection points as determined by the algebraic solution.

Independent Practice

▶ **Ex. 13–16** These exercises should be assigned as a group. Students should justify their answers.

Answers

29. $(-1, 0), (0, -1)$
30. $(-5\sqrt{2}, 5\sqrt{2}), (5\sqrt{2}, -5\sqrt{2})$
31. $(6, 8), (8, 6)$
32. $\left(-\frac{12}{5}, -\frac{9}{5}\right), (3, 0)$
33. $\left(\frac{1 + \sqrt{7}}{2}, \frac{-1 + \sqrt{7}}{2}\right), \left(\frac{1 - \sqrt{7}}{2}\right), \left(\frac{-1 - \sqrt{7}}{2}\right)$
34. $(-30, -75), (6, -3)$
35. $(-8, 8), (0, 0)$
36. No points of intersection

▶ **Ex. 17–28** Students could use a compass to sketch neat graphs of the circles.

▶ **Ex. 29–36** Students should use Example 2 as a solution model. Encourage students to sketch the graphs of the equations on grid paper before finding an algebraic solution.

▶ **Ex. 39–40** Assign these exercises as a pair. Students should use Example 3 as a solution model.

Answers

37. $x^2 + y^2 = \frac{2,400,000}{\pi}$

41. The points that lie on the line segment with endpoints $\left(\frac{4}{5}, -\frac{3}{5}\right)$ and $\left(-\frac{3}{5}, -\frac{4}{5}\right)$, but not including the endpoints.

42. Vertices: $(-10, 0)$, $(10, 0)$, $(6, 8)$; side lengths: 20, $\sqrt{80}$, $\sqrt{320}$; $(\sqrt{80})^2 + (\sqrt{320})^2 = 20^2$

37. *Desert Irrigation* Each of the circular fields shown at the right has an area of about 2,400,000 square yards. Write an equation that represents the boundary of one of the fields. Let (0, 0) represent the center. See margin.

This irrigation project in Libya enables farmers to raise crops in the desert. Water from deep wells is pumped to sprinklers that rotate in circular patterns.

38. *Communications Satellites* Most communications satellites have *synchronous* orbits, which means they circle the earth at exactly the same speed as Earth is rotating. For a viewer on Earth, such a satellite appears to be stationary. Find an equation that represents the circular orbit of a satellite placed 22,240 miles above Earth. Let (0, 0) represent the center. (The radius of Earth is about 3960 miles.) $x^2 + y^2 = 26,200^2$

Two Lights in a Room **In Exercises 39 and 40, use the following information.**

A room contains two ceiling lights. In a coordinate system in which the center of the ceiling is the origin, the lights are located at (0, 8) and (0, 16). The light that is farthest from the center of the room is twice as bright as the other light.

39. Find an equation that represents all points on the ceiling that receive the same illumination of light from each source. (*Hint:* Recall from Lesson 10.2 that the illumination of light at a point varies inversely with the square of the distance between the point and the source.) $x^2 + y^2 = 128$

40. Which of the lights is brighter at the point (8, 8)? Neither

41. *Busing Boundary* To be eligible to ride the school bus to East High School, a student must live at least 1 mile from the school. Describe the portion of Clark Street for which the residents are *not* eligible to ride the school bus. Use a coordinate system in which the school is at (0, 0) and each unit represents 1 mile. See margin.

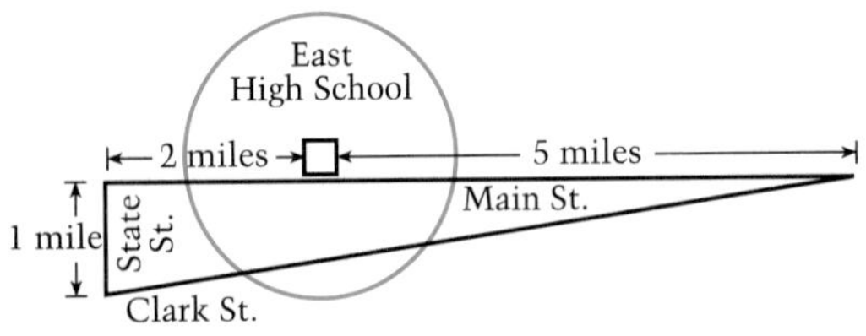

42. *Geometry* A theorem from geometry states that if a triangle is inscribed in a circle so that one side of the triangle is a diameter of the circle, then the triangle is a right triangle. Show this theorem is true for the circle $x^2 + y^2 = 100$ and the triangle formed by the lines $y = 0$, $y = \frac{1}{2}x + 5$, and $y = -2x + 20$. (Find the vertices of the triangle and verify that it is a right triangle.) See margin.

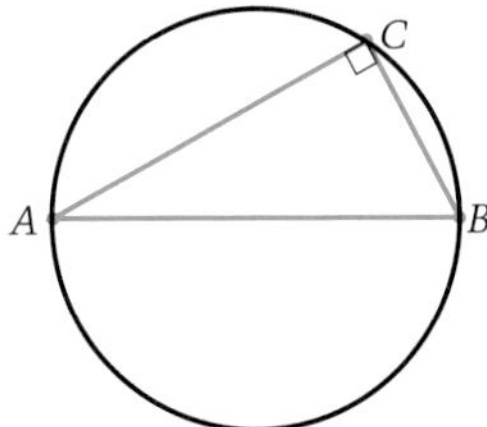

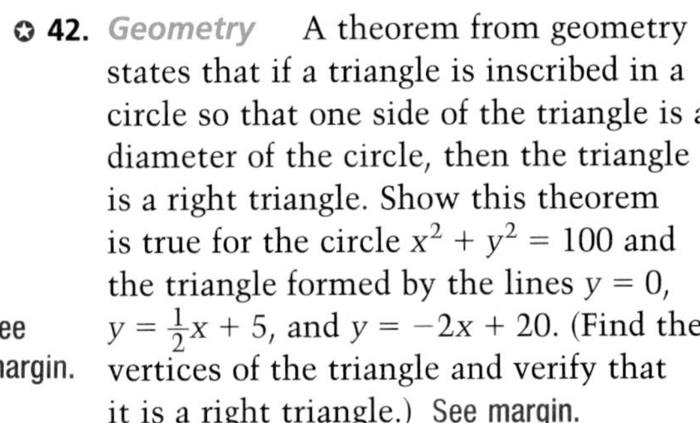

Integrated Review

In Exercises 43–48, find the distance between the two points.

43. (1, 2), (6, 3) $\sqrt{26}$

44. (3, 0), (−1, −2) $\sqrt{20}$

45. (−3, 2), (1, 3) $\sqrt{17}$

46. (5, −5), (1, 3) $\sqrt{80}$

47. (0, 0), (9, 1) $\sqrt{82}$

48. (4, 3), (−1, 6) $\sqrt{34}$

In Exercises 49–54, find the point of intersection of the lines.

49. $\begin{cases} x + 2y = 1 \\ 2x - 3y = 16 \end{cases}$ (5, −2)

50. $\begin{cases} -x + 4y = 11 \\ 3x + 5y = 18 \end{cases}$ (1, 3)

51. $\begin{cases} 4x + y = 4 \\ 9x - 5y = -20 \end{cases}$ (0, 4)

52. $\begin{cases} 3x + y = 3 \\ -x - 5y = 13 \end{cases}$ (2, −3)

53. $\begin{cases} x + 5y = -9 \\ -2x + 3y = -8 \end{cases}$ (1, −2)

54. $\begin{cases} 2x + y = -1 \\ x - 3y = 10 \end{cases}$ (1, −3)

55. *Powers of Ten* Which best estimates the diameter of a bicycle tire? **d**

a. 2.5×10^{-2} in. **b.** 2.5×10^{-1} in.
c. 2.5×10^{0} in. **d.** 2.5×10^{1} in.

56. *Powers of Ten* Which best estimates the radius of a penny? **c**

a. 1.0×10^{-1} mm **b.** 1.0×10^{0} mm
c. 1.0×10^{1} mm **d.** 1.0×10^{2} mm

Exploration and Extension

In Exercises 57–60, match the equation with its graph.

57. $|x|^{1/2} + |y|^{1/2} = 1$ **b** **58.** $|x|^{1} + |y|^{1} = 1$ **d** **59.** $|x|^{2} + |y|^{2} = 1$ **c** **60.** $|x|^{4} + |y|^{4} = 1$ **a**

a.
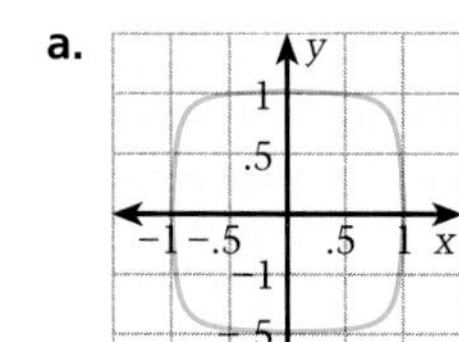

b.
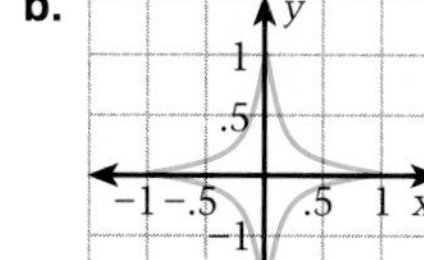

c.
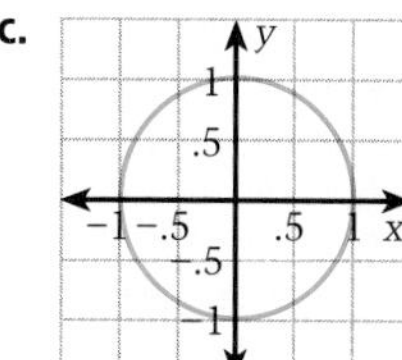

d.
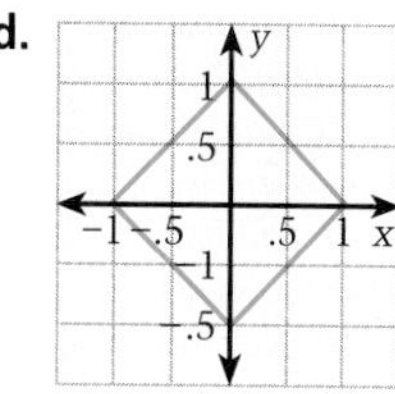

61. *Water Lily* In his novel, *Kavenaugh,* Henry Wadsworth Longfellow stated the following puzzle about a water lily. **17.05 cm**

When the stem of the water lily is vertical, the blossom is 10 centimeters above the surface of a lake. If you pull the lily to one side, keeping the stem straight, the blossom touches the water at a spot 21 centimeters from where the stem formerly cut the surface. How deep is the water?

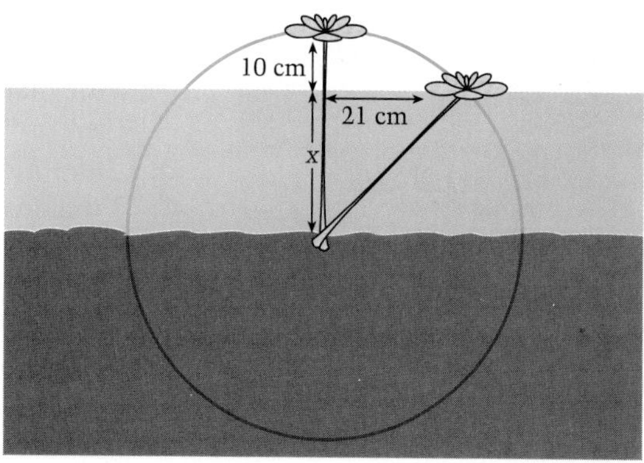

EXTEND Ex. 43–48 Ask students to explain the connection between the Distance Formula and the equation of a circle. The Distance Formula can be used to find the radius of a circle.

▶ **Ex. 57–60** These exercises should be assigned as a group. Students should justify their answers.

▶ **Ex. 61** Remind students that the stem that subsequently touches the water surface represents a radius of the circle and has a length of $x + 10$.

Enrichment Activity

This activity requires students to go beyond lesson goals.

CONNECTIONS Students who have completed a year of geometry have already studied the properties of a circle.

List any properties of circles that you can recall. Draw a diagram to accompany your properties. Share your list with a classmate. How many properties did you have in common?

COOPERATIVE LEARNING Work with your partner. How many more properties can you recall?

Some properties that students might recall include:

- Three points that are not in a straight line determine a circle.
- The radius of a circle that is perpendicular to a chord bisects the chord.
- The measure of an inscribed angle of a circle is half the measure of its intercepted arc.
- The diameter of a circle is the longest chord of a circle.
- An inscribed triangle with one side a diameter is a right triangle. (See Ex. 42.)

Students might also recall the relationship between a circle and constructions.

Mixed REVIEW

1., 6., 13., 14. See Additional Answers. 4. −4 18. $\frac{2x^2 - 2x - 6}{(x-1)(x-3)(x+3)}$

1. Sketch $-4 < x \le 2$. **(1.6)**
2. Solve $|2x + 3| < 4$. **(1.7)** $-3\frac{1}{2} < x < \frac{1}{2}$
3. Write 3.72×10^5 in standard form. 372,000
4. Evaluate $f(x) = -x^2$ when $x = -2$. **(6.1)**
5. 15 is $\frac{5}{9}$ of what? 27
6. Sketch $y = -2x^2 + 4x - 3$. **(5.2)**
7. Find the intercepts of $2x + 4 = x - 4y$. **(5.2)** (0, −1) and (−4, 0)
8. Find the vertex of $y = 3x^2 - 9x + 5$. **(5.2)** $\left(\frac{3}{2}, -\frac{7}{4}\right)$
9. Write the inverse of $f(x) = \frac{2}{3}x - 5$. **(6.3)** $g(x) = \frac{3}{2}x + 7\frac{1}{2}$
10. Find the slope of $5x + 2y - 7 = 8x - 3y$. **(2.2)** $\frac{3}{5}$
11. Find $\frac{5}{8}$ of $\frac{4}{15}$. $\frac{1}{6}$
12. Simplify: $\left(\frac{1}{3}\right)^{-2}$. **(7.1)** 9
13. Sketch $y = \sqrt[3]{x+1} - 4$. **(7.6)**
14. Sketch $f(x) = e^{x+5}$. **(8.1)**
15. Evaluate $\log_3 9 - \log_2 8$. **(8.2)** −1
16. Factor $4x^2 - 20x + 25$. **(9.3)** $(2x - 5)^2$
17. Solve $\frac{x+6}{3} = \frac{6}{5-x}$. **(10.4)** −4, 3
18. Add: $\frac{x+2}{x^2 + 2x - 3} + \frac{x}{x^2 - 9}$. **(10.5)**
19. If 100 is increased by 10%, and then decreased by 10%, what remains? 99
20. Solve for y: $\frac{x+y}{y} = \frac{2x}{5}$. **(1.5)** $y = \frac{5x}{2x-5}$

Career Interview

Sales Coordinator

Teresa Carrera-Hanley is a sales coordinator in the foreign languages department of a publishing company. "About 60% of my time is spent in schools, talking to teachers and students."

Q: ***What lead you into this career?***

A: I grew up in Ecuador, so Spanish was my first language. In this country, I taught Spanish before moving into publishing. This career combines my business and multi-lingual background.

Q: ***Did you like math in school?***

A: I have always liked math. My grandfather who was Chinese used the abacus to teach me math and to play math games. In school, I earned a lot of respect from my friends because I was good in math. Whenever there was a difficult math problem, everyone would say, "Let's ask Teresa," then we would sit down and solve it together.

Q: ***Does having a math background help you on the job?***

A: I majored in economics in college. My math background was very helpful to me. On the job, I use a lot of statistics. We have many variables to consider when reviewing data. To plan well, I need to interpret the data accurately.

Q: ***What would you like to tell kids about math?***

A: If you don't understand, never hesitate to ask for help.

11.3 Ellipses

What you should learn:

Goal 1 How to write an equation of an ellipse and sketch its graph

Goal 2 How to use ellipses to solve real-life problems

Why you should learn it:

You can model many shapes, such as archways, with elliptical models.

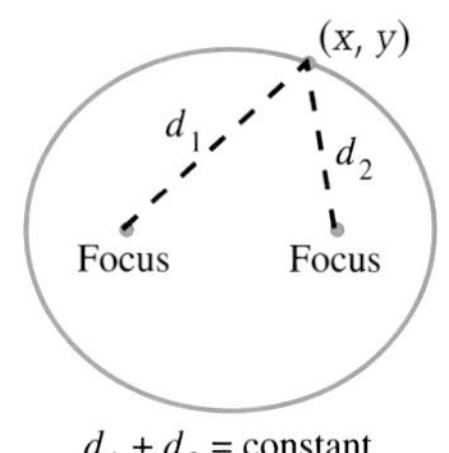

$d_1 + d_2 =$ constant

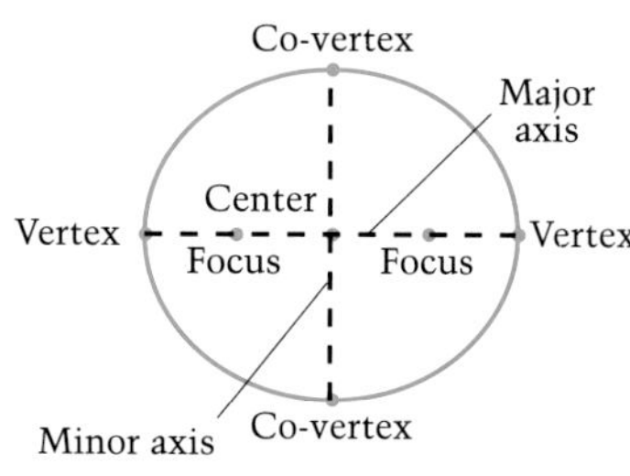

Goal 1 Writing an Equation of an Ellipse

An **ellipse** is the set of all points (x, y) such that the sum of the distances between (x, y) and two distinct fixed points, the **focuses** or **foci,** is a constant. The line through the foci intersects the ellipse at two points, the **vertices.** The line segment joining the vertices is the **major axis,** and its midpoint is the **center** of the ellipse. The line segment perpendicular to the major axis at the center is the **minor axis** of the ellipse. The endpoints of the minor axis are the **co-vertices.**

Standard Equation of an Ellipse (Center at Origin)

The **standard form of the equation of an ellipse** with center at (0, 0) and major and minor axes of lengths $2a$ and $2b$, where $a > b$, is as follows.

$$\frac{x^2}{a^2} + \frac{y^2}{b^2} = 1 \quad \textit{Horizontal major axis}$$

$$\frac{x^2}{b^2} + \frac{y^2}{a^2} = 1 \quad \textit{Vertical major axis}$$

The vertices lie on the major axis, a units from the center. The co-vertices lie on the minor axis, b units from the center.

The foci of the ellipse lie on the major axis, c units from the center, where $c^2 = a^2 - b^2$.

The two orientations of an ellipse are shown below.

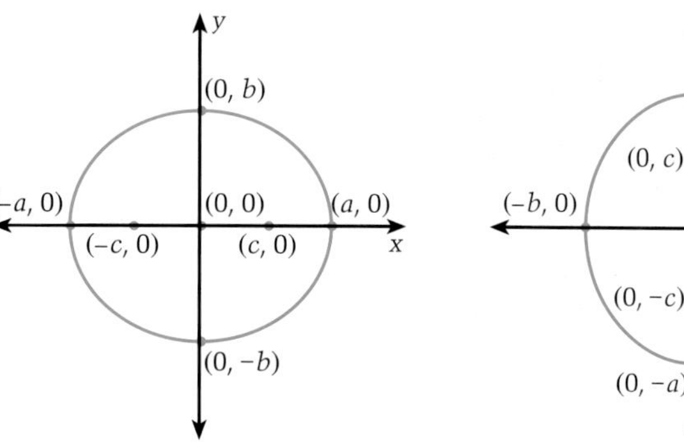

Problem of the Day

A train goes from Boston to New York averaging 40 km per hour and returns averaging 60 km per hour. What is the average speed (it is not 50 km per hour) for the round trip?
48 km per hour

ORGANIZER

Warm-Up Exercises

Find the missing value of the relationship $c^2 = a^2 - b^2$ for given values of a, b, or c.

a. Find c, given $a = 9$, $b = 5$.
b. Find c, given $a = 6$, $b = 2$.
c. Find b, given $a = 13$, $c = 8$.

a. $c = \sqrt{56}$, **b.** $c = \sqrt{32}$, **c.** $b = \sqrt{105}$

Lesson Resources

Teaching Tools
- Transparencies: 7, 8
- Problem of the Day: 11.3
- Warm-up Exercises: 11.3
- Answer Masters: 11.3

Extra Practice: 11.3
Technology Handbook: p. 79

LESSON Notes

Example 1

Ask students to state the lengths of the major and minor axes (major axis: 6 units; minor axis: 4 units).

Suppose the major and minor axes were allowed to be the same length. Ask students to describe the "ellipse" if $a = b$. You would have a circle!

Examples 1–2

Be sure that students understand how to determine which axis of a coordinate plane is the major axis for an ellipse with its center at the origin. In Example 1, point out that, because the positive x-coordinate is larger than the positive y-coordinate, the major axis is on the x-axis. Similarly, in Example 2, because one-half of the width of the gateway (on the x-axis) is larger than the height of the gateway (on the y-axis), the major axis is on the x-axis of the superimposed coordinate plane.

Example 3

Ask students to explain how the system of linear equations was determined. The Sun is located at one of the foci of the ellipse. Therefore, it is "c" units (in millions of kilometers) from the center. Mercury is closest to the Sun when it is on the same side of the center as the Sun. But then it is "a" units from the center. Because the Sun is then between the center and Mercury, the distance from the Sun to Mercury is $a - c$. Or $a - c = 46.04$. Similarly, the larger distance is $a + c = 69.86$.

Extra Examples

Here are additional examples similar to **Examples 1–2.**

1. Write an equation of the ellipse whose vertices are $(-5, 0)$ and $(5, 0)$ and whose co-vertices are $(0, -3)$ and $(0, 3)$. Find the foci of the ellipse.
   $\frac{x^2}{25} + \frac{y^2}{9} = 1$; foci: $(-4, 0)$ and $(4, 0)$

Example 1 *Writing an Equation of an Ellipse*

Write an equation of the ellipse whose vertices are $(-3, 0)$ and $(3, 0)$ and whose co-vertices are $(0, -2)$ and $(0, 2)$. Find the foci of the ellipse.

Solution Begin by graphing the four vertices and co-vertices. The center of the ellipse is $(0, 0)$ because that is the point that lies halfway between the vertices (or halfway between the co-vertices). The equation of the ellipse has the form

$$\frac{x^2}{a^2} + \frac{y^2}{b^2} = 1. \quad \textit{Horizontal major axis}$$

The distance between the center and either vertex is $a = 3$. The distance between the center and either co-vertex is $b = 2$. Thus the standard equation of the ellipse is

$$\frac{x^2}{3^2} + \frac{y^2}{2^2} = 1 \quad \text{or} \quad \frac{x^2}{9} + \frac{y^2}{4} = 1.$$

The foci lie on the major axis, c units from the center, where $c^2 = 3^2 - 2^2 = 5$. The coordinates of the foci are $(-\sqrt{5}, 0)$ and $(\sqrt{5}, 0)$. ■

Goal 2 Using Ellipses in Real-Life Problems

Example 2 *Building an Archway*

You are responsible for designing an archway whose opening is an ellipse, as shown at the left. The height of the archway is 10 feet and the width is 30 feet. Find an equation of the ellipse and use the equation to sketch an accurate drawing of the archway.

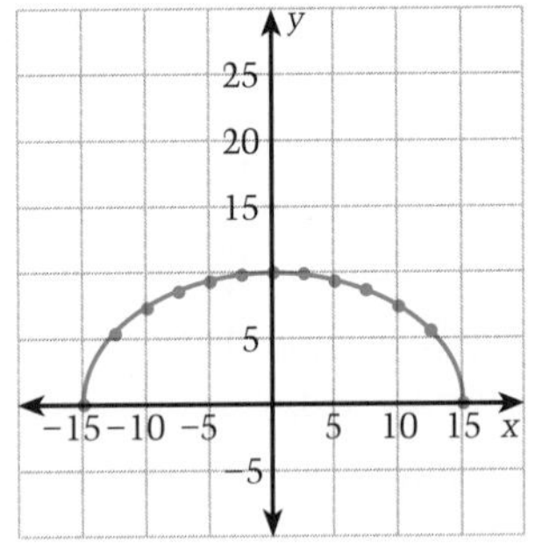

Solution Locate the origin at the center of the base of the archway. Because the major axis is horizontal, you know that $a = 15$ and $b = 10$, which implies that the equation is

$$\frac{x^2}{15^2} + \frac{y^2}{10^2} = 1 \quad \text{or} \quad \frac{x^2}{225} + \frac{y^2}{100} = 1.$$

To make an accurate sketch of the ellipse, solve this equation for (positive) y, and construct a table of values.

$$y = 10\sqrt{1 - \frac{x^2}{225}}$$

***x*-value**	±15	±12.5	±10	±7.5	±5	±2.5	0
***y*-value**	0	5.53	7.45	8.66	9.43	9.86	10

Using the values in the table, you can sketch the graph shown at the left. ■

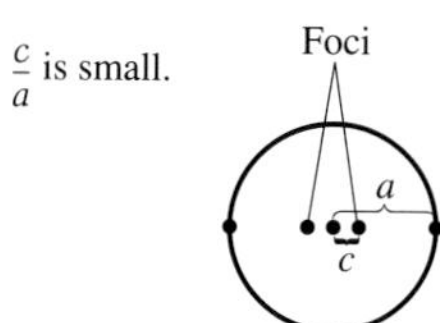

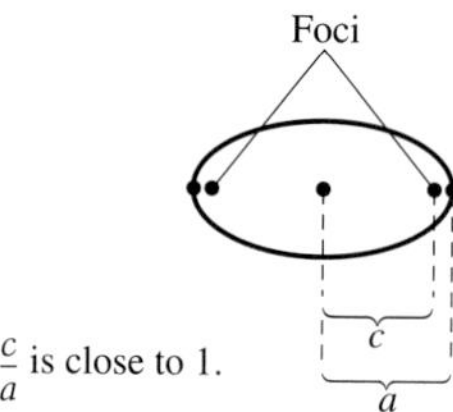

The **eccentricity** of an ellipse is a measure of its ovalness. The eccentricity is given by $\frac{c}{a}$. If the ellipse is nearly circular, then the foci are close to the center and the eccentricity $\frac{c}{a}$ is close to 0. If the ellipse is very elongated, then the foci are close to the vertices and the eccentricity is close to 1.

All planets have elliptical orbits about the sun. (The sun is a focus of the elliptical orbit, *not* the center.) Some orbits, however, are nearly circular. The eccentricity of each is given below. Notice that the orbit of Venus is the most circular and the orbit of Pluto is the most elongated.

Mercury: 0.2055	Venus: 0.0068	Earth: 0.0167
Mars: 0.0934	Jupiter: 0.0484	Saturn: 0.0543
Uranus: 0.0460	Neptune: 0.0082	Pluto: 0.2481

Connections
Astronomy

Example 3 *The Orbit of Mercury*

In its orbit, Mercury ranges between 46.04 million kilometers and 69.86 million kilometers from the sun. Use this information to write an equation for the orbit of Mercury.

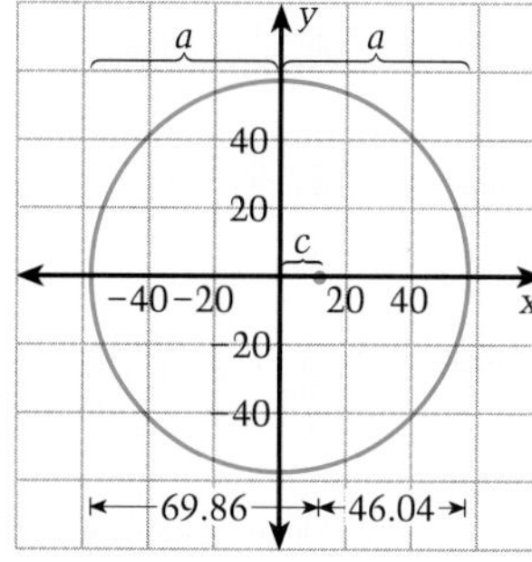

Solution Notice in the diagram at the left that the greatest and least distances occur at the x-intercepts. Thus you can write a system of linear equations involving a and c.

$$\begin{cases} a - c = 46.04 \\ a + c = 69.86 \end{cases}$$

The solution of this system is $a = 57.95$ and $c = 11.91$. From the relationship $c^2 = a^2 - b^2$, you can conclude that

$$b = \sqrt{a^2 - c^2} \approx 56.71.$$

Thus an equation of the elliptical orbit is

$$\frac{x^2}{(57.95)^2} + \frac{y^2}{(56.71)^2} = 1,$$

where the units for x and y are millions of kilometers. ■

Communicating about ALGEBRA

SHARING IDEAS about the Lesson

A. The closest that Pluto gets to the sun is 4.443 billion kilometers. Use this and the eccentricity of Pluto to find the farthest that Pluto gets from the sun. 7.375 billion km

B. Find an equation for the orbit of Pluto. See below.

B. $\frac{x^2}{5.909^2} + \frac{y^2}{5.724^2} = 1$ where x and y are billions of kilometers

2. Write an equation of the ellipse whose vertices are $(0, -13)$ and $(0, 13)$ and whose co-vertices are $(-12, 0)$ and $(12, 0)$. Find the foci of the ellipse.
$\frac{x^2}{144} + \frac{y^2}{169} = 1$; foci: $(0, -5)$ and $(0, 5)$

3. In its orbit, the earth ranges between 147.09 million kilometers and 152.24 million kilometers from the sun. Write an equation of the orbit of the earth.
$\frac{x^2}{(149.67)^2} + \frac{y^2}{(149.64)^2} = 1$

Check Understanding

1. What do you know about the sum of the distances between the foci of an ellipse and any point on the ellipse? The sum of the distances is a constant.
2. What is the major axis of an ellipse? How does it differ from the minor axis? See page 583.
3. What is the standard form of the equation of an ellipse? See page 583.
4. What is the relationship between the foci and the center of an ellipse? The center of an ellipse is the midpoint of the segment joining the foci.

Communicating about ALGEBRA

The farthest that Uranus gets from the sun is 2.993 billion kilometers. Use the eccentricity of Uranus to find the closest that Uranus gets to the sun. Find an equation for the orbit of Uranus.
The closest that Uranus gets to the sun is about 2.730 billion kilometers. An equation for the orbit of Uranus is
$\frac{x^2}{2.861^2} + \frac{y^2}{2.858^2} = 1.$

EXERCISE Notes

ASSIGNMENT GUIDE

Basic/Average: Ex. 1–6, 11, 17, 21, 25–28, 31–39 odd, 47–51, 56, 63–67

Above Average: Ex. 1–6, 11, 17, 21, 25–28, 31–39 odd, 45–54, 56, 66–68

Advanced: Ex. 1–6, 15–25 odd, 26–29, 31–43 odd, 47–56, 68

Selected Answers
Exercises 1–6, 7–67 odd

✪ More Difficult Exercises
Exercises 17–18, 49–56, 68

Guided Practice

▶ **Ex. 1–6** You can use the Guided Practice exercises as an in-class readiness check.

▶ **Ex. 1–2** Encourage students to use their own words, not to just quote the statements given in the text.

▶ **Ex. 5** Remind students that in order to find the foci, they need to use the fact that $c^2 = a^2 - b^2$ to find the value of c.

Independent Practice

▶ **Ex. 13–18** Students need to write the equations in standard form in order to properly locate the values of a, b, and c. Remind students that the right-hand side of the standard-form equation is equal to 1. Also, the coefficients of x^2 and of y^2 in the standard-form equation are both 1. Thus, in Exercise 15, an equation such as $\frac{4x^2}{25} + \frac{9y^2}{25} = 1$ needs to be written as $\frac{x^2}{\frac{25}{4}} + \frac{y^2}{\frac{25}{9}} = 1.$

▶ **Ex. 22–24** Remind students that they need to use the fact that $c^2 = a^2 - b^2$ to find the value of b.

EXERCISES

Guided Practice

CRITICAL THINKING about the Lesson

In Exercises 1–6, use the graph at the right.

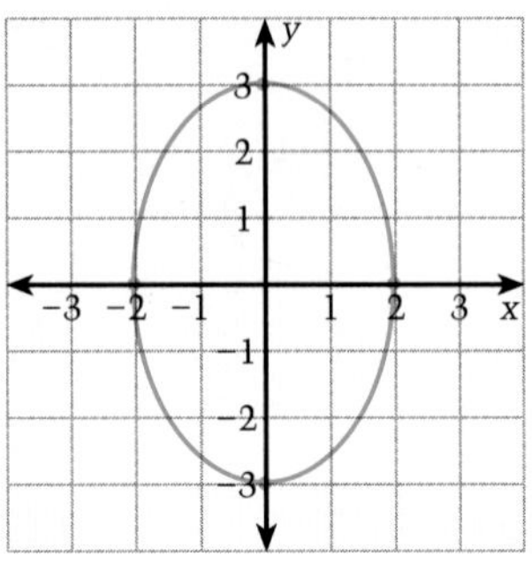

1. Describe the major axis of the ellipse. See below.
2. Describe the minor axis of the ellipse. See below.
3. What are the coordinates of the vertices?
4. What are the coordinates of the co-vertices? See below.
5. What are the coordinates of the foci?
6. Write an equation of the ellipse. $\frac{x^2}{4} + \frac{y^2}{9} = 1$

3. (0, −3) and (0, 3)

5. $(0, -\sqrt{5})$ and $(0, \sqrt{5})$

Independent Practice

In Exercises 7–12, find the foci and vertices of the ellipse. See Additional Answers.

7. $\frac{x^2}{25} + \frac{y^2}{16} = 1$
8. $\frac{x^2}{144} + \frac{y^2}{169} = 1$
9. $\frac{x^2}{4} + \frac{y^2}{36} = 1$
10. $\frac{x^2}{9} + \frac{y^2}{25} = 1$
11. $\frac{x^2}{1} + \frac{y^2}{9} = 1$
12. $\frac{x^2}{81} + \frac{y^2}{9} = 1$

In Exercises 13–18, write the equation in standard form. Find the foci and vertices. See Additional Answers.

13. $49x^2 + 16y^2 = 784$
14. $49x^2 + 64y^2 = 3136$
15. $4x^2 + 9y^2 = 25$
16. $x^2 + 36y^2 = 4$
✪ 17. $\frac{x^2}{600} + \frac{y^2}{2100} = 150$
✪ 18. $\frac{x^2}{288} + \frac{y^2}{200} = 2$

In Exercises 19–24, find an equation of the ellipse. The center is (0, 0). See Additional Answers.

19. Vertex: (0, 6)
Co-vertex: (5, 0)
20. Vertex: (0, 5)
Co-vertex: (−4, 0)
21. Vertex: (−4, 0)
Co-vertex: (0, 3)
22. Vertex: (2, 0)
Focus: (1, 0)
23. Vertex: (4, 0)
Focus: (−3, 0)
24. Vertex: (0, −6)
Focus: (0, 2)

In Exercises 25–28, match the equation with its graph.

25. $\frac{x^2}{225} + \frac{y^2}{100} = 1$ c
26. $\frac{x^2}{121} + \frac{y^2}{25} = 1$ a
27. $\frac{x^2}{225} + \frac{y^2}{400} = 1$ d
28. $\frac{x^2}{1} + \frac{y^2}{25} = 1$ b

a. 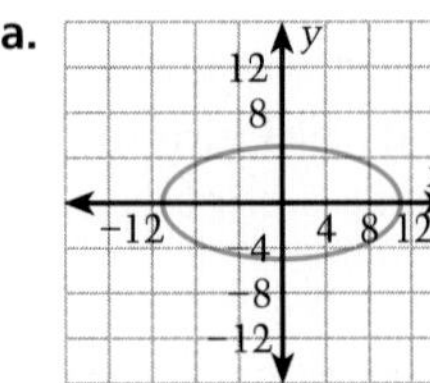

b. 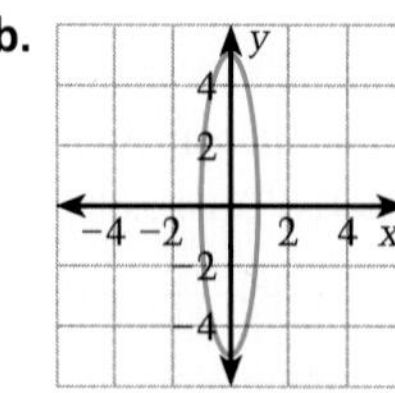

c. 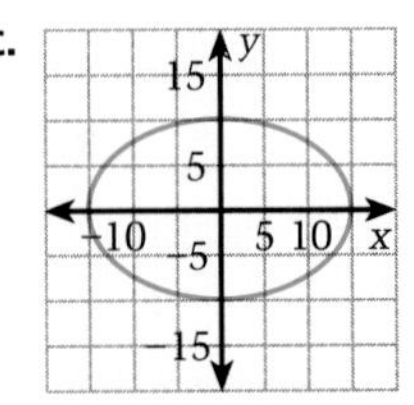

d. 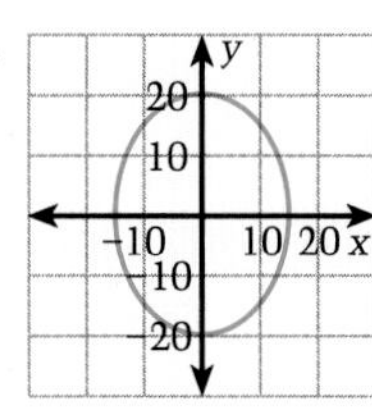

1. Segment with (0, −3) and (0, 3) as its endpoints
2. Segment with (−2, 0) and (2, 0) as its endpoints
4. (−2, 0) and (2, 0)

In Exercises 29–40, sketch the ellipse and find its eccentricity. See Additional Answers for graphs.

29. $\frac{x^2}{16} + \frac{y^2}{36} = 1$ ≈0.7454

30. $\frac{x^2}{4} + \frac{y^2}{25} = 1$ ≈0.9165

31. $\frac{x^2}{36} + \frac{y^2}{81} = 1$ ≈0.7454

32. $\frac{x^2}{49} + \frac{y^2}{121} = 1$ ≈0.7714

33. $\frac{x^2}{4} + y^2 = 100$ ≈0.8660

34. $\frac{x^2}{16} + \frac{y^2}{25} = 4$ 0.6

35. $\frac{x^2}{4} + \frac{y^2}{6} = 1$ ≈0.5774

36. $\frac{x^2}{9} + \frac{y^2}{5} = 1$ ≈0.6667

37. $64x^2 + 25y^2 = 1600$ ≈0.7806

38. $400x^2 + 441y^2 = 420^2$ ≈0.3049

39. $\frac{x^2}{4} + \frac{4y^2}{9} = 1$ ≈0.6614

40. $\frac{25x^2}{4} + \frac{25y^2}{9} = 1$ ≈0.7454

In Exercises 41–46, find the points of intersection of the graphs. 46. (0, −5), (−8, 3)

41. $\frac{x^2}{4} + \frac{y^2}{9} = 1$ $(-2, 0), \left(\frac{10}{13}, -\frac{36}{13}\right)$
$y = -x - 2$

42. $\frac{x^2}{16} + y^2 = 1$
$y = \frac{1}{2}(x + 1)$

42., 43. See Additional Answers.

43. $\frac{x^2}{25} + y^2 = 1$
$y = 2x$

44. $\frac{x^2}{9} + \frac{y^2}{4} = 1$ $\left(\frac{6}{\sqrt{13}}, \frac{6}{\sqrt{13}}\right), \left(-\frac{6}{\sqrt{13}}, -\frac{6}{\sqrt{13}}\right)$
$y = x$

45. $\frac{x^2}{16} + \frac{y^2}{4} = 1$
$y = x + 2$ $(0, 2), \left(-\frac{16}{5}, -\frac{6}{5}\right)$

46. $\frac{x^2}{100} + \frac{y^2}{25} = 1$ (0, −5),
$y = -x - 5$ (−8, 3)

47. *Drawing an Ellipse* You can draw an ellipse with a pencil, paper, string, and two thumbtacks. Place the two thumbtacks at the foci. Tie the string in a loop that is $2(a + c)$ units around. Loop the string around the thumbtacks, pull the string taut with a pencil, and trace a path as shown in the diagram at the right. Explain why the path is an ellipse. See below.

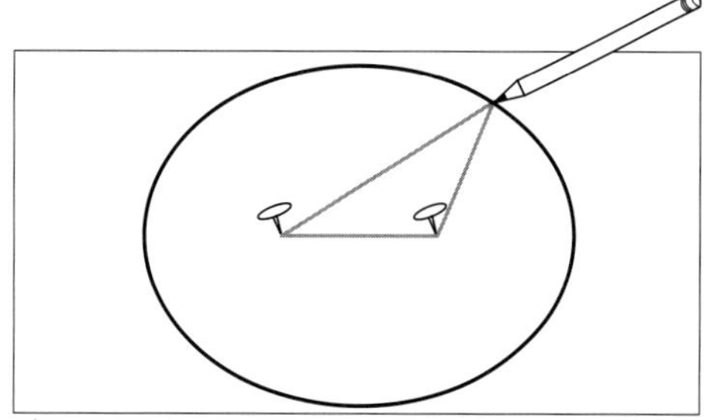

48. *Eccentricity* Suppose you are holding a soda pop can. When the top of the can is facing you, its rim is a circle. When you tip the can, however, the rim of the can appears to become elliptical. What happens to the eccentricity of this "ellipse" the farther you tip the top of the can away from you? It increases, getting closer to 1.

Bicycle Chainwheel **In Exercises 49–51, use the following information.**

The pedals of a bicycle drive a *chainwheel,* which drives a smaller *sprocket wheel* on the rear axle. Many chainwheels are circular. Some, however, are slightly elliptical, which tends to make pedaling easier. The front chainwheel on the bicycle shown at the right is 8 inches at its widest and $7\frac{1}{2}$ inches at its narrowest.

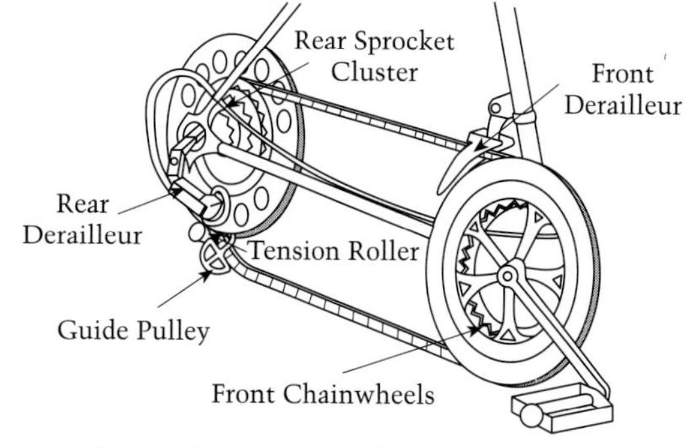

49. Find an equation for the outline of this elliptical chainwheel.

50. What is its eccentricity? ≈0.3480

51. What is the area of the chainwheel? (Use $A = \pi ab$ for an ellipse.) 15π in.^{2}

49. $\frac{x^2}{4^2} + \frac{y^2}{\left(\frac{15}{4}\right)^2} = 1$ or $\frac{x^2}{\left(\frac{15}{4}\right)^2} + \frac{y^2}{4^2} = 1$

47. The sum of the distances between any point on the path and the two thumbstacks is a constant.

▶ **Ex. 25–28** Assign these exercises as a group. Have students state a justification in terms of the orientation of the graph.

▶ **Ex. 29–40** The centers are at the point (0, 0). Refer students to the lesson presentation at the top of page 585 to find the eccentricity $e = \frac{c}{a}$.

▶ **Ex. 41–46** You may have to model an algebraic solution for these exercises before assigning them.

▶ **Ex. 47** **GROUP ACTIVITY** Use this exercise as an in-class activity. Be sure that all students write an explanation as to why the path is an ellipse.

▶ **Ex. 48** Students should use a soda can to act out the exercise.

▶ **Ex. 49–51** Assign these exercises as a group. Students might find the measurements for their own bicycles.

▶ **Ex. 52–54** Assign these exercises as a group. Ask students who have been to Statuary Hall, or any other building that has a "whispering gallery," to describe the experience.

▶ **EXTEND Ex. 56** **RESEARCH** Ask students when Halley's comet will next be seen on earth.
2062–2064

Enrichment Activities

These activities require students to go beyond lesson goals.

TECHNOLOGY Remember, the graphics calculator is designed to graph functions in the form $y = f(x)$. In order to graph equations of circles such as $x^2 + y^2 = 25$, or equations of ellipses such as $\frac{x^2}{4} + \frac{y^2}{25} = 1$ (Exercise 30) using a graphics calculator, you must separate them into halves. For example, consider

$$\frac{x^2}{4} + \frac{y^2}{25} = 1.$$

Solve for y as follows.

$$25x^2 + 4y^2 = 100$$
$$4y^2 = 100 - 25x^2$$
$$4y^2 = 25(4 - x^2)$$

Take the square root of both sides.

$$2y = \pm 5\sqrt{4 - x^2}$$
$$y = \pm\frac{5\sqrt{4 - x^2}}{2}$$

To graph the entire ellipse you need to graph its top half,

$$y_1 = \frac{5\sqrt{4 - x^2}}{2}$$

and then graph its bottom half,

$$y_2 = -\frac{5\sqrt{4 - x^2}}{2}.$$

Use the Trace feature of a graphics calculator to find the points of intersection in Exercises 41–46. (*Hint:* The linear equations would be labeled y_3 and graphed on the same coordinate grid.)

41. (0.769, −2.769), (−2, 0)
42. (0.944, 0.972), (−2.544, −0.772)
43. (0.498, 0.995), (−0.498, −0.995)
44. (1.664, 1.664), (−1.664, −1.664)
45. (0, 2), (−3.2, −1.2)
46. (0, −5), (−8, 3)

Whispering Gallery **In Exercises 52–54, use the following information.**

Statuary Hall is an elliptical room in the United States Capitol in Washington, D.C. The room is 46 feet wide and 96 feet long. Because of a reflective property of an ellipse, a person standing at one focus can hear even a whisper spoken by a person standing at the other focus. (John Quincy Adams is said to have used this feature of the room to overhear conversations.)

52. Find an equation of the ellipse.

53. How far apart are the two foci? ≈84.26 ft

54. What is the floor area of the room? 1104π ft^2

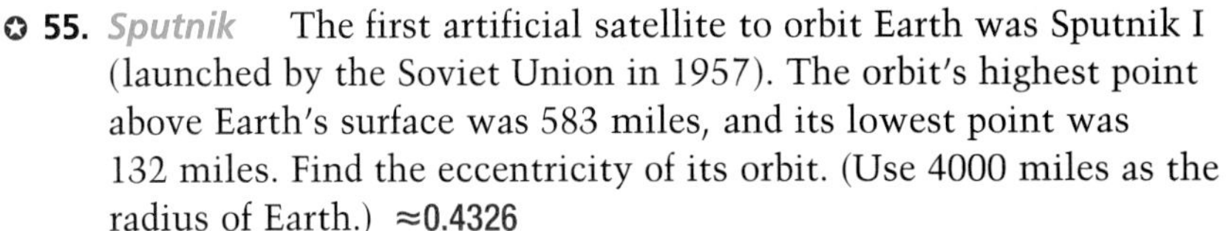

55. *Sputnik* The first artificial satellite to orbit Earth was Sputnik I (launched by the Soviet Union in 1957). The orbit's highest point above Earth's surface was 583 miles, and its lowest point was 132 miles. Find the eccentricity of its orbit. (Use 4000 miles as the radius of Earth.) ≈0.4326

52. $\frac{x^2}{48^2} + \frac{y^2}{23^2} = 1$ or $\frac{x^2}{23^2} + \frac{y^2}{48^2} = 1$

56. *Comet's Orbit* Halley's Comet has an elliptical orbit with the sun at a focus. Its maximum distance from the sun is approximately 35.64 astronomical units and its minimum distance is approximately 0.594 astronomical unit. Find the eccentricity of its orbit. (An astronomical unit is about 93×10^6 miles.) ≈0.9672

Integrated Review

In Exercises 57–62, solve the equation for *y*. 59., 62. See below.

57. $\frac{x^2}{1} + \frac{y^2}{4} = 1$ $y = \pm 2\sqrt{1 - x^2}$

58. $\frac{x^2}{36} + \frac{y^2}{64} = 1$ $y = \pm\frac{4}{3}\sqrt{36 - x^2}$

59. $\frac{x^2}{4} + \frac{y^2}{9} = 1$

60. $\frac{x^2}{4} + \frac{y^2}{100} = 1$ $y = \pm 5\sqrt{4 - x^2}$

61. $\frac{x^2}{100} + \frac{y^2}{81} = 1$ $y = \pm\frac{9}{10}\sqrt{100 - x^2}$

62. $\frac{x^2}{25} + \frac{y^2}{144} = 1$

In Exercises 63–67, sketch the graph of the equation. See Additional Answers.

63. $x^2 + y^2 = 81$ **64.** $x^2 = -16y$ **65.** $y^2 = 4x$ **66.** $x^2 + y^2 = 100$ **67.** $y^2 = 20x$

Exploration and Extension

68. *Geometry* For the ellipse at the right, a is half the length of the major axis. Write an expression, in terms of a, for the perimeter of each triangle in which one vertex is a focus and the opposite side contains the other focus. $4a$

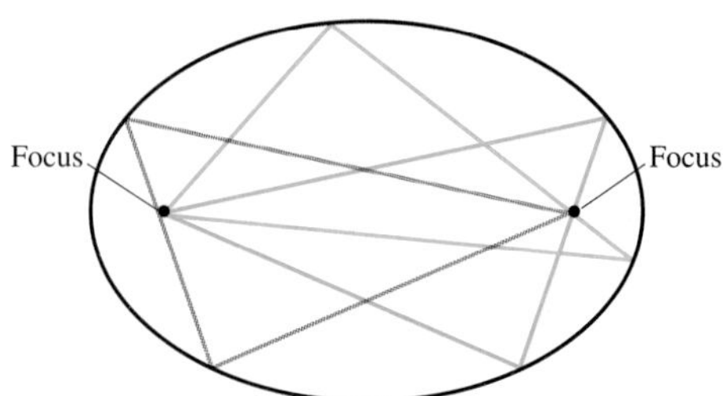

59. $y = \pm\frac{3}{2}\sqrt{4 - x^2}$

62. $y = \pm\frac{12}{5}\sqrt{25 - x^2}$

Mid-Chapter SELF-TEST

Take this test as you would take a test in class. The answers to the exercises are given in the back of the book.

In Exercises 1 and 2, find the focus and the directrix of the parabola. (11.1)

1. $16y = x^2$ (0, 4), $y = -4$

2. $-12x = y^2$ (−3, 0), $x = 3$

In Exercises 3 and 4, write an equation of the parabola whose vertex is at the origin and has the given focus. (11.1)

3. $(0, -4)$ $-16y = x^2$

4. $\left(\frac{1}{2}, 0\right)$ $2x = y^2$

In Exercises 5–10, sketch the graph of the equation. (11.1, 11.2, 11.3) 9.–10. See Additional Answers.

5. $20y = x^2$ See margin.

6. $-4x = y^2$ See margin.

7. $x^2 + y^2 = 49$ See margin.

8. $4x^2 + 4y^2 = 36$ See margin.

9. $\dfrac{x^2}{16} + \dfrac{y^2}{4} = 1$

10. $9x^2 + 4y^2 = 36$

11. Write an equation of the circle of radius 5 with center at the origin. **(11.2)** $x^2 + y^2 = 25$

12. Write an equation of the circle whose center is the origin and passes through the point $(4, -1)$. **(11.2)** $x^2 + y^2 = 17$

In Exercises 13 and 14, find the foci and the vertices of the ellipse. (11.3)

13. $\dfrac{x^2}{9} + \dfrac{y^2}{25} = 1$ Foci (0, 4), (0, −4), vertices (0, 5), (0, −5)

14. $9x^2 + 16y^2 = 144$ Foci $(-\sqrt{7}, 0)$, $(\sqrt{7}, 0)$, vertices (−4, 0), (4, 0)

In Exercises 15–17, find all intersection points. (11.1, 11.2, 11.3)

15. $y = \frac{1}{12}x^2$, $y = -\frac{1}{6}x + \frac{2}{3}$ $\left(-4, \frac{4}{3}\right)$, $\left(2, \frac{1}{3}\right)$

16. $\dfrac{x^2}{4} + \dfrac{y^2}{1} = 1$, $y = x + 1$ $\left(-\frac{8}{5}, -\frac{3}{5}\right)$, (0, 1)

17. $x^2 + y^2 = 25$, $y = 10 - 2x$ (3, 4), (5, 0)

18. Write the equation in standard form for the ellipse whose co-vertices are (0, 4) and (0, −4) and whose vertices are (6, 0) and (−6, 0). **(11.3)** $\dfrac{x^2}{36} + \dfrac{y^2}{16} = 1$

19. The moon's orbit about Earth is elliptical. The distance between the moon and Earth varies from 363,000 kilometers to 406,000 kilometers. Find the eccentricity of the moon's orbit. **(11.3)** 0.0559

20. The jet of water from a fountain forms a parabola. Write an equation for the path the water follows. **(11.1)** $y = -\frac{2}{5}x^2$

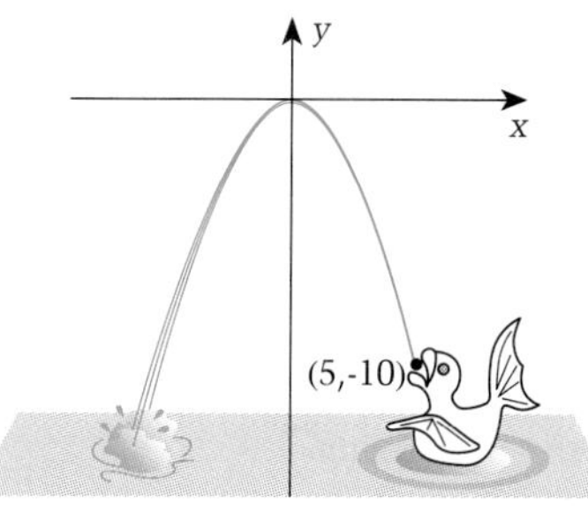

21. A planetarium is sponsoring a craft show in the circular corridor surrounding the auditorium. It is 10 yards from the center of the auditorium to the inside wall of the corridor, and the corridor is 6 yards wide. Write equations for the inside and outside walls of the corridor. **(11.2)** $x^2 + y^2 = 100$, $x^2 + y^2 = 256$

Answers

5.

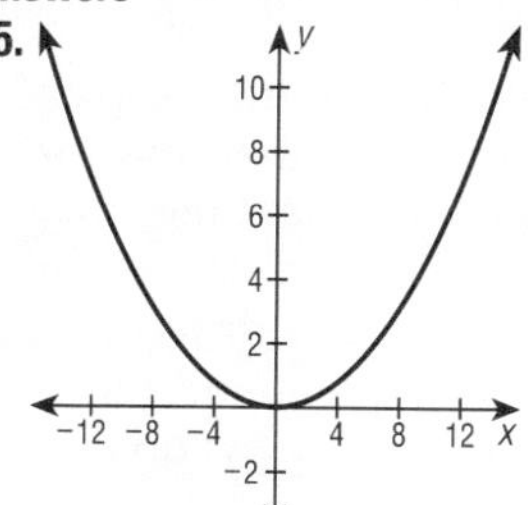

6.

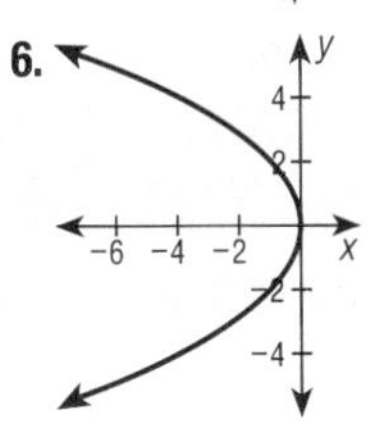

7.

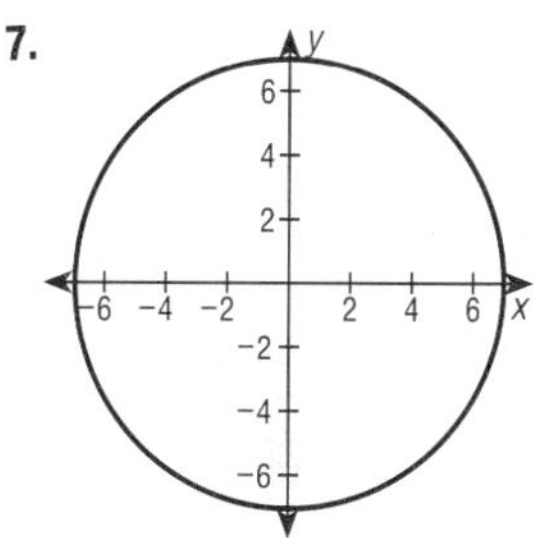

8.

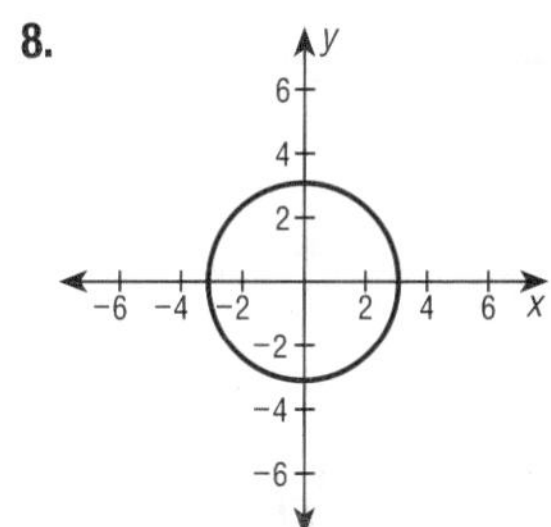

9.

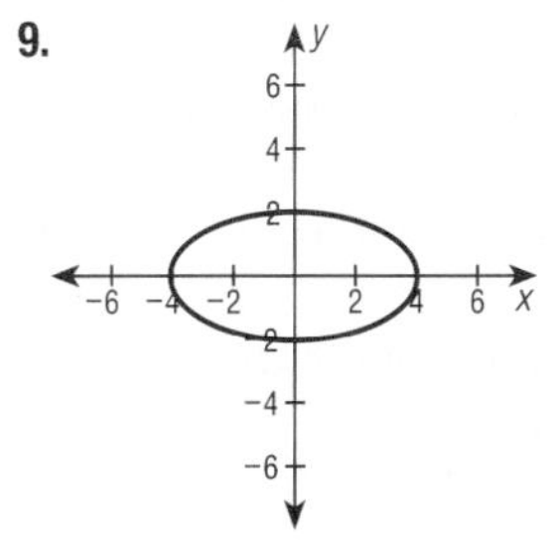

10.

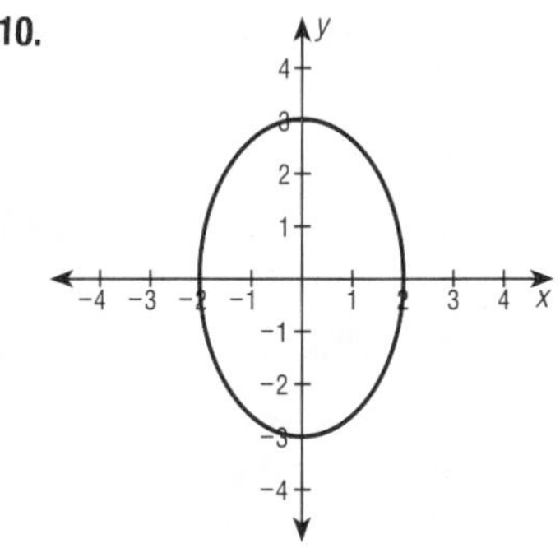

Problem of the Day

Ron and Sam run a race. Ron runs half the time and walks half the time. Sam runs half the distance and walks half the distance. If they both walk and run at the same rates, who wins?
Ron

ORGANIZER

Warm-Up Exercises

1. Find the missing value of the relationship $c^2 = a^2 + b^2$ for given values of a, b, or c.
   a. Find c, given $a = 6$, $b = 5$.
   b. Find c, given $a = 3$, $b = 2$.
   c. Find b, given $a = 4$, $c = 8$.
   d. Find a, given $b = 14$, $c = 17$.

   a. $\sqrt{61}$, b. $\sqrt{13}$, c. $\sqrt{48}$, d. $\sqrt{93}$
2. Rewrite the given equation of an ellipse in standard form.
   a. $4x^2 + 9y^2 = 36$
   b. $16x^2 + 25y^2 = 400$

   a. $\frac{x^2}{9} + \frac{y^2}{4} = 1$
   b. $\frac{x^2}{25} + \frac{y^2}{16} = 1$

Lesson Resources

Teaching Tools
- Transparencies: 7, 8
- Problem of the Day: 11.4
- Warm-up Exercises: 11.4
- Answer Masters: 11.4

Extra Practice: 11.4
Color Transparency: 59

LESSON Notes

Example 1

Be sure students recognize that the location of the vertices indicates whether the transverse axis is on the x-axis or on the

(continued)

11.4 Hyperbolas

What you should learn:

Goal 1 How to write the equation of a hyperbola and graph it using asymptotes

Goal 2 How to use hyperbolas to solve real-life problems

Why you should learn it:

You can use hyperbolic models in real-life situations that require a locus of points for which the difference of the distances to two fixed points is constant.

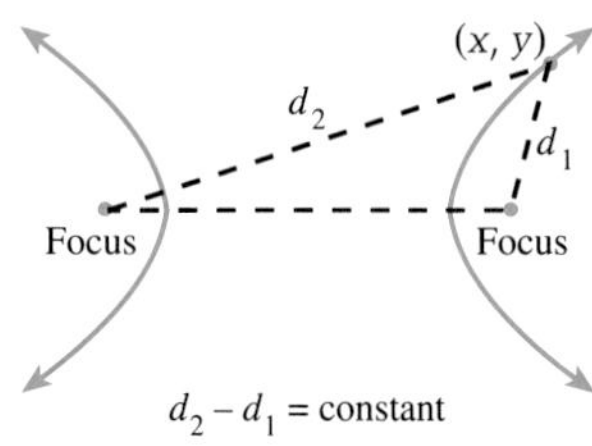

Goal 1 Writing the Equation of a Hyperbola

The definition of a hyperbola is similar to that of an ellipse. The distinction is that, for an ellipse, the *sum* of the distances between a point on the ellipse and the two fixed points is constant, while, for a hyperbola, the *difference* of these distances is constant.

The line through the *foci* (the two fixed points) intersects the hyperbola at two points, the **vertices.** The line segment joining the vertices is the **transverse axis,** and its midpoint is the **center** of the hyperbola. The graph of a hyperbola has two branches.

Standard Equation of a Hyperbola (Center at Origin)

The **standard form of the equation of a hyperbola** with center at (0, 0) is as follows.

$$\frac{x^2}{a^2} - \frac{y^2}{b^2} = 1 \quad \textit{Horizontal transverse axis}$$

$$\frac{y^2}{a^2} - \frac{x^2}{b^2} = 1 \quad \textit{Vertical transverse axis}$$

The vertices and foci are, respectively, a and c units from the center, and $c^2 = a^2 + b^2$.

The graphs show the two orientations of a hyperbola. Note that a hyperbola has two asymptotes.

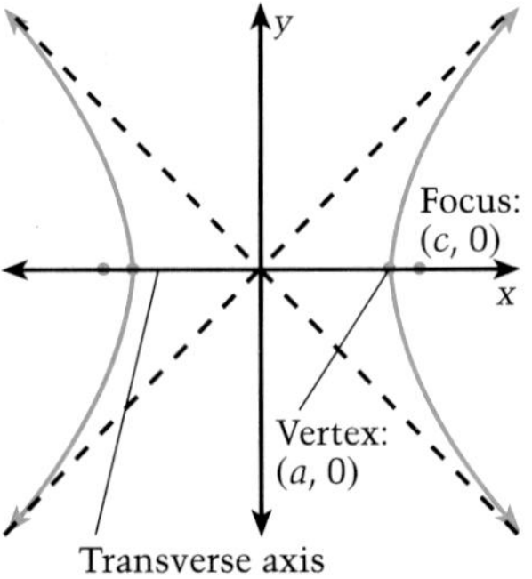

Horizontal transverse axis

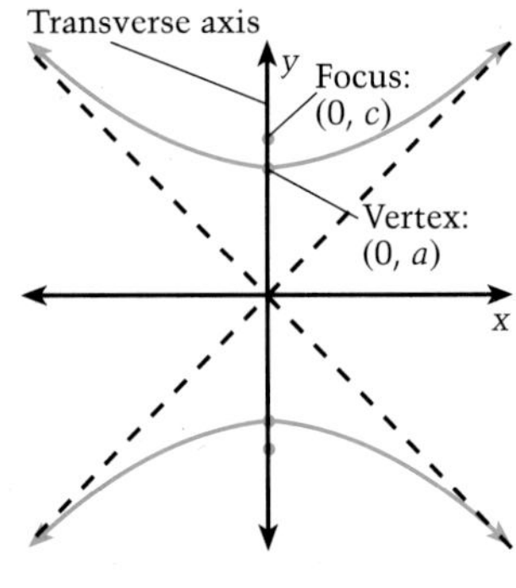

Vertical transverse axis

Example 1 Writing an Equation of a Hyperbola

Write an equation of the hyperbola with foci at $(-3, 0)$ and $(3, 0)$ and vertices at $(-2, 0)$ and $(2, 0)$.

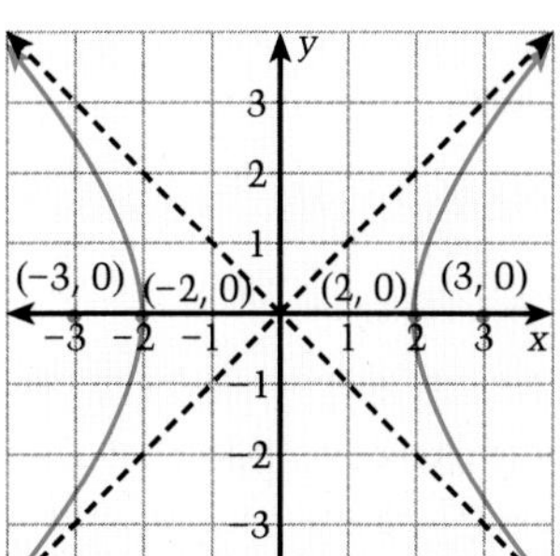

Solution The center of the hyperbola is the origin and the transverse axis is horizontal. Because the foci are each three units from the center, you can conclude that $c = 3$. Similarly, because the vertices are each two units from the center, you can conclude that $a = 2$. Using these values of a and c, you can write

$$b^2 = c^2 - a^2$$
$$= 3^2 - 2^2 = 9 - 4 = 5$$

which implies that $b = \sqrt{5}$. Because the transverse axis is horizontal, the standard form of the equation is as follows.

$$\frac{x^2}{a^2} - \frac{y^2}{b^2} = 1 \quad \textit{Horizontal transverse axis}$$

$$\frac{x^2}{2^2} - \frac{y^2}{(\sqrt{5})^2} = 1 \quad \textit{Substitute } a = 2 \textit{ and } b = \sqrt{5}.$$

$$\frac{x^2}{4} - \frac{y^2}{5} = 1 \quad \textit{Simplify.}$$

■

The asymptotes intersect at the center of the hyperbola and pass through the corners of a rectangle with corners $(\pm a, \pm b)$.

Asymptotes of a Hyperbola (Center at Origin)

The asymptotes of a hyperbola with center at (0, 0) are as follows.

$y = \frac{b}{a}x$ and $y = -\frac{b}{a}x$ *Horizontal transverse axis*

$y = \frac{a}{b}x$ and $y = -\frac{a}{b}x$ *Vertical transverse axis*

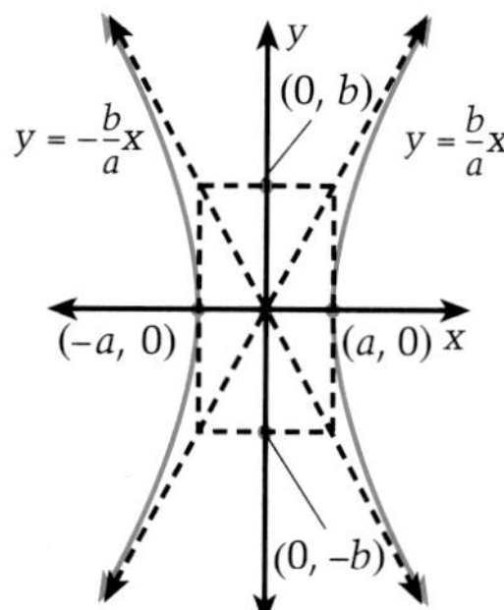

Horizontal transverse axis

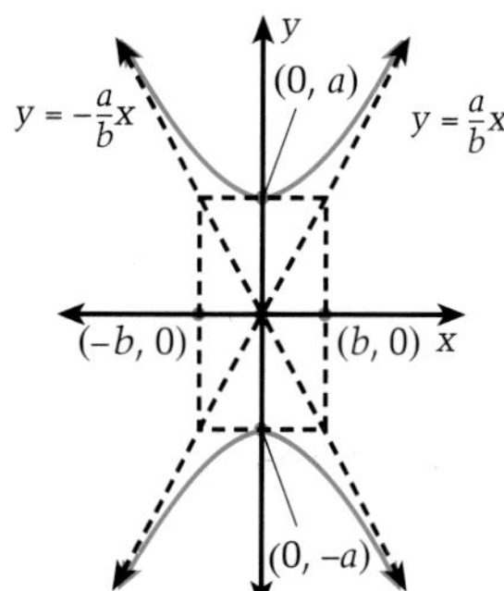

Vertical transverse axis

(continued)

y-axis. In this case, the transverse axis is on the x-axis because the x-coordinates of the vertices are non-zero. Consequently, $\frac{x^2}{a^2}$ is the leading term $\left(\text{instead of } \frac{y^2}{a^2}\right)$ because there is a horizontal transverse axis.

Point out that the b-value, $\sqrt{5}$, does not occur on the graph of the hyperbola, but as the next example shows, it is used to derive the asymptotes of the hyperbola.

Example 2

Ask students to find the foci of the hyperbola. (The foci can be obtained from the relationship, $c^2 = a^2 + b^2 = 2^2 + 4^2 = 4 + 16 = 20$. So $c = \sqrt{20} = 2\sqrt{5}$.)

Ask students to verify that the hyperbola drawn in the example has asymptotes that satisfy the relationships $y = \frac{b}{a}x$ and $y = -\frac{b}{a}x$. (The asymptote of positive slope through the origin contains the points (0, 0) and (2, 4). The slope is $m = \frac{4-0}{2-0} = 2$. Therefore, the equation of the line is $y = 2x$, which satisfies the relationship $y = \left(\frac{b}{a}\right)x$. The equation of the negatively sloped asymptote can be obtained similarly.)

Example 3

Ask students to write an equation of the hyperbola.

$$\frac{y^2}{9} - \frac{x^2}{\frac{9}{4}} = 1 \text{ or } y^2 - 4x^2 = 9$$

Example 4

Be sure students recognize that the situation described in the example is modeled by a hyperbola. In particular, reiterate that the distance from B to the explosion is always 2200 feet more than the distance from A

(continued)

(continued)
to the explosion. This means, then, that the difference between the "*B* distance" and the "*A* distance" is constant. Therefore, the location of the explosion lies on a hyperbola.

Furthermore, points *A* and *B* are the foci of the hyperbola.

Extra Examples

Here are additonal examples similar to **Examples 1–3.**

1. Write an equation of a hyperbola with the given foci and vertices.
   a. foci: $(-5, 0)$ and $(5, 0)$ vertices: $(-3, 0)$ and $(3, 0)$
   b. foci: $(0, -6)$ and $(0, 6)$ vertices: $(0, -4)$ and $(0, 4)$

a. $\dfrac{x^2}{9} - \dfrac{y^2}{16} = 1$

b. $\dfrac{y^2}{16} - \dfrac{x^2}{20} = 1$

2. Sketch the graphs of the following hyperbolas.
   a. $4x^2 - 9y^2 = 36$
   b. $2.25y^2 - 1.44x^2 = 3.24$

a.

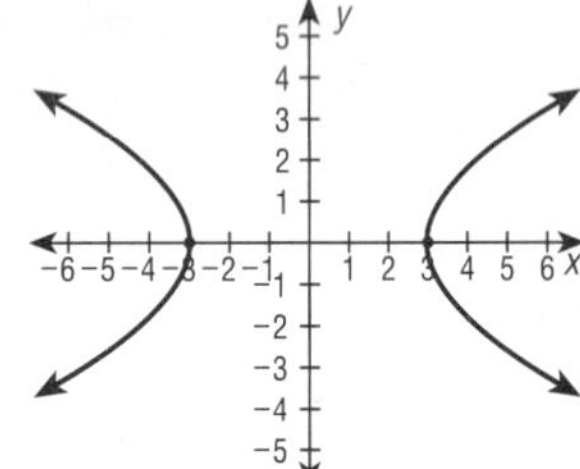

b.

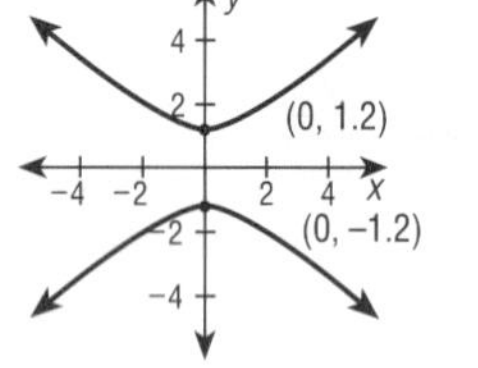

592

Example 2 Sketching a Hyperbola

Sketch the hyperbola $4x^2 - y^2 = 16$.

Solution Begin by rewriting the equation in standard form.

$$4x^2 - y^2 = 16 \qquad \textit{Given equation}$$

$$\frac{4x^2}{16} - \frac{y^2}{16} = \frac{16}{16} \qquad \textit{Divide both sides by 16.}$$

$$\frac{x^2}{2^2} - \frac{y^2}{4^2} = 1 \qquad \textit{Standard form}$$

Because the x^2-term is positive, you can conclude that the transverse axis is horizontal. The vertices are $(-2, 0)$ and $(2, 0)$. To sketch the hyperbola, first sketch a rectangle that is centered at the origin, $2a = 4$ units wide and $2b = 8$ units high. Then sketch asymptotes by drawing the lines that pass through opposite corners of the rectangle. Finally, sketch the hyperbola as shown below at the right.

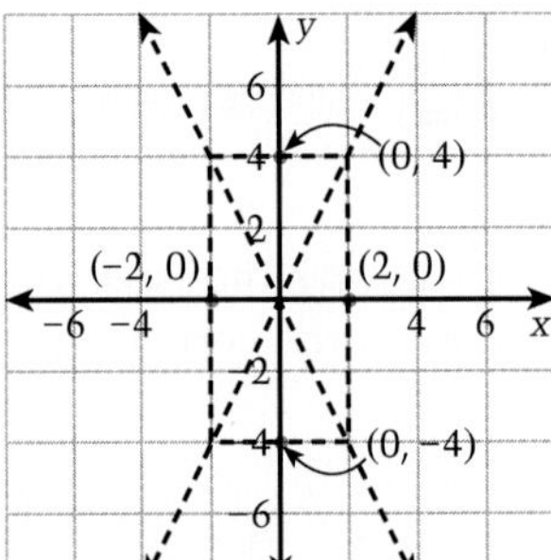

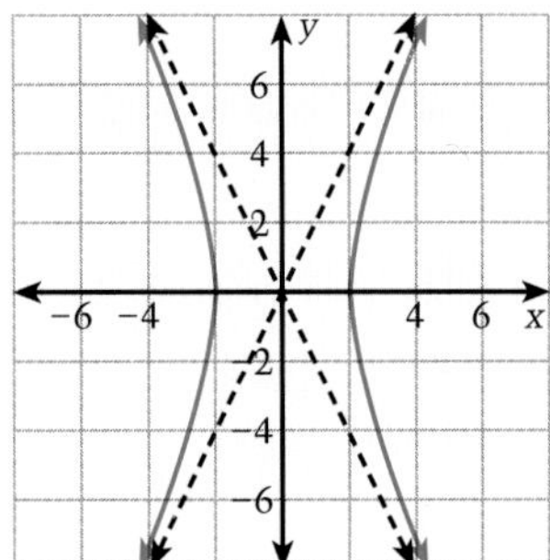

Example 3 Finding the Foci of a Hyperbola

Find the foci of the hyperbola having vertices at $(0, -3)$ and $(0, 3)$ and with asymptotes $y = 2x$ and $y = -2x$.

Solution Because the transverse axis is vertical, the asymptotes are of the form

$$y = \tfrac{a}{b}x \quad \text{and} \quad y = -\tfrac{a}{b}x.$$

From the given equations of the asymptotes, it follows that $\frac{a}{b} = 2$, and since $a = 3$, you can conclude that $b = \frac{3}{2}$. Using the equation $c^2 = a^2 + b^2$, you can conclude that

$$c = \sqrt{a^2 + b^2}$$

$$= \sqrt{3^2 + \left(\tfrac{3}{2}\right)^2} = \sqrt{\tfrac{45}{4}} = \tfrac{3}{2}\sqrt{5}.$$

Thus the foci are $\left(0, -\frac{3}{2}\sqrt{5}\right)$ and $\left(0, \frac{3}{2}\sqrt{5}\right)$.

y = −2x
(0, 3)
y = 2x
(0, −3)

Goal 2 Using Hyperbolas in Real-Life Problems

Example 4 *Locating an Explosion*

Two microphones, 1 mile apart, recorded an explosion. Microphone A received the sound 2 seconds before Microphone B. Is this enough information to decide where the sound came from?

Solution Because sound travels at 1100 feet per second, you know that the explosion took place 2200 feet farther from B than from A. The set of *all* points that are 2200 feet farther from B than from A is one branch of a hyperbola. Let $(c, 0)$ be the coordinates of A and let $(-c, 0)$ be the coordinates of B. Because A and B are 5280 feet apart, it follows that $c = 2640$. To find the vertex $(a, 0)$, use the fact that

$$(c + a) - (c - a) = 2200.$$

Solving this equation produces $a = 1100$. Finally, you can use the equation $b^2 = c^2 - a^2 = 5{,}759{,}600$ to conclude that the explosion occurred somewhere on the right branch of the hyperbola given by

$$\frac{x^2}{a^2} - \frac{y^2}{b^2} = 1$$

$$\frac{x^2}{1{,}210{,}000} - \frac{y^2}{5{,}759{,}600} = 1.$$

Thus you were not given enough information to determine the exact location of the explosion—only that it came from some point on the hyperbola. ■

Communicating about ALGEBRA

SHARING IDEAS about the Lesson

Match the equation with its graph. Explain your reasoning.

A. $x^2 - \frac{y^2}{4} = 1$ **B.** $\frac{x^2}{4} - y^2 = 1$ **C.** $y^2 - \frac{x^2}{4} = 1$

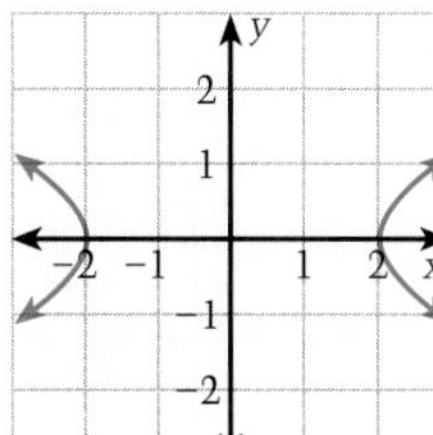

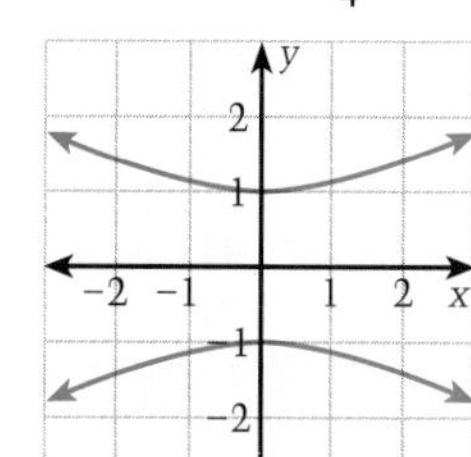

A. 1, (1, 0) is only on graph 1 **B.** 2, (2, 0) is only on graph 2 **C.** 3, (0, 1) is only on graph 3

3. Find the foci of the hyperbola having the vertices $(-2, 0)$ and $(2, 0)$ and asymptotes described by $y = 3x$ and $y = -3x$.
foci: $(-2\sqrt{10}, 0)$ and $(2\sqrt{10}, 0)$

Check Understanding

1. How are the definitions of ellipse and hyperbola the same? How are they different? See the top of page 590.
2. What information do the asymptotes of a hyperbola provide? The asymptotes of a hyperbola are two intersecting lines to which the hyperbola gets closer and closer; they determine the shape of the hyperbola.
3. Describe how the asymptotes of a hyperbola with a horizontal transverse axis may be graphed without using the algebraic relationships, $y = \frac{b}{a}x$ and $y = -\frac{b}{a}x$. See Examples 1 and 2.

Communicating about ALGEBRA

Write an equation of the hyperbola that satisfies the conditions indicated in the following graph. Also provide the foci of the hyperbola.

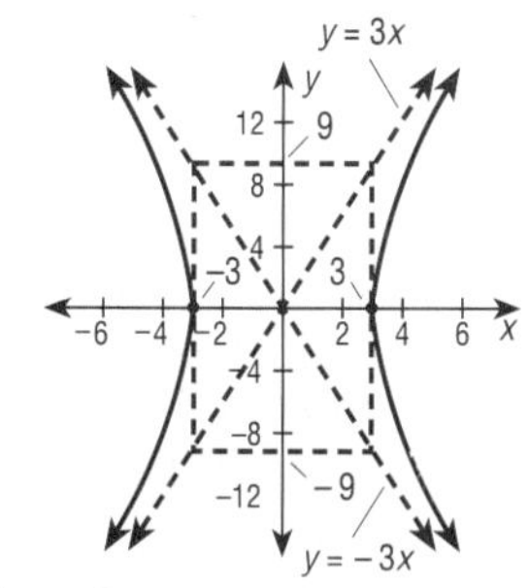

$\frac{x^2}{9} - \frac{y^2}{81} = 1$,
foci: $(-3\sqrt{10}, 0)$ and $(3\sqrt{10}, 0)$

EXERCISE Notes

ASSIGNMENT GUIDE

Basic/Average: Ex. 1–6, 11–19 odd, 25–28, 33, 37, 43, 49, 52–62

Above Average: Ex. 1–6, 9–19 odd, 24–28, 31–43 odd, 47, 44–59, 51–52, 57–63

Advanced: Ex. 1–6, 9–19 odd, 24–28, 31–43 odd, 47, 49–53, 60–64

Selected Answers
Exercises 1–6, 7–61 odd

More Difficult Exercises
Exercises 48–49, 51–52, 63–64

Guided Practice

Ex. 1–6 Use these Guided Practice exercises as a final student readiness check of the lesson. Make sure that students understand how to find the value of b.

Ex. 5 Check the students' understanding of why the graph would have a vertical transverse axis.

Independent Practice

Ex. 19–24 Make sure students understand why they must use $c^2 = a^2 + b^2$ to find the value of b.

Ex. 25–28 Assign these exercises as a group. Ask students to connect the horizontal or vertical orientation of a graph to a positive coefficient for x^2 or y^2.

EXERCISES

Guided Practice

CRITICAL THINKING about the Lesson

In Exercises 1–4, use the graph at the right.

1. State the coordinates of the vertices and foci.
2. Is the transverse axis $x = 0$ or $y = 0$? $y = 0$
3. Write an equation of the hyperbola.
4. Write equations for the asymptotes.
5. Write $y^2 - 4x^2 = 36$ in standard form.
6. Which a, b, c relationship is true for hyperbolas? **b**
   a. $c^2 = a^2 - b^2$ **b.** $c^2 = a^2 + b^2$

1. $(-4, 0), (4, 0); (-6, 0), (6, 0)$ 3. $\frac{x^2}{16} - \frac{y^2}{20} = 1$ 4. $y = \pm\frac{\sqrt{5}}{2}x$ 5. $\frac{y^2}{36} - \frac{x^2}{9} = 1$

Independent Practice

In Exercises 7–12, find the vertices and foci of the hyperbola. See Additional Answers.

7. $\frac{x^2}{9} - \frac{y^2}{49} = 1$
8. $\frac{x^2}{144} - \frac{y^2}{9} = 1$
9. $\frac{x^2}{16} - \frac{y^2}{4} = 1$
10. $\frac{y^2}{36} - \frac{x^2}{4} = 1$
11. $\frac{y^2}{25} - x^2 = 1$
12. $x^2 - \frac{y^2}{49} = 1$

In Exercises 13–18, write the equation in standard form. Find the foci and vertices. See Additional Answers.

13. $25x^2 - 4y^2 = 100$
14. $y^2 - 64x^2 = 64$
15. $36y^2 - 4x^2 = 9$
16. $16y^2 - 36x^2 + 9 = 0$
17. $y^2 - \frac{x^2}{25} = 4$
18. $\frac{x^2}{9} - \frac{4y^2}{9} = 9$

In Exercises 19–24, write an equation of the hyperbola. Its center is (0, 0). See below.

19. Foci: $(-2, 0), (2, 0)$
    Vertices: $(-1, 0), (1, 0)$
20. Foci: $(-4, 0), (4, 0)$
    Vertices: $(-3, 0), (3, 0)$
21. Foci: $(-9, 0), (9, 0)$
    Vertices: $(-2, 0), (2, 0)$
22. Foci: $(-6, 0), (6, 0)$
    Vertices: $(-4, 0), (4, 0)$
23. Foci: $(0, -9), (0, 9)$
    Vertices: $(0, -7), (0, 7)$
24. Foci: $(0, -2), (0, 2)$
    Vertices: $(0, -1), (0, 1)$

In Exercises 25–28, match the equation with its graph.

25. $\frac{x^2}{16} - \frac{y^2}{4} = 1$ **b**
26. $\frac{y^2}{4} - \frac{x^2}{2} = 1$ **a**
27. $\frac{y^2}{16} - \frac{x^2}{4} = 1$ **d**
28. $\frac{x^2}{4} - \frac{y^2}{2} = 1$ **c**

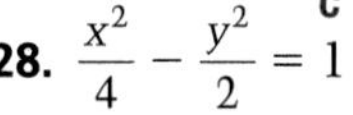

a.
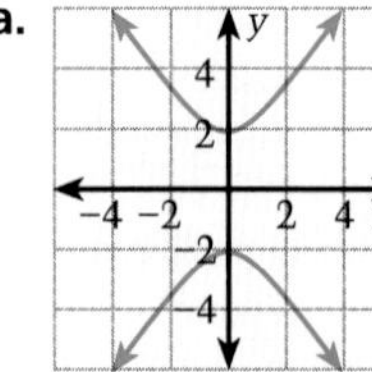

b.
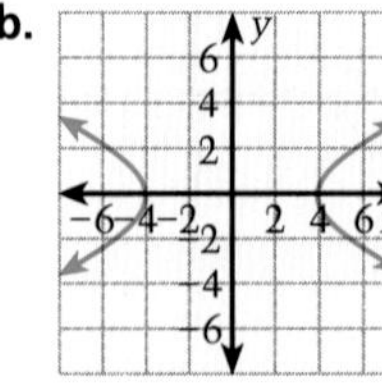

c.
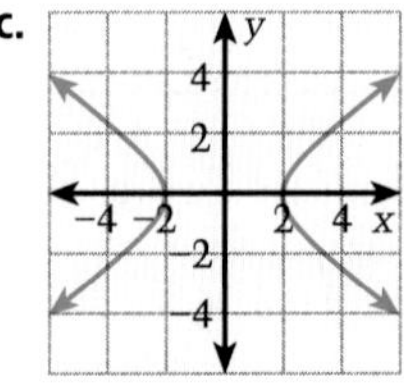

d.
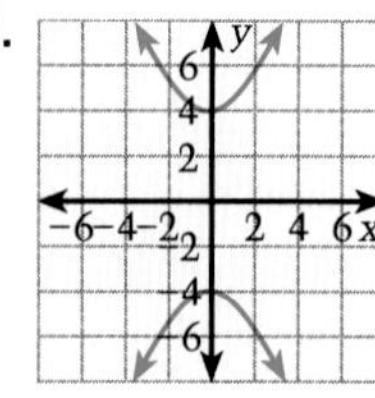

19. $\frac{x^2}{1} - \frac{y^2}{3} = 1$ 20. $\frac{x^2}{9} - \frac{y^2}{7} = 1$ 21. $\frac{x^2}{4} - \frac{y^2}{77} = 1$ 22. $\frac{x^2}{16} - \frac{y^2}{20} = 1$

23. $\frac{y^2}{49} - \frac{x^2}{32} = 1$ 24. $\frac{y^2}{1} - \frac{x^2}{3} = 1$

In Exercises 29–40, find the asymptotes of the hyperbola. Then sketch its graph.

29. $\dfrac{x^2}{25} - \dfrac{y^2}{144} = 1$ $y = \pm\frac{12}{5}x$

30. $\dfrac{x^2}{49} - y^2 = 1$ $y = \pm\frac{1}{7}x$

31. $\dfrac{y^2}{16} - \dfrac{x^2}{9} = 1$ $y = \pm\frac{4}{3}x$

32. $\dfrac{y^2}{9} - \dfrac{x^2}{64} = 1$ $y = \pm\frac{3}{8}x$

33. $\dfrac{x^2}{169} - \dfrac{y^2}{9} = 1$ $y = \pm\frac{3}{13}x$

34. $\dfrac{y^2}{81} - x^2 = 1$ $y = \pm 9x$

35. $\dfrac{25x^2}{36} - \dfrac{y^2}{121} = 1$ $y = \pm\frac{55}{6}x$

36. $\dfrac{x^2}{49} - \dfrac{9y^2}{4} = 1$ $y = \pm\frac{2}{21}x$

37. $4y^2 - x^2 = 4$ $y = \pm\frac{1}{2}x$

38. $100x^2 - 81y^2 = 8100$ $y = \pm\frac{10}{9}x$

39. $\dfrac{y^2}{25} - \dfrac{x^2}{16} = 16$ $y = \pm\frac{5}{4}x$

40. $x^2 - 25y^2 = 25$ $y = \pm\frac{1}{5}x$

See Additional Answers for graphs.

In Exercises 41–46, find the foci of the hyperbola.

41. Vertices: $(-1, 0), (1, 0)$ $(-\sqrt{10}, 0), (\sqrt{10}, 0)$
Asymptotes: $y = 3x,\ y = -3x$

42. Vertices: $(0, -2), (0, 2)$ $(0, -\sqrt{5}), (0, \sqrt{5})$
Asymptotes: $y = 2x,\ y = -2x$

43. Vertices: $(0, -5), (0, 5)$ $(0, -\sqrt{26}), (0, \sqrt{26})$
Asymptotes: $y = 5x,\ y = -5x$

44. Vertices: $(-4, 0), (4, 0)$ $(-\sqrt{80}, 0), (\sqrt{80}, 0)$
Asymptotes: $y = 2x,\ y = -2x$

45. Vertices: $(0, -6), (0, 6)$ $(0, -10), (0, 10)$
Asymptotes: $y = \dfrac{3}{4}x,\ y = -\dfrac{3}{4}x$

46. Vertices: $(-3, 0), (3, 0)$ $(-5, 0), (5, 0)$
Asymptotes: $y = \dfrac{4}{3}x,\ y = -\dfrac{4}{3}x$

Geometry **In Exercises 47 and 48, use the following information.**

The *annular ring* at the right has an inside radius of x and an outside radius of y. Its area is 8π.

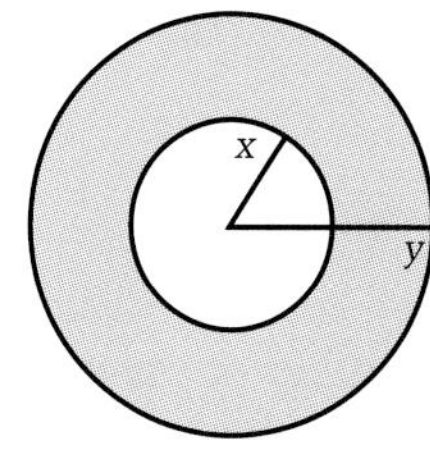

47. Find an equation that represents the relationship between x and y. Sketch the first-quadrant portion of the graph of the equation. $\dfrac{y^2}{8} - \dfrac{x^2}{8} = 1$, See margin.

✪ **48.** Copy and complete the table. What do you notice about the values of y as x increases? Can an annular ring have an area of 8π and an outside radius of 1 million units?

x	0	1	2	3	4	5	10	15	20
y	?	?	?	?	?	?	?	?	?

≈ 2.83, 3, ≈ 3.46, ≈ 4.12, ≈ 4.90, ≈ 5.74, ≈ 10.39, ≈ 15.26, ≈ 20.20; the value of y approaches the value of x; no

✪ **49.** *Thunder* You and a friend live 4 miles apart and are talking on the phone. You hear a crack of thunder, and 18 seconds later your friend hears the crack. Find an equation that gives the possible places where the lightning could have occurred. See margin.

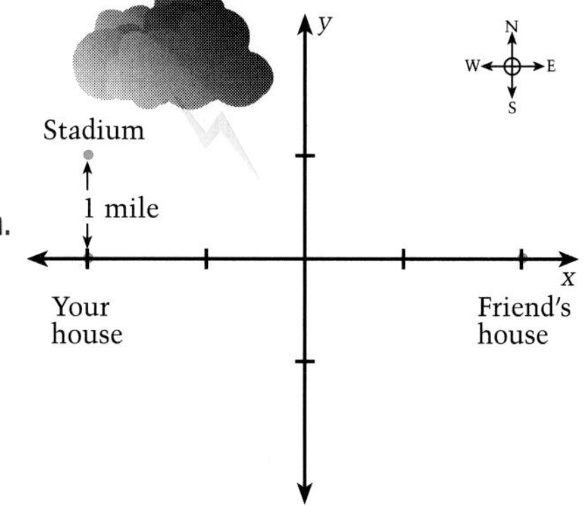

50. *Where Did It Hit?* You are still on the phone with your friend. After hearing the thunder, your friend says, "I hope that didn't hit the baseball stadium." Knowing that the baseball stadium is 1 mile directly north of you, how can you be sure that lightning didn't hit the stadium? The stadium is not located on a branch of the hyperbola.

▶ **Ex. 29–40** Encourage students to use a format and process similar to that modeled by Example 2.

▶ **Ex. 41–46** Encourage students to use a format and process similar to that modeled by Example 3.

▶ **Ex. 47** Suggest to the students that the area of the ring equals the area of the outer circle minus the area of the inner circle. So, $8\pi = \pi y^2 - \pi x^2$.

▶ **Ex. 49–50** Encourage students to use a format and process similar to that modeled by Example 4.

Answers

47.

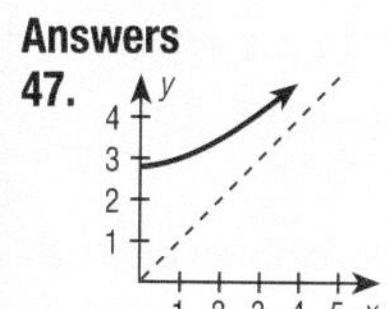

49.

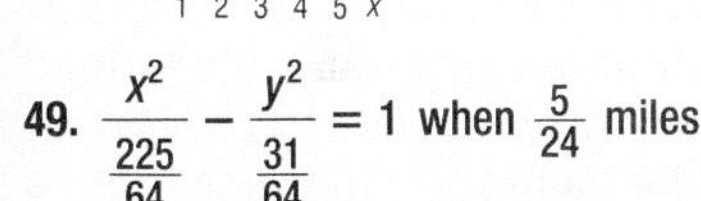

$\dfrac{x^2}{\frac{225}{64}} - \dfrac{y^2}{\frac{31}{64}} = 1$ when $\frac{5}{24}$ miles per second is used for the speed of sound.

▶ **Ex. 53–60** Assign these exercises as a group to help students define specific characteristics of each equation as related to their graphs.

▶ **Ex. 63–64** **TECHNOLOGY** Students might use graphics calculators to find the points of intersection. (See the Enrichment Activities for Lessons 11.1 and 11.3 for a technique that can be used to graph conic sections using graphics calculators.)

Enrichment Activity

This activity requires students to go beyond lesson goals.

COOPERATIVE LEARNING

Work with a partner to complete the activities.

Parabolas, hyperbolas, and ellipses are know as conic sections because they are formed as cross sections of a cone when the cone is intersected by a plane. A circle can be considered a special type of ellipse.

The light from a flashlight held in the air is a cone. When the light hits a flat surface, like a wall, a conic section can be formed by the boundaries of the light.

In a darkened room, shine a flashlight against a wall so that the edge of light forms a circle, an ellipse, a parabola, and a hyperbola. Describe how you can do this.
A flashlight held perpendicular to the wall forms a circle. If you move it up a little bit, you form an ellipse; if you move it up about 45° more, it forms a parabola. A flashlight held up parallel to the wall forms one branch of a hyperbola.

51. *Hyperbolic Mirror* In a hyperbolic mirror, light rays directed to one focus will be reflected to the other focus. The mirror below has the equation $\frac{x^2}{36} - \frac{y^2}{64} = 1$. (≈6.54, ≈3.46)
At which point on the mirror will light from the point (0, 10) reflect to the focus?

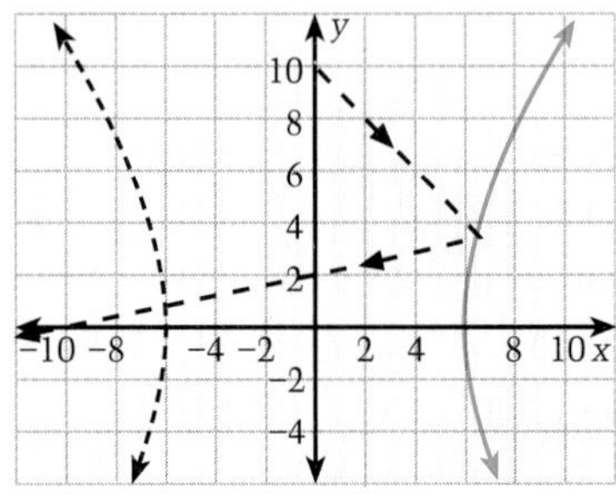

52. *Miniature Golf* You are playing miniature golf. Your golf ball is at (−15, 25). The wall at the end of the enclosed area is part of a hyperbola whose equation is $\frac{x^2}{19} - \frac{y^2}{81} = 1$. (≈4.99, ≈5.01)
At which point on the wall must your ball hit for it to go into the hole? (The ball bounces off a wall only once.)

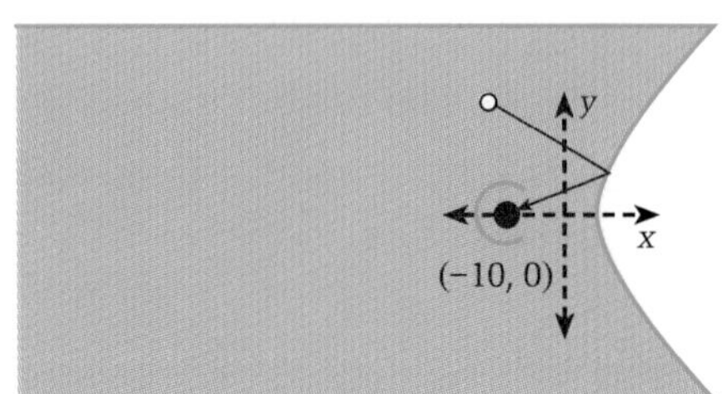

Integrated Review

In Exercises 53–60, identify the graph of the equation as a circle, an ellipse, or a hyperbola. Then sketch the graph. See Additional Answers.

53. $\frac{x^2}{25} + \frac{y^2}{4} = 1$ Ellipse **54.** $x^2 + y^2 = 81$ Circle **55.** $x^2 - y^2 = 16$ Hyperbola **56.** $\frac{x^2}{4} + \frac{y^2}{4} = 1$ Circle

57. $\frac{x^2}{9} + \frac{y^2}{49} = 1$ Ellipse **58.** $3x^2 - y^2 = 27$ Hyperbola **59.** $x^2 + y^2 = 25$ Circle **60.** $x^2 - 4y^2 = 16$ Hyperbola

61. *Powers of Ten* Which best estimates the number of maple leaves that will fill a 32-gallon plastic bag? d
a. 6.0×10^1 **b.** 6.0×10^2
c. 6.0×10^3 **d.** 6.0×10^4

62. *Powers of Ten* Which best estimates the number of cubic feet in a 32-gallon plastic bag? a
a. 4.3×10^0 **b.** 4.3×10^1
c. 4.3×10^2 **d.** 4.3×10^3

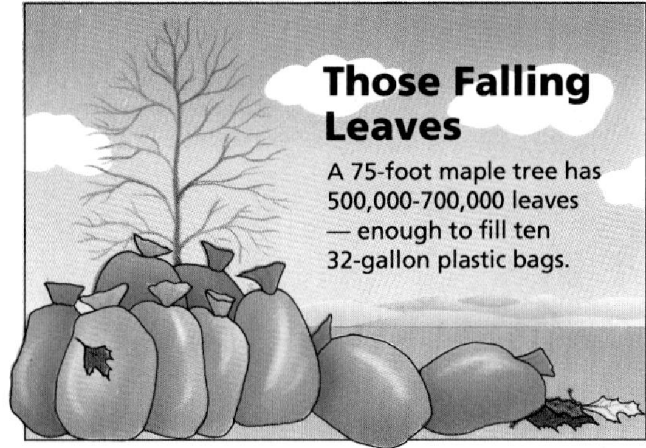

Those Falling Leaves
A 75-foot maple tree has 500,000-700,000 leaves — enough to fill ten 32-gallon plastic bags.

Exploration and Extension

In Exercises 63 and 64, sketch the graphs of the equations. Find all points of intersection of the graphs. See Additional Answers.

63. $\frac{x^2}{25} + \frac{y^2}{9} = 1$ and $\frac{x^2}{25} - \frac{y^2}{9} = 1$ (−5, 0), (5, 0)

64. $x^2 + y^2 = 16$ and $-x^2 + \frac{y^2}{16} = 1$ (0, −4), (0, 4)

11.5 Translations of Conics

What you should learn:

Goal 1 How to write an equation of a parabola with a vertex at (h, k) and an equation of a circle, an ellipse, or a hyperbola with its center at (h, k)

Goal 2 How to use translations to solve real-life problems

Why you should learn it:

If two conics do not share the same center, such as the cam and camshaft of an electric motor, you need to model one of these figures with an equation that is translated from the origin.

Goal 1 Writing Equations of Conics

Parabolas, circles, ellipses, and hyperbolas are called **conic sections** or simply **conics.** In Lessons 11.1 through 11.4, you studied equations of parabolas whose vertex is the origin, and equations of circles, ellipses, and hyperbolas whose center is the origin. In this lesson, you will study equations of conics that have been translated in the coordinate plane.

Standard Form of Equations of Translated Conics

The point (h, k) is the *vertex* of the parabola and the *center* of the other conics.

Circle: $(x-h)^2 + (y-k)^2 = r^2$

	Horizontal axis	Vertical axis
Parabola:	$(y-k)^2 = 4p(x-h)$	$(x-h)^2 = 4p(y-k)$
Ellipse:	$\frac{(x-h)^2}{a^2} + \frac{(y-k)^2}{b^2} = 1$	$\frac{(x-h)^2}{b^2} + \frac{(y-k)^2}{a^2} = 1$
Hyperbola:	$\frac{(x-h)^2}{a^2} - \frac{(y-k)^2}{b^2} = 1$	$\frac{(y-k)^2}{a^2} - \frac{(x-h)^2}{b^2} = 1$

Example 1 *Equations of Translated Conics*

a. $(x-1)^2 + (y+2)^2 = 3^2$ *Circle*

b. $\frac{(x-2)^2}{3^2} + \frac{(y-1)^2}{2^2} = 1$ *Ellipse*

c. $\frac{(x-3)^2}{1^2} - \frac{(y-2)^2}{3^2} = 1$ *Hyperbola*

d. $(x-2)^2 = 4(-1)(y-3)$ *Parabola*

a. 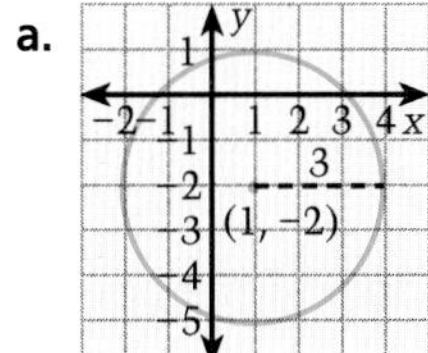

b. 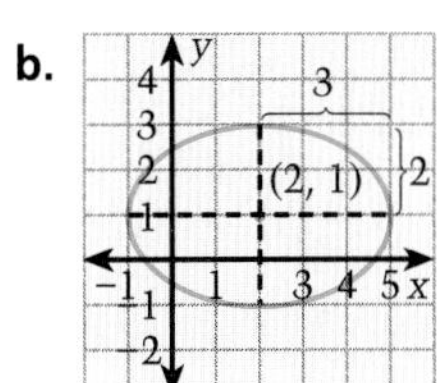

c. 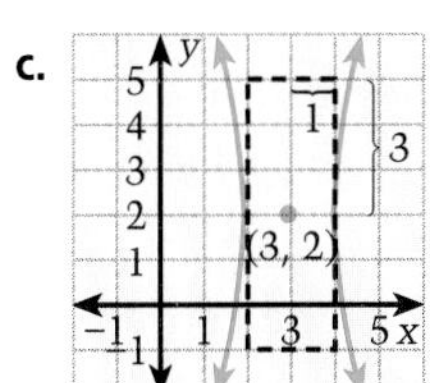

d. 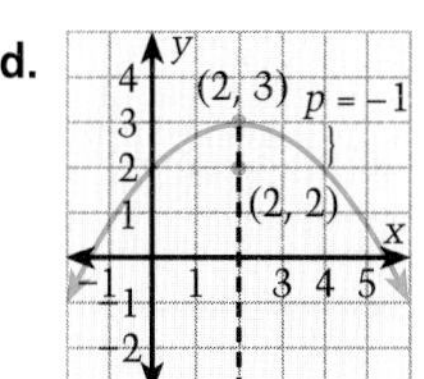

■

Problem of the Day

A circular dart board consists of a bull's-eye with radius 1 in., surrounded by 4 rings, each 1 in. wide. Which area is greater, the area of the bull's-eye together with the first two rings, or the area of the outer ring?

The areas are the same.

ORGANIZER

Warm-Up Exercises

1. Explain how the graph of $g(x) = (x-3)^2 + 4$ is different from the basic graph of $f(x) = x^2$.
The graph of $g(x)$ is a translation of the graph of $f(x)$ shifted 3 units to the right and 4 units up.
2. Solve by completing the square.
a. $x^2 - 4x + 1 = 2$
b. $4x^2 + 3x - 11 = 1$
a. $x = 2 \pm \sqrt{5}$
b. $x = \left(\frac{-3 \pm \sqrt{201}}{8}\right)$

Lesson Resources

Teaching Tools
- Transparencies: 2, 3, 7, 8
- Problem of the Day: 11.5
- Warm-up Exercises: 11.5
- Answer Masters: 11.5

Extra Practice: 11.5
Color Transparency: 60
Technology Handbook: p. 81

LESSON Notes

It is good to remind students that their work involving the effects of constants on algebraic functions will be important in the study of the translation of conic sections.

Example 1

Remind students that the horizontal and vertical shifts exhibited in the graphs of the equations of this example are exactly the effects of constants discussed in earlier lessons. Specifically, the expression $(x - 1)$ in the equation of the circle indicates that the circle whose center is at the origin should be shifted 1 unit to the right. The expression $(y + 2)$ in the equation of the circle indicates that the circle should then be shifted 2 units down. The new center, then, is located at the point $(1, -2)$.

The other equations can be graphed in a similar manner.

Example 2

Ask students why they should think that the equation is of a parabola. (One explanation is that only the x-variable is squared.) Help students to see that $p < 0$ indicates that the focus point is "below" the vertex of the parabola; consequently, the parabola opens downward. Emphasize that a factor of 4 is required on the side of the equation that has the linear factor.

Example 3

Both the x- and y-variables are squared. Therefore, you could have one of three conic sections. However, note that the coefficients of both the x^2 and y^2 terms are the same. This indicates that upon completing the square, you will get a circle. (Ask students to state why.)

To write the equation of a (shifted) conic in standard form, complete the square as demonstrated in Examples 2 through 5.

Example 2 *Writing the Standard Form of a Parabola*

Find the vertex and focus of the parabola given by

$$x^2 - 2x + 4y - 3 = 0.$$

Solution

$x^2 - 2x + 4y - 3 = 0$	*Given equation*
$x^2 - 2x = -4y + 3$	*Group terms.*
$x^2 - 2x + 1 = -4y + 3 + 1$	*Add 1 to both sides.*
$(x - 1)^2 = -4y + 4$	*Completed-square form*
$(x - 1)^2 = 4(-1)(y - 1)$	*Standard-form vertical axis*
$(x - h)^2 = 4p(y - k)$	

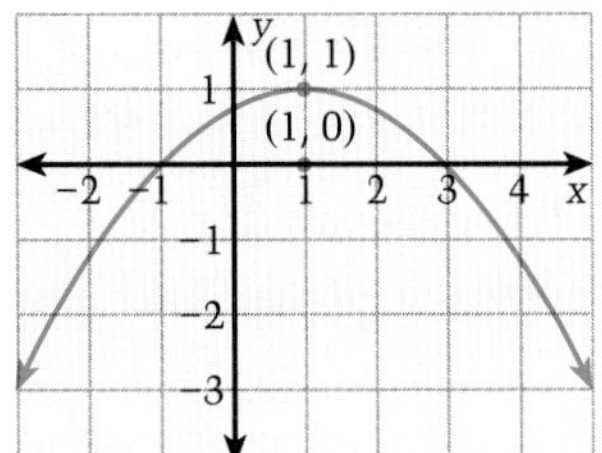

From the standard form, you can see that $h = 1$, $k = 1$, and $p = -1$. Because the axis is vertical and p is negative, the parabola opens downward. The vertex is $(h, k) = (1, 1)$ and the focus is $(h, k + p) = (1, 0)$. The graph is shown at the left. ■

Example 3 *Sketching a Circle*

Identify the center and radius of the circle given by the equation, and sketch the circle.

$$4x^2 + 4y^2 + 20x - 16y + 37 = 0$$

Solution Begin by writing the given equation in standard form. To do this, you must complete the square for both the x-terms *and* the y-terms.

$$4x^2 + 4y^2 + 20x - 16y + 37 = 0$$

$$x^2 + y^2 + 5x - 4y + \frac{37}{4} = 0$$

$$(x^2 + 5x + \quad) + (y^2 - 4y + \quad) = -\frac{37}{4}$$

$$\left(x^2 + 5x + \left(\frac{5}{2}\right)^2\right) + (y^2 - 4y + (-2)^2) = -\frac{37}{4} + \frac{25}{4} + 4$$

$$\left[\tfrac{1}{2}(5)\right]^2 \qquad \left[\tfrac{1}{2}(-4)\right]^2$$

$$\left(x + \frac{5}{2}\right)^2 + (y - 2)^2 = 1$$

$$(x - h)^2 + (y - k)^2 = r^2$$

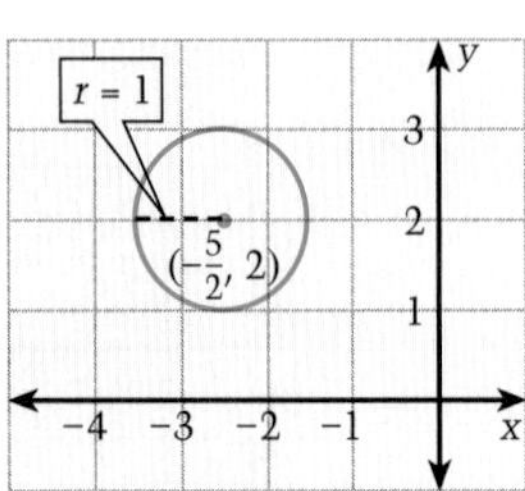

The center of the circle is $\left(-\frac{5}{2}, 2\right)$ and the radius of the circle is 1. The graph is shown at the left. ■

Example 4 Sketching an Ellipse

Sketch the ellipse: $x^2 + 4y^2 + 6x - 8y + 9 = 0$.

Solution

$$x^2 + 4y^2 + 6x - 8y + 9 = 0$$
$$(x^2 + 6x + \quad) + (4y^2 - 8y + \quad) = -9$$
$$(x^2 + 6x + \quad) + 4(y^2 - 2y + \quad) = -9$$
$$(x^2 + 6x + 9) + 4(y^2 - 2y + 1) = -9 + 9 + 4(1)$$
$$(x + 3)^2 + 4(y - 1)^2 = 4$$
$$\frac{(x + 3)^2}{4} + \frac{(y - 1)^2}{1} = 1$$
$$\frac{(x - h)^2}{a^2} + \frac{(y - k)^2}{b^2} = 1$$

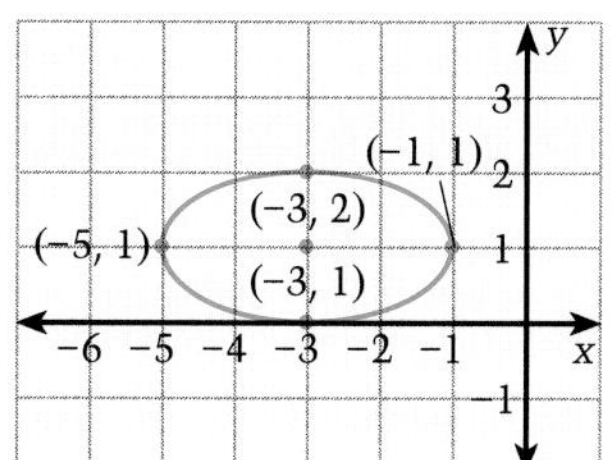

The center is $(h, k) = (-3, 1)$. Because the denominator of the x^2-term is $a^2 = 2^2$, the vertices lie two units to the right and left of the center. Because the denominator of the y^2-term is $b^2 = 1^2$, the co-vertices lie one unit up and down from the center. The graph is shown at the left. ■

Example 5 Sketching a Hyperbola

Sketch the hyperbola: $y^2 - 4x^2 + 4y + 24x - 41 = 0$. Also show the asymptotes.

Solution

$$y^2 - 4x^2 + 4y + 24x - 41 = 0$$
$$(y^2 + 4y + \quad) - (4x^2 - 24x + \quad) = 41$$
$$(y^2 + 4y + \quad) - 4(x^2 - 6x + \quad) = 41$$
$$(y^2 + 4y + 4) - 4(x^2 - 6x + 9) = 41 + 4 - 4(9)$$
$$(y + 2)^2 - 4(x - 3)^2 = 9$$
$$\frac{(y + 2)^2}{9} - \frac{4(x - 3)^2}{9} = 1$$
$$\frac{(y + 2)^2}{9} - \frac{(x - 3)^2}{\frac{9}{4}} = 1$$
$$\frac{(y - k)^2}{a^2} - \frac{(x - h)^2}{b^2} = 1$$

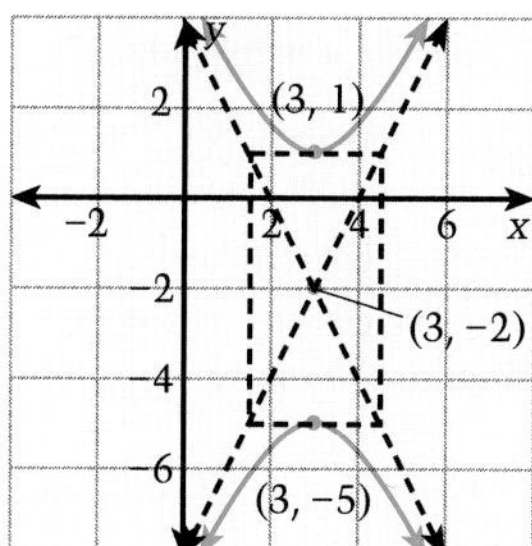

The transverse axis is vertical and the center is $(h, k) = (3, -2)$. Because the denominator of the y^2-term is $a^2 = 3^2$, the vertices lie three units above and below the center. To sketch the hyperbola, draw a rectangle centered at $(3, -2)$ of width $2b = 3$ and height $2a = 6$. Sketch the asymptotes and draw the hyperbola, as shown at the left. ■

Example 4

As in Example 3, both the x- and y-variables are squared. However, in this equation, the x^2 and y^2 terms are not the same. You may reason that the equation models either an ellipse (if terms are added) or a hyperbola (if terms are subtracted).

Remind students that because the denominator of the x term is larger, the vertices lie on a horizontal major axis.

Example 5

Point out that because the coefficient of x^2 is negative, the equation most likely models a hyperbola.

Ask students to derive the equation of the asymptotes.
$y = 2x - 8$, $y = -2x + 4$

Extra Examples

Here are additional examples similar to **Examples 1–5.**
Write the standard form of each equation of a conic section.

a. $9x^2 - 4y^2 + 18x + 16y - 43 = 0$
b. $y^2 - 8x - 6y + 9 = 0$
c. $x^2 + y^2 + 10x - 6y + 18 = 0$
d. $4x^2 + y^2 + 64x - 12y + 288 = 0$

a. $\frac{(x + 1)^2}{4} - \frac{(y - 2)^2}{9} = 1$
b. $(y - 3)^2 = 4(2)x$
c. $(x + 5)^2 + (y - 3)^2 = 16$
d. $\frac{(x + 8)^2}{1} + \frac{(y - 6)^2}{4} = 1$

Example 6

Make sure students recognize the effect of using an elliptical cam rather than a circular one. Would the follower move back and forth if the cam were circular and the camshaft were centered at the center of the circle?

Ask students to express the equation of the ellipse using integer coefficients.
$12x^2 + 16y^2 - 12x - 9 = 0$

Check Understanding

1. What is at least one advantage of expressing the equations of conic sections in standard form?
   Answers may include: to identify the type of conic section or to identify its center, its vertex, or its asymptotes.
2. How can you tell if a given equation models a parabola or a circle?
   By examining its standard form.
3. Why is completing the square a useful algebraic technique when working with conics?
   Completing the square is used to write the equation in standard form.

Communicating about ALGEBRA

Write each equation found in Parts A, B, and C using integer coefficients.

A. $3x^2 + 4y^2 - 4y - 2 = 0$

B. $12x^2 + 16y^2 + 12x - 9 = 0$

C. $3x^2 + 4y^2 + 4y - 2 = 0$

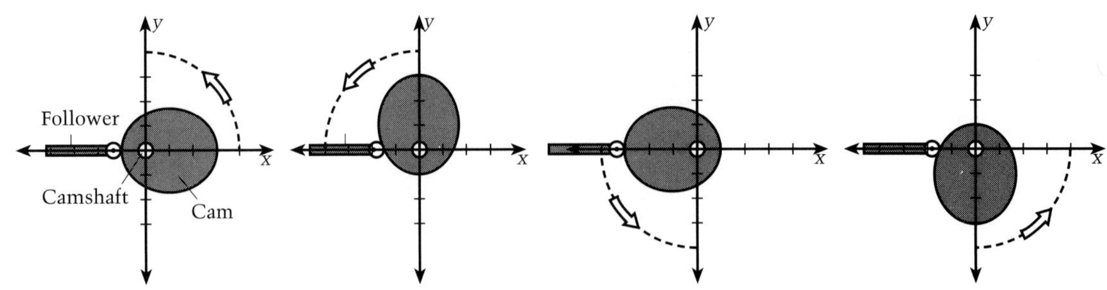

In the above diagram, the camshaft is rotating counterclockwise. An elliptical cam is attached to the camshaft. As the cam rotates, the follower moves back and forth. Thus this cam has changed the direction of the motion from rotational to oscillating (as occurs in a sewing machine).

Goal 2 Using Translations in Real-Life Problems

Real Life
Engineering Design

Example 6 *Finding the Equation of an Ellipse*

In the above diagram, the center of the camshaft is the focus of the elliptical cam. As the follower moves back and forth, it varies between $\frac{1}{2}$ inch and $\frac{3}{2}$ inches from the center of the camshaft. Find an equation of the elliptical cam in its starting position (on the left).

Center of Cam $(\frac{1}{2}, 0)$
$(-\frac{1}{2}, 0)$
$(\frac{3}{2}, 0)$
Focus
c a

Solution From the diagram at the left, you can see that the length of the major axis of the cam is 2 inches, which implies that $a = 1$. The center of the cam is $\left(\frac{1}{2}, 0\right)$. Because the focus is $\frac{1}{2}$ inch from the center of the cam, you can conclude that $c = \frac{1}{2}$. Using the equation $b^2 = a^2 - c^2$, it follows that $b^2 = \frac{3}{4}$ and $b = \frac{\sqrt{3}}{2}$. Thus the equation of the elliptical cam is:

$$\frac{\left(x - \frac{1}{2}\right)^2}{1^2} + \frac{(y - 0)^2}{\left(\frac{\sqrt{3}}{2}\right)^2} = 1 \text{ or } \left(x - \frac{1}{2}\right)^2 + \frac{y^2}{\frac{3}{4}} = 1.$$

A. $\frac{x^2}{\frac{3}{4}} + \left(y - \frac{1}{2}\right)^2 = 1$ B. $\left(x + \frac{1}{2}\right)^2 + \frac{y^2}{\frac{3}{4}} = 1$

Communicating about ALGEBRA

SHARING IDEAS about the Lesson

C. $\frac{x^2}{\frac{3}{4}} + \left(y + \frac{1}{2}\right)^2 = 1$

In Example 6, an equation was found for the elliptical cam in its starting position. Find an equation for the cam in the other three positions shown at the top of this page.

A. Rotated 90° **B.** Rotated 180° **C.** Rotated 270°

EXERCISES

Guided Practice

CRITICAL THINKING about the Lesson

1. What does it mean to "translate a conic in the coordinate plane?" To graph a conic whose only vertex or center is not at the origin.

In Exercises 2–5, identify the type of conic. If the conic is a circle, an ellipse, or a hyperbola, find its center. If the conic is a parabola, find its vertex.

2. $(x-1)^2 + (y-2)^2 = 1$ Circle, (1, 2)
3. $\frac{(y-1)^2}{30} - \frac{(x+1)^2}{25} = 1$ Hyperbola, (−1, 1)
4. $\frac{(x-10)^2}{9} + \frac{(y+2)^2}{4} = 1$ Ellipse, (10, −2)
5. $(y-4)^2 = 4(2)(x-1)$ Parabola, (1, 4)
6. Write $x^2 + y^2 - 10x + 12y + 37 = 0$ in standard form. Explain your steps. Identify the type of conic. $(x-5)^2 + (y+6)^2 = 24$, complete the square for the x- and y-terms, circle

Independent Practice

In Exercises 7–14, match the equation with its graph.

7. $\frac{(x-2)^2}{4} + \frac{(y+3)^2}{9} = 1$ g
8. $\frac{(x-2)^2}{9} + \frac{(y+3)^2}{4} = 1$ d
9. $\frac{(x-2)^2}{9} - \frac{(y+3)^2}{4} = 1$ c
10. $\frac{(x+2)^2}{4} - \frac{(y-3)^2}{9} = 1$ h
11. $\frac{(x+2)^2}{9} - \frac{(y-3)^2}{4} = 1$ f
12. $\frac{(x+2)^2}{9} + \frac{(y-3)^2}{4} = 1$ b
13. $\frac{(x-2)^2}{4} - \frac{(y+3)^2}{9} = 1$ a
14. $\frac{(x+2)^2}{4} + \frac{(y-3)^2}{9} = 1$ e

a. 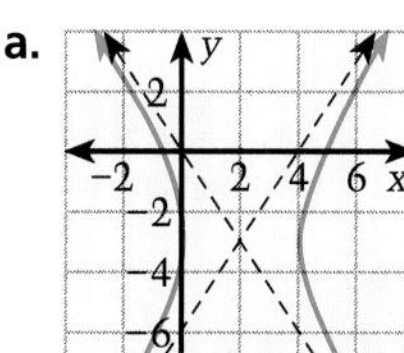

b. 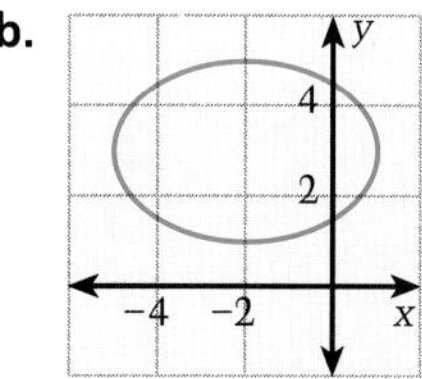

c. 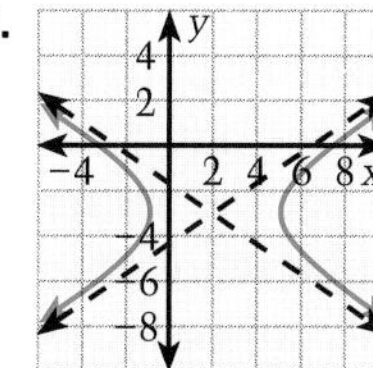

d. 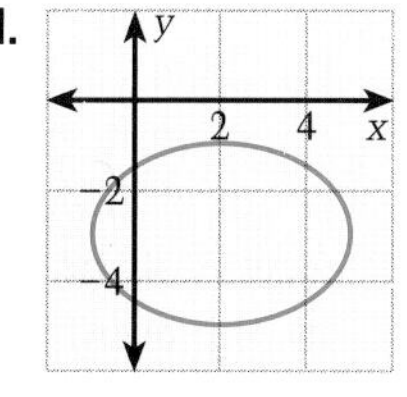

e. 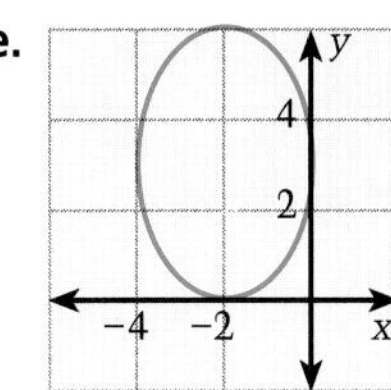

f. 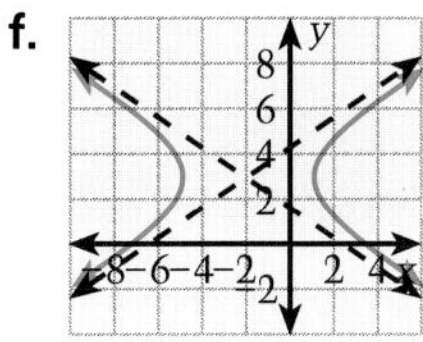

g. 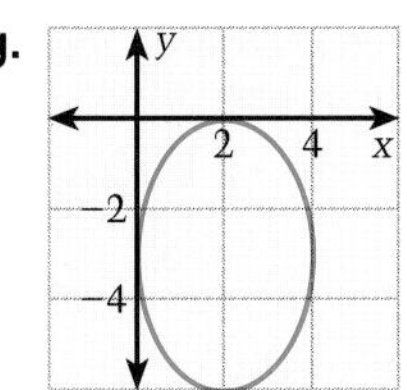

h. 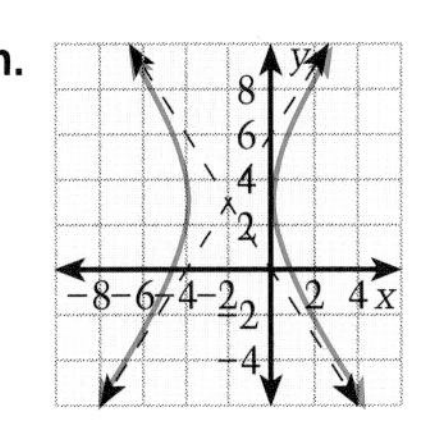

EXERCISE Notes

ASSIGNMENT GUIDE

Basic/Average: Ex. 1–6, 7–14, 15–23 odd, 27, 31, 35, 39-42, 43–51 odd

Above Average: Ex. 1–6, 7–14, 15–25 odd, 29, 33, 37, 39–42, 52

Advanced: Ex. 1–6, 7–14, 15–25 odd, 29, 33, 37, 39–42, 50–52

Selected Answers
Exercises 1–6, 7–49 odd

Use **Mixed Review** as needed.

More Difficult Exercises
Exercises 39–42, 51–52

Guided Practice

Ex. 1–6 COOPERATIVE LEARNING Have students work in pairs to complete these Guided Practice exercises. Encourage students to use their own words for Exercise 1. In Exercises 2–5, students should give reasons for their choices.

Independent Practice

Ex. 7–14 WRITING Assign these exercises as a group. Ask students to write a one-sentence justification for their choices.

Ex. 15–22 Refer students to the list of equations stated on page 597.

Ex. 23–38 These exercises are modeled by Examples 2–5. Encourage students to use quick-graph techniques to sketch the graphs.

Ex. 39 Remind students that the formula for the area of an ellipse is $A = \pi ab$.

Answers

19. $\frac{(x+2)^2}{8} + \frac{\left(y+\frac{1}{2}\right)^2}{\frac{81}{4}} = 1$

21. $\frac{y^2}{16} - \frac{(x-8)^2}{20} = 1$

39. $\frac{x^2}{100^2} + \frac{(y-85)^2}{85^2} = 1$, 8500π yd^2

Ex. 41 The center of the circle is $(-1, 3)$ and the point of tangency is $(1, 1)$. Remind students that the line tangent to a circle at point $(1, 1)$ is perpendicular to the line containing the radius of the circle.

Enrichment Activities

These activities require students to go beyond lesson goals.

Consider the following equation $x^2 + y^2 + 4x + 10y + 29 = 0$ in the form $Ax^2 + Bxy + Cy^2 + Dx + Ey + F = 0$ where $B = 0$.

1. State the values of A, B, C, D, E, and F. In order, they are 1, 0, 1, 4, 10, 29.

In Exercises 15–22, find an equation for the conic.

15. *Circle* Center: $(9, 3)$ Radius: 4 $(x-9)^2 + (y-3)^2 = 16$
16. *Circle* Center: $(-3, 1)$ Radius: 2 $(x+3)^2 + (y-1)^2 = 4$
17. *Parabola* Vertex: $(1, -2)$ Focus: $(1, 1)$ $(x-1)^2 = 12(y+2)$
18. *Parabola* Vertex: $(3, -2)$ Focus: $(5, -2)$ $(y+2)^2 = 8(x-3)$
19. *Ellipse* Vertices: $(-2, -5)$, $(-2, 4)$ Foci: $(-2, -4)$, $(-2, 3)$ See margin.
20. *Ellipse* Vertices: $(-2, 2)$, $(4, 2)$ Co-vertices: $(1, 1)$, $(1, 3)$ $\frac{(x-1)^2}{9} + \frac{(y-2)^2}{1} = 1$
21. *Hyperbola* Vertices: $(8, -4)$, $(8, 4)$ Foci: $(8, -6)$, $(8, 6)$ See margin.
22. *Hyperbola* Vertices: $(-10, 3)$, $(10, 3)$ Foci: $(-12, 3)$, $(12, 3)$ $\frac{x^2}{100} - \frac{(y-3)^2}{44} = 1$

In Exercises 23–26, write the equation of the circle in standard form. Identify the radius and center and sketch the graph. See Additional Answers.

23. $x^2 + y^2 - 2x + 6y + 9 = 0$
24. $x^2 + y^2 - 8x - 20y + 115 = 0$
25. $x^2 + y^2 - 6x - 8y + 24 = 0$
26. $x^2 + y^2 + 10x - 6y + 33 = 0$

In Exercises 27–30, write the equation of the parabola in standard form. Identify the vertex and focus and sketch the graph. See Additional Answers.

27. $y^2 - 4y + 12x - 8 = 0$
28. $y^2 - 12y + 4x + 4 = 0$
29. $x^2 + 4x - 8y + 12 = 0$
30. $x^2 + 6x - 2y + 13 = 0$

In Exercises 31–34, write the equation of the ellipse in standard form. Identify the vertices and foci and sketch the graph. See Additional Answers.

31. $25x^2 + y^2 - 100x - 2y + 76 = 0$
32. $x^2 + 4y^2 - 10x - 40y + 121 = 0$
33. $x^2 + 36y^2 - 16x - 72y + 64 = 0$
34. $4x^2 + y^2 - 48x - 4y + 48 = 0$

In Exercises 35–38, write the equation of the hyperbola in standard form. Identify the vertices and foci and sketch the graph. See Additional Answers.

35. $x^2 - 25y^2 - 14x + 100y - 76 = 0$
36. $9x^2 - y^2 - 72x + 8y + 119 = 0$
37. $-9x^2 + 4y^2 - 36x - 16y - 164 = 0$
38. $-x^2 + y^2 - 2x - 12y + 31 = 0$

39. **Aussie Football** In Australia, football by *Australian Rules* (or rugby) is played on elliptical fields. The field can be a maximum of 170 yards wide and a maximum of 200 yards long. Let the center of a field of maximum size be represented by the point $(0, 85)$. Write an equation of the ellipse that represents this field. Find the area of the field. ***(Source: Oxford Companion to World Sports and Games)*** See margin.

40. **Geometry** Find the points of intersection of the circle $(x-2)^2 + (y-1)^2 = 100$ and the line $y = -\frac{3}{4}x + \frac{5}{2}$. Show that the line contains a diameter of the circle. See below.

This rugby match is being played on an elliptical field in Melbourne, Australia.

40. $(10, -5)$, $(-6, 7)$; $\sqrt{(10-(-6))^2 + (-5-7)^2} = 20 = 2\sqrt{100}$, the length of the diameter

41. *Geometry* Find an equation of a line that is tangent to the circle $(x+1)^2 + (y-3)^2 = 8$ at the point (1, 1). $y = x$

42. *Designing a Menu* You are opening a restaurant called the Treetop Restaurant and are using a computer program to design the menu cover. The equation for the tree trunk is $25x^2 - 4y^2 + 80y - 800 = 0$. Write this equation in standard form and then sketch the graph. $\frac{x^2}{16} - \frac{(y-10)^2}{100} = 1$

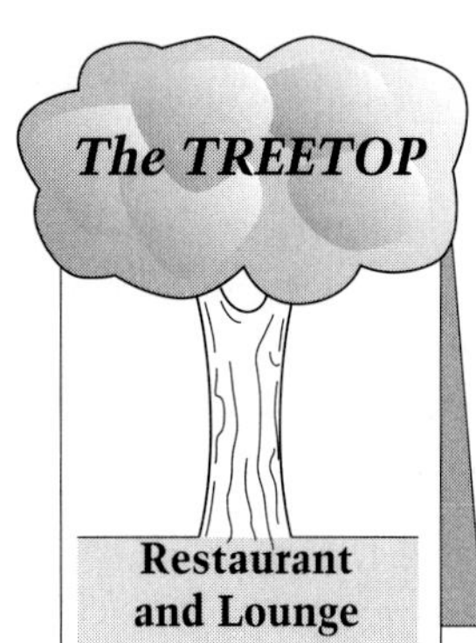

Integrated Review

In Exercises 43–46, solve the equation by completing the square.

43. $x^2 + 4x - 1 = 0$ $-2 \pm \sqrt{5}$

44. $x^2 - 2x - 2 = 0$ $1 \pm \sqrt{3}$

45. $x^2 - 6x + 4 = 0$ $3 \pm \sqrt{5}$

46. $x^2 - 4x - 8 = 0$ $2 \pm \sqrt{12}$

In Exercises 47–50, sketch the graph of the equation. See Additional Answers.

47. $x^2 + y^2 = 225$

48. $\frac{x^2}{25} - \frac{y^2}{4} = 1$

49. $\frac{x^2}{16} + \frac{y^2}{144} = 1$

50. $\frac{y^2}{49} - x^2 = 1$

Exploration and Extension

In Exercises 51 and 52, write the equation in standard form. Then sketch the graph. See Additional Answers.

51. $9x^2 + y^2 + 36x + 4y + 4 = 0$ $\frac{(x+2)^2}{4} + \frac{(y+2)^2}{36} = 1$

52. $4x^2 - y^2 - 8x - 6y - 9 = 0$ $\frac{(x-1)^2}{1} - \frac{(y+3)^2}{4} = 1$

Mixed REVIEW

4. $16x^2$ 9. $2x^2 - 2x + 10$ 11. $\frac{2 \pm \sqrt{4-3c}}{3}$

16. See margin.

1. Divide: $\frac{7}{9} \div \frac{5}{6}$. $\frac{14}{15}$
2. Find all intercepts of $y = 3x^2 - 7x + 4$. **(2.3)** (0, 4), (1, 0), and $\left(\frac{4}{3}, 0\right)$
3. Simplify: $\sqrt{72x^2y^3}$. **(7.4)** $6|x|y\sqrt{2y}$
4. Find $g(f(x))$ for $f(x) = 4x$, $g(x) = x^2$. **(6.2)**
5. Complete the square: $3x^2 - 12x + 7$. **(5.3)** See margin.
6. Find the geometric mean of 3 and 48. **(7.5)** 12
7. Sketch $f(x) = \ln(x-5) - 2$. **(8.2)** See margin.
8. Expand $\ln\left(\frac{5}{8}\right)^3$. **(8.3)** $3(\ln 5 - \ln 8)$
9. Divide: $\frac{2x^3 - 8x^2 + 16x - 30}{x-3}$. **(9.4)**
10. Find the range of $f(x) = \frac{|x|}{\sqrt{x^2}}$. **(6.2)** 1
11. Solve for x: $3x^2 - 4x + c = 0$. **(5.4)**
12. Solve $3x^4 - 14x^2 + 8 = 0$. **(9.5)** $\pm 2, \pm\sqrt{\frac{2}{3}}$
13. Solve $|4x + 2| > 6$. **(1.7)** $x < -2$ or $x > 1$
14. Sketch $3y - 2x \le 15$. **(2.5)** See margin.
15. Multiply: $\sqrt{-6} \cdot \sqrt{-15}$. **(5.5)** $-3\sqrt{10}$
16. Factor $x^4 + 2x^3 - 13x^2 - 14x + 24$. **(9.4)**
17. Find the focus of $y^2 = x$. **(11.1)** $\left(\frac{1}{4}, 0\right)$
18. Find the mean: 2, 4, 8, 8, 10, 12. **(9.7)** $\frac{22}{3}$
19. Solve $\frac{x+4}{3x} + \frac{x-1}{x+6} = 1$. **(10.4)** 3, 8
20. Find the center of $x^2 + 4y^2 = 5$. **(11.3)** (0, 0)

2. Complete the square to express the equation in standard form. $(x+2)^2 + (y+5)^2 = 0$
3. What conic is represented by the equation? a circle
4. Find the value of $D^2 + E^2 - 4AF$. 0
5. What are the real number solutions for the standard-form equation you found? What do you notice? The only real number solution is the point $(-2, -5)$. This equation defines a point, not a circle.
6. What generalization can you conclude from the results? An equation in the form $Ax^2 + Bxy + Cy^2 + Dx + Ey + F = 0$, $B = 0$, defines a point when $D^2 + E^2 - 4AF = 0$.

Answers

5. $3(x-2)^2 - 5$

7.
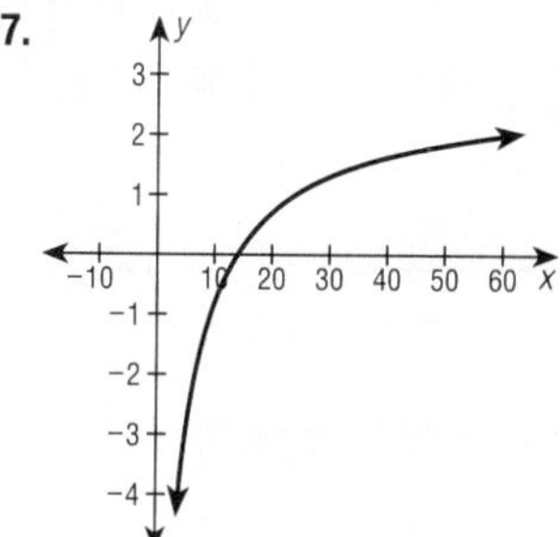

14.
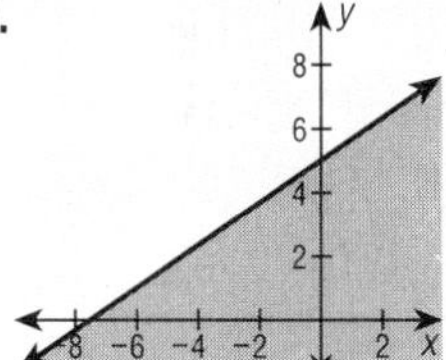

16. $(x+4)(x-3)(x-1)(x+2)$

TECHNOLOGY Graphics calculators are powerful graphing tools. However, emphasize that graphing relations can only be achieved if students have good algebraic skills that allow them to rewrite equations of conics as functional relationships. In Example 1, students must be able to solve for y in the equation of a circle. In Example 2, not only must students be able to solve for y in the equation of a hyperbola; they must also be able to write the equation of a straight line given a point and its slope.

Answers

1.

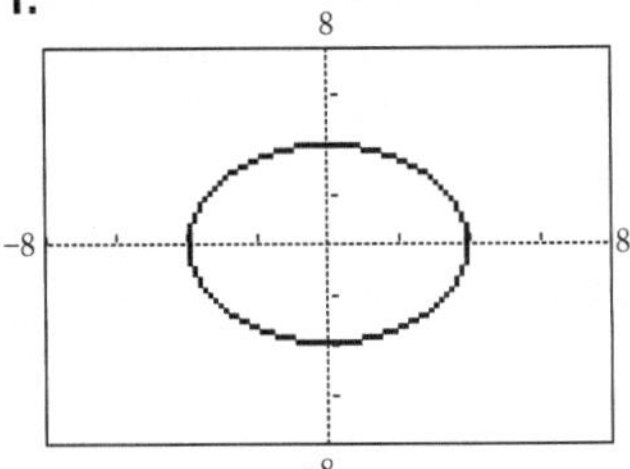

2.

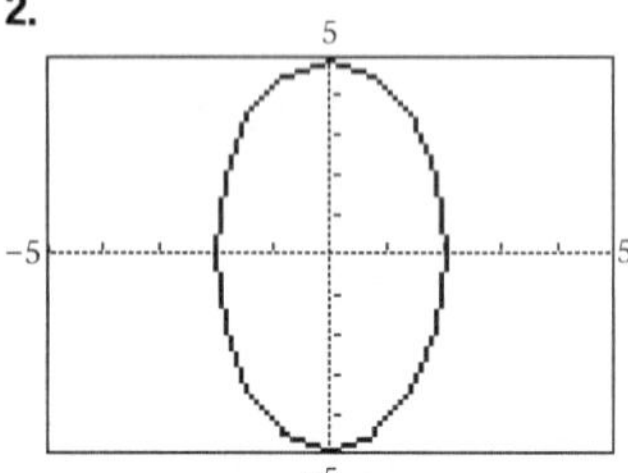

3.

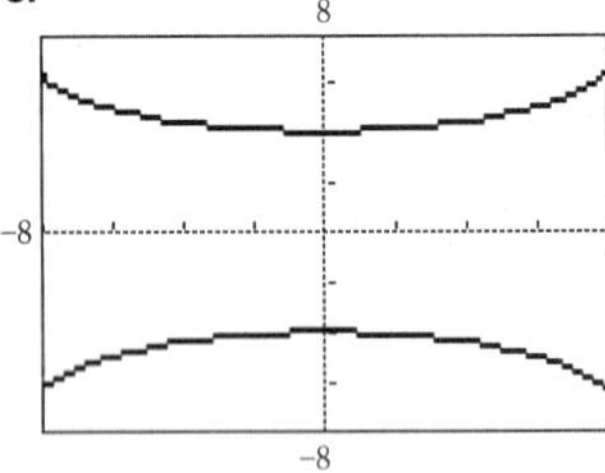

USING A GRAPHING CALCULATOR

When sketching a conic with a graphing calculator, you must remember two things. First, most graphing calculators are set up to sketch the graphs of functions, not relations—so you must first write the relation as one *or more* functions. Second, to obtain graphs with true perspectives (in which circles look like circles), you must use a "square setting." For a graphing calculator whose display screen's height is two thirds of its width, you can obtain a square setting by using

$$\frac{\text{Ymax} - \text{Ymin}}{\text{Xmax} - \text{Xmin}} = \frac{2}{3}. \quad \textit{Square setting}$$

Example 1: Sketching a Circle with a Graphing Calculator

Use a graphing calculator to sketch the graph of $x^2 + y^2 = 4$.

Solution Begin by solving the given equation for y.

$$x^2 + y^2 = 4$$
$$y^2 = 4 - x^2$$
$$y = \pm\sqrt{4 - x^2}$$

Next, set the calculator's range setting to $-3 \le x \le 3$ and $-2 \le y \le 2$. Finally, enter the following two equations—one to sketch the top half of the circle and the other to sketch the bottom half of the circle.

$$y = \sqrt{4 - x^2} \quad \textit{Top half of circle}$$

$$y = -\sqrt{4 - x^2} \quad \textit{Bottom half of circle}$$

The graph is shown at the right. ■

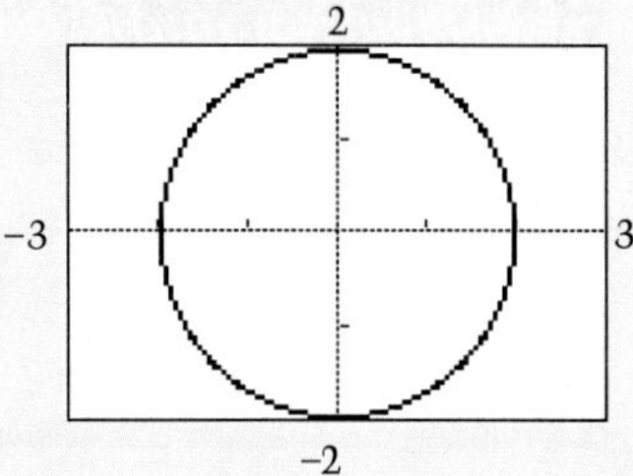

A graphing calculator can be used to estimate the points of intersection of two graphs. For instance, to find the points of intersection of the line $y = 2x - 2$ and the circle $x^2 + y^2 = 4$, you can show the graph of the line with the graph of Example 1. Using the trace feature, you can approximate the points of intersection to be $(0, -2)$ and $(1.6, 1.2)$. Both points can be verified algebraically as follows.

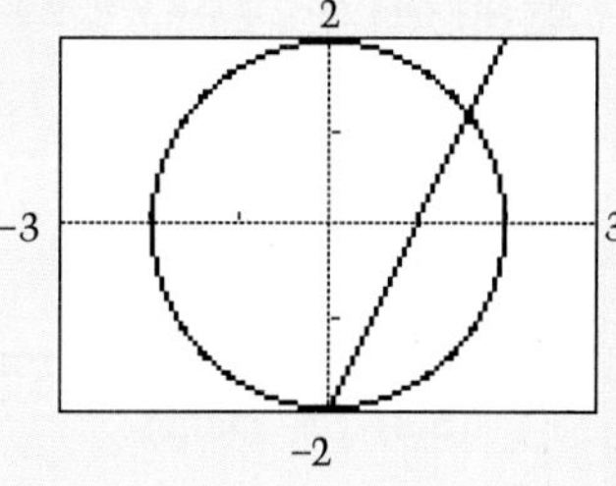

$$x^2 + y^2 = 4 \quad \textit{Given equation of circle}$$

$$x^2 + (2x - 2)^2 = 4 \quad \textit{Substitute } 2x - 2 \textit{ for } y.$$

$$5x^2 - 8x = 0 \quad \textit{Standard form}$$

$$x(5x - 8) = 0 \quad \textit{Factor.}$$

$$\left.\begin{array}{r} x = 0 \\ \text{or } x = \frac{8}{5} \end{array}\right\} \quad \textit{Zero Factor Property}$$

Example 2: Investigating Asymptotic Behavior

Sketch the hyperbola $\frac{(x-2)^2}{3^2} - \frac{(y-3)^2}{1^2} = 1$ and its asymptotes on the same display screen.

Solution To sketch the hyperbola, solve the given equation for y.

$$\frac{(x-2)^2}{3^2} - \frac{(y-3)^2}{1^2} = 1$$

$$\frac{1}{9}(x-2)^2 - 1 = (y-3)^2$$

$$\pm\sqrt{\frac{1}{9}(x-2)^2 - 1} = y - 3$$

$$3 \pm \sqrt{\frac{1}{9}(x-2)^2 - 1} = y$$

The "plus-version" is the equation of the top half of each branch and the "minus-version" is the equation of the bottom half of each branch. The asymptotes of the hyperbola pass through the point (2, 3) with slopes of $\pm\frac{b}{a} = \pm\frac{1}{3}$.

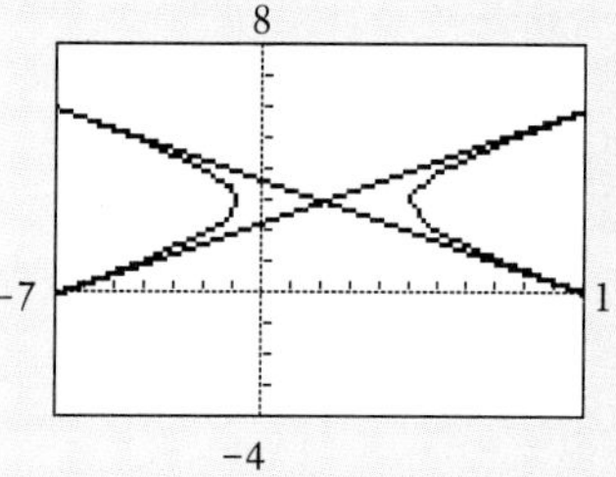

$$y = \frac{1}{3}(x+7) \quad \text{and} \quad y = \frac{1}{3}(-x+11) \quad \textit{Asymptotes}$$

By sketching the hyperbola and the two asymptotes on the same screen, you can see that the hyperbola gets closer to its asymptotes as you move to the right side or left side of the screen. ■

Exercises

In Exercises 1–6, use a graphing calculator to sketch the graph of the equation. See margin.

1. $x^2 + y^2 = 16$
2. $\frac{x^2}{2^2} + \frac{y^2}{5^2} = 1$
3. $\frac{y^2}{4^2} - \frac{x^2}{6^2} = 1$
4. $(x-1)^2 + (y-2)^2 = 10$
5. $\frac{(x+1)^2}{12} + \frac{(y-3)^2}{9} = 1$
6. $\frac{(x-2)^2}{16} - \frac{(y+4)^2}{10} = 1$

In Exercises 7–9, find the points of intersection of the graphs. Use a graphing calculator to verify your result.

7. $(x-4)^2 + (y+2)^2 = 10$
   $x + 2y = -5$ (5, −5), (1, −3)
8. $\frac{4(x+1)^2}{9} + (y-5)^2 = 5$
   $x + y = 6$ (2, 4), $\left(\frac{-16}{13}, \frac{94}{13}\right)$
9. $y^2 = \frac{1}{2}x$
   $x - 2y = 0$ (0, 0), (2, 1)

10. Find the equations of the asymptotes of the hyperbola. Then use a graphing calculator to sketch the hyperbola and the asymptotes on the same display screen.
    $\frac{(x-4)^2}{5^2} - \frac{(y+2)^2}{4^2} = 1$ $y = \frac{4}{5}x - \frac{26}{5}$, $y = -\frac{4}{5}x + \frac{6}{5}$, See margin.

Answers

4.

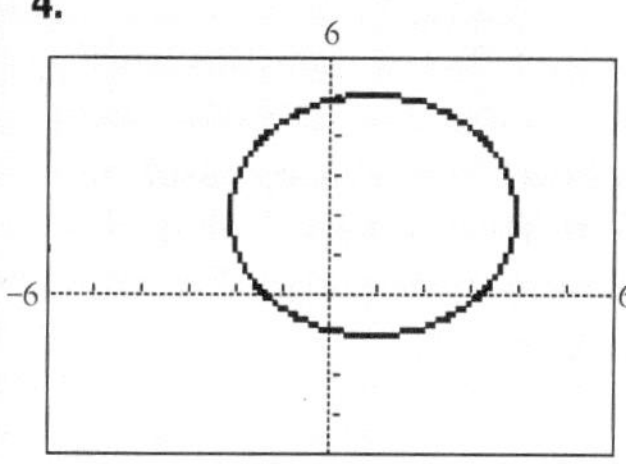

5.

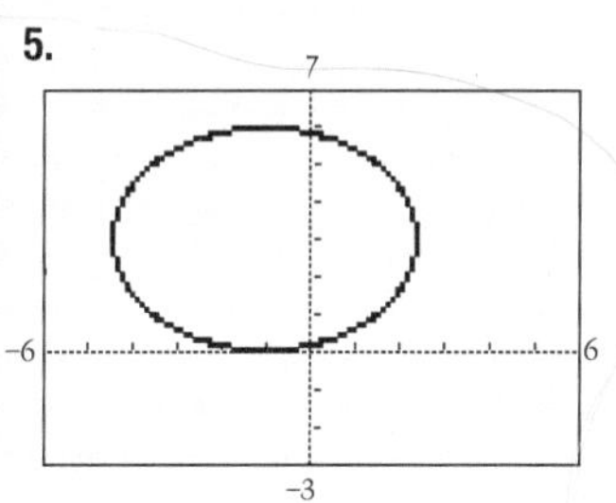

6.

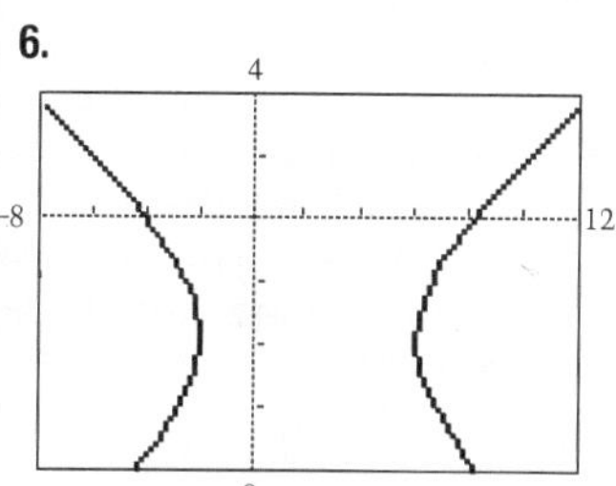

10.

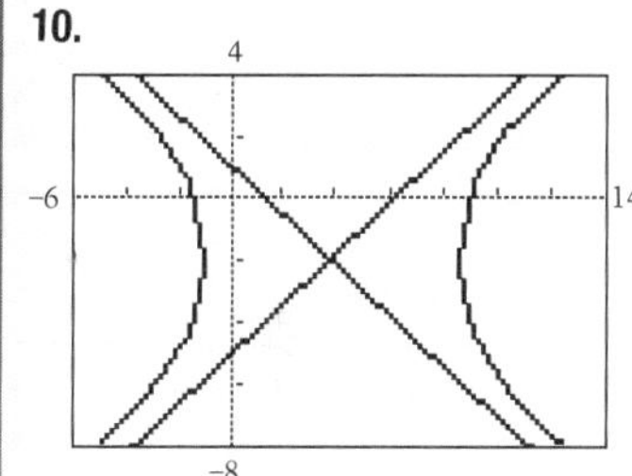

Problem of the Day

Each side of triangle ABC is 1. Arcs are drawn with centers at A, B, and C. (See diagram.) What is the perimeter of the "curved triangle"?

π

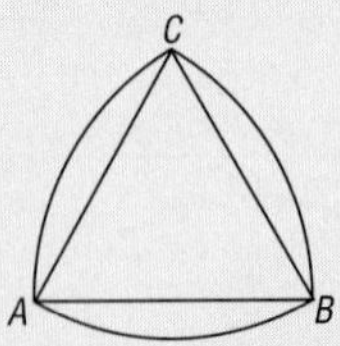

ORGANIZER

TIME SCHEDULE: All levels, two days

Warm-Up Exercises

1. Compute the discriminant of the following quadratic equations.
   a. $y = 3x^2 - 2x + 8$
   b. $y = 4 + 9x - 5x^2$
   a. -92 **b.** 161
2. Solve for y in terms of x and graph each "branch" of the relation $x^2 + y^2 = 0$.
   $y = \pm x$

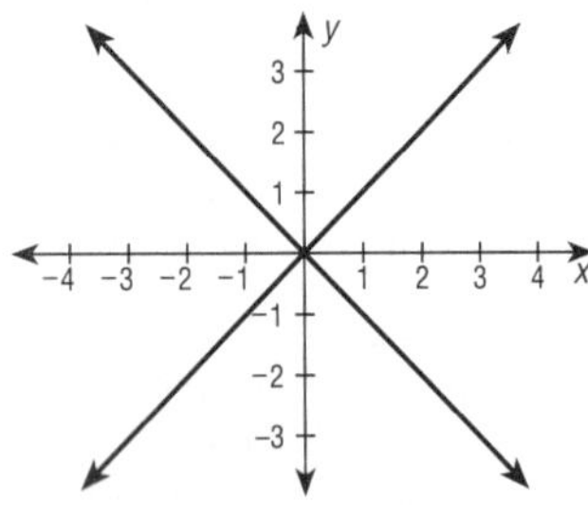

Lesson Resources

Teaching Tools
- Transparencies: 7, 8, 23
- Problem of the Day: 11.6
- Warm-up Exercises: 11.6
- Answer Masters: 11.6

Extra Practice: 11.6

Color Transparencies: 60, 61

Technology Handbook: p. 34, 35

11.6 Classifying Conics

What you should learn:

Goal 1 How to classify a conic from its equation

Goal 2 How to use equations of conics to solve mathematical and real-life problems

Why you should learn it:

You can use pairs of quadratic equations to solve many real-life problems, such as locating a ship at sea using the LORAN system of navigation.

Goal 1 Classifying a Conic from Its Equation

The reason that parabolas, circles, ellipses, and hyperbolas are called *conics* or *conic sections* is that each can be formed by the intersection of the plane and a double-napped *cone,* as shown below.

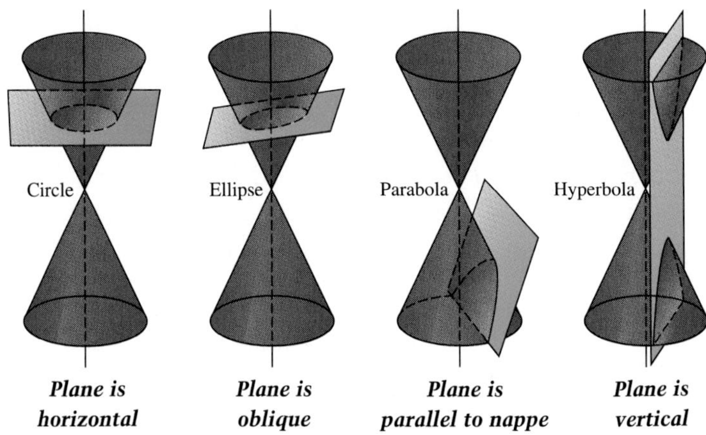

Plane is horizontal	***Plane is oblique***	***Plane is parallel to nappe***	***Plane is vertical***

Classifying Conics

If the graph of $Ax^2 + Bxy + Cy^2 + Dx + Ey + F = 0$ is a conic, then the type of conic is determined by the **discriminant,** $B^2 - 4AC$, as follows.

1. Ellipse or circle: $B^2 - 4AC < 0$
2. Parabola: $B^2 - 4AC = 0$
3. Hyperbola: $B^2 - 4AC > 0$

In case 1, the graph is a circle if $A = C$.

The equation $Ax^2 + Bxy + Cy^2 + Dx + Ey + F = 0$ is a **general second-degree equation** in x and y. Some second-degree equations in x and y have graphs that are not conics. For instance, the graph of $x^2 + y^2 = 0$ is a single point, and the graph of $x^2 - y^2 = 0$ is a pair of intersecting lines. Graphs of second-degree equations in x and y that are not one of the four "standard" conics are **degenerate conics.**

In a second-degree equation in x and y, if $B \neq 0$ and the graph is a parabola, an ellipse, or a hyperbola, then the axis of the graph in the coordinate plane is neither horizontal nor vertical.

Example 1 ***Classifying a Conic***

Classify the conic given by $xy - 1 = 0$.

Solution For this equation, $B = 1$ and $F = -1$. The other coefficients, A, C, D, and E, are zero. The discriminant is

$$B^2 - 4AC = 1^2 - 4(0)(0) = 1 > 0$$

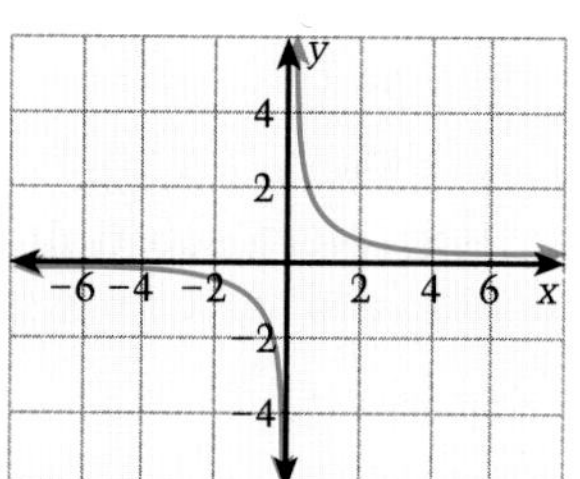

which implies that the graph is a hyperbola. To sketch the graph, write the equation in the form

$$y = \frac{1}{x}$$

and use the strategies described in Lesson 10.1 for sketching the graph of a rational function. The graph is shown at the left. The transverse axis lies on the line $y = x$, and its center is at the origin. ■

Example 2 ***Classifying a Conic***

Classify the conic given by $2x^2 + y^2 - 4x - 4 = 0$.

Solution For this equation, $A = 2$, $B = 0$, and $C = 1$. The discriminant is

$$B^2 - 4AC = 0^2 - 4(2)(1) = -8 < 0 \text{ and } A \neq C$$

which implies that the graph is an ellipse. To sketch the graph of the ellipse, complete the square as follows.

$$2x^2 + y^2 - 4x - 4 = 0$$
$$(2x^2 - 4x) + y^2 = 4$$
$$2(x^2 - 2x) + y^2 = 4$$
$$2(x^2 - 2x + 1) + y^2 = 4 + 2(1)$$
$$2(x - 1)^2 + y^2 = 6$$
$$\frac{(x - 1)^2}{3} + \frac{y^2}{6} = 1$$
$$\frac{(x - h)^2}{b^2} + \frac{(y - k)^2}{a^2} = 1$$

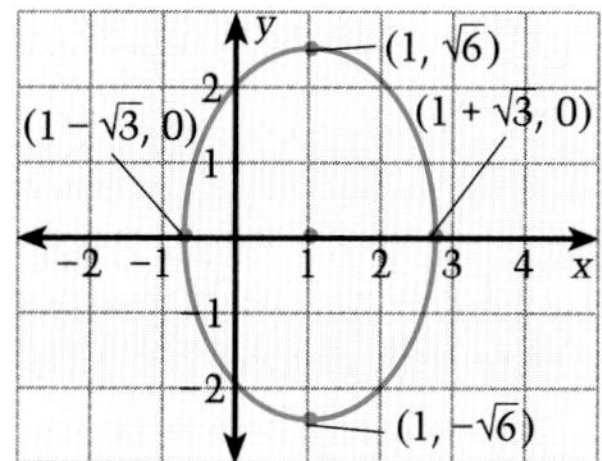

The center is $(h, k) = (1, 0)$. Because the denominator of the y^2-term is $a^2 = 6$, the vertices lie $\sqrt{6}$ units above and below the center. Because the denominator of the x^2-term is $b^2 = 3$, the co-vertices lie $\sqrt{3}$ units to the left and right of the center. The graph is shown at the left. ■

LESSON Notes

Note: Since the material contained in this lesson requires considerable student time, it is suggested that you present Goal 1, the classification of conics (Examples 1–2 along with Exercises 1–4 and 7–42), on the first day; Goal 2, applications of conics (Examples 3–5 along with Exercises 5–6 and 43–68), on the second day.

Example 1

Ask students to identify the vertices of the hyperbola. $(-1, -1)$ and $(1, 1)$

Example 2

Ask students whether the major axis is horizontal or vertical, and ask them to find the foci of the ellipse. vertical; $(1, -\sqrt{3})$ and $(1, \sqrt{3})$

Examples 3–4

The third method of solving systems of equations, graphing, can also be used to solve a system of quadratic equations. However, graphing the quadratic system is a good way of checking that the solutions obtained algebraically are reasonable.

Example 5

Ask students to solve the system using the substitution $x = \frac{24}{y}$. (You get $\left(\frac{24}{y}^2\right) - 25y^2 + 100 = 0$
$25y^4 - 100y^2 - 576 = 0$
$(5y^2 - 36)(5y^2 + 16) = 0$
The real solutions are $y = \pm\frac{6}{\sqrt{5}}$. From the diagram, the y-coordinate of the ship's location is $\frac{6}{\sqrt{5}} \approx 2.68$. The x-coordinate is ≈ 8.94.)

Extra Examples

Here are additional examples similar to **Examples 1–4.**

1. Classify the following conics.
   a. $5x^2 + 2y^2 - 20x + 4y + 24 = 0$
   b. $y^2 + 8x - 12y = 0$
   c. $-24x^2 + 18y^2 + 18 = 0$
   d. $16 - xy = 0$

   a. ellipse b. parabola
   c. hyperbola d. hyperbola
2. Solve the following quadratic systems.
   a. $x^2 + y^2 + 4x - 6y - 12 = 0$
   $3x^2 + 12x - 16y + 60 = 0$
   b. $25x^2 + 9y^2 - 100x - 18y - 116 = 0$
   $x^2 + 9y^2 - 4x - 18y + 4 = 0$

   a. $(-6, 6)$ and $(2, 6)$
   b. $(-1, 1)$ and $(5, 1)$

Goal 2 Using Conics to Solve Problems

In Lesson 11.2, you learned how to find points of intersection of a line and a conic by solving a *linear-quadratic system.* The next two examples show how to find the points of intersection of two conics by solving a *quadratic system.*

Example 3 Solving a Quadratic System by Substitution

Find the points of intersection of the graphs of the system.

$$x^2 + 4y^2 - 4x - 8y + 4 = 0 \quad \textit{Equation 1}$$
$$x^2 + 4y - 4 = 0 \quad \textit{Equation 2}$$

Solution Because the second equation has no y^2-term, solve that equation for y to obtain $y = 1 - \frac{1}{4}x^2$. Next, substitute this expression for y in the first equation and solve for x.

$$x^2 + 4y^2 - 4x - 8y + 4 = 0$$
$$x^2 + 4\left(1 - \tfrac{1}{4}x^2\right)^2 - 4x - 8\left(1 - \tfrac{1}{4}x^2\right) + 4 = 0$$
$$x^2 + 4 - 2x^2 + \frac{1}{4}x^4 - 4x - 8 + 2x^2 + 4 = 0$$
$$\tfrac{1}{4}x^4 + x^2 - 4x = 0$$
$$x^4 + 4x^2 - 16x = 0$$
$$x(x - 2)(x^2 + 2x + 8) = 0$$

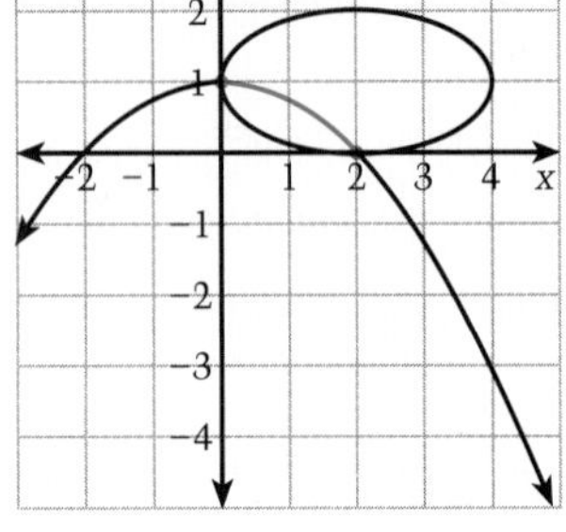

There are two real solutions: $x = 0$ and $x = 2$. The corresponding y-values are $y = 1$ and $y = 0$. Thus the graphs have two points of intersection, $(0, 1)$ and $(2, 0)$, as shown at the left. ■

Example 4 Solving a Quadratic System by Elimination

Find the points of intersection of the graphs of the system.

$$x^2 + y^2 - 16x + 39 = 0 \quad \textit{Equation 1}$$
$$x^2 - y^2 - 9 = 0 \quad \textit{Equation 2}$$

Solution For this system, eliminate the y^2-term by adding the two equations. The resulting equation can be solved for x.

$$2x^2 - 16x + 30 = 0$$
$$2(x - 3)(x - 5) = 0$$

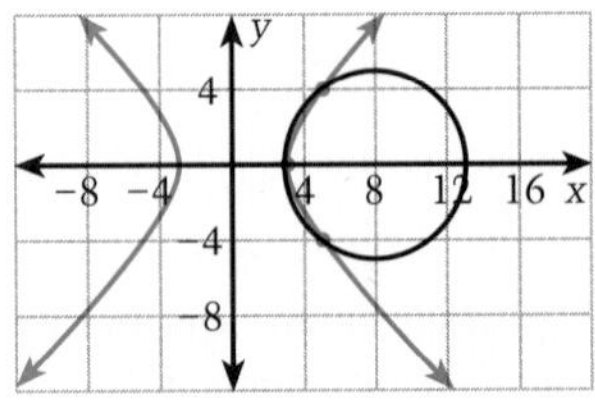

There are two real solutions: $x = 3$ and $x = 5$. The corresponding y-values are $y = 0$ and $y = \pm 4$. Thus the graphs have three points in common, $(3, 0)$, $(5, 4)$, and $(5, -4)$, as shown at the left. ■

LORAN (long distance radio navigation) uses synchronized pulses sent out by pairs of transmitting stations. By calculating the difference in the times of arrival of the pulses from two stations, the LORAN equipment locates the ship on a hyperbola. By doing the same thing with a second pair of stations, the ship's location will be the intersection of two hyperbolas.

Real Life Navigation

Example 5 *Finding the Point of Intersection of Two Hyperbolas*

LORAN equipment has calculated a ship's position to be the point of intersection of the graphs of the following hyperbolas. Find the coordinates of the point of intersection.

$$xy - 24 = 0 \quad \textit{Equation 1}$$

$$x^2 - 25y^2 + 100 = 0 \quad \textit{Equation 2}$$

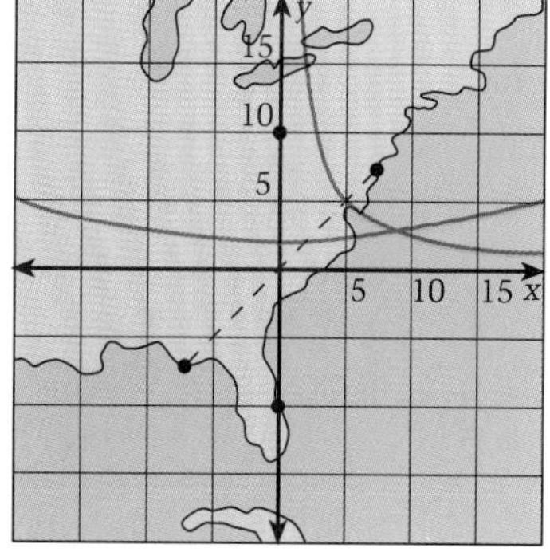

Solution Solve the first equation for y to obtain $y = \frac{24}{x}$. Then substitute this expression for y in the second equation.

$$x^2 - 25y^2 + 100 = 0$$

$$x^2 - 25\left(\frac{24}{x}\right)^2 + 100 = 0$$

$$x^4 + 100x^2 - 14{,}400 = 0$$

$$(x^2 - 80)(x^2 + 180) = 0$$

There are two real solutions: $x = \pm\sqrt{80} \approx \pm 8.94$. From the diagram, the positive solution must represent the x-coordinate of the ship's location. The y-coordinate is $\frac{24}{\sqrt{80}} \approx 2.68$. Thus the ship's position is (8.94, 2.68). ■

Communicating about ALGEBRA

SHARING IDEAS about the Lesson

A. Classify the conic $4x^2 + 9y^2 - 16x - 18y - 11 = 0$.
B. Classify the conic $4x^2 - 9y^2 - 16x + 18y - 29 = 0$.
C. Sketch the conics in parts **A** and **B** and find all points of intersection. **A.** Ellipse **B.** Hyperbola **C.** (−1, 1), (5, 1)

See margin.

Check Understanding

1. Why are parabolas, circles, ellipses, and hyperbolas called conic sections? See page 606.
2. What is a degenerate conic? The graphs of second degree equations in x and y that are not one of the four "standard conics."
3. Can the methods of solving systems of linear equations be used to solve systems of quadratic equations? Yes; see Examples 3–5.

Communicating about ALGEBRA

Answer

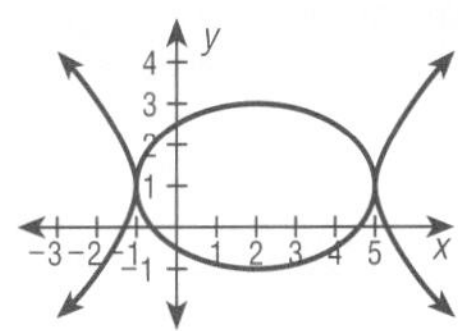

EXTEND *Communicating*
Show that the following system has the four real solutions: (0, 2), (0, −2), (2, 2), and (2, −2).

(1) $2x^2 + 5y^2 - 4x - 20 = 0$
(2) $3x^2 - y^2 - 6x + 4 = 0$

One method is to multiply the second equation by 5 to obtain the equation

(3) $15x^2 - 5y^2 - 30x + 20 = 0$

Add equations (1) and (3) to eliminate the y^2-term, then solve for x.

$17x^2 - 34x = 0$
$17x(x - 2) = 0$
$x = 0$ or $x = 2$

Substitute the values of x in one of the original equations to find y. When $x = 0$, $y = \pm 2$. When $x = 2$, $y = \pm 2$. So, the four real solutions are (0, 2), (0, −2), (2, 2), and (2, −2).

EXERCISE Notes

ASSIGNMENT GUIDE

Day 1

Basic/Average: Ex. 1–4, 7–21 odd, 23–30, 31, 37, 41

Above Average: Ex. 1–4, 7–21 odd, 23–30, 31–39 odd

Advanced: Ex. 1–4, 7–21 odd, 23–30, 31–39 odd

Day 2

Basic/Average: Ex. 5–6, 43–53 odd, 55, 59, 61–64

Above Average: Ex. 5–6, 43–53 odd, 55–66

Advanced: Ex. 5–6, 43–53 odd, 55–68

Selected Answers
Exercises 1–6, 7–63 odd

More Difficult Exercises
Exercises 55–60, 65–68

Guided Practice

Ex. 1 It might be helpful to bring a clear plastic model to class.

Ex. 3 Consider the equation of a point given in the Enrichment Activities of Lesson 11.5 on page 603.

Ex. 5–6 Encourage students to solve the equations intuitively by inspecting their graphs. The graphs will at least confirm the number of solutions.

Answers

5. Substitute $-y$ for x^2 in the first equation and solve for y; then solve for x by using $x^2 = -y$.
6. Add the equations and solve for x; then solve for y by substituting for x.

Independent Practice

Ex. 7–22 Be sure students illustrate their use of the discriminant $B^2 - 4AC$ as justification.

EXERCISES

Guided Practice

CRITICAL THINKING about the Lesson

1. How must a plane intersect a double-napped cone to form a circle? an ellipse? a parabola? a hyperbola? See page 606.
2. How can the discriminant $B^2 - 4AC$ be used to classify the graph of $Ax^2 + Bxy + Cy^2 + Dx + Ey + F = 0$? See page 606.
3. Match the equation with one of the figures below. In each case, describe the relationship between the plane and the cone. See below.

   a. $x^2 - y^2 = 0$ **b.** $x + y = 0$ **c.** $x^2 + y^2 = 0$

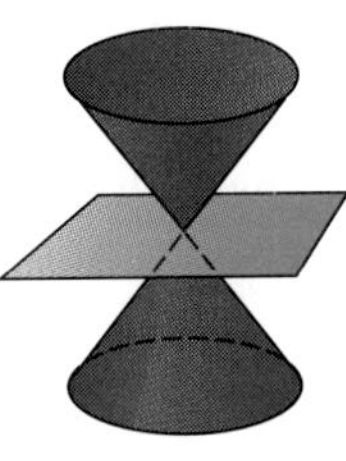

Point

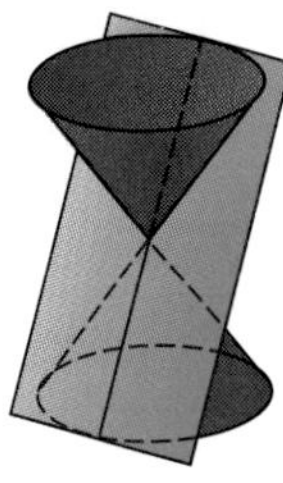

Line

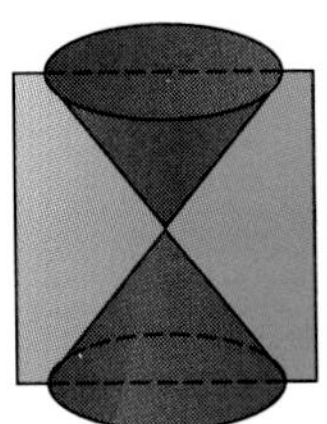

Intersecting Lines

4. If the axis of a parabola, an ellipse, or a hyperbola is horizontal or vertical, what must be true of B in the following equation? $B = 0$

   $Ax^2 + Bxy + Cy^2 + Dx + Ey + F = 0$

In Exercises 5 and 6, solve the system. Explain your steps. See margin.

5. $x^2 + y^2 - 4y = 0$ (0, 0)
   $x^2 + y = 0$
6. $x^2 + y^2 - 4x = 0$ (0, 0), $(3, \sqrt{3})$, $(3, -\sqrt{3})$
   $x^2 - y^2 - 2x = 0$

Independent Practice

In Exercises 7–22, classify the conic.

7. $9x^2 + 4y^2 + 36x - 24y + 36 = 0$ Ellipse
8. $x^2 - 9y^2 + 2x - 54y - 80 = 0$ Hyperbola
9. $x^2 - 9y^2 + 36y - 72 = 0$ Hyperbola
10. $x^2 + y^2 - 10x - 2y + 10 = 0$ Circle
11. $16x^2 + 25y^2 - 36x - 50y + 61 = 0$ Ellipse
12. $x^2 + y^2 - 16x + 4y + 67 = 0$ Circle
13. $9y^2 - x^2 + 2x + 54y + 62 = 0$ Hyperbola
14. $9x^2 + 25y^2 - 36x - 50y + 61 = 0$ Ellipse
15. $x^2 - 2x + 8y + 9 = 0$ Parabola
16. $y^2 - 4y - 4x = 0$ Parabola
17. $12x^2 + 20y^2 - 12x + 40y - 37 = 0$ Ellipse
18. $9x^2 - y^2 + 54y + 10y + 55 = 0$ Hyperbola
19. $x^2 + y^2 - 4x - 2y - 4 = 0$ Circle
20. $36x^2 + 9y^2 + 48x - 36y + 43 = 0$ Ellipse
21. $16y^2 - x^2 + 2x + 64y + 63 = 0$ Hyperbola
22. $x^2 - 2x + 8y + 17 = 0$ Parabola

3. a: Intersecting lines, a vertical plane contains the axis of the cone; b: Line, a plane is tangent to the cone; c: Point, a plane intersects the cone only at its vertex.

In Exercises 23–30, match the equation with its graph.

23. $x^2 + 4y^2 - 4x + 8y - 8 = 0$ **b**

24. $y^2 - x^2 + 12y + 4x + 31 = 0$ **f**

25. $x^2 + 6x + 2y + 5 = 0$ **d**

26. $x^2 + y^2 - 4x - 6y - 3 = 0$ **a**

27. $x^2 + y^2 + 2x - 2y - 7 = 0$ **c**

28. $y^2 - 2y - 4x + 9 = 0$ **e**

29. $4x^2 - 9y^2 - 16x - 20 = 0$ **g**

30. $25x^2 + 4y^2 + 100x + 16y + 16 = 0$ **h**

a.
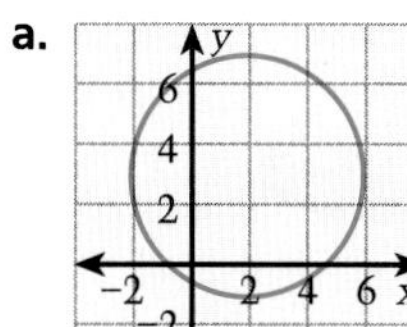

b.
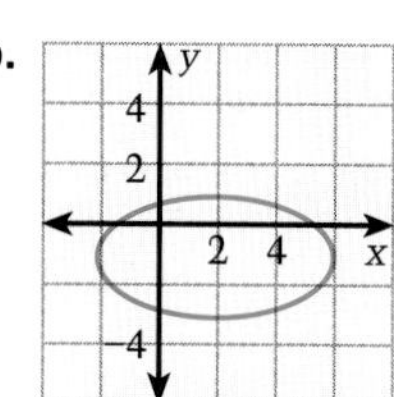
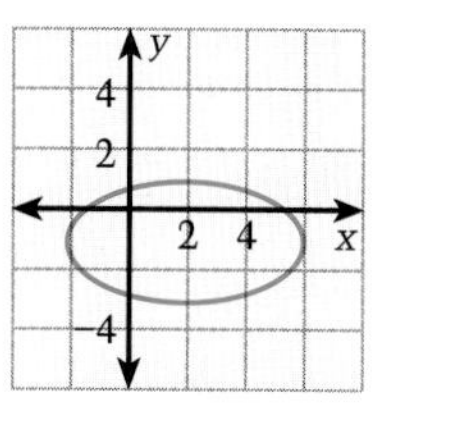

c.
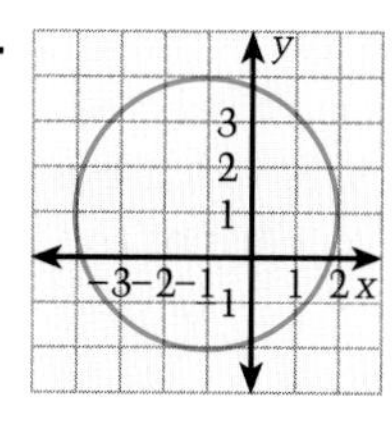

d.
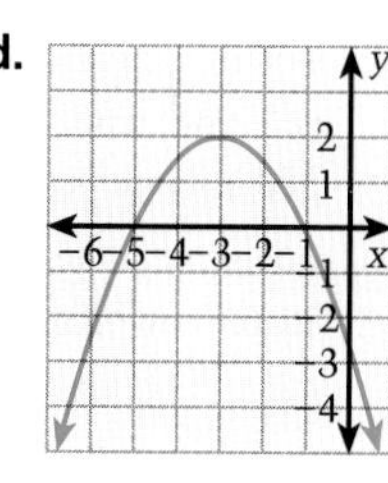

e.
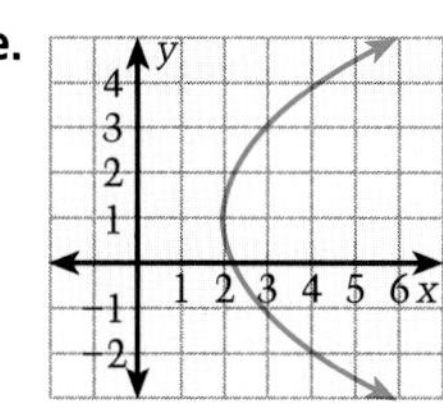

f.
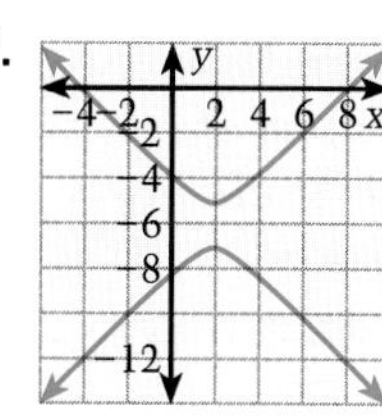

g.
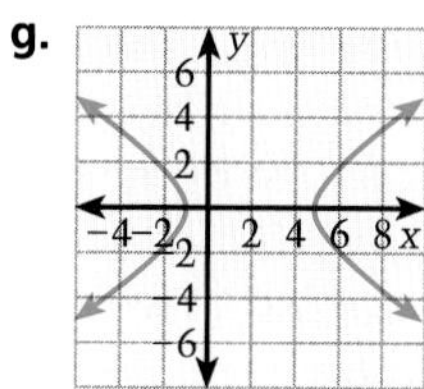

h.
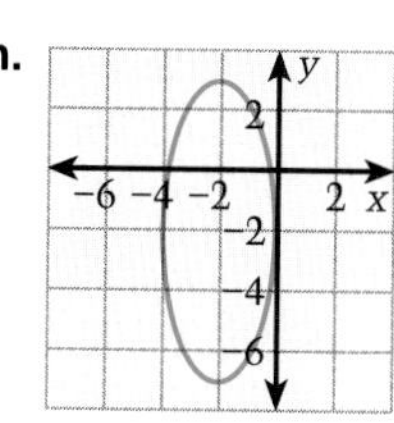

In Exercises 31–42, sketch the graph of the equation. See Additional Answers.

31. $x^2 + 4x + 4y - 16 = 0$

32. $y^2 - 2x - 20y + 94 = 0$

33. $15x^2 + 15y^2 - 150x - 30y + 330 = 0$

34. $x^2 + y^2 - 20x - 16y + 64 = 0$

35. $16x^2 + y^2 + 160x - 22y + 505 = 0$

36. $4x^2 - 9y^2 + 32x - 144y - 548 = 0$

37. $y^2 - 16x - 14y + 17 = 0$

38. $x^2 + y^2 - 12x - 24y + 36 = 0$

39. $25x^2 + 16y^2 + 50x - 320y + 1225 = 0$

40. $x^2 + 9y^2 - 28x - 36y + 151 = 0$

41. $-25x^2 + y^2 + 50x + 20y + 50 = 0$

42. $4x^2 - y^2 - 8x + 4y - 9 = 0$

In Exercises 43–54, find the points of intersection of the graphs.

43. $16x^2 - y^2 + 16y - 128 = 0$
$y^2 - 48x - 16y - 32 = 0$

44. $16x^2 + y^2 - 24y + 80 = 0$ (0, 4)
$16x^2 + 25y^2 - 400 = 0$

45. $-x^2 + y^2 + 4x - 6y + 4 = 0$ (2, 4), (2, 2)
$x^2 + y^2 - 4x - 6y + 12 = 0$

46. $x^2 + y^2 + 8x - 20y + 7 = 0$ (−7, 0), (−1, 0)
$x^2 + 9y^2 + 8x + 4y + 7 = 0$

47. $4x^2 + y^2 + 32x - 24y + 64 = 0$ (−10, 12)
$4x^2 + y^2 + 56x - 24y + 304 = 0$

48. $x^2 - 4y^2 - 20x - 64y - 172 = 0$ (6, −8),
$4x^2 + y^2 - 80x + 16y + 400 = 0$ (14, −8)

49. $x^2 + y^2 + 2x + 2y = 0$ (0, 0)
$x^2 + y^2 - 2x - 2y = 0$

50. $x^2 + y^2 + 4x + 3 = 0$ (−1, 0)
$64x^2 - y^2 - 64 = 0$

51. $x^2 - y^2 - 12x + 12y - 36 = 0$ (0, 6), (12, 6)
$x^2 + y^2 - 12x - 12y + 36 = 0$

52. $x^2 + y^2 + 36x - 10y + 324 = 0$ (−18, 0)
$x^2 + y^2 + 36x - 20y + 324 = 0$

53. $x^2 + 4y^2 - 2x - 8y + 1 = 0$ (1, 0)
$x^2 - 2x + 4y + 1 = 0$

54. $x^2 + y^2 + 4y - 12 = 0$ (4, −2), (−4, −2)
$x^2 - 16y^2 - 64y - 80 = 0$

43. $(-2, 8), (5, 8 \pm 4\sqrt{21})$

Ex. 23–30 **WRITING** Assign these exercises as a group. Have students write a one-sentence justification for their choices.

Ex. 31–42 Students should use quick-graph techniques to sketch the equations. If graphics calculators or computer graphing programs are available, you might have students work with partners to graph the equations.

Ex. 43–54 Refer students to Examples 3 and 4 for solution models. If the technology is available, have students graph the equations in order to visualize the solutions.

Ex. 57–58 **COOPERATIVE LEARNING** To encourage creativity in the method of solution and connections with the lesson, have students work with partners to complete the exercises in class.

Answers

57. The following table shows how the tangent circle is related to the three circles.

	Small	Medium	Large
	outside	inside	outside
	outside	inside	inside
1	inside	outside	outside
2	outside	outside	inside
3	inside	inside	outside
4	inside	outside	inside
5	inside	inside	inside
6	outside	outside	outside

58.

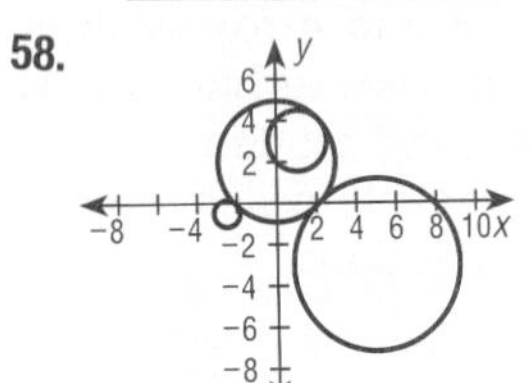

EXTEND **Ex. 59** Ask students to interview the school's earth science teacher to verify the data given in the exercise.

▶ **Ex. 61–62** **CONNECTIONS** You may want to compare the trends illustrated in these real-life applications to those of your local area. Has your county or state gotten new area codes? What apples are grown in your area?

▶ **Ex. 65–68** Conic graph paper can be found in *Teaching Tools: Transparencies and Copy Masters.*

55. *Sharpening a Pencil* The sharpened part of the pencil at the right has a conical shape. What type of conic occurs at the intersection of the sharpened cone and each flat side of the pencil? Explain.

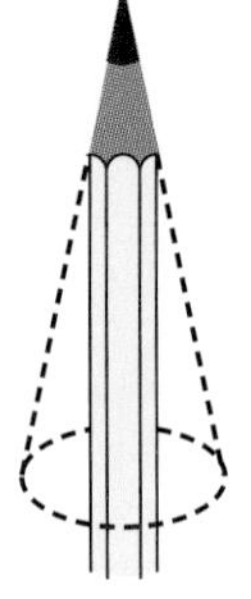

56. *Tipping an Hourglass* The hourglass at the right has the shape of a double-napped cone. The hourglass has the same amount of sand inside each nappe. When the hourglass is laid on its side, what type of conic is traced by the top edge of the sand? Explain.

57. *Circles of Apollonius* Apollonius (*circa* 225 B.C.) lived during the classical period of Greece. He proposed the following problem.

Suppose you are given any three nonintersecting circles in a plane. The circles can have different radii, but none of the three should be inside one of the other two. Is it always possible to find a fourth circle that is tangent to each of the three given circles? (Two circles are tangent if their intersection is exactly one point.)

Apollonius showed that it *is* always possible to find a fourth "tangent circle." In fact, he showed that it is always possible to find eight such tangent circles! Two of the solutions are shown below. Find the other six. See margin.

55. Hyperbola; plane through cone is vertical
56. Hyperbola; plane through cone is vertical

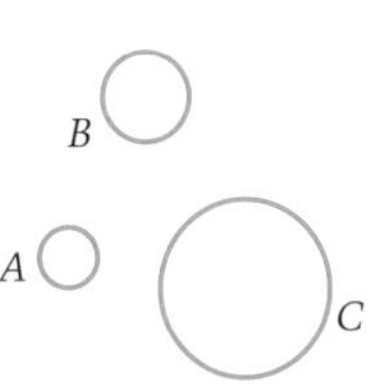

Given Three Circles

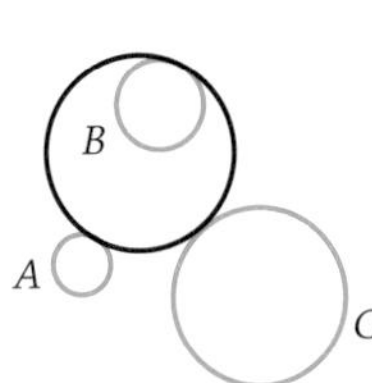

One Possible Solution

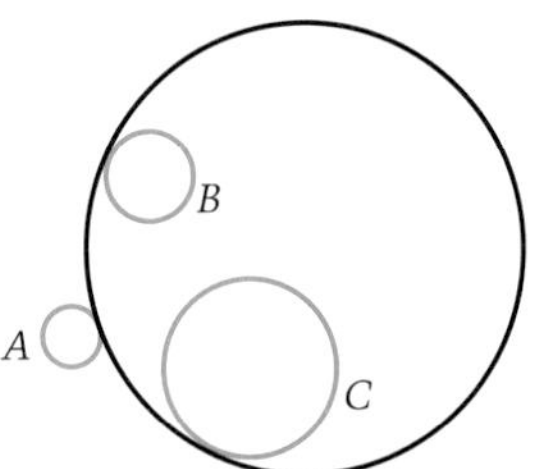

Another Possible Solution

58. *More About Circles* Which of the two solutions shown in Exercise 57 could be represented by the following system of equations? Justify your answer graphically. First solution, see margin.

$$x^2 + y^2 - 4y - 4 = 0$$
$$x^2 + y^2 + 5x + y + 6 = 0$$
$$x^2 + y^2 - 2x - 6y + 8 = 0$$
$$x^2 + y^2 - 10x + 6y + 16 = 0$$

59. *Seismographs and Earthquakes* Three seismographs are located at the points (0, 0), (8, −8), and (11, 10), where each unit in the coordinate plane represents 1 mile. An earthquake is recorded by each of the seismographs. The epicenter of the earthquake is 5 miles from the first seismograph, 13 miles from the second, and 10 miles from the third. What are the coordinates of the epicenter? Explain your answer graphically. See below.

59. (3, 4); draw a circle from each seismograph's location, then write the coordinates of the point where all three circles intersect.

60. *Locating an Explosion* Two of your friends live 4 miles apart and on the same "east-west" street, and you live halfway between them. You are all talking on a three-way phone call when you hear an explosion. Six seconds later your friend to the east hears the explosion, and your friend to the north hears it 8 seconds after you do. Find the equations of two hyperbolas that would locate the explosion. (Remember sound travels at a rate of 1100 ft/sec.)

Integrated Review

61. *Telephone Area Codes* What percentage of the 144 possible area codes were in use in 1991? ≈95%

62. *Delicious Apples* Of the top five varieties of apples picked in 1990, what percent were of the red delicious variety? ≈58%

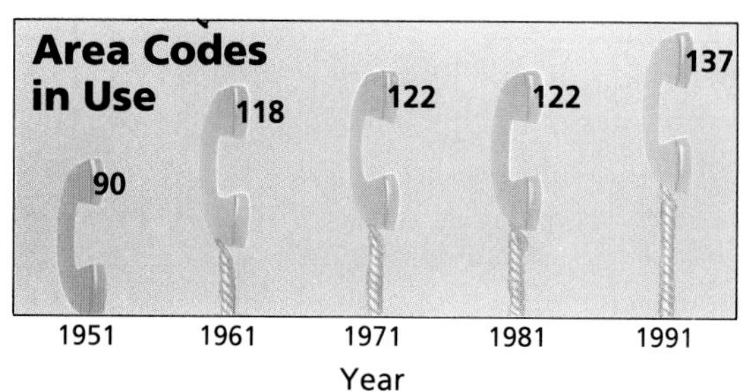

63. *College Entrance Exam Sample* If x and y are two different real numbers and $rx = ry$, then $r = \boxed{?}$. a

a. 0 **b.** 1 **c.** $\frac{x}{y}$ **d.** $\frac{y}{x}$ **e.** $x - y$

64. *College Entrance Exam Sample* A cube has 4-inch edges. If each edge is increased by 25%, then the volume is increased by about d

a. 25% **b.** 48% **c.** 73% **d.** 95% **e.** 122%

Exploration and Extension

Conic Graph Paper **In Exercises 65 and 66, use conic graph paper (shown below).**

See Additional Answers.

65. Sketch a hyperbola whose foci are the centers of the concentric circles. The difference of the distances between each point on the hyperbola and the foci is 10.

66. Sketch an ellipse whose foci are the centers of the concentric circles. The sum of the distances between each point on the ellipse and the foci is 14.

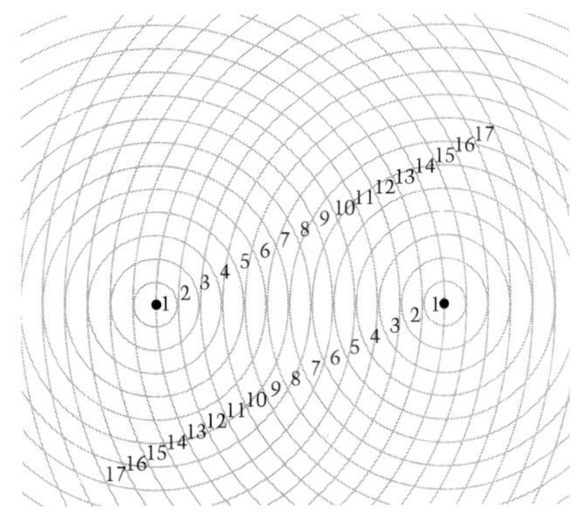

67. Sketch the graph of $x^2 - y^2 + 18x + 8y + 65 = 0$.

68. Sketch the graph of $x^2 + y^2 - 2x + 4y + 5 = 0$.

67. and **68.** See Additional Answers.

60. $\frac{x^2}{10{,}890{,}000} - \frac{y^2}{16{,}988{,}400} = 1$, $\frac{(x + 10{,}560)^2}{19{,}360{,}000} - \frac{y^2}{8{,}518{,}400} = 1$

Enrichment Activity

This activity requires students to go beyond lesson goals.

An Alternate Approach
Using Technology
The following BASIC program can be used to determine the type of conic section from its general-form equation. If you have access to a computer, use the program to check your answers to Exercises 7–22. (If you do not have a computer, but you have a graphics calculator, you should consult the manual for directions on how to program the calculator.)

```
10   PRINT "A*X^2 + C*Y^2
       + D*X + E*Y + F = 0
20   PRINT "ENTER A, C, D, E, F"
25   INPUT A, C, D, E, F
27   IF (A = 0) AND (C = 0)
       THEN 100
30   IF A < >0 THEN 60
40   IF D = 0 THEN 100
50   GOTO 120
60   IF C < >0 THEN 80
70   GOTO 120
80   LET F1 = D^2/(4*A)
       + E^2/(4*A)
90   IF F1 < >F THEN 120
100  PRINT "DEGENERATE
       CASE": GOTO 190
120  IF A = C THEN PRINT
       "CIRCLE": GOTO 190
140  IF A*C < 0 THEN PRINT
       "HYPERBOLA": GOTO 190
160  IF A*C > 0 THEN PRINT
       "ELLIPSE": GOTO 190
180  PRINT "PARABOLA"
190  END
```

ORGANIZER

TIME SCHEDULE: All levels, two days.

The Chapter Summary helps students organize the main ideas of the chapter. In this chapter, we studied the four basic conic sections: circles, parabolas, ellipses, and hyperbolas. Attention was given to expressing the equations of conic sections in a standard form that makes them easier to graph.

Work with students to review the skills, strategies, and concepts of the chapter. The first day's homework assignment can be used as the basis for the second day of review.

COOPERATIVE LEARNING Encourage students to study together. Emphasize the importance of "teaching" a classmate how to perform a skill or how to recall a strategy. When students work together, everyone wins. The students receiving help get additional instruction and the students giving help gain a deeper understanding of the skills and concepts involved.

Chapter SUMMARY

Review the chapter by asking students, "What did you learn? Why did you learn it? How does it fit with the other algebra skills and concepts you have learned?"

11 Chapter Summary

What did you learn?

Skills

1. Write an equation of a conic.
   - Write an equation of a parabola. **(11.1, 11.5)**
   - Write an equation of a circle. **(11.2, 11.5)**
   - Write an equation of an ellipse. **(11.3, 11.5)**
   - Write an equation of a hyperbola. **(11.4, 11.5)**
2. Find characteristics of conics.
   - Find vertex, focus, and directrix of a parabola. **(11.1, 11.5)**
   - Find center and radius of a circle. **(11.2, 11.5)**
   - Find vertices, foci, and co-vertices of an ellipse. **(11.3, 11.5)**
   - Find vertices, foci, and asymptotes of a hyperbola. **(11.4, 11.5)**
3. Find points of intersection of two graphs.
   - Linear-quadratic systems **(11.2–11.4)**
   - Quadratic systems **(11.6)**
4. Classify the graph of a general second-degree equation. **(11.6)**

Strategies

5. Use conics to solve real-life problems. **(11.1–11.6)**

Why did you learn it?

Conic sections are used in many different types of construction problems. From earlier work with parabolas, you know that suspension bridges include parabolic shapes. In this chapter, you learned that parabolas have reflective properties that can be used in the construction of antennas and light reflectors. You also studied many applications of circles, ellipses, and hyperbolas. For instance, you learned that the navigational system LORAN involves finding the point of intersection of two hyperbolas.

How does it fit into the bigger picture of algebra?

In this chapter, you studied the graphs of quadratic relations of the form

$$Ax^2 + Bxy + Cy^2 + Dx + Ey + F = 0.$$

The graph of a quadratic relation can be a *parabola*, a *circle*, an *ellipse*, or a *hyperbola.* Each of these graphs are called *conic sections* or simply *conics.*

In the first four lessons, you studied properties of the four types of conics. To keep the equations simple, the conics in the first four lessons have their centers (or in the case of a parabola, its vertex) at the origin. In the fifth lesson, you studied equations of conics that have been translated.

Chapter REVIEW

In Exercises 1–9, write an equation of the specified conic. (11.1–11.5) 6. – 8. See below.

1. *Circle* $x^2 + y^2 = 25$
   Center: (0, 0)
   Radius: 5

2. *Circle* $(x - 3)^2 + (y + 1)^2 = 4$
   Center: (3, −1)
   Radius: 2

3. *Parabola* $(y - 2)^2 = -8(x - 3)$
   Vertex: (3, 2)
   Focus: (1, 2)

4. *Parabola*
   Vertex: (−1, 2)
   Focus: (−1, 0)
   $(x + 1)^2 = -8(y - 2)$

5. *Ellipse* $\frac{(x-2)^2}{4} + \frac{(y-2)^2}{1} = 1$
   Vertices: (0, 2), (4, 2)
   Co-vertices: (2, 3), (2, 1)

6. *Ellipse*
   Vertices: (−1, 6), (−1, −2)
   Co-vertices: (1, 2), (−3, 2)

7. *Ellipse*
   Vertices: (2, 0), (2, 4)
   Foci: (2, 1), (2, 3)

8. *Hyperbola*
   Vertices: (2, 0), (6, 0)
   Foci: (0, 0), (8, 0)

9. *Hyperbola*
   Vertices: (2, 3), (2, −3)
   Foci: (2, 5), (2, −5)
   $\frac{(y)^2}{9} - \frac{(x-2)^2}{16} = 1$

In Exercises 10–15, find the vertex, focus, and directrix of the parabola. (11.1, 11.5) See margin.

10. $y = 4x^2$
11. $y = 2x^2$
12. $y^2 = -6x$
13. $y^2 = 3x$
14. $(x - 1) + 8(y + 2)^2 = 0$
15. $(x + 3) + (y - 2)^2 = 0$

In Exercises 16–21, find the center, vertices, and foci of the ellipse. (11.3, 11.5) See margin.

16. $\frac{x^2}{25} + \frac{y^2}{16} = 1$
17. $\frac{x^2}{144} + \frac{y^2}{169} = 1$
18. $(x + 2)^2 + 4(y + 4)^2 = 1$
19. $\frac{(x-5)^2}{16} + \frac{y^2}{9} = 1$
20. $\frac{(x-1)^2}{9} + \frac{(y-5)^2}{25} = 1$
21. $\frac{(x+3)^2}{25} + \frac{(y-1)^2}{36} = 1$

In Exercises 22–27, find the center, vertices, foci, and asymptotes of the hyperbola. (11.4, 11.5) See margin.

22. $\frac{y^2}{25} - \frac{x^2}{144} = 1$
23. $\frac{x^2}{36} - \frac{y^2}{4} = 1$
24. $\frac{(x-1)^2}{4} - \frac{(y+2)^2}{1} = 1$
25. $\frac{(x+1)^2}{144} - \frac{(y-4)^2}{25} = 1$
26. $(y + 6)^2 - (x - 2)^2 = 1$
27. $\frac{(y-1)^2}{1/4} - \frac{(x+3)^2}{1/9} = 1$

In Exercises 28–31, match the equation with its graph. (11.5)

28. $\frac{(x-3)^2}{9} + \frac{(y+5)^2}{4} = 1$ b
29. $(x - 5)^2 + (y - 3)^2 = 4$ d
30. $\frac{(x+3)^2}{9} - \frac{(y-5)^2}{4} = 1$ c
31. $(y - 3)^2 - (x + 5) = 0$ a

a.
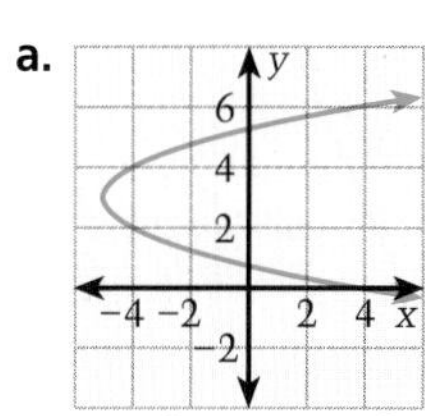

b.
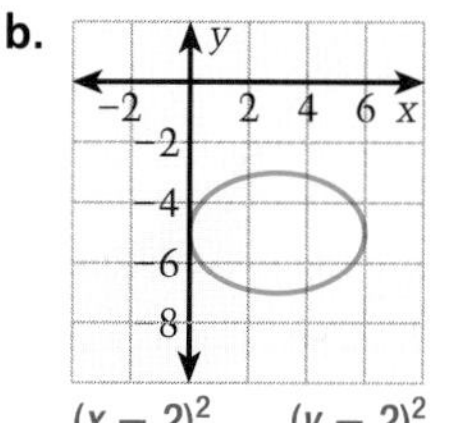

c.
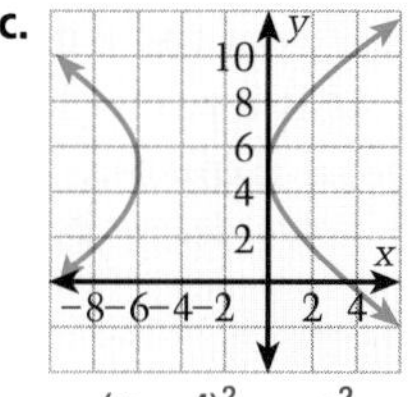

d.
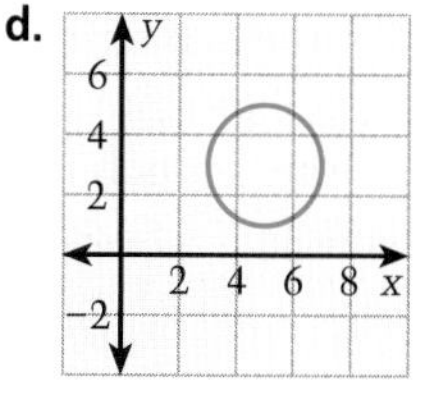

6. $\frac{(x+1)^2}{16} + \frac{(y-2)^2}{4} = 1$
7. $\frac{(x-2)^2}{3} + \frac{(y-2)^2}{4} = 1$
8. 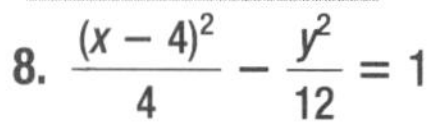 $\frac{(x-4)^2}{4} - \frac{y^2}{12} = 1$

Chapter REVIEW

ASSIGNMENT GUIDE

Basic/Average: Ex. 3–72 multiples of 3, 79

Above Average: Ex. 3–72 multiples of 3, 75–77, 81–82

Advanced: Ex. 3–72 multiples of 3, 75–77, 81–82

✪ More Difficult Exercises
Exercises 75–82

▶ **Ex. 1–82** The Chapter Review exercises include a reference to the lesson in which the skill or strategy was first presented. Students may use these references as a study aid if they are uncertain about how to do a given exercise in the Chapter Review.

Answers

10. (0, 0), $\left(0, \frac{1}{16}\right)$, $y = -\frac{1}{16}$
11. (0, 0), $\left(0, \frac{1}{8}\right)$, $y = -\frac{1}{8}$
12. (0, 0), $\left(-\frac{3}{2}, 0\right)$, $x = \frac{3}{2}$
13. (0, 0), $\left(\frac{3}{4}, 0\right)$, $x = -\frac{3}{4}$
14. (1, −2), $\left(\frac{31}{32}, -2\right)$, $x = \frac{33}{32}$
15. (−3, 2), $\left(-\frac{13}{4}, 2\right)$, $x = -\frac{11}{4}$
16. (0, 0); (−5, 0), (5, 0); (−3, 0), (3, 0)
17. (0, 0); (0, −13), (0, 13); (0, −5), (0, 5)
18. (−2, −4); (−3, −4), (−1, −4); $\left(-2 - \frac{\sqrt{3}}{2}, -4\right)$, $\left(-2 + \frac{\sqrt{3}}{2}, -4\right)$
19. (5, 0); (1, 0), (9, 0); $(5 - \sqrt{7}, 0)$, $(5 + \sqrt{7}, 0)$
20. (1, 5); (1, 0), (1, 10); (1, 1), (1, 9)
21. (−3, 1); (−3, −5), (−3, 7); $(-3, 1 - \sqrt{11})$, $(-3, 1 + \sqrt{11})$
22. (0, 0); (0, −5), (0, 5); (0, −13), (0, 13); $y = \pm\frac{5}{12}x$
23. (0, 0); (−6, 0), (6, 0); $(-\sqrt{40}, 0)$, $(\sqrt{40}, 0)$; $y = \pm\frac{1}{3}x$
24. (1, −2); (−1, −2), (3, −2); $(1 - \sqrt{5}, -2)$, $(1 + \sqrt{5}, -2)$; $y = \frac{x}{2} - \frac{5}{2}$, $y = -\frac{x}{2} - \frac{3}{2}$

Answers

25. $(-1, 4)$; $(-13, 4)$, $(11, 4)$; $(-14, 4)$, $(12, 4)$; $y = \frac{5}{12}x + \frac{53}{12}$, $y = -\frac{5}{12}x + \frac{43}{12}$

26. $(2, -6)$; $(2, -7)$, $(2, -5)$; $(2, -6 - \sqrt{2})$, $(2, -6 + \sqrt{2})$; $y = x - 8$, $y = -x - 4$

27. $(-3, 1)$; $\left(-3, \frac{1}{2}\right)$, $\left(-3, \frac{3}{2}\right)$; $\left(-3, 1 - \frac{\sqrt{13}}{6}\right)$, $\left(-3, 1 + \frac{\sqrt{13}}{6}\right)$; $y = \frac{3}{2}x + \frac{11}{2}$, $y = -\frac{3}{2}x - \frac{7}{2}$

41. $(x + 4)^2 = 4(2)y$

42. $(y + 3)^2 = 4\left(\frac{1}{2}\right)x$

43. $(y - 5)^2 = 4(4)(x + 2)$

44. $(x + 1)^2 = 4\left(\frac{1}{4}\right)(y + 5)$

45. $(x - 5)^2 + (y + 1)^2 = 100$

46. $(x + 4)^2 + (y - 1)^2 = 4$

47. $(x + 2)^2 + (y + 4)^2 = 1$

48. $(x + 6)^2 + (y + 2)^2 = 1$

49. $\frac{(x + 4)^2}{1} + \frac{(y - 1)^2}{9} = 1$

50. $\frac{(x - 2)^2}{4} + \frac{(y + 1)^2}{16} = 1$

51. $\frac{(x - 1)^2}{4} + \frac{(y - 5)^2}{\frac{25}{4}} = 1$

52. $\frac{(x + 2)^2}{\frac{1}{9}} + \frac{(y - 2)^2}{16} = 1$

53. $\frac{(y - 9)^2}{1} - \frac{(x + 1)^2}{\frac{1}{4}} = 1$

54. $\frac{(x - 2)^2}{3} - \frac{(y + 2)^2}{9} = 1$

55. $\frac{(x + 1)^2}{25} - \frac{(y + 1)^2}{\frac{49}{4}} = 1$

56. $\frac{(y - 2)^2}{\frac{4}{9}} - (x - 3)^2 = 1$

In Exercises 32–40, sketch the graph of the equation. See Additional Answers.

32. $\frac{x^2}{9} + \frac{y^2}{49} = 1$

33. $\frac{(x + 1)^2}{25} - \frac{(y + 3)^2}{49} = 1$

34. $x^2 + (y - 5) = 0$

35. $(y - 1)^2 - (x + 3) = 0$

36. $(x - 3)^2 + (y + 7)^2 = 36$

37. $\frac{(x - 5)^2}{9} + \frac{(y + 2)^2}{25} = 1$

38. $\frac{(x + 9)^2}{25} - \frac{(y - 1)^2}{4} = 1$

39. $\frac{(y - 4)^2}{16} - \frac{(x - 1)^2}{4} = 1$

40. $(x + 5)^2 + (y - 4)^2 = 81$

In Exercises 41–56, write the equation of the conic in standard form. (11.5) See margin.

41. $x^2 + 8x - 8y + 16 = 0$

42. $y^2 + 6y - 2x + 9 = 0$

43. $y^2 - 10y - 16x - 7 = 0$

44. $x^2 + 2x - y - 4 = 0$

45. $x^2 + y^2 - 10x + 2y - 74 = 0$

46. $x^2 + y^2 + 8x - 2y + 13 = 0$

47. $x^2 + y^2 + 4x + 8y + 19 = 0$

48. $x^2 + y^2 + 12x + 4y + 39 = 0$

49. $9x^2 + y^2 + 72x - 2y + 136 = 0$

50. $4x^2 + y^2 - 16x + 2y + 1 = 0$

51. $25x^2 + 16y^2 - 50x - 160y + 325 = 0$

52. $144x^2 + y^2 + 576x - 4y + 564 = 0$

53. $y^2 - 4x^2 - 18y - 8x + 76 = 0$

54. $3x^2 - y^2 - 12x - 4y - 1 = 0$

55. $49x^2 - 100y^2 + 98x - 200y - 1276 = 0$

56. $-16x^2 + 36y^2 + 96x - 144y - 16 = 0$

In Exercises 57–64, classify the conic. (11.6)

57. $x^2 + y^2 + 2x - 12y + 28 = 0$ Circle

58. $4y^2 - x^2 - 16y - 2x + 3 = 0$ Hyperbola

59. $3x^2 - y^2 - 12x - 2y + 5 = 0$ Hyperbola

60. $x^2 + 8x - 14y + 16 = 0$ Parabola

61. $2x^2 + 3y^2 - 12x + 24y + 36 = 0$ Ellipse

62. $x^2 + y^2 - 6x - 2y - 6 = 0$ Circle

63. $y^2 - 9y + 2x + 81 = 0$ Parabola

64. $4x^2 + y^2 + 8x - 4y + 7 = 0$ Ellipse

In Exercises 65–70, find the points of intersection. (11.6)

65. $x^2 + y^2 - 18x + 24y + 200 = 0$
$4x + 3y = 0$ $(6, -8)$, $(12, -16)$

66. $y^2 - 2x - 10y + 31 = 0$
$x - y + 2 = 0$ $(5, 7)$, $(3, 5)$

67. $x^2 + y^2 + 2x - 12y + 12 = 0$
$y^2 + x = 0$ $(-1, 1)$, $(-4, 2)$

68. $x^2 + 2y^2 + 4x + 8y + 11 = 0$
$y^2 - 40x + 4y - 36 = 0$ $(-1, -2)$

69. $4x^2 + y^2 - 48x - 2y + 129 = 0$
$x^2 + y^2 - 2x - 2y - 7 = 0$ $(4, 1)$

70. $9x^2 - 16y^2 + 18x + 153 = 0$
$9x^2 + 16y^2 + 18x - 135 = 0$ $(-1, 3)$, $(-1, -3)$

71. *Paper Folding* Draw a dot a few inches from the side of a sheet of paper. Then crease the paper 20 to 30 times, as shown in the diagram. The creases of the paper are tangents of a conic. Is it a parabola or one branch of a hyperbola? Explain.
Parabola; the distance between a point on the curve and the focus is the same as the distance between the point and the bottom edge of the paper.

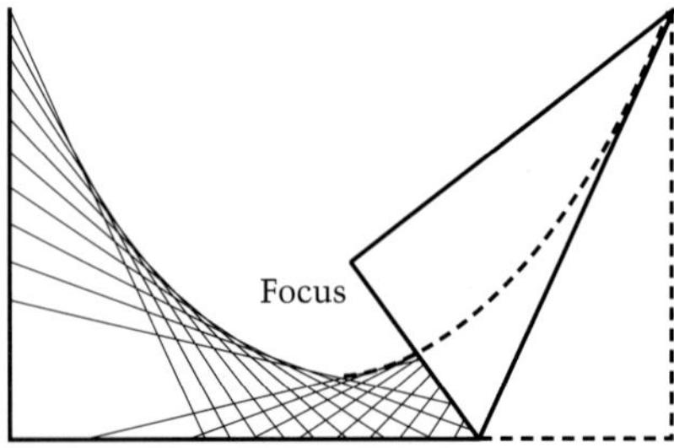

72. *Purifying Water* To detoxify the ground water at Livermore National Laboratory in California, water is pumped through tubes that are at the foci of parabolic mirrors. Cross sections of the parabolic mirrors can be modeled by the equation

$$y = \frac{1}{15}x^2$$

where x and y are measured in feet. How far is the tube above the vertex of the mirror? $\frac{1}{60}$ ft

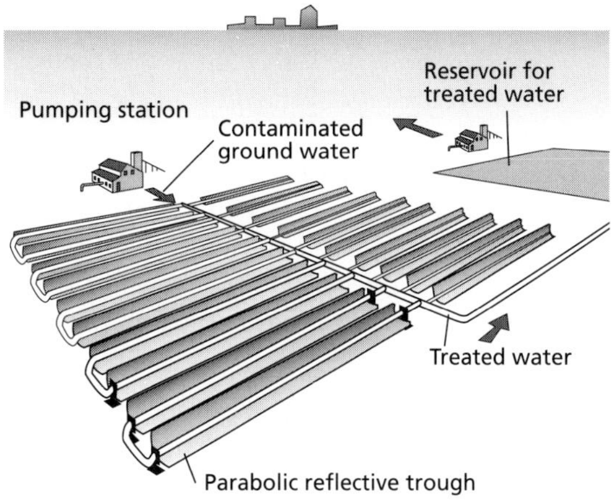

73. *Archway* A semi-elliptical arch is to be formed over the entrance to an estate. The arch is to set on pillars that are 10 feet apart and is to have a height atop the pillars of 4 feet. Make an accurate sketch of the arch. See margin.

74. *One Man on an Island* A man lives on an island in a large freshwater lake. Each morning he rides his horse to his gold mine—one mile away. Each evening he rides back to his house. Because he has no well, he always rides to the island's shore to get water on each trip to and from the gold mine. For several weeks, he has been hunting for the shortest path, but no matter how he travels, the distance is always the same. Why? Because the island has the shape of an ellipse and his home and goldmine are at the foci.

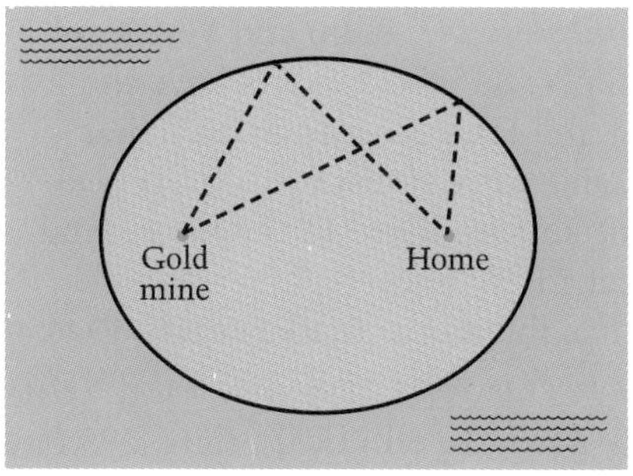

Keeping Sheep **In Exercises 75 and 76, use the following information.**

The wheel-shaped pen at the right is used in Iceland to hold sheep at shearing time. The inner circle can be approximated by the equation

$$x^2 + y^2 - 250y + 13{,}224 = 0$$

where x and y represent feet. Each individual section outside the inner circle is about 76 feet long.

75. Find the center and radius of the inner circle. (0, 125), 49 ft

76. Write an equation that represents the outer circle. $x^2 + (y - 125)^2 = 125^2$

Answer

73.

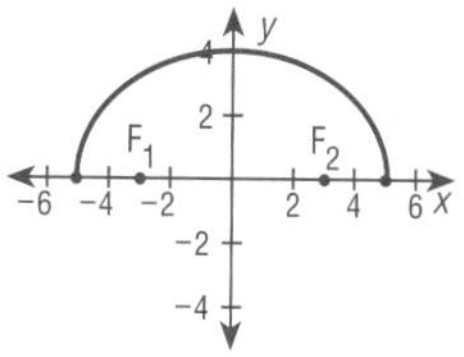

Answer

80. Those with hyperbolic and parabolic orbits; those with elliptical orbits can enter and leave our solar system many times.

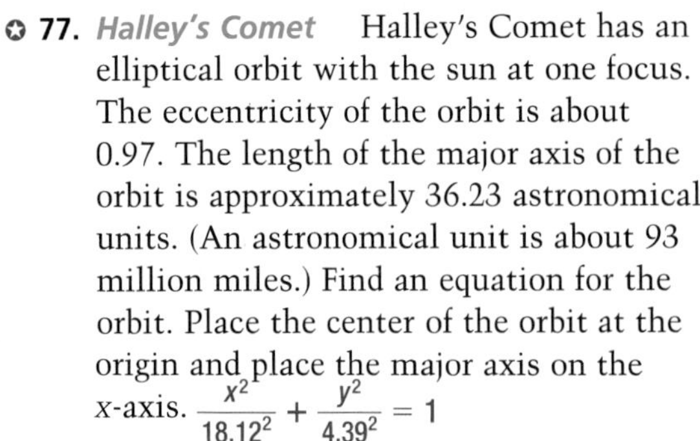

77. *Halley's Comet* Halley's Comet has an elliptical orbit with the sun at one focus. The eccentricity of the orbit is about 0.97. The length of the major axis of the orbit is approximately 36.23 astronomical units. (An astronomical unit is about 93 million miles.) Find an equation for the orbit. Place the center of the orbit at the origin and place the major axis on the x-axis. $\frac{x^2}{18.12^2} + \frac{y^2}{4.39^2} = 1$

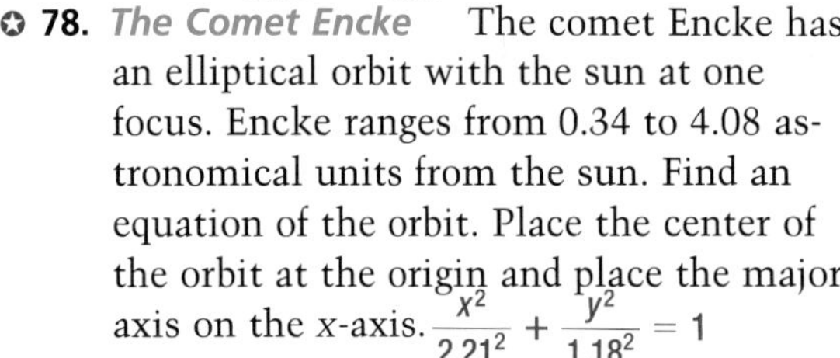

78. *The Comet Encke* The comet Encke has an elliptical orbit with the sun at one focus. Encke ranges from 0.34 to 4.08 astronomical units from the sun. Find an equation of the orbit. Place the center of the orbit at the origin and place the major axis on the x-axis. $\frac{x^2}{2.21^2} + \frac{y^2}{1.18^2} = 1$

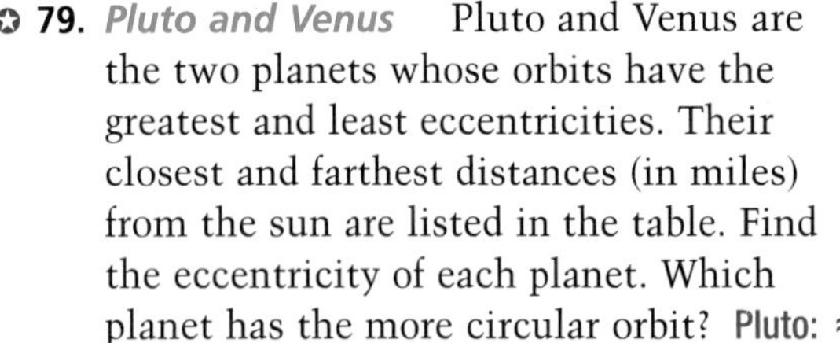

79. *Pluto and Venus* Pluto and Venus are the two planets whose orbits have the greatest and least eccentricities. Their closest and farthest distances (in miles) from the sun are listed in the table. Find the eccentricity of each planet. Which planet has the more circular orbit? Pluto: ≈ 0.2500, Venus: ≈ 0.0067, Venus

	Closest	Farthest
Pluto	2,749,600,000	4,582,700,000
Venus	66,800,000	67,700,000

80. *Comets in Our Solar System* Of the 610 comets identified prior to 1970, 245 have elliptical orbits, 295 have parabolic orbits, and 70 have hyperbolic orbits. In each case, the sun lies at the focus of the orbit. Which types of comets pass through our solar system only once? Explain. See margin.

Classifying Comets **In Exercises 81 and 82, classify the type of orbit of the comet from its equation. What is the closest distance that the comet gets to the sun? (Each unit represents 1 million miles.)**

81. $9y^2 - 7x^2 + 1170y + 31{,}725 = 0$
Hyperbola, ≈ 13.5 million miles

82. $x^2 - 80y = 1600$ Parabolic, 20 million miles

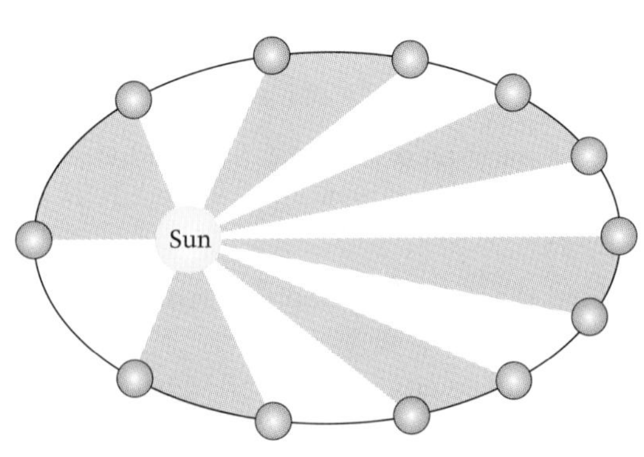

Johannes Kepler, a 17th century German astronomer, discovered that if a ray is extended from the sun to a comet or a planet with an elliptical orbit, the ray sweeps out equal areas in equal times. This principal, called* Kepler's Second Law, *implies that the comet or planet travels faster as it approaches the sun.

Chapter TEST

In Exercises 1–4, write the equation of the indicated conic in standard form. (11.1–11.4)

1. **Parabola** Vertex: (2, 8) Directrix: $y = 4$ $(x-2)^2 = 16(y-8)$
2. **Circle** Center: (4, −2) Radius: 11 $(x-4)^2 + (y+2)^2 = 121$
3. **Ellipse** Vertices: (−4, 6), (8, 6) Co-vertices: (2, 8), (2, 4) See margin.
4. **Hyperbola** Vertices: (−3, 2), (7, 2) Foci: (−5, 2), (9, 2) See margin.

In Exercises 5–8, classify the graph as a parabola, circle, ellipse, or hyperbola. Then sketch its graph. (11.5) 5.–8. See Additional Answers for graphs.

5. $8y = (x-5)^2$ Parabola
6. $(x-3)^2 + (y+2)^2 = 4$ Circle
7. $\frac{(x+4)^2}{4} + \frac{(y+1)^2}{16} = 1$ Ellipse
8. $\frac{(x+4)^2}{9} - \frac{(y+3)^2}{25} = 1$ Hyperbola

In Exercises 9–12, classify the graph as a parabola, circle, ellipse, or hyperbola. Then write the equation in standard form. (11.6)

9. $4x^2 + y = 3$ Parabola, $x^2 = 4\left(-\frac{1}{16}\right)(y-3)$
10. $x^2 - 4x + y^2 + 2y - 4 = 0$ Circle, $(x-2)^2 + (y+1)^2 = 9$
11. $x^2 + 6x + 2y^2 - 8y - 1 = 0$ Ellipse, see margin.
12. $3x^2 + 6x - 6y^2 + 24y - 33 = 0$ Hyperbola, See margin.
13. Find the vertex and directrix of $x^2 = -16y$. **(11.1)** (0, 0), $y = 4$
14. Find the center and radius of $(x+4)^2 + (y-6)^2 = 36$. **(11.5)** (−4, 6), 6
15. Find the vertices of $\frac{(x-3)^2}{12} + \frac{(y-2)^2}{16} = 1$. **(11.5)** (3, 6), (3, −2)
16. Write equations of the asymptotes of $\frac{y^2}{4} - \frac{x^2}{9} = 1$. **(11.4)** $y = \frac{2}{3}x$, $y = -\frac{2}{3}x$
17. Find the points of intersection of $x^2 - y = 0$ and $x^2 + y^2 - 6y + 4 = 0$. **(11.6)** (−1, 1), (1, 1), (−2, 4), (2, 4)
18. An elliptical cam has a shaft mounted at one focus. As it rotates, it lifts a follower from half an inch to two inches above the camshaft center. What is the eccentricity of the cam? **(11.5)** 0.6
19. A bowl has cross-sections that are parabolic. The equation of a parabolic cross-section is $y = \frac{1}{2}x^2$ where x varies between −4 and 4. Is the focus "inside" or "outside" the bowl? Explain. **(11.1)**
 19. Inside, $\left(0, \frac{1}{2}\right)$
20. A light ray from the point (5, 0) is directed toward the focus of a hyperbolic mirror. The mirror has the equation $\frac{1}{9}y^2 - \frac{1}{16}x^2 = 1$. At which point on the mirror will light reflect back to the other focus? (1.73, −3.27)

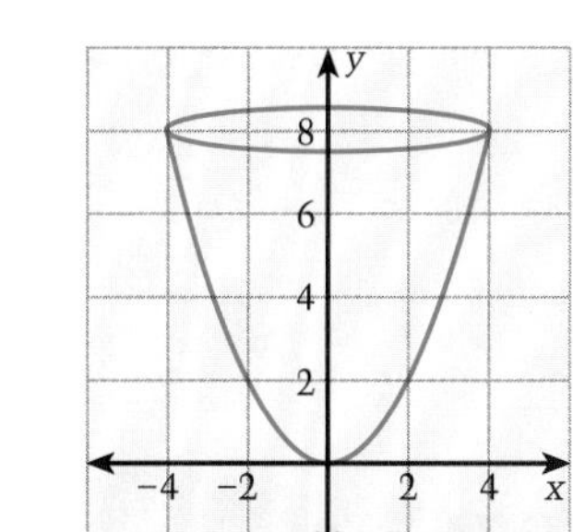

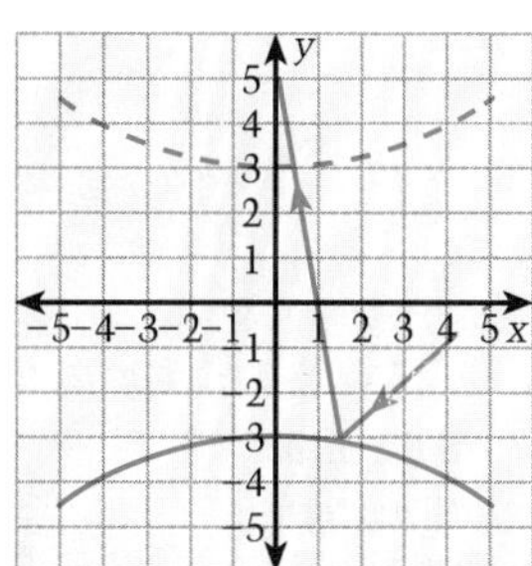

Answers

3. $\frac{(x-2)^2}{36} + \frac{(y-6)^2}{4} = 1$
4. $\frac{(x-2)^2}{25} - \frac{(y-2)^2}{24} = 1$
11. $\frac{(x+3)^2}{18} + \frac{(y-2)^2}{9} = 1$
12. $\frac{(x+1)^2}{4} - \frac{(y-2)^2}{2} = 1$

An important concept used in both engineering and physics applications is "limit." This concept is also an integral part of such branches of mathematics as calculus and statistics. In addition to the uses of sequences and series in modeling real-life situations, the study of sequences and series will provide students with experiences that facilitate their understanding of "limit."

The Chapter Summary on page 666 provides you and the students with a synopsis of the chapter. It identifies key skills and concepts. You may want to have students look at the Chapter Summary as an overview before beginning the chapter.

CHAPTER 12

Sequences and Series

LESSONS

The close-up of a daisy head shows the spiraling patterns formed by its florets.

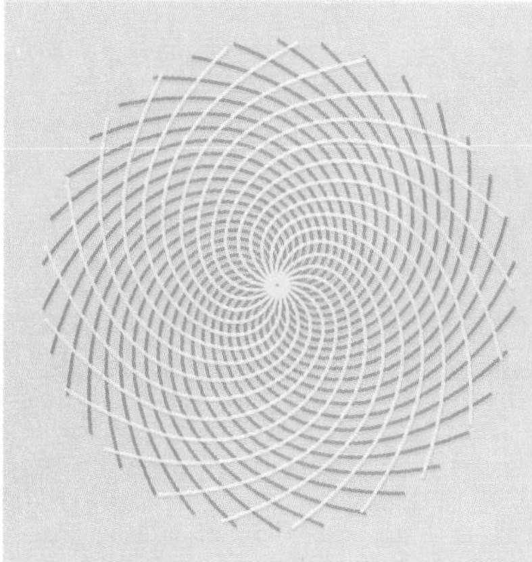

Looking directly into the head of a daisy, you see florets spiraling outward from its center in different directions, forming an intricate pattern. The number of spirals that swirl outward in a clockwise direction is 21. The number of spirals that swirl outward in a counterclockwise direction is 34. When the Fibonacci numbers (see page 323) are listed in order to form the Fibonacci sequence, you see that 21 and 34 are successive numbers:

1, 1, 2, 3, 5, 8, 13, 21, 34, . . .

If you look at the spirals in sunflower heads, pineapple scales, and pinecone brachs, you will again find that their numbers are successive numbers of the Fibonacci sequence.

Fibonacci (Leonardo of Pisa) first introduced the sequence of numbers that was to bear his name in *Liber abaci* in 1202. The sequence was generated from his now famous rabbit population problem, which is explained in Example 3 on page 323.

In addition to its occurrence in nature, Fibonacci's sequence has a stunning variety of other mathematical applications. The Scottish mathematician Robert Simson related it to the "golden ratio" of classical Greek architecture. The English mathematician Charles Lutwidge Dodgson, who wrote *Through the Looking Glass* under the pseudonym Lewis Carroll, used it to solve a famous mathematical paradox. It is also related to the entries along the diagonals of Pascal's Triangle (see Exercise 42, page 656).

12 Sequences and Series

PACING CHART

The basic course concludes with this chapter on sequences and series. However, after completing this chapter, you may still have some further teaching days available. If so, you may wish to consult the introductory notes on pages 677A, 729A, and 781A.

DAY	FULL COURSE	BASIC COURSE
1	12.1	12.1
2	12.2	12.2
3	12.3	12.2
4	Mid-Chapter Self-Test	12.3
5	12.4	12.3
6	12.5	Mid-Chapter Self-Test
7	12.6 & Using a Programmable or Graphing Calculator	12.4
8	Chapter Review	12.4
9	Chapter Test	12.5
10	Cumulative Review 7–12	12.6
11		12.6 & Using a Programmable or Graphing Calculator
12		Chapter Review
		Chapter Test
		Cumulative Review 7–12

CHAPTER ORGANIZATION

LESSON	PAGES	GOALS	MEETING THE NCTM STANDARDS
12.1	622–628	1. Write and use sequences and series 2. Use sequences and series as models of real-life situations	Problem Solving, Communication, Reasoning, Connections, Calculus concepts, Structure
12.2	629–635	1. Find the *n*th term and the sum of an arithmetic series 2. Use arithmetic series in real-life problems	Problem Solving, Communication, Connections, Calculus concepts, Structure
Mixed Review	636	Review algebraic and arithmetic skills	
Milestones	636	Godel's proof	Connections
12.3	637–643	1. Find the *n*th term and the sum of a geometric series 2. Use geometric series in real-life problems	Problem Solving, Communication, Connections, Calculus concepts, Structure
Mid-Chapter Self-Test	644	Diagnose student weaknesses and remediate with correlated Reteach worksheets	
12.4	645–650	1. Find the sum of an infinite geometric series 2. Use infinite geometric series as models of real-life problems	Problem Solving, Communication, Connections, Calculus concepts, Structure
12.5	651–657	1. Use the Binomial Theorem to expand a binomial that is raised to a power 2. Use the Binomial Theorem in real-life situations	Problem Solving, Communication, Reasoning, Connections, Structure
Mixed Review	657	Review algebraic and arithmetic skills	
12.6	658–663	1. Find the balance of an increasing annuity 2. Find the monthly deposit to reach a specified balance in an annuity	Problem Solving, Communication, Connections, Discrete Math
Using a Calculator	664–665	Find and graph the balance of an increasing annuity	Technology
Chapter Summary	666	Restate for students what they have learned, why they have learned it, and how it fits into the structure of algebra	Structure, Connections
Chapter Review	667–670	Review concepts and skills learned in the chapter	
Chapter Test	671	Diagnose student weaknesses and remediate with correlated Reteaching worksheets	
Cumulative Review 7–12	672–675	Review concepts and skills from chapters 7–12	

LESSON RESOURCES

MEETING INDIVIDUAL NEEDS

RETEACHING For students who need to spend more time on basics:

If a mid-chapter self-test or chapter test indicates a deficiency, teachers can help students with the appropriate ***Reteaching Copymaster.***

PRACTICE For students who need more practice:

Additional exercises like those in the Pupil's Edition are provided for each lesson in ***Extra Practice Copymasters.***

ENRICHMENT For enriching and broadening students' experiences:

Problem of the Day copymasters in ***Teaching Tools*** provide a daily opportunity to use logical reasoning, looking for a pattern, writing an equation, and other routine and non-routine problem-solving strategies.

Math Log copymasters in ***Alternative Assessment*** provide opportunities to report on investigations, research, and open-ended problems.

Technology: Using Calculators and Computers provides enriching activities with graphing and scientific calculators and computers.

The ***Applications Handbook*** provides additional information about the cross-curriculum topics such as astronomy, chemistry, physics, sports, economics, genetics, and music that are integrated into the Pupil's Edition.

LESSON	12.1	12.2	12.3	12.4	12.5	12.6
PAGES	622-628	629-635	637-643	645-650	651-657	658-663
Teaching Tools						
Transparencies					✓	
Problem of the Day	✓	✓	✓	✓	✓	✓
Warm-up Exercises	✓	✓	✓	✓	✓	✓
Answer Masters	✓	✓	✓	✓	✓	✓
Extra Practice Copymasters	✓	✓	✓	✓	✓	✓
Reteaching Copymasters	Teacher-directed and independent activities tied to results on the Mid-Chapter Self-Tests and Chapter Tests					
Color Transparencies	✓	✓	✓		✓	✓
Applications Handbook	Additional background information for many real-life applications					
Technology	Calculator and computer worksheets for appropriate lessons					
Complete Solutions Manual	✓	✓	✓	✓	✓	✓
Alternative Assessment	Assess student's ability to reason, analyze, solve problems, and communicate using mathematical language.					
Formal Assessment	Mid-Chapter Self-Tests, Chapter Tests, Cumulative Tests, and Practice for College Entrance Tests					
Computer Test Bank	Customized tests can be created by choosing from over 2500 items.					

INSIGHTS

12.1 Sequences and Series

A number of special functions and relations have been investigated in our study of algebra. To the list of special functions, this lesson adds sequences and series. Students explore real-life situations that can be modeled by this new class of algebraic functions. In the process, students encounter many new mathematical patterns that can be used to describe real-life phenomena.

12.2 Arithmetic Sequences and Series

Students continue to explore sequences and series. Arithmetic sequences and series are characterized by an algebraic pattern that has a constant increment from one term to the next. In this lesson, students learn how to recognize arithmetic sequences and to find the nth term of a sequence. A formula for evaluating an arithmetic series is also developed.

12.3 Geometric Sequences and Series

Geometric sequences and series are characterized by an algebraic pattern in which successive terms are multiplied by the same factor. In this lesson, students learn how to recognize geometric sequences and find the nth term of a sequence. A formula for evaluating a geometric series is also evaluated.

12.4 Infinite Geometric Series

That an infinite geometric series can have a finite sum is a powerful mathematical phenomenon, and involves the concept of "limit." The limit concept is useful in many real-life applications in such areas as economics, physics, calculus, and statistics.

12.5 The Binomial Theorem

The expansion of a binomial raised to a given power of n generates a special finite series that is described by the Binomial Theorem. In this lesson, students explore several ways to generate the coefficients of a binomial expansion. Probability is a branch of mathematics that frequently utilizes binomial expansions.

12.6 Exploring Data: The Algebra of Finance

This is the second lesson dealing with a topic likely to affect the financial "health" of every average citizen. The balance of an ordinary increasing (or decreasing) annuity is an important application of the sum of a geometric series. In this lesson, students will have an opportunity to use the formula for the balance of an ordinary annuity in real-life finance problems.

Problem of the Day

Ellen made a scalene triangle from matches. Later she tried to remember how long each side was. She thought *AB* was one match long, *BC* two matches long, and *CA* three matches long. She was wrong about the length of only one side. Which side was she wrong about, and how many matches long was it actually?

She was wrong about *AB*, which was 4 matches long.

ORGANIZER

Warm-Up Exercises

1. Evaluate $f(x)$ at the indicated values of x.
   a. $f(x) = 2x + 3$
      $x = 1, 2, 3, 4, 5, 6, 7$
   b. $f(x) = 5x - 1$
      $x = 0, 1, 2, 3, 4, 5, 6, 7, 8$

   a. 5, 7, 9, 11, 13, 15, 17
   b. −1, 4, 9, 14, 19, 24, 29, 34, 39
2. Evaluate $g(n)$ at the indicated values of n.
   a. $g(n) = 3n - 2$
      $n = 1, 2, 3, 4, 5$
   b. $g(n) = 3^n - 6$
      $n = 0, 1, 2, 3, 4, 5, 6$

   a. 1, 4, 7, 10, 13
   b. −5, −3, 3, 21, 75, 237, 723

Lesson Resources

Teaching Tools
- Problem of the Day: 12.1
- Warm-up Exercises: 12.1
- Answer Masters: 12.1

Extra Practice: 12.1
Color Transparency: 63

12.1 Sequences and Series

What you should learn:

Goal 1 How to write and use sequences and series

Goal 2 How to use sequences and series as models of real-life situations

Why you should learn it:

You can model many real-life situations with sequences and series, such as geometric arrangements of items in a display.

Goal 1 Working with Sequences and Series

When we say that a collection of objects is listed "in sequence," we usually mean that the collection is ordered so that it has a first member, a second member, a third member, and so on. In mathematics, the word "sequence" is used in much the same way. Here are two examples.

$$1, 3, 5, 7, \ldots \quad \text{and} \quad \frac{1}{2}, \frac{1}{4}, \frac{1}{8}, \frac{1}{16}, \ldots$$

A Sequence

A **sequence** is a function whose domain is the set of positive integers. The function values

$$a_1, a_2, a_3, a_4, \ldots, a_n, \ldots$$

are the **terms** of the sequence. If the domain consists of the first n positive integers only, then the sequence is a **finite sequence.**

Sometimes it is convenient to include zero in the domain of a sequence and list the terms as

$$a_0, a_1, a_2, a_3, a_4 \ldots.$$

Unless otherwise stated, however, you may assume a sequence begins with a_1.

Example 1 *Writing Terms in a Sequence*

When you are given a formula for the nth term of a sequence, you can find the terms by substituting 1, 2, 3, and so on, for n.

a. Here are the first four terms of the sequence whose nth term is $a_n = 4n - 3$.

$a_1 = 4(1) - 3 = 1$
$a_2 = 4(2) - 3 = 5$
$a_3 = 4(3) - 3 = 9$
$a_4 = 4(4) - 3 = 13$

b. Here are the first four terms of the sequence whose nth term is $a_n = 2 + (-1)^n$.

$a_1 = 2 + (-1)^1 = 2 - 1 = 1$
$a_2 = 2 + (-1)^2 = 2 + 1 = 3$
$a_3 = 2 + (-1)^3 = 2 - 1 = 1$
$a_4 = 2 + (-1)^4 = 2 + 1 = 3$ ■

Note that simply listing the first few terms is not sufficient to define a sequence—the nth term *must be given.* To see this, consider the following sequences, both of which have the same first three terms.

$$\frac{1}{2}, \frac{1}{4}, \frac{1}{8}, \frac{1}{16}, \ldots, \frac{1}{2^n}, \ldots$$

$$\frac{1}{2}, \frac{1}{4}, \frac{1}{8}, \frac{1}{15}, \ldots, \frac{6}{(n+1)(n^2-n+6)}, \ldots$$

Some very important sequences in mathematics involve factorials. Recall from Lesson 6.6 that $0! = 1$ and

$$n! = 1 \cdot 2 \cdot 3 \cdot 4 \cdots (n-1) \cdot n.$$

Example 2 *Writing Terms of a Sequence Involving Factorials*

Write the first six terms of the sequence whose nth term is $a_n = \frac{30}{n!}$. Begin with $n = 0$.

Solution

$$a_0 = \frac{30}{0!} = \frac{30}{1} = 30 \qquad a_1 = \frac{30}{1!} = \frac{30}{1} = 30$$

$$a_2 = \frac{30}{2!} = \frac{30}{2} = 15 \qquad a_3 = \frac{30}{3!} = \frac{30}{6} = 5$$

$$a_4 = \frac{30}{4!} = \frac{30}{24} = \frac{5}{4} \qquad a_5 = \frac{30}{5!} = \frac{30}{120} = \frac{1}{4}$$

■

There is a convenient notation for the sum of the terms of a *finite sequence.* It is called **summation notation** or **sigma notation** because it uses the uppercase Greek letter *sigma,* written as Σ.

Summation Notation

The expression formed by adding the first n terms of a sequence is called a **series** and is represented by

$$\sum_{i=1}^{n} a_i = a_1 + a_2 + a_3 + a_4 + \cdots + a_n$$

where i is called the **index of summation,** n is the **upper limit of summation,** and 1 is the **lower limit of summation.** It is read as "the sum from $i = 1$ to n of a_i."

The index of summation does not have to be i—any letter can be used. Also, the index does not have to begin at 1. For instance, in part b of the next example, the index begins with 3.

LESSON Notes

Give students an opportunity to ask questions about sequences and how they are expressed. Many of the difficulties that students experience with sequences and series can be traced back to an unclear understanding of what the symbols convey.

Example 1

It is important to point out that sequences are simply functions. Consequently, in Part a, the sequence whose nth term is $a_n = 4n - 3$, can be rewritten in standard function notation as $f(n) = 4n - 3$.

Students often lose sight of what they already know about functions when they study sequences and series because the notation is new to them.

Ask students to express the sequence in Part b in standard function notation.

$f(n) = 2 + (-1)^n$

Example 2

Ask students to express the first six terms of the sequence in standard function notation. For example, since $f(n) = \frac{30}{n!}$, $f(0) = \frac{30}{0!}$.

Example 3

The expressions $3i$, $1 + k^2$, and $\frac{1}{j!}$, are called the arguments of the summation notation.

Be sure students understand that each indicated sum in Parts a–c is a series. Part c suggests the possibility of using rational relationships to approximate irrational values.

Example 4

This is an example of a sequence that does not begin with either 0 or 1. Ask students to express this sequence in standard function notation.

Example 5

Ask students to evaluate the following series.

a. $\sum_{i=1}^{7} i$ b. $\sum_{i=1}^{8} 1$ c. $\sum_{i=1}^{12} i^2$

a. 28, b. 8, c. 650

Extra Examples

Here are additional examples similar to **Examples 1–4.**

1. Write the first 5 terms of the sequence whose nth term is $a_n = \frac{(n-2)!}{n!}$, $n \geq 2$.
   $\frac{1}{2}, \frac{1}{6}, \frac{1}{12}, \frac{1}{20}, \frac{1}{30}$
2. Write the first eight terms of the sequence whose nth term is $a_n = 2^n - n^2$.
   1, 0, −1, 0, 7, 28, 79, 192
3. Evaluate the following series.
   a. $\sum_{i=1}^{8} 2i$ 72
   b. $\sum_{m=1}^{4} (1 + m)^2$ 54
   c. $\sum_{i=1}^{12} \frac{1}{i^2}$ 1.565

Example 3 *Using Summation Notation*

a. Let i take on values from 1 to 5 to evaluate the sum.

$$\begin{aligned}\sum_{i=1}^{5} 3i &= 3(1) + 3(2) + 3(3) + 3(4) + 3(5)\\ &= 3(1 + 2 + 3 + 4 + 5)\\ &= 3(15)\\ &= 45\end{aligned}$$

b. Let k take on values from 3 to 6 to evaluate the sum.

$$\begin{aligned}\sum_{k=3}^{6} (1 + k^2) &= (1 + 3^2) + (1 + 4^2) + (1 + 5^2) + (1 + 6^2)\\ &= 10 + 17 + 26 + 37\\ &= 90\end{aligned}$$

c. Let j take on values from 0 to 8 to evaluate the sum.

$$\begin{aligned}\sum_{j=0}^{8} \frac{1}{j!} &= \frac{1}{0!} + \frac{1}{1!} + \frac{1}{2!} + \frac{1}{3!} + \frac{1}{4!} + \frac{1}{5!} + \frac{1}{6!} + \frac{1}{7!} + \frac{1}{8!}\\ &= 1 + 1 + \frac{1}{2} + \frac{1}{6} + \frac{1}{24} + \frac{1}{120} + \frac{1}{720} + \frac{1}{5040} + \frac{1}{40{,}320}\\ &\approx 2.71828\end{aligned}$$

This sum is very close to the natural base $e \approx 2.71828$. It can be shown that as more terms of the sequence are added, the sum becomes closer and closer to e. ■

Goal 2 Using Sequences and Series in Real Life

Connections
Geometry

Example 4 *Finding the Terms of a Sequence*

The number of degrees, d_n, in each angle at the tip of the six n-pointed stars shown on this page is given by

$$d_n = \frac{180(n-4)}{n}, \qquad n \geq 5.$$

Write the first six terms of this sequence.

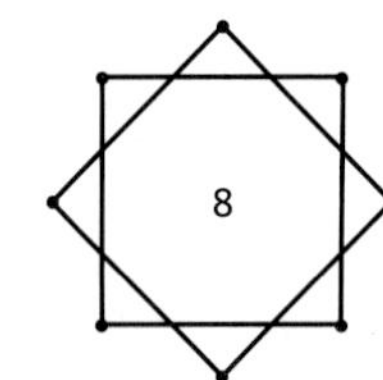

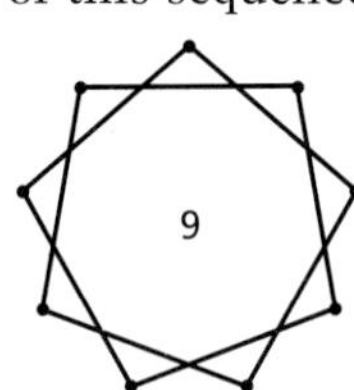

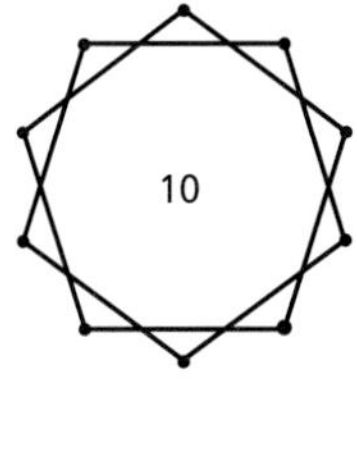

Solution The solution is shown in the table.

n	5	6	7	8	9	10
$\frac{180(n-4)}{n}$	36°	60°	$\frac{540°}{7} \approx 77.14°$	90°	100°	108°

■

The sum of the terms of a finite sequence can be found by simply adding the terms. For sequences with many terms, however, adding the terms can be tedious. It is often more efficient to use a formula for the sum of special types of sequences.

Formulas for Special Series

1. $\sum_{i=1}^{n} 1 = n$ 2. $\sum_{i=1}^{n} i = \frac{1}{2}(n)(n+1)$

3. $\sum_{i=1}^{n} i^2 = \frac{1}{6}(n)(n+1)(2n+1)$

Real Life
Retail Sales

Example 5 ***Using a Formula to Find the Sum of a Series***

You are working in the produce department of a grocery store. The manager of the produce department asks you to create a display in which oranges are stacked in a pyramid as shown at the left. The pyramid is to have 16 layers. How many oranges do you need?

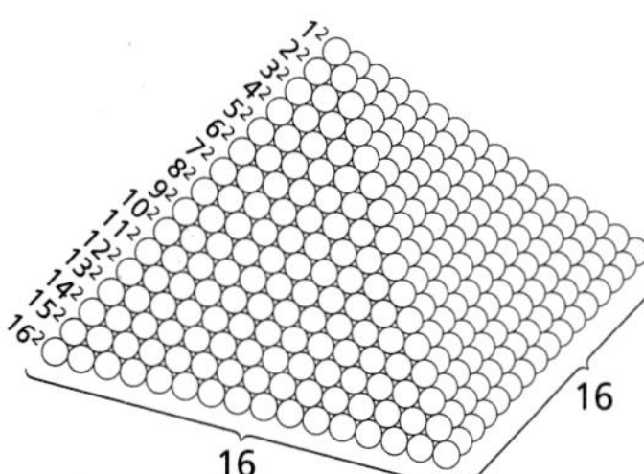

Solution To solve the problem, you first need to recognize that the bottom layer has 16^2 oranges, the next layer has 15^2, and so on. This means that the total number of oranges is

$$\sum_{i=1}^{16} i^2 = 1^2 + 2^2 + \cdots + 16^2.$$

One way to find the sum is to add the 16 terms. Another way is to use the formula 3 (shown above).

$$\sum_{i=1}^{16} i^2 = \frac{1}{6}(16)(16+1)(2\cdot 16+1) = \frac{1}{6}(16)(17)(33) = 1496$$

There are 1496 oranges in the stack. Check this by actually adding the number of oranges in each of the 16 layers. ■

Communicating about ALGEBRA

SHARING IDEAS about the Lesson

Find the sum in two ways. First add the terms. Then use one of the formulas shown at the top of this page.

A. $\sum_{i=1}^{15} i$ **B.** $\sum_{i=1}^{10} i^2$ A. 120 B. 385

Check Understanding

1. How are sequences related to algebraic functions? They are special functions.
2. What are the terms of a sequence? See page 622.
3. What is the domain of a finite sequence? a finite set of integers
4. What information is provided by the upper and lower limits of summation? first and last elements of the domain

Communicating about ALGEBRA

Have students evaluate each series using the formulas for the special series (page 625) and the following property of sigma notation:

$$\sum_{i=1}^{n} (a_i + b_i) = \sum_{i=1}^{n} a_i + \sum_{i=1}^{n} b_i$$

a. $\sum_{i=1}^{n} 3i^2$ b. $\sum_{i=1}^{n} (2i^2 + i)$

c. $\sum_{i=1}^{n} (5i^2 + 2i + 1)$

a. $\frac{3}{6}(n)(n+1)(2n+1)$
$= \frac{1}{2}(n)(n+1)(2n+1)$

b. $\sum_{i=1}^{n} 2i^2 + \sum_{i=1}^{n} i = 2\sum_{i=1}^{n} i^2 + \sum_{i=1}^{n} i$

$\frac{1}{3}(n)(n+1)(2n+1) + \frac{1}{2}n(n+1)$

c. $\frac{5}{6}(n)(n+1)(2n+1) +$
$n(n+1) + n$

2. **COOPERATIVE LEARNING** Have students work in pairs to show that the following property is true.

$$\sum_{i=1}^{n} (a_i + b_i) = \sum_{i=1}^{n} a_i + \sum_{i=1}^{n} b_i$$

Expand the left-hand side as $a_1 + b_1 + a_2 + b_2 + \cdots + a_n + b_n$. Use the associative and commutative properties of real numbers to rearrange the terms appropriately. $a_1 + b_1 + a_2 + b_2 + \cdots + a_n + b_n = a_1 + a_2 + a_3 + \cdots + a_n + b_1 + b_2 + b_3 + \cdots + b_n$.

EXERCISE Notes

ASSIGNMENT GUIDE

Basic/Average: Ex. 1–4, 9–21 odd, 29, 35–41 odd, 52, 55–58, 63

Above Average: Ex. 1–4, 9–23 odd, 29, 35–43 odd, 47–50, 52, 54, 61–67 odd

Advanced: Ex. 1–4, 9–23 odd, 35–43 odd, 47–50, 52, 54, 60–73, 66–68

Selected Answers
Exercises 1–4, 5–63 odd

More Difficult Exercises
Exercises 51–54, 60, 64, 65–68

Guided Practice

Ex. 1–4 Use the Guided Practice exercises as a readiness check of students' understanding of the lesson. Students' responses to Exercise 1 will provide you with feedback about their perception of the lesson.

Independent Practice

Ex. 5–28 Ask students to use a solution format as modeled in Example 1.

EXERCISES

Guided Practice

CRITICAL THINKING about the Lesson

1. Which of the following is a sequence? Which is a series? Explain the difference. a is a sequence, b is a series; a series is the sum of the terms of a sequence.
   a. 5, 12, 19, 26, 33, 40 **b.** 5 + 12 + 19 + 26 + 33 + 40
2. Write the first six terms of the sequence whose nth term is $a_n = 2n$. 2, 4, 6, 8, 10, 12
3. Write the first six terms of the sequence whose nth term is $a_n = (-1)^{n+1}$. 1, −1, 1, −1, 1, −1
4. Evaluate the sum $\sum_{i=1}^{4} 2(i-1)$. 12

Independent Practice

In Exercises 5–28, write the first five terms of the sequence. Begin with $n = 1$. See Additional Answers.

5. $a_n = 3n - 5$
6. $a_n = 4 - n$
7. $a_n = n^2 + 2$
8. $a_n = (n+1)^2$
9. $a_n = 2^n$
10. $a_n = \left(\frac{1}{2}\right)^n$
11. $a_n = \left(-\frac{1}{2}\right)^n$
12. $a_n = (-1)^{n-1}$
13. $a_n = \frac{n}{n+1}$
14. $a_n = \frac{n^2}{2n}$
15. $a_n = (-2)^n$
16. $a_n = (-5)^n$
17. $a_n = n! - 2$
18. $a_n = \frac{n!}{2}$
19. $a_n = \frac{3^n}{n!}$
20. $a_n = (2n+1)!$
21. $a_n = 2(n!)$
22. $a_n = 2(n+2)!$
23. $a_n = (n-1)!$
24. $a_n = \frac{n^2}{n!}$
25. $a_n = 4 + (-1)^n$
26. $a_n = 2(-1)^n$
27. $a_n = 4 - (-1)^{n+1}$
28. $a_n = 2(-1)^{n+1}$

In Exercises 29 and 30, the sequence is defined recursively. Write the first six terms of the sequence. Begin with $n = 0$.

29. $a_0 = 1$, $a_1 = 1$, $a_n = a_{n-2} + a_{n-1}$ 1, 1, 2, 3, 5, 8
30. $a_0 = 1$, $a_n = 3a_{n-1} + 2$ 1, 5, 17, 53, 161, 485

The sequence given by the recursive definition in Exercise 29 is called the Fibonacci sequence (see pages 323 and 621). The terms of the Fibonacci sequence occur as patterns in nature. For instance, from the center of the pine cone shown at the right, there are 8 clockwise spirals and 13 counterclockwise spirals.

In Exercises 31–42, write the series represented by the summation notation. Then evaluate the sum. See margin.

31. $\sum_{n=0}^{4} n^2$ **32.** $\sum_{i=0}^{6} (2i+5)$ **33.** $\sum_{k=1}^{5} (k^2+1)$ **34.** $\sum_{j=3}^{7} (6j-10)$

35. $\sum_{n=1}^{4} n(n+1)$ **36.** $\sum_{n=0}^{5} 2n^2$ **37.** $\sum_{k=2}^{6} (k!-k)$ **38.** $\sum_{j=0}^{4} \frac{6}{j!}$

39. $\sum_{m=2}^{6} \frac{2m}{2(m-1)}$ **40.** $\sum_{i=2}^{5} -2i!$ **41.** $\sum_{n=0}^{6} (n!+10)$ **42.** $\sum_{k=1}^{5} \frac{10k}{k+2}$

In Exercises 43–46, use summation notation to represent the sum. Use *i* as the index and begin with *i* = 1. See Additional Answers.

43. $2+4+6+8+10+12$ **44.** $3+5+7+9+11+13$

45. $3+9+27+81+243$ **46.** $\frac{1}{4}+\frac{1}{16}+\frac{1}{64}+\frac{1}{256}$

In Exercises 47–50, use one of the formulas for special series to evaluate the sum.

47. $\sum_{i=1}^{24} 1$ 24 **48.** $\sum_{i=1}^{54} i$ 1485 **49.** $\sum_{i=1}^{42} i^2$ 25585 **50.** $\sum_{i=1}^{20} i^2$ 2870

51. *Towers of Hanoi* In the puzzle called the Towers of Hanoi, you are asked to move the rings from one peg and stack them *in order* on another peg. You can make as many moves as you want, but each move must consist of moving exactly one ring. Moreover, no ring may be placed on top of a smaller ring. The minimum number of moves required to move *n* rings is 1 for 1 ring, 3 for 2 rings, 7 for 3 rings, 15 for 4 rings, and 31 for 5 rings. Find a formula for this sequence. What is the minimum number of moves required to move 6 rings? $\sum_{i=1}^{n} (2^i - 1)$, 63

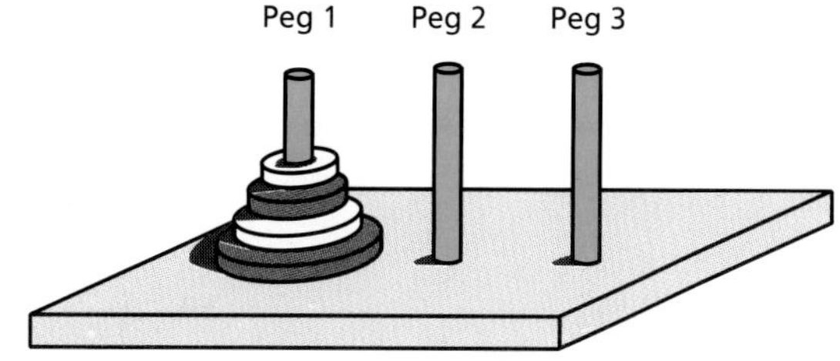

52. *Soccer Ball* The number of degrees, a_n, in each angle of a regular *n*-sided polygon is

$$a_n = \frac{180(n-2)}{n}, \qquad n \geq 3.$$

The surface of a soccer ball is made of regular hexagons and pentagons. If a soccer ball is taken apart and flattened, as shown at the right, the sides of the hexagons don't meet each other. Use the terms a_5 and a_6 to explain why there are gaps between adjacent hexagons. See below.

52. $a_5 = 108$, $a_6 = 120$ A pentagon has 108° in each angle, a hexagon has 120° in each angle; around any point there are 360°, but around each point that is common to two hexagons and a pentagon there are only 348° (108° + 120° + 120°); so there is a gap of 12°.

▶ **Ex. 29–30** *Explicit* definitions for sequences state how the *n*th term of a sequence is related to the number *n*. *Recursive* definitions for sequences state how the *n*th term is related to the term or terms preceding it. The Fibonacci sequence is a sequence that is usually defined only recursively. Students may wish to review recursive functions in Lesson 6.6.

▶ **Ex. 31–42** Students should use Example 3 as a solution model.

Answers

31. $0+1+4+9+16=30$
32. $5+7+9+11+13+15+17=77$
33. $2+5+10+17+26=60$
34. $8+14+20+26+32=100$
35. $2+6+12+20=40$
36. $0+2+8+18+32+50=110$
37. $0+3+20+115+714=852$
38. $6+6+3+1+\frac{1}{4}=\frac{65}{4}$
39. $2+\frac{3}{2}+\frac{4}{3}+\frac{5}{4}+\frac{6}{5}=\frac{437}{60}$
40. $-4+(-12)+(-48)+(-240)=-304$
41. $11+11+12+16+34+130+730=944$
42. $\frac{10}{3}+5+6+\frac{20}{3}+\frac{50}{7}=\frac{197}{7}$

▶ **Ex. 47–50** Students should review the patterns shown at the top of page 625.

▶ **Ex. 51** COOPERATIVE LEARNING You might wish to bring in a Tower of Hanoi puzzle and have students work in pairs to complete the exercise.

Ex. 66 Ask students to apply function notation.

Ex. 67–68 These exercises lead to Lessons 12.2 and 12.3. Encourage students to think about linear and exponential models.

Enrichment Activities

These activities require students to go beyond lesson goals.

1. Rewrite the following statement using Σ- notation. The square of the sum of integers from 1 to n is the sum of the cubes of the integers from 1 to n.
   $\left(\sum_{i=1}^{n} i\right)^2 = \sum_{i=1}^{n} i^3$
2. Verify that the statement is true for $n = 5$.
   $(1 + 2 + 3 + 4 + 5)^2 = (15)^2 = 225$;
   $1^3 + 2^3 + 3^3 + 4^3 + 5^3 = 1 + 8 + 27 + 64 + 125 = 225$
3. Using Formula 2 on page 625, write a formula for $\sum_{i=1}^{n} i^3$ and test it for $n = 5$.
   $\sum_{i=1}^{n} i^3 = \left(\sum_{i=1}^{n} i\right)^2 = \left[\frac{1}{2}n(n+1)\right]^2$
   $= \left[\frac{1}{4}n^2(n^2 + 2n + 1)\right]$
   When $n = 5$,
   $\left[\frac{1}{4}(5)^2((5)^2 + 2(5) + 1)\right]$
   $= \left[\frac{1}{4}(25)(25 + 10 + 1)\right] = 225$

53. *Stars* The stars in Example 4 were formed by placing n equally spaced points on a circle and connecting each point with the second point from it on the circle. The stars below are formed in a similar way except each point is connected with the third point from it. For these stars, the number of degrees in a tip is given by
$$d_n = \frac{180(n-6)}{n}, \qquad n \geq 7.$$
$\frac{180}{7}$, 45, 60, 72, $\frac{900}{11}$
Write the first five terms of this sequence.

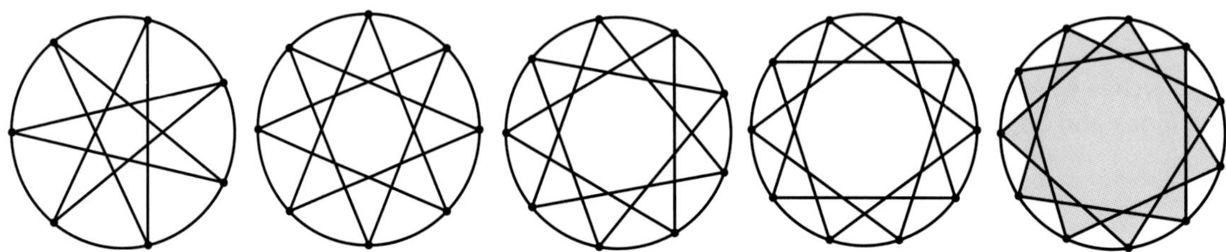

54. *More Stars* If you form the stars by connecting each point with the fourth point from it, you obtain stars with the following numbers of tips and degrees: 9 tips (20°), 10 tips (36°), 11 tips ($49\frac{1}{11}$°), 12 tips (60°). Find a formula for the number of degrees in an n-pointed star. $d_n = \frac{180(n-8)}{n}$, $n \geq 9$

Integrated Review

In Exercises 55–60, evaluate the expression.

55. $8!$ 40,320
56. $10!$ 3,628,800
57. $\frac{18}{3!} - 3$ 0
58. $\frac{6!}{80} + 4$ 13
59. $2\left(\frac{48}{4!} - \frac{60}{5!}\right)$ 3
60. $6!\left(\frac{1260}{7!} + \frac{3(2!)}{4!}\right)$ 360

In Exercises 61–64, find the first six terms of the sequence defined by the given recursive function.

61. $f(0) = 1$ 1, 2, 4, 8, 16, 32
    $f(n) = 2f(n-1)$
62. $f(0) = 2$ 2, −1, 3, 0, 4, 1
    $f(n) = n - f(n-1)$
63. $f(0) = 0$ 0, 1, 1, 0, −1, −1
    $f(1) = 1$
    $f(n) = f(n-1) - f(n-2)$
64. $f(0) = 2$ 2, 5, 1, 14, −11, 53
    $f(1) = 5$
    $f(n) = 3f(n-2) - f(n-1)$

Exploration and Extension

In Exercises 65–68, find a formula for the nth term of the sequence. See below.

65. $2, 1 + \frac{1}{2}, 1 + \frac{1}{3}, 1 + \frac{1}{4}, 1 + \frac{1}{5}, 1 + \frac{1}{6}, \ldots$
66. $1, \frac{1}{2}, \frac{1}{6}, \frac{1}{24}, \frac{1}{120}, \frac{1}{720}, \ldots$
67. 2, 6, 10, 14, 18, . . .
68. 3, 6, 12, 24, 48, 96, . . .

65. $f(n) = 1 + \frac{1}{n}, n \geq 1$
66. $f(n) = \frac{1}{n!}, n \geq 1$
67. $f(n) = 4n - 2, n \geq 1$
68. $f(n) = 3 \cdot 2^{n-1}, n \geq 1$

12.2

Arithmetic Sequences and Series

What you should learn:

Goal 1 How to find the nth term or the sum of an arithmetic series

Goal 2 How to use arithmetic sequences in real-life problems

Why you should learn it:

You can model many real-life situations with arithmetic sequences, such as rows of increasing lengths in some concert halls.

Goal 1 Finding the Terms and the Sum

A sequence whose consecutive terms have a common difference is an **arithmetic sequence.**

An Arithmetic Sequence

The sequence $a_1, a_2, a_3, a_4, \ldots, a_n, \ldots$ is **arithmetic** if there is a number d such that

$$a_2 - a_1 = d, \quad a_3 - a_2 = d, \quad a_4 - a_3 = d$$

and so on. The number d is the common difference of the arithmetic sequence.

Example 1 *Examples of Arithmetic Sequences*

a. The sequence whose nth term is $4n + 3$ is arithmetic. The common difference between consecutive terms is 4.

$$\underbrace{7, 11,}_{11 - 7 = 4} 15, 19, \ldots, 4n + 3, \ldots$$

b. The sequence whose nth term is $7 - 5n$ is arithmetic. The common difference between consecutive terms is -5.

$$\underbrace{2, -3,}_{-3 - 2 = -5} -8, -13, \ldots, 7 - 5n, \ldots$$

c. The sequence whose nth term is $\frac{1}{4}(n + 3)$ is arithmetic. The common difference between consecutive terms is $\frac{1}{4}$.

$$\underbrace{1, \tfrac{5}{4},}_{\frac{5}{4} - 1 = \frac{1}{4}} \tfrac{3}{2}, \tfrac{7}{4}, \ldots, \tfrac{1}{4}(n + 3), \ldots$$

■

Every arithmetic sequence has an nth term that is of the form

$$a_n = a_1 + (n - 1)d \qquad \textit{nth term of an arithmetic sequence}$$

where d is the common difference of the sequence.

Problem of the Day

Four men made the following statements.

Archie: Dave did it.

Dave: Tom did it.

Geraldo: I didn't do it.

Tom: Dave lied when he said I did it.

If only one of the statements is true, who is guilty?
Geraldo

ORGANIZER

Warm-Up Exercises

1. Identify the sequences that have a common difference or a common factor between consecutive terms in the sequence. Identify the sequences that have neither.
   a. 4, 2, 6, 1, 7, 12, 9, 5, 24, . . .
   b. 1.5, 1.8, 2.16, 2.592, 3.1104, . . .
   c. 8.4, 8.27, 8.14, 8.01, 7.88, . . .
   d. 4, 5.5, 7, 8.5, 10, 11.5, 13, . . .
   a. neither
   b. common factor of 1.2
   c. common difference of -0.13
   d. common difference of 1.5
2. Evaluate each term for the indicated value of n.
   a. $a_n = 3n + 11, n = 6$ 29
   b. $a_n = 13 - 2n, n = 3$ 7
   c. $a_n = -2n^2 + n, n = 4$ -28

Lesson Resources

Teaching Tools
Problem of the Day: 12.2
Warm-up Exercises: 12.2
Answer Masters: 12.2
Extra Practice: 12.2
Color Transparency: 63

LESSON Notes

Example 1

Ask students to verify that the nth term of each sequence is equivalent to $a_n = a_1 + (n - 1)d$ where d is the common difference of the sequence. In Part a, for example, a_n is $7 + (n - 1)4 = 7 + 4n - 4 = 4n + 3$.

Example 2

Be sure students understand that, although the points lie on the straight line given by the equation $y = 3x - 1$, the graph of the sequence is not a straight line! The graph is a collection of points associated with the positive integers on the x-axis.

Example 3

Ask students to graph the sequence. Is it a straight line or a discrete collection of points that lie on a straight line? It is a discrete collection of points.

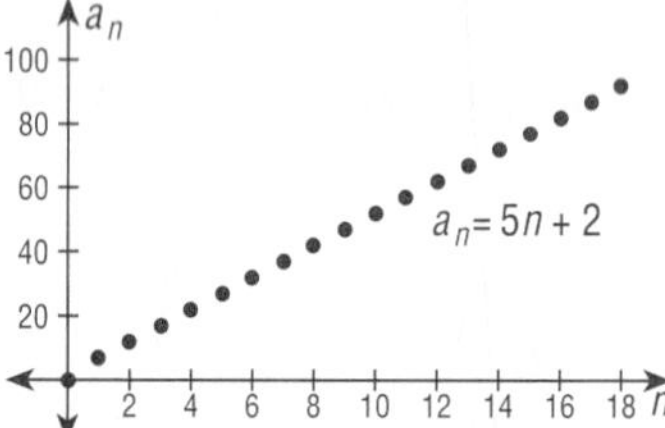

Example 4

Some students may reason in the following manner to find the common difference, d. The difference between the thirteenth and fourth terms is $43 - 16 = 27$. The difference is spread over the 9 terms from the fourth to the thirteenth term, so the common difference, d, is $d = \frac{27}{9} = 3$.

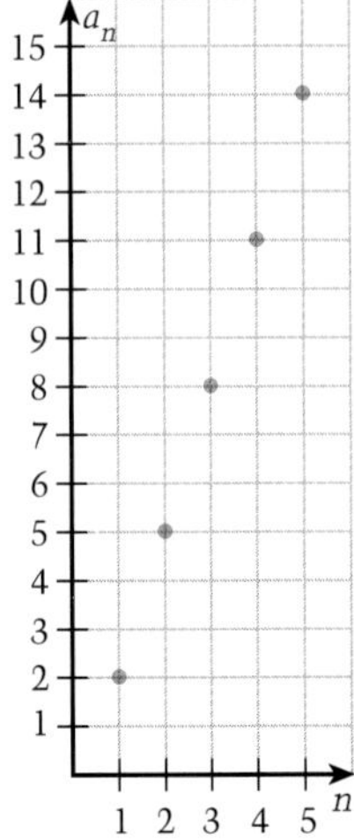

Example 2 Finding the nth Term

Find a formula for the nth term of the arithmetic sequence whose first term is 2 and whose common difference is 3.

Solution

$$a_n = a_1 + (n - 1)d \quad \textit{nth term formula}$$
$$= 2 + (n - 1)3 \quad \textit{Substitute 2 for } a_1 \textit{ and 3 for } d.$$
$$= 3n - 1 \quad \textit{Simplify.}$$

The terms of the sequence are as follows.

$2, 5, 8, 11, 14, \ldots, 3n - 1, \ldots$

A graph of the sequence is shown at the left. Notice that the points lie on a line. This is true of the graph of *any* arithmetic sequence. ■

Example 3 Finding the nth Term

Find a formula for the nth term of the arithmetic sequence whose common difference is 5 and whose *second* term is 12. What is the eighteenth term of this sequence?

Solution Because the second term is 12 and the common difference is 5, you know that the first term is 7.

$$a_n = a_1 + (n - 1)d \quad \textit{nth term formula}$$
$$= 7 + (n - 1)5 \quad \textit{Substitute 7 for } a_1 \textit{ and 5 for } d.$$
$$= 5n + 2 \quad \textit{Simplify.}$$

The eighteenth term of the sequence is $a_{18} = 5(18) + 2 = 92$. ■

Example 4 Finding the nth Term

The fourth term of an arithmetic sequence is 16. The thirteenth term is 43. Write a formula for the nth term.

Solution To obtain the thirteenth term from the fourth term, add the common difference d to the fourth term 9 times.

$$a_{13} = a_4 + 9d$$
$$43 = 16 + 9d$$
$$27 = 9d$$
$$3 = d$$

To obtain the first term, subtract the common difference d from the fourth term 3 times: $a_1 = a_4 - 3(d) = 16 - 9 = 7$. The formula for the nth term is

$$a_n = a_1 + (n - 1)d = 7 + (n - 1)3 = 3n + 4.$$ ■

The expression formed by adding the terms of a finite arithmetic sequence is called an **arithmetic series.** To find the sum of n terms of an arithmetic series, find the average of the first and last terms, $\frac{1}{2}(a_1 + a_n)$, and multiply the result by n.

Here is why the procedure works. Let S represent the sum of the first n terms. You can write the sum S in two ways, and add the results.

$$\begin{array}{rl} S = & a_1 + (a_1 + d) + (a_1 + 2d) + \cdots + a_n \\ S = & a_n + (a_n - d) + (a_n - 2d) + \cdots + a_1 \\ \hline 2S = & (a_1 + a_n) + (a_1 + a_n) + (a_1 + a_n) + \cdots + (a_1 + a_n) \end{array}$$

From this you can conclude that $2S = n(a_1 + a_n)$, which implies that the sum is $S = \frac{1}{2}n(a_1 + a_n)$.

The Sum of an Arithmetic Series

The sum of an arithmetic series with n terms is

$$\sum_{i=1}^{n} a_i = n\left(\frac{a_1 + a_n}{2}\right).$$

This sum can also be written as $\frac{1}{2}n(2a_1 + (n - 1)d)$.

Before using this formula, you should check to be sure that the series is arithmetic.

Example 5 *Evaluating the Sum of an Arithmetic Series*

Find the following sum.

$$1 + 3 + 5 + 7 + 9 + 11 + 13 + 15 + 17 + 19 + 21$$

Solution Because the terms have a common difference of 2, the series is arithmetic. Because the series has 11 terms, its sum is as follows.

$$\begin{aligned} \sum_{i=1}^{n} a_i &= 1 + 3 + 5 + 7 + 9 + 11 + 13 + 15 + 17 + 19 + 21 \\ &= n\left(\frac{a_1 + a_n}{2}\right) \\ &= 11\left(\frac{1 + 21}{2}\right) \\ &= 11(11) \\ &= 121 \end{aligned}$$

The sum of the series is 121. ■

Example 5

It is a good idea to demonstrate the meaning of the formula for the Sum of an Arithmetic Series. In this example, show that twice the sum is the sum of the first and last terms times the number of terms. That is,

$$\begin{array}{l} S = 1 + 3 + \cdots + 19 + 21 \\ S = 21 + 19 + \cdots + 3 + 1 \\ \hline 2S = 22 + 22 + \cdots + 22 + 22 \end{array}$$

So, $2S = 11(22)$ or $S = \frac{11(22)}{2}$.

Refer students to the *Handbook of Mathematical Connections* for a geometric interpretation of the sum of an arithmetic series.

Extra Examples

Here are additional examples similar to **Examples 1–5.**

1. Find the nth term of each arithmetic sequence. Write the first 5 terms of each.

a. $a_1 = 3$, $d = 2$
b. $a_1 = 1.4$, $d = -3.2$
c. $a_3 = 14$, $d = 4$
d. $a_8 = 26$, $a_{12} = 36$
e. $a_7 = -15$, $a_{13} = 18$

a. $2n + 1$; 3, 5, 7, 9, 11
b. $4.6 - 3.2n$; 1.4, −1.8, −5, −8.2, −11.4
c. $4n + 2$; 6, 10, 14, 18, 22
d. $2.5n + 6$; 8.5, 11, 13.5, 16, 18.5
e. $5.5n - 53.5$; −48, −42.5, −37, −31.5, −26

2. Find the sum of each finite series.

a. $3 + 8 + 13 + 18 + 23 + 28 + 33 + 38$
b. $-3 + (-1.5) + 0 + 1.5 + 3 + 4.5 + 6 + 7.5 + 9 + 10.5$
c. $26 + 25.77 + 25.54 + 25.31 + 25.08 + 24.85 + 24.62$

a. 164, b. 37.5, c. 177.17

Example 6

Ask students to rewrite the nth term of the arithmetic sequence in terms of the common difference, d.
$87.5 + 1.5n = 89 + (n - 1)(1.5)$

Check Understanding

1. What is the difference between an arithmetic sequence and an arithmetic series?
A series is an indicated sum of the terms of a sequence.
2. Describe the graph of an arithmetic sequence. How does it compare to the graph of a straight line?
The graph of an arithmetic sequence is a discrete set of points that lie on a straight line.

***Communicating* about ALGEBRA**

EXTEND *Communicating*
COOPERATIVE LEARNING
You might have students work in pairs to show the equivalence of the following formulas for the Sum of an Arithmetic Series. The sum of an arithmetic series with n terms is
$n\left(\frac{a_1 + a_n}{2}\right)$ or
$\frac{1}{2}n(2a_1 + (n - 1)d)$.
Substitute $a_1 + (n - 1)d$ for a_n in the first formula and simplify.

***Red Rocks Park** is an open-air amphitheater carved out of rock, near Denver, Colorado. The amphitheater has 69 rows of seats. Rows 46 through 69 have seats for 3318 people. The number of seats in the first 45 rows can be modeled by the arithmetic sequence whose nth term is $87\frac{1}{2} + \frac{3}{2}n$.*

Goal 2 Using Arithmetic Series in Real Life

Real Life **Business**

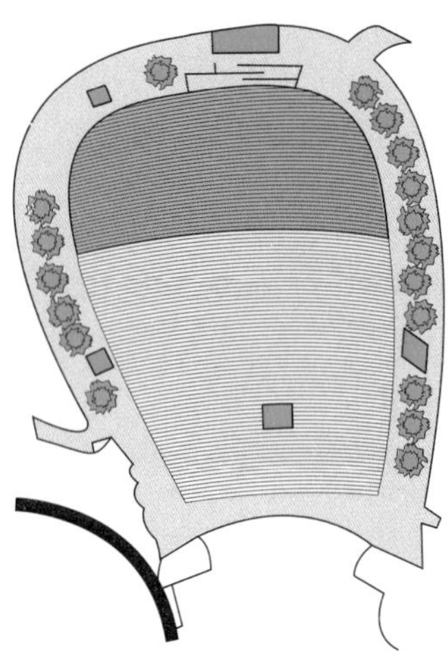

Example 6 *Seating Capacity*

You are organizing a concert at Red Rocks Park. How much should you charge per ticket in order to receive $50,000 for the ticket sales of a performance that is sold out?

Solution From the information given above, you know that there are 3318 seats in the last 24 rows. To approximate the number of seats in the first 45 rows, you can use the formula for the sum of an arithmetic series.

$$\sum_{i=1}^{45} a_i = 89 + 90\tfrac{1}{2} + 92 + 93\tfrac{1}{2} + \cdots + 155$$
$$= 45\left(\frac{89 + 155}{2}\right)$$
$$= 5490$$

The total number of seats is about $3318 + 5490 = 8808$. To bring in $50,000, you should charge about $\frac{1}{8808}(50{,}000) \approx$ $5.68 per ticket. ■

Communicating about ALGEBRA

SHARING IDEAS about the Lesson

A. Your job pays $6.50 an hour. After each of the first six years, you receive a raise of $0.50 an hour. Let $a_1 = 6.50$ represent your hourly wage during the first year. Write a formula for your hourly wage during the nth year.

B. Explain how to use the formula for the sum of an arithmetic series to find the sum of the first 100 odd positive integers.

A. $a_n = 6 + 0.5n$ **B.** $S = \frac{1}{2}n(a_1 + a_n) = \frac{1}{2} \cdot 100(1 + 199) = 50(200) = 10{,}000$

EXERCISES

Guided Practice

CRITICAL THINKING about the Lesson

1. What makes a sequence arithmetic? A common difference between consecutive terms.
2. The second and third terms of an arithmetic sequence are 7 and 12, respectively. What is the first term? 2
3. The fourth term of an arithmetic sequence is 15. The common difference is 2. Write a formula for the nth term of the sequence. $a_n = 2n + 7$
4. The second term of an arithmetic sequence is 10. The eleventh term is 37. What is the sixth term? 22
5. What is the common difference of the sequence? 5

   3, 8, 13, 18, 23, . . .

6. Evaluate the sum: $\sum_{i=1}^{4} 5(2i + 3)$. 160

Independent Practice

In Exercises 7–12, decide whether the sequence is arithmetic.

7. 10, 8, 6, 4, 2, . . . It is.
8. 1, 2, 4, 8, 16, . . . It is not.
9. $3, \frac{5}{2}, 2, \frac{3}{2}, 1, \ldots$ It is.
10. $\frac{9}{4}, 2, \frac{7}{4}, \frac{3}{2}, \frac{5}{4}, \ldots$ It is.
11. −12, −8, −4, 0, 4, . . . It is.
12. $\frac{1}{3}, \frac{2}{3}, \frac{4}{3}, \frac{8}{3}, \frac{16}{3}, \ldots$ It is not.

In Exercises 13–18, find the common difference of the arithmetic sequence. Write the next term in the sequence.

13. 5, 13, 21, 29, 37, . . . 8, 45
14. 11, 23, 35, 47, 59, . . . 12, 71
15. $\frac{1}{3}, \frac{1}{2}, \frac{2}{3}, \frac{5}{6}, 1, \ldots$ $\frac{1}{6}, \frac{7}{6}$
16. $3, \frac{11}{3}, \frac{13}{3}, 5, \frac{17}{3}, \ldots$ $\frac{2}{3}, \frac{19}{3}$
17. $\frac{7}{2}, \frac{9}{4}, 1, -\frac{1}{4}, -\frac{3}{2}, \ldots$ $-\frac{5}{4}, -\frac{11}{4}$
18. $\frac{5}{2}, \frac{11}{6}, \frac{7}{6}, \frac{1}{2}, -\frac{1}{6}, \ldots$ $-\frac{2}{3}, -\frac{5}{6}$

In Exercises 19–21, find a formula for the *n*th term of the arithmetic sequence.

19. Common difference: 4
    First term: 1 $a_n = 4n - 3$
20. Common difference: 8
    First term: −2
    $a_n = 8n - 10$
21. Common difference: −2
    Second term: 9
    $a_n = -2n + 13$

In Exercises 22–27, answer the question about the arithmetic sequence.

22. Common difference: −5
    Second term: 12
    What is the first term? 17
23. Common difference: 3
    Third term: 6 15
    What is the sixth term?
24. Common difference: −2
    Fourth term: 1
    What is the tenth term? −11
25. Common difference: −10
    Sixth term: −20
    What are the first five terms? 30, 20, 10, 0, −10
26. Fifth term: 6
    Eleventh term: 36
    What are the first four terms? −14, −9, −4, 1
27. Eighth term: 2
    Nineteenth term: 101
    What are the first four terms?
    −61, −52, −43, −34

EXERCISE Notes

ASSIGNMENT GUIDE

Basic/Average: Ex. 1–6, 9–27 odd, 28–31, 35–47 odd, 51–61 odd, 62

Above Average: Ex. 1–6, 9–27 odd, 28–31, 35–45 odd, 46–48, 55–62

Advanced: Ex. 1–6, 9–27 odd, 28–31, 35–43 odd, 44–48, 57–62

Selected Answers
Exercises 1–6, 7–61 odd

Use **Mixed Review** as needed.

More Difficult Exercises
Exercises 43–48, 57–60, 63

Guided Practice

Ex. 1–6 Use these exercises to review the necessary skills for success in the Independent Practice. Exercises 3 and 4 are modeled by Examples 3 and 4, respectively.

Independent Practice

Ex. 7–12 WRITING Have students write a one-sentence justification that defends their decision.

Ex. 19–27 Encourage students to review Examples 2–4 for solution models for these exercises.

▶ **Ex. 28–31** Assign these exercises as a group.

▶ **Ex. 32–39** Students can use Example 5 as a solution model.

▶ **Ex. 43–46** Students can use Example 6 as a solution model.

▶ **Ex. 47** You can connect this real-life application to a linear equation. The salary a_n for each year is $a_n = 26{,}000 + 1500n$ or $27{,}500 + (n - 1)1500$.

▶ **Ex. 49–56** Ask students to observe the pattern of the terms as well as the description of a_n to identify those sequences that are arithmetic. Exercises 49, 52, and 53

In Exercises 28–31, match the arithmetic sequence with its graph.

28. $a_n = -\frac{1}{3}n + 2$ **d**

29. $a_n = 5n - 4$ **b**

30. $a_n = 2n - 3$ **c**

31. $a_n = -\frac{1}{2}n + 4$ **a**

a. 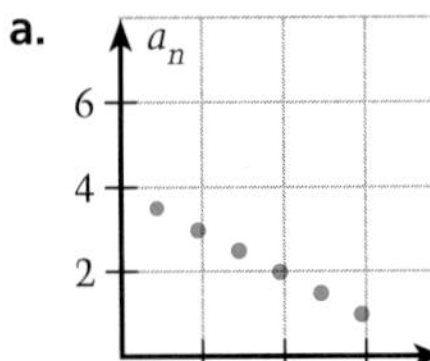

b. 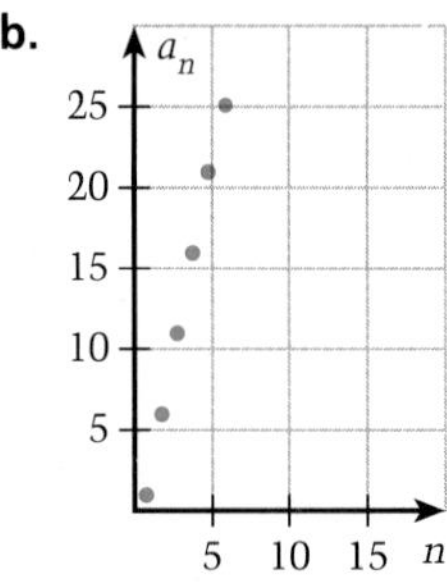

c. 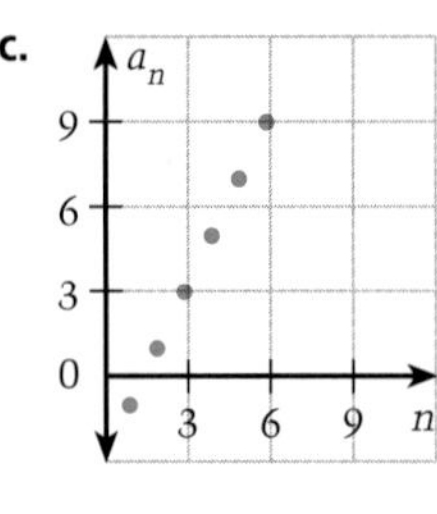

d. 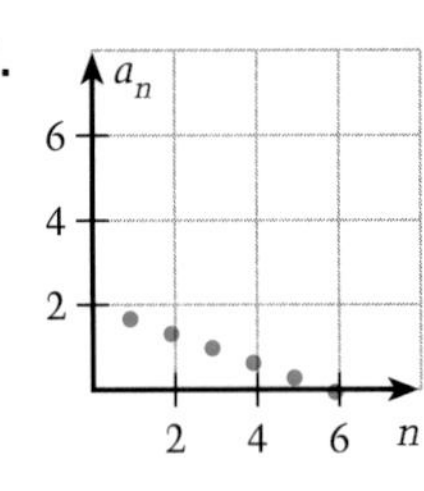

In Exercises 32–39, find the sum of the first n terms of the arithmetic sequence.

32. 8, 20, 32, 44, . . . ; $n = 10$ **620**

33. 2, 8, 14, 20, . . . ; $n = 25$ **1850**

34. −6, −2, 2, 6, . . . ; $n = 50$ **4600**

35. 0.5, 0.9, 1.3, 1.7, . . . ; $n = 10$ **23**

36. 40, 37, 34, 31, . . . ; $n = 10$ **265**

37. 1.50, 1.45, 1.40, 1.35, . . . ; $n = 20$ **20.5**

38. $a_1 = 100$, $a_{25} = 220$, $n = 25$ **4000**

39. $a_1 = 15$, $a_{100} = 307$, $n = 100$ **16,100**

In Exercises 40–42, evaluate the sum.

40. $\sum_{i=1}^{20} (3i + 2)$ **670**

41. $\sum_{i=1}^{32} (2i + 8)$ **1312**

42. $\sum_{i=1}^{45} (4i + 9)$ **4545**

43. ***Ticket Prices*** There are 20 rows of seats on the main floor of a concert hall: 20 seats in the first row, 21 seats in the second row, 22 seats in the third row, and so on. How much should you charge per ticket in order to obtain $15,000 for the sale of all of the seats on the main floor? **$25.42**

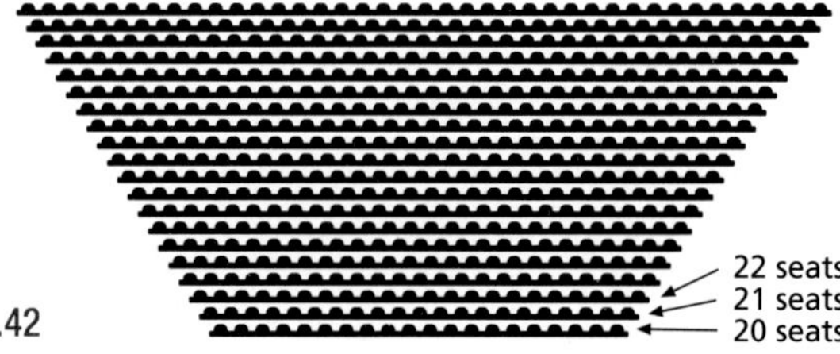

44. ***Drilling a Well*** A well-drilling company charges $15 for drilling the first foot of a well, $15.25 for drilling the second foot, $15.50 for drilling the third foot, and so on. How much would it cost to have this company drill a 100-foot well? **$2737.50**

45. ***Pile of Logs*** Logs are stacked in a pile as shown below. The top row has 15 logs and the bottom row has 21 logs. How many logs are in the stack? **126**

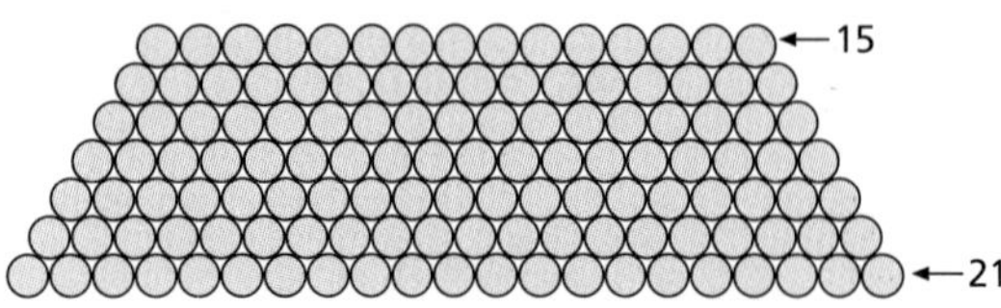

46. ***Brick Pattern*** A brick patio has 20 rows of bricks. The first row has 14 bricks and the twentieth row has 33 bricks. How many bricks are in the patio? **470**

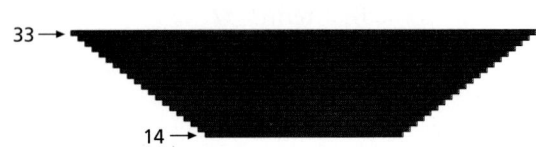

47. *Salary Increase* You have accepted a job with a salary of \$27,500 the first year. At the end of each year for five years, you receive a \$1500 raise. Write a formula for your salary, S_n, during the nth year. What is your total salary during the first six years? $S_n = 1500n + 26{,}000$, \$187,500

48. *Clock Chimes* A clock chimes once at 1:00, twice at 2:00, three times at 3:00, and so on. The clock also chimes once at 15-minute intervals that are not on the hour. How many times does the clock chime in a 12-hour period? 114

Integrated Review

In Exercises 49–56, write the first five terms of the sequence.

49. $a_n = 5 + 3n$ 8, 11, 14, 17, 20

50. $a_n = (2^n)n$ 2, 8, 24, 64, 160

51. $a_n = \frac{1}{n+1}$ $\frac{1}{2}, \frac{1}{3}, \frac{1}{4}, \frac{1}{5}, \frac{1}{6}$

52. $a_n = 1 + 4(n - 1)$ 1, 5, 9, 13, 17

53. $a_n = 100 - 3n$ 97, 94, 91, 88, 85

54. $a_n = 2^{n-1}$ 1, 2, 4, 8, 16

55. $a_n = (2 + n) - (1 + n)$ 1, 1, 1, 1, 1

56. $a_n = (-1)^n$ −1, 1, −1, 1, −1

In Exercises 57–60, decide whether the recursive function is linear, quadratic, or neither.

57. $f(0) = 3$, $f(n) = 5 + f(n - 1)$ Linear

58. $f(1) = 2$, $f(n) = f(n - 1) + 2n$ Quadratic

59. $f(1) = 0$, $f(n) = f(n - 1) - n^2$ Neither

60. $f(0) = 1$, $f(n) = f(n - 1) + 13$ Linear

61. *Powers of Ten* Which best estimates the height of a 10-story building? **b**

a. 1.2×10^1 ft **b.** 1.2×10^2 ft **c.** 1.2×10^3 ft **d.** 1.2×10^4 ft

62. *Powers of Ten* Which best estimates the volume of this book? **d**

a. 1.2×10^{-1} in.^{3} **b.** 1.2×10^0 in.^{3} **c.** 1.2×10^1 in.^{3} **d.** 1.2×10^2 in.^{3}

Exploration and Extension

63. *Japanese Circle* The Japanese "magic circle" shown at the right is the creation of Seki Kowa, a 17th-century Japanese mathematician. Each set of numbers on a diameter has the same total. Use the magic circle to find the sum of the first 51 positive integers. Check your result using techniques that you studied in this lesson. Can you explain why the magic circle works? 5.53 + 1 = 266

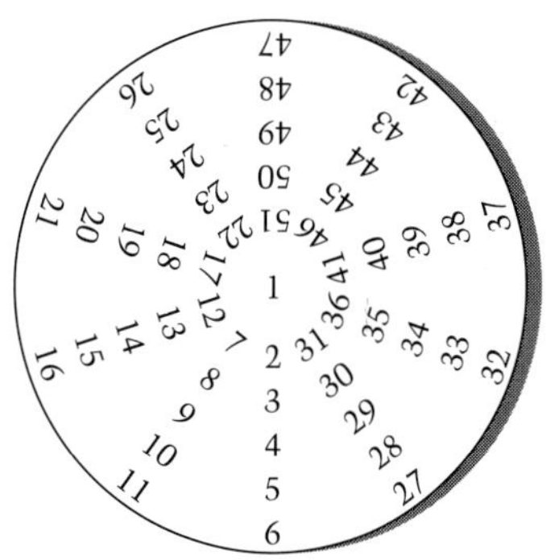

▶ **Ex. 63 ALTERNATIVE ASSESSMENT** As an assessment opportunity in the last five or ten minutes of class, you could have students work in pairs to complete this activity.

Enrichment Activities

These activities require students to go beyond lesson goals.

TECHNOLOGY The following BASIC program will generate the terms of an arithmetic sequence recursively and the sum of the terms of the sequence.

```
10 REM PROGRAM TO PRINT
   TERMS OF ARITHMETIC
   SEQUENCE AND SUM OF
   SERIES
15 LET N = 1
20 LET TERM = 2
25 LET SUM = 0
30 LET SUM = SUM + TERM
35 PRINT "N", "TERM", "SUM"
40 PRINT N, TERM, SUM
45 FOR N = 2 TO 20
50 TERM = TERM + 5
55 SUM = SUM + TERM
60 PRINT N, TERM, SUM
65 NEXT N
70 END
```

Run the program.

1. List the first three terms of the sequence. 2, 7, 12
2. List the last three terms of the sequence. 87, 92, 97
3. List the last line of data. 20, 97, 990
4. Rewrite the program to generate the problem modeled in Exercise 44.

```
20 LET TERM = 15
45 FOR N = 2 TO 100
50 TERM = TERM + 0.25
```

Answers.
1. $g(x) = \sqrt[3]{x - 2}$

14.
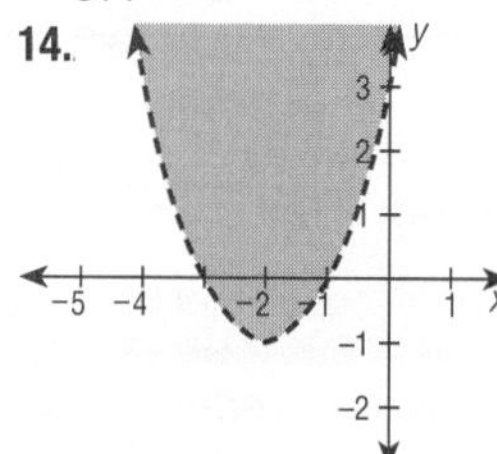

Mixed REVIEW

5. $(x + 2)(x + 6)(x - 3)$
6. $4x^2 + 7x + 7$
9. $\begin{bmatrix} -2 & 13 \\ 3 & 7 \end{bmatrix}$
10. $\begin{bmatrix} -4 & \frac{3}{2} \\ 2 & 9 \end{bmatrix}$

1. Find the inverse $f(x) = x^3 + 2$. **(6.3)** See margin.
2. Classify the conic: $6x^2 - 4y^2 = 24$. **(11.6)** Hyperbola
3. Simplify: $\frac{3x^4y^{-1}}{x^2} \cdot \frac{y^2}{9x^3}$. **(7.1)** $\frac{y}{3x}$
4. Solve $\begin{cases} x + 2y = 8 \\ 4x - y = 5 \end{cases}$. **(3.2)** (2, 3)
5. Factor $x^3 + 5x^2 - 12x - 36$. **(9.3)**
6. Divide $(4x^3 + 3x^2 - 7)$ by $(x - 1)$. **(9.4)**
7. Solve $|4x - 6| \le 3$. **(1.7)** $\frac{3}{4} \le x \le \frac{9}{4}$
8. Evaluate $36^{-3/2}$. **(7.3)** $\frac{1}{216}$
9. Add: $\begin{bmatrix} 2 & 5 \\ -6 & 0 \end{bmatrix} + \begin{bmatrix} -4 & 8 \\ 9 & 7 \end{bmatrix}$. **(4.1)**
10. Multiply: $\begin{bmatrix} \frac{1}{2} & 2 \\ 3 & -1 \end{bmatrix}\begin{bmatrix} 0 & 3 \\ -2 & 0 \end{bmatrix}$. **(4.2)**
11. Solve $\sqrt{x} - \sqrt[4]{2x - 1} = 0$. **(7.5)** 1
12. Write in logarithmic form $5^{-2} = \frac{1}{25}$. **(8.2)**
13. Expand $\log_6 \frac{3x^2}{4}$. **(8.3)**
14. Sketch $y > x^2 + 4x + 3$. **(5.7)** See margin.
15. Solve $3mn + 2 = m + n$ for n. **(1.5)** $\frac{m - 2}{3m - 1}$
16. Solve $3x^2 + 4x = 2$. **(5.3)** $\frac{-2}{3} \pm \frac{\sqrt{10}}{3}$
17. Write an equation of the line passing through $(-3, 2)$ and $(3, -4)$. **(2.4)**
18. Write an equation of the line that passes through $(1, 1)$ with a slope of $-\frac{1}{4}$. **(2.2)**
19. Find a polynomial equation with leading coefficient of 1 and solutions $2i$, $-2i$, and 1. **(9.6)** $x^3 - x^2 + 4x - 4 = 0$
20. Find the mean and median of the numbers: 12, 13, 22, 28, 30, 18, 12, 13, 17, 18. **(6.7)** 18.3, 17.5

12. $\log_5\left(\frac{1}{25}\right) = -2$ 13. $\log_6 3 + 2\log_6 x - \log_6 4$ 17. $x + y = -1$ 18. $x + 4y = 5$

Milestones GÖDEL'S PROOF

1880 Gottlob Frege's work published — 1910 First volume of *Principia Mathematica* published — 1930 Gödel's proof published

Kurt Gödel

In the late 1800's, a movement began to organize all branches of mathematics similar to the way it is organized in geometry. Efforts were made to identify a complete set of axioms for each branch of mathematics from which all of its theorems and corollaries could be deduced. In 1879, the German mathematician Gottlob Frege published his efforts in two volumes. Alfred North Whitehead and Bertrand Russell made a more complete attempt in the three volumes of *Principia Mathematica*, published between 1910 and 1913. Then, in 1931, a faculty member at the University of Vienna, Kurt Gödel, showed that their goal was unattainable. He proved that a complete set of axioms could never be identified for a branch of mathematics such that all of its propositions could be proven or disproven on the basis of these axioms. Although Gödel effectively closed one avenue of mathematical research, he also pointed out new directions for the future.

In the 1930's and 1940's many mathematicians immigrated to the United States to work and study. Why do you think this was so?

Answers may include the following: many mathematicians came to escape the turmoil in their countries caused by World War II and the number of graduate programs in mathematics at universities and colleges in the U.S. was increasing.

Milestones

Language Arts: ***Critical Reading***

1. Find out who David Hilbert was. What role did he play in the movement to organize mathematical knowledge? David Hilbert (1862–1943) was a German mathematics professor. He attempted, through imposing the language and rules of logic on mathematical knowledge, to show the internal consistency of all mathematics. Gödel's proof showed that this could not be done.
2. Bertrand Russell wrote to Gottlob Frege concerning Frege's work. Find out Russell's reason for writing. Russell had discovered a contradiction in Frege's writing just as the second volume was about to be published. Russell later worked out the contradiction in *Principia Mathematica.*

12.3 Geometric Sequences and Series

What you should learn:

Goal 1 How to find the nth term and the sum of a finite geometric series

Goal 2 How to use geometric series in real-life problems

Why you should learn it:

You can model many real-life situations with geometric sequences, such as revenues that increase at a constant percent.

Goal 1 Finding the Terms and the Sum

A sequence whose consecutive terms have a common ratio is a **geometric sequence.**

Definition of a Geometric Sequence

The sequence $a_1, a_2, a_3, a_4, \ldots, a_n, \ldots$ is **geometric** if there is a nonzero number r such that

$$\frac{a_2}{a_1} = r, \quad \frac{a_3}{a_2} = r, \quad \frac{a_4}{a_3} = r$$

and so on. The number r is the *common ratio* of the sequence.

Example 1 *Examples of Geometric Sequences*

a. The first 4 terms of the sequence whose nth term is 2^n are 2, 4, 8, and 16. The common ratio between consecutive terms is 2. Therefore, the sequence is geometric.

$$\underbrace{2, 4,}_{\frac{4}{2} = 2} 8, 16, \ldots, 2^n, \ldots$$

b. The first 4 terms of the sequence whose nth term is $4(3^n)$ are 12, 36, 108, and 324. The common ratio between consecutive terms is 3. Therefore, the sequence is geometric.

$$\underbrace{12, 36,}_{\frac{36}{12} = 3} 108, 324, \ldots, 4(3^n), \ldots$$

c. The first 4 terms of the sequence whose nth term is $9\left(-\frac{1}{3}\right)^n$ are -3, 1, $-\frac{1}{3}$, and $\frac{1}{9}$. The common ratio between consecutive terms is $-\frac{1}{3}$. Thus, the sequence is geometric.

$$\underbrace{-3, 1,}_{\frac{1}{-3} = -\frac{1}{3}} -\frac{1}{3}, \frac{1}{9}, \ldots, 9\left(-\frac{1}{3}\right)^n, \ldots$$

■

If r is the common ratio of a geometric sequence, then the formula for the nth term of the sequence is

$$a_n = a_1 r^{n-1} \qquad \textit{nth term of a geometric sequence}$$

Problem of the Day

Find the sum of all the numbers that can be obtained by rearranging the digits in the three-digit number *abc*.

$222(a + b + c)$

ORGANIZER

Warm-Up Exercises

1. Identify the sequences that have a common difference or a common factor between consecutive terms in the sequence. Identify the sequences that have neither.
   a. 3, 2, 7, 1, 9, 12, 8, 5, 18, . . .
   b. 2.3, 6.9, 20.7, 62.1, 186.3, . . .
   c. 8.4, 8.27, 8.14, 8.01, 7.88, . . .
   d. 4, 6, 9, 13.5, 20.25, 30.375, . . .

   a. neither
   b. common factor of 3
   c. common difference of -0.13
   d. common factor of 1.5
2. Evaluate each term for the indicated value of n.
   a. $a_n = 3(1.5)^{n-1}$, $n = 4$
   b. $a_n = 24(3)^{n-1}$, $n = 5$
   c. $a_n = 0.2\left(\frac{3}{4}\right)^{n-1}$, $n = 3$

   a. 10.125, b. 1944, c. 0.1125

Lesson Resources

Teaching Tools
- Problem of the Day: 12.3
- Warm-up Exercises: 12.3
- Answer Masters: 12.3

Extra Practice: 12.3
Color Transparencies: 64, 65

LESSON Notes

Example 1

The graphs of arithmetic sequences lie on the graphs of related linear equations. Ask students to describe the graphs of geometric sequences in a similar manner. If $r > 0$, but not equal to 1, then the graph of the geometric sequence is similar to the graph of a related exponential equation. If $r = 1$, then the graph of the geometric sequence is related to the graph of a horizontal line; if $r < 0$, the graph of the geometric sequence consists of points that move back and forth from below the x-axis to above the x-axis.

Examples 2–3

If you have access to an overhead graphics calculator or an overhead computer plotter, you might sketch the graph of each of the geometric sequences to show their relationship to the corresponding exponential functions. (Be sure students understand that the graphs consist of discrete points plotted on a coordinate plane.)

Common-Error ALERT!

In **Example 5,** students often confuse the assignment values of a_1, r, and n in the formula for the Sum of a Finite Geometric Series. Specifically, the last term, a_1r^{n-1}, may appear to be the $(n - 1)$th term, when, in fact, it's the nth term. Carefully check that students make the correct assignments for the first few examples by writing out the first terms and labeling them.

Example 2 Finding a Term of a Geometric Sequence

Find the fifteenth term of the geometric sequence whose first term is 20 and whose common ratio is 1.05.

Solution Use the formula for the nth term of the sequence.

$$a_n = a_1r^{n-1} \quad \textit{nth term of geometric sequence}$$

$$= 20(1.05)^{n-1} \quad \textit{Substitute 20 for } a_1 \textit{ and 1.05 for } r.$$

The fifteenth term is $a_{15} = 20(1.05)^{14} \approx 39.6$. ■

Example 3 Finding a Term of a Geometric Sequence

Find the twelfth term of the geometric sequence whose first three terms are 5, 15, and 45.

Solution The common ratio of this sequence is $r = \frac{15}{5} = 3$. Because the first term is $a_1 = 5$, the formula for the nth term is as follows.

$$a_n = a_1r^{n-1} \quad \textit{nth term of geometric sequence}$$

$$= 5(3)^{n-1} \quad \textit{Substitute 5 for } a_1 \textit{ and 3 for } r.$$

The twelfth term is $a_{12} = 5(3)^{11} = 885{,}735$. ■

Example 4 Finding a Term of a Geometric Sequence

The fourth term of a geometric sequence with positive terms is 125. The tenth term is $\frac{125}{64}$. Find the first term.

Solution You can obtain the tenth term by multiplying the fourth term by r^6. Because $a_{10} = \frac{125}{64}$ and $a_4 = 125$, you can solve for r as follows.

$$a_{10} = a_4r^6 \quad \textit{Multiply } a_4 \textit{ by } r^6 \textit{ to obtain } a_{10}.$$

$$\frac{125}{64} = 125r^6 \quad \textit{Substitute } \tfrac{125}{64} \textit{ for } a_{10} \textit{ and 125 for } a_4.$$

$$\frac{1}{64} = r^6 \quad \textit{Divide both sides by 125.}$$

$$\frac{1}{2} = r \quad \textit{Take sixth root of both sides.}$$

Now, using $r = \frac{1}{2}$, you can solve for the first term of the sequence as follows.

$$a_4 = a_1r^{4-1} \quad \textit{Formula for fourth term.}$$

$$125 = a_1\left(\frac{1}{2}\right)^3 \quad \textit{Substitute 125 for } a_4 \textit{ and } \tfrac{1}{2} \textit{ for } r.$$

$$125 = a_1\left(\frac{1}{8}\right) \quad \textit{Simplify.}$$

$$1000 = a_1 \quad \textit{Multiply both sides by 8.}$$

■

The expression formed by adding the terms of a geometric sequence is called a **geometric series.** You can develop a formula for the sum, S, of a (finite) geometric series as follows.

$$\begin{array}{rl} S = & a_1 + a_1r + a_1r^2 + a_1r^3 + \cdots + a_1r^{n-1} \\ -rS = & \quad - a_1r - a_1r^2 - a_1r^3 - \cdots - a_1r^{n-1} - a_1r^n \\ \hline S(1 - r) = a_1(1 - r^n) & \qquad\qquad\qquad\qquad - a_1r^n \end{array}$$

By dividing both sides of this equation by $(1 - r)$, you can obtain the following formula for S.

The Sum of a Finite Geometric Series

The sum of the geometric series $a_1 + a_1r + a_1r^2 + \cdots + a_1r^{n-1}$ with common ratio $r \neq 1$ is

$$\sum_{i=1}^{n} a_1r^{i-1} = a_1\left(\frac{1 - r^n}{1 - r}\right).$$

Example 5 Finding the Sum of a Geometric Series

Evaluate the sum.

$$\sum_{i=1}^{12} 4(0.3)^i = 4(0.3) + 4(0.3)^2 + 4(0.3)^3 + \cdots + 4(0.3)^{12}$$

Solution Because $a_1 = 4(0.3)$, $r = 0.3$, and $n = 12$, you can apply the formula for the sum of a geometric series.

$$\sum_{i=1}^{12} 4(0.3)^i = a_1\left(\frac{1 - r^n}{1 - r}\right) = 4(0.3)\left(\frac{1 - (0.3)^{12}}{1 - 0.3}\right)$$
$$\approx 1.714$$

■

Example 6 Finding the Sum of a Geometric Series

Evaluate the sum.

$$\sum_{i=0}^{10} 10\left(-\frac{1}{2}\right)^i = 10 + 10\left(-\frac{1}{2}\right) + 10\left(-\frac{1}{2}\right)^2 + \cdots + 10\left(-\frac{1}{2}\right)^{10}$$

Solution By examining the given terms, you can see that the *first term* is $a_1 = 10$, and $r = -\frac{1}{2}$. Moreover, by starting with $i = 0$ and ending with $i = 10$, you are adding $n = 11$ terms.

$$\sum_{i=0}^{10} 10\left(-\frac{1}{2}\right)^i = a_1\left(\frac{1 - r^n}{1 - r}\right) = 10\left(\frac{1 - \left(-\frac{1}{2}\right)^{11}}{1 - \left(-\frac{1}{2}\right)}\right)$$
$$\approx 6.670$$

■

Example 5

Refer students to the *Handbook of Mathematical Connections* for a proof of the formula for the sum of a finite geometric series.

Example 6

Take the time to let students sort out the potentially confusing use of subscripts. The index i runs from 0 to 10, but the first term, a_1, uses the subscript 1! The reason for this is that the formula for the sum of a geometric series begins with a_1. However, the index $i = 0$ is convenient because the first term of the geometric series can then be expressed without a factor of r (as in Example 6).

Extra Examples

Here are additional examples similar to **Examples 1–6.**

1. Find the indicated term and the nth term of each geometric sequence.

a. $a_1 = 4$, $r = 2$, $a_7 = ?$

b. $a_1 = 1.2$, $a_2 = 3.6$, $a_3 = 10.8$, $a_9 = ?$

c. $a_4 = 10$, $a_9 = 0.0032$, $a_3 = ?$

d. $a_6 = 54$, $a_{10} = 3.375$, $a_8 = ?$

a. $a_n = 4(2)^{n-1}$, $a_7 = 256$
b. $a_n = 1.2(3)^{n-1}$, $a_9 = 7873.2$
c. $a_n = 1250(0.2)^{n-1}$, $a_3 = 50$
d. $a_n = 1728(0.5)^{n-1}$, $a_8 = 13.5$

2. Find the sum of each finite series.

a. $1.4 + 2.8 + 5.6 + 11.2 + \cdots + 179.2$

b. $-12 + (-18) + (-27) + \cdots + (-91.125)$

c. $\sum_{i=1}^{6} 30(0.2)^i$ **d.** $\sum_{i=0}^{12} 0.01(3)^i$

a. 357, b. −249.375, c. 7.49952, d. 7971.61

Example 7

Observe the use of the index i as an exponent in the argument of the summation notation. The effect is the same as beginning the index with $i = 0$ for the argument $361(1.175)^i$.

Check Understanding

1. What is the difference between a geometric sequence and a geometric series? A geometric series is an indicated sum of the terms of a geometric sequence.
2. Describe the graph of a geometric sequence. How does it compare to the graph of an exponential function? A discrete set of points that lie on the graph of an exponential function; see the Lesson Notes for Example 1.
3. How many terms are there in the finite series $\sum_{i=0}^{6} 30(0.2)^i$? $6 + 1 = 7$

Communicating about ALGEBRA

EXTEND *Communicating*

Have students construct an algebraic model and a bar graph that shows the annual revenue, R (in thousands of dollars), over a 10-year period, of a company whose annual revenue in the first year was $320,000 and whose annual revenue, R, in thousands of dollars

a. increased by an average of $100,000 each year. (Let n = number of years.)

b. decreased at an average rate of 1.5% per year. (Let n = number of years.)

a. $R = 320 + (n - 1)100$
The bars are 320, 420, 520, etc. Each bar of the graph is 100 greater than the previous bar.

b. $R = 320(0.985)^{n-1}$
The bars are 320, 315.2, 310.47, etc. Each bar is 0.985 of the previous bar.

Goal 2 Using Geometric Series in Real Life

Revenues for Rubbermaid

Revenue (in millions of dollars): 500, 1000, 1500

Year (1 ↔ 1981): 1 2 3 4 5 6 7 8 9 10

Example 7 *Finding the Sum of a Geometric Series*

For 1981 through 1990, the annual revenue for Rubbermaid, Inc. increased at an average rate of 17.5% per year. Over this 10-year period, the annual revenue, R (in millions of dollars), can be modeled by $R = 361(1.175)^{n-1}$ where $n = 1$ represents 1981. Use this model to approximate the total revenue for Rubbermaid from 1981 to 1990. ***(Source: Rubbermaid, Inc.)***

Solution Because the annual revenues are modeled by a geometric sequence, you can approximate the total revenue by using the formula for the sum of a geometric series.

$$\sum_{i=1}^{10} 361(1.175)^{i-1} = 361 + 361(1.175) + \cdots + 361(1.175)^9$$
$$= 361\left(\frac{1 - 1.175^{10}}{1 - 1.175}\right)$$
$$\approx 361\left(\frac{-4.016}{-0.175}\right) \approx 8284$$

The total revenue over the 10-year period was about $8284 million or $8,284,000,000. (Rubbermaid's actual total income over the 10-year period was $8,289,000,000.) ■

Communicating about ALGEBRA

SHARING IDEAS about the Lesson

The bar graphs show the annual revenue, R (in millions of dollars), for two companies over a 10-year period. One company's revenue growth followed an arithmetic pattern; the other company's revenue growth followed a geometric pattern. Find a model for each company's revenue and find the total revenue over the 10-year period.

A.

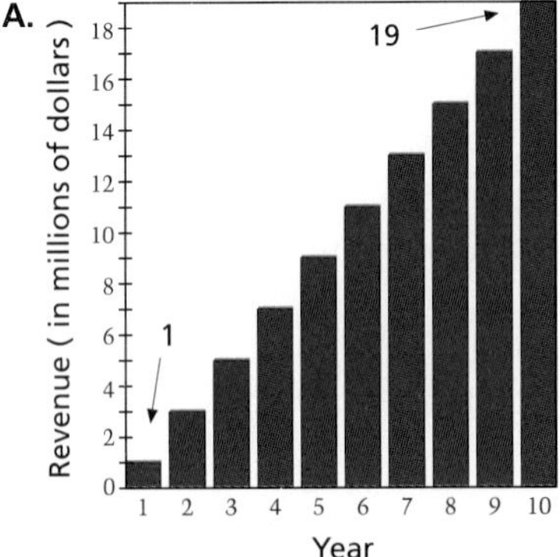

B.

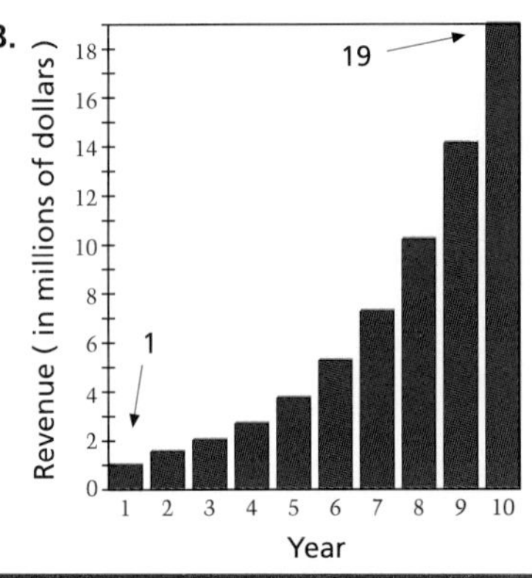

A. $a_n = 2n - 1$, $100 million **B.** $a_n \approx 1.387^{n-1}$, ≈$65.5 million

EXERCISES

Guided Practice

CRITICAL THINKING about the Lesson

1. What makes a sequence geometric? A common ratio of consecutive terms
2. What is the general formula for the nth term of a geometric sequence? $a_n = a_1r^{n-1}$
3. Find the common ratio of the sequence 2, 6, 18, 54, 162, ... 3
4. Which is the sixth term of the geometric sequence whose third term is 64 and whose common ratio is 4? a
   a. $64(4^3)$ **b.** $64(4^4)$ **c.** $64(4^5)$ **d.** $64(4^6)$
5. Find the sum: $\sum_{i=1}^{10} 6\left(\frac{1}{2}\right)^i$. ≈5.99
6. Find the sum: $\sum_{i=0}^{10} 6\left(\frac{1}{2}\right)^i$. ≈11.99

Independent Practice

In Exercises 7–12, decide whether the sequence is arithmetic, geometric, or neither.

7. 5, 15, 45, 135, . . . Geometric
8. 1, 4, 9, 16, . . . Neither
9. 3, 12, 21, 30, . . . Arithmetic
10. 1, −2, −5, −8, . . . Arithmetic
11. $\frac{1}{2}, \frac{3}{2}, \frac{9}{2}, \frac{27}{2}, \ldots$ Geometric
12. −3, −1, 1, 3, . . . Arithmetic

In Exercises 13–18, find the common ratio of the geometric sequence. Then write the next term.

13. 1, −3, 9, −27, . . . −3, 81
14. 3, 9, 27, 81, . . . 3, 243
15. 2, 10, 50, 250, . . . 5, 1250
16. 6, −24, 96, −384, . . . −4, 1536
17. $5, -\frac{5}{2}, \frac{5}{4}, -\frac{5}{8}, \ldots$ $-\frac{1}{2}, \frac{5}{16}$
18. $2, \frac{4}{3}, \frac{8}{9}, \frac{16}{27}, \ldots$ $\frac{2}{3}, \frac{32}{81}$

In Exercises 19–24, write the first five terms of the geometric sequence. See below.

19. $a_1 = 2, r = 3$
20. $a_1 = 1, r = \frac{1}{2}$
21. $a_1 = 6, r = 2$
22. $a_1 = 5, r = -\frac{1}{10}$
23. $a_1 = 1, r = -\frac{1}{4}$
24. $a_1 = 1, r = \frac{1}{3}$

In Exercises 25–30, find the indicated term of the geometric sequence.

25. $a_1 = 4, r = \frac{1}{2}, \quad a_{10} = \boxed{?}$ $\frac{1}{128}$
26. $a_1 = 5, r = \frac{3}{2}, \quad a_8 = \boxed{?}$ $\frac{10,935}{128}$
27. $a_1 = 6, r = -\frac{1}{3}, \quad a_{12} = \boxed{?}$ $-\frac{2}{59,049}$
28. $a_1 = 1, r = -\frac{2}{3}, \quad a_7 = \boxed{?}$ $\frac{64}{729}$
29. $a_1 = 100, r = 2, \quad a_9 = \boxed{?}$ 25,600
30. $a_1 = 8, r = 4, \quad a_4 = \boxed{?}$ 512

In Exercises 31–34, write a formula for the nth term of the geometric sequence.

31. $1, \frac{3}{2}, \frac{9}{4}, \frac{27}{8}, \ldots$ $a_n = \left(\frac{3}{2}\right)^{n-1}$
32. $1, \frac{2}{3}, \frac{4}{9}, \frac{8}{27}, \ldots$ $a_n = \left(\frac{2}{3}\right)^{n-1}$
33. $4, 6, 9, \frac{27}{2}, \ldots$ $a_n = 4\left(\frac{3}{2}\right)^{n-1}$
34. 24, 12, 6, 3, . . . $a_n = 24\left(\frac{1}{2}\right)^{n-1}$

19. 2, 6, 18, 54, 162 20. $1, \frac{1}{2}, \frac{1}{4}, \frac{1}{8}, \frac{1}{16}$ 21. 6, 12, 24, 48, 96
22. $5, -\frac{1}{2}, \frac{1}{20}, -\frac{1}{200}, \frac{1}{2000}$ 23. $1, -\frac{1}{4}, \frac{1}{16}, -\frac{1}{64}, \frac{1}{256}$ 24. $1, \frac{1}{3}, \frac{1}{9}, \frac{1}{27}, \frac{1}{81}$

For each company, find the total revenue over the 10-year period.

(*Hint:* a. the nth term = $100n + 220$, b. the nth term = $320(0.985)^{n-1}$)

a. \$7,700,000 **b.** ≈\$2,992,417

EXERCISE Notes

ASSIGNMENT GUIDE

Basic/Average: Ex. 1–6, 9–13 odd, 21–29 odd, 33–41 odd, 47–53 odd, 65–71

Above Average: Ex. 1–6, 9–17 odd, 21–27 odd, 21–43 odd, 47–53 odd, 54, 68–71

Advanced: Ex. 1–6, 9–17 odd, 23–43 odd, 45–51 odd, 52–54, 68–71

Selected Answers
Exercises 1–6, 7–69 odd

✪ More Difficult Exercises
Exercises 52–54, 71

Guided Practice

▶ **Ex. 1–6 COOPERATIVE LEARNING** Have students work with partners to complete these exercises in class. In Exercise 1, students should relate the common ratio r to the constant percent increase/decrease concepts associated with exponential functions.

Independent Practice

▶ **Ex. 25–43** Encourage students to refer to Examples 2–4 for solution models.

Answers

35. $a_n = 3\left(\frac{1}{2}\right)^{n-1}$

36. $a_n = \frac{1}{2}(4)^{n-1}$

37. $-\frac{1}{177{,}147}$

38. 12,288

39. 8748 or -8748

40. $\frac{1}{2187}$ or $-\frac{1}{2187}$

41. $\frac{1}{4}$ or $-\frac{1}{4}$

42. 3

43. $a_n = 32\left(\frac{1}{2}\right)^{n-1}$ or $a_n = 32\left(-\frac{1}{2}\right)^{n-1}$

Ex. 44–51 You may need to help students transform the summation notation to the summation formula

$$S_n = a_1\left(\frac{1-r^n}{1-r}\right).$$

Ex. 52 Students should refer to Example 7 for a solution model.

In Exercises 35–43, give the missing information about the geometric sequence. See margin.

35. Common ratio: $\frac{1}{2}$
First term: 3
Formula for a_n: ?

36. Common ratio: 4
First term: $\frac{1}{2}$
Formula for a_n: ?

37. Common ratio: $-\frac{1}{3}$
First term: 1
Twelfth term: ?

38. Common ratio: 2
First term: 3
Thirteenth term: ?

39. First term: 4
Fifth term: 324
Eighth term: ?

40. First term: 9
Third term: 1
Tenth term: ?

41. Third term: 64
Seventh term: $\frac{1}{4}$
Common ratio: ?

42. Eighth term: $\frac{1}{9}$
Fifteenth term: 243
Common ratio: ?

43. Third term: 8
Ninth term: $\frac{1}{8}$
Formula for a_n: ?

In Exercises 44–51, evaluate the sum of the finite geometric series.

44. $\sum_{n=1}^{10} 8(2^{n-1})$ 8184

45. $\sum_{n=1}^{6} 3(4^{n-1})$ 4095

46. $\sum_{i=1}^{10} 8\left(\frac{1}{4}\right)^{i-1}$ ≈10.67

47. $\sum_{i=1}^{10} 5\left(\frac{1}{3}\right)^{i-1}$ ≈7.50

48. $\sum_{n=1}^{8} 2^{n-1}$ 255

49. $\sum_{n=1}^{8} (-2)^{n-1}$ −85

50. $\sum_{n=0}^{5} 3\left(\frac{3}{2}\right)^{n}$ $\frac{1995}{32}$

51. $\sum_{n=0}^{6} 2\left(\frac{4}{3}\right)^{n}$ $\frac{28{,}394}{729}$

52. *Student Guaranteed Loans* For 1982 through 1990, the amount, a_n (in millions of dollars), of guaranteed student loans can be modeled by $a_n = 5650.2(1.2)^{n-1}$ where $n = 2$ represents 1982. Use this model to approximate the total amount of guaranteed student loans for 1982 through 1990. ***(Source: Student Loan Marketing Association)*** \$141,021.55 million

53. *Geometry* A square has 16-inch sides. It is partitioned to form a new square by connecting the midpoints of the sides of the original square. Then two of the corner triangles are shaded. This process is repeated five more times. Each time, two of the corner triangles are shaded. The portion of the original square that is shaded on the nth partitioning is $\frac{1}{4}\left(\frac{1}{2}\right)^{n-1}$. Find the total shaded area. 126 in.²

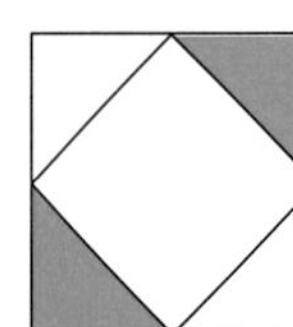
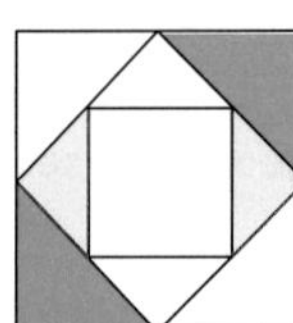
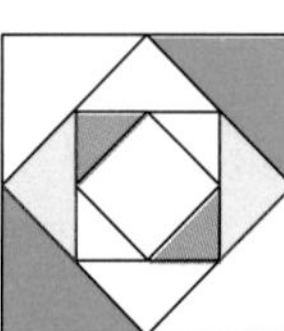
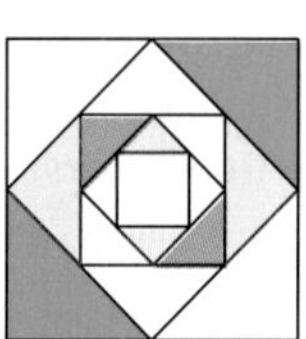

54. *Geometry* A 27-inch by 27-inch square is partitioned into nine squares and the center square is shaded. Each of the unshaded squares is then partitioned into nine squares and their center squares are shaded. The process is repeated three more times. The portion of the original square that is shaded on the nth partitioning is $\frac{1}{9}\left(\frac{8}{9}\right)^{n-1}$. Find the total shaded area. $\frac{26{,}281}{81}$ in.²

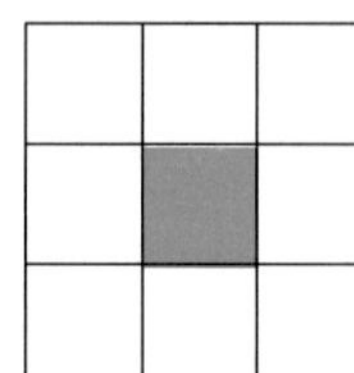
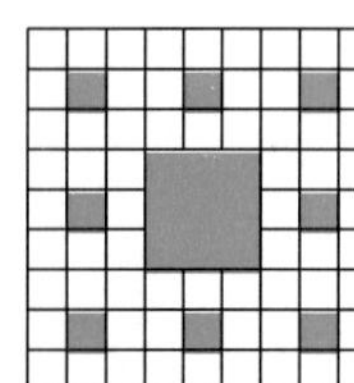
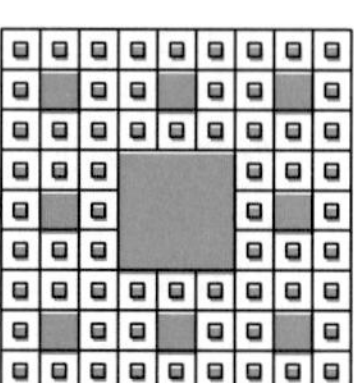

Integrated Review

In Exercises 55–63, is the sequence arithmetic, geometric, or neither?

55. $\frac{1}{3}, 1, 3, 9, \ldots$ Geometric

56. $\frac{1}{4}, \frac{1}{2}, \frac{3}{2}, 2, \ldots$ Neither

57. $-5, -3, -1, 1, \ldots$ Arithmetic

58. $2, 1, \frac{1}{2}, \frac{1}{4}, \ldots$ Geometric

59. $\frac{1}{2}, 1, \frac{7}{2}, 5, \ldots$ Neither

60. $\frac{9}{2}, 1, \frac{2}{9}, \frac{4}{81}, \ldots$

60. Geometric

61. $-5, 1, 3, 6, \ldots$ Neither

62. $-4, -2, 2, 4, \ldots$ Neither

63. $0, 1, 2, 4, 8, \ldots$ Neither

In Exercises 64–67, evaluate the sum of the series.

64. $\sum_{i=1}^{4} (6i + 5)$ 80

65. $\sum_{i=1}^{5} 2(0.8)^{i-1}$ $\frac{4202}{625}$

66. $\sum_{i=1}^{10} i^2$ 385

67. $\sum_{i=1}^{5} (4 - 2i)$ -10

Native Americans **In Exercises 68–70, use the following data from 1990.**

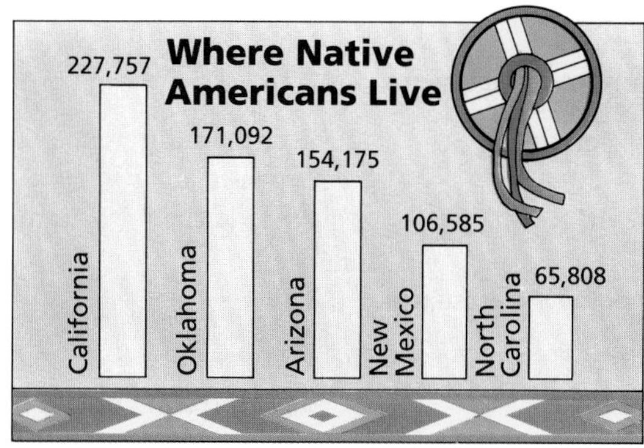

Recently actor Lou Diamond Phillips, who is part Cherokee, produced a documentary on the problems of the 1.5 million Native Americans living in the United States.

68. Of all Native Americans living in the U.S. in 1990, what percent lived in California? ≈15%

69. In 1990, the population of New Mexico was 1,515,000. What percent of the population were Native Americans? ≈7%

70. In 1990, Native Americans represented 4.2% of the population of Arizona. What was the population of Arizona in 1990? ≈3,670,833

Exploration and Extension

71. *Finding a Pattern* In the large square at the right, four corner squares were drawn and shaded. The sides of each corner square are one-fourth the length of a side of the original square. This process is then repeated in the center square. What portion of the original square is shaded after performing the process five times? $\frac{341}{1024}$

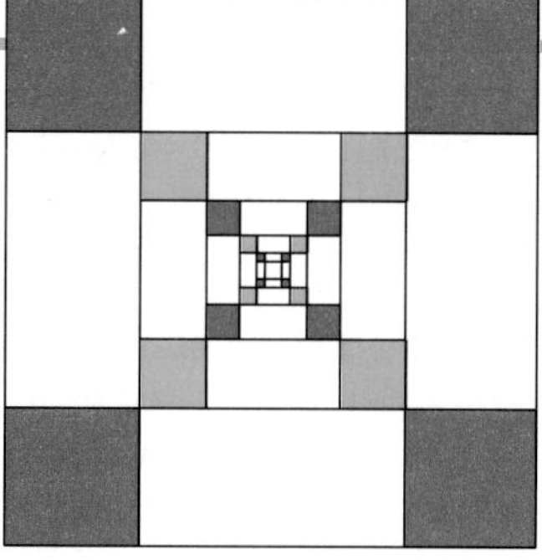

EXTEND Ex. 64–67 Have students identify those series that are arithmetic and those that are geometric.
Ex. 64 and 67 are arithmetic; Ex. 65 is geometric.

Ex. 68–70 MULTICULTURAL These exercises provide an opportunity to discuss Native American issues in your local community. Students can be asked to find the Native American population in your area. Is the population growing? Is it shrinking?

Ex. 71 PROBLEM SOLVING If Exercises 53 and 54 are assigned as part of a homework assignment, then you should use Exercise 71 as an in-class problem-solving opportunity.

Enrichment Activities

These activities require students to go beyond lesson goals.

The sales, S, in millions of dollars, for the years 1980 to 1990 of a toy store chain can be modeled by the following geometric sequence: $S = 640(1.125)^{t-1}$, where $t = 1$ corresponds to the year 1980.

1. Find the first term of the sequence. What does this term represent?
   640; the total sales in 1980 were \$640 million.
2. Find the last term of the sequence. What does this term represent?
   2078.3; the total sales in 1990 were \$2078.3 million.
3. Find the sum of the terms of the sequence. What does this sum represent?
   $S = \frac{640(1 - 1.125^{11})}{1 - 1.125}$
   $\approx 13,584.6$
   The total sales from 1980 through 1990 were \$13,584.6 million.

Mid-Chapter SELF-TEST

Take this test as you would take a test in class. The answers to the exercises are given in the back of the book.

In Exercises 1–3, decide whether the sequence is arithmetic, geometric, or neither. (12.2, 12.3)

1. 4, 10, 20, 34, 52, . . .
Neither

2. 11, 7, 3, −1, −5, . . .
Arithmetic

3. $-1, \frac{1}{2}, -\frac{1}{4}, \frac{1}{8}, -\frac{1}{16}, \ldots$
Geometric

In Exercises 4–6, write the first 5 terms of the sequence. Begin with $n = 1$. (12.1)

4. $a_n = 3n + 2$
5, 8, 11, 14, 17

5. $a_n = -3n^2 + 2$
−1, −10, −25, −46, −73

6. $a_n = \frac{n!}{n^2}$
$1, \frac{1}{2}, \frac{2}{3}, \frac{3}{2}, \frac{24}{5}$

In Exercises 7–9, find a formula for the nth term of the sequence. (12.2, 12.3)

7. 4, 10, 16, 22, 28, . . .
$a_n = 6n - 2$

8. 6, 12, 24, 48, 96, . . .
$a_n = 6(2)^{n-1}$

9. $\frac{10}{3}, \frac{11}{3}, 4, \frac{13}{3}, \frac{14}{3}, \ldots$
$a_n = \frac{n+9}{3}$

In Exercises 10–12, answer the question about the arithmetic sequence. (12.2)

10. Common difference: −4
Second term: 8
What is the 5th term?
−4

11. Common difference: 6
First term: −20
What is the 4th term?
−2

12. Common difference: 2
Tenth term: 15
What is the 15th term?
25

In Exercises 13–15, answer the question about the geometric sequence. (12.3)

13. Common ratio: $\frac{1}{5}$
First term: 10
Write a formula for a_n.
$a_n = 10\left(\frac{1}{5}\right)^{n-1}$

14. Second term: 18
Fifth term: 486
Find the 7th term.
4374

15. Third term: −24
Eighth term: 768
Find the common ratio.
−2

In Exercises 16–18, evaluate the sum of the series. (12.1–12.3)

16. $\sum_{n=1}^{44} n$ 990

17. $\sum_{i=0}^{4} \left(\frac{1}{3}\right)^i$ $\frac{121}{81}$

18. $\sum_{n=1}^{16} (4n - 3)$ 496

19. For 1989 through 1994, the amount, a_n (in billions of dollars), of retail sales for general merchandise stores can be modeled by $a_n = 15.66n + 60.28$ where $n = 9$ represents 1989. Use this model to approximate the total retail sales for general merchandise stores for 1989 through 1994. **(12.2)** ***(Source: U.S. Bureau of Census)*** $1442.22 billion

20. For 1985 through 1994, the sales, a_n (in millions of dollars), of compact disks can be modeled by $a_n = 251(1.286)^n$ where $n = 5$ represents 1985. Use this model to approximate the total sales of compact disks for 1985 through 1994. **(12.3)**
(Source: U.S. Bureau of Census) $35,100.76 million

21. For 1989 through 1994, the amount, a_n (in billions of dollars), of sales for bookstores can be modeled by $a_n = 1.2 + 0.6n$ where $n = 9$ represents 1989. Use this model to approximate the total sales for 1989 through 1994. **(12.2)** ***(Source: U.S. Bureau of Census)*** $48.6 billion

12.4 Infinite Geometric Series

What you should learn:

Goal 1 How to find the sum of an infinite geometric series

Goal 2 How to use infinite geometric series as models of real-life problems

Why you should learn it:

You can model many real-life situations with geometric series, such as the total distance traveled by a bouncing ball.

Goal 1 Summing an Infinite Geometric Series

It should not surprise you that all *finite* geometric series have a finite sum. You may, however, be surprised that some *infinite* geometric series have a finite sum. Here is an example.

$$\sum_{i=0}^{\infty} 0.3\left(\frac{1}{10}\right)^i = 0.3 + 0.03 + 0.003 + 0.0003 + \cdots$$
$$= 0.3333\ldots$$
$$= \frac{1}{3}$$

To indicate that an **infinite series** has no upper limit of summation, we write the symbol for *infinity*, ∞, in place of the upper limit. This symbol is never used to represent a number. Instead, it is used to indicate a type of unbounded or unending situation. For an infinite series, the infinity symbol is used to indicate that the series is composed of infinitely many terms.

Sum of an Infinite Geometric Series

If $|r| < 1$, then the infinite geometric series

$$a_1 + a_1 r + a_1 r^2 + a_1 r^3 + \cdots + a_1 r^{n-1} + \cdots$$

has the sum $\sum_{n=1}^{\infty} a_1 r^{n-1} = \frac{a_1}{1-r}$.

If $|r| \geq 1$, then the series has no sum.

Example 1 *Summing an Infinite Geometric Series*

Find the sum.

$$\sum_{n=1}^{\infty} 4(0.6)^{n-1} = 4 + 4(0.6) + 4(0.6)^2 + 4(0.6)^3 + \cdots$$

Solution For this series, $a_1 = 4$ and $r = 0.6$. Because $|r| < 1$, you can apply the formula for an infinite geometric series.

$$\sum_{n=1}^{\infty} 4(0.6)^{n-1} = \frac{a_1}{1-r} = \frac{4}{1-(0.6)} = 10$$ ■

Problem of the Day

The longer base of an isosceles trapezoid has a length equal to one of its diagonals, and the shorter base has a length equal to the altitude. If these lengths are integers, what is the smallest possible area of the trapezoid?
12 square units

ORGANIZER

Warm-Up Exercises

1. Find the sum.

a. $\sum_{i=1}^{7} 100(0.2)^i$ **b.** $\sum_{i=1}^{9} 5(2)^i$

a. 24.99968, **b.** 5110

2. Compare the following finite sums.

a. $\sum_{i=1}^{5} \left(\frac{1}{2}\right)^i$ **b.** $\sum_{i=1}^{10} \left(\frac{1}{2}\right)^i$

c. $\sum_{i=1}^{20} \left(\frac{1}{2}\right)^i$ **d.** $\sum_{i=1}^{50} \left(\frac{1}{2}\right)^i$

As i gets larger, the sum of the geometric series gets closer to 1. The sums are (approximately) **a.** 0.96875, **b.** 0.999023438, **c.** 0.999999046, **d.** 1.0000000

Lesson Resources

Teaching Tools
- Problem of the Day: 12.4
- Warm-up Exercises: 12.4
- Answer Masters: 12.4

Extra Practice: 12.4

Technology Handbook: p. 83

LESSON Notes

Example 1

Ask students to compute the sums of the following finite series and compare their sums to the infinite series given in the example.

a. $\sum_{i=1}^{10} 4(0.6)^{i-1}$ **b.** $\sum_{i=1}^{20} 4(0.6)^{i-1}$

c. $\sum_{i=1}^{30} 4(0.6)^{i-1}$

The sums, ≈ 9.939534, ≈ 9.999634, and ≈ 9.999998, approach the sum of the infinite series, 10.

Example 2

According to Homer's classic tale, *The Illiad,* Achilles was a Greek warrior and leader of the Trojan War. In one legend, his mother Thetis wishing to make him immortal, held the baby Achilles by his heel and dipped him in the river Styx for divine protection. He was killed by an arrow of the Trojan warrior Paris. The arrow struck Achilles in his heel, the only place on his body that was not protected. Hence, your Achilles' heel is considered to be your most vulnerable spot.

Example 3

Some students may describe the activity of the bouncing ball in the following way: $10 + 2\left[10\left(\frac{4}{5}\right)\right] + 2\left[10\left(\frac{4}{5}\right)^2\right] + 2\left[10\left(\frac{4}{5}\right)^3\right] + \ldots$ where the distance of the bouncing ball in both directions is expressed in one geometric series, $2\left[10\left(\frac{4}{5}\right)\right] + 2\left[10\left(\frac{4}{5}\right)^2\right] + 2\left[10\left(\frac{4}{5}\right)^3\right] + \ldots$. Here, $a_1 = 16$ and $r = \frac{4}{5}$. The distance of the ball, 10, is added to the sum of this geometric series.

Goal 2 Using Infinite Geometric Series

A mathematical *paradox* is an argument in which an apparent untruth has been proved. Many famous paradoxes in mathematics involve infinite geometric series. One is the paradox of *Achilles and the Tortoise.* The paradox was first proposed by the Greek mathematician Zeno (*circa* 480 B.C.).

Connections
Paradoxes

Example 2 *Finding the Sum of an Infinite Series*

Achilles is a heroic character in Greek mythology, thought to run faster than any other living creature. One day, as Achilles is walking in a field, he sees a tortoise that is 100 feet away. The tortoise starts running in a straight line at a speed of 50 feet per second. Achilles starts running at a speed of 100 feet per second. By common sense, you know that Achilles can catch the tortoise—in fact, it should take only two seconds. In the paradox, however, Zeno argues that Achilles can *never* catch the tortoise. His argument is as follows.

Achilles starts running, but when he gets to the position where the tortoise began, the tortoise is no longer there—it is in a new position. So Achilles keeps running, but again when he gets to the new position, the tortoise is no longer there—it is in yet a new position. By this reasoning, Achilles can never catch the tortoise because when Achilles gets to where the tortoise was, the tortoise will have moved to a new position!

How can you resolve this paradox?

Solution The "solution" to this paradox is that it is possible for Achilles, or anyone, to occupy infinitely many positions during a finite length of time. In Achilles' pursuit, the distances traveled between consecutive positions of the tortoise and the times required to reach those positions form geometric series that have finite sums.

Distance (ft)	100	50	25	12.5	6.25	3.125	. . .
Time (sec)	1	0.5	0.25	0.125	0.0625	0.03125	. . .

The total distance traveled and the total time elapsed are as follows.

$$\text{Distance} = \sum_{n=1}^{\infty} 100\left(\frac{1}{2}\right)^{n-1} = \frac{100}{1-\frac{1}{2}} = 200 \text{ feet}$$

$$\text{Time} = \sum_{n=1}^{\infty} 1\left(\frac{1}{2}\right)^{n-1} = \frac{1}{1-\frac{1}{2}} = 2 \text{ seconds}$$

■

The bouncing ball at the right was dropped from a height of 10 feet. Each time it hits the ground, it bounces $\frac{4}{5}$ of its previous height. (Although the ball's path is shown as if it were moving horizontally to the right, its motion is actually straight up and down.)

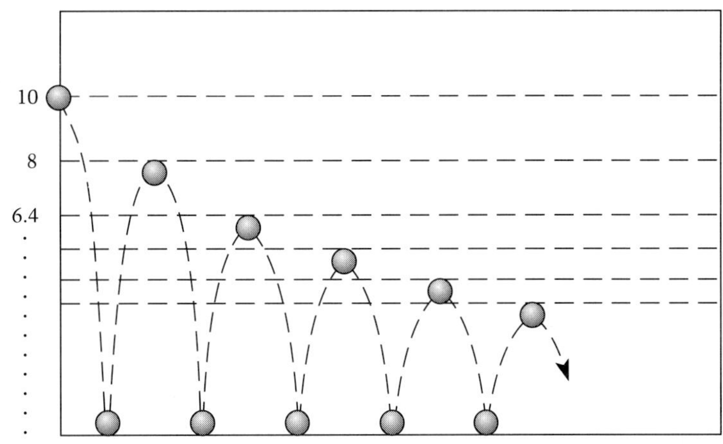

Connections
Physics

Example 3 *Finding the Sum of an Infinite Series*

Find the total distance traveled by the bouncing ball above.

Solution The ball first fell 10 feet. It then bounced to a height of $\frac{4}{5}(10) = 8$ feet. After falling 8 feet back to the ground, it bounced to a height of $\frac{4}{5}(8) = \frac{32}{5}$ feet, and so on. The total distance the ball fell is

$$\text{Falling distance} = 10 + 10\left(\frac{4}{5}\right) + 10\left(\frac{4}{5}\right)^2 + \cdots = \frac{10}{1 - \frac{4}{5}} = 50 \text{ feet.}$$

The total distance the ball bounced upwards is

$$\text{Bouncing distance} = 8 + 8\left(\frac{4}{5}\right) + 8\left(\frac{4}{5}\right)^2 + \cdots = \frac{8}{1 - \frac{4}{5}} = 40 \text{ feet.}$$

Thus the total distance traveled by the ball (in both directions) is $50 + 40 = 90$ feet. Did the ball ever stop bouncing? Why? ■

SHARING IDEAS about the Lesson

A frog at the bottom of a 20-foot well is jumping up the wall, climbing its way to the top. The first jump is 10 feet, the second jump is 5 feet, the third jump is $2\frac{1}{2}$ feet, and so on. At each jump, the frog jumps half of the remaining distance. Each jump takes the frog 1 second. Will the frog ever make it to the top of the well? Explain.

No, see below.

No; although the total distance traveled has a finite sum, the total time elapsed does not.

Extra Examples

Here are additional examples similar to **Examples 1–3.**

1. Find the sum.

   a. $\sum_{i=1}^{\infty} 8(0.4)^{i-1}$ b. $\sum_{i=1}^{\infty} 12\left(\frac{2}{3}\right)^i$

   a. $\frac{40}{3}$, b. 24

2. Suppose the ball in Example 3 is dropped from a height of 12 feet and bounces $\frac{5}{6}$ of its previous height each time it hits the ground. What is the total distance traveled by this bouncing ball? 132 feet

Check Understanding

1. Express the fraction $\frac{2}{3}$ as a geometric series.

   $\sum_{i=0}^{\infty} 0.6\left(\frac{1}{10}\right)^i$

2. What is a mathematical paradox?
   See the top of page 646.

3. Does the symbol ∞ represent a number value?
   No, the symbol indicates a type of unbounded or unending situation. See page 645.

Communicating about ALGEBRA

Consider the same jumping frog. This time, however, suppose that the first jump takes 1 second, the second jump takes $\frac{1}{2}$ second, the third jump takes $\frac{1}{4}$ second, and so on. Each jump takes half the time the previous jump took. Now can the frog make it to the top of the well? If so, how long will it take?
Yes; 2 seconds

EXTEND *Communicating*

WRITING PROBLEMS Ask students to describe one or more problem situations in which a geometric series can be used to find the solution(s). The problems can be collected to use as extra practice later, or students can exchange problems with their classmates for an additional homework assignment.

EXERCISE Notes

ASSIGNMENT GUIDE

Basic/Average: Ex. 1–4, 5–23 odd, 29, 30, 33, 34, 35–45 odd, 47

Above Average: Ex. 1–4, 5–29 odd, 30–34, 35–51 odd

Advanced: Ex. 1–4, 5–29 odd, 30–35, 42–47, 50, 51

Selected Answers
Exercises 1–4, 5–45 odd

More Difficult Exercises
Exercises 29–34, 47–51

Guided Practice

Ex. 1–4 You can use these exercises as an in-class check for understanding. In Exercise 3, ask students what happens to the sum if $r = 0$.
The sum is the first term.

Independent Practice

Ex. 5–12 **WRITING** Ask students to write a one-sentence justification for their decision.

Ex. 13–28 Students should first state the summation formula and then substitute the appropriate values for a and r.

Ex. 16, 19, 26 Ask students why these series have no sum.
$|r| \geq 1$

EXERCISES

Guided Practice

CRITICAL THINKING about the Lesson

1. Write the symbol for *infinity*. ∞
2. Use summation notation to write $a_1 + a_2 + a_3 + \cdots + a_n + \cdots$. $\sum_{i=1}^{\infty} a_i$
3. Under what condition will $\sum_{i=1}^{\infty} a_1 r^{i-1}$ have a sum? $|r| < 1$
4. Find the sum of $\sum_{i=1}^{\infty} 4\left(\frac{1}{3}\right)^{i-1}$. 6

Independent Practice

In Exercises 5–12, decide whether the infinite geometric series has a sum.

5. $\sum_{n=1}^{\infty} 3\left(\frac{3}{2}\right)^{n-1}$ It does not.
6. $\sum_{n=0}^{\infty} -4\left(\frac{1}{4}\right)^{n}$ It does.
7. $\sum_{n=1}^{\infty} \frac{3}{2}\left(\frac{1}{3}\right)^{n-1}$ It does.
8. $\sum_{n=1}^{\infty} \frac{2}{3}\left(\frac{4}{3}\right)^{n-1}$ It does not.
9. $\sum_{n=0}^{\infty} 3\left(-\frac{4}{5}\right)^{n}$ It does.
10. $\sum_{n=1}^{\infty} \left(\frac{2}{9}\right)^{n-1}$ It does.
11. $\sum_{n=0}^{\infty} 6(2)^{n}$ It does not.
12. $\sum_{n=1}^{\infty} \left(\frac{6}{5}\right)^{n-1}$ It does not.

In Exercises 13–28, find the sum (if it has one).

13. $\sum_{n=0}^{\infty} \left(\frac{1}{2}\right)^{n}$ 2
14. $\sum_{n=0}^{\infty} 2\left(\frac{2}{3}\right)^{n}$ 6
15. $\sum_{n=1}^{\infty} \left(-\frac{1}{2}\right)^{n-1}$ $\frac{2}{3}$
16. $\sum_{n=1}^{\infty} \frac{1}{2}(2)^{n-1}$ It has none.
17. $\sum_{n=0}^{\infty} 4\left(\frac{1}{4}\right)^{n}$ $\frac{16}{3}$
18. $\sum_{n=0}^{\infty} \left(\frac{1}{10}\right)^{n}$ $\frac{10}{9}$
19. $\sum_{n=1}^{\infty} 2\left(\frac{7}{5}\right)^{n-1}$ It has none.
20. $\sum_{n=0}^{\infty} 4\left(\frac{3}{7}\right)^{n}$ 7
21. $\sum_{n=1}^{\infty} -\frac{1}{6}\left(-\frac{1}{2}\right)^{n-1}$ $-\frac{1}{9}$
22. $\sum_{n=1}^{\infty} \frac{1}{2}\left(-\frac{2}{5}\right)^{n-1}$ $\frac{5}{14}$
23. $\sum_{n=0}^{\infty} \frac{1}{12}(-0.12)^{n}$ $\frac{25}{336}$
24. $\sum_{n=1}^{\infty} -\left(-\frac{2}{11}\right)^{n-1}$ $-\frac{11}{13}$
25. $\sum_{n=1}^{\infty} 20\left(\frac{4}{5}\right)^{n-1}$ 100
26. $\sum_{n=0}^{\infty} \frac{1}{2}(2)^{n}$ It has none.
27. $\sum_{n=1}^{\infty} \frac{1}{4}\left(\frac{7}{8}\right)^{n-1}$ 2
28. $\sum_{n=0}^{\infty} -5\left(\frac{1}{5}\right)^{n}$ $-\frac{25}{4}$

29. ***Stacking Paper*** Start with a piece of paper that is 8 inches by 10 inches. It has a surface area (both sides) of $2(8)(10) = 160$ square inches. Cut a second piece of paper whose length and width are each $\frac{4}{5}$ of the original. It has a surface area of $2\left(\frac{4}{5}\right)(8)\left(\frac{4}{5}\right)(10)$ square inches. Suppose that you could continue this process forever, each time reducing the length and width of the next piece of paper to $\frac{4}{5}$ of the length and width of the previous piece. Would the total surface area of the infinitely many pieces of paper in the stack be finite or infinite? Would the height be finite or infinite? Explain.

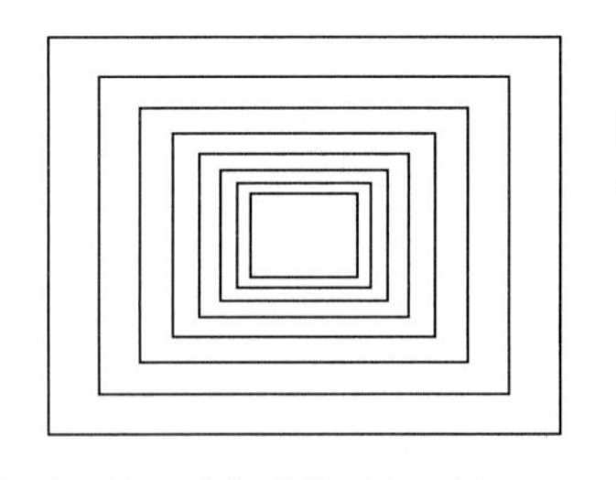

Finite, $\frac{a_1}{1-r} = \frac{160}{1 - \frac{16}{25}} = \frac{4000}{9}$; infinite, $r = 1$ so that $\frac{a_1}{1-r}$ is undefined for the series of heights.

30. *Length of a Spring* The length of the first loop of a spring is 20 inches. The length of the second loop is $\frac{9}{10}$ the length of the first. The length of the third loop is $\frac{9}{10}$ the length of the second, and so on. Suppose the spring has infinitely many loops. Does it have a finite or infinite length? Explain. If it has a finite length, find the length.

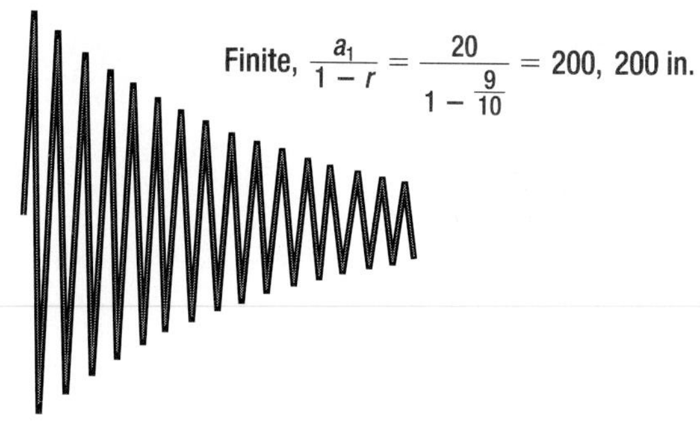

Finite, $\frac{a_1}{1-r} = \frac{20}{1-\frac{9}{10}} = 200$, 200 in.

31. *Superball* You drop a superball from a height of 5 feet. After each time it hits the ground, it bounces to $\frac{8}{9}$ of its previous height. What is the total distance that the superball travels before it comes to rest? **85 ft**

32. *A Jumping Flea* A flea named Argus jumps 12 inches toward another flea, Almo. On the second jump, Argus jumps 6 more inches toward Almo. On the third jump, Argus jumps 3 more inches toward Almo, and so on. Each jump takes one second. Argus keeps getting closer and closer, but cannot quite reach Almo. How far apart were they at the time of Argus's first jump? **24 in.**

33. **$\frac{I}{3}$, $\frac{a_1}{1-r} = \frac{\frac{I}{4}}{1-\frac{1}{4}} = \frac{I}{3}$**

33. *Thermopane Window* A thermopane window is constructed with two parallel pieces of glass, each of which reflects half of the sunlight that hits it from either side. The other half of the sunlight passes through the pane. How much of the sunlight will pass through *both* panes? Explain.

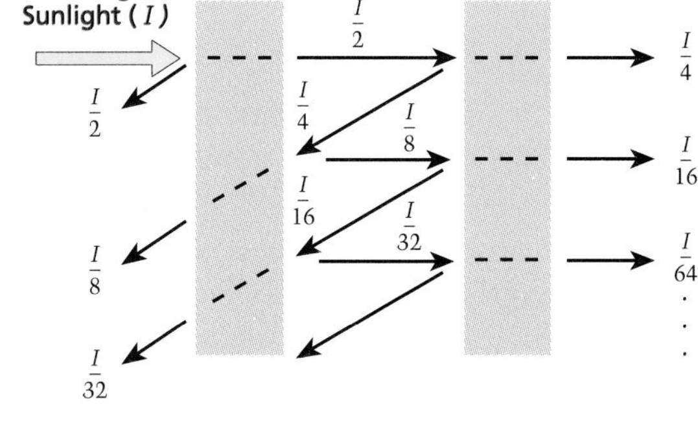

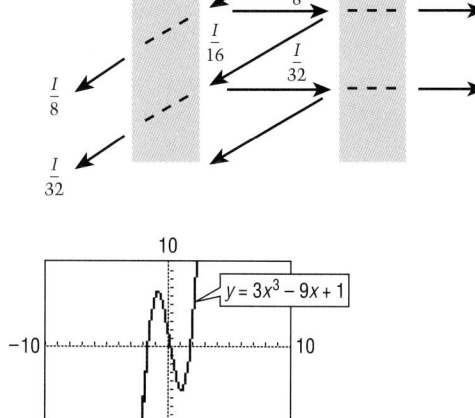

34. *Zooming-In* The zoom feature on most graphing calculators crops 75% of the width (and 75% of the height) of the previous viewing screen. Let W represent the width of the original screen. The amount of the previous screen's width that is cropped on the nth zoom-in is $(0.75W)(0.25)^{n-1}$. If the zooming could continue forever, the total width that is cropped from the original viewing screen would be given by the sum

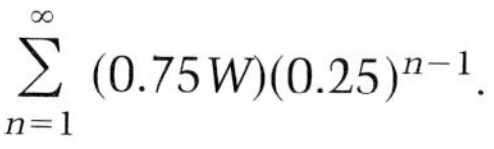

$$\sum_{n=1}^{\infty} (0.75W)(0.25)^{n-1}.$$

Find this sum and interpret the result. Then use a graphing calculator to find the number of times it will actually allow you to zoom in. Use the top graph at the right as the original screen.

10
$y = 3x^3 - 9x + 1$
−10 10
−10
Original Screen

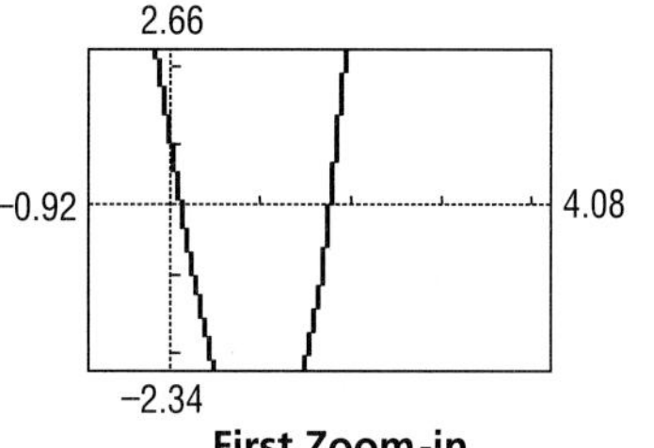

First Zoom-in

▶ **Ex. 29 COOPERATIVE LEARNING** Have students work in small groups to complete this exercise in class. If you have a copying machine that reduces, you could create a series of 80% reduction models for each group.

▶ **Ex. 33** Help students realize that they need to sum the right-hand column of numbers.

Answer

34. $S = W$; this would imply that the entire screen would be cropped, which is impossible; answers will vary with the calculator model, but the *TI-81* will allow you to zoom 13 times.

EXTEND Ex. 35–42 Ask students to identify the type of series shown in each exercise. Ex. 35 and 38 are arithmetic; Ex. 36, 37, and 39–42 are geometric.

Enrichment Activities

This activity requires students to go beyond lesson goals.

RESEARCH Discover more about Zeno of Elea and his paradoxes. Write a one-page report to share with your classmates.
Zeno of Elea (circa 490–435 B.C.) tried to prove that reality is made up of many changing things. He argued that if a thing moves, it must move either where it is, or where it is not. If it moves where it is, there is a contradiction, because it is at rest and thus not moving. It cannot move to where it is not because it does not exist there. In the arrow paradox, Zeno states that a moving arrow will never reach its target because it travels half the remaining distance to the target over time.

Integrated Review

In Exercises 35–42, find the sum (if it has one).

35. $\sum_{n=1}^{5} (2n + 3)$ 45

36. $\sum_{n=0}^{4} (2)^n$ 31

37. $\sum_{n=1}^{6} (5)^n$ 19,530

38. $\sum_{i=1}^{4} (5i - 2)$ 42

39. $\sum_{n=1}^{6} 4(3)^{n-1}$ 1456

40. $\sum_{n=1}^{20} (2)^{n-1}$ 1,048,575

41. $\sum_{n=1}^{10} \frac{1}{2}\left(\frac{1}{2}\right)^{n-1}$ $\frac{1023}{1024}$

42. $\sum_{n=0}^{\infty} 2(3)^n$ It has none.

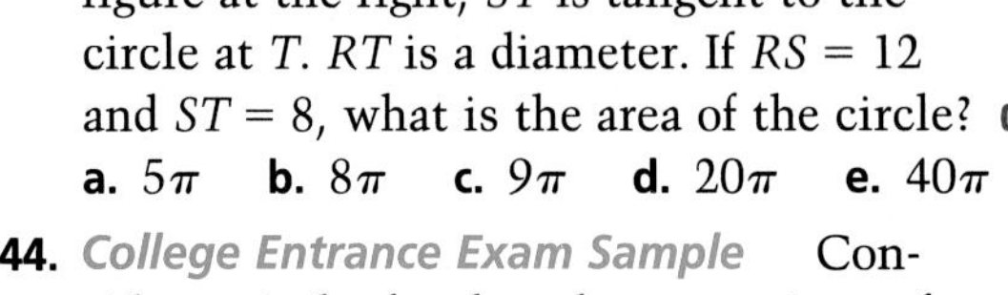

43. *College Entrance Exam Sample* In the figure at the right, ST is tangent to the circle at T. RT is a diameter. If $RS = 12$ and $ST = 8$, what is the area of the circle? d

a. 5π **b.** 8π **c.** 9π **d.** 20π **e.** 40π

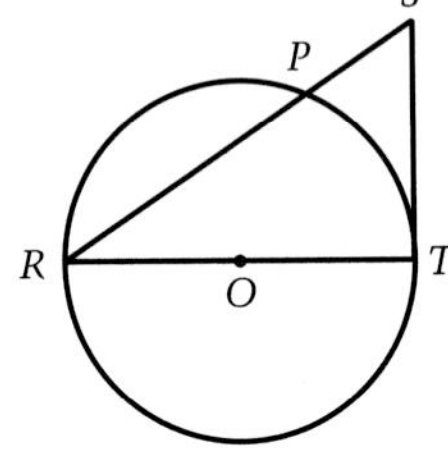

44. *College Entrance Exam Sample* Consider a circle that has the same circumference as the perimeter of a regular hexagon. Each side of the hexagon has a length of 22 inches. Which of the following is the best estimate of the radius of the circle? c

a. 7 **b.** 14 **c.** 21 **d.** 24 **e.** 28

45. *Powers of Ten* Which best estimates the temperature (in degrees Fahrenheit) of a hot tub? c

a. 1.0×10^0 **b.** 1.0×10^1 **c.** 1.0×10^2 **d.** 1.0×10^3

46. *Powers of Ten* Which best estimates the diameter (in millimeters) of a high school class ring? b

a. 2.0×10^0 **b.** 2.0×10^1 **c.** 2.0×10^2 **d.** 2.0×10^3

Exploration and Extension

In Exercises 47–50, write the repeated decimal as an infinite series using summation notation. Evaluate the sum and write the sum as the ratio of two integers.

47. 0.11111 . . . $\sum_{n=1}^{\infty} (0.1)^n = \frac{1}{9}$

48. 0.4444 . . . $\sum_{n=1}^{\infty} 4(0.1)^n = \frac{4}{9}$

49. 0.363636 . . . $\sum_{n=1}^{\infty} 36(0.01)^n = \frac{4}{11}$

50. 0.212121 . . . $\sum_{n=1}^{\infty} 21(0.01)^n = \frac{7}{33}$

51. You and your friend Lisa begin walking around a $\frac{1}{4}$-mile track. Because Lisa is preparing for a race, she is walking twice as fast as you are. As you start, the two of you are walking side by side. How far will each of you have walked when you are again walking side by side? You: $\frac{1}{4}$ mi, Lisa: $\frac{1}{2}$ mi

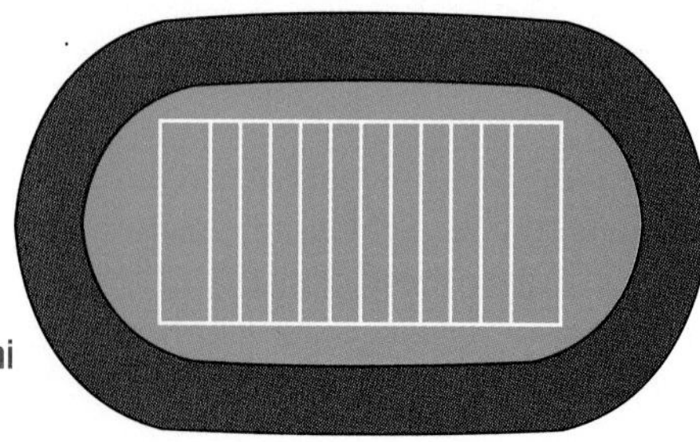

12.5 The Binomial Theorem

What you should learn:

Goal 1 How to use the Binomial Theorem to expand a binomial that is raised to a power

Goal 2 How to use the Binomial Theorem in real-life situations

Why you should learn it:

You can use binomial expansions for many real-life problems, such as the probability of a marble rolling into a particular slot of a certain game board.

Goal 1 Using the Binomial Theorem

Recall that a *binomial* is a polynomial that has two terms. In this lesson, you will study a quick method for raising a binomial to a power. To begin, consider the following examples.

$$(x + y)^0 = 1$$
$$(x + y)^1 = x + y$$
$$(x + y)^2 = x^2 + 2xy + y^2$$
$$(x + y)^3 = x^3 + 3x^2y + 3xy^2 + y^3$$
$$(x + y)^4 = x^4 + 4x^3y + 6x^2y^2 + 4xy^3 + y^4$$
$$(x + y)^5 = x^5 + 5x^4y + 10x^3y^2 + 10x^2y^3 + 5xy^4 + y^5$$

For *each* of these expansions, notice the following.

1. $(x + y)^n$ expands to $n + 1$ terms.
2. The first term is x^n and the last term is y^n. Each of these terms has a coefficient of 1.
3. From term to term, the exponents with base x decrease by 1. The exponents with base y increase by 1.
4. The sum of the exponents of each term is n. For instance, in the expansion of $(x + y)^5$, the sum of the exponents of each term is 5.

$$(x + y)^5 = x^5 + 5\overbrace{x^4y^1}^{4+1=5} + 10\overbrace{x^3y^2}^{3+2=5} + 10x^2y^3 + 5xy^4 + y^5$$

5. The **binomial expansion coefficients** or simply **binomial coefficients** increase and then decrease in a symmetrical pattern. For $(x + y)^5$, the pattern is

   1 5 10 10 5 1.

The Binomial Theorem

The binomial expansion of $(x + y)^n$ is

$$(x + y)^n = x^n + nx^{n-1}y + \cdots + \frac{n!}{(n-m)!m!}x^{n-m}y^m + \cdots + nxy^{n-1} + y^n.$$

The coefficient of $x^{n-m}y^m$ is denoted by $\binom{n}{m}$.

Problem of the Day

Let P be any point in the interior of an equilateral triangle. Prove that the sum of the lengths of the three perpendicular segments from P to the sides of the triangle is equal to the altitude of the triangle.

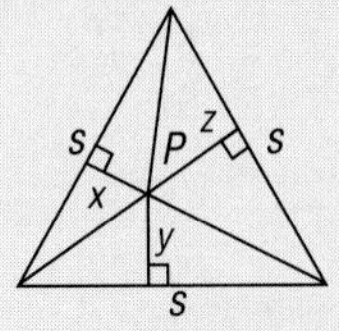

$$\frac{1}{2}xs + \frac{1}{2}ys + \frac{1}{2}zs = \frac{1}{2}hs$$
$$x + y + z = h$$

ORGANIZER

Warm-Up Exercises

1. Compute the following mathematical expressions.
   a. $(5 - 2)!$ b. $\frac{5!}{3!}$
   c. $\frac{4!3!}{6!}$ d. $\frac{3!}{0!2!}$
   a. 6, b. 20, c. 0.2, d. 3
2. Expand $(p + q)^4$.
   $p^4 + 4p^3q + 6p^2q^2 + 4pq^3 + q^4$

Lesson Resources

Teaching Tools
- Transparency: 24
- Problem of the Day: 12.5
- Warm-up Exercises: 12.5
- Answer Masters: 12.5

Extra Practice: 12.5
Color Transparency: 65
Technology Handbook: p. 87

LESSON Notes

Remind students that 0! equals 1. Show students that $n!$ can be rewritten in a variety of ways. In particular,

$$\begin{aligned}8! &= 8 \cdot 7 \cdot 6 \cdot 5 \cdot 4 \cdot 3 \cdot 2 \cdot 1\\ &= 8 \cdot 7 \cdot 6 \cdot 5 \cdot 4 \cdot 3 \cdot 2!\\ &= 8 \cdot 7 \cdot 6 \cdot 5 \cdot 4 \cdot 3!\\ &= 8 \cdot 7 \cdot 6 \cdot 5 \cdot 4!\\ &= 8 \cdot 7 \cdot 6 \cdot 5!\\ &= 8 \cdot 7 \cdot 6!\\ &= 8 \cdot 7!\end{aligned}$$

Example 1

The ability to rewrite a factorial in several different ways can be helpful in simplifying expressions with binomial coefficients.

Example 2

Be sure students understand that if Pascal's Triangle has already been computed for the desired expansion, they need only read from the appropriate row. That is, no computation is required.

Example 3

Make sure students recognize that the five observations made at the beginning of this lesson are also being used to generate the appropriate expansion.

Many students might find it helpful to expand the basic binomial $(x + y)^3$ and then substitute the values "x" for x and "1" for y in the binomial.

Examples 4–5

As suggested for Example 3, one can first expand $(x + y)^3$ as $x^3 + 3x^2y + 3xy^2 + y^3$. Then, substituting "$2x$" for x and "$-y$" for y (because $2x - y = 2x + (-y)$), $(2x)^3 + 3(2x)^2(-y) + 3(2x)(-y)^2 + (-y)^3 = 8x^3 - 12x^2y + 6xy^2 - y^3$.

Ask students what the substitution would be in Example 5. It would be "x" for x and "3" for y in the basic expansion for $(x + y)^4$.

Study Tip

When m and n are not zero, as in parts **a** and **b** in Example 1, the formula for binomial coefficients can be simplified as follows.

$$\binom{8}{2} = \frac{\overbrace{8 \cdot 7}^{\text{2 factors}}}{\underbrace{2 \cdot 1}_{\text{2 factorial}}} = 28$$

$$\binom{10}{3} = \frac{\overbrace{10 \cdot 9 \cdot 8}^{\text{3 factors}}}{\underbrace{3 \cdot 2 \cdot 1}_{\text{3 factorial}}} = 120$$

Example 1 *Finding Binomial Coefficients*

a. $\binom{8}{2} = \frac{8!}{6!2!} = \frac{8 \cdot 7 \cdot 6!}{6! \cdot 2!} = \frac{8 \cdot 7}{2 \cdot 1} = 28$

b. $\binom{10}{3} = \frac{10!}{7!3!} = \frac{10 \cdot 9 \cdot 8 \cdot 7!}{7! \cdot 3!} = \frac{10 \cdot 9 \cdot 8}{3 \cdot 2 \cdot 1} = 120$

c. $\binom{7}{0} = \frac{7!}{7! \cdot 0!} = 1$

d. $\binom{6}{6} = \frac{6!}{0! \cdot 6!} = 1$ ■

LESSON INVESTIGATION

■ Investigating Pascal's Triangle

Partner Activity A convenient pattern for writing binomial coefficients is called **Pascal's Triangle,** after the famous French mathematician Blaise Pascal (1623–1662). He was the first to notice that the numbers in this triangle are precisely the same numbers that occur as coefficients in the expansion of $(x + y)^n$ as listed on page 651.

```
                    1                    ←— Row 0
                  1   1                  ←— Row 1
                1   2   1                ←— Row 2
              1   3   3   1              ←— Row 3
            1   4   6   4   1            ←— Row 4
          1   5  10  10   5   1          ←— Row 5
        1   6  15  20  15   6   1        ←— Row 6
      1   7  21  35  35  21   7   1      ←— Row 7
```

a. With your partner, discuss any patterns you see in the triangle. Write a summary of your discussion.

b. Use your results to write the next two rows of Pascal's Triangle.

c. Find the sum of each of the first six rows of Pascal's Triangle. What is the pattern?

a. Answers will vary.
b. 1 8 28 56 70 56 28 8 1; 1 9 36 84 126 126 84 36 9 1.
c. 1, 2, 4, 8, 16, 32; each sum is twice the previous sum.

The top row in Pascal's Triangle is **Row 0** because it corresponds to the expansion of $(x + y)^0$. Similarly, the next row is **Row 1,** because it corresponds to the expansion of $(x + y)^1$. In general, **Row n** in Pascal's Triangle gives the coefficients of $(x + y)^n$.

Example 2 Using Pascal's Triangle

Use Pascal's Triangle to find the following nine binomial coefficients for the expansion of $(x + y)^8$.

$$\binom{8}{0}, \binom{8}{1}, \binom{8}{2}, \binom{8}{3}, \binom{8}{4}, \binom{8}{5}, \binom{8}{6}, \binom{8}{7}, \binom{8}{8}$$

Solution These nine binomial coefficients represent Row 8 of Pascal's Triangle. Using Row 7 of the triangle, you can calculate the numbers in Row 8, as follows.

$$\begin{array}{ccccccccccccccccc} & 1 & & 7 & & 21 & & 35 & & 35 & & 21 & & 7 & & 1 & \\ 1 & & 8 & & 28 & & 56 & & 70 & & 56 & & 28 & & 8 & & 1 \\ \binom{8}{0} & & \binom{8}{1} & & \binom{8}{2} & & \binom{8}{3} & & \binom{8}{4} & & \binom{8}{5} & & \binom{8}{6} & & \binom{8}{7} & & \binom{8}{8} \end{array}$$

■

Example 3 Expanding a Binomial Sum

Expand the binomial $(x + 1)^3$.

Solution The binomial coefficients in Row 3 of Pascal's Triangle are 1, 3, 3, 1. Therefore, the expansion is as follows.

$$\begin{aligned}(x + 1)^3 &= (1)x^3 + (3)x^2(1) + (3)x(1^2) + (1)(1^3) \\ &= x^3 + 3x^2 + 3x + 1\end{aligned}$$

■

In the next examples, notice how any coefficient other than 1 must be raised to a power when writing the expansion.

Example 4 Expanding a Binomial Difference

Write the expansion for $(2x - y)^3$.

Solution The binomial coefficients in Row 4 of Pascal's Triangle are 1, 3, 3, 1. Note the alternating pattern of signs when the binomial is a difference.

$$\begin{aligned}(2x - y)^3 &= (1)(2x)^3 + 3(2x)^2(-y) + 3(2x)(-y)^2 + (1)(-y)^3 \\ &= 8x^3 - 12x^2y + 6xy^2 - y^3\end{aligned}$$

■

Example 5 Expanding a Binomial

Write the expansion for $(x + 3)^4$.

Solution The binomial coefficients in Row 4 of Pascal's Triangle are 1, 4, 6, 4, 1. Therefore, the expansion is as follows.

$$\begin{aligned}(x + 3)^4 &= (1)x^4 + (4)x^3(3) + (6)x^2(3^2) + (4)x(3^3) + (1)(3^4) \\ &= x^4 + 12x^3 + 54x^2 + 108x + 81\end{aligned}$$

■

Extra Examples

Here are additional examples similar to **Examples 1–5.**

1. Find the binomial coefficients.

a. $\binom{4}{3}$ 4

b. $\binom{12}{8}$ 495

c. $\binom{7}{5}$ 21

2. Use Pascal's Triangle to find the 10 binomial coefficients for the expansion of $(x + y)^9$.
1, 9, 36, 84, 126, 126, 84, 36, 9, 1

3. Expand the following binomials.

a. $(x + 2)^4$ **b.** $(3x - 2)^3$
c. $(2x - 3y)^4$ **d.** $(x + 5)^5$

a. $x^4 + 8x^3 + 24x^2 + 32x + 16$
b. $27x^3 - 54x^2 + 36x - 8$
c. $16x^4 - 96x^3y + 216x^2y^2 - 216xy^3 + 81y^4$
d. $x^5 + 25x^4 + 250x^3 + 1250x^2 + 3125x + 3125$

Check Understanding

1. How many terms are in the expansion of a binomial raised to the 7th power? 8

2. Explain how Pascal's Triangle is related to chance.
See Example 6.

3. Describe how the binomial $(3x - 6y)^6$ can be expanded using the expansion of $(x + y)^6$.
See Lesson Notes for Examples 3–5.

Example 6

Why do you start with 128 marbles in the experiment? The sum of the coefficients in Row 7 of Pascal's Triangle totals to 128.

This example certainly suggests why binomial expansions are frequently used in the branch of mathematics called probability.

Communicating *about* **ALGEBRA**

One disadvantage in using Pascal's Triangle is that each of the preceding rows in the display must be computed before the row you need for an expansion can be obtained. Using factorials to obtain the binomial coefficients can be time-consuming and error-prone.

Ask students to verify that the following technique for obtaining the binomial coefficients works. (Some students may wish to provide a rationale for why it works.)

An Alternate Approach

Use the preceding coefficient to determine binomial coefficients.

"The kth coefficient of the binomial expansion $(x + y)^n$ is:

a. 1 if $k = 1$;

b. $\frac{CE}{(k-1)}$, where C is the coefficient of the preceding term and E is the exponent of x in the preceding term."

The five observations at the beginning of the lesson are used to determine the exponents of x and y. For example, the expansion of $(x + y)^4$ becomes

$x^4 + \frac{(1)(4)}{1}x^3y + \frac{(4)(3)}{2}x^2y^2 + \frac{(6)(2)}{3}xy^3 + \frac{(4)(1)}{4}y^4 = x^4 + 4x^3y + 6x^2y^2 + 4xy^3 + y^4$.

The model shown at the right is designed to experimentally produce the numbers in Row 7 of Pascal's Triangle. To conduct the experiment, 128 marbles are poured into the funnel at the top. As each marble hits a hexagonal divider, the marble is equally likely to fall to the left or to the right.

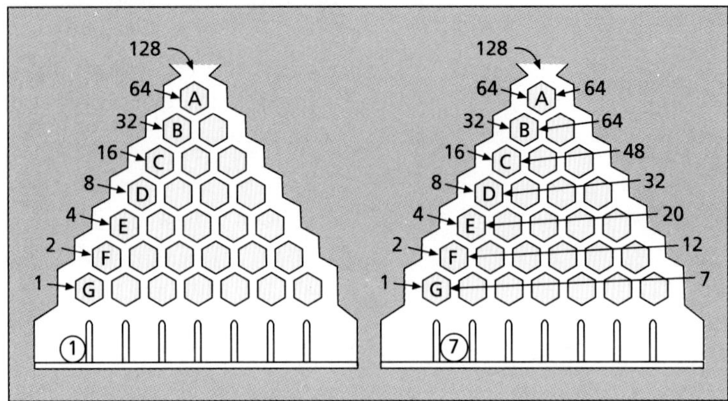

Goal 2 Using Binomial Coefficients in Real Life

Connections
Experimental Mathematics

Example 6 *An Experiment Involving Pascal's Triangle*

How many marbles would you expect to fall into the two leftmost compartments at the bottom of the apparatus shown above?

Solution Begin by considering how many should fall into the first compartment. Half (or 64) of the 128 marbles that hit hexagon A will fall to the left. Of those, half (or 32) that hit hexagon B will fall to the left. By continuing this pattern, you can conclude that only one marble will fall into the first compartment.

To find the number of marbles that should fall into the second compartment, refer to the second diagram above. The number of marbles falling through each channel is shown in the diagram. From the diagram, you can see that 7 marbles should fall into the second compartment. ■

Communicating *about* **ALGEBRA**

SHARING IDEAS about the Lesson See Additional Answers.

A. Copy the model above. Then use it to show how many marbles should fall into the third compartment. Explain your answer.

B. Use a model to show how many marbles should fall into the fourth compartment.

EXERCISES

Guided Practice

CRITICAL THINKING about the Lesson

1. In the expansion of $(x + y)^n$, describe the pattern for the exponents with base x. The exponents with base x decrease by 1 from term to term.
2. Describe two methods you could use to expand the binomial $(x + 5)^4$. Binomial Theorem or Pascal's Triangle
3. Identify the binomial coefficients of the expansion of $(x + y)^7$ shown below. 1, 7, 21, 35, 35, 21, 7, 1

$$x^7 + 7x^6y + 21x^5y^2 + 35x^4y^3 + 35x^3y^4 + 21x^2y^5 + 7xy^6 + y^7$$

4. Which of the following is equal to $\binom{11}{5}$? Explain. a, $\binom{11}{5} = \frac{11!}{6!5!} = \frac{11 \cdot 10 \cdot 9 \cdot 8 \cdot 7}{5!}$

   a. $\frac{11 \cdot 10 \cdot 9 \cdot 8 \cdot 7}{5 \cdot 4 \cdot 3 \cdot 2 \cdot 1}$ **b.** $\frac{11 \cdot 10 \cdot 9 \cdot 8 \cdot 7}{6 \cdot 5 \cdot 4 \cdot 3 \cdot 2 \cdot 1}$

5. Which row in Pascal's Triangle is given by the following binomial coefficients? Row 6

$$\binom{6}{0} \quad \binom{6}{1} \quad \binom{6}{2} \quad \binom{6}{3} \quad \binom{6}{4} \quad \binom{6}{5} \quad \binom{6}{6}$$

6. Write the expansions for $(x + 7)^4$ and $(x - 7)^4$. How are they similar? How are they different?

$x^4 + 28x^3 + 294x^2 + 1372x + 2401$, $x^4 - 28x^3 + 294x^2 - 1372x + 2401$; absolute values of the coefficients are the same for like terms, the even-numbered terms have opposite signs.

Independent Practice

In Exercises 7–14, evaluate the binomial coefficient.

7. $\binom{5}{2}$ 10	**8.** $\binom{5}{3}$ 10	**9.** $\binom{7}{7}$ 1	**10.** $\binom{8}{6}$ 28
11. $\binom{12}{5}$ 792	**12.** $\binom{12}{0}$ 1	**13.** $\binom{12}{7}$ 792	**14.** $\binom{14}{8}$ 3003

In Exercises 15–22, use the Binomial Theorem to expand the binomial. See Additional Answers.

15. $(a + 2)^3$	**16.** $(x - 5)^4$	**17.** $(x + 2)^4$	**18.** $(2x - 1)^4$
19. $(x - 3)^5$	**20.** $(x - 8)^6$	**21.** $(2x + 1)^6$	**22.** $(x + y)^7$

23. Find the coefficient of x^4 in the expansion of $(x + 2)^8$. 1120

24. Find the coefficient of x^6 in the expansion of $(x - 3)^7$. -21

In Exercises 25–28, use Pascal's Triangle to find the binomial coefficient.

25. $\binom{3}{2}$ 3	**26.** $\binom{6}{0}$ 1	**27.** $\binom{7}{5}$ 21	**28.** $\binom{9}{9}$ 1

In Exercises 29–36, use Pascal's Triangle to expand the binomial. See Additional Answers.

29. $(4x - y)^3$	**30.** $(3s + 2)^3$	**31.** $(x + 4)^4$	**32.** $(4t - 1)^4$
33. $(2t - s)^5$	**34.** $(x + 7)^5$	**35.** $(2x + y)^6$	**36.** $(x + 2)^6$

EXERCISE Notes

ASSIGNMENT GUIDE

Basic/Average: Ex. 1–6, 9–33 odd, 40–42, 47, 48, 51, 52

Above Average: Ex. 1–6, 9–33 odd, 37–42, 47–52

Advanced: Ex. 1–6, 9–35 odd, 37–42, 47–52

Selected Answers
Exercises 1–6, 7–47

Use **Mixed Review** as needed.

✪ More Difficult Exercises
Exercises 49–52

Guided Practice

▶ **Ex. 1–6** This lesson contains many new terms and notations. You can use these Guided Practice exercises as an in-class check for understanding of the connection between Pascal's Triangle and the terms of binomial expansions.

▶ **Ex. 4** Ask students why $\binom{11}{5} = \frac{11!}{6!5!}$.

Independent Practice

▶ **Ex. 7–14 and 25–28** TECHNOLOGY Some scientific calculators have a factorial function and an $_nC_r$ function that will compute the combination of n things taken r at a time. Students who have these calculators could use them to check computed values.

▶ **Ex. 15–22 and 29–36** WRITING After students complete these exercises, have them write a paragraph describing the method they think is the most efficient.

▶ **Ex. 37–39** Assign these exercises as a group.

EXTEND EX. 42 Ask students if they recognize the sequence of diagonal sums, 1, 1, 2, 3, 5, 8, ... It is the Fibonacci sequence.

Stacking Cubes **In Exercises 37–39, use the following information.**

The picture at the right shows the different ways to stack four cubes.

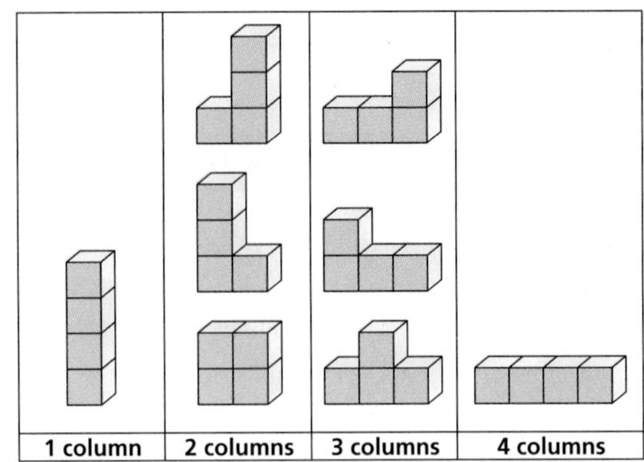

37. Sketch the different ways to stack three cubes. See Additional Answers.

38. Sketch the different ways to stack five cubes. Describe the pattern in the number of ways to stack three, four, and five cubes. See Additional Answers.

39. In how many different ways could you stack ten cubes? Explain your reasoning.

40. What is the sum of the numbers in Row n of Pascal's Triangle? Explain. 39.–40. See below.

41. What is the sum of the numbers in rows 0 through 20 of Pascal's Triangle? 2,097,151

42. *Finding a Pattern* Describe the pattern formed by the sums of the numbers along the diagonal segments (shown at the right) of Pascal's Triangle. The Fibonacci numbers: 1, 1, 2, 3, 5, 8, . . .

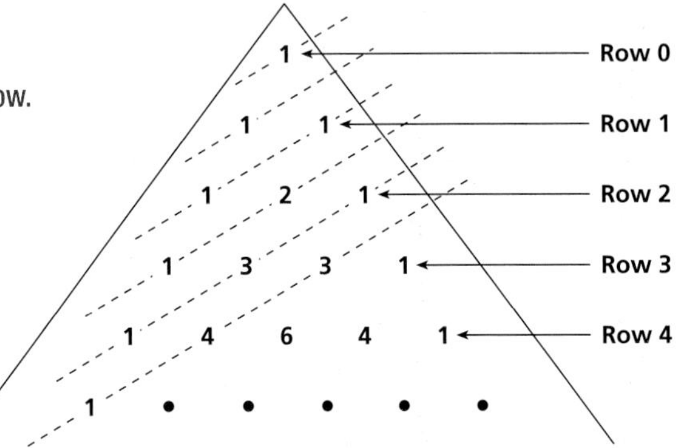

Integrated Review

In Exercises 43–46, evaluate the expression.

43. $\frac{5!}{1! \cdot 4!}$ 5

44. $\frac{7!}{4! \cdot 3!}$ 35

45. $\frac{9!}{2! \cdot 7!}$ 36

46. $\frac{6!}{2! \cdot 4!}$ 15

47. *Wedding Costs* The average cost of a wedding in 1996 was about \$15,000. How much, on the average, was spent on the engagement ring? *(Source: USA Today)* \$3000

Percent of Wedding Expenses

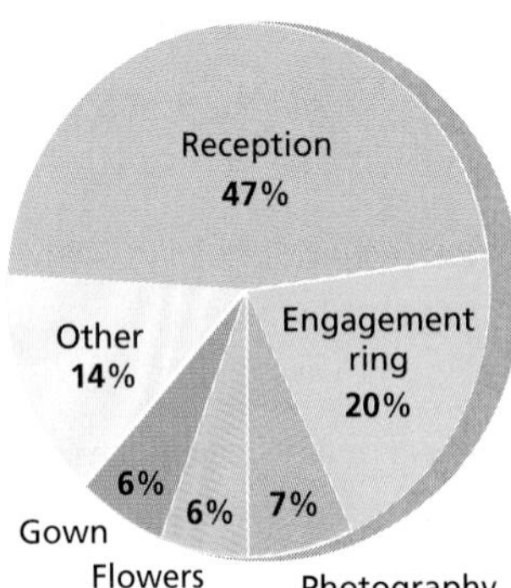

48. *Basketball* What was the percent increase in attendance at Women's NCAA college basketball games from 1990 to 1993? *(Source: Women's NCAA College Basketball)* 50%

Women's College Basketball (NCAA)

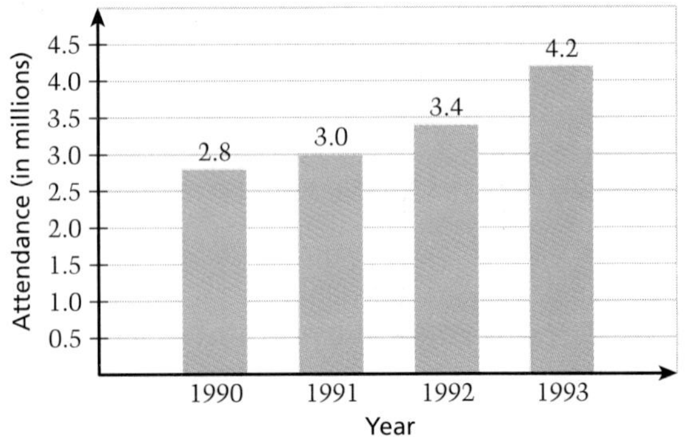

48. ≈3.2%, ≈2.7%, ≈2.2%

39. 512, add the numbers in Row 9 of Pascal's Triangle

40. 2^n; in row 3 the sum is $2^3 = 8$, in row 5 the sum is $2^5 = 32$, etc.

Exploration and Extension

True or False? **In Exercises 49 and 50, decide whether the formula is true or false for all integers m and n such that $1 \le m \le n$. Explain your reasoning.**

49. $\binom{n}{m} = \binom{n}{n-m}$ See Additional Answers.

50. $\binom{n+1}{m} = \binom{n}{m-1} + \binom{n}{m}$ See Additional Answers.

51. *Patterns in Pascal's Triangle* Use each encircled group of numbers to form a 2×2 matrix. Find the determinant of each matrix. Describe the pattern.

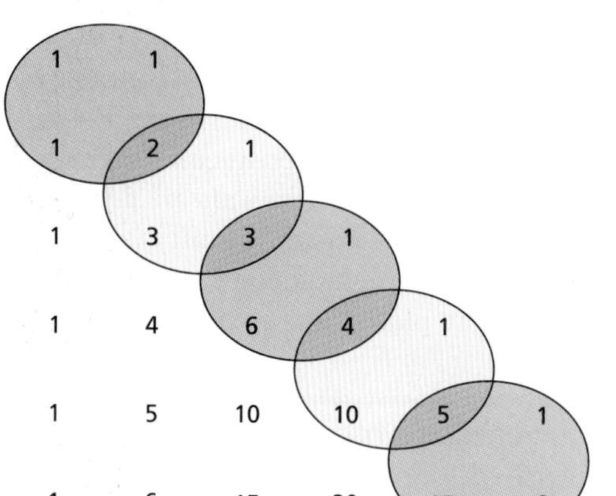

1, 3, 6, 10, 15; $f(1) = 1$, $f(n) = f(n-1) + n$

52. *Patterns in Pascal's Triangle* In the "flowers" below, taken from Pascal's Triangle, fill in the missing numbers. Which rows do the flowers come from?

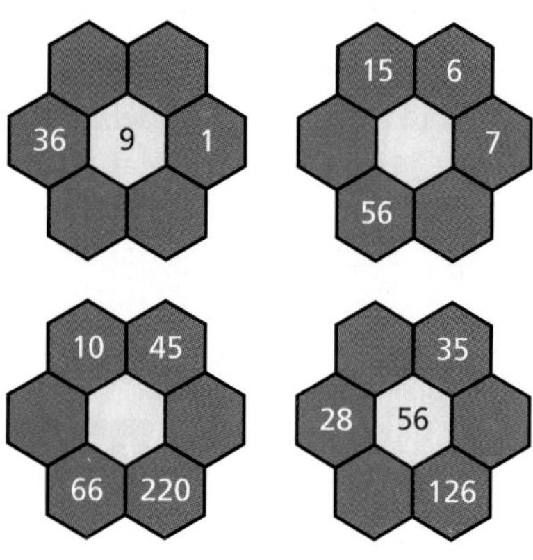

See Additional Answers.

Mixed REVIEW

5., 11. and 16. See Additional Answers.

12. $\frac{2x(x-3)}{x+4}$

1. Find the inverse of $\begin{bmatrix} 2 & 0 \\ 5 & 1 \end{bmatrix}$. **(4.4)** $\begin{bmatrix} \frac{1}{2} & 0 \\ -\frac{5}{2} & 1 \end{bmatrix}$

2. Evaluate the determinant of $\begin{bmatrix} 4 & -3 \\ 2 & 6 \end{bmatrix}$. **(4.3)** 30

3. Solve $5x^2 = 245$. **(7.3)** ± 7

4. Condense $\log_4 3x - \log_4 y$. **(8.3)** $\log_4\left(\frac{3x}{y}\right)$

5. Sketch $y = |3x - 2|$. **(2.6)**

6. Simplify: $\sqrt[3]{27x^3} + 2x$. **(7.4)** $5x$

7. Write in exponential form: $\log_4 64 = 3$. **(8.2)** $4^3 = 64$

8. Find the asymptotes of $y = \frac{3x^2}{x^2 - 9}$. **(10.1)** $x = \pm 3$, $y = 3$

9. Solve $\begin{cases} 5x - 2y = -11 \\ 2x + 5y = -16 \end{cases}$. **(3.2)** $(-3, -2)$

10. Solve $\begin{cases} 3x + 2y - z = -3 \\ 4y + 2z = -12 \\ 5z = 10 \end{cases}$. **(3.6)** $\left(\frac{7}{3}, -4, 2\right)$

11. Sketch $f(x) = \frac{3x + 5}{2x - 3}$. **(10.1)**

12. Multiply: $\frac{6x^3 - 18x^2}{x^2 + 5x + 4} \cdot \frac{x^2 - 2x - 3}{3x^2 - 9x}$. **(10.3)**

13. Write an equation of the horizontal line through $(-1, 2)$. **(2.2)** $y = 2$

14. Find the focus and directrix of the parabola $x = 8y^2$. **(11.1)** $\left(\frac{1}{32}, 0\right)$, $x = -\frac{1}{32}$

15. Find the slope of $6x + 2y = 3$. **(2.4)** -3

16. Sketch $(x + 3)^2 + (y - 4)^2 = 9$. **(11.5)**

17. Classify the conic whose equation is $4x^2 + 24x + y^2 - 4y + 36 = 0$. **(11.5)** Ellipse

18. Identify the vertices in Exercise 17. **(11.5)** $(-3, 0)$, $(-3, 4)$

Ex. 49–50 Ask students to test the formulas by using specific values of m and n. For example, let $m = 3$ and $n = 10$.

Ex. 51–52 COOPERATIVE LEARNING Have students work in pairs to complete this activity in class.

Enrichment Activities

These activities require students to go beyond lesson goals.

Look at the seven rows of Pascal's triangle given on page 652.

1. Write the next three rows for the triangle.
   1, 8, 28, 56, 70, 56, 28, 8, 1
   1, 9, 36, 84, 126, 126, 84, 36, 9, 1
   1, 10, 45, 120, 210, 252, 210, 120, 45, 10, 1
2. Where can you find the sequence of natural numbers?
   The second diagonal row, that is, 1, 2, 3, 4, etc.
3. What do you notice about the numbers in the *third* diagonal row, that is, 1, 3, 6, 10, 15, etc.?
   They are the triangular numbers.
4. Look at the fourth diagonal row. What relationship can you find between these numbers and the numbers in the third diagonal row?
   The numbers 1, 4, 10, etc. form a sequence of partial sums of the triangular numbers in the third diagonal row. That is, 1, 4, and 10 are 1, 1 + 3, and 1 + 3 + 6, respectively.
5. What other patterns do you notice?
   Answers will vary.

Problem of the Day

In a string of 33 pearls, the middle one is the most expensive. Starting at one end, each pearl is worth \$100 more than the previous one, to the middle pearl. Starting from the other end, each is worth \$150 more than the previous one, to the middle pearl. The whole string is worth \$65,000. What is the value of the middle pearl?
\$3000

ORGANIZER

Warm-Up Exercises

1. Evaluate to two decimal places.
   a. $(1 + i)^k$, $i = 0.01$, $k = 5$
   b. $(1 + i)^k$, $i = 0.02$, $k = 12$
   a. 1.05, b. 1.27
2. Find the nth term of each sequence.
   a. 5, 13, 21, 29, 37, . . .
   b. 3, 9, 27, 81, . . .
   c. $2, \frac{4}{3}, \frac{8}{9}, \frac{16}{27}, \ldots$
   d. $\frac{5}{2}, \frac{11}{6}, \frac{7}{6}, \frac{1}{2}, -\frac{1}{6}, \ldots$
   a. $8n - 3$ b. 3^n
   c. $2\left(\frac{2}{3}\right)^{n-1}$ d. $\frac{19 - 4n}{6}$
3. Use sigma notation to rewrite each geometric series in Exercise 2 above.
   2b. $\sum_{i=1}^{\infty} 3^i$ 2c. $\sum_{i=1}^{\infty} 2\left(\frac{2}{3}\right)^{i-1}$

Lesson Resources

Teaching Tools
- Transparencies: 2, 3
- Problem of the Day: 12.6
- Warm-up Exercises: 12.6
- Answer Masters: 12.6

Extra Practice: 12.6
Color Transparency: 66
Technology Handbook: p. 89

12.6 Exploring Data: The Algebra of Finance

What you should learn:

Goal 1 How to find the balance of an increasing annuity

Goal 2 How to find the monthly deposit to reach a specified balance in an annuity

Why you should learn it:

You may in the future want to save money using an increasing annuity. It is the most common form of savings plan.

Goal 1 Finding the Balance of an Annuity

In Lesson 7.2, you studied savings plans that involved a *single* deposit. Most people, however, prefer to make several deposits over a long period of time. **Annuities** are savings accounts to which many equal deposits (or withdrawals) are made at regular intervals. Annuities involving equal deposits are **increasing annuities,** and those involving equal withdrawals are **decreasing annuities.** One example of an increasing annuity is a payroll savings plan in which part of an employee's paycheck is deducted each pay period and placed in a savings account.

Example 1 *Constructing an Increasing Annuity Table*

You deposit \$50 at the end of each month into a credit union savings plan that pays annual interest of 6%, compounded monthly. What is the balance at the end of six months?

Solution On the last day of the first month, the balance is \$50. Then, at the end of the second month, when the second deposit is made, the account will have the following balance.

$$\text{Balance} = \underbrace{50.00}_{\text{1st Deposit}} + \underbrace{\tfrac{1}{12}(0.06)(50.00)}_{\text{Interest on 1st Deposit}} + \underbrace{50.00}_{\text{2nd Deposit}} = \$100.25$$

Similarly, at the end of the third month, when the third deposit is made, the balance will be the sum of \$100.25 (previous balance), $\frac{1}{12}(0.06)(100.25)$ (interest on previous balance), and \$50 (new deposit). Continuing, you obtain:

End of Month	Previous Balance	Interest	Deposit	New Balance
1	\$ 0.00	\$0.00	\$50.00	\$ 50.00
2	\$ 50.00	\$0.25	\$50.00	\$100.25
3	\$100.25	\$0.50	\$50.00	\$150.75
4	\$150.75	\$0.75	\$50.00	\$201.50
5	\$201.50	\$1.01	\$50.00	\$252.51
6	\$252.51	\$1.26	\$50.00	\$303.77

The balance after 6 months (and 6 deposits) is \$303.77. ■

Balance of an Increasing Annuity

Consider an increasing annuity with equal periodic deposits of D dollars made n times per year. The annual interest rate is r, and the interest is compounded n times per year. The balance, A, after t years is

$$A = D\left[\frac{(1+i)^{nt}-1}{i}\right]$$

where $i = \frac{r}{n}$.

For this formula, the total number of deposits is nt. Deposits are made at the *end* of consecutive compounding periods, and the last deposit is made on the last day of the time period (and therefore receives no interest). This type of increasing annuity is called an ordinary *increasing annuity.*

Real Life
Savings

Example 2 Finding the Balance of an Increasing Annuity

You deposit $100 each month into an increasing annuity that pays an annual interest rate of 7.5%, compounded monthly. What is the balance after 40 years? How much of this balance is interest?

Solution For this increasing annuity, you have $D = 100$, $r = 0.075$, $n = 12$, $i = \frac{0.075}{12} = 0.00625$, and $t = 40$. The balance after 40 years will be

$$\begin{aligned} A &= D\left[\frac{(1+i)^{nt}-1}{i}\right] \\ &= 100\left[\frac{(1+0.00625)^{480}-1}{0.00625}\right] \\ &= \$302{,}382.22. \end{aligned}$$

Since the total amount of money deposited is $480(100) = \$48{,}000$, you can conclude that the total amount of interest earned by the annuity is

$$\begin{aligned} \text{Interest} &= 302{,}382.22 - 48{,}000.00 \\ &= \$254{,}382.22. \end{aligned}$$ ■

From Example 2, you can see the primary benefit of a long-term annuity—relatively small deposits can produce very large balances. Because of this, increasing annuities are frequently used as retirement accounts.

Banks take elaborate precautions to safeguard the money and valuables of their customers.

LESSON Notes

Example 1

Find the balance of the savings using the ordinary annuity formula. Ask students to identify the values of the variables D, r, n, i, and t in the formula.
$D = \$50$, $r = 6\%$, $n = 12$, $i = 0.5\%$, and $t = 0.5$; $A = 303.78$. The difference of 1 cent is due to round-off error in the computation of Example 1.

Example 2

This impressive growth of $100 deposited monthly for 40 years has led to compound interest being called "the eighth wonder of the world."

Example 3

Ask students to describe situations in their lives for which it is necessary (or just a good idea) to save a fixed amount of money each month to achieve a financial goal. Then, have them compute the monthly deposit for that objective. (Some students may wish to save money for a car, a summer vacation, special articles of clothing, etc.)

Extra Examples

Here are additional examples similar to **Examples 1–3.**

1. You deposit $150 every three months into an ordinary annuity that pays 8.5% annually, compounded quarterly. What is the balance after 5 years? How much of the balance is interest?
   $3690.32; $690.32
2. Suppose you want to save $3500 to buy a used car in two years. If the annual savings rate is 9.25% compounded monthly, what monthly deposit will be required?
   $133.32

Check Understanding

1. What is an annuity?
   See page 658.
2. What is an example of an increasing annuity?
   See Examples 1–2.
   A decreasing annuity?
   See Exercises 27 and 28.

Communicating about ALGEBRA

Students may be wondering why the annuity formula is in a chapter about sequences and series. To address this concern, ask students to explain the connection between the following sequence and Example 1 of this lesson.

$50,\ 50\left(1+\frac{0.06}{12}\right),$
$50\left(1+\frac{0.06}{12}\right)^2,\ 50\left(1+\frac{0.06}{12}\right)^3,$
$50\left(1+\frac{0.06}{12}\right)^4,\ 50\left(1+\frac{0.06}{12}\right)^5$

Each term in the sequence represents a $50 deposit that has earned interest for the number of months indicated by the exponent.

The kth term of the sequence is $a_k = 50\left(1+\frac{0.06}{12}\right)^{k-1}$. For the variables D, n, i (equal to $\frac{r}{n}$), and t, you can express the sum of the series as $\sum_{k=1}^{nt} D(1+i)^{k-1}$.

If A is the sum of a finite geometric series, then for $a_1 = D$ and $r = (1+i)$ (this "r" is from the geometric series formula),

$$A = D\left[\frac{1-(1+i)^{nt}}{1-(1+i)}\right]$$
$$= D\left[\frac{1-(1+i)^{nt}}{-i}\right]$$
$$= D\left[\frac{(1+i)^{nt}-1}{i}\right]$$

As an algebraic exercise, ask students to show how to rewrite the formula for the balance of an ordinary annuity as a formula for the deposit of an ordinary annuity.

Goal 2 Finding the Deposits Needed

The formula for the balance of an increasing annuity can also be used to determine equal deposits that will produce a given balance.

$$D = A\left[\frac{i}{(1+i)^{nt}-1}\right]$$

Real Life
Retirement Planning

Example 3 How to Save to Become a Millionaire

What monthly deposits will produce a balance of $1,000,000 after 45 years? Assume that the annual percentage rate is 6.6%, compounded monthly. What is the total amount deposited over the 45-year period?

Solution Using $A = 1{,}000{,}000$, $r = 0.066$, $n = 12$, $i = \frac{0.066}{12} = 0.0055$, and $t = 45$, you can compute the monthly deposit as follows.

$$D = A\left[\frac{i}{(1+i)^{nt}-1}\right]$$
$$= 1{,}000{,}000\left[\frac{0.0055}{(1+0.0055)^{540}-1}\right]$$
$$= \$299.99$$

The total amount deposited over the 45-year period is

$$540(299.99) = \$161{,}994.60.$$

The growth of this annuity is shown graphically at the left. ■

How to Become a Millionaire

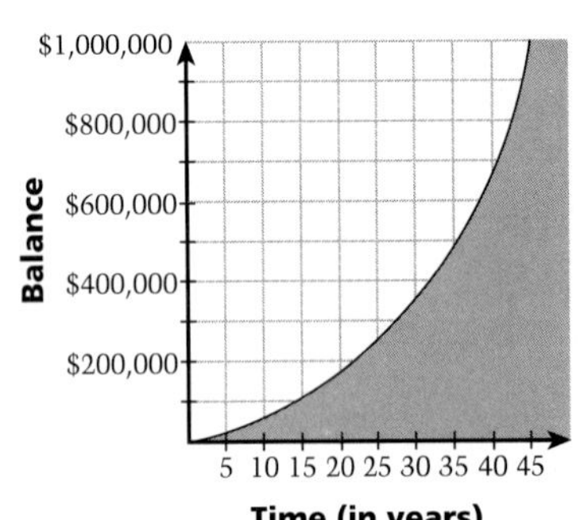

Communicating about ALGEBRA

SHARING IDEAS about the Lesson

The annual interest rate and the length of time can greatly affect the balance of an increasing annuity. Find the following balances.

	Monthly Deposit	Annual Interest	Number of Years	
A.	$100	5%	30	$83,225.86
B.	$100	8%	30	$149,035.94
C.	$200	7.5%	35	$406,152.40
D.	$200	7.5%	45	$893,406.19

EXERCISES

Guided Practice

CRITICAL THINKING about the Lesson

1. What is an increasing annuity? Give an example of how an increasing annuity is used. See page 658; as a retirement account

2. *Savings Plan* You deposit $80 into a savings plan that pays 6% annual interest, compounded monthly. How much interest will your deposit earn in the first month? $0.40

3. *Annuity Table* The table below represents an increasing annuity that pays 7.5% annual interest, compounded monthly. Find the missing entries in the table. $1.25; $301.88; $301.88; $403.77

End of Month	Previous Balance	Interest	Deposit	New Balance
1	$0.00	$0.00	$100.00	$100.00
2	$100.00	$0.63	$100.00	$200.63
3	$200.63	?	$100.00	?
4	?	$1.89	$100.00	?

4. In an ordinary increasing annuity, when are the monthly deposits made: at the end of the month or at the beginning of the month? at the end of the month

5. Over a 10-year period, you make monthly deposits of $200 into an increasing annuity that pays 6% annual interest, compounded monthly. Identify the values of D, r, n, i, and t. $D = 200$, $r = 0.06$, $n = 12$, $i = 0.005$, $t = 10$

6. Describe a situation in which you could use the formula

$$D = A\left[\frac{i}{(1+i)^{nt} - 1}\right].$$ To determine equal deposits that will produce a given balance

Independent Practice

Constructing an Annuity Table **In Exercises 7–10, construct an annuity table (similar to the one in Exercise 3) showing the deposits and balances for the given savings plan.** See Additional Answers.

7. At the end of each year for five years, $1000 is invested in an increasing annuity. The annuity pays 8% annual interest, compounded annually.

8. At the end of each quarter for one year, $300 is invested in an increasing annuity. The annuity pays $6\frac{1}{2}$% annual interest, compounded quarterly.

9. Twice a year for five years, $500 is invested in an increasing annuity. The annuity pays 6% annual interest, compounded semiannually.

10. At the end of each month for one year, $100 is invested in an increasing annuity. The annuity pays $6\frac{1}{2}$% annual interest, compounded monthly.

EXERCISE Notes

ASSIGNMENT GUIDE

Basic/Average: Ex. 1–6, 9, 13–21 odd, 25–27

Above Average: Ex. 1–6, 9, 13, 15–20, 24–28

Advanced: Ex. 1–6, 7, 11, 15–21, 24–28

Selected Answers
Exercises 1–6, 7–25 odd

More Difficult Exercises
Exercises 15–24, 27, 28

Guided Practice

Ex. 1–6 Use these exercises as an in-class readiness check for the Independent Practice. In Exercise 1, discuss the use of annuities for Individual Retirement Accounts or college savings plans. In Exercise 6, one example of a "sinking fund" situation is to suppose you want $50,000 for a college education 18 years from now. What is the required monthly deposit?

Independent Practice

▶ **Ex. 7–14** Students should use Example 1 on page 658 as a solution model. Students may also use computer spreadsheet software for constructing annuity tables.

EXTEND **Ex. 15–18** Have students compare the balances to the actual money paid into the fund.

Answers

19. Monthly deposits of $80 at 6% interest compounded monthly
20. Monthly deposits of $45 at 9% interest compounded monthly

▶ **Ex. 21–24** Students should use Example 3 on page 660 as a solution model.

In Exercises 11–14, find the balance of the increasing annuity.

	Periodic Deposit	Number of Deposits per Year	Number of Years	Annual Interest Rate	Compounding Period
11.	$2000	1	15	12%	Annually $74,559.43
12.	$50	26	2	$6\frac{1}{2}$%	Biweekly $2772.87
13.	$40	52	5	6%	Weekly $12,120.35
14.	$75	12	3	$6\frac{1}{2}$%	Monthly $2972.38

15. *Saving for College* The parents of a newborn child decide to deposit $40 a month into a savings account. The account earns 6%, compounded monthly. What is the balance after 15 years? $11,632.75

16. *Saving for a Home* A newly-married couple are both working and decide to save $1000 a month for a down payment on a home. The account earns $8\frac{1}{2}$% annually, compounded monthly. How large a down payment will they have saved in three years? In four years? $40,842.66; $56,931.49

17. *Two-Account Savings Plan* You deposit $30 a month for 20 years into an account that pays 6%, compounded monthly. After 20 years, you leave the money in the account for five more years, but make no more deposits. What is the balance after the 25-year period? $18,696.72

18. *Two-Account Savings Plan* You deposit $500 a year for 10 years into an account that pays 6.36%, compounded annually. After 10 years, you transfer the money into another account that pays 7%, compounded continuously. The money is left in the second account for eight years. What is the balance after the 18-year period? $11,734.55

19. *Comparing Two Savings Plans* Which would give you a higher balance in 10 years: monthly deposits of $50 at 10% interest, compounded monthly, or monthly deposits of $80 at 6% interest, compounded monthly? See margin.

20. *Comparing Two Savings Plans* Which would give you a higher balance in 15 years: monthly deposits of $45 at 9% interest, compounded monthly, or monthly deposits of $60 at 5% interest, compounded monthly? See margin.

21. *How Much to Deposit?* How much must be deposited each month into an ordinary annuity paying $7\frac{1}{2}$%, compounded monthly, to attain a balance of $25,000 after five years? $344.70

22. *How Much to Deposit?* How much must be deposited every other week into an ordinary annuity paying $6\frac{1}{2}$%, compounded biweekly, to attain a balance of $8000 after three years? (Use 52-week years.) $93.02

23. *Comparing Two Savings Plans* You want to save \$50,000. Which plan would require a smaller *total* deposit: monthly payments with 5% interest, compounded monthly for 15 years, or monthly payments with $7\frac{1}{2}\%$ interest, compounded monthly for 10 years? See margin.

24. *Comparing Two Savings Plans* You want to save \$20,000. Which plan would require a smaller *total* deposit: monthly payments with 8% interest, compounded monthly for 10 years, or monthly payments with 6% interest, compounded monthly for 8 years? Monthly payments with 8% interest compounded monthly for 10 years

Integrated Review

Retirement Age **In Exercises 25 and 26, use the bar graph at the right, which shows the results of a survey taken in 1991.**

25. What is the ratio of the number of people who said they would retire after age 65 to the number who said they would retire at age 65 or earlier? $\frac{1}{15}$

26. Of the total number of people who said they would retire at age 65 or earlier, what percent said they would retire at age 54 or earlier? 10%

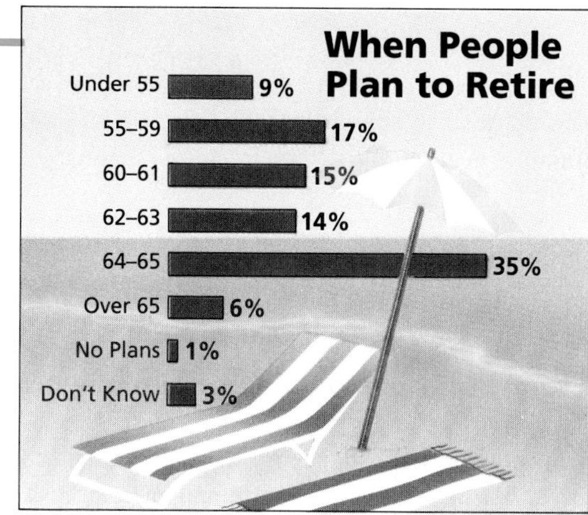

Exploration and Extension

Decreasing Annuity **In Exercises 27 and 28, use the following information.**

A **decreasing annuity** is a savings account that has an initial deposit followed by several equal withdrawals until the account is depleted. A decreasing annuity of n equal withdrawals of W per year for t years is described by the equation

$$W = P\left[\frac{i}{1 - \left(\frac{1}{1+i}\right)^{nt}}\right]$$

where P is the initial deposit and r is the annual interest rate and $i = \frac{r}{n}$.

27. Your grandparents deposit \$50,000 in an account for you to use during college. The account pays $7\frac{1}{2}\%$ annual interest and is set up to send you equal monthly withdrawals for five years. How much will be sent to you each month? How much will be sent over the five-year period? How much interest will the account have earned? \$1001.90; \$60,114.00; \$10,114.00

28. Your grandparents have just retired. They have a balance of \$500,000 in an increasing annuity. How much can they withdraw each month in a decreasing annuity that pays 8% annual interest if they plan to deplete the account over a 30-year period? \$3668.82

Answer

23. Monthly payments with 5% interest compounded monthly for 15 years

▶ **Ex. 25–26** CONNECTIONS
You can use these exercises to discuss the financial issues that adults face upon retirement. Students might research the inflation rates, housing costs, food costs, and medical costs over many years to understand why the money put aside by their grandparents in the 1960s through 1980s for retirement is inadequate for the 1990s.

▶ **Ex. 27–28** TECHNOLOGY
You might wish to model how to use the calculator to perform the computations.

Enrichment Activities

These activities require students to go beyond lesson goals.

1. RESEARCH Work with a partner. Use the financial pages of a newspaper to look for financial institutions that offer annuity programs. Contact the institutions and gather information on the various types of annuities. Write a one-paragraph report on the programs.
2. One important consideration in accepting employment at one company rather than another is often the company's retirement plan or profit-sharing plan. Contact some businesses in your area to determine which types of plans are available to employees who work for those companies. Write a one-page report on the plans. What percentage of employees take advantage of the plans? Are figures available on how the money grows?

TECHNOLOGY The programmable feature of most calculators can be used to calculate many different algebraic relationships. Ask students to identify formulas and other relationships that they would like to program into their graphing calculators. A useful and fun class project involves students working in small groups to write the programs for the identified collection of formulas and relationships. The programs can then be shared with all the members of the class.

USING A PROGRAMMABLE CALCULATOR

A programmable calculator can be programmed to calculate the balance of an increasing annuity. To do this, you can program the calculator so that it accepts four inputs.

$D =$ Periodic deposit

$r =$ Annual interest rate

$n =$ Number of deposits (and compoundings) per year

$t =$ Number of years

After these amounts are input, the program can calculate and display the value of A, which is the balance in the annuity after t years. This balance is given by

$$A = D\left[\frac{(1 + i)^{nt} - 1}{i}\right]$$

where $i = \frac{r}{n}$.

Example 1: Finding the Balance of an Increasing Annuity

Complete a table that compares the balances for 30-year increasing annuities in which the annual interest rates are 5%, 6%, 7%, 8%, 9%, and 10%, all compounded monthly. Use monthly deposits of $50, $100, and $200.

Solution

Monthly Deposit	$50.00	$100.00	$200.00
Balance at 5%	$41,612.93	$83,225.86	$166,451.73
Balance at 6%	$50,225.75	$100,451.50	$200,903.01
Balance at 7%	$60,998.55	$121,997.10	$243,994.20
Balance at 8%	$74,517.97	$149,035.94	$298,071.89
Balance at 9%	$91,537.17	$183,074.35	$366,148.70
Balance at 10%	$113,024.40	$226,048.79	$452,097.58

From the table, you can see that for a given interest rate, the balance varies directly with the monthly deposit. For instance, at a 5% interest rate, you can double your balance by doubling your monthly deposit.

From the table, you can also see that for a given monthly deposit, the balance *does not* vary directly with the annual interest rate. For instance, for a given monthly deposit, the balance at 10%, is more than twice the balance at 5%. ■

Example 2: Sketching a Graph of the Balance

Use a graphing calculator to sketch a graph that shows the balance in a 30-year increasing annuity. Use a monthly deposit of \$50 and an annual interest rate of 6%, compounded monthly. How does the balance after 15 years compare to the balance after 30 years? Is it half?

Solution For this increasing annuity, the monthly deposit is $D = 50$, the annual interest rate is $r = 0.06$, and the number of compoundings per year is $n = 12$. This implies that $i = 0.005$. To show how the balance, A, increases over time, t, you can sketch the graph of the function

$$A = D\left[\frac{(1+i)^{nt}-1}{i}\right] = 50\left[\frac{(1.005)^{12t}-1}{0.005}\right].$$

The graph is shown at the right. From the graph, you can see that the balance after 15 years is not nearly half of the balance after 30 years. In fact, after 15 years, the annuity has built up a balance that is only about 29% of the 30-year balance. ■

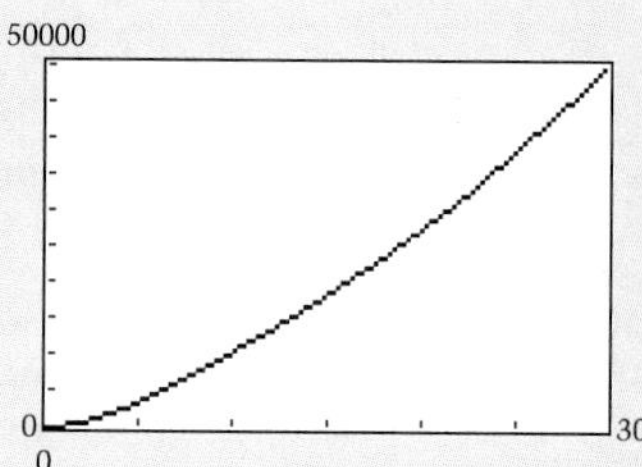

Exercises

In Exercises 1 and 2, use a programmable calculator to complete the table. See margin.

1. *20-Year Increasing Annuity*

Monthly Deposit	\$50	\$100	\$200
Balance at 6%	?	?	?
Balance at 8%	?	?	?
Balance at 10%	?	?	?

2. *40-Year Increasing Annuity*

Monthly Deposit	\$50	\$100	\$200
Balance at 6%	?	?	?
Balance at 8%	?	?	?
Balance at 10%	?	?	?

3. *Finding the Periodic Deposit* Write a program (for a programmable calculator or a computer) that will calculate the periodic deposit that is required to produce a given balance in an increasing annuity. See margin.

Finding the Monthly Deposit **In Exercises 4–6, use the program from Exercise 3 to find the monthly deposit required to produce the balance.** See margin.

4. Balance: \$100,000
   Term: 15 years
   Annual interest: 6%
   Compounded monthly

5. Balance: \$10,000
   Term: 5 years
   Annual interest: 8%
   Compounded monthly

6. Balance: \$25,000
   Term: 10 years
   Annual Interest: 7%
   Compounded monthly

7. Use a graphing calculator to sketch a graph that shows the balance in a 40-year increasing annuity. Use a monthly deposit of \$100 and an annual interest rate of 8%, compounded monthly. What percent of the balance has accrued after 20 years? See margin.

Answers

1. \$23,102.04; \$46,204.09; \$92,408.18.
   \$29,451.02; \$58,902.04; \$117,804.08.
   \$37,968.44; \$75,936.88; \$151,873.77.
2. \$99,574.54; \$199,149.07; \$398,298.15.
   \$174,550.39; \$349,100.78; \$698,201.57.
   \$316,203.98; \$632,407.96; \$1,264,815.92.
3. Program 2: DEPOSIT
```
Display "ENTER, A"
Input A
Display "ENTER, R"
Input R
Display "ENTER, N"
Input N
Display "ENTER, T"
Input T
R/N → I
A(I / ((1 + I) ^ (NT)−1) → D
Display D
```
4. \$343.86
5. \$136.10
6. \$144.44
7.

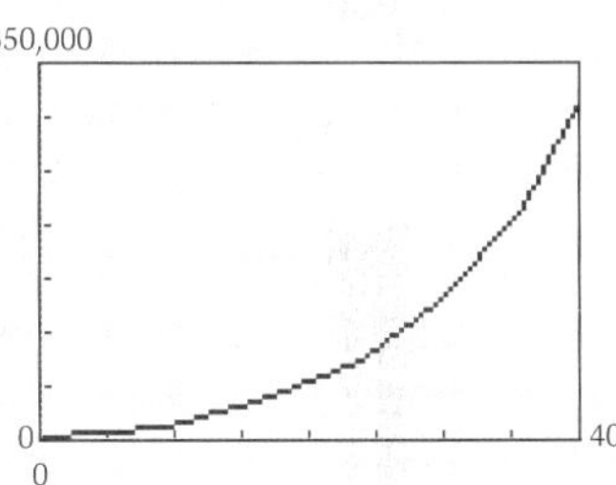

≈ 16.9%

ORGANIZER

TIME SCHEDULE: All levels, two days

The Chapter Summary helps students organize the main ideas of the chapter. In this chapter, we investigated arithmetic and geometric sequences and series. Binomial expansions and ordinary annuities were examined as examples of series, and summation notation for sums of series was also introduced.

Work with students to review the skills, strategies, and concepts of the chapter. The first day's homework assignment can be used as the basis for the second day of review.

COOPERATIVE LEARNING

Encourage students to study together. Emphasize the importance of "teaching" a classmate how to perform a skill or how to recall a procedure. When students work together, everyone wins. The students receiving help get additional instruction and the students giving help gain a deeper understanding of the skills and concepts involved.

Chapter SUMMARY

Review the chapter by asking students, "What did you learn? Why did you learn it? How does it fit with the other algebra skills and concepts you have learned?"

12 Chapter Summary

What did you learn?

Skills

1. Use summation notation to write a series. **(12.1)**
2. Write a formula for the nth term
   - of an arithmetic sequence. **(12.2)**
   - of a geometric sequence. **(12.3)**
3. Find the sum of a series. **(12.1)**
   - Find the sum of a finite arithmetic series. **(12.2)**
   - Find the sum of a finite geometric series. **(12.3)**
   - Find the sum of an infinite geometric series. **(12.4)**
4. Find binomial coefficients. **(12.5)**
5. Expand a binomial
   - using the Binomial Theorem. **(12.5)**
   - using Pascal's Triangle. **(12.5)**

Strategies

6. Use sequences to solve real-life problems. **(12.2–12.3)**
7. Use series to solve real-life problems. **(12.1–12.4)**

Exploring Data

8. Solve increasing annuity problems. **(12.6)**

Why did you learn it?

Many real-life problems can be modeled with sequences. Often these sequences form patterns that enable you to write a formula for the nth term of the sequence. For instance, suppose your property tax was $2000 in 1991. If your tax increased by 5% each year for 10 years, then a formula for the tax during the nth year would be $a_n = 2000(1.05)^{n-1}$. Because this sequence is geometric, you could find the total property tax paid during the 10 years by using the formula for the sum of a geometric series.

How does it fit into the bigger picture of algebra?

In this chapter, you studied *sequences* of numbers such as

$2, 5, 8, 11, 14, \ldots$ and $2, 4, 8, 16, 32, \ldots$.

The first of these is an arithmetic sequence, and the second is a geometric sequence. The expression formed by adding the terms in a sequence is the series associated with a sequence. All finite series have a sum, and some infinite series have a sum.

In the chapter, you also studied the Binomial Theorem. This theorem describes patterns for the coefficients of the terms in the expansion of a binomial that is raised to a power.

21. $\sum_{n=1}^{7}(6n-5) = 133$ **22.** $\sum_{n=1}^{7}(2n+2) = 70$ **23.** $\sum_{n=1}^{8}(28-6n) = 8$

Chapter REVIEW

In Exercises 1–8, write the first five terms of the sequence. Begin with $n = 1$. See margin.

1. $a_n = 6 - 2n$
2. $a_n = 10n - n^2$
3. $a_n = \dfrac{10}{n!}$
4. $a_n = \dfrac{n!}{10}$
5. $a_n = (-0.2)^{n-1}$
6. $a_n = \left(-\dfrac{1}{4}\right)^n$
7. $a_n = n^2 + 5$
8. $a_n = (n + 1)^3$

In Exercises 9–12, find the sum.

9. $\sum_{i=0}^{4} (i! + 2)$ 44
10. $\sum_{n=5}^{8} n(2n - 1)$ 322
11. $\sum_{n=1}^{25} n^2$ 5525
12. $\sum_{i=1}^{12} i$ 78

In Exercises 13–18, decide whether the sequence is arithmetic, geometric, or neither. If the sequence is arithmetic, state its common difference. If the sequence is geometric, state its common ratio.

13. 22, 28, 34, 40, 46, . . . Arithmetic, 6
14. $\frac{1}{10}, \frac{1}{5}, \frac{3}{5}, \frac{12}{5}, 12, \ldots$ Neither
15. $1, \frac{1}{2}, 0, -\frac{1}{2}, -1, \ldots$ Arithmetic, $-\frac{1}{2}$
16. $\frac{1}{2}, 2, 8, 32, 128, \ldots$ Geometric, 4
17. 1, 1, 2, 6, 24, . . . Neither
18. $3, \frac{11}{3}, \frac{13}{3}, 5, \frac{17}{3}, \ldots$ Arithmetic, $\frac{2}{3}$
19. Write a formula for the nth term of the arithmetic sequence whose common difference is 5 and whose first term is 13. $a_n = 5n + 8$
20. Write the first five terms of the arithmetic sequence whose common difference is 14 and whose third term is 36. 8, 22, 36, 50, 64

In Exercises 21–26, use sigma notation to write the arithmetic series. Then find the sum of the series. See below.

21. $1 + 7 + 13 + 19 + 25 + 31 + 37$
22. $4 + 6 + 8 + 10 + 12 + 14 + 16$
23. $22 + 16 + 10 + 4 - 2 - 8 - 14 - 20$
24. $\frac{3}{2} + 2 + \frac{5}{2} + 3 + \frac{7}{2} + 4 + \frac{9}{2} + 5$
25. $\frac{1}{3} + \frac{2}{3} + 1 + \frac{4}{3} + \frac{5}{3} + 2 + \frac{7}{3} + \frac{8}{3}$
26. $-\frac{3}{8} + 0 + \frac{3}{8} + \frac{3}{4} + \frac{9}{8} + \frac{3}{2} + \frac{15}{8} + \frac{9}{4}$
27. Write a formula for the nth term of the geometric sequence whose common ratio is 6 and whose first term is $\frac{1}{4}$. $a_n = \frac{1}{4}(6)^{n-1}$
28. Write the first five terms of the geometric sequence whose common ratio is 3 and whose first term is 2. 2, 6, 18, 54, 162

In Exercises 29–40, find the sum of the series.

29. $\sum_{i=1}^{20} 16(4)^{i-1}$ $\approx 5.864 \times 10^{12}$
30. $\sum_{n=1}^{10} \frac{1}{2}(3)^{n-1}$ 14,762
31. $\sum_{n=0}^{15} 6\left(-\frac{1}{4}\right)^n$ ≈ 4.800
32. $\sum_{n=0}^{12} 10\left(\frac{1}{2}\right)^n$ ≈ 20.00
33. $\sum_{n=1}^{9} \frac{1}{10}(2)^{n-1}$ $\frac{511}{10}$
34. $\sum_{i=1}^{20} 2(-2)^{i-1}$
35. $\sum_{i=1}^{\infty} 2\left(\frac{3}{5}\right)^{i-1}$ 5
36. $\sum_{i=0}^{\infty} 15\left(\frac{2}{9}\right)^i$ $\frac{135}{7}$
37. $\sum_{n=1}^{\infty} 4(-0.2)^{n-1}$ $\frac{10}{3}$
38. $\sum_{n=0}^{\infty} 8(0.15)^n$ $\frac{160}{17}$
39. $\sum_{i=0}^{\infty} 1.4(0.8)^i$ 7
40. $\sum_{i=1}^{\infty} 25(0.5)^{i-1}$ 50

34. −699,050

24. $\sum_{n=1}^{8} \left(\frac{1}{2}n + 1\right) = 26$
25. $\sum_{n=1}^{8} \frac{1}{3}n = 12$
26. $\sum_{n=1}^{8} \left(\frac{3}{8}n - \frac{3}{4}\right) = \frac{15}{2}$

Chapter Review

ASSIGNMENT GUIDE

Basic/Average: Ex. 1–27 odd, 30–54 multiples of 3, 57–58, 61–63 odd, 69–71

Above Average: Ex. 1–27 odd, 30–54 multiples of 3, 59–60, 63–64, 68–71

Advanced: Ex. 1–27 odd, 30–54 multiples of 3, 59–60, 63–64, 68–71

More Difficult Exercises

Exercises 59–60, 69–72

Ex. 1–72 The Chapter Review exercises include a reference to the lesson in which the skill or strategy was first presented. Students may use these references as a study aid if they are uncertain about how to do a given exercise in the Chapter Review.

Answers

1. 4, 2, 0, −2, −4
2. 9, 16, 21, 24, 25
3. $10, 5, \frac{5}{3}, \frac{5}{12}, \frac{1}{12}$
4. $\frac{1}{10}, \frac{1}{5}, \frac{3}{5}, \frac{12}{5}, 12$
5. 1, −0.2, 0.04, −0.008, 0.0016
6. $-\frac{1}{4}, \frac{1}{16}, -\frac{1}{64}, \frac{1}{256}, -\frac{1}{1024}$
7. 6, 9, 14, 21, 30
8. 8, 27, 64, 125, 216

Answers

45. $x^3 + 12x^2 + 48x + 64$
46. $x^4 + 40x^3 + 600x^2 + 4000x + 10{,}000$
47. $x^5 + 5x^4y + 10x^3y^2 + 10x^2y^3 + 5xy^4 + y^5$
48. $y^6 + 18y^5 + 135y^4 + 540y^3 + 1215y^2 + 1458y + 729$
49. $y^7 - 7y^6 + 21y^5 - 35y^4 + 35y^3 - 21y^2 + 7y - 1$
50. $x^8 + 16x^7 + 112x^6 + 448x^5 + 1120x^4 + 1792x^3 + 1792x^2 + 1024x + 256$
51. $x^3 + 6x^2y + 12xy^2 + 8y^3$
52. $16x^4 + 32x^3y + 24x^2y^2 + 8xy^3 + y^4$
59. 1, 1, 2, 3, 5, 8, 13, 21, 34, 55; 1, 2, 1.5, ≈ 1.667, 1.6, 1.625, ≈ 1.615, ≈ 1.619, ≈ 1.618; the ratios get closer and closer to the Golden Ratio.

668

In Exercises 41–44, evaluate the binomial coefficient.

41. $\binom{9}{2}$ 36 **42.** $\binom{12}{3}$ 220 **43.** $\binom{11}{11}$ 1 **44.** $\binom{6}{0}$ 1

In Exercises 45–52, expand the binomial. See margin.

45. $(x + 4)^3$ **46.** $(x + 10)^4$ **47.** $(x + y)^5$ **48.** $(y + 3)^6$

49. $(y - 1)^7$ **50.** $(x + 2)^8$ **51.** $(x + 2y)^3$ **52.** $(2x + y)^4$

Annuity Table **In Exercises 53 and 54, construct an annuity table showing interest earned, deposits, and balances for the first six months of the savings plan.**

53. You deposit \$50 at the end of each month in an increasing annuity that pays 8% annual interest, compounded monthly. See Additional Answers.

54. You deposit \$85 at the end of each month in an increasing annuity that pays 6.5% annual interest, compounded monthly. See Additional Answers

55. *Monthly Deposits* An increasing annuity pays 7.5% annual interest, compounded monthly. What monthly deposits will produce a \$5000 balance in 10 years? \$28.10

56. *Comparing Two Savings Plans* You want to build a balance of \$10,000 in an increasing annuity. Find the monthly deposits for the following plans.

a. A 16-year annuity that earns 6% annual interest, compounded monthly \$31.14

b. A 14-year annuity that earns 8% annual interest, compounded monthly \$32.47

Cell Division **In Exercises 57 and 58, use the following information.**

In early embryological growth, a human cell divides into two cells, each of which divides into two cells, and so on. The number, a_n, of *new* cells formed after the nth division is $a_n = 2^{n-1}$.

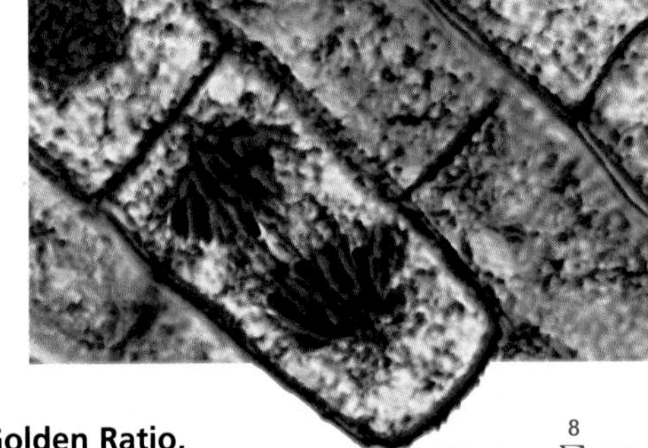

57. Use summation notation to write the series that represents the total number of cells after eight divisions.

58. Find the sum of the series in Exercise 57. 256

57. $1 + \sum_{n=1}^{8} 2^{n-1}$

Fibonacci Sequence **In Exercises 59 and 60, use the Golden Ratio, which is $\frac{1}{2}(1 + \sqrt{5})$.**

59. Write the first 10 terms of the Fibonacci sequence. Then calculate the ratios of a_{n+1} to a_n for these terms. What do you notice? See margin.

60. Calculate the sum:

$$\sum_{n=1}^{10} \left[\frac{1}{2}(1 + \sqrt{5})\right]^{n-1}.$$ ≈197.39, yes

Is it greater than the sum of the first 10 Fibonacci numbers?

Answer

64. For $n > 1$, the nth triangular number is equal to the sum of n consecutive positive integers beginning with 1; the diagonal rows beginning with 1 in row 2

61. *Stack of Cannonballs* The stack of cannonballs shown below is part of a monument in Gettysburg National Park in Gettysburg, Pennsylvania. The stack has the shape of a pyramid with a square base. How many cannonballs are in the stack? 55

62. *Bouncing Ball* A bouncing ball is dropped from a height of 10 feet. Each time the ball hits the ground, it bounces to three-fifths of its previous height. Find the total *vertical* distance traveled by the bouncing ball before it comes to rest. 40 ft

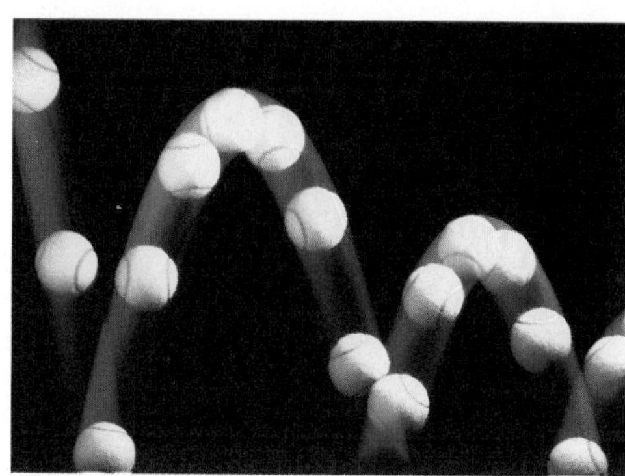

63. *Turntable* The first second after the power is turned off, a turnable makes 0.8 revolution. Each second after that, the turntable makes $\frac{4}{5}$ as much of a revolution as it did the second before. After a total of 15 seconds, the turntable stops. How many revolutions did it make? ≈ 3.86

64. *Triangular Numbers* The nth triangular number is $\frac{1}{2}(n)(n + 1)$ (see page 326). What is the relationship between the triangular numbers and the arithmetic sequence whose nth term is n? The triangular numbers occur as two diagonal rows in Pascal's Triangle. Which diagonal rows are they? See margin.

Multiples of Integers **In Exercises 65–67, each of the triangles represents the Rows 0 through 31 of Pascal's Triangle. In one of the triangles, the blue hexagons represent multiples of 2, in another, they represent multiples of 3 and in the other, multiples of 4. Identify the triangles.**

65. Multiples of 4

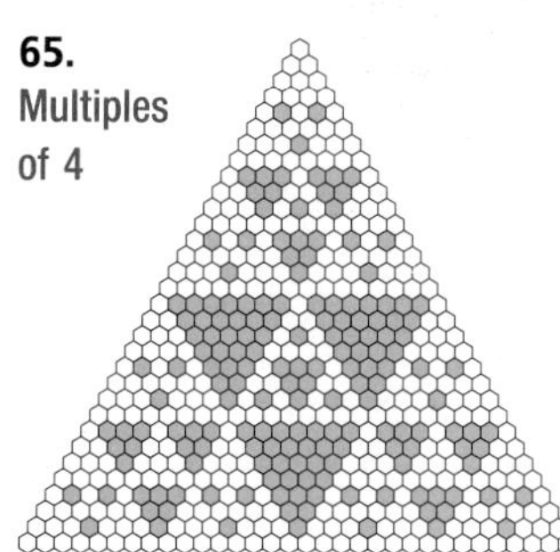

66. Multiples of 2

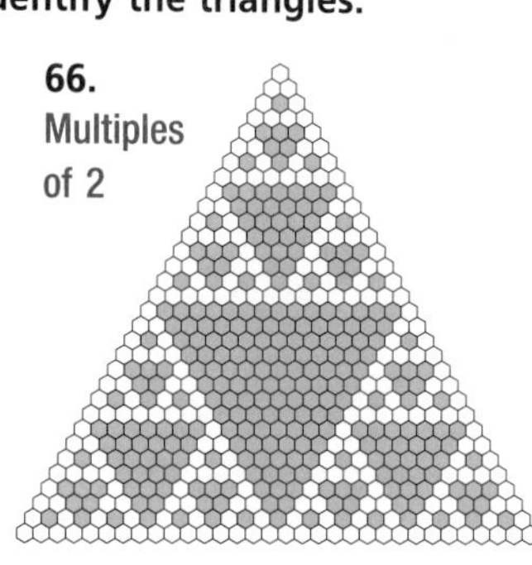

67. Multiples of 3

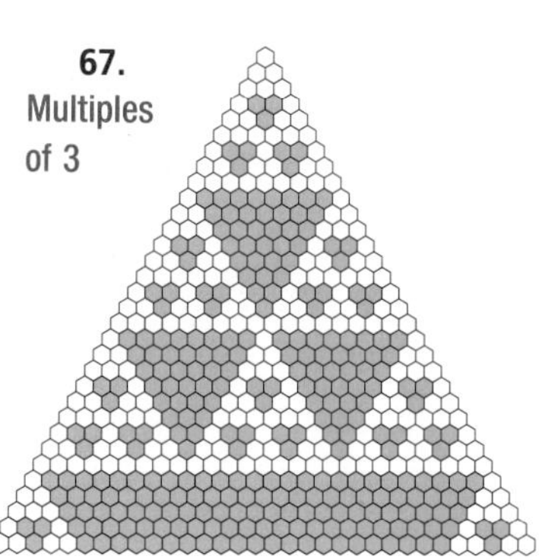

68. Using Exercises 65–67 as models, create the Pascal's Triangle pattern for multiples of 5. See Additional Answers.

Find the Treasure! **In Exercises 69–71, your goal is to move from the circle marked *Start* to the circle on the bottom. To make each move, you must find the sum of the series and move to the circle represented by the sum. *Only one of the three paths leads to the treasure!* Which path is it?** Path in Exercise 71

69.

$\sum_{n=0}^{5} (n! - 2n^2)$	44
$\sum_{i=0}^{\infty} 21\left(\frac{3}{10}\right)^i$	30
$\sum_{k=1}^{5} \frac{1}{20}(6)^{k-1}$	$\frac{311}{4}$
$\sum_{j=8}^{15} (2j - 15)$	64
$\sum_{n=1}^{\infty} \frac{1}{6}\left(\frac{5}{6}\right)^{n-1}$	1
$\sum_{i=1}^{8} 4(1.1)^{i-1}$	45.74
$\sum_{j=0}^{4} (10j - 5)$	75
$\sum_{k=5}^{9} k(k - 5)$	80

70.

$\sum_{n=1}^{21} (n - 10)$	21
$\sum_{n=0}^{5} (6n - 5)$	60
$\sum_{i=1}^{\infty} 8\left(\frac{7}{9}\right)^{i-1}$	36
$\sum_{n=1}^{10} 25\left(\frac{1}{8}\right)^{n-1}$	28.6
$\sum_{i=1}^{9} \frac{1}{7}(2)^{i-1}$	73
$\sum_{n=0}^{\infty} 16(0.2)^n$	20
$\sum_{n=2}^{6} \frac{n^2 + 80}{n!}$	61.9
$\sum_{n=6}^{9} (n^2 - 33)$	98

71.

$\sum_{i=1}^{4} i^2$	30
$\sum_{j=1}^{8} j$	36
$\sum_{n=1}^{4} 2\left(\frac{8}{3}\right)^{n-1}$	59.5
$\sum_{n=0}^{6} 2\left(\frac{1}{2}\right)^n$	3.97
$\sum_{i=0}^{6} (i + 4)$	49
$\sum_{n=1}^{\infty} 20\left(\frac{2}{3}\right)^{n-1}$	60
$\sum_{n=0}^{\infty} 3\left(\frac{4}{5}\right)^n$	15
$\sum_{i=0}^{4} \frac{i!}{2}$	17

72. *How Much Treasure?* The amount of the treasure is the larger of the following amounts. Which amount is it? a

a. Each month for 100 years, \$10 was deposited in an increasing annuity that earned 8% annual interest, compounded monthly.

b. The sum of all of the numbers in Row 0 through Row 9 of Pascal's Triangle is multiplied by \$1000.

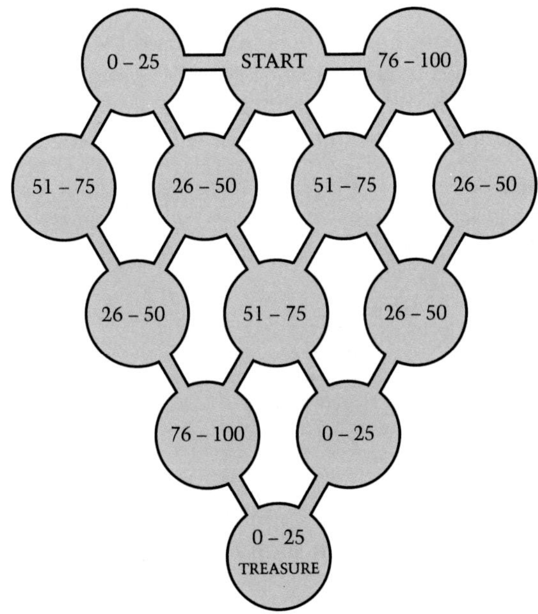

Chapter TEST

In Exercises 1 and 2, write the formula for the *n*th term of the sequence. (12.2, 12.3)

1. $\frac{1}{2}, 2, \frac{7}{2}, 5, \frac{13}{2}, \ldots$ $a_n = \frac{3n-2}{2}$

2. 3.6, 3.24, 2.916, 2.6244, 2.36196, . . . $a_n = 3.6(0.9)^{n-1}$

In Exercises 3–10, find the sum. (12.1–12.4)

3. $\sum_{i=1}^{3} \frac{i!}{i^2}$ $\frac{13}{6}$

4. $\sum_{n=1}^{5} (2n+3)$ 45

5. $\sum_{i=1}^{10} i$ 55

6. $\sum_{i=1}^{14} i^2$ 1015

7. $\sum_{n=1}^{20} (-4n+6)$ −720

8. $\sum_{i=0}^{11} (1.5)^i$ ≈257.49

9. $\sum_{n=1}^{\infty} \frac{1}{4}\left(\frac{1}{2}\right)^{n-1}$ $\frac{1}{2}$

10. $\sum_{n=1}^{\infty} 5(-0.75)^{n-1}$ $\frac{20}{7}$

In Exercises 11–14, evaluate the binomial coefficient. (12.5)

11. $\binom{4}{3}$ 4

12. $\binom{10}{8}$ 45

13. $\binom{9}{4}$ 126

14. $\binom{15}{0}$ 1

In Exercises 15–18, use the Binomial Theorem to expand the binomial. (12.5) 15.–17. See margin.

15. $(2x+3y)^4$

16. $(s+3t)^3$

17. $(x-6)^5$

18. $(x-2y)^2$ $x^2 - 4xy + 4y^2$

In Exercises 19–22, use Pascal's Triangle to expand the binomial. (12.5) See margin.

19. $(2r+4)^6$

20. $(s+2t)^3$

21. $(2x-y)^3$

22. $(y-2)^7$

23. Complete the table below for an increasing annuity that pays 6% annual interest compounded quarterly. **(12.6)**

End of Quarter	Previous Balance	Interest	Deposit	New Balance
1	$0.00	$0.00	$250.00	$250.00
2	$250.00	$3.75	$250.00	$503.75
3	$503.75	$7.56 ?	$250.00	? $761.31
4	? $761.31	$11.42	$250.00	? $1022.73

24. Which savings plan has a higher balance after 20 years: monthly deposits of $35 at 5% interest compounded monthly, or quarterly deposits of $100 at 6% interest compounded quarterly? **(12.6)** Quarterly deposits of $100 at 6%

25. For 1985 through 1990 the amount, a_n (in millions of dollars), of food and drink sales at hotel and motel restaurants can be modeled by $a_n = 10{,}688.1(1.072)^n$ where $n = 0$ represents 1985. Use the model to approximate the total food and drink sales for 1985 through 1990. ***(Source: National Restaurant Association)*** **(12.3)** ≈$76,841.47 million

26. You release a yo-yo attached to a string whose length is 3 feet. With no further movement of your hand, the yo-yo returns $\frac{4}{5}$ of the way up the string. If this pattern continues, what is the total vertical distance the yo-yo travels before coming to a stop? **(12.3)** 27 ft

Answers

15. $16x^4 + 96x^3y + 216x^2y^2 + 216xy^3 + 81y^4$
16. $s^3 + 9s^2t + 27st^2 + 27t^3$
17. $x^5 - 30x^4 + 360x^3 - 2160x^2 + 6480x - 7776$
19. $64r^6 + 768r^5 + 3840r^4 + 10{,}240r^3 + 15{,}360r^2 + 12{,}288r + 4096$
20. $s^3 + 6s^2t + 12st^2 + 8t^3$
21. $8x^3 - 12x^2y + 6xy^2 - y^3$
22. $y^7 - 14y^6 + 84y^5 - 280y^4 + 560y^3 - 672y^2 + 448y - 128$

Answers

12. $\dfrac{-3x^3 + 5x^2 + 58x + 30}{(x-5)(x+5)(x-1)}$

19. $\dfrac{(x-1)^2}{4} - (y+1)^2 = 1$

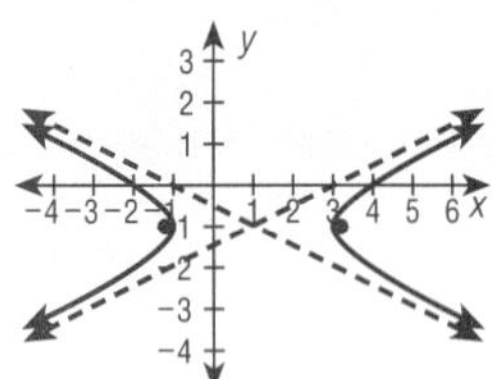

20. $\dfrac{(x-1)^2}{4} + \dfrac{(y+3)^2}{9} = 1$

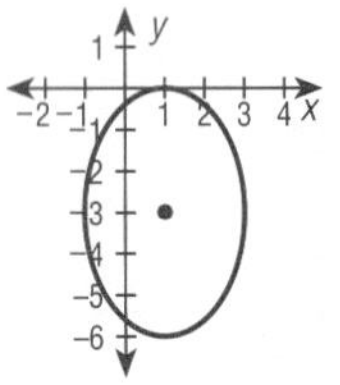

21. $(x-3)^2 = 4(y-1)$

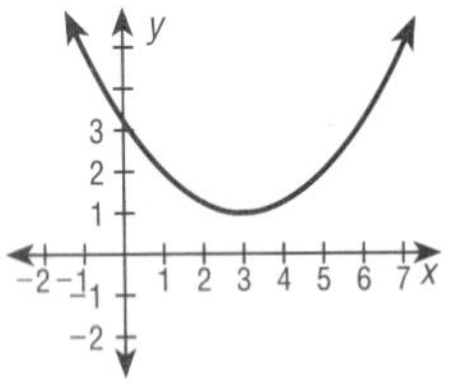

22. $(x-4)^2 + y^2 = 1$

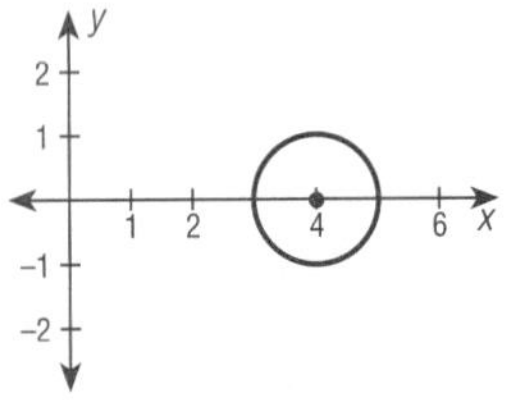

23. $\dfrac{(x+2)^2}{16} + \dfrac{(y+2)^2}{9} = 1$

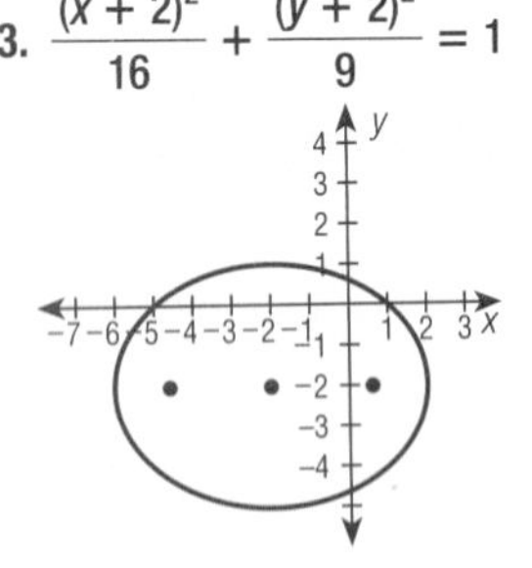

Cumulative REVIEW ■ Chapters 7–12

In Exercises 1–4, match the function with its graph.

1. $f(x) = \dfrac{-2}{x^2+2}$ c

2. $f(x) = \dfrac{3x}{x^2-2x+1}$ d

3. $f(x) = \dfrac{x+6}{x-3}$ a

4. $f(x) = \dfrac{6}{4-x}$ b

a. 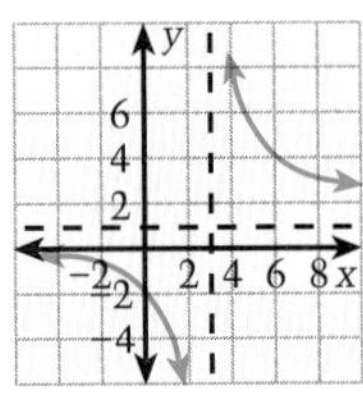

b. 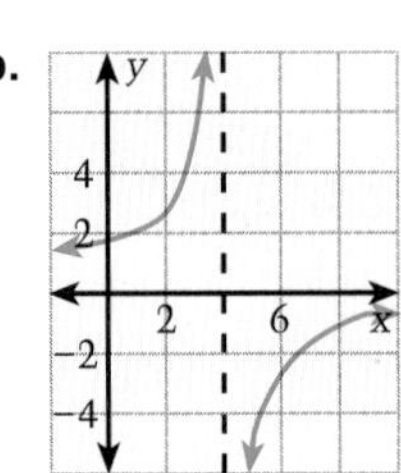

c. 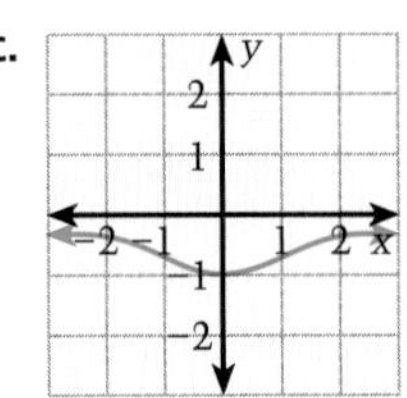

d. 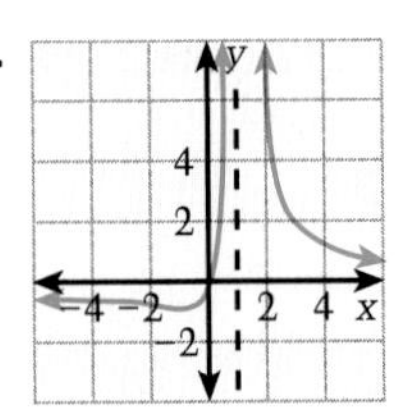

In Exercises 5 and 6, the variables *x* and *y* vary inversely. Use the given values to find an equation that relates the variables.

5. $x = 6,\ y = 12$ $xy = 72$

6. $x = 2,\ y = 11$ $xy = 22$

In Exercises 7 and 8, the variable *z* varies jointly with the product of *x* and *y*. Use the given values to find an equation that relates the variables.

7. $z = 4,\ x = 2,\ y = 3$ $\dfrac{z}{xy} = \dfrac{2}{3}$

8. $z = 3,\ x = 12,\ y = 7$ $\dfrac{z}{xy} = \dfrac{1}{28}$

In Exercises 9–14, perform the operations and simplify.

9. $\dfrac{3x^2y}{x-2} \cdot \dfrac{x^2+x-6}{3x-6} \div (x^2-4)$ $\dfrac{x^2y(x+3)}{(x-2)^2(x+2)}$

10. $\dfrac{x-1}{x^3+5x^2-6x} \div \dfrac{x^2-36}{x+3} \cdot \dfrac{x^2}{x+3}$ $\dfrac{x}{(x-6)(x+6)^2}$

11. $-3\left(\dfrac{x}{x+4} + \dfrac{6x}{x^2-16}\right) + \dfrac{5}{x-4}$ $\dfrac{-3x^2-x+20}{(x+4)(x-4)}$

12. $\dfrac{x+2}{x^2-6x+5} - \dfrac{3x}{x-1} + \dfrac{4}{x+5}$ See margin.

13. $\dfrac{\left(\dfrac{6}{x} + \dfrac{3x}{x+4}\right)}{\left(\dfrac{x+1}{x-4} + \dfrac{x-2}{x+4}\right)}$ $\dfrac{3x^3-6x^2-96}{2x^3-x^2+12x}$

14. $\dfrac{\left(\dfrac{x-1}{2} - \dfrac{4-x}{2x}\right)}{\left(\dfrac{6}{x+3} + \dfrac{3}{x}\right)}$ $\dfrac{x^3+3x^2-4x-12}{18x+18}$

In Exercises 15–18, match the equation with its graph.

15. $(x+2)^2 + (y-4)^2 = 1$ d

16. $(y+3)^2 = -4(x+1)$ c

17. $\dfrac{(x-1)^2}{4} + (y+2)^2 = 1$ b

18. $\dfrac{(x-3)^2}{9} - \dfrac{(y-1)^2}{4} = 1$ a

a. 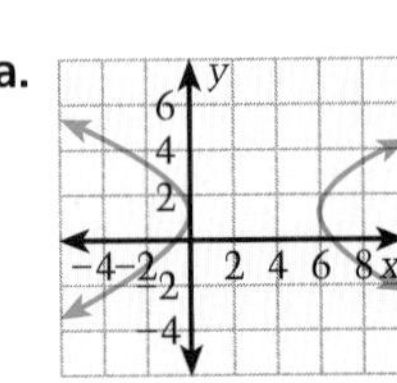

b. 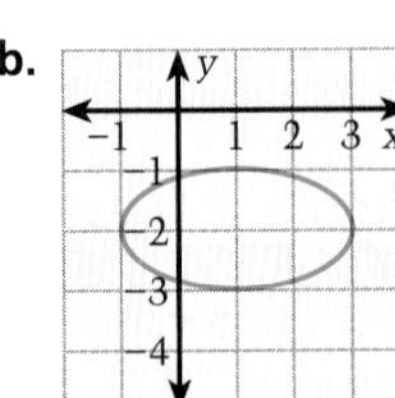

c. 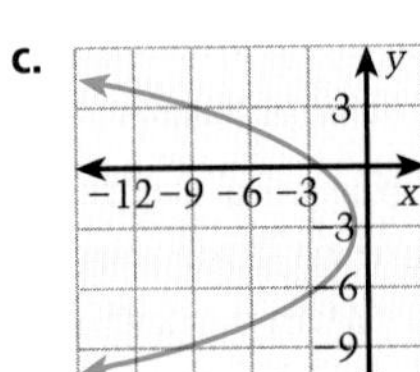

d. 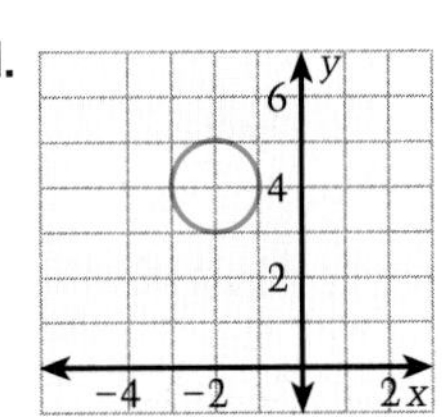

In Exercises 19–26, classify the conic. Write the equation in standard form and then sketch its graph. See margins for graphs and equations.

19. $x^2 - 4y^2 - 2x - 8y - 7 = 0$ hyperbola

20. $9x^2 + 4y^2 - 18x + 24y + 9 = 0$ ellipse

21. $x^2 - 6x - 4y + 13 = 0$ parabola

22. $x^2 + y^2 - 8x + 15 = 0$ circle

23. $9x^2 + 16y^2 + 36x + 64y - 44 = 0$ ellipse

24. $y^2 + 8x + 2y + 17 = 0$ parabola

25. $x^2 + y^2 + 4x - 2y - 4 = 0$ circle

26. $-9x^2 + y^2 + 36x - 45 = 0$ hyperbola

In Exercises 27–30, find the points of intersection of the graphs.

27. $-9x^2 + 4y^2 + 18x - 45 = 0$ (1, 3), (1, −3)
$x^2 + y^2 - 2x - 8 = 0$

28. $x^2 - y^2 - 2x + 2y = 0$ (0, 0), (2, 0),
$4x^2 + y^2 - 8x = 0$ $\left(\frac{2}{5}, \frac{8}{5}\right), \left(\frac{8}{5}, \frac{8}{5}\right)$

29. $9x^2 + 25y^2 + 36x - 200y + 211 = 0$ (−2, 1),
$x^2 + y^2 + 4x - 8y + 11 = 0$ (−2, 7)

30. $36x^2 - y^2 + 216x + 4y + 284 = 0$
$2x^2 + 12x + y + 14 = 0$
$(-3 \pm \sqrt{10}, -16), (-4, 2), (-2, 2)$

In Exercises 31–34, write the first five terms of the sequence. Begin with *n* = 1. See margin.

31. $a_n = 2n^2 + 3$

32. $a_n = (-1)^n \cdot \frac{1}{n}$

33. $a_n = n^3 - n^2$

34. $a_n = \frac{1}{4}(n!)$

In Exercises 35–42, evaluate the sum.

35. $\sum_{i=1}^{10} (4i + 2)$ 240

36. $\sum_{n=1}^{20} (10n - 5)$ 2000

37. $\sum_{n=1}^{13} n^2$ 819

38. $\sum_{i=1}^{14} i$ 105

39. $\sum_{i=1}^{10} 6\left(\frac{1}{3}\right)^{i-1}$ ≈9.000

40. $\sum_{n=0}^{12} \left(-\frac{3}{4}\right)^n$ ≈0.585

41. $\sum_{n=1}^{\infty} 4(0.2)^{n-1}$ 5

42. $\sum_{i=0}^{\infty} 3\left(-\frac{1}{5}\right)^i$ 2.5

43. Write a formula for the *n*th term of the arithmetic sequence whose common difference is 7 and whose third term is 25. $a_n = 7n + 4$

44. Write a formula for the *n*th term of the geometric sequence whose common ratio is 4 and whose fourth term is 64. $a_n = 4^{n-1}$

In Exercises 45–48, evaluate the binomial coefficient.

45. $\binom{4}{2}$ 6

46. $\binom{3}{1}$ 3

47. $\binom{5}{0}$ 1

48. $\binom{14}{12}$ 91

49. Expand the binomials: $(2x + 3)^3$, $(x - y)^5$. See below.

In Exercises 50 and 51, find the ratio of the area of the blue region to the total area. Write your result in simplified form.

50.

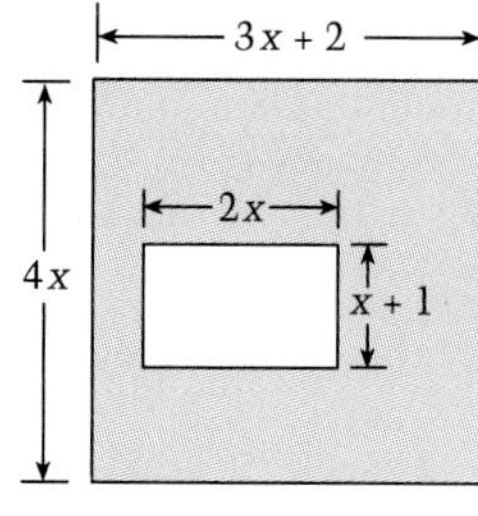

$\frac{5x + 3}{6x + 4}$

51.

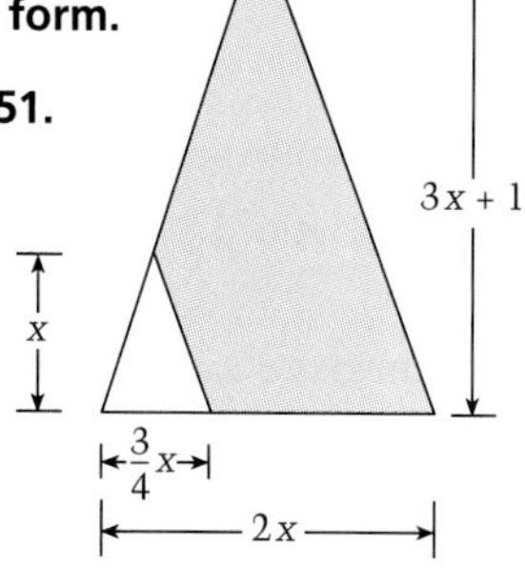

$\frac{21x + 8}{24x + 8}$

49. $8x^3 + 36x^2 + 54x + 27$, $x^5 - 5x^4y + 10x^3y^2 - 10x^2y^3 + 5xy^4 - y^5$

24. $(y + 1)^2 = -8(x + 2)$

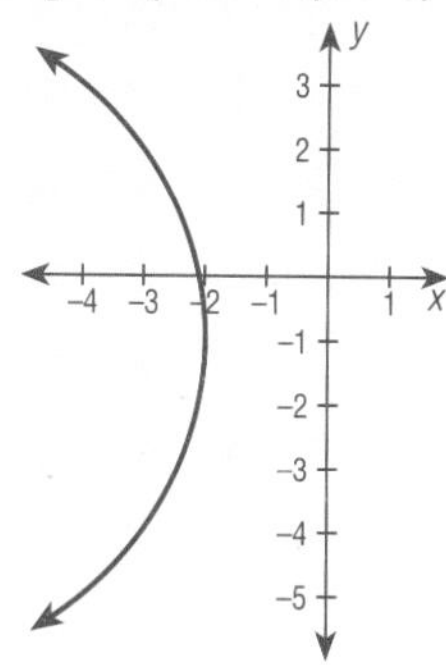

25. $(x + 2)^2 + (y - 1)^2 = 9$

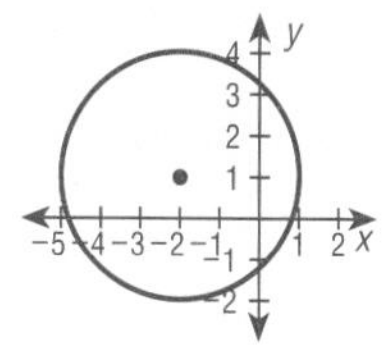

26. $\frac{y^2}{9} - (x - 2)^2 = 1$

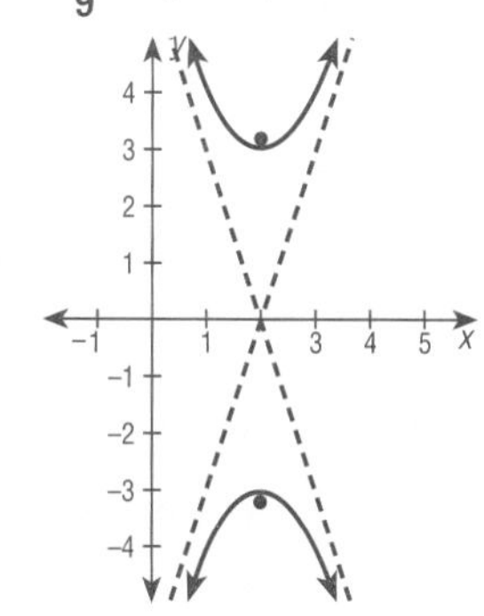

31. 5, 11, 21, 35, 53

32. $-1, \frac{1}{2}, -\frac{1}{3}, \frac{1}{4}, -\frac{1}{5}$

33. 0, 4, 18, 48, 100

34. $\frac{1}{4}, \frac{1}{2}, \frac{3}{2}, 6, 30$

Ski Membership Fees **In Exercises 52 and 53, use the following information.**

The number, M, of memberships for the Mountain Top Ski Club for 1980 through 1992 can be modeled by

$$M = \frac{100(3 + 2t)}{3 + 0.5t}$$

where $t = 0$ represents 1980. The cost, C, of a membership can be modeled by

$$C = \frac{225 + 262.5t}{3 + 2t}.$$

52. $A = \frac{100(225 + 262.5t)}{3 + 0.5t}$

52. Write a model for the total amount, A, received in membership fees.

53. Approximate the amount the club received in 1987 from membership fees. $31,730.77

Suzanne's Savings Plan **In Exercises 54 and 55, use the following information.**

Your friend, Suzanne, wants to buy a car. Her parents agree, *if* for one month she can follow a savings plan they have designed. On the first day, Suzanne must save \$0.01. The second day, she must save \$0.02, the third day \$0.04, and so forth for 30 days.

54. Write a geometric series that represents the amount Suzanne would save in 30 days if she followed the savings plan. $\sum_{n=1}^{30} 0.01(2)^{n-1}$

55. Do you think Suzanne can follow her parents' plan? Explain. See below.

Your Savings Plan **In Exercises 56–59, use the following information.**

You decide to save \$25 per month in an increasing annuity that pays an annual interest rate of 6%, compounded monthly.

56. Copy and complete the following table showing the balances and earned interest during the first six months of the annuity. See Additional Answers.

End of Month	Previous Balance	Interest	Deposit	New Balance
1	\$0.00	\$0.00	\$25.00	\$25.00
2	?	?	\$25.00	?
3	?	?	\$25.00	?
4	?	?	\$25.00	?
5	?	?	\$25.00	?
6	?	?	\$25.00	?

57. What is the balance of your account after two years? \$635.80

58. What monthly deposit would produce an amount of \$1000 after two years? \$39.33

59. After three years, you have saved enough for the down payment on a car. To purchase the car, however, you need to take out a 3-year loan of \$4000 at 11% annual interest. What are your monthly payments? Find the total interest you will pay over the life of the loan. \$130.95, \$714.20

55. No, since $\sum_{i=1}^{30} (0.01)(2)^{i-1} = 10{,}737{,}418.23$, Suzanne would have to save over \$10 million in 30 days.

Solar Oven **In Exercises 60 and 61, use the following information.**

Located in the Pyrenees mountains of France, the Odieollo Solar Oven can produce temperatures up to 6870°F. Sixty-three flat mirrors, called heliostats, track the sun and reflect its light waves onto the central parabolic mirror, which is composed of 9500 smaller mirrors. The central mirror focuses the light into a small region on the tower that houses the furnace. The resulting solar radiation on the tower is up to 12,000 times more powerful than normal solar radiation.

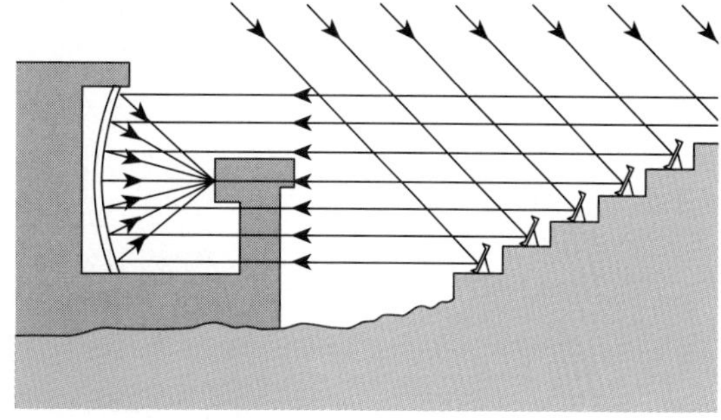

60. A cross-section of the parabolic mirror can be modeled by $y^2 = 322x$ where x and y are measured in feet. What is the distance between the surface of the mirror and the focus? 80.5 ft

61. The surface area, S (in square inches), of each of the heliostats can be found by solving the equation

$$\frac{30{,}240}{S} = \sum_{n=1}^{63} n.$$

What is the surface area of each heliostat? 15 in.^{2}

Galaxies **In Exercises 62 and 63, use the following information.**

In the 1920's, Edwin P. Hubble classified the galaxies known at that time according to their structural formation: spiral, irregular, or elliptical. Elliptical galaxies contain mostly dying stars. Below are graphs of the structural form of three elliptical galaxies.

62. Write an equation for each of the graphs. See below.

63. Which galaxy has the greatest eccentricity? c, the galaxy NGC205

a.

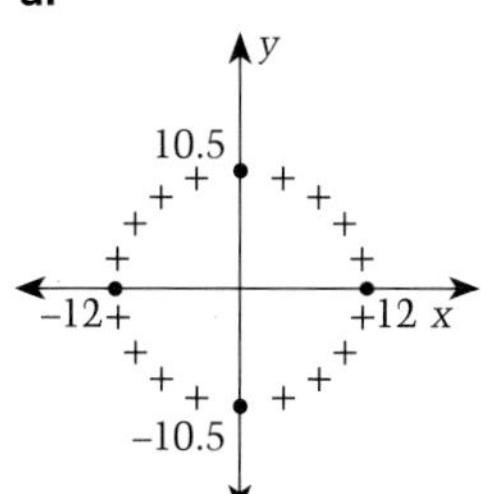

The galaxy NGC4486 is in the Virgo cluster.

b.

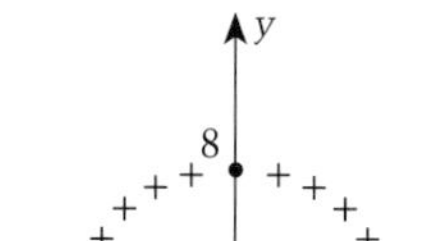

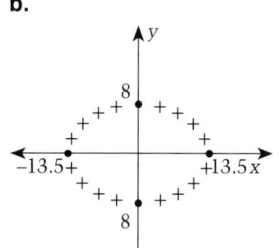

The galaxy NGC147 is in the constellation Cassiopeia.

c.

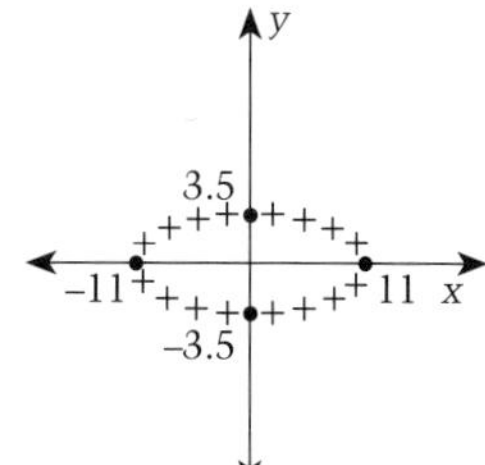

The galaxy NGC205 is in the constellation Andromeda.

62. a. $\frac{x^2}{144} + \frac{y^2}{110.25} = 1$, b. $\frac{x^2}{182.25} + \frac{y^2}{64} = 1$, c. $\frac{x^2}{121} + \frac{y^2}{12.25} = 1$

Trigonometric functions represent another class of functions that students may not have encountered previously.

In this chapter many of the algebraic foundations of trigonometric functions will be developed. Because trigonometric functions have many real-life applications, they are an important collection of functions that should be studied carefully.

The Chapter Summary on page 722 provides you and the students with a synopsis of the chapter. It identifies key skills and concepts. You may want to have students look at the Chapter Summary as an overview before beginning the chapter.

CHAPTER 13

Trigonometric Functions

LESSONS

Molikpaq, a state-of-the-art oil drilling platform, lies in Alaska's Beaufort Sea, just inside the Arctic Circle.

Trigonometry is a branch of mathematics that deals with the relationships found in the angles and sides of triangles. It has many applications in such fields as surveying, navigation, and engineering. Trigonometric functions are also important in the study of another branch of mathematics—calculus.

What is the circumference of the earth at the equator? Compare the length of the equator to that of the Arctic Circle.

The circumference of the earth at the equator is about 24,881 miles. The equator is about 16 times as long as the Arctic Circle.

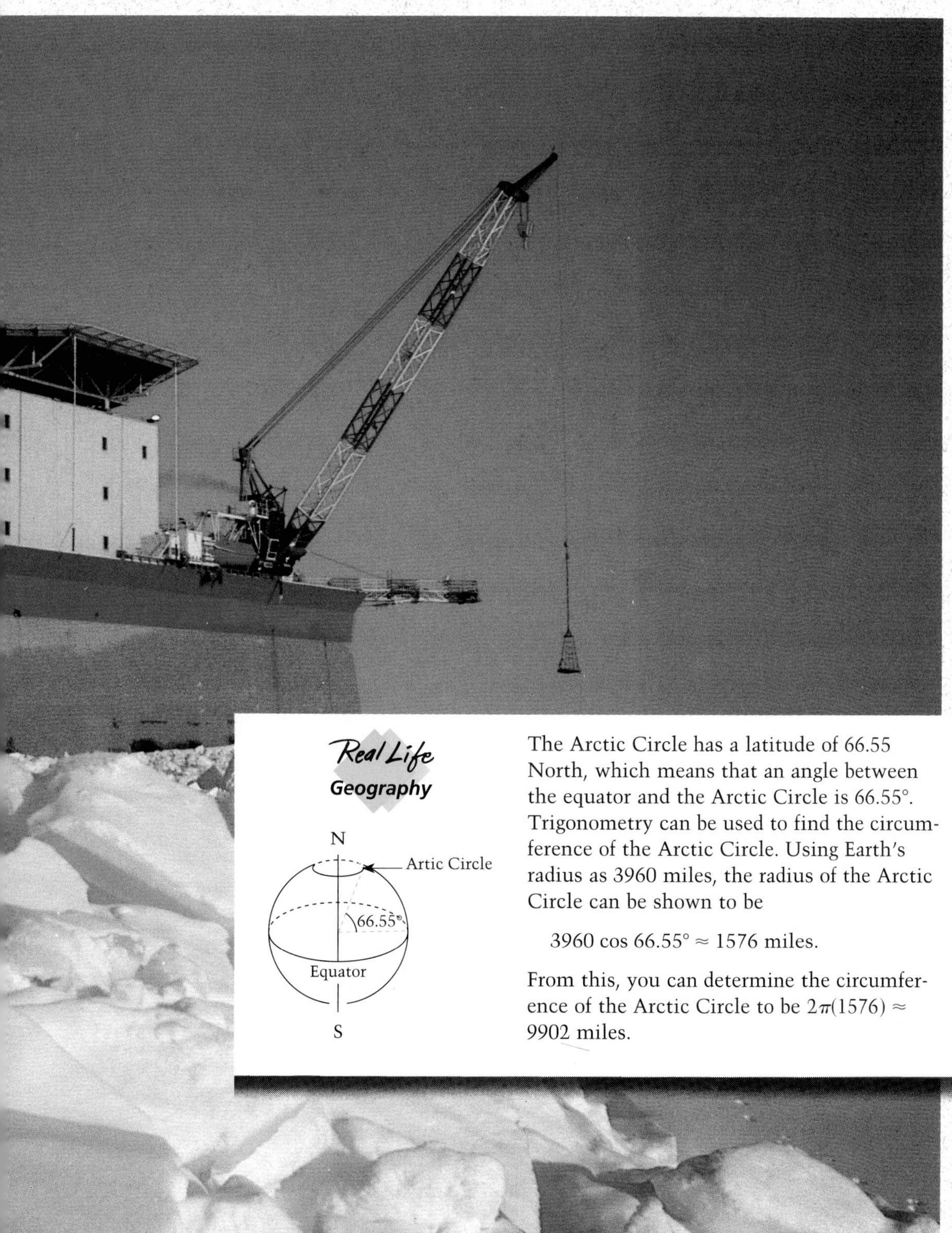

The Arctic Circle has a latitude of 66.55 North, which means that an angle between the equator and the Arctic Circle is 66.55°. Trigonometry can be used to find the circumference of the Arctic Circle. Using Earth's radius as 3960 miles, the radius of the Arctic Circle can be shown to be

$$3960 \cos 66.55° \approx 1576 \text{ miles.}$$

From this, you can determine the circumference of the Arctic Circle to be $2\pi(1576) \approx 9902$ miles.

Trigonometric Ratios and Functions

PACING CHART

This chapter is an integral part of the full course for Algebra 2. However, if your class is taking the basic course, and you still have teaching days available for further topics, you may decide to teach some basic lessons on trigonometry. Lessons 13.1 through 13.4 are recommended for this purpose. You may also wish to consult the introductory notes on pages 729A and 781A.

DAY	FULL COURSE
1	13.1
2	13.2
3	13.2
4	13.3
5	13.3 & Using a Graphing Calculator
6	Mid-Chapter Self-Test
7	13.4
8	13.4
9	13.5
10	13.5
11	13.6
12	13.6
13	Chapter Review
14	Chapter Test

CHAPTER ORGANIZATION

LESSON	PAGES	GOALS	MEETING THE NCTM STANDARDS
13.1	678–684	1. Use trigonometric relationships to evaluate the trigonometric functions of acute angles 2. Use trigonometric ratios to solve real-life problems	Problem Solving, Communication, Reasoning, Connections, Functions, Trigonometry
13.2	685–691	1. Measure angles in standard position using degree measure and radian measure 2. Use radian measure in real-life problems	Problem Solving, Communication, Connections, Geometry, Trigonometry, Structure
Mixed Review	692	Review algebraic and arithmetic skills	
Career Interview	692	Dermatologist	Connections
13.3	693–699	1. Evaluate trigonometric functions of any angle 2. Use trigonometric functions to solve real-life problems	Problem Solving, Communication, Reasoning, Connections, Functions, Trigonometry
Using a Calculator	700–701	Draw angles and the unit circle	Technology, Trigonometry
Mid-Chapter Self-Test	702	Diagnose student weaknesses and remediate with correlated Reteach worksheets	
13.4	703–708	1. Evaluate inverse trigonometric functions 2. Use inverse trigonometric functions to solve real-life problems	Problem Solving, Communication, Connections, Functions, Trigonometry
13.5	709–715	1. Use the Law of Sines to find the sides and angles of any triangle 2. Use the Law of Sines to solve real-life problems	Problem Solving, Communication, Connections, Geometry (synthetic), Trigonometry
Mixed Review	715	Review algebraic and arithmetic skills	
13.6	716–721	1. Use the Law of Cosines to find the sides and angles of any triangle 2. Use the Law of Cosines to solve real-life problems	Problem Solving, Communication, Connections, Geometry, Trigonometry
Chapter Summary	722	Restate for students what they have learned, why they have learned it, and how it fits into the structure of algebra	Structure, Connections
Chapter Review	723–726	Review concepts and skills learned in the chapter	
Chapter Test	727	Diagnose student weaknesses and remediate with correlated Reteaching worksheets	

LESSON RESOURCES

MEETING INDIVIDUAL NEEDS

RETEACHING For students who need to spend more time on basics:

If a mid-chapter self-test or chapter test indicates a deficiency, teachers can help students with the appropriate ***Reteaching Copymaster.***

PRACTICE For students who need more practice:

Additional exercises like those in the Pupil's Edition are provided for each lesson in ***Extra Practice Copymasters.***

ENRICHMENT For enriching and broadening students' experiences:

Problem of the Day copymasters in ***Teaching Tools*** provide a daily opportunity to use logical reasoning, looking for a pattern, writing an equation, and other routine and non-routine problem-solving strategies.

Math Log copymasters in ***Alternative Assessment*** provide opportunities to report on investigations, research, and open-ended problems.

Technology: Using Calculators and Computers provides enriching activities with graphing and scientific calculators and computers.

The ***Applications Handbook*** provides additional information about the cross-curriculum topics such as astronomy, chemistry, physics, sports, economics, genetics, and music that are integrated into the Pupil's Edition.

LESSON	13.1	13.2	13.3	13.4	13.5	13.6
PAGES	678-684	685-691	693-699	703-708	709-715	716-721
Teaching Tools						
Transparencies		✓	✓			
Problem of the Day	✓	✓	✓	✓	✓	✓
Warm-up Exercises	✓	✓	✓	✓	✓	✓
Answer Masters	✓	✓	✓	✓	✓	✓
Extra Practice Copymasters	✓	✓	✓	✓	✓	✓
Reteaching Copymasters	Teacher-directed and independent activities tied to results on the Mid-Chapter Self-Tests and Chapter Tests					
Color Transparencies			✓	✓	✓	✓
Applications Handbook	Additional background information for many real-life applications					
Technology	Calculator and computer worksheets for appropriate lessons					
Complete Solutions Manual						
Alternative Assessment	Assess student's ability to reason, analyze, solve problems, and communicate using mathematical language.					
Formal Assessment	Mid-Chapter Self-Tests, Chapter Tests, Cumulative Tests, and Practice for College Entrance Tests					
Computer Test Bank	Customized tests can be created by choosing from over 2500 items.					

INSIGHTS

13.1 Right Triangle Trigonometry

The right triangle is the basis for the definitions of the six basic trigonometric ratios. In this lesson, these ratios are defined as the ratios of the sides of a right triangle. These definitions can then be extended more broadly to situations not directly involving right triangles.

13.2 General Angles and Radian Measures

Not all angles are part of a triangle, nor are they necessarily acute angles. Many angles that describe real-life situations are greater than 90° and exist on a coordinate plane. To evaluate the trigonometric functions of such angles, it is necessary to first define angles and angle measurement in a context beyond that of right triangles. In this lesson, definitions and rules by which this can be accomplished are introduced.

13.3 Trigonometric Functions of Any Angle

Frequently, real-life situations are modeled by relationships that involve obtuse and negative angles. In this lesson, students see how angles of rotation can be evaluated using acute angles.

13.4 Inverse Trigonometric Functions

Like other algebraic functions, trigonometric functions have inverses, but these inverses are relations, not functions. However, by suitably restricting the domain of each trigonometric function, an inverse can be devised that is a function. Trigonometric inverses can be used to model many real-life situations in physics, manufacturing, and engineering.

13.5 The Law of Sines

Not all triangles used to model real-life situations are right triangles, and so the Pythagorean Theorem cannot necessarily be applied to find unknown parts of a triangle. Missing information in an oblique triangle can be found using either the Law of Sines or the Law of Cosines. In this lesson, students examine the use of the Law of Sines.

13.6 The Law of Cosines

Sometimes, missing information in oblique triangles cannot be found using the Law of Sines, for example, if none of the angle measures is known. In this lesson, students examine the use of the Law of Cosines, which can be applied to such cases.

Problem of the Day

How many times will the hands of a clock exactly line up *between* noon and midnight?
10 times

ORGANIZER

Warm-Up Exercises

1. Identify the acute angles in the following triangles.

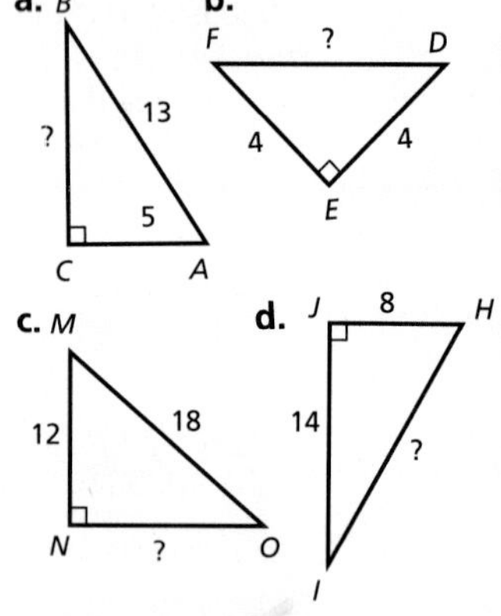

a. ∠A, ∠B b. ∠D, ∠F
c. ∠M, ∠O d. ∠H, ∠I

2. Use the Pythagorean Theorem to find the unknown lengths of the right triangles shown in Warm-Up Exercise 1.

a. 12, b. $4\sqrt{2}$, c. $2\sqrt{65}$, d. $6\sqrt{5}$

Lesson Resources

Teaching Tools
- Problem of the Day: 13.1
- Warm-up Exercises: 13.1
- Answer Masters: 13.1

Extra Practice: 13.1

13.1 Right Triangle Trigonometry

What you should learn:

Goal 1 How to use trigonometric relationships to evaluate the trigonometric functions of acute angles

Goal 2 How to use trigonometric functions to solve real-life problems

Why you should learn it:

You can use trigonometric functions to solve many real-life problems, such as finding the height of a mountain.

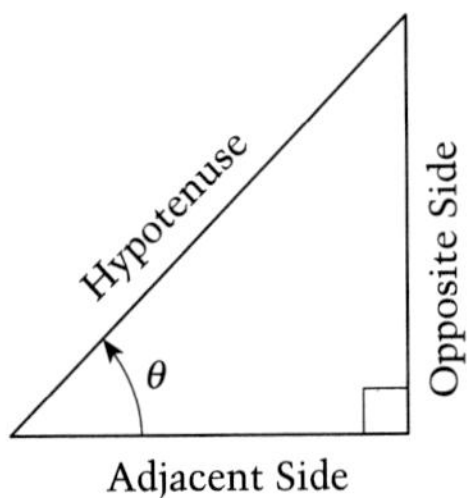

Goal 1 Using Trigonometric Relationships

As derived from the Greek language, the word **trigonometry** means "measurement of triangles." To begin our study of trigonometry, consider a right triangle, one of whose acute angles is θ (the Greek letter *theta*). The three sides of the right triangle are the **hypotenuse,** the **opposite side** (the side opposite the angle θ), and the **adjacent side** (the side adjacent to the angle θ).

Using these three sides, you can form the six **trigonometric functions: sine, cosine, tangent, cosecant, secant,** and **cotangent** (abbreviated as sin, cos, tan, csc, sec, and cot, respectively).

Right Triangle Definition of Trigonometric Functions

Let θ be an *acute* angle of a right triangle.

$$\sin\theta = \frac{\text{opp}}{\text{hyp}} \qquad \cos\theta = \frac{\text{adj}}{\text{hyp}} \qquad \tan\theta = \frac{\text{opp}}{\text{adj}}$$

$$\csc\theta = \frac{\text{hyp}}{\text{opp}} \qquad \sec\theta = \frac{\text{hyp}}{\text{adj}} \qquad \cot\theta = \frac{\text{adj}}{\text{opp}}$$

The abbreviations *opp, adj,* and *hyp* represent the lengths of the three sides of the right triangle. Note that the ratios in the second row are the *reciprocals* of the ratios in the first row.

Example 1 *Evaluating Trigonometric Functions*

Evaluate the six trigonometric functions of the angle θ shown in the right triangle at the left.

Solution From the Pythagorean Theorem, the length of the hypotenuse is $\sqrt{4^2 + 3^2} = \sqrt{25} = 5$. Using adj = 3, opp = 4, and hyp = 5, you can write the following statements.

$$\sin\theta = \frac{\text{opp}}{\text{hyp}} = \frac{4}{5} \qquad \cos\theta = \frac{\text{adj}}{\text{hyp}} = \frac{3}{5} \qquad \tan\theta = \frac{\text{opp}}{\text{adj}} = \frac{4}{3}$$

$$\csc\theta = \frac{\text{hyp}}{\text{opp}} = \frac{5}{4} \qquad \sec\theta = \frac{\text{hyp}}{\text{adj}} = \frac{5}{3} \qquad \cot\theta = \frac{\text{adj}}{\text{opp}} = \frac{3}{4}$$ ■

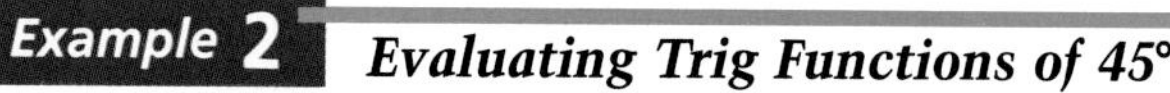

Example 2 Evaluating Trig Functions of 45°

Evaluate sin 45°, cos 45°, and tan 45°.

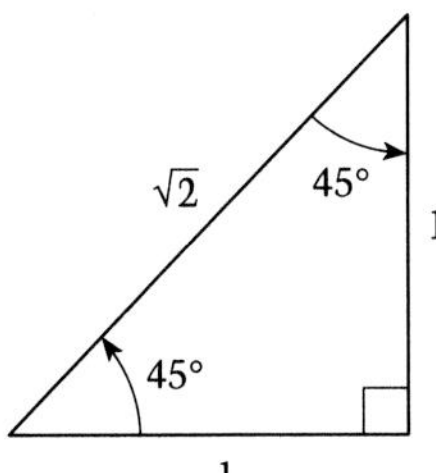

Solution Sketch a right triangle having 45° as one of its acute angles. All such triangles are similar. Thus to make the calculations simple, you can choose 1 as the length of the adjacent side. Because the triangle is isosceles, the length of the opposite side is also 1. From the Pythagorean Theorem, the length of the hypotenuse is $\sqrt{1^2 + 1^2} = \sqrt{2}$. Thus you can write the following.

$$\sin 45^\circ = \frac{\text{opp}}{\text{hyp}} = \frac{1}{\sqrt{2}} = \frac{\sqrt{2}}{2} \approx 0.707$$

$$\cos 45^\circ = \frac{\text{adj}}{\text{hyp}} = \frac{1}{\sqrt{2}} = \frac{\sqrt{2}}{2} \approx 0.707$$

$$\tan 45^\circ = \frac{\text{opp}}{\text{adj}} = \frac{1}{1} = 1$$

■

Example 3 Evaluating Trig Functions of 30°and 60°

Use the equilateral triangle shown at the left to help you evaluate sin 60°, cos 60°, sin 30°, and cos 30°.

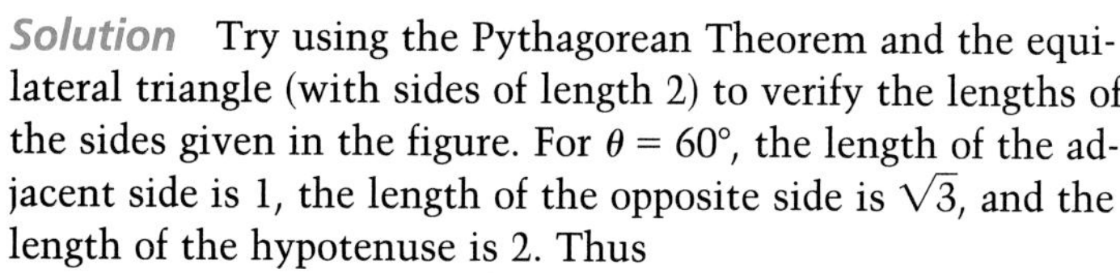

Solution Try using the Pythagorean Theorem and the equilateral triangle (with sides of length 2) to verify the lengths of the sides given in the figure. For $\theta = 60^\circ$, the length of the adjacent side is 1, the length of the opposite side is $\sqrt{3}$, and the length of the hypotenuse is 2. Thus

$$\sin 60^\circ = \frac{\text{opp}}{\text{hyp}} = \frac{\sqrt{3}}{2} \approx 0.866 \quad \text{and} \quad \cos 60^\circ = \frac{\text{adj}}{\text{hyp}} = \frac{1}{2}.$$

For $\theta = 30^\circ$, the length of the adjacent side is $\sqrt{3}$, the length of the opposite side is 1, and the length of the hypotenuse is 2. Thus

$$\sin 30^\circ = \frac{\text{opp}}{\text{hyp}} = \frac{1}{2} \quad \text{and} \quad \cos 30^\circ = \frac{\text{adj}}{\text{hyp}} = \frac{\sqrt{3}}{2} \approx 0.866.$$

■

Because the angles 30°, 45°, and 60° occur frequently in trigonometry, you should learn to sketch the triangles shown in Examples 2 and 3. The values of the sines, cosines, and tangents of these common angles are summarized below.

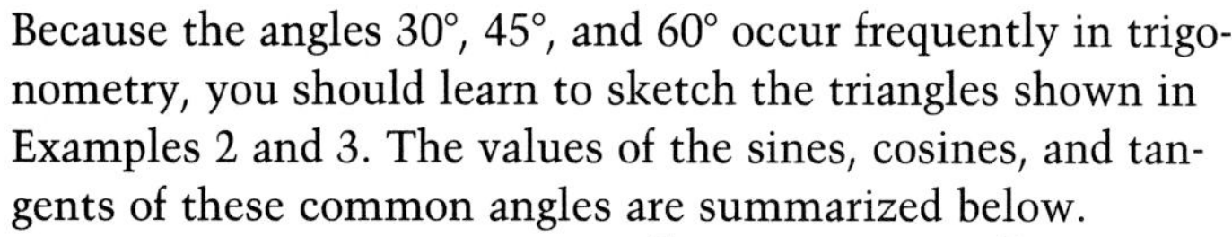

$$\sin 30^\circ = \frac{1}{2} \qquad \cos 30^\circ = \frac{\sqrt{3}}{2} \qquad \tan 30^\circ = \frac{\sqrt{3}}{3}$$

$$\sin 45^\circ = \frac{\sqrt{2}}{2} \qquad \cos 45^\circ = \frac{\sqrt{2}}{2} \qquad \tan 45^\circ = 1$$

$$\sin 60^\circ = \frac{\sqrt{3}}{2} \qquad \cos 60^\circ = \frac{1}{2} \qquad \tan 60^\circ = \sqrt{3}$$

LESSON Notes

Example 1

Students may note that each acute angle of a right triangle has two adjacent sides! But one of the sides is always the hypotenuse. Hence, the term "adjacent side" always refers to a leg of the triangle, not the hypotenuse.

Emphasize that the definitions of the trigonometric functions depend on the lengths of the sides of a given triangle. Students will later see that the corresponding ratios of similar right triangles will be the same. Encourage students to memorize the definitions of the six trigonometric functions.

Example 2

Have students compare csc 45° and sin 45°. Their values, $\sqrt{2}$ and $\frac{\sqrt{2}}{2}$, are reciprocals of each other. Have students find a similar relationship for sec 45° and cos 45° and for cot 45° and tan 45°.

Example 3

Ask students to evaluate the functions csc θ, sec θ, and cot θ at $\theta = 30^\circ$ and $\theta = 60^\circ$.

$\csc 30^\circ = 2$, $\sec 30^\circ = \frac{2\sqrt{3}}{3}$,

$\cot 30^\circ = \sqrt{3}$, $\csc 60^\circ = \frac{2\sqrt{3}}{3}$,

$\sec 60^\circ = 2$, $\cot 60^\circ = \frac{\sqrt{3}}{3}$

All of the trigonometric functions of the commonly occurring angles can be evaluated using the information contained in the sketches shown in Examples 2 and 3. If students learn to sketch the triangles , it will not be necessary for them to memorize the values of the sines, cosines, and tangents of the common angles.

Example 4

Be sure students understand that the expression $\sin^2\theta$ is shorthand notation for $(\sin \theta)^2$. Stated differently, mathematicians have agreed to represent the square of the sine function in this manner.

Refer students to the *Handbook of Mathematical Connections* for proofs of the Trigonometric Pythagorean Identities.

Example 5

Students may find it interesting that $\cos 69.5° = \frac{a}{\text{hyp}}$ can also be used to determine a. That is,

$a = (\cos 69.5°)(9388)$
$\approx (0.350207)(9388) \approx 3288.$

Extra Examples

Here are additional examples similar to **Examples 1–4.**

1. Evaluate the six trigonometric functions of the angle θ shown.

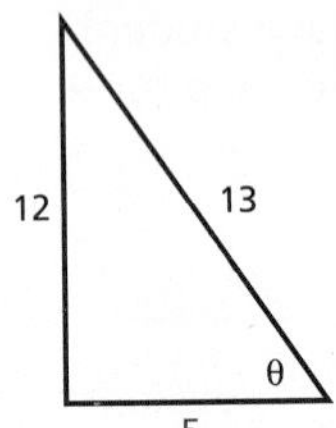

$\sin \theta = \frac{12}{13}$, $\cos \theta = \frac{5}{13}$, $\tan \theta = \frac{12}{5}$, $\csc \theta = \frac{13}{12}$, $\sec \theta = \frac{13}{5}$, $\cot \theta = \frac{5}{12}$

2. Evaluate.
   a. $\frac{\sin 45°}{\cos 60°}$ $\quad \sqrt{2}$
   b. $(\tan 60°)(\csc 30°)$ $\quad 2\sqrt{3}$
   c. $\sec 30° + \cot 30°$ $\quad \frac{5\sqrt{3}}{3}$
3. Let θ be the acute angle such that $\cos \theta = 0.5$. Find $\sin \theta$.
   $\sin \theta = 0.5\sqrt{3} \approx 0.866$
4. Let θ be the acute angle such that $\tan \theta = \sqrt{3}$. Find $\sec \theta$.
   $\sec \theta = 2$

Relationships between trigonometric functions are called **trigonometric identities.**

Fundamental Trigonometric Identities

Reciprocal

$\sin \theta = \frac{1}{\csc \theta}$ $\qquad \cos \theta = \frac{1}{\sec \theta}$ $\qquad \tan \theta = \frac{1}{\cot \theta}$

$\csc \theta = \frac{1}{\sin \theta}$ $\qquad \sec \theta = \frac{1}{\cos \theta}$ $\qquad \cot \theta = \frac{1}{\tan \theta}$

Tangent and Cotangent

$\tan \theta = \frac{\sin \theta}{\cos \theta}$

$\cot \theta = \frac{\cos \theta}{\sin \theta}$

Pythagorean

$\sin^2 \theta + \cos^2 \theta = 1$

$1 + \tan^2 \theta = \sec^2 \theta$

$1 + \cot^2 \theta = \csc^2 \theta$

Note that $\sin^2 \theta$ is used to represent $(\sin \theta)^2$.

Example 4 *Using Trigonometric Identities*

Let θ be the acute angle such that $\sin \theta = 0.6$. Find $\cos \theta$.

Solution To find the value of $\cos \theta$, use the Pythagorean Identity that relates the sine and cosine of an angle.

1 0.6 θ 0.8

sin θ = 0.6

$\sin^2 \theta + \cos^2 \theta = 1$ *Pythagorean Identity*

$(0.6)^2 + \cos^2 \theta = 1$ *Substitute 0.6 for sin θ.*

$\cos^2 \theta = 1 - (0.6)^2 = 0.64$

$\cos \theta = \sqrt{0.64} = 0.8$

Note that you choose the positive square root because the cosine of an acute angle is positive. ■

A calculator can be used to obtain decimal approximations of the values of trigonometric functions. When doing this, *be sure to set the calculator to the correct mode:* degrees or radians. For instance, using a calculator set to degree mode, you can approximate the value of cos 20° to be 0.940.

Most calculators do not have keys for cosecant, secant, or cotangent. To evaluate these, use the reciprocal key $\boxed{1/x}$ or $\boxed{x^{-1}}$ and the appropriate reciprocal identity. For instance,

$$\sec 20° = \frac{1}{\cos 20°} \approx 1.064.$$

Mt. Everest is in two countries, Nepal and Tibet. Its height first was determined in the Great Trigonometric Survey of India (1849–1850). *Modern techniques have verified that the results of the 1850 calculations were correct to within several feet.*

Goal 2 Using Trigonometric Functions in Real Life

Connections **Topography**

Example 5 *Determining the Height of a Mountain*

You are at an elevation of 25,740 feet. Your position is below the peak of Mt. Everest. Using surveying instruments, you measure the *angle of elevation* to the peak to be 20.5°. The distance between you and the peak is 9388 feet. What is the elevation of the peak?

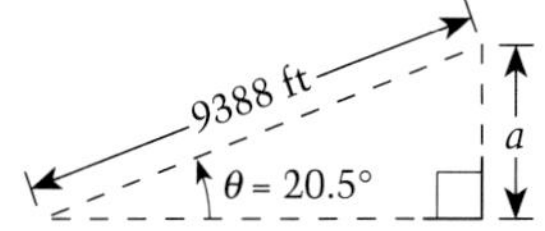

Solution Using the right triangle at the left, let a represent the difference in elevations of the peak and your position.

$\sin\theta = \frac{\text{opp}}{\text{hyp}}$ *Definition of sine function*

$\sin 20.5° = \frac{a}{9388}$ *Let $\theta = 20.5°$ and hyp = 9388.*

$(\sin 20.5°)(9388) = a$ *Multiply both sides by 9388.*

$3288 \approx a$ *Use a calculator.*

Since your elevation is 25,740 feet, the elevation of the peak is 25,740 + 3288 = 29,028 feet. This elevation, established in 1954, is only 26 feet more than that established in 1850. ■

Communicating about ALGEBRA

SHARING IDEAS about the Lesson

A. The sides of a right triangle are 5, 12, and 13. Sketch the triangle. Let θ represent the angle that is opposite the side whose length is 5. Evaluate the six trigonometric functions of θ. See Additional Answers.

B. One of the acute angles in a right triangle is $\theta = 32°$. The length of the hypotenuse is 14 meters. Solve the right triangle by making a sketch and labeling all sides and angles. See Additional Answers.

5. Use a calculator to evaluate the following to three decimal places.

a. sin 87° 0.999

b. tan 23° 0.424

c. sec 42° 1.346

d. cos 55° 0.574

Check Understanding

1. What is the hypotenuse of a right triangle?
The side opposite the right angle; the longest side

2. What does "trigonometry" literally mean?
Measurement of triangles

3. How is the Pythagorean Theorem useful in evaluating trigonometric functions?
It can be used to find a missing leg length or hypotenuse length in a right triangle; in developing some of the trigonometric identities at the top of page 680

4. Sketch and label the appropriate triangle that can be used to evaluate the sines and cosines of 30° and 60°.
See Example 3.

EXTEND *Communicating*
Ask students to show that if $\sin^2\theta + \cos^2\theta = 1$ then

a. $1 + \tan^2\theta = \sec^2\theta$ and

b. $1 + \cot^2\theta = \csc^2\theta$.

(*Hint:* Rewrite one side of the equations in Parts a and b in terms of sine and cosine, and use the result to derive the other side.)

a. Here is a sample derivation.

$$1 + \tan^2\theta = 1 + (\tan\theta)^2 = 1 + \left(\frac{\sin\theta}{\cos\theta}\right)^2 = 1 + \frac{\sin^2\theta}{\cos^2\theta} = \frac{\cos^2\theta + \sin^2\theta}{\cos^2\theta} = \frac{1}{\cos^2\theta} = \sec^2\theta$$

b. A similar derivation can be used in Part b.

EXERCISE Notes

ASSIGNMENT GUIDE

Basic/Average: Ex. 1–6, 7–19 odd, 25–35 odd, 45–55 odd, 57, 61–73

Above Average: Ex. 1–6, 9–21 odd, 27–37 odd, 47–59 odd, 64–73

Advanced: Ex.1–6, 9–21 odd, 27–37 odd, 43–59 odd, 64–73

Selected Answers
Exercises 1–6, 7–65 odd

✪ More Difficult Exercises
Exercises 56–59

Guided Practice

To Rationalize or Not?

In Example 2 on page 679, the trigonometric values for 45° are given in nonrationalized form and in rationalized form. If you have a preference, be sure to tell your students. Our preference is to rationalize simpler fractions, such as those in Example 2, and to leave more complicated fractions, such as $5/\sqrt{41}$ in Exercise 2, in nonrationalized form. Our reasoning is that $5/\sqrt{41}$ is easier to enter in a calculator than $5\sqrt{41}/41$.

▶ **Ex. 1–6** Students may have studied some right triangle trigonometry in previous courses. You might wish to use these exercises before the lesson as a way of determining what students already know.

▶ **Ex. 3–4** Ask students to draw a right triangle that reflects the information given in these exercises.

▶ **Ex. 6** TECHNOLOGY Remind students that calculators need to be in degree mode. Note also that scientific and graphics calculators do not have built-in cot, sec, or csc functions. To find the value of cot, sec, or csc of an angle by

(continued)

EXERCISES

Guided Practice

▶ **CRITICAL THINKING about the Lesson**

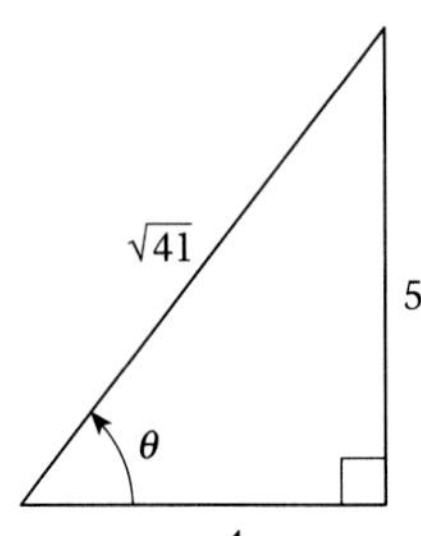

1. State the full name and the abbreviation of each of the six trigonometric functions.

2. Evaluate the six trigonometric functions of the angle θ in the triangle at the right. See below.

3. Consider a right triangle that has θ as one of its acute angles. The length of the adjacent side is 2, and the tangent of θ is $\frac{1}{4}$. What is the length of the opposite side? What is the length of the hypotenuse? $\frac{1}{2}$, $\frac{\sqrt{17}}{2}$

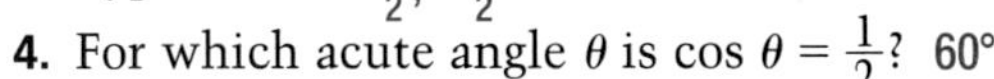

4. For which acute angle θ is $\cos\theta = \frac{1}{2}$? 60°

5. You are given $\cos\theta = 0.55$, where θ is an acute angle. Explain how to find $\sin\theta$.

6. Use a calculator to approximate the value of sec 32° to three decimal places. 1.179

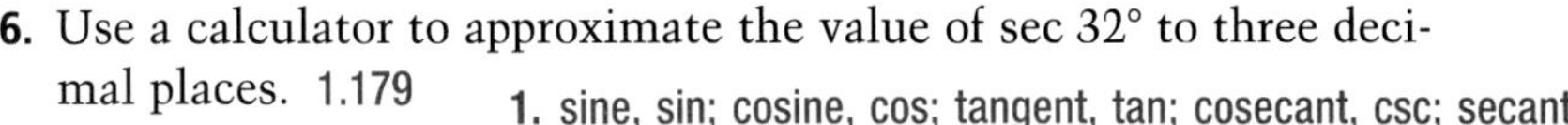

1. sine, sin; cosine, cos; tangent, tan; cosecant, csc; secant, sec; cotangent, cot
5. $\sin^2\theta + \cos^2\theta = 1$, $\sin^2\theta + (0.55)^2 = 1$, $\sin\theta \approx 0.8352$

Independent Practice

In Exercises 7–14, evaluate the six trigonometric functions of θ. See Additional Answers.

7.
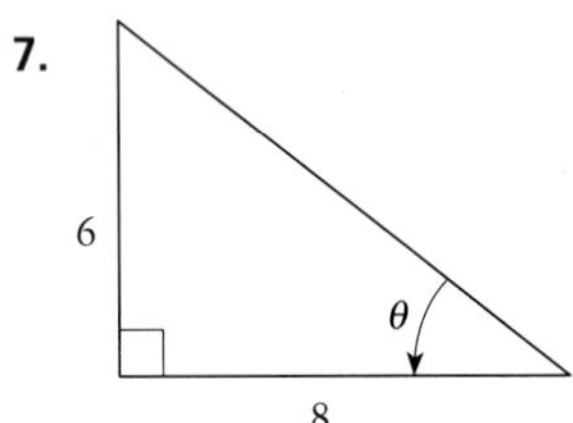

8.
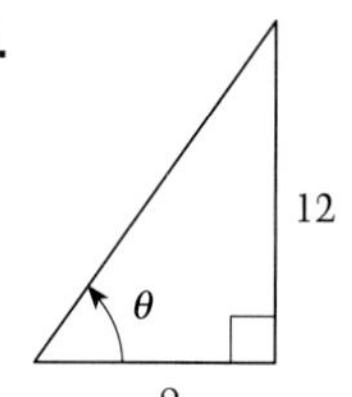

9.
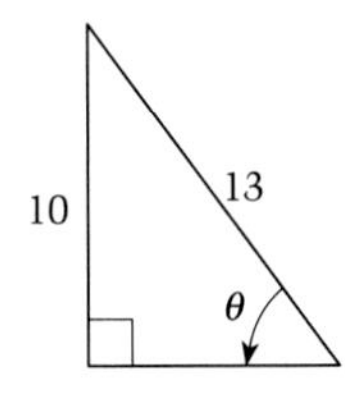

10.
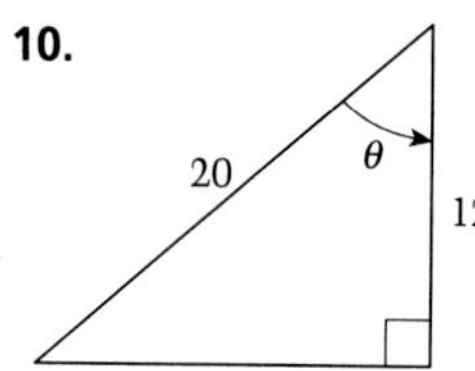

11.
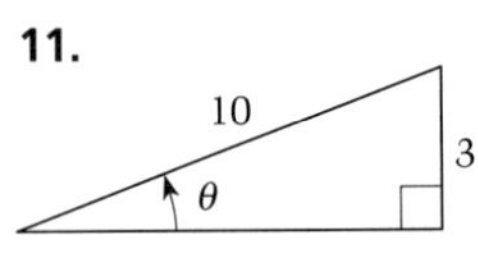

12.
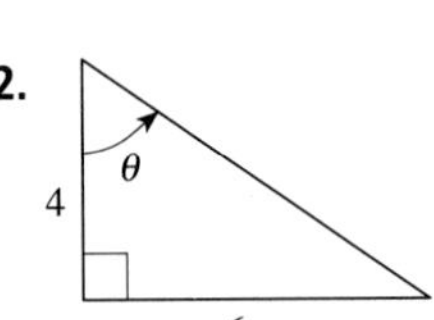

13.
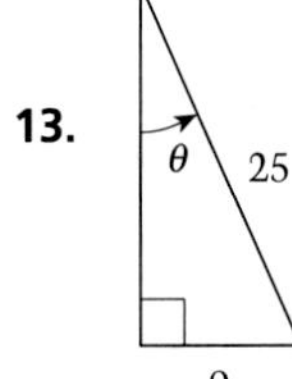

14.
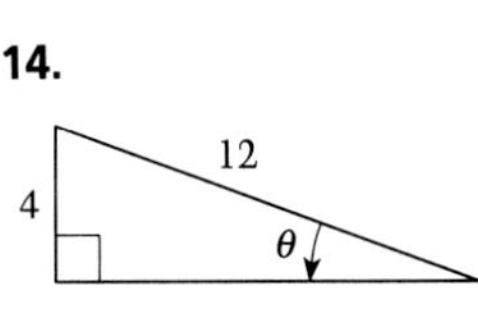

In Exercises 15–22, sketch a right triangle that has θ as one of its acute angles. Then find the values of the five trigonometric functions that are not given. See Additional Answers.

15. $\sin\theta = \frac{2}{3}$ **16.** $\cot\theta = 5$ **17.** $\sec\theta = 2$ **18.** $\cos\theta = \frac{5}{7}$

19. $\tan\theta = 3$ **20.** $\csc\theta = \frac{17}{4}$ **21.** $\cot\theta = \frac{3}{2}$ **22.** $\sin\theta = \frac{3}{8}$

2. $\sin\theta = \frac{5}{\sqrt{41}}$, $\cos\theta = \frac{4}{\sqrt{41}}$, $\tan\theta = \frac{5}{4}$, $\csc\theta = \frac{\sqrt{41}}{5}$, $\sec\theta = \frac{\sqrt{41}}{4}$, $\cot\theta = \frac{4}{5}$

In Exercises 23–30, evaluate the trigonometric function *without* using a calculator.

23. sec 45° $\sqrt{2}$ **24.** sec 30° $\frac{2}{\sqrt{3}}$ **25.** csc 60° $\frac{2}{\sqrt{3}}$ **26.** csc 45° $\sqrt{2}$

27. cot 45° 1 **28.** csc 30° 2 **29.** cot 60° $\frac{1}{\sqrt{3}}$ **30.** sec 60° 2

31. Given $\sin \theta = \frac{1}{2}$, find $\cos \theta$. $\frac{\sqrt{3}}{2}$

32. Given $\tan \theta = \sqrt{3}$, find $\sin \theta$. $\frac{\sqrt{3}}{2}$

33. Given $\cos \theta = \frac{1}{9}$, find $\sec \theta$. 9

34. Given $\cot \theta = 3$, find $\tan \theta$. $\frac{1}{3}$

35. Given $\cos \theta = \frac{1}{2}$, find $\cot \theta$. $\frac{1}{\sqrt{3}}$

36. Given $\sin \theta = \frac{4}{5}$, find $\cot \theta$. $\frac{3}{4}$

37. Given $\sin \theta = \frac{1}{3}$, find $\sec \theta$. $\frac{3}{2\sqrt{2}}$

38. Given $\tan \theta = \frac{1}{3}$, find $\cos \theta$. $\frac{3\sqrt{10}}{10}$

39. Given $\cot \theta = \frac{4}{3}$, find $\sin \theta$. $\frac{3}{5}$

40. Given $\sec \theta = 5$, find $\sin \theta$. $\frac{2\sqrt{6}}{5}$

In Exercises 41–55, use a calculator to approximate the value of the trigonometric function. Round your result to three decimal places.

41. sin 10° 0.174 **42.** csc 80° 1.015 **43.** tan 73° 3.271

44. cos 29° 0.875 **45.** cot 48° 0.900 **46.** sec 16° 1.040

47. cos 8° 0.990 **48.** tan 66° 2.246 **49.** sec 31° 1.167

50. cot 39° 1.235 **51.** csc 50° 1.305 **52.** sin 52° 0.788

53. sec 62° 2.130 **54.** cot 25° 2.145 **55.** csc 2° 28.654

56. *The Corinth Canal* The tugboat in the photograph is pulling a ship through the Corinth Canal, which is located near the city of Corinth in southern Greece. Between the points where the two tow lines are fastened to the ship, the ship is 50 feet wide. The angle between the tow lines is 22°. Approximate the length of the tow lines. 128.6 ft

The Corinth Canal links the Aegean and Ionian Seas. The canal is 4 miles long and has an average depth of 190 feet.

57. *Grouse Mountain* The Grouse Mountain ski slope in Avon, Colorado, rises at an angle of about 25.2°. How long is the ski slope? ≈4246.3 ft

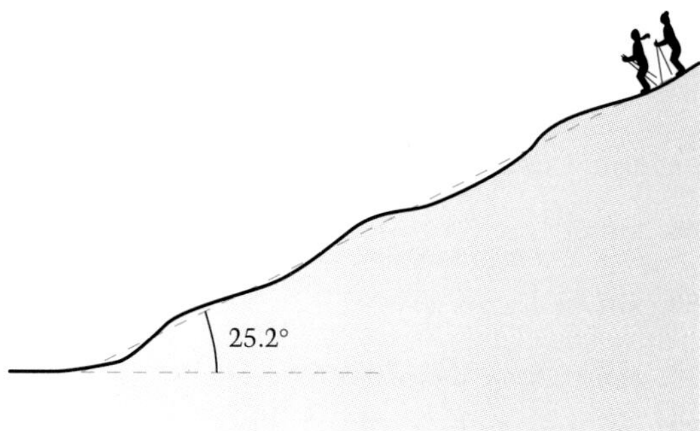

Grouse Mountain is a ski slope at Beaver Creek in Avon, Colorado. The vertical height of the slope is 1808 feet. Beaver Creek Ski Resort is only ten miles from Vail Ski Resort. Between the two resorts, skiers have a choice of 31 ski lifts and 178 trails.

(continued)
using a calculator, students must evaluate the reciprocals of the tan, cos, or sin of a given angle, respectively. (Be sure to tell students that keys labeled $\sin^{-1}$ do not execute the reciprocal of the sine function. They represent the inverse of the sine function.)

Unless otherwise stated, students should round their answers to three decimal places.

Independent Practice

▶ **Ex. 7–14** Students should be able to recognize some common Pythagorean triples and their multiples.

▶ **Ex. 15–22** These exercises require the use of the Pythagorean Theorem.

▶ **Ex. 23–40** GEOMETRY Students should draw right triangles to visualize the answers.

▶ **Ex. 41–55** See the Exercise Notes for Exercise 6.

▶ **Ex. 56** If students need help in solving this problem, explain this diagram and suggest that they use it.

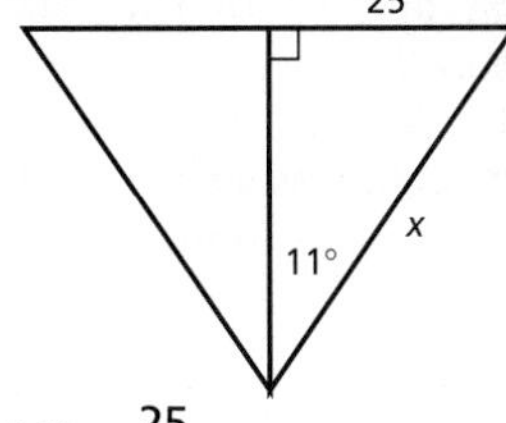

$\sin 11° = \frac{25}{x}$

EXTEND Ex. 56 Ask students to find out more about the Corinth Canal.

▶ **Ex. 58–59** TECHNOLOGY Assign these exercises as a pair. Encourage students to use their calculators to find the missing measures.

▶ **Ex. 65** It is a good idea to memorize some of the most common Pythagorean triples because questions on Pythagorean triples commonly appear on standardized tests.

▶ **Ex. 66–73** GROUP ACTIVITY You may choose to use these exercises as an in-class group activity.

Enrichment Activity

This activity requires students to go beyond lesson goals.

EXTEND Ex. 56–59 WRITING PROBLEMS Write your own right-triangle application problems. Provide a complete solution for each problem, including a diagram. Share your problems with your classmates.

Sample problem:
Last Saturday, Angela walked 10,000 feet from the bottom to the top of a hill in San Francisco. Her friend Walter stayed at the bottom of the hill. He measured the angle of elevation of the hill to be 18°. How much higher was Angela at the top of the hill than Walter?

$\sin 18° = \frac{x}{10{,}000} \approx 3090$

3090 feet

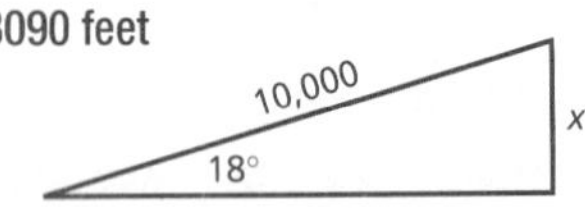

Note: The problems could be collected and included on tests or used as extra-credit problems. You might wish to credit the "author."

Duquesne Incline **In Exercises 58 and 59, use the following information.**

The Duquesne Incline, built in Pittsburgh in 1877, transports people up and down the side of a mountain in cable cars. The original cable cars are still in use, each holding about 20 people. The track is 800 feet long. The top of the incline is 400 feet above the base.

58. What is the angle of inclination of the Duquesne Incline? 30°

59. What is the slope of the track? ≈0.577

Integrated Review

In Exercises 60–63, find the length of the unlabeled side of the triangle.

60. $4\sqrt{5}$

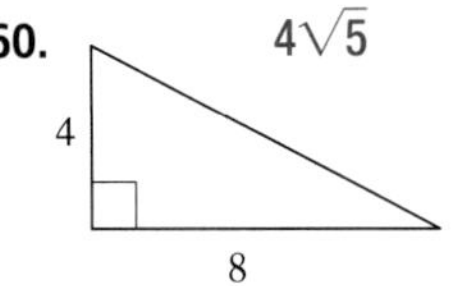

61. $3\sqrt{10}$

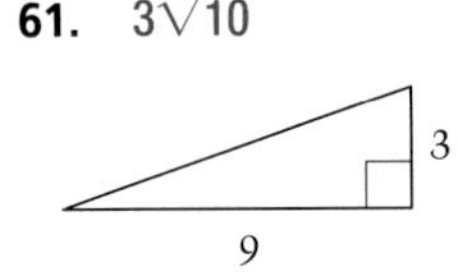

62. 4

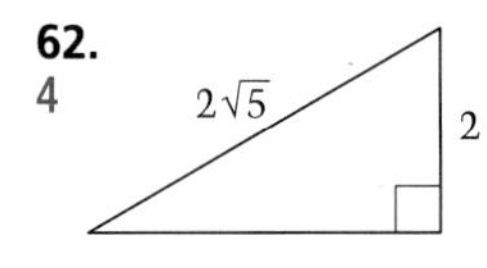

63. 5

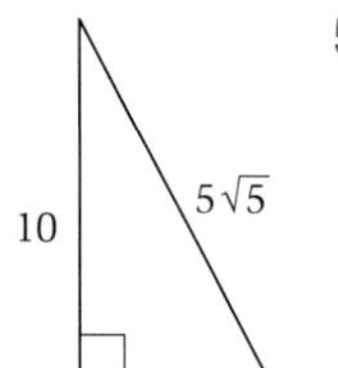

64. *College Entrance Exam Sample* Use the diagram to determine the relationship between the lengths of sides AC and AB. **b**

a. $AC < AB$ **b.** $AC = AB$
c. $AC > AB$

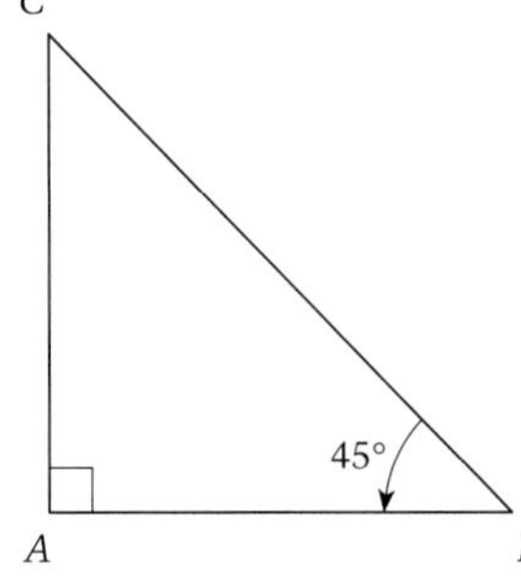

65. *College Entrance Exam Sample* The lengths of the sides of five triangles are given. Which is *not* a right triangle? **e**

a. 5, 12, 13 **b.** 3, 4, 5 **c.** 8, 15, 17
d. 9, 40, 41 **e.** 12, 15, 18

Exploration and Extension

True or False? **In Exercises 66–73, decide whether the statement is true or false. Explain your reasoning.** See Additional Answers.

66. $\sin 60° \csc 60° = 1$

67. $\sec 30° = \csc 60°$

68. $\sin 45° + \cos 45° = 1$

69. $\frac{\sin 60°}{\sin 30°} = \sin 2°$

70. $\tan[(0.8)^2] = \tan^2(0.8)$

71. $\tan \theta \csc \theta = \sec \theta$

72. $\cot \theta \sec \theta \sin \theta = 1$

73. $\csc \theta \sec \theta = \tan \theta$

13.2 Angles of Rotation and Radian Measure

What you should learn:

Goal 1 How to measure angles in standard position using degree measure and radian measure

Goal 2 How to use radian measure in real-life problems

Why you should learn it:

You will use radian measure for angles in most courses in science, engineering, and advanced mathematics.

Goal 1 Angles in Standard Position

In Lesson 13.1, the angle θ was always an acute angle—an angle whose measure was between 0° and 90°. In this lesson, you will study *angles of rotation,* whose measures can be any real number. Then, in Lesson 13.3, you will learn how to evaluate trigonometric functions of these angles.

An **angle of rotation** is determined by rotating a ray about its endpoint, or **vertex.** The starting position of the ray is the **initial side** of the angle. The position after rotation is the **terminal side.** In **standard position,** the vertex of the angle in the coordinate plane is at the origin, and the initial side is the positive x-axis.

Because *angles* and *angles of rotation* can be distinguished by how they are used, from now on we will refer to an angle of either type as simply an angle.

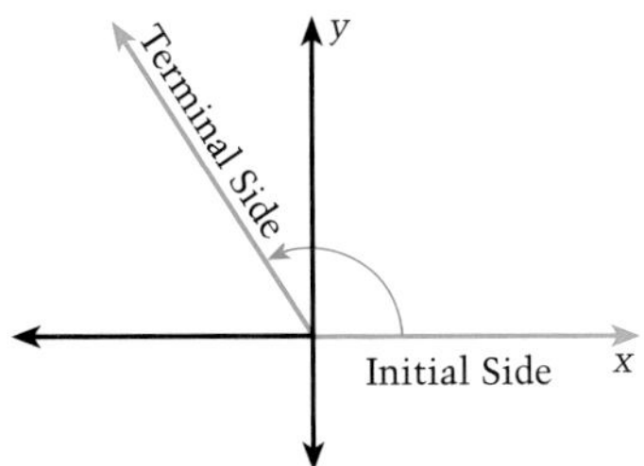

Standard Position of an Angle

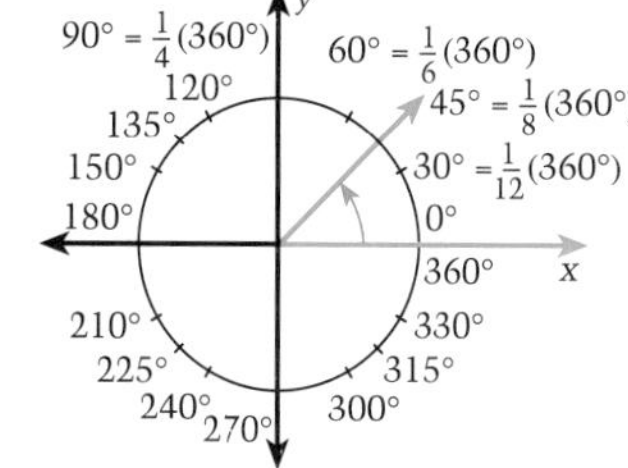

Degree Measure

The **measure of an angle** is determined by the amount of rotation from the initial side to the terminal side. A measure of **one degree (1°)** is equivalent to $\frac{1}{360}$ of a full revolution. To measure angles, it is convenient to mark degrees on a circle whose center is the origin. A full revolution (counterclockwise) corresponds to 360°, a half revolution to 180°, a quarter revolution to 90°, and so on.

The measure of an **acute** angle is between 0° and 90°. The measure of an **obtuse** angle is between 90° and 180°. **Positive angles** are generated by counterclockwise rotation, and **negative angles** by clockwise rotation.

Problem of the Day

While driving at a constant speed, Amir passed a milepost with a two-digit number. An hour later he passed another milepost with the same two digits, but in reverse order. An hour later he passed a third milepost with the same two digits, but with a zero between them. What was the number on the first milepost?
16

ORGANIZER

Warm-Up Exercises

1. Combine the terms. Express the results in exact form.
   a. $\frac{3\pi}{2} + \frac{4\pi}{6}$ $\quad \frac{13\pi}{6}$
   b. $\frac{\pi}{6} - \frac{5\pi}{12}$ $\quad -\frac{\pi}{4}$
   c. $\frac{23\pi}{5} - \frac{11\pi}{3}$ $\quad \frac{14\pi}{15}$
2. Which angles are
   a. complementary?
   b. supplementary?

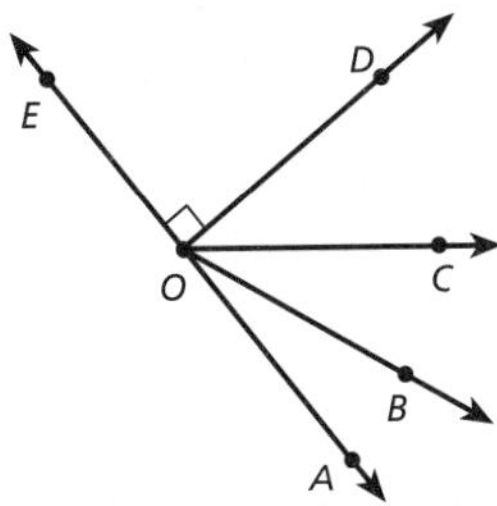

a. ∠AOB and ∠BOD; ∠AOC and ∠COD

b. ∠AOB and ∠BOE; ∠AOC and ∠COE; ∠AOD and ∠DOE

Lesson Resources

Teaching Tools
- Transparencies: 6, 25
- Problem of the Day: 13.2
- Warm-up Exercises: 13.2
- Answer Masters: 13.2

Extra Practice: 13.2
Technology Handbook: p. 91

LESSON Notes

Example 1

If the x-axis is considered the "west(−)-east(+)" line and the y-axis is considered the "south(−)-north(+)" line, then counterclockwise rotation from the x-axis can be considered "naturally" positive.

Examples 2–3

Radian measure is convenient because it is real-valued. That is, the sine of 30 radians is expressed as the sine of the number 30. Consequently, when trigonometric functions are defined for radian measure, the real number line can be used as the domain.

Be sure students understand how to use the relationship $180° = \pi$ radians. Since $180° = \pi$ radians, you get $180(1°) = \pi$ radians or $1° = \frac{\pi \text{ radians}}{180}$. So, to rewrite 135° as radians, carefully show the following.

$$135° = 135(1°)$$
$$= 135 \cdot \frac{\pi \text{ radians}}{180}$$
$$= \frac{135\pi}{180} \text{ radians}$$
$$= \frac{3\pi}{4} \text{ radians}$$

Similarly, using $180° = \pi$ radians, you find that $\pi \cdot 1$ radian $= 180°$ or $1 \text{ radian} = \frac{180°}{\pi}$.

So, to rewrite $\frac{9\pi}{2}$ radians as degrees, show the following.

$$\frac{9\pi}{2} \text{ radians} = \frac{9\pi}{2} \cdot 1 \text{ radian}$$
$$= \frac{9\pi}{2} \cdot \frac{180°}{\pi}$$
$$= \frac{9}{2}(180°) = 810°$$

Example 4

Ask students to determine how many feet you would move each minute. What is the circumference of the restaurant? about 4 feet per minute; about 251 feet

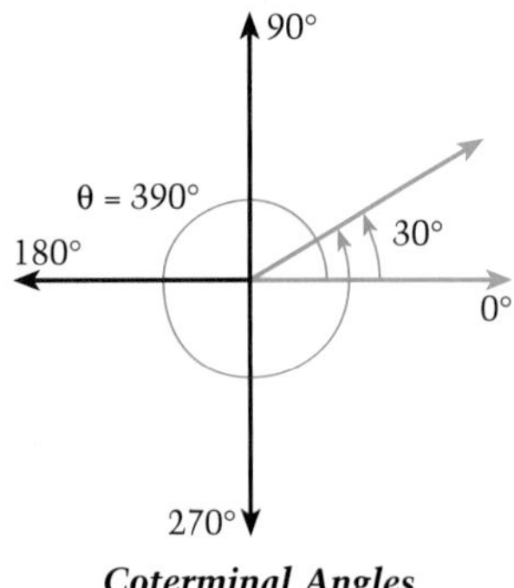

Coterminal Angles

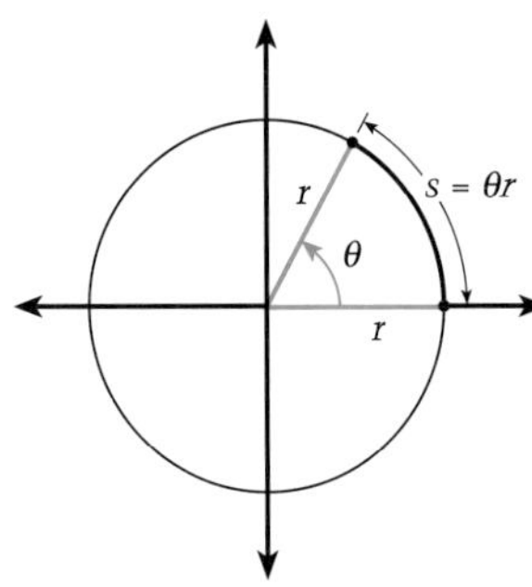

If s = r, then θ = 1 radian.

Two angles are **coterminal** if they have the same initial and terminal sides. For instance, the angles 30° and 390° (shown at the left) are coterminal, as are the angles 240° and −120° (a clockwise rotation of 120°). Coterminal angles have different measures, such as 240° and −120°, but describe the same angle.

The second way to measure angles is in terms of radians. To define a radian, consider a circle of radius r whose center is the origin. One **radian** is the measure of an angle θ whose terminal side intercepts an arc of length r. The *radian measure* of θ, the arc length s, and the radius r are related by the equation $s = \theta r$.

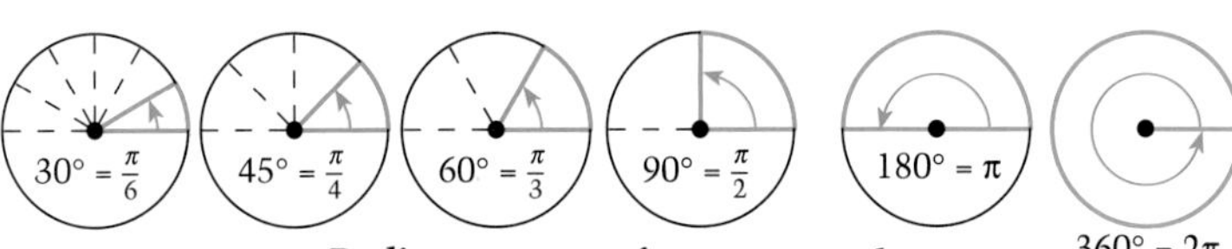

Radian measure of common angles

Two positive angles are **complementary** if the sum of their measures is $\frac{\pi}{2}$ radians (or 90°). Two positive angles are **supplementary** if the sum of their measures is π radians (or 180°).

Example 1 *Complementary, Supplementary, and Coterminal Angles*

a. The complement of $\theta = \frac{\pi}{12}$ is

$\frac{\pi}{2} - \frac{\pi}{12} = \frac{6\pi}{12} - \frac{\pi}{12} = \frac{5\pi}{12}$. (See diagram **a.** below.)

b. The supplement of $\theta = \frac{5\pi}{6}$ is

$\pi - \frac{5\pi}{6} = \frac{6\pi}{6} - \frac{5\pi}{6} = \frac{\pi}{6}$. (See diagram **b.** below.)

c. In radian measure, a coterminal angle is found by adding or subtracting 2π. For $\theta = \frac{17\pi}{6}$, you can find a coterminal angle by subtracting 2π to obtain

$\frac{17\pi}{6} - 2\pi = \frac{17\pi}{6} - \frac{12\pi}{6} = \frac{5\pi}{6}$.

Thus $\frac{17\pi}{6}$ and $\frac{5\pi}{6}$ are coterminal.

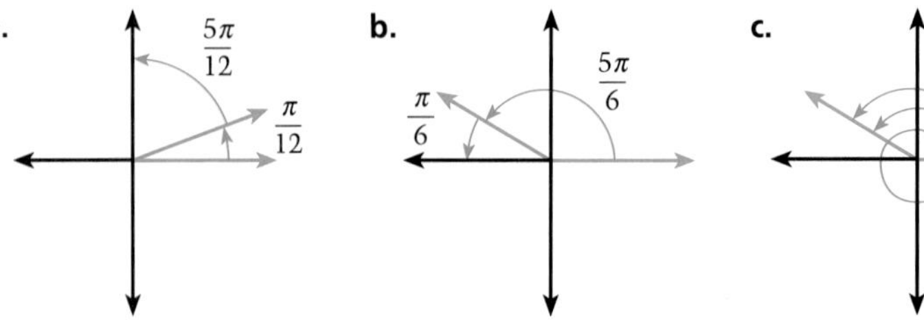

■

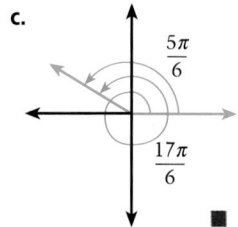

Degree measure and radian measure are related by the equation $360° = 2\pi$ radians, or

$$180° = \pi \text{ radians.}$$

From the latter equation, you can obtain the following conversions between degree and radian measure.

> **Conversions Between Degrees and Radians**
>
> **1.** To rewrite degrees as radians, multiply by $\dfrac{\pi \text{ radians}}{180°}$.
>
> **2.** To rewrite radians as degrees, multiply by $\dfrac{180°}{\pi \text{ radians}}$.
>
> *Note:* When no units of angle measure are specified, *radian measure is implied.* For instance, $\theta = 2$ means $\theta = 2$ radians.

To apply either of these two conversion rules, you simply need to remember the basic relationship $180° = \pi$ radians.

Example 2 *Rewriting Degrees as Radians*

a. $135° = (135 \text{ degrees})\left(\dfrac{\pi \text{ radians}}{180 \text{ degrees}}\right) = \dfrac{3\pi}{4}$ radians

b. $540° = (540 \text{ degrees})\left(\dfrac{\pi \text{ radians}}{180 \text{ degrees}}\right) = 3\pi$ radians

c. $-270° = (-270 \text{ degrees})\left(\dfrac{\pi \text{ radians}}{180 \text{ degrees}}\right) = -\dfrac{3\pi}{2}$ radians

a.

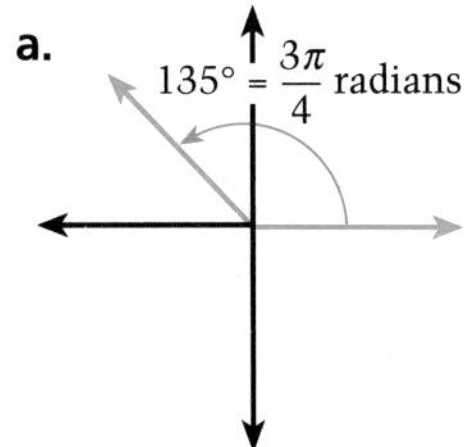

b.

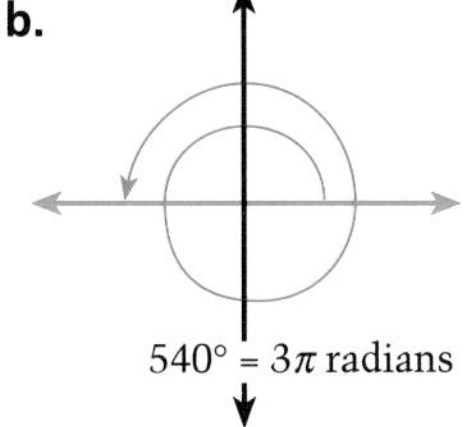

c.

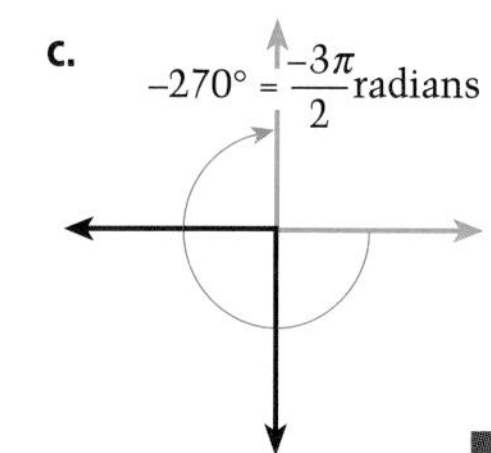

■

Example 3 *Rewriting Radians as Degrees*

a. $-\dfrac{\pi}{2}$ radians $= \left(-\dfrac{\pi}{2} \text{ radians}\right)\left(\dfrac{180 \text{ degrees}}{\pi \text{ radians}}\right) = -90°$

b. $\dfrac{9\pi}{2}$ radians $= \left(\dfrac{9\pi}{2} \text{ radians}\right)\left(\dfrac{180 \text{ degrees}}{\pi \text{ radians}}\right) = 810°$

c. $2 \text{ radians} = (2 \text{ radians})\left(\dfrac{180 \text{ degrees}}{\pi \text{ radians}}\right) = \dfrac{360}{\pi} \approx 114.59°$ ■

Extra Examples

Here are additional examples similar to **Examples 1–3.**

1. Find the complement of each angle.

a. $\theta = \dfrac{\pi}{8}$ **b.** $\theta = \dfrac{2\pi}{5}$

c. $\theta = \dfrac{3\pi}{10}$ **d.** $\theta = \dfrac{\pi}{6}$

a. $\frac{3\pi}{8}$, b. $\frac{\pi}{10}$, c. $\frac{\pi}{5}$, d. $\frac{\pi}{3}$

2. Find the supplement of each angle.

a. $\theta = \dfrac{3\pi}{8}$ **b.** $\theta = \dfrac{2\pi}{5}$

c. $\theta = \dfrac{7\pi}{10}$ **d.** $\theta = \dfrac{5\pi}{12}$

a. $\frac{5\pi}{8}$, b. $\frac{3\pi}{5}$, c. $\frac{3\pi}{10}$, d. $\frac{7\pi}{12}$

3. Find a coterminal angle for each given angle.

a. $\dfrac{19\pi}{6}$ **b.** $\dfrac{23\pi}{8}$

c. $\dfrac{26\pi}{10}$ **d.** $\dfrac{11\pi}{4}$

a. $\frac{7\pi}{6}$, b. $\frac{7\pi}{8}$, c. $\frac{3\pi}{5}$, d. $\frac{3\pi}{4}$

4. Rewrite each angle measure in radians.

a. 240° **b.** −80°

c. 340° **d.** 135°

a. $\frac{4\pi}{3}$, b. $-\frac{4\pi}{9}$, c. $\frac{17\pi}{9}$, d. $\frac{3\pi}{4}$

5. Rewrite each angle measure in degrees.

a. $\dfrac{5\pi}{8}$ radians 112.5°

b. 5 radians ≈ 286.48°

c. $-\dfrac{24\pi}{25}$ radians −172.8°

d. $\dfrac{16\pi}{5}$ radians 576°

Check Understanding

Describe when an angle is in standard position.
Vertex at origin; initial side on positive x-axis

Can a triangle contain two obtuse angles? Two acute angles?
No; yes

3. What is a measure of 1°?
$\frac{1}{360}$ of a revolution

4. How many radians are in a circle?
2π

5. How are negative angles represented?
Using clockwise rotation

Communicating about ALGEBRA

EXTEND Communicating
COOPERATIVE LEARNING

Suppose angles θ_1 and θ_2 are complementary angles, and angles θ_3 and θ_4 are supplementary. Ask students to work in small groups to investigate and form conjectures about the relationships among the six trigonometric functions evaluated at complementary and supplementary angles.

For any complementary angles θ_1 and θ_2, it can be shown that $\cos\theta_1 = \sin\theta_2$ and $\sin\theta_1 = \cos\theta_2$.

For any supplementary angles θ_3 and θ_4, it can be shown that $\cos\theta_3 = -\cos\theta_4$ and $\sin\theta_3 = \sin\theta_4$.

The other relationships can be derived in terms of the sine and cosine of the angles. For example:

For any complementary angles θ_1 and θ_2, it can be shown that $\cot\theta_1 = \tan\theta_2$ and $\csc\theta_1 = \sec\theta_2$.

For any supplementary angles θ_3 and θ_4, it can be shown that $\cot\theta_3 = -\cot\theta_4$ and $\sec\theta_3 = -\sec\theta_4$.

*The Seattle **Space Needle** was built in 1962 for the Seattle World's Fair. The restaurant at the top of the needle is circular, with a radius of 40 feet. The dining part of the restaurant (by the windows) revolves, making one complete revolution per hour.*

Goal 2 Using Radian Measure in Real Life

Example 4 *Finding an Angle and Arc Length*

You go to dinner with your family at the Seattle Space Needle. At 6:42 P.M., you take a seat at a window table. You finish dinner at 8:18 P.M. Through what angle did your position rotate during your stay? How many feet did your position revolve?

Solution You spent a total of 96 minutes at dinner. Because the Space Needle makes one revolution every 60 minutes, your angle of rotation was $\theta = \frac{96}{60}(2\pi) \approx \frac{16}{5}\pi$ radians. Because the radius of the restaurant is 40 feet, your position moved through an arc length of $s = \theta r = \frac{16}{5}\pi\,(40) \approx 402$ feet.

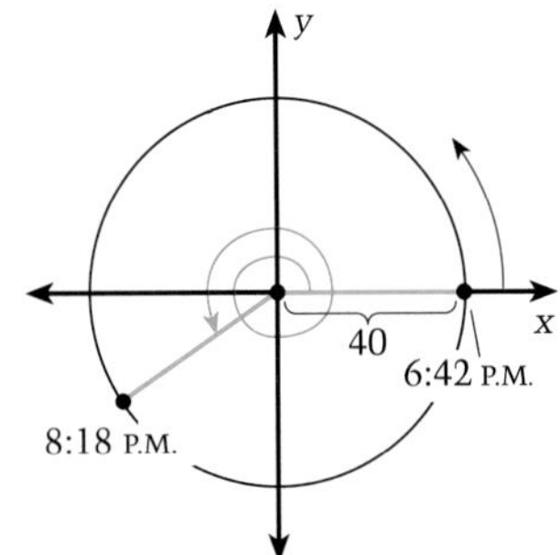

■

Communicating about ALGEBRA

SHARING IDEAS about the Lesson

Match the degree angle measure **(A–D)** with its radian angle measure **(E–H).** Then match the measure with the angle **(I–L).**

A. 20° **B.** 420° **C.** −60° **D.** −320°

E. $-\frac{\pi}{3}$ **F.** $\frac{\pi}{9}$ **G.** $\frac{7\pi}{3}$ **H.** $-\frac{16\pi}{9}$

I. **J.**

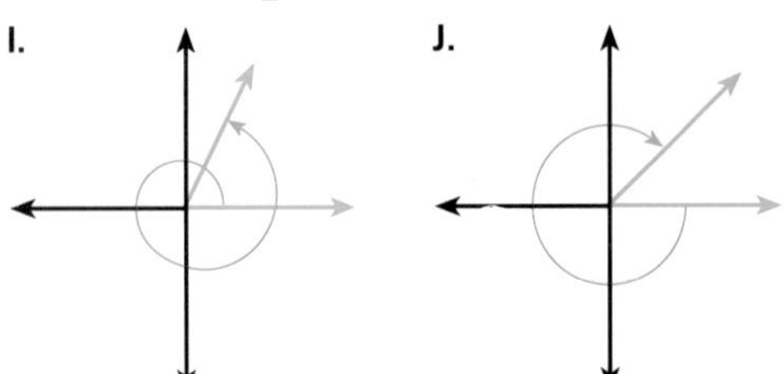

K.

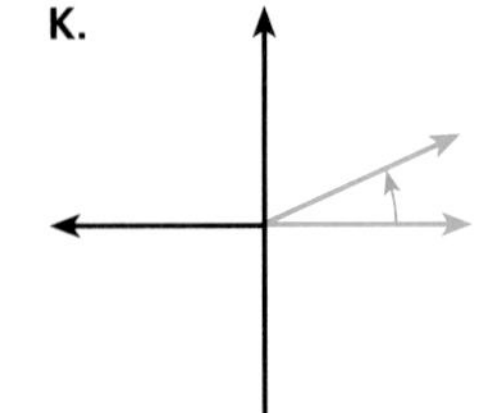

L.

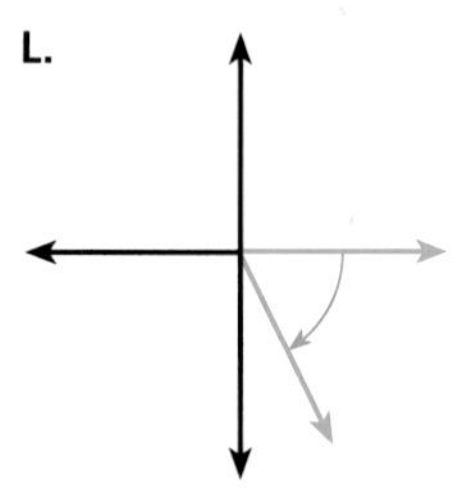

A, F, K; B, G, I; C, E, L; D, H, J

EXERCISES

Guided Practice

CRITICAL THINKING about the Lesson

1. What are the locations of the vertex and the initial side of an angle that is in standard position? Origin, positive x-axis
2. Match the angle θ with its terminal side. a, B; b, C; c, D; d, A
   a. $\theta = \frac{\pi}{2}$ **b.** $\theta = \pi$ **c.** $\theta = \frac{3\pi}{2}$ **d.** $\theta = 2\pi$
   A. Positive x-axis **B.** Positive y-axis **C.** Negative x-axis
   D. Negative y-axis
3. Find two angles, one with positive measure and the other with negative measure, that are coterminal with the angle 60°. 420°, −300°
4. A circle has a radius of r. What is the length of the are corresponding to a central angle of π radians? πr
5. Find the supplement of $\theta = \frac{\pi}{4}$. $\frac{3\pi}{4}$
6. Rewrite $\frac{3\pi}{2}$ radians as degrees. 270°

Independent Practice

In Exercises 7–14, decide whether the angles are coterminal.

7. 30°, 390° They are. **8.** 30°, −390° They are not. **9.** 280°, 640° They are. **10.** 35°, −335° They are not.

11. $\frac{\pi}{3}, -\frac{5\pi}{3}$ They are. **12.** $-\frac{11\pi}{6}, \frac{13\pi}{6}$ They are. **13.** $\frac{3\pi}{4}, -\frac{13\pi}{4}$ They are. **14.** $\frac{2\pi}{3}, -\frac{5\pi}{6}$ They are not.

In Exercises 15–22, find the complement of the angle.

15. 30° 60° **16.** 45° 45° **17.** 52° 38° **18.** 4° 86°

19. $\frac{2\pi}{9}$ $\frac{5\pi}{18}$ **20.** $\frac{5\pi}{12}$ $\frac{\pi}{12}$ **21.** $\frac{\pi}{6}$ $\frac{\pi}{3}$ **22.** $\frac{2\pi}{5}$ $\frac{\pi}{10}$

In Exercises 23–30, find the supplement of the angle.

23. 100° 80° **24.** 150° 30° **25.** 24° 156° **26.** 36° 144°

27. $\frac{2\pi}{9}$ $\frac{7\pi}{9}$ **28.** $\frac{4\pi}{5}$ $\frac{\pi}{5}$ **29.** $\frac{\pi}{3}$ $\frac{2\pi}{3}$ **30.** $\frac{3\pi}{16}$ $\frac{13\pi}{16}$

In Exercises 31–38, find two angles, one with positive measure and the other with negative measure, that are coterminal with the given angle.

31. 50° 410°, −310° **32.** 210° 570°, −150° **33.** 420° 60°, −300° **34.** 840° 120°, −240°

35. $\frac{13\pi}{3}$ $\frac{\pi}{3}, -\frac{5\pi}{3}$ **36.** $\frac{17\pi}{4}$ $\frac{\pi}{4}, -\frac{7\pi}{4}$ **37.** $\frac{24\pi}{7}$ $\frac{10\pi}{7}, -\frac{4\pi}{7}$ **38.** $\frac{26\pi}{3}$ $\frac{2\pi}{3}, -\frac{4\pi}{3}$

In Exercises 39–46, rewrite the degree measure as radians.

39. 20° $\frac{\pi}{9}$ **40.** 225° $\frac{5\pi}{4}$ **41.** 150° $\frac{5\pi}{6}$ **42.** 45° $\frac{\pi}{4}$

43. −110° $-\frac{11\pi}{18}$ **44.** 315° $\frac{7\pi}{4}$ **45.** 320° $\frac{16\pi}{9}$ **46.** 190° $\frac{19\pi}{18}$

EXERCISE Notes

ASSIGNMENT GUIDE

Basic/Average: Ex. 1–6, 9–19 odd, 27–53 odd, 55–62, 63, 65, 67–70

Above Average: Ex. 1–6, 9–19 odd, 25–37 odd, 43–51 odd, 55–62, 63, 65, 67–71

Above Average: Ex. 1–6, 11–21 odd, 25–37 odd, 43–51 odd, 55–62, 63, 65, 67–72

Selected Answers
Exercises 1–6, 7–69 odd

Use **Mixed Review** as needed.

More Difficult Exercises
Exercises 63–64, 71–72

Guided Practice

Ex. 1–6 You may use these exercises as a final check for understanding before assigning the Independent Practice. This lesson contains a lot of new terminology and procedures for students. If students have difficulty with these exercises, you may wish to assign the Independent Practice exercises over two days.

Independent Practice

Ex. 7–30 Be sure that students have a clear understanding of the meanings of the terms *coterminal*, *complement*, and *supplement*.

▶ **Ex. 39–54** Students can use Examples 2 and 3 as solution models.

▶ **Ex. 55–62** Assign these exercises as a group.

▶ **Ex. 63 and 65** Remind students to use the $s = \theta r$ formula that relates the radius r and radian measure of an angle θ to the intercepted arc of length s.

In Exercises 47–54, rewrite the radian measure as degrees.

47. $\frac{5\pi}{3}$ 300°
48. $-\frac{3\pi}{2}$ −270°
49. $\frac{\pi}{10}$ 18°
50. $\frac{\pi}{12}$ 15°

51. $\frac{7\pi}{15}$ 84°
52. $\frac{9\pi}{2}$ 810°
53. $-\frac{7\pi}{6}$ −210°
54. $\frac{8\pi}{5}$ 288°

In Exercises 55–62, match the measure with the angle.

55. $\frac{5\pi}{4}$ a
56. $460°$ c
57. $\frac{7\pi}{3}$ b
58. $-\frac{13\pi}{3}$ d

a. 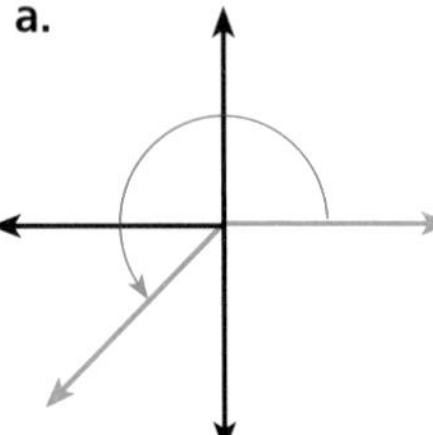
b. 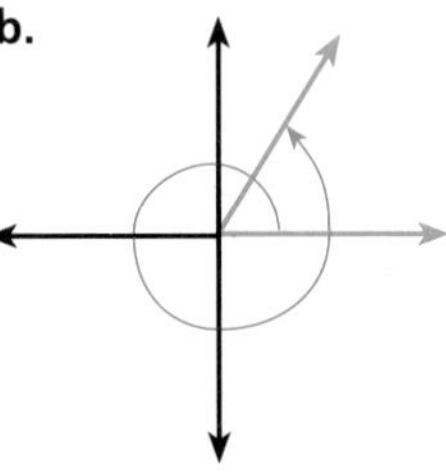
c. 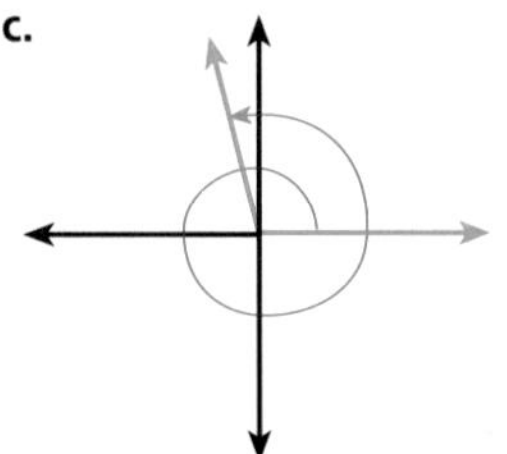
d. 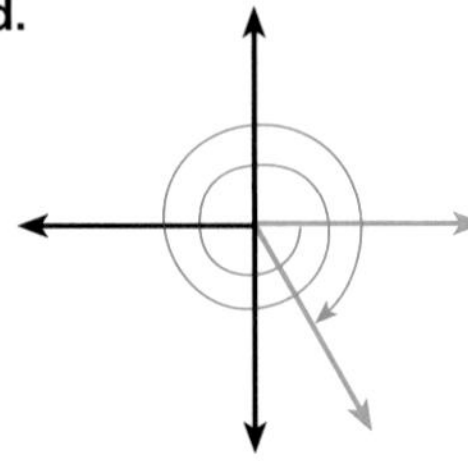

59. $\frac{5\pi}{3}$ f
60. $-210°$ h
61. $150°$ e
62. $-\frac{\pi}{2}$ g

e. 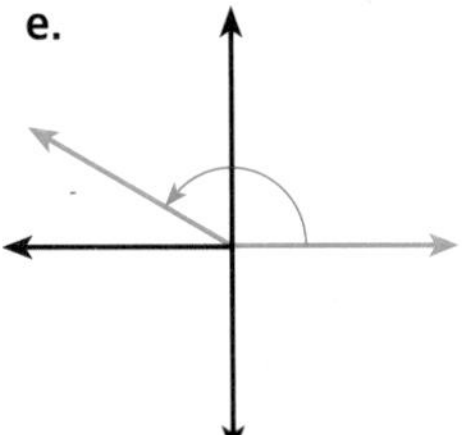
f. 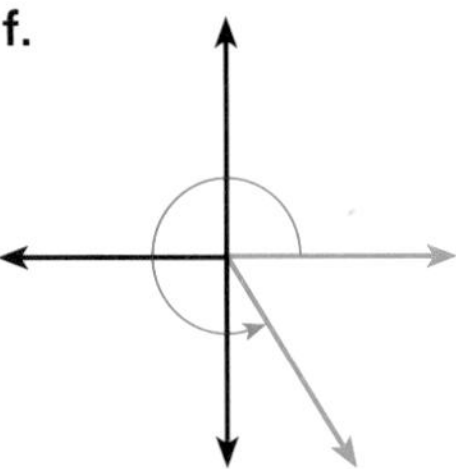
g. 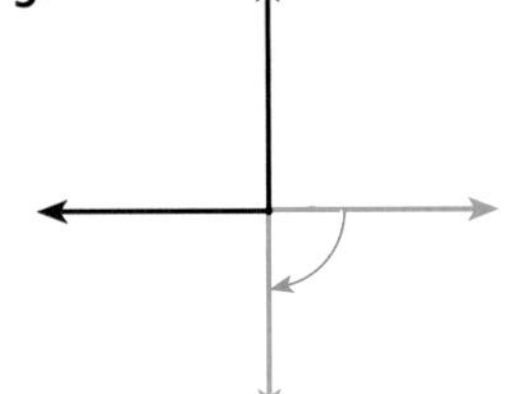
h. 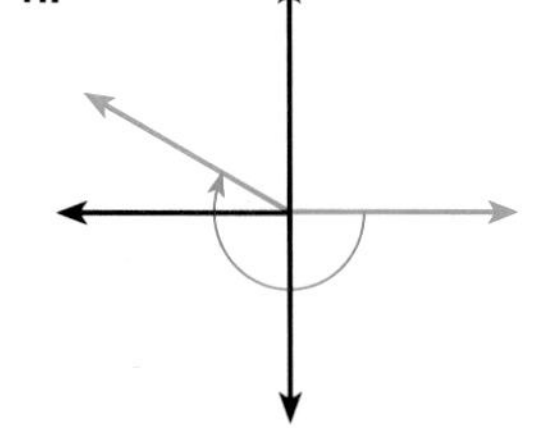

63. *Fire Truck Ladder* For the ladder on a fire truck to operate properly, the base of the ladder must be almost level. The diagram at the right shows part of a leveling device that is used to determine whether the ladder's base is level enough—the ball of mercury must be within the allowable range. Find the length of the arc that describes the allowable range. $\frac{7\pi}{8}$

Top View

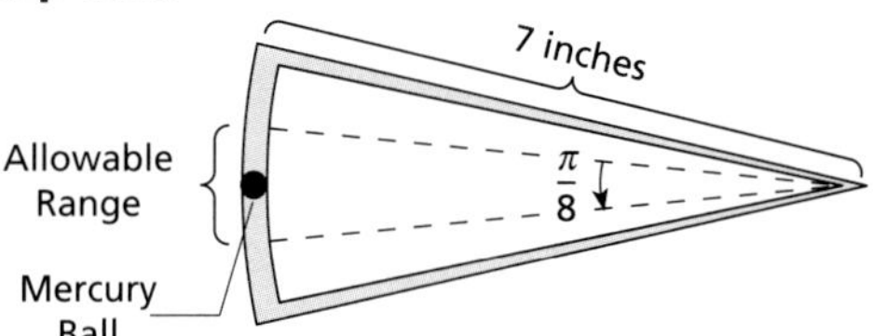

64. *Rotating a Bridge* A rotating bridge is on an island in the middle of a river. Six circular gears, each with a $2\frac{1}{2}$-foot diameter, are attached to the underside of the bridge. As the gears revolve, they roll around the circumference of a stationary circular plate on the island. The circular plate has a diameter of 40 feet. How many revolutions must each of the smaller gears make to rotate the bridge 90°? 4

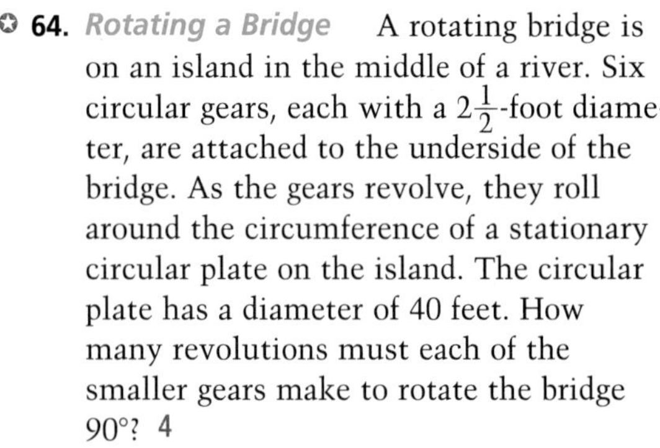

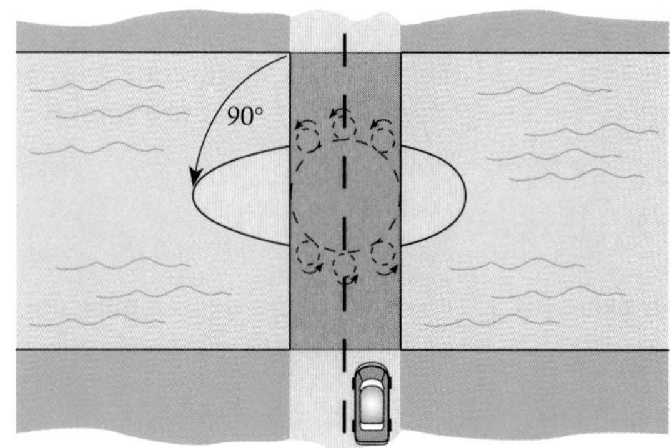

Spiral Stairs **In Exercises 65 and 66, use the following information.**

The freestanding spiral stairs at the right have 13 steps, including the top landing step. Each "pie-shaped" step has a radius of 36 inches and has a central angle of $\frac{\pi}{7}$.

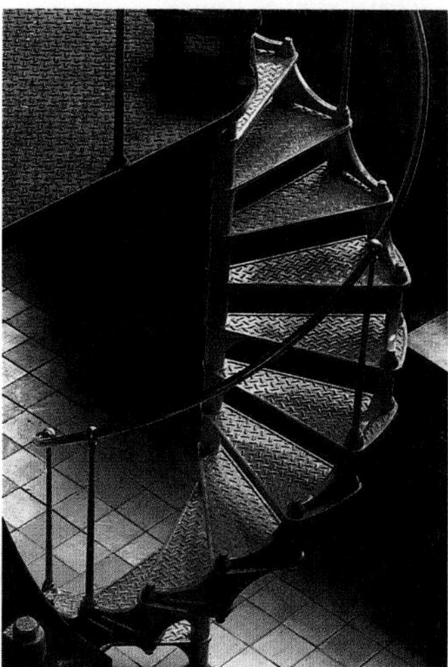

65. What is the arc length of the outer edge of each step? $\frac{36\pi}{7}$ in.

66. Through what angle would you turn by climbing the stairs? Assume that you start on the ground floor and end on the top landing step. $\frac{13\pi}{7}$

Integrated Review

67. Complete the table.

θ	$\frac{\pi}{6}$	$\frac{\pi}{4}$	$\frac{\pi}{3}$
sin θ	?	?	?
cos θ	?	?	?
tan θ	?	?	?

$\frac{1}{2}, \frac{\sqrt{2}}{2}, \frac{\sqrt{3}}{2}$; $\frac{\sqrt{3}}{2}, \frac{\sqrt{2}}{2}, \frac{1}{2}$; $\frac{\sqrt{3}}{3}, 1, \sqrt{3}$

68. Complete the table.

θ	$\frac{\pi}{6}$	$\frac{\pi}{4}$	$\frac{\pi}{3}$
csc θ	?	?	?
sec θ	?	?	?
cot θ	?	?	?

$2, \sqrt{2}, \frac{2\sqrt{3}}{3}$; $\frac{2\sqrt{3}}{3}, \sqrt{2}, 2$; $\sqrt{3}, 1, \frac{\sqrt{3}}{3}$

69. *Powers of Ten* Which best estimates the distance (in feet) traveled by a bicycle tire that makes 10 revolutions? **c**

a. 6.5×10^{-1} **b.** 6.5×10^{0} **c.** 6.5×10^{1} **d.** 6.5×10^{2}

70. *Powers of Ten* Which best estimates the angle (in radians) through which Earth rotates on its axis in a week? **b**

a. 4.4×10^{0} **b.** 4.4×10^{1} **c.** 4.4×10^{2} **d.** 4.4×10^{3}

Exploration and Extension

✪ **71.** *Geometry* Find the angle θ. 18°

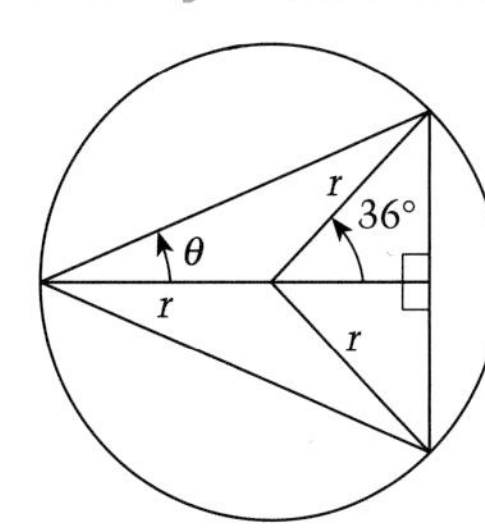

✪ **72.** *Geometry* Explain how the result of Exercise 71 can be used to find the angle measure of a point on the star below.

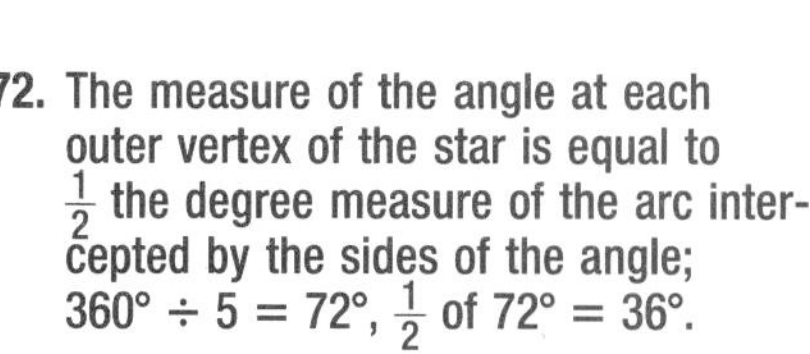

72. The measure of the angle at each outer vertex of the star is equal to $\frac{1}{2}$ the degree measure of the arc intercepted by the sides of the angle; $360° \div 5 = 72°$, $\frac{1}{2}$ of $72° = 36°$.

▶ **Ex. 67–68** These exercises should be done using mental math techniques.

▶ **Ex. 71** GEOMETRY Give students the diagram shown below. Ask them if they recall some properties from geometry that could be used to find the measure of angle 1.

Given: Triangle *ABC* is an isosceles triangle. Exterior angle *BCD* = 36°.

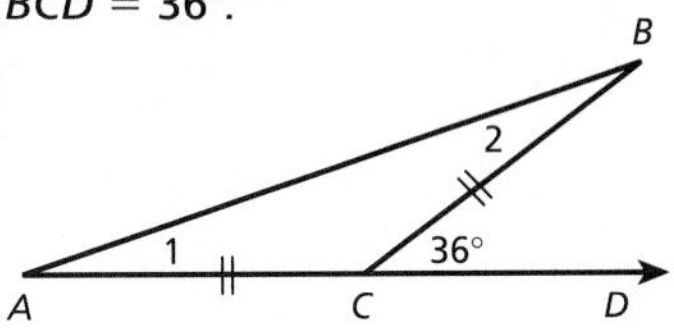

By the exterior angle theorem, $\angle 1 + \angle 2 = 36$.

Because triangle *ABC* is isosceles, $\angle 1 = \angle 2$.

So, $\angle 1 + \angle 2 = 36$
$\angle 1 + \angle 1 = 36$
$2(\angle 1) = 36$
$\angle 1 = 18$

The measure of angle 1 is 18°.

Enrichment Activities

These activities require students to go beyond lesson goals.

An Alternate Approach
Using Technology
The following program can be used on a graphics calculator to convert between degrees and radians.

```
Prgm #:DEGTORAD
:ClrHome
:Disp "DEGREES"
:Input D
:D π/180 → R
:Disp "RADIANS = "
:Disp R
```

1. How would you change the program to express radians as degrees?

```
Prgm #:RADTODEG
:ClrHome
:Disp "RADIANS"
:Input R
:180R/π → D
:Disp "DEGREES = "
:Disp D
```

2. Use the programs to check your answers to Exercises 39–54. (If you do not have access to graphics calculators, students could write similar BASIC programs to use on a computer.)

Answers

3.
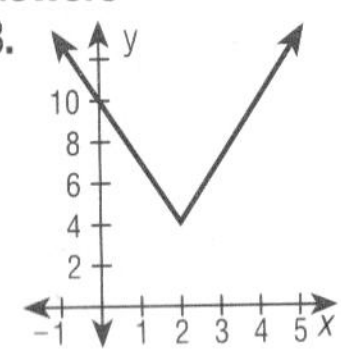

15.
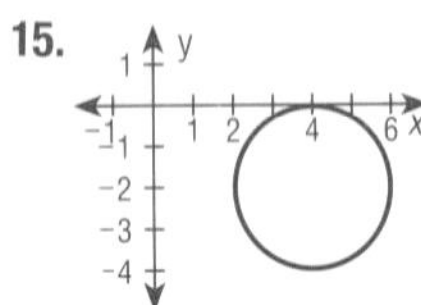

Mixed REVIEW

9. $x^3 - 4x^2 + 16x - 64$ 11. $(x - 1)(x + 1)(x + 2)$
15. See margin. 17. $xy = 3$ 18. $y = -3$

1. Solve $|x + 2| \geq 7$. **(1.7)** $x \leq -9$ or $x \geq 5$
2. Evaluate $\left(\frac{1}{6}\right)^y$ when $y = 4$. **(1.2)** $\frac{1}{1296}$
3. Sketch $y = 3|x - 2| + 4$. **(2.6)** See margin.
4. Solve $3x^2 + 8x + 9 = 0$. **(5.4)** $\frac{-4 \pm i\sqrt{11}}{3}$
5. Evaluate the determinant: $\begin{vmatrix} 6 & -10 \\ 14 & 12 \end{vmatrix}$. **(4.3)** 212
6. Solve $\begin{cases} 3x + \frac{1}{4}y = 5 \\ \frac{1}{4}x - 6y = -\frac{93}{8}. \end{cases}$ **(3.2)** $\left(\frac{3}{2}, 2\right)$
7. Find the median: 12, 3, 9, 14, 27, 36. **(6.7)** 13
8. Simplify: $3 \log_5 10 + 7 \log_5 4$. **(8.3)**
9. Divide: $x^4 - 256$ by $x + 4$. **(9.4)**
10. Solve $\log_2 x = 6$. **(8.2)** 64
11. Factor $x^3 + 2x^2 - x - 2$. **(9.3)**
12. Evaluate $\binom{6}{4}$. **(12.5)** 15
13. Evaluate $\sum_{n=1}^{5} (3n + 2)$. **(12.1)** 55
14. Solve $\frac{3x}{16} - \frac{3}{2} = -\frac{24}{x}$. **(10.4)** $4 \pm 4i\sqrt{7}$
15. Sketch $(x - 4)^2 + (y + 2)^2 = 4$. **(11.5)**
16. Name the conic: $4x^2 + 9y^2 = 36$. **(11.3)** Ellipse
17. Write an equation: x and y vary inversely when $x = 6$ and $y = \frac{1}{2}$. **(10.2)**
18. Write an equation for the line through the point $(4, -3)$ with a slope of 0. **(2.2)**
19. Is the inverse of $f(x) = 3x^2 + 2x + 1$ a function? **(6.3)** No
20. For $f(1) = 6$ and $f(n + 1) = f(n) + 3n$, find $f(4)$. **(6.6)** 24

8. $\log_5(10^3 \cdot 4^7)$ or $\log_5 16{,}384{,}000$

Career Interview

Dermatologist

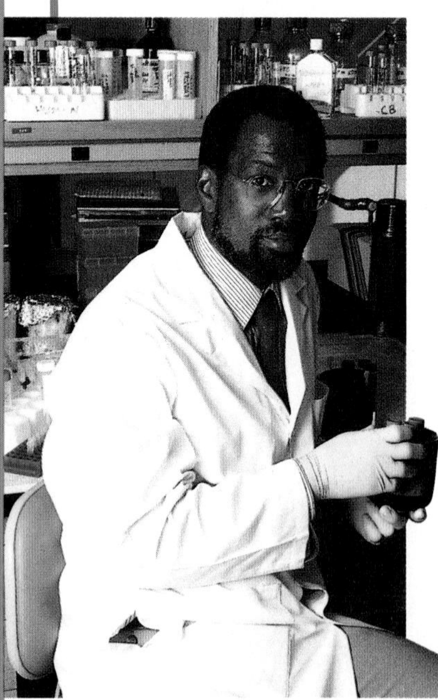

Michael Bigby is a dermatologist. Dr. Bigby spends 80% of his time on research related to the skin and divides his remaining time teaching resident dermatologists and treating patients.

Q: ***What is the focus of your research?***
A: I study T-lymphocytes, which are a group of white blood cells, and their effects on skin diseases, such as poison ivy.

Q: ***Does having a math background help you on the job?***
A: It's a tremendous help in my work. The application of math is basic to data collection. My research involves different measurements and calculations. For example, measuring the amount of light absorbed in a solution, or counting cells in a special volume.

Q: ***How much of your work is done by computers?***
A: In the technology of research, everything is computerized from counters to spectrophotometers to scanners. Easily 90% of the actual math is done by computers.

Q: ***Does this mean you do not need to understand math?***
A: The opposite! You have to understand the math if you want to know what to do with the computers and how to work with the numbers.

13.3 Evaluating Trigonometric Functions

What you should learn:

Goal 1 How to evaluate trigonometric functions of any angle

Goal 2 How to use trigonometric functions to solve real-life problems

Why you should learn it:

You can model many real-life problems with trigonometric functions, such as the horizontal distance that a projectile travels.

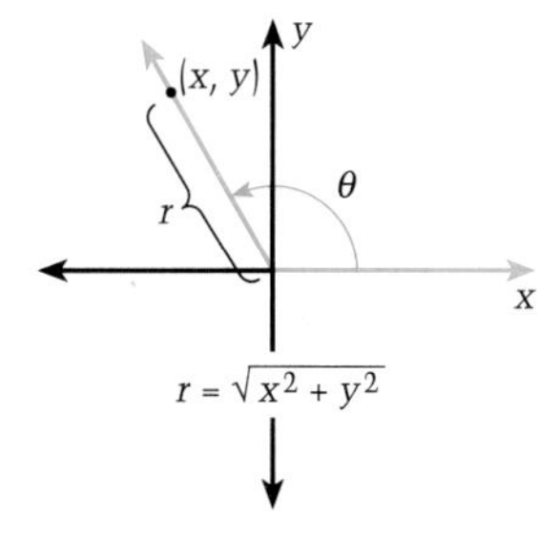

Goal 1 Evaluating Functions of Any Angle

In Lesson 13.1, you learned how to evaluate trigonometric functions of an acute angle. In this lesson, you will learn how to evaluate trigonometric functions of *any* angle.

Evaluating Trigonometric Functions

Let θ be an angle in standard position with any point (x, y) (except the origin) on the terminal side of θ. Let $r = \sqrt{x^2 + y^2}$.

$$\sin\theta = \frac{y}{r} \qquad \cos\theta = \frac{x}{r} \qquad \tan\theta = \frac{y}{x},\quad x \neq 0$$

$$\csc\theta = \frac{r}{y},\quad y \neq 0 \qquad \sec\theta = \frac{r}{x},\quad x \neq 0 \qquad \cot\theta = \frac{x}{y},\quad y \neq 0$$

For acute angles, these definitions give the same values as those given by the definitions in Lesson 13.1.

Because $r = \sqrt{x^2 + y^2}$ cannot be zero, the sine and cosine functions are defined for any angle θ. If $x = 0$, the tangent and secant of θ are undefined. For example, the tangent of $90°$ $\left(\text{or } \frac{\pi}{2}\right)$ is undefined. Similarly, if $y = 0$, the cotangent and cosecant of θ are undefined.

Example 1 *Evaluating Trigonometric Functions*

Let $(-3, 4)$ be a point on the terminal side of θ. Find the sine, cosine, and tangent of θ.

Solution Using $x = -3$ and $y = 4$, you can determine that

$$r = \sqrt{x^2 + y^2} = \sqrt{(-3)^2 + 4^2} = \sqrt{25} = 5.$$

From these values of x, y, and r, you can write the following.

$$\sin\theta = \frac{y}{r} = \frac{4}{5}$$

$$\cos\theta = \frac{x}{r} = \frac{-3}{5} = -\frac{3}{5}$$

$$\tan\theta = \frac{y}{x} = \frac{4}{-3} = -\frac{4}{3}$$ ■

Problem of the Day

One angle of the 4-inch square is trisected. What is the area of the shaded region?

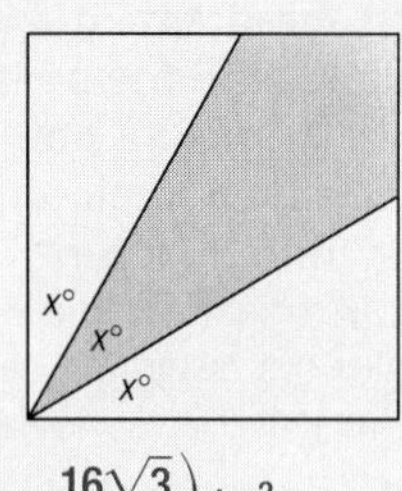

$\left(16 - \frac{16\sqrt{3}}{3}\right)$ in.², or ≈ 6.76 in.²

ORGANIZER

Warm-Up Exercises

Find $c = \sqrt{a^2 + b^2}$.

a. $a = 4, b = 6$ $\quad c \approx 7.21$

b. $a = 8, b = 12$ $\quad c \approx 14.42$

c. $a = 11, b = 23$ $\quad c \approx 25.50$

d. $a = 3, b = 5$ $\quad c \approx 5.83$

Lesson Resources

Teaching Tools
- Transparencies: 2, 3, 6, 25
- Problem of the Day: 13.3
- Warm-up Exercises: 13.3
- Answer Masters: 13.3

Extra Practice: 13.3

Color Transparencies: 67, 68, 69

Applications Handbook: p. 3, 46–48

Technology Handbook: p. 93

LESSON Notes

Example 1

Have students draw an angle in the first quadrant of a coordinate plane with the point (3, 4) on the terminal side. Have them compare their angle to the second-quadrant angle shown in this example.
The angles are congruent mirror images in the *y*-axis.

Example 2

Remind students that $\theta = 2.3$ is measured in radians and not degrees. If it were measured in degrees, then it would read: $\theta = 2.3°$. Students may find it helpful to know that 1 radian is about 57°.

Point out that before the reference angle of a negative angle is determined, you should rewrite the negative angle as a positive coterminal angle.

Example 3

Students should observe that each of the given angles has one of the special angles, 60°, 30°, or 45°, as reference angles.

Ask students why reference angles are important.
Reference angles are important because they facilitate the evaluation of trigonometric functions without the aid of a calculator or other computational tool.

It is often helpful to sketch the given angle on a coordinate plane when deciding what its reference angle is.

The values of the trigonometric functions of angles greater than 90° (or less than 0°) can be determined from their values at corresponding acute angles called **reference angles.**

The Reference Angle of an Angle

Let θ be an angle in standard position. Its **reference angle** is the acute angle θ' (read *theta prime*) formed by the terminal side of θ and the *x*-axis.

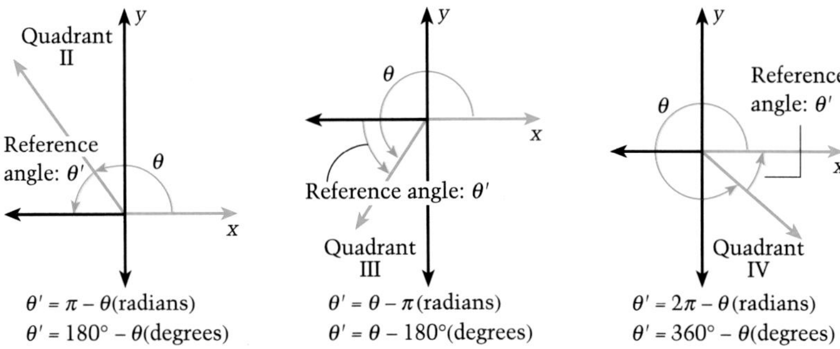

Example 2 *Finding Reference Angles*

Find the reference angle θ' for each of the following.

a. $\theta = 300°$ **b.** $\theta = -\frac{3\pi}{4}$ **c.** $\theta = 2.3$

Solution

a. Because $\theta = 300°$ lies in Quadrant IV, the reference angle is

$\theta' = 360° - 300° = 60°.$

b. First determine that $\theta = -\frac{3\pi}{4}$ is coterminal with $\frac{5\pi}{4}$, which lies in Quadrant III. The reference angle is

$\theta' = \frac{5\pi}{4} - \pi = \frac{\pi}{4}.$

c. Because $\theta = 2.3$ lies between $\frac{\pi}{2} \approx 1.57$ and $\pi \approx 3.14$, it follows that θ is in Quadrant II. The reference angle is

$\theta' = \pi - 2.3 \approx 0.8416.$

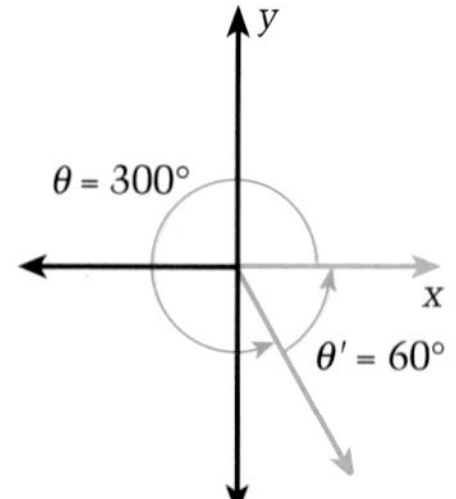

a. θ in Quadrant IV

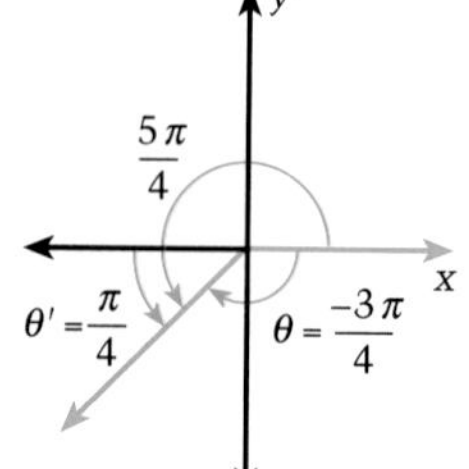

b. θ in Quadrant III

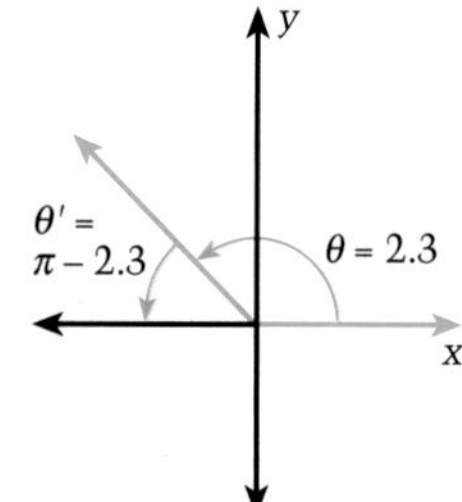

c. θ in Quadrant II ■

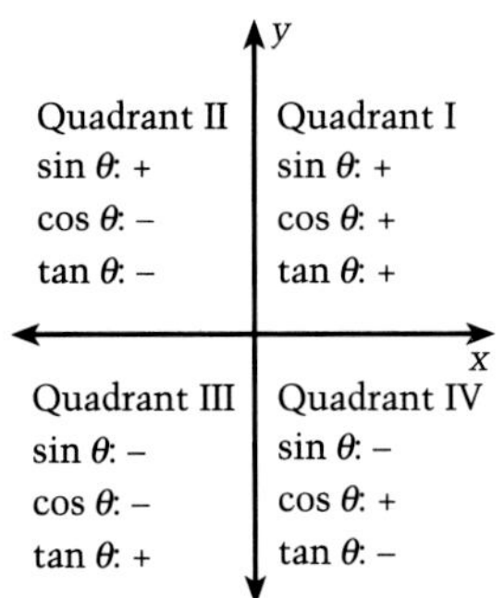

Quadrant Signs

The *signs* of the trigonometric function values in the four quadrants can be determined from the definitions of the functions. For instance, because $\cos\theta = \frac{x}{r}$, it follows that $\cos\theta$ is positive wherever $x > 0$, which is in Quadrants I and IV. (Remember, r is always positive.)

Evaluating Trigonometric Functions of Any Angle

Use the following steps to evaluate a trigonometric function of any angle θ.

1. Find the reference angle θ'.
2. Evaluate the trigonometric function for the angle θ'.
3. Use the quadrant in which θ lies to determine the sign of the trigonometric function of θ.

Example 3 ***Evaluating Trigonometric Functions of Any Angle***

Evaluate the following.

a. $\cos\frac{4\pi}{3}$ **b.** $\tan(-210°)$ **c.** $\csc\frac{11\pi}{4}$

Solution

a. Because $\theta = \frac{4\pi}{3}$ lies in Quadrant III, the reference angle is

$\theta' = \frac{4\pi}{3} - \pi = \frac{\pi}{3}.$

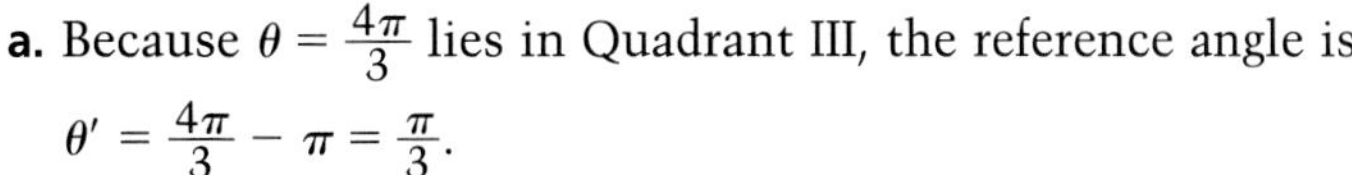

The cosine is negative in Quadrant III, so you can write

$\cos\frac{4\pi}{3} = -\cos\frac{\pi}{3} = -\frac{1}{2}.$

b. Because $-210° + 360° = 150°$, it follows that $-210°$ is coterminal with the second-quadrant angle 150°. The reference angle is

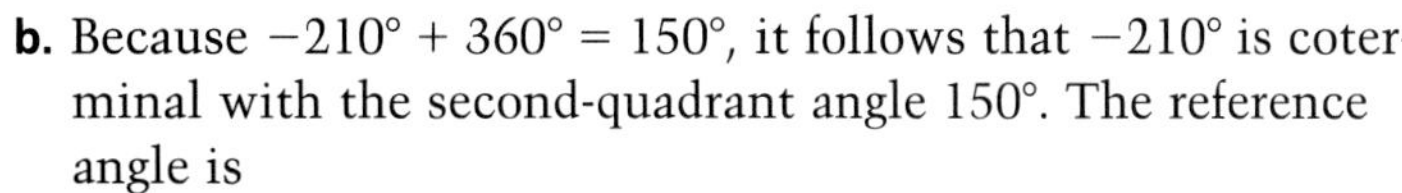

$\theta' = 180° - 150° = 30°.$

The tangent is negative in Quadrant II, so you can write

$\tan(-210°) = -\tan 30° = -\frac{\sqrt{3}}{3}.$

c. Because $\frac{11\pi}{4} - 2\pi = \frac{3\pi}{4}$, it follows that $\frac{11\pi}{4}$ is coterminal with the second-quadrant angle $\frac{3\pi}{4}$. Therefore, the reference angle is

$\theta' = \pi - \frac{3\pi}{4} = \frac{\pi}{4}.$

The cosecant is positive in Quadrant II, so you can write

$\csc\frac{11\pi}{4} = +\csc\frac{\pi}{4} = \sqrt{2}.$ ■

a. (figure: $\theta = \frac{4\pi}{3}$, $\theta' = \frac{\pi}{3}$)

b. (figure: $\theta' = 30°$, $\theta = -210°$)

c.

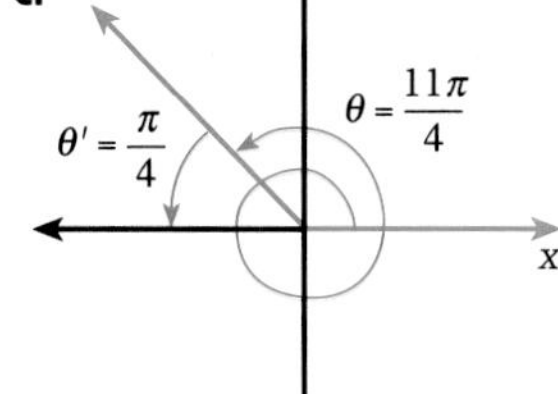

Extra Examples

Here are additional examples similar to **Examples 1–3.**

1. Find the sine, cosine, and tangent of the angle θ if the given point lies on its terminal side.
   a. $(-5, 12)$ **b.** $(8, -6)$
   a. $\sin\theta = \frac{12}{13}$
   $\cos\theta = -\frac{5}{13}$
   $\tan\theta = -\frac{12}{5}$
   b. $\sin\theta = -\frac{6}{10}$
   $\cos\theta = \frac{8}{10}$
   $\tan\theta = -\frac{6}{8}$
2. Find the reference angle θ' for each of the following.
   a. $\theta = 340°$ $\theta' = 20°$
   b. $\theta = 3.5$ $\theta' \approx 0.358$
   c. $\theta = \frac{4\pi}{5}$ $\theta' = \frac{\pi}{5}$
   d. $\theta = \frac{3\pi}{5}$ $\theta' = \frac{2\pi}{5}$
3. Evaluate the following.
   a. $\cot\left(\frac{5\pi}{3}\right)$ $-\frac{\sqrt{3}}{3}$
   b. $\sin(-310°)$ ≈ 0.766
   c. $\sec\left(\frac{13\pi}{6}\right)$ $\frac{2\sqrt{3}}{3}$
   d. $\cos(315°)$ $\frac{\sqrt{2}}{2}$

Check Understanding

1. What are reference angles?
   See page 694.
2. Describe how you would find the sine of any angle.
   See page 695.
3. In which quadrants is the cotangent positive?
   First and third

Communicating about ALGEBRA

It might be interesting to analyze the swings of some hitters to see if their homerun swings are at a 45° angle with the ground.

What is the minimal initial velocity in miles per hour?
113.14 feet per second ≈ 77.14 miles per hour.

EXTEND *Communicating*
COOPERATIVE LEARNING
Have groups of students evaluate the following to look for relationships among angle measures and among values of their trigonometric functions.

1. a. $\cos \pi$ b. $\cos 3\pi$
   c. $\cos 5\pi$
2. a. $\cos 2\pi$ b. $\cos 4\pi$
   c. $\cos 6\pi$
3. a. $\sin \frac{7\pi}{3}$ b. $\sin \frac{\pi}{3}$
   c. $\sin 420°$ d. $\sin 60°$
4. a. $\sec 540°$ b. $\frac{1}{\cos 540°}$
   c. $\sec 180°$

From 1a–c, trigonometric functions of odd multiples of π are equivalent (all cosines equal -1).

From 2a–c, trigonometric functions of even multiples of π are equivalent (all cosines equal 0).

From 3a–d, trigonometric functions of angle measures given in degrees are equivalent to trigonometric functions of angle measures given in radians $\left(\text{all sines equal } \frac{\sqrt{3}}{2}\right)$.

From 4a–b, the cosine and secant are reciprocal trigonometric functions (both equal -1).

From 3a–d and from 4a and c, trigonometric functions of θ, $\theta + 2\pi$, and $\theta + 360°$ are all equivalent (all sines equal $\frac{\sqrt{3}}{2}$; both secants equal -1).

The horizontal distance, d (in feet), traveled by a projectile that is launched from ground level with an initial speed, v (in feet per second), is given by $d = \frac{v^2}{32} \sin 2\theta$ *where θ is the angle at which the projectile is launched.*

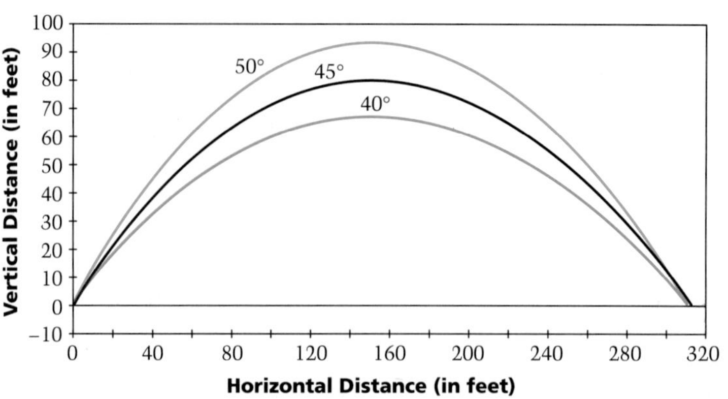

Goal 2 Using Trig Functions in Real Life

Real Life
Baseball

Example 4 *Finding How Far a Projectile Travels*

You are at bat and hit the baseball so that it has an initial velocity of 100 feet per second. Approximately how far will the ball travel horizontally for the following angles of elevation?

a. 40° **b.** 45° **c.** 50°

Solution

a. For $\theta = 40°$, the distance traveled horizontally is

$$d = \frac{100^2}{32} \sin(2 \cdot 40°) \quad \textit{Substitute 100 for v and 40° for } \theta.$$
$$= 312.5 \sin 80° \quad \textit{Simplify.}$$
$$\approx 307.75 \text{ feet.} \quad \textit{Use a calculator.}$$

b. For $\theta = 45°$, the distance traveled horizontally is

$$d = \frac{100^2}{32} \sin(2 \cdot 45°) \quad \textit{Substitute 100 for v and 45° for } \theta.$$
$$= 312.5 \sin 90° \quad \textit{Simplify.}$$
$$= 312.5 \text{ feet.} \quad \textit{Use } \sin 90° = 1.$$

c. For $\theta = 50°$, the distance traveled horizontally is

$$d = \frac{100^2}{32} \sin(2 \cdot 50°) = 312.5 \sin 100° \approx 307.75 \text{ feet.}$$

■

Communicating about ALGEBRA

SHARING IDEAS about the Lesson

What is the minimum initial velocity that you would need to hit a baseball a horizontal distance of at least 400 feet? Explain your reasoning. ≈113.14 feet/second

The baseball goes farthest when the angle of launching is 45°, so substitute 45° for θ in the formula, and substitute 400 for d; then solve for v.

EXERCISES

Guided Practice

CRITICAL THINKING about the Lesson

1. The point $(x, y) = (-4, -5)$ lies on the terminal side of the angle θ. What is r? $\sqrt{41}$
2. The point $(x, y) \neq (0, 0)$ lies on the terminal side of the angle θ. What is $\tan\theta$? For which values of x is $\tan\theta$ undefined? For which values of θ is $\tan\theta$ undefined? $\frac{y}{x}$; 0; 90°, 270°
3. What is the maximum value of $\sin\theta$? Explain your reasoning. 1; $\sin\theta = \frac{y}{r}$, $y \leq r$
4. In which quadrants must θ lie so that $\cos\theta$ is positive? I, IV
5. Find the reference angle for $\theta = \frac{7\pi}{4}$. $\frac{\pi}{4}$
6. Find the reference angle for $\theta = -120°$. 60°
7. Evaluate $\cos\left(-\frac{4\pi}{3}\right)$. $-\frac{1}{2}$
8. Evaluate $\tan 240°$. $\sqrt{3}$

Independent Practice

In Exercises 9–20, you are given a point on the terminal side of an angle θ. Find $\sin\theta$, $\cos\theta$, and $\tan\theta$.

9. $\frac{5}{13}, -\frac{12}{13}, -\frac{5}{12}$ 10. $-\frac{3}{5}, \frac{4}{5}, -\frac{3}{4}$ 11. $-\frac{4}{5}, -\frac{3}{5}, \frac{4}{3}$ 12. $\frac{4}{5}, -\frac{3}{5}, -\frac{4}{3}$

9.
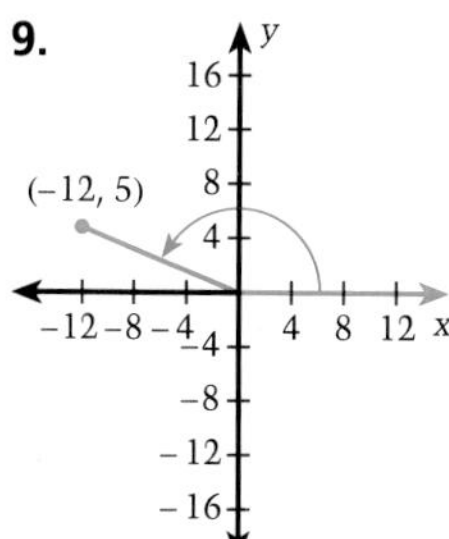

10.
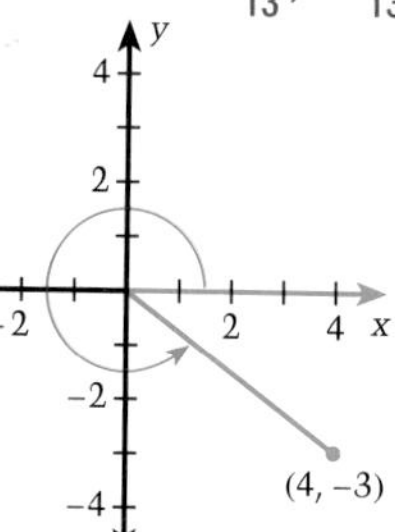

11.
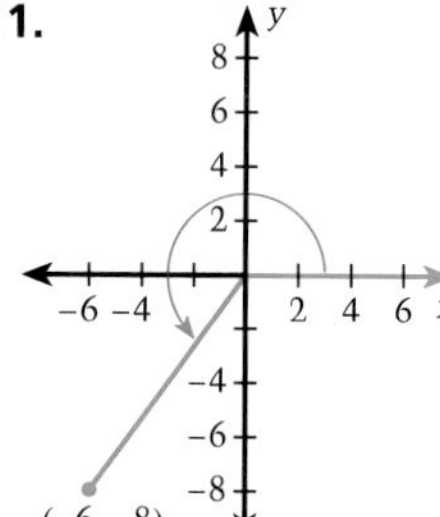

12.
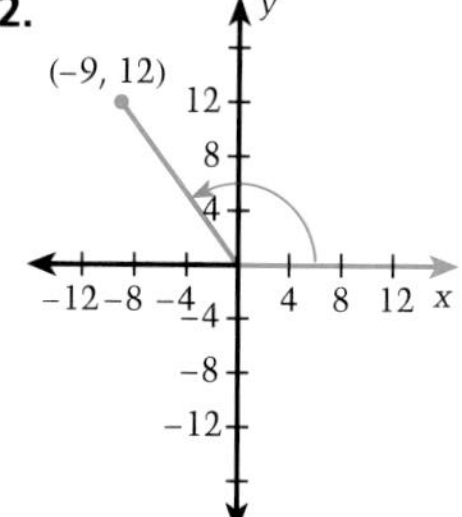

13. $(-12, -5)$
14. $(1, -1)$ $-\frac{1}{\sqrt{2}}, \frac{1}{\sqrt{2}}, -1$
15. $(-\sqrt{3}, 1)$
16. $(-2, 2)$
17. $(8, -15)$
18. $(-3, -4)$ $-\frac{4}{5}, -\frac{3}{5}, \frac{4}{3}$
19. $(-1, 1)$
20. $(9, 12)$ $\frac{4}{5}, \frac{3}{5}, \frac{4}{3}$

In Exercises 21–32, sketch the angle. Then find its reference angle. See Additional Answers.

21. 220° 40°
22. −510° 30°
23. −155° 25°
24. 345° 15°
25. $-\frac{8\pi}{3}$ $\frac{\pi}{3}$
26. $\frac{35\pi}{3}$ $\frac{\pi}{3}$
27. $\frac{12\pi}{5}$ $\frac{2\pi}{5}$
28. $\frac{21\pi}{4}$ $\frac{\pi}{4}$
29. −4.3 ≈1.16
30. 6.9 ≈0.62
31. 10.6 ≈1.18
32. −1.9 ≈1.24

In Exercises 33–40, without using a calculator, decide whether the equation is true or false. Then use a calculator to verify your decision. 33., 35., 36., 38. True 34., 37., 39.–40. False

33. $\sin 174° = \sin 6°$
34. $\cos 200° = \cos 20°$
35. $\tan 220° = \tan 40°$
36. $\cos 300° = \cos 60°$
37. $\tan\frac{5\pi}{3} = \tan\frac{\pi}{3}$
38. $\tan\frac{7\pi}{4} = \tan\frac{3\pi}{4}$
39. $\cos\frac{5\pi}{3} = -\cos\frac{\pi}{3}$
40. $\sin\frac{5\pi}{4} = \sin\frac{3\pi}{4}$

13. $-\frac{5}{13}, -\frac{12}{13}, \frac{5}{12}$ 15. $\frac{1}{2}, -\frac{\sqrt{3}}{2}, -\frac{1}{\sqrt{3}}$ 16. $\frac{1}{\sqrt{2}}, -\frac{1}{\sqrt{2}}, -1$ 17. $-\frac{15}{17}, \frac{8}{17}, -\frac{15}{8}$ 19. $\frac{1}{\sqrt{2}}, -\frac{1}{\sqrt{2}}, -1$

EXERCISE Notes

ASSIGNMENT GUIDE

Basic/Average: Ex. 1–8, 11–21 odd, 27–37 odd, 41–71 odd, 77–83 odd

Above Average: Ex. 1–8, 11–21 odd, 25–37 odd, 42–66 multiples of 3, 70–74, 77–83 odd

Advanced: Ex. 1–8, 11–21 odd, 27–37 odd, 39–66 multiples of 3, 70–76, 83

Selected Answers
Exercises 1–8, 9–81 odd

More Difficult Exercises
Exercises 70–76

Guided Practice

Ex. 1–8 COOPERATIVE LEARNING Have students work together in pairs to complete these exercises in class. Students should draw appropriate diagrams for Exercises 1–3.

Independent Practice

Ex. 9–20 Remind students that the value of $r = \sqrt{x^2 + y^2}$.

Ex. 21–32 Students should use Example 2 as a solution model.

EXTEND Ex. 33–40 WRITING Have students write a one- or two-sentence justification for their true-false choices.

Ex. 41–56 Students could use a calculator to verify their answers. However, these exercises should be completed using mental-math techniques.

EXTEND Ex. 81–82 Have students find the correct answers.
sin 10° ≈ 0.174; csc 3 ≈ 7.086

Ex. 83 Remind students that corresponding angles of similar triangles are congruent and corresponding sides of similar triangles are proportional.

Enrichment Activities

These activities require students to go beyond lesson goals.

The Unit Circle
A circle with its center at the origin of a coordinate system and with radius 1 unit is called the unit circle. If x and y are the coordinates of any point (x, y) on the circle, then $x^2 + y^2 = 1$.

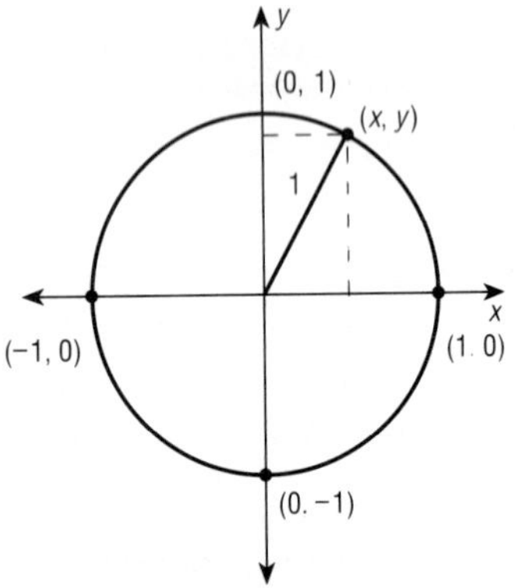

Consider a 30° angle in standard position. The initial side of the angle intercepts the unit circle at the point (1, 0). The terminal side intercepts the unit circle at point (x, y). Consider the angle as part of a 30°-60°-90° right triangle as shown.

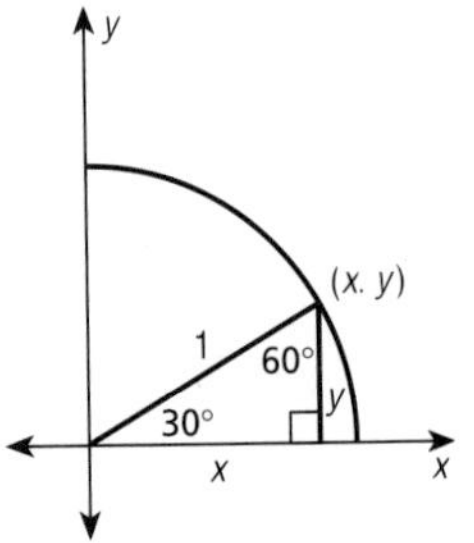

In Exercises 41–56, evaluate the function without using a calculator.

41. $\cos 300°$ $\frac{1}{2}$
42. $\cos(-225°)$ $-\frac{\sqrt{2}}{2}$
43. $\csc(-120°)$ $-\frac{2}{\sqrt{3}}$
44. $\tan 225°$ 1
45. $\sec 750°$ $\frac{2}{\sqrt{3}}$
46. $\sin 240°$ $-\frac{\sqrt{3}}{2}$
47. $\cos(-240°)$ $-\frac{1}{2}$
48. $\tan(-120°)$ $\sqrt{3}$
49. $\cot \frac{5\pi}{6}$ $-\sqrt{3}$
50. $\sec \frac{11\pi}{4}$ $-\sqrt{2}$
51. $\sin\left(-\frac{7\pi}{6}\right)$ $\frac{1}{2}$
52. $\cos \frac{2\pi}{3}$ $-\frac{1}{2}$
53. $\sin\left(-\frac{13\pi}{6}\right)$ $-\frac{1}{2}$
54. $\sin \frac{5\pi}{4}$ $-\frac{\sqrt{2}}{2}$
55. $\cos \frac{10\pi}{3}$ $-\frac{1}{2}$
56. $\csc \frac{17\pi}{3}$ $-\frac{2\sqrt{3}}{3}$

In Exercises 57–68, use a calculator to evaluate the function.

57. $\sin 18°$ ≈0.3090
58. $\sin 390°$ 0.5
59. $\sec 29°$ ≈1.14
60. $\cot 420°$ ≈0.577
61. $\cos \frac{20\pi}{3}$ −0.5
62. $\cot \frac{10\pi}{3}$ ≈0.5774
63. $\sec \frac{9\pi}{2}$ Undefined
64. $\csc \frac{18\pi}{5}$
65. $\tan 3.2$ ≈0.0585
66. $\csc 2.9$ ≈4.1797
67. $\cos 5.4$ ≈0.635
68. $\tan 2.3°$ ≈0.04

64. ≈−1.0515

69. *Fishing* You and a friend are fishing. Each of you casts with an initial velocity of 40 feet per second. Your cast was projected at an angle of 45° and your friend's at an angle of 60°. About how much further will your fishing tackle go than your friend's? ≈6.70 ft

70. *Ferris Wheel* The largest Ferris wheel in operation is the Cosmolock 21 at Yokohama City, Japan. It has a diameter of 328 feet. Passengers board the cars at the bottom of the wheel, about 16.5 feet above ground. Imagine that you have boarded the Cosmolock 21. The wheel rotates 300° before it stops. How far above ground are you when your ride on the Cosmolock 21 stops? 98.5 ft

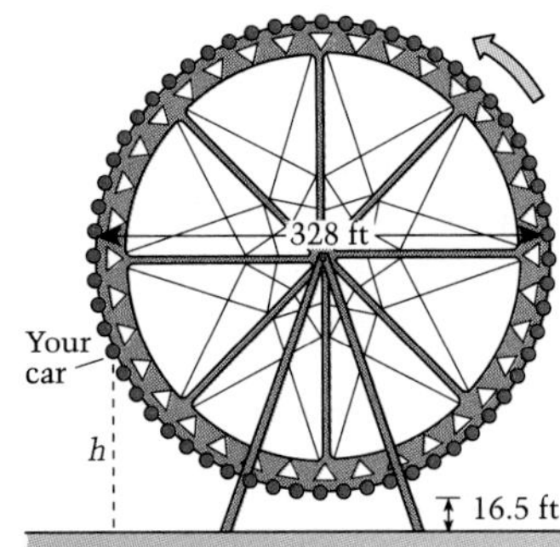

71. *Volleyball* While playing a game of volleyball, you set the ball to your teammate, who sets it to the spiker. You hit the ball with an initial velocity of 24 feet per second at an angle of 70°. About how far away should your teammate be to receive your set? ≈11.57 ft

Tropic of Cancer **In Exercises 72 and 73, use the following information.**

The Tropic of Cancer is the latitude farthest north of the equator where the sun appears directly overhead and is shown at the right.

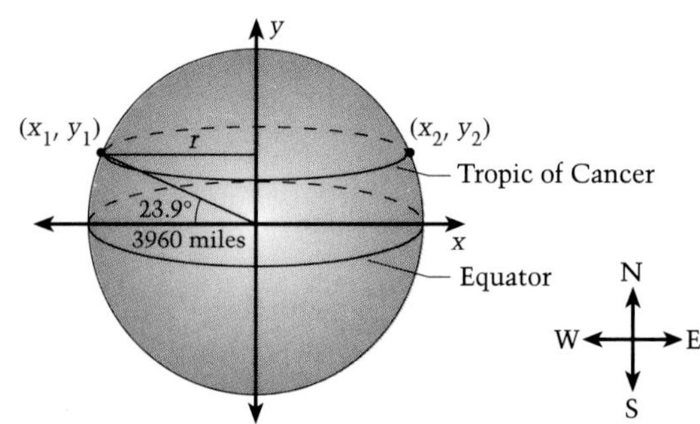

72. Find the circumference of the Tropic of Cancer. ≈22,748 mi

73. Imagine that a coordinate plane cuts Earth through its center, as shown in the diagram. If each unit in the coordinate system represents one mile, what are the coordinates of the two points that lie on the Tropic of Cancer? ≈(−3620.4, 1604.3), ≈(−3620.4, 1604.3)

74. *Football Halftime* Your school's marching band is performing at halftime. At one point in the performance, the band members form a 100-foot-wide circle in the center of the field. As the band begins to march around the circle, you are 100 feet from a goal line. How far from the same goal line will you be after you have marched 300° around the circle? 125 ft

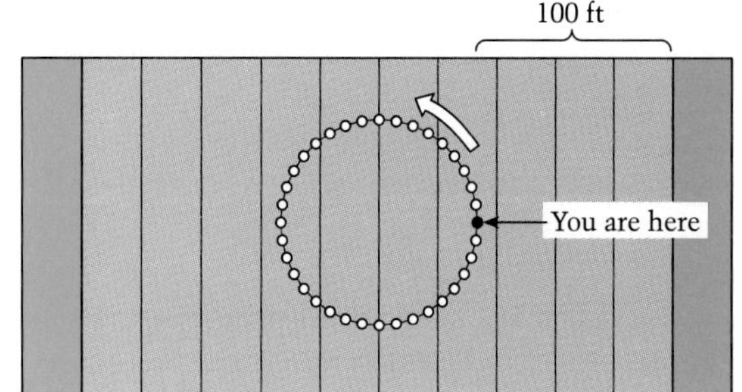

Tunneling Through Earth **In Exercises 75 and 76, use the following information.**

The diagram at the right shows the "top view" of the circle of 30° north latitude. Imagine that you travel along this circle from the point at 0° longitude in the Sahara Desert to the point at 120° east longitude, just south of Shanghai, China.

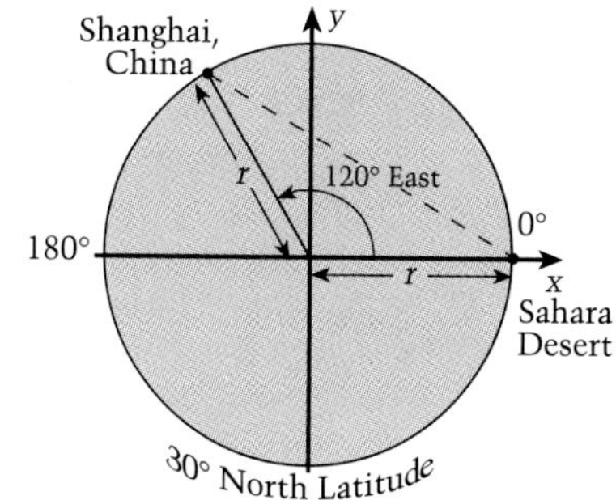

75. Find the distance (in miles) that you travel. ≈7182.7 mi

76. Find the coordinates of the point (x, y) just south of Shanghai. Imagine that, instead of traveling around a circular arc, you tunnel through Earth. How many miles long would the tunnel be? ≈(−1714.8, 2970.0), ≈5940.1 mi

Integrated Review

In Exercises 77–80, use the given value to find the values of the other five trigonometric functions. (Assume θ is an acute angle.) See Additional Answers.

77. $\tan \theta = \frac{3}{2}$ **78.** $\cos \theta = \frac{2}{3}$ **79.** $\sin \theta = \frac{1}{4}$ **80.** $\cot \theta = 2$

81. *What Went Wrong?* Your friend used a calculator to evaluate sin 10° and obtained −0.544. How can you tell this is wrong? What did your friend do wrong? sin 10° is positive, the calculator is in radian mode.

82. *What Went Wrong?* Your friend used a calculator to evaluate csc 3 and obtained 0.327. What did your friend do wrong?
In radian mode, your friend did the following: $\csc 3 = \sin \frac{1}{3} \approx 0.327$. The correct sequence is $\csc 3 = \frac{1}{\sin 3} \approx 7.086$

Exploration and Extension

83. *Geometry* The two triangles are similar. Find the six trigonometric functions of θ. See Additional Answers.

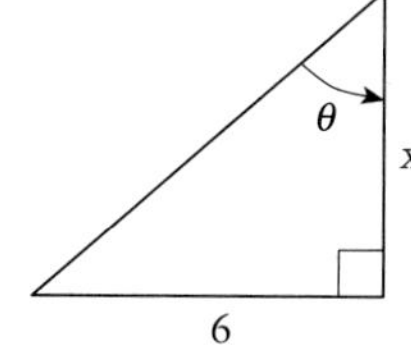

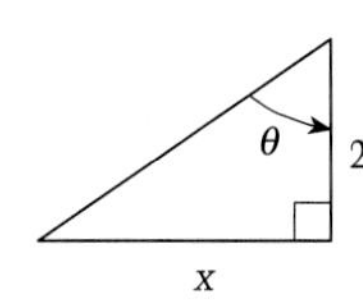

The length of the side opposite the 30° angle corresponds to the y-coordinate of the point. The length of the side opposite the 60° angle corresponds to the x-coordinate of the point.

1. Consider the ratio of the lengths of the sides of a 30°-60°-90° right triangle. If the triangle has a hypotenuse of 1 unit, then what are the x- and y-coordinates of the point?
The lengths of the sides of the triangle are 1, $\frac{1}{2}$, and $\frac{\sqrt{3}}{2}$. Thus, the 30° angle intercepts the unit circle at point $\left(\frac{\sqrt{3}}{2}, \frac{1}{2}\right)$.
2. Identify the x- and y-coordinates of points on the unit circle that correspond to these angles in standard position: 45° and 60°.
It can be shown that a 45° angle in standard position corresponds to point $\left(\frac{\sqrt{2}}{2}, \frac{\sqrt{2}}{2}\right)$; and a 60° angle in standard position corresponds to point $\left(\frac{1}{2}, \frac{\sqrt{3}}{2}\right)$.
3. Evaluate the sine and cosine of these same 30°, 45°, and 60° angles. Compare the results obtained to the results in Activities 1 and 2 above.
The x- and y-coordinates of the point on the unit circle correspond to the cosine and sine of the angle, respectively.
4. Identify the x- and y-coordinates of points on the unit circle that correspond to these angles in standard position: −30°, 210°, −270°, 495°, and 630°.
$\left(\frac{\sqrt{3}}{2}, -\frac{1}{2}\right)$, $\left(-\frac{\sqrt{3}}{2}, -\frac{1}{2}\right)$, $(0, 1)$, $\left(-\frac{\sqrt{2}}{2}, \frac{\sqrt{2}}{2}\right)$, $(0, -1)$

TECHNOLOGY The exploration in the "Enrichment Activities" section in the previous lesson will be informative if you choose to use a graphing calculator to graph angles and circles as described in this section.

USING A GRAPHING CALCULATOR

Graphing calculators, such as the *TI-82* or the Casio *fx-9700GE*, can be programmed to perform relatively complicated tasks. The program listed on page 863 takes time to enter, but the result is both fun and instructive.

When the program is run, it prompts you to enter "0" for radians or "1" for degrees. Next, you are prompted to enter an angle—any angle θ can be used. After entering the angle, press ENTER. The calculator then draws the angle in standard position. The point of the terminal side of the angle lies on the *unit circle*. Moreover, the coordinates of the endpoint are $(x, y) = (\cos\theta, \sin\theta)$. *To clear the screen,* press ENTER.

Cooperative Learning

Partner Activity Without showing your partner, use the angle drawing program listed on page 863 to draw an angle. Then, give the graphing calculator to your partner and ask him or her to estimate the value of the angle within 5 degrees. Ask your partner how he or she made the estimate.

After your partner has found the angle, switch roles and repeat the activity.

Two examples of the results of this program are shown below. The screen on the left shows the angle $\theta = 420°$. The screen on the right shows the angle $\theta = -\frac{13\pi}{4}$.

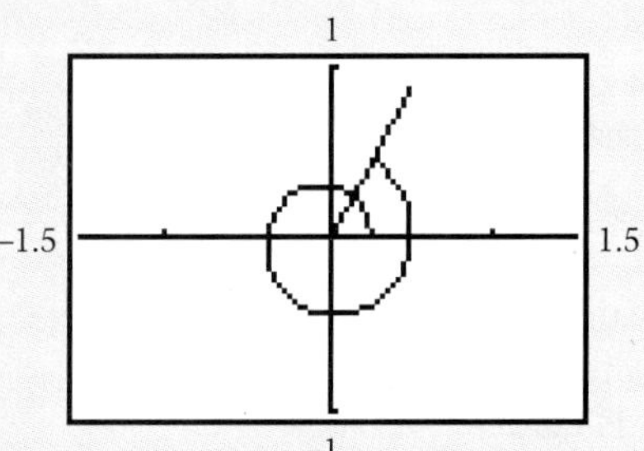

The screen shows an angle of 420°. The "terminal point" is $(\cos 420°, \sin 420°)$ or $\left(\frac{1}{2}, \frac{\sqrt{3}}{2}\right)$.

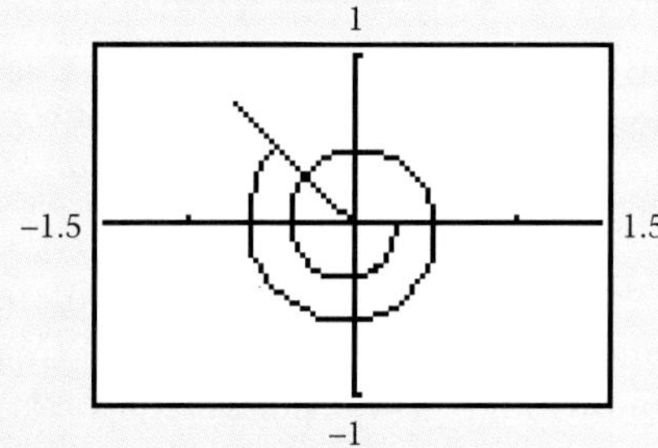

The screen shows an angle of $-\frac{13\pi}{4}$. Its "terminal point" is $\left(\cos\left(-\frac{13\pi}{4}\right), \sin\left(-\frac{13\pi}{4}\right)\right)$ or $\left(-\frac{\sqrt{2}}{2}, \frac{\sqrt{2}}{2}\right)$.

Another nice use of a graphing calculator is to sketch a graph of the unit circle using the calculator's *parametric mode.* On the *TI-82,* you can begin by pressing MODE, cursoring to PAR, and pressing ENTER. Next, press WINDOW and set the bounds for X, Y, and T.

Tmin = 0	Xmin = −1.5	Ymin = −1
Tmax = 6.3	Xmax = 1.5	Ymax = 1
Tstep = .031416	Xscl = 1	Yscl = 1

You are now ready to enter the x-coordinates and y-coordinates that will trace out the unit circle. Press Y= and enter

$:X_{1T} = \cos T$ and $:Y_{1T} = \sin T.$

Check that the calculator is set to radian mode, then press GRAPH. The screen will appear as shown below. Using the *trace* feature of the calculator, you can move the cursor around the unit circle. The displayed value of T represents the measure of θ in radians. The displayed value of x represents the cosine of θ, and the displayed value of y represents the sine of θ. (When you are done using this program, remember to change the mode back to FUNCTION.)

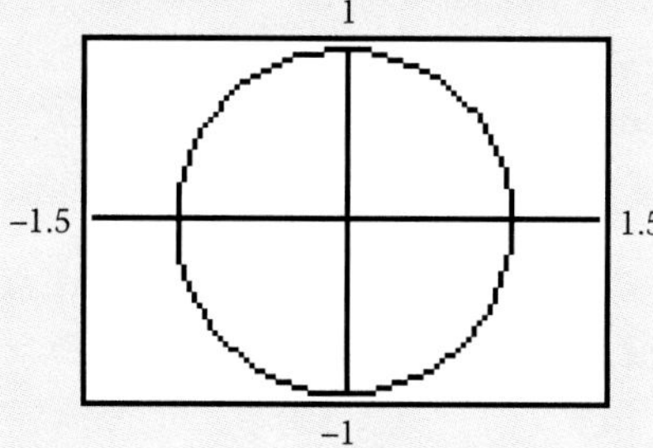

Exercises

In Exercises 1–4, use the angle-drawing program to sketch the angle. See margin.

1. $\theta = 500°$ **2.** $\theta = -600°$ **3.** $\theta = \frac{11\pi}{3}$ **4.** $\theta = -\frac{17\pi}{4}$

In Exercises 5–10, use a graphing calculator set in parametric mode to sketch the unit circle. Then use the trace key to approximate the radian measure of the angle whose cosine and sine are given.

5. $\cos\theta \approx -0.6846$, $\sin\theta \approx 0.7290$ ≈ 2.32

6. $\cos\theta \approx 0.6846$, $\sin\theta \approx -0.7290$ ≈ 5.47

7. $\cos\theta \approx -0.1879$, $\sin\theta \approx 0.9823$ ≈ 1.76

8. $\cos\theta \approx 0.9409$, $\sin\theta \approx 0.3387$ ≈ 0.35

9. $\cos\theta \approx -0.8607$, $\sin\theta \approx -0.5090$ ≈ 3.68

10. $\cos\theta \approx 0.9956$, $\sin\theta \approx -0.0941$ ≈ 6.20

Answers

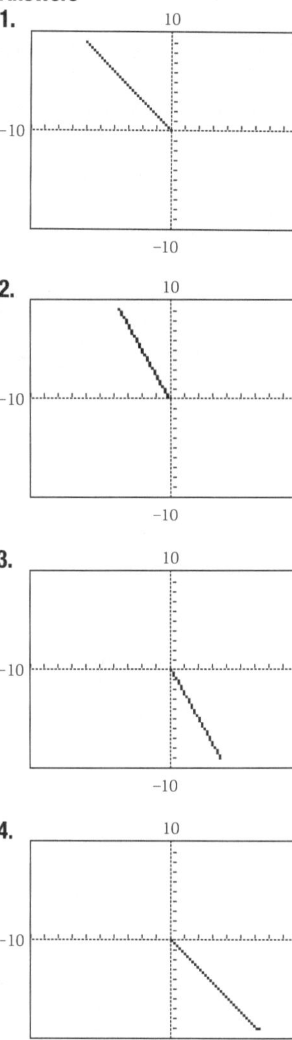

Answers

1. $\sin\theta = \frac{3}{\sqrt{13}}, \cos\theta = \frac{2}{\sqrt{13}},$
$\tan\theta = \frac{3}{2}, \csc\theta = \frac{\sqrt{13}}{3},$
$\sec\theta = \frac{\sqrt{13}}{2}, \cot\theta = \frac{2}{3}$

2. $\sin\theta = \frac{1}{\sqrt{5}}, \cos\theta = \frac{2}{\sqrt{5}},$
$\tan\theta = \frac{1}{2}, \csc\theta = \sqrt{5},$
$\sec\theta = \frac{\sqrt{5}}{2}, \cot\theta = 2$

3. $\sin\theta = \frac{1}{2}, \cos\theta = \frac{\sqrt{3}}{2},$
$\tan\theta = \frac{1}{\sqrt{3}}, \csc\theta = 2,$
$\sec\theta = \frac{2}{\sqrt{3}}, \cot\theta = \sqrt{3}$

Mid-Chapter SELF-TEST

Take this test as you would take a test in class. The answers to the exercises are given in the back of the book.

In Exercises 1–3, evaluate the six trigonometric functions of θ. (13.1) See margin.

1.

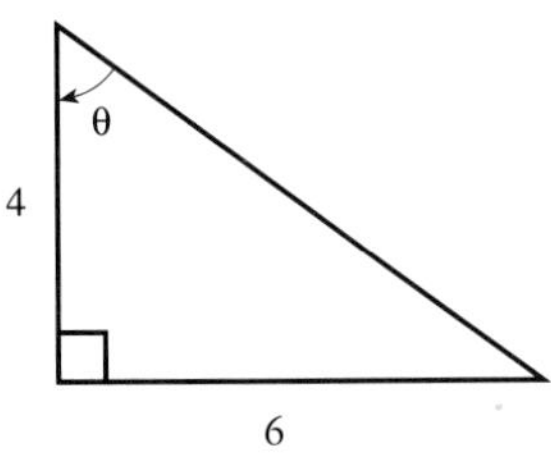

2.

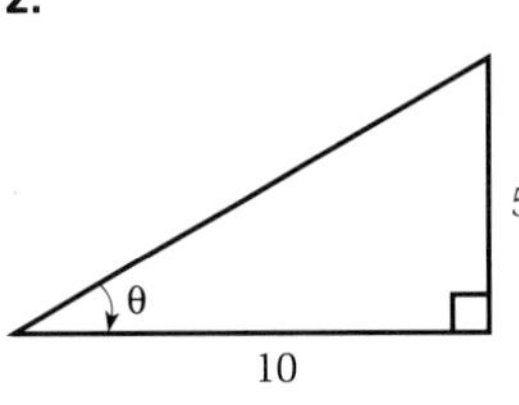

3.

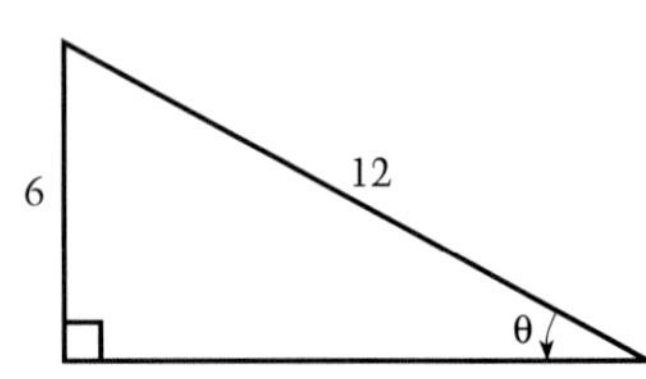

In Exercises 4–6, find the complement of the angle. (13.2)

4. $37°$ 53°

5. $15°$ 75°

6. $\frac{3\pi}{8}$ $\frac{\pi}{8}$

In Exercises 7–9, find the supplement of the angle. (13.2)

7. $120°$ 60°

8. $\frac{7\pi}{8}$ $\frac{\pi}{8}$

9. $\frac{5\pi}{12}$ $\frac{7\pi}{12}$

In Exercises 10–12, find two angles, one with positive measure and the other with negative measure, that are coterminal with the given angle. (13.2) For Exercises 10–12, other answers are possible.

10. $560°$ 200°, −160°

11. $-75°$ 285°, −435°

12. $-\frac{5\pi}{8}$ $\frac{11\pi}{8}, \frac{-21\pi}{8}$

In Exercises 13–15, you are given a point on the terminal side of an angle θ. Find $\sin\theta$, $\cos\theta$, and $\tan\theta$. (13.3)

13. $(3, 4)$ $\sin\theta = \frac{4}{5}, \cos\theta = \frac{3}{5}, \tan\theta = \frac{4}{3}$

14. $(-8, 6)$ $\sin\theta = \frac{3}{5}, \cos\theta = \frac{-4}{5}, \tan\theta = \frac{-3}{4}$

15. $(-5, -12)$ $\sin\theta = \frac{-12}{13}, \cos\theta = \frac{-5}{13}, \tan\theta = \frac{12}{5}$

In Exercises 16–19, decide whether the equation is true or false. (13.3)

16. $\sin 47° = \sin 227°$ False

17. $\cos 110° = \cos 290°$ False

18. $\tan\frac{5\pi}{4} = -\tan\frac{\pi}{4}$ False

19. $\sin 155° = -\sin 35°$ False

20. The carpenter's square at the right is marked at 12 on the short arm and $3\frac{1}{4}$ and 12 on the long arm. To which angles do these correspond? **(13.1)** ≈17.35°, 45°

21. The carpenter's square at the right is marked at 12 on the short arm. Where should you place marks on the long arm to correspond to 30° and 60°? **(13.1)** $6\frac{7}{8}''$, $20\frac{3}{4}''$

22. How far will a ball travel horizontally if hit with an initial velocity of 80 feet per second and an angle of elevation of 40°? Use the formula $d = \frac{1}{32}v^2 \sin 2\theta$. **(13.3)** ≈197 ft

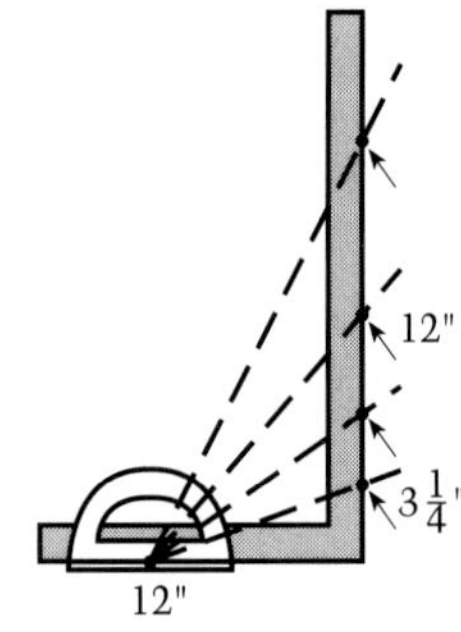

13.4 Inverse Trigonometric Functions

What you should learn:

Goal 1 How to evaluate inverse trigonometric functions

Goal 2 How to use inverse trigonometric functions to solve real-life problems

Why you should learn it:

You can solve many problems from real life, such as finding the length of a pulley belt, by using inverse trigonometric functions and trigonometric equations.

Goal 1 Evaluating Inverse Trig Functions

LESSON INVESTIGATION

Investigating Inverse Trig Functions

Partner Activity A function f is **one-to-one** if no two values of x correspond to the same value of y.

a. The function $f(x) = \sin x$ is not one-to-one. Why?

b. For a function to have an inverse, it must be one-to-one. Can you find an interval on the real number line in which the sine function is one-to-one? If so, describe the interval.

a. $\sin(0) = \sin(2\pi)$

b. $-\frac{\pi}{2} \le x \le \frac{\pi}{2}$

In this investigation, you may have discovered that the sine function does not have an inverse unless its domain is restricted to an interval over which the sine function is one-to-one. This is also true of the other trigonometric functions.

Inverse Trigonometric Functions

1. If $-1 \le a \le 1$, then the **inverse sine** of a is

$$\arcsin a = \theta$$

where $\sin \theta = a$ and $-90° \le \theta \le 90°$.

2. If $-1 \le a \le 1$, then the **inverse cosine** of a is

$$\arccos a = \theta$$

where $\cos \theta = a$ and $0° \le \theta \le 180°$.

3. If a is any real number, then the **inverse tangent** of a is

$$\arctan a = \theta$$

where $\tan \theta = a$ and $-90° < \theta < 90°$.

Study Tip

Not all mathematics texts agree on the definitions of the inverse trigonometric functions. Moreover, other notations such as Arcsin a or $\sin^{-1} a$ are often used. Because of this, whenever you encounter inverse trigonometric functions, you should check the definitions and notations that are being used.

The expression arcsin a is read as "the arcsine of a" or as "the angle whose sine is a." The expressions arccos a and arctan a are read in similar ways. Inverse trigonometric functions corresponding to the secant, cosecant, and cotangent functions can also be defined, but we will not do that in this book.

Problem of the Day

The average age of a group of doctors and lawyers is 40 years. If the average age of the doctors is 35, and the average age of the lawyers is 50, what is the ratio of doctors to lawyers?
2:1

ORGANIZER

Warm-Up Exercises

1. Using angle degree measure, find three angles (for θ) that satisfy the given equation.
   a. $\sin \theta = \frac{\sqrt{3}}{2}$
   b. $\cos \theta = \frac{\sqrt{2}}{2}$

   a. 60°, 120°, −300°
   b. 45°, 315°, −45°
2. Using angle radian measure, find three angles (for θ) that satisfy the given equation.
   a. $\sin \theta = 0.5$
   b. $\cos \theta = -\frac{\sqrt{3}}{2}$

   a. $\frac{\pi}{6}, \frac{5\pi}{6}, -\frac{7\pi}{6}$
   b. $\frac{5\pi}{6}, \frac{7\pi}{6}, -\frac{5\pi}{6}$

Lesson Resources

Teaching Tools
- Problem of the Day: 13.4
- Warm-up Exercises: 13.4
- Answer Masters: 13.4

Extra Practice: 13.4
Color Transparency: 70
Applications Handbook: p. 38, 39

LESSON Notes

In the first three lessons of this chapter, students learned how to evaluate trigonometric functions of a given angle. In this lesson, students will learn how to find angles that correspond to a given value of a trigonometric function.

Note: Some texts use arcsin or $\sin^{-1}$ to refer to the relation and use Arcsin and Sin^{-1} to refer to the function with its restricted range.

Example 1

Evaluating the arcsine function is like answering an "I am an angle whose sine is. . . . Who am I?" riddle.

Observe in Part a that there are many angles whose sine is $-\frac{1}{2}$. For example, consider $\theta = 210°$ and $\theta = 330°$. However, neither of these angles is in the indicated range of arcsine values, $-90°$ to $90°$.

It is often necessary to move clockwise on the coordinate plane to obtain an angle in the specified range that is a coterminal angle of 330°.

Example 2

TECHNOLOGY Scientific calculators have an inverse function key for the sine, cosine, and tangent functions. For example, to find arccos 0.3, in degree mode, you would enter

2nd | cos | 0.3 | ENTER

to obtain the value, 72.54.

Although inverse function values can always be obtained using the scientific calculator, it is instructive for students to verify in which quadrants the (restricted) ranges occur.

Example 1 Evaluating the Inverse Sine Function

Evaluate the following, if possible. List your responses in degree measure *and* in radian measure.

a. $\arcsin\left(-\frac{1}{2}\right)$ **b.** $\arcsin \frac{\sqrt{3}}{2}$ **c.** arcsin 2

Solution

a. The value of $\arcsin\left(-\frac{1}{2}\right)$ is the angle θ between $-90°$ and $90°$ whose sine is $-\frac{1}{2}$. Because $\sin(-30°) = -\frac{1}{2}$, you can write

$$\arcsin\left(-\frac{1}{2}\right) = -30°.$$

In radian measure, the angle is $-\frac{\pi}{6}$.

b. The value of $\arcsin \frac{\sqrt{3}}{2}$ is the angle θ between $-90°$ and $90°$ whose sine is $\frac{\sqrt{3}}{2}$. Because $\sin 60° = \frac{\sqrt{3}}{2}$, you can write

$$\arcsin \frac{\sqrt{3}}{2} = 60°.$$

In radian measure, the angle is $\frac{\pi}{3}$.

c. It is not possible to evaluate arcsin 2 because there is no angle whose sine is 2. ■

Example 2 Evaluating Inverse Trigonometric Functions

Evaluate the following, if possible. List your responses in degree measure and in radian measure.

a. $\arccos(-1)$ **b.** arctan 0 **c.** arccos 0.3

Solution

a. Because $\cos 180° = -1$, you can write

$$\arccos(-1) = 180°.$$

In radian measure, the angle is π.

b. Because $\tan 0° = 0$, it follows that

$$\arctan 0 = 0°.$$

In radian measure, the angle is also 0.

c. There is no common angle whose cosine is 0.3. Using a calculator set to *degree mode,* you can approximate the angle to be

$$\arccos 0.3 \approx 72.54°.$$

Using a calculator set to *radian mode,* you can approximate the angle to be

$$\arccos 0.3 \approx 1.27 \text{ radians}.$$ ■

Goal 2 Using Inverse Trigonometric Functions

Connections
Mechanics

Example 3 *Finding the Length of a Pulley Belt*

Find the length of the pulley belt shown at the left.

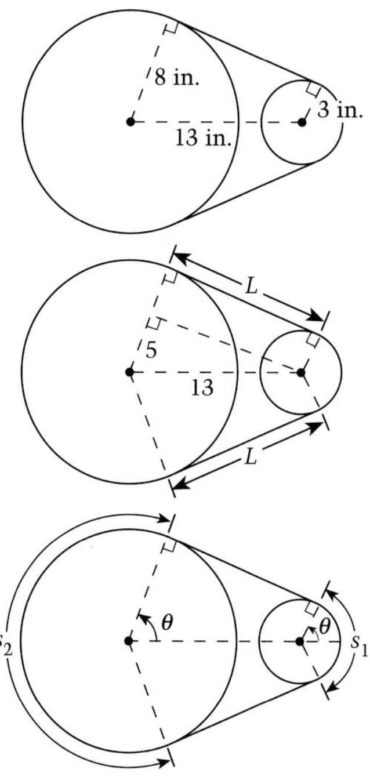

Solution To find the length, partition the pulley belt into four parts: the two straight segments, the arc around the small wheel, and the arc around the large wheel.

Two Straight Segments: Each of the straight segments is the side, L, of a right triangle whose hypotenuse is 13 and whose other side is 5. Using the Pythagorean Theorem, you can find the length of each straight segment to be

$$L = \sqrt{13^2 - 5^2} = \sqrt{144} = 12 \text{ inches.}$$

Arc around the Small Wheel: The angle θ is part of a right triangle whose adjacent side has a length of 5 and whose hypotenuse has a length of 13. Therefore, θ is the angle whose cosine is $\frac{5}{13}$. Using the formula for the length of a circular arc, $s = r\theta$, the length of the segment around the small wheel is

$$s_1 = 2\left(3 \text{ arccos } \frac{5}{13}\right) = 6 \text{ arccos } \frac{5}{13} \text{ inches.}$$

Arc around the Large Wheel: The circumference of the large wheel is $2\pi r = 16\pi$ inches. Therefore, the length of the segment around the large wheel is

$$s_2 = 16\pi - 2\left(8 \text{ arccos } \frac{5}{13}\right) = 16\pi - 16 \text{ arccos } \frac{5}{13} \text{ inches.}$$

By adding the lengths of the four parts, you can find the length of the pulley.

$$\begin{aligned}\text{Pulley length} &= 2L + s_1 + s_2 \\ &= 2(12) + 6 \text{ arccos } \frac{5}{13} + \left(16\pi - 16 \text{ arccos } \frac{5}{13}\right) \\ &= 24 + 16\pi - 10 \text{ arccos } \frac{5}{13} \approx 62.5 \text{ inches.}\end{aligned}$$

Note that when using the formula, $s = r\theta$, for the length of a circular arc, you must use *radian mode.* ■

Communicating about ALGEBRA

SHARING IDEAS about the Lesson

Evaluate the following in degree and radian measure.

A. $\text{arcsin}\left(-\frac{\sqrt{2}}{2}\right)$ **B.** $\text{arccos}\left(-\frac{\sqrt{2}}{2}\right)$ **C.** $\text{arctan}\left(-\frac{\sqrt{2}}{2}\right)$

A. $-45°, -\frac{\pi}{4}$ **B.** $135°, \frac{3\pi}{4}$ **C.** $\approx -35.3°, \approx -0.615$

Example 3

Ask students to explain why each straight segment is the same length as the long leg of the right triangle whose hypotenuse and short leg are 13 and 5 inches. Ask students to explain why the length of the segment around the small wheel is twice the value of $3 \text{ arccos}\left(\frac{5}{13}\right)$.

Extra Examples

Here are additional examples similar to **Examples 1–2.**

1. Evaluate the following, if possible, in degree measure.
   a. $\text{arcsin}\left(\frac{\sqrt{3}}{2}\right)$ 60°
   b. arccos (−1) 180°
   c. arctan (5) ≈ 78.69°
   d. $\text{arcsin}\left(-\frac{3}{2}\right)$ not possible
2. Evaluate the following, if possible, in radian measure.
   a. $\text{arccos}\left(\frac{\sqrt{2}}{2}\right)$ $\frac{\pi}{4}$
   b. $\text{arctan}\left(-\frac{1}{2}\right)$ ≈ −0.464
   c. $\text{arcsin}\left(\frac{\sqrt{3}}{2}\right)$ $\frac{\pi}{3}$
   d. $\text{arccos}\left(\frac{3}{4}\right)$ ≈ 0.723

Check Understanding

1. What is an inverse trigonometric function?
   A set of angles that correspond to a given value of a trigonometric function
2. Why must the angle θ be restricted in the definition of arcsin θ?
   To insure that arcsin is a function
3. Is the $\sin^{-1}\theta$ function equivalent to the arcsin θ function?
   yes

Communicating about ALGEBRA

COOPERATIVE LEARNING

Have students work with a partner and evaluate the following. Then, students should discuss the observed relationships and make conjectures about the basic trigonometric functions and their inverses.

a. arcsin (sin 45°) 45°

b. arccos (cos 30°) 30°

c. arctan (tan 60°) 60°

d. $\cos\left(\arccos -\frac{1}{2}\right)$ $-\frac{1}{2}$

e. sin (arcsin 0) 0

f. tan (arctan 12) 12

g. arcsin (cos 45°) 45°

h. arccos (tan 30°) ≈ 54.74°

i. arctan (sin 150°) ≈ 26.57°

Note in Part h that the arccos and tan are not directly related. Similarly, in Part i, the arctan and sin are not directly related.

EXERCISE Notes

ASSIGNMENT GUIDE

Basic/Average: Ex. 1–4, 7–19 odd, 31–37 odd, 40, 41, 45–48

Above Average: Ex. 1–4, 7–19 odd, 27–37 odd, 39–42, 45–50

Advanced: Ex. 1–4, 7–19 odd, 27–37 odd, 39–45, 47, 49–52

Selected Answers

Exercises 1–4, 5–47 odd

✪ More Difficult Exercises

Exercises 8, 37–44, 49–52

Guided Practice

EXTEND Ex. 1–2 Ask students to explain their reasoning.

EXERCISES

Guided Practice

CRITICAL THINKING about the Lesson

1. Because $\sin 45° = \frac{\sqrt{2}}{2}$, $\arcsin \frac{\sqrt{2}}{2} = \boxed{?}$. 45°

2. Because $\tan \frac{\pi}{3} = \sqrt{3}$, $\arctan \boxed{?} = \frac{\pi}{3}$. $\sqrt{3}$

3. Find the degree *and* radian measure of the angle θ in the triangle at the right. ≈39.8°, ≈0.695

4. Use a calculator to evaluate $\arccos \frac{9}{10}$. Write your result in radians and degrees. ≈25.8°, ≈0.451

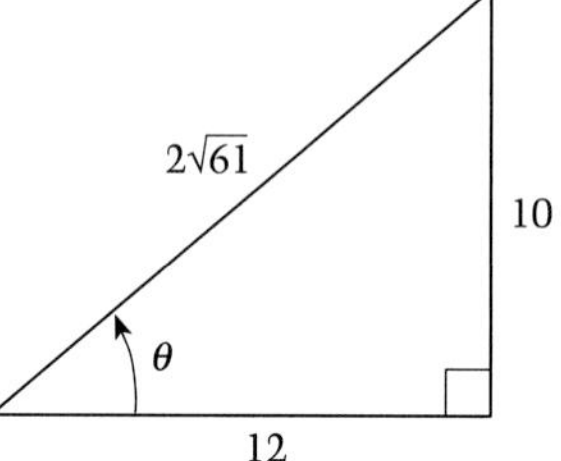

Independent Practice

7. 30°, ≈0.524 8. ≈126.9°, ≈2.214

In Exercises 5–12, write the measure of the angle θ in degrees and in radians.

5. ≈33.7°, ≈0.588

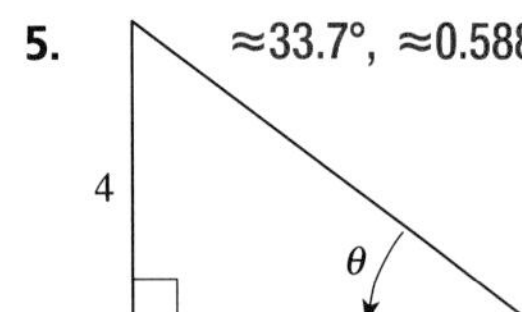

6.

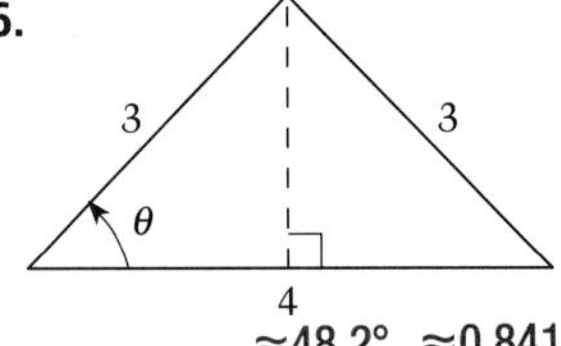

≈48.2°, ≈0.841

7.

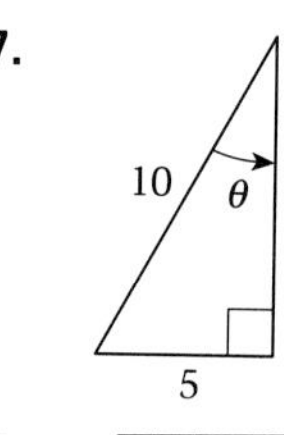

✪ 8.

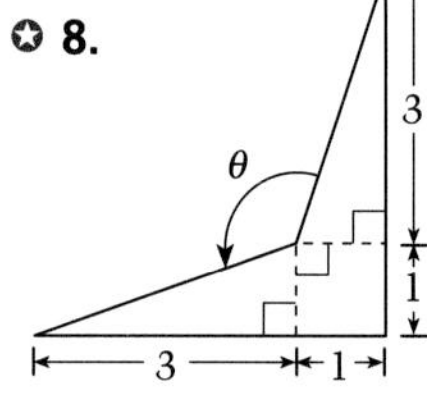

9.

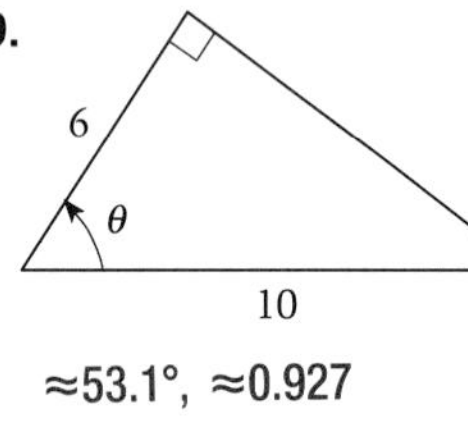

≈53.1°, ≈0.927

10.

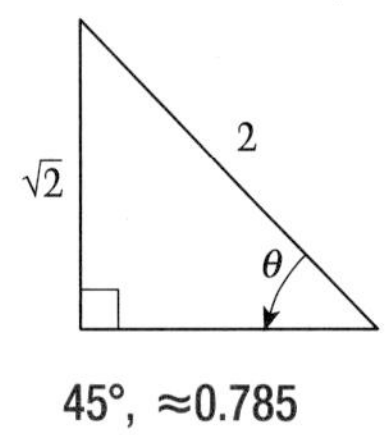

45°, ≈0.785

11.

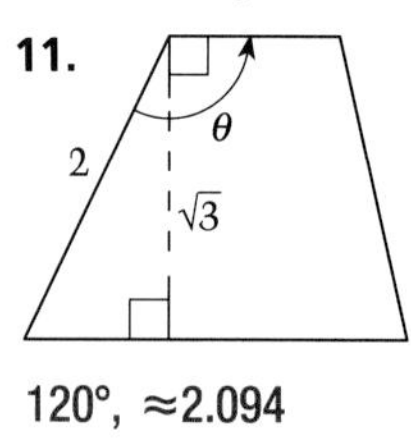

120°, ≈2.094

12.

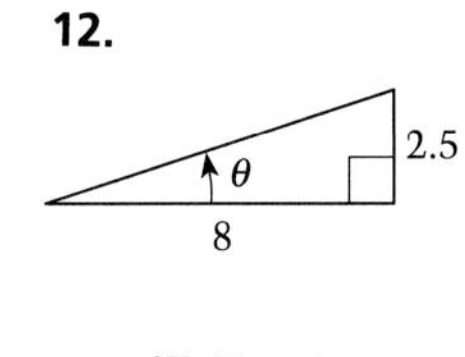

≈17.4°, ≈0.303

In Exercises 13–28, evaluate the expression without using a calculator. Write the result in degrees and in radians.

13. $\arccos\left(\frac{\sqrt{2}}{2}\right)$ 45°, $\frac{\pi}{4}$

14. $\arcsin\left(\frac{1}{2}\right)$ 30°, $\frac{\pi}{6}$

15. $\arcsin(1)$ 90°, $\frac{\pi}{2}$

16. $\arccos(0)$ 90°, $\frac{\pi}{2}$

17. $\arctan(\sqrt{3})$ 60°, $\frac{\pi}{3}$

18. $\arctan\left(\frac{\sqrt{3}}{3}\right)$ 30°, $\frac{\pi}{6}$

19. $\arccos\left(-\frac{\sqrt{3}}{2}\right)$ 150°, $\frac{5\pi}{6}$

20. $\arcsin\left(-\frac{\sqrt{2}}{2}\right)$ −45°, $-\frac{\pi}{4}$

21. $\arcsin(-1)$ −90°, $-\frac{\pi}{2}$

22. $\arctan(0)$ 0°, 0

23. $\arctan(-1)$ −45°, $-\frac{\pi}{4}$

24. $\arccos\left(-\frac{1}{2}\right)$

25. $\arcsin\left(-\frac{\sqrt{3}}{2}\right)$ −60°, $-\frac{\pi}{3}$

26. $\arccos\left(-\frac{\sqrt{2}}{2}\right)$ 135°, $\frac{3\pi}{4}$

27. $\arcsin(0)$ 0°, 0

28. $\arctan(-\sqrt{3})$ −60°, $-\frac{\pi}{3}$

24. 120°, $\frac{2\pi}{3}$

In Exercises 29–36, use a calculator to approximate the value. Write the result in degrees and in radians. Round to three decimal places. See below.

29. $\arctan(3.9)$

30. $\arccos(0.14)$

31. $\arccos(0.22)$

32. $\arcsin(0.75)$

33. $\arcsin(-0.6)$

34. $\arccos(-0.9)$

35. $\arctan(-0.1)$

36. $\arctan(2.25)$

29. 75.619°, 1.320 **30.** 81.952°, 1.430 **31.** 77.291°, 1.349

32. 48.590°, 0.848 **33.** −36.870°, −0.644 **34.** 154.158°, 2.691

35. −5.711°, −0.100 **36.** 66.038°, 1.153

37. *What Went Wrong?* Suppose you used a calculator to evaluate arctan 45, and the calculator displayed 1.000. What did you do wrong? Write the correct answer in degree measure and in radian measure. **See margin.**

38. *What Went Wrong?* Suppose you used a calculator to evaluate arcsin $\frac{1}{2}$, and the calculator displayed 0.479°. What did you do wrong? Write the correct answer in degree measure and in radian measure. **See margin.**

39. *Swimming Pool* The swimming pool shown below is 24 feet long. The pool depth ranges from 2 feet to 10 feet. Find the *angle of depression* of the bottom of the pool. **≈18.4°**

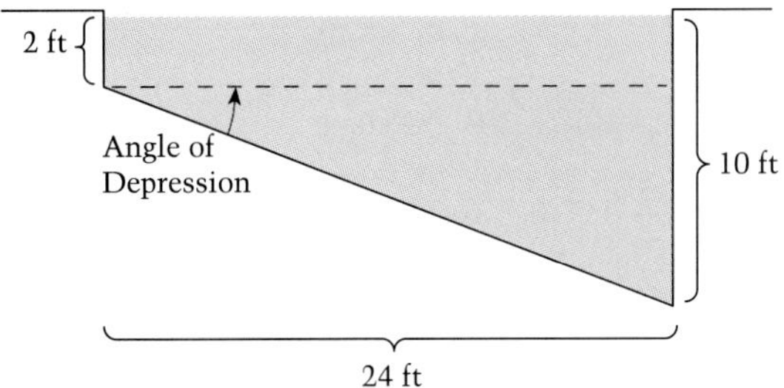

40. *Dump Truck* The dump truck shown below has a 10-foot bed. When tilted at its maximum angle, the bed reaches a height of 7 feet above its original position. What is the truck bed's maximum angle? **≈44.4°**

41. *Refraction of Light* When a light ray enters glass, it bends towards a line perpendicular to the surface of the glass. The *angle of incidence,* θ_1, is the angle between the ray and the perpendicular line before entering the glass (and after leaving the glass). The *angle of refraction,* θ_2, is the angle between the ray and the perpendicular line in the glass. The *index of refraction, n,* of the glass is

$$n = \frac{\sin \theta_1}{\sin \theta_2}.$$

A typical index of refraction for glass is $n = 1.5$. For such glass, if $\theta_1 = 59°$, what is θ_2? **≈34.9°**

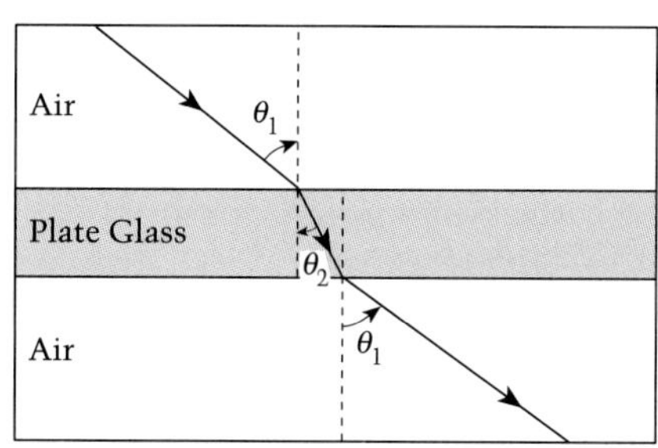

42. *Civil Engineering* Curves in roads are often constructed as arcs of circles. In the diagram shown below, θ is the central angle of the arc. The radius of the circle is 225 feet, and the length of each "tangent" is 158 feet. Find the degree measure of θ. **≈70.2°**

Aerial view of the Pan-American Highway near Arequipa, Peru

▶ Ex. 4 COOPERATIVE LEARNING Have students using the same brand of calculator work together to build confidence in using the calculator to evaluate the inverse trigonometric functions.

Independent Practice

▶ Ex. 5–12 These exercises provide practice in the right triangle trigonometric ratios.

▶ Ex. 13–28 It is critical that students have a clear understanding of the range of possible answers as stated on page 703. The arcsin and arctan functions are defined for angles $-90° < \theta < 90°$. The arccos function is defined for angles $0° \le \theta \le 180°$.

▶ Ex. 29–36 Ask students to check their answers by applying the trigonometric function to the value. For example, in Exercise 29, if arctan (3.9) = 75.619 then tan 75.619 = 3.9.

Answers

37. You didn't push the second function key in the degree mode and found tan 45 instead of arctan 45; ≈88.7°, ≈1.549

38. You didn't push the second function key in the radian mode and found sin $\frac{1}{2}$ instead of arcsin $\frac{1}{2}$; 30°, ≈0.524

▶ Ex. 39 Remind students how to view the angle of depression.

▶ Ex. 42 Some students may need to review the idea from geometry of a line tangent to a circle. You might suggest that students imagine there is an auxiliary line through the nonright angles of the quadrilateral in order to find $\frac{1}{2}\theta$.

Ex. 45 All students should complete the table from memory.

Ex. 49–52 You may need to review composition of functions with your students.

Enrichment Activities

These activities require students to go beyond lesson goals.

Trigonometric Equations

The inverse trigonometric functions can be used to solve some trigonometric equations for their angles.

For example, solve the equation for angle θ.

$$2 \sin \theta + 1 = 0$$
$$2 \sin \theta = -1$$
$$\sin \theta = -\frac{1}{2}$$
$$\arcsin (\sin \theta) = \arcsin \left(-\frac{1}{2}\right)$$
$$\theta = -30^\circ$$

Other possible solutions include 210° and 330°.

1. Find at least two solutions for each equation. You may have to use a calculator.
   a. $2 \tan \theta = 3$
   b. $4 \cos \theta + 1 = 0$
   c. $2 \cos \theta = \sec \theta$

   a. $56.3^\circ, 236.3^\circ$
   b. $104.5^\circ, 255.5^\circ$
   c. $45^\circ, 135^\circ$

2. Recall from Lesson 13.3 that the horizontal distance, d, in feet, traveled by a projectile launched at an angle θ with an initial velocity, v, in feet per second, is given by the formula $d = \frac{v^2}{32} \sin 2\theta$. A projectile launched at a velocity of 64 feet per second traveled a horizontal distance of 100 feet. Find the launching angle of the projectile.
   $\approx 25.7^\circ$

Bicycle Chain Length **In Exercises 43 and 44, use the following information.**

The chain on a one-gear bicycle is turned by two circular sprocket wheels whose centers are 18 inches apart. The front sprocket wheel has 48 teeth, and the rear sprocket wheel has 24 teeth. Each tooth accounts for $\frac{\pi}{6}$ of an inch of its sprocket wheel's circumference.

43. Find the length of the chain. ≈ 53.29 in.

44. The pedals of the bicycle are attached to the center of the large front sprocket. The diameter of each bicycle wheel is 26 inches. How many inches along the ground does the bike travel with one full revolution of the pedals? ≈ 163.4 in.

Integrated Review

45. Complete the table.

θ	0	$\frac{\pi}{6}$	$\frac{\pi}{4}$	$\frac{\pi}{3}$	$\frac{\pi}{2}$	π	$\frac{3\pi}{2}$	2π
sin θ	?	?	?	?	?	?	?	?
cos θ	?	?	?	?	?	?	?	?
tan θ	?	?	?	?	?	?	?	?
csc θ	?	?	?	?	?	?	?	?
sec θ	?	?	?	?	?	?	?	?
cot θ	?	?	?	?	?	?	?	?

$0, \frac{1}{2}, \frac{\sqrt{2}}{2}, \frac{\sqrt{3}}{2}, 1, 0, -1, 0;$

$1, \frac{\sqrt{3}}{2}, \frac{\sqrt{2}}{2}, \frac{1}{2}, 0, -1, 0, 1;$

$0, \frac{\sqrt{3}}{3}, 1, \sqrt{3}, *, 0, *, 0;$

$*, 2, \sqrt{2}, \frac{2\sqrt{3}}{3}, 1, *, -1, *;$

$1, \frac{2\sqrt{3}}{3}, \sqrt{2}, 2, *, -1, *, 1;$

$*, \sqrt{3}, 1, \frac{\sqrt{3}}{3}, 0, *, 0, *$

* means function is not defined

46. Find the radian measure of two angles that are coterminal with $\arcsin(-1)$. $-\frac{\pi}{2}; \frac{3\pi}{2}$

47. *College Entrance Exam Sample* In the figure below, the area of square $OABC$ is 2. What is the area of the circle? **c**

a. $\frac{\pi}{4}$ **b.** $\pi\sqrt{2}$ **c.** 2π **d.** 4π **e.** 16π

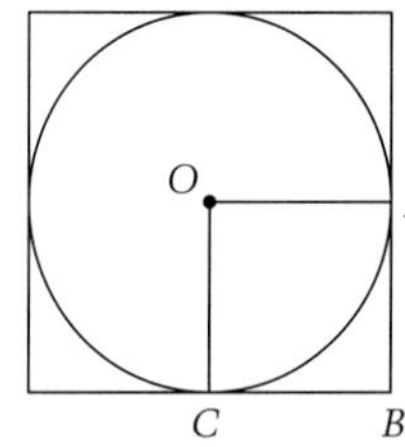

48. *College Entrance Exam Sample* In the figure below, $AB = 5$, $AC = 3$, and $BC = CD$. What is the length of BE? **a**

a. 1 **b.** 2 **c.** 3 **d.** 4 **e.** 5

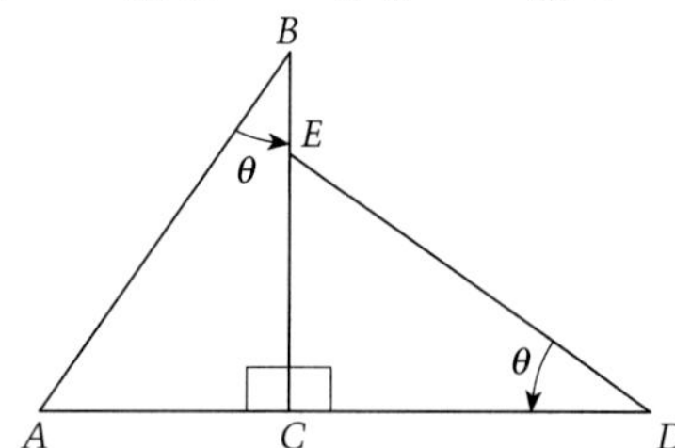

Exploration and Extension

In Exercises 49–52, evaluate the expression. Use a calculator to verify your result.

49. $\arccos\left(\cos \frac{\pi}{3}\right)$ $\frac{\pi}{3}$ **50.** $\tan\left(\arctan \frac{4\pi}{3}\right)$ $\frac{4\pi}{3}$ **51.** $\cos\left(\arcsin \frac{5}{13}\right)$ $\frac{12}{13}$ **52.** $\sin\left(\arctan \frac{3}{4}\right)$ $\frac{3}{5}$

13.5 The Law of Sines

What you should learn:

Goal 1 How to use the Law of Sines to find the sides and angles of any triangle

Goal 2 How to use the Law of Sines to solve real-life problems

Why you should learn it:

You can use the Law of Sines to solve for many distances that cannot be measured directly.

Goal 1 Using the Law of Sines

In Lesson 13.1, you learned how to solve right triangles. In this lesson and the next, you will learn to solve **oblique triangles**—triangles that have no right angles.

To solve an oblique triangle, you need to know the measure of at least one side and any two other parts of the triangle—either two sides, two angles, or one angle and one side. This breaks down into four possible cases.

1. Two angles and any side (AAS or ASA)
2. Two sides and an angle opposite one of them (SSA)
3. Three sides (SSS)
4. Two sides and their included angle (SAS)

The first two cases can be solved using the **Law of Sines.** The last two cases require the **Law of Cosines** (Lesson 13.6).

C, b, a, A, c, B

Law of Sines

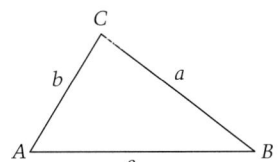

If ABC is a triangle with sides a, b, and c, then

$$\frac{a}{\sin A} = \frac{b}{\sin B} = \frac{c}{\sin C} \quad \text{or} \quad \frac{\sin A}{a} = \frac{\sin B}{b} = \frac{\sin C}{c}.$$

Example 1 *Given Two Angles and One Side—AAS*

Given a triangle with $C = 102°$, $B = 28°$, and $b = 27$ feet, find the measures of the remaining angle and sides.

C, a, b = 27 ft, 102°, 28°, A, c, B

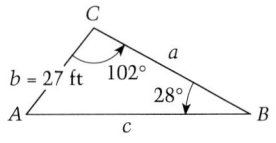

Solution Because the sum of the angles of a triangle is 180°, you know that the third angle $A = 180° - B - C = 50°$. By the Law of Sines, you can write

$$\frac{a}{\sin 50°} = \frac{b}{\sin 28°} = \frac{c}{\sin 102°}.$$

Because $b = 27$, you can solve for a and c as follows.

$$a = \frac{27}{\sin 28°}(\sin 50°) \approx 44.06 \text{ feet}$$

$$c = \frac{27}{\sin 28°}(\sin 102°) \approx 56.25 \text{ feet}$$ ■

Problem of the Day

A man born in the early part of the 19th century was x years old in the year x^2. When was he born? 1806

ORGANIZER

Warm-Up Exercises

1. In an isosceles triangle, one angle is known to be 50°. Find the measures of the other two angles. The measures are 65° each, or 80° and 50°.
2. In an isosceles triangle, one angle is known to be 100°. Find the measures of the other two angles. The measure of each angle is 40°.
3. Solve the following equations.
   a. $\frac{3x}{36} = \frac{9}{24}$ $x = 4.5$
   b. $\frac{6x}{25} = \frac{18}{35}$ $x = \frac{15}{7}$

Lesson Resources

Teaching Tools
- Problem of the Day: 13.5
- Warm-up Exercises: 13.5
- Answer Masters: 13.5

Extra Practice: 13.5

Color Transparency: 71

LESSON Notes

You might wish to use Exercises 71–72 on page 715 as an in-class activity to introduce the lesson.

Refer students to the *Handbook of Mathematical Connections* for a proof of the Law of Sines.

Example 1

Call attention to the way in which sides and angles of a triangle are labeled. In particular, make sure students see that an angle is denoted by a capital letter (A, B, or C, for example) and that the side opposite a given angle is denoted by the lower case of the letter used to denote the angle.

Example 2

Be sure students recognize the difference between the two cases, SSA and SAS. (The latter case is solved using the Law of Cosines.) Point out that if

$$\sin B \approx 0.365$$

then

$$\arcsin(\sin B) = \arcsin(0.365)$$

or

$$B = \arcsin(0.365)$$
$$B \approx 21.4°.$$

Example 3

Observe that a different relationship could be used to determine the height of the ceiling of the cave, as follows: $\frac{a}{\sin A} = \frac{c}{\sin C}$, $\frac{a}{\sin 71} = \frac{50}{\sin 47°}$,

$a = \frac{50 \sin 71°}{\sin 47°} \approx 64.64.$

The height h can be found by solving the following equation.

$\sin B = \frac{h}{a}$ or $\sin 62° = \frac{h}{64.64}$.

So, $h = 64.64 \sin 62° \approx 57.1$.

Extra Examples

Here are additional examples similar to **Examples 1–3.**

1. Given a triangle with $A = 135°$, $C = 23°$, and $b = 10$ meters, find the remaining angle and sides.
   $B = 22°$, $a \approx 18.88$ m, $c \approx 10.43$ m
2. Given a triangle with $B = 56°$, $C = 84°$, and $c = 15$ feet, find the remaining angle and sides.
   $A = 40°$, $b \approx 12.50$ ft, $a \approx 9.69$ ft

When solving triangles, a careful sketch is useful as a quick check. Remember that the longest side lies opposite the angle with the greatest measure, and the shortest side lies opposite the angle with the least measure.

Two angles and one side (AAS or ASA) determine exactly one triangle. Two sides and an opposite angle (SSA) may determine no triangle, exactly one triangle, or two triangles.

Example 2 Single Solution Case—SSA

Given a triangle with sides $a = 22$ meters and $b = 12$ meters and angle $A = 42°$, find the measure of the remaining side and angles.

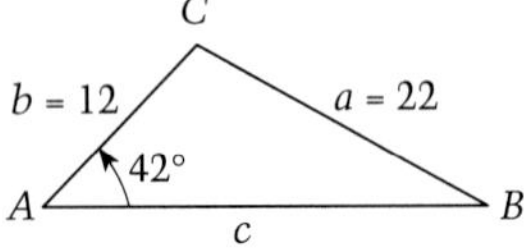

Solution Begin by using the given information to make a careful sketch. Then, using the Law of Sines, you can solve for B.

$$\frac{\sin B}{12} = \frac{\sin 42°}{22} \quad \textit{Law of Sines}$$

$$\sin B = 12\left(\frac{\sin 42°}{22}\right) \quad \textit{Multiply both sides by 12.}$$

$$\sin B \approx 0.365 \quad \textit{Use a calculator in degree mode.}$$

$$B \approx \arcsin 0.365 \quad \textit{Use arcsine to solve for B.}$$

$$B \approx 21.4° \quad \textit{Use a calculator in degree mode.}$$

Now you know that $C \approx 180° - 42° - 21.4° = 116.6°$, and you can use the Law of Sines again to find the remaining side of the triangle.

$$\frac{c}{\sin 116.6°} = \frac{22}{\sin 42°}$$

$$c = \sin 116.6°\left(\frac{22}{\sin 42°}\right)$$

$$\approx 29.4 \text{ meters}$$

■

The Law of Sines can be used to derive a formula for the area of triangle when you know the lengths of two sides and the measure of the included angle.

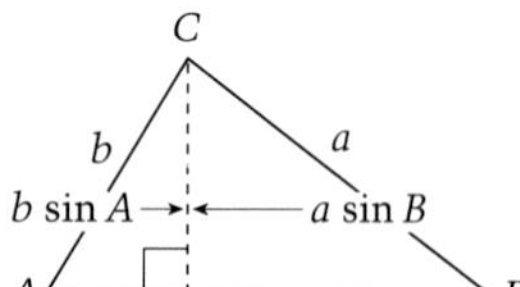

Area of a Triangle

The area of any triangle is given by one half the product of the lengths of two sides times the sine of their included angle.

$$\text{Area} = \frac{1}{2}bc \sin A$$

$$\text{Area} = \frac{1}{2}ab \sin C$$

$$\text{Area} = \frac{1}{2}ac \sin B$$

Goal 2 Using the Law of Sines in Real Life

Real Life
Surveying

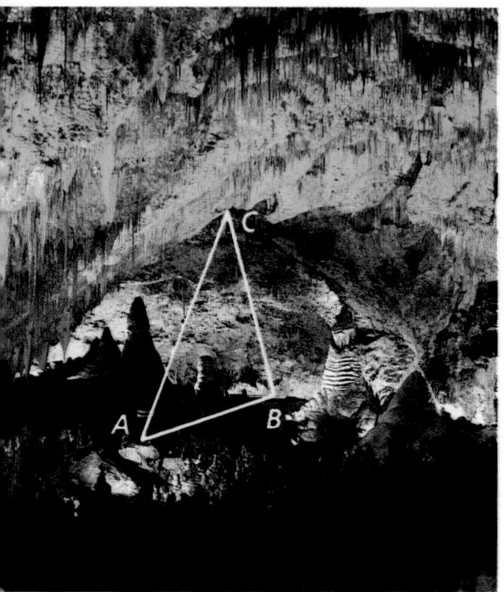

Calcium deposits have formed beautiful stalactites and stalagmites in the Carlsbad Caverns of New Mexico.

Example 3 *Exploring a Cave*

You are at position A and another surveyor is at position B. The distance between A and B is 50 feet. You each locate a point C on the ceiling of a cave and measure the angles to be 71° at A and 62° at B, as shown at the left. Point C lies directly above the horizontal line connecting A and B. Find the height of C.

Solution The angle C is given by $C = 180 - 71 - 62 = 47°$. Using the Law of Sines, you can find b.

$$\frac{b}{\sin B} = \frac{c}{\sin C}$$

$$\frac{b}{\sin 62°} = \frac{50}{\sin 47°}$$

$$b = \sin 62°\left(\frac{50}{\sin 47°}\right)$$

$$b \approx 60.4 \text{ feet}$$

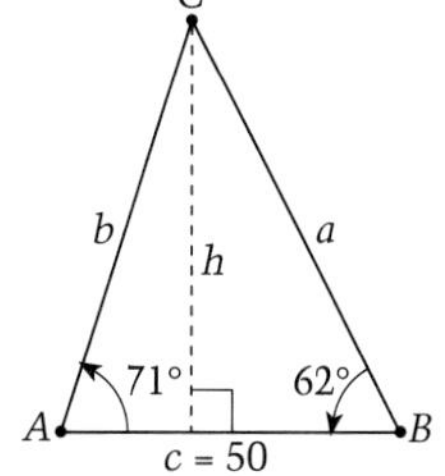

Now, because the height, h, is the side of a right triangle, you can use the sine function to solve for h.

$$\sin A = \frac{h}{b}$$

$$\sin 71° = \frac{h}{60.4}$$

$$60.4 \sin 71° = h$$

$$57.1 \approx h$$

Therefore, C is about 57.1 feet above the horizontal line connecting A and B. ■

Communicating about ALGEBRA

SHARING IDEAS about the Lesson

Use the Law of Sines to solve the triangle.

A.

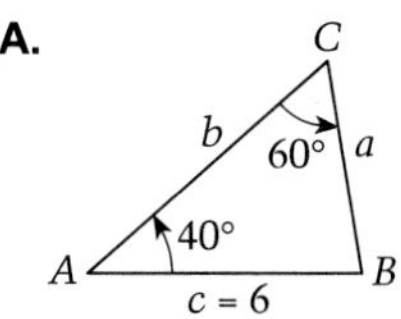

B.

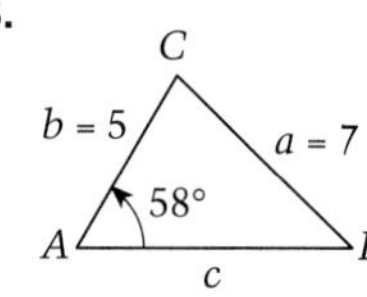

A. $\angle B = 80°$, $a \approx 4.45$, $b \approx 6.82$ B. $\angle B = 37°$, $\angle C = 85°$, $c = 8.3$

3. Given a triangle with $B = 65°$, $a = 20$, and $b = 30$, find the remaining side and angles.
   $A \approx 37.2°$, $C \approx 77.8°$, $c \approx 32.4$
4. Given a triangle with $B = 100°$, $c = 15$, and $b = 20$, find the remaining side and angles.
   $A \approx 32.4°$, $C \approx 47.6°$, $a \approx 10.88$
5. Find the area of each triangle given the following.
   a. $B = 55°$, $a = 10$, $c = 15$
   b. $A = 135°$, $b = 8$, $c = 20$
   a. 61.44 square units
   b. 56.57 square units

Check Understanding

1. What is an oblique triangle?
   A triangle that has no right angles
2. Can the Law of Sines be used for right triangles?
   Yes, but it reduces to $\frac{b}{\sin b} = c$ (when the right angle is at C). This is equivalent to the definition of sin b.
3. Can the remaining angles and side be found for every triangle described by SSA?
   Not necessarily; 0, 1, or 2 triangles can be determined by the given information.
4. Can the remaining angle and sides be found for every triangle described by ASA or AAS?
   Yes, only 1 triangle can be determined by the given information.

Communicating about ALGEBRA

COOPERATIVE LEARNING

Have students work in small groups to complete the following investigations.

For each triangle described below, make a careful sketch of the information and find the remaining side and angles, if possible. If two triangles can be found that fit the given information, include both solutions. If there is no solution, explain why.

a. Given a triangle with $B = 30°$, $a = 20$, and $b = 15$.
   A sketch indicates that two triangles are possible.
   One solution is:
   $A \approx 41.8°$, $C \approx 108.2°$, $c \approx 28.5$
   Other solution is:
   $A \approx 138.2°$, $C \approx 11.8°$, $c \approx 6.1$

b. Given a triangle with $A = 20°$, $a = 20$, and $b = 7$.
   $B \approx 6.9°$, $C \approx 153.1°$, $c \approx 26.4$

c. Given a triangle with $C = 57°$, $a = 12$, and $c = 7$.
   Values do not form a triangle.

EXERCISE Notes

ASSIGNMENT GUIDE

Basic/Average: Ex. 1–6, 13–25 odd, 43–53 odd, 57, 61, 62, 63–71 odd, 72

Above Average: Ex. 1–6, 9–23 odd, 29–39 odd, 45–59 odd, 62–64, 69–72

Advanced: Ex. 1–6, 9–23 odd, 29–39 odd, 45–57 odd, 58–64, 71–72

Selected Answers
Exercises 1–6, 7–69 odd

Use **Mixed Review** as needed.

More Difficult Exercises
Exercises 57–62

EXERCISES

Guided Practice

CRITICAL THINKING about the Lesson

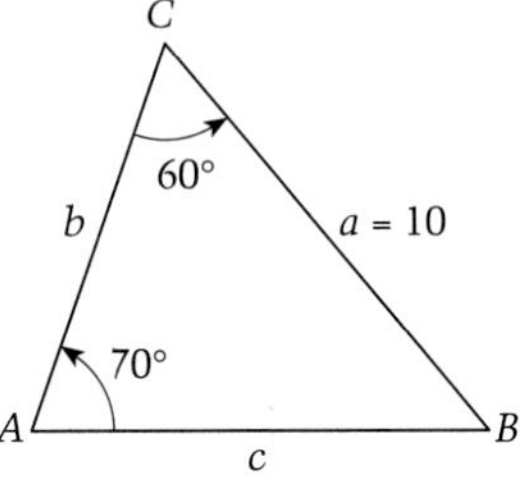

1. State the Law of Sines. See page 709.
2. Which of the following cannot be solved with the Law of Sines? d
   a. AAS **b.** ASA **c.** SSA **d.** SAS
3. Explain why there is no triangle that has sides of $a = 3$, $b = 4$, and $c = 8$. See below.
4. If you are given all three angles of a triangle, is it possible to determine the lengths of its sides? Explain.
5. Find c in the triangle at the right. ≈ 9.22
6. What is the area of the triangle at the right? ≈ 35.31

Independent Practice

In Exercises 7–30, solve the triangle. (*Hint:* Some of the "triangles" have no solution and some have two solutions.) 7.–14. See margin. 15.–30. See Additional Answers.

7.
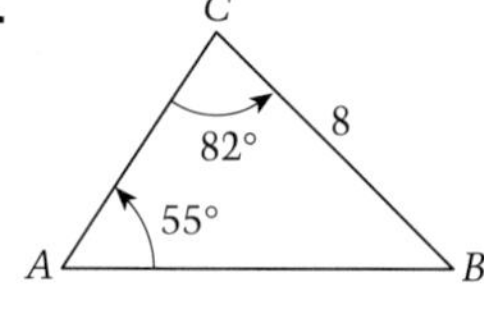

8.
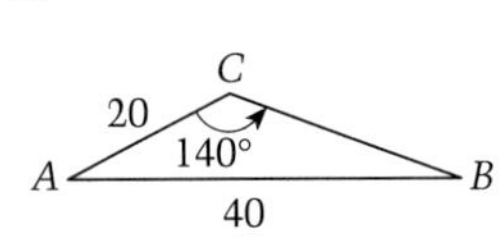

9.
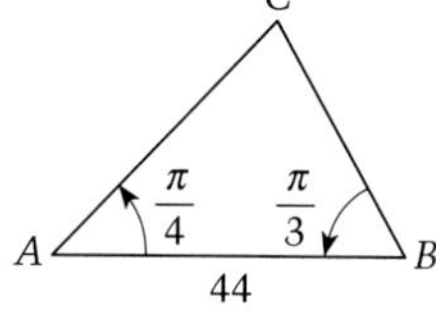

10.
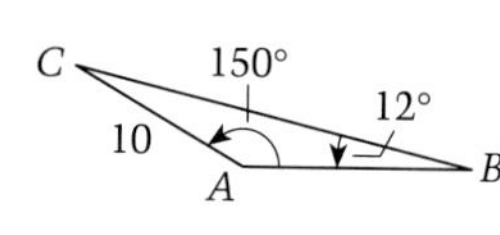

11.
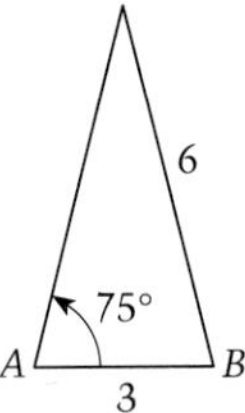

12.
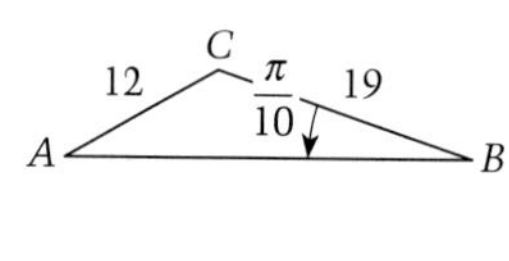

13.
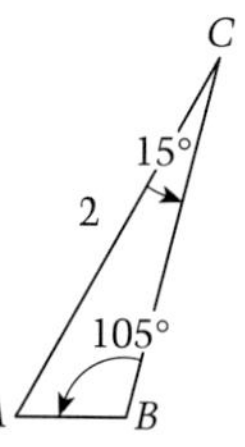

14.
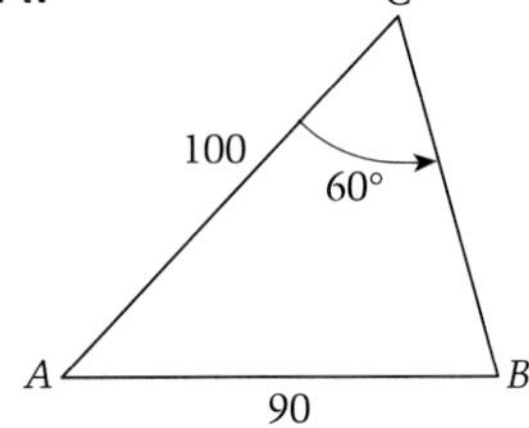

15. $A = 72°$, $C = 50°$, $a = 40$
16. $A = 18°$, $B = 28°$, $b = 100$
17. $B = \frac{\pi}{3}$, $b = 30$, $c = 20$
18. $B = 130°$, $a = 10$, $b = 8$
19. $B = 80°$, $C = 53°$, $a = 2$
20. $B = 18°$, $C = 152°$, $b = 4$
21. $A = 15°$, $C = 120°$, $b = 3$
22. $A = 45°$, $B = 100°$, $c = 15$
23. $C = 16°$, $b = 92$, $c = 32$
24. $C = 95°$, $a = 5$, $c = 6$
25. $C = \frac{3\pi}{4}$, $b = 48$, $c = 12$
26. $B = \frac{\pi}{4}$, $b = 27$, $c = 26$
27. $B = 110°$, $C = 30°$, $a = 12$
28. $A = 70°$, $B = 60°$, $c = 25$
29. $A = 85°$, $a = 23$, $c = 24$
30. $A = 20°$, $a = 8$, $c = 9$

3. The length of the longest side must be less than the sum of the lengths of the other two sides.

In Exercises 31–46, decide whether there are no solutions, exactly one solution, or two solutions. (You do not need to solve the triangle.) See margin.

31. 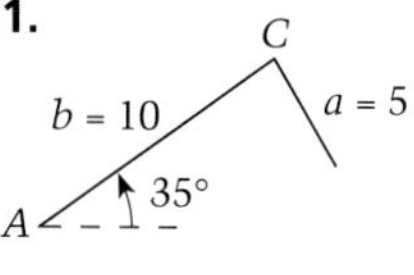

32. 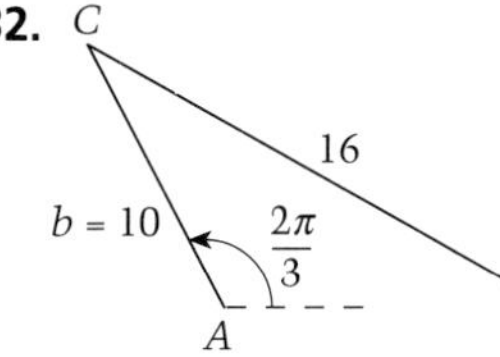

33. 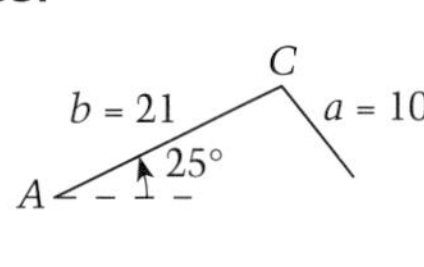

34. 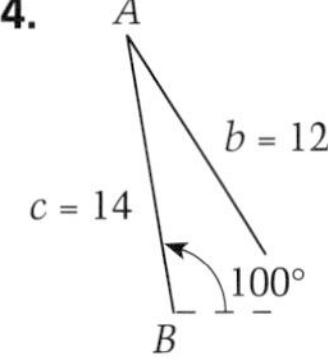

35. $C = 65°, c = 46, b = 30$

36. $B = 105°, b = 11, a = 5$

37. $A = 18°, a = 38, c = 50$

38. $A = 70°, a = 155, c = 160$

39. $B = \frac{4\pi}{9}, b = 90, a = 100$

40. $C = \frac{11\pi}{36}, c = 12, b = 16$

41. $A = 140°, a = 6, c = 7$

42. $A = 16°, a = 14, c = 10$

43. $C = 80°, c = 34, a = 20$

44. $B = \frac{\pi}{3}, b = 14, c = 15$

45. $B = \frac{\pi}{6}, b = 12, a = 9$

46. $C = 170°, c = 10, b = 15$

In Exercises 47–56, find the area of the triangle.

47. ≈54.38 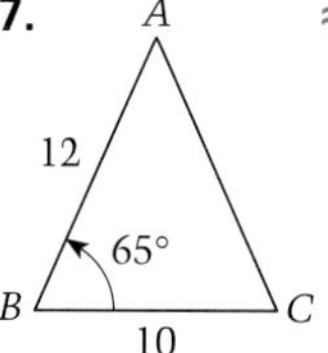

48. ≈108.19 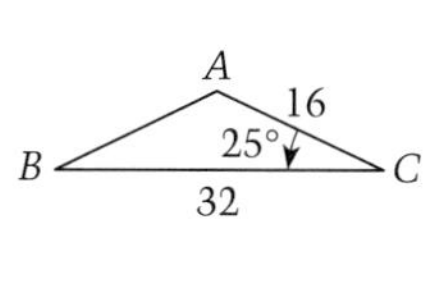

49. ≈12.99 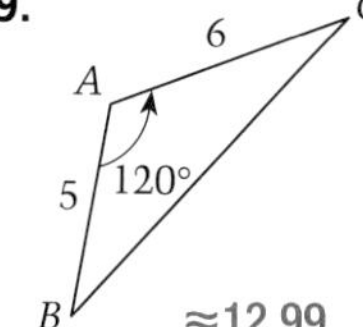

50. 7 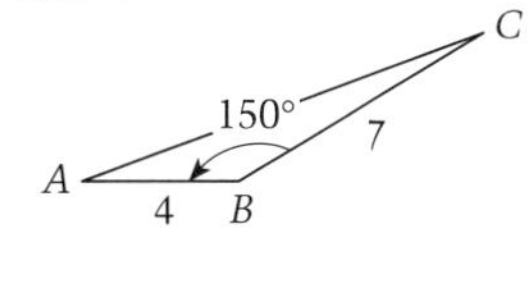

51. $A = 60°, b = 9, c = 12$ ≈46.77

52. $B = 25°, a = 15, c = 31$ ≈98.26

53. $C = 130°, a = 21, b = 17$ ≈136.74

54. $A = 85°, b = 11, c = 18$ ≈98.62

55. $B = \frac{\pi}{4}, a = 4, c = 1$ ≈1.41

56. $C = 140°, a = 6, b = 4$ ≈7.71

57. *Aqueduct* A reservoir supplies water through an aqueduct to Springfield, which is 15 miles southeast of the reservoir. A pumping station at Springfield pumps water 12 miles to Centerville, which is due east of the reservoir. Plans have been made to build an aqueduct directly from the reservoir to Centerville. How long will the aqueduct be? ≈16.22 mi

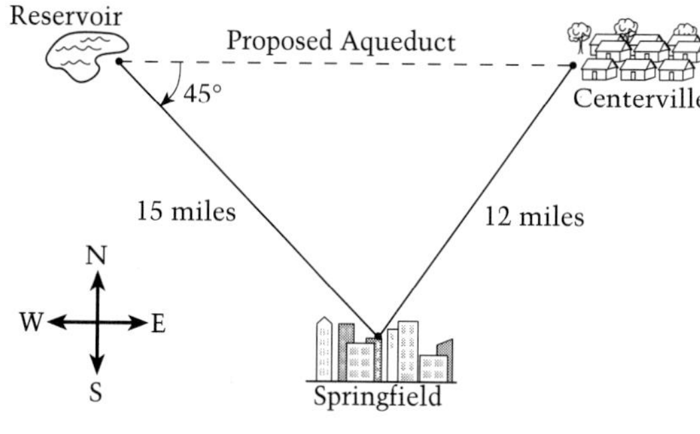

Hoover Dam, on the Colorado River in Nevada, supplies water for much of southern California through a 240-mile aqueduct.

Guided Practice

Ex. 1–6 COOPERATIVE LEARNING Have students work with partners to complete these exercises.

Answer

4. No; for example: if you are given that the three angles are 40°, 60°, and 80°, then many triangles of different sizes are possible.

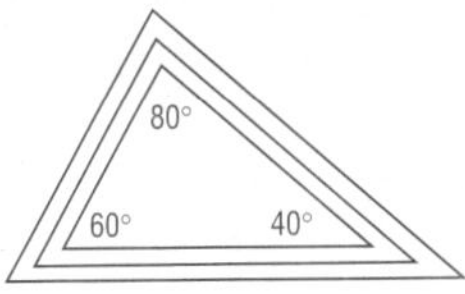

Independent Practice

Answers

7. $B = 43°, b \approx 6.66, c \approx 9.67$
8. $B \approx 18.7°, A \approx 21.3°, a \approx 22.60$
9. $C = \frac{5\pi}{12}, a \approx 32.21, b \approx 39.45$
10. $C = 18°, a \approx 24.05, c \approx 14.86$
11. $C \approx 28.9°, B \approx 76.1°, b \approx 6.03$
12. $A \approx 0.511, C \approx 2.316, c \approx 28.54$
13. $A = 60°, a \approx 1.79, c \approx 0.54$
14. $B \approx 74.2°, A \approx 45.8°, a \approx 74.49$

Ex. 15–30 Encourage students to draw a diagram for each problem to identify the type of information (ASA, AAS, SSA, or SSS) given in each exercise.

EXTEND Ex. 31–46 WRITING Ask students to write a one-sentence justification for their choice.

Answers

31., 34., 39–41., 46. No solutions **32., 35., 36., 42., 43., 45.** Exactly one solution **33., 37., 38., 44.** Two solutions

Ex. 47–56 Refer students to the area formula given at the bottom of page 710.

▶ **Ex. 58** Students will need to realize that the angle of depression to point *B* is congruent to the angle of elevation from point *B* to the airplane.

Answer

60. $A = a^2 + b^2 + c^2 + 2\left(\frac{1}{2}ab\right) + \frac{1}{2}bc \sin 150° + \frac{1}{2}ac \sin 120°$
$= 5^2 + (5\sqrt{3})^2 + 10^2 + 2\left(\frac{1}{2} \cdot 5 \cdot 5\sqrt{3}\right) + \frac{1}{2} \cdot 5\sqrt{3} \cdot 10 \cdot \frac{1}{2} + \frac{1}{2} \cdot 5 \cdot 10 \cdot \frac{\sqrt{3}}{2}$
$= 200 + 50\sqrt{3}$
$= 50\sqrt{3} + 200$
$= 2 \cdot 5 \cdot 5\sqrt{3} + 2 \cdot 10^2$
$= 2ab + 2c^2$

RESEARCH Students might be asked to find out more about Sophie Germaine.

58. *Measuring an Island* What is the width of the island in the diagram below? **≈5495.58 ft**

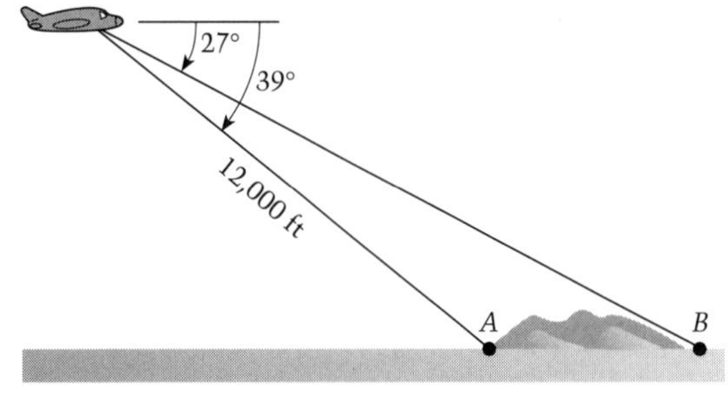

59. *Tracking Stations* What is the altitude of the airplane in the diagram below? **≈11.2 mi**

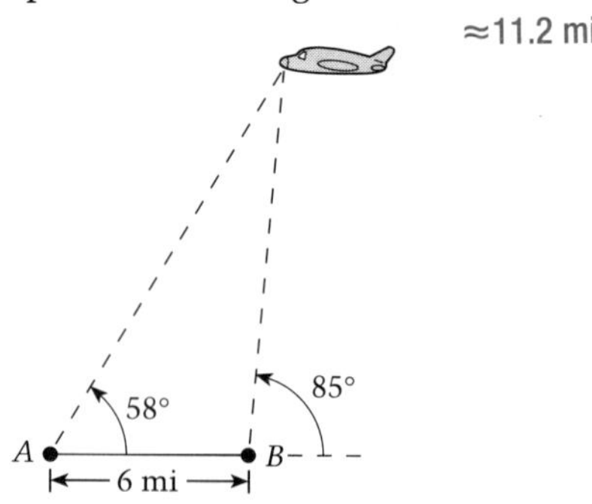

60. *Pythagorean Theorem* The area of the entire enclosed region is given by $A = 2ab + 2c^2$. Verify this formula for the values $a = 5$, $b = 5\sqrt{3}$, and $c = 10$. **See margin.**

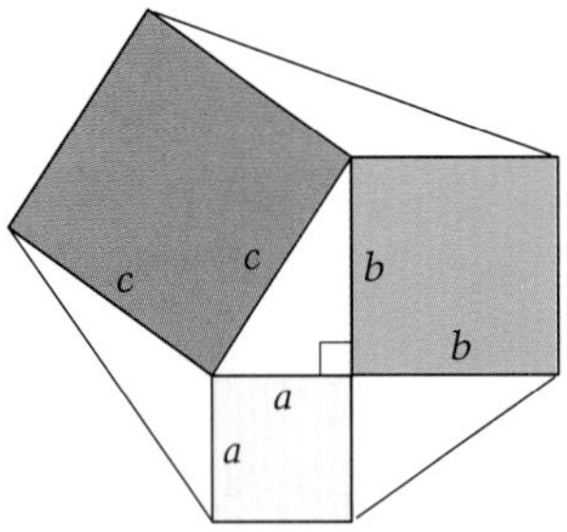

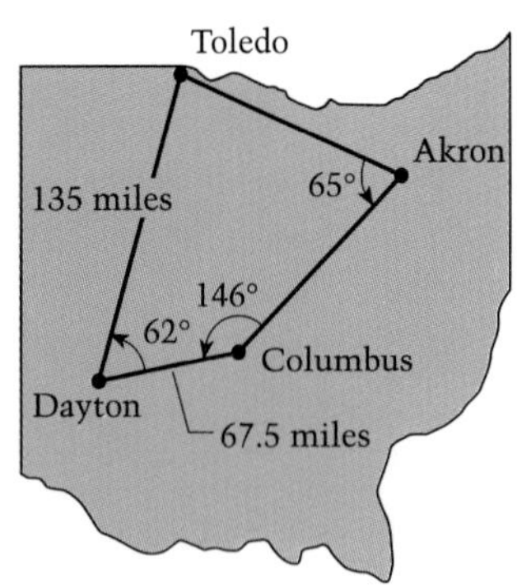

61. *Grand Piano* Use $C = 62°$, $A = 66°$, and $c = 4.5$ feet to find the length, b (in feet), of the longest string of the piano shown below. **≈4.0 ft**

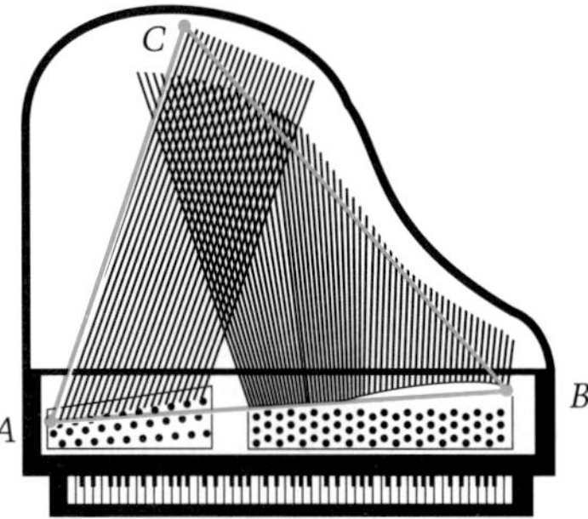

62. *Ohio Cities* Use the map below to find the straight-path distance between Toledo and Akron. Then find the straight-path distance between Akron and Columbus. (The straight-path distance between Dayton and Akron is 175 miles.) **≈118.6 mi, ≈104.0 mi**

Sophie Germain (1776–1831) was a French mathematician who worked in number theory and the theory of elasticity. This statue of her now stands in the courtyard of the École Sophie Germain, a lycée in Paris.

Integrated Review

In Exercises 63–66, find *x*.

63. ≈ 13.05

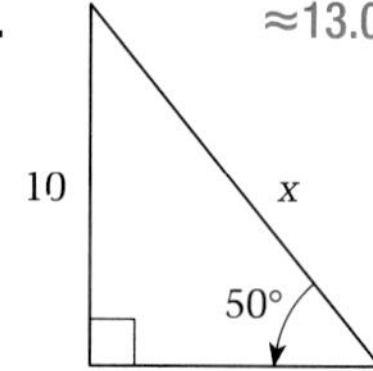

64. $\approx 71.8°$

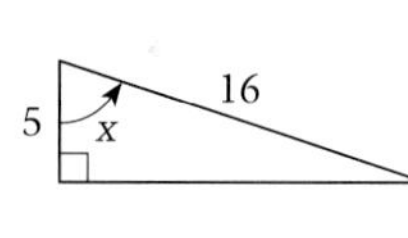

65. $\sqrt{3}$

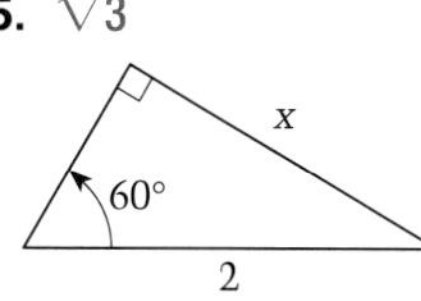

66. ≈ 28.98

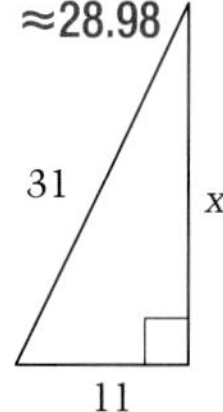

In Exercises 67–70, solve for *x*.

67. $\frac{14}{\sin 40°} = \frac{6}{\sin x}$ $\approx 16.0°$

68. $\frac{12}{\sin x} = \frac{14}{\sin 110°}$ $\approx 53.7°$

69. $\frac{10}{\sin x} = \frac{12}{\sin \frac{2\pi}{3}}$ ≈ 0.806

70. $\frac{4}{\sin 135°} = \frac{3}{\sin x}$ $\approx 32°$

Exploration and Extension

71. *Geometry* Measure the sides of the triangle at the right (in millimeters). Then use a protractor to measure the angles of the triangle. $\angle A \approx 56°$, $\angle B \approx 36°$, $\angle C \approx 88°$

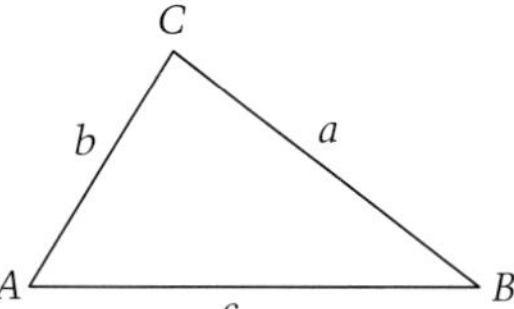

$a \approx 26$ mm, $b \approx 18$ mm
$c \approx 30$ mm

72. *Geometry* Use the Law of Sines to check the measurements you obtained in Exercise 71. How closely do your measurements conform to the Law of Sines? $\frac{\sin A}{a} \approx 0.030$, $\frac{\sin B}{b} \approx 0.026$, $\frac{\sin C}{c} \approx 0.030$; to the hundredths place

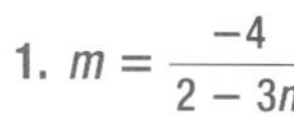

Mixed REVIEW

1. $m = \frac{-4}{2 - 3n}$ 8., 9. See Additional Answers.

1. Solve $4n + 2mn = 3mn^2$ for m. **(1.5)**

2. Graph $3x - 2y = 6$. **(2.3)** See margin.

3. Find the slope: $3y - 2x = 17$. **(2.3)** $\frac{2}{3}$

4. Solve $|3t - 8| \le 2$. **(1.7)** $2 \le t \le \frac{10}{3}$

5. Find the inverse of $\begin{bmatrix} 3 & 1 \\ 4 & 2 \end{bmatrix}$. **(4.4)** $\begin{bmatrix} 1 & -\frac{1}{2} \\ -2 & \frac{3}{2} \end{bmatrix}$

6. Evaluate $\sum_{n=1}^{\infty} 3\left(\frac{1}{4}\right)^{n-1}$. **(12.4)** 4

7. Solve $\begin{cases} 3x + y = -11 \\ 2x + 4y = 16 \end{cases}$. **(3.2)** $(-6, 7)$

8. Sketch the graph of $y = \sqrt{x + 2} - 3$. **(7.6)**

9. Sketch the graph of $y = 4^x - 5$. **(8.1)**

10. Expand $\ln\left(\frac{3x^2}{4y}\right)$. **(8.8)**

11. Multiply: $\frac{3x + 3}{4x} \cdot \frac{x^2}{x + 1}$. **(10.3)**

12. Solve $x^2 - 10x + 20 = 0$. **(5.4)** $5 \pm \sqrt{5}$

13. Find the vertex of $y = x^2 + 3x - 2$. **(5.2)**

14. Factor $4x^3 - 8x^2 - 5x$. **(9.5)**

15. Find $f(-2)$ for $f(x) = -3x^2$. **(6.1)** -12

16. Evaluate $\sin^2 30°$. **(13.1)** $\frac{1}{4}$

17. Find the standard deviation of 3, 6, 7, 4, 3, 9, 10, 11, 12. **(9.7)** 3.258

18. Find the *n*th term of the sequence $1, \frac{1}{8}, \frac{1}{27}, \frac{1}{64}, \frac{1}{125}, \ldots$ **(12.1)** $\frac{1}{n^3}$

10. $\ln 3 + 2\ln x - \ln 4 - \ln y$ 11. $\frac{3x}{4}$ 13. $\left(-\frac{3}{2}, -\frac{17}{4}\right)$ 14. $x(2x - 5)(2x + 1)$

Enrichment Activity

The following activity requires students to go beyond lesson goals.

TECHNOLOGY The following graphics calculator program uses the Law of Sines to solve a triangle when two angles and their common side are given. Use the program to verify the appropriate answers from Exercises 7–30.

```
Prgm #: LAWSIN
: ClrHome
: Disp "1ST ANGLE"
: Input A
: Disp "2ND ANGLE"
: Input B
: Disp "COMMON SIDE"
: Input F
: 180 − A − B → C
: F sin A / sin C → D
: F sin B / sin C → E
: ClrHome
: Disp "3 ANGLES = "
: Disp A
: Disp B
: Disp C
: Pause (Wait for ENTER key to be
  pressed)
: Disp "OPPOSITE SIDES = "
: Disp D
: Disp E
: Disp F
```

Notes:

- Set calculator to Degree mode.
- For Radian mode change $180 - A - B \rightarrow C$ in program line 8 to $\pi - A - B \rightarrow C$.

Answer

Mixed Review

2.

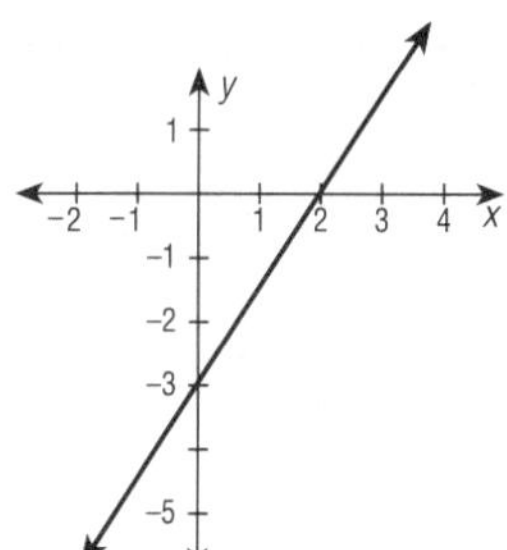

Problem of the Day

All of the curves in this square are semicircles. Find the area of the shaded region.

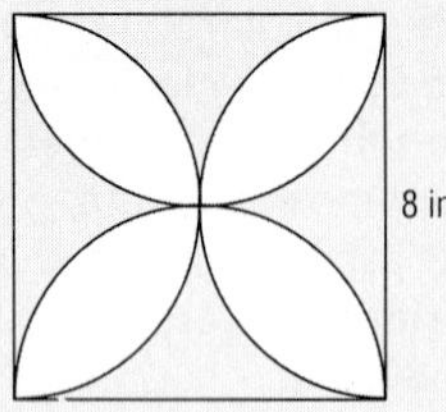

$(32\pi - 64)$ in.^{2}, or ≈ 36.53 in.^{2}

ORGANIZER

Warm-Up Exercises

1. Use the Law of Sines to find the unknown parts of the given triangle.
   a. $A = 35°$, $B = 120°$, and $a = 36$ feet.
   b. $C = 115°$, $b = 10$ meters, and $c = 27$ meters.

   a. $C = 25°$, $b \approx 54.36$ ft, $c \approx 26.53$ ft
   b. $A \approx 45.4°$, $B \approx 19.6°$, $a \approx 21.21$ m

Lesson Resources

Teaching Tools
- Problem of the Day: 13.6
- Warm-up Exercises: 13.6
- Answer Masters: 13.6

Extra Practice: 13.6
Color Transparency: 71
Applications Handbook: p. 38

13.6 The Law of Cosines

What you should learn:

Goal 1 How to use the Law of Cosines to find the sides and angles of any triangle

Goal 2 How to use the Law of Cosines to solve problems in real life

Why you should learn it:

You can use the Law of Cosines to solve for many distances that cannot be measured directly, such as the distance from a starting point when a change of direction is involved.

Goal 1 Using the Law of Cosines

Two cases of oblique triangles left to solve are SSS and SAS. Both can be solved with the **Law of Cosines.**

> **Law of Cosines**
>
> Let ABC be a triangle with sides a, b, and c.
>
> | **Standard Form** | **Alternative Form** |
> |---|---|
> | $a^2 = b^2 + c^2 - 2bc \cos A$ | $\cos A = \dfrac{b^2 + c^2 - a^2}{2bc}$ |
> | $b^2 = a^2 + c^2 - 2ac \cos B$ | $\cos B = \dfrac{a^2 + c^2 - b^2}{2ac}$ |
> | $c^2 = a^2 + b^2 - 2ab \cos C$ | $\cos C = \dfrac{a^2 + b^2 - c^2}{2ab}$ |

Example 1 Given Three Sides of a Triangle—SSS

Find the three angles of the triangle whose sides have lengths $a = 8$ feet, $b = 19$ feet, and $c = 14$ feet.

B; a = 8; c = 14; C; A; b = 19

Solution First find the angle opposite the longest side—side b in this case. Using the Law of Cosines, you can write

$$\cos B = \frac{a^2 + c^2 - b^2}{2ac} = \frac{8^2 + 14^2 - 19^2}{2(8)(14)} \approx -0.451.$$

Using the inverse cosine function, you can find B to be the obtuse angle given by $B \approx \arccos(-0.451) \approx 116.8°$. Now use the Law of Sines to solve for A.

$$\frac{\sin A}{a} = \frac{\sin B}{b}$$

$$\sin A = a\left(\frac{\sin B}{b}\right) \approx 8\left(\frac{\sin 116.8°}{19}\right) \approx 0.376$$

Thus $A \approx \arcsin 0.376 \approx 22.1°$. Finally, you can find angle C.

$$C \approx 180° - 22.1° - 116.8° = 41.1°$$ ■

In Example 1, notice that we solved for the largest angle *first* because then you know whether the triangle is acute or obtuse. After you have found the largest angle, you know that both of the other angles must be acute.

The Law of Cosines can be used to establish the following formula for the area of a triangle. This formula is credited to the Greek mathematician Hero (*circa* 100 B.C.).

Hero's Area Formula

The area of the triangle with sides of length a, b, and c is

$$\text{Area} = \sqrt{s(s-a)(s-b)(s-c)}$$

where $s = \frac{1}{2}(a + b + c)$.

Goal 2 Using the Law of Cosines in Real Life

Example 2 *Given Two Sides and the Included Angle—SAS*

The pitcher's mound on a softball field is 46 feet from home plate. The distance between the bases is 60 feet. How far is the pitcher's mound from first base? How much closer is the pitcher's mound to second base than it is to first base?

Solution Begin by forming the triangle HPF. In this triangle, you know that $H = 45°$ because the line HP bisects the right angle at home plate. From the given information, you know that $f = 46$ and $p = 60$. Using the Law of Cosines, you can solve for h.

$$\begin{aligned} h^2 &= f^2 + p^2 - 2fp \cos H \\ &= 46^2 + 60^2 - 2(46)(60) \cos 45° \\ &\approx 1812.8 \end{aligned}$$

60 ft, 60 ft, 60 ft, P, h, F, f = 46 ft, 45°, p = 60 ft, H

Softball Diamond

The approximate distance from the pitcher's mound to first base is

$$h \approx \sqrt{1812.8} \approx 42.58 \text{ feet.}$$

The distance between home plate and second base is $\sqrt{60^2 + 60^2} \approx 84.85$ feet, which implies that the distance between the pitcher's mound and second base is

$$84.85 - 46 \approx 38.85 \text{ feet.}$$

Thus the pitcher's mound is about 4 feet closer to second base than it is to first base. ■

LESSON Notes

Refer students to the *Handbook of Mathematical Connections* for a proof of the Law of Cosines.

Example 1

Observe that it is acceptable to solve for, say, angle A first. However, it is important to sketch the triangle so that the relative sizes of the angles can be determined, in which case, you would have a sketch similar to the one shown in the text. Then $\cos A = \frac{19^2 + 14^2 - 8^2}{2(19)(14)} \approx 0.92669$, so that $A \approx 22.1°$.

In a manner similar to the approach used in this example, and noting the sketch of the triangle, you can use the Law of Sines to find angle B.

$$\begin{aligned} \sin B &= 19\left(\frac{\sin 22.1°}{8}\right) \\ &\approx 0.89353 \end{aligned}$$

and $B \approx 63.3°$ or $116.7°$. But $B \approx 116.7°$ is more appropriate for the sketch. Finally, $C \approx 41.2°$. Also note that the measures of the angles in this solution are slightly different from those of Example 1.

Example 2

Ask students to determine the remaining angles of triangle HPF.
$P \approx 85.2°$, $F \approx 49.8°$

Example 3

Have students determine the other angles of triangle ABC.
$A \approx 8.6°$, $C \approx 6.4°$

Extra Examples

Here are additional examples similar to **Examples 1–3.**

1. Find the three angles of the triangle whose sides have lengths $a = 9$, $b = 14$, $c = 21$.
   $A \approx 19.0°$, $B \approx 30.4°$, $C \approx 130.6°$

2. Find the remaining angles and side of the triangle whose sides have lengths $a = 34$ and $c = 28$, and whose angle $B = 92°$.
$A \approx 49.3°$, $C \approx 38.7°$, $b \approx 44.8$
3. Find the three angles of the triangle whose sides have lengths $a = 11$, $b = 8$, $c = 15$.
$A \approx 45.6°$, $B \approx 31.3°$, $C \approx 103.1°$
4. Use Hero's Formula to find the areas of the triangles in Extra Examples 1–3 above.
   1. ≈47.8 square units
   2. ≈475.7 square units
   3. ≈42.8 square units

Check Understanding

1. In the standard form of the Law of Cosines given by $a^2 = b^2 + c^2 - 2bc \cos A$, describe the situation in which the term, $-2bc \cos A$, equals zero.
When angle A is 90°
2. Must the Law of Sines also be used after initially applying the Law of Cosines as shown in Example 1?
No
3. Why is it a good idea when using the Law of Cosines to identify the angle with the greatest measure first?
See the top of page 717.

Communicating about ALGEBRA

Ask students to find the area of each of the following.

a. Parallelogram $MNOP$ whose adjacent sides are 18 and 12 feet, with an angle of 135°.
b. Rhombus $ABCD$ whose sides measure 15 cm, with one diagonal 20 cm long.
c. Parallelogram $HIJK$ whose diagonals are 50 and 60 inches long, and form an angle of 60°.

a. $\approx 152.74\text{ ft}^2$,
b. $\approx 223.61\text{ cm}^2$,
c. $\approx 1299.04\text{ in.}^2$

Real Life Navigation

The Amoco Procyon won one of Popular Science's "Best of What's New" awards in 1991. The yacht has many new features including a computer navigation system.

Example 3 Given Two Sides and the Included Angle—SAS

A boat travels 60 miles due east. Then it adjusts its course 15° northward and travels 80 miles, as shown below. How far is the boat from its point of departure?

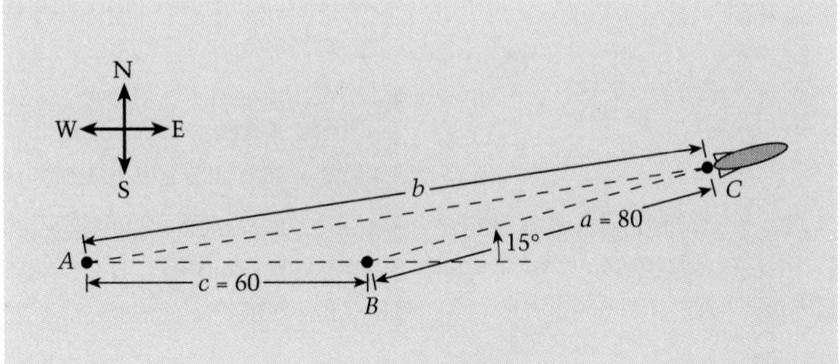

Solution In the above triangle, you know that $a = 80$, $c = 60$, and

$$B = 180° - 15° = 165°.$$

Using the Law of Cosines, you can find b.

$$\begin{aligned} b^2 &= a^2 + c^2 - 2ac \cos B \\ &= 80^2 + 60^2 - 2(80)(60) \cos 165° \\ &\approx 19{,}273 \end{aligned}$$

Therefore, the distance from the point of departure is

$$b \approx \sqrt{19{,}273} \approx 138.8 \text{ miles.}$$

■

Communicating about ALGEBRA

SHARING IDEAS about the Lesson

You have studied several formulas for the area of a triangle. Find the area of each of the following triangles. In each case, use the formula that you think is most appropriate and explain your choice.

A. $\approx 3.06\text{ ft}^2$ **B.** $\approx 2.90\text{ ft}^2$ **C.** $\approx 4\text{ ft}^2$

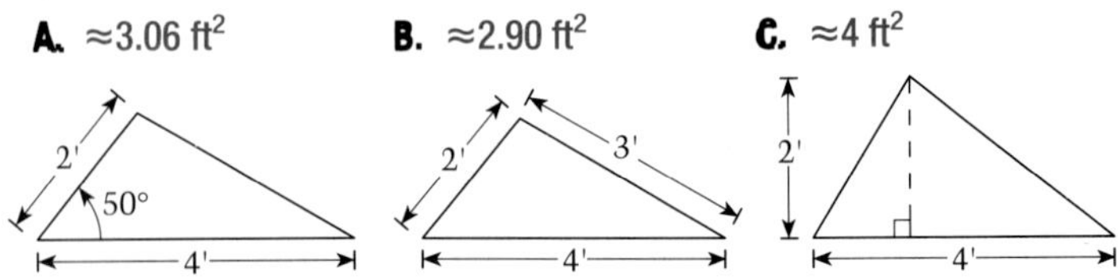

A. $\approx 3.06\text{ ft}^2$, use the formula derived from the Law of Sines because two sides and the included angle are given.
B. $\approx 2.90\text{ ft}^2$, use Hero's Formula because three sides are given.
C. 4 ft^2, use the formula involving the height and the base because they are given.

EXERCISES

Guided Practice

CRITICAL THINKING about the Lesson

1. Of the triangles represented by AAS, ASA, SSA, SAS, and SSS, which are best solved with the Law of Cosines? SAS and SSS
2. Describe how to find angle B in the triangle at the right. See margin.
3. Which form of the Law of Cosines would you use to find a, given b, c, and A? $a^2 = b^2 + c^2 - 2bc \cos A$
4. A triangle has two sides of lengths 12 feet and 16 feet. The angle between these two sides is 30°. What is the length of the other side? ≈8.21 ft

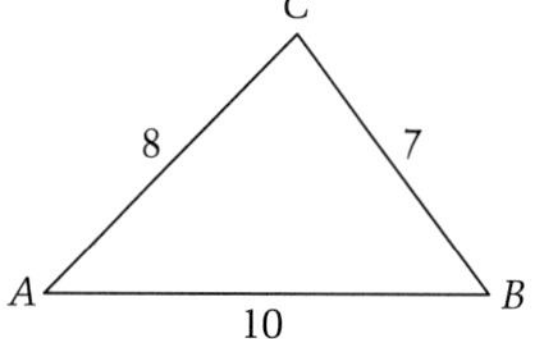

Independent Practice

In Exercises 5–30, find the remaining angles and sides. 5.–12. See margin. 13.–30. See Additional Answers.

5. 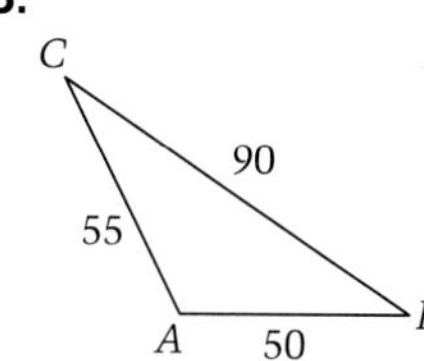

6. 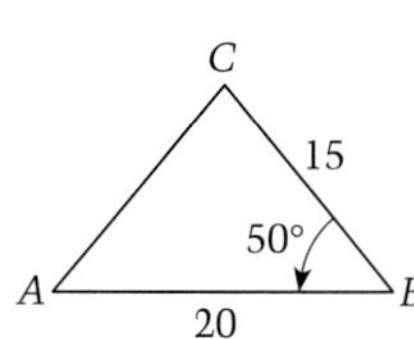

7. 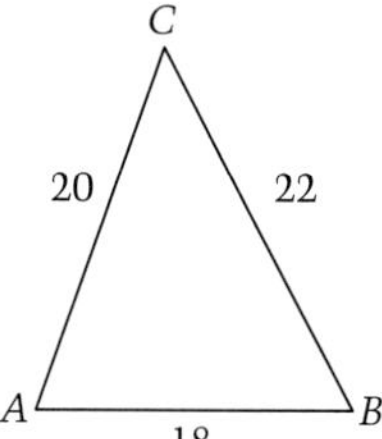

8. 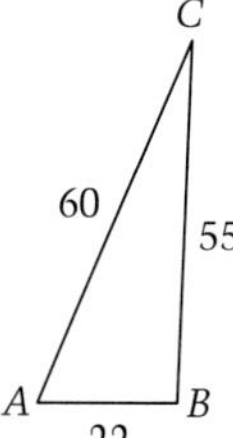

9. 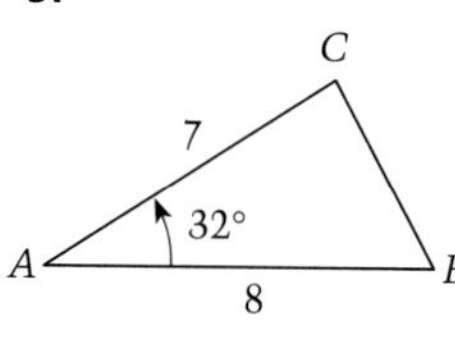

10. 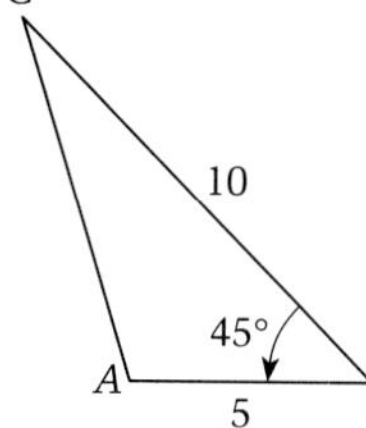

11. 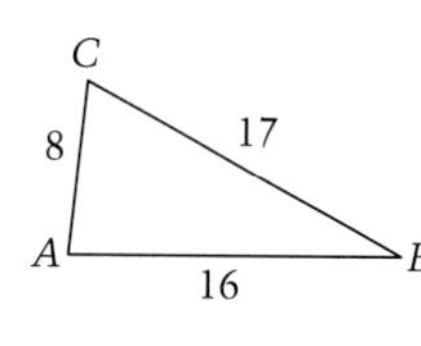

12. 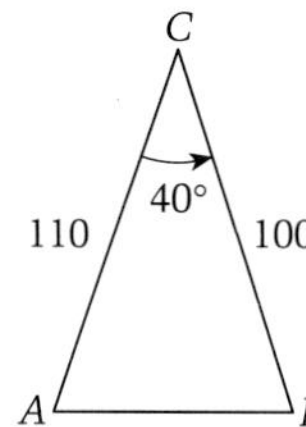

13. $B = 20°, a = 120, c = 100$
14. $C = 95°, a = 5, b = 6$
15. $a = 30, b = 15, c = 29$
16. $a = 2, b = 4, c = 5$
17. $A = 78°, b = 2, c = 4$
18. $A = 60°, b = 50, c = 48$
19. $B = 45°, a = 9, c = 19$
20. $C = 30°, a = 15, b = 15$
21. $a = 9, b = 3, c = 11$
22. $B = 15°, a = 12, c = 6$
23. $a = 25, b = 26, c = 5$
24. $a = 39, b = 20, c = 54$
25. $C = \frac{3\pi}{4}, a = 1, b = 5$
26. $B = 100°, a = 25, c = 33$
27. $A = 98°, b = 11, c = 19$
28. $a = 1, b = 1, c = \sqrt{2}$
29. $a = 3, b = 4, c = 5$
30. $a = 6, b = 10, c = 6$

EXERCISE Notes

ASSIGNMENT GUIDE

Basic/Average: Ex. 1–4, 7–23 odd, 33–41 odd, 44, 47, 48, 51, 55

Above Average: Ex. 1–4, 7–23 odd, 31–39 odd, 43–48, 55, 56

Advanced: Ex. 1–4, 7–23 odd, 29–39 odd, 43–48, 56, 57

Selected Answers
Exercises 1–4, 5–55 odd

✪ More Difficult Exercises
Exercises 43–48, 57

Guided Practice

▶ **Ex. 1** Refer students to Examples 1 and 2 for solution models.

▶ **Ex. 2** Remind students that they must use the inverse trigonometric functions from Lesson 13.4 to find angle B.

Answer

2. Use the formula $\cos B = \frac{a^2 + c^2 - b^2}{2ac}$, substituting for a, b, and c, and solving for B.

▶ **Ex. 4** Encourage students to draw a diagram.

Answers

5. $A \approx 117.9°$, $B \approx 32.7°$, $C \approx 29.4°$
6. $b \approx 15.47$, $A \approx 48.0°$, $C \approx 82.0°$
7. $A \approx 70.5$, $B \approx 59.0°$, $C \approx 50.5°$
8. $A \approx 66.4°$, $B \approx 88.5°$, $C \approx 25.1°$
9. $a \approx 4.24$, $B \approx 61.0°$, $C \approx 87.1°$
10. $b \approx 7.37$, $A \approx 106.3°$, $C \approx 28.7°$
11. $A \approx 83.0°$, $B \approx 27.8°$, $C \approx 69.2°$
12. $b \approx 72.44$, $A \approx 62.5°$, $B \approx 77.5°$

Independent Practice

▶ **Ex. 13–30, 43, 48** Encourage students to draw a diagram for each problem.

▶ **Ex. 31–42** Students should use Hero's Formula listed on page 717.

EXTEND Ex. 46 Ask students who have been to Epcot Center to describe the *Spaceship Earth* exhibit.

EXTEND Ex. 48 PROJECT Ask students to write a report on the connections between architectural designs for buildings and geometric figures as shown in this exercise.

▶ **Ex. 49–56** Assign these exercises for practice in either method.

Answer

43. Distance between the pitcher's mound and second base is $(\sqrt{90^2 + 90^2} - 60.5)$, or ≈ 66.78, ft. Distance between the pitcher's mound and first base is $(\sqrt{60.5^2 + 90^2 - 2(60.5)(90)\cos 45°})$, or ≈ 63.72, ft. 66.78 ft − 63.72 ft = 3.06 ft

In Exercises 31–42, find the area of the triangle.

31. ≈116.31

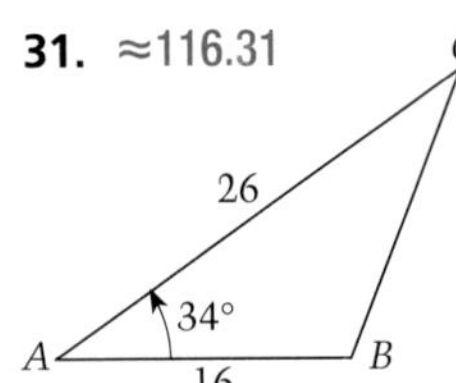

32. ≈158.75

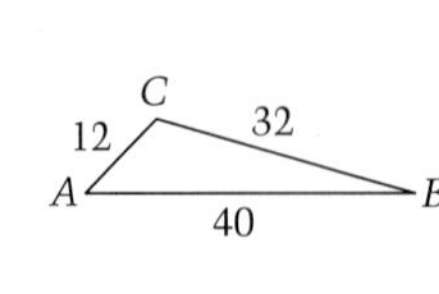

33. ≈404.54

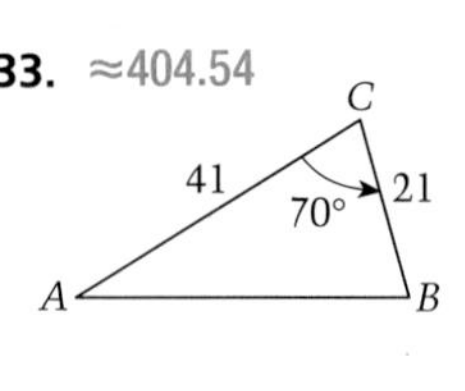

34. ≈56.92

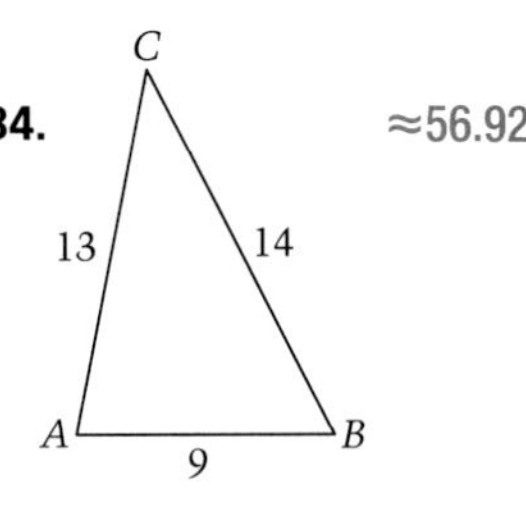

35. $a = 25, b = 60, c = 45$ ≈509.90

36. $a = 11, b = 2, c = 12$ ≈9.92

37. $a = 100, b = 55, c = 61$ ≈1467.05

38. $a = 4, b = 24, c = 26$ ≈43.16

39. $a = 2, b = \frac{1}{2}, c = 2$ ≈0.50

40. $a = 20, b = 21, c = 37$ ≈163.33

41. $a = 8, b = 8, c = 8$ ≈27.71

42. $a = 12, b = 9, c = 5$ ≈20.40

43. *Baseball Diamond* The pitcher's mound on a baseball field is 60.5 feet from home plate. The distance between the bases is 90 feet. How much closer is the pitcher's mound to first base than it is to second base? Explain. 3.06 ft

44. *Measuring a Pond* How long is the pond shown in the diagram at the right? ≈110.02 ft

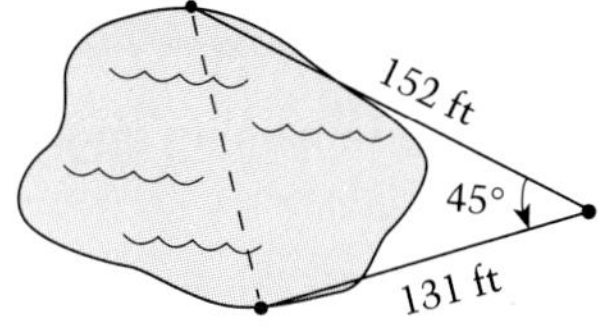

45. *Distance Between Two Airplanes* Two airplanes leave the same airport in opposite directions. At 1:00 P.M., the angle of elevation from the airport to the first airplane is 25° and to the second airplane is 16°. The elevation of the first airplane is 8 miles. The elevation of the second airplane is 6 miles. Find the air distance between the two airplanes. ≈38.14 mi

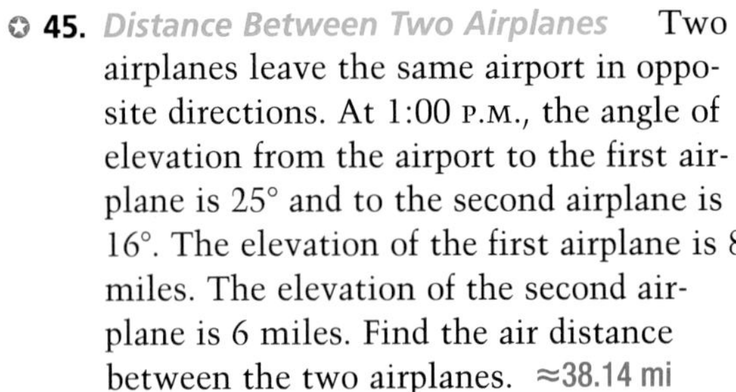

46. *Spaceship Earth* One of the seven different sizes of panels on *Spaceship Earth* is shown at the right. What are the angles of the outer triangle of the panel? What is the area? ≈54.03°, ≈54.03°, ≈71.94°; ≈13,099.06 in.^{2}

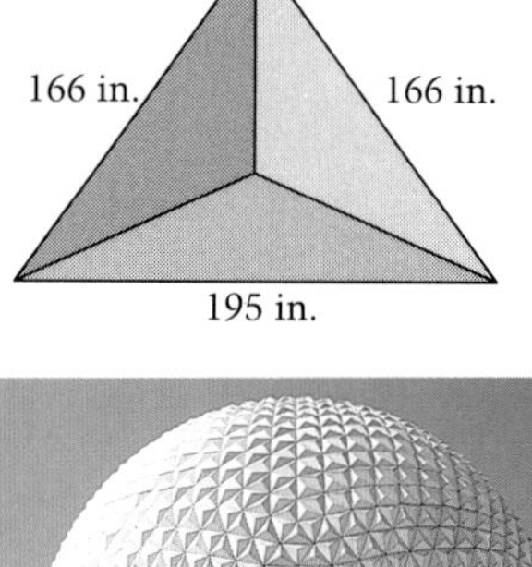

47. *Flying to Orlando* As your flight is descending to Orlando, the pilot announces that the airplane's elevation is 18,000 feet and the ground distance to the airport is 20 miles. Later the pilot announces that the airplane's elevation is 11,000 feet and the ground distance to the airport is 8 miles. How far (in a straight air path) did the airplane travel between the two announcements? ≈12.07 mi

Spaceship Earth, ***at Disney World's Epcot Center, is the first completely spherical geodesic dome ever built.***

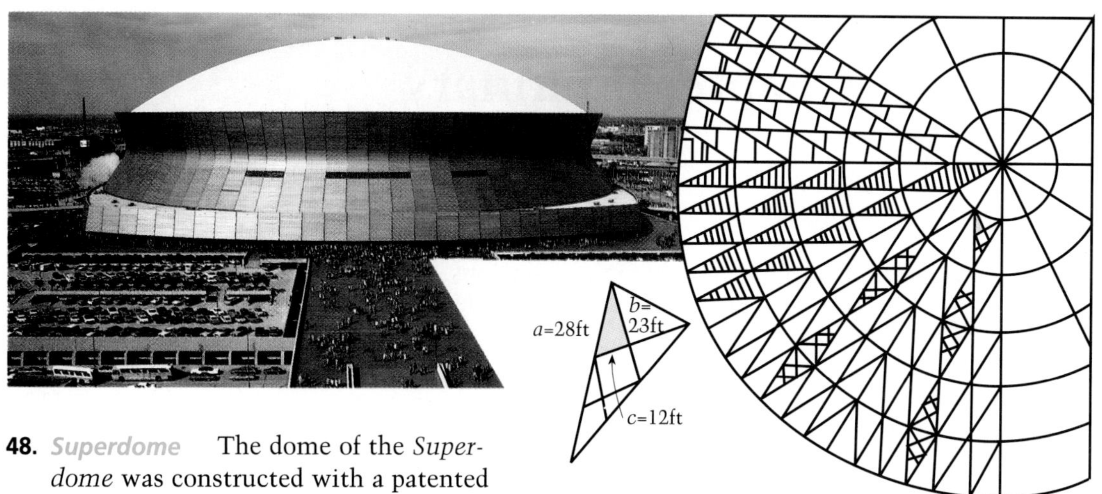

48. *Superdome* The dome of the *Superdome* was constructed with a patented configuration composed of hundreds of triangles of various shapes and sizes. Find the angles of the blue triangle at the right.
≈24.8°, ≈53.6°, ≈101.6°

The Superdome in New Orleans, Louisiana, is the world's largest indoor covered stadium. It can seat over 75,000 people. The dome has a diameter of 680 feet and is 273 feet high at its center.

Integrated Review

In Exercises 49–56, use the Law of Sines or the Law of Cosines to solve the triangle. See Additional Answers.

49.

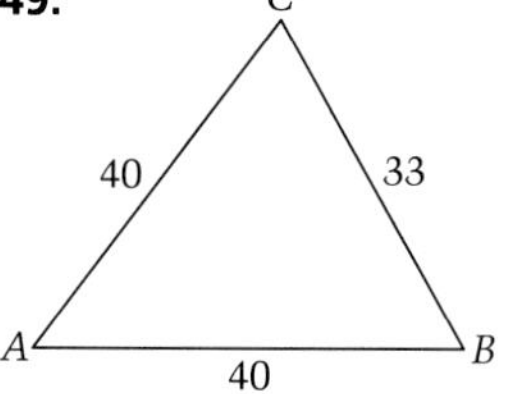

50.

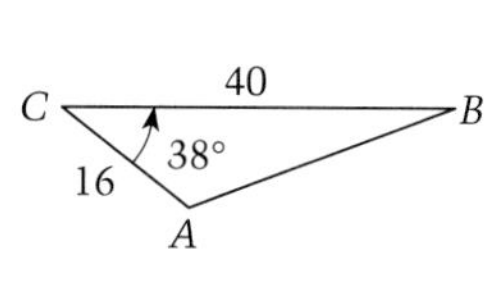

51.

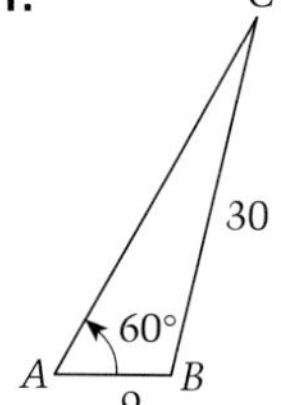

52.

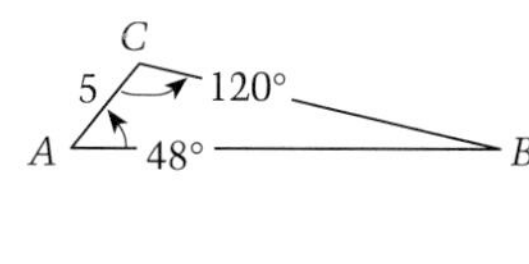

53. $A = \frac{\pi}{4}$, $B = \frac{\pi}{6}$, $a = 26$

54. $B = 62°$, $a = 4$, $c = 5$

55. $B = 15°$, $C = 135°$, $b = 15$

56. $A = 91°$, $C = 10°$, $b = 100$

Exploration and Extension

57. *Mirrors* In the diagram at the right, a beam of light is directed at the blue mirror, reflected to the red mirror, then reflected back to the blue mirror. Find the length (PT) that the light travels from the red mirror back to the blue mirror given that $OQ = 6$ feet and $OP = 4.7$ feet. ≈2.01 ft

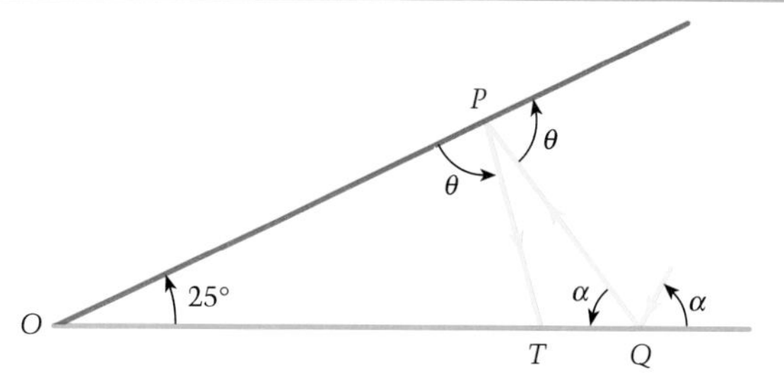

Enrichment Activity

The following activity requires students to go beyond lesson goals.

TECHNOLOGY The following graphics calculator program uses the Law of Cosines to solve a triangle when two sides and their included angle are given. Use the program and that shown on page 715 to verify the appropriate answers for Exercises 5–30.

```
Prgm #: LAWCOS
: ClrHome
: Disp "1ST SIDE"
: Input A
: Disp "2ND SIDE"
: Input B
: Disp "INCLUDED ANGLE"
: Input F
: √(A² + B² − 2AB cos F) → C
: cos⁻¹ ((B² + C² − A²) / 2BC) → D
: cos⁻¹ ((A² + C² − B²) / 2AC) → E
: ClrHome
: Disp "3 ANGLES = "
: Disp D
: Disp E
: Disp F
: Pause (Wait for ENTER key to be pressed)
: Disp "OPPOSITE SIDES = "
: Disp A
: Disp B
: Disp C
```

Note: You may want to set the calculator to Degree mode.

ORGANIZER

TIME SCHEDULE: All levels, two days

The Chapter Summary helps students organize the main ideas of the chapter. In this chapter, trigonometric functions were introduced and examined. Real-life situations and problems involving trigonometric relationships were presented and solved. Two important formulas for solving problems that can be modeled by trigonometric ideas were the Law of Sines and the Law of Cosines.

Work with students to review the formulas, strategies, and concepts of the chapter. The first day's homework assignment can be used as the basis for the second day of review.

COOPERATIVE LEARNING

Encourage students to study together. Emphasize the importance of "teaching" a classmate how to perform a skill or how to recall a strategy. When students work together, everyone wins. The students receiving help get additional instruction and the students giving help gain a deeper understanding of the strategies and concepts involved.

Chapter SUMMARY

Review the chapter by asking students "What did you learn? Why did you learn it? How does it fit with the other algebra skills and concepts you have learned?"

13 Chapter Summary

What did you learn?

Skills

1. Evaluate trigonometric functions
   - of an acute angle. **(13.1)**
   - of any angle. **(13.3)**
2. Use degree and radian measure to measure any angle.
   - Write equivalent degree and radian measures. **(13.2)**
3. Evaluate inverse trigonometric functions. **(13.4)**
4. Solve a triangle.
   - Use trigonometric functions to solve a right triangle. **(13.1)**
   - Use Law of Sines to solve an oblique triangle. **(13.5)**
   - Use Law of Cosines to solve an oblique triangle. **(13.6)**

Strategies

5. Use trigonometric functions to solve real-life problems. **(13.1–13.6)**
6. Use inverse trigonometric functions to solve real-life problems. **(13.4–13.6)**

Why did you learn it?

Many real-life problems can be solved with trigonometry. The most common type of problem involves a triangle in which some of the sides and angles are *known* and others need to be *determined.* This use of trigonometry is common in surveying, engineering, astronomy, and navigation. For instance, in Example 5 on page 681, you saw how trigonometry could be used to measure the elevation of a mountain.

You also learned how trigonometry can be used to measure circular arcs. For instance, on a tire with a 12-inch radius, a central angle of $\frac{\pi}{3}$ radians corresponds to an arc length of $s = r\theta = 12\left(\frac{\pi}{3}\right) = 8\pi$ inches.

How does it fit into the bigger picture of algebra?

There are two basic types of trigonometry. The type you studied in this chapter deals with trigonometric functions of *angles.* Most of the applications of this type of trigonometry involve triangles or arcs of circles. In the next chapter, you will study the second basic type of trigonometry, which deals with trigonometric functions of *real numbers.*

Almost all practical uses of trigonometry require algebra. To solve the problems in this chapter, you used many algebraic skills, including simplification techniques and techniques for solving equations.

Chapter REVIEW

In Exercises 1–4, find the six trigonometric functions of θ. (13.1) See margin.

1. 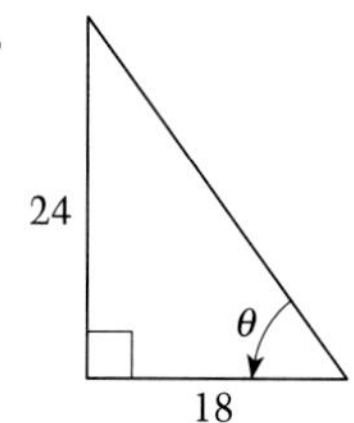

2. 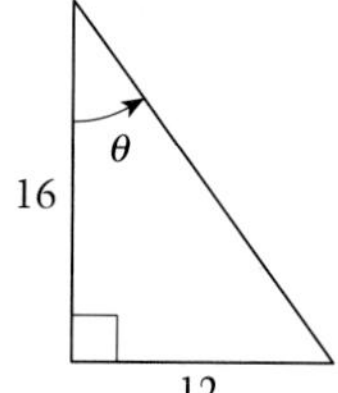

3. 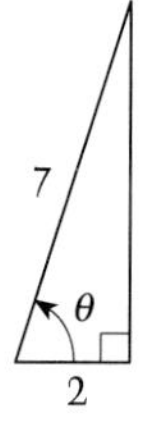

4. 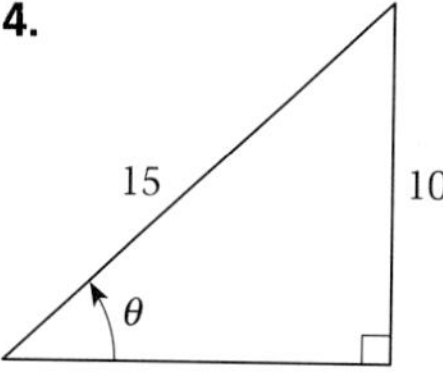

5. Given $\cos\theta = \frac{5}{13}$, find the other five trigonometric functions of θ. (Assume that θ is acute.) **(13.1)** $\sin\theta = \frac{12}{13}$, $\tan\theta = \frac{12}{5}$, $\csc\theta = \frac{13}{12}$, $\sec\theta = \frac{13}{5}$, $\cot\theta = \frac{5}{12}$

6. Given $\tan\theta = \frac{3}{4}$, find the other five trigonometric functions of θ. (Assume that θ is acute.) **(13.1)** $\sin\theta = \frac{3}{5}$, $\cos\theta = \frac{4}{5}$, $\csc\theta = \frac{5}{3}$, $\sec\theta = \frac{5}{4}$, $\cot\theta = \frac{4}{3}$

In Exercises 7–12, find the complement of the angle. (13.2)

7. 40° 50°
8. 29° 61°
9. $\frac{\pi}{3}$ $\frac{\pi}{6}$
10. $\frac{2\pi}{11}$ $\frac{7\pi}{22}$
11. $\frac{4\pi}{9}$ $\frac{\pi}{18}$
12. $\frac{\pi}{7}$ $\frac{5\pi}{14}$

In Exercises 13–18, find the supplement of the angle. (13.2)

13. 95° 85°
14. 163° 17°
15. 80° 100°
16. $\frac{6\pi}{7}$ $\frac{\pi}{7}$
17. $\frac{3\pi}{4}$ $\frac{\pi}{4}$
18. $\frac{2\pi}{3}$ $\frac{\pi}{3}$

In Exercises 19–24, find two angles, one with positive measure and the other with negative measure, that are coterminal with the given angle. (13.2)

19. 355° 715°, −5°
20. $\frac{3\pi}{8}$ $\frac{19\pi}{8}$, $-\frac{13\pi}{8}$
21. $-\pi$ π, -3π
22. 114° 474°, −246°
23. 66° 426°, −294°
24. −32° 328°, −392°

In Exercises 25–30, rewrite the degree measure as radians. (13.2)

25. −15° $-\frac{\pi}{12}$
26. 20° $\frac{\pi}{9}$
27. 60° $\frac{\pi}{3}$
28. 315° $\frac{7\pi}{4}$
29. 260° $\frac{13\pi}{9}$
30. 109° $\frac{109\pi}{180}$

In Exercises 31–36, rewrite the radian measure as degrees. (13.2)

31. $\frac{7\pi}{6}$ 210°
32. $\frac{5\pi}{6}$ 150°
33. -3π −540°
34. $\frac{4\pi}{15}$ 48°
35. $\frac{4\pi}{5}$ 144°
36. $\frac{8\pi}{9}$ 160°

In Exercises 37–42, sketch the angle. Then find its reference angle. (13.3) See Additional Answers.

37. $\frac{11\pi}{4}$ $\frac{\pi}{4}$
38. 190° 10°
39. −115° 65°
40. $-\frac{17\pi}{15}$ $\frac{2\pi}{15}$
41. 460° 80°
42. 208° 28°

Chapter REVIEW

ASSIGNMENT GUIDE

Basic/Average: Ex. 3–90 multiples of 3, 91–95 odd, 96–98

Above Average: Ex. 3–90 multiples of 3, 94–95, 99–102

Advanced: Ex. 3–90 multiples of 3, 94–95, 99–102

✪ More Difficult Exercises
Exercises 94–102

▶ **Ex.1–102** The Chapter Review exercises include a reference to the lesson in which the skill or strategy was first presented. Students may use these references as a study aid if they are uncertain about how to do a given exercise in the Chapter Review.

Answers

1. $\sin\theta = \frac{4}{5}$, $\cos\theta = \frac{3}{5}$, $\tan\theta = \frac{4}{3}$, $\csc\theta = \frac{5}{4}$, $\sec\theta = \frac{5}{3}$, $\cot\theta = \frac{3}{4}$

2. $\sin\theta = \frac{3}{5}$, $\cos\theta = \frac{4}{5}$, $\tan\theta = \frac{3}{4}$, $\csc\theta = \frac{5}{3}$, $\sec\theta = \frac{5}{4}$, $\cot\theta = \frac{4}{3}$

3. $\sin\theta = \frac{3\sqrt{5}}{7}$, $\cos\theta = \frac{2}{7}$, $\tan\theta = \frac{3\sqrt{5}}{3}$, $\csc\theta = \frac{7}{3\sqrt{5}}$, $\sec\theta = \frac{7}{2}$, $\cot\theta = -\frac{2}{3\sqrt{5}}$

4. $\sin\theta = \frac{2}{3}$, $\cos\theta = \frac{\sqrt{5}}{3}$, $\tan\theta = \frac{2}{\sqrt{5}}$, $\csc\theta = \frac{3}{2}$, $\sec\theta = \frac{3}{\sqrt{5}}$, $\cot\theta = \frac{\sqrt{5}}{2}$

Answers

71. $C = 98°$, $a \approx 9.68$, $c \approx 18.10$
72. $a \approx 33.05$, $B \approx 0.433$, $C \approx 0.615$
73. $A \approx 25.8°$, $B \approx 28.2°$, $a \approx 6.46$
74. $A \approx 86.6°$, $B \approx 70.5°$, $C \approx 22.9°$
75. $B \approx 17.4°$, $C \approx 109.6°$, $c \approx 9.44°$
76. $A \approx 84.0°$, $C \approx 56.0°$, $c \approx 7.76$
77. $A = \frac{\pi}{12}$, $a \approx 3.66$, $b \approx 12.25$
78. $A \approx 0.198$, $B \approx 0.588$, $b \approx 14.11$
79. $A \approx 81.8°$, $B = 60°$, $C \approx 38.2°$
80. $A \approx 34.0°$, $B \approx 122.9°$, $C \approx 23.1°$

In Exercises 43–58, evaluate the expression without using a calculator. For Exercises 53–58, write the result in degrees and in radians. (13.3, 13.4)

43. $\sec 60°$ 2 **44.** $\csc \frac{\pi}{6}$ 2 **45.** $\cot \frac{\pi}{4}$ 1 **46.** $\sin 60°$ $\frac{\sqrt{3}}{2}$

47. $\sin 45°$ $\frac{1}{\sqrt{2}}$ **48.** $\cot 45°$ 1 **49.** $\tan 330°$ $-\frac{1}{\sqrt{3}}$ **50.** $\sin 225°$

51. $\csc 120°$ $\frac{2}{\sqrt{3}}$ **52.** $\cos \frac{11\pi}{6}$ $\frac{\sqrt{3}}{2}$ **53.** $\arccos 0.5$ $60°, \frac{\pi}{3}$ **54.** $\arcsin \frac{\sqrt{2}}{2}$

55. $\arcsin \frac{1}{2}$ $30°, \frac{\pi}{6}$ **56.** $\arctan \frac{\sqrt{3}}{3}$ $30°, \frac{\pi}{6}$ **57.** $\arctan \sqrt{3}$ $60°, \frac{\pi}{3}$ **58.** $\arccos \frac{\sqrt{2}}{2}$

50. $-\frac{1}{\sqrt{2}}$ **54.** $45°, \frac{\pi}{4}$ **58.** $45°, \frac{\pi}{4}$

In Exercises 59–70, use a calculator to evaluate the expression. Round the result to three decimal places. For Exercises 65–70, write the result in degrees and in radians. (13.3, 13.4)

59. $\cos 43°$ 0.731 **60.** $\sin 26°$ 0.438 **61.** $\csc \frac{10\pi}{11}$ 3.549 **62.** $\cot \frac{\pi}{9}$ 2.747

63. $\tan 160°$ −0.364 **64.** $\sec 220°$ −1.305 **65.** $\arcsin 0.9$ 64.158°, 1.120 **66.** $\arctan 51$

67. $\arccos \frac{3}{4}$ 41.410°, 0.723 **68.** $\arccos 0.42$ 65.165°, 1.137 **69.** $\arctan 3$ 71.565°, 1.249 **70.** $\arcsin \frac{1}{8}$

66. 88.877°, 1.551 **70.** 7.181°, 0.125

In Exercises 71–80, solve the triangle. (13.5, 13.6) See margin.

71.

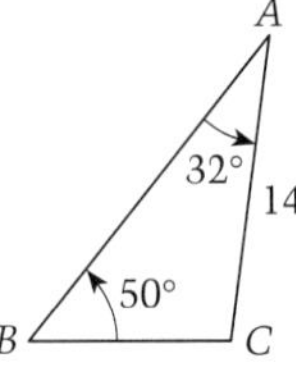

72.

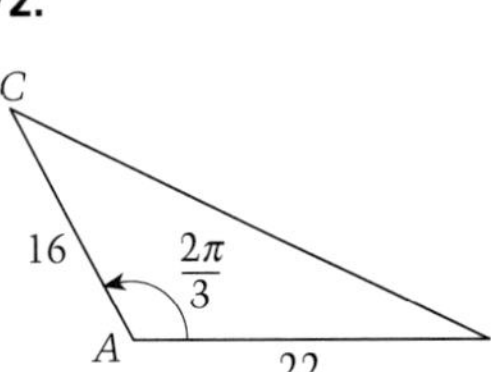

73.

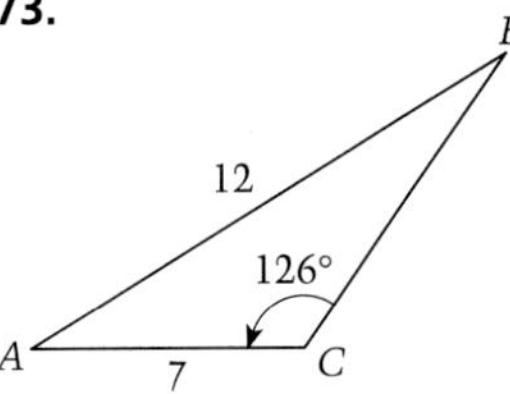

74.

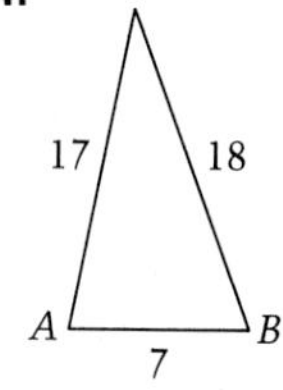

75. $A = 53°, a = 8, b = 3$ **76.** $B = 40°, a = 12, c = 10$

77. $B = \frac{\pi}{4}, C = \frac{2\pi}{3}, c = 10$ **78.** $C = \frac{3\pi}{4}, c = 18, a = 5$

79. $a = 8, b = 7, c = 5$ **80.** $a = 10, b = 15, c = 7$

In Exercises 81–90, find the area of the triangle. (13.5)

81. ≈164.40

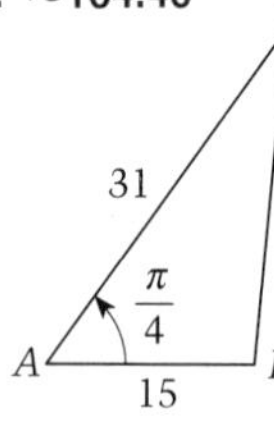

82. ≈10809.69

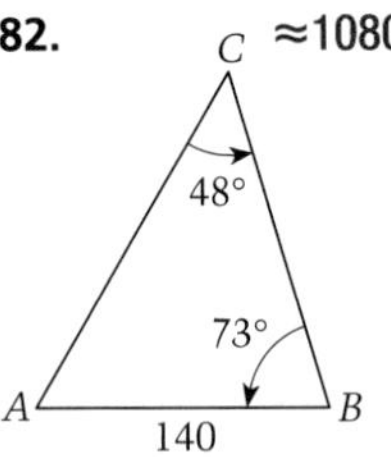

83. ≈26.83

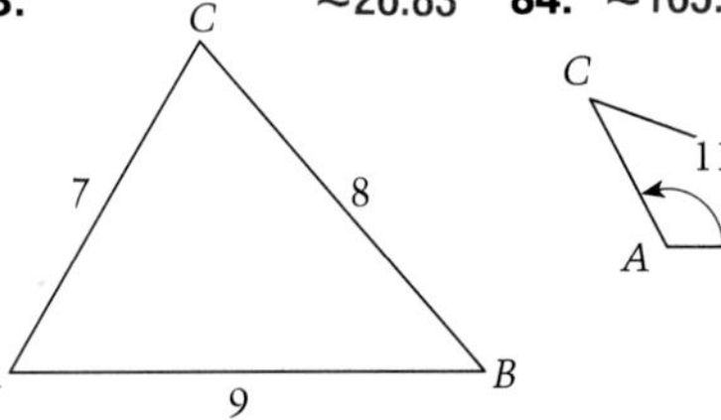

84. ≈165.27

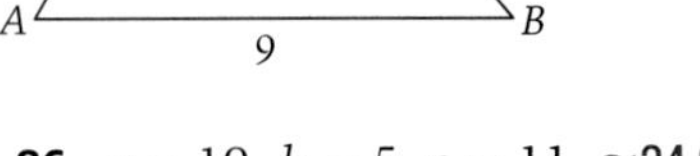

85. $A = 66°, b = 9, c = 5$ ≈20.55 **86.** $a = 10, b = 5, c = 11$ ≈24.98

87. $a = 6, b = 14, c = 9$ ≈18.41 **88.** $C = 28°, a = 12, b = 8$ ≈22.53

89. $a = 5, b = 4, c = 7$ ≈9.80 **90.** $B = 82°, a = 21, c = 29$ ≈301.54

91. *Fountain Place Tower* The Bank Tower shown at the right is about 700 feet tall and has 60 floors. The figure below shows dimensions of the section between the 13th and 45th floors. Find the angle θ. 13.26°

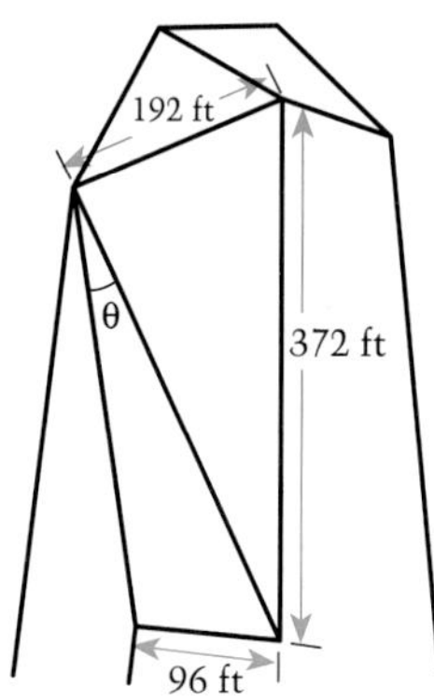

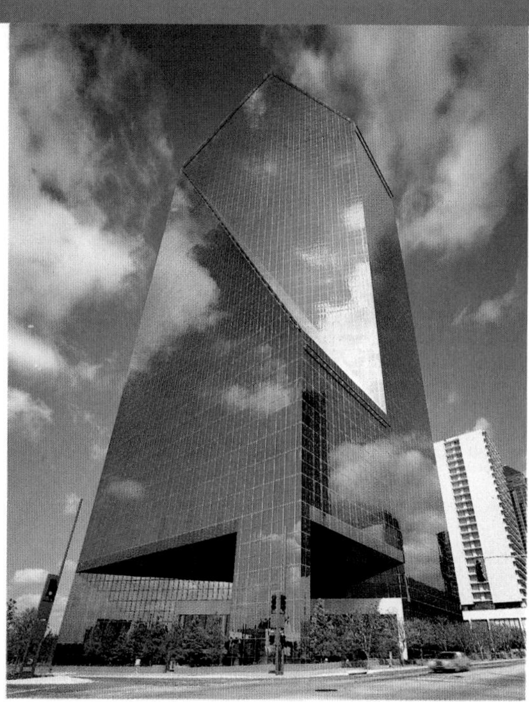

The Bank Tower at Fountain Place in Dallas

92. *Finding a Distance* You live 4 miles east of your school, and your best friend lives 5 miles northwest of the school. How far does your friend live from you? ≈ 8.32 mi

93. *Circular Pipe* Find the inner diameter of the narrowest circular pipe that will fit over the hexagonal nut shown at the right. ≈ 1.27 cm

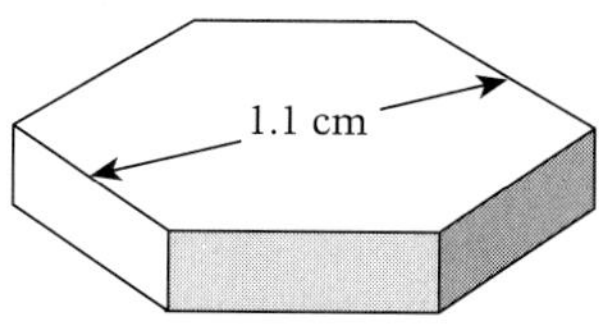

94. *The Star Sirius* The *parallax* of a star can be measured by comparing pairs of photographs taken several months apart. The parallax for Sirius, the brightest star that can be seen from Earth, is

$\theta = 1.04 \times 10^{-4}$ degree.

What is the distance between Earth and Sirius in astronomical units? (An *astronomical unit* is the mean distance between Earth and the sun.) ≈ 550921 a.u.

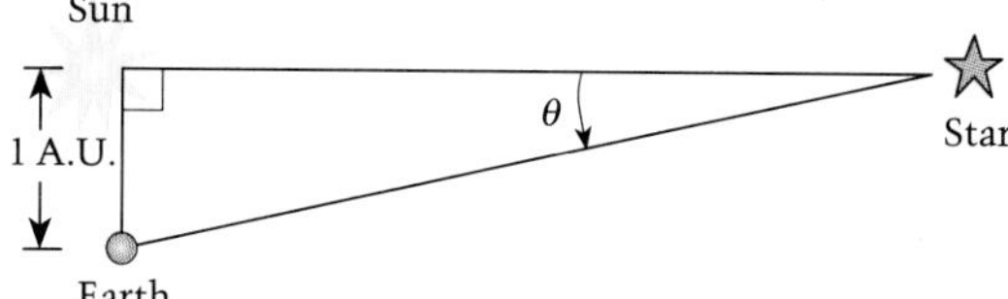

95. *Submarine Torpedo* In World War II, a submarine tracked a ship on a course 10° west of north. As the ship passed directly west of the submarine, the submarine fired a torpedo that traveled three times as fast as the ship. At what angle did the torpedo have to be fired to hit the ship?

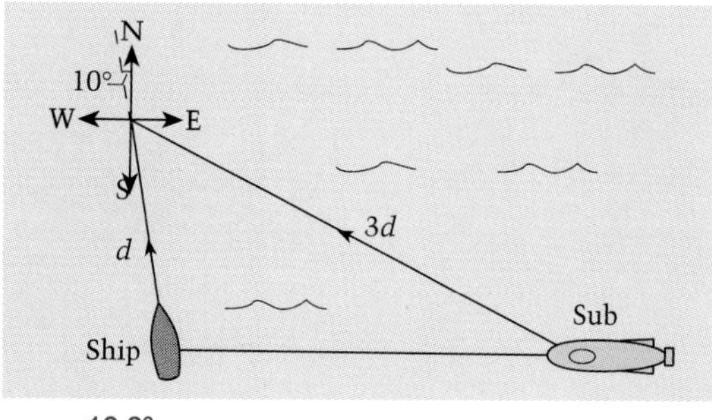

$\approx 19.2°$

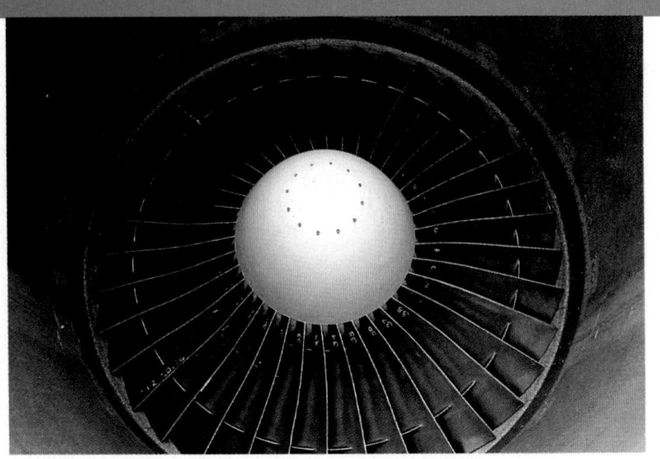

The fan casing for the jet engine's fan blades has a diameter of 6 feet. (Each of the fan blades extends to the end of the fan casing.)

96. *Jet Engine* Use the above photo to find the angle between each pair of consecutive fan blades. ≈9.5°

97. *Jet Engine* Approximate the distance between the outer tips of two consecutive fan blades. ≈0.50 ft

98. *Jet Engine* The tip of one of the fan blades starts at the lowest position. The fan blade then rotates 75° counterclockwise. How far did the tip move horizontally? How far did it move vertically? ≈2.90 ft, 2.22 ft

Jet Airplane **In Exercises 99–102, use the following information.**

The diagram below shows a top view of a DC-10 jet airplane. The length of the airplane is 182 feet, and the length from one wingtip to the other is 161 feet.

99. Find AB, the distance between a wingtip and the nose of the airplane. 148.7 ft

100. Find the length of the front edge of each wing. 89 ft

101. Find the angle, θ, that the wing makes with the airplane. 127.2°

102. Find CD, the distance between a wingtip and the tail of the airplane. ≈92.2 ft

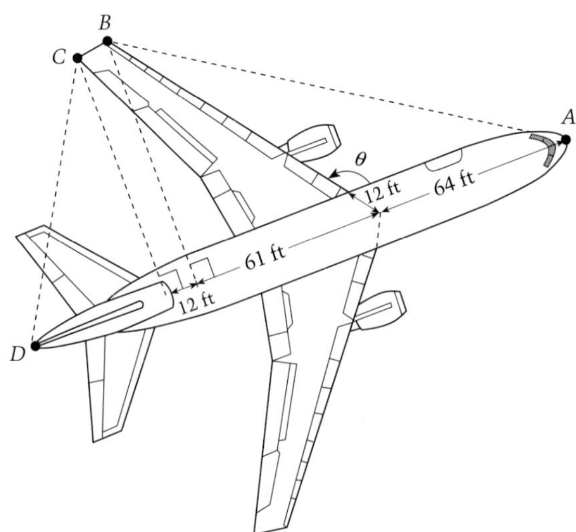

Chapter TEST

In Exercises 1–3, evaluate the expression. (13.1, 13.3)

1. $\csc \frac{\pi}{4}$ $\sqrt{2}$

2. $\tan 135°$ -1

3. $\cos \frac{11\pi}{6}$ $\frac{\sqrt{3}}{2}$

In Exercises 4–6, evaluate the expression. Write the result in degrees and in radians. (13.4)

4. $\arccos 0.5$ $60°, \frac{\pi}{3}$

5. $\arcsin\left(-\frac{\sqrt{2}}{2}\right)$ $-45°, \frac{-\pi}{4}$

6. $\arctan(-1)$ $-45°, \frac{-\pi}{4}$

In Exercises 7–12, solve the triangle. (13.1–13.6) **8.** $B \approx 30.7°$, $C \approx 99.3°$, $c \approx 3.9$

7.

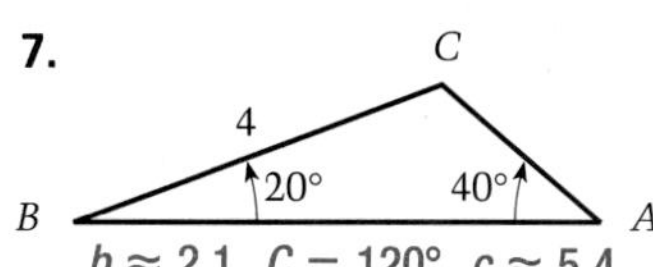

$b \approx 2.1$, $C = 120°$, $c \approx 5.4$

8.

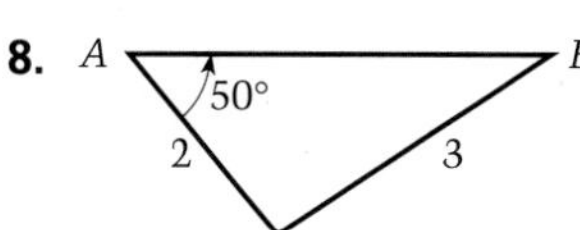

9.

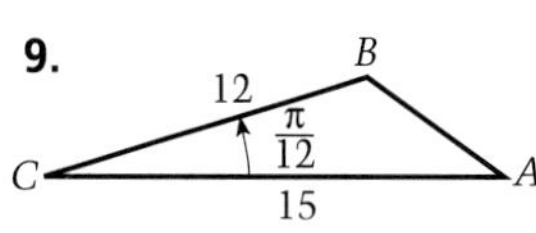

9. $c \approx 4.6$, $A \approx 0.74$ radians, $B \approx 1.00$ radian

10. $a = 7,\ C = 35°,\ b = 14$ See margin.

11. $A = 29°,\ B = 130°,\ a = 5$ See margin.

12. $a = 6,\ b = 12,\ c = 9$ See margin.

In Exercises 13–18, find the area of the triangle. (13.1–13.6)

13. ≈ 19.8

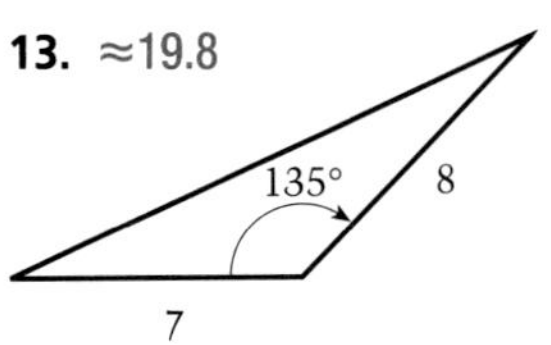

14. ≈ 11.8

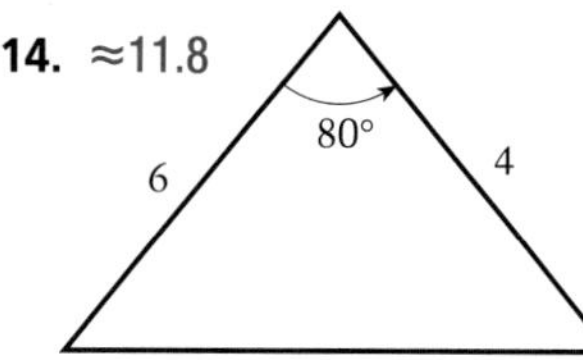

15. $4\sqrt{15}$

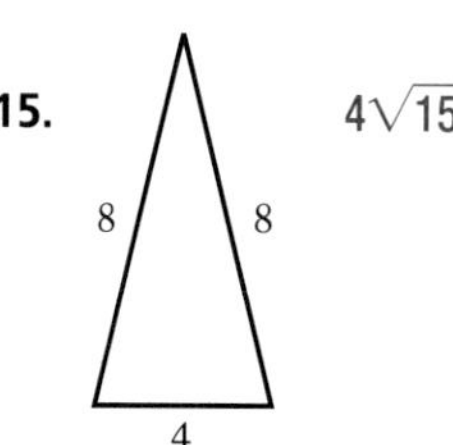

16. $b = 4,\ c = 9,\ A = 36°$ ≈ 10.6

17. $a = 5,\ c = 10,\ B = \frac{3\pi}{4}$ ≈ 17.7

18. $a = 3,\ b = 6,\ C = \frac{3\pi}{8}$ ≈ 8.3

19. A boat travels 50 miles due west before adjusting its course 30° northward and traveling an additional 25 miles. How far is the boat from its point of departure? **(13.1)** ≈ 72.7 mi

20. A bike has tires with radii of 13 inches. How many degrees of revolution does each tire go through when the bike travels 10 feet? **(13.2)** $\approx 529°$

21. The kitchen work triangle at the right has acute angles of 20° each. The longest side of the triangle measures 10 feet. What are the lengths of the two legs of the triangle? **(13.1)** ≈ 5.3 ft

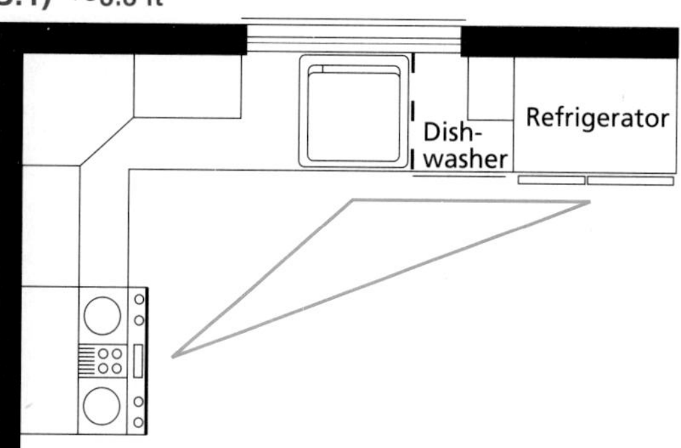

22. Find the total length of the sides of a work triangle that contains an angle of 67° with adjoining sides of 5 feet and 9 feet. For the greatest efficiency, the total length of the sides of the work triangle should be less than 26 feet. Is this kitchen work triangle efficient? **(13.5)** Yes

Answers

10. $c \approx 9.2$, $A \approx 25.9°$, $B \approx 119.1°$

11. $C = 21°$, $b \approx 7.9$, $c \approx 3.7$

12. $A \approx 29.0°$, $B \approx 104.5°$, $C \approx 46.5°$

Once the basic trigonometric functions have been defined, the next logical step is to describe their graphs on a coordinate plane. As in the case of other algebraic functions, the graphs of trigonometric functions can be used to describe the relationships modeled by the functions and can be used to solve related real-life problems.

In this chapter, students will explore trigonometric relationships that always hold and others that hold for given values of the angle θ. In this context, students will have the opportunity to derive proofs of trigonometric relationships and find algebraic solutions of trigonometric equations.

The Chapter Summary on page 772 provides you and the students with a synopsis of the chapter. It identifies key skills and concepts. You may want to have students look at the Chapter Summary as an overview before beginning the chapter.

CHAPTER 14

Trigonometric Graphs, Identities, and Equations

LESSONS

Exercising vigorously, as these cross-country runners are, increases your pulse rate. One effect of this on your blood pressure is shown in the graphs on the opposite page.

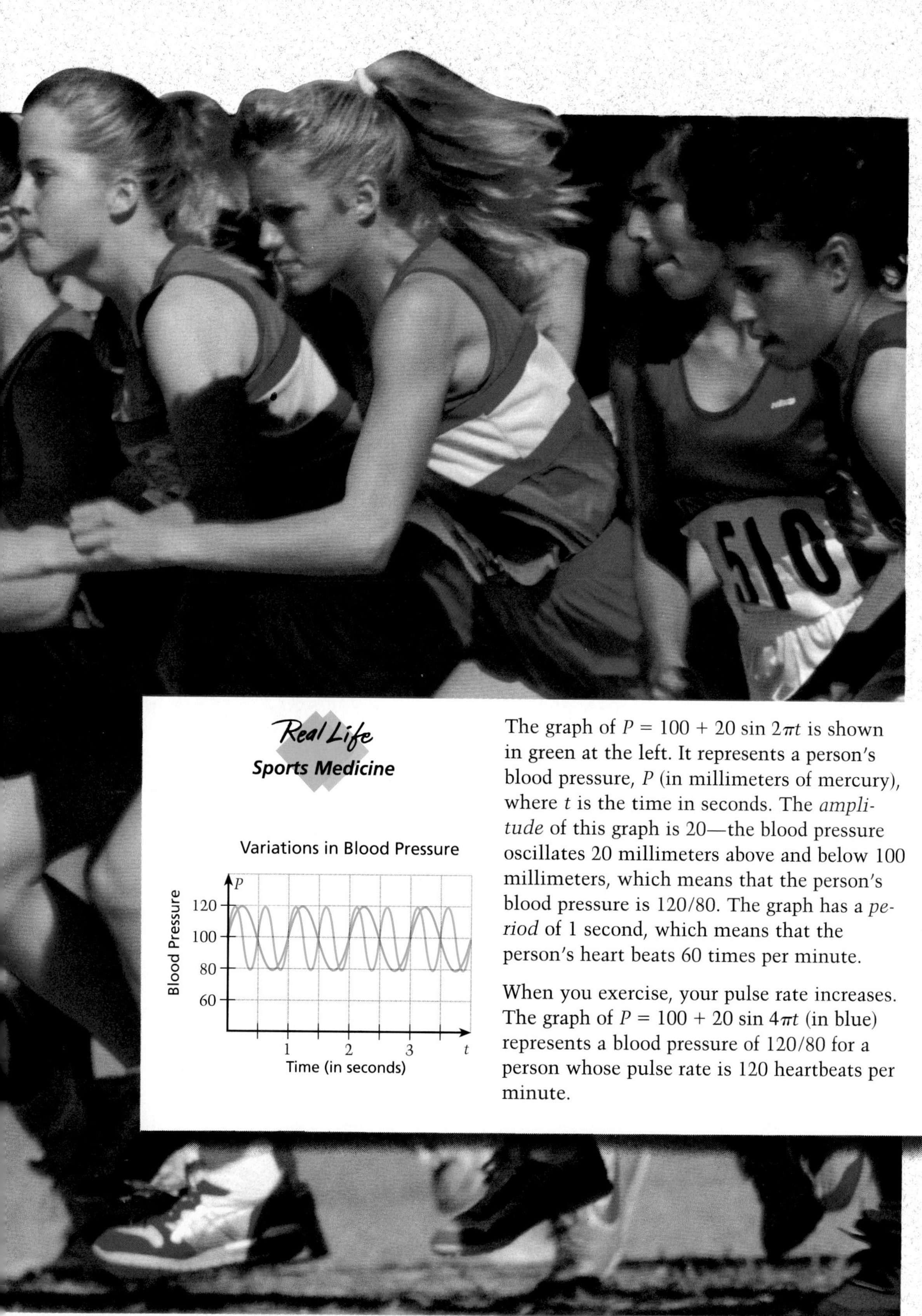

Real Life
Sports Medicine

The graph of $P = 100 + 20 \sin 2\pi t$ is shown in green at the left. It represents a person's blood pressure, P (in millimeters of mercury), where t is the time in seconds. The *amplitude* of this graph is 20—the blood pressure oscillates 20 millimeters above and below 100 millimeters, which means that the person's blood pressure is 120/80. The graph has a *period* of 1 second, which means that the person's heart beats 60 times per minute.

When you exercise, your pulse rate increases. The graph of $P = 100 + 20 \sin 4\pi t$ (in blue) represents a blood pressure of 120/80 for a person whose pulse rate is 120 heartbeats per minute.

Trigonometric relationships are used to represent many different real-life situations. The trigonometric model of a person's blood pressure is a good example of this. It shows the important characteristics of a graph of a trigonometric relationship.

Based upon the information and the graphs in the opening discussion, ask students what the amplitude and period of the second graph (in green) are. Can students identify characteristics of the graphs that indicate how the period is shown on a graph?

amplitude: 20, period: $\frac{1}{2}$ second; The period is shown on the horizontal axis of the graph. It represents the distance on the x-axis for the graph to complete one full pattern.

Trigonometric Identities and Equations

PACING CHART

This chapter is an integral part of the full course for Algebra 2. However, it is not recommended for a limited treatment of trigonometry such as that suggested on page 677A for classes taking the basic course. You may wish to consult also the introductory note on page 781A.

DAY	FULL COURSE
1	14.1
2	14.1
3	14.2
4	14.2
5	14.3
6	14.3 & Using a Graphing Calculator
7	Mid-Chapter Self-Test
8	14.4
9	14.4
10	14.5
11	14.5
12	14.6
13	14.6
14	Chapter Review
15	Chapter Test
16	Cumulative Review 13–14

CHAPTER ORGANIZATION

LESSON	PAGES	GOALS	MEETING THE NCTM STANDARDS
14.1	730–736	1. Sketch the graphs of the sine and cosine functions 2. Use sine and cosine functions as models of real-life problems	Problem Solving, Communication, Connections, Functions, Trigonometry
14.2	737–742	1. Graph vertical and horizontal shifts and reflections of the graphs of the sine and cosine functions 2. Use sine and cosine functions to solve real-life problems	Problem Solving, Communication, Reasoning, Connections, Functions, Geometry (algebraic), Trigonometry
Mixed Review	743	Review algebraic and arithmetic skills	
Milestones	743	Unsolved Problems	Connections
14.3	744–749	1. Use trigonometric identities to simplify trigonometric expressions, and verify trigonometric identities 2. Use trigonometric identities to solve real-life problems	Problem Solving, Communication, Connections, Functions, Trigonometry
Using a Calculator	750–751	Discover trigonometric identities graphically	Technology
Mid-Chapter Self-Test	752	Diagnose student weaknesses and remediate with correlated Reteach worksheets	
14.4	753–759	1. Solve trigonometric equations 2. Solve trigonometric equations that model real-life problems	Problem Solving, Communication, Reasoning, Connections, Functions, Geometry (algebraic), Trigonometry
14.5	760–765	1. Use the sum and difference formulas to evaluate trigonometric functions of the sum and difference of two angles 2. Use sum and difference formulas to solve real-life problems	Problem Solving, Communication, Reasoning, Connections, Functions, Trigonometry
Mixed Review	765	Review algebraic and arithmetic skills	
14.6	766–771	1. Use double-angle and half-angle formulas 2. Use trigonometry to solve real-life problems	Problem Solving, Communication, Connections, Functions, Geometry (algebraic), Trigonometry
Chapter Summary	772	A restatement of what has been learned, why it has been learned, and how it fits into the structure of algebra	Structure, Connections
Chapter Review	773–776	Review concepts and skills learned in the chapter	
Chapter Test	777	Diagnose student weaknesses and remediate with correlated Reteaching worksheets	
Cumulative Review 13, 14	778–779	Review concepts and skills from chapters 13, 14	

LESSON RESOURCES

MEETING INDIVIDUAL NEEDS

RETEACHING For students who need to spend more time on basics:

If a mid-chapter self-test or chapter test indicates a deficiency, teachers can help students with the appropriate ***Reteaching Copymaster.***

PRACTICE For students who need more practice:

Additional exercises like those in the Pupil's Edition are provided for each lesson in ***Extra Practice Copymasters.***

ENRICHMENT For enriching and broadening students' experiences:

Problem of the Day copymasters in ***Teaching Tools*** provide a daily opportunity to use logical reasoning, looking for a pattern, writing an equation, and other routine and non-routine problem-solving strategies.

Math Log copymasters in ***Alternative Assessment*** provide opportunities to report on investigations, research, and open-ended problems.

Technology: Using Calculators and Computers provides enriching activities with graphing and scientific calculators and computers.

The ***Applications Handbook*** provides additional information about the cross-curriculum topics such as astronomy, chemistry, physics, sports, economics, genetics, and music that are integrated into the Pupil's Edition.

LESSON	14.1	14.2	14.3	14.4	14.5	14.6
PAGES	730-736	737-742	744-749	753-759	760-765	766-771
Teaching Tools						
Transparencies	✓	✓		✓		
Problem of the Day	✓	✓	✓	✓	✓	✓
Warm-up Exercises	✓	✓	✓	✓	✓	✓
Answer Masters	✓	✓	✓	✓	✓	✓
Extra Practice Copymasters	✓	✓	✓	✓	✓	✓
Reteaching Copymasters	Teacher-directed and independent activities tied to results on the Mid-Chapter Self-Tests and Chapter Tests					
Color Transparencies	✓	✓		✓	✓	
Applications Handbook	Additional background information for many real-life applications					
Technology	Calculator and computer worksheets for appropriate lessons					
Complete Solutions Manual						
Alternative Assessment	Assess student's ability to reason, analyze, solve problems, and communicate using mathematical language.					
Formal Assessment	Mid-Chapter Self-Tests, Chapter Tests, Cumulative Tests, and Practice for College Entrance Tests					
Computer Test Bank	Customized tests can be created by choosing from over 2500 items.					

INSIGHTS

14.1 Graphs of Sine and Cosine Functions

The sine and cosine functions are used extensively in physics, engineering, economics, and biology. The behavior of these functions is perhaps best described by their graphs. Because the graph of a trigonometric function can be affected by several factors, students should be taught to carefully explore the effects of constants on these special algebraic functions.

14.2 Translations of Graphs

The effects of constants on the sine and cosine functions are similar to those observed for other algebraic functions in previous lessons. Real-life situations are seldom modeled by the basic trigonometric functions alone. Instead, problem-solving situations usually involve applications in which the basic functions have been shifted in vertical or horizontal directions, or reflected about the x-axis. In this lesson, students explore the effects of constants on the sine and cosine functions.

14.3 Trigonometric Identities

Trigonometric identities are frequently used to simplify trigonometric expressions in applications and in proving theorems. Verifying trigonometric identities provides practice in recognizing trigonometric expressions that may be simplified.

14.4 Solving Trigonometric Equations

Determining the values of x for which a statement, such as $\cos x = 1$, is true is called solving a trigonometric equation. In this lesson, students examine solution techniques for trigonometric equations. Since trigonometric functions are periodic, there are an infinite number of solutions to the equation $\cos x = 1$. Students should learn to express such solution sets precisely and comprehensively so as to "cover all the bases".

14.5 Sum and Difference Formulas

In this lesson, students continue to extend their ability to use trigonometric functions effectively as models of real-life situations. Frequently, trigonometric relationships involve angle measures that are related as either the sum or difference of other angles. In such cases, special trigonometric formulas presented in this lesson can be used to evaluate trigonometric functions. The advantage in using the sum and difference formulas is that resulting evaluations can be expressed as exact values (rather than as approximate decimal values).

14.6 Double- and Half-Angle Formulas

Students continue to develop the skills and understanding needed to use trigonometric functions effectively in modeling real-life situations. In this lesson, they examine trigonometric relationships that involve angle measures that are related as either the double- or half-angle of other angles. In such cases, special trigonometric formulas presented in this lesson can be used to evaluate trigonometric functions. Again, one of the advantages in using the formulas is that resulting evaluations can be expressed as exact values (rather than as approximate decimal values).

Problem of the Day

In how many ways can a panel of four on-off switches in a row be set so that no two adjacent switches are off?
8

ORGANIZER

Warm-Up Exercises

1. Evaluate the following functions for the given angle measures in radians. What do their angles have in common?

   a. $\sin\frac{\pi}{2}$ b. $\sin\frac{5\pi}{2}$

   c. $\sin\frac{9\pi}{2}$ d. $\sin\frac{13\pi}{2}$

   a. 1, b. 1, c. 1, d. 1; All the angles are coterminal.

2. For each set of values, x, y, z, for which $z = \frac{x}{y}$, find the unknown.

   a. $x = 2\pi, y = 4, z = ?$
   b. $x = 3\pi, y = 60, z = ?$
   c. $x = 4\pi, y = 30, z = ?$
   d. $x = ?, y = 4, z = \frac{3\pi}{4}$
   e. $x = 2\pi, y = \frac{3}{2}, z = ?$
   f. $x = 3\pi, y = \frac{3}{5}, z = ?$

   a. $z = \frac{\pi}{2}$ b. $z = \frac{\pi}{20}$

   c. $z = \frac{2\pi}{15}$ d. $x = 3\pi$

   e. $z = \frac{4\pi}{3}$ f. $z = 5\pi$

Lesson Resources

Teaching Tools
- Transparencies: 26, 27, 28
- Problem of the Day: 14.1
- Warm-up Exercises: 14.1
- Answer Masters: 14.1

Extra Practice: 14.1
Color Transparency: 73
Applications Handbook: p. 49–51, 56, 57
Technology Handbook: p. 95

14.1 Graphs of Sine and Cosine Functions

What you should learn:

Goal **1** How to sketch the graphs of the sine and cosine functions

Goal **2** How to use sine and cosine functions as models of real-life problems

Why you should learn it:

You can model wave patterns, such as sound waves, with the sine and cosine functions.

Domain: $-\infty < x < \infty$,
Range: $-1 \le y \le 1$;
Domain: $-\infty < x < \infty$,
Range: $-1 \le y \le 1$

Study Tip

A function is ***periodic*** *if its graph has a pattern that repeats indefinitely. The shortest repeating portion is a* ***cycle.*** *The horizontal length of each cycle is the* ***period.***

Goal 1 Graphing Sine and Cosine Functions

LESSON INVESTIGATION

■ Investigating Sine and Cosine Graphs

Partner Activity Use a graphing calculator to sketch the graph of the sine function, $y = \sin x$. Set the calculator to radian mode and choose an appropriate viewing rectangle. With your partner, describe the graph. What is its domain? Its range? Repeat this activity for the cosine function. Then compare the graphs of the two functions.

In this investigation, you may have discovered the following.

1. The *domain* of each function is all real numbers.
2. The *range* of each function is $-1 \le y \le 1$.
3. Each of the graphs below has a period of 2π.
4. The maximum value of $y = \sin x$ occurs when $x = \frac{\pi}{2} + 2n\pi$, where n is an integer. The maximum value of $y = \cos x$ occurs when $x = 2n\pi$.
5. The minimum value of $y = \sin x$ occurs when $x = \frac{3\pi}{2} + 2n\pi$, where n is an integer. The minimum value of $y = \cos x$ occurs when $x = (2n + 1)\pi$.

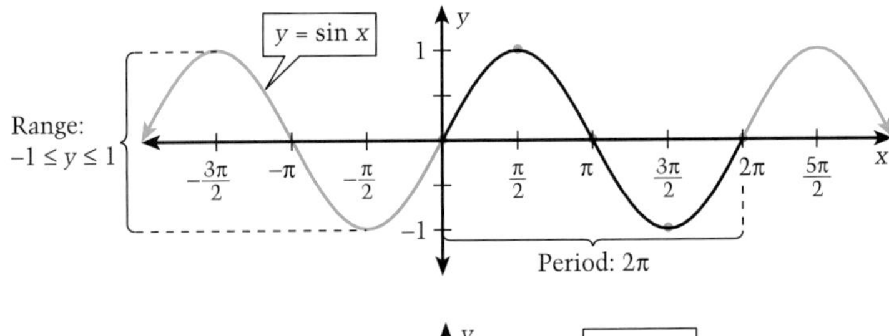

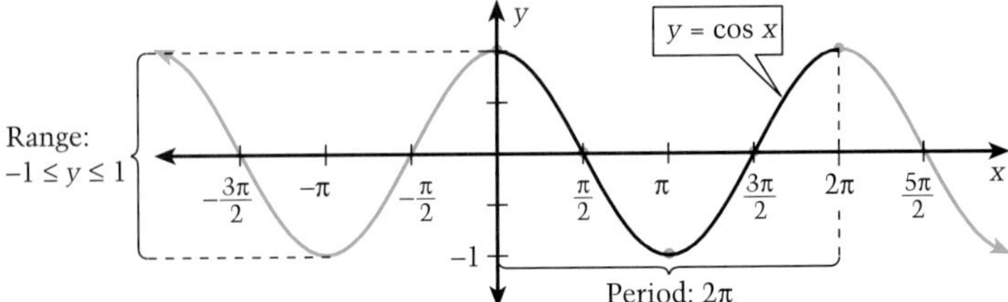

On the interval from 0 to 2π, the graphs of the basic sine and cosine functions have five key points: the x-intercepts, the maximum, and the minimum. These are shown in the graphs below.

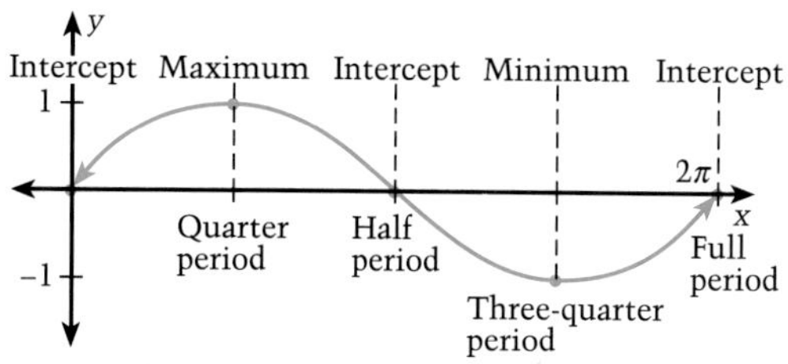

Five key points on graph of y = sin x

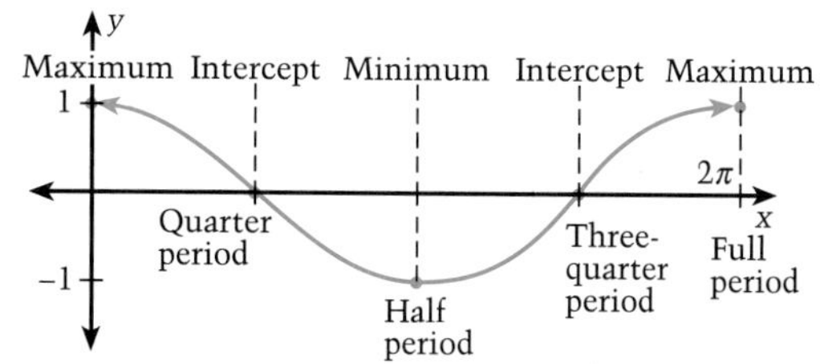

Five key points on graph of y = cos x

Amplitude and Period of Sine and Cosine Graphs

The **amplitude** of the graph of $y = a \sin bx$ or $y = a \cos bx$ is the amount by which it varies above and below the x-axis. The **period** of each graph is the length of the shortest interval on the x-axis over which the graph repeats.

Amplitude $= |a|$ and Period $= \dfrac{2\pi}{|b|}$

Example 1 *Using Amplitude and Period to Sketch a Graph*

Sketch the graph of $y = 2 \sin x$.

Solution Because $a = 2$, the graph has an amplitude of 2. Because $b = 1$, the graph has a period of 2π. Thus, it cycles once from 0 to 2π with one maximum of 2, one minimum of -2, and three x-intercepts.

Intercept	Maximum	Intercept	Minimum	Intercept
$(0, 0)$,	$\left(\frac{\pi}{2}, 2\right)$,	$(\pi, 0)$,	$\left(\frac{3\pi}{2}, -2\right)$,	$(2\pi, 0)$

Note how the graph of $y = 2 \sin x$ is like the graph of $y = \sin x$. Each has the same x-intercepts, but $y = 2 \sin x$ has an amplitude that is twice the amplitude of $y = \sin x$. ■

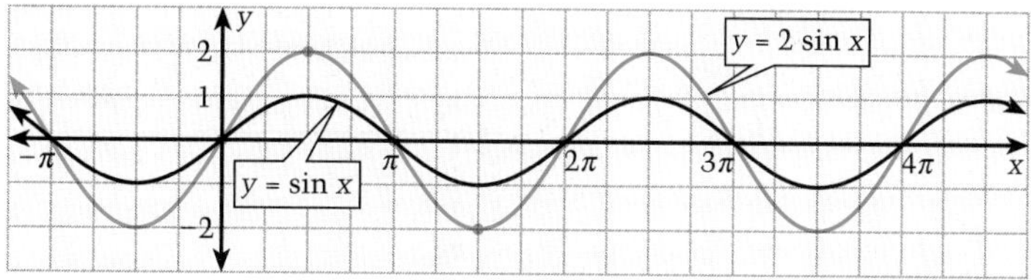

LESSON Notes

Examples 1–2

Be sure that students understand that the graphs of sine and cosine functions are not constructed of straight line segments joining the five identified key points.

Example 3

Ask students to explain why the trigonometric function with the greatest period oscillates the least.

Answers will vary, but students should respond that a function with a large period has a larger interval over which to complete one cycle. Hence there are fewer oscillations for a fixed interval.

Example 4

A hertz is a unit of measure of the frequency of waves in cycles per second, used especially for electromagnetic and acoustic waves. It is named for a pioneer in the study of electromagnetic waves, the German physisist Heinrich Rudolf Hertz (1857–1894).

Extra Examples

Here are additional examples similar to **Examples 1–3.**

1. Sketch the following graphs.

a. $y = 3 \sin x$ **b.** $y = 4 \cos x$

a. The graph is similar to the graph of $y = \sin x$ with a period of 2π and an amplitude of 3.

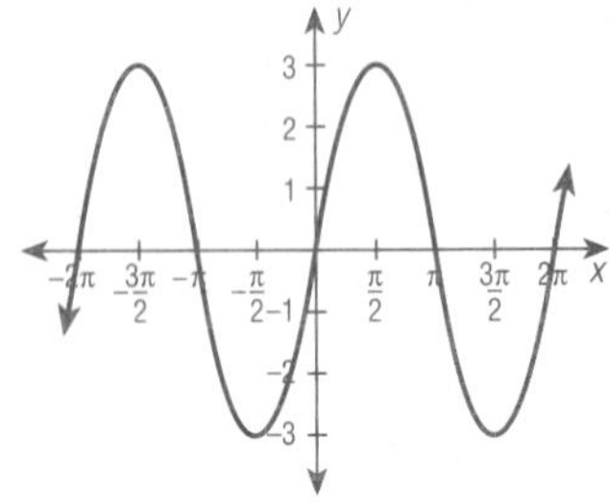

b. The graph is similar to the graph of $y = \cos x$ with a period of 2π and an amplitude of 4.

2. Sketch the following graphs.

a. $y = \sin 4x$ **b.** $y = \cos 2x$

a. The graph is similar to the graph of $y = \sin x$ with a period of $\frac{\pi}{2}$.

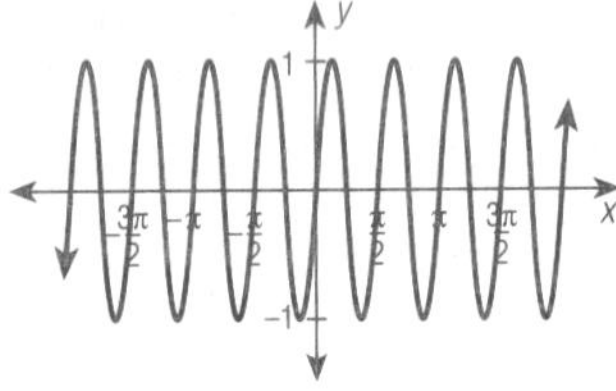

b. The graph is similar to the graph of $y = \cos x$ with a period of π.

3. Sketch the following graphs.

a. $y = \frac{1}{3} \cos \frac{x}{3}$

b. $y = 5 \sin 3x$

a.

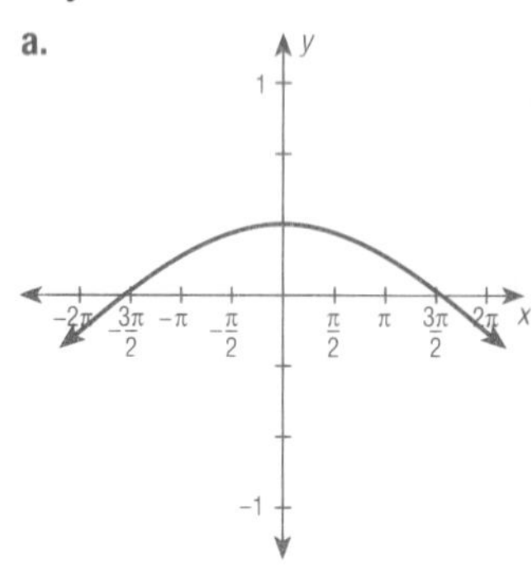

Example 2 Using Amplitude and Period

Sketch the graph of $y = \frac{1}{2} \cos x$ and $y = 3 \cos x$.

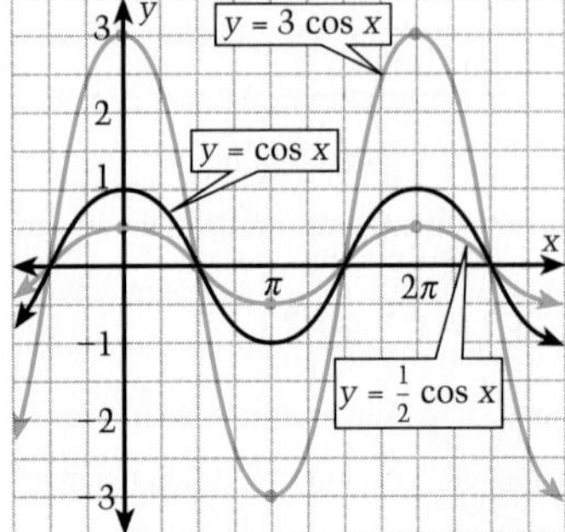

Solution For $y = \frac{1}{2} \cos x$, the amplitude is $\frac{1}{2}$ and the period is 2π. Between 0 and 2π, the five key points on the graph are

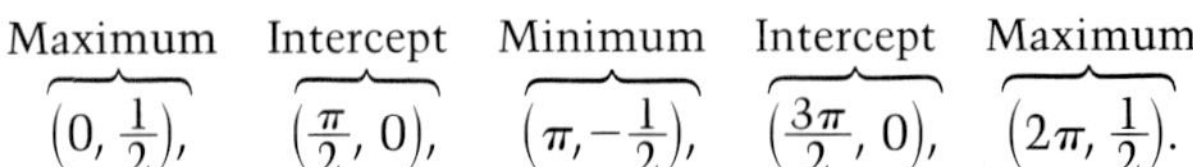

Maximum $\left(0, \frac{1}{2}\right)$,
Intercept $\left(\frac{\pi}{2}, 0\right)$,
Minimum $\left(\pi, -\frac{1}{2}\right)$,
Intercept $\left(\frac{3\pi}{2}, 0\right)$,
Maximum $\left(2\pi, \frac{1}{2}\right)$.

For $y = 3 \cos x$, the amplitude is 3 and the period is 2π. Between 0 and 2π, the five key points on the graph are

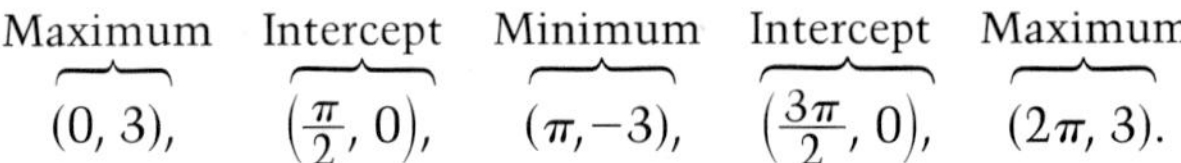

Maximum $(0, 3)$,
Intercept $\left(\frac{\pi}{2}, 0\right)$,
Minimum $(\pi, -3)$,
Intercept $\left(\frac{3\pi}{2}, 0\right)$,
Maximum $(2\pi, 3)$.

The graphs of both functions are shown at the left with the graph $y = \cos x$. ■

Example 3 Comparing Graphs with Other Periods

In the graphs below, note that the one with the least period oscillates most frequently, and the one with the greatest period oscillates least frequently (over intervals of the same length).

a. The graph of $y = \sin 2x$ has a period of $\frac{2\pi}{2} = \pi$.

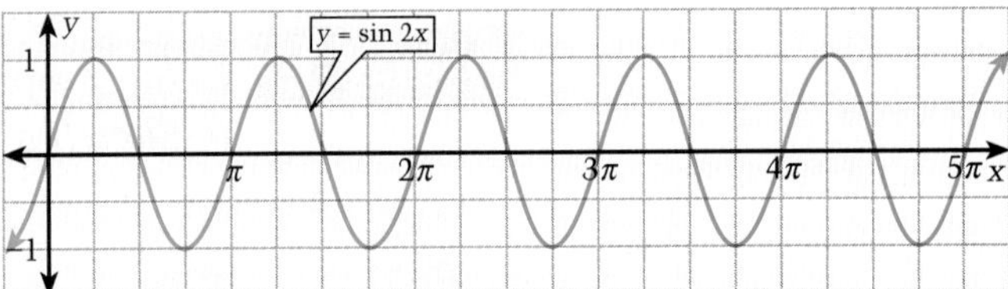

b. The graph of $y = \sin x$ has a period of 2π.

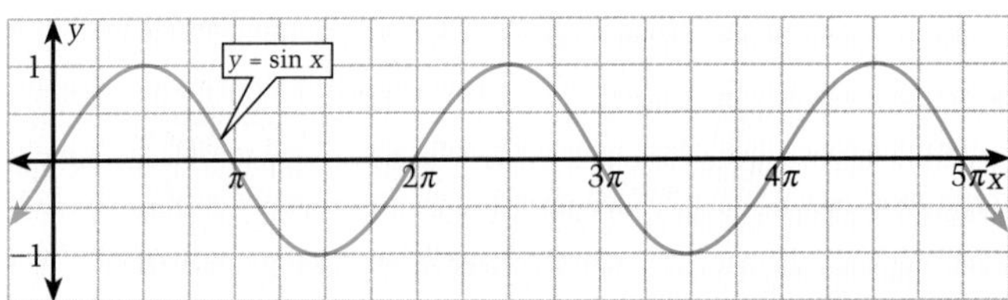

c. The graph of $y = \sin \frac{x}{2}$ has a period of $\frac{2\pi}{1/2} = 4\pi$.

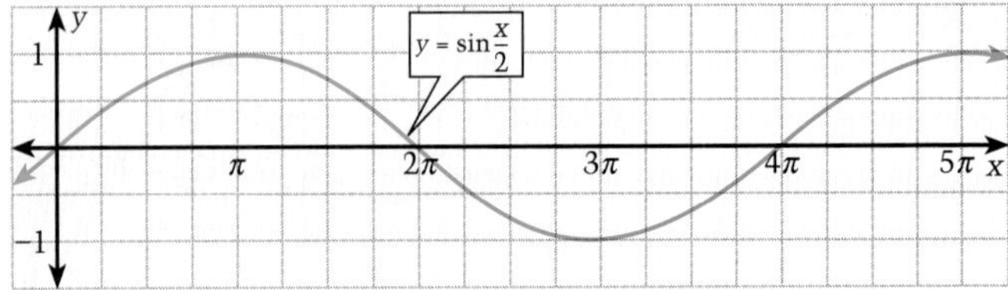

■

Musical notes are classified by their frequency. For instance, the musical note A above middle C has a frequency of 440 **hertz** ***(cycles per second), and the A above A-440 has a frequency of 880 hertz. The sound waves of musical notes can be modeled by sine functions. The frequency is the reciprocal of the period.***

Goal 2 Using Trig Functions in Real Life

Example 4 *Amplitude and Frequency of Musical Sound*

The amplitude of a sound wave is a measure of its *intensity.* Which of the following sine functions, where t is the time in seconds, models the musical note A-220? A-440? A-880? Which note has the greatest intensity?

a. $y = 3 \sin 440\pi t$ **b.** $y = 4 \sin 880\pi t$ **c.** $y = 2 \sin 1660\pi t$

Solution

a. The amplitude of this sound wave is 3. Its period is
$\frac{2\pi}{|b|} = \frac{2\pi}{440\pi} = \frac{1}{220}$ second.
The frequency is 220 hertz. Thus the note is A-220.

b. The amplitude of this sound wave is 4. Its period is
$\frac{2\pi}{|b|} = \frac{2\pi}{880\pi} = \frac{1}{440}$ second.
The frequency is 440 hertz. Thus the note is A-440.

c. The amplitude of this sound wave is 2. Its period is
$\frac{2\pi}{|b|} = \frac{2\pi}{1660\pi} = \frac{1}{880}$ second.
The frequency is 880 hertz. Thus the note is A-880.

The second note has the greatest intensity. ■

Communicating about ALGEBRA

SHARING IDEAS about the Lesson

Sketch the graphs of the functions. Explain your steps.

A. $y = 2 \sin 3x$ **B.** $y = \cos \frac{x}{3}$ **C.** $y = \sin 2\pi x$

b.

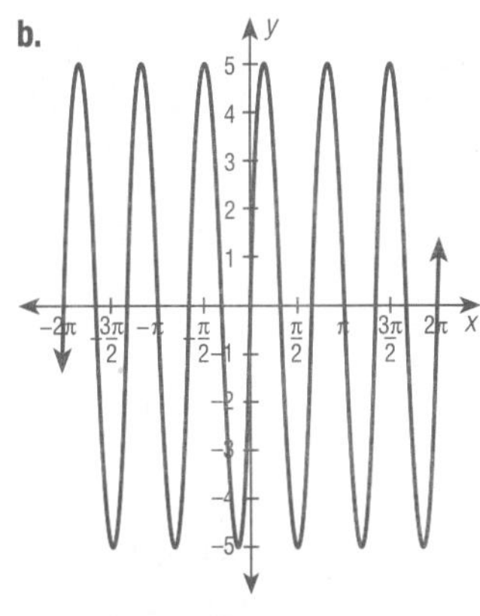

Check Understanding

1. Name the five key points of the sine and cosine functions. See the top of page 731.
2. Why is radian measure used in graphing the sine and cosine functions instead of degree measure? The domains of the sine and cosine functions are all real numbers. Angles in radian measure are real numbers.
3. What is the distinction between a cycle and a period of a trigonometric function? See the third feature listed on page 730.

Communicating about ALGEBRA

EXTEND *Communicating*
COOPERATIVE LEARNING
Have students work in small groups to identify the x-intercepts of the sine and cosine functions. Write an algebraic expression that represents all x-intercepts for each function.
For the sine function, the x-intercepts occur at $0 + 2n\pi$ and $\pi + 2n\pi$ or, written in one expression, $0 + n\pi$, where n is an integer. For the cosine function, the x-intercepts occur at $\frac{\pi}{2} + n\pi$, where n is an integer.
A.–C. See Additional Answers.

EXERCISE Notes

ASSIGNMENT GUIDE

Basic/Average: Ex. 1–5, 7, 8, 10, 12, 16–20, 25–41 odd, 44, 48–51

Above Average: Ex. 1–5, 7, 8, 10, 12, 16–20, 25–41 odd, 42, 43–52

Advanced: Ex. 1–5, 7, 8, 10, 12, 16–20, 25–41 odd, 42–47, 50–52

Selected Answers
Exercises 1–5, 7–51 odd

✪ More Difficult Exercises
Exercises 40–45, 52

Guided Practice

▶ **Ex. 1–5** These exercises provide a final readiness check for students' understanding of key information from the lesson.

▶ **Ex. 1 WRITING** Ask students to write a one-sentence explanation for their answer.

Independent Practice

▶ **Ex. 6–7** Help students to "visualize" a trace of one period for each graph.

▶ **Ex. 10** Warning! Some students will write $\frac{2\pi}{|b|}$ as $\frac{2\pi}{\frac{1}{2}}$ and then simplify to π instead of 4π.

▶ **Ex. 16–19** Assign these exercises as a group. Ask students to justify their choices. (Just stating the "answer" as a, b, c, or d isn't sufficient.)

Ask students to compare the value of b when the graph oscillates quickly over an interval to the value of b when the graph completes a period more slowly.

If $|b| < 1$, the curve stretches out. If $|b| > 1$, the curve oscillates (completes a period) more quickly.

EXERCISES

Guided Practice

CRITICAL THINKING about the Lesson

1. Which x-value does *not* give a maximum of $y = \sin x$? **b**

   a. $x = -\frac{3\pi}{2}$ b. $x = -\frac{\pi}{2}$ c. $x = \frac{\pi}{2}$

2. What is the domain and range of the sine and cosine functions? **All real numbers, $-1 \le y \le 1$**

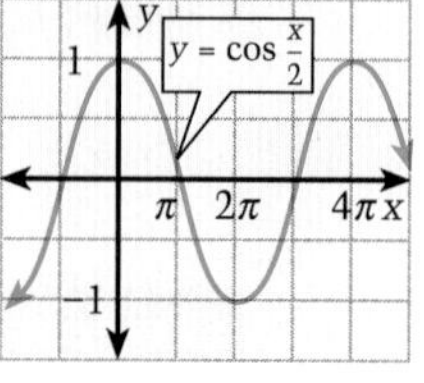

3. What is the period of the graph at the right? **4π**
4. What is the period of $y = \cos \pi x$? **2**
5. What is the amplitude of $y = 5 \sin x$? **5**

Independent Practice

In Exercises 6–15, find the amplitude and period of the graph. **7. $\frac{1}{2}$, $\frac{2\pi}{3}$**

6. 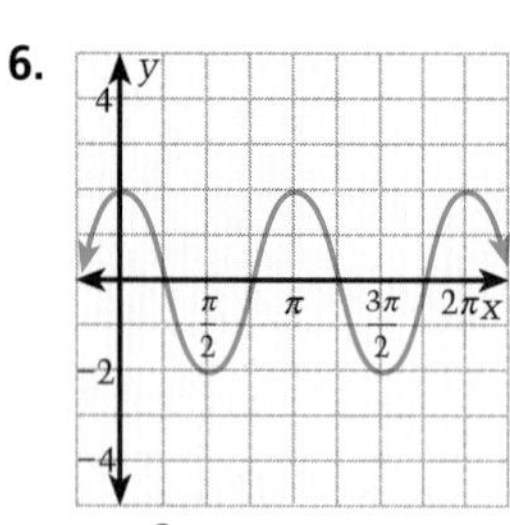

2, π

7. 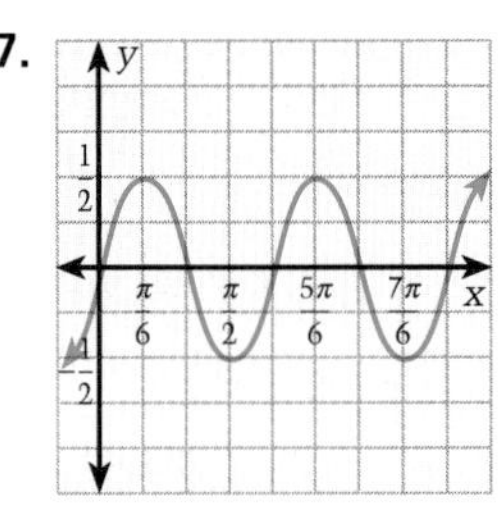

8. $y = 2 \sin 2x$ **2, π**
9. $y = \frac{1}{2} \cos 2x$
10. $y = 3 \cos \frac{1}{2}x$ **3, 4π**
11. $y = \sin 4x$
12. $y = \frac{1}{4} \cos \pi x$ **$\frac{1}{4}$, 2**
13. $y = 4 \cos 2\pi x$
14. $y = 3 \sin \frac{1}{2}\pi x$ **3, 4**
15. $y = \frac{1}{3} \sin 4\pi x$

9. $\frac{1}{2}$, π 11. 1, $\frac{\pi}{2}$ 13. 4, 1 15. $\frac{1}{3}$, $\frac{1}{2}$

In Exercises 16–19, match the equation with its graph.

16. $y = 2 \cos 2x$ **b**
17. $y = 2 \sin 2x$ **d**

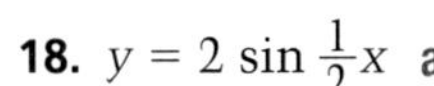

18. $y = 2 \sin \frac{1}{2}x$ **a**
19. $y = 2 \cos \frac{1}{2}x$ **c**

a. 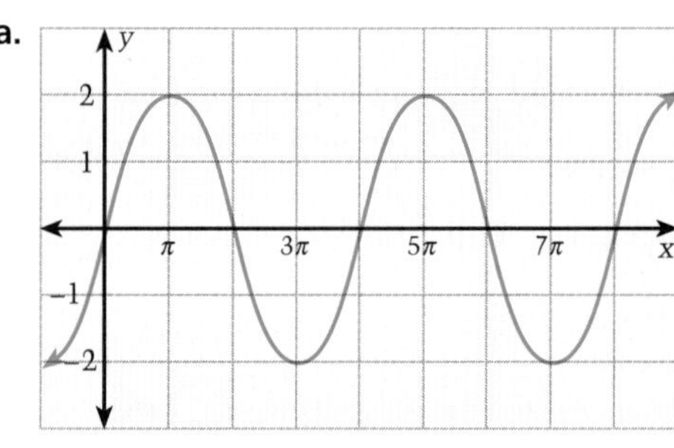

b. 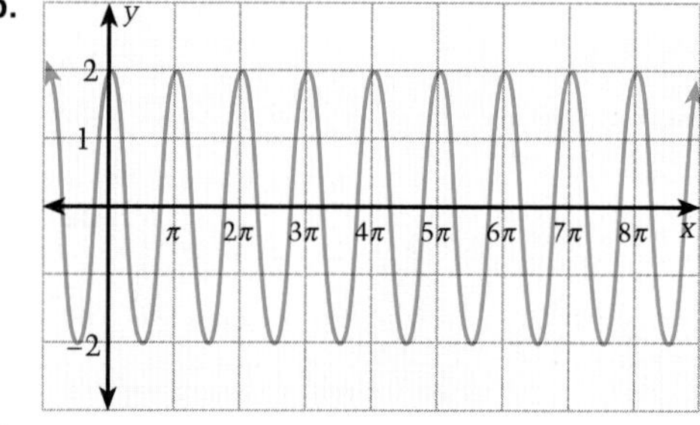

c. 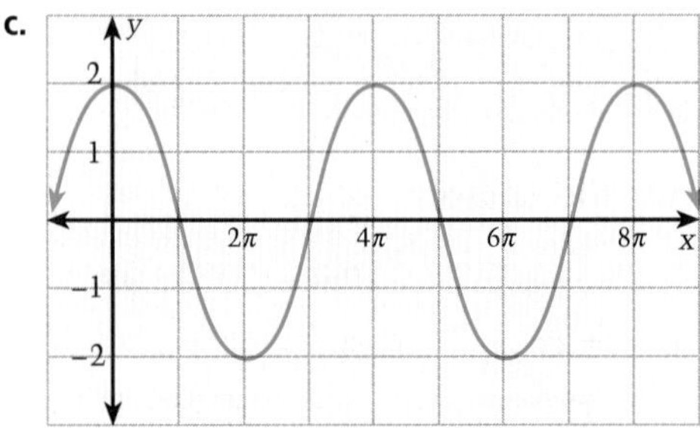

d. 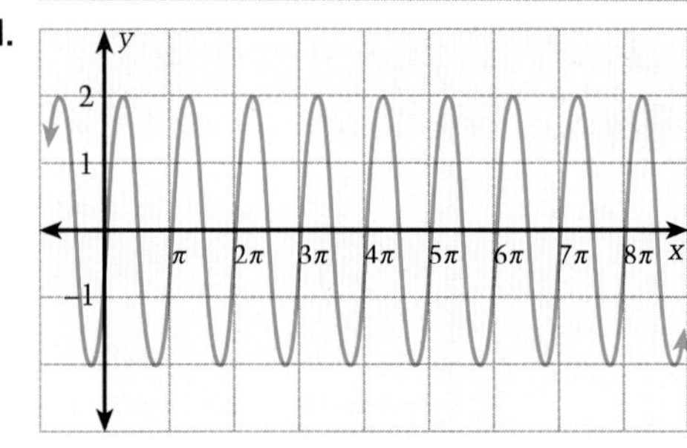

In Exercises 20–31, sketch two cycles of the graph of the function.

20. $y = \sin \frac{1}{3}x$ **21.** $y = \cos \frac{1}{5}x$ **22.** $y = \frac{1}{2} \sin x$ **23.** $y = 6 \cos x$

24. $y = 5 \cos x$ **25.** $y = 10 \sin x$ **26.** $y = 4 \cos 2x$ **27.** $y = \frac{1}{3} \sin 4x$

28. $y = \frac{1}{3} \sin \pi x$ **29.** $y = 3 \sin 2\pi x$ **30.** $y = 2 \cos 6\pi x$ **31.** $y = \frac{1}{2} \cos \frac{1}{4}\pi x$

Technology **In Exercises 32 and 33, the graph was sketched with a graphing calculator, but it is *incorrect*. Can you see what went wrong?**

32. $y = 3 \cos(2x)$ Calculator was in degree mode

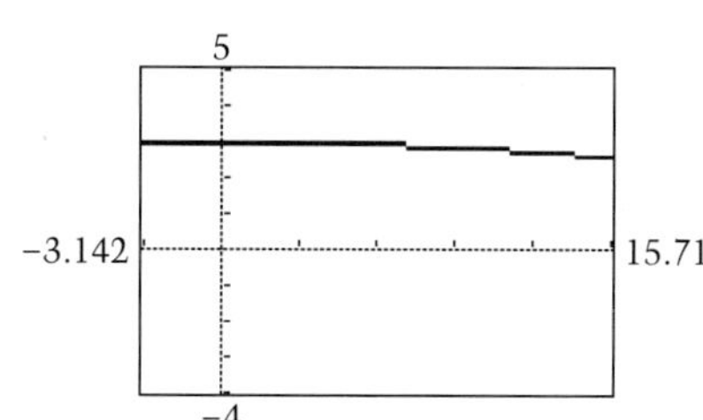

33. $y = \frac{1}{2} \sin x$ "cos x" was entered instead of "sin x"

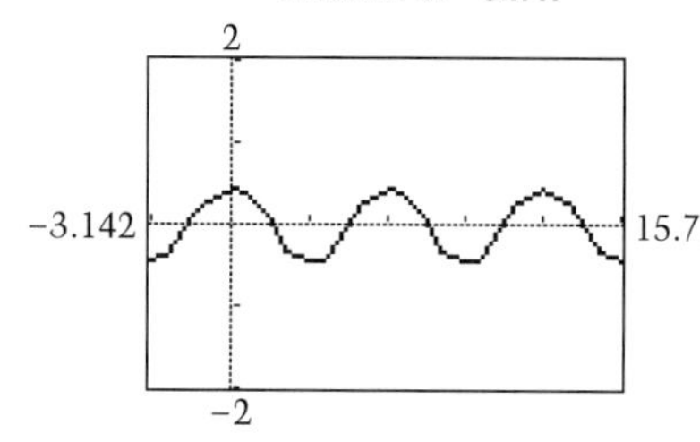

In Exercises 34–37, write an equation of the indicated function. (Use $a > 0$ and $b > 0$.)

34. $y = a \sin bx$
Amplitude: 4
Period: 3π $y = 4 \sin \frac{2}{3}x$

35. $y = a \sin bx$
Amplitude: 3
Period: $\frac{\pi}{3}$ $y = 3 \sin 6x$

36. $y = a \cos bx$
Amplitude: 1
Period: 4 $y = \cos \frac{\pi}{2}x$

37. $y = a \cos bx$
Amplitude: $\frac{5}{2}$
Period: π
$y = \frac{5}{2} \cos 2x$

In Exercises 38 and 39, write two x-values at which the function has a minimum. Write two x-values at which it has a maximum.

38. $y = 2 \cos 3x$ $\frac{\pi}{3}, \pi$; $0, \frac{2\pi}{3}$

39. $y = \frac{1}{2} \sin 4x$ $\frac{3\pi}{8}, \frac{7\pi}{8}$; $\frac{\pi}{8}, \frac{5\pi}{8}$

In Exercises 40 and 41, which graph oscillates most frequently?

40. **a.** $y = \sin \frac{1}{4}x$ **b.** $y = 2 \sin \frac{1}{2}x$ **c.** $y = \frac{1}{4} \sin x$ **d.** $y = \sin 3x$ d

41. **a.** $y = \frac{1}{4} \cos 2x$ **b.** $y = 2 \cos \frac{1}{3}x$ **c.** $y = 4 \cos \frac{5}{2}x$ **d.** $y = \cos \frac{1}{2}x$ c

Sound Waves **In Exercises 42 and 43, use the following information.**

Plucking or striking a stretched string such as a violin string causes sound waves. Sound waves can be modeled on the screen of an **oscilloscope** as shown, and by sine functions of the form $y = a \sin bx$, where x is measured in seconds.

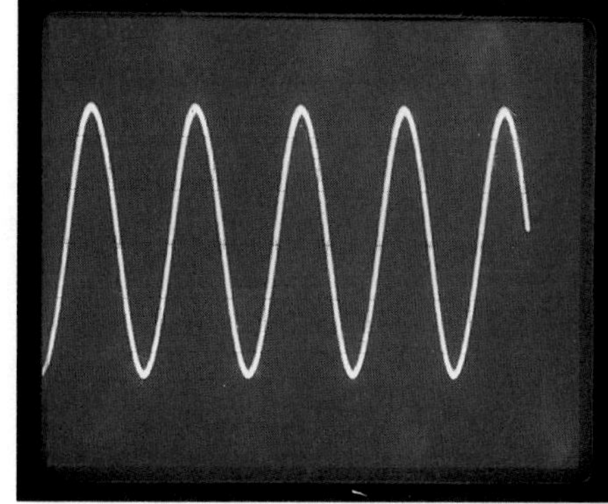

42. Write an equation of a sound wave whose amplitude is 2 and whose period is $\frac{1}{264}$ second. $y = 2 \sin 528\pi x$

43. What is the frequency of the sound wave described in Exercise 42? 264 hertz (cycles per second)

▶ **Ex. 20–31** Encourage students to use the techniques described in Examples 1, 2, and 3. Also encourage students to check their graphs by using computer graphing programs or graphing calculators (see the Enrichment Activities that follow). See Additional Answers.

▶ **Ex. 34–37** Remind students that the period $P = \frac{2\pi}{|b|}$, or $|b| = \frac{2\pi}{P}$.

▶ **Ex. 40–41** Ask students to justify their choices.

EXTEND Ex. 50–51 Use these exercises as an in-class experiment for your location in the United States.

Enrichment Activities

These activities require students to go beyond lesson goals.

TECHNOLOGY When using a graphing calculator, the sine and cosine functions can be graphed using either radians or degrees. For this activity, make sure your calculator is in the radians mode.

Example 1: Graph $y = \sin x$.

Note: Refer to your user's manual for a ZOOM feature that sets the x-range as a multiple of π and sets the Xscl as $\pi/2$.

Example 2: Graph $y = 3\cos(2x)$.

The graph of this function is given below.

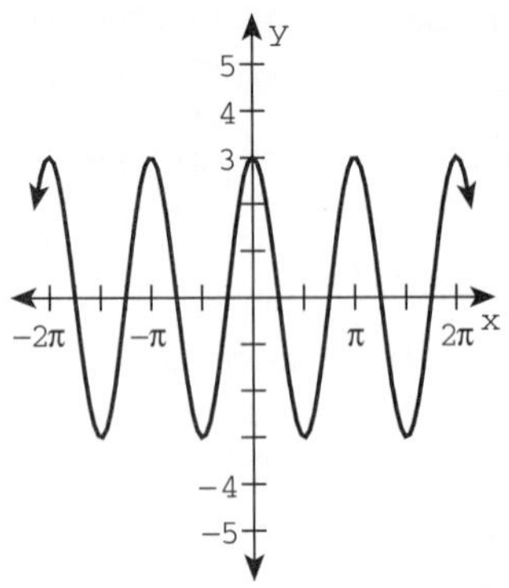

Use a graphing calculator to graph the functions in Exercises 20–31. Use the TRACE feature to find the following:

a. The x-intercepts, maximum and minimum values, and fundamental period of each function;

b. Decimal equivalents of values such as $\frac{\pi}{2}$. $\frac{\pi}{2} \approx 1.57$ on the x-axis.

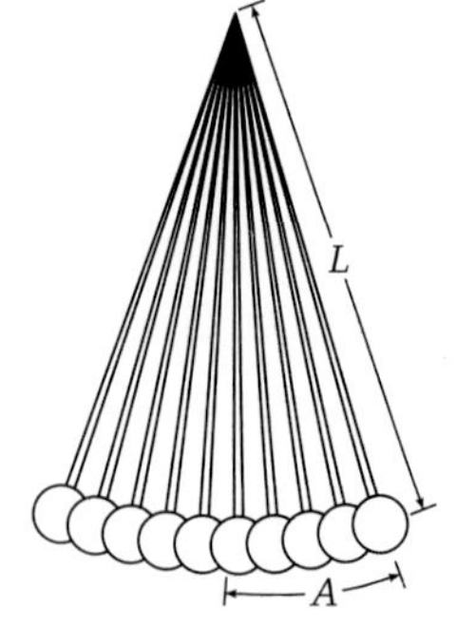

44. *Pendulum Motion* The motion of a pendulum can be modeled by

$$y = A\cos\frac{\sqrt{32}t}{L}.$$

In this model, y is the *directed* length (in feet) of the arc, A is the length (in feet) of the arc from which the pendulum is released, L is the length (in feet) of the pendulum, and t is the time in seconds. How many seconds does it take a 2-foot-long pendulum that is released with an initial arc of 4 inches to swing through one complete cycle? ≈ 2.22

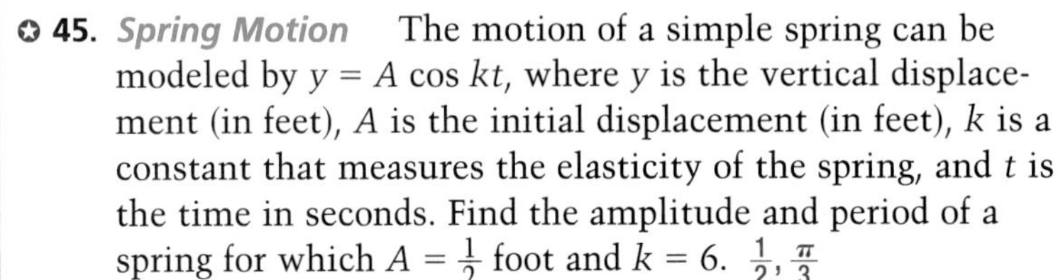

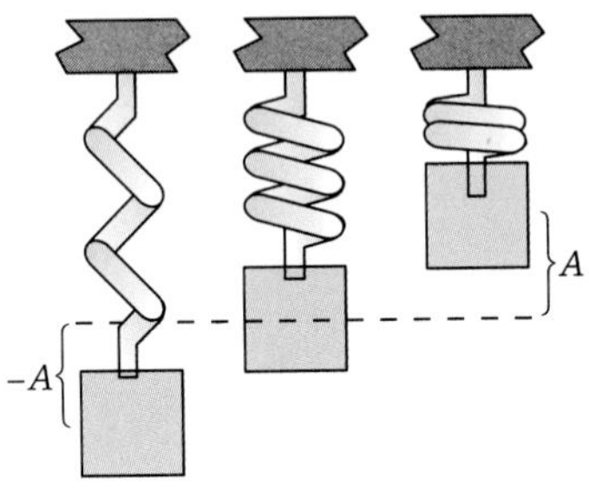

45. *Spring Motion* The motion of a simple spring can be modeled by $y = A\cos kt$, where y is the vertical displacement (in feet), A is the initial displacement (in feet), k is a constant that measures the elasticity of the spring, and t is the time in seconds. Find the amplitude and period of a spring for which $A = \frac{1}{2}$ foot and $k = 6$. $\frac{1}{2}, \frac{\pi}{3}$

Integrated Review

In Exercises 46–49, evaluate the expression.

46. $\sin\frac{3\pi}{4}$ $\frac{1}{\sqrt{2}}$ or ≈ 0.707 **47.** $\cos\frac{7\pi}{3}$ $\frac{1}{2}$ or 0.5 **48.** $\tan\frac{7\pi}{4}$ -1 **49.** $\sec\frac{5\pi}{6}$ $-\frac{2}{\sqrt{3}}$ or ≈ -1.155

Estimating Latitude **In Exercises 50 and 51, use the following information.**

You can estimate the latitude of your location on Earth by pointing one arm at the North Star and the other at the horizon directly below the North Star. The latitude of your location is approximately equal to the angle (in degrees) between your two arms.

50. One of the people pictured is in Portland, Maine. The other is in San Antonio, Texas. Which is which? Explain.

51. Use a protractor to estimate the latitudes of Portland and San Antonio.

Exploration and Extension

52. Copy and complete the table. Round each value to two decimal places. Sketch the graph of $y = \tan x$ for $-\frac{\pi}{2} < x < \frac{\pi}{2}$. What is the period of the graph?

x	$-\frac{11\pi}{24}$	$-\frac{5\pi}{12}$	$-\frac{\pi}{3}$	$-\frac{\pi}{4}$	$-\frac{\pi}{6}$	0	$\frac{\pi}{6}$	$\frac{\pi}{4}$	$\frac{\pi}{3}$	$\frac{5\pi}{12}$	$\frac{11\pi}{24}$
tan x	?	?	?	?	?	?	?	?	?	?	?

−7.60, −3.73, −1.73, −1, −0.58, 0, 0.58, 1, 1.73, 3.73, 7.60; see Additional Answers for graph; π

50. The girl in the picture on the left is in San Antonio, Texas.

51. San Antonio: $\approx 29°$; Portland: $\approx 43°$

14.2 Translations and Reflections of Graphs

What you should learn:

Goal 1 How to graph vertical and horizontal shifts and reflections of the graphs of the sine and cosine functions

Goal 2 How to use sine and cosine functions to solve real-life problems

Why you should learn it:

You can model many cyclical relationships, such as average daily temperature, using translations or reflections of the sine and cosine function.

Goal 1 Graphing Shifts and Reflections

In Lesson 14.1, you learned how to sketch the graphs of functions of the form $y = a \sin bx$ and $y = a \cos bx$ where a is positive. In this lesson, you will study the graphs of functions of the form $y = a \sin(bx + c) + d$ and $y = a \cos(bx + c) + d$ where a is any real number.

Shifts of Sine and Cosine Graphs

To obtain the graph of

$$y = a \sin(bx + c) + d \quad \text{or} \quad y = a \cos(bx + c) + d$$

shift the graph of $y = a \sin bx$ or $y = a \cos bx$ as follows.

1. If $d > 0$, shift d units up.
2. If $d < 0$, shift $|d|$ units down.
3. If $\frac{c}{b} > 0$, shift $\frac{c}{b}$ units to the left.
4. If $\frac{c}{b} < 0$, shift $\left|\frac{c}{b}\right|$ units to the right.

The *amplitude* of the resulting graph is $|a|$ and the *period* is $\frac{2\pi}{|b|}$.

Example 1 *Using a Vertical Shift to Sketch a Graph*

Sketch the graph of $y = 2 + 3 \sin 2x$.

Solution The amplitude of the graph is 3, and the period is π. Because the graph is shifted two units up, it oscillates about the line $y = 2$ rather than about the x-axis. One way to sketch the graph is to sketch the graph of $y = 3 \sin 2x$ first, and then shift the graph up two units. Another way is to locate five key points within one period (between 0 and π). At three of the points, the graph intersects the line $y = 2$. The other two represent the maximum and minimum values of the graph.

$$\underbrace{(0, 2)}_{y=2},\quad \underbrace{\left(\frac{\pi}{4}, 5\right)}_{\text{Maximum}},\quad \underbrace{\left(\frac{\pi}{2}, 2\right)}_{y=2},\quad \underbrace{\left(\frac{3\pi}{4}, -1\right)}_{\text{Minimum}},\quad \underbrace{(\pi, 2)}_{y=2}$$

The graph of the function is shown at the left. ■

$y = 2 + 3 \sin 2x$

$y = 2$

Period: π

Problem of the Day

Ten Ping-Pong balls are numbered 1 to 10. If a pair is drawn at random, what sum is most likely to be obtained? 11

ORGANIZER

Warm-Up Exercises

1. Given the function, $g(x) = x^2$, how would you graph $h(x) = (x + 3)^2 + 2$? Shift the graph of the basic function $g(x)$ 3 units to the left and then 2 units up.
2. Given the function, $f(x) = |x|$, how would you graph $j(x) = |x - 1| - 5$? Shift the graph of the basic function $f(x)$ 1 unit to the right and then 5 units down.
3. Given the function, $p(x) = e^x$, how would you graph $q(x) = -e^{x-2}$? Reflect the graph of the basic function $p(x)$ in the x-axis and then shift the reflection 2 units to the right.

Lesson Resources

Teaching Tools
- Transparencies: 26, 27, 28
- Problem of the Day: 14.2
- Warm-up Exercises: 14.2
- Answer Masters: 14.2

Extra Practice: 14.2
Color Transparency: 74
Applications Handbook: p. 3
Technology Handbook: p. 97

LESSON Notes

Example 1

The period of π was found by solving the equation, period $= \frac{2\pi}{|b|}$ where $b = 2$.

Examples 1–3

To identify the five key points, students should evaluate a function at the start of the period, at $\frac{1}{4}$ of the period, at $\frac{1}{2}$ of the period, at $\frac{3}{4}$ of the period, and at the end of the period. So, if the period is 2π, the function should be evaluated at $0 \cdot 2\pi = 0$, at $\frac{1}{4} \cdot 2\pi = \frac{\pi}{2}$, at $\frac{1}{2} \cdot 2\pi = \pi$, at $\frac{3}{4} \cdot 2\pi = \frac{3\pi}{2}$, and at $1 \cdot 2\pi = 2\pi$; if the period is π, the function should be evaluated at $0 \cdot \pi = 0$, at $\frac{1}{4} \cdot \pi = \frac{\pi}{4}$, at $\frac{1}{2} \cdot \pi = \frac{\pi}{2}$, at $\frac{3}{4} \cdot \pi = \frac{3\pi}{4}$, and at $1 \cdot \pi = \pi$; etc.

Since the period of the function in **Example 2** is 2π, it should be evaluated at 0, at $\frac{\pi}{2}$, at π, at $\frac{3\pi}{2}$, and at 2π. Thus:

$$\cos\left(0 - \frac{\pi}{2}\right) = \cos\left(-\frac{\pi}{2}\right) = 0,$$

$$\cos\left(\frac{\pi}{2} - \frac{\pi}{2}\right) = \cos 0 = 1,$$

$$\cos\left(\pi - \frac{\pi}{2}\right) = \cos\left(\frac{\pi}{2}\right) = 0,$$

$$\cos\left(\frac{3\pi}{2} - \frac{\pi}{2}\right) = \cos(\pi) = -1,$$

$$\cos\left(2\pi - \frac{\pi}{2}\right) = \cos\left(\frac{3\pi}{2}\right) = 0.$$

Example 3

Ask students to describe how to sketch the graph of this function, $y = -3 \cos x$, by graphing the reflection of $y = 3 \cos x$.

Example 4

Have students identify how the key points of the period are related to the following x-values: $0 \cdot 12$, $\frac{1}{4} \cdot 12$, $\frac{1}{2} \cdot 12$, $\frac{3}{4} \cdot 12$, and $1 \cdot 12$.

Example 2 *Graphing Horizontal Shifts*

Sketch the graph of $y = \cos\left(x - \frac{\pi}{2}\right)$.

Solution The amplitude of the graph is 1, and the period is 2π. One way to sketch the graph is to first sketch the graph of $y = \cos x$, and then shift that graph $\frac{\pi}{2}$ units to the right. Another way is to locate five key points within one period (between 0 and 2π).

Intercept	Maximum	Intercept	Minimum	Intercept
$(0, 0)$,	$\left(\frac{\pi}{2}, 1\right)$,	$(\pi, 0)$,	$\left(\frac{3\pi}{2}, -1\right)$,	$(2\pi, 0)$

The graph is shown below in blue. The black curve is the graph $y = \cos x$.

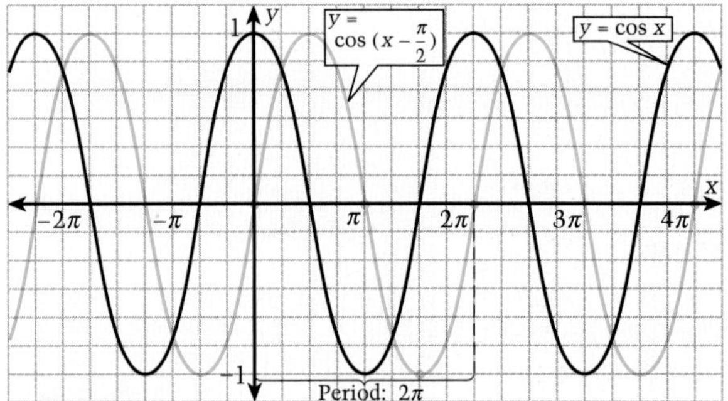

■

Example 3 *Using a Reflection to Sketch a Graph*

Sketch the graph of $y = -3 \cos x$.

Solution The amplitude of the graph is 3, and the period is 2π. One way to sketch the graph is to graph $y = 3 \cos x$ first, and then reflect it in the x-axis. Another way is to locate five key points in the interval between 0 and 2π.

Minimum	Intercept	Maximum	Intercept	Minimum
$(0, -3)$,	$\left(\frac{\pi}{2}, 0\right)$,	$(\pi, 3)$,	$\left(\frac{3\pi}{2}, 0\right)$,	$(2\pi, -3)$

The graph is shown below in blue. The black curve is the graph $y = 3 \cos x$.

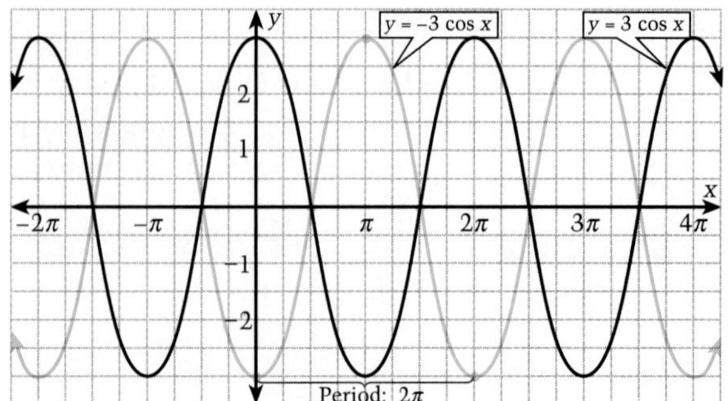

■

Goal 2 Using Sine and Cosine Functions

Real Life
Meteorology

The Riverfront Plaza in downtown Louisville

Example 4 *Using a Temperature Model*

A model for the average daily temperature, T (in degrees Fahrenheit), in Louisville, Kentucky, is given by

$$T = 55.2 + 22.8 \sin\left(\tfrac{1}{6}\pi t + 4.2\right)$$

where $t = 0$ represents January 1, $t = 1$ represents February 1, and so on. What is the period of this function? Which months have the highest and lowest average daily temperatures?

Solution The period of the function is

$$\frac{2\pi}{|b|} = \frac{2\pi}{\frac{1}{6}\pi} = 12.$$

Thus the temperature cycle repeats every 12 months. By calculating temperatures, you can sketch the graph shown below. The highest and lowest average temperatures are

$$55.2 + 22.8 = 78°\text{F} \quad \text{and} \quad 55.2 - 22.8 = 32.4°\text{F}$$

which occur around the first of July and the first of February, respectively.

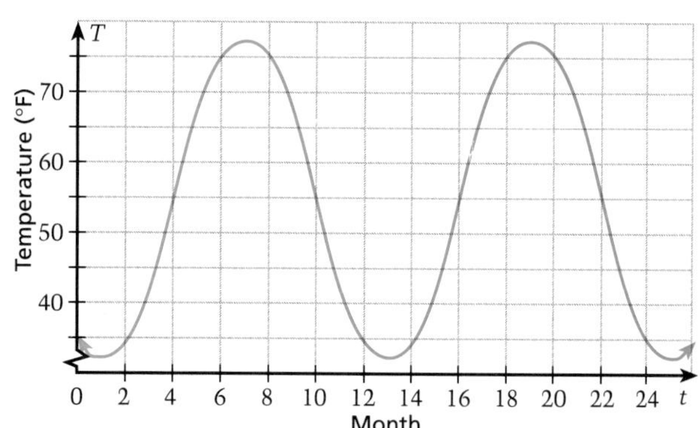

■

Communicating about ALGEBRA

SHARING IDEAS about the Lesson See Additional Answers.

Sketch the graph of each function from 0 to 4π. Explain the effects of the constants a, b, c, and d in $y = a \sin(bx + c) + d$.

A. $y = 3 \sin x$
B. $y = 3 \sin 2x$
C. $y = 3 \sin(2x - 2\pi)$
D. $y = -2 + 3 \sin(2x - 2\pi)$

Extra Examples

Here are additional examples similar to **Examples 1–4.**

1. Sketch the graph.
   $y = \sin\left(2x + \frac{\pi}{4}\right)$

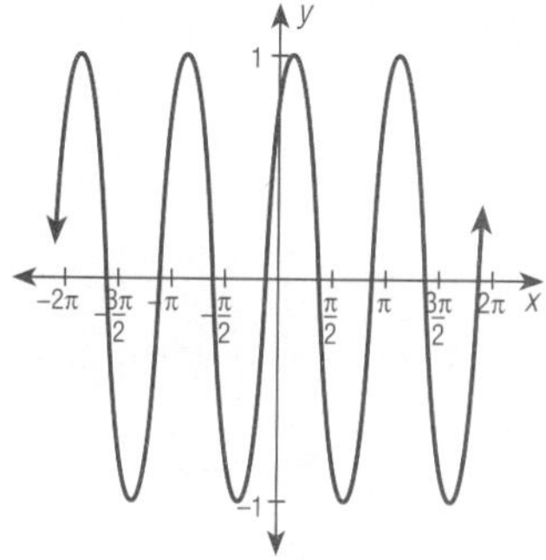

2. How do the following graphs differ from the graph of $y = \cos x$?
   a. $y = \frac{2}{3}\pi \cos x$
   b. $y = \frac{2}{3}\pi + \cos x$
   c. $y = \cos\left(\frac{2}{3}\pi\right)x$
   d. $y = \cos\left(x - \frac{2}{3}\pi\right)$

   a. the amplitude is $\frac{2}{3}\pi$; b. the graph is a vertical shift $\frac{2}{3}\pi$ units up; c. the period is 3; d. the graph is a horizontal shift $\frac{2}{3}\pi$ units to the right.

Check Understanding

1. In terms of its amplitude, how far apart (in vertical distance) are the maximum and minimum values of a cosine function? $|2a|$
2. When graphing the general sine function,
   $y = a \sin(bx + c) + d,$
   why would you first graph the basic sine function? Answers may vary. To obtain the shape of the graph.

Communicating about ALGEBRA

MATH JOURNAL You might have students enter the effects of the constants in the graphing section of their math journals.
A.–D. See Additional Answers.

EXERCISE Notes

ASSIGNMENT GUIDE

Basic/Average: Ex. 1–6, 11–17 odd, 19–24, 31–39 odd, 45–51, 59–62

Above Average: Ex. 1–6, 9–19 odd, 20–24, 29–39 odd, 45–51, 59–64

Advanced: Ex. 1–6, 9–19 odd, 20–24, 29–39 odd, 45–54, 61–64

Selected Answers
Exercises 1–6, 7–61 odd

Use **Mixed Review** as needed.

More Difficult Exercises
Exercises 47–54, 61–64

Guided Practice

Ex. 1–6 Use these exercises to help students systematically work through to the final graph. For example, in Exercise 6, students can find the fundamental period, $\frac{2\pi}{|b|} = 2\pi$, the horizontal shift c, π units to the left, the amplitude a, 3, and the vertical shift d, 2 units up.

Independent Practice

Ex. 7–18 Be sure students indicate that the shift is so many units up/down or to the left/right.

EXERCISES

Guided Practice

CRITICAL THINKING about the Lesson

In Exercises 1–4, state whether the graph of the function is a vertical shift, a horizontal shift, or a reflection of the graph of $y = 6 \cos 2x$.

1. $y = 1 + 6 \cos 2x$ Vertical shift
2. $y = -6 \cos 2x$ Reflection
3. $y = 6 \cos(2x - 3)$ Horizontal shift
4. $y = 6 \cos(2x + 1)$ Horizontal shift
5. Sketch the graph of $y = \sin\left(x + \frac{\pi}{2}\right)$. See Additional Answers.
6. Sketch the graph of $y = 2 + 3 \cos(x + \pi)$. See Additional Answers.

Independent Practice

In Exercises 7–18, describe how the graphs of $y = \sin x$ or $y = \cos x$ can be shifted to produce the graph of the function.

7. $y = 1 + \sin x$ 1 unit up
8. $y = 6 + \cos x$ 6 units up
9. $y = -3 + \cos x$ 3 units down
10. $y = -5 + \sin x$ 5 units down
11. $y = \cos\left(x + \frac{\pi}{2}\right)$ $\frac{\pi}{2}$ units left
12. $y = \sin(x + \pi)$ π units left
13. $y = \sin(x - \pi)$ π units right
14. $y = \cos\left(x - \frac{\pi}{2}\right)$ $\frac{\pi}{2}$ units right
15. $y = 4 + \cos\left(x - \frac{\pi}{4}\right)$ $\frac{\pi}{4}$ units right, 4 units up
16. $y = -2 + \cos\left(x + \frac{3\pi}{4}\right)$ $\frac{3\pi}{4}$ units left, 2 units down
17. $y = -1 + \cos(x - \pi)$ π units right, 1 unit down
18. $y = 3 + \cos(x + \pi)$ π units left, 3 units up

In Exercises 19–24, match the equation to its graph.

19. $y = \cos(x + \pi)$ e
20. $y = 1 + \sin \frac{1}{2}x$ d
21. $y = -2 + \cos x$ f
22. $y = -2 + \sin(2x + \pi)$ a
23. $y = -\cos(x + \pi)$ c
24. $y = 1 + 2 \cos\left(\frac{1}{2}x + \frac{\pi}{2}\right)$ b

a.
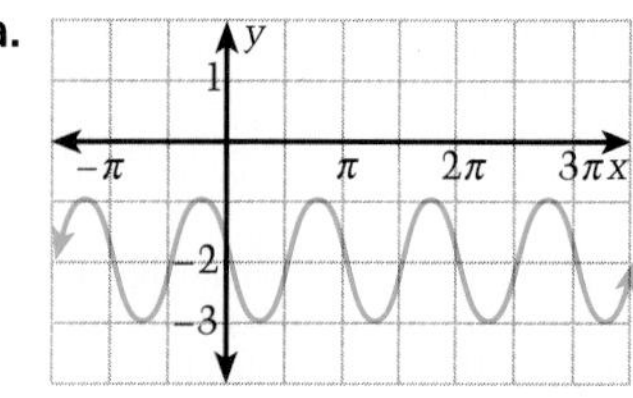

b.
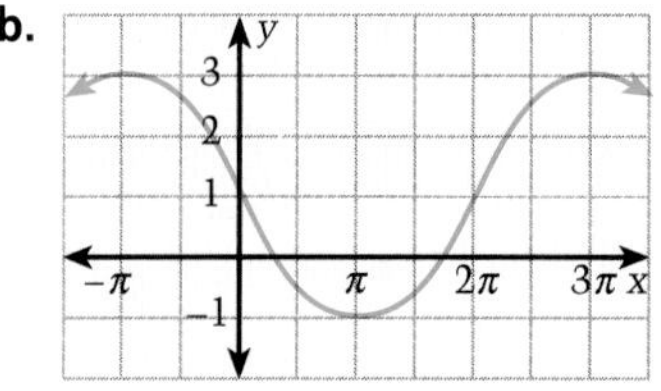

c.
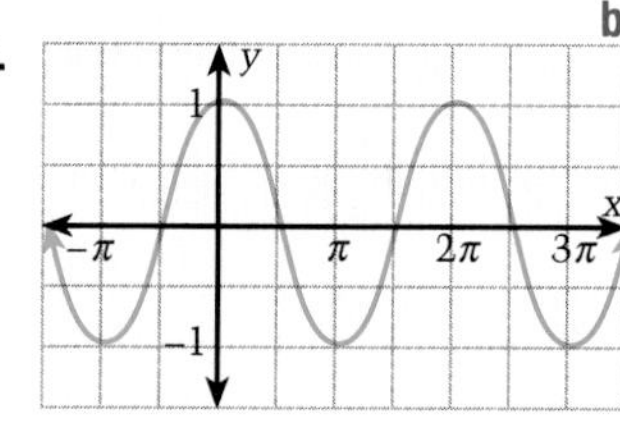

d.
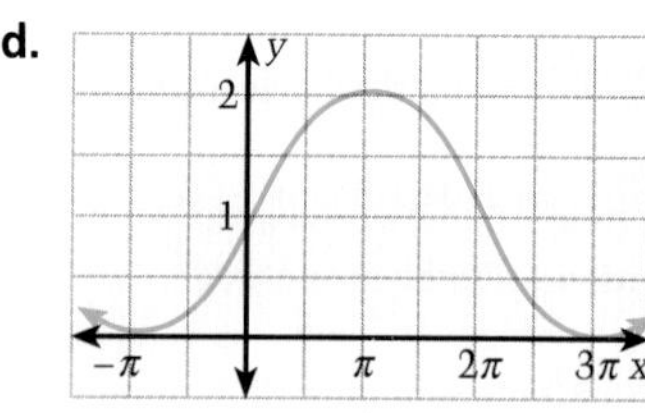

e.
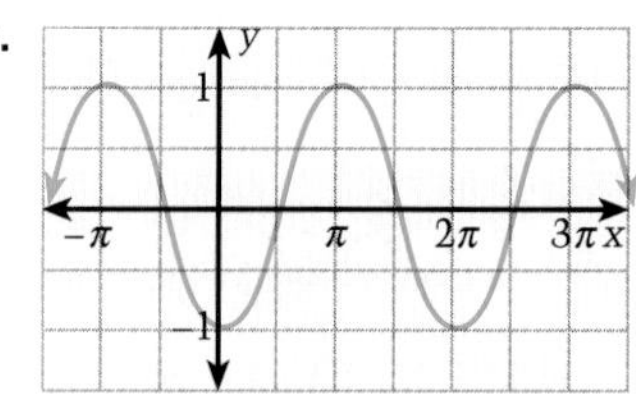

f.
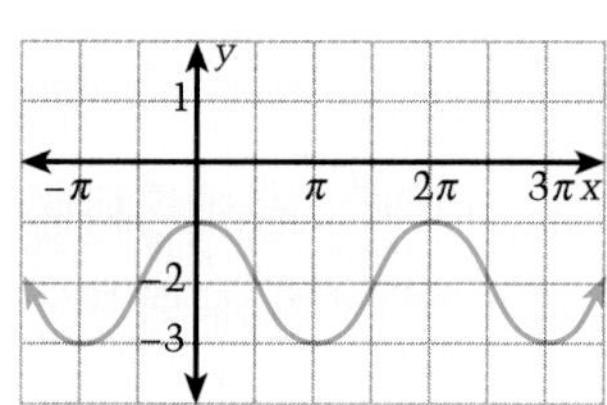

25. The graph of $y = 2 \sin 3x$ is drawn and then shifted up two units. Write an equation of the resulting graph. $y = 2 + 2 \sin 3x$
26. The graph of $y = 2 \sin x$ is drawn and then shifted π units to the right. Write an equation of the resulting graph. $y = 2 \sin(x - \pi)$

In Exercises 27–44, sketch the graph of the function.

27. $y = 4 + \sin \frac{1}{2}x$
28. $y = \cos\left(x - \frac{\pi}{2}\right)$
29. $y = -2 + \cos 3x$
30. $y = 3 + \sin 2x$
31. $y = 1 + \sin(x + \pi)$
32. $y = -1 + \cos(x - \pi)$
33. $y = 3 + 2\cos(2x - \pi)$
34. $y = 2 + 3\sin(4x + \pi)$
35. $y = -4\cos \frac{1}{4}x$
36. $y = -\frac{1}{2}\sin 4x$
37. $y = -1 + 2\sin(x - \pi)$
38. $y = -8 + 6\cos\left(2x - \frac{\pi}{2}\right)$
39. $y = 10 + \sin\left(\frac{1}{4}x + \pi\right)$
40. $y = -2 + \cos\left(2x + \frac{\pi}{2}\right)$
41. $y = 2 - \sin x$
42. $y = 1 - \cos x$
43. $y = 1 + 5\cos(x - \pi)$
44. $y = 5 + 10\sin\left(x + \frac{3\pi}{2}\right)$

45. What are the minimum and maximum values of $y = -1 + 2\sin 6x$? Write two x-values at which the minimum occurs. Write two x-values at which the maximum occurs. **−3, 1; at −3: $\frac{\pi}{4}, \frac{7\pi}{12}$; at 1: $\frac{\pi}{12}, \frac{5\pi}{12}$**

46. What are the minimum and maximum values of the function $y = 4 - 3\cos\left(x + \frac{\pi}{2}\right)$? Write two x-values at which the minimum occurs. Write two x-values at which the maximum occurs. **1, 7; at 1: $-\frac{\pi}{2}, \frac{3\pi}{2}$; at 7: $\frac{\pi}{2}, \frac{5\pi}{2}$**

In Exercises 47 and 48, write an equation of the graph.

47.

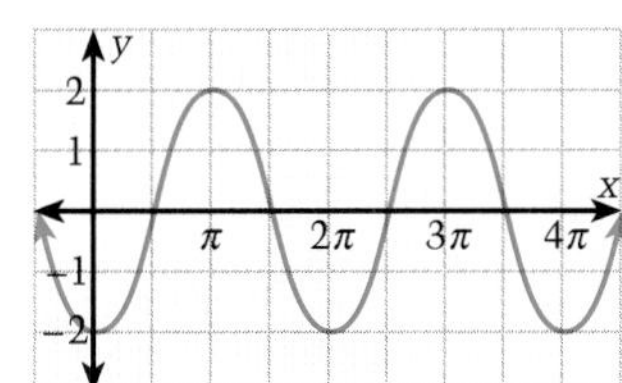

48.

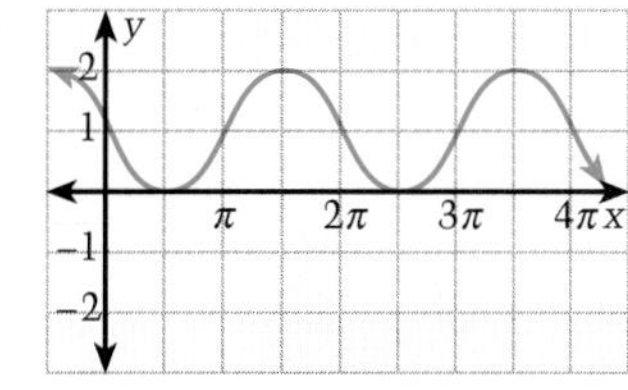

Coyotes and Rabbits **In Exercises 49 and 50, use the following information.**

Sine functions are used in biology to model oscillations in predator and prey populations. The population, R, of rabbits and the population, C, of coyotes in a region are modeled by

$$R = 25{,}000 + 15{,}000 \cos \frac{2\pi t}{24}$$

$$C = 5000 + 2000 \sin \frac{2\pi t}{24}$$

where t is the time in months.

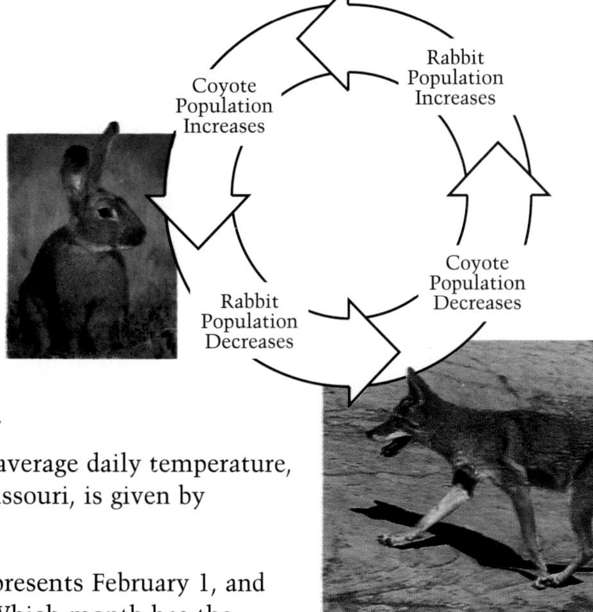

49. Find the amplitude and period of each graph. **R: 15,000; 24; C: 2000; 24**

50. Sketch both of these functions on the same coordinate plane and explain the oscillations in the size of each population.

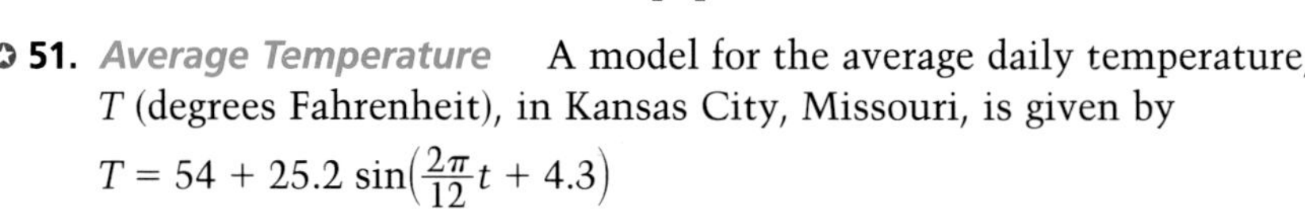

51. *Average Temperature* A model for the average daily temperature, T (degrees Fahrenheit), in Kansas City, Missouri, is given by

$$T = 54 + 25.2 \sin\left(\frac{2\pi}{12}t + 4.3\right)$$

where $t = 0$ represents January 1, $t = 1$ represents February 1, and so on. Sketch the graph of this function. Which month has the highest average temperature? The lowest average temperature? **July, January**

47. $y = 2\sin\left(x - \frac{\pi}{2}\right)$ or $y = 2\cos(x - \pi)$

48. $y = 1 + \cos\left(x + \frac{\pi}{2}\right)$ or $y = 1 + \cos\left(x - \frac{3\pi}{2}\right)$ or $y = 1 + \sin(x - \pi)$ or $y = 1 - \sin x$, and so on

▶ **Ex. 19–24** Assign these exercises as a group.

WRITING Have students write a one-sentence explanation of their reasoning.

▶ **Ex. 27–44** Be sure to have students discuss a general strategy for completing the paper-and-pencil graphs in class.

TECHNOLOGY If computer graphing programs or graphics calculators are available, you might assign some students to use them to graph the equations to provide a visual check.

Answers
See Additional Answers for exercises 27–44 and 50–51.

▶ **Ex. 49–54** **TECHNOLOGY** Students could use graphics-calculator graphs to solve these problems. You may need to discuss how to locate appropriate range values in order to view the pattern of the graph. In Exercise 50, for example, a range of

Xmin = 0 Ymin = 0
Xmax = 48 Ymax = 42,000
Xscl = 3 Yscl = 5000
Xres = 1

yields a visual verification of the oscillation pattern.

EXTEND Ex. 49–54
PROJECT The "predator-prey" relationship between two animals is but one real-life example of cyclical or periodic relationships. Ask students to interview science teachers to determine other places in the schools' curriculum where periodic or cyclical functions are used. (For example, electrical waves; light waves; radioactive emission patterns; biological or circadian rhythms such as sleep-wake cycles, sound wave patterns, brain wave patterns, economic cycles, etc.)

Ex. 61 This problem is often referred to in geometry as "crooks problem." Suggest that students draw a line parallel to $\overline{PQ}$ and $\overline{RS}$ through angle θ. Then apply the theorem from geometry that the alternate interior angles, formed by a transversal that intersects parallel lines, are congruent.

Ex. 63–64 TECHNOLOGY Using a graphics calculator, try $c = \pi, \frac{\pi}{4}, \frac{\pi}{3}$, and $\frac{\pi}{2}$.

Answers
See Additional Answers for exercises 55–60.

Enrichment Activities

These activities require students to go beyond lesson goals.

1. Sketch the graph of the basic trigonometric function, $y = \tan x$. (*Hint:* See Exercise 52 on page 736. This graph has asymptotes.)

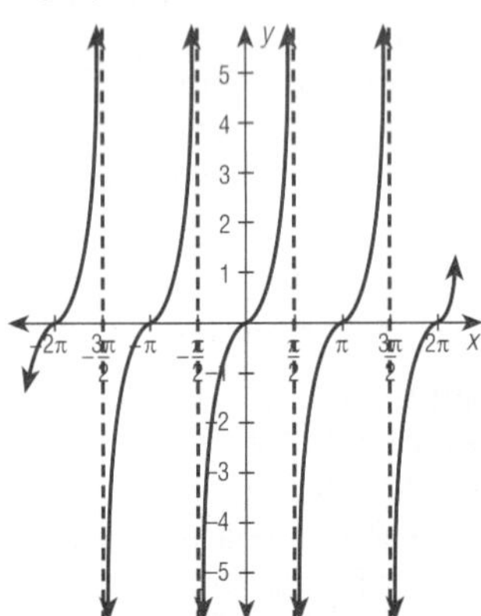

2. Use the basic graph of $y = \tan x$, and your knowledge of the effects of constants on the graphs of algebraic functions to graph the following.
   a. $y = 4 + \tan x$
   b. $y = -\tan x$
   c. $y = \tan\left(x - \frac{\pi}{2}\right)$

Sunrise, Sunset **In Exercises 52 and 53, use the following information.**

Cheyenne (Wyoming), DuBois (Pennsylvania), Ft. Wayne (Indiana), and Omaha (Nebraska) all have a latitude of 41° north. On this latitude, the position of the sun at sunrise (and sunset) can be modeled by

$$T = 31\sin\left(\frac{2\pi}{365}t - 1.4\right)$$

where t is the time in days with $t = 1$ representing January 1. In this model, T represents the number of degrees north or south of due east (and west) that the sun rises (and sets).

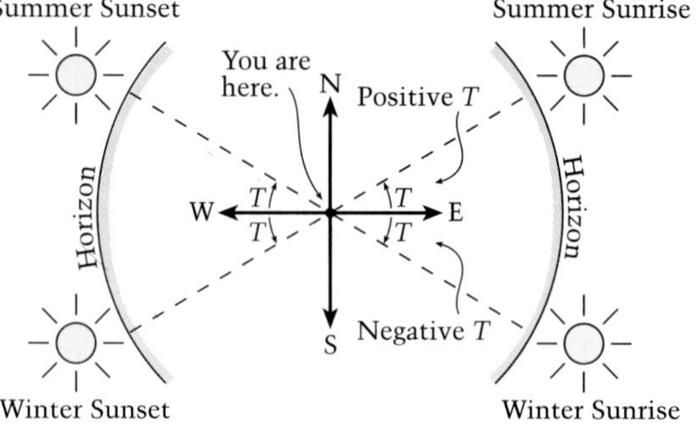

52. Sketch the graph of this function. **See Additional Answers.**

53. Locate the spring equinox, the summer solstice, the fall equinox, and the winter solstice on the graph. Which of these four days has the most hours of sunlight? Which has the least? **See below.**

54. *Blood Pressure* The blood pressure, P (in millimeters of mercury), of a certain person is given by $P = 100 - 20\cos\frac{8\pi}{3}t$, where t is the time in seconds. Sketch the graph of this function. What is this person's pulse rate in heartbeats per minute? **80; see Additional Answers for graph.**

Integrated Review

In Exercises 55–60, sketch the graph of the function.

55. $y = 6\sin 4x$ **56.** $y = 5\cos 6x$ **57.** $y = \frac{1}{3}\cos 5x$

58. $y = \frac{1}{2}\sin 2x$ **59.** $y = 3\cos\left(\frac{1}{4}x\right)$ **60.** $y = \cos\left(\frac{1}{3}x\right)$

61. *College Entrance Exam Sample* In the figure below, line PQ is parallel to line RS. What is the degree measure of θ? **c**

a. 90° **b.** 100° **c.** 110° **d.** 120° **e.** 130°

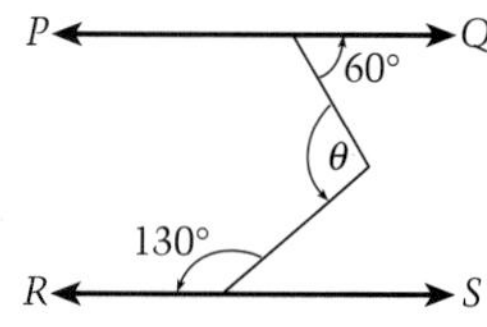

62. *College Entrance Exam Sample* In the figure below, what is the length of $\overline{NP}$? **e**

a. 3 **b.** 8 **c.** 9 **d.** 12 **e.** 15

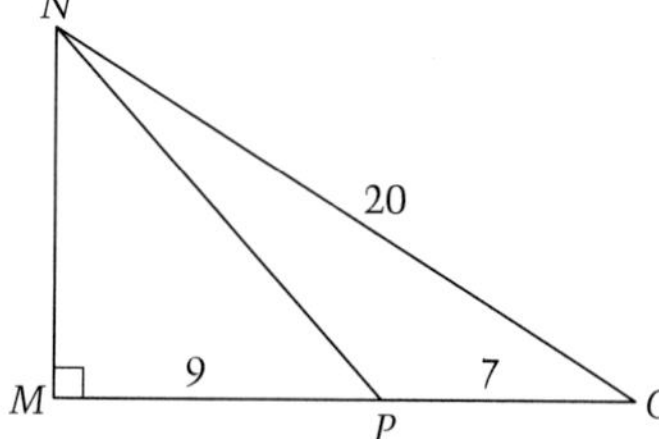

Exploration and Extension

63. Find a value of c for which the graphs of $y = \sin x$ and $y = \cos(x - c)$ are the same. **Answers vary; example: $\frac{\pi}{2}$**

64. Find a value of c for which the graphs of $y = \cos x$ and $y = \sin(x - c)$ are the same. **Answers vary; example: $-\frac{\pi}{2}$**

53. Spring Equinox: day 79, March; Summer Solstice: day 172, June—most sunlight; Fall Equinox: day 265, September; Winter Solstice: day 355, December—least sunlight.

Mixed REVIEW

4. $y = \sqrt[3]{3x + 9}$ 6., 8., 9., and 11. See Additional Answers.

1. Solve $\frac{3}{4}x^2 - 6 = 8$. **(5.2)** $\pm 2\sqrt{\frac{14}{3}}$
2. Solve $3x + 2 \leq -3$. **(1.6)** $x \leq \frac{-5}{3}$
3. Find the mean: 0, 2, 3, 6, 5, 4, 2. **(6.7)** $\frac{22}{7}$
4. Find the inverse of $y = \frac{1}{3}x^3 - 3$. **(6.3)**
5. Simplify: $6^2x^2y^{-3} \cdot 2^{-2}xy^4$. $9x^3y$
6. Complete the square: $3x^2 + 3x - 4$. **(5.3)**
7. Condense $2 \log_4 x + 4 \log_4 y$. **(8.3)** $\log_4 x^2y^4$
8. Sketch the graph of $y = e^{-2x}$. **(8.1)**
9. Sketch the graph of $y^2 - 4x = 0$. **(11.1)**
10. Find the complement of $\frac{3\pi}{13}$. **(13.2)** $\frac{7\pi}{26}$
11. Sketch the graph of $4x^2 + 6y^2 = 24$. **(11.3)**
12. Sketch the graph of $y = x^2 + 5x + 6$. **(11.1)** See Additional Answers.
13. What is the reciprocal of the $\cos \theta$? **(13.1)** $\sec \theta$
14. Sketch the graph of $y \leq 2x^2 + 5x + 3$. **(5.7)** See Additional Answers.
15. Factor $4x^5 + 3x^2 + x$. **(9.3)** $x(4x^4 + 3x + 1)$
16. True or False? $1 + \tan^2 \theta = \sec^2 \theta$ **(13.1)** True
17. Solve $\frac{3y}{x} + 2y = x^2$ for y. **(1.5)** $y = \frac{x^3}{3 + 2x}$
18. Solve for a: $\frac{5}{\sin 30°} = \frac{a}{\sin 60°}$. **(13.5)** $5\sqrt{3}$ or ≈ 8.66
19. Find the area of a triangle whose sides have lengths of 2, 4, and 5. **(13.6)** $\frac{\sqrt{231}}{4}$
20. Identify the conic whose equation is $4x^2 + 8x - 2y^2 + 8y = 6$. **(11.6)** Hyperbola

Milestones UNSOLVED PROBLEMS

400 | 200 | B.C. | A.D. | 200 | 400 | 600 | 800 | 1000 | 1200 | 1400 | 1600 | 1800 | 2000 | 2200

Euclid's proof of the number of primes — Fermat's last theorem — Topology becomes a division of mathematics

Some of the questions raised in the study of mathematics are answered, while others raise new questions. Around 300 B.C., Euclid proved that the number of primes is infinite, but no one has found a formula for the next larger prime number for any given prime. Other questions arise from new areas of research, such as the topological question, How many ways can a rubber sphere be wrapped around another sphere?

In the 1600's, Pierre de Fermat made the assertion that the equations $x^n + y^n = z^n$ has no solution in the positive integers when $n > 2$. "Fermat's last theorem," as it is called, had been proven true for all values of $n < 100{,}000$, and certain other values of n. Only recently did Andrew Wiles of Princeton University prove it for *all* the positive integers greater than 2. In his notes describing the theorem, Fermat asserted, ". . . I have assuredly found an admirable proof of this, but the margin is too narrow to contain it." The proof has not been found in any of his other writings, leading many people to believe that he later found a flaw in it.

Pierre de Fermat was a lawyer by profession and a mathematician in his spare time.

Find out what topology is.

Topology is the study of geometric shapes and the properties that do not change when the shapes are stretched, twisted, or in some way manipulated.

a. Vertical shift up 4 units

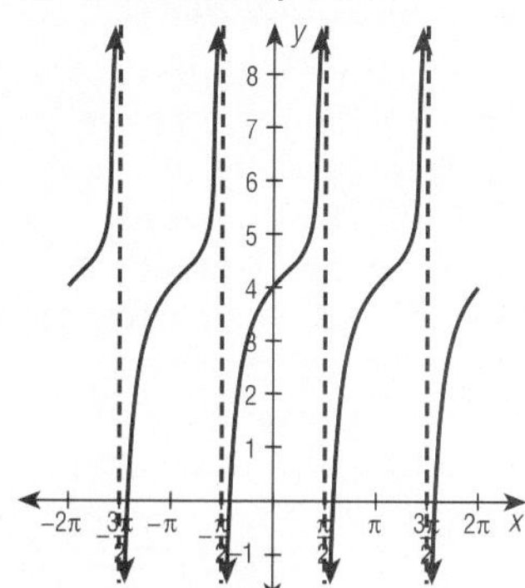

b. Reflection in the x-axis

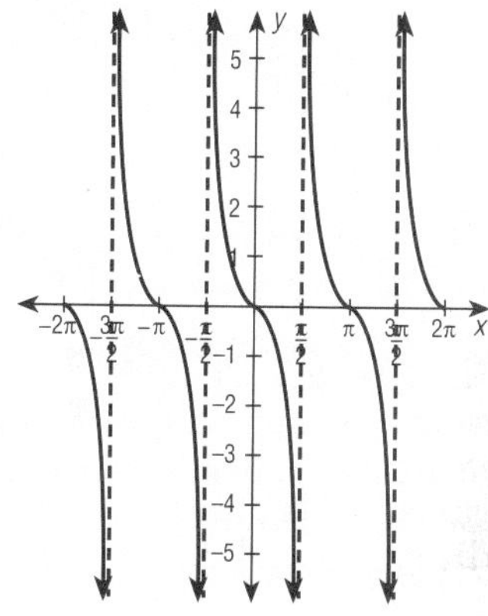

c. Horizontal shift $\frac{\pi}{2}$ units to the right of $y = \tan x$

Milestones

Language Arts: *Critical Reading*

1. Find a solution in the positive integers for $x^n + y^n = z^n$ when $n \leq 2$. Answers may include the following: $3^2 + 4^2 = 5^2$, $5^2 + 12^2 = 13^2$, $1^1 + 2^1 = 3^1$
2. Find and describe the two parts to Goldbach's Conjecture. 1. Every even number ≥ 6 is the sum of two odd prime numbers. 2. Every odd number ≥ 9 is the sum of three odd prime numbers.

Problem of the Day

In a certain country, everyone is a Knight, a Knave, or a Squire. Knights always tell the truth, Knaves always lie, and Squires sometimes lie. A stranger approaches a group of three consisting of a Knight, a Knave, and a Squire. A says, "I am a Squire." B says, "That is true." C says, "I am not a Squire." What are they in fact?

A is a Knave, B is a Squire, C is a Knight.

ORGANIZER

Warm-up Exercises

Fill in the blank with an expression that completes the trigonometric identity.

a. $\cos^2 x + \sin^2 x = \underline{\ ?\ }$

b. $\dfrac{\sin x}{\cos x} = \underline{\ ?\ }$

c. $\dfrac{1}{\sec x} = \underline{\ ?\ }$

d. $1 + \tan^2 x = \underline{\ ?\ }$

a. 1 b. $\tan x$

c. $\cos x$ d. $\sec^2 x$

Lesson Resources

Teaching Tools

Transparencies: 2, 3

Problem of the Day: 14.3

Warm-up Exercises: 14.3

Answer Masters: 14.3

Extra Practice: 14.3

Color Transparency: 74

Technology Handbook: p. 99

14.3 Trigonometric Identities

What you should learn:

Goal 1 How to use trigonometric identities to simplify trigonometric expressions and how to verify trigonometric identities

Goal 2 How to use trigonometric identities to solve real-life problems

Why you should learn it:

In real-life problems, trigonometric expressions often occur in unsimplified forms. You can use trigonometric identities to simplify these expressions.

Goal 1 Using Trigonometric Identities

In Lesson 13.1 (page 680), you studied three of the five types of fundamental trigonometric identities: reciprocal identities, tangent and cotangent identities, and Pythagorean identities. Two other types are *cofunction identities* and *negative angle identities.*

Cofunction Identities:

$\sin\left(\frac{\pi}{2} - x\right) = \cos x \qquad \tan\left(\frac{\pi}{2} - x\right) = \cot x \qquad \sec\left(\frac{\pi}{2} - x\right) = \csc x$

$\cos\left(\frac{\pi}{2} - x\right) = \sin x \qquad \cot\left(\frac{\pi}{2} - x\right) = \tan x \qquad \csc\left(\frac{\pi}{2} - x\right) = \sec x$

Negative Angle Identities:

$\sin(-x) = -\sin x \qquad \cos(-x) = \cos x \qquad \tan(-x) = -\tan x$

$\csc(-x) = -\csc x \qquad \sec(-x) = \sec x \qquad \cot(-x) = -\cot x$

Example 1 *Simplifying a Trigonometric Expression*

Simplify $\cos x - \cos x \sin^2 x$.

Solution

$\cos x - \cos x \sin^2 x = \cos x(1 - \sin^2 x)$ *Factor.*

$= \cos x(\cos^2 x)$ *Pythagorean identity: $\sin^2 x + \cos^2 x = 1$*

$= \cos^3 x$ *Simplify.* ■

Example 2 *Simplifying a Trigonometric Expression*

Simplify $\sin\left(\frac{\pi}{2} - x\right) \tan x$ for $x \neq \frac{\pi}{2}, \frac{3\pi}{2}$.

Solution

$\sin\left(\frac{\pi}{2} - x\right) \tan x = \cos x \tan x$ *Cofunction identity*

$= \cos x \dfrac{\sin x}{\cos x}$ *Tangent identity*

$= \cancel{\cos x} \dfrac{\sin x}{\cancel{\cos x}}$ *Divide by common factor.*

$= \sin x$ *Simplify.* ■

A *verification* of a statement is a logical argument that the statement is true. The fundamental trigonometric identities can be used to verify other trigonometric identities.

Example 3 *Verifying a Trigonometric Identity*

Verify the identity $\frac{\cot^2 x}{\csc x} = \csc x - \sin x$.

Solution

$$\frac{\cot^2 x}{\csc x} = \frac{\csc^2 x - 1}{\csc x} \quad \textit{Pythagorean identity}$$

$$= \frac{\csc^2 x}{\csc x} - \frac{1}{\csc x} \quad \textit{Write as separate fractions.}$$

$$= \frac{\csc^{\not 2} x}{\not{\csc x}} - \frac{1}{\csc x} \quad \textit{Divide by common factor.}$$

$$= \csc x - \frac{1}{\csc x} \quad \textit{Simplify.}$$

$$= \csc x - \sin x \quad \textit{Reciprocal identity}$$

■

Example 4 *Verifying a Trigonometric Identity*

Verify the identity $\cos^2 x - \sin^2 x = 2\cos^2 x - 1$.

Solution

$$\cos^2 x - \sin^2 x = \cos^2 x - (1 - \cos^2 x) \quad \textit{Pythagorean identity}$$

$$= \cos^2 x - 1 + \cos^2 x \quad \textit{Simplify.}$$

$$= 2\cos^2 x - 1 \quad \textit{Simplify.}$$

■

Example 5 *Verifying a Trigonometric Identity*

Verify the identity $\frac{\sin x}{1 - \cos x} = \frac{1 + \cos x}{\sin x}$.

Solution

$$\frac{\sin x}{1 - \cos x} = \frac{\sin x(1 + \cos x)}{(1 - \cos x)(1 + \cos x)} \quad \textit{Multiply numerator and denominator by } (1 + \cos x).$$

$$= \frac{\sin x(1 + \cos x)}{1 - \cos^2 x} \quad \textit{Simplify denominator.}$$

$$= \frac{\sin x(1 + \cos x)}{\sin^2 x} \quad \textit{Pythagorean identity}$$

$$= \frac{\not{\sin x}(1 + \cos x)}{\sin^{\not 2} x} \quad \textit{Divide by common factor.}$$

$$= \frac{1 + \cos x}{\sin x} \quad \textit{Simplify.}$$

■

LESSON Notes

Example 1

Students might find it useful to express the given trigonometric functions in terms of a single trigonometric function, or in terms of sines and cosines. In this case, the sine function is expressed in terms of the cosine.

Example 2

Using the cofunction identity and the tangent identity, you can express the given trigonometric expression in terms of sines and cosines of x.

Ask students why x cannot equal $\frac{\pi}{2}$ and $\frac{3\pi}{2}$. Recall that the domain of the function $y = \cos x$ is restricted to the interval $0 \le x \le \pi$, so that the arccos y function can be well-defined. The tan function is undefined when $\cos x$ is 0. So, $\tan x$ is undefined when $x = \frac{\pi}{2}$ and $x = \frac{3\pi}{2}$.

Examples 3–5

When verifying trigonometric identities, students must understand that they rewrite one side of the identity in terms of the functions found on the other side.

Call students' attention to the technique used in Example 5. This technique of multiplying a rational expression by "1," written in a special form, is a common step used in many algebraic proofs.

Example 6

The equations $x = 2\cos t$ and $y = 3\sin t$ imply (upon solving for $\cos t$ and $\sin t$) that $\cos t = \frac{x}{2}$ and $\sin t = \frac{y}{3}$.

TECHNOLOGY If the parametric equations are graphed on a graphics calculator, be sure to use the trigonometric setting for the coordinate plane.

Extra Examples

Here are additional examples similar to **Examples 1–6.**

1. Simplify the following.
   a. $\dfrac{1-\cos^2 x}{\sec^2 x-\tan^2 x}$ $\sin^2 x$
   b. $\dfrac{\cos^2 x}{\cos\left(\frac{\pi}{2}-x\right)}+\sin x$ $\csc x$
2. Verify the following identities.
   a. $\dfrac{\sin x}{\csc x-1}+\dfrac{\sin x}{\csc x+1}=2\tan^2 x$
   b. $\dfrac{\sin^2 x}{\cos x}+\cos x=\sec x$
   c. $(\sec x-\tan x)\cdot(\sec x+\tan x)=1$

   Verifications may vary.

   a. $\dfrac{\sin x(\csc x+1)+\sin x(\csc x-1)}{(\csc x-1)(\csc x+1)}=\dfrac{1+\sin x+1-\sin x}{\csc^2 x-1}=\dfrac{2}{\cot^2 x}=2\tan^2 x$

   b. $\dfrac{\sin^2 x+\cos^2 x}{\cos x}=\dfrac{1}{\cos x}=\sec x$

   c. $\sec^2 x-\tan^2 x=1+\tan^2 x-\tan^2 x=1$
3. Show that the graph of the parametric equations $x = 4\cos t$ and $y = 4\sin t$ is a circle.
   The standard equation associated with the parametric equations is $x^2+y^2=16$.

Check Understanding

1. What is a verification of an algebraic statement? See the top of page 745.
2. How are parametric relationships different from the standard function relationships? Parametric relationships in x and y both depend on another variable, t. See the top of page 746.
3. What are the Pythagorean identities? $\sin^2 x+\cos^2 x=1$, $1+\cot^2 x=\csc^2 x$, and $\tan^2 x+1=\sec^2 x$.

Goal 2 Using Trig Identities in Technology

Most graphing calculators have two graphing modes: *function* mode and *parametric* mode. In function mode, you must enter y as a function of x. In parametric mode, you enter x and y as functions of a *parameter* t. In either case, the resulting graph is in the standard xy-coordinate plane.

Connections **Technology**

Example 6 *Sketching an Ellipse*

Show that the graph of these parametric equations is an ellipse: $x = 2\cos t$ and $y = 3\sin t$.

Solution To get an idea of the shape of the graph, you can substitute several values of t and plot the resulting points (x, y).

t	0	$\frac{\pi}{4}$	$\frac{\pi}{2}$	$\frac{3\pi}{4}$	π	$\frac{5\pi}{4}$	$\frac{3\pi}{2}$	$\frac{7\pi}{4}$
(x, y)	$(2, 0)$	$\left(\sqrt{2}, \frac{3\sqrt{2}}{2}\right)$	$(0, 3)$	$\left(-\sqrt{2}, \frac{3\sqrt{2}}{2}\right)$	$(-2, 0)$	$\left(-\sqrt{2}, -\frac{3\sqrt{2}}{2}\right)$	$(0, -3)$	$\left(\sqrt{2}, -\frac{3\sqrt{2}}{2}\right)$

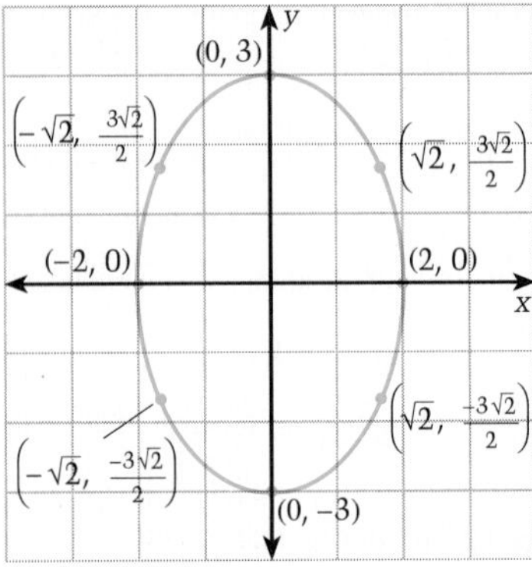

By plotting the points in the table, it appears that each point lies on an ellipse. You can use a trig identity to verify this.

$\sin^2 t+\cos^2 t=1$ *Pythagorean identity*

$\left(\frac{y}{3}\right)^2+\left(\frac{x}{2}\right)^2=1$ *Substitute for sin t and cos t.*

$\frac{y^2}{3^2}+\frac{x^2}{2^2}=1$ *Simplify.*

From the last equation, you can recognize that the graph of the equations is an ellipse. If you have a graphing calculator, try using it in parametric mode to sketch this ellipse. ■

Communicating about ALGEBRA

SHARING IDEAS about the Lesson

You are taking a course in calculus. For one of the homework problems, the (correct) answer in the back of the book is $\sec x\csc x$. You and two classmates all get different answers. Which of them are correct? Explain.

A. $\dfrac{\sec^2 x}{\tan x}$ **B.** $\cot x+\tan x$ **C.** $\dfrac{1}{\sin x\cos x}$

EXERCISES

Guided Practice

CRITICAL THINKING about the Lesson

1. Write the six cofunction identities. See page 744.
2. A function is *odd* if $f(-x) = -f(x)$. A function is *even* if $f(-x) = f(x)$. Which of the six trigonometric functions are odd? Which are even?
3. Simplify $\cos\left(\frac{\pi}{2} - x\right) \cot x$. Which identities did you use?
4. Complete the verification shown at the right.

2. odd: $\sin x, \tan x, \cot x, \csc x$; even: $\cos x, \sec x$

3. $\cos x$, cofunction and cotangent 4. $1 - \cos^2 x$, $\frac{1}{\cos^2 x}$

$$\begin{aligned}\tan^2 x &= \frac{\sin^2 x}{\cos^2 x}\\ &= \frac{\boxed{?}}{\cos^2 x}\\ &= \boxed{?} - 1\\ &= \sec^2 x - 1\end{aligned}$$

Independent Practice

In Exercises 5–22, simplify the expression. See page 680 for the Fundamental Trigonometric Identities.

5. $\sec x \cos(-x) - \sin^2 x$ $\cos^2 x$
6. $\frac{\cot(-x)}{\csc(-x)}$ $\cos x$
7. $\cos\left(\frac{\pi}{2} - x\right) \csc x$ 1
8. $\cos x(1 + \tan^2 x)$ $\sec x$
9. $\sec^2\left(\frac{\pi}{2} - x\right) - 1$ $\cot^2 x$
10. $\frac{\cos^2 x \cot^2\left(\frac{\pi}{2} - x\right) - 1}{\cos^2 x}$ -1
11. $-1 + \sin^2 x + \sec^2(-x)$ $\sec^2 x - \cos^2 x$
12. $\cot\left(\frac{\pi}{2} - x\right) \cot x$ 1
13. $(1 - \cos^2 x)(\cot^2(-x) + 1)$ 1
14. $\frac{\csc x \sin^2 x - \csc x}{\sec\left(\frac{\pi}{2} - x\right)}$ $-\cos^2 x$
15. $\frac{\sin(-x)}{\csc x} + \cos^2(-x)$ $1 - 2\sin^2 x$ or $2\cos^2 x - 1$
16. $\sec(-x) - \sec(-x)\sin^2 x$ $\cos x$
17. $\csc^2(-x)\tan^2\left(\frac{\pi}{2} - x\right)$ $\csc^2 x \cot^2 x$ or $\frac{\cos^2 x}{\sin^4 x}$
18. $\frac{\cos^2(-x)}{\cot^2 x}$ $\sin^2 x$
19. $\frac{\cos\left(\frac{\pi}{2} - x\right) - 1}{1 + \sin(-x)}$ -1
20. $\frac{\sec x \tan x}{\sin(-x)\cot\left(\frac{\pi}{2} - x\right)}$ $-\csc x \sec x$
21. $\frac{\cos x \cot x + \tan\left(\frac{\pi}{2} - x\right)}{1 + \cos x}$ $\cot x$
22. $\tan x \cot\left(\frac{\pi}{2} - x\right) - \sec^2 x$ -1

In Exercises 23–38, verify the identity. See Additional Answers.

23. $\cos^2 x - \sin^2 x = 1 - 2\sin^2 x$
24. $\csc^4 x - \cot^4 x = 2\csc^2 x - 1$
25. $\cot^2 x(\sec^2 x - 1) = 1$
26. $4\tan^4 x + \tan^2 x - 3 = \sec^2 x(4\tan^2 x - 3)$
27. $\sin x \csc x = 1$
28. $\tan y \cot y = 1$
29. $\tan^2 x + 4 = \sec^2 x + 3$
30. $2 - \sec^2 x = 1 - \tan^2 x$
31. $\sin^2 x - \sin^4 x = \cos^2 x - \cos^4 x$
32. $\cos x + \sin x \tan x = \sec x$
33. $\frac{1 + \sin x}{\cos x} + \frac{\cos x}{1 + \sin x} = 2\sec x$
34. $\frac{1}{\cot x + 1} + \frac{1}{\tan x + 1} = 1$
35. $\frac{\cos(-x)}{1 + \sin(-x)} = \sec x + \tan x$
36. $\frac{1 + \sec(-x)}{\sin(-x) + \tan(-x)} = -\csc x$
37. $\cos\left(\frac{\pi}{2} - x\right) \csc x = 1$
38. $\frac{\cos\left(\frac{\pi}{2} - x\right)}{\sin\left(\frac{\pi}{2} - x\right)} = \tan x$

Communicating about ALGEBRA

Identify the graph of the following parametric equations.

a. $x = 4\cos t$, $y = 3\sin t$
b. $x = \pi\cos t$, $y = \pi\sin t$
c. $x = 3\cos t$, $y = 8\sin t$

a. an ellipse, $\left(\frac{x}{4}\right)^2 + \left(\frac{y}{3}\right)^2 = 1$
b. a circle, $x^2 + y^2 = \pi^2$
c. an ellipse, $\left(\frac{x}{3}\right)^2 + \left(\frac{y}{8}\right)^2 = 1$

A.–C. See Additional Answers.

EXERCISE Notes

ASSIGNMENT GUIDE

Basic/Average: Ex. 1–4, 9–17 odd, 27–35 odd, 41, 45, 49–55 odd

Above Average: Ex. 1–4, 9–19 odd, 25–35 odd, 39–45 odd, 46, 54–57

Advanced: Ex. 1–4, 9–19 odd, 25–35 odd, 39–43 odd, 44–47, 56–58

Selected Answers
Exercises 1–4, 5–55 odd

More Difficult Exercises
Exercises 41–46, 57–60

Guided Practice

Ex. 2 If an overhead graphing calculator or computer is available, you can graph $y = \cos x$ to help students view the "y-axis symmetry" associated with an even function. The function $y = \sin x$ will have the "origin symmetry" associated with odd functions.

Independent Practice

Ex. 5–22 GROUP ACTIVITY In class, have students suggest specific strategies that they

(continued)

2. Odd: sine, tangent, cosecant, cotangent; even: cosine, secant

(continued)
might use to begin the simplifications. For example,

1. Replace any cofunction or negative angle terms with their equivalent identities.
2. Look to replace a "squared" term with a Pythagorean identity equivalent (see Example 3 on page 745).
3. Write the expression in terms of sine or cosine and try to simplify.

Ex. 23–38 TECHNOLOGY After completing the graphics calculator activity on pages 750 and 751, students may verify the identities. For example, in Exercise 29, let $y_1 = \tan^2 x + 4$ and $y_2 = \sec^2 x + 3$. The two graphs should coincide. In Exercise 34, you could graph $y_1 = \frac{1}{\cot x + 1}$, $y_2 = \frac{1}{\tan x + 1}$, and $y_3 = y_1 + y_2$. The graph $y_1 + y_2$ should be the horizontal line $y = 1$.

Answers
See Additional Answers for exercises 23–38.

Ex. 44 Ask students if $\cot \theta = \tan 2\theta$ for all values of θ. No

Answers

41. Circle,
$$x^2 + y^2 = 16\cos^2 t + 16\sin^2 t = 16(\cos^2 t + \sin^2 t) = 16$$

42. Hyperbola,
$$\frac{x^2}{9} - y^2 = \sec^2 t - \tan^2 t = 1$$

43. Ellipse,
$$\frac{x^2}{4} + \frac{y^2}{9} = \cos^2 t + \sin^2 t = 1$$

44.
$$\cot\theta = \cot\frac{\pi}{6} = \tan\left(\frac{\pi}{2} - \frac{\pi}{6}\right) = \tan\left(\frac{3\pi}{6} - \frac{\pi}{6}\right) = \tan\left(\frac{2\pi}{6}\right) = \tan\frac{\pi}{3} = \tan 2\theta$$

In Exercises 39 and 40, show that the equation is not an identity (for *all* values of *x*) by finding a value of *x* for which the equation is not true.

39. $\sin x = \sqrt{1 - \cos^2 x}$
$x = \frac{3\pi}{2}$, $\sin x = -1 \neq \sqrt{1 - \cos^2 x}$

40. $\sqrt{\tan^2 x} = \tan x$
$x = \frac{3\pi}{4}$, $\tan x = -1 \neq \sqrt{\tan^2 x}$

Technology **In Exercises 41–43, use a graphing calculator in parametric mode to sketch the graph represented by the parametric equations. Use a trigonometric identity to verify that the graph is a circle, an ellipse, or a hyperbola. (Use a calculator setting of $-6 \le x \le 6$ and $-4 \le y \le 4$.)**

41. $x = 4\cos t$, $y = 4\sin t$

42. $x = 3\sec t$, $y = \tan t$

43. $x = 2\cos t$, $y = 3\sin t$

44. *Wheel Spokes* The diagram below shows a wheel with 12 equally-spaced spokes. Each pair of consecutive spokes meets at an angle of $\theta = \frac{\pi}{6}$. Use the trigonometric identity $\tan\left(\frac{\pi}{2} - x\right) = \cot x$ to show that $\cot\theta = \tan 2\theta$.

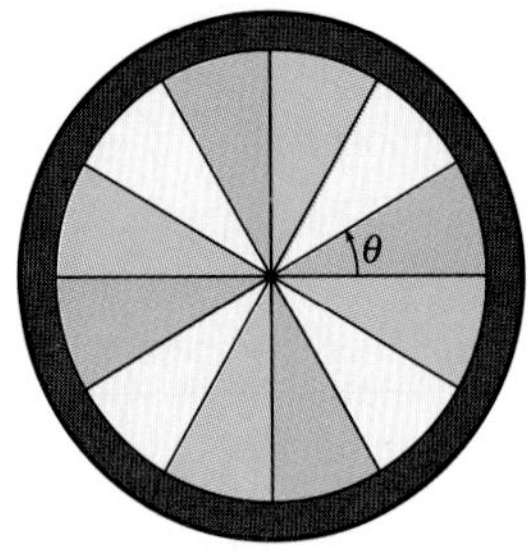

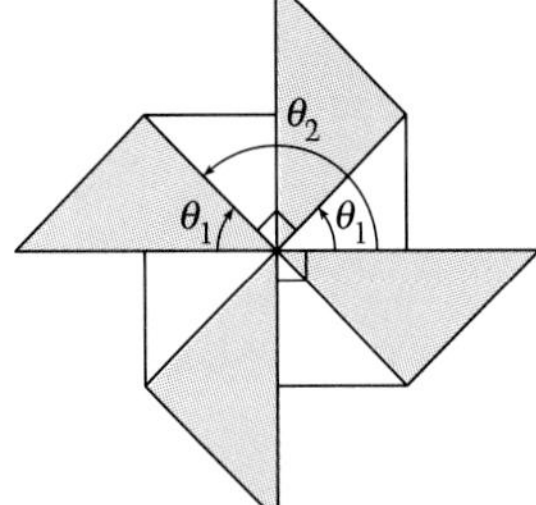

45. *Propeller* The diagram above shows a toy propeller. Use the identity $\sec\left(\frac{\pi}{2} - x\right) = \csc x$ and a negative angle identity to show that $\sec\theta_1 = \csc\theta_2$.

46. *Up the Down Staircase* Each of the handrails on the up and down spiral stairs of the Statue of Liberty has the shape of a *helix.* The *projection* of the *down* handrail onto an *xy*-plane can be modeled by $y = 3.5\sin\frac{\pi}{7}x$, where *x* and *y* are measured in feet. Which of the following models represents the projection of the *up* handrail? Explain. b

a. $y = 3.5\cos\left(-\frac{\pi}{7}x\right)$ b. $y = 3.5\sin\left(-\frac{\pi}{7}x\right)$

c. $y = 3.5\cos\left(\frac{\pi}{2} - \frac{\pi}{7}x\right)$

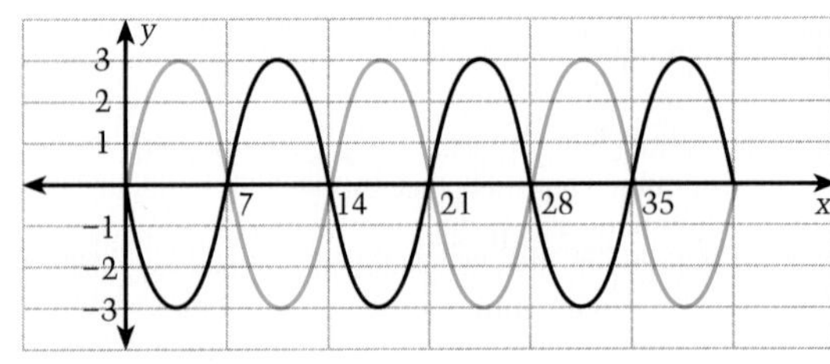

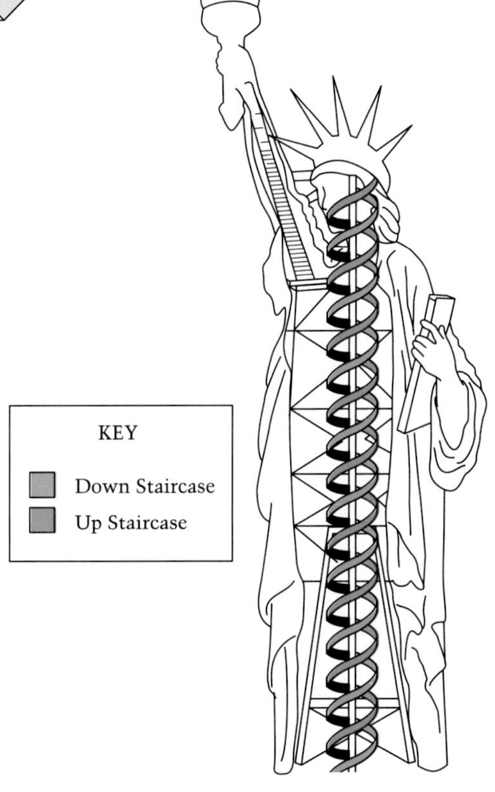

Inside the Statue of Liberty there are two 112-foot spiral stairs—one for people walking up and the other for people walking down.

Integrated Review

In Exercises 47–54, evaluate the indicated trigonometric function. (Assume $0 \le x \le \frac{\pi}{2}$.)

47. Given $\cos x = \frac{4}{5}$, find $\sin x$.

48. Given $\sin x = \frac{2}{3}$, find $\csc x$.

49. Given $\tan x = 10$, find $\sec x$.

50. Given $\cot x = 1$, find $\cos\left(\frac{\pi}{2} - x\right)$.

51. Given $\cos x = \frac{1}{2}$, find $\csc\left(\frac{\pi}{2} - x\right)$.

52. Given $\tan x = 3$, find $\sec\left(\frac{\pi}{2} - x\right)$.

53. Given $\sin x = \frac{1}{3}$, find $\csc x$.

54. Given $\cos x = \frac{1}{5}$, find $\tan x$.

55. *Archerfish* The archerfish at the right is shooting a jet of water into the air to catch the beetle. The beetle is 12 in. above and 11 in. to one side of the fish. At what angle must the archerfish aim to hit the beetle?

56. *Angelfish* Approximate the length of the angelfish shown below.

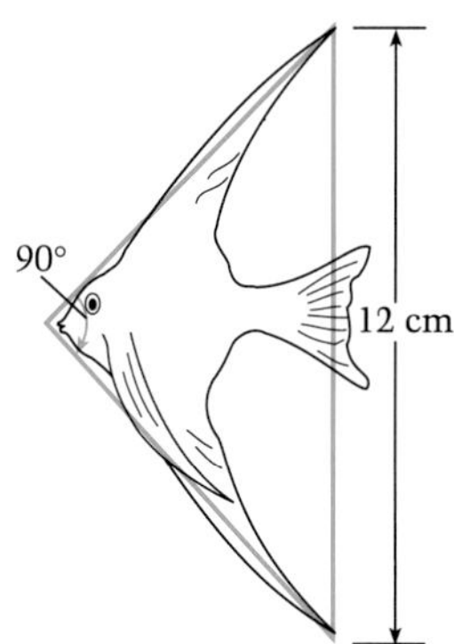

Exploration and Extension

In Exercises 57–60, substitute the given value of *x* into the radical expression. Then use a trigonometric identity to rewrite the expression so that it contains no radical.

57. $\sqrt{25 - x^2}$
$x = 5 \sin \theta$

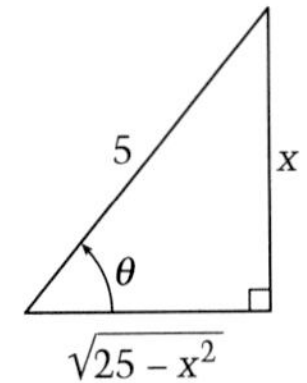

58. $\sqrt{16 - 4x^2}$
$x = 2 \sin \theta$

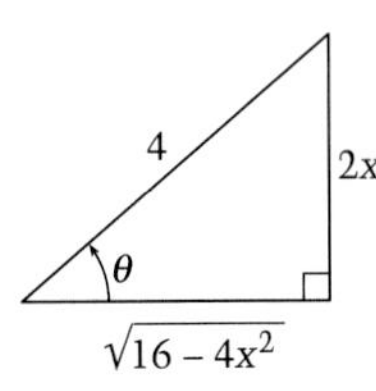

59. $\sqrt{1 - (x - 1)^2}$
$x = 1 + \cos \theta$

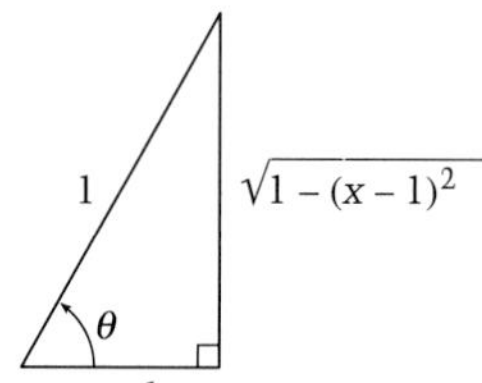

60. $\sqrt{x^2 + 9}$
$x = 3 \tan \theta$

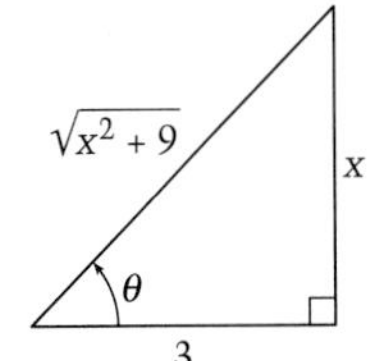

45. Because $\frac{\pi}{2} + \theta_1 = \theta_2$,

$$\sec \theta_1 = \sec\left(\theta_2 - \frac{\pi}{2}\right) = \sec\left(\frac{\pi}{2} - \theta_2\right) = \csc \theta_2$$

▶ **Ex. 46** Help students use either the negative angle identity and cofunction identity, or the visual connection of the horizontal shift of $y = 3.5 \cos\left(\frac{\pi}{7}x\right)$.

Answer

46. The projection of the up handrail is the reflection of the down handrail in the *x*-axis.

Enrichment Activity

This activity requires students to go beyond lesson goals.

Use the identities from this lesson to create your own list of trigonometric identities for all the basic trigonometric functions. For example, $1 + \sec^2 x \sin^2 x = \sec^2 x$.

Provide a complete verification of the identity as well. Exchange lists with a classmate. Verify each others' identities before turning them in for credit.

TECHNOLOGY Using a graphics calculator to verify trigonometric identities is an excellent way to convince students of the plausibility of the relationships.

After completing the activity, have students use the graphics calculator to verify the trigonometric identities of Lesson 14.3.

USING A GRAPHING CALCULATOR

A graphing calculator can be used to verify trigonometric identities graphically. For instance, to verify the Pythagorean identity $\sin^2 x + \cos^2 x = 1$, you can sketch the graph of

$$y = \sin^2 x + \cos^2 x$$

over the interval $0 \le x < 2\pi$. From the graph, you can see that $y = 1$ for all values of x in the interval. Then, using the fact that the sine and cosine functions each have a period of 2π, you can conclude that $y = 1$ for *any* value of x.

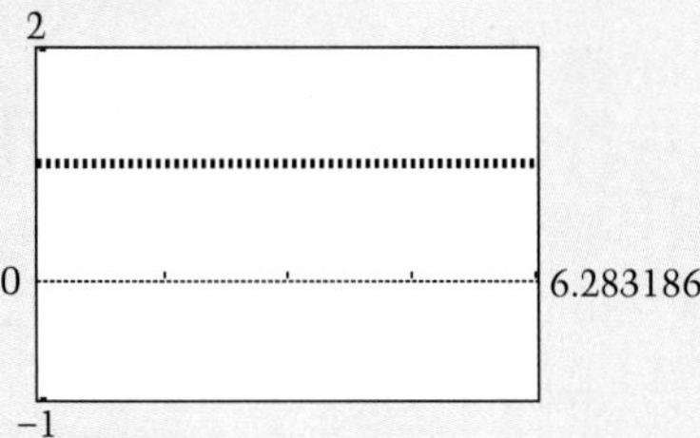

Example 1: Discovering Trigonometric Identities

Sketch the graph of

$$y = \sin x + \cos x.$$

From the graph, can you see how to represent this function as a single sine or cosine function?

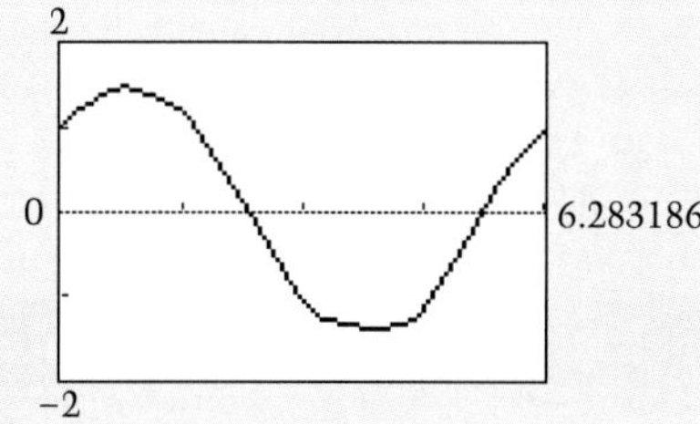

Solution From the graph, as shown at the right, it looks like the sum of $\sin x$ and $\cos x$ is itself a sine or cosine function. Suppose you guess that the function is of the form $y = a\cos(bx + c)$. Using the trace key, the minimum and maximum values appear to be $-1.414 \approx -\sqrt{2}$ and $1.414 \approx \sqrt{2}$, which would imply that $a = \sqrt{2}$. Because the period of the function is 2π, you can conclude that $b = 1$. Finally, because the graph appears to have been shifted $\frac{\pi}{4}$ units to the right, you can conclude that $\frac{c}{b} = -\frac{\pi}{4}$, which implies that $c = -\frac{\pi}{4}$. Thus an equation for the graph appears to be

$$y = \sqrt{2}\cos\left(x - \frac{\pi}{4}\right).$$

You can verify this by sketching both graphs on the same screen. Because both equations have the same graph, you have discovered the following trigonometric identity.

$$\sin x + \cos x = \sqrt{2}\cos\left(x - \frac{\pi}{4}\right) \quad \blacksquare$$

Another important use of a graphing calculator in trigonometry is to sketch graphs of functions that would be very tedious to sketch by hand.

Example 2: Sketching a Graph of a Function

Since 1957, the Mauna Loa Climate Observatory in Hawaii has been collecting data on the carbon dioxide level of Earth's atmosphere. A model that represents the data is given by

$$y = (316 + 0.654t + 0.0216t^2) + 2.5 \sin 2\pi t$$

where y represents the carbon dioxide concentration (in parts per million) and $t = 0$ represents 1960 (January 1). Sketch a graph of this function and explain the oscillations in the graph.

Solution The graph of the function is shown below. From the graph, you can see that the carbon dioxide level fluctuates each year. The low level each year, which occurs toward the end of summer in the northern hemisphere, is caused by the intake of carbon dioxide by growing plants. ■

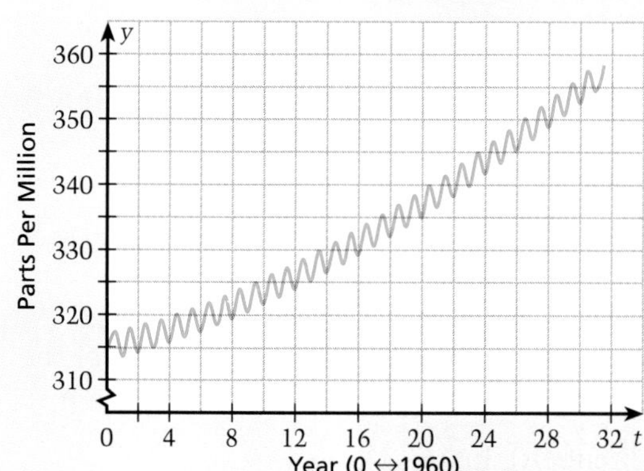

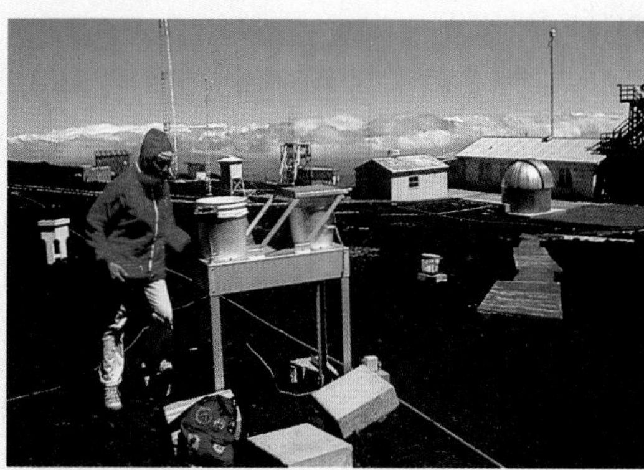

Mauna Loa Climate Observatory

Exercises

In Exercises 1–4, sketch the graphs of all three functions on the same display screen. Which two graphs are the same? What trigonometric identity have you discovered?

1. **a.** $y = \sin^2 x$ **b.** $y = \frac{1}{2}(1 - \cos 2x)$ **c.** $y = \frac{1}{2}(1 + \cos 2x)$

2. **a.** $y = 2\cos^2 x$ **b.** $y = 1 - \cos 2x$ **c.** $y = 1 + \cos 2x$

3. **a.** $y = 2 \sin x \cos 2x$ **b.** $y = \sin 3x + \sin x$ **c.** $y = \sin 3x - \sin x$

4. **a.** $y = \sin 2x$ **b.** $y = 2 \sin x$ **c.** $y = 2 \sin x \cos x$

5. *Carbon Dioxide* Sketch the graph of the model from Example 2 for $28 \le t \le 30$. Between January 1, 1988 and January 1, 1990, what were the highest and lowest levels of carbon dioxide? When did each occur? ≈356.1 in 1989, ≈350.1 in 1988

Answers

1. a and b, $\sin^2 x = \frac{1}{2}(1 - \cos 2x)$
2. a and c, $2\cos^2 x = 1 + \cos 2x$
3. a and c, $2 \sin x \cos 2x = \sin 3x - \sin x$
4. a and c, $\sin 2x = 2 \sin x \cos x$

Answers

4.
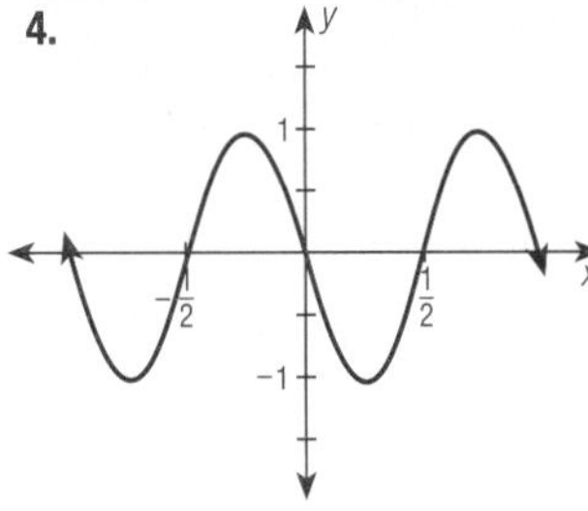

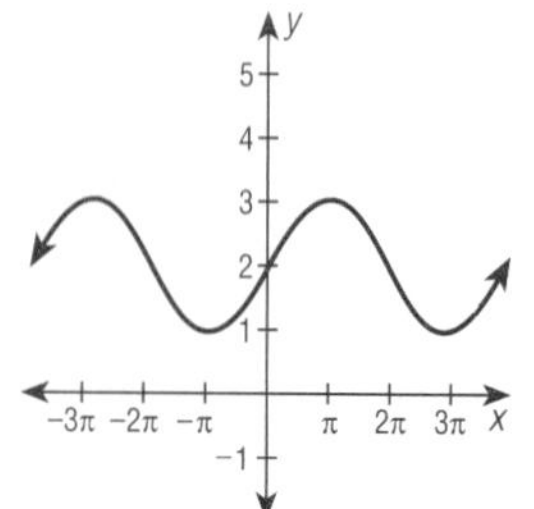

6.
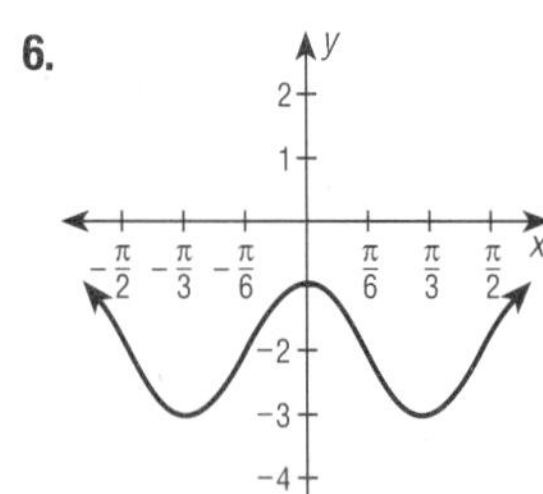

18.
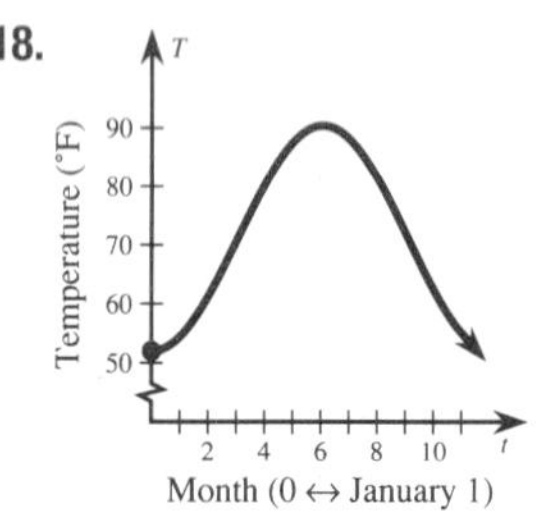

Mid-Chapter SELF-TEST

Take this test as you would take a test in class. The answers to the exercises are given in the back of the book.

In Exercises 1–3, find the amplitude and period of the graph. (14.1, 14.2)

1. $y = \frac{5}{2} \sin 2x$ $\frac{5}{2}, \pi$
2. $y = 3 + 4 \cos \pi x$ 4, 2
3. $y = \frac{1}{2} - 3 \sin 4x$ $3, \frac{\pi}{2}$

In Exercises 4–6, sketch two cycles of the graph of the function. (14.1, 14.2) See margin.

4. $y = -\sin 2\pi x$
5. $y = 2 + \sin \frac{x}{2}$
6. $y = -2 + \cos 3x$

In Exercises 7–9, write an equation of the function, where *a* > 0 and *b* > 0. (14.1)

7. $y = a \sin bx$
   Amplitude: 6
   Period: $\frac{1}{2}$ $y = 6 \sin 4\pi x$
8. $y = a \sin bx$
   Amplitude: $\frac{4}{3}$
   Period: 3π $y = \frac{4}{3} \sin \frac{2}{3}x$
9. $y = a \cos bx$
   Amplitude: $\frac{2}{5}$
   Period: π $y = \frac{2}{5} \cos 2x$

In Exercises 10–12, simplify the expression. (14.3)

10. $\frac{-\sec(-x)}{\csc(-x)}$ $\tan x$
11. $\sin x + \cos x \cot x$ $\csc x$
12. $\frac{\sin x \cos x - \tan x}{\tan(-x) \sin^2 x}$ 1

In Exercises 13–15, verify the identity. (14.3) See Additional Answers.

13. $\frac{\cos^2(-x)}{\sin(-x)} = \frac{\cos x}{\tan(-x)}$
14. $\sin^2 x + \frac{1}{\sec^2 x} = 1$
15. $\cos x + \sin x \tan x = \sec x$

16. The graph of $y = 3 \sin 4x$ is shifted down 3 units. Write an equation of the resulting graph. **(14.2)** $y = -3 + 3 \sin 4x$

17. The graph of $y = \frac{1}{4} \cos 2x$ is shifted $\frac{\pi}{2}$ units to the left. Write an equation of the resulting graph. **(14.2)** $y = \frac{1}{4} \cos 2\left(x + \frac{\pi}{2}\right)$

18. The average daily temperature, T (in degrees Fahrenheit), in Spartanburg, South Carolina, can be modeled by $T = 71.2 + 19.2 \sin\left(\frac{1}{6}\pi t + 4.7\right)$, where $t = 0$ represents 1, $t = 1$ represents February 1, and so on. Sketch the graph of this model. **(14.2)** See margin.

19. In Exercise 18, which month has the highest average daily temperature? Which has the lowest? **(14.2)** July, January

20. Which of the following equations represents the graph of a person's blood pressure? Simplify that equation. **(14.2)**

    a. $P = 120 + 40 \sin\left(2\pi t - \frac{\pi}{2}\right)$

    b. $P = 100 - 20 \cos\left(2\pi t - \frac{\pi}{2}\right)$

    c. $P = 100 + 20 \sin\left(\frac{\pi}{2} - 2\pi t\right)$

    d. $P = 120 + 20 \sin\left(\pi t - \frac{\pi}{2}\right)$

    c, $P = 100 + 20 \cos(2\pi t)$

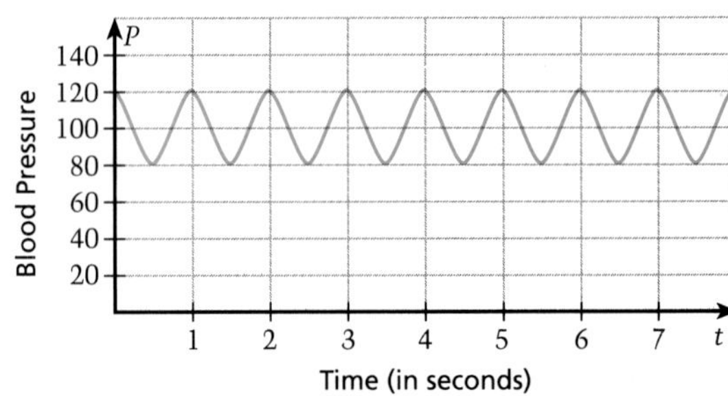

14.4 Solving Trigonometric Equations

What you should learn:

Goal 1 How to solve a trigonometric equation

Goal 2 How to solve trigonometric equations that model real-life problems

Why you should learn it:

You can use trigonometric equations to find the measure of the angles in many complex geometric figures.

Goal 1 Solving a Trigonometric Equation

In Lesson 14.3, you learned to *verify* a trigonometric identity. In this lesson, you will learn to solve a trigonometric equation. To see the difference, consider the following two equations: $\sin^2 x + \cos^2 x = 1$ and $\sin x = 1$. The first equation is an identity because it is true for all real values of x. The second equation, however, is true only for some values of x. When you find these values, you are solving the equation.

Example 1 *Solving a Trigonometric Equation*

Solve $2 \sin x - 1 = 0$.

Solution

$2 \sin x - 1 = 0$ *Rewrite original equation.*

$2 \sin x = 1$ *Add 1 to both sides.*

$\sin x = \frac{1}{2}$ *Divide both sides by 2.*

The equation $\sin x = \frac{1}{2}$ has solutions $x = \frac{\pi}{6}$ and $x = \frac{5\pi}{6}$ in the interval $0 \le x < 2\pi$. Moreover, because $\sin x$ has a period of 2π, there are infinitely many other solutions, which can be written as

$$x = \frac{\pi}{6} + 2n\pi \quad \text{and} \quad x = \frac{5\pi}{6} + 2n\pi \qquad \textit{General solution}$$

where n is an integer. ■

To see that the equation $\sin x = \frac{1}{2}$ has infinitely many solutions, consider the graph. Each x-value at which the graph intersects the line $y = \frac{1}{2}$ is a solution.

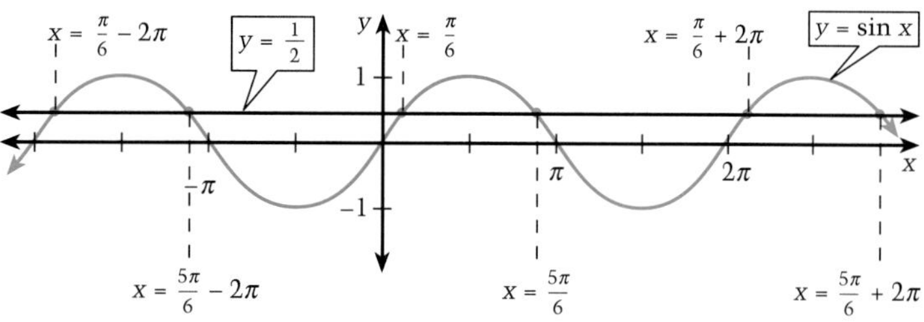

Problem of the Day

A cross is formed from 5 congruent squares, as shown. If $x = 10$, find the area of the cross.
100

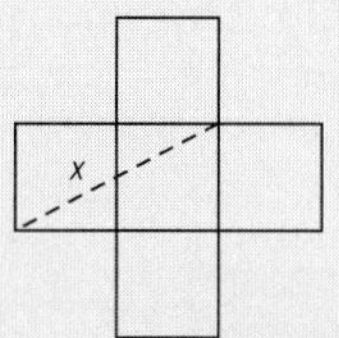

ORGANIZER

Warm-up Exercises:

1. Solve for x.
   a. $2x - 3 = 5$
   b. $3x^2 - 4 = 0$
   c. $4x^3 = 9x$
   d. $6x^2 + x = 12$

   a. $x = 4$ b. $x = \pm\frac{2\sqrt{3}}{3}$
   c. $x = 0, \pm\frac{3}{2}$ d. $x = -\frac{3}{2}, \frac{4}{3}$

2. Describe all values of x for which the following trigonometric equations are true.
   a. $\cos x = 0.5$
   b. $\sin x = \frac{\sqrt{2}}{2}$

   a. $x = \frac{\pi}{3} + 2n\pi$ and $x = \frac{5\pi}{3} + 2n\pi$, where n is an integer;
   b. $x = \frac{\pi}{4} + 2n\pi$ and $x = \frac{3\pi}{4} + 2n\pi$, where n is an integer

Lesson Resources

Teaching Tools
- Problem of the Day: 14.4
- Warm-up Exercises: 14.4
- Answer Masters: 14.4

Extra Practice: 14.4
Color Transparencies: 75, 76

LESSON Notes

Example 1

Students might find it easier to solve a parallel problem. If you substitute u for $\sin x$, then solving $2 \sin x - 1 = 0$ is similar to solving the equation $2u - 1 = 0$ for u.

Examples 1–2

Ask students to show on the graphs why adding $2n\pi$ to the solutions in the interval $0 \le x \le 2\pi$ produces the general solutions to the trigonometric equations.

Example 2

Be sure students recognize that $x = \frac{\pi}{4}$ is not a solution of Example 2 because $\cos \frac{\pi}{4}$ is positive.

Example 3

Refer students to "Enrichment Activities" in Lesson 14.2 for a graph of $y = \tan x$. Students can see from the graph of the basic tangent function that it has a cycle of π units, and so its period is π.

Example 4

Ask students why the interval for the solution is $0 \le x < \pi$. Because the cotangent has a period of π, the expressions on either side of the equation also have a period of π.

Example 5

As illustrated in Example 1, many students will find it helpful to make an initial substitution, say u, for $\sin x$. The original equation then becomes $2u^2 - u = 1$.
Solving gives

$$2u^2 - u - 1 = 0$$
$$(2u + 1)(u - 1) = 0 \text{ and}$$
$$2u + 1 = 0 \text{ or}$$
$$u - 1 = 0.$$

Replacing $\sin x$ for u gives $2 \sin x = -1$ or $\sin x = 1$.

Example 2 *Collecting Like Terms*

Solve $\cos x + \sqrt{2} = -\cos x$.

Solution

$\cos x + \sqrt{2} = -\cos x$	*Rewrite original equation.*
$\cos x + \cos x = -\sqrt{2}$	*Add cos x to both sides.*
$2 \cos x = -\sqrt{2}$	*Collect like terms.*
$\cos x = -\frac{\sqrt{2}}{2}$	*Divide both sides by 2.*

Because $\cos x$ has a period of 2π, first find all solutions in the interval $0 \le x < 2\pi$. These are $x = \frac{3\pi}{4}$ and $x = \frac{5\pi}{4}$. Adding $2n\pi$ to each of these solutions produces the general solution

$$x = \frac{3\pi}{4} + 2n\pi \quad \text{and} \quad x = \frac{5\pi}{4} + 2n\pi$$

where n is an integer. ■

Examples 1 and 2 used the fact that the sine and cosine functions have a period of 2π. The periods of the other four trigonometric functions are as follows.

Tangent function:	Period is π.
Cotangent function:	Period is π.
Secant function:	Period is 2π.
Cosecant function:	Period is 2π.

Example 3 *Finding Square Roots*

Solve $3 \tan^2 x - 1 = 0$.

Solution

$3 \tan^2 x - 1 = 0$	*Rewrite original equation.*
$3 \tan^2 x = 1$	*Add 1 to both sides.*
$\tan^2 x = \frac{1}{3}$	*Divide both sides by 3.*
$\tan x = \pm\frac{1}{\sqrt{3}}$	*Find square roots.*
$\tan x = \pm\frac{\sqrt{3}}{3}$	*Simplify.*

Because $\tan x$ has a period of π, first find all solutions in the interval $0 \le x < \pi$. These are $x = \frac{\pi}{6}$ and $x = \frac{5\pi}{6}$. Adding $n\pi$ to each produces the general solution

$$x = \frac{\pi}{6} + n\pi \quad \text{and} \quad x = \frac{5\pi}{6} + n\pi$$

where n is an integer. ■

Example 4 Factoring

Solve $\cot x \csc^2 x = 2 \cot x$ in the interval $0 \le x < \pi$.

Solution

$\cot x \csc^2 x = 2 \cot x$ *Rewrite original equation.*

$\cot x \csc^2 x - 2 \cot x = 0$ *Subtract 2 cot x from both sides.*

$\cot x(\csc^2 x - 2) = 0$ *Factor.*

Use the Zero Product Property to obtain the solutions. From $\cot x = 0$, you obtain $x = \frac{\pi}{2}$. From $\csc^2 x - 2 = 0$, you obtain $\csc x = \pm\sqrt{2}$, which has $x = \frac{\pi}{4}$ and $x = \frac{3\pi}{4}$ as solutions. Check each of these solutions in the original equation. ■

If you have a graphing calculator, try using it to sketch the graphs of $y = \cot x \csc^2 x$ and $y = 2 \cot x$, as shown at the left. If your calculator does not have cotangent and cosecant keys, you can enter these functions as

$$y = \frac{1}{\tan x \sin^2 x} \quad \text{and} \quad y = \frac{2}{\tan x}.$$

Note that the graphs intersect at three points between 0 and π.

Example 5 Solving an Equation of Quadratic Type

Solve $2 \sin^2 x - \sin x = 1$ in the interval $0 \le x < 2\pi$.

Solution Treat the equation as a quadratic in $\sin x$.

$2 \sin^2 x - \sin x = 1$ *Rewrite original equation.*

$2 \sin^2 x - \sin x - 1 = 0$ *Subtract 1 from both sides.*

$(2 \sin x + 1)(\sin x - 1) = 0$ *Factor.*

Setting each factor equal to zero produces the following.

$2 \sin x + 1 = 0$ and $\sin x - 1 = 0$

$\sin x = -\frac{1}{2}$ $\quad$ $\sin x = 1$

$x = \frac{7\pi}{6}, \frac{11\pi}{6}$ $\quad$ $x = \frac{\pi}{2}$ ■

When an equation contains more than one trigonometric function, you may want to begin by rewriting the equation in terms of a single trigonometric function. Here is an example.

$-2 \cos^2 x - \sin x + 1 = 0$ *Original equation*

$-2(1 - \sin^2 x) - \sin x + 1 = 0$ *Pythagorean identity*

$2 \sin^2 x - \sin x - 1 = 0$ *Simplify.*

This equation is the same as that solved in Example 5.

Example 6

Emphasize that this is an excellent example of how mathematics in general, and trigonometry in particular, is used to explain real-life phenomena.

Extra Examples

Here are additional examples similar to **Examples 1–5.**

1. Solve the equations.

a. $4 \cos x + 2 = 0$

b. $4 \tan x - \sqrt{3} = \tan x$

a. $\frac{2\pi}{3} + 2n\pi, \frac{4\pi}{3} + 2n\pi$;

b. $\frac{\pi}{6} + n\pi$

(where n is an integer)

2. Solve the equations in the interval $0 \le x < 2\pi$.

a. $2 \csc^2 x - 4 = 0$

b. $2 \sin^2 x - \cos x = 1$

c. $\sin^2 x - \cos^2 x = 0$

a. $\frac{\pi}{4}, \frac{3\pi}{4}, \frac{5\pi}{4}, \frac{7\pi}{4}$

b. $\pi, \frac{\pi}{3}, \frac{5\pi}{3}$

c. $\frac{\pi}{4}, \frac{3\pi}{4}, \frac{5\pi}{4}, \frac{7\pi}{4}$

Check Understanding

1. How are solving trigonometric equations similar to solving other algebraic equations? Solving trigonometric equations means to find all the values of the variable that make the equation true.
2. What is the difference between an identity and an equation? See the top of page 753.
3. Why do many trigonometric equations have infinitely many solutions? The functions are periodic.
4. What is the period of the basic tangent and cotangent functions? π

Communicating about ALGEBRA

See Additional Answers.

EXTEND Communicating
Have students use the inverse-function relationships between the sin and csc functions, the cos and sec functions, and the tan and cot functions to graph the cosecant, secant, and cotangent functions.

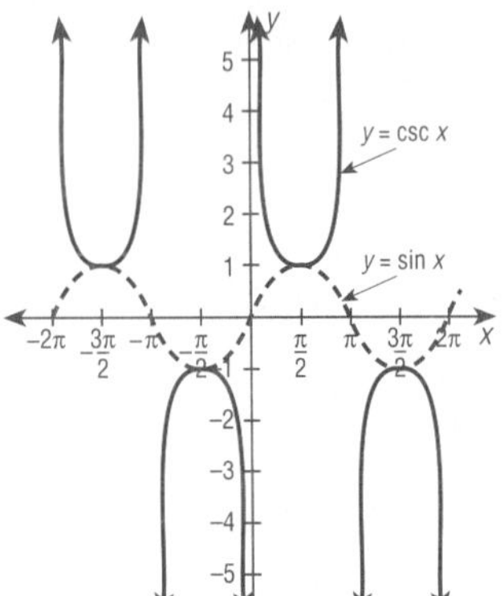

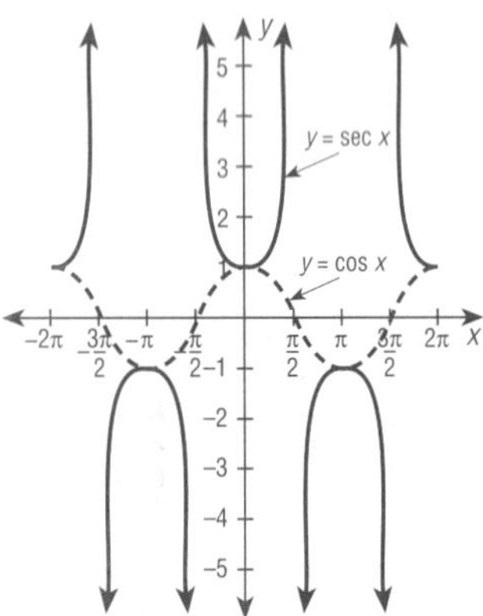

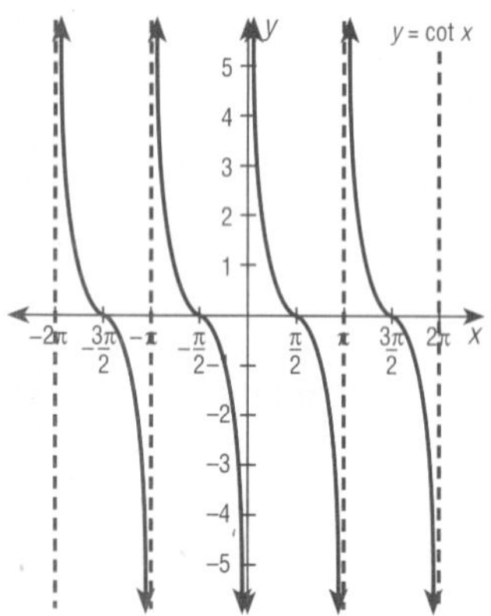

Top view of honeycomb (left) Cross section of honeycomb (right)

Goal 2 Solving Trig Equations in Real Life

Real Life Biology

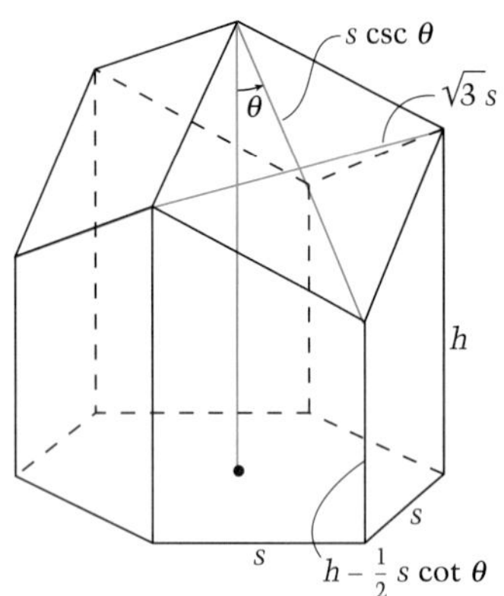

The top (or bottom) of a honeycomb cell is composed of three congruent rhombi. The sides of the cell are trapezoids. Each rhombus has an area of $\frac{\sqrt{3}}{2}s^2 \csc \theta$. Each trapezoid has an area of $hs - \frac{1}{4}s^2 \cot \theta$. ($s$, θ, and h are shown in the diagram above.)

Example 6 *Surface Area of a Honeycomb Cell*

The total surface area of a honeycomb cell is $6hs + \frac{3}{\sqrt{2}}s^2$.

Find the angle, θ, at which the face of each rhombus meets the vertical line passing through the center of the cell.

Solution Because the total surface area consists of the area of three rhombi and six trapezoids, you can use the areas given above to write the following.

$$6hs + \frac{3}{\sqrt{2}}s^2 = 3\left(\frac{\sqrt{3}}{2}s^2 \csc \theta\right) + 6\left(hs - \frac{1}{4}s^2 \cot \theta\right)$$

This equation simplifies to $\sqrt{2} \sin \theta = \sqrt{3} - \cos \theta$, which can be solved as follows.

$\sqrt{2} \sin \theta = \sqrt{3} - \cos \theta$ *Simplified equation*

$2 \sin^2 \theta = 3 - 2\sqrt{3} \cos \theta + \cos^2 \theta$ *Square both sides.*

$2(1 - \cos^2 \theta) = 3 - 2\sqrt{3} \cos \theta + \cos^2 \theta$ *Pythagorean identity*

$0 = 1 - 2\sqrt{3} \cos \theta + 3 \cos^2 \theta$ *Simplify.*

$0 = (1 - \sqrt{3} \cos \theta)^2$ *Factor.*

This means that the angle θ is as follows.

$\cos \theta = \frac{1}{\sqrt{3}}$ *Zero Product Property*

$\theta = \arccos \frac{1}{\sqrt{3}} \approx 54.7°$

■

Communicating about ALGEBRA

SHARING IDEAS about the Lesson

Show that the first equation in the solution to Example 6 simplifies to $\sqrt{2} \sin \theta = \sqrt{3} - \cos \theta$.

4. sin, cos, sec, and csc: 2π; tan and cot: π

5. $x = 0$; get all terms on one side, factor, set each factor equal to zero, and solve for x with $0 \leq x < \pi$.

EXERCISES

Guided Practice

CRITICAL THINKING about the Lesson 2. $\frac{\pi}{3}$ and $\frac{5\pi}{3}$; see Additional Answers for graph.

1. What makes a trigonometric equation an identity? It is true for all real values of the variable.
2. Solve $2 \cos x + 4 = 5$ for $0 \le x < 2\pi$. Show your result graphically.
3. Write the general solution of the equation in Exercise 2. $x = \frac{\pi}{3} + 2n\pi$ or $x = \frac{5\pi}{3} + 2n\pi$
4. State the period of each of the six trigonometric functions.
5. Solve the equation $\tan^2 x = \sec x \tan^2 x$ for $0 \le x < \pi$. Explain.
6. Use a trigonometric identity to rewrite $1 + \cot^2 x = 2 \csc x - 1$ in terms of a single trigonometric function. Explain how to solve the resulting equation.

Independent Practice

In Exercises 7–12, verify that the *x*-value is a solution of the equation.

7. $3 - 2 \cot x - 1 = 0,\ x = \frac{\pi}{4}$ $\cot x = 1$
8. $\csc x - 2 = 0,\ x = \frac{5\pi}{6}$ $\csc x = 2$
9. $4 \cos^2 2x - 3 = 0,\ x = \frac{\pi}{12}$ $\cos 2x = \frac{\sqrt{3}}{2}$, $2x = \frac{\pi}{6}$
10. $3 \tan^3 4x - 3 = 0,\ x = \frac{\pi}{16}$ $\tan 4x = 1$, $4x = \frac{\pi}{4}$
11. $2 \sin^4 x - \sin^2 x = 0,\ x = \frac{5\pi}{4}$
12. $2 \cot^4 x - \cot^2 x - 15 = 0,\ x = \frac{13\pi}{6}$

$\cot x = \pm\sqrt{3}$, $\cot \frac{13\pi}{6} = \sqrt{3}$

11. $\sin x = 0, \pm\frac{1}{\sqrt{2}}$, $\sin \frac{5\pi}{4} = -\frac{1}{\sqrt{2}}$

In Exercises 13–34, solve the equation for $0 \le x < 2\pi$.

13. $\sqrt{2} \cos x - 1 = 0$ $\frac{\pi}{4}, \frac{7\pi}{4}$
14. $7 \sec x - 7 = 0$ 0
15. $3 \sin x = \sin x + 1$ $\frac{\pi}{6}, \frac{5\pi}{6}$
16. $5 \cos x - \sqrt{3} = 3 \cos x$ $\frac{\pi}{6}, \frac{11\pi}{6}$
17. $2 \csc x + 17 = 15 + \csc x$ $\frac{7\pi}{6}, \frac{11\pi}{6}$
18. $\cos x - 1 = -\cos x$ $\frac{\pi}{3}, \frac{5\pi}{3}$
19. $4 \sin^2 x - 2 = 0$ $\frac{\pi}{4}, \frac{3\pi}{4}, \frac{5\pi}{4}, \frac{7\pi}{4}$
20. $\tan^2 x - 3 = 0$ $\frac{\pi}{3}, \frac{2\pi}{3}, \frac{4\pi}{3}, \frac{5\pi}{3}$
21. $3 \tan^2 x - 1 = 0$ $\frac{\pi}{6}, \frac{5\pi}{6}, \frac{7\pi}{6}, \frac{11\pi}{6}$
22. $3 \sec^2 x - 4 = 0$ $\frac{\pi}{6}, \frac{5\pi}{6}, \frac{7\pi}{6}, \frac{11\pi}{6}$
23. $\sec x \csc x - 2 \csc x = 0$ $\frac{\pi}{3}, \frac{5\pi}{3}$
24. $\sqrt{2} \cos x \sin x - \cos x = 0$ $\frac{\pi}{4}, \frac{\pi}{2}, \frac{3\pi}{4}, \frac{3\pi}{2}$
25. $4 \sin^2 x - 3 = 0$ $\frac{\pi}{3}, \frac{2\pi}{3}, \frac{4\pi}{3}, \frac{5\pi}{3}$
26. $3 \tan^3 x - \tan x = 0$ $0, \frac{\pi}{6}, \frac{5\pi}{6}, \pi, \frac{7\pi}{6}, \frac{11\pi}{6}$
27. $\cos^3 x = \cos x$ $0, \frac{\pi}{2}, \pi, \frac{3\pi}{2}$
28. $3 \sin x \sec x - 2\sqrt{3} \sin x = 0$ $0, \frac{\pi}{6}, \pi, \frac{11\pi}{6}$
29. $\sqrt{3} \tan x = 3 \tan x \csc x - \sqrt{3} \tan x$

29. $0, \frac{\pi}{3}, \frac{2\pi}{3}, \pi$

30. $2 \csc x - \sqrt{2} \cos x \csc x = 3 \csc x$ $\frac{3\pi}{4}, \frac{5\pi}{4}$
31. $\sqrt{3} \tan^2 x + 4 \tan x + \sqrt{3} = 0$ $\frac{2\pi}{3}, \frac{5\pi}{6}, \frac{5\pi}{3}, \frac{11\pi}{6}$
32. $\csc^2 x - \csc x - 2 = 0$ $\frac{\pi}{6}, \frac{5\pi}{6}, \frac{3\pi}{2}$
33. $2 \cos^2 x - \sin x - 1 = 0$ $\frac{\pi}{6}, \frac{5\pi}{6}, \frac{3\pi}{2}$
34. $2 \cot x + \sec^2 x = 0$ $\frac{3\pi}{4}, \frac{7\pi}{4}$

In Exercises 35–40, use a graphing calculator to approximate the solutions in the interval $0 \le x < 2\pi$.

35. $3 \tan x + 1 = 13$ $\approx 1.33, \approx 4.47$
36. $9 \cos x + 2 = 3$ $\approx 1.46, \approx 4.82$
37. $\csc^2 x + \cot x = 7$ $\approx 0.46, \approx 2.82, \approx 3.61, \approx 6.00$
38. $\tan^2 x + 6 = 4 \sec x$ No solution
39. $\sin x + \cos x \tan^2 x = 30 \cos x$ $\approx 1.37, \approx 1.74, \approx 4.51, \approx 4.88$
40. $2 \cos^2 x + \sin^2 x \cot x = \sin^2 x$ $\approx 1.11, \approx 2.36, \approx 4.25, \approx 5.50$

EXERCISE Notes

ASSIGNMENT GUIDE

Basic/Average: Ex. 1–6, 11–25 odd, 39–45 odd, 52–54, 55–67 odd, 68

Above Average: Ex. 1–6, 9–27 odd, 33–45 odd, 49–54, 65–69

Advanced: Ex. 1–6, 9–27 odd, 33–45 odd, 49–54, 67–70

Selected Answers
Exercises 1–6, 7–69 odd

More Difficult Exercises
Exercises 51–54, 69–70

Guided Practice

Ex. 1–6 Use these exercises as a final in-class readiness check before assigning the Independent Practice. Make sure students understand some of the basic paper-and-pencil strategies such as collecting like terms and finding square roots or fractions as outlined in Examples 2–5.

Ex. 5 Ask students why you should not first divide both sides by $\tan^2 x$ to obtain $1 = \sec x$.
You would lose a possible root. See Example 4 for a solution model.

Answers
See Additional Answers for exercises 4 and 5.

Independent Practice

Ex. 7–12 These exercises provide review practice for evaluating trigonometric expressions.

Ex. 13–34 TECHNOLOGY Students can use the Trace feature of graphics calculators to verify solutions. They can locate *x*-intercepts in exercises such as Exercise 23, or examine intersections of the graphs of both sides of the equation.

Ex. 37–40 TECHNOLOGY If students are using scientific calculators they will be required to use the reciprocal functions in order to evaluate cot x, sec x, or csc x. (See Example 4 on page 755.) Note that the text does not formally introduce the graphs of tan x, csc x, cot x, or sec x.

Ex. 43–48 Have some students do these exercises using paper-and-pencil algorithms while other students use graphics calculators. Compare the solutions.

Answer

45. $\left(\frac{\pi}{6}, \frac{8}{3}\right)$, $\left(\frac{5\pi}{6}, \frac{8}{3}\right)$, $\left(\frac{2\pi}{3}, \frac{8}{3}\right)$, $\left(\frac{11\pi}{6}, \frac{8}{3}\right)$

Ex. 49–50 You can use an overhead graph generator, if available, to show that the solutions of one equation occur at the maximum/ minimum values of the other equation.

Ex. 54 Demonstrate the graph using the following range.

X min = 0 Y min = 0
X max = 24 Y max = 10
X scl = 2 Y scl = 2

Ex. 55–66 Use these exercises for in-class review questions.

Ex. 52 To see that the diameter equals the ratio of the length of any side to the sine of the opposite angle, use this figure.

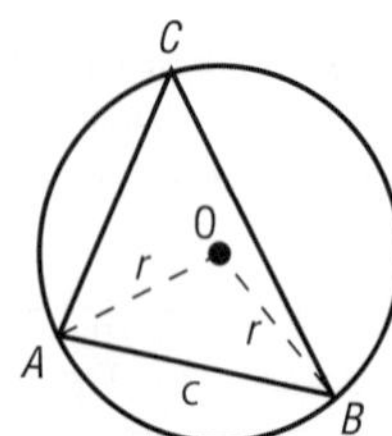

In $\triangle AOB$,

$$c^2 = r^2 + r^2 - 2r^2 \cos O$$
$$= 2r^2 - 2r^2 \cos 2C$$
$$= 2r^2 - 2r^2(1 - 2\sin^2 C)$$
$$= 4r^2 \sin^2 C$$

so $c = 2r \sin C$

and $2r = \frac{c}{\sin C}$.

In Exercises 41 and 42, find the x-intercepts of the graph in the interval $0 \le x \le 2\pi$.

41. $y = 2\cos x + 1$

42. $y = 2\tan^2 x - 6$

In Exercises 43–48, find the intersection points of the two graphs in the interval $0 \le x \le 2\pi$.

43. $y = \sqrt{3}\tan^2 x$
$y = \sqrt{3} - 2\tan x$

44. $y = 9\cos^2 x$
$y = \cos^2 x + 8\cos x - 2$

45. $y = 2\sec^2 x$
$y = 3 - \tan^2 x$

46. $y = \sin^2 x$
$y = 2\sin x - 1$

47. $y = \sin x \tan x$
$y = 2 - \cos x$

48. $y = 4\cos^2 x$
$y = 4\cos x - 1$

49. In calculus, it can be shown that the function $y = \sin x - \cos x$ has minimum or maximum values when

$\cos x + \sin x = 0.$ $\frac{3\pi}{4}, \frac{7\pi}{4}$

Find all solutions of $\cos x + \sin x = 0$ in the interval $0 \le x < 2\pi$. Verify your solutions with a graphing calculator.

50. In calculus, it can be shown that the function $y = 2\sin x + \cos 2x$ has minimum or maximum values when

$2\cos x - 4\sin x\cos x = 0.$ $\frac{\pi}{2}, \frac{3\pi}{2}, \frac{\pi}{6}, \frac{5\pi}{6}$

Find all solutions of $2\cos x - 4\sin x\cos x = 0$ in the interval $0 \le x < 2\pi$. Verify your solutions with a graphing calculator.

51. *Soap Film* A wire tetrahedron has sides of x inches each. When the tetrahedron is dipped into soapy water, six triangular soap films form within the tetrahedron as shown at the right. The six triangular soap films meet at four liquid edges, and all have a common vertex. Each triangular soap film has an area of $\frac{\sqrt{2}}{8}x^2$, and the base of each of these triangles is an edge of the tetrahedron frame. The height of each triangular soap film is equal to $\frac{x}{2}\tan\theta$. Find θ.

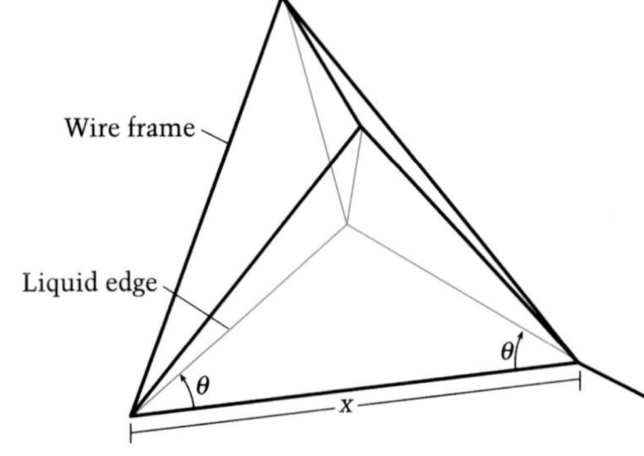

52. *Geometry* The diameter of a circle can be found by drawing any triangle whose vertices all lie on the circle. The diameter is then equal to the ratio of any side of the triangle to the sine of its opposite angle. For a triangle inscribed in a 5-meter radius circle, find the angle, θ, whose opposite side is 6 meters. ≈36.9° or ≈0.64

53. *Kite* Find the angle θ in the kite shown at the right by solving the equation $\cot\theta = 16\tan\theta$. ≈14.0° or ≈0.24

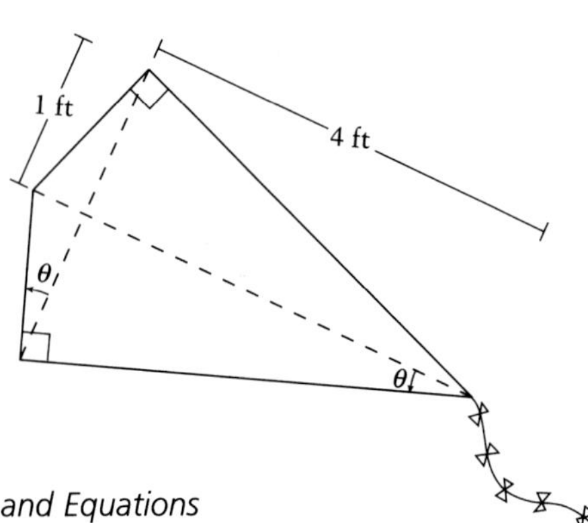

41. ≈2.09, ≈4.19 **42.** ≈ 1.05, ≈2.09, ≈4.19, ≈5.24

43. $\left(\frac{\pi}{6}, \frac{\sqrt{3}}{3}\right)$, $\left(\frac{7\pi}{6}, \frac{\sqrt{3}}{3}\right)$, $\left(\frac{2\sqrt{3}}{3}, 3\sqrt{3}\right)$, $\left(\frac{5\pi}{3}, 3\sqrt{3}\right)$ **44.** $\left(\frac{\pi}{3}, \frac{9}{4}\right)$, $\left(\frac{5\pi}{3}, \frac{9}{4}\right)$

54. *High Tide* The depth of the Atlantic Ocean at a channel buoy off the coast of Maine can be modeled by

$y = 5 - 3\cos\left(\frac{\pi}{6}t\right)$

where y is the water depth in feet and t is the time in hours. Consider a day in which $t = 0$ represents 12:00 midnight. For that day, when do the high and low tides occur? 6 A.M. and 6 P.M., noon and midnight

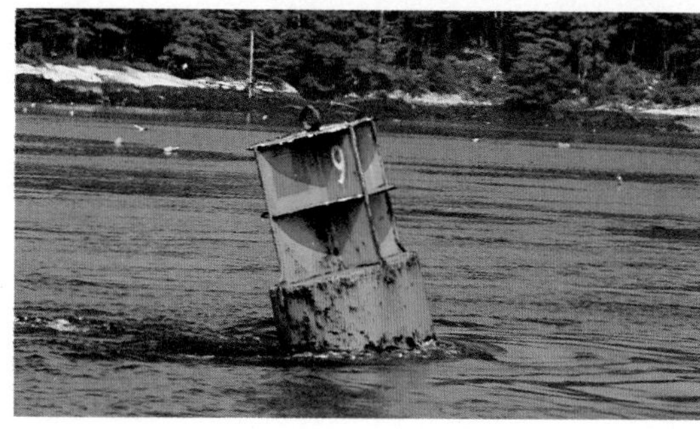

Channel marker off the Maine coast

Integrated Review

In Exercises 55–60, find the period and amplitude of the function.

55. $y = 10 \sin \frac{1}{2}x$ 4π, 10

56. $y = 6 \cos 3x$ $\frac{2\pi}{3}$, 6

57. $y = -4 \cos(x - \pi)$ 2π, 4

58. $y = -5 + \sin\left(\frac{1}{3}x + \pi\right)$ 6π, 1

59. $y = 8 + 3\cos\left(3x - \frac{\pi}{2}\right)$ $\frac{2\pi}{3}$, 3

60. $y = -12 \sin(2x + \pi)$ π, 12

In Exercises 61–66, sketch the graph of the function.

61. $y = 3 \cos 3x$

62. $y = -2 \sin(-2x + \pi)$

63. $y = 5 + 2 \sin 4x$

64. $y = -2 + \cos\left(\frac{1}{2}x + \pi\right)$

65. $y = -4\cos\left(x - \frac{\pi}{2}\right)$

66. $y = 1 - 3\sin\left(x - \frac{\pi}{2}\right)$

67. *Powers of Ten* Which best estimates the angle (in radian measure) of the tip of a sharpened pencil? a

a. 2.7×10^{-1} **b.** 2.7×10^{0} **c.** 2.7×10^{1} **d.** 2.7×10^{2}

68. *Powers of Ten* Which best estimates the number of times a ten-pin bowling ball revolves as it rolls down a bowling lane? c

a. 2.7×10^{-1} **b.** 2.7×10^{0} **c.** 2.7×10^{1} **d.** 2.7×10^{2}

Exploration and Extension

Computer Graphics **In Exercises 69 and 70, use the following information.**

Matrix multiplication can be used to rotate a point (x, y) an angle of θ around the origin. The coordinates of the resulting point, (x', y'), are determined by the following matrix equation.

$$\begin{bmatrix} \cos\theta & -\sin\theta \\ \sin\theta & \cos\theta \end{bmatrix}\begin{bmatrix} x \\ y \end{bmatrix} = \begin{bmatrix} x' \\ y' \end{bmatrix}$$

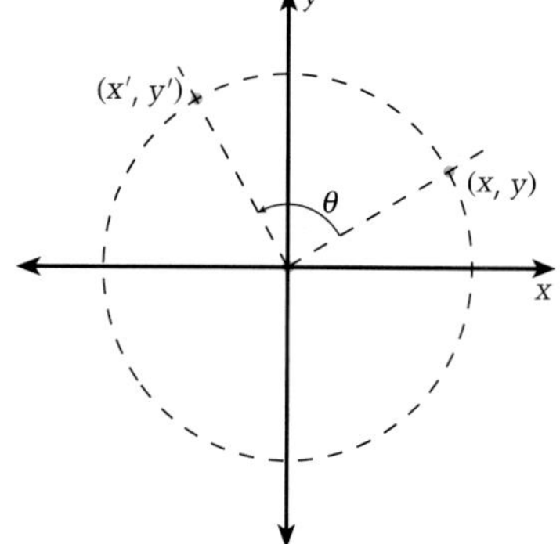

69. The point (4, 1) is rotated an angle of $\frac{\pi}{6}$ about the origin. What are the coordinates of the "resulting point"? $\left(2\sqrt{3} - \frac{1}{2}, 2 + \frac{\sqrt{3}}{2}\right)$

70. Through what angle θ must the point (2, 4) be rotated to produce $(x', y') = (-2 + \sqrt{3}, 1 + 2\sqrt{3})$? $\frac{\pi}{6}$

46. $\left(\frac{\pi}{2}, 1\right)$ 47. $\left(\frac{\pi}{3}, \frac{3}{2}\right), \left(\frac{5\pi}{3}, \frac{3}{2}\right)$ 48. $\left(\frac{\pi}{3}, 1\right), \left(\frac{5\pi}{3}, 1\right)$ 51. $\approx 35.26°$

Answers

See Additional Answers for exercises 61–66.

▶ **Ex. 69–70** You may need to review matrix multiplication.

Enrichment Activities

These activities require students to go beyond lesson goals.

When x is in radians, $\sin x$ can be estimated by the following series.

$$\sin x \approx x - \underset{(3!)}{\frac{x^3}{6}} + \underset{(5!)}{\frac{x^5}{120}} - \underset{(7!)}{\frac{x^7}{5040}}$$

1. Evaluate the function at $x = \frac{\pi}{3}$.

$$\sin\frac{\pi}{3} \approx +\frac{\pi}{3} - \frac{\left(\frac{\pi}{3}\right)^3}{6} + \frac{\left(\frac{\pi}{3}\right)^5}{120} - \frac{\left(\frac{\pi}{3}\right)^7}{5040} \approx 0.8660$$

2. Graph $y = \sin x$ and $y = x - \frac{x^3}{6} + \frac{x^5}{120} - \frac{x^7}{5040}$ on the same set of axes. What do you notice?

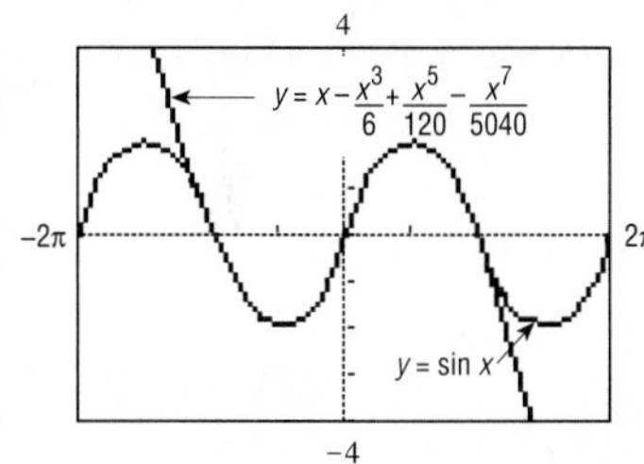

For $-\pi \le x < \pi$, the graphs match.

3. Look for a pattern in the series. Add more terms to it. Then repeat Activities 1 and 2. Do you get a more accurate value for $\sin\frac{\pi}{3}$? Do the graphs match over a greater domain? yes; yes

Problem of the Day

A man has two children. One child is a boy. What is the probability that the other child is a girl?

$\frac{2}{3}$

ORGANIZER

Warm-Up Exercises

1. Evaluate the expressions.
   a. sin 30° b. tan 45°
   c. cos 60° d. csc 90°

   a. $\frac{1}{2}$ b. 1 c. $\frac{1}{2}$ d. 1

2. Multiply.
   a. $\frac{\sqrt{3}}{2} \cdot \frac{\sqrt{3}}{3}$
   b. $(\sqrt{2})\left(1 - \frac{\sqrt{3}}{2}\right)$
   c. $\left(\frac{\sqrt{3}}{2}\right)\left(\sqrt{2} + \frac{\sqrt{2}}{2}\right)$

   a. $\frac{1}{2}$ b. $\frac{2\sqrt{2} - \sqrt{6}}{2}$ c. $\frac{3\sqrt{6}}{4}$

Lesson Resources

Teaching Tools
- Problem of the Day: 14.5
- Warm-up Exercises: 14.5
- Answer Masters: 14.5

Extra Practice: 14.5

Color Transparencies: 76, 77

Applications Handbook: p. 38, 39, 56, 57

LESSON Notes

Derivations for some of the sum and difference formulas are given in the *Handbook of Mathematical Connections.*

Example 1

Because there are many other ways to express 75° as a sum, ask students to explain why 45° and 30° are used.

14.5 Sum and Difference Formulas

What you should learn:

Goal 1 How to use the sum and difference formulas to evaluate trigonometric functions of the sum or difference of two angles

Goal 2 How to use sum and difference formulas to solve real-life problems

Why you should learn it:

You can model many real-life situations, such as the angle of refraction of water or glass, using the sum or difference of two angles.

Goal 1 Using Sum and Difference Formulas

In this lesson, you will study formulas that will allow you to evaluate trigonometric functions of the sum or difference of two angles.

Sum and Difference Formulas

$$\sin(u + v) = \sin u \cos v + \cos u \sin v$$
$$\sin(u - v) = \sin u \cos v - \cos u \sin v$$
$$\cos(u + v) = \cos u \cos v - \sin u \sin v$$
$$\cos(u - v) = \cos u \cos v + \sin u \sin v$$
$$\tan(u + v) = \frac{\tan u + \tan v}{1 - \tan u \tan v}$$
$$\tan(u - v) = \frac{\tan u - \tan v}{1 + \tan u \tan v}$$

Note that $\sin(u + v) \neq \sin u + \sin v$. Similar statements can be made for $\cos(u + v)$ and $\tan(u + v)$.

Example 1 *Evaluating a Trigonometric Function*

Find the exact value of sin 75° and sin 15°.

Solution Use the fact that $75° = 45° + 30°$ and $15° = 45° - 30°$.

$$\begin{aligned}\sin 75° &= \sin(45° + 30°)\\ &= \sin 45° \cos 30° + \cos 45° \sin 30°\\ &= \frac{\sqrt{2}}{2}\left(\frac{\sqrt{3}}{2}\right) + \frac{\sqrt{2}}{2}\left(\frac{1}{2}\right) = \frac{\sqrt{6} + \sqrt{2}}{4}\end{aligned}$$

$$\begin{aligned}\sin 15° &= \sin(45° - 30°)\\ &= \sin 45° \cos 30° - \cos 45° \sin 30°\\ &= \frac{\sqrt{2}}{2}\left(\frac{\sqrt{3}}{2}\right) - \frac{\sqrt{2}}{2}\left(\frac{1}{2}\right) = \frac{\sqrt{6} - \sqrt{2}}{4}\end{aligned}$$

Try checking these results with a calculator. For instance, evaluate sin 75° and $\frac{\sqrt{6} + \sqrt{2}}{4}$ to check that both have the same value. ■

Example 2 Evaluating a Trigonometric Function

Find the exact value of $\cos \frac{\pi}{12}$.

Solution Use the fact that $\frac{\pi}{12} = \frac{\pi}{3} - \frac{\pi}{4}$.

$$\begin{aligned}\cos \frac{\pi}{12} &= \cos\left(\frac{\pi}{3} - \frac{\pi}{4}\right)\\ &= \cos \frac{\pi}{3} \cos \frac{\pi}{4} + \sin \frac{\pi}{3} \sin \frac{\pi}{4}\\ &= \frac{1}{2}\left(\frac{\sqrt{2}}{2}\right) + \frac{\sqrt{3}}{2}\left(\frac{\sqrt{2}}{2}\right)\\ &= \frac{\sqrt{2} + \sqrt{6}}{4}\end{aligned}$$

■

Example 3 Evaluating a Trigonometric Expression

Find the exact value of $\frac{\tan 80° + \tan 55°}{1 - \tan 80° \tan 55°}$.

Solution Use the fact that $55° + 80° = 135°$.

$$\begin{aligned}\frac{\tan 80° + \tan 55°}{1 - \tan 80° \tan 55°} &= \tan(80° + 55°)\\ &= \tan 135°\\ &= -\tan 45°\\ &= -1\end{aligned}$$

■

Example 4 Solving a Trigonometric Equation

Solve $\sin\left(x + \frac{\pi}{4}\right) + 1 = \sin\left(\frac{\pi}{4} - x\right)$ for $0 \le x < 2\pi$.

Solution

$$\sin\left(x + \frac{\pi}{4}\right) + 1 = \sin\left(\frac{\pi}{4} - x\right)$$

$$\sin x \cos \frac{\pi}{4} + \cos x \sin \frac{\pi}{4} + 1 = \sin \frac{\pi}{4} \cos x - \cos \frac{\pi}{4} \sin x$$

$$2 \sin x \cos \frac{\pi}{4} = -1$$

$$2(\sin x)\left(\frac{\sqrt{2}}{2}\right) = -1$$

$$\sin x = -\frac{1}{\sqrt{2}}$$

$$\sin x = -\frac{\sqrt{2}}{2}$$

In the interval $0 \le x < 2\pi$, the solutions are

$$x = \frac{5\pi}{4} \quad \text{and} \quad x = \frac{7\pi}{4}.$$

■

The graphs at the left are a visual check for the solutions obtained in Example 4.

$y = \sin\left(x + \frac{\pi}{4}\right) + 1$

$y = \sin\left(\frac{\pi}{4} - x\right)$

Example 2

Most students will recognize how to express the value $\frac{\pi}{12}$ as a difference of $\frac{\pi}{3}$ and $\frac{\pi}{4}$ if these radian measures are first expressed in degree measures as 15°, 60°, and 45°, respectively.

Example 3

This is an example of how the sum and difference formulas can be applied in either direction. Students often ignore the fact that the sum and difference formulas are equally true when read from right to left.

Example 4

The sum and difference formulas are used here to simplify the expressions in the trigonometric equation. In a similar way, sum and difference formulas can also be used to verify trigonometric identities.

Example 5

Be sure students recognize that before you can express the quotient, $\frac{\cos \frac{\theta}{2}}{\sin \frac{\theta}{2}}$, in the next to last line as the cotangent function, you must be certain that the argument (the angle measure) of the cosine and sine is the same. In the example, both the sine and cosine functions are expressed in terms of an angle measure of $\frac{\theta}{2}$.

Extra Examples

Here are additional examples similar to **Examples 1–4.**

1. Find the exact value of the expression.

   a. $\sin 105°$ **b.** $\tan \frac{5\pi}{12}$

   a. $\frac{\sqrt{6} + \sqrt{2}}{4}$ **b.** $2 + \sqrt{3}$

2. Find the exact value of the expression.

   a. $\cos 50° \cos 100° - \sin 50° \sin 100°$

   b. $\dfrac{\tan 128° - \tan 98°}{1 + \tan 128° \tan 98°}$

   a. $-\dfrac{\sqrt{3}}{2}$ b. $\dfrac{\sqrt{3}}{3}$

3. Solve the equation $\cos\left(x + \frac{\pi}{4}\right) + \cos\left(x - \frac{\pi}{4}\right) = 1$ in the interval $0 \leq x < 2\pi$. $x = \frac{\pi}{4}, x = \frac{7\pi}{4}$

Check Understanding

1. Use numerical examples to show that $\cos(u + v) \neq \cos u + \cos v$. Suppose that $u = \frac{\pi}{4}$ and $v = \frac{\pi}{4}$. Then $\cos u = \cos v = \frac{\sqrt{2}}{2}$, and $\cos(u + v) = \cos\frac{\pi}{2} = 0$; but $\cos u + \cos v = \sqrt{2}$. Since $0 \neq \sqrt{2}$, $\cos(u + v) \neq \cos u + \cos v$.
2. Indicate how the tan 105° can be evaluated using either the sum or difference formulas. Substitute for u and v in the $\tan(u + v)$ formula. Let $u = 45°$ and $v = 60°$.

***Communicating* about ALGEBRA**

EXTEND *Communicating*
COOPERATIVE LEARNING
The sum and difference formulas can be used to verify trigonometric identities. Have students work with a partner to verify each of the following identities.

a. $\cos\left(x + \frac{\pi}{2}\right) = -\sin x$

b. $\cos\left(x + \frac{3\pi}{2}\right) = \sin x$

c. $\sin\left(\frac{3\pi}{2} - x\right) = -\cos x$

a. $\cos\left(x + \frac{\pi}{2}\right) = \cos x \cos\frac{\pi}{2} - \sin x \sin\frac{\pi}{2} = \cos x(0) - \sin x(1) = -\sin x$

b. Use a solution similar to that for Part a.

When light passes from one medium to another, such as from water to air, it is refracted, or bent. To demonstrate this, place a penny at the same position in two cups. Fill one cup with water and place the cups side by side. By looking down at the right angle, you will see the penny in the water but not the other penny.

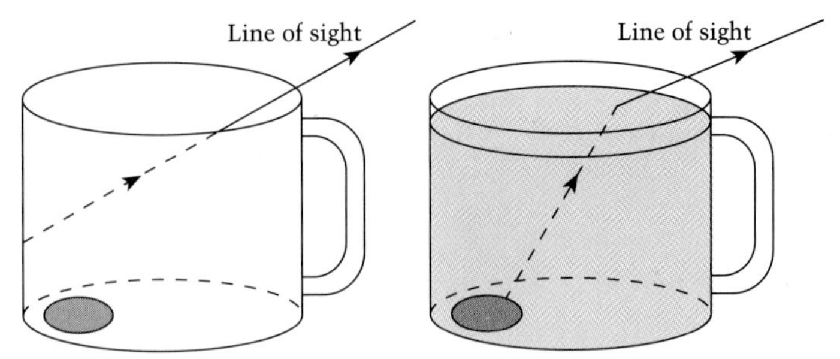

Goal 2 Using Sum and Difference Formulas

Example 5 *Measuring the Index of Refraction*

The *index of refraction, n,* of a transparent material is the ratio of the speed of light in a vacuum to the speed of light in the material. Some common indexes are air (1.00), water (1.33), glass (1.5), and diamond (2.4). Triangular prisms are often used to measure the index of refraction using the formula

$$n = \frac{\sin\left(\frac{\theta}{2} + \frac{\alpha}{2}\right)}{\sin\frac{\theta}{2}}.$$

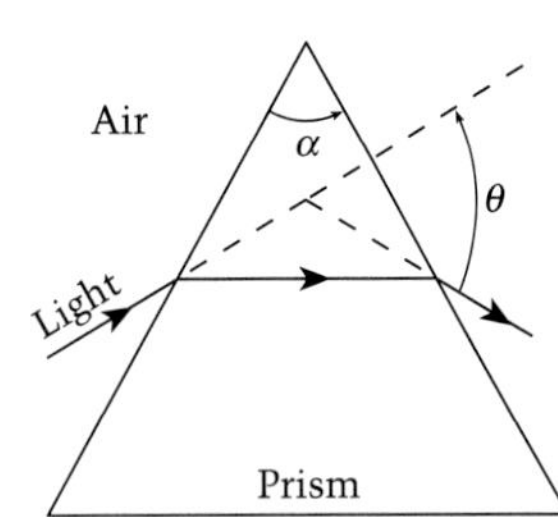

For the prism at the left, $\alpha = 60°$. Write the index of refraction as a function of $\cot\frac{\theta}{2}$.

Solution

$$n = \frac{\sin\left(\frac{\theta}{2} + 30°\right)}{\sin\frac{\theta}{2}} \qquad \textit{Substitute } \alpha = 60°.$$

$$= \frac{\sin\frac{\theta}{2}\cos 30° + \cos\frac{\theta}{2}\sin 30°}{\sin\frac{\theta}{2}} \qquad \textit{Sine of sum of two angles}$$

$$= \frac{\left(\sin\frac{\theta}{2}\right)\left(\frac{\sqrt{3}}{2}\right) + \left(\cos\frac{\theta}{2}\right)\left(\frac{1}{2}\right)}{\sin\frac{\theta}{2}} \qquad \textit{Evaluate sin 30° and cos 30°.}$$

$$= \frac{\sqrt{3}}{2} + \frac{1}{2}\cot\frac{\theta}{2} \qquad \textit{Simplify.}$$

■

Communicating about ALGEBRA

SHARING IDEAS about the Lesson

For the diagram in Example 5, find the angle θ for each material: water, glass, diamond. See below.

Water: ≈94.3°; glass: ≈76.5°; diamond: ≈36.1°

EXERCISES

Guided Practice

CRITICAL THINKING about the Lesson

1. Explain how to use a sum or difference formula to find the exact value of tan 240°.

2. The sum and difference formulas can be written as follows. Explain why "∓" rather than "±" occurs in the cosine and tangent formulas. The top sign on the left side corresponds to the top sign on the right side.

$\sin(u \pm v) = \sin u \cos v \pm \cos u \sin v$

$\cos(u \pm v) = \cos u \cos v \mp \sin u \sin v$

$\tan(u \pm v) = \dfrac{\tan u \pm \tan v}{1 \mp \tan u \tan v}$

3. Find the exact value of $\sin(-15°)$. $\frac{1-\sqrt{3}}{2\sqrt{2}}$

4. Solve $\cos\left(x - \frac{\pi}{6}\right) = 1 + \cos\left(x + \frac{\pi}{6}\right)$ for $0 \le x < 2\pi$. $\frac{\pi}{2}$

Independent Practice

In Exercises 5–16, find the exact value of the expression.

5. cos 105° $\frac{1-\sqrt{3}}{2\sqrt{2}}$
6. tan 195° $\frac{-1+\sqrt{3}}{1+\sqrt{3}}$
7. tan 75° $\frac{1+\sqrt{3}}{-1+\sqrt{3}}$
8. sin 345° $\frac{1-\sqrt{3}}{2\sqrt{2}}$
9. cos 255° $\frac{1-\sqrt{3}}{2\sqrt{2}}$
10. cos 375° $\frac{1+\sqrt{3}}{2\sqrt{2}}$
11. $\tan\frac{11\pi}{12}$ $\frac{1-\sqrt{3}}{1+\sqrt{3}}$
12. $\sin\frac{13\pi}{12}$ 12. $\frac{1-\sqrt{3}}{2\sqrt{2}}$
13. $\sec\frac{19\pi}{12}$ $\frac{2\sqrt{2}}{-1+\sqrt{3}}$
14. $\cos\frac{17\pi}{12}$ $\frac{1-\sqrt{3}}{2\sqrt{2}}$
15. $\cot\frac{5\pi}{12}$ $\frac{-1+\sqrt{3}}{1+\sqrt{3}}$
16. $\cot\frac{7\pi}{12}$ $\frac{1-\sqrt{3}}{1+\sqrt{3}}$

In Exercises 17–20, simplify the expression. Do not evaluate it.

17. cos 35° cos 15° − sin 35° sin 15° cos 50°
18. sin 110° cos 50° + cos 110° sin 50° sin 160°
19. sin 4 cos 2.2 − cos 4 sin 2.2 sin 1.8
20. $\sin\frac{\pi}{9}\sin\frac{\pi}{3} + \cos\frac{\pi}{9}\cos\frac{\pi}{3}$ $\cos\frac{2\pi}{9}$ or $\cos\frac{-2\pi}{9}$

In Exercises 21–26, evaluate the expression.

21. sin 30° cos 45° + cos 30° sin 45° $\frac{1+\sqrt{3}}{2\sqrt{2}}$
22. $\cos\frac{\pi}{4}\cos\frac{\pi}{3} - \sin\frac{\pi}{4}\sin\frac{\pi}{3}$ $\frac{1-\sqrt{3}}{2\sqrt{2}}$
23. sin 240° cos 45° − cos 240° sin 45° $\frac{1-\sqrt{3}}{2\sqrt{2}}$
24. $\dfrac{\tan 60° - \tan 45°}{1 + \tan 60° \tan 45°}$ $\frac{-1+\sqrt{3}}{1+\sqrt{3}}$
25. $\cos\frac{5\pi}{3}\cos\frac{\pi}{4} + \sin\frac{5\pi}{3}\sin\frac{\pi}{4}$ $\frac{1-\sqrt{3}}{2\sqrt{2}}$
26. sin 420° cos 45° + cos 420° sin 45° $\frac{1+\sqrt{3}}{2\sqrt{2}}$

In Exercises 27–32, solve the equation for $0 \le x < 2\pi$.

27. $\sin\left(x + \frac{\pi}{3}\right) + \sin\left(x - \frac{\pi}{3}\right) = 1$ $\frac{\pi}{2}$
28. $\sin\left(x + \frac{\pi}{4}\right) - \sin\left(x - \frac{\pi}{4}\right) = 1$ $\frac{\pi}{4}, \frac{7\pi}{4}$
29. $\cos\left(x + \frac{\pi}{4}\right) + \cos\left(x - \frac{\pi}{4}\right) = 1$ $\frac{\pi}{4}, \frac{7\pi}{4}$
30. $\cos\left(x + \frac{\pi}{6}\right) - \cos\left(x - \frac{\pi}{6}\right) = 1$ $\frac{3\pi}{2}$
31. $\tan(x + \pi) + 2\sin(x + \pi) = 0$ $0, \frac{\pi}{3}, \pi, \frac{5\pi}{3}$
32. $\tan(x + \pi) - \cos\left(x + \frac{\pi}{2}\right) = 0$ $0, \pi$

c. $\sin\left(\frac{3\pi}{2} - x\right) = \sin\frac{3\pi}{2}\cos x - \cos\frac{3\pi}{2}\sin x = (-1)\cos x - (0)\sin x = -\cos x$

EXERCISE Notes

ASSIGNMENT GUIDE

Basic/Average: Ex. 1–4, 9–17 odd, 21–29 odd, 33, 37, 38, 41–44, 49, 50

Above Average: Ex. 1–4, 7–17 odd, 19–29 odd, 35, 37–40, 44, 49, 51

Advanced: Ex. 1–4, 7–17 odd, 19–29 odd, 31, 35, 37–40, 43, 47, 51

Selected Answers
Exercises 1–4, 5–49 odd

Use **Mixed Review** as needed.

✪ More Difficult Exercises
Ex. 37–40, 51

Guided Practice

▶ **Ex. 1–4** Use these exercises as a summary of the lesson. In Exercise 4, refer students to Example 4 for a solution method. Or, students could use a graphics calculator.

Answer

1. $\tan 240° = \tan(180° + 60°) = \dfrac{\tan 180° + \tan 60°}{1 - \tan 180° \tan 60°} = \dfrac{0 + \sqrt{3}}{1 - 0 \cdot \sqrt{3}} = \sqrt{3}$

Independent Practice

▶ **Ex. 5–16** Refer students to the solution models shown in Examples 1 and 2.

▶ **Ex. 21–26** These exercises are modeled by Example 3 on page 761.

▶ **Ex. 27–32** Students can either use the solution model of Example 4 or use a graphics calculator.

Ex. 37–38 These exercises provide an interesting connection of tangents to slopes of lines.

Ex. 39–40 Use these exercises in class as a real-life connection of the sum and difference formulas.

In Exercises 33–36, evaluate the expression. Use the fact that $\sin u = -\frac{12}{13}$ with $\pi < u < \frac{3\pi}{2}$ and $\cos v = \frac{3}{5}$ with $\frac{3\pi}{2} < v < 2\pi$.

33. $\cos(u + v)$ $\frac{-63}{65}$ **34.** $\sin(u + v)$ $\frac{-16}{65}$ **35.** $\sin(v - u)$ $\frac{56}{65}$ **36.** $\cos(u - v)$ $\frac{33}{65}$

Angle Between Two Lines **In Exercises 37 and 38, use the following information.**

In the figure at the right, the acute angle of intersection, $\theta_2 - \theta_1$, of two lines with slopes m_1 and m_2 is given by the formula

$$\tan(\theta_2 - \theta_1) = \frac{m_2 - m_1}{1 + m_1 m_2}.$$

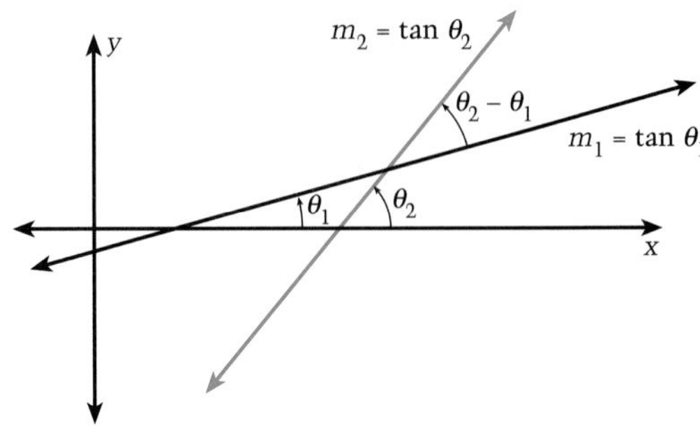

37. Find the acute angle of intersection between the lines $y = \frac{1}{2}x + 3$ and $y = 2x - 3$. ≈36.9°

38. Find the acute angle of intersection between the lines $y = x + 2$ and $y = 3x + 1$. ≈26.6°

Tuning a Guitar **In Exercises 39 and 40, use the following information.**

A guitar is being tuned by using a piano that is known to be in tune. The note A-440 is struck on the piano producing sound waves given by the model

$$y = \cos 880\pi t$$

where t is measured in seconds. Then the note that is supposed to be A-440 is struck on the guitar. If the guitar note has a frequency of a, then the guitar produces sound waves given by the model

$$y = \cos 2a\pi t.$$

If the guitar is not in tune, then the sound from both notes has a beat (or pulse) that is represented by the graph of

$$y = \cos 880\pi t + \cos 2a\pi t.$$

As a gets closer and closer to 440, the beat occurs less frequently. Finally, when the beat disappears, you know that a is equal to 440 and the guitar is in tune.

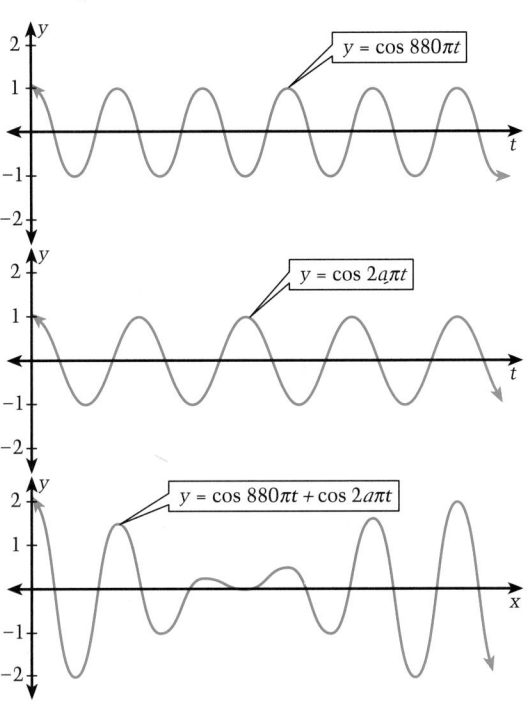

39. Use the trigonometric formula

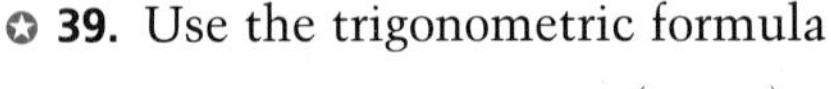

$$\cos u + \cos v = 2\cos\left(\frac{u + v}{2}\right)\cos\left(\frac{u - v}{2}\right)$$

to rewrite $y = \cos 880\pi t + \cos 2a\pi t$.

40. Use the result of Exercise 39 to rewrite the sound from *both* notes as

$$y = 2(\cos^2 440\pi t \cos^2 a\pi t - \sin^2 440\pi t \sin^2 a\pi t).$$

39. $y = 2\cos(440\pi t + a\pi t)\cos(440\pi t - a\pi t)$

40. $y = 2(\cos 440\pi t \cos a\pi t - \sin 440\pi t \sin a\pi t)(\cos 440\pi t \cos a\pi t + \sin 440\pi t \sin a\pi t)$ = given expression

Integrated Review

In Exercises 41–44, find the values of the other five trigonometric functions of θ. $\left(\text{Assume } 0 < \theta < \frac{\pi}{2}.\right)$

41. $\cot\theta = \sqrt{6}$ **42.** $\csc\theta = \frac{2\sqrt{3}}{3}$ **43.** $\sin\theta = \frac{\sqrt{2}}{2}$ **44.** $\cos\theta = \frac{1}{2}$

In Exercises 45–50, simplify the expression.

45. $\sec x - \sec x\tan^2 x$ **46.** $\sin^2\left(\frac{\pi}{2} - x\right) + \sin^2 x$ 1 **47.** $\frac{\sin x}{\cos x} + \sec x\tan x$

48. $\left(\frac{1}{\sin x}\right)^2 - \left(\frac{1}{\tan x}\right)^2$ 1 **49.** $\frac{\cos x\csc x}{\tan x}$ $\cot^2 x$ **50.** $\frac{\cos x}{1 - \sin x} - \tan x$ $\sec x$

Exploration and Extension

51. *Diamond or Zircon?* Diamond has an index of refraction of 2.4. Zircon has an index of refraction of 1.9. One of the prisms shown below is diamond; the other is zircon. Which is which? Zircon is on the left.

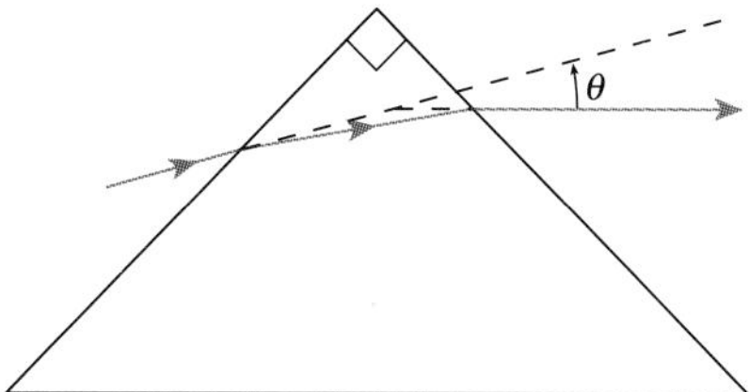

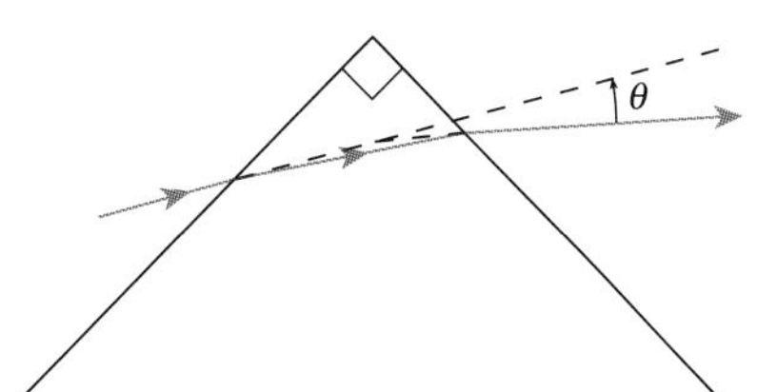

Mixed REVIEW

4., 15. See Additional Answers. 10. $y = \frac{1}{3}(x - 2)$ 12. $x^2 + 2x - 3$

1. Solve $8x^3 = -27$. **(7.3)** $-\frac{3}{2}$

2. Write the nth term: $\frac{1}{2}, 0, -\frac{1}{2}, -1, \ldots$ **(12.1)** $1 - \frac{1}{2}n$

3. Evaluate $\binom{7}{5}$. **(12.5)** 21

4. Sketch $(x - 3)^2 + (y + 3)^2 = 16$. **(11.5)**

5. Write in terms of radians: 75°. **(13.2)** $\frac{5\pi}{12}$

6. Write in terms of degrees: $\frac{5\pi}{4}$. **(13.2)** 225°

7. Solve $\log_a 16 = 4$ for a. **(8.3)** $a = 2$

8. Evaluate $e^{\ln 5}$. **(8.5)** 5

9. Solve $\sqrt{x - 2} = 5$. **(7.5)** 27

10. Write the inverse of $y = 3x + 2$. **(6.3)**

11. Find the reference angle of $-\frac{5\pi}{12}$. **(13.3)** $\frac{5\pi}{12}$

12. Divide $(x^3 + 3x^2 - x - 3)$ by $(x + 1)$. **(9.4)**

13. Evaluate $f(-1)$ for $f(x) = 3^x - 2$. **(7.1)** $-\frac{5}{3}$

14. True or False? $\sec\theta = \frac{1}{\tan\theta}$ **(13.1)** False

15. Sketch $y \le |3x - 2| + 3$. **(2.5)**

16. Classify the conic: $9x^2 - 4y^2 = 36$. **(11.6)** Hyperbola

17. Solve $|4 - t| > 10$. **(1.7)** $t < -6$ or $t > 14$

18. Evaluate $\begin{vmatrix} 4 & -4 \\ 6 & 3 \end{vmatrix}$. **(4.3)** 36

19. Write an equation of the line passing through $\left(-3, -\frac{1}{2}\right)$ and $(-1, 6)$. **(2.2)** $-13x + 4y = 37$

20. Write a quadratic model that represents the points $(1, 3)$, $(0, 0)$, and $(-1, 3)$. **(6.6)** $y = 3x^2$

Ex. 41–50 These exercises review key concepts from Lessons 14.1–14.4

Answers

41. $\sin\theta = \frac{1}{\sqrt{7}}$, $\cos\theta = \frac{\sqrt{6}}{\sqrt{7}}$, $\tan\theta = \frac{1}{\sqrt{6}}$, $\csc\theta = \sqrt{7}$, $\sec\theta = \frac{\sqrt{7}}{\sqrt{6}}$

42. $\sin\theta = \frac{3}{2\sqrt{3}}$, $\cos\theta = \frac{1}{2}$, $\tan\theta = \frac{3}{\sqrt{3}}$, $\sec\theta = 2$, $\cot\theta = \frac{\sqrt{3}}{3}$

43. $\cos\theta = \frac{\sqrt{2}}{2}$, $\tan\theta = 1$, $\csc\theta = \frac{2}{\sqrt{2}}$, $\sec\theta = \frac{2}{\sqrt{2}}$, $\cot\theta = 1$

44. $\sin\theta = \frac{\sqrt{3}}{2}$, $\tan\theta = \sqrt{3}$, $\csc\theta = \frac{2}{\sqrt{3}}$, $\sec\theta = 2$, $\cot\theta = \frac{1}{\sqrt{3}}$

45. $2\sec x - \sec^3 x$

47. $\tan x(1 + \sec x)$

Ex. 51 Refer students to Example 5 on page 762 for help with this exercise.

Enrichment Activity

This activity requires students to go beyond lesson goals.

Use the formulas for $\sin(u + v)$ and $\sin(u - v)$ and $\cos(u + v)$ and $\cos(u - v)$ to derive the formulas for $\tan(u + v)$ and $\tan(u - v)$.

(*Hint:* You want a 1 as the first term of the denominator expression.)

$$\tan(u + v) = \frac{\sin(u + v)}{\cos(u + v)}$$
$$= \frac{\sin u\cos v + \cos u\sin v}{\cos u\cos v - \sin u\sin v}$$
$$= \frac{\frac{\sin u\cos v}{\cos u\cos v} + \frac{\cos u\sin v}{\cos u\cos v}}{\frac{\cos u\cos v}{\cos u\cos v} - \frac{\sin u\sin v}{\cos u\cos v}}$$
$$= \frac{\tan u + \tan v}{1 - \tan u\tan v}$$

The derivation for $\tan(u - v)$ is similar.

Problem of the Day

On a balance scale, 3 green balls balance 6 blue balls, 2 yellow balls balance 5 blue balls, and 6 blue balls balance 4 white balls. How many yellow balls are needed to balance 5 green and 10 white balls?
10

ORGANIZER

Warm-Up Exercises

1. Solve the equation for x and for y: $2x + 2yx = 0$.
   $x = 0, y = -1$
2. Simplify the radical expressions.
   a. $\frac{3 + \sqrt{3}}{\sqrt{3}}$ b. $\sqrt{\frac{3 + \sqrt{2}}{6}}$
   a. $\sqrt{3} + 1$ b. $\frac{\sqrt{18 + 6\sqrt{2}}}{6}$

Lesson Resources

Teaching Tools
  Problem of the Day: 14.6
  Warm-up Exercises: 14.6
  Answer Masters: 14.6
Extra Practice: 14.6
Applications Handbook: p. 55, 56

LESSON Notes

Derivatives of some of the double-angle and half-angle formulas are given in the *Handbook of Mathematical Connections,* pages 854–855.

Example 1

The double-angle formula is used here to simplify the expression on the left-hand side of the trigonometric equation. Remind students that a solution of a trigonometric equation is an x-intercept of its graph.

766

14.6 Double- and Half-Angle Formulas

What you should learn:

Goal 1 How to use double-angle and half-angle formulas

Goal 2 How to use trigonometry to solve real-life problems

Why you should learn it:

You can model many real-life situations with double-angle and half-angle relationships, such as the path of a projectile launched at an initial angle θ.

Goal 1 Double- and Half-Angle Formulas

In this lesson, you will use formulas for double angles (angles of measure 2θ) and half angles (angles of measure $\frac{1}{2}\theta$).

Three formulas for $\cos 2u$ are listed below. All three formulas are equivalent, and you can use whichever one is most convenient in a problem.

Double-Angle Formulas

$\sin 2u = 2 \sin u \cos u$

$\tan 2u = \frac{2 \tan u}{1 - \tan^2 u}$

$\cos 2u = \cos^2 u - \sin^2 u$

$\cos 2u = 2 \cos^2 u - 1$

$\cos 2u = 1 - 2 \sin^2 u$

Note that $\sin 2u \neq 2 \sin u$. Similar statements can be made for $\cos 2u$ and $\tan 2u$.

Example 1 *Solving a Trigonometric Equation*

Solve $2 \cos x + \sin 2x = 0$ for $0 \le x < 2\pi$.

Solution Begin by rewriting the equation using functions of x rather than $2x$. Then factor and solve as usual.

$2 \cos x + \sin 2x = 0$ *Rewrite original equation.*

$2 \cos x + 2 \sin x \cos x = 0$ *Double-angle formula*

$2 \cos x(1 + \sin x) = 0$ *Factor.*

Using the Zero Product Property, set each factor equal to zero and solve for x.

$\cos x = 0$
$x = \frac{\pi}{2}, x = \frac{3\pi}{2}$

$1 + \sin x = 0$
$\sin x = -1$
$x = \frac{3\pi}{2}$

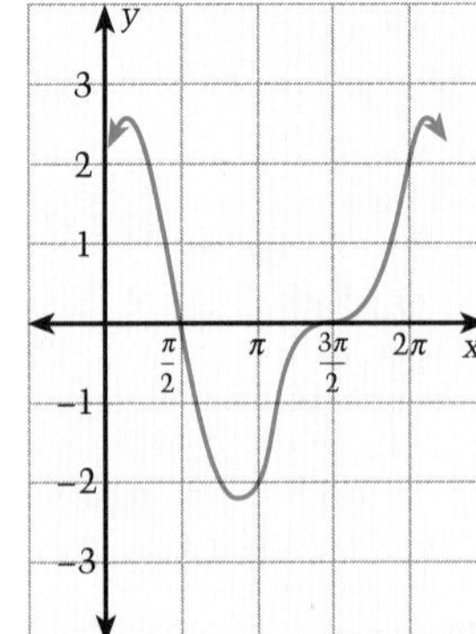

In the interval $0 \le x < 2\pi$, the equation has two solutions, $\frac{\pi}{2}$ and $\frac{3\pi}{2}$. Check these in the original equation. Or, if you have a graphing calculator, use a graphing check, as shown at the left. ■

Half-Angle Formulas

$$\sin\frac{u}{2} = \pm\sqrt{\frac{1-\cos u}{2}} \qquad \cos\frac{u}{2} = \pm\sqrt{\frac{1+\cos u}{2}}$$

$$\tan\frac{u}{2} = \frac{1-\cos u}{\sin u} \qquad \tan\frac{u}{2} = \frac{\sin u}{1+\cos u}$$

The signs of $\sin\frac{u}{2}$ and $\cos\frac{u}{2}$ depend on the quadrant in which $\frac{u}{2}$ lies.

Example 2 *Using a Half-Angle Formula*

Find the exact value of $\tan\frac{\pi}{8}$.

Solution Use the fact that $\frac{\pi}{8}$ is half of $\frac{\pi}{4}$.

$$\tan\frac{\pi}{8} = \frac{1-\cos\frac{\pi}{4}}{\sin\frac{\pi}{4}} = \frac{1-\frac{\sqrt{2}}{2}}{\frac{\sqrt{2}}{2}} = \frac{2-\sqrt{2}}{\sqrt{2}} = \sqrt{2}-1$$

Try checking this result with a calculator by evaluating $\tan\frac{\pi}{8}$ and $\sqrt{2}-1$. ■

Example 3 *Using a Half-Angle Formula*

Use $\cos 72° = \frac{-1+\sqrt{5}}{4}$ to find the exact value of cos 36°.

Solution Because 36° lies in the first quadrant, cos 36° is positive.

$$\cos 36° = \sqrt{\frac{1+\cos 72°}{2}} = \sqrt{\frac{1+\frac{-1+\sqrt{5}}{4}}{2}} = \sqrt{\frac{\frac{3+\sqrt{5}}{4}}{2}} = \sqrt{\frac{3+\sqrt{5}}{8}}$$

Try checking this result with a calculator. ■

Example 2

Observe that there are two formulas for the tangent of a half angle. Ask students to check whether the result obtained in this example is the same if the other formula is used.

The formula is

$\tan\frac{u}{2} = \frac{\sin u}{1+\cos u}$.

Since $\frac{\pi}{8}$ is half of $\frac{\pi}{4}$, $\tan\frac{\pi}{8} =$

$$\frac{\sin\frac{\pi}{4}}{1+\cos\frac{\pi}{4}} = \frac{\sqrt{2}}{2+\sqrt{2}} = \frac{\sqrt{2}(2-\sqrt{2})}{(2+\sqrt{2})(2-\sqrt{2})} = \frac{2\sqrt{2}-2}{4-2} = \frac{2\sqrt{2}-2}{2} = \sqrt{2}-1$$

Example 3

The decimal value of the complex radical expression is ≈0.809017, which is equal to cos 36°.

Example 4

Point out that the last step uses the double-angle formula for $\sin 2\theta$ by reading the formula from right to left. Ask students to justify each step in the verification of the relationship, $y = \frac{v^2 \sin 2\theta}{32}$.

Extra Examples

Here are additional examples similar to **Examples 1–3.**

1. Solve each equation for x in the interval $0 \le x < 2\pi$.

a. $\cos x + \cos 2x = 0$

b. $\sin 2x = 2\sin x$

a. $x = \frac{\pi}{3}, x = \pi, x = \frac{5\pi}{3}$

b. $x = 0, x = \pi$

2. Use the fact that $\cos 72° = \frac{-1+\sqrt{5}}{4}$ to find the exact value of cos 144°.

$-\frac{1+\sqrt{5}}{4}$

3. Use the fact that $\cos 105° = \frac{\sqrt{2} - \sqrt{6}}{4}$ to find the exact value of sin 52.5°. $\sqrt{\frac{4 - \sqrt{2} + \sqrt{6}}{8}}$

Check Understanding

1. Is $\sin 2x = 2 \sin x$ an identity? No; $\sin 2x = 2 \sin x \cos x$
2. Why do the half-angle formulas for sine and cosine have "±" in front of each radical expression? The signs of the sine and cosine depend on the interval in which the angle $\frac{u}{2}$ lies.
3. What identity can you use to show the equivalence of the three versions of the double-angle formula for cosine? You can use the Pythagorean identity $\sin^2 u + \cos^2 u = 1$. For example, $\cos^2 u - \sin^2 u = (1 - \sin^2 u) - \sin^2 u = 1 - 2\sin^2 u$. $\cos^2 u - \sin^2 u = \cos^2 u - (1 - \cos^2 u) = 2\cos^2 u - 1$.

Communicating about ALGEBRA

COOPERATIVE LEARNING

Have students work with a partner and use the sum formulas shown on page 760 to derive the double-angle formulas for the sine, cosine, and tangent functions.
(*Hint:* $\sin 2\theta = \sin(\theta + \theta)$)

a. $\sin 2\theta = \sin(\theta + \theta)$
$= \sin\theta\cos\theta + \cos\theta\sin\theta$
$= 2\sin\theta\cos\theta$

b. $\cos 2\theta = \cos(\theta + \theta)$
$= \cos\theta\cos\theta - \sin\theta\sin\theta$
$= \cos^2\theta - \sin^2\theta$

c. $\tan 2\theta = \tan(\theta + \theta)$
$= \frac{\tan\theta + \tan\theta}{1 - \tan\theta\tan\theta}$
$= \frac{2\tan\theta}{1 - \tan^2\theta}$

Goal 2 Using Trigonometry in Real-Life Problems

The path traveled by an object that is projected at an initial height of h feet, an initial velocity of v (in feet per second), and an initial angle of θ is given by

$$y = -\frac{16}{v^2 \cos^2\theta}x^2 + (\tan\theta)x + h$$

where y is the height of the object (in feet), and x is the horizontal distance (in feet).

Connections: Physics

Example 4 Path of a Projectile

In Lesson 13.3 (page 696), the horizontal distance traveled by an object that is projected from ground level at an angle of θ is given as $\frac{1}{32}v^2 \sin 2\theta$. Use the equation above to verify this.

Solution Because the object starts at ground level, you know that $h = 0$. You can find the horizontal distance that the object travels by letting y be equal to 0 and solving for x.

$$-\frac{16}{v^2\cos^2\theta}x^2 + (\tan\theta)x = 0 \quad \textit{Let } y = 0.$$

$$(-x)\left(\frac{16}{v^2\cos^2\theta}x - \tan\theta\right) = 0 \quad \textit{Factor.}$$

$$\frac{16}{v^2\cos^2\theta}x - \tan\theta = 0 \quad \textit{Zero Product Property}$$

$$\frac{16}{v^2\cos^2\theta}x = \tan\theta$$

$$x = \frac{1}{16}v^2\cos^2\theta\tan\theta$$

$$x = \frac{1}{16}v^2\cos\theta\sin\theta$$

$$x = \frac{1}{32}v^2(2\cos\theta\sin\theta)$$

$$x = \frac{1}{32}v^2\sin 2\theta$$

The last step uses the double-angle formula for $\sin 2\theta$. ■

y, θ, x, $\frac{1}{32}v^2\sin 2\theta$

Communicating about ALGEBRA

SHARING IDEAS about the Lesson

In Example 3, $\cos 72° = \frac{-1 + \sqrt{5}}{4}$ was used to find the exact value of cos 36°. Use $\cos 72° = \frac{-1 + \sqrt{5}}{4}$ to find the exact value of sin 36°. $\sqrt{\frac{5 - \sqrt{5}}{8}}$

13. $\frac{\pi}{2}, \frac{3\pi}{2}$ 14. $0, \frac{2\pi}{3}, \frac{4\pi}{3}$ 15. $\frac{\pi}{4}, \frac{3\pi}{4}, \frac{5\pi}{4}, \frac{7\pi}{4}, \frac{\pi}{2}, \frac{3\pi}{2}$ 16. $\frac{\pi}{3}, \frac{2\pi}{3}, \frac{4\pi}{3}, \frac{5\pi}{3}$

17. $\frac{\pi}{6}, \frac{5\pi}{6}, \frac{7\pi}{6}, \frac{11\pi}{6}, \frac{\pi}{2}, \frac{3\pi}{2}$

EXERCISES

Guided Practice

CRITICAL THINKING about the Lesson

1. Rewrite $\tan x$ as a function of $\tan \frac{x}{2}$. $\tan x = \dfrac{2\tan\frac{x}{2}}{1-\tan^2\frac{x}{2}}$
2. Use $\cos\frac{4\pi}{9} \approx 0.174$ to approximate $\cos\frac{8\pi}{9}$. $2(0.174)^2 - 1 \approx 0.939$
3. Use $\sin\frac{4\pi}{9} \approx 0.985$ to approximate $\cos\frac{8\pi}{9}$. Do your answers for Exercises 2 and 3 agree? $1 - 2(0.985)^2 \approx -0.940$; agree to 0.001
4. Use $\sin 105° = \frac{1}{4}(\sqrt{2}+\sqrt{6})$ and $\cos 105° = \frac{1}{4}(\sqrt{2}-\sqrt{6})$ to find the exact value of $\tan 52.5°$. $\dfrac{\sqrt{2}+\sqrt{6}}{4+\sqrt{2}-\sqrt{6}}$ or $\dfrac{4-\sqrt{2}+\sqrt{6}}{\sqrt{2}+\sqrt{6}}$ or $\sqrt{\dfrac{4-\sqrt{2}+\sqrt{6}}{4+\sqrt{2}-\sqrt{6}}}$
5. Solve $\cos 2x + 3\sin^2 x - 2 = 0$ for $0 \le x < 2\pi$.
6. Which of the following is the graph of $y = \cos 2x + 3\sin^2 x - 2$? Explain. b, The graph is the same as the graph of $y = -\frac{1}{2}\cos 2x - \frac{1}{2}$

a.

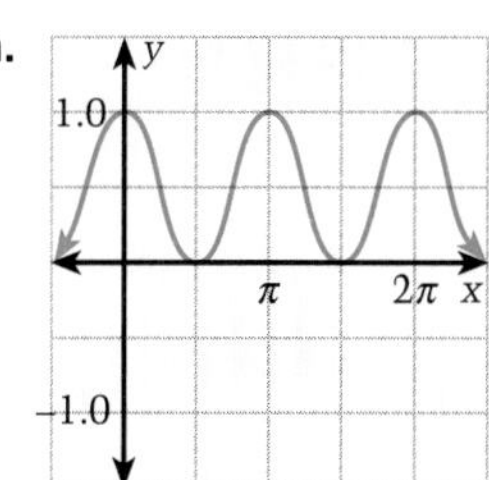

b.

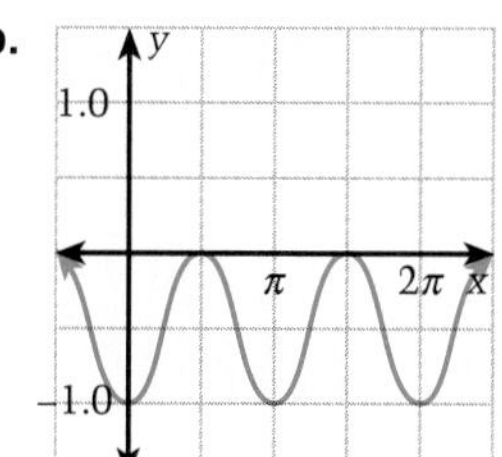

c.

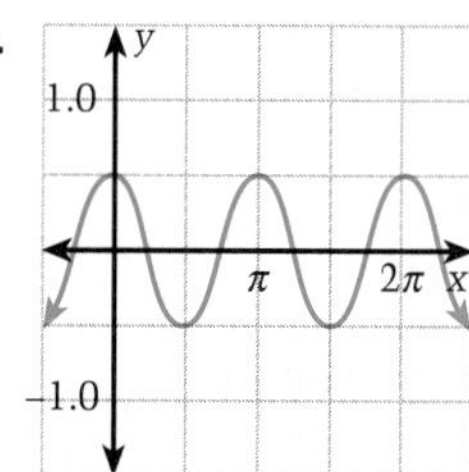

Independent Practice

In Exercises 7–12, rewrite each expression without double angles. Simplify the expression.

7. $\tan 2x(1 + \tan x)$ $\dfrac{2\tan x}{1-\tan x}$
8. $\cos 2x - 4\sin^4 x$ $1 - 2\sin^2 x - 4\sin^4 x$
9. $\cot x \tan 2x$ $\dfrac{2}{1-\tan^2 x}$
10. $\dfrac{\sin 2x}{\sin x}$ $2\cos x$
11. $\dfrac{\cos 2x}{\cos^2 x}$ $2 - \sec^2 x$
12. $\dfrac{-1+\sqrt{2}\cos x}{\cos 2x}$ $\dfrac{1}{1+\sqrt{2}\cos x}$

In Exercises 13–24, solve the equation for $0 \le x < 2\pi$.

13. $\cos 2x = -1$
14. $\cos 2x - \cos x = 0$
15. $\sin 2x \sin x = \cos x$
16. $\cos 2x = -2\cos^2 x$
17. $\cot x = \tan 2x$
18. $\sin 2x + \sin x = 0$
19. $2\tan x = \tan 2x$
20. $\sin 2x = \tan x$
21. $\cos 2x - 3\sin x = 2$
22. $\dfrac{3}{2}\tan 2x = 2$
23. $\dfrac{\cos 2x + \sqrt{2}\cos x}{\cos^2 x} = 1$
24. $\dfrac{\tan x \sin 2x + 2\sin^2 x}{-1 + 4\sin x} = 1$

In Exercises 25–30, find the exact value of the expression.

25. $\tan\frac{\pi}{12}$ $2-\sqrt{3}$
26. $\sin\frac{7\pi}{12}$ $\sqrt{\dfrac{2+\sqrt{3}}{4}}$
27. $\cos(-22.5°)$ $\sqrt{\dfrac{\sqrt{2}+1}{2\sqrt{2}}}$
28. $\cos\frac{7\pi}{8}$ $-\sqrt{\dfrac{\sqrt{2}+1}{2\sqrt{2}}}$
29. $\tan\frac{3\pi}{8}$ $\sqrt{2}+1$
30. $\tan\left(-\frac{7\pi}{8}\right)$ $\sqrt{2}-1$

18. $0, \pi, \frac{2\pi}{3}, \frac{4\pi}{3}$ 19. $0, \pi$ 20. $\frac{\pi}{4}, \frac{3\pi}{4}, \frac{5\pi}{4}, \frac{7\pi}{4}, 0, \pi$ 21. $\frac{7\pi}{6}, \frac{11\pi}{6}, \frac{3\pi}{2}$

22. $\approx 0.464, \approx 3.605, \approx 2.034, \approx 5.176$ 23. $\approx 1.026, \approx 5.256$

EXERCISE Notes

ASSIGNMENT GUIDE

Basic/Average: Ex. 1–6, 7–19 odd, 25–31 odd, 36–40 even, 41–43, 46, 48–50

Above Average: Ex. 1–6, 7–21 odd, 27–33 odd, 36–40 even, 41–46, 48–50

Advanced: Ex. 1–6, 7–21 odd, 27–33 odd, 38–46, 47, 51, 52

Selected Answers
Exercises 1–6, 7–49 odd

More Difficult Exercises
Exercises 44–47, 51, 52

Guided Practice

Ex. 1–6 COOPERATIVE LEARNING Have students work in small groups to complete these Guided Practice exercises. Discuss the answers in class to be sure students understand the application of the formula. Help students connect Exercises 5 and 6.

Answer

5. $\frac{\pi}{2}, \frac{3\pi}{2}$

Independent Practice

Ex. 7–24 A common student error is to express $\tan 2x$ as $2\tan x$ or $\cos 2x$ as $2\cos x$. Make sure students understand the expressions are not equivalent.

Ex. 13–24 TECHNOLOGY Encourage students to check the exact solutions by using the Trace function of graphics calculators, if they are available.

Answer

24. $\frac{\pi}{6}, \frac{5\pi}{6}$

Ex. 25–30 Remind students that $\frac{\pi}{12}$ is "half of" $\frac{\pi}{6}$, etc.

Ex. 34–39 Students may need to practice simplifying with radicals and fractions.

Answers

34. $\frac{1}{\sqrt{5}}, \frac{2}{\sqrt{5}}, \frac{1}{2}$

35. $\sqrt{\frac{\sqrt{5}-1}{2\sqrt{5}}}, \sqrt{\frac{\sqrt{5}+1}{2\sqrt{5}}}, \frac{\sqrt{5}-1}{2}$ or $\frac{2}{\sqrt{5}+1}$ or $\sqrt{\frac{\sqrt{5}-1}{\sqrt{5}+1}}$

36. $\sqrt{\frac{3-2\sqrt{2}}{6}}, \sqrt{\frac{3+2\sqrt{2}}{6}}, 3-2\sqrt{2}$ or $\frac{1}{3+2\sqrt{2}}$ or $\sqrt{\frac{3-2\sqrt{2}}{3+2\sqrt{2}}}$

37. $\sqrt{\frac{10-\sqrt{19}}{2\sqrt{5}}}, -\sqrt{\frac{10-\sqrt{19}}{2\sqrt{5}}}, \frac{10-\sqrt{19}}{-9}$ or $\frac{-9}{10+\sqrt{19}}$ or $-\sqrt{\frac{10-\sqrt{19}}{10+\sqrt{19}}}$

38. $\sqrt{\frac{\sqrt{5}-2}{2\sqrt{5}}}, \sqrt{\frac{\sqrt{5}+2}{2\sqrt{5}}}, \sqrt{5}-2$ or $\frac{1}{\sqrt{5}+2}$ or $\sqrt{\frac{\sqrt{5}-2}{\sqrt{5}+2}}$

39. $\sqrt{\frac{1}{5}}, \sqrt{\frac{4}{5}}, \frac{1}{2}$

Ex. 40–43 Assign these exercises as a group. Have students justify their answers.

Ex. 47 **COOPERATIVE LEARNING** Have students work with partners to complete this exercise in class. It will allow for student exploration of the different formulas.

Ex. 51–52 These can be used as an in-class demonstration of graphing parametric equations and as an extension of the chapter.

In Exercises 31–33, use the given value to find the exact value of the indicated expression. Check your result with a calculator.

31. $\cos 195° = -\frac{1}{2}\sqrt{2+\sqrt{3}}$

$\sin 97.5° = \boxed{?}$ $\sqrt{\frac{2+\sqrt{2+\sqrt{3}}}{4}}$

32. $\sin 105° = \frac{\sqrt{6}+\sqrt{2}}{4}$

$\cos 105° = \frac{\sqrt{2}-\sqrt{6}}{4}$

$\tan 52.5° = \boxed{?}$

32. $\frac{4-\sqrt{2}+\sqrt{6}}{\sqrt{6}+\sqrt{2}}$ or $\frac{\sqrt{6}+\sqrt{2}}{4+\sqrt{2}-\sqrt{6}}$

33. $\sin \frac{\pi}{12} = \frac{\sqrt{6}-\sqrt{2}}{4}$

$\cos \frac{\pi}{12} = \frac{\sqrt{6}+\sqrt{2}}{4}$

$\tan \frac{\pi}{24} = \boxed{?}$

33. $\frac{4-\sqrt{6}-\sqrt{2}}{\sqrt{6}-\sqrt{2}}$ or $\frac{\sqrt{6}-\sqrt{2}}{4+\sqrt{6}+\sqrt{2}}$

In Exercises 34–39, find the exact values of $\sin \frac{u}{2}$, $\cos \frac{u}{2}$, and $\tan \frac{u}{2}$.

34. $\cos u = \frac{3}{5},\ 0 \le u < \frac{\pi}{2}$

35. $\tan u = 2,\ 0 \le u < \frac{\pi}{2}$

36. $\sin u = \frac{1}{3},\ \frac{\pi}{2} \le u < \pi$

37. $\sin u = -\frac{9}{10},\ \frac{3\pi}{2} \le u < 2\pi$

38. $\tan u = \frac{1}{2},\ 0 \le u < \frac{\pi}{2}$

39. $\sin u = \frac{4}{5},\ 0 \le u < \frac{\pi}{2}$

In Exercises 40–43, match the equation with its graph.

40. $y = \sin x \cos x$ **c**

41. $y = \cos^2 2x - \sin^2 2x$ **a**

42. $y = 1 - 2\sin^2 x$ **d**

43. $y = \sin \frac{1}{2}x \cos \frac{1}{2}x$ **b**

a.

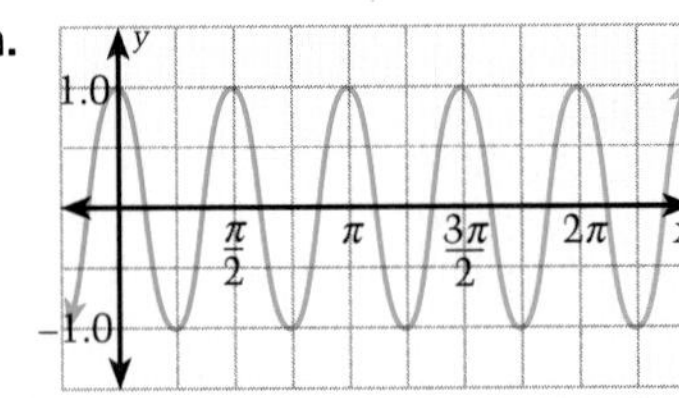

b.

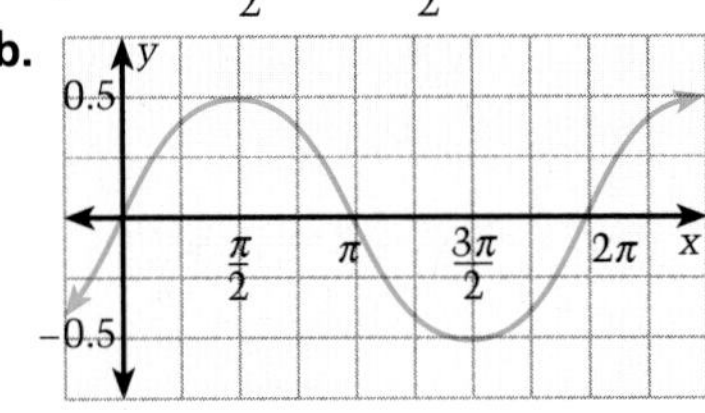

c.

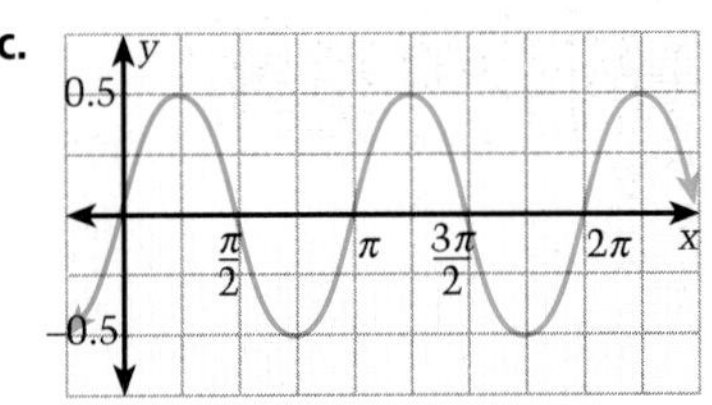

d.

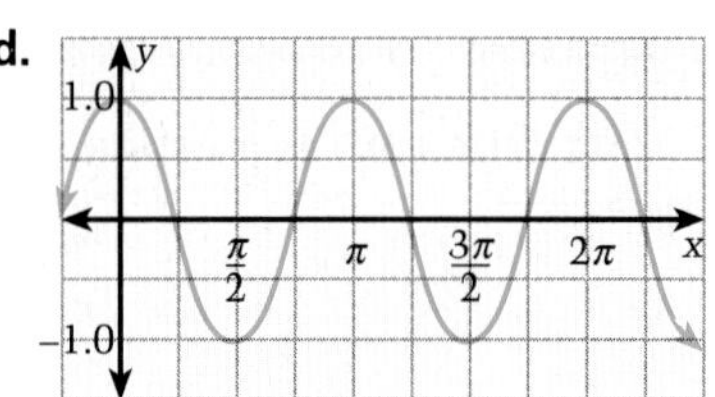

Inca Dwelling **In Exercises 44 and 45, use the following information.**

The picture at the right is a drawing of an Inca dwelling. This dwelling was in the Inca city of Machu Picchu, about 50 miles northwest of Cusco, Peru. All that remains of the ancient city today are stone ruins.

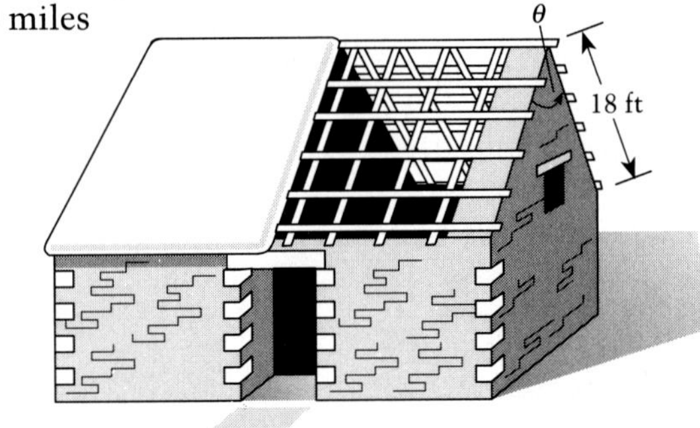

44. Express the area of the triangular portion of the right side of the dwelling as a function of $\sin \frac{\theta}{2}$ and $\cos \frac{\theta}{2}$. $324 \sin \frac{\theta}{2} \cos \frac{\theta}{2}$ ft^2

45. Express the area found in Exercise 44 as a function of $\sin \theta$. Then solve for θ using an area of 132 square feet. $162 \sin \theta$, $\approx 54.6°$

46. *Mach Number* The *mach number, M,* of an airplane is the ratio of its speed to the speed of sound. When an airplane travels faster than the speed of sound, the sound waves form a cone behind the airplane. The mach number is related to the apex angle of the cone by $\sin \frac{\theta}{2} = \frac{1}{M}$. Find the angle θ that corresponds to a mach number of 4.5. $\approx 25.7°$

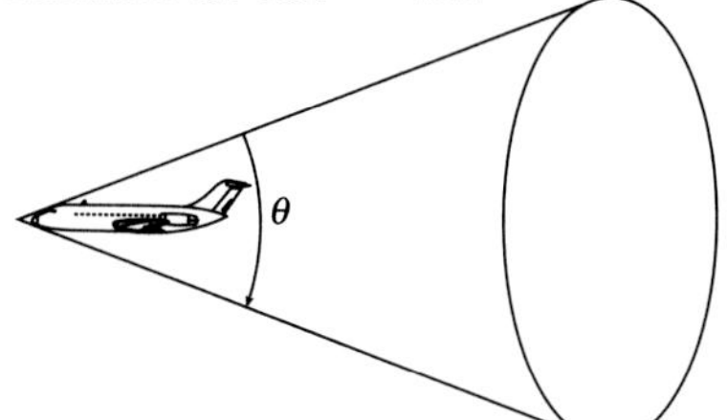

47. *Railroad Track* When two railroad tracks merge, the overlapping portions of the tracks are in the shape of circular arcs. The radius of the arcs, r (in feet), the width of the track, x (in feet), and the angle, θ, are related by $\frac{x}{2} = 2r \sin^2 \frac{\theta}{2}$. Write a formula for x in terms of $\cos \theta$. $x = 2r(1 - \cos \theta)$

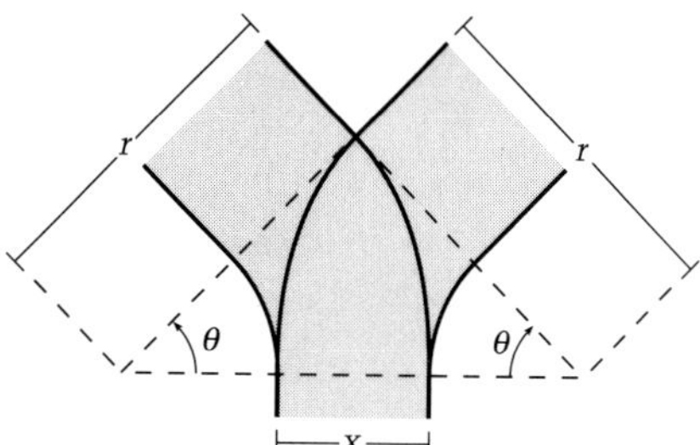

Integrated Review

In Exercises 48–50, solve the equation for $0 \le x < 2\pi$.

48. $2 \sin^2 x = 1$ $\frac{\pi}{4}, \frac{3\pi}{4}, \frac{5\pi}{4}, \frac{7\pi}{4}$

49. $2 \cos^2 x - 9 \cos x = 5$ $\frac{2\pi}{3}, \frac{4\pi}{3}$

50. $2 \sin^2 x - \sin x = 1$ $\frac{7\pi}{6}, \frac{11\pi}{6}, \frac{\pi}{2}$

Exploration and Extension

Carpet Area **In Exercises 51 and 52, use the following information.**

The following parametric equations can be used to model a cross section of one roll of carpeting.

$$x = \left(1 + \frac{1}{2\pi}t\right) \cos t \qquad y = \left(1 + \frac{1}{2\pi}t\right) \sin t$$

In this model, x and y are measured in inches.

51. Use a graphing calculator to sketch the graph given by the parametric equations. Let the parameter, t, vary between 0 and 20π and use a calculator setting of $-15 \le x \le 15$ and $-10 \le y \le 10$.

52. Using calculus, the length of the rug in feet (when unrolled) can be shown to be

$$\frac{1}{4\pi}(11a - b) - \frac{1}{4\pi^2} \ln\left(\frac{22\pi + a}{2\pi + b}\right) \text{ where}$$

$a = \sqrt{1 + (22\pi)^2}$ and $b = \sqrt{1 + (2\pi)^2}$. What is the length of the rug? $\approx$60 ft

Fernando Mateo owns two successful carpet stores in Manhattan. He has won several civic awards for his work in helping young people help themselves. "I rose from nowhere—so can others," says Mateo.

Answer
51. See Additional Answers.

Enrichment Activity

This activity requires students to go beyond lesson goals.

TECHNOLOGY This BASIC program finds the degree solutions to the trigonometric equation

$$2 \cos x - \sin 2x = 0$$

within a given interval. By modifying lines 20 and 120, you can solve the equations in Exercises 13–24 and 48–50 within the interval $0° \le x < 360°$. Check your answers by comparing the degree solutions with their radian equivalents.

```
10  PRINT "FIND THE SOLU-
    TIONS OF"
20  PRINT "2 * COS (X) −
    SIN (2X) = 0"
30  INPUT "ENTER THE LEAST
    VALUE TO BE TESTED.";X1
40  INPUT "ENTER THE GREAT-
    EST VALUE TO BE TESTED.";
    X2
50  PRINT
60  FOR J = X1 TO X2
70  LET Y = J/57.2957795;
    GOSUB 120
80  NEXT J
90  IF X = 1 THEN 180
100 PRINT "NO SOLUTIONS
    BETWEEN" ;X1; "AND" ;X2;
    "DEGREES."
110 GOTO 180
120 LET F = 2 * COS (Y)
    − SIN (2Y)
130 IF ABS(F) > 0.0000001
    THEN 160
140 LET X = 1
150 PRINT "ONE SOLUTION IS"
    ;J; "DEGREES."
160 RETURN
170 PRINT "THESE ARE THE
    SOLUTIONS BETWEEN"
    ;X1; "AND" ;X2
180 END
```

(If computers are not available, then consult the programming section of the graphics calculator manual to rewrite the program to the correct TI or Casio syntax.)

ORGANIZER

TIME SCHEDULE: All levels, two days

The Chapter Summary helps students organize the main ideas of the chapter. In this chapter, the basic graphs of the sine and cosine functions were explored, including the effects of constants on the graphs of the basic sine and cosine functions. The distinction between trigonometric identities and equations was made and students had the opportunity to work with each. The sum and difference formulas and the double- and half-angle formulas were also presented.

Work with students to review the concepts, graphs, and strategies of the chapter. The first day's homework assignment can be used as the basis for the second day of review.

COOPERATIVE LEARNING
Encourage students to study together. Emphasize the importance of "teaching" a classmate how to perform a skill or how to recall a strategy. When students work together, everyone wins. The students receiving help get additional instruction and the students giving help gain a deeper understanding of the concepts and strategies involved.

Chapter SUMMARY

Review the chapter by asking students "What did you learn? Why did you learn it? How does it fit with the other algebra skills and concepts you have learned?"

14 Chapter Summary

What did you learn?

Skills

1. Sketch graphs of sine and cosine functions. **(14.1)**
   - Find the amplitude of sine and cosine graphs. **(14.1)**
   - Find the period of sine and cosine graphs. **(14.1)**
   - Use horizontal shifts to sketch sine and cosine graphs. **(14.2)**
   - Use vertical shifts to sketch sine and cosine graphs. **(14.2)**
   - Use reflections to sketch sine and cosine graphs. **(14.2)**
2. Use trigonometric identities to simplify expressions. **(14.3)**
3. Verify trigonometric identities. **(14.3)**
4. Solve trigonometric equations. **(14.4)**
5. Use sum and difference formulas. **(14.5)**
6. Use double- and half-angle formulas. **(14.6)**

Strategies

7. Use trigonometric functions to solve real-life problems. **(14.1–14.6)**

Why did you learn it?

In Chapters 13 and 14, you have studied many types of real-life problems that can be modeled with trigonometry. Most of these fall into two categories: those dealing with angles and those dealing with periodic behavior. Many of the applications in this chapter have been of the second type.

For instance, in this chapter, you learned that heartbeats, sound waves, and temperatures are periodic and can therefore be modeled with trigonometric functions.

How does it fit into the bigger picture of algebra?

In this chapter, you have seen many examples of the use of algebra in trigonometry. For instance, to solve the trigonometric equation in Example 5 on page 755,

$$2\sin^2 x - \sin x = 1$$

you used the same basic procedure used to solve the algebraic equation,

$$2u^2 - u = 1.$$

That is, you subtracted 1 from both sides, factored the left side, and used the Zero Product Property to solve for the variable.

If you go on to study higher-level algebra, you will see other connections between trigonometry and algebra. For instance, trigonometry is used in algebra to find the complex nth roots of a real number.

Chapter REVIEW

In Exercises 1–9, find the amplitude and period of the function. (14.1)

1. $y = 4 \sin \frac{1}{2}x$ 4, 4π

2. $y = \sin \frac{2\pi}{3}x$ 1, 3

3. $y = \cos \frac{\pi}{4}x$ 1, 8

4. $y = 10 \cos 8x$ 10, $\frac{\pi}{4}$

5. $y = \frac{3}{2} \sin 2\pi x$ $\frac{3}{2}$, 1

6. $y = 1.2 \cos \pi x$ 1.2, 2

7. $y = 1 - 2 \sin 2x$ 2, π

8. $y = 6 \sin 4\pi x - 2$ 6, $\frac{1}{2}$

9. $y = 1 + \frac{1}{3} \cos\left(2x - \frac{\pi}{2}\right)$ $\frac{1}{3}$, π

In Exercises 10–18, match the equation with its graph. (14.1, 14.2)

10. $y = 1 - 3 \cos(x + 2\pi)$ i

11. $y = 3 \sin \frac{1}{2}x$ a

12. $y = 1 + 3 \cos \frac{1}{2}x$ f

13. $y = 3 \cos\left(x - \frac{\pi}{2}\right)$ c

14. $y = 3 \sin(x + \pi)$ h

15. $y = 1 + 3 \sin\left(2x + \frac{\pi}{4}\right)$ g

16. $y = 3 \sin \frac{\pi}{2}x$ d

17. $y = -3 \cos(x + \pi)$ e

18. $y = -3 \sin(2x - 3\pi)$ b

a.

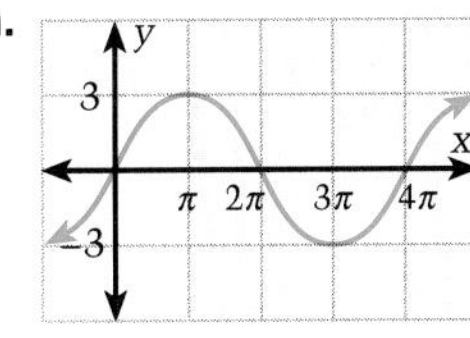

b.

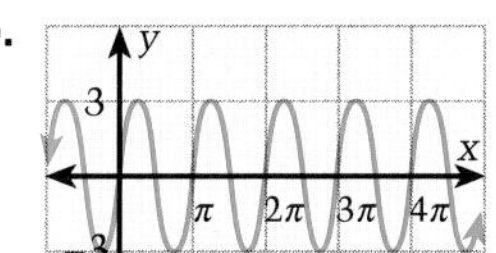

c.

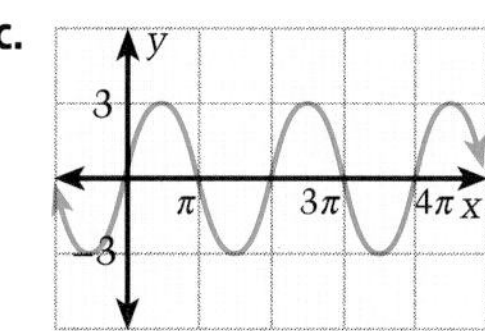

d.

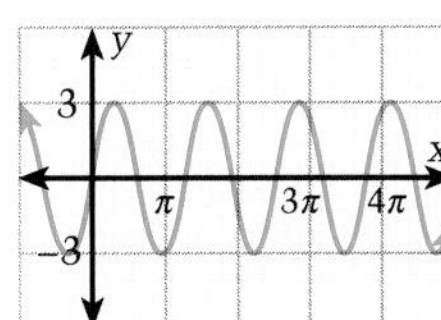

e.

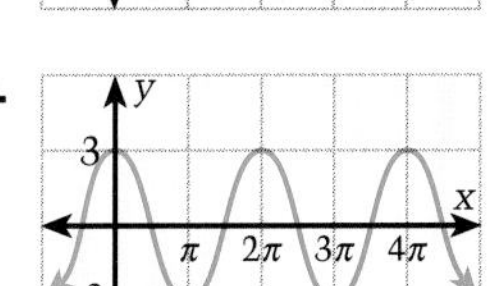

f.

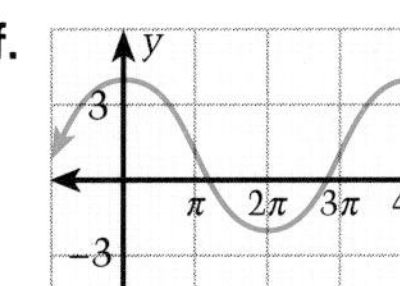

g.

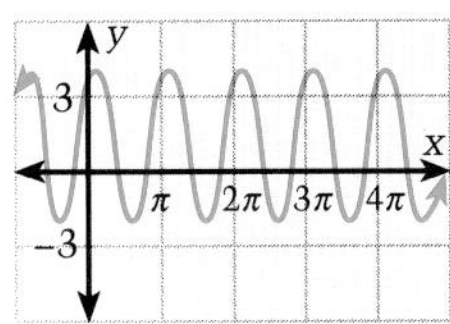

h.

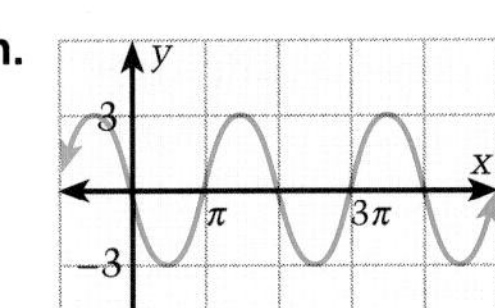

i.

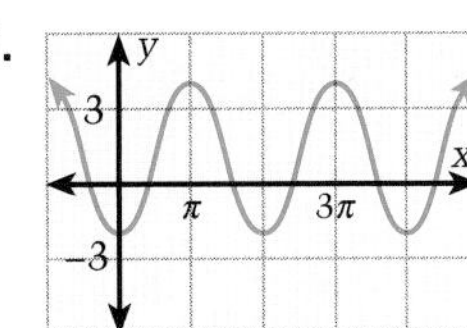

In Exercises 19–27, sketch the graph of the function. (14.1, 14.2) See Additional Answers.

19. $y = 4 - \sin x$

20. $y = -1 + \cos 4x$

21. $y = 2 \cos \pi x$

22. $y = \frac{1}{2} \sin 8x$

23. $y = 6 \cos\left(x - \frac{\pi}{2}\right)$

24. $y = 5 \sin(2x + \pi)$

25. $y = 2 + \sin\left(\frac{1}{2}x + 2\pi\right)$

26. $y = 3 + 3 \cos\left(\frac{1}{2}x + 2\pi\right)$

27. $y = -4 \cos(x - \pi)$

In Exercises 28–33, simplify the expression. (14.3)

28. $\cos^2 x + \sin(-x)$ $\cos^2 x - \sin x$

29. $\tan(-x) \cos(-x)$ $-\sin x$

30. $\sin^2\left(\frac{\pi}{2} - x\right) - 2 \sin^2 x + 1$ $-1 + 3\cos^2 x$ or $2 - 3\sin^2 x$

31. $\cot^2\left(\frac{\pi}{2} - x\right) - \sec^2(-x)$ -1

32. $\csc^2(-x) \cos^2\left(\frac{\pi}{2} - x\right)$ 1

33. $-1 + \tan(-x)\dfrac{\cos\left(\frac{\pi}{2} - x\right)}{\sin\left(\frac{\pi}{2} - x\right)}$ $-\sec^2 x$

ASSIGNMENT GUIDE

Basic/Average: Ex.1–9 odd, 10–18, 21–33 odd, 34–37, 39–54 multiples of 3, 56, 61–63

Above Average: Ex.1–9 odd, 10–18, 21–33 multiples of 3, 34–37, 39–57 odd, 58–60, 62, 64–75

Advanced: Ex.1–9 odd, 10–18, 21–33 multiples of 3, 34–37, 39–57 odd, 58–60, 62, 64–78

More Difficult Exercises
Exercises 56–60, 64–78

EXERCISE Notes

Exercises 1–78 The Chapter Review exercises include a reference to the lesson in which the skill or strategy was first presented. Students may use these references as a study aid if they are uncertain about how to do a given exercise in the Chapter Review.

Answers
See Additional Answers for Exercises 19–27.

Answers

34. $\cot^2 x(\tan^2 x + 1)$
$= \cot^2 x\left(\frac{1}{\cot^2 x} + 1\right)$
$= 1 + \cot^2 x$

35. $\cot x + \tan x = \frac{\cos x}{\sin x} + \frac{\sin x}{\cos x}$
$= \frac{\cos^2 x + \sin^2 x}{\cos x \sin x}$
$= \frac{1}{\cos x \sin x}$
$= \sec x \csc x$

36. $\sin^2(-x) = \sin^2 x$
$= 1 - \cos^2 x$
$= 1 - \frac{1}{\sec^2 x}$
$= \frac{\sec^2 x - 1}{\sec^2 x}$
$= \frac{\tan^2 x}{\tan^2 x + 1}$

37. $\tan^2(-x)\cos^2 x = \tan^2 x \cos^2 x$
$= \frac{\sin^2 x}{\cos^2 x} \cdot \cos^2 x$
$= \sin^2 x$
$= 1 - \cos^2 x$

38. $0, \pi, \frac{\pi}{4}, \frac{3\pi}{4}, \frac{5\pi}{4}, \frac{7\pi}{4}$

39. $\frac{\pi}{2}, \frac{3\pi}{2}, \frac{\pi}{6}, \frac{5\pi}{6}, \frac{7\pi}{6}, \frac{11\pi}{6}$

40. $\frac{7\pi}{6}, \frac{11\pi}{6}, \frac{\pi}{2}$

41. $\frac{\pi}{4}, \frac{3\pi}{4}, \frac{5\pi}{4}, \frac{7\pi}{4}$

42. $\frac{\pi}{6}, \frac{11\pi}{6}$

43. $\frac{\pi}{2}$

44. $\frac{\pi}{3}, \frac{2\pi}{3}, \frac{4\pi}{3}, \frac{5\pi}{3}$

45. $0, \pi, \frac{3\pi}{4}, \frac{7\pi}{4}$

46. $\frac{\pi}{2}$

47. $0, \pi, \frac{\pi}{2}, \frac{3\pi}{2}, \frac{\pi}{4}, \frac{3\pi}{4}, \frac{5\pi}{4}, \frac{7\pi}{4}$

56.

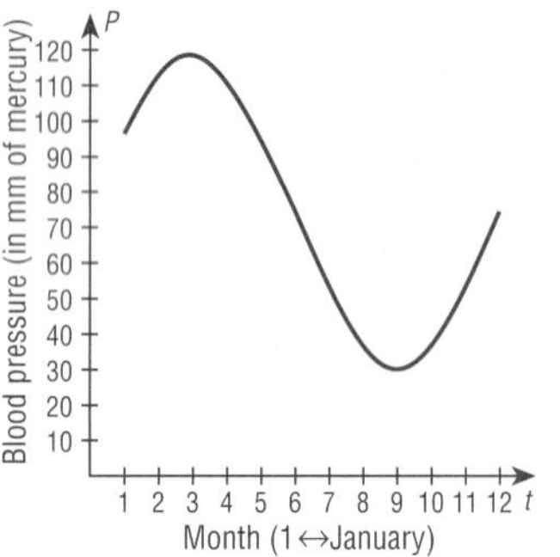

In Exercises 34–37, verify the identity. (14.3)

34. $\cot^2 x(\tan^2 x + 1) = 1 + \cot^2 x$

35. $\cot x + \tan x = \sec x \csc x$

36. $\sin^2(-x) = \frac{\tan^2 x}{\tan^2 x + 1}$

37. $1 - \cos^2 x = \tan^2(-x) \cos^2 x$

In Exercises 38–47, solve the equation for $0 \le x < 2\pi$. (14.4)

38. $2 \sin^2 x \tan x - \tan x = 0$

39. $\cos x \csc^2 x + 3 \cos x = 7 \cos x$

40. $2 \sin^2 x - \sin x - 1 = 0$

41. $\sec^2 x - 2 = 0$

42. $\sin\left(x + \frac{\pi}{2}\right) - \sqrt{3} = \sin\left(x - \frac{\pi}{2}\right)$

43. $\sin\left(x - \frac{\pi}{3}\right) + \sqrt{3} \cos x = \sin\left(\frac{\pi}{3} - x\right) + 1$

44. $\tan\left(x + \frac{\pi}{4}\right) + \tan x + 2 = 0$

45. $\sin x \tan(x + \pi) - \sin(x + \pi) = 0$

46. $\cos 2x + 2 \sin^2 x - \sin x = 0$

47. $(\sin 2x + \cos 2x)^2 - 1 = 0$

In Exercises 48–55, find the exact value of the expression. (14.5)

48. $\sin 285°$ $\frac{-1 - \sqrt{3}}{2\sqrt{2}}$

49. $\cos 195°$ $\frac{-1 - \sqrt{3}}{2\sqrt{2}}$

50. $\tan \frac{7\pi}{12}$ $\frac{1 + \sqrt{3}}{1 - \sqrt{3}}$

51. $\cot\left(-\frac{\pi}{12}\right)$ $\frac{1 + \sqrt{3}}{1 - \sqrt{3}}$

52. $\frac{\tan 145° - \tan 55°}{1 + \tan 145° \tan 55°}$ Undefined

53. $\cos 226° \cos 136° + \sin 226° \sin 136°$ 0

54. $\sin \frac{\pi}{5} \cos \frac{\pi}{30} - \sin \frac{\pi}{30} \cos \frac{\pi}{5}$ $\frac{1}{2}$

55. $\frac{\tan \frac{7\pi}{12} - \tan \frac{\pi}{4}}{1 + \tan \frac{7\pi}{12} \tan \frac{\pi}{4}}$ $\sqrt{3}$

56. *Seasonal Sales* For 1990, the national monthly sales, S (in thousands of units), of *Easy Glide* lawn furniture can be modeled by $S = 74.50 + 43.75 \sin \frac{\pi}{6}t$ where t represents the time in months, with $t = 1$ corresponding to January. In which months did sales exceed 100,000 units? Sketch the graph of this function for one year. February, March, April

57. *Biorhythms* A popular but unproven theory called *biorhythms* explains the ups and downs of everyday life with three cycles that begin at birth.

Physical: $P = \sin \frac{2\pi}{23}t$

Emotional: $E = \sin \frac{2\pi}{28}t$

Intellectual: $I = \sin \frac{2\pi}{33}t$

Positive values indicate high-energy days, and negative values indicate low-energy days. Consider a person who was born on August 16, 1967. Describe this person's three energy levels on May 3, 1995 (the 10,122nd day of the person's life). Physical: high, emotional: neutral, intellectual: low

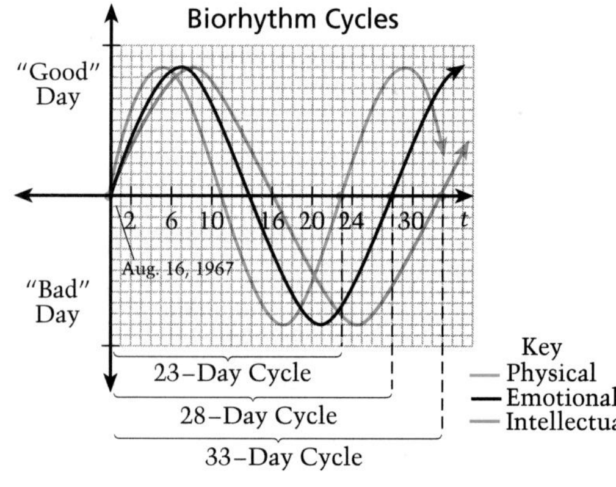

Answer

61. $\sin\left(\frac{\pi}{2} - \theta_1\right) = \sin\theta_2 = \frac{b}{c} = \cos\theta_1$

$\cos\left(\frac{\pi}{2} - \theta_1\right) = \cos\theta_2 = \frac{a}{c} = \sin\theta_1$

$\tan\left(\frac{\pi}{2} - \theta_1\right) = \tan\theta_2 = \frac{b}{a} = \cot\theta_1$

$\cot\left(\frac{\pi}{2} - \theta_1\right) = \cot\theta_2 = \frac{a}{b} = \tan\theta_1$

$\sec\left(\frac{\pi}{2} - \theta_1\right) = \sec\theta_2 = \frac{c}{a} = \csc\theta_1$

$\csc\left(\frac{\pi}{2} - \theta_1\right) = \csc\theta_2 = \frac{c}{b} = \sec\theta_1$

Calculating Pi **In Exercises 58–60, use the following information.**

58. 3.220250731, within 0.08
59. 2.358328174, within 0.8
60. 3.141592652, within 0.000000002

Many mathematicians have worked at finding ways to compute the decimal form of π. Here are three well-known series.

James Gregory (1638–1675):

$$\pi \approx 4\left(1 - \frac{1}{3} + \frac{1}{5} - \frac{1}{7} + \cdots + \frac{(-1)^{n+1}}{2n-1}\right)$$

John Machin (1680–1751): (For n = 4)

$$\pi \approx 16\left(\frac{1}{5} - \frac{1}{3 \cdot 5^3} + \frac{1}{5 \cdot 5^5} - \frac{1}{7 \cdot 5^7}\right) - 4\left(\frac{1}{5} - \frac{1}{3 \cdot 239^3} + \frac{1}{5 \cdot 239^5} - \frac{1}{7 \cdot 239^7}\right)$$

Srinivasa Ramanujan (1877–1920): (For n = 2)

$$\pi \approx \frac{9801}{\sqrt{8}}\left(\frac{1}{1103 + 24(1103 + 26{,}390)394^{-4}}\right)$$

58. Find the sum of Gregory's series for $n = 11$ and $n = 15$. Average the two results. How close is the answer to π?

59. Evaluate Machin's approximation for $n = 4$. How close is the answer to π?

60. Evaluate Ramanujan's approximation for $n = 2$. How close is the answer to π?

Srinivasa Ramanujan developed methods of calculating π with extraordinary efficiency. His approach is now incorporated in computer algorithms that yield millions of digits of π.

61. *Cofunction Identities* Use the triangle below to verify the six cofunction identities for an *acute* angle.

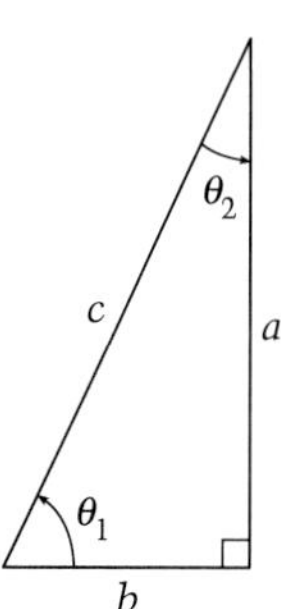

62. *x-Intercepts of a Graph* Find the x-intercepts of the graph of $y = \sin 2x + \sin x$ in the interval $0 \le x < 2\pi$. 0, π, $\frac{2\pi}{3}$, $\frac{4\pi}{3}$

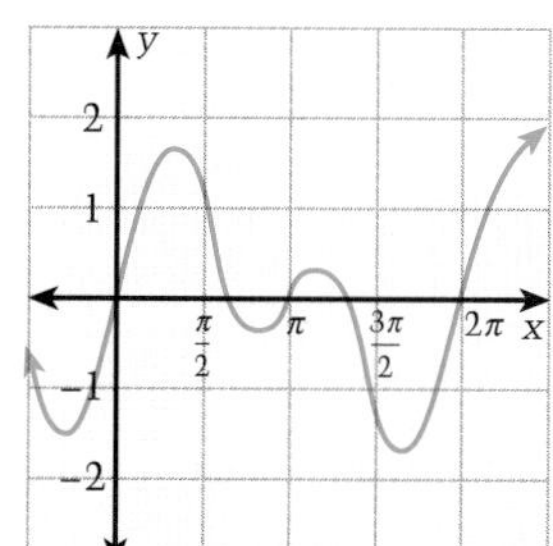

63. *Canal Construction* A canal is being constructed with a cross section in the shape of an isosceles trapezoid, as shown at the right. What angle θ is required for the cross-sectional area of the canal to be $75\sqrt{3}$ square yards? 60°

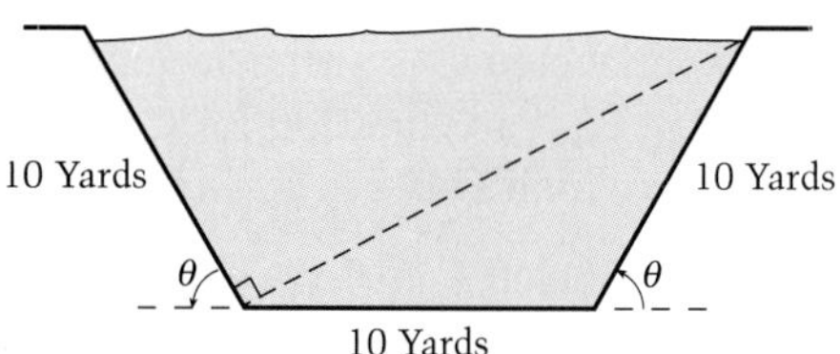

Blanket Pattern **In Exercises 64–73, match the equation with its graphs in the blanket. Then state the amplitude and period of each "matched graph." (The equations match all 22 graphs in the blanket except C, H, K, L, O, and T.)**

64. $y = d + \frac{1}{4}\cos 2x$ F, G, R, S; $\frac{1}{4}$, π

65. $y = d - \frac{1}{4}\cos 2x$ D, E, P, Q; $\frac{1}{4}$, π

66. $y = d + 2\sin x$ B, 2, 2π

67. $y = d - \sin x$ V, 1, 2π

68. $y = d + 2\cos\left(x + \frac{3\pi}{4}\right)$ J, 2, 2π

69. $y = d + 2\cos\left(x + \frac{\pi}{4}\right)$ I, 2, 2π

70. $y = d - 2\sin x$ U, 2, 2π

71. $y = d + \sin x$ A, 1, 2π

72. $y = d + 2\cos\left(x + \frac{5\pi}{4}\right)$ N, 2, 2π

73. $y = d + 2\cos\left(x + \frac{7\pi}{4}\right)$ M, 2, 2π

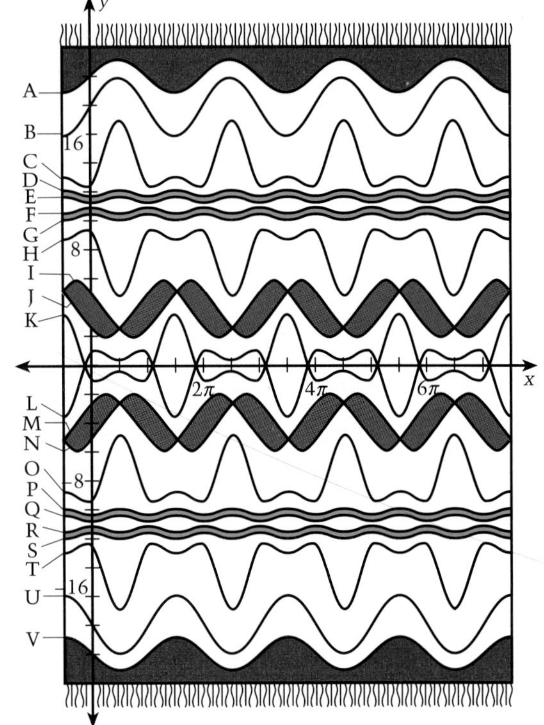

74. Which of the following is *not* an equation of the graph labeled U? d

a. $y = -18 + 2\sin(x + \pi)$

b. $y = -18 + 2\sin(-x)$

c. $y = -18 + 2\cos\left(x + \frac{\pi}{2}\right)$

d. $y = -18 + 2\cos\left(x + \frac{3\pi}{2}\right)$

75. Which of the following is *not* an equation of the graph labeled D? a

a. $y = 12 + \frac{1}{4}\cos(2x + 2\pi)$

b. $y = 12 - \frac{1}{4}\cos(2x + 2\pi)$

c. $y = 12 - \frac{1}{4}\sin\left(2x + \frac{\pi}{2}\right)$

d. $y = 12 + \frac{1}{4}\sin\left(2x + \frac{3\pi}{2}\right)$

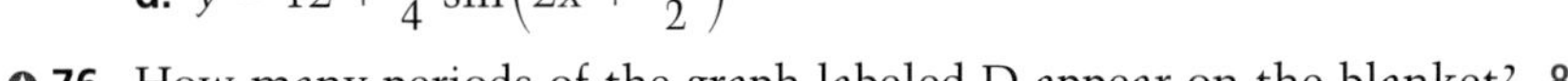

76. How many periods of the graph labeled D appear on the blanket? 8

77. The graph labeled K has the following equation.

$$y = \frac{1}{2} + 2\sin(x - \pi) - \cos(2x - 2\pi)$$

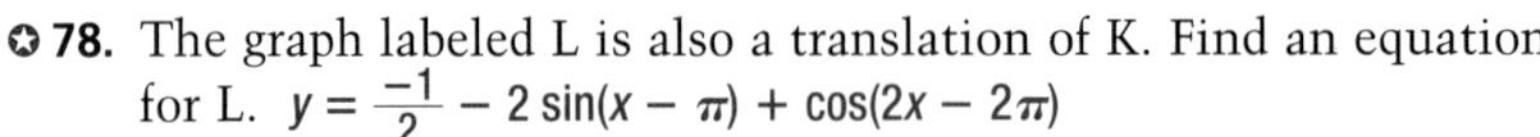

The graph labeled C is a translation of K: 13.5 units up and π units to the left. Find an equation for C. Then find all x-values on the blanket at which C is 17. $y = 14 + 2\sin x - \cos 2x$; $\frac{\pi}{2}, \frac{5\pi}{2}, \frac{9\pi}{2}, \frac{13\pi}{2}$

78. The graph labeled L is also a translation of K. Find an equation for L. $y = \frac{-1}{2} - 2\sin(x - \pi) + \cos(2x - 2\pi)$

Chapter TEST

In Exercises 1–3, sketch the graph of the function. (14.2) See margin.

1. $y = 4 + 2\sin x$
2. $y = -1 - \sin\left(x + \frac{\pi}{2}\right)$
3. $y = 3 + 6\cos(x - \pi)$

In Exercises 4–6, simplify the expression. (14.3)

4. $2\sin^2 u \cot u$ $2\sin u \cos u$ or $\sin 2u$
5. $\dfrac{4\sin x \cos x - 2\sin x \sec x}{2\tan x}$ $2\cos^2 x - 1$ or $\cos 2x$
6. $\dfrac{1}{\csc x + \cot x}$ $\dfrac{\sin x}{1 + \cos x}$ or $\tan \dfrac{x}{2}$

In Exercises 7–9, verify the identity. (14.3) See Additional Answers.

7. $\csc u - \cot u = \tan \dfrac{u}{2}$
8. $\cos^2 u - \sin^2 u = 1 - 2\sin^2 u$
9. $-2\cos^2 u \tan(-u) = \sin 2u$

In Exercises 10–13, solve the equation for $0 \le x \le 2\pi$. (14.4, 14.5)

10. $-6 + 10\cos x = -1$ $\dfrac{\pi}{3}, \dfrac{5\pi}{3}$
11. $\tan^2 x - 2\tan x + 1 = 0$ $\dfrac{\pi}{4}, \dfrac{5\pi}{4}$
12. $\cos x - \sin x \sin 2x = 0$
13. $4 - 3\csc^2 x = 0$ $\dfrac{\pi}{3}, \dfrac{2\pi}{3}, \dfrac{4\pi}{3}, \dfrac{5\pi}{3}$

12. $\dfrac{\pi}{4}, \dfrac{\pi}{2}, \dfrac{3\pi}{4}, \dfrac{5\pi}{4}, \dfrac{3\pi}{2}, \dfrac{7\pi}{4}$

In Exercises 14–17, find the exact value of the expression. (14.5, 14.6)

14. $\dfrac{\tan 175° + \tan 140°}{1 - \tan 175° \tan 140°}$ -1
15. $\dfrac{\tan 160° - \tan 40°}{1 + \tan 160° \tan 40°}$ $-\sqrt{3}$
16. $\sin \frac{2\pi}{3} \cos \frac{5\pi}{3} + \cos \frac{2\pi}{3} \sin \frac{5\pi}{3}$ $\frac{\sqrt{3}}{2}$
17. $\cos \frac{13\pi}{12} \cos \frac{11\pi}{12} + \sin \frac{13\pi}{12} \sin \frac{11\pi}{12}$ $\frac{\sqrt{3}}{2}$

18. On a horseshoe-pitching court, the two stakes tilt towards each other, 40 feet apart. Use the diagram at the right to find the angle, θ, that a horseshoe stake makes with the ground. **(14.4)** $\approx 78.7°$
19. Find the length of the stake above the ground. Round the result to one decimal place. **(14.4)** 15.3 in.

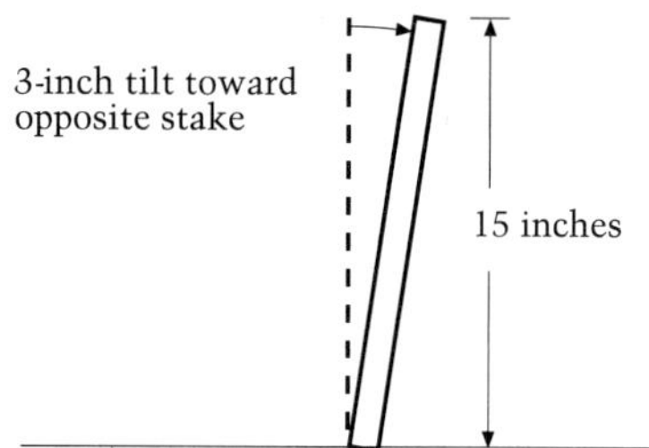

20. The depth of the ocean at a swim buoy can be modeled by

$$y = 4 + 2\sin\left(\frac{\pi}{6}t\right)$$

where y is the water depth in feet and t is the time in hours. Consider a day in which $t = 0$ represents 12:00 midnight. For that day, when do high and low tides occur? **(14.2)** High tides: 3 A.M., 3 P.M., low tides: 9 A.M., 9 P.M.

21. A ball thrown and caught at a height of 4 feet with an initial velocity of 30 feet per second traveled 14 feet. Use the formula

$$x = \frac{1}{16}v^2 \cos\theta \sin\theta$$

to determine the angle of the throw. **(14.6)** ≈ 0.26 radians or $\approx 14.9°$

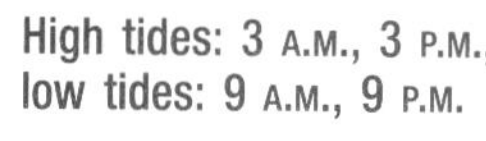

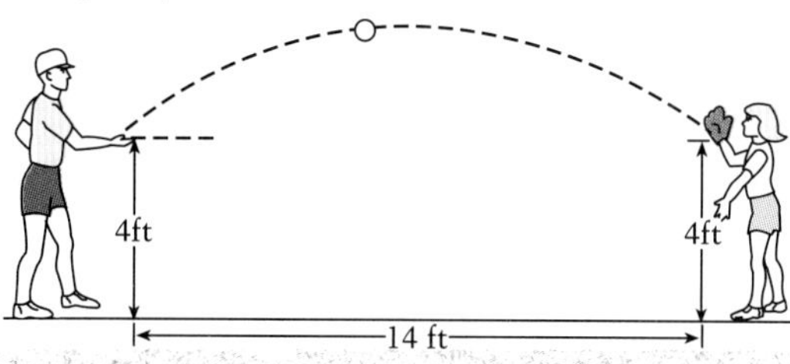

Answers

1.

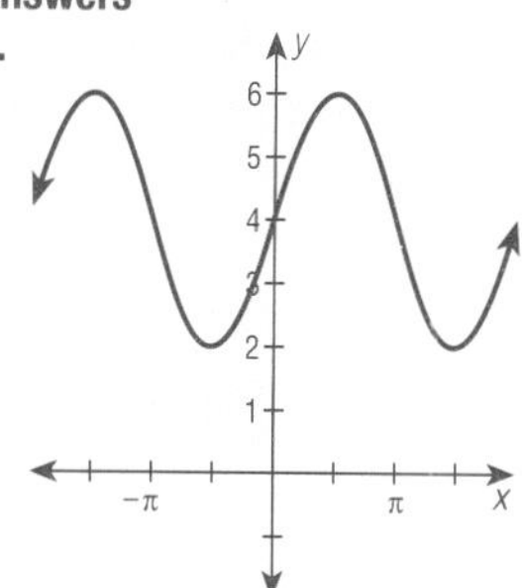

2.

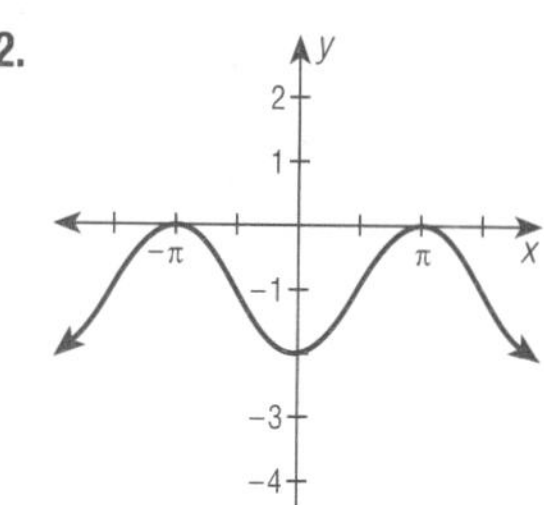

3.

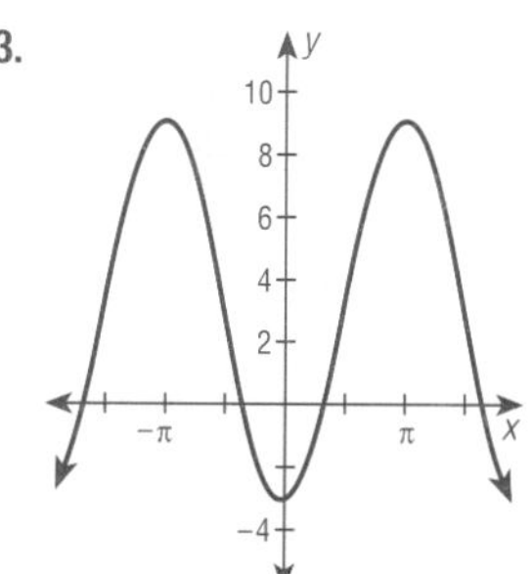

Answers

13. $C = 113°$, $a \approx 3.18$, $b \approx 1.49$

14. $B \approx 48.7°$, $C \approx 21.3°$; $a \approx 1.93$

15. $A \approx 36.18°$, $B \approx 100.28°$, $C \approx 43.54°$

16. $A \approx 30.5°$, $C \approx 49.5°$, $b \approx 7.77$

19.

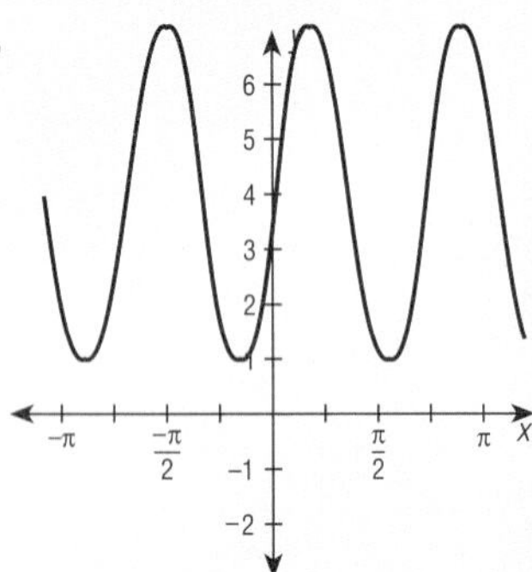

20.

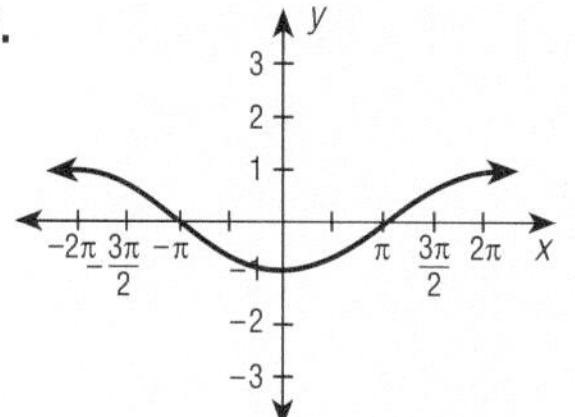

21.

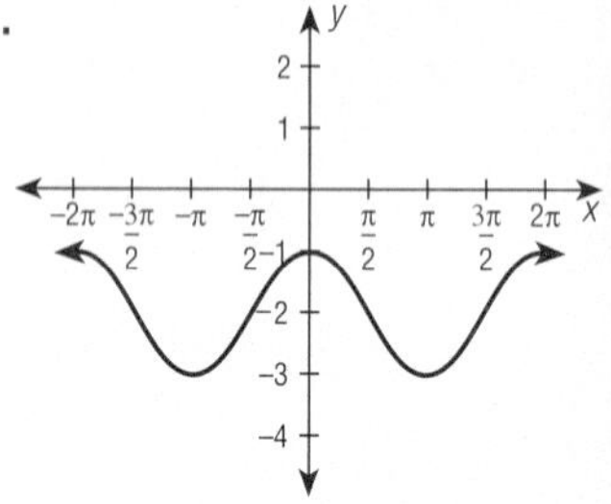

Cumulative REVIEW ▪ Chapters 13–14

In Exercises 1–4, find the complement and supplement of the angle.

1. $25°$ 65°, 155°
2. $89°$ 1°, 91°
3. $\frac{\pi}{12}$ $\frac{5\pi}{12}, \frac{11\pi}{12}$
4. $\frac{3\pi}{8}$ $\frac{\pi}{8}, \frac{5\pi}{8}$

In Exercises 5–8, find two angles, one with positive measure and the other with negative measure that are coterminal with the given angle.

5. $195°$ 555°, −165°
6. $-47°$ 313°, −407°
7. $\frac{4\pi}{3}$ $\frac{10\pi}{3}, -\frac{2\pi}{3}$
8. $-\frac{9\pi}{2}$ $\frac{3\pi}{2}, -\frac{5\pi}{2}$

In Exercises 9–12, evaluate the expression without using a calculator. For Exercises 11 and 12, write the answer in degrees and radians. Accept all equivalent answers.

9. $\csc \frac{13\pi}{4}$ $-\sqrt{2}$
10. $\cot\left(-\frac{\pi}{4}\right)$ −1
11. $\arccos \frac{\sqrt{3}}{2}$ 30°, $\frac{\pi}{6}$
12. $\arcsin\left(-\frac{1}{2}\right)$ −30°, $-\frac{\pi}{6}$

In Exercises 13–16, solve the triangle.

13. $A = 47°$, $B = 20°$, $c = 4$
14. $b = 4$, $c = 5$, $C = 110°$
15. $a = 6$, $b = 10$, $c = 7$
16. $a = 4$, $B = 100°$, $c = 6$

In Exercises 17 and 18, find the area of the triangle.

17. $a = 3$, $B = \frac{\pi}{6}$, $C = \frac{2\pi}{3}$ $\frac{9\sqrt{3}}{4}$
18. $a = 7$, $b = 4$, $c = 6$ ≈11.98

In Exercises 19–24, sketch the graph of the function.

19. $y = 4 + 3 \sin 3x$
20. $y = -\cos \frac{1}{2}x$
21. $y = -2 - \cos(x + \pi)$
22. $y = -4 \sin\left(x + \frac{\pi}{2}\right)$
23. $y = 3 + \cos(x - \pi)$
24. $y = -\sin 3x$

In Exercises 25–30, simplify the expression.

25. $\sin^2 x \sec^2\left(\frac{\pi}{2} - x\right) - \sin^2(-x)$ $\cos^2 x$
26. $2 \sin\left(\frac{x}{2}\right) \sin\left(\frac{\pi}{2} - x\right) \cos \frac{1}{2}x$ $\sin x \cos x$
27. $\left(\frac{\sin 4x}{1 + \cos 4x}\right)^2 + 1$ $\sec^2 2x$
28. $\tan(-x) + \tan x \sec^2 x$ $\tan^3 x$
29. $2 \sin x \cos^3 x - 2 \sin^3 x \cos x$ $\sin 2x \cos 2x$
30. $\frac{\cos^2 x}{\cos^2\left(\frac{\pi}{2} - x\right)} + 1 + \sin^2 x$ $\csc^2 x + \sin^2 x$

In Exercises 31–34, verify the identity.

31. $\frac{\sin\left(\frac{\pi}{2} - x\right)}{\sin x} = \cot x$
32. $2 \cos^2 x - 1 = 1 - 2 \sin^2 x$
33. $\csc 2x - \cot 2x = \tan x$
34. $\tan x \sec x - \csc x \sec^2 x = -\csc x$

In Exercises 35–38, solve the equation for $0 \le x < 2\pi$.

35. $3 \sin x = \sqrt{3} + 5 \sin x$ $\frac{4\pi}{3}, \frac{5\pi}{3}$
36. $\tan^2 x - \tan x - \sqrt{3} \tan x + \sqrt{3} = 0$ $\frac{\pi}{4}, \frac{\pi}{3}, \frac{5\pi}{4}, \frac{4\pi}{3}$
37. $4 \cos^2 x - 1 = 0$ $\frac{\pi}{3}, \frac{2\pi}{3}, \frac{4\pi}{3}, \frac{5\pi}{3}$
38. $4 \sin\left(\frac{\pi}{2} - x\right) = 3 \sec x$ $\frac{\pi}{6}, \frac{5\pi}{6}, \frac{7\pi}{6}, \frac{11\pi}{6}$

In Exercises 39–42, find the exact value of the expression. $\frac{(\sqrt{3}-1)^2}{2}$ or $\frac{\sqrt{3}-1}{\sqrt{3}+1}$

39. $\cos \frac{5\pi}{12}$ $\frac{\sqrt{6}-\sqrt{2}}{4}$

40. $\tan\left(\frac{\pi}{12}\right)$

41. $\sin 65° \cos 355° + \sin 355° \cos 65°$ $\frac{\sqrt{3}}{2}$

42. $\cos \frac{5\pi}{3} \cos \frac{3\pi}{2} - \sin \frac{5\pi}{3} \sin \frac{3\pi}{2}$ $\frac{-\sqrt{3}}{2}$

Shuffleboard **In Exercises 43 and 44, use the shuffleboard diagram.**

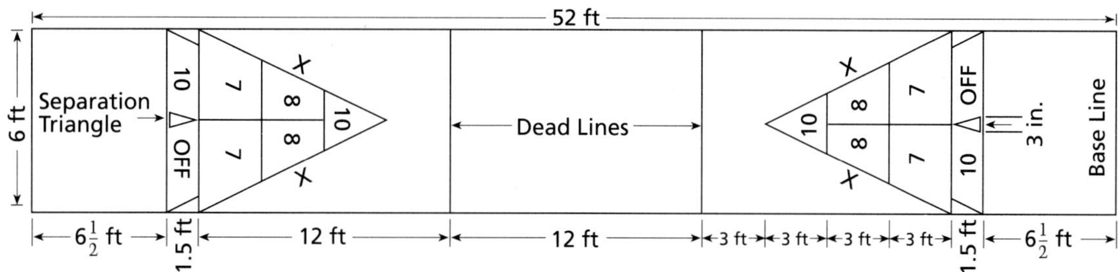

43. Determine the degree measure of the angles of the scoring triangles. ≈71.57°, ≈71.57°, ≈36.86°

44. Find the length, x (in feet), of the scoring triangles. ≈9.49 ft

Cloverleaf **In Exercises 45 and 46, use the following information.**

A car is traveling on a circular portion of a cloverleaf at a highway interchange. The radius of one ramp of the cloverleaf is 150 feet.

45. Through what angle has the car driven after the "inside tires" of the car have traveled 600 feet on the cloverleaf? (List your result in degrees.) 229.2°

46. The inside and outside tires of the car are 5 feet apart. How much farther did the outside tires of the car travel? 20 ft

Throwing a Softball **In Exercises 47 and 48, use the following equation for the path of a projectile. (In this equation, *y* is the height in feet, *x* is the horizontal distance in feet, *v* is the initial velocity, *h* is the initial height, and *θ* is the initial angle at which the projectile is launched.)**

$$y = -\frac{16}{v^2 \cos^2 \theta}x^2 + (\tan \theta)x + h$$

47. You and a teammate are warming up for a softball game. You throw the ball at an initial height of 6 feet, an initial velocity of 30 feet per second, and an angle of 30°. The ball lands at your teammate's feet. How far away is your teammate? ≈32 ft

48. Your teammate returns the ball, releasing it at a height of 6 feet with an initial velocity of 40 feet per second. You catch the ball 6 feet above the ground. At what angle was the ball thrown? ≈19.9°

Answers

22.

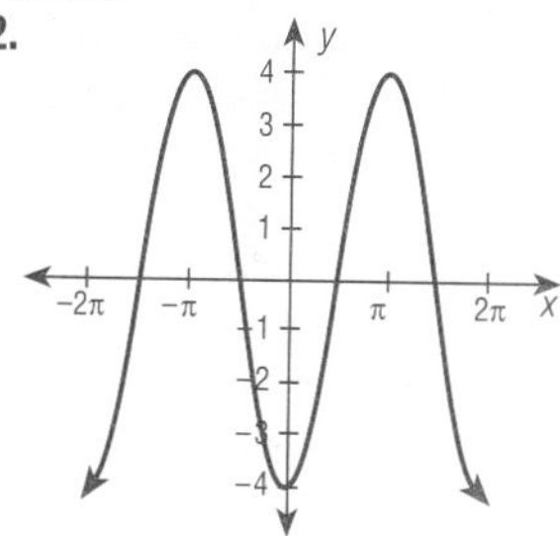

23.

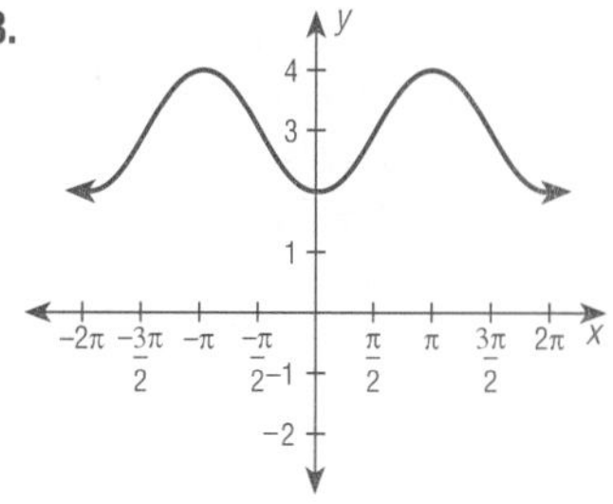

24.

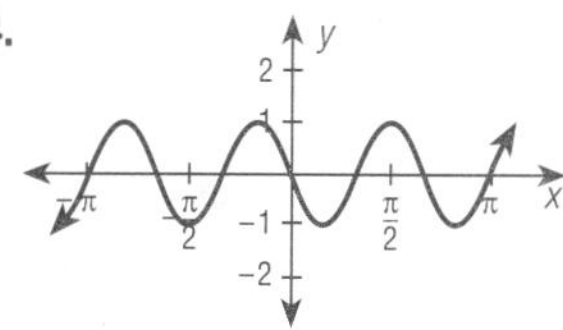

See Additional Answers for exercises 31–34.

In this chapter, students will learn to systematically find permutations and combinations of events in order to apply the theories of probability to real-life situations. Many real-life decisions can be expressed in terms of probabilities rather than certainties. Based on past performance, what are the odds that a certain team will win the game? Based on the polls, how likely is it that a candidate will win the election? What are the chances that a new drug will cure cancer or AIDS? How likely is the drug to have side effects? What are the chances that a machine will have a defective part?

COOPERATIVE LEARNING
Many of the probability applications in this chapter involve situations that are open to interpretation. You may wish to have students work with partners to consider the appropriate solution models.

The Chapter Summary on page 823 provides you and the students with a synopsis of the chapter. It identifies key skills and concepts. You may want to have students look at the Chapter Summary as an overview before beginning the chapter.

CHAPTER 15 Probability

LESSONS

The gleaming roof of white ceramic tiles belongs to the Sydney Opera House of Sydney, Australia, which was designed by the Danish architect, Jorn Utzon. Completed in 1973, it houses five auditoriums for the performing arts in a complex that covers $4\frac{1}{2}$ acres.

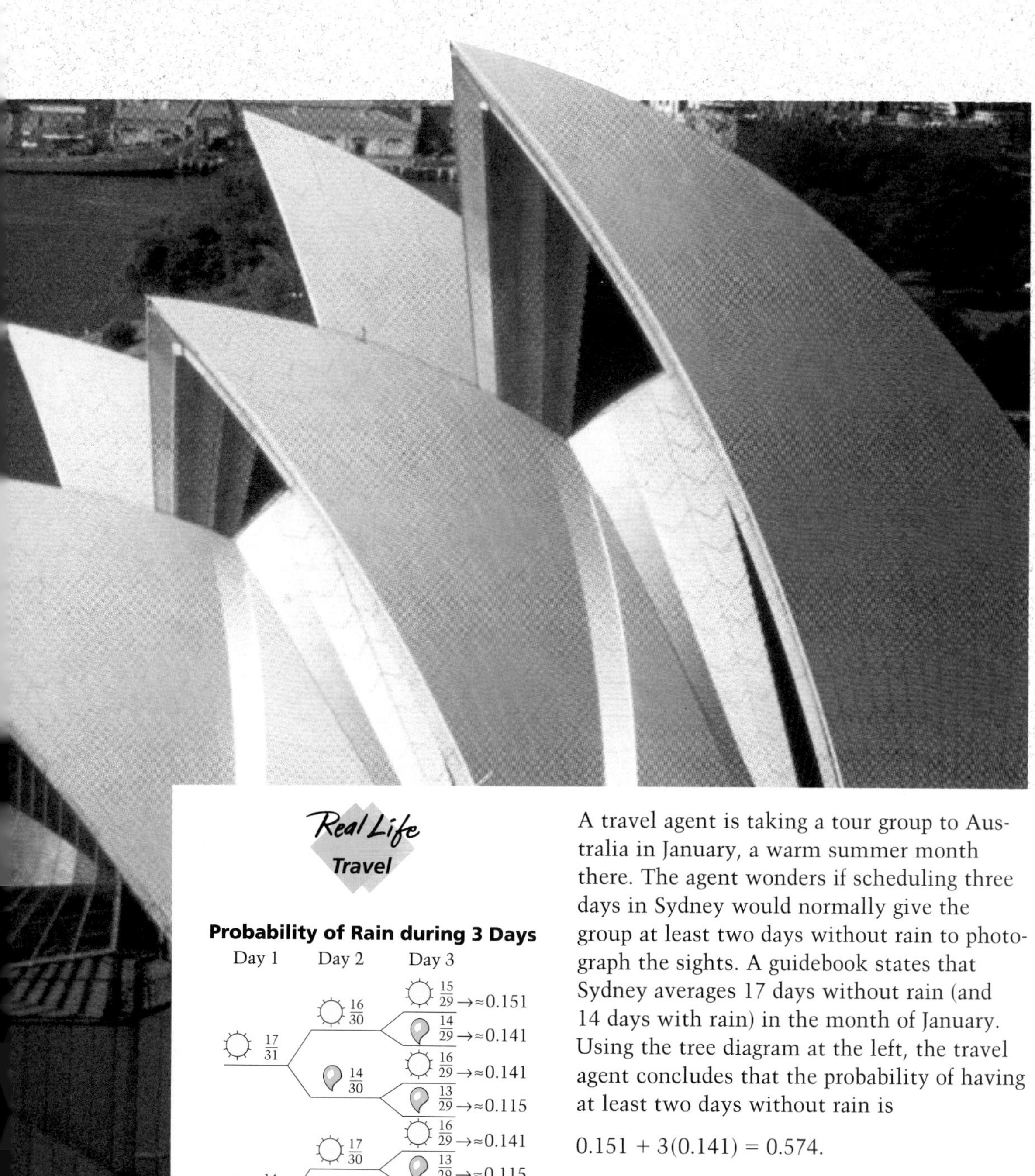

Real Life
Travel

Probability of Rain during 3 Days

A travel agent is taking a tour group to Australia in January, a warm summer month there. The agent wonders if scheduling three days in Sydney would normally give the group at least two days without rain to photograph the sights. A guidebook states that Sydney averages 17 days without rain (and 14 days with rain) in the month of January. Using the tree diagram at the left, the travel agent concludes that the probability of having at least two days without rain is

$$0.151 + 3(0.141) = 0.574.$$

Because this is greater than one half, the group should be able to expect at least two days without rain during their stay.

Games of chance have been played throughout history. However, it wasn't until the fifteenth century that mathematicians began to be concerned with the question of a fair distribution of winnings and with the odds of success in games of chance. Games of chance were very popular in the French court of King Louis XIV. In 1654, a member of the court asked the mathematician Blaise Pascal to consider the odds for a game of dice. The works of Pascal and his contemporary Pierre deFermat are considered to be the foundations of the mathematical theory of probability. Probability has many uses beyond games of chance. It has contributed to the solution of many problems in science and industry.

Students will probably have encountered tree diagrams in previous courses. To prepare them for some of the terminology that will be used in the chapter, have them use the tree diagram shown to answer questions such as the following.

a. What is the probability of having *exactly* two days without rain?
$0.141 + 0.141 + 0.141 = 0.423$

b. Is the probability of having no rain on the first two days the same as having no rain on the last two days?
yes; $0.151 + 0.141 = 0.292$; $0.151 + 0.141 = 0.292$

Probability and Statistics

PACING CHART

This chapter is an integral part of the full course for Algebra 2. However, if your class is taking the Basic Course, and you still have teaching days available for further topics, you may decide to teach some basic lessons on probability. Lessons 15.1 and 15.4 are recommended for this purpose.

DAY	FULL COURSE
1	15.1
2	15.2
3	15.2
4	15.3
5	15.3 & Using a Graphing Calculator
6	Mid-Chapter Self-Test
7	15.4
8	15.4
9	15.5
10	15.5
11	15.6
12	15.6
13	Chapter Review
14	Chapter Test

CHAPTER ORGANIZATION

LESSON	PAGES	GOALS	MEETING THE NCTM STANDARDS
15.1	782–787	1. Find the probability of an event 2. Use probability to answer questions about real-life	Problem Solving, Communication, Connections, Probability, Discrete Math
15.2	788–794	1. Use the Fundamental Counting Principle to find probabilities 2. Use permutations to find probabilities	Problem Solving, Communication, Reasoning, Connections, Probability, Discrete Math
Mixed Review	795	Review algebraic and arithmetic skills	
Career Interview	795	Illustrator	Connections
15.3	796–801	1. Use combinations to count the number of ways an event can happen 2. Use combinations to find probabilities	Problem Solving, Communication, Reasoning, Connections, Probability, Discrete Math
Using a Calculator	802–803	Experiment with probability	Technology
Mid-Chapter Self-Test	804	Diagnose student weaknesses and remediate with correlated Reteach worksheets	
15.4	805–810	1. Use unions to find probabilities 2. Use complements and intersections to find probabilities	Problem Solving, Communication, Reasoning, Connections, Probability, Discrete Math
15.5	811–816	1. Find the probability of independent events 2. Use the complement of the event to find the probability of the event	Problem Solving, Communication, Reasoning, Connections, Probability, Discrete Math
Mixed Review	816	Review algebraic and arithmetic skills	
15.6	817–822	1. Find the expected value of a sample space 2. Use expected value to answer questions about real-life situations	Problem Solving, Communication, Reasoning, Connections, Probability, Discrete Math
Chapter Summary	823	Restate for students what they have learned, why they have learned it, and how it fits into the structure of algebra	Structure, Connections
Chapter Review	824–827	Review concepts and skills learned in the chapter	
Chapter Test	828	Diagnose student weaknesses and remediate with correlated Reteaching worksheets	

LESSON RESOURCES

MEETING INDIVIDUAL NEEDS

RETEACHING For students who need to spend more time on basics:

If a mid-chapter self-test or chapter test indicates a deficiency, teachers can help students with the appropriate ***Reteaching Copymaster.***

PRACTICE For students who need more practice:

Additional exercises like those in the Pupil's Edition are provided for each lesson in ***Extra Practice Copymasters.***

ENRICHMENT For enriching and broadening students' experiences:

Problem of the Day copymasters in ***Teaching Tools*** provide a daily opportunity to use logical reasoning, looking for a pattern, writing an equation, and other routine and non-routine problem-solving strategies.

Math Log copymasters in ***Alternative Assessment*** provide opportunities to report on investigations, research, and open-ended problems.

Technology: Using Calculators and Computers provides enriching activities with graphing and scientific calculators and computers.

The ***Applications Handbook*** provides additional information about the cross-curriculum topics such as astronomy, chemistry, physics, sports, economics, genetics, and music that are integrated into the Pupil's Edition.

LESSON	15.1	15.2	15.3	15.4	15.5	15.6
PAGES	782-787	788-794	796-801	805-810	811-816	817-822
Teaching Tools						
Transparencies			✓			
Problem of the Day	✓	✓	✓	✓	✓	✓
Warm-up Exercises	✓	✓	✓	✓	✓	✓
Answer Masters	✓	✓	✓	✓	✓	✓
Extra Practice Copymasters	✓	✓	✓	✓	✓	✓
Reteaching Copymasters	Teacher-directed and independent activities tied to results on the Mid-Chapter Self-Tests and Chapter Tests					
Color Transparencies	✓	✓		✓		
Applications Handbook	Additional background information for many real-life applications					
Technology	Calculator and computer worksheets for appropriate lessons					
Complete Solutions Manual						
Alternative Assessment	Assess student's ability to reason, analyze, solve problems, and communicate using mathematical language.					
Formal Assessment	Mid-Chapter Self-Tests, Chapter Tests, Cumulative Tests, and Practice for College Entrance Tests					
Computer Test Bank	Customized tests can be created by choosing from over 2500 items.					

INSIGHTS

15.1 Introduction to Probability

It might be helpful, as you begin this chapter, to encourage students to think of probability formulas as another example of how algebra models the real world. They have already learned that models differ in complexity—linear models are relatively simple to identify and symbolize, as compared with rational models, for example. The first lesson explains how the probability of an event can be expressed as a decimal or as a fraction. Students are introduced to the concept of sample space, and to the basic probability ratio. A number of examples from real-life contexts explain informally how a probability ratio can be put together by interpreting the data. Note that in Example 3, and in several exercises, the sample space and the event are interpreted geometrically, in terms of area.

15.2 Counting Methods: Permutations

This lesson introduces students formally to the role of the Fundamental Counting Principle in computing probability, to the concept of counting with permutations (where the arrangement of elements in the '"event" is important), and to the role and significance of $n!$ ("n factorial"). These basic ideas will help students understand, for example, the constraints that apply to license-plate numbers, lock combinations, phone numbers, etc.

15.3 Counting Methods: Combinations

The process of counting events where the arrangement of elements is not important requires a different approach. This lesson introduces students to counting with combinations, and to the use and significance of the symbol $\binom{n}{m}$. Just as the use of graphing often gives a geometric interpretation to an algebraic concept, Exercises 9 and 10 should be a reminder that algebra often offers a quick solution to a geometry problem!

15.4 Probability, Unions, and Intersections

In this lesson, the probability formulas become more complex, as more complex sets of outcomes are modeled. This lesson makes use of set notation to model the probability of the union of two events occurring, such as the drawing of a heart *or* a face card from a deck of playing cards. Students are introduced to the algebraic model for this situation, and to the concepts of mutually exclusive events and the complement of an event. Finally the lesson explains how to model the probability of the intersection of two events occurring, such as the drawing of a card that is *both* a face card and a heart.

15.5 Independent Events

A common error, as the Chapter Summary points out, is to think that probabilities have "memories"—that if a coin has landed heads up, it must land tails up on the next toss! This lesson explains the significance of independent events such as successive coin tosses. In Example 1, independent events are compared with dependent events, and students are shown how to model both situations. Example 2 explains how the complement of an event is used in modeling probability.

15.6 Expected Value

Many real-life contexts, including sales, insurance, and contract bidding, make constant use of probability and expected value as an integral part of their financial management. This lesson explains the concept of the expected value of a sample space, and how to model it algebraically.

Problem of the Day

Eight steel balls, each with a radius of one inch, are melted down and remolded into one ball. What is its radius?
2 in.

ORGANIZER

Warm-Up Exercises

1. Write each fraction as a decimal and as a percent.
   a. $\frac{5}{6}$ ≈ 0.8333, $\approx 83.33\%$
   b. $\frac{5}{11}$ ≈ 0.4545, $\approx 45.45\%$
2. Find the area of the shaded region.

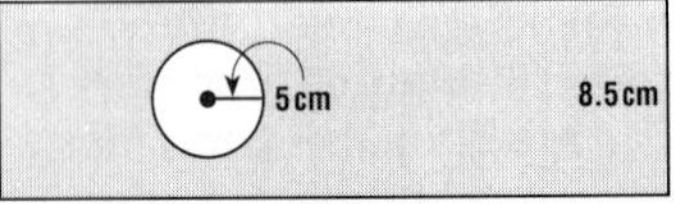

$(255 - 25\pi)$ cm^2, or ≈ 176.46 cm^2

Lesson Resources

Teaching Tools
- Problem of the Day: 15.1
- Warm-up Exercises: 15.1
- Answer Masters: 15.1

Extra Practice: 15.1
Color Transparency: 79
Applications Handbook: p. 13, 14

LESSON Notes

GROUP ACTIVITY Before presenting the theoretical model in Example 1, you might wish to perform this activity in class to provide students with an experimental model.

15.1 Introduction to Probability

What you should learn:

Goal 1 How to find the probability of an event

Goal 2 How to use probability to answer questions about real-life situations

Why you should learn it:

You can use probability to solve many real-life problems, such as those involving survey results or Punnett squares.

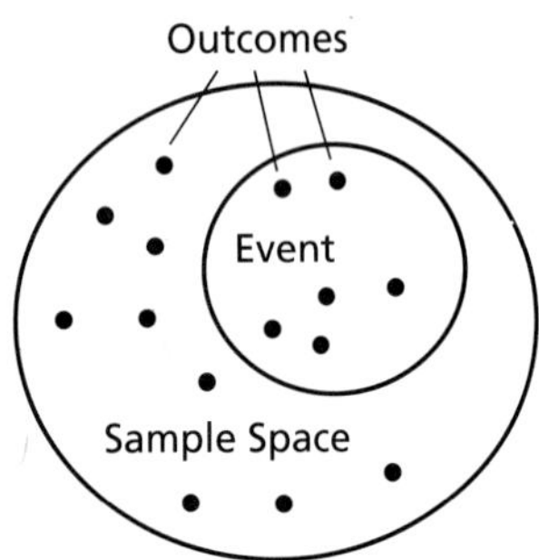

This event has six outcomes, and the sample space has 15 outcomes.

Goal 1 Finding the Probability of an Event

The **probability of an event** is a number between 0 and 1 that indicates the likelihood the event will occur. An event that is certain to occur has a probability of 1. An event that *cannot* occur has a probability of 0. An event that is *equally likely* to occur or not occur has a probability of $\frac{1}{2}$, or 0.5.

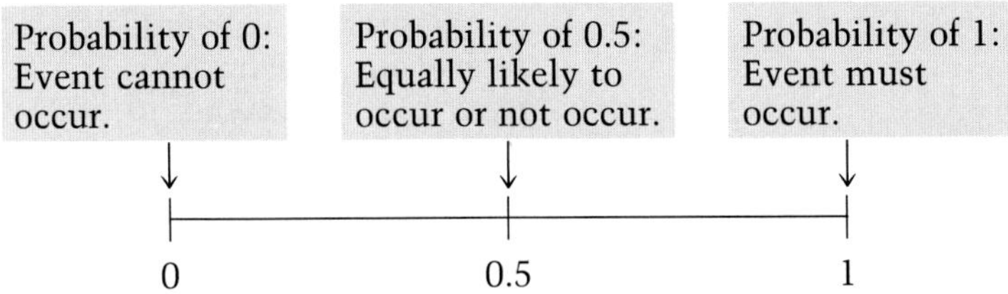

Finding the Probability of an Event

Consider a set, S, called a **sample space,** that is composed of a finite number of outcomes, each of which is equally likely to occur. A subset, E, of the sample space is an **event.** The probability, P, that an outcome in E will occur is the ratio of the number of outcomes in E to the number of outcomes in S.

$$P = \frac{\text{Number of outcomes in event}}{\text{Number of outcomes in sample space}}$$

Example 1 ***Finding the Probability of an Event***

a. You are dialing a friend's phone number, but can't remember the last digit. If you choose a digit at random, what is the probability that you dial the correct number?

$$P = \frac{\text{Number of correct digits}}{\text{Number of possible digits}} = \frac{1}{10}$$

b. On a multiple-choice test, you know the answer to Question 8 is not **a** or **d,** but you are not sure about **b, c,** and **e.** If you guess, what is the probability that you are *wrong?*

$$P = \frac{\text{Number of wrong answers}}{\text{Number of possible answers}} = \frac{2}{3}$$

■

Goal 2 Using Probability in Real-Life Situations

Real Life
Polling

Example 2 *Conducting a Poll*

In 1990, the Centers for Disease Control took a survey of 11,631 high school students. The students were asked whether they considered themselves to be at a good weight, underweight, or overweight. The results of the survey are shown at the left.

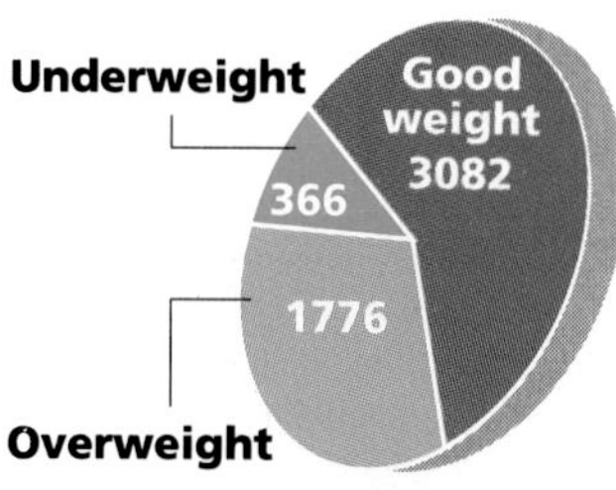

Female

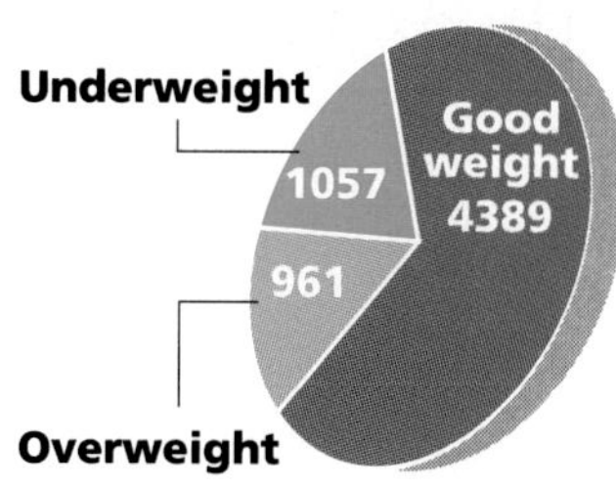

Male

a. If you choose a female at random from those surveyed, the probability that she said she was underweight is

$$P = \frac{\text{Number who answered "underweight"}}{\text{Number of females in survey}}$$

$$= \frac{366}{3082 + 366 + 1776}$$

$$= \frac{366}{5224} \approx 0.07.$$

b. If you choose a person who answered "underweight" from those surveyed, the probability that the person is female is

$$P = \frac{\text{Number of females who answered "underweight"}}{\text{Number in survey who answered "underweight"}}$$

$$= \frac{366}{366 + 1057} = \frac{366}{1423} \approx 0.26.$$ ■

Polls like the one described in Example 2 are often used to make inferences about a population that is larger than the sample. For instance, from Example 2, you might infer that 7% of *all* high school girls consider themselves to be underweight. When you make such an inference, it is important that those surveyed are representative of the entire population.

Connections
Geometry

Example 3 *Using Area to Find Probability*

You have just stepped into the tub to take a shower when one of your contact lenses falls out. (You have not yet turned on the shower.) Assuming that the lens is equally likely to land anywhere on the bottom of the tub, what is the probability that it landed in the drain? Use the dimensions shown at the left to answer the question.

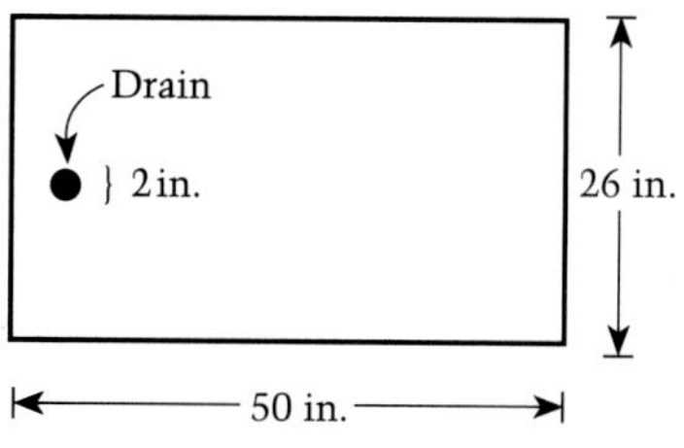

Solution Because the area of the tub bottom is $(26)(50) = 1300$ square inches and the area of the drain is $\pi(1^2) = \pi$ square inches, the probability that the lens landed in the drain is about

$$P = \frac{\pi}{1300} \approx 0.0024.$$ ■

An Experment

Have students form small groups. Ask students to discuss whether they think the ten digits, which might be the last digit of a telephone number, occur with the same frequency.

1. You or a student should record the groups' responses on the blackboard or overhead projector.
2. Then distribute pages of a telephone directory to each group. Have them pick a page at random and record the frequencies of the ten digits that occur as the last digit in a column of telephone numbers.
3. Have each group of students present their information to the class. Record the frequency of each digit for every group of students.
4. Discuss the terms *sample space*, *events*, and *outcomes*, and how to find the probability of an event. Ask students to find the probability for the occurrence of each digit. Use the results to answer these questions. Are there any digits that are more likely to occur than others? What is the sum of the probabilities in the sample space?
5. Summarize this experiment by checking the hypotheses that the students made before the experiment took place. How accurate were their initial guesses? Are any students surprised by the results of the experiment?

Example 1

When finding the probability of an event, students may find it helpful to write what the outcomes in the event are and what the outcomes in the sample space are, in the context of the problem. In Part a discuss with students why the number of correct digits is 1 and the number of possible digits is 10.

EXTEND Example 2 Have pairs of students decide on a topic and conduct class polls of their own. If students do not come up with topics, you might suggest that students determine the class favorite sport and athlete, or favorite type of music and recording artist. To motivate the students, you might wish to tell students that these polls could be used as extra credit for the students. Or you could display the polls on the bulletin board, or use them for test and quiz questions later.

Example 3

This example illustrates a probability application for geometric concepts such as area, surface area, and volume.

Example 4

COOPERATIVE LEARNING

Some students may have studied Punnett squares in biology. Pair those students with students who have not. Encourage the students to copy and complete the Punnett square for Example 4 before finding the probability asked for in this example.

Extra Examples

Here are additional examples similar to **Examples 1–4.**

1. You are dialing a friend's phone number. You know that the last digit is not a 0 or 1 but otherwise you can't remember the last digit. If you choose a digit at random, what is the probability that you dial the correct number? $\frac{1}{8}$
2. In Example 2, if you choose a male at random from those surveyed, find the probability that he said he was "overweight." $\frac{961}{6407}$

Example 4 *The Probability of Inheriting Genes*

Common parakeets have genes that can produce four feather colors: green (BBCC, BBCc, BbCC, or BbCc), blue (BBcc or Bbcc), yellow (bbCC or bbCc), or white (bbcc). Use the *Punnett square* below to find the probability that an offspring of two green parents (both with BbCc feather genes) will be yellow. Note that each parent passes along a B or b gene and a C or c gene.

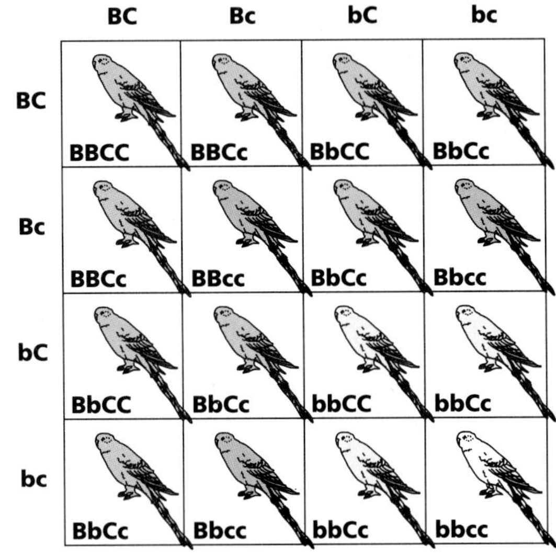

Solution $P = \frac{\text{Number of yellow possibilities}}{\text{Number of possibilities}} = \frac{3}{16}$ ■

Communicating about ALGEBRA

SHARING IDEAS about the Lesson See Additional Answers.

	Bc	bc	Bc	bc
bC	?	?	?	?
bc	?	?	?	?
bC	?	?	?	?
bc	?	?	?	?

A. In Example 1a, suppose you remembered that the last digit is even. If you choose an even digit at random, what is the probability that you dial the correct number? $\frac{1}{5}$

B. In Example 2, if you choose a person who answered "overweight" from those surveyed, what is the probability that the person is male? $\frac{961}{2737} \approx 0.35$

C. Use Example 4 as a model to copy and complete the Punnett square at the left for a blue parent (Bbcc) and a yellow parent (bbCc). For these parents, what is the probability that an offspring will be yellow?

EXERCISES

Guided Practice

CRITICAL THINKING about the Lesson 3. 7, any one of the 7 marbles may be drawn at random.

1. One of your friends says that there is a 110% chance of snow tomorrow. Why is this statement not mathematically correct?
2. Is it possible that the probability of an event is negative? No
3. *Marbles* A jar contains three red marbles and four blue marbles. One of the marbles is drawn at random. How many possible outcomes are in the sample space: two or seven? Explain.
4. *Picking a Pencil* There are six pencils, eraser end up, in a pencil can. Two are newly sharpened. Because all have the same length, you can't tell which are sharpened. If you choose one of the pencils, what is the probability that it is sharpened? $\frac{1}{3}$
5. *Punnett Square* Draw a Punnett square that represents the possible combinations of color genes for the offspring of a green parakeet (BBCc) and a yellow parakeet (bbCc). What is the probability that one of the offspring is green? $\frac{3}{4}$; see Additional Answers for Punnett square.

Independent Practice

In Exercises 6–9, use the sample space {1, 2, . . . , 20} to find the probability of the event.

6. An even number is chosen. $\frac{1}{2}$
7. A prime number is chosen. $\frac{2}{5}$
8. A perfect square is chosen. $\frac{1}{5}$
9. A number with at least four factors is chosen. $\frac{9}{20}$

Geometry **In Exercises 10–15, find the probability that the marble will come to rest in the indicated region.**

A marble is dropped into a large box whose base is painted different colors, as shown at the right. Because the base a is a horizontal surface, the marble has an equal likelihood of coming to rest at any point on the base.

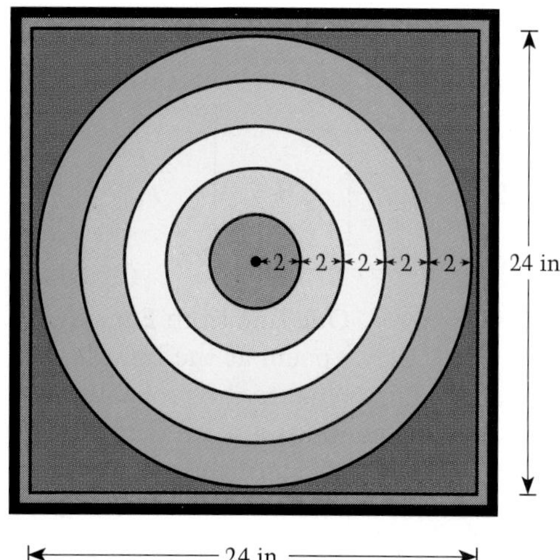

10. The red center $\frac{\pi}{144}$
11. The orange ring $\frac{\pi}{48}$
12. The yellow ring $\frac{5\pi}{144}$
13. The green ring $\frac{7\pi}{144}$
14. The blue ring $\frac{\pi}{16}$
15. The purple border $1 - \frac{25\pi}{144}$

1. The probability of an event always occurring is 1, or 100%.

3. In Example 3, what is the probability that the contact lens landed on the bottom of the tub? ≈0.9976
4. In Example 4, find the probability that an offspring of two green parents is green. $\frac{9}{16}$

Check Understanding

1. Define sample space, event, and probability of an event. See page 782.
2. What is the probability of an event that is equally as likely to occur as not to occur? Give an example of such an event.

$\frac{1}{2}$; answers may vary. One example is tossing a coin with the probability that it will land "heads up."

Communicating about ALGEBRA

These problems are a good summary of the lesson goals.

COOPERATIVE LEARNING

For Part C, pair students who have not studied Punnett squares in biology with those who have. Encourage students to work together as they copy and complete the Punnett square.

EXERCISE Notes

ASSIGNMENT GUIDE

Basic/Average: Ex. 1–5, 7, 10–21, 22–25, 28–30

Above Average: Ex. 1–5, 8–23, 26–30

Advanced: Ex. 1–5, 7–23, 25–30

Selected Answers
Exercises 1–5, 7–29 odd

✪ More Difficult Exercises
Exercises 16–23

Guided Practice

▶ **Ex. 1–5** Use these exercises for an in-class summary of the lesson. Many students will have taken biology. Ask them to write about or discuss other Punnett squares they may have encountered before.

Independent Practice

▶ **Ex. 10–23** Since answers can be written as fractions, rounded decimals, or percents, it is important that students know your expected format for their written answers. For example, the answer to Exercise 11 can be written as either $\frac{12\pi}{576}$, $\frac{\pi}{48}$, ≈0.0654, or ≈6.54%.

EXTEND **Ex. 19–20** Have students prepare interview questions for a biology teacher in your school to obtain uses of Punnett squares in human genetics.

Answers

19.

XX	XX
XY	XY

Each probability is $\frac{1}{2}$.

20.

AB	Bo
Ao	oo

A, B; $\frac{1}{4}$, $\frac{1}{4}$, $\frac{1}{4}$, $\frac{1}{4}$

16. *Picking a Penguin* Estimate the probability that a penguin that is chosen randomly in the photo will be an adult penguin. (The baby penguins are brown.) ≈0.19

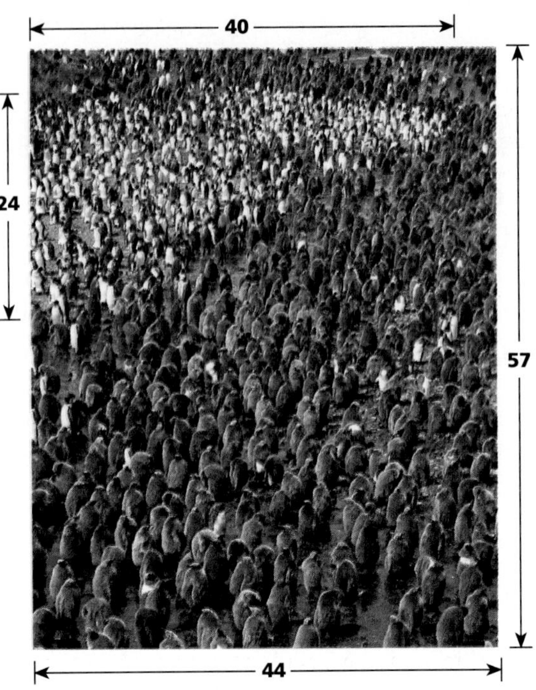

17. *Unemployed Workers* In October, 1991, of the 5,711,000 American workers who had recently left their jobs, 4,722,000 left because they had been laid off. If you randomly choose one of the workers who had recently left his or her job, what is the probability that the person left because he or she was laid off? ***(Source: Bureau of Labor Statistics)*** $\frac{4722}{5711}$

18. *Giving a Speech* You are sitting in English class, and your teacher is drawing names to see which student will give the first speech. There are 31 students in your class. What is the probability that the first name drawn will be yours? Four speeches are given each day. What is the probability that you will have to give your speech on the first day? $\frac{1}{31}$; $\frac{4}{31}$

19. *Girl or Boy?* The genes that determine the sex of humans are denoted as XX (female) and XY (male). Copy and complete the Punnett square below. Then use the result to explain why the probability that a newborn baby will be a boy or a girl is equally likely.

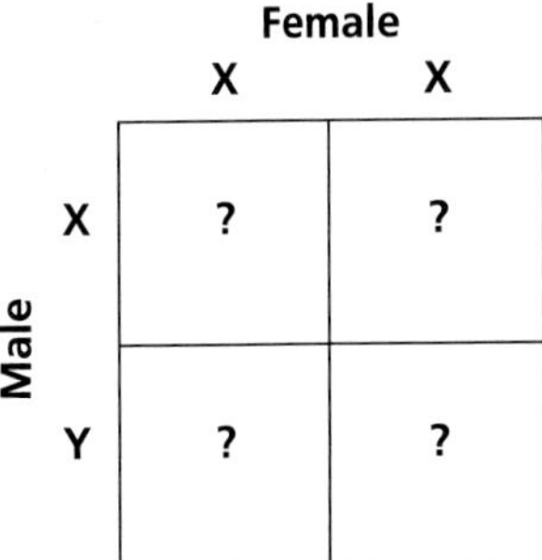

20. *Blood Types* There are four basic human blood types: A (AA or Ao), B (BB or Bo), AB (AB), and O (oo). Copy and complete the Punnett square below. What is the blood type of each parent? What is the probability that their offspring will have blood type A? B? AB? O?

	A	o
B	?	?
o	?	?

21. *Paris, France* One line from Edna St. Vincent Millay's poem at the right is chosen at random. What is the probability that the line contains a word with the letter "i"? What is the probability that the line contains a word with the letter "p"? $\frac{7}{8}$, $\frac{1}{4}$

Look, Edwin! Do you see that boy
Talking to the other boy?
No, over there by those two men—
Wait, don't look now—now look again.
No, not the one in navy-blue:
That's the one he's talking to.
Sure you see him? Stripéd pants?
Well, he was born in Paris, France.

United States Blood Types **In Exercises 22 and 23, use the following information.**

The circle graph at the right shows the number of people in the United States in 1990 with each blood type. ***(Source: American Association of Blood Banks)***

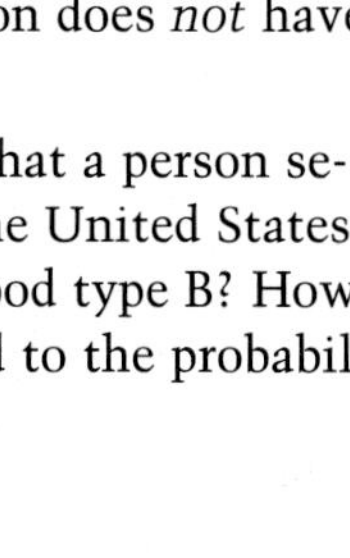

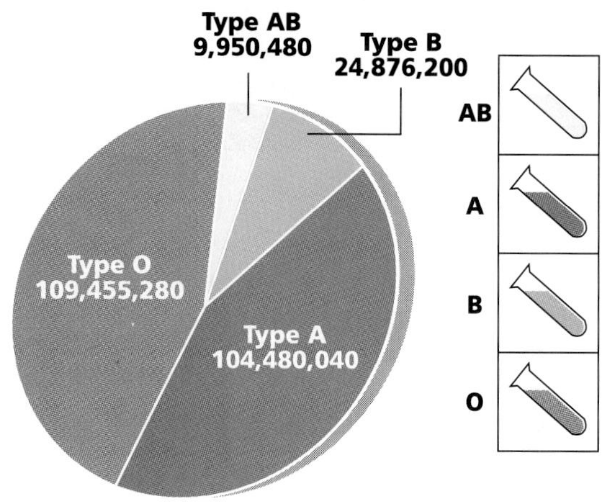

22. A person is selected at random from the United States population. What is the probability that the person does *not* have blood type B? $\frac{9}{10}$

23. What is the probability that a person selected at random from the United States population *does* have blood type B? How is this probability related to the probability found in Exercise 22?
$\frac{1}{10}$; $\frac{1}{10} = 1 - \frac{9}{10}$

Integrated Review

In Exercises 24–27, find the area of the shaded region.

24.

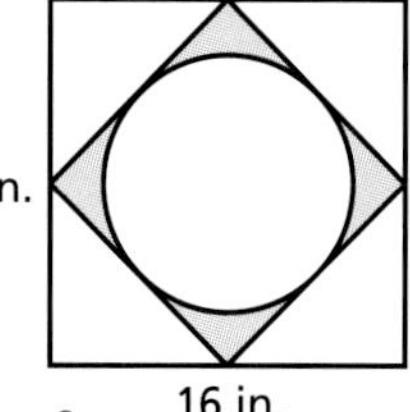

$100\sqrt{3} - 25\pi\ \text{cm}^2$

25.

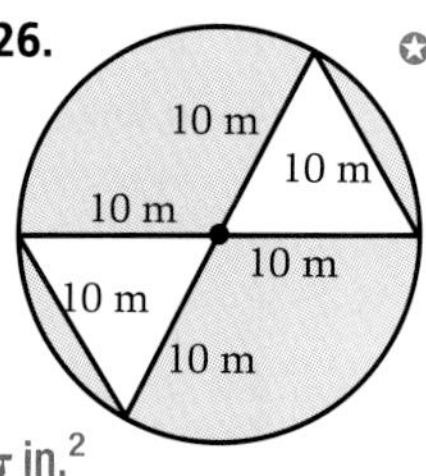

$128 - 32\pi\ \text{in.}^2$

26.

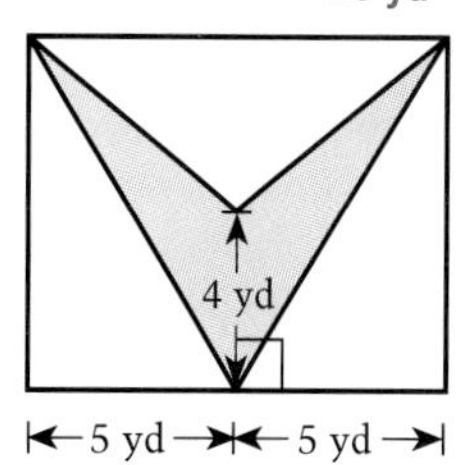

$100\pi - 50\sqrt{3}$

27. 4 yd; 5 yd; 5 yd

20 yd²

28. *College Entrance Exam Sample* In the figure at the right, what percent of the area of rectangle $PQRS$ is shaded blue? b

a. 20% **b.** 25% **c.** 30% **d.** $33\frac{1}{3}$% **e.** 40%

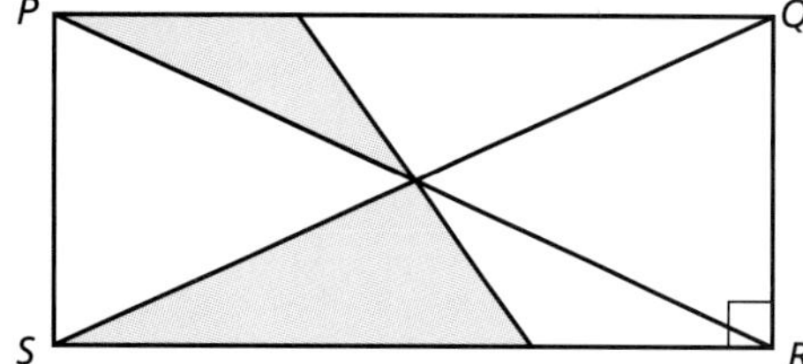

29. *College Entrance Exam Sample* How many of the whole numbers between 100 and 300 begin or end with 2? d

a. 20 **b.** 40 **c.** 100 **d.** 110 **e.** 180

Exploration and Extension

30. *Choose a Number* Studies have shown that when people are asked to choose one of the integers 1, 2, 3, or 4, more than half of the people choose the same number. Conduct this experiment by asking several people to choose one of the four integers. Which number was chosen most often? Use this result to explain why asking a person to choose a number is *not* the same as randomly selecting a number. (Experiment)

30. "Asking a person to choose a number" allows the person to know before the choice what the number will be; "randomly selecting a number" does not allow the person to know before the choice what the number will be.

▶ **Ex. 24–27** These exercises provide a review of finding areas of regions.

▶ **Ex. 28** This is an example of a probability question from the ACT exam.

▶ **Ex. 30** Overall, recent experiments have shown that the number 3 is chosen most often.

Enrichment Activities

These activities require students to go beyond lesson goals.

1. **EXTEND Example 2** How would you design a questionnaire that can be used to gather data in your school similar to that shown in Example 2 on page 783? Since teenagers can be very sensitive about their weight and attractiveness, students should agree that an anonymous survey would produce the best results. Students should also include questions that ask about the gender of the participant as well as age or grade.
2. How can you distribute the survey in order to obtain a fair sample of the entire population of the school? Students can distribute the questionnaire during the lunch periods or before school; or they can position tables in corridors or place a folder on each bulletin board of the school.
3. Conduct the survey. Compile the results. Compare the results of your survey with the results issued by the Center for Disease Control's 1990 survey. How alike are the results? Are there any significant differences? Can you think of any reasons for these differences?
4. The survey shown in the text was completed in 1990. Based on your data, could you predict that the results would be any different if the Center for Disease Control took the survey today? Why?

Problem of the Day

Each letter represents a different digit. Find the digits.

twenty	546250
twenty	546250
+ thirty	593750
sev enty	1686250

ORGANIZER

Warm-Up Exercises

1. Simplify.
   a. 0! b. $(2 \cdot 3)!$ c. 2!3!
   d. 2! + 3! e. $(2 + 3)!$
   a. 1, b. 720, c. 12, d. 8, e. 120
2. Stirling's Formula is given by $n! \approx \left(\frac{n}{e}\right)^n \cdot \sqrt{2\pi n}$ for large values of n. Use a calculator and Stirling's Formula to estimate the factorials listed below.
   a. 12! 475687486.5
   b. 30! $2.645 \cdot 10^{32}$
   c. 52! $8.053 \cdot 10^{67}$
3. TECHNOLOGY Use a calculator to compute the factorials. Compare the estimates you get with the computed values.

Lesson Resources

Teaching Tools
- Problem of the Day: 15.2
- Warm-up Exercises: 15.2
- Answer Masters: 15.2

Extra Practice: 15.2

Color Transparency: 80

15.2 Counting Methods: Permutations

What you should learn:

Goal 1 How to use the Fundamental Counting Principle to find probabilities

Goal 2 How to use permutations to find probabilities

Why you should learn it:

You can use the Fundamental Counting Principle to find many real-life probabilities, such as the probability of guessing a number correctly.

Goal 1 The Fundamental Counting Principle

From Lesson 15.1, you know that a basic skill needed to find probabilities is to be able to count the number of outcomes in an event and in a sample space. Sometimes this can be done by simply listing all possible outcomes. At other times, when there are too many outcomes to list, it is more efficient to use a formula to count the number of outcomes.

Almost all "counting formulas" are based on the following **Fundamental Counting Principle.**

Fundamental Counting Principle

If one event can occur in m ways and another event can occur in n ways, then the number of ways that both events can occur is equal to $m \cdot n$.

This principle can be extended to three or more events. For instance, if three events can occur in m, n, and k ways, then the number of ways that all three can occur is $m \cdot n \cdot k$.

Example 1 *Using the Fundamental Counting Principle*

How many different three-letter "words" from the English alphabet are possible?

Solution This experiment has three events. The first event is the choice of the first letter, the second event is the choice of the second letter, and the third event is the choice of the third letter. Because the English alphabet contains 26 letters, it follows that each event can occur in 26 ways.

Three-letter "words": □ □ □
26 26 26

Using the Fundamental Counting Principle, it follows that the number of three-letter "words" is $26 \cdot 26 \cdot 26 = 17{,}576$. ■

Real Life
Word Processing

A gnu, or wildebeest

Example 2 *Finding a Probability*

You are using a word processor to prepare a term paper. The word processor has a spelling checker that contains all commonly used English words, including 522 three-letter words. You are unsure of the correct spelling for *gnu*, so you type three letters at random. What is the probability that the spelling checker will identify this "word" as misspelled?

Solution From Example 1, you know there are 17,576 possible three-letter "words." Of these, $17{,}576 - 522 = 17{,}054$ would be flagged as incorrect by the spelling checker. Therefore, the probability that your word will be flagged is

$$\frac{\text{Number of incorrect words}}{\text{Number of possible words}} = \frac{17{,}054}{17{,}576} \approx 0.970 \text{ or } 97\%.$$

■

Real Life
Phone Systems

Example 3 *Applying the Fundamental Counting Principle*

Telephone numbers in the United States have ten digits. The first three form the *area code* and the next seven form the *local telephone number.* What is the maximum number of local telephone numbers that are possible within each of the states shown in blue on the map? (A local telephone number cannot have 0 or 1 as its first or second digit.)

Solution There are only eight choices for the first two digits, because neither can be 0 or 1. For each of the other digits, there are ten choices.

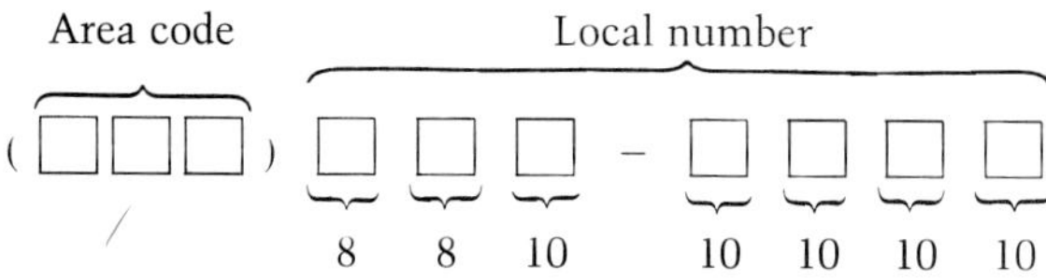

Using the Fundamental Counting Principle, the number of telephone numbers possible within each area code is

$$8 \cdot 8 \cdot 10 \cdot 10 \cdot 10 \cdot 10 \cdot 10 = 6{,}400{,}000.$$

Each of these states had, at most, 6.4 million phone numbers in 1990. ■

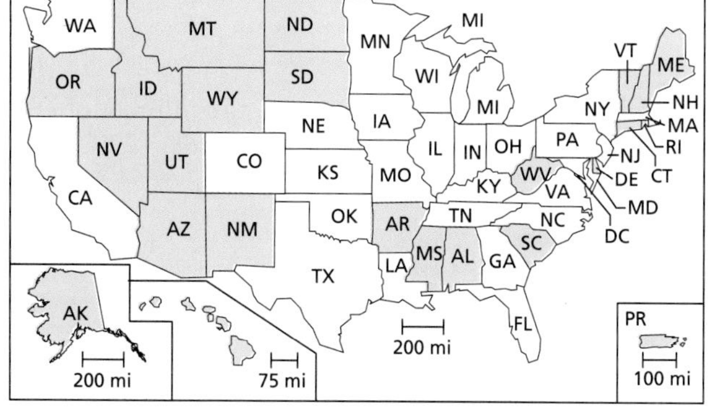

In 1990, the 23 states that are shown in blue on the map had a single area code for all telephone numbers in the state. The District of Columbia and Puerto Rico also have a single area code.

LESSON Notes

Example 1

This example provides students with a solution model for applying the Fundamental Counting Principle. Students will find it helpful to draw and label the boxes as a concrete representation of the problems in this lesson.

Example 2

In this example, the first step of the solution is given in Example 1. Students should note that the probability expression given on page 782 was rewritten in terms that are more meaningful for the sample space in this particular problem. For example, the "Number of outcomes in event" is written as "Number of incorrect words."

Example 3

As in Example 1, students are being given a concrete representation for the problem.

EXTEND Example 3 You could use Extra Example 2 shown on page 791 as an extension for this Example. Students may be interested in the fact that if the number of choices for the second digit is increased to 9, then 800,000 new numbers can be added to the phone system.

Goal 2 Using Permutations to Find Probabilities

The Fundamental Counting Principle can be used to determine the number of ways that n elements can be arranged *in order.* Such an ordering of n elements is a **permutation** of the elements.

Example 4 *Listing Permutations*

Write all the permutations of the letters A, B, and C.

Solution The three letters can be arranged in six ways.

ABC, ACB, BAC, BCA, CAB, CBA ■

In Example 4, the number of permutations could also have been found using the Fundamental Counting Principle. To do this, note that there are three choices for the first letter, two choices for the second, and only one choice for the third.

Permutations of three letters

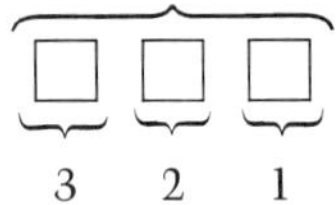

Thus the number of permutations is $3 \cdot 2 \cdot 1 = 3! = 6$.

Example 5 *Finding the Number of Permutations of n Different Elements*

How many permutations of the letters A, B, C, D, E, and F are possible?

Solution There are six choices for the first letter, five for the second, four for the third, and so on.

Permutations of six letters

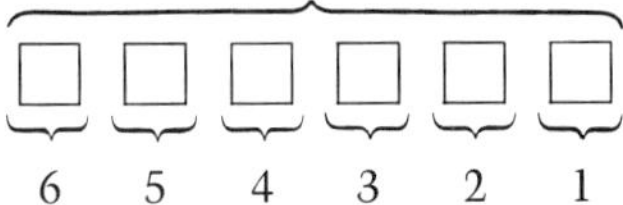

Using the Fundamental Counting Principle, the number of permutations of the six letters is $6 \cdot 5 \cdot 4 \cdot 3 \cdot 2 \cdot 1 = 6! = 720$. ■

Number of Permutations of n Elements

The number of permutations of n elements is given by

$$n! = n \cdot (n-1) \cdot \cdots \cdot 4 \cdot 3 \cdot 2 \cdot 1.$$

Examples 4–5

In a permutation, there is a definite order to the elements in the sample space.

Before presenting these examples and the formula for the Number of Permutations of n Elements at the bottom of the page, you could conduct an experiment as shown below. After completing the experiment, students should be able to state the formula for the number of ways n elements can be arranged in order.

GROUP ACTIVITY Have four volunteers stand together where everyone in the room can see them. Ask the class, "In how many ways can these students stand in a straight line?" Through discussion, lead the students to understand that the order Amy, Charles, Jose, Tawanda is different from the order Jose, Tawanda, Charles, Amy. There are 4 ways in which to pick a person for the head of the line. Once the first place has been filled, then there are 3 ways to pick a person to be second, 2 ways to pick a person to be third, and so on. In all, there are $4 \cdot 3 \cdot 2 \cdot 1$, or 24, ways in which the four students can stand in a straight line.

Example 6

This is a good illustration of how permutations are used to find the number of outcomes in a sample space. Students may be surprised by the answer to this problem.

Extra Examples

Here are additional examples similar to **Examples 1–6.**

1. Suppose that you decide to buy a car. Among the options you may choose from are 6 paint colors (red, yellow, blue, brown, black, white), 3 interior colors (brown, black, white), and 2 transmissions (manual, automatic).

Real Life
Test Taking

First Ten Elements of the Periodic Table

Name	Symbol
Beryllium	Be
Boron	B
Carbon	C
Fluorine	F
Helium	He
Hydrogen	H
Lithium	Li
Neon	Ne
Nitrogen	N
Oxygen	O

Example 6 *Probability of Guessing Correctly*

You are taking a chemistry test and are asked to list the first ten elements *in order* as they appear in the periodic table of elements. Suppose you have no idea of the correct order and simply guess. What is the probability that you guess correctly?

Solution You have ten choices for the first element, nine choices for the second, eight choices for the third, and so on.

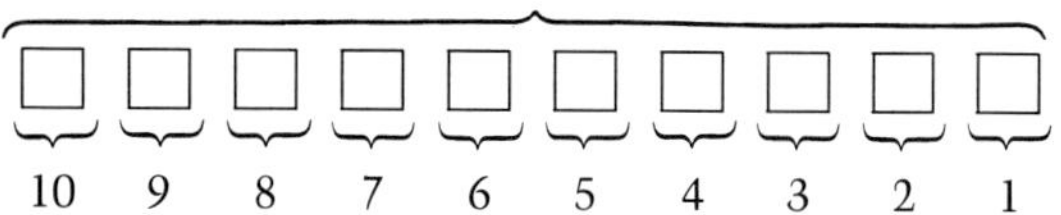

The number of different orders is $10! = 3,628,800$, so your probability of guessing correctly is $P = \frac{1}{3,628,800}$. ■

Communicating about ALGEBRA

SHARING IDEAS about the Lesson

The diagram below represents the 36 possible outcomes when rolling two six-sided dice.

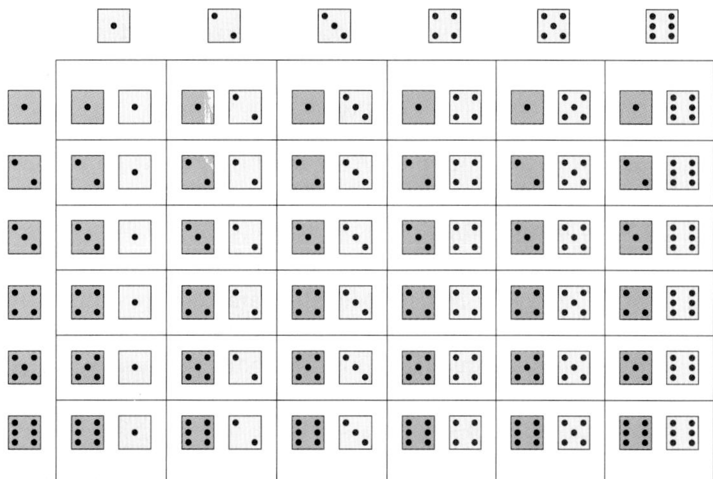

A. What is the probability that the total is 7? $\frac{1}{6}$

B. Roll a pair of dice 90 times and record each total. Use the results to estimate the probability of rolling a 7. How closely does the probability agree with the probability you found in part A? Depends on experiment.

a. If a car dealer has, in the parking lot, one of every type of car made using these options, how many cars are in the parking lot? 36

b. How many cars of each color are in the parking lot? 6

c. What is the probability that if you choose a car at random, the car will be red? $\frac{1}{6}$

2. The Commonwealth of Massachusetts has three area codes: one for the Boston metropolitan area, one for the rest of eastern Massachusetts, and one for western Massachusetts. Using the information about telephone numbers in Example 3, how many telephone numbers can the phone system accommodate in the state of Massachusetts? 19,200,000

3. Suppose you go out to a movie with six of your friends. In how many ways can you form a line at the ticket booth? 7! or 5040 ways

Check Understanding

1. What is the Fundamental Counting Principle? See page 788.
2. How do you find the number of permutations of n elements? See the bottom of page 790.

Communicating about ALGEBRA

This activity is a good way to illustrate the difference between the theoretical and the empirical (experimental) studies of probability. To find the solution to Part A, the theoretical definition of probability must be used. In Part B, an experiment is used to estimate the probability of rolling a 7.

EXERCISE Notes

ASSIGNMENT GUIDE

Basic/Average: Ex. 1–6, 7, 11, 14, 15, 16, 17, 23, 24, 26–34

Above Average: Ex. 1–6, 9–13 odd, 14–20, 23–26, 32–34

Advanced: Ex. 1–6, 9–13 odd, 14–26, 32–34

Selected Answers
Exercises 1–6, 7–33 odd

Use **Mixed Review** as needed.

✪ More Difficult Exercises
Exercises 13, 18–22, 25, 34

Guided Practice

▶ **Ex. 1–6** COOPERATIVE LEARNING Have students work with partners to complete these Guided Practice exercises. Use the exercises to make the connection between the Fundamental Counting Principle and the concept of permutation.

▶ **Ex. 2** Encourage the students to think of 001, 002, 003, etc. as three-digit numbers.

Answers

1. The number of ways 2 or more events can occur.
2. There are 10 choices for the first digit, 10 choices for the second digit, and 10 choices for the third digit. So, $10 \cdot 10 \cdot 10 = 1000$.
3. Consider each toss an event that has 2 possible choices—heads or tails. Then the number of ways the coin can land is $2 \cdot 2 \cdot 2 \cdot 2 \cdot 2 \cdot 2 = 2^6$. No.

EXERCISES

Guided Practice

CRITICAL THINKING about the Lesson

1. When you use the Fundamental Counting Principle, what are you counting?
2. Use the Fundamental Counting Principle to explain why there are 1000 three-digit numbers from 000 to 999.
3. A coin is tossed six times. Explain why the number of ways the coin can land heads up or tails up is 2^6. Is this an example of a permutation? (*Hint:* Three ways the coin can land are HHHHHH, THHHHH, and HTHHHH.)
4. Find the number of permutations of the letters ARST. How many of these are English words? If one of the permutations is chosen at random, what is the probability that it is an English word? 24; 5, STAR, ARTS, RATS, TARS, and TSAR; $\frac{5}{24}$
5. In how many orders can six people be arranged in a line? 720
6. How many three-digit numbers have only even digits? 125

Independent Practice

In Exercises 7–10, find the number of permutations of the given letters. Then, find the number of "words" of the same length that can be formed with the given letters if you are allowed to repeat letters. (For instance, from STOP, you could form SSSS, SSST, SSTT, and so on.)

7. STOP 24; 256
8. SMILE 120; 3125
9. PENCIL 720; 46,656
10. FORMULA 5040; 823,543

11. *Taking a Trip* Four people are taking a long trip in a car. Two sit in the front seat and two in the back seat. Three of the people agree to share the driving. In how many different arrangements can the four people sit? 18
12. *Morse Code* In Morse code, all characters are transmitted using a sequence of *dots* and *dashes.* How many different characters can be formed with a sequence of four symbols, each of which is a dot or a dash? How many can be formed with a sequence of one, two, or three symbols? 16; 2, 4, 8

✪ 13. *Holiday Lights* You are decorating the living room window with a string of holiday lights. You have four red, four blue, four green, four yellow, and four white lights. In how many different patterns can you string the lights so that matching colors are five lights apart? 120

14. *Batting Order* A baseball coach is determining the batting order for the team. The team has nine members, but the coach does not want the pitcher to be one of the first four to bat. How many batting orders are possible? 201,600

15. *School Lunch* Your school cafeteria offers three salads, four main courses, two vegetables, and three desserts. How many different combinations of a salad, main course, vegetable, and dessert are possible? 72

16. *Combination Lock* You have forgotten the combination of the lock on your school locker. There are 40 numbers on the lock, and the correct combination is "R☐ – L☐ – R☐." How many combinations may it be necessary to try in order to get the correct combination? (Numbers may be repeated, such as 8-8-8.) 64,000

17. *Combination Lock* In Exercise 16, suppose you remember that the first part of the combination is "Right 16." Now how many combinations may it be necessary to try in order to get the correct combination? 1600

18. *Scheduling Classes* Next year you are taking math, English, history, keyboarding, chemistry, physics, and physical education. Each class is offered during each of the seven periods in the day. In how many different orders can you schedule your classes? 5040

19. *Scheduling Classes* Your best friend is taking four of the seven courses you are taking. She has already scheduled her classes. In how many ways can you schedule your classes so that you and your friend share four classes? (Assume that each of the nonshared classes is offered during each of the periods in the day.) 6

Boston Computer Museum **In Exercises 20–22, use the following information.**

You visited the Boston Computer Museum and bought four different robots—one for each of your four cousins. When you bought the robots, you intended a specific one for each cousin. After wrapping the packages, however, you realize that you don't know which robot is in which package. You don't want to unwrap the packages, so you randomly assign names to the packages.

20. What is the probability that all four cousins receive the robot you intended? $\frac{1}{24}$

21. What is the probability that exactly three receive the robot you intended? 0

22. What is the probability that exactly two receive the robot you intended? Exactly one? None? $\frac{1}{4}, \frac{1}{3}, \frac{3}{8}$

Independent Practice

Ex. 7–19 Students should use the solution format modeled in Examples 3–5. The number of available elements can be listed in each box before applying the Fundamental Counting Principle. For example, in Exercise 7, the solution would look like the following.

4	3	2	1	= 24

4	4	4	4	= 256

Ex. 21–22 The phrases "exactly three" or "exactly two" may create difficulty for students when finding the probability. Help students think through the choices before you assign these problems for homework.

Ex. 34 This exercise requires students to find the product of three numbers of permutations.

Enrichment Activities

These activities require students to go beyond lesson goals.

The following example shows the number of permutations, or orders, that are possible when all the elements of a set are not used.

Suppose you own a bookstore. From a box of nine new titles, you want to select three to arrange in a window display. The number of possible arrangements is $\boxed{9}\boxed{8}\boxed{7} =$ 504. Another way to express this product is $\frac{9 \cdot 8 \cdot 7 \cdot 6 \cdot 5 \cdot 4 \cdot 3 \cdot 2 \cdot 1}{6 \cdot 5 \cdot 4 \cdot 3 \cdot 2 \cdot 1} = \frac{9!}{6!}$. This is an example of the permutation of 9 objects taken 3 at a time.

In general, the number of permutations of n objects taken m at a time is given by the expression

$$\frac{n!}{(n-m)!}.$$

1. In how many ways can 3 playing cards be laid in a row in order face up from a deck of 52 cards?
   $52 \cdot 51 \cdot 50 = 132{,}600$
2. In how many ways can the letters of the word MATH be arranged in order using only two letters?
   $4 \cdot 3 = 12$
3. In how many ways can three cheerleaders be arranged in order for a picture from a group of twelve cheerleaders?
   $12 \cdot 11 \cdot 10 = 1320$

23. *Sydney Opera House* You are attending a concert at the Sydney Opera House. As you are walking in, you see a friend you haven't seen in years. Your friend is attending the same concert. What is the probability that your friend will be sitting in one of the two seats beside you? $\frac{1}{773}$

The Opera House in Sydney, Australia, has a seating capacity of 1547.

24. *Car License Number* You witnessed a hit-and-run automobile accident. You remember the license number of the automobile, except for the last two digits. You do remember that the license is from your state, that the last two symbols are digits (not letters of the alphabet), and that they are *not* the same. With this information, how many automobile license numbers would the police have to check? 90

✪ **25.** *Mobile Home* You are buying a new mobile home. You have seven choices of carpet and four choices of wall paneling for each of six rooms. In how many different ways can you choose carpeting and paneling for the six rooms? 28^6 or 481,890,304

Integrated Review

In Exercises 26–31, evaluate the expression.

26. 4! 24	**27.** 3! 6	**28.** 6! 720
29. 7! 5040	**30.** 5! 120	**31.** 8! 40320

Geometry **In Exercises 32 and 33, use the figure at the right.**

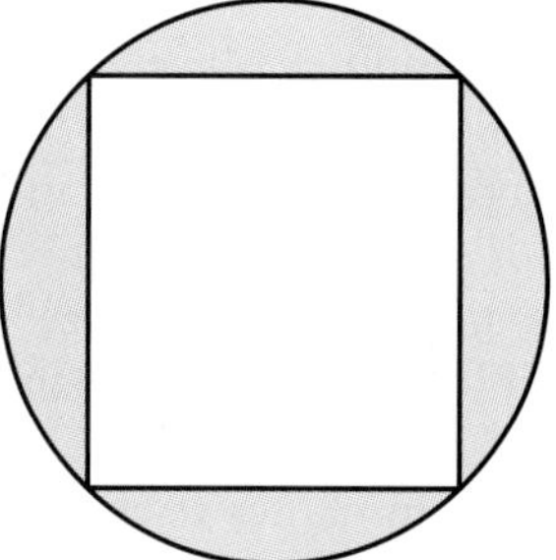

32. What is the probability that a point chosen randomly inside the circle will also be inside the square? $\frac{2}{\pi}$

33. What is the probability that a point chosen randomly inside the square will also be inside the circle? 1

Exploration and Extension

✪ **34.** *Dance Recital* For a dance recital, five beginner groups, eight intermediate groups, and six advanced groups are to perform. The director wants to set up the program so that all the beginner groups perform first, then the intermediate groups, then the advanced groups. How many orders are possible? 3,483,648,000

Mixed REVIEW

1. $a = \pm\sqrt{\frac{bc - c}{3b}}$ 3. $\frac{3\pi}{8}$ 4. $(x - 3)(x^2 + 4)$ 10. $x^3 + x + 2$
6., 18. See Additional Answers.

1. Solve $3a^2b + c = bc$ for a. **(1.5)**
2. Solve $25 - 3t < 20$. **(1.6)** $t > \frac{5}{3}$
3. Find the reference angle of $\frac{11}{8}\pi$. **(13.2)**
4. Factor $x^3 - 3x^2 + 4x - 12$. **(9.3)**
5. Solve $|4 - t| \geq 10$. **(1.7)** $t \leq -6$ or $t \geq 14$
6. Sketch $y = |4x| + 7$. **(2.6)**
7. Complete the square: $x^2 - x + 9$. **(5.3)** 7., 9. See margin.
8. Solve $\begin{cases} 5x - y = -14 \\ -5x + 2y = 13 \end{cases}$. **(3.2)** $(-3, -1)$
9. Find the inverse of $y = 4x^2 - \frac{1}{4}$. **(6.3)**
10. Add: $(x^2 + x) + (x^3 - x^2 + 2)$. **(9.1)**
11. Solve $\sqrt[3]{27r} + 15 = 0$. **(7.5)** -125
12. Find the amplitude of $y = 3 \sin 2x$. **(14.1)** 3
13. Classify: $4x^2 - 4x + 4y^2 + 2y = 8$. **(11.5)** Circle
14. Find the mean: 17, 21, 6, 15, 17, 19. **(6.7)** $15\frac{5}{6}$
15. Simplify: $(3x^0b^{-1}) \cdot \left(\frac{1}{6}x^2y^{-2}\right) \div b^2$. **(7.1)** $\frac{x^2}{2b^3y^2}$
16. Find the period of $y = \cos 2\pi\theta$. **(14.1)** 1
17. Solve $\log_2 16 = x$. **(8.6)** 4
18. Sketch $y = 2 \sin(3x - 2)$. **(14.1)**
19. Write an equation of the vertical line passing through the point $\left(\frac{3}{4}, \frac{1}{2}\right)$. **(2.2)** $x = \frac{3}{4}$
20. Find $g(f(x))$ for $g(x) = x - 2$ and $f(x) = x^3$. **(6.2)** $x^3 - 2$

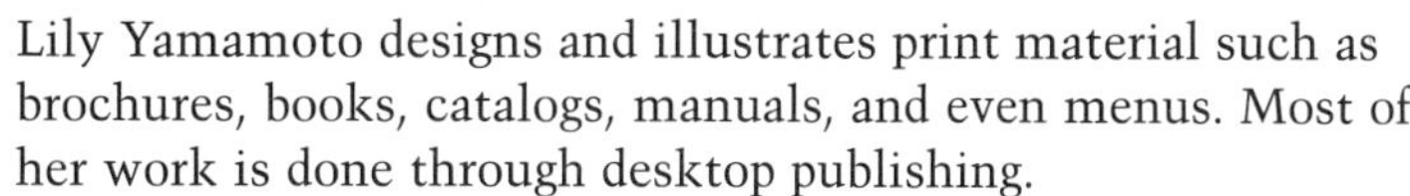

Career Interview

Illustrator

Lily Yamamoto designs and illustrates print material such as brochures, books, catalogs, manuals, and even menus. Most of her work is done through desktop publishing.

Q: ***What led you into this career?***

A: I was attracted to my career by my love for art, particularly with drawing, painting, visual communication, and problem solving.

Q: ***Does having a math background help you on the job?***

A: To create different colors, I need to combine percentages of cyan, magenta, yellow, and black. To position art, I need to consider proportion in relation to space and text. I am constantly converting between picas, a unit of measure used in publishing, and inches. I knew math was needed to do daily tasks, like banking and shopping, but was I surprised to discover how much math can be applied to graphic design.

Q: ***Do you use calculators?***

A: Complicated resizing of art is usually done first with a calculator. However, you still need to understand math in order to do it correctly. Also, understanding the math allows me to compute mentally when I do not have the luxury of time.

Answers

Mixed Review

7. $\left(x - \frac{1}{2}\right)^2 + \frac{35}{4}$

9. $y = \pm\frac{\sqrt{4x + 1}}{4}$

Problem of the Day

You have 50,000 names in a list and are trying to locate one of them. You cut the list in half as nearly as possible and keep the half that contains the name. If you continue this process, what is the greatest number of times you will need to halve the list?
16

ORGANIZER

Warm-Up Exercises

1. How many permutations are there of 3 people selected from a group of 10 people?
$10 \cdot 9 \cdot 8 = 720$
2. Does the situation involve a permutation?
   a. Selecting 4 students to complete a 9-member committee
   b. Selecting 4 students to fill the vacant positions on a baseball team
   c. Selecting the answers to a true-false test
   d. Selecting a hand of 7 cards from a deck of playing cards

   a. no, b. yes, c. yes, d. no

Lesson Resources

Teaching Tools
- Transparencies: 2, 3, 24
- Problem of the Day: 15.3
- Warm-up Exercises: 15.3
- Answer Masters: 15.3

Extra Practice: 15.3
Technology Handbook: p. 101

15.3 Counting Methods: Combinations

What you should learn:

Goal 1 How to use combinations to count the number of ways an event can happen

Goal 2 How to use combinations to find probabilities

Why you should learn it:

You can solve many real-life probability problems using combinations, such as the probability that you will be dealt a particular five-card hand.

Goal 1 Counting Combinations

For some counting problems, order is important. For instance, when counting phone numbers, 454-5526 and 454-5562 are not the same. These are called permutations. For other counting problems, order is not important. For instance, in most card games, the order in which your cards are dealt is not important. Reordering the cards that are dealt does not change your "card hand." These are called combinations.

Choosing m elements from a set that has n elements is called forming a **combination of n elements taken m at a time.**

Example 1 ***Combination of n Elements Taken m at a Time***

In how many ways can three letters be chosen from the letters A, B, C, D, and E? Their order is not important.

Solution The following subsets represent the combinations of three letters that can be chosen from five letters.

{A, B, C} {A, B, D} {A, B, E} {A, C, D} {A, C, E}
{A, D, E} {B, C, D} {B, C, E} {B, D, E} {C, D, E}

Thus there are ten ways that three letters can be chosen from five letters. ■

Combinations of n Elements Taken m at a Time

The number of combinations of n elements taken m at a time is

$$\binom{n}{m} = \frac{n!}{(n-m)!m!}.$$

Note that this formula is the same one given for binomial coefficients in Lesson 12.5. Applying the formula to Example 1 produces the original result

$$\binom{5}{3} = \frac{5!}{2! \cdot 3!} = \frac{5 \cdot 4}{2} = 10.$$

Goal 2 Using Combinations to Find Probabilities

Connections **Game Theory**

Standard 52 - Card Deck

A ♠	A ♥	A ♦	A ♣
K ♠	K ♥	K ♦	K ♣
Q ♠	Q ♥	Q ♦	Q ♣
J ♠	J ♥	J ♦	J ♣
10 ♠	10 ♥	10 ♦	10 ♣
9 ♠	9 ♥	9 ♦	9 ♣
8 ♠	8 ♥	8 ♦	8 ♣
7 ♠	7 ♥	7 ♦	7 ♣
6 ♠	6 ♥	6 ♦	6 ♣
5 ♠	5 ♥	5 ♦	5 ♣
4 ♠	4 ♥	4 ♦	4 ♣
3 ♠	3 ♥	3 ♦	3 ♣
2 ♠	2 ♥	2 ♦	2 ♣

Example 2 *The Probability of a Run of Cards*

Five cards are dealt at random from a standard deck of 52 playing cards. What is the probability that the cards are 10–J–Q–K–A of the same suit?

Solution The number of five-card hands possible from a deck of 52 cards is given by the number of combinations of 52 elements taken five at a time.

$$\binom{52}{5} = \frac{52!}{47! \cdot 5!} = \frac{52 \cdot 51 \cdot 50 \cdot 49 \cdot 48}{5 \cdot 4 \cdot 3 \cdot 2 \cdot 1} = 2{,}598{,}960$$

Because only four of these five-card hands are 10–J–Q–K–A of the same suit, the probability that the hand contains these cards is

$$P = \frac{4}{2{,}598{,}960} = \frac{1}{649{,}740}.$$

■

Real Life **Polling**

Example 3 *Conducting a Survey*

In 1990, a survey was conducted of 500 adults who have worn Halloween costumes. Each person was asked how he or she acquired a Halloween costume: created it, rented it, bought it, or borrowed it. The results are shown at the left. What is the probability that the first four people who were polled all created their costumes?

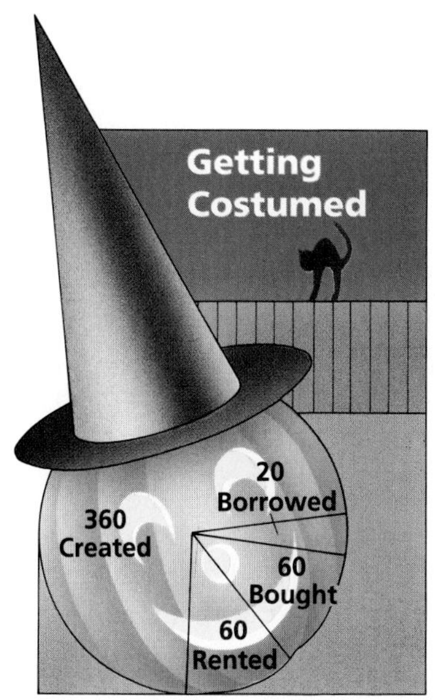

Solution To answer this question you need to use the formula for the number of combinations *twice*. First, find the number of ways to choose four people from the 360 who created their own costumes.

$$\binom{360}{4} = \frac{360 \cdot 359 \cdot 358 \cdot 357}{4 \cdot 3 \cdot 2 \cdot 1} = 688{,}235{,}310$$

Next, find the number of ways to choose four people from the 500 who were surveyed.

$$\binom{500}{4} = \frac{500 \cdot 499 \cdot 498 \cdot 497}{4 \cdot 3 \cdot 2 \cdot 1} = 2{,}573{,}031{,}125$$

The probability that all of the first four people surveyed created their own costumes is the ratio of these two numbers.

$$P = \frac{\text{Number of ways to choose 4 from 360}}{\text{Number of ways to choose 4 from 500}}$$

$$= \frac{688{,}235{,}310}{2{,}573{,}031{,}125} \approx 0.267$$

■

LESSON Notes

For many students, the difference between the concepts of *combination* and *permutation* is very subtle. Encourage students to read the first two paragraphs at the top of page 796 for examples of situations that use a combination or a permutation when counting the number of ways an event can occur.

Example 1

When selecting combinations of the three letters, stress to students that order does not matter. For example, the combination of letters $\{A, C, E\}$ is the same as $\{E, A, C\}$.

The Fundamental Counting Principle can be used to derive the formula for the number of combinations of n elements taken m at a time, as follows. The number of permutations of n objects taken m at a time is given by the expression $\frac{n!}{(n-m)!}$. To find the number of permutations of n objects taken m at a time, you have to select the m objects and then arrange them. You can select the m objects in $\binom{n}{m}$ ways and arrange them in $m!$ ways. By the Fundamental Counting Principle, the number of permutations of n objects taken m at a time is

$\binom{n}{m} \cdot m! = \frac{n!}{(n-m)!}$.

Dividing both sides by $m!$, you get $\binom{n}{m} = \frac{n!}{(n-m)!m!}$, which is the formula for the number of combinations of n elements taken m at a time.

Examples 2–4

These examples show how combinations can be used when computing probabilities in game theory, conducting a survey, and committee representation. Using Example 4 as a model, students may want to investigate the representation of diverse cultural groups on committees for clubs and organizations in the school.

Extra Examples

Here are additional examples similar to **Examples 1–4.**

1. In how many ways can three letters be chosen from the alphabet? The order of the three letters is not important. 2600
2. Five cards are dealt at random from a standard deck of 52 playing cards. What is the probability that all five cards are face cards or aces? $\frac{1}{595}$
3. Using the information in Example 3, what is the probability that the first ten people who were polled all borrowed their costumes? $\approx 7.5 \cdot 10^{-16}$
4. Using the information in Example 4, what is the probability that all five committee members were selected from minority groups? ≈ 0.0015

Check Understanding

1. Explain the difference between counting combinations and permutations by giving examples of where you need to count the number of combinations and where you need to count the number of permutations. See the top of page 796.
2. State the formula for the Number of Combinations of n Elements Taken m at a Time. See the bottom of page 796.

Real Life
Minority Representation

Example 4 ***Forming a Committee***

To obtain input from 200 company employees, the management of a company selected a committee of 5. Of the 200 employees, 56 are from minority groups. None of the 56, however, were selected to be on the committee. Does this indicate that the management's selection was biased?

Solution Part of the solution is similar to that of Example 3. If the 5 committee members were selected at random, the probability that all 5 would be nonminorities is

$$P = \frac{\text{Number of ways to choose 5 from 144 nonminorities}}{\text{Number of ways to choose 5 from 200 employees}}$$

$$= \frac{\binom{144}{5}}{\binom{200}{5}}$$

$$= \frac{481{,}008{,}528}{2{,}535{,}650{,}040}$$

$$\approx 0.19.$$

Thus if the committee were chosen at random (that is, without bias), then the likelihood that it would have no minority members is about 0.19. While this does not *prove* there was bias, it does suggest it. ■

Communicating about ALGEBRA

SHARING IDEAS about the Lesson

A. Six pennies are tossed in the air and land on a table—either heads up or tails up. The $\binom{6}{3}$ ways that exactly three of the pennies can land heads up are as follows.

HHHTTT	HHTHTT	HHTTHT	HHTTTH	HTHHTT
HTHTHT	HTHTTH	HTTHHT	HTTHTH	HTTTHH
TTTHHH	TTHTHH	TTHHTH	TTHHHT	THTTHH
THTHTH	THTHHT	THHTTH	THHTHT	THHHTT

Explain how this number could be found using the formula for the number of combinations.

B. Eight pennies are tossed in the air and land on a table. What is the probability that exactly four of them land heads up? $\frac{35}{128}$ **A.** $\binom{6}{3} = \frac{6!}{(6-3)!3!} = 20$

EXERCISES

Guided Practice

CRITICAL THINKING about the Lesson

1. How many different *combinations* of four letters can be chosen from the letters A, B, C, D, E, F, and G? 35

2. In the portion of Pascal's Triangle shown below, which entry is $\binom{7}{2}$? 21

$$\begin{array}{ccccccccccccccc} & & & & & & & 1 & & & & & & & \\ & & & & & & 1 & & 1 & & & & & & \\ & & & & & 1 & & 2 & & 1 & & & & & \\ & & & & 1 & & 3 & & 3 & & 1 & & & & \\ & & & 1 & & 4 & & 6 & & 4 & & 1 & & & \\ & & 1 & & 5 & & 10 & & 10 & & 5 & & 1 & & \\ & 1 & & 6 & & 15 & & 20 & & 15 & & 6 & & 1 & \\ 1 & & 7 & & 21 & & 35 & & 35 & & 21 & & 7 & & 1 \end{array}$$

Disc Jockey **In Exercises 3 and 4, use the following information.**

By winning a contest on a local radio station, you get to be the disc jockey for two hours. During your time, you must play 20 songs, 17 of which are selected by the station and 3 of which are your own choice.

3. Six of your favorite songs are not among the 17 chosen by the station. In how many ways can you choose 3 songs from the 6? 20

4. If the 20 songs are played in random order, what is the probability that your 3 choices will be played first? $\frac{1}{1140}$

Independent Practice

5. *Study Questions* Your history teacher gives your class a list of eight study questions. Five of the eight will be on the next test. How many different combinations of five test questions can your teacher choose from the eight study questions? 56

6. *Basketball Lineup* A high school basketball team has 15 players. In how many different ways can the coach choose the starting lineup? (Assume each player can play each position.) 3003

7. *First-String Team* A high school football team has 2 centers, 9 linemen (who can play either guard or tackle), 2 quarterbacks, 5 halfbacks, 5 ends, and 6 fullbacks. The coach uses 1 center, 4 linemen, 2 ends, 2 halfbacks, 1 quarterback, and 1 fullback to form an offensive unit. In how many ways can the offensive unit be selected? 302,400

8. *Pizza Toppings* A pizza shop offers nine toppings. How many different "three-topping pizzas" can be formed with the nine toppings? (Assume no topping is used twice.) 84

Communicating about ALGEBRA

Discuss the number of events in the sample space in Part B. Since a penny can land either heads up or tails up, there are two outcomes in each event. By the Fundamental Counting Principle, there are

$\boxed{2}\,\boxed{2}\ldots\boxed{2}$ (8 factors)

or $2^8 = 256$ ways in which the eight pennies can land heads up or tails up.

EXERCISE Notes

ASSIGNMENT GUIDE

Basic/Average: Ex. 1–4, 5, 8, 9, 12, 14, 15, 21–28

Above Average: Ex. 1–4, 5–9, 12, 15–22, 27–29

Advanced: Ex. 1–4, 5–9, 12, 14–21, 27–30

Selected Answers
Exercises 1–4, 5–27 odd

✪ More Difficult Exercises
Exercises 15–20, 29–30

Guided Practice

▶ **Ex. 2** Remind students that the last row represents the coefficients for the expansion of $(a + b)^7$. Thus the third coefficient for that row is $\binom{7}{2}$ or $\binom{n}{m-1}$ where n is the power of the binomial expansion at the mth term.

Ex. 4 Students should use Example 2 as a solution model. The number of song orders for the first three songs is given by the number of combinations of 20 elements taken three at a time, or $\frac{20!}{3!17!} = 1140$. Because there is only 1 possibility that the first 3 choices will be your songs, the probability that your 3 choices will be played first is $\frac{1}{1140}$.

Independent Practice

Ex. 7 You might have students who play football give an explanation of what an "offensive unit" in football is. Remind students to apply the Fundamental Counting Principle to find the total number of possible units.

Ex. 11, 13, 14 Students should express answers as reduced fractions.

Ex. 15–20 Make sure you remind students of your expected answer format (rounded decimals, factorial form, exponential form).

Ex. 17–18 and 19–20 Assign these exercises in pairs.

Ex. 27–28 These exercises check students' fundamental understanding of the difference between permutations and combinations. Combinations are used when the order of elements in an event does not matter; permutations are used when order does matter.

Ex. 30 Ask students to notice the connection between the numerator of the probability that a marble will land in each of the 8 compartments and the eighth row of Pascal's triangle.

9. *Geometry* Three points that are not on a line determine three lines. How many lines are determined by seven points, no three of which are on a line? 21

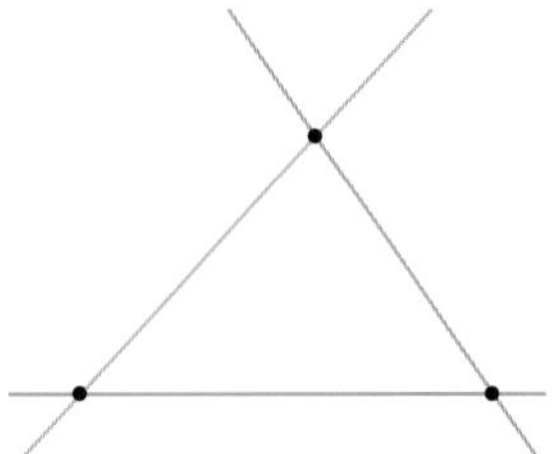

10. *Geometry* How many circles are determined by seven points, no four of which are on a circle? (Assume no three of the points lie on a line.) 35

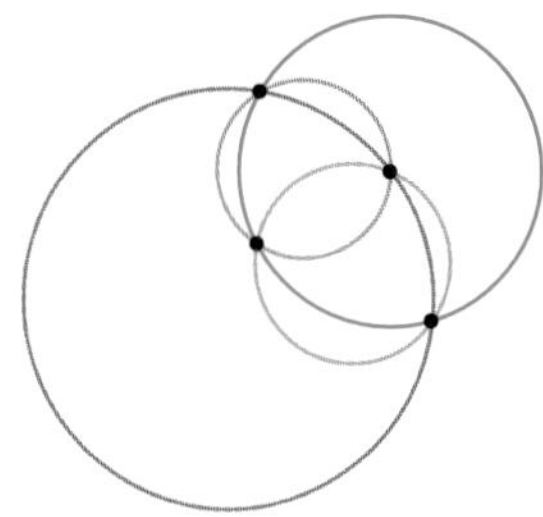

11. *Microwave Shipment* Fifty microwave ovens were shipped to an appliance store. Four of the fifty were defective. What is the probability that exactly two of the first five sold are defective? $\frac{99}{2303}$

12. *100-Yard Dash* You, your friend, and five other high school students are running in the 100-yard dash at the state finals. In how many different orders can the seven runners finish? In how many of these are you and your friend in first and second place, respectively? (Assume there are no ties.) 5040; 120

13. *Television Game Show* A contestant on a game show must choose 2 of 16 boxes. Of the 16 boxes, 3 contain money, 5 contain prizes, and 8 are empty. What is the probability that both boxes chosen by a contestant contain money? What is the probability that both are empty? $\frac{1}{40}, \frac{7}{30}$

14. *Selecting a Card* Two cards are selected at random from a standard deck of 52 playing cards. What is the probability that both are hearts? $\frac{1}{17}$

15. *Pizza* The graph below shows the percents of different types of pizza sold in 1990. Suppose 3 pizzas are randomly chosen from 100 pizzas. What is the probability that all 3 have thin crusts? (Assume the 100 pizzas are represented by the circle graph.) ***(Source: National Restaurant Association)*** $\frac{11,713}{80,850} \approx 0.145$

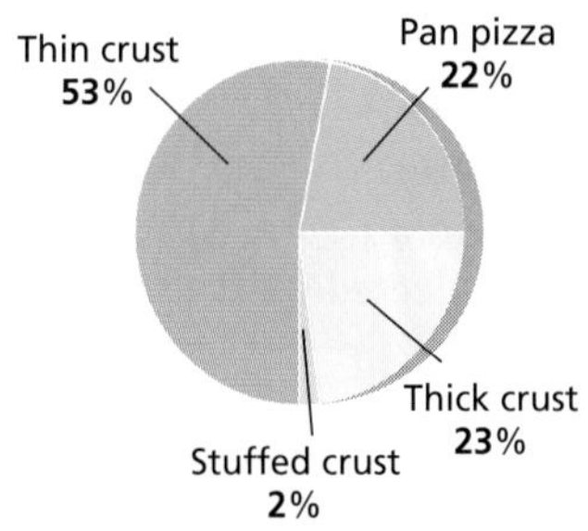

16. *Bank Loans* The graph below shows the percents of different types of bank loans in 1992. Suppose 4 loans are randomly chosen from 1000 loans. What is the probability that all 4 are auto loans? (Assume the 1000 loans are represented by the circle graph.) ***(Source: Federal Reserve System)*** ≈ 0.008

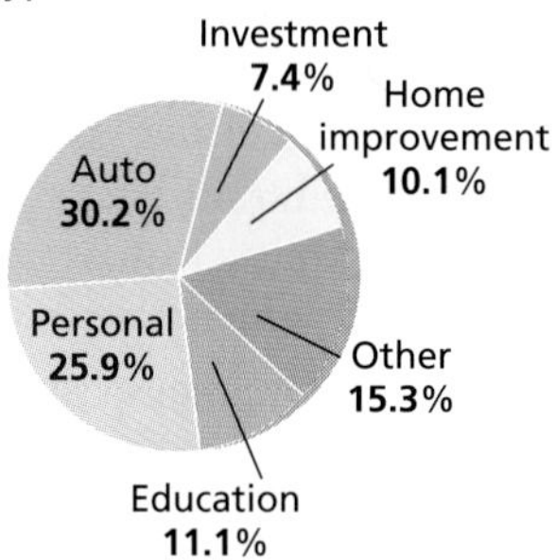

Hospital Inspection **In Exercises 17 and 18, use the following information.**

As part of a monthly inspection at a hospital, the inspection team randomly selects reports from 8 of the 84 nurses who are on duty.

17. What is the probability that none of the reports selected will be from the 10 most experienced nurses on duty? ≈ 0.346

18. What is the probability that all of the reports selected will be from the 20 least experienced nurses on duty? ≈ 0.0000029

19. *Jelly Beans* A bag of 4340 jelly beans contains equal numbers of red, green, purple, yellow, black, white, and orange. You pick 7 jelly beans at random. What is the probability that at least 1 is orange? 0.6604

20. *Jelly Beans* In Exercise 19, what is the probability that none of the 7 you choose are orange? ≈ 0.3396

Integrated Review

In Exercises 21–26, evaluate the expression.

21. $\binom{8}{2}$ 28

22. $\binom{5}{4}$ 5

23. $\binom{30}{26}$ 27405

24. $\binom{1000}{2}$ 499500

25. $\binom{55}{6}$ 28989675

26. $\binom{103}{3}$ 176851

In Exercises 27 and 28, identify each situation as involving permutations, combinations, or neither.

27. The number of ways a 5-person volleyball team can be chosen from 15 people Combinations

28. The number of orders in which five people can stand in line Permutations

Exploration and Extension

29. *Probability Demonstrator* What is the probability that the marble at the right will land in each of the four compartments at the bottom of the "probability demonstrator"? $\frac{1}{8}, \frac{3}{8}, \frac{3}{8}, \frac{1}{8}$

30. *Probability Demonstrator* You are constructing a probability demonstrator similar to the one shown at the right, except that your probability demonstrator has eight bottom compartments. What is the probability that a marble will land in each of the eight compartments?

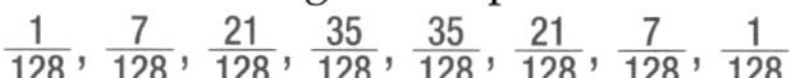

$\frac{1}{128}, \frac{7}{128}, \frac{21}{128}, \frac{35}{128}, \frac{35}{128}, \frac{21}{128}, \frac{7}{128}, \frac{1}{128}$

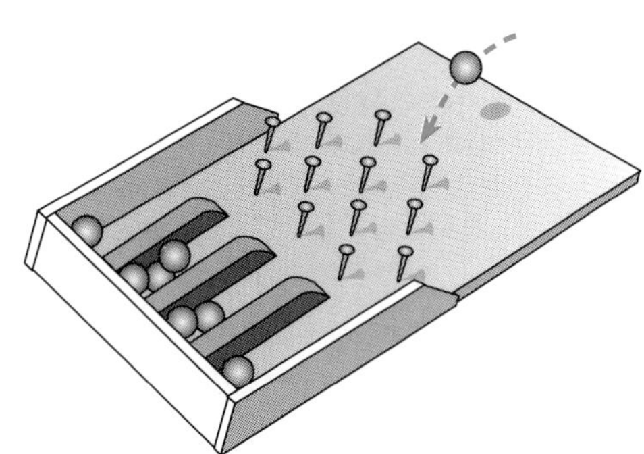

Enrichment Activity

The following activity can be used as an alternate assessment project.

Probability Project-
Creating a Survey

COOPERATIVE LEARNING

Work with a partner. Consider the survey that was shown in Example 3 on page 797.

1. What are some issues to consider in a survey design? The topic, the number of people to poll, how to tally the results, whether it is a whole group or small group project, etc.
2. Create a survey question that has three or four possible responses. For example, as listed in Exercise 15, you could poll students on their pizza choices.
3. **WRITING** Write some probability questions similar to those shown in the lesson, which can be answered using the results of your survey.
4. Conduct your survey. Tabulate and display the results. Use the real data to answer your probability questions. Compare these experimental probabilities with expected theoretical probabilities.

USING A GRAPHING CALCULATOR

Most graphing calculators have a random number generator that can be used to perform probability experiments. Each time the random number generator is used, it selects a real number between 0 and 1. (The number selected is greater than 0 and less than 1.) If you want the calculator to randomly select an *integer* x, such that $n \le x \le m$, you can enter the following keystrokes.

[iPart] $((m - n + 1)$ [rand] $+ n)$

Choose the integer part of the number — Select a random number between 0 and 1.

Specific keystroke instructions for a *TI-82*, Casio *fx-9700GE*, and Sharp *EL-9300C* are listed on page 865.

Example: Tossing a Die with a Graphing Calculator

Use one of the calculator programs on page 865 to simulate the tossing of a six-sided die 120 times. Record the number of times you obtain 1, 2, 3, 4, 5, and 6.

Solution Theoretically, you would expect each number to occur 20 times. *Experimentally*, however, you shouldn't expect this exact result. Here are the results we obtained.

Number	1	2	3	4	5	6
Tally	20	20	22	23	25	10

From these results, it appears that the probability of obtaining each number is as follows.

1: $\frac{20}{120}$ 2: $\frac{20}{120}$ 3: $\frac{22}{120}$ 4: $\frac{23}{120}$ 5: $\frac{25}{120}$ 6: $\frac{10}{120}$

A probability that is determined by this method is called an **experimental probability.** As you can see from this example, experimental probabilities can differ from theoretical probabilities. By conducting an experiment with a larger number of trials, you will obtain experimental results that are closer to the theoretical results. For instance, running the dice-rolling program 10,000 times produced the results shown in the following table.

Number	1	2	3	4	5	6
Tally	1633	1670	1663	1644	1698	1692

From this table, you obtain experimental probabilities that differ from the theoretical ones by less than 0.004. ■

In 1978, 32 students in a mathematics class at the Pennsylvania State University at Erie each tossed a standard six-sided playing die 1000 times. The results of the 32,000 tosses are shown in the table.

A six-sided die with recessed indicia is slightly unbalanced. Higher numbers are more likely to turn up than lower numbers.

Number	1	2	3	4	5	6
Tally	5057	5186	5292	5401	5491	5573

The experimental probabilities for each of the numbers from 1 through 6 can be obtained by dividing each tally by 32,000.

1: 0.158, 2: 0.162, 3: 0.165
4: 0.169, 5: 0.172, 6: 0.174

With such a large number of trials, why were the results not closer to the theoretical probability of $\frac{1}{6}$ (0.167)? The answer is that standard six-sided playing dice are made with recessed *indicia* (or dots), causing them to be unbalanced. The side with only one recessed dot is the heaviest, the side with only two recessed dots is the next heaviest, and so on. The lightest sides, those with six and five recessed dots, tend to land up more often than the sides with one and two dots.

To overcome this problem, Ron Larson (one of the authors of this book) devised a new type of die, called *Duo-Dice.** These dice also have recessed dots, but because each die is 12-sided, it is perfectly balanced—a 1 is opposite a 1, a 2 is opposite a 2, and so on. Using *Duo-Dice,* the experiment described above was repeated with the following results.

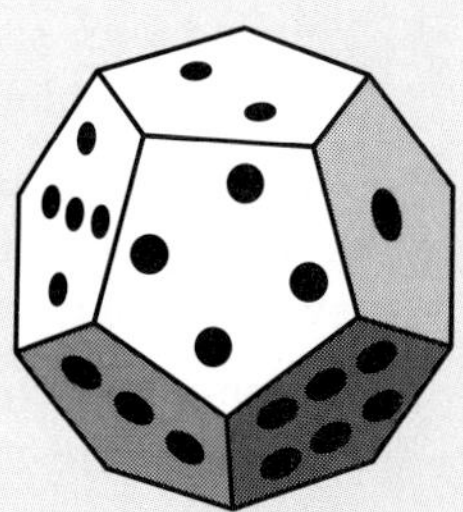

A 12-sided Duo-Die *is perfectly balanced. Each number is equally likely to turn up.*

Number	1	2	3	4	5	6
Tally	5288	5316	5373	5271	5402	5350

* *Duo-Dice* are patented by Roland E. Larson, Patent #4,465,279.

Exercises

1. *Experimental Probability* Use a graphing calculator to perform the experiment described in the example on page 802. How closely do your experimental results agree with the theoretical results? **Answers vary.**

2. *Duo-Dice* If a single *Duo-Die* is tossed, what is the probability that a given number turns up? $\frac{1}{6}$

3. *Duo-Dice* The game *Monopoly* requires two standard dice. Explain why this game would also require two *Duo-Dice.* **The standard dice are unbalanced.**

4. *Duo-Dice* In the experiment described above in which *Duo-Dice* were tossed 32,000 times, what is the experimental probability for each number? How much does it differ from the theoretical probability of $\frac{1}{6}$?

Answers

4. 0.16525, 0.166125, 0.16790625, 0.16471875, 0.1688125, 0.1671875; $\approx$0.0014, $\approx$0.0005, $\approx$0.0012, $\approx$0.0019, $\approx$0.0021, $\approx$0.0005

Mid-Chapter SELF-TEST

Take this test as you would take a test in class. The answers to the exercises are given in the back of the book.

1. A bunch of balloons contains 1 red, 2 blue, 2 white, and 2 yellow balloons. You randomly choose one of the strings that are attached to the balloons. What is the probability that you choose a red one? A yellow one? **(15.1)** $\frac{1}{7}, \frac{2}{7}$
2. An archer shoots an arrow at the target shown at the right. $\frac{1}{25}, \frac{7}{25}$ Suppose that the arrow is equally likely to hit any point on the target. What is the probability that the arrow hits the bull's eye? What is the probability that the arrow hits the blue ring? **(15.1)**
3. The notepapers in a memo cube alternate from pink to yellow to blue to green. (The cube contains the same number of each color.) You take a piece of notepaper from the cube. What is the probability that it is pink? **(15.1)** $\frac{1}{4}$

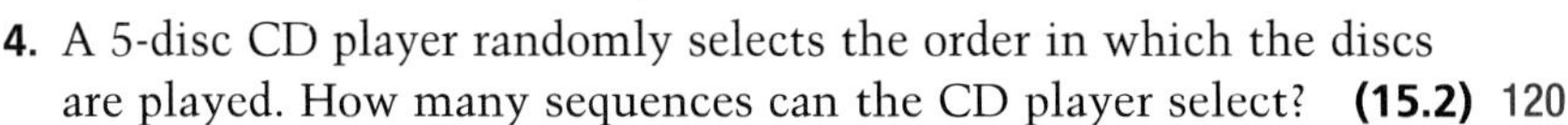

4. A 5-disc CD player randomly selects the order in which the discs are played. How many sequences can the CD player select? **(15.2)** 120
5. In how many ways can the letters A, E, L, M, S be arranged? **(15.2)** 120
6. You and your sister rent 3 VCR movies from the corner store. In how many orders can the three tapes be viewed? **(15.2)** 6
7. The softball coach is determining the batting order for the nine-member team. The catcher and pitcher will bat 5th and 9th, respectively. How many batting orders can the coach form? **(15.2)** 5040
8. A committee consisting of 12 girls and 8 boys is planning the senior prom. The committee randomly selects six members to form a subcommittee to be in charge of decorating. What is the probability that the subcommittee will have 2 girls and 4 boys? **(15.3)** ≈ 0.12
9. A deck of UNO® cards is made up of 108 cards. (Twenty-five each are red, yellow, blue, and green, and 8 are wild cards.) Each player is randomly dealt a seven-card hand. What is the probability that a hand will have 2 wild cards? What is the probability that a hand will have 2 wild cards, 2 red cards, and 3 blue cards? **(15.3)** ≈ 0.076, ≈ 0.00069
10. A sales clerk in a shoe store is setting up a display using 6 different colors of socks. The clerk wants to display all 2-color combinations of socks (a 2-color combination consists of two socks, each of a *different* color). How many *pairs* of socks are needed to form all possible 2-color combinations? **(15.3)** 30
11. You have time to listen to 3 selections from a CD that contains 12 songs. How many combinations of songs do you have to choose from? **(15.3)** 220
12. Twenty-five television sets are delivered to the appliance store. Three of these sets are defective. What is the probability that exactly one of the first five sold is defective? **(15.3)** ≈ 0.413

15.4

Probability, Unions, and Intersections

What you should learn:

Goal 1 How to use unions to find probabilities

Goal 2 How to use complements and intersections to find probabilities

Why you should learn it:

You can solve many probability problems using unions and intersections, such as the probability that one event or another will not happen.

Goal 1 Using Unions to Find Probabilities

The **union** of two events, A and B, written as $A \cup B$, consists of all outcomes that are in A *or* in B (or in both A and B). The **intersection** of A and B, written as $A \cap B$, consists of all outcomes that are in A *and* B.

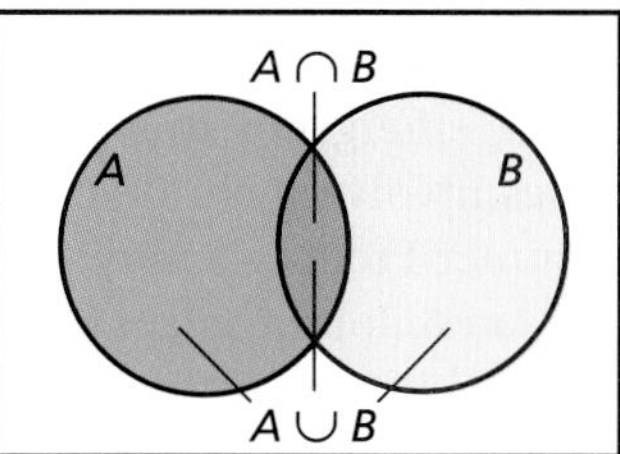

Union and Intersection

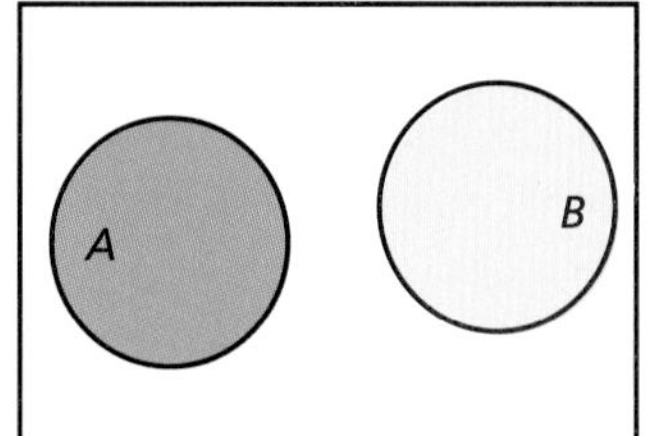

Mutually exclusive events

The Probability of the Union of Two Events

If A and B are events in the same sample space, then the probability of A *or* B occurring is $P(A \cup B) = P(A) + P(B) - P(A \cap B)$. If $A \cap B$ is empty, then A and B are called **mutually exclusive events** and $P(A \cup B) = P(A) + P(B)$.

Connections
Game Theory

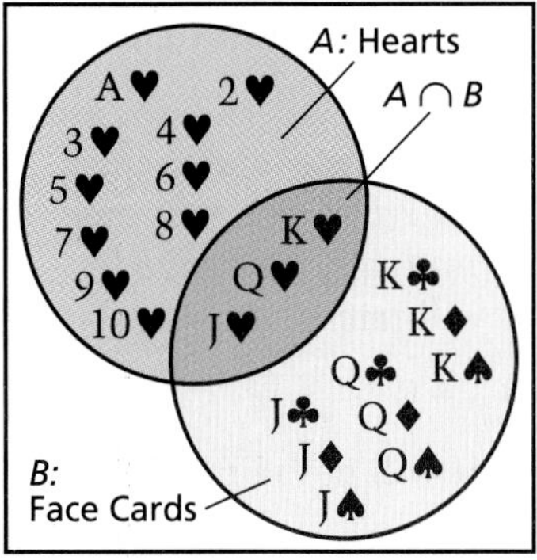

Example 1 *Using Unions to Find Probability*

One card is selected from a standard deck of 52 playing cards. What is the probability that the card is a heart *or* a face card?

Solution Let event A represent selecting a heart and event B represent selecting a face card. Because there are 13 hearts and 12 face cards, you can conclude that $P(A) = \frac{13}{52}$ and $P(B) = \frac{12}{52}$. Also, because three of the cards, J♡, Q♡, and K♡, are both hearts *and* face cards, you know that $P(A \cap B) = \frac{3}{52}$. Using this information, you can find the probability of selecting a heart *or* a face card.

$$P(A \cup B) = P(A) + P(B) - P(A \cap B)$$
$$= \frac{13}{52} + \frac{12}{52} - \frac{3}{52} = \frac{22}{52} \approx 0.423$$ ■

Problem of the Day

A clock chimes on the hour, chiming the same number of times as the hour, and also chimes once every quarter hour. If the face of the clock cannot be seen and the clock has just chimed once, what is the longest time you might have to wait to be sure of knowing the time?

$1\frac{1}{2}$ hours

ORGANIZER

Warm-Up Exercises

One card is drawn at random from a standard deck of 52 playing cards. What is the probability that the card is:

1. a black card $\frac{1}{2}$
2. a face card $\frac{3}{13}$
3. neither an ace nor a two $\frac{11}{13}$

Lesson Resources

Teaching Tools
- Problem of the Day: 15.4
- Warm-up Exercises: 15.4
- Answer Masters: 15.4

Extra Practice: 15.4
Color Transparency: 81
Technology Handbook: p. 103

LESSON Notes

Example 1

The use of set theory is fundamental to the study of probability theory. Give special attention to the use of the terms *and* and *or* in the context of working with sets of events as described on top of page 805. Venn diagrams are an ideal model for illustrating unions and intersections of sets. Encourage students to draw Venn diagrams when solving this type of problem.

The Venn diagram shown at the top of page 806 illustrates the complement of an event. You can use numerical examples with the students to discuss why the sum of the probability of an event and the probability of its complement must equal 1. For example, if the probability that you will pick the ace of diamonds from an ordinary deck of 52 cards is $\frac{1}{52}$, then the probability that you will not pick the ace of diamonds is $\frac{51}{52}$. $\frac{1}{52} + \frac{51}{52} = 1$.

Solving the equation $P(A) + P(A') = 1$ for $P(A')$, you find that $P(A') = 1 - P(A)$.

Example 2

Discuss with students the events that make up this situation. A is the event that an injury involved swings, monkey bars, slides, or seesaws. Its complement, A', is the event that an injury occurred on another type of equipment.

EXTEND Example 2 Have students contact insurance companies to find similar accident statistics in order to write probability questions to share with the class.

Goal 2 Using Complements and Intersections

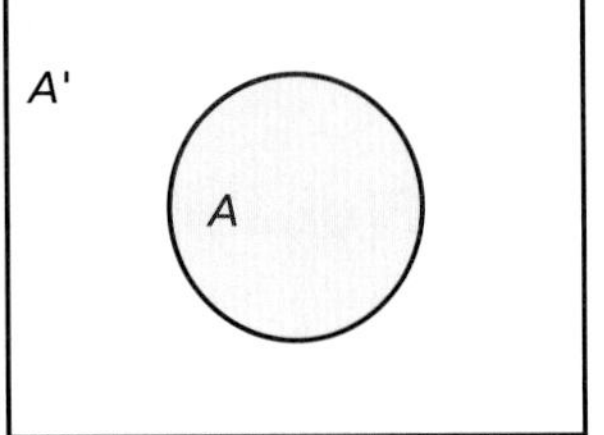

If A is an event in a sample space S, then the event consisting of all outcomes in S that are not in A is called the **complement** of A. The complement of A is written as A'.

> **The Probability of the Complement of an Event**
>
> The probability of the complement of A is
>
> $$P(A') = 1 - P(A).$$

Example 2 *Finding the Probability of a Complement*

In 1990, emergency rooms treated 237,484 cases of playground injuries involving children. You work for the United States Consumer Product Safety Commission and have used a computer to randomly select one of the cases for study. What is the probability that the injury involved playground equipment other than swings, monkey bars, slides, and seesaws?

(Source: U.S. Consumer Product Safety Commission)

Playground Injuries Needing Emergency Room Treatment

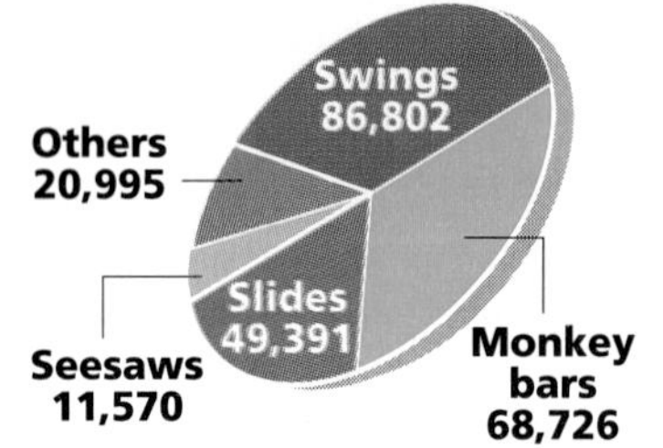

Solution From the information given at the left, the probability that the case involved swings, monkey bars, slides, or seesaws is

$$P(A) = \frac{86{,}802 + 68{,}726 + 49{,}391 + 11{,}570}{237{,}484} \approx 0.9116.$$

The probability that the case involved playground equipment other than swings, monkey bars, slides, and seesaws is

$$P(A') = 1 - P(A) \approx 1 - 0.9116 = 0.0884.$$ ■

Real Life
Meteorology

Example 3 *Finding the Probability of a Complement*

A television meteorologist announces that your area has a 50% chance of rain tomorrow and a 20% chance of snow. What is the probability that it will not rain or snow tomorrow? (Assume that it will not rain *and* snow.)

Solution Let event A represent rain tomorrow and event B represent snow. Because $P(A) = 0.5$ and $P(B) = 0.2$, you can use the assumption that A and B are mutually exclusive to conclude that the probability that it will rain *or* snow is

$$P(A \cup B) = P(A) + P(B) = 0.5 + 0.2 = 0.7.$$

This means that the probability that it will not rain or snow is

$$P((A \cup B)') = 1 - 0.7 = 0.3.$$ ■

Example 4 Using Intersection to Find Probability

In 1992, your company paid overtime wages *or* hired temporary help during 27 weeks. Overtime wages were paid during 21 weeks, and temporary help was hired during 12 weeks. At the end of the year, an auditor checks your accounting records and randomly selects one week to check your company's payroll. What is the probability that the auditor will select a week in which you paid overtime wages *and* temporary help?

Solution Let event A represent the weeks during which overtime wages were paid and let event B represent the weeks during which temporary help was hired. From the given information, you know that

$$P(A) = \frac{21}{52}, \quad P(B) = \frac{12}{52}, \quad \text{and} \quad P(A \cup B) = \frac{27}{52}.$$

The probability that the selected week comes from event A *and* event B is $P(A \cap B)$.

$$\begin{aligned} P(A \cup B) &= P(A) + P(B) - P(A \cap B) \\ \frac{27}{52} &= \frac{21}{52} + \frac{12}{52} - P(A \cap B) \\ P(A \cap B) &= \frac{21}{52} + \frac{12}{52} - \frac{27}{52} \\ P(A \cap B) &= \frac{6}{52} \end{aligned}$$

Therefore, the probability that the auditor will select a week in which you paid both overtime and temporary help is $P(A \cap B) = \frac{6}{52} \approx 0.1154$. ■

Communicating about ALGEBRA

SHARING IDEAS about the Lesson

A section of forest is composed of three species of trees as shown in the matrix below. One of the trees is selected at random to be cut down.

Number of Trees	Birch	Pine	Oak
Less than 50 feet	123	165	214
50 feet or more	56	189	78

A. What is the probability that the selected tree is an oak or is less than 50 feet? $\frac{116}{165}$

B. What is the probability that the selected tree is not a pine that is 50 feet or more? $\frac{212}{275}$

C. What is the probability that the selected tree is an oak or a birch? $\frac{157}{275}$

Examples 3–4

These examples are direct applications of the formula given on page 805.

After making the appropriate substitutions into the formula in Example 4, solving for $P(A \cap B)$ will give the solution to the problem.

Extra Examples

Here are additional examples similar to **Examples 1–4.**

1. One card is selected from a standard deck of 52 playing cards. What is the probability that the card is a club or not a face card? $\frac{43}{52}$
2. Using the information in Example 2, what is the probability that the injury involved playground equipment other than slides? ≈ 0.792
3. For the families chosen for a survey in a certain metropolitan area, 93% own a color television set, 31% own a black and white television set, and 28% own both kinds of television sets. What is the probability that a family will not own either kind of television set? $\frac{1}{25}$
4. The probability that a car stopped at a roadblock will have a burnt-out headlight is 0.35; worn-out muffler, 0.23; and burnt-out headlight and/or worn-out muffler, 0.43. What is the probability that such a car will have both a burnt-out headlight and a worn-out muffler? 0.15

Check Understanding

1. Define union and intersection of two events. See the top of page 805.
2. Define mutually exclusive events. See page 805.
3. Define the complement of an event. See page 806.

Communicating *about* ALGEBRA

For many problems, data can be collected into a matrix like the one shown here. When finding the probabilities, ask students questions about the sample space, such as the following. "What is the number of elements in the sample space?" 825 "What species of tree occurs most often?" Pine

EXERCISE Notes

ASSIGNMENT GUIDE

Basic/Average: Ex. 1–6, 7–15 odd, 18, 20, 21, 23, 24, 27–31

Above Average: Ex. 1–6, 7–15 odd, 17–23 odd, 24–31

Advanced: Ex. 1–6, 7–15 odd, 18–24, 26–31

Selected Answers

Exercises 1–6, 7–29 odd

More Difficult Exercises

Exercises 25–29

EXERCISES

Guided Practice

CRITICAL THINKING about the Lesson

In Exercises 1 and 2, consider the integers from 1 through 12. One of the integers is chosen at random. Let event *A* represent choosing an even integer. Let event *B* represent choosing a multiple of three.

1. Describe the event $A \cup B$. What is the probability of this event? 1. $A \cup B$ is the event of choosing an integer that is either even or a multiple of three; $\frac{2}{3}$
2. Describe the event $A \cap B$. What is the probability of this event? 2. $A \cap B$ is the event of choosing an integer that is even and a multiple of three; $\frac{1}{6}$
3. State an example of two events that are mutually exclusive. Answers vary
4. *Doing Your Homework* Let event A represent completing your homework tonight. What is A'? If $P(A) = 0.95$, what is $P(A')$? A' represents not completing your homework tonight, 0.05
5. A and B are mutually exclusive events such that $P(A) = 0.1$ and $P(B) = 0.2$. What is $P(A \cup B)$? What is $P(A \cap B)$? 0.3, 0
6. A and B are events in the same sample space such that $P(A) = \frac{1}{4}$, $P(B) = \frac{2}{5}$, and $P(A \cup B) = \frac{1}{2}$. What is $P(A \cap B)$? 0.15

Independent Practice

In Exercises 7–12, find the indicated probability. State whether *A* and *B* are mutually exclusive.

7. $P(A) = 0.55$
   $P(B) = 0.30$
   $P(A \cup B) = 0.65$
   $P(A \cap B) = \boxed{?}$ 0.20, No
8. $P(A) = 0.5$
   $P(B) = 0.30$
   $P(A \cup B) = \boxed{?}$ 0.6, No
   $P(A \cap B) = 0.2$
9. $P(A) = 0.25$
   $P(B) = \boxed{?}$ 0.65, Yes
   $P(A \cup B) = 0.90$
   $P(A \cap B) = 0$
10. $P(A) = 0.75$
    $P(B) = 0.20$
    $P(A \cup B) = \boxed{?}$ 0.95, Yes
    $P(A \cap B) = 0$
11. $P(A) = \frac{11}{15}$
    $P(B) = \frac{8}{15}$
    $P(A \cup B) = \boxed{?}$ 1, No
    $P(A \cap B) = \frac{4}{15}$
12. $P(A) = \frac{1}{3}$
    $P(B) = \frac{1}{4}$
    $P(A \cup B) = \frac{1}{3}$
    $P(A \cap B) = \boxed{?}$ $\frac{1}{4}$, No
13. If $P(A) = 0.64$, what is $P(A')$? 0.36
14. If $P(A) = 0.19$, what is $P(A')$? 0.81
15. Find $P(A')$ using the figure below. (The sample space has 60 outcomes.) $\frac{7}{12}$
16. Find $P(A \cup B)$ using the figure below. (The sample space has 800 outcomes.) $\frac{9}{16}$

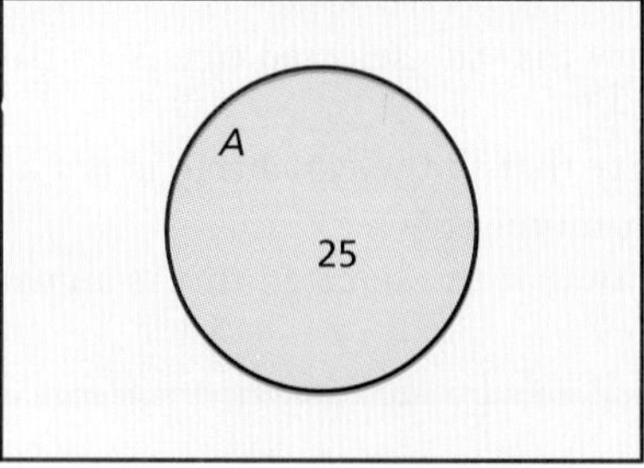

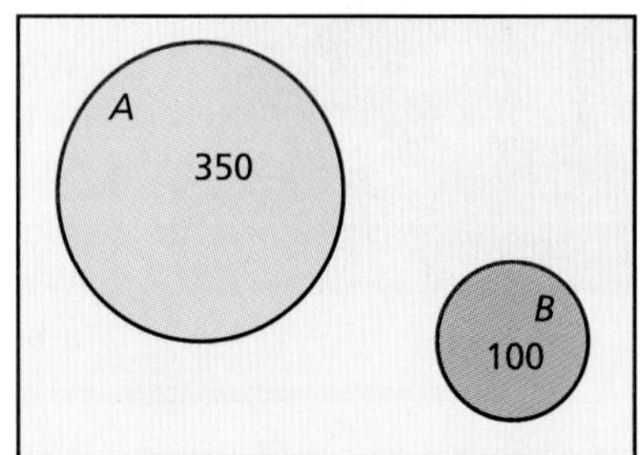

17. *Hockey or Swimming* The probability that Mick will make the hockey team is $\frac{2}{3}$. The probability that he will make the swimming team is $\frac{3}{4}$. If the probability that he makes both teams is $\frac{1}{2}$, what is the probability that he makes at least one of the teams? That he makes neither team?

18. *Aquarium* An aquarium at a pet store contains 30 orange swordtails (12 females and 18 males) and 48 green swordtails (32 females and 16 males). You randomly net one of the fish. What is the probability that it is a female or a green swordtail? $\frac{20}{39}$

19. *Class Election* Three people are running for class president. From a poll that was taken, the probability that Candidate A will win is estimated to be 0.3. The probability that Candidate B will win is estimated to be 0.55. Given a reasonable estimate of the probability that Candidate C will win. 0.15

17. $\frac{11}{12}, \frac{1}{12}$

20. *Probability of Rain* The probability that it will rain today is 0.4, and the probability that it will rain tomorrow is 0.2. The probability that it will rain both days is 0.1. What is the probability that it will rain today *or* tomorrow? 0.5

21. *Picking a Pie* The freezer case at a grocery store contains several frozen pies: 10 apple, 4 peach, 6 blueberry, 5 pumpkin, 3 peanut butter, and some pecan. You are shopping in a hurry and pick one of the pies from the freezer without looking at the type. The probability that the pie you picked is apple or pecan is $\frac{13}{22}$. How many pecan pies are in the freezer case? 16 pecan pies

22. *Picking a Cola* A picnic cooler contains several types of colas: 12 regular, 8 cherry, 10 diet, 6 diet cherry, 8 caffeine-free, and some caffeine-free diet. You pick one of the colas without looking at the type. The probability that it is a diet drink is $\frac{5}{12}$. How many caffeine free diet colas are in the picnic cooler? 4

23. *Holiday Shopping* In 1995 a survey was conducted to find when people do their holiday shopping. The results are shown in the graph at the right. If one of those indicated in the graph is chosen at random, what is the probability that the person said that he or she shops on the day before Christmas? ***(Source: USA Today)*** $\frac{56}{800} = 0.07$

24. *Holiday Shopping* If two of those surveyed in Exercise 23 are chosen at random, what is the probability that both said that they shop in the last two weeks of November? 0.0194

When People Shop

Category	Number
Already done / don't know	40
Last 2 weeks of November	112
Day after Thanksgiving	32
First 2 weeks of December	280
Evenly through the season	176
Last 2 weeks of December	104
Day before Christmas	56

Guided Practice

Ex. 1–6 Use these exercises as a student readiness check. It is essential that students understand that $P(A \cup B) = P(A) + P(B) - P(A \cap B)$. If $P(A \cap B) = 0$ then A and B are said to be *mutually exclusive.*

Independent Practice

Ex. 7–12 Ask students to use a solution format that first shows the equation $P(A \cup B) = P(A) + P(B) - P(A \cap B)$ and then shows the appropriate substitutions.

Ex. 13–15 Apply $P(A') = 1 - P(A)$ as illustrated by Goal 2 at the top of page 806.

Ex. 17–26 Encourage students to follow this solution format.
(1) Show the formula: $P(A \cup B) = P(A) + P(B) - P(A \cap B)$.
(2) Decide what they are asked to determine.
(3) Substitute the appropriate information.
(4) Solve for the appropriate quantity.

Ex. 27–29 Assign these exercises as a group. Extend the survey to your class. Do your students' responses match the probability results of the survey?

Ex. 28 This exercise will require students to use combinations.

Enrichment Activity

This activity requires students to go beyond lesson goals.

WRITING PROBLEMS Find data in newspapers or other sources similar to those shown below for recreational sports. From these data sources, generate some probability questions. The questions should have solutions that involve unions and intersections. Include the solutions with the questions.

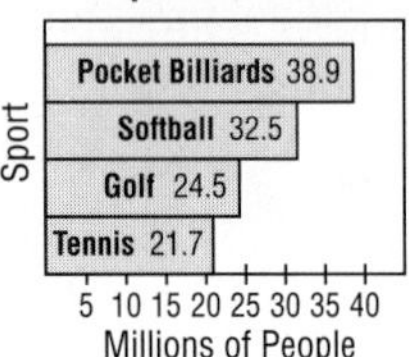

Sample questions:

1. Use the graph to find the probability that a person surveyed plays tennis. $\frac{21.7}{117.6} \approx 0.185$
2. If a person is chosen at random, find the probability that the person played softball or played golf. (You would need to know that $P(S \cap G) = 0$ or assign a value to it.) ≈ 0.485

25. *Honors Banquet* Of 148 students honored at an academic awards banquet, 40 won awards for mathematics and 82 for English. Twelve of these students won awards in both mathematics and English. One of the 148 students is chosen at random to be interviewed for a newspaper article. What is the probability that the student won an award in mathematics or English? $\frac{55}{74}$

26. *Art Gallery* An art gallery has scheduled 38 shows during the year: 18 that feature painters and 20 that feature sculptors. Some of the shows are scheduled for one week, and the others are scheduled for two weeks. Each week, only one painter or sculptor is shown. If you attend one show at random, the probability that it is a two-week show of paintings is $\frac{2}{13}$. What is the probability that it is a one-week show of sculptures? $\frac{5}{26}$

Integrated Review

Salary Expectations **In Exercises 27–29, use the following information.**

In 1991, a survey was conducted to determine salary expectations of high school students. In the survey, 1879 "A" and "B" students were asked what they expected to earn in their first job after completing their education. The salary expectations of the 1728 who responded are shown in the graph at the right.

27. One of the 1728 students who responded is chosen at random. What is the probability that this student expects to earn between \$30,000 and \$44,999? $\frac{545}{1728}$

28. Five of the 1728 students who responded are chosen at random. What is the probability that all five students expect to earn between \$30,000 and \$44,999? ≈ 0.003

29. One of the 1728 students who responded is chosen at random. What is the probability that this student expects to earn \$60,000 or more? \$12,000 or more? $\frac{47}{864}, \frac{209}{216}$

Exploration and Extension

Odds **In Exercises 30 and 31, use the following information.**

The odds of an event A occurring are the ratio of the probability that the event will occur to the probability that the event will not occur.

30. A jar contains three red marbles and five green marbles. What are the odds that a randomly chosen marble is green? $\frac{5}{3}$

31. A jar contains three red marbles and some green marbles. The odds are "3 to 1" that a randomly chosen marble is green. How many green marbles are in the jar? 9

15.5 Independent Events

What you should learn:

Goal 1 How to find the probability of independent events

Goal 2 How to use the complement of an event to find the probability of the event

Why you should learn it:

You can solve many probability problems in real life using independent events, such as finding the probability that in a group of 35 people, at least two will have the same birthday.

Goal 1 Independent Events

Two events are **independent** if the occurrence of one has no effect on the occurrence of the other. For instance, if two coins are tossed, the outcomes (heads or tails) are independent. Two events are **dependent** if the occurrence of one *does* affect the occurrence of the other.

Probability of Independent Events

If A and B are independent events, then the probability that both A and B occur is $P(A \cap B) = P(A) \cdot P(B)$.

Example 1 *Random Number Generator*

You use a graphing calculator to randomly select two integers between 1 and 20. What is the probability that both integers are less than 6?

Solution Let event A represent selecting a first number that is less than 6. Let event B represent selecting a second number than is less than 6. Each of these outcomes has a probability of $\frac{5}{20} = \frac{1}{4}$. Because the program randomly selects numbers, the two events are independent. Therefore, the probability that both numbers are less than 6 is

$$P(A \cap B) = P(A) \cdot P(B) = \frac{1}{4} \cdot \frac{1}{4} = \frac{1}{16}.$$

Study Tip

In Example 1, the events are independent. Here is a similar experiment in which the events are dependent. A box contains 20 pieces of paper numbered from 1 to 20. Two pieces are drawn from the box (the second is drawn without replacing the first). The probability that both numbers are less than 6 is

$$P = \frac{5}{20} \cdot \frac{4}{19} = \frac{1}{19}.$$

Can you see why?

LESSON INVESTIGATION

Investigating Random Numbers

Partner Activity Use a graphing calculator to randomly select two integers between 1 and 20. (See page 865 for keystrokes.) Repeat the experiment several times and record your results. How many times were both numbers less than 6? Compare your result with the result obtained in Example 1 above.

Answers will vary, but with a large sample space should approach the theoretical value.

Problem of the Day

Each letter represents a different digit. Find the digits.

$$\begin{array}{r} ABCDE \\ \times\ F \\ \hline EDCBA \end{array} \qquad \begin{array}{r} 21978 \\ \times\ 4 \\ \hline 87912 \end{array}$$

ORGANIZER

Warm-Up Exercises

For each event A, write the event A' and find $P(A')$.

1. The probability that a light in a shipment of 10,000 lights is defective is 2%.
   a light in a shipment of 10,000 lights is not defective; 0.98
2. The probability of getting 2 inches of snow on Monday is 0.35.
   not getting two inches of snow; 0.65

Lesson Resources

Teaching Tools
  Problem of the Day: 15.5
  Warm-up Exercises: 15.5
  Answer Masters: 15.5
Extra Practice: 15.5

LESSON Notes

Example 1

TECHNOLOGY Students may use the graphing calculator keystrokes on page 865 to estimate the reasonableness of the theoretical probability. The number 6 in the keystrokes should be replaced by 20 to generate integers from 1 to 20. Stress that since the program will only generate numbers greater than or equal to 1, there are only five numbers less than 6 in the sample space. Discuss with students why the two events are independent.

Example 2

In many instances it is easier to work with complements of events. This example provides students with a model for solving such problems.

Example 3

This classic problem uses a solution technique similar to that used in Example 2. It is easier to solve the problem by assuming that no two people have the same birthday, which is the complement of the event that at least two people have the same birthday. As an in-class experiment, determine whether there are at least two people in your class or grade who have the same birthday. Does the theoretical probability of 0.813 mean that at least two of any 35 people must have the same birthday 81% of the time?

Extra Examples

Here are additional examples similar to **Examples 1–3.**

1. Suppose that you use the program from Example 1 to generate random numbers between 1 and 20. Using this program to randomly select two integers between 1 and 20, what is the probability that one of the integers is between 9 and 18, and the other integer is even?
   $\frac{1}{5}$
2. In Example 2, suppose that 3% of the 10,000 lights are defective. If you make 100 random tests of the lights, what is the probability that at least one of the tests reveals a defective light?
   ≈ 0.952

Goal 2 Using Complements to Find Probabilities

Sometimes it is easier to find the probability that an event A *does not* occur than it is to find the probability that it *does* occur. In such cases, you can still find the probability of the event occurring by using the formula $P(A') = 1 - P(A)$.

Real Life
Quality Control

Example 2 *Using the Complement to Find a Probability*

You work in the quality control department of a company that manufactures fluorescent lights. Suppose that 1% of the lights in a shipment of 10,000 lights are defective. If you make 100 random tests of the lights (each time choosing a light, testing it, and replacing it), what is the probability that at least one of the tests reveals a defective light?

Solution Let event A represent selecting a light that is defective. Then event A' represents selecting a good light. Because there are 9900 good lights, the probability of A' is

$$P(A') = \frac{9900}{10{,}000} = 0.99. \qquad \textit{Select 1 good light.}$$

Because each light is replaced after it is tested, the selection of the next light is independent of the selection of the first. This means that the probability that the first two selections are good lights is $[P(A')]^2$. By similar reasoning, you can conclude that the probability of selecting 100 good lights is

$$[P(A')]^{100} = (0.99)^{100} \approx 0.366. \qquad \textit{Select 100 good lights.}$$

Finally, you can conclude that the probability of selecting at least one defective light is

$$P(A) = 1 - (0.99)^{100} \approx 0.634. \qquad \textit{Select at least 1 bad light.}$$ ■

Connections
Birthday Problem

Of the approximately 35 people shown in the square in the photo, the probability that at least 2 have the same birthday is 0.813.

Example 3 Using the Complement

A famous probability problem, called the *Birthday Problem,* can be stated as follows.

Thirty-five people are chosen at random. What is the probability that at least two of them have the same birthday?

Solve this problem. (Assume there are 366 different birthdays in a year.)

Solution Let event A represent each of the 35 people having *different* birthdays. The probability that the first 2 people have different birthdays is

$$\frac{366}{366} \cdot \frac{365}{366}, \qquad \textit{First 2 birthdays are different.}$$

that the first 3 people have different birthdays is

$$\frac{366}{366} \cdot \frac{365}{366} \cdot \frac{364}{366}, \qquad \textit{First 3 birthdays are different.}$$

and that the first 4 people have different birthdays is

$$\frac{366}{366} \cdot \frac{365}{366} \cdot \frac{364}{366} \cdot \frac{363}{366}. \qquad \textit{First 4 birthdays are different.}$$

By continuing this pattern, you can reason that the probability that all 35 people have different birthdays is

$$\frac{366}{366} \cdot \frac{365}{366} \cdot \cdots \cdot \frac{332}{366} \approx 0.187. \qquad \textit{All 35 birthdays are different.}$$

The probability that at least 2 of the 35 people have the same birthday is $P(A') = 1 - P(A) \approx 1 - 0.187 = 0.813$. ■

Communicating about ALGEBRA

SHARING IDEAS about the Lesson

A. You are tossing a coin five times. What is the probability that the coin lands heads up all five times? $\frac{1}{32}$

B. You are tossing a coin five times. What is the probability that the coin lands heads up at least once? $\frac{31}{32}$

C. A bookbinding machine at a printing company has a probability of 0.005 of producing a defective book. Your school has ordered 300 books that were bound on this machine. What is the probability that at least 1 of the books is defective? Explain.

The probability that there are no defective books in the order is $(0.995)^{300} \approx 0.222$. Thus, the probability that there is at least one defective book is $1 - (0.995)^{300} \approx 0.778$.

3. In Example 3, suppose that 20 people are chosen at random instead of 35. Do you think the probability of at least two of them having the same birthday will increase or decrease? What is this probability?
Decrease, ≈0.411

Check Understanding

In your own words, define what it means for two events to be independent.
See the top of page 811.

Communicating about ALGEBRA

You can use these problems as a summary and review of the types of problems that are applications of independent events.

EXTEND *Communicating*
BINOMIAL TRIALS

You might wish to show students how binomial expansions are related to "binomial trials," as illustrated in Parts A and B. Binomial trials are experiments of independent events that have two possible outcomes, such as heads or tails, win or lose, or success and failure. Coin-tossing experiments are perfect examples of binomial trials, since the outcome of one toss does not affect the outcome of another toss.

In tossing a coin five times, if the expression H^2T^3 represents the outcome two heads (H) and three tails (T), then the expansion $(H + T)^5 = H^5 + 5H^4T + 10H^3T^2 + 10H^2T^3 + 5HT^4 + T^5$ describes all the possible outcomes of the five coin tosses. To find the probability of exactly two heads, you would use the $10H^2T^3$ term as follows.

$$P(\text{exactly 2 H}) = 10(P(H))^2(P(T))^3 = 10\left(\frac{1}{2}\right)^2\left(\frac{1}{2}\right)^3 = \frac{10}{32} = \frac{5}{16}$$

Have students use the expansions to illustrate their solutions to Parts A and B.
For example, in Part A, you would use the term H^5 to find $\left(\frac{1}{2}\right)^5 = \frac{1}{32}$. In Part B, you would need to use any term in which at least 1 head appeared. So, $H^5 + 5\,H^4T + 10\,H^3T^2 + 10\,H^2T^3 + 5\,HT^4 = \left(\frac{1}{2}\right)^5 + 5\left(\frac{1}{2}\right)^4\left(\frac{1}{2}\right) + 10\left(\frac{1}{2}\right)^3\left(\frac{1}{2}\right)^2 + 10\left(\frac{1}{2}\right)^2\left(\frac{1}{2}\right)^3 + 5\left(\frac{1}{2}\right)\left(\frac{1}{2}\right)^4 = \frac{1}{32} + \frac{5}{32} + \frac{10}{32} + \frac{10}{32} + \frac{5}{32} = \frac{31}{32}$.

Coin-tossing experiments can be used to simulate the playing of "win-lose" games. For example, suppose your team (H) wins 3 out of 5 games against an opponent (T) in the regular season. What is the probability that your team will win exactly 3 out of 5 playoff games in post-season play? Substitute in the term $10\,H^3T^2$. So, the probability that your team will win exactly 3 games is $10\,H^3T^2 = 10\left(\frac{3}{5}\right)^3\left(\frac{2}{5}\right)^2 = \frac{216}{625} \approx 0.35$

EXERCISE Notes

ASSIGNMENT GUIDE

Basic/Average: Ex. 1–6, 7–13 odd, 16–19, 21–23

Above Average: Ex. 1–6, 7–15 odd, 16–22, 24

Advanced: Ex. 1–6, 7–15 odd, 16–24

Selected Answers
Exercises 1–6, 7–23 odd

Use **Mixed Review** as needed.

✪ More Difficult Exercises
Exercises 12, 14–21

EXERCISES

Guided Practice

CRITICAL THINKING about the Lesson

In Exercises 1–4, state whether events *A* and *B* are independent or dependent.

1. Event A is rolling a 5 on a die.
Event B is rolling a 3 on the same die. Independent
2. Event A is drawing a queen from a standard deck of cards.
Event B is drawing a king from the remaining cards in the same deck. Dependent
3. Event A is getting a red gumball from a gumball machine.
Event B is getting a yellow gumball from the remaining gumballs in the same machine. Dependent
4. Event A is getting 0.25 inches of snow on Monday.
Event B is getting 0.45 inches of snow on Tuesday. Independent

In Exercises 5 and 6, *A* and *B* are independent events. Evaluate *P(A ∩ B)*.

5. $P(A) = 0.29,\ P(B) = 0.80$ 0.232
6. $P(A) = \frac{1}{2},\ P(B) = \frac{5}{6}$ $\frac{5}{12}$

Independent Practice

In Exercises 7–10, *A* and *B* are independent events. Find the indicated probability.

7. $P(A) = \frac{2}{3}$, $P(B) = \frac{7}{8}$, $P(A \cap B) = \boxed{?}$ $\frac{7}{12}$
8. $P(A) = 0.1$, $P(B) = 0.1$, $P(A \cap B) = \boxed{?}$ 0.01
9. $P(A) = 0.5$, $P(B) = \boxed{?}$ 0.02, $P(A \cap B) = 0.01$
10. $P(A) = \boxed{?}$ $\frac{5}{8}$, $P(B) = \frac{2}{5}$, $P(A \cap B) = \frac{1}{4}$

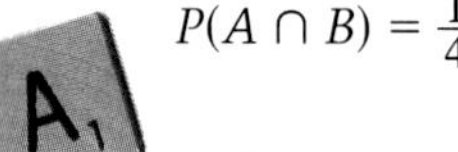

11. *Six Girls* Consider all families that have exactly six children. If one of these families is chosen at random, what is the probability that all six children are girls? $\frac{1}{64}$

✪ 12. *SCRABBLE^R* You and your friend are playing SCRABBLE^R. The distribution of letters is shown at the right. If you pick your seven letters first, what is the probability that you will pick three vowels and four consonants? (Count "Y" as a vowel.) ≈0.262

A: 9	G: 3	M: 2	S: 4	Y: 2
B: 2	H: 2	N: 6	T: 6	Z: 1
C: 2	I: 9	O: 8	U: 4	Blank: 2
D: 4	J: 1	P: 2	V: 2	
E: 12	K: 1	Q: 1	W: 2	
F: 2	L: 4	R: 6	X: 1	

13. *Numbers in a Hat* There are 25 pieces of paper, numbered 1 to 25, in a hat. You pick a piece of paper, replace it, and then pick another piece of paper. What is the probability that each number is greater than 20 or less than 4? $\frac{64}{625}$

12. 0.262

14. *Pick a Prize* There are 30 gifts for 30 people at a party. To determine who gets which gift, numbers from 1 to 30 are drawn. Starting with the person who drew the number 1, each person opens a gift. As each person opens a gift, he or she can retain the gift or switch with any previously opened gift. (At the end, the person who drew the number 30 has a choice of any of the 30 gifts.) What is the probability that you and your best friend will draw numbers 29 and 30 (in either order). $\frac{1}{435}$

15. *Baby Shower* There are 7 prizes at a baby shower. Each person at the shower places a tag by 3 prizes. You choose the first 3 prizes and, including your tags, there are 10 tags by the first prize, 8 tags by the second prize, and 18 tags by the third prize. To determine the winner of each prize, a tag beside the prize is chosen at random. What is the probability that you will win all 3 prizes? What is the probability that you won't win any of the prizes? $\frac{1}{1440}, \frac{119}{160}$

16. *Wrong Number* A friend of yours works as a telephone operator. From past records, your friend has a probability of about 0.001 of connecting a caller to an incorrect number. Last Tuesday, your friend connected 906 callers. What is the probability that at least 1 was connected to an incorrect number? ≈ 0.596

17. *American Travelers* In 1990, three percent of all Americans who took trips traveled outside the United States. Twenty Americans who took a trip in 1990 are chosen at random. What is the probability that at least one traveled out of the United States? ≈ 0.456

18. *Birthdays* If there are 26 people in your algebra class, what is the probability that at least 2 of you have the same birthday? 0.597

19. *Technology* Modify the graphing calculator program shown on page 811 to choose random numbers from 1 through 366. Use the program to choose 26 numbers. What is the probability that at least 2 of the numbers are the same? Are at least 2 of the numbers the same? Answers will vary

20. *Popping Balloons* Your school is having a fair. At the sundae bar, the price of a sundae is determined by choosing a balloon with a price inside. When you get to the sundae bar, there are 20 balloons left, 3 of which contain tags marked "FREE." If you and two of your friends get sundaes, what is the probability that at least one of you gets a free sundae? ≈ 0.404

21. *Ping Pong Balls* Marcus had 4 table tennis balls with small cracks. His friend, Sasha, accidentally mixed them in with 16 good ones. If Marcus randomly picks 3 table tennis balls, what is the probability that at least 1 is cracked? ≈ 0.509

Guided Practice

▶ **Ex. 1–6** Use these Guided Practice exercises as a check for understanding before assigning the Independent Practice for homework. Students will need to reflect on models of dependent versus independent events.

Independent Practice

▶ **Ex. 7–10** Ask students to write the formula $P(A \cap B) = P(A) \cdot P(B)$ and then evaluate the formula using the appropriate values.

▶ **Ex. 13** This exercise is an example of independent events due to the replacement. It is similar to Example 1 on page 811.

▶ **Ex. 16–21** Examples 2 and 3 can be used as solution models for these exercises. Encourage students to use the "complement" of the event to help find the solution. Calculators can be helpful with the required repeated multiplications.

▶ **Ex. 22–23** You might need to have students discuss these exercises in class. In Exercise 22 2 of the 66 remaining numbers will allow you to win. So the probability is $\frac{2}{66} = \frac{1}{33}$.

EXTEND Ex. 24 RESEARCH Have students who are interested find out more about conditional probability in a statistics text. Ask students to look up Bayes' Formula, a general rule for conditional probability.

Enrichment Activities

These activities require students to go beyond lesson goals.

1. Choose two movies that are currently popular and create a table like that shown below by surveying students in the school.

Saw movie	Grade 9	10	11	12	Total
X	4	11	28	16	59
Y	8	3	9	13	33
Total	12	14	37	29	92

2. Create probability questions similar to those shown in the lesson, which can be solved by using the table. Include the solutions with your questions.

   Sample questions:

   a. If a student is randomly selected, find the probability that the student saw movie X.

   $\frac{59}{92}$

   b. What is the probability that the student saw movie X given that the student is an 11th grader?

   $\frac{28}{37}$

3. **EXTEND** Ex. 24 How can you use the table to verify the formula for dependent events:

   $P(B|A) = \frac{P(A \cap B)}{P(A)}$?

Consider the two sample questions. If you let event A be the selecting of a student who saw Movie X, then the condition that the student must be an 11th grader is event B. Using the table, a total of 59 students saw Movie X, 28 of whom are in grade 11. So,

$P(X|\text{11th grade}) = \frac{P(X \cap \text{11th grade})}{P(X)}$

$= \frac{28}{59}$.

Integrated Review

In Exercises 22 and 23, use the following information.

You and your friends are playing a game. Each player receives a card that has 25 different numbers from 1 through 75. Your card is shown at the right. One player randomly draws numbers from 1 through 75 from a basket (without replacement). As your number is called, you mark the appropriate space on your card. The goal is to mark five squares in a row on your card, either horizontally, vertically, or diagonally.

5	16	41	48	72
1	30	44	49	69
10	28	39	50	65
4	19	42	55	73
2	20	35	59	75

22. The numbers that have been drawn so far are 5, 39, 55, 2, 72, 1, 73, 19, and 30. What is the probability that you will win the game with the next number called? $\frac{1}{33}$

23. The numbers that have been drawn so far are 49, 39, 4, 42, 72, 69, 55, and 19. What is the probability that you will win the game with the next number called? $\frac{2}{67}$

Exploration and Extension

24. *Choosing Marbles* If A and B are *dependent* events, then the probability that event B will occur given that event A has occurred is given by $P(B|A)$ and read as "the probability of B given A." This probability is

$$P(B|A) = \frac{P(A \cap B)}{P(A)}$$

where $P(A \cap B)$ is the probability that *both* A and B occur. A jar contains eight white marbles numbered 1 through 8 and six blue marbles numbered 1 through 6. What is the probability that a randomly chosen marble is white, given that the number on the marble is 4? $\frac{1}{2}$

Mixed REVIEW

3., 6., 10., 12. See Additional Answers. 4. $\frac{7}{3y-3}$

1. Does $\sin(-x) = -\sin x$? **(14.3)** Yes

2. Evaluate $\frac{1}{2} \cdot 4 \cdot 3 \cdot \sin \frac{\pi}{4}$. **(13.6)** $3\sqrt{2}$

3. Sketch $y = (x-3)^2 + 2$. **(11.5)**

4. Solve $3xy^2 - 3xy = 7y$ for x. **(1.5)**

5. Solve $3s + 4 \le 5s - 6$. **(1.6)** $s \ge 5$

6. Sketch $|x - 4| < 5$. **(1.6)**

7. Solve for θ: $\tan \theta = \sqrt{3}$. **(13.4)** $\frac{\pi}{3}, \frac{4\pi}{3}$

8. Solve $|3x - 2| = 12$. **(1.7)** $\frac{14}{3}, \frac{-10}{3}$

9. Solve $0 = \frac{3}{2}x^2 - \frac{3}{4}x + 1$. **(5.6)** $\frac{3 \pm i\sqrt{87}}{12}$

10. Sketch $y > x^2 + x - 12$. **(5.7)**

11. Solve $\frac{1}{5}x^2 = 180$. **(7.1)** ± 30

12. Sketch $y = \sqrt{x} + 2$. **(7.6)**

13. Find the cofunction of $\cot\left(\frac{\pi}{2} - x\right)$. **(14.3)** $\tan x$

14. Find the x-intercepts of $y = 3x^2 + 4$. **(5.5)** None

15. Multiply: $(3x^2 - x + 2) \cdot (x + 1)$. **(9.1)** $3x^3 + 2x^2 + x + 2$

16. Evaluate $f(-3)$ for $f(x) = -x^2 + x^3$. **(6.1)** -36

17. Find $f(g(x))$ for $f(x) = 2x^2$ and $g(x) = x + 1$. $2(x+1)^2$

18. Write in standard form: $x^2 + 25y^2 = 25$. **(11.3)** $\frac{x^2}{5^2} + \frac{y^2}{1^2} = 1$

15.6 Expected Value

What you should learn:

Goal 1 How to find the expected value of a sample space

Goal 2 How to use expected value to answer questions about real-life situations

Why you should learn it:

You can use expected value to calculate how much you will win in games in which winners are determined randomly, such as random drawing.

Goal 1 Finding an Expected Value

Suppose you and a friend are playing a game. You flip a coin. If the coin lands heads up, your friend scores 1 point, and you lose 1 point. If the coin lands tails up, you score 1 point, and your friend loses 1 point. After playing the game many times, would you expect to have more, fewer, or the same number of points as when you started? The answer is that you should expect to end up with about the same number. You can expect to lose a point about half the time and win a point about half the time. Your *expected value* for this game is 0.

Expected Value of a Sample Space

A sample space is partitioned into n events, no two of which have any outcomes in common. The probabilities of the n events occurring are $P_1, P_2, P_3, \ldots, P_n$ and the **payoffs** for the n events are $x_1, x_2, x_3, \ldots, x_n$. The **expected value,** V, of the sample space is

$$V = P_1x_1 + P_2x_2 + P_3x_3 + \cdots + P_nx_n$$

where $P_1 + P_2 + P_3 + \cdots + P_n = 1$.

Connections
Game Theory

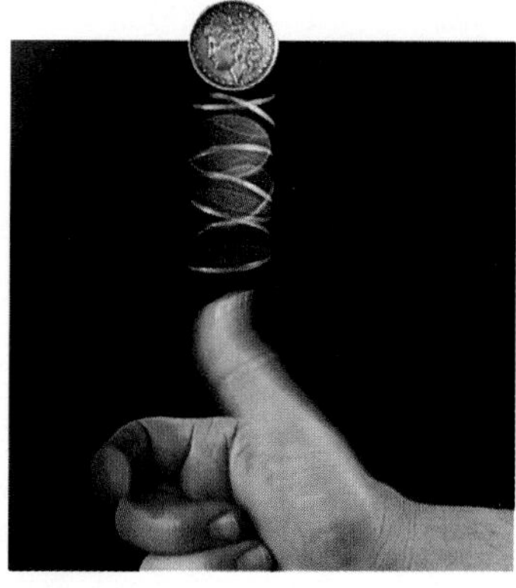

Example 1 *Finding the Expected Value of a Game*

You and a friend each flip a coin. If both coins land heads up, then your friend scores 3 points, and you lose 3 points. If one or two coins land tails up, then you score 1 point, and your friend loses 1 point. What is the expected value of this game?

Solution When two coins are tossed, four outcomes are possible: HH, HT, TH, TT.

Let event A be HH. Then event A' is HT, TH, *or* TT. The probability of A is $\frac{1}{4}$, and the probability of A' is $\frac{3}{4}$. *From your point of view,* the payoff for event A is -3 and the payoff for event A' is 1. Therefore, the expected value of the game is

$$V = P(A)\cdot(-3) + P(A')\cdot(1) = \tfrac{1}{4}(-3) + \tfrac{3}{4}(1) = 0.$$

The game is called **fair** because its expected value is 0. ■

Problem of the Day

Of the homeowners in a certain town, $\frac{4}{5}$ are women and $\frac{2}{3}$ of all women who own homes are married. Three-fifths of all women have no property, although half of them are married. What fraction of the women are unmarried homeowners?
$\frac{2}{15}$

ORGANIZER

Warm-Up Exercises

An experiment is performed where a die is rolled and then a coin is tossed. Find each probability.

1. $P(6)$ $\frac{1}{6}$
2. $P(H)$ $\frac{1}{2}$
3. $P(5, H)$ $\frac{1}{12}$

Lesson Resources

Teaching Tools
- Problem of the Day: 15.6
- Warm-up Exercises: 15.6
- Answer Masters: 15.6

Extra Practice: 15.6

LESSON Notes

Goal 1 Present the concept of the Expected Value of a Sample Space carefully. The sample space must be partitioned into events that have no outcomes in common. Reinforce the idea that the expected value is the sum of the products of the payoffs for each event times the probability that each event will occur.

Example 1

A game is called "fair" if its expected value is 0. As a class project, students can pick games of chance and compute the expected values of the games they choose.

Example 2

The students must compute what the person receives from the insurance company when he or she has an accident. Students who are interested in buying a car may want to do some research about car insurance, car loans, and other costs of operating a car. Some of this information can be found in a consumer mathematics textbook.

Example 3

Fast food chains are always running contests. Students can use this example as a model to compute the expected value of raffle tickets sold when fund raising for school clubs and organizations.

Real Life
Insurance

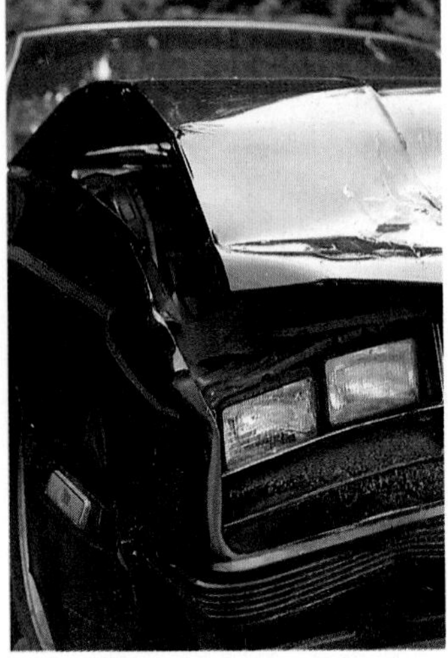

In 1993, there were 121,100,000 cars in use in the United States. That year, there were 14,100,000 automobile accidents.

Connections
Game Theory

Example 2 *Finding an Expected Value*

The average premium paid in 1993 for automobile collision insurance was \$638, and the average automobile collision claim paid by insurance companies was \$1796. What was the expected value of the insurance coverage for a person who paid \$638 for collision insurance in 1993?

Solution Let event A represent having an automobile accident. Then event A' represents not having an automobile accident. In 1993, the probability of having an automobile accident was

$$P(A) = \frac{14{,}100{,}000}{121{,}100{,}000} \approx 0.116$$

which means $P(A') \approx 0.884$. You can calculate the payoffs for A and A' as follows. If the person had an accident, he or she paid \$638 and received \$1796, so the person received \$1796 − \$638 = \$1158. If the person didn't have an accident, he or she paid \$638. Thus the expected value was

$$\begin{aligned} V &= P(A) \cdot (1158) + P(A') \cdot (-638) \\ &= 0.116(1158) - 0.884(638) \approx -\$429.66. \end{aligned}$$

The \$429.66 represents an average amount paid by each policy-holder to cover insurance company expenses (and produce a profit for the insurance companies). ■

Example 3 *Entering a Contest*

A fast-food restaurant chain is having a contest. No purchase is necessary to enter. The contest has five prizes. What is the expected value for a contest ticket?

Prize	Value	Probability of Winning
Soda Pop	\$0.69	0.01
Hamburger	\$1.29	0.005
Double Cheeseburger	\$2.59	0.002
Scholarship	\$1,000.00	0.000005
New Home	\$100,000.00	0.00000004

Solution Let event A represent not winning a prize. Because you paid nothing for the ticket, it is not necessary to compute the probability of A. The expected value is as follows.

$$\begin{aligned} V &= 0.01(0.69) + 0.005(1.29) + 0.002(2.59) + 0.000005(1000) \\ &\quad + 0.00000004(100{,}000) \\ &= 0.0069 + 0.00645 + 0.00518 + 0.005 + 0.004 \\ &= 0.02753 \end{aligned}$$

Thus the average amount won per ticket is about $2\frac{3}{4}$ cents. ■

Goal 2 Using Expected Value in Real Life

Real Life **Sales**

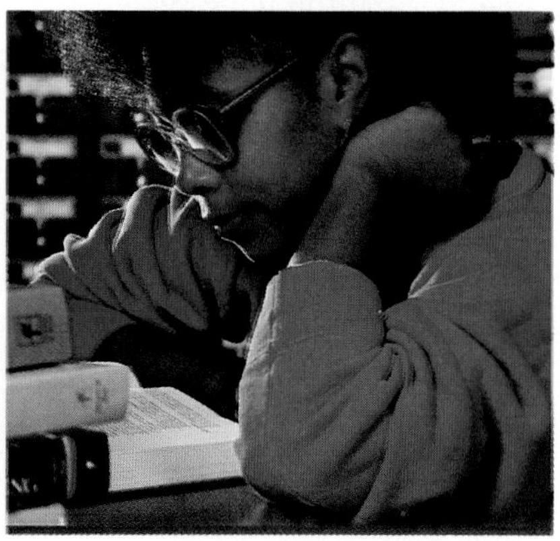

Example 4 *Estimating Sales*

You are a sales representative for a publisher that sells college textbooks. It is March, and your regional manager has asked you to prepare a "top-20 list." (A top-20 list contains your 20 largest potential sales for next fall.) With each potential sale, you are asked to estimate the probability that you will get the sale. Your top-20 list is shown below. What is the expected value of each entry on the list? What is the total expected value of the list?

Text	Price	Copies	Prob.	Text	Price	Copies	Prob.
B	\$39.95	3000	0.40	G	\$45.95	900	0.70
A	\$35.95	2500	0.90	A	\$35.95	900	0.95
C	\$46.95	2000	0.75	C	\$46.95	800	0.25
C	\$46.95	2000	0.25	D	\$42.95	750	0.40
E	\$28.95	1750	0.50	A	\$35.95	700	0.30
A	\$35.95	1500	0.50	E	\$28.95	600	0.80
D	\$42.95	1500	0 60	E	\$28.95	500	0.75
D	\$42.95	1200	0.75	G	\$45.95	500	0.90
E	\$28.95	1100	0.70	B	\$39.95	400	0.60
F	\$37.95	1000	0.90	A	\$35.95	400	0.50

Expected Values

\$47,940.00	\$28,948.50
\$80,887.50	\$30,737.25
\$70,425.00	\$9,390.00
\$23,475.00	\$12,885.00
\$25,331.25	\$7,549.50
\$26,962.50	\$13,896.00
\$38,655.00	\$10,856.25
\$38,655.00	\$20,677.50
\$22,291.50	\$9,588.00
\$34,155.00	\$7,190.00

Solution The expected value of each potential sale can be found by multiplying the price by the number of copies by the probability. For instance, the expected value of the first text on the list is $V = (39.95)(3000)(0.40) = \$47,940$. The expected values of all 20 potential sales are shown at the left. The expected value of the list is the total of these 20 expected values, or \$560,495.75. ■

Communicating about ALGEBRA

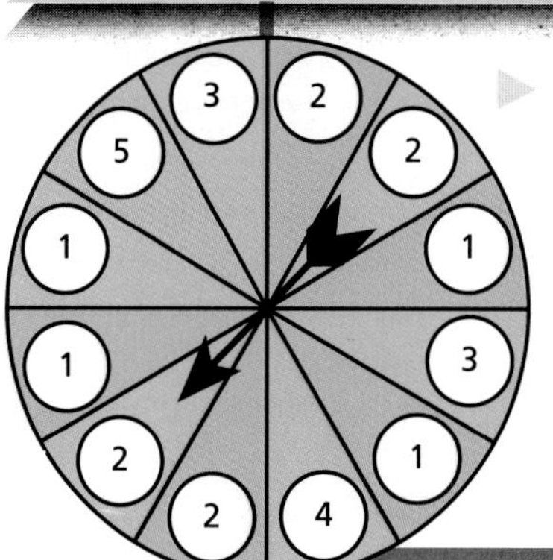

SHARING IDEAS about the Lesson

The spinner at the left is used in a game for two people: the Blue Player and the Green Player. Whichever color the spinner points to is the player who receives the indicated number of points from the other player.

A. What is the expected value for the Blue Player? $\frac{1}{4}$

B. What is the expected value for the Green Player? $-\frac{1}{4}$

C. What number of points should be in the "southeast" region to make the game fair? 4

Example 4

The expected value of each potential sale is given by the following verbal model.

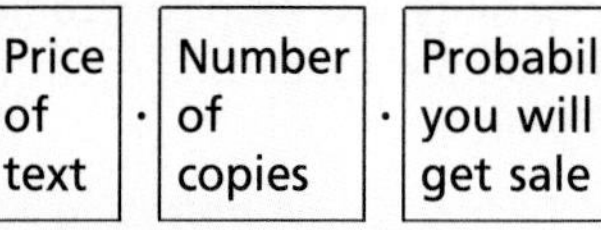

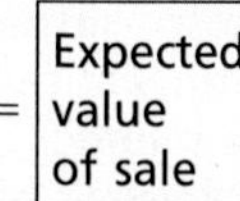

Extra Examples

Here are additional examples similar to **Examples 1–4.**

1. You and a friend flip a coin 3 times. If one or two coins land tails up, then you score 3 points. Otherwise, your friend scores 1 point. What is your expected value of this game? Is the game fair? 2; No
2. Using the information in Example 3, what is the probability of not winning a prize? ≈0.983
3. A school sells 500 raffle tickets for \$1 each. The prize is a \$200 racing bicycle. What is the expected value if you buy
   - **a.** 1 ticket?
   - **b.** 2 tickets?
   - **c.** 100 tickets?
   - **d.** 500 tickets?

   a. −\$0.60 b. −\$1.20 c. −\$60 d. −\$300

Check Understanding

1. In your own words, define the Expected Value of a Sample Space. See page 817.
2. What does it mean for a game to be "fair"? See page 817.

Communicating about ALGEBRA

Help the students compute the probability of each outcome. The sample space is the entire spinner.

EXTEND *Communicating*
Even though most lottery games are not mathematically fair, a considerable number of people play them anyway because of the potentially large payoffs. If your state has a lottery, have students investigate what the expected value of a certain kind of lottery ticket is. In some daily games, people have the choice of playing numbers in *exact* order or in *any* order. Why are the payoffs for the exact-order tickets so much greater than for the any-order tickets? Other games require you to pick 5 or more numbers in any order from a set of 36 to 50 numbers. These games are cumulative: the jackpot keeps growing until someone wins. Does the expected value keep growing from game to game?

EXERCISE Notes

ASSIGNMENT GUIDE

Basic/Average: Ex. 1–4, 5–11, 13–15, 18–21

Above Average: Ex. 1–4, 5–15, 17–21

Advanced: Ex. 1–4, 5–21

Selected Answers
Exercises 1–4, 5–19 odd

✪ More Difficult Exercises
Exercises 12–15

EXERCISES

Guided Practice

CRITICAL THINKING about the Lesson

1. A sample space has five equally likely outcomes. The payoffs for the outcomes are 1, 2, 3, 4, and 5. What is the expected value of the sample space? **3**

2. A sample space has five equally likely outcomes. The payoffs for four of the outcomes are 1, 2, 3, and 4. The expected value of the sample space is 5. What is the payoff for the fifth outcome? **15**

3. Two people are playing a *fair* game. What is the expected value for each player? Why is the game called "fair"? **Zero, neither person has an advantage.**

4. Two people are playing a game in which they exchange points, depending on the outcomes in the game. The expected value for one player is 3 points. What is the expected value for the other player? **−3**

Independent Practice

In Exercises 5 and 6, a sample space has five outcomes, each with the indicated payoff and probability. Find the expected value.

5.

Payoff	\$3	\$5	\$1	−\$3	−\$2
Probability	0.10	0.12	0.50	0.10	0.18

\$0.74

6.

Payoff	\$0	\$10	\$20	\$30	\$40
Probability	0.9955	0.0030	0.0010	0.0004	0.0001

\$0.066

Game Theory **In Exercises 7–10, consider a game in which each of two people simultaneously chooses an integer: 1, 2, or 3. Find the expected value for player *A* and the expected value for player *B*. Is each game fair?**

7. If the sum of the two numbers is odd, then player *A* loses that sum of points, and player *B* wins that sum. If the sum of the two numbers is even, player *B* loses 4 points, and player *A* wins 4 points.

8. If both players choose the same number, then no points are exchanged. Otherwise, the player with the higher number wins a point, and the other player loses a point. **A: 0, B: 0; Yes**

9. If both players choose the same number, then no points are exchanged. If the two numbers differ by one, then the player with the higher number wins two points and the other player loses two points. If the two numbers differ by two, then the player with the higher number wins one point and the other player loses one point.

10. If both players choose the same number, then no points are exchanged. If the sum of the two numbers is even, the player with the higher number wins two points and the other player loses two points. If the sum of the numbers is odd, the player with the lower number wins two points and the other player loses two points.

9. A: 0, B: 0, Yes **7. A: $\frac{4}{9}$, B: $-\frac{4}{9}$; No** **10. A: 0, B: 0, Yes**

11. *Fair Game?* Two players play *Stone, Scissors, Paper,* a game in which each player simultaneously represents either a stone, a pair of scissors, or a sheet of paper using one hand. If there is a tie, then neither player receives any points. When stone "breaks" scissors, scissors must give stone two points. When paper "covers" stone, stone must give paper three points. When scissors "cut" paper, paper must give scissors four points. Find the expected value to determine whether this game is fair. **0, the game is fair.**

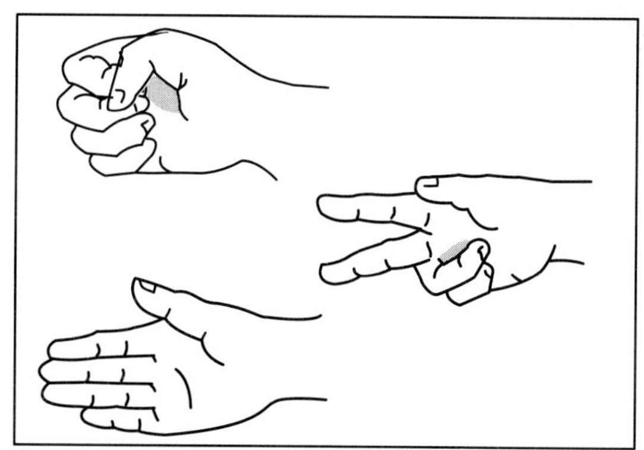

12. *Board Game* You are playing a board game. To advance around the board, you must first draw a card, roll a die three times, and add up the points you receive for each number rolled. (The card tells you the number of points each roll receives.) If the sum of the points is positive, you must advance that many spaces around the board; if it is negative, you must move backwards. Find the expected value for each of the four cards below.

a. 1

Number Rolled	Points
1	4
2	3
3	–5
4	2
5	1
6	–3

b. 1

Number Rolled	Points
1	3
2	–3
3	2
4	2
5	–3
6	1

c. $\frac{5}{2}$

Number Rolled	Points
1	6
2	–5
3	–1
4	2
5	0
6	3

d. $\frac{1}{2}$

Number Rolled	Points
1	–3
2	4
3	–1
4	–1
5	3
6	–1

13. *Buying Insurance* You have purchased a one-year policy to insure your contact lenses. The insurance fee is \$18. If you have to replace a pair of contact lenses, the insurance will pay \$30 toward the cost. Your optician told you that of 1000 people who purchase your type of contact lens, about 400 need to replace their lenses within a year. What is the expected value for this insurance policy? **–\$6**

14. *Roofing Jobs* You are a roofer and have six potential customers. The list below shows the price per square of shingles, the number of squares of shingles, and the probability that you will get the job. What is the expected income for these jobs? **\$3650**

Job	Price Installed	Number of Squares	Probability
A	\$90	12	0.9
B	\$130	9	0.5
C	\$125	10	0.3
D	\$100	7	0.1
E	\$140	10	0.8
F	\$110	8	0.6

Guided Practice

▶ **Ex. 1–4** These exercises can be used to reinforce the idea that the expected value of a game is the sum of the products of the payoffs for each event times the probability that each event will occur.

Independent Practice

▶ **Ex. 7** The game is not a "fair" game since the expected value is not zero. Ask students which player they would prefer to be and why?
Player A. On the average, Player A gains $\frac{4}{9}$ points per trial.

EXTEND Ex. 11 Have students play the game and record the results. Does this seem to be a "fair" game?
The game is "fair." Students will see this if they perform enough trials.

▶ **Ex. 13** The students can use Example 2 on page 818 as a solution model for this exercise. Ask students to discuss or explore the job of an actuary. This type of problem relates to establishing insurance premiums, which is part of the job of an actuary.

▶ **Ex. 14–15** The students can use Examples 3 and 4 on pages 818 and 819 as solution models for these exercises.

▶ **Ex. 21** Ask students if the expected value of 2.5 makes sense for a tossing of 5 coins.

Enrichment Activity

This activity requires students to go beyond lesson goals.

WRITING Game theory is a branch of mathematics that is relatively new and generates extensions and applications in probability theory.

Use library resources to write a one-page report on the origin of game theory, especially exploring the idea of a "zero-sum" game.

Game theory is a mathematical theory founded by John von Neumann (1903-1957), a Hungarian-born American mathematician and economist. A zero-sum game is one in which the sum of the winnings (and losings) of the various players is zero.

15. *Selling Yearbooks* You are a sales representative for a company that prints high school yearbooks. The following is a list of 16 potential customers, the prices of the books, the estimated sales, and the probability that you will make the sale. What is the expected value of your sales this year? **$46182.31**

School	Price	Copies	Probability	School	Price	Copies	Probability
1	12.95	225	0.60	**9**	15.95	25	0.95
2	11.00	450	0.85	**10**	14.00	175	0.80
3	10.50	100	0.80	**11**	18.00	500	0.75
4	13.00	200	0.65	**12**	16.00	400	0.80
5	15.00	100	0.70	**13**	12.00	750	0.80
6	11.00	600	0.90	**14**	12.50	450	0.65
7	11.50	50	0.60	**15**	14.50	80	0.90
8	14.00	75	0.85	**16**	10.50	400	0.80

Integrated Review

16. *Contract Bidding* Ten construction companies are bidding for a job. Three are local companies, three have state-wide operations, and four have national operations. Each company is equally likely to win the contract. What is the probability that the contract will be awarded to a local company? $\frac{3}{10}$

17. *Boy or Girl?* There are two children in a family. At least one of them is a boy. What is the probability that both children are boys? (*Hint:* The sample space is BB, BG, and GB.) $\frac{1}{3}$

18. *College Entrance Exam Sample* In the figure, what is the value of z? **b**

a. 80 **b.** 100 **c.** 110 **d.** 120 **e.** 130

19. *College Entrance Exam Sample* In the figure, what is the value of $x + y$? **c**

a. 140 **b.** 130 **c.** 120 **d.** 110 **e.** 100

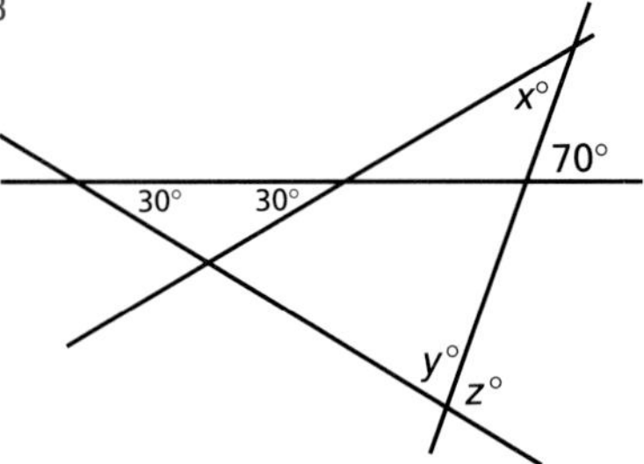

Exploration and Extension

Tossing Coins **In Exercises 20 and 21, consider an experiment in which five coins are tossed.**

20. Copy and complete the table that gives the probabilities for 0, 1, 2, 3, 4, or 5 heads turning up. $\frac{1}{32}, \frac{5}{32}, \frac{5}{16}, \frac{5}{16}, \frac{5}{32}, \frac{1}{32}$

Number of Heads	0	1	2	3	4	5
Probability	?	?	?	?	?	?

21. Find the expected number of heads that will turn up. $\frac{5}{2}$

15

Chapter Summary

What did you learn?

Skills

1. Find the probability of an event. **(15.1)**
2. Count the number of outcomes in an event
   - by making a list. **(15.1)**
   - by using the Fundamental Counting Principle. **(15.2)**
   - by using permutations. **(15.2)**
   - by using combinations. **(15.3)**
3. Find the probability of the union of two events. **(15.4)**
4. Find the probability of the complement of an event. **(15.5)**
5. Find the probability of the intersection of two events. **(15.4)**
6. Find the probability of independent events. **(15.5)**
7. Find the expected value of a sample space. **(15.6)**

Strategies

8. Use probability to solve real-life problems. **(15.1–15.6)**

Why did you learn it?

We all have to cope with the uncertainties of life. When you plan a picnic, you consider the probability that it will rain. When you drive a car, you consider the probability that you will be involved in an accident. When you enter a contest, you consider the probability that you will win.

Understanding the basic principles of probability can help you make decisions. For instance, in Lesson 15.6 you saw how a sales representative can use probability to forecast sales. This can help the sales representative decide whether to expect a bonus.

Using probability to make decisions can be difficult. A common error is to think that probabilities have "memories." For instance, a person who has just tossed a coin that lands heads up might reason that the next toss must be tails up, because heads and tails occur half the time. This reasoning is not correct. Regardless of how many times the coin has been tossed, the probability that the next toss will be heads is still one half.

How does it fit into the bigger picture of algebra?

In this chapter, and for that matter, in the entire book, you have learned that there are many connections between the many areas of mathematics. Algebra is connected to geometry, to trigonometry, to statistics, and to probability. For instance, in Chapter 12 you studied Pascal's Triangle as it related to binomial coefficients. In this chapter, you saw that Pascal's Triangle is connected to probability.

ORGANIZER

TIME SCHEDULE All levels, two days

The Chapter Summary helps students organize the main ideas of the chapter. In this chapter, students learned how to systematically count outcomes using lists, permutations, and combinations, in order to find their probabilities. An important application of probability, expected values, was also presented.

Work with students to review the skills, strategies, and concepts of the chapter. The first day's homework assignment can be used as the basis for the second day of review.

COOPERATIVE LEARNING Encourage students to study together. Emphasize the importance of "teaching" a classmate how to perform a skill or how to recall a strategy. When students work together, everyone wins. The students receiving help get additional instruction and the students giving help gain a deeper understanding of the skills and concepts involved.

Chapter SUMMARY

Review the chapter by asking students, "What did you learn? Why did you learn it? How does it fit with the other algebra skills and concepts you have learned?"

Chapter REVIEW

ASSIGNMENT GUIDE

Basic/Average: Ex. 1–2, 3–23 odd, 24–29

Above Average: Ex. 3–23 odd, 24–31

Advanced: Ex. 5–23 odd, 24–34

✪ **More Difficult Exercises:** Exercises 30–34

▶ **Ex. 1–34** The Chapter Review exercises include a reference to the lesson in which the skill or strategy was first presented. Students may use these references as a study aid if they are uncertain about how to do a given exercise in the Chapter Review.

Chapter REVIEW

In Exercises 1 and 2, find the probability of a successful event given the number of successes and the number of possibilities.

1. Number of successes: 16
   Number of possibilities: 360 $\frac{2}{45}$
2. Number of successes: 3
   Number of possibilities: 135 $\frac{1}{45}$
3. *Choosing a Calculator* A calculator is chosen randomly from a display case that contains four solar calculators and eight battery-operated calculators. What is the probability that the calculator is solar? $\frac{1}{3}$
4. *Geometry* The two larger squares in the figure on the right have the same dimensions. The centers of these squares are labeled A and B. What is the probability that a randomly chosen point inside the figure will lie in the shaded region? $\frac{1}{7}$

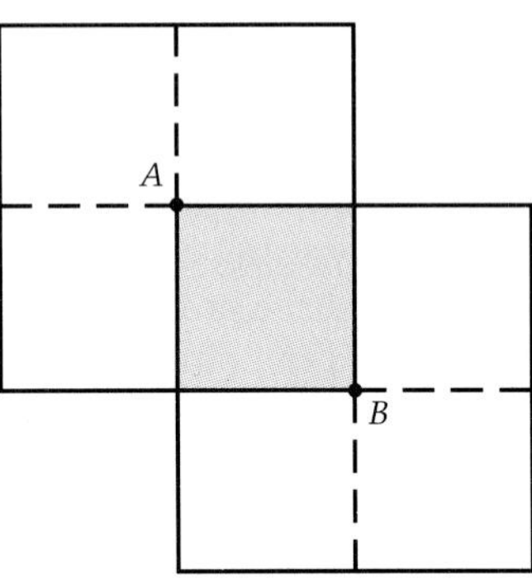

5. *Payroll Error* The employees of a company are in six departments: 31 are in sales, 54 are in research, 42 are in marketing, 20 are in engineering, 47 are in finance, and 58 are in production. If one employee's paycheck is lost, what is the probability that the employee is in the research department? $\frac{3}{14}$
6. *Defective Units* A shipment of 12 microwave ovens contains 3 defective units. A vending company has ordered 4 of these units, and since each is identically packaged, the selection will be at random. What is the probability that all 4 units are good? Exactly 2 units are good? At least 2 units are good? $\frac{14}{55}$, $\frac{12}{55}$, $\frac{54}{55}$
7. *Permuting Letters* In how many ways can you order the letters A, B, C, and D? 24
8. *Permuting Letters* How many permutations are possible for the letters in the word HOUSE? 120
9. *Making a Key* The key blank at the right has five notches, each of which has four possible heights. How many different keys can be made using this key blank? 1024
10. *Trying a Key* You have found a key that was made from a key blank like the one at the right. You recognize that the key to your apartment was made from the same type of blank. What is the probability that the key you found will open your door? $\frac{1}{1024}$

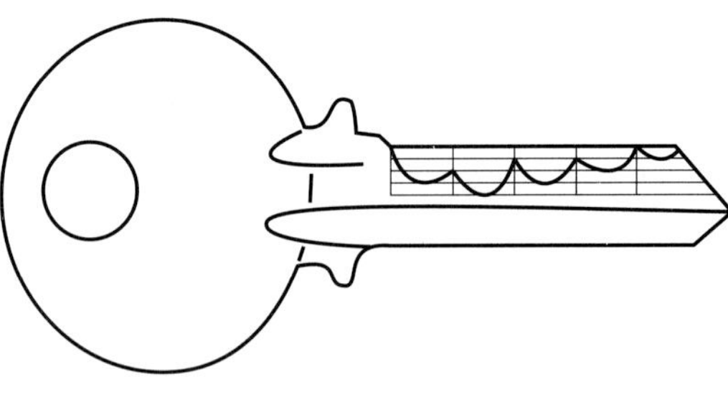

11. *Random Number Generator* A computer generates random integers from 1 through 20. What is the probability that the first two numbers are 5 or less? $\frac{1}{16}$

12. *School Pictures* There are 34 students in your homeroom. Numbers from 1 through 34 are randomly assigned to the students to determine the order in which your school pictures will be taken. What is the probability that you and your best friend will be the first and second to get your pictures taken (in either order)? ≈0.002

13. *Supreme Court* The Supreme Court of the United States has nine justices. On a certain case, the justices voted 5 to 4 in favor of the defendant. In how many ways could this have occurred? 126

14. *Supreme Court* On a certain case, the justices of the Supreme Court voted 7 to 2 in favor of the plaintiff. In how many ways could this have occurred? In which row of Pascal's Triangle does the number of ways occur? 36, Row 9

In 1981, Sandra Day O'Connor became the first woman to serve as a justice of the United States Supreme Court.

In Exercises 15 and 16, find the indicated probability.

15. $P(A) = \frac{1}{4}$

$P(B) = \frac{1}{5}$

$P(A \cap B) = \frac{2}{15}$

$P(A \cup B) = \boxed{?}$ $\frac{19}{60}$

16. $P(A) = \frac{2}{5}$

$P(B) = \frac{1}{10}$

$P(A \cap B) = \boxed{?}$ 0

$P(A \cup B) = \frac{1}{2}$

In Exercises 17 and 18, you are given the probability that an event will happen. Find the probability that the event will not happen.

17. $P = 0.04$ 0.96

18. $P = 0.63$ 0.37

19. *Asian-Americans* The circle graph at the right is representative of the relative percents of Asian-Americans living in the United States in 1990. If 2 people are chosen randomly from this group of 219 people, what is the probability that both are Asian-Indians? ***(Source: United States Bureau of Census)*** ≈0.022

20. *Asian-Americans* If a random sample of 219 Asian-Americans are polled, what is the probability that the first 3 people are Japanese-Americans? ≈0.003

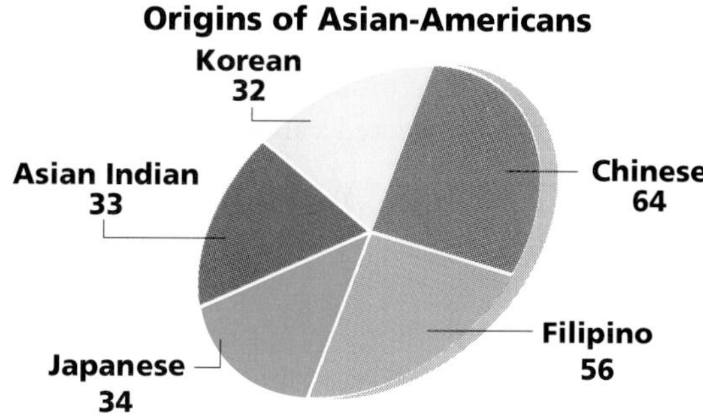

Answers

29. Use the ninth row of Pascal's Triangle. The first entry is the number of ways 0 heads and 8 tails can occur, the second entry is the number of ways 1 head and 7 tails can occur, etc. The largest entry is 70, representing the number of ways 4 heads and 4 tails can occur.

30.

	SM	Sm	sM	sm
Sm	SSMm	SSmm	SsMm	Ssmm
Sm	SSMm	SSmm	SsMm	Ssmm
sm	SsMm	Ssmm	ssMm	ssmm
sm	SsMm	Ssmm	ssMm	ssmm

31. Sable: $\frac{1}{8}$, tricolor: $\frac{1}{8}$, trifactored sable: $\frac{1}{4}$, blue merle: $\frac{1}{8}$, sable merle: $\frac{1}{8}$, trifactored sable merle: $\frac{1}{4}$

21. *Soccer or Baseball* You play soccer and baseball during the summer (June 1 through August 15). On 22 days you play soccer, and on 26 days you play baseball. On 36 days, you play soccer or baseball. On how many days do you play both? 12

22. *YAHTZEE®* In the game of *YAHTZEE®*, 5 six-sided dice are rolled. A "yahtzee" is when all 5 dice show the same number. What is the probability of this happening on a single roll of the 5 dice? $\frac{1}{1296}$

23. *Family Trip* Your family decides to visit the local amusement park twice during the summer (June 1 through August 15). During the summer, your city averages 32 days with rain and 10 days when the temperature is under 70°F. There are 37 days with rain *or* a temperature under 70°F. What is the probability that it will rain and be under 70°F on at least 1 of the days planned for your visits? ≈ 0.127

The World's Children **In Exercises 24 and 25, use the following information.**

In 1990, 33% of the world's population were under the age of 15 and 6% were over 65. Suppose that 20 people are chosen at random.

24. What is the probability that none of the 20 chosen are under the age of 15? 0.0003

25. What is the probability that none of the 20 chosen are over the age of 65? 0.290

26. *The World's Languages* The graph at the right shows the percents of people in the world who speak the world's most-used languages as their primary language. (The "other" includes, in descending order of frequency, Bengali, Portuguese, Indonesian, Japanese, German, French, and 200 other languages.) In a random sample of two people, what is the probability that both people speak Spanish *or* English as their primary language? 0.0225

Languages of the World's Population

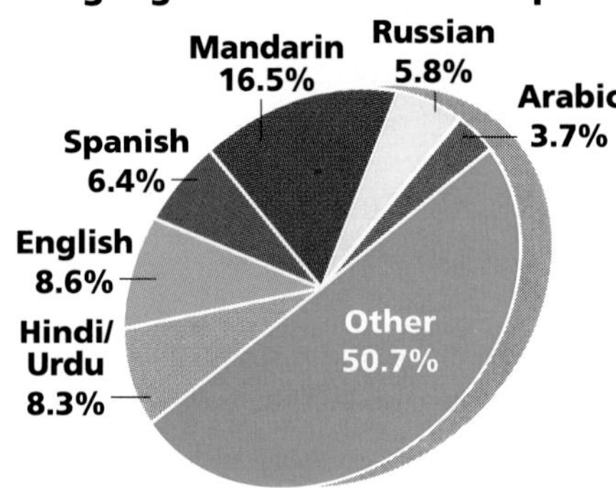

27. *Care Package* Your mother is sending a "care package" to your brother at college. She insures delivery of the package by paying \$1.60 extra. If the package is lost in the mail, your mother will collect \$60.00. The probability that the package will be lost is 0.001. What is the expected value of this insurance? −\$1.54

28. *Pascal's Triangle* Write the entire row of Pascal's triangle in which $\binom{8}{4}$ appears. 1 8 28 56 70 56 28 8 1

29. *Tossing Coins* Eight coins are tossed. Explain how to use Pascal's triangle to decide which of the following is most likely. e

a. 0 heads and 8 tails	**b.** 1 head and 7 tails	**c.** 2 heads and 6 tails
d. 3 heads and 5 tails	**e.** 4 heads and 4 tails	**f.** 5 heads and 3 tails
g. 6 heads and 2 tails	**h.** 7 heads and 1 tail	**i.** 8 heads and 0 tails

Collie Colors **In Exercises 30 and 31, use the following information.**

There are six basic types of coloring in registered collies: sable (SSmm), tricolor (ssmm), trifactored sable (Ssmm), blue merle (ssMm), sable merle (SSMm), and trifactored sable merle (SsMm).

Trifactored Sable

Trifactored Sable Merle

Blue Merle

Tri color

Sable Merle

Sable

30. Copy and complete the following Punnett square, which shows the possible colorings of offspring of a trifactored sable merle collie and a trifactored sable collie.

	SM	**Sm**	**sM**	**sm**
Sm				
Sm				
sm				
sm				

31. Use the results of Exercise 30 to find the probability of obtaining each of the six basic collie colorings in the offspring of the two collies described by the Punnett square.

32. *Choosing a Dog* Two breeds of dogs are randomly chosen from the following list. What is the probability that both breeds are spaniels? $\frac{1}{45}$

Boston Terrier	Collie
Cocker Spaniel	Beagle
Golden Retriever	Great Dane
Doberman Pinscher	King Charles Spaniel
Scottish Terrier	
Labrador Retriever	

33. *Dog Show* The fee for entering a dog in a dog show is \$50. The owner of the winning dog receives \$1000. Forty dogs are entered in the show. What is the expected value for each contestant? (Assume each contestant is equally likely to win.) −\$25

34. *Dog Show* In how many different orders can the 40 dogs in Exercise 33 be shown? $40! \approx 8.16 \times 10^{47}$

Chapter TEST

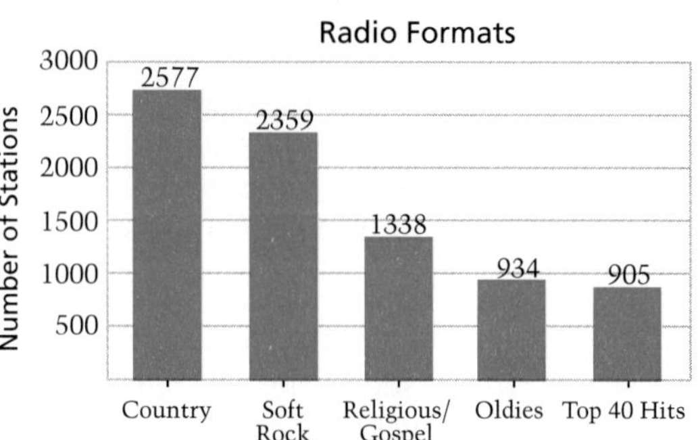

There are 10,924 AM and FM radio stations in the USA.

1. A radio is tuned to one of the radio stations in the United States. What is the probability that the radio station is an "oldies" station? What is the probability that it is a "country" station? **(15.1)**

2. A quiz consists of ten true or false questions. Each question is answered true or false. In how many ways can this be done? **(15.2)** **1024**

3. How many two-person teams can be chosen from eight tennis players? **(15.3)** **28**

1. ≈ 0.085, ≈ 0.236

4. Of the 15 players on a basketball team, six play guard, six play forward, and three play center. How many starting lineups can be formed from 15 players? (A starting lineup has two guards, two forwards, and one center.) **(15.3)** **675**

5. The directions on a snack cake mix instruct you to add the cake mix, water, oil, and eggs in a bowl. In how many orders can you put the ingredients into the bowl? **(15.2)** **24**

In Exercises 6–8, find the indicated probability. (15.4)

6.
$P(A) = 0.61$
$P(B) = 0.39$
$P(A \cup B) = 0.75$
$P(A \cap B) = \boxed{?}$ **0.25**

7.
$P(A) = \boxed{?}$
$P(B) = \frac{3}{8}$
$P(A \cup B) = \frac{7}{16}$
$P(A \cap B) = \frac{1}{16}$ $\frac{1}{8}$

8.
$P(A) = \frac{5}{16}$
$P(B) = \boxed{?}$
$P(A \cup B) = \frac{3}{4}$
$P(A \cap B) = 0$
$\frac{7}{16}$

9. Your mom claims that 90% of incoming phone calls are for you. Assuming that this claim is true, what is the probability that all of the next three phone calls are for you? **(15.5)** **0.729**

10. You want to invest \$500 in the stock of a new company. There is a 0.25 probability of receiving a profit of \$1000 in a year and a 0.75 probability of losing the \$500. What is the expected value of this investment? **(15.6)** **−\$125**

11. If $P(A) = 0.71$, what is $P(A')$? **(15.4)** **0.29**

12. If $P(A) = \frac{7}{12}$, what is $P(A')$? **(15.4)** $\frac{5}{12}$

In Exercise 13, use the following information.

One of the students from a high school will be chosen to attend a statewide meeting of high school students. The table at the right shows the composition of the high school. **(15.4)**

	Boys	Girls
Freshmen	178	150
Sophomores	188	185
Juniors	150	147
Seniors	143	166

13. What is the probability that the student chosen is a junior or senior? not a sophomore boy? a junior boy or a senior girl? ≈ 0.464, ≈ 0.856, ≈ 0.242

End-of-Book CONTENTS

PROPERTIES

		Addition	Multiplication
Basic Properties	**Closure**	$a + b$ is a real number	$a \cdot b$ is a real number
	Commutative	$a + b = b + a$	$ab = ba$
	Associative	$(a + b) + c = a + (b + c)$	$(ab)c = a(bc)$
	Identity	$a + 0 = a,\ 0 + a = a$	$a(1) = a,\ 1(a) = a$
	Inverse	$a + (-a) = 0$	$a \cdot \frac{1}{a} = 1,\ a \neq 0$
	Distributive	$a(b + c) = ab + ac$ or $(a + b)c = ac + bc$	
Properties of Exponents	**Product of Powers**	$a^m \cdot a^n = a^{m+n}$	
	Power of a Power	$(a^m)^n = a^{m \cdot n}$	
	Power of a Product	$(a \cdot b)^m = a^m \cdot b^m$	
	Quotient of Powers	$\frac{a^m}{a^n} = a^{m-n},\ a \neq 0$	
	Power of a Quotient	$\left(\frac{a}{b}\right)^m = \frac{a^m}{b^m},\ b \neq 0$	
	Negative Exponent	$a^{-n} = \frac{1}{a^n},\ a \neq 0$	
	Zero Exponent	$a^0 = 1,\ a \neq 0$	
Properties of Radicals	**Product Property**	$\sqrt{ab} = \sqrt{a} \cdot \sqrt{b}$	
	Quotient Property	$\sqrt{\frac{a}{b}} = \frac{\sqrt{a}}{\sqrt{b}}$	
Properties of Matrices	**Equal Matrices**	$\begin{bmatrix} a & b \\ c & d \end{bmatrix} = \begin{bmatrix} e & f \\ g & h \end{bmatrix}$ then $a = e,\ b = f,\ c = g,$ and $d = h.$	
	Adding Matrices	$\begin{bmatrix} a & b \\ c & d \end{bmatrix} + \begin{bmatrix} e & f \\ g & h \end{bmatrix} = \begin{bmatrix} a+e & b+f \\ c+g & d+h \end{bmatrix}$	
	Multiplying Matrices	$\begin{bmatrix} a & b \\ c & d \end{bmatrix} \cdot \begin{bmatrix} e & f \\ g & h \end{bmatrix} = \begin{bmatrix} ae+bg & af+bh \\ ce+dg & cf+dh \end{bmatrix}$	
Properties of Matrix Multiplication	**Associative Property**	$A(BC) = (AB)C$	A, B, and C are matrices.
	Distributive Property	$A(B + C) = AB + AC$	
	Distributive Property	$(A + B)C = AC + BC$	
	Scalar Multiplication	$c(AB) = (cA)B = A(cB)$	
Special Products and their Factors	**Sum and Difference Pattern**	$(a + b)(a - b) \longleftrightarrow a^2 - b^2$	
	Square of a Binomial Pattern	$(a + b)^2 \longleftrightarrow a^2 + 2ab + b^2$	
		$(a - b)^2 \longleftrightarrow a^2 - 2ab + b^2$	
	Sum of Cubes	$a^3 + b^3 \longleftrightarrow (a + b)(a^2 - ab + b^2)$	
	Difference of Cubes	$a^3 - b^3 \longleftrightarrow (a - b)(a^2 + ab + b^2)$	
Properties of Logarithms	Let a, u, and v be positive numbers such that $a \neq 1$.		
	Product Property	$\log_a (uv) = \log_a u + \log_a v$	
	Quotient Property	$\log_a \frac{u}{v} = \log_a u - \log_a v$	
	Power Property	$\log_a u^n = n \log_a u$	

MODELS

A Problem-Solving Plan	**Ask** yourself what you need to know to solve the problem. Then **write a verbal model** that will give you what you need to know. **Assign labels** to each part of your verbal model. Use the labels to **write an algebraic model** based on your verbal model. **Solve** the algebraic model. **Answer** the original question. **Check** that your answer is reasonable.
Summary of Equations of Lines in Two Variables	**Slope of a Line through Two Points** $m = \frac{(y_2 - y_1)}{(x_2 - x_1)}$ **Standard Form** $Ax + By = C$ **Intercept Form** $\frac{x}{a} + \frac{y}{b} = 1$ **Slope-intercept Form** $y = mx + b$ **Point-slope Form** $y - y_1 = m(x - x_1)$
Algebraic Models	**Linear** $f(x) = mx + b$ **Quadratic** $f(x) = ax^2 + bx + c$ or $f(x) = a(x - h)^2 + k$ **Rational** $f(x) = \frac{a}{x - h} + k$ **Exponential Growth Model, $a > 1$, $C > 0$** $y = Ca^x$ **Exponential Decay Model, $0 < a < 1$, $C > 0$** $y = Ca^x$
Models for Conic Sections	In a parabola the point (h, k) is the vertex, and p is the distance from the focus to the vertex. For the other conics, (h, k) is the center. **Circle** $(x - h)^2 + (y - k)^2 = r^2$ Focus (or foci) on $x = h$ / Focus (or foci) on $y = k$ **Parabola** $(y - k)^2 = 4p(x - h)$ / $(x - h)^2 = 4p(y - k)$ **Ellipse** $\frac{(x - h)^2}{a^2} + \frac{(y - k)^2}{b^2} = 1$ / $\frac{(x - h)^2}{b^2} + \frac{(y - k)^2}{a^2} = 1$ **Hyperbola** $\frac{(x - h)^2}{a^2} - \frac{(y - k)^2}{b^2} = 1$ / $\frac{(y - k)^2}{a^2} - \frac{(x - h)^2}{b^2} = 1$
Models for Trigonometric Functions	Let θ be an angle in standard position with any point (x, y) (except the origin) on the terminal side of θ. Let $r = \sqrt{x^2 + y^2}$. $\sin \theta = \frac{y}{r}$ $\cos \theta = \frac{x}{r}$ $\tan \theta = \frac{y}{x}, x \neq 0$ $\csc \theta = \frac{r}{y}, y \neq 0$ $\sec \theta = \frac{r}{x}, x \neq 0$ $\cot \theta = \frac{x}{y}, y \neq 0$
Models for Direct and Inverse Variation	**1.** The variables x and y vary directly if, for a constant k, $\frac{y}{x} = k$ or $y = kx$ **2.** The variables x and y vary inversely if, for a constant k, $yx = k$ or $y = \frac{k}{x}$ In both cases, the number k is the constant of variation.

TRANSFORMATIONS

Using Transformations to Solve Equations, Inequalities, and Systems

Transformations that Produce Equivalent Equations

1. Add or subtract the same number to *both* sides.
2. Multiply or divide *both* sides by the same nonzero number.
3. Interchange the two sides.
4. Raise *both* sides to the same power. (Be careful, this may give an extraneous solution.)

Transformations that Produce Equivalent Inequalities

1. Add or subtract the same number to *both* sides.
2. Multiply or divide *both* sides by the same *positive* number.
3. Multiply or divide *both* sides by the same *negative* number and reverse the inequality sign.

Transformations of Absolute Value Inequalities

1. The inequality $|ax + b| < c$ means that $ax + b$ is between $-c$ and c. This is equivalent to $-c < ax + b < c$.
2. The inequality $|ax + b| > c$ means that $ax + b$ is not between $-c$ and c inclusive. This is equivalent to $ax + b < -c$ or $ax + b > c$.

Equivalent Systems of Equations

Each of the following **row operations** on a system of linear equations produces an equivalent system of linear equations.

1. Interchange two equations.
2. Multiply one of the equations by a nonzero constant.
3. Add a multiple of one of the equations to another equation to replace an equation.

Using Transformations to Graph Related Functions

Vertical and Horizontal Shifts

Let c be a positive real number. **Vertical** and **horizontal** shifts of the graph of f are represented as follows.

1. Vertical shift of c units upward: $g(x) = f(x) + c$
2. Vertical shift of c units downward: $g(x) = f(x) - c$
3. Horizontal shift of c units to the right: $g(x) = f(x - c)$
4. Horizontal shift of c units to the left: $g(x) = f(x + c)$

Reflection in the *x*-Axis

If $g(x) = -f(x)$, then the graph of g is a reflection in the x-axis of the graph of f.

Shifts of Sine and Cosine Graphs

To graph $y = a \sin(bx + c) + d$ or $y = a \cos(bx + c) + d$, shift the graph of $y = a \sin bx$ or $y = a \cos bx$ as follows.

1. If $d > 0$, shift d units up.
2. If $d < 0$, shift $|d|$ units down.
3. If $\frac{c}{b} > 0$, shift, $\frac{c}{b}$ units to the left.
4. If $\frac{c}{b} < 0$, shift, $\left|\frac{c}{b}\right|$ units to the right.

FORMULAS

The Inverse of a 2 × 2 Matrix	The inverse of the matrix $A = \begin{bmatrix} a & b \\ c & d \end{bmatrix}$ is $A^{-1} = \frac{1}{\lvert A \rvert}\begin{bmatrix} d & -b \\ -c & a \end{bmatrix} = \frac{1}{ad - cb}\begin{bmatrix} d & -b \\ -c & a \end{bmatrix}$, $ad - cb \neq 0$.
Cramer's Rule	Consider a linear system whose coefficient matrix has a determinant D. If $D \neq 0$, then the system has exactly one solution. **1.** The solution of $\begin{cases} ax + by = c \\ dx + ey = f \end{cases}$ is $x = \dfrac{\begin{vmatrix} c & b \\ f & e \end{vmatrix}}{\begin{vmatrix} a & b \\ d & e \end{vmatrix}}$ and $y = \dfrac{\begin{vmatrix} a & c \\ d & f \end{vmatrix}}{\begin{vmatrix} a & b \\ d & e \end{vmatrix}}$ **2.** The solution of $\begin{cases} ax + by + cz = d \\ ex + fy + gz = h \\ ix + jy + kz = l \end{cases}$ is $x = \dfrac{\begin{vmatrix} d & b & c \\ h & f & g \\ l & j & k \end{vmatrix}}{\begin{vmatrix} a & b & c \\ e & f & g \\ i & j & k \end{vmatrix}}$, $y = \dfrac{\begin{vmatrix} a & d & c \\ e & h & g \\ i & l & k \end{vmatrix}}{\begin{vmatrix} a & b & c \\ e & f & g \\ i & j & k \end{vmatrix}}$, and $z = \dfrac{\begin{vmatrix} a & b & d \\ e & f & h \\ i & j & l \end{vmatrix}}{\begin{vmatrix} a & b & c \\ e & f & g \\ i & j & k \end{vmatrix}}$
The Quadratic Formula	Let a, b, and c be real numbers such that $a \neq 0$. The solutions of the quadratic equation $ax^2 + bx + c = 0$ are $x = \dfrac{-b \pm \sqrt{b^2 - 4ac}}{2a}$
Interest	An initial principal of P is deposited in an account that pays interest at an annual rate of r. The balance, A, in the account after t years is as follows. **Compounded *n* times per year:** $A = P\left(1 + \frac{r}{n}\right)^{nt}$. **Continuously compounded:** $A = Pe^{rt}$
The Distance Formula	The distance d between the points (x_1, y_1) and (x_2, y_2) is $d = \sqrt{(x_2 - x_1)^2 + (y_2 - y_1)^2}$
Standard Deviation of a Collection of Numbers	The standard deviation of $x_1, x_2, x_3, \ldots, x_n$ is $s = \sqrt{\dfrac{(x_1 - \overline{x})^2 + (x_2 - \overline{x})^2 + \cdots + (x_n - \overline{x})^2}{n}}$ $= \sqrt{\dfrac{x_1{}^2 + x_2{}^2 + \cdots + x_n{}^2}{n} - \overline{x}^2}$ where $\overline{x}$ is the mean of $x_1, x_2, \ldots,$ and x_n.

Conic Sections	**Ellipse** If a, b, and c represent the distances from the center of an ellipse to a vertex, covertex, and a focus, respectively, then: $a^2 = b^2 + c^2$. **Hyperbola** If a and c are the distances from the center (h, k) to a vertex and a focus, respectively, then b is related to the hyperbola by the equation $a^2 + b^2 = c^2$, and the equations for the asymptotes are: $y - k = \pm\frac{b}{a}x - h$ when the transverse axis is horizontal, or $y - k = \pm\frac{a}{b}x - h$ when the transverse axis is vertical.
Formulas for Special Series	**1.** $\sum_{i=1}^{n} 1 = n$ **2.** $\sum_{i=1}^{n} i = \frac{1}{2}(n)(n+1)$ **3.** $\sum_{i=1}^{n} i^2 = \frac{1}{6}(n)(n+1)(2n+1)$
Sum of a Series	▪ The sum of an arithmetic series with n terms is $\sum_{i=1}^{n} a_i = n\left(\frac{a_1 + a_n}{2}\right)$ or $\frac{1}{2}n(2a_1 + (n-1)d)$. ▪ The sum of the geometric series $a_1 + a_1r + a_1r^2 + a_1r^3 + \cdots + a_1r^{n-1}$ with common ratio $r \neq 1$ is $\sum_{i=1}^{n} a_1r^{i-1} = a_1\left(\frac{1 - r^n}{1 - r}\right)$. ▪ If $\lvert r\rvert < 1$, then the sum of the infinite geometric series $a_1 + a_1r + a_1r^2 + a_1r^3 + \cdots + a_1r^{n-1} + \cdots$ is $\sum_{i=1}^{\infty} a_1r^{i-1} = \frac{a_1}{1 - r}$. If $\lvert r\rvert \geq 1$, then the series has no sum.

FORMULAS

Fundamental Trigonometric Identities	**Reciprocal** $\sin\theta = \frac{1}{\csc\theta}$ $\cos\theta = \frac{1}{\sec\theta}$ $\tan\theta = \frac{1}{\cot\theta}$ $\csc\theta = \frac{1}{\sin\theta}$ $\sec\theta = \frac{1}{\cos\theta}$ $\cot\theta = \frac{1}{\tan\theta}$ **Tangent and Cotangent** $\tan\theta = \frac{\sin\theta}{\cos\theta}$ $\cot\theta = \frac{\cos\theta}{\sin\theta}$ **Pythagorean** $\sin^2\theta + \cos^2\theta = 1$ $1 + \tan^2\theta = \sec^2\theta$ $1 + \cot^2\theta = \csc^2\theta$
Sum and Difference Formulas	$\sin(u+v) = \sin u \cos v + \cos u \sin v$ $\tan(u+v) = \frac{\tan u + \tan v}{1 - \tan u \tan v}$ $\sin(u-v) = \sin u \cos v - \cos u \sin v$ $\tan(u-v) = \frac{\tan u - \tan v}{1 + \tan u \tan v}$ $\cos(u+v) = \cos u \cos v - \sin u \sin v$ $\cos(u-v) = \cos u \cos v + \sin u \sin v$
Double- and Half-Angle Formulas	$\sin 2u = 2 \sin u \cos u$ $\cos 2u = \cos^2 u - \sin^2 u$ $\tan 2u = \frac{2\tan u}{1 - \tan^2 u}$ $= 2\cos^2 u - 1$ $= 1 - 2\sin^2 u$ $\sin\frac{u}{2} = \pm\sqrt{\frac{1-\cos u}{2}}$ $\cos\frac{u}{2} = \pm\sqrt{\frac{1+\cos u}{2}}$ $\tan\frac{u}{2} = \frac{1-\cos u}{\sin u}$ $\tan\frac{u}{2} = \frac{\sin u}{1+\cos u}$ The signs of $\sin\frac{u}{2}$ and $\cos\frac{u}{2}$ depend on the quadrant in which $\frac{u}{2}$ lies.
Law of Sines	If ABC is a triangle with sides a, b, and c, then $\frac{a}{\sin A} = \frac{b}{\sin B} = \frac{c}{\sin C}$ or $\frac{\sin A}{a} = \frac{\sin B}{b} = \frac{\sin C}{c}$.
Law of Cosines	Let ABC be a triangle with sides a, b, and c, then $a^2 = b^2 + c^2 - 2bc\cos A$, $b^2 = a^2 + c^2 - 2ac\cos B$, and $c^2 = a^2 + b^2 - 2ab\cos C$.
The Binomial Theorem	The binomial expansion of $(x+y)^n$ is $(x+y)^n = x^n + nx^{n-1}y + \cdots + \binom{n}{m}x^{n-m}y^m + \cdots + nxy^{n-1} + y^n$ where $\binom{n}{m} = \frac{n!}{(n-m)!m!}$

absolute value of a real number (44, 101) The distance between the origin and the point representing the real number. For example: $|3| = 3$, $|-3| = 3$, or $|0| = 0$.

acute angle (678) An angle whose measure is greater than 0° and less than 90°.

additive identity (4) See Properties page 830.

algebraic expression (9) A collection of numbers, variables, operations, and grouping symbols.

algebraic model (9) An algebraic expression or equation used to represent a real-life situation.

amortization (555) The process of repaying an installment loan.

amplitude (730, 731) The amount by which a sine or cosine graph varies above and below its axis.

annuities (658) Savings accounts that involve many equal deposits (or withdrawals) at regular intervals over a long period of time.

associative property (4, 259) See Properties on page 830.

associative property of matrices (175, 181) See Properties on page 830.

asymptote (421) A line to which a graph becomes arbitrarily close as $|x|$ or $|y|$ increases without bound.

augmented matrix (210) A matrix containing the coefficients and constants of a system of linear equations.

axis of symmetry of a parabola (236) The vertical line passing through the vertex of a parabola and dividing the parabola into two parts that are mirror images of each other.

bar graph (51) A graph that organizes a collection of data by using horizontal or vertical bars to display how many times each number occurs in the collection.

base of a logarithm (405) In logarithmic notation, $\log_a x$, a is the base.

base of an exponent (10, 346) The number or variable that is used as a factor in repeated multiplication. For example, in the expression 4^6, 4 is the base.

best-fitting line (107) A line that best fits the data points on a scatter plot.

binomial (456) A polynomial that has only two terms.

box-and-whisker plot (328) A plot used to present a graphic summary of information. The quartiles and extreme values of a set of data displayed using a number line.

branches (518) Two portions of a graph that are not connected.

catenary (249) A U-shaped curve that resembles a parabola.

centroid (33) The point at which a plane figure will balance.

change of base formula (406) A formula which makes it possible to use a calculator to evaluate logarithms in bases other than base 10.

$$\log_a x = \frac{\log_{10} x}{\log_{10} a}$$

circle graph (51) A circle partitioned into sectors used to show the relative size of different portions of a whole.

closure property (4) See Properties on page 830.

coefficient (456) In a term that is the product of a number and a variable, the number is the coefficient of the variable.

coefficient matrix (216) A matrix representation of the coefficients of the variables in a system of equations.

collinearity test (193) A test which involves finding the area of the "triangle" formed by three points to determine if those points are collinear (lie on the same line). If the area is zero, the points are collinear.

column matrix (174) A matrix that has only one column.

combinations (796) Any choice of m elements from a set of n elements.

common difference (629) The difference between successive terms in an arithmetic sequence.

common logarithmic function (406) A logarithmic function with base 10.

common ratio (637) The ratio of successive terms in a geometric sequence.

commutative property (4, 259) See Properties on page 830.

commutative property for adding matrices (175) If A and B are matrices, $A + B = B + A$.

complement of an event (806) If an event A is in a sample space S, then the complement of A consists of all outcomes in S that are not in A.

completed-square form (250) A quadratic equation in the form "perfect square = k," $k \geq 0$; a quadratic function in the form $f(x) = a(x - h)^2 + k$. The vertex of the parabola is the point (h, k).

completing the square (245) A process which can be used to solve any quadratic equation.

complex conjugate (496) A complex number of the form $a + bi$ is $a - bi$.

complex fraction (549) A fraction that contains at least one fraction in its numerator and/or denominator.

complex number (259) A number that is the sum of a real number and an imaginary number. Each complex number can be written in the standard form, $a + bi$.

complex plane (260) A plane formed by a real number line and an imaginary number line intersecting at a right angle. The horizontal axis is the real axis and the vertical axis is the imaginary axis.

composition of two functions (292) The composition of the function f with the function g is given by $h(x) = f(g(x))$.

compound function (306) A function that may be represented by a combination of equations.

compound inequality (38) Two inequalities connected by *and* or *or*.

compound interest (35, 352) Interest paid on the original principal and on interest that becomes part of the account.

constant of variation (88, 526) The constant k in the relationships of variation.

constant term (456) A monomial without a variable. For example, in $x^2 + 2x + 5$ the number 5 is the constant term.

constraints (151) The system of linear inequalities in a linear programming problem.

continuous function (463) A function whose graph has no breaks.

coordinate plane (64) A plane formed by two real number lines intersecting at a right angle.

correlation coefficient (110) A statistical measure used to determine how well a collection of data points can be modeled by a line. Correlation coefficients range between -1 and $+1$.

cosecant (678) See Formulas on page 835.

cosine (678) See Formulas on page 835.

cotangent (678) See Formulas on page 835.

Cramer's rule (216) See Formulas on page 833.

cycle (730) The shortest repeating portion of a periodic function.

data (50) The facts, or numbers, that describe something.

definition of division (4) For all real numbers a and b, division is defined as multiplying by the reciprocal, $\frac{a}{b} = a \cdot \frac{1}{b}$.

definition of subtraction (4) For all real numbers a and b, subtraction is defined as adding the opposite, $a - b = a + (-b)$.

degree of a monomial (456) The integer k is the degree of ax^k.

denominator (9) The divisor in a quotient expressed in fraction form. For example, 2 is the denominator of $\frac{5x}{2}$.

dependent events (811) Two events where the occurrence of one affects the occurrence of the other.

determinant (187) A real number associated with each square matrix. For example, $ad - cb$ is the determinant of the matrix.

diagonal method (189) A method used to evaluate the determinant of a 3×3 matrix, where the first and second columns of the matrix form fourth and fifth columns. The determinant is then obtained by adding the products of three diagonals and subtracting the products of three diagonals.

difference (4) The result obtained when numbers or expressions are subtracted. The difference of a and b is defined as $a - b$.

difference of two cubes (474) See Properties on page 830.

difference of two squares (458) See Properties on page 830.

direct variation (88) A function whose equation has the form $y = kx$ or $\frac{x}{y} = k$.

directrix (570) A fixed line from which all the points of a parabola are equidistant.

discriminant (253) The expression $b^2 - 4ac$ is for a quadratic equation $ax^2 + bx + c = 0$.

distance formula (376) The distance, d, between any two points, (x_1, y_1) and (x_2, y_2), is given by $d = \sqrt{(x_2 - x_1)^2 + (y_2 - y_1)^2}$.

distributive property (4, 259) See Properties on page 830.

distributive property for matrices (175, 181) See Properties on page 830.

dividing rational expressions (535) To divide one rational expression by another, multiply the first by the reciprocal of the second.
$\frac{a}{b} \div \frac{c}{d} = \frac{a}{b} \cdot \frac{d}{c}$

domain, or input, of a function (284) The x-values of a function $y = f(x)$.

domain of a rational expression (518) The set of all real numbers, except those for which the denominator is zero.

eccentricity of an ellipse (585) The measure of the ovalness of an ellipse.

elementary row operations (210) Row operations which when performed on an augmented matrix produce a matrix that is row-equivalent to the original matrix.

ellipse (583) The set of all points (x, y) such that the sum of the distances between (x, y) and two distinct fixed points, the focuses or foci, is a constant.

equal complex numbers (259) Two complex numbers, $a + bi$ and $c + di$, are equal if $a = c$ and $b = d$.

equal matrices (174) Matrices are equal if the entries in corresponding positions are equal.

equally-likely outcome (782) An event that is as likely to happen as not to happen.

equation (18) A statement formed when an equality symbol is placed between two equal expressions.

equivalent equations (18) Equations with the same solution set.

equivalent inequalities (37) Inequalities with the same solution set.

equivalent systems of equations (158) Two linear systems are equivalent if they have the same set of solutions.

Euler number (420) The natural base e.

evaluate a function (286) To find $f(x)$ for a given value of x is to evaluate the function f.

evaluate an algebraic expression (9) To find the value of an algebraic expression by replacing each variable in the expression by a number.

expansion by minors (188) A method that can be used to find the value of any third or higher order determinant. It expresses the determinant of a 3×3 matrix in terms of three 2×2 determinants.

expected value (817) The sum of the probabilities of n events occurring, each multiplied by its payoff.

experimental probability (802) A probability determined by performing a number of trials.

exponent (10, 346) The number or variable that represents the number of times the base is used as a factor. For example, in the expression 4^6, 6 is the exponent.

exponential decay (354) A situation in which the original amount is repeatedly multiplied by a change factor less than one.

exponential form of a power (346) The expression 4^6 is the exponential form of $4 \cdot 4 \cdot 4 \cdot 4 \cdot 4 \cdot 4$.

exponential growth (354) A situation in which the original amount is repeatedly multiplied by a change factor greater than one.

extraneous solutions (375, 542) A solution of a transformed equation that is not a solution of the original equation.

Factor Theorem (482) A polynomial $f(x)$ has a factor $(x - k)$ if and only if $f(k) = 0$.

factorial function (321) The function defined by the equation $f(n) = n!$ The product of the positive integers from n to 1. 0! is defined to be 1.

factors (473) The numbers and variables that are multiplied in an expression. For example, 4 and x are factors of $4x$.

Fibonacci sequence (323) The sequence 1, 1, 2, 3, 5, 8, 13, ... This special sequence, discovered by Leonardo Fibonacci, is often found in nature.

finite sequence (622) A sequence with a limited number of terms.

first differences of a recursive function (322) The first differences of a recursive function are found by subtracting consecutive values of the function.

FOIL pattern (457) A method used to multiply two binomials in a single step. Find the sum of the products of the **F**irst terms, **O**uter terms, **I**nner terms, and **L**ast terms.

formula (31) An algebraic expression that represents a relationship in real-life situations.

fractal (261) Pictures that are drawn using the complex plane.

frequency distribution (50) A table that organizes a collection of data by counting how many times each number occurs in the collection.

function (285) A special type of relationship between two values in which each input value corresponds to exactly one output value. A relation in which no two ordered pairs (x, y) have the same x-value.

Fundamental Counting Principle (788) If one event can occur in m ways and another in n ways, then the number of ways that both can occur is $m \cdot n$.

Fundamental Theorem of Algebra (495) Counting complex and repeated solutions, an nth-degree polynomial equation has exactly n solutions.

Gaussian elimination (158) The process of rewriting a system of linear equations into triangular form.

general form of a linear equation (78) See Models on page 831.

general form of a second-degree equation (606) See Models on page 831.

geometric mean (375) The geometric mean of a and b is $\sqrt{ab}$.

geometric sequence (637) A sequence in which each term after the first is the product of the preceding term and a constant, r. The constant is called the common ratio.

geometric series (639) The expression formed by adding the terms of a geometric sequence.

golden ratio (377, 668) The golden ratio is $\frac{\sqrt{5} + 1}{2}$.

graph of an equation (64) A visual model of an equation.

greatest common monomial factor (473) The greatest monomial factor that can be factored from the terms of a polynomial.

half-plane (94) In a plane, the region on either side of a boundary.

horizontal shift of a graph (314) See Transformations on page 832.

Hero's Area Formula (717) The area of the triangle with sides of length a, b, and c, is Area $= \sqrt{s(s - a)(s - b)(s - c)}$ where $s = \frac{1}{2}(a + b + c)$.

hyperbola (590) The graph of a rational function $f(x) = \frac{a}{(x-h)} + k$, whose center is (h, k).

identity matrix (196) For a 2×2 matrix the multiplicative identity is $I = \begin{bmatrix} 1 & 0 \\ 0 & 1 \end{bmatrix}$.

identity property (4) See Properties on page 830.

imaginary number (258) If a is a positive number, then $\sqrt{-a}$ is an imaginary number and $\sqrt{-a} = i\sqrt{a}$.

independent events (811) Two events where the occurrence of one has no effect on the occurrence of the other.

index of a radical (369) In the radical expression $c\sqrt[n]{a}$, the number n is the index.

index of summation (623) In the summation $\sum_{i=1}^{n} a_i$, i is the index of summation.

inequality (37) An open sentence formed when an inequality symbol is placed between two expressions.

infinite series (645) A series that has no upper limit of summation.

integers (2) The set of numbers $\{..., -3, -2, -1, 0, 1, 2, 3, ...\}$.

intercept form of an equation of a line (79) See Models on page 831.

intersection of two events (805) The intersection of two events consists of all outcomes that are in both A and B.

inverse of a matrix (196) Two 2×2 matrices are inverses of each other if their product (in both orders) is the identity matrix.

inverse of a relation (298) The set of ordered pairs obtained by switching the coordinates of each ordered pair in the relation.

inverse property (4) See Properties on page 830.

inverse property of exponents and logarithms (434) Let a be a positive real number. $a^x = a^y$ if and only if $x = y$. $\log_a x = \log_a y$ if and only if $x = y$.

inverse variation (526) A function defined by an equation of the form $xy = k$ or $y = \frac{k}{x}$.

irrational number (2) A real number that cannot be expressed as the quotient of two integers. For example, the square roots of numbers that are not perfect squares ($\sqrt{2}$, $\sqrt{3}$, $\sqrt{5}$, and $\sqrt{6}$). When expressed as a decimal, it neither repeats nor terminates.

joint variation (528) A joint variation occurs when a quantity varies directly as the product of two or more other quantities. An equation of the form $z = kxy$ where k is a nonzero constant.

Law of Cosines (716) If ABC is a triangle with sides a, b, and c, then $a^2 = b^2 + c^2 - 2bc \cos A$, $b^2 = a^2 + c^2 - 2ac \cos B$ and $c^2 = a^2 + b^2 - 2ab \cos C$.

Law of Sines (709) If ABC is a triangle with sides a, b, and c, then $\frac{a}{\sin A} = \frac{b}{\sin B} = \frac{c}{\sin C}$ or $\frac{\sin A}{a} = \frac{\sin B}{b} = \frac{\sin C}{c}$.

leading coefficient (230, 456) The coefficient of the first term in a polynomial written in standard form.

least common multiple of polynomials (541) The simplest polynomial that is a multiple of each of the original polynomials.

like radicals (369) Two radical expressions are like radicals if they have the same index and the same radicand.

linear combination method (131) A method of solving systems of equations that involves adding multiples of the given equations.

Linear Factorization Theorem (496) A polynomial of degree n has exactly n linear factors $f(x) = a_n(x - k_1)(x - k_2) \ldots (x - k_n)$, where $k_1, k_2, \ldots, k_n$ are complex numbers and a_n is the leading coefficient of $f(x)$.

linear function (285) A function defined by the equation $f(x) = mx + b$.

linear model (24) An equation of a line that is used to represent a real-life situation.

linear programming (151) A type of optimization process for finding the maximum or minimum value of an objective quantity by using a system of constraints.

literal equation (31) An equation which uses more than one letter as a variable. For example, $3x + y = 4$ is a literal equation.

logarithm (405) If $a > 0$ and $x > 0$, $y = \log_a x$ if and only if $x = a^y$.

logarithmic function (405) The inverse of an exponential function.

logistics growth models (441) Models used to represent populations or quantities whose rate of growth changes from an increasing rapid growth to a decreasing rapid growth.

major axis of an ellipse (31, 583) The major axis is the longer of the two line segments that form the axes of symmetry for an ellipse.

Mandelbrot set (261) A set of famous fractal pictures.

mapping diagram (284) A diagram that pictures a correspondence between two sets.

matrix (174) A rectangular arrangement of numbers into rows and columns.

matrix multiplication (180) If A is an $m \times n$ matrix and B is an $n \times p$ matrix, then the product AB is an $m \times p$ matrix.

mean (327) The sum of a set of numbers divided by the number of numbers in the set.

measures of central tendency (327) The mean, median, and mode of a distribution.

median (327) The middle number of an odd collection of numbers; the average of the two middle numbers in an even collection of numbers.

minor axis of an ellipse (31, 583) The line segment perpendicular to the major axis at the center.

mode (327) The number that occurs most frequently in a collection of numbers. If two numbers tie for most frequent occurrence, the collection is said to be bimodal.

modeling (9) Writing algebraic expressions or equations to represent real-life situations.

monomial (456) A polynomial with only one term.

multiplicative identity (4) See Properties on page 830.

multiplying rational expressions (534) To multiply two rational expressions, multiply the numerators, multiply the denominators, and write the new fraction in simplest form.

mutually exclusive events (805) Events which have no outcomes in common.

natural base (404, 426) The number $e \approx 2.71828$ is called the natural base.

negative correlation (110) Data points on a scatter plot that approximate a line with a negative slope.

negative numbers (2) Numbers represented by points to the left of the origin on the real number line. Numbers less than 0.

negative power property (346) See Properties on page 830.

numerator (9) The dividend in a quotient expressed in fraction form. For example, $5x$ is the numerator of $\frac{5x}{2}$.

objective quantity (151) In linear programming, the quantity or expression to be maximized or minimized.

obtuse angle (685) An angle whose measure is between 90° and 180°.

odds (810) The ratio of the number of ways an event can occur to the number of ways the event cannot occur.

opposites (3) Two points on the real number line that are the same distance from the origin but on opposite sides of the origin. For example, 3 and -3 are opposites.

optimization (151) Finding the minimum or maximum value of a quantity.

order of a matrix (174) The number of rows of the matrix "by" the number of columns of the matrix. For example, a 2×3 matrix has 2 rows and 3 columns.

ordered pair (64) A pair of real numbers used to locate each point in a coordinate plane.

origin of the coordinate plane (64) The point in the coordinate plane at which the horizontal axis intersects the vertical axis. The point (0, 0).

origin of the real number line (2) The point that represents 0 on the real number line.

parabola (236, 570) The graph of a quadratic equation.

paraboloid (574) A shape that is made up of vertical cross sections that are parabolas and horizontal cross sections that are circles.

parallelogram rule (261) To add (or subtract) two complex numbers, add (or subtract) the real and imaginary parts of the numbers.

Pascal's Triangle (652) A convenient triangular pattern for writing binomial coefficients. The first and last number in each row is 1. Each of the other entries in any row is formed by adding the two numbers immediately above it.

perfect square trinomial (245) A trinomial that factors into the square of a binomial.

perimeter of a polygon (35) The sum of the lengths of the sides of the polygon.

period (731) The horizontal length of each cycle of a periodic function.

periodic function (730) A function whose graph has a repeating pattern that continues indefinitely.

permutation (790) The number of ways that n elements can be arranged in order.

point-slope form of the equation of a line (86) See Models on page 831.

polynomial (456) An expression whose terms are of the form $a_k x^k$, where k is a nonnegative integer, is a polynomial in x.

polynomial function (463) A function of the form $f(x) = a_n x^n + a_{n-1}x^{n-1} + \cdots + a_1 x + a_0,\ a_n \neq 0$.

positive correlation (110) Data points on a scatter plot that approximate a line with a positive slope.

positive numbers (2) Numbers represented by the points to the right of the origin on the real number line. Numbers greater than 0.

power (346) The result of repeated multiplication. For example, in the expression $4^2 = 16$, 16 is the power.

power of a power property (346) See Properties on page 830.

power of a product property (346) See Properties on page 830.

power of a quotient property (346) See Properties on page 830.

power property of logarithms (413) See Properties on page 830.

principal (352, 555) The initial amount of money invested or borrowed.

probability (782) The ratio of successful outcomes to the total number of outcomes of the event.

probability of an event (782) A number between 0 and 1 that indicates the likelihood the event will occur.

probability of independent events (811) If A and B are independent events, then the probability that both A and B occur is $P(A \cap B) = P(A) \cdot P(B)$.

product (4) The result obtained when numbers or expressions are multiplied. The product of a and b is $a \cdot b$, or ab.

properties See page 830.

Pythagorean Identity (680) See Formulas on page 835.

Pythagorean Theorem (232) For any right triangle, the sum of the squares of the lengths of the legs, a and b, equals the square of the length of the hypotenuse, c. For example, $a^2 + b^2 = c^2$.

quadrant (64) In the coordinate plane, one of the four parts into which the axes divide the plane.

quadratic equation (230) An equation in x that can be written in the standard form $ax^2 + bx + c = 0$, $a \neq 0$.

quadratic formula (252) See Formulas on page 833.

quartile (328) The values in a collection of numbers that separate the data into four equal parts.

quotient (4) The result obtained when numbers or expressions are divided. The quotient of a and b is $\frac{a}{b}$, where a and b are real numbers, and $b \neq 0$.

radian (686) A unit of angle, arc, or rotation measure such that π radians = 180°.

radicand (230, 369) The number or expression under a radical symbol.

range, or output, of a function (284) The y-values of a function $y = f(x)$.

rate of change (73) A relationship such as distance over time, often described by using a slope.

rational equation (541) An equation that involves rational expressions with the variable occurring in the denominator(s).

rational expressions (548) Fractions whose numerators and denominators are polynomials.

rational function (518) A function of the form $f(x) = \frac{p(x)}{q(x)}$, $q(x) \neq 0$.

rational numbers (2) Any number that can be written in the form $\frac{a}{b}$, where a and b are integers and $b \neq 0$.

real number line (2) A horizontal line that pictures real numbers as points.

real numbers (2) The set of numbers consisting of the positive numbers, the negative numbers, and zero.

reciprocal (4) If $\frac{a}{b}$ is a non-zero number then its reciprocal is $\frac{b}{a}$.

recursive function (321) A function whose domain is the set of nonnegative integers and which can be defined by stating the value of the function at 0 and a formula relating each of the other values to the preceding value.

reference angle (694) The acute angle formed by the terminal side of an angle and the x-axis.

reflection (316) See Transformations on page 832.

reflection of the graph (298) See Transformations on page 832.

relation (284) A set of ordered pairs (x, y).

Remainder Theorem (482) If a polynomial $f(x)$ is divided by $(x - k)$, then the remainder is $r = f(k)$.

round-off error (23) The error produced when a decimal result is rounded in order to provide a meaningful answer.

row-equivalent matrices (210) Two matrices are row equivalent if one can be obtained from the other by a sequence of elementary row operations.

row matrix (174) A matrix that has only one row.

sample space (782) The set of all possible outcomes of an experiment.

scalar (175) In matrix algebra, a real number is called a scalar.

scatter plot (108) The plot, or graph, of the points that correspond to the ordered pairs in a collection.

secant (678) See Formulas on page 835.

second differences of a recursive function (322) The second differences of a recursive function are found by subtracting consecutive first differences.

sequence (622) A function whose domain is the set of positive integers. Each value is a term of the sequence.

series (623) The expression formed by adding the first n terms of a sequence.

set (2) A well-defined collection of objects.

sigma notation (623) The Greek letter Σ.

simple interest (352) The amount paid or earned for the use of money for a unit of time.

sine (678) See Formulas on page 835.

slope of a line (70) The ratio of the vertical change to the horizontal change; the ratio of the change in y to the corresponding change in x.

slope-intercept form (79) See Models on page 831.

solution (18, 37) A replacement for a variable that makes an equation or inequality true.

solution of a system (122, 145, 204) An ordered pair that is a solution of each of the equations in the system.

solution of a system of equations in x, y, z (157) An ordered triple (x, y, z) that is a solution of each of the equations in the system.

speed (254) The absolute value of the velocity of an object; the rate of motion given by distance traveled over time.

square matrix (174) A matrix that has the same number of rows and columns.

square root (2, 230) If $b^2 = a$, then b is a square root of a.

standard deviation (502) For any set of data, x_1, x_2, x_3, ..., x_n, the standard deviation is

$$s = \sqrt{\frac{(x_1 - \overline{x})^2 + (x_2 - \overline{x})^2 + \dots + (x_n - \overline{x})^2}{n}}$$

where $\overline{x}$ is the mean of x_1, x_2, ..., and x_n.

standard form of a polynomial (456) A polynomial in which the terms are written in descending order, from the greatest degree to the smallest degree.

standard form of an equation (576) See Models on page 831.

statistics (107) A field of mathematics that organizes large collections of data in ways that can be used to understand trends and make predictions.

substitution method (130) A method of solving a system of equations in which one variable is written in terms of other variables in one equation and the resulting expression is substituted into the other equation.

sum (4) The result obtained when numbers or expressions are added. The sum of a and b is $a + b$.

summation notation (623) The Greek letter Σ.

supplementary (686) Two positive angles are supplementary if the sum of their measures is π radians (or 180°).

synthetic division (480) A method that uses only coefficients to display the process of dividing a polynomial in x by $x \pm k$.

system of equations (122, 157) A set of equations in the same variables.

system of inequalities (147) A set of inequalities in the same variables.

tangent (678) See Formulas on page 835.

terms of an expression (9) In an expression such as $3x + 2$, $3x$ and 2 are called terms.

three-dimensional coordinate system (164) A system that begins with the xy-coordinate plane in a horizontal position and then adds the z-axis as a vertical line through the origin.

time-line graph (52) A graph used to represent events that occur in a given sequence in time.

transformations (314) See Transformations on page 832.

transverse axis (590) The line segment joining the vertices of a hyperbola.

triangular form (157) A linear system in which the equations follow a stair-step pattern and each equation has 1 as the coefficient of its first nonzero term.

trigonometric identities (680) See Formulas on page 835.

trigonometry (678) The study of the measurements of triangles.

trinomial (456) A polynomial that has only three terms.

union of two events (805) The union of two events, A and B, consists of all outcomes that are in A or B (or in both A and B).

variable (9) A letter that is used to represent one or more numbers.

velocity (254) The speed and direction in which an object is traveling.

vertex of a parabola (236) The lowest point of a parabola that opens up and the highest point of a parabola that opens down.

vertical shift of a graph (314) See Transformations on page 832.

vertices of a hyperbola (590) The line through the foci intersects the hyperbola at two points, the vertices.

vertices of an ellipse (583) The line through the foci intersects the ellipse at two points, the vertices.

whole numbers (2) The set of numbers $\{0, 1, 2, 3, 4, \ldots\}$.

***x*-axis** (64) The horizontal axis in the coordinate plane.

***x*-coordinate** (64) The first number of an ordered pair, the horizontal coordinate.

***x*-intercept** (78) The value of x when $y = 0$.

***y*-axis** (64) The vertical axis in the coordinate plane.

***y*-coordinate** (64) The second number of an ordered pair, the vertical coordinate.

***y*-intercept** (78) The value of y when $x = 0$.

zero matrix (174) A matrix whose entries are all zero.

zero power property (346) See Properties on page 830.

zero product property (475) Let a and b be real numbers. If $ab = 0$, then $a = 0$ or $b = 0$.

All answers to exercises in the Students Handbook will be found in the Additional Answers, pages 923–924.

Handbook of Mathematical Connections

You may have taken a geometry course in which some geometric properties (called theorems) were proved using other theorems and basic assumptions (called postulates). The proofs show how the geometric properties are interrelated and help convince us that the proved properties are correct.

In this handbook, we will show how various properties of algebra are related to other properties in algebra, in geometry, and in trigonometry.

THE QUADRATIC FORMULA

Lesson 5.4, page 252

The quadratic formula can be derived by completing the square of the general quadratic equation written in standard form. In completing the square, we use properties of addition and multiplication (see page 4) and transformations that produce equivalent equations (see page 18).

$$ax^2 + bx + c = 0$$ *Standard form*

$$x^2 + \frac{b}{a}x + \frac{c}{a} = 0$$ *Divide both sides by a, $(a \neq 0)$.*

$$x^2 + \frac{b}{a}x = -\frac{c}{a}$$ *Subtract $\frac{c}{a}$ from both sides.*

$$x^2 + \frac{b}{a}x + \left(\frac{b}{2a}\right)^2 = -\frac{c}{a} + \left(\frac{b}{2a}\right)^2$$ *Complete the square by adding the square of half the coefficient of x to both sides.*

$$\left(x + \frac{b}{2a}\right)^2 = \frac{b^2 - 4ac}{4a^2}$$ *Simplify.*

$$x + \frac{b}{2a} = \pm\sqrt{\frac{b^2 - 4ac}{4a^2}}$$ *Take the square root of both sides.*

$$x = -\frac{b}{2a} \pm \sqrt{\frac{b^2 - 4ac}{4a^2}}$$ *Subtract $\frac{b}{2a}$ from both sides.*

$$x = \frac{-b \pm \sqrt{b^2 - 4ac}}{2a}$$ *Simplify.*

Try this. Solve the quadratic equation $3x^2 + 5x + 2 = 0$ by completing the square. Compare each step in your solution with the steps in the above proof.

PROPERTIES OF LOGARITHMS

Lesson 8.3, page 413

Since logarithms are defined in terms of exponents:

$\log_a x = y$ if and only if $a^y = x$,

the properties of logarithms can be derived directly from the properties of exponents which were developed in Lesson 7.1.

Properties of Logarithms	***Properties of Exponents***
1. $\log_a(uv) = \log_a u + \log_a v$	**1.** $a^n a^m = a^{n+m}$
2. $\log_a \frac{u}{v} = \log_a u - \log_a v$	**2.** $\frac{a^n}{a^m} = a^{n-m}$
3. $\log_a(u^n) = n \log_a u$	**3.** $(a^n)^m = a^{nm}$

For example, to derive the first property of logarithms, let

$n = \log_a u$ and $m = \log_a v$.

Then,

$u = a^n$ and $v = a^m$. *Definition of logarithm*

Multiplying, we obtain

$uv = a^{n+m}$ *First property of exponents*

Taking the logarithm of both sides, we obtain

$\log_a(uv) = \log_a(a^{n+m})$

$= n + m$ *Definition of logarithm*

$= \log_a u + \log_a v$. *Substitution*

Try this. The second property of logarithms would be derived in a very similar fashion. The difference would be in dividing u and v instead of multiplying. Write the derivation. Which property of exponents is used?

The third property of logarithms depends on the third property of exponents as follows.

Let $m = \log_a u$.

Then $u = a^m$. *Definition of logarithm*

$u^n = (a^m)^n$ *Raise both sides to the same power.*

$\log_a u^n = \log_a(a^m)^n$ *Take the logarithm of both sides.*

$\log_a u^n = \log_a(a^{mn})$ *Third property of exponents*

$= mn$ *Definition of logarithm*

$= n \log_a u$ *Substitution*

CHANGE OF BASE FORMULAS

Pages 406 and 427

The Change of Base Formula can be stated in terms of logarithms to the base 10, base e, or, in general, any base b:

$$\log_a x = \frac{\log_{10} x}{\log_{10} a} \qquad \log_a x = \frac{\ln x}{\ln a} \qquad \log_a x = \frac{\log_b x}{\log_b a}$$

The formulas can be derived from the properties of logarithms. Here is a derivation of the first formula.

Let $u = \log_a x$.

$a^u = x$ *Definition of logarithm*

$\log_{10} a^u = \log_{10} x$ *Take the logarithm of both sides.*

$u \log_{10} a = \log_{10} x$ *3rd property of logarithms*

$u = \dfrac{\log_{10} x}{\log_{10} a}$ *Solve for u.*

$\log_a x = \dfrac{\log_{10} x}{\log_{10} a}$ *Substitution*

Try this. What changes would you make in the above derivation to derive the other two change of base formulas?

EQUATION OF A PARABOLA

Lesson 11.1, page 570

Using the geometric definition of a parabola and the distance formula, we can derive the equation of a parabola.

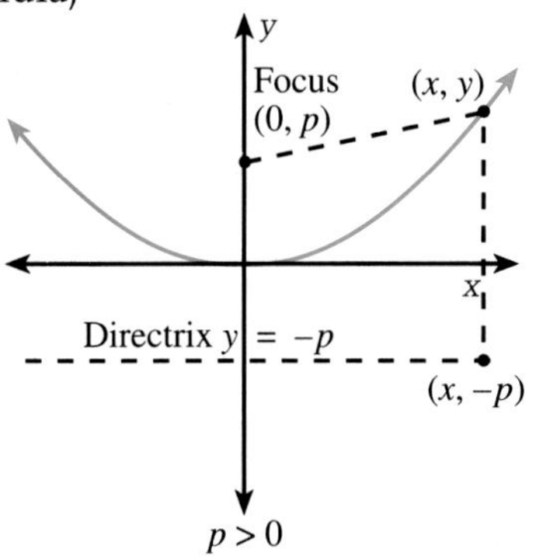

Definition: A parabola is the set of all points (x, y) that are equidistant from a fixed line, the directrix, and a fixed point, the focus.

y
Focus
(0, p)
(x, y)
x
Directrix y = −p
(x, −p)
p > 0

Let the coordinates of the focus be $(0, p)$ and the equation of the directrix be $y = -p$ as in this diagram. Notice that $(0, 0)$ is p units from the focus and also p units from the directrix. Therefore $(0, 0)$ is on the parabola.

For any point (x, y) on the parabola,

distance from (x, y) to $(0, p)$ = distance from (x, y) to $y = -p$ *Definition of a parabola*

$\sqrt{(x-0)^2 + (y-p)^2} = \sqrt{(x-x)^2 + (y-(-p))^2}$ *Distance formula*

$\sqrt{(x^2 + (y-p)^2)} = \sqrt{(y+p)^2}$ *Simplify.*

$x^2 + (y-p)^2 = (y+p)^2$ *Square both sides.*

$x^2 + y^2 - 2yp + p^2 = y^2 + 2yp + p^2$ *Multiply.*

$x^2 = 4yp$ *Subtract $y^2 - 2yp + p^2$ from both sides.*

In the diagram on the previous page, we assumed that p was positive. If p were negative, the focus would be below the x-axis and the directrix would be above it. However, there would be no change in the proof.

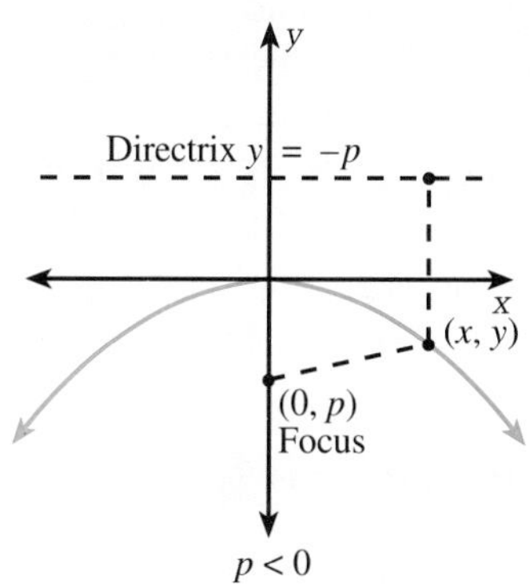

So, if p is positive in the equation $x^2 = 4py$, the parabola opens upward; if p is negative, it opens downward.

The proof that $y^2 = 4px$ is the equation of a parabola that opens to the left or right is similar to the proof on page 848.

Try this. Derive the formula for the set of all points (x, y) that are the same distance from $(0, 5)$ as they are from $y = -5$. Compare your derivation with each step in the proof on page 848.

EQUATION OF AN ELLIPSE

Lesson 11.3, page 583

Using the geometric definition of an ellipse and the distance formula, we can derive the equation of an ellipse.

Definition: An ellipse is the set of all points (x, y) such that the sum of the distances between (x, y) and two distinct points, the foci, is a constant.

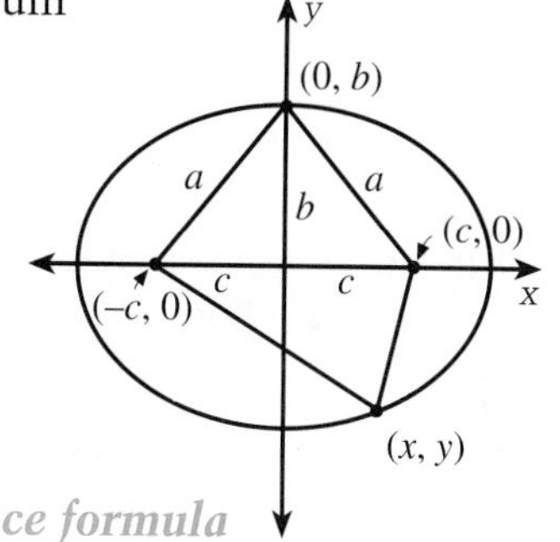

Let the two foci have coordinates $(-c, 0)$ and $(c, 0)$. Let $(0, b)$ be the point where the ellipse intersects the positive y-axis. Let the distance from $(0, b)$ to each focus be a. Then the sum of the distances from any point (x, y) on the ellipse to the two foci must be $2a$. So,

$$\sqrt{(x-(-c))^2+(y-0)^2}+\sqrt{(x-c)^2+(y-0)^2}=2a. \qquad \textit{Distance formula}$$

By squaring both sides and simplifying, we can obtain

$$4a\sqrt{(x-c)^2+y^2}=4a^2-4cx.$$

Again, squaring both sides and simplifying, we can obtain

$$\frac{(a^2-c^2)x^2}{a^2}+y^2=a^2-c^2.$$

Since a, b, and c are sides of a right triangle, $a^2 - c^2 = b^2$. Substituting, we obtain

$$\frac{b^2x^2}{a^2}+y^2=b^2 \quad \text{or} \quad \frac{x^2}{a^2}+\frac{y^2}{b^2}=1.$$

STUDENT HANDBOOK

We now have the equation of an ellipse whose center is (0, 0). We know that the length of the vertical axis of the ellipse, the distance from $(0, b)$ to $(0, -b)$, is $2b$. But what is the length of the other axis? Let $(d, 0)$ be the intersection of the ellipse with the positive x-axis.

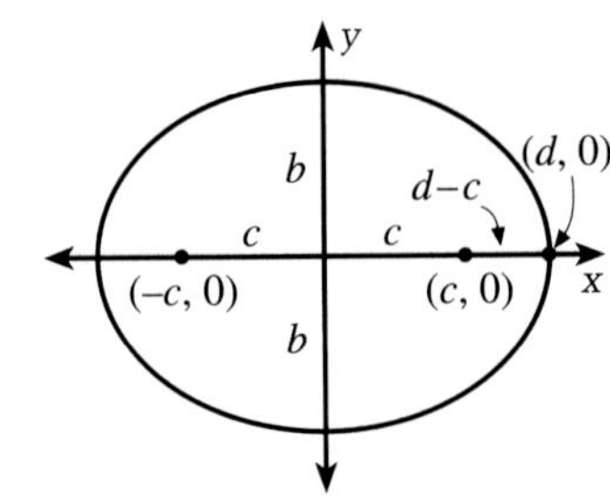

Then, distance from $(d, 0)$ to $(-c, 0)$ plus distance from $(d, 0)$ to $(c, 0)$ is $2a$.

$$\begin{aligned}[c + c + (d - c)] + (d - c) &= 2a \\ 2d &= 2a \\ d &= a\end{aligned}$$

Therefore, the ellipse intersects the x-axis at $(-a, 0)$ and $(a, 0)$ and the length of the horizontal axis of the ellipse is $2a$.

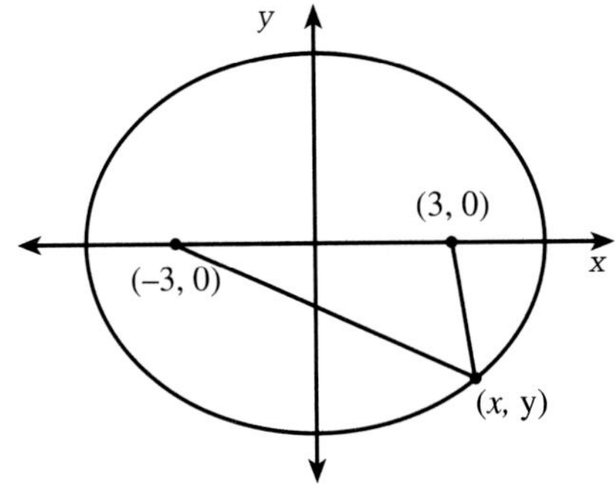

Try this. Derive the formula for the set of all points (x, y) such that the sum of the distances between (x, y) and the two points $(-3, 0)$ and $(3, 0)$ is 10. Compare your derivation with each step in the above proof.

EQUATION OF A HYPERBOLA

Lesson 11.4, page 590

Using the geometric definition of a hyperbola and the distance formula, we can derive the equation of a hyperbola.

Definition: A hyperbola is the set of all points (x, y) such that the difference of the distances between (x, y) and two distinct points, the foci, is a constant.

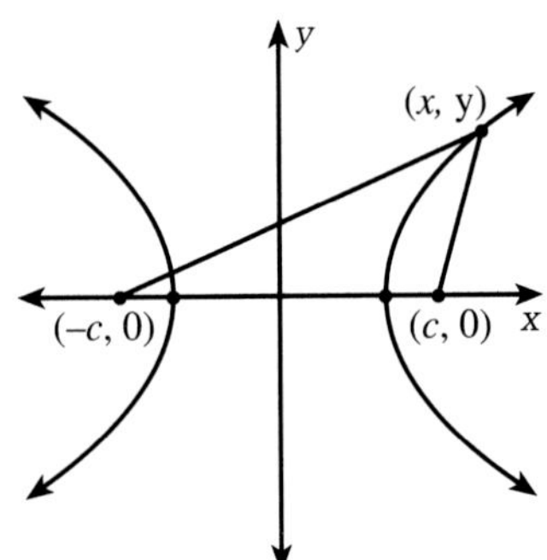

Let the two foci have coordinates $(-c, 0)$ and $(c, 0)$. Let $2a$ be the constant difference in distances.

Then,

$$\sqrt{(x - (-c))^2 + (y - 0)^2} - \sqrt{(x - c)^2 + (y - 0)^2} = \pm 2a.$$

Squaring both sides and simplifying, we can obtain

$$\frac{cx}{a} - a = \pm\sqrt{(x - c)^2 + y^2}.$$

Again, squaring both sides and simplifying we can obtain

$$\frac{(c^2 - a^2)x^2}{a^2} - y^2 = c^2 - a^2.$$

Let $b^2 = c^2 - a^2$ and substitute to obtain

$$\frac{b^2x^2}{a^2} - y^2 = b^2 \quad \text{or} \quad \frac{x^2}{a^2} - \frac{y^2}{b^2} = 1.$$

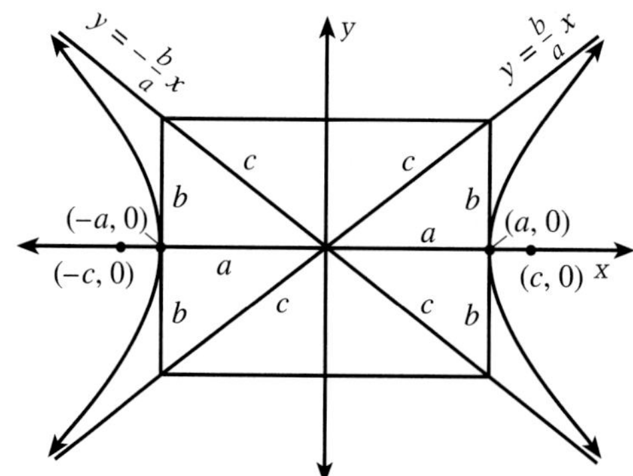

This diagram shows how a, b, and c are related to the hyperbola. The vertices of the hyperbola are at $(a, 0)$ and $(-a, 0)$. The lines $y = \frac{b}{a}x$ and $y = -\frac{b}{a}x$ are the asymptotes of the hyperbola—the lines which the branches of the hyperbola approach.

Try this. Derive the formula for the set of all points (x, y) such that the difference of the distances from (x, y) to the points $(5, 0)$ and $(-5, 0)$ is ± 10. Compare your derivation with each step in the above proof.

FORMULA FOR THE SUM OF AN ARITHMETIC SERIES

Lesson 12.2, page 631

Think of an arithmetic series in geometric terms and think of a duplicate arithmetic series stacked on top as in the diagram. The height of each bar is $a_1 + a_n$ and there are n of them.

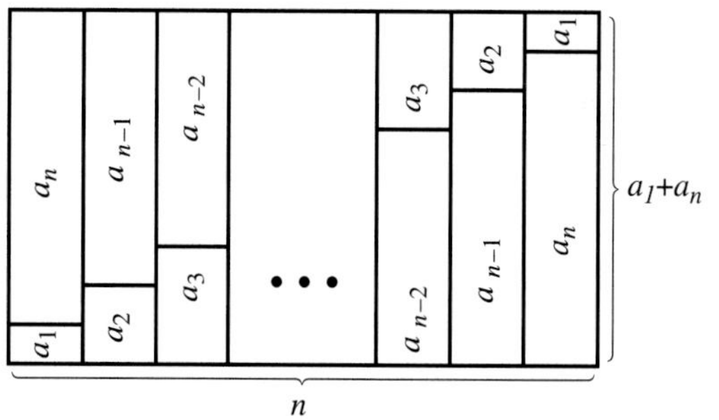

Algebraically we can write this as

$$S_n = a_1 + (a_1 + d) + (a_1 + 2d) + \ldots + (a_1 + (n - 1)d).$$

Each term is d larger than the previous one.

$$S_n = a_n + (a_n - d) + (a_n - 2d) + \ldots + (a_n - (n - 1)d)$$

In reverse order, each term is d less than the previous one.

Adding we get

$$2S_n = (a_1 + a_n) + (a_1 + a_n) + (a_1 + a_n) + \ldots + (a_1 + a_n).$$

Where the term $(a_1 + a_n)$ occurs n times.

Therefore, $2S_n = n(a_1 + a_n)$ or $S_n = \frac{n}{2}(a_1 + a_n)$.

Try this. Find the sum of this arithmetic series by using the techniques in the above derivation; that is by adding the series to itself with the order reversed.

$$7 + 10 + 13 + \ldots + 178 + 181 + 184 \text{ (60 terms)}$$

FORMULA FOR THE SUM OF A GEOMETRIC SERIES

Lesson 12.3, page 639

This derivation is similar to the one for an arithmetic series in that it adds two related series. However, it does not lend itself to a geometric representation.

In expanded form, the sum of a geometric series, S_n, is

$$S_n = a_1 + a_1r + a_1r^2 + \ldots + a_1r^{n-2} + a_1r^{n-1}.$$

Multiply both sides by $-r$ to obtain:

$$-rS_n = -a_1r - a_1r^2 - \ldots - a_1r^{n-2} - a_1r^{n-1} - a_1r^n.$$

Adding gives:

$$S_n - rS_n = a_1 - a_1r^n$$
$$S_n(1 - r) = a_1(1 - r^n)$$
$$S_n = a_1\left(\frac{1 - r^n}{1 - r}\right) \qquad (r \neq 1).$$

Try this. Find the sum of the geometric series by using the techniques used in the above derivation; that is, multiply each term by -3, add the two series, etc.

$2 + 6 + 18 + \ldots + 1458 + 4374$ (8 terms)

TRIGONOMETRIC PYTHAGOREAN IDENTITIES

Lesson 13.1, page 680

The trigonometric Pythagorean Identities can be derived directly from the Pythagorean Theorem.

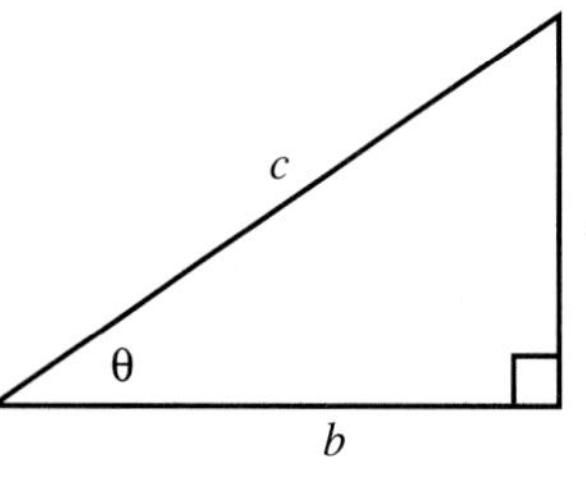

Since $\sin\theta = \frac{a}{c}$ and $\cos\theta = \frac{b}{c}$,

$$\sin^2\theta + \cos^2\theta = \left(\frac{a}{c}\right)^2 + \left(\frac{b}{c}\right)^2 = \frac{a^2 + b^2}{c^2} = \frac{c^2}{c^2} = 1.$$

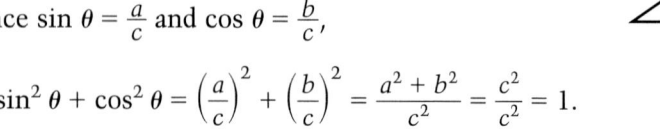

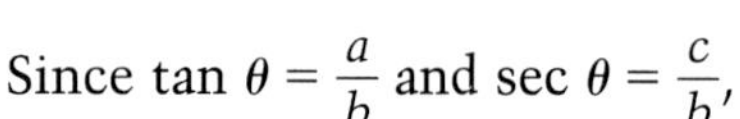

Since $\tan\theta = \frac{a}{b}$ and $\sec\theta = \frac{c}{b}$,

$$1 + \tan^2\theta = 1 + \left(\frac{a}{b}\right)^2 = \frac{b^2 + a^2}{b^2} = \frac{c^2}{b^2} = \left(\frac{c}{b}\right)^2 = \sec^2\theta.$$

Try this. Use the definitions of $\cot\theta$ and $\csc\theta$ and the Pythagorean Theorem to show that $1 + \cot^2\theta = \csc^2\theta$.

LAW OF SINES

Lesson 13.5, page 709

This law can be derived directly from the formula for the area of a triangle.

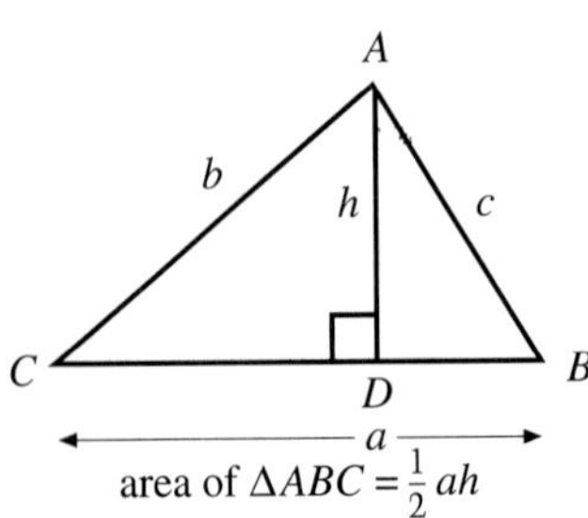

In $\triangle CAD$, $\sin C = \frac{h}{b}$. So, $h = b \sin C$. Therefore, area of $\triangle ABC = \frac{1}{2}ab \sin C$. In a similar manner, area of $\triangle ABC = \frac{1}{2}ac \sin B$ and area of $\triangle ABC = \frac{1}{2}bc \sin A$. Therefore, $\frac{1}{2}ab \sin C = \frac{1}{2}ac \sin B = \frac{1}{2}bc \sin A$. Multiplying by $\frac{2}{abc}$ gives $\frac{\sin C}{c} = \frac{\sin B}{b} = \frac{\sin A}{a}$.

Try this. Use the ideas in the proof on page 852 to find the area of this triangle.

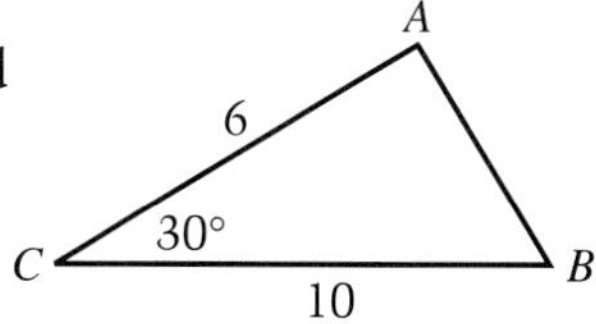

LAW OF COSINES

Lesson 13.6, page 716

The Law of Cosines can be derived from the Pythagorean Theorem and the trigonometric Pythagorean Identity, $\sin^2 \theta + \cos^2 \theta = 1$.

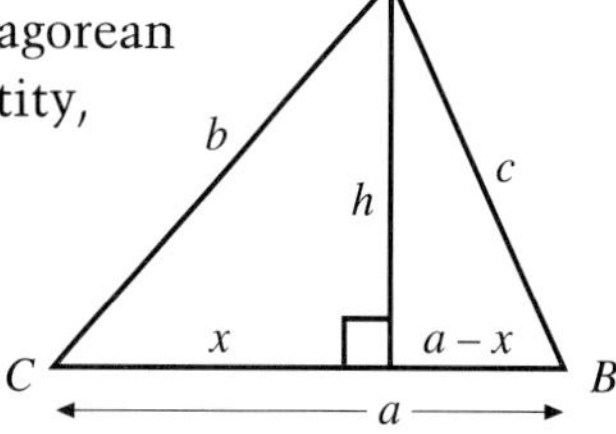

Suppose that C is an acute angle as shown.

$$c^2 = h^2 + (a - x)^2$$

Since $\sin C = \frac{h}{b}$ and $\cos C = \frac{x}{b}$, $h = b \sin C$ and $x = b \cos C$. Substitute into the first equation to get $c^2 = (b \sin C)^2 + (a - b \cos C)^2$. Simplifying and replacing $\sin^2 C + \cos^2 C$ with 1 gives $c^2 = a^2 + b^2 - 2ab \cos C$.

The same result is obtained if C is an obtuse angle, and similar arguments establish that

$$a^2 = b^2 + c^2 - 2bc \cos A \quad \text{and} \quad b^2 = a^2 + c^2 - 2ac \cos B.$$

Try this. Given ΔABC with $a = 7$, $b = 6$, and $C = 40°$, use the Pythagorean Theorem, $\sin^2 \theta + \cos^2 \theta = 1$, and the procedure used in the above derivation to show that

$$c^2 = 36 + 49 - 84 \cos 40°.$$

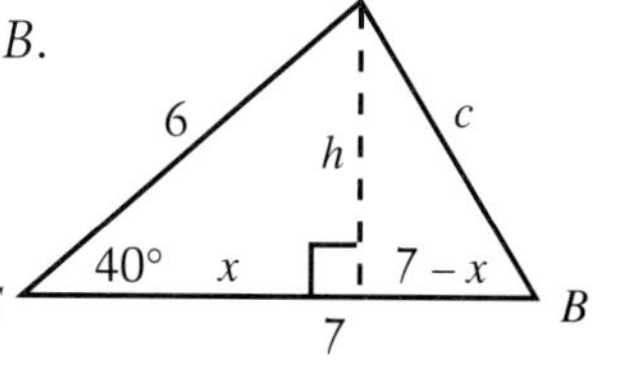

SUM AND DIFFERENCE FORMULAS

Lesson 14.5, page 760

The formula for the cosine of the difference of two angles can be derived using the Law of Cosines, the distance formula, and the identity $\sin^2 \theta + \cos^2 \theta = 1$. The other sum and difference formulas can be derived from the formula for the cosine of the difference.

Draw two angles whose sides are the x-axis, as shown in the diagram. Let v be the measure of the smaller angle and u be the measure of the larger angle. Choose points P and Q on the sides of the angles so that they are each 1 unit from the origin. Then the sides of the right triangles have length $\sin v$, $\cos v$, $\sin u$ and $\cos u$ as labeled. Therefore, the coordinates of P are $(\cos v, \sin v)$ and the coordinates of Q are $(\cos u, \sin u)$.

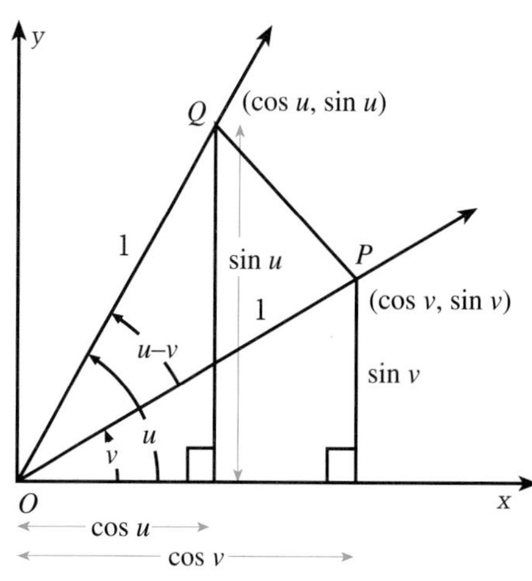

STUDENT HANDBOOK

By the distance formula,

$$PQ = \sqrt{(\cos u - \cos v)^2 + (\sin u - \sin v)^2}.$$

By the Law of Cosines applied to $\triangle QOP$,

$$PQ^2 = 1^2 + 1^2 - 2(1)(1)\cos(u - v).$$

By squaring the first formula, substituting into the second one, and simplifying we can obtain

$$\cos(u - v) = \cos u \cos v + \sin u \sin v.$$

The same argument can be used for P and Q in any quadrants.

Try this. Derive the formula for the cosine of a sum of two angles by substituting $-v$ for v in the above formula and using the fact that $\cos(-v) = \cos v$ and $\sin(-v) = -\sin v$.

STUDENT HANDBOOK

SINE OF A SUM

Lesson 14.5, page 760

Since sine and cosine are cofunctions, $\sin\left(\frac{\pi}{2} - u\right) = \cos u$ and $\cos\left(\frac{\pi}{2} - u\right) = \sin u$.

$$\begin{aligned}
\text{So, } \sin(u + v) &= \cos\left(\frac{\pi}{2} - (u + v)\right) \\
&= \cos\left(\left(\frac{\pi}{2} - u\right) - v\right) \\
&= \cos\left(\frac{\pi}{2} - u\right)\cos v + \sin\left(\frac{\pi}{2} - u\right)\sin v \quad \textit{Cosine of a difference} \\
&= \sin u \cos v + \cos u \sin v.
\end{aligned}$$

Try this.

1. Derive the formula for the sine of a difference of two angles by replacing v in the above formula by $-v$.
2. Derive the formulas for the tangent of a sum and difference of two angles from the sine and cosine formulas and the fact that $\tan\theta = \frac{\sin\theta}{\cos\theta}$.

DOUBLE-ANGLE FORMULAS

Lesson 14.6, page 766

The double-angle formulas are obtained from the sum formulas. In the formula for the cosine of the sum of two angles:

$$\cos(u + v) = \cos u \cos v - \sin u \sin v,$$

replace v by u:

$$\begin{aligned}
\cos(u + u) &= \cos u \cos u - \sin u \sin u \\
\cos 2u &= \cos^2 u - \sin^2 u.
\end{aligned}$$

Try this. Use the formula on page 854 and $\sin^2 u + \cos^2 u = 1$ to derive the other two formulas for $\cos 2u$. Derive the formulas for $\sin 2u$ and $\tan 2u$ from the sum formulas.

HALF-ANGLE FORMULAS

Lesson 14.6, page 767

The half-angle formulas are obtained from the double-angle formulas. In the formula for $\cos 2u$:

$$\cos 2u = 2\cos^2 u - 1$$

replace $2u$ by u and replace u by $\frac{u}{2}$.

$$\cos u = 2\cos^2 \frac{u}{2} - 1$$

$$\cos^2 \frac{u}{2} = \frac{\cos u + 1}{2}$$

$$\cos \frac{u}{2} = \pm\sqrt{\frac{\cos u + 1}{2}}$$

Try this.

1. Derive the formula for $\sin \frac{u}{2}$ from the formula $\cos 2u = 1 - 2\sin^2 u$.

2. When deriving $\tan \frac{u}{2}$ from $\sin \frac{u}{2}$ and $\cos \frac{u}{2}$ we get

$$\tan \frac{u}{2} = \pm\sqrt{\frac{1 - \cos u}{1 + \cos u}}.$$

First, simplify this radical by multiplying numerator and denominator of the fraction by $1 + \cos u$. Then, simplify this radical by multiplying the numerator and denominator of the fraction by $1 - \cos u$.

STUDENT HANDBOOK

TECHNOLOGY

Keystrokes for Other Graphing Calculators

Keystrokes for page 17

You can use your graphing calculator as a programmable calculator. This means you can *write, store,* and *execute* programs using your calculator. For example, you can write a program that calculates the volume of a right circular cylinder. The formula for the volume is

$V = \pi r^2 h$

where r is the radius and h is the height.

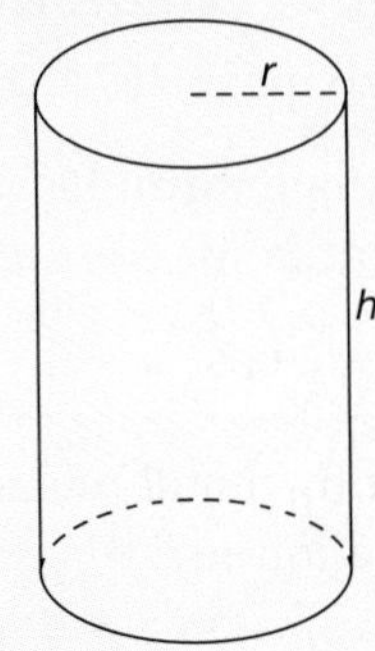

The following programs use a Texas Instruments *TI-82,* a Casio *fx-9700GE,* and a Sharp *EL-9300C* to find the volume of a right circular cylinder. When you execute, or run, the programs, you will be prompted to enter the radius and height of a cylinder. The program will then calculate the cylinder's volume.

TI-82

```
PROGRAM:VOLUME
:Disp "ENTER RADIUS"
:Input R
:Disp "ENTER HEIGHT"
:Input H
:π * R² * H→V
:Disp "THE VOLUME IS"
:Disp V
```

Casio *fx-9700GE*

```
VOLUME
"ENTER RADIUS"
?→R
"ENTER HEIGHT"
?→H
π * R² * H→V
"THE VOLUME IS"
V
```

Sharp *EL-9300C*

```
volume
——————————REAL
Print "enter radius"
Input r
Print "enter height"
Input h
v=πr²h
Print "the volume is"
Print v
```

Exercises

1. Enter a program listed above on your graphing calculator. (See your owner's manual to learn how to enter a program.) Then use the program to find the volume of several cylinders, including the one shown below.

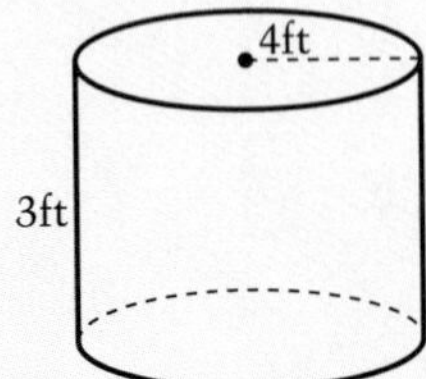

2. Write a graphing calculator program that calculates the volume of a rectangular prism. Then use the program to find the volume of several prisms, including the one shown below.

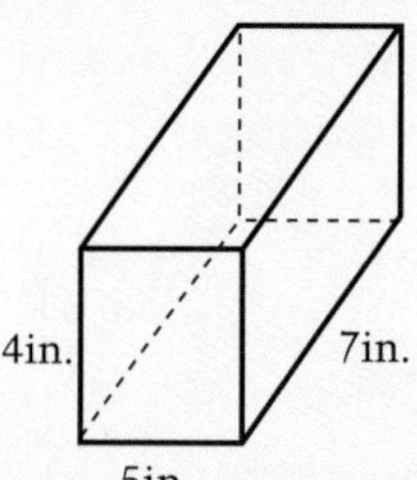

TECHNOLOGY

Keystrokes for Other Graphing Calculators

Keystrokes for page 84

A graphing calculator can be used to sketch the graph of an equation. Here we show how to use a Texas Instruments *TI-82,* a Casio *fx-9700GE,* and a Sharp *EL-9300C* to sketch the graph of $x + 2y = 4$.

Using a graphing calculator is fairly easy, but you must remember four things:

- Set the calculator mode to rectangular.
- Solve the equation for y in terms of x.
- Set the viewing window (the range) by entering the least and greatest x- and y-values and the scale (units per tick mark).
- Use parentheses if you are unsure of the calculator's order of operations.

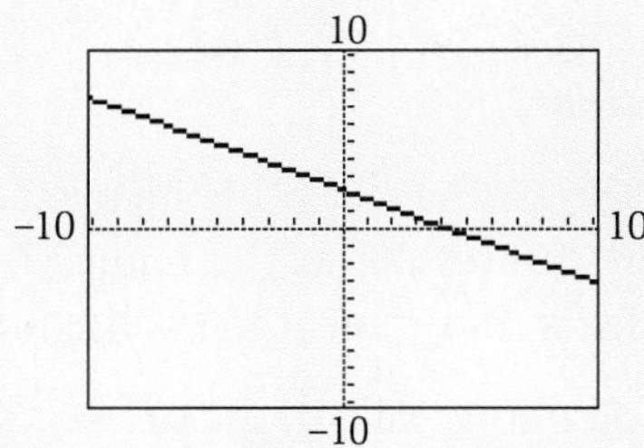

Graph of x + 2y = 4

Before sketching the graph of the equation, solve the equation for y: $y = -\frac{1}{2}x + 2$.

Xmin=−10	Ymin=−10
Xmax=10	Ymax=10
Xscl=1	Yscl=1

TI-82

[WINDOW] [▽] (Set window.)

[Y=] [CLEAR]

[(] [(−)] 1 [÷] 2 [)] [X,T,θ] [+] 2

[GRAPH] [CLEAR] (Clear screen.)

Sharp *EL-9300C*

[graph key] [RANGE] (Set range.)

[QUIT] [CL]

[(] [(−)] 1 [÷] 2 [)] [X/θ/T] [+] 2 [graph key]

[QUIT] (Clear screen.)

Casio *fx-9700GE*

In MAIN MENU, enter 1.

[Range] (Set range.) [EXIT]

[SHIFT] [F5] [EXE]

[Graph]

[(] [(−)] 1 [÷] 2 [)] [X,θ,T] [+] 2

[EXE]

[SHIFT] [F5] [EXE] (Clear screen.)

If you want a closeup view of the point where the line $y = -\frac{1}{2}x + 2$ crosses the x-axis, you could enter a new range setting to obtain the graph shown at the right.

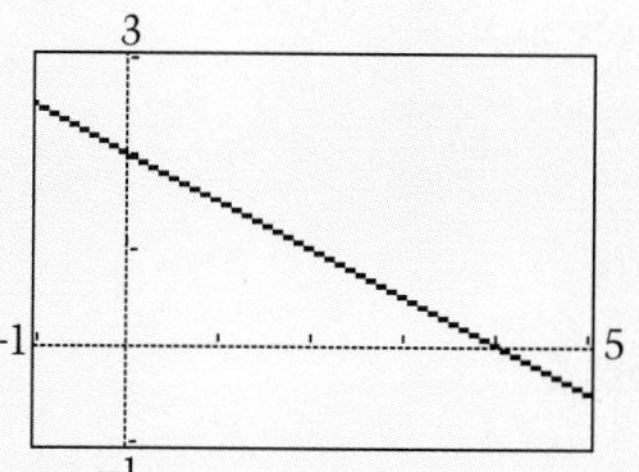

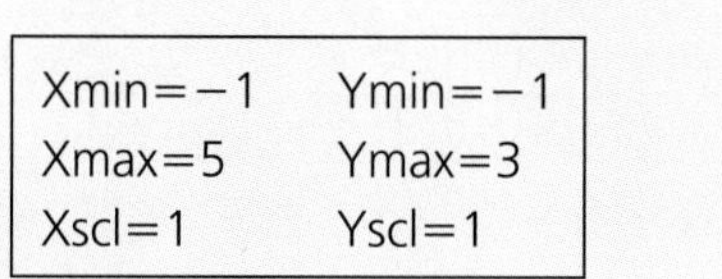

Xmin=−1	Ymin=−1
Xmax=5	Ymax=3
Xscl=1	Yscl=1

Refer to page 85 for an example of using a graphing calculator to approximate the solution of an equation and for related exercises.

TECHNOLOGY APPENDIX

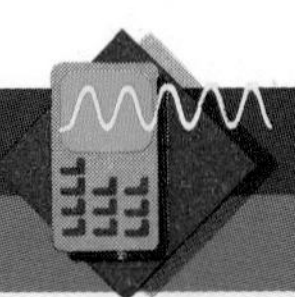

Keystrokes for page 128

A graphing calculator can be used to graph a system of equations. Here we show how to use a *TI-82,* a Casio *fx-9700GE,* and a Sharp *EL-9300C* to graph the following system.

$$\begin{cases} 3x - 4y = 2 & \textbf{Equation 1} \\ 5x + 6y = 16 & \textbf{Equation 2} \end{cases}$$

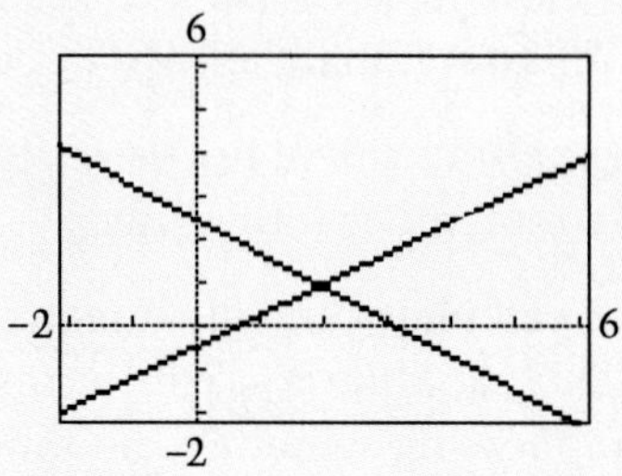

The display at the right shows the graph of the system. Note that the two lines appear to intersect at the point (2, 1). To check this, substitute 2 for x and 1 for y in each equation.

The first step in using a graphing calculator to sketch the graph of a linear system is to solve each equation for y.

$$\begin{cases} y = \frac{3}{4}x - \frac{1}{2} & \textbf{Equation 1} \\ y = -\frac{5}{6}x + \frac{8}{3} & \textbf{Equation 2} \end{cases}$$

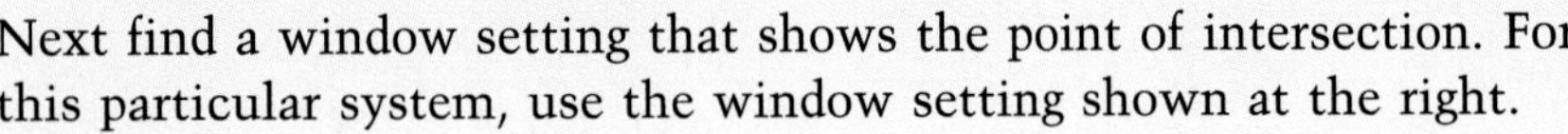

Next find a window setting that shows the point of intersection. For this particular system, use the window setting shown at the right.

Xmin=−2
Xmax=6
Xscl=1
Ymin=−2
Ymax=6
Yscl=1

TI-82

[WINDOW] [▽] (Set window.)

[Y=] [CLEAR]

[(] 3 [÷] 4 [)] [X,T,θ] [−] [(] 1 [÷] 2 [)]

[ENTER] [CLEAR]

[(] [(−)] 5 [÷] 6 [)] [X,T,θ] [+] [(] 8 [÷] 3 [)]

[GRAPH]

[CLEAR] (Clear screen.)

Casio *fx-9700GE*

In MAIN MENU, enter 1.

[Range] (Set range.) [EXIT]

[SHIFT] [F5] [EXE]

[Graph]

[(] 3 [÷] 4 [)] [X,θ,T] [−] [(] 1 [÷] 2 [)] [EXE]

[Graph]

[(] [(−)] 5 [÷] 6 [)] [X,θ,T] [+] [(] 8 [÷] 3 [)] [EXE]

[SHIFT] [F5] [EXE] (Clear screen.)

Sharp *EL-9300C*

[graph key] [RANGE] (Set range.)

[QUIT] [CL]

[(] 3 [÷] 4 [)] [X/θ/T] [−] [(] 1 [÷] 2 [)]

[ENTER] [CL]

[(] [(−)] 5 [÷] 6 [)] [X/θ/T] [+] [(] 8 [÷] 3 [)]

[graph key] [QUIT] (Clear screen.)

Refer to page 129 for an example of using a graphing calculator to approximate the solution of a linear system and for related exercises.

Keystrokes for page 194

Most graphing calculators can perform matrix algebra. That is, they can add, subtract, and multiply matrices, multiply a matrix by a scalar, find the determinant of a matrix, and find the inverse of a matrix. (The inverse of a matrix is introduced in Lesson 4.4.) Here we show how to use a Texas Instruments *TI-82,* a Casio *fx-9700GE,* and a Sharp *EL-9300C* to perform the following matrix operation.

$$\begin{bmatrix} 5 & -1 & 0 \\ 7 & 2 & -1 \end{bmatrix} + \begin{bmatrix} 2 & 0 & -2 \\ 1 & -2 & -4 \end{bmatrix} = \begin{bmatrix} 7 & -1 & -2 \\ 8 & 0 & -5 \end{bmatrix}$$

TI-82

[MATRX]

Cursor to EDIT, [1]

2 [ENTER] 3 [ENTER]

Input 1st matrix in [A].

[A] 2 × 3
1,1=5 2,1=7
1,2=−1 2,2=2
1,3=0 2,3=−1

[MATRX]
Cursor to edit, [2]

2 [ENTER] 3 [ENTER]

Input 2nd matrix in [B].

[B] 2 × 3
1,1=2 2,1=1
1,2=0 2,2=−2
1,3=−2 2,3=−4

[2nd] [QUIT]

[MATRX] [1] [+] [MATRX] [2]

[ENTER]

Display will be
[[7 −1 −2]
[8 0 −5]]

Casio *fx-9700GE*

In MAIN MENU, enter 5.

[F4] [F2] 2 [EXE] 3

[EXIT] [F1]

Input 1st matrix in [A].

A	1	2	3
1	5	−1	0
2	7	2	−1

[EXIT] Cursor to Mat B.

[F2] 2 [EXE] 3

[EXIT] [F1]

Input 2nd matrix in [B].

B	1	2	3
1	2	0	−2
2	1	−2	−4

[EXIT] [EXIT]

[F1] [ALPHA] [A] [+]

[F1] [ALPHA] [B] [EXE]

Display will be

Ans	1	2	3
1	7	−1	−2
2	8	0	−5

Sharp *EL-9300C*

[⊞⊟⊠⊞] [MENU] [3]

[MENU] Cursor to C DIM.

[ENTER] [ENTER]

2 [ENTER] 3 [ENTER]

Input 1st matrix in [A].

A[1,1]=5 A[2,1]=7
A[1,2]=−1 A[2,2]=2
A[1,3]=0 A[2,3]=−1

[MENU] [ENTER]

Cursor to matrix B. [ENTER]

2 [ENTER] 3 [ENTER]

Input 2nd matrix in [B].

B[1,1]=2 B[2,1]=1
B[1,2]=0 B[2,2]=−2
B[1,3]=−2 B[2,3]=−4

[QUIT]

[2nd F] [MAT] [A] [+]

[2nd F] [MAT] [B] [ENTER]

Display will be
Ans [1, 1]=7
Ans [2, 1]=8

Cursor to see entire display.

Refer to page 195 for an example of using a graphing calculator to solve problems involving matrices and for related graphing exercises.

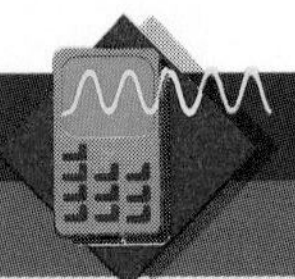

Keystrokes for page 205

Example 2

Use a matrix equation and a graphing calculator to solve the linear system.

$$\begin{cases} 2x + 3y + z = -1 \\ 3x + 3y + z = 1 \\ 2x + 4y + z = -2 \end{cases}$$

Solution The matrix equation that represents the system is

$$\underbrace{\begin{bmatrix} 2 & 3 & 1 \\ 3 & 3 & 1 \\ 2 & 4 & 1 \end{bmatrix}}_{A}\underbrace{\begin{bmatrix} x \\ y \\ z \end{bmatrix}}_{X} = \underbrace{\begin{bmatrix} -1 \\ 1 \\ -2 \end{bmatrix}}_{B}.$$

Using a graphing calculator, you can find the inverse of the matrix A to be

$$A^{-1} = \begin{bmatrix} -1 & 1 & 0 \\ -1 & 0 & 1 \\ 6 & -2 & -3 \end{bmatrix}.$$

To find the solution, multiply B (on the left) by A^{-1}.

$$X = A^{-1}B = \begin{bmatrix} -1 & 1 & 0 \\ -1 & 0 & 1 \\ 6 & -2 & -3 \end{bmatrix}\begin{bmatrix} -1 \\ 1 \\ -2 \end{bmatrix} = \begin{bmatrix} 2 \\ -1 \\ -2 \end{bmatrix} = \begin{bmatrix} x \\ y \\ z \end{bmatrix}$$

The solution $x = 2$, $y = -1$, and $z = -2$. ■

The keystrokes for the *TI-82*, the Casio *fx-9700GE*, and the Sharp *EL-9300C* for Example 2 are given below.

TI-82

Enter matrix A in [A] and matrix B in [B]. (See p. 859.)

[MATRX] [1] [x^{-1}]

[MATRX] [2] [ENTER]

Casio *fx-9700GE*

In matrix mode, enter matrix A in [A] and matrix B in [B]. (See p. 859.)

[F1] [ALPHA] [A] [SHIFT] [x^{-1}]

[F1] [ALPHA] [B] [EXE]

Sharp *EL-9300C*

In matrix mode, enter matrix A in [A] and matrix B in [B]. (See p. 859.)

[2nd F] [MAT] [A] [2nd F] [x^{-1}]

[2nd F] [MAT] [B] [ENTER]

Refer to Exercises on page 207 and 208 for practice using a graphing calculator to solve linear systems.

Keystrokes for page 266

If you have a graphing calculator, you can write a program that will find the real solutions of a quadratic equation. The following programs for the *TI-82*, the Casio *fx-9700GE*, and the Sharp *EL-9300C* will find the real solutions of the quadratic equation $ax^2 + bx + c = 0$.

TI-82

```
PROGRAM:QUADRTIC
:Disp "Enter A"
:Input A
:Disp "ENTER B"
:Input B
:Disp "ENTER C"
:Input C
:B²−4AC→D
:If D<0
:Goto 1
:(−B+√D)/(2A)→S
:Disp S
:(−B−√D)/(2A)→S
:Disp S
:Stop
:Lbl 1
:Disp "NO REAL SOLUTION"
:Stop
```

Casio *fx-9700GE*

```
QUADRATIC
"A="?→A
"B="?→B
"C="?→C
B²−4AC→D
D<0⇒Goto 1
(−B+√D)÷(2A)→S
S◢
(−B−√D)÷(2A)→S
S
Goto 2
Lbl 1
"NO REAL SOLUTION"
Lbl 2
```

Sharp *EL-9300C*

```
quadratic
—————————REAL
Input a
Input b
Input c
d=b²−4a∗c
If d<0 Goto 1
s=(−b+√d)/(2a)
Print s
s=(−b−√d)/(2a)
Print s
End
Label 1
Print "no real solutions"
```

Refer to page 266 for an example of using a graphing calculator to find real solutions of a quadratic equation. Also, refer to exercises on pages 268 and 269 for practice.

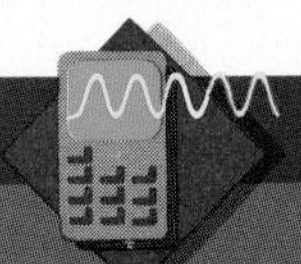

Keystrokes for page 389

Example 2: Finding a Mathematical Model

Most graphing calculators can fit power models of the form $y = ax^b$ to data. For instance, the *TI-82*, the Casio *fx-9700GE*, and the Sharp *EL-9300C* all have this capability programmed into their statistical features. Use a graphing calculator to find a model that relates the diameter, y (in inches), to the length, x (in inches), of a common nail.

Length, *x*	1.00	2.00	3.00	4.00	5.00	6.00
Diameter, *y*	0.070	0.111	0.146	0.176	0.204	0.231

Solution To solve this problem, enter the keystrokes shown below.

TI-82

[STAT] [4] [2nd] [L1] [,]

[2nd] [L2] [ENTER]

[STAT] [1]

Enter data.

L1(1)=1 L2(1)=.070
L1(2)=2 L2(2)=.111
L1(3)=3 L2(3)=.146
L1(4)=4 L2(4)=.176
L1(5)=5 L2(5)=.204
L1(6)=6 L2(6)=.231

[STAT] Cursor to CALC.

[ALPHA] [B]

[2nd] [L1] [,] [2nd] [L2]

[ENTER]

Casio *fx-9700GE*

In MAIN MENU, enter 4.

[F2] [F3] [F1]

[SHIFT] [SET UP]

Cursor to STAT GRAPH. [F2]

Cursor to REG MODEL. [F4]

[EXIT]

Enter data.

1 [F3] .070 [F1]
2 [F3] .111 [F1]
3 [F3] .146 [F1]
4 [F3] .176 [F1]
5 [F3] .204 [F1]
6 [F3] .231 [F1]

[F6] [F6]

[F1] [EXE]

[F2] [EXE]

Sharp *EL-9300C*

[▤] [MENU]

[D] [2]

[ENTER] [3]

Enter data.

X1=1 Y1=.070
X2=2 Y2=.111
X3=3 Y3=.146
X4=4 Y4=.176
X5=5 Y5=.204
X6=6 Y6=.231

[2nd] [⁂]

[F] [5]

[MENU] [ENTER]

From each display, you obtain $a \approx 0.07$ and $b \approx 0.66$. Because $2/3 \approx 0.66$, you can write the following model.

$$y = 0.07x^{2/3}$$

Refer to Exercise 5 on page 389 for a related graphing exercise.

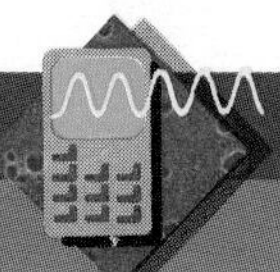

Keystrokes for page 700

The following programs for the *TI-82* and the Casio *fx-9700GE* are angle drawing programs. When you run the program, you are prompted to enter an angle. The calculator then draws the angle in standard position. The point of the terminal side of the angle lies on the unit circle. The coordinates of the endpoint are $(x, y) = (\cos \theta, \sin \theta)$.

TI-82

```
PROGRAM:ANGLE
:Disp "ENTER MODE"
:Disp "0, RADIAN"
:Disp "1, DEGREE"
:Input M
:Disp "ENTER ANGLE"
:Input T
:If M=1
:πT/180→T
:Radian
:ClrDraw
:FnOff
:−1.5→Xmin
:1.5→Xmax
:1→Xscl
:−1→Ymin
:1→Ymax
:1→Yscl
:0→Tmin
:abs T→Tmax
:.1→Tstep
:cos T→A
:sin T→B
:Param
:1→S
:If T<0
:−1→S
:"(.25+.04T)cos T"→X1T
:"S(.25+0.4T)sin T"→Y1T
:DispGraph
:Line(0,0,A,B)
:Pause
:Func
:FnOn
:Stop
```

Casio *fx-9700GE*

```
ANGLE
Lbl 1
"ENTER MODE"
"0=RADIAN"
"1=DEGREE"
?→M
"ENTER ANGLE"
?→T
M=1⇒πT÷180→T
Rad
Cls
Range −1.5,1.5,1,−1,1,1,0,Abs T,.15
cosT→A
sinT→B
1→S
T<0⇒−1→S
Graph(X,Y)=((.25+.04T)cos T,S (.25+.04T)sin T)
Plot 0,0
Plot A,B
Line
Goto 1
```

Note: Before running the program, set the calculator in parametric mode. To do so, enter the following keystrokes.

In MAIN MENU, enter 1. [*SHIFT*] [*SET UP*] [*F3*] [*EXIT*]

Refer to Exercises 1–4 on page 701 for practice using this program.

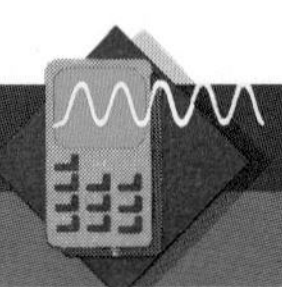

TECHNOLOGY

Keystrokes for Other Graphing Calculators

Keystrokes for page 701

Another nice use of a graphing calculator is to sketch a graph of the unit circle using the calculator's *parametric* mode. The keystrokes listed below show how to sketch the graph of the unit circle using a *TI-82*, a Casio *fx-9700GE*, and a Sharp *EL-9300C*.

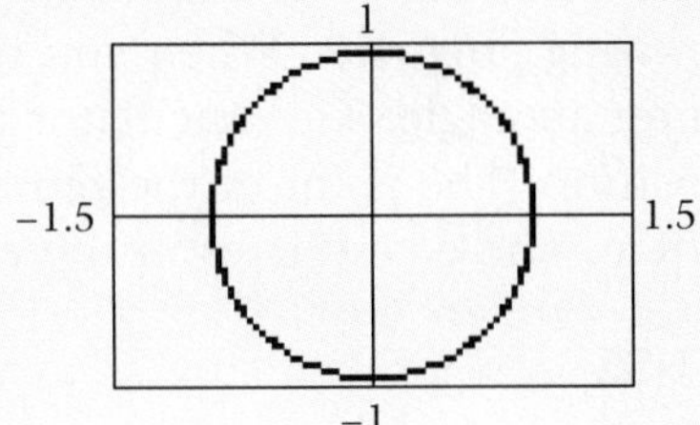

The graph of the unit circle is shown at the right. Remember, after graphing the unit circle, change the calculator back to *function* mode.

TI-82

[MODE] Cursor to Radian. [ENTER]

Cursor to Par. [ENTER]

[WINDOW] [▽] (Set window.)

Tmin=0	Xmin=−1.5	Ymin=−1
Tmax=6.3	Xmax=1.5	Ymax=1
Tstep=.031416	Xscl=1	Yscl=1

[Y=] [CLEAR]

[COS] [X,T,θ]

[ENTER] [CLEAR]

[SIN] [X,T,θ]

[GRAPH]

Sharp *EL-9300C*

[⍭] [SET UP]

[ALPHA] [B] [2]

[ALPHA] [E] [3] [QUIT]

[RANGE] (Set range.)

Tmin=0	Xmin=−1.5	Ymin=−1
Tmax=6.3	Xmax=1.5	Ymax=1
Tstp=Do not set.	Xscl=1	Yscl=1

[QUIT] [MENU] [1] [CL]

[cos] [X/θ/T] [ENTER]

[CL] [sin] [X/θ/T]

[⍭]

Casio *fx-9700GE*

In MAIN MENU, enter 1.

[SHIFT] [SET UP] [F3] [EXIT]

[SHIFT] [DRG] [F2] [EXE]

[Range] (Set range.)

Xmin: −1.7	Ymin: −1
Xmax: 1.7	Ymax: 1
Xscale: 1	Yscale: 1

[EXE] (or cursor down)

T,θ

min:0

max:6.3

pitch:0.031416

[EXIT]

[Graph]

[cos] [X,θ,T] [SHIFT] [,] [sin] [X,θ,T] [)] [EXE]

Refer to Exercises 5–10 on page 701 for practice using this feature.

TECHNOLOGY APPENDIX

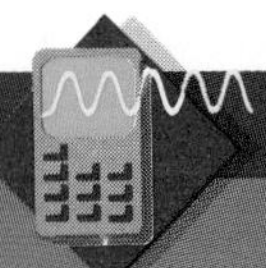

Keystrokes for page 802

Most graphing calculators, such as the *TI-82,* the Casio *fx-9700GE,* and the Sharp *EL-9300C,* have random number generator that can be used to perform probability experiments. Each time the random number generator is used, it selects a number between 0 and 1. (The number selected is greater than 0 and less than 1.) If you want the calculator to randomly select an integer x, such that $n \leq x \leq m$, you must follow an algorithm similar to the one shown below.

Integer Part of $[(m - n + 1) \times \text{(random number)} + 1]$

For example, to generate a random number between 1 and 6, enter the following keystrokes on your calculator.

TI-82: [MATH] [▷] [2] [(] 6 [MATH] [◁] [1] [+] 1 [)] [ENTER]

Casio *fx-9700GE:* [SHIFT] [MATH] [F3] [F2] [(] 6 [SHIFT] [MATH] [F2] [F4] [+] 1 [)] [EXE]

Sharp *EL-9300C:* [MATH] [3] [(] 6 [MATH] [8] [+] 1 [)] [ENTER]

Often when performing probability experiments, you want to generate many random numbers. This is a simple task when using the programming features of a graphing calculator. For example, each program listed below simulates the roll of a six-sided die. Every time you press enter, the program generates a number between 1 and 6.

TI-82

```
PROGRAM:RANDOM
:Lbl 1
:iPart (6rand+1)→X
:Disp X
:Pause
:Goto 1
```

To exit the program enter [ON] 2.

Casio *fx-9700GE*

```
RANDOM
Lbl 1
Int (6Ran#+1)→X
X◢
Goto 1
```

To exit the program, enter [AC/ON] [AC/ON].

Sharp *EL-9300C*

```
random
————————REAL
Label 1
X=ipart (6random+1)
Print X
Wait
Goto 1
```

To exit the program, enter [ON] [CL].

Refer to pages 802 and 803 for an example of using these programs to generate random numbers and for related exercises.

USING THE COMMUNICATION LINK

The *TI-82,* the Casio *fx-9700GE*, and the Sharp *EL-9300C* all have a port that allows you to communicate, or link, with a similar calculator. That is, this feature allows you to send and receive programs, data, and other information to and from a calculator of the same type. Instructions for using the communication link to send and receive a program are given below.

TI-82

The link port is located at the bottom center of the calculator. Insert the link cable in each calculator port.

To send a program:
[2nd] [LINK] [2] Cursor to program name. [▷] [ENTER]

To receive a program:
[2nd] [LINK] [▷] [ENTER]

Casio *fx-9700GE*

The link port is located at the bottom right of the calculator. Turn both calculators off and insert the link port in each calculator. Turn the calculators on.

To send a program:
In **MAIN MENU**, enter [B].
[F1] Cursor to program. [EXE]
[F2] Cursor to program name. [EXE]

To receive a program:
In **MAIN MENU**, enter [ALPHA] [B].
[F2] Cursor to program. [EXE]
[F2] Cursor to empty program location.
[EXE]

Note: The parameters of both calculators must be identical. To determine the parameters, press [F6]. Change as necessary.

Sharp *EL-9300C*

The link port is located at the bottom center of the calculator. Insert the link cable in each calculator port.

To send a program:
[2ndF] [OPTION]
[G] [ENTER]
[ENTER] Cursor to program name.
[ENTER]

To receive a program:
[2ndF] [OPTION]
[G] [ENTER]
[C] [ENTER]

TECHNOLOGY

Keystrokes for Other Graphing Calculators

Use with Lesson 5.4, pp. 252–257

Some graphing calculators can be used to solve equations with tables. The following shows how to use the table features of a Texas Instruments *TI-82* and a Casio *fx-9700GE* to solve the equation

$$-2x^2 + 4x + 7 = 0.$$

X	Y1	
-5	-63	
-4	-41	
-3	-23	
-2	-9	
-1	1	
0	7	
1	9	

X= -5

TI-82

[Y=] [CLEAR]

[(–)] 2 [X,T,θ] [x^2] [+] 4 [X,T,θ] [+] 7

[2nd] [TblSet]

TblMin = –5

△Tbl = 1

Indpnt: Auto

Depend: Auto

[2nd] [TABLE]

A possible table is shown above. Use [△] and [▽] to scroll through the table. A solution occurs when the y-value is 0. From the table, it appears that one solution occurs when x is between -2 and -1. Alter the table by changing TblMin to -2 and △Tbl to 0.1. Continue to alter TblMin and △Tbl until you can approximate the solution $x = -1.121$ (accurate to three decimal places).

Casio *fx-9700GE*

In MAIN MENU, enter 8.

[F1] [F1] [F1] (Clears any functions.)

[(–)] 2 [X,θ,T] [x^2] [+] 4 [X,θ,T] [+] 7

[F2] (RANGE)

Start: –5

End : 10

Pitch: 1

[F1]

A possible table is shown above. Use [△] and [▽] to scroll through the table. A solution occurs when the y-value is 0. From the table, it appears that one solution occurs when x is between -2 and -1. Alter the table by changing the Start to -2, End to -1, and Pitch to 0.1. (To do so, enter [F2] [F2].) Continue to alter the Start, End, and Pitch until you can approximate the solution $x = -1.121$ (accurate to three decimal places).

Exercises

1. Find the second solution to the equation $-2x^2 + 4x + 7 = 0$.

In Exercises 2–4, use a table to find the solutions to the equation. Each equation has two solutions. Your solution should be accurate to three decimal places.

2. $3x^2 + 4x - 5$

3. $-5x^2 - 7x + 3$

4. $0.25x^2 - 2x - 1$

CHAPTER 1

1.6 Independent Practice p. 41

33. 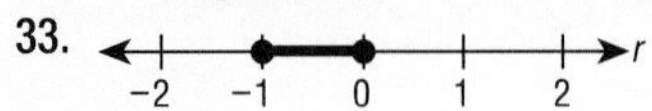

34. 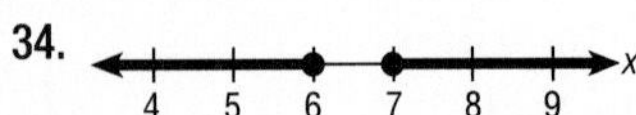

35.

36.

37.

38.

1.7 Independent Practice p. 48

37. 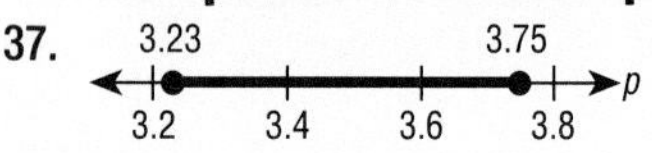

38. 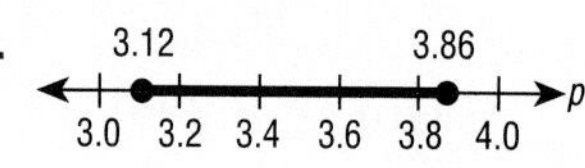

1.8 Independent Practice p. 53

5.

Number	Tally	Frequency
1	||	2
2	|||	3
3	||||	4
4	||||	4
5	|||	3
6	|||	3
7	||	2

6.

Number	Tally	Frequency
295	|	1
296	|	1
297	|	1
298	|	1
299	|||	3
300	~~||||~~	5

CHAPTER 2

2.1 Communicating About Algebra p. 66

A. 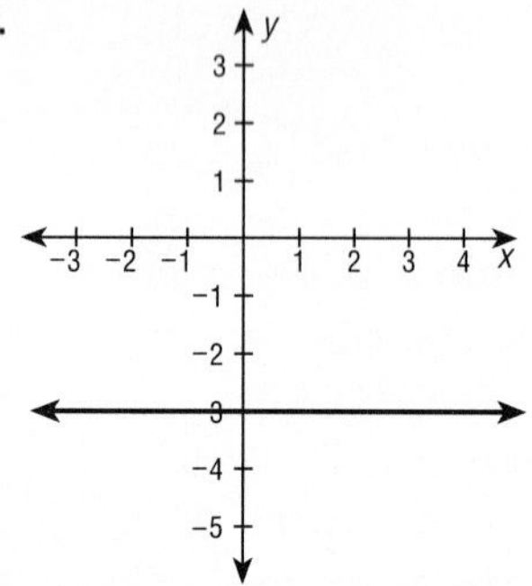

B. 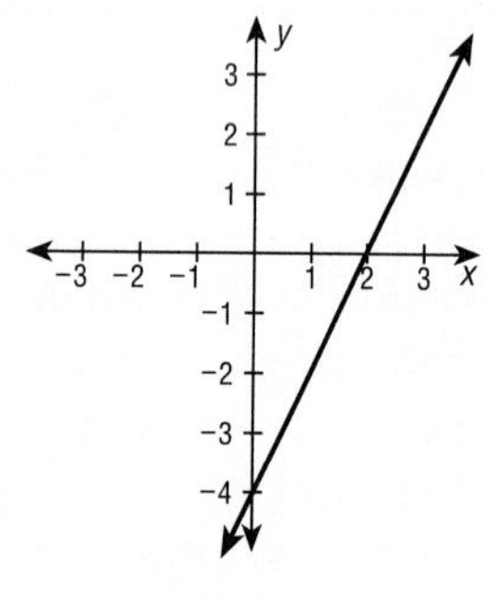

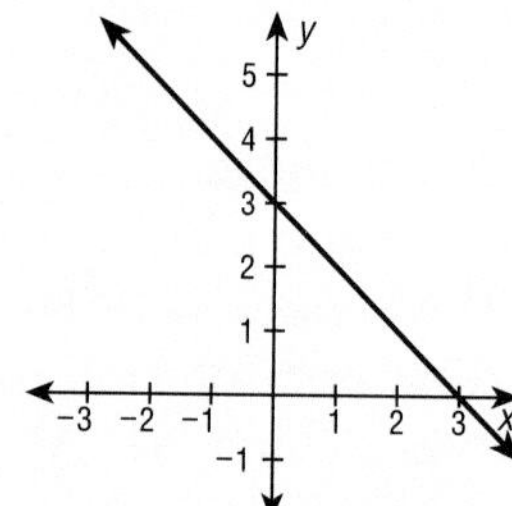

2.1 Guided Practice p. 67

1.

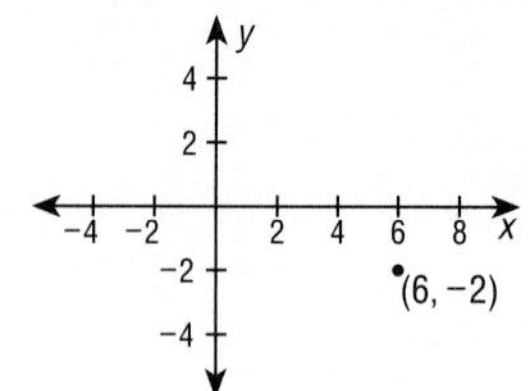

3. No, because it is endless. One can choose any value for x, and find a value for y.

2.1 Independent Practice p. 68

25. 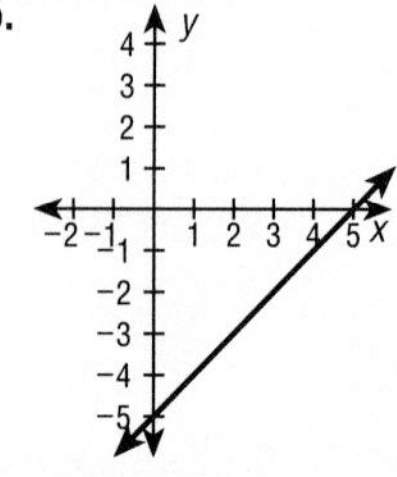

26. 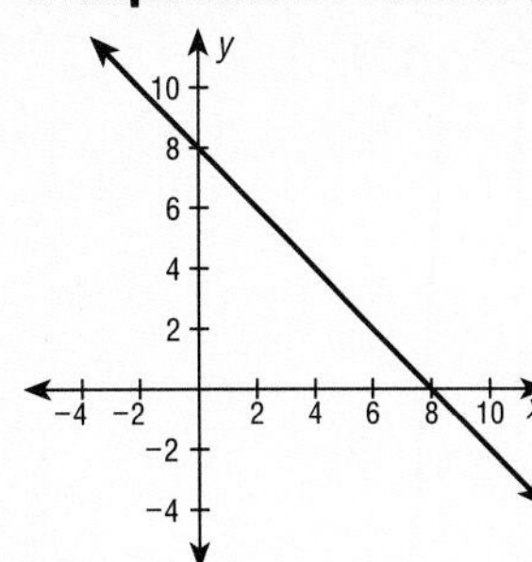

27. 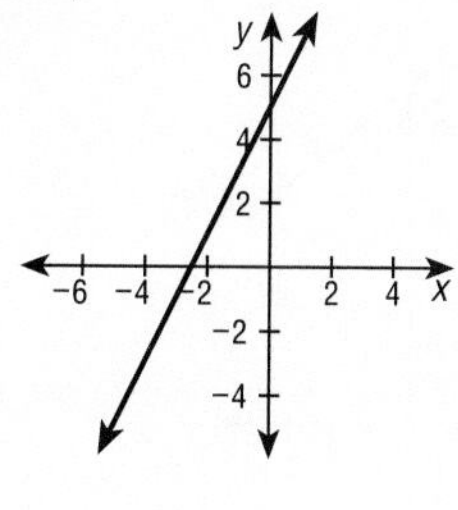

28. 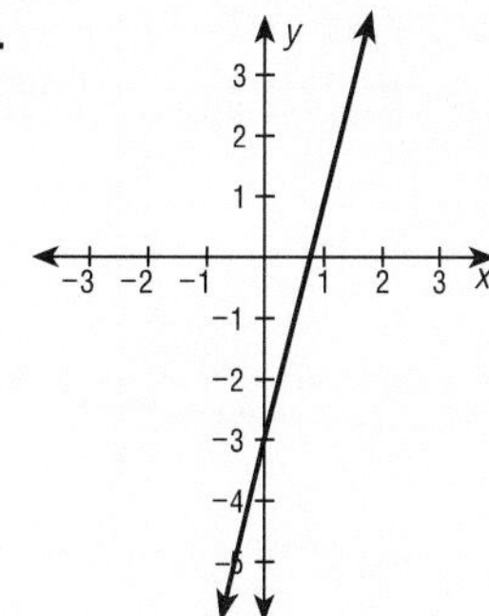

29. 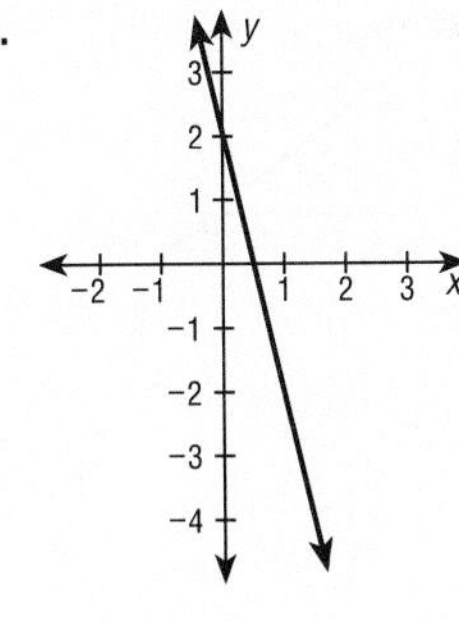

30. 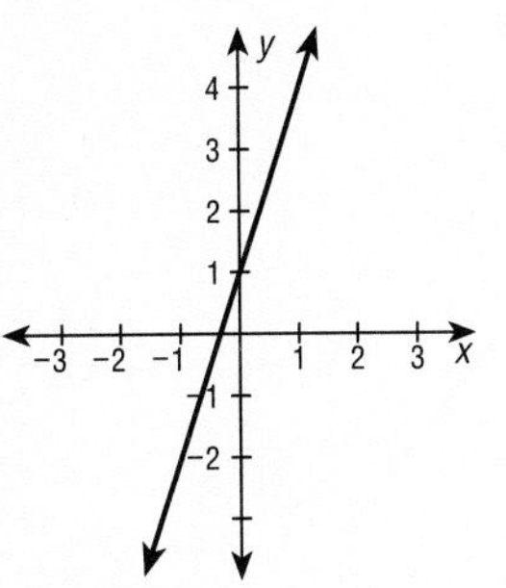

31. 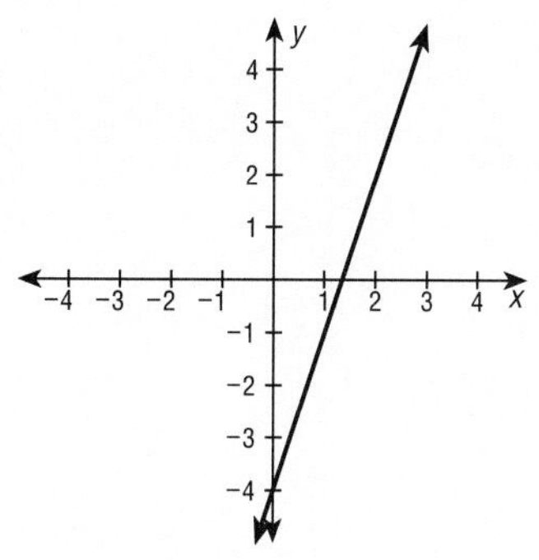

32. 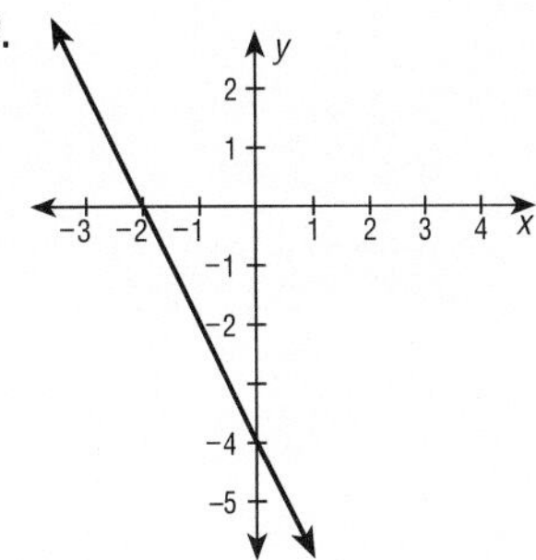

33. 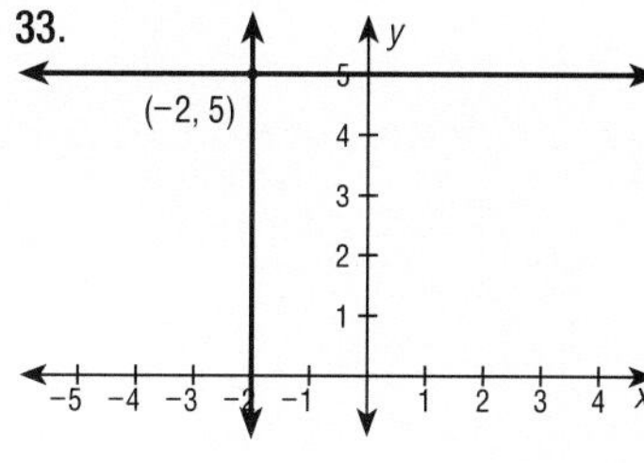

34. 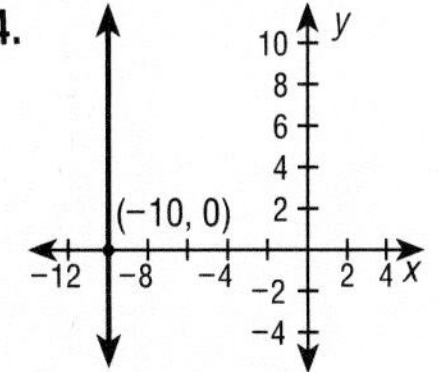

35. 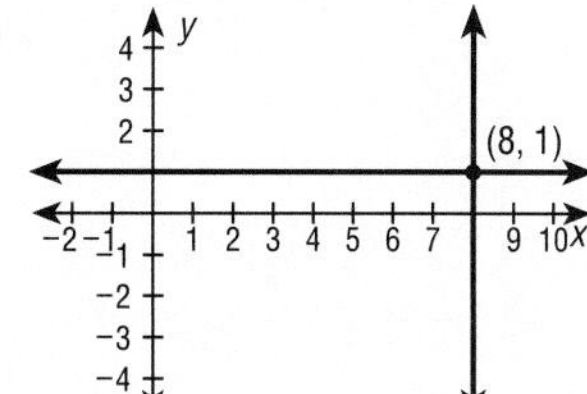

36. 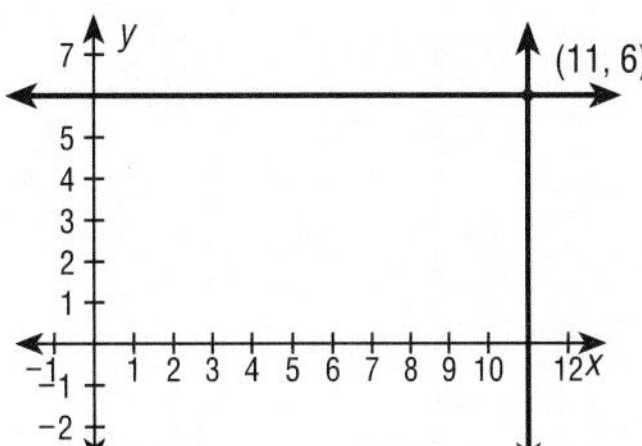

2.2 Independent Practice p. 74

9. 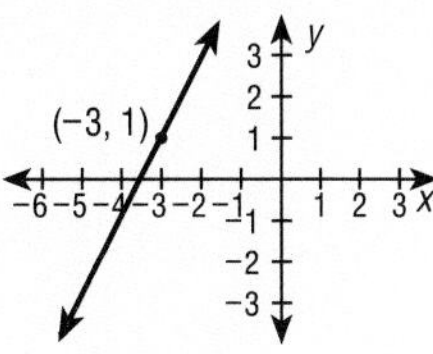

10. 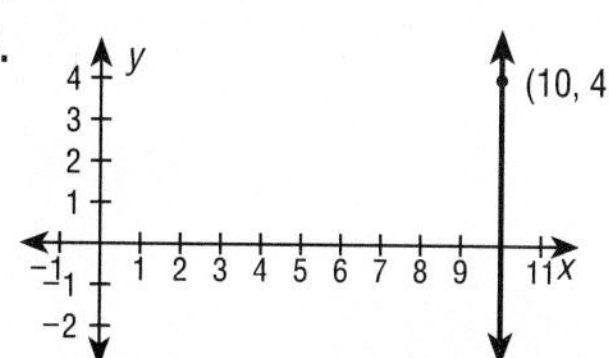

11. 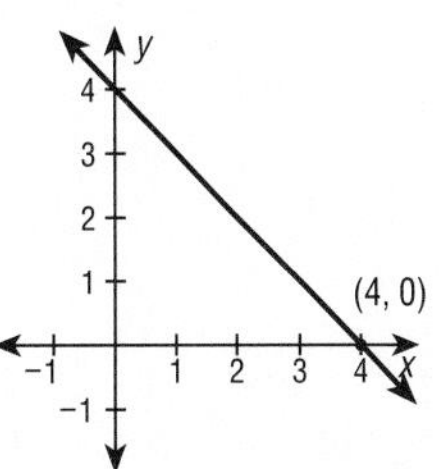

12. 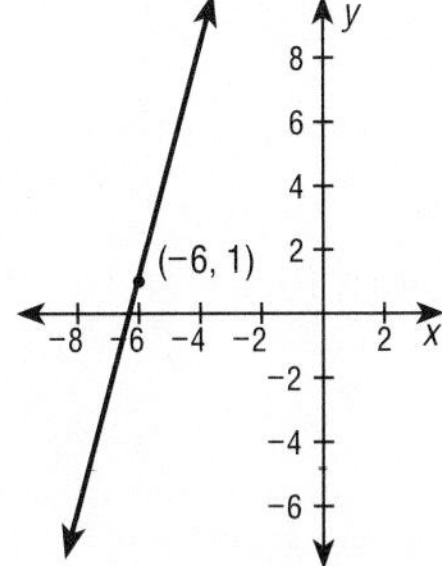

13. 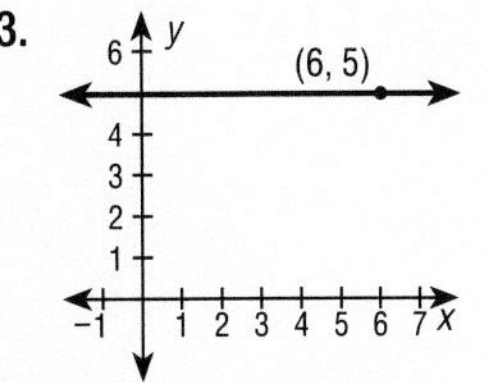

14. 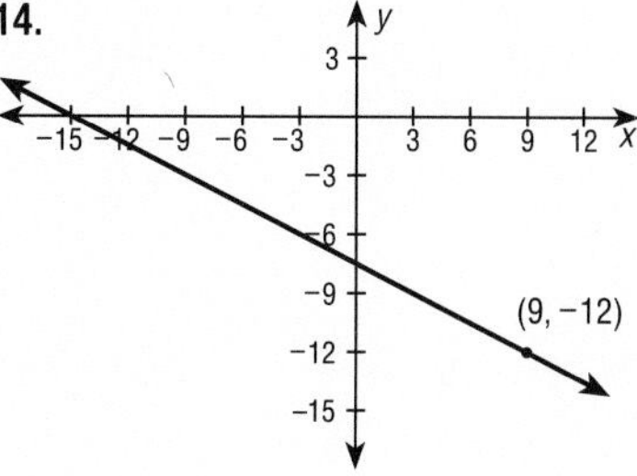

15. 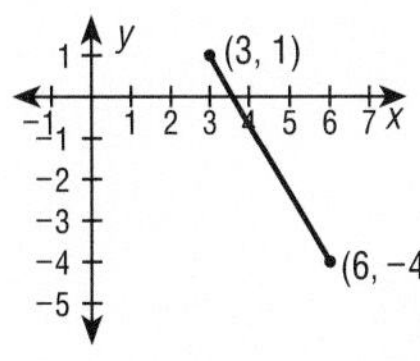

16. 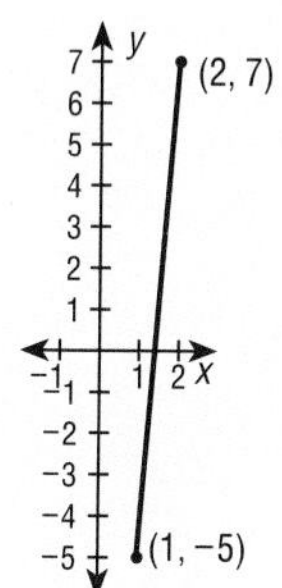

2.2 Independent Practice *(continued)*

17.
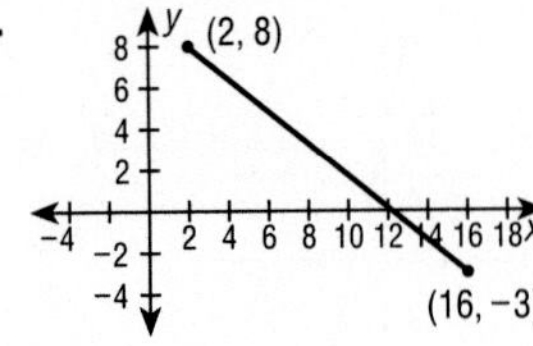

18.
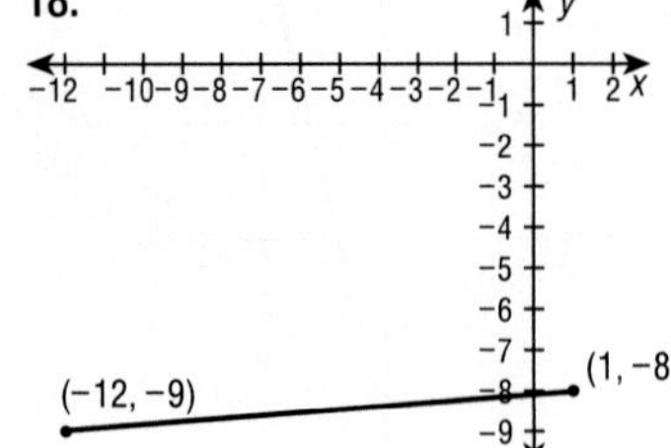

19.
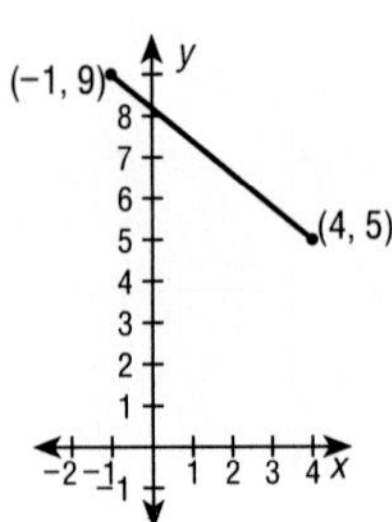

20.
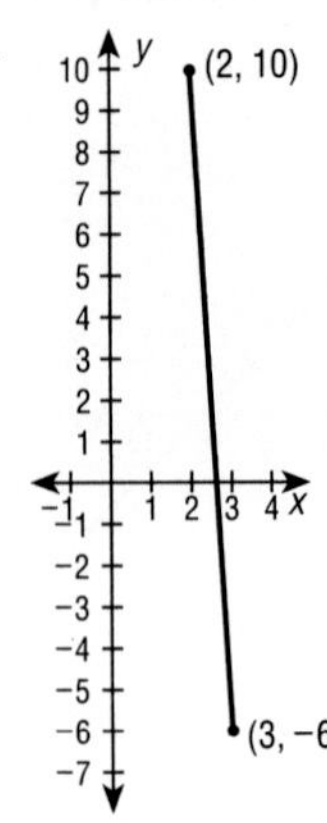

21.
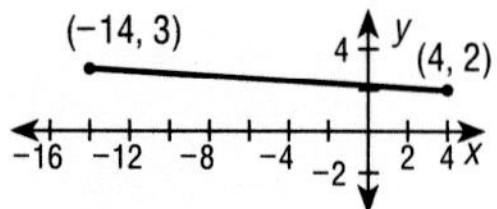

22.
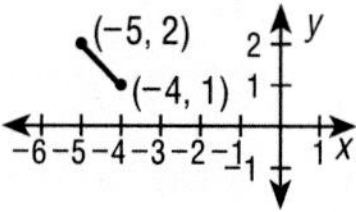

23.
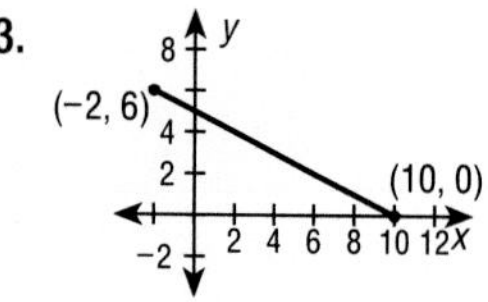

24.
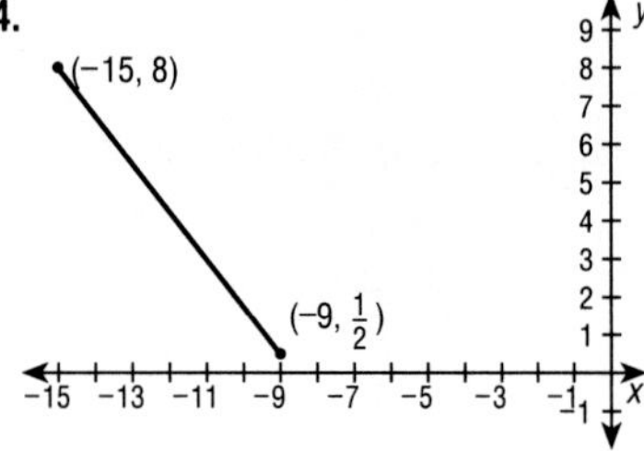

25.
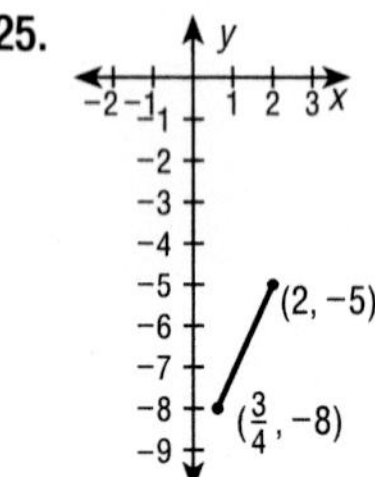

26.
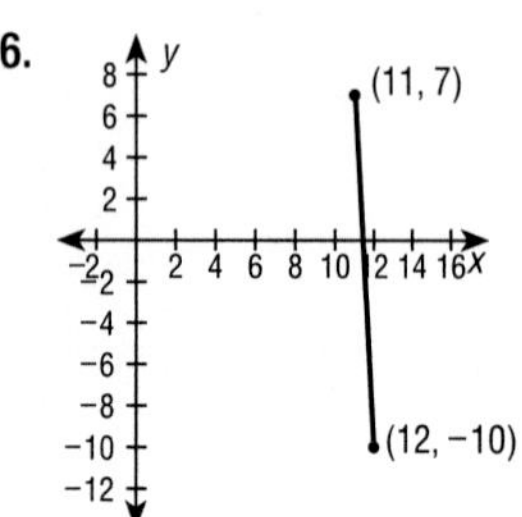

2.2 Integrated Review p. 76

45.
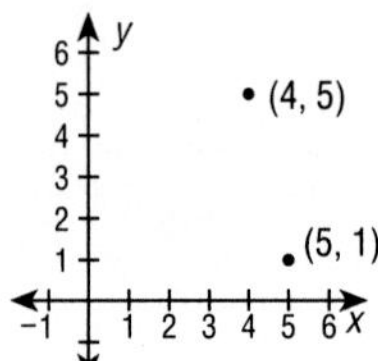

46.
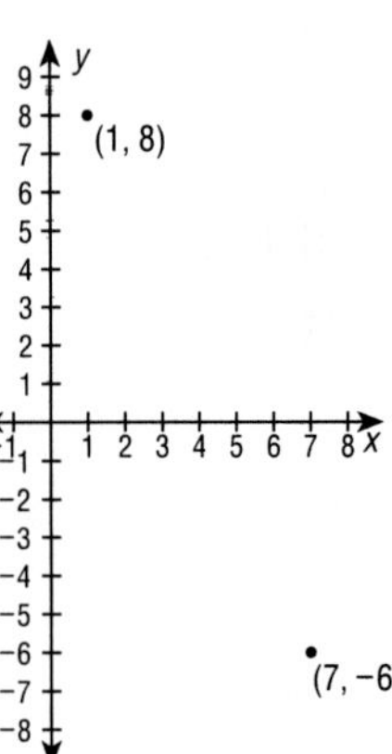

47.
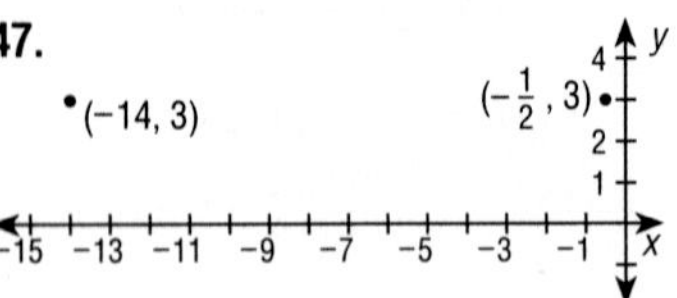

48.
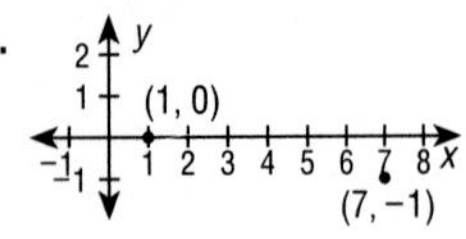

49.
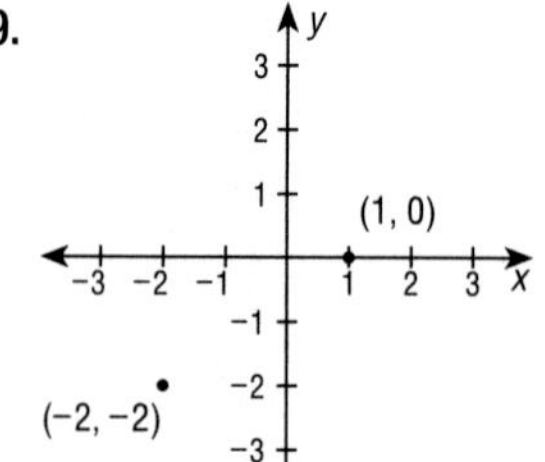

50.
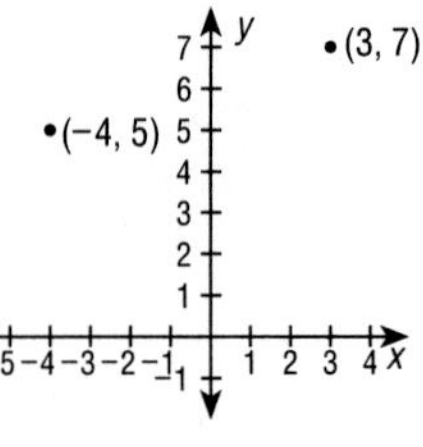

ADDITIONAL ANSWERS

2.3 Independent Practice p. 81

19.
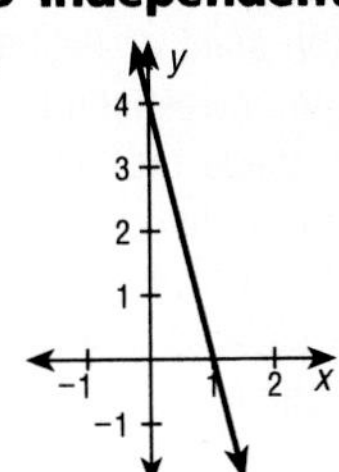

20.
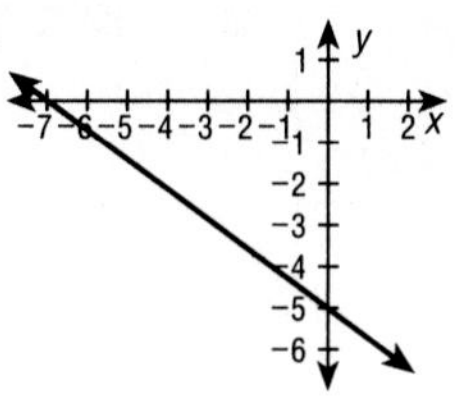

21.
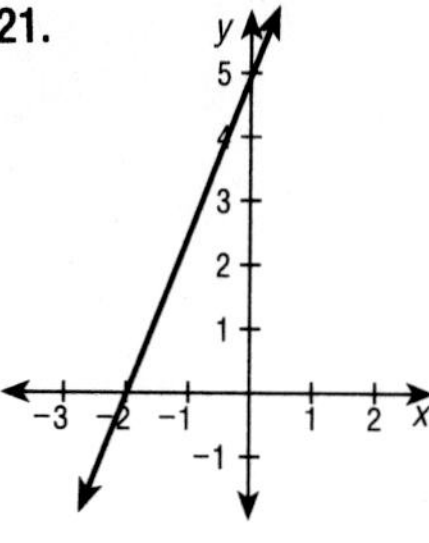

22.
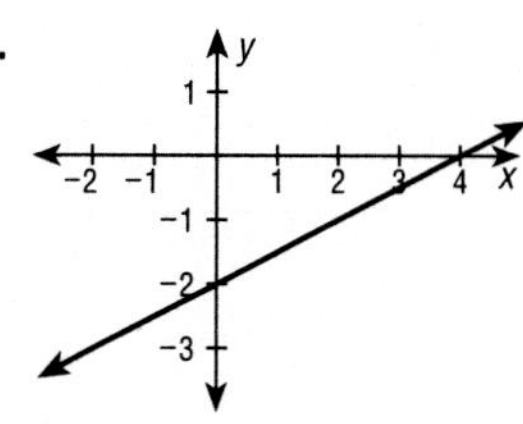

23.
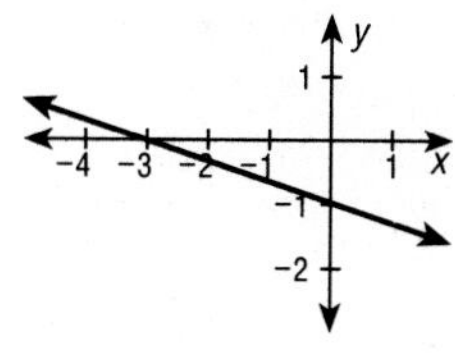

40.
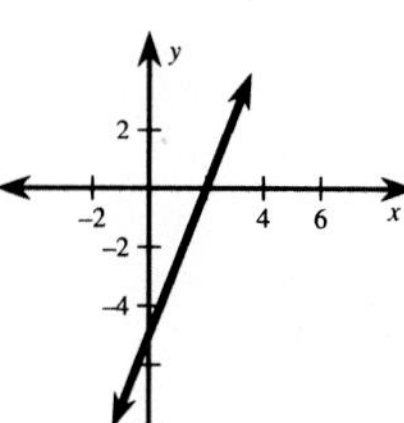

41.
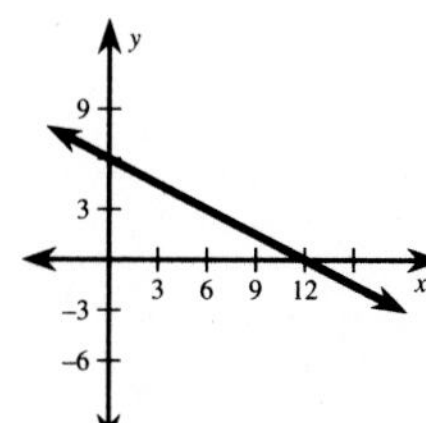

42.
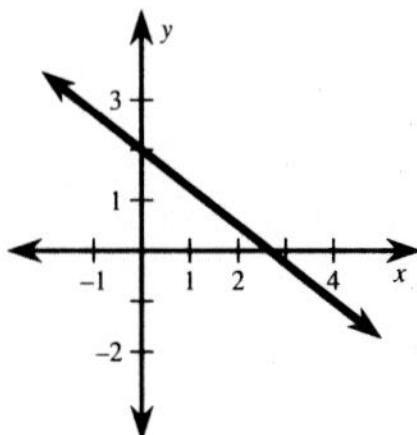

2.3 Integrated Review p. 83

61.
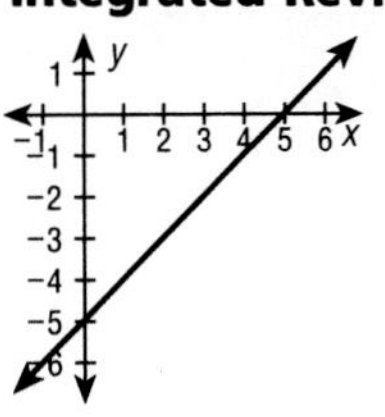

62.
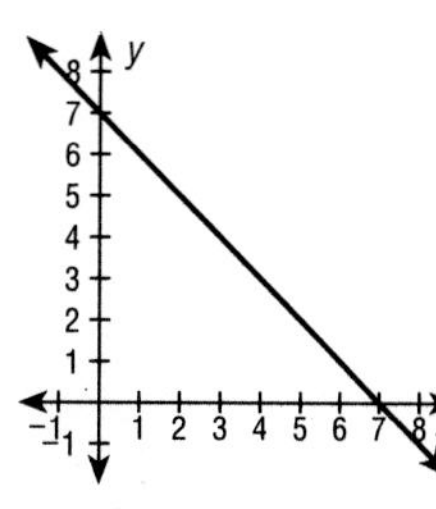

63.
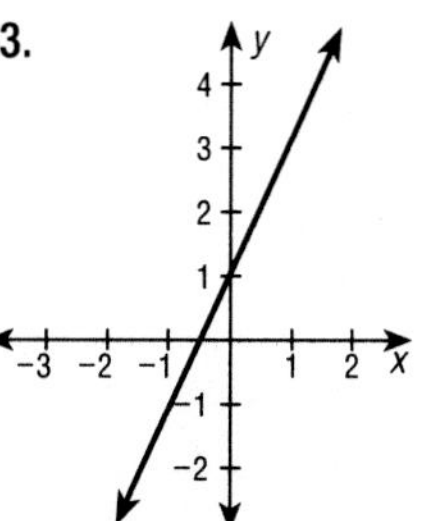

64.
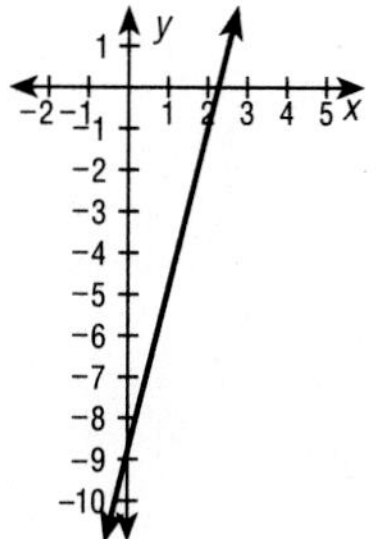

65.
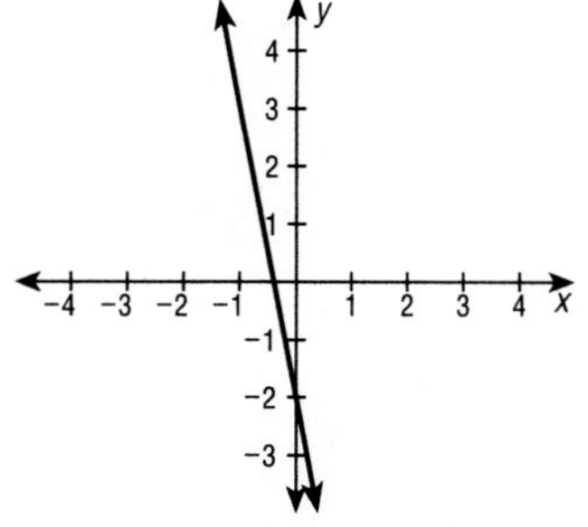

66.
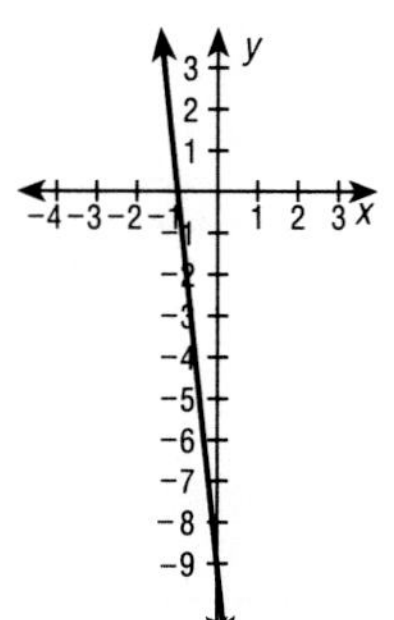

2.3 Using a Graphing Calculator p. 85

1.
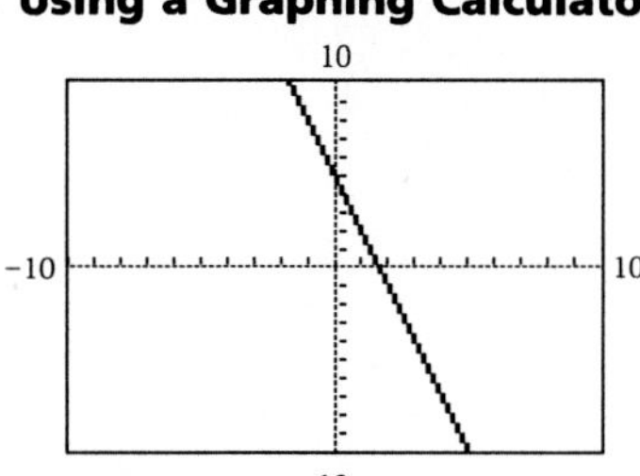

2.
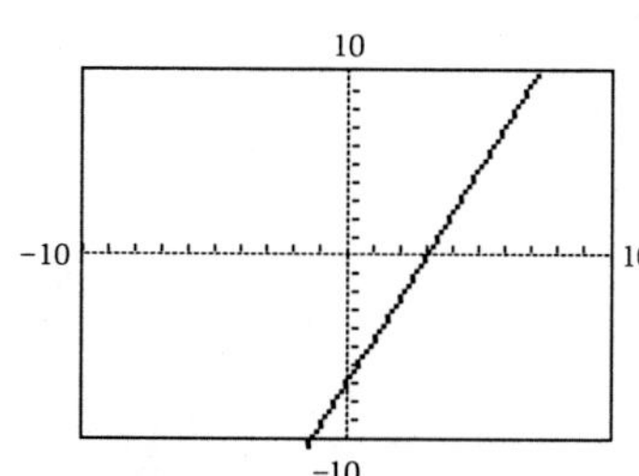

3.
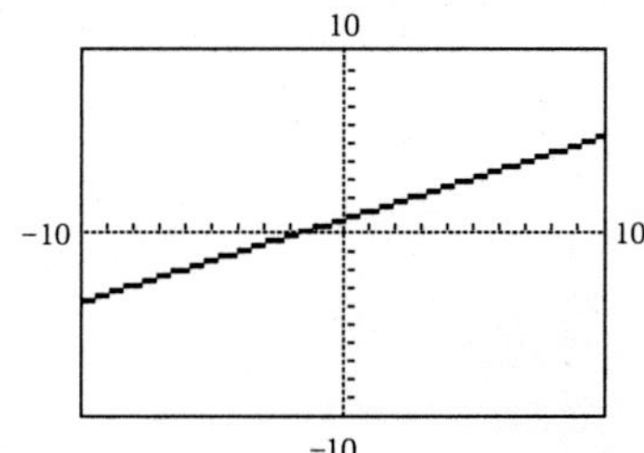

4.
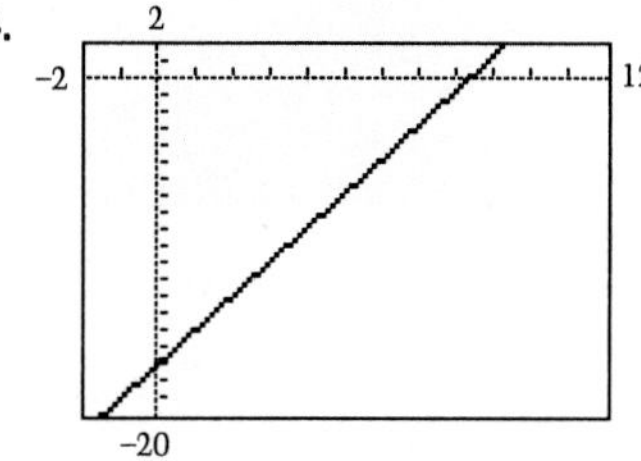

5.
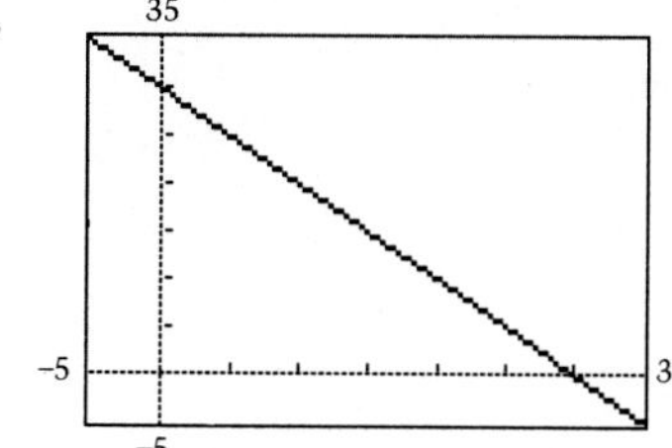

6.
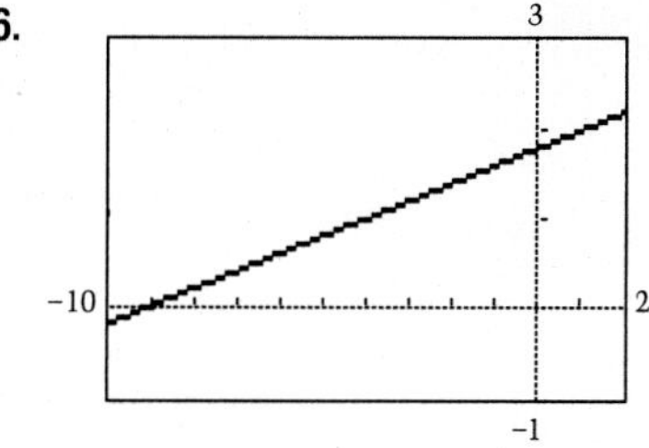

ADDITIONAL ANSWERS

2.5 Communicating About Algebra p. 96

A.
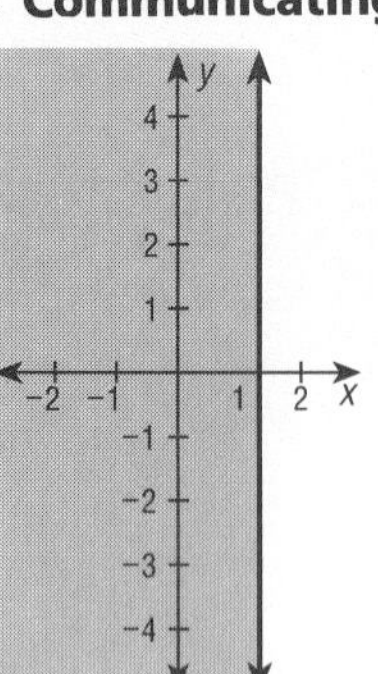

B.
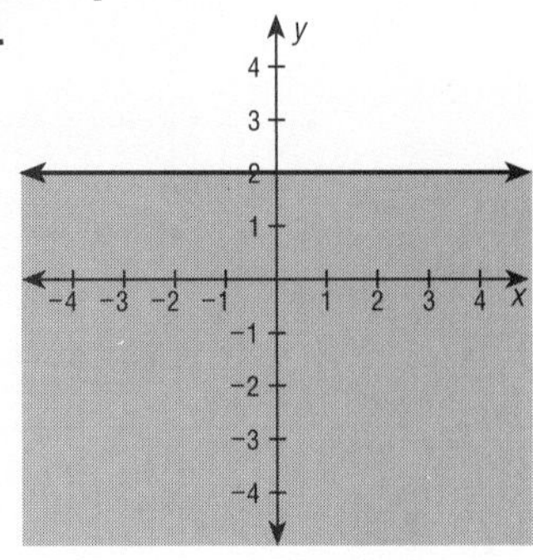

C.
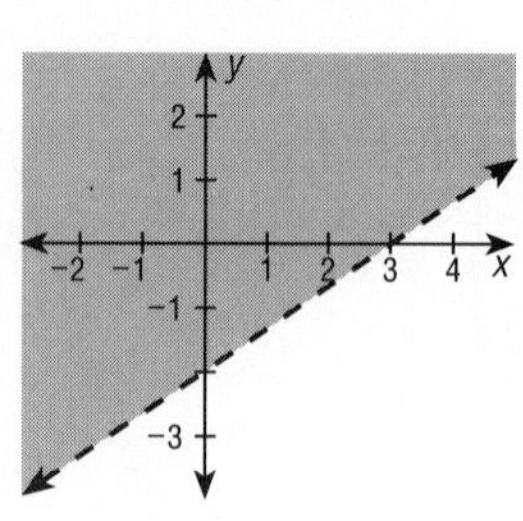

A.–C. Sketch the solid or dashed line of the corresponding equality; shade the half-plane that contains solutions after one point in that half-plane is shown to be a solution.

2.5 Independent Practice p. 97

17.
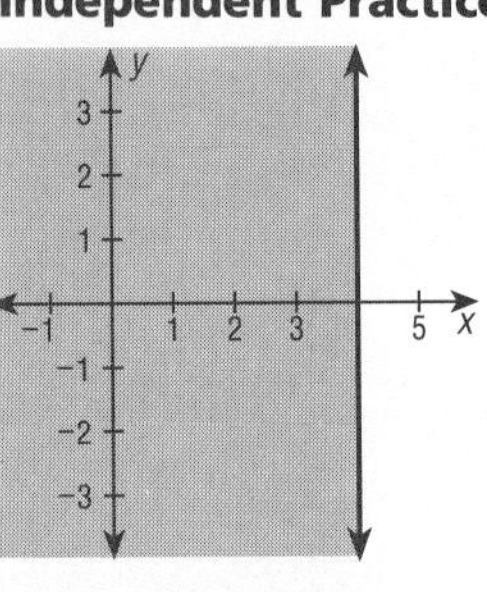

18.
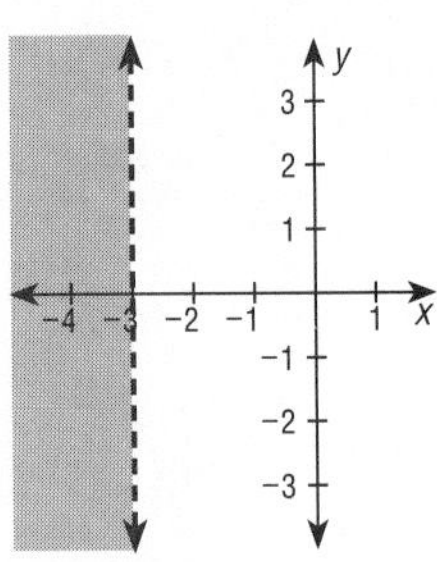

19.
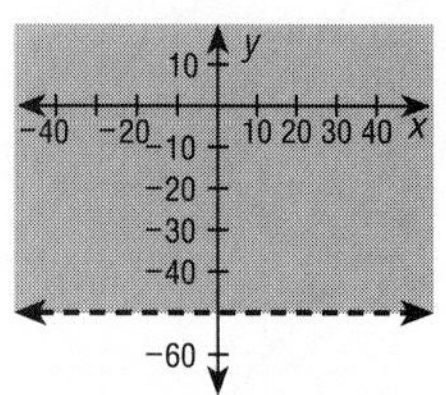

20.
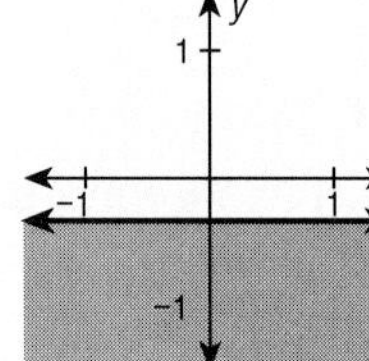

21.
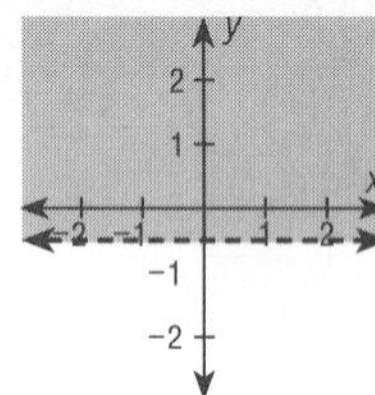

22.
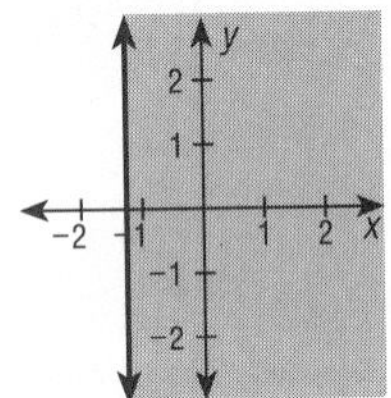

23.
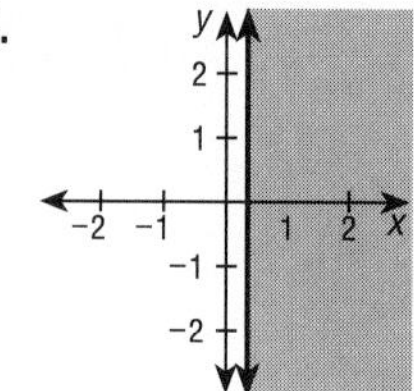

24.
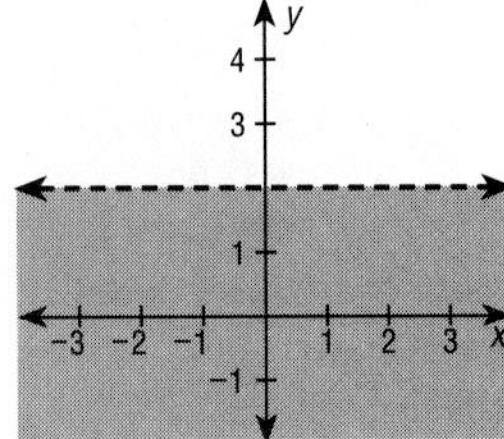

29.
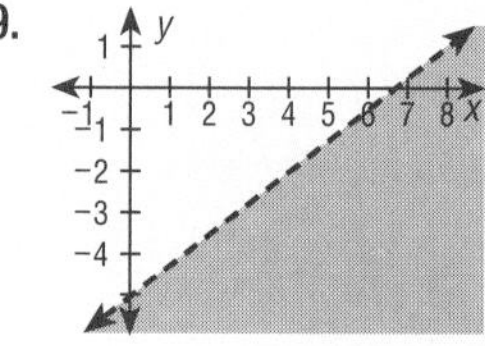

30.
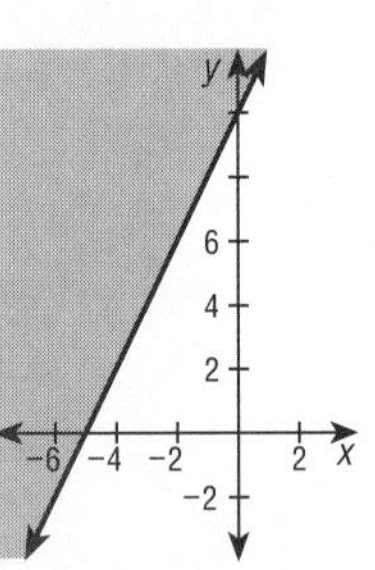

31.
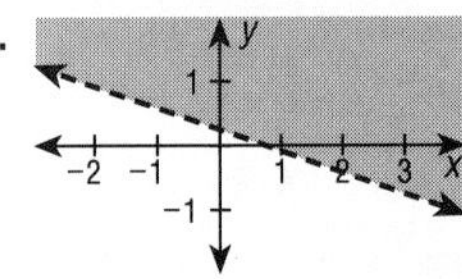

32.
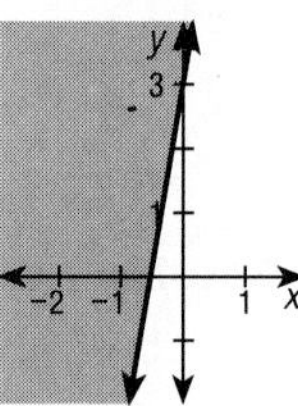

33.
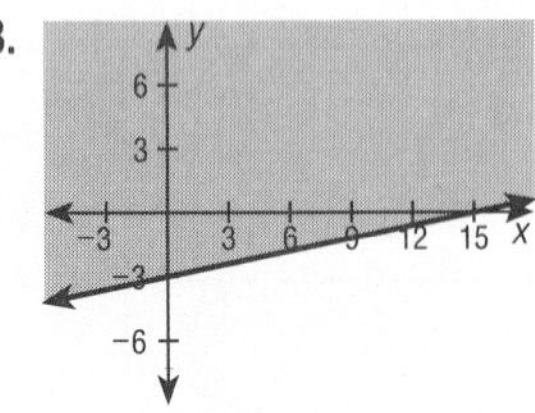

34.
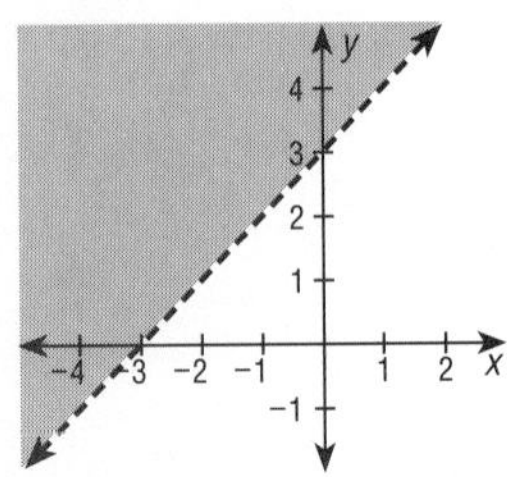

35.
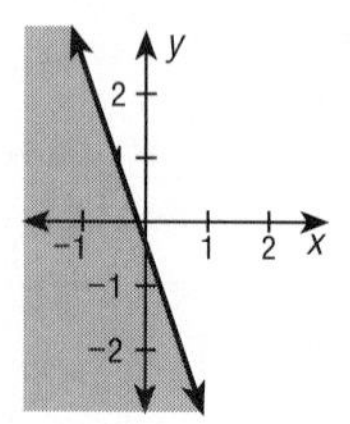

36.
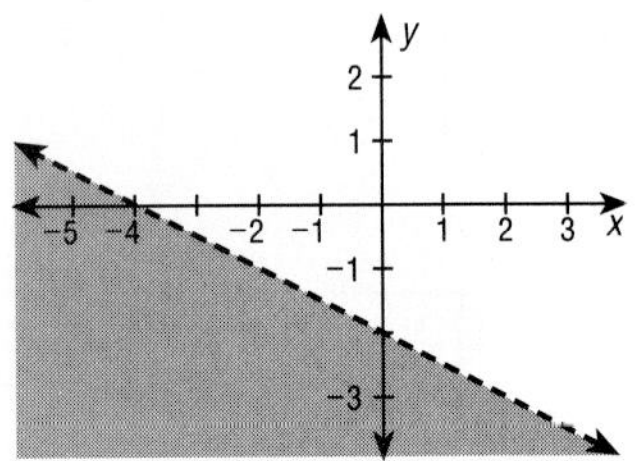

2.5 Integrated Review p. 99

51. 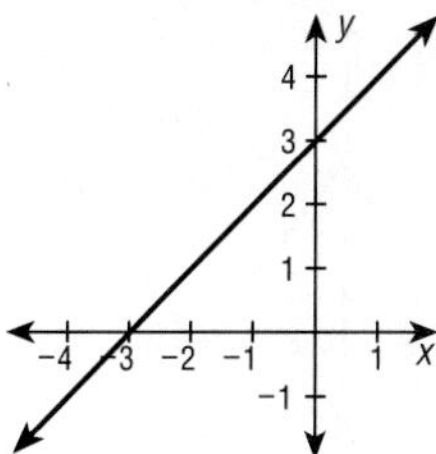

52. 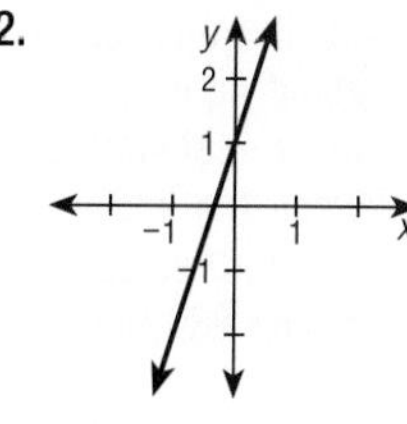

53. 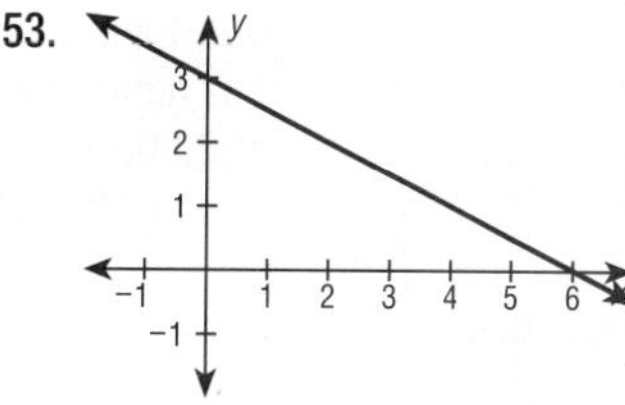

54. 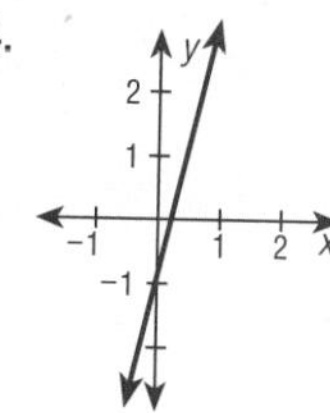

2.6 Communicating About Algebra p. 103

A., B., C. 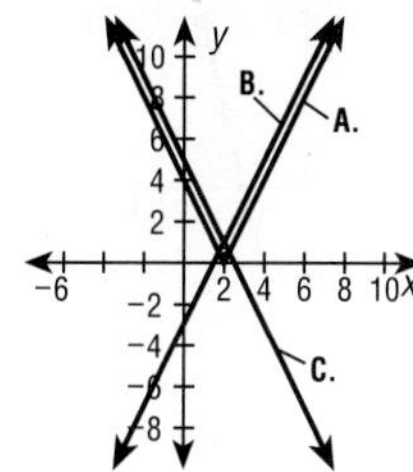

2.6 Guided Practice p. 104

4. 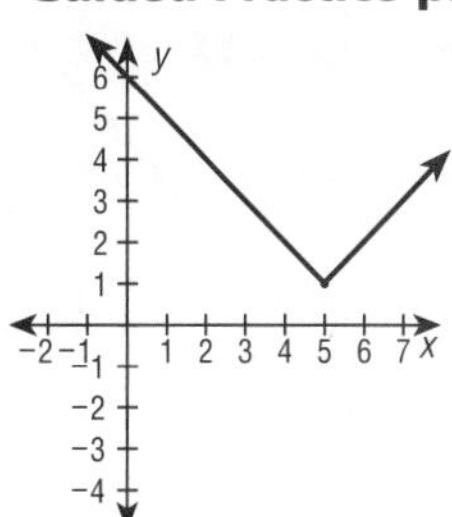

2.6 Independent Practice p. 105

29. 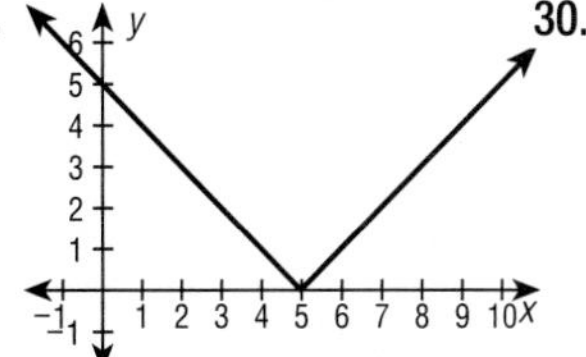

30. 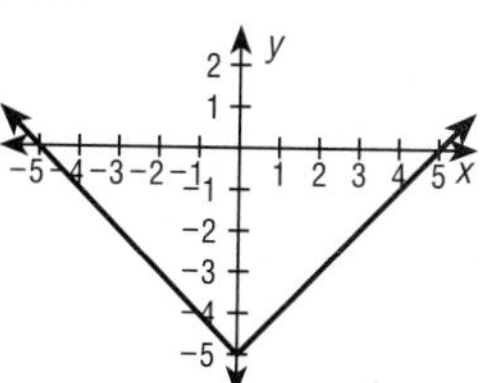

31. 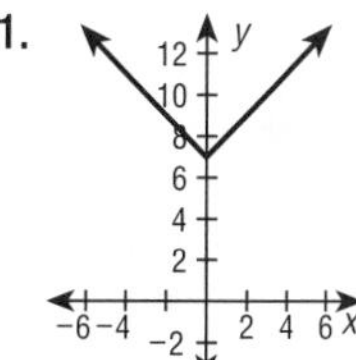

32. 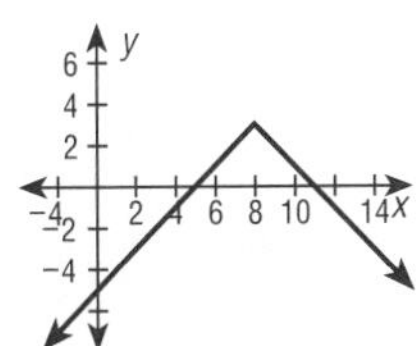

33. 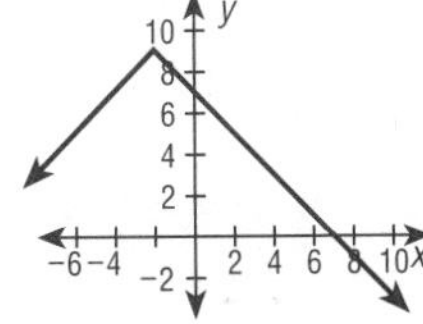

34. 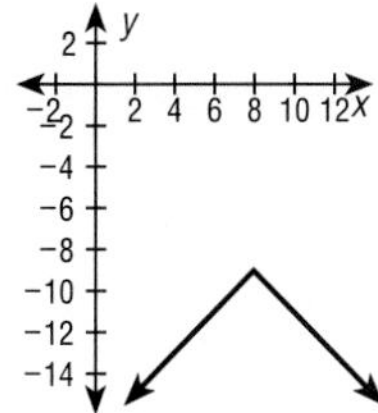

35. 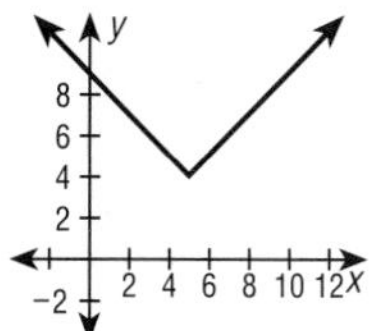

36. 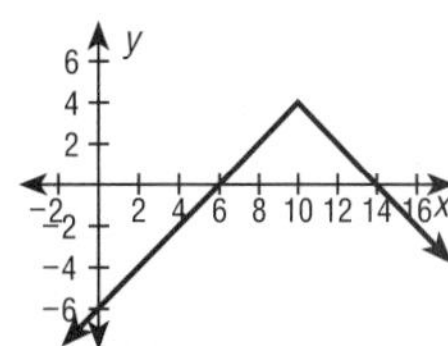

37. 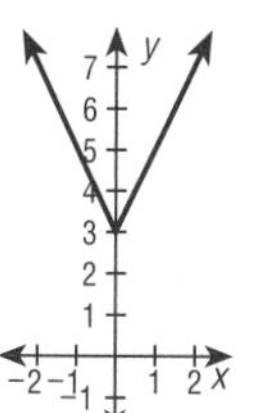

38. 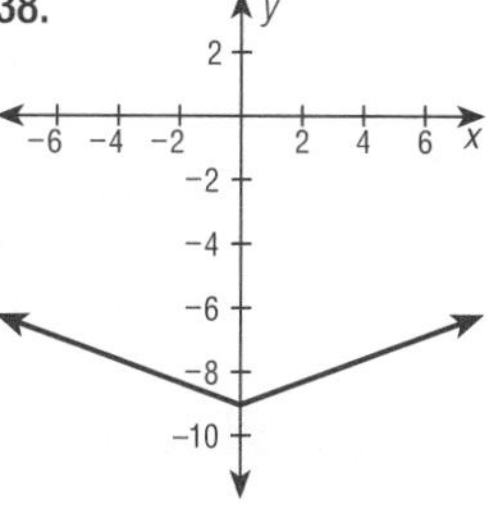

39. 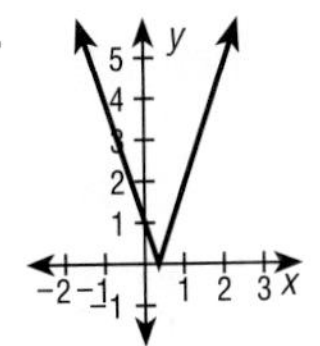

40. 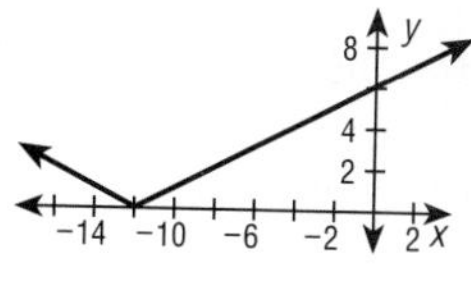

41. 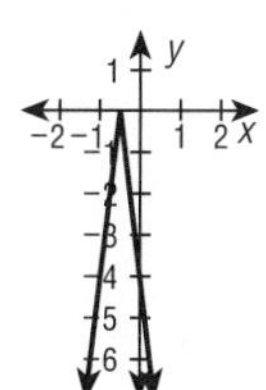

42. 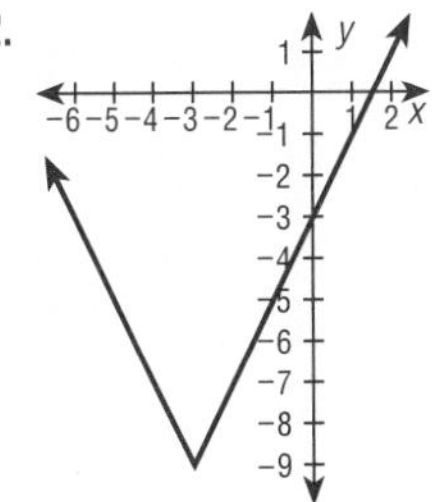

43. 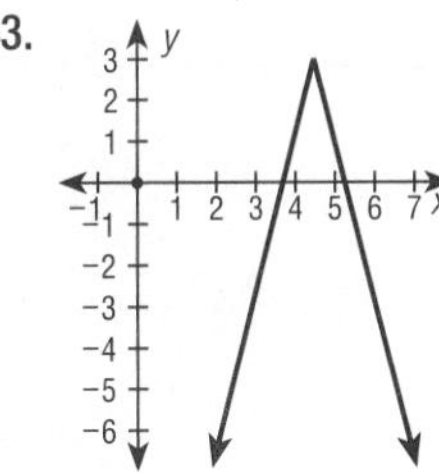

2.6 Independent Practice *(continued)*

44.

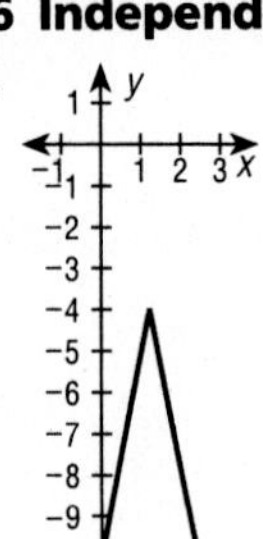

45.

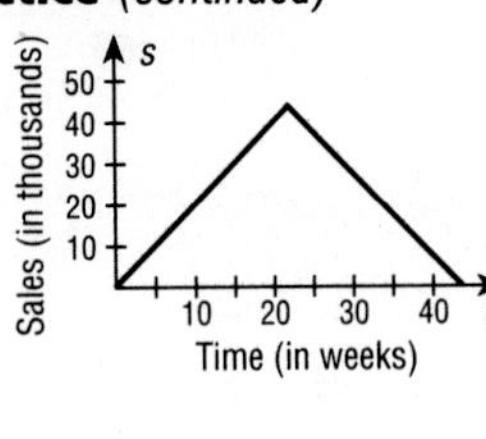

46. 968 thousand, yes, no; number is over 500,000 but less than 1,000,000

47. No; the total sales would equal the area of the rectangle between the line $s = 20$ and the t-axis, which is only 880,000.

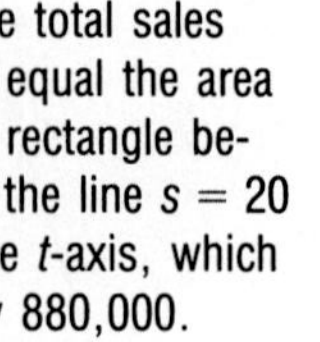

48.

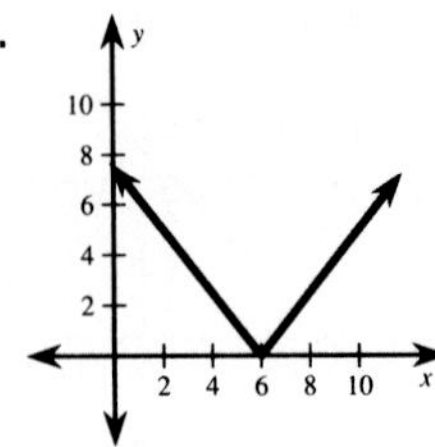

2.6 Integrated Review p. 106

49.

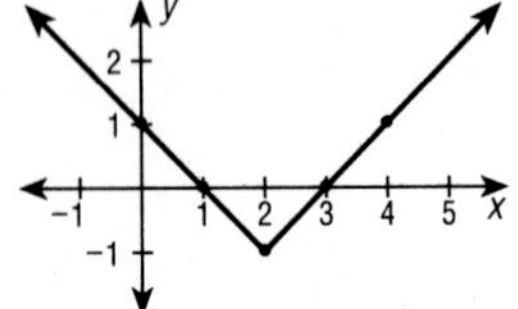

50.

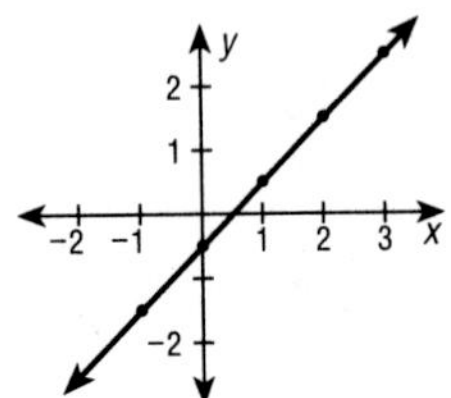

51.

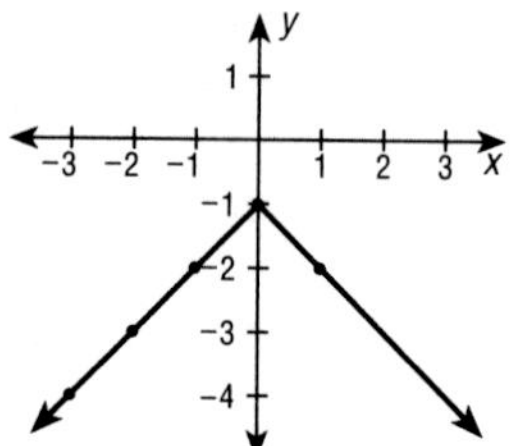

52.

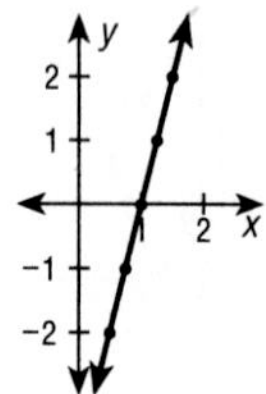

2.6 Exploration and Extension p. 106

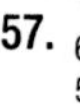

57.

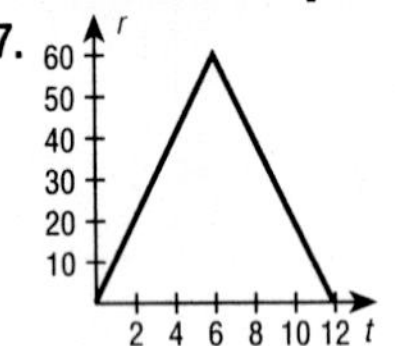

2.7 Chapter Review p. 115

1.

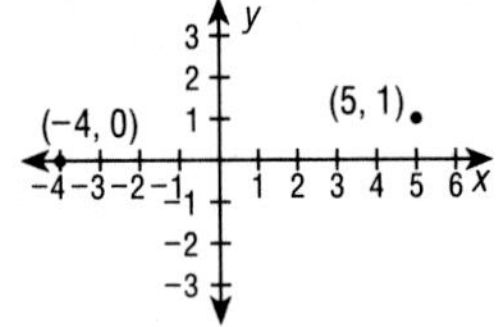

2.

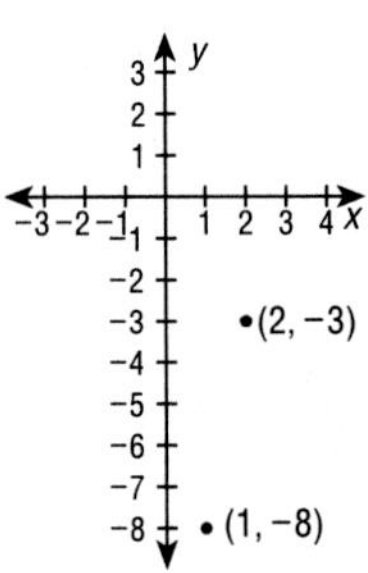

3.

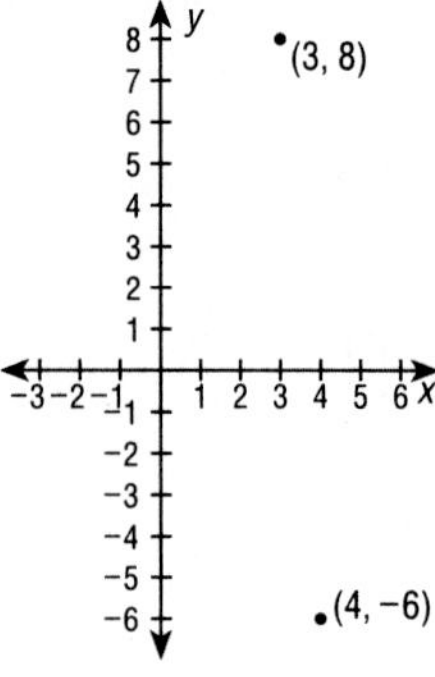

4.

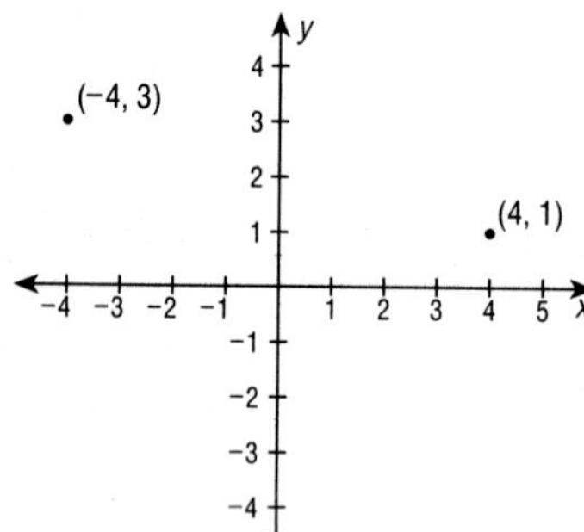

5.

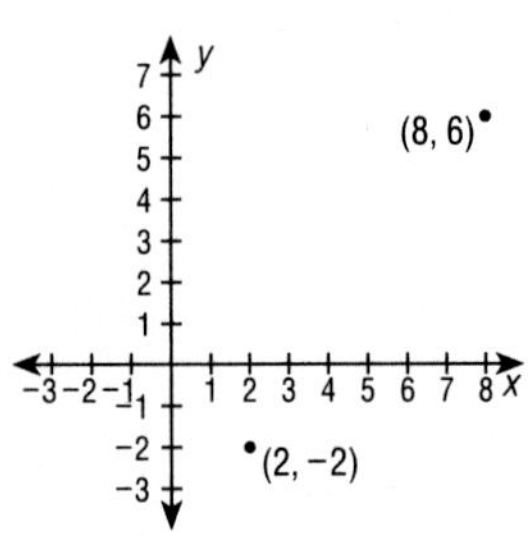

6.

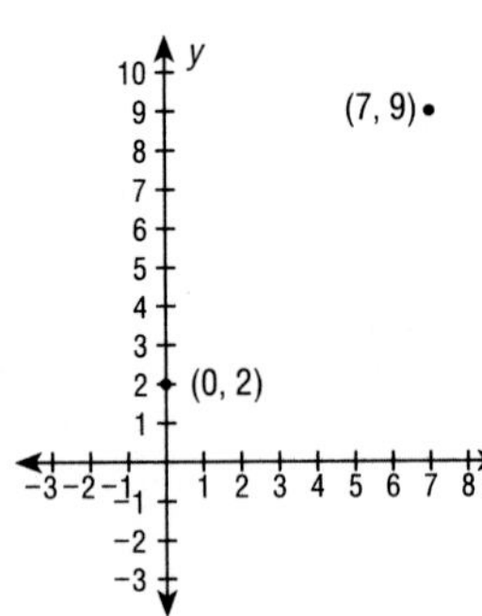

2.7 Chapter Test p. 119

3.

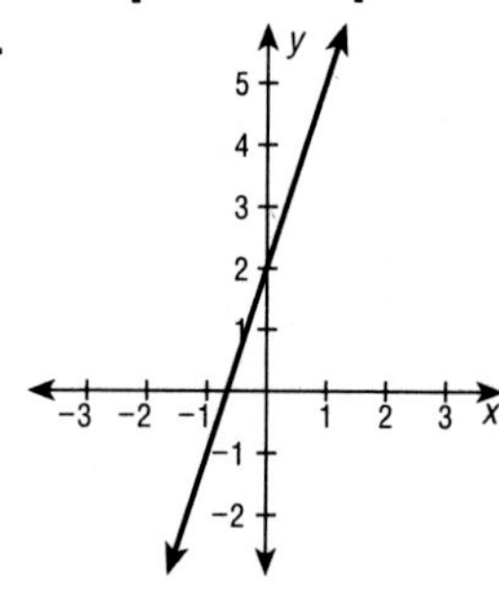

11.

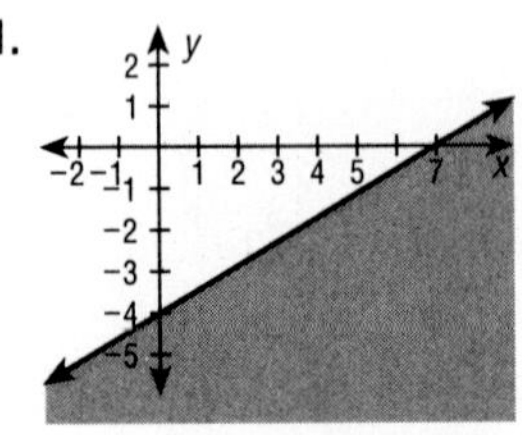

12.

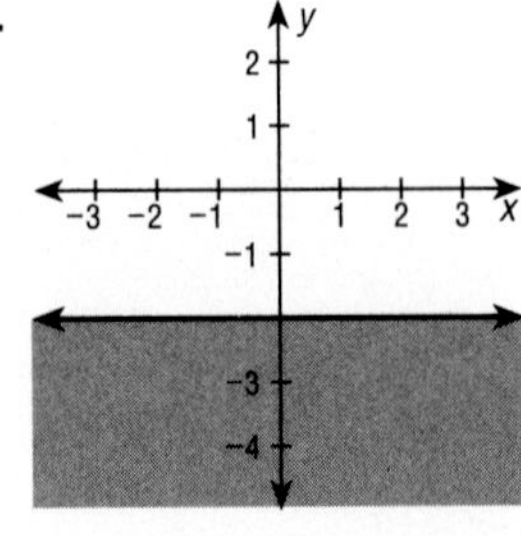

13.

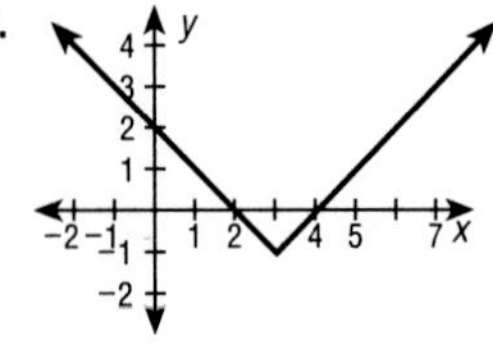

14.

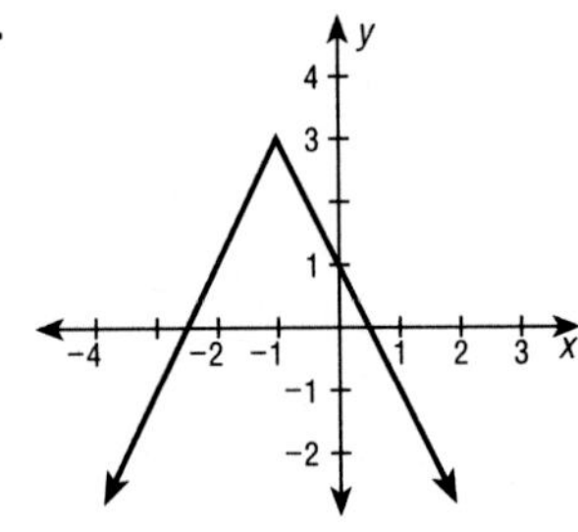

ADDITIONAL ANSWERS

3.1 Communicating About Algebra p. 124

A. (3, 1); if the two graphs intersect, the point of intersection is the solution.

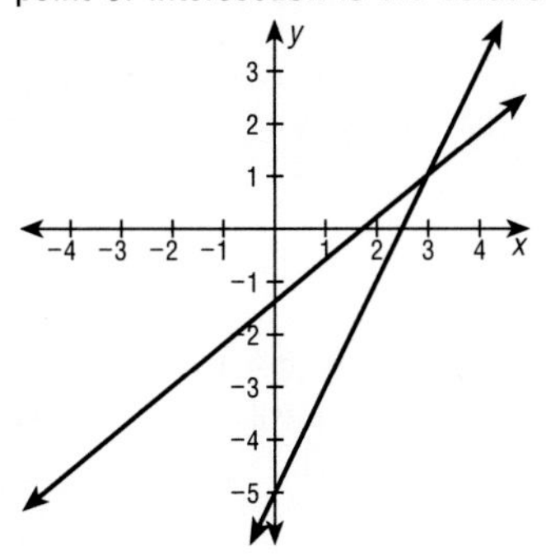

B. Any point on the line $3x - 4y = 5$; if the two graphs coincide, any point on the line is a solution.

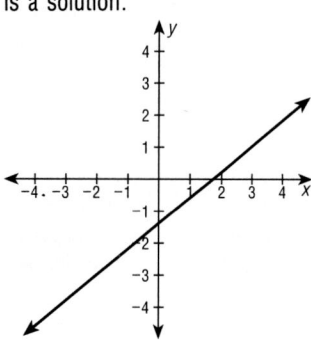

C. No solution; if the two graphs are parallel, there is no solution.

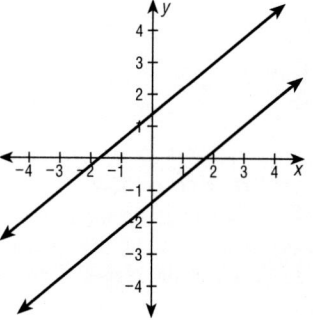

3.1 Guided Practice p. 125

1. If the graphs of the equations intersect, the system has one solution; if the graphs of the equations do not intersect, the system has no solutions; if the graphs of the equations coincide, the system has infinitely many solutions.

2. It is impossible for two lines to intersect in exactly two points.

3. $(4, -1); 4 + 5(-1) = -1, 4 + 4(-1) = 0$

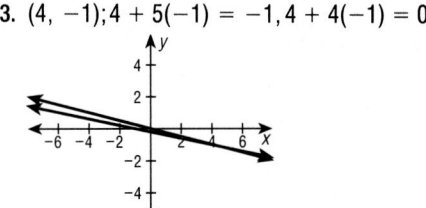

3.1 Independent Practice p. 126

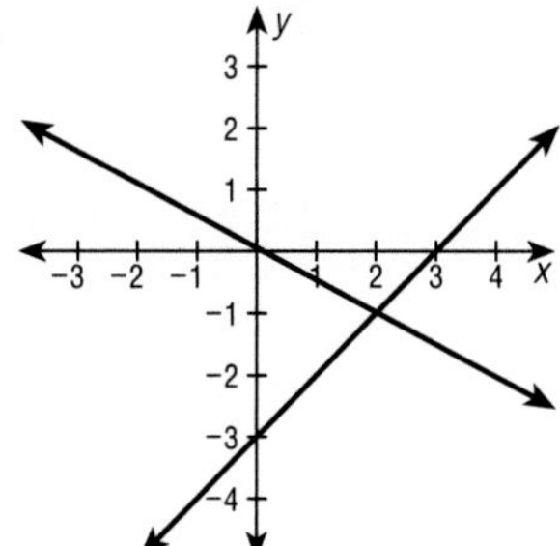

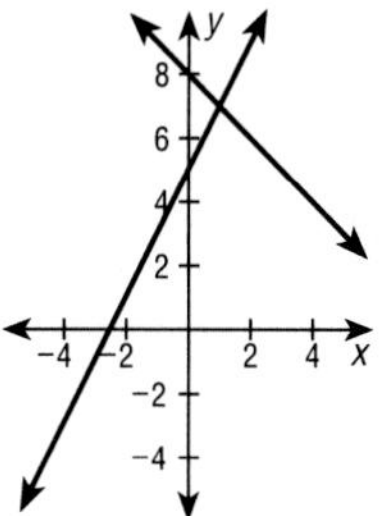

27.

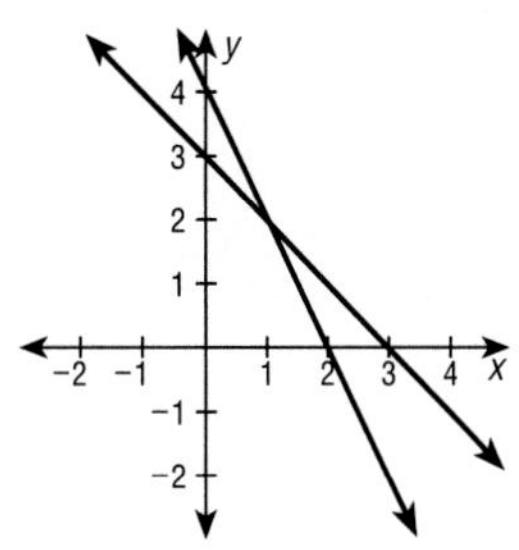

28.

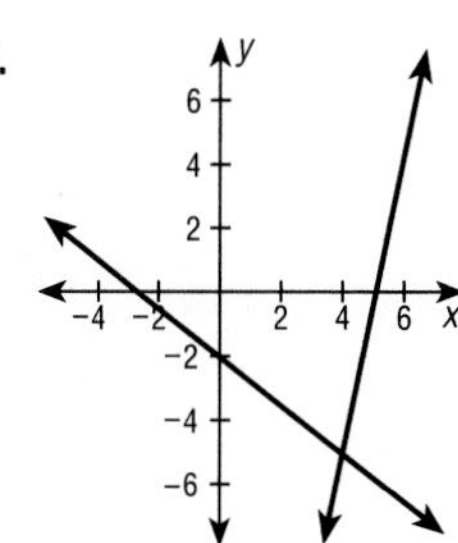

29.

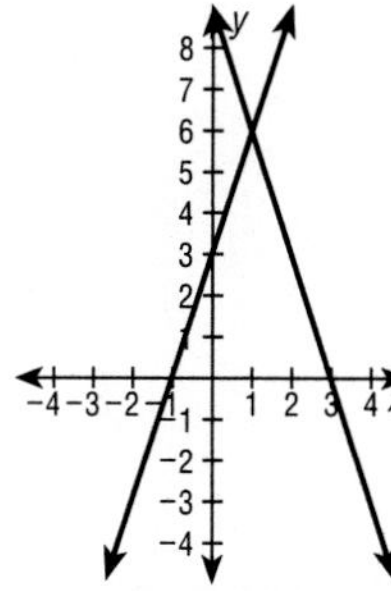

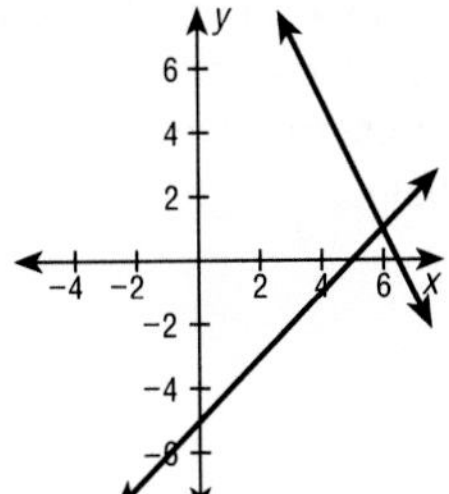

31.

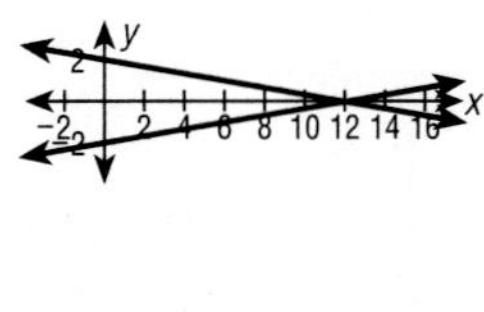

32.

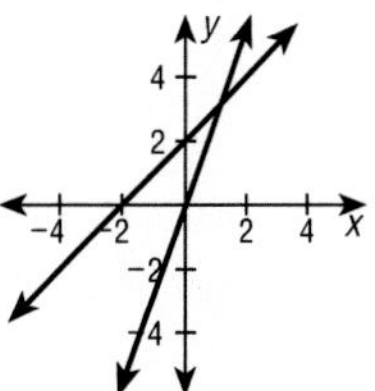

33.

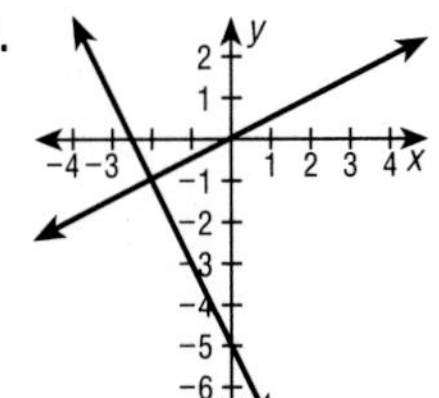

34.

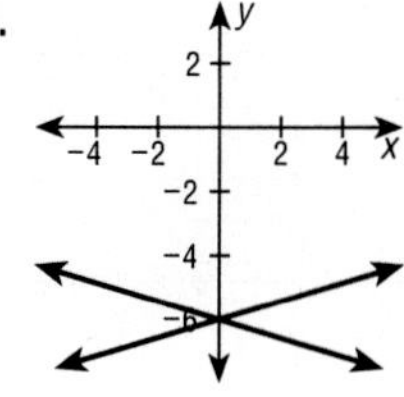

35.

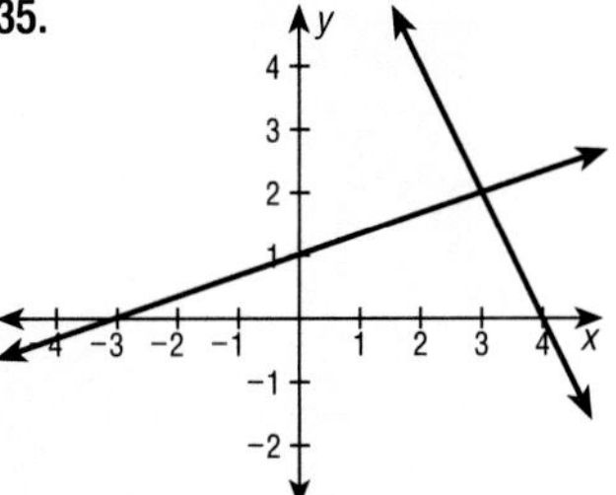

36.

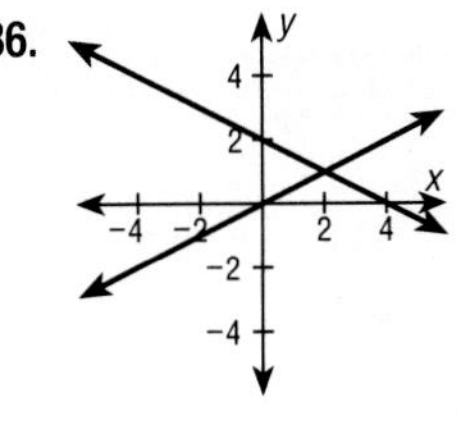

41. $y = 2x - 4,\ y = \frac{2}{3}x - 2;\ \left(\frac{3}{2}, -1\right)$

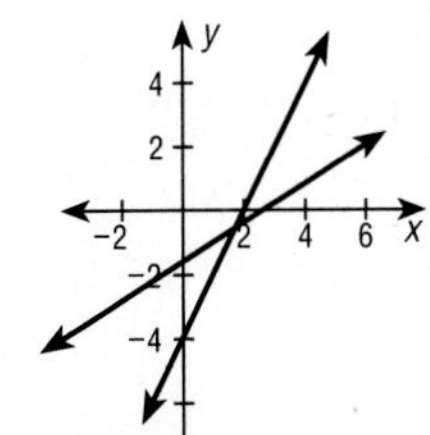

3.1 Integrated Review p. 127

54. 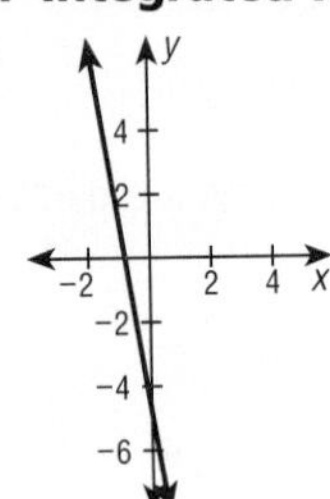

55. 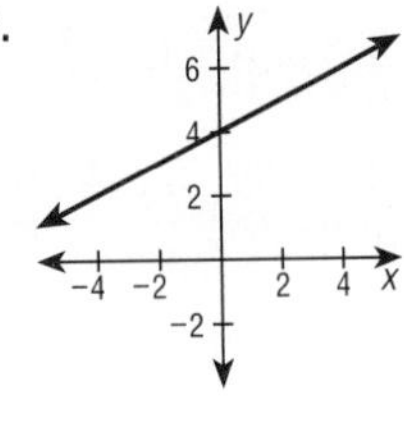

56. 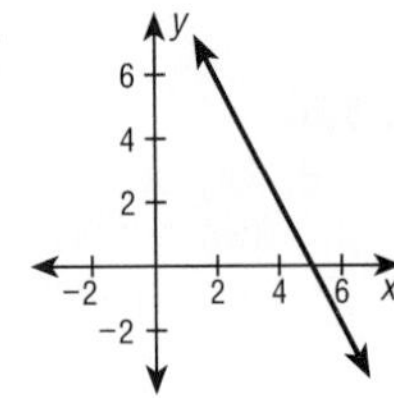

3.2 Guided Practice p. 134

1. Solve the second equation for x in terms of y, substitute the resulting expression into the first equation and solve for y; substitute the value for y into the revised second equation and solve for x; check.
2. Multiply the first equation by 2 and the second equation by 3, add the equations, and solve for y; substitute the value for y into either equation and solve for x; check.

3.2 Independent Practice p. 134

19. 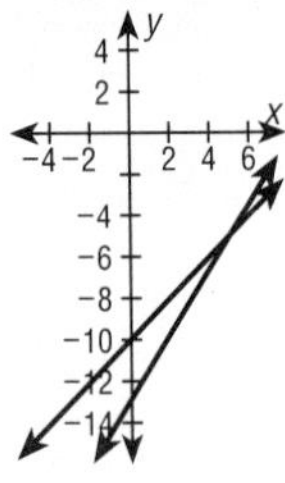

20. 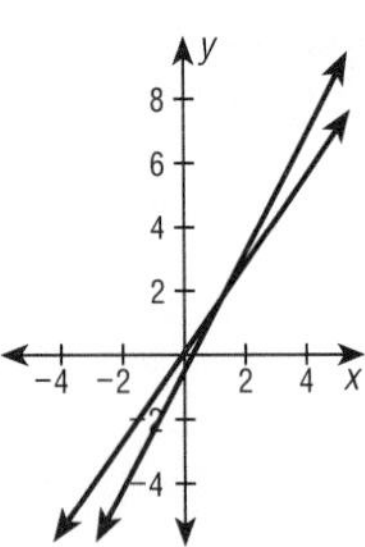

21. 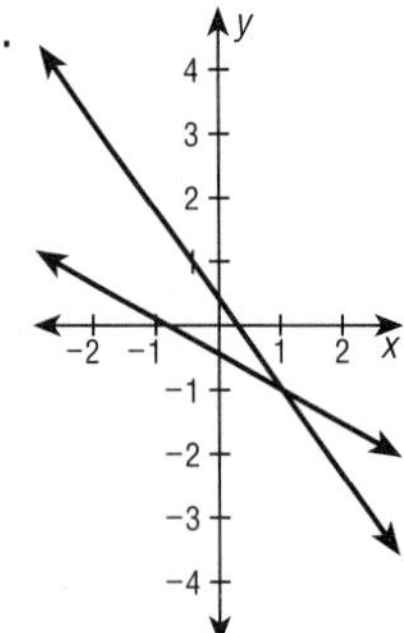

22. 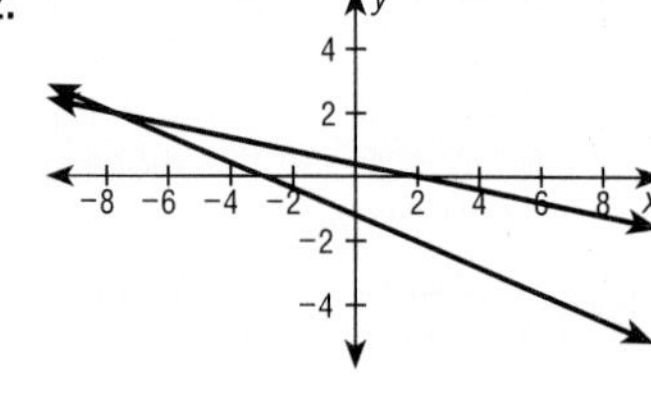

23. 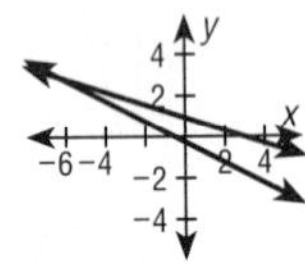

24. 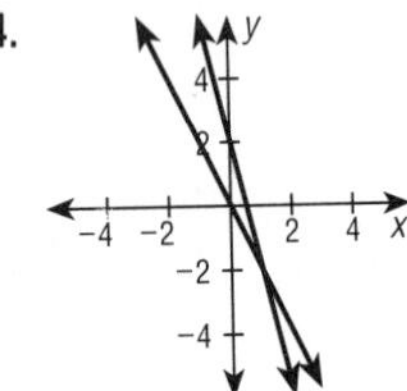

3.4 Communicating About Algebra p. 147

A. 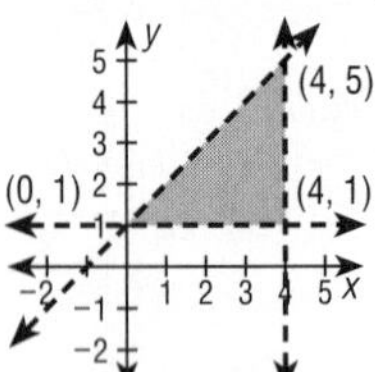

B. 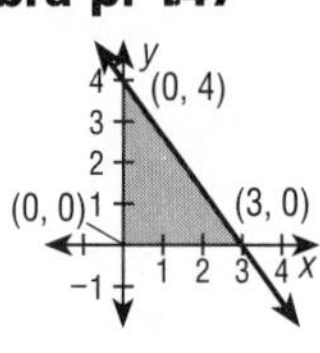

C. 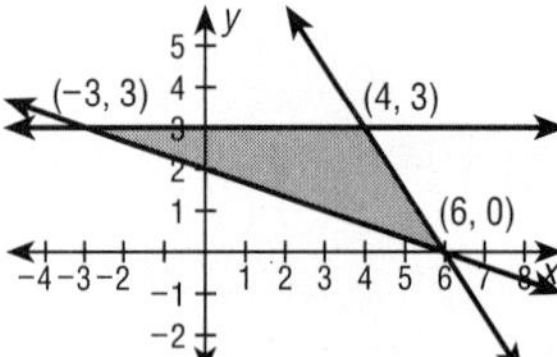

A.–C. Graph the inequalities on the same coordinate plane to get three half-planes that intersect in a common region.

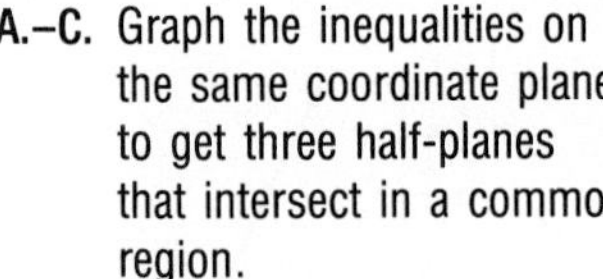

3.4 Independent Practice p. 149

13. 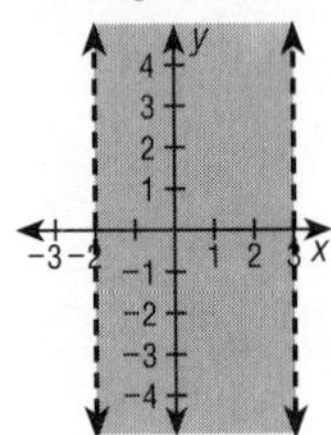

14. 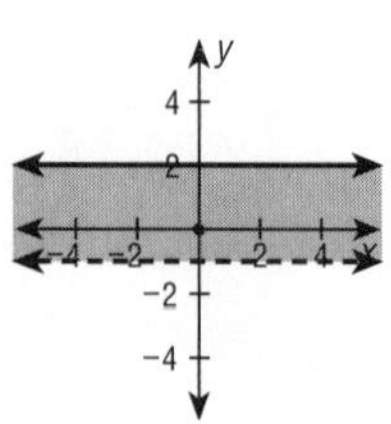

15. 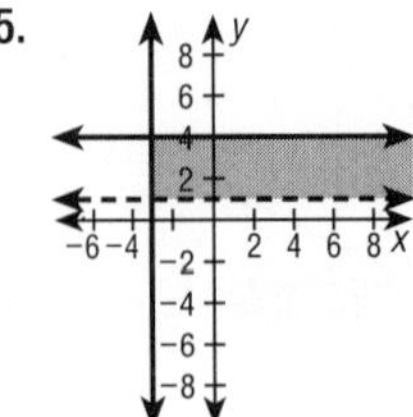

16. 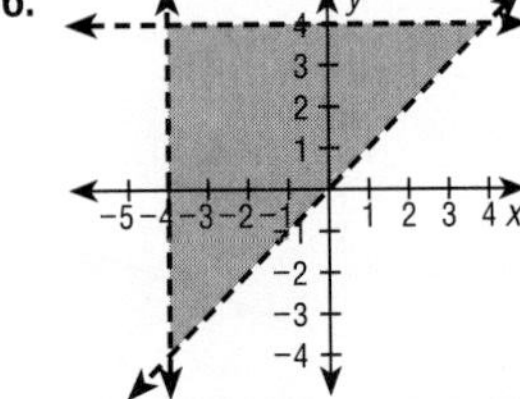

3.4 Independent Practice *(continued)*

17.

18.

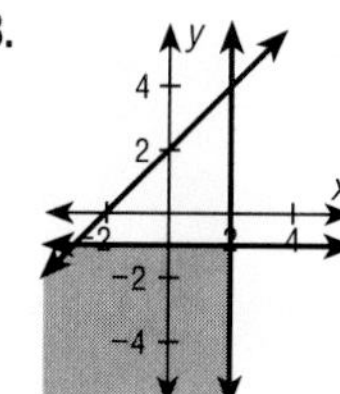

19.

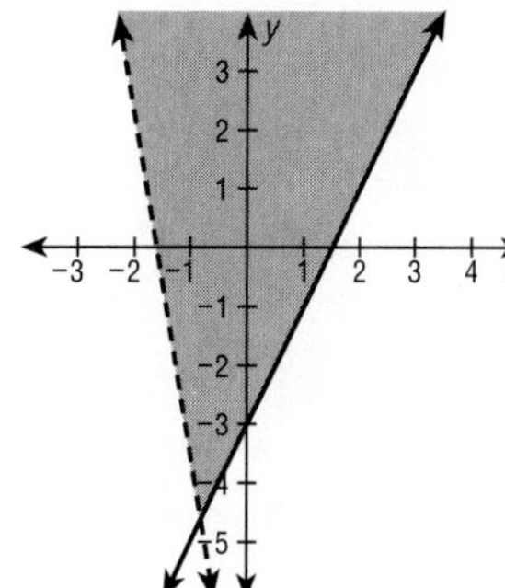

20.

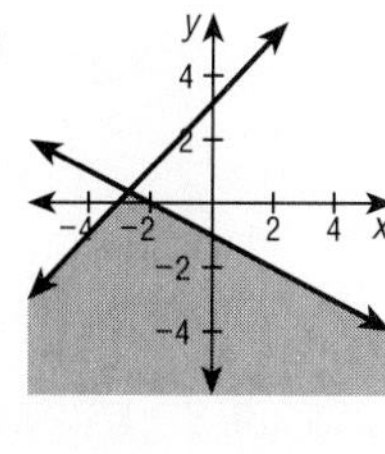

21.

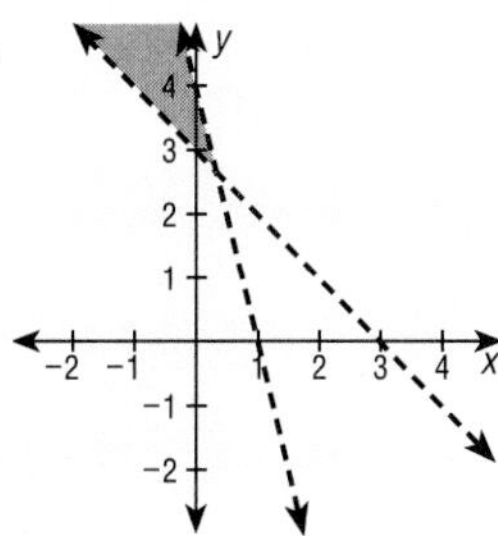

22.

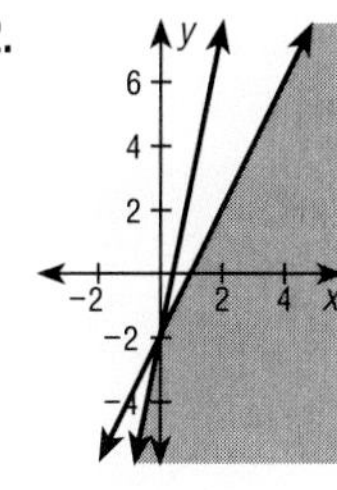

23.

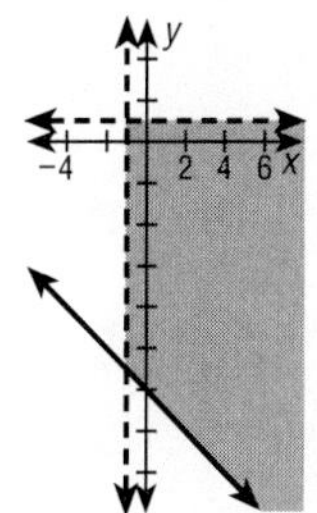

24.

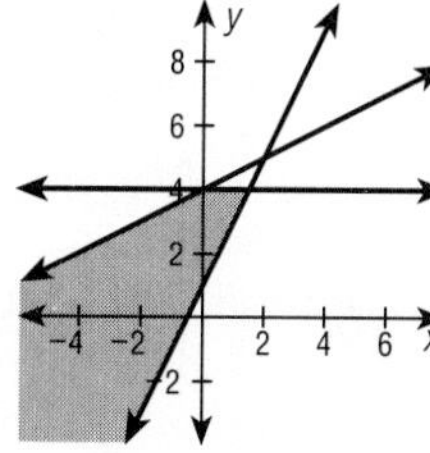

25.

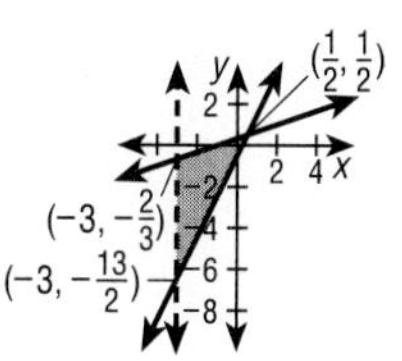

26.

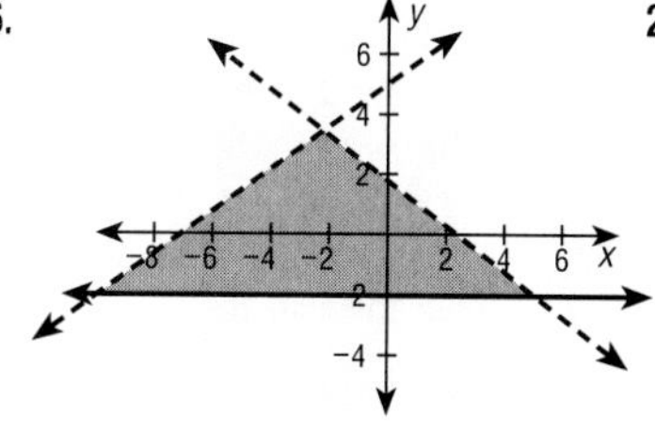

27.

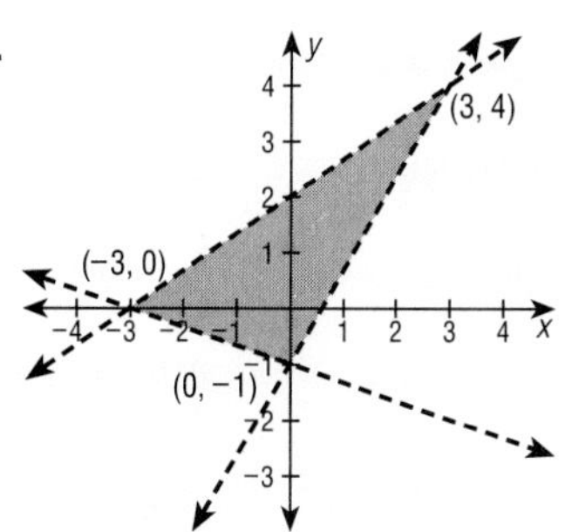

28.

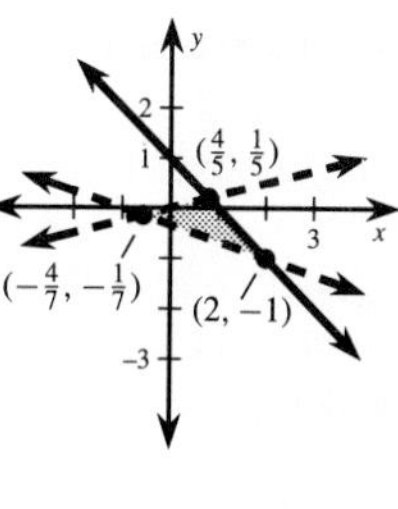

29.

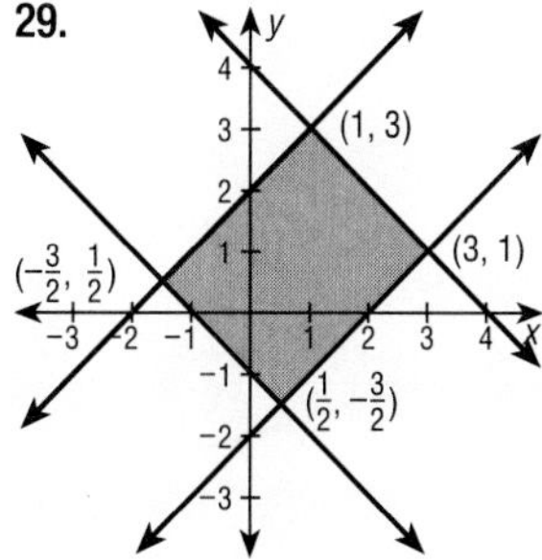

30.

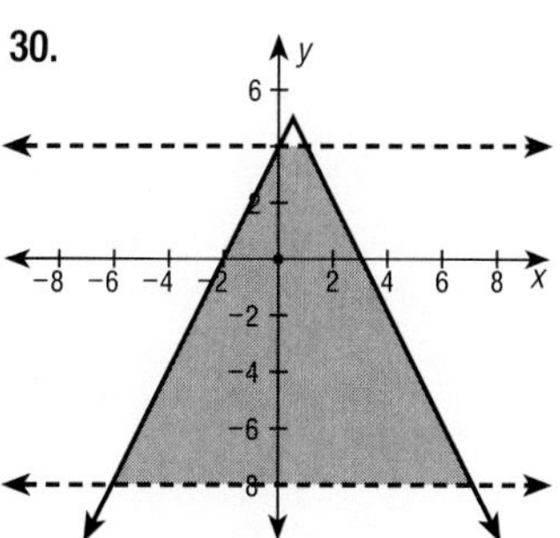

3.4 Integrated Review p. 150

39.

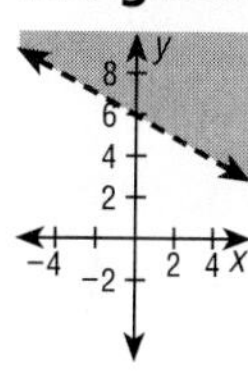

40.

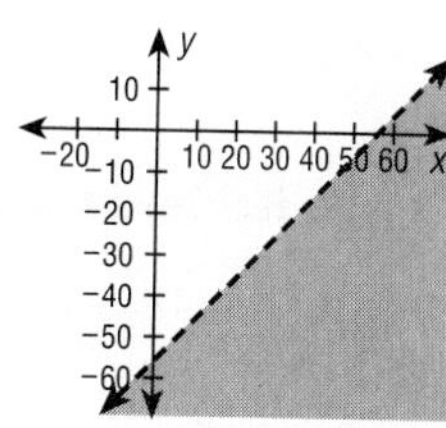

41.

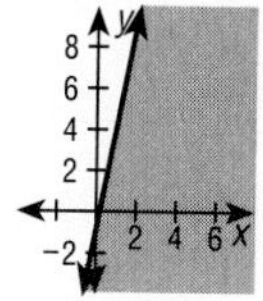

42.

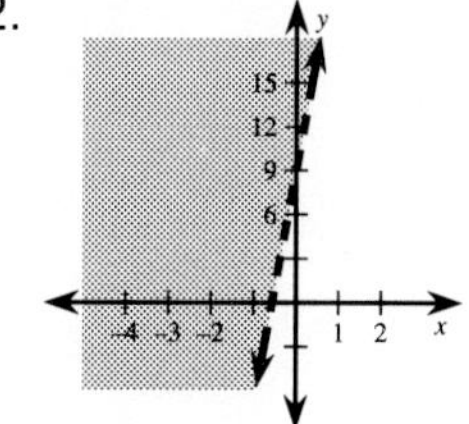

43.

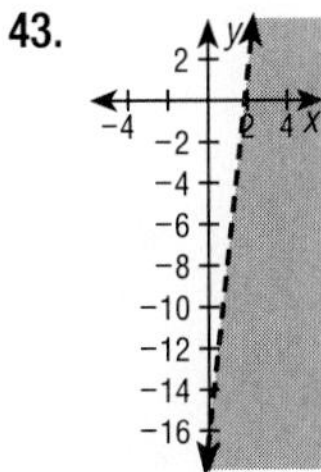

44.

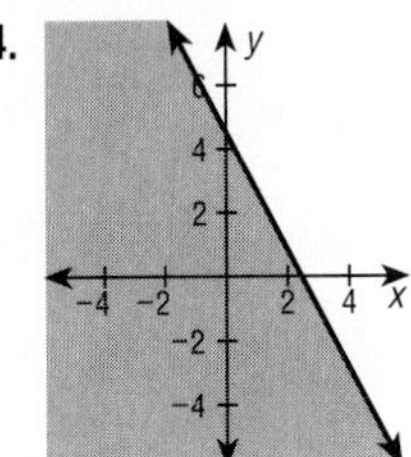

3.4 Exploration and Extension p. 150

53.

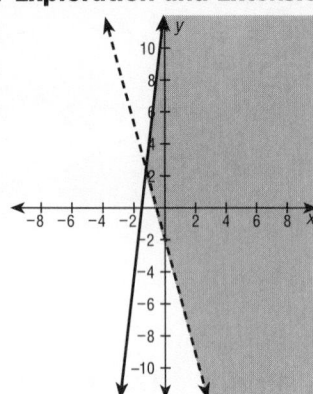

54.

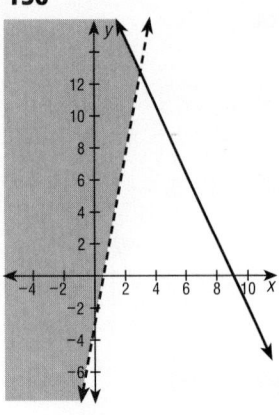

3.5 Integrated Review p. 156

24.

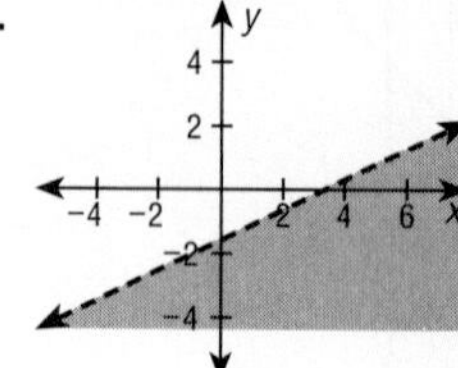

25.

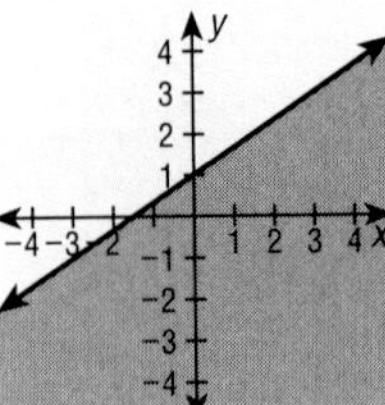

26.

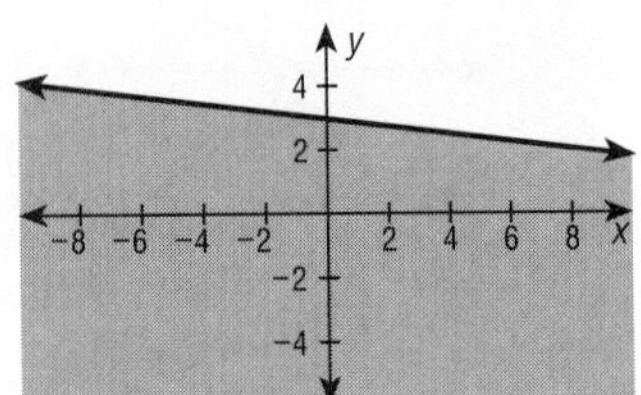

27.

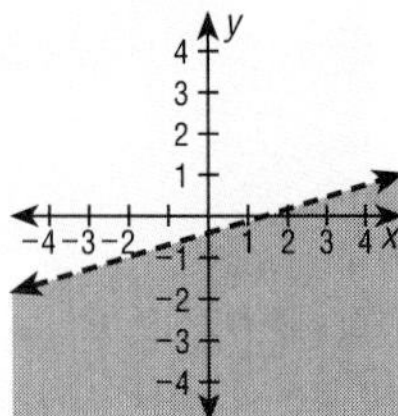

28.

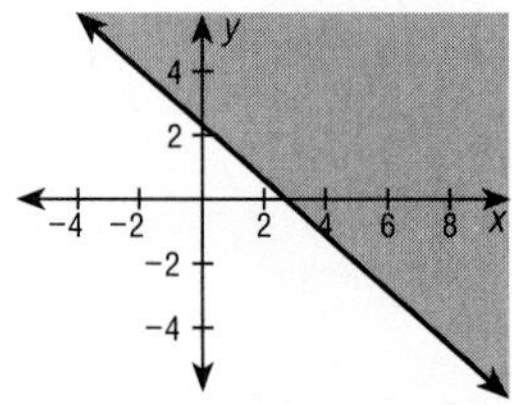

29.

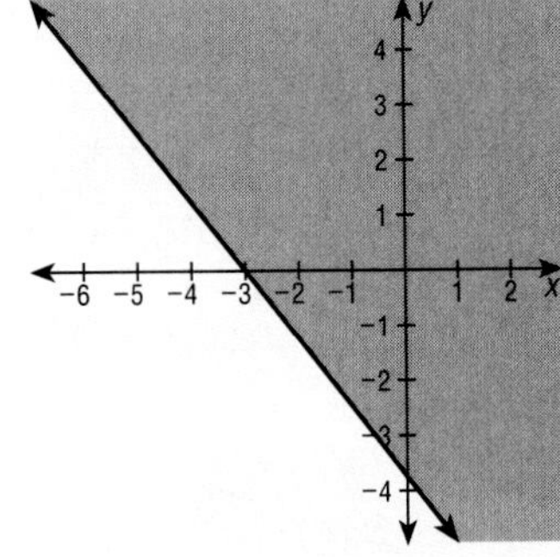

3.6 Communicating About Algebra p. 160

A. (4, −3, 2) Add −3 times $x - 3z = -2$ to $3x + y - 2z = 5$, to get $y + 7z = 11$; add −2 times $x - 3z = -2$ to $2x + 2y + z = 4$, to get $2y + 7z = 8$; solve the linear system in y and z; solve for x in the original system.

B. (1, 2, 3) Add −2 times $x + y + z = 6$ to $2x - y + z = 3$, to get $-3y - z = -9$; add −3 times $x + y + z = 6$ to $3x - z = 0$, to get $-3y - 4z = -18$; solve the linear system in y and z; solve for x in the original system.

3.6 Independent Practice p. 161

13. $$\begin{aligned} x + y - z &= 39 \\ -2x + 2y - z &= 12 \\ 4x + y - 3z &= 19 \end{aligned}$$

14. $$\begin{aligned} x + \frac{3}{2}y - \frac{1}{2}z &= \frac{9}{2} \\ -3x + y - 2z &= 14 \\ 4x - 3y + 9z &= 36 \end{aligned}$$

15. $$\begin{aligned} x - 3y + 4z &= 24 \\ 7y - 10z &= -18 \\ -3x - 2y + z &= 46 \end{aligned}$$

16. $$\begin{aligned} x + 3y - 4z &= 12 \\ 11y - 13z &= 74 \\ 2x + y + z &= 46 \end{aligned}$$

Cumulative Review for Chapters 1–3 pp. 170–171

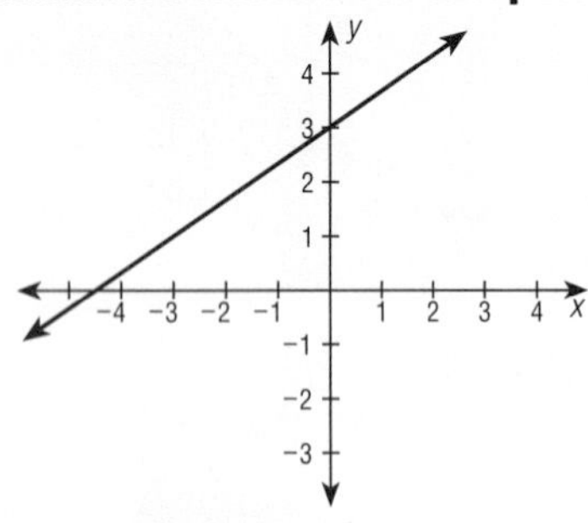

10.

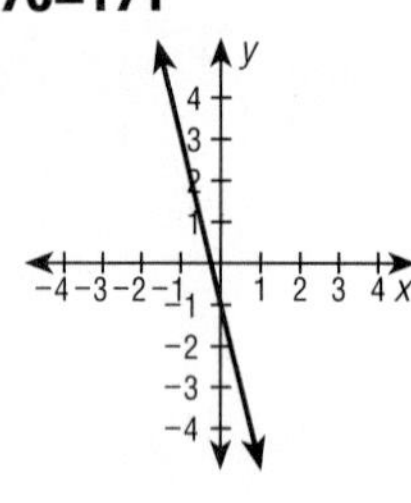

11.

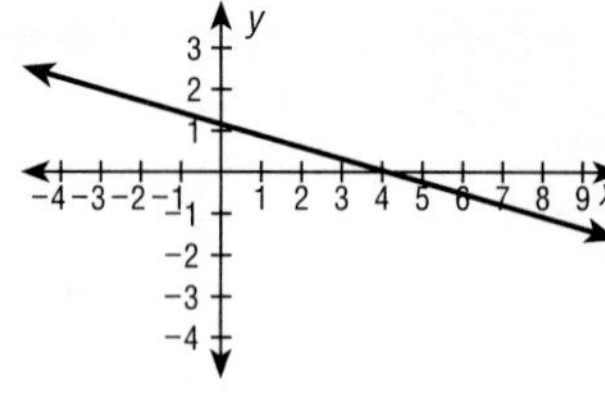

12.

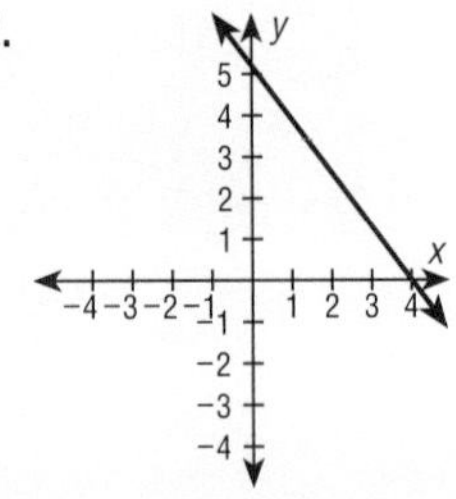

28., 30.

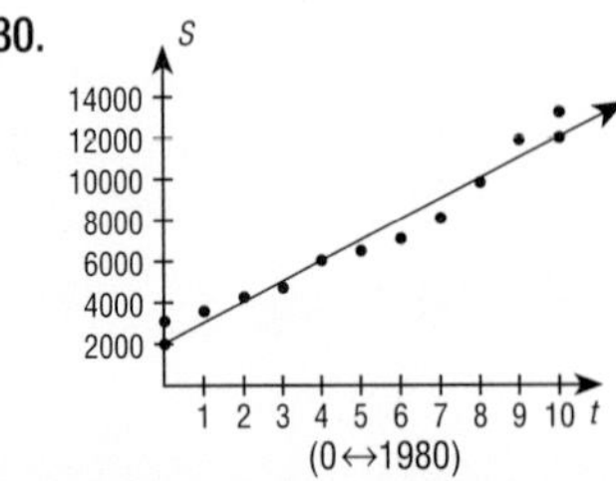

32.

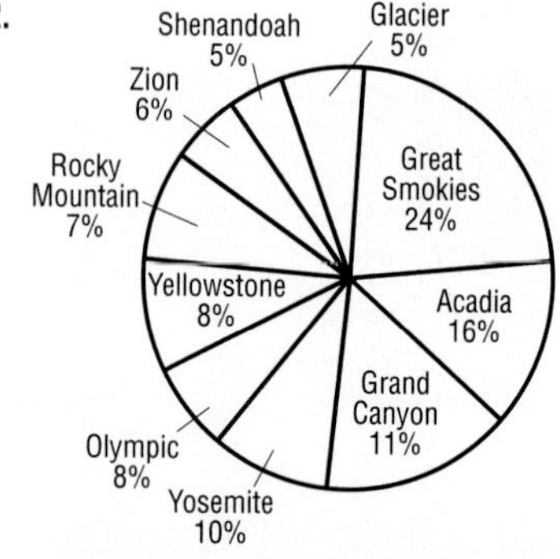

4.1 Communicating About Algebra p. 176

A. $2\left(\begin{bmatrix} 4 & -5 \\ -2 & 8 \end{bmatrix} + \begin{bmatrix} -3 & 7 \\ 8 & 2 \end{bmatrix}\right) \stackrel{?}{=} 2\begin{bmatrix} 4 & -5 \\ -2 & 8 \end{bmatrix} + 2\begin{bmatrix} -3 & 7 \\ 8 & 2 \end{bmatrix}$

$2\begin{bmatrix} 1 & 2 \\ 6 & 10 \end{bmatrix} \stackrel{?}{=} \begin{bmatrix} 8 & -10 \\ -4 & 16 \end{bmatrix} + \begin{bmatrix} -6 & 14 \\ 16 & 4 \end{bmatrix}$

$\begin{bmatrix} 2 & 4 \\ 12 & 20 \end{bmatrix} = \begin{bmatrix} 2 & 4 \\ 12 & 20 \end{bmatrix}$

B. $c\left(\begin{bmatrix} d_{11} & d_{12} \\ d_{21} & d_{22} \end{bmatrix} + \begin{bmatrix} e_{11} & e_{12} \\ e_{21} & e_{22} \end{bmatrix}\right) \stackrel{?}{=} c\begin{bmatrix} d_{11} & d_{12} \\ d_{21} & d_{22} \end{bmatrix} + c\begin{bmatrix} e_{11} & e_{12} \\ e_{21} & e_{22} \end{bmatrix}$

For any d and e: $c(d + e) = cd + ce$.

4.1 Exploration and Extension p. 179

46.

Hourly Wage ($)	3 months	6 months	12 months
Regular employee	5.25	5.355	5.62
Assistant employee	6.04	6.14	6.405

4.2 Communicating About Algebra p. 182

Multiply the given matrix by $\begin{bmatrix} 25{,}000 \\ 30{,}000 \\ 45{,}000 \end{bmatrix}$ to get: $\begin{bmatrix} 28{,}700 \\ 35{,}750 \\ 35{,}550 \end{bmatrix}$

4.2 Independent Practice p. 183

15. $\begin{bmatrix} 2 & 6 & 21 \\ -8 & 2 & -46 \\ -13 & 8 & 2 \end{bmatrix}$

16. $\begin{bmatrix} -12 & 0 & 6 \\ 4 & 2 & 4 \\ -2 & -1 & -1 \end{bmatrix}$

17. Not defined, the number of columns in the first matrix is not equal to the number of rows in the second.

4.2 Integrated Review p. 185

38. Not possible, orders do not match.

39. Not possible, orders do not match.

40. Not possible, orders do not match.

41. Possible, 2×3.

42. Possible, 6×6.

43. Possible, 1×4.

4.2 Exploration and Extension p. 185

44. $BC = \begin{bmatrix} 24 & -6 \\ 16 & -4 \end{bmatrix}$, $A(BC) = \begin{bmatrix} 72 & -18 \\ 200 & -50 \end{bmatrix}$;

$AB = \begin{bmatrix} 18 & 9 \\ 50 & 25 \end{bmatrix}$, $(AB)C = \begin{bmatrix} 72 & -18 \\ 200 & -50 \end{bmatrix}$

45. $B + C = \begin{bmatrix} 8 & 3 \\ 8 & 0 \end{bmatrix}$, $A(B + C) = \begin{bmatrix} 32 & 3 \\ 64 & 27 \end{bmatrix}$;

$AB = \begin{bmatrix} 18 & 9 \\ 50 & 25 \end{bmatrix}$, $AC = \begin{bmatrix} 14 & -6 \\ 14 & 2 \end{bmatrix}$, $AB + AC = \begin{bmatrix} 32 & 3 \\ 64 & 27 \end{bmatrix}$

46. $A + B = \begin{bmatrix} 7 & 6 \\ 13 & 1 \end{bmatrix}$, $(A + B)C = \begin{bmatrix} 38 & -12 \\ 30 & -2 \end{bmatrix}$;

$AC = \begin{bmatrix} 14 & -6 \\ 14 & 2 \end{bmatrix}$, $BC = \begin{bmatrix} 24 & -6 \\ 16 & -4 \end{bmatrix}$;

$AC + BC = \begin{bmatrix} 38 & -12 \\ 30 & -2 \end{bmatrix}$

47. $AB = \begin{bmatrix} 18 & 9 \\ 50 & 25 \end{bmatrix}$, $c(AB) = \begin{bmatrix} 72 & 36 \\ 200 & 100 \end{bmatrix}$;

$cA = \begin{bmatrix} 4 & 12 \\ 36 & -4 \end{bmatrix}$, $(cA)B = \begin{bmatrix} 72 & 36 \\ 200 & 100 \end{bmatrix}$;

$cB = \begin{bmatrix} 24 & 12 \\ 16 & 8 \end{bmatrix}$, $A(cB) = \begin{bmatrix} 72 & 36 \\ 200 & 100 \end{bmatrix}$

4.3 *Guided Practice* p. 191

2. $\begin{bmatrix} 8 & 7 \\ 2 & 3 \end{bmatrix}$ and $\begin{bmatrix} 6 & -4 \\ -1 & 1 \end{bmatrix}$

30. $(0)\begin{vmatrix} 3 & -1 \\ -2 & 7 \end{vmatrix} - (0)\begin{vmatrix} 4 & -5 \\ -2 & 7 \end{vmatrix} + (0)\begin{vmatrix} 4 & -5 \\ 3 & -1 \end{vmatrix} = 0 + 0 + 0 = 0,$ the matrix has a column of zeros.

32. $(3)\begin{vmatrix} 1 & 3 \\ 3 & 9 \end{vmatrix} - (-4)\begin{vmatrix} -2 & -6 \\ 3 & 9 \end{vmatrix} + (1)\begin{vmatrix} -2 & -6 \\ 1 & 3 \end{vmatrix} = (3)(0) - (-4)(0) + (1)(0) = 0,$ one column is a multiple of another.

4.3 Independent Practice p. 192

29. $(-1)\begin{vmatrix} 2 & 3 \\ 4 & 6 \end{vmatrix} - (0)\begin{vmatrix} 1 & 3 \\ 2 & 6 \end{vmatrix} + 8\begin{vmatrix} 1 & 2 \\ 2 & 4 \end{vmatrix} = (-1)(0) - (0)(0) + (8)(0) = 0,$ one row is a multiple of another.

31. $(5)\begin{vmatrix} -4 & -4 \\ 3 & 3 \end{vmatrix} - (-3)\begin{vmatrix} 2 & 2 \\ 3 & 3 \end{vmatrix} + (8)\begin{vmatrix} 2 & 2 \\ -4 & -4 \end{vmatrix} = (5)(0) - (-3)(0) + (8)(0) = 0,$ two columns are the same.

4.4 Independent Practice p. 201

31. $\begin{bmatrix} 0 & 1 & 0 \\ 7 & 66 & -9 \\ -3 & -29 & 4 \end{bmatrix}$

32. $\begin{bmatrix} -1 & 18 & 6 \\ -7 & 136 & 45 \\ 1 & -21 & -7 \end{bmatrix}$

33. $\begin{bmatrix} -4 & -\frac{7}{2} & 2 \\ -5 & -5 & 3 \\ 2 & 2 & -1 \end{bmatrix}$

34. $\begin{bmatrix} 12 & -7 & 3 \\ -20 & 12 & -5 \\ \frac{3}{2} & -1 & \frac{1}{2} \end{bmatrix}$

4.4 Integrated Review p. 202

44. $\begin{bmatrix} -7 & -4 & 3 \\ 2 & 2 & -4 \\ 0 & -6 & -3 \end{bmatrix}$

46. $\begin{bmatrix} 5 & 16 & 13 \\ -4 & -6 & -6 \\ -5 & 26 & 13 \end{bmatrix}$

4.5 Communicating About Algebra p. 206

A. $\begin{bmatrix} \frac{3}{7} & -\frac{2}{7} \\ -\frac{4}{7} & \frac{5}{7} \end{bmatrix}\begin{bmatrix} 21 \\ 35 \end{bmatrix} = \begin{bmatrix} -1 \\ 13 \end{bmatrix}$, (−1, 13)

B. $\begin{bmatrix} -0.25 & 0.6 & 0.65 \\ 0 & -0.2 & 0.2 \\ -0.25 & 0 & 0.25 \end{bmatrix}\begin{bmatrix} -3 \\ 12 \\ 18 \end{bmatrix} = \begin{bmatrix} 19.65 \\ 1.2 \\ 5.25 \end{bmatrix}$, (19.65, 1.2, 5.25)

4.5 Independent Practice p. 207

5. $\begin{bmatrix} 2 & -4 \\ -3 & 1 \end{bmatrix}\begin{bmatrix} x \\ y \end{bmatrix} = \begin{bmatrix} 7 \\ 12 \end{bmatrix}$

6. $\begin{bmatrix} 1 & 9 \\ 2 & -4 \end{bmatrix}\begin{bmatrix} x \\ y \end{bmatrix} = \begin{bmatrix} 20 \\ 15 \end{bmatrix}$

7. $\begin{bmatrix} 2 & -5 \\ 1 & -3 \end{bmatrix}\begin{bmatrix} x \\ y \end{bmatrix} = \begin{bmatrix} 12 \\ -3 \end{bmatrix}$

8. $\begin{bmatrix} 1 & -2 & 3 \\ 2 & 1 & -2 \\ -3 & -5 & 9 \end{bmatrix}\begin{bmatrix} x \\ y \\ z \end{bmatrix} = \begin{bmatrix} 14 \\ 16 \\ 36 \end{bmatrix}$

9. $\begin{bmatrix} 3 & -1 & 4 \\ 2 & 4 & -1 \\ 1 & -1 & 3 \end{bmatrix}\begin{bmatrix} x \\ y \\ z \end{bmatrix} = \begin{bmatrix} 16 \\ 10 \\ 31 \end{bmatrix}$

10. $\begin{bmatrix} 1 & 1 & -1 \\ 2 & 0 & -1 \\ 0 & 1 & 1 \end{bmatrix}\begin{bmatrix} x \\ y \\ z \end{bmatrix} = \begin{bmatrix} 0 \\ 1 \\ 2 \end{bmatrix}$

4.5 Mixed Review p. 209

8. −20

9. 

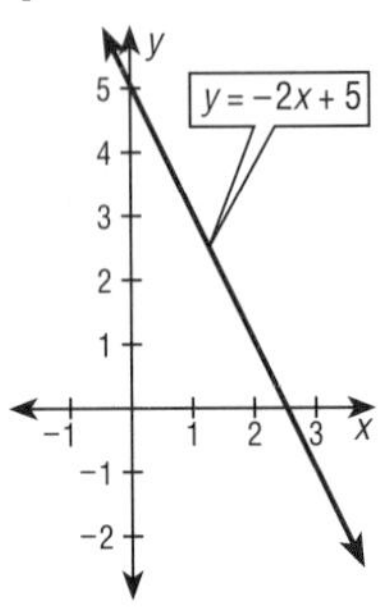

10.

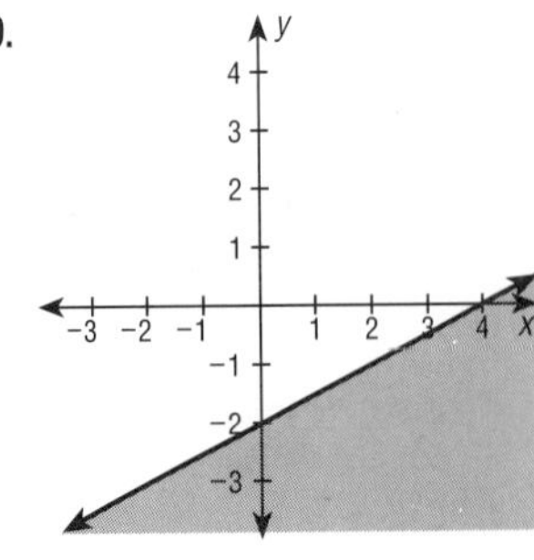

11. 0 1 2 3 4 5 6 x

12. 50.27 ft²

4.5 Mixed Review *(continued)*

15. $\begin{bmatrix} -6 & 3 & -2 \\ 2 & -9 & 3 \end{bmatrix}$ 16. $\begin{bmatrix} -3 & 4 \\ 3 & -10 \end{bmatrix}$ 17. $\begin{bmatrix} 1 & 0 \\ 0 & 1 \end{bmatrix}$ 18. $\begin{bmatrix} 0 & 0 \\ 0 & 0 \end{bmatrix}$

4.6 Independent Practice p. 213

5. $\left[\begin{array}{ccc|c} 1 & 3 & -1 & 16 \\ 4 & -6 & 0 & 9 \\ 0 & 2 & -3 & 12 \end{array}\right]$ 6. $\left[\begin{array}{ccc|c} -1 & -3 & 2 & 8 \\ 2 & 1 & -3 & 9 \\ 1 & 0 & -3 & 10 \end{array}\right]$ 7. $\left[\begin{array}{ccc|c} 2 & -1 & 3 & 4 \\ -3 & 0 & -1 & 1 \\ 1 & -3 & 1 & 5 \end{array}\right]$

4.7 Independent Practice p. 219

14. $x = \dfrac{\begin{vmatrix} 2 & -1 & 10 \\ 4 & 0 & 1 \\ 0 & 2 & 1 \end{vmatrix}}{\begin{vmatrix} 1 & -1 & 10 \\ 3 & 0 & 1 \\ 7 & 2 & 1 \end{vmatrix}}, y = \dfrac{\begin{vmatrix} 1 & 2 & 10 \\ 3 & 4 & 1 \\ 7 & 0 & 1 \end{vmatrix}}{\begin{vmatrix} 1 & -1 & 10 \\ 3 & 0 & 1 \\ 7 & 2 & 1 \end{vmatrix}}, z = \dfrac{\begin{vmatrix} 1 & -1 & 2 \\ 3 & 0 & 4 \\ 7 & 2 & 0 \end{vmatrix}}{\begin{vmatrix} 1 & -1 & 10 \\ 3 & 0 & 1 \\ 7 & 2 & 1 \end{vmatrix}}$

15. $x = \dfrac{\begin{vmatrix} 3 & 0 & 2 \\ -1 & 2 & 5 \\ 10 & 1 & -7 \end{vmatrix}}{\begin{vmatrix} 4 & 0 & 2 \\ -1 & 2 & 5 \\ 3 & 1 & -7 \end{vmatrix}}, y = \dfrac{\begin{vmatrix} 4 & 3 & 2 \\ -1 & -1 & 5 \\ 3 & 10 & -7 \end{vmatrix}}{\begin{vmatrix} 4 & 0 & 2 \\ -1 & 2 & 5 \\ 3 & 1 & -7 \end{vmatrix}}, z = \dfrac{\begin{vmatrix} 4 & 0 & 3 \\ -1 & 2 & -1 \\ 3 & 1 & 10 \end{vmatrix}}{\begin{vmatrix} 4 & 0 & 2 \\ -1 & 2 & 5 \\ 3 & 1 & -7 \end{vmatrix}}$

16. $x = \dfrac{\begin{vmatrix} 4 & 6 & -1 \\ -4 & 1 & 2 \\ 1 & -1 & 3 \end{vmatrix}}{\begin{vmatrix} 2 & 6 & -1 \\ 3 & 1 & 2 \\ 6 & -1 & 3 \end{vmatrix}}, y = \dfrac{\begin{vmatrix} 2 & 4 & -1 \\ 3 & -4 & 2 \\ 6 & 1 & 3 \end{vmatrix}}{\begin{vmatrix} 2 & 6 & -1 \\ 3 & 1 & 2 \\ 6 & -1 & 3 \end{vmatrix}}, z = \dfrac{\begin{vmatrix} 2 & 6 & 4 \\ 3 & 1 & -4 \\ 6 & -1 & 1 \end{vmatrix}}{\begin{vmatrix} 2 & 6 & -1 \\ 3 & 1 & 2 \\ 6 & -1 & 3 \end{vmatrix}}$

4.7 Chapter Test p. 227

5. $\begin{bmatrix} \frac{2}{11} & \frac{1}{11} \\ \frac{3}{22} & \frac{7}{22} \end{bmatrix}$ 6. does not exist 7. $\begin{bmatrix} -\frac{1}{3} & \frac{1}{3} \\ -\frac{1}{3} & -\frac{2}{3} \end{bmatrix}$

ADDITIONAL ANSWERS

CHAPTER 5

5.2 Independent Practice pp. 240–241

29.

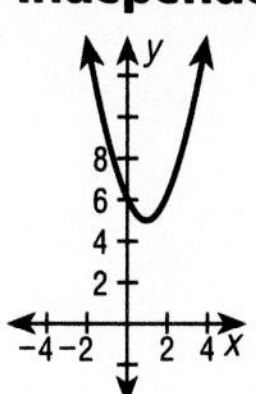

30.

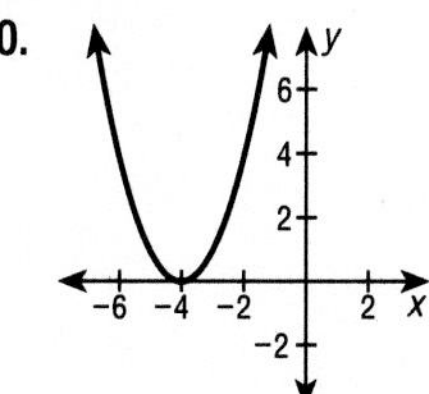

31.

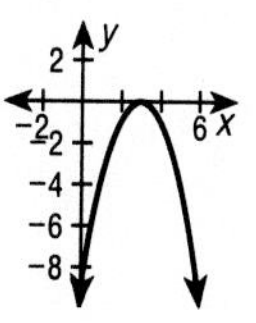

32.

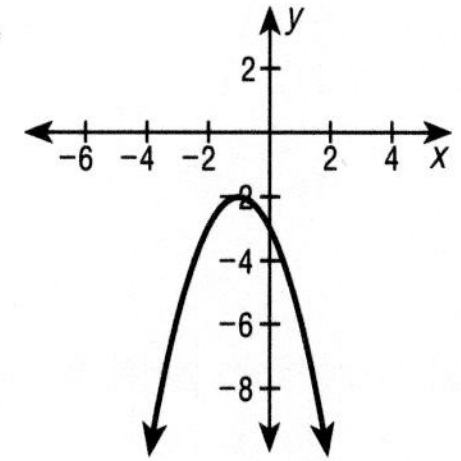

33.

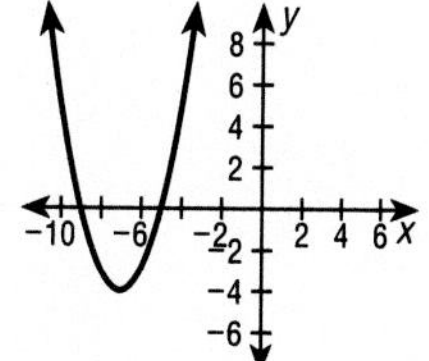

34.

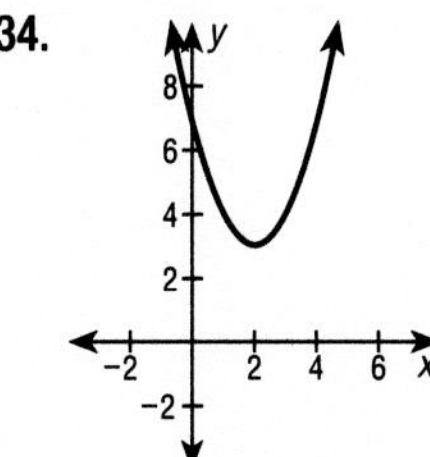

35.

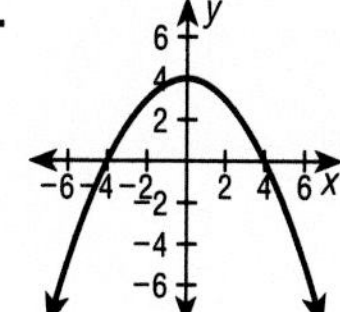

36.

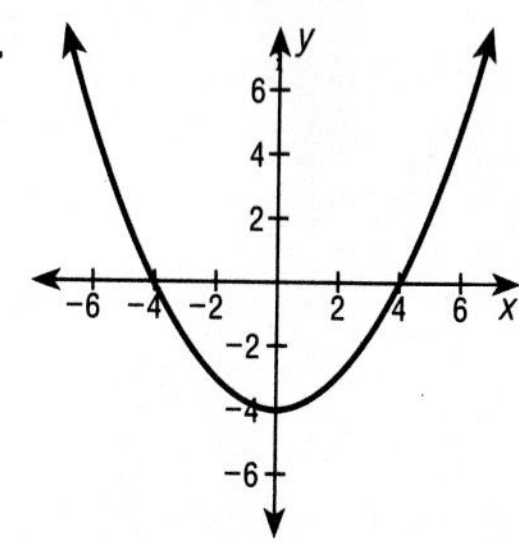

5.2 Independent Practice *(continued)*

37.

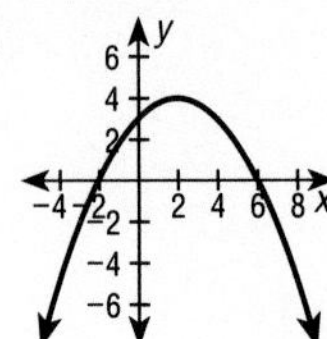

38.

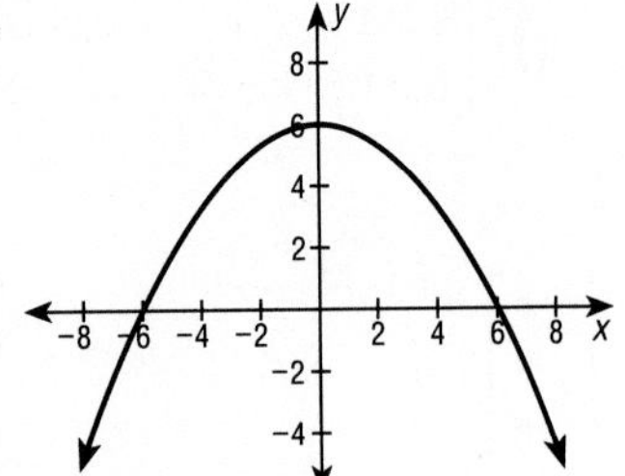

39.

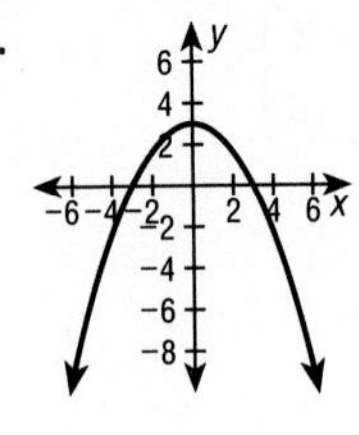

40.

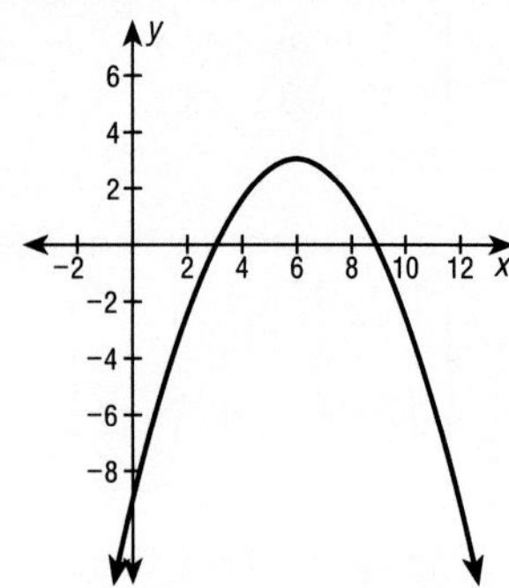

44.

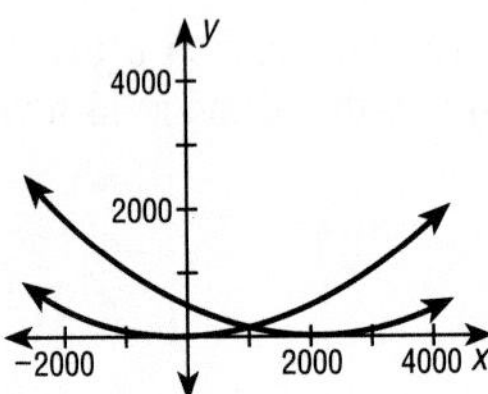

5.2 Integrated Review p. 241

47.

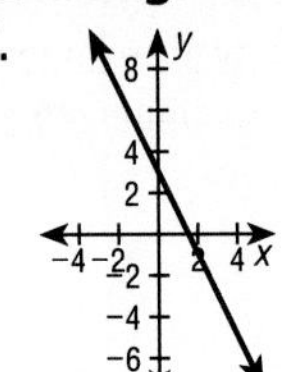

48.

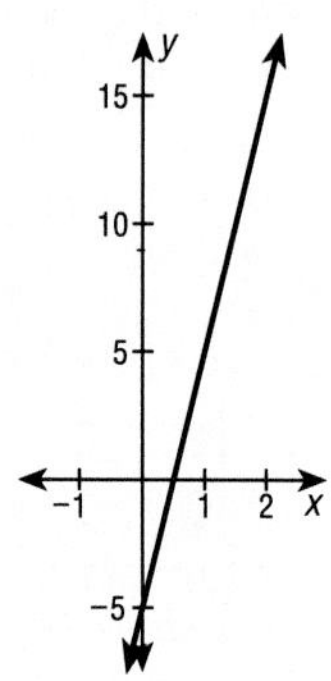

49.

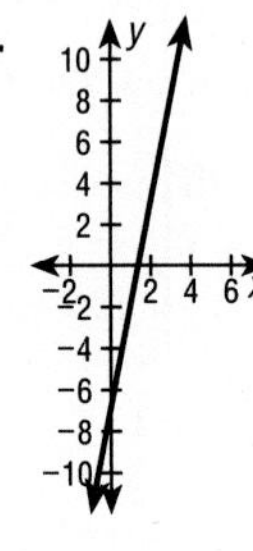

50.

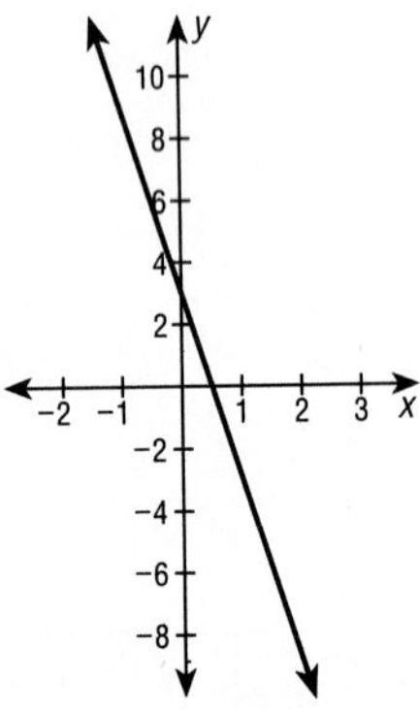

51.

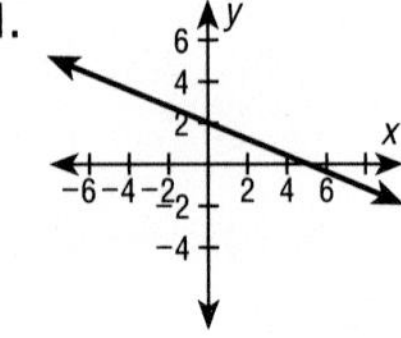

52.

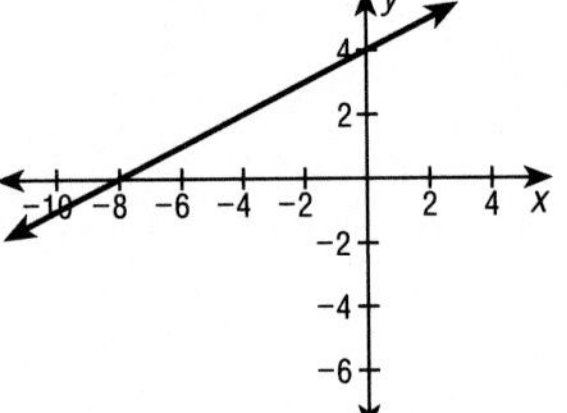

5.2 Exploration and Extension p. 241

61.

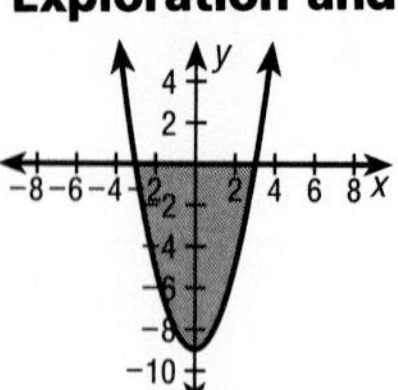

62.

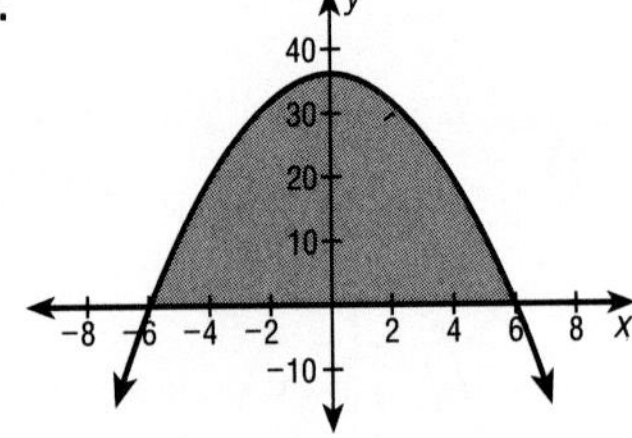

5.3 Communicating About Algebra p. 247

A. $4 \pm \sqrt{6}$; subtract 10 from both sides, add the square of half of -4 to both sides, find the square roots.

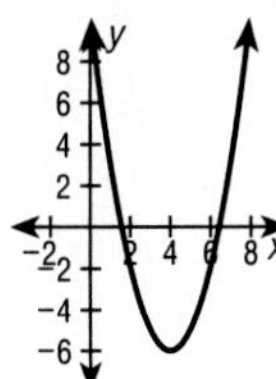

B. $-1, -3$; add 6 to both sides, divide both sides by -2, add the square of half 4 to both sides, find the square roots.

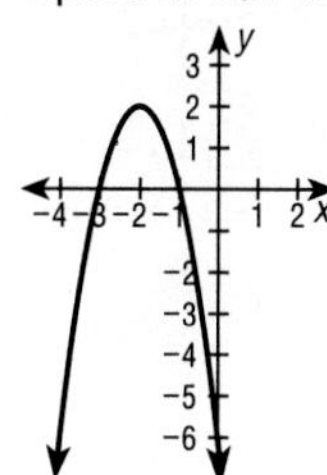

5.3 Guided Practice p. 248

5. $\left(x + \frac{3}{2}\right)^2 = \frac{13}{4}$; divide both sides by 2, add the square of half of 3 to both sides, write the binomial squared.

6. Algebraic method: substitute the values found for x into the equation and simplify; geometric method: set the equation equal to zero, then substitute y for zero and graph the equation to find the x-intercepts.

5.3 Independent Practice p. 248

19. 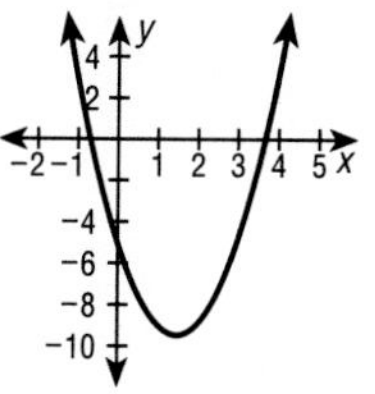

20. 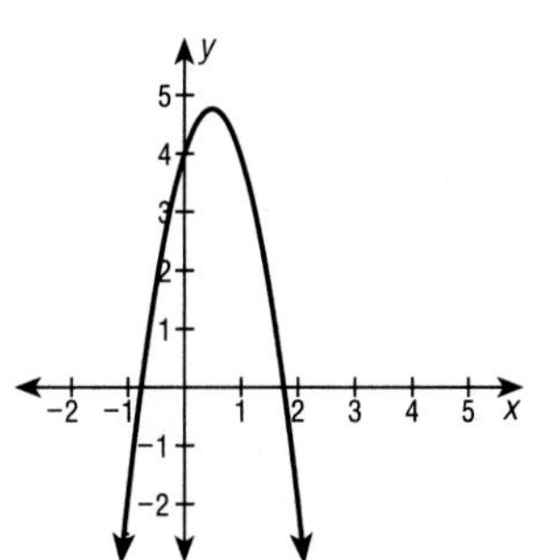

21. 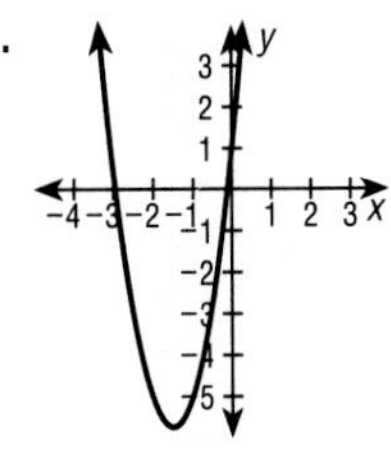

22. 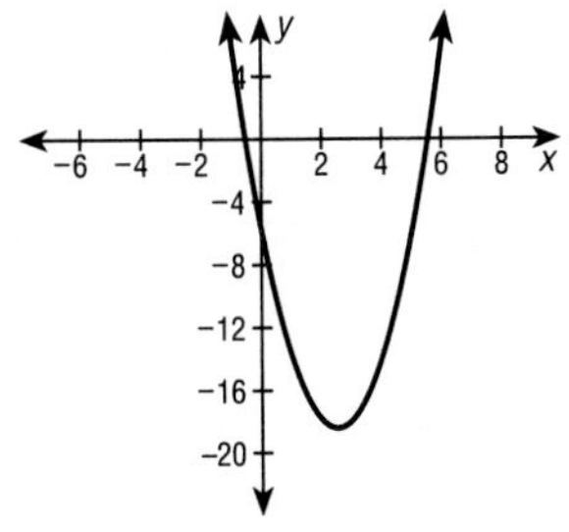

23. 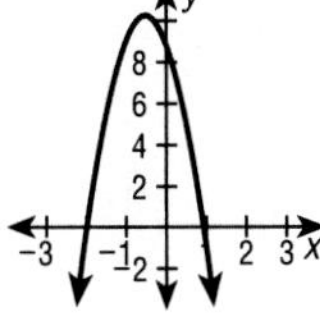

24. 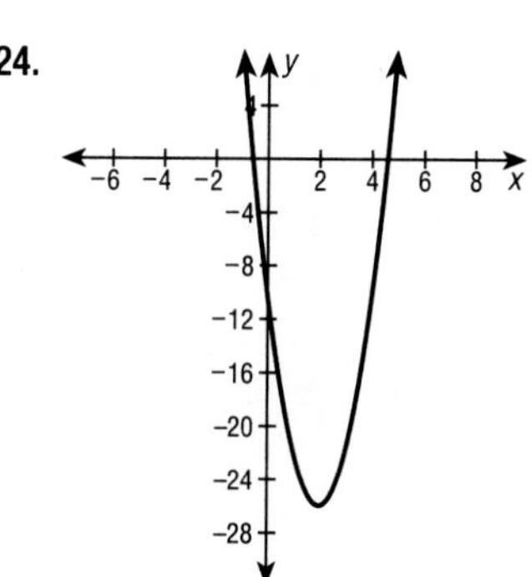

5.3 Integrated Review p. 250

45. 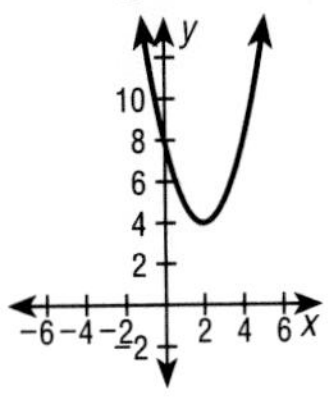

46. 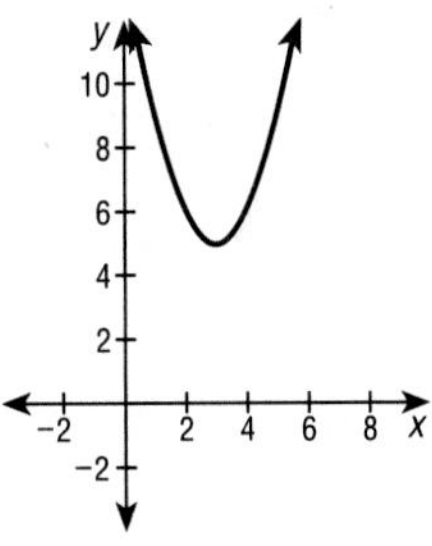

47. 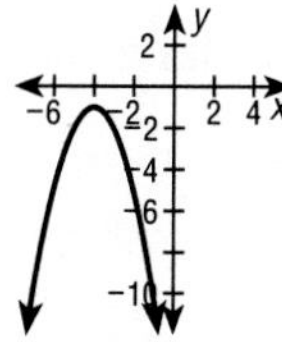

48. 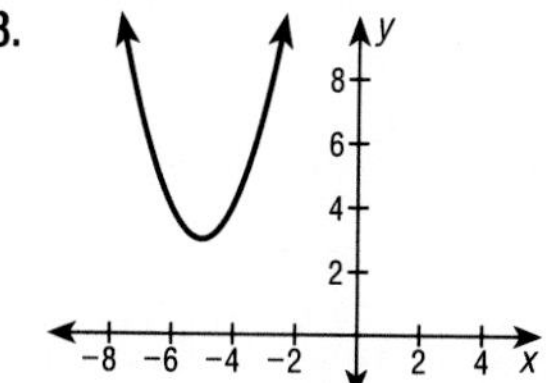

49. 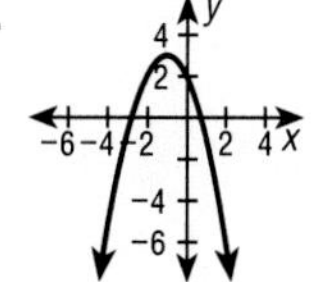

50. 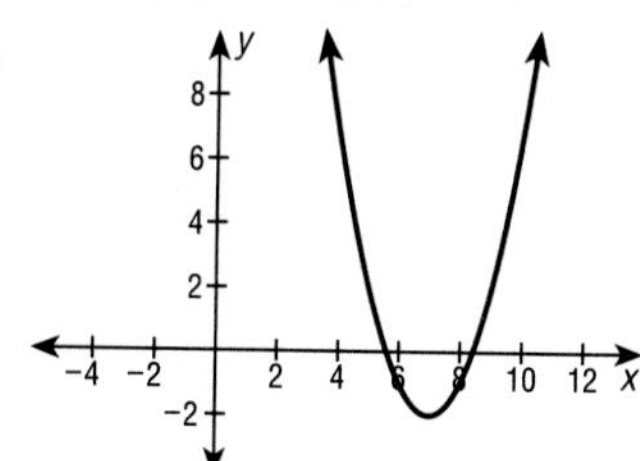

5.4 Guided Practice p. 255

1. Solutions are the opposite of b, plus or minus the square root of b squared minus $4ac$, all divided by $2a$.

2. $b^2 - 4ac$
If positive, there are two solutions;
if zero, there is one solution;
if negative, there are no solutions.

4. The first h is the height of an object after it is dropped, the second h is the height of an object after it is thrown.

ADDITIONAL ANSWERS

5.4 Exploration and Extension p. 257

59. 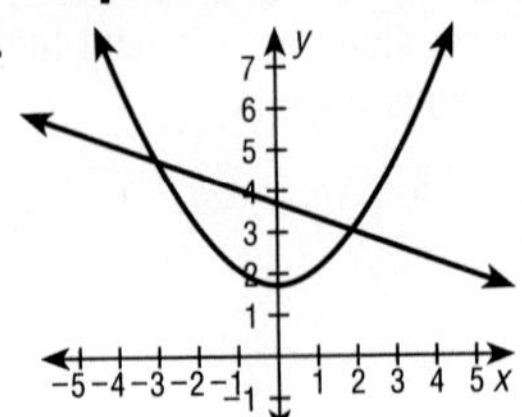

60. 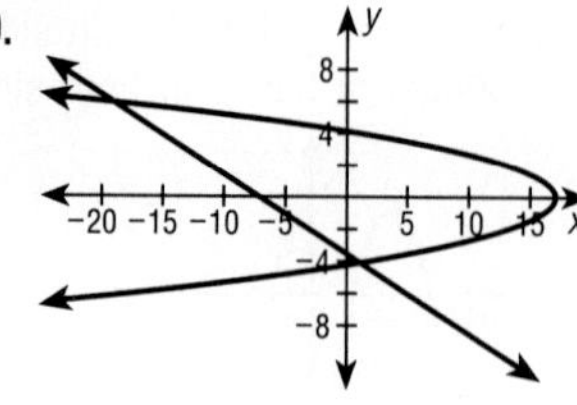

5.5 Independent Practice p. 263

63. 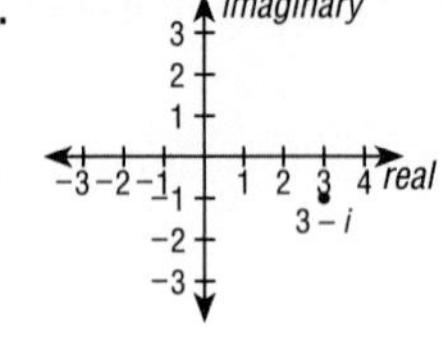

64. 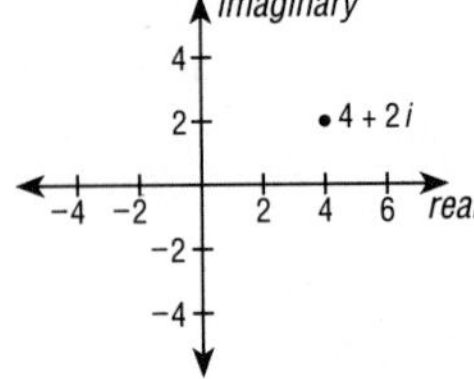

65. 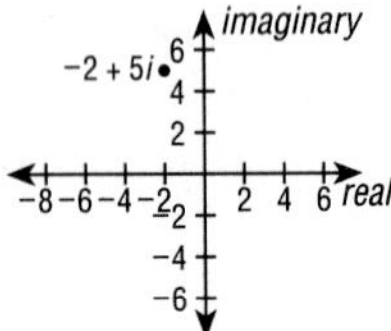

66. 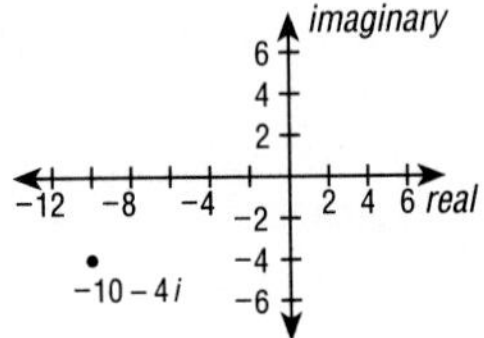

67. 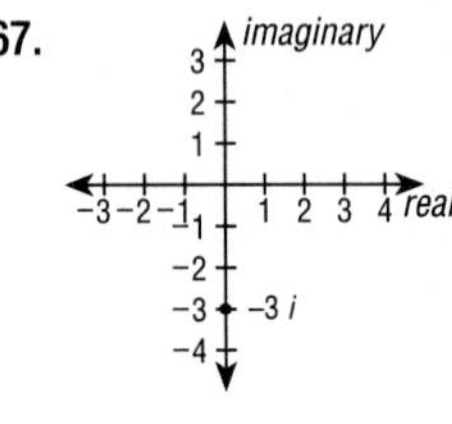

68. 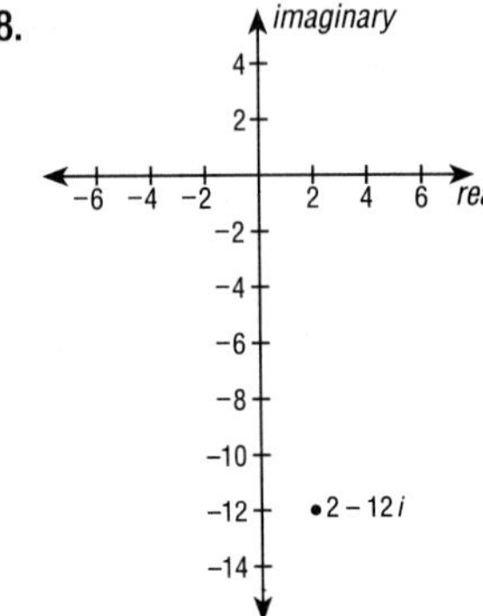

69.–72. 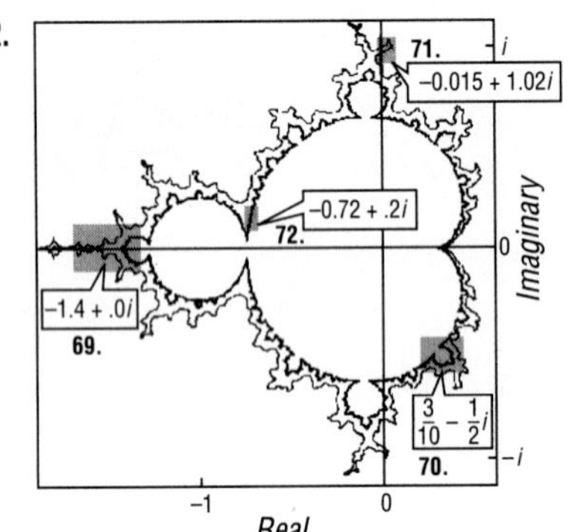

5.6 Independent Practice p. 268

11. $\frac{5}{2} \pm \frac{\sqrt{11}}{2}i$

12. $\frac{7}{12} \pm \frac{\sqrt{23}}{12}i$

13. $1 \pm \frac{\sqrt{24}}{2}i$

14. $1 \pm \frac{\sqrt{20}}{10}i$

15. $\frac{3}{4} \pm \frac{\sqrt{12}}{8}i$

16. $\frac{1}{2} \pm \frac{\sqrt{3}}{2}i$

17. $-\frac{5}{6} \pm \frac{\sqrt{11}}{6}i$

18. $-3 \pm \frac{\sqrt{8}}{4}i$

19. $-\frac{8}{21} \pm \frac{\sqrt{80}}{42}i$

20. $-\frac{1}{2} \pm \frac{\sqrt{12}}{4}i$

21. $\frac{1}{6} \pm \frac{\sqrt{35}}{6}i$

22. $\frac{1}{2} \pm \frac{\sqrt{23}}{2}i$

29. No, area $= \frac{1}{2}(b_1 + b_2)h = \frac{1}{2}[x + (500 - 2x)]x = 43{,}560$. The equation has no real solutions.

30. No, area $= l \cdot w = x(50 - x) = 630$. The equation has no real solutions.

31. It was never down to 300; there are no real-number solutions because the discriminant is negative, and the minimum point on the graph is at (20, 312).

32. 1977 and 1987, it was never \$2.75 million; for \$2.3 million, $t \approx 7.1$ and ≈ 16.9, and the graph crosses the t-axis twice; for \$2.75 million, there are no real solutions because the discriminant is negative, and the graph does not cross the t-axis.

5.6 Independent Practice *(continued)*

33.

Year	1980	1981	1982	1983	1984	1985	1986	1987	1988	1989	1990	1991	1992	1993	1994
R	822.51	790.99	769.79	758.91	758.35	768.11	788.19	818.59	859.31	910.35	971.71	1043.39	1125.39	1217.71	1320.35

No, there are no real number solutions.

5.6 Integrated Review p. 270

45. 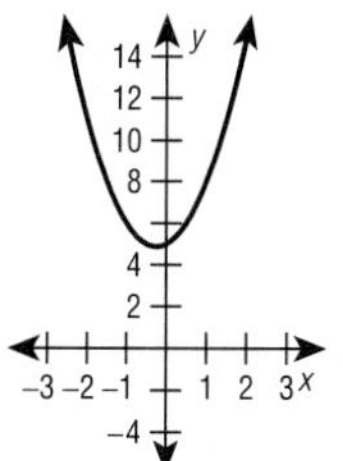

46. 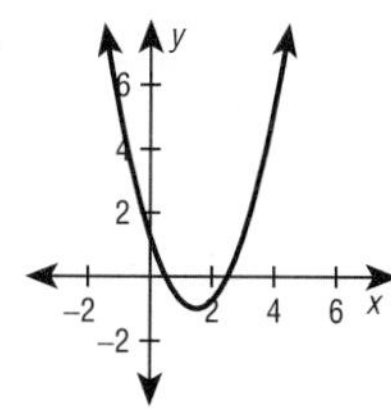

47. 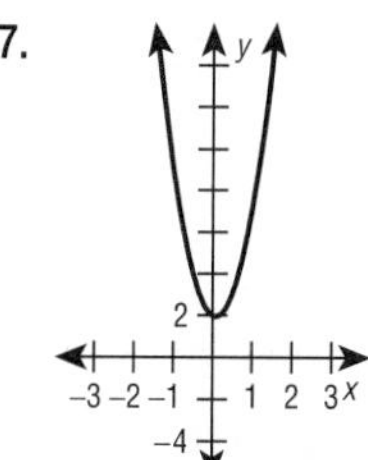

48. 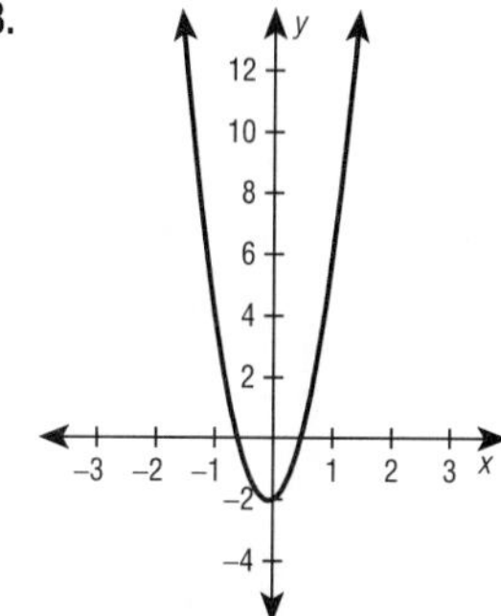

49. 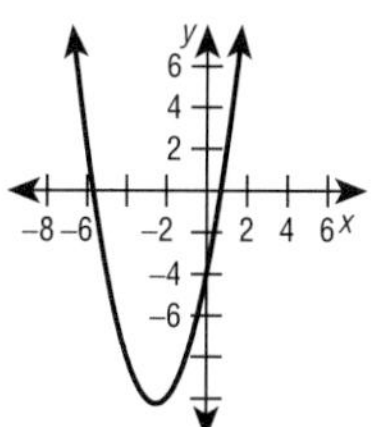

50. 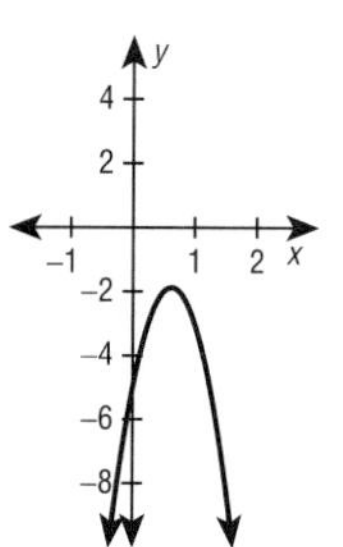

5.7 Independent Practice pp. 274–275

17. 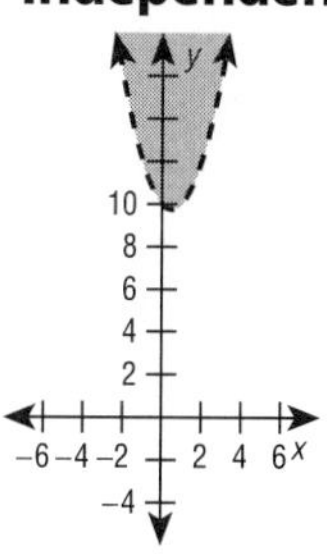

18. 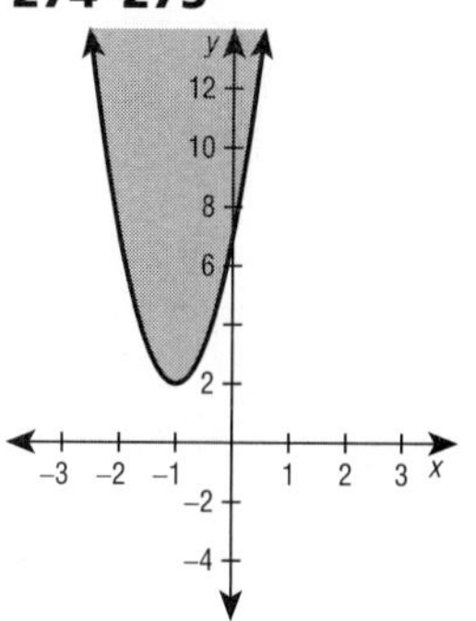

19. 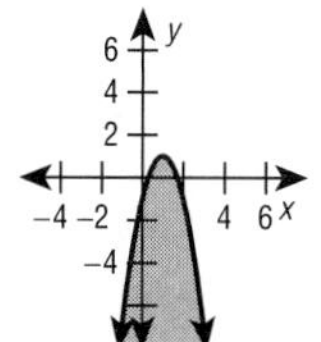

20. 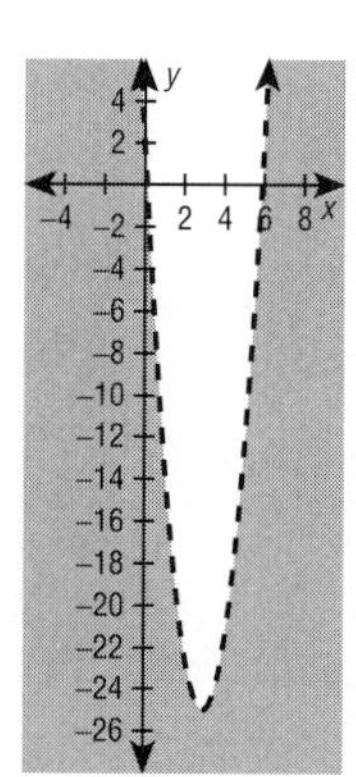

21. 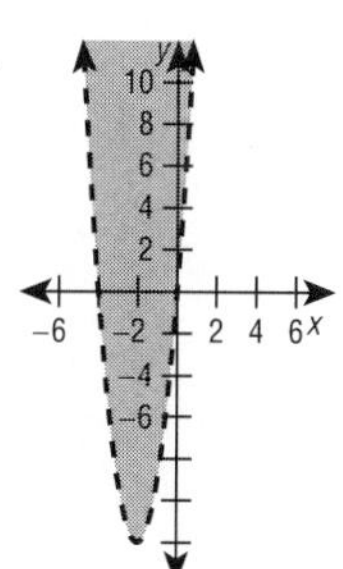

22. 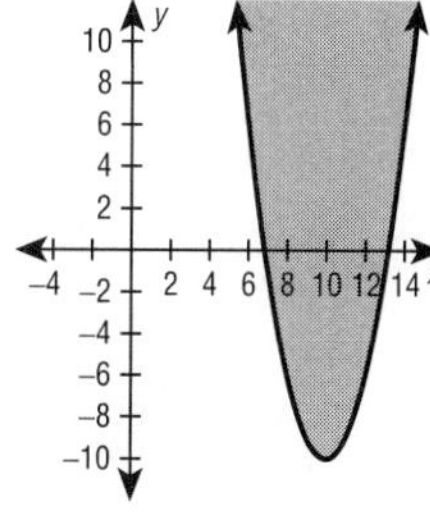

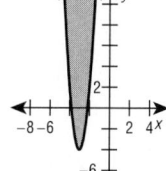

24. 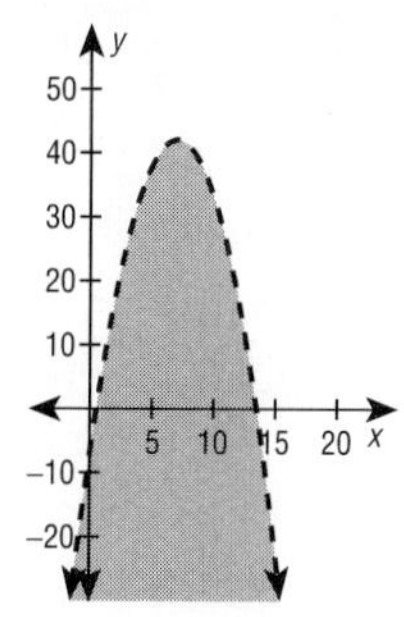

25. 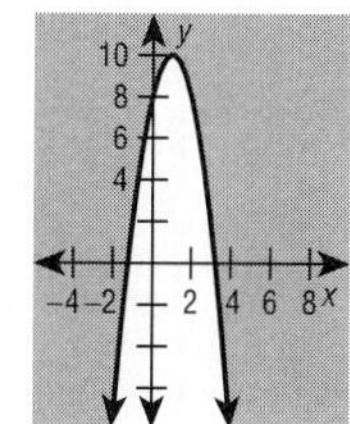

26. 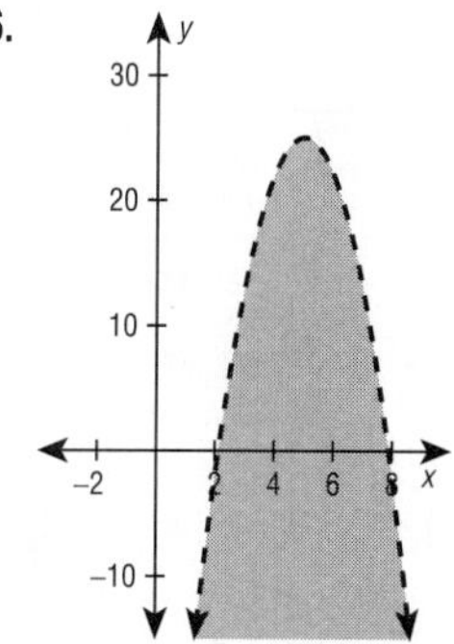

ADDITIONAL ANSWERS

5.7 Independent Practice *(continued)*

27.

28.

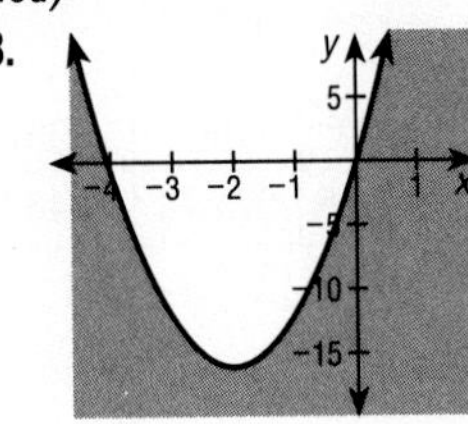

29.

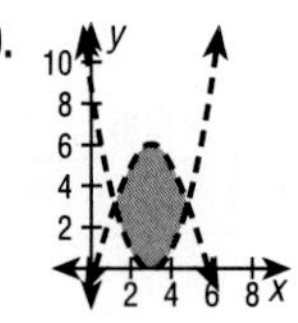

30.

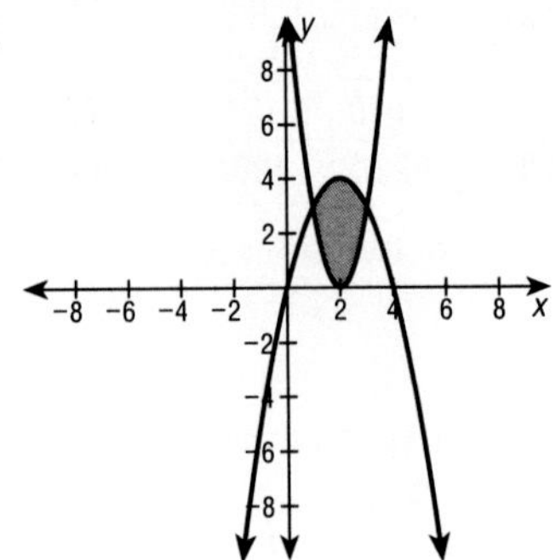

31.

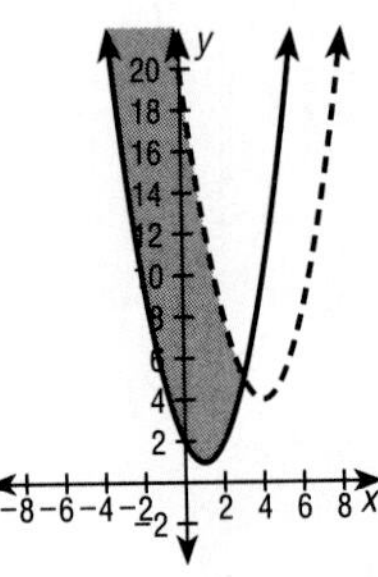

32.

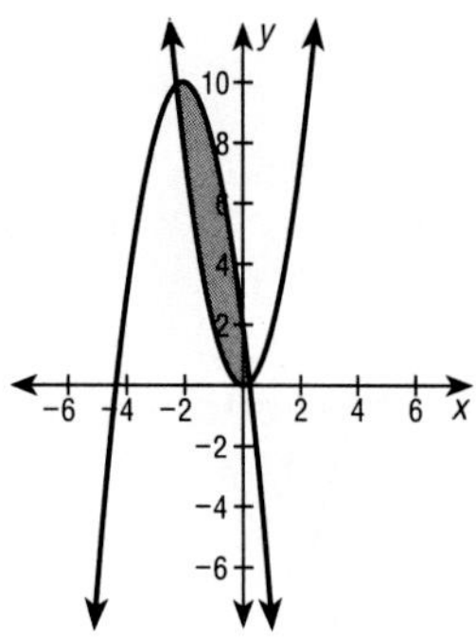

33.

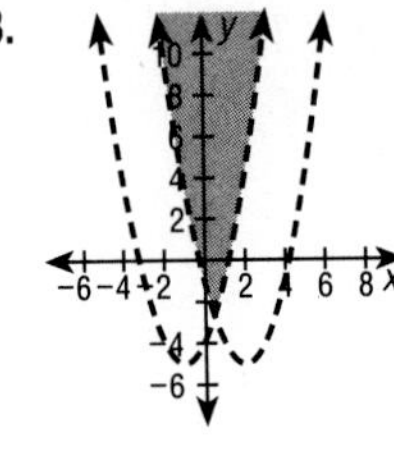

34.

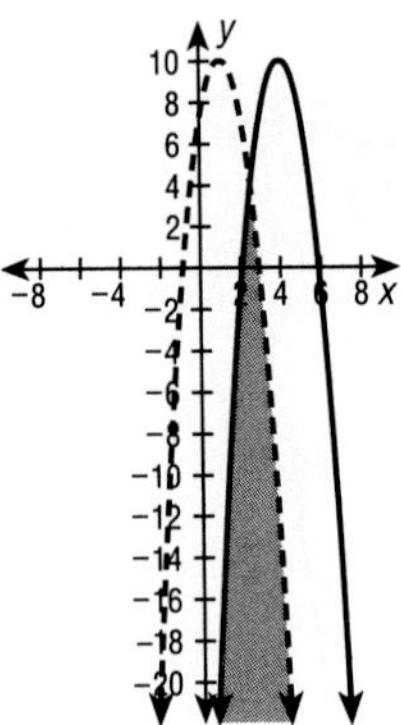

35.

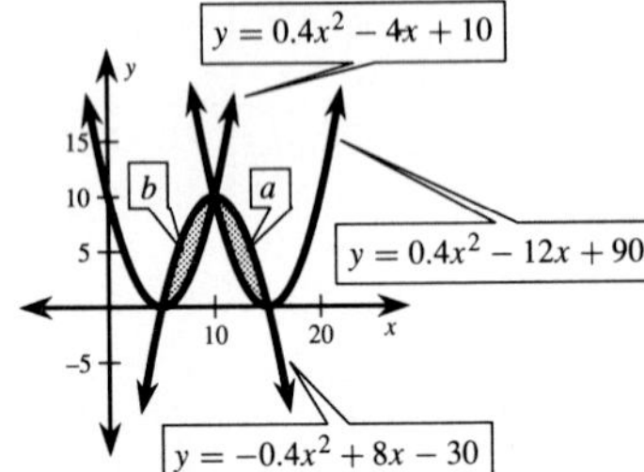

36.

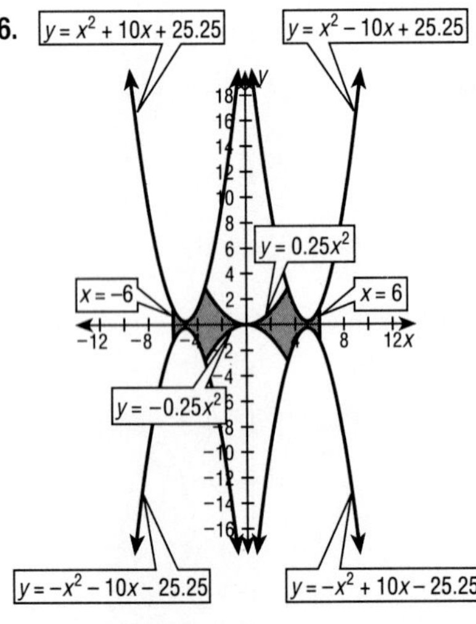

$(-3.4, 2.89)$, $(3.4, 2.89)$,
$(-3.4, -2.89)$, $(3.4, -2.89)$

5.7 Integrated Review p. 276

39.

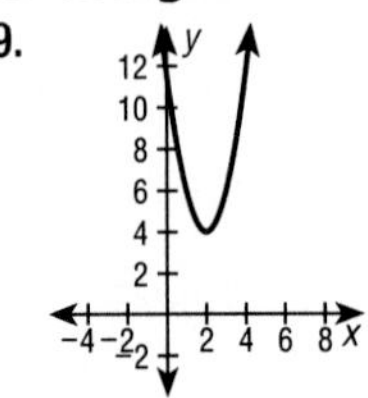

40.

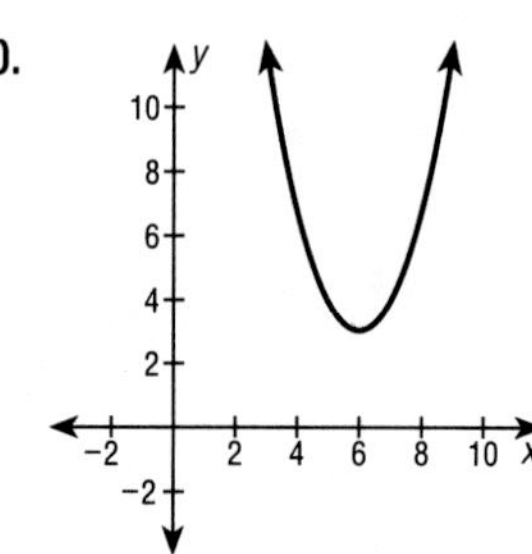

41.

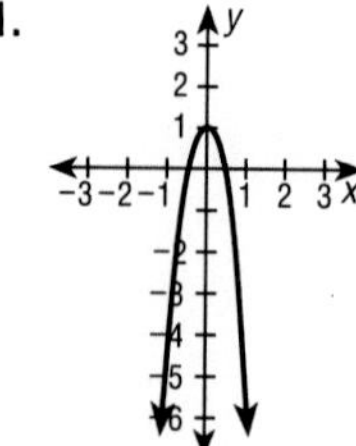

42.

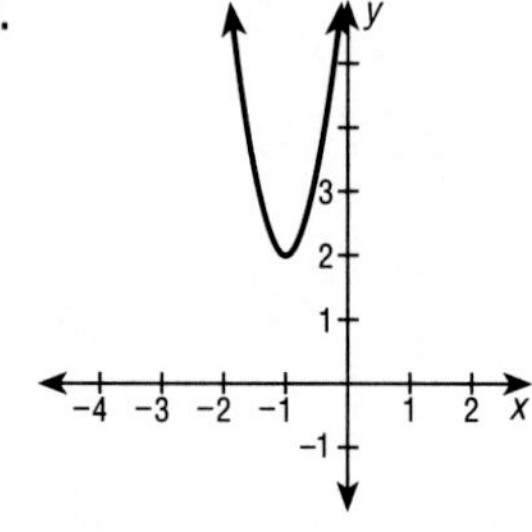

5.7 Integrated Review *(continued)*

49. 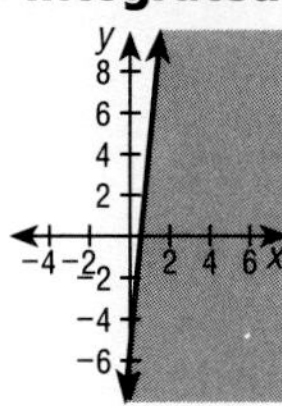

50. 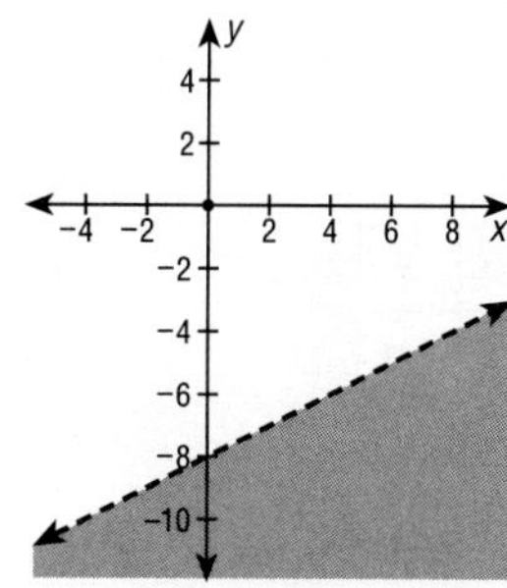

51. 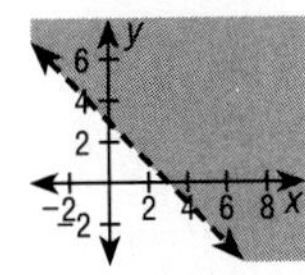

52. 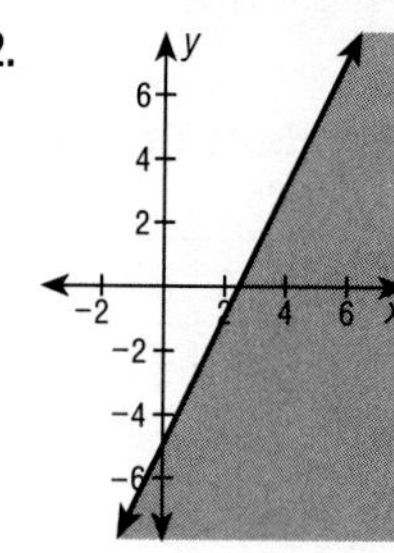

54. 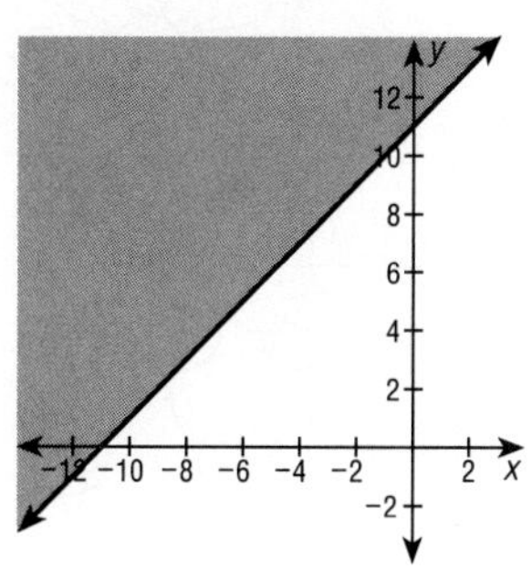

5.7 Chapter Review p. 278

4. 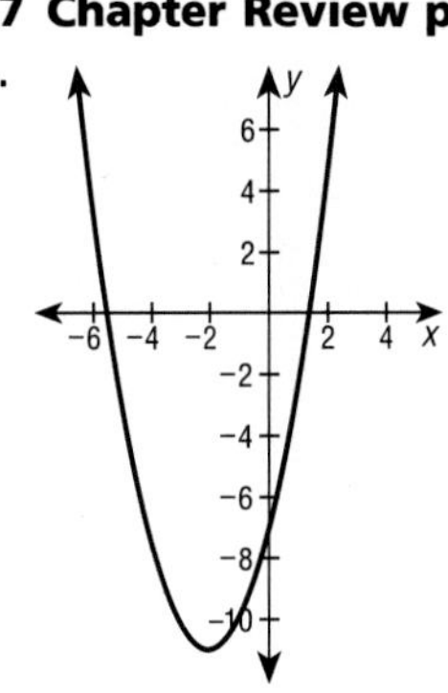

5. 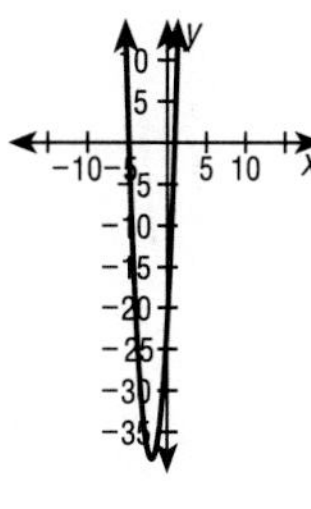

6. 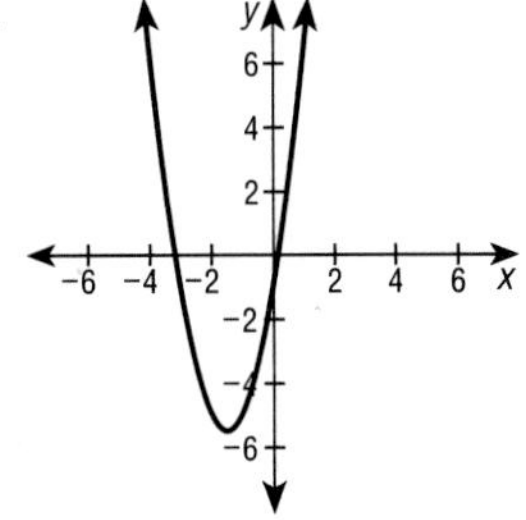

7. 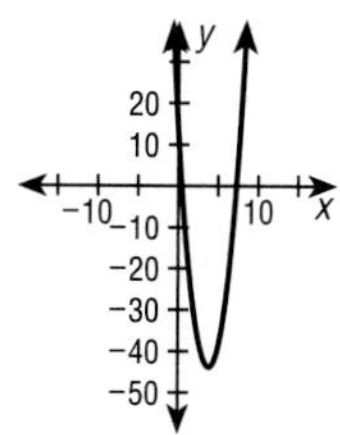

8. 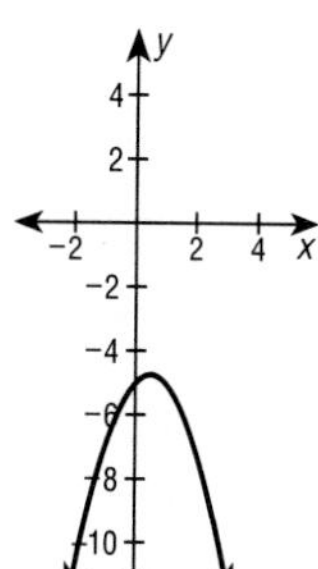

9. 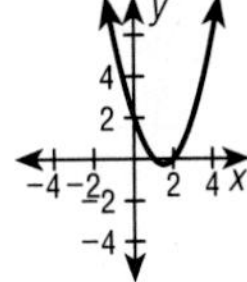

40. 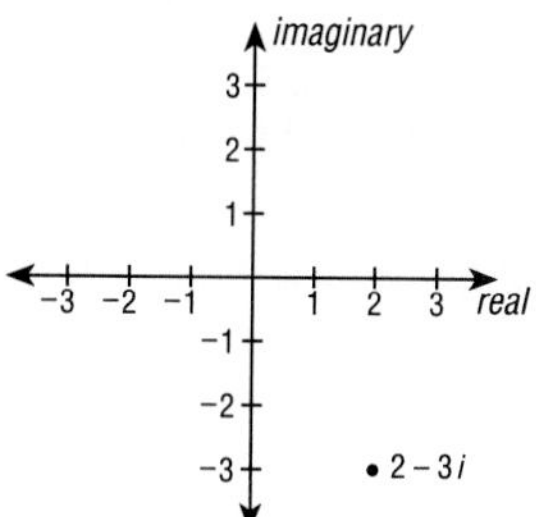

41. 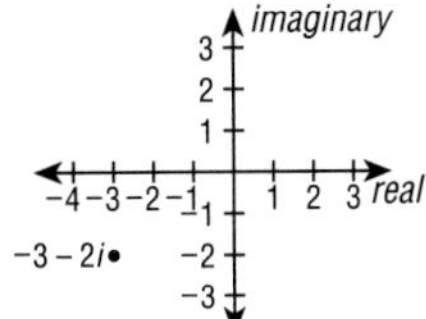

42. 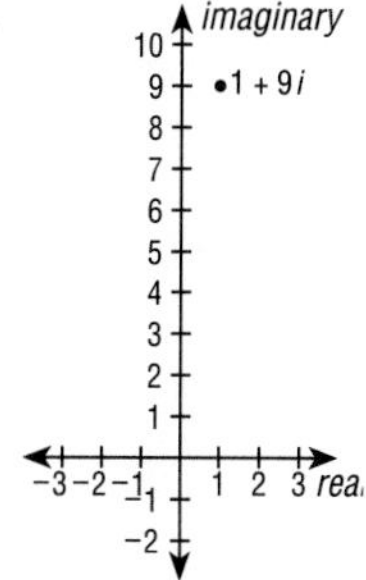

6.1 Independent Practice p. 289

13.

Input, x	−4	−3	0	3	4
Output, y	−3	0	3	0	−3

14.

Input, x	0	0	2	2	4	4
Output, y	4	−4	3	−3	1	−1

15.

Input, x	−6	−4	−4	0	0
Output, y	0	2	−2	3	−3

16.

Input, x	−5	−4	−3	0	3	4	5
Output, y	−6	−4	−2	−1	−2	−4	−6

17.

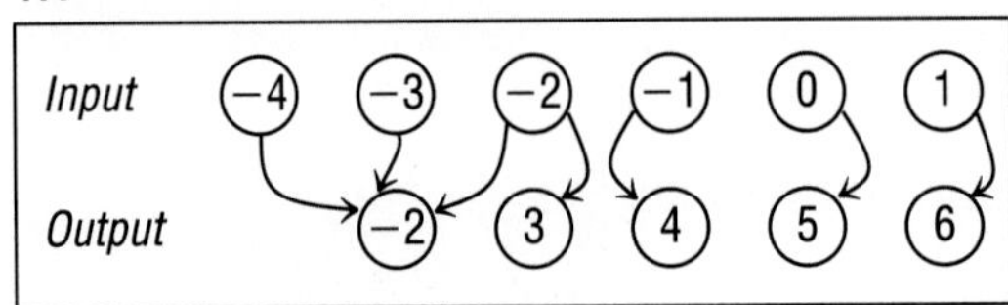

18.

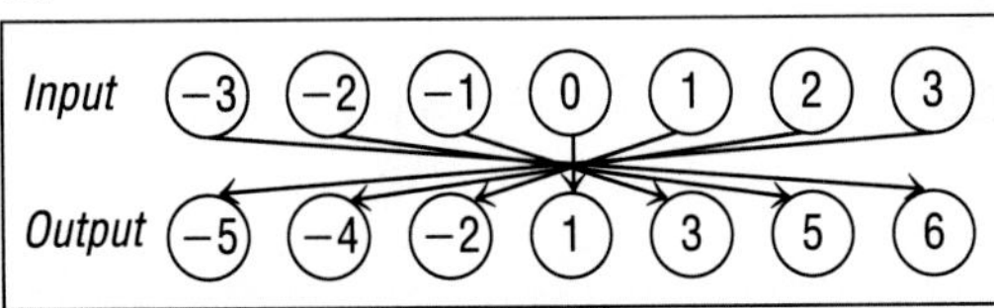

19.

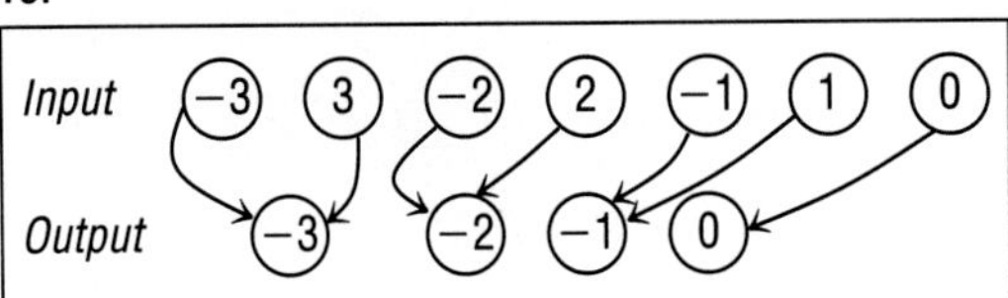

20.

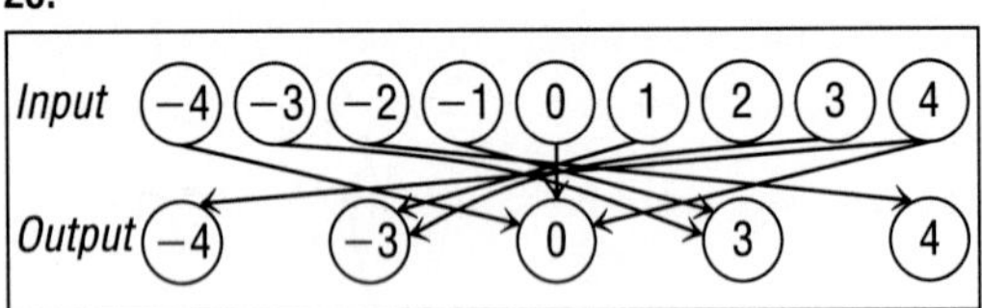

6.1 Integrated Review p. 290

43.

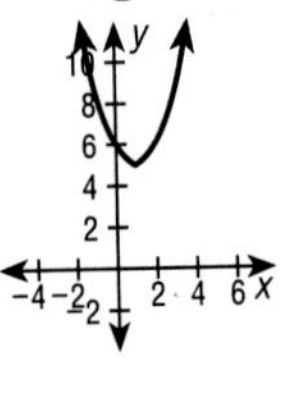

44.

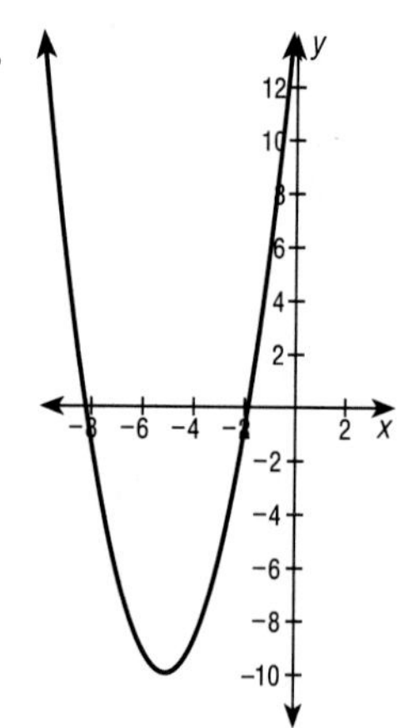

45.

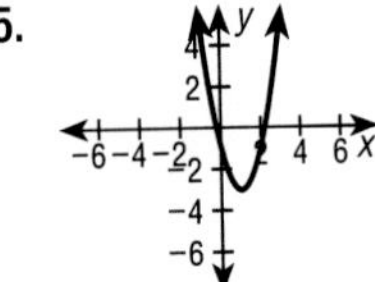

46.

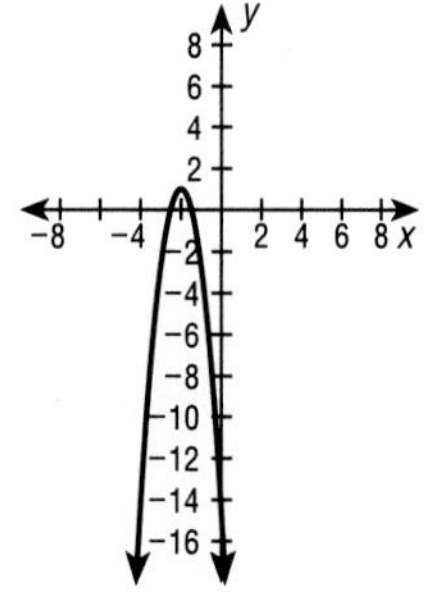

6.2 Communicating About Algebra p. 293

Sale price with rebate of \$1200: $f(x) = x - 1200$,

Sale price with discount of 15%: $g(x) = 0.85x$;

Rebate first: $g(f(x)) = g(x - 1200) = 0.85(x - 1200)$,

Discount first: $f(g(x)) = f(0.85x) = 0.85x - 1200$;

Rebate first: \$14,620; discount first: \$14,440; \$180

6.2 Independent Practice pp. 294–295

21. $f(g(x)) = 2x + 1$, $g(f(x)) = 2x + 2$

22. $f(g(x)) = 2x - 5$, $g(f(x)) = 2x - 2$

23. $f(g(x)) = x^2 - 2x + 1$, $g(f(x)) = x^2 - 1$

24. $f(g(x)) = x^2 + 4x + 3$, $g(f(x)) = x^2 + 1$

6.2 Independent Practice *(continued)*

25. $f(g(x)) = x,\ g(f(x)) = x$

26. $f(g(x)) = -x^2 - 10x - 26,\ g(f(x)) = -x^2 + 4$

27. $h(x) = 4x - 16$, all real numbers

28. $h(x) = x^2 + 2x - 8$, all real numbers

29. $h(x) = x^2 - 2x + 8$, all real numbers

30. $h(x) = 6x^2$, all real numbers

31. $h(x) = \dfrac{2x - 8}{x^2}$, all real numbers except 0

32. $h(x) = 2x^3 - 8x^2$, all real numbers

33. $h(x) = 2x^2 - 8$, all real numbers

34. $h(x) = 4x^2 - 32x + 64$, all real numbers

6.2 Mixed Review

5.

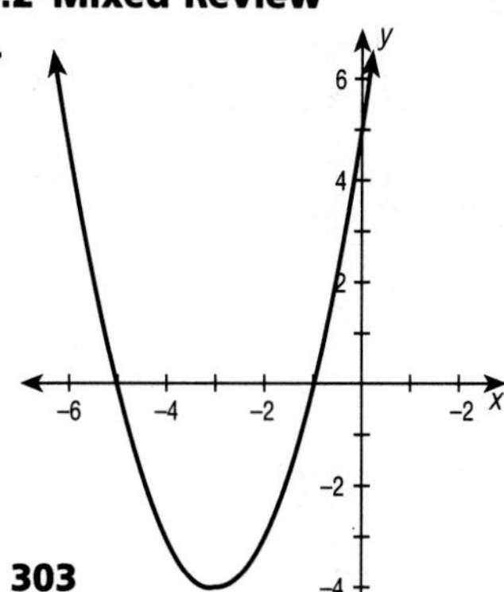

6.3 Communicating About Algebra p. 301

A. $x = \frac{2}{3}y - 4$, yes

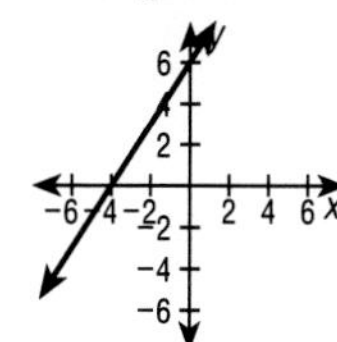

B. $x = \frac{1}{2}y^2 + 3$, no

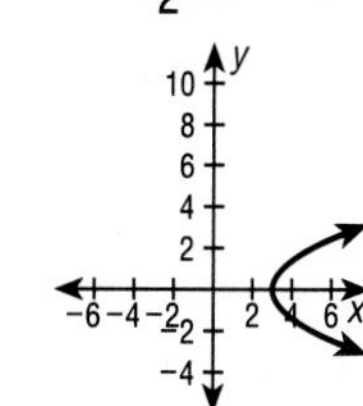

6.3 Independent Practice p. 303

23.

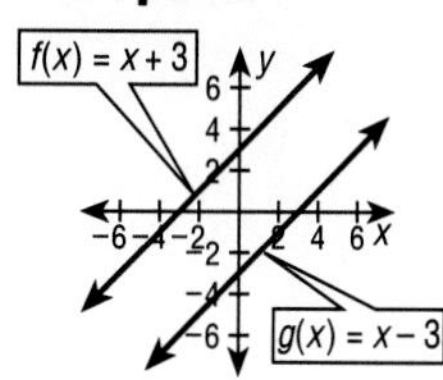

24.

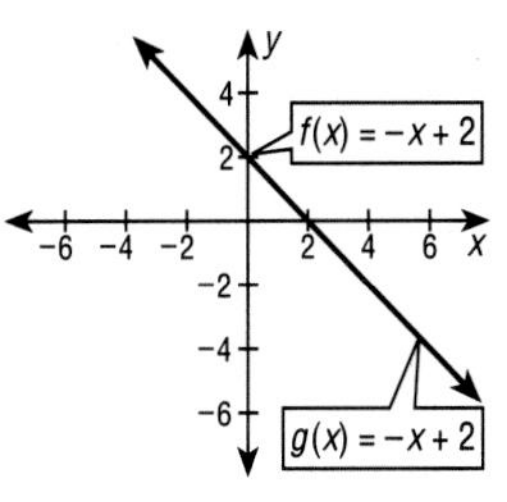

25.

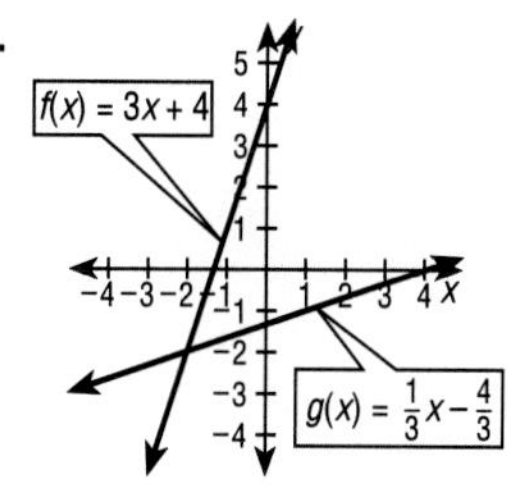

26.

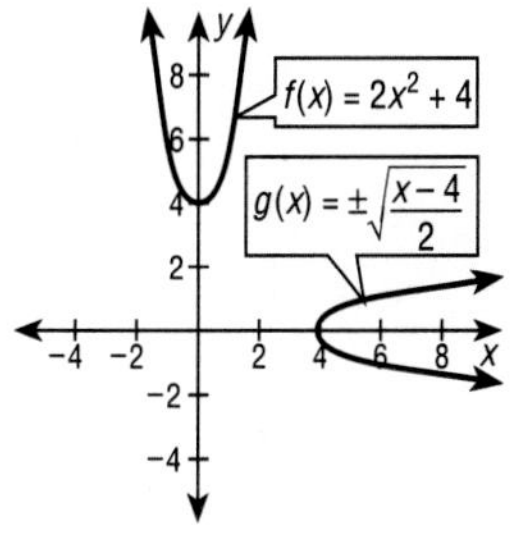

27.

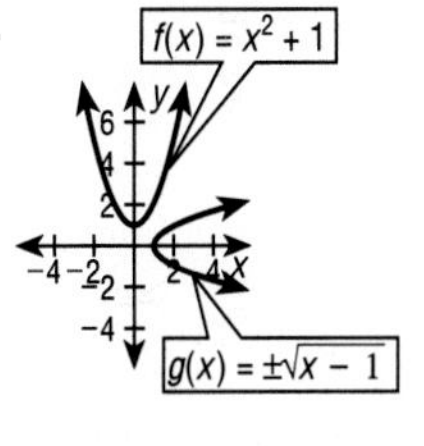

28.

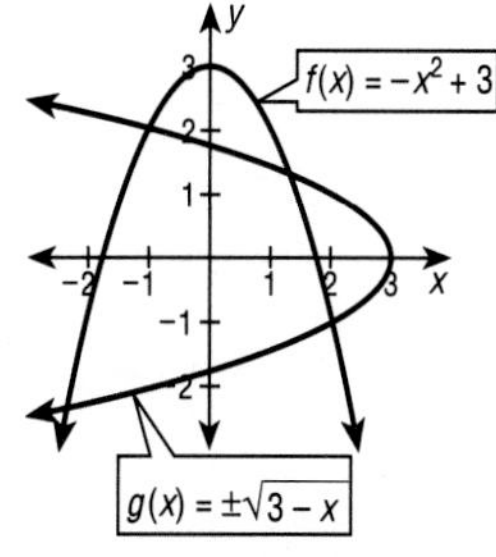

29.

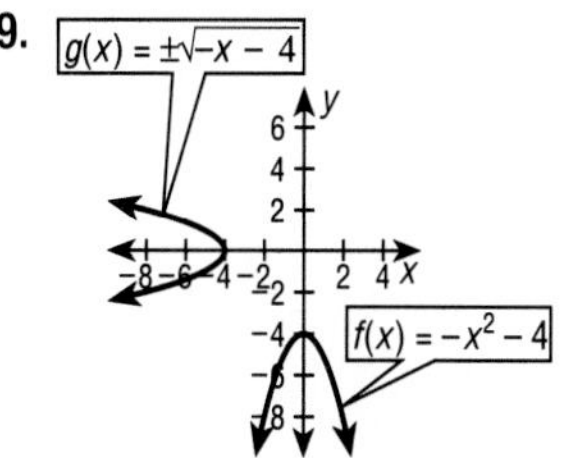

30.

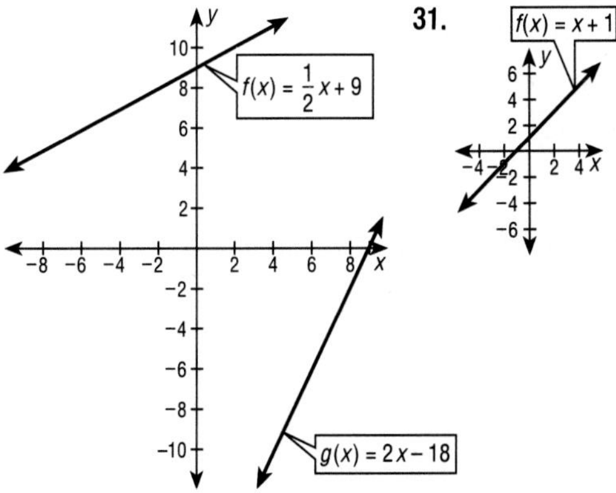

31. f(x) = x + 1

32.

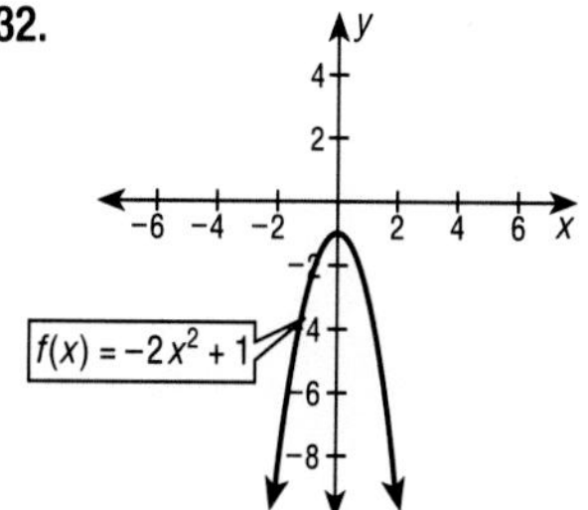

33.

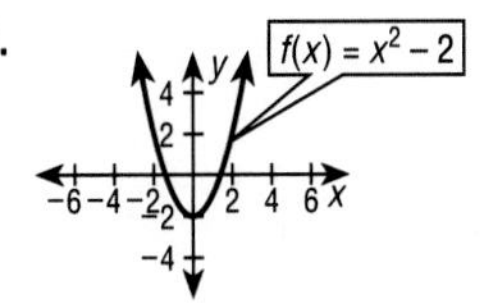

34.

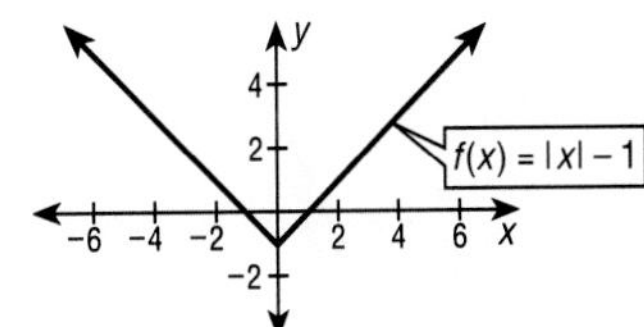

6.3 Mid-Chapter Self-Test p. 305

1.

Input: 2, 4, 5, 6

Output: 3, 7, 8, 12

2. 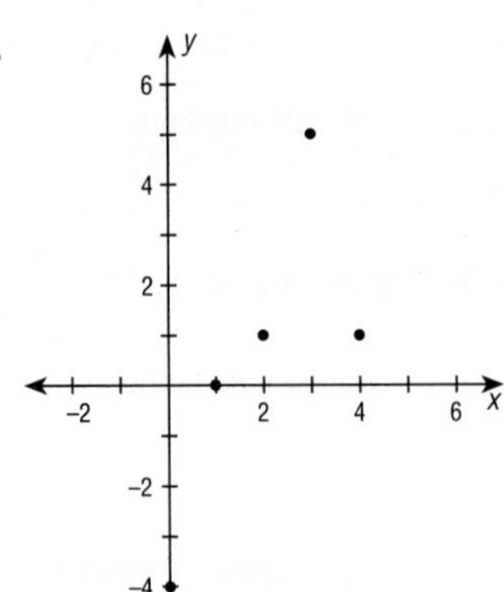

6.4 Communicating About Algebra p. 308

A. 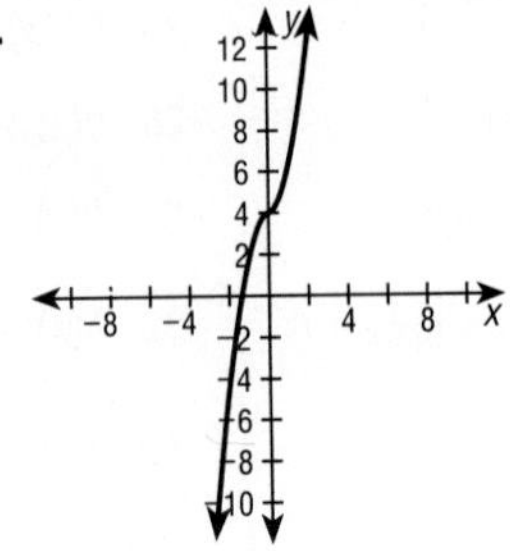

Lightly draw each graph for all values of x, then shade the portion that is the graph of the compound function.

B. 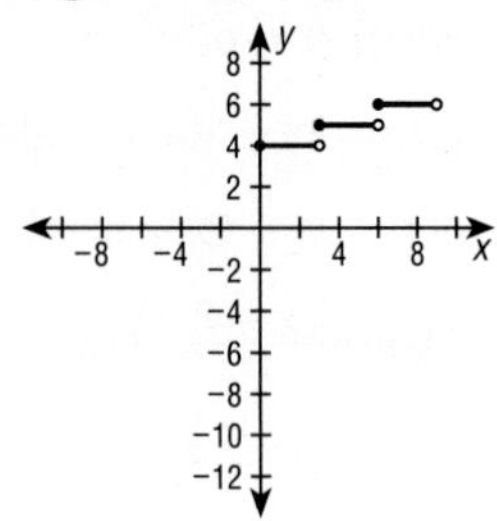

Draw horizontal line segments, each with a solid dot on the left and an open dot on the right.

6.4 Guided Practice p. 309

1. Answers vary. $f(x) = \begin{cases} x, & x \geq 0 \\ -x, & x < 0 \end{cases}$

2. $f(x) = \begin{cases} 10 - 2x, & x < 5 \\ -(10 - 2x), & x \geq 5 \end{cases}$ or $f(x) = \begin{cases} 10 - 2x, & x \leq 5 \\ -(10 - 2x), & x > 5 \end{cases}$

3. 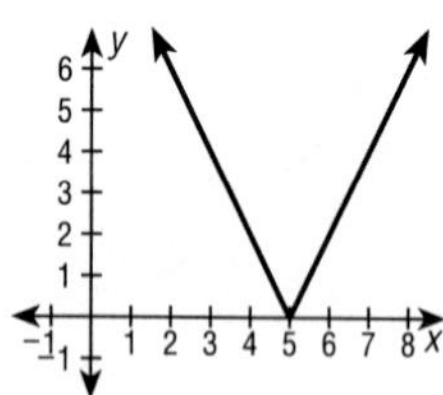

4. Answers vary. $f(x) = \begin{cases} -1, & 0 \leq x < 2 \\ 0, & 2 \leq x < 4 \\ 1, & 4 \leq x < 6 \end{cases}$

6.4 Independent Practice pp. 309–310

13. 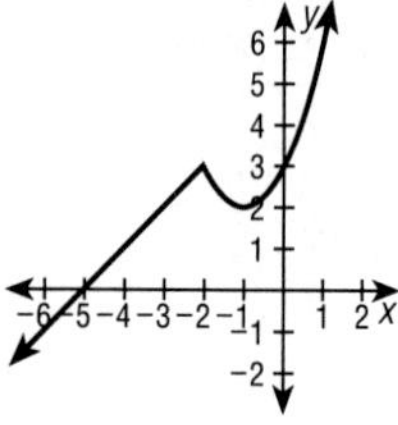

14. 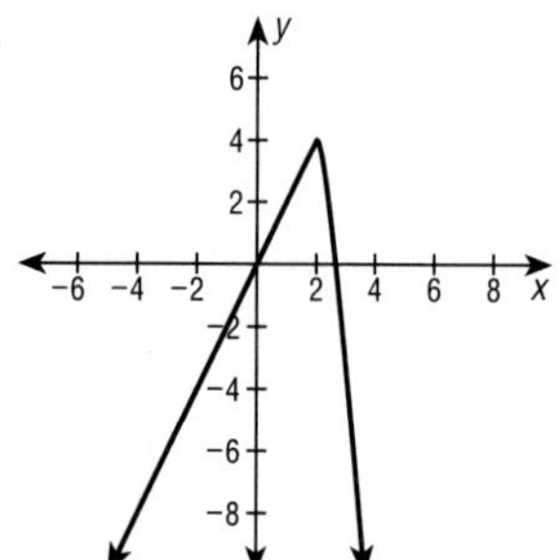

15. 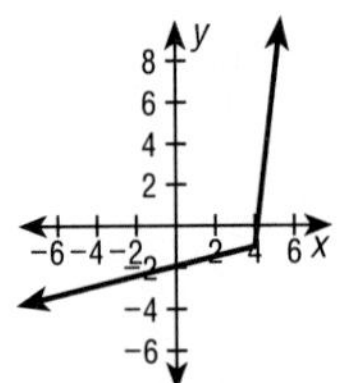

16. 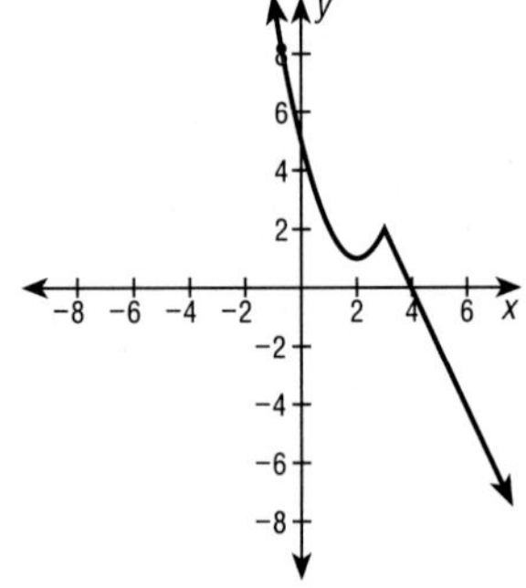

17. 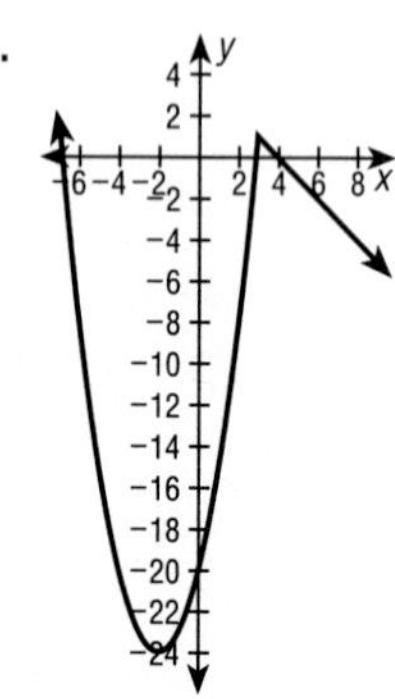

18. 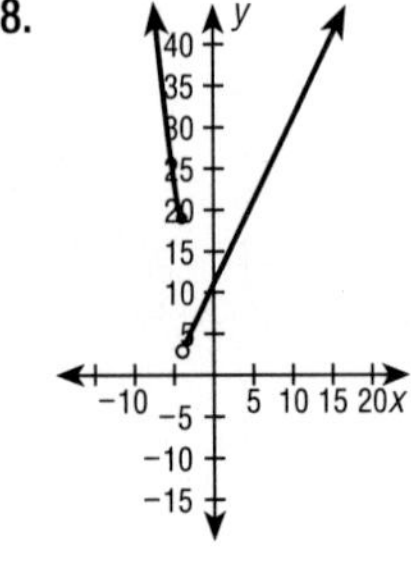

6.4 Independent Practice *(continued)*

23. $f(x) = \begin{cases} -(x+2), & x < -2 \\ x+2, & x \geq -2 \end{cases}$ or $\begin{cases} -(x+2), & x \leq -2 \\ x+2, & x > -2 \end{cases}$

24. $f(x) = \begin{cases} -x+5, & x < 5 \\ -(-x+5), & x \geq -5 \end{cases}$ or $\begin{cases} -x+5, & x \leq 5 \\ -(-x+5), & x > 5 \end{cases}$

25. $f(x) = \begin{cases} -2x-6, & x < -3 \\ -(-2x-6), & x \geq -3 \end{cases}$ or $\begin{cases} -2x-6, & x \leq -3 \\ -(-2x-6), & x > -3 \end{cases}$

26. $f(x) = \begin{cases} -\left(\frac{1}{2}x-3\right), & x < 6 \\ \frac{1}{2}x-3, & x \geq 6 \end{cases}$ or $\begin{cases} -\left(\frac{1}{2}x-3\right), & x \leq 6 \\ \frac{1}{2}x-3, & x > 6 \end{cases}$

27. $f(x) = \begin{cases} -\left(\frac{1}{3}x-1\right), & x < 3 \\ \frac{1}{3}x-1, & x \geq 3 \end{cases}$ or $\begin{cases} -\left(\frac{1}{3}x-1\right), & x \leq 3 \\ \frac{1}{3}x-1, & x > 3 \end{cases}$

28. $f(x) = \begin{cases} -3x+6, & x < 2 \\ -(-3x+6), & x \geq 2 \end{cases}$ or $\begin{cases} -3x+6, & x \leq 2 \\ -(-3x+6), & x > 2 \end{cases}$

29.
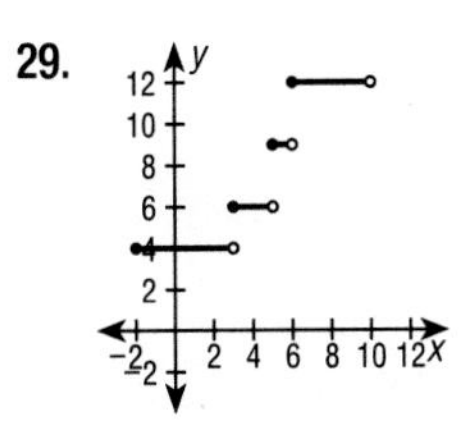

30.
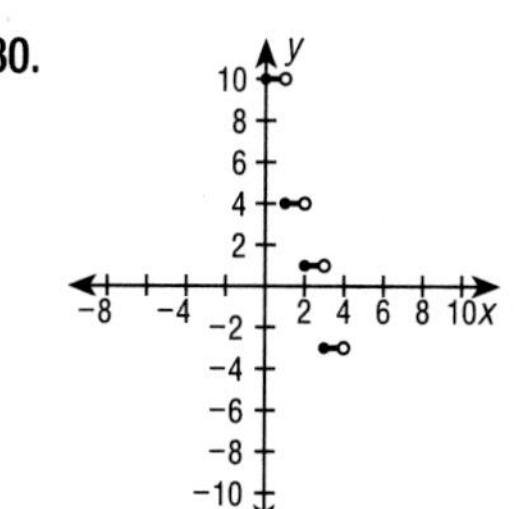

31.
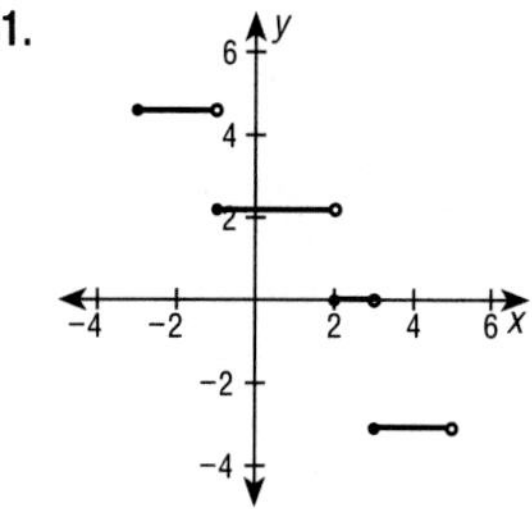

33. $C = \begin{cases} 10 + 8.5x, & x \leq 25 \\ 10 + 7.75x, & x > 25 \end{cases}$
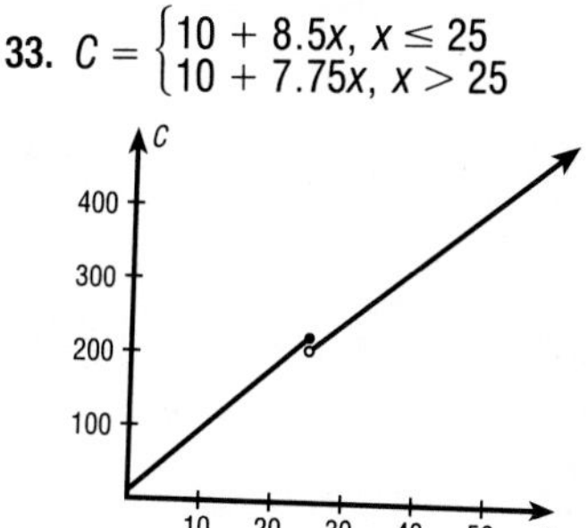

34.
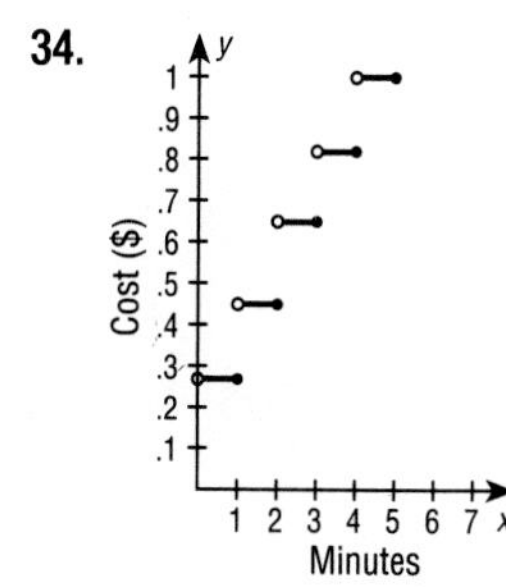

35.
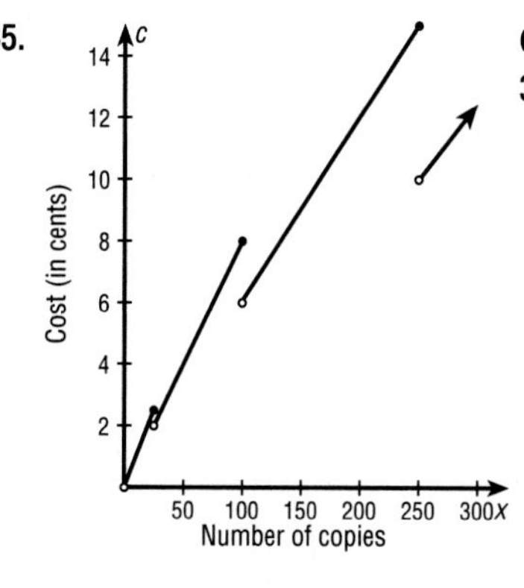

6.4 Integrated Review p. 311

39.
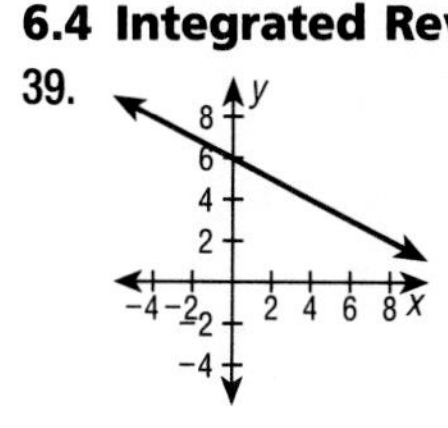

40.
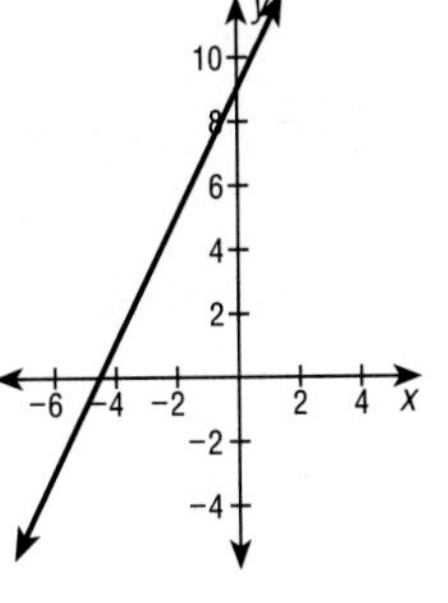

41.
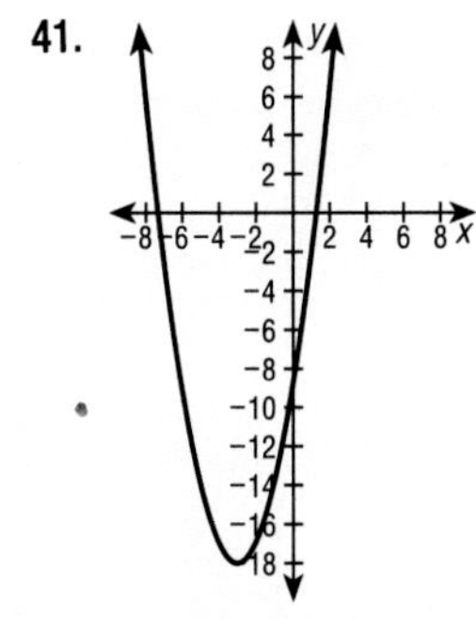

42.
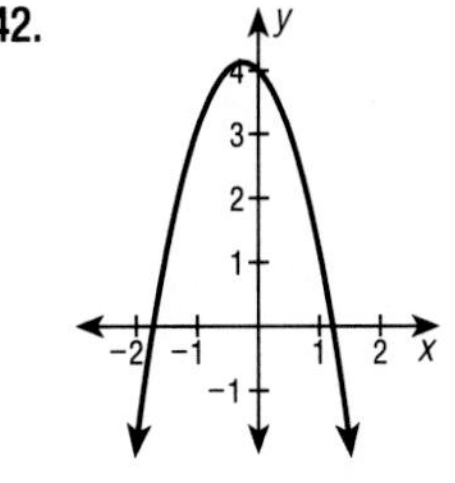

43.
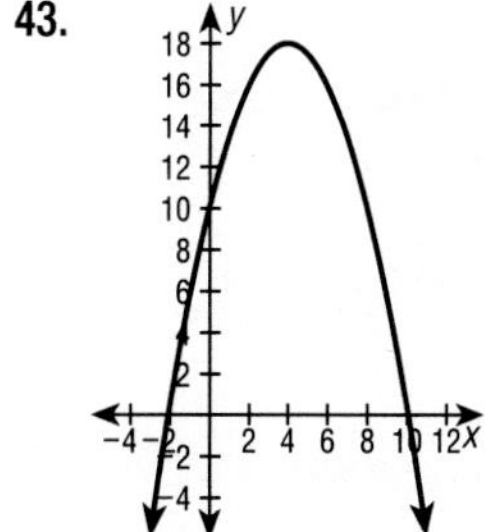

44.
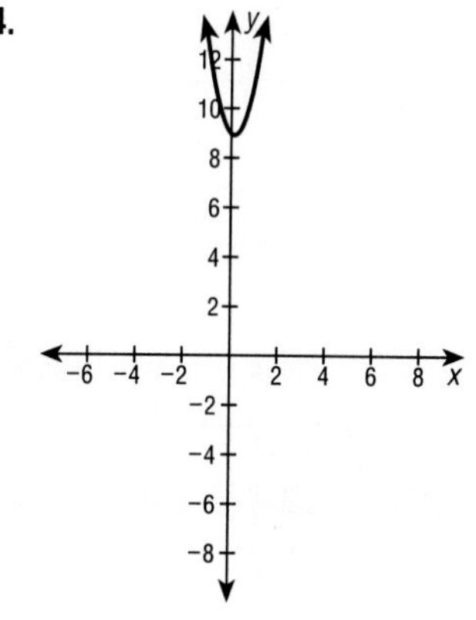

6.4 Exploration and Extension p. 311

51. $f(x) = \begin{cases} -\frac{1}{4}x^2 + 2x + 4, & 0 \le x < 8 \\ 2x - 12, & 8 \le x < 12 \\ -2x + 36, & 12 \le x < 16 \\ -\frac{1}{4}x^2 + 10x - 92, & 16 \le x \le 24 \end{cases}$

Variations of $\le$ and $<$ are possible

6.5 Guided Practice p. 318

3.

4.

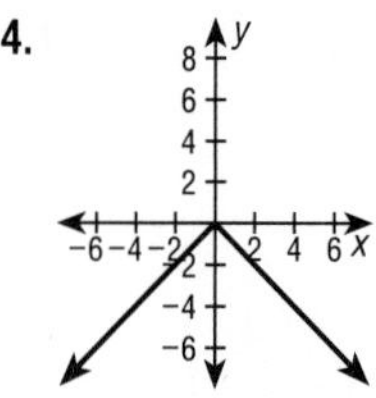

6.5 Independent Practice pp. 318–319

9. $g(x)$ is $f(x)$ shifted up 3 units.
10. $g(x)$ is $f(x)$ shifted up 9 units.
11. $g(x)$ is $f(x)$ shifted down 1 unit.
12. $g(x)$ is $f(x)$ shifted down 3 units.
13. $g(x)$ is $f(x)$ shifted to the right 8 units.
14. $g(x)$ is $f(x)$ shifted to the left 4 units.
15. $g(x)$ is $f(x)$ shifted to the left 5 units.
16. $g(x)$ is $f(x)$ shifted to the right 7 units.
17. $g(x) = f(x + 1)$
18. $g(x) = f(x - 9)$
19. $g(x) = f(x) - 3$
20. $g(x) = f(x) + 4$

25.

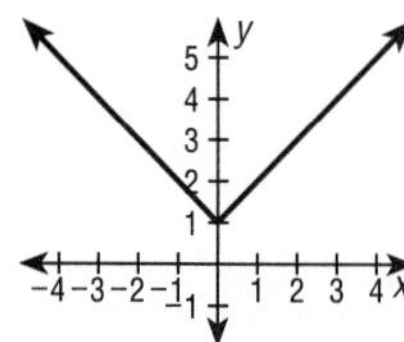

26.

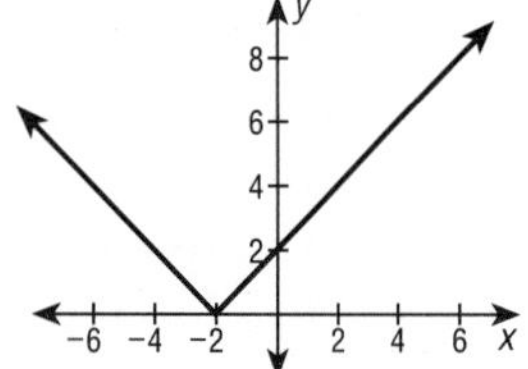

27.

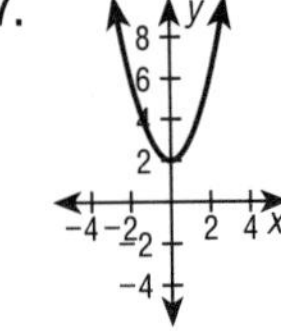

28.

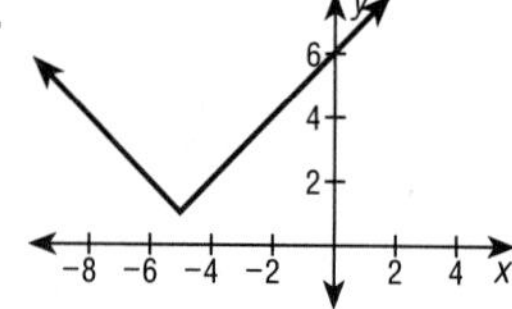

29.

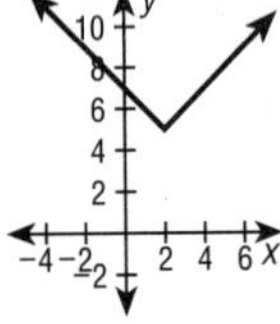

30.

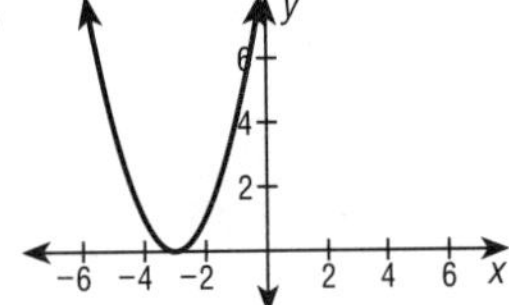

31.

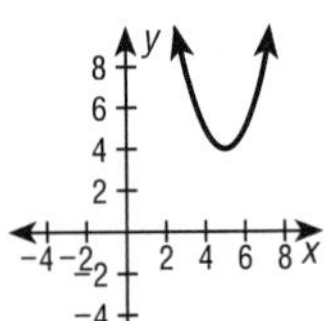

32.

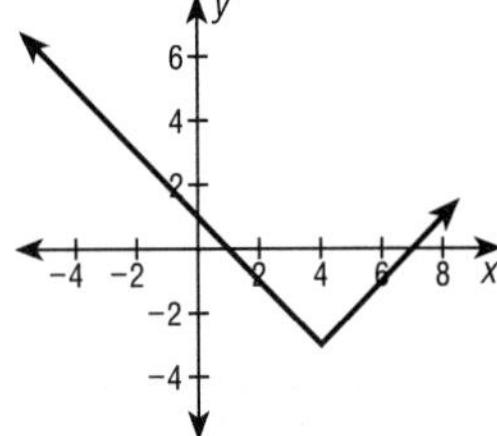

33.

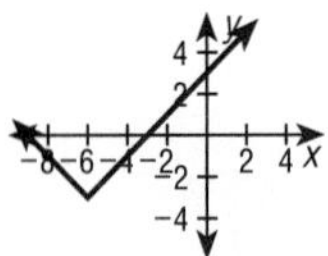

34.

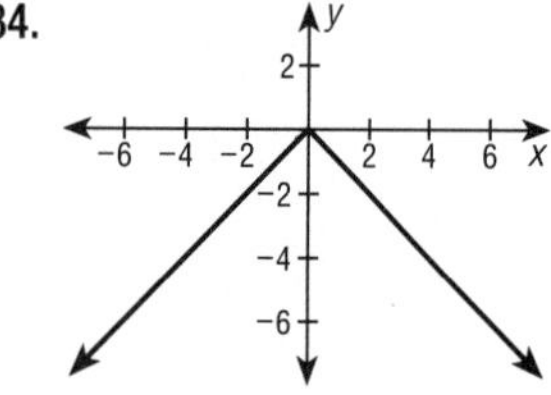

35.

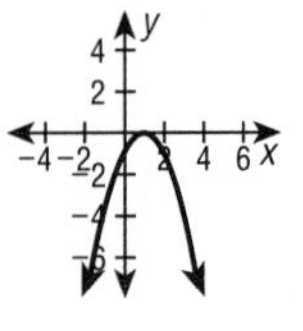

36.

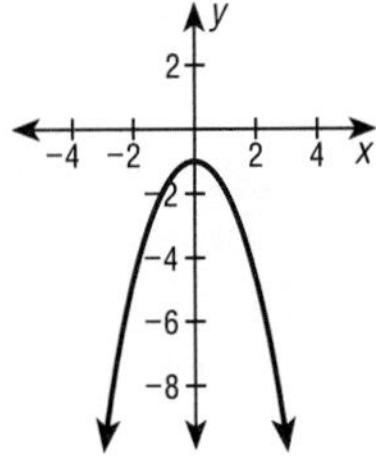

37.

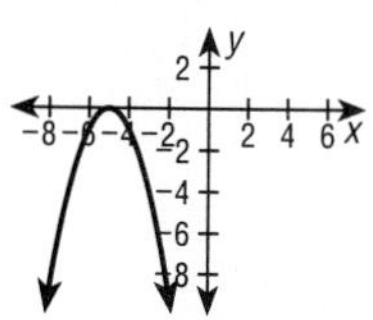

38.

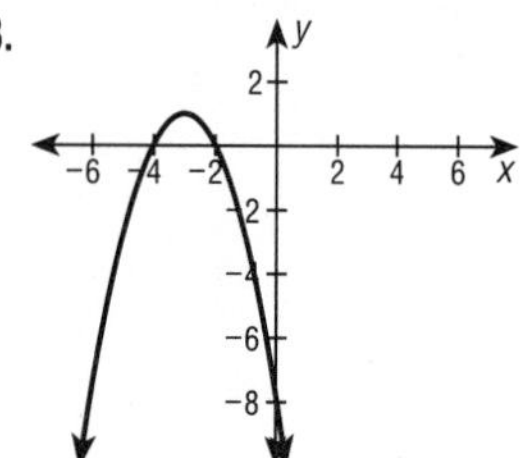

39.

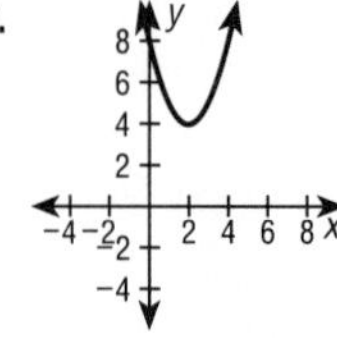

40.

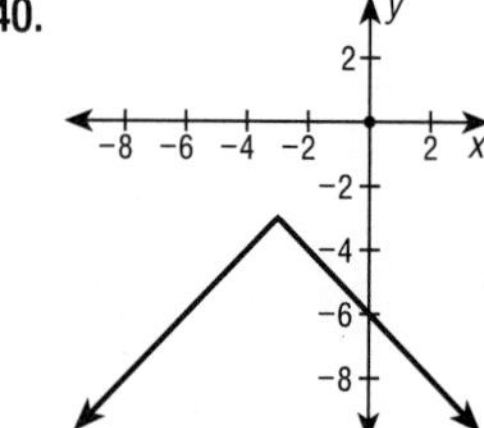

6.5 Independent Practice *(continued)*

41.

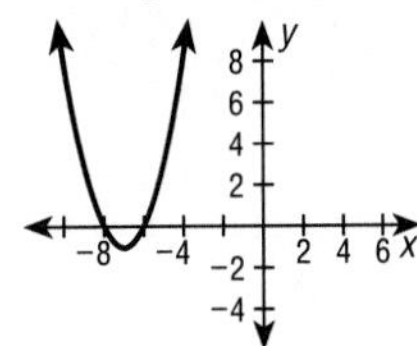

42.

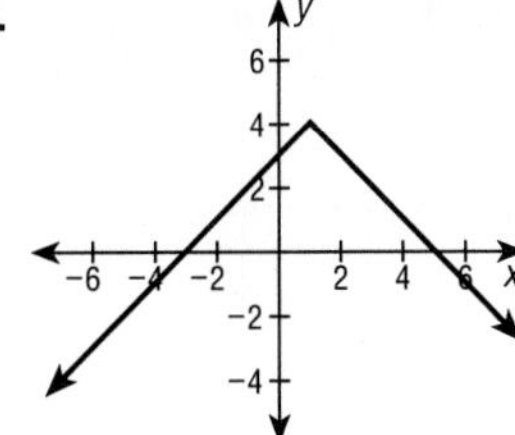

6.5 Integrated Review p. 320

51.

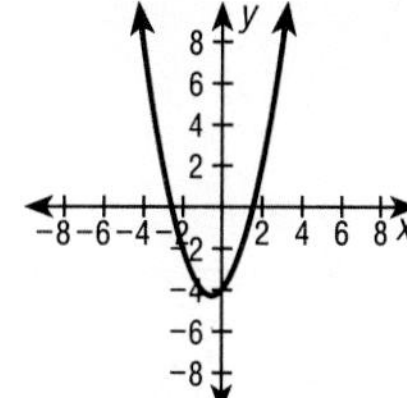

52.

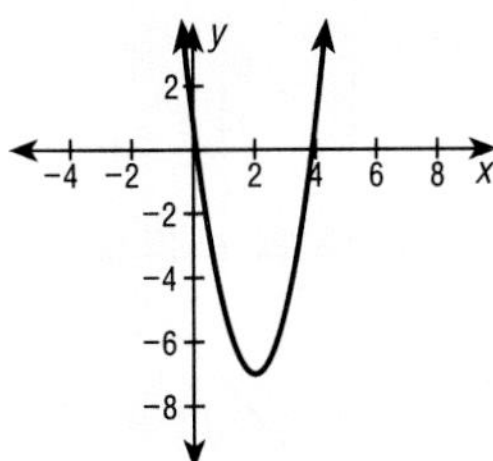

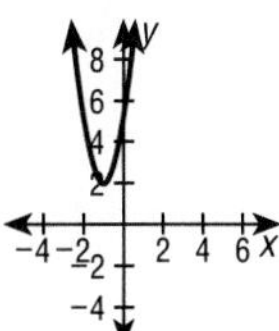

54.

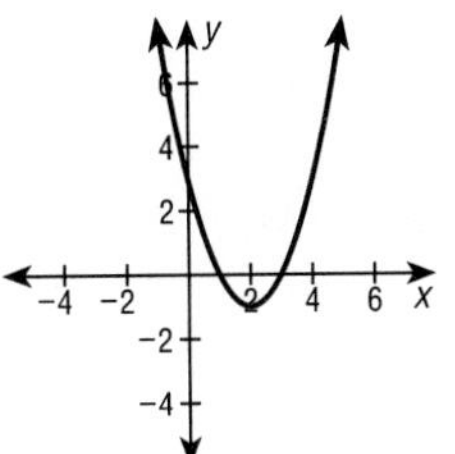

55.

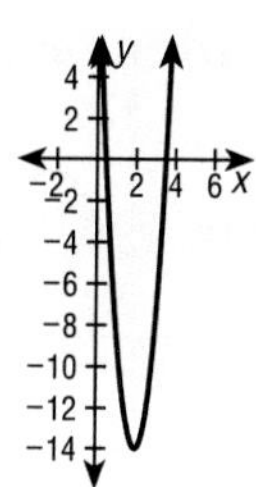

56.

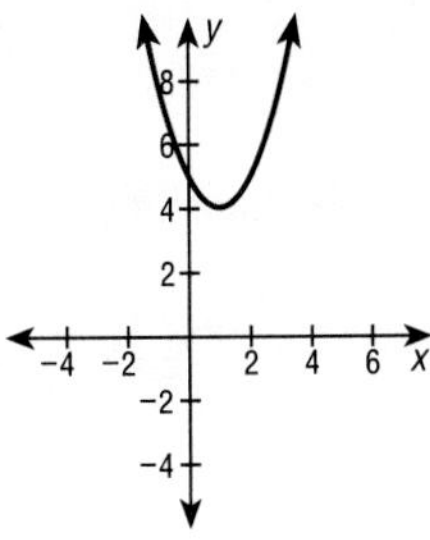

6.5 Exploration and Extension p. 320

59.

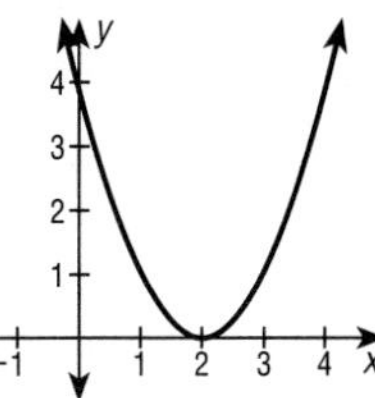

60.

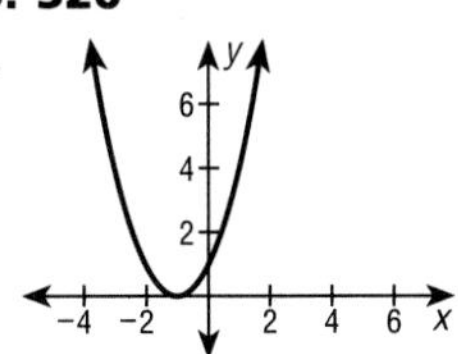

6.5 Mixed Review p. 320

7.

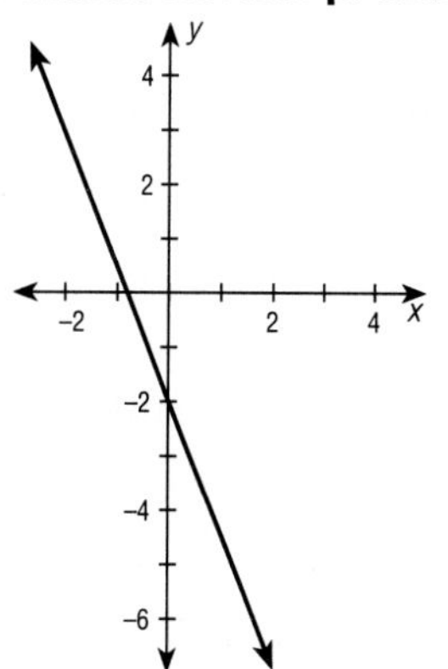

8.

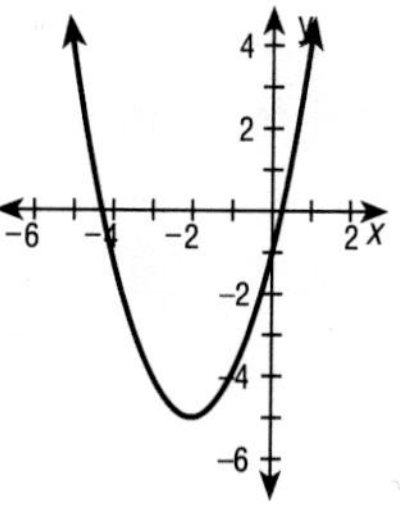

6.7 Chapter Review p. 335

47.

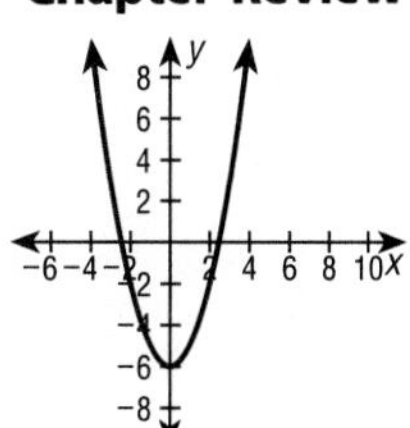

48.

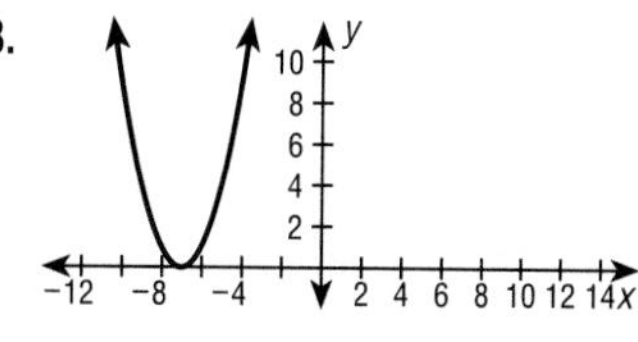

49.

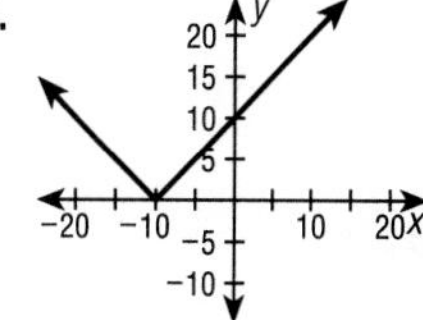

50.

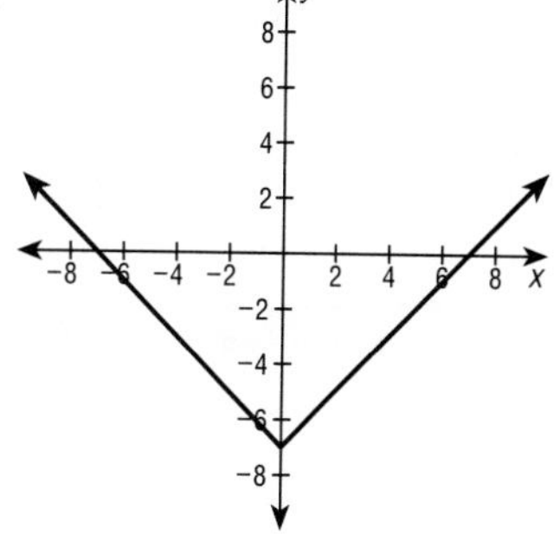

51.

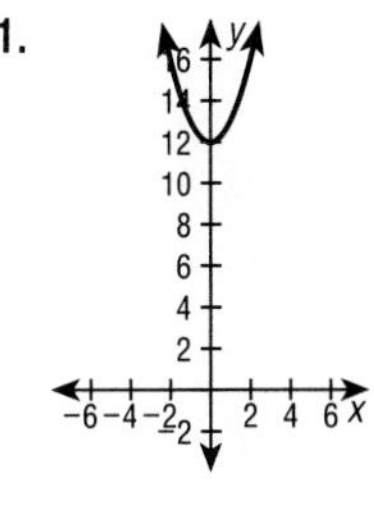

52.

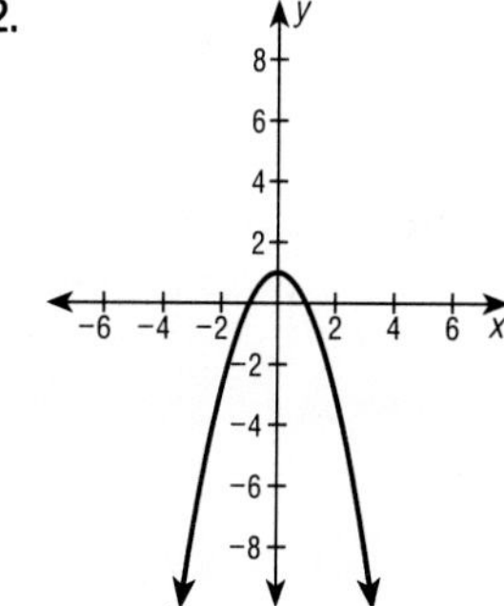

53.

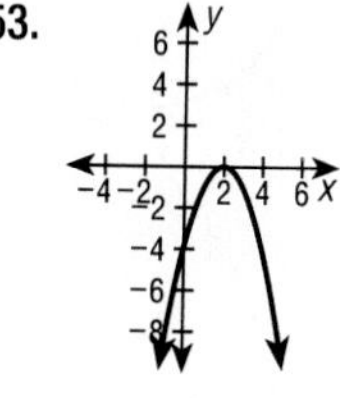

ADDITIONAL ANSWERS

6.7 Chapter Review *(continued)*

54.

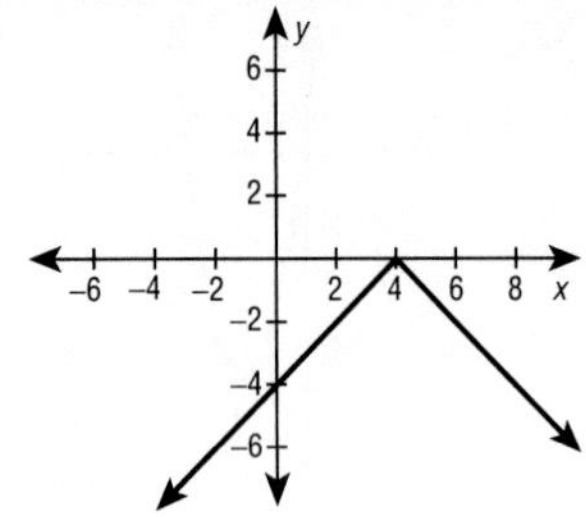

55.

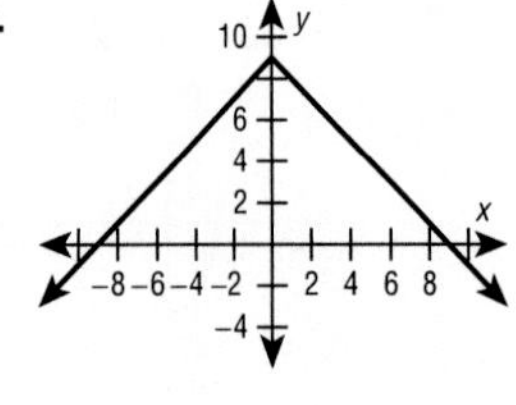

56.

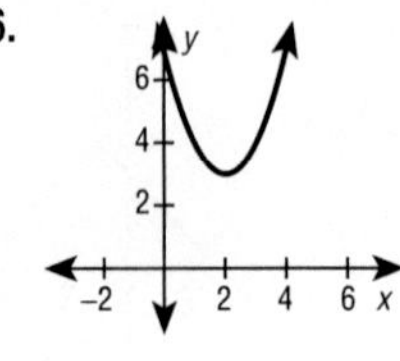

57.

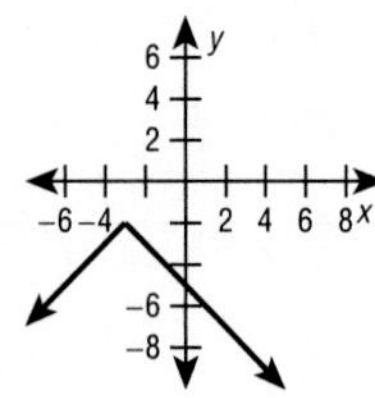

58.

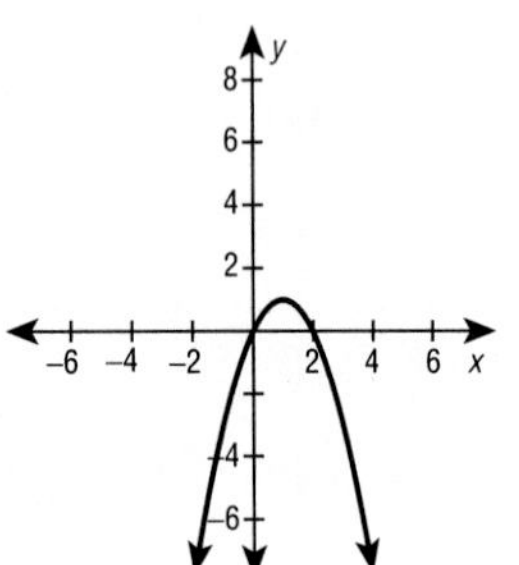

6.7 Chapter Test p. 339

12.

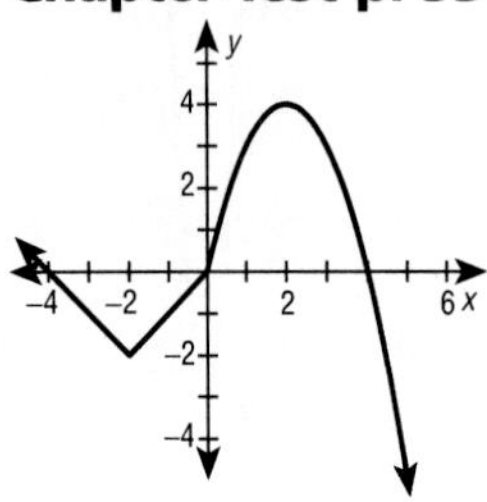

Cumulative Review for Chapters 1–6 p. 343

58.

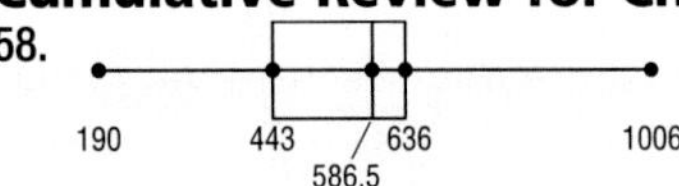

7.1 Communicating About Algebra p. 348

A. $$\frac{180 \cdot \frac{4}{3}\pi\left(\frac{3}{10}\right)^3}{\pi\left(\frac{3}{2}\right)^2\left(\frac{11}{2}\right)} = \frac{180 \cdot \frac{4}{3}\pi\left(\frac{27}{1000}\right)}{\pi\left(\frac{9}{4}\right)\left(\frac{11}{2}\right)}$$ Power of a Quotient Property

$$= \frac{180 \cdot 4 \cdot 27 \cdot 4 \cdot 2}{3 \cdot 1000 \cdot 9 \cdot 11} = \frac{144}{275}$$ Simplify

B. 8000, $\frac{1000}{n} = \frac{6^3}{12^3}$, $n = 8000$

7.2 Independent Practice p. 357

23.

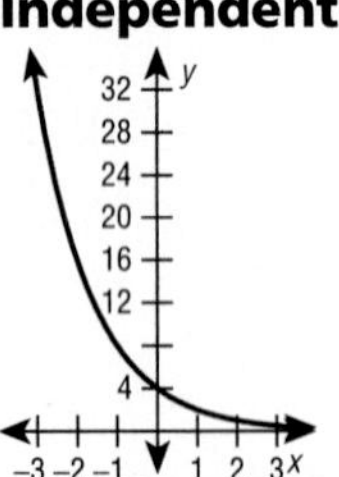

24.

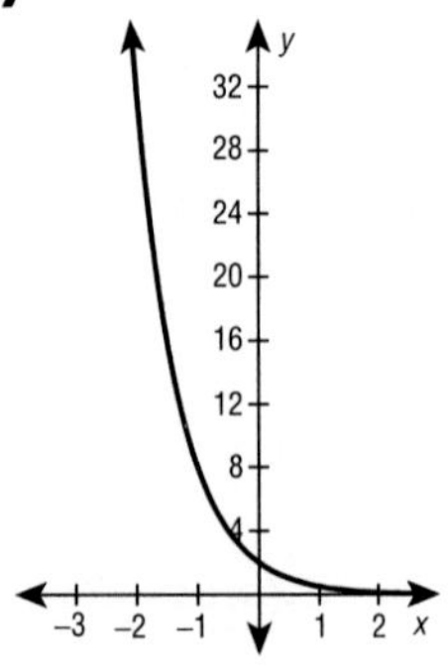

25.

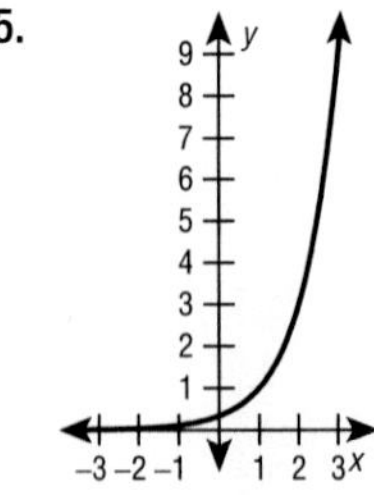

26.

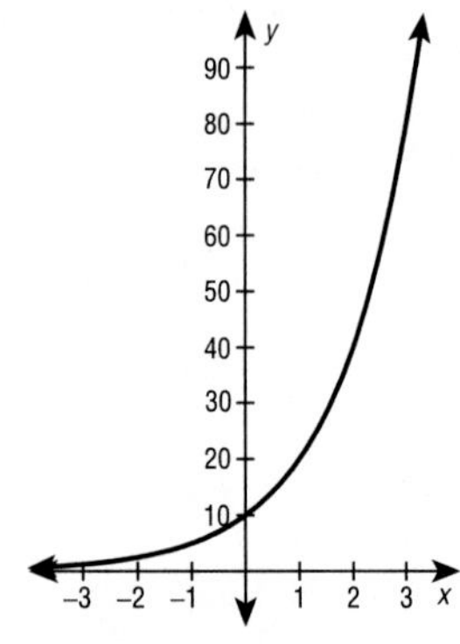

7.2 Independent Practice *(continued)*

27. 2146.86, 2343.32, 2536.48, 2745.57, 2971.89, 3216.87

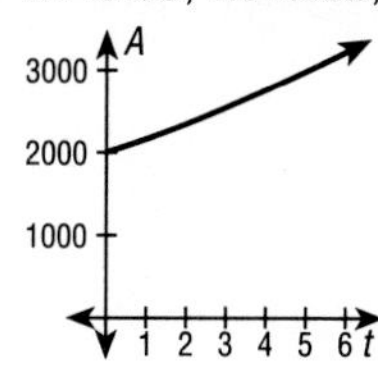

28. 2160, $\approx$2333, $\approx$2519, $\approx$2721, $\approx$2939, $\approx$3174

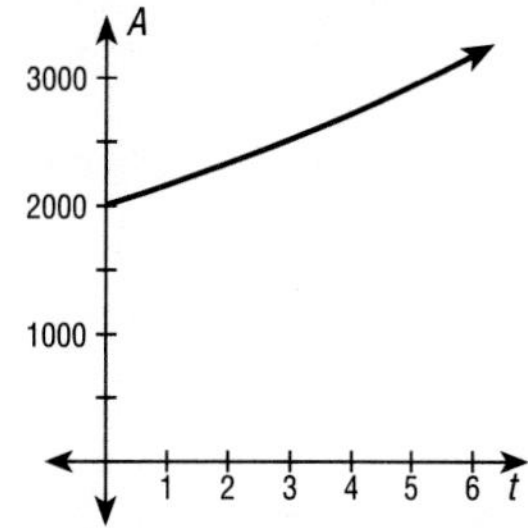

7.5 Communicating About Algebra p. 377

A. $b = (\sqrt{5} - 2) - (-1) = \sqrt{5} - 1$

$a = 1 - (\sqrt{5} - 2) = -\sqrt{5} + 3$

$\frac{\sqrt{5} - 1}{-\sqrt{5} + 3} \stackrel{?}{=} \frac{\sqrt{5} + 1}{2}$

$2\sqrt{5} - 2 \stackrel{?}{=} -5 + 3\sqrt{5} - \sqrt{5} + 3$

$2\sqrt{5} - 2 = 2\sqrt{5} - 2$

B. $\frac{2}{\sqrt{5} - 1} \cdot \frac{\sqrt{5} + 1}{\sqrt{5} + 1} = \frac{2(\sqrt{5} + 1)}{5 - 1} = \frac{2(\sqrt{5} + 1)}{4} = \frac{\sqrt{5} + 1}{2}$

7.5 Guided Practice p. 378

1. Isolate the radical on one side, raise both sides to the same power to eliminate the radical, solve the resulting equation.

2. $\sqrt[3]{2x + 10} = 2$ Isolate the radical by subtracting 1 from both sides.

$2x + 10 = 8$ Raise both sides to the third power to eliminate the radical.

$x = -1$ Solve the resulting equation.

4. $d = \sqrt{(6 - (-2))^2 + (1 - 5)^2} = \sqrt{80}$;
$d^2 = (6 - (-2))^2 + (1 - 5)^2$, $d^2 = 80$, $d = \sqrt{80}$

7.5 Independent Practice p. 379

49.

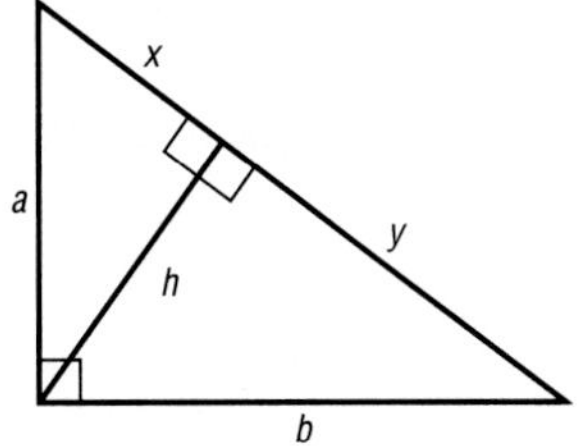

h is the geometric mean of x and y.

$(x + y)^2 = a^2 + b^2$

$x^2 + 2xy + y^2 = x^2 + h^2 + y^2 + h^2$

$2xy = 2h^2$

$xy = h^2$

$h = \sqrt{xy}$

7.5 Mixed Review p. 380

12.

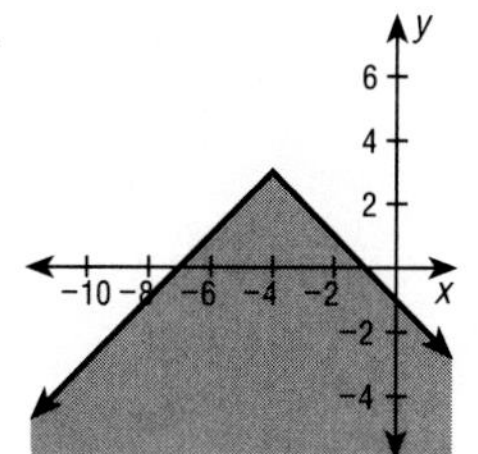

15.

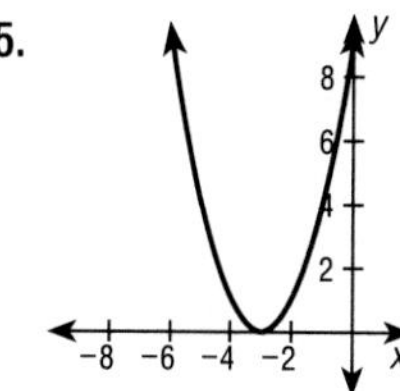

16.

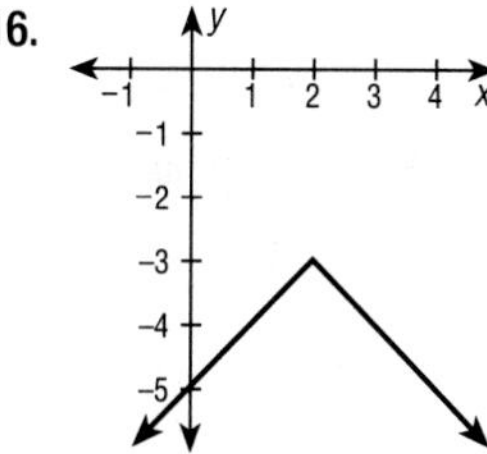

7.6 Communicating About Algebra p. 383

A. Translate the graph of $\sqrt{x}$ down 2 units and to the right 5 units.

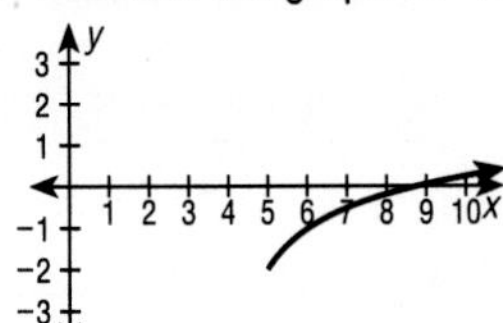

B. Translate the graph of $\sqrt[3]{x}$ up 3 units and to the left 2 units.

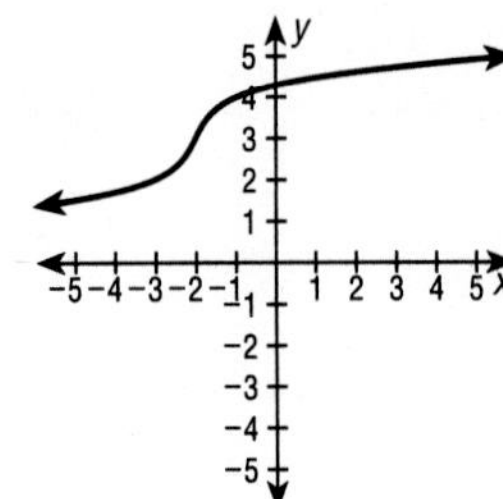

7.6 Independent Practice pp. 385–386

17.

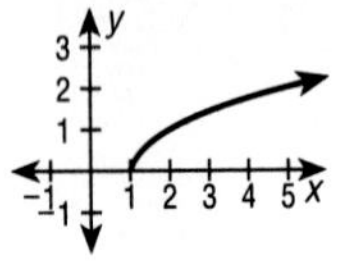

$x \geq 1$, $y \geq 0$

18.

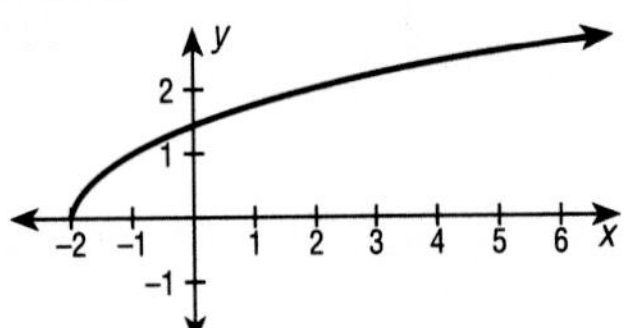

$x \geq -2$, $y \geq 0$

19.

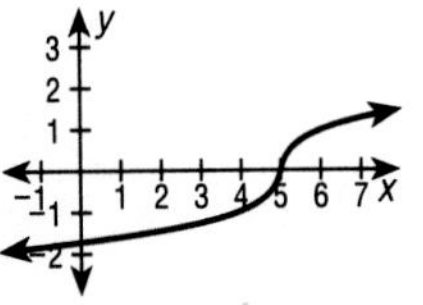

All real numbers, all real numbers

20.

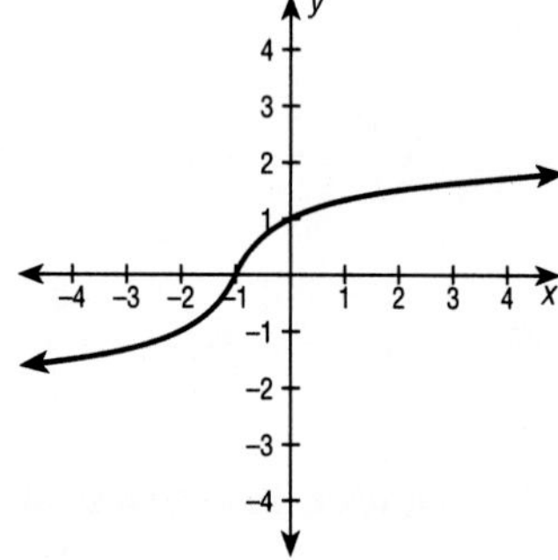

All real numbers, all real numbers

21.

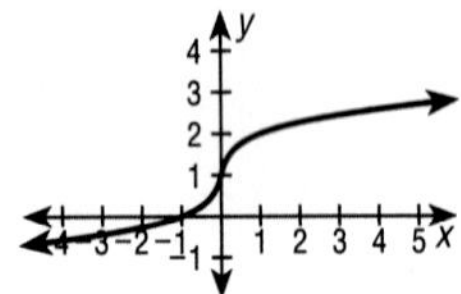

All real numbers, all real numbers

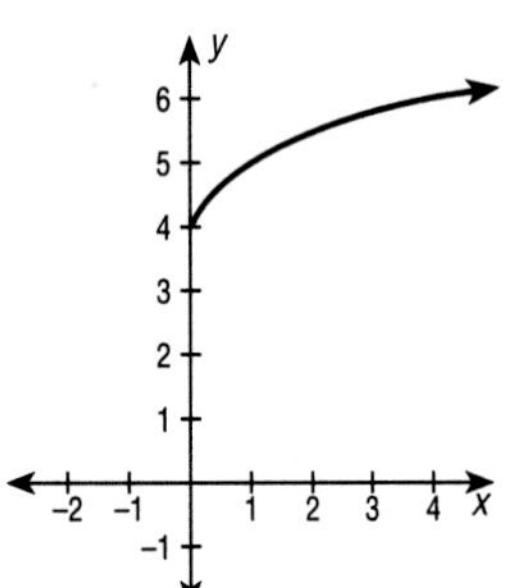

$x \geq 0$, $y \geq 4$

23.

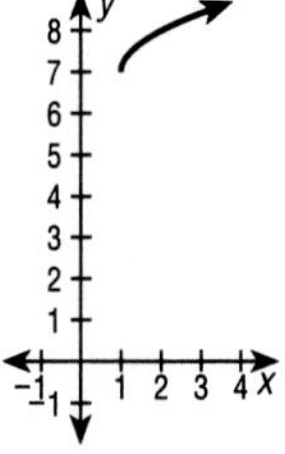

$x \geq 1$, $y \geq 7$

24.

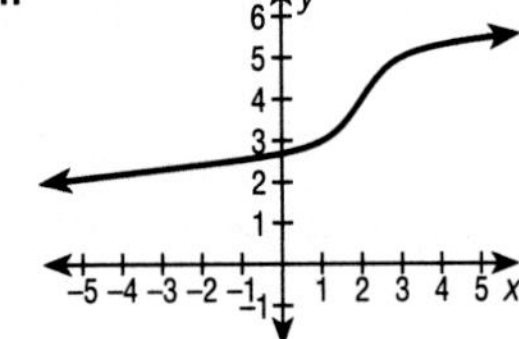

All real numbers, all real numbers

25.

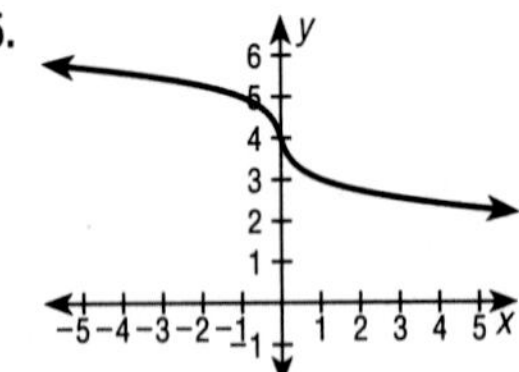

All real numbers, all real numbers

26.

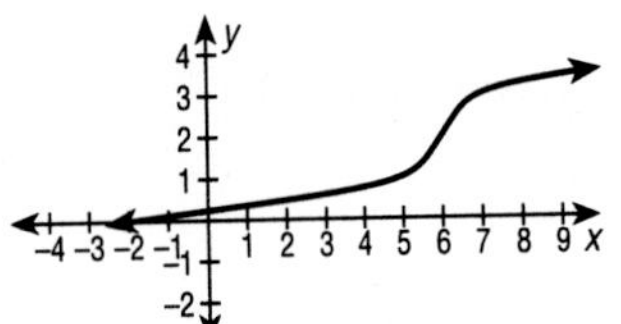

All real numbers, all real numbers

27.

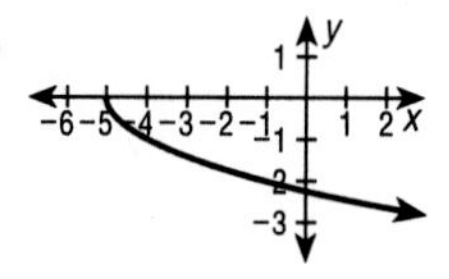

$x \geq -5$, $y \leq 0$

28.

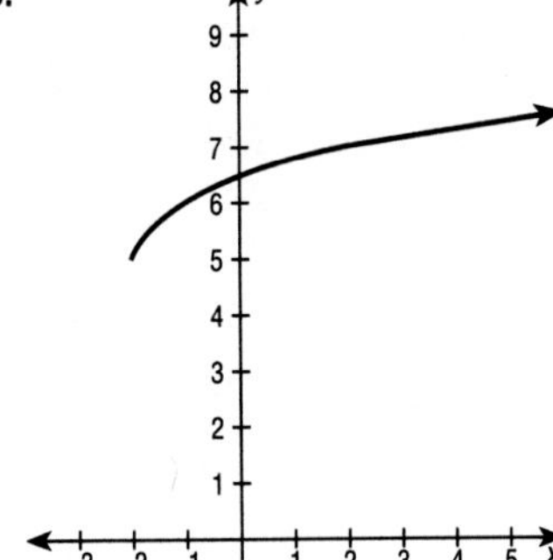

$x \geq -2$, $y \geq 5$

7.6 Independent Practice *(continued)*

29.

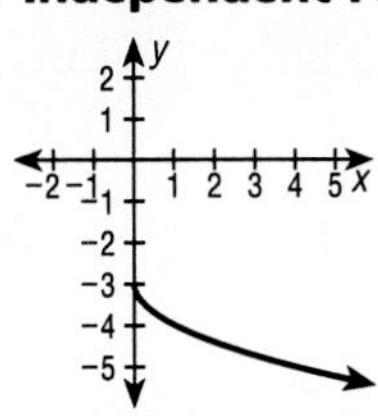

$x \geq 0,\ y \leq -3$

30.

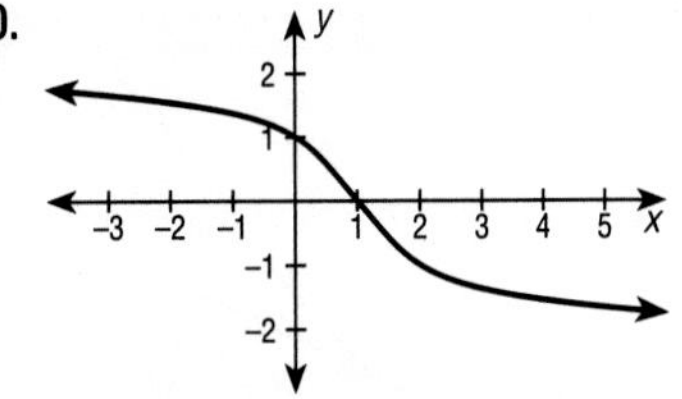

All real numbers, all real numbers

31.

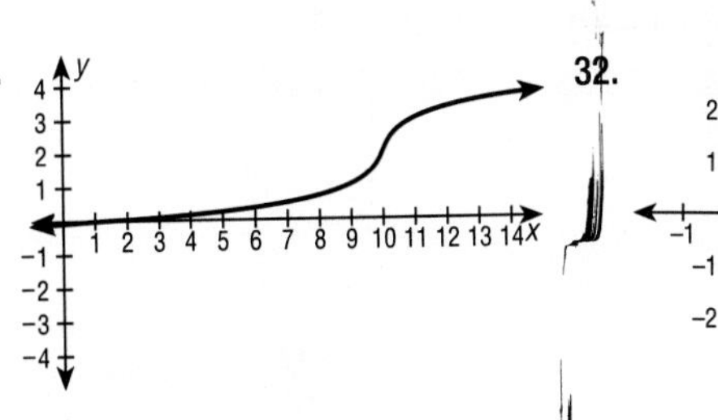

All real numbers, all real numbers

32.

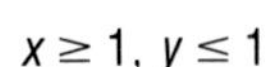

$x \geq 1,\ y \leq 1$

33.

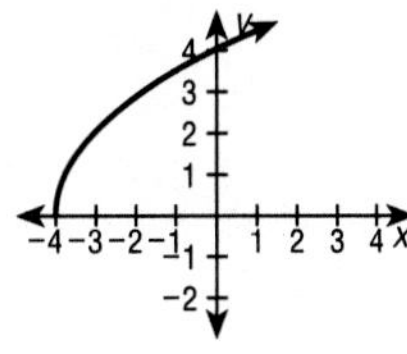

$x \geq -4,\ y \geq 0$

34.

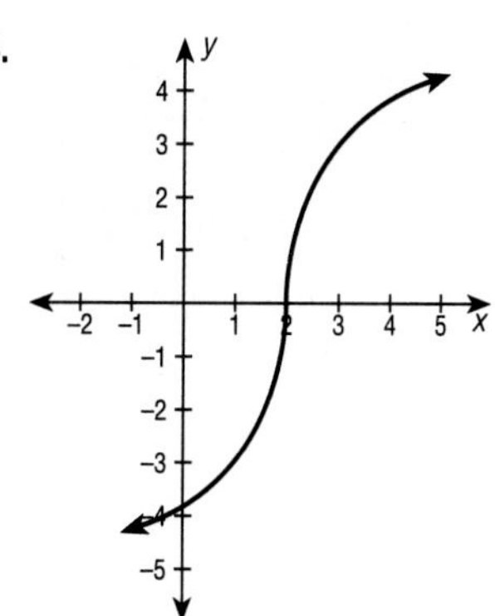

All real numbers, all real numbers

37.

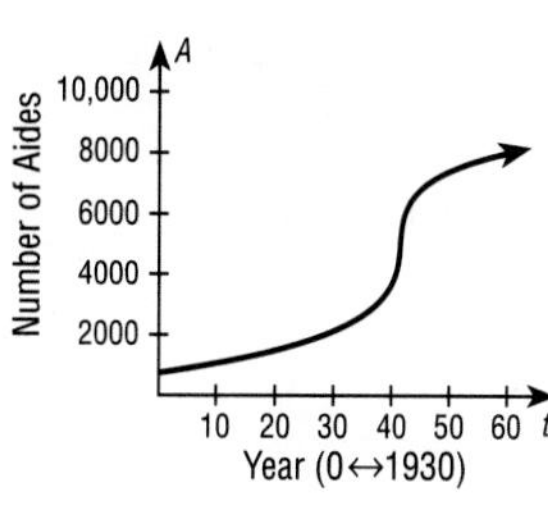

38. $N = \dfrac{244\sqrt[3]{t-42} + 980}{87}$

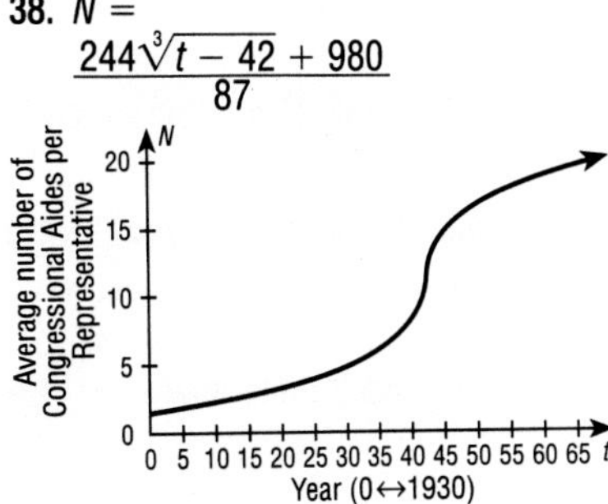

39.

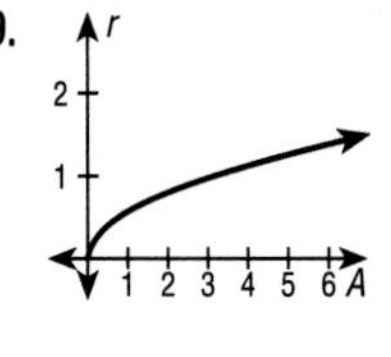

40.

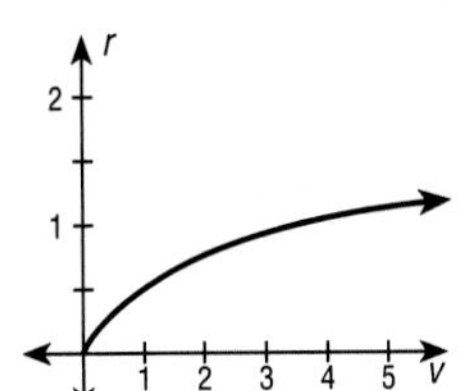

41.

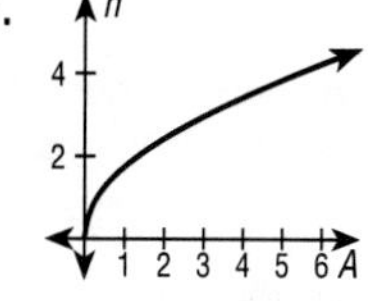

42.

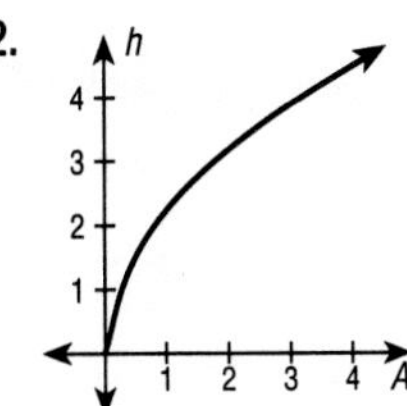

43.

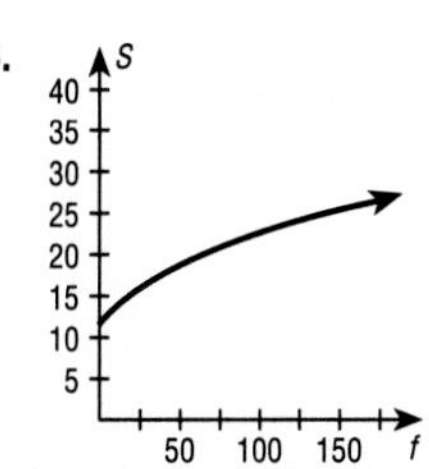

7.6 Exploration and Extension p. 387

57.

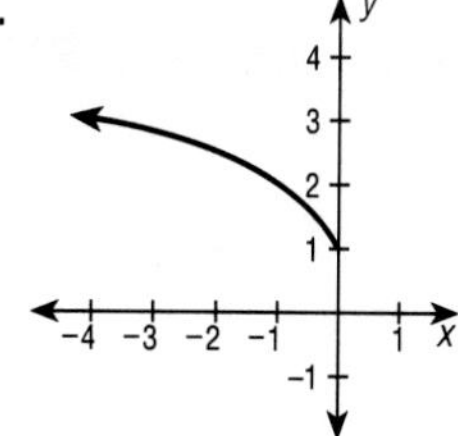

$x \leq 0,\ y \geq 1$

58.

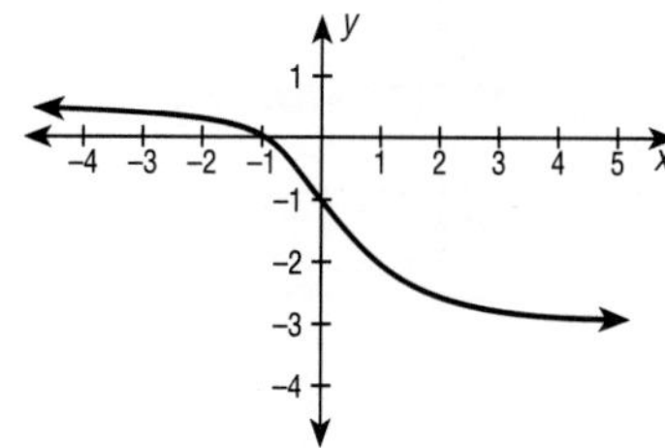

All real numbers, all real numbers

7.6 Using a Graphing Calculator p. 389

1.–4.

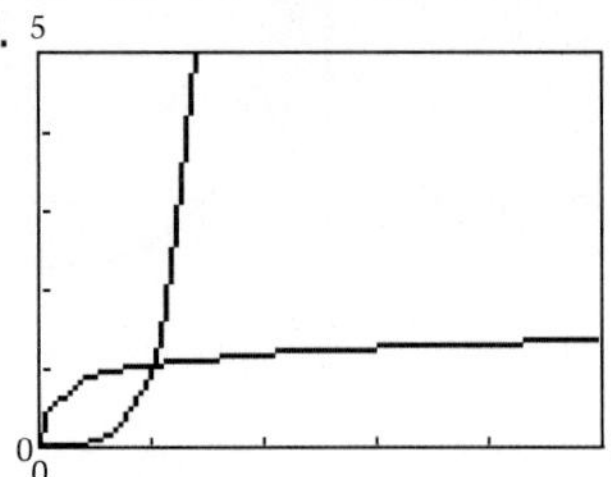

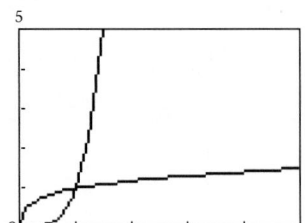

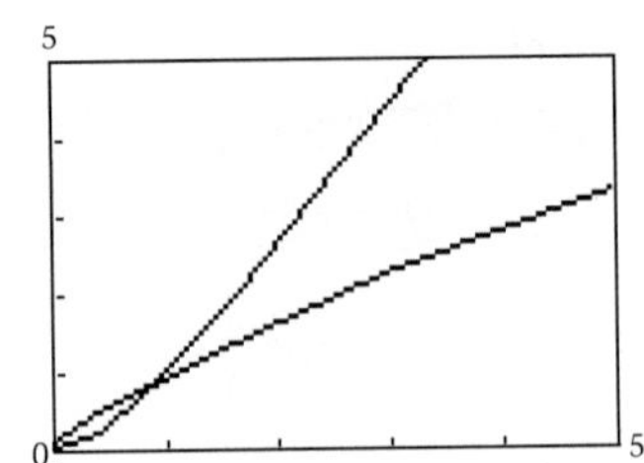

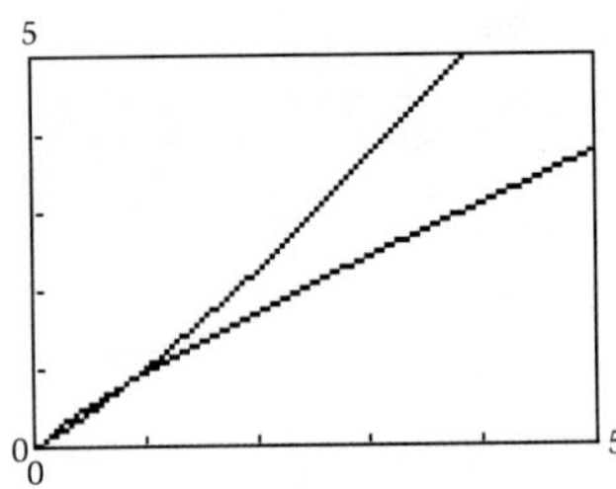

CHAPTER 8

8.1 Communicating About Algebra p. 401

A.

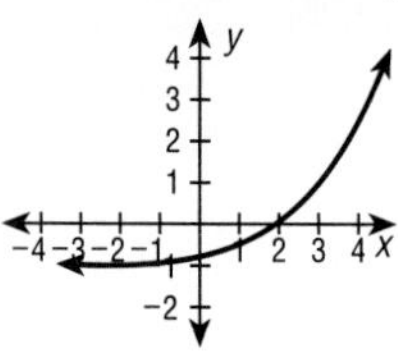

Shift the graph of 2^x two units to the right and one unit down.

B.

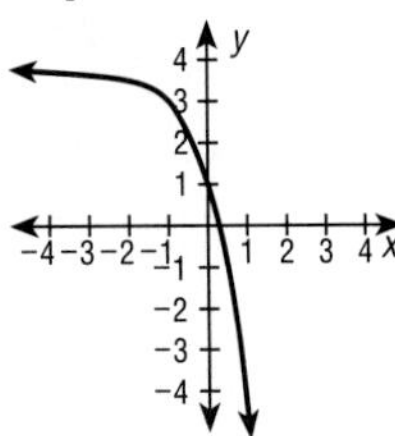

Reflect the graph of 3^x in the x-axis, then shift the graph 1 unit to the left and 4 units up.

8.1 Independent Practice pp. 402–404

5. The graph of g is the graph of f shifted 2 units down.
6. The graph of g is the graph of f shifted 1 unit to the right.
7. The graph of g is the graph of f shifted 6 units up.
8. The graph of g is the graph of f shifted 10 units up.
9. The graph of g is the graph of f reflected in the x-axis and shifted 4 units to the left.
10. The graph of g is the graph of f reflected in the x-axis and shifted 9 units up.
11. The graph of g is the graph of f shifted 4 units to the right and 5 units up.
12. The graph of g is the graph of f shifted 2 units to the left and 2 units down.

21.

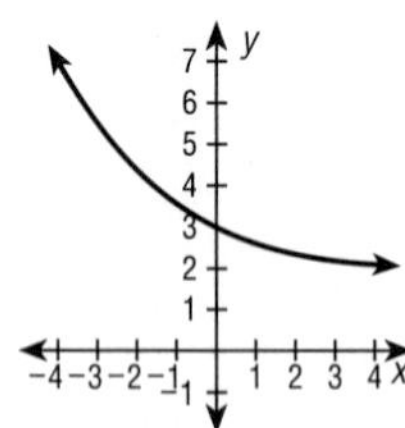

22.

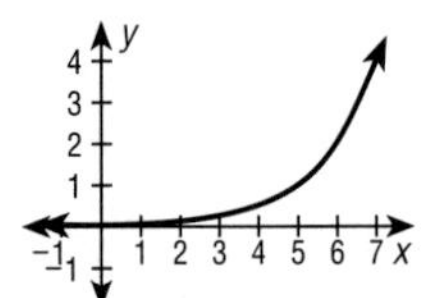

23.

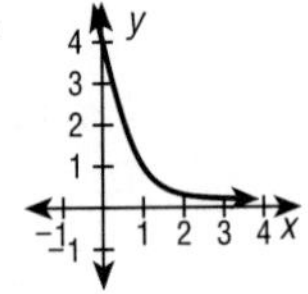

24.

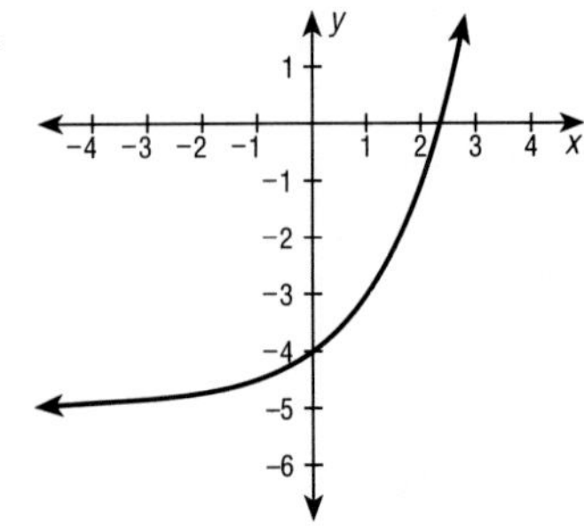

25.

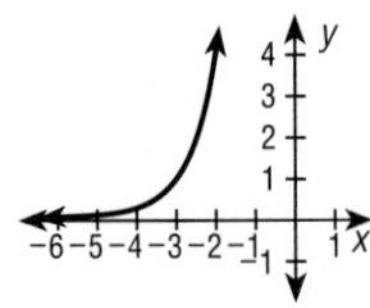

26.

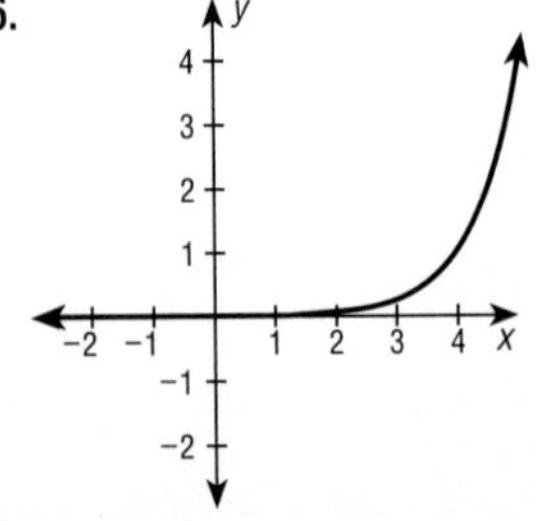

27.

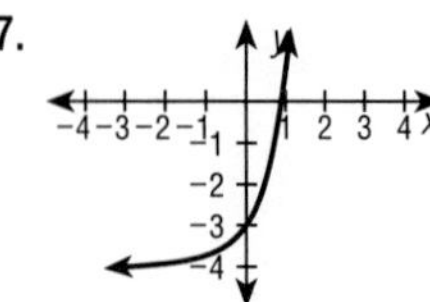

28.

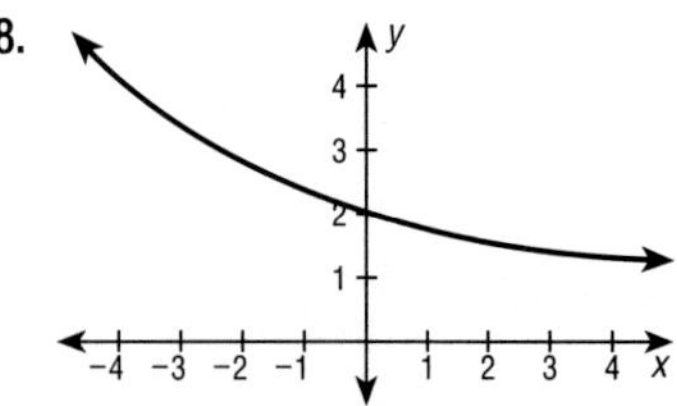

8.1 Independent Practice *(continued)*

29.
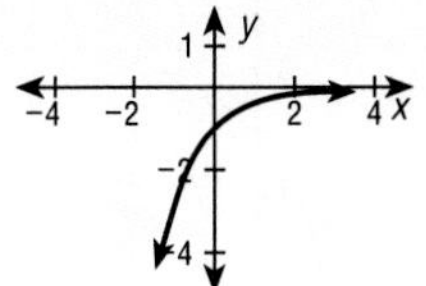

30.
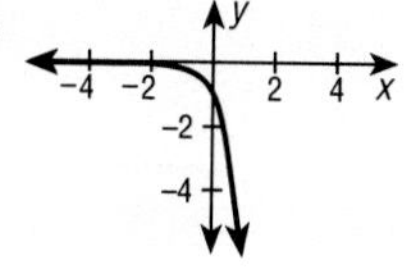

31.
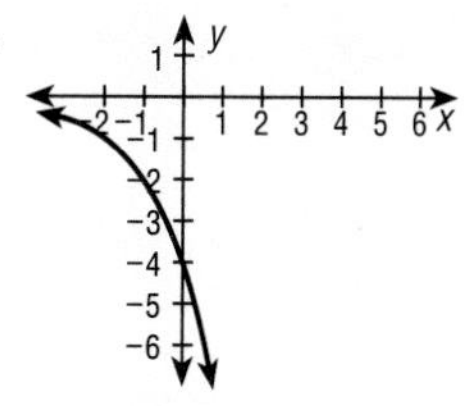

32.
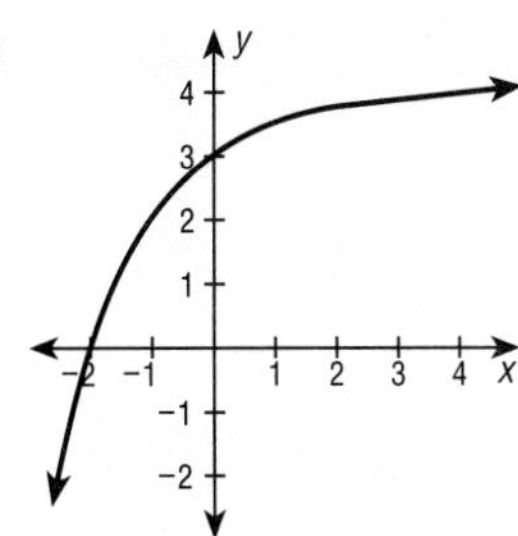

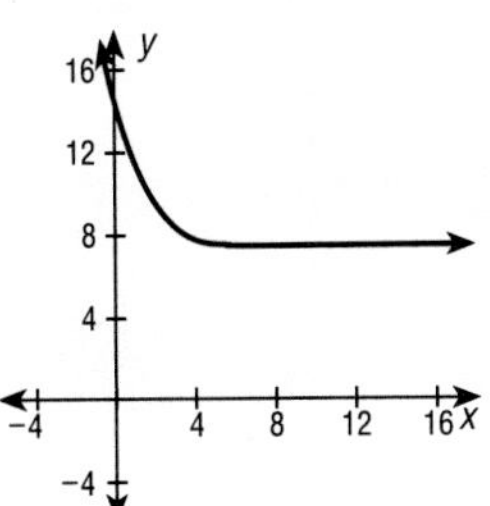

34.
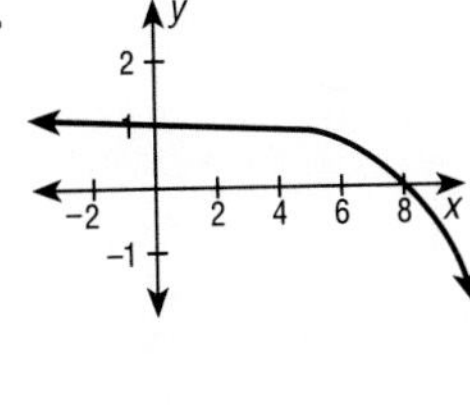

35.
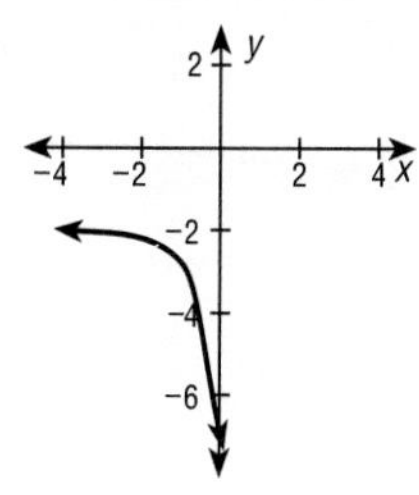

36.
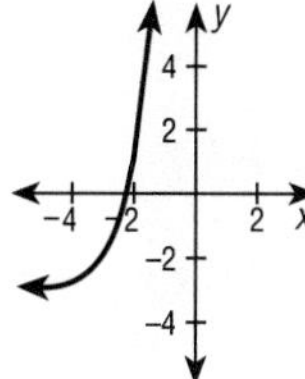

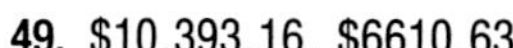

49. \$10,393.16, \$6610.63

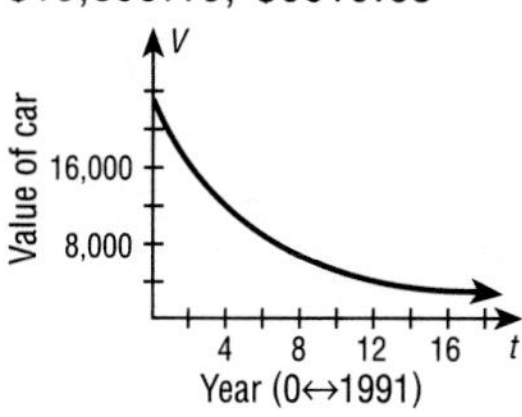

50. \$121,366.73, \$160,507.50

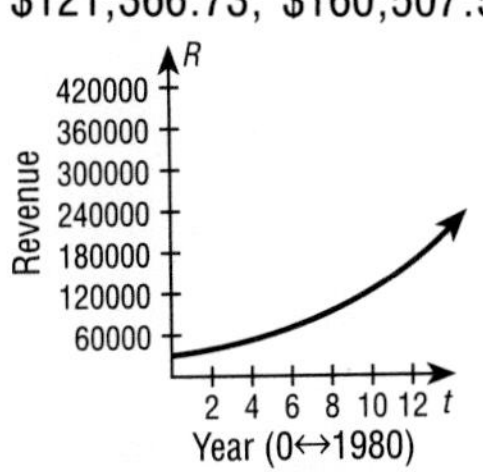

52.
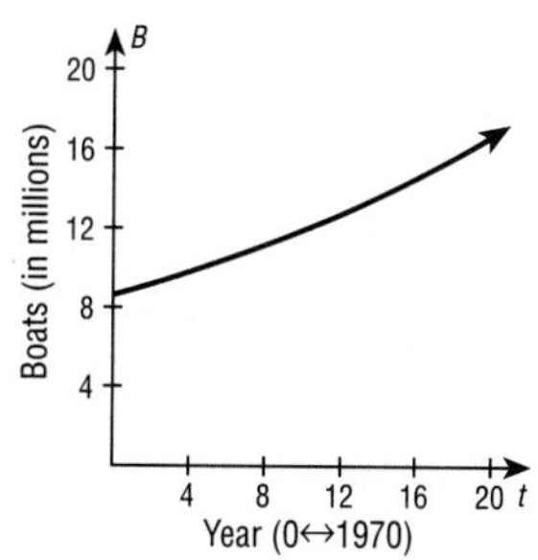

8.1 Integrated Review p. 404

53.
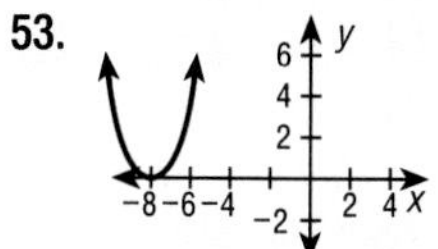

54.
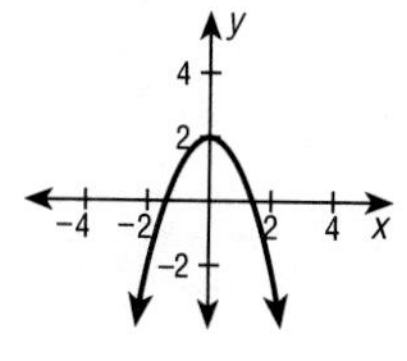

55.
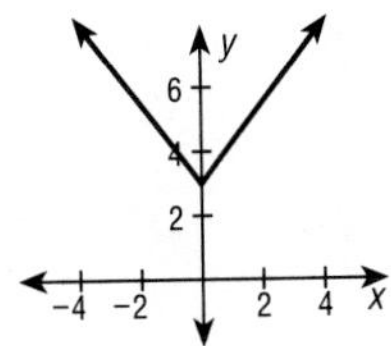

56.
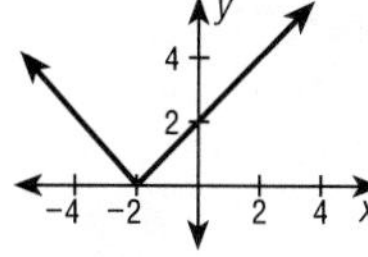

57.
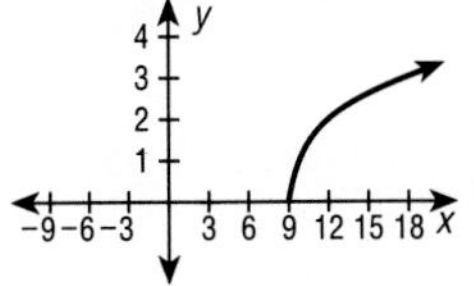

58.
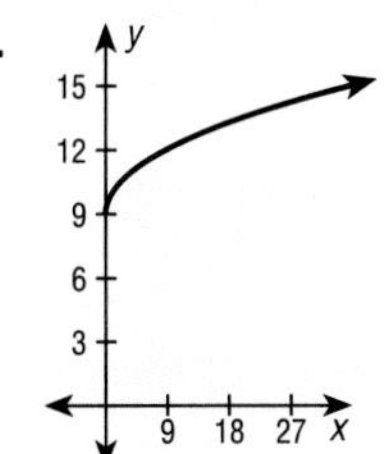

59.
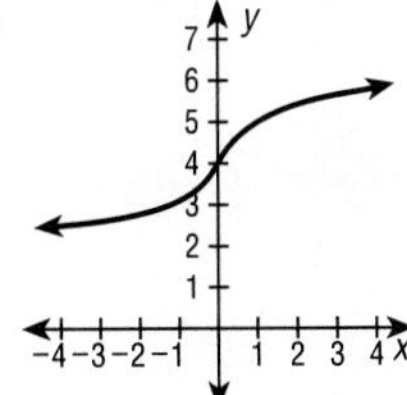

60.
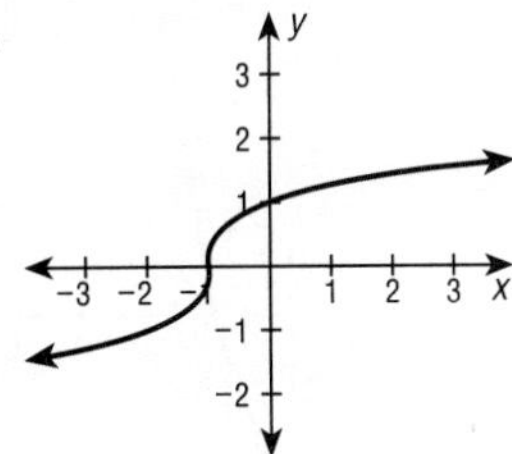

61. exponential decay

62. exponential growth

63. exponential growth

64. exponential decay

ADDITIONAL ANSWERS

8.1 Exploration and Extension p. 404

72.

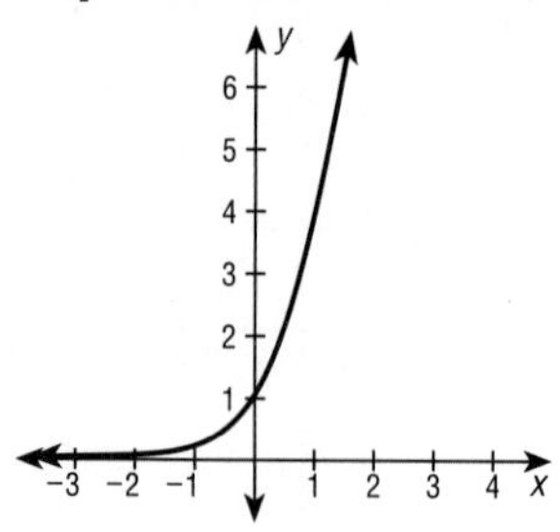

8.2 Independent Practice p. 410

67.

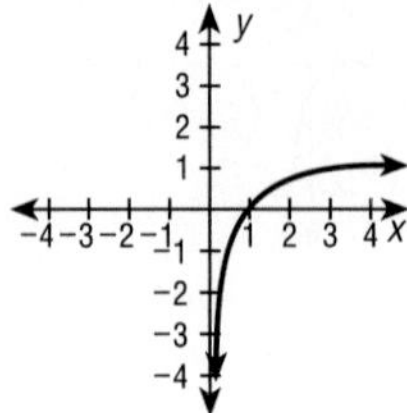

68.

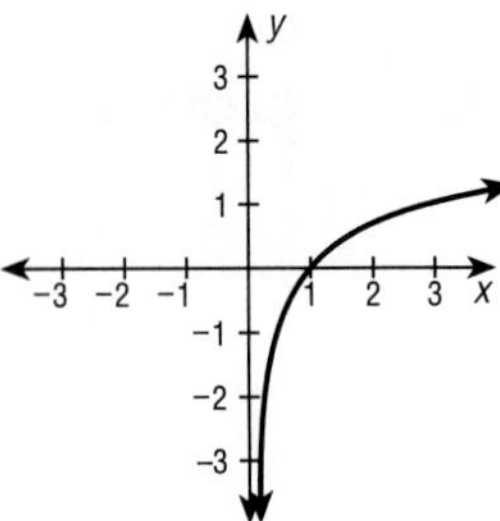

69.

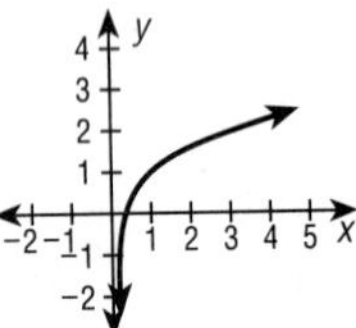

70.

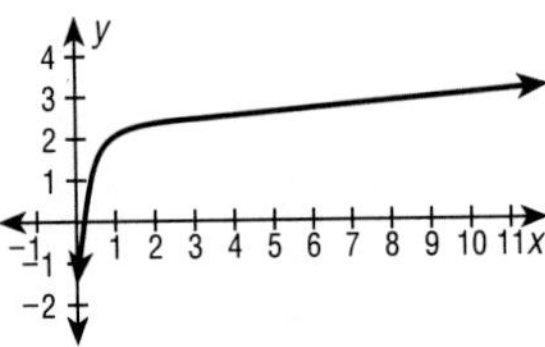

71.

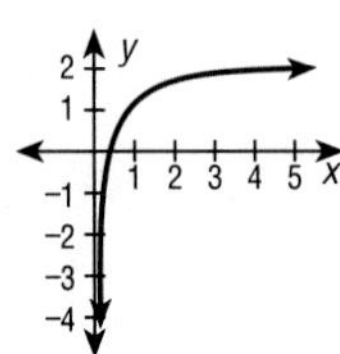

72.

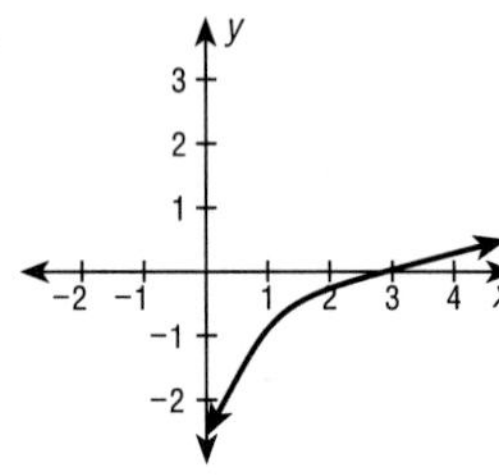

73.

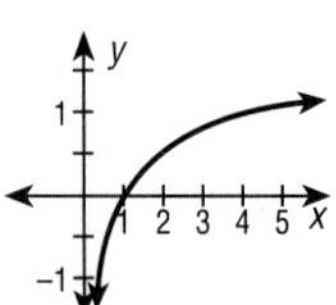

74.

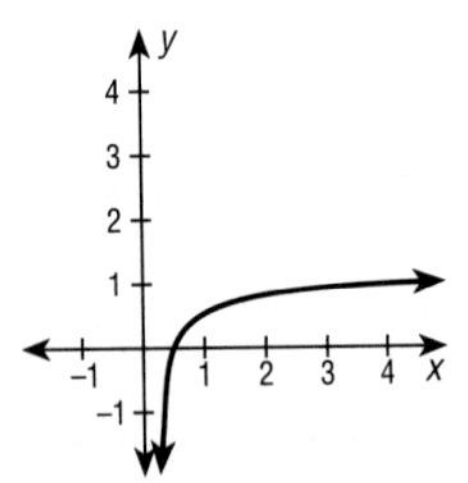

75.

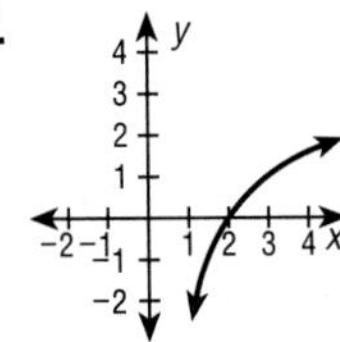

76.

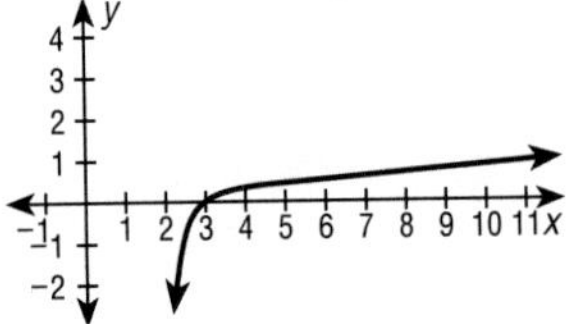

77.

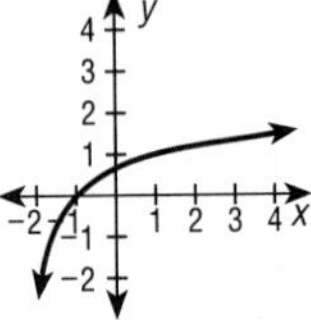

78.

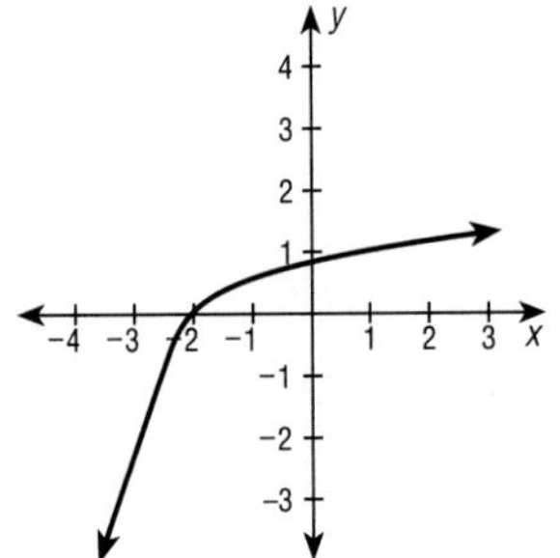

8.2 Integrated Review p. 411

84.

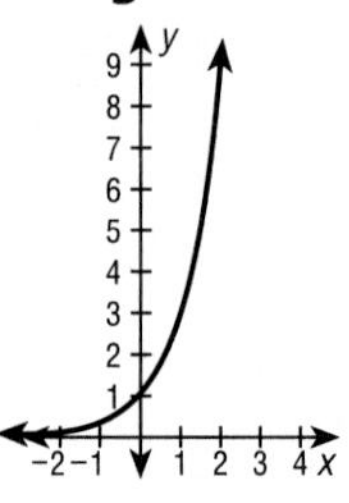

85.

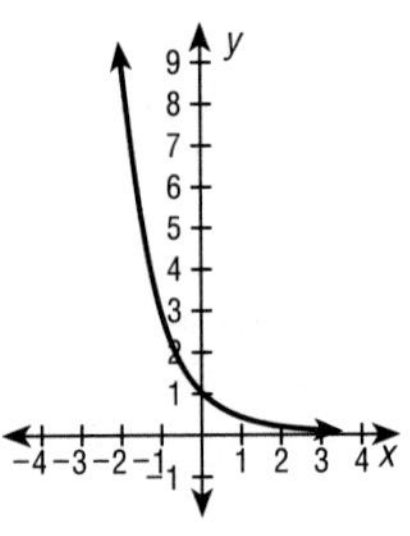

86.

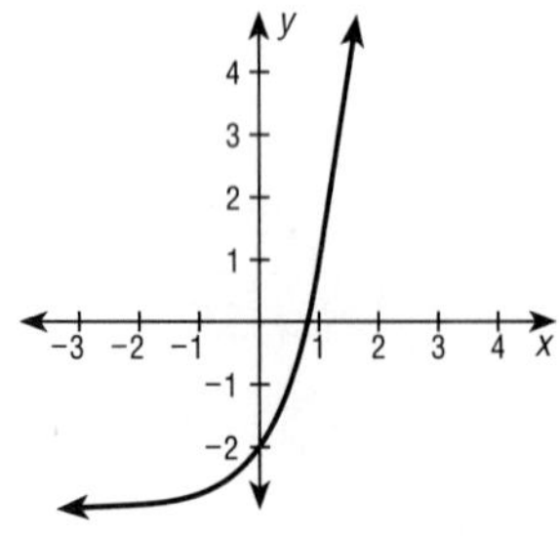

87.

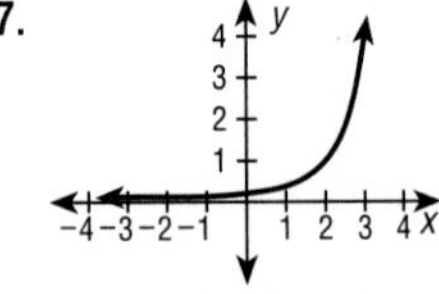

88.

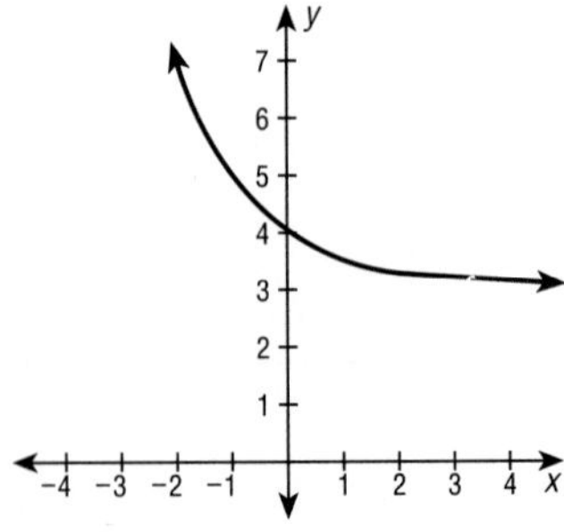

ADDITIONAL ANSWERS

8.2 Exploration and Extension p. 411

91. $\log_a b = \dfrac{\log_{10} b}{\log_{10} a} = \dfrac{1}{\dfrac{\log_{10} a}{\log_{10} b}} = \dfrac{1}{\log_b a}$

8.4 Communicating About Algebra p. 422

A.

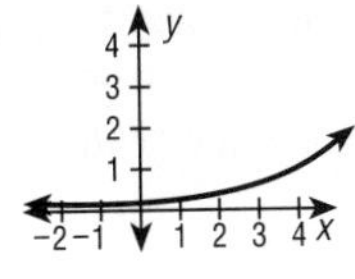

Exponential growth

B.

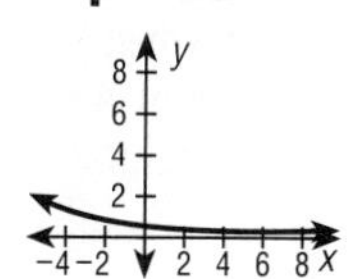

Exponential decay

8.4 Guided Practice p. 423

3.

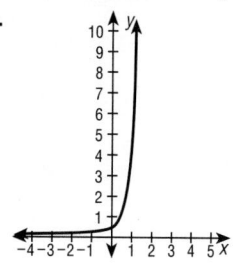

Exponential growth

8.4 Independent Practice p. 424

33.

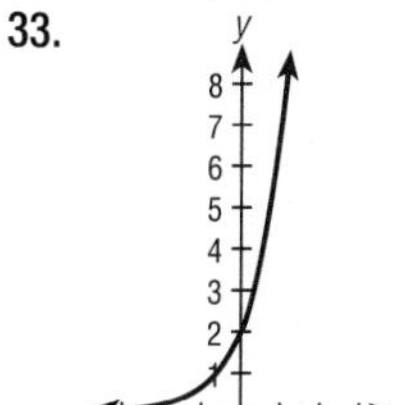

34.

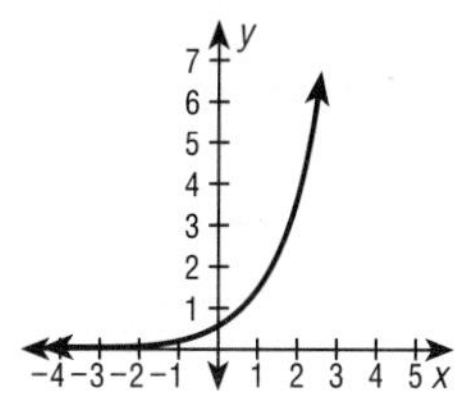

35.

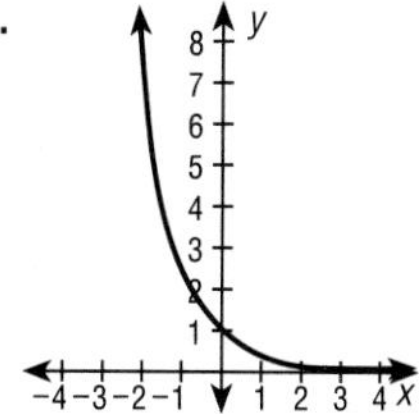

36.

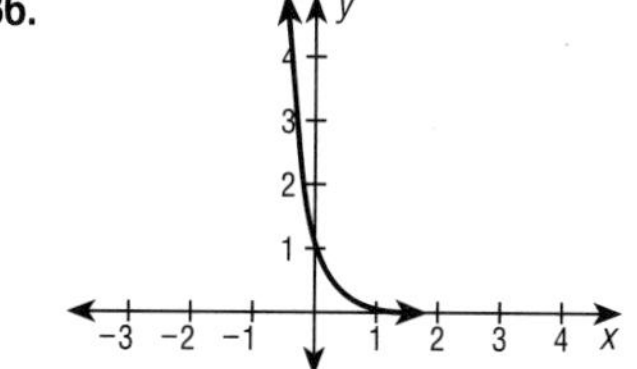

37.

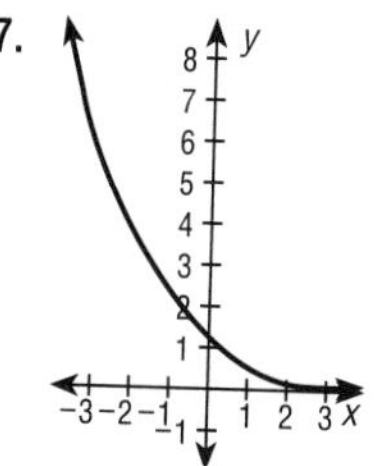

38.

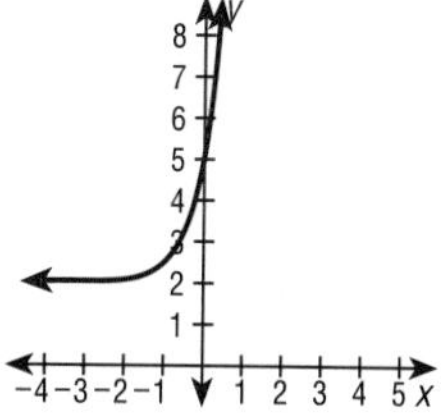

39.

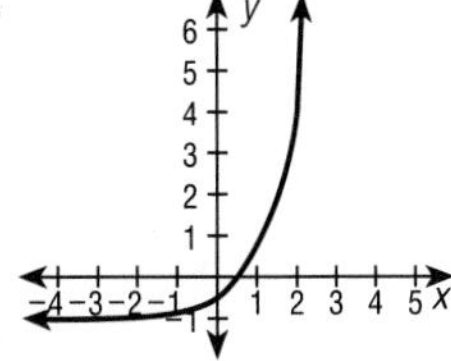

40.

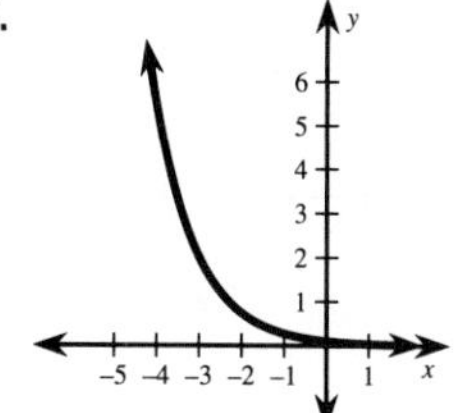

41.

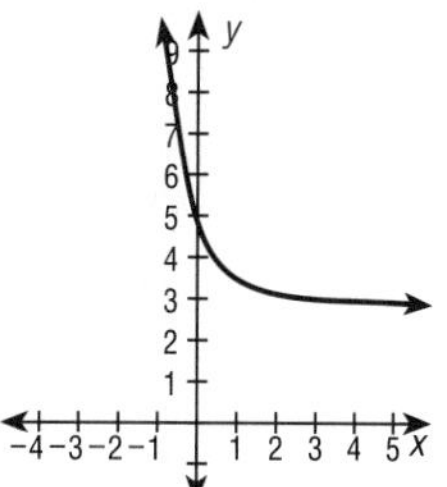

42.

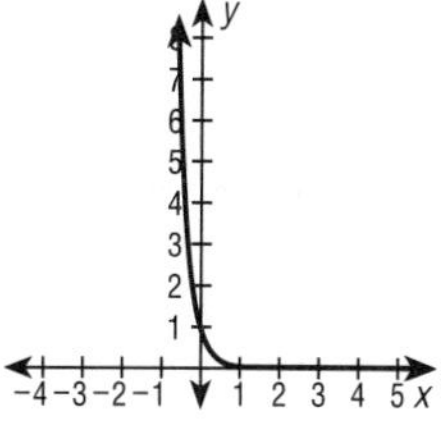

43.

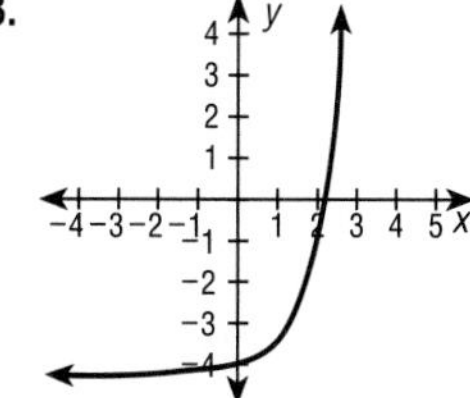

44.

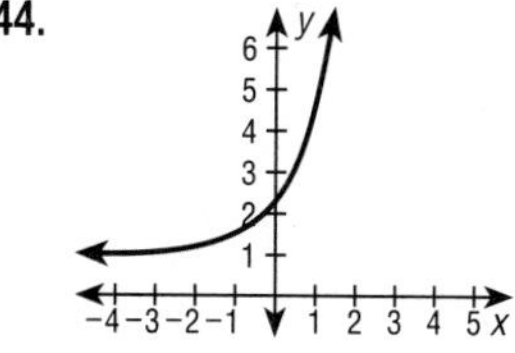

51.

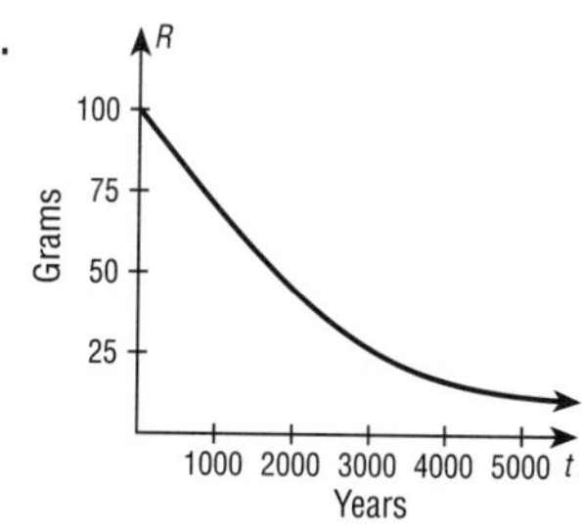

8.4 Integrated Review p. 425

59. The value of A when $t = 0$ is the same for the model and the graph; the value of A when $t = 10$ is 1,149,254 for the model and 1,149,159 for the graph; the model does not show the slight decline from 1981 to 1982 shown on the graph.

60. The value of A when $t = -15$ is the same for the model and the graph; the value of A when $t = 20$ is 124,975 for the model and 124,973 for the graph; the model does not show the declines from 1955 to 1960 and from 1965 to 1970 shown on the graph; the model does not show the regularity from 1975 to 1980 shown on the graph.

8.5 Guided Practice p. 429

3. 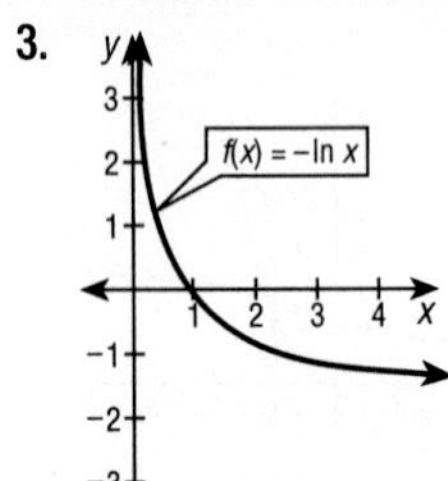

It is the reflection of ln x in the x-axis.

8.5 Independent Practice p. 430

33. 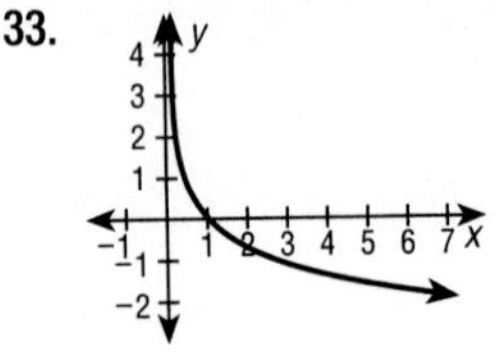

34. 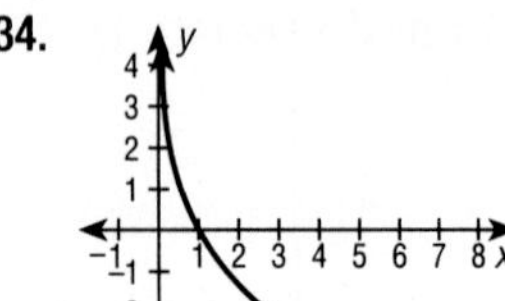

35. 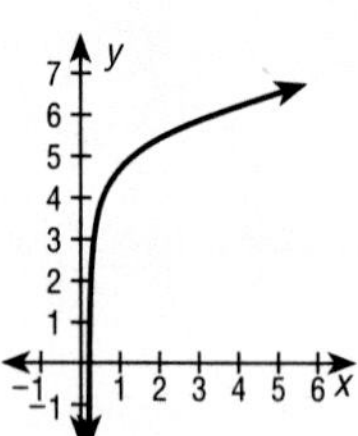

36. 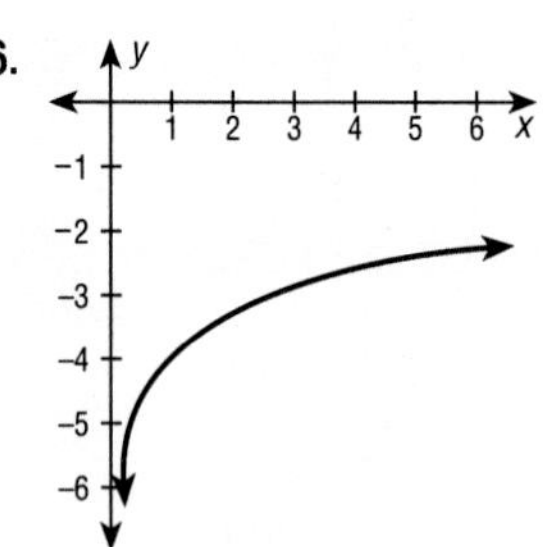

37. 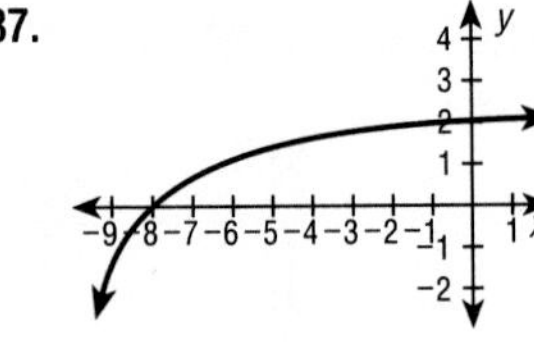

38. 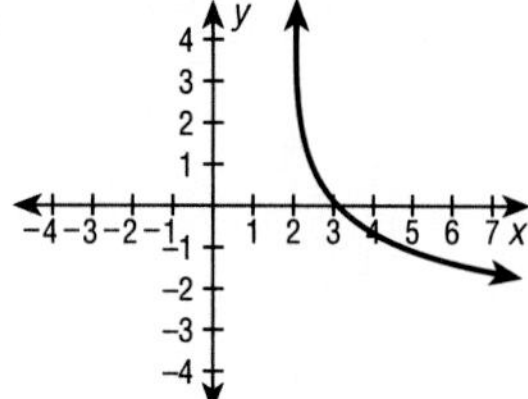

39. 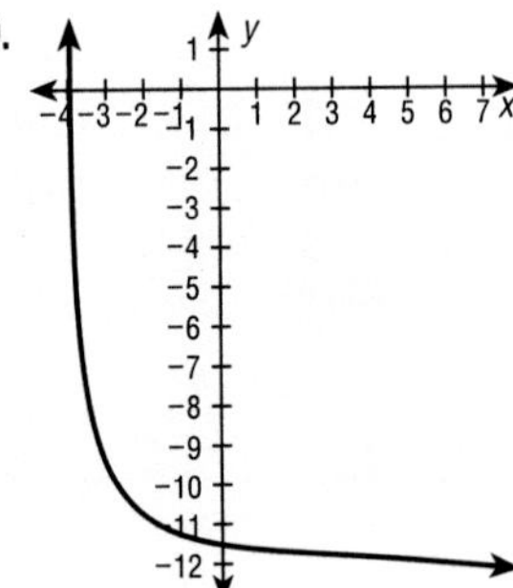

40. 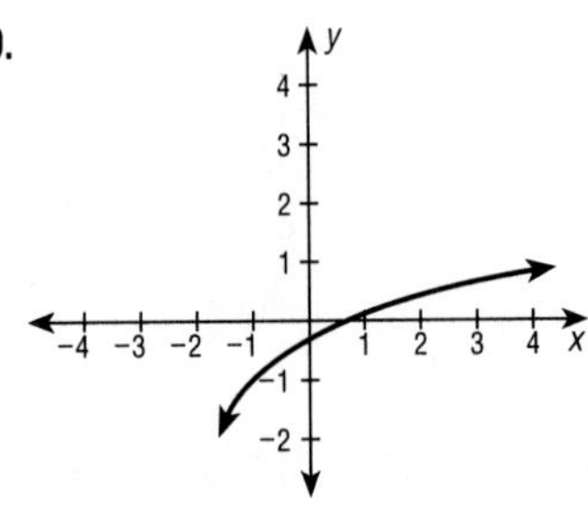

41. 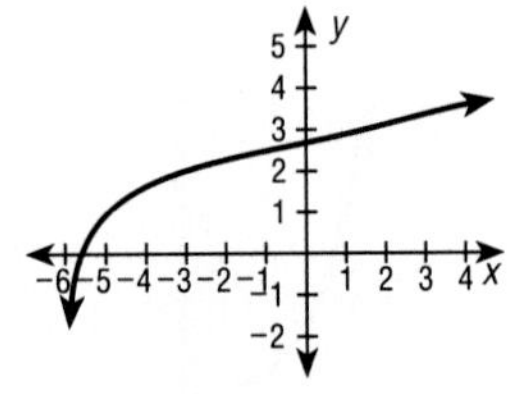

42. 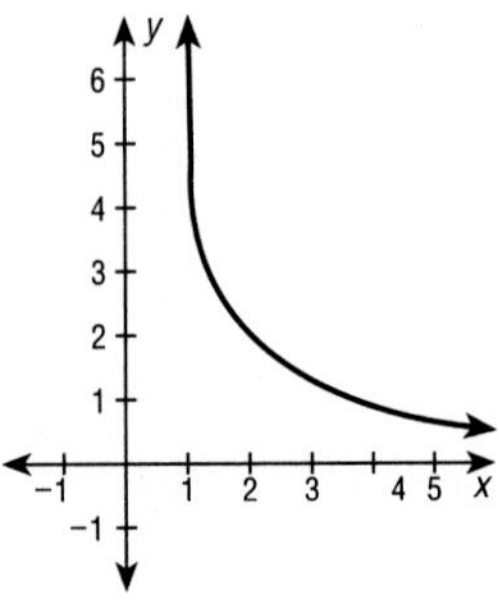

43. 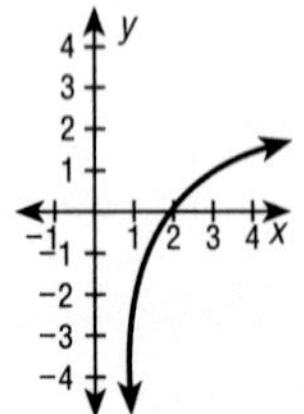

44. 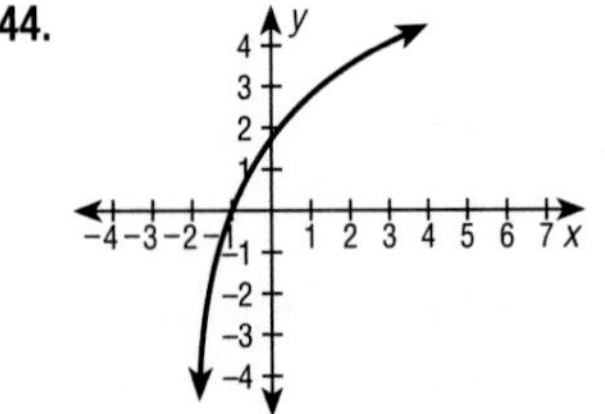

8.5 Using a Graphing Calculator p. 433

1. 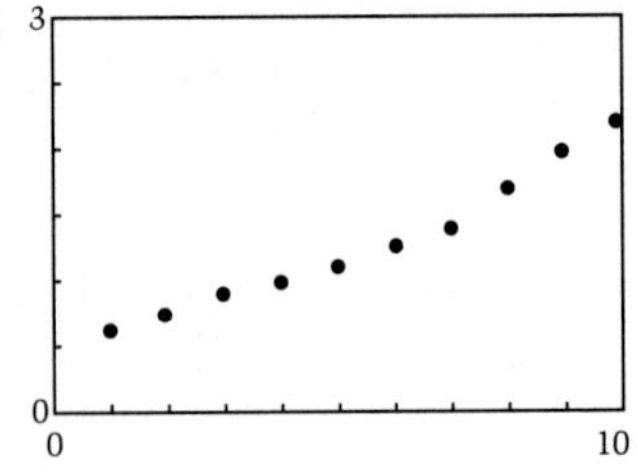

8.5 Using a Graphing Calculator *(continued)*

2.

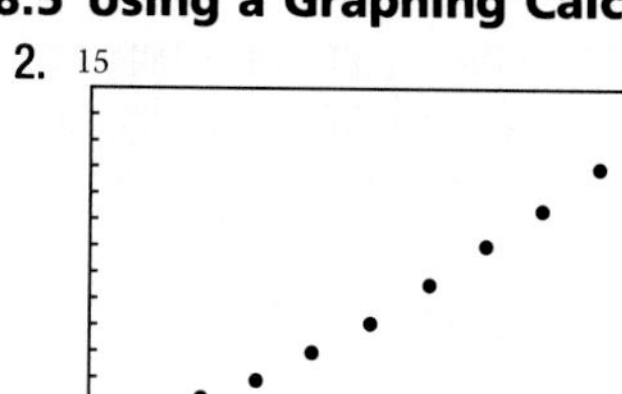

3. 4

0 10

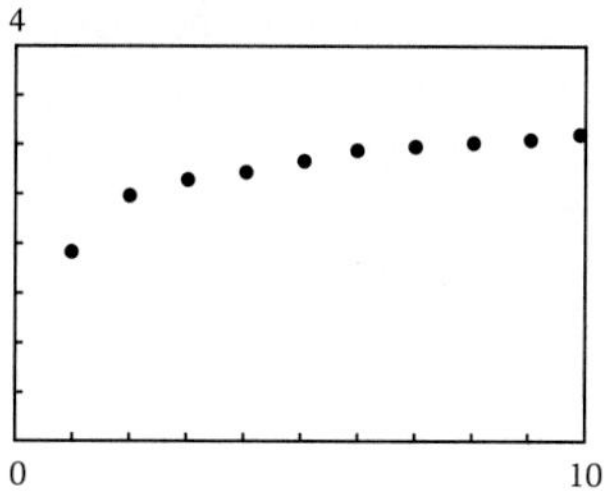

4.

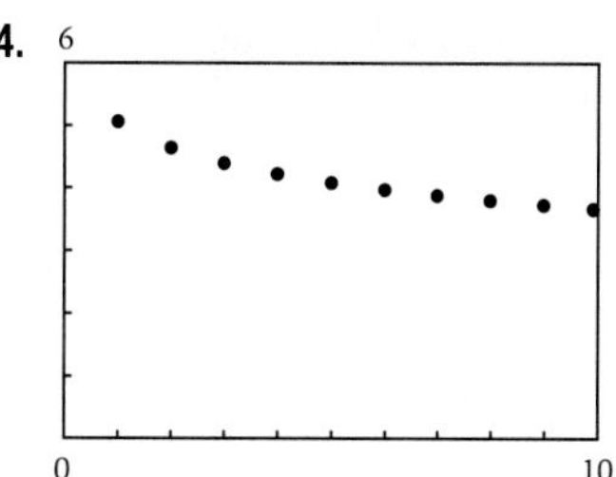

5.

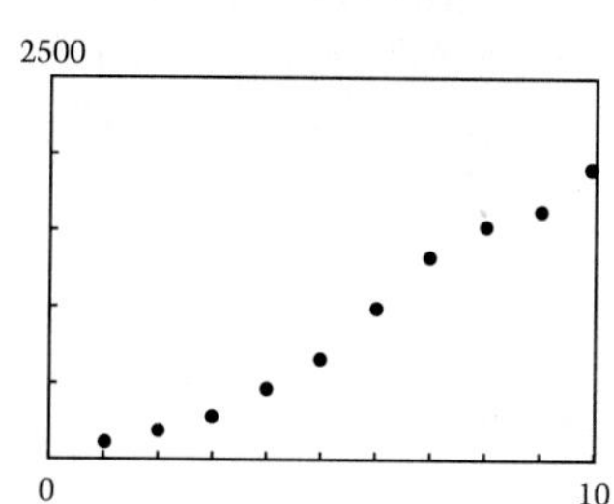

8.6 Exploration and Extension p. 440

63. Each function is a straight line through the origin with slope of 1. Conclusion: $f(x) = g(x) = h(x) = x$

8.6 Mixed Review p. 440

12.

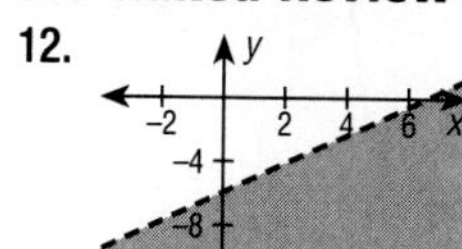

15.

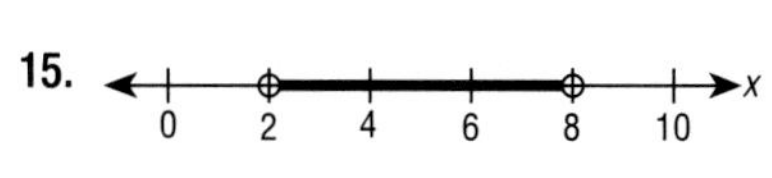

8.7 Communicating About Algebra p. 443

A.

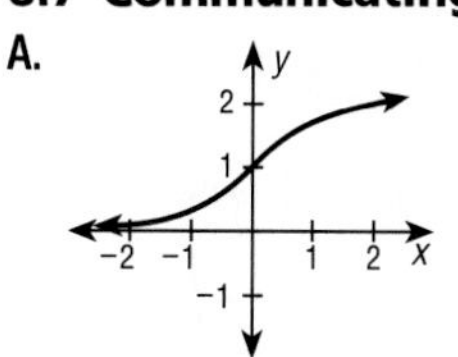

$y = 0$, $y = 2$

B.

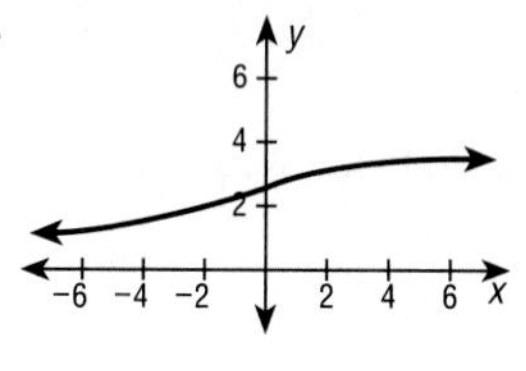

$y = 0$, $y = 5$

8.7 Independent Practice, pp. 444–445

22.

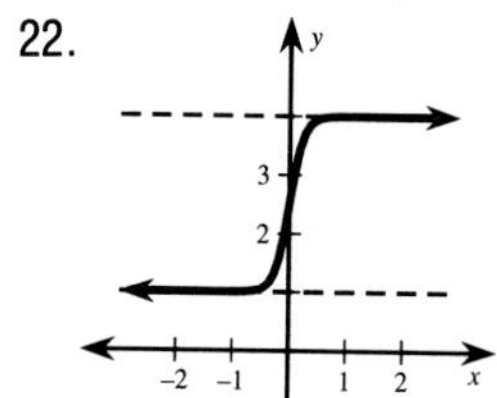

23.

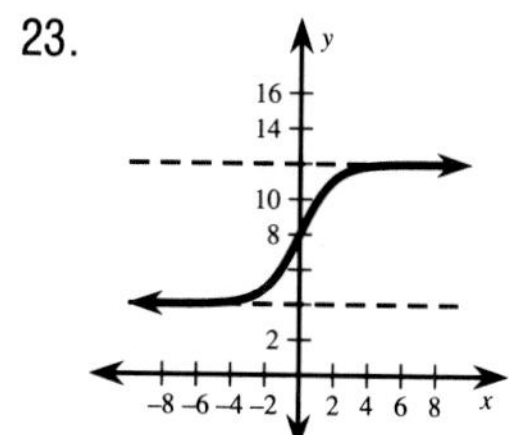

24.

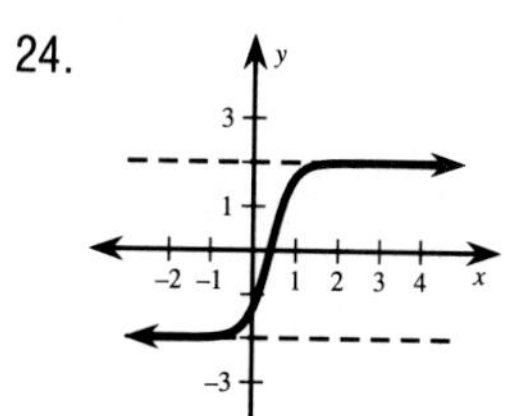

25.

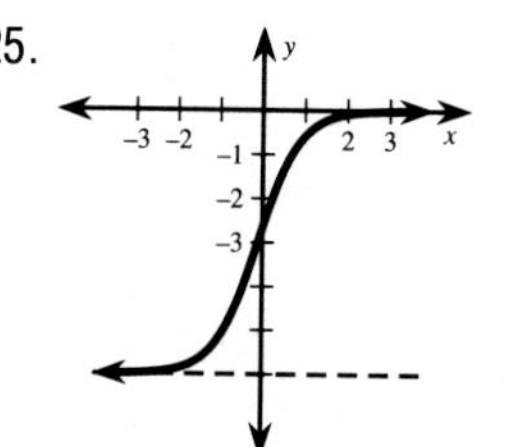

26.

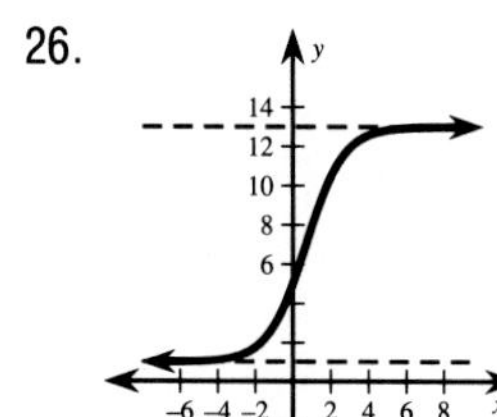

27.

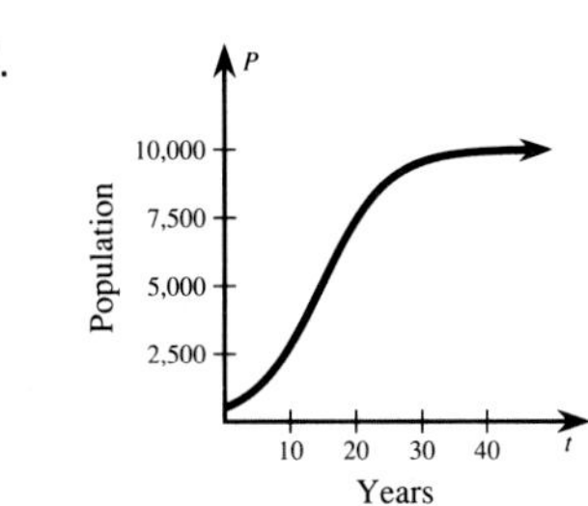

30.

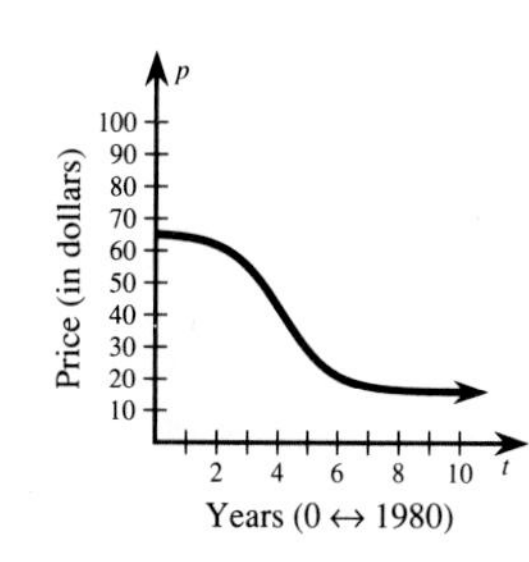

Competition, reduced marketing cost, possibility of renting them

CHAPTER 9

9.1 Independent Practice p. 460

17. $13x^2 + x + 14$, 2

18. $-\frac{5}{3}x^2 + \frac{2}{3}x$, 2

19. $6x^3 + x^2 - 1$, 3

20. $x^3 - 2x^2 + 5x + 2$, 3

21. $-x^3 - 3x^2 + 3$, 3

22. $-4x^3 + 6x^2 - 19x + 18$, 3

23. $11x^2 - 3$

24. $x^3 + 13x - 5$

25. $-4x + 4$

26. $-6x + 9$

27. $-12x^3 - 2x^2 + x + 4$

28. $-x^3 + 36x^2 - 16x - 3$

29. $4x^2 - 13x - 4$

30. $3x^3 + x^2 + 13x - 12$

31. $-5x^3 - x^2 + 5x - 9$

32. $-8x^2 - 6x - 6$

33. $-10x^3 - 3x^2 - 6x + 15$

34. $-8x^3 - 9x^2 + 48x - 7$

35. $20x^3 + 9x^2 - 15x - 30$

36. $-5x^3 - 5x^2 - 14x + 1$

37. $x^3 + 9x^2 - 5x$

38. $12x^3 - 96x^2$

39. $-2x^2 - 8x$

40. $6x^3 - 2x^2 + 12x$

41. $x^2 - 6x + 8$

42. $x^2 + 7x - 8$

43. $x^3 + 2x^2 - x + 6$

44. $x^3 + 3x^2 - 50x + 36$

45. $6x^4 - 3x^3 - 2x^2 + 7x - 3$

46. $12x^3 + 10x^2 + 8x + 2$

47. $x^3 + 7x^2 - 12x + 54$

48. $8x^3 - 18x^2 + 15x - 9$

ADDITIONAL ANSWERS

9.2 Independent Practice pp. 467–469

7. a shift 5 units to the right
8. a shift 3 units up
9. a reflection in the x-axis
10. a shift 1 unit to the left
15. falls to the left and falls to the right.
16. falls to the left and falls to the right.
17. rises to the left and rises to the right.
18. falls to the left and rises to the right.
19. falls to the left and falls to the right.
20. falls to the left and rises to the right.
21. falls to the left and rises to the right.
22. rises to the left and falls to the right.
23. falls to the left and rises to the right.
24. rises to the left and rises to the right.

31. 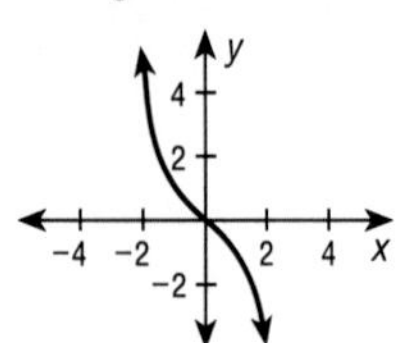

32. 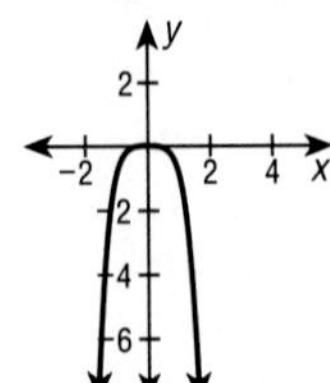

33.

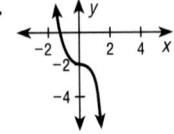

34.

35.

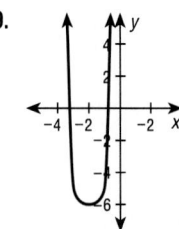

36. 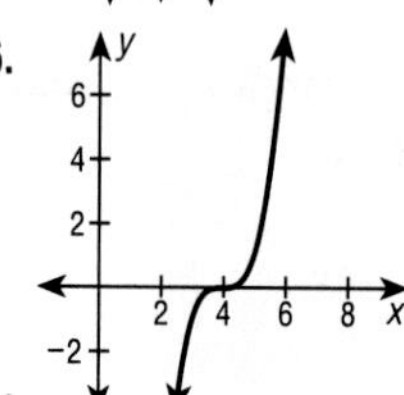

37.

38. 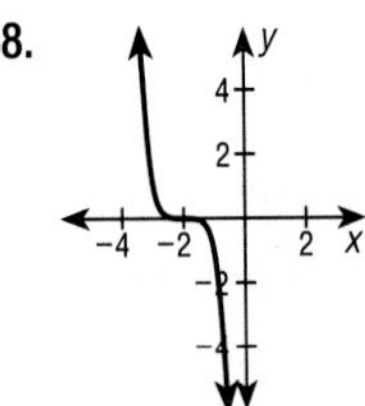

39.

40. 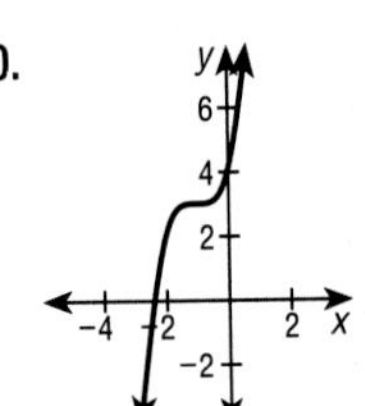

41.

42. 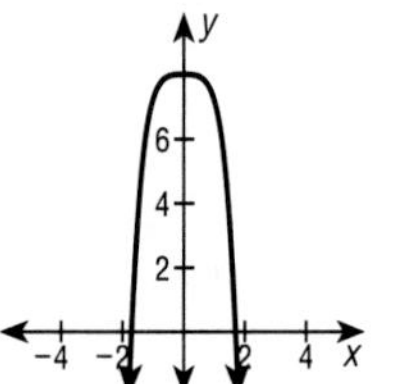

43. 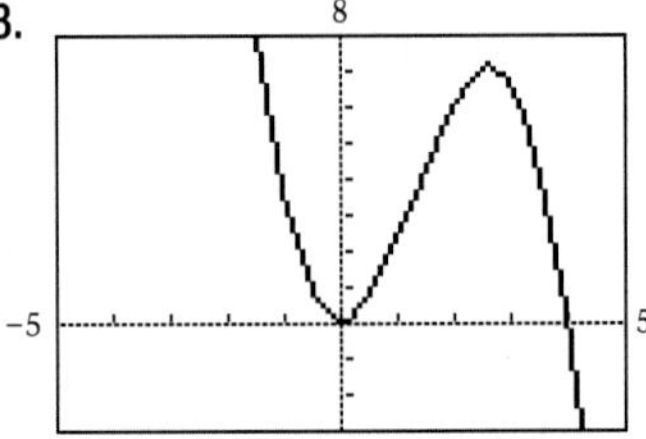

44. 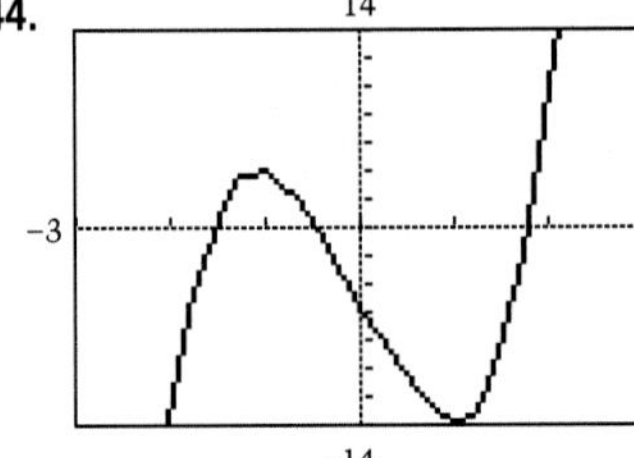

45. 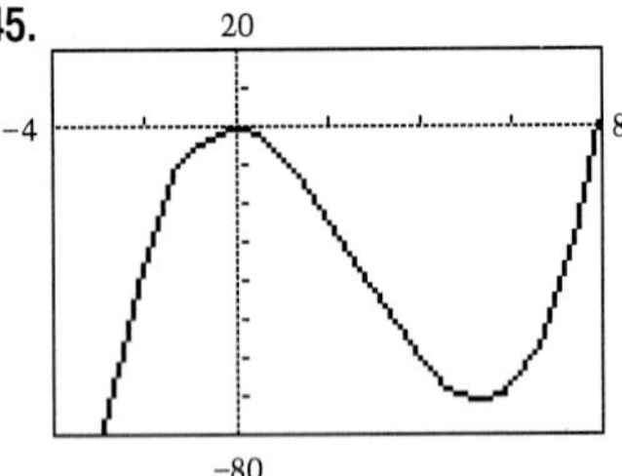

46. 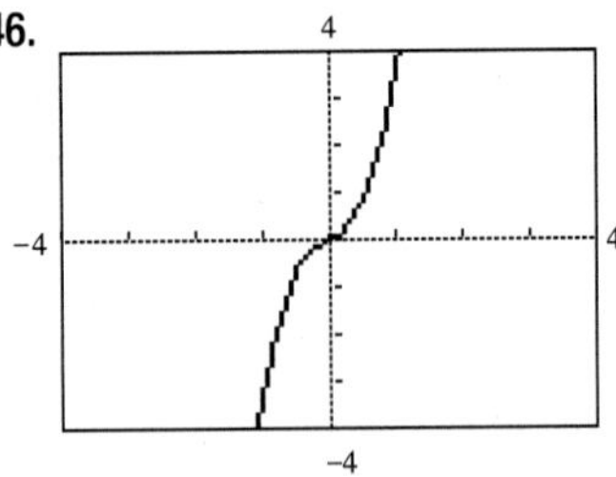

47. 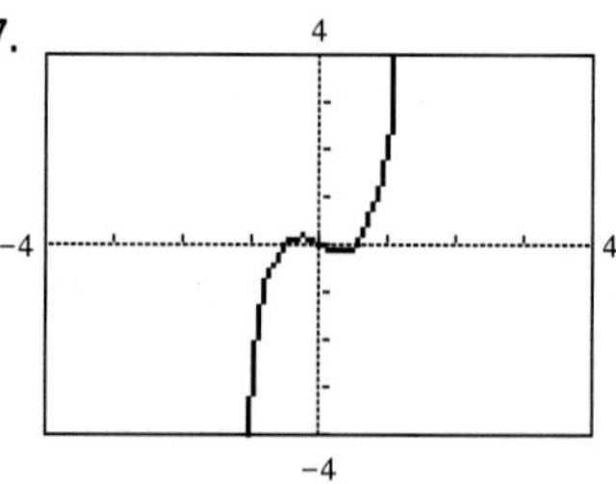

48. 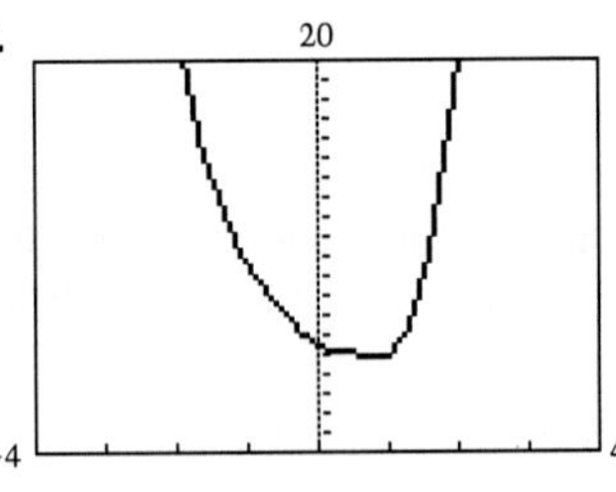

56. 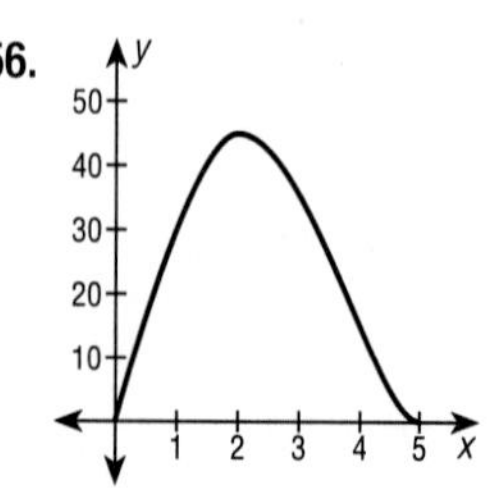

9.2 Integrated Review p. 469

57. 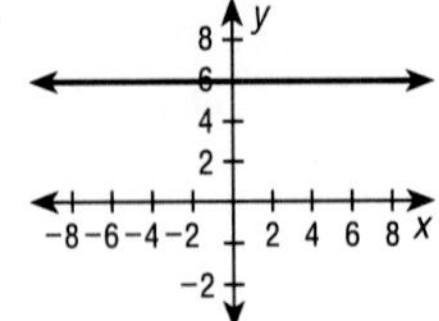

58. 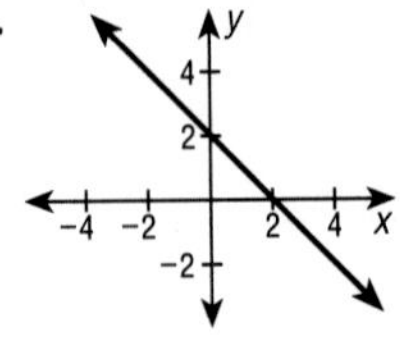

9.2 Integrated Review *(continued)*

59. 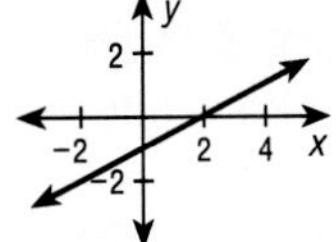

60. 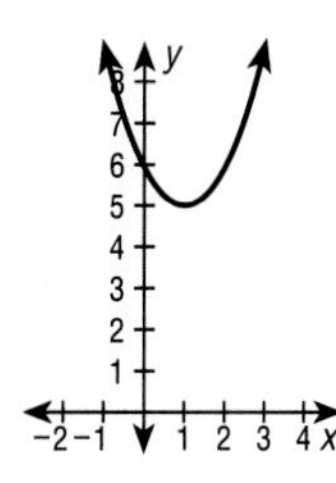

61. 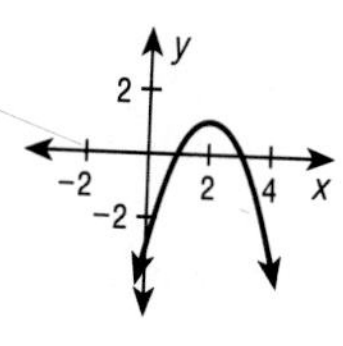

62. 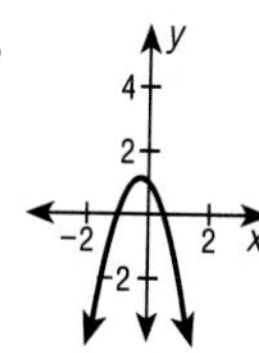

9.3 Communicating About Algebra p. 476

A. $-3, 2, 3$; write the equation in standard form, factor by grouping, factor the difference of two squares, use the Zero-Product Property

B. $-5, -2, 2, 5$; factor the trinomial, factor the difference of two squares twice, use the Zero-Product Property

9.3 Independent Practice p. 477

15. $(3x + 2)(3x - 2)$
16. $3(x + 4)(x - 4)$
17. $(x - 2)(x^2 + 2x + 4)$
18. $(x + 4)(x^2 - 4x + 16)$
19. $(6x + 1)(36x^2 - 6x + 1)$
20. $(5x - 1)(25x^2 + 5x + 1)$
21. $50(2x + 1)(2x - 1)$
22. $4(2x - 1)(4x^2 + 2x + 1)$
23. $2(4x - 1)(16x^2 + 4x + 1)$
24. $8(x - 2)(x^2 + 2x + 4)$
25. $3(x + 3)(x^2 - 3x + 9)$
26. $5(2x - 1)(4x^2 + 2x + 1)$
27. $(x^2 + 1)(x + 1)$
28. $(10x^2 + 1)(3x + 4)$
29. $(x^2 + 5)(x + 2)$
30. $(x^2 + 4)(x - 2)$
31. $(9x^2 + 7)(x + 2)$
32. $-1(2x^2 + 3)(x + 2)$
33. $2(x^2 + 2)(x + 2)$
34. $(6x^2 + 1)(3x + 5)$
35. $(2x^2 + 5)(x - 1)$
36. $(x + 4)(x - 4)(2x + 3)$
37. $(5x^2 + 3)(x - 4)$
38. $(2x^2 + 3)(9x - 1)$
39. $7x(x + 1)^2$
40. $3(x - 4)^2$
41. $(2x^2 - 3)(x - 2)$
42. $2(3x^2 - 1)(x - 3)$
43. $3(x - 2)(x^2 + 2x + 4)$
44. $(x + 5)(x^2 - 5x + 25)$
45. $3x^2(x + 10)(x - 10)$
46. $7x(2x + 1)(2x - 1)$
47. $3x(x^2 + 2)(x + 1)$
48. $x(x^2 + 4)(x + 12)$
49. $2(5x^2 - 1)(x - 2)$
50. $9(x + 1)(x - 1)(2x - 1)$

9.3 Integrated Review p. 479

69. $x^2 + 8x + 16$
70. $x^2 - 4x + 4$
71. $x^3 - 9x^2 + 27x - 27$
72. $x^3 + 3x^2 + 3x + 1$
73. $x^3 + 27x^2 + 243x + 729$
74. $x^3 - 15x^2 + 75x - 125$

9.3 Exploration and Extension p. 479

83. a^3 represents the volume of the larger cube and b^3 represents the volume of the smaller cube; when you take away the smaller volume from the larger, the volumes of the remaining solids are $(a - b)a^2$, $(a - b)ab$, and $(a - b)b^2$.

9.4 Independent Practice p. 484

7. $x + 9 + \frac{16}{x - 2}$
8. $2x - 7 + \frac{27}{x + 4}$
9. $3x + \frac{1}{x + 3}$
10. $x - 4$
11. $4x^2 - 4 + \frac{9}{x^2 + 1}$
12. $x + 6 + \frac{-10x - 10}{x^2 + 5}$
13. $2x^2 + 3 + \frac{6}{3x^2 + x - 2}$
14. $10x + 7 + \frac{5}{x^2 + 2x}$
15. $2x + 3 - \frac{1}{x^2 - 2}$
16. $2x - \frac{9}{x^3 + x^2 - 5}$
17. $5x^2 + 3x + 1$
18. $x^2 + x - 1$

9.4 Independent Practice *(continued)*

19. $x^2 + 2x - 3$

20. $x^2 - 4x - 12$

21. $4x - 1 - \frac{2}{x + 1}$

22. $7x - 10 + \frac{23}{x + 2}$

23. $2x + 11 + \frac{30}{x - 2}$

24. $4x + 2 + \frac{4}{x - 2}$

25. $x - 3 + \frac{18}{x + 3}$

26. $x - 3 + \frac{12}{x + 3}$

27. $10x^3 - 5x^2 + 9x - 9$

28. $x^3 + 3x^2 + 15x + 5 + \frac{45}{x - 5}$

29. $2x^3 + x - \frac{3}{x - 3}$

30. $4x^3 + x^2 + x - 1$

9.5 Integrated Review p. 492

35.

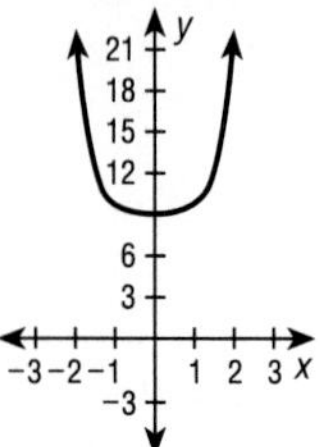

36.

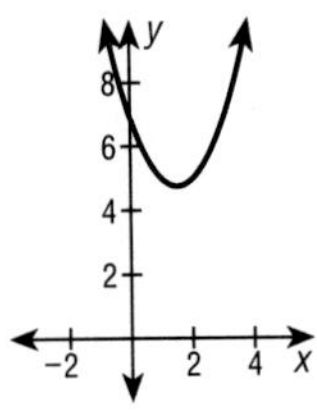

37.

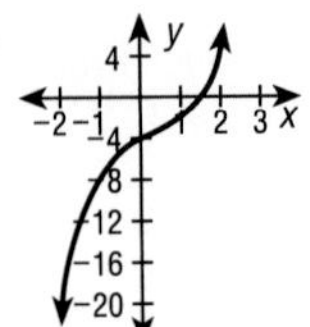

38.

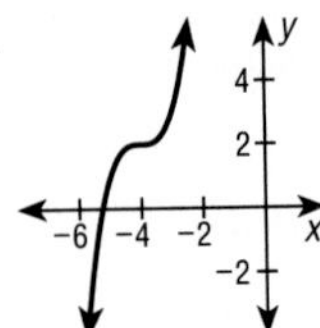

39.

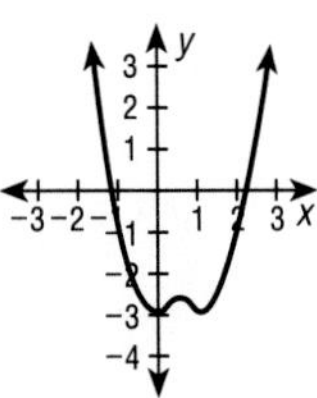

40.

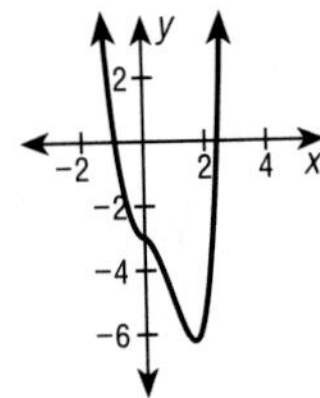

9.5 Mixed Review p. 493

4.

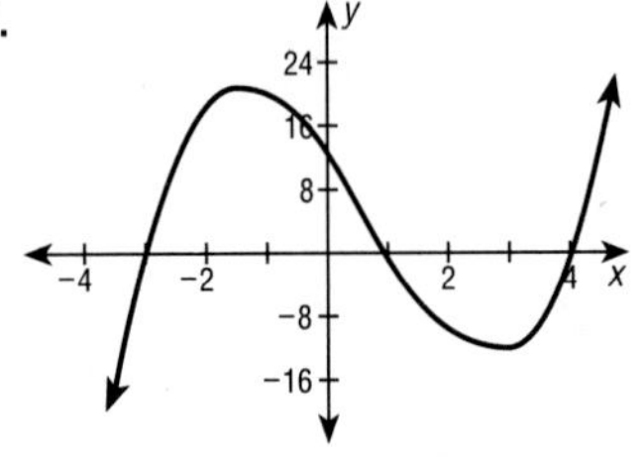

6.

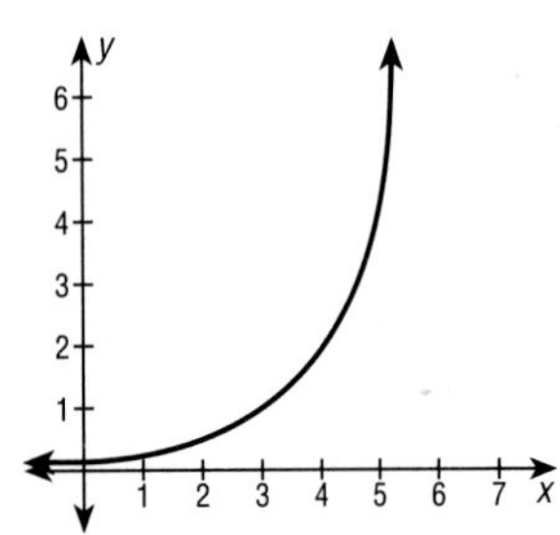

7.

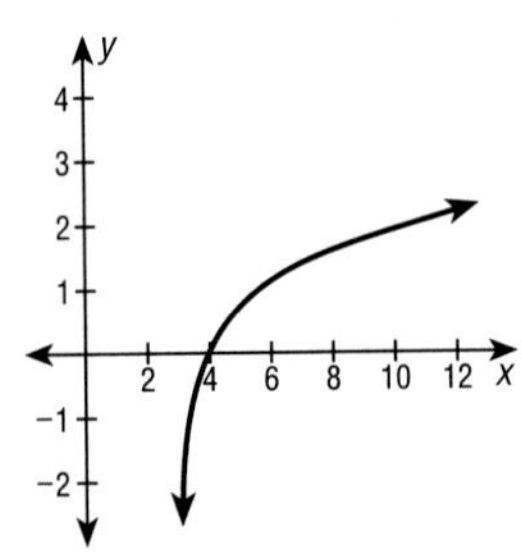

20.

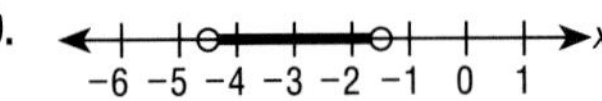

9.6 Independent Practice p. 498

12. $f(x) = (x - 2)(x - 1)(x - 5)$

13. $f(x) = (x - 1)(x + 3)(x - 4)$

14. $f(x) = (x + 6)(x - 4)(x - 2)$

15. $f(x) = (x - 6)(x + 2)(x - 2)$

16. $f(x) = (x - 5)(x^2 + 1)$

17. $f(x) = (x^2 + 9)(x - 4)$

18. $f(x) = (x - 3)(x + 3)(x^2 + 4)$

19. $f(x) = (x^2 + 1)(x^2 + 4)$

20. $f(x) = (x - 4)(x - 1)$

21. $f(x) = (x + 2)(x - 5)$

22. $f(x) = (x + 3)(x - 2)$

23. $f(x) = (x - 9)(x + 1)$

24. $(x + 1)(x - 2)(x + 5)(x - 3)$

25. $(x + 1)(x - 2)(x + 2)(x + 2)$

26. $x(x - 2)(x + 1)$

27. $(x + 3)(x - 3)(x + 5)$

28. $(x + 3)(x - 3)(x + 2i)(x - 2i)$

29. $(x - 1)(x + 3i)(x - 3i)$

30. $(x - 1)(x + 4i)(x - 4i)$

31. $(x + 3)(x + 5i)(x - 5i)$

32. $(x + 2)(x - 3)(x + i)(x - i)$

33. $(x + 1)(x - 2)(x + 2i)(x - 2i)$

34. $(x + 3)(x - 3)(2x - 9)(x + 1)$

35. $(x - 2)(x + 5i)(x - 5i)$

9.7 Chapter Review p. 509

1. $4x^3 + 3x^2 + x - 4$
2. $x^4 + x^3 - x^2 - 4x + 7$
3. $8x^2 - 5x - 3$
4. $-4x^3 + 2x^2 - 2x - 2$
5. $2x^3 + 6x^2 - 8x$
6. $x^5 + 2x^3 - 3x$
7. $x^2 - 36$
8. $x^2 - 196$
9. $x^3 - 2x^2 - 10x + 21$
10. $x^3 + 4x^2 + 4x + 1$
11. $x^2 + 18x + 81$
12. $x^2 - 20x + 100$
13. $x^3 - 21x^2 + 147x - 343$
14. $x^3 + 15x^2 + 75x + 125$

21.

22.

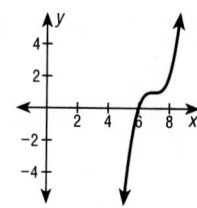

23. 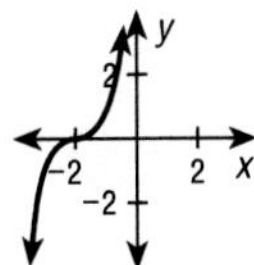

24. 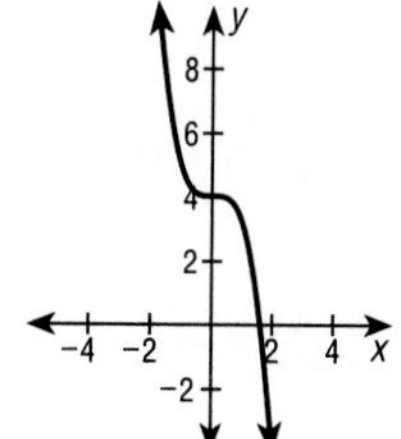

25. 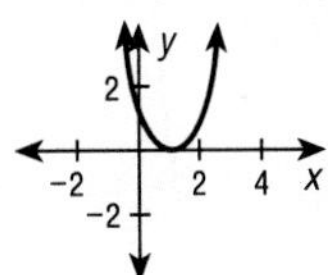

26. 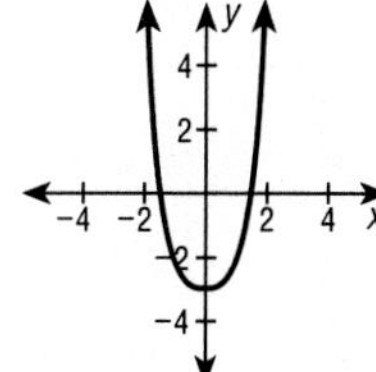

27. 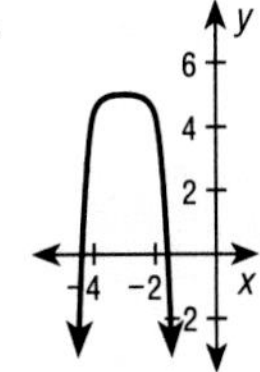

28. 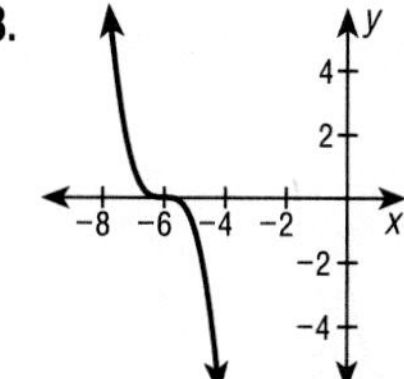

29. 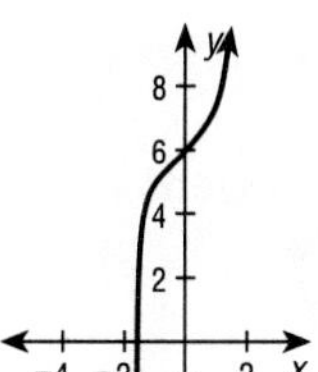

30. 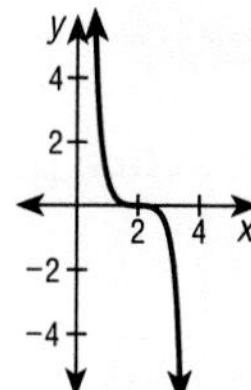

31. 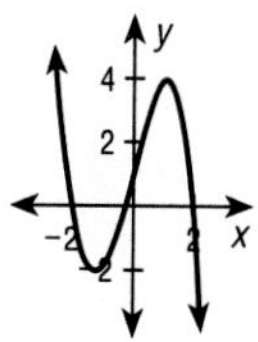

32. 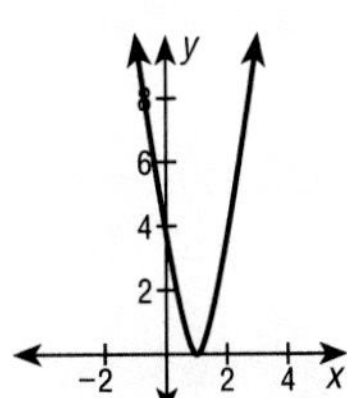

33. $16(x - 3)(x + 3)$
34. $(2x + 7)(2x - 7)$
35. $(5x - 2)(25x^2 + 10x + 4)$
36. $(4x + 7)(16x^2 - 28x + 49)$
37. $(x - 11)(x + 2)$
38. $(x + 9)(x - 1)$
39. $(2x - 1)(x - 2)$
40. $(3x + 4)(x - 5)$
41. $(x^2 + 5)(x - 1)$
42. $(x^2 - 5)(x - 3)$
43. $(x^2 + 2)(2x - 3)$
44. $(x^2 + 1)(2x + 1)$

Cumulative Review for Chapters 7–9 p. 514

27.
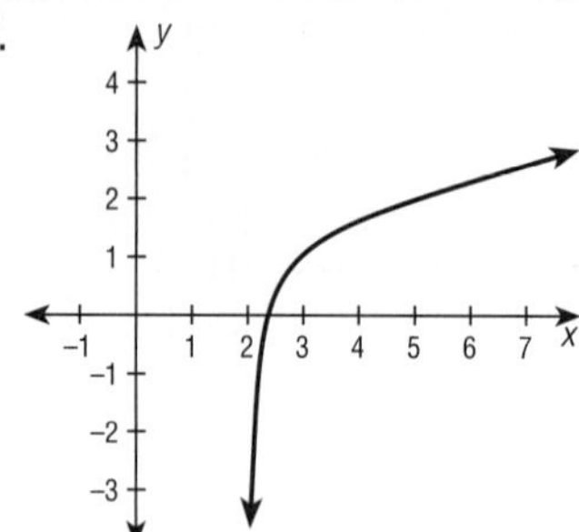

28.
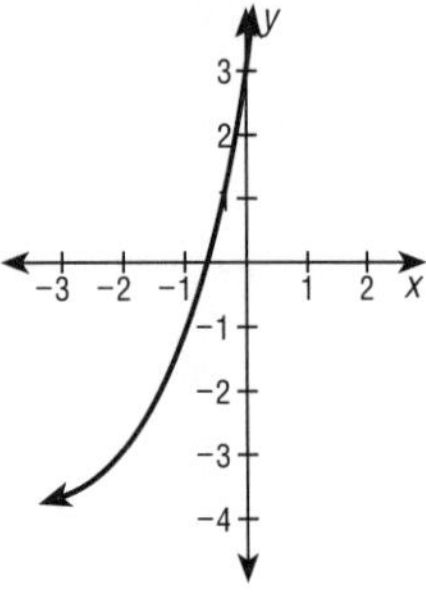

29.
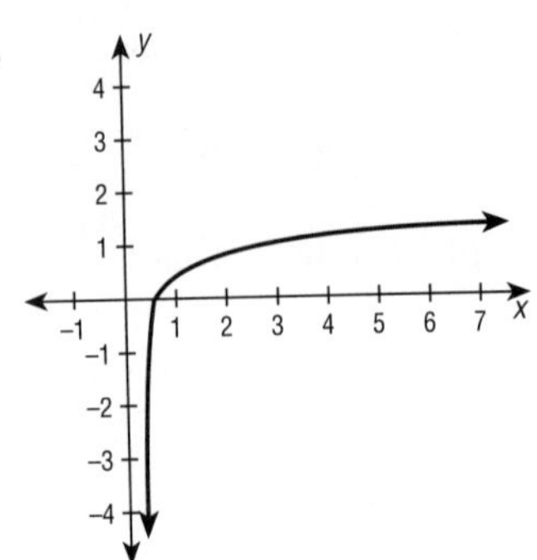

30.
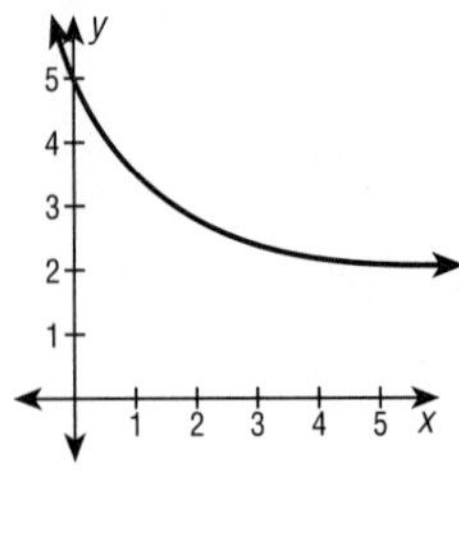

31.
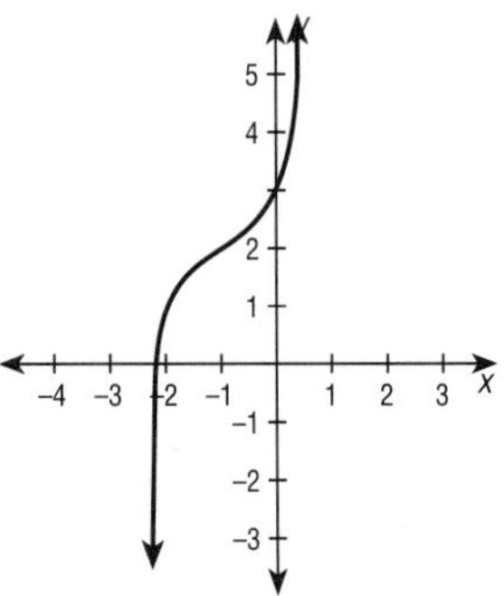

32.
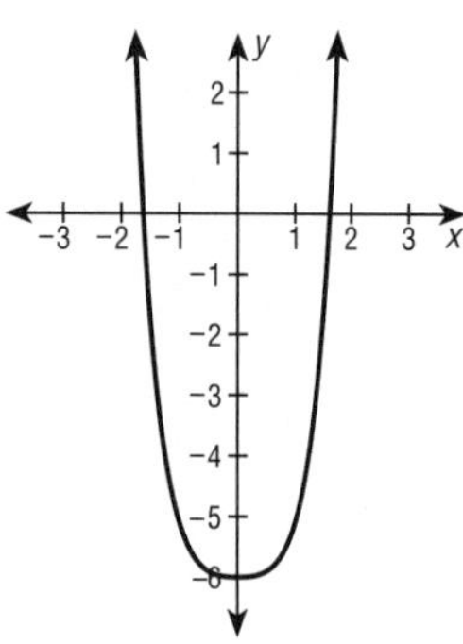

ADDITIONAL ANSWERS

CHAPTER 10

10.1 Guided Practice p. 521

2.
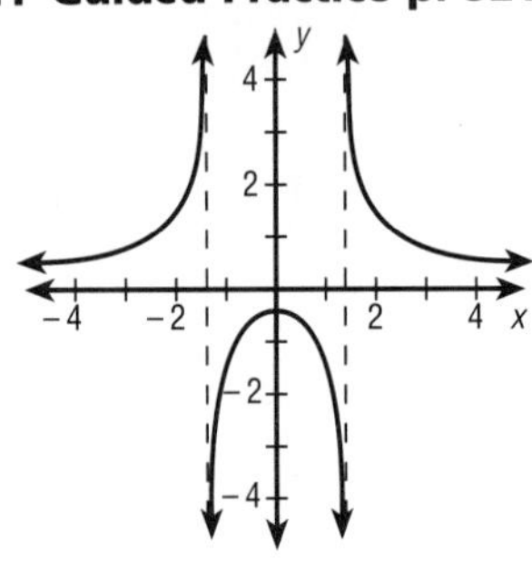

10.1 Independent Practice pp. 522–523

31.
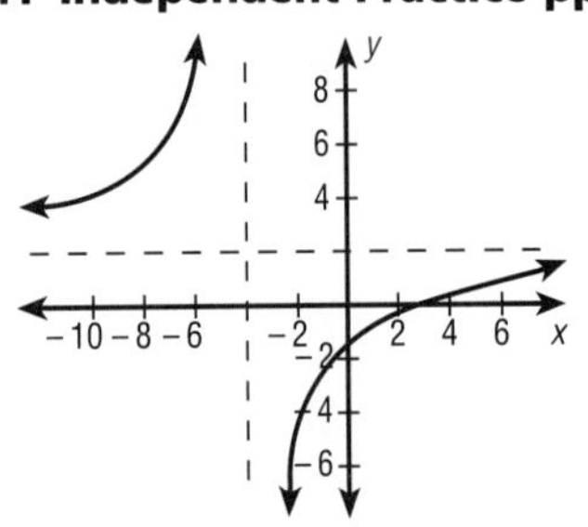

32.
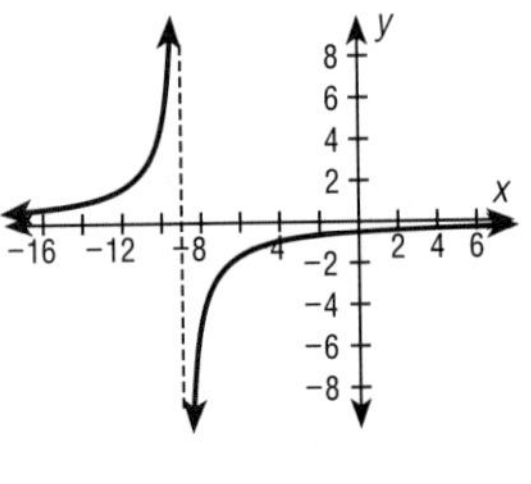

33.
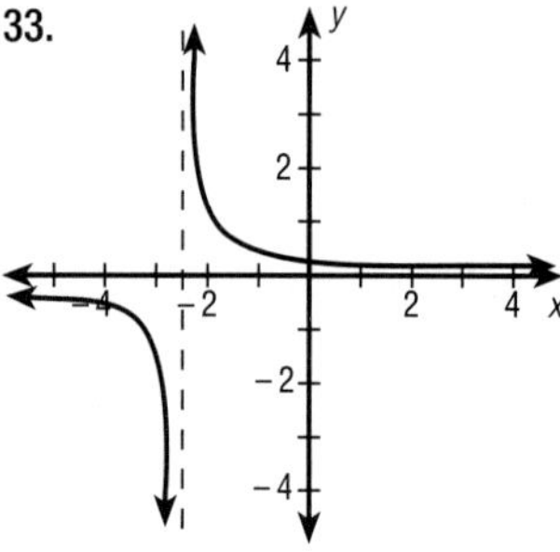

34.
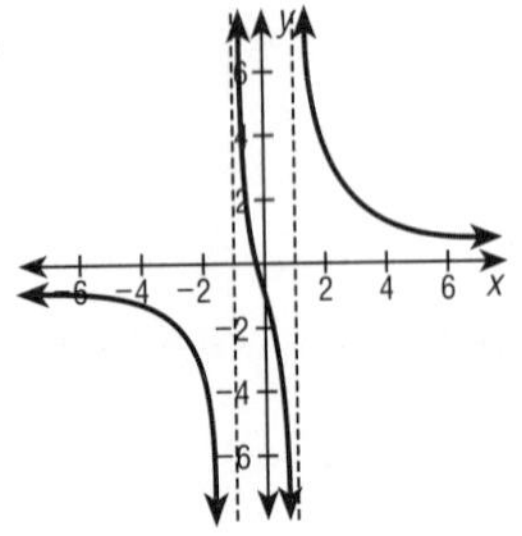

35.
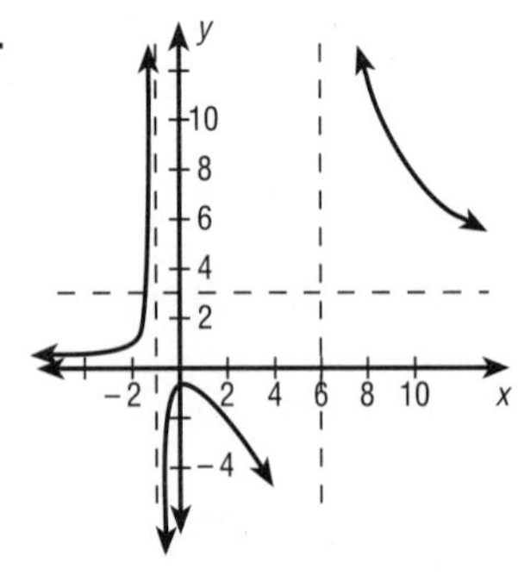

36.
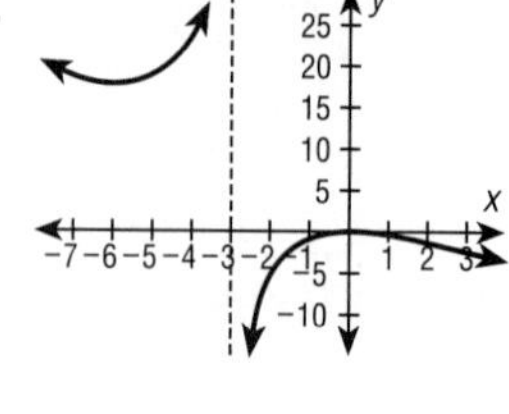

37.
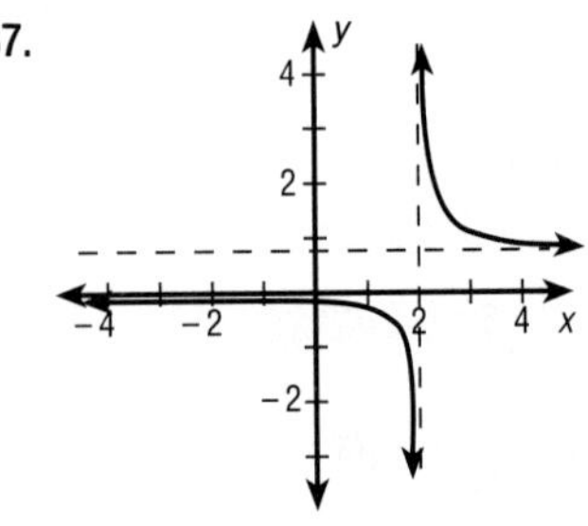

10.1 Independent Practice *(continued)*

38.

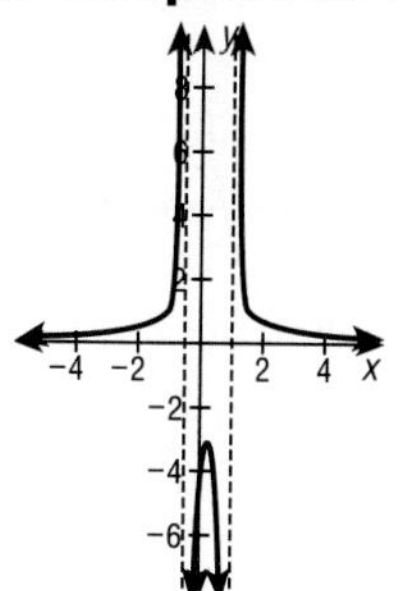

39.

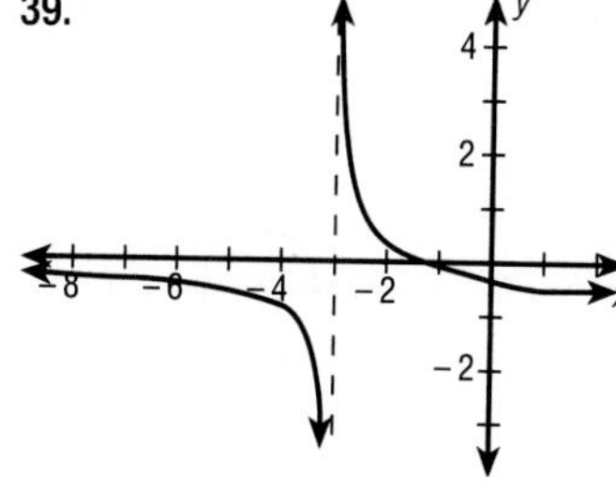

40.

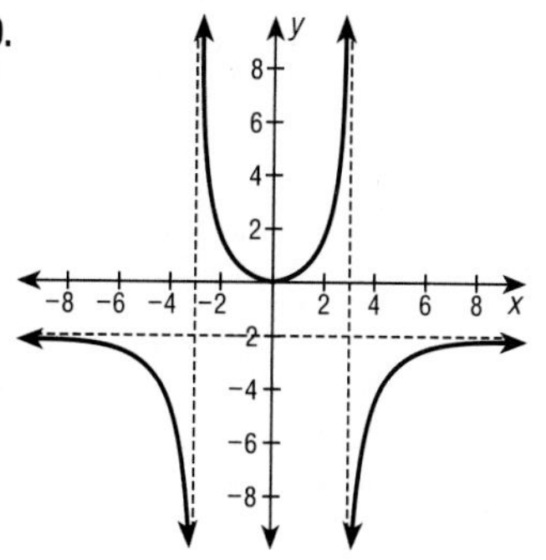

41.

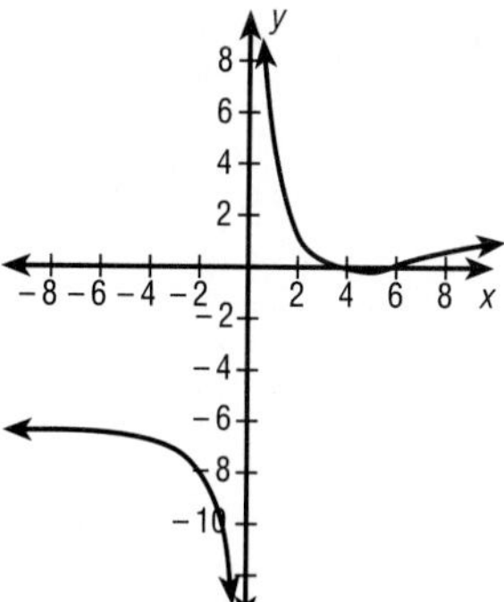

42.

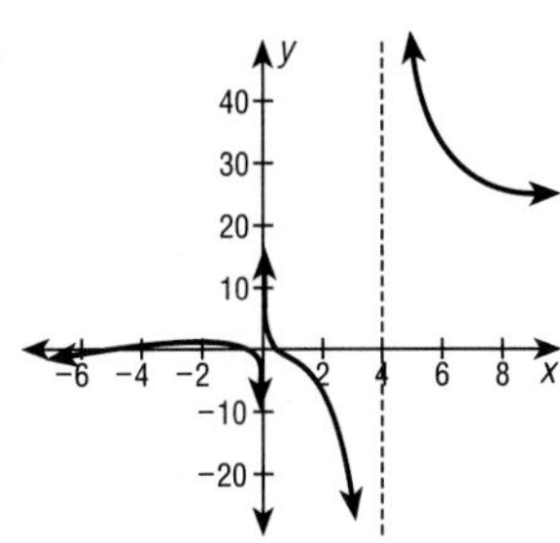

43. $A = \dfrac{3m + 180}{m}$; $A = 3$; when the VCR is paid for, \$3 will be spent for each movie.

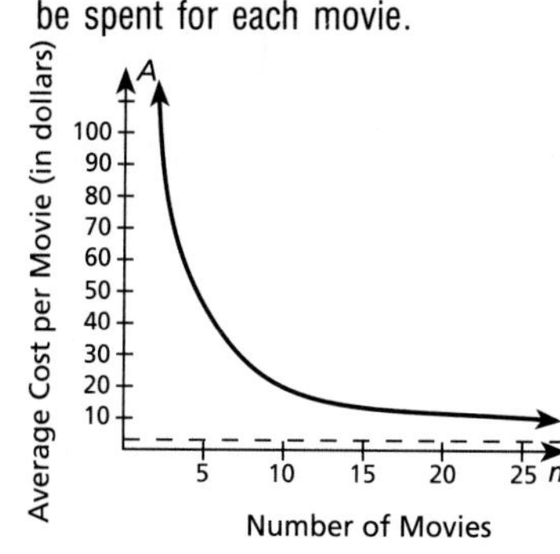

45.

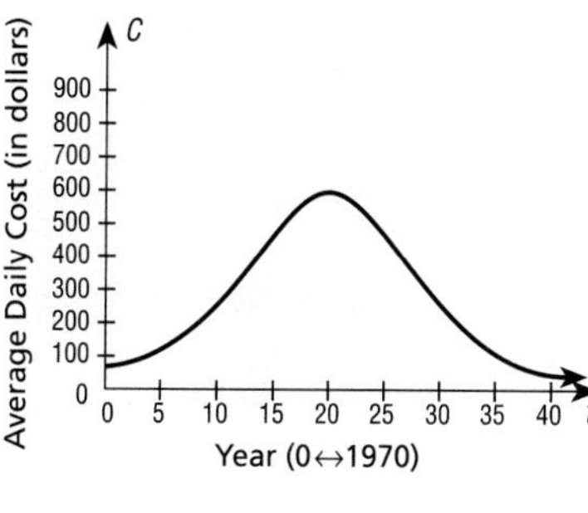

46.

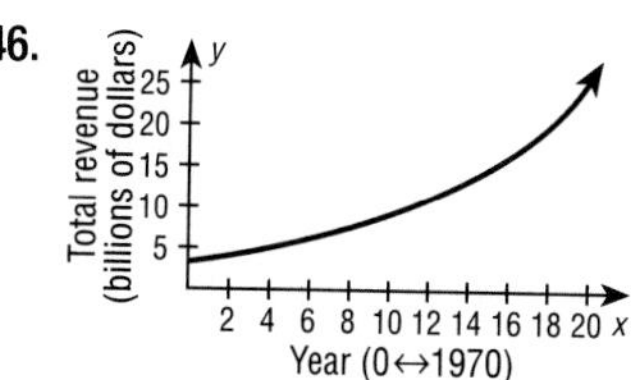

10.1 Exploration and Extension p. 523

56.

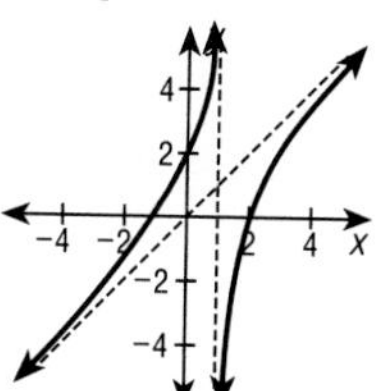

57.

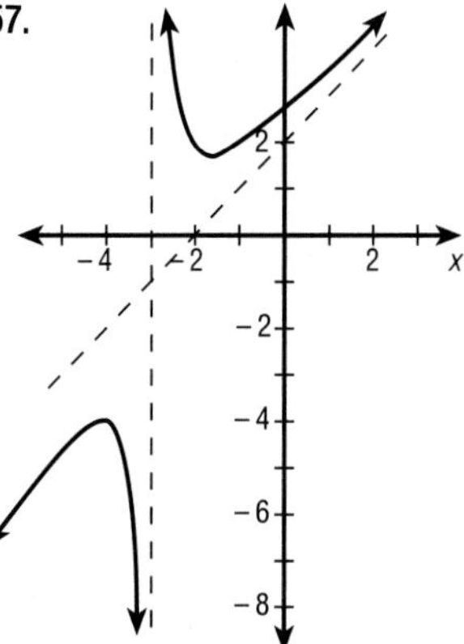

10.2 Integrated Review p. 531

24.

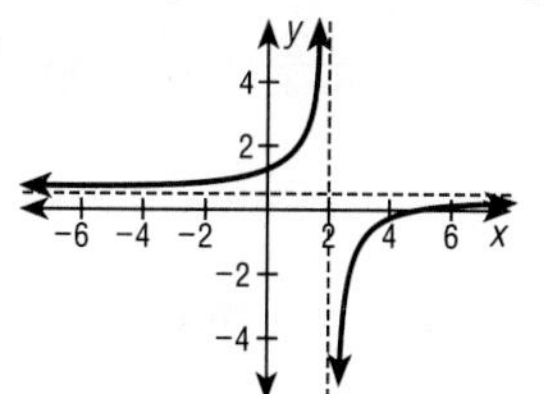

25.

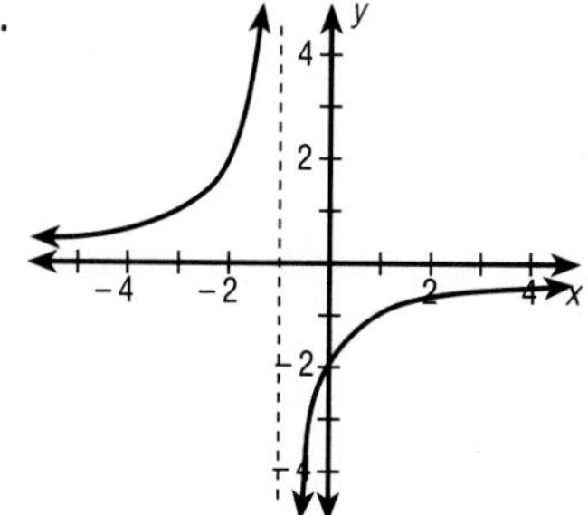

26.

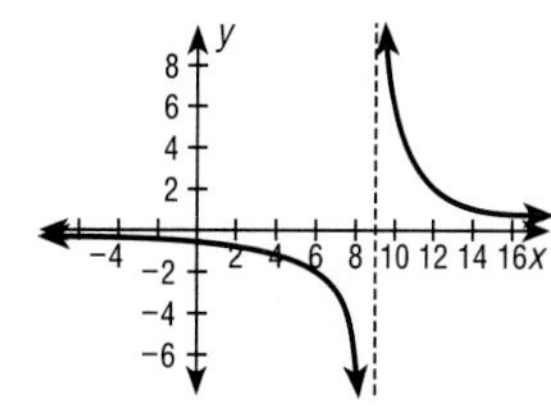

27.

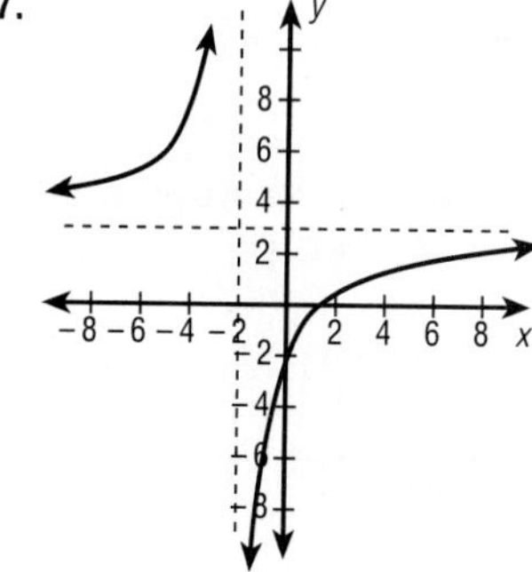

10.2 Integrated Review *(continued)*

28. 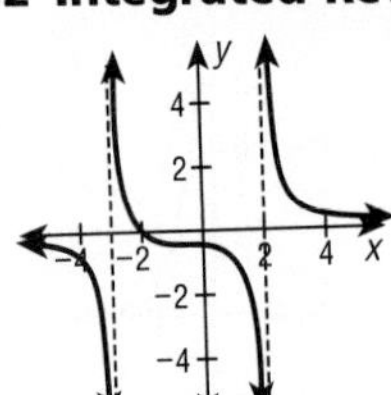

29. 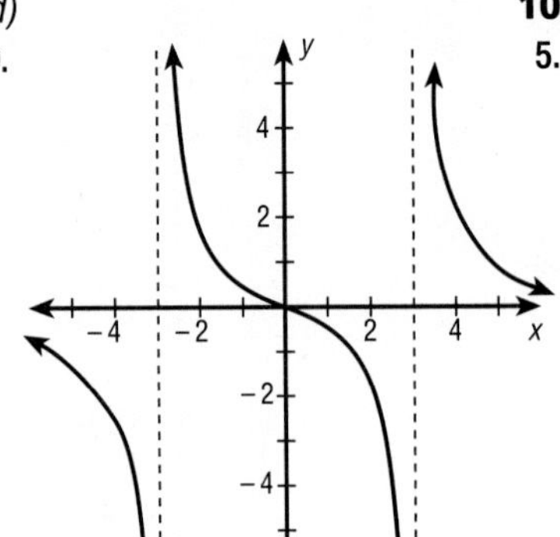

10.2 Mixed Review p. 532

5. 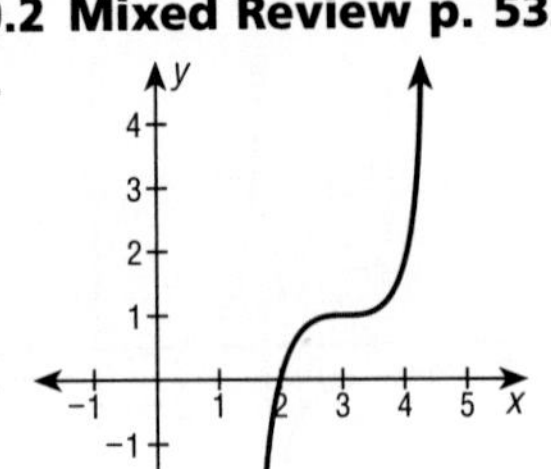

6. 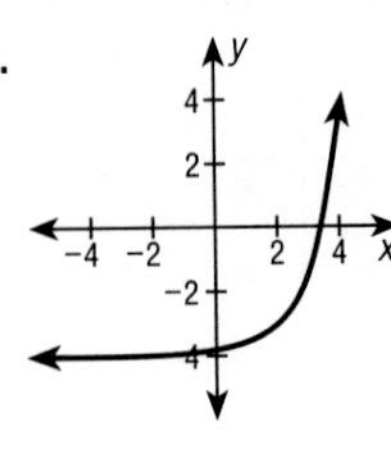

7. 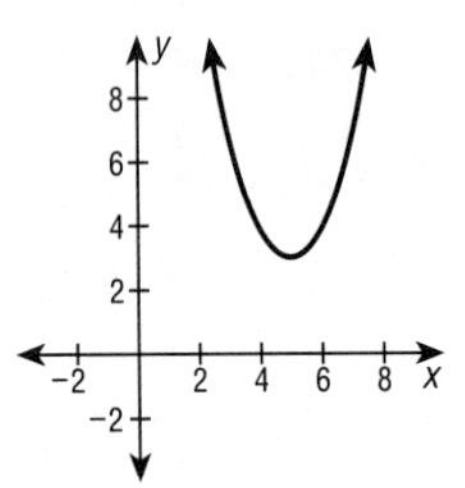

8. 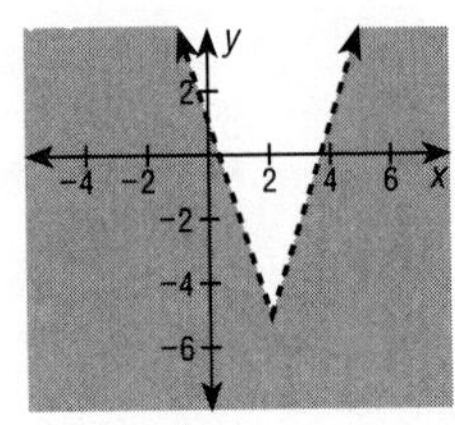

10.3 Mid-Chapter Self-Test p. 540

1. 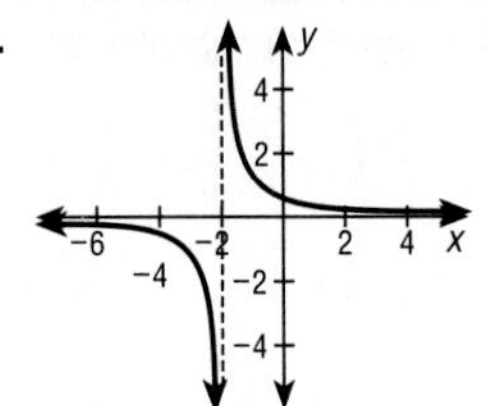

2. 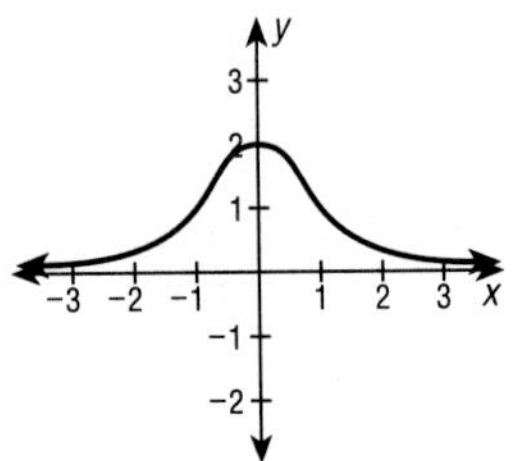

3. 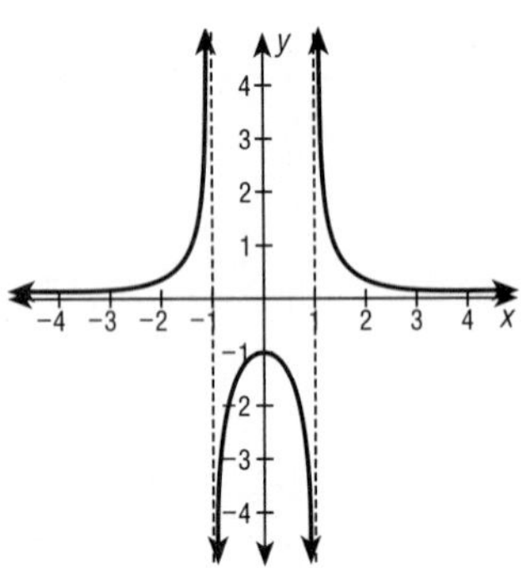

4. 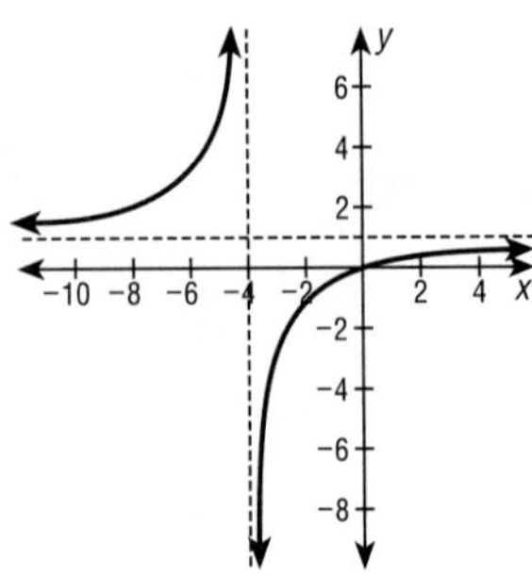

5. 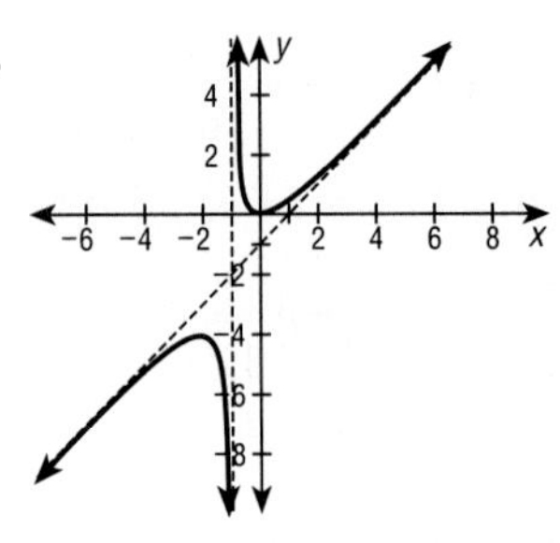

6. 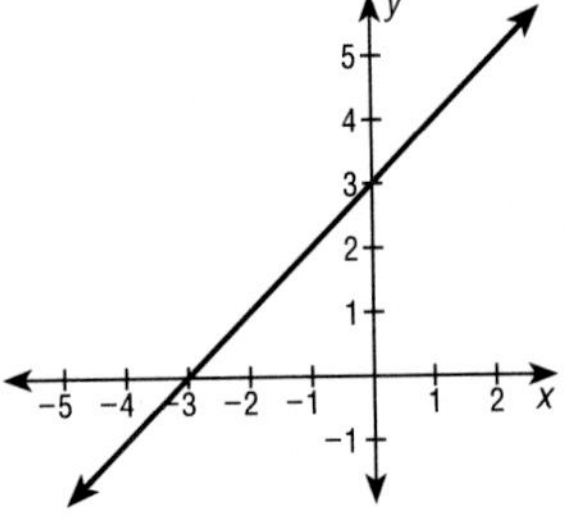

10.5 Independent Practice p. 551

8. $\frac{6 + x}{3x^2}$

9. $-\frac{43}{28x}$

10. $\frac{-x^2 + 15x + 5}{3x(x - 5)}$

11. $\frac{x - 5}{(x + 2)(x - 2)}$

12. $\frac{-3x^3 - x^2 + 13x + 12}{7x(2x + 3)}$

13. $\frac{-20x - 25}{(x - 6)(x + 5)}$

14. $\frac{2x^3 + 14x^2 - 42x - 70}{(3x + 5)(x + 7)}$

15. $\frac{12x^2 - x - 39}{3(x + 11)}$

16. $\frac{-6x^2 + x - 11}{(x - 6)(2x + 1)}$

17. $\frac{8x^3 - 15x^2 + 9x + 12}{x(x + 1)(2x - 3)}$

18. $\frac{-x^3 - 5x - 42}{3(x + 2)(x - 2)}$

19. $\frac{-4x^2 + 3x - 1}{(x + 10)(x - 1)}$

20. $\frac{4x^2 - 11x - 3}{(x - 8)(x - 4)}$

21. $\frac{-6x + 7}{(x + 5)(x - 2)}$

22. $\frac{49x^2 + 24x - 5}{6x(x + 1)(x - 1)}$

10.5 Mixed Review p. 554

3.

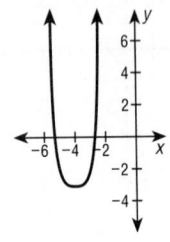

4. 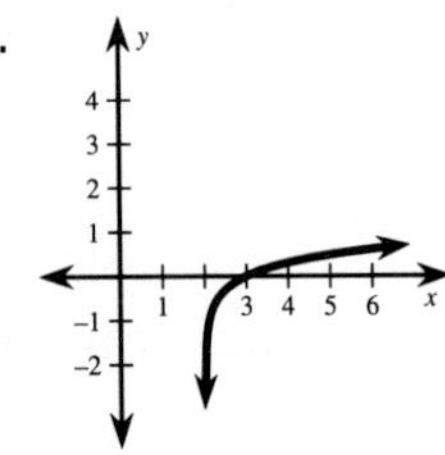

5. 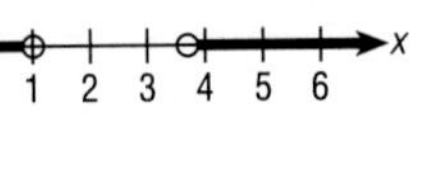

6. 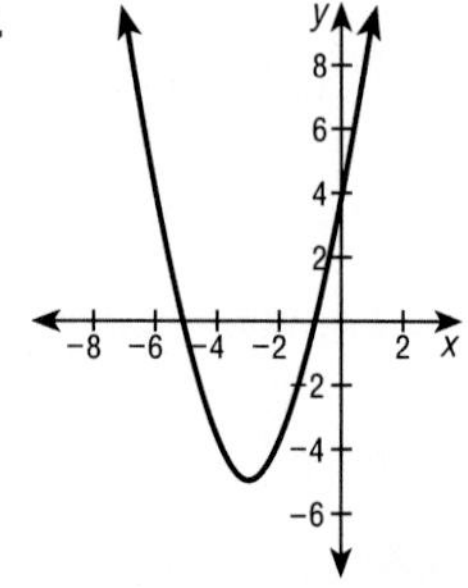

10.5 Mixed Review *(continued)*

7.
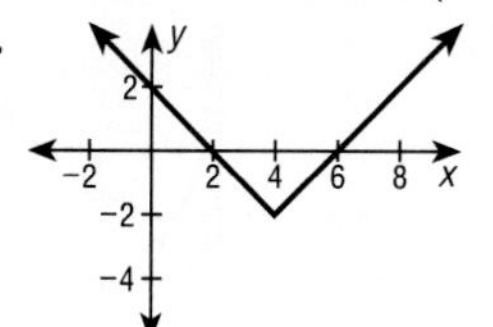

8.
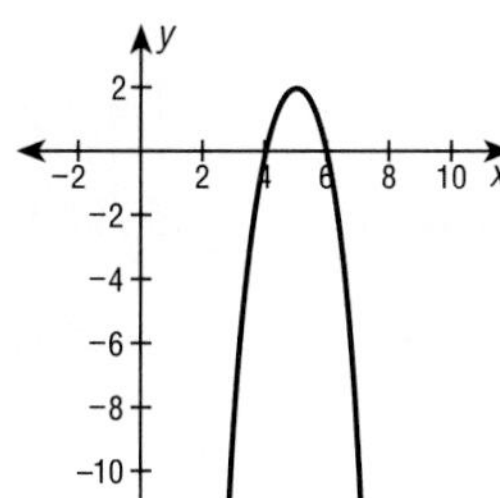

10.6 Independent Practice p. 558

5.

Payment number	Balance before payment	Payment	Interest payment	Principal payment	Balance after payment
1	$1200.00	$139.23	$10.50	$128.73	$1071.27
2	$1071.27	$139.23	$ 9.37	$129.86	$ 941.41
3	$ 941.41	$139.23	$ 8.24	$130.99	$ 810.42
4	$ 810.42	$139.23	$ 7.09	$132.14	$ 678.28
5	$ 678.28	$139.23	$ 5.93	$133.30	$ 544.98
6	$ 544.98	$139.23	$ 4.77	$134.46	$ 410.52
7	$ 410.52	$139.23	$ 3.59	$135.64	$ 274.88
8	$ 274.88	$139.23	$ 2.41	$136.82	$ 138.06
9	$ 138.06	$139.27	$ 1.21	$138.06	$ 0.00

6.

Payment number	Balance before payment	Payment	Interest payment	Principal payment	Balance after payment
1	$750.00	$88.27	$8.75	$79.52	$670.48
2	$670.48	$88.27	$7.82	$80.45	$590.03
3	$590.03	$88.27	$6.88	$81.39	$508.64
4	$508.64	$88.27	$5.93	$82.34	$426.30
5	$426.30	$88.27	$4.97	$83.30	$343.00
6	$343.00	$88.27	$4.00	$84.27	$258.73
7	$258.73	$88.27	$3.02	$85.25	$173.48
8	$173.48	$88.27	$2.02	$86.25	$ 87.23
9	$ 87.23	$88.25	$1.02	$87.23	$ 0.00

7.

Payment number	Balance before payment	Payment	Interest payment	Principal payment	Balance after payment
1	$5000.00	$434.94	$33.33	$401.61	$4598.39
2	$4598.39	$434.94	$30.66	$404.28	$4194.11
3	$4194.11	$434.94	$27.96	$406.98	$3787.13
4	$3787.13	$434.94	$25.25	$409.69	$3377.44
5	$3377.44	$434.94	$22.52	$412.42	$2965.02
6	$2965.02	$434.94	$19.77	$415.17	$2549.85
7	$2549.85	$434.94	$17.00	$417.94	$2131.91
8	$2131.91	$434.94	$14.21	$420.73	$1711.18
9	$1711.18	$434.94	$11.41	$423.53	$1287.65
10	$1287.65	$434.94	$ 8.58	$426.36	$ 861.29
11	$ 861.29	$434.94	$ 5.74	$429.20	$ 432.09
12	$ 432.09	$434.97	$ 2.88	$432.09	$ 0.00

8.

Payment number	Balance before payment	Payment	Interest payment	Principal payment	Balance after payment
1	$400.00	$69.61	$5.00	$64.61	$335.39
2	$335.39	$69.61	$4.19	$65.42	$269.97
3	$269.97	$69.61	$3.37	$66.24	$203.73
4	$203.73	$69.61	$2.55	$67.06	$136.67
5	$136.67	$69.61	$1.71	$67.90	$ 68.77
6	$ 68.77	$69.63	$0.86	$68.77	$ 0.00

10.6 Chapter Review pp. 562–566

11.
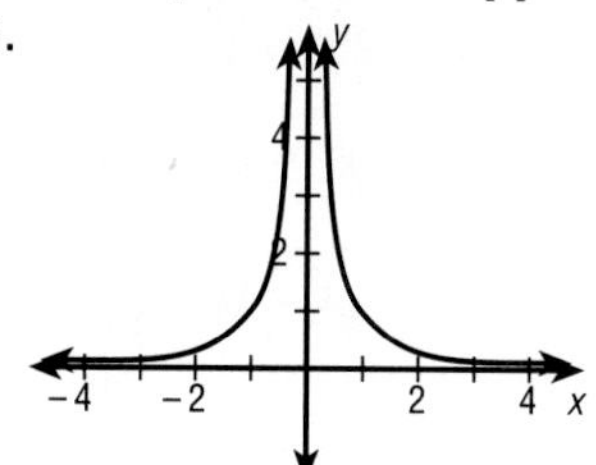

12.
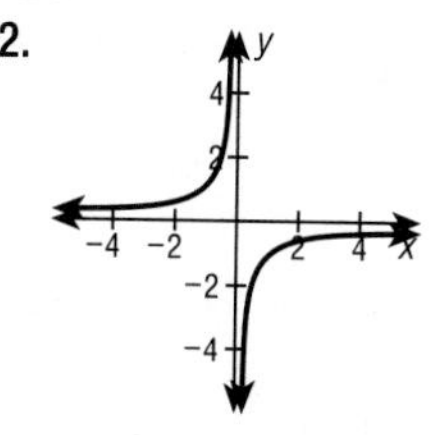

13.
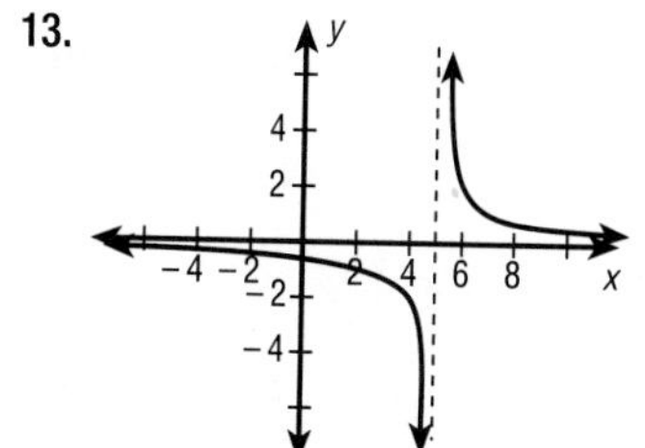

14.
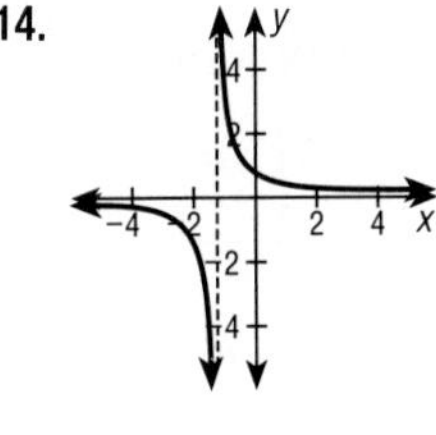

10.6 Chapter Review *(continued)*

15. **16.**

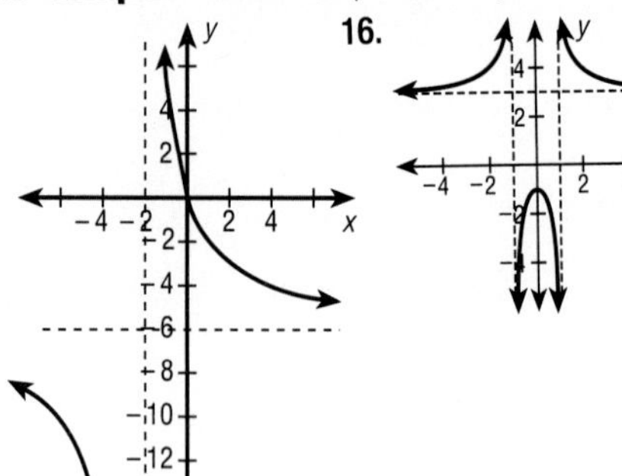

17.

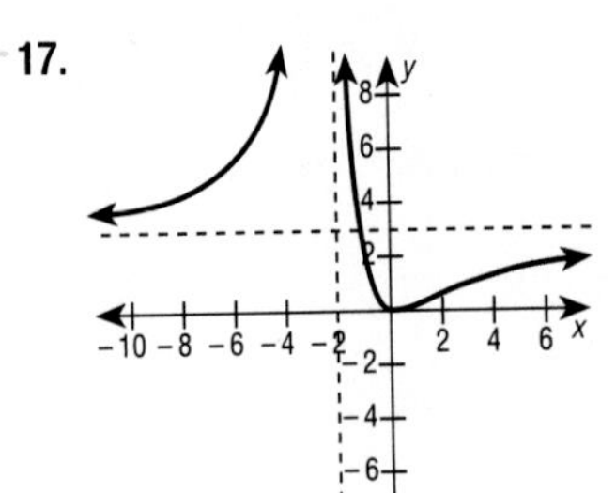

18.

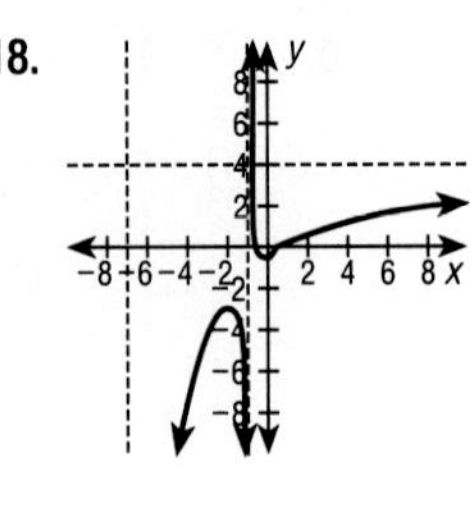

19.

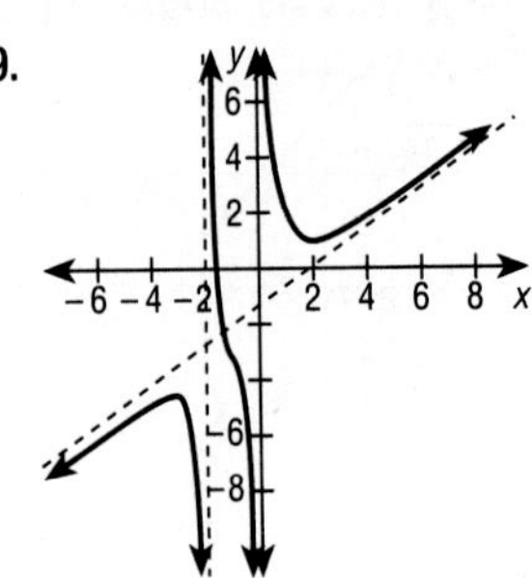

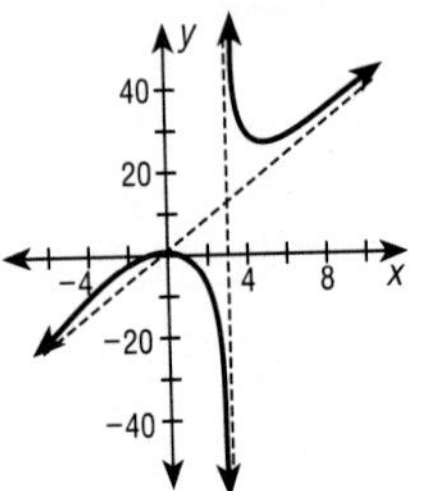

21.

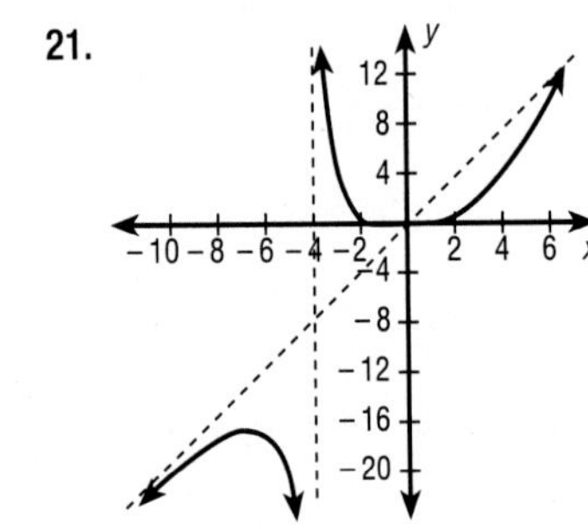

22.

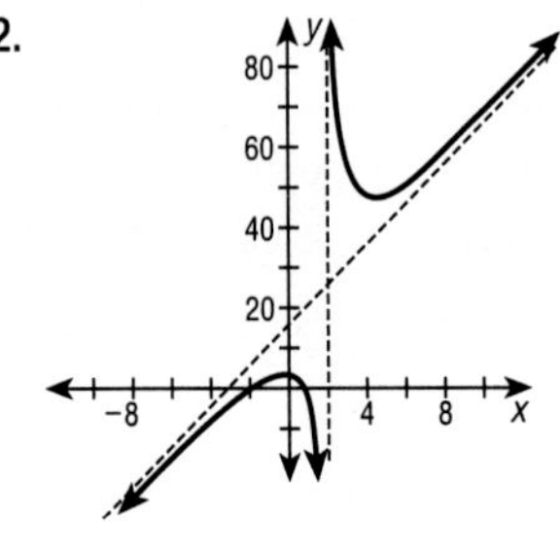

75.

Payment number	Balance before payment	Payment	Interest payment	Principal payment	Balance after payment
1	\$900.00	\$79.12	\$7.50	\$71.62	\$828.38
2	\$828.38	\$79.12	\$6.90	\$72.22	\$756.16
3	\$756.16	\$79.12	\$6.30	\$72.82	\$683.34
4	\$683.34	\$79.12	\$5.69	\$73.43	\$609.91
5	\$609.91	\$79.12	\$5.08	\$74.04	\$535.87
6	\$535.87	\$79.12	\$4.47	\$74.65	\$461.22
7	\$461.22	\$79.12	\$3.84	\$75.28	\$385.94
8	\$385.94	\$79.12	\$3.22	\$75.90	\$310.04
9	\$310.04	\$79.12	\$2.58	\$76.54	\$233.50
10	\$233.50	\$79.12	\$1.95	\$77.17	\$156.33
11	\$156.33	\$79.12	\$1.30	\$77.82	\$ 78.51
12	\$ 78.51	\$79.16	\$0.65	\$78.51	\$ 0.00

83.

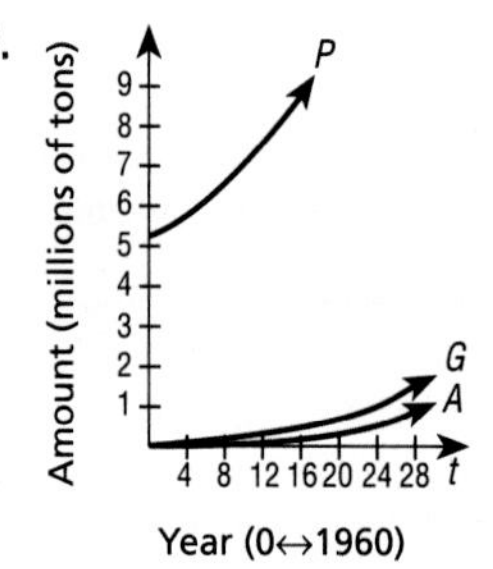

CHAPTER 11

11.1 Guided Practice p. 573

6.

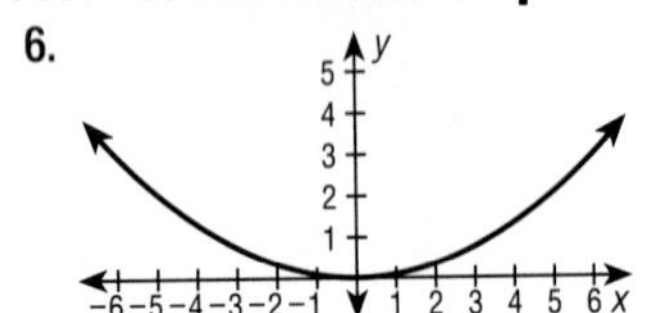

11.1 Independent Practice pp. 573–574

7. $x^2 = 4\left(\frac{1}{20}\right)y$ **8.** $x^2 = 4\left(-\frac{1}{12}\right)y$ **9.** $x^2 = 4\left(-\frac{1}{8}\right)y$ **10.** $y^2 = 4\left(\frac{1}{36}\right)x$

11. $y^2 = 4\left(\frac{1}{6}\right)x$ **12.** $y^2 = 4\left(\frac{9}{2}\right)x$ **13.** $y^2 = 4\left(\frac{1}{56}\right)x$ **14.** $y^2 = 4\left(-\frac{1}{8}\right)x$

19. $\left(0, -\frac{1}{8}\right)$, $y = \frac{1}{8}$ **20.** $\left(\frac{1}{16}, 0\right)$, $x = -\frac{1}{16}$ **21.** $(0, 2)$, $y = -2$ **22.** $(5, 0)$, $x = -5$

23. $(-4, 0)$, $x = 4$ **24.** $(0, -9)$, $y = 9$ **25.** $(0, -3)$, $y = 3$ **26.** $(0, 10)$, $y = -10$

27. $(1, 0)$, $x = -1$ **28.** $\left(0, \frac{1}{3}\right)$, $y = -\frac{1}{3}$ **29.** $(0, 6)$, $y = -6$ **30.** $\left(\frac{1}{9}, 0\right)$, $x = -\frac{1}{9}$

11.1 Independent Practice *(continued)*

43.
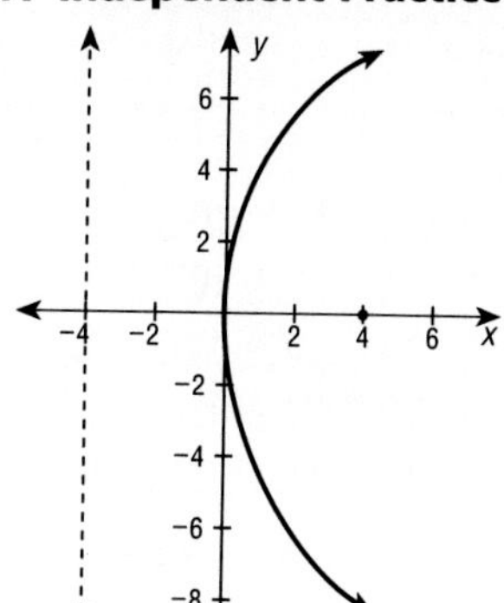

44.
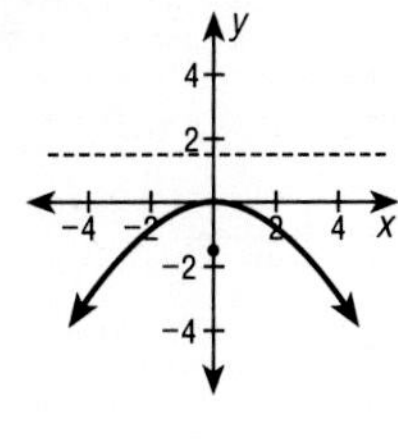

45.
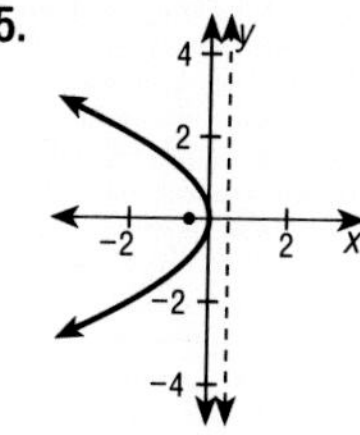

46.
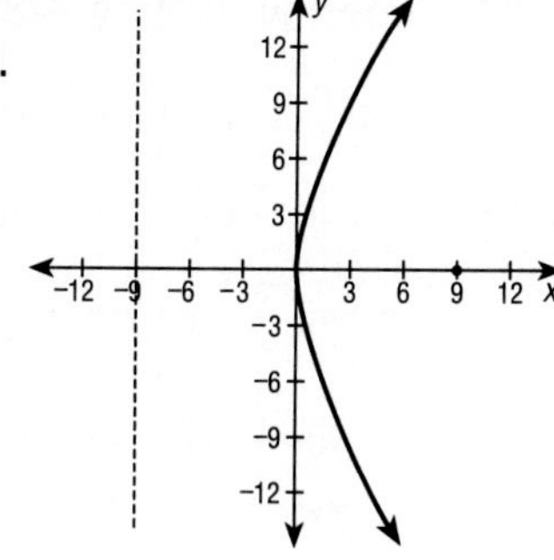

47.
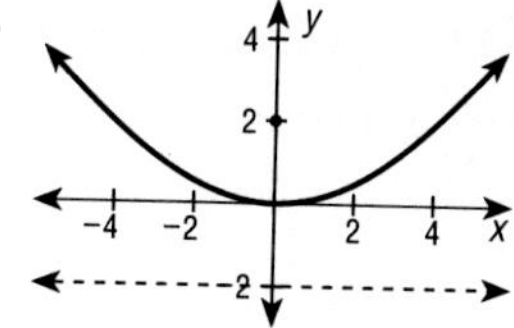

48.
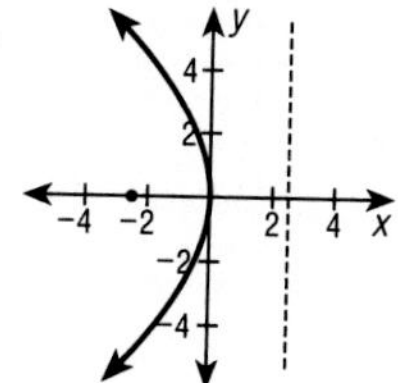

49.
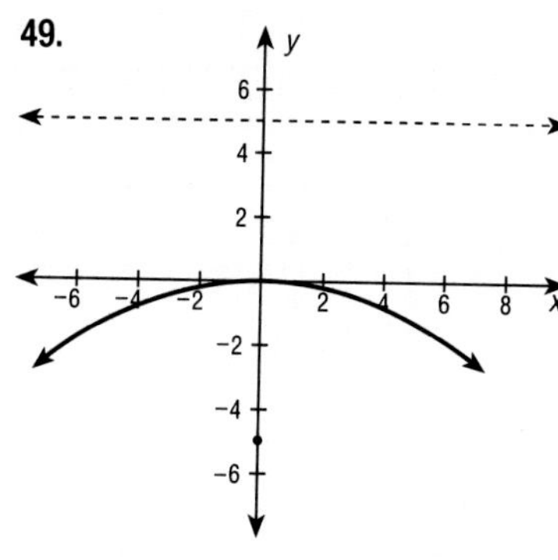

50.
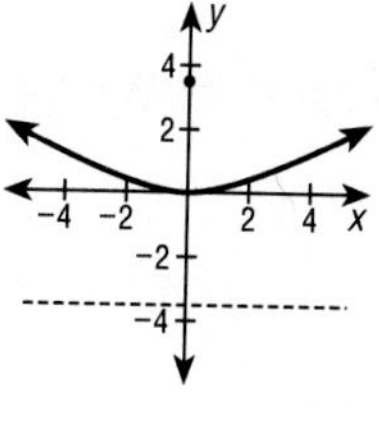

51.
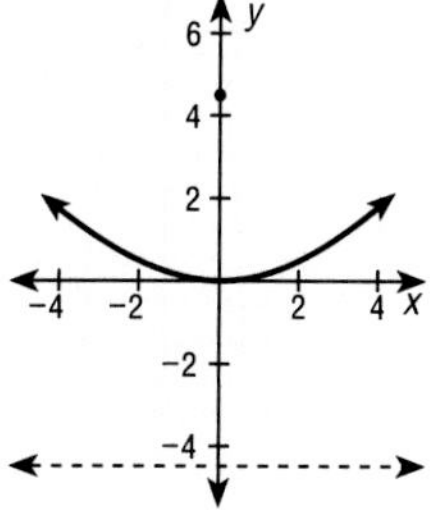

52.
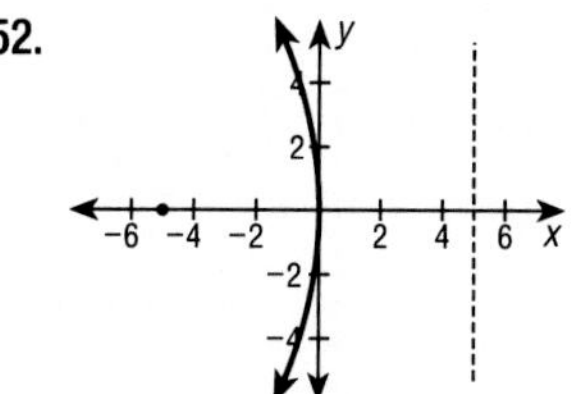

53.
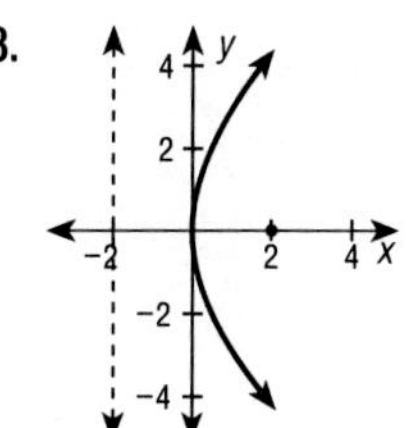

54.
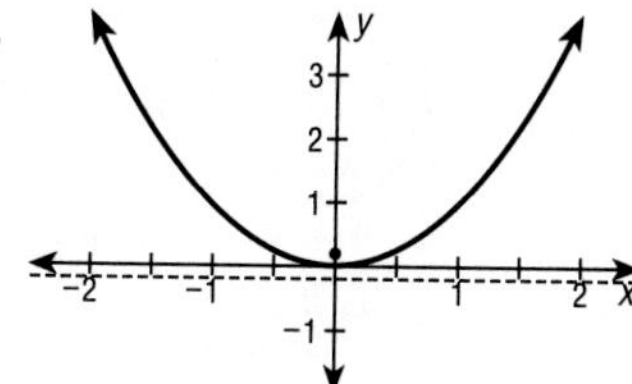

ADDITIONAL ANSWERS

11.1 Integrated Review p. 575

61.
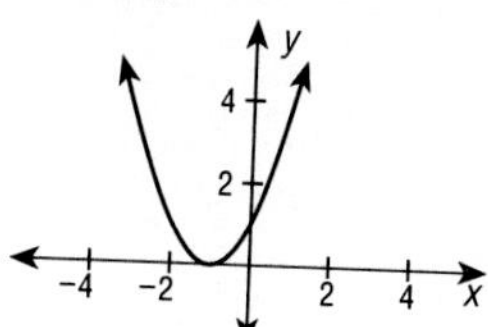

62.
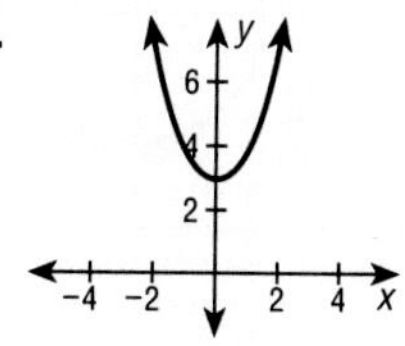

63.
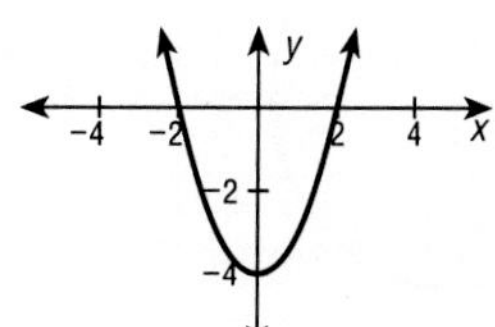

64.
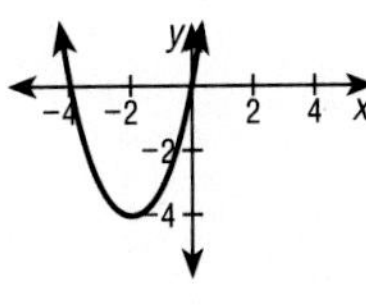

65.
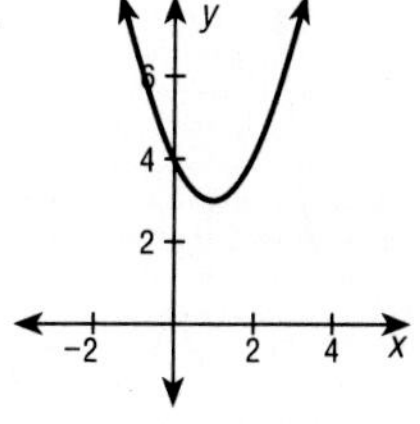

66.
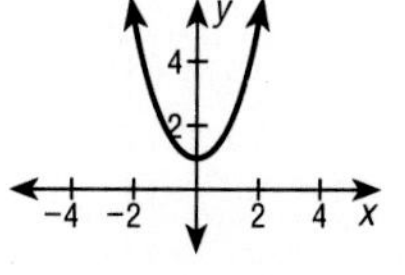

11.2 Communicating About Algebra p. 578

$$\frac{I}{(\sqrt{(x-10)^2+(y-0)^2})^2} = \frac{2I}{(\sqrt{(x-20)^2+(y-0)^2})^2}$$

$$\frac{I}{(x-10)^2+y^2} = \frac{2I}{(x-20)^2+y^2}$$

$$\frac{I}{x^2-20x+100+y^2} = \frac{2I}{x^2-40x+400+y^2}$$

$$\frac{1}{x^2-20x+100+y^2} = \frac{2}{x^2-40x+400+y^2}$$

$$x^2-40x+400+y^2 = 2x^2-40x+200+2y^2$$

$$-x^2-y^2 = -200$$

$$x^2+y^2 = 200$$

11.2 Guided Practice p. 579

3. 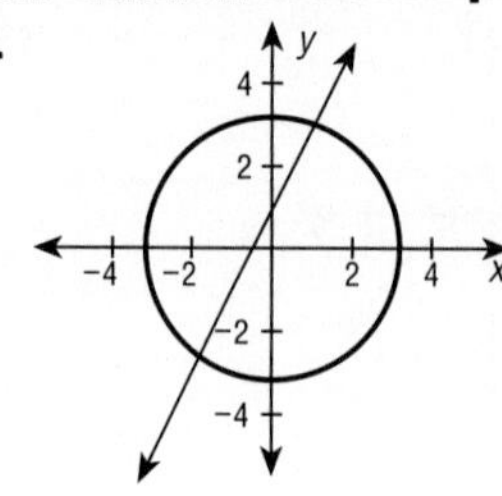

4. $y = 2x + 1$
$x^2 + y^2 = 10$

$$x^2 + (2x+1)^2 = 10$$
$$x^2 + 4x^2 + 4x + 1 = 10$$
$$5x^2 + 4x - 9 = 0$$
$$(5x+9)(x-1) = 0$$
$$5x + 9 = 0 \quad \text{or} \quad x - 1 = 0$$
$$x = -\frac{9}{5} \quad \text{or} \quad x = 1$$

Since $y = 2x + 1$, either

$y = 2\left(-\frac{9}{5}\right) + 1$ or $y = 2(1) + 1$

Therefore the graphs intersect at $\left(-\frac{9}{5}, -\frac{13}{5}\right)$ and (1, 3)

11.2 Independent Practice p. 579

17. 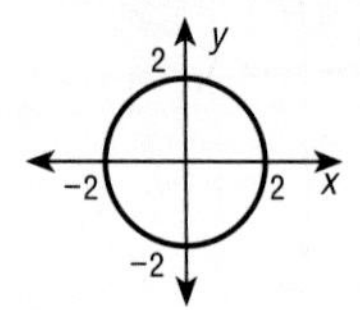

18. 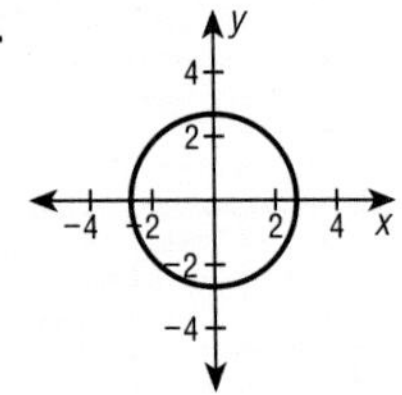

19. 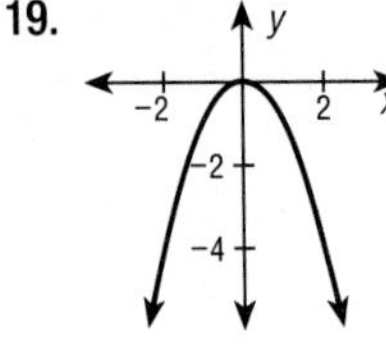

20. 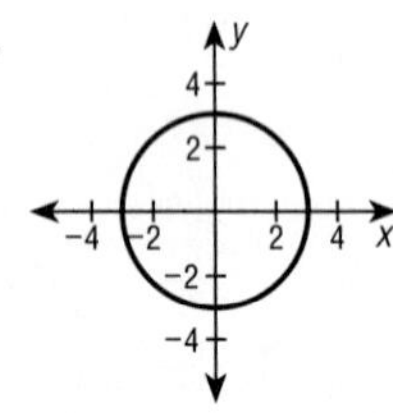

21. 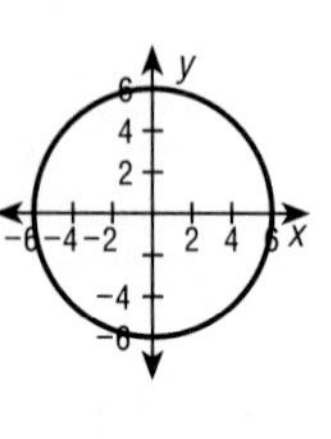

22. 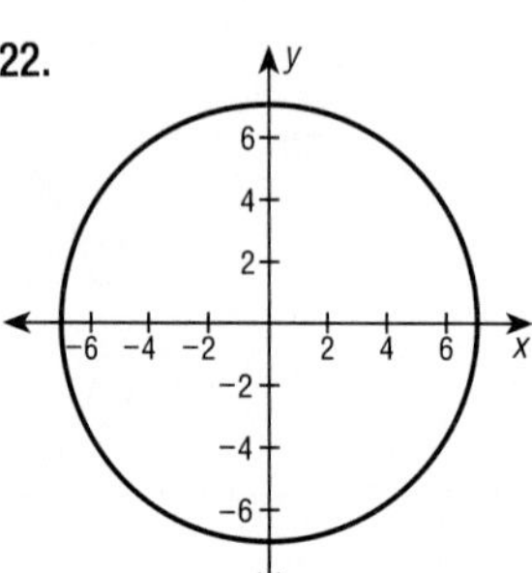

23. 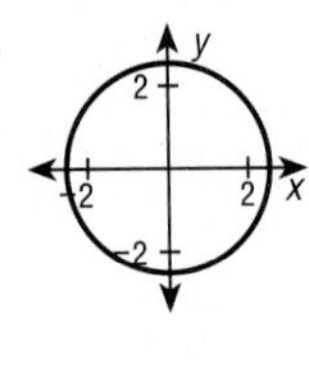

24. 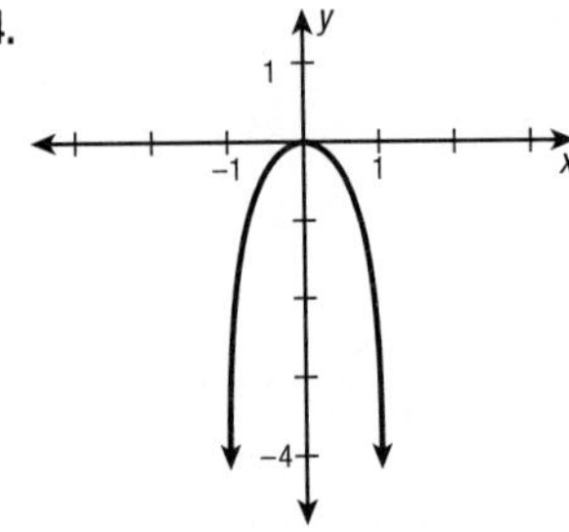

25. 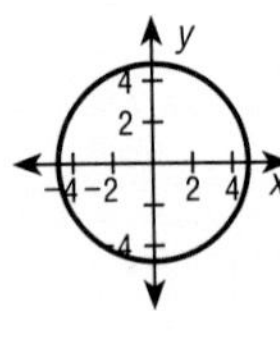

26. 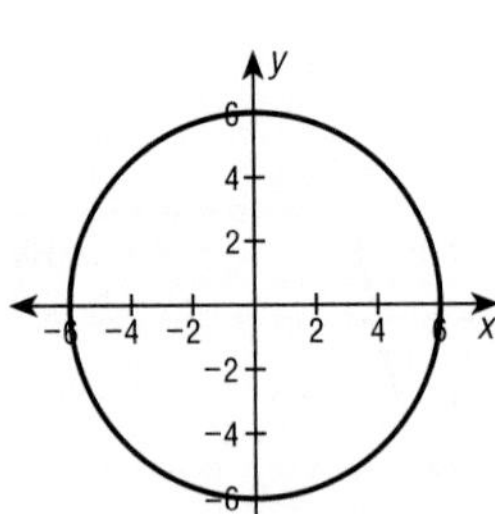

27. 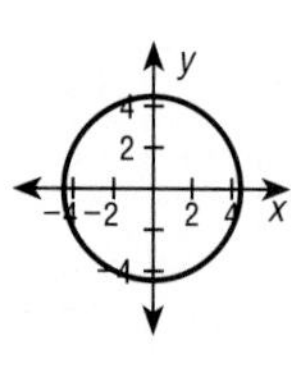

28. 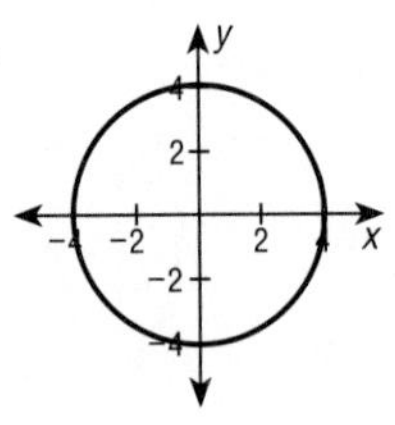

11.2 Mixed Review p. 582

1. 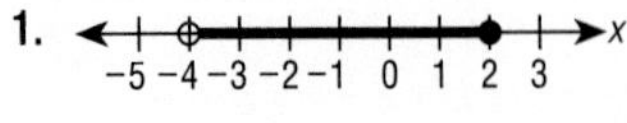

6. 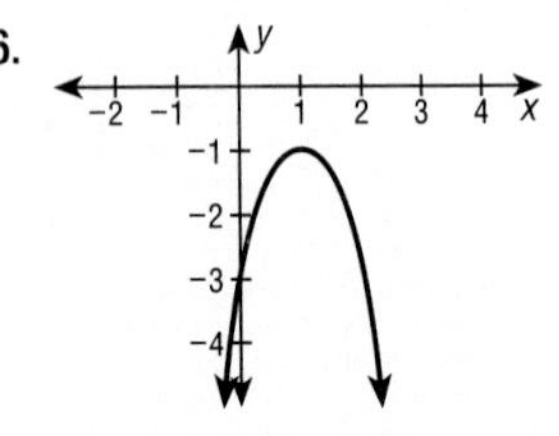

13. 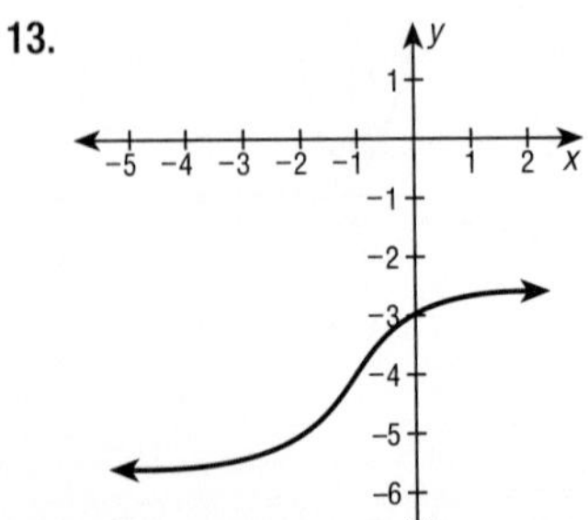

14. 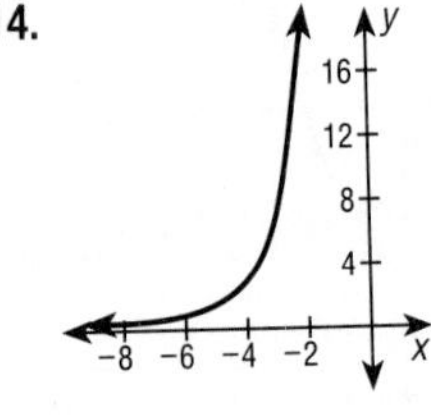

ADDITIONAL ANSWERS

11.3 Independent Practice pp. 586–587

7. $(-3, 0), (3, 0); (-5, 0), (5, 0)$ **8.** $(0, -5), (0, 5); (0, -13), (0, 13)$ **9.** $(0, -\sqrt{32}), (0, \sqrt{32}); (0, -6), (0, 6)$ **10.** $(0, -4), (0, 4);$ $(0, -5), (0, 5)$

11. $(0, -\sqrt{8}), (0, \sqrt{8}); (0, -3), (0, 3)$ **12.** $(-\sqrt{72}, 0), (\sqrt{72}, 0);$ $(-9, 0), (9, 0)$ **13.** $\frac{x^2}{16} + \frac{y^2}{49} = 1; (0, -\sqrt{33}),$ $(0, \sqrt{33}); (0, -7), (0, 7)$ **14.** $\frac{x^2}{64} + \frac{y^2}{49} = 1, (-\sqrt{15}, 0),$ $(\sqrt{15}, 0); (-8, 0), (8, 0)$

15. $\frac{x^2}{\frac{25}{4}} + \frac{y^2}{\frac{25}{9}} = 1;$ $\left(-\frac{\sqrt{125}}{6}, 0\right),$ $\left(\frac{\sqrt{125}}{6}, 0\right); \left(-\frac{5}{2}, 0\right),$ $\left(\frac{5}{2}, 0\right)$

16. $\frac{x^2}{4} + \frac{y^2}{\frac{1}{9}} = 1;$ $\left(-\frac{\sqrt{35}}{3}, 0\right), \left(\frac{\sqrt{35}}{3}, 0\right);$ $(-2, 0), (2, 0)$

17. $\frac{x^2}{90{,}000} + \frac{y^2}{315{,}000} = 1;$ $(0, -\sqrt{225{,}000}),$ $(0, \sqrt{225{,}000});$ $(0, -\sqrt{315{,}000}),$ $(0, \sqrt{315{,}000})$

18. $\frac{x^2}{576} + \frac{y^2}{400} = 1;$ $(-\sqrt{176}, 0), (\sqrt{176}, 0);$ $(-24, 0), (24, 0)$

19. $\frac{x^2}{25} + \frac{y^2}{36} = 1$ **20.** $\frac{x^2}{16} + \frac{y^2}{25} = 1$ **21.** $\frac{x^2}{16} + \frac{y^2}{9} = 1$ **22.** $\frac{x^2}{4} + \frac{y^2}{3} = 1$ **23.** $\frac{x^2}{16} + \frac{y^2}{7} = 1$ **24.** $\frac{x^2}{32} + \frac{y^2}{36} = 1$

29. **30.**

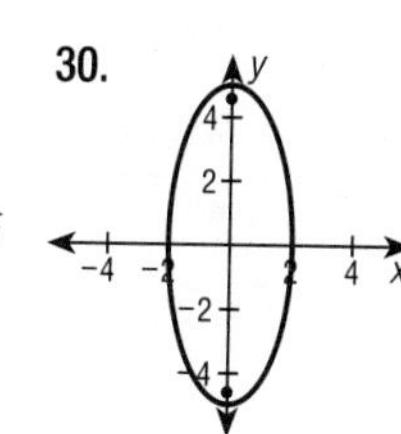

31. **32.**

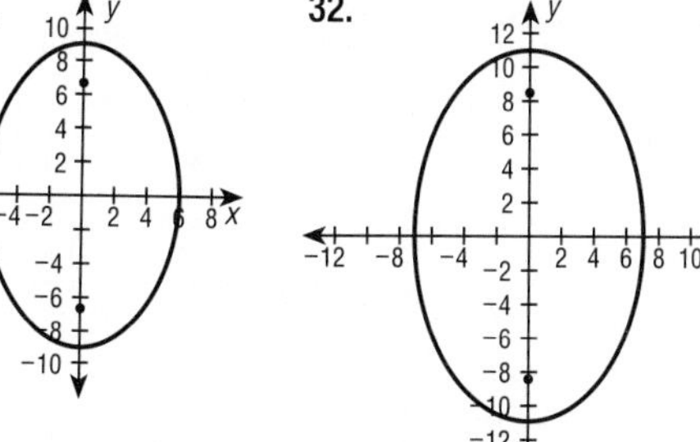

33.

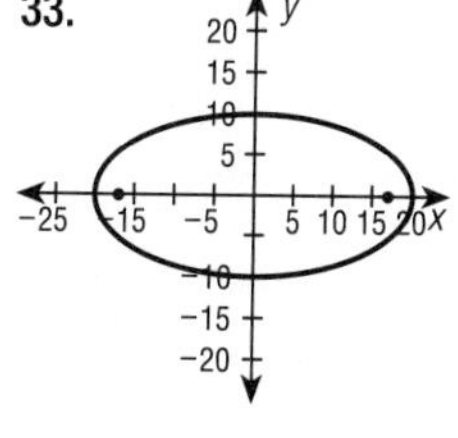

34.

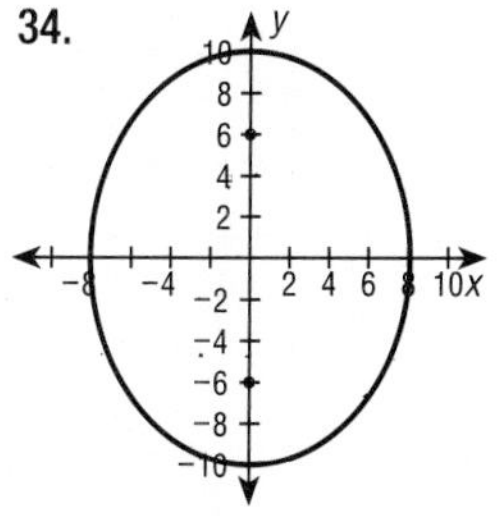

35. **36.**

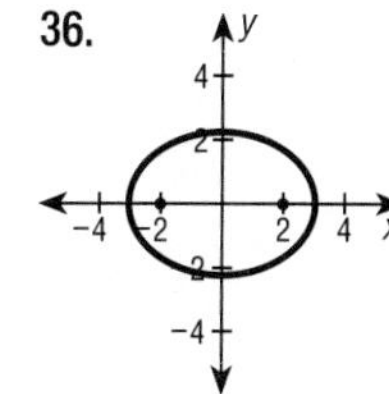

37.

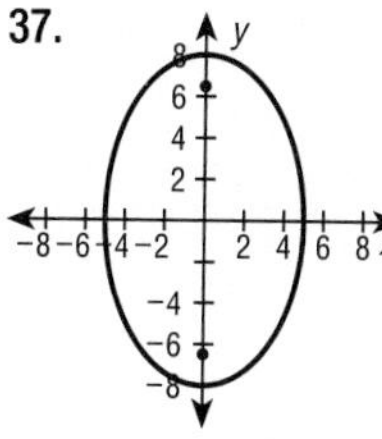

38.

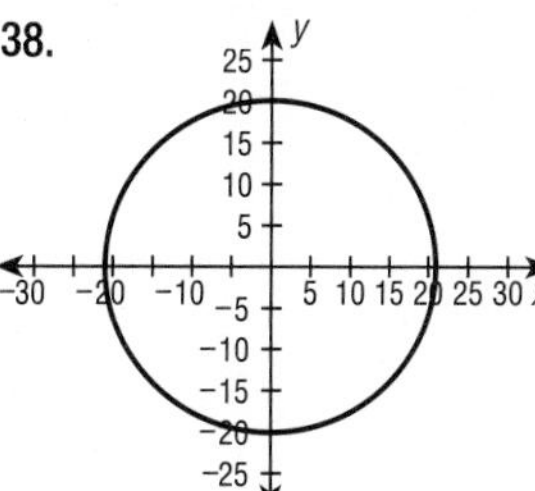

39.

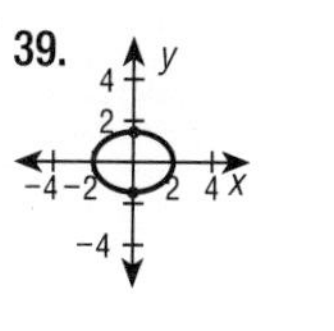

40.

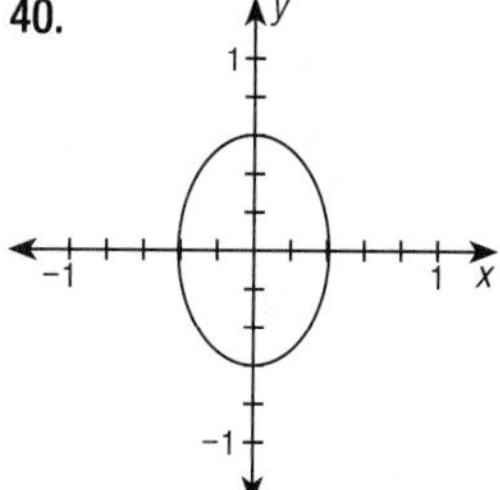

42. $\left(-\frac{4}{5} + \frac{2\sqrt{19}}{5}, \frac{1}{10} + \frac{\sqrt{19}}{5}\right), \left(-\frac{4}{5} - \frac{2\sqrt{19}}{5}, \frac{1}{10} - \frac{\sqrt{19}}{5}\right)$

43. $\left(\frac{5}{\sqrt{101}}, \frac{10}{\sqrt{101}}\right), \left(-\frac{5}{\sqrt{101}}, -\frac{10}{\sqrt{101}}\right)$

11.3 Integrated Review p. 588

63.

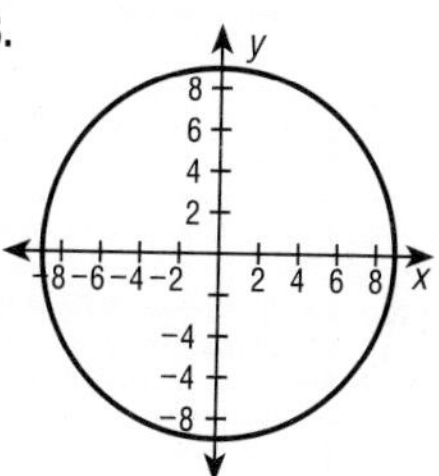

64.

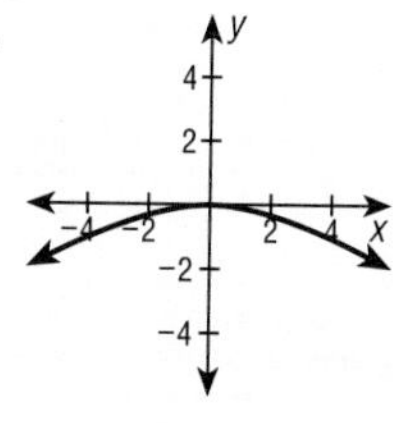

65.

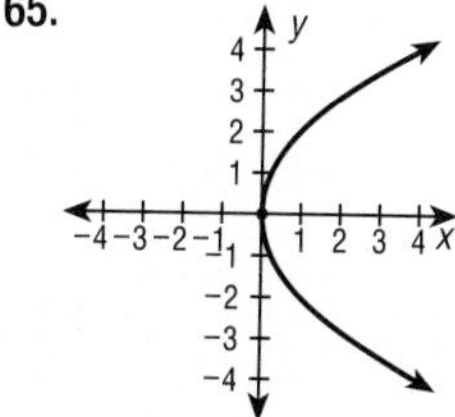

66.

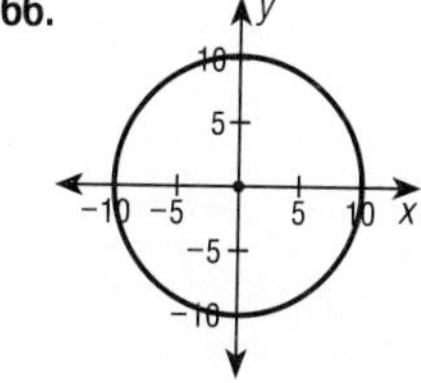

67.

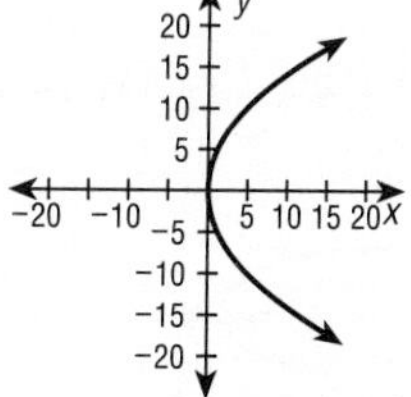

ADDITIONAL ANSWERS

11.4 Independent Practice pp. 594–595

7. $(-3, 0), (3, 0); (-\sqrt{58}, 0), (\sqrt{58}, 0)$

8. $(-12, 0), (12, 0); (-3\sqrt{17}, 0), (3\sqrt{17}, 0)$

9. $(-4, 0), (4, 0); (-2\sqrt{5}, 0), (\sqrt{5}, 0)$

10. $(0, -6), (0, 6); (0, -2\sqrt{10}), (0, \sqrt{10})$

11. $(0, -5), (0, 5); (0, -\sqrt{26}), (0, \sqrt{26})$

12. $(-1, 0), (1, 0); (-5\sqrt{2}, 0), (5\sqrt{2}, 0)$

13. $\frac{x^2}{4} - \frac{y^2}{25} = 1; (-\sqrt{29}, 0), (\sqrt{29}, 0); (-2, 0), (2, 0)$

14. $\frac{y^2}{64} - \frac{x^2}{1} = 1; (0, -\sqrt{65}), (0, \sqrt{65}); (0, -8), (0, 8)$

15. $\frac{y^2}{\frac{1}{4}} - \frac{x^2}{\frac{9}{4}} = 1; \left(0, -\frac{\sqrt{10}}{2}\right), \left(0, \frac{\sqrt{10}}{2}\right); \left(0, -\frac{1}{2}\right), \left(0, \frac{1}{2}\right)$

16. $\frac{x^2}{\frac{1}{4}} - \frac{y^2}{\frac{9}{16}} = 1; \left(-\frac{\sqrt{13}}{4}, 0\right), \left(\frac{\sqrt{13}}{4}, 0\right); \left(-\frac{1}{2}, 0\right), \left(\frac{1}{2}, 0\right)$

17. $\frac{y^2}{4} - \frac{x^2}{100} = 1; (0, -2\sqrt{26}), (0, 2\sqrt{26}); (0, -2), (0, 2)$

18. $\frac{x^2}{81} - \frac{y^2}{\frac{81}{4}} = 1; \left(-\frac{9\sqrt{5}}{2}, 0\right), \left(\frac{9\sqrt{5}}{2}, 0\right); (-9, 0), (9, 0)$

29.
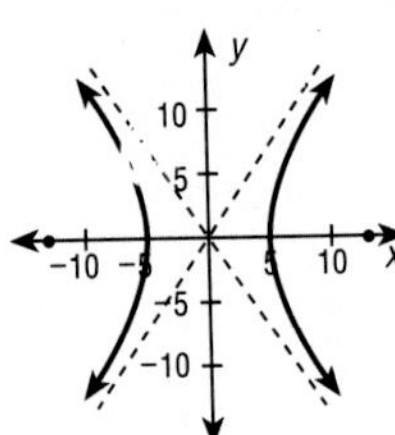

30.
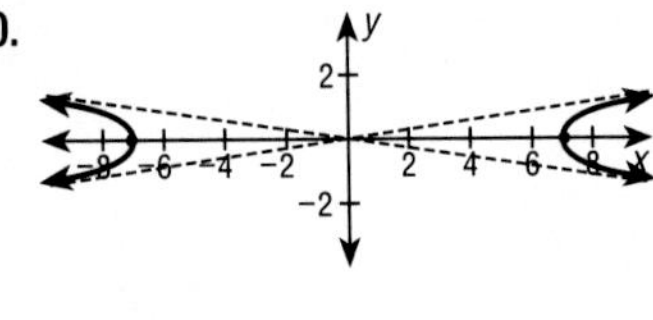

31.
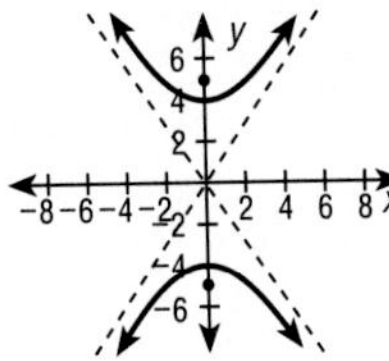

32.
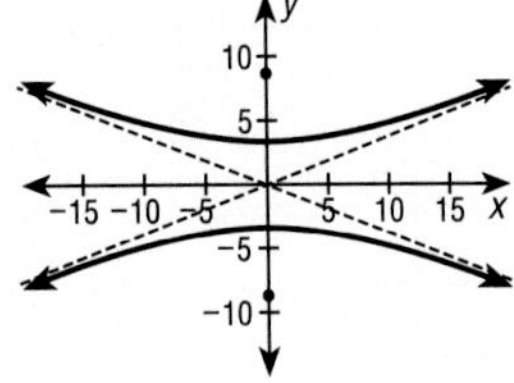

33.
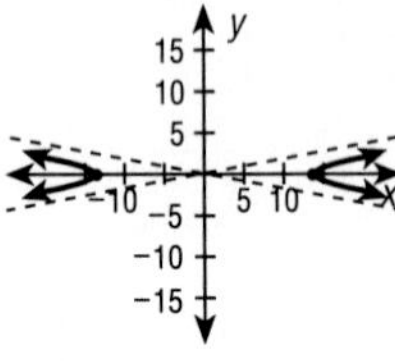

34.
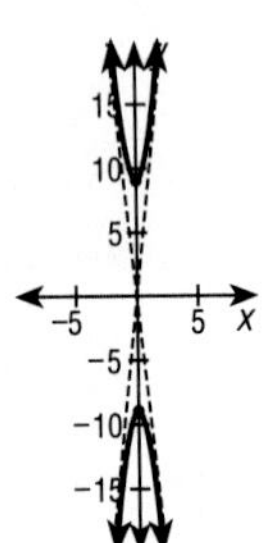

35.
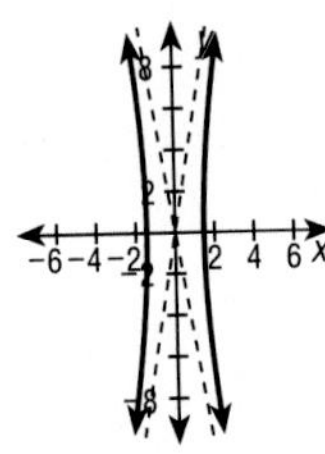

36.
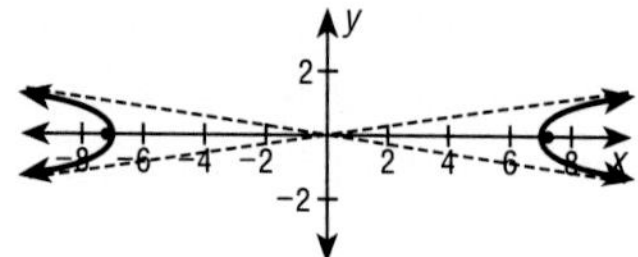

37.
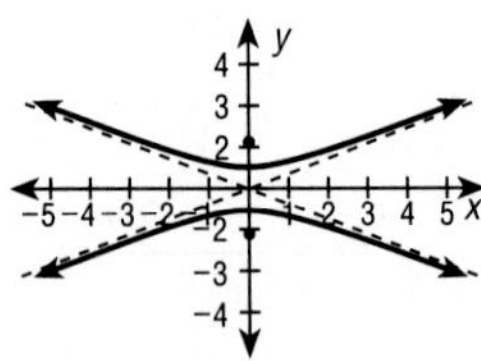

38.
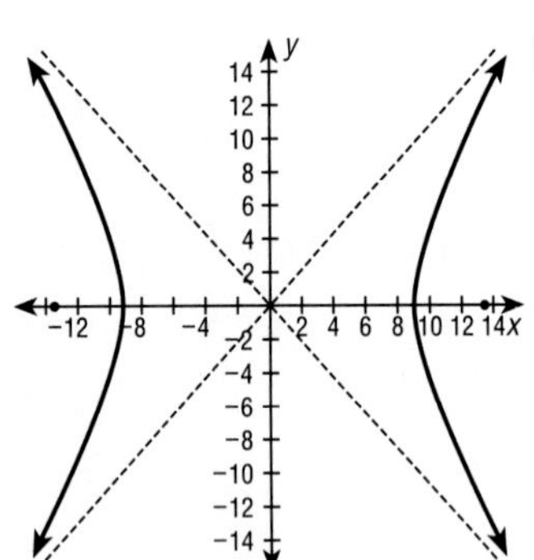

39.
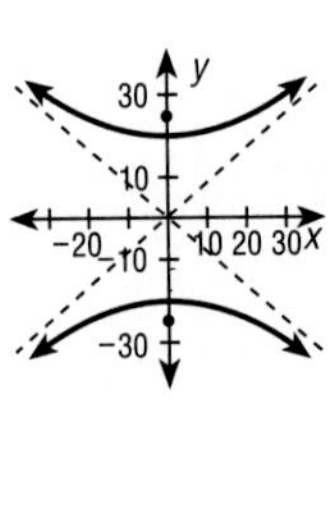

40.
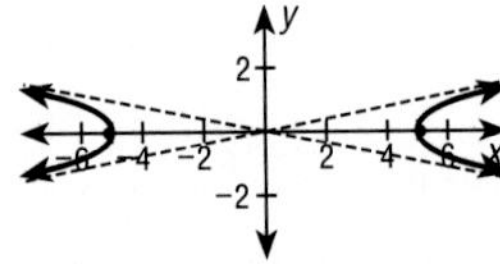

11.4 Integrated Review p. 596

53.
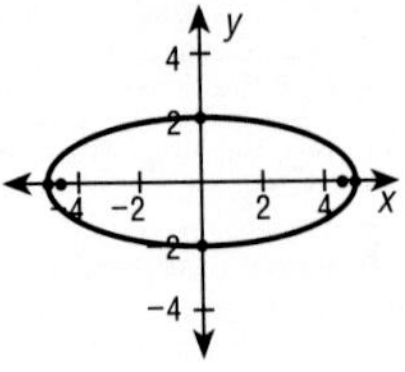

54.
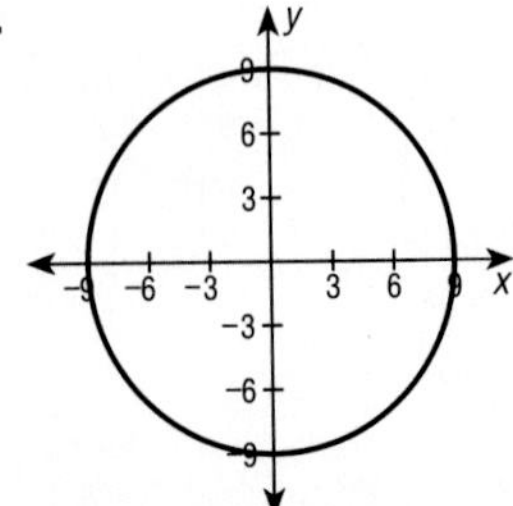

55.
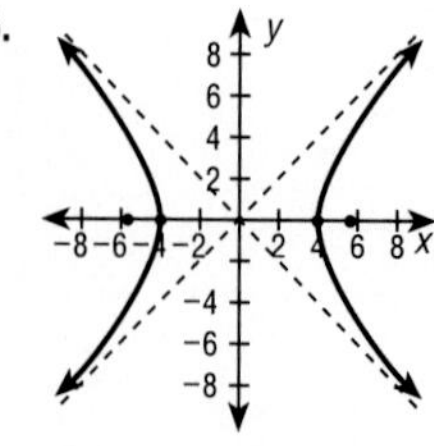

56.
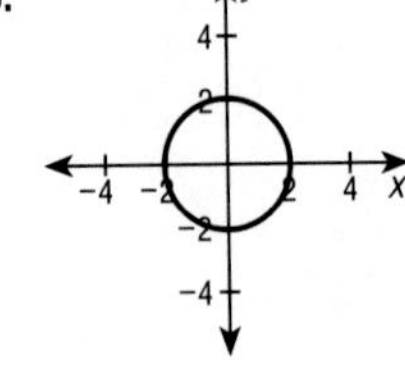

11.4 Integrated Review *(continued)*

57.

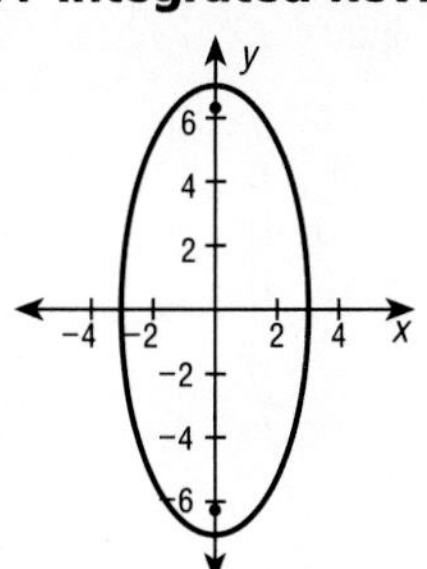

58.

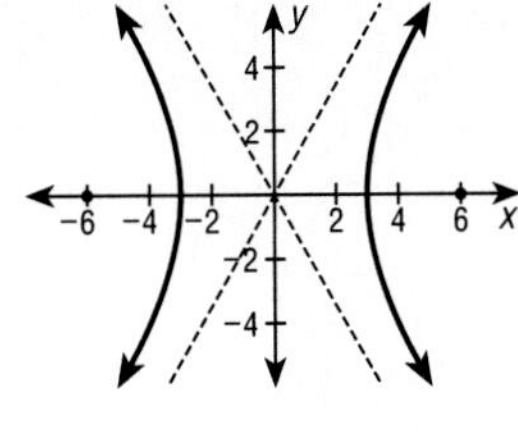

59.

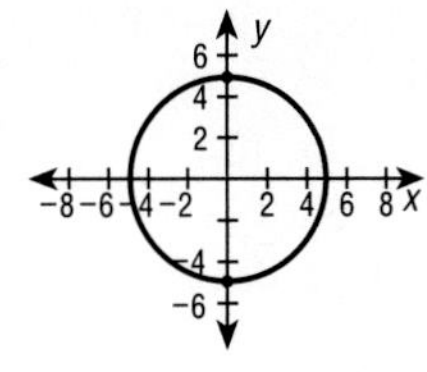

60.

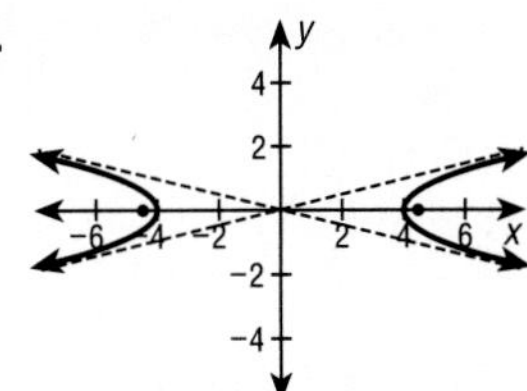

11.4 Exploration and Extension p. 596

63.

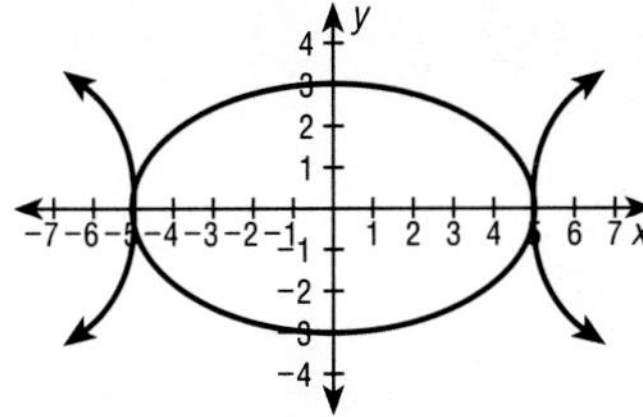

64.

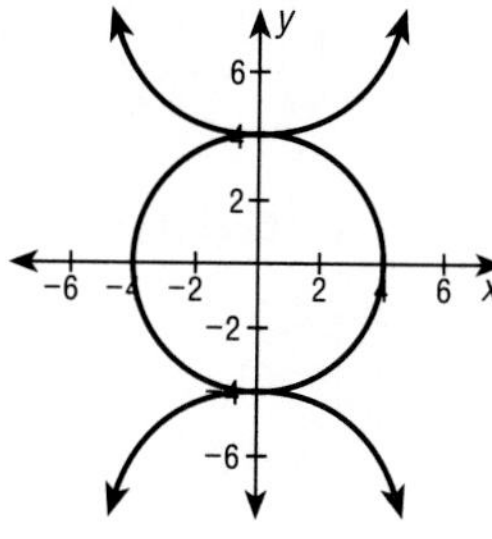

11.5 Independent Practice p. 602

23. $(x - 1)^2 + (y + 3)^2 = 1$; 1, $(1, -3)$

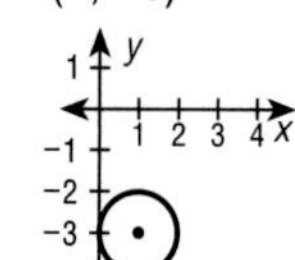

24. $(x - 4)^2 + (y - 10)^2 = 1$; 1, $(4, 10)$

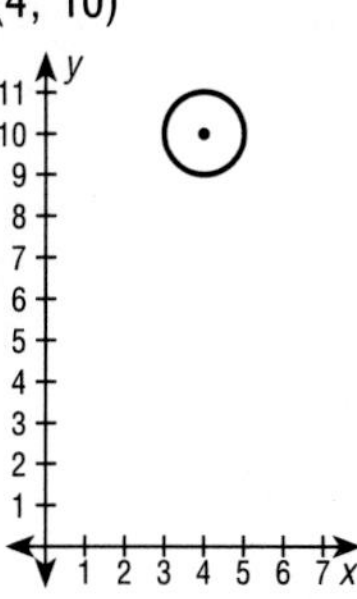

25. $(x - 3)^2 + (y - 4)^2 = 1$; 1, $(3, 4)$

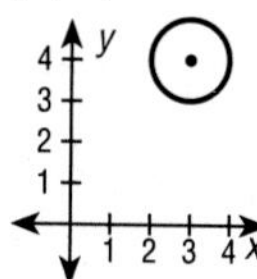

26. $(x + 5)^2 + (y - 3)^2 = 1$; 1, $(-5, 3)$

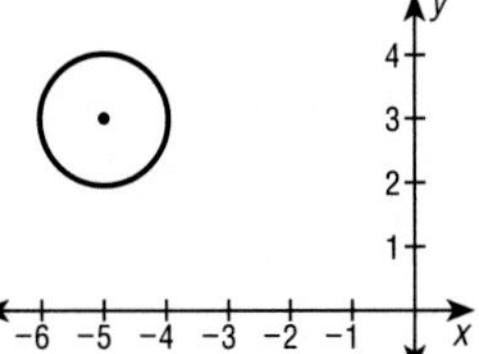

27. $(y - 2)^2 = 4(-3)(x - 1)$; $(1, 2)$, $(-2, 2)$

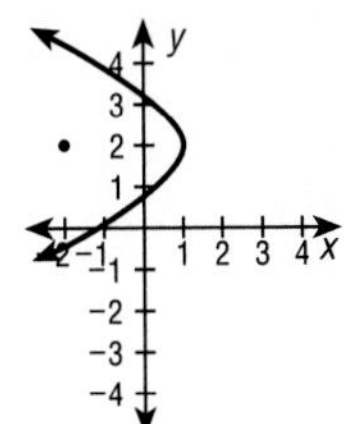

28. $(y - 6)^2 = 4(-1)(x - 8)$; $(8, 6)$, $(7, 6)$

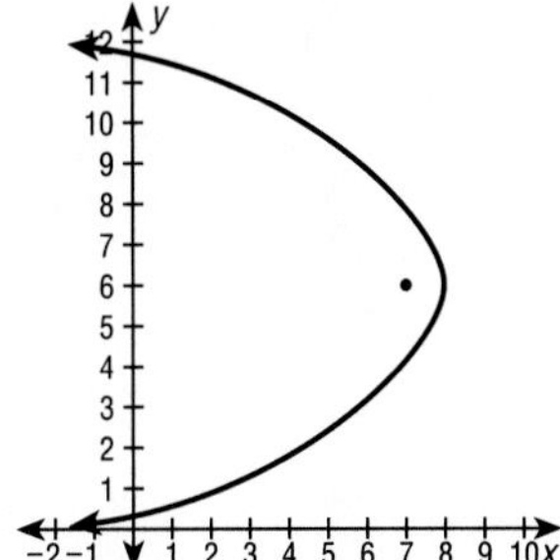

29. $(x + 2)^2 = 4(2)(y - 1)$; $(-2, 1)$, $(-2, 3)$

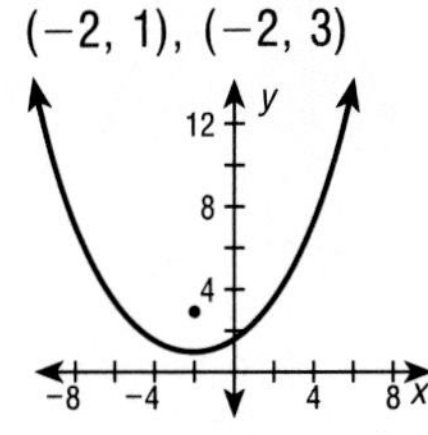

30. $(x + 3)^2 = 4\left(\frac{1}{2}\right)(y - 2)$; $(-3, 2)$, $\left(-3, 2\frac{1}{2}\right)$

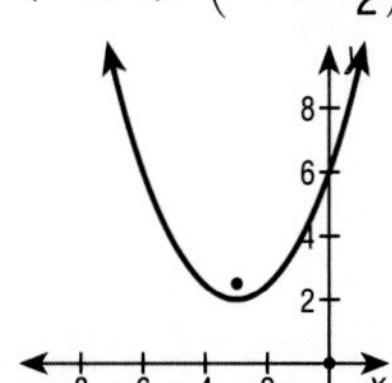

31. $\frac{(x - 2)^2}{1} + \frac{(y - 1)^2}{25} = 1$; $(2, 6)$, $(2, -4)$; $(2, 1 + 2\sqrt{6})$, $(2, 1 - 2\sqrt{6})$

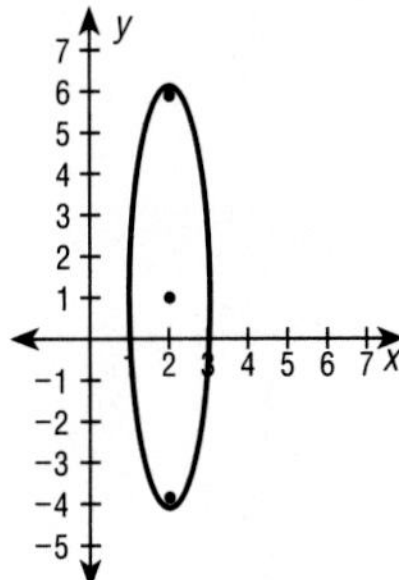

32. $\frac{(x - 5)^2}{4} + \frac{(y - 5)^2}{1} = 1$; $(7, 5)$, $(3, 5)$; $(5 + \sqrt{3}, 5)$, $(5 - \sqrt{3}, 5)$

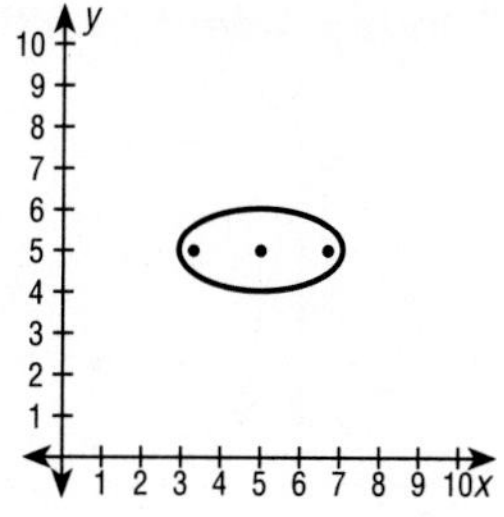

ADDITIONAL ANSWERS

11.5 Independent Practice *(continued)*

33. $\dfrac{(x-8)^2}{36}+\dfrac{(y-1)^2}{1}=1$; $(14, 1), (2, 1)$; $(8+\sqrt{35}, 1)$, $(8-\sqrt{35}, 1)$

34. $\dfrac{(x-6)^2}{25}+\dfrac{(y-2)^2}{100}=1$; $(6, -8), (6, 12)$; $(6, 2-5\sqrt{3})$, $(6, 2+5\sqrt{3})$

35. $\dfrac{(x-7)^2}{25}-\dfrac{(y-2)^2}{1}=1$; $(12, 2), (2, 2)$; $(7+\sqrt{26}, 2)$, $(7-\sqrt{26}, 2)$

36. $\dfrac{(x-4)^2}{1}-\dfrac{(y-4)^2}{9}=1$; $(5, 4), (3, 4)$; $(4+\sqrt{10}, 4)$, $(4-\sqrt{10}, 4)$

37. $\dfrac{(y-2)^2}{36}-\dfrac{(x+2)^2}{16}=1$; $(-2, -4), (-2, 8)$; $(-2, 2+\sqrt{52})$, $(-2, 2-\sqrt{52})$

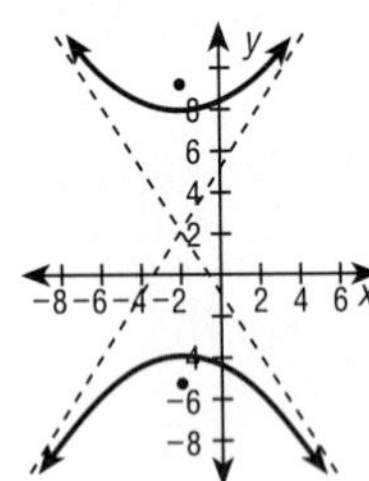

38. $\dfrac{(y-6)^2}{4}-\dfrac{(x+1)^2}{4}=1$; $(-1, 4), (-1, 8)$; $(-1, 6+\sqrt{8})$, $(-1, 6-\sqrt{8})$

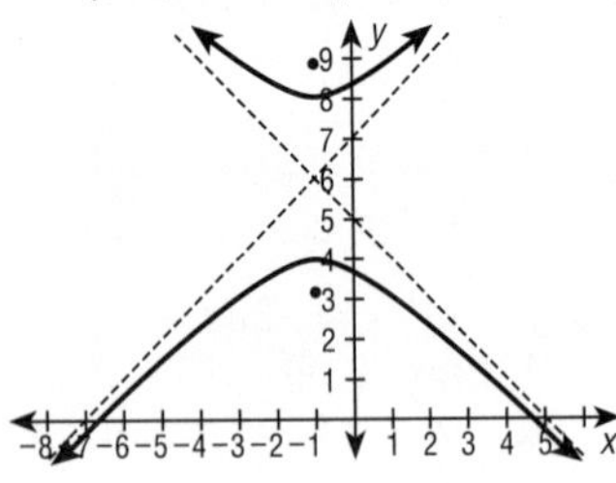

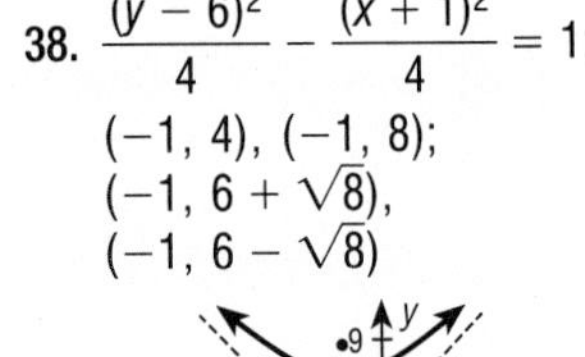

11.5 Integrated Review p. 603

47.

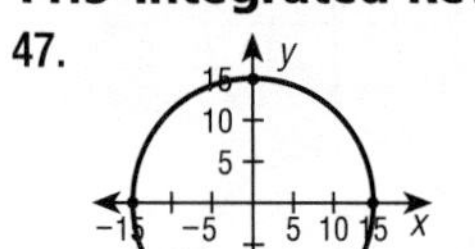

48.

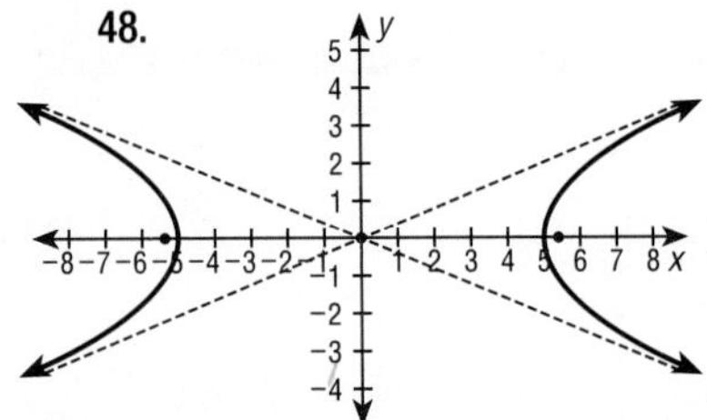

49.

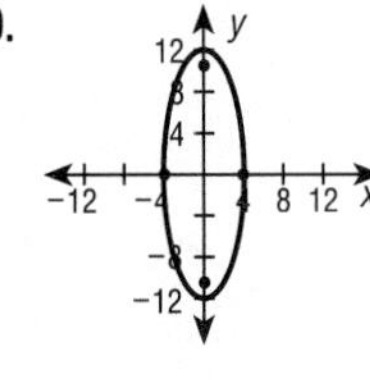

50.

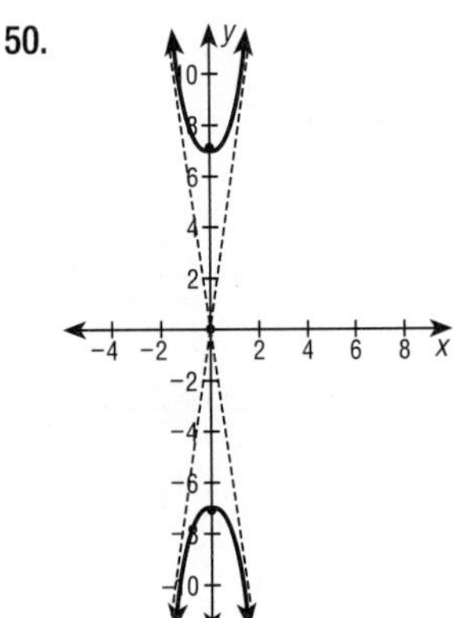

11.5 Exploration and Extension p. 603

51. $\dfrac{(x+2)^2}{4}+\dfrac{(y+2)^2}{36}=1$

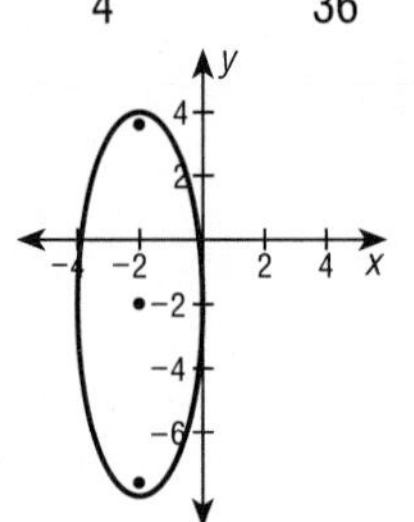

52. $\dfrac{(x-1)^2}{1}-\dfrac{(y+3)^2}{4}=1$

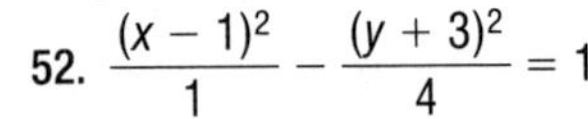

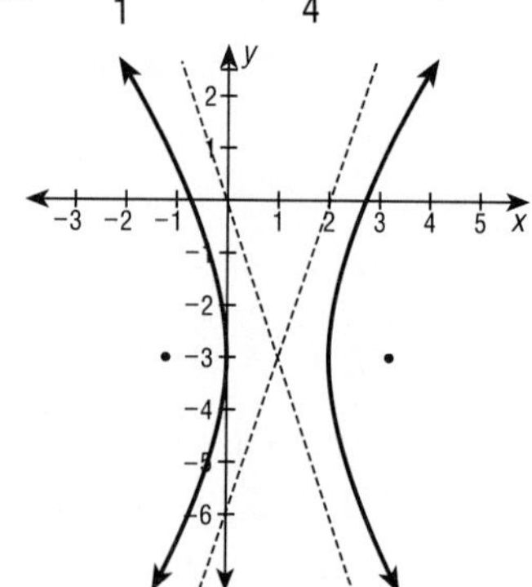

11.6 Independent Practice p. 611

31.

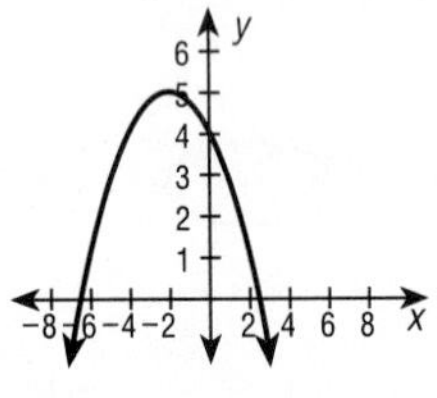

32.

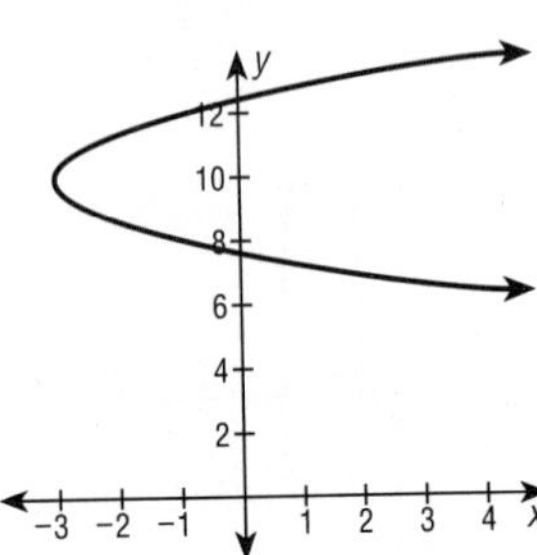

33.

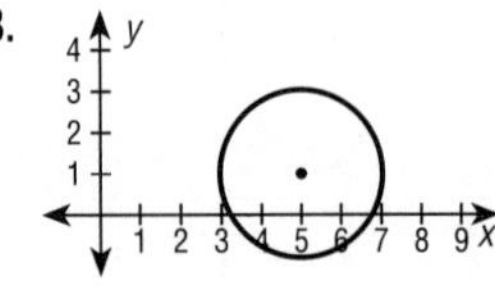

34.

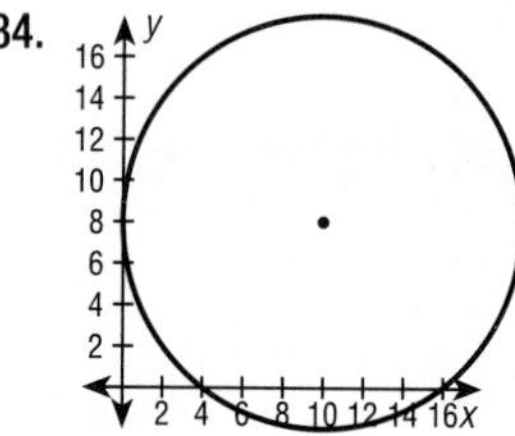

11.6 Independent Practice *(continued)*

35.

36.

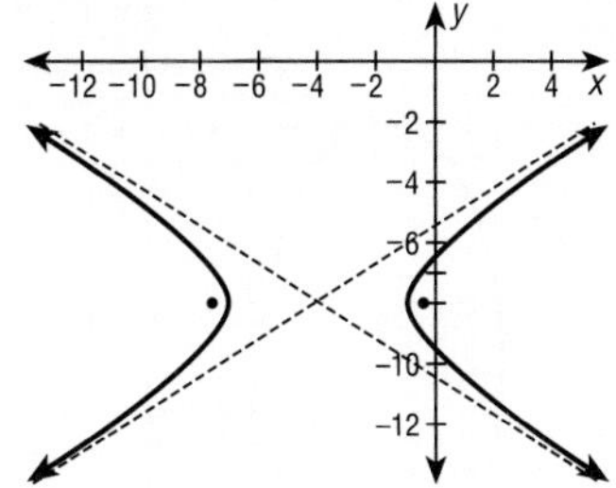

37.

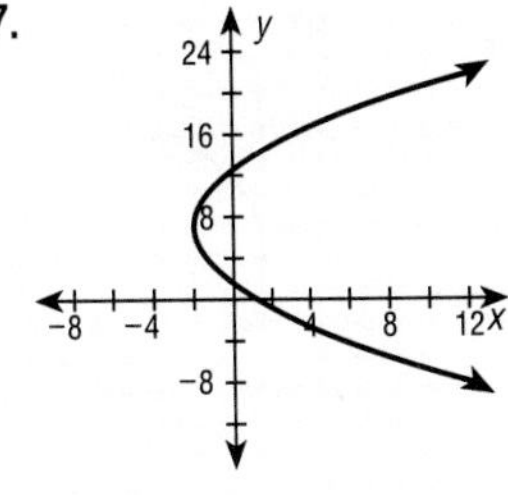

38.

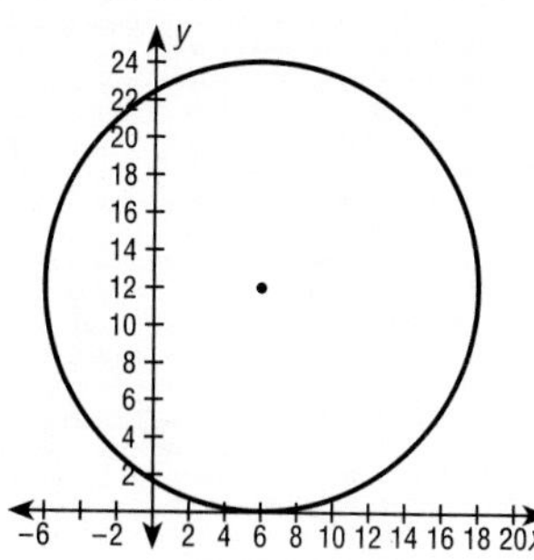

39.

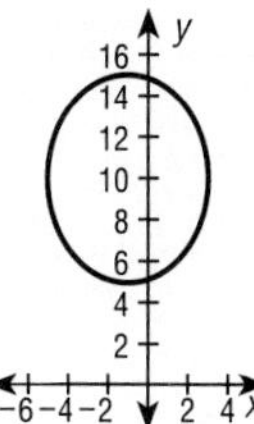

40.

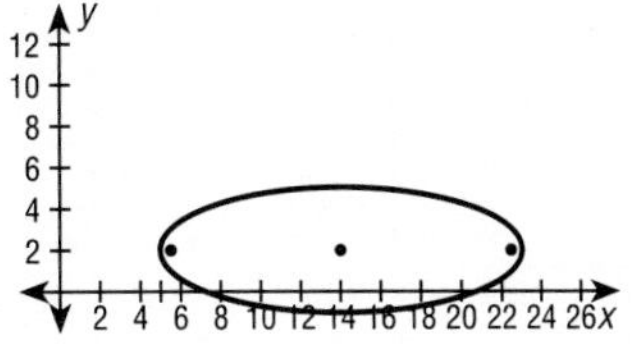

41.

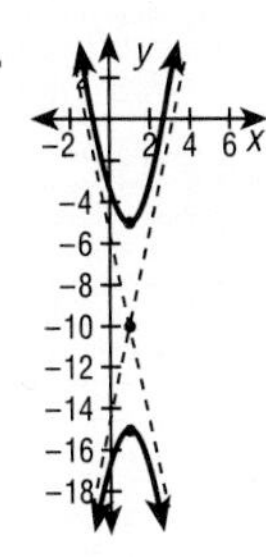

42.

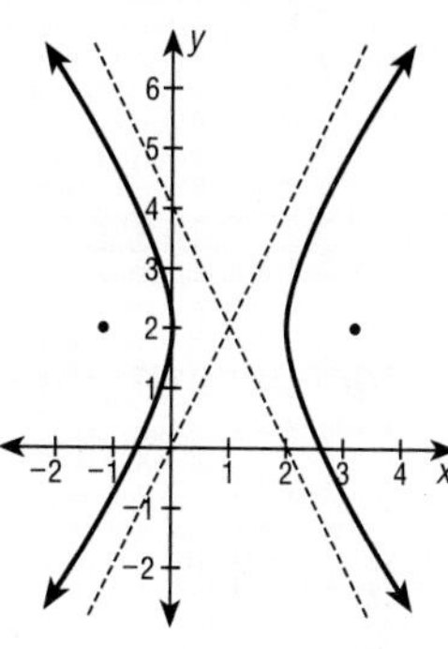

11.6 Exploration and Extension p. 613

65.

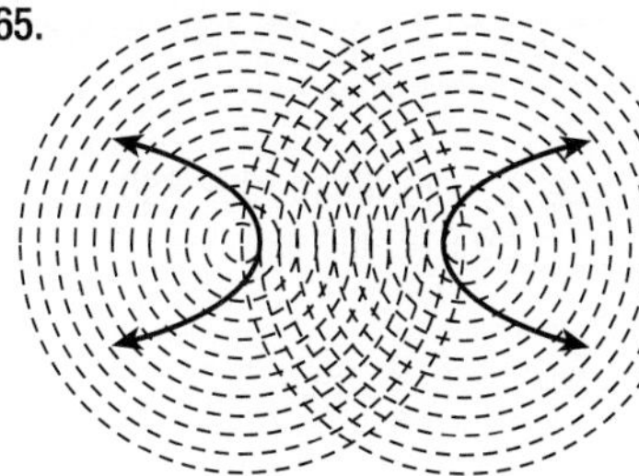

66.

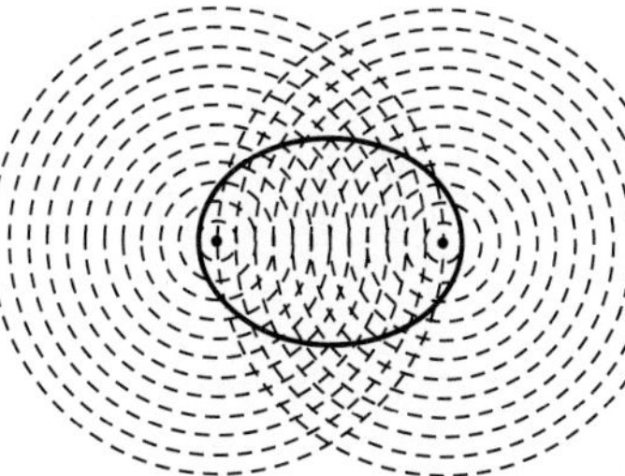

67.

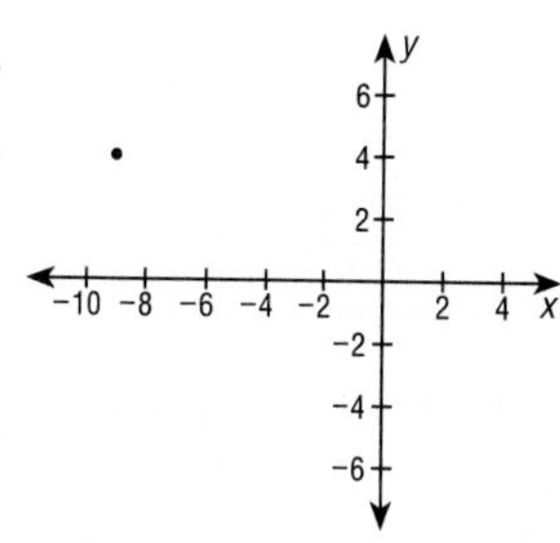

68.

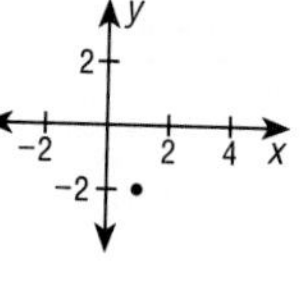

11.6 Chapter Review p. 616

32.

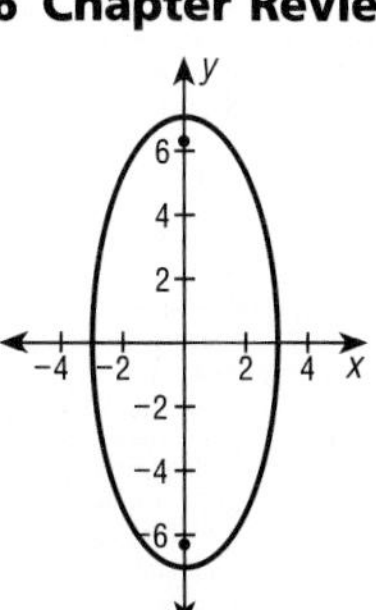

33.

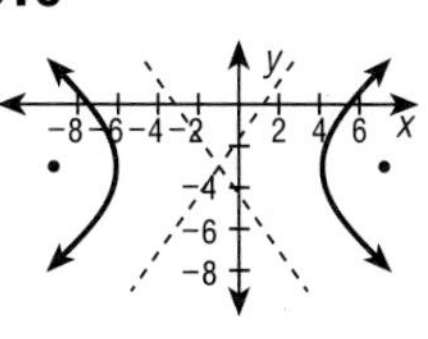

34.

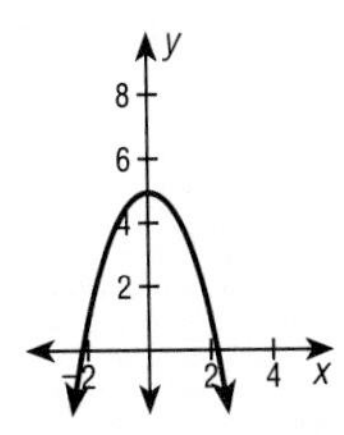

35.

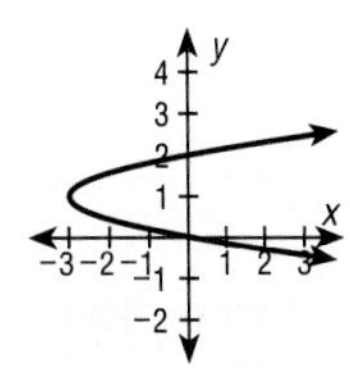

36.

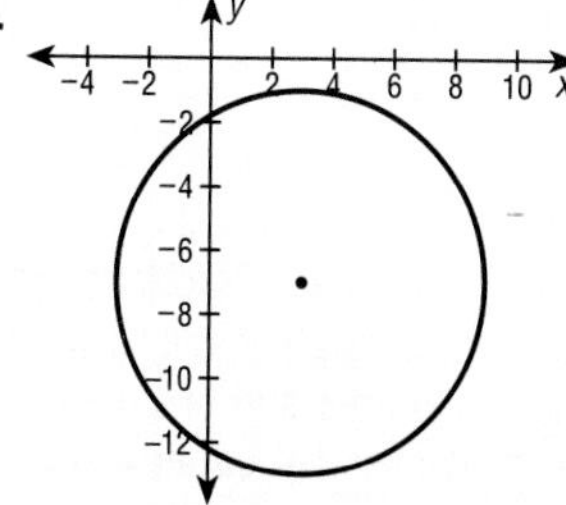

37.

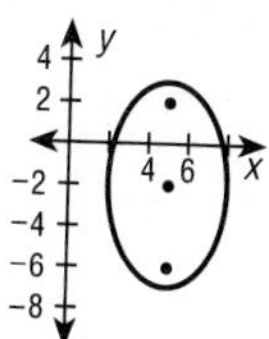

38.

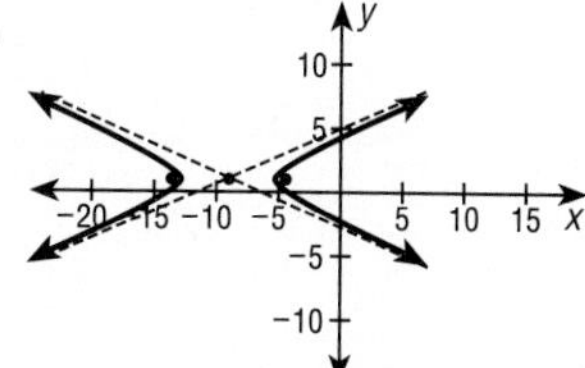

39.

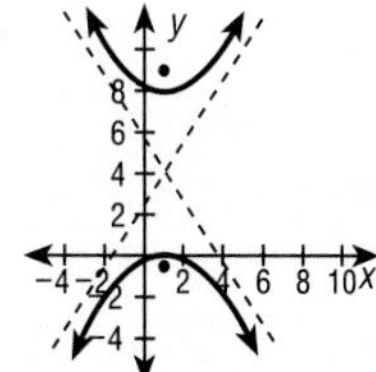

40.

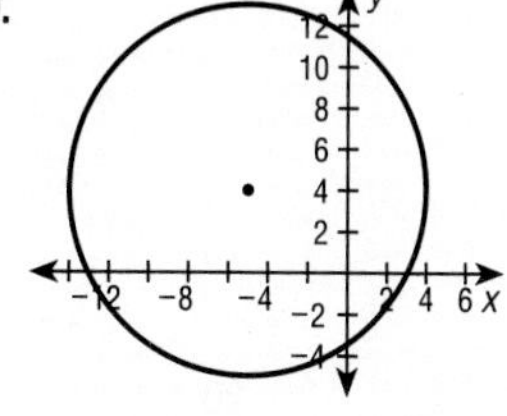

11.6 Chapter Test p. 619

5. 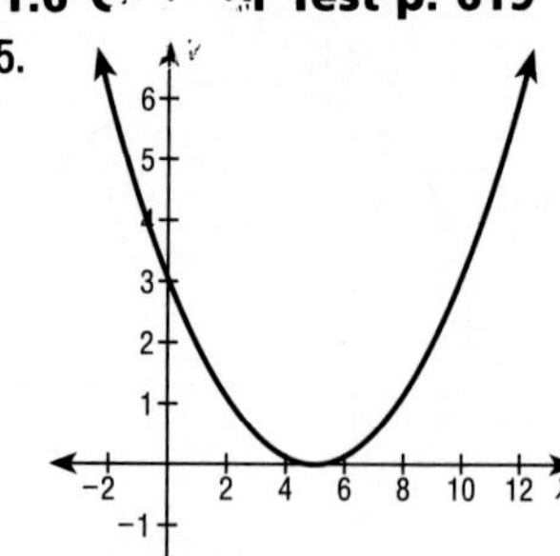

6. 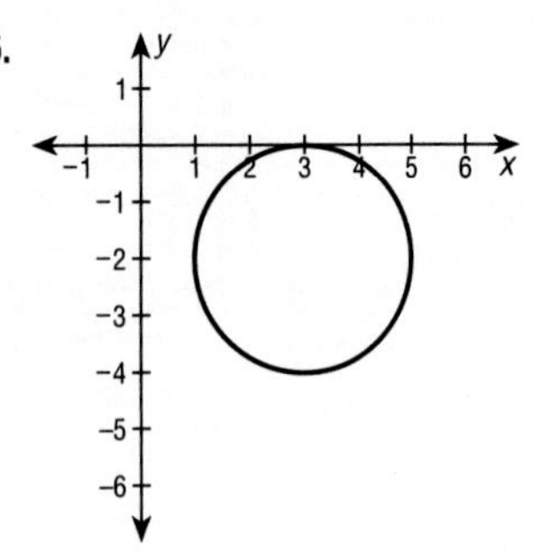

7. 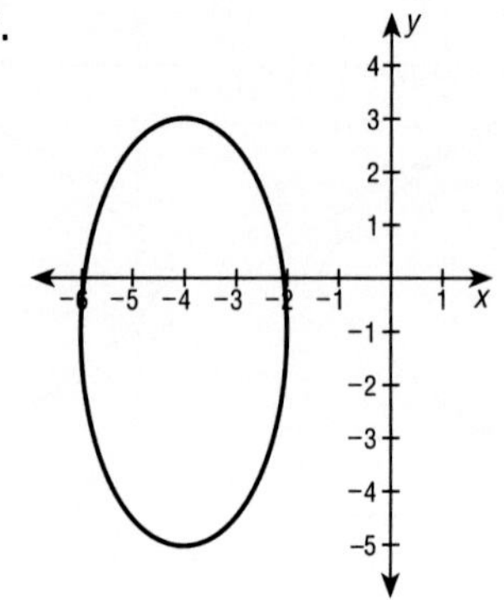

8. 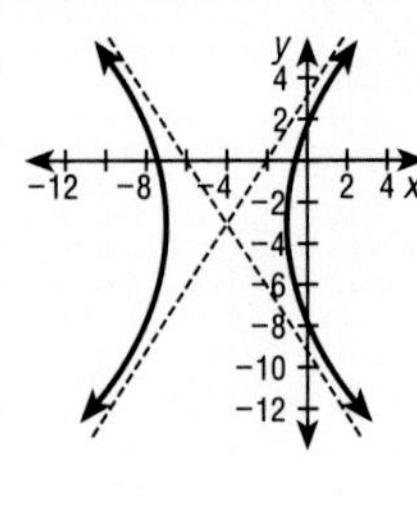

CHAPTER 12

12.1 Independent Practice p. 626

5. $-2, 1, 4, 7, 10$

6. $3, 2, 1, 0, -1$

7. $3, 6, 11, 18, 27$

8. $4, 9, 16, 25, 36$

9. $2, 4, 8, 16, 32$

10. $\frac{1}{2}, \frac{1}{4}, \frac{1}{8}, \frac{1}{16}, \frac{1}{32}$

11. $-\frac{1}{2}, \frac{1}{4}, -\frac{1}{8}, \frac{1}{16}, -\frac{1}{32}$

12. $1, -1, 1, -1, 1$

13. $\frac{1}{2}, \frac{2}{3}, \frac{3}{4}, \frac{4}{5}, \frac{5}{6}$

14. $\frac{1}{2}, 1, \frac{3}{2}, 2, \frac{5}{2}$

15. $-2, 4, -8, 16, -32$

16. $-5, 25, -125, 625, -3125$

17. $-1, 0, 4, 22, 118$

18. $\frac{1}{2}, 1, 3, 12, 60$

19. $3, \frac{9}{2}, \frac{9}{2}, \frac{27}{8}, \frac{81}{40}$

20. 6, 120, 5040, 362,880, 39,916,800

21. 2, 4, 12, 48, 240

22. 12, 48, 240, 1440, 10,080

23. 1, 1, 2, 6, 24

24. $1, 2, \frac{3}{2}, \frac{2}{3}, \frac{5}{24}$

25. 3, 5, 3, 5, 3

26. $-2, 2, -2, 2, -2$

27. 3, 5, 3, 5, 3

28. $2, -2, 2, -2, 2$

43. $\sum_{i=1}^{6} 2i$

44. $\sum_{i=1}^{6} (2i + 1)$

45. $\sum_{i=1}^{5} 3^i$

46. $\sum_{i=1}^{4} \left(\frac{1}{4}\right)^i$

12.5 Communicating About Algebra p. 654

Half of the marbles fall to the left and half fall to the right; then each number on the end of a row is half the number diagonally above it, and each number not on an end of a row is the sum of half of each of the two numbers diagonally above it.

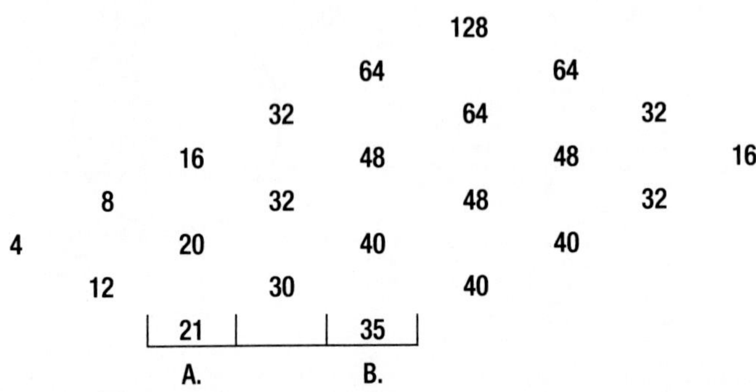

12.5 Independent Practice pp. 655–656

15. $a^3 + 6a^2 + 12a + 8$

16. $x^4 - 20x^3 + 150x^2 - 500x + 625$

17. $x^4 + 8x^3 + 24x^2 + 32x + 16$

18. $16x^4 - 32x^3 + 24x^2 - 8x + 1$

19. $x^5 - 15x^4 + 90x^3 - 270x^2 + 405x - 243$

20. $x^6 - 48x^5 + 960x^4 - 10{,}240x^3 + 61{,}440x^2 - 196{,}608x + 262{,}144$

21. $64x^6 + 192x^5 + 240x^4 + 160x^3 + 60x^2 + 12x + 1$

22. $x^7 + 7x^6y + 21x^5y^2 + 35x^4y^3 + 35x^3y^4 + 21x^2y^5 + 7xy^6 + y^7$

29. $64x^3 - 48x^2y + 12xy^2 - y^3$

30. $27s^3 + 54s^2 + 36s + 8$

31. $x^4 + 16x^3 + 96x^2 + 256x + 256$

32. $256t^4 - 256t^3 + 96t^2 - 16t + 1$

33. $32t^5 - 80t^4s + 80t^3s^2 - 40t^2s^3 + 10ts^4 - s^5$

34. $x^5 + 35x^4 + 490x^3 + 3430x^2 + 12{,}005x + 16{,}807$

35. $64x^6 + 192x^5y + 240x^4y^2 + 160x^3y^3 + 60x^2y^4 + 12xy^5 + y^6$

36. $x^6 + 12x^5 + 60x^4 + 160x^3 + 240x^2 + 192x + 64$

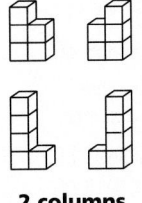

37.

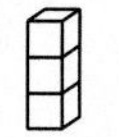

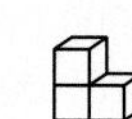
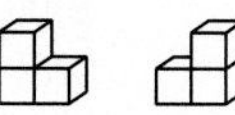

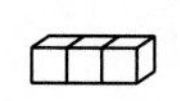

38.

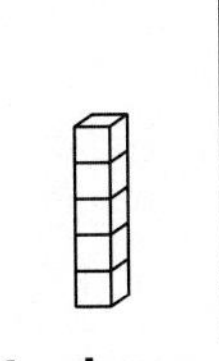

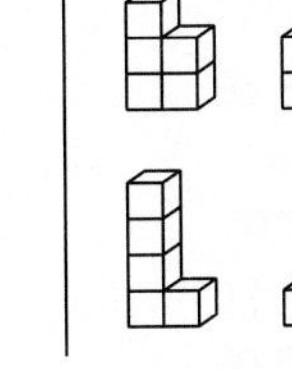

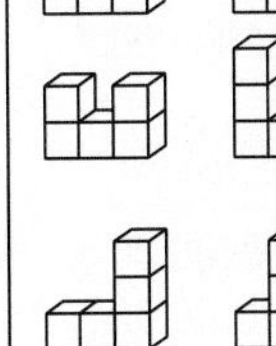
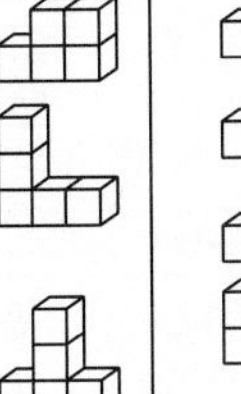

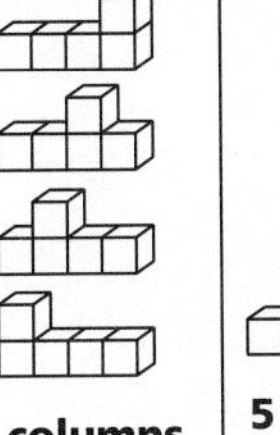

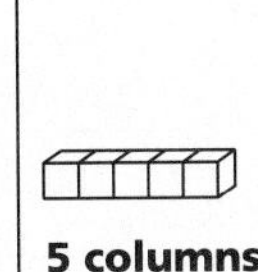

3 cubes: 1, 2, 1;
4 cubes: 1, 3, 3, 1;
5 cubes: 1, 4, 6, 4, 1—the pattern in Pascal's triangle

12.5 Exploration and Extension p. 657

49.

$$\binom{n}{m} \stackrel{?}{=} \binom{n}{n-m}$$

$$\frac{n!}{(n-m)!m!} \stackrel{?}{=} \frac{n!}{(n-(n-m))!(n-m)!}$$

$$\frac{n!}{(n-m)!m!} = \frac{n!}{m!(n-m)!}$$

50.

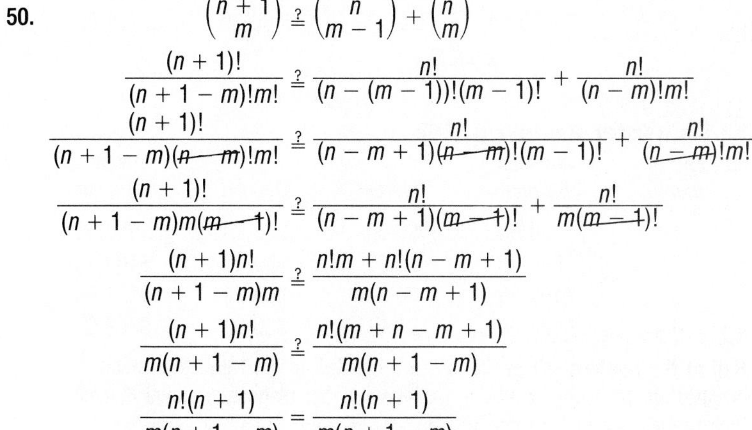

$$\binom{n+1}{m} \stackrel{?}{=} \binom{n}{m-1} + \binom{n}{m}$$

$$\frac{(n+1)!}{(n+1-m)!m!} \stackrel{?}{=} \frac{n!}{(n-(m-1))!(m-1)!} + \frac{n!}{(n-m)!m!}$$

$$\frac{(n+1)!}{(n+1-m)(\cancel{n-m})!m!} \stackrel{?}{=} \frac{n!}{(n-m+1)(\cancel{n-m})!(m-1)!} + \frac{n!}{\cancel{(n-m)}!m!}$$

$$\frac{(n+1)!}{(n+1-m)m(\cancel{m-1})!} \stackrel{?}{=} \frac{n!}{(n-m+1)(\cancel{m-1})!} + \frac{n!}{m(\cancel{m-1})!}$$

$$\frac{(n+1)n!}{(n+1-m)m} \stackrel{?}{=} \frac{n!m + n!(n-m+1)}{m(n-m+1)}$$

$$\frac{(n+1)n!}{m(n+1-m)} \stackrel{?}{=} \frac{n!(m+n-m+1)}{m(n+1-m)}$$

$$\frac{n!(n+1)}{m(n+1-m)} = \frac{n!(n+1)}{m(n+1-m)}$$

52.

8 1 ← row 8
36 9 1 ← row 9
45 10 ← row 10

15 6 ← row 6
35 21 7 ← row 7
56 28 ← row 8

10 45 ← row 10
11 55 70 ← row 11
66 220 ← row 12

21 35 ← row 7
28 56 70 ← row 8
84 126 ← row 9

12.5 Mixed Review p. 657

5.

11. 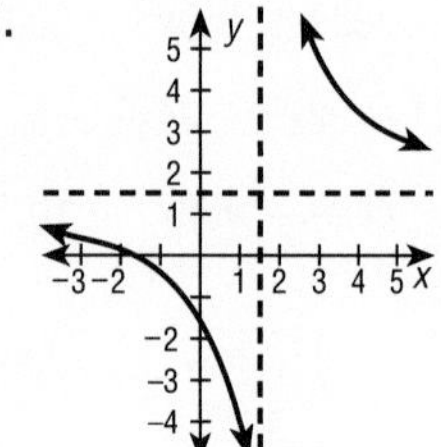

16. 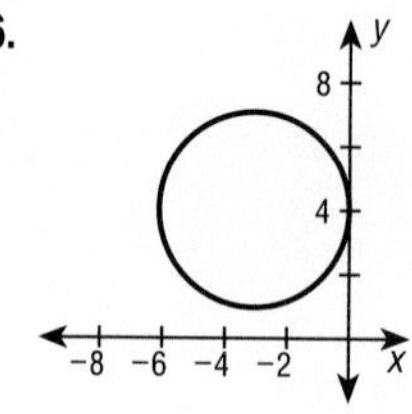

12.6 Independent Practice p. 661

7.

End of year	*Previous balance*	*Interest*	*Deposit*	*New balance*
1	$0.00	$0.00	$1000.00	$1000.00
2	$1000.00	$80.00	$1000.00	$2080.00
3	$2080.00	$166.40	$1000.00	$3246.40
4	$3246.40	$259.71	$1000.00	$4506.11
5	$4506.11	$360.49	$1000.00	$5866.60

8.

End of year	*Previous balance*	*Interest*	*Deposit*	*New balance*
1	$0.00	$0.00	$300.00	$300.00
2	$300.00	$4.88	$300.00	$604.88
3	$604.88	$9.83	$300.00	$914.71
4	$914.71	$14.86	$300.00	$1229.57

9.

End of halfyear	*Previous balance*	*Interest*	*Deposit*	*New balance*
1	$0.00	$0.00	$500.00	$500.00
2	$500.00	$15.00	$500.00	$1015.00
3	$1015.00	$30.45	$500.00	$1545.45
4	$1545.45	$46.36	$500.00	$2091.81
5	$2091.81	$62.75	$500.00	$2654.56
6	$2654.56	$79.64	$500.00	$3234.20
7	$3234.20	$97.03	$500.00	$3831.23
8	$3831.23	$114.94	$500.00	$4446.17
9	$4446.17	$133.39	$500.00	$5079.56
10	$5079.56	$152.39	$500.00	$5731.95

10.

End of month	*Previous balance*	*Interest*	*Deposit*	*New balance*
1	$0.00	$0.00	$100.00	$100.00
2	$100.00	$0.54	$100.00	$200.54
3	$200.54	$1.09	$100.00	$301.63
4	$301.63	$1.63	$100.00	$403.26
5	$403.26	$2.18	$100.00	$505.44
6	$505.44	$2.74	$100.00	$608.18
7	$608.18	$3.29	$100.00	$711.47
8	$711.47	$3.85	$100.00	$815.32
9	$815.32	$4.42	$100.00	$919.74
10	$919.74	$4.98	$100.00	$1024.72
11	$1024.72	$5.55	$100.00	$1130.27
12	$1130.27	$6.12	$100.00	$1236.39

12.6 Chapter Review p. 668

53.

End of month	*Previous balance*	*Interest*	*Deposit*	*New balance*
1	$0.00	$0.00	$50.00	$50.00
2	$50.00	$0.33	$50.00	$100.33
3	$100.33	$0.67	$50.00	$151.00
4	$151.00	$1.01	$50.00	$202.01
5	$202.01	$1.35	$50.00	$253.36
6	$253.36	$1.69	$50.00	$305.05

54.

End of month	*Previous balance*	*Interest*	*Deposit*	*New balance*
1	$0.00	$0.00	$85.00	$85.00
2	$85.00	$0.46	$85.00	$170.46
3	$170.46	$0.92	$85.00	$256.38
4	$256.38	$1.39	$85.00	$342.77
5	$342.77	$1.86	$85.00	$429.63
6	$429.63	$2.33	$85.00	$516.96

68. 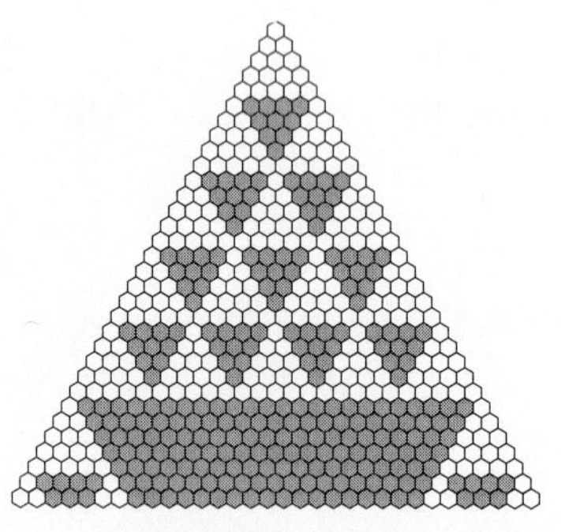

Cumulative Review for Chapters 7–12 p. 674

56.

End of month	*Previous balance*	*Interest*	*Deposit*	*New balance*
1	$0.00	$0.00	$25.00	$25.00
2	$25.00	$0.13	$25.00	$50.13
3	$50.13	$0.25	$25.00	$75.38
4	$75.38	$0.38	$25.00	$100.76
5	$100.76	$0.50	$25.00	$126.26
6	$126.26	$0.63	$25.00	$151.89

CHAPTER 13

13.1 Communicating About Algebra p. 681

A.

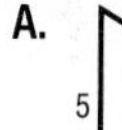
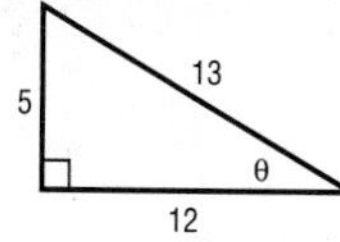

B.

58°
14m
7.42m
32°
11.87m

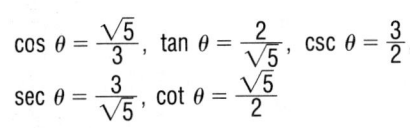

$\sin\theta = \frac{5}{13}$, $\cos\theta = \frac{12}{13}$, $\tan\theta = \frac{5}{12}$, $\csc\theta = \frac{13}{5}$, $\sec\theta = \frac{13}{12}$, $\cot\theta = \frac{12}{5}$

13.1 Independent Practice p. 682

7. $\sin\theta = \frac{3}{5}$, $\cos\theta = \frac{4}{5}$, $\tan\theta = \frac{3}{4}$, $\csc\theta = \frac{5}{3}$, $\sec\theta = \frac{5}{4}$, $\cot\theta = \frac{4}{3}$

8. $\sin\theta = \frac{4}{5}$, $\cos\theta = \frac{3}{5}$, $\tan\theta = \frac{4}{3}$, $\csc\theta = \frac{5}{4}$, $\sec\theta = \frac{5}{3}$, $\cot\theta = \frac{3}{4}$

9. $\sin\theta = \frac{10}{13}$, $\cos\theta = \frac{\sqrt{69}}{13}$, $\tan\theta = \frac{10}{\sqrt{69}}$, $\csc\theta = \frac{13}{10}$, $\sec\theta = \frac{13}{\sqrt{69}}$, $\cot\theta = \frac{\sqrt{69}}{10}$

10. $\sin\theta = \frac{4}{5}$, $\cos\theta = \frac{3}{5}$, $\tan\theta = \frac{4}{3}$, $\csc\theta = \frac{5}{4}$, $\sec\theta = \frac{5}{3}$, $\cot\theta = \frac{3}{4}$

11. $\sin\theta = \frac{3}{10}$, $\cos\theta = \frac{\sqrt{91}}{10}$, $\tan\theta = \frac{3}{\sqrt{91}}$, $\csc\theta = \frac{10}{3}$, $\sec\theta = \frac{10}{\sqrt{91}}$, $\cot\theta = \frac{\sqrt{91}}{3}$

12. $\sin\theta = \frac{3}{\sqrt{13}}$, $\cos\theta = \frac{2}{\sqrt{13}}$, $\tan\theta = \frac{3}{2}$, $\csc\theta = \frac{\sqrt{13}}{3}$, $\sec\theta = \frac{\sqrt{13}}{2}$, $\cot\theta = \frac{2}{3}$

13. $\sin\theta = \frac{9}{25}$, $\cos\theta = \frac{4\sqrt{34}}{25}$, $\tan\theta = \frac{9}{4\sqrt{34}}$, $\csc\theta = \frac{25}{9}$, $\sec\theta = \frac{25}{4\sqrt{34}}$, $\cot\theta = \frac{4\sqrt{34}}{9}$

14. $\sin\theta = \frac{1}{3}$, $\cos\theta = \frac{2\sqrt{2}}{3}$, $\tan\theta = \frac{1}{2\sqrt{2}}$, $\csc\theta = 3$, $\sec\theta = \frac{3}{2\sqrt{2}}$, $\cot\theta = 2\sqrt{2}$

15.

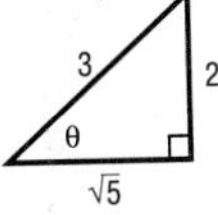

$\cos\theta = \frac{\sqrt{5}}{3}$, $\tan\theta = \frac{2}{\sqrt{5}}$, $\csc\theta = \frac{3}{2}$, $\sec\theta = \frac{3}{\sqrt{5}}$, $\cot\theta = \frac{\sqrt{5}}{2}$

16.

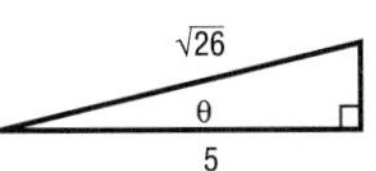

$\sin\theta = \frac{1}{\sqrt{26}}$, $\cos\theta = \frac{5}{\sqrt{26}}$, $\tan\theta = \frac{1}{5}$, $\csc\theta = \sqrt{26}$, $\sec\theta = \frac{\sqrt{26}}{5}$

17.

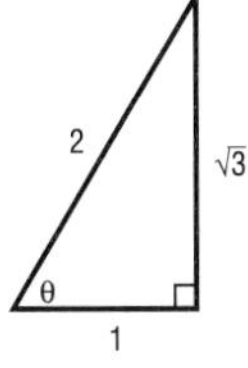

$\sin\theta = \frac{\sqrt{3}}{2}$, $\cos\theta = \frac{1}{2}$, $\tan\theta = \sqrt{3}$, $\csc\theta = \frac{2}{\sqrt{3}}$, $\cot\theta = \frac{1}{\sqrt{3}}$

18.

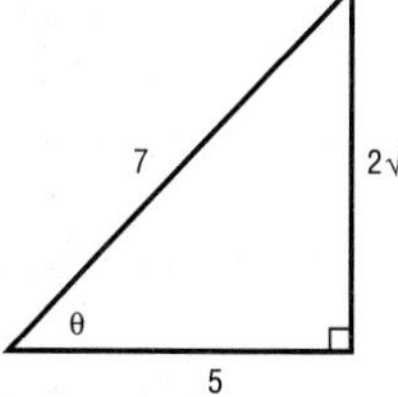

$\sin\theta = \frac{2\sqrt{6}}{7}$, $\tan\theta = \frac{2\sqrt{6}}{5}$, $\csc\theta = \frac{7}{2\sqrt{6}}$, $\sec\theta = \frac{7}{5}$, $\cot\theta = \frac{5}{2\sqrt{6}}$

19.

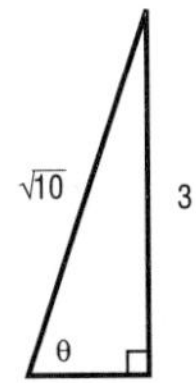

$\sin\theta = \frac{3}{\sqrt{10}}$, $\cos\theta = \frac{1}{\sqrt{10}}$, $\csc\theta = \frac{\sqrt{10}}{3}$, $\sec\theta = \sqrt{10}$, $\cot\theta = \frac{1}{3}$

13.1 Independent Practice *(continued)*

20.

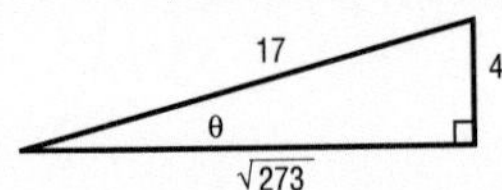

$\sin\theta = \frac{4}{17}$, $\cos\theta = \frac{\sqrt{273}}{17}$, $\tan\theta = \frac{4}{\sqrt{273}}$, $\sec\theta = \frac{17}{\sqrt{273}}$, $\cot\theta = \frac{\sqrt{273}}{4}$

21.

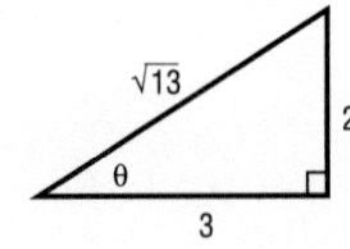

$\sin\theta = \frac{2}{\sqrt{13}}$, $\cos\theta = \frac{3}{\sqrt{13}}$, $\tan\theta = \frac{2}{3}$, $\csc\theta = \frac{\sqrt{13}}{2}$, $\sec\theta = \frac{\sqrt{13}}{3}$

22.

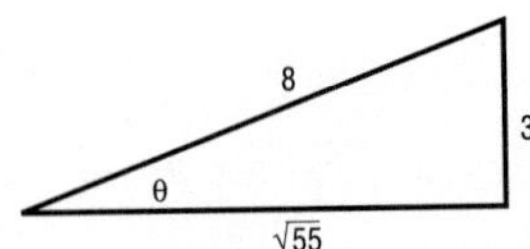

$\cos\theta = \frac{\sqrt{55}}{8}$, $\tan\theta = \frac{3}{\sqrt{55}}$, $\csc\theta = \frac{8}{3}$, $\sec\theta = \frac{8}{\sqrt{55}}$, $\cot\theta = \frac{\sqrt{55}}{3}$

ADDITIONAL ANSWERS

13.1 Exploration and Extension p. 684

66. True, $\sin 60^\circ \csc 60^\circ = \sin 60^\circ\left(\frac{1}{\sin 60^\circ}\right) = 1$

67. True, $\sec 30^\circ = \frac{1}{\cos 30^\circ} = \frac{1}{\frac{\sqrt{3}}{2}} = \frac{1}{\sin 60^\circ} = \csc 60^\circ$

68. False, $\sin 45^\circ + \cos 45^\circ = \frac{1}{\sqrt{2}} + \frac{1}{\sqrt{2}} = \frac{2}{\sqrt{2}} \neq 1$

69. False, $\frac{\sin 60^\circ}{\sin 30^\circ} \approx \frac{0.8660}{0.5} = 1.7320 \not\approx 0.0349 \approx \sin 2^\circ$

70. False, $\tan (0.8)^2 \approx 0.7445 \not\approx 1.0602 \approx \tan^2 (0.8)$

71. True, $\tan\theta \csc\theta = \frac{\sin\theta}{\cos\theta} \cdot \frac{1}{\sin\theta} = \frac{1}{\cos\theta} = \sec\theta$

72. True, $\cot\theta \sec\theta \sin\theta = \frac{\cos\theta}{\sin\theta} \cdot \frac{1}{\cos\theta} \cdot \sin\theta = 1$

73. False, $\csc\theta \sec\theta = \frac{1}{\sin\theta} \cdot \frac{1}{\cos\theta} \neq \frac{\sin\theta}{\cos\theta} = \tan\theta$

13.3 Independent Practice p. 697

21.

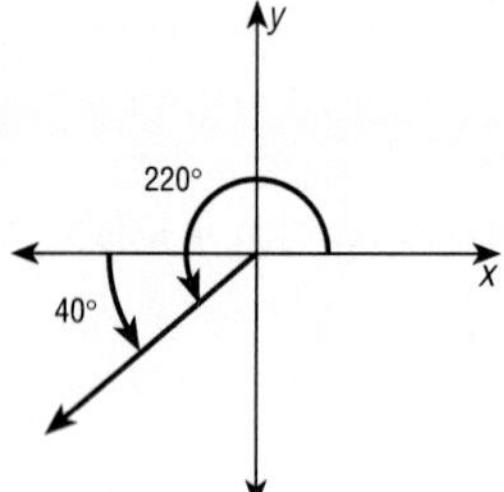

22.

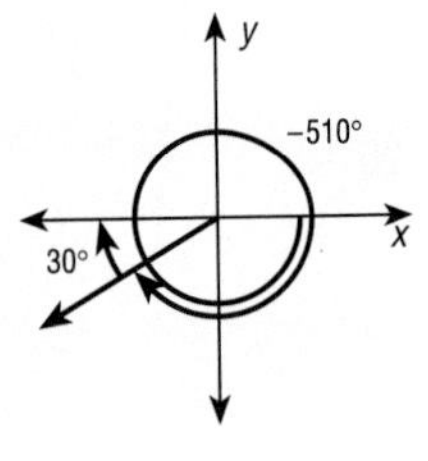

23.

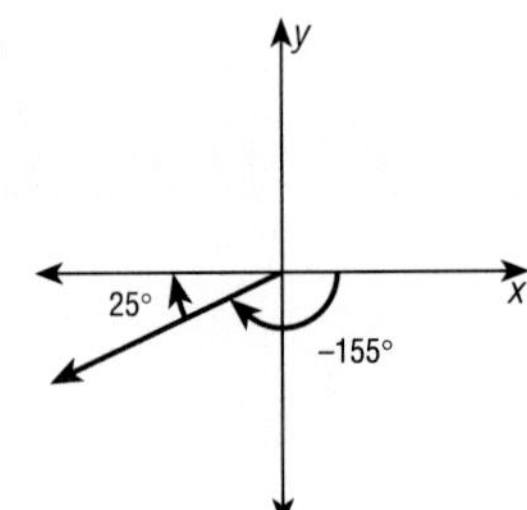

24.

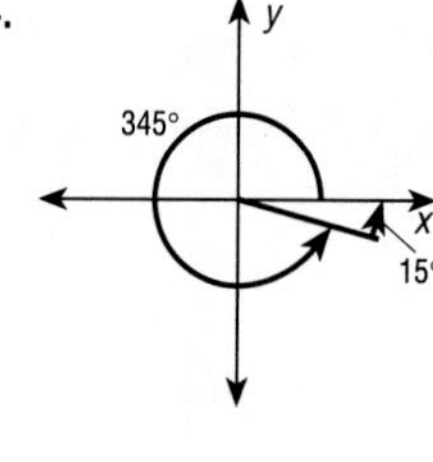

25.

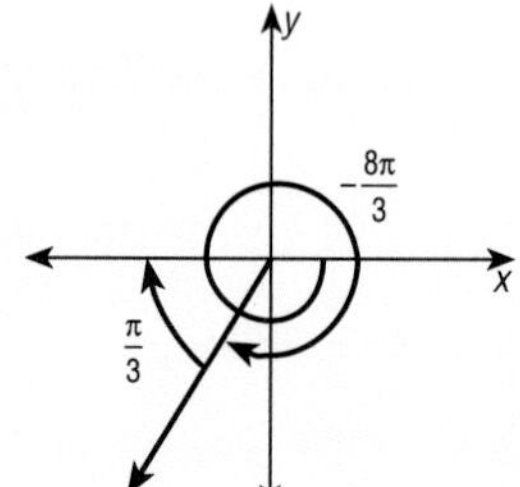

26.

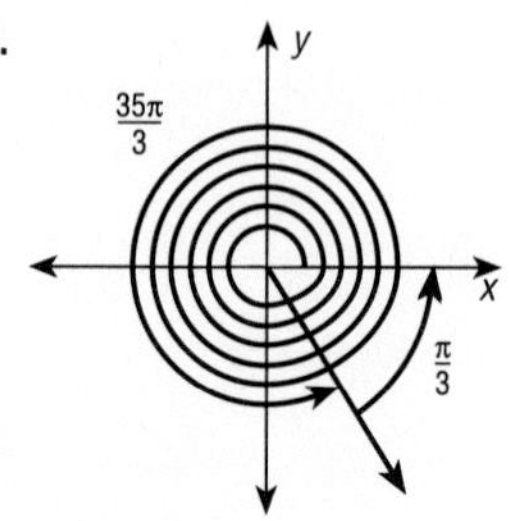

27.

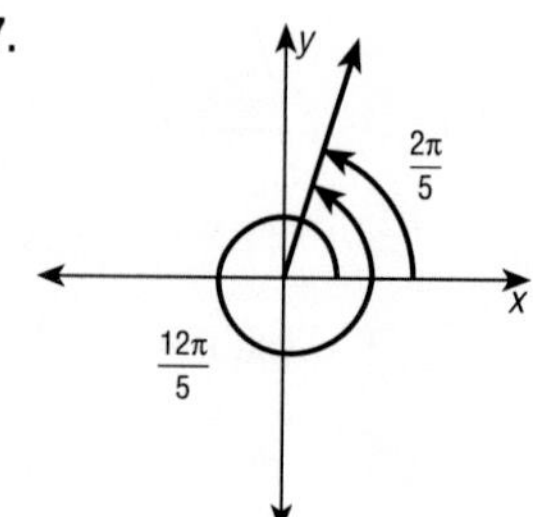

28.

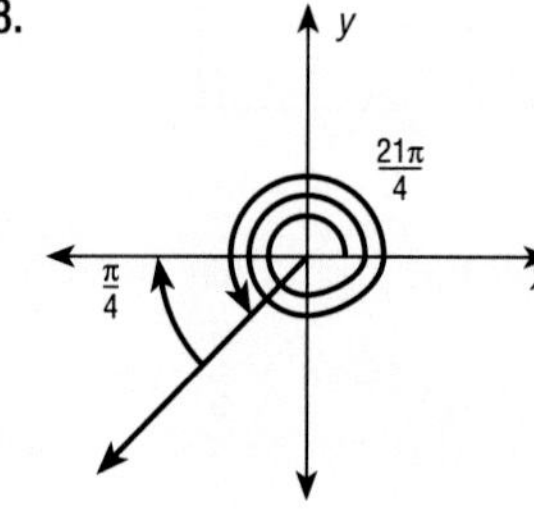

13.3 Independent Practice *(continued)*

29. 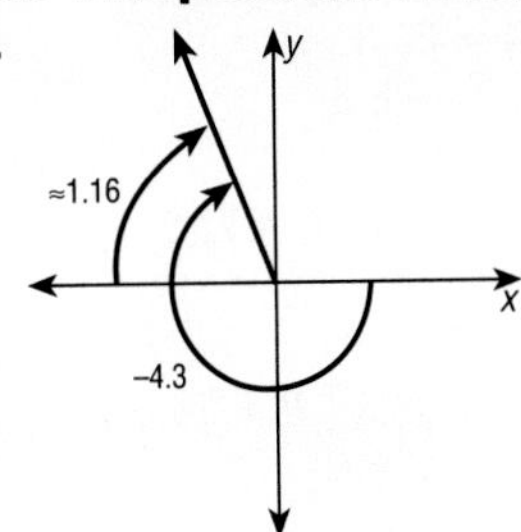

30. 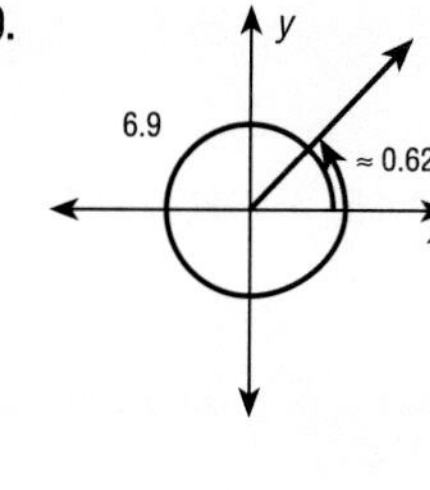

31. 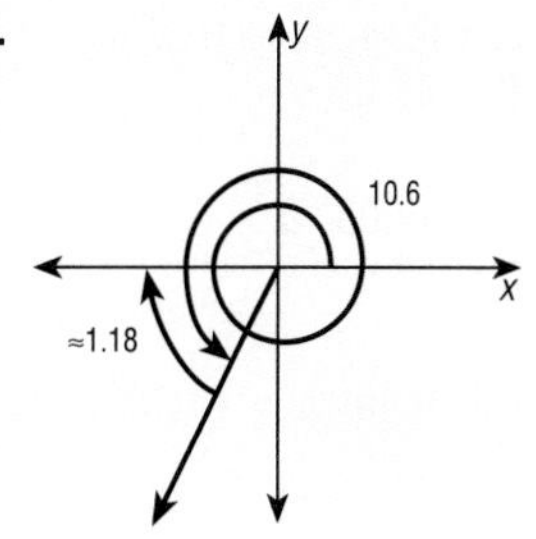

32. 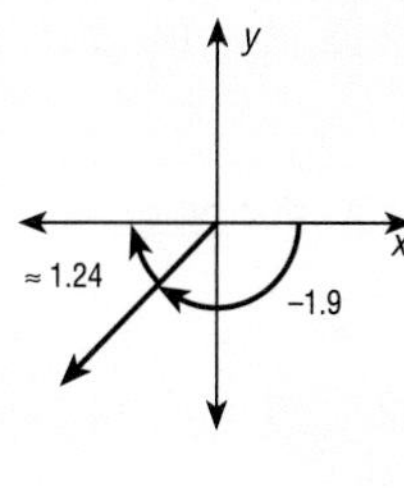

13.3 Integrated Review p. 699

77. $\sin\theta = \frac{3}{\sqrt{13}}$, $\cos\theta = \frac{2}{\sqrt{13}}$, $\csc\theta = \frac{\sqrt{13}}{3}$, $\sec\theta = \frac{\sqrt{13}}{2}$, $\cot\theta = \frac{2}{3}$

78. $\sin\theta = \frac{\sqrt{5}}{3}$, $\tan\theta = \frac{\sqrt{5}}{2}$, $\csc\theta = \frac{3}{\sqrt{5}}$, $\sec\theta = \frac{3}{2}$, $\cot\theta = \frac{2}{\sqrt{5}}$

79. $\cos\theta = \frac{\sqrt{15}}{4}$, $\tan\theta = \frac{1}{\sqrt{15}}$, $\csc\theta = 4$, $\sec\theta = \frac{4}{\sqrt{15}}$, $\cot\theta = \sqrt{15}$

80. $\sin\theta = \frac{1}{\sqrt{5}}$, $\cos\theta = \frac{2}{\sqrt{5}}$, $\tan\theta = \frac{1}{2}$, $\csc\theta = \sqrt{5}$, $\sec\theta = \frac{\sqrt{5}}{2}$

13.3 Exploration and Extension p. 699

83. $\sin\theta = \frac{3}{2\sqrt{3}}$, $\cos\theta = \frac{1}{2}$, $\tan\theta = \frac{3}{\sqrt{3}}$, $\csc\theta = \frac{2\sqrt{3}}{3}$, $\sec\theta = 2$, $\cot\theta = \frac{\sqrt{3}}{3}$

13.5 Independent Practice p. 712

15. $B = 58°$, $b \approx 35.67$, $c \approx 32.22$

16. $C = 134°$, $a \approx 65.82$, $c \approx 153.22$

17. $A = 1.479$, $C \approx 0.615$, $a \approx 34.495$

18. No solution

19. $A = 47°$, $b \approx 2.69$, $c \approx 2.18$

20. $A = 10°$, $a \approx 2.25$, $c \approx 6.08$

21. $B = 45°$, $a \approx 1.10$, $c \approx 3.67$

22. $C = 35°$, $a \approx 18.49$, $b \approx 25.75$

23. $A = 111.6°$, $B \approx 52.4°$, $a \approx 107.95$; $A \approx 36.4°$, $B \approx 127.6°$, $a \approx 68.89$

24. $A \approx 56.1°$, $B \approx 28.9°$, $b \approx 2.91$

25. No solution

26. $A \approx 1.607$, $C \approx 0.749$, $a \approx 38.16$

27. $A = 40°$, $b \approx 17.54$, $c \approx 9.33$

28. $C = 50°$, $a \approx 30.67$, $b \approx 28.26$

29. No solution

30. $B \approx 137.4°$, $C \approx 22.6°$, $b \approx 15.83$; $B \approx 2.6°$, $C \approx 157.4°$, $b \approx 1.06$

13.5 Mixed Review p. 715

8. 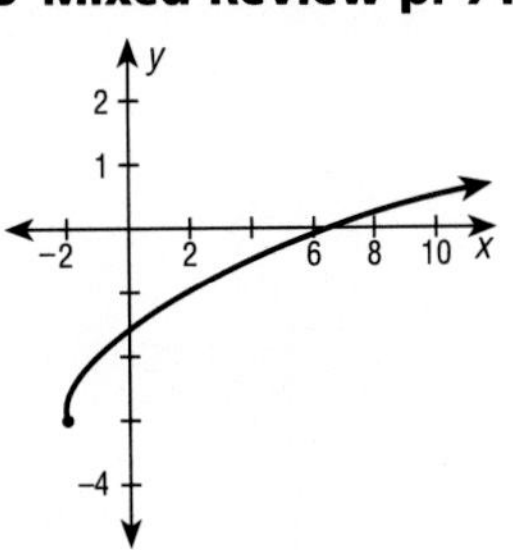

9. 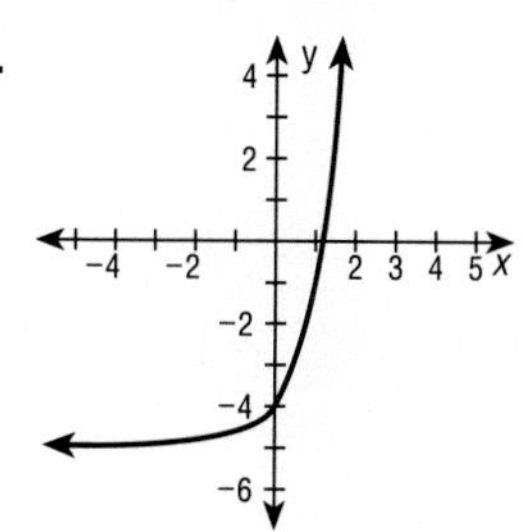

ADDITIONAL ANSWERS

13.6 Independent Practice p. 719

13. $b \approx 42.98$, $A \approx 107.3°$, $C \approx 52.7°$

14. $c \approx 8.14$, $A \approx 37.7°$, $B \approx 47.3°$

15. $A \approx 79.0°$, $B \approx 29.4°$, $C \approx 71.6°$

16. $A \approx 22.3°$, $B \approx 49.4°$, $C \approx 108.3°$

17. $a \approx 4.08$, $B \approx 28.6°$, $C \approx 73.4°$

18. $a \approx 49.03$, $B \approx 62.0°$, $C \approx 58.0°$

19. $b \approx 14.15$, $A \approx 26.7°$, $C \approx 108.3°$

20. $c \approx 7.76$, $A \approx 75°$, $B \approx 75°$

21. $A \approx 42.1°$, $B \approx 12.9°$, $C \approx 125.0°$

22. $b \approx 6.40$, $A \approx 151.0°$, $C \approx 14.0°$

23. $A \approx 73.0°$, $B \approx 96.0°$, $C \approx 11.0°$

24. $A \approx 33.8°$, $B \approx 16.6°$, $C \approx 129.6°$

25. $c \approx 5.75$, $A \approx 0.123$, $B \approx 0.662$

26. $b \approx 44.73$, $A \approx 33.4°$, $C \approx 46.6°$

27. $a \approx 23.2$, $B \approx 27.9°$, $C \approx 54.1°$

28. $A = 45°$, $B = 45°$, $C = 90°$

29. $A \approx 36.9°$, $B \approx 53.2°$, $C = 89.9°$

30. $A \approx 33.6°$, $B \approx 112.9°$, $C \approx 33.5°$

13.6 Integrated Review p. 721

49. $A \approx 48.72°$, $B \approx 65.64°$, $C \approx 65.64°$

50. $c \approx 29.11$, $A \approx 122.2°$, $B \approx 19.8°$

51. $b \approx 33.48$, $B \approx 104.9°$, $C \approx 15.1°$

52. $B = 12°$, $a \approx 17.87$, $c \approx 20.83$

53. $C = \frac{7\pi}{12}$, $b \approx 18.38$, $c \approx 35.52$

54. $b \approx 4.71$, $A \approx 48.6°$, $C \approx 69.4°$

55. $A = 30°$, $a \approx 28.98$, $c \approx 40.98$

56. $B = 79°$, $a \approx 101.86$, $c \approx 17.69$

13.6 Chapter Review p. 723

37.

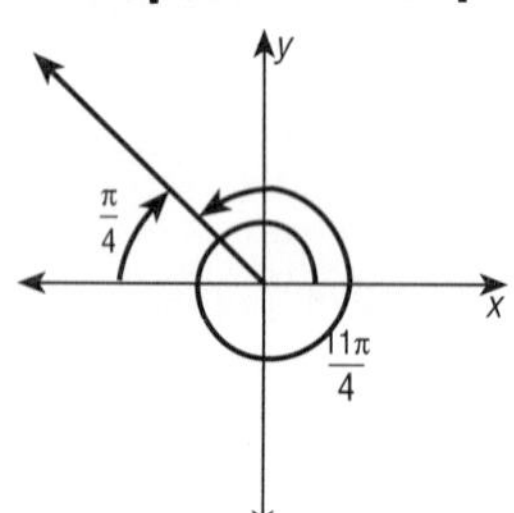

38.

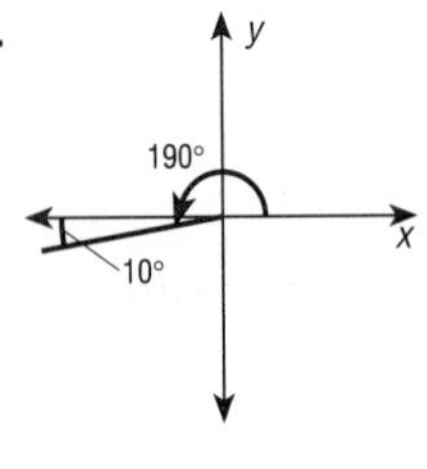

39.

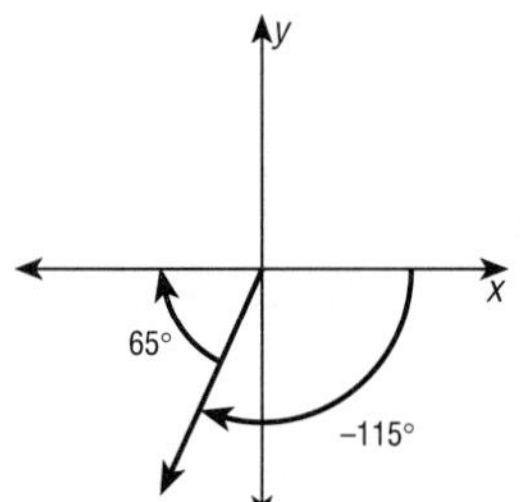

40.

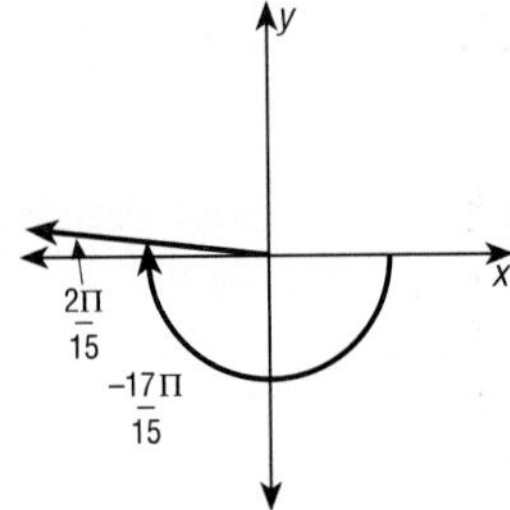

41.

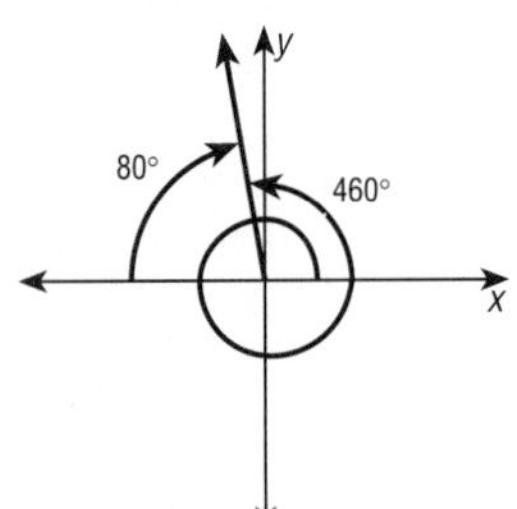

42.

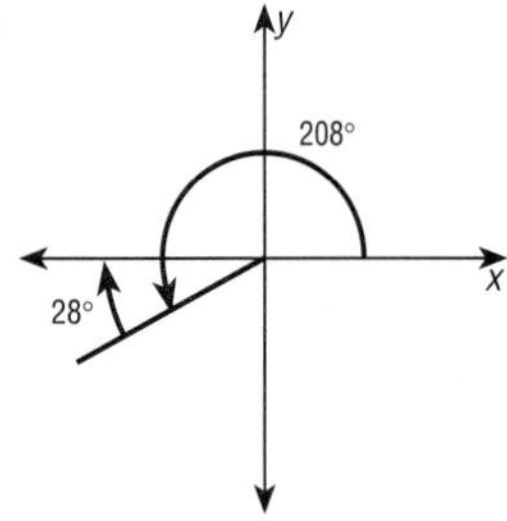

14.1 Communicating About Algebra p. 733

A.

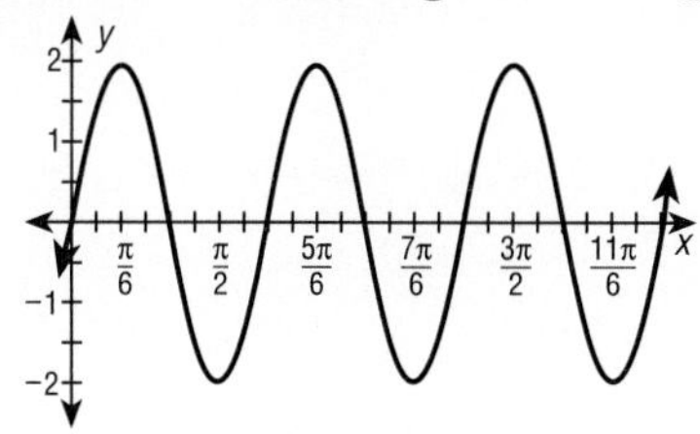

Amplitude: 2, period: $\frac{2\pi}{3}$

B.

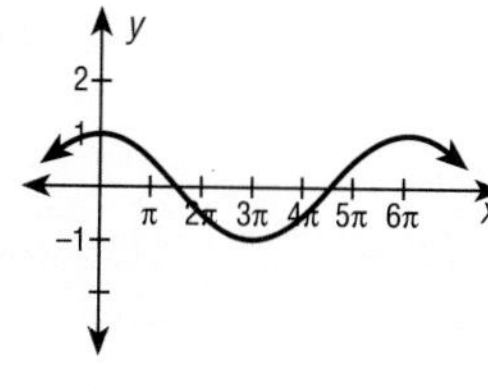

Amplitude: 1, period: 6π

C.

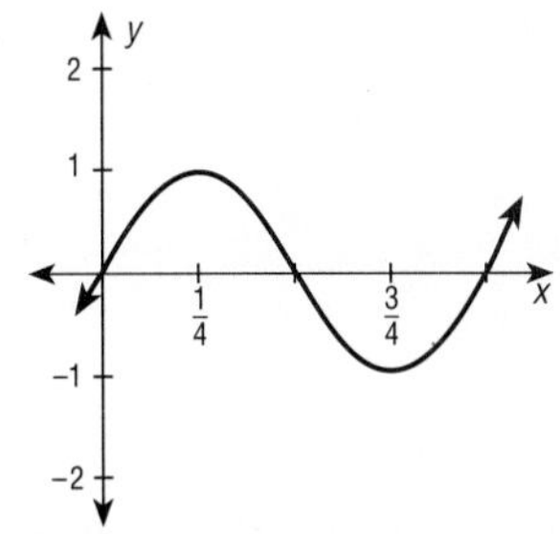

Amplitude: 1, period: 1

14.1 Independent Practice p. 735

20.

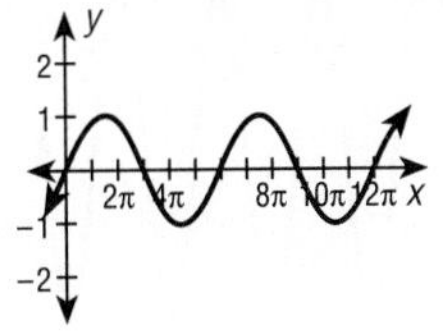

21.

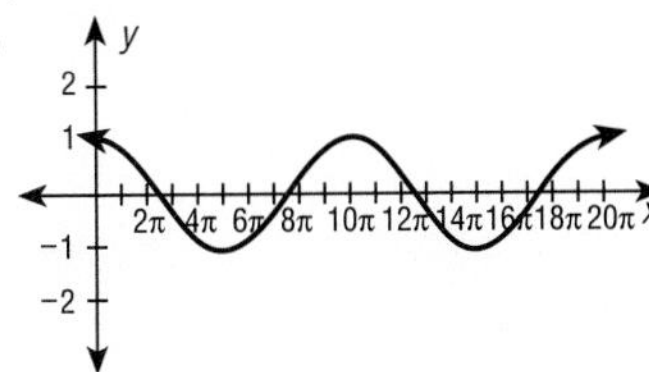

22.

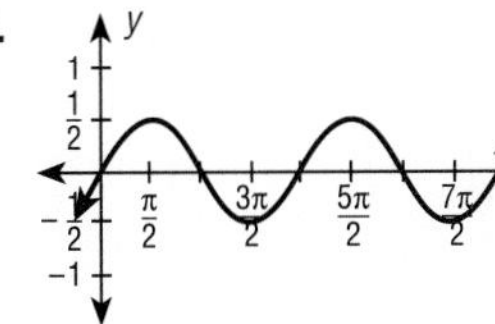

23.

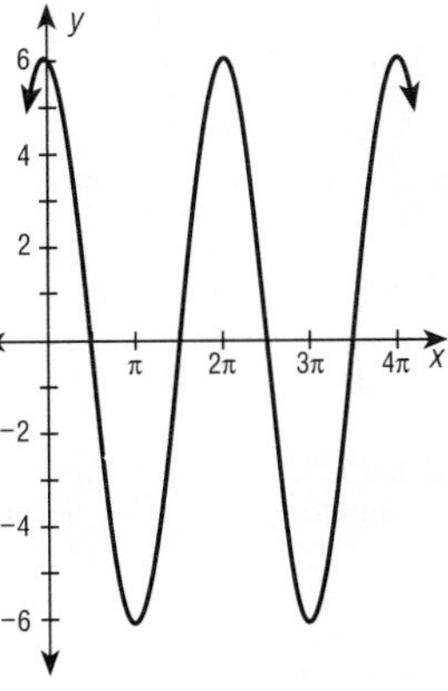

24.

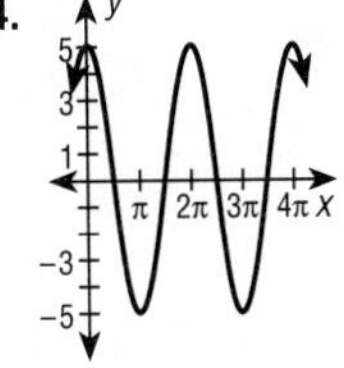

25.

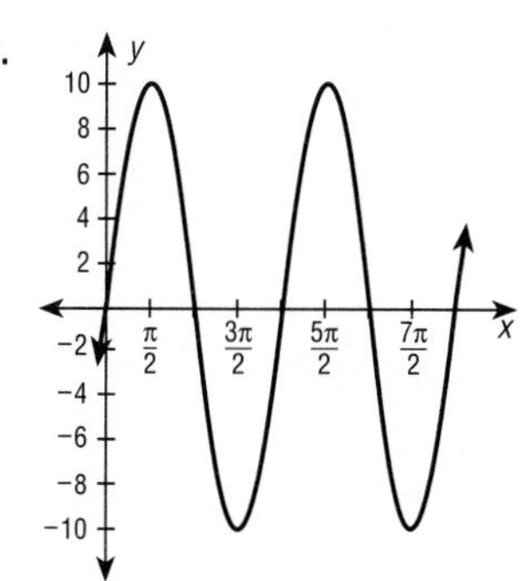

26.

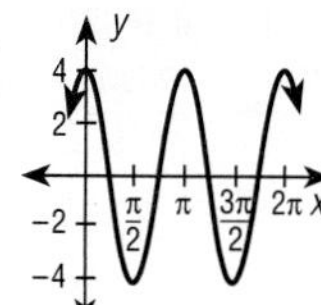

27.

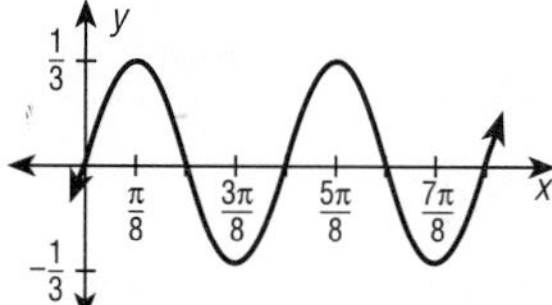

28.

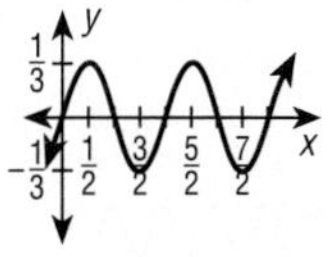

29.

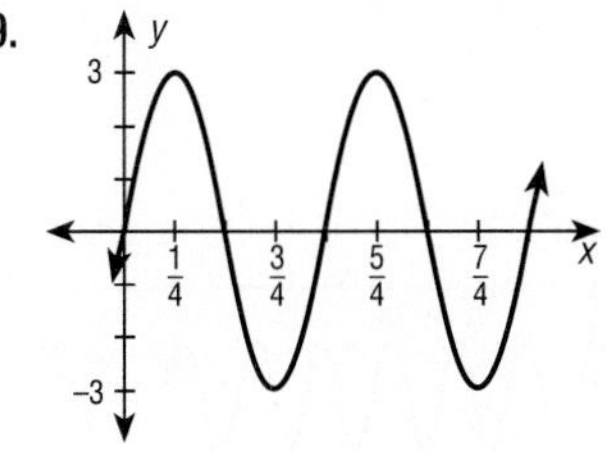

30.

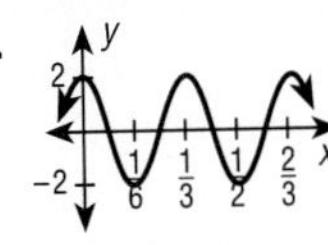

31.

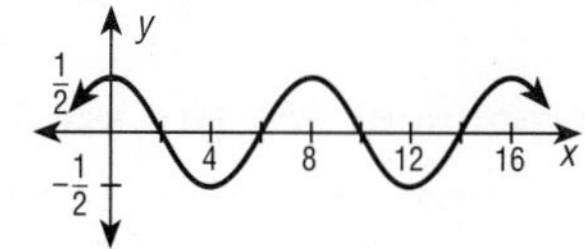

14.1 Exploration and Extension p. 736

52.

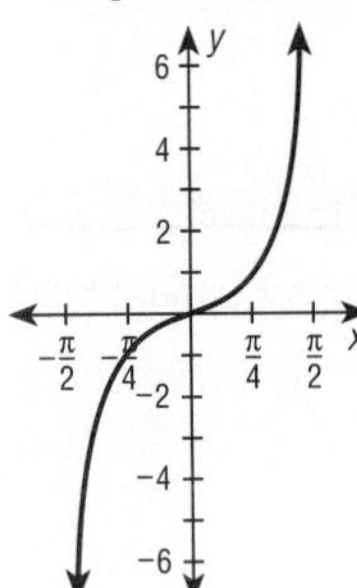

14.2 Communicating About Algebra p. 739

A.

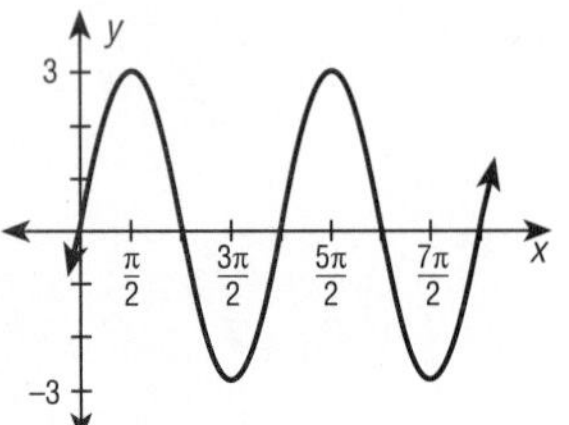

B.

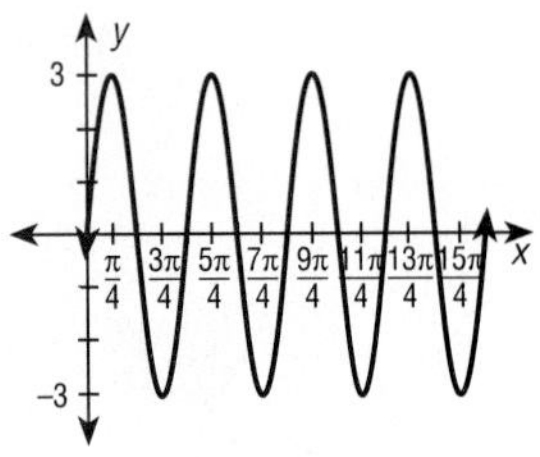

C.

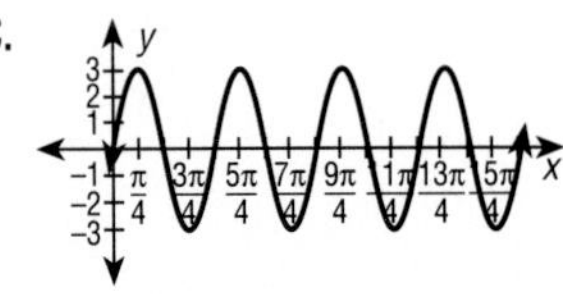

D.

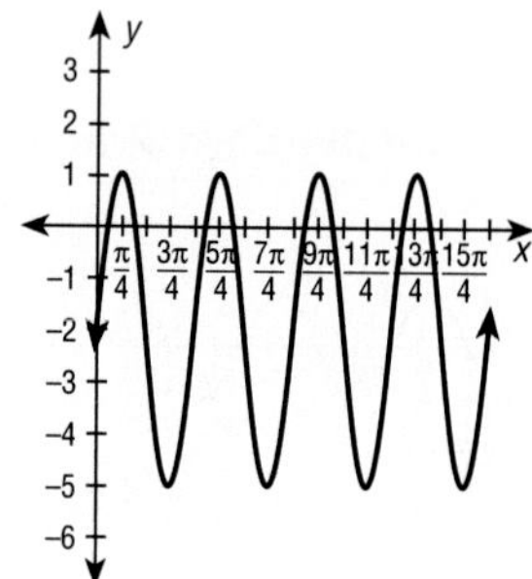

A.–D.	a indicates the amplitude is	b indicates the period is	$\frac{c}{b}$ indicates a horizontal shift of	d indicates a vertical shift of
A.	3	2π	none	none
B.	3	π	none	none
C.	3	π	π units to the right	none
D.	3	π	π units to the right	2 units down

14.2 Guided Practice p. 740

5.

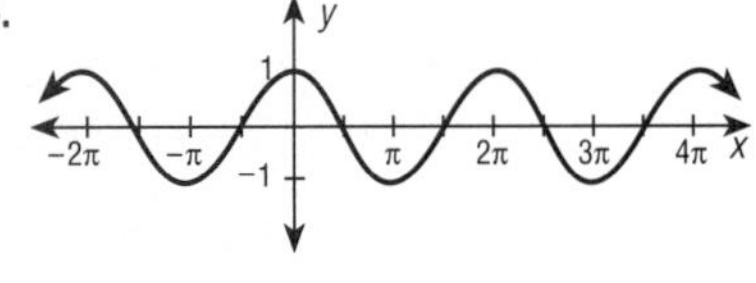

6.

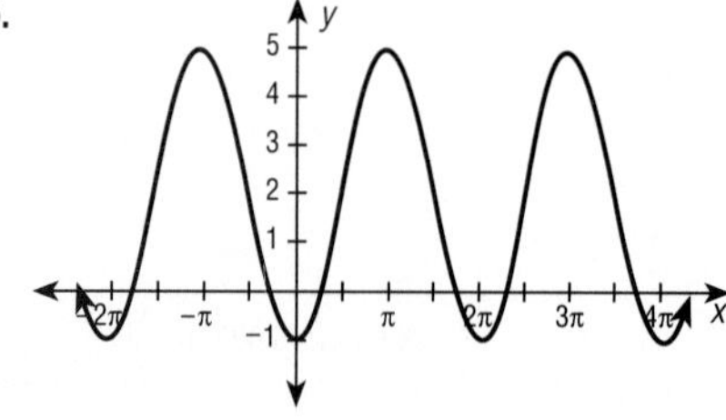

14.2 Independent Practice pp. 741–742

27.

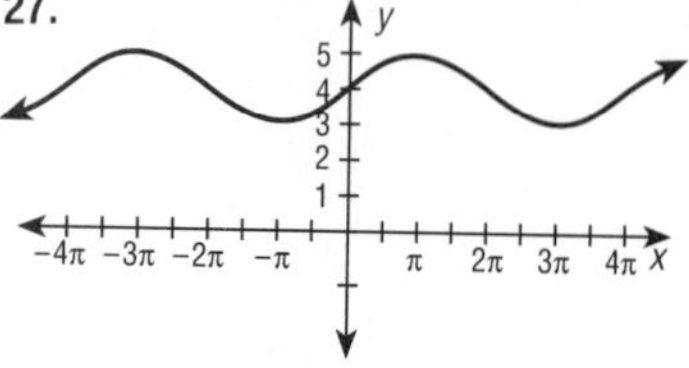

28.

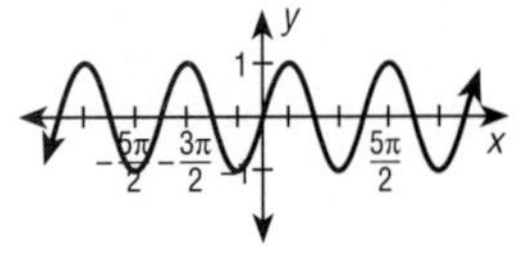

29.

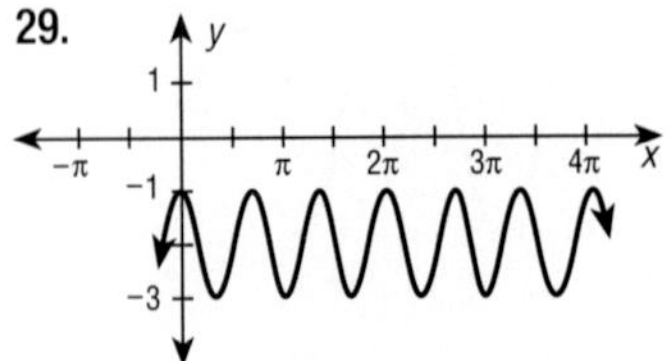

30.

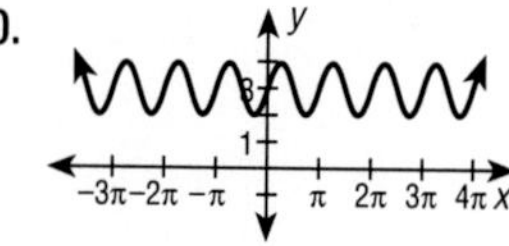

14.2 Independent Practice *(continued)*

31.

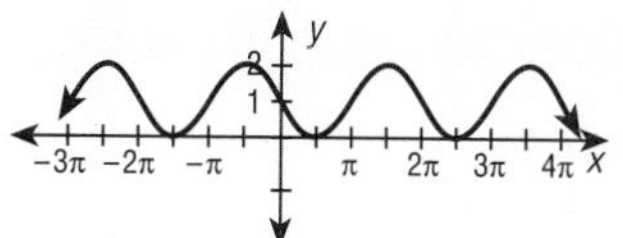

32.

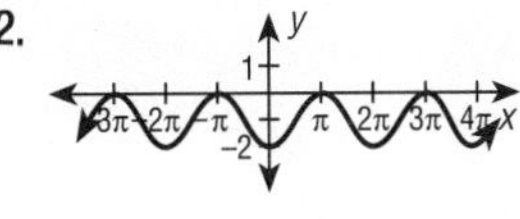

33.

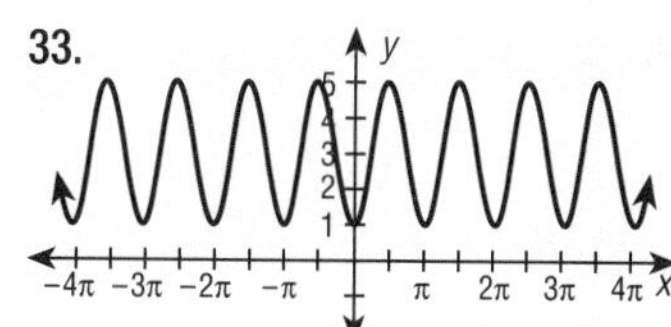

34.

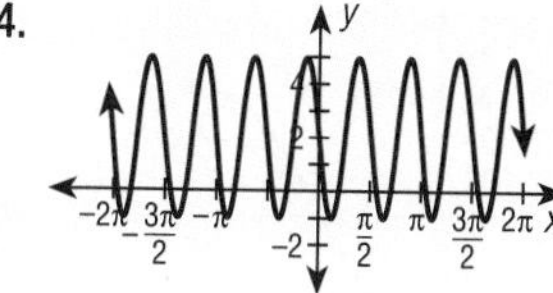

35.

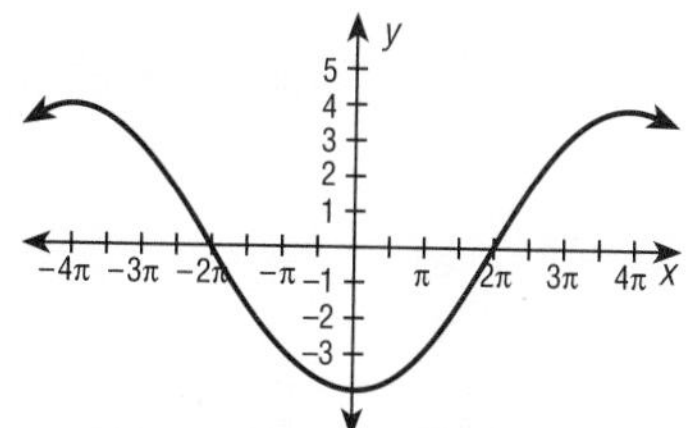

36.

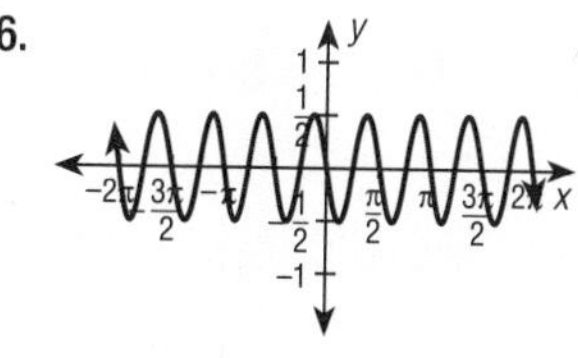

37.

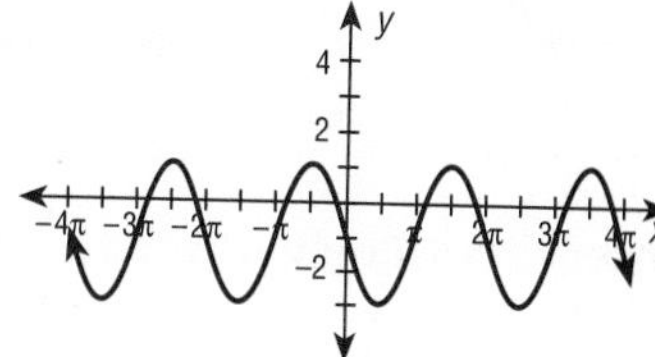

38.

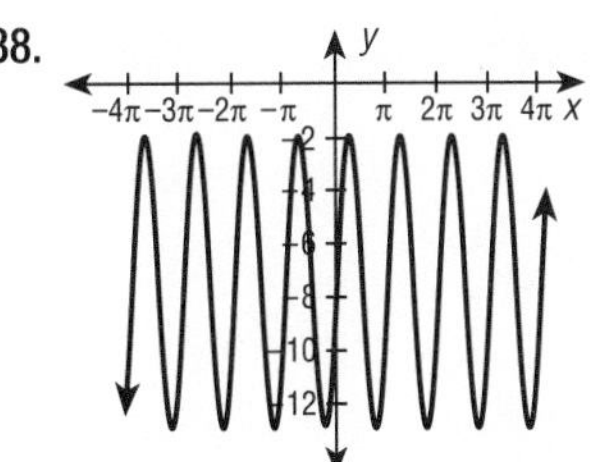

39.

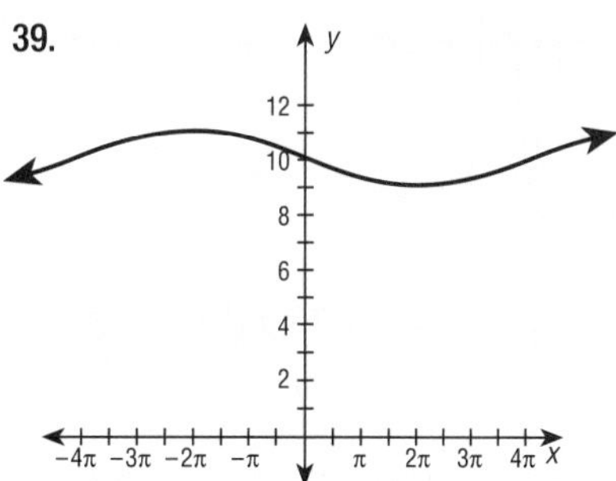

40.

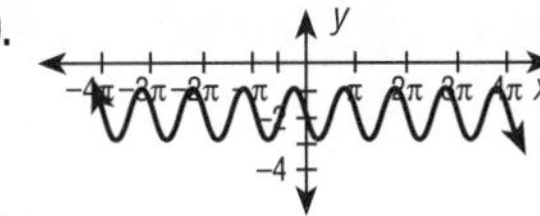

41.

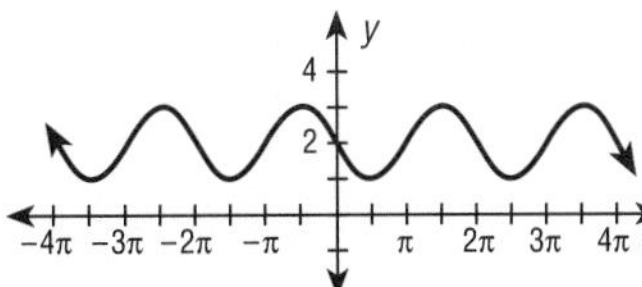

42.

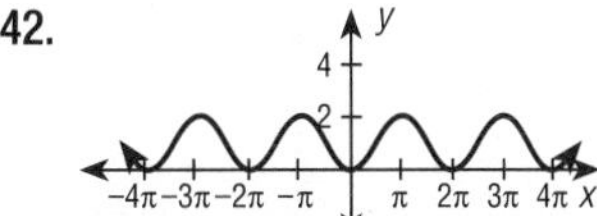

43.

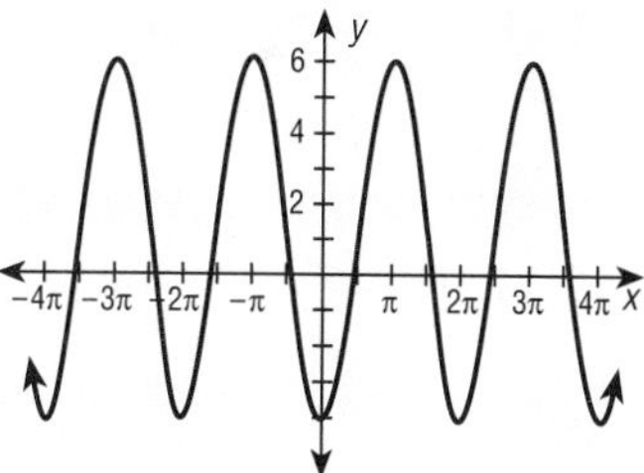

44.

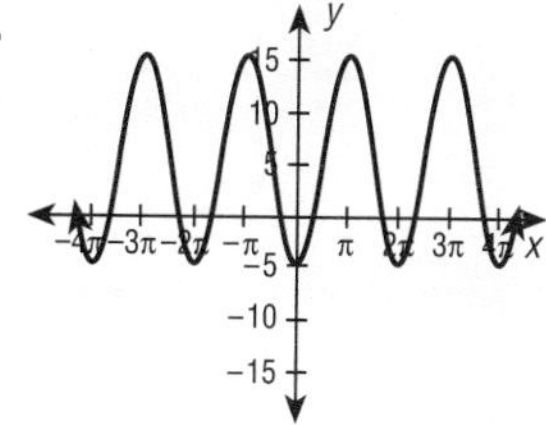

50.

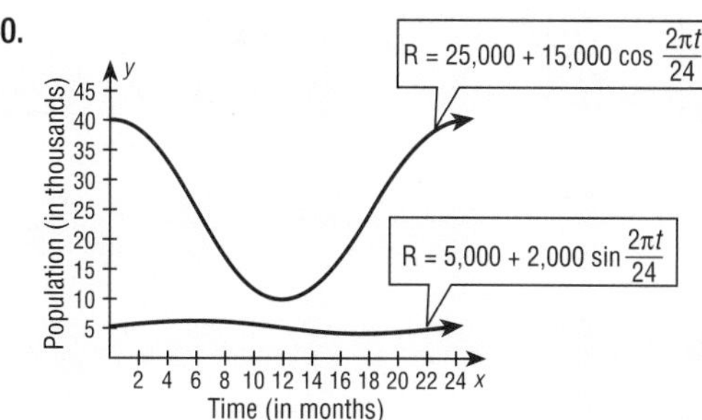

51.

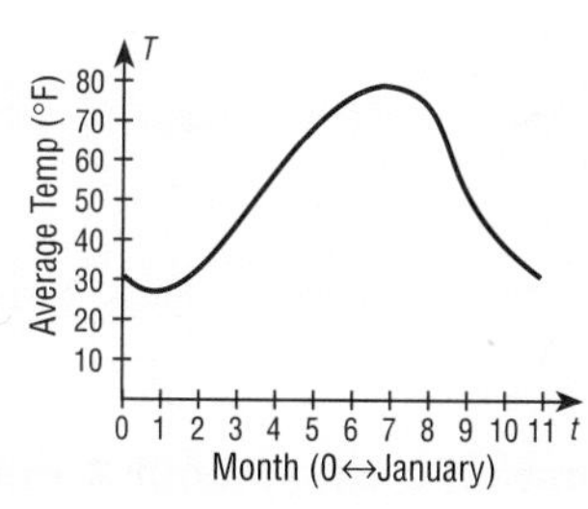

14.2 Independent Practice *(continued)*

52.

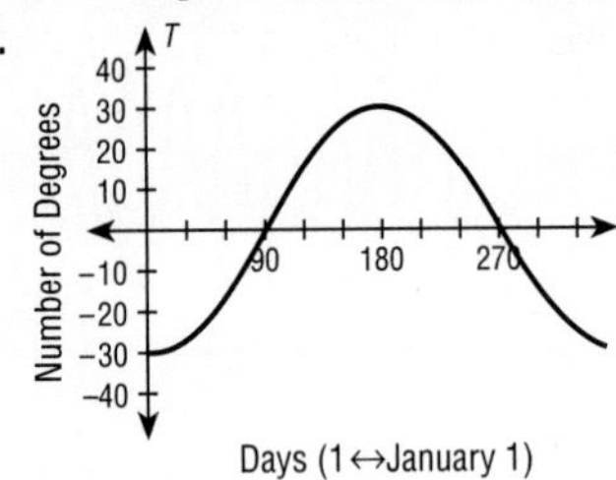

54.

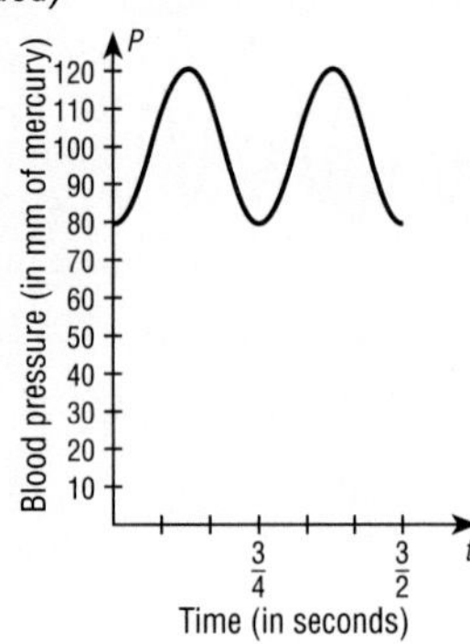

14.2 Integrated Review p. 742

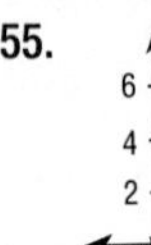

55.

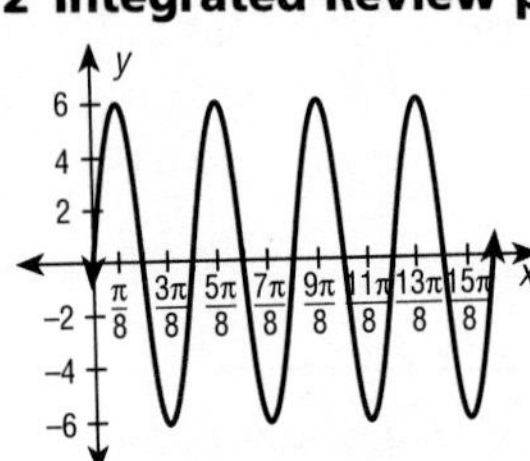

56.

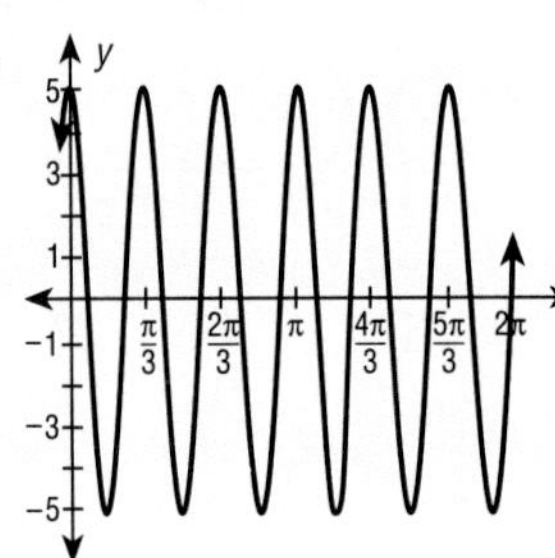

57.

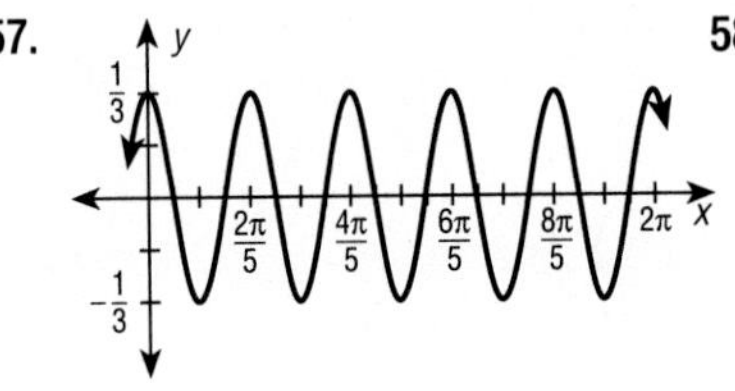

58.

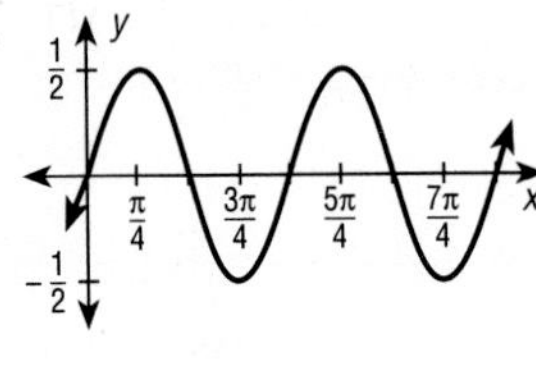

59.

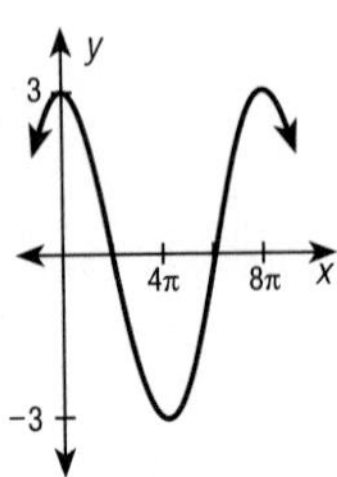

60.

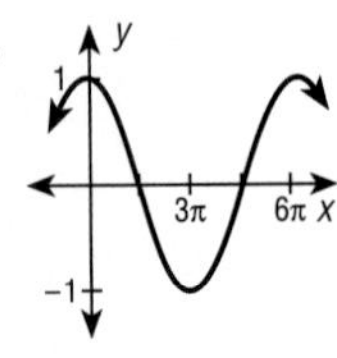

14.2 Mixed Review p. 743

6. $3\left(x + \frac{1}{2}\right)^2 - \frac{19}{4}$

8.

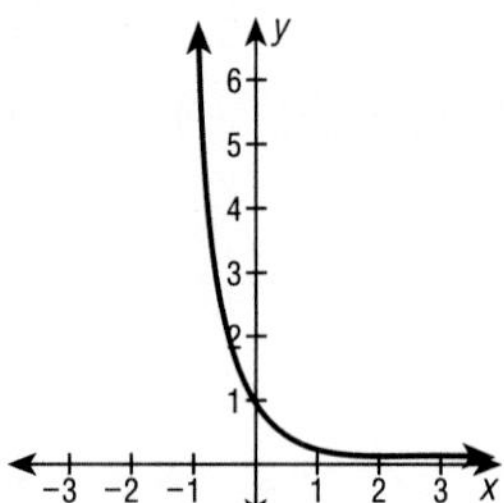

9.

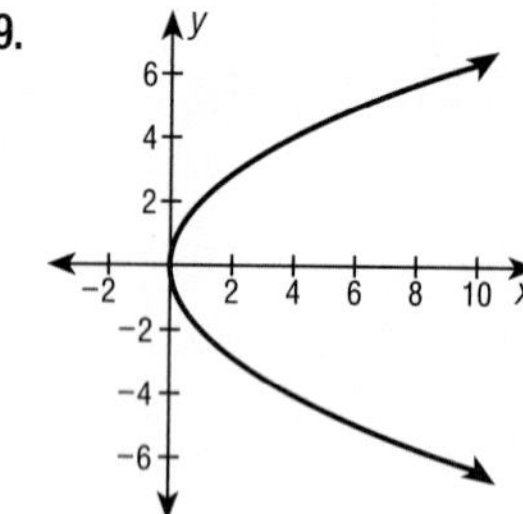

11.

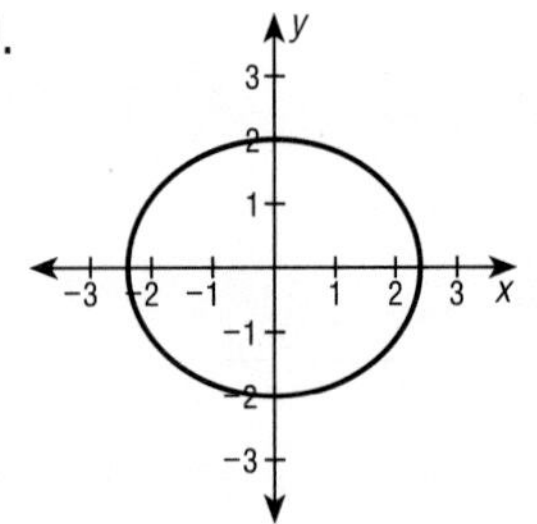

12.

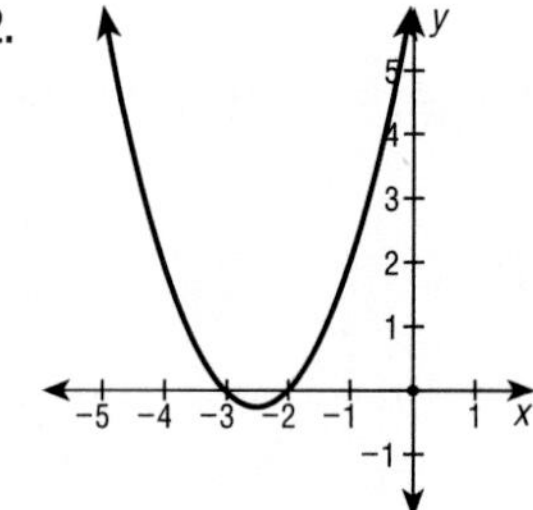

14.

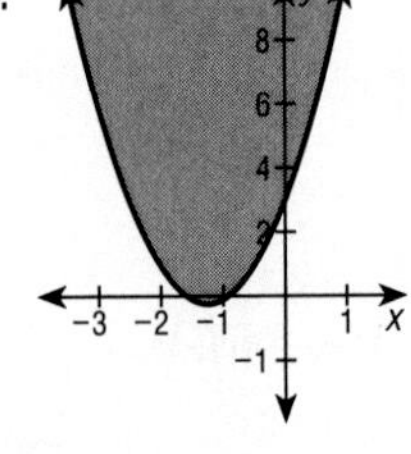

14.3 Communicating About Algebra p. 746

A. $\frac{\sec^2 x}{\tan x} = \frac{1}{\cos^2 x} \cdot \frac{\cos x}{\sin x} = \frac{1}{\cos x \sin x} = \sec x \csc x$, A is correct

B. $\cot x + \tan x = \frac{\cos x}{\sin x} + \frac{\sin x}{\cos x} = \frac{\cos^2 x + \sin^2 x}{\sin x \cos x} = \frac{1}{\sin x \cos x} =$ $\csc x \sec x$

C. $\frac{1}{\sin x \cos x} = \csc x \sec x$, C is correct

14.3 Independent Practice p. 747

23. $\cos^2 x - \sin^2 x = 1 - \sin^2 x - \sin^2 x = 1 - 2\sin^2 x$

24. $\csc^4 x - \cot^4 x = (\csc^2 x - \cot^2 x)(\csc^2 x + \cot^2 x)$
$= \csc^2 x + \cot^2 x$
$= \csc^2 x + (\csc^2 x - 1)$
$= 2\csc^2 x - 1$

25. $\cot^2 x(\sec^2 x - 1) = \cot^2 x(\tan^2 x) = \frac{1}{\tan^2 x}(\tan^2 x) = 1$

26. $4\tan^4 x + \tan^2 x - 3 = (\tan^2 x + 1)(4\tan^2 x - 3) =$
$\sec^2 x(4\tan^2 x - 3)$

27. $\sin x \csc x = \sin x \frac{1}{\sin x} = 1$

28. $\tan y \cot y = \tan y \frac{1}{\tan y} = 1$

29. $\tan^2 x + 4 = (\sec^2 x - 1) + 4 = \sec^2 x + 3$

30. $2 - \sec^2 x = 2 - (\tan^2 x + 1) = 2 - \tan^2 x - 1 = 1 - \tan^2 x$

31. $\sin^2 x - \sin^4 x = \sin^2 x(1 - \sin^2 x) = (1 - \cos^2 x)\cos^2 x =$
$\cos^2 x - \cos^4 x$

32. $\cos x + \sin x \tan x = \cos x + \sin x \frac{\sin x}{\cos x} = \frac{\cos^2 x + \sin^2 x}{\cos x} =$
$\frac{1}{\cos x} = \sec x$

33. $\frac{1 + \sin x}{\cos x} + \frac{\cos x}{1 + \sin x} = \frac{(1 + \sin x)^2 + \cos^2 x}{\cos x(1 + \sin x)} =$
$\frac{1 + 2\sin x + \sin^2 x + \cos^2 x}{\cos x(1 + \sin x)} = \frac{2 + 2\sin x}{\cos x(1 + \sin x)} =$
$\frac{2(1 + \sin x)}{\cos x(1 + \sin x)} = 2\sec x$

34. $\frac{1}{\cot x + 1} + \frac{1}{\tan x + 1} = \frac{(\tan x + 1) + (\cot x + 1)}{2 + \cot x + \tan x} =$
$\frac{2 + \cot x + \tan x}{2 + \cot x + \tan x} = 1$

35. $\frac{\cos(-x)}{1 + \sin(-x)} = \frac{\cos x}{1 - \sin x} = \frac{\cos x}{1 - \sin x} \cdot \frac{1 + \sin x}{1 + \sin x} =$
$\frac{\cos x + \cos x \sin x}{1 - \sin^2 x} = \frac{1}{\cos x} + \frac{\sin x}{\cos x} = \sec x + \tan x$

36. $\frac{1 + \sec(-x)}{\sin(-x) + \tan(-x)} = \frac{1 + \sec x}{-\sin x - \tan x}$
$= \frac{1 + \frac{1}{\cos x}}{-\sin x - \frac{\sin x}{\cos x}} \cdot \frac{\cos x}{\cos x}$
$= \frac{\cos x + 1}{-\sin x \cos x - \sin x}$
$= \frac{\cos x + 1}{-\sin x(\cos x + 1)}$
$= -\frac{1}{\sin x} = -\csc x$

37. $\cos\left(\frac{\pi}{2} - x\right)\csc x = \sin x \csc x = \sin x \frac{1}{\sin x} = 1$

38. $\frac{\cos\left(\frac{\pi}{2} - x\right)}{\sin\left(\frac{\pi}{2} - x\right)} = \frac{\sin x}{\cos x} = \tan x$

14.3 Mid-Chapter Self-Test p. 752

13. $\frac{\cos x}{\tan(-x)} = \frac{\cos(-x)}{\frac{\sin(-x)}{\cos(-x)}} = \frac{\cos^2(-x)}{\sin(-x)}$

14. $\sin^2 x + \frac{1}{\sec^2 x} = \sin^2 x + \cos^2 x = 1$

15. $\cos x + \sin x \tan x = \cos x + \sin x \cdot \frac{\sin x}{\cos x}$
$= \cos x + \frac{\sin^2 x}{\cos x} = \frac{\cos^2 x + \sin^2 x}{\cos x} = \frac{1}{\cos x} = \sec x$

14.4 Communicating About Algebra p. 756

$$6hs + \frac{3}{\sqrt{2}}s^2 = 3\left(\frac{\sqrt{3}}{2}s^2 \csc \theta\right) + 6\left(hs - \frac{1}{4}s^2 \cot \theta\right)$$

$$6hs + \frac{3}{\sqrt{2}}s^2 = 3\left(\frac{\sqrt{3}}{2}s^2 \csc \theta\right) + 6hs - \frac{3}{2}s^2 \cot \theta$$

$$\frac{3}{\sqrt{2}}s^2 = \frac{3\sqrt{3}}{2}s^2 \csc \theta - \frac{3}{2}s^2 \cot \theta$$

$$\frac{3}{\sqrt{2}}s^2 = \frac{3}{2}s^2(\sqrt{3} \csc \theta - \cot \theta)$$

$$\frac{2}{\sqrt{2}} = \sqrt{3} \csc \theta - \cot \theta$$

$$\sqrt{2} = \sqrt{3} \cdot \frac{1}{\sin \theta} - \frac{\cos \theta}{\sin \theta}$$

$$\sqrt{2} \sin \theta = \sqrt{3} - \cos \theta$$

14.4 Guided Practice p. 757

2.

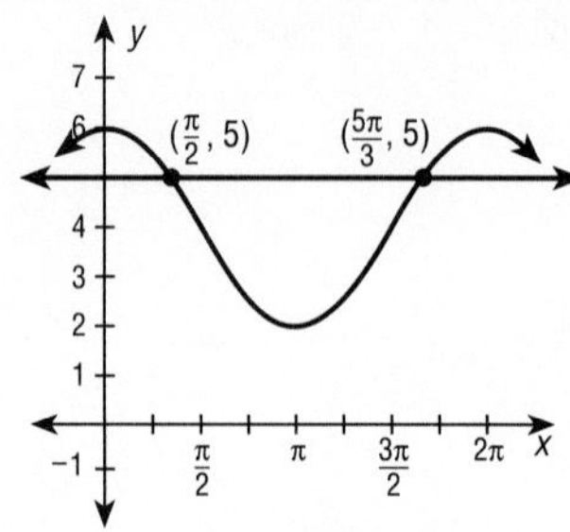

4. Sine, cosine, secant, and cosecant have a period of 2π; tangent and cotangent have a period of π.
5. Solutions occur when $\tan x = 0$ or $\sec x = 1$, i.e., $x = 0$ or $x = \frac{\pi}{2}$. However, at $\frac{\pi}{2}$, $\tan x$ is undefined, so that $\frac{\pi}{2}$ is an extraneous solution.

14.4 Integrated Review p. 759

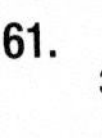

61.

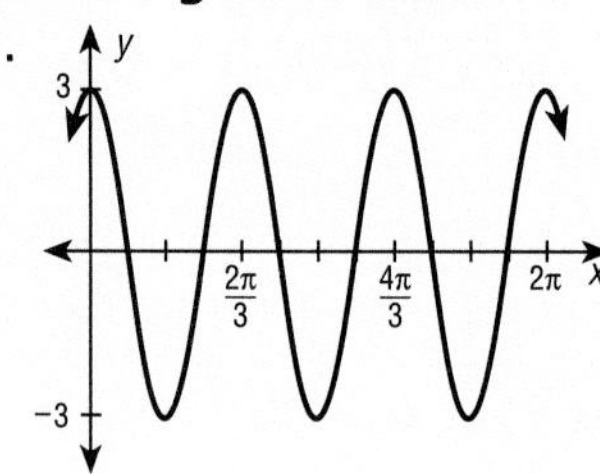

62.

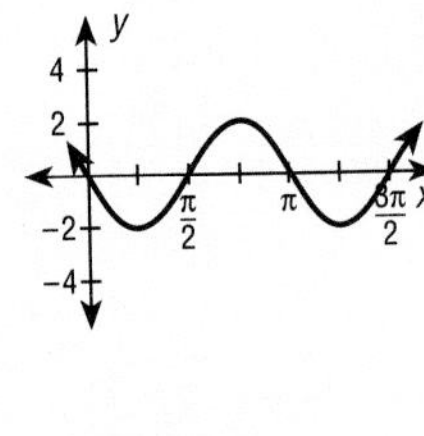

63.

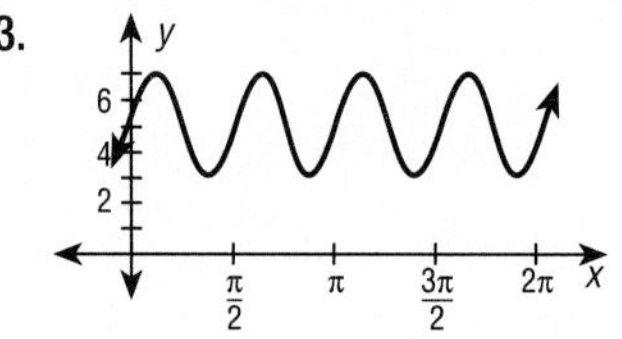

64.

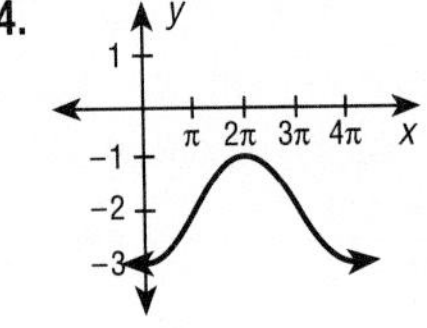

65.

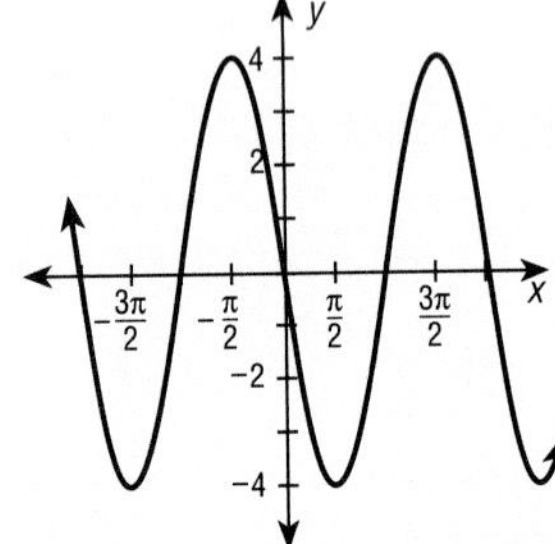

66.

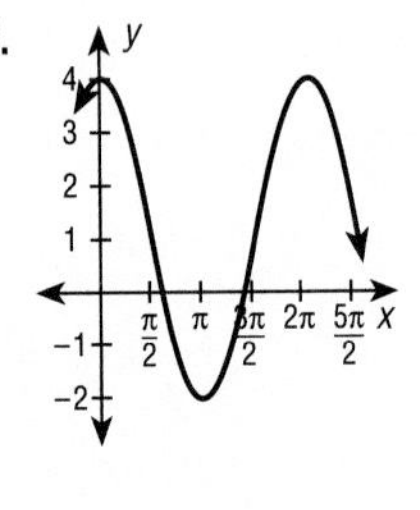

14.5 Mixed Review p. 765

4.

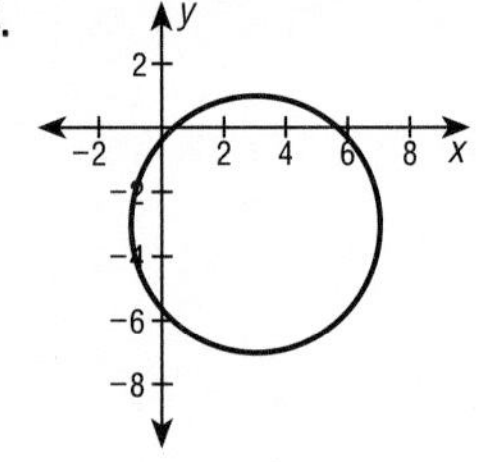

15.

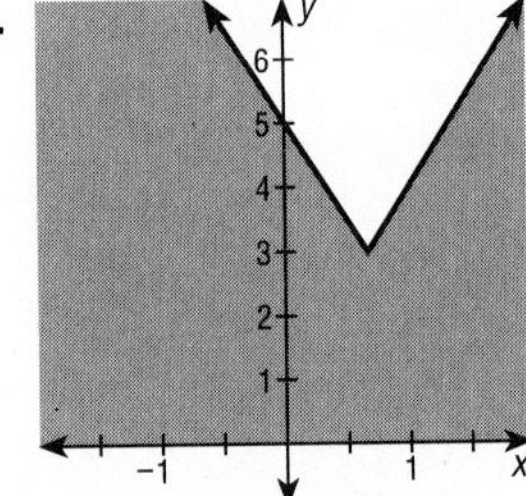

14.6 Exploration and Extension p. 771

51.

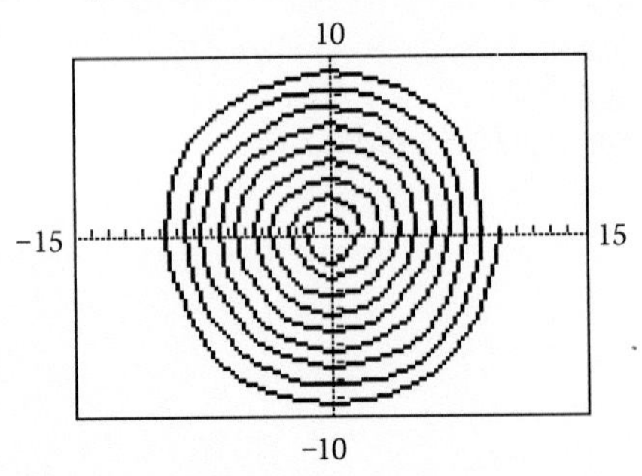

14.6 Chapter Review p. 773

19.

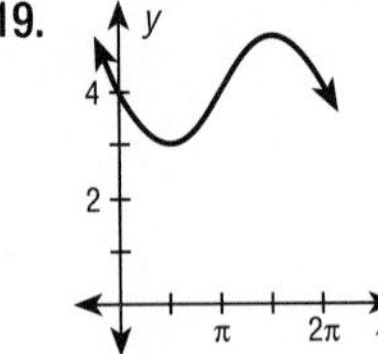

20.

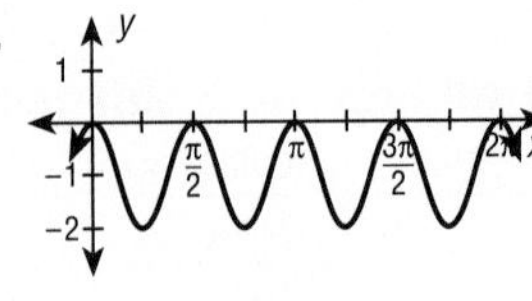

14.6 Chapter Review *(continued)*

21. 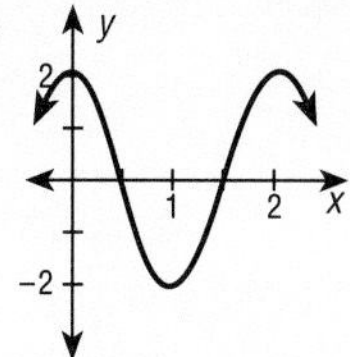

22. 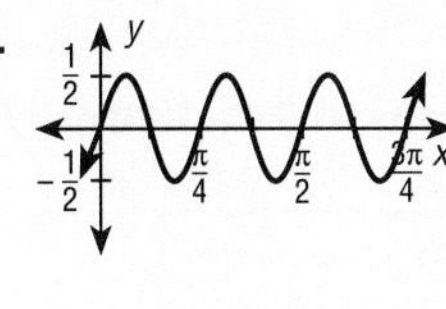

23. 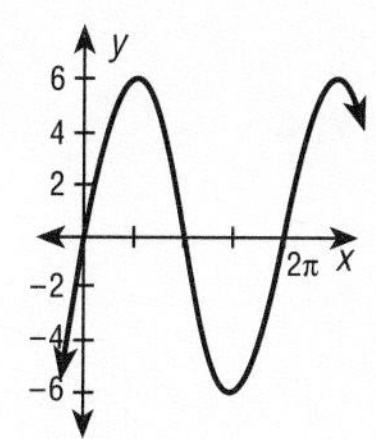

24. 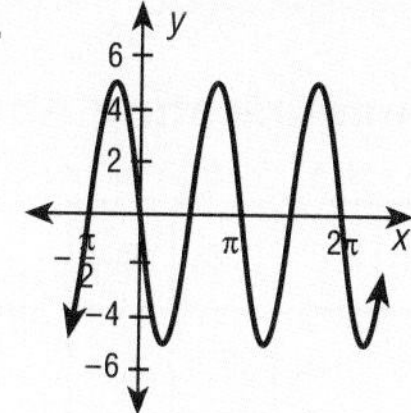

25. 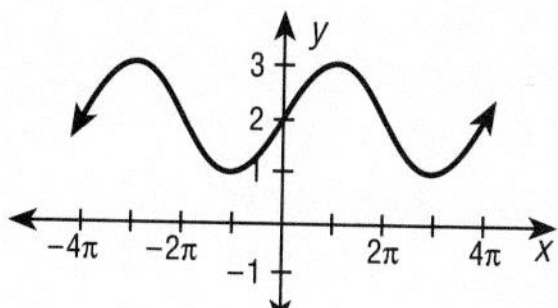

26. 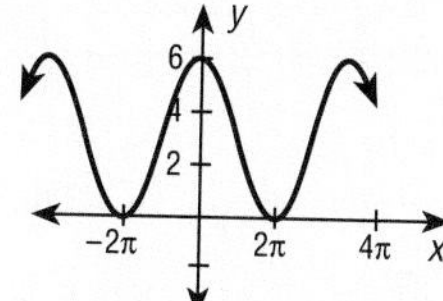

27. 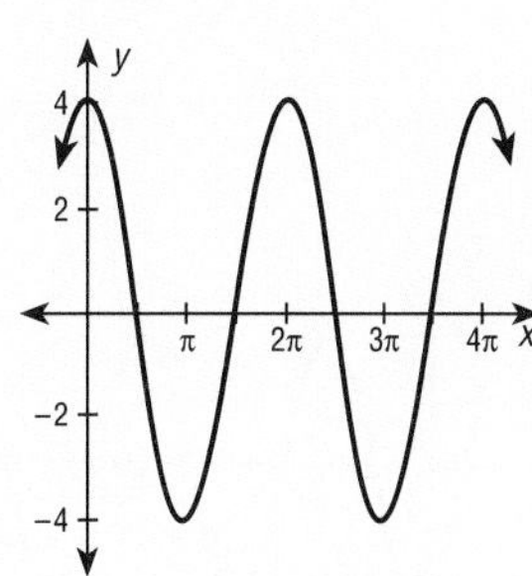

14.6 Chapter Test p. 777

7. $\csc u - \cot u = \dfrac{1}{\sin u} - \dfrac{\cos u}{\sin u} = \dfrac{1 - \cos u}{\sin u} = \tan \dfrac{u}{2}$

8. $\cos^2 u - \sin^2 u = 1 - \sin^2 u - \sin^2 u = 1 - 2 \sin^2 u$

9. $-2 \cos^2 u \tan(-u) = 2 \cos^2 u \tan u = 2 \cos^2 u \cdot \dfrac{\sin u}{\cos u} = 2 \cos u \sin u = \sin 2u$

Cumulative Review for Chapters 13–14 p. 778

31. $\dfrac{\sin\left(\dfrac{\pi}{2} - x\right)}{\sin x} = \dfrac{\cos x}{\sin x} = \cot x$

32. $2 \cos^2 x - 1 = 2(1 - \sin^2 x) - 1 = 2 - 2 \sin^2 x - 1 = 1 - 2 \sin^2 x$

33. $\csc 2x - \cot 2x = \dfrac{1}{\sin 2x} - \dfrac{\cos 2x}{\sin 2x} = \dfrac{1 - \cos 2x}{\sin 2x}$

$= \dfrac{1 - (\cos^2 x - \sin^2 x)}{\sin 2x} = \dfrac{1 - \cos^2 x + \sin^2 x}{\sin 2x}$

$= \dfrac{\sin^2 x + \sin^2 x}{\sin 2x} = \dfrac{2 \sin^2 x}{2 \sin x \cos x} = \dfrac{\sin x}{\cos x} = \tan x$

34. $\tan x \sec x - \csc x \sec^2 x$

$= \dfrac{\sin x}{\cos x} \cdot \dfrac{1}{\cos x} - \dfrac{1}{\sin x} \cdot \dfrac{1}{\cos^2 x}$

$= \dfrac{\sin x}{\cos^2 x} - \dfrac{1}{\sin x \cos^2 x}$

$= \dfrac{\sin^2 x}{\sin x \cos^2 x} - \dfrac{1}{\sin x \cos^2 x}$

$= \dfrac{\sin^2 x - 1}{\sin x \cos^2 x} = -\dfrac{\cos^2 x}{\sin x \cos^2 x} = -\dfrac{1}{\sin x} = -\csc x$

ADDITIONAL ANSWERS

CHAPTER 15

15.1 Communicating About Algebra p. 784

C.

	Bc	bc	Bc	bc
bC	BbCc	bbCc	BbCc	bbCc
bc	Bbcc	bbcc	Bbcc	bbcc
bC	BbCc	bbCc	BbCc	bbCc
bc	Bbcc	bbcc	Bbcc	bbcc

15.1 Guided Practice p. 785

5.

	BC	Bc	BC	Bc
bC	BbCC	BbCc	BbCC	BbCc
bc	BbCc	Bbcc	BbCc	Bbcc
bC	BbCC	BbCc	BbCC	BbCc
bc	BbCc	Bbcc	BbCc	Bbcc

15.2 Mixed Review p. 795

6.

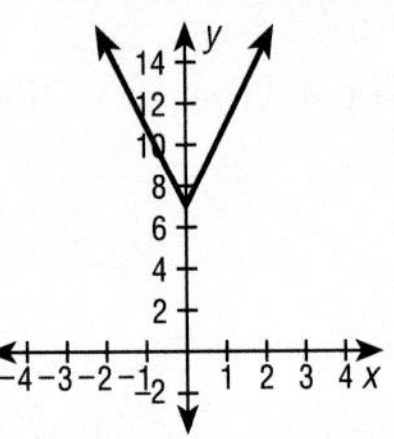

18.

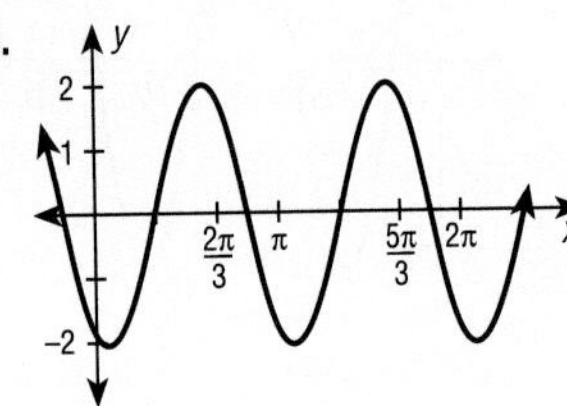

15.5 Mixed Review p. 816

3.

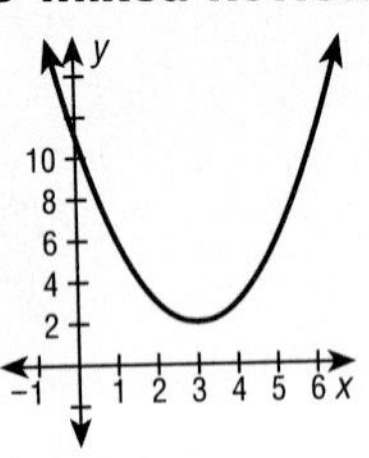

6.

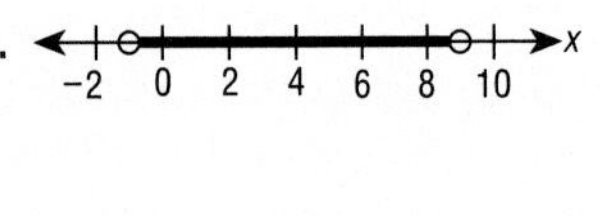

10.

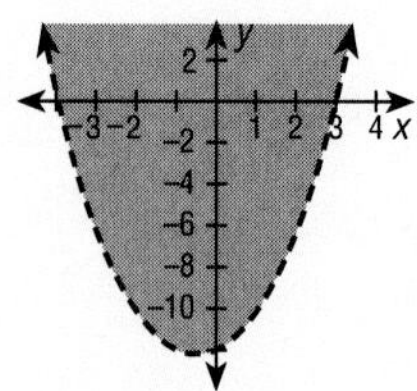

12.

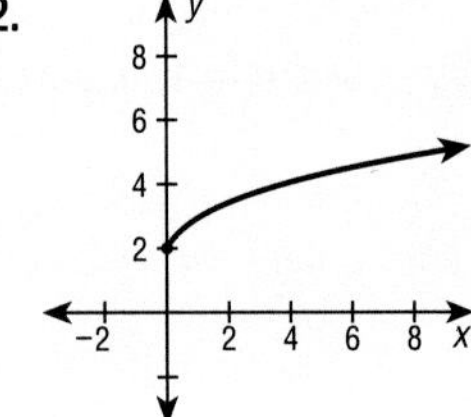

ADDITIONAL ANSWERS

Page 846 (The Quadratic Formula)

$3x^2 + 5x + 2 = 0 \Rightarrow x^2 + \frac{5}{3}x + \frac{2}{3} = 0 \Rightarrow x^2 + \frac{5}{3}x = \frac{-2}{3} \Rightarrow x^2 + \frac{5}{3}x + \left(\frac{5}{6}\right)^2 = \frac{-2}{3} + \left(\frac{5}{6}\right)^2 \Rightarrow \left(x + \frac{5}{6}\right)^2 = \frac{1}{36} \Rightarrow x + \frac{5}{6} = \pm\frac{1}{6} \Rightarrow x = \frac{-5}{6} \pm \frac{1}{6} \Rightarrow x = \frac{-2}{3}$ or -1

Page 847 (Properties of Logarithms)

Let $n = \log_a u$ and $m = \log_a v$. Then $u = a^n$ and $v = a^m$. Divide to obtain $\frac{u}{v} = a^{n-m}$. Take the logarithm of both sides to obtain $\log_a \left(\frac{u}{v}\right) = \log_a (a^{n-m}) = n - m = \log_a u - \log_a v$.

Page 848 (Change-of-Base Formulas)

To derive the second change-of-base formula, substitute ln for $\log_{10}$. To derive the third change-of-base formula, substitute $\log_b$ for $\log_{10}$.

Page 848 (Equation of a Parabola)

For any point (x, y) on the parabola, the distance from (x, y) to $(0, 5)$ equals the distance from (x, y) to the line $y = -5$. So, $\sqrt{(x-0)^2 + (y-5)^2} = \sqrt{(x-x)^2 + (y-(-5))^2} \Rightarrow \sqrt{x^2 + (y-5)^2} = \sqrt{(y+5)^2} \Rightarrow x^2 + y^2 - 10y + 25 = y^2 + 10y + 25 \Rightarrow x^2 = 4(5)y$

Page 849 (Equation of an Ellipse)

For any point (x, y) on the ellipse, the sum of the distances from (x, y) to $(-3, 0)$ and to $(3, 0)$ is 10. So, $c = 3$, $a = 5$, and $\sqrt{(x-(-3))^2 + (y-0)^2} + \sqrt{(x-3)^2 + (y-0)^2} = 2(5) = 10$. Square both sides and simplify to obtain $5\sqrt{(x-3)^2 + y^2} = -3x + 25$. Again, square both sides and simplify to obtain $\frac{16}{25}x^2 + y^2 = 16$ or $\frac{x^2}{25} + \frac{y^2}{16} = 1$, where $a^2 - c^2 = 25 - 9 = 16 = b^2$.

Page 851 (Equation of a Hyperbola)

For any point (x, y) on the hyperbola, the difference of the distances from (x, y) to $(5, 0)$ and to $(-5, 0)$ is ± 10. So, $c = 5$, $a = 5$, and $\sqrt{(x-(-5))^2 + (y-0)^2} - \sqrt{(x-5)^2 + (y-0)^2} = \pm 10$. Square both sides and simplify to obtain $x - 5 = \pm\sqrt{(x-5)^2 + y^2}$. Again, square both sides and simplify to obtain $y = 0$. This is a degenerate form of a hyperbola consisting of all points on the x-axis. A degenerate hyperbola was obtained because $a = c$ and $b^2 = a^2 - c^2 = 0$.

Page 851 (Formula for the Sum of an Arithmetic Series)

The sum of the series S_{60} is $7 + 10 + 13 + \cdots + 178 + 181 + 184$. In the reverse order, $S_{60} = 184 + 181 + 178 + \cdots + 13, +10 + 7$. Add to get $2S_{60} = 191 + 191 + 191 + \cdots + 191 + 191 + 191$ (60 terms). Therefore $2S_{60} = 60(191)$ or $S_{60} = \frac{60}{2}(191) = 5730$.

Page 851 (Formula for the Sum of a Geometric Series)

The sum S_8 of the geometric series $2 + 6 + 18 + \cdots + 1458 + 4374$ can be written in expanded form as $S_8 = 2 + 2(3) + 2(3)^2 + \cdots + 2(3)^6 + 2(3)^7$. Multiply both sides by -3 to obtain $-3S_8 = -2(3) - 2(3)^2 - \cdots - 2(3)^6 - 2(3)^7 - 2(3)^8$. Add the two forms of the series to get $S_8 - 3S_8 = 2 - 2(3)^8 \Rightarrow S_8(1-3) = 2(1-3^8) \Rightarrow S_8 = 2\left(\frac{1-3^8}{1-3}\right) = 6560$.

Page 852 (Trigonometric Pythogorean Identities)

Since $\cot\theta = \frac{b}{a}$ and $\csc\theta = \frac{c}{a}$, then $1 + \cot^2\theta = 1 + \left(\frac{b}{a}\right)^2 = \frac{a^2 + b^2}{a^2} = \frac{c^2}{a^2} = \left(\frac{c}{a}\right)^2 = \csc^2\theta$.

Page 852 (Law of Sines)

To find the length of the altitude from A, use $\sin 30° = \frac{h}{6}$, so $h = 6 \sin 30°$. Then the area of triangle ABC is $\frac{1}{2}bh = \frac{1}{2}(10)(6)\sin 30° = 30\left(\frac{1}{2}\right) = 15$.

Page 853 (Law of Cosines)

By the Pythagorean Theorem, $c^2 + h^2 + (7-x)^2$. Since $\sin 40° = \frac{h}{6}$ and $\cos 40° = \frac{x}{6}$, then $h = 6 \sin 40°$ and $x = 6 \cos 40°$. Substituting into the first equation gives $c^2 = (6 \sin 40°)^2 + (7 - 6\cos 40°)^2$. Simplifying and replacing $\sin^2 40° + \cos^2 40°$ by 1 gives $c^2 = 36 + 49 - 84 \cos 40°$.

Page 853 (Sum and Difference Formulas)

$$\begin{aligned}\cos(u+v) &= \cos(u-(-v)) \\ &= \cos u \cos(-v) + \sin u \sin(-v) \\ &= \cos u \cos v + \sin u(-\sin v) \\ &= \cos u \cos v - \sin u \sin v\end{aligned}$$

Page 854 (Sine of a Sum)

1. $\sin(u - v) = \sin(u + (-v))$
$= \sin u \cos(-v) + \cos u \sin(-v)$
$= \sin u \cos v + \cos u(-\sin v)$
$= \sin u \cos v - \cos u \sin v$

2. $$\tan(u + v) = \frac{\sin(u + v)}{\cos(u + v)} = \frac{\sin u \cos v + \cos u \sin v}{\cos u \cos v - \sin u \sin v} = \frac{\frac{\sin u \cos v}{\cos u \cos v} + \frac{\cos u \sin v}{\cos u \cos v}}{\frac{\cos u \cos v}{\cos u \cos v} - \frac{\sin u \sin v}{\cos u \cos v}} = \frac{\tan u + \tan v}{1 - \tan u \tan v}$$

$$\tan(u - v) = \frac{\sin(u - v)}{\cos(u - v)} = \frac{\sin u \cos v - \cos u \sin v}{\cos u \cos v + \sin u \sin v} = \frac{\frac{\sin u \cos v}{\cos u \cos v} - \frac{\cos u \sin v}{\cos u \cos v}}{\frac{\cos u \cos v}{\cos u \cos v} + \frac{\sin u \sin v}{\cos u \cos v}} = \frac{\tan u - \tan v}{1 + \tan u \tan v}$$

Page 854 (Double-Angle Formulas)

$\cos 2u = \cos^2 u - \sin^2 u$
$= \cos^2 u - (1 - \cos^2 u)$
$= \cos^2 u + \cos^2 u - 1$
$= 2\cos^2 u - 1$

Similarly,

$\cos 2u = \cos^2 u - \sin^2 u$
$= (1 - \sin^2 u) - \sin^2 u$
$= 1 - \sin^2 u - \sin^2 u$
$= 1 - 2\sin^2 u$

Sine formula:

$\sin 2u = \sin(u + u)$
$= \sin u \cos u + \sin u \cos u$
$= 2 \sin u \cos u$

Tangent formula:

$$\tan 2u = \tan(u + u) = \frac{\tan u + \tan u}{1 - \tan u \tan u} = \frac{2 \tan u}{1 - \tan^2 u}$$

ADDITIONAL ANSWERS

Page 855 (Half-Angle Formulas)

1. Substitute $\frac{u}{2}$ for x in the formula $\cos 2x = 1 - 2\sin^2 x$ and solve for $\sin \frac{u}{2}$:

$$\cos 2\left(\frac{u}{2}\right) = 1 - 2\sin^2 \frac{u}{2}$$

$$\sin^2\left(\frac{u}{2}\right) = \frac{1 - \cos u}{2}$$

$$\sin \frac{u}{2} = \pm\sqrt{\frac{1 - \cos u}{2}}.$$

2. $$\tan \frac{u}{2} = \pm\sqrt{\frac{1 - \cos u}{1 + \cos u}} = \pm\sqrt{\frac{(1 - \cos u)}{(1 + \cos u)} \cdot \frac{(1 + \cos u)}{(1 + \cos u)}} = \pm\sqrt{\frac{1 - \cos^2 u}{(1 + \cos u)^2}} = \pm\sqrt{\frac{\sin^2 u}{(1 + \cos u)^2}} = \frac{\sin u}{1 + \cos u}$$

Similarly,

$$\tan \frac{u}{2} = \pm\sqrt{\frac{1 - \cos u}{1 + \cos u}} = \pm\sqrt{\frac{(1 - \cos u)}{(1 + \cos u)} \cdot \frac{(1 - \cos u)}{(1 - \cos u)}} = \pm\sqrt{\frac{(1 - \cos u)^2}{1 - \cos^2 u}} = \pm\sqrt{\frac{(1 - \cos u)^2}{\sin^2 u}} = \frac{1 - \cos u}{\sin u}$$

A

B

INDEX

C

D

INDEX

E

F

G

INDEX

M

N

O

INDEX

R

S

T

V

W

X

Y

Z

Appreciation to the following art/photo production staff:

Leslie Concannon, Pam Daly, Irene Elios, Julie Fair, Martha Friedman, Susan Geer, Aimee Good, Carmen Johnson, Judy Kelly, Maureen Lauran, Mark MacKay, Helen McDermott, Penny Peters, Nina Whitney, Bonnie Yousefian.

COVER DESIGN

Linda Fishborne (Hememway design)
Background Photo: (COMSTOCK, INC.) Russ Kinne
Inset Photo: COMSTOCK, INC.

ILLUSTRATION CREDITS

Calligraphy by **Jean Evans**
Illustration by **Pat Rossi and Associates**
Technical Illustration by **Tech-Graphics**

PHOTO CREDITS

viii *t*: NASA. viii *b*: Ken Levine (Allsport). ix *t*: Robert Frerck (Odyssey). ix *b*: Superstock. x *t*: Marc Romanelli (The Image Bank). x *b*: William Meyer (Third Coast Stock Source). xi *t*: Pixar. xi *b*: George Zimbel (Monkmeyer Press). xii *t*: George Loehr (The Image Bank). xii *b*: Carleton Ray (Photo Researchers, Inc.). xiii *t*: Peter Pearson (Tony Stone Worldwide). xiii *b*: Thomas W. Martin, APSA (Photo Researchers, Inc.). xiv *t*: Dan Guravich (Photo Researchers, Inc.). xiv *b*: Bob Daemmerich (The Image Works). xv: Peter Hendrie (The Image Bank).

CHAPTER 1 0–1: NASA. 5: Phil Degginger. 7 *t*: Focus on Sports. 7 *b*: The Bettman Archive. 8: V.J. Anderson (Animals, Animals). 11: Edward L. Miller (Stock Boston). 14: Courtesy IBM, Almaden Research Center. 15: Susan Doheny/© D.C. Heath. 20: Brian Smith. 23: FPG. 24: Jack Fields (Photo Researchers, Inc.). 25: The Bettmann Archive. 26: Lee Boltin Picture Library. 27: Luiz Claudio Marigo (Peter Arnold, Inc.). 28: Seares (Photo Researchers, Inc.). 32: Jerry Wachter (Photo Researchers, Inc.). 35: Scott Camazine (Photo Researchers, Inc.). 36: Secchi-Lecaque-Roussel-UCLAF (Photo Researchers, Inc.). 40: David Frazier. 42 *t*: Historical Pictures Services. 42 *b*: Stephen Dalton (Photo Researchers, Inc.). 46: Joe MacDonald (Animals, Animals). 48: Ralph Reinhold (Animals, Animals). 50: John Mahar (Stock Boston). 51: The Image Works. 52: The Bettmann Archive. 53: Tom & Pat Leeson (Photo Researchers, Inc.). 55: Willard Luce (Animals, Animals). 60: Ben Osborne (Animals, Animals). 61: Martha Swope Photography.

CHAPTER 2 62–63: Ken Levine (Allsport). 68: Dan Suzio. 75 *t*: J.M. Charles-Rapho (Photo Researchers, Inc.). 75 *b*: Stacy Pick (Stock Boston). 80: David Frazier. 83 *t*: R.F. Head (Animals, Animals). 83 *b*: Jerome Wexler (Photo Researchers, Inc.). 88: Nuridsany et Perennou (Photo Researchers, Inc.). 89: The Bettmann Archive. 96: Al Grotell. 98 *t*: Springer/Bettmann. 98 *b*: Vandystadt (Photo Researchers, Inc.). 99: Bonnie Rauch (Photo Researchers, Inc.). 105: Arthur D'Amario (FPG). 106: Frank Oezas (Tony Stone Worldwide). 108: IFA (Peter Arnold, Inc.). 109: Andrew Cox (Tony Stone Worldwide). 112: Ken Levine (Allsport). 117 *t*: Lee Boltin Picture Library. 117 *b*: Rod Planck (Tony Stone Worldwide). 118: David Frazier.

CHAPTER 3 120–121: Robert Frerck (Odyssey). 126: John Scowgn (FPG). 133: E.R. Degginger. 135: Nathan Benn (Stock Boston). 136: Paul Conklin (Uniphoto). 137: John Henebry/© D.C. Heath, Inc. 139: David H. Wells (The Image Works). 141: Superstock. 142 *b*: The Image Works. 143: Tom McHugh (Photo Researchers, Inc.). 147: David Frazier. 153: Paulo Curto (The Image Bank). 155: Jon Feingers (Tom Stack and Assoc.). 160: Ellis Herwig (The Picture Cube). 162: Bob Dammerich (The Image Works). 163: E. Nagel (FPG). 168 *t*: John Cancalosi (Tom Stack & Assoc.). 168: The Image Works.

CHAPTER 4 172–173: Superstock. 176: Larry Lawfer. 178: J.D. Cuban (Allsport). 182: Shelby Thorner (David Madison). 184: Dorothy Littell (Stock Boston). 185: E.R. Degginger. 186: Topham (The Image Works). 190l: Oxford Scientific Films (Animals, Animals). 190 *r*: D.R. Specker (Animals, Animals). 193: Alan G. Nelson (Animals, Animals). 198: British Museum. 201: all Ira Block. 202: The Image Works. 206: Superstock. 208 *t*: Focus on Sports. 208 *b*: John Gichigi (Allsport). 212: FPG. 214: Superstock. 220: Captain Meriwether Lewis Meeting the Shoshonees, 1864–1926, American; photo from Superstock. 225: Focus on Sports.

CHAPTER 5 228–229: Marc Romannelli (The Image Bank). 231: The Bettmann Archive. 234: NASA. 238: Paul Logsdon. 241: David Weintraub (Photo Researchers, Inc.). 244: Susan Doheny/© D.C. Heath. 247: TV: courtesy Sony Corp. 247 *inset:* John Eastcott (The Image Works). 251: Carini (The Image Works). 254: Donald Graham (Leo deWys). 261: Courtesy Westinghouse Electric Corporation. 261: Dr. Seth Shostak (Photo Researchers, Inc.). 263 *tl*: Gregory Sams (Photo Researchers, Inc.). 263 *tr*: Gregory Sams (Photo Researchers, Inc.). 263 *bl*: Gregory Sams (Photo Researchers, Inc.). 263 *br*: Dr. Fred Espenar (Photo Researchers, Inc.). 267: Nancy Kaye (Leo de Wys). 269: Arthur D'Arazien (The Image Bank). 270: The Bettmann Archive. 272: European Space Agency (Photo Researchers, Inc.). 279: Jay Freis (The Image Bank). 280 *t*: Wolfgang Hille (Leo deWys). 280 *b*: Jacana (Photo Researchers, Inc.).

CHAPTER 6 282–283: William Meyer (Third Coast Stock Source). 289: Ken O'Donoghue/© D.C. Heath. 293: Mark Antman (The Image Works). 295: Martin Dohrn (Photo Researchers, Inc.). 297: by R.F. Voss;

from the Fractal Geometry of Nature by Benoit B. Mandelbrot, 1982. 301: Arthur Hustwitt (Leo deWys). 303: Antman (The Image Works). 307: Nicholas Foster (The Image Bank). 308: Steve Proehl (The Image Bank). 310: The Image Works. 316: Don Landwehrle (The Image Bank). 319: Benn Mitchell (The Image Bank). 323: Paulette Crowley. 328: Bob Daemmerich (The Image Works). 329: Paul Conklin (Monkmeyer Press). 331: Mimi Forsyth (Monkmeyer Press). 332: Dan Guravich (Photo Researchers, Inc.). 337: John P. Kelly (The Image Bank). 338: The Bettmann Archive. 342: Gazuit (Photo Researchers, Inc.). 342: Paolo Koch (Photo Researchers, Inc.). 343: Grant Heilman (Grant Heilman Photography).

CHAPTER 7 344–345: Pixar. 348: Margot Granitsas (The Image Works). 350: Jet Propulsion Lab, NASA. 351: Renee Lynn (David Madison). 357: Dan Helme (Duomo). 359: Susan Doheny/© D.C. Health. 362: Breck Kent (Animals, Animals). 363 *t*: Christy Hill. 363 *b*: Christy Hill. 364: Henryk T. Kaiser (The Picture Cube). 365 *t*: Ulrike Welsch (Photo Researchers, Inc.). 365 *b*: Stuart Dee (The Image Bank). 366: Del Mulkey (Photo Researchers, Inc.). 370 *t*: NASA. 370 *b*: David Madison. 373: David Madison. 379: Courtesy Children's Hospital of Michigan. 382: E.R. Degginger. 385: Everett C. Johnson (Leo de Wys). 388: Tom McHugh (Photo Researchers, Inc.). 389: Anup & Manoj Shah (Animals, Animals). 391: Margot Granitsas (Photo Researchers, Inc.). 393: Superstock. 394: Churchill & Klehr (Tony Stone Worldwide).

CHAPTER 8 396–397: George Zimbel (Monkmeyer Press). 403: The Bettmann Archive. 408: W. Gregory Brown (Animals, Animals). 410: The Image Works. 411: Harry Casey. 412: Culver Pictures. 415: Superstock. 417all: Susan A. Anderson/© D.C. Heath. 418 *l*: Howard Dratch (Leo deWys). 418 *cr*: Read D. Brugger (The Picture Cube). 418 *r*: The Image Works. 418 *lc*: Hal Clason (Tom Stack & Assoc.). 419: AP/Wide World. 422: Will & Demi McIntyre (Photo Researchers, Inc.). 424: Culver Pictures. 428: Richard Wood (The Picture Cube). 430: Renee Lynn (David Madison). 431: Thomas R. Taylor (Photo Researchers, Inc.). 433: Grant LeDuc (Monkmeyer Press). 437: Frank Siteman (The Picture Cube). 442: The American Red Cross. 443: Audrey Gottlieb (Monkmeyer Press). 445: Treat Davidson (Photo Researchers, Inc.). 451 *t*: Don Carl Steffen (Photo Researchers, Inc.). 451 *b*: Chet Seymore (The Picture Cube). 452: Susan A. Anderson/ © D.C. Heath.

CHAPTER 9 456–457: George Loehr (The Image Bank). 459: Julian Calder (Tony Stone Worldwide). 462: Steve Elmore (The Stock Market). 472: Susan Doheny/© D.C. Heath. 476: Robert Isear (Photo Researchers, Inc.). 478: Susan A. Anderson/© D.C. Heath. 479 *i*: Adam Woolfitt/Daily Telegraph Magazine (Woodfin Camp). 479: Bill Bachmann (The Image Works). 483: Gale Zucker (Stock Boston). 490: Michael L. Abramson (Woodfin Cam). 493: Walter E. Harvey (Photo Researchers, Inc.). 495: Historical Pictures Service. 497: William Curtsinger © National Geographic Society. 499 *l*: Cary Wolinski (Stock Boston). 499 *r*: Cary Wolinski (Stock Boston). 500: Vandystadt (Photo Researchers, Inc.). 501: Davies/Pashko (Envision). 503: John Cancalosi (Stock Boston). 504: William Johnson (Stock Boston). 505: ASP/Science Source (Photo Researchers, Inc.). 506 *at*: Joseph Nettis (Stock Boston). 506 *b*: Robert Frerck (Woodfin Camp). 511: Tom & Pat Leeson (Photo Researchers, Inc.). 512 *t*: Culver Pictures. 512 *l*: North Winds Picture Archive. 512 *lt*: George E. Jones (Photo Researchers, Inc.). 512 *c*: UPI/Bettmann. 512 *cb*: George Holton (Photo Researchers, Inc.). 512 *cr*: Bettmann Archive.

CHAPTER 10 516–517: Carleton Ray (Photo Researchers, Inc.). 520: National Gallery of Art, Washington, Cornelius Van S. Roosevelt Collection. 522: Sybil Shackman (Monkmeyer Press). 525: William Curtsinger (Photo Researchers, Inc.). 527: John Coletti (Tony Stone Worldwide). 528: Courtesy GM Hughes Electronics. 529: Diane M. Lowe (Stock Boston). 530: Francoise Sauze (Photo Researchers, Inc.). 531: Bob Daemmerich (Stock Boston). 532 *t*: The Bettmann Archive. 532 *b*: The Bettmann Archive. 536: Ellis Herwig (Stock Boston). 539: Bob Daemmerich (The Image Works). 543: Dennis J. Cipnic (Photo Researchers, Inc.). 544: Bob Daemmerich (Stock Boston). 547: Stephen G. Maka. 550: Blair Seitz (Photo Researchers, Inc.). 552: Bob Daemmerich (Stock Boston). 553 *t*: Crandall (The Image Works). 553 *b*: Walt Anderson (Visuals Unlimited). 557: The Image Works. 559: The Image Works. 560: Paul Barton (The Stock Market). 560: John Coletti (Stock Boston). 565: Science Source (Photo Researchers, Inc.). 566: The Image Works.

CHAPTER 11 568–569: Peter Pearson (Tony Stone Worldwide). 572: Hank Morgan (Photo Researchers, Inc.). 575: The Bettman Archive. 580: Derek Bayes (Tony Stone Worldwide). 582: Susan Doheny/© D.C. Heath. 588: Superstock. 602: Gatha Ashvin (Leo DeWys). 609: Robert Kristofik (The Image Bank). 617: J.B. Bidner. 618: NASA.

CHAPTER 12 620–621: Thomas W. Martin, APSA (Photo Researchers, Inc.). 626: Scott Camazine (Photo Researchers, Inc.). 627: David Madison. 632: Eric Lars Bakke. 635: Ken Kaminsky (The Picture Cube). 636: Alfred Eisenstaedt, Life Magazine © Time Warner Inc. 643: PIP/LGI. 647: Mary Clay (Tom Stack & Assoc.). 650: Ken O'Donoghue/© D.C. Heath. 659: Robert Essel (The Stock Market). 662: Bob Daemmerich (Stock Boston). 668: C.W. Brown (Photo Researchers, Inc.). 669 *l*: Phil Degginger (Tony Stone Worldwide). 669 *r*: Richard Megna (Fundamental Photos). 675: Gazuit (Photo Researchers, Inc.).

CHAPTER 13 676–677: Dan Guravich (Photo Researchers, Inc.). 681: Ira Kirschenbaum (Stock Boston). 683: Isreal Talby (Woodfin Camp). 684: Walt Urbina. 688: Alain Thomas (Photo Researchers, Inc.). 691: R.P. Kingston (Stock Boston). 692: Susan Doheny/© D.C. Heath. 696: David Madison. 707: Geoff Tompkinson (Aspect Picture Library). 711: Dave Brown (Stock Market). 713: D. & J. Heaton (Stock Boston). 714: Steve Murez (Black Star). 718: Courtesy of Amoco Chemical Company. 720: Brian Parker (Tom Stack & Assoc.). 721: Walter Looss (The Image Bank). 725: Wes Thompson (The Stock Market). 726: Bill Gallery (Stock Boston).

CHAPTER 14 728–729: Bob Daemmerich (The Image Works). 733: Steve Dunwell (The Image Bank). 735: Alvis Upitis (The Image Bank). 741 *l*: Charles Mahaux (The Image Bank). 741 *b*: Joe Van Os (The Image Bank). 743: The Bettmann Archive. 749: Roy Pinney (Photo Researchers, Inc.). 751: Anne La Bastille (Photo Researchers, Inc.). 765: Treat Davidson (Photo Researchers, Inc.). 759: Jim Hamilton (The Picture Cube). 756: Michel Tcherevkoff (The Image Bank). 771: People Weekly © Mario Ruiz. 775: John Moss, Courtesy of the Royal Society of London. 779: Beringer/Dratch (The Image Bank).

CHAPTER 15 780–781: Peter Hendrie (The Image Bank). 784: Treat Davidson (Photo Researchers, Inc.). 786: Mark Jones (Hedgehog House of New Zealand). 789: Tim Davis (Photo Researchers, Inc.). 792: Hank deLespinasse (The Image Bank). 793: Susan A. Anderson/© D.C. Heath. 794: Matt Kelly. 795: Susan Doheny/© D.C. Heath. 798: *t*: Stephen Wilkes (The Image Bank)., 798 *b*: Murray Alcosser (The Image Bank). 801: Hank deLespinasse (The Image Bank). 807: Dr. Wm. M. Harlow (Photo Researchers, Inc.). 809: Robert Frerck (The Stock Market). 812: Gary Gladstone (The Image Bank). 813: Steve Dunwell (The Image Bank). 814: Susan A. Anderson/© D.C. Heath. 815: Richard Gross (The Stock Market). 817: Fred Lyon (Photo Researchers, Inc.). 818: Gabe Palmer/Mug Shots (The Stock Market). 819: Jeff Smith (The Image Bank). 825: AP/Wide World Photos. 827: Ralph A. Reinhold (Animals, Animals).

Appreciation to the following editorial staff at Larson Texts:

Richard Bambauer, Linda Bollinger, Laurie Brooks, Patti Jo Campbell, Jordan Feidler, Linda Kifer, Deanna Larson, Patricia Larson, Timothy Larson, Amy Marshall, John Musser, Scott O'Neil, Louis Rieger, Paula Sibeto Steinhart, Kristin Smith, and Evelyn Wedzikowski

TEACHER'S EDITION DESIGN AND PRODUCTION

Christine Reynolds (Reynolds Design & Management)